Bones and Cartilage: Developmental and Evolutionary Skeletal Biology

Brian K. Hall

AMSTERDAM • BOSTON • HEIDELBERG • LONDON • NEW YORK • OXFORD
PARIS • SAN DIEGO • SAN FRANCISCO • SINGAPORE • SYDNEY • TOKYO

ELSEVIER
ACADEMIC
PRESS

Permissions may be sought directly from Elsevier's Science & Technology Rights
Department in Oxford, UK: phone: (+44) 1865 843830, fax: (+44) 1865 853333,
e-mail: permissions@elsevier.co.uk. You may also complete your request on-line via
the Elsevier homepage (http://www.elsevier.com), by selecting 'Customer Support'
and then 'Obtaining Permissions'

Elsevier Academic Press
525 B Street, Suite 1900, San Diego, California 92101-4495, USA
http://www.elsevier.com

Elsevier Academic Press
84 Theobald's Road, London WC1X 8RR, UK
http://www.elsevier.com

British Library Cataloguing in Publication Data
A catalogue record for this book is available from the British Library

Library of Congress Control Number: 2005922522
A catalogue record for this book is available from the Library of Congress

ISBN 0-12-31906-06

For information on all Elsevier Academic Press publications
visit our website at http://www.books.elsevier.com

Typeset by Newgen Imaging Systems (P) Ltd., Chennai, India
Printed and bound in Great Britain
05 06 07 08 9 8 7 6 5 4 3 2 1

Bones and Cartilage:
Developmental and Evolutionary
Skeletal Biology

Epigraph

Problems of the developmental physiology of bone have always had a great fascination for those who have studied them. That fascination is likewise easily understood when we reflect how notable an example is provided by bone of the way in which, from the apparently simple, more complex problems arise, and how the pursuit of these further problems leads the investigator, if [s]he be so minded, straight to fundamental problems of the nature of life itself.

(Brash, 1934, p. 306)

The facts to which I wish to call your attention, and which are confirmatory of the views here developed, have been brought to light by different observers, at different times.

(Hubrecht, 1897, p. 37)

Contents

Part III Unusual Modes of Skeletogenesis

Part IV Stem Cells

Part V Skeletogenic Cells

Part VI Embryonic Origins

Part VII Getting Started

Part VIII Similarity and Diversity

Part IX Maintaining Cartilage in Good Times and Bad

Part X Growing Together

Part XI Staying Apart

Part XII Limb Buds

Preface

The skeleton has fascinated humankind ever since it was realized that, aside from one or several sets of genes, bare bones are our only bequest to posterity. But the skeleton is more than an articulated set of bones: its three-dimensional conformation establishes the basis of our physical appearance; its formation and rate of differentiation determine our shape and size at birth; its postnatal growth orders us among our contemporaries and sets our final stature; while its decline in later life is among the primary causes of loss of the swiftness and agility of youth. Not surprisingly, the skeleton is a central focus of many scientific and biomedical disciplines and investigations.

For the developmental or cell biologist, the skeleton provides an excellent model for studies of gene action, cell differentiation, morphogenesis, polarized growth, epithelial–mesenchymal interactions, programmed cell death, and the role of the extracellular matrix. The skeleton supplies the geneticist with a permanent record of the vicissitudes of its growth, whereby the phenotypic expression of genetic abnormalities can be studied. The orthopaedic surgeon earns a livelihood from correcting abnormalities and breaks, while the orthodontist corrects the position of teeth displaced consequent to alveolar bone dysfunction. Physiologists, biochemists and nutritionists are concerned with the skeleton's store of calcium and phosphorus and its response to vitamins and hormones. Haematologists, on the other hand, find that the skeleton houses the progenitors of the blood cells. Pathologists endeavour to understand the disease states that result from abnormalities in skeletal cellular differentiation or function; surgeons want to prevent formation of skeletal tissues in the wounds that bear witness to their work. Vertebrate palaeontologists make their living from the analysis of the skeletons of extinct taxa. Veterinarians, physical anthropologists, radiographers, forensic scientists – the list goes on.

Bones come in all shapes and sizes. There are long bones, flat bones, curved bones, bones of irregular and geometrically indefinable shapes, large bones and small. Bones exhibit bumps, ridges, grooves, holes and depressions where they articulate with other bones, attach to tendons and ligaments, and where nerves and blood vessels course through them. Some bones and cartilages arise within the skeleton and are integral parts of it. Others arise outside the skeleton, some as sesamoids or ossifications within tendons or ligaments, others as pathological ossifications in what otherwise would be benign soft tissues. Bones and cartilages may develop during embryonic or foetal life, in larval stages or in adulthood – often late in adulthood – during normal ontogeny,

wound repair, or regeneration. Bones modify themselves in response to injury, disease or parasitic infection, in the aftermath of surgery, as a defensive response to predators, as a consequence of domestication or hibernation, and through evolutionary adaptations.

My previous book on the skeleton – *Developmental and Cellular Skeletal Biology* – was published in 1978. That book concerned itself with how bones and cartilages are made and how these tissues, organs and systems evolved. So too does the present book, which includes and updates the earlier treatment. With respect to *skeletal development*, I address such questions as the following.

- Is bone always bone, no matter where and under what conditions it forms?
- Do bones that develop *indirectly* by replacing another tissue – be it cartilage, marrow, connective tissue, fat, tendon or ligament – differ from one another, and/or from bone that develops *directly* (intramembranously)?
- Is fast-growing the same as slow-growing bone?
- Is fish bone the same as human bone?
- Does bone form continuously or in cycles?
- Do bears make new bone during hibernation?
- Can sharks make bone?
- If cartilage does not contain type II (cartilage-type) collagen, is it still cartilage?
- Does the body contain cells that can differentiate as chondrocytes *or* as osteocytes and, if so, what factors allow cells to choose their fate?
- Are progenitor (stem) cells for bone and cartilage only found within the skeleton? If not, how do we recognize such cells and activate them for skeletogenesis?
- Why is aggregation (condensation) of cells so important for the initiation of the skeleton?
- Does the skeleton display daily or circadian rhythms?
- Do similar genes/growth factors regulate the differentiation of osteoblasts and chondroblasts?
- Can mononucleated cells resorb bone?
- How do joints form and remain patent?
- How does activating FGF receptors cause cranial sutures to fuse?
- What can mutants tell us about normal skeletogenesis?
- Does Wolff's law really govern the structure of bone?
- How do chondroid, chondroid bone, osteoid and bone differ from one another?
- How do antlers, horns and knobs (ossicones) differ one from the other?
- Can we restart cell division in articular cartilage to effect repair?

With respect to the *evolution* of skeletal tissues, organs and systems, I ask such questions as the following.

- What are the evolutionary relationships between cartilage and bone and between acellular and cellular bone?
- How did novel features such as tetrapod limbs arise from fish fins?
- Can fossilized bone reveal patterns of growth, metabolism or physiology?
- Why are so few aware of the extensive cartilaginous skeletons found in many invertebrates?
- Is five the canonical number of tetrapod digits?
- If tetrapods are vertebrates with limbs, then how can limbless snakes be tetrapods?
- How did snakes lose their limbs?
- How did whales lose their hind limbs and transform their forelimbs into flippers?
- How do we recognize the diverse range of tissues in fossilized skeletons that are intermediate between connective tissues and cartilage, cartilage and bone, bone and dentine, or dentine and enamel?
- Why can some vertebrates regenerate their limbs or tails and others not?
- How does reduction in body size (miniaturization) affect the skeleton?

The answers to the above and many other questions may be found in this book. Sometimes the 'answers' are limited to descriptions. In other cases we have an extensive knowledge of the molecular, cellular, developmental and evolutionary processes involved. Some transitions (fins → limbs, for example) are understood in considerable detail, with paleontology, paleobiology, paleohistology, paleopathology, and the study of extant forms through molecular, cell and developmental biology contributing to our understanding. Other transitions – the origin of the turtle shell, for example – are much less well understood, with fossils contributing little and developmental information only beginning to appear.

Discussion of the mechanisms of skeletal development and evolution is organized into 15 parts to enable you to select with ease a topic of special interest. The range of skeletal tissues covered by the book is outlined in Part I. Although primarily devoted to bone and cartilage, Part I introduces dentine and enamel and four skeletal tissues that I call 'intermediate' because they display features of two or more of cartilage, bone, dentine and enamel. The four are chondroid, chondroid bone, cementum and enameloid. Discussion of these intermediate tissues is expanded in Part II in the context of what I refer to as 'natural experiments,' a category that includes invertebrate cartilages and an examination of the evolution of skeletal tissues.

Unusual tissues are followed in Part III by unusual modes of skeletogenesis, namely, horns, antlers, intratendinous ossifications and sesamoids, and the ossicones (knobs) of

giraffes. Parts IV and V deal with the origin of skeletogenic cells, either as stem cells in embryos or adults (Chapters 10 and 11) or as more definitive skeletogenic cells (Part V). Here the emphasis is on those cells that can differentiate *either* as chondro- or osteoblasts (Chapter 12), on dedifferentiation as a source of skeletogenic cells in normally developing long bones and jaws and in regenerating urodele limbs (Chapters 13 and 14), and on the relationship(s) between the cells that make and the cells that break bone – osteoblasts and osteoclasts (Chapter 15).

I move explicitly into embryonic development in the three chapters in Part VI through examination of the embryonic origins of skeletogenic cells in somitic mesoderm and the neural crest, and an evaluation of the roles of epithelial–mesenchymal interactions in initiating skeletogenesis. The developmental processes that underpin skeletal formation – differentiation, morphogenesis and growth – are mediated through modification of cell division, movement, death (apoptosis) and/or specialization. To our amazement, similar genes and gene networks or pathways may be involved in all these developmental processes, and in vertebrates as diverse as sharks, salamanders, shrews and shore birds; *pathways are conserved*. In other instances, genetic pathways have diverged in different taxa; *development evolves*. Consequently, constancy, variation, change and novelty all confound our analyses. A major challenge for the future is to understand how conservation and/or modification at molecular, cellular and developmental levels produces and has produced the diversity of skeletal tissues, elements and systems seen in bone, cartilage and chondroid (tissues), the clavicle, humerus and ribs (elements), and in the endo- and exo-, appendicular, axial and craniofacial skeletal systems.

Skeletal development comprises a stepwise set of events, each depending on the step before, but each involving different cellular processes – migration, adhesion, proliferation, growth – and each subject to different genetic control. Condensation, the pivotal stage in the development of skeletal and other mesenchymal tissues, when a previously dispersed population of cells gathers together to differentiate into a single cell/tissue type such as cartilage, bone, muscle, tendon, kidney or lung, is the earliest stage during organ formation when tissue-specific genes are up-regulated. Condensation itself is a multistep process, involving initiation, establishment of boundary conditions, cell adhesion, proliferation, growth and cessation of growth. Chapters 19 and 20 in Part VII discuss condensation and the transition from condensation to overt differentiation of cartilage and bone, processes that are elaborated and related to epithelial–mesenchymal interactions in the context of a discussion of the development of the skull and optic and nasal capsules in Chapter 21.

Because cartilage is not cartilage is not cartilage, and an osteoblast is not an osteoblast is not an osteoblast, I devote Part VIII to an evaluation of similarity and diversity of cartilages and bones and of chondro- and osteogenic cells. Included in these chapters are discussions of the role of type X collagen, chondrocyte hypertrophy, vascularity and/or resistance to vascular invasion in different cartilages, and gender-based differences in the skeleton.

Part IX takes us to how chondrocytes are maintained as differentiated cells (Chapter 26), how chondrogenesis gone awry can lead to achondroplasia (Chapter 27), and to discussions of restarting chondrogenesis in articular cartilages or during bone repair and regeneration (Chapters 28 and 29).

Skeletal organs do not exist in isolation but are articulated at joints or sutures to form skeletal systems. How skeletal growth, especially long-bone growth, is initiated, how skeletal shape is maintained, and that perennial topic, Wolff's law and the response of bone (and cartilage) to mechanical stimulation, are the topics of Part X; movement, maintenance of joints and abnormal fusion of sutures (craniosynostosis) are the topics of Part XI.

Although the search has been long, evolutionary developmental mechanisms are more elusive. Almost a century and a half ago, Thomas Henry Huxley recognized but three processes in the generation of animal morphology: (i) excess or (ii) suppression of one or more parts with respect to other parts, and (iii) the coalescence of parts. In essence, modification in a descendant of any of the developmental processes noted above and discussed in the text could lead to evolutionary changes in the skeleton. Reduce the size of a limb bud and digit number falls below five. Expand the width of the limb bud and an extra digit (polydactyly) can result. Limblessness and taillessness often result, not from failure to initiate a limb or tail bud, but from the inability to maintain the bud for long enough for skeletal tissues/organs to form. A mutation that allows limb buds to persist in an individual limbless tetrapod can result in the formation of an atavistic skeletal element. Add natural selection and a limbless taxon can result.

The six chapters in Parts XII and XIII discuss is some details limb buds, the limb skeleton, limbless tetrapods, fins, and the transformation of fins into limbs with the evolutionary origin of the tetrapods. Many are familiar with the apical ectoderm ridge or AER. Fewer will know that developing tails possess a ventral ectodermal ridge or VER. Part XIV is taken up with development of the vertebrae, including those of the tail, and including a discussion of tailless vertebrates in which the VER fails to function normally. Part XV – the last – explicitly treats evolutionary skeletal biology with discussions of skeletal variation, heterochrony, miniaturization, the evolution of novel skeletal elements (neomorphs) and atavisms.

Skeletal biology is a vast field; the research of thousands of individuals is included in the close to 6800 references cited. Nonetheless, and although I have endeavoured to be comprehensive, I have not been inclusive. I would appreciate important references I have omitted being brought to my attention.

Important conclusions are highlighted in the text. Topics that apply throughout the book and to which you the reader may want to refer frequently – the major proteins of bone, cartilage mineralization, Bmps and their receptors, as three examples – along with unexpected findings (chondrocytes with cilia, cartilage inside the notochord, the evolutionary consequences of hunting big sheep, cartilages and bones in the heart, how hibernation affects the skeleton, coal trimmers and shoemakers, fish without tails) are placed in boxes, which can be read as part of the text or in isolation. To make the text as user friendly as possible I have placed references, comments, elaborations, asides and tit(tid)bits[a] in endnotes, gathered together at the end of the text.

A device I have used to show how our understanding of a topic has evolved and how research programmes develop is to outline the research from a research group in a single list in chronological sequence. Furthermore, individual studies often pertain with equal utility to several aspects of, or approaches to, skeletal biology. Because of this, and because I want each section to be as self-contained as possible, you will occasionally find that I have repeated an example, explanation or mechanism. This does not reflect sloppy copy-editing. Rather, it is a deliberate way of demonstrating how interrelated the apparently separate topics are.

Although I want you to read the book, you should be able to access information readily through the index. Consequently, I have provided a detailed index, which serves as both subject and taxonomic index. The annotated list of abbreviations provided also can be used as a glossary of the many genes and growth factors discussed.

Inclusion of a list of abbreviations raises the issue of gene nomenclature.

Because of the rampant confusion and variation in the literature, a comment on the notation used for genes and their products has become a standard element in the preface of many recent books – Weiss and Buchanan (2004) for example, or been extended to an extensive appendix – Appendix 1 in Wilkins (2002) for example. Gene, growth factor and receptor nomenclature are in considerable disarray – Older names have been supplanted by newer names. And, that's a good thing; it reflects the recognition of families of growth factors and genes and a consequent rationalization of nomenclature. So GHox-4.6 and GHox-8 isolated from chick embryos are now known to be a member of the Hoxd group (Hoxd-11) and an Msx gene (Msx-2), respectively. Independent and often contemporaneous discovery of the same gene or gene product in different laboratories introduced into the literature different names for the same entity. Examples are osteogenin, Bmp2B and rhBmp2B (human recombinant Bmp2B) for what is now known as Bmp-3. Even so, one has to be on one's guard;

---

[a] Tidbit, US variant of titbit. Titbit, a dainty morsel, a piquant item of news. 'The book is chock-full of colorful titbits about theater and theater people' (Alec Guinness) is an example of the use of the term found on many web sites.

older names remain in use in recent literature. I have endeavoured to provide current nomenclature throughout, including older names where appropriate.[b] Genes are in italics (*Fgf-4*), gene products in plain typeface (Fgf-4). Following convention, human genes or their products are capitalized (*FGF-4*, FGF-4).

My own interest in the skeleton was kindled by my Ph.D. supervisor, the late P. D. F. Murray, whose 1936 monograph *Bones, A Study of the Development and Structure of the Vertebrate Skeleton* remains one of the most lucid and, paradoxically, most modern treatments of the developing skeleton. The present book was meant to be a post-retirement project, one that I could dawdle over into my declining years. Charles (Chuck) Crumly, then of Academic Press, convinced me that I should make an earlier start on the project by providing a contract with a long lead-time and then believed me when I said I was working on the book and not dawdling…My thanks Chuck, for your faith, encouragement, and friendship. Completion of the book was aided enormously by the granting of a Killam Research Fellowship by the Canada Council for the Arts in 2003. Without the consequent release-time from teaching and administration I would be even further behind the generous deadline given when I signed the contract.

I am grateful that Tim Fedak applied his considerable artistic skills to the preparation of the figures, which enhance the text enormously. My thanks to Hollie Knoll, who prepared the majority of the tables (often from quite raw starting material), and to Patricia (Paty) Avendaño for assistance with some of the tables and for providing Figures 40.6 to 40.8. A number of other friends and colleagues kindly made figures available. My thanks to Michael Locke for Figure 2.3, Eckhard Witten for Figure 2.11, Tamara Franz-Odendaal and Andrew Gillis for the figures of scleral ossicles in salmon (Fig. 21.1), David Precious for Figure 33.2, and Ulrich Zeller for Figure 44.1.

[b] M. P. Smith (1992) and S. Stein *et al.* (1996) provide guides and checklists to the vertebrate homeobox (Hox) genes, Duboule (1994) a guide to those homeobox genes known a decade ago. The rationale for gene nomenclature outlined by Wilkins (2000, pp. 525–526) is eminently sensible and rational; Wendy Olson and I followed this convention for *Keywords & Concepts in Evolutionary Developmental Biology* (Hall and Olson, 2003, p. xvi) and I use it here.

Although I hope it is not too obvious from the book, it is not possible for one person to be expert in all the areas of skeletal biology. Skeletal biology is a vast field; the research of thousands of individuals is included in the close to 6800 reference cited. I am grateful to a number of experts in specific areas who took the time and trouble to respond to my request to review individual chapters. With the chapter they reviewed in parentheses, my thanks to Mike Benjamin (9), George Bernard (2), John Bertram (32), Bobo Christ (16), Nelly Farnum (31), Benedikt Hallgrímsson (44), Greg Handrigan (43), Tuomo Kantomaa (33), Gillian Morriss-Kay (34), Lynne Opperman (34), Pertti Pirttiniemi (33), Robin Poole (3), Cheryll Tickle (38) and Eckhard Witten (1–6). Tim Fedak, Tamara Franz-Odendaal and Matt Vickaryous, three current members of my laboratory, each provided comments on several chapters and engaged in a scavenger hunt for elusive titbits. Allison Cole commented on Chapter 4. Annie Burke, Andy Horn and Marty Leonard, Peggy Kirby, David Precious and Marvalee Wake provided information concerning somite organization, the identification of the species in Figure 7.7, cells migrating from the ventral neural tube, nasal septal growth, and chondroid in caecilians, respectively. Tom MacRae and Vett Lloyd kindly checked the list of abbreviations. It is a pleasure to thank Priscilla Goldby for expert copyediting and Pauline Sones and Andy Richford of Elsevier for their professionalism in bringing the book into production. June Hall read two drafts of the entire book – each of which I thought was the penultimate draft – and provided much excellent advice, more of which, as always, I should have taken.

Since 1968 my research has been supported continuously by the National Research Council (NRC) and then the Natural Sciences and Engineering Research Council (NSERC) of Canada (grants A5056 and 257447-02), with additional support from time to time from the Research Development Fund and Killam Trust of Dalhousie University and from funds associated with the George S. Campbell Chair in Biology and a University Research Professorship, the Victoria General Hospital (Halifax), the Medical Research Council (Canada), the Canadian Institutes of Health Research, the Killam Trust of the Canada Council for the Arts, and the US National Institutes of Health (NIH; grant 45344). To all these agencies, my heartfelt thanks.

Abbreviations

Because acronyms and abbreviations abound, especially in molecular biology, metabolic pathways and enzymology, I provide a list of the abbreviations used in the book, including abbreviations used in tables but excluding those used in figures; many of the latter pertain only to a single figure legend. Where I felt it would be helpful I have annotated the explanation of an abbreviation. Entries marked * (e.g. *Fgf) represent families of molecules. In such entries, individual family members are not listed, but their identity in the text should be obvious. For example Fgf-1 and Fgf-2 – fibroblast growth factor one and fibroblast growth factor two – are two members of the Fgf family, BmpR-1B is a BmpR (bone morphogenetic protein receptor) and so forth. When only a single member of a family is mentioned in the book I only list that member. *Bapx-1* and *Gli-3* are two examples.

Genes and gene products are listed under the same abbreviation with the gene name in italics and the gene product in plain text as per convention, e.g. *Fgf*, Fgf. Gene names in humans are capitalized (*FGF-2*) in the text, but not listed separately in these abbreviations. Gene names for Hox genes were regularized a little over a decade ago. Nevertheless, and for ease of reference, I have included older names in this list and cross-listed them to the newer name, e.g. Hox 1.1 = *Hoxa-7*, Quox 7 = *Msx-2*.

A

A23187,	a monocarboxylic acid extracted from the actinomycete *Streptomyces chartreusensis*; an ionophore that binds Ca^{++}. Also known as calcimycin
Ac,	the *achondroplasia* mutant rabbit
AEMF,	apical ectodermal maintenance factor
AER,	apical ectodermal ridge of limb bud (should more properly be termed the apical *epithelial* ridge, as it arises from epithelium long after the ectoderm has become epithelial)
aFGF,	acidic fibroblast growth factor, a synonym for Fgf-1
Alk-6,	activin receptor-like kinase (=BmpRIB)
Alx-3,	a mouse paired, *Aristaless*-like homeobox gene
AmphiWnt,	the *Wnt* gene family in amphioxus (*Branchiostoma*), e.g. *AmphiWnt-3*
6-AN,	6-aminonicotinamide ($C_6H_7N_3O$, MW 137.14), a teratogenic vitamin antagonist
6-AN-NAD,	6-aminonicotinamide adenine dinucleotide
ANZ,	anterior necrotic zone of limb buds; a zone of apoptosis

A-P,	antero-posterior axis/polarity of organ rudiments (e.g. limb bud), organs (e.g. limbs) or organism
ATCD-5,	a chondrogenic murine embryonic carcinoma cell line
ATP,	adenosine 5′-triphosphate

B

BAC library,	a DNA library based on the cloning of large fragments of genomic DNA into vectors known as <u>B</u>acterial <u>A</u>rtificial <u>C</u>hromosomes
BAG-75,	bone acidic glycoprotein
BALB/c,	an inbred strain of mice that develops numerous tumours in later life
Bapx-1,	*Bagpipe homeobox gene-*, a member of the *Nkx* family of transcription factors. Also known as *Nkx3.2*
Barx-1,	*BarH-like homeobox 1*, a gene encoding a mouse homeodomain transcription factor; orthologue of the Bar subclass of homeodomain proteins of *Drosophila*
B&C score,	Boone & Crockett score; official measure for the sizes of antlers and horns
bFGF,	basic fibroblast growth factor, a synonym for Fgf-2
BGJ$_b$,	a culture medium for skeletal tissues/cells devised by Biggers, Gwatkin and Judah in the 1960s
BGP,	bone Gla-protein, a synonym for osteocalcin
bHLH,	basic helix-loop-helix gene family
Bm,	*Brachymorphic* mutant mice which displays disproportionate dwarfism
Bmp, Bmp,	bone morphogenetic protein gene and gene product
BmpR,	bone morphogenetic protein receptor, e.g. BmpRIB
bpH,	the symbol for the *brachypodism* mouse mutant
Br,	*Brachyrrhine* mouse mutant
BrdU,	bromodeoxyuridine, an analogue of thymidine
BSP,	bone sialoprotein

C

C-1-1,	a transcription factor variant of the *ets* DNA-binding transcription factor *ch-egr*, expressed in articular chondrocytes
C1–C7,	cervical vertebrae one to seven

C342Y, a cysteine to tyrosine mutation of the human *FGFR-2* gene resulting in Crouzon syndrome

C3H10T1/2, a murine multipotential mesenchymal cell line

C-4-S, C-6-S, chondroitin-4 and chondroitin-6 sulphates

C57BL/6, an inbred strain of mice derived in 1921 and widely used in research; often shortened to C57

CaCl₂, calcium chloride

CAM, chorioallantoic membrane; one of the extraembryonic membranes of amniote embryos, formed by the fusion of the chorion and allantois

cAMP, cyclic adenosine 3′,5′-monophosphate, a second messenger

Can, the *Cartilage anomaly* achondroplastic mouse mutant

Cart-1, the gene for cartilage homeo protein-1, a 326-amino acid homeoprotein containing a paired-like domain

CBA/Ca an inbred strain of mice with low spontaneous but high inducible incidence of leukemia

Cbfa-1, core binding factor alpha 1; also known as osf-2; see *Runx-2*

*CD-44, a family of cell surface transmembrane glycoproteins

CD-57, see HNK-1

Cdmp-1, the gene for cartilage-derived morphogenetic protein-1, a synonym for *Bmp-14*

Cdx-1, a homeobox gene encoding a transcription factor related to *caudal* in *Drosophila*

Cfkh-1, chicken forkhead (winged)-Helix transcription factor gene; regulates TGFβ and interacts with Smad transcription factors

C-fms, receptor protein of the CSF-1 proto-oncogene

CFUs, colony-forming units

cGMP, guanosine 3′,5′-cyclic monophosphate, a second messenger for peptide hormones and nitric oxide that activates cGMP-dependent protein kinases

Ch, the *congenital hydrocephalus* mutant mouse

CHL, chordin-like protein

Cho, *Chondrodysplasia*, a recessive lethal mutant in C57 inbred mice resulting from a cysteine deletion in the gene for collagen type X

CHO, Chinese hamster ovarian established cell line

CHox-4.6, older name for *Hoxd-11*

*ClC-7, member of a family of chloride-channel proteins found in the ruffled border of osteoclasts; mutated in osteopetrotic mice

Ck-erg, the chicken *erg* gene, an Ets transcription factor

Clim-2, the gene for carboxyl-terminal lim domain protein-2

Cmd, the *cartilage matrix deficient* chondrodystrophic mouse mutant

Cn, an *achondroplasia* mutant mouse

CNC, cranial neural crest or cranial neural crest cells

Col1a1, the gene for the procollagen type I alpha 1 chain

Col2a1, the gene for the procollagen type II alpha 1 chain. Older symbols were COLLII and CG2A1A

COS-7, an SV40 transformed cell line from the kidney of the African green (velvet) monkey, *Cercopithecus aethiops*

CP22-/1, a G-protein-coupled extracellular matrix receptor for ATP

CPCs, chondroprogenitor cells

CPZ, carboxypeptidase Z, a zinc-dependent enzyme involved in *Wnt* signaling

*CRABP, cellular retinoic acid binding protein

CS, chondroitin sulfate

CSF-1, colony stimulating factor-1

CTgf, connective tissue growth factor, a secreted protein that interacts with other growth factors

D

dak, *dackel*, a gene in zebrafish that acts with *Shh* to activate *Fgf-4* and *Fgf-8*

Dbf, *doublefoot*, a polydactylous mouse limb mutant

Dh, *dominant hemimelia*, a mouse limb mutant

Diam, diameter

DiI, 1,1-Didodecyl-3,3,3′,3′-tetramethyl indocarbocyanine; a lipophylic dye used as a cell tracer

*Dl, a family of genes (e.g. *Dl-1*) that encode Delta proteins, which are type-1 cytokine receptor family protein

Dll, the *distalless* gene in *Drosophila*

*Dlx, *distalless* gene family in vertebrates, e.g. *Dlx-1*

DMEM, Dulbecco-Vogt minimal essential medium for cell culture

Dmm, the symbol for *Disproportionate micromelia*, a chondrodysplasia mouse mutant

DON, 6-diazo-5-oxo-L-norleucine, an amino acid that interferes with glutamine metabolism

DPNH, reduced diphosphopyridine nucleotide; also known as reduced nicotinamide adenine dinucleotide (NADH)

Dpp, Dpp decapentaplegic gene and protein in *Drosophila*

DRG, dorsal root ganglia

Ds, the *Disorganization* polydactylous mouse mutant

D-V, dorso-ventral axis/polarity of organ rudiments (e.g. limb bud), organs (e.g. limbs) or organism

Dy, Dy²⁵, symbol for a mouse muscular dystrophy mutant

E

ECM, extracellular matrix

Edn-1, see Suc

EDTA, ethylenediamine tetraacetic acid, a chelating agent used in decalcification of mineralized tissues

eHand a basic helix-loop-helix protein

*Egf, Egf, epithelial growth factor proteins, genes, e.g. Egf-1, Egf-1

EGTA, ethylenebis (oxyethylenenitrilo) tetraacetic acid; a chelating agent used to decalcify mineralized tissues

EMILIN-5, elastin microfibril interface protein-5

EMSP-1, enamel matrix serine proteinase-1, a proteolytic enzyme found in enamel

*En, engrailed gene family, e.g. En-1

Eph-A4, eph receptor tyrosine kinase-A4, a member of a large subfamily of receptor protein-tyrosine kinases consisting of receptors related to Eph, a receptor expressed in an erythropoietin-producing human hepatocellular carcinoma cell line

E.R., endoplasmic reticulum

Et-1, endothelin-1, a vasoactive peptide regulating blood pressure and craniofacial development

EphA-7, the gene for the receptor for members of the ephrin-A subfamily of receptor tyrosine kinases

*Ext, a family of genes (Ext-1, Ext-2, Ext-3) that code for glycosyltransferases (heparan sulphate co-polymerases)

F

F12, see Ham's F12

Far, first arch murine craniofacial mutation

FBJ virus, Finkel, Biskis, and Jinkins murine osteosarcoma virus

FCFU, fibroblast colony-forming unit

*Fgf, Fgf, fibroblast growth factor protein, gene, e.g. Fgf-2, Fgf-2

*FgfR, fibroblast growth factor receptor

*Fhf-1, -2, fibroblast growth factor homologous factor-1 and -2 genes (synonyms for Fgf-12 and Fgf-13)

Frzb-1, frizzled-related protein precursor-1 gene, a secreted antagonist of Wnt signaling

Ft, fused toes; a mouse mutant which displays fusion of distal phalanges

G

GAG, glycosaminoglycan

gas-2, growth-arrest-specific-2, a gene that acts as a substrate for caspase enzymes and regulates apoptosis

GCMJ, growth cartilage-metaphyseal junction

*Gdf, Gdf, a family of growth and differentiation factor genes and products, e.g. Gdf-5, Gdf-5

GFP, green fluorescent protein

Ghox-4.6, older name for Hoxd-11

GHox-8, older name for Msx-2

Gli-3, one of a family of mouse genes whose members share a zinc-finger domain with the Drosophila gene cubitus interruptus

GMP, guanosine 5'-monophosphate

GPC-3, glypican-linked heparan sulphate modified proteoglycan (Glypican)

GPI, glycosylphosphatidylinositol, a cell surface lipid involved in linking proteins to cell membranes

Gy, gray unit, the unit of absorbed radiation (named after a British physician, Hal Gray), one Gy depositing one joule of energy in a kilogram of irradiated tissue. One Gy (equal to 100 rad) of ionizing radiation is sufficient to sterilize an individual human

H

HA, hyaluronate

Ham's F12, a culture medium for vertebrate cells, tissues or organs devised by R. G. Ham in 1975

HCCs, hypertrophic chondrocytes

Hd, hypodactyly, a mutant chick with defective limb development

HGF/SC, hepatocyte growth factor/scatter factor

H.H., Hamilton-Hamburger stage of chick embryonic development

*Hh, Hedgehog gene family, e.g. Indian hedgehog (Ihh)

Hmˣ, the mouse mutant Hemimelia-extra toes. Also known as Hx

HNK-1, a cell surface carbohydrate (known as CD-57 in immunology) used as a marker for neural crest cells

*Hox, homeotic gene classes in vertebrate, e.g. Hoxd-10

Hox1.1, older name for Hoxa-7

Hox1.6, older name for Hoxa-1

Hox3.3, older name for Hoxc-6

Hox3.6, older name for Hoxc-10

Hox4.2, older name for Hoxd-4

Hox4.5, older name for Hoxd-10

Hox4.6, older name for Hoxd-11

Hox7.1, older name for Msx-1

HPLC, high performance liquid chromatography

Hx, see Hmˣ

I

ia, symbol for incisor absent osteopetrotic mutant mice

*Igf, Igf, insulin-like growth factor genes and products

Ihh, Indian hedgehog gene

*IL, interleukins (e.g. IL-1), a family of cytokines discovered by their mediation of interactions between leukocytes

int-2, murine mammary tumor virus integration site oncogene; synonym for *Fgf-3*

Islet-1, a homeobox gene expressed during mouse tooth development

K

KB cells, human laryngeal carcinoma cell line

KC cells, a human tumour cell line derived from keratinocytes

kDa, kilo daltons

Knox-1, *knotted-like homeobox gene-1*

Krox-1, gene encoding a zinc-finger transcription factor

L

LACA, L-azetidine-2-carboxylic acid, a proline analogue that disrupt collagen synthesis

LAG, lines of arrested growth (in bone)

LDH, lactate (lactic acid) dehydrogenase, an enzyme that catalyzes the interconversion of pyruvate and lactate

LDL, low-density lipoproteins

*Lef, lymphoid enhancer-binding factor gene, a family of transcription factor genes (e.g. *Lef-1*) for members of the Wnt gene family

lof, *long fin*, a zebrafish mutant

Lp, the *Loop-tail* mutant mouse, in which hindbrain and spinal cord fail to close and so neural crest cells fail to migrate

LM, light microscopy

LPM, lateral plate mesoderm

M

M1, M2, the first and second molar teeth of mammals

MAP kinase, mitogen-activated protein kinase, an enzyme involved in the phosphorylation of target molecules such as transcription factors and other kinases in the cytoplasm and nucleus

MDH, malate dehydrogenase, an enzyme that catalyzes the conversion of (S)-malate and NAD+ to oxaloacetate and NADH

Meis-1, Meis-2, hox-DNA-binding co-factor genes

*Mfh, *mesenchyme fork head* transcription factor genes (e.g. *Mfh-1*)

Mhox, older name for the gene *Prx-1*

*MMP, a family of matrix metalloproteinases (e.g. MMP-20), of which collagenases are major members

MW, molecular weight

Mrf-4, a gene in the *MyoD* gene family encoding a muscle-specific basic helix-loop-helix transcription factor. Also known as *Myf-6* (myogenic factor-6 or Herculin)

Mrf-5, a gene in the *MyoD* gene family encoding a muscle-specific basic helix-loop-helix transcription factor

Ms-1, *myocyte stress-1*, a stress-responsive gene involved in response of muscle cells to pressure

Msh, the gene for melanocyte stimulating hormone in *Drosophila*

*Msx, homeobox genes (e.g. *Msx-1*, *Msx-2*) of vertebrates; orthologue of the *Drosophila* msh (muscle segment homeobox) gene family

Mtsh-1 mouse orthologue of *Drosophila teashirt* (*tsh*) gene

mya, million years ago

Myf-6, *see Mrf-4*

MyoD, the *myogenic differentiation* gene for a muscle-determining transcription factor

N

N-2, the receptor for the *Dl-1* gene

NAD, nicotinamide adenine dinucleotide

NADH, reduced form of nicotinamide adenine dinucleotide

NaF, sodium fluoride

N-CAM, neural cell adhesion molecule

NCC, neural crest cells

NDST-2, N-deacetylase N-sulfotransferase-2; a heparan sulphate modifying enzyme

NEM, nutrient-enriched medium, used in cell culture

Nkx3.2, *see Bapx-1*

NLK, neuroleukin (also known as phosphoglucose isomerase)

Nm, the *nanomelia* chick mutant resulting from marked depression of chondroitin sulphate synthesis

O

1,25 (OH)$_2$-D$_3$, 1,25 dihydroxyvitamin D$_3$

OAF, osteoclast activating factor

OB1, a murine osteoblastic cell line

OB 7.3, a monoclonal antibody against chicken (*Gallus domesticus*) osteocytes

obk, the *obake* mutation in *Drosophila*, which results in the formation of multiple antenna morphogenetic fields

OIP, osteogenic inhibitory protein

op, symbol for *osteopetrotic* mutant mice

OP-1, osteogenin protein-1, a synonym for Bmp-7

OPCs, osteoprogenitor cells

osf-2, synonym for *cbfa-1*

*Otx, *orthodenticle* family of genes in vertebrates, e.g. *otx-1*

ozd, *oligozeugodactyly*, a mutant chick with defective limbs

P

p53, a protein involved in transcription of DNA and in cell proliferation

P-107, P-130, proteins in the retinoblastoma family of *pRB* genes involved in regulation of the cell cycle

PAS, periodic acid Schiff reagent; used in histology especially to visualize glycogen

*Pax a family of nine mammalian genes containing a paired-type homeodomain as a DNA-binding motif; e.g. *Pax-1, Pax-9*

P-D, proximo-distal axis/polarity of organ rudiments (e.g. limb bud) or organs (e.g. limbs)

PDL, periodontal ligament

*Pdgf, Pdgf, platelet-derived growth factor genes and gene products, e.g. *Pdgf-1*, Pdgf-1

PEMF, pulsed electromagnetic field

*PGE, prostaglandins, e.g. PGE-1, PGE-2

PG-Lb, proteoglycan-LB (also known as epiphycan), a small chondroitin/dermatan sulphate proteoglycan expressed in epiphyseal cartilage

*Pitx, pituitary homeobox gene in mouse (e.g. *Pitx-1*), related to *bicoid* in *Drosophila*. Also known as *Ptx*

PKA, a cyclic AMP-dependent protein kinase

PKC, protein kinase-C, an 80 kDa enzyme activated as a result of membrane lipid hydrolysis of lipids and involved in the regulation of N-cadherin

PNA, peanut agglutinin lectin; visualizes skeletogenic cells at the condensation stage

PNZ, posterior necrotic zone of limb buds; a zone of apoptosis

pRB, see P-107

PRE, pigmented retinal epithelium

*Prx-1, a family of paired *Aristaless*-like homeobox genes, e.g. *Prx-1*, the older name for which was *Mhox*

Ptc-1, patched one, a binding protein for hedgehog gene products

PTH, parathyroid hormone

PTHrP, parathyroid hormone related protein

*Ptx, *see Pitx*

Q

QJ, quadratojugal, a bone of the upper jaw

QTL, quantitative trait loci; a region of a chromosome that carries many genes and influences quantitative characters such as shape and size

R

r-1 to r-7, rhombomeres 1–7 in the vertebrate hindbrain

RA, retinoic acid, a biologically active form of vitamin A

RA-2, retinaldehyde dehydrogenase-2 gene

*RAR, retinoic acid receptor family, e.g. RAR-γ

RAW 264.7, a cell line derived from ascites fluid of leukemic mice and related to the monocytes-macrophage lineage

RCJ3.1, a cell line derived from foetal rat calvariae

*Runx-2, *runt-related transcriptional factor-2* gene, a transcriptional activator of osteoblast differentiation; other names are *cbfa-1* and *osf-2*

S

S1, sacral vertebra number one

S252W, a mutation of the human *FGFR-2* gene resulting in Apert syndrome

SAOS-2, human osteosarcoma-derived cell line

SEM, scanning electron microscopy

SF/HGF, scatter factor/hepatocyte growth factor, a large polypeptide involved in patterning limb buds

SGBD, Simpson-Golabi-Benmel dysmorphia syndrome in humans

Shh, *Sonic hedgehog* gene

Six-2, a gene that produces a homeodomain protein containing the Six domain; orthologue of *sine oculis* in *Drosophlia*

*Slit, a family of genes that encode large glycoproteins found in cell membranes or extracellular matrix

smc, *Spondylometaphyseal chondrodysplasia*, a mouse mutant resulting from a mutation in the gene for collagen type X, in which columns fail to form in the growth plates of long bones

sof, *short fin*, a zebrafish mutant

*Sox, a multi-gene family (e.g. *Sox-9*) that encode proteins with high mobility group DNA-binding domains

S phase, the phase of the cell cycle during which DNA is synthesized

Spp-1, the gene for secreted phosphoprotein-1; also known as bone sialoprotein and osteopontin

Stm, *Stumpy*, a chondrodystrophic mouse mutant

Suc, the gene *sucker* in zebrafish, *Danio rerio*, which disrupts Endothelin-1. Also known as *Endothelin-1, Edn-1*

SVL, snout-vent length, a standard measure of length of tetrapods

T

T3, T4, triiodothyronine (T3) and thyroxine (T4); two thyroid hormones, T3 being the active hormone

T6, T13 the sixth and thirteenth thoracic vertebrae

*ta, the symbol for the *Talpid* gene family in domestic fowl, e.g. *talpid², talpid³*

*T-box, a family of genes encoding transcription factors; e.g. *Brachyury, Tbx-5*

T/btm, symbol for a tailless mouse mutant

*tbx, a class of genes within the T-box family of transcription factors, e.g. *Tbx-6*

$T_c,$ total cell cycle time from one cell division to the next

Tcf-1, *transcription factor-1,* a transcription factor in the *Wnt* pathway

Tcof-1, the gene for the nuclear protein Treacle; when mutated elicits Treacher Collins syndrome

Te 85, a cell line derived from a human osteosarcoma

TEM, transmission electron microscopy

*Tgfβ, transforming growth factor beta, e.g. Tgfβ-1

*TgfβR transforming growth factor beta receptor

tl, symbol for *toothless* osteopetrotic mutant mice

TMJ, temporomandibular joint of mammals connecting the lower jaw to the skull

TNC, trunk neural crest or trunk neural crest cells

TNF-α, tumour necrosis factor-alpha

TRα-1, thyroid hormone receptor alpha-1

TRAP, tartrate-resistant acid phosphatase, an enzyme produced by osteoclasts and their precursors and involved in resorption of bone

TRβ-1, thyroid hormone receptor beta-1

TUNEL, terminal deoxynucleotidyl transferase Biotin-dUTP nick end labeling, used to detect broken strands of DNA cleaved by Ca++- and Mg++-dependent endonucleases, and indicative of apoptosis or programmed cell death

U

UDP, uridine 5′-diphosphate

UDPG, uridine 5′-diphosphoglucose

UDP-NaGal, uridine diphosphate-N-acetylgalactosamine

UMR-106, a cell line derived from rat osteosarcoma

Uncx4.1, the gene for a paired-type homeodomain transcription factor

V

V-ATPase, vacuolar adenosine triphosphatase; protein complexes of more than 700 kDa that act as proton pumps and acidify membrane-bound compartments within eukaryotic cells, especially osteoclasts and their precursors

Vegf, Vegf, vascular epithelial growth factor gene and protein

VER, ventral ectodermal (epithelial) ridge on developing tail buds

Vgo, the *van Gogh* mutant in zebrafish, *Danio rerio,* caused by a mutation in *tbx-1.* The pharyngeal region fails to develop

Vgr-1, a synonym for *Bmp-6*

Vit-C, vitamin C

v-myc, a proto-oncogene from retrovirus-associated DNA sequences originally isolated from an avian myelocytomatosis virus

VNT, ventral neural tube

vt, the *vestigial tail* tailless mouse mutant

W

W-20-17, mouse bone-marrow stromal cell line

wl, a *wingless* chick mutant

ws, a *wingless* chick mutant

*Wnt, a large gene family orthologous to *wingless* in *Drosophila* that produces secreted molecules involved in intercellular signaling. *Wnt* is a combination of *Drosophila wingless* and *int-1* for *Wnt-1,* the first vertebrate (mouse) family member discovered

X

Xt, the *extra toes* mouse mutant

Z

Zfgf, zebrafish (*Danio rerio*) fibroblast growth factor gene

ZfRAR-γ, zebrafish retinoic acid receptor-gamma

ZPA, zone of polarizing activity at base of limb buds

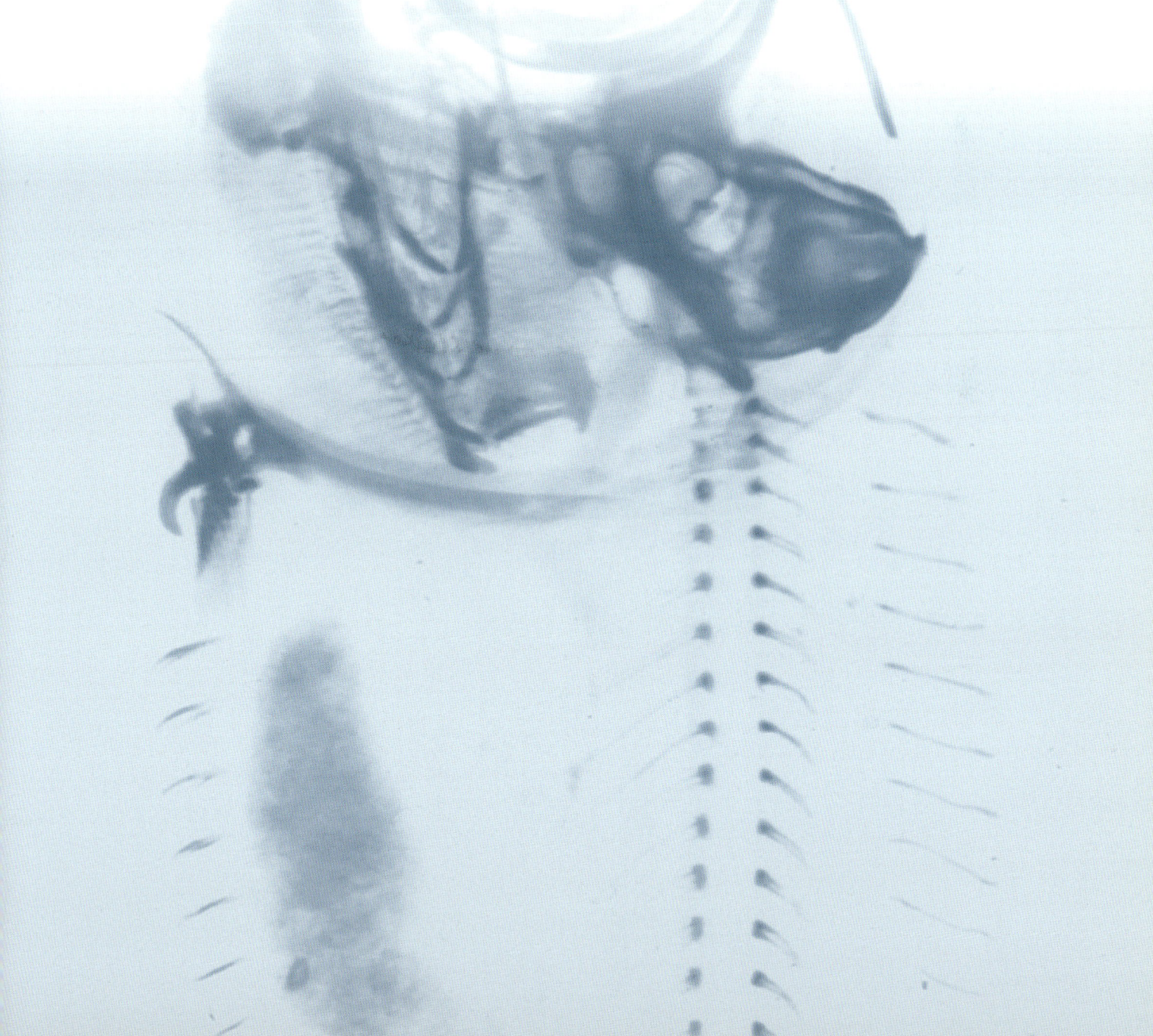

Part I

Skeletal Tissues

'When *Tess* made me too weepy, I turned to the timeless serenity of the frontal bones'

(Bellairs, 1989, p. 93).

Acellular bone links evolutionary and developmental studies on the one hand, and normality and pathology on the other.

Types of Skeletal Tissues

'In the pioneering stages of Natural Science we recognize the work of collecting, describing and classifying the typical units as a fundamental necessity. The study of their morphology and their history belongs to a more advanced period.'[1]

Skeletal tissues are ancient, their origins reaching back perhaps three-quarters of a billion years. The number of skeletal tissues or organs is far more limited. In 2000, Thomas and colleagues evaluated 182 characters of skeletal design as possible design options in morphospace, a 3D representation of the distribution of all known morphologies. Of these 182 characters, 146 were already in use in animals of the Burgess Shale fauna 530 mya. Indeed, within 15 million years of the appearance of the crown groups of the major phyla, 80 per cent of the design elements were already in use.[2]

Four classes of mineralized tissues are found in vertebrates. The four are *bone, cartilage, dentine* and *enamel*. This may seem an unlikely list. Normally, we think of cartilage and bone as *skeletal tissues*, enamel and dentine as *dental tissues*. But enamel and dentine arose evolutionarily as skeletal tissues in the *exoskeleton* (dermal, dermoskeleton) of early vertebrates (Fig. 1.1, Table 1.1). Bone is also a primitive tissue of that exoskeleton. Cartilage, on the other hand, provided the basis for the second vertebrate skeletal system, the *endoskeleton* (Table 1.1), and has an even more diverse distribution and, potentially, a longer evolutionary history than bone.[3] This is because cartilage and cartilage-like tissues form endoskeletal elements in many *invertebrates*. Although most invertebrates have non-cartilaginous endoskeletons the diversity of taxa with cartilage is astonishing. Where there is an exoskeleton or cuticle in invertebrates – and there almost always is – it is composed of chitin (sometimes with glycoproteins, sometimes mineralized) or calcium carbonate,

but not calcium phosphate, which is the major component of bone.[4] Neither bone nor mineralized cartilage is ever found in invertebrates, although some invertebrate cartilages have surprisingly 'bone-like' features (Chapter 4).

Mineralization is not the exclusive property of the four vertebrate mineralized tissues. Mineralization is ubiquitous within metazoans as well as being a property of many single-celled organisms. In Table 1.2, I summarize the range and diversity of mineralized biological tissues in various groups and the major organic component(s) associated with mineralization in each. Mineralization can place enormous demands on an organism, for example when deer regrow their annual set of antlers (Chapter 8) or hens lay eggs (Chapter 25).

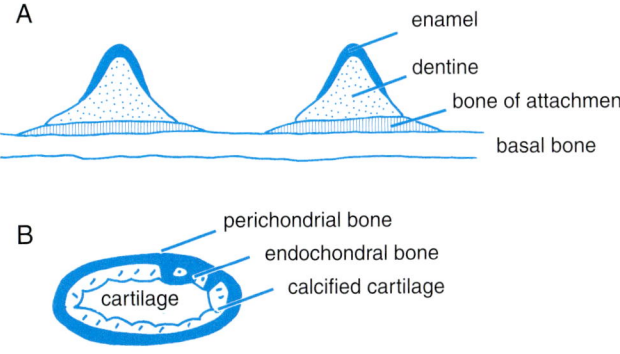

Figure 1.1 Diagrammatic representations of the tissues that comprise (A) the dermal (exo) and (B) the endoskeletons of vertebrates. The major difference is that the dermal skeleton is based on dentine and associated bone, while the endoskeleton is based on cartilage. (A) Enamel caps the dentine in the dermal skeleton. Individual exoskeletal units (odontodes), which produce dentine and bone of attachment, fuse to adjacent basal bone. (B) The cartilage of the endoskeleton may (i) remain unmineralized, (ii) mineralize and remain as a permanent mineralized (calcified) cartilage, (iii) be surrounded by perichondral bone, or (iv) be invaded and replaced by endochondral bone.

Table 1.1 Definitions of terms for skeletal systems and modes of ossification[a]

Structure	Definition	Mode of ossification
Exoskeleton	Skeletal system that forms in contact with the ectoderm or endoderm; dermal bone, scales, fin rays, gill rakers, teeth	Intramembranous ossification
Endoskeleton	Skeletal system that does not form in contact with ectoderm or endoderm; large or chondral bone	All ossification types
Chondral bone	Type of bone that develops in a preformed cartilage model; the majority of the endoskeleton	Endochondral ossification
Perichondrial bone	Subtype of chondral bone that ossifies from the perichondral connective tissue	Perichondrial ossification
Membrane bone	Type of bone that develops without contact with the ectoderm or endoderm, but which is not preformed in cartilage	Intramembranous ossification
Subdermal bone	Neoformations that develop without contact with the ectoderm or endoderm; sesamoid bones, pathological ossifications	Intramembranous ossification
Perichordal bone	Type of bone that forms in the connective tissue sheath surrounding the notochord	Perichordal ossification
Dermal bone	Type of bone that is not preformed in cartilage and that develops in contact with ectoderm or endoderm; the majority of the exoskeleton	Intramembranous ossification

[a] Based on Hilton and William (1999) to which I have added modes of ossification. Although these terms were outlined in the context of a comparative analysis of fish skeletons and skeletal development they apply across the vertebrates.

These four broad classes of vertebrate skeletal tissues may be further subdivided in ways that reflect the interests of individual skeletal biologists and the scope of their fields. Embryologists subdivide skeletal tissues by developmental process, anatomists by structure, pathologists on the basis of deviation from the norm, and so on.

For vertebrate skeletons, I add a fifth 'catch-all' category to encompass those tissues that, on the basis of one or more criteria, are *intermediate* between two of the four mineralized tissues.[5] Examples are tissues intermediate between bone and cartilage (chondroid, chondroid bone), bone and dentine (osteodentine, cementum), enamel and dentine (enameloid). The usefulness of this category for understanding the dynamic development of skeletal tissues will become apparent as we proceed. But first, I introduce the players.

BONE

Bone is a vascularized, supporting skeletal tissue – although it may also arise ectopically[6] outside the skeleton – consisting of cells and a mineralized extracellular matrix (ECM). Bone is deposited by bone-forming cells (*osteoblasts*) and by *osteocytes*, some of which are ciliated (Box 1.1). Osteoblasts cease dividing when they transform into osteocytes. Bone is modeled, remodeled and/or removed by mono- or multinucleated osteoclasts (and sometimes by osteocytes). Type I collagen, composed of two αI chains and one αII chain (products of the Col1a1 and Col2a1 genes, respectively) and depicted as $\alpha1$ $(II)_3$, is the major extracellular matrix component. Osteocalcin (bone gla protein), osteopontin and

Table 1.2 The diversity of mineralization of biological tissues[a]

Group	Mineralized tissues	Chemical formula	Major organic components
Plant	Cell wall	$CaCO_3$	Cellulose, pectin, lignin
Diatoms	Exoskeleton	Si	Pectin
Radiolaria	Exoskeleton	$SrSO_4$	–
Coelenterates, sponges, and molluscs	Exoskeleton	$CaCO_3$	Proteins, conchiolin
Arthropoda	Exoskeleton	$CaCO_3$	Chitin, proteins
Vertebrates	Exo- and endoskeleton	$Ca_{10}(PO_4)_6OH_2$, $CaCO_3$	Collagen, mucopolysaccharides

[a] Based on data from Dacke (1979).

osteonectin are the major non-collagenous proteins (Box 1.2). Bone matrix is permeated by canals (*canaliculi*) which contain osteocyte processes, and which connect to other osteocytes, to osteoblasts and to osteogenic cells on the surface via gap junctions, forming a syncytium. The fluid content of bone is low. Embryologically, bone arises from mesoderm and from neural crest, two of the four germ layers of vertebrate embryos.

The first bone matrix deposited is unmineralized and known as *osteoid*. Subsequently, osteoid is impregnated with hydroxyapatite to form bone, the mineralized tissue. An organic layer – the *lamina limitans* – separates osteoid from already mineralized bone (Scherft, 1978).

Bone is an aerobic tissue with high oxygen consumption. Bone functions to support the body, protect major organs such as the brain and spinal cord, and as a site of attachment for ligaments and muscles. Bone is also a

Box 1.1 Cilia and skeletal cells

Chondroblasts/cytes and osteoblasts/cytes are embedded within matrices of varying degrees of fluidity, viscosity or solidity. Cilia are organelles that project from cells or single-celled organisms. Projecting into the external environment or into a lumen such as the gut, cilia function to move currents or particles (including food) or to move the entire organism. The notion of chondrogenic or osteogenic cells possessing cilia is the histologist's version of an oxymoron. Nevertheless, cilia have been described on cartilage and bone cells.

Cells that clearly are *osteocytes* in rat and chick calvariae possess cilia, usually one per cell, one speculation being that cilia function to move fluids through the canaliculi (Fig. 1.2). Given that osteocytes reside within lacunae (Figs 1.3 and 1.4) this is not an implausible function.

Cells that clearly are *chondrocytes* in mouse radii, and rat and canine articular chondrocytes, also each possesses a single cilium (Fig. 1.5).[a]

[a] See Federman and Nichols (1974) and Tenenbaum *et al.* (1986) for cilia on osteocytes, and Scherft and Daems (1967), Wilsman and Fletcher (1978) and Vidinov and Vasilev (1985) for cilia on chondrocytes. See the chapters in *Volume 1: The Osteoblast and Osteoclast* of Hall (1990–1994) for overviews of the basic structure and function of osteoblasts and osteocytes.

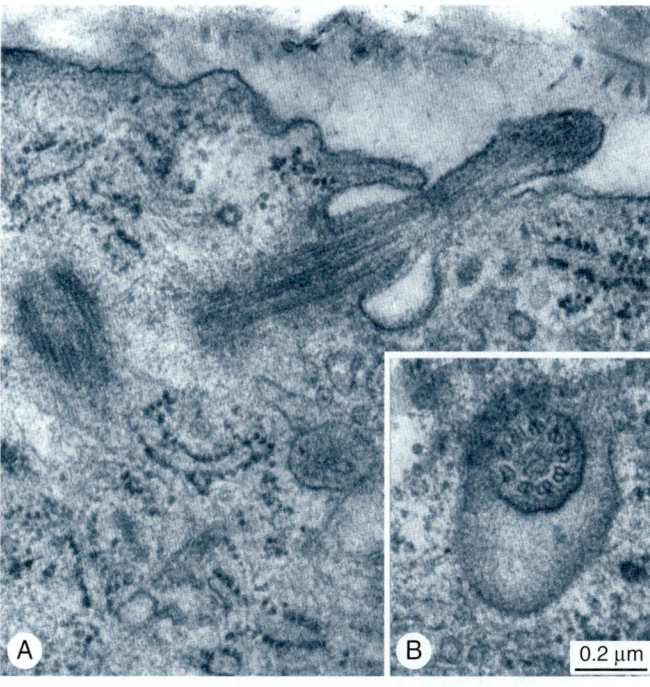

Figure 1.2 Cilia and osteocytes. Occasionally, osteocytes display cilia. (A) A transmission electron micrograph of a cilium projecting from a rat osteocyte. (B) A cross section of the basal body shows the typical 9 + 0 fibril arrangement of cilia. Modified from Federman and Nichols (1974).

storehouse for calcium and phosphorus, and a major site for the metabolic regulation of mineral homeostasis. Bone houses the haematopoietic tissues of adult mammals.

Bone is found only in vertebrates. Extant jawless vertebrates (lampreys), craniates (hagfishes) and all invertebrates lack bone. Conodonts possess bone; their phylogenetic status is not resolved to everyone's satisfaction, however.

CARTILAGE

Cartilage is an avascular, supporting and articular skeletal tissue consisting of cells in an extracellular matrix (ECM), which may or may not mineralize depending on cartilage type. Like bone, cartilage can arise ectopically outside the skeleton, for example, in connective tissue, muscle and the heart. Cartilage is deposited by cartilage-forming cells (*chondroblasts, chondrocytes*; Fig. 1.7) and removed by mono- and multinucleated chondroclasts. Cartilage cells are separated from one another by pericellular and extracellular matrices. Unlike bone cells, chondrocytes lack connecting cell processes. Some chondrocytes are ciliated (see Box 1.1, Fig. 1.5). Most chondrocytes continue to divide throughout life, although in some cartilages (mammalian articular cartilages, for example), the number of dividing cells may be less than one per cent of the chondrocyte population.

The hydrated ECM of vertebrate cartilage is primarily composed of glycosaminoglycans (GAGs), notably chondroitin sulphates and proteoglycans. The major collagen is type II, composed of three αII chains, depicted as $\alpha 1(I)_2\alpha II$. Some types of vertebrate cartilages contain additional collagens; for example, type I in articular, fibro- and secondary cartilages, and type X in hypertrophic cartilage (Fig. 1.8).

Lamprey and hagfish cartilages lack collagen type II and have different classes of fibrous proteins (lamprins, myxin) in place of collagen. Cartilages of living agnathans do not mineralize *in vivo*. Within invertebrates, cartilaginous extracellular matrix is composed of GAGs and a modified form of type I collagen. No invertebrate cartilages mineralize. Vertebrate cartilages can arise from mesodermal and from neural crest-derived mesenchyme. The origins of invertebrate cartilages are less well known, although some may be epithelial.

Cartilage resists pressure, has a high fluid content, is anaerobic as a tissue, and so has low oxygen consumption. Cartilage functions as the embryonic endoskeletal tissue in vertebrate embryos (primary cartilage) and as an endoskeleton in many invertebrates. In vertebrates, remnants of primary cartilage transform to articular cartilage and function as the articular tissue at joints in endochondral bones (Figs 1.8 and 1.9; see Fig. 3.11), while secondary cartilage forms on many membrane bones in birds and mammals.

DENTINE

Dentine, a tubular, mineralized, dental and skeletal tissue, comprises the bulk of true teeth, true teeth being those composed of enamel and dentine, in contrast to keratinized structures that function as 'teeth' in, for example, anuran amphibian tadpoles. Dentine is the primary tissue of the vertebrate exoskeleton, and thus is both an odontogenic (tooth-forming) *and* a skeletogenic (skeleton-forming)

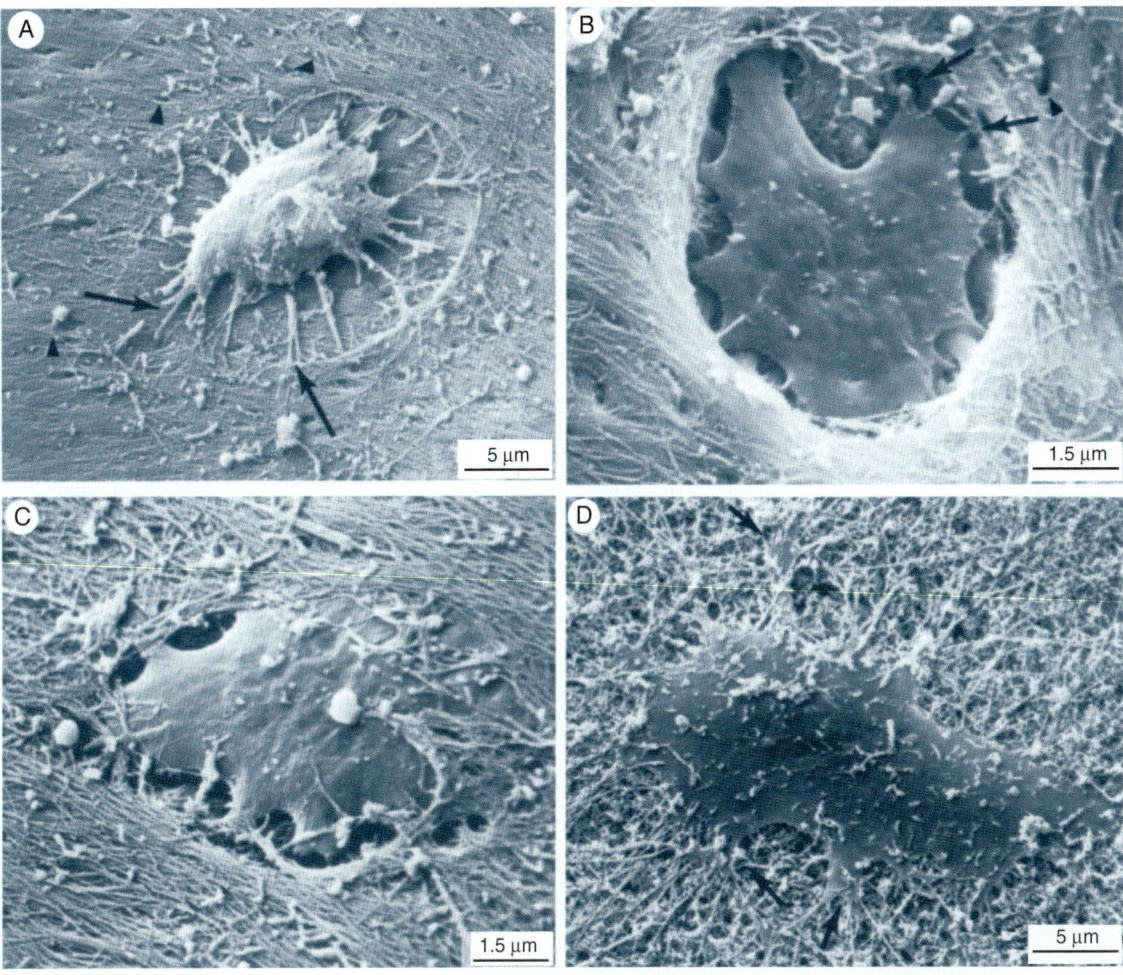

Figure 1.3 Osteocyte formation, as seen on the endosteal surface of a rat femur visualized with scanning electron microscopy. (A) Surface osteoblasts in the earliest stage of lacuna formation. Note the extensive and branched cell processes (arrows). (B, C) Further development of the lacuna with collagen fibrils in the lacunar wall and cell processes entering the pores (future canaliculi; arrows in B) in the wall of the lacuna. (D) An osteoblast depositing and being incorporated into osteoid. Modified from Menton *et al.* (1984).

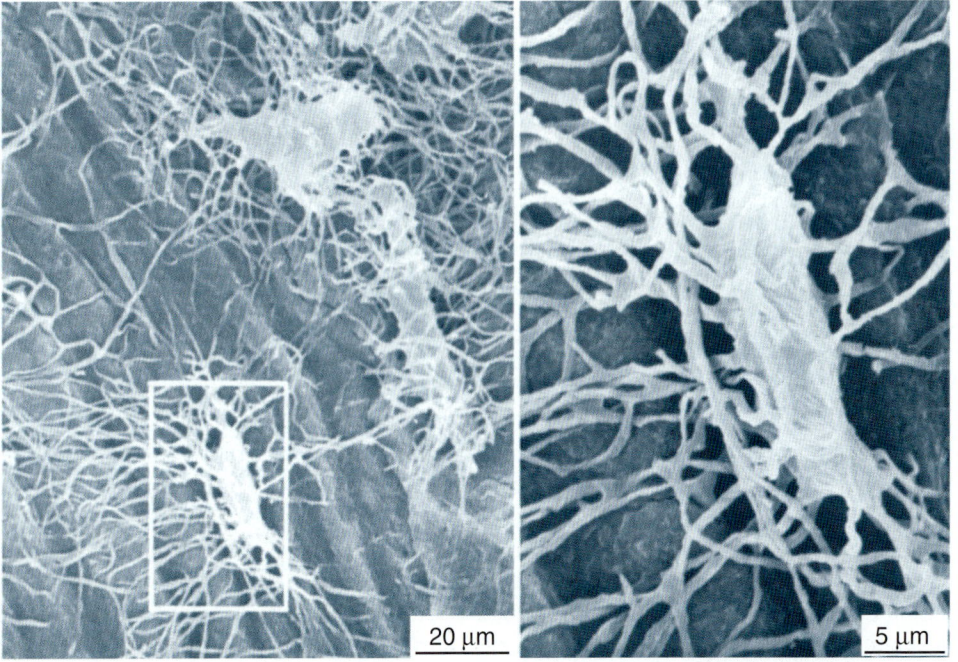

Figure 1.4 Scanning electron micrographs of osteocytes from the dentary of a 70-year-old human. These methacrylate replicas show the lacunae in which the osteocytes resided and the canaliculi through which osteocyte cell processes ran. The osteocyte in the inset on the left is enlarged on the right and shown on the cover. Modified from Atkinson and Hallsworth (1983).

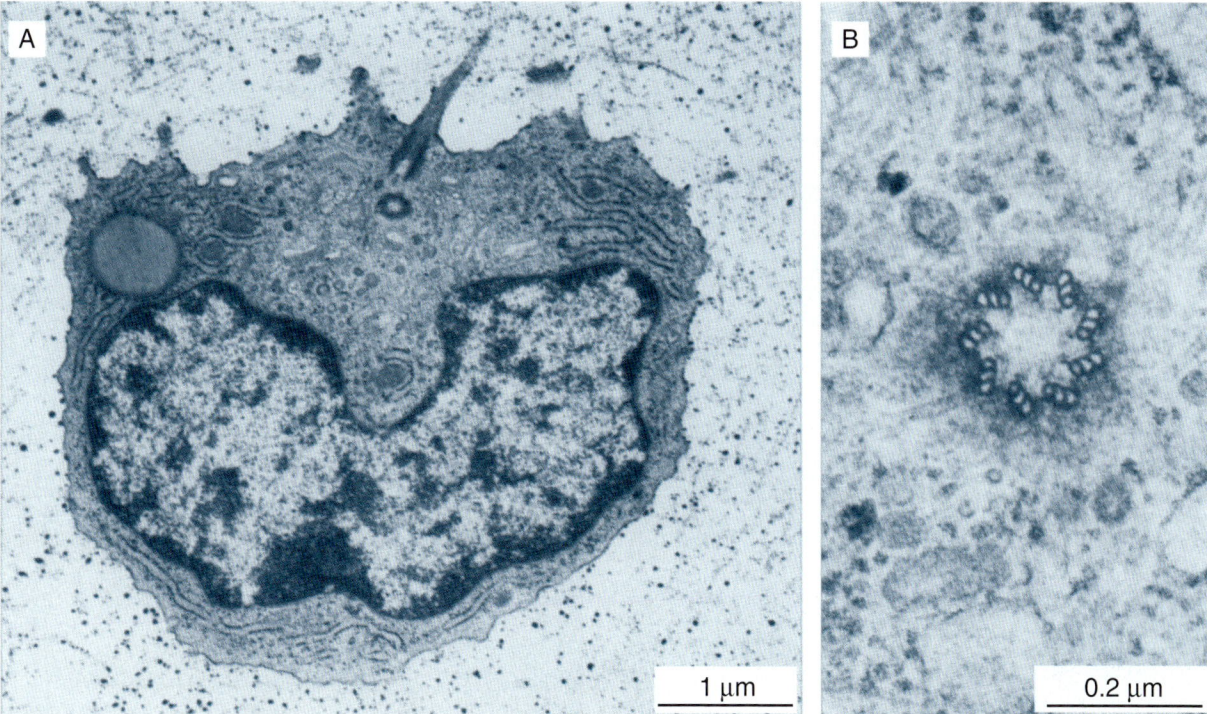

Figure 1.5 Cilia in cartilage. A transmission electron micrograph of an articular chondrocyte from the humerus of a two-day-old Labrador pup, with cilium and basal body attached to a centriole (A) and sets of microtubule triplets in the basal body (B). Modified from Wilsman and Fletcher (1978).

Box 1.2 Major non-collagenous proteins of bone matrix

Osteocalcin

Osteocalcin (bone γ-carboxyglutamic acid, Gla; bone Gla protein, BGP) is a 5800 MW, vitamin K-dependent, γ-carboxyglutamic acid-containing, Ca^{++}-binding protein. The seventh most abundant protein in human bone, osteocalcin is the most abundant non-collagenous protein in bone, accounting for 10–20 per cent of the non-collagenous protein. Levels in human serum are 7.0 ± 2.5 ng/ml. Osteocalcin recruits osteoclasts or osteoclast precursors to bone for resorption. See Box 24.1 for further details.

Osteopontin

Osteopontin, a 66-kDa glycosylated phosphoprotein synthesized by osteoblasts, osteoclasts and macrophages, is found at active sites of bone metabolism: endochondral and membrane bone, osteoid, preosteoblasts, osteoblasts and osteocytes (Fig. 1.6). Osteopontin enhances cell survival and migration but inhibits mineralization. Positive and negative regulation of transcription, and regulation of osteocalcin and osteopontin protein levels have been described.[a] See Box 24.2 for further details.

Osteonectin

Osteonectin (or SPARC; secreted protein, acidic, cysteine-rich), a 32 000 MW extracellular matrix protein comprises some ten per cent of the protein in bone. Appearing with mineralization (Fig. 1.6), osteonectin links collagen to hydroxyapatite, serves as a nucleus for mineralization, and regulates the formation and growth of hydroxyapatite crystals. See Box 24.3 for further details.

[a] See T. A. Owens *et al.* (1991), Ayad *et al.* (1994) and Yagami *et al.* (1999) for osteocalcin and osteopontin.

tissue. Seventy-five per cent of the mass of dentine is inorganic, 20 per cent organic (cf. enamel below), and five per cent liquid.

Dentine is produced by dentine-forming cells known as *odontoblasts*. Odontoblasts (and therefore dentine) are always derived from neural crest cells (Box 1.3). Dentine, like enamel, is produced in two sequential phases. Synthesis and deposition of *predentine* – an organic matrix composed of GAGs and type I collagen – comprises the first phase. Mineralization of predentine by hydroxyapatite to form *dentine* is the second phase. Odontoblasts interact with enamel-forming cells at all stages of mammalian tooth formation. Dentine is resorbed by odontoclasts. Dentine is not found in invertebrates.

ENAMEL

Enamel is the highly mineralized, very hard, prismatic, avascular outer layer of vertebrate teeth and of some scales. Like dentine, enamel is an odontogenic and skeletogenic tissue of the exoskeleton. Enamel is much more resistant to wear than dentine, some 96 per cent of mammalian enamel being inorganic, and only 0.5 per cent organic. Enamel is produced by enamel-forming cells (*ameloblasts*), which are derived from ectodermal or endodermal epithelia (Box 1.3). The epithelium specifies the shapes of mammalian teeth into broad classes, such as incisiform or molariform, and participates in the initiation

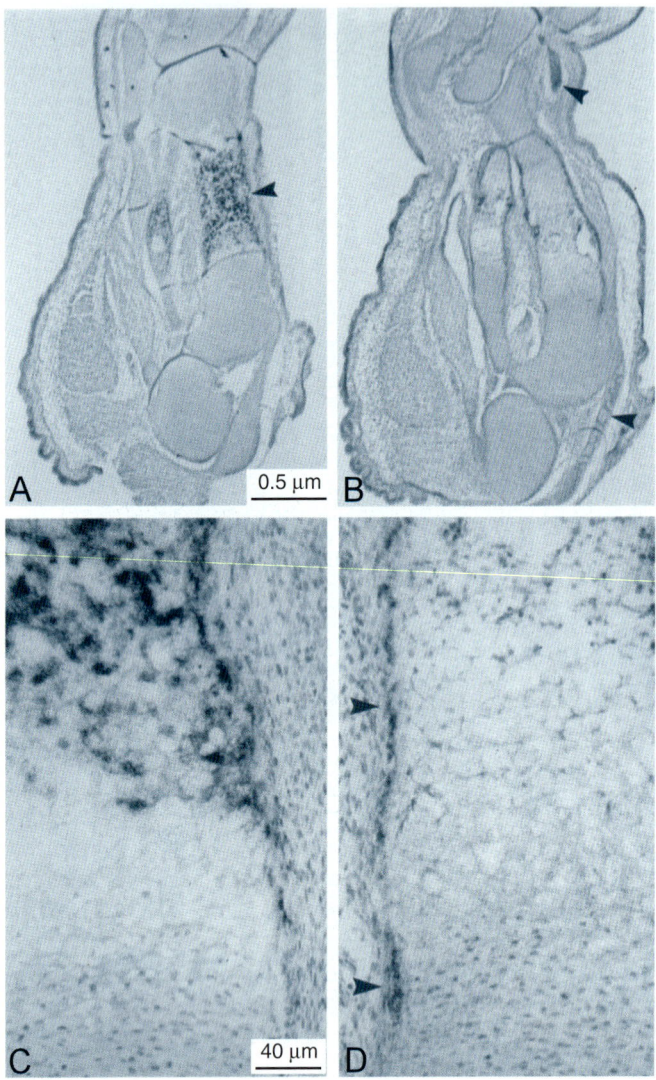

Figure 1.6 Patterns of expression of osteopontin and osteonectin during murine limb development. (A) A longitudinal section of the forelimb from a 15.5-day-old embryo showing osteopontin (black) within the epiphyseal cartilage (arrowhead). (B) A section near that of A showing osteonectin transcripts (black) in the periosteum, subperiosteal bone, and in a developing tendon (arrowhead). (C) A higher magnification of part of A showing osteopontin transcripts (black), where endochondral ossification has been initiated, and absence of expression in the other chondrocyte zones. (D) A higher magnification of part of B showing osteonectin transcripts (black) in the periosteum (arrowheads). Modified from Nomura *et al.* (1988).

of dentinogenesis through inductive interactions with associated ectomesenchymal cells. As with the formation of all mineralized tissues in vertebrates, enamel formation (*amelogenesis*) consists of two phases: deposition of organic matrix and mineralization of that matrix. Enamel is not found in invertebrates.[7]

INTERMEDIATE TISSUES

Intermediate tissues, my fifth category of skeletal tissues, possess characteristics of two or more of the four skeletal

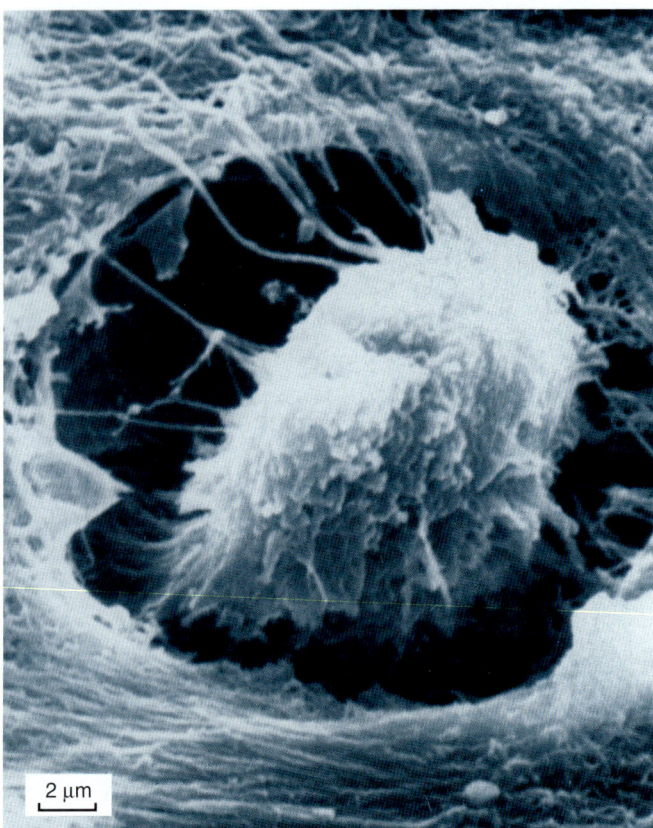

Figure 1.7 A scanning electron micrograph of a chondrocyte in the articular cartilage of a human patella. The lacuna and cell processes are shown especially well. From Hall (1978a).

tissues. The existence of such permanent tissues suggests that the four skeletal tissues represent a continuum or perhaps continua, and not discrete, bounded categories. The most recognizable intermediate tissues are cementum, enameloid, chondroid and chondroid bone. Some tissues are 'so intermediate' that they cannot be classified as bone, cartilage, dentine, enamel, or as any one of the more readily recognizable intermediate tissues. Some intermediate tissues, but by no means all, are pathological (Chapter 5).

Cementum

Cementum, the supporting tissue that anchors the teeth of many vertebrates into their sockets, has features of dentine and of calcified cartilage, as well as unique features (Fig. 1.11). Cementum is deposited by *cemento-blasts*. Cementoblasts, along with the odontoblasts that deposit dentine, the osteoblasts that deposit supporting (alveolar) bone, and the fibroblasts that deposit pulp and periodontal ligament (Fig. 1.11), all arise from the dental papilla, which has its origin in the neural crest. *Cementum is deposited onto existing dentine.* Consequently, cementum has features of dentine as well as features peculiar to itself.

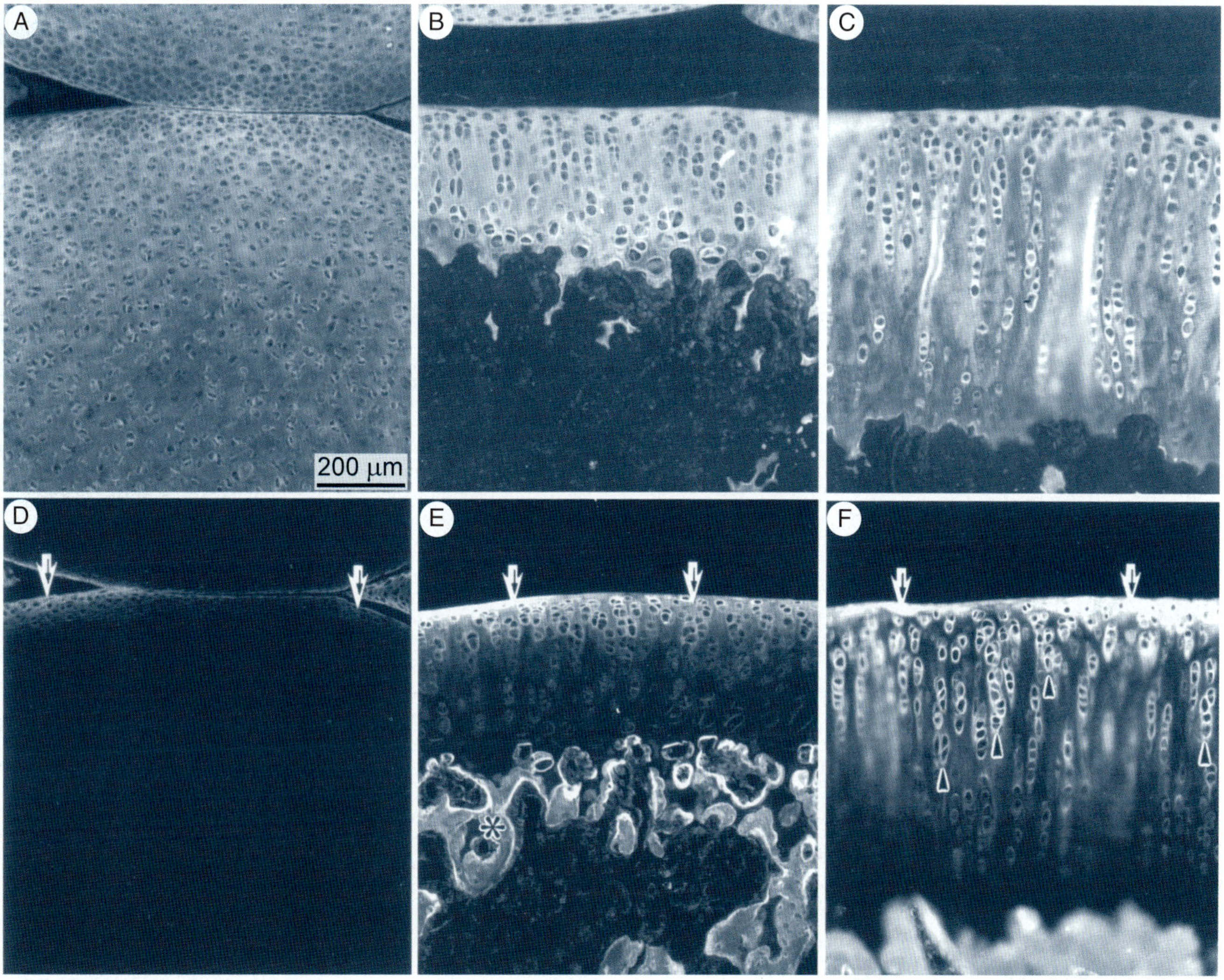

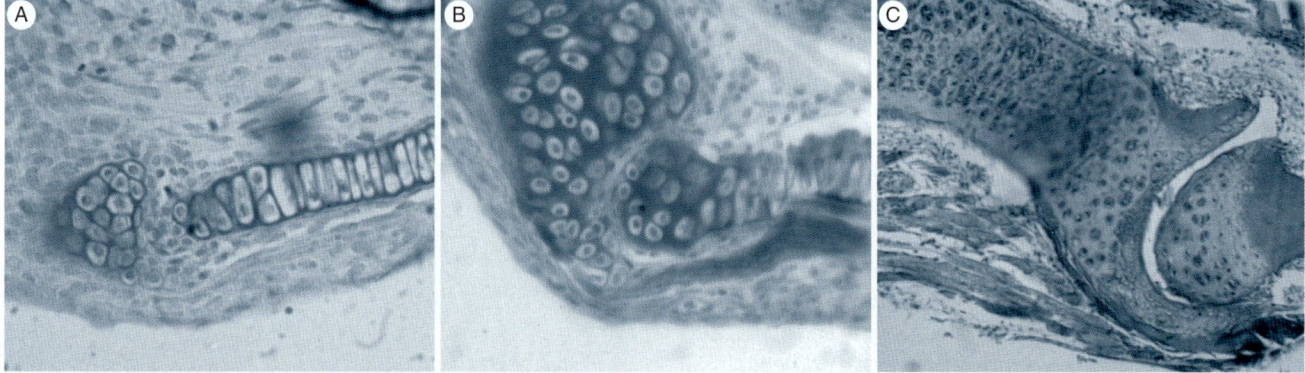

Figure 1.8 Histological sections of articular cartilages from rat tibiae at one (A, D), five (B, E) and 11 (C, F) weeks after birth to show the distribution of type II collagen (A–C) and of type I collagen (D–F) as visualized with antibodies. Type II collagen is distributed throughout the cartilage matrix (A–C). Type I collagen is preferentially expressed in the superficial chondrocytes at the articular surface (D–F, arrows). Adapted from Sasano *et al.* (1996).

Figure 1.9 Development of the mandibular joint in the Japanese medaka, *Oryzias latipes*. Two stages post-fertilization (pf) and an adult are shown as longitudinal histological sections. Meckel's cartilage is the element on the right in each. (A) six days pf; (B) 49 days pf; (C) adult. The position of the joint cavity is evident at six days pf (A) and is well developed by 49 days (B). Photographed by Tom Miyake.

Box 1.3 Ectodermal and endodermal teeth

The neural crest origin of the dentine and pulp of the teeth of the spotted salamander, *Ambystoma punctatum*, was determined in an insightful experimental study by Adams (1924). The same study established that endoderm *and* ectoderm participate in tooth formation and both can and do form enamel. Enamel of the teeth on the palatine and splenial bones of *Ambystoma* is derived from endoderm, enamel of the teeth on the maxillary, vomer and dentary from ectoderm; see Figure 1.10 for teeth on the splenial cartilage of another species of *Ambystoma*, the Mexican axolotl, *A. mexicanum*.

Transplanting mouth ectoderm to the gills prevents tooth formation, indicating that interactions between mouth ectoderm and mesenchyme are required for tooth formation. 'Foreign' (i.e. non-mouth) ectoderm transplanted to the mouth cavity elicits tooth formation, indicating that the ectodermal requirement is not specific. The latter finding is relevant to current efforts to overturn the homology of teeth and dermal denticles – with a consequent 'inside-out' theory of the origin of teeth – in which teeth are argued to have arisen in association with endoderm (inside), not from dermal denticles and an association with ectoderm – the traditional 'outside-in' theory, according to which teeth arose from dermal denticles.

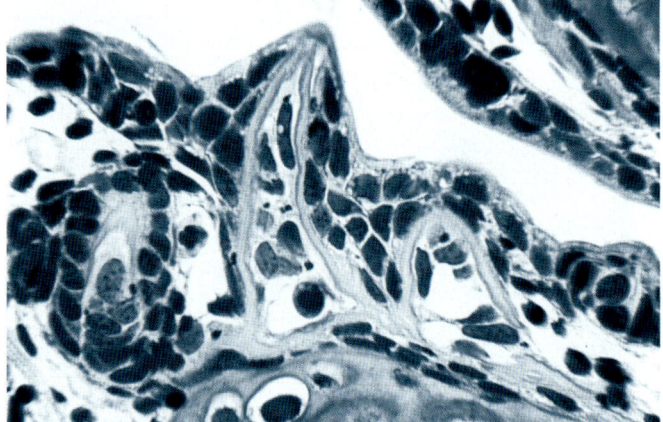

Figure 1.10 A histological section of three teeth and the splenial cartilage from an adult Mexican axolotl, *Ambystoma mexicanum*. Compare with Fig. 18.4, which shows teeth evoked from trunk neural crest. Image provided by Moya Smith.

Three types of mammalian cementum are recognized: crown cementum, and primary root and secondary root cementum. Cementum equivalent to mammalian root cementum is found in alligators and crocodiles, in primitively toothed birds such as *Hesperornis* from the Late Cretaceous, and in various other groups. The cementum on the coronal surface of the molar teeth in guinea pigs, *Cavia porcellus*, has some of the features of cementum, some of cartilage, and some that are unique. Indeed, it has been argued that in all rodents and ruminants, cementum is a form of calcified cartilage.[8]

Enameloid

Enameloid is a mineralized aprismatic tissue that coats the teeth of teleosts and larval urodeles, and is found as a component of the scales of some groups. Enameloid is also variously known as mesoenamel, vitrodentine and durodentine, names that reflect the presence of components typical of enamel and dentine. Enameloid is deposited by the combined and orchestrated action of ameloblasts and odontoblasts. Whether in teeth or in scales, the bulk of the enameloid is deposited by ameloblasts but penetrated by fibres and cellular processes from adjacent dentine. The dual contribution of these two cells types is demonstrated nicely in the ballan wrasse, *Labrus bergylta*, the eel, *Anguilla anguilla*, and the checkered puffer fish, *Sphaeroides testudineus*, in all of which odontoblasts produce and deposit collagen and ameloblasts produce and deposit the other proteins of the enameloid matrix. Uniquely among vertebrate mineralized tissues, collagen is removed from enameloid as mineralization occurs.[9]

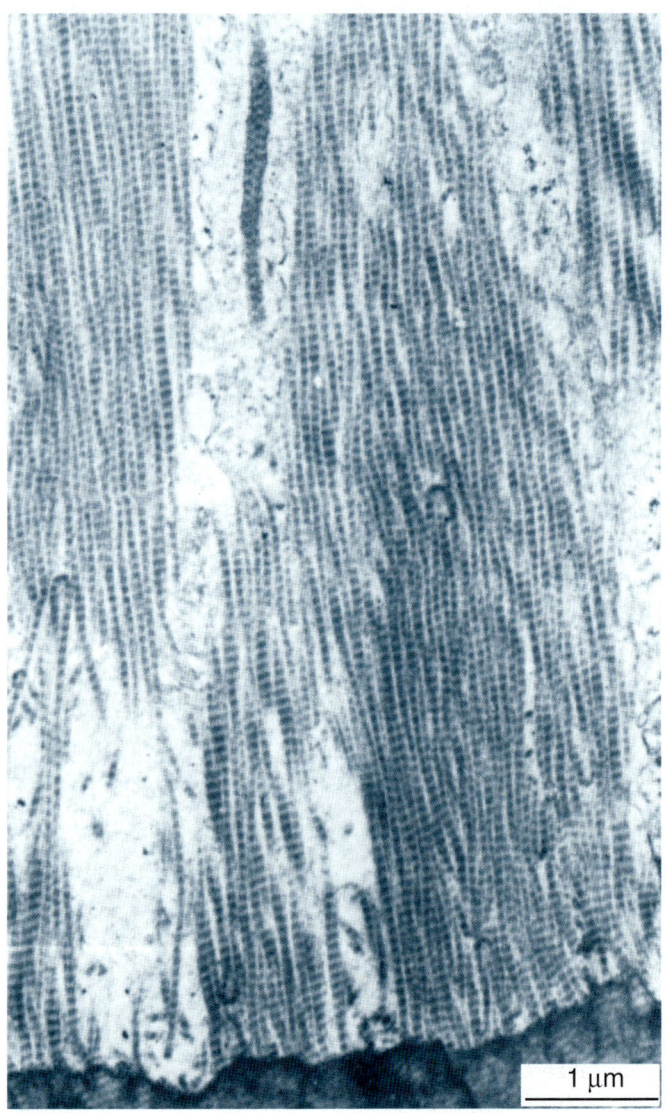

1 μm

Figure 1.11 The collagenous fibres of a murine periodontal ligament insert into the cementum (bottom) of the first molar tooth. From Ten Cate *et al.* (1976).

Chondroid and chondroid bone

Chondroid and chondroid bone possess, to different extents, features of cartilage and bone, and resemble, to different degrees, cartilage or bone in their development. Typically, chondroid and chondroid bone have collagen types I and II in their matrices, although this is not sufficient to identify either tissue; articular and secondary cartilages also possess both collagen types, as demonstrated in the articular cartilage of rat tibiae over the first 11 weeks of postnatal life (see Fig. 1.8).

The matrix of *chondroid bone* is basophilic (i.e. it stains with basic dyes in light microscopical histological analyses), as is bone, but has nests of cells in lacunae resembling chondrocytes. In mammals, especially, chondroid bone is often a permanent skeletal tissue. In this sense, chondroid bone is a stable intermediate skeletal tissue, and not a transitional one. Primary cartilage, being a transitional tissue replaced by bone, is a less permanent part of the skeleton than chondroid bone.

Chondroid is a less permanent tissue. In the antlers of white-tailed deer, *Odocoileus virginianus*, chondroid is replaced by bone in what has been called 'chondroidal bone formation,' a process intermediate between intramembranous and endochondral bone formation (see Chapter 5). In other situations, chondroid is remodeled and transformed to bone. Examples of the formation of chondroid and chondroid bone are discussed in various parts of this text.[10]

The existence of such tissues, whether transitional or permanent (Chapter 5), raises fundamental questions about how we *identify* bone and cartilage, and osteoblasts and chondrocytes; and therefore how we *define* and/or *distinguish* bone and cartilage, and osteoblasts and chondroblasts, in embryonic development and through evolutionary history. Such issues are central to my discussions.

BONE OR CARTILAGE

Historically, the distinction between cartilage and bone has been recognized since at least the time of Aristotle (384–322 BC), who recognized and separated the cartilaginous fishes, or Chondrichthyes, from the bony fishes, or Osteichthyes, by the presence of a cartilaginous or osseous skeleton. Two thousand and fifty years later, the oldest articulated chondrichthyan, the 50–75 cm long *Doliodus problematicus*, was described from the Early Devonian (409 mya) of New Brunswick, Canada. Previously known only from isolated dermal scales and teeth, this specimen includes the braincase, prismatic mineralized cartilage, and paired spines associated with the pectoral fins, previously unknown as a feature of cartilaginous fishes (R. F. Miller *et al.*, 2003).

Until the 17th century, cartilage was thought to transform into bone. Such a transformation *can* occur but, as

we shall see, it is difficult to detect. The Belgian and Danish anatomists Adrian Spigelius (1631) and Thomas Bartholinus (1676) distinguished bone that forms from cartilage – what we now refer to as *endochondral bone* arising from *endochondral ossification* – from bone that forms directly – what we now refer to as *membrane bone* arising by *intramembranous ossification* (Table 1.1). All the bone of an endochondral bone is not the result of endochondral ossification; primary ossification of the cartilaginous rudiment occurs subperiosteally around the shaft (Fig 1.12), but by endochondral replacement of cartilage of the shaft or metaphysis. So, endochondral can be a misnomer. Now we also know that endochondral is too limited a term; bone can replace marrow, tendon or ligament without cartilage being involved. *Replacement bone* and *indirect ossification* may be better terms.[11]

Volcher Coiter, a pupil of the anatomist Gabriele Falloppio at the University of Padua, described the complete human skeleton in 1559. In 1736, Robert Nesbitt set

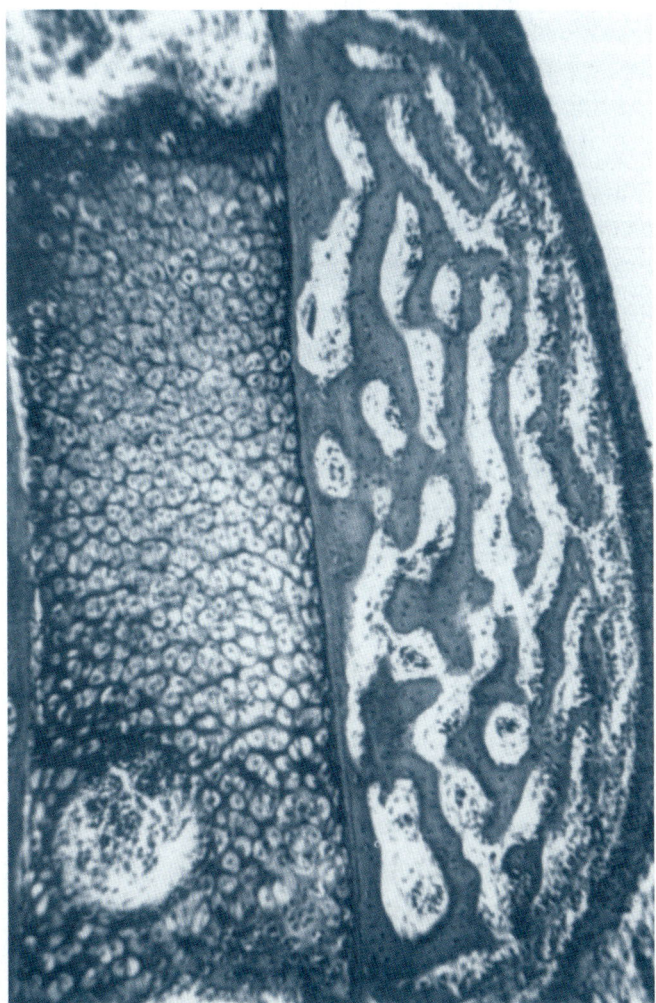

Figure 1.12 Bony exostosis. Extensive subperiosteal bone (right) has formed in this embryonic chick tibia forming a subperiosteal exostosis. Note the absence of any endochondral ossification or replacement of the primary cartilage (left), a situation typical of long bones in avian embryos. From Hall (1991a).

himself to 'shew the ancient and common notion of all bones being originally cartilaginous to be a vulgar error,' and that 'bony particles in foetuses begin to be deposited or to shoot between membranes or within cartilages.' Although Portal (1770) argued that the *embryonic skeleton* is first formed in cartilage and only later in bone, almost 70 years elapsed before Johannes Müller demonstrated that cartilage is destroyed and replaced by bone. Meischer, also in 1836, was the first to refer to the stages in endochondral ossification in growing bone. Twelve years later, W. S. Sharpey refined the views of Nesbitt and Müller with his *Theory of Substitution*, which states that bone replaces cartilage in endochondral bones.

Today, the difficulty of distinguishing cartilage from bone poses problems for at least three professional groups. Palaeohistologists have to identify these tissues and determine their relationships during the origin, evolution and radiation of the vertebrates. Pathologists often find skeletal tissues they cannot classify into any definitive category. Developmental biologists who study cell differentiation and tissue formation – and especially those who study bones other than mammalian long bones, which are the 'model' animals, and 'model' bones from which most of our knowledge is derived and extrapolated – are frustrated by classes of tissues in the non-pathological skeleton that appear to be intermediate between cartilage and bone.

In the two chapters that follow, I will examine bone and cartilage in their various guises, and at various time periods and situations in different taxa, before going on to evaluate the evidence for cartilage in many invertebrates (Chapter 4), and for tissues intermediate between bone and cartilage among the vertebrates (Chapter 5).

NOTES

1. From J. M. Petrie (1914), *Handbook for New South Wales*, British Association for the Advancement of Science, Sydney, Australia, 1914, cited from L. Gilbert (1997), p. 283.

2. See Hall (1998a, 2002a) for analyses of morphospace and the Burgess Shale fauna, and Dzik (2000) on the origin of the mineral skeleton in chordates.

3. For overviews of the evolution of skeletal issues, for the delineation of exo- and endoskeletons, and for models of exo- and endoskeletal evolution see Patterson (1977), Hanken and Hall (1983), Smith and Hall (1990, 1993), Janvier (1996), Donoghue et al. (2000) and references therein, Franz-Odendaal et al. (2003), Vickaryous et al. (2003) and Witten et al. (2004). See Prothero and Schoch (1994) for excellent reviews of the major features of vertebrate evolution from craniata, through gnathostomes to mammals.

4. Throughout the book, I have tried to use mineralized/mineralization rather than calcified/calcification. I do this to emphasize that mineralization is much more than the deposition of calcium phosphates or calcium carbonates. Brachiopods deposit calcium phosphate along with calcium carbonate into the inner layer of their shells (see Chapter 4).

5. From the studies Alison Cole has recently completed in my laboratory it became clear that many of the invertebrate cartilages also fit into a category of intermediate tissues, but I will not preempt her studies here. See Cole and Hall (2004a,b) for the beginnings of this story.

6. By ectopic I mean 'out of place,' or 'extraskeletal,' in the sense that the element develops outside the primary skeleton and so is extraskeletal in origin and location. Some elements such as sesamoids are ectopic in origin and initial location but then join with the primary skeleton. Ectopic osteogenesis is sometimes referred to as heterotopic osteogenesis. I use ectopic osteogenesis throughout. Ectopic mineralization – deposition of mineral without bone formation – differs from ectopic osteogenesis (Connor, 1983; Russell and Kanis, 1984).

7. See Glimcher et al. (1965), Miles (1967), Scott and Symons (1974) and Linde (1984) for the structure and function of enamel and dentine, and see the papers in Diekwisch (2002) for an overview of the development and mineralization of dentine and enamel. For reviews of mineralized skeletal tissues, see the chapters in Bourne (1971–1976), J. G. Carter (1990), in the nine-volume series bone edited by Hall (1990–1994), and in Bilezikian et al. (1996).

8. See Chai et al. (2000) and Jiang et al. (2000) for neural-crest origin of dental tissues in mouse embryos, and see Miles (1967), Listgarten (1973), Osborn (1981) and Sperber (2001) for types of mammalian cementum and the distribution of cementum within the vertebrates.

9. See Andreucci and Blumen (1971), Balatz (1974), Shellis and Miles (1974), Moss (1968b, 1977) and Sasagawa (2002) for enameloid formation in teleost fishes, and Berkovitz and Moore (1974), Bergot (1975), Berkovitz (1978), Huyssaune (1983, 2000), Smith and Hall (1990, 1993), Sire et al. (1998), Smith and Coates (1998), Teaford et al. (2000) and Sire and Huyssaune (2003) for basic studies on tooth formation in teleosts.

10. See Wright and Moffett (1974) and Beresford (1981) for chondroid bone, Goret-Nicaise (1984b), Goret-Nicaise and Dhem (1985) and Mizoguchi et al. (1992, 1993) for chondroid in mammals, Wislocki et al. (1947) for chondroid in antler development, and Enlow (1962a) for remodeling of chondroid as it transforms to bone, including a good discussion of the early literature.

11. See Hall (1988) for a discussion of the various types of direct and indirect osteogenesis. The journal *Nature* published a *Nature Insight on Bone and Cartilage* in May 2003 (Vol. 423, pp. 315–361), consisting of eight overviews on various aspects of the biology of both tissues. Individual reviews are cited where appropriate in the text.

Bone

Bone crops up in the oddest places. The ability of elasmobranchs (cartilaginous fishes) to produce bone is a good example of progenitor cells of the skeleton retaining, over long periods of evolutionary time, the ability to modulate to either cartilage or bone, cartilaginous fishes having had bony ancestors.

DISCOVERY OF THE BASIC STRUCTURE OF BONE

A rudimentary (in modern terms) 'histology' of bone, consisting of analyses of bone slices viewed with the newly invented microscope, allowed Havers (1691/1692) and Leeuwenhoek (1693) to describe the microstructure of bone. Both described tubules coursing through an otherwise solid structure. As Leeuwenhoek wrote: 'I reviewed the shin bone of a calf, in which I found several little holes, passing from without inwards; and I then imagined, that this bone had divers small pipes going long ways.'[1] The insight and ability to visualize structure in three dimensions, that allowed Leeuwenhoek to recognize these holes as the openings of pipes or canals over 300 years ago, is a skill that eludes many present-day students when they are first introduced to bone histology.

At the turn of the 18th century, Duverney (1700) showed that some of the holes identified by Leeuwenhoek and Havers contain nerves, evidence that bone must not be an inactive, inert tissue. The majority of the holes, however, represent tubules – now known as *Haversian canals* after Clopton Havers (1665–1702) – or canaliculi, through which osteocyte processes course (Figs 1.4 and 2.1). Observing irregular canals in pathological bone, Howship (1817) correctly inferred that they were lacunae – now

known as *Howship's lacunae* – representing sites of bone absorption (resorption), a conclusion confirmed by Tomes. Such integrated units of bone structure are now known as *Haversian systems* (Figs 2.2 and 2.3).[2]

When John Quekett (1846) set out to identify the structural peculiarities of bone in a variety of species, he ushered in the field of comparative osteology, which continues today. In using muriatic acid (hydrochloric acid) to remove the mineralized matrix, Quekett created a fundamental methodology – *demineralization* or *decalcification* – which we still use to help us visualize organic structure.

In 1845, John Goodsir demonstrated a special class of cells within the substance of bone. His proposal that these cells deposited the bone was not universally accepted. Goodsir identified what he also thought were bone-forming cells within the *periosteum*, which he recognized as a cellular covering of bone: 'On the surface of young and vigorous bones I have observed numerous cells, flattened, elongated and more or less turgid, belonging doubtless to the systems of Haversian cells.'[3] Carl Gegenbaur (1864) who also recognized these cells as bone forming, gave them the name *osteoblasts* (Fig. 2.4). Observations of resorption and formation laid the foundation of our understanding that bone is replaced as it ages, that replacement rate declines with age, and that old bone is replaced by newly deposited bone. Bone is dynamic despite its stony appearance.

The years 1872 and 1873 were a watershed for identification and analysis of the functions of osteoblasts and osteoclasts (Fig. 2.4).[4] *Osteoclasts*, the bone-resorbing cells named by Kölliker in 1873, were described and distinguished from giant cells of bone marrow by Robin (1849), although Robin had no inkling of their function. Robin also described the lining-up of *osteoblasts* along 'walls' of bone at sites of bone formation. He was describing the

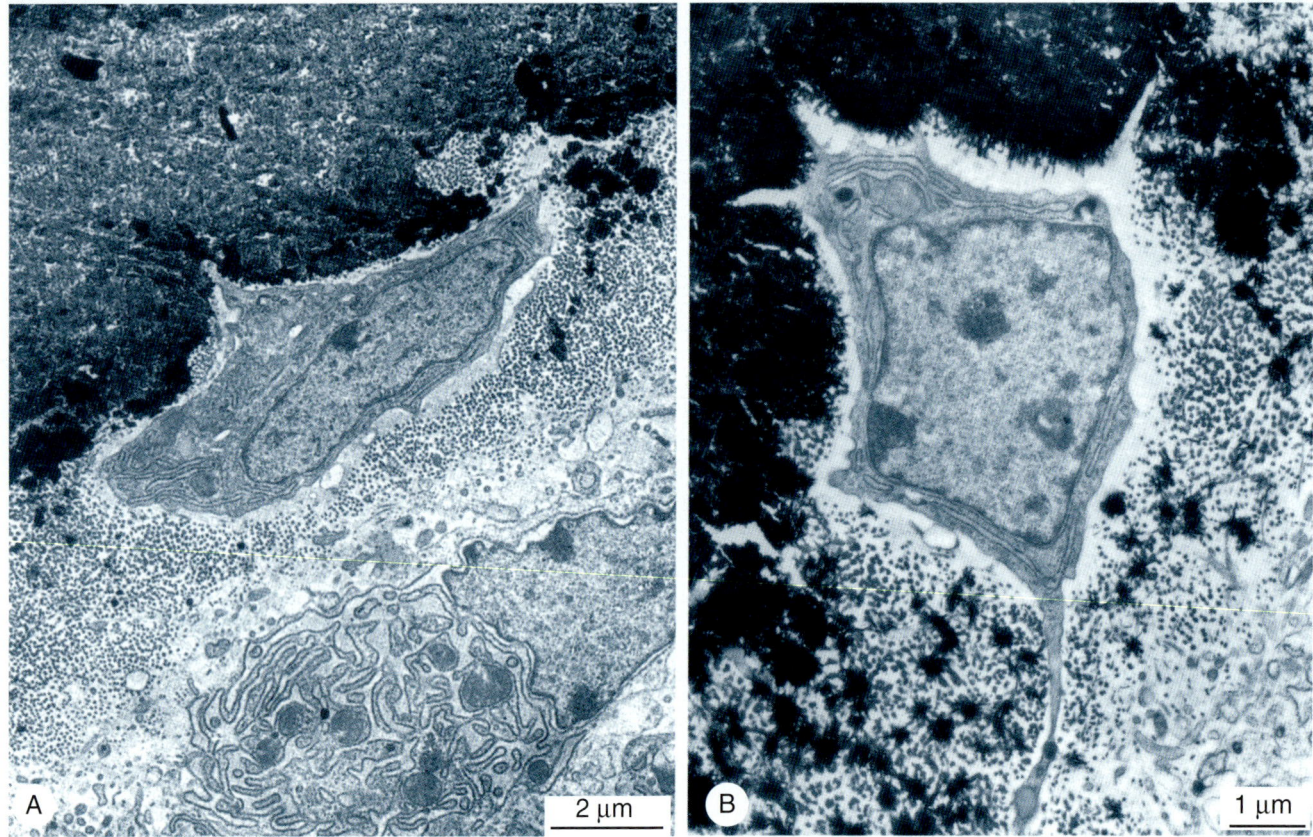

Figure 2.1 From osteoblast to osteocytes and from osteoid to bone as visualized in these transmission electron micrographs of subperiosteal bone from five-day-old chick embryos. (A) An active site of bone formation with mineralized matrix (top); an active osteoblast within the collagenous osteoid matrix – note the cell processes within the matrix – and additional osteoblasts further from the osteogenic front (bottom). (B) An osteoblast is shown transforming into an osteocyte through the deposition of mineralized matrix that progressively buries the cell in matrix, except immediately around the cell and its cytoplasmic processes. From Palumbo (1986).

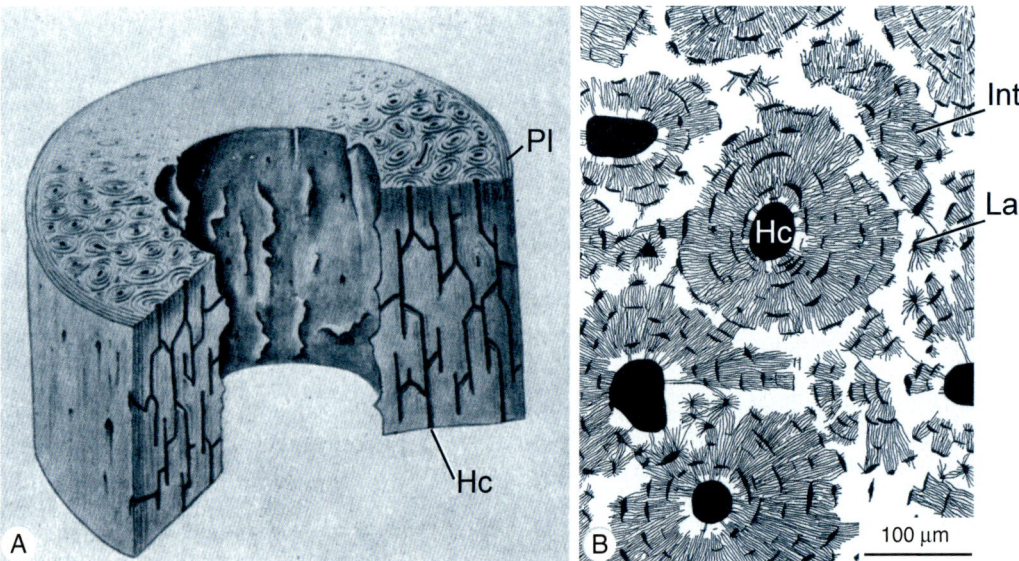

Figure 2.2 The lamellar structure of mammalian long bone (3D representation). (A) Haversian canals (Hc), whose scale is exaggerated, make up the bulk of the lamellar bone. Peripheral lamellae (Pl) are found at the periosteal surface. (B) A drawing of a histological section of a human femur showing Haversian canals in cross section. Interstitial lamellae (Int) between more fully developed Haversian systems, and lacunae (La) in which osteocytes would have been housed. From Le Gros Clark (1958).

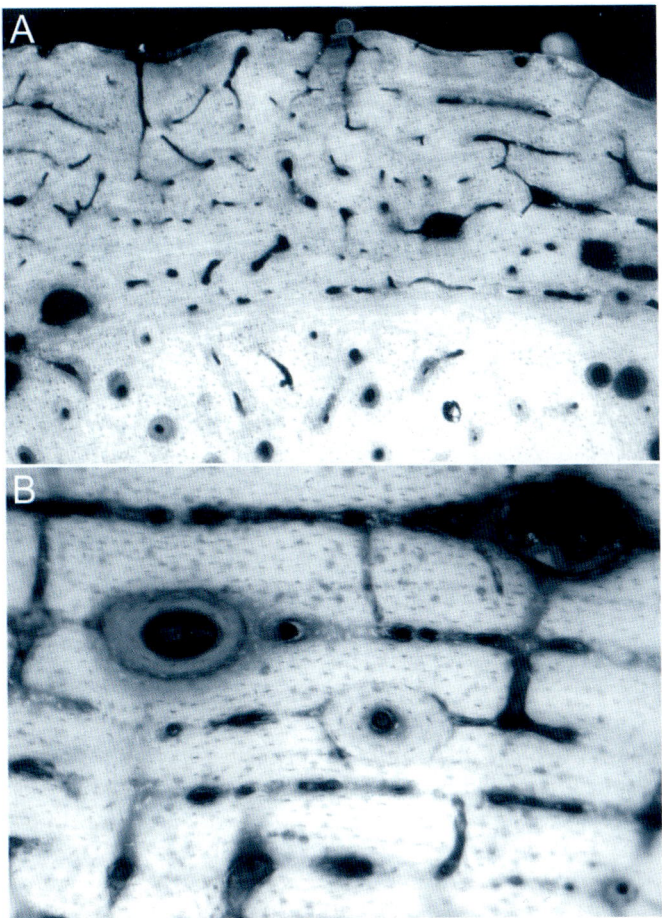

Figure 2.3 The laminar structure of human bone. (A) Superficial bone and medial osteonic bone seen in a transverse ground section of the femur from a 52-year-old woman, the femur having been impregnated with osmium black. (B) Several stages in osteone formation in a silver-stained ground section of the femur of a 3-year-old buffalo, *Bison bison*. See Locke (2005) for details of sample preparation. Images kindly provided by Michael Locke.

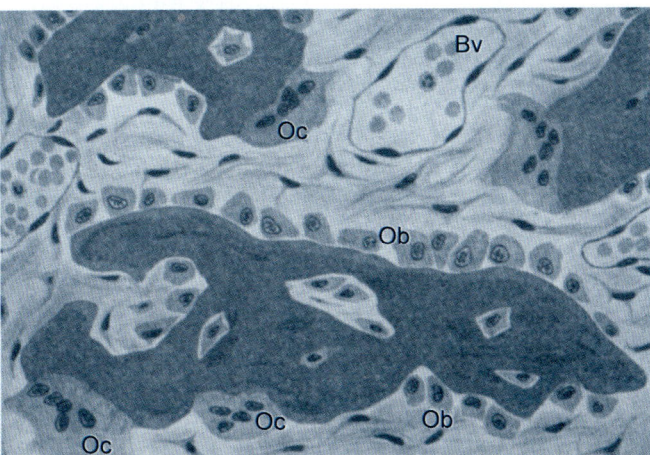

Figure 2.4 An elegantly drafted representation of osteoblasts (Ob), forming a cellular layer (endosteum) along bone surfaces; isolated multinucleated osteoclasts (Oc), associated with regions of resorption; and blood vessels (Bv), indicating the vascularity of bone. From Le Gros Clark (1958).

inner membrane of bone now known as the *endosteum* or endosteal surface (Fig. 2.4).

Now we know that resorption is multistep. From a synthesis of 351 papers on osteoclast biology, Vaes (1988) enumerated three steps:

 (i) formation of resorbing cells from precursors;
 (ii) activation of the resorbing cells by osteoblasts;
 (iii) resorption of bone matrix.

As late as 1910, however, some skilled pathologists and surgeons, including the eminent German pathologist Friedrich von Recklinghausen, still believed that osteoclasts were consequential to, rather than the cause of, bone resorption.[5]

CELLULAR BONE

On the basis of developmental processes, bone has been classified for a century and a half as *endochondral* (developing by the replacement of a cartilaginous model) or *intramembranous* (developing by the replacement of a fibrous or fibrocellular model). Growth of bone in tendons or ligaments, or the growth of ectopic bone, is typical of either long bone growth (if cartilage forms first) or of membrane bone growth (if no cartilage phase is present).

Although these terms primarily apply to the *processes of ossification*, often they are used to specify the organs – the bones – that result from these processes. Cellular or acellular bone, woven (cancellous) or lamellar bone, coarse-fibre or fine-fibre bone: these are the classifications of the histologist. Such classifications need not align with phylogenetic relationships, nor is the same skeletal organ always composed of similar tissue. Thus, while the skeleton of the paired fins in the Queensland lungfish, *Neoceratodus forsteri*, is composed of cellular bone, the fin skeleton in the African lungfish, *Protopterus annectens*, consists of acellular bone.[6]

Locations where particular types of bone are found in birds and mammals include bone with:

- coarse bundles of parallel fibres – at sites of attachment of tendons and ligaments in birds and mammals, and in the ossified tendons of birds;
- coarse bundles of woven fibres – in foetal mammals and in the early stages of fracture repair in tetrapods;
- fine bundles of parallel fibres – in the long bones of birds and young mammals, and around blood vessels in ossified tendons;
- fine lamellar fibres – in adult mammals; and
- bone with coarse *and* fine fibrous bundles – near attachment sites of tendons and ligaments, and where coarse bundles are removed and replaced.

The processes that produce these various types of bone differ. Coarse, woven bone is deposited rapidly.

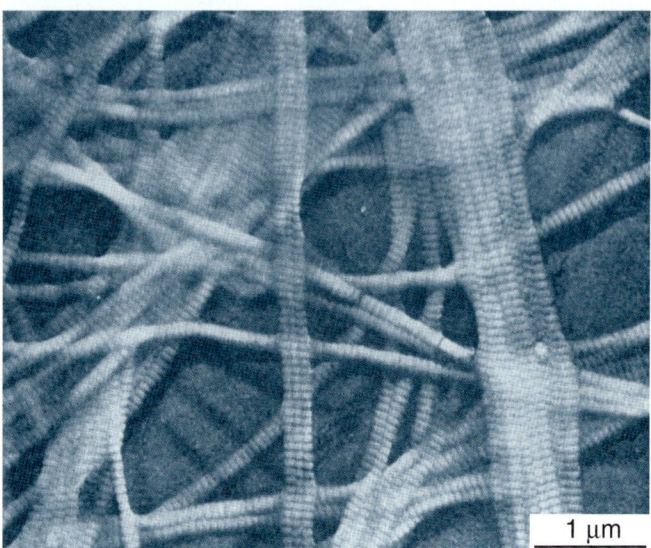

Figure 2.5 Banded collagen fibrils, shadowed with chromium and visualized with electron microscopy. From Le Gros Clark (1958).

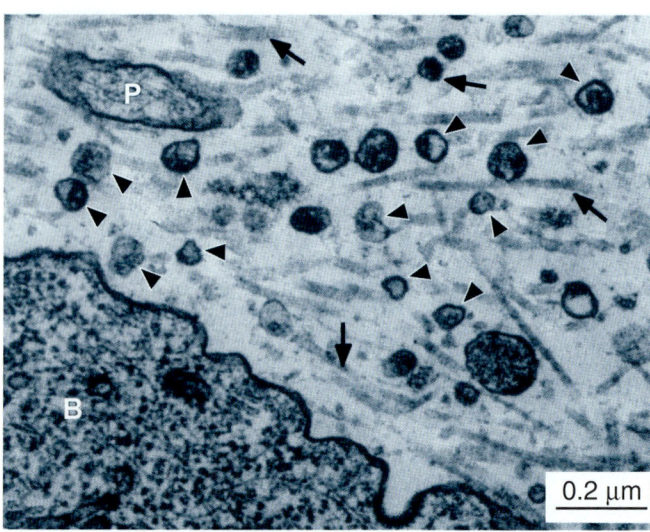

Figure 2.6 A transmission electron micrograph of part of the extracellular matrix (osteoid) of the calvarium of a rat at 17 days of gestation, showing collagen fibrils (arrows) membrane-bound bodies (arrowheads), cell processes (P) and a bone cell (B). From Katchburian and Severs (1983).

As a result, the osteocytes and fibres are haphazardly arranged (Fig. 2.5). Replacement by lamellar bone is considerably more orderly and predictably progressive with the formation of primary and then secondary *osteones* (Fig. 2.3 and see below).

In the past, it has been argued that the specialized histology of bone represents either an adaptation to mechanical stresses placed upon it, or an adaptation to the metabolic requirement for calcium and/or phosphorus. Armand de Ricqlès related bone histology to the pattern of growth and to the general metabolism exhibited by individual species or bones. He contrasted the bones of the slow-growing salamander, which might achieve a body weight of 20 g after four years, with the many mammals that achieve weights of several hundred kilograms in as little as two years. De Ricqlès correlated types of periosteal bone – which comprises most of the bulk of the long bones – with vascularity, species growth rate, rate of mineralization, organization of matrix fibres and periodicity of bone deposition, all of which are a function of the rate of bone deposition. These studies achieved some prominence in discussions on the possible warm-bloodedness of dinosaurs, and they merit the close attention of all those interested in the functional significance of bone histology.[7]

The most recently deposited, unmineralized, metabolically active bone (*osteoid*) (a term and concept developed by Virchow, 1853; see Figs 1.3, 2.1 and 2.6) is found adjacent to periosteal (outer) or endosteal (inner) bone surfaces, which are lined by formative cells (*osteoblasts*), resorptive cells (*osteoclasts*) and precursor cells (*osteoprogenitor cells*). These bone surfaces are of prime importance in metabolic function, in reactions to vitamins and hormones, and in the initiation of pathological

changes. Endosteal surfaces are especially important in bones that exhibit secondary remodeling of Haversian systems (Figs 2.2 and 2.3). *In 40 cm³ of a human pelvis there are 80 cm² of periosteal surface, but 1,600 cm² of endosteal surface.* Maureen Owen from Oxford pioneered studies of the dynamics of the cells on these surfaces (see Chapter 12). All the cells of a bone communicate in a most astounding way. Ultrastructural evidence shows junctions between adjacent osteocytes, between osteocytic processes in canaliculi, between osteocytes and osteoblasts, and between osteoblasts. As a result, signals can be transported rapidly throughout the bone; one need only witness the appearance of radioisotope in cortical bone minutes after intraperitoneal injection.[8]

OSTEOCYTES

Once osteoblasts are surrounded by unmineralized osteoid or mineralized bone, they are known as osteocytes. Osteocytes are distinguished by their characteristic morphology; by the synthesis of type I collagen (Fig. 2.5) and other specialized 'bone' proteins, such as osteocalcin, osteonectin and osteopontin (see Box 1.2, Fig. 1.6 and Chapter 23); and by the deposition of a mineralized matrix (see Figs 1.4 and 2.1).

It would be daunting to review all the available data on factors maintaining the osteocytic differentiated state. Vitamins (A, C and D; see Box 27.1), hormones (calcitonin, parathyroid hormone growth hormones, thyroxine, oestrogen), growth factors too numerous to enumerate, nutrition, and biomechanical and bioelectrical factors all play their roles, often interactively and/or differentially with respect to age and skeletal region. The large number of texts available on the physiology of bone in comparison to the paucity of volumes on cartilage attests to the depth of our knowledge of bone physiology and to the importance of the bony skeleton. The 25–30 million North Americans who are edentulous in one or both jaws because of inadequate knowledge of how to prevent excessive breakdown of alveolar bone (see below) would no doubt argue that even more should be learned.[9]

The physiological activities of osteoblasts and osteocytes are usually high in growing animals. Rat parietal bone is deposited at a rate of up to 470 μm³ matrix per osteoblast per day. In American elk, *Cervus canadensis*, osteoblastic and osteocytic activities are sufficiently high to increase the length of the antlers by up to 2.7 cm per day during the March to August growing season when growth hormone reaches its annual peak.[10]

Unlike many chondrocytes, osteocytes survive mineralization of their ECM – which, indeed, they initiate and control – obtaining metabolites and nutrients via an extensive canalicular system (see Figs 1.3, 1.4 and 2.1). Later in this chapter we will see that osteoblasts in teleost acellular bone either do not become enclosed within ECM (and hence do not become osteocytes) or become so walled in by matrix that they can no longer survive. An understanding of why these osteocytes do not survive might provide valuable clues as to how osteocytes of cellular bone do survive.[11]

INTRAMEMBRANOUS VERSUS ENDOCHONDRAL BONE

Kölliker (1859) – the first to report the absence of osteocytes from the bone of teleosts, thereby identifying the subclass of *acellular bone* – elaborated and discussed extensively the concept of bones formed in cartilage (*cartilage bones*) versus bones formed in membrane (*membrane bones*). As introduced in Chapter 1, bone develops either (i) directly (*de novo*) in mesenchyme (*intramembranous ossification*), or (ii) indirectly. The most common form of indirect ossification is within and by replacement of cartilage, the process of *endochondral ossification*. Other modes are possible (see Chapters 8 and 9).[12]

Embryonic origins

Because most bones that develop intramembranously have a neural crest origin (the major exceptions being the subperiosteal bone that forms around long bones or ribs) and most endochondral bones have their origin in embryonic mesoderm (Chapter 16), it is often thought (assumed, hoped?) that the two types of bone tissue would be (might be, should be?) metabolically distinct. Although they are not, some interesting physiologically differences merit discussion. Membrane and endochondral bones differ in germ layer of origin, mode of histogenesis

and in *evolutionary history*. Most membrane bones are part of the exoskeleton. All endochondral bones are endoskeletal (see Table 1.1). These are not the same differences as the *site-specific regulation of osteogenesis* seen when comparing periosteal and endosteal surfaces of long bones, or the ecto- and endocranial periosteal surfaces of skull bones; ectocranial surfaces show higher levels of alkaline phosphatase and more osteoblasts. Their cells also produce more bone *in vitro* than do those on the endocranial surface (McCulloch *et al.*, 1989, 1990).

Other modes

Intramembranous and endochondral ossification are usually regarded as separate and very different modes of ossification, which they are. This is so, even though much of the ossification in the long bones of many tetrapods occurs *subperiosteally* and so is essentially intramembranous, the membrane in this case being the perichondrium, which transforms into a periosteum. Despite the strong stance taken by Patterson (1977), for whom exo- and endoskeletons were absolutely separate, *mixed bones* consisting of fused dermal and chondral bone do exist; for example, in the skeleton of the striped bass, *Morone saxatilis*; in fusion between tooth-bearing dermal bones and the perichondral and endoskeletal bone of the visceral arches in zebrafish and cichlids; and in fusion of the endochondral bone that replaces Meckel's cartilage (endoskeleton) and dermal mandibular bones in hamsters.[13]

We expect these patterns of ossification to be conserved phylogenetically – once an endochondral bone, always an endochondral bone – and, indeed, phylogenetic conservation is what we see almost all the time. Examples are also known, however, where an endochondral bone in an ancestral lineage has been replaced by a membrane bone in a descendant lineage. The simplest mechanisms would be suppression of the cartilaginous model and formation of bone *de novo*. Of course, in any such example, we have to be sure that we are looking at the same elements in ancestor and descendant, and that one bone has not been replaced by another from a different position within the embryo and/or with a different phylogenetic history.

An oft-cited example is the orbitosphenoid of a South American limbless 'worm-lizard' (amphisbaenians), *Leposternon microcephalum*, studied by Bellairs and Gans (1983). The orbitosphenoid is an endochondral bone in all other species studied. In this species it is a membrane bone with an associated cartilage nodule. This nodule is probably not a secondary cartilage; it does not lie in the periosteum of the membrane bone, nor have secondary cartilages been reported from reptiles (see Fig. 5.4).[14] It could be a remnant:

- of the orbital cartilage, which is present but exceedingly small in this species;

- of the cartilaginous model that forms the orbitosphenoid in other species; or
- a neomorph.

A developmental series is required to resolve this issue.

Metabolic differences

Numerous studies demonstrate physiological differences in rat bones that appear to relate to whether they develop intramembranously or endochondrally. Membrane bones ossify faster than long bones, as can be seen from the rate at which the collagen of rat calvariae and long bones mineralize (Table 2.1). Somatostatin receptors are restricted to a subpopulation of osteoblasts during endochondral bone formation but are not found in membrane bone at all (Mackie et al., 1990).

Dentary bones have a higher calcium content and are more cellular than endochondral bones in the same individual. In rats raised on a quarter of the normal protein levels and thus suffering from protein-energy malnutrition, more calcium is lost from the dentaries than from other bones, bringing the mandibular calcium levels to those normally seen in long bones. Ca^{++} concentration declines by 51 per cent in the dentary but shows no change in the long bones. DNA levels rise by 48 per cent in the dentary but drop by 20 per cent in the long bones (Table 2.2). Such differences may reflect the higher resorption rates of

alveolar bone and the large amounts of alveolar bone in rodent dentaries; the long root of the incisor tooth means that alveolar bone occupies much of the body of the dentary.

The bones of rats given diets of 6, 12 or 20 per cent protein respond differently to caffeine: calcium increases in the mandibles but not in the long bones, collagen increases in the long bones but not in the mandibles, and so forth. From an examination of the impact of weaning it is clear that early growth is under tight genetic control and later growth is influenced by nutrition, and hence also by functional demands of mastication (see Box 12.2).[15]

Differences in the kinetics of mineralization between human membrane and endochondral bones (parietals and femora between six and 41 weeks of gestation) indicate that such differences are present from early on in skeletal development. A similar conclusion can be drawn from a radiographic study of the development of human parietal bone and interparietal sutures from 15 individuals, a study that shows this region to be a primary ossification centre. The lower mineral content in the non-weight-bearing parietals could have a functional explanation independent of the origin or mode of histogenesis. Cellularity could also be a factor.[16]

Morphogenetic differences

Morphogenetic differences have also been reported. When transplanted intramuscularly, osteoblasts derived from scapulae form ossicles of bone with abundant marrow, while osteoblasts derived from calvariae form islands of bone with little marrow. Moskalewski and colleagues were unsure whether these differences reflected inherent differences between osteoblasts from membrane and endochondral bones, the normal presence of marrow – and hence marrow precursors – in endochondral but not in membranes bones, or differential adaptation/response to mechanical conditions. Isolated scapular or vertebral osteoblasts grafted onto devitalized cranial vaults form similar bone tissues, which may indicate that morphogenetic specificity is lost when osteoblasts are isolated before being grafted.[17]

Portions of a single bone can also differ in their morphogenetic properties, as reflected in phylogenetically conserved shapes. Monteiro and Abe (1999) used the components approach to skeletal morphogenesis developed by Atchley and Hall (1991) to demonstrate phylogenetic influences on scapular morphology of members of the mammalian order Xenarthra – armadillos, anteaters, sloths and the armadillo-like, 100-kg glyptodonts – differences that reflect differential behaviour of osteoblasts in the two major components of the scapula.

Modes of ossification can also vary. From an analysis of four species of marsupials, Sánchez-Villagra and Maier (2002) determined that the spine of the scapula displays two modes of osteogenesis. Only the acromial portion of the scapular spine is endochondral. The major dorsal

Table 2.1 Approximate composition (%) of mineralized and unmineralized collagen in rat long bones and calvariae between birth and two months after birth[a]

Age (weeks)	Long bones (femur)		Calvariae (frontal and parietal)	
	Unmineralized collagen	Mineralized collagen	Unmineralized collagen	Mineralized collagen
0	75	25	58	42
1	–	–	30	70
2	55	45	25	75
4	35	65	10	90
8	15	85	7	93

[a] Based on data from Zika and Klein (1975).

Table 2.2 Protein, DNA and calcium contents of mandibles and long bones from control and malnourished newborn rats[a]

	Control 250 g protein/kg body weight		Malnourished 60 g protein/kg body weight	
	Mandible	Long bone	Mandible	Long bone
Protein (mg)	33.7	55.9	44.3	59.0
DNA (μg)	12.5	168.2	18.5	135.0
Calcium (mg)	130.6	68.2	66.7	69.6

[a] Based on data in Nakamoto et al. (1983).

portion is appositional bone deposited intramembranously. In a subsequent comparative analysis of scapular development in marsupials, these same researchers concluded that the dorsal portion of the scapular spine is a neomorph in therian mammals.[18]

Clinical experience with humans and transplant experience with other mammals also demonstrate physiological differences. Craniofacial defects accompany one in every 600 live births in the US of which 65 per cent have cleft lip or cleft palate. When used to repair cleft palate in children, grafts of dentary bone are better than iliac grafts; dentary bone is incorporated more completely and undergoes less resorption (Box 2.1). Koole and colleagues (1989) thought the embryonic origin of the two bones (neural crest versus mesoderm) explained the difference.

It is difficult, however, to separate embryonic origin from such site-specific differences as local functional adaptations that are independent of embryonic origin. Kasperk et al. (1995) attribute the differing properties of human osteoblasts from different sites of four individuals to site-specific adaptations. mRNA levels for *Fgf-2* are higher and mRNA for alkaline phosphatase is lower in mandibular osteoblasts than in osteoblasts from the iliac crest. Levels of mRNA for *Tgfβ* are higher in iliac crest osteoblasts. Another difference is that mandibular osteoblasts divide more rapidly than do osteoblasts from the iliac crest.

In a study using membrane or endochondral bone for grafts in craniofacial onlays in rabbits, Hardesty and Marsh (1990) demonstrated that membrane bone grafts grew and increased in size but that endochondral grafts were resorbed. Their explanation: the three-dimensional architecture of the two bone types, not embryological origin. Using similar logic, oral surgeons use clavicular secondary cartilage to replace condylar cartilage of the human dentary in situations where the latter is damaged or malformed, essentially replacing the temporomandibular joint with a sternoclavicular joint. As discussed in Chapter 16, the rationale is the similarity of the two secondary cartilages in responding to movement by growth, and their organization into similar layers; growth capacity of the costochondral graft correlates with the amount of cartilage in the graft.[19]

Box 2.1 Grafting bone and cartilage

"No bone graft, no matrix, no growth factor, no cytokine can contribute to the generation or integration of new tissue, except through the influence it has on the behavior of cells. The efficacy of all current clinical tools depends *entirely* on the cells in the grafted site, particularly the small subset of stem cells and progenitor cells that are capable of generating new tissue."[a]

The first bone-grafting procedure – bone from a dog's skull grafted into a battle wound in a soldier's skull – was reported in 1668 by Job Janszoon van Meekeren. Evidently, the graft was successful; the Church excommunicated the soldier as a result of the treatment. When he asked van Meekeren to remove the graft so that he could return to his church, the surgeon was unable to grant his request; the graft had taken too well.

Grafting of cartilage was a much later undertaking; Paul Bert (1865) was among the first to graft cartilages into experimental animals. Nowadays we know that homografts of cartilage, even large grafts such as half-joints consisting of femoral articular cartilage and subjacent bone, survive longer than any other tissue except cornea (Campbell, 1972).

Two hundred years after van Meekeren, Ollier's (1867) experimentation on dogs and rabbits became the benchmark for mechanisms of osteogenesis, especially the viability of *autografts*, which are taken from and grafted into the same individual. The first autograft was performed in Germany by Walter (1821) to replace skull bone. The first allograft (graft from one individual to another), was carried out by William Macewen, in Scotland in 1881 when he replaced humeral with tibial bone. The first cartilage grafts in humans may have been as late as 1900 with the grafting of rib cartilages.[b]

Phelps (1891) suggested that the most desirable form of bone grafts was with living, vascularized bone, a technique pioneered by Curtis (1893). Senn (1889) was among the first to implant antiseptically treated bone.[c]

[a] Muschler and Midura (2002), p. 66, emphasis mine.
[b] See Trueta (1968) and de Boer (1988) for a history of bone grafting, Bert (1865) for cartilage for the first cartilage graft, and Meijer and Walia (1983) for a history of cartilage grafting.
[c] Foetal bone was used as a transplant to replace massive bone loss by Siegal et al. (1977). An and Draughn (2000) evaluate the mechanical properties of bone, especially at the bone–implant interface. See Brooks (1971) and Singh et al. (1991) for vascularity of bone.

OSTEONES

The basic structural elements of mineralized cortical bone matrix in mammals and larger tetrapods are known as *osteones*, which come in different types. Groups of osteones form a Haversian system.

A *primary osteone* (Fig. 2.3) has a central canal that is less than 100 μm diameter and contains two or more blood vessels, but lacks a delimiting cement line or interstitial lamellae. A *secondary osteone* has a larger canal, a single central blood vessel, is limited externally by a cement line, and is wedged between interstitial lamellae. The lamellae may be concentrically arranged as in a Haversian system (Fig. 2.2), circumferential as in near-surface bone, or interstitial as in remnants of old osteones. The bones of many fish lack osteones or have only primary osteones.

Osteone life spans and the time it takes to produce an osteone vary from species to species and group to group. It takes a two-year-old cat (Fig. 2.8) 50 days to make an osteone; a 45-year-old man needs 100 days. And, since the average life span of human osteones is 15 years, only 0.05 per cent of our skeletons are turned over per day. Nor is the rate of osteone mineralization uniform; 70 per cent of the mineralization occurs within one to two days of the deposition of osteoid, the remaining 30 per cent can take many months. These rate differences become important when assessing pathological states, particularly in metabolic bone diseases.[20]

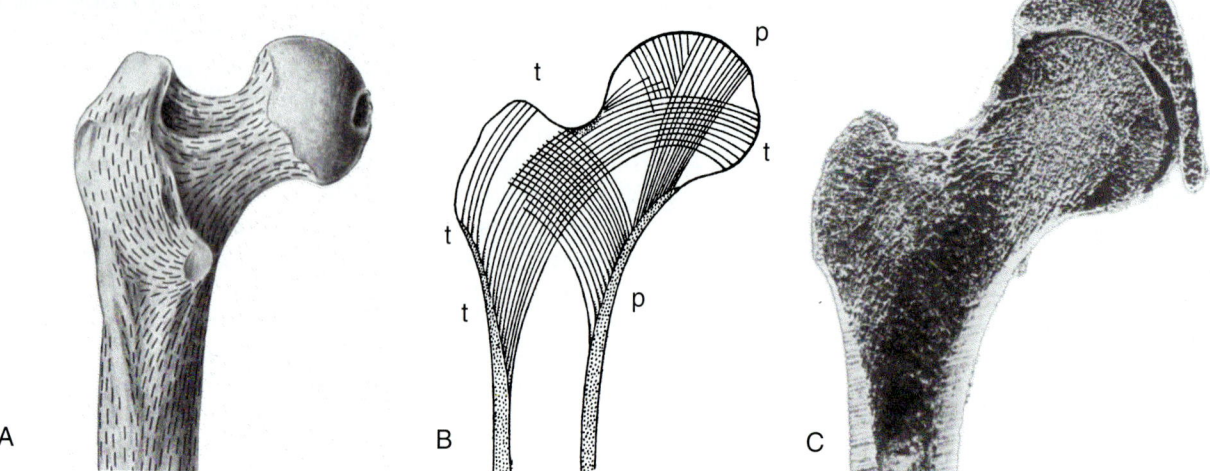

Figure 2.7 Various ways of visualizing the distribution and orientation of trabeculae, illustrated using the proximal end of the human femur. (A) The bone is pricked with an awl and stain rubbed into the surface. Removal of excess stain leaves residual stain in 'splint lines' that reveal fibre bundle and osteone orientation. Method developed by Benninghoff in the mid-1920s. See also Fig. 31.1. (B) A diagrammatic representation of pressure (p) and tension (t) trabeculae as depicted by Meyer in 1867. (C) A longitudinal section of dried bone. Compare the trabecular organization with (B) and with Figure 31.2. Modified from Murray (1936).

Figure 2.8 A typical cat, relaxed yet alert. Photographed by the author.

Table 2.3 Some of the densest mammalian bone[a]

Species	Bone	Density (g/cm³)
Mesoplodon densirostris (Blainville's beaked whale)	Mesorostral ossification	2.3–2.6
Mesoplodon carlhubbsi (beaked whale)	Mesorostral ossification	2.40
Physeter macrocephalus (sperm whale)	Tympanic bulla	2.16
Trichechus inunguis, T. manatus, T. senegalensis (Amazon, West Indies and West African manatees)	Rib	1.81
Delphinus delphis (common dolphin)	Rostrum	0.79
Elephas maximus (Asian elephant), *Loxodonta africana* (African elephant)	Ivory	1.71

[a] From data assembled by MacLeod (2002).

Numerous secondary osteones develop in dense bones. Among the densest is the bone that replaces the mesethmoid cartilage in the rostrum of mature (especially male) Blainville's beaked whales, *Mesoplodon densirostris*. The estimated density of 2.3–2.6 g/cm³ is three times that of the bone in the rostrum of the common dolphin, and 40 per cent higher than the density of the ivory that forms elephant tusks (Table 2.3). For comparisons, the density of rat bone is of the order of 0.6 g and human bone 0.35–0.55 g of Ca^{++}/cm^3 (Jowsey, 1968).

Human cortical bone contains three types of secondary osteones, which have similar biological lifetimes. Some 90 per cent of the human osteone population is made up of unremodeled secondary osteones, with age and gender-related changes seen in the cortical bone of the ribs. Osteones 'drift' through cortical bone matrix as they – the osteones – are remodeled. Examination of the kinetics of secondary Haversian systems in dogs reveals 9.2 osteoclasts per secondary osteone, 9.1 nuclei per osteoclast, 11.5 days as the average time nuclei spend in osteoclasts, and seven nuclei per day as the turnover rate.[21]

The standard histology textbook gives the impression that all mammalian bone is fine lamellar bone containing numerous secondary osteones. In fact, this picture is true only for human bone, and even then not for all human bone. Surveys of mammalian bone are available, notably those of Enlow and Brown (1958), Enlow (1966a) and Singh *et al.* (1974), the latter including a valiant attempt at quantification based on the number and size of primary longitudinal canals and the number of lacunae and empty lacunae.

Variation includes areas of bone lacking primary or secondary osteones, acellular or avascular areas, and necrotic (dead) areas. Moreover, presence or extent of such regions varies from bone to bone within an individual, with age for a given bone, and between individuals. One side of a bone may be highly vascularized, the other avascular. Osteones change with age of the osteone and of the individual (see Ageing below). A highly sophisticated knowledge of the microenvironment in which bone and bones develop is required before we can interpret this diversity of expression of the differentiated state of osteocytes. One of the aims of the book is to marshal and analyse some of that evidence.

Osteones can be detected in ground sections of fossilized bone. A histological analysis of tibiae from an early bird was used by Houde (1987) to argue that *Hesperornis* had osteones similar to those of most living (neognathous) birds. The Upper Permian amphibian (temnospondyl) *Australerpeton cosgriffi*, had abdominal scales of primary bone, with prominent vascular canals and osteones, the latter providing evidence for bone remodeling in adults but not in juveniles.

GROWTH

While cartilage grows internally (interstitial growth) and appositionally (on a preexisting surface), bone only grows by apposition of new osteoid onto existing surfaces. When Duhamel (1742) placed a circle of silver wire subperiosteally against the cortex, the wire became imbedded in the shaft of the bone. Duhamel concluded that bone grew by apposition from the periosteum 'as an exogenous stem grows from the inner layer of the bark.' We now know that the periosteum is not the only surface involved in the deposition and growth of bone (see Table 1.1). In long bones deposition takes place on:

- the outside of bone – *periosteal ossification* – by osteogenic cells in the outer layer or periosteum;
- an inner surface – *endosteal ossification* – by osteogenic cells in an inner layer or endosteum;
- mineralized cartilage – *perichondral ossification* – by cells brought into the marrow cavity by invading blood vessels; or
- following removal of mineralized cartilage – *endochondral ossification* – by cells brought into the marrow cavity by invading blood vessels.

Postnatally, the skeleton grows intramembranously as in membrane bones and at sutural margins of skull bones, subperiosteally as along the shafts of long bones, and by endochondral replacement as in long bones, vertebrae and ribs. In many situations, extensive remodeling accompanies growth. Indeed, remodeling is a vital part of the growth processes.

Several lines of evidence document the osteogenic activity of endosteal lining cells and osteoblasts as well as differences in potential between them and periosteal cells:

- Endosteal osteoblasts isolated from tibiae of four- to nine-month-old rats retain their differentiated features *in vitro* and deposit a mineralized matrix.
- Systemic injection of rats with Fgf-2 (100 μg/kg body weight for seven days) stimulates endosteal bone formation, as evidenced by an increase in the numbers of preosteoblasts and their transformation to osteoblasts.
- The distinction between delayed healing and lack of healing (non-union) of fractures is evident at the cellular level (Chapter 29). In delayed union, the block to timely repair is cessation of the periosteal response. In non-unions, both periosteal *and* endosteal responses cease or are not initiated. Interestingly, if the digits of mice or rhesus monkeys are amputated, the periosteal cells produce cartilage and bone, while the endosteal cells only produce bone, reflecting the general non-chondrogenic potential of endosteal progenitor cells.[22]

In most vertebrates, bone grows throughout life. Growth may not be continuous or uninterrupted but bone continues to grow throughout the life of even small mammals. Bernard Sarnat (1986), who has spent an enormously long career analyzing bone growth, listed seven issues to be resolved whenever the growth of bone is considered. These are: (i) *pattern*, (ii) *site*, (iii) *amount*, (iv) *rate*, (v) *direction*, (vi) *proportions*, and (vii) *factors influencing*.[23]

REGIONAL REMODELING

Remodeling turns bone from an inert mineralized supporting tissue to a vital and metabolically active component of an organism's mineral metabolism and ionic balance. Typically, remodeling is taken to refer to the ongoing and usually constant turnover of bone matrix as old matrix is replaced by new. But neither ongoing nor constant are true, as illustrated by the following four diverse examples:

- remodeling and resorption of the breast-bone (sternum) in female blackbirds, *Turdus merula*, during the breeding season but not at other times of the year;
- remodeling of the ribs and long bones of deer peaks in June when antlers are at their highest growth rates and demand for calcium is at its highest;
- remodeling of the fracture callus in repairing long bones follows a diurnal rhythm (which is abolished following hypophysectomy); while
- remodeling of the bony abdominal scales of the Upper Permian amphibian *Australerpeton cosgriffi* occurs in adults but not in juveniles.[24]

Table 2.4 Differences between remodeling rates in proximal, mid-shaft and distal portions of the ribs of beagle dogs[a]

Parameter	Proximal	Mid-shaft	Distal
Ratio of cortical area to total area	0.67	0.66	0.76
Number of osteoid seams/mm^2	6.55	4.69	4.00
Number of resorption spaces/mm^2	2.18	1.30	0.90
Rate of bone formation (mm^2/mm^2/year)	0.61	0.43	0.35

[a] Based on Anderson and Danylchuk (1978).

Because bone is not bone is not bone, remodeling is not remodeling is not remodeling. Rates of remodeling are not constant even within different sites on the same bone:

- A study of the proximal, mid-shaft and distal portions of the ribs of beagle dogs revealed differences in rate of bone formation, numbers of resorption spaces and osteoid seams, and the ratio of cortical bone to total bone area (Table 2.4).
- The alveolar bone that supports the teeth is more metabolically active than other types of bone and so is remodeled more rapidly than is other bone. Remodeling takes place during the normal physiological drifting of teeth in their sockets, and even more rapidly when orthodontic appliances are applied (Chapter 25).
- As is well known to oral surgeons who reposition the lower jaw to counteract an imbalance in growth between upper and lower jaws, resorption takes place in some but not other sites along the jaw. Insert the pin into resorption sites and the surgery fails.[25]

Remodeling also responds to environmental change or to surgical intervention. 'Suckling' and swallowing *in utero* stimulate remodeling in the mandibular condyle of human foetuses as early as the fifth month of gestation. In rhesus monkeys, *Macaca mulatta*, as in many other mammals, a soft diet results in low rates of remodeling, a hard diet promotes higher rates of remodeling of the dentary bone. Altered remodeling of the condylar process and zygomatic arch follows destruction of the motor nucleus of the trigeminal nerve. Drilling a hole in a bone is a sufficient stimulus to accelerate remodeling and vascular changes around the hole, as shown in beagle dogs of various ages.[26]

AGEING

At whatever level analyses are undertaken – whether bone cells, bone as a tissue, bones as organs, bone as a skeletal system – the passage of time influences bone.[27] A tiny sample of alterations of bone with age at the four levels (cells → tissue → organs → skeletal system) follows.

Ageing at the cellular level: Rats show a generalized and decreased capacity to form bone with age, whether

Table 2.5 Comparison of calcium accumulation, alkaline phosphatase activity, and bone Gla-protein in bone induced by demineralized bone matrix implanted into rats of different ages[a]

Age (months)	Total Ca^{++}/mg[b]	Alkaline phosphatase activity (units/g)[c]	Appearance of BPG (ng BPG/mg)[d]
1	110	35	45
3	65	35	23
10	30	30	15
16	15	13	0

[a] Based on data from Nishimoto *et al.* (1985)
[b] Total calcium present six weeks after implantation.
[c] Activity three weeks after implantation.
[d] BPG, bone Gla-protein; total BPG four weeks after implantation.

Table 2.6 Average number of osteoclasts per unit area of metaphyseal bone in the distal epiphysis of the mouse femur at different ages[a]

Age (weeks)	Number of osteoclasts/unit area
1	1.28
2	1.85
3	1.35
5	1.55
6	1.17
7	0.88
8	0.49
9	0.38
10	0.30
26	0.15
27	0.05
28	0.04
52	0.02
53	0.02
54	0

[a] Based on data from Tonna (1960).

assessed by total calcium accumulation, alkaline phosphatase or osteocalcin (bone Gla protein), as can be seen from a study in which bone was induced in rats of different ages (Table 2.5; see Box 14.3). Studies using osteoprogenitor cells from young (1.5-month-old) and old (17–26-month-old) rats maintained *in vitro* demonstrate that cells from the older rats have a reduced capacity for self-renewal (Bellows *et al.*, 2003).

Ageing at cellular and tissue levels: Osteoclast populations decrease with age as, for example, in both periosteal and metaphyseal bone of rat radii (Table 2.6). Indeed, metabolic evidence of skeletal ageing can be detected in mice as early as eight weeks after birth, when the activity of respiratory enzymes begins to decline. The number of progenitor cells that can be isolated from the cambial layer of periostea of mammalian bones declines with age.

Ageing at the organ level: Ageing affects the ability to repair fractures. Comparison of the mechanical properties of healed fractures after 40 and 80 days in rats aged three months and two years shows that, although there is no significant difference in mechanical properties after 40 days, maximal load, maximal stress, stiffness, and energy absorption in three-point bending are all lower in older animals after 80 days, indicating delayed healing (Table 2.7).

Ageing of the skeletal system: The physiological bone atrophy that accompanies age in mice, and that may contribute to ageing, can be reduced with 30 minutes of daily exercise; dry weight, Ca^{++}, PO_4^{3-} and Mg^{++} content, and femoral volume all increase in exercised mice.[28]

Table 2.7 Differences in the mechanical properties of healing tibial fractures measured 80 days after fracture in rats of two different ages[a]

Mechanical property	Three months old	Two years old
Maximum load (N)[b]	122.0 ± 6.00	53.9 ± 8.2[c]
Maximum stress (N/mm^2)	103.6 ± 13.5	33.8 ± 6.7[c]
Energy as maximum load (/mm)	19.2 ± 1.8	82.3 ± 1.0[c]
Relative energy at maximum load (10^3 Newtons/mm)	18.15 ± 3.9	5.2 ± 0.09[c]
Deflection at maximum load (mm)	0.32 ± 0.02	0.36 ± 0.05

[a] Based on data in Bak and Andreassen (1989).
[b] N, Newtons.
[c] P < 0.001 in comparison of three-month-old with two-year-old rats.

Osteones over time

Because of the importance for human health (including the desirability of prolonged activity into old age) and for fisheries and wildlife, tens of thousands of studies have been undertaken on the ageing of mammalian and fish bones. Changes in osteones are one reflection of the ageing of bone. For example, although osteone size decreases, the number of osteones in the cortical bone of femoral and tibial shafts from humans between 36 and 75 years increases by some 20 per cent. All the standard measures of the mechanical strength of bone – breaking load, strength, strain, modulus of elasticity and bone density – *decrease with age* (Table 2.8). As seen in a study comparing rodents, dogs, cows and humans, the number of secondary osteones – an indicator of regulation of calcium metabolism – correlates with life span (Table 2.9). As expected, the number of primary osteones decreases with age in the non-weight-bearing frontal bones of rabbits.[29]

The pore structure of human mandibular bone changes with age, as determined using a combination of microradiography, scanning electron microscopy and methacrylate replication (see Fig. 1.4). The age of an Alaskan mummy was estimated as 50–59 years based on osteone density, and 48–58 years based on racemization of aspartic acid, an excellent fit of the two ageing methods.[30]

Table 2.8 The average mechanical and physical properties, number and size of osteones in femoral cortical bone of younger and older human males[a]

Age group	Breaking load (kg)	Ultimate tensile strength (kg/mm^2)	Tensile strain (% elongation)	Modulus of elasticity (kg/mm^2)	Density (g/cm^2)	No. osteones/mm^2	Area/Osteone (mm^2)
Younger[b]	73.03	10.27	1.73	1766	1.93	11.06	0.04
Older[c]	61.87	7.73	1.27	1500	1.80	13.37	0.03

[a] Based on data from Evans (1976).
[b] Average age, 41.5 years.
[c] Average age, 71 years.

Table 2.9 Species and average life span compared to the presence of secondary osteones and developmental time at which they appear[a]

Species	Average lifespan (years)	Presence of secondary osteones	Duration of epiphyseal growth	Time at which secondary osteones develop
Rat (*Rattus rattus*)	2–4	No	9 months	–
Rabbit (*Oryctolagus cuniculus*)	8–10	Yes	6 months	Before closure of epiphyseal plates
Human (*Homo sapiens*)	75	Yes		Two-thirds of the cortical bone has secondary osteones at 2.5 years

[a] Based on Jowsey (1968).

ACELLULAR BONE

Bone in humans that is acellular is usually regarded as degenerated, pathological or an artifactual consequence of poor histological processing. However, large areas of acellular or avascular bone do exist in mammalian bone; hence the sampling problems and difficulty in assigning causation encountered by palaeontologists, forensic scientists and anthropologists. In two major groups of vertebrates, however, bone acellularity is the rule, not the exception. The two are extant *teleost fishes* and a group of Ordovician jawless vertebrates, the *heterostracans* (see Box 6.2). Acellular bone links evolutionary and developmental studies on the one hand, normality and pathology on the other.[31]

I use Caisson disease to illustrate abnormal acellular bone in humans, introduce the normal acellular bone of extant teleosts, and comment on acellular bone (aspidine) in extinct agnathans and on bone in cartilaginous fishes.[32]

Caisson disease and abnormal acellular bone in mammals

Caisson disease of bone is the development of avascular necrosis as a result of diving or tunneling under increased atmospheric pressure, the degree of necrosis in divers correlating with diving depth, frequency of dives and recovery period between dives. Necrosis is not seen in those who dive at shallow depths. As pressure increases by one atmosphere for every 10 m of depth, the deeper and more frequent the dives, the more necrosis. Curiously, although necrosis is seen in the knee joints, necrosis is not seen in the ankles, wrists, elbows or vertebrae of commercial divers.[33]

Acellular areas occur in articular cartilage and underlying subchondral bone in compression-induced arthritis in rabbits. Reduced atmospheric pressure also affects healing of bone fractures. At half normal atmosphere, cartilage proliferation and ossification are reduced, as are accumulation of collagen, Ca^{++}, and the tensile strength of any bone formed (Table 2.10). Mucopolysaccharides (glycosaminoglycans) appear unaffected.[34]

Acellular bone in teleost fishes

Acellular bone is the norm in that most speciose group of vertebrates, teleost fishes, which at somewhere between 24 000 and 28 000 species exceed the 20 000 avian and mammalian species combined.

Kölliker (1859) first described acellular bone in an analysis of teleost fishes. Unfortunately, he called it osteoid. I say unfortunately because, as discussed earlier, Virchow (1853) had already used the term osteoid for the initial unmineralized but mineralizable matrix deposited by osteogenic cells (see Fig. 2.1). The term osteoid should be reserved for recently deposited, unmineralized bone, whether cellular or acellular.

In her analysis of acellular bone in higher teleosts, Parenti (1986) presented convincing arguments that acellular bone should be used as a character in phylogenetic analysis. She notes that only one teleost has both cellular and acellular bone, that species being *Albula vulpes*, which has the very appropriate common name of bonefish. Maisey (1988) included acellular bone in his phylogenetic analysis of vertebrate skeletal tissues, while Smith and Hall (1990) analyzed the phylogeny of acellular bone and dealt with the perennial question of whether cellular or acellular bone arose first in evolution (see Chapter 6). To reflect the presence of cells on the surfaces of acellular bones it has been argued that acellular bone should be renamed *anosteocytic* bone and cellular bone renamed *osteocytic* bone. I sympathize with this view, but because of recognized usage, will continue to use cellular and acellular.[35]

Of course, we cannot say that *only one* species of fish has both cellular and acellular bone; Parenti did not examine every species known, nor could she. The pike, *Esox*, is described as having acellular bone, but nests phylogenetically within a group of fishes, all of which have cellular bone. Is the pike an exception, and/or should its bone be reexamined? Hughes *et al.* (1994a,b) identified osteocytes in the alleged acellular bone of three species of sea bream off the east coast of Australia, the yellow fin bream, *Acanthopagrus australis*, snapper, *Pagrus auratus* and Tarwhine or gold lined seabream, *Rhabdosargus sarba*. Furthermore, bony exostoses (hyperostotic bone) that develop in fish bone are comprised of cellular bone,

Table 2.10 The effects of reduced atmospheric pressure on collagen[a] and Ca^{++} accumulation, and tensile strength in healing bone fractures of rat tibiae[b]

	7 Days post-fracture		14 Days post-fracture		Difference (%)[c]
	Hypoxia	Control	Hypoxia	Control	
Hydroxyproline (mg)	1.32 ± 0.12	1.39 ± 0.10	1.79 ± 0.15	2.41 ± 0.18	−25
Ca^{++} (mg)	1.45 ± 0.23	2.45 ± 0.41	4.76 ± 0.63	8.78 ± 0.74	−12
Tensile strength (kg)	1.36 ± 0.10	1.65 ± 0.08	1.95 ± 0.18	2.22 ± 0.09	−46

[a] Collagen accumulation was measured as hydroxyproline.
[b] Based on data from Penttinen *et al.* (1972).
[c] Difference as seen 14 days post-fracture.

even when the individual and species has cellular bone (see Box 2.2). As elaborated below, acellular bone:

- as might be expected, displays an unusual mode of development involving polarized secretion of osteoid as osteoblasts retain their position on the bone surface;
- plays a role in Ca^{++} regulation;
- can be resorbed;
- repairs fractures; and
- is found in the dermal skeleton of cartilaginous fishes (sharks and rays).

Development

Although especially interesting, the development of acellular bone has received little study. Extant higher teleosts with acellular bone arose from ancestors with cellular bone. Similarly, during teleost ontogeny, acellular bone

arises secondarily from a cellular tissue. In his pioneering studies carried out in the early 1960s, Melvin Moss from Columbia University, New York, showed that acellular bone arises by osteogenesis from one or a combination of more than one cell type: (i) periosteal osteoprogenitor cells (OPCs), (ii) cells in tendons (tendinous osteogenesis), and/or (iii) metaplastic transformation from cartilage.

In each of the situations – and the three types of osteogenesis may be found within one individual at different skeletal sites – one of two fates befalls an osteoblast. Some become trapped as osteocytes within the matrix, after which they become pyknotic and mineralize, much as in the mineralization of lamprey cartilages *in vitro* or in the lignification of plants. Other osteoblasts remain on the surface of the bone, a location that implies *polarized secretion of osteoid*. That implication was made real by Sunetra Ekanayake in her studies on vertebral osteogenesis in the Japanese medaka, *Oryzias latipes*, studies that led to the model shown in Figs 2.9 and 2.10. Osteoblasts avoid becoming entrapped in bone because osteoid is deposited only from the surface of the osteoblasts facing the bone matrix, i.e. deposition is polarized.[36]

Resorption

Surprisingly, teleosts can resorb acellular bone. How do they do it?

When bone resorption was initiated by removing scales from the African mouth-breeding cichlid, *Tilapia macrocephala*, and in two other species, one with cellular and one with acellular bone (see Box 2.3 for scales), osteoblasts could be seen retreating from the mineralization front of the acellular bone. Ekanayake and Hall (1987, 1988) saw a similar phenomenon in vertebral bone of the Japanese medaka, *Oryzias latipes*, and identified two types of mononucleated resorptive cells, both of which lack ruffled borders. With our human orientation, we are accustomed to thinking of *multinucleated* mammalian osteoclasts, which adhere to bone through a specialized cytoplasmic region, the ruffled border. Many teleost osteoclasts are multinucleated with ruffled borders and sealing zones (Fig. 2.11).[37]

Subsequently, it was shown that osteoclast formation could be induced in a teleost, the kelp bass, *Paralabrax clathratus*, and in the leopard shark, *Triakis semifasciata*, after bone particles had been implanted (Fig. 2.12). This is surprising. Cartilaginous fishes would not be expected to be able to produce active osteoclasts, unless osteoclasts and the chondroclasts that resorb cartilage in these fishes are closely related, which indeed they may be (Chapter 15).[38]

Takagi and Kaneko (1995) used intramuscular transplantation of mineral-containing bone particles into rainbow trout, *Oncorhynchus mykiss*, to investigate the temporal sequence of activation of bone-resorbing cells. After two weeks, small multinucleated cells were found on the surfaces of the implanted matrix. After four to eight weeks,

Box 2.2 Hyperostosis

Hyperostosis is the formation of excess bone (exostoses) in teleost fish. I know now that the rounded 'galls' I saw on the vertebral spines of fish skeletons washed up on the beaches off Wollongong, Australia, were examples of hyperostotic bone. Palaeontologists have similarly been confused; fossils from the Miocene of Tunisia, previously identified as the mollusc *Sepia*, were shown by Hewitt (1983) to be hyperostotic bones from teleost fishes, as revealed by the characteristic growth rings discussed in Box 6.3.[a]

Hyperostosis has a broad distribution within fishes, having arisen multiple times even within single lineages; in a study of 92 species from 22 families of teleosts, half the hyperostotic bones found came from members of one family, the jacks (Carangidae), the condition having arisen many times within that family. Interestingly and inexplicably, in species with acellular bone, hyperostotic bone is often cellular. Therein rests a rather nice Ph.D. project.

Although the term hyperostosis is usually applied to fish, other vertebrates develop hyperostotic bone. The ribs of two species of whales show histological evidence of hyperostosis in periostea and bone cortices, abnormally thickened bone (pachyostosis), absence of free medullary cavities, mineralized cartilage, and globular ossification with prolonged subperiosteal osteogenesis, all of which were interpreted by de Buffrènil and colleagues as adaptations to aquatic life.[b]

Hereditary multiple exostoses in humans have been linked to three genes (*Ext-1, Ext-2, Ext-3*) that code for glycosyltransferases (heparan sulphate co-polymerases) for the synthesis of heparan sulphate. Homologues of *Ext-1* and *Ext-2* in mice have identical patterns of expression in long bone primordia, which suggested to Stickens et al. (2000) that they code for a glycosyltransferase complex (Table 27.4, and see Box 18.1). These workers model *Ext* genes as interacting with *Ihh* and parathyroid hormone-related protein in regulating chondrocyte differentiation (see Box 31.2).

[a] A reversal and correction in the opposite direction is the case of the infilled burrows of a shrimp previously identified as plesiosaur embryos (Thulborn, 1982).
[b] See Smith-Vaniz et al. (1995) for the distribution in Carangidae, Meunier and Desse (1986) for an overview of hyperostosis, and de Buffrènil et al. (1990) for whales.

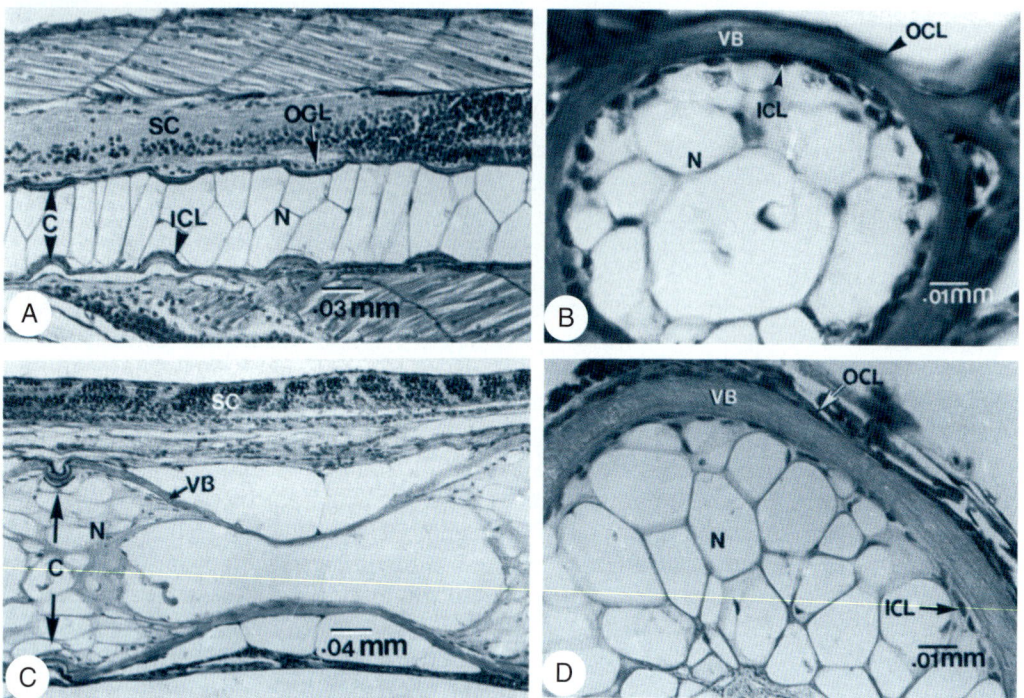

Figure 2.9 Stages in the development of acellular vertebral bone in the Japanese medaka, *Oryzias latipes*, shown as light microscopic histological sections. (A) A longitudinal section one day before hatching shows the spinal cord (SC), the inner and outer cell layers (ICL, OCL) and the constrictions (C) of the notochord (N). (B) A transverse section five weeks after hatching to show the relationship of the acellular vertebral bone (VB) to the inner and outer cell layers of the notochord. (C) A longitudinal section of the vertebral column of an adult medaka, showing the constricted notochord (N) between two vertebrae, which are composed of vertebral bone (VB). SC, spinal cord. (D) A transverse section 20 weeks after hatching. Labels as in (B). Adopted from Ekanayake and Hall (1987).

multinucleated osteoclasts were resorbing the matrix. And, as noted earlier, devitalized bone from eels, *Anguilla anguilla*, implanted into dogfish, *Scyllium canicula*, evokes formation of mono- and multinucleated cells that resorb the implanted bone.

Ultrastructural analysis of osteoclasts in several species of teleost fish has been used to demonstrate resorption of perichondrial, dermal and chondroid bone, as well as elasmoid, the presence of Howship's lacunae, ruffled borders, many mononucleated cells and resorption of unmineralized matrix; Fig. 2.11 shows the ruffled border and ATPase in a multinucleated osteoclast from a juvenile carp, *Cyprinus carpio*. Like their mammalian counterparts, teleost osteoclasts contain vacuolar-ATPase which they use in a vacuolar proton pump.[39]

Repair of fractures

Another approach to the development and physiology of acellular bone is to explore its ability to repair fractures.

Fractured acellular opercular bones and bones of the lower jaws of *Tilapia macrocephala* produce a callus of mineralized cartilage and bone. Fibroblasts from connective tissue surrounding the bone modulate to osteo- and chondroprogenitor cells, which differentiate into the chondroblasts and osteoblasts that deposit the cartilage and bone of the callus. Because adjacent connective tissue cells can

modulate into skeletogenic cells, the loss of bone cells during ontogeny does not impede the ability to repair fractures (in this species). The concept of lineages of cells with skeletogenic potential is explored in Chapters 11 and 12.

Ability to form a fracture callus *is* diminished when fish with acellular bone are maintained in water with lower than normal calcium levels. Under similar acalcemic conditions, species with cellular bone such as the goldfish, *Carassius auratus*, can and do initiate fracture repair (Moss, 1962a).

In a parallel study, Moss (1962b) implanted acellular and cellular bones either subcutaneously or into defects in the femora or crania of adult rats to further assess the potential of the two bone types. When implanted subcutaneously, cellular and acellular bones produced an immune response, following which they were resorbed and removed by host cells. On the other hand, in the intraskeletal sites both bone types were incorporated into the host bones, and then slowly removed by resorption. The implication is that components of the matrix of acellular bone or adherent periostea possess a species specificity that elicits a rejection response.

Ca^{++} regulation

Acellular bone is found in both freshwater and marine teleosts, raising the question of whether it plays a role in regulating Ca^{++} and/or PO$_4^{3-}$. Eighty per cent of the Ca^{++} in

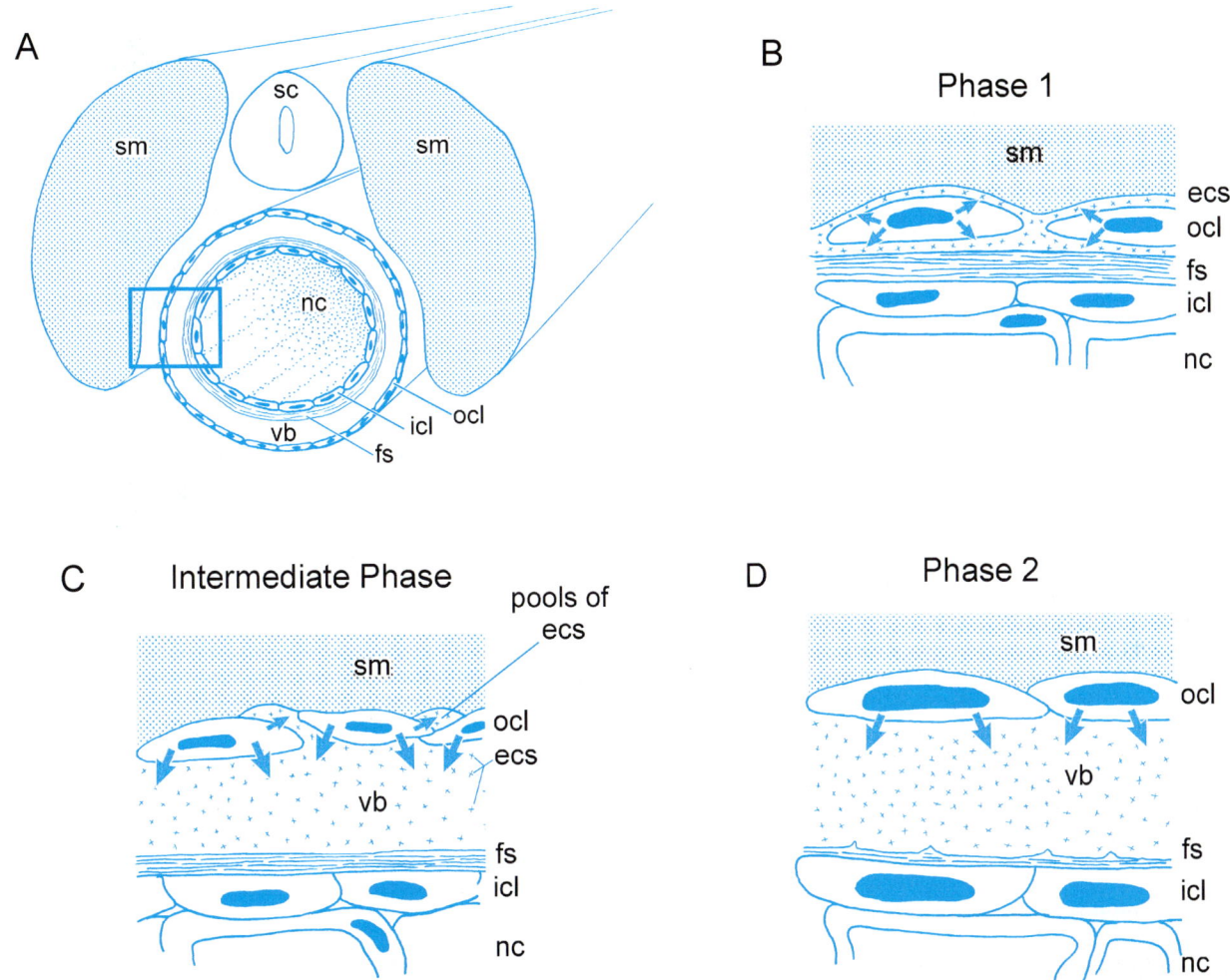

Figure 2.10 The model for the development of acellular vertebral bone proposed by Ekanayake and Hall (1988) on the basis of studies of the Japanese medaka, *Oryzias latipes*. (A) A diagrammatic cross section of a medaka to show the topographical relationships between the notochord (nc) with its inner and outer cell layers and fibrous sheath (icl, ocl, fs), the somites (sm), spinal cord (sc) and developing vertebral bone (vb) (B) An enlargement of the region in the box in A showing extracellular secretion (ecs, arrows) of matrix by osteocytes (ocl) immediately adjacent to the fibrous sheath (fs) deposited by cells of the inner cell layer (icl) of the notochord (nc), as seen in embryonic and larval medaka. (C) In the intermediate phase seen in juvenile medaka, secretion of matrix (ecs) is becoming polarized (arrows) as vertebral bone (vb) is deposited toward the fibrous sheath (fs) of the notochord. (D) In mature (adult) medaka, secretion of bone matrix is fully polarized (arrows), leaving osteocytes (ocl) on the surface and creating acellular bone. Modified from Ekanakaye and Hall (1988).

the killifish, *Fundulus kansae*, is stored in acellular bone in diffusible form and can be mobilized in response to seasonal needs or induced stress, as seen after hypophysectomy (Brehe and Fleming, 1976). This may not, however, be typical.

Since neither Ca^{++} nor PO_4^{3-} is in limited supply in the aquatic environment, the presence of acellular bone has been assumed not to be correlated with retention of Ca^{++} or PO_4^{3-}. The availability of Ca^{++} stored in acellular bone, however, varies from species to species (Table 2.12). Furthermore, bone is often not the site of the most accessible calcium. The highest concentrations were found in the viscera and muscles in a study of eight species, whether inhabitants of fresh or salt water. The lowest concentrations in salt-water species tend to be in the gut (Table 2.12).[40]

Examination of Ca^{45} uptake in the acellular bone of the oyster toadfish, *Opsanus tau*, demonstrated little, which led the researchers to conclude that little osteogenic activity was taking place. However, tetracycline is incorporated into acellular bone within as little as three hours after being administered intraperitoneally to the northern pike, *Esox lucius*, indicating osteogenesis and metabolic activity.[41]

Ca^{++} from the skeletons of species with cellular bone such as the American eel, *Anguilla rostrata*, is similarly mobile. Skeletal cells respond to exogenous calcitonin, pituitary extract or 1,25-dihydroxycholecalciferol by diminished bone resorption, as do osteoblasts and osteocytes of mammalian bone. Interestingly, osteoclastic resorption of scales in freshwater (goldfish) and marine (nibbler fish) teleosts is suppressed by calcitonin.[42]

Box 2.3 Fish scales

Scales, which come in an abundance of types and compositions, form an important part of the dermal skeleton of teleost fishes. They provide protection for the body, especially for such superficial organs as the lateral lines; a barrier between organism and environment, and against infection; and a source of Ca^{++} and PO_4^{3-} that can be mobilized by resorption. In some species they even serve as a source of food for other fishes – *Haplochromis welcommei*, a cichlid from the Mwanza Gulf, feeds exclusively by scraping the scales off another cichlid, obtaining as much protein from one scale as from a planktonic crustacean. The scales of the prey species regenerate, creating a happy and balanced situation all round.[a]

The most extensive analyses of scale development, especially at the ultrastructural level, are those by Jean-Yves Sire and his colleagues in Paris, a research programme that shows no sign of slowing down. Since 1990 they have:

- described scales in actinopterygian fishes, including the range of types from ganoid to elasmoid, the presence of dentine, the underlying osseous plate and the role of paedomorphosis in scale development (Sire, 1990);
- provided one of the first analyses of the development of scale patterns (squamation) in four teleost fish, development usually starting at the peduncle (but with six other patterns described), and discussed the role of epithelial–mesenchymal interactions and the lateral line (Sire and Arnulf, 1990);
- examined the development of elasmoid scales in the zebrafish, *Danio rerio*, and the Japanese medaka, *Oryzias latipes*, and postulated that *Shh* may be involved in scale morphogenesis in zebrafish (Sire and Akimenko, 2004);
- evaluated epidermal–dermal and fibronectin–cell interactions during scale regeneration in the jewel cichlid, *Hemichromis bimaculatus*, after 48 hours of healing, with emphasis on the role of the basement membrane and epithelial attachment in interactions (Sire *et al.*, 1990a);
- provided ultrastructural evidence that the epidermis participates in the deposition of the superficial layer of scales in the zebrafish, *Danio rerio*, during initial development and in regeneration (Sire *et al.*, 1997a,b);
- shown, in only the second ultrastructural study, that osteoclasts resorb elasmoid, perichondrial, dermal and chondroid bone, and that Howship's lacunae and ruffled borders are present, although many mononucleated cells are involved, especially in resorbing unmineralized matrix (Sire *et al.*, 1990b);
- described the bony scales and developing odontodes of armoured catfish, *Corydoras arcuatus* (Sire, 1993; Sire and Huysseune, 1996);
- in analyzing the developing frontal bones and scales in the jewel cichlid, *Hemichromis bimaculatus*, described two condensation types, one well delimited with tightly packed cells (scales and lepidotrichia), one with a loose meshwork of cells (bones and scutes; Sire and Huysseune, 1993);
- provided information on regenerating scales in an ultrastructural analysis of the canaliculi of Williamson in Holostean bone in extant and fossil forms (Sire and Meunier, 1994);
- compared teeth and dermal denticles in the West African herring, *Denticeps clupeoides*, in which teeth cover most dermal bones in the head, and argued for homology of teeth and dermal denticles as neural crest derived and developing under the same genetic programme (Sire *et al.*, 1998); and
- described the development and structure of ctenoid spines on the scales in the Central American convict cichlid, *Cichlasoma nigrofasciatum*. These unusual scales are entirely collagenous, lack an enameloid covering, are connected by ligamentous connections, and resemble but are not homologous with odontodes (Sire and Arnulf, 2000).

Table 2.11 Comparison of the histochemical composition of the osseous layer, osteoid margin, and matrix of trout scales[a]

| Test for: | Chemical reaction | | |
	Osseous layer	Growing osteoid margin	Scale matrix
Calcium salts	+	−	−
Elastic fibres	+ +	+	−
Collagen fibres	+	+	+ +
Mucopolysaccharides	+ +	−	+

[a] Based on data from Maekawa and Yamada (1970).

Other studies that contain valuable information on the nature of the tissues or collagen in teleost scales include: developmental studies of catfish scales as dermal skeleton; a comparative study examining scale composition as acellular or cellular bone; a study that considered *Tilapia mossambica* scales as mineralized derivatives of dermal collagen; and an examination of two types of type I collagen in the scales and bone of the common carp, *Cyprinus carpio*, the two being 'standard' type I collagen with two α1 and an α2 chain [(α1)$_2$ α2] and a collagen with three different α1 chains, designated α1α2α3.[b]

Using histochemical and ultrastructural techniques to investigate aspects of growing scales in rainbow trout, Maekawa and Yamada (1970) found that (Table 2.11):

- the upper layer of the scale contains neutral mucopolysaccharides, sulphated acid mucopolysaccharides and collagen;
- the growing osteoid margin is negatively metachromatic and negative after alcian blue staining, indicating absence of mucopolysaccharides, but it contains collagen as 300Å fibres; while
- the scale matrix is mineralized and contains periodic acid Schiff (PAS)-positive material.

As with all structures, scale development is affected by mutations. A mutation in a member of the tumour necrosis gene family in the Japanese medaka, *Oryzias latipes*, results in lack of scale formation. A mutation in members of the same gene family elicits *ectodermal dysplasia* (a syndrome in which hair and teeth fail to form) in humans, which suggested to Sharpe (2001) that scales, hair and teeth are homologous, although he is careful to note that homology of genes, especially regulatory genes, is a risky basis on which to homologize organs.

The gene for apolipoprotein E is expressed in zebrafish in the epidermis of developing fins and scales, and during fin regeneration, but not in trunk scales. Monnot *et al.* (1999) discuss its potential role in epithelial–mesenchymal interactions, while Quilhac and Sire (1998) describe restoration of subepidermal tissues during scale regeneration in the jewel cichlid, *Hemichromis bimaculatus*.

[a] See J. R. Dunn (1984) for an overview of scale development, and see Miyake and Hall (1994) and Koumans and Sire (1996) for methods of culturing the dermal skeleton, including fish scales and teeth. See Goldschmidt (1996) for *Darwin's Dreampond*, an engaging account of the cichlids of Lake Victoria.

[b] See Ballantyne (1930) and Meunier and Huysseune (1992) for cellular or acellular bone, Lanzing and Higginbotham (1974) and Lanzing and Wright (1976) for *Tilapia*, and S. Kimura *et al.* (1991) for the two types of collagen. Yamada (1971) described the ultrastructure of developing scales in chum salmon fry; Ikeda *et al.* (1973) studied goldfish scale growth.

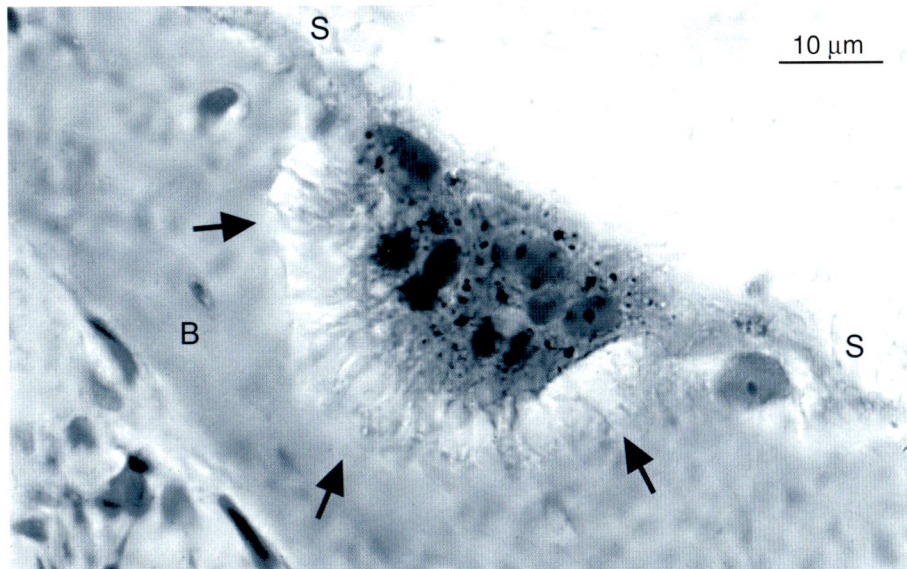

Figure 2.11 A multinucleated osteoclast from a juvenile carp, *Cyprinus carpio*, actively resorbing bone matrix (B). A prominent ruffled border (arrows) and sealing zone (S) can be seen. The granules within the cytoplasm are sites of ATPase. Image courtesy of P. Eckhard Witten.

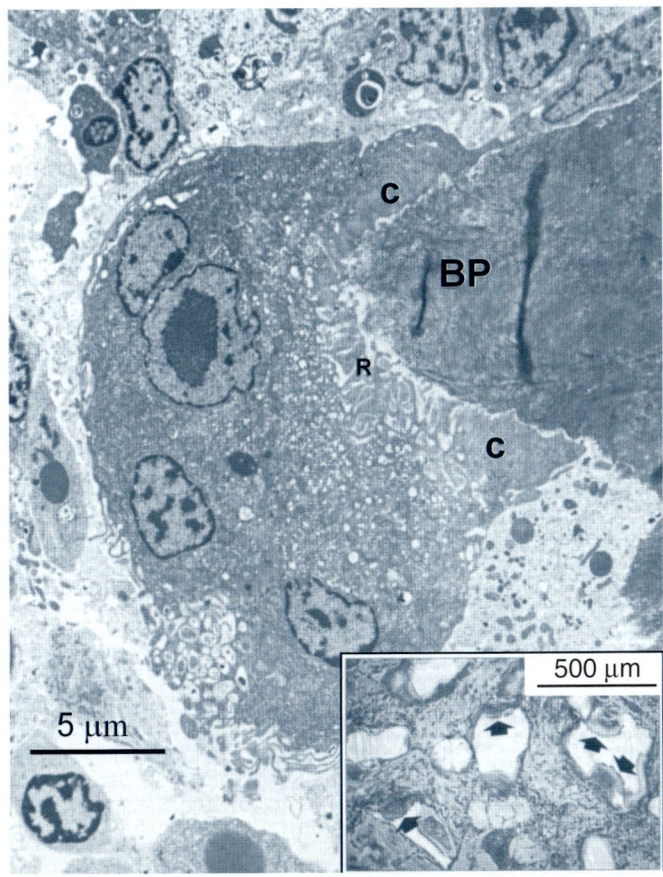

Figure 2.12 Fish produce osteoclasts. A transmission electron micrograph of multinucleated cells with ruffled borders and clear zones (c) elicited on bone particles (BP) derived from the acellular vertebral bone of the kelp bass, *Paralabrax clathratus*, and implanted intramuscularly into another kelp bass. The insert shows osteoclasts (arrows) on the implanted bone. Modified from Glowacki *et al.* (1986).

Hormones – especially estradiol-17β – and nutritional status are important elements in mobilization and/or utilization of Ca^{++} via scale resorption, as shown in juvenile and maturing rainbow trout, *Oncorhynchus mykiss*, by Persson, who found tartrate-resistant acid phosphatase (TRAP)-staining to be an effective marker for scale resorption. Estradiol-17β induces osteoclastic activity associated with the scales whose cells have the E2 estradiol receptor. This receptor also binds testosterone, which may explain why scales are resorbed in both sexes. Ca^{++} levels are regulated during migration and sexual maturation, with osteoclastic activity increasing at these life-history stages in response to the rising levels of estradiol-17β. Persson found what she interpreted as two osteoclast populations. Bone resorption is also influenced by estradiol-17β and by nutritional status; fasting decreasing pharyngeal bone formation.[43]

Resorption of scales in precocious male parr of the Masu salmon, *Oncorhynchus masou*, occurs in autumn and winter, beginning at the posterior margin of each scale. In this early phase resorbing cells are difficult to distinguish from those depositing scale material, which led Ouchi *et al.* (1972) to suggest that forming cells transformed into mononuclear resorbing cells. Once much of the scale margin is resorbed, multinucleated cells are involved in resorption. The availability of Ca^{++} stored within acellular bone might be explained by the presence of vascular canals in the bone; perivascular connective tissue cells may be able to mobilize Ca^{++} (Moss, 1963). However, the basis of calcium utilization from acellular bone remains to be elucidated, and might well provide useful information for the treatment of prevention of pathological necrosis of cellular bone.

Table 2.12 Rank of concentration of radioactive Ca^{++} as stored in several species of teleost fish[a]

Rank of concentration[b]	Fresh water fish				Salt water fish			
	Goldfish[c]	Guppy[d]	L. h[e]	S. j[f]	S. i[g]	S. m[h]	S. n[i]	T. m[j]
1	–	Spine	Fin	Gut	Gut	Scale	Gut	Gut
2	Gill	Head	Scale	Fin	Scale	Fin	Vertebrae	Bone
3	Scale	–	Gill	Gill	Fin	Gill	Fin/Gill	–
4	Skin	–	–	–	–	–	–	–
5	–	–	Gut	Vertebrae	Gill	Viscera	–	–
6	–	–	Skin	Skin	Vertebrae	Vertebrae	–	–
7	–	–	Vertebrae	–	Skin	Skin	Skin	Skin
8	–	Muscle	Viscera	Viscera	Viscera	–	Viscera	Gill
9	Viscera	Viscera	Muscle	Muscle	Muscle	Muscle	–	Muscle

[a] Based on data from Simmons (1971).
[b] 1–9: concentration of stored Ca^{++} from lowest to highest.
[c] Goldfish (*Carassius auratus*); [d] Guppy (*Lebistes*); [e] *L. h*, flatfish (*Limanada herzensteini*); [f] *S. j*, chub mackerel (*Scomber japonicus*); [g] *S. i*, black rockfish (*Sebastes inermis*); [h] *S. m*, Japanese pilchard (*Sardinia melanosticta*); [i] *S. n*, pufferfish (*Sphaeroides niphobles*); [j] *T. m*, Mozambique tilapia (*Tilapia mossambica*).

Aspidine

Aspidine (sometimes *aspidin*), the type of acellular bone found in some of the earliest fossil vertebrates, the jawless heterostracans, is discussed in Chapter 6. The oft-debated relationship between cellular and acellular bone during evolution takes on an element of rationality when viewed in light of knowledge of the ontogenetic development of acellular bone from cellular tissues. Understanding how acellular bone develops in extant teleosts may provide clues as to how aspidine developed.[44]

BONE IN CARTILAGINOUS FISHES (SHARKS AND RAYS)

Acellular bone crops up in the oddest places. Sharks and rays and batoid fishes are cartilaginous fishes, defined as lacking a bony skeleton. The ability of elasmobranchs to produce bone is a good example of progenitor cells of the skeleton retaining, over long periods of evolutionary time, the ability to modulate to either cartilage or bone, cartilaginous fishes having had bony ancestors.[45]

Numerous reports in the literature document bone in sharks and rays, especially in association with the vertebrae, for example at the edges of the neural arches in the small spotted catshark, *Scyliorhinus canicula*. Even within the cartilaginous portion that makes up the bulk of each vertebra, one third of the collagen in the milk shark, *Rhizoprionodon* (*Carcharias*) *acutus*, is type I (bone-type) collagen. The basal tissue supporting the teeth in sharks is also acellular bone, while the pattern of mineralization of shark dentine has also been documented as bone-like.[46]

The most detailed studies on bone in shark vertebrae were undertaken in the 1980s by Jacquelyn Peignoux-Deville. The neural arches of dogfish (*Scyllium canicula*) vertebrae contain hypertrophic cartilage with globular mineralization. Bone forms by perichondral ossification adjacent to the hypertrophic cartilage. The bone that forms is perichondral lamellar bone containing osteoblasts but with few osteocytes. The level of mineralization in cartilage and bone is of the order of 0.83–1.3 g mineral/cm^3, levels comparable to those seen in tetrapod bone.

A less direct line of evidence for osteogenetic activity in cartilaginous fish is that sharks can resorb implanted foreign bone. Implants of devitalized eel bone transplanted into dogfish for two and a half months are resorbed by mono- and multinucleated bone-resorbing cells. We do not know whether these are the same cells that remove cartilage.[47]

NOTES

1. Leeuwenhoek (1693) cited in translation by De Boer, 1988, p. 19.
2. For the structure and function of the Haversian system of bone, including excellent reviews of the literature, see Enlow (1962b) and Jaworski (1992).
3. Goodsir (1845) cited by Brand (1979, p. 290). See Aarden *et al.* (1994) for a review of osteocyte function.
4. Following the pioneering work of Kölliker (1849), workers such as Bassini (1872), Wegner (1872), Morrison (1873) and Kölliker himself (1873) made contributions. See the chapters in *Volume 2: The Osteoclast* of Hall (1990–1994) for overviews of osteoclast origins, structure and function.
5. The form of neurofibromatosis from which Joseph Merrick – the 'elephant man' – was thought to have suffered, was named von Recklinghausen neurofibromatosis after Friedrich Daniel von Recklinghausen (1833–1910), who attracted postgraduate students from all over the world, but rarely bothered even to speak to them; see Hall and Hörstadius (1988) and Hall (1999a) for some details.
6. See H. M. Smith (1947), Ørvig (1967), Gardner (1971), Hancox (1972a), Pritchard (1972a), Ham (1974), Patterson (1977), Ham and Cormack (1979) and Smith and Hall (1990, 1993) for in-depth discussions of bone type and bone development from the perspectives of histologists, palaeontologists and

bone biologists. See Géraudie and Meunier (1982, 1984) for lungfish bone. See volumes 6 and 7 of *Bone* (Hall, 1990–1994) for bone growth. See Hall (2004d) for an evaluation of Arthur Ham's contributions to skeletal biology.

7. See de Ricqlès (1973, 1974, 1976), de Buffrènil *et al.* (1986, 1990) and de Ricqlès *et al.* (1990, 2000) for bone histology as indicator of metabolic state, Bennett and Dalzell (1973) and Desmond (1976) for warm-blooded dinosaurs, and Roy *et al.* (2004) for alteration in bone histology in response to phorphorous deficiency. See Brooks (1971) and Singh *et al.* (1991) for vascularity of bone, and see the chapters in *Volume 4: Bone Metabolism and Mineralization* of Hall (1990–1994) for overviews of bone metabolism.

8. See Neuman (1969) and Ramp (1975) for the reactivity of bone surfaces, Owen (1970, 1971) for the dynamics of surface cells, and Holtrop and Weinger (1971), Furseth (1973), Weinger and Holtrop (1974), Stanka (1975) and Holtrop (1990) for junctional connections.

9. For excellent reviews of the physiology and maintenance of bone and bones, see McLean and Urist (1968), Bourne (1971–1976), Hancox (1972a), Little (1973), Rasmussen and Bordier (1974), Vaughan (1975), Cruess (1982), Urist (1980) and Hall (1985a, 1990–1994). For approaches to skeletal research, see Simmons and Kunin (1979) and Kunin and Simmons (1983). Serafini-Fracassini and Smith (1974), Stockwell (1979), the three volumes edited by Hall (1983a), Kuettner *et al.* (1986) and Hall and Newman (1991) are basic reference sources on the biology of cartilage.

10. See Jones (1974) for parietal bones, and Goss (1970) and Ryg and Langvatn (1982) for antlers.

11. See Pawlicki (1974, 1975b) for an ultrastructural analysis of canaliculi, especially with respect to branching and the arrangement of collagen fibres, and termination of canaliculi in osteocyte lacunae.

12. See Russell (1916) for a discussion and evaluation of the early work on intramembranous and endochondral ossification, and see Smith (1947) for the classification of bone. For basic reviews of the embryonic development of bone see Hall (1988, 1991a, 1997a, 1998b), Cohen (2000b,c) and Olsen *et al.* (2000).

13. See Groman (1982) for the striped bass, Huysseune (1986, 1989, and pers. comm.) for cichlids, and Kanazawa and Takano (1979) for Meckel's cartilage. Also unusual, in the rat, unmineralized cartilage is resorbed by macrophages and fibroblasts unassociated with ossification, vascularization or mineralization (Muhlhauser, 1986).

14. Irwin and Ferguson (1986) investigated whether reptiles could form secondary cartilage by making incisions in parietal bones of three species of lizards and two species of snakes. Bony union typically occurred by 18 days after incision. Secondary cartilage was never seen. See Fig. 5.4 and Hall (1984e) for an overview of skeletogenesis in reptiles.

15. The physiological distinctiveness of neural crest-derived alveolar bone is discussed in Chapter 25. See Nakamoto *et al.* (1983) for calcium content, Nakamoto and Shaye (1984) for the response to caffeine, Helm and German (1996) for weaning, and Atchley and Hall (1991) and Herring (1993) for functional influences on facial development.

16. See Dziedzic-Goclawska *et al.* (1988) for mineralization, Silau *et al.* (1995) for the radiographic study, and the chapters in Volume 4: *Bone Metabolism and Mineralization* of Hall (1990–1994) for overviews.

17. See Moskalewski *et al.* (1986, 1988, 1990) for these transplantation studies.

18. See Sánchez-Villagra and Maier (2003) for the comparative analysis, which includes embryos, neonates and postnatal specimens of the gray short-tailed possum, *Monodelphis domestica*, the four-eyed opossum, *Philander opossum*, the bare-tailed woolly opossum, *Caluromys philander*, and the red-cheeked dunnart, *Sminthopsis virginae*. See Young (2004) for an analysis of modularity of the scapulae in hominoids (apes).

19. Hardesty and Marsh (1990) contains a good discussion of the relevant literature on grafting different types of bone. See Ellis and Carlson (1986), Hennig *et al.* (1992), Peltomäki (1992, 1993), Daniels *et al.* (1987), and the discussion and literature in Hall (2001a).

20. See Bordier *et al.* (1969), Baylink *et al.* (1972), Rasmussen and Bordier (1974), Fornasier (1977), Rodan and Martin (2000) and An (2003) for osteone deposition and mineralization in metabolic bone diseases.

21. See Ascenzi *et al.* (1965) for a basic study of the ultrastructure of osteone mineralization, and Ascenzi *et al.* (1966) for age- and gender-related differences. Using ox bone, Ascenzi *et al.* (1967) showed that not all collagen fibres are totally mineralized. See Takahashi and Frost (1965, 1966) for types of secondary osteones in human bone, and Jaworski and Hooper (1980) and Jaworski *et al.* (1981) for the kinetics of osteones in dogs. Currey (1970, 1982, 1984a,b), Yeager *et al.* (1975) and Frost (1990a–d) discuss how osteones are treated in the biomechanical literature. Robling and Stout (1999) is an excellent review of the morphology of drifting osteones in the cortical bone of humans and baboons. See MacLeod (2002) for osteones in the whale rostrum.

22. See Modrowski and Marie (1993) for the study *in vitro*, T. Nakamura *et al.* (1995) for Fgf-2, Marsh (1998) for delayed vs. non-union, and Trueta (1968) for an excellent history of bone under normal and diseased conditions. Of the estimated 6.25 million fractures/year in the USA some 5–10 per cent are delayed or non-unions. In discussing work on non-union of calvarial defects, Schmitz *et al.* (1990) comment on 186 different causes of non-union!

23. Sarnat (1968, 1986, 1992) and Sarnat and Selman (1982) review the use of markers to study bone growth, with special emphasis on mandibular and craniofacial growth. Sarnat (1992, 2001) reviews his extensive work on postnatal craniofacial growth. See also Dixon and Sarnat (1982, 1985) and Dixon *et al.* (1991) for the results of three symposia on bone growth, and the chapters in volumes 6 and 7: Bone Metabolism and Mineralization of Hall (1990–1994) for bone growth.

24. See Chapter 8 for antlers, Lidauer *et al.* (1985) for blackbirds, and Simmons and Cohen (1980) for fracture callus. Hypophysectomy also retards fracture repair in adult chicks (Negulesco, 1971). See Dias and Richter (2002) for *Australerpeton*. Dias and Richter regard the abdominal scales as equivalent to the gastralia ('stomach stones') seen in modern-day reptiles such as alligators.

25. See Atkinson *et al.* (1977) and Bouvier *et al.* (1991) for alveolar bone, Enlow (1962a, 1968b, 1975) for resorption and remodeling sites in the human facial skeleton, Hinrichsen (1985) for morphogenesis of the human face, P. A. Johnson *et al.* (1976) for sites of growth and remodeling of chimpanzee, *Pan troglodytes*, mandibles, and the papers in Glowacki (2004) for repair and regeneration of oral tissues.

26. See Goret-Nicaise and Dhem (1984) for remodeling *in utero*, Bouvier and Hylander (1981) and Hylander (1981) for diet, Shih and Norrdin (1985) for drilling, and Phillips *et al.* (1982) for nerve ablation and a review of past studies on muscle or nerve ablation in the rhesus monkey. Remodeling of condylar and glenoid fossae of the temporomandibular joint is a major contributor to mandibular prognathism in humans (Ruf and Pancherz, 1998). See Vig and Burdi (1988) and Vig (1990) for an integrative analysis of clinical and basic mechanisms in understanding craniofacial dysmorphogenesis.

27. See Sharpe (1979), Radin and Lanyon (1982), Falkner and Tanner (1986), Snow (1986), Carlson (1990), Lampl *et al.* (1992) and Hoppa and Fitzgerald (1999) for overviews of age-related changes in human bone.

28. See Nishimoto *et al.* (1985) for decreased capacity to form bone with age, Tonna (1960, 1965, 1966) for decreased osteoclast numbers, Simon *et al.* (2003) for reduced numbers of progenitor cells, Bak and Andreassen (1989) for decreased ability to repair fractures, and Ringe and Steinhagen-Thiessen (1985) for the positive affects of exercise on mice, which we hope extrapolates to humans.

29. For an overview of age changes in human bone, see Sharpe (1979), Radin and Lanyon (1982) and Falkner and Tanner (1986). See Jowsey (1968) and Copping (1978) for changing numbers of primary and secondary osteones with age. Fish bone also can be aged using information laid down in otoliths, scales and fin rays (Box 2.3).

30. See Atkinson and Hallsworth (1983) for pore structure; Masters and Zimmerman (1978) for the Alaskan mummy.

31. See Carroll (1997), Hall (1998a, 2002a), Fedak *et al.* (2002a,b), Vickaryous *et al.* (2002, 2003), Cole *et al.* (2003), Franz-Odendaal *et al.* (2003) and Schoch and Carroll (2003) for recent commentaries on bone from the perspectives of development, palaeontology and other fields.

32. See Enlow (1966a,b), Wells (1973) and the chapters in Iscan and Kennedy (1989) for discussions of acellular bone, and Brookes (1971) for an excellent analysis of the blood supply and avascularity of bone.

33. See Jaffe (1972), Ohta and Matsunaga (1974), Walder (1976) and Nixon (1983) for Caisson disease.

34. See Gritzka *et al.* (1973) for acellularity, the chapters in Kuettner *et al.* (1986) for overviews of articular cartilage, Mow *et al.* (1980) for the biomechanics of articular cartilages, and Penttinen *et al.* (1971, 1972) for fracture healing under reduced atmospheric pressure.

35. See Weiss and Watabe (1978a,b) for the terms anosteocytic and osteocytic. Moss (1961a) and Enlow and Brown (1956, 1957, 1958) contain the bulk of the classic histological studies on acellular bone in the teleosts. For recent analyses see the studies by Ann Huysseune, Jean-Yves Sire and P. Eckhard Witten.

36. See Moss (1961a,b, 1963) for basic studies on the formation and mineralization of acellular bone, Person and Philpott (1963) and Person (1983a) for lignification in plants, and Langille and Hall (1993a) for mineralization of lamprey cartilage *in vitro*. Meunier (1989) also provides an overview of the acellularization process. See Ekanayake and Hall (1987, 1988) for the model of how osteoblasts remain on the periosteal surface.

37. See Weiss and Watabe (1978a,b) for the studies based on scale removal, Weiss and Watabe (1979) for the retreating osteoblasts, Ryder *et al.* (1981) for the ruffled border, and

38. Witten *et al.* (2000) for resorption of teleost bone by both mono- and multinucleated osteoclasts.

38. See Glowacki *et al.* (1986) and Peignoux-Deville *et al.* (1989) for these studies. Analysis of a vertebral deformity in the tope shark, *Galeorhinus galeus*, provides circumstantial evidence that mineralized *cartilage* can also be resorbed (Officer *et al.*, 1995).

39. See Sire *et al.* (1990b) for the ultrastructural analysis and see Witten (1997) and Witten *et al.* (1999a,b) for V-ATPase and other histochemical features of teleost osteoclasts.

40. See Moss (1961a, 1963, 1965) and Dacke (1979) for the lack of correlation with Ca^{++} or PO_4^{3-}, and Simmons (1971) for the concentrations of Ca^{++} in various organs in the eight teleost species.

41. See Simmons and Marshall (1970) for the study with the toadfish, and Meunier and Boivin (1974) for that with the northern pike.

42. See Fenwick (1974) for the eel study, and Lopez (1970a,b), Lopez and Martelly-Bagot (1971), Lopez and Deville (1973) and Lopez *et al.* (1980) for the response of cellular bone to hormonal stimulation. A calcitonin gene-related peptide receptor (CGRPR) cloned from the gills of the Japanese flounder, *Paralichthys olivaceus*, but also in found in bone, is the first non-mammalian CGRPR isolated (N. Suzuki *et al.*, 2000a,b). Calcitonin is synthesized by cells of the ultimobranchial gland in bonnethead sharks, *Sphyrna tiburo*, serum concentrations rising fivefold to some 400 ng/ml during early pregnancy (Nichols *et al.*, 2003).

43. See Persson *et al.* (1994) for the initial study, Persson *et al.* (1995) for TRAP as a marker, Persson (1997) and Persson *et al.* (1998) for osteoclastic activity, Persson *et al.* (1997) for nutritional status, and Persson *et al.* (2000) for testosterone binding to the estrogen receptor oestradiol.

44. Moss (1968a–c), Hall (1975a, 1995a, 1998a,b, 2002a,b), Maderson (1975), Schaeffer (1977), Patterson (1977) and Smith and Hall (1990, 1993) address the interplay between knowledge of developmental processes and evolutionary mechanisms with respect to skeletal evolution. See Matthew (1925), Romer (1969), Smith and Hall (1990), Conway Morris (1994, 1998), Carroll (1997), Shubin and Marshall (2000), Hall (2002a) and Hadly (2003) for 80 years of trying to integrate palaeontology with studies using extant organisms.

45. See Miyake and McEachran (1991) and Miyake *et al.* (1992) for comparative analyses of the gill arch skeleton and rostral cartilages in batoid fishes. Moss *et al.* (1964) demonstrated that ectodermal and mesodermal collagen can mineralize in shark teeth and scales, using the term *ichthylepedin* for collagen from fish scales.

46. See Applegate (1967), Moss (1970, 1977), Kemp and Westrin (1979), Hall (1982a), Peignoux-Deville and Janvier (1984) and Bordat (1987) for bone at the base of teeth and in the neural arches of elasmobranchs, Rama and Chandrakasan (1984) for type I collagen in vertebrae, and Kemp and Smith (1976) for bone-like mineralization of dentine.

47. See Peignoux-Deville *et al.* (1981, 1982, 1985) for bone formation, and Kemp and Westrin (1979) and Peignoux-Deville *et al.* (1989) for resorption of implanted bone. Romer (1942) discusses the value of cartilage as an embryonic adaptation and the phylogenetic trend away from bone, i.e. of reduced ossification, while Moss (1977) discusses possible inhibition of osteoblast activity by mineralized cartilage in other parts of the shark skeleton.

Cartilage

Chapter 3

Chondrones are to cartilage as osteones are to bone.

TYPES

Vertebrate cartilage(s) is subdivided into types primarily according to histology. On histological grounds, cartilage is:

- *hyaline*, if the matrix is composed predominantly of glycosaminoglycans;
- *elastic*, if elastic fibres are present in the extracellular matrix (ECM); or
- *fibrous* (fibrocartilage), if the matrix has an enriched collagenous fibre content, usually accompanied by deposition of type I collagen.[1]

In mammals, *hyaline* cartilage is found in the embryonic models of endochondral bones and in portions of the laryngeal cartilage (Fig. 3.1); *elastic* cartilage in the pinna, larynx, epiglottis and intervertebral discs (Figs 3.2 and 3.3); and *fibrous* cartilage where ligaments and tendons attach to bone (Fig. 3.4), in intraarticular discs of joints and as articular cartilages at joint surfaces.[2]

This superficially simple and apparently 'universal' classification applies with any accuracy only to mammalian cartilages. The types of cartilage recognized using histological criteria expand considerably when other tetrapods and fish are considered, and expand enormously when lamprey and hagfish skeletons are taken into account.

Those cartilages that are models for endochondral bones are *primary* and form the primary cartilaginous skeleton. Cartilages that arise from the periostea of membrane bones are *secondary cartilages* because they arise after (secondary to) bone formation. Sesamoid cartilages

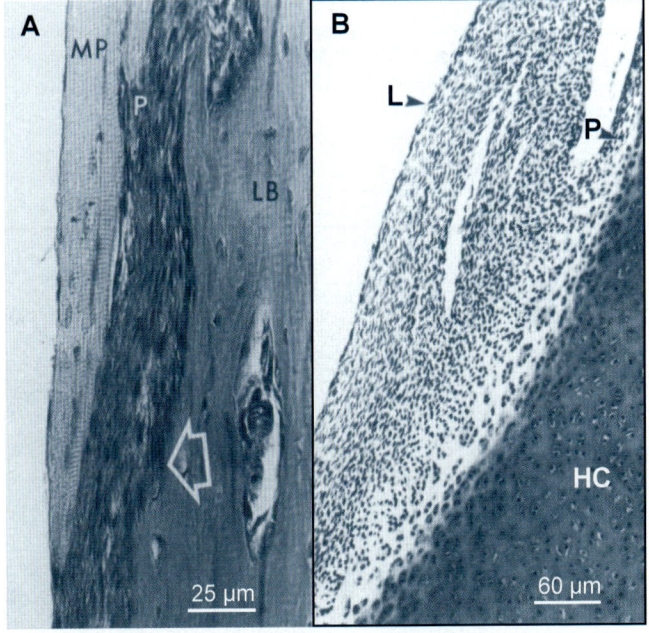

Figure 3.1 Muscles and ligaments insert onto bone. (A) A longitudinal section of the insertion of a muscle (MP) onto lamellar bone (LB) via collagen fibres (arrowhead) that cross the periosteum (P) in a 49-day-old rabbit. (B) A longitudinal section of the attachment of the medial collateral ligament (L) to hyaline cartilage (HC) of the femur in a two-day-old rabbit. Adapted from Hurov (1986).*

or *sesamoids* arise outside the skeleton, and so are extraskeletal (*ectopic*) in origin and initial location, although many later fuse with the primary skeleton.

Many cartilages develop from mesoderm (Chapter 16), but many craniofacial cartilages – Meckel's cartilage (Figs 3.5 and 3.6), much of the chondrocranium and the visceral arch (gill) cartilages (Figs 3.7 and 3.8) – are of neural crest (ectomesenchymal) origin (see Chapter 17).

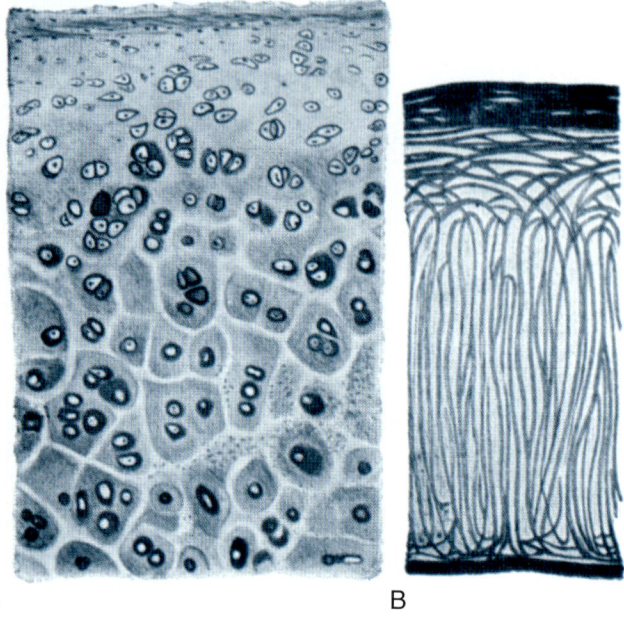

A   B

Figure 3.2 (A) A drawing of a histological section of human tracheal (elastic) cartilage from the perichondrium at the top to mature chondrocytes at the bottom, the latter in groups identified as chondrones (see Fig. 3.9). (B) A diagram of fibril arrangement in tracheal cartilage from a cow, as depicted by Benninghoff (1925), oriented as in (A) but with perichondria at top and bottom. Modified from Murray (1936).

Neural crest-derived cartilages thus constitute much of the endoskeleton of the head. Secondary cartilage on the clavicle, and cartilages that arise in the valves or septa of the heart also arise from neural crest cells, as do ectopic cartilages in the heart and ectopic cartilages or sesamoids in the jaws. Cartilages derived from mesoderm constitute what we normally consider as the endoskeleton, but also include cartilages that form as sesamoids and become associated with the endoskeleton (the patella or knee cap), as well as cartilages that arise in tendons, ligaments or soft connective tissues and that are not normally included as components of the endoskeleton.[3]

CHONDRONES

Chondrones are to cartilage as osteones are to bone.

It had been known for centuries that osteones are the basic structural unit of compact bone (Chapter 2), yet although the notion of similar units – chondrones – in cartilage has been around for a century (Figs 3.2 and 3.9), extensive analysis of chondrones, for the most part carried out by C. Anthony Poole in Auckland, New Zealand, is less than 20 years old. A chondrone consists of a nest of chondrocytes and their collective pericellular matrix

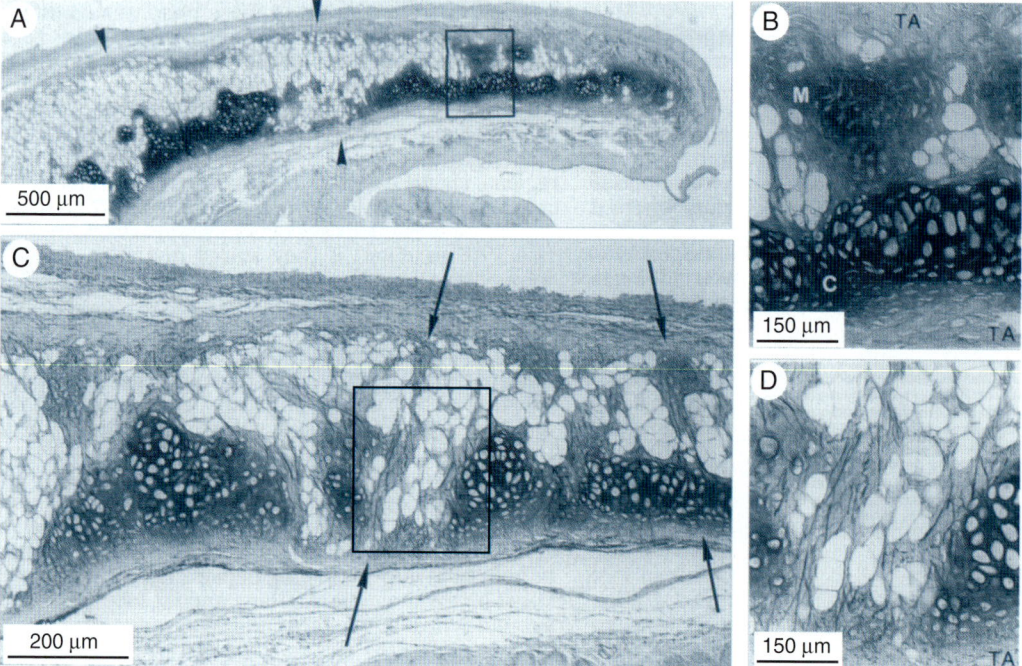

Figure 3.3 Elastic cartilage, as seen in canine epiglottal cartilages. (A) A longitudinal section of adult cat epiglottal cartilage and myxoid surrounded by a connective tissue sheath (arrowheads). The box shows the location of (B), which is a higher power image showing cartilage (C), myxoid tissue (M) and the sheath of the tunica albuginea (TA). (C) Connective tissue strands (septa) cross the tissue and connect to the connective sheath shown in A. (D) A higher power image of the box in C showing elastic fibres in the connective tissue septa, primarily between fat lobules. Modified from Egerbacher *et al.* (1995).

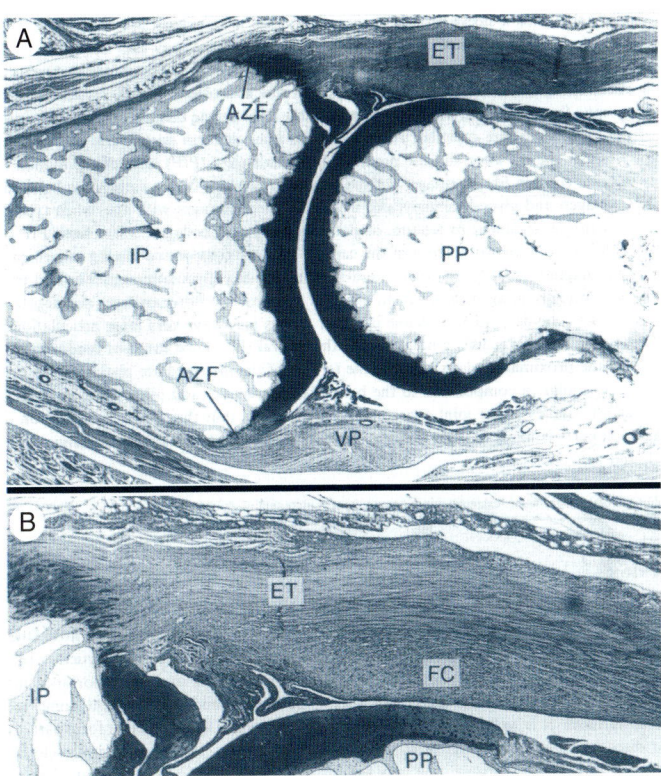

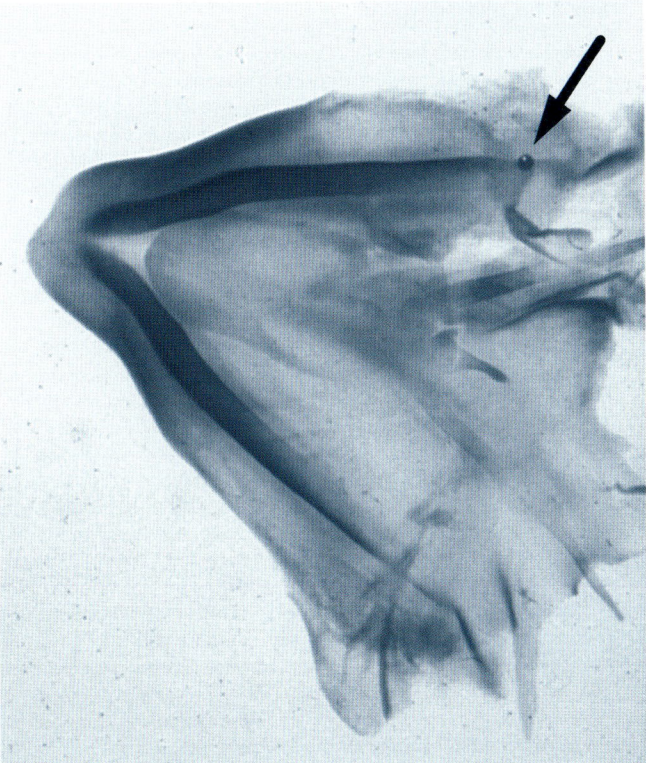

Figure 3.4 The synovial joint between two phalanges of a human finger, as seen in longitudinal histological section. (A) The extensor tendon (ET) and a ventral fibrocartilage (VP) form the dorsal and ventral portions of the joint capsule. Both tendon and fibrocartilage attach to the intermediate phalanx (IP) via attachment zone fibrocartilage (AZF). (B) A higher magnification of the attachment of the extensor tendon shown in A to show fibrocartilage (FC) on the deep surface of the tendon. Modified from Ralphs and Benjamin (1994).

Figure 3.5 Ventral view of the head of a chick embryo of Hamburger–Hamilton (H.H.) stage 36 showing a bead that was implanted at H.H. 22, now located at the end of Meckel's cartilage. Also see Figs 3.6 and 12.3, and see Ekanayake and Hall (1997) for details.

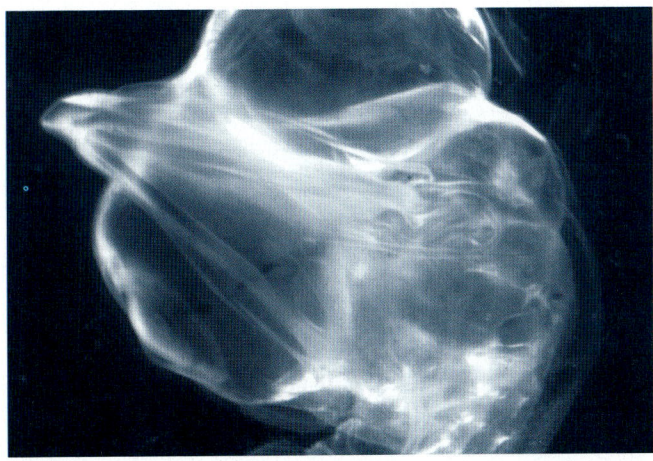

Figure 3.6 Ventral view of the head of a H.H. 36 chick embryo showing a bead that was implanted at H.H. 22. Illuminated to reveal details of skeletal development. Also see Figs 3.5 and 12.3, and see Ekanayake and Hall (1997) for details.

(Fig. 3.10). Poole and his colleagues have demonstrated that:

- pericellular capsules of fine fibrils characterize deep but not superficial canine tibial articular cartilage;
- adult human articular cartilage contains similar chondrones;
- chondrones can be physically isolated as intact structural units;
- chondrones are rich in proteoglycans (Fig. 3.10);
- type VI collagen is found in the pericellular capsule of isolated canine tibial chondrones, suggesting that chondrones resist loading in dynamic compression (Box 3.1); and
- chondrones isolated from human or canine osteoarthritic cartilage are defective in collagen and proteoglycan.[4]

CARTILAGE GROWTH

In contrast to the solely appositional growth of bone, cartilage growth is both internal (*interstitial*) by chondroblast/cyte proliferation and deposition of ECM and *appositional* by addition to cartilage surfaces. This diversity of growth mechanisms is thought to explain the success of cartilage as an embryonic skeletal tissue. The large volumes of fluid bound by cartilaginous matrices make cartilage an ideal tissue to resist compression. In addition, the resistance of cartilage to compression enhances its usefulness in growing organisms and makes it an ideal tissue for joint surfaces (Figs 1.8 and 3.11). Romer (1942) contains a classic and oft-cited discussion of the value of

Box 3.1 Collagen type VI

Fibrillar collagens (see Fig. 2.5) occupy a key position in verte-brate evolution for several important reasons, among which is the structural role collagen plays in the Metazoa. This is reflected in part in the fact that collagen arose before the genome duplication event at the base of vertebrate evolution.[a]

Collagen type VI, a glycoprotein with a short collagenous central domain, is a ubiquitous constituent of animal connec-tive tissues, existing as 5-nm-diameter microfibrils with a perio-dicity of 100 nm. Within the specialized connective tissues that we know as skeletal tissues, collagen type VI is up-regulated with chondrocyte differentiation and plays a mechanical role.

In ovo, type VI collagen is expressed in articular tibial chon-drocytes from 13 days of incubation onwards. From cultures of tibial chondrocytes of chick embryos we know that expression of type VI collagen is down-regulated with dedifferentiation and attachment. Chondrocytes also divide and synthesize type I collagen under these conditions. If placed in suspen-sion culture, dedifferentiated chondrocytes cease dividing, rapidly up-regulate type II and VI collagen, and hypertrophy (Quarto *et al.*, 1993).

Consistent with playing a role in cartilage that is under mechanical stress, type VI collagen is found in the pericellular capsule of isolated canine tibial chondrones (see text, Chapter 3), suggesting that chondrones resist loading in dynamic compression. Site-specific fibres composed of colla-gen types III and VI and fibrillin are found where tendons and ligaments attach to human mineralized bone cortex and to periostea.[b]

[a] See Hall (1998a), Boot-Handford and Tuckwell (2003) and Cole and Hall (2004a,b) for discussions, and see Ayad *et al.* (1994) for collagen type VI.
[b] See Poole *et al.* (1988a) for type VI collagen in chondrone capsules, Keene *et al.* (1991) for collagen type VI, and Liu *et al.* (1995) for an overview of collagen in tendon, ligament and bone healing.

- *superficial canals*, found in unmineralized cartilage, and containing fibroblasts and macrophages;
- *intermediate canals*, found in hypertrophic cartilage, and containing fibroblasts with lysosomes; and
- *deep canals*, found in mineralized cartilage, and con-taining chondroclasts.[5]

The idea that canals act as a conduit for stem cells is not new. Canals have been identified using a variety of techniques. Lutfi (1970a) identified stem cells in perichon-dria and cartilage canals in the growth plates of nine-day-old chicks using ^{3}H-thymidine over three decades ago. Cole and Wezeman (1985) analyzed the basement mem-brane composition of cartilage canals during cartilage differentiation and epiphyseal ossification; type IV colla-gen is not present at initial perichondrial invasion of the canals but laminin is present at all stages (and laminin is mitogenic; Box 9.3). They, and Ganey *et al.* (1995), saw cartilage canals as a conduit for stem cells. Nerves can course through cartilage canals as, for example, unmyeli-nated axons in cartilage canals associated with the initia-tion of secondary ossification in rat knee joint (Fig. 3.12).[6]

In the course of studies on vitamin D-enhanced osteo-clastic bone resorption (see Box 27.1), mercury porosime-try was proposed as a method to measure the volume of vascular canals. An experimental approach to the study of cartilage canals in rat tibiae, involving culturing tibiae in the presence of puromycin or thyroxine, demonstrates that programmed cell death, stimulated by thyroxine, contributes to canal growth.[7] Cartilage canals, therefore, play several important roles in development, mainte-nance, growth and removal of cartilage. They also facili-tate development of secondary ossification centres.

cartilage as an embryonic adaptation in the context of the evolutionary origins of bone. Cartilage growth is taken up in depth in Chapters 26 and 30.

CARTILAGE CANALS

While cartilage is normally avascular, non-chondrified channels may be present within cartilaginous extracellular matrices. These channels – *cartilage canals* – carry blood vessels some distance into the body of a cartilage. Cartilage canals also supply progenitor cells for bone formation and for resorption of mineralized cartilage and osteoid. The sequence of events that replace some carti-lages with bone depends upon blood vessels being brought into cartilage via these canals. In developing human long bones, cartilage canals appear during the 12th week of gestation just before the first breakdown (chondrolysis) of unmineralized cartilage and invasion of the cartilage by blood vessels. Mineralization of the cartilage follows quickly (see Box 4.3).

Cole and Wezeman (1985) describe three types of cartilage canals, occurring at three stages during the development of mouse epiphyseal cartilages:

SECONDARY CENTRES OF OSSIFICATION

As already introduced, *primary ossification* of the carti-laginous rudiment of a long bone occurs subperiosteally around the shaft and by endochondral replacement of cartilage of the shaft or metaphysis. Subsequently, a *sec-ondary ossification centre* may form distally within the epiphysis; the time of appearance of such centres during the life of a long bone or individual varies greatly from group to group. Both primary and secondary ossification centres require that osteogenic cells be brought into the cartilage model. For primary ossification, those cells are brought in by the vascular system as it invades the miner-alized matrix of the hypertrophic chondrocytes. A sec-ondary ossification centre normally requires that canals penetrate the epiphyseal cartilage to bring osteogenic cells into the centre of the epiphysis. Haines (1939) described the primitive secondary centres of ossification of the New Zealand tuatara, *Sphenodon punctatus*, emphasizing that most early tetrapods lacked secondary ossification centres. In *Sphenodon*, cartilage mineralizes and is replaced by bone and marrow, processes that

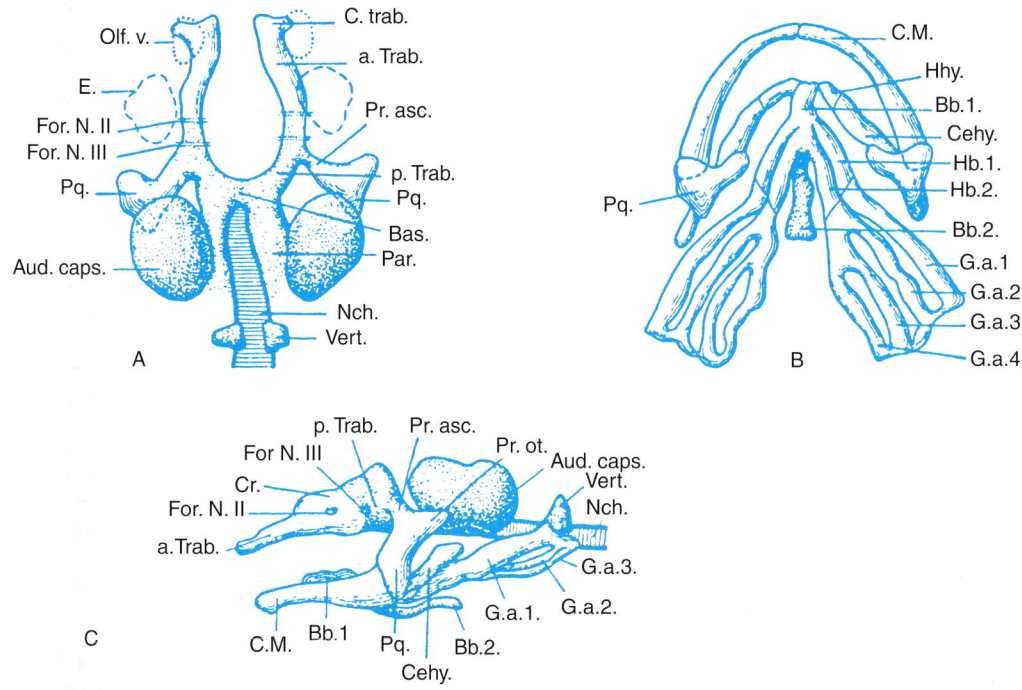

Figure 3.7 The developing cartilaginous skull (chondrocranium) of the Mexican axolotl, *Ambystoma mexicanum*. (A) The neurocranium, palatoquadrates (Pq.), auditory capsule (Aud. caps.) and first vertebra (vert.) in dorsal view. (B) The viscerocranium, including Meckel's cartilages (C.M.), palatoquadrates (Pq.), hyobranchials (Hb.1., Hb.2.), basibranchials (Bb.1, Bb.2) and gill arch cartilages (G.a.1–4), as seen in dorsal view. (C) A lateral view of the chondrocranium. Other abbreviations for skeletal elements: a.Trab., anterior trabeculae; Bas., basal plate of neurocranium; Cehy., ceratohyal; C.M., Meckel's cartilage; Hhy., hypohyal; Nch., notochord; Par., parachordals; p. Trab., posterior trabeculae. Modified from Hörstadius (1950).

provide a centre of growth separate from the articular cartilage. As the osteogenic cells for secondary centres in mammals arise from cells within cartilage canals (Kugler *et al.*, 1979), reptiles may lack such a source of cells.[8]

Epiphyses in such marsupials as the opossum, *Didelphis*, kangaroo, *Macropus*, and tree kangaroos, *Dendrolagus*, lack cartilage canals. Although a secondary centre of ossification forms, it ossifies directly from the perichondrium, an arrangement that Haines (1941) assumed was primitive for mammals (see Fig. 19.2).

Haines assumed a progressive evolution of secondary centres from a single origin. Representing another view, Walter (1985) thought that secondary centres of ossification evolved independently in mammals, snakes and lizards. Carter *et al.* (1998b) revisited the evolution of long bone epiphyses, especially the role of mechanical factors in their development, finding that:

- *periosteal ossification* dominates in fish, amphibians, birds and dinosaurs, all of which are taxa in which mechanical stresses decrease the importance of secondary ossification centres; and
- *periosteal and endochondral ossification* play equivalent roles in mammals and lizards, taxa in which mechanical stress increases the need for secondary ossification centres.[9]

In marine leatherbacks there is an extensive system of vascular cartilage canals (Rhodin *et al.*, 1981), and Carroll

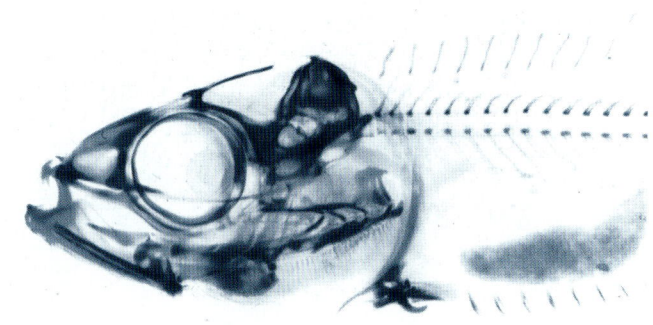

Figure 3.8 A specimen of the alewife, *Alosa pseudoharengus*, cleared and stained with alcian blue and alizarin red to reveal the craniofacial and viscerocranial skeletons. Photographed by Tom Miyake.

(1977) describes secondary ossification centres in lizards in the context of the evolutionary origin of lizards. Most long bones in domestic fowl do not contain secondary ossification centres, the exceptions being the proximal end of the metatarsus, and the proximal and distal ends of the tibiae. Consequently, growth rates vary among and between bones in fowl; see Table 3.1 for growth of different long bones.

Fibrocartilage often forms at the insertion site of a tendon onto a bone, a site known as the *enthesis*

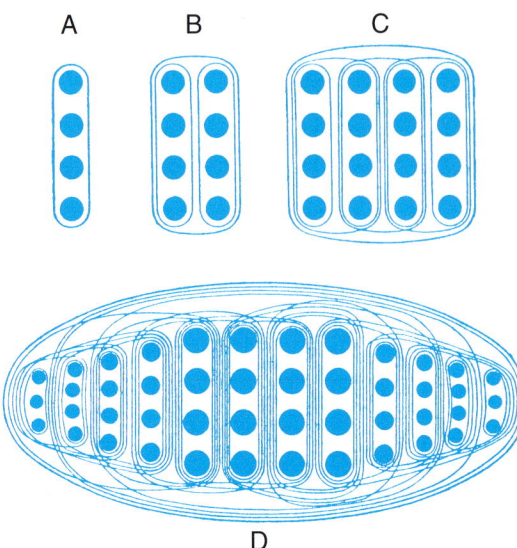

Figure 3.9 Benninghoff's depiction of increasing numbers of chondrones and the arrangement of fibrils around chondrones (A–C) and within hyaline cartilage (D). One (A), two (B), three (C) and 12 (D) chondrones are shown. The outer layer in D is the perichondrium. Each black circle represents a chondrocyte and its ECM (see Figure 3.2). From Murray (1936).

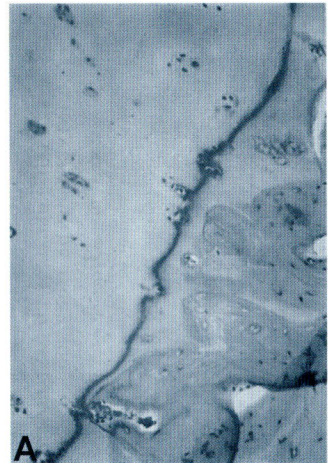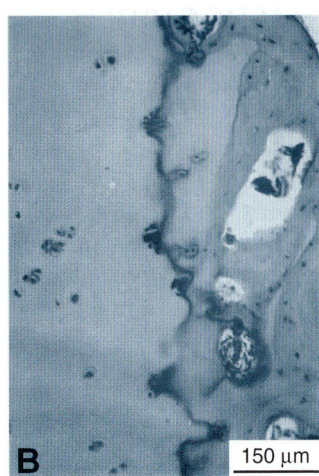

150 μm

Figure 3.11 The tidemark: where mineralized cartilage meets bone. (A, B) The junction of articular cartilage (left) and subchondral bone (right) in the femur of a 21-year-old human male to show the tidemark, which is basophilic after staining with haematoxylin and eosin. Modified from Havelka *et al.* (1984).

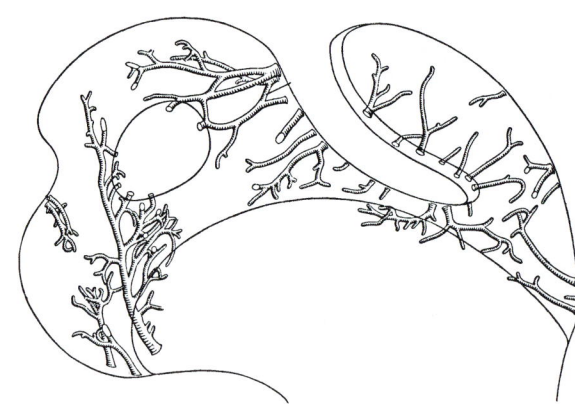

Figure 3.12 The distribution of cartilage canals in the proximal end of the femur from an 18-month-old child. From Le Gros Clark (1958).

Table 3.1 The rate of growth in length of the long bones of chicks of domestic fowl between 3 and 14 weeks after hatching[a]

Long bone	Rate of growth (mm/week)
Humerus	5
Femur	5
Tibia	8
Metatarsal	5

[a] Based on data from Church and Johnson (1964) from a cross of a New Hampshire male with a Barrier Rock female.

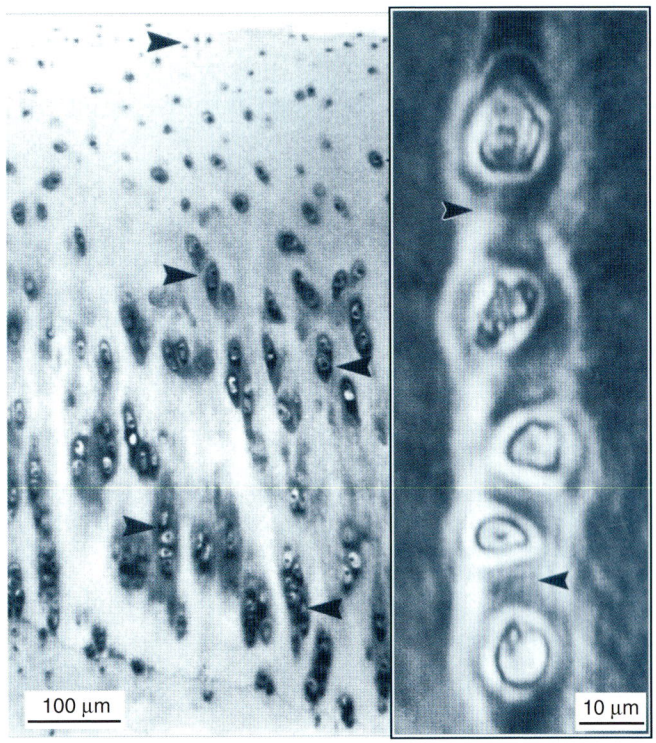

100 μm

10 μm

Figure 3.10 Chondrones – the structural units of cartilage. On the left is a longitudinal section of a growth plate cartilage from the tibia of a dog stained with haematoxylin and eosin to show the organization of chondroblasts/cytes into longitudinal groups of chondrones (arrowheads). On the right is a phase-contrast image of five chondrones enclosed in a common fibrous capsule (arrowheads). Modified from C. A. Poole *et al.* (1988a).

(see Fig. 3.4). The fibrocartilage arises, *not from within the tendon, but from primary cartilage of the long bone rudiment*, which is eroded rapidly, forming what is essentially a secondary ossification centre within a fibrocartilage.

I now turn to elastic cartilage as an example of how to identify a specific type of vertebrate cartilage.

ELASTIC CARTILAGE

The availability of specific histochemical procedures and, more recently, of immunohistochemistry, means that elastic fibres and elastin have now clearly been shown to be associated with cartilages, bones and joints. In some cartilages, elastic fibres are in sufficient quantity that the cartilage is known histologically as *elastic cartilage*, and the cartilaginous elements as elastic cartilages. Elastic cartilage of the mammalian external ear – the pinna – is perhaps the best studied (see Figs 3.2 and 3.3).[10]

Elastic fibres

Elastic fibres are also prominent in joints; we have descriptions of elastic fibres in murine temporomandibular joints from six days prenatally to two months after birth. Rabbit meniscal cartilage contains thick collagen fibres, elastin and fibronectin deposited by cells called *fibrochondrocytes*. Elastic fibres are found in the articular fibrous tissue of various joints, and in the fibrous perichondria of hyaline and fibrous cartilages. Within bones, elastic fibres are found in the outer fibrous layer of the periosteum, but not in the inner cellular (cambial) layer. We also know that elastin can form in tumours; chondroid containing elastin and type II collagen is present in salivary gland adenomas.[11]

The cells

Just as we can recognize elastic cartilage as a subset of cartilages, so the cells that deposit elastin-rich extracellular matrices form a recognizable subset. *Auricular*

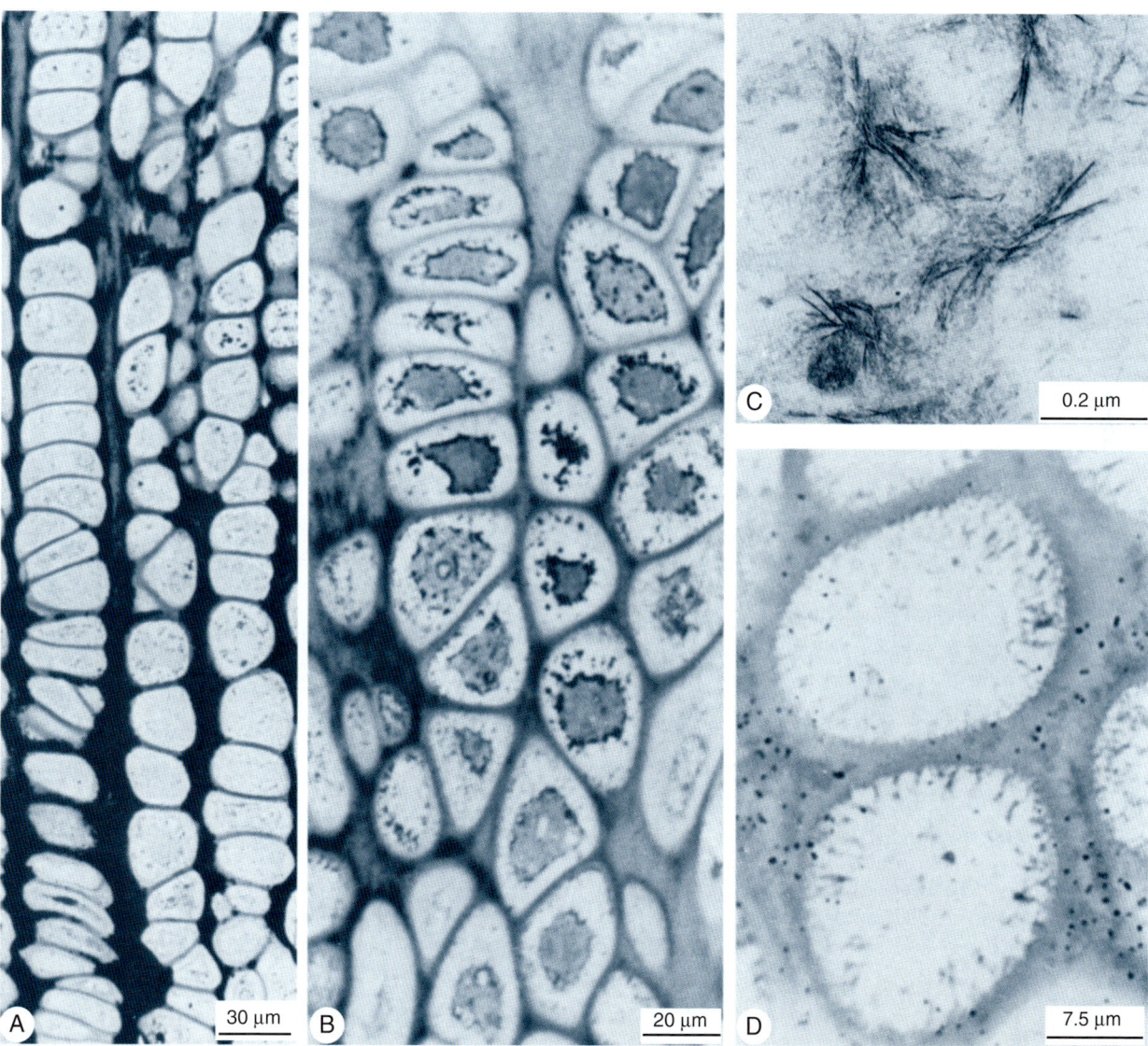

Figure 3.13 Mineralization of rachitic albino rat cartilage is effected by matrix vesicles *in vitro*. (A) The hypertrophic (upper) and mineralizing (lower) zones of a growth plate 48 hours after intraperitoneal injection of phosphate shows extensive mineralization (black), especially in the longitudinal septa. (B) Mineralization of rachitic cartilage, cultured in a medium that promotes mineralization, and revealed by Kossa staining. (C) After three hours in a mineralizing solution, matrix vesicles from rachitic hypertrophic cartilage accumulate apatite crystals, as seen in this electron micrograph. (D) Hypertrophic chondrocytes cultured as in (C) accumulate ^{45}Ca, especially into the longitudinal septa. Adapted from Anderson *et al.* (1975).

Box 3.2 Epidermal and fibroblast growth factor families

The first growth factor to be discovered (in the early 1960s) was named *epidermal growth factor* (EGF, now designated Egf) because its action was thought to be specific to epidermal epithelial cells. Similarly, fibroblast growth factor (FGF, now designated Fgf) was thought to act specifically on fibroblastic (mesenchymal) cells. Despite its name and acronym, Egf also acts on mesenchymal cells; for example, promoting proliferation of osteoprogenitor cells from foetal calvariae, stimulating growth of long bones and calvariae, and suppressing proline incorporation into collagen. Subsequently, Egf was shown to influence cartilage morphogenesis.

The growth factor family has grown. Eighteen Fgfs are now known, probably more by the time you read this. Other classes of growth factors have also been discovered – transforming growth factor-β, bone morphogenetic proteins, insulin-like growth factors, and so forth – each a family with as many as 20 members. In turn, the families can be grouped on the bases of shared sequences into superfamilies. Receptors for most have been isolated. As we now know, these 'growth' factors are even more important as differentiation factors during embryonic development. Had the first study been undertaken with embryonic tissue rather than the eyelids of adult mice, I am sure Egf would have been named epidermal differentiation factor (Edf).[a]

Fibroblast growth factor (Fgf), the second growth factor to be discovered, came in acidic and basic forms, designated aFGF and bFGF, now but two of 18 members of a fibroblast growth family with important roles in embryonic development, including skeletal development.[b]

Initially named for their chemical characteristics, basic fibroblast growth factor (bFGF) is now Fgf-2, acidic fibroblast growth factor (aFGF) now Fgf-1. Because bFGF is the name used in so many of the studies discussed throughout the text, I occasionally remind you the reader that Fgf-2 is bFGF. aFGF has far fewer actions on the skeleton and so I use Fgf-1 for aFGF without a reminder. To further clarify (confuse?) terminology I follow the convention of italicizing and placing the gene in lower case, while capitalizing the gene product without italics. Hence:

● *Fgf-2* = bFGF = the gene *fibroblast growth factor-2* (gene names being in italics);
● *Fgf-1* = aFGF = the gene *fibroblast growth factor-1*;
● *Fgf-2* = bFGF = Fibroblast growth factor-2 (= basic fibroblast growth factor); while
● *Fgf-1* = aFGF = Fibroblast growth factor-1 (= acidic fibroblast growth factor).

Discussion of the skeletogenic roles of Fgf-2 abounds in the text. Box 23.3 provides an overview. Fgf-2 is produced by skeletal cells and can initiate or enhance osteogenesis and chondrogenesis in a dosage-dependent manner over a wide range of concentrations (see Table 15.1). More recently, Fgf receptors have been shown to have major roles in skeletogenesis, especially in maintaining the sutures in growing mammalian skulls. Receptor action is primarily discussed in Chapters 14 and 34.

[a] See Canalis and Raisz (1979) and Raisz *et al.* (1980) for early studies of the action of Egf on osteogenic cells; Pratt (1987) for a review of early studies; Slavkin *et al.* (1992) and Shum *et al.* (1993) for Egf and morphogenesis of mouse Meckel's cartilages; and Coffin-Collins and Hall (1989) and Hall and Coffin-Collins (1990) for Egf, proliferation and cartilage induction. *In vitro*, murine Meckelian chondrocytes respond to Egf with enhanced DNA synthesis and cell division, inhibition of chondrogenic differentiation, and transformation to fibroblasts. See Isheseki *et al.* (2001) for the role of Egf.
[b] For a major review of Fgfs as multifunctional signaling factors see Szebenyi and Fallon (1999).

---

chondrocytes deposit the elastic cartilages of mammalian external ears. Rabbit auricular chondrocytes continue to synthesize, produce and deposit elastic fibres for up to 28 days in organ or monolayer culture, indicating the stability of this particular specialized cell state. Auricular chondrocytes contain matrix vesicles whose function relates to the secretion of elastic fibres, rather than to mineralization, which is the role played by matrix vesicles in other cartilages (Fig. 3.13; see Chapter 22), raising the question of whether elastic cartilage can mineralize. It can and it does.[12]

Rabbit auricular cartilage is an excellent model for wound healing and cartilage repair. As an example of one study, holes were made in rabbit ear cartilages and expression of growth factors examined during subsequent regeneration of the cartilage. Maximal expression of *Igf-1* and *-2*, *Fgf-2*, and *Tgfβ-3* was seen seven days after surgery, after which expression declined. This can be compared with healing of a fractured long bone, where Tgfβ-1 is present at the equivalent stage of repair (Bos *et al.*, 2001). See Box 3.2 for epidermal and fibroblast growth factors, Box 13.2 for interactions between Egf and Tgfβ, and Box 15.1 for an overview of the Tgfβ growth factor superfamily and their receptors.

The elastic cartilages of mice pinnae are also interesting because of the lag phase that occurs between condensation and deposition of elastic fibres. In other cartilages, condensation of mesenchymal cells occurs no earlier than a day before chondrogenesis is initiated and the major ECM molecules are deposited (see Figs 13.8, 19.1 and 19.4, and Chapter 19). According to Mallinger and Böck (1985), condensation of mesenchymal cells for mouse elastic cartilages occurs four days postnatally, glycosaminoglycans are deposited at eight days, and elastin not until 11–13 days.[13]

Table 3.2 Degree of chondrogenesis elicited in defective articular cartilage of miniature pigs by members of the transforming growth factor-β superfamily or other growth factor families[a]

Growth factor		Degree of chondrogenesis (%)
Type	Concentration/ml	
Tgfβ-1	4 ng	2
Tgfβ-1	600 ng	70
Tgfβ-2	600 ng	65
Tgfβ-3	600 ng	50
Bmp-2	1 μg	90
Bmp-13	6 μg	85
Bmp-13	320 μg	90
Insulin-like growth factor-1	400 μg	2
Epidermal growth factor	1.6 μg	3
Tgf-α	2.6 μg	1
Tenascin-C	400 ng	0

[a] Based on data from Hunziker *et al.* (2001).

Elastic cartilage intermediates

Cartilages intermediate between elastic and other cartilage types are known:

- foetal hyaline chondrocytes can be converted to elastogenic chondrocytes that express and deposit elastin-related macromolecules;
- chondrocytes in zones of appositional growth immediately beneath the perichondrium of human elastic cartilage contain free elastic fibres which produce a region that is intermediate between elastic and hyaline cartilage;
- cat epiglottic cartilage is a composite of elastic cartilage, fibrous cartilage, myxoid and fat. Egerbacher and colleagues speculate that the fat cells (adipocytes) could be dedifferentiated chondrocytes, which could make myxoid a precursor of cartilage in this instance (see Fig. 3.3).[14]

The suggested chondrocyte → adipocyte transformation is interesting in light of the secretion of *leptin* by adipocytes. Agents such as hormones or vitamins that regulate the skeleton, even though they are produced outside the skeleton, act locally. Leptin, which binds to a leptin receptor in the hypothalamus, regulates fat deposition. Rodents and mice lacking leptin are obese *and have much enhanced bone masses* because, surprisingly, leptin provides a central, essentially hypothalamic control over bone mass. The reverse transformation (adipocytes → chondrocytes) might provide a use for adipocytes obtained as a result of liposuction to remove excess body fat; micromass cultures of mesenchymal stem cells isolated from 'lipoaspirates' differentiate as chondroblasts when grown with Tgfβ-1, insulin and transferrin.[15]

Lamprey cartilages are fascinating because of their phylogenetic position as living agnathans and because their long-independent evolution allowed novel cartilages, with novel components, to evolve. More fully discussed below, *mucocartilage* of the sea lamprey, *Petromyzon marinus*, contains many elastic-like microfibrils, 11–13 nm in diameter, and associated with the branched fibrous protein *lamprin*, which bears structural similarity to human elastin and cross-reacts with elastin antibodies. Other lamprey cartilages have elastin in the perichondria and peripheral ECM. Spotted spiny 'eels', *Macrognathus siamensis* – which although spiny are not eels but member of the perch (Perciformes) family – have a characteristic mobile rostral tentacle on the heads. The tentacles contain an elastic cellular cartilage, the rostral cartilage. Like the cellular elastic cartilages of lampreys, the rostral cartilage contains many elastic fibres and is enclosed in a thick perichondrium.[16]

Skeletal elements that do not contain elastic fibres can influence the development of structures that do. For instance, the amphibian tympanic membrane fails to thin or form the elastic fibres characteristic of the membrane if the columella is removed (See Box 13.1).[17]

SHARK CARTILAGE

Cartilaginous fishes or *elasmobranchs* have to do with cartilage what other vertebrates do with bone. As a consequence, sharks make an interesting case study for how to be an active, fast-swimming, manoeuverable predator with 'only' a cartilaginous skeleton.

Cartilaginous fishes are classified, in large part, on the basis of having an entirely cartilaginous skeleton, albeit (and as discussed below) a skeleton that may be as densely mineralized as the bone of other vertebrates. That said, cartilage makes up only six per cent of the body weight of basking sharks, *Cetorhinus maximus*, and spiny dogfish, *Squalus acanthias*. Shark cartilage, which is resistant to extraction of matrix products, contains lysozymes, a trypsin inhibitor, mitogens and inhibitors of tumour angiogenesis, all of which may explain why neoplasia is only rarely seen in sharks.[18]

Development and mineralization

Because the bulk of the skeleton of cartilaginous fishes is composed of mineralized cartilage and not bone, it is of interest to know whether mineralization of the cartilaginous skeleton is similar to mineralization of cartilage in those vertebrates with an extensive bony skeleton.

Mineralization of extracellular matrices is not uncommon in sharks. In a series of papers published during the 1970s, based on his analyses of three species of sharks – the grey reef shark, *Carcharhinus menisorrah*, the white tip reef shark, *Triaenodon obesus*, and the lemon shark, *Negaprion brevirostris* – and several species of skates (*Raja* spp), Norman Kemp described two routes to mineralization of the endoskeleton – one by chondroblasts and the other by perichondral fibroblasts that produce a mineralized cartilage matrix *without themselves becoming chondroblasts*. The former is cellular metaplasia, the latter tissue metaplasia. I will not use these terms, however, as the distinctions are not always so clear-cut.

The mineralized jaw cartilage of the lemon shark is a hyaline cartilage, but an unusual jaw cartilage for it contains perichondrial *tesserae* (Fig. 3.14), named after the small square blocks used to make Roman and other mosaics; a tessera in Greek and Roman antiquity was also a small square of bone used as a ticket or token.

Kemp described endoskeletal tesserae that 'grow not by apposition of mineralized cartilage around their inner and lateral margins but by addition of layers of mineralized connective tissue subperichondrially.' Kemp speculated that the perichondrial fibroblasts that deposit the layers of mineralized connective tissues could be considered osteoblasts, and the elasmobranch endoskeleton "considered to consist of calcified cartilage with a veneer of bone as in some placoderms and possibly ostracoderms. This relationship may be primitive rather than regressive."[19]

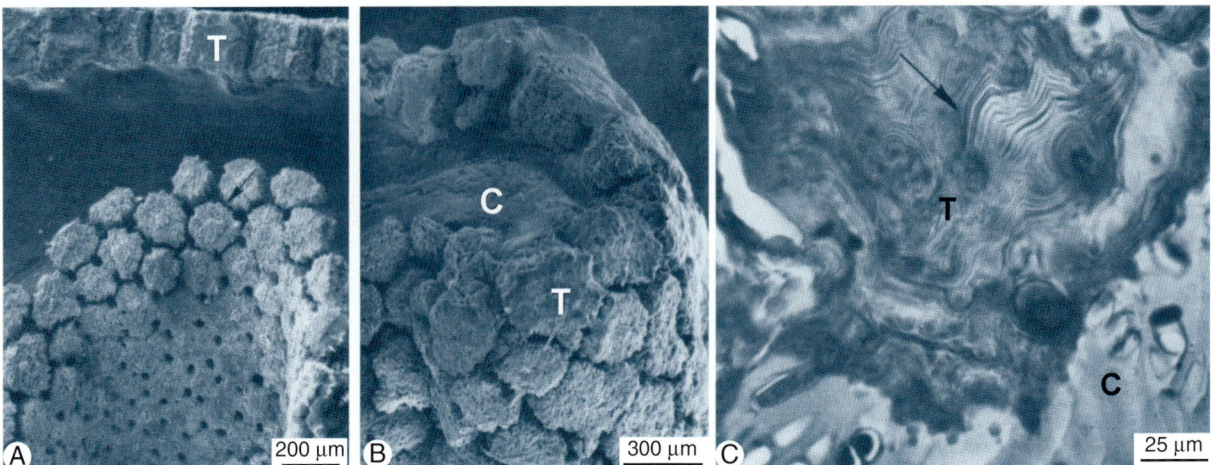

Figure 3.14 Tesserae in the endoskeleton of sharks as seen with scanning electron microscopy. (A) Tesserae (T) from the gill arch of a 60-cm-long gray reef shark, *Carcharhinus menisorrah*. (B) Tesserae from the jaw skeleton of a 40-cm-long lemon shark, *Negaprion brevirostris*. The tesserae lie on the surface of Meckel's cartilage (C). (C) An interior view of tesserae bordering cartilage (C) with waves of Liesegang lines (arrows) indicating mineralization. Modified from Kemp and Westrin (1979).

Subsequently, Kemp and Westrin (1979) looked at the ultrastructure of shark tesserae, identifying an inner zone of mineralized cartilage lacking coarse collagen, and an outer zone of 'osteoblasts' with coarse collagen (type I?) fibres. Mineralized globules enlarge and fuse to form mineralized prisms or tesserae (Fig. 3.14), matrix glycosaminoglycans are removed, collagen fibres exposed, and hydroxyapatite deposited in close association with collagen fibres. Individual tesserae are separated by connective tissue strands. Bordat (1987) described ultrastructural features of vertebral mineralization in the small spotted cat shark, *Scyliorhinus canicula*, finding fibrous cartilages at the edges of each vertebral body, mineralized cartilage in the centre, and bone at the periphery of the neural arches. Layers of prismatic calcium phosphate are deposited as tesserae to strengthen the jaw cartilages. Tesserae continue to grow as long as the skeleton continues to grow, which for sharks effectively means the entire life span, sharks having indeterminate growth.[20]

Growth

Fluorescent markers administered to growing sharks have helped to elucidate the dynamics of skeletal growth. If oxytetracycline is administered to young sharks, two bands are deposited in the vertebral centra per year and three bands are deposited in the cartilage of the upper jaws. The growth rate of the vertebral centra is calculated as 0.31–3.43 μm/day in the wobbegong or carpet shark, *Orectolobus maculates*, and 1.61–4.51 μm/day in the swell shark, *Cephaloscyllium ventriosum*. Similarly, tetracycline-labeled bands in the vertebrae of Pacific angel sharks, *Squatina californica*, are not deposited as a single band once a year throughout the skeleton but reflect different growth cycles in different parts of the skeleton.[21]

The patterns of growth of extinct sharks have also been determined, but from spines rather than cartilage. The Upper Carboniferous xenacanthid shark, *Orthacanthus*, shows growth patterns consistent with the spines arising from dermal papillae (Soler-Gijon, 1999).

As seen in cartilages in other vertebrates, shark cartilage contains canals. Indeed, those in sharks are among the largest known – 120–200 μm diameter in the centra of large sharks.[22]

Inhibition of vascular invasion

It has been argued that mineralization of shark cartilages inhibits osteogenesis, the argument relating to degree and speed of cartilage mineralization. Whether this is a real inhibition is debatable. What is not debatable, however, is that shark cartilage is resistant to the extraction of products from the ECM, inhibits vascular invasion – which in itself may be sufficient to prevent ossification – and resists invasion by tumours. Indeed, neoplasia is only rarely seen in sharks. In addition to resisting the extraction of matrix products, shark cartilages contain lysozymes, a trypsin inhibitor, and mitogens, and inhibit tumour angiogenesis. No mechanistic relationship between lack of vascular invasion and lack of osteogenesis in sharks has been uncovered, although vascular invasion is a prerequisite for endochondral osteogenesis in bony vertebrates (see Chapter 23).[23]

LAMPREYS

As the only extant jawless vertebrates or agnathans, lampreys and hagfishes have attracted much interest. Views differ markedly on the closeness of their phylogenetic relationship to one another, some retaining the classic view of a single agnathan clade, the cyclostomes, others regarding the two as independent. One major difference is that lampreys have a vertebral column, i.e. are vertebrates, but hagfishes do not and are not. Consequently, some view hagfish as craniates but not vertebrates, or as degenerate vertebrates. Both lampreys and hagfishes have cartilaginous skeletons reflecting hundreds of millions of years of evolution independent from the jawed vertebrates. Recent knowledge of lamprey cartilages comes primarily from studies on the sea lamprey, *Petromyzon marinus*, by Glenda Wright and her colleagues in Canada. I introduced these cartilages in discussing elastic cartilage above.[24]

Fate-mapping of skeletal elements of the sea lamprey (Fig. 3.15) demonstrates a major contribution of neural crest cells to the head skeleton. Indeed the patterns of neural crest contributions are remarkably similar to the crest contributions in gnathostomes, indicating that a spatially patterned neural crest arose before lampreys separated from the vertebrate lineage.[25]

With knowledge of cranial cartilages well established, researchers turned to experimental approaches and genetic analysis to further understand lamprey skeletal development and evolution. In a study of epithelial–mesenchymal interactions in lamprey and chick lower jaws, Shigetani *et al.* (2002), following up the neural crest extirpations of Langille and Hall (1986, 1988b, 1989b), saw similar genes acting in both arches but acting in different regions of the arch mesenchyme. They saw this as a heterochronic (timing) shift associated with the origin of the jaws. It is also a heterotopic (placement) shift. We showed that such heterochronic shifts can occur quite readily; they are evident in the timing of the interactions that initiate mandibular development in different inbred strains of mice, whose 'evolutionary' origins are very recent (MacDonald and Hall, 2001; see Green and Witham, 1991, for descriptions and the origins of inbred strains of mice and see Chapter 44 for further discussion of heterochrony).

Tomsa and Langeland (1999) cloned a single *Otx* gene from the sea lamprey, *Petromyzon marinus*. The gene is expressed in cranial neural crest cells, mesenchyme and epithelia of the first pharyngeal arches (which is consistent with a role in chondrogenesis) and in fore- and midbrain. This single gene is equivalent to the two mammalian genes *Otx-1* and *Otx-2*, which are the vertebrate homologues of the gene *Orthodenticle* in *Drosophila*. A single copy of *Otx* in the lamprey means that the major gene duplication event early in vertebrate evolution occurred after the origin of the lampreys.

Cohn (2002) cloned a homeobox gene, *HoxL-6*, from the European river lamprey, *Lampetra fluviatilis*. Expression co-localized with *distalless* (*Dlx*) in the mandibular arch but in a mutually exclusive pattern (see Box 4.1). This study identifies the lamprey as the only 'vertebrate' with a *Hox* gene in the mandibular arch. Cohn also compared expression patterns with the same (orthologous) gene, in amphioxus (*Branchiostoma*), which is designated *AmphiHox-6* and expressed throughout the neural tube caudal to the cerebral vesicle, and in endoderm to the level of the first gill slit.

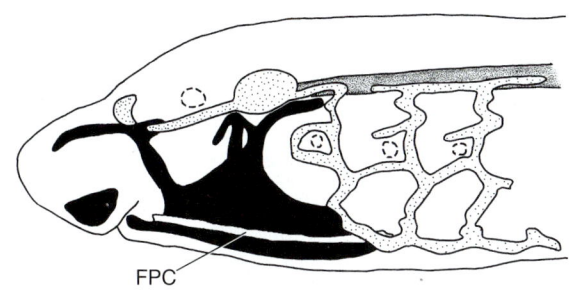

Figure 3.15 The head skeleton of the ammocoete larva of the sea lamprey, *Petromyzon marinus*. Cartilages of the head and three branchial arches are stippled. Mucocartilage is shown in black with the future piston cartilage (FPC) marked. Modified from Armstrong *et al.* (1987).

A reminder of the caution that must be exercised in extrapolating data from one species comes from the isolation of 11 Hox genes from a second species, the Japanese lamprey, *Lethenteron japonicum*, and the finding that none of these genes is expressed in the mandibular arch (Takio *et al.*, 2004).

Mucocartilage

One further peculiarity of lampreys is a tissue known as *mucocartilage*, a temporary cartilage found in the heads of ammocoete larvae, introduced when discussing elastic cartilage above (Fig. 3.15). Ultrastructural analysis reveals that mucocartilage is an avascular cartilage surrounded by a perichondrium, with a few 'fibroblasts' scattered throughout an ECM that contains elastic-like microfibrils, 11–13 nm in diameter.

Among the suggested fates of mucocartilage are reduction, loss or transformation to a connective tissue or adult cartilage. Transformation of mucocartilage to a definitive adult cartilage in *Petromyzon marinus* takes place at metamorphosis. The ECM is degraded and the mucocartilage dedifferentiates to mesenchyme, producing an aggregation of cells (a blastema) within the core of the mucocartilage. Blastemal cells then differentiate to chondroblasts, secrete proteoglycans and an ECM containing 20-nm fibrils composed of the protein *lamprin*. The resulting cartilage is a specific adult head cartilage, the piston or tongue cartilage (Figs 3.15 and 3.16). Because of this structure and these changes, mucocartilage is not always regarded as a true cartilage. Cartilage without collagen but with the novel protein lamprin also expands our horizons.[26]

Lamprin

Collagen is only a minor constituent of the annular cartilages of lampreys. In 1983, Wright and her colleagues described a protein that is the major structural protein of adult lamprey cartilage, making up 44–51 per cent of the

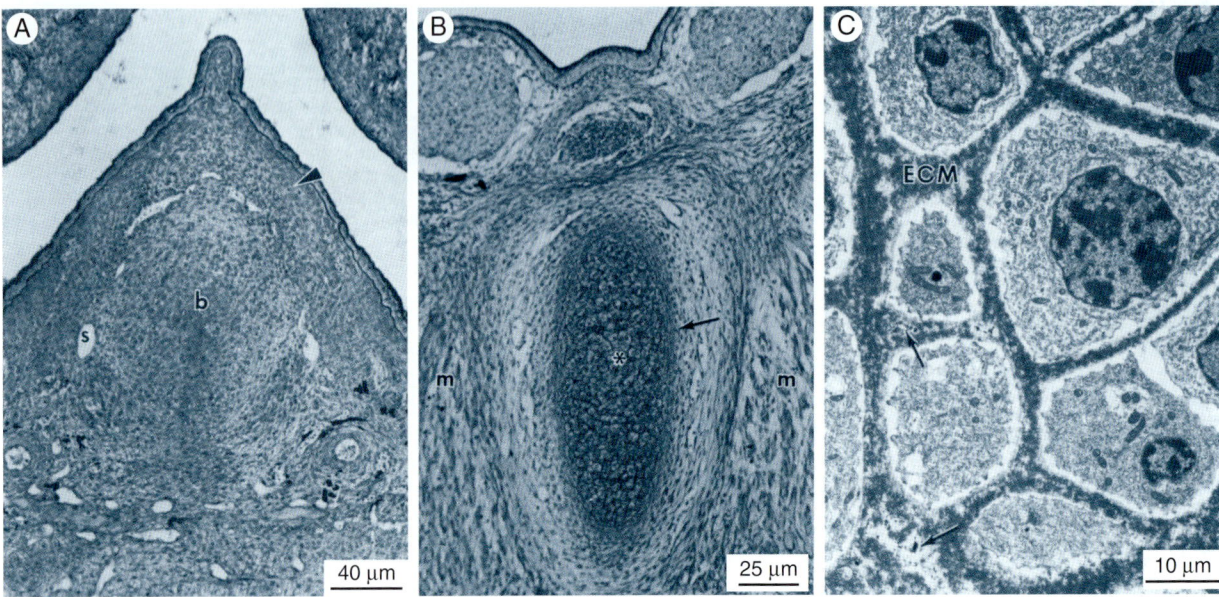

Figure 3.16 Stages in the development of the adult piston cartilage from larval mucocartilage of the sea lamprey, *Petromyzon marinus.* (A) A cross section through mucocartilage (arrowhead) of an ammocoete larva showing the blastema (b) of the piston cartilage surrounded by blood sinuses (s). (B) A cross section through the elongate piston cartilage (*) of an older larva. The arrow marks the perichondrium and m the musculature of the future piston. (C) A transmission electron micrograph of the piston cartilage illustrating the seams of ECM and cell debris (arrows). Modified from Armstrong *et al.* (1987).

dry weight of the annular cartilage. They named this new protein *lamprin*. Unlike collagen, lamprin has only traces of hydroxyproline.

Like collagen, however, lamprin is a fibrous protein consisting of branched, 150–400 Å diameter fibrils (collagen fibrils/fibres are not branched). Ultrastructural analysis of piston cartilage reveals a highly cellular cartilage with a vascularized perichondrium, hypertrophic cells, branched lamprin fibrils and many elastic fibres, collagen being confined to the peripheral ECM (Fig. 3.16). Subsequent comparison of trabecular, nasal, branchial and pericardial cartilages using transmission electron microscopy and immunohistochemistry, showed that they resemble adult annular and piston cartilages with elastin in the perichondria and peripheral ECM. Recent studies of the mechanical properties of pericardial and annular cartilages of the sea lamprey, *Petromyzon marinus*, show that lamprin confers similar mechanical properties to these cartilages as collagen does to bovine auricular cartilage.[27]

The single protein lamprin is now known to be one of a family of proteins whose structure is similar to human elastin; lamprin antibodies cross-react with elastin antibodies. In further studies using *Petromyzon marinus*, the annular and neurocranial cartilages were shown to contain lamprin, and the branchial and pericardial cartilages to contain a second member of the lamprin family. Lamprin mRNA can be localized in trabecular cartilage using *in situ* hybridization. As with mRNA for type II collagen in avian or mammalian cartilages, which appears with

condensation as a marker for initial chondrogenesis (see Chapter 19), lamprin mRNA first appears with chondrogenic condensations at day 19.[28]

Chondrogenesis of the branchial skeleton of the sea lamprey begins with condensation of mesenchymal cells at 13 days postfertilization and chondrogenesis at 14 days in one-cell-wide stacks of chondroblasts, reminiscent of those described in the zebrafish pharyngeal skeleton. Three condensations fuse to form a single element. In jawed vertebrates the usual situation is for each condensation to give rise to a single skeletal element, although more than one skeletal element can arise from a single condensation.[29]

Mineralization

Lamprey cartilage is not mineralized and probably cannot mineralize *in vivo*. Cartilage can, however, mineralize *in vitro*, provided that it is cultured between 20 and 37°C, and a source of phosphate (β-glycerophosphate) is provided (Fig. 3.17). Almost 20 years after these *in vitro* studies by Robert Langille in my laboratory, the discovery was announced of a 370-million year-old ostracoderm, *Euphanerops longaevus*, with a fully mineralized cartilaginous skeleton.[30] The pattern of mineralization bears an amazing resemblance to that seen when cartilage from extant lampreys mineralizes *in vitro*, indicating parallel evolution in early vertebrates, and long-term conservation of the pattern of cartilage mineralization in agnathans.

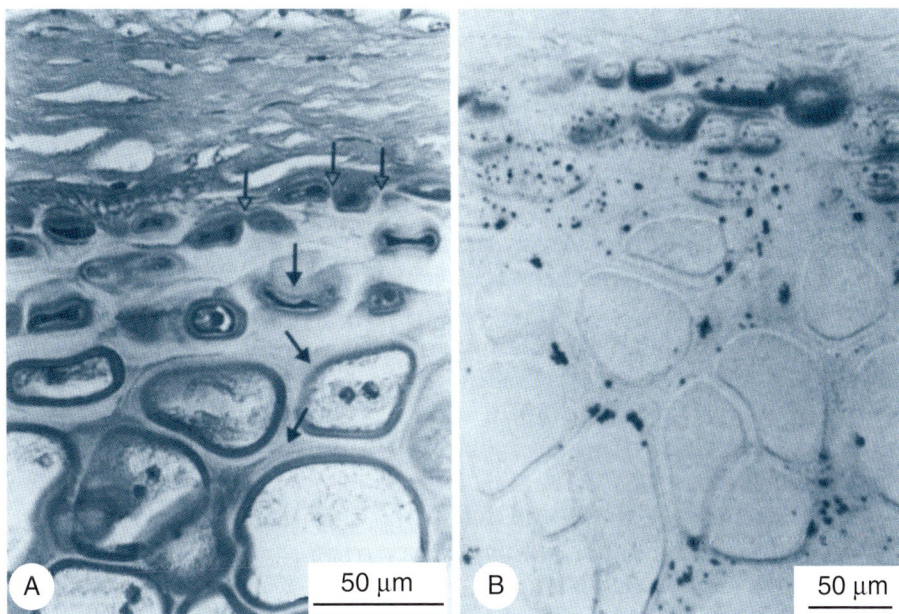

Figure 3.17 Lamprey cartilage can mineralize. Histological sections of cranial cartilage from an adult sea lamprey, *Petromyzon marinus*. (A) Note the connective tissue (top), perichondrium (open arrows), and increasing chondroblast/cyte size away from the perichondrium. Rings of ECM (solid arrows) are evident. Haematoxylin and eosin. (B) Adult lamprey cartilage, incubated in a hydroxyapatite metastable solution at 30°C for six days, and stained with the Kossa technique; same orientation as A. Mineral granules (black) can be seen in the ECM of all cartilage zones, but not in the connective tissue (top). From Langille and Hall (1993a).

HAGFISH

As with lampreys, the several types of hagfish head cartilages have also been most intensively examined by Glenda Wright and her colleagues. Their model 'hag' is the Atlantic hagfish, *Myxine glutinosa*.

Two types of lingual cartilages are present:

- type 1, with the unusual protein *myxine*, and
- type II, which resembles invertebrate cartilages (see the following chapter) but has some superficial similarities to notochord (see Box 4.4).

Subsequent studies identified a family of endoskeletal cartilages. Unlike lamprey cartilages, collagen is found in some hagfish cartilages. Two distinct types of type I collagen are present in cartilage of the inshore hagfish, *Eptatretus burgeri*, one equivalent to shark type I and found in the skin, the other equivalent to invertebrate type I collagen and found in the peritoneum, muscles and intestine. The notochord of the inshore hagfish has an over-sulphated chondroitin sulphate, chondroitin sulphate H, and a unique dermatan sulphate. The Japanese lamprey, *Entosphenus* (*Lampetra*) *japonicus*, has a skin collagen equivalent to vertebrate type I collagen, and a collagen in muscle, intestine and cartilage equivalent to invertebrate cartilage type I.[31]

NOTES

1. For basic references on the cytology and histology of vertebrate cartilage, see Fell (1925), Gardner (1971), Ham (1974), Ham and Cormack (1979), Stockwell (1979), and Moss and Moss-Salentijn (1983). See Serafini-Fracassini and Smith (1974), Stockwell (1983) and the chapters in Urist (1980), Kuettner *et al.* (1986), Hall (1983a), and Hall and Newman (1991) for the biochemistry, metabolism, physiology and physical structure of cartilage, and Glenister (1976) for an embryological view of cartilage. For now-classic studies on the ultrastructure of cartilage – using transmission and scanning electron microscopy – see Godman and Porter (1960), Sheldon and Kimball (1962), Sheldon (1964b, 1983), Palfrey and Davies (1966) and Boyde and Jones (1983).

2. See Haines (1942a, 1947), Gardner (1950), Barnett (1954), Barnett and Lewis (1958), Moffett (1965) and the chapters in Sokoloff (1978, 1980), Sledge (1981), Hall (1985a) and Bilezikian *et al.* (1996) for joint structure and development, and see Mow *et al.* (1980) for the biomechanics of joint cartilage.

3. See Chapter 17, Maderson (1987), Hall and Hörstadius (1988) and Hall (1999a) for overviews of neural crest, and see Northcutt and Gans (1983), Northcutt (1996), Hall (1999a, 2005a) and Shimeld (2003) for the importance of neural crest in the evolutionary origin of the vertebrate head. See Langille (1993) for the formation of the vertebrate face, and Hall (1995, 1999a), Kuratani *et al.* (1997, 2001), Morriss-Kay (2001) and Shimeld (2003) for evolutionary origin of the viscerocranium and mammalian skull. See Hall (1998c) for a bibliography of connective tissue research, and see de Beer (1937) and Hall and Hanken (1993) for bibliographies on skull development spanning 1937–1992.

4. See C. A. Poole *et al.* (1984, 1987, 1988a,b, 1991a,b) and C. A. Poole (1997) for recent studies and Murray (1936) and Hall (1985a) for analyses of earlier studies.

5. See Novak (1964), Moss-Salentijn (1975) and Blumer *et al.* (2004) for analyses of cartilage canals, Kugler *et al.* (1979), Agrawal *et al.* (1983, 1984), Cole and Wezeman (1985), Chandraraj and Briggs (1988) and Cole and Cole (1989) for the role(s) of perivascular connective tissue cells, and Chappard *et al.* (1983, 1986) for the human data.

6. See Nozawa-Inoue et al. (1999) for the data on laminin, and Hedberg et al. (1995) for the axon data. See Strong (1925) and Wright et al. (1958) detailed studies of the order, time and rate of ossification of the skeletons of the albino rat (Mus [Rattus] norvegicus albinus) and the Long-Evans rat. Laminin, is also found in the lining cells of the synovial membrane of the rat TMJ, and in blood vessels. Laminin, which is a normal constituent of basement membrane, appears within one hour of chick facial epithelia and mesenchyme being separated and recombined (Xu et al., 1990a,b).

7. See Liu et al. (1974) for mercury porosimetry, Delgado-Baeza et al. (1992) for information on programmed cell death, Hall (1973a) for thyroxine and chick tibial development, and Hirata and Hall (2000) for patterns of cell death in chick embryos.

8. See Haines (1969) for review of epiphyses and sesamoids with emphasis on secondary ossification centres in epiphyses, and Mottershead (1988) for review of sesamoids in relation to their function.

9. See Haines (1941, 1942b, 1969) for marsupials and for the evolution of epiphyses from fish to marsupials.

10. See Sheldon (1964a), Bradamante et al. (1975, 1991), Keith et al. (1977), Bradamante and Svajger (1981) and Kostovic-Knezevic et al. (1981, 1986) for the histology of elastic cartilages, and see Ayad et al. (1994) for elastin.

11. See O'Dell et al. (1989) for the distribution of elastic fibres in the articular disc of the temporomandibular joint (TMJ) in rabbits, Frommer and Monroe (1966) for elastic fibres in and for development of the TMJ in mice, McDevitt and Webber (1990) for fibrochondrocytes, Taylor and Yeager (1966), Chong et al. (1982) and Wlodarski (1989) for elastic fibres in periostea, Cotta-Pereira et al. (1984) for elastic fibres in perichondria, and Landini (1991) for elastin in tumours. Recently, Doi et al. (2004) isolated elastin microfibril interface protein-5 (EMILIN-5) from human stem cells in which osteogenesis had been elicited, and demonstrated expression in other osteoblast cell lines and in the perichondria of murine long-bone primordia at 13.5 days of gestation.

12. See Moskalewski (1976) for auricular chondrocytes in vitro, and Gabrovska and Vancov (1982) for matrix vesicles.

13. See Svajger (1970, 1971) for the development of rat external ear (elastic) cartilage in vitro.

14. See Jurié-Lekic et al. (1982) for human, Egerbacher et al. (1995) for cat intermediate tissues, and Lee et al. (1994) for myxoid.

15. See Ducy et al. (2000) and references therein, Amling et al. (2000) and Harada and Rodan (2003) for overviews of leptins and osteogenesis, and J. I. Huang et al. (2004) for the lipoaspirates.

16. See Wright and Youson (1982, 1983) and Wright et al. (1988) for elastic fibres, Robson et al. (1993) for the lamprin-elastin study, and Benjamin and Sandhu (1990) for the spotted spiny eel.

17. See Helff (1928, 1949) for studies on the tympanic membrane, and Chapter 21 for studies on the columella. According to Luther's studies with the edible frog, Rana esculenta, the columella itself forms without any signal from the otic vesicle.

18. See Lee and Langer (1983) and Lee et al. (1984) for these data. Ashhurst (2004) has now found that dogfish (Scyliorhinus spp.) cannot repair fractured fin rays by initiation of chondrogenesis at the cut surface, although some chondrocyte-like cells do accumulate and deposit sparse ECM.

19. See Kemp et al. (1975), Kemp (1977a, 1979), and Dingerkus et al. (1991). The two quotations are from p. 932 of Kemp (1977a). Recent studies by Summers et al. (1998) and Summers (2000) have added to our knowledge of mineralized tesserae in stingrays.

20. See Kemp (1984) for a review of his views on mineralization and tesserae. As seen in the nurse shark, Gingylmostoma cirratum, and in the lemon shark, Negaprion brevirostris, the specialized fibres (ceratotrichia) of shark fins are composed of large collagen fibres that grow by apposition of collagen to their surface (Kemp, 1977b). See Clemen (1992) for the fine structure of mineralization of the elasmobranch skeleton, based on a review of 53 species.

21. See Tanaka (1990) and Nathanson and Calliet (1990) for these tetracycline labeling studies, and see Fig. 5.7 for tetracycline administered to a teleost to label forming ossicles.

22. See Applegate (1967) for an overview of mineralization of shark cartilages, Hoenig and Walsh (1982) for cartilage canals, and Alluchon-Gérard (1982) for involvement of thyroid hormones in the development of dogfish, Scyllium canicula, cartilages. PTHrP is expressed in sharks, rays and lampreys, e.g. in the dermal denticles of elasmobranchs (Trivett et al., 1999).

23. See Lee and Langer (1983) and Lee et al. (1984) for the nature of the ECM of shark cartilages.

24. For the biology of lamprey skeletons see Hardisty (1979, 1981), Wright and Youson (1982, 1983), Wright et al. (1983, 1988), Langille and Hall (1986, 1988b, 1989a,b), Matsui et al. (1990), Robson et al. (1993, 1997), McBurney and Wright (1996), Hall (1999a), Tomsa and Langeland (1999), Morrison et al. (2000) and Wright et al. (2001). For hagfish skeletons, see Hardisty (1979), Wright et al. (1984), Armstrong et al. (1987), Kimura and Matsui (1990), Jørgensen et al. (1998), Robson et al. (2000) and Wright et al. (2001).

25. See Langille and Hall (1986, 1988a), and the literature discussed in Hall (1995a, 1999a), Kuratani et al. (1998, 2001, 2002), Ogasawara et al. (2000) and Shimeld (2003). Mallatt (1996) argued for ventilation (respiration for feeding) as the driving force for the evolution of the jaws, a scenario that involves evolution of a new mouth.

26. See Johnels (1949) for literature on the fate of mucocartilage, Wright and Youson (1982) for its ultrastructure, Armstrong et al. (1987) for transformation to an adult cartilage, and Youson (2004) for a synthesis of lamprey larval biology and metamorphosis.

27. See Wright and Youson (1983) and McBurney and Wright (1996) for features of adult cartilage and chondrogenesis, and see Courtland et al. (2003) for the mechanical properties of lamprey cartilages.

28. See Robson et al. (1993) for lamprin as a family of proteins. Robson et al. (1997) grouped the cartilages on the basis of the type of lamprin present. McBurney et al. (1996) did the in-situ labeling. Benjamin and Sandhu (1990) describe the rostral cartilages in the tentacles on the head of the spiny eel, Macrognathus siamensis, as consisting of elastic cellular cartilage with a thick perichondrium akin to that of lamprey cartilage.

29. See Morrison et al. (2000) for chondrogenesis in the lamprey, Kimmel et al. (1998) for zebrafish, Miyake and Hall (1994) for chondrogenesis in teleost fishes, and Hall and Miyake (1992, 1995, 2000) and Miyake et al. (1996a,b) for condensation in gnathostomes.

30. See Langille and Hall (1983a) for mineralization of lamprey cartilage *in vitro*, Langille and Hall (1988b) for how to organ culture and graft lamprey cartilages and teeth, Janvier and Arsenault (2002) for *Euphanerops longaevus*, and Janvier (2001) for the classic French recipe *Matelotte de Lamproie à la Bordelaise/à la Nantaise*.

31. See Wright *et al.* (1984) for the two types of lingual cartilage, Robson *et al.* (2000) and Wright *et al.* (2001) for the family of cartilages, Ueoka *et al.* (1999) for notochordal chondroitin sulphate, and Kimura and Matsui (1990) and Matsui *et al.* (1990) for the two collagen types in hagfish and the lamprey.

Part II

Natural Experiments

A new type of *acellular cartilage* was discovered as part of a valve-like system in the conus arteriosus of the freshwater ray, *Potamotrygon laticeps* from the Amazon. This 'cartilage' has neither chondrocytes nor perichondrium and is surrounded by and attached directly to a basement membrane-like structure. The collagenous extracellular matrix is perforated by ramifying vascular canals that form 'canaliculi,' akin to those seen in bone. Unlike most cartilages, the glycosaminoglycans of the extracellular matrix are not bound to core protein (Junqueira *et al.*, 1983). Such a tissue, reminiscent of vertebrate cartilage, acellular bone and invertebrate cartilage, challenges our neat compartmentalization of cellular cartilage and cellular/acellular bone.

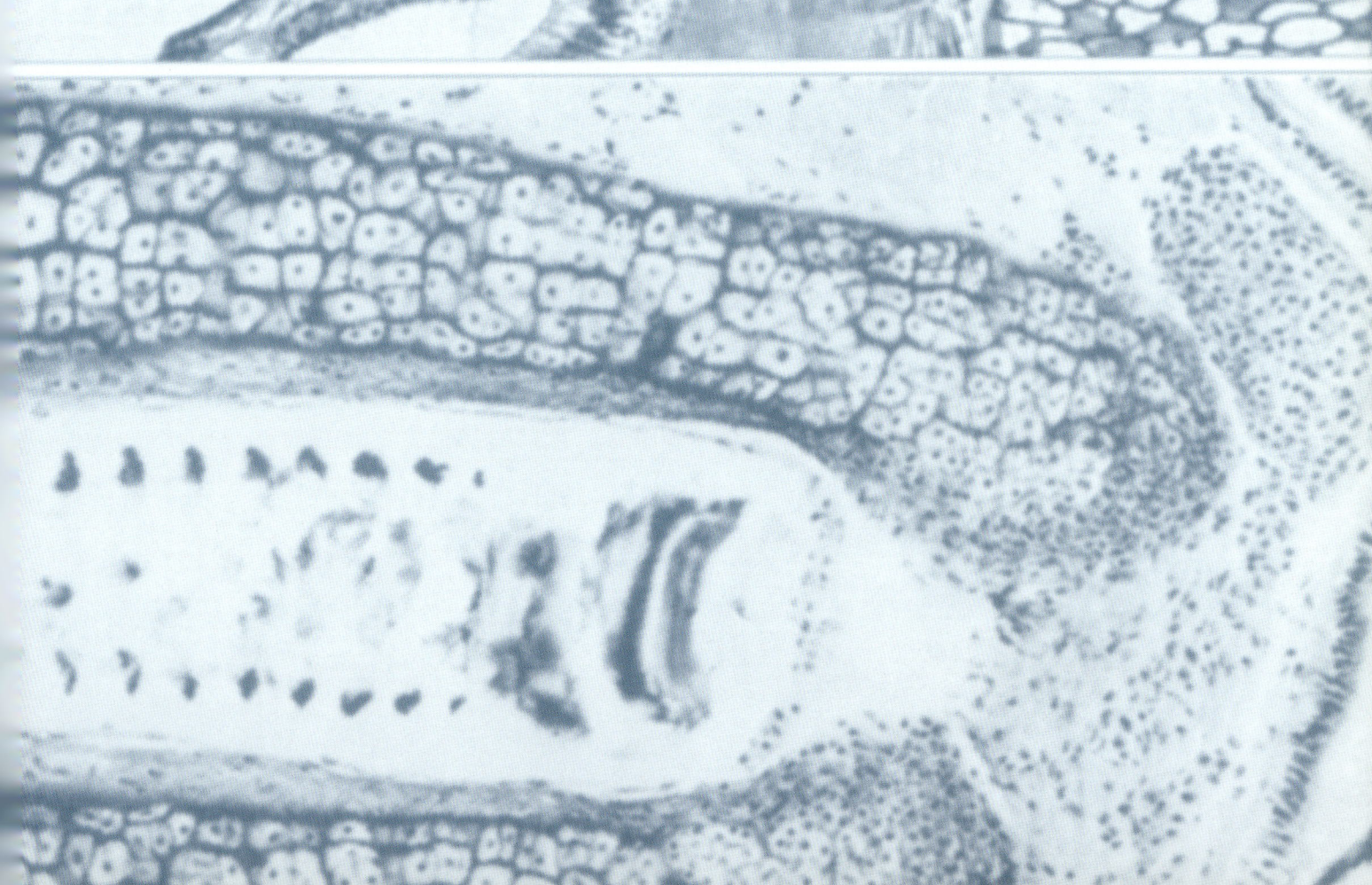

Invertebrate Cartilages

One of the generalities that led Schwann and Schleiden to the formulation of the Cell Theory in 1838–1839 was the recognition that many invertebrates possess a cartilage-like – chondroid, chordoid, mucoid – connective tissue, structurally similar to the parenchymal tissues of plants.

CHONDROID, CARTILAGE OR NEITHER

In 1930, Schaffer exhaustively summarized and reviewed the 19th and early 20th century studies on these tissues. He did not, however, regard them as true cartilages. Their immature histology, combined with a scant extracellular matrix, the inability to detect glycosaminoglycans or collagen in their matrices – both of which are diagnostic of vertebrate cartilage – and their failure to mineralize, led Schaffer to regard these invertebrate tissues as *chondroid tissues*. By 'chondroid' Schaffer meant cartilage-like; by implication, a less fully evolved skeletal tissue than cartilage. Schaffer classified chondroid tissues into grades of greater or lesser resemblance to cartilage without, however, implying any phylogenetic trends or affiliations.[1]

Schaffer's view prevailed, and skeletal biologists turned to other problems. Indeed, no research publications on invertebrate cartilages appeared until 1959. Then in the 1960s and '70s, primarily through the work of one man, Philip Person, several structural studies appeared. Person was thorough. He used light and electron microscopy, biochemical characterization of extracellular matricial products, histochemistry and mineralization, and provided some results on the regenerative ability of invertebrate cartilage.[2] With the exception of the last study, no information on the development of invertebrate cartilages was available until Alison Cole initiated studies in my laboratory this century.[3]

By 1969, sufficient data had accumulated that Person and Philpott could review the work of the previous decade to present arguments in favour of regarding these invertebrate tissues as true cartilages, arguments that included the presence of cells in lacunae embedded in a matrix containing collagen and chondroitin sulphate-A or -C (chondroitin-4- and chondroitin-6-sulphate, respectively) producing a tissue providing mechanical support and a site for muscle and ligament attachment.[4]

A deep homology of extracellular matrices, and so a monophyletic origin of the animal kingdom, has been proposed on what seem to me to be entirely reasonable grounds. In an analysis placing fibrillar collagen in a key position in vertebrate evolution – fibrillar collagen, it is argued, having arisen before the genome duplication at the base of vertebrate evolution – Boot-Handford and Tuckwell (2003) discuss invertebrate collagens only briefly. Older studies saw collagen first appearing in the sponges.[5] Although type I is the most widespread collagen, invertebrates contain several collagen genes. Type II collagen in cartilage arose with the vertebrates, perhaps providing an argument for the primitiveness of collagen with three identical chains during evolution.[6]

The distribution of elastin in elastic cartilages and in lamprey mucocartilage was discussed in Chapter 3. Previously thought to be absent from all invertebrates but present in all craniates other than hagfishes and lampreys, elastin appeared phylogenetically with the evolution of closed, high-pressure circulatory systems, but is arranged differently from vertebrate to vertebrate.[7] The nature of invertebrate cartilages can, perhaps, best be reviewed by summarizing our knowledge of the cartilages in different taxa (Fig. 4.1). I discuss:

- odontophore cartilages in a gastropod, the channeled whelk, *Busycon canaliculatum*;

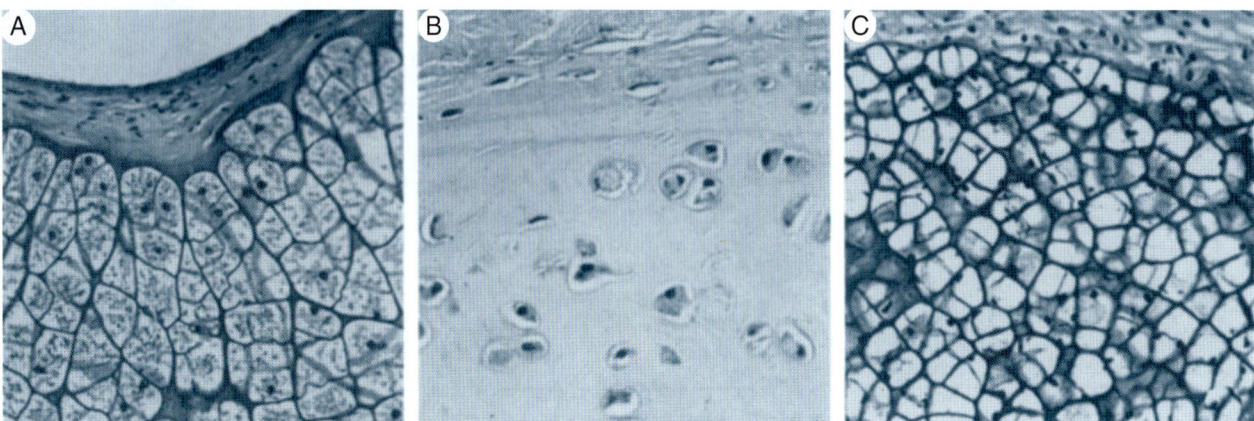

Figure 4.1 The histology of invertebrate cartilages. (A) Odontophoral cartilage of the marine snail, *Busycon canaliculatum*, with large cells and scant extracellular matrix. (B) Cranial cartilage of the longfin squid, *Loligo pealii*, with abundant extracellular matrix. (C) Branchial cartilage of the horseshoe crab, *Limulus polyphemus*, again with scant extracellular matrix. All stained with haematoxylin and eosin. Adapted from Person (1960).

- branchial (gill book) cartilage in an arthropod, the horseshoe crab, *Limulus polyphemus*;
- cranial cartilages of several cephalopods – the longfin squid, *Loligo pealii*, the common cuttlefish, *Sepia officinalis*, and the common octopus, *Octopus vulgaris*;
- tentacular cartilages of two polychaete annelids, the feather duster worms *Eudistylia polymorpha* and *Sabella melanostigma*; and
- lophophore cartilage in the lampshell, *Terebratalia transversa*, an articulate brachiopod.

ODONTOPHORE CARTILAGE IN THE CHANNELED WHELK, *BUSYCON CANALICULATUM*

In one of the few functional morphological studies on invertebrate cartilage, Guralnick and Smith (1999) examined the phylogenetic distribution of cartilage in gastropods – including criteria to classify cartilages – and looked at correlations between the evolution of gastropod cartilages and the radula.

The channeled whelk has a large *odontophore cartilage* consisting of hypertrophic cells in capsules separated by scant amounts of extracellular matrix (ECM) (Fig. 4.1). While it is true that invertebrate cartilage with scant ECM presented a stumbling block to acceptance of these tissues as true cartilages, a number of vertebrate cartilages are characterized by scant amounts of ECM, notably xiphisternal, secondary and callus cartilages in jawed vertebrates and lampreys, and cell-rich cartilage in teleosts.

In general, the glycosaminoglycans (GAGs) of invertebrate cartilages tend to be over-sulphated in comparison to those from vertebrate cartilages. Fitting this pattern, the odontophore cartilage lacks chondroitin sulphate, having instead a polyglucose sulphate akin to chitin.[8] Echinoderm connective tissues have a similar polyfucose sulphate, but also have chondroitin sulphate. Odontophore cartilages

have collagen fibres with a 640 Å periodicity and a typical X-ray diffraction pattern. *Busycon* cartilage is unusual in that it contains myoglobin within its cells, unusual because the blood pigment in molluscs is haemocyanin and cartilages are normally not pigmented.[9]

Busycon cartilage was the first cartilage from any taxon – vertebrate or invertebrate – in which cytochrome oxidase was found. Indeed, the low utilization of oxygen by vertebrate cartilage prompted Person and colleagues (1959) to examine invertebrate cartilages in the first place. They were prescient. Paradoxically, low O_2 utilization was coupled with high levels of dehydrogenases requiring cytochrome oxidase and considerable synthesis of sugars requiring oxygen, an unexpected mix of metabolic parameters.

Little else is known about mollusc cartilages outside the cephalopods, for which, see below. After amputating the proboscis, members of a number of genera of boring prosobranch molluscs can regenerate a new proboscis. A single aggregation of cells gives rise to cartilage and muscle of the new proboscis. A single aggregation suggests a single developmental origin and a homogeneous population of cells. Studies of vertebrate limb development and of the regeneration of adult urodele limbs (Chapter 38), tell us that neither need be so.[10]

BRANCHIAL (GILL BOOK) CARTILAGE IN THE HORSESHOE CRAB, *LIMULUS POLYPHEMUS*

The gill books of the horseshoe crab, *Limulus polyphemus*, contain a cartilaginous endoskeleton (Fig. 4.1). The scant extracellular matrix contains chondroitin-4-sulphate – which the whelk does not – and reacts as does vertebrate cartilage to histochemical tests for GAGs, although a novel sulphated oligosaccharide containing 3–*o*-sulphated glucuronic acid was isolated from chondroitin sulphate K from *Limulus* cartilage.[11]

Hydroxyproline and hydroxylysine are found in amounts typical of vertebrate cartilage, and with X-ray diffraction patterns typical of collagen. Lipid and glycogen are found as storage products, while the presence of cytochrome oxidase indicates potential aerobic metabolism. As with all invertebrate cartilages, gill book cartilages are unmineralized. The presence of lipids, which in vertebrate skeletal tissues play a role in mineralization, is of interest (see the section on mineralization below).

As gill book chondrocytes divide, a phragmosome-like structure[12] – reminiscent of the membrane formed during the division of supporting tissue in plants – is deposited between the daughter cells and provides the basis for new ECM. As the cells mature, the entire chondrocyte is 'chondrified' as its cytoplasm fills in with matrix products and the cell dies, a process reminiscent of lignification in plants. This unusual pattern of chondrification may explain, in part, *the entirely appositional growth of gill book cartilage*, which contrasts with interstitial growth of cartilages in other cephalopods (below).[13]

CRANIAL CARTILAGES IN SQUID, CUTTLEFISH AND OCTOPUSES

The longfin squid, *Loligo pealii*, the common cuttlefish, *Sepia officinalis*, and the common octopus, *Octopus vulgaris*, contain an extensive cranial cartilage that protects the brain and a scleral cartilage protecting the eye (Fig. 4.1). Unlike cartilages in the whelk and horseshoe crab, these cranial cartilages have extensive matrix and sparsely scattered cells, cell densities being more typical of vertebrate bone than cartilages. The matrix may be quite fibrous and often has a canalicular system coursing through it. Cell processes lie within the canals, connecting cells one to another, giving the appearance of osteocytes. Some squid have been reported to possess cartilaginous epidermal scales.[14]

Composition of the extracellular matrix

Much diversity of matrix products is seen in cephalopods. Whether more than in other invertebrate group is hard to say. Although a number of cephalopod species have been examined, only one or two species have been studied in most other groups. Histochemical evidence shows less GAG, and a different spectrum of matrix products in cephalopod cartilage than in *Limulus*, emphasizing the need for more biochemical characterization of matricial products.

Glycosaminoglycans (GAGs)

The predominant GAG is chondroitin-4-sulphate, as in *Limulus*. No hyaluronan is found in squid cartilage, and the core protein appears to differ from that in vertebrate cartilage. Novel tetrasaccharide sequences are found

in squid chondroitin sulphate; the common squid, *Ommastrephes sloani pacificus*, has a glucuronic acid-containing glycopeptide distinct from any other GAGs.[15]

Cartilages of the brown or short-finned squid, *Ilex illecebrosus coidentii*, contain hyaluronan (Box 4.1), a previously undescribed heavily over-sulphated chondroitin sulphate, and a previously undescribed 39 200 molecular weight polysaccharide mainly composed of glucuronic acid, galactose and mannose in the ratio of 1:2:1. Extraction and characterization of proteoglycans from *Todarodes pacificus* show that one-quarter of the proteoglycans from cranial cartilages do not aggregate with hyaluronan. Cranial cartilage is low in proteoglycan content, proteoglycans do not interact with hyaluronan, and keratan sulphate is absent (Box 4.1).[16]

Tsilemov *et al.* (1998) showed that a 35-kDa protein from squid cartilage – species not specified, but the cartilage lacks hyaluronan – reacts to antiserum against sheep link protein and binds to the same G1 domain of aggrecan (the globular domain at the N-terminus that functions as the hyaluronan-binding region), suggesting a similar role for this protein to that carried out by *aggrecan* in vertebrates, which is to bind to hyaluronan to modulate the osmotic properties of cartilage (Box 4.2). In vertebrates, synthesis of aggregan is enhanced in mesenchyme subjected to compression, while expressing aggrecan or aggrecan link protein in cultured fibroblasts or chondrocytes disrupts cell–substrate interaction.[17]

Cranial cartilage from the arrow squid, *Nototodarus gouldi*, contains highly sulphated chondroitin sulphate E. The epidermis contains unsulphated, but highly glycosylated, glycosaminoglycans (Falshaw *et al.*, 2000).

Collagens

Cranial cartilage collagen in the Japanese flying or common squid, *Todarodes pacificus*, resembles vertebrate type I collagen (Table 4.1).

A combined light, polarizing and ultrastructural microscopical study of the head cartilages of *Sepia officinalis* and *Octopus vulgaris* revealed 10–25-nm-diameter collagen fibres similar to vertebrate type I (albeit with somewhat different banding patterns), polymeric aggregate and a defined perichondrium – features typical of vertebrate cartilages – but with chondrocytes connected to one another by cell processes and blood vessels within the cartilage – features of vertebrate bone, not cartilage (Fig. 4.1). Bairati *et al.* (1987) argued that vascularization of these cartilages provides a high level of nutrition, eliminating the 'need' for mineralization. This, and subsequent studies by Bairati and colleagues, provide the most detailed information on any squid or octopus cartilage. They undertook:

- a transmission electron microscopical (TEM) study of the perichondrium of octopus cartilage, demonstrating a fibrous layer that was discontinuous at connective tissue or muscle attachments, and remarkably vertebrate-like;

Box 4.1 Hyaluronan

Hyaluronan, a high-molecular-weight polyanion with a highly negative charge, occurs in low concentrations in many animal tissues. A single molecule may have a molecular weight of 10 million. Consisting of an alternating polymer of glucuronic acid and N-acetylglucosamine joined by a β1–3 linkage, hyaluronan expands enormously in solution, forming a net of extended, random-coiled molecules. Hydrated, it typically occupies the volume of a sphere 400 nm in diameter – a thousand times greater than the volume of the non-hydrated chain.

Hyaluronan – then known as hyaluronic acid – was isolated in 1934. Hyaluronan is degraded by hyaluronidase. Much of the research over the 40 years after its discovery was biochemical, with some suggestions of an association of hyaluronan with cell proliferation, migration and aggregation, and of hyaluronidase – and therefore removal of hyaluronan – with subsequent cell differentiation. Perhaps the first to demonstrate these links with cell behaviour were Maurer and Hudack (1952), who showed that a large amount of hyaluronan is synthesized in the early stages of callus formation during repair of fractured long bones. Since then, hyaluronan has been shown to be involved in vertebral and limb chondrogenesis, migration of neural crest cells, regeneration of amphibian limbs and ectopic chondrogenesis and osteogenesis (Chapter 14), and corneal development, to name only a few examples.

Mouse embryonic calvarial mesenchyme produces more membrane bone nodules when cultured with low-molecular-weight hyaluronan (30–40 kDa), high-molecular-weight hyaluronan (600–1300 kDa) having no effect (Pilloni and Bernard, 1998).

Although some hyaluronan is intracellular, most lies within ECM, where it forms the backbone to which proteochondroitin sulphate is attached, forming large aggregates with aggrecan within cartilaginous ECM (see Box 4.2). Both hyaluronan and the aggregate impede transport and help to maintain the greatly hydrated nature of cartilaginous ECM. Prechondrogenic cells produce only monomeric proteoglycan. With progressive differentiation, larger and more aggregated forms of proteoglycan are deposited. Interestingly, as they age *in vivo* or *in vitro*, chondrocytes revert to synthesizing small monomer proteoglycans, a change that may be a form of dedifferentiation as, to some extent, is true of some tumours. With chondrocyte differentiation, hyaluronan is localized preferentially in the hypertrophic zone of the growth plate, concentration correlating with lacuna size, reflecting water-binding properties; lacunae fail to enlarge if chondroblasts are cultured in hyaluronidase.[a]

We have some indications of how levels of hyaluronan and chondroitin sulphate (CS) are correlated (Fig. 4.2). Hexosamine synthesis is enhanced when tibial cartilage is treated for two days with hyaluronidase and then organ-cultured for four days in the absence of hyaluronidase. Less CS is synthesized than in untreated cartilage; the CS has a shorter chain length than normal; and CS is uncoupled from the hyaluronan backbone of the hyaluronan–proteochondroitin sulphate aggregate. Synthesis of CS is depressed when hyaluronan is added to cultures of chondrocytes from embryonic or adult mammals or birds, suggesting that synthesis of CS is controlled – at least in part – by the level of hyaluronan. The effect of hyaluronan is specific. Other GAGs – chondroitin, keratan or heparin sulphates – do not stimulate synthesis of CS. Oligosaccharides derived from hyaluronan do.[b]

As hyaluronan is primarily extracellular, little attention has been paid to how it might be internalized by cells. It appears that mammalian articular chondrocytes use the CD-44 receptor. Using rat cranial base synchondroses as a model, Gakunga and colleagues localized hyaluronan in the hypertrophic but not the other chondroblast zones. (Autoradiographic and histological analyses of growth of the cartilages in the midline cranial base of the Norwegian rat, *Rattus norvegicus*, to 16 months after birth, show that proliferation is specific to individual cartilages.) Hyaluronidase and CD-44 co-localize with hyaluronan. These researchers argue that hyaluronan is internalized by hypertrophic chondrocytes and degraded via a CD-44-based mechanism. Synchondroses cultured in *Streptomyces* hyaluronidase show reduced lacunar expansion, suggesting that hyaluronan-mediated hydrostatic pressure plays a role in lacunar expansion during chondrocyte hypertrophy.[c]

[a] Miner (1950) provided a comprehensive review of the early work on hyaluronan and on hyaluronidase. See Larsson and Kuettner (1974) for intracellular hyaluronan, Wiebkin and Muir (1973), Goldberg and Toole (1984), Knudson and Knudson (1991), Pavasant et al. (1996) and Toole (2001) for aggregate formation, and Ovadia et al. (1980) and Pacifici et al. (1981) for synthesis of monomers by prechondroblasts and ageing chondrocytes. See Pavasant et al. (1996) for lacunar enlargement, and Toole (1990, 2001) for a review of hyaluronan and its binding proteins, the hyaladherins. The 2001 paper deals especially well with the role of hyaluronan in the pericellular matrix surrounding chondrocytes. Spicer and Tien (2004) review hyaluronan's role in morphogenesis, especially condensation and joint cavitation during skeletogenesis. They (see their Table 1) identify 30 genes involved in hyaluronon synthesis, function and/or degradation.
[b] See Hardingham et al. (1972) for hexosamine synthesis in tibial cartilage, and Wiebkin and Muir (1973, 1975) and Solursh et al. (1974) for addition of hyaluronan to cartilage cultures.
[c] See Jiang et al. (2001) for the CD-44 receptor, Gakunga et al. (2000) for the hyaluronan study, and Roberts and Blackwood (1983, 1984) for the study with the Norwegian rat. CD44 is expressed in mouse development (9.5–12.5 days) in high levels in the somites, heart, limb condensations, AER, tooth primordia and upper and lower jaw mesenchyme (Wheatley et al., 1993).

- a TEM study of chondrocytes from the cranial cartilages of both species, demonstrating many long processes connecting cells – reminiscent of vertebrate osteocyte processes – an absence of secretory granules, the presence of rough endoplasmic reticulum cisternae and vesicles opening to the cell surface, haemocyanin in many chondrocytes and gap junctions; and
- an immunological study of collagens from *Sepia officinalis* using vertebrate and cuttlefish collagen antibodies. Response to antibodies against cuttlefish type I and rat type V was intense, but reactions to chick anti-type I and calf anti-type II were weaker. Indeed, and surprisingly, unless *Sepia* collagen is not at all highly derived, except for

antibodies against mammalian type I collagen, all antibodies tested cross-react with *Sepia* cartilage, implicating type I collagen at an early step in skeletal evolution.[18]

Sepia cranial cartilage contains a novel collagen with three distinct chains – named C1–C3 – consisting of 105, 115 and 130 kDa, respectively. This collagen is more cross-linked than shark cartilage type I collagen. Interestingly, an antibody against shark cartilage cross-reacts with *Sepia* cartilage but a *Sepia* cartilage antibody does not cross-react with shark cartilage. Based on rotary shadowing electron microscopy, the matrix collagen fibres have been compared with those of chick sternal cartilage.[19]

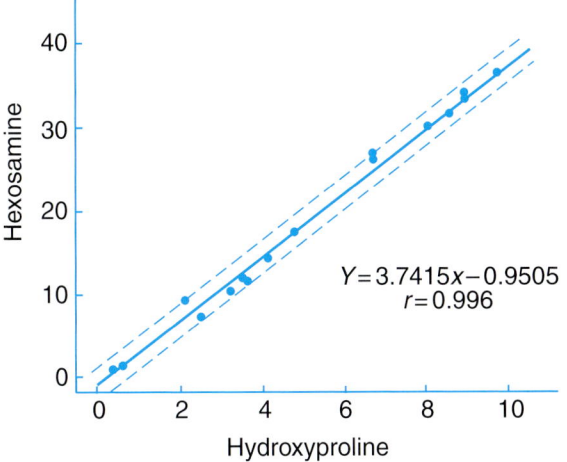

Figure 4.2 Concentrations of hexosamine and hydroxyproline as μm/pair of tibiae from chick embryos incubated for 12 days are significantly correlated (r = 0.996) in this study comparing results from control embryos and from embryos treated with vitamin C or with the metal thallium. See the text and Hall (1973b) for details of treatments.

The equation shown in the figure:

$$Y = 3.7415x - 0.9505$$
$$r = 0.996$$

Box 4.2 Aggrecan

Aggrecan (aggregating chondroitin sulphate proteoglycan) comprises as much as ten per cent of the dry weight of some cartilages. As many as 100 aggrecan monomers bind to a single hyaluronan chain via a link protein to form a high-molecular-weight aggregate, with a globular domain (G1, the hyaluronan-binding region) at the N-terminus where the keratan sulphate chains are preferentially located. Aggrecan shares numerous sequence similarities with versican (Box 21.3) and CD-44, a hyaluronan receptor (Ayad et al., 1994).

Aggrecan is a structural proteoglycan, by which I mean that the aggregation to which aggrecan contributes provides an important structural property to cartilage, namely to maintain an osmotic swelling pressure and the highly hydrated nature of cartilaginous extracellular matrix. Compressive forces, such as those found on articular cartilages, stimulate synthesis of aggrecan. The core protein of aggrecan is broken down by proteases, a breakdown that facilitates vascular invasion of cartilage.

Aggrecan mRNA is expressed in periosteal cells of membrane bones immediately before they begin to differentiate as chondroblasts (see Evidence for bipotentiality, Chapter 12).

Table 4.1 The amino acid composition of collagen and collagen α chains from cranial cartilage of the Japanese flying squid, *Todarides pacificus* compared with type I collagen from rat tail tendon and rat skin[a]

Amino acid	Squid			Rat tail tendon	Rat skin
	Collagen	α1(I)	α1(II)	Collagen	Collagen
3-Hydroxyproline	1	1	1	4.2	1
4-Hydroxyproline	90	92	88	90	92
Alanine	94	102	76	107	106
Arginine	59	58	60	50	51
Aspartic acid	58	53	65	45	46
Glutamic acid	91	92	90	71	71
Glycine	338	341	339	331	331
Histidine	6	6	8	4	5
Hydroxylysine	14	11	24	7	6
Isoleucine	15	11	22	10	11
Leucine	27	25	31	24	24
Lysine	10	12	9	27	28
Methionine	11	11	8	8	8
Phenylalanine	9	10	8	12	11
Proline	90	88	90	122	121
Serine	42	44	36	43	43
Threonine	23	24	20	20	20
Tyrosine	2	3	1	4	2
Valine	20	16	24	23	25

[a] From data in Kimura and Karasawa (1985) for squid and in Serafini-Fracassini and Smith (1974) for rat tail tendon and skin. Based on limited pepsin proteolysis and expressed as residues/1000 amino acid residues.

TENTACULAR CARTILAGE IN POLYCHAETE ANNELIDS

The feather duster worms *Eudistylia polymorpha* and *Sabella melanostigma* have extensive, branched and sub-divided (segmented?) cartilages at the base of the tentacle complex, within the tentacles, and in the pinnae that branch from the tentacles (Fig. 4.3). These cartilages (presumably) give considerable support and flexibility to the tentacles as a food-gathering organ. The cartilaginous matrix is sparse and vascularized, causing it superficially to resemble osteoid. The glycosaminoglycan is a highly sulphated chondroitin-6-sulphate. Ultrastructural studies of *Sabella* cartilage show heterogeneity of matrix structure.[20]

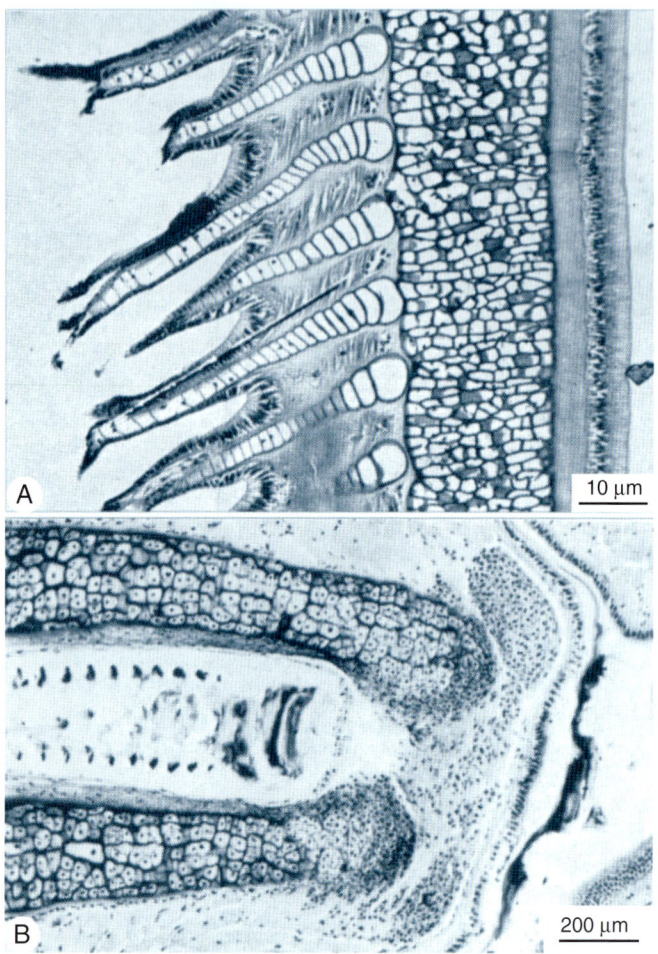

Figure 4.3 Polychaete and gastropod cartilages. (A) A longitudinal section of a tentacle of the marine polychaete, the feather duster worm, *Eudistylia polymorpha*, showing the central cartilage and the cartilages of the pinnae. (B) Regenerating odontophoral cartilage of the gastropod, the marine oyster drill, *Urosalpinx cinerea follyensis*, 11 days after amputation of the proboscis. Note the proliferating, blastema-like cartilage on the right. Adapted from Person and Mathews (1967) and Carriker *et al.* (1972).

LOPHOPHORE CARTILAGE IN AN ARTICULATE BRACHIOPOD, *TEREBRATALIA TRANSVERSA*

Cartilage in the pedicle of an articulate (two-valve) brachiopod, *Terebratalia transversa*, was discovered serendipitously during an ultrastructural study of pedicle development, the pedicle being the fleshy stalk that attaches the animal to the substrate. It consists of a weakly metachromatic connective tissue that develops in the pedicle after metamorphosis. The connective tissues of the tentacles are *acellular* and not metachromatic after toluidine blue, but at the ultrastructural level they resemble (indeed, appear equivalent to) vertebrate fibrous cartilage. The connective tissue in the remainder of the lophophore is metachromatic and clearly cartilaginous.[21]

MINERALIZATION OF INVERTEBRATE CARTILAGES

An overview of the mineralization of vertebrate cartilage may be found in Box 4.3. Chapter 22 contains a discussion of mineralization in relation to chondrocyte hypertrophy and the synthesis of type X collagen.

Although many invertebrates possess mineralized skeletal matrices, no instance of *in vivo* mineralization has been recorded. Nor has osseous tissue or bone been found in any invertebrate, the vascularized cartilage of polychaetes and decapods being morphologically closest to osteoid. Interestingly, many invertebrate taxa with *unmineralized* cartilages do have *mineralized* non-cartilaginous endoskeletons, raising the question of why the cartilages do not mineralize. Do they lack the ability to mineralize *in vivo*, or are they inhibited from mineralizing? Would they mineralize *in vitro* under conditions shown to permit lamprey cartilage to mineralize (Chapter 3)?[22]

Preparations of cartilage from *Loligo*, *Limulus* and *Busycon* can mineralize *in vitro* if incubated in a solution metastable to hydroxyapatite. Mineralization only occurs at a temperature (37°C) well above normal environmental temperatures of 20°C. Mineralization is in the form of hydroxyapatite, which is the normal form of calcium phosphate in vertebrate cartilages. In the initial abstract, the authors reported accumulation of 18 g Ca^{++}/g cartilage in *Loligo*. Ca^{++} accumulation follows formation of perichondrial and matrix granules and progressive intracellular mineralization of the chondrocytes (Fig. 4.10). Such a pattern of intracellular mineralization is normally associated with cell death in vertebrate chondrocytes (Hall, 1972a), but is seen when lamprey cartilage mineralizes *in vitro*.[23] The rate of mineralization *in vitro* is positively correlated with the concentration of phosphatidyl-serine within cartilage, consistent with involvement of lipids in mineralization of vertebrate cartilage and bone.[24]

Mineralization of cartilage in invertebrates appears to be effectively blocked by temperature, perhaps coupled with the absence of sufficient concentrations of calcium and phosphorus in seawater. Hall (1975a) discusses how the mineralization of invertebrate skeletal tissues relates to the evolution of vertebrate mineralizing skeletal tissues.[25]

CARTILAGE ORIGINS

Although theories of the evolutionary origin(s) of cartilage abound, *all* start from the premise that cartilage is exclusively a vertebrate tissue, with the exceptions of the analyses Alison Cole has initiated (Cole and Hall, 2004a,b). Berrill (1955) supported Garstang's earlier view that vertebrates arose from free-swimming larvae of Precambrian tunicates. He thought cartilage evolved as an extracellular secretion by tunicate connective tissue cells to serve a mechanical function, and that bone

Box 4.3 Cartilage mineralization

Interest in mineralization goes back to Honor Fell's pioneering studies 80 years ago. Mineralization is so regular, dynamic and precisely controlled and can be approached from so many different levels and backgrounds – crystallography, mineralogy, endocrinology, cell biology, pharmacology, pathology – that the literature is enormous. Throughout the book I touch on mineralization as a developmental process. Here, I survey mineralization using the mineralization of cartilage to highlight some important features.[a]

The players

Robin Poole from Montreal has contributed much of our basic understanding of cartilage mineralization and its relationship to growth of long bones at growth plates. In 1989 he and his colleagues summarized and reviewed the roles of alkaline phosphatase, collagen, chondrocalcin and proteoglycans in cartilage mineralization. Here, I outline calmodulin, chondrocalcin and osteonectin.

Calmodulin is so named because this abundant cytoplasmic constituent is a *calcium-modul*ated prote*in*. It functions by sensing Ca^{++} levels and then signaling calcium-sensitive enzymes or calcium-sensitive ion channels. Calmodulin is found in chondrocytes (detected in mature and mineralizing chondrocytes of rodent primary and condylar cartilages) but is absent from the perichondrial and periosteal cells (Lewinson and Boskey, 1984).

Chondrocalcin (the c-propeptide of type II procollagen) is a calcium-binding extracellular protein, surprisingly giving the c-propeptide two different roles: assembly of the triple helix of type II collagen, and binding calcium in mineralizing cartilage. Not found in the proliferative layers of bovine foetal epiphyseal or growth plate cartilages, chondrocalcin is present in the hypertrophic zone and in the longitudinal septa of the extracellular matrix (Fig 4.4), where its distribution coincides with mineralization. The interface between non-mineralized and mineralized cartilage is known as the *tidemark* (Fig. 3.11). Tidemarks occur in hyaline and fibrocartilage, provided that there is a junction between mineralized and non-mineralized cartilage (Fig 3.11, and see Havelka *et al.*, 1984, for a review).[b]

Osteonectin is a 32 000-MW protein that makes up around one 10th of bone protein. Linking collagen to hydroxyapatite and acting as a nucleus for mineralization, osteonectin is found in osteoblasts, osteoprogenitor cells, young but not old osteocytes, chondrocytes of mineralizing cartilage, fracture callus, and ectopic ossifications such as *myositis ossificans* (Fig. 1.6). Osteonectin is expressed in hypertrophic chondroblasts, mineralized ECM and in the perichondrium of rat mandibular condylar cartilage. In chicken tibial growth cartilage, osteonectin is found in the cells of all zones but only in the matrix in the mineralizing zone.[c]

The role of *proteoglycans* in the mineralization of cartilage is complex, little understood and cartilage or taxon-specific. Acridine orange preserves proteoglycans, allowing their visualization in rosettes in the longitudinal septa of the rat growth plate at sites where mineralization occurs. Proteoglycans, monomers and link protein also are retained during the mineralization of mammalian and avian epiphyseal growth plates (Fig. 4.5).[d]

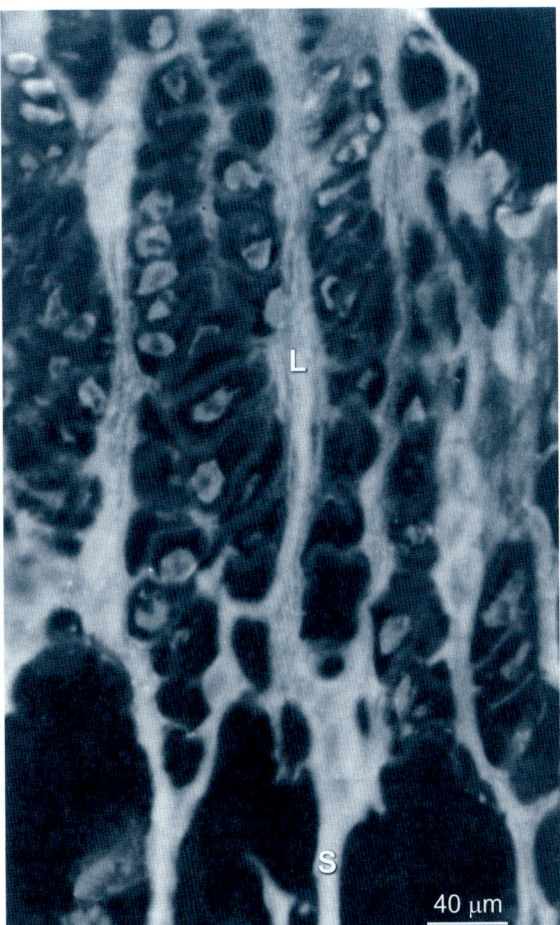

Figure 4.5 Link protein. This longitudinal section of a portion of the hypertrophic zone of a bovine foetal growth plate shows link protein (green in original) localized to the longitudinal septa (L) and to spicules of mineralized cartilage matrix (S). Little link protein can be visualized in the transverse septa. Cytoplasmic staining (orange in original) reflects a counter stain, not link protein. Modified from A. R. Poole *et al.* (1982a).

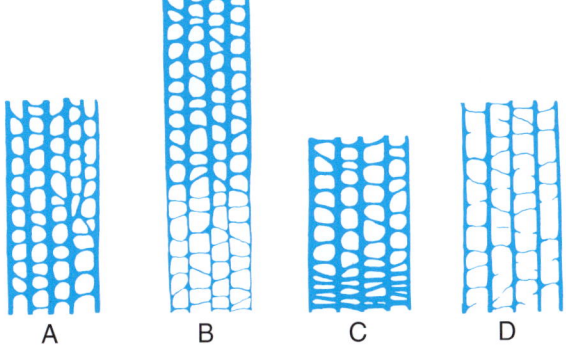

Figure 4.4 The relative amounts of longitudinal and transverse septa (trabeculae) – depicted by the thickness of the lines – in normal bone (A), under conditions of rapid growth in length (B), in arrested growth (C), and with atrophy (D). The lengths of the bone segments in A–D are proportional to each physiological situation. From Harris (1933).

Box 4.3 (*Continued*)

Cartilage specificity

Mineralization of cartilage varies in a cartilage-specific manner. Mammalian small-celled cartilage normally does not mineralize, mineralization being the prerogative of hypertrophic chondrocytes. Secondary cartilages on avian membrane bones mineralize but Meckel's cartilage (a small-celled cartilage) does not. Mineralization and alkaline phosphatase are virtual synonyms but mineralization of rat tracheal cartilages is not associated with deposition of alkaline phosphatase. Guinea pig mandibular condylar, tibial epiphyseal and articular cartilages exhibit different patterns of mineralization in response to scurvy. The jaw cartilages of clam-crushing Myliobatid stingrays are stiffened internally with mineralized struts and rows of tesserae in a manner only found in bone in other vertebrates.[e]

Mineralization *in vitro*

Many instances in which cartilage mineralizes *in vitro* are discussed throughout the text (Figs 4.6 and 4.7). For example, in what Zimmermann *et al.* (1990) termed 'organoid cultures,' explanted mouse limb cells had become almost completely mineralized after 21 days of culture in the presence of 10 nM β-glycerophosphate. Examination at the ultrastructural level revealed that almost all chondrocytes in the mineralized areas had died. In a later study, addition of osteoblastic calvarial cells enhanced mineralization, while addition of fibroblasts decreased mineralization, presumably by the release of soluble factors or the disruption of interactions between mineralizing cells; addition of dexamethasone induces these cells to chondrify (Zimmermann and Cristea, 1993).

Vitamins

Vitamin D plays a critical role in the mineralization of cartilage and bone; vitamin D elevates plasma Ca^{++} and PO_4^- to levels high enough for bone to mineralize. Of the two major hormones that mediate calcium metabolism, parathyroid hormone is vitamin D-dependent and calcitonin is not. The first direct evidence that vitamin D_3 stimulates movement of calcium into bone came from studies on calvariae from Japanese quail maintained *in vitro* by Gunasekeran *et al.* (1986a,b), who paradoxically found that Ca^{++} uptake into mouse bone was reduced by vitamin D_3.[f]

Vitamin D deficiency can result in an inability to mineralize cartilage because hypertrophic chondrocytes fail to mature, stunt growth and lead to *rickets* in children and *osteomalacia* in adults. Rickets and osteomalacia are two terms for the same condition of 'softening' of bones, although the consequences differ the earlier the onset. Collagen synthesis and calcium nucleation are both enhanced when rachitic bone is exposed to vitamin D. Some of the action of vitamin D on bone growth and mineralization may be indirect through enhancement of plasma Ca^{++} and PO_4^{3-}. Demineralized bone matrix implanted into vitamin D-deficient rats induces normal amounts of cartilage, but less bone is produced and the bone is abnormally mineralized unless vitamin D is added.[g]

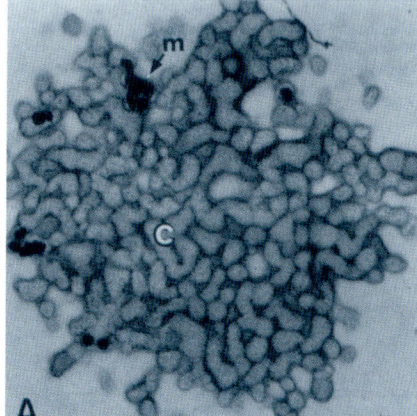 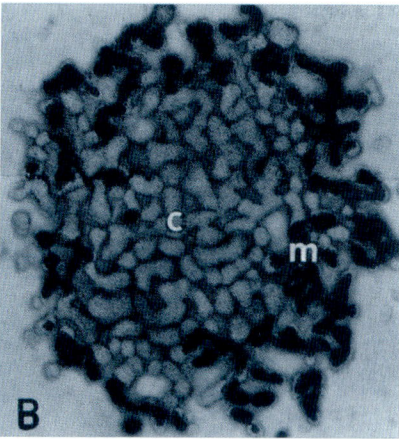 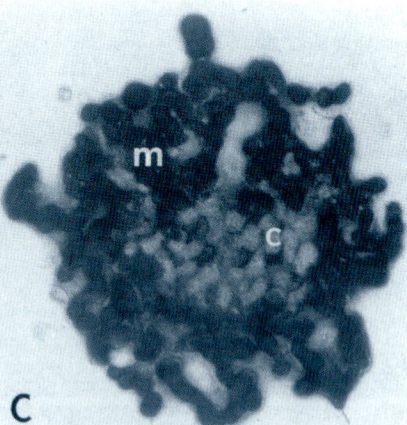

Figure 4.6 Mineralization of cartilage *in vitro*. Mandibular mesenchyme removed from chick embryos of H.H. stage 21 (3.5 days of incubation), dissociated and established in micromass culture (10 μl drops of 2×10^7 cells/ml) undergoes chondrogenesis and matrix mineralization (as seen with Kossa staining) when cultured in Ham's F12:BGJ$_b$ [3:1] + 10% fetal calf serum + ascorbic acid (150 μg/ml). (A) Cartilage nodules (c) with initial signs of mineralization (m) after 12 days of culture. (B) Increasing mineralization (m) after 14 days *in vitro*. Also see Figure 4.7. (C) After 20 days *in vitro* only the central cartilage (c) is unmineralized. Adapted from Ekanayake and Hall (1994b).

Box 4.3 (*Continued*)

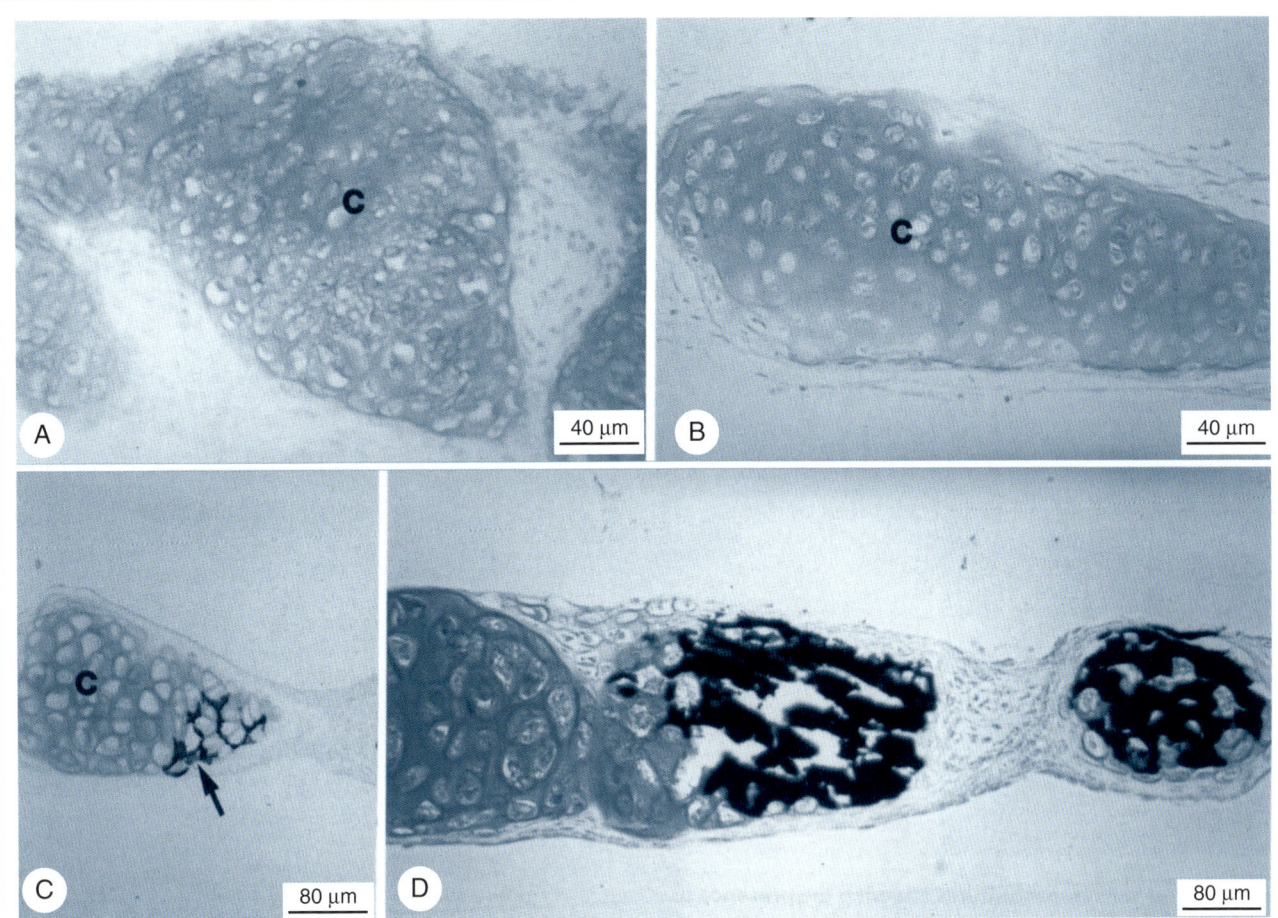

Figure 4.7 Mineralization of cartilage *in vitro*. Mandibular mesenchyme, removed from chick embryos of H.H. stage 21 (3.5 days of incubation), dissociated and established in micromass culture (10 μl drops of 2 × 10⁷ cells/ml) undergoes chondrogenesis with matrix mineralization. Culture media were Ham's F12:BGJ$_b$ [3:1] + 10% fetal calf serum (A); Ham's F12: BGJ$_b$ [3:1] + 10% fetal calf serum + 150 μg/ml ascorbic acid and 10^{-7}M dexamethasone (B–D). Culture period was 12 or 14 days; see also Figure 4.6. Mineralization visualized with the Kossa reaction. (A, B) Histological sections after 12 days of culture. Cartilage (c) is present in both, mineralization in neither (cf. Fig. 4.6). 12-day (C) and 14-day (D) cultures in test medium. Onset of mineralization can be seen in one region of the cartilage (c) after 12 days of culture with ascorbic acid and dexamethasone (C), and more extensive and heavier mineralization after 14 days (D). Adapted from Ekanayake and Hall (1994b).

[a] For reviews of the microstructure and mineralization of skeletal tissues, see Carter (1990), Francillon-Vieillot *et al.* (1990) and de Ricqlès *et al.* (1991).
[b] See A. R. Poole *et al.* (1984) for chondrocalcin and Havelka *et al.* (1984) for the tidemark.
[c] See Termine *et al.* (1981), Stenner *et al.* (1986), Jundt *et al.* (1987) and Nomura *et al.* (1988) for osteonectin, and Copray *et al.* (1989) and Pacifici *et al.* (1990) for expression in condylar cartilage and tibial growth cartilage; and see Figs 4.8 and 4.9 for proximal tibial growth plates in 16- and 25-day-old mice.
[d] See Shepard and Mitchell (1985) for acridine orange, and Davis *et al.* (1982) and A. R. Poole *et al.* 1982c) for retention of proteoglycans during mineralization.
[e] See Hall (1971b) and Sasano *et al.* (1993a) for mineralization of avian and tracheal cartilages; Irving and Durkin (1965) and Durkin *et al.* (1969a,b) for guinea pig; and Summers (2000) and Summers *et al.* (1998) for myliobatid

stingrays, which include such species as the cow nose ray, *Rhinoptera bonasus*, eagle ray, *Aetobatus narinari*, and bat ray, *Myliobatis californica*. *R. bonasus* also develops a fibrocartilage where the ventral intermandibular tendon inserts onto the mineralized jaw cartilage (Summers *et al.*, 2003). Clement (1992) analyzed the fine structure of mineralization of elasmobranch skeletons on the basis of a review of 53 species, while Summers *et al.* (2004) correlated CT-scanning images with measures of stiffness in mineralized cartilages of the California horn shark, *Heterodontus francisci*.
[f] See Wong (1982, 1984) for the skeletal actions of PTH, and see Pacifici *et al.* (1980), Ornoy and Zusman (1983), Narbaitz (1992) and Boskey *et al.* (2001) for vitamins and cartilage.
[g] See Cousins and DeLuca (1972) and Stern (1980) for vitamin D and bone, Underwood and DeLuca (1984) for plasma regulation, and Van der Steenhoven *et al.* (1988) for bone implanted into vitamin D-deficient rats.

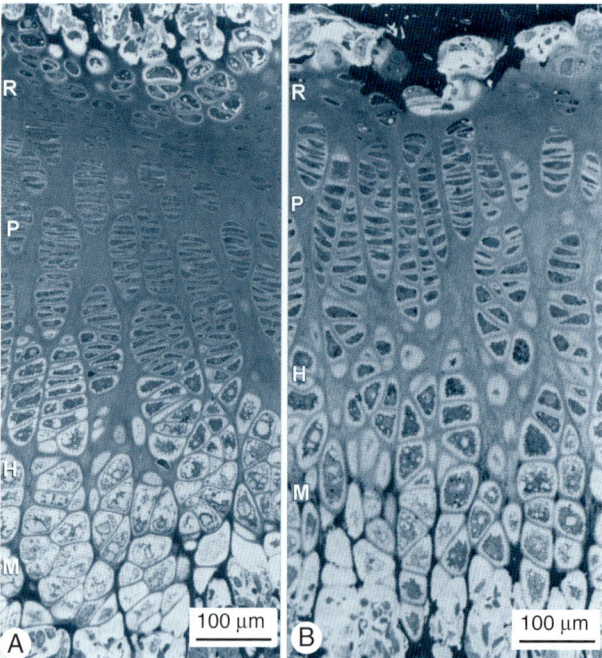

Figure 4.8 The organization of growth plates. Proximal tibial growth plates from 16-(A) and 25- (B) day-old mice showing the organization of the cell columns. The articular surface is at the top. P, proliferative, H, hypertrophic, and R, resting cartilage cell zones. M, mineralization front. Modified from Wikström *et al.* (1984a).

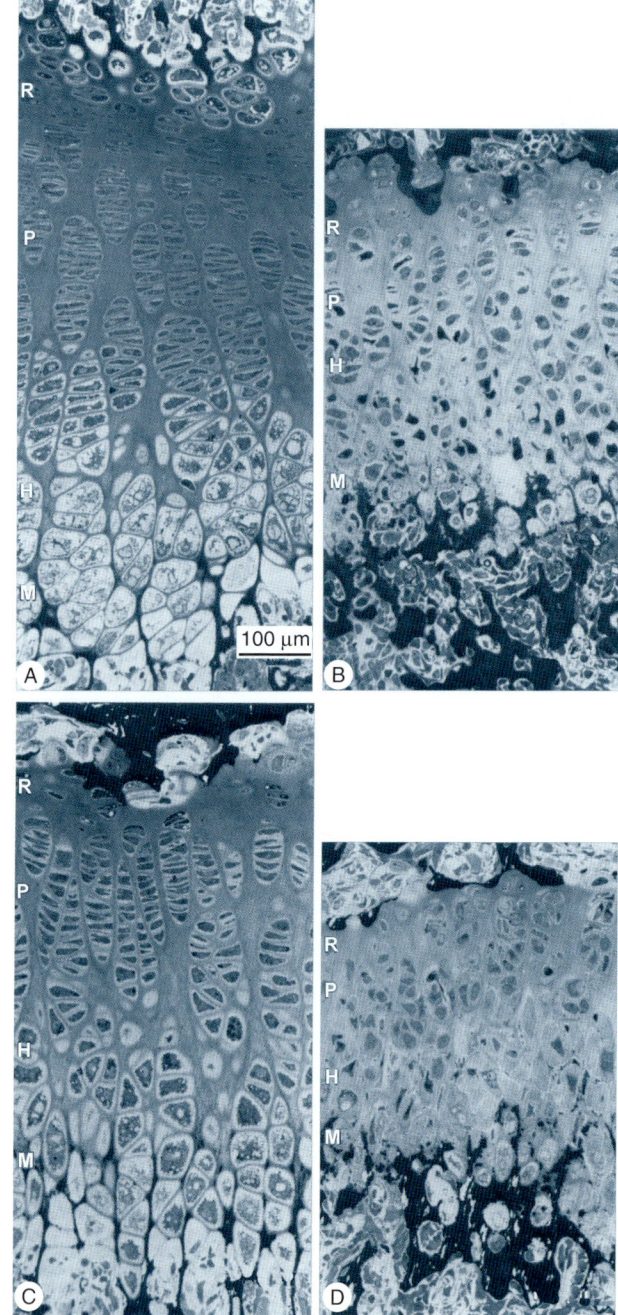

Figure 4.9 Reduced development of the proximal tibial growth plates is seen in *Brachymorphic (Bm/Bm)* mice. (A, B) Growth plates from 16-day-old wild-type (A), and *bm/bm* (B) mice; (C, D) growth plates from 25-day-old wild-type (C) and *Bm/Bm* (D) mice. Cartilage and mineralization (black) is disorganized in the growth plates from the *Bm/Bm* embryos. The entire growth plate is reduced in height in mutant embryos of both ages (all at same scale). P, proliferative, H, hypertrophic, and R, resting cartilage cell zones. M, mineralization front. Modified from Wikström *et al.* (1984a).

formed an outer, impermeable layer because of the tendency for calcium phosphates to be excreted from the skin. In teleost fish, in fact, calcium is excreted via the gills, and phosphorus is transported in the intestine with little if any skeletal involvement.[26]

Such a premise is now either untenable or, if tenable, relates to the origin of vertebrate cartilage, not to the origin of cartilage *per se*. Consequently, several important questions remain. When and how many times did cartilage evolve as a tissue? What is the precursor of cartilage among the earliest Metazoans? Does the evolution of vertebrate cartilage(s) relate in any way to the evolution of invertebrate cartilage(s)? Can we separate the evolution of cartilage as a tissue from the evolution of type II (cartilage-type) collagen? Cole and Hall (2004a,b) think we can.

Often taken as a surrogate vertebrate ancestor, amphioxus, *Branchiostoma lanceolatum*, has gill bars that contain skeletal rods. Only occasionally have these been regarded as cartilaginous. It also has been suggested that cartilage or a cartilage-like tissue exists in the tentacle-like oral cirri around the oral hood. Amazing recent fossil finds, however, mean that we can now place less emphasis on modern organisms as surrogate vertebrate ancestors. *Haikouella lanceolata* (over 300 specimens) and *H. jianshanensis*, from the Chengjiang formation in China (530 mya), possessed gills, a tail, six branchial arches – presumed to be cartilaginous – and pharyngeal teeth, which may represent the earliest mineralized tissues.[27]

No invertebrate has bone, although some invertebrate cartilages contain such bone-cell markers as osteonectin mRNA (Ringuette *et al.*, 1991), a finding that raises fascinating issues: Do invertebrates do with cartilage all that vertebrates do with cartilage *and* bone? Cartilaginous fishes

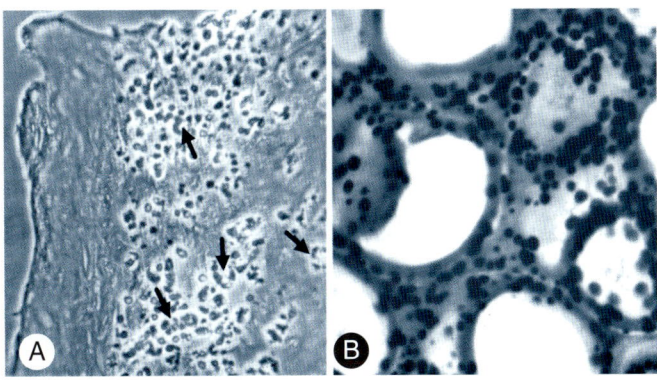

Figure 4.10 Squid cartilage can mineralize. Head cartilage of the longfin squid, *Loligo pealii*, after incubation for 32 (A) or 40 (B) hours in a hydroxyapatite mineral solution at 37°C. (A) Phase-contrast microscopy reveals accumulation of mineral (arrows) beneath the perichondrium. (B) Kossa staining with toluidine blue counter stain to show accumulation of mineral granules (black) over the ECM. Modified from Libbin *et al.* (1976), which contains details of the incubation solution.

certainly do with cartilage what teleosts and tetrapods do with bone. Whatever invertebrate cartilages possess that brings them into the cartilage family – perhaps superfamily would be a better term – we must project the origin(s) of cartilage much further back than tunicates.

We have seen the enormous differences between lamprey, hagfish, gnathostome and invertebrate cartilages. Indeed, just as extant vertebrates are subdivided into the three groups lampreys, hagfishes and gnathostomes, invertebrates should be subdivided into groups – cephalopods, brachiopods, annelids – to reflect the diversity of invertebrate cartilaginous tissues. Extant agnathans and invertebrates have cartilage but not collagen type II, a *sine qua non* for the identification of cartilage in jawed vertebrates but not connected to the first evolution of cartilage as a tissue. Consequently, we must widen our definition and concepts of what constitutes cartilage (Box 4.4). This is not true for bone.

Box 4.4 Notochordal cartilage

Notochord is an unusual tissue with large vacuolated cells connected by desmosomes and surrounded by an expanded basement membrane as a perinotochordal sheath. Notochordal cells synthesize and deposit type II collagen, which we think of as cartilage-type collagen. I believe we may have this reversed and that type II collagen is notochordal collagen, a notochordal molecule that was taken over by cartilage and elaborated into an ECM when cartilage arose evolutionarily.[a]

Even more intriguing issues are whether notochord should be included in the cartilage 'superfamily,' how notochord as a tissue relates to connective tissues, and how it relates to tissues such as chondroid or myxoid, typically regarded as intermediates between two other tissue types (Chapters 1 and 5).

It is well known that the notochord induces sclerotomal mesenchyme around it to chondrify as vertebral cartilage (Fig. 4.11 and Chapter 42). Metachromatic, GAG-rich matrix accumulates around the notochord *before* chondrogenesis of the adjacent mesenchyme, a timing that led to the suggestion that these *notochordal* matrix products progressively *become* the matrix of the vertebral cartilage.[b]

More intriguing, cartilage can arise from the notochord itself. Although not common, notochordal chondrogenesis is part of normal vertebral development in some taxa and is a regenerative response in at least one taxon of teleost fishes.

Lawson (1966) described the formation of notochordal (intravertebral) cartilage in the centra of an apodan, *Hypogeophis rostratus*, the cartilage apparently being formed by cells of the notochordal epithelial sheath and not from invading mesenchymal cells. As Lawson points out, urodele centra ossify from connective tissue, while anuran centra ossify in cartilage.[c]

Absence of a distinct sclerotome but plentiful notochordal cartilage characterizes vertebral development in the northern two-lined lungless salamander, *Eurycea bislineata*. Wake and Lawson (1973) considered this a larval adaptation to stress.

During regeneration of the tail of the glass knifefish, *Eigenmannia virescens*, cartilage develops at the end of the notochord. The chondrocytes do not hypertrophy, nor does the matrix mineralize. Cartilage on the ventral surface is removed by multinucleated cells and replaced by perichondral bone (Kirschbaum and Meunier, 1981, 1988).

A further fascinating example, which comes to us from the fossil record, illustrates the advantages to be had when an informed

palaeontologist incorporates a developmental approach in his/her analysis. The example is a group of extinct, limbless, enormously elongate reptiles, the aystopods, with many hundreds of vertebrae. The challenges in forming so many vertebrae without greatly extending ontogeny are considerable. Carroll (1986, 1989) identified two patterns of vertebral development in these and Palaeozoic tetrapods, both of which reveal unusual involvement of the notochord. Vertebrae developed either from a perichondral tube around the notochord or from a medial notochordal fibrous sheath, a mechanism that allowed more rapid ossification than does perichondral ossification.

A final example is the prolacertiform reptile *Tanystropheus longobardicus* from the Middle Triassic (Wild, 1988). At least three-quarters of the total body length of these six-metre-long ancient reptiles was neck and tail (Fig. 4.11). The enormously elongate neck, which has been described as the longest neck possible within the laws of physics, amazingly consists of no more than

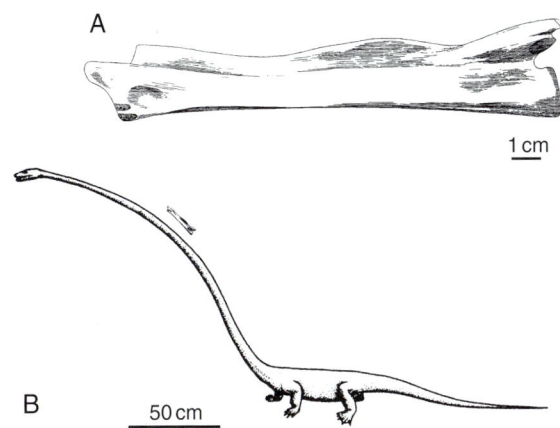

Figure 4.11 The longest cervical vertebrae (A) may have belonged to the middle Triassic reptile, *Tanystropheus longobardicus* (B). A cervical vertebra is shown to scale beside the neck in (B). Based on information in Wild (1973) and Tschanz (1988).

Box 4.4 (*Continued*)

12 cervical vertebrae (often only 10), each of which is enormously elongated (Fig. 4.11). As determined by Tschanz (1988) from an analysis of eight specimens from Switzerland that comprise an ontogenetic series, the largest individuals had cervical vertebrae as long as 26 cm, the average length of cervical vertebrae 3–11 being 17.5 cm in the largest specimen and 3.5 cm in the smallest, a five-fold increase in size.

[a] See Ekanayake and Hall (1991) for the perinotochordal sheath, and see Hall (1998a) and Cole and Hall (2004a,b) for notochord–cartilage affinities.
[b] See Frederickson and Low (1971), Strudel (1971) and Corsin (1974) for discussions of studies on the notochordal extracellular matrix.
[c] See Carroll et al. (1999) and Hall (2003b) for overviews of evolution and development, respectively, and see Carroll (1997), Hall (1998a) and Hall and Olson (2003) for books that integrate both approaches.

The enormous diversity of cartilages in invertebrates and vertebrates demonstrates the lability of skeletogenic cells, a lability that is further reinforced when we consider the range of intermediate tissues in vertebrates, tissues that possess features of two or more of the four skeletogenic and odontogenic tissues – bone, cartilage, dentine and enamel.

NOTES

1. See Beresford (1981) and Dhem *et al.* (1989) for analyses of chondroid, its contribution to skeletal growth, and various uses of the term.
2. See Person and Fine (1959), Person *et al.* (1959), Person (1960, 1983a), Person and Philpott (1963, 1969a,b, 1983), Philpott and Person (1970), and Person in Slavkin (1972, pp. 140–151, the evolution of cartilage section of a workshop held on Santa Catalina Island, CA).
3. See Cole and Hall (2002, 2003, 2004a,b).
4. See Jeanloz (1960) for a discussion of the nomenclature of the glycosaminoglycans (GAGs) or acid mucopolysaccharides, and Anderson (1976b), who surveyed the distribution and composition of GAGs and glycoproteins.
5. See P. J. Morris (1993) for the origin of extracellular matrix, and Garrone and Pottu (1973), Expisoto and Garrone (1993), Har-El and Tanzer (1993) and Garrone (1998) for collagen appearing with sponges. See Hall (2003a) for an analysis of such issues as deep homology and the complex relationships between homology at various levels of the biological hierarchy.
6. Mathews (1968, 1975) and E. Adams (1978) review the composition of invertebrate collagens, Kobayashi (1971) and Finerty (1981) the homology of collagens, and Garrone (1978, 1998) the evolution of metazoan collagens, while Cole and Hall (2004) discuss the collagens of invertebrate cartilages and the evolution of type II collagen.
7. See Sage and Gray (1979, 1980) and a more recent analysis by Wright *et al.* (1988).
8. See Person's (1960) emphasis that invertebrate cartilages possess chitin but lack chondroitin sulphate.
9. See Person (1960, 1983a) and Person and Philpott (1963) for light and transmission microscope cytology/histology, Lash and Whitehouse (1960) and Katzman and Jeanloz (1969) for polyglucose and polyfucose sulphates, and Lash (1959), Person *et al.* (1959) and Person and Philpott (1963) for myoglobin. See Cole and Hall (2004a,b) for overviews of gastropod cartilages, Ponder and Lindberg (1997) for a phylogeny of gastropod molluscs (117 characters, 40 taxa, mostly prosobranch), including the use of cartilage as character,

and see Guralnick and Smith (1999) and references therein, for gastropod cartilages.
10. Information on regeneration of mollusc cartilage is from Carriker *et al.* (1972); also see Cole and Hall (2004a,b).
11. Sugahara *et al.* (1996), who did this study, use the common name 'king crab', an older name for the horseshoe crab, but it is clear that they were examining *Limulus*. The common name 'king crab' (red, blue or golden king crab) is more commonly used for the three species of Alaska king crab. The red king crab, *Paralithodes camtschaticus*, is the most widespread, its range extending from British Columbia to Japan. The golden king crab, *Lithodes aequispinus*, also is distributed from British Columbia to Japan, while the blue king crab, *P. platypus*, is found from southeastern Alaska to Japan.
12. Descriptions of phragmosomes differ, in part reflecting differences in different plant taxa. The phragmosome (partition body) is variously described as: (i) a *structure* within plant cells – a flat membranous vesicle composed of cell-wall components; microbodies of the cell plate that form after nuclear division; a raft-like structure composed of actin filaments (Box 5.1) and microtubules formed immediately before mitosis at the pre-prophase stage – (ii) a *region* within plant cells – the cytoplasmic region containing the nucleus during mitosis – or (iii) by *function* – cytoplasmic strands that aid the nucleus in its migrate to a central position before mitosis. All these descriptions fit the membrane formed in dividing gill-book chondrocytes.
13. See Person (1960), Person and Philpott (1963) and Cowden (1967) for light and transmission microscope cytology/histology, Mathews *et al.* (1962), Cowden (1967) and Sugahara *et al.* (1996) for the GAGs, Person and Philpott (1969b) for collagen, Person and Fine (1959) and Philpott and Person (1966, 1970) for cytochrome oxidase, and Person and Philpott (1963, 1969b) for the pattern of chondrification and its resemblance to lignification.
14. See Cowden (1967) and Person (1969, 1983a) for cranial, scleral and epidermal cartilages, Person (1960), Person and Philpott (1963) and Philpott and Person (1970) for GAG composition, and the chapters in Wight and Mecham (1987) for the biology of proteoglycans.
15. See Kinoshita *et al.* (1997) for the novel sequences and Habuchi *et al.* (1983) for the distinct glycopeptide.
16. See Hjerpe *et al.* (1983), Vynios *et al.* (1985) and Vynios and Tsiganos (1990) for these studies. Two mannose 6-phosphate receptors are expressed in cartilage during mouse development (Matzner *et al.*, 1992).
17. See Ayad *et al.* (1994) and Knudson and Knudson (2001) for detailed reviews of aggrecan and its functions, and see Yang *et al.* (1998) for expression in cultured cells.
18. See Bairati *et al.* (1989, 1995, 1998, 1999).

19. See Sivakumar and Chandrakasan (1998) for C1–C3 collagens, and Rigo and Bairati (1998) for the rotary shadowing.
20. See Person and Mathews (1967) for structure and biochemistry, and Cowden and Fitzharris (1975) for ultrastructure.
21. See Stricker and Reed (1985, and references therein) and Reed and Cloney (1997) for these tissues and the arguments for and against them being cartilaginous.
22. For descriptions of mineralized tissues in invertebrates see Watabe (1965), Travis et al. (1967), Travis (1968) and Schrarer (1970) and J. G. Carter (1990).
23. See Milberg et al. (1974) for mineralization of invertebrate cartilage, Eilberg et al. (1974, 1975), Libbin et al. (1976) and Hall (1972a) for intracellular mineralization of invertebrate chondrocytes, and Langille and Hall (1988a, 1993a) and Janvier and Arsenault (2002) for mineralization of lamprey cartilage.
24. Genes that regulate lipid biosynthesis or transport can act differentially on cartilage and on lipiogenesis. The mutation gonzo (goz) in zebrafish, a mutation in the gene site-1 protease, which regulates key enzymes involved in lipid synthesis and transport, acts independently to produce cartilage and lipid defects (Schlombs et al., 2003).
25. See Libbin et al. (1976) and Rabinowitz et al. (1976) for lipid and in-vitro mineralization of invertebrate cartilages, Irving and Wuthier (1968) Irving (1973), Schuster et al. (1975) and Rogers (2000) for lipids and mineralization of bone and vertebrate cartilages, and Roy et al. (2004) for the influence of dietary phosphorus on bone mineralization.
26. See Romer (1942), Bone (1972), Hall (1975a, 1999a), Gans and Northcutt (1983), Northcutt and Gans (1983), Hall and Hörstadius (1988), Smith and Hall (1990, 1993), Janvier (1996), Northcutt (1996) and Cole and Hall (2004a,b) for theories of the origin of cartilage.
27. See Rahr (1981) and de Beer (1937) for gill and oral hood cartilages, Chen et al. (1999) for Haikouella lanceolata, Shu et al. (2003a) for H. jianshanensis, and Holland and Holland (2001) for how adequate amphioxus is as a surrogate for a (the) ancestral vertebrate. Haikouichthys ercaicunensis from the same Lower Cambrian formation in China has been identified as a stem group craniate by Shu et al. (2003b).

Intermediate Tissues

Chapter 5

Normally there is no difficulty...but life is not always so simple.

Normally, there is no difficulty in assigning vertebrate skeletal tissues and cells to one of the four categories of mineralized tissues. Similarly, invertebrate tissues can be identified as one of a variety of cartilaginous or connective tissues. Recognizing the stability of differentiating or differentiated cell phenotypes and the tissues/organs they inhabit within the vertebrates, we readily identify four cell types and the tissues/organs they create: (i) osteoblasts (bone), (ii) chondroblasts (cartilage), (iii) ameloblasts (enamel), and (iv) odontoblasts (dentine).

Nevertheless, all is not so simple. Life becomes troublesome if we attempt to classify tissues on the basis of a specific molecule (even specific molecules) they produce:

- The type of collagen produced by odontoblasts depositing predentine is the same as that produced by osteoblasts depositing osteoid, viz. type I collagen $[\alpha1(I)_3]$, but odontoblasts are not osteoblasts.
- Notochord, cartilage and early stages of membrane bone produce type II collagen $[\alpha1(I)_2\alpha(II)]$, but notochord is not cartilage is not bone.
- Lamprey cartilages do not deposit type II collagen but can be classified as cartilages on the basis of many other criteria.

This list could go on. Gorski (1998) argues that not all bone is the same. Specifically, rapidly deposited woven bone is not the same as lamellar bone, which is deposited much more slowly (Table 5.1). This conclusion was arrived at after analysis of the presence or absence (or relative amounts) of non-collagenous proteins in bone matrices, and whether deposition or resorption is influenced by mechanical factors or zero gravity (Table 5.1, and see the discussion of the affects of gravity in Chapter 32.[1] So bone is not necessarily bone.

Such observations suggest diversity among cells (osteoblasts) of a single skeletal tissue (bone), and relationships between cells of different skeletal tissues (bone and dentine). Moss (1964a) grouped the cells that produce the four mineralized tissues and named them *scleroblasts*. By grouping them in this way, Moss identified a family or race of related cells. Synthesis of the same type of collagen by cells producing two quite different mineralized tissues – bone and dentine – is one sign of affinity. We now realize, however, as discussed above, that sharing the production of a single molecule – even an important 'luxury' molecule – is not a sufficient criterion to unite cell types. I illustrate this in Box 5.1 with a discussion of actin and myosin.[2]

Although we do not understand the processes fully, sufficient examples are available to indicate that, at a variety of sites within embryonic and adult skeletons, and *in vitro*, skeletal cells can produce tissues whose characteristic features are intermediate between cartilage and bone.

Table 5.1 Some features separating woven from lamellar bone[a]

Woven bone	Lamellar bone
Rapid deposition	Slow deposition
BAG-75 and BSP[b]	BAG-75 and BSP not found
Low osteocalcin levels	Enriched in osteocalcin
Enriched in acid phosphoproteins	
Deposition enhanced at high levels of mechanical strain	
More susceptible to resorption in Spaceflight	

[a] Based on data from Gorski (1998).
[b] BAG-75, bone acidic glycoprotein; BSP, bone sialoprotein.

Box 5.1 'Luxury molecules': distribution of actin and myosin

Until the late 1970s it was thought that actin and myosin were unique biochemical and morphological markers for myoblasts as cell types and muscle as a tissue. Production of such 'luxury molecules' was thought to be diagnostic for contractile function – a function that was the prerogative of cells within the myogenic lineage. The continued synthesis of actin and myosin while cartilage proteoglycans are being produced, and as cells derived from muscle transform to chondroblasts, was explained away by arguments that the myogenic cells were caught in their transformation and so had not yet switched off actin and myosin synthesis.[a]

Reports of identification of actin and/or myosin in non-myogenic cells began to appear in the late 1970s, accompanied by controversy over methodology and interpretation. The controversy centered on whether actin and myosin were really present in non-myogenic cells, and, if they were, whether they functioned in the same way as actin and myosin function in myoblasts. If they were serving a contractile function, was contractility a more general property of cells than previously thought? Alternatively, did actin and myosin play a role in the intracellular transport of materials, or in such major cellular events as cytokinesis? And what of notions of the early separation of myogenic from other cell lines during development?

By the early 1970s it had been shown that actin and/or myosin are present in leukocytes, fibroblasts, the myofibroblasts that surround implants, the brush borders of intestinal cells, the primitive streak, and many tissues of early chick embryos including sources of skeletal mesenchyme such as somites, lateral plate mesoderm, and wing buds.[b]

Myofibroblasts that form the capsule surrounding implanted blood clots are especially interesting. As determined using ultrastructural and functional data, they have characteristics of both fibroblasts *and* smooth muscle cells. Like fibroblasts, they contain intracellular accumulations of collagen and bundles of microfilaments but, like smooth muscle cells, they bind anti-smooth-muscle serum and respond to pharmacological stimulation. Myofibroblasts provide an excellent example of the ability of differentiated cells to express functions normally thought to be the exclusive properties of separate cell populations. They reaffirm the need for **caution** when identifying a cell type solely on the basis of one differentiated feature or function; actin and myosin continue to be produced during the production of proteoglycans typical of chondroblasts, as cells within muscles transform to chondroblasts when cultured on bone matrix.[c]

Myosin is now known to be present in many skeletal and odontogenic cell types, including dental pulp fibroblasts, odontoblast cytoplasmic processes, and at the periphery of osteoclasts derived from rat femora and maintained *in vitro* in the presence of calcitonin, and resorbing bone. So too is actin. When cultured for three days in the presence of parathyroid extract (1 IU/ml) mouse calvariae bind actin-like filaments to heavy meromyosin in 50–70 Å-diameter filaments localized beneath the plasma membrane, and inside cell processes of osteoblasts, osteocytes and osteoclasts. Myosin-like and actin-like structures composed of myosin and actin indistinguishable from the proteins in muscle form in cartilage exposed to acute ischemia. Although similar protein products are found in various cell types, the structural genes specifying actin differ from tissue to tissue, indicating specificity at the genomic level. Many isoforms of actin have been identified within invertebrates, while six actin proteins are known from mammals: α-skeletal and α-cardiac actin in striated and cardiac muscle respectively, α- and β-actin in smooth muscle, and α- and β-cytoskeletal actins.[d]

These examples of the distribution of actin and myosin raise questions of the role contractility plays in developmental and

physiological processes, including cell division. The chondroblast actin cytoskeleton is important for maintaining cell shape and initiating chondrogenesis; disrupting actin with cytochalasin (2 μg/ml) 'causes' chondroblasts to round up and differentiate, an effect that is reversible with fibronectin. Molecules of the hyaluronan-binding chondroitin sulphate proteoglycan versican that lack the two Egf-like repeats in the G3 domain enhance chondrocyte differentiation by acting via *actin assembly in the cytoskeleton* (Box 9.3). Dedifferentiation of sternal chondrocytes *in vitro* is accompanied by substantial increase in versican expression, while transfecting chondrocytes accelerates and exposure to antisense versican probes decreases dedifferentiation, both of which raise interesting possibilities of the role of the cytoskeleton in differentiation and dedifferentiation; a far cry from its exclusive presence in myogenic cells.[e]

From a variety of approaches, a number of which are discussed in subsequent chapters, we know that the cortical cytoskeleton is linked to the extracellular matrix by transmembrane links. Cytoplasm, cell membrane and extracellular matrix form a coordinated functional unit. Norman Maclean and I stressed this when we spoke of the cell and its pericellular matrix as the essential cellular functional unit. The notion of ECM as a central regulator of tissue homeostasis captures the same idea (Fig. 5.1).[f]

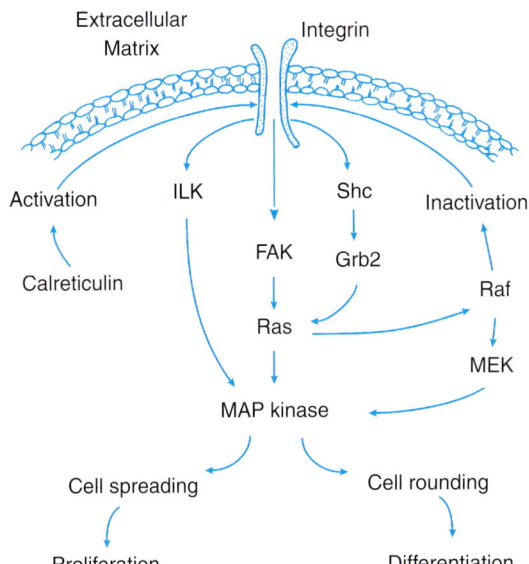

Figure 5.1 Integrin signaling pathways involved in the regulation of cell proliferation and/or differentiation, as depicted with integrin inserted into a cell membrane. The flow diagram includes a well-established pathway from focal adhesion kinase (FAK) to Ras and mitogen-activated protein (MAP-kinase) in the regulation of cell shape, proliferation, and differentiation. Alternate pathways independent of FAK (Shc → Grb2 → Ras) and independent of Ras (ILK → Map-kinase) and from Ras → Raf → MEK → MAP-kinase also are shown. Integrin signaling could be blocked by (i) blocking ligand binding via Ras/RAF and/or via the Ca^{++}-binding protein calreticulin, or (ii) by blocking integrin cytoplasmic domains to activate ligand binding. Based on the scheme proposed by Weaver and Roskelley (1997).

Box 5.1 *Continued*

[a] See Nathanson (1983a,b, 1985, 1986) and Nathanson *et al.* (1986) for these studies.
[b] See Adelstein and Conti (1971), Adelstein *et al.* (1971, 1972) and Shibata *et al.* (1972) for leukocytes and fibroblasts, Ryan *et al.* (1973) for myofibroblasts, Tilney and Mooseker (1971) for the intestinal cells, and Orkin *et al.* (1973) for embryonic chick cells.
[c] See Ryan *et al.* (1973) and Beresford (1981) for myofibroblasts, and Nathanson (1983a,b, 1985, 1986) and Nathanson *et al.* (1986) for cells transforming to chondroblasts.
[d] See Leonard and Sharawy (1974) for the dental cells, King and Holtrop (1975) for culture with parathyroid extract, Warshafsky *et al.* (1980, 1985) for resorption, Aziz and Dhem (1987) for ischemia, Storti and Rich (1976) for

different structural genes, Kedes *et al.* (1985) for mammalian actins, and Moody (1998) for an overview of actin and myosin.
[e] See Zanetti and Solursh (1984) for cytochalasin, and Zhang *et al.* (2001) for versican. Versican also is expressed in condensing limb mesenchyme, the perinotochordal sheath and the neuroepithelial basement membrane in chick embryos (Yamagata *et al.* (1993).
[f] See Daniels and Solursh (1991) for cortical cytoplasmic links, Maclean and Hall (1987), Ettinger and Doljanski (1992), Weaver and Roskelley (1997) and Behonick and Werb (2003) for peri- and extracellular matrices as functionally part of the cell. See Selleck (2000) and Table 27.4 for cell surface proteoglycans and their role in morphogenesis, growth and tumour suppression.

Such tissues represent for developmental biologists what mutants are for geneticists and disease states are for pathologists – a way of analyzing differentiation, not only within the intermediate tissues, but in the more readily identifiable classes of skeletal tissues. Recognizing the mineralized tissues that I refer to as intermediate tissues takes Moss's scleroblasts to the next level. On the basis of one or more – often quite a few more – of the criteria outlined in the definitions of the four classes provided in Chapter 1, these intermediate tissues or cells show features of two of the four classes.

I discuss the most common intermediate tissues, which I consider to be those intermediate in one or more way between *cartilage and bone, dentine and bone,* or *dentine and enamel*. As cartilages are fresh in our collective minds, I treat the connective tissue → cartilage → bone intermediates first.

CHONDROID AND CHONDROID BONE

The range of cartilages in invertebrates and their relationships to notochord and connective tissues in vertebrates demonstrates the continuity that exists between three tissues that I recognize as endpoints in a continuum, the three being notochord, cartilage and connective tissue (see Box 4.4). The existence of a category of skeletal tissues named *chondroid* reflects our ability to recognize a permanent tissue type that has sufficient properties of cartilage and connective tissue that it cannot be classified as one or the other; see above, Table 5.2 and Chapter 1.

Numerous examples of tissues intermediate between cartilage and bone are known. *Chondroid bone*, perhaps the most recognized class, was introduced in Chapter 1. Many tumours, sarcomas, and skeletal neoplasia fall into this category.[3] Others are normal skeletal tissues in one taxon that resemble pathological situations in another taxon. The skeletal tissue(s) in deer antlers (Chapter 8) or in the kype of salmon (this chapter) can be compared to neoplastic bone; certainly they are unusual modes of ossification. Below, I briefly discuss cartilage from fibrous tissue and metaplasia of chondrogenic to osteogenic cells before expanding on chondroid and chondroid bone, the major intermediate tissues in this category. But first,

we need to distinguish between:

- *modulation of cellular activity*, for example, a cell switching from the synthesis of type I to the synthesis of type II collagen;
- *transformation of cell identity*, for example, a fibroblast becoming a chondroblast or a hypertrophic chondrocyte becoming an osteoprogenitor cell (Chapter 13); and
- *formation of permanent intermediate tissues*, for example, the differentiation of cementum, enameloid or chondroid bone.

MODULATION AND INTERMEDIATE TISSUES

Modulation, a temporary change in cell behaviour, structure and/or the type of matrix product(s) produced, occurs in response to altered environmental conditions. Maintenance of the new state depends on the continued presence of the environmental stimulus. Remove the stimulus and the cell reverts to its original structure, function or behaviour. Modulation, therefore, is a physiological response.

Transformation to a new cell identity is usually permanent; the new state remains after the stimulus is removed. Intermediate tissues are permanent. A metaphor would be the difference between growth of a foetus during pregnancy (a permanent change) and the weight gained by the mother, a temporary modulation (although it may seem permanent to the mother).

Many studies document the fact that altering microenvironments *in vivo* or *in vitro* can modulate the synthetic activity of skeletal cells. Indeed, the same environmental change – e.g. oxygen tension – may have different effects on different cell populations. Bone cell metabolism – especially glucose and glycine metabolism – is exquisitely sensitive to oxygen levels. This does not mean invariance of responses to altered levels:

- The neotropical teleost fish, the pacu, *Piaractus mesopotamicus*, develops extensions of dermal bone in its jaws when in oxygen-deficient waters. The extensions regress when the water is aerated.
- Undifferentiated mesenchymal cells and prechondroblasts of prenatal mouse mandibular condylar

Table 5.2 Comparison of the developmental patterns and matrix composition of cartilage, bone, chondroid bone and related skeletal tissues[a]

Tissue	Development pattern		Matrix
	Initial	Late	
Mucochondroid	Endosteal	Endosteal	Collagen, microfibrils, granules, nerves, blood vessels
Hyaline cell cartilage	Endosteal Replaced by hyaline cartilage Metaplasia to bone on some teleost dentary	Endosteal	Collagen, microfibrils, granules, nerves, blood vessels
Yellow perch cartilage	Endosteal Replaced by endochondral bone Chondrocytes show degenerative features when stained	Endosteal	Collagen
Secondary cartilage	Endosteal Replaced by fibrocartilage Replaced by endochondral bone or other tissue	Endosteal	Collagen
Anomalous bone	Periosteal Replaced by chondroblasts	Endosteal	Collagen Immunofluorescence (−) type II collagen (+) cartilage control non-lamellar
Chondroid bone	Periosteal Not replaced Occurs in area of fast growth Frequently shows degenerative features	Endosteal	Collagen: random Matrix identical histologically to bone
Mammalian chondroid bone	Endosteal and periosteal Not a transitional tissue	Periosteal	Interstitial: type I collagen Perilacunar: type I and II Collagens Branched microfilaments Matrix granules
Acellular bone	Periosteal	Periosteal	Collagen lamellar
Japanese Medaka acellular bone	Periosteal Not a secondary response	Periosteal	Collagen layer over cells → pool Extracellular space → disappears lamellar

[a] For more information see Taylor et al. (1994).

cartilages (Fig. 5.2) are especially sensitive to increased oxygen tension; 28 per cent O_2 is mitogenic, increasing the undifferentiated and prechondroblastic cells layers while decreases glycosaminoglycan synthesis. Reflecting stages of maturity or regional differences in cell populations, different regions of the condyle respond differentially (Fig. 5.2).[4]

- As many studies have shown, variation in oxygen tension influences osteogenesis in vitro. The optimal oxygen tension for osteogenesis is 35 per cent. Invariably, no osteogenesis occurs at 5 per cent and 95 per cent initiates resorption. In one study, rabbit periostea were cultured for six weeks in agarose in oxygen tensions ranging from 1 to 90 per cent. Aerobic conditions (12–15% O_2) were optimal for osteogenesis. While no bone forms in 5 per cent O_2, 90 per cent of the cultures in 1 to 5 per cent produced cartilage, showing how variation in the same environmental signal can mediate the modulation of synthetic activity and differentiation.[5]

Such results support other studies that show that oxygen tension plays a role in the switch to chondrogenesis (Chapter 28). For example, when fresh, crushed or lyophilized bone matrix is placed into diffusion chambers with marrow or muscle, and implanted into host mice or rabbits, host cells are induced to form mature bone outside the chamber, while chondro-osteoid or cellular bone forms inside. The differences in tissue types produced are attributed to reduced oxygen tension within the chamber.[6]

The substrata on which skeletal tissues are maintained invitro also modulate differentiative pathways. For example, rabbit articular chondrocytes cultured on demineralized bone produce a chondroid matrix with features of cartilage and bone. When cultured on gelfoam sponge, however, these same chondrocytes produce a chondromyxoid tissue. Such transformations are modulations, not permanent changes. As assessed by histological analysis, culturing embryonic chick tibiae and mandibular bones – endochondral and membrane bones, respectively – in complement-sufficient antiserum results in the production of extensive amounts of a chondro-osteoid tissue and secondary cartilage from periosteal cells, plus release of chondrocytes from their matrices and their modulation to fibroblasts and osteoblasts.[7]

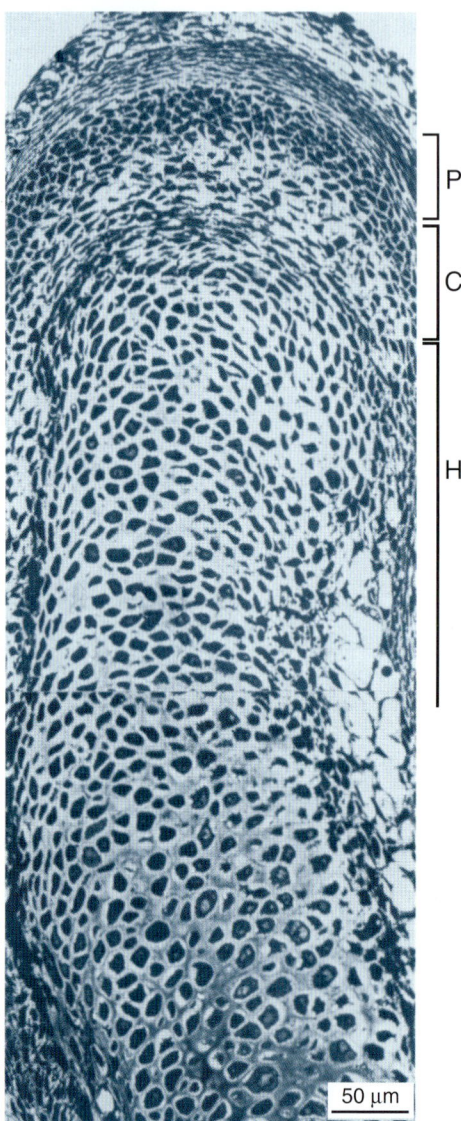

Figure 5.2 Condylar cartilage from the mandible of a 20-day-old mouse embryo cultured for two days retains the organization it had *in vivo*. C, chondrogenic zone; H, hypertrophic zone; P, proliferative zone. Modified from Silbermann *et al.* (1983).

CARTILAGE FROM FIBROUS TISSUE AND METAPLASIA

An unusual fibrocartilage is produced by chondrogenic replacement of the patellar ligament during development of the epiphyseal tubercle of the tibiae in rats. The ligamentous tissue undergoes metaplasia to a fibrocartilage, which is then mineralized and replaced by bone. Mineralization occurs, however, *without* the chondrocytes undergoing hypertrophy or accumulating stores of glycogen, both of which are normally associated with the maturational changes in cartilage that precede mineralization (Chapter 22). Connective tissue around the tail vertebrae of four-day-old rats undergoes metaplasia to a chondroid tissue after the vertebrae are transplanted. As growth

pressure compresses the annulus fibrosus, the chondroid tissue is replaced by metaplastic bone, not unlike changes seen in late stages of *ankylosing spondylitis* in humans.

Metaplasia is quite common outside the mammals. Turtle femora contain substantial amounts of metaplastic bone, which forms as hypertrophic chondrocytes transform into osteoblasts. Lizards too, contain much metaplastic bone.[8]

A further example comes from bullfrogs, *Rana catesbiana*, which possess an osteochondral ligament as a fibrous attachment between bone and articular cartilage. Osteoblasts are found on the inner area of the ligament near the bone. Mineralization of the growth cartilage and formation of bony trabeculae occur late in ontogeny and are dissociated. At one year of age there is neither mineralization of cartilage nor replacement of cartilage by bone, although the chondrocytes are alkaline-phosphatase positive, which usually is a sign of incipient mineralization or a marker for preosteoblasts (Box 5.2). By two years, mineralization and replacement of cartilage are underway, but being so late in ontogeny, neither process contributes to long bone growth, giving the unusual situation of long bone growth without cartilage mineralization.[9]

METAPLASIA OF EPITHELIAL CELLS TO CHONDROBLASTS OR OSTEOBLASTS

A body of literature provides *circumstantial* evidence for *metaplasia* of epithelial cells to chondroblasts or osteoblasts, as seen in the osteoid that forms in human pancreatic carcinomas or in ectopic cartilage in the thyroid. The altered epithelium was regarded as the source of metaplastic bone found in a stomach polyp after transformation of the stomach epithelium to a cancerous state. Evidence for the epithelial origin of these and other clinical cases is indirect at best, although, as we saw in Chapter 4, evidence for the ectodermal (epithelial) origin of some invertebrate cartilages is more secure.[10]

The demonstration that epithelial cells can produce sulphated glycosaminoglycans (GAGs) and collagen, both of which are major and characteristic macromolecules of skeletal extracellular matrices (ECMs), is certainly consistent with potential transformation to chondrogenesis, although as emphasized throughout the text, the chief attribute of chondrogenic cells is their ability to deposit such products into a *stable* ECM. The epithelia analyzed include embryonic chick amniotic cells inhibited by 5-bromo-2′-deoxyuridine (BrdU, a thymidine analogue that substitutes for thymidine and blocks terminal differentiation of a wide variety of cell types), chick corneal epithelium, and human amnion AU-2 epithelial-like cells *in vitro*.[11]

In fact, *most* cell types synthesize GAGs when maintained in culture, and chondrocytes and quite dissimilar fibroblasts (skin and gingiva) produce similar GAGs

Box 5.2 Alkaline phosphatase

An important prerequisite for cells within skeletal, connective or other tissues to ossify and/or mineralize an extracellular matrix is the ability to synthesis alkaline phosphatase. Although this statement could have been made in 1930 – the pioneering work of Fell and Robison – we still do not understand why alkaline phosphatase is so important.

Over sixty years ago, Gomori (1943) established a method – known simply as 'the Gomori method' – that became the gold standard for determining the presence of alkaline phosphatase, an enzyme that is universally used as a marker for osteogenic cells

and/or osteogenesis *in vivo* and *in vitro*. Lorch published important findings in the late 1940s on the association of alkaline phosphatase with osteogenesis in mammals and developing trout, and with cartilage mineralization in the small spotted catshark, *Scyliorhinus canicula*. A decade later, Henrichsen showed the utility of alkaline phosphatase as a marker for osteogenesis and mineralization *in vitro*.[a]

With the advent of electron microscopy alkaline (and acid) phosphatases were localized to collagen-containing vesicles in fibroblasts, while in 1978 George Bernard published the ultrastructural localization of alkaline phosphatase in the initial phases of intramembranous ossification. This and two other papers from his laboratory remain the best analyses of intramembranous ossification. A bone-specific alkaline phosphatase has been described in cultured chick limb mesenchymal cells, and stage-specific expression of alkaline phosphatase used to map the skeletal derivatives of the first arch skeleton in inbred strains of mice (Fig. 5.3).[b]

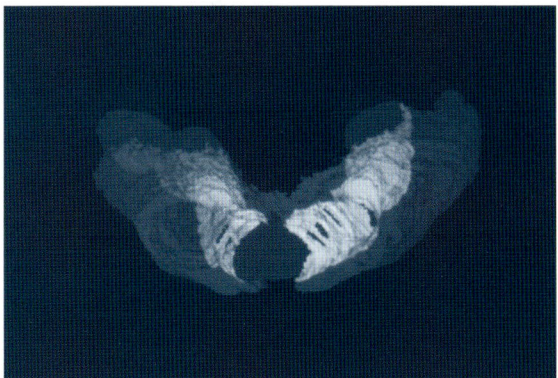

Figure 5.3 A 3D-reconstruction of the skeletal elements of the developing mandible of a C57BL/6 inbred mouse strain at Theiler stage 21. The reconstruction was based on an embryo in which alkaline phosphatase was used to visualize osteogenic cells, and alcian blue to visualize Meckel's cartilage (Miyake *et al.*, 1997a).

[a] Of the many papers on the requirement for alkaline phosphatase synthesis, see Fell and Robison (1929, 1930), Gomori (1943), Lorch (1947, 1949a–c), Henrichsen (1956a,b, 1958a,b) and Johnson and McMinn (1956). For alkaline phosphatase as a marker for osteogenic cells, see Farley and Baylink (1986), Wlodarski and Reddi (1986), and Miyake *et al.* (1997a) and the literature cited therein, and virtually any paper in which osteogenesis is analyzed *in vitro*.

[b] See Deporter and Ten Cate (1973) for localization of phosphatases to vesicles, Bernard and Pease (1969), Marvaso and Bernard (1977) and Bernard (1978) for initial intramembranous ossification, Osdoby and Caplan (1981) for the bone-specific antibody, and Miyake *et al.* (1997a) for alkaline phosphatase and mapping skeletal primordia. Alkaline phosphatase has long been known to be associated with the initial mineralization of cartilage and bone, perhaps because it functions as a Ca^{++}-binding glycoprotein (de Bernard *et al.*, 1986). Alkaline phosphatase also is active in the metabolism of mandibular condylar cartilage cells at all stages of maturation (Lewinson *et al.*, 1982). See Fig. 5.2 for the cell zones in murine condylar cartilage.

in vitro. Such results may reflect a heterogeneity of many cell populations that is only revealed when they are cultured, a heterogeneity that may extend to collagen synthesis. Clonal epithelial cells, cells from melanomas, liver, pituitary and kidney, neuroblastomas and HeLa cells, all produce collagen *in vitro*, although few do so *in vivo*.

Chondroid

We tend to expect to find chondroid in what were once called 'lower vertebrates,' although Darwin **cautioned** that we should never speak of higher and lower in these terms. Fish, amphibians and reptiles – the ectothermic (hence lower?) vertebrates – fall into this derogatory category.[12]

A cartilaginous tissue in the cloacae of male caecilians has been described as unmineralized chondroid, although the points of the connective tissue sheath surrounding the tissue are mineralized. Another unusual and highly derived caecilian feature is the presence of a pair of chemosensory retractable tentacles located beside the eyes. The single study on tentacle development in the Mexican burrowing caecilian, *Dermophismexicanus*, suggests that caecilian skeletal development would profitably bear closer study (Table 5.3).[13]

Table 5.3 Caecilians: an annotated list of studies on skeletal development and skeletal tissues

Allometric analysis of the skull in *Dermophis mexicanus* (Lessa and Wake, 1992)

Chondrocranial development in the French Guyana caecilian, *Typhlonectes compressicaudus*, and comparison with other species (M. H. Wake *et al.*, 1985)

Cloacal *chondroid*/cartilage in male caecilians (M. H. Wake, 1985, 1998)

Early Jurassic caecilian (*Eocaecilia*) with limbs which shares many features with salamanders (Jenkins and Walsh, 1993)

Larval *hyobranchial* of bone is 'replaced' by a cartilage in adult *Epicrionops bicolor*, reversing the normal sequence (M. H. Wake, 1989)

Osteoderms and scales in caecilians (Casey and Lawson, 1977; Zylberberg *et al.*, 1980; Zylberberg and Wake, 1990)

The *skull* in caecilians (Kesteven, 1957a; M. H. Wake, 1985, 1998)

Skull development in *Dermophis mexicanus*, which has a kinetic skull and early ossification of the jaw suspension associated with intraovidual feeding (M. H. Wake and Hanken, 1982)

The *tentacle*, a unique chemosensory structure in caecilians (M. H. Wake, 1985, 1998)

Vertebral development in caecilians (D. B. Wake and M. H. Wake, 1986)

Teleosts

Chondroid is frequently found in teleost fishes (Table 5.2). Much of the research leading to this statement and much of our knowledge of the diversity of cartilage types found in teleost fishes comes from Michael Benjamin in Cardiff, Wales. Chondroid tissue is found as part of the maxilla, premaxilla and dentary, three membrane bones of a cypriniform fish, the lemon (Chinese) algae eater, *Gyrinocheilus aymonieri*. Identification of this tissue as chondroid was based on light, transmission and scanning electron microscopy, histochemistry and x-ray microprobe analysis. Determining the origin of the tissue was more difficult; Benjamin (1986) could not decide whether these tissues were secondarily attached to the bones or secondary cartilages arising from periosteal cells. A developmental analysis would be required to settle this issue, especially in light of secondary cartilage having only been reported from one teleost, the black molly, *Poecilia sphenops* (Fig. 5.4), an identification that required a close developmental analysis that was part of an examination of cartilaginous and mucocartilaginous tissues in some 56 species of teleosts representing 26 families (Benjamin, 1988, 1989a,b). The intermediate tissue mucocartilage defined by Benjamin (1988) as a mucous connective tissue, was especially common beneath the skin and in the opercular valves.

A detailed description of chondroid – *hyaline cell cartilage* to use Benjamin's terminology – in teleost heads followed, as did an analysis of the development of chondroid on the maxilla, dentary and cleithrum of the black molly, *Poecilia sphenops*. Subsequently, on the basis of a sample of 45 species from 24 families, Benjamin (1990) undertook a detailed analysis of cranial cartilages from which he identified six types: hyaline-cell, cell-rich hyaline, fibro/cell-rich, fibrohyaline-cell and matrix-rich hyaline cartilages, and *Zellknorpel*. To enable further evaluation of the cartilaginous nature of these tissues, Benjamin included an analysis of the ECMs of the connective tissues of teleosts heads in his 1990 paper. Five of the 12 species studied in this regard had collagen type I in their bones and type II in the connective tissues, illustrating where the potential for

transformation to chondroid or cartilage resides. A parallel study identified similar classes of cartilage in the trunk, although *Zellknorpel* was not found in the trunk.[14]

Mammals

Chondrogenic tissues, including chondroid (as in the nasofrontal suture) are frequently found in the sutures of mammalian skulls (Chapter 34). Goret-Nicaise published a series of studies during the 1980s identifying chondroid in the human skeleton:

- on the coronoid process in infants, where the temporal muscle inserts to allow migration of the muscle insertion;
- in the mandibular symphysis of newborns;
- at the ventral border of the mandible;
- at the sutural margins of the cranial vault, where chondroid and chondroid bone were thought to be neural crest-derived, the balance of the bone mesodermal; and
- in the midline suture between the frontal bones along with secondary cartilage.[15]

Goret-Nicaise demonstrated collagen types I and II in human and feline mandibular chondroid tissues. Importantly, she demonstrated that chondroid bone differs from bone and cartilage in not being a transitional tissue, but being a permanent intermediate tissue in the sense that I define the term.

Studies of infant mandibles one or two months after birth demonstrated similar calcium levels in chondroid and woven bone. The levels are lower than in enamel or mineralized cartilage but higher than in lamellar bone and dentine (Table 5.4). Furthermore, and illustrating how locally adaptive such tissues are, there is a 60 per cent increase in mRNA for osteonectin in chondroid of mandibles subjected to loading.[16]

Chondroid is often a component of novel skeletal tissues that form outside the bounds of what we traditionally consider to be the 'normal' skeleton. As one example, cartilage is present in the muscle at the base of the adipose fin in many teleost fish.[17] In some species, the tissue is cartilage, in others chondroid (Fig. 5.5). One interpretation of the

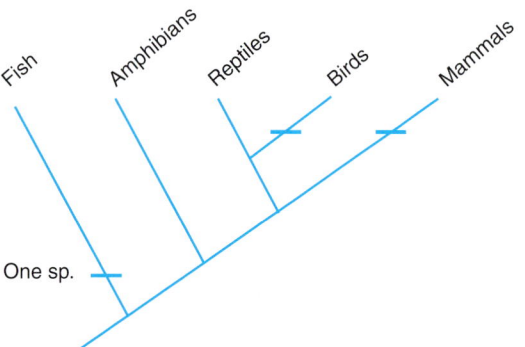

Figure 5.4 The distribution of secondary cartilage in vertebrates. Secondary cartilage has been described in a single species of teleost, the black molly, *Poecilia sphenops,* in birds and in mammals. Independent origin of secondary cartilage in the three groups is suggested, a conclusion that is supported by the differences between avian and mammalian secondary cartilages described in the text. The implication is that avian and mammalian secondary cartilages are not homologous tissues. Modified from Hall (2000a).

Table 5.4 A comparison of the relative calcium content of the mineralized tissues of the mandible of a two-month-old human child[a]

Tissue	Average calcium content %	Maximum calcium content %
Enamel	100	100
Mineralized cartilage	95	87
Woven bone	83	83
Chondroid	80	77
Dentine	67	60
Lamellar bone	63	63

[a] From data in Goret-Nicaise and Dhem (1985), based on photometric analysis of radiographs. As the most mineralized tissue, enamel was given the value of 100 per cent.

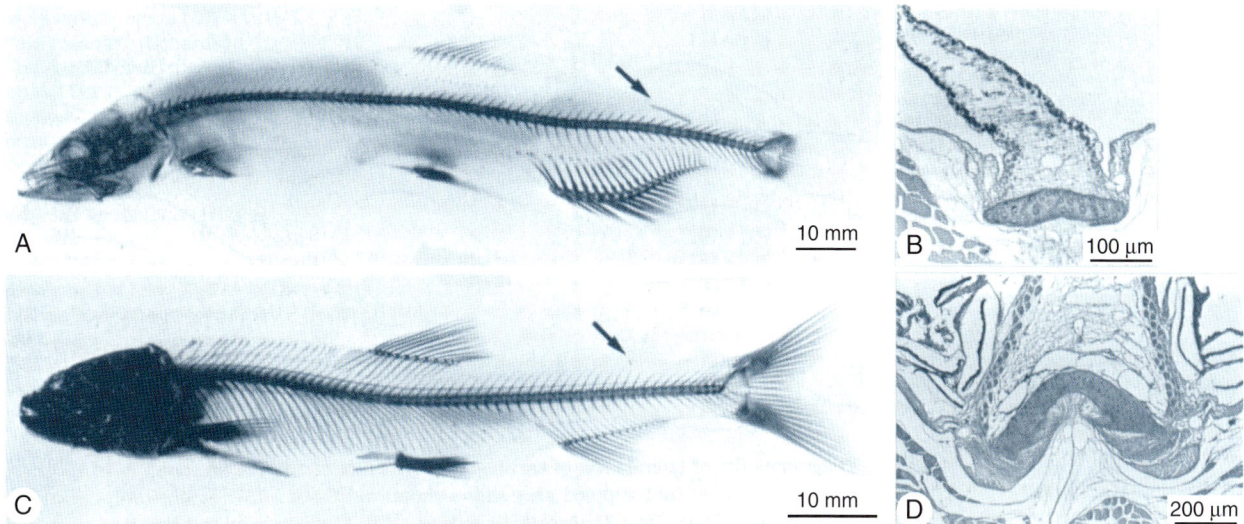

Figure 5.5 Teleost fish adipose fins. (A, C) Cleared and stained specimens of the shishamo smelt, *Spirinchus lanceolatus* (A), and the ayu fish, *Plecoglossus altivelis* (C), both from Japan, showing the adipose fin between dorsal and caudal fins, and the cartilage → in the base of the adipose fins. The adipose fins of most teleosts lack cartilage. Transverse sections of the base of the fleshy adipose fins of the shirauo or Japanese icefish, *Salangichthys microdon* (B), and *S. lanceolatus* (D) show the location of cartilage at the base of the fins. Modified from Matsuoka and Iwai (1983).

origin of such elements is that they represent undeveloped pterygiophores, which are the endoskeletal cartilaginous elements that support the bony fin rays. Pterygiophores are not normally visible externally, being embedded in the body at the base of the fin. If these cartilages at the adipose fin base are pterygiophores as Matsuoka and Iwai (1983) advocate, they could arise because of a truncation of pterygiophore development through a developmental mechanism such as paedomorphosis.

CHONDROID BONE

Chondrocytes secrete an osteoid matrix to form chondroid bone (Table 5.2).

Teleosts

As early as 1915 reports appeared that yellow perch, *Perca flavescens*, infected with metacercaria of the trematode, *Apophallus brevis*, develop mineralized 'bony' cysts in the muscle.[18] These doughnut-shaped hollow cysts have pores plugged with connective tissue at two of the poles, apparently escape ports for the parasite that lives within (Figs 5.6, 5.7 and 34.1). Lawrence Taylor investigated these 'bone cysts' in my laboratory, only to find that their structure is far more complex and interesting than originally thought; they contain collagen but little GAGs, and are organized into outer and inner layers of chondroid bone. Chondroid bone associated with the pharyngeal jaws of cichlid fishes is discussed below.[19]

'Kype' is the term given to the hook that develops on the distal tip of the lower jaw in Atlantic salmon, *Salmo salar*, as they migrate upstream to breed. Although the

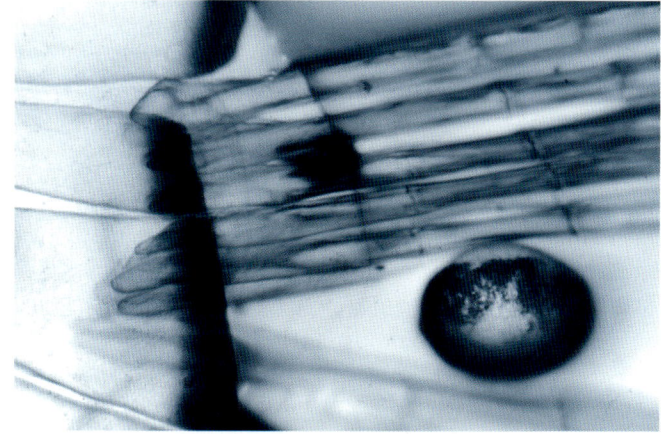

Figure 5.6 An ectopic ossicle of bone and chondroid bone between the fin rays of a yellow perch, *Perca flavescens*. The ossicle has arisen in response to infection with metacercaria of the trematode parasite, *Apophallus brevis*. A band of mineralized encircles the ossicle. Alizarin red-stained whole mount. Also see Figs 5.7 and 34.1, and see Taylor *et al.* (1994) for details.

kype forms from the periosteum of the dentary bone, the needle-like bone matrix, rapid growth and presence of chondrocytes are all features that distinguish kype bone from regular dentary bone (Witten and Hall, 2002, 2003).

Mammals

The cartilaginous articular surface of the temporal component of the human temporomandibular joint *transforms into chondroid bone*. Transformation involves the formation of a predominantly collagenous ECM with

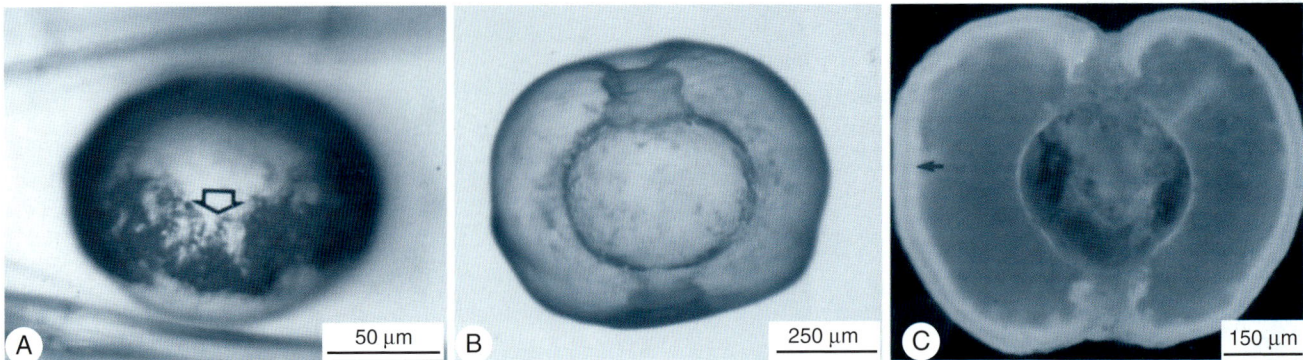

Figure 5.7 Ectopic ossicles composed of bone and chondroid bone form in yellow perch, *Perca flavescens*, infected with metacercaria of the trematode parasite, *Apophallus brevis*. (A) An early ossicle stained with alizarin red shows a band of mineralized matrix → encircling the ossicle. (B) A more mature ossicle shows the central chamber that houses the metacercaria and two channels that allow the metacercaria to move in and out of the ossicle. (C) A cross section of an ossicle recovered from a yellow perch previously given three intraperitoneal injections of tetracycline to show lamellar bone deposited peripherally ←; when viewed under fluorescent light. Also see Figs 5.6 and 34.1. Modified from Taylor *et al.* (1994).

the basophilic characteristic histology of bone but with nests of cells resembling chondrocytes in lacunae.[20] I presented evidence for a similar transformation of secondary cartilage to chondroid bone in paralyzed embryonic chicks, and analyzed the then-available literature on such transformations (Hall, 1972b). Matrix changes are the result of trapped chondroblasts and chondrocytes assuming osteoblastic activity and modifying the ECM toward an osseous type. Such alterations – true metaplasia – are also seen in the repair of fractured long bones in lizards and frogs, at the distal tip of penile bones (Chapter 25), in human jaws in response to ill-fitting dentures, and in the formation of bone and cartilage in mammary tumours.[21]

Evidence of metaplasia is often primarily cytological. It would be nice to have an analysis of the changes in collagen type produced by these cells; some transformations are not as readily interpreted as these and those I reviewed in 1972 (Hall, 1972b). Lutfi (1971) described what he termed 'cartilage bone' lining marrow spaces of the tibiae of one- and two-day-old chicks as the first step in endochondral ossification; however, his figures are reminiscent of extension of staining from the surface periosteum.[22]

Although Dodds and Cameron (1939) described osteoblasts in open chondrocyte lacunae in old cartilage in rachitic rat metaphyses, these could be interpreted as locally transformed chondrocytes, or transformed migratory marrow cells, either of which could deposit the surface osteoid. A similar phenomenon is seen in articular cartilage exposed to excess vitamin A in organ culture (see Box 27.1),[23] a phenomenon regarded, not as 'true osteogenesis', but as "selective collagenous regeneration by the liberated chondrocytes" (Barratt, 1973, p. 212). The necessity for such semantic refinements illustrates the difficulty encountered in interpreting the cellular activity underlying such changes, although Boyde and Shapiro (1987) used high resolution microscopical analysis of the growth cartilages of rachitic chicks to demon-

strate that chondrocytes do survive to become osteoblasts. Unraveling the mechanisms of such transformations will be a major step toward understanding interrelations between skeletal cells and how they modulate their activity (see Chapter 12).

If mandibular condyles from one- or 20-day-old foetuses are organ-cultured for 10 days, differentiated hypertrophic chondrocytes influence chondroprogenitor cells to transform to osteoblasts and deposit a mineralized, collagen type I-containing matrix, containing osteonectin and bone sialoprotein and interpreted as chondroid bone (Figs 5.8 and 5.9). Chondroid bone identified in the rat glenoid fossa contains type I and type II and osteocalcin, i.e. it has features of hypertrophic chondrocytes and of osteoblasts/cytes.[24]

In addition to chondroid bone, a variety of types of cartilage also have both type I and type II collagen in their ECMs. Aspects of development that predispose cartilages to synthesize and deposit type I collagen include periosteal origin of the cartilage (secondary cartilages), function as an articular cartilage, mineralization and/or hypertrophy and replacement by bone (Table 5.5).

Chondroid bone and pharyngeal jaws

An adaptation found in some teleost fishes is the development of *a second set of jaws known as pharyngeal jaws*, which develop from branchial arch elements that support the gills in other fishes. Pharyngeal jaws come equipped with powerful muscles and very plastic teeth, by which I mean that the morphology of the teeth can be modified in response to the prey items being eaten. Pharyngeal jaws allow the 'regular' jaws to specialize for prey capture, the pharyngeal jaws doing the breaking up of food. As you might expect, such a division of labour opens up evolutionary opportunities, which the cichlid fishes in the African rift lakes have exploited to an amazing degree. Speciation in these cichlids is almost as rapid

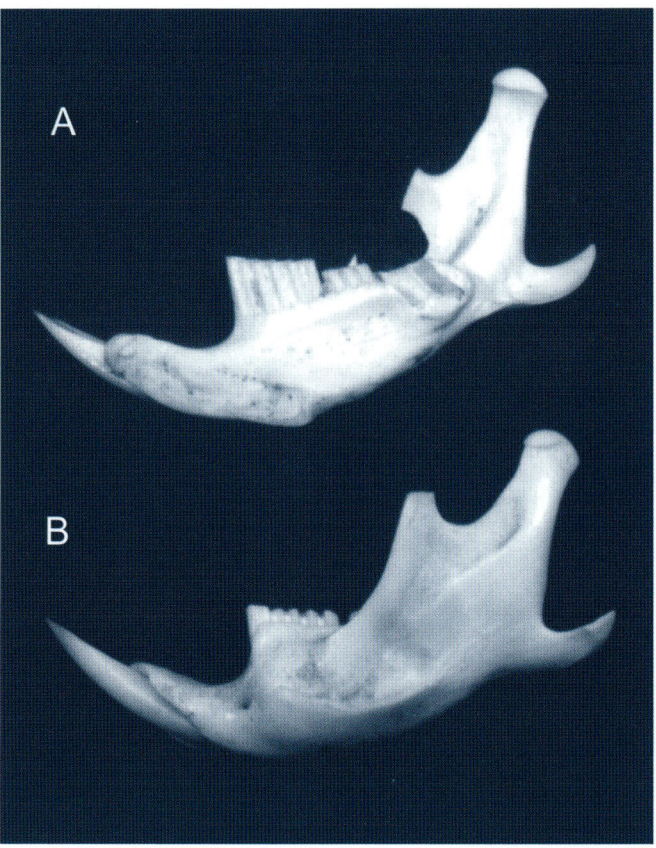

Figure 5.8 The dentary of the Magdalen Island subspecies of the meadow vole, *Microtus pennsylvanicus magdalensis,* in medial (A) and lateral (B) views to show the prominent condylar process with its articulating surface for the temporomandibular joint, and the smaller coronoid and angular processes anterior and posterior to the condylar process.

as has ever been measured (Hall, 1998a). Liem (1973) developed the thesis that pharyngeal jaws are an *evolutionary innovation* and studied them extensively, especially from the points of view of functional morphology and musculoskeletal adaptation. A shift in the insertion of the fourth levator externus muscle coupled with developmental flexibility of periostea to form cartilage when exposed to mechanical forces allows secondary fusion of the lower pharyngeal jaws with the basipharyngeal joint.[25]

The development of pharyngeal jaws and their associated skeletal apparatus in the pharynx and branchial arches of the cichlid *Astatotilapia (Haplochromis) elegans* is known in some detail, as is the microstructure of the bone of the pharyngeal jaws in two *morphotypes* (morphologically distinguishable stable variants) of the cichlid *Astatoreochromis alluaudi.*[26]

Unusual tissues develop in these regions of cichlid fishes.

(i) In a series of studies Anker (1986, 1987, 1989) described in some detail the morphology of the cichlid head skeleton, including joints and presence of cartilage grading into bone in the hyoid, branchial basket jaw and skeletons of *A. elegans.* See Fig. 1.9 for three stages in the development of the mandibular joint in the Japanese medaka, *Oryzias latipes.*

(ii) Chondroid bone develops late in ontogeny at the articulation of the upper pharyngeal jaws with the base of the neurocranium. This chondroid bone resembles secondary cartilage but is formed by osteoblast-like cells rather than chondroblasts, has histological staining properties of bone rather than cartilage, forms on bone and on cartilage, and does not contain type II collagen.

(iii) Huysseune (1985) investigated the cartilage that develops on the opercular bone at its articulation with the hyosymplectic in the cichlid, *Astatotilapia elegans.* Thus unusual cartilage does not hypertrophy, plays a limited role in growth of the opercular bone, and arises from the hyosymplectic, which is an endochondral bone. Huysseune identified the tissue as chondroid bone.

Chondroid bone persists in adults on infrapharyngobranchials III and IV, and on the parasphenoid and basioccipital. In further studies of the contribution of chondroid bone to bone growth incorporation of cells into chondroid bone on the parasphenoid was shown to correlate with local demand, chondroid bone being deposited whenever and wherever there was rapid growth. No cement lines (lines of arrested bone growth; Fig. 5.10) develop between chondroid bone and the underlying acellular bone, a feature also suggestive of a matricial switch.[27]

Chondroid bone in the jewel cichlid, *Hemichromis bimaculatus,* is described as having a mineralized collagenous ECM equivalent to bone, with osteocyte-like cells and matrix vesicles, and as permanent, neither transforming into acellular or into cellular bone (Table 5.2). The carbohydrate histochemistry of this chondroid bone was analyzed and compared with acellular bone and cartilage. The development and resorption of bone and cartilage of the lower jaws – especially as both processes relate to tooth development – were analyzed using transmission electron and light microscopy.[28]

A related issue is whether teleost fish can develop secondary cartilage (see Chondroid above). Nigrelli and Gordon (1946) described spontaneous neoplasms – osteochondromas of the maxilla and opercular bones – in *H. bimaculatus.* These consisted of hyaline cartilage and were evaluated in the context of the chondrogenic potential of the periosteum of these dermal bones, a potential that implies the ability to form secondary cartilage. Norman (1926) includes in the discussion of his paper that the posterior portion of the premaxilla is cartilaginous in *Salmo,* arising late in development. This could reflect secondary chondrogenesis or it could reflect a separate cartilaginous element associated with the premaxilla.

Pharyngeal jaws are toothed. Distinctive larval and adult pharyngeal dentition was described for 13 Japanese cyprinidid fish species with emphasis on correlating tooth growth with rates of growth of the pharyngeal bone. A further study of several species emphasized ankylosis of teeth to bone.[29]

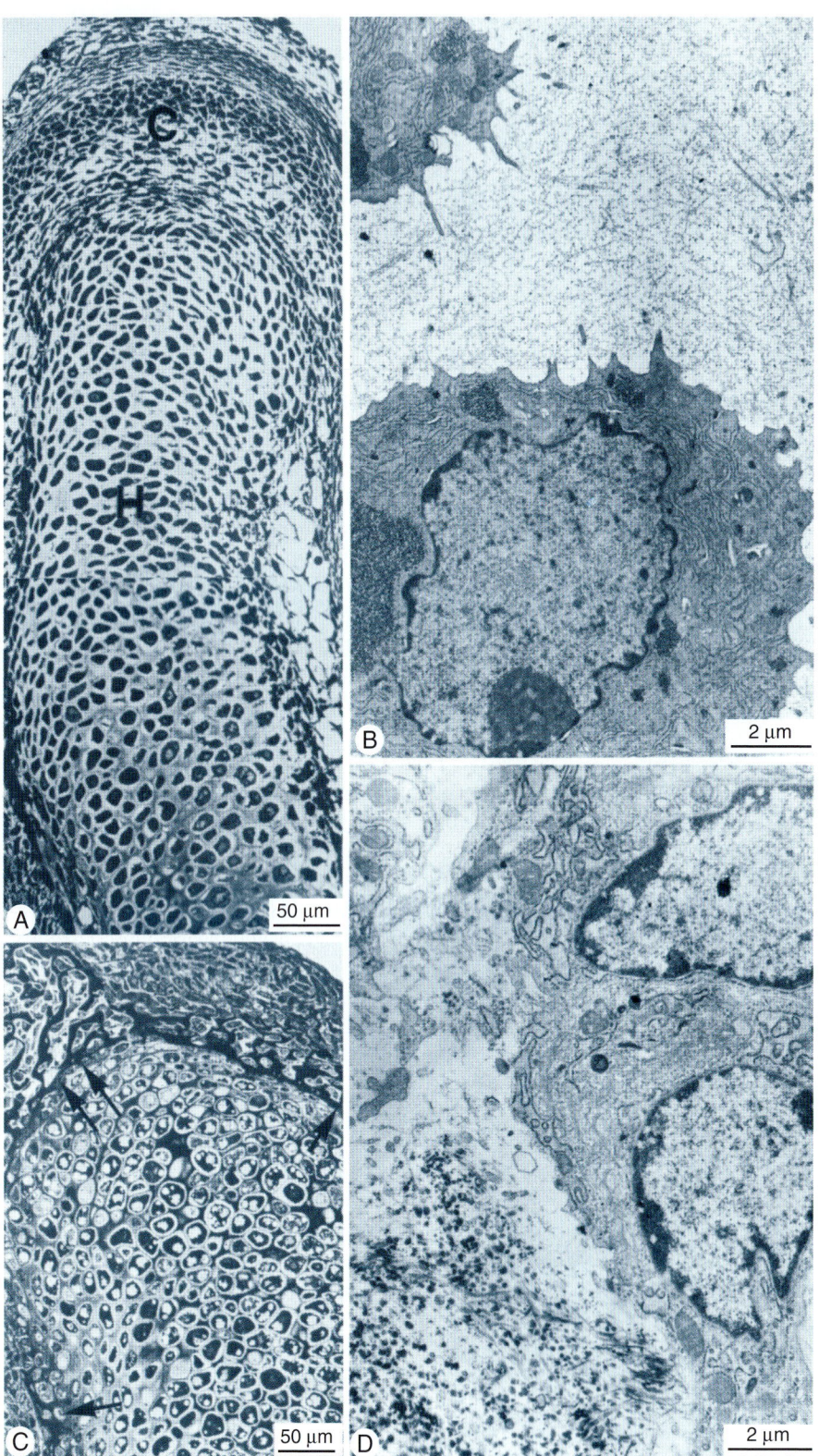

Figure 5.9 Chondroprogenitor cells can become osteoblasts and deposit bone. (A) A longitudinal section of condylar cartilage from the mandible of 20-day-old mouse embryos cultured for two days retains the organization it had *in vivo*, with articular surface and perichondrium (top), chondroprogenitor cell (C), chondroblast, and hypertrophic chondrocyte (H) zones. (B) A transmission electron micrograph of mature chondroblasts. Note the glycogen 'lakes' (granular at left) and the fine collagenous fibres in the ECM. (C) In condylar cartilage cultured for five days, almost all the cells are hypertrophic chondrocytes – compare with zone H in (A) – many in mineralized matrix. New bone has formed along the perichondrium (arrows), especially towards the articular surface (double arrows). (D) A transmission electron micrograph of an osteoblast (right) and collagenous osteoid matrix formed after condylar cartilage had been cultured for five days; cf. (B). Modified from Silbermann *et al.* (1983).

The processes of skeletal formation in the development and regrowth of antlers shed considerable light on how the presence of a tissue such as chondroid bone modifies the process of endochondral ossification. This topic is taken up in Chapter 8.

TISSUES INTERMEDIATE BETWEEN BONE AND DENTINE

The two tissues I introduce briefly in this category are dentine in its various guises, and cementum. Smith and

Table 5.5 Some examples of cartilages that contain collagen types I and II

As part of their normal differentiation, a number of different cartilages synthesize and deposit type I *and* type II collagen into their ECMs. Most of the examples below are discussed more fully in other parts of the text. Absence of type I collagen from many avian cartilages reflects the presence of a cartilage-specific promoter within intron two of the chick Colα2(I) collagen gene. As a consequence, cartilage has type I collagen mRNA but does not produce the protein product (Bennett and Adams, 1990).

Type I and type II collagens are found in the following cartilage types:

Articular, fibro- and *secondary* cartilages of the rat (Fig. 1.8; Sasano et al., 1996).

Cartilage of the *hagfish, Eptatretus burgeri,* has two distinct types of type I collagen, neither of which are equivalent to type I in vertebrate cartilages.[a]

Chondroid bone of the rat glenoid fossa has type I and type II and osteocalcin, i.e. this intermediate tissue has features of hypertrophic chondrocytes and of osteoblasts/cytes.[b]

Human *thyroid cartilages* contain substantial quantities of type I collagen (Claasen Kirsch, 1994).

Mammalian *hypertrophic chondrocytes* contain type I collagen at the stage when endochondral osteoid is being deposited onto them, although some of the type I may not have been synthesized by the chondrocytes.[c]

Mandibular condylar cartilages of rodents, calves and monkeys.[d] Distribution of types I and II within condylar cartilage correlates with mechanical loading; type I is depressed and type II enhanced at sites of compression sites, type I increased and type II depressed at sites of tension in tensile sites (Mizoguchi et al., 1996).

Osteoarthritic human articular cartilage synthesizes and deposits type I collagen contains considerably more type I collagen and so is less able to bind GAGs and therefore more liable to destruction from abrasion.[e]

Tesserae of mineralized cartilage in sharks (Kemp and Westrin (1979).

[a] Kimura and Matsui (1990), Matsui et al. (1990).
[b] Tuominen et al. (1996), Mizoguchi et al. (1997a).
[c] von der Mark and von der Mark (1977a), Yasui et al. (1984).
[d] Hirschmann and Shuttleworth (1976), Mizoguchi et al. (1990, 1992, 1997b), Silbermann and von der Mark (1990), Silbermann et al. (1990), Ishii et al. (1998).
[e] Nimni and Deshmukh (1973), Gay et al. (1976b), Müller and Kühn (1977).

Sansom (2000) review the evolutionary origin of dentine back to Pander's observations in 1857 of the presence of dentine in dermal armour, i.e. dentine as a primary tissue of the exoskeleton (Chapter 1).

Dentine

As also introduced in Chapter 1, various types of dentine are known. *Tubular dentine* is acellular with odontoblast processes persisting as tubules. *Vasodentine* or vascularized dentine is found on the lingual side of the incisors of the little pocket mouse, *Perognathus longimembris*. Cellular dentine forms the basal tissue of the teeth of the bowfin, *Amia*.[30]

When the coelacanth, *Latimeria chalmunae*, was discovered the nature of the scales provided important clues to its likely affinities. The scales are cosmoid, composed of isopedine and collagens organized into tissues that are tubular dentine and enameloid. Using a combination of

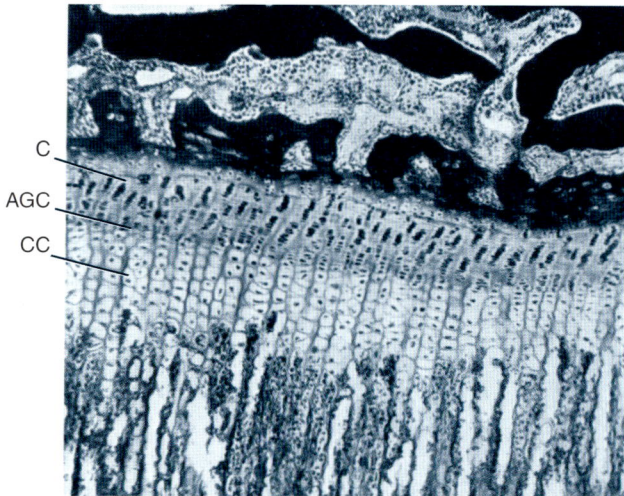

Figure 5.10 Arrested growth of bone. A histological section of the distal end of the radius of a growing dog showing formation of a line of arrested growth (AGC), evident as the layer of shrunken chondrocytes with heavily mineralized ECM between the proliferating chondroblasts (C) and the mineralized (calcified) cartilage zone (cc). Compare with growth arrest following three periods of illness shown in Figure 16.3. From Harris (1933).

radiographic and histological analysis, Miller (1979) demonstrated remodeling in the jaw bones, and identified primary and secondary Haversian systems.[31]

Scanning electron microscopical analysis of *odontodes* in the scales of a coelacanth embryo demonstrated that each *odontode* has a bone of attachment independent of the osseous scale plate. An odontode is the development unit from which a dermal scale develops; essentially equivalent to (homologous with) a tooth bud. Wolfgang Reif, the author of the *odontode regulation theory* of independent units comparable to those that form teeth, reviewed the theory and contrasted it with the *lepidomorial* theory – fusion of denticles to form the dermal skeleton – of development and morphogenesis of the dermal skeleton.

Bhatti (1938) provided one of the most detailed descriptions of the integument, including the dermal skeleton, describing bony scutes, with an outer covering of osteoid and with one or more denticles, but which develop independently of the denticles. The denticles are tooth-like with enamel caps overlying dentine. Bhatti describes the denticle, the pediment on which it rests and the associated connective tissue as arising in a single denticle papilla. Three species of Andean catfishes (astroblepids) form two types of odontodes, one on fin rays associated with dermal bone (the typical odontode situation), the second on the skin of the head, nose and lips, associated, not with dermal bone, but with epithelial structures to form mechanosensory organs. Schaefer and Buitrago-Suârez (2002) interpret these as a specialized adaptation to the fast-flowing mountain streams in which these species live.[32]

Dentine with various degrees of contribution from ameloblasts or osteoblasts has been described in Chapter 1. In the bowfin, *Amia calva*, a holostean fish with *cellular dentine*, developmental processes that result in the persistence of odontoblasts after dentine matrix is deposited are identical to those that allow osteocytes to persist within bone.

In the same study in which he reported acellular bone in teleosts, Kölliker (1859) described four teleost mineralized tissues. The four types recognized 150 years later (of which the first three are mineralized) are:

- cellular bone;
- osteodentine – bone with dentinal tubules but containing osteocytes;
- acellular bone (Kölliker's osteoid), lacking cells or tubules; and
- osteoid with occasional tubules.

The evident gradation between these four types indicates that teleost fishes can modulate cellular activity between odontogenesis and osteogenesis, and/or that these two processes had an early common evolutionary origin and have been conserved with little change.

In his extensive studies on fossil fishes, Jarvik (1959) saw dentine, bone, cartilage and connective tissues, as well as tissues he thought were intermediate between two or more of these tissues, as evidence of transformation between categories of dental, skeletal and connective tissues. As a further example illustrating the difficulty of separating dentine and bone, scanning electron microscopy was used to demonstrate that the 'bone' of attachment of the teeth in several species of sea bream (yellow fin bream, snapper, and the Tarwhine) is dentine, specifically circumpolar tubular dentine. These authors also identified osteocytes in the alleged acellular bone of these species (see Chapter 2).[33]

Cementum

In a recent and thorough review, Diekwisch (2001) defined tooth cementum as "a bone-like mineralized tissue secreted by cementoblasts on the surface of root dentine or, in some animals, crown enamel" (p. 695). Because epithelial cells are separated from the root surface by a basal lamina when initial cementum is deposited, and because cementum proteins did not cross-react with amelogenic (enamel) antibodies on Western blots, Diekwisch concluded that rodent cementum is a derivative of the dental follicle. Given this analysis, in what sense is cementum said to be intermediate between bone and dentine?

Cementum is often ignored by skeletal biologists, probably because it is associated in our minds and in the organism with teeth, rather than the skeleton (see Chapter 1). However, like dentine, cementum exists in a variety of forms, some of which reflect deposition of odontoblast products into the cementum matrix. Consequently, cementum can stand alone as a tissue, or present as an admixture of tissues – an intermediate tissue. Variation exists even within individuals; cementum can be acellular distally but cellular proximally.

Box 5.3 Dentine and bone induction

As discussed in Chapter 5, dentine, bone and cementum form a family of tissues producing inductive agents that promote chondrogenesis, osteogenesis and/or cementogenesis. Another indication of the close relationship are the findings that demineralized dentine can induce *chondrogenesis* in muscle, skin, bone marrow and periodontal ligament *in vivo* and *in vitro*, and that extracts of bovine or human dentine stimulate chondrogenesis from mesenchyme of H.H. 24 chick limb buds.[a]

Within five to 10 days of implanting *demineralized dentine* into the abdominal muscles of rats or guinea pigs, proliferation of quiescent host progenitor cells is enhanced as these cells accumulate within the implanted matrix and differentiate into chondroblasts. By 14 days the cartilage is replaced by bone and marrow, both of which persist for several months but are then resorbed; why such ectopic bone is not self-sustaining still remains to be understood.

The ectopic bone that forms in guinea pig muscle contains 5–7 nm microfilaments and matrix vesicles, although initial mineralization is associated with collagen fibres. With repeated implantation of demineralized dentine an accelerated immune response is provoked and bone inducing capacity reduced. Irradiation, ultrasound, heat treatment and proteolytic enzymes, such as collagenase or trypsin, all reduce the bone-inducing capability of dentine.[b]

Demineralized dentine along with alkaline phosphatase has also been implanted over the skulls of rats. After four weeks there was an 80 per cent increase in Ca^{++} and a 60 per cent increase in phosphorus in the implants, implicating alkaline phosphatase in mineralization of the dentine collagen (Beersten and van den Bos, 1992).

Butler *et al.* (1977) started the search for the active agent in dentine by isolating non-collagenous proteins with bone inductive activity from rat dentine. A 6000–10 000-MW polypeptide isolated from rat incisor dentine is an active chondrogenesis inducer at concentrations between one and 10 ng/ml, but is not mitogenic (Amar *et al.*, 1991). Now we know that Bmp-2 is the active agent (Box 12.2).

[a] See Inoue *et al.* (1986) and Somerman *et al.* (1987) for induction from various tissues, and Rabinowitz *et al.* (1990) from chick limb mesenchyme.
[b] Huggins and Urist (1970) and Bang (1973c) published the initial studies on demineralized dentine. See Bang (1972, 1973a,b) for the immune response, Nilsen (1977, 1980a,b) and Nilsen and Magnusson (1979) for the nature of the induced bone, and Bang and Johannessen (1972a,b) for the various treatments of dentine. Nilsen (1977) and Nilsen and Magnusson (1979) analyzed dentine-induced bone in guinea pigs muscle using electron microscopical and enzyme histochemical approaches, including descriptions of dentinoclasts resorbing the dentine.

Interestingly, cells other than cementoblasts and odontoblasts derived from dental papillae can deposit cementum. Calvarial cells – which, like cementoblasts, are of neural crest origin – can produce a cementum-like tissue when co-cultured with slices of tooth root, presumably as a result of an inductive interaction (Melcher *et al.*, 1987). And cementum is deposited onto existing dentine.

Another indication of the close relationship between cementum and skeletal tissues is a protein isolated from cementum that promotes chondrogenesis and mineralization. Similarly, demineralized dentine and bone matrices *both* induce bone formation when implanted *in vivo* (Box 5.3), while cementogenesis in the Chacma baboon,

Table 5.6 Teleost fishes:an annotated list of skeletal development and skeletal tissues

Development of the entire skeleton

African catfish, *Clarias gariepinus*, correlated with functional demand with opercular, and tooth-bearing bones developing early (Adriaens and Verraes, 1998).

Anemone fish, *Amphiprion melanopus*; early mineralization of the jaws and mineralization of the fins at metamorphosis by Green and McCormick (2001).

Atlantic cod, *Gadus morhua*, histological atlas, including skeleton (Morrison, 1993); ontogeny of skeleton (Hunt von Herbing *et al.*, 1996a–c).

Basic developmental osteology, including brief discussions of the major skeletal regions, sequences of ossification (feeding and respiration first with references), patterns of chondrification and scale development (J. R. Dunn, 1984).

Guppy, *Poecilia reticulata* (Weisel, 1967).

Lantern fishes (Family Myctophidae); ossification sequence of 14 species (Moser and Ahlstrom, 1970). See Table 5.7 for the ossification sequence of bones in the mandibular and hyoid arches, Table 5.8 for the branchial arches, and Table 5.9 for the skull. The sequence in the mandibular arches is highly conserved (Table 5.7). The most obvious variation in the branchial arches is early or late ossification of the ceratobranchial gill rakers, the order in which the ceratobranchials develop, and that hypobranchials develop either early or late in the sequence. The parasphenoid is always the first and the vomer the last of the skull bones to ossify. Ossification of the circumorbital bones is very variable (Table 5.9).

Large-scale sucker, *Catostomus macrocheilus* (Weisel, 1967).

Nile tilapa, *Oreochromis niloticus*, embryonic and early larvae (Morrison *et al.*, 2001).

Paddlefishes (Polyodontidae), ontogeny of skeleton in recent and fossil (Grande and Bemis, 1991).

Red sea bream, *Pagrus major*, skeletal development tables from pre-larva to late juvenile.

Red snapper, *Lutjanus campechanus* (Potthoff *et al.*, 1988).

Head skeleton

Sturgeon (*Acipenser*) including reduction in ossification of the endocranium, the unique jaw suspension associated, enormous variation including anamestic bones cannot homologize skull bones (Jollie, 1980; Hilton and Bemis, 1999).

Clarias gariepinus (African catfish), cranial lateral-line bones (Adriaens *et al.*, 1997).

Bowfin (*Amia*) including the relation of lateral-line canals to bones (Jollie, 1984e).

Eel (*Anguis vulgaris*), chondrocranium and comparison with other bony fishes (Norman (1926).

Bagrid catfish (*Chrysichthys auratus*), chondrocranium from hatching to 18-days post-hatching (Vandewalle *et al.*, 1999).

Barbel (*Barbus barbus*), chondrocranium from hatching to 24 days post-hatching; (Vandewalle *et al.*, 1992).

Siamese fighting fish (*Betta splendens*) including intraspecific variation and interspecific comparisons (Mabee and Trendler, 1996).

Catfish (*Corydoras paleatus*), ontogeny and homology of pterygoid bones (Howes and Teugels, 1989).

Wrestling halfbeak (*Dermogenys pusillus*), tooth-bearing bones in which pharyngeal tooth plates exist, the anterior growth of the jaw coming from cartilages connected to Meckel's cartilage (Clemen *et al.*, 1997).

Pike (*Esox*), lack of invasion of endoskeleton by dermal bones but secondary fusions between the two (Jollie, 1975).

Flatfish (*Solea solea*), chondrocranium from post-embryonic to juvenile (post-metamorphic) (Wagemans and Vandewalle, 1999).

Cichlid (*Astatotilapia [Haplochromis] elegans*), development and function of frontal bones (Verraes and Ismail, 1980).

Longnose gar (*Lepisosteus*), Jollie (1984d).

Bichir (*Polypterus*) especially for common blastema forming cartilage and bone, membranous vs. chondral bones and membranous extensions from chondral bones (Tchernavin, 1937a,b, 1938a–c, 1944; Jollie, 1984a,c).

Salmon, *Oncorhynchus kisutch* (coho), *O. nerka* (sockeye) and *Salmo*, Jollie (1984a).

Sea bream (*Sparus aurata*), viscerocranium from one to 90 days post-hatch; dermal bones ossify before endochondral; different ossification strategies in higher vs. lower teleosts (Faustinoi and Power, 2001).

Papio ursinus, is promoted by recombinant human osteogenic protein-1 (Bmp-7; see Box 28.2) following single applications in bovine collagenous matrix at concentrations between 100 and 500 μg/g matrix.[34]

ENAMELOID: A TISSUE INTERMEDIATE BETWEEN DENTINE AND ENAMEL

As noted above when discussing dentine, coelacanth scales are capped with enameloid, an enamel-like tissue into which odontoblasts deposit matrix products. In another study using an ultrastructural approach, Prostak *et al.* (1990) showed that enameloid in the little skate, *Raja erinacea*, has giant fibres unique to sharks, and that odontoblasts deposit the non-collagenous fibres of the enameloid.[35]

The teeth of adult urodeles are 'true teeth' capped by enamel (see Fig. 1.10). The teeth of larval urodeles con-tain enameloid not enamel. In two urodele amphibians, the Mexican axolotl, *Ambystoma mexicanum*, and the smooth newt, *Triturus vulgaris*, ameloblasts produce and deposit enameloid from the *larval* dental epithelium but produce and deposit enamel from the dental epithelium in *adults*, one of the few examples of a switch in ameloblast activity between two life history stages, reflecting differential activation of regulatory genes pre- and post-metamorphosis. Similar regulatory shifts have been proposed as underlying changes in the production of mineralized tissues that occurred during the evolution of the vertebrates.[36]

In part, such shifts are possible because of the linking and conservation of genes made possible by gene duplication events that occurred early in metazoan and vertebrate evolution. So we find that enamel matrix proteins, salivary gland proteins and milk caseins form a linked family, amelogenin being the only enamel protein not in this family. All these genes produce secreted

Table 5.7 Sequence of ossification of hyoid and mandibular arch skeletons in larvae of 14 species of lanternfishes (Family Myctophidae)[a]

1	2	3	4	5	6	7	8	9	10	11	12	13	14
Hyoid arch skeleton[b]													
opercular	operc	operc	operc	operc	operc	operc	operc	operc	operc	operc	operc	operc	operc
preopercular	urohy	urohy	urohy	urohy	urohy	urohy	urohy	urohy	preop	urohy	urohy	urohy	urohy
urohyal	preop	preop	preop	preop	preop	preop	preop	preop	urohy	preop	preop	preop	preop
ceratohyal	subop	subop	subop	subop	subop	subop	subop	subop	subop	subop	subop	subop	subop
subopercular	interop	interop	interop	interop	interop	interop	interop	interop	interop	interop	interop	interop	interop
interopercular	cerato	hyo	cerato	cerato	cerato	hyo	cerato	cerato	cerato	cerato	cerato	hyo	hyo
hyomandibular	hyo	cerato	hyo	hyo	hyo	cerato	hyo	hyo	hyo	hyo	hyo	cerato	cerato
hypohyal	hypohy	hypohy	hypohy	hypohy	hypohy	hypohy	hypohy	hypohy	hypohy	hypohy	hypohy	interhy	hypohy
interhyal	interhy	interhy	interhy	interhy	interhy	interhy	interhy	interhy	interhy	interhy	Interhy	hypohy	interhy
Mandibular arch skeleton[c]													
maxillary	max	max	max	max	max	max	max	max	max	max	maxi	max	max
dentary	dentary	dentary	dentary	dentary	dentary	dentary	dentary	dentary	premax	dentary	dentary	dentary	dentary
articular	artic	premax	premax	premax	premax	premax	premax	premax	dentary	premax	premax	premax	premax
angular	angular	artic	angular	artic	artic	artic	artic	artic	artic	artic	artic	artic	artic
premaxillary	premax	angular	artic	angular	angular	angular	angular	angular	angular	angular	angular	angular	angular

[a] Based on data in Moser and Ahlstrom (1970). The fourteen species are: 1, the bigeye lanternfish, *Protomyctophum thompsoni*; 2, the California flashlightfish, *P. crockeri*; 3, the chubby flashlightfish, *Electrona rissoi*; 4, the longfin lanternfish, *Diogenichthys atlanticus*; 5, Diogenes lanternfish, *D. lanternatus*; 6, the lamp fish, *Benthosema panamense*; 7, the thickhead lanternfish, *Hygophum atratum*; 8, Reinhardt's lanternfish, *H. reinhardti*; 9, the bigfin lanternfish, *Symbolophorus californiensis*; 10, the pearly lanternfish, *Myctophum nitidulum*; 11, Laura's lanternfish, *Loweina rara*; 12, the blue lanternfish, *Tarletonbeania crenularis*; 13, the slendertail lanternfish, *Gonichthys tenuiculus*; 14, *Centrobranchus choerocephalus*.

[b] Hyoid arch skeletal elements are written in full in the first column showing the sequence for *Protomyctophum thompsoni*, and abbreviated in the other columns. Abbreviations: cerato, ceratohyal; hyo, hyomandibular; hypohy, hypohyal; interhy, interhyal; interop, interopercular; operc, opercular; preop, preopercular; subop, subopercular; urohy, urohyal.

[c] Mandibular arch skeletal elements are written in full in the first column showing the sequence for *Protomyctophum thompsoni*. Abbreviated in other columns are: artic, articular; max, maxillary; premax, premaxillary.

Table 5.8 Sequence of ossification of branchial arch bones in larvae of 14 species of lanternfishes (Family Myctophidae)[a]

Protomyctophu thompsoni	Protomyctophum crocken	Electrona rissoi	Diogenichthys atlanticus	Diogenichthy lanternatus	Bentosema panamense	Hygosema atratum	Myctophum nitidulum	Hygophum reinhardti	Loweina rara	Symbolophorus californiensis	Tarletonbeania crenularis	Gonichthys tenuiculus	Centrobranchus choerocephalus
Pharyngeobranchial teeth (4th arch)	P-4	P-4	P-4	P-4	P-4	P-4	P-4	P-4	P-4	P-4	P-4	P-4	P-4
Ceratobranchial gill rakers (1st arch)	CGR-1	CGR-1	CGR-1	CGR-1	CGR-1	CGR-1	CGR-1	B	E-4	B	CGR-1	C-1	B
Ceratobranchial gill rakers (2nd arch)	EGR-1	EGR-1	EGR-1	EGR-1	B	EGR-1	EGR-1	CGR-1	CGR-1	H-1	EGR-1	C-4	CGR-1
Epibranchial gill rakers (1st arch)	B	B	CGR-2	B	CGR-2	B	B	EGR-1	EGR-1	CGR-1	B	E-4	C-1
Basibranchials	C-1	CGR-2	C-1	CGR-2	C-4	C-4	CGR-2	C-1	B	H-2	C-4	CGR-1	C-2
Ceratobranchial (1st arch)	C-2	C-1	C-4	C-1	E-4	E-4	C-1	C-2	CGR-2	C-5	C-5	EGR-1	C-3
Ceratobranchial (2nd arch)	C-3	C-2	B	C-2	H-1	H-1	C-2	C-3	C-1	C-1	E-4	B	C-4
Ceratobranchial (3rd arch)	C-4	C-3	C-2	C-3	H-2	H-2	C-3	E-3	C-2	C-2	H-1	C-2	C-5
Ceratobranchial (4th arch)	H-1	C-4	C-3	C-4	EGR-1	CGR-2	C-4	E-4	C-3	C-3	H-2	C-3	E-1
Epibranchial (1st arch)	E-1	C-5	C-5	C-5	C-1	C-1	C-5	H-1	C-4	C-4	C-1	C-5	E-2
Epibranchial (4th arch)	E-4	E-1	E-1	E-1	C-2	C-2	E-1	H-2	C-5	E-4	C-2	E-1	E-3
Ceratobranchial (5th arch)	H-2	E-2	E-2	E-2	C-3	C-3	E-2	C-4	E-1	EGR-1	C-3	E-2	E-4
Epibranchial (2nd arch)	CGR-2	E-3	E-3	E-3	C-5	C-5	E-3	CGR-2	E-2	CGR-2	E-1	E-3	H-1
Epibranchial (3rd arch)	C-5	E-4	E-4	E-4	E-1	E-1	E-4	C-5	E-3	E-1	E-2	H-1	H-2
Hypobranchial (1st arch)	E-3	H-1	H-1	H-1	E-2	E-2	H-1	E-1	H-1	E-2	E-3	H-2	EGR-1
Hypobranchial (2nd arch)	E-3	H-2	H-2	H-2	E-3	E-3	H-2	E-2	H-2	E-3	CGR-2	CGR-2	CGR-2

[a] Based on data in Moser and Ahlstrom (1970). Skeletal elements written in full in first column showing sequence for *Protomyctophum thompsoni*, and abbreviated in other columns. Abbreviations: B, basibranchials; C-1 to C-5, ceratobranchials of the first to fifth gill arches; CGR-1 and CGR-2, ceratobranchial gill rakers on the first and second arches; E-1 to E-4, epibranchials of the first to fourth gill arches; EGR-1, epibranchial gill rakers on the first and second arches; H-1 and H-2, hypobranchials of the first and second arches; P-4, pharyngobranchial teeth on the fourth arch. See Table 5.7 for common names for the 14 species.

Table 5.9 Sequence of ossification of 18 skull bones in larvae of 14 species of lanternfishes (Family Myctophidae)[a]

Protomyctophum thompsoni	Protomyctophum crockeri	Electrona rissoi	Diogenichthys atlanticus	Diogenichthys laternatus	Bentosema panamense	Hygophum atratum	Hygophum reinhardti	Symbolophorus californiensis	Myctophum nitidulum	Loweina rara	Tarletonbeania crenularis	Gonichthys tenuiculus	Centrobranchus choerocephalus
1 Parasphenoid	paras	paras	paras	paras	paras	paras	paras	paras	paras	paras	paras	paras	paras
2 Frontal	frontal	frontal	frontal	frontal	frontal	frontal	frontal	frontal	frontal	frontal	frontal	frontal	frontal
3 Nasal	exoc	exoc	pter	nasal	pter	nasal	pter	pter	pter	pter	pter	pter	pter
4 Exoccipital	nasal	nasal	exoc	pter	exoc	pter	exoc	nasal	nasal	nasal	exoc	exoc	exoc
5 Parietal	par	par	nasal	exoc	nasal	exoc	nasal	antorb	exoc	exoc	nasal	nasal	par
6 Pterotic	pter	pter	par	par	par	par	sphen	exoc	antorb	C-5	antorb	C-1	nasal
7 Circumorbital 1	C-3	C-1	C-1	C-5	supra	epit	par	par	par	epit	C-5	epit	C-1
8 Circumorbital 3	basi	C-5	C-5	sphen	antorb	antorb	C-1	C-6	C-1	antorb	epit	antorb	C-5
9 Basioccipital	supra	C-6	epit	antorb	C-4	C-4	C-5	sphen	C-5	C-4	C-4	C-4	epit
10 Circumorbital 2	antob	epit	supra	C-2	C-5	sphen	epit	C-1	epit	par	par	par	supra
11 Circumorbital 6	C-1	sphen	antorb	C-6	epit	C-1	supra	C-2	supra	C-1	supra	C-5	antorb
12 Vomer	C-2	C-3	C-6	supra	C-6	C-5	antorb	C-4	C-6	basio	sphen	supra	C-4
13 Epitotic	C-4	basio	C-2	C-1	C-2	supra	C-6	epit	C-2	supra	C-1	C-6	sphen
14 Sphenotic	C-5	supra	C-4	epit	sphen	C-6	C-2	C-3	C-4	C-6	C-6	C-2	C-3
15 Antorbital	C-6	antorb	sphen	C-3	basio	C-2	C-4	C-5	sphen	C-2	C-2	sphen	basio
16 Circumorbital 4	epit	C-2	basio	basio	C-1	C-3	C-3	supra	C-3	Sphen	C-3	C-3	C-6
17 Circumorbital 5	sphen	C-4	C-3	C-4	C-3	basio	basio	basio	basio	C-3	basio	basio	C-2
18 Supraoccipital	vomer	vomer	vomer	vomer	vomer	vomer	vomer	Vomer	vomer	vomer	vomer	vomer	vomer

[a] Based on data in Moser and Ahlstrom (1970). Skeletal elements written in full in first column showing sequence for *Protomyctophum thompsoni*. Longer names of bones abbreviated in other columns. Abbreviations: antorb, antorbital; basio, basioccipital; C-1 to C-6, circumorbital bones 1–6; epit, epitotic; exoc, exoccipital; paras, parasphenoid; par, parietal; pter, pterygoid; sphen, sphenotic; supra, supraoccipital. See Table 5.7 for common names for the 14 species.

products that are calcium-binding phosphoproteins, many of which share the function of regulating extracellular calcium phosphate concentrations. A set of genes for dentine and bone ECM products forms a second such family.[37]

NOTES

1. See Veis (1987) for an attempt to standardize the nomenclature of bone proteins and for proposals for standards. See Mori et al. (2003) for an elegant comparison of lamellar and Haversian bone.

2. See Bassett (1964) for a discussion of osteoblasts as a separate race of mesenchymal cells, and Hall (1970a) for the now-outdated view that differential synthesis of collagens, GAGs or patterns of enzymes is sufficient to separate chondrogenic from osteogenic cells.

3. See Willis (1962), Jaffe (1968), Ashley (1970) and Wright and Cohen (1983) for intermediate tissues associated with pathological skeletal conditions, and Sanerkin (1980) for definitions and distinctions between osteosarcoma, chondrosarcoma and fibrosarcomas of bone. A rare mesenchymal chondrosarcoma diagnosed in 19 patients by Dabska and Huvos (1983) is of developmental interest because the chondrosarcoma mimics endochondral ossification, even to the production of what appear to be growth plates (Fig. 23.1).

4. See Vaes and Nichols (1962) for sensitivity to oxygen levels, Saint Paul and Bernardino (1988) for the study on the pacu, and Kantomaa (1986a,c) for condylar cartilage.

5. See Shaw and Bassett (1967) and O'Driscoll et al. (1977) for the two studies.

6. See Koskinen et al. (1972), Upton (1972), Simmons et al. (1973), Buring (1974), Kantorova (1976, 1981a,b), Ashton et al. (1980), Bab et al. (1986), Friedenstein et al. (1987), Rooney et al. (1993) and C.-Y. Li et al. (1995) for studies utilizing diffusion chambers in vivo.

7. See Green (1971) and Green and Ferguson (1975) for the studies with rabbit chondrocytes, and Fell et al. (1968) for the chick studies.

8. See Badi (1972b) for the rat and Haines (1969) for the reptile study, and Feik and Storey (1982) for the transplanted vertebrae. Ankylosing spondylitis and polyarthritis can be induced in mice immunized with chondroitinase ABC, proteoglycan and adjuvant (Glant et al., 1987).

9. See Felisbino and Carvalho (2000, 2001) for these studies.

10. See Kay and Harrison (1969), Finkle and Goldman (1973) and Ohtsuki et al. (1987) for the three clinical examples, and see Person and Philpott (1969a), Moss (1972a), Hall (1975a) and Cole and Hall (2004a,b) for epithelial origin of some invertebrate cartilages.

11. See Willis (1962), Jaffe (1968) and W. A. Beresford (1981) for the nature of the evidence, Bischoff (1971) for amniotic cells, Hay and Dodson (1973) and Meier and Hay (1973, 1974a) for corneal epithelium, and Thonard and Wiebkin (1973) for the human amniotic cells.

12. See Beresford (1981) and Dhem et al. (1989) for analyses of chondroid, its contribution to skeletal growth, and use of the term. I use ectothermic rather than cold-blooded or poikilothermic as the latter terms imply mammalian body temperatures as a universal norm against which other animals should be compared. Ectotherms regulate their body temperatures behaviourally in response to external conditions. In so doing many ectotherms can maintain quite a 'high' body temperature (again, using mammals as the standard for what is high). Two examples are the side-blotched lizard, Utastansburiana hesperis, from California with a cloacal temperature of 38°C when the air temperature was 13°C (Cowles, 1947), and a high-altitude lizard that emerged from its burrow into air temperature of −5°C and within a short time had a body temperature of 26°C (Pearson, 1954).

13. See M. H. Wake (1985, 1998) for caecilian tentacles, and see Billo and Wake (1987) for the developmental study.

14. See Benjamin (1988, 1989a,b) for the detailed analyses, Benjamin (1990) and Benjamin and Ralphs (1991) for the five types of cranial cartilages and the connective tissue analysis, and Benjamin et al. (1992) for trunk cartilages. Table 5.6 provides an entrée into studies on the cranial skeleton and skeletal tissues of teleost fishes, Table 5.7 evaluates some of the variation in sequence of ossification, and Hall (2000a) evaluates the data on secondary cartilage in teleosts and the homology of secondary cartilages across the vertebrates.

15. See Rafferty and Herring (1999) for the nasofrontal suture and for a general review of craniofacial sutures, masticatory strains and a discussion of chondroid, Goret-Nicaise (1981, 1982, 1984a, 1986), Goret-Nicaise and Dhem (1982), Goret-Nicaise et al. (1988) and Manzanares et al. (1988) for chondroid in the human skeleton, and de Beer (1937) and Hall and Hanken (1993) for bibliographies on skull development.

16. See Goret-Nicaise (1984b) and Goret-Nicaise and Dhem (1985) for studies on mandibular chondroid, including calcium levels, and Haas and Holick (1996) for changes in chondroid after loading.

17. Members of the Salmonidae (salmon, trout), Osmeridae (smelts), Myctophidae (lanternfishes), and various catfishes and characins (cichlids, cyprinids, killifish, piranhas) possess adipose fins which, unlike median fins (Box 40.1), lack skeletal elements and are covered by dermis and epidermis, not by scales.

18. Apophallus brevis produces the condition known as 'black spot' in brook trout (Salvelinus fontinalis) and brown trout (Salmo trutta), with sand-grain-sized black spots on the skin. Only yellow perch produce ossicles. Frogs infected with cysts of the trematode Ribeiroia ondatrae develop limb deformities, especially limb duplications. Whether Apophallus and Ribeiroia utilize similar mechanisms to evoke skeletogenesis is worth exploring.

19. See Pike and Burt (1983) for the original descriptions, and Taylor et al. (1993, 1994) for analyses of the nature of the tissues.

20. See Wright and Moffett (1974) for the human temporomandibular joint (TMJ) study, Beresford (1981) for chondroid bone, du Brul (1964) for the evolution of the TMJ, and Enlow (1962a) for remodeling of chondroid as it transforms to bone (remodeling that differs from cartilage or bone remodeling), including a good discussion of the early literature.

21. For metaplasia in fractures and penile bones, see Ruth (1934), Pritchard and Ruzicka (1950), Beresford (1970, 1975a) and Beresford and Clayton (1977). For metaplasia in response to dentures, see Cutright (1972), and in tumours see B. H. Smith and Taylor (1969).

22. For the contribution of periosteal cells to marrow formation, as demonstrated using mouse limb buds in vitro, see D. R. Johnson, 1980. Neither marrow cavity nor endochondral

bone forms unless the rudiments contain periostea perforated by blood vessels. See Menton *et al.* (1982) for the presence of a 'marrow sac' – an epithelium-like covering over the myeloid tissue with osteoprogenitor cells exiting through holes in the sac.

23. See Hall (1983a), Ornoy and Zusman (1983), Narbaitz (1992) and Bilezikian *et al.* (1996) for overviews of the role of vitamins in cartilage. During mouse embryogenesis retinoic acid receptor-gamma (RAR-γ) is expressed at eight days in presomitic mesoderm, from 9.5 to 11.5 days in frontonasal processes, pharyngeal arches and limb mesoderm sclerotome, at 12.5 days in all precartilage tissue, and at 13.5 days in *all cartilages*, keratinized epithelium, teeth and whiskers (Ruberte *et al.*, 1990, 1992; and see Box 20.2).

24. See Silbermann *et al.* (1978a,b, 1983) for chondroid bone from the condylar cartilage, and Tuominen *et al.* (1996) and Mizoguchi *et al.* (1992, 1993, 1997a) for chondroid bone on the glenoid fossa.

25. Lauder (1983), a student of Liem's, reviewed the functional design and evolution of pharyngeal jaws and associated muscles and ligaments in teleosts fishes, while Lauder and Liem (1989) provided a six-point test to use when searching for key innovations or developmental constraints. For innovations (novelties) involving skeletal tissues, see Hanken (1985), Müller and Streicher (1989), Müller and Wagner (1991), Stiassny (1992), Smirnov (1997), Hall (1998a, 2005a), Jernvall (2000) and Hall and Stone (2004).

26. See Verraes *et al.* (1979) and Huysseune *et al.* (1994) for these cichlid studies.

27. See Ismail *et al.* (1982) and Huysseune (1986, 1989) for chondroid bone on pharyngeal jaws, Huysseune *et al.* (1986, 1994) for growth, Huysseune and Verraes (1986) for persistence of chondroid bone into adult life, and Huysseune (2000) for an excellent overview of fish skeletal systems including a discussion of the endo-exoskeleton (or endoskeleton–dermal skeleton as she prefers; p. 307), dermal membrane bone and intermediate tissues.

28. See Huysseune and Sire (1990, 1992a,b) and Huysseune and Verraes (1990) for chondroid bone

29. See Nakajima (1984, 1987) and Nakajima and Yue (1989) for pharyngeal dentition.

30. See Moss (1964b) for cellular dentine in *Amia*, Meredith-Smith *et al.* (1972) and W. A. Miller (1979) for tubular dentine in *Latimeria* scales, Moss-Salentijn and Moss (1975) for vasodentine, and Meinke (1986) for the dermal skeleton in lungfish, including cosmine and the timing of tissue interactions.

For syntheses of various approaches and for a bibliography of papers on lungfishes, see Bemis *et al.* (1987). In a surprising discovery, hatchling lungfish, whether from the three living genera or 360-mya extinct species have identical patterns of tooth development, although adults patterns differ. This finding points not only to the long-term conservation of the basic developmental pattern, but also to the adaptability of the adult pattern (Reisz and Smith, 2001; Smith and Krupina, 2001; Smith *et al.*, 2002).

31. An entrée into the coelacanth skeletal literature includes: for the fin skeleton (Géraudie and Meunier, 1980, 1982; Géraudie and Landis, 1982), the medial fins (Ahlberg, 1992), odontodes and scales (Meredith-Smith *et al.*, 1972; W. A. Miller, 1979), patterns of mineralization (Francillon *et al.*, 1975), remodeling (W. A. Miller, 1979) and for scales that are cosmoid (Meredith-Smith *et al.*, 1972). Danke *et al.* (2004) generated a BAC library from genomic DNA isolated from the heart of the Indonesian coelacanth, *Latimeria menadoensis*, providing a resource that will facilitate studies at the genomic level.

32. See Meredith-Smith *et al.* (1972) and W. A. Miller (1979) for a basic description of scales and remodeling, M. M. Smith (1979) for the odontode study, Reif (1978, 1980a,b, 1982) and Reif and Richter (2001) for morphogenesis of the dermal skeleton as based on odontodes in recent and fossil sharks, and Smith and Hall (1990, 1993) and Donoghue (2002) for evaluation of Reif's odontode theory and syntheses of the tissues in the dermal skeleton. For an overview on evolutionary changes in tooth attachment, especially the histological changes, between reptiles and mammals, see Osborn (1984).

33. See Hughes *et al.* (1994a,b) for sea bream bone of attachment.

34. See Amar *et al.* (1991) and Arzate *et al.* (1996) for the chondrogenesis-promoting ability of cementum, and for a 6000–10 000 MW polypeptide isolated from rat incisor dentine that promotes chondrogenesis. See Ripamonti *et al.* (1996) for the action of Bmp-7.

35. For ultrastructure of shark tooth enamel, and a discussion of enamel versus enameloid in shark teeth, see Kemp (1985).

36. See Meredith Smith and Miles (1977) for enameloid in the teeth of adult axolotls, and Hall (1975a, 1978a) and Maderson (1975) for the developmental processes underlying the switch from larval to adult tissues.

37. See Kawasaki and Weiss (2003) for the study of calcium-binding phosphoproteins. The enamel proteins amelogenin, enamelin, tuftelin, MMP-20 and EMSP-1 form another highly conserved family of matrix proteins (Satchell *et al.*, 2002).

An Evolutionary Perspective

Although conodonts are not 'little whatzits' all is not complete tranquility.

Much information is now available on the structure of skeletal tissues from extinct vertebrates. From histological studies of fossilized skeletal tissues we can conclude that these tissues in the earliest vertebrates were no less specialized than those of today's vertebrates, such as ourselves. Rather than involving major changes in cell or tissue type, evolution worked with adaptive modulation of a highly plastic series of scleroblasts as dictated by local conditions. Further, it is now possible to interpret developmental processes and the functional significance of skeletal tissues during vertebrate evolution. An examination of the skeletal tissues found in ancestral vertebrates, even in the earliest chordates, indicates a surprisingly high degree of homology with skeletal tissues in present-day vertebrates.

FOSSILIZED SKELETAL TISSUES

It continues to amaze me that fossilized skeletal tissues can be processed for light and electron microscopy as well as for chemical analyses (Table 6.1). On the latter point, osteocalcin was extracted from fossil bovid bones and teeth that are between 12 000 and 13 million years old, and from rodents that are 30 million years old. Hydroxyproline and hydroxylysine, identified in fossil crocodiles and rhinos, are taken as evidence of the presence of collagen (Table 6.1). Basic structural elements are conserved with surprising fidelity. Mathews (1975) exhaustively summarized the then-available data on the macromolecular structure of connective and skeletal tissues during evolution, while Fietzek et al. (1977) provided data

on conservation of the cross-linking region of the collagen molecule during evolution.[1] Pawlicki's extensive studies of dinosaur bone serve as an example of the methods used and cellular detail that can be obtained from 'fossil histology' (palaeohistology):

- scanning electron microscopy to visualize vascular canals with collagenous walls;
- histochemistry to demonstrate lipids and acid mucopolysaccharides in perivascular spaces;
- light, transmission and scanning electron microscopy to identify two cell types, as a result Pawlicki identified;
  - a vascular communication system similar to that found in contemporary animals;
  - osteocytic processes linking osteocytes to one another; and
  - two types of lacunae and associated collagen.

Pawlicki showed that less mature osteocytes are distributed near blood vessels, an observation with implications for the approaches we should take to understanding metabolic pathways. These studies do not 'stand alone' (although some difficulties with how such histological data are used are discussed under the heading Problematica below).

Collagen in the bone and dentine of an 11 200-year-old mammoth from the late Pleistocene of central Utah examined by light and transmission electron microscopy, and by cyanogen bromide digestion after controlled desiccation for nine months, was shown to be type I. Cell processes, presumed to be odontoblast, were identified in the dentine. An ultrastructural examination of a variety of fossil taxa, spanning the wide time frame 1300 BC to 400 mya, revealed little collagen in Tertiary, Mesozoic or Palaeozoic fossils, 50-nm banded collagen in Pleistocene

Table 6.1 Summary of techniques applied to and findings obtained from fossilized skeletal tissues

Subject	Findings	References
Teleost fish	Described and compared acellular with cellular bone	Moss (1961b)
Ordovician eriptychiids and astraspids	SEM[a] and light microscopy to determine hard tissue histology; determined aspidine may not be present in heterostracans like eriptychiids	Ørvig (1989)
Ordovician heterostrachi	Reviewed works on chemical analysis of fossilized skeleton	Halstead (1969a)
Jurassic to Devonian fossils	Automatic amino acid analyzer to assay amino acid contents; micro quantities were found in higher concentrations than in rock	Armstrong and Halstead Tarlo (1966)
Heterostracan ostracoderms	Decalcified dentine to analyze organic content	Halstead Tarlo and Mercer (1966)
Pleistocene mammals	Used a linear relationship between body temperature and imino acid content in the collagen of extant mammals to estimate the body temperatures	Ho (1967a,b)
Invertebrates and vertebrates	Macromolecular evidence and phylogeny to analyze evolutionary relationships between species	Mathews (1967a)
Pliocene *Gampthotherium*, Miocene *Brontosaurus* and *Uintatherium*	Fossilized bone organic material is partially depleted of organic material and replaced by calcium carbonate	Biltz and Pellegrino (1969)
Dinosaur *Tarbozauru bataar*	SEM to visualize vascular canals with collagenous walls (1975a)	Pawlicki (1975a, 1976, 1977a,b, 1978, 1983, 1984a,b, 1985)
	Staining procedures on ground unfixed sections to find lipids in the perivascular space of vascular canals (1976)	
	Histochemistry to demonstrate lipids and acid mucopolysaccharides in perivascular spaces (1977a,b)	
	Light microscopy, SEM, TEM[b] to identify two cell types (1978)	
	Identified a vascular communication system similar to that found in contemporary animals (1983)	
	Determined that intermediary osteocytes were distributed near blood vessels, with implications for metabolic pathways (1984a)	
	Identified osteocytic processes linking osteocytes to one another (1984b)	
	Identified two lacunar types and associated collagen (1985)	
Jurassic sauropod	Light microscopy discovered lamellar-zonal compact bone characteristic of ectotherms	Reid (1981, 1984)
Fossilized antler	Electrophoresis and chondroitinase digestion to isolate chondroitin sulphate	Scott and Hughes (1981)
Pleistocene fossils	TEM to determine that fossils with collagen profiles typical of modern collagen showed significant peptide fingerprints differences when exposed to proteolytic enzymes	Armstrong *et al.* (1983)
	Changes in collagen may have occurred due to fossilization	
	Collagen does not stay unchanged upon fossilization	
Early hominids	SEM/replica technique for microscopical taphonomy, bone damage, bone growth and remodeling analysis	Bromage (1987)
Bovids between 12 000 and 13 million years old; rodents 30 million years old	Detected osteocalcin using HPLC[c], ion exchange and size exclusion columns	Ulrich *et al.* (1987)
Fossilized crocodiles and rhinos	Amino acid analysis to identify hydroxyproline and hydroxylysine and took the presence of these amino acids as evidence of collagen	Glimcher *et al.* (1990)

SEM, scanning electron microscopy; TEM, transmission electron microscopy; HPLC, high performance liquid chromatography.

taxa (perhaps showing evolution of collagen fibrillogenesis; Fig. 2.4), and evaluated the literature on the electron microscopical analysis of fossil tissues.[2]

Insights into life style also are recorded in features of skeletons such as fractures; some seven per cent of Lower Jurassic 'early crocodilian' (eosuchian) reptiles have fractures of their jaws with deposition of cancellous bone in the healing site. Evans (1983) attributed the high frequency and localization to the jaws to intraspecific territorial conflicts.

ALL FOUR SKELETAL TISSUES ARE ANCIENT

All four mineralized tissues are present in representatives of the Ordovician agnatha (which, despite its name, is not a monophyletic group) as fossilized head shields (Fig. 6.1) or elements of the endoskeleton (Table 6.2).

Of the best samples are early late Ordovician vertebrates from Colorado with cellular bone and dentine. Indeed, these specimens form important evidence for the early association of dentine and bone (M. M. Smith, 1991). Reanalysis of such specimens is furthering our understanding. A specimen, previously identified as *Vert. Indet. A* (a vertebrate of indeterminate affinity) has been restudied,

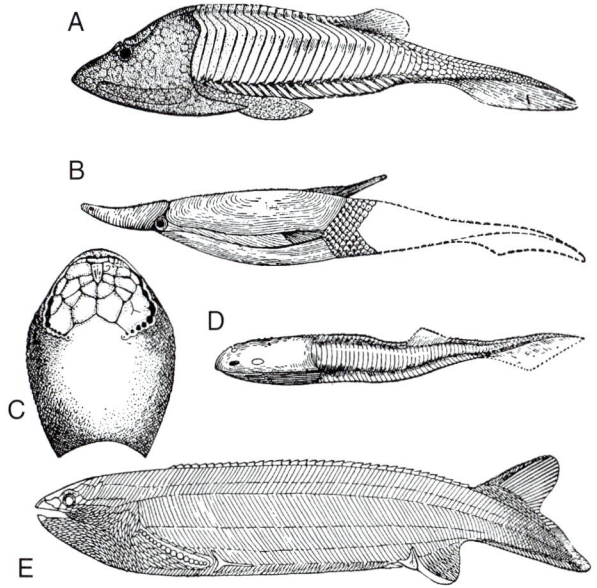

Figure 6.1 Ostracoderms from the Upper Silurian and Devonian. All are shown in lateral view, except C, which is a dorsal view of the head shield. (A) *Pterolepis*. (B,C) *Tremataspis*. (D) *Pteraspis*. (E) *Cephalaspis*. Note the extent of development of the dermal skeleton in the different species. From Gregory (1929).

Table 6.2 A summary of mineralized tissues and the classes of vertebrates within the fossil record

Geological period	Class	Mineralized tissue
Quaternary		
Cretaceous		
Jurassic	Mammals	
Triassic	Birds[a]	Enamel, dentine,
Permian		cartilage, membrane
Pennsylvanian	Reptiles	bone, aspidine
Mississippian		
	Amphibians	Endochondral bone
Devonian	Chondrichthyes[d]	
Silurian	Placoderms	
	Osteichthyes[b]	
Ordovician	Agnatha[c]	
Cambrian		
Pre-Cambrian		

[a] Enamel and dentine were secondarily lost in the birds.
[b] Cartilage was secondarily lost in Osteichthyes.
[c] Aspidine is only found in the heterostracan agnathan, acellular bone arose again in teleost fish.
[d] Endochondral bone was secondarily lost in the Chondrichthyes.

named *Skiichthys halsteadia*, and identified as an acanthodian or placoderm with scales of enameloid, mesodentine, odontocytes and basal cellular bone.[3]

Whether these jawless vertebrates also possessed unmineralized tissues such as cartilage is a matter of inference. Given what we know of modern-day marine fishes, the Ordovician skeletal tissues may not have served as a major reservoir for calcium or phosphorus; other organs contain much higher levels of both elements. Given the enormous surface area of gills, and loss of calcium and phosphorus to the environment from the gills, it is unlikely that exoskeletal denticles or plates acted as a barrier against loss of calcium and phosphorus to the aquatic environment. They may have functioned as protective armour against large and predatory eurypterids (sea scorpions), marine arthropods that ranged in size from 10 cm to over two metres, and are most closely related to modern-day scorpions and spiders. *Eurypterus remipes*, a scavenging predatory eurypterid from New York State, was adopted as the state fossil in 1984. Mechanical support was probably of lesser importance for these early marine vertebrates than for later terrestrial taxa.[4]

With further vertebrate evolution, especially with the evolution of jaws and the origin of the jawed vertebrates (gnathostomes), adaptations of skeletal tissues arose in response to new and local environmental demands, utilizing the ability of skeletal cells (scleroblasts) to modulate their synthetic activities and differentiative pathways. There is, however, no evidence of a generalized trend toward progressive specialization during evolution. As examples, the degree of skeletal mineralization and bone density in jawless fish that existed over 500 mya (ostracoderms; Fig. 6.1), in amphibians of the Carboniferous and Permian (stegocephalians), and in extinct and extant whales, dolphins and porpoises (cetaceans), reflects adaptations for such features as equilibrium, buoyancy and conservation of temperature and energy, rather than progressive evolution within the groups.

The skeletons of aquatic vertebrates do not have to support the weight of the body. Consequently, de Ricqlès (1989) argued that there is no selective advantage to one bony structure over another in the non-weight-bearing skeletons of aquatic tetrapods; tetrapods that have reverted to an aquatic mode of life show convergence of bone histology, which de Ricqlès attributes to heterochrony (see Box 6.1). While this is an accurate generalization when comparing skeletal tissues, variation can and does exist at the organ level (from skeletal element to skeletal element) and from taxon to taxon. Such variation includes bone that is very dense, elements that lack cortical bone, various degrees of loss of bone marrow, and so forth. The spongy bone of the vertebrae of bony fishes is deposited in response to mechanical stresses, although the biconid shape of each vertebra results from alternating constrictions and dilations of the notochord and sclerotome rather than response to mechanical stresses.[5]

Box 6.1 Heterochrony

Heterochrony has a long history. The term was coined by Ernst Haeckel in the 19th century for *similar features that arise at different time* in different organisms. A more modern view of heterochrony goes back to Gavin de Beer in the 1920s. de Beer took the knowledge that genes affect the timing of physiological or metabolic processes (rate genes, discovered by Richard Goldschmid) and reasoned that changes in timing must be an important mechanism of evolutionary change, explicitly *linking heterochrony to timing differences in a phylogenetic context*. Gould (1977) resurrected heterochrony as a mechanism to compare change in the timing of a feature – usually some aspect of size or shape – in a descendant with respect to an ancestor.

An enormous literature has grown up since 1977.[a] Nowadays heterochrony is used in two ways:

- de Beer and Gould's way of comparison with an ancestral condition; or
- to make comparisons within an individual, usually between two different systems, more or less Haeckel's original view of heterochrony.

Alteration in the timing of developmental, cellular or genetic pathways – one aspect of *heterochrony* – influences various aspects of skeletogenesis over both developmental and evolutionary time scales. Examples of both usages are discussed in the final chapter.

[a] A sample of the heterochrony literature, especially as it relates to skeletal systems, includes Shea (1983b), Hall (1984c, 1990a, 1998a, 2001c, 2002b), Zelditch and Fink (1996), Hall and Miyake (1997), Cubo (2000), K. K. Smith (2003), and the chapters in Zelditch (2001) and Hall and Olson (2003).

EVOLUTIONARY EXPERIMENTATION

Given such lability of skeletal tissues, and given the existence of intermediate tissues in recent vertebrates, one might expect to find many intermediate tissues in fossil vertebrates as different tissue organizations were tried out in response to changing environmental conditions.

Intermediate tissues in fossil agnatha

Tor Ørvig was one of the first to undertake detailed studies of the three types of dentine, and of those tissues intermediate between dentine and bone and between bone and mineralized cartilage in extinct agnathans. The presence of intermediate tissues in early vertebrates suggests that osteoblasts, odontoblasts and chondroblasts could modulate their synthetic and differentiative activity much as they do today. Acquisition of such adaptable cells so early in their evolution was a major reason for the later evolutionary success of the vertebrates.[6]

The stance I took in 1978 on modulation of scleroblast differentiative activity during evolution in response to such epigenetic factors as growth rates, muscle attachments, vascularity, habitat, feeding cycles, and so forth (Hall, 1978a) was in accord with then-current notions on the genetic changes necessary for directing evolution, which, following King and Wilson (1975), I saw to *be regulatory rather than changes in structural genes*.

Dermal skeletal tissues of extant agnathans fall into two classes found, respectively, in two major groups, the *osteostracans* (cephalaspids; Fig. 6.1) and the *heterostracans* (Box 6.2). Both share dentine, enamel and mineralized cartilage as skeletal tissues but the skeletal similarities stop there. Ostracoderms have *cellular bone*, heterostracans *acellular bone* or *aspidine*, both aspidine and cellular bone having arisen contemporaneously during the Ordovician (Table 6.2). Once this was known, much debate ensued over whether aspidine was bone or dentine, individuals holding different views at different times. For example:

- aspidine and cellular bone evolved independently (Ørvig, 1965; Moss, 1968a);
- aspidine evolved from cellular bone (Ørvig, 1957, 1968, 1989); or
- cellular bone arose from aspidine (Denison, 1963).

Of course, there is no shame in changing your view as new data/theories emerge, and, in fairness, Ørvig, Moss, and Denison as often as not posed their theories as questions such as: Did aspidine evolve from cellular bone?[7]

In 1978 I argued that development of acellular bone from cellular tissues during the ontogeny of modern teleosts provides a necessary clue to understanding the evolutionary problem. A subsequent study of the development of acellular bone revealed the inadequacy of this view. Acellular bone is not acellular at the outset. It becomes acellular because of the polarized secretion of extracellular matrix by cells that are not incorporated into the bone matrix (Chapter 2). Similarly, I argued that as bone and mineralized cartilage arose contemporaneously during the Ordovician (Table 6.2), the conundrum of

Box 6.2 Silurian and Devonian jawless vertebrates

Osteostracans (Osteostraci) were a group of some 200 extinct species of near shore marine or freshwater armoured agnathans, 20–40 cm in length (range 4 cm–1 metre) that existed for some 60 million years between the Early Silurian and the Late Devonian (about 370 million years). Their large blunt heads and skulls were covered with a shield (head shield, cephalic shield) of dermal bone (Fig. 6.1); hence the other common name, *cephalaspids*. The remainder of the body was encased in interlocking scales that gave rise to the analogy to a mediaeval suit of armour. Paired anterior fins connected to the head.

Heterostracans (Heterostraci) were a group of some 300 extinct species of armoured bottom-dwelling agnathan inhabitants of lagoons, deltas and freshwater. Heterostracans were contemporaneous with osteostracans. Their funnel-shaped heads were armoured. Most were 5–30 cm long, the longest perhaps 1.5 metres. Like osteostracans they had armoured heads, in this case consisting of separate ventral and dorsal shields with small elements laterally and around the mouth. A central disc in the dorsal shield is characteristic. Unlike osteostracans, heterostracans lacked a mineralized endoskeleton. Their internal anatomy is now well preserved or understood.

whether cartilage arose first during evolution, as it does during ontogeny, had finally been put to rest (Hall, 1978a, 1988).[8]

Given our present state of knowledge, both views may now be regarded as entirely specious, inadequate or inappropriate. Twenty-seven years ago, my concern was that 'our inability to identify *unmineralized* cartilage in the fossil record is an obstacle to settling the time of origin of endochondral ossification.' I had discussed this question earlier (Hall, 1975a), relating it to phylogenies of the vertebrates, and modulation of scleroblast activity. Part of my argument went as follows: 'Because the embryonic skeleton is initially cartilaginous in many areas, because cartilage is the forerunner of bone in endochondral ossification, and because the Chondrichthyes (cartilaginous fishes) were assumed to be primitive, the logical extension to evolution was that cartilage arose before bone.'

As a result of a highly productive collaboration in the late 1980s and early 1990s with Moya Smith from what was then Guy's and St Thomas's Medical and Dental Schools, it is now possible to say that cartilage was the primary tissue of the endoskeleton and that bone (and dentine) were the primary skeletal tissues of the exoskeleton. It may still be useful to know which came first, cartilage or bone. The question, however, no longer relates to which came first within *the* skeleton, for vertebrates have *two* skeletal systems. Cartilage came first in the endoskeleton, bone in the exoskeleton. Any cartilage in the exoskeleton is secondary evolutionarily and secondary developmentally.

DINOSAUR BONE

We have gleaned a great deal of information about comparative skeletal biology from analyses of the histology of dinosaur bone. Many of these studies came from John Horner's laboratory at Montana State University. In one study, Horner compared different-aged embryos of two ornithischian dinosaurs and argued that bony processes and protuberances, equivalent to those onto which muscles attach in extant taxa, are not present until after the muscles develop. Function drove skeletal morphology then as it does now.

The histology of long bones of embryonic and perinatal archosaurs was compared with that of birds, dinosaurs and other long bones. As the avian and dinosaurian histology overlapped, Horner and colleagues concluded that particular types of bone histology can be used as a unique shared character (a synapomorphy) of birds and dinosaurs. Furthermore, the arrangement of the embryos indicated that this was a nest site, which, if nesting meant for dinosaurs what it does for extant birds, implies parental behaviour. Bone histology in context reveals life history.

Histological features of long bones from six growth stages of the hadrosaurian dinosaur *Maiasaura peeblesorum*

indicate fast growth, a growth pattern that differs greatly from the slow growth of reptilian bone. Earlier studies of variation in 12 hadrosaur bones demonstrated lines of arrested growth, which, although they provide some indication of age, have no special physiological meaning, at least according to Horner and colleagues.[9] Histology reveals patterns of growth.

Palaeohistology has been used to assess patterns of long bone growth in juvenile and adult dinosaurs from the Upper Cretaceous and mechanisms of growth thought to be similar to (homologous with) patterns seen in the growth plates of avian long bones have been identified and described. However, this study by Barretto and colleagues (1993) was criticized on two grounds: the confounding effect of the irregular nature of the growth plate–bone boundary in dinosaurs, and the argument that crocodilians are a more appropriate taxon (a better outgroup) than birds to use for such comparisons.

A third criticism would be that bird bones are not bird bone are not bird bones, and that growth is not growth is not growth. Birds that can feed as soon as they hatch and those that cannot (praecocial and altricial birds) have very different amounts of cartilage in their long bones at hatching, praecocial birds having ossified their long bones more fully, a feature that reflects differential growth rates. Starck (1993) was able to construct a classification of ontogenies and timing of skeletal development – the development and evolution of avian life histories – for many avian species, from among which appropriate taxa could be chosen to compare with dinosaurs. Uniformity is not the watchword.[10]

When de Ricqlès *et al.* (2001) used palaeohistology to examine embryos of theropod dinosaurs from the Upper Jurassic of Portugal they found extensive mineralized cartilage, with canals and marrow buds – indicative of endochondral ossification in the long bones – and an avian-like medullary core (Figs 6.2 and 25.4, and see Chapter 25). They estimated the rate of periosteal ossification to have been at least 20 µm/day, a rate that in extant taxa requires a high bone growth rate (and therefore a high body temperature?).

Armand de Ricqlès and his colleagues used a palaeohistological analysis of *pterosaur* bone to inform anatomy, development and biomechanics. The histology of pterosaur bones is closer to what is found in birds than in dinosaurs or crocodilians. A high growth rate early in ontogeny allows rapid deposition of long bone growth through subperiosteal ossification. Subsequent remodeling, deposition of secondary endosteal parallel-fibre bone, and the development of internal trabecular struts provided the strength such flying creatures required, yet the question of whether pterosaurs flew or glided remains controversial. Kevin Padian, a leading advocate and student of the flying school, demonstrated that their inferred locomotory abilities are consistent with flying not gliding and resemble the features seen in birds.[11]

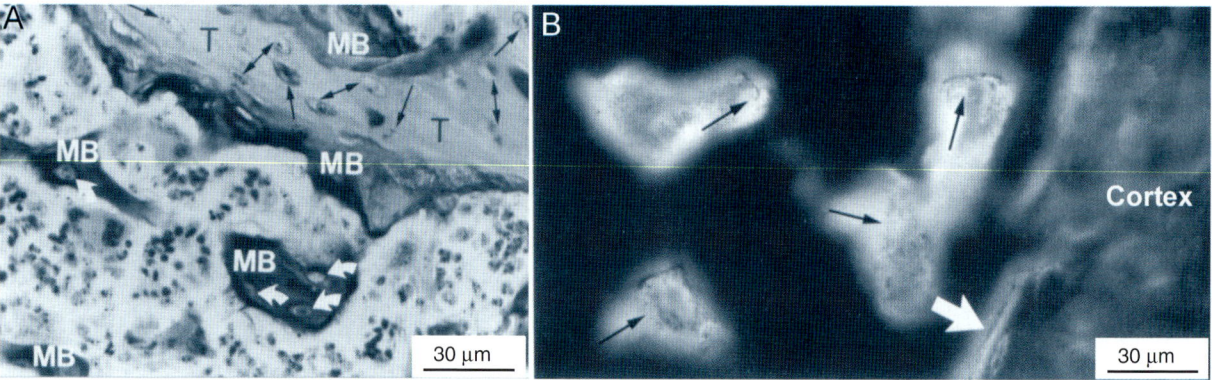

Figure 6.2 Medullary bone formation in a female Japanese quail, *Coturnix coturnix japonica*. (A) Considerable medullary bone (MB, black arrows) has formed, both on existing trabecular bone (T) and in the marrow cavity (MB, white arrows). (B) An autoradiograph to show limited uptake of technetium[99] into the mineralized front of cortical trabecular bone (white arrow) and much more widespread uptake into the medullary bone (black arrows). Adapted from Van de Velde *et al.* (1985). Also see Fig. 25.3.

DEVELOPING FOSSILS

The fortuitous preservation of a size series of fossilized individuals from a single species creates an opportunity to study skeletal development. Histology of individuals of the same size, age or stage of ontogeny, can also provide developmental information, provided that there is a variable component in the skeletal tissue or organ such as a range of developmental stages of scales or teeth, more and less differentiated stages in a bone, and so forth.

A record in stone of the evolution of bone development in Late Cretaceous plesiosaurs from New Zealand is a particularly fine example – a unique ontogenetic trajectory was discovered by Wiffen *et al.* (1995). Their description is that juveniles are pachyosteosclerotic (the bones are large and dense), adults osteoporotic (bones weakened by imbalance of resorption over deposition, Box 8.2) as a consequence of what was interpreted as a heterochronic shift (an acceleration in the timing of bone development; Box 6.1 and Chapter 44). Patterns of limb ossification in the ichthyosaur, *Stenopterygius*, and the ontogeny and phylogeny of the mesopodium in mososaurs (extinct giant marine lizards related to snakes) have proven amenable to analysis in Mike Caldwell's hands. From these studies, a modified pattern of perichondral ossification in the evolution of paddle-like limbs in marine reptiles was uncovered; loss of perichondral bone is correlated with evolution of a paddle for aquatic locomotion. Additional endochondral ossifications on the pre- and post-axial margins of the limbs would have given rise to the polydactylous limbs these animals possessed.[12] There is a **caution**, however. Not all evolutionary trends are toward increase in size. As discussed in Chapter 44, reduction in size (miniaturization) characterizes many lineages, especially frogs and burrowing lizards.

PROBLEMATICA

All is not complete tranquility, however.

The ability to 'do' histology on fossils ignited heated debates, some of which continue to smolder. Some revolve around patterns of bone growth, others around bone resorption or remodeling, including whether remodeling occurred.

For example, as introduced above, controversy surrounds interpretation of the histology of dinosaur bone, especially whether bone structure provides evidence that dinosaurs were warm blooded, could maintain a constant body temperature, and/or how rapid their growth rates would have been. The degree of ossification of the pelvic girdle in perinatal specimens of the duck-billed dinosaur *Maiasaura* has been used to conclude that dinosaurs were precocial, i.e. able to fend for themselves from hatching. The girdles were well ossified as they are in precocial extant taxa such as ducks, emu, alligators, caiman, and in contrast to the poorly ossified pelves seen at hatching in extant altricial taxa such as doves and starlings. Now we have eggs, clutches and evidence of maternal care to fuel (end?) the debates.

It has been argued that the pelvic bones of sauropod dinosaurs have lamellar-zonal bone with zones and annuli, and that dinosaur bone, in general, is fibrolamellar and exhibits growth rings (Box 6.3), endochondral ossification, Haversian systems, secondary cancellous bone and metaplasia. An examination of the histology of the long bones of four genera of sauropod dinosaurs from the Upper Jurassic Tendaguru beds in Tanzania identified fibrolamellar bone, prompted arguments that remodeling and growth lines can be used to distinguish the four genera, and fueled speculation that two types of bone histology in *Barosaurus* might reflect sexual dimorphism. Sexual dimorphism in bone structure is well documented in mammals, but much less evident in reptiles.[13]

Box 6.3 Growth rings in bone

Bone (and dentine) are often deposited in sufficiently precise layers that age or other aspects of the life history of an individual, population or species can be determined with surprising accuracy. Here are two examples from *mammals*:

- Layers in the dentary bone of the mandible and in the teeth have been used to age the white whale or beluga, *Delphinapterus leucas*, there being twice as many annual layers in the teeth as in the dentary for any given animal.
- Annual lines also are deposited as periosteal growth lines in the mandibles of European hares.[a]

Reptilian bone has been the subject of considerable study, in part because of our fascination over whether dinosaurs were warm blooded and/or homeothermic, homeotherms having a constant body temperature, which is not necessarily the same as being warm blooded. A sample of studies includes:

- Bone is deposited in concentric annular rings in turtles, snakes and lizards.
- Dinosaur bone is fibrolamellar, with growth rings that can be used to distinguish four genera of sauropod dinosaurs.
- Teeth of Late Cretaceous dinosaurs have growth rings similar to those seen in extant crocodiles, and unlike those in contemporary mammals. Dinosaurs were therefore said to be ectothermic.
- Skeletal growth lines in the skull of an Eocene crocodile, *Crocodylus affinus*, were used as indicators of ontogenetic change in the skeleton and of paleoclimatic conditions in the Eocene.[b]

Amphibian bone also is deposited in concentric annular rings that show considerable stability, and so remain suitable for age determination long after they are deposited:

- Layers of bone can appear up to two years after completion of metamorphosis in the phalangeal diaphyses of the European common frog, *Rana temporaria*, survive subsequent endosteal resorption, and be used in age determinations.
- Administering four fluorescent labels to the edible frog, *Rana esculenta*, reveals periodicity in the annual growth rings within long bones; lines of arrested growth (LAGs) develop annually in winter, and are not fully destroyed by remodeling.[c]

The bones, scales and otoliths of *teleost fish* provide perhaps one of the most accurate recording systems of growth there is. Identification of growth rings to show that fossils purported to be molluscs were actually teleost hyperostotic bone was noted in Box 2.2.

As with growth studies in tetrapods, analysis of ageing can be augmented by incorporating a label into bone or scales; oxytetracycline administered to Atlantic salmon, *Salmo salar*, or to juvenile Atlantic cod, *Gadus morhua*, persists in the centra and ribs of salmon for two years, and in cod for three years, with no obvious effects on growth or survival. Oxytetracycline has also been used to show that cartilage growth in *sharks* is regulated locally; two bands per year in vertebral centra, three bands per year in cartilages of the upper jaw.[d]

Otoliths retain excellent records of age in teleost fishes. Because an otolith records *daily* growth increments in fibroprotein, otoliths can be used to determine age, rate of growth, spawning time, and even – from the size of bands and distances between them – provide indications of environmental conditions such as diet, periodicity of feeding, and so forth.[e]

[a] See Brodie (1969) for whales, and Frylestam and von Schantz (1997) for hares.
[b] See Castanet and Cheylan (1979) for annual rings, P. A. Johnston (1979) and de Buffrènil and Buffetaut (1981) for fossil crocodiles, Reid (1981, 1984) and Sander (2000) for dinosaurs, and Castanet *et al.* (1993) for individual ageing using bone. Closure of vertebral sutures in crocodiles proceeds from caudal to cranial, the most caudal sutures being closed at hatching and the closure of the cervical sutures occurring at morphological maturity, which may or may not correspond to the end of growth. Brochu (1996) used this approach to assess the maturity of fossil archosaurs.
[c] See Castanet and Cheylan (1979) for annual rings, Guyetant *et al.* (1984) for phalanges in *Rana*, and Francillon and Castanet (1985) for periodicity of growth in *Rana* long bones, in which increase in length is by intramembranous ossification, the role of the cartilage being to increase width only, there being minimal endochondral ossification (Felisbino and Carvalho, 1999). Histochemistry and TEM of the epiphyseal apparatus of the ranid frog *Rana esculenta* showing lack of vascular invasion or mineralization of the metaphyseal cartilage, absence of trabecular bone and periosteal rather than metaphyseal ossification (Dell'Orbo *et al.*, 1992).
[d] See Meunier *et al.* (1979) and Summers (2000) for overviews, Odense and Logan (1974) for salmon, Nordeide *et al.* (1992) for cod, and Nathanson and Calliet (1990) and Tanaka (1990) for the shark studies.
[e] See Campana (1983, 2004), Brothers (1984), Maisey (1987) and Moser (1984) for otoliths.

Histology of dental tissues also provides important information. Erickson (1996a) used the incremental lines of von Ebner – the daily growth lines seen in dentine and cementum first described by Richard Owen ('Old Bones' as *Vanity Fair* depicted him in a cartoon published in 1873, Fig. 6.3) – in dinosaur teeth to assess rates of tooth replacement, arguing that the lines in dentine indicate a pattern of daily matrix formation. Three different types of evidence show how conclusions regarding thermoregulation have been sought from palaeohistology.

- Histological analyses of long bones from a prosauropod dinosaur reveal laminar bone unlike that in recent reptiles, and more vascularization than in recent reptiles; the patterns of vascularization resemble those in recent mammals, evoking speculation about their physiological significance for presence or absence of endothermy.
- The bony plates along the backs of *Stegosaurus* are composed of a thin wall of incompletely remodeled bone around large cancellous regions with 'pipes,' a pattern that is consistent with but not conclusive proof of thermoregulation.
- Ultrastructural analysis of avian and dinosaur bones demonstrates that they can be separated on the basis of canaliculi and collagen fibre bundles, ornithischian dinosaurs being more like mammals than they are like other dinosaurs.[14]

PALAEOPATHOLOGY

These problems are not unique to palaeohistologists. Pathologists must also deal with individual variation in bone structure, some caused by the demands that particular life styles place on skeletal structure – Wolff's law in action – some by gender, age, disease or nutritional status.

Bone lesions are sufficiently common in the fossil record that *palaeopathology* is now a subfield of

Figure 6.3 Richard Owens or 'Old Bones' as he was depicted in this cartoon from *Vanity Fair*, March 1, 1873, one of the more flattering portrayals of Owens.

palaeontology. Among the bone lesions recorded in fossils is a case of osteopetrosis in a dinosaur that lived 70 mya, a lesion with remarkable similarity to osteopetrosis in humans. Pathologies of the humerus and fibula have been reported from *Tyrannosaurus rex*. Osteoarthritis, periostitis, osteomyelitis, lumpy jaw and fractures in various stages of healing have all been described in a Pleistocene population of kangaroos.[15]

CONODONTS

'Little whatzits'

Conodonts, tooth-like conical elements (*conodont elements*), a conundrum of the fossil record since the first described by Pander in 1856, have in recent years joined the chordate family and have done so because of the skeletal tissues that comprise them. Conodont elements first appeared in the geological record in the Late Cambrian, persisting until the Triassic.

Only recently has it become clear that these tooth-like elements are part of a complex feeding apparatus in an organism (*the conodont animal*) known from some 15 specimens, eight of which have the feeding apparatus *in situ*. Studies, largely by Phil Donoghue, on wear patterns and growth of conodont elements provide strong evidence for retention of the elements throughout life, periods of use followed by periods of growth, and a pattern of occlusion of pairs of elements as complex as the occlusion patterns of mammalian teeth.

Conodonts appear to have been active swimmers, some 40–60 mm long, with V-shaped muscle blocks, a rayed caudal fin (Box 43.1) and (possibly) scleral cartilage supporting large eyes. One of the largest conodonts known (length of the anterior part, 109 mm), *Promissum pulchrum*, from the Upper Ordovician of South Africa, is also one of the only extant agnathans to show details of muscle fibre organization.[16]

Phylogenetic reconstruction places the conodonts as more derived than (*crownwards of*) hagfishes (myxinoids) and as a sister group of the heterostracan fishes introduced in Chapter 2. It is critical in phylogenetic reconstructions based on fossils to distinguish *stem* from *crown taxa*, a stem taxon being an ancestral taxon and a crown taxon a derived taxon, normally one that has not left further descendants. Crown taxa are, if you like, at the top of the tree, stem taxa at the bottom. The term *crownwards* indicates that a taxon is further toward the top of the tree than the taxon with which it is being compared.

Several workers and lines of evidence address the nature of the tissues that comprise conodont elements. In an early analysis of the amino acids found in fossil mineralized tissues, Armstrong and Halstead Tarlo (1966) failed to detect hydroxyproline in Devonian conodonts. Whether the fact that they found amino acids in the rocks as well as the fossils makes such analyses more or less convincing is hard to say. Two decades later, Dzik (1986), who pioneered the use of histological analysis of conodont elements, proposed the now widely accepted view that the tissues within conodont elements are vertebrate tissues (dentine, enamel), and conodonts not 'little whatzits.'[17]

In what is the most thorough and sophisticated analysis, Sansom *et al.* (1992, 1994) identified cellular bone, dentine, an enamel-like tissue, and (more debatably) spherulitic mineralized cartilage in conodont elements. An excellent review of conodont affinities and chordate phylogeny, based on 17 taxa using 103 characters, was published by Donoghue *et al.* in 2000. In all the phylogenetic reconstructions tested, conodonts came out crownward of hagfishes and lampreys. The review includes an extensive discussion of skeletal tissues in fossil vertebrates, encompassing cellular and acellular bone and dentine, an important conclusion being that the first bone was acellular.

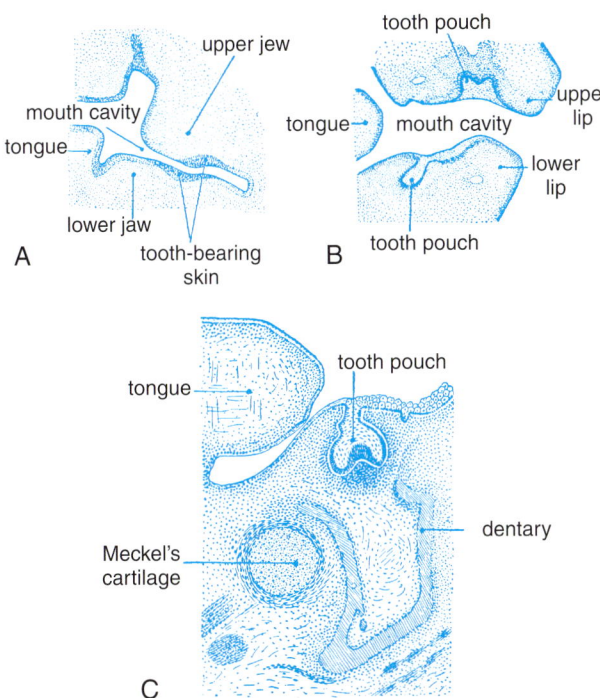

Figure 6.4 Early stages in the establishment of human teeth shown as drawings of histological sections. (A) An 11 mm long embroy (40 days post-ovulation) showing the location of the dental epithelium ('tooth-bearing skin'). (B) A 16 mm long embryo (45 days post-ovulation) showing onset of the dental lamina ('tooth pouch'). (C) The lower jaw of a 32.5 mm long embryo (58 days post-ovulation) illustrating the bell stage ('tooth pouch'), differentiated Meckel's cartilage and the dentary bone. Modified from Gregory (1929).

Donoghue (2001) and Donoghue and Aldridge (2001) regard conodont elements as the earliest chordates with a mineralized skeleton, and discuss the (unsettled) issue of homology of the hard tissues in conodont elements. Following upon their studies, M. M. Smith and Coates (1998) used the internal location of the conodont apparatus to argue that conodont elements are not homologous with dermal denticles or teeth and that, consequently, teeth did not arise evolutionarily from dermal denticles associated with ectoderm (Fig. 6.4), but rather arose internally in association with endoderm. One of the difficulties with this scheme is the phylogenetic position of conodonts, which places them later than the origin of dermal denticles. Sire (2001) described teeth *outside* the mouth in the western Pacific reef fish, the bearded silverside, *Atherion elymus*, and showed them to be developmentally and structurally equivalent to teeth. Furthermore, examination of a more basal group of agnathans, the *thelodonts*, reveals external exoskeletal denticles over the mouth and pharynx, entirely consistent with denticles → teeth.[18]

NOTES

1. See Ulrich *et al.* (1987) for osteocalcin, and Glimcher *et al.* (1990) for hydroxyproline and hydroxylysine. In addition to the references in Table 6.1, see Ørvig (1965) and Kobayashi (1971). For preservation of key biomolecules in fossils, see Bada *et al.* (1999). See Gould (2002) for a discussion of the first fossil ever figured in print, a drawing of a striated bivalve shell in a treatise on metals by Christopher Encelius, published in 1557.

2. For references to Pawlicki's work, see Table 2.1. See Schaedler *et al.* (1992) for the study of mammoth collagen, and Armstrong *et al.* (1983) for the ultrastructural analysis.

3. See M. M. Smith (1991) for the Ordovician fauna, and M. M. Smith and Sansom (1997) for *Skiichthys halsteadia* (named to honour the late Beverly Halstead, a pioneer in the study of fossil skeletal tissues), which they differentiate from two well-known genera, *Astraspis* and *Eriptychius*, both of which have acellular bone and tubular dentine. M. P. Smith *et al.* (1996) described dentine in *Anatolepis*, which they regard as a vertebrate and as the oldest fish. Recently, Smith and Johanson (2003a) identified what they regard as dentine in the teeth of representative arthrodires (derived placoderms). The significance of this finding lies in their contention that more basal placoderms lack teeth. *If* the identification of dentine holds up, and *if* the phylogenetic relationships among the placoderms are robust, and *if* more basal placoderms really lack teeth, the implication would be that teeth evolved twice; see Stokstad (2003) and Burrow (2003) for commentaries and see Smith and Johanson (2003b) for a response to the latter.

4. Eastoe (1970) is a good source for earlier discussions on the place of cartilage and bone among vertebrate mineralized tissues. See Forey and Janvier (1993) for agnathans and the origin of jawed vertebrates, and Denison (1963), Romer (1963), Ørvig (1968), Dacke (1979), Smith and Hall (1990) and Janvier (1996) for the likely function of these skeletal shields.

5. For skeletal mineralization and bone density see Schaeffer (1961), Felts and Spurrell (1965; 1966, and Figure 44.3 therein), Bryden and Felts (1974), de Ricqlès (1975), Laerm (1976, 1979) and Stein (1989). See Moss (1963) for bone structure in aquatic vertebrates. For a detailed discussion of the significance of histological variation in fossil vertebrates, see Enlow and Brown (1956, 1957, 1958), Enlow (1969), de Ricqlès (1973, 1974, 1976, 2000), Hall (1975a, 1978a), Smith and Hall (1990, 1993) and Horner *et al.* (1999, 2000, 2001). For the importance of the marrow cavity for the mechanical strength of bones and why most bones are not solid, see Pauwels (1974) and Currey (1984a,b).

6. See Ørvig (1951, 1989), Halstead (1969a–c, 1973), Hall (1975a, 1977a), Maisey (1988) and Smith and Hall (1990, 1993) for detailed discussion of the significance of intermediate tissues for vertebrate evolution.

7. On a visit to faculty X at university Y in Canada to present some seminars and lead a workshop, I was delighted to have a local faculty member greet me by saying that he had spent the past three days reading my papers, only to find that his reason for doing so was to discuss every instance in which I had presented (apparently?) different views or interpretations in different publications, some of them far in the past.

8. See pp. 4–6 in Hall (1978a). For discussions on the adequacy of such evidence see Ørvig (1989), Hall (1975a), Schaeffer (1977) and Smith and Hall (1990, 1993). See Ekanayake and Hall (1987, 1988) for the development of acellular bone in the Japanese medaka, *Oryzias latipes*.

9. See Horner and Weishampel (1988) for bony processes and muscle attachment, and Horner *et al.* (1999, 2000, 2001) for hadrosaur bone histology. The histology of the tibiae of the

great western marine bird, *Hesperornis*, was used by Houde (1987) to argue that this species had osteones like neognathous birds. Palaeognathus birds have a bone histology resembling dinosaurs.

10. See Barretto *et al.* (1993) for the palaeohistology, Fischman (1993) for the comment, and Starck (1993, 1994, 1996) and Starck and Ricklefs (1998) for an enormously useful analysis of avian long-bone growth.

11. See de Ricqlès *et al.* (2000) for palaeohistology, de Ricqlès (2000) for growth rate, and Padian (1983), Padian and Rayner (1993) and Unwin (2003) for pterosaur flight.

12. See Caldwell (1996, 1997a,b) for these studies, Caldwell (2003) for an overview of limb loss, and Fedak and Hall (2004) and literature therein for studies on polydactyly in secondarily aquatic mammals.

13. See Reid (1981, 1984) and Geist and Jones (1996) for the pelvic bone studies, and Sander (2000) for the study of the four genera of sauropods: *Brachiosaurus* and *Barosaurus* (each with growth series), fewer samples of *Dicraeosaurus* and a single specimen of *Janenschia*.

14. See Currey (1962) for the prosauropod study, de Buffrénil *et al.* (1986) for *Stegosaurus*, and Rensberger and Watabe (2000) for the ultrastructural analysis. From studies on mallard ducks, *Anas platyrhynchos*, de Margerie (2002) proposed that laminar bone in birds may be an adaption to the torsional loads associated with flapping flight.

15. See Campbell (1966) for the dinosaur study, Brochu (2003) for *T. rex*, which was described 100 years ago (1905) from a specimen now on display in the Carnegie Museum of Natural History in Pittsburg, PA; Wells (1973) for the palaeopathology of bone disease, Ortner and Putschar (1982) for pathological conditions in human skeletal remains, Horton and Samuel (1978) for Pleistocene kangaroos, and Brandt (2001) for an atlas of osteoarthritis in humans.

16. See Aldridge *et al.* (1993, 1994) for the conodont animal, and Gabbott *et al.* (1995) for *Promissum pulchrum*. For conodont elements functioning as cutting teeth and for the basic model of the conodont apparatus, see Purnell and von Bitter (1992), Purnell and Donoghue (1997), Donoghue (1998, 2001), Donoghue and Purnell (1999a,b) and Purnell (1999). For the phylogenetic position of conodonts, see Aldridge *et al.* (1993), Aldridge and Purnell (1996), Aldridge and Donoghue (1998), Donoghue *et al.* (1998), Sweet and Donoghue (2001) and Donoghue and Sansom (2002). Donoghue (1998) provides an excellent historical analysis of studies on conodonts.

17. See Sweet (1985) for conodont elements as 'little whatzits,' and Benton (1987), Sansom *et al.* (1992), and Donoghue and Chauffe (1998) for the nature of the tissues. Sweet and Donoghue (2001) discuss the phosphatic skeleton of conodont elements. For reviews of the evolution of skeletal tissues, especially those derived from the neural crest, see M. M. Smith and Hall (1990), Hall (2000a), Donoghue *et al.* (2000) and especially Donoghue and Sansom (2002), which provides an excellent review of the origins of the vertebrate skeleton using developmental, molecular and fossil data.

18. See van der Brugghen and Janvier (1993) for thelodont denticles and Wilson and Caldwell (1993) for scales of Silurian and Devonian thelodonts. As an example of dermal denticles and their development, in this case in the little skate, *Leucoraja erinacea*, see Miyake *et al.* (1999). See Franz-Odendaal *et al.* (2003) for conodonts and M. M. Smith (2003) for the origins of vertebrate dentition and jaws.

Part III

Unusual Modes of Skeletogenesis

Just as the existence of intermediate tissues – chondroid, chondroid bone, cementum, enameloid – has confounded students of the skeleton for a century and a half, so the existence of unusual modes of skeletogenesis has confounded anatomists, histologists, and comparative biologists.

I am not thinking of pathological conditions or responses to parasitic infections or diseases. Rather, I am thinking of the skeletal tissues and processes of skeletogenesis associated with mammalian horns and antlers, and sesamoids and ossifications in tendons and ligaments. In deer, the skeletal organ – antlers – develops after birth, is shed (usually annually) and grows back – regenerates – annually. So, rather than discussing the standard stories of endochondral and intramembranous ossification – which can be found in every textbook of histology or anatomy, and in most first year university biology textbooks – I use antlers, which I contrast with the development of horns and ossicones, to introduce modes of osteogenesis and chondrogenesis, the topic of Part III. This approach also provides the opportunity to place the skeletal organs in the context of organismal life history, rather than treating them, the skeletal organs, as inanimate objects, bones.

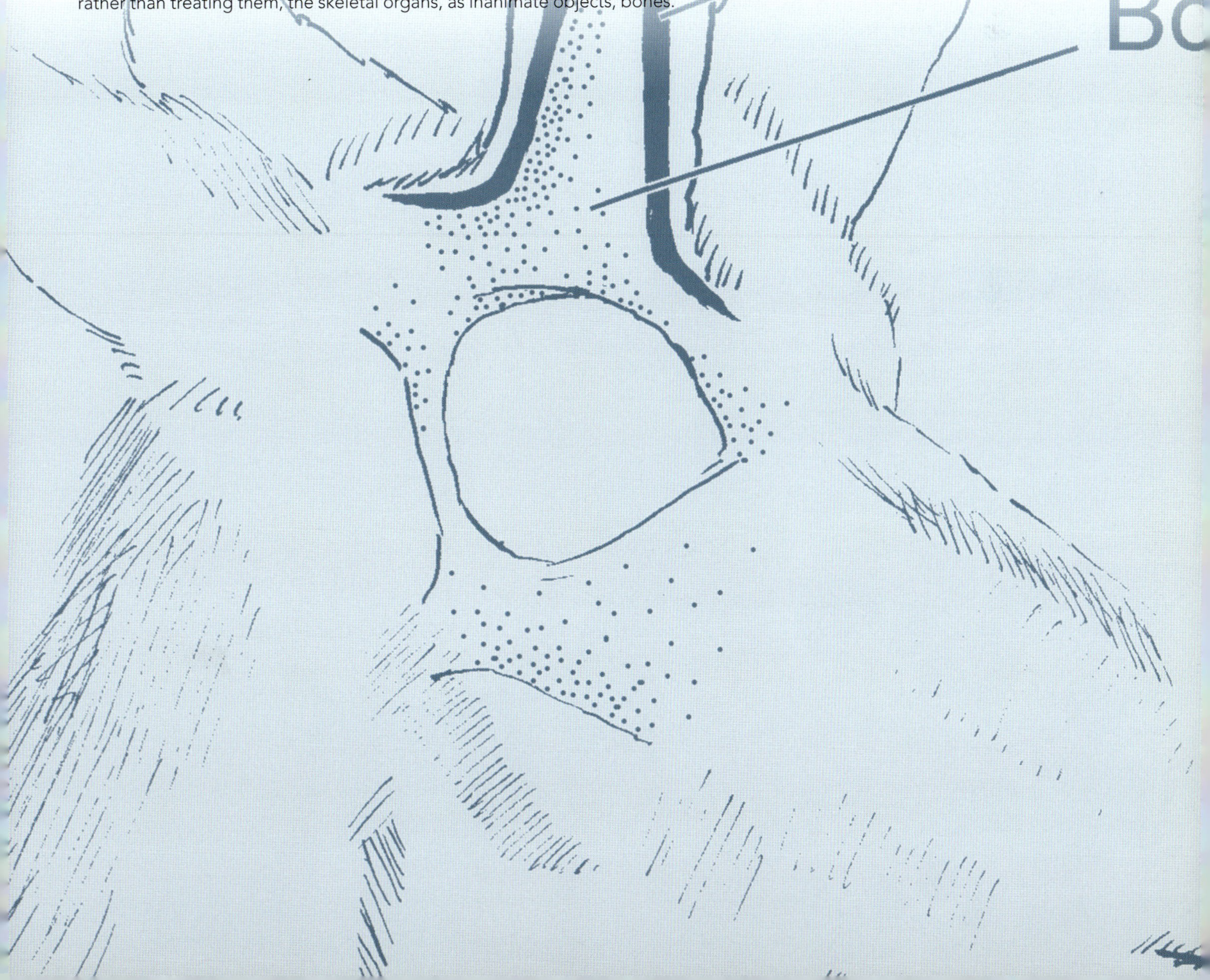

Horns and Ossicones

'In the frequent fits of anger to which males especially are subject, the efforts of their inner feelings cause the fluids to flow more strongly towards that part of their head; in some there is hence deposited a secretion of horny matter, and in others of bony matter mixed with horny matter, which gives rise to solid protuberances: thus we have the origin of horns and antlers.'

(Lamarck, 1809, p. 122)

To the layperson, the prongs on the heads of hoofed animals such as cows, deer and giraffes appear to be more or less branched variations on the same theme or variants of the same structures. Not so. Fundamental differences exist between cow horns and deer antlers, the chief of which are summarized in Table 7.1; the differences are substantial. Giraffes, it turns out, have neither horns nor antlers, but ossicones (knobs). Five families of hoofed animals (ungulates) possess horns, antlers or ossicones:

- Rhinocerotidae (rhinos), with one or two *midline nasal horns*;
- Bovidae (cattle, sheep, goats and antelopes), with *paired, symmetrical horns*;
- Antilocapridae (pronghorn antelope), with *paired, symmetrical, deciduous horns*;
- Cervidae (moose, deer, elk), with *paired antlers* in the males, although female reindeer and caribou are antlered; and
- Giraffidae (giraffe, okapi), with permanent *paired ossicones*.[1]

In this chapter I examine horns and ossicones. Chapter 8 deals with antlers.

Table 7.1 The major features distinguishing horns from antlers

Horns	Antlers
Living	Dead
Cornified (±osseous core)	Ossified
Unbranched	Branched
Permanent (except pronghorn antelope)	Deciduous (usually annually)
Cannot regenerate	Regenerate
Basal growth	Apical growth

HORNS

'Truly was produced a K'i-lin [giraffe] whose shape was high 15 feet With the body of a deer and the tail of an ox, and a fleshy boneless horn.'[2]

The Oxford English Dictionary (OED) defines a horn as 'a non-deciduous excrescence, often curved and pointed, on [the] head of cattle, sheep, goats and other mammals, found in pairs, single, or one in front of the other'. This definition applies to horns as *organs* in mammals, but does not embrace horn as a *tissue*, or those organs called horns in animals other than mammals. The evolutionary significance of horns relates to their use as weapons, shields and for determining and/or maintaining dominance ranks or hierarchies.[3]

The OED also defines horns as 'hornlike projections on the head of other animals, e.g., a snail's tentacle, the crest of a horned owl, etc.' This use of the term 'horn' covers projections from the heads of non-mammalian vertebrates such as chameleons or dinosaurs, and analogous projections on the heads of beetles. So, we have such common names as the Amazonian horned frog *Ceratophrys cornuta*; the horned screamer *Anhima cornuta*, a marsh-dwelling bird; the great horned owl

Bubo virginianus; the Texas horned lizard *Phrynosoma cornutum*; and the Asian long-horned beetle *Anoplophora glabripennis*.[4]

Used in this second way, the word describes an anatomical structure or organ, irrespective of whether that structure arises on the head of a vertebrate or an invertebrate, and irrespective of the nature of the tissue or tissues that comprise the horn. As a further complication, as with the term 'bone,' the term horn is used – in vertebrates at least – for both organ and tissue. This double usage is so embedded in the scientific literature and lay usage that it is impossible to avoid.

We are all familiar with the horns of cattle, sheep and goats – those animals that together constitute the family Bovidae. These organs are all composed of the tissue horn, which the OED defines accurately as 'the structure of a horn [as organ], consisting of a core of bone encased in keratinized skin.' Although they may have a bony core, horns are keratinized *epidermal*, not skeletal, structures. Horn is produced by epithelial cells, known as *keratinocytes*. Both elements – bony core and keratinized covering – are required to fully describe horn as an organ. The following organs are all composed of keratin, but not all are horns:

- the *horns* of some chameleons, rhinos and such extinct animals as ceratopsian dinosaurs and titanotheres;
- *horny scales* and feathers of birds, reptiles, and on the tails of beavers and opossums;
- the *spurs* of game cocks;
- *dew claws* of ruminants, swine and dogs;
- *chestnuts, ergots* and *hooves* of horses;
- fingernails and toe *nails* of many tetrapods; and
- the *baleen* or whale bone of baleen whales.

DISTRIBUTION OF HORNS AS ORGANS

Bovidae

The more than 100 members of the family Bovidae exhibit tremendous diversity in the shapes and sizes of their horns. Bovid horns may be straight, curved or spiral – the spirals being either simple or complex – and may project forward over the head or back toward the tail (Figs 7.1 and 7.2). Horns may be enormous; the horns of the Indian water buffalo extend 2.5–2.7 metres from tip to tip. Measuring horns or antlers to establish world records is arcane (Box 7.1). Some current world records for horns are listed in Table 7.2.

Cattle horns are hollow. The bony core is continuous with the cavity of the frontal bone. In domesticated Nepalese 'unicorn' sheep, the two horns are joined artificially by transplanting horn buds to form a single midline horn. Both male and female wild sheep and goats possess horns, although horns of the males are always larger.

Antelope horns also exhibit a great diversity of shapes and sizes. Depending on the species, horns may be present in both sexes, as in the eland, *Antelope oreas*, or only in males, as in the greater kudu, *Tragelaphus strepsiceros*. Horns in male African antelopes probably evolved between three and five times, essentially whenever body weight exceeded 20 kg; horns are found in females in half the genera of African antelopes, and then only in heavier species. In female bovids that do possess them, horns or hornlike organs develop later than in males, and horn ontogeny is less tightly controlled. Castration precipitates loss of the horns, demonstrating their dependence on male sex hormones for their maintenance.[5]

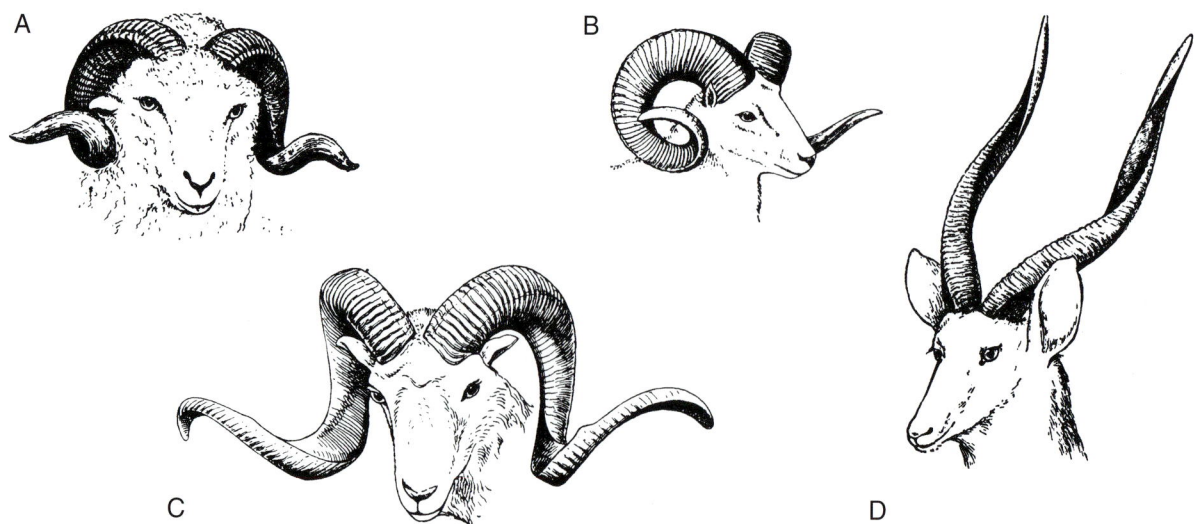

Figure 7.1 Spiral horns of wild sheep and goats. (A) The horns of a male (ram) merino sheep, a breed that is prized for the quality of the wool. (B) The Tibetan Argali, *Ovis ammon hodgsoni*, found in China, India and Nepal. (C) Marco Polo's Argali, *Ovis ammon poli*, the largest of the wild sheep, with body weight up to 250 kg and horns up to 170 cm in length. (D) The Nyala, *Tragelaphus angasi*, the spiral horned antelope of South Africa. Compiled from Cook (1914).

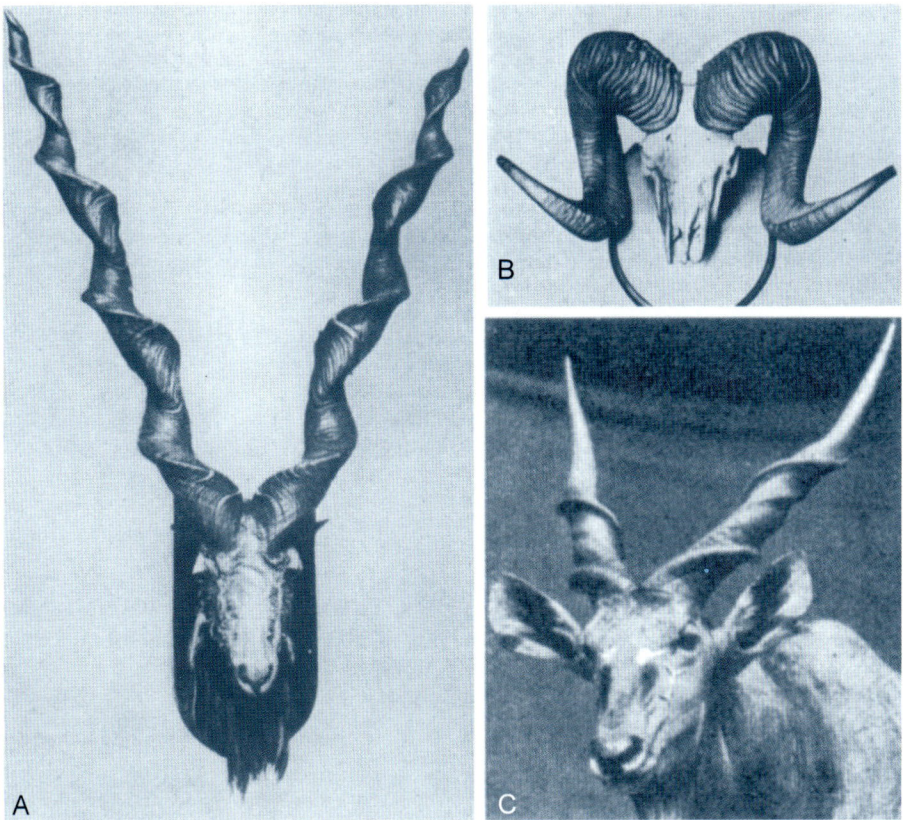

Figure 7.2 The tightly spiraled horns of eland and sheep. (A) An impressive specimen of the, now endangered, straight-horned (Suleman) markhor, *Capra falconeri jerdoni*, from the Suleman Range, Punjab frontier and Afghanistan. (B) The twisted curves of the horns of a wild sheep of the Gobi Desert, the Hangai agali, *Ovis ammon mongolica*. (C) Ninety-two centimetre horns of a giant (Derby, the Hon. Walter Rothschild's Senegambian) eland, *Taurotragus derbianus*, in which horns are present in both sexes, reaching lengths of 120 cm in males. Compiled from Cook (1914).

Box 7.1 How to measure horns and antlers

The sizes of horns and antlers are measured using a complex formulation that results in the assignment of 'Boone & Crockett points,' a system established by the Boone & Crockett Club in 1932 as a consistent means to establish world records. To be considered eligible the official scoring forms from B&C must be used and measurements taken using the following equipment:

- one 1/4″ wide by 6′ long flexible steel tape measure with a lip end;
- one 1/4″ wide by 6′ long flexible steel tape measure with a ring end;
- one roll of half-inch-wide masking tape;
- one carpenter's folding ruler, with slide end.
- a thin, flexible bicycle brake or gear selector type steel cable, three to four feet long; and
- two alligator clips.

Having taken the measurements, one (a) adds up the inside spread, both main beams, all typical points and all four circumferences from each beam; (b) compares the corresponding measurements – beams, points and circumferences – from the right antler to the left and subtracts the smaller number from the larger; (c) adds up all the differences (and any non-typical points); and (d) subtracts them from the grant total to obtain the final score.[a]

[a] Source: HTTP://WWW.GAMECALLS.NET/HUNTINGTIPS/ THEBOONEANDCROCKETT MEASURINGSYSTEM.HTML

Table 7.2 World records for horns[a]

Species	B&C score	Length of horn[b]		Tip-to-tip spread[b]
		Right	Left	
Bighorn sheep (*Ovis canadensis*)	208 3/8	47 4/8	46 5/8	23 1/8
Desert sheep (*Ovis canadensis mexicana*)	205 1/8	43 5/8	43 6/8	25 5/8
Stone's sheep (*Ovis dalli stonei*)	196 6/8	50 1/8	51 5/8	31
Dall's sheep (*Ovis dalli*)	189 6/8	48 5/8	47 7/8	34 3/8
American bison (*Bison bison*)	136 4/8	21 2/8	23 2/8	27
Musk ox (*Ovibos moschatus*)	127 2/8	29 2/8	29 7/8	28 6/8
Pronghorn antelope (*Antilocapra americana*)[c]	95	17 2/8	17/2/8	4 3/8
Rocky Mountain goat (*Oreamnos americanus*)	56 6/8	11 7/8	10 6/8	8 2/8

[a] Source: Boone & Crockett Club web site (HTTP://WWW.BOONE-CROCKETT.ORG) and see Table 8.1 for comparable data for antlers.
[b] Being official B&C point records I have not translated these Imperial measures (inches) to metric units.
[c] See text for the distinction between true horns and the horns of pronghorns.

Domestication is followed by changes ranging from reduction in horn size to total loss (see Box 7.2 for the effects of domestication on the skeleton and skeletal tissues). Domesticated cattle have smaller horns, or have lost the horns possessed by their wild relatives. Reduced competition, fighting over resources or fighting for mates are the likely selective forces underlying such changes.

One flipside of domestication is the hunting of sheep and antelope and the removal of their horns as trophies. Trophy hunting has evolutionary consequences (Box 7.3).

Rhinos

Rhino horns, which are composed of a fused mass of long, hair-like strands of keratin, lack a bony core – and are therefore sometimes called *keratin-fibre horns* – resting

instead on a bony knob on a much-strengthened nasal bone. Since antlers always form on the frontal bones, rhino 'horns' therefore share some features of horns and antlers. Horns can form in ectopic locations; in 1515, Albrecht Dürer produced a famous woodcut of an Indian rhino (Fig. 7.3), with a small ectopic, midline horn on the shoulder. Homage to the unicorn?

Male and female Indian rhinos, *Rhinoceros unicornis*, have horns of equal size. The front curve of these horns can be as much as 0.6 metres. A female of the southern race of the white rhino, *Ceratotherium simum simum*,

---

Box 7.2 Domestication

Given that various attributes of the skeleton can be modified by natural or artificial selection, we might expect the skeletons of domesticated animals to deviate from their wild relatives. We might expect it, but few have studied or even considered the issue. This is a shame. Domestication does affect bone structure, as known for centuries.

Domestication of sheep in an area that is now modern-day Turkey was investigated at two archaeological sites, one dated at 5780 BC, the other at 6570 BC, one with wild, the other with domesticated sheep. In comparison with wild sheep, bone from domesticated sheep shows a higher preferential alignment of hydroxyapatite crystals, significant increase in lacunar size, thicker trabeculae, and a sharp transition between compact and spongy bone. Whether such changes are the direct result of domestication, or whether they flow from the inbreeding that often accompanies domestication, is hard to determine. Inbreeding does affect skeletal development; inbred plague rats, *Rattus villosissimus*, for instance, have smaller skull lengths and widths, and increased skeletal fluctuating asymmetry in comparison with random-bred rats.[a]

Some of the most detailed studies of the skeletons in domesticated mammals have been carried out by Robert Wayne, who compared changes in cranial and limb skeletons in domestic and wild dogs, concluding that:

- differences in cranial morphology between breeds of domestic dogs are as great as those between genera of wild dogs;
- small breeds are paedomorphic;
- a developmental basis for the diversity in skeletal morphology among breeds can be found in changes in shape during development, usually involving simple changes in the timing of postnatal growth rates.[b]

Because of the resulting inbreeding and reduced gene pool, it was argued that domestication leads to increased dental or skeletal abnormalities, although domesticated populations could be interbred. No evidence of this was found in an analysis of skull and tooth development in wild and domestic foxes. Captivity – which is not the same as domestication – leads to loss of teeth in alligators, but one can imagine many potential causes, including diet, confined space, metal enclosures, and so forth.[c]

Domestication does stabilize some life history features; maintaining a laboratory strain Japanese quail, *Coturnix coturnix japonica*, results in selection for females with more stable egg-laying routines than found in quail in the wild (Houdelier *et al.*, 2004).

Another aspect of what could be regarded as domestication is the analysis of strontium, lead, oxygen and carbon isotopes in the bones and teeth of the 'alpine iceman,' an exceptionally well preserved circa 46-year-old human male who lived some 5200 years ago in what archaeologists call the Neolithic-Copper Age of Europe. Comparison of the isotopic ratios in enamel, bone and intestine enabled Wolfgang Müller and his colleagues to determine that this individual spent his entire life no more than 60 km southeast of the discovery site, lived on gneissic soils during his later life, and used plant and animal materials from a wider range, an amazing series of deductions (Müller *et al.*, 2003).

[a] See Drew *et al.* (1971) for the archaeological sites, and Lacy and Horner (1996) for the study on inbred rats.
[b] See Wayne (1986a,b) and Wayne and Ruff (1993). Also see Alberch (1985) for paedomorphosis and digit number in dogs.
[c] See McKeown (1975) for the fox, and Erickson (1996b) for the alligator study.

Box 7.3 Evolutionary consequences of taking horns as trophies

A fascinating analysis of a 30-year record of the hunting of bighorn sheep, *Ovis canadensis*, in Alberta, Canada, revealed undesirable evolution consequences of hunters selectively killing rams with rapidly growing horns. Hunters will go to considerable lengths and expense to secure a world-class trophy; hunting permits exchange hands for hundreds of thousands of dollars, one hunter having paid more than one million Canadian dollars over two years for special permits.[a] The current world record for bighorn sheep (which may be the record for horns of any species) is held by a ram with a Boone and Crockett score (Box 7.1) of 208 3/8", lengths of right and left horns of 47 4/8" and 46 5/8" and a tip-to-tip spread of 23 1/8" (Table 7.2).

Both body weight and horn size are highly heritable. Body weight has a heritability of 0.69, horn length a value of 0.41, on a scale where 1.0 would be a trait that was 100 per cent heritable with no environmental or other influences. Selective culling (by shooting) of larger rams might be expected to influence body

and horn size in subsequent generations. A surprising result that emerged from this analysis was the short time period over which such culling affected the population. Most horn growth occurs during the second to fourth years of ram life. Eighty per cent of the rams shot were under eight years of age, a fifth as young as in their fourth year of life.

Over the 30-year period studied (1972–2002), mean ram weight declined from 85 to 65 kg (a 24 per cent drop), mean horn length from 70 to 50 cm (a 28 per cent drop), and population size (ewes two years old or older plus yearlings) from 65 to 20 individuals (a 70 per cent decline). Clearly, the population has responded rapidly and dramatically to selective shooting of big-horned big rams.

[a] The study by Coltman *et al.* (2003), from which the statistics are taken, utilized quantitative genetic analysis on a pedigree reconstructed using genetic criteria. See Box 43.3 for a discussion of quantitative trait loci (QTLs).

Figure 7.3 Albrecht Dürer produced this woodcut of an Indian rhinoceros, *Rhinoceros unicornis,* in 1515. The ectopic horn (arrow) is usually interpreted as illustrating the influence of the unicorn.

holds the record for the longest horn recorded – a front horn of 1.6 metres and a posterior horn of 0.56 metres. Males of the lesser one-horned Java rhino, *Rhinoceros sondaicus,* of Burma, Bengal and Java have a larger single horn than do females. The anterior horn of the Sumatran (Asiatic) two-horned rhino, *Ceratorhinos sumatrensis,* is always larger than the posterior horn.

Titanotheres

Titanotheres are an extinct group of ungulates of the order Perissodactyla, the order that includes horses, tapirs and rhinos. The short evolutionary history of titanotheres was restricted geographically to North America and temporally to the Eocene and Oligocene. Their evolution was characterized by a rapid increase in body size and by the acquisition and enlargement of 'horns.' *Brontotherium* (thunder-beast), a N. American titanothere and part of the mythology of the Sioux Indians, had a pair of forked horns on the tip of the snout, in much the same position as the horns of rhinos. *Brontotherium* represents the end of a lineage in which the horns increased in size. The horns and their bony attachment to the skulls becoming more and more elaborate (Osborn, 1929).

Pronghorn antelopes

Antilocapra americana, the pronghorn antelope, was once one of the most common native mammals in North America. Their numbers were enormous – as many as 40 million in 1800 – before excess hunting and disease reduced their numbers dramatically. Although *A. americana* is the most well known pronghorn, as many as a dozen species populated the prairies of North America as little as a million years ago, each with a distinctive set of 'horns,' some – of the spiraled pronghorn, *Ilingoceros* – resembling the spiraled horns of wild sheep and goats (Figs 7.1, 7.4 and 7.5), others – of *Paramoceros* and *Ramoceros* from New Mexico and *Paracosoryx* from Nebraska – remarkably convergent on antlers. Phylogenetically, pronghorns are close to bovids, although their horns are much smaller than the true horns of bovids (Table 7.2).[6]

Unlike antlers, which are usually shed annually, horns are normally permanent. Male pronghorn antelopes, however, have *deciduous* 'horns' consisting of a pair of permanent bony projections on the skull (Fig. 7.6) that are sharp-edged in section, solid, unbranched, and covered with a layer of epidermal cells that forms horn (keratin) but not velvet. In pronghorns:

- the epidermal layer is shed annually and then replaced, becoming subdivided into the prongs that give the animal its common name;
- like antlers, pronghorn horns harden and are shed in synchrony with an annual testosterone cycle; and
- female pronghorns sometimes develop a bony knob on the skull from which a horn may or may not develop.

The horns of the Saiga antelope, *Saiga tatarica,* also have a permanent bony core and a horny sheath that is shed periodically. Interestingly, at least one species of most major lineages within the Bovidae sheds its horns (Figs 7.1 and 7.5).[7]

Figure 7.4 The corkscrew spiral horns of a male Cabul markhor, *Capra falconeri*, from Northern India. These horns attain lengths of 160 cm in males but rarely more than 25 cm in females. From Cook (1914).

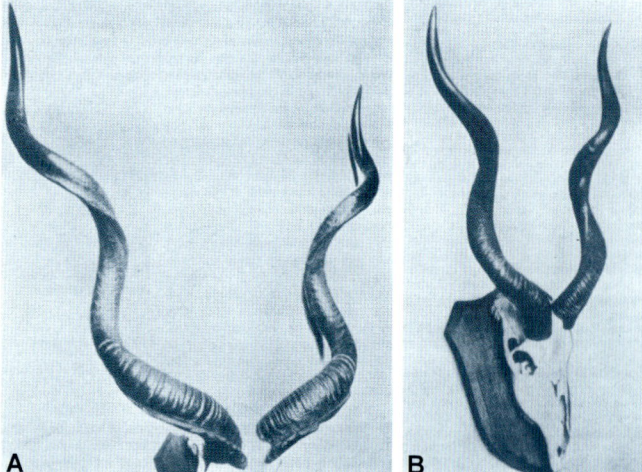

Figure 7.5 Antelope horns. (A) The elegant horns of a male greater kudu, *Tragelaphus strepsiceros,* the second tallest of the antelopes and the species with the most spectacular horns. The 2.5 turns in each horn is typical of horns that are usually around 120 cm in length, the record being 180 cm. (B) The no less impressive horns of a male lesser kudu, *Strepsiceros imberbis*. Compiled from Cook (1914).

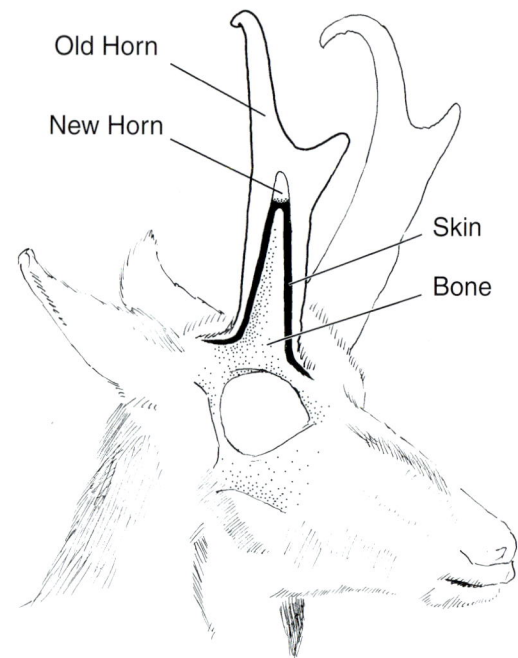

Figure 7.6 Pronghorn 'horns'. A pronghorn antelope, *Antilocapra americana,* just before the old horn is cast, which is in November in Montana, USA, showing the position of the new horn with its bony core and skin covering. Based on data in O'Gara and Matson (1975).

Giraffes

Giraffes (*Giraffa camelopardalis*), as the tallest animals in the world, are among the most well known of *all* mammals. A male standing close to 5.5 metres (16–18 feet), weighing almost 2000 kg and consuming over 60 kg of leaves per day is an amazing sight in the wild and a challenge to keep in captivity. Nine subspecies are recognized, the reticulated giraffe, *Giraffa camelopardalis reticulata,* and Rothschild's giraffe, *Giraffa camelopardalis rothschildi,* being the most frequent subspecies seen in zoos. The specific epithet comes from cameleopard, one of the early names for this animal thought to be part camel and part leopard. The Chinese named it 'K'i-lin' after a mythical animal resembling a unicorn. In the Koran it is the 'serafe' or lovely one.

The paired structures on the heads of giraffes and okapi are neither horns nor antlers, but *ossicones,* sometimes referred to as *knobs*. Ossicones, permanent bony outgrowths covered with vascularized skin, have fascinated biologists ever since Richard Owen (1841) published his paper on the anatomy of the Nubian giraffe, *Giraffa camelopardali typica*. Lydekker (1904) thought he could identify subspecies in part based on the horns.

According to Ganey *et al.* (1990) ossicones are not true horns because they lack a keratin sheath. Another complication when making comparisons is that giraffes have several sets of ossicones, the most prominent of which are those on the parietal bones, but with, in addition, a more anterior median horn on the frontal bone of all males and a few females – antlers develop on the frontal bones – an occipital ossicone on the occipital bone, orbital ossicones associated with the eyes, and an azygous ossicone just posterior to one of the orbital ossicones. Sexual dimorphism in ossicones relates to the mode of fighting; males use their heads as clubs, females kick with their legs. Constant awareness of their environment is achieved by sleeping for no more than half an an hour a day, the half hour spread over some five-minute 'naps' throughout the day.

Ossicones begin as nodules of bone beneath the skin, quite separate from the bone of the skull – like the human patella, they are sesamoid bones – but each ossicone later fuses with the underlying bone. The overlying skin is not shed, in contrast to the annual shedding of the velvet from antlers. Again, in contrast to antlers, ossicones are permanent and present in both sexes; in the Miocene moose-like Asian and African giraffids, *Sivatherium giganteum* and *S. maurusium*, the ossicones (horns) were almost a metre in length and palmate.

In one of the few recent studies, Spinage (1968) claimed that ossicones are present in foetal giraffes as cartilaginous precursors, basing this claim on Ray Lankester's study published in 1907. The line drawing Spinage provides as Figure 4 is not informative, although he describes ossification of the cartilage commencing after birth, beginning apically and progressing proximally, leaving a pad of cartilage at the base for subsequent growth – the latter based on Blumenbach's 1805 *Handbuch* and an 1872 paper by Murie, both of whom thought the ossicones developed endochondrally from this cartilage.

HORN AS A TISSUE

The principal component of horn, the one that gives it its characteristic physiological and physical properties, is *keratin*, or more properly, a keratin, for there is a whole family of these fibrous proteins. The two basic types of keratin – α and β – are differentiated from one another by the periodicity of their fibres: β-keratins are stretched with a periodicity of 0.33 nm; α-keratins have a periodicity of 0.51 nm. Mammals produce only α-keratins, birds and reptiles can produce both types.

The sulphur content of keratins can be very high, as for example in wool, and is responsible for the cross-linking of individual molecules by disulphide bonds; hence the flexibility of particular keratins. Keratin may also mineralize either lightly – as in human fingernails and rhinoceros horn – or heavily, as in the beaks of birds (Fig. 7.7) and baleen. Although horn is produced by specialized epidermal cells, *keratinocytes*, most epithelial cells can produce keratin. In fact, many epithelia that do not normally keratinize will do so if separated from the deeper layers of the skin. I have experienced this when CAM-grafting; exposing epithelia for only a few minutes initiates keratinization. Horns do to excess what many other epithelial cells do to a limited degree.

Figure 7.7 Rainbow lorikeets (*Trichoglossus haematodus*), photographed by the author in Currumbin, Queensland, in May 1965, illustrating keratinized beaks in action.

Baleen, one of the most unusual keratinized structures in the animal kingdom, consists of amorphous and α-keratin and hydroxyapatite. Baleen grows continuously throughout life as pairs of plates – as many as 350 pairs up to 4.5 metres long in the bowhead whale, *Balaena mysticetus*. Despite its fragile appearance, baleen has a breaking strength of 2–$9 \times 10^6 \, N^{-2}$, equal to the breaking strength of buffalo horn. Horses hooves, which are designed to withstand fractures, have tubular keratin with a 'work to fracture' of around $10\,000 \, J \, m^{-2}$, which compares with 1000–$3000 \, J \, m^{-2}$ for fresh bone and $10\,000$ for wood.[8]

DEVELOPMENT AND GROWTH OF HORNS

As non-deciduous, unbranched structures, incapable of regenerating, horns stand in stark contrast to antlers. The nature of the epithelial product that encases antlers is also fundamentally different from keratin. Horns are fundamentally epithelial, with a bony core. Antlers are fundamentally bony structures that for part of each year are covered with an epithelial covering, velvet. Although possessing a bony core and covered by an epidermal tissue, horns develop very differently from antlers.

The primordium of the bony core of a horn does not arise as an outgrowth from the frontal bone – as is the case with antlers – but rather as a separate bony ossicle, the *os cornu*, that develops within dermal connective tissue above but distinct from the nasal bone (Fig. 7.8). Extirpation of the bony ossicle does not prevent horn development, in contrast to the effect on antler development of the same operation. Instead, the epidermis stops forming hair and begins to synthesize horn, *after which* a new bony nodule forms. The implication is that primary control of horn development resides within the skin, not in bone, an implication that was demonstrated experimentally

Figure 7.8 Initiation of horn development. (A) A longitudinal section through the *os cornu* of a young goat. It is separate from the frontal bone below. (B) A single horn in an Ayreshire bull produced by fusing the two horn buds of the calf. Modified from Goss (1983).

in at least three ways:

- Removing the skin from horn-forming regions prevents horn development.
- Grafting an *os cornu* under the epidermis but elsewhere on the head does not initiate horn development.
- Grafting skin from the horn-forming region elsewhere on the body initiates horn development in that ectopic site.

Once initiated, the epidermis lays down layers of keratin fibres that accumulate and elongate as the bony core grows. Subsequent growth is *basal*, the older layers being pushed ahead of the new layers beneath. Basal growth means that a horn cannot change its shape during life, nor can it branch. Spiral growth results from asymmetrical growth on opposite sides of the horn (Figs 7.1, 7.2 and 7.4). Before a horn begins its development skin at the site is producing hair follicles. How hair follicle formation ceases and development of horn begins remains unclear.

No systematic information is available on histogenesis of the *os cornu* (Fig. 7.8). Whether osteogenesis is direct or endochondral is not known, although over a century ago Hans Gadow discussed the presence of cartilage (possible chondroid bone) in horn, and 80 years ago Atzenkern described islands of trabecular cartilage in goat and sheep horns, and cartilage surrounded by osteoblasts

in cow, sheep and goat horns.[9] Nor do we know the details of how the *os cornu* unites with the bone of the skull.

Being formed of layer upon layer of keratin fibres, horn is a dead tissue, but because of the underlying layer of proliferative keratinocytes, a horn can continue to grow throughout the lifetime of an individual. As horns are not shed, older or larger individuals have thicker or heavier horns than younger or lighter animals; an exception to the rule that horns are not shed and do not branch are the horns of the pronghorn antelope discussed above. In practice, horn growth is interrupted and episodic, being speeded up or slowed down by seasonal and annual cycles, sickness, pregnancy, and any trauma that initiates damage.

NOTES

1. See Goss (1983) and Lincoln (1992) for antlers, Gadow (1902), Modell (1969), Mitchell (1980), Kiltie (1985), Bubenik and Bubenik (1990) and Janis (1982) for horns, pronghorns and antlers, and Ganey *et al.* (1990) for giraffe ossicones.
2. From an address by the Imperial Academy to the Emperor of China, penned in 1413, one year before the first k'i-lin arrived at the Imperial Court. Cited from Boorstin (1983), p. 198. K'i-lin was the Chinese name for a legendary animal resembling a unicorn. See Lagueux (2003) for an account of the giraffe that the Pasha of Egypt sent to King Charles X of France 413 years later.
3. See Gadow and Geist (1966a,b) and Janis (1982, 1986) for evolutionary, ecological, behavioural and life history aspects of horns.
4. See Arrow (1951) for horned beetles, Geist (1966a) for the evolution of horn-like structures, and Emlen and Nijhout (1999, 2000) and Emlen (2000) for analyses of integrated developmental and evolution studies on beetle horns, especially in the dung beetle, *Onthophagus*, which is polymorphic, populations consisting of large horned males and small sneaker males that lack horns. One of the contexts of these studies, known to Charles Darwin, but long forgotten, is the allocation of developmental resources from one feature to another, enlargement of horns (or wings) coming at the expense of other tissues (Klingenberg and Nijhout, 1998; Nijhout and Emlen, 1998). We have here a potential developmental mechanism for the phenomenon of scaling discussed in Chapter 32. In the hyperossified horned frog, *Ceratophrys cornuta*, the chondrocranium lacks the frontoparietal fontanelle (which may be unique among frogs) and a plate of cartilage covers the skull (Wild, 1997).
5. See Packer (1983) and Kiltie (1985) for horns in females and relationship to body size.
6. See Mitchell (1980), Bubenik and Bubenik (1990) and Prothero and Schocho (2002) for the general biology of pronghorns.
7. See O'Gara *et al.* (1971) and Mitchell (1980) for the annual shedding cycle, Mitchell (1980) and Kiltie (1985) for bony knobs in female pronghorns, and O'Gara and Matson (1975) for phylogenetic relationships.
8. See D. J. St. Aubin *et al.* (1984) for baleen, and Bertram and Gosline (1986) for horse hooves.
9. See Gadow (1902) and Altzenkern (1923).

Antlers

Loss of vasculature when an antler is shed is comparable to the loss associated with shedding the placenta at birth. The growing antler bud is so highly vascularized that it may be hot to the touch.

Antlers are the many-branched, paired, cranial appendages of deer, moose, elk, caribou and reindeer of the family Cervidae.

ANTLERS

Antlers are used in combative interactions between males. These are not casual encounters; the frequency of wounding in such interactions is high. In male caribou and reindeer, *brow tines* (literally, the points or prongs on the forehead) are asymmetrical – left-dominant animals having a shovel-shaped tine, right-dominant animals a spiked tine – an adaptation though to permit avoiding excessive contact during interactions. Like horns, antlers establish social status, as evinced by several lines of evidence, notably the:

- correlation between antler weight and dominance hierarchy;
- negative correlation between social rank and the order in which antlers are cast, dominant males retaining their antlers longer than submissive male white-tailed deer, *Odocoileus virginianus*; and that
- removing antlers from farmed red deer, *Cervus elaphus*, produces only slight alterations in the dominance hierarchy, although farmed deer are not the ideal animals upon which to examine dominance hierarchies.[1]

Antlers are bony structures covered for part of the year by velvet, a specialized skin of fine velvety hairs. Once the antler is fully formed each year, the velvet is lost and the bony antler dies, as it is no longer vascularized. Antlers bud and branch as they grow. Antler development is progressive and incremental, each passing year bringing a larger and more branched sets of antlers (Fig. 8.1). Antler growth, unlike that of horns, is apical (terminal); surgical removal of the apex of an antler prevents further growth (Goss, 1961). Most temperate deer shed their antlers annually. Other deer, including the Indian sambar, *Rusa unicolor*, can carry a single set of antlers for several years. Deer at the equator carry one set of antlers for life. How they manage not to shed their antlers is discussed below.

Except for reindeer and caribou (*Rangifer tarandus platyrhynchus* and *R. t. caribou*), antlers are found only on male deer. Male reindeer shed their antlers in late November. The antlers of female reindeer are smaller, less branched, retained over the winter, and shed in the spring. Female wapiti, *Cervus canadensis*, and female muntjak, *Cervus moschatus*, have sharp bony knobs on their skulls where the males have antlers.

Size and absence

The ice age Irish elk, *Megaloceros giganteus*, had truly massive antlers, as much as 4.3 metres across, and weighing 50 kg or more, yet its antlers were not the largest relative to body size; that distinction goes to the European fallow deer, *Dama dama*, and caribou, *Rangifer tarandus caribou*, each of which has eight grams of antler/kg body weight. Various subspecies of caribou also have the largest antlers in absolute as well as relative terms (Table 8.1). The smallest antlers in terms of absolute size are the single spikes of the tufted deer, *Elaphodus cephalophus* (Fig. 8.2). At 0.6 g/kg body weight, the pudu has the smallest relative antler size (Table 8.2).

Loss of the upper canine teeth – which can develop into tusks – accompanied the evolution of antlers. An exception is the muntjak or Barking deer, *Muntiacus muntjak*, of India,

China and Borneo, which has both tusks and small antlers, usually no more than a spike or little branched (Table 8.2). Evolution of the Cervidae is characterized by reduction in chromosome number from the basal number of 70/diploid nucleus – the number still found in the subfamily Odocoileinae – to 68 in the Cervinae, 46 in the Muntiacinae, and an extreme reduction to six in female Indian muntjak, *Muntiacus muntjak vaginalis* (Fontana and Rubini, 1990).

Two species of extant deer lack antlers altogether, the males retaining their upper canines, which are modified into permanently growing tusks. The two are the musk deer, *Moschus moschiferus* (Fig. 8.2), from central Asia and the Himalayas, and the Chinese water deer, *Hydropotes inermis*, from China and Korea. Both are reminiscent of Miocene deer, which were tusked and lacked antlers.[2]

INITIATION OF ANTLER FORMATION

In their first year, fawns develop a bony platform or *pedicle* on the frontal bones of the skull. In the second year, a spike-like antler develops on the pedicle. This antler is shed and replaced in the third year by a branched antler that in turn is lost and replaced by a set of even more branched antlers in subsequent years (Fig. 8.1).

Pedicle formation

A variety of experimental approaches allow us to conclude that the ability to initiate pedicle development lies within the frontal bone, from which the skeletal tissue of the antler arises, rather than within the overlying epidermis, from which the velvet develops:

- Neither pedicle nor antler forms if the bony protuberance on the frontal is removed from a fawn that has not produced its first set of antlers.
- Transplanting the frontal bone to an ectopic site under the epidermis – either elsewhere on the head, or onto a

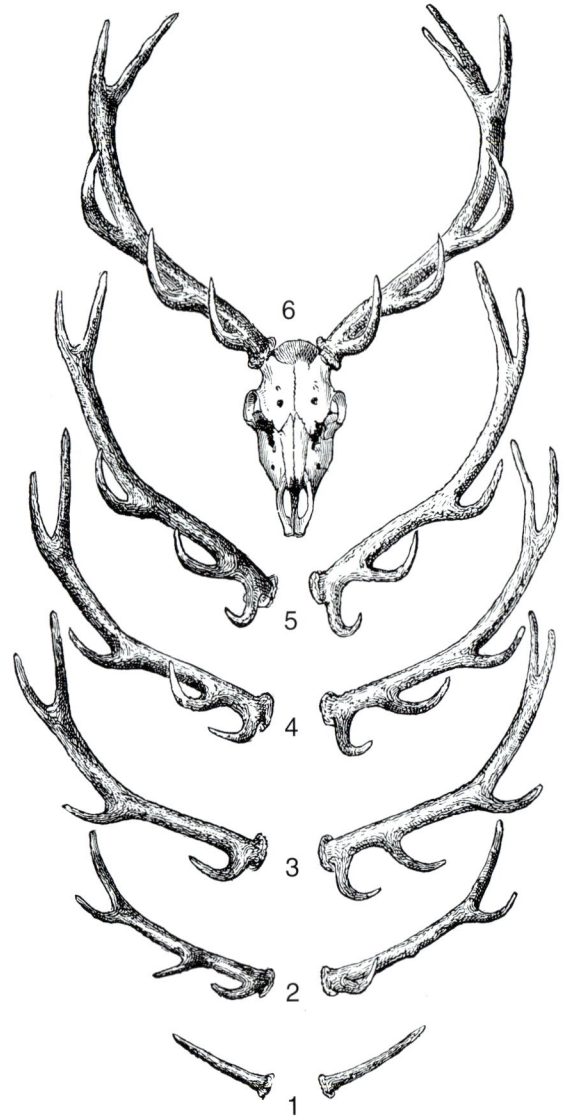

Figure 8.1 Growth and branching of the antlers of a stag shown over the first six years of life (1–6). From Romanes (1901).

Table 8.1 World records for antlers[a]

Species	B&C score[b]	Length of main beam[c]		Number of points[c]	
		Right	Left	Right	Left
Caribou					
Barren ground (Grant's) caribou (*Rangifer tarandus granti*)	477	55 5/8	57 7/8	17	25
Quebec–Labrador caribou (*Rangifer tarandus caribou*)	474 6/8	60 4/8	61 1/8	22	30
Woodland (mountain) caribou (*Rangifer tarandus caribou*)	453	48 2/8	49 5/8	23	21
Central Canada barren ground caribou (*Rangifer tarandus caribou*)	433 4/8	48 7/8	49 4/8	29	26
Moose/Elk					
American elk (*Cervus canadensis*)	442 5/8	56 2/8	56 2/8	6	7
Alaska–Yukon moose (*Alces alces gigas*)	261 5/8	54 4/8	53 6/8	19	15
Canada moose (*Alces alces americanus*)	242	44 5/8	45	15	16
Wyoming moose (*Alces alces andersoni*)	205 4/8	38 6/8	38 5/8	15	15
Deer					
Rocky Mountain mule deer (*Odocoileus hemionus*)	213 5/8	28 4/8	28 4/8	8	6
Mule deer (*Odocoileus hemionus*)	226 4/8	30 1/8	28 6/8	6	5

[a] Source, Boone & Crockett Club web site (HTTP://WWW.BOONE-CROCKETT.ORG) and see Table 7.2 for comparable data for horns.
[b] See Box 7.2 for how B&C scores are determined.
[c] Being official B&C point records I have not translated these Imperial measures (inches) to metric units. For moose, the measure is of the right and left palm.

Figure 8.2 Antlers and tusked canines do not coexist in deer as shown in (A) the red deer, *Cervus elaphus*, which has antlers and no tusks; (B) the tufted deer, *Elaphodus cephalophus*, which has small tusks and short antlers; and (C) the musk deer, *Moschus moschiferus*, which has prominent tusks and no antlers. Modified from Goss (1983).

Table 8.2 A comparison of relative antler mass among deer[a]

Species	Antler mass (g/kg)[b]
Dama dama (fallow deer)	8.0
Rangifer tarandus tarandus (caribou/reindeer)	8.0
Cervus elaphus hippelaphus (wapitis/red deer)	6.3
Axis axis (Indian chital/spotted deer)	5.6
Elaphurus davidianus (Pere David's deer)	5.6
Rusa timorensis (Timor deer)	5.1
Rucervus eldi (Eld's deer)	4.9
Rucervus duvauceli (Barasingha deer)	4.9
Przewalskium (white-lipped deer)	4.9
Cervus elaphus canadensis (elk)	4.5
Capreolus capreolus (roe deer)	4.4
Blastocerus dichotomus (marsh deer)	4.4
Alces gigas (Alaska) (moose)	4.4
Capreolus pygarus (Siberian roe deer)	4.2
Odocoileus hemionus (mule deer)	4.2
Przewalskium albirostris (Thorold's deer)	4.2
Cervus nippon (sika deer)	3.5
Rusa unicolor (sambar deer)	3.2
Odocoileus virginianus (white-tailed deer)	3.0
Alces alces (Europe) (moose)	3.0
Muntiacus reevesi (Reeve's muntjac)	2.4
Odocoileus bezoarticus (pampas deer)	2.3
Mazama americana (Brocket deer)	1.0
Pudu mephistopheles (pudu deer)	0.6

[a] Based on data in Geist (1998).
[b] Grams of antler mass per kilogram of body mass raised to the power of 1.35.

metacarpal – results in initiation of pedicle formation in that ectopic site.

- Discs of frontal-bone periosteum transplanted under leg skin induce pedicle bone and antlers, which are subsequently shed.

- Grafting periostea with or without the overlying skin from frontal bones of red deer, *Cervus elaphus*, into nude mice shows that a skin – periosteal interaction is required for antlerogenesis.[3]

The periosteum of the frontal bone imposes polarity on the developing antler, as demonstrated by Goss (1991) who rotated, inverted or 'minced' periostea to show that antero-posterior (A-P) polarity is determined by the periosteum and proximo-distal (P-D) polarity by the overlying skin.

Testosterone is essential for pedicle formation. The bony knob on the frontal that presages development of the pedicle is present in male but not female foetal red deer, *Cervus elaphus*. The knob's appearance and development over gestation days 60–100 parallel testicular development and testosterone accumulation. No further development occurs until the fawn receives a photoperiodic cue. If fawns are castrated before antlerogenesis commences, the pedicle fails to form and the first set of antlers do not develop; a pedicle begins to form, however, if such castrated fawns are given testosterone. If in order to provide a wound stimulus, this pedicle is amputated, an antler will start to develop.[4]

The pedicle arises during the first year of life. Pedicles are not seen in foetal or newborn reindeer. Rather, the epidermis becomes folded over the flat plate of the frontal bone during late foetal life and a bony knob begins to grow several days after birth.[5]

The antler bud and dermal–epidermal interactions

During the formation of the first set of antlers – and in subsequent antler regeneration/shedding – the position of the pedicle depends upon properties of the frontal bone. Formation of the antler bud on the pedicle depends upon interactions between pedicle and epidermis, and perhaps also on interactions with the dermis.

If the pedicle itself is removed, formation of an antler bud, and hence of antlers, is prevented. If only the bone of the pedicle is removed, however, the wound heals, and antlers begin to develop. Thus, while pedicle formation is initiated by the frontal bone, antler formation is not initiated by bone of the pedicle. Instead, however, the antler develops from dermal connective tissue immediately overlying the pedicle rather than from the pedicle itself. Antler bud development is initiated as a result of interaction between the epidermis and the dermal covering of the pedicle.[6]

Thickened epidermis and extensive epithelial down growths are seen, beginning at day five into the regeneration. Both features are reminiscent of changes seen in regenerating rabbit ears (Box 25.1), and after holes are punched in the external ears of mice of the MRL strain, the one strain that can regenerate holes in its ears, including regenerating the cartilage.[7]

Mammalian dermal tissue normally responds to wounding by forming scar tissue.[8] Absence of this response on the pedicle allows antlers to form and regenerate. If head, limb or ear skin is transplanted over the pedicle, an antler covered with velvet will develop, i.e. the epidermal requirement for initiation of antler development is not specific. Regional specificity of the *dermal* component has not been thoroughly tested. Trauma to the frontal bones of female deer, followed by electrical stimulation of the antler nerve or periosteum, results in a 70 per cent increase in antler length and a 40 per cent increase in antler weight. Antlers in control animals grew to 278 mm with four 'points'; stimulated antlers grew to 338 mm and had six points.[9]

HORMONAL CONTROL OF PEDICLE DEVELOPMENT AND GROWTH

Hormones stimulate division, accumulation and differentiation of populations of skeletal cells. Since many hormones influence skeletogenesis, an analysis of the hormonal control of antler growth might shed (no pun) light on the nature of these tissues.

Sexually dimorphic elements of the skeleton are particularly dependent upon gonadal hormones for their differentiation, morphogenesis and growth. Control of antler growth by testosterone (below), transformation of female murine symphyseal cartilages into an interpubic ligament under oestrogen control (Chapter 23), resorption of the pelvic girdle under the influence of oestrogen, and deposition of medullary bone in egg-laying hens (Chapter 25), are four examples of hormones regulating site- and gender-specific responses from skeletal tissues. A major subject for future investigation is the basis for such regional specificity.

Furthermore, hormones and vitamins interact. Thus, while vitamin C increases osteoblast formation, deposition of osteoid and proliferation of cartilage, and parathyroid hormone decreases osteoblast function, when used in combination they stimulate deposition of large amounts of bone within marrow cavities of explanted mouse radii (Gaillard, 1974). This balance between deposition and resorption controls bone turnover, and determines whether bone will accumulate (grow) or be lost (atrophy).

Evidence of testosterone control of pedicle growth comes from direct experimental analysis: pedicles fail to form in male fawn castrated before the onset of development of the first set of antlers. This is not from trauma associated with the surgery but through lack of testosterone; administering testosterone to castrated fawn elicits a normal pedicle and antler development. Further evidence for the directive role of testosterone comes from experiments in which female deer are induced to produce small pedicles in response to exogenous testosterone. The normal inability of these animals to form antlers is related to inhibition from the circulating female sex hormone oestrogen; administering oestrogen to males stops pedicle and antler development.

ANTLER REGENERATION

Examination of the initiation of the first or subsequent set of antlers shows how precise interactions involving specific attributes of the frontal bone, its bony pedicle, and the overlying epidermis all participate in localizing and initiating antlers on the head.

The shedding cycle

Mature antlers are solid, osseous appendages lacking vascularization and innervation. Early in the spring, after the mating season, and over the course of just a few days, osteoclasts erode the base of the antlers at their junction with the pedicle (Fig. 8.3). Whether the osteoclasts are local in origin or from migratory cells that 'home' to the base of the antler is unclear. Similarly, we do not know what stimulates osteoclasts or osteoclast precursors to accumulate at the base of the antler or what causes them to act, although testosterone may trigger resorption. As determined in male red deer, *Cervus elaphus atlanticus* (Fig. 8.2), an annual peak of growth hormone follows the casting of the old set of antlers, and may trigger development of the new set.[10]

Shedding the antlers leaves an exquisitely sensitive wound on the pedicle, with bone, blood vessels and nerve endings all exposed (Fig. 8.4). Indeed, antlers have been described as a 'disposable vascular bed,' the loss of vasculature at shedding comparable to loss of the umbilical-vascular bed at birth in placental mammals. The growing antler bud is so highly vascularized that the bud may be hot to the touch.[11]

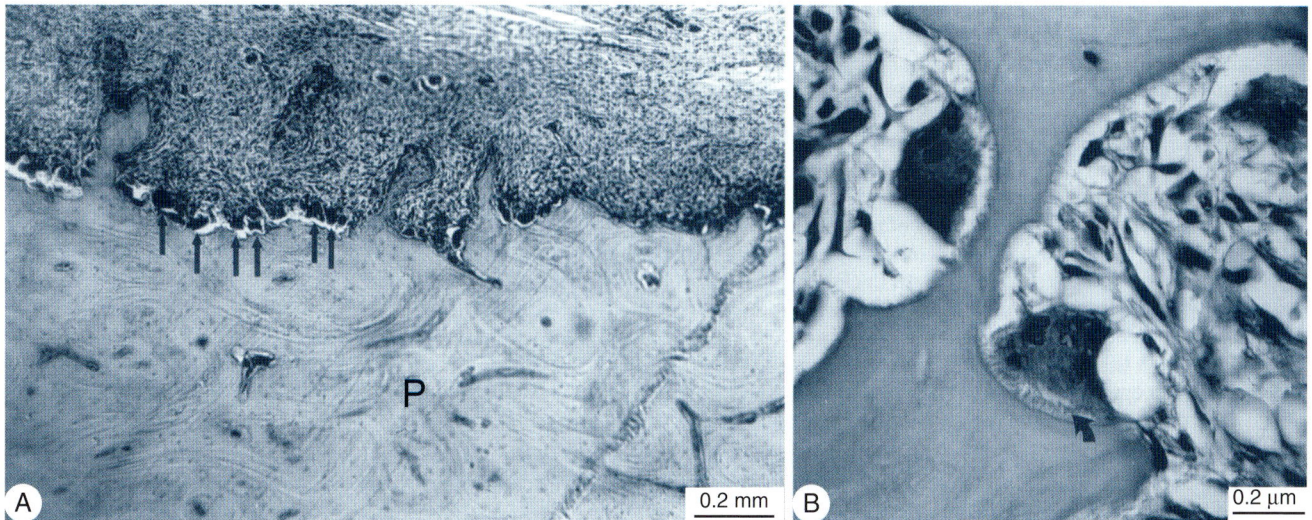

Figure 8.3 Shedding of the antlers in the European fallow deer, *Dama dama*. (A) A histological section of the pedicle (P) 14 days after castration, showing that the pedicle is lined by osteoclasts (arrows) resorbing the bone. (B) A higher magnification of osteoclasts on the surface of the pedicle 18 days after castration. The ruffled border of one osteoclast is marked with the arrow. Modified from Goss *et al.* (1992).

Regeneration in the animal kingdom always requires a *wound stimulus*. Regrowth of antlers is no exception, the wounded pedicle exposed by the casting of the previous set of antlers serving as a vital stimulus for renewal of antler growth. This was convincingly demonstrated in congenitally polled ('hummel') red deer, which lack antlers, have pedicles, and can be induced to form a set of antlers if the tips of the pedicles are amputated. Skin overgrows the pedicle to form a flat antler bud that is both highly vascularized and heavily innervated. Rapid growth of this antler bud over the spring and summer produces the next season's antlers (Fig. 8.4). Similarly, castrated deer, which form pedicles in response to injections of testosterone, will only complete the cycle and form antlers if the pedicles are amputated. This response by the dermal tissues over the pedicle to the wound stimulus is unusual; if other parts of the dermis are wounded, the dermal connective tissue forms scar tissue, which inhibits regeneration.[12]

HISTOGENESIS OF ANTLERS

Histogenesis of antlers illustrates the ability of cells to produce a tissue having features of cartilage and bone. It also indicates that the process(es) of ossification need not be the strict endochondral or intramembranous sequence described in most textbooks.

The sequence of cellular events that results in the production of such a massive portion of the skeleton is of special interest for several reasons.

Antler development is rapid, antlers being among the fastest growing – if not the fastest growing – of skeletal tissues (Table 8.3). Antlers of American elk, *Cervus canadensis*, can grow as much as 2.75 cm/day during the March to August growing season when growth hormone reaches an annual peak, a staggering level of osteoblastic and osteocytic activities.[13]

Mobilization of Ca^{++} and PO_4^{3-} from other parts of the skeleton for antler mineralization is another cause for interest.

Thirdly, the sequence of cellular events in antlerogenesis includes *chondrogenesis and osteogenesis*, as well as some replacement of cartilage by bone, as occurs in long bones. However, the cartilage of developing antlers is so highly vascularized that the *process of endochondral ossification is unusual*; some bone forms directly (*intramembranously*) without replacement of cartilage. In other regions of the antler, bone arises by *metaplasia* of cartilage.

When implanted *in vivo* in diffusion chambers, both the pedicle and the first antler tissue will form trabecular bone irrespective of whether the tissue was chondrogenic or osteogenic when implanted. Consequently, Li *et al.* (1995) suggest that the default pathway is osteogenesis, the cartilage in antlers essentially being secondary cartilage and extrinsically induced. A cartilaginous tissue with osteogenesis as the default pathway is unusual indeed. I do not think, however, that this is unusual in the animal kingdom. It seems unusual to us because we have studied so little of what constitutes skeletogenesis in the animal kingdom, expecting all vertebrates to conform to the mammalian pattern discussed in most textbooks.

The nature of the cellular processes associated with antler development has been a source of debate for over 150 years.[14] Debated questions include: does a cartilaginous phase characterize antler formation? Is the ossification process typically endochondral or does metaplasia of cartilage to bone occur? Or is the tissue only 'cartilage-like' and not true cartilage? Finally, does neither cartilage nor chondroid form, in which case ossification would be

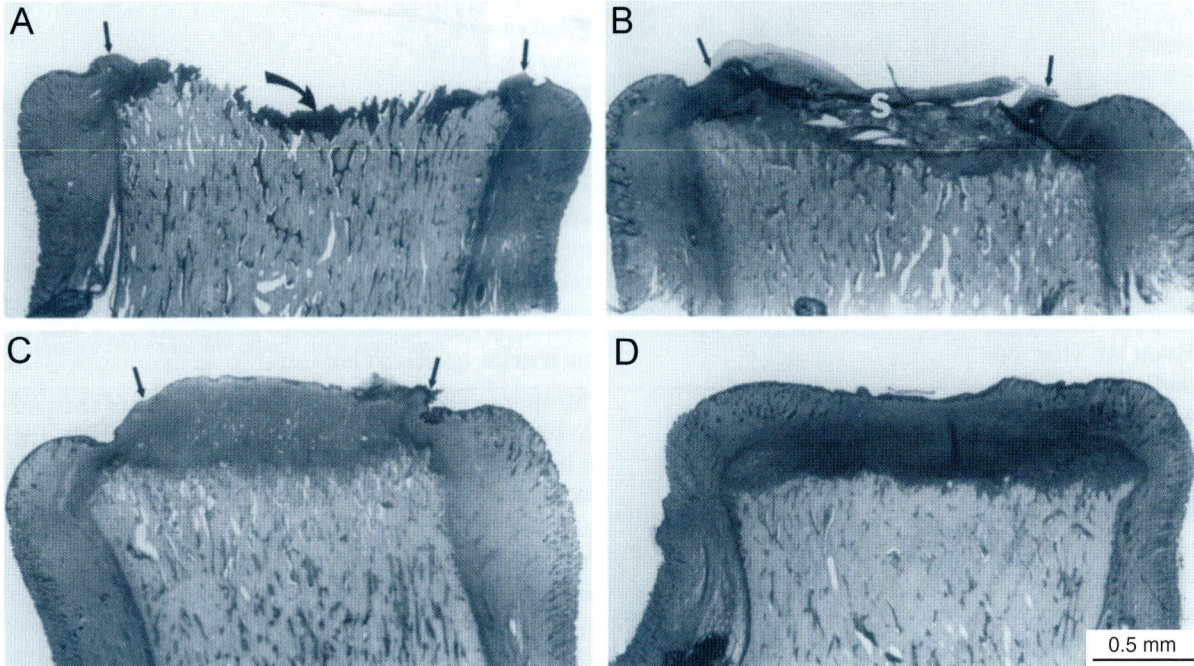

0.5 mm

Figure 8.4 Shedding of the antlers in the European fallow deer, *Dama dama*, shown as histological sections through the pedicle. (A) The pedicle immediately after the antler was cast is partly covered with dense connective tissue (curved arrow). The epidermis (arrows) has begun to migrate around the edges of the pedicle. (B) One or two days after the antler was cast a 'scab' (S) covers the wound surface and the epidermis (arrows) now covers the periphery of the pedicle. (C) Four days after the antler was cast, mesenchymal cells now cover the pedicle and hair follicles are differentiating in the peripheral epidermis. The arrows mark the extent of epidermal migration over the surface. (D) One week after the old antler was cast epidermis covers what is now an antler bud of mesenchymal cells. Modified from Goss *et al.* (1992).

Table 8.3 Maximum daily growth rates (cm/day) and final sizes (cm) of antlers from various species of deer[a, b]

Species	Final antler length (cm)	Maximum growth rate (cm/day)	Growth period (months)
Cervus canadensis (American elk)	129.5	2.75	March to August
Cervus dama (fallow deer)	64.0	0.68	April to September
Cervus nippon (sika deer)	42.2	0.37	April to September

[a] Antler length is variable between species, but the growth period (months) is constant.
[b] Based on data in Goss (1970).

entirely intramembranous? Consequently, tissues of developing antlers are an elegant model for testing the veracity of distinctions between endochondral and intramembranous ossification and between vascular bone and avascular cartilage, and for comparing and contrasting the appositional growth of bone with the combined interstitial and appositional growth of cartilage.

Such basic issues as whether antler ossification is intramembranous or endochondral may have lain unresolved for so long because the processes of antlerogenesis vary from species to species. Consequently, I discuss the histogenesis of antler formation species by species, commenting on similarities and differences.

White-tailed deer, American elk, European fallow and roe deer

Modell and Noback (1931) described various cellular zones in the growing antlers of three species, the Virginia or white-tailed deer, *Odocoileus virginianus*, the American elk or wapiti, *Cervus canadensis*, and the European fallow deer, *Dama dama*. Remembering that antlers *grow apically* (Fig. 8.5), the zones comprise:

- mesenchymal cells at the antler tip followed, progressively more basally, by
- osseous tissue,
- a zone in which preosseous tissue is degenerating, and,
- most basally, a zone of ossification (Fig. 8.5).

You will note that cartilage or a cartilaginous zone are neither listed nor illustrated in Fig. 8.5. Modell and Noback used the term *preosseous* to denote a zone of tissue that *precedes* the laying down of bone rather than a zone of bone-forming tissue. They described the tissue as somewhat resembling hyaline cartilage when analyzed histologically and as a supporting tissue that played the role of cartilage but was not itself cartilaginous.

These cartilage-like cells degenerate further toward the base of the antler, leaving behind intact longitudinal columns of matrix. Modell and Noback employed the term 'fibrils.' These fibrils provide the scaffolding upon which osteoblasts settle before depositing osteoid and

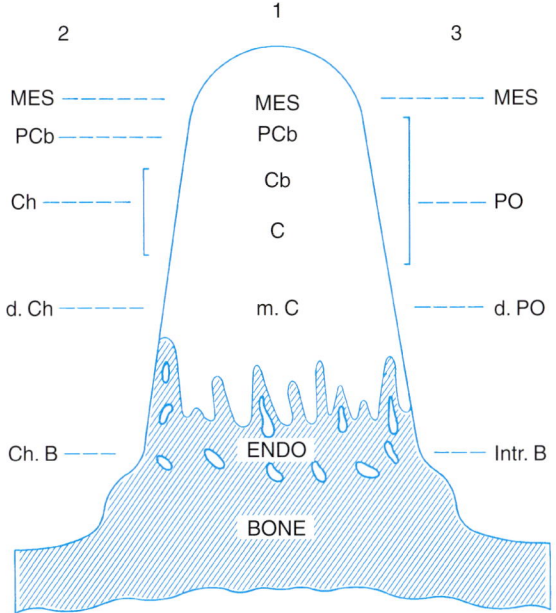

Figure 8.5 Three versions of the major cellular zones of an antler, shown as a longitudinal section of a developing antler bud. C, chondrocytes; Cb, chondroblasts; Ch, chondroid; Ch. B, chondroidal bone formation; d. Ch, zone of degenerating chondroid; d. PO, degenerating preosseous tissue; ENDO, endochondral bone formation; Intr. B, intramembranous bone formation; m. C, mineralizing chondrocytes; MES, mesenchymal tissue; PCb, prechondroblast zone; PO, preosseous zone. The three interpretations are by (1) Banks (1974), (2) Wislocki et al. (1947), and (3) Modell and Noback (1931). Reproduced from Hall (1978a).

bone. Because this scaffold was not regarded as cartilage, the ossification process was regarded as intramembranous, even though the bone was deposited within a fibrillar extracellular matrix rather than in a membrane. An interesting parallel is seen in foetal rat parietal bones *in vitro*. If cultured in a chemically defined medium in the presence of 1 mM phosphate, osteoid is deposited. Addition of 3 mM phosphate is sufficient for the osteoid to be mineralized. In the presence of 6 mM phosphate, mineralization extends to the periosteum in what is, to all extents, ectopic mineralization.[15]

Wislocki et al. (1947) examined the antlers of the white-tailed deer using histochemical procedures to show that the capsules of the cells in the preosseous zone are strongly metachromatic, with pronounced affinity for basic histological stains (basophilia), usually indicative of the high acid mucopolysaccharide content of cartilage. Histologically, the tissue was assessed as neither hyaline nor elastic cartilage, but as fibrocartilage or chondroid. Because the preosseous zone was cartilage-like and not cartilaginous, Wislocki and colleagues thought that the process of ossification was neither typical of long bones nor of intramembranous ossification (Fig. 8.5). They coined the term '*chondroidal bone formation*,' which they regarded as *intermediate between intramembranous and endochondral bone formation*.

Kuhlman et al. (1963), on the other hand, described ossification as occurring on preosseous fibrocellular columns.

The process is intermediate only in the sense that the tissue being replaced is not the cartilage typically seen in long bones; for example, protein–polysaccharides decline substantially as cartilage is replaced during endochondral ossification in mammals. This particular type of cartilage is similar to the secondary cartilage found on such membrane bones as the mammalian dentary or the avian quadratojugal (Chapter 18). Furthermore, early phases of intramembranous ossification are normally associated with an *increase* in glycosaminoglycans (GAGs), molecules normally thought to be associated with cartilage rather than bone. This is why bone and osteoid stain with 'cartilage stains' such as methylene blue in whole mount preparations.[16]

Bone of the antlers of the American elk, *Cervus canadensis*, was described by Modell (1969) as equivalent in its histology and rapid development to malignant bone sarcoma. He denied the existence of a cartilaginous phase, describing instead ossification on a fibrous framework, much as Modell and Noback had done 38 years before.

Studies undertaken on the histogenesis of the antlers of the white-tailed deer in the 1970s, using light and electron microscopy and biochemical techniques, supported the cartilaginous nature of the preosseous tissue. Banks (1974) described the replacement of cartilage by bone (and growth of the antler) as a process of modified endochondral ossification. Figure 8.5 shows the arrangement of the major zones as described by Banks (1974), along with the interpretations of Wislocki and Modell. The cartilage is arranged in columnar fashion, as in the growth plate of a long bone.

As with the secondary cartilages found on membrane bones, there is scant extracellular matrix, and the cartilaginous columns are separated by highly vascularized spaces. The highly vascularized nature of the cartilage was a stumbling block to its correct identification. The cartilaginous columns mineralize and are replaced by columns of mineralized bone occasionally containing islands of chondrocytes embedded within the osteoid. Mineralization of the cartilage is associated with matrix vesicles and microfibrils with a hyaluronan core.[17]

Subsequent light microscopical analysis of the antlers of the European fallow deer, *Dama dama*, confirmed the distinct zones described 60 years earlier, and documented a modified form of endochondral ossification that begins apically and can be contrasted with the intramembranous ossification that ensues from the sides of the antler bud. The endochondral ossification does not involve any metaplasia of cartilage to bone, as Modell and Noback's interpretation required. Indicative of species differences, ossification of Lapland reindeer antlers involves invasion of cartilage lacunae by osteoblasts at the edge of the cartilage, and metaplasia of cartilage to bone centrally within the cartilage.[18]

Most recently, Kierdorf and colleagues (2003) reexamined histogenesis immediately after the old antlers has been cast in both roe and European fallow deer (*Capreolus capreolus* and *Dama dama*), and described initial intramembranous ossification, *followed by* formation of cartilage at the distal tips and replacement of the cartilage endochondrally. They thought the thickened cambial periosteal layer of ther pedicle contained osteo- and chondroprogenitor cells and was the source of the bone and cartilage that formed, as is typical of secondary chondrogenesis from membrane bones (Hall, 2004a, and see Chapter 12).

Rocky Mountain mule deer

Examination of antler histogenesis in the Rocky Mountain mule deer, *Odocoileus hemionus*, by light and electron microscopy, shows *cartilage as the primary skeletal tissue*. Mineral accumulates within mitochondria of those chondrocytes immediately distal to the zone of matrix mineralization, i.e. mineralization is initiated intracellularly, as is the mineralization of the cartilages of long bones, in contrast to the matrix vesicles and microfibrils in white-tailed deer described above.[19]

Histological and biochemical analysis of mature antlers during the rapid growth phase demonstrated chondroitin-4- and chondroitin-6-sulphates and hyaluronan in the terminal cap of reserve cells, and synthesis and accumulation of GAGs in the subjacent cartilaginous zone. Hyaluronan (Box 4.1) and GAGs interact in cartilage; chondrocytes have cell-surface receptors for hyaluronan that mediate GAG synthesis and form part of the chondrocyte pericellular coat. Indeed, various cell types can deposit an organized chondrocyte-like pericellular matrix following activation of cell-surface hyaluronan receptors.[20]

Sika deer

A proliferative zone has been described at the tip of the antlers of sika deer, *Cervus nippon*. Cells exit this zone to differentiate as chondrocytes, which then undergo metaplasia into the bone of the mature antlers (Goss, 1969a, 1970). Further studies on this species are required.

HORMONES, PHOTOPERIOD AND ANTLER GROWTH

Even if symmetrical, any paired structures will show slight deviation between left and right members of the pair. Such variation may be genetically controlled, as in left and right directions of coiling in snail shells, or it may be random, in which case it is known as *fluctuating asymmetry* (Box 8.1). The lower the level of fluctuating asymmetry between developing paired structures – antlers, hind limbs, humeri – the more constrained and the less variable development is thought to be. The left and right antlers in fallow deer, *Dama dama*, show fluctuating asymmetry, correlated not with population density or

individual body size, but with maximal cohort body size. Fluctuating asymmetry decreases in fallow deer older than two years, a decrease that has been interpreted as resulting from increased sexual selection with age.[21]

In red deer, antler size correlates with an increase in lifetime breeding success, antler mass having a heritability component of 0.33. Despite this finding – which was based on almost 30 years' data, and which is one of the more robust correlations between a secondary sexual character and lifetime breeding success obtained for any mammal – and because of environmental influences such as nutrition, there is no positive genetic correlation between antler size and fitness (Kruuk *et al.*, 2002).

In June, coincident with peak antler growth in fallow deer, remodeling of the ribs, metacarpals, metatarsals and tibial Haversian systems also peaks. Antler growth comes at a cost, as the needs of the antlers take precedence. In reindeer, for example, bone is lost from the skeleton during antler growth, though reindeer chew on the cast antlers so that at least some of the mineral is not wasted. (Paradoxically, female reindeer probably evolved the ability to form antlers for access to food – antlers are used to scrape away snow to reveal the vegetation beneath.) Even when deer are kept on a high calcium diet, cortical bone is mobilized from the ribs; bone density, and calcium, phosphorus and magnesium levels all decrease in ribs sampled when antler mineralization is taking place. This *cyclic physiological osteoporosis* is thought to provide calcium and phosphorus for antler mineralization; indeed, mineralization of antlers may be a good model for the metabolic bone disease osteoporosis (Box 8.2).[22]

Photoperiod and testosterone

Variations in the endogenous level of testosterone mediate pedicle and antler growth. As amply demonstrated by Richard Goss, these levels are led by local exogenous photoperiod, alternating periods of dark and light being critical for initiation of antler replacement. Shedding is controlled by photoperiod – progressive, seasonal changes in day length – as demonstrated by Goss in studies in which the photoperiod was reversed, or day length altered.[23]

What about tropical deer, which normally experience far less variation in day length? Even after being moved to a temperate climate, tropical deer such as the axis or Indian deer, *Axis axis*, still fail to display seasonal synchrony of antler shedding. Instead, shedding is synchronized to the testis cycle, which varies from individual to individual, meaning that on a population level shedding is asynchronous. Essentially, the photoperiodic control coupled with the annual testosterone cycle in temperate deer is missing from such tropical deer (Loudon and Curlewis, 1988).

Two tropical deer, the southern pudu, *Puda puda*, and the roe deer, *Capreolus capreolus*, show two peaks of

Box 8.1 Fluctuating asymmetry

Fluctuating asymmetry (FA in the trade) is random deviation from the expected symmetry of paired left and right features of an organism.

Normally we expect paired organs (eyes, antlers, tibiae, digits) to be equal in size and/or shape, attributing any deviation from equality to instability of development, or environmental disturbances (which may cause deviations in development). Deviations with a functional basis (the serving arm of a tennis player) or a genetic basis (eye rotation in flatfish, flounder and halibut; the asymmetry of the ears in Tengmalm's owl, *Aegolius funereus*, that facilitates discriminating sound frequencies between 10 000 and 16 000 cycles a second; the asymmetrical and single tusk of the narwhal, *Monodon monoceros*; the crushing claw of a lobster; left- or right-handed coiling of snail shells) are examples of biased or directed, not fluctuating asymmetry. So too is the growth of the left and right portions of the syrinx (the lower portion of the larynx) in ducks. M. M. Takahashi and Noumura (1993) co-cultured left and right halves of the duck syrinx with and without oestrogen treatment and found that a soluble factor from the left syrinx from male ducks inhibits growth and chondrogenesis of the right syrinx.

Although much is known about variation and variability of skeletal elements (see Chapters 32 and 44), few *reports* document FA in skeletal elements. Ferrario *et al.* (1997) reported FA in the shape of the condyle between left and right mandibles in a sample of 20 men and 20 women. That even fewer detailed *analyses* of FA exist is due in part, I am sure, to the far-from-trivial morphometrics and statistics involved. FA is not for the faint of heart. If you are statistically maladroit, collaborate with someone morphometrically and statistically adroit![a]

On examining the FA of a large sample of mammalian mandibles, as well as rhesus monkey, *Macaca mulatta*, and human skeletons, Hallgrímsson (1998, 1999) found that FA variance increases as development or ontogeny proceeds, increasing to a greater extent in slow-growing than in rapidly growing animals. Despite analyses directed to this issue, it is not clear whether this finding is explained by (i) the cumulative effect of asymmetrical mechanical factors, (ii) an accumulation of variation in local growth regulation, (iii) inherent features of bone morphology, or (iv) some combination of all three, the precise combination being specific to individual circumstances.

Sexual dimorphism of facial growth in rhesus monkeys certainly has to be considered, male faces growing at twice the rate of female faces (a consequence of hormonal status), with growth ceasing at the same age in both sexes. Consequently, the faces of adult females are like those of juvenile males.[b]

Adams (1992a,b, 1998) examined the stages and sequences of forearm and postcranial skeletal formation in the little brown bat, *Myotis lucifugus*, and also looked at the functional integration of bat wing skeleton and wing membrane. Growth compensation occurs in the wing skeleton by alterations in morphogenesis. Wing shape is conserved across bats but there is plasticity in development of the hand skeleton, which Adams interpreted as functional integration with a proximo-distal gradient in FA.

In an important analysis of FA in nine mandibular characters in random-bred mice, Leamy (1993) concluded that one would expect morphological integration at the population but not the individual level. In another population-level analysis Pertoldi *et al.* (2000) examined FA in characters of 234 skulls and lower jaws of adult Eurasian otters, *Lutra lutra*, from nine populations from seven countries, to compare historical changes in healthy populations with those in populations considered endangered. They found increased FA and decreased size of skull traits over time in some endangered populations but no similar changes in healthy populations. Sexual dimorphism was also reduced in the endangered populations.

Inbred rats (*Rattus villosissimus*) have smaller skull lengths and widths and increased skeletal FA in comparison with random-bred rats (Lacy and Horner, 1996), in which FA is best detected at the population rather than the individual level, as Leamy found.

Again using mandibular characters in random-bred mice, Leamy (1999) determined that directional asymmetry (which he considers genetic) and fluctuating asymmetry (which he considers environmental) have extremely low heritabilities, respectively 0.06 and 0.03. Sixteen quantitative trait loci affect directional asymmetry, while eleven influence fluctuating asymmetry. Somewhat surprisingly, they found less genetic variability for fluctuating than for directional asymmetry (Leamy *et al.*, 1997).

[a] Benedikt Hallgrímsson is my adroit collaborator in such studies; see Hallgrímsson *et al.* (2002, 2003) for our approach using the mammalian appendicular skeleton, and see Hallgrímsson (1998, 1999).
[b] See Cheverud and Richtsmeier (1986) for sexual dimorphism, and McNamara (1980), McNamara *et al.* (1982) and Hallgrímsson (1999) for the functional determinants of craniofacial size and shape in rhesus monkeys. Klingenberg and Nijhout (1998) used measures of FA in the sizes of the forewings of the butterfly *Precis coenia*, the caterpillars of which had one or both hindwing imaginal discs removed. Their interpretation of competition between organ rudiments for developmental resources – in this case for haemolymph-derived resources – points to likely developmental, metabolic and physiological mechanisms underlying FA as well as positive and negative allometry. Also see Nijhout and Emlen (1998) and Emlen and Nijhout (2000) on this point.

---

testosterone and luteinizing hormone per year. Antlers are shed and begin to regrow after animals are injected intramuscularly with an anti-androgen (cyproterone acetate), demonstrating that the cycle in these deer is sensitive to androgens, as is true for deer in temperate latitudes (Bubenik *et al.*, 2002).

In adult males, testosterone controls ossification of the antlers so that growth and shedding are coordinated. In reindeer and caribou, the level of plasma testosterone is normally about 1 ng/ml plasma, but during the period of maximal antler growth, levels rise to between 30 and 60 ng/ml plasma. *The time of antler shedding* is thus under hormone control. If adults are castrated in the fall and winter, when the antlers have stopped growing and consist of bony tissue, the antlers are not shed but remain

as permanent structures. If castration is performed in the summer, when the antlers are still growing, velvet is retained and the antlers not shed. Administering a testosterone inhibitor, cyproterone acetate, to adult white-tailed deer at 3.5 mg/kg body weight/week blocks antler mineralization, fewer Haversian systems form, and the antlers are retained. Castrated fallow deer develop a tumour-like growth on their antlers (an antleroma), representing an abnormal regenerative response.[24]

Antlers are thus target organs for testosterone, which accumulates in the prechondroblastic blastema as the growth season progresses (Bubenik *et al.*, 1974). The absence of testosterone either prevents antlers from forming in the first place, or stops their growth and prevents normal shedding. Excess testosterone leads to

Box 8.2 Osteoporosis and statins

Essentially, three types of metabolic bone disease afflict the human skeleton:

- *hypervitaminosis-D* and *hyperthyroidism*, which result from an excess of resorption of extracellular matrix over deposition;
- *rickets* and *osteomalacia*, which result from deposition of an extracellular matrix that cannot mineralize as a result of *hypovitaminosis-D*; and
- *osteoporosis*, which results from lack of synthesis of or an inability to deposit extracellular matrix.

Estimates by Mundy *et al.* (1999) of those at risk for osteoporosis are 100 million world wide, and 30 million in the United States alone. In introducing a special insert on Bone and Cartilage in the journal *Nature* in May 2003, DeWitt estimated 200 million affected women world wide, and annual expenditures of US$ 14 billion treating fractures resulting from osteoporosis. Such estimates may be no more that 'guestimates,' since determining rates of bone loss depends enormously on the population under consideration, on its nutritional and reproductive status, and so forth.

Given these qualifications, bone loss associated with osteoporosis in Western human populations typically begins around age 40. As a consequence of declining oestrogen levels, postmenopausal women lose 0.5 to 1.5 per cent of their bone/year, requiring 1500 mg Ca^{++}/day, combined with physical activity and oestrogen to counteract this bone loss. Because bone mass is lost, osteoporosis is a major factor predisposing ageing humans to fractures; 25 per cent of white females in the USA will have one or more fractures by age 65, including 100 000 wrist fractures/year. Risk of osteoporosis increases with mild alcohol intake or heavy smoking in men aged 35 to 50 (men whom the authors of the study regarded as middle-aged), apparently because of decreased bone formation, bone resorption being within normal limits. Rapid ageing syndromes (*progerias*) also have osteological complications, including osteoporosis.[a]

Osteoporosis – more properly osteoporeses – comes in various forms. *Immobilization osteoporosis* sets in three weeks after canine limbs are immobilized. An initial increase in bone resorption is followed by decreased resorption and increased formation, bringing resorption back into equilibrium. Interestingly, osteoporosis fails to appear in thyroidectomized or parathyroidectomized dogs, indicating a balance between hormonal and mechanical factors in the etiology of osteoporosis. In dolphin humeri a form of what is essentially osteoporosis occurs as periosteal bone transforms to cancellous bone; the long bones lack a medullary cavity, and resorption exceeds deposition.[b]

It has been claimed that *disuse osteoporosis* – and congenital pseudarthroses – in rats can be prevented using pulsing electromagnetic fields (PEMF Box, 32.1). A pulse train of PEMF increases cellular Ca^{++} and promotes mineralization, while a single pulse decreases cellular Ca^{++} but promotes bone formation, a desirable outcome in patients with osteoporosis or avascular necrosis.[c]

Experimental studies indicate that osteoporosis may influence osteoblasts as well as osteoclasts; Bmp-induced bone formation is slowed in osteoporotic rats in comparison with control rats, an effect ascribed to suppression of osteoblast differentiation in osteoporosis. Drug-induced hypothyroidism leads to increased bone formation in the mandibular condyle and to osteoporosis, with recovery mediated by administration of growth hormone and thyroxine.[d]

Osteoblasts are not always functioning maximally; parathyroid hormone (PTH) produces a rapid increase in bone formation when administered subcutaneously to rats. Consequently, administration of PTH or of PTHrP – the ligand for the PTH receptor – is an effective means of treating osteoporosis.[e]

Statins

Statins would seem to be an unlikely source of drugs to combat osteoporosis, for statins are used to lower serum cholesterol levels, especially low-density lipoprotein (LDL) levels. Statins are popular; five are available in the USA: lovastatin, simvastatin, pravastatin, fluvastatin and atorvastatin. Surprisingly, those taking statins are reported to be only half as likely to fracture bones. The unexpected finding that statins enhance bone formation emerged from a large screening study of 30 000 natural products conducted in the late 1990s in which osteogenesis was assessed in mouse calvarial bones *in vitro* and *in vivo*.[f]

A second *in-vivo* model utilized oral administration of a statin to female rats – some of which were ovariectomized to mimic postmenopausal osteoporosis – and then assessed trabecular bone. Statins enhanced *in-vitro* bone formation two- to threefold and increased new calvarial bone formation by 50 per cent following five days of thrice-daily injections. After oral administration, trabecular bone volume rose 40 to 94 per cent as a result of enhanced bone formation and declining osteoclast numbers.

[a] See An (2003) for an overview of osteoporosis, Teitelbaum (2000) for an overview of osteoclast control in relation to osteoporosis, Mundy *et al.* (1999) for the estimates of the incidence of osteoporosis, Marx (1980) for figures on incidence of fractures, de Vernejoul *et al.* (1983) for the alcohol/smoking study, and Moen (1982) for skeletal aspects of progeria. See Owen and MacPherson (1963) for the periosteal study, and Roberts and Jee (1974) and Box 28.3 for studies using the periodontal ligament.
[b] See Burkhart and Jowsey (1967) for the hormonal studies, and de Buffrènil and Schoevaert (1988) for dolphin humeri.
[c] See L. S. Bassett *et al.* (1979), C. A. L. Bassett *et al.* (1981a,b), Goodman *et al.* (1983) and Rubin *et al.* (1989).
[d] See Iida *et al.* (1994) for the study on Bmp-induced bone, and Lewinson *et al.* (1994) for induced hypothyroidism.
[e] See Manolagas (2000) for an overview.
[f] See Mundy *et al.* (1999) for the screening, Ferber (2000) for the data on fracture repair, and Rodan and Martin (2000), Rogers (2000), Garrett and Mundy (2002) and Karsenty (2003) for statins and the skeleton.

increased rates of ossification, premature cessation of antler growth, premature shedding of the velvet, and retention of the set of antlers.

Oestrogen opposes the action of testosterone, while hypophysectomy prevents antler growth altogether, implicating pituitary hormones. Although growth hormone may be involved, the exact role of the pituitary hormones remains to be established.

Pituitary hormone involvement in fracture repair may provide some clues. Fracture repair in adult domestic fowl is also retarded following hypophysectomy, while healing of fractures is delayed substantially in mice with hereditary pituitary insufficiencies. Normally, mice heal fractured long bones in 13 weeks. Pituitary insufficient mice show no signs of healing at this age, having formed only a soft unmineralized callus.[25]

Alkaline phosphatase accumulates within the preosseous tissue of antlers. Levels of testosterone, androstenedione, thyroxine, calcium, phosphorus, and alkaline phosphatase – all of which are associated with the annual cycle of antler development – were investigated in white-tailed and sika deer. Levels of alkaline phosphatase and 1,25, dihydroxyvitamin D_3 increase sevenfold in fallow deer, *Dama dama*, in July as the antlers are actively forming. Presumably changes are similar in other species.[26]

Graham *et al.* (1962) investigated some aspects of blood chemistry during the cycle of antler growth in Virginia deer. In *females*, blood levels of calcium, phosphorus and alkaline phosphatase are stable when antlers are growing in males. Serum levels of calcium and phosphorus remain stable in *males*, but the level of alkaline phosphatase rises. The increase in alkaline phosphatase causes premature cessation of antler growth by clamping off the base of the antler to constrict blood flow, a condition that is followed by lowered levels of alkaline phosphatase.

Parathyroid hormone and calcitonin

In white-tailed deer, levels of parathyroid hormone (PTH) increase at the velvet stage (April to May) and then decrease. PTH levels increase again after the velvet is shed, decreasing once the antlers are cast – the increased PTH regulating calcium absorption from the gut and mineralization of the antlers – while *calcitonin*, which inhibits bone resorption, increases during antler growth (Chao and Brown, 1984). PTH stimulates osteoclasts and osteocytes to resorb bone, and inhibits osteoblast function. Calcitonin, while inhibiting osteoclastic resorption, stimulates osteoblastic function. Both hormones influence the differentiation of precursor cells; osteoprogenitor cells possess calcitonin-binding sites, and precursor or osteogenic cell populations can be separated *in vitro* on the basis of their differential sensitivity to PTH and to calcitonin.[27]

NOTES

1. See Geist (1986) for wounding in combat, Goss (190a) for antler asymmetry, Forand *et al.* (1985) for social rank and order of casting antlers, and see Suttie (1980) for the study with farmed red deer.
2. In *The Name of the Rose* Umberto Eco provides an informed discussion on horns, tusks and the uses to which horns are put, as William seeks to instruct his charge in Aristotle's causes: '… all animals with horns are without teeth in the upper jaw … You then try to imagine a material cause for horns – say, the lack of teeth provides the animal with an excess of osseous matter that must emerge somewhere else … And you must also imagine a final cause. The osseous matter emerges in horns only in animals without other means of defense' (Eco, 1984, pp. 365–366).
3. See Goss *et al.* (1964), Hartwig (1968), Goss (1970), Hartwig and Schrudde (1974) and Goss and Powell (1985) for the first three sets of studies, and C. Y. Li *et al.* (2001) for grafting into nude mice. In the last study, antlerogenic and non-antlerogenic periosteum from the frontal bone and/or scalp skin from red deer, *Cervus elaphus*, calves was grafted.
4. See Lincoln (1973) for changes in testosterone levels during gestation, Lincoln (1992) for an overview of hormonal regulation of the antler cycle, Goss (1963) for effects of castration before antlerogenesis begins, and Jaczewski and Krzywinska (1974) for effects of pedicle amputation.
5. See Wika (1980, 1982b) for foetal changes in reindeer.
6. See Goss (1961) and Goss *et al.* (1964) for removal of the pedicle, and Goss (1964, 1970) for removal of only the bone of the pedicle.
7. See Wika Clark *et al.* (1998) and Heber-Katz (1999) for the epithelial down growth and regeneration of holes punched in mouse ears.
8. Healing in embryos, even late-stage embryos, is usually accomplished rapidly and without any scar tissue; see Nodder and Martin (1997) for a discussion of the role of growth factors in 'scarless' healing. Powdered cartilage accelerates wound healing in rats with a 20 per cent increase in tensile strength of sutured wounds seven days after wounding (Prudden *et al.*, 1957).
9. See Goss (1964) and Hartwig and Schrudde (1974) for transplantation of ectopic skin to the pedicle, Goss (1987a) for the response of ectopic epidermis to periosteum, and Bubenik *et al.* (1982) for response to electrical stimulation.
10. See Goss (1963) and Goss *et al.* (1992) for basic descriptions of antler resorption, Ryg and Langvatn (1982) for growth hormone, and Herskovits *et al.* (1991) for innervation of bone.
11. See Wika and Krog (1980) for vasculature. Initially, perfusion was the tool used to analyze bone vasculature (Langer, 1876). Later perfusion was replaced by radiopaque dyes and X-rays (Lexer *et al.*, 1904; Deleskamp, 1906).
12. See Lincoln and Fletcher (1976) for the study with hummel red deer, and Li and Suttie (1994) for pedicle formation in red deer.
13. See Goss (1970) and Ryg and Langvatn (1982) for antler growth rates, and Bak *et al.* (1990) and Bak and Andreassen (1991) for growth hormone and the strength of repair tissue. In an analogous situation of rapid skeletal growth – fracture repair – a response to 0–10 mg of growth hormone/kg body weight influences the mechanical properties of repairing rat tibial fractures. Growth hormone administered daily for the first 40 days of repair results in enhanced mechanical properties at 80 days and, interestingly, increased mechanical properties in the contralateral control. Growth hormone may play similar roles in antler growth.
14. For reviews of the early literature on the histogenesis of antlers, see Modell and Noback (1931), Goss (1969a, 1970, 1983), Bubenik and Bubenik (1990), Rönning *et al.* (1990) and C.-Y. Li *et al.* (2001).
15. See Gronowicz *et al.* (1989) for the studies on addition of phosphate. Pioneering studies on the development and growth of embryonic bones in chemically defined media by Kieny (1958), and a series of important studies by Biggers and his colleagues (Biggers, 1960, 1965; Biggers and Heyner, 1963; Biggers *et al.*, 1957, 1966), culminated in the development of the Biggers–Gwatkin–Jones medium (BGJ_b) now in standard use for skeletal cell and tissue culture.
16. See Hirschman and Dziewiatkowski (1966) and de Bernard *et al.* (1977) for the association between GAGs and stages of endochondral ossification. See Ralis and Watkins (1991) for a specific tetrachrome method for staining osteoid (and defectively mineralized bone), and see Scott-Savage and Hall (1979) and Miyake *et al.* (1996b) for its utilization in chick and mouse embryos.

17. See Newbrey and Banks (1983) for matrix vesicles. Banks (1974) provides a critical review of the literature pertaining to the existence and fate of cartilage in the antler and published the best micrographs of its structure.
18. See Kierdorf *et al.* (1994, 1995) for ultrastructural studies on the European fallow deer, and Rönning *et al.* (1990) for the Lapland reindeer.
19. See Sayegh *et al.* (1974) and Newbry and Banks (1975) for the electron microscopy, and Ali (1983, 1987) for reviews of mineralization within mitochondria.
20. See Frazier *et al.* (1975) for the matrix composition of the antlers of Rocky Mountain mule deer, Hascall (1977), Goldberg and Toole (1984) and Knudson and Toole (1985, 1987) for chondrocyte hyaluronan receptors, and Knudson and Knudson (1991) and Knudson *et al.* (1995) for hyaluronan receptor-mediated deposition of chondrocyte-like pericellular matrices.
21. See Hallgrìmsson *et al.* (2002, 2003) for fluctuating asymmetry (FA) as an index of developmental stability, and see Putman *et al.* (2000) for sexual selection and antler FA.
22. See Banks *et al.* (1968a,b), Hillman *et al.* (1973) and Baksi and Newbrey (1989) for seasonal and regional resorption, and Wika (1982a) for reutilization of mineral from cast antlers.

Similarly, phosphorous deficiency in teleost fish increases bone resorption and then decreases mineralization and bone formation, as demonstrated in haddock, *Mellanogrammus aeglefinus*, fed phosphorus-deficient diets (Roy *et al.*, 2004).

23. See Goss (1968, 1969b,c, 1970, 1976, 1977, 1980a, 1984, 1987b) for the correlation of antler shedding with photoperiod.
24. See Goss (1968) for testosterone control of antler growth and shedding, Whitehead and McEwan (1973) for plasma testosterone levels, Goss (1963) for seasonal effects of castration, Bubenik *et al.* (1975) for administration of the testosterone inhibitor, and Goss (1990b) for tumours.
25. See Goss (1969a, 1970) and Bubenik *et al.* (1975) for effects of oestrogen, hypophysectomy and growth hormone, and Negulesco (1971) and Hsu and Robinson (1969) for fracture repair.
26. See Kuhlman *et al.* (1963), Chao and Brown (1984) and Szuwart *et al.* (1994a,b) for levels of alkaline phosphatase in preosseous tissue, and R. D. Brown *et al.* (1983a,b) for the hormone, ion and enzymes studies.
27. See Warshawsky *et al.* (1980) for calcitonin binding, and Wong and Cohn (1974, 1975), Luben *et al.* (1977) and Cohn and Wong (1979) for cell separation using hormone sensitivity.

Tendons and Sesamoids

Chapter 9

Chickens are a joy to eat, turkeys take some effort.

From the turkeys some of us consume at festival dinners, many of us are familiar with mineralized tendons. *Mineralization* – the deposition of calcium phosphates into a tendon – is not my concern in this book: I treat mineralization only when it occurs in connection with a skeletal tissue. Chondrogenesis or osteogenesis often occurs in tendons and in ligaments, the result being the formation of mineralized cartilage or bone. Tendinous chondrification and ossification are the topics of the first half of this chapter. Formation of *sesamoids*, which represents one type of skeletogenesis within tendons, is the topic of the other half of the chapter. Sesamoids may begin and remain as cartilage, or begin as cartilage and be replaced by bone. Cartilage fits the mechanical needs of the environment in which sesamoids develop. I am unaware of any sesamoids that arise intramembranously.

TENDONS AND SKELETOGENESIS

Tendons may mineralize or ossify, either as an occasional event in some individuals in response to trauma or, as in birds and dinosaurs, as a normal part of the life history of all individuals within a species. To take but four examples:

- Most dermal ossifications in reptiles arise by *intratendinous ossification* or by ossification of other dense connective tissue.
- Ossification of particular tendons in woodcreepers is sufficiently common to be a synapomorphy for the woodcreeper family, the Dendrocolaptinae.
- Urodermal bones develop in tendons of the epaxial caudal muscles in lanternfish (family Myctophidae).[1]

- Ossification of the human larynx affects the structure of associated tendinous insertions, which have their own osteogenic potential. Macroscopic and histological techniques demonstrate tendinous insertions into the thyroid cartilages. The mechanical advantages that accrue are self-evident.[2]

Tendon insertion into thyroid cartilages is interesting. Thyroid cartilages contain substantial quantities of type I collagen as a normal constituent along with collagen type II. Some chondrocytes switch from synthesis of type II collagen to synthesis and deposition of type I collagen in a mode of ossification that is endochondral rather than subperiosteal (Claasen and Kirsch, 1994).

'Tendon bones' should be considered part of the normal skeleton in woodcreepers and lanternfishes, although it is not always easy to determine whether or not an element is a tendon bone. *Urohyal 'bones'* in bony fishes may be cartilage, tendon or bone, being preformed in cartilage in sarcopterygians, arising as an ossified tendon in teleosts, and from the ossification of three tendons in *Polypterus*.[3]

The nature of the *pteroid 'bone'* of pterosaurs – whether a true bone, an ossified cartilage or a tendon – has long been debated.

The pteroid is an extra preaxial skeletal element in the wrists of more derived pterosaurs. I say extra rather than duplicated as the pteroid is clearly not a bifurcation of digit I. And that's the rub. Digits should not develop on the preaxial side of digit I. Although only a slender splint of bone, the pteroid articulates by a joint with one or more carpal elements. It runs toward the shoulder supporting a basal portion of the wing membrane. The most parsimonious explanation is that the pteroid develops in the insertion of a tendon, as indeed bones and cartilages do in many species today. Contraction of the tendon would enable the pteroid to play a role in upward movement or rotation of the wing. Unwin et al. (1996), who review pterosaur biology, regard the pteroid as a true bone.

Since tendon development depends on local tension, any bone that occurs in a tendon is also under tension. Fibrous tissue in tendons can modulate to bone, as happens for example when defects in the radii of dogs are isolated by inserting Millipore filters. I consider two examples in more depth: development of *fibrocartilages* during normal tendon development or ageing, and the *Achilles tendon*, which we can injure in a variety of sporting and other activities. Mineralization, chondrogenesis, and/or osteogenesis commonly accompany the repair process, most extensively studied in rodent Achilles tendons.[4]

Fibrocartilage in tendons

Hurov (1986) investigated muscle, tendon and ligament attachment to the periostea of rabbit long bones, finding cartilage, fibrocartilage and chondroid bone in the fibrous periosteum and in the bone at sites of attachment (Fig. 3.1). Over the past decade, Michael Benjamin and his colleagues in Cardiff, Wales, have provided a sustained body of knowledge on the structure and basic nature of the fibrocartilages that arise in mammalian tendons. Benjamin and Ralphs (1995) surveyed 38 regions of tendons in elderly humans, concentrating on where the tendons wrap around bony pulleys or pass beneath connective tissue fascia. Twenty-two of the 38 sites contain fibrocartilage, especially where tendons press against bone. Depending on the site, the extracellular matrix consists of randomly oriented collagen fibres or rows of parallel collagen fibres with intervening fibrocartilaginous cells. Interestingly, and as shown in many of their studies:

- a single tendon can be variably modified along its length, variability including whether fibrocartilage is present and the nature of any fibrocartilage present; while
- the degree of modification of the periosteum of the adjacent bone (e.g. the formation of periosteal fibrocartilage) often mirrors fibrocartilage development in the tendon.

Two specialized regions exist along the length of tendons and ligaments. One is the region where tendons change direction as they wrap around bony pulleys. The second is the *enthesis*, the insertion site of the tendon onto the bone. We associate such insertion sites with Sharpey's fibres (Box 9.1).

Fibrocartilage is commonly found in the wrap-around region, a region often subject to degeneration with age or with overuse. The enthesis may be fibrocartilaginous or fibrous, depending on site-, ligament- and/or tendon-specific factors. Enthesis fibrocartilage appears to arise not from within the tendon, but rather from the cartilaginous model, which is eroded rapidly and replaced by fibrocartilage, essentially forming a secondary ossification centre in the tendon.

Tendon sheaths also show regions of specialization related to mechanical loads, the theory being that fibrocartilages arise in tendons by metaplasia where epiphyseal tendons attach to bone. Fibrocartilaginous cells in

---

Box 9.1 Sharpey's fibres and Sharpey fibre bone

Some tendons with fibrous entheses attach to bone by collagenous fibres that extend from the insertion of the tendon into the bone, effectively imbedding the tendon in the bone. Such fibres are known as *Sharpey's fibres* and the associated bone as *Sharpey fibre bone*. In mouse periodontia, synthesis of Sharpey's fibres is coordinated with synthesis and deposition of new bone (Fig. 1.11).[a]

William Sharpey, who discovered these fibres, added a postscript to his 1848 paper on the theory of substitution of cartilage by bone in which he expressed surprise that Clementi claimed priority over the discovery of the fibres that bear his (Sharpey's) name. The postscript captures beautifully the importance attached to structural studies in the mid- to late-19th century: "… and in conclusion, I cannot help saying that when I first observed these fibres I had no idea that they had been recognized before, still less did I imagine that the subject of my observations would ever acquire such importance as to lead to a formal claim of priority on the part of Italian science."[b]

A difficulty concerning Sharpey's fibres is to determine how tendons or muscles maintain attachment sites while bones and muscles are growing, especially if growth of skeleton and muscular systems is differential. Bones employ two tricks: insert fibres only into the periosteum and/or deposit cartilage at the junction, two strategies that either allow the tendon to stay in place as the periosteum drifts or to modulate the cartilaginous connection. In chicks, the relative position of 20 muscles on long bones from pre-hatching to maturity is unchanging, the muscles being attached only to the periosteum and so maintaining their relative position during bone growth. In rabbits, cartilage, fibrocartilage or chondroid bone form at the site of attachment/insertion of muscle, tendon and ligament to the periostea of long bones.[c]

Distinguishing Sharpey's fibres from other fibres inserting into bone is not always straightforward, even if one restricts the definition of Sharpey's fibres to those found only at tendinous insertions. Ham and Cormack (1979) use a broader definition to include collagenous fibres anchoring tendons, muscles or the periodontal ligament to bone, emphasizing that the insertion effectively acts as a periosteum as the fibres are buried in bone. Figure 1.11 shows fibres anchoring cementum. A second approach is to restrict the definition to collagen fibre insertions into bone, an approach used by Simmons et al. (1993) in their study of periosteal attachment fibres in rat calvariae. These fibres are unmineralized, pass between individual osteoblasts or small groups of osteoblasts, and are sensitive to mechanical stresses; Simmons and colleagues did not regard these as Sharpey's fibres.

[a] See Witten and Hall (2002) for Sharpey fibre bone and rapid bone growth, and see Garant and Cho (1979) for deposition in mouse periodontia.
[b] See Müller (1836), Miescher (1836) and Sharpey (1848), and Schäfer (1878) for an overview of knowledge of bone organization in the late 1870s. The quotation from Sharpey is taken from Schäfer (1878), p. 144.
[c] See Grant et al. (1980) and Hurov (1986) for the chick rabbit studies, and see Witten and Hall (2002, 2003) for how Sharpey fibre bone is to be distinguished.

tendons and ligaments have elaborate cell processes forming a three-dimensional, almost canalicular network throughout the matrix. Cell processes are joined by gap junctions, suggesting extensive cell-to-cell communication and a potential load-sensing system. Fibrocartilage cells are often packed with intermediate filaments, providing another potential mechanical transducing system.[5]

Rodent Achilles tendons

Achilles tendons in rats are a favoured site for investigating metaplastic changes associated with mineralization or ossification. Bony spurs (*enthesophytes*) form in the Achilles tendon, as do three fibrocartilages associated with the insertion of the tendon:

- *enthesial* fibrocartilage at the interface of tendon and bone;
- *sesamoid* fibrocartilage deeper in the tendon, where the tendon presses on the calcaneus[6]; and
- *periosteal* fibrocartilage on the surface of the bone opposing the tendon.

All three appear around birth, which, of course, is a time when mechanical and other lifestyle conditions change dramatically.

All three have differing arrangements of collagen fibres, each mirroring local mechanical loads. For the tendon fibrocartilage itself, blood vessels invade rows of fibres at the tendon–bone junction (the enthesis), fibrocartilaginous cells form and endochondral ossification ensues.

Changes in the sagittal suture in mouse skulls in the first months after birth have been interpreted as transitional between bone forming and bone–tendon junctions (implying that the open suture is transitional between bone formation and the processes occurring at a bone–tendon junction), displaying as it does vascularization before mineralization and apoptosis accompanying osteoid mineralization. In a further example of regional specificity and adaptability, in PTHrP-deficient mice, a prominent bony crest develops on the mandibular ramus, attached to the masseter muscles by a tendon, with bone formation along the tendon (Box 31.2).[7]

Counter intuitively, tension is required to initiate chondrogenesis but not to initiate osteogenesis in a tendon. In part, this reflects the fact that typical endochondral ossification ensues as the enthesial cartilage – or cartilage during repair – is replaced by bone, osteogenesis depending on all those factors and conditions that allow cartilage to be replaced, rather than conditions required for bone to form directly.[8]

Rat Achilles tendons also make good models because cartilage and bone formation – including formation of mineralized cartilage, osteoid and mineralized bone – can be induced in diffusion chambers. If a tendon is cut mid-point along its length, and the reparative tissue placed in Millipore chambers and implanted intraperitoneally, first cartilage and then bone forms (Rooney et al., 1993).

Table 9.1 Effect of zinc deficiency on formation of cartilage and bone in rat Achilles tendons at various days post-tenotomy[a]

Days post-tenotomy	Animals showing ossification (%)	
	Zn-sufficient animals	Zn-deficient animals
30	0	0
60	75	25
90	100	45

[a] From data in Calhoun et al. (1974).

An interesting effect of zinc on ectopic bone formation within rat Achilles tendons has been documented; ossification following tenotomy is dramatically reduced in Zn-deficient rats (Table 9.1). The concentration of Zn is higher in bone than in any other tissue, the daily requirement for humans being 15 mg. Zn deficiency in pregnancy is associated with congenital bone deformation, because of either abnormal function of Zn-dependent enzymes or abnormal vitamin action.

Ossification of avian tendons

Tendons begin to mineralize in the common fowl, *Gallus domesticus*, around four months after hatching, as fibroblasts transform into osteoblasts and deposit bone. Such tendon bone can be sufficiently mature to contain osteones, except around the periphery.

As in rat Achilles tendons, fibrocartilage forms in response to tension; the synovial cavity and fibrocartilage associated with tendons are lost in chick embryos paralyzed with D-tubocurarine from eight days of incubation onwards. Tendons themselves can fail to form with paralysis and can adapt to bone shortening (Bertram et al., 1997).

Different fibrocartilages respond differently to the absence of movement: tendon fibrocartilage fails to form, the plantar tarsal sesamoid fails to form – see below – while the meniscus of the tibiofemoral joint forms, fails to mature and then regresses.[9]

Formation and composition of tendon fibrocartilages

Tendons of the axial skeleton and vertebral cartilages have their origins in the sclerotomal portion of somitic mesoderm. Tendons associated with the appendicular skeleton arise from lateral plate mesoderm. Tendons are not cartilages and so express different genes from those expressed by chondrocytes (Table 9.2). However, tendons and cartilage also express a number of common genes for extracellular matrix molecules and contain some similar growth factors (Table 9.2). Chondrogenesis in tendons, therefore, should not come as a surprise.

Condensation

Several studies suggest that intratendinous chondrogenesis is not initiated by individual cells acting alone, but

Table 9.2 Extracellular matrix molecules, growth factors, transmembrane proteins and transcription factors expressed in avian (chick) and mammalian (mouse, human) tendons[a]

	Chick	Mouse	Human
Extracellular matrix molecules			
Collagens			
Type I	+	+	+
Type III	+	+	
Type V	+	+	
Type VI	+	+	
Type XII	+	+	
Type XIV	+		
Proteoglycans			
Biglycan		+	+
Decorin	+	+	+
Fibromodulin	+	+	+
Lumican		+	+
Elastin	+		
Emilin	+		
Fibrillin-1	+		
Tenascin-C	+		
Growth factors			
Bmp-4	+		
Fgf-8	+		
Fgf-18	+		
TGFβ-2	+	+	
Follistatin (binds activin)	+		
Transmembrane proteins			
Eph-A4	+		
Teneurin-2	+		
Tenomodulin		+	
Transcription factors			
Scleraxis	+	+	
Six-2		+	

[a] Based on studies accumulated by Edom-Vovard and Duprez (2004).

by aggregations of cells. Aggregation–condensation – is a critical step for chondrogenesis during embryogenesis. I think condensation so important that I have devoted three reviews and three chapters (19–21) to it. Condensation was an important theme in the introduction I wrote for the 1985 reprinting of *Bones* by my Ph.D. supervisor P. D. F. Murray (Murray, 1936; Hall, 1985a). The question whether condensation is *a necessary precondition* for fibrocartilage to form in tendons is less clear than the finding that condensation *occurs*; a conclusion based on placing a positive spin on a number of observations.[10]

Murine thrombospondin-5 (cartilage oligomeric matrix protein) cloned and sequenced in 2000 (Box 9.2) is expressed in tendon and in cartilage, in both of which it first appears within the condensation of prechondrogenic cells.

The homeodomain-containing protein, Cart-1, is expressed in mouse chondrogenic mesenchyme, tendon, kidney and lung. Zhao and colleagues interpreted this distribution pattern as indicating tissues with chondrogenic potential. A second common link may be the ability of mesenchyme to condense, a link that would implicate condensation in the initiation of tendon chondrogenesis.

Box 9.2 Thrombospondins

Thrombospondins constitute a superfamily of multifunctional extracellular matrix adhesive glycoproteins that are designated 1 through 5.[a] With primary functions in coagulation and anti-coagulation, thrombospondins also mediate cell proliferation and adhesion, angiogenesis and growth of blood vessels, and tissue repair and metastasis. A variety of roles and locations have been demonstrated for thrombospondins in developing skeletal tissues.

For instance, articular chondrocytes from pig ankle joints synthesize thrombospondin. Splice variants of thrombospondin-2 mRNAs are localized in prechondrogenic condensations in chick embryos. Thrombospondin-4 is expressed in early osteogenic tissue; R. P. Tucker and his colleagues, who cloned chick thrombospondin-4, found transient expression in osteogenic mesenchyme but not in osteoblasts. Thrombospondin-5 (also known as cartilage oligomeric matrix protein) also is expressed in cartilage and prechondrogenic mesenchyme. Mouse thrombospondin-5, which was cloned and sequenced by C. Fang and colleagues, first appears in condensing prechondroblasts at 10 days of gestation, and is expressed in all cartilages and muscles by 13 days and in hypertrophic cartilage, perichondria and periostea by 19 days of gestation. Thrombospondin-5 also is expressed in tendons, trachea, bone, muscle, eye, heart and placenta.[b]

[a] See Ayad *et al.* (1994) and Adams and Tucker (2000) for reviews of the thrombospondin type 1 repeat superfamily.
[b] See Miller and McDevitt (1988) for expression in joints, R. P. Tucker (1993) and R. P. Tucker *et al.* (1995) for thrombospondin-2 and -4, and C. Fang *et al.* (2000) for thrombospondin-5.

Third, tenosynovial and synovial cells respond to Bmps by transforming (transdifferentiating?) into chondroblasts. Tendon morphogenesis is altered when the Bmp-antagonist noggin is delivered to chick limb buds cells in a retroviral vector. Chondrogenesis is suppressed at the condensation stage in the same specimens, suggesting a molecular approach to inducing chondrogenesis within tendons and a link with condensation.[11]

Scleraxis

The role of *scleraxis* in chondrogenesis and tendon development is consistent with condensation occurring in tendon fibrocartilage development (Table 9.2). The product of the gene *scleraxis*, the basic helix–loop–helix protein scleraxis is expressed in tendons, ligaments and in their respective precursor cells. Schweitzer *et al.* (2001) took advantage of this to show that tendon cells are induced by epithelial signals. The Bmp antagonist noggin allows ectopic tendon cells to be specified but not new tendons to form, demonstrating that the epithelial signal is regulated or restricted by Bmp but that other factors regulate tendon maturation.

Scleraxis is neither diagnostic for tendons nor for ligaments. Scleraxis is expressed throughout mouse embryos at gastrulation but restricted to prechondrogenic sites at 9.5 days of gestation, being expressed at condensations in the axial, appendicular and cranial skeletons. Expression is down-regulated as chondrogenic cells differentiate.

Scleraxis-null mice created by gene targeting fail to gastrulate because of a failure to form mesoderm. In chimaeric mice, scleraxis-null cells are excluded from the sclerotome but contribute to other embryonic regions. Thus, scleraxis has at least three major roles: formation of mesoderm; formation of sclerotomal mesenchyme; and initiation of chondrogenesis.[12]

Composition

Normal tendons, such as bovine flexor tendons, were shown in the early 1980s to contain an aggregating proteoglycan similar to that found in cartilage. Subsequent studies identify this proteoglycan as *decorin* (Table 9.2), a chondroitin–dermatan sulphate proteoglycan that binds to and thus inactivates Tgfβ. Decorin is expressed in mesenchymal cells as they transform into chondrocytes during fracture repair in rats. Such cells lack Tgfβ; hypertrophic chondrocytes possess Tgfβ but not decorin.[13]

Formation of fibrocartilage is associated with deposition of other GAGs, especially *aggregan* and *biglycan* – the latter also found in tendons (Table 9.2) – allowing the fibrocartilage to imbibe water and withstand compressive forces. Foetal bovine tendon exposed to cyclic compression *in vitro* shows selective synthesis of biglycan without alteration in decorin. Keratan sulphate is found at the enthesis fibrocartilages in the medial collateral ligament of rat knee joints.[14]

Collagen fibrillogenesis also varies depending on whether tendons are exposed to tension or pressure: tendons at tension zones contain collagen with a periodicity of 63 nm, those at pressure zones collagen with 53 nm periodicity (plus microfilaments and lipid). Sites of attachment of tendons and ligaments to human mineralized bone cortex and to periostea display site-specific fibres containing collagen types III and VI and fibrillin. Cartilage-specific collagens (types I, II, V, IX, XI) are found at the ligament–bone interface in bovine medial collateral and anterior cruciate ligaments. Type XII collagen is found in developing mouse tendon (and in connective tissue, membrane bone, fibrocartilage and perichondria). With heavy loading of a tendon, type II collagen is deposited, even if no cartilage forms. Finally, there is a cartilage-specific promoter within intron two of the chick α2(I) collagen gene. Consequently, the cartilage produced has type I collagen mRNA but no protein. Bone and tendon share a promoter.[15]

Tenascin is found in high amounts in the osteotendinous junction of bone–tendon attachment in the quadriceps muscle in rats. Mechanical loading regulates tenascin-C levels at this junction; immobilization in a cast for three weeks results in removal of almost all tenascin. Eight weeks of free cage activity following removal of the cast only results in a slight increase in tenascin, but eight weeks of low or high intensity treadmill running restores tenascin levels to those in controls. High levels of tenascin are deposited around chondrocytes and fibroblasts of the

---

Box 9.3 Egf-repeats in components of extracellular matrices

A number of the major molecules found within extracellular matrices (ECMs) possess Egf or Egf-like repeats. Even more importantly from the perspective of developmental regulation, the mitogenic properties of these ECM components can be attributed to mitogenic properties of the Egf repeats. Levesque *et al.* (1991) argued that this mitogenic activity reflects common ancestral genes underlying growth factors, cell adhesion molecules, proteins of the ECM, and immunoglobulins; the greatest similarity of the repeats in the ECM protein tenascin is with the cluster of fibronectin type III repeats found in the class III region of the human major histocompatability complex (Matsumoto *et al.*, 1992).

Laminin, an integral component of epithelial basement membranes, many of which are involved in skeletogenic interactions (Fig. 10.7; Chapter 18) contains Egf domains with growth factor activity; cells with Egf receptors can use laminin or peptides derived from laminin to promote mitosis. A large proteoglycan from the articular cartilages of humans and baboons has an Egf-like domain[a]

Tenascin, which has 13 *Egf* repeats and is induced by *Tgfβ*, is mitogenic for several cell lines, in part because of the *Egf* repeats and in part because it enhances the mitogenic activity of Egf. Tenascin is found in skeletogenic locations where proliferation is high: chondrogenic condensations, perichondria, osteogenic cells and differentiated cartilage. Engel (1991) emphasized the common structural motifs in proteins of the ECM and that similar domains serve related functions in apparently unrelated proteins (see Table 21.9).

The hyaluronan-binding chondroitin sulphate proteoglycan versican contains two Egf-like repeats in the G3 domain. Molecules that lack these two repeats enhance chondrocyte differentiation, acting via actin assembly in the cytoskeleton.

[a] See Panayotou *et al.* (1989) and Kubota *et al.* (1992) for laminin, Colognato and Yurchenco (2000) for a review of the structure and function of members of the laminin family, and Stanescu *et al.* (1991) and Ayad *et al.* (1994) for the ECM proteoglycan.

---

tendon, and around collagen fibres within the body of the tendon. Tenascin, like laminin, contains Egf-repeats, and so exerts a proliferative effect (Box 9.3 and see Box 21.3 and Chapter 25 for tenascin).[16]

Teneurin-2, a homologue of the *Drosophila ten-m* gene, is expressed transiently in tendons; there is more prolonged expression in the mesenchyme of chick limbs, somites and the craniofacial region. *Teneurin-2* is co-expressed with *Fgf-8*; implanting Fgf-8-coated beads leads to ectopic expression of *teneurin-2*, which is therefore downstream of *Fgf-8* (Tucker *et al.*, 2001). Whether *teneurin-2* plays a role in tendon development or tendon skeletogenesis remains to be determined.

SESAMOIDS

Sesamoids – so named because of their resemblance to a sesame seed – are nodules of cartilage or bone formed in tendons or ligaments, especially where a tendon passes over an angulation of the skeleton. The patella (knee cap) in humans and its association with the knee joint is perhaps the most familiar sesamoid. In humans, sesamoids or sesamoid bones also occur in the elbow and heel as intratendinous chondrifications.[17]

Sesamoids, which may be cartilages or bones, develop intratendinously. Fibrocartilaginous sesamoids, as seen in the capsule of the proximal interphalangeal joint of human fingers and toes, show a high degree of differentiation, prominent GAGs, and type II collagen.[18]

Although ectopic – they arise outside the skeleton – sesamoids are not occasional ossifications found only in some individuals. As Al Capone knew well, we all have kneecaps. Sesamoids are also distinct from *traction epiphyses* (apophyses), projections from long bones that insert onto tendons, which have their own ossification centres, and which are associated with secondary ossification centres, although traction epiphyses – which are distinct from the pressure epiphyses found at the ends of endochondral bones – may have started out (evolutionarily) as sesamoids.[19]

Sesamoids develop independently of the skeletal element, only later becoming associated with the primary skeleton. Interestingly, using this definition, the 'secondary' cartilage that develops on the dentary bone in some mammals is a sesamoid – it arises *outside* the periosteum of the dentary (and so is not a secondary cartilage) – while in other mammals and in birds, these cartilages arise *within* the periosteum, and so are not sesamoids, but are secondary cartilages (Chapter 12).

Over the span of almost three decades from the 1940s to the 1960s, R. Wheeler Haines evaluated whether sesamoid bones give rise to epiphyses or whether sesamoids arise from isolated parts of long bones, concluding that neither is true but that sesamoids are independent, stable centres of ossification. Also distinct from sesamoids – but part of the same functional matrix – are fibrocartilages such as those found in the attachment zones of the human quadriceps tendon and patellar ligaments, and discussed earlier in the chapter.[20]

The ossicones on the heads of giraffes and okapi arise as nodules of bone beneath the skin, separate from the bone of the skull (Chapter 7). Ganey *et al.* (1990) regard ossicones as sesamoids, which is fine with me, unless we restrict the definition of sesamoids to independent ossifications/chondrifications within tendons. The same reasoning applies to the bone at the base of cattle horns, which also arise independently of the skull, but in very close association with it (Fig. 7.8 and see Chapter 7).

Sesamoids are sensitive to Hox gene knockout. Sesamoids develop abnormally in the forelimbs, and an additional sesamoid forms beside the tibiale in mice following disruption of *Hoxa-11* (Small and Potter, 1993). The latter is interesting. It is present in 10 per cent of wild-type individuals but in 100 per cent of homozygous mutant individuals; knocking out *Hoxa-11* brings the abnormal varient to full expression in all individuals. Other skeletal elements affected include the ulna/radius, tibia/fibula, and the 13th thoracic vertebra, which is transformed into a first lumbar vertebra as the sacral region is transformed into a lumbar region.

A brief survey of sesamoids in several vertebrate taxa follows.

Amphibians

Little work has been undertaken on sesamoids in amphibians. Olson (2000) described seven sesamoids in each hind limb of the South African dwarf frog, *Hymenochirus boettheri*, more than reported for any other frog, and most associated with tendons and muscles. Some develop during metamorphosis before the limbs are fully developed, some later. At least some of these sesamoids depend for their formation on normal hind-limb function; the proximal *os sesamoides tarsale*, *tibialis anticus* and *cartilago sesamoides* failed to form in the right hind limbs of individuals in which the the sciatic nerve had been

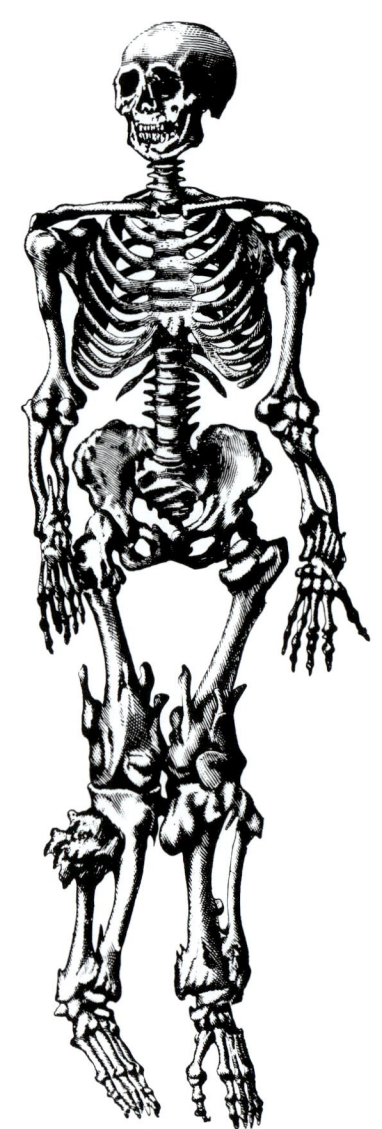

Figure 9.1 Multiple exostoses, especially on the long bones, most prominently associated with the knees and ankles. From Gould and Pyle (1896).

severed. Other sesamoids – *distal os sesamoides tarsale, cartilago plantares* – were present but their ossification was delayed in comparison to the elements in the contralateral control limb, implicating function in formation of some and maturation of other sesamoids (Kim *et al.*, 2002).

Reptiles

Sesamoids are either much more common in reptiles than in amphibians, and/or reptiles have been studied more intensively.

Ankylosaurs (fused lizards), the heavily armoured, herbivorous, ornithischian dinosaurs of the late Cretaceous, had extensive cranial ornamentation. The skull was bossed and decorated with bumps and humps that are a combination of exostoses of dermal cranial bones, sesamoids and ectopic extracranial bones (Vickaryous *et al.*, 2001). Multiple exostoses, especially associated with the knees

and ankles in a human skeleton, are illustrated in Fig. 9.1. Within extant taxa, a description of chondrogenesis and mineralization of the limb skeleton in the bloodsucker or garden lizard, *Calotes versicolor*, documents many sesamoids and examples of metaplastic mineralization (Mathur and Goel, 1976). Extra cartilages – known as paraphalanges because of their association with phalanges – were described in 57 species of 16 genera of gecko lizards by Russell and Bauer (1988), who interpreted them as sesamoids (Fig. 9.2). Paraphalanges, which support the footpad, appear to have arisen multiple times.

Birds

Within birds, sesamoids can develop in association with, and form an integral part of, the articulation between the quadrate and Meckel's cartilage or the retroarticular of the lower jaw. In the North Island kokako, *Callaeas*

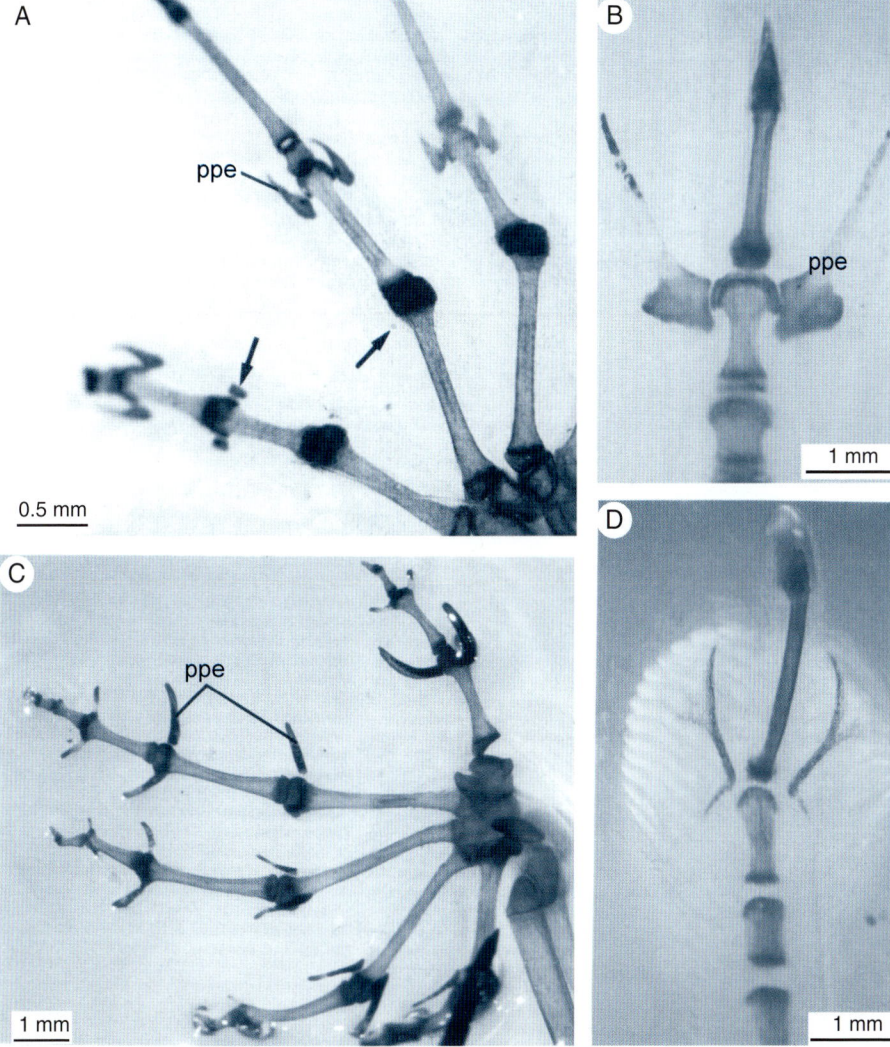

Figure 9.2 Paraphalangeal bones in gecko lizards. (A) A cleared and stained left forelimb of Brook's house gecko, *Hemidactylus brookii*, showing three digits, large paraphalanges (ppe) distally and smaller ones (arrows) proximally. (B) Digit IV from the left forelimb of the turnip-tailed gecko, *Thecadactylus rapicauda*, has paraphalanges (ppe) that are broad proximally and taper distally. (C) The left hind foot of the Namibian web footed gecko, *Palmatogecko rangei*, showing five digits and the extensive formation of paraphalanges (ppe). (D) Digit IV from the left hind limb of the Micronesian gecko, *Perochirus ateles*, has branched paraphalanges. All specimens stained with Alizarin red and cleared. Adapted from Russell and Bauer (1988).

cinerea, from New Zealand, two sesamoids occur as ossifications within the internal jugomandibular ligament (Fig. 9.3). Another lies between the quadrate and the mandible, with cartilaginous pads facing each skeletal element. Clearly, this sesamoid is an integral part of the functional joint (Fig. 9.3). The sclera of the common potoo, *Nyctibius griseus*, and of 10 species of owl, contains an additional sesamoid bone that forms in cartilage, lies on the trajectory of the tendon of the pyramidal muscle, and plays a role in contracting the nictitating membrane.[21]

When we consider sesamoids in birds, we have the advantage that experimental work can be and has been undertaken, especially paralysis to determine whether sesamoids arise in response to mechanical forces, as seems likely given their origin in tendons and ligaments.

The plantar tarsal sesamoid fails to form in paralyzed chick embryos, its formation being dependent on movement. This is a different response from the meniscus of the tibio-femoral joint, which is a derivative of the primary cartilaginous skeleton. The meniscus forms in paralyzed embryos, fails to mature and regresses. A cartilaginous sesamoid arises on the tibiae of embryonic chicks. The cartilage ossifies and is incorporated into the tibiofibularis as a crest, the *syndesmosis tibiofibularis*. This element requires movement to form; in one sample of paralyzed chick embryos, none (0/47) formed the tibial cartilage and only 8.5 per cent (4/47) formed a patella.[22]

Teleosts

Teleost fish are notorious for possessing multiple small *Wormian bones* in their skulls. Using a combination of morphological, developmental, palaeontological and comparative data to homologize skeletal elements in several dozen catfish species, Diogo *et al.* (2001) concluded that, because many of these bones ossify in ligaments or in association with a ligament, they are sesamoids.

NOTES

1. The terms epaxial and hypaxial were coined for somitic muscles innervated by dorsal or ventral branches of the spinal nerves, respectively. Although transferred to somite development, Burke and Nowicki (2003) argue that the division is not appropriate developmentally. Epaxial and hypaxial somitic domains are consistent with the fate map of epaxial and hypaxial muscles, but do not reflect key developmental events associated with muscle differentiation. They propose *primaxial* and *abaxial* to distinguish somitic contributions on the basis of the origin of the cell lineages. The primaxial domain comprises those structures that differentiate entirely within the somitic environment – the vertebral column, vertebral ribs, periaxial and intercostal muscles and associated connective tissue – and that are all generated from somitic cells. The abaxial domain comprises all structures that differentiate within the lateral-plate environment, including the migrating somite-derived myoblasts that migrate into and differentiate within lateral-plate mesoderm – limb and abdominal muscles, sternal and sternum.

2. The evolutionary and functional significance of tendon ossification in woodcreepers is discussed by Bledsoe *et al.* (1993). Moss (1969) compares dermal ossifications among reptiles, demonstrating that most are tendinous ossifications rather than periosteal, and that they produce a range of structures from mineralized tendons to bone. See Miyake and Uyeno (1987) for urodermals in lanternfish, Fischer and Tillmann (1991) for thyroid cartilage tendons, Urist *et al.* (1964), Rooney *et al.* (1992) and Rooney (1994) for reviews of mechanisms of tendon mineralization, and Koob and Summers (2002) and Summers and Koob (2002) for the introduction to a recent symposium on tendons and for the evolution of tendons.

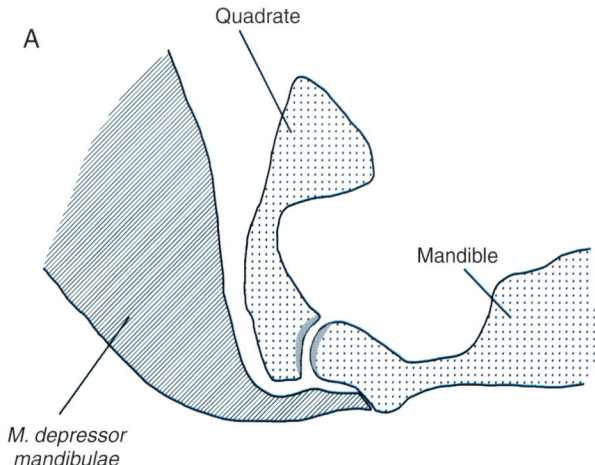

A

Quadrate

Mandible

M. depressor mandibulae

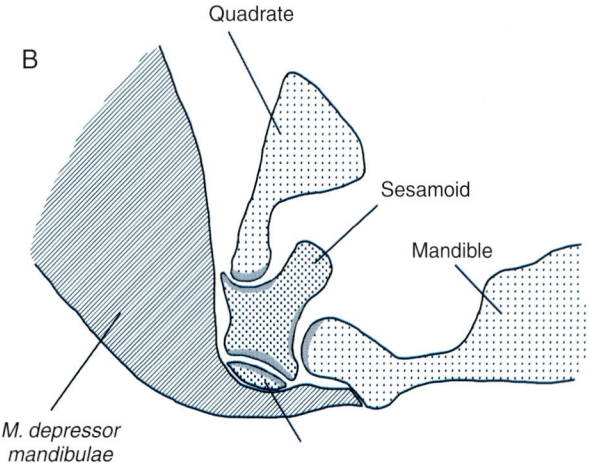

B

Quadrate

Sesamoid

Mandible

M. depressor mandibulae

Figure 9.3 Sesamoids in an avian jaw joint. (A) A typical avian jaw joint in lateral view (anterior to the right) with articulation between the quadrate and the mandible via a synovial joint. The insertion of *M. depressor mandibulae* is shown. (B) The jaw joint in the New Zealand wattlebird, the kokako (*Callaeas cinerea*), showing a sesamoid associated with *M. depressor mandibulae* and articulating with a second sesamoid, which in turn articulates with the mandible and quadrate. Each articulation is facilitated through the development of joints with articular cartilages (shaded). (B) modified from Burton (1973).

3. *Amia* and *Lepisosteus* lack urohyal bones. See Arratia and Schultze (1990) for a detailed analysis. The different modes of development of urohyals caused Arratia and Schultze to debate and then question whether cartilaginous and tendinous urohyals are homologues. See Hall (1994a, 1995c,e, 1998a, 2003a) for homology as based on developmental processes.

4. See Murray (1936), Krompecher (1937), Altmann (1964) and Currey (1984a) for extensive treatments of the literature on tension and skeletogenesis, and see Bassett and Ruedi (1966) for the dog study.

5. See Ralphs *et al.* (1992) and Benjamin and Ralphs (1997a,b, 1998) for development of fibrocartilages – the suprapatella and the attachment of the patella – at two sites in quadriceps tendon of rats.

6. For the onset of ossification of the human calcaneus as visualized using radiographic and histological analyses, see Meyer and O'Rahilly (1976). Sixteen per cent of individuals have an intramembranous parachordal site that appears *before* the primary centre.

7. See Rufae *et al.* (1992, 1996) and Benjamin *et al.* (2000) for development and ageing of these fibrocartilages, Zimmermann *et al.* (1998) for sagittal sutures, and Shibata *et al.* (2000) for the PTHrP-deficient mice.

8. See Wong and Buck (1972) for the role of tension, and Rooney *et al.* (1992) and Rooney (1994) for basic descriptions of endochondral ossification in repair of rat Achilles tendon.

9. See Abdalla (1979) for tendon mineralization and structure, Beckham *et al.* (1977) for the paralysis study, Mikic *et al.* (2000b) for the differential responses, Tardieu *et al.* (1983) for how tendons adapt to bone shortening, and Whedon and Heaney (1993) for paralysis and bone growth.

10. See Hall and Miyake (1992, 1995, 2000) for the three reviews on condensation, Hall (1995b) for an overview of the major stages and processes of vertebrate embryogenesis, and Hall (1997a,b, 2001b) for the major stages and processes of skeletogenesis.

11. See C. Fang *et al.* (2000) for thrombospondin-5, Zhao *et al.* (1994) for Cart-1, and Sato *et al.* (1988) and Pizette and Niswander (2000) for Bmp and *noggin*.

12. See Cserjesi *et al.* (1995) and Brown *et al.* (1999) for *scleraxis* expression.

13. See J. C. Anderson (1982), Y. Yamaguchi *et al.* (1990) and Ayad *et al.* (1994) for the aggregating proteoglycan decorin, and Matsumoto *et al.* (1994) for its role in fracture repair. For the role of proteoglycans such as decorin, syndecan and heparan sulphate in binding growth factors and/or fibronectin, see Ruoshahti and Yamaguchi (1991). In general, cartilage proteoglycans from different species cross-react (Wieslander and Heinegard, 1981); and see Wight and Mecham (1987) for the biology of proteoglycans.

14. See Benjamin and Ralphs (1998) for aggrecan, Gao *et al.* (1996) for keratan sulphate and Ayad *et al.* (1994) for biglycan. In their study of the postnatal (birth – 120 days) development of the insertion of the medial collateral ligaments in the rat knee, Wei and Messner (1996) described perichondrial transformation into fibrocartilage at eight days and secondary ossification at 15 days.

15. See Merrilees and Flint (1980) for the data on GAGs and collagen, Evanko and Vogel (1993) for the study with cyclic compression, Benjamin and Ralphs (1998), Oh *et al.* (1993), Keene *et al.* (1991) and Visconti *et al.* (1996) for the collagen types, and Bennett and Adams (1990) for the cartilage-specific promoter. For an overview of collagen in tendon, ligament and bone healing, see Liu *et al.* (1995). Collagenous bone matrix but not tail tendon collagen is mitogenic for potentially osteogenic cells (Rath and Reddi, 1979).

16. See Järvinen *et al.* (1999) for this tenascin study, and see Ayad *et al.* (1994) for tenascin structure and function. Tucker *et al.* (1994) identified a novel tenascin in tendons, ligaments and mesenchyme at sites of epithelial–mesenchymal interactions in chick embryos.

17. See Parsons and Keith (1897), Barnett and Lewis (1958), Mottershead (1988), Benjamin *et al.* (1995) and Sarin *et al.* (1999) for sesamoids in humans, and see Rooney (1994) for intratendinous ossification.

18. See Lewis *et al.* (1998) and Milz *et al.* (1998) for the interphalangeal joint, and see Ralphs and Benjamin (1994) for an overview of the joint capsule in normal development, ageing and disease states.

19. In a review of the functional role of sesamoids Mottershead (1988) noted that no animal before the Jurassic had secondary ossification centres within sesamoids.

20. See Barnett and Lewis (1958) for the distinction between sesamoids and traction epiphyses, Haines (1940, 1969) and Sarin *et al.* (1999) for sesamoids and epiphyses, and Evans *et al.* (1990) for the fibrocartilages. In their evaluation of ontogenetic and phylogenetic transformation of ear ossicles in marsupials Sánchez *et al.* (2002) classified the cartilage of Paauw as functionally equivalent to a sesamoid.

21. See Burton (1973) for the kokako, and Bohúrquez Mahecha *et al.* (1998) for the data on the potoo and owls.

22. See Hogg and Hosseini (1992) for data on the tibial crest and patella formation, and Mikic *et al.* (200b) for the tibiofemoral meniscus. The latter authors kindly refer to such mechanical effects as "the Hall effect transducer." See Hall and Herring (1990), Hosseini and Hogg (1991a,b) Hogg and Hosseini (1992) and Wu *et al.* (2001) for growth of the skeleton in paralyzed chick embryos. For another analysis using paralysis, and for discussion of the connection to theropod dinosaurs (putative avian ancestors), which also possess the tibiofibularis crest, see Müller and Streicher (1989).

Part IV

Stem Cells

The examples of the range of tissues and modes of skeletogenesis intermediate between two generally recognized classes of skeletal tissues and processes of skeletogenesis – intramembranous and endochondral – discussed in Parts II and III set the stage for the two major questions addressed in Parts IV and V:

How do skeletogenic cells arise?
What are the relationships between the various classes of skeletogenic cells?

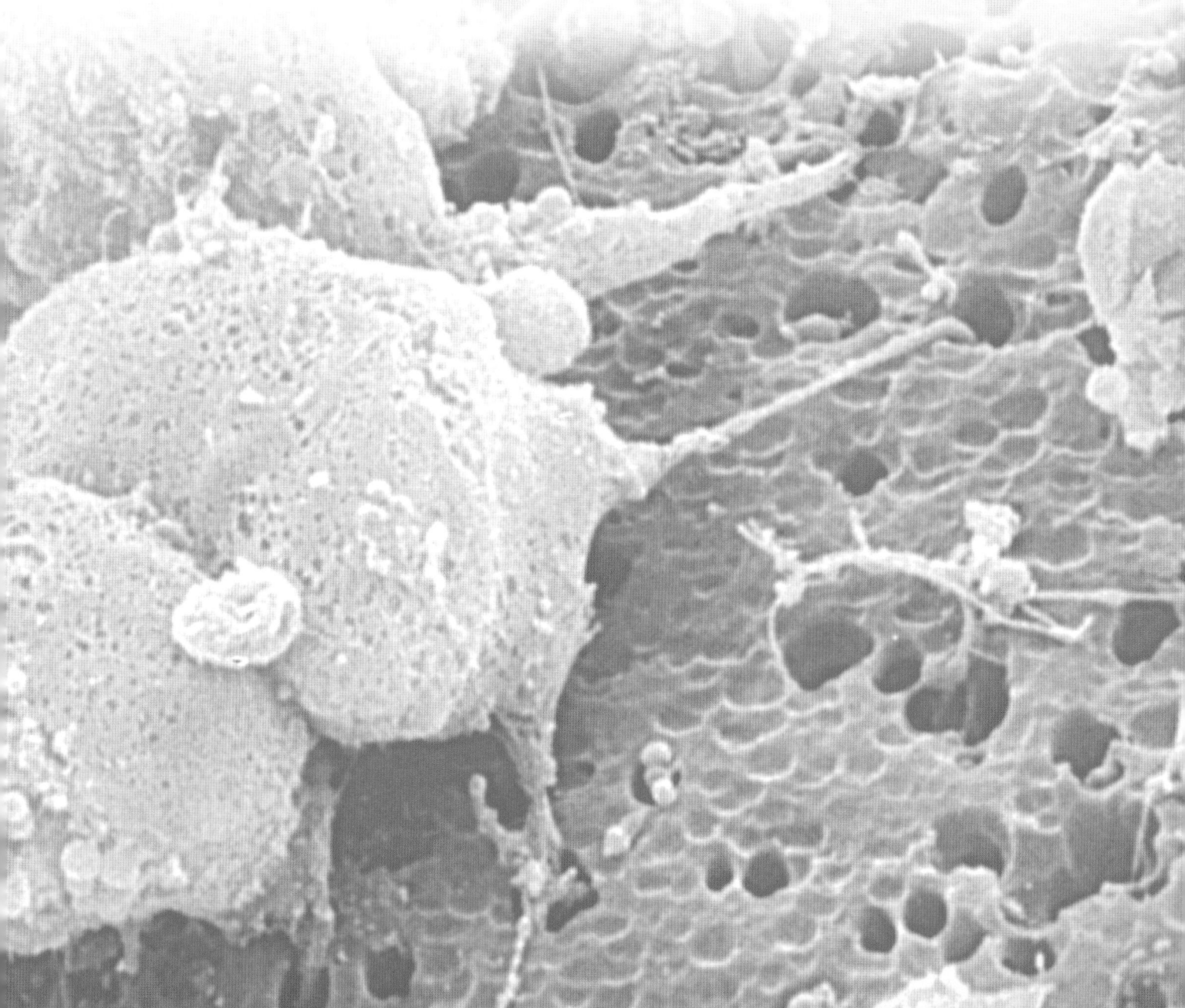

Embryonic Stem Cells

The specification of stem cells varies from location to location and from time to time during development.

A vital issue when considering skeletogenesis is the origin and nature of the cells capable of forming cartilage or bone *in vivo* and *in vitro*, in embryo and in adults, within the skeleton and without, and under normal or the abnormal conditions associated with pathology, disease, trauma, parasitism, temperature shocks and environmental insults. The issue is major, the literature vast, and the approaches many and varied. I approach the topic on several fronts through examination of the:

- embryonic origins of the skeletal cells that form the normal endoskeleton and exoskeleton found in all individuals of a species (this chapter and Chapter 12);
- identification and localization of skeletogenic stem cells outside the skeleton in adult mammals (Chapter 11);
- origins of the cells that form skeletal tissues (usually cartilage) during repair and regeneration (Chapters 14 and 29); and
- isolation and identification of skeletogenic cells *in vitro* (Chapter 24).

In this chapter I place the search for skeletogenic cells into a framework proposed by the late E. Neville Willmer, a framework that anticipated much of the current interest in stem cells.

STEM CELLS

In *Cytology and Evolution*, the second edition of which was published in 1970, E. Neville Willmer of Cambridge University, a pioneer in cell and tissue culture, and for many years editor of *The Biological Reviews of the*
Cambridge Philosophical Society, provided one of the most thought-provoking analyses of the origin of cell lineages during development and evolution. Ths span of the book is from the prototype of the first organism – which Willmer though would have been nemertine-like – to the 'sophisticated organs and tissues of man' which he saw 'as the results of perpetual modification and selection of cellular form and function' (1970, p. 5). The opening sentence of the introduction sets the tone, style and approach: 'The transition from the Victorian carriage to the racing car of today [1969] or from the Wright brothers' flying machine to the 'concorde' are object lessons in evolution. Step by step, some features have been modified and adapted; others have been discarded. Some entirely new characters have appeared, while certain basic structures have been maintained relatively unchanged throughout' (Willmer, 1970, p. 1).

Willmer was seeking the most fundamental cell types. Figure 10.1 summarizes the pedigree of stem cell types Willmer envisaged as leading to skeletal and connective tissues, and myogenic and haematopoietic cells. From this listing alone it is clear that Willmer envisaged fundamental relationships between these four great classes of tissues. From Fig. 10.1 you will see that there is an initial broad subdivision of zygotic blastomeres into *amoeboblasts* – forerunners of the haematopoietic and osteoclast cells – and *mechanoblasts* – forerunners of skeletal, connective tissue and myogenic cells. Willmer based the identification of these two classes of cells, and of a third class – *epitheliocytes*, on the properties displayed by cells *in vitro*, in particular, the cells that grow out from explant (organ) cultures; i.e. he was endeavouring to bridge or unify the first and fourth categories listed above.[1]

The three classes of cells differ from one another morphologically, physiologically and in metabolic requirements, are stable, and reproduce true to type. These distinguishing

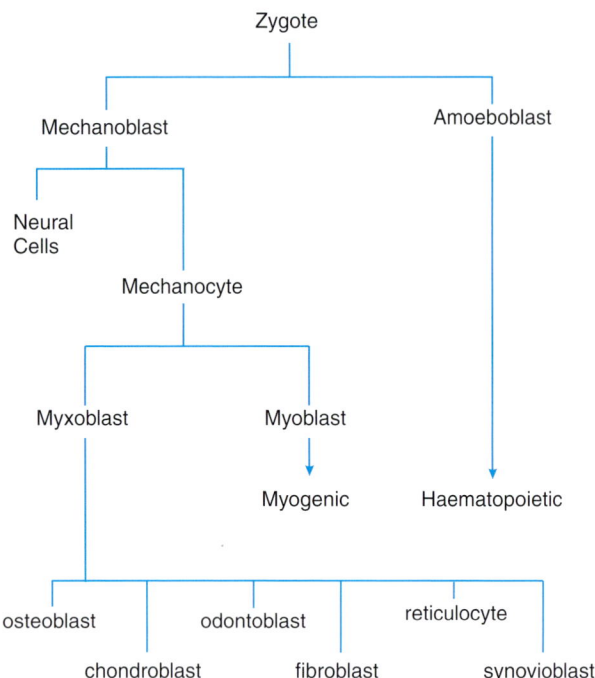

Figure 10.1 Cell lineages. The origin of skeletal, dental and connective tissue-forming cells as progressive segregation from precursor cells during embryonic development. 'Clasts (osteo-, chondro- and odonto-) arise from the amoeboblast line. The precursor cell terminology (amoeboblast, mechanoblast, mechanocyte) is from Willmer (1970). Modified from Hall (1978a).

Table 10.1 A summary of the features distinguishing epitheliocytes, mechanocytes, and amoebocytes from one another[a]

Epitheliocytes	Mechanocytes	Amoebocytes
1. Grow as a sheet	Grow as a net	Isolated
2. Close cell-to-cell contact	Contact inhibition	No contact
3. ±[b] Desmosomes	± Desmosomes	No desmosomes
4. Cells move as a sheet	Cells move with polarity	Amoeboid movement
5. Grow at surfaces	Some ability to penetrate the substrate	Penetrate Substrate
6. ± Phagocytic	Not phagocytic	Phagocytic
7. Do not require embryo extract to divide	Require embryo extract	Fatty degeneration in presence of embryo extract
8. –	Do not support Rous chick sarcoma virus	Support Rous chick sarcoma virus
9. Rough and smooth E.R.[c]	Rough E.R.	Smooth E.R.
10. Intra- and extracellular mucosubstances	Extracellular mucosubstances	Intracellular mucosubstances
11. Produce keratin	Produce collagen	Produce albumin and fibrinogen
12. ± Acid phosphatase	No acid phosphatase	Acid phosphatase in lysosomes
13. ± Alkaline phosphatase	Alkaline phosphatase	No alkaline phosphatase
14. –	Cause epithelial differentiation	Do not cause epithelial differentiation

[a] Adapted from Willmer (1970).
[b] ±, presence or absence.
[c] E.R., endoplasmic recticulum.

features embrace fundamental attributes of cellular organization including the presence of desmosomes and contact inhibition, the production of specialized molecules such as keratins and collagens, and modes of growth (Table 10.1).

The schema proposed by Willmer postulates a dichotomous branching of stem cells, with *progressive determination* (we would now say *specification* or *restriction*) at each step. At each stage the stem-cell population subdivides into two or more stem-cell pools, each of which is able to differentiate along a particular pathway.

For example, two stem cells arise within the mechanoblast series: (i) the *myxoblast* from which scleroblasts, fibroblasts and reticulocytes arise, and (ii) the *amoeboblast* from which the haematopoietic cells arise (Fig. 10.1). Myxoblasts and myoblasts each have a more restricted potential than do their parent mechanoblast. Each specific type of myoblast is more restricted than was the parent myoblast, and so on.

A note of **caution**: While the lines joining cell types in the pedigree in Fig. 10.1 are shown as simple, one-way paths, in actual fact few data are available on the sequence or number of steps involved in proceeding from one cell to another within the same stem-cell pool. As Willmer noted, cells, by reversing their development, appear to be able, to some extent, to start again on another line; i.e. cells can *dedifferentiate* and then *redifferentiate* along another pathway (Chapters 13 and 14). We know, for example, that

the various skeletal cells have more in common with one another than they do with myoblasts (Hall, 1970a). Unstimulated mechanocytes from bone can form either bone or cartilage *in vitro*; thus osteoblasts and chondroblasts can be modulations of the one mechanocyte determined for osteo- and/or chondrogenesis (see Chapter 12).[2]

On the other hand, mechanocytes isolated from muscle either cannot form bone or cartilage, or else they require stimulation – induction – to do so. Whether this observation justifies the establishment of two stem-cell pools (myxo- and myoblasts), or whether skeletal, connective and myogenic cells can form from one stem cell, depends on the skeletal region examined; specification of stem cells varies from location to location and from time to time during development (Chapter 12).

The ability of scleroblasts to modulate their differentiation, and/or to undergo metaplasia, as seen in tissues intermediate between various mineralized tissues described in Chapter 1, and the transformations described in Chapters 10 and 11, suggest that the existence of many classes of multipotential myxoblasts is unlikely. An excellent recent review by Muschler and Midura (2002) of progenitor cells in connective tissues, especially with respect to tissue engineering and clinical applications, places much emphasis on the central role of the cells. Other elements – growth factors, glues and mechanical applicators – are necessary, but without competent cells, there is nothing to activate.[3]

Similarly, whether embryonic myxoblasts initially possess six potential fates, to form bone, cartilage, dentine, connective tissue, reticular tissue[4] or synovium, the cell type produced depending on local environmental factors and conditions, or whether myxoblasts represent six pools of stem cells, each possessing one of the above fates as an intrinsic feature, depends on skeletal region and embryonic origin. These two options do not exhaust the possibilities. Rather, they represent two extremes.

There may be six pools of stem cells, some or all of which require environmental interactions before they differentiate. Or there may be three pools of stem cells, each with two potential fates, with local environmental conditions directing cells into one of the two possible alternatives. In relation to differences between 'similar' cell types later in development, Conrad et al. (1977) showed that fibroblasts from embryonic chick cornea, heart and skin differ morphologically and in their responses to enzymes or to chelating agents. Lineages of fibroblasts exist. Later approaches to these cells, including the one that Norman Maclean and I adopted, used a stochastic compartmental model, with mesenchyme, osteoclasts, osteoblasts and osteocytes as four compartments that exchange cells; but we now know that osteoblasts and osteoclasts are in non-exchangeable compartments (see Chapter 15).[5]

Local differences in progenitor cells means that they have different potentials depending on their location. This statement assumes that we can distinguish the differential potential of stem cells from the differential environments acting on equivalent stem cells, Several lines of evidence support the concept of progenitor cells having dual potentials: (i) as already mentioned, and as discussed in depth by Hall (1970a, 1972b) and in Chapter 12, osteoblasts and chondroblasts can be modulations of a single mechanocyte; (ii) Willmer (1970) presented evidence that fibroblasts, chondroblasts, osteoblasts, synovioblasts and odontoblasts represent 'a little subfamily of cells' (Fig. 10.1).

The observation that fibroblasts can show a wide range of synthetic activities (i.e. can synthesize many different cell products), supports alternative pathways being open to stem cells (see the following section), and that progenitor cells are likely to possess at least two potential fates. Examples are progenitor cells that can form cartilage or bone, bone or fibrous tissue, bone or ligament, dentine or bone, and fibroblasts or reticulocytes.

Modulations of stem cell behaviour can occur after the stem-cell line has been established. A classic example is regeneration, which, more often than not, is initiated following dedifferentiation of cells and their redifferentiation into the cells from which the regeneration arises (Chapter 14).

SET-ASIDE CELLS

Some major groups of invertebrates display indirect development in which the embryos forms a larva (tadpole, imago), which is then either transformed into an adult via metamorphosis or which contains stem cells from which the adult arises, again by metamorphosis. In the latter case, few if any of the larval cells may contribute to the adult. The stem cells from which the adult arises are set aside (hence, set-aside cells) in the embryo but make no contribution to the larva. Their role is to form the adult.

The most well known example of set-aside cells are those in the imaginal discs of flies such as the Hawaiian fruit fly genus Drosophila (of which there are hundreds of species). Cells within imaginal discs are 'specified' for a particular cell fate – and so we have eye discs, antennal discs, wing discs, and so forth – but will only begin to differentiate when exposed to a hormonal cue associated with metamorphosis. A less well known example are set-aside cells in echinoderms (sand dollars, sea urchins, sea stars) known as the 'juvenile rudiment,' and from which the adult arises.

Anuran and urodele amphibians (frogs, toads, urodeles, salamanders) have a larval stage – the tadpole – and undergo metamorphosis to produce the adult frog or salamander (but see Box 10.1 for loss of the tadpole stage). One example of set-aside cells has been described. It is therefore an unusual (unique in the real meaning of the word) stem cell situation for vertebrates. The species is the northern two-lined salamander, Eurycea bislineata. The cells are those that give rise to the adult epibranchial cartilage during metamorphosis (Fig. 10.2).[6]

The epibranchial is an element of the branchial arch cartilage complex of larval and adult salamanders. The normal situation is for larval cartilages to be resorbed and replaced by cartilages (bones) of the adult that arise from cells with no connection to the larval elements. For example, during development of the Surinam toad, Pipa pipa and the South African clawed frog, Xenopus laevis, the larval ethmoid cartilage disappears at metamorphosis and the adult ethmoid (an endochondral bone of the skull) develops from a new cartilage that arises dorsal to the larval ethmoid. This is the typical situation for adult bones.[7]

In E. bislineata, as in other salamanders, larval chondrocytes die under the phagocytic action of lysosomal enzymes. Consequently, the three larval epibranchial cartilages degenerate and play no part in formation of the adult elements. The single adult epibranchial forms from cells within a portion of the perichondrium of the surviving larval cartilage, which proliferate in response to rising levels of thyroxine (Alberch et al., 1985; Alberch and Gale, 1986; Fig. 10.2). This can be likened to the imaginal

Box 10.1 Frogs without tadpoles

Of the many frogs that have lost or reduced the tadpole stage in their life cycles, perhaps the best studied is the Puerto Rican coqui, *Eleutherodactylous coqui*. Such direct-developing frogs (direct developers) advance the stage when the limb buds develop, presenting the bizarre situation of a fully yolked amphibian embryo with well-developed limb buds, even skeletal elements. The hind limbs are more advanced than the forelimbs; neither possesses an apical epithelial ridge (AER). Removal of limb bud epithelium does not disturb limb outgrowth but does result in deficient distal limb elements.[a]

Coqui also lack neuromasts and lateral line ganglia – two more deviations from the developmental patterns found in indirectly developing (biphasic) frogs, i.e. frogs with a tadpole stage and metamorphosis in their life cycle. Ectodermal placodes for these structures fail to develop, although other placodes are present and develop normally. Experimental studies by Schlosser *et al.* (1999) demonstrate that coqui embryonic head ectoderm is not competent to respond to inductive signals, although coqui form placodes in response to signals from grafted axolotl ectoderm.[b]

Most cranial cartilages typical of anuran tadpoles *fail to form in coqui*. Indeed, in many regions of the head, a mid-metamorphic or adult morphology is evident from the onset of skeletal development. Comparable accelerations are seen in development of the skulls of two species of salamanders, the black-bellied salamander, *Desmognathus quadramaculatus*, and the seepage salamander, *D. aeneus*, which lie at extremes of the life history strategies of

indirect and direct development. The direct developer, *D. aeneus*, has lost such ancestral larval structures as the palatopterygoid and altered the time of appearance of other elements, as seen in precocious development of the maxilla.

On the other hand, in a direct-developing, lung-less (plethodontid) salamander, the arboreal salamander, *Aneides lugubris* – which has the most prolonged ontogeny of any plethodontid – ossification of the long bones continues throughout life, indeed is never completed, a great deal of cartilage remaining even at the end of life. Similarly, in the Surinam toad, *Pipa pipa*, a direct-developing anuran that is the flattest toad in the world, major skeletal changes, including hyperossification of the skull, occur late in metamorphosis after the metamorphic climax. Such extremes of patterning variation in direct-developing urodeles and anurans cry out for further study.[c]

[a] See S. C. Smith *et al.* (1994) for developmental interactions between placodal ectoderm and neural crest-derived mesenchyme, and see Webb and Noden (1993) for overview of placodal development.
[b] See Townsend and Stewart (1985) for a staging table for coqui, Richardson *et al.* (1998) and Shi (2000) for an overview of metamorphosis, Callery and Elinson (2000) and Callery *et al.* (2001) for thyroid-hormone-dependent metamorphosis before hatching in coqui, and Hanken *et al.* (2001) for the lack of an AER and the role of limb bud ectoderm.
[c] See Hanken *et al.* (1992, 2001) and Olsson *et al.* (2001) for cranial development in coqui, and S. B. Marks (2000), Wake *et al.* (1983) and Trueb *et al.* (2000) for direct development in *Desmognathus aeneus*, *Aneides lugubris* and *Pipa pipa*.

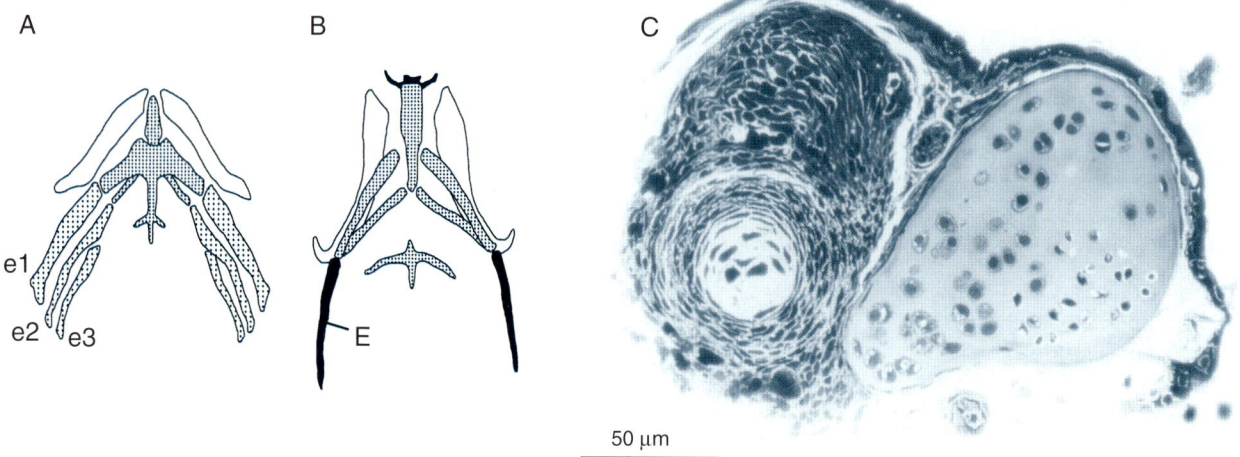

Figure 10.2 Larval elements lost and an adult element gained from set-aside cells. (A, B). The hypobranchial skeletons in larval (A) and adult (B) lungless northern two-lined salamanders, *Eurycea bislineata*. The three larval epibranchials (e1–e3, stippled in A) disappear during metamorphosis and their place is taken by a single epibranchial (E, black in B). Unshaded elements in A and B are the ceratohyals, which are retained but remodeled during metamorphosis. (C) During metamorphosis, chondrocytes in the larval epibranchial (right) show vacuole formation as they degenerate. The future adult epibranchial (left) is undergoing active chondrogenesis; compare the thickened perichondrium with the virtual absence of a perichondrium in the degenerating larval element. Modified from Alberch and Gale (1986).

discs of *Drosophila* in which adult cells are set aside in the larva to await hormonal cues that trigger differentiation.

Presumably, the rising level of thyroxine during metamorphosis is the cue used by the salamander cells to begin to differentiate. Thyroxine, the 'metamorphic hormone' in amphibians, certainly influences ossification in anuran amphibians. The sequence in which the skeleton ossifies, and whether ossification of individual elements is

initiated before or after metamorphosis, is controlled by thyroxine. This has been studied, for example, in the northern leopard frog, *Rana pipiens*, the major features of skeletal development being: ossification in a cephalocaudal sequence; some ossifying before, some after metamorphosis; and the timing of ossification and skeletal maturation being controlled by thyroxine (Figs 10.3 and 10.4). The sequence of ossification of skull but not the

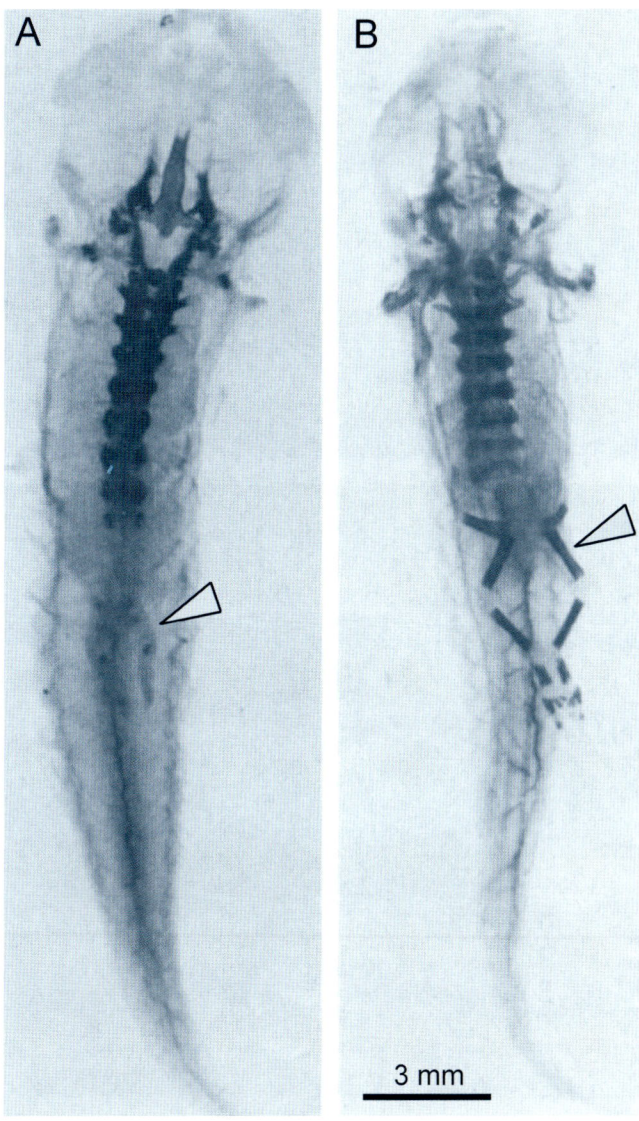

Figure 10.3 Alizarin red-stained tadpoles of the northern leopard frog, *Rana pipiens,* viewed from the dorsal surface to show the extent of skeletal development. (A) Control with ossifying vertebrae but minimal ossification of the femora (arrowhead). (B) A tadpole exposed to 6.25×10^{-8} M thyroxine shows precocious development of both fore- and hind limbs. Again, the right femur is identified (arrowhead). Modified from Kemp and Hoyt (1969).

limb bones can be manipulated by administering exogenous thyroxine early in larval life. In the Oriental fire-bellied toad, *Bombina orientalis,* in response to implants of thyroxine, chondrogenesis is initiated earlier, but osteogenesis is not, uncoupling chondrogenesis from osteogenesis.[8]

STEM CELLS FOR PERIOSTEAL OSTEOGENESIS IN LONG BONES

Following initial studies by Stump (1925) on the histogenesis of bone, Dame Honor B. Fell – who spent her wonderfully productive career at the Strangeways

Research Laboratory in Cambridge (UK) – pioneered the analysis of skeletal tissues *in vitro* with studies on the nature of cell populations within the periostea and endostea of bones from embryonic chicks.[9]

Periostea are bi-layered, consisting of an outer fibrous and an inner cambial (cellular) layer. Bone formation was initiated when Fell stripped periostea from the tibiae of six-day-old embryonic chicks and cultured the intact periostea for 10 days in a medium consisting of embryo extract in a plasma clot. Only fibrous tissue forms if the *outer fibrous* layer of the periosteum is cultured alone, demonstrating that osteogenic capability – and therefore the location of osteogenic stem cells – is restricted to the *inner layer* of the periosteum, which became known as the *cellular or cambial layer* to reflect that it is more cellular than the fibrous layer, and – by analogy to the cambium of plants – the progenitor layer from which new bone forms. Fell concluded that the outer fibrous and inner cellular layers consisted of two separate populations of cells, each with different potentials. She considered the fibrous layer to contain fibroblast progenitor cells (myxoblasts, to use Willmer's terminology) unable to form bone, while the inner layer consisted of osteoblast progenitor cells (mechanocytes) able to form bone. The dual nature of the periostea of endochondral and membrane bones (Fig. 10.5) has been confirmed and reinforced in many later studies investigating a number of properties of periostea or periosteal progenitor cells. For example:

- Using periostea from chick and murine calvariae, Burger *et al.* (1986) confirmed that osteoblasts arose only from the inner layer, the outer layer forming fibroblasts.
- Cell cultures established from periostea from two-month-old rabbits produce cartilage, and so contain chondrogenic precursors, but these are limited to the cambial layer (Ito *et al.,* 2001, and see below).
- Scott-Savage and Hall (1980) evaluated the differentiative ability of the cambial and fibrous periosteum from chick embryos.[10]
- Similarly, Syftestad *et al.* (1985) isolated and characterized osteogenic cells associated with initial osteogenesis of the embryonic chick tibia.
- Cambial periosteal cells can be used as grafts to repair articular cartilage, to repair subchondral bone and for studies of perichondria *in vitro.*[11]

The cambial layer of a periosteum can arise from superficial cartilage cells, as demonstrated when Kahn and Simmons (1977) CAM-grafted tibial epiphyses free of perichondrium and observed the development of a periosteum containing osteoblasts and osteocytes. The vascularized conditions of the CAM-graft (Box 12.3) permitted the cartilage matrix in the graft to undergo metaplasia to a more bone-like matrix. I have seen the same transformation of the matrix of secondary cartilage in paralyzed chick embryos (Hall, 1972b).

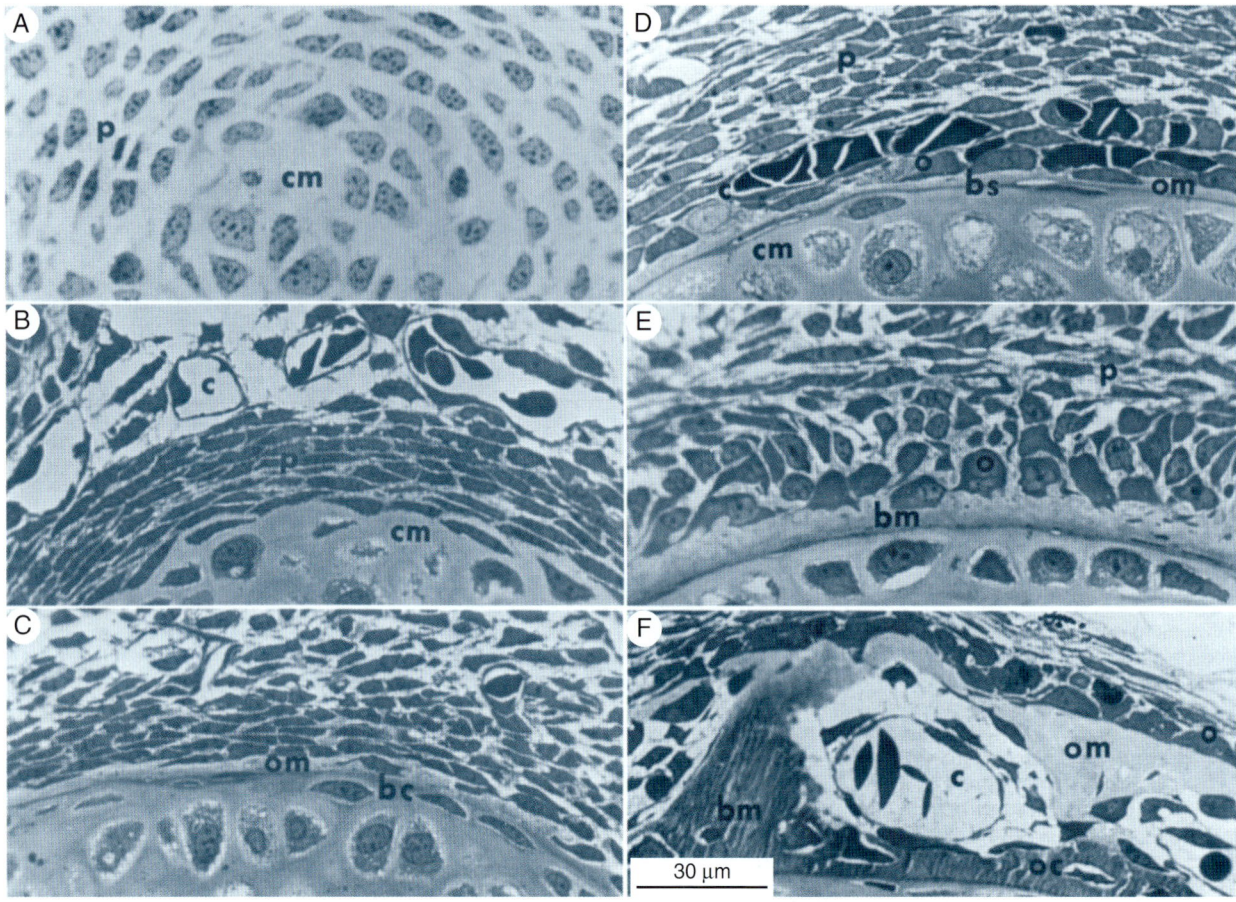

Figure 10.4 Histological cross sections of femora from control tadpoles of the northern leopard frog, *Rana pipiens* (A–C) and from tadpoles treated with thyroxine (D–F) to show the extent to which precocious exposure to thyroxine enhances skeletal development. (A) Condensing mesenchyme (cm) and initial perichondrium (p). (B) Cartilage matrix (cm) has been deposited, the perichondrium (p) is multilayered and capillaries (c) course through the connective tissues adjacent to the perichondrium. (C) Osteoid (om) has been deposited and the peripheral chondrocytes ('border cells', bc) have flattened. (D) After four days of exposure to L-thyroxine osteoblasts (o), osteoid (om), border cells (bs) and a thickened perichondrium (p) are all established; compare with A. (E) After seven days of exposure to L-thyroxine an extensive layer of osteoblasts (o) and bone matrix (bm) are present; compare with B. (F) After nine days of exposure to L-thyroxine mineralized bone matrix (bm), osteoid (om), osteoblasts (o) osteocytes (oc) and capillary invasion (c) are evident; compare with C. Adapted from Kemp and Hoyt (1969).

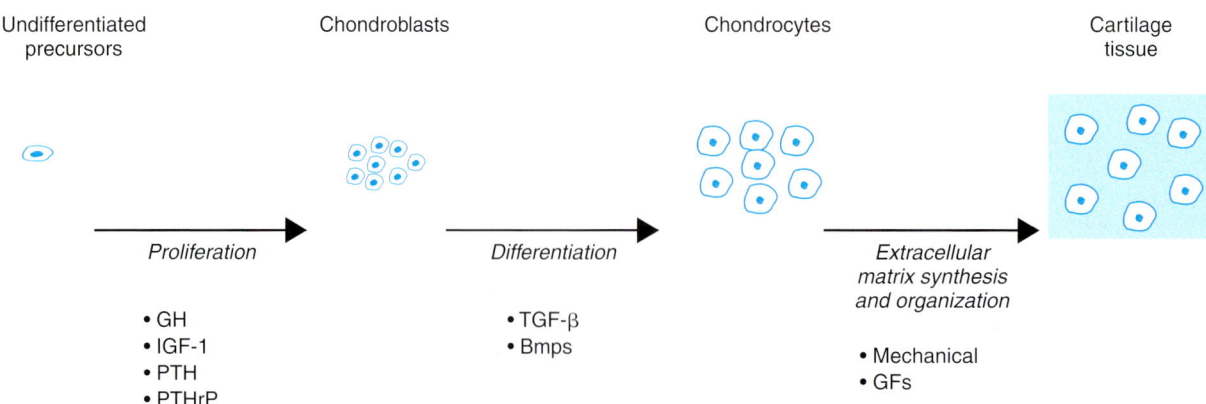

Figure 10.5 A three-step model of chondrogenesis, based on periostea transplanted into defects in mammalian articular cartilage. The three steps are: (1) proliferation of undifferentiated precursors in the periosteum to form chondroblasts; (2) differentiation of chondroblasts into chondrocytes; and (3) deposition of ECM to produce cartilage. Growth factors active at steps 1 and 2 are shown. Mechanical factors act at step 3. Bmps, bone morphogenetic proteins; GFs, growth factors; GH, growth hormone; IGF-1, insulin-like growth factor-1; PTH, parathyroid hormone; PTHrP, parathyroid hormone related peptide; TGF-β, transforming growth factor-beta. Modified from O'Driscoll (1999). I have modified the cell terminology from that in the original model.

The properties and behaviour of periosteal cells can vary with position along the periosteum. Thus, any progenitor cell in the periosteum of an avian membrane bone can differentiate as a chondroblast to form secondary cartilage, but only those at joints or sites of muscle attachment or ligament insertion receive the necessary mechanical stimuli to initiate chondrogenesis (Chapter 12). Production of osteoprogenitor cells is controlled locally within the periostea of the mandibular rami in pigs, apparently via differential rates of cell replication in the cellular layer of the periosteum. No such regional replication is seen in the fibrous layer.[12]

MODULATION OF SYNTHETIC ACTIVITY AND DIFFERENTIATIVE PATHWAYS OF CELL POPULATIONS

When investigating stem or progenitor cells, it is critically important to know when a change is a short-term or temporary response to altered conditions (a modulation or physiological change, modulation having been introduced in Chapter 5), and when it is a more permanent alteration in cell state associated with switching cell fate and/or differentiating along a particular pathway. I address this issue by examining the ability of cells to modulate between fibroblast and chondroblast, to modulate the synthesis of matrix products, or to simultaneously deposit and degrade extracellular matrix (ECM). In Box 5.1, I used the synthesis of actin and myosin to show how reliance on individual 'luxury' molecules to identify cell types can lead us astray.

Fibroblast–chondroblast modulation

A large literature exists on the modulation of cell states between chondroblasts and fibroblasts via modulation of chondrogenesis and fibrogenesis. As one example from cell culture: bovine articular chondrocytes treated with liver lysosomes or lysosomal enzyme, or cultured for a prolonged time, acquire a fibroblastic morphology, and switch to the synthesis of type I collagen. As one in-vivo example: in disease conditions or syndromes such as chondromalacia of the human patella, chondrocytes modulate to fibroblasts and modify their patterns of protein synthesis. Although much evidence shows that the synthetic activity in cultured chondrocytes does change from typically chondrocytic to typically fibroblastic, the transformation is often regarded as a dedifferentiation.[13]

Modulation of glycosaminoglycan synthesis

Many other studies have shown that synthesis, secretion and/or degradation of glycosaminoglycans (GAGs) and collagen alter when cells are perturbed in vivo or in vitro. Examples of GAG synthesis follow.

Populations of fibroblasts from a variety of sources synthesize collagen and either hyaluronan or chondroitin sulphate (CS) when maintained in vitro, but can vary their synthesis if culture conditions change. For example, fibroblasts from embryonic chick amnion incorporate glucosamine exclusively into hyaluronan when cultured at low cell densities, but they incorporate glucosamine one-to-one into hyaluronan and CS when maintained at high cell densities. Other cells show morphological transformation along with the switch in synthetic activity: under the influence of cAMP and/or testosterone, Chinese hamster ovarian cells transform to fibroblasts with concomitant initiation of collagen synthesis.[14]

Of course, the synthesis, secretion and deposition of hyaluronan and CS are subject to other controls; interactions among ECM products are complex. Synthesis of hyaluronan by cultured chondrocytes from avian embryos is unaffected by BrdU while synthesis of CS is reduced to below half control values (Table 10.2). Synthesis of different chondroitin sulphates may be controlled independently: when cells from rabbit aortae are stretched on elastic membranes, synthesis of hyaluronan and chondroitin-6-sulphate increases by 300 per cent; synthesis of chondroitin-4-sulphate does not (Table 10.2). Both the epithelium and the sub-epithelial mesenchyme of embryonic chick limb buds are rich in hyaluronan. Epithelial hyaluronan is inducible by Fgf-2, and mesenchymal hyaluronan by Tgfβ, and epithelium can induce a hyaluronan-dependent pericellular matrix from mesenchymal cells (Knudson et al., 1995). Control is multi-factorial.

In any event, the level of hyaluronan exerts some control on the synthesis of CS, rather than the reverse; high levels of hyaluronan associated with condensation depress synthesis of CS, while removal of hyaluronan by hyaluronidase augments synthesis of CS.

Matalon and Dorfman's (1966) study on fibroblasts from patients with Hurler syndrome[15] heralded the use of biochemical analyses of cell lines to assess synthetic shifts in metabolic bone diseases; see Box 10.2 for the C3H10T1/2 cell line as an example. In fact, most cell types synthesize GAGs when maintained in vitro, and quite

Table 10.2 Effects of BrdU and mechanical stretching on the synthesis of hyaluronan and chondroitin sulphates in embryonic chick chondrocytes and rabbit aorta cells[a]

Glycosaminoglycan	Chick embryo chondrocyte cultures		Rabbit aorta cells	
	Control	BrdU	Control	Stretched
Chondroitin-4-sulphate	–	–	140	140
Chondroitin-6-sulphate	–	–	530	1540
Chondroitin sulphate	7874	3580	–	–
Hyaluronan	784	702	4800	15 730

[a] Based on data from Daniel et al. (1973) and Leung et al. (1976). Data expressed as counts of radioactivity/min/µg DNA.
[b] BrdU: 5-bromodeoxyuridine.

Box 10.2 C3H10T1/2 cell line

The wonders of and cautionary tales about cell culture…

The C3H10T1/2 cell line of multipotential fibroblast-like cells was established from mice of the C3H strain (hence the C3H designation at the start of the cell-line name) and made available in 1993. Easily cultured in standard media (Eagle's minimal essential medium + 10 per cent foetal bovine serum), the cells will proliferate indefinitely as fibroblast-like, contact-inhibited (they do not aggregate) undifferentiated cells. The cell line, pronounced C3H ten two and a half cells, can be switched into skeletogenesis with ease and so is a favourite among skeletal cell biologists. A sample of uses is outlined here.

As evidenced by synthesis of alkaline phosphatase and responsiveness to parathyroid hormone, C3H10T1/2 cells differentiate into the osteoblastic lineage in response to Bmp-2. As in any such study, the nature of the evidence, in this case two of the chief characteristics of osteogenic cells, does not demonstrate that the cells can (or have) become osteoblasts, nor whether they can synthesize and deposit osteoid and mineralize the resulting osteoid to produce bone matrix. Tgfβ-1 did not elicit the same response, nor were any features of chondrogenesis elicited (Katagiri et al., 1990b).

More recently, Denker et al. (1999) elicited chondrogenesis from C3H101/2 cells cultured at high cell density as micromass cultures in the presence of Bmp-2. Poly-L-lysine, which is known to promote cell–cell interactions, enhanced the effect of Bmp-2.

So culture conditions are critical if cells are to reveal their full potential. C3H10T1/2 cells normally are contact inhibited. Overcome that inhibition (poly-L-lysine in this case) and culture the cells at high density (micromass) and they can respond to a growth factor (in this case Bmp-2) by differentiating along a pathway (chondrogenesis) which it was thought this growth factor could not elicit and/or which the cells were thought not capable of entering.

There are more than enough salutary lessons in this study about the limitations of cell culture analysis, and more importantly, about assumptions we make about the differentiative or lack of differentiative capability of cells. Illustrating all these items, chondrogenesis has been elicited from C3H10T1/2 cells in response to Tgfβ-1 or to combinations of Bmps. As these new chondrogenic cells synthesize *Bmp-2* and *Bmp-5*, the potential for positive feedback and maintenance of the chondrogenic state exists.[a]

The heparan sulphate proteoglycan *perlecan* is found in basement membranes, cartilage condensations, but not in condensations of membrane bone in mouse embryos between 12.5 and 15.5 days of gestation. Expression of perlecan precedes type II collagen, as expected of a proteoglycan present at the condensation stage (Chapter 19). A role for percelan in chondrogenesis is suggested by the formation of cartilage nodules containing type II collagen in C3H10T1/2 cells exposed to perlecan.[b]

[a] See Denker et al. (1995) and Atkinson et al. (1997) for the Tgfβ and Bmp studies, and Atkinson et al. (1997) for synthesis of Bmps by the now-chondrogenic cells. Bächner et al. (1998) used subtractive cloning to identify 21 cDNAs expressed by C3H10T1/2 cells after exposure to Bmp-2. One of the major issues in developmental biology is how to detect when a cell has entered a new differentiation pathway. See *Cell Commitment and Differentiation* (Maclean and Hall, 1987) and *Cell Lineage and Fate Determination* (Moody, 1998) in both of which this is a major theme.
[b] Dedifferentiated human chondrocytes also form cartilage nodules when exposed to perlecan (French et al., 1999; Kirn-Safran et al., 2004). See Ayad et al. (1994) and the paper by Gomes et al. in Glowacki (2004) for perlecan, which is one of the four heparan sulphate proteoglycans found in cartilage, the other two being syndecan-3, glypican-3 and glypican-5 (Kirn-Safran et al., 2004).

dissimilar fibroblasts (e.g. those from skin and gingival) and chondrocytes produce similar GAGs *in vitro*. Such results probably do not reflect simple heterogeneity of cell populations; clonal cultures of epithelial, melanoma, liver, HeLa, pituitary, kidney and neuroblastoma cells all produce collagen *in vitro*. With the exception of these clonal studies – which, you will note, do not include any skeletogenic cells – such findings illustrate the effect of *in vitro* perturbations on *populations* of cells, and similar synthetic activity in apparently dissimilar cell types. Whether *single* cells can synthesize more than one ECM product and, if so, whether such syntheses take place simultaneously or sequentially must also be addressed.[16]

MODULATION OF SYNTHETIC ACTIVITY AND DIFFERENTIATIVE PATHWAYS IN SINGLE CELLS

The studies just discussed only pertain to modulation if it can be shown that they occur in the progeny of single cells. Otherwise, the results may reflect differential survival or selection of one population over another, or heterogeneity arising from a mixed-cell population. Compelling evidence for modulation of synthetic ability *in single cells* was provided in the late 1970s, when two groups took advantage of the distinctive primary structure of collagen of types I and II to develop an immunofluorescence technique that could visualize tagged antibodies against them.

Both collagen types I and type II are found in human dermal fibroblasts. Type I collagen is normally regarded as a product of fibroblasts, bone, ligament, cornea and 'dedifferentiated' chondrocytes, type II a product of cartilage and notochord, distributions exemplified by the sobriquets bone-type and cartilage-type collagen. Such delineation of collagen types to specific cell types is by no means absolute. Indeed, if skeletogenic cells could modulate their differentiation, they would be expected to be able to produce both collagen types, and they do: differentiated chondrocytes contain mRNA for type I and type II collagens, as do articular chondrocytes at joint surfaces and secondary chondrocytes that arise from periosteal cells, and as does chondroid bone.[17]

Dual synthetic activity is also demonstrated by cells that can synthesize and degrade ECM products simultaneously.

Degradative activity

Especially pertinent to a discussion of the synthetic abilities of skeletogenic cells are the studies of Richard Ten Cate and his colleagues in Toronto, and Wouter Beertsen and colleagues in Amsterdam. Both research groups used transmission electron microscopy to show that individual fibroblasts synthesize and degrade collagen simultaneously. This remarkable conclusion is based on the identification of intracellular, membrane-bound, banded collagen fibrils within fibroblasts of mouse, rat and guinea pig periodontal ligaments (PDLs), and in healing skin wounds in mice *at the same time* as these cells are synthesizing collagen (Fig. 10.6). These particular fibroblast populations contain alkaline phosphatase, as expected of cells synthesizing collagen. Indicative of degradative activity, acid phosphatase is also present within membrane-bound fibrils. In newborn mice, fibroblasts of mouse PDLs synthesize and degrade collagen. In 20-month-old mice, some 17 per cent of *all* PDL cells are multinucleated 'fibroblasts,' with two to seven nuclei/histological section, i.e. they are functioning as the PDL equivalent of osteoclasts.[18]

We are accustomed to cells displaying a single functional role, the separation of cells as epithelial or mesenchymal (Fig. 10.7) and cells 'breeding true' in clonal culture (Fig. 10.8) being two examples. Simultaneous synthesis and degradation of collagen may occur in these fibroblasts because of the multiple roles they play, and/or the speed with which fibroblasts of the PDL have to respond to mechanical stimuli, remove old tissue or create new cells during wound healing. However, foetal lung fibroblasts also degrade as much as 30 per cent of the collagen they synthesize, they do so intracellularly, and within minutes of its synthesis. This is an active process; lung fibroblasts are sensitive to agents such as L-azetidine-2-carboxylic acid (LACA), which increases collagen degradation twofold. These fibroblasts can adjust the concentration of collagen *before* it is secreted into the ECM.[19]

These data on the synthetic and degradative activities of fibroblasts are consistent with a search for precursor cells with more than one potential, and/or for specific precursor cells in specific regions. The findings are of especial interest for osteogenic cell lineage for *osteoclasts*

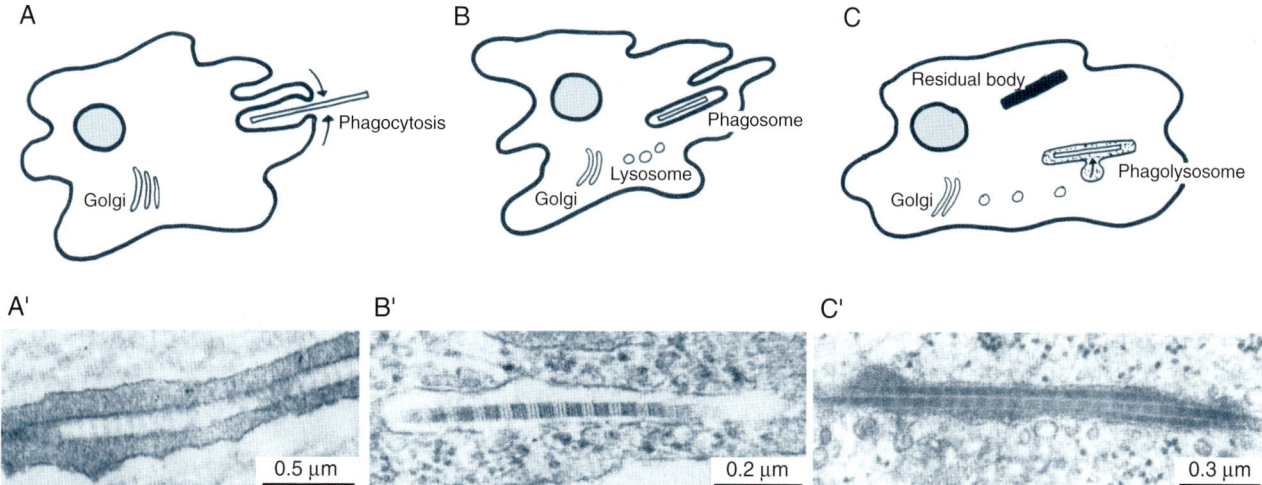

Figure 10.6 The correspondence between ingestion of a foreign body by a macrophage and ingestion of collagen by fibroblasts of the periodontal ligament. (A–C) Ingestion by a macrophage in which pseudopodia surround the foreign body (A) to enclose the foreign body in a phagosome (B) which then fuses with lysosomes to form a phagolysosome (C). (A′–C′) Phagocytosis of a collagen fibre by a fibroblast. The three stages correspond to A–C above. (A′) Phagocytosis. (B′) Collagen within a phagosome. (C′) Development of the phagolysosome. Modified from Ten Cate *et al.* (1976).

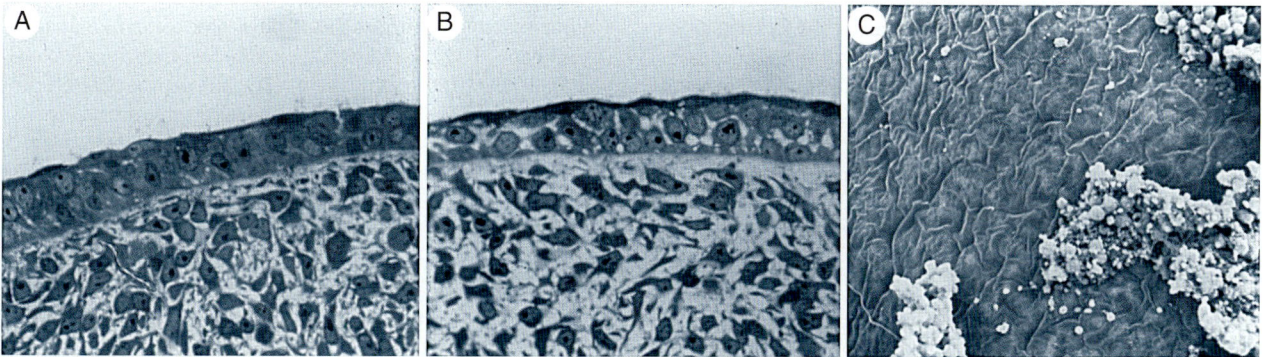

Figure 10.7 The basement membrane of the mandibular arch of chick embryos of H.H. 22, which is when the epithelium is required for osteogenesis to be initiated. (A, B) Two 0.5 μm sections stained with toluidine blue show the epithelium–mesenchyme interface on the side of the mandibular arch where epithelium induces bone: (A) where peripheral mesenchyme is apposed to the basement membrane, and (B) the lack of such contact on the medial face of the arch, where the epithelium is not inductively active. (C) A scanning electron micrograph shows the appearance of the basal lamina (and a few epithelial cells) after the epithelium has been removed following digestion with a solution of EDTA.

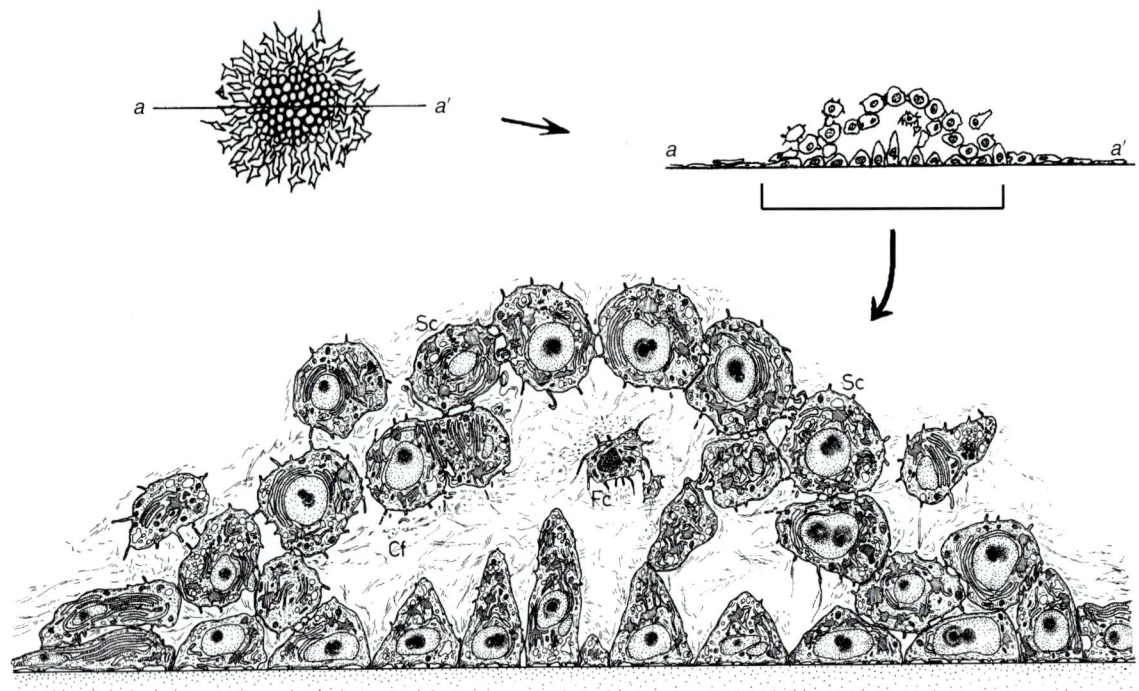

Figure 10.8 Clonal culture of cartilage cells. A clonal culture established using a single sternal chondrocyte from a 13-day-old chick embryo shows (top left) a surface view of the colony with central cartilage and peripheral fibroblasts, and (top right) a cross section of the colony along the line a-a'. Below is the colony in cross section as seen at higher magnification. Cf, collagen fibres in the lumen of the colony; Sc, fully differentiated chondrocytes at the surface of the colony and along the plastic substrate. Note how cell morphology adapts to attachment to the plastic. Modified from Eguchi and Okada (1971).

contain acid phosphatase, osteoblasts do not. Osteoclasts are now usually identified as TRAP-positive cells, a technique introduced in the early 1980s by Cedric Minkin (using mouse calvariae; see Chapter 15) for lineage relationships – or lack of a lineage relationship – between osteoblasts and osteoclasts.[20]

NOTES

1. Willmer's interests in evolutionary cytology arose from his seminal studies on the behaviour of cells in tissue culture, studies that he summarized in his book *Tissue Culture*, published in 1935 (2nd Edn, 1954) and in the influential two-volume treatise *Cells and Tissues in Cultures: Methods, Biology and Physiology*, published in 1965. Some 25 years' experience in tissue culture culminated in a 1951 review on evolutionary cytology and in *Cytology and Evolution*. The term ameloblasts as used by Willmer should not be confused with the same term used for those epithelial cells that deposit enamel (Chapter 1).

2. See Maclean and Hall (1987), Moody (1998) and Gilbert (2003) for analyses of dedifferentiation and redifferentiation. Both phenomena have been studied extensively in amphibian limb regeneration (Ferretti and Géraudie, 1998; and see Chapter 14).

3. See Hall (1970a) for common features among skeletal cells, and Haines and Mohiuddin (1968), Hall (1970a, 1972b) and Beresford (1981, 1983) for modulation and metaplasia.

4. Reticular tissue is a connective tissue with a high content of reticular fibres which are thin fibres composed of collagen type III (Miller, 1985).

5. See Maclean and Hall (1987) and Reddy and Joshi (1987) for stochastic models, and see the chapters in Moody (1998) for current views on cell lineage in many invertebrate and vertebrate organ systems.

6. I use the phrase 'adult' epibranchial (and 'adult' ethmoid in the example that follows) as shorthand for the epibranchial (ethmoid) that forms during metamorphosis to replace the larval cartilage, which is lost at metamorphosis. See Box 10.1 for skeletal development in frogs that lack a tadpole stage and consequently either lack metamorphosis or have a very abbreviated metamorphosis (Callery and Elinson, 2000; Callery et al., 2001). See Hall and Wake (1999), Hall (2004b) and Hall et al. (2004) for developmental approaches to larval stages and metamorphosis. See Schoch (2002) and Schoch and Carroll (2003) for how analysis of larval stages in exitinct taxa (Palaeozoic salamanders) can reveal both the life history of individual taxa and inform knowledge of relationships between higher taxa.

7. See Roček and Vesely (1989) for the ethmoid, Heatwole and Carroll (2000), Hall (2003b) and Heatwole and Davies (2003) for amphibian skeletons, Duellman and Trueb (1986) for amphibian biology, and Tinsley and Kobel (1996) for the biology of *Xenopus*.
Given that we have little information concerning the origin, identification or localization of cells producing the adult craniofacial skeleton in anurans, I have initiated a study with Neely Vincent to identify such cells in metamorphosing

tadpoles of the South African dwarf frog, *Hymenochirus boettgeri*, and to investigate how differentiation of the adult elements is initiated (Vincent and Hall, 2004).

8. See Kemp and Hoyt (1965a–c, 1969a,b) for studies on *Rana pipiens*, Hanken and Hall (1988b), Hanken and Summers (1988) and Hanken *et al.* (1989) for those on *Bombina*, C. S. Rose (1995a–c, 1996, 2004) for thyroxine-mediated resorption of urodele skulls, and see Shi (2000), Youson (2004) and the other chapters in Hall *et al.* (2004) for thyroxine and metamorphosis. See Tables 5.7–5.9 for data on sequence of ossification in lanternfishes. Daily injection of tadpoles of the northern leopard frog with 1 IU of parathyroid for two to three weeks slows metamorphosis and fails to reverse the inhibition of hind-limb development that occurs in animals treated with the thyroid hormone inhibitor thiourea, suggesting an indirect action of PTH via the thyroid. Hind limbs are bent as a result of failure of initiation of mineralization of long bones such as the femur (Kemp *et al.*, 1970).

9. See Fell (1931, 1932, 1933) for periostea and endostea in culture, Fell (1956, 1959) for overviews of the first 25 years of her studies, and Murray (1936) and Wlodarski (1989) for overviews of periostea. See Mathur (1979) for humerus and femur histogenesis in the crested tree lizard, *Calotes versicolor*, especially the condensation stage, formation of inner and outer layers of the periosteum, and transformation of perichondria to periostea during development.

10. If the perichondrium is stripped from Meckel's cartilage, it reforms from superficial chondrocytes (Kavumpurath and Hall, 1989).

11. O'Driscoll (1999) contains a good review of periostea as grafts to repair articular cartilage and subchondral bone. Figure 10.5 summarizes their model. O'Driscoll and Fitzsimmons (2001) is an excellent review of the role of periostea in cartilage repair, including a discussion of the nature of the periosteum, its ability to produce chondroblasts, and the roles of growth factors and mechanical influences.

12. See Decker *et al.* (1996) for porcine mandibular periostea and Chapter 12 for secondary chondrogenesis.

13. See Mayne *et al.* (1973) and von der Mark *et al.* (1977a) for transformation of chondroblasts to fibroblasts *without*

dedifferentiation, Deshmukh (1974) for bovine articular chondrocytes, Simny and Redler (1969, 1972) for chondromalacia, and Hsie *et al.* (1971) for transformation to fibroblasts. See Chapters 13 and 14 for dedifferentiation as a source of skeletogenic progenitor cells.

14. See Daniel *et al.* (1973) for general synthesis of GAGs, Hsie *et al.* (1971), Goggins *et al.* (1972) and Kulonen and Pikkarairen (1973) for modulation of synthesis and deposition of matricial products, Matalon and Dorfman (1966), Green and Goldberg (1968), Conrad (1970) and Levene and Bates (1970) for studies on the synthesis of ECM products by fibroblasts *in vitro*, and Mayne *et al.* (1971) for modulation of glucosamine.

15. Hurler syndrome (mucopolysaccharidosis type I), an autosomal recessive present in 1 in 10 000 births is normally lethal at age 13–15 years. Mental retardation and dwarfism are two of the major alterations that result from storage of abnormal amounts of mucopolysaccharides in the tissues as a result of not being able to produce the enzyme lysosomal alpha-L-iduronate.

16. See Goggins and Billups (1972) and Mayne *et al.* (1973) for production of similar GAGs by different cell types, and see Green and Goldberg (1968) and Langness and Udenfriend (1974) for the behaviour of clonal cells *in vitro*.

17. See Gay *et al.* (1976a) and Müller and Kühn (1977) for the immunofluorescence technique, Miller and Matukas (1969), Kosher *et al.* (1986a) and McDonald and Tuan (1989) for data on dedifferentiated chondrocytes, and Silbermann and von der Mark (1990) and Buxton *et al.* (2003) for mammalian and avian secondary cartilages.

18. See Ten Cate (1972), Ten Cate and Freeman (1974) and Beertsen *et al.* (1974a,b) for simultaneous synthesis and degradation of collagen, and Deporter and Ten Cate (1973) and Ten Cate and Syrbu (1973) for simultaneous localization of acid and alkaline phosphatases.

19. Ten Cate and Deporter (1975) and Ten Cate *et al.* (1976) reviewed the available data on response to mechanical stimuli. See Bienkowski *et al.* (1978) for the study with LACA.

20. See Burstone (1959), Lucht (1971, 1980) and Hall (1975b) for acid phosphatase and osteoclasts/blasts, and Minkin (1982) for TRAP staining.

Stem Cells in Adults

Self-perpetuating populations of determined osteogenic precursor cells can be isolated by cloning bone-marrow cells.

Aside from Neville Willmer's synthesis, the most extensive contribution to the resolution of the origin and diversity of mechanocyte (skeletogenic) precursors is that of Alexander Friedenstein and his colleagues at the Gamelaya Institute of the Academy of Medical Sciences in Moscow. Whereas Willmer concentrated on the origin of stem cells in embryos and in culture, the Russian work was on precursors of mechanocytes in postnatal and adult mammals. Friedenstein and Lalykina's 1973 monograph – *Bone Induction and Osteogenic Precursor Cells* – provides the most detailed analysis of their work. Unfortunately, it is in Russian. Their summary, reproduced below, captures the essence of the monograph, approach and major conclusions.

> The self-perpetuating population of determined osteogenc [sic] precursor cells can be isolated by cloning in monolayer cultures of bone-marrow cells. The determined osteogenic precursor cells have the characteristics of stem cells. Their progeny is [sic] capable of bone formation without any heterogenic inductive stimuli being required. Typing of determined osteogenic precursor cells in heterotopic bone-marrow transplants and in radiation chimaeras indicates that they are histogenetically independent of the haematopoietic stem cells. Another category of osteogenic precursors is present in the connective tissue and in populations of lymphoid cells (thymus cells, spleen cells, peritoneal macrophages, blood cells). These inducible osteogenic precursor cells require heterogenic inductive stimuli for bone formation to start. Bone induction by the transitional epithelium [of the bladder] and by decalcified bonn [sic] matrix is discussed in this book in detail using materiae [sic] largely collected by the authors. For both inductive systems the nature of the inductive stimuli, the composition of the reactive system and the properties of the induced bone are analised [sic]. (Friedenstein and Lalykina, 1973, p. 223)

FIBROBLAST COLONY-FORMING CELLS

One particular class of mechanocytes – stromal reticular cells of haematopoietic and lymphoid organs – was examined closely. Friedenstein's approach utilized monolayer conditions to produce clones of fibroblasts from large, mononuclear stem cells, which they termed *fibroblast colony-forming cells* (FCFCs). Under these conditions, cells cultured from bone marrow, spleen, thymus and lymph nodes form fibroblast colonies in quantities specific to each tissue, and with properties that vary depending on the tissue of origin. For example, while FCFCs from bone marrow initiate osteogenesis, those from spleen or thymus do not. Such differing potentials (determinations, in their minds) of apparently similar mechanocytes form the basis of the thesis that mechanocytes or their precursors differ in different haematopoietic organs.[1]

Friedenstein also produced evidence that mechanocytes or their precursors circulate within the blood vessels of adult mammals. FCFCs can be isolated from blood, albeit in low concentrations – one FCFC/100 000 leukocytes in guinea pig blood. This finding is relevant for ectopic osteogenesis after surgery, which seems to occur in more than half of those who have hip replacements and 17 per cent of those of us with fractured acetabulae (Box 11.1). Age-related differences exist between the osteogenic and haematopoietic potential of bone-marrow stem cells of young and old rabbits and rats. Cells from young animals produce

Box 11.1 Ectopic bone following surgery

The statistics in Table 11.1 show the incidence of ectopic osteogenesis following hip replacement in humans based on 12 data sets spanning two decades. The incidence can be alarmingly high.

Following their analysis of 237 cemented, total hip replacements, Hierton *et al.* (1983) concluded that high body weight, postoperative fever, and/or trauma are the major factors predisposing to ectopic bone formation. Orzel and Rudd (Table 11.1) examined 50 patients and found that ectopic bone formation was most common after muscle trauma or paralysis following total hip replacement, and recommend motion, steroids, aspiration, irradiation or

Table 11.1 The incidence of ectopic osteogenesis following hip replacement in humans[a]

Incidence (%)	Surgery	Source
21	Six months after total hip replacement	Brooker *et al.* (1973)
12–50	Following total hip replacement	Ayers *et al.* (1991) summarizing eight studies
50	A minimum of four months after total hip arthroplasty	Riegler and Harris (1976)
53	An average of 22.5 months after total hip arthroplasty	Orzel and Rudd (1985)
17	One year post-surgical treatment of acetabular fractures	Ghalambor *et al.* (1994)

[a] Some of these sources summarize several clinical studies.

diphosphonates to inhibit mineralization. Ayers *et al.* (1991) and Deflitch and Stryker (1993) found radiation to be an effective therapy in high-risk patients, the latter study commenting on reasons for treatment failure.[a]

With respect to thoracic surgery, Eke and Warrington (1981) described three cases of ectopic ossification in abdominal scars, each of which was characterized by deposition of lamellar and cancellous bone and formation of marrow. They discuss periosteal contamination, metaplasia of connective tissue and venous insufficiency as potential causes.

Ectopic ossification occurs as part of a small number of genetic syndromes, the most well known of which in humans is *myositis ossificans progressiva* (*fibrodysplasia ossificans progressiva*) in which the connective tissues progressively ossify. Another is *hereditary osteoma cutis* associated with Albright's hereditary osteodystrophy, which was studied in three generations of one family. Membrane bones were normal in these individuals but ectopic bones formed in the skin. Whether the ectopic bone formed from aberrant osteoblasts or from pluripotential cells could not be determined. A similar syndrome, *familial ectopic ossification*, is characterized by 'multifocal subcutaneous ossifications,' which in one individual extended along the muscle planes of one limb.[b]

[a] Radiotherapy can influence bone growth adversely. A dose of 15 Gy given to 15 children with early-stage Hodgkin's disease reduced growth of the clavicle by 30 per cent in comparison with controls (Merchant *et al.*, 2004). A Gy (Gray unit) named after a British physician, Hal Gray, is the unit of absorbed radiation, one Gy depositing one joule of energy in a kilogram of irradiated tissue. One Gy (equal to 100 rad) of ionizing radiation is sufficient to sterilize an individual.

[b] See Bona *et al.* (1967) and Maxwell *et al.* (1977) for *myositis ossificans progressiva*, Fawcett and Marsden (1983) for *hereditary osteoma cutis*, and Gardner *et al.* (1988) for *familial ectopic ossification*.

1.2 ± 0.7 colonies/10^6 cells, those from old animals, 0.15 ± 0.23 colonies/10^6 cells.[2]

Rat marrow-derived mesenchymal stem cells proliferate in a serum-free medium and retain the ability to form bone and cartilage when transplanted *in vivo*. Human mesenchymal stem cells with osteogenic and chondrogenic potential can be diluted with dermal fibroblasts such that, even in the presence of 50 per cent fibroblasts, expression of alkaline phosphatase and calcium increases *in vitro*, and bone still forms if the stem cells are implanted. This may be of practical importance. An implant of even 100 per cent mesenchymal cells is likely to be infiltrated with connective and vascular elements, diluting the effective skeletogenic population; marrow stem cells are now being used to engineer bone, in association with appropriate matrices in which they are implanted.[3]

OSTEOGENIC PRECURSOR CELLS

Because the studies carried out by Friedenstein and his colleagues on the origin of osteogenic precursor cells in postnatal mammals is such an excellent example of how to search for lineages of cells or determined precursor stem cells, their experiments are discussed in some depth. Three basic approaches were used.

(1) They first utilized *urinary bladder epithelium*, which Neuhof as early as 1917 had shown could induce bone formation (see later in this chapter). Eighty years later, Urist *et al.* (1998) demonstrated expression of Bmp in urinary bladder epithelium, implicating Bmp in osteogenic induction (Box 14.2). In Friedenstein's approach, epithelium of the urinary bladder was placed into a diffusion chamber, implanted *in vivo*, and normally *non-osteogenic host cells* were induced to undergo ectopic osteogenesis. Bone formation was induced by transitional epithelium at sites locally irradiated with 2000 to 5000 rads, confirming that inducible osteogenic precursor cells can migrate into the implant site.[4]

Lymphocytes from thymus, spleen, peritoneal fluid and blood – but not from lymphatic nodes – responded to signals from transitional epithelium implanted in diffusion chambers by producing ectopic bone. Each of the responsive organs possesses a population(s) of cells able to respond to a signal(s) from transitional epithelium by producing bone.[5] As discussed above, these organs possess populations of FCFCs. Whether FCFCs and the osteogenic cell populations are one and the same was the next question investigated.

(2) Bone marrow carrying a chromosomal marker was transplanted under the kidney capsule of host animals that had a different chromosomal marker. Osteogenesis was evoked from the implanted marrow cells. Therefore,

bone marrow contains a line of cells that can be induced to form bone ectopically. Do these cells arise from the stem cells of the haematopoietic tissue, or are they a separate stem cell line?

Reverse transplantation of repopulated marrow back to donor bone marrow demonstrated that ectopic bone is not resorbed by donor cells, i.e. it is immunologically the same as donor tissues, and therefore derived from host osteogenic marrow cells. Host marrow replaces the graft-cell population and is resorbed when transplanted back to the original donor. Therefore, marrow contains a line of osteogenic stem cells, distinct from the stem-cell line producing haematopoietic tissue. This was confirmed by irradiation experiments in which osteogenic precursors were maintained but haematopoietic precursors destroyed. Ectopic bone forms in such transplants, haematopoietic tissue does not. Storage at 0°C for one to two weeks inactivates the potential for ectopic osteogenesis, but does not destroy the population of determined precursor cells. Subsequently, a common precursor of osteo- and chondrogenesis was obtained from bone marrow.[6]

(3) In their third approach, cultures of embryonic spleen, liver and marrow, established from clonal cultures of FCFCs, were reimplanted *in vivo*. When bone marrow was placed into diffusion chambers, it routinely formed reticular tissue unless cell density was high, in which case bone formed. As one example, 10^7 marrow cells/chamber implanted intraperitoneally into athymic mice are actively dividing three days later; form alkaline phosphatase-positive cells by 13 days (Box 5.2); and form fibrous tissue, mineralized cartilage, bone, and tissue transitional between cartilage and bone after 20 days.[7]

Clonal analysis

In another pioneering study, Friedenstein *et al.* (1970) *cloned marrow progenitor cells*. Fibroblasts from these monolayer clonal cultures routinely formed bone when implanted. Fibroblasts from cloned spleen cells formed reticular tissue and not bone when implanted under similar conditions, but did form bone when transplanted in contact with the transitional epithelium of the urinary bladder. Confirmation that the bone arose from FCFC came when eliminating FCFCs from suspensions of spleen or thymic cells was shown to eliminate the osteogenic response. This result, when combined with formation of bone from clonally derived FTFCs, provides strong evidence for the dual potential of FTFCs.[8]

Osteogenic cell populations isolated from the marrow of skeletal elements such as mammalian tibiae or radii, undergo osteogenesis in the *absence* of an inductive stimulus. A greater *amount* of bone is produced if an inducer is present, but *initiation* of osteogenesis is not dependent on an inducer. This is not true for induced ectopic bone. Ectopic bone produced *in vivo* becomes populated with bone marrow, but the progenitor cells within the marrow must be induced to produce bone. Even though this

difference in osteogenic potential exists, the fibroblast colony-forming abilities of the two populations are similar (Friedenstein, 1976). Therefore, normal bone marrow either contains some inducible cells, or the activity of determined osteogenic precursor cells is increased by an inducer; recall that splenic FCFCs cannot initiate osteogenesis without an inductive stimulus. The conclusions from these important studies as they apply to osteogenesis are that adult mammals (or at least those species tested) possess two categories of osteogenic precursor cell populations:

- *determined osteogenic precursor cells* within bone and bone marrow; and
- *inducible osteogenic precursor cells* – precursor cells requiring induction to form bone – located within the blood, spleen, thymus and peritoneal fluids.

Independent studies undertaken in the 1970s and '80s by Maureen Owen and her group in Oxford support these conclusions. In a summary of the first decade of their work, Owen (1980) described bone cells arising in postnatal mammals from the soft connective tissues of periostea, endostea, Haversian canals and marrow stroma, and emphasized cellular interactions between haematopoietic, stromal marrow and bone cells. Determined osteogenic precursor cells of marrow are separate from the haematopoietic precursor cells of the bone marrow. In light of Willmer and Friedenstein's studies, these cells must be regarded as a lineage of determined cells, separate from those of the marrow, and separate from those stromal mechanocytes of connective tissues that also undergo osteogenesis.[9]

Lineages of cells

A further line of evidence for separate lineages of mechanocytes is the differential effect mesenchyme of different origins has on epithelial proliferation, differentiation and maintenance. For epithelia to continue to proliferate and grow they must interact with mesenchymal cells (Fig. 10.7). One aspect of such dependence is that culturing mesenchymal cells on epithelial basement lamellae alters mesenchymal cell behaviour (Overton, 1977; and see Chapter 18).

The reciprocal of this interaction – the effects of epithelia on mesenchyme – which operates during and after the stage at which mesenchymal cells are specified for skeletogenesis, and which provides an essential clue to one of the mechanisms controlling the sites within embryos where individual skeletal elements form, is discussed in Chapters 18 and 21.

Dexamethasone

Skeletogenic cells can be elicited from marrow stroma using a variety of agents, chiefly the synthetic adrenocorticoid *dexamethasone* or natural glucocorticoids (catabolic steroids) produced by the adrenal glands. The anti-inflammatory, immunosuppressive and blood calcium-regulating properties of glucocorticoids make them incredibly useful agents. With respect to identifying responsive cells in marrow, Wurtz et al. (2001) isolated a protein (RP59) with 2.6 Kb and 2.8 Kb components from a rat femur cDNA library. The protein is expressed strongly in the femur, spleen and cytoplasm of marrow cells (subsequently shown to be osteoblastic) and weakly in osteoblast nuclei.[10]

The synthetic adrenocortical steroid dexamethasone, which is effective at extremely low concentrations, is the glucocorticoid of choice to *evoke osteogenesis from stromal cells*. In the presence of 10 nM dexamethasone (and without any added phosphate source such as β-glycerophosphatase)[11] rat bone-marrow stromal cells produce mineralized nodules *in vitro*, mineralization being initiated in matrix vesicles. Chondrogenesis may also be initiated. As one might expect, markers of osteogenesis such as osteonectin mRNA are up-regulated in the presence of dexamethasone, as they are by retinoic acid and vitamin D$_3$. Osteoprogenitor cells recruited from chick bone-marrow stromal cell cultures by dexamethasone are self-renewing (proliferative) and alkaline phosphatase positive.[12]

When dexamethasone is administered in combination with other agents such as growth factors, later phases of osteogenesis can be targeted; the combined synergistic action of dexamethasone and Bmp-2 on rat stromal cells leads to rapid osteoblast differentiation by enhancing osteoblastic activity rather than by stimulating the proliferation of preosteoblasts (Rickard *et al.*, 1994). A mouse stromal cell line containing a virus expressing *Bmp-2* that induces ectopic bone in rodent muscle and speeds healing of femoral defects is a most promising gene therapy.[13]

Chondrogenesis can also be elicited from marrow stroma in response to dexamethasone – at least it can from rabbit bone-marrow stroma established in monolayer culture before being cultured as aggregates in the presence of 10^{-7} M dexamethasone; 25 per cent of the aggregates produce cartilage. Add Tgfβ and all cultures form cartilage (Box 15.1).[14]

Which cell types respond to dexamethasone? A monoclonal antibody against bone alkaline phosphatase (AP) labels osteogenic and preosteoblastic cells of calvariae and long bones, and so was used to immunoselect calvarial cells into AP-positive and AP-negative populations. Osteoprogenitor cells came out in the AP-positive population. Treating this population with dexamethasone led to a fivefold increase in the formation of bony nodules. However, treating the AP-negative population leads to a 30-fold increase. Counter-intuitively, the cells that responded to dexamethasone with enhanced osteogenesis (dexamethasone-dependent cells) were in the AP-negative population.[15] (Also see calvarial cells in Chapter 24.)

EPITHELIAL INDUCTION OF ECTOPIC BONE

Transitional epithelium of the urinary bladder

Transitional epithelium of the urinary bladder can induce ectopic osteogenesis from inducible osteogenic progenitor cells from a variety of tissues in postnatal mammals (above), demonstrating that epithelial–connective tissue interactions can elicit skeletogenesis *outside* the normal bounds of the skeleton from cells that would normally become neither chondroblasts nor osteoblasts. The site of the epithelium dictates the location of the ectopic skeletal nodule.

Charles Huggins carried out the classic studies inducing bone with transitional epithelium, studies that he began in the 1930s and continued to the late '60s; Huggins shared the 1996 Nobel Prize in Medicine with Peyton Rous for their work on hormonal regulation of cancer cells. Huggins implanted transitional epithelium intramuscularly or subcutaneously and showed that host connective tissue adjacent to the epithelium formed ectopic bone. No bone arose when transitional epithelium was transplanted into kidney, liver or spleen, unless fascia or connective tissue was included. Huggins and his colleagues concluded that only some tissues contain cells capable of responding to transitional epithelium.[16]

Huggins' original model for ectopic osteogenesis by transitional epithelium of the urinary bladder has been modified in various ways. Different epithelia have been tested. Different transplantation sites have been used in different species. And, of course, the molecular basis of the induction has been sought by analysis of epithelia, and by using individual molecules or molecular cocktails as inducers. By the late 1990s we knew that Bmp is expressed in urinary bladder, and that Bmp embedded within a fibrous glass membrane and implanted subcutaneously into rats induces cartilage, which is replaced by bone.[17] A small sample of studies demonstrating the efficacy of urinary bladder epithelium follows.

Ectopic bone forms in the bladder *in situ*; when transplanted subcutaneously, pelvic transitional epithelium of the pelvis induces ectopic bone (Tavassoli and Crosby, 1971).

Tumours of the urinary bladder evoke ectopic osteoid, bone and cartilage, while cells within some tumours possess or acquire skeletogenic ability themselves.[18] A correlation between tumours and ectopic skeletogenesis is a matter of interest in light of the relationship between epithelial proliferation and inductive ability discussed in the following section and in Chapter 18. Many tumours proliferate rapidly. Huggins observed that only actively growing epithelia were inductively active, Ioseliani (1972) correlated heightened epithelial mitotic activity with inductive ability, and we demonstrated that skeletogenetically active embryonic epithelia are mitotically active.[19]

Transitional epithelium can act on osteogenic progenitor cells located *intra*skeletally, as demonstrated in two studies on repair of fractured bones. Pieces of bladder dispersed in polyurethane sponge or in protein-free bone matrix accelerate repair of fractured cranial bones in guinea pigs and rats, while urinary bladder mucosa accelerates fracture repair in dogs. Whether the epithelium commits more cells to osteogenesis or accelerates the rate of bone formation of already committed cells is not known. Friedenstein's work suggests the latter.[20]

Skeletal tissues that form part of the normal skeleton become *independent* of the action of the inducer soon after induction. This is true for the apical ectodermal ridge (AER) and limb chondrogenesis, notochord/spinal cord and somitic chondrogenesis, mandibular epithelium and mandibular osteogenesis, otic capsular cartilages and otic epithelium, and so forth (Chapter 21). Formation of epithelially induced ectopic bone, however, *continues only* while the epithelium is present. This is also true of somitic cartilage induced *in vitro* (Chapter 41), and may be a general feature of induced skeletal tissues.[21]

Callis (1982) described ultrastructural events associated with bone induced by urinary bladder epithelium in guinea pigs. Bone appears 10 days following epithelial implantation, and is undergoing osteoclastic resorption by 17 days. Bone formation is inhibited by smooth muscle in the cyst, a finding consistent with the *inability* of cells from visceral organs to respond to induction. In guinea pigs, bone is induced *directly* without prior induction of cartilage, i.e. the mode of bone formation is intramembranous. In other species, a cartilage nodule is induced and replaced by bone, usually (always?) by cells from outside the ectopic cartilage.

This was an important statement. You may wish to read the last sentence again.

So often, when we speak of *bone induction* we really mean *induction of cartilage*, and subsequent *replacement of the cartilage by bone*. The induced cells and tissue are chondroblasts and cartilage, respectively. Bone comes later as cartilage is replaced by bone. The final skeletal element is *a bone* but cartilage is induced.

At least one epithelium actively stimulates *resorption* of bone: human gingival epithelia secrete a 10 000–70 000 kDa factor that stimulates resorption of adjacent alveolar bone, especially in periodontal disease. Human gingival cells contain collagenases that can cleave type I but not type II collagen; see Chapter 15 for specificity of resorption of skeletal matrices.[22]

Epithelial cell lines

A variety of epithelial cell lines induce ectopic cartilage or bone when transplanted *in vivo*, as first demonstrated by H. Clark Anderson and his colleagues in Kansas City, USA and then by Krzystof Wlodarski in Warsaw, Poland.[23]

After cultured human amniotic cells are injected intramuscularly into mice treated with cortisone to slow immunological rejection, tumours arise in which *host* fibroblasts modulate to chondroblasts and deposit cartilage, which is replaced by mineralized bone. Injection of ^{3}H-thymidine-labeled amniotic cells shows that only one 10th of the fibroblasts within muscle exhibit light labeling three days post-injection, a finding interpreted by Anderson and Coulter (1967) as reutilization of label from amniotic cells rather than transformation of amniotic cells to cartilage. It was suggested that this would be a good model for the progressive and debilitating disease, *myositis ossificans progressiva*, which has an incidence of one in a million in the human population, and in which intramembranous and endochondral ossification occur ectopically and progressively within muscles. Death usually comes from ossification of the muscles of respiration.[24]

Wlodarski published his first studies in 1969. He showed that a variety of established normal and neoplastic, murine and non-murine 'epithelial' cell lines – HeLa, WISH, CLV-J3, CLV-4, CLV-X, CLV-14 – evoke chondrogenesis and osteogenesis, but that primary cultures do not. Some of these cell lines are *not epithelial in origin* but are transformed fibroblasts – the CLV (cell line vaccine) line of human fibroblasts, for example – that exhibit an *epithelial pattern of growth* (see Box 11.2). Epithelial behaviour not epithelial origin 'gives' these cells their inductive ability. Wlodarski showed that: induction of ectopic cartilage is neither genus nor species specific; neonatal or X-irradiated mice do not form cartilage; and subcutaneous injection is ineffective, presumably because of a lack of inducible fibroblasts in such sites.[25]

The active cell lines Wlodarski tested all have one or more additional chromosomes – they are *heteroploid* – and agglutinate with concanavalin A; non-inductive cell lines are diploid and do not agglutinate.[26] In addition to being a useful marker for inductively active cells, the signaling mechanisms on host cells may be membrane based, which could imply that contact of injected cells with host fibroblasts is necessary for modulation to occur. Release of a diffusible signal by injected epithelial cells is also a possibility, but as the response is localized to injected muscle, any diffusible signal would have to be short range, and/or degraded rapidly. Furthermore, transfilter studies carried out by Clarke Anderson establish that induction fails to cross the filter barrier, further implicating cell-to-cell contact; see Fig. 11.1 for a transfilter culture from one of our studies. Nor will co-culture or co-CAM-grafting thigh muscle with epithelial cells elicit cartilage or bone formation, although both skeletal tissues form when muscle and epithelial cells are grafted intracerebrally.[27]

Wlodarski continued his studies throughout the 1980s, demonstrating that:

- when transplanted intramuscularly into lethally irradiated mice, transitional epithelium evokes cartilage, bone and marrow;
- Malony murine sarcoma virus and various tumour cells injected into the shank muscles of mice elicit periosteal

Box 11.2 Viruses, Paget's disease and the skeleton

Viruses influence osteogenesis in a number of ways.

Osteogenesis

The rubella (German measles) virus has a direct action on bone, inhibiting the growth of embryonic human and foetal rat bones *in vitro* (Heggie, 1977). Any potential relevance to virally transformed epithelia and bone induction (see text) has not been explored.

Infecting chick periosteal cells with Fujinami sarcoma virus transforms osteoblasts into cells resembling those found in an osteosarcoma, with the deposition of cellular bone and formation of multinucleated cells (Cogliano *et al.*, 1987).

Viruses can elicit bony tumours in mice. For example, the FBJ murine osteosarcoma virus induces tumours that contain chondroid, osteoid and mineralized bone; virally transformed human tumour KB cells will induce periosteal and ectopic bone; and Maloney murine sarcoma virus evokes periosteal (but not ectopic) osteogenesis when injected into mouse thighs. An interesting sidebar to these studies is that the ectopic bone does not develop a periosteum.[a]

Paget's disease (*osteitis deformans*), described by Sir James Paget in 1876 and characterized by abnormal bone remodeling, is manifest in the skull, vertebrae, and bones of the legs and hip, usually in men over age 40, afflicting some three per cent of men in that age range. Normal bone is replaced with soft, porous, weak bone, often so rapidly that excess bone forms, producing bone liable to fracture. Pagetic bone is excessively ossified. Although there is evidence of viral involvement, perhaps associated with rubella, a causative virus has not been identified.

Ultrastructural abnormalities in osteoblasts characterize Paget's disease. Whether viral particles are always present is less clear. Barbara Mills and her colleagues cultured cells from patients with Paget's disease through seven passages for eight months and found crystalline Ca^{++} in cytoplasm and extracellular matrix. Nuclear inclusions similar to those described in Pagetic osteoclasts were retained and the cells synthesized acid and alkaline phosphatase. Respiratory syncytial virus antigens were detected as inclusions in nuclei and cytoplasm of Pagetic osteoclasts. A viral antigen-bearing cell line, established from cells maintained for 3.5 years through 185 subcultures, consists of cells that are epith-

elioid, contain acid and alkaline phosphatase, and are responsive to vitamin D$_3$. Although these cells do not form bone *in vitro*, they do form osteosarcomas when implanted into athymic mice.[b]

Chondrogenesis

Viruses also affect cartilage and chondrogenesis.

Transformation of chondrocytes by a temperature-sensitive mutant of Rous sarcoma virus results in the synthesis of copious amounts of fibronectin, an extracellular matrix molecule not normally produced by chondrocytes. The mechanism is via translational control of type I collagen mRNA, with concomitant down-regulation of type II collagen mRNA synthesis (Adams *et al.*, 1982, 1987).

Fibroblasts transformed by a temperature-sensitive mutant (*ts-LA24A*) Rous sarcoma virus synthesize similar glycoproteins to those produced by *embryonic* chick vertebral chondroblasts (Cossu *et al.*, 1982), indicating at least a partial reversion to an embryonic state, a situation not true for many tumours (Maclean and Hall, 1987).

Murine chondrocytes can be immortalized by the introduction of a temperature-sensitive Simian virus-40 (SV-40). The treatment elicits features of hypertrophic chondrocytes, including synthesis of type X collagen, osteocalcin, and osteopontin, but not synthesis of type II collagen or deposition of extracellular matrix (Lefebvre *et al.*, 1995).

[a] See Yumoto *et al.* (1970) for the FBJ (Finkel, Biskis, and Jinkins) virus, P. K. Wlodarski *et al.* (1995) for the KB (human laryngeal carcinoma) cells, and Wlodarski (1985, 1989), Wlodarski *et al.* (1981, 1984) and Wlodarski and Reddi (1987) for Maloney virus.
[b] See Basle *et al.* (1978) for the ultrastructure, Mills *et al.* (1979, 1980, 1981, 1985) for the cell lines, and Mills (1991) for an overview. Clonal cell lines derived from murine osteosarcomas retain their tumourigenicity for many years in culture, and secrete a protein that induces bone when the cells are implanted *in vivo*. Although the osteoinductive factor is lost with prolonged culture, it is regained after implantation. See Kuettner *et al.* (1977b, 1978) and Kuettner and Pauli (1983) for the cancer cells, Amitani *et al.* (1974) and Amitani and Nakata (1975a,b, 1977) for the clonal cell lines, and see Sanerkin (1980) for definitions and distinctions between osteosarcomas, chondrosarcomas and fibrosarcomas of bone.

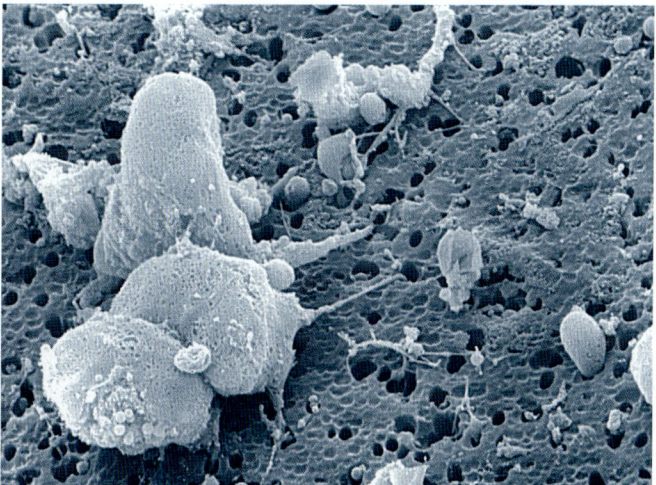

Figure 11.1 A scanning electron micrograph of embryonic chick mandibular mesenchymal cells cultured on the surface of a Nuclepore filter of 0.8 μm porosity, on the opposite side of which, epithelial cells have been cultured. The large mesenchymal cells are sending processes down through pores in the filter. The 'blebs' projecting up through some of the pores are epithelial cell processes.

proliferation and periosteal osteogenesis – viral particles are seen in the newly deposited periosteal bone but not in ectopic bone (see Box 11.2); and

- ectopic bone produced in response to neoplastic cells or viral injection fails to develop a true periosteum.[28]

The demonstration that a periosteum fails to form around ectopic bone has important implications; it says that a renewing cambial layer fails to form. This may be important when we consider that, as with most (all?) ectopic bone, bone induced by established cell lines is only maintained as long as the epithelial cells survive, in practice, several months, after which the bone is resorbed.[29] This finding may be compared with the studies in which transitional epithelium is used to evoke ectopic skeletogenesis, and to studies of epithelial–mesenchymal interactions during skeletal development (Chapters 18 and 21). Bone induced within the skeleton continues to form for decades after induction is over. Induction during normal development generates determined and proliferative skeletogenic stem cells.

Induction of ectopic bone induces cells that differentiate as osteoblasts or chondroblasts, not stem cells that persist to differentiate at a later date.

NOTES

1. See Friedenstein et al. (1976) for tissue-specific FCFCs, and Bab et al. (1984a) for their absence from spleen and thymus. Thymectomy of rats between one and five days postnatally decreases bone growth over the following three months with a concomitant decrease in monocytes and osteoclasts, because of reduced production of bone marrow (Gunther et al., 1980).

2. See Friedenstein (1976) for FCFCs from blood, Russell et al. (1986), Urist (1980) and Chapter 11 for ectopic (sometimes called heterotopic) osteogenesis, and Bernard and Dodson (1992) for age affects. Co-culturing rabbit marrow stem cells with rat osteoblasts increases the number of alkaline phosphatase-positive cells over that obtained from marrow cells cultured alone, but not as regards the numbers obtained from calvarial osteoblasts cultured alone (Kim et al., 2003). Eke and Warrington (1981) described three cases of (ectopic) ossification in abdominal scars, each of which was characterized by deposition of lamellar and cancellous bone and formation of marrow. As potential origins they discussed periosteal contamination, metaplasia of connective tissue and venous insufficiency. For a review of ectopic ossification in soft tissues, see Connor (1983).

3. See Lennon et al. (1995, 2000) for the stem-cell data, Service (2000) for an overview of 'engineered' bone and Bruder et al. (1994) for the mesenchymal stem-cell concept and cell lineages. Budenz and Bernard (1980) isolated two marrow populations that each produced more cartilage, bone and connective tissue than whole marrow cultures; see Tibone and Bernard (1983) for their model for intramembranous ossification from adult marrow stem cells.

4. For induced ectopic osteogenesis from normally non-osteogenic cells, see Friedenstein (1962), Petrakova and Friedenstein (1965), Friedenstein et al. (1967) and Abdin and Friedenstein (1972). Friedenstein (1973) did the irradiation study. Although inducible osteogenic precursor cells can migrate, they need not (Hall, 1975b).

5. See Lalykina and Friedenstein (1969) and Friedenstein and Lalykina (1970, 1972).

6. See Friedenstein and Kuralesova (1971) and Kuralesova (1971) for the irradiation experiments, which were confirmed by Amsel et al. (1969), Amsel and Dell (1971, 1972) and Elves and Pratt (1975). See Bierley et al. (1975) and Nelson (1975) for the effects of storage at 0°C, and see Friedenstein et al. (1987) for the common precursor for osteo- and chondrogenesis.

7. See Friedenstein et al. (1966) and Ashton et al. (1980, 1984) for the effect of cell density, and Bab et al. (1984a,b, 1986) for the time-course study.

8. See Friedenstein et al. (1968, 1970), Friedenstein (1973), Friedenstein and Lalykina (1973) and Bab et al. (1984a) for the responsiveness of cloned splenic fibroblasts in diffusion chambers.

9. The major studies from Maureen Owen's group are Owen (1970, 1971, 1985, 1988), Ashton et al. (1980), Bab et al. (1984a) and Owen et al. (1987). See Hall (1970a, 1975b) for ameloblast stem cells forming bone.

10. Kohyama et al. (2001) claimed to demonstrate that marrow stroma and isolated osteoblasts could transform ('meta-differentiate') into neurons in culture in response to the Bmp antagonist noggin or to a demethylating agent. Stimulated by two reports of formation of neurons from marrow cells (Brazelton et al., 2000; Mezey et al., 2000), Castro et al. (2002) found that they could not elicit neural cells from transplanted bone-marrow cells.

11. Benayahu et al. (1989) and J. N. Beresford (1989) isolated a stromal cell line that becomes osteoblastic in vitro, mineralizes in the presence of β-glycerophosphate and continues to form bone when implanted in vivo.

12. See Satomura and Nagayama (1991) and Leboy et al. (1991) for rat marrow stromal cells, and Kamalia et al. (1992) for chick.

13. See Rickard et al. (1994) for dexamethasone and Bmp-2, and J. R. Lieberman et al. (1998) for the gene therapy.

14. See Johnstone et al. (1998) and Cancedda et al. (1995, 2000) for chondrogenesis from bone-marrow stromal cells, and also for cell-culture models, condensation and the fate of hypertrophic chondrocytes.

15. See Turksen and Aubin (1991) and Turksen et al. (1992) for immunoselection, and Bellows et al. (1989) for chondrogenesis. These calvarial cells undergo chondrogenesis after 100 nM dexamethasone is added to the medium.

16. See Huggins (1931a,b, 1969), Huggins and Sammett (1933) and Huggins et al. (1936) for the pioneering work on transitional epithelial induction of bone. By the mid 1950s, alkaline phosphatase was being used as a marker for induced osteogenesis (Johnson and McMinn, 1956, and see Box 6.2).

17. See Urist et al. (1998) and Sasano et al. (1997).

18. Meningiomas are one example (Ball et al., 1975). See Box 23.2 for skeletogenesis in association with breast and prostate cancers.

19. See Willis (1962), Damjanov and Urbanke (1969), Delides (1972) and Urist et al. (1998) for proliferative activity of bladder tumours. See Hall (1980a,b, 1981a), Coffin-Collins and Hall (1989) and Hall and Coffin-Collins (1990) for embryonic epithelia and for regulation of mitotic activity by Egf.

20. See Beresford and Hancox (1965, 1967) for cranial bone repair, and Gilbert and Gorman (1971) for repair in dogs.

21. See Keilisborok et al. (1982) for induced bone, and Hall (1977a) for induced cartilage.

22. See Goldhaber et al. (1973) for resorption of bone by gingival epithelia, Robertson and Miller (1972) for collagenase activity, and Woessner (1973) for a review of mammalian collagenases. Gingival and skin fibroblasts produce similar GAGs when cultured (Goggins and Billups, 1972). The collagenases are subtypes of matrix metalloproteinases; see Box 13.3.

23. For overviews of interactions involving epithelial cell lines, see Bridges and Pritchard (1958), Bridges (1959), Ostrowski and Wlodarski (1971), Anderson (1976a,b) and Wlodarski (1985, 1991).

24. See Anderson et al. (1964) and Anderson (1967) for the initial studies, Anderson and Coulter (1967) for transplantation of labeled cells, and Anderson (1976a) for evidence that the host cells that modulate most readily in response to injection of amniotic cells are fibroblasts of skeletal muscle. See Bona et al. (1967), Maxwell et al. (1977) and R. Smith and Triffitt (1986) for myositis ossificans progressiva.

25. See Ostrowski and Wlodarski (1971) for epithelial growth not origin, Wlodarski *et al.* (1970, 1971a,b) for lack of species specificity and the effects of irradiation, and Hancox and Wlodarski (1972) for the data on subcutaneous injection.

26. Concanavalin A is a lectin isolated from the jack bean, *Canavalia ensiformis*. Depending on their cell-surface characteristics, some cells will bind to (agglutinate with) a specific lectin, others will not, providing a means of characterizing (and separating) cells.

27. See Wlodarski *et al.* (1974) for heteroploidy and agglutination, and Anderson (1976a) for the transfilter study.

28. See Wlodarski and Jakobisiak (1981) for intramuscular transplants, Wlodarski *et al.* (1981, 1984) and Wlodarski and Reddi (1987) for periosteal osteogenesis, Wlodarski (1985, 1989) and Wlodarski and Reddi (1987) for lack of periosteum formation, and Wlodarski (1991) for an overview of bone formation in soft tissues.

29. See Pritchard (1962), Wlodarski *et al.* (1970, 1971b) and Urist (1980).

Part V

Skeletogenic Cells

Regulation of the differentiative pathway entered by skeletogenic cells (scleroblasts) and whether or not that pathway can change with time and/or with variation in scleroblast microenvironment fascinated Charles Darwin, as can be seen from the following.

'The most interesting point with respect to supernumerary digits is their occasional regrowth after amputation. Mr. White [a physician] described a child, three years old, with a thumb double from the first joint. He removed the lesser thumb, which was furnished with a nail; but to his astonishment, it grew again, and reproduced a nail. The child was then taken to an eminent London surgeon, and the newly-grown thumb was wholly removed by its socket-joint, but again it grew and reproduced a nail. Dr. Struthers mentions a case of partial regrowth of an additional thumb, amputated when the child was three years old … [the] gentleman, who first called my attention to this subject … had fourteen children, of whom three have inherited additional digits; and one of them, when about six weeks old, was operated on by an eminent surgeon. The additional finger … was removed at the joint; the wound healed, but immediately the degit began growing; and in about three months' time the stump was removed for the second time by the root. But it has since grown again …'

(Darwin, 1861, pp. 14–15)

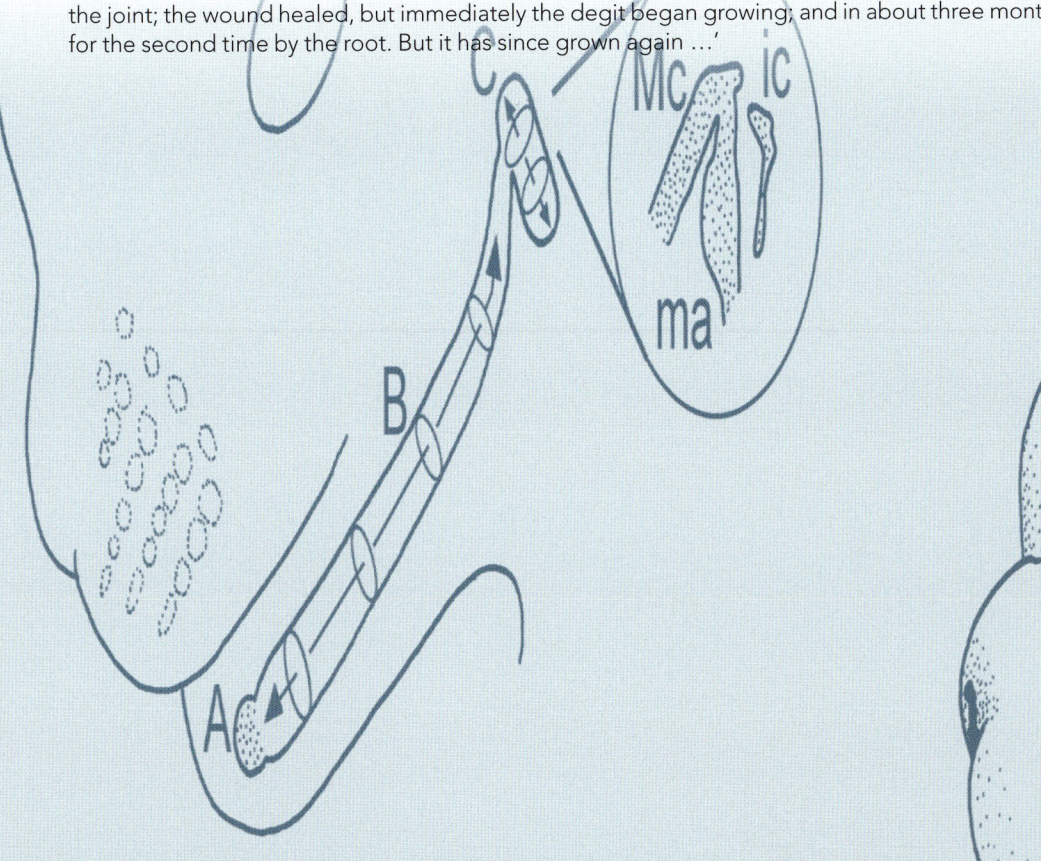

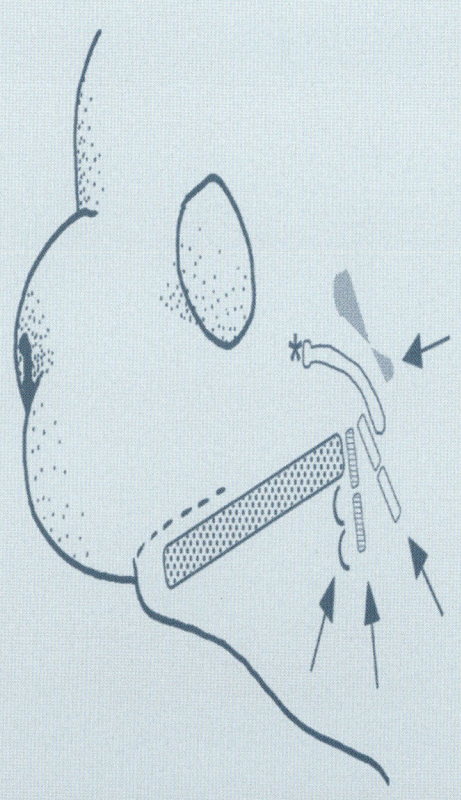

Osteo- and Chondroprogenitor Cells

Chapter 12

Not in rats and pigs, but in mice and men.

As are all stem cells, skeletal stem cells are set aside (determined) as separate lineages of progenitor cells some time before the onset of cytodifferentiation, or when they are found in the skeleton, whichever comes first. We can pinpoint the exact time when some lineages are set aside. For others, timing is much less precise. In either situation, we have substantial information on the conditions that must be met for cells to be or become skeletogenic. Contact with an epithelium or an implanted matrix and condensation are preconditions for differentiation (Chapters 18 and 19). Furthermore, the data and model presented and proposed in Chapter 11 show that specification is progressive; cells become more and more restricted in potential with each successive stage.

By the time they can be recognized topographically, progenitor cells are restricted in their potential. Whether particular populations of progenitor cell are restricted to only one fate – for example, differentiating as a chondroblast – or whether they possess two or more possible fates – differentiating as an chondroblast or an osteoblast – cannot be determined from their morphology or from their normal fate, except under unusually fortuitous circumstances; expressing one fate does not mean that the cells could not have expressed another under different circumstances.

The existence of tissues intermediate between cartilage and bone (Chapter 5) implies lability within particular scleroblasts for production of differentiated products normally produced by different progenitor cells and presenting as separate tissues. This, however, does not prove that particular scleroblasts are capable of responding to different environmental conditions by differentiating into one or another cell type. Happily, there is much evidence for bipotential progenitor cells, even in adults.

The literature on the ability of skeletal progenitor cells – more strictly, the progeny of progenitor cells – to differentiate into more than one differentiated cell type is vast, diverse, overwhelming, and intimidating. Inducible osteogenic progenitor cells, which can form fibroblasts or osteoblasts, are one such population (Chapter 11).

In this chapter I work through two more examples that illustrate that progenitor cells with ongoing skeletogenesis still possess the potential for forming more than one skeletal cell type and tissue. This is true in embryos and adults; in normal development and in repair and pathology. Admittedly, this potential declines with age, repeated trauma, or disease states, but it declines; it is not lost.

I discuss two case studies, on the basis of which it has been proposed that cartilage arises from periosteal (osteogenic?) cells, to evaluate whether the evidence allows the conclusion that progenitor cells in these systems are bipotential for chondrogenesis and osteogenesis. The two are (i) cartilage associated with the condylar process of the mammalian dentary; and (ii) secondary cartilage on craniofacial membrane bones of domestic fowl.

I then investigate *dedifferentiation* of existing cells as another way of deriving skeletogenic cells (Chapters 13 and 14). Limb regeneration is the classic example (Chapter 14) but the same phenomenon occurs during normal endochondral bone formation in mammalian long bones (Chapter 13). I round out the discussion of skeletogenic cells by turning to another aspect of bipotentiality, namely whether osteoblasts as bone-forming cells and osteoclasts as bone-resorbing cells arise from the same precursors (Chapter 15). This question has had a checkered history in studies of mammalian osteoclasts – which are multinucleated, by definition – but is even more complex, when we consider that *mononucleated cells* resorb bone in other taxa such as teleost fishes.

IDENTIFYING OSTEO- AND CHONDROPROGENITOR CELLS

In the previous chapter, I discussed the ingenious experiments of Alexander Friedenstein and his associates, which revealed osteoprogenitor cells within a variety of tissues in adult mammals. These experiments are even more remarkable when we recall that there were no morphological or genetic criteria to identify the cells within connective tissues or marrow *in vivo*. Identification was based on the product produced. Progenitor cells *can* be identified and visualized when they are associated with the surfaces of preexisting bone or cartilage, or when they are located in places in embryos – such as condensations (Chapter 19) – where we know osteogenesis or chondrogenesis will occur; location and knowledge of subsequent fate allow identification.

Execrable terminology

The terms *skeletogenic cells* or *scleroblasts* are generic and apply to cells that produce skeletal tissues, whether cartilage, bone, dentine or enamel. As with terms such as 'myogenic cells,' they reveal little about relationships between the cells (how do cardiac, striated and smooth myoblasts arise from myogenic cells?), embryonic origins, or how differentiation is regulated.

A variety of more useful terms are used to describe proliferative progenitor cells. Those associated with osteogenic activity have been termed spindle cells, reticulum cells, mesenchymal cells, *preosteoblasts*, or *osteoprogenitor cells* (OPCs). I prefer *OPCs* and the companion term *chondroprogenitor cells* (CPCs) for progenitors of osteoblasts and chondroblasts, respectively; although, as discussed below, the many sites where precursor cells can become chondrogenic or osteogenic makes the clumsier term *osteochondroprogenitor cells* (OCPCs) more appropriate. I reemphasize that these cells are committed as progenitor cells long before they can first be identified by virtue of their association with normal sites of skeletogenesis.[1]

Features

Histochemically, OPCs can be identified because of their strong staining for DNA, RNA, alkaline phosphatase and glycogen. Ultrastructurally, they are spindle-shaped, with a large nucleus and an extensive endoplasmic reticulum,[2] small Golgi bodies, many free ribosomes, and numerous mitochondria. These features are characteristic, whether the cells are associated with endochondral or membrane bone, cortical or endosteal bone, or at fracture or ectopic sites. Uptake of isotopically labeled precursors, response to hormones, pattern of isozymes and distinctive gene products are additional means of identifying these progenitor cells.[3]

Cell cycle dynamics

Population kinetics of progenitor cells were examined in detail by several workers, in particular Norman Kember and Maureen Owen.[4] Bear in mind that progenitor cells vary from site to site and time to time. Given this caveat, I estimate cell cycle data for an 'average' progenitor cell for mammalian cartilage or bone as follows:

S phase of DNA synthesis = 7 hr;
G_2 phase of growth = 1.5 hr; and
M phase (mitosis) = 1.25 hr.

In a study using rat proximal tibial metaphyses that contains excellent criteria for distinguishing osteoprogenitor cells, osteoblasts and osteoclasts, Donald Kimmel and Webster Jee identified two populations of osteoprogenitor cells. One has a high turnover and resides in primary spongiosa. The second has a low rate of turnover and is located in secondary spongiosa (Table 12.1). Bone in primary spongiosa is added at a rate of 41.2 μm^2/day, bone in the secondary spongiosa at only a fifth that rate or 9.0 μm^2/day (Table 12.1). Loss of cartilage and addition of bone are balanced, the balance being especially close in primary spongiosa (Table 12.1). T_c for osteoprogenitor cells of the primary spongiosa is 39 ± 18 hours, indicating variation of as much as three quarters of a day in either direction in a cycle that is one and two thirds days long.[5]

Although I have described OPCs and CPCs with uniform morphology, as if a single category of cells exists in all species and classes of vertebrates, lineages of progenitor

Table 12.1 Per cent of the total amount of osteoprogenitor cells, osteoblasts and osteoclasts, and rates of cartilage and bone loss and addition, in primary and secondary spongiosa of growing long bone metaphyses in rats[a]

Distance from GCMJ[b] (mm)	Age (days)	% osteoprogenitor cells	% osteoblasts	% osteoclasts
Primary spongiosa[c]				
0.1	0.5	35	43.3	21.5
0.3	2	30	29.4	15
0.8	4.5	10	8.6	16
Secondary spongiosa[d]				
1.2	7	4.7	2.8	13.6
1.6	9.5	4.9	4.1	8.6
2.0	12	4.2	3.2	7.1
2.5	14.5	3.1	3	4.6
3.0	17	3.4	2.4	5.6
3.3	20	2.7	1.6	3.8
3.8	22	1.6	1.3	3.2
4.2	25	0.6	0.4	1.1

[a] Based on data from Kimmel and Jee (1980).
[b] GCMJ: growth cartilage–metaphyseal junction.
[c] Cartilage is lost from the primary spongiosa at the rate of 0.036 mm^2/day, and bone added at the rate of 0.041 mm^2/day.
[d] Cartilage is lost from the secondary spongiosa at the rate of 0.0014 mm^2/day, and bone added at the rate of 0.001 mm^2/day.

cells do exist. Determined and inducible osteogenic progenitor cells found outside the skeleton (Chapter 11) are two such lineages.

BIPOTENTIAL PROGENITOR CELLS FOR OSTEOGENESIS AND CHONDROGENESIS

Bipotential cell populations or bipotential cells?

We can think of bipotentiality at two levels. The first is bipotentiality of *individual cells*; the progeny of an individual cell have the potential to become one or other of two different cell types. Here we seek the basis for bipotentiality in the 'switch' that sends a cell down one pathway rather than another, and seek to understand the genetic and cellular machinery allowing such bipotentiality to be expressed. The second is a bipotential *population of cells*. This may be a misnomer, but is often what is meant when bipotentiality is investigated. A bipotential population often consists of two (or more) unipotential populations. Bipotentiality in this instance – seen in limb-bud and neural crest-derived mesenchyme – involves a 'switch' selecting one population of cells to differentiate (perhaps also to proliferate) over another. In the first, we could seek to isolate bipotential stem cells. In the second, we could seek to expand one cell population at the expense of another neither of which need be stem cells or bipotential.

Uncovering bipotentiality

An enormous body of literature supports the conclusion that, in many skeletal sites, progenitor cells for osteoblasts (OPCs) are also progenitors for chondroblasts (CPCs). Because OPCs reside within periostea of membrane bones, we should not assume that their potential is limited to differentiating as osteoblasts. Such cells may differentiate into chondroblasts or fibroblasts, depending on local conditions and circumstances. Contrary findings can often be attributed to local conditions that are unsuitable for chondrogenesis: Herold *et al.* (1971) found that cartilage extracts in gelfoam implanted into calvarial defects accelerated osteogenesis repair without any formation of cartilage, and so concluded, erroneously, that only endochondral bones formed cartilage in repair. In reality, conditions were not appropriate to evoke chondrogenesis, and membrane bones can form cartilage during repair.[6] Alternatively, bones in some taxa have minimal periostea (Fig. 12.1) and so have limited ability for periosteal chondro- or even osteogenesis. Although Meckelian chondrocytes are released from their capsules in spotted salamanders whose lower jaws have been fractured, the released cells have little potential to reinitiate division or to repair the cartilage (Fig. 12.2).

Alternatively, without a time series (longitudinal study), potential to differentiate along a particular pathway may not be revealed. A study testing the osteoinductive properties of implanted bovine frontal bone matrix will serve as an example. Demineralized endochondral bone was shown to elicit cartilage in two-thirds of implants when demineralized membrane bone elicited none (0/18) under the same conditions. It was therefore suggested that membrane bone be chosen as the graft material when treating craniofacial fractures where cartilage differentiation may be disadvantageous. However, the claim of no cartilage phase in the repair was based on examining graft sites no earlier than 39 days after grafting, by which time we expect any cartilage present to have been replaced by bone.[7]

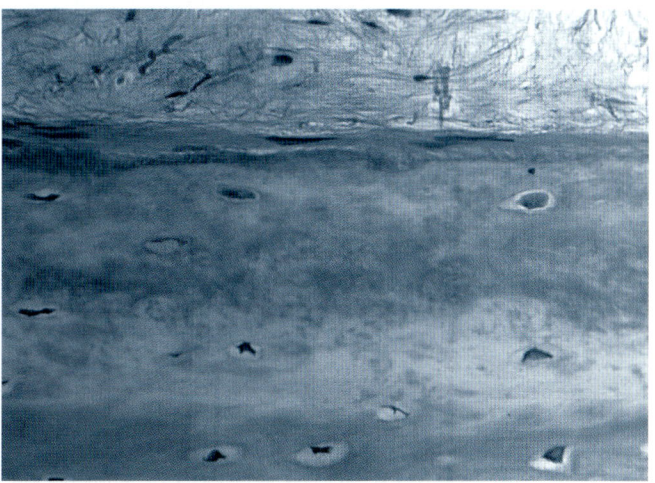

Figure 12.1 A histological section of a dermal bone of the Mexican axolotl, *Ambystoma mexicanum*, to illustrate the low cellularity and virtual absence of a periosteum.

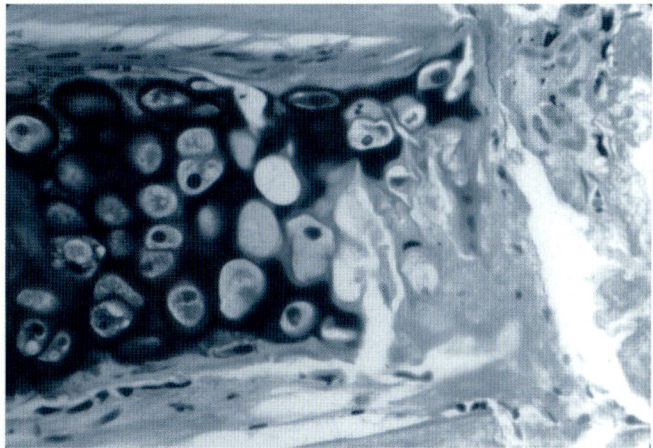

Figure 12.2 Three weeks after fracturing the jaw of an adult spotted salamander, *Ambystoma maculatum*, Meckel's cartilage at the fracture site is undergoing matrix erosion and releasing chondrocytes, but there is minimal evidence of reinitiation of cell division. See Hall and Hanken (1985a) for details.

Discovering bipotentiality

Stating that progenitor cells *can* differentiate into osteoblasts or chondroblasts describes their potentialities. In situations such as the formation of condylar cartilage on the mammalian dentary or of secondary cartilages on avian membrane bones (discussed later in this chapter) or in cartilage formation during repair of fractured bones (Chapter 29), we know that mechanical stimulation is the environmental signal to which the progenitor cells respond. While this describes the *conditions* under which a particular cytodifferentiation occurs, it says nothing about *how* progenitor cells receive and translate the environmental signal, nor which physiological and metabolic activities must be altered or set in motion for chondrogenesis to be initiated. Potential to differentiate as an osteoblast or chondroblast exists in progenitor cells *before* they are exposed to the environmental signal(s), such exposure being a prerequisite to *expression* of a determined state established earlier. At least three important areas concerning the differentiation of progenitor cells must be addressed:

- identification of the environmental factors that direct progenitor cells to differentiate along a particular pathway;
- understanding how the progenitor cells receive these messages; and
- knowing which genetic, biochemical, metabolic and structural changes are set in motion as progenitor cells differentiate along one pathway rather than another.

Biochemical and metabolic markers

I spent much time at the beginning of my research career and into the 1970s searching the literature for biochemical (largely histochemical) features of chondro- and osteogenesis that would indicate how differentiation was controlled, and continued then to gather information on how the two cell states can be distinguished using enzyme profiles. The rationale was that cartilage and bone differ metabolically, and so seeking metabolic changes associated with early differentiation should provide clues to how differentiative decisions are established. Although I once thought otherwise, cataloging the differences between chondroblasts and osteoblasts is not a useful way to understand how they differentiate; by the time they are 'blasts' the decisions are history.[8]

Another approach was metabolic, using isozymes of lactate dehydrogenase (LDH) and the ratio of LDH to MDH (malate dehydrogenase) as measures of anaerobic chondroblast and aerobic osteoblast precursors. Among approaches to distinguishing CPCs from OPCs, or to determining how chondroblasts or osteoblasts differentiate from common progenitor cells, was to determine the ratio of collagen to acid mucopolysaccharides produced by cells. A close correlation between the concentration of collagen and acid mucopolysaccharides in developing long bones in chick embryos holds even after administration of various disruptive agents (Table 12.2; Fig 4.2).[9] I made these attempts because, until the late 1970s, adequate criteria to unequivocally separate osteogenic from chondrogenic cell types at early stages in their differentiation were not available, although differentiated cell types could be identified and separated. For example, many used differential synthesis and accumulation of sulphated glycosaminoglycans (GAGs) as the marker for chondroblast–chondrocyte differentiation, but osteoblasts and fibroblasts also synthesize sulphated GAGs, albeit in lower amounts.[10]

An observation that has not been explored in any depth is that *chondrocytes are thermoresistant but osteogenic cells thermosensitive,* at least as based on studies showing

Table 12.2 Concentrations of hydroxyproline and hexosamine, and molar ratios of hexosamine to hydroxyproline, in developing tibiae of embryonic chicks treated *in ovo* with thallium, tetracycline, vitamin C or cortisone[a]

Treatment	Hexosamine (μmol/mg wet wt)	Hydroxyproline (μmol/mg wet wt)	Molar ratio
Control	2.75 ± 0.12	0.76 ± 0.05	3.65
Thallium (0.6 mg)	2.79 ± 0.7	0.78 ± 0.04	3.58
Tetracycline (1 mg)	2.87 ± 0.18	0.79 ± 0.04	3.62
Tetracycline (2 mg)	2.64 ± 0.20	0.72 ± 0.05	3.61
Tetracycline (2.5 mg)	3.38 ± 0.14	0.98 ± 0.03	3.44
Vitamin C (2 mg)	3.02 ± 0.17	0.83 ± 0.04	3.63
Vitamin C (4 mg)	1.55 ± 0.13	0.55 ± 0.13	2.82
Vitamin C (6 mg)	1.76 ± 0.26	0.41 ± 0.09	4.32
Thallium (0.6 mg) + cortisone (2 mg)	3.10 ± 0.11	0.96 ± 0.02	3.21
Thallium (0.6 mg) + vitamin C (2 mg)	2.37 ± 0.04	0.70 ± 0.2	3.39
Thallium (0.6 mg)+ vitamin C (4 mg)	1.88 ± 0.18	0.16 ± 0.01	3.08
Thallium (0.6 mg) + vitamin C (6 mg)	2.16 ± 0.14	0.67 ± 0.1	3.23

[a]Based on data from Hall (1973b). Agents all administered at seven days of incubation and embryos recovered at 14 days of incubation.

that after four days at 4°C chondrocytes show no change in proliferation or mitochondrial activity, but enhanced protein synthesis and cell volume; while osteogenic cell lines (MC3T3.E1 and Ros 17/2.8) show reduced proliferation, and enhanced protein synthesis, cell volumes and mitochondrial activity (Flour et al., 1992).

Collagen types

The discovery that collagen differs in its primary structure in a tissue-specific manner, the characterization of procollagens, and translation of collagen mRNA in cell-free systems provided sensitive markers for osteogenic and chondrogenic cells. Development of immunohistochemical methods to localize procollagen in osteoblasts and osteoid allowed collagens to be visualized within cells and tissues.[11]

But becoming a chondroblast rather than an osteoblast involves more than the synthesis of type II rather than type I collagen,[12] as the following examples show.

- Type II collagen is regulated transcriptional but type I translationally in mesenchyme of chick limb buds.
- mRNA for Col 2a1 – the $\alpha1(I)$ subunit of type II collagen – is 'switched off' before mRNA for $\alpha2(I)$ and $\alpha1(II)$ subunits, and there is two-step synthesis of procollagen type I when chondrocytes dedifferentiate as, for example, if cultured under low cell density.
- Human foetal epiphyseal chondrocytes dedifferentiate under monolayer culture conditions, with down-regulation of mRNA for type II collagen and up-regulation of message for types I and III. Placing these chondrocytes into agarose culture promotes redifferentiation with concomitant up-regulation of type II collagen mRNA.
- The pro $\alpha1(II)$ chain of chondroblasts is under-methylated when compared with pro $\alpha1(II)$ in fibroblasts or red blood cells. Degree of methylation is not affected when chondroblasts dedifferentiate.
- Type II collagen mRNA is transcriptionally regulated in prechondrocytes with transcription increasing substantially at condensation (Chapters 19 and 20). Although type I is regulated at the translational level, type I mRNA is found in differentiated chondrocytes.
- Mesenchymal cells of H.H. 24 chick limb buds contain 20 copies of type II procollagen mRNA – as assessed using a cDNA probe – a number that rises to 2000 copies by H.H. 31. Sternal chondrocytes from 17-day-old embryos contain 10 000 copies, indicative of differential expression in the two cartilages.
- Alternatively spliced mRNA encoding type II procollagen with a cysteine-rich amino-propeptide is preferential expression in early cartilage and epithelium.
- Two forms of alternatively spliced type II procollagen mRNA are differentially distributed during vertebral development in humans, IIB in the chondrocyte extracellular matrix (ECM), and IIA in prechondrogenic mesenchyme around cartilage and in spinal ganglia.[13]

Type IIA procollagen is an alternatively spliced product of the collagen type II gene. In chick embryos between five and nine days of incubation, type IIA is found in the ECM of condensations and in early differentiating cartilage. Subsequently it is found in perichondria and then in early hypertrophic chondrocytes. Interestingly, type IIA is not found in normal human articular cartilage but it is found in osteoarthritic cartilage. Type IIA is not found in the human intervertebral disc, evidently because of regulated expression of the NH_2-propeptide domain.[14]

As you can see, specific collagen types provide excellent markers. Even so, we only have a glimpse of the biochemical events required for a progenitor cell to differentiate along one pathway rather than the other. An example using a tumour suppressor gene illustrates this further.

The tumor suppressor gene p53

Whether a chondrogenic line established from mice containing the tumor suppressor gene p53 differentiate all the way to hypertrophic chondrocytes or only to prehypertrophic chondrocytes depends on the signals to which they are exposed.

p53 targets genes that regulate cell proliferation and apoptosis. Newborn $p53^{(-/-)}$ mice have no skulls (exencephaly). Minimal TUNEL (terminal deoxynucleotidyl transferase biotin-dUTP nick end labeling) activity exists within hypertrophic chondrocytes of the tibial growth plates (Figs 4.8 and 4.9), the level of cell death resembling that seen in 15-day-old wild-type embryos, indicating delayed cartilage maturation. Similarly, TUNEL analysis was used to examine the balance of apoptosis and decreased cell proliferation in digital growth of mouse limb buds.[15]

In order to investigate chondrocyte dynamics and response to growth factors and/or hormones, Kamiya et al. (2002) established a chondrogenic cell line (N1511) from the rib cartilage of a p53 null mouse. Once confluence was reached, the cells differentiated, the pathway of differentiation depending on the cocktail of growth factors and hormones provided.

In the presence of parathyroid hormone (1×10^{-7} M) and dexamethasone (1×10^{-6} M) the cells form alcian blue-positive nodules and are negative for alizarin red and alkaline phophatase, indicating a chondrogenic phenotype. Collagen type II appears slowly in the cultures, followed by type IX, but no type X, indicating that chondrogenesis is arrested before chondrocyte hypertrophy.

No cartilaginous nodules develop in the presence of Bmp-2 (50 ng/ml; Box 12.1) and bovine insulin (1×10^{-6} M). Instead the cells are positive for alizarin red and alkaline phosphatase, indicating an osteogenic phenotype. Collagen types II, IX and X are deposited in a temporal sequence as the cells mature. Type X, of course, is indicative of hypertrophic chondrocytes not of osteogenic cells (Chapter 22).

I now discuss the two case studies: the condylar process of the mammalian dentary, and secondary cartilage on

Box 12.1 Bmp-2 and Bmp-4

Bmp-2 (*BMP2A*) and *Bmp-4* have been cloned from various vertebrates, including zebrafish, *Danio rerio* (*swirl* is a nonsense mutation of *Bmp-2*), *Xenopus* and mice.[a] Anderson *et al.* (2000) localized seven Bmps (*Bmp-1* to *Bmp-7*) in developing human long bones, their major findings being that:

- Bmp is cytoplasmic in localization;
- the highest levels are in hypertrophic and mineralizing growth plate (relating either to terminal differentiation or cell death in the growth plate), osteoblasts and osteoprogenitor cells; and, surprisingly,
- there are high levels in osteoclasts related, they speculate, to coupling of osteoclast–osteoblasts required to activate osteoclasts.

The chromosomal location of several human *BMP* genes has been mapped; *BMP-1* to chromosome 8, *BMP-2* to chromosome 20 and *BMP-3* to chromosome 4. *BMP-2* and *BMP-3* are conserved between mice and humans and map to regions of loci associated with disorders of cartilage and bone development (Tabas *et al.*, 1991). Human *BMP-2* was purified and characterized by Wang *et al.*

(1990), and a dose and time-course study undertaken to show that it induces cartilage (day 7) and bone (day 14).

Over-expressing *Bmp-2* and *Bmp-4* increases the volume and alters the shape of chick limb mesenchyme *in vitro* by a mechanism involving enhanced synthesis and deposition of ECM and recruitment of cells rather than via enhanced proliferation. Hypertrophy and periosteal ossification are delayed (Duprez *et al.*, 1996).

Bmp-2 and *Bmp-4* both are expressed in chick facial epithelia, while Bmp in facial mesenchyme controls the outgrowth of facial processes. Bmp-2 implanted on beads into embryonic chick mandibular processes affects mandibular skeletogenesis, in part by initiating local cell death (Fig. 12.3). Important interactions between *Bmp-4* and *Msx* genes are discussed in Box 13.4[b]

Four 70-amino-acid, cysteine-rich domains allow *chordin* to bind to *Bmp-4*, making chordin an inactivator of *Bmp-4*. Similar cysteine-rich domains are found in other molecules, such as procollagens and procollagen-IIA mRNA, which like chordin can function as a dorsalizing agent in specifying the dorsal region of early embryos. Chordin begins to appear early in development; high concentrations are found in murine chondrogenic condensations,

Figure 12.3 Bmp-2 affects both proliferation and cell death, demonstrated by implanting Bmp-2-soaked beads into chick embryos. (A) An embryo 24 hours after an unloaded (control) bead had been implanted into the right mandibular arch (arrow) at Hamburger–Hamilton (H.H.) stage 22. Both developing mandibular arches (m) are of equal size and stage of development. (B) An embryo 24 hours after a bead loaded with 75 ng/μl Bmp-2 had been implanted into the right mandibular arch (arrow) at H.H. 22. The implanted (right) mandibular arch (m) is substantially smaller than the left arch. Also compare with A. (C) The mandibular arches (m) from an embryo of H.H. 22 cultured for 24 hours after an unloaded (control) bead (b) had been implanted into the right mandibular arch. (D) Cell death has been initiated in the mandibular arch into which a bead (b) loaded with 75 ng/μl Bmp-2 was implanted 24 hours earlier. Compare with C. (E) The mandibular arches of an embryo similar to that in D to show cell death around the bead (b) and the much smaller size of the mandibular arch (m) into which Bmp-2 had been implanted. (F) Little cell death has occurred in the mandibular arch into which a bead (b) loaded with 25 ng/μl Bmp-2 was implanted 24 hours earlier. Compare with D and E. Modified from Ekanayake and Hall (1997).

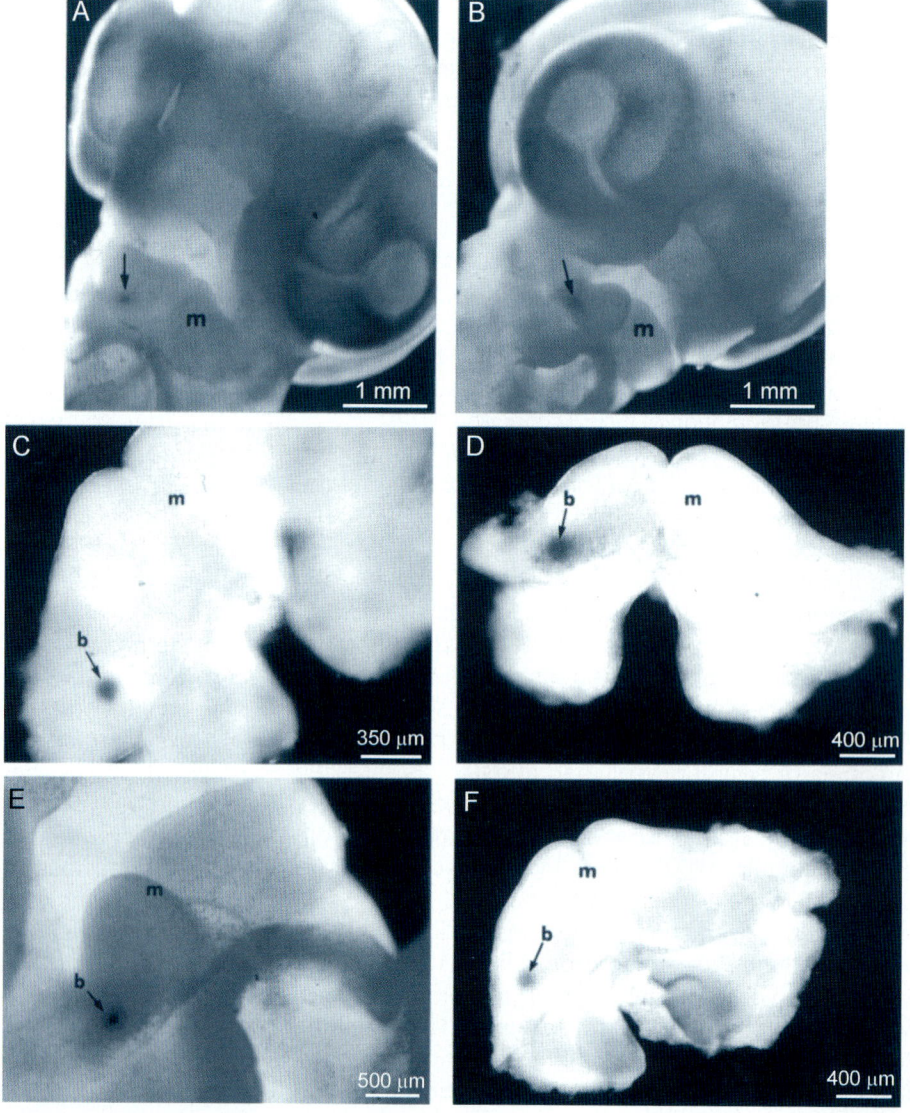

Box 12.1 (*Continued*)

differentiating cartilages and the central nervous system from 11.5 days onwards. Homozygous *chordin* mutant mice die perinatally with extensive craniofacial defects, some of which are similar to human DiGeorge syndrome. Chordin up-regulates *Tbx-1* and other transcription factors required for pharyngeal development.[c]

Bmp dimers

Growth factors usually act in combinations or in concert, either by *dimerization* or by joint action such as binding to the same receptors or being activated by the same signals; Bmp-2 and Bmp-4 act independently but also form dimers with enhanced activity. To take but two examples:

- a point mutation in *Bmp-14* (*Cdmp-1; cartilage-derived morphogenetic protein-1*) that blocks Bmp secretion from cells, influences limb morphogenesis, primarily because Bmp-14 regulates the secretion of other Bmps by forming heterodimers with them;
- using chambers implanted into the proximal metaphyses of the tibiae of five rabbits, it was shown that decreased bone formation was greater after implanting bovine collagen sponges containing both Bmp-2 (0.6 μg/implant) and Fgf-2 (0.1 μg/implant) than after implanting sponges with either growth factor alone.[d]

Xenopus

In *Xenopus* embryos, *Bmp-2* and *Bmp-4* are expressed in the neural crest, fin mesenchyme, olfactory placodes and craniofacial primordia. Bmps-2, -4 and -7 all are differentially transcribed in early embryos. The functional equivalent of *Xenopus* and mammalian Bmps was shown by demonstrating that *Xenopus* Bmp-4 up-regulates alkaline phosphatase in mammalian cells and cell lines.[e]

Action on cell differentiation

Mouse xiphoid cartilage transdifferentiates to adipocytes, and clonal myoblasts (C2C12) are converted into osteoblasts, in response to Bmp-2 *in vitro*. Ninety to ninety-five per cent of C3H10T1/2 cells associated with *Bmp-2* in a retroviral construct in micromass culture differentiate into chondrocytes, but direct infection is needed.[f]

Osteogenesis

Bmp-2 promotes osteogenesis under a variety of conditions and from a variety of cell types, as outlined below.

Bmp-2, -4 and -6 activate osteoprogenitor cells from rat calvariae *in vitro*, Bmp-2 being the least potent. Conditioned medium from rat calvarial osteoblast cultures stimulates osteogenesis (both

proliferation and differentiation) from rat bone marrow stromal cells via a 10–30 kDa molecule that is not Tgfβ. Osteogenesis also is promoted in foetal rat calvarial cells exposed to recombinant Bmp-2 (as measured by increased alkaline phosphatase and osteocalcin mRNAs), as is *Bmp-2* mRNA expression and *Bmp-2* promoter activity.[g]

Recombinant human BMP-2 (RhBMP-2) induces ectopic bone formation from muscle cells, both *in vivo* and *in vitro*, and induces osteoblastic differentiation from W-20-17 – which is a bone marrow stromal cell line – as evidenced by increased alkaline phosphatase, development of a cAMP response to PTH, and synthesis of osteocalcin.[h]

In three separate osteogenic cell populations (calvarial cells, C3H10T1/2 and MC3T3-E1) Ogata *et al.* (1993) showed that addition of Bmp-2 increased expression of the gene *Id* (*inhibitor of differentiation*) that encodes a helix–loop–helix molecule, increased expression being evident after 24 hours and for as long as 96 hours in culture. The Bmp-2 effect was not blocked by actinomycin D but was blocked by cyclohexamide, indicative of an effect on post-translational control. If mRNA for *Id* was enhanced by exposing cells to dexamethasone, or when mRNA synthesis resumed after exposure to 1,25-dihydroxyvitamin D$_3$, Bmp-2 no longer altered mRNA levels, indicating a correlation of the effect with osteoblastic cell differentiation.

Bmp-2 inhibits retinoic acid-induced expression of alkaline phosphatase in osteoblastic cell lines, indicating antagonistic interaction between Bmp-2 and retinoic acid (Ogata *et al.*, 1994).

[a] See K.-H. Lee *et al.* (1998) for zebrafish and the *swirl* mutant, Feng *et al.* (1994) for *Xenopus*, and Plessow *et al.* (1991) and Nishimatsu *et al.* (1992, 1993) for mice.
[b] See Francis-West *et al.* (1994) and Ekanayake and Hall (1997).
[c] See Larraín *et al.* (2000) for the cysteine-rich domain, I. C. Scott *et al.* (2000) for chordin distribution, Bachiller *et al.* (2003) for *chordin* knockout, Hall (1999a) and Gorlin *et al.* (2001) for DiGeorge syndrome, and Garg *et al.* (2001) for *Tbx-1*, which is up-regulated by *Shh*. *Tbx-1* is localized in pharyngeal endoderm and the mesodermal but not neural crest-derived mesenchyme of the pharyngeal arches. *Shh* is expressed in neural crest-derived mesenchyme.
[d] See J. T. Thomas *et al.* (1997) for limb morphogenesis, and Vonau *et al.* (2001) for the implant study.
[e] See Nishimatsu *et al.* (1992, 1993), Suzuki *et al.* (1993), Yamagishi *et al.* (1995) and Dale and Jones (1999) for *Xenopus* Bmps.
[f] See Heermeier *et al.* (1994) and Katahiri *et al.* (1994) for xiphoid cartilage and C2C12 cells, and Carlsberg *et al.* (2000) C3H10T1/2 cells.
[g] See Hughes and McCulloch (1991), Harris *et al.* (1995) and Hughes *et al.* (1995) for studies with calvarial cells.
[h] See Thies *et al.* (1992) and Volek-Smith and Urist (1996) for *RhBMP-2*.

---

craniofacial membrane bones of domestic fowl. Until quite recently, *I thought both were excellent examples of bipotentiality of periosteal cells*. I now realize that condylar cartilage is not always derived from periosteal cells. The basis for reaching that conclusion is elaborated below. This does not disprove the concept of bipotentiality but does remove condylar cartilage in some mammals as an example of cartilage arising from bipotential cells.

CONDYLAR CARTILAGE ON THE CONDYLAR PROCESS OF THE MAMMALIAN DENTARY

The condylar process of the mammalian dentary articulates with the temporal portion of the skull to form the

temporomandibular joint (Figs 5.8 and 5.9). The body of the dentary and the basal portion of the condylar process develop by intramembranous ossification. However, the distal portion of the condylar process forms by the development of a cartilage known as the *condylar cartilage* (Fig. 12.4). Similar, though more, transitory cartilage is found on the coronoid and angular processes.

Histodifferentiation and scurvy

Numerous studies have been published on the histodifferentiation of the condylar cartilage, especially in rodent development. Some examine condylar chondrogenesis in unperturbed embryos, some after perturbations as diverse as the application of orthodontic devices, hard or

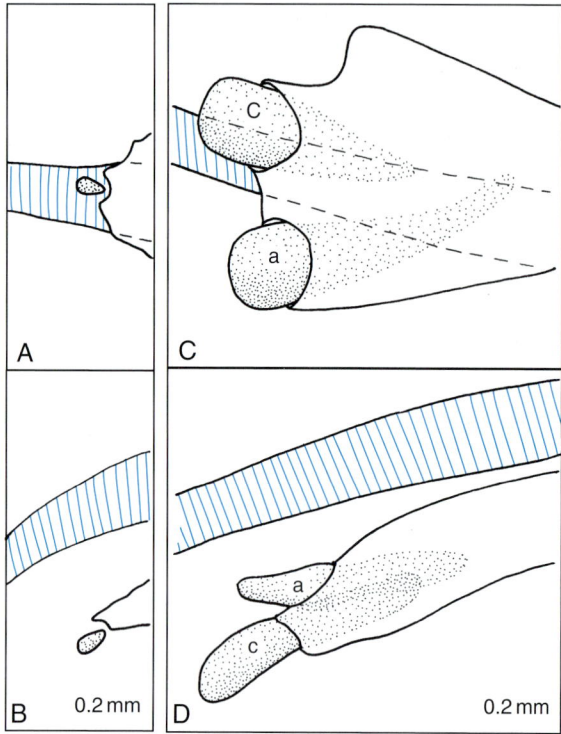

Figure 12.4 The condylar cartilage on the condylar process of the rat dentary bone arises from a condensation that is separate from the bone. (A, B) Lateral (A) and dorsal (B) views of the posterior end of the dentary at 17 days of gestation show Meckel's cartilage (hatched) lying medial to both the bone of the dentary (unshaded) and the separate condensation for the condylar cartilage (stippled). (C, D) Lateral (C) and dorsal (D) views of the posterior of the dentary at 18 days of gestation show Meckel's cartilage (hatched) and the condensations for the condylar (c) and angular (a) cartilages (stippled), now associated with the bone of the condylar and angular processes; the condensation for the angular cartilage arises during the 18th day and so is not seen in A and B. Adapted from Duterloo and Jansen (1969).

One or two cell populations

For condylar chondrogenesis to qualify as an example of skeletogenesis involving bipotential cells, the cartilage and subjacent bone have to arise from the same population of cells, *ideally* from progeny of the same single cells. You would think that even the former would be straightforward – the latter is much more difficult, only rarely having been achieved – especially given the interest in this cartilage from the clinical dental community, temporomandiblar joint pain being a major source of pain in many ageing humans.

It turns out not to be straightforward, in part because of species or strain differences, and in part because of the need to examine developmental series to resolve the issue. To give a flavour of the literature available, some of which is discussed in the following sections:

- Baume (1962c), Duterloo and Jansen (1969), Keith (1982) and Vinkka-Puhakka and Thesleff (1993) thought that the condylar cartilage originates *from a separate condensation adjacent to the periosteum*; Baume (1962c) spoke of a condylar blastema arising in human embryos at 24 mm crown–rump length and contributing to both muscle and dentary. Keith (1982) documented development of the temporomandibular joint (TMJ), initial ossification of the dentary around the alveolar nerve, and independent condensation of what he refers to as the articular disc. It appears that the periosteal cells may not be bipotential. They could be but, if they are, their bipotentiality is not utilized in forming the condylar cartilage.
- On the other hand, Meikle (1973b), Tengan (1990) and Shibata *et al.* (1996) thought the condylar cartilage originates *from periosteal cells*, i.e. that the periosteal cells are bipotential, *assuming* that the cartilage arises from the same cells that produce the bone of the condylar process and not from chondrogenic (and nonosteogenic) cells in the periosteum.

These conclusions were not reached on the basis of studies of the same species. This list encompasses studies on rats (Duterloo, Meikle), mice (Shibata, Tengan), humans (Baume) and the Syrian hamster. Not that species differences tell the whole story. You will see below that different individuals working on the same species come to opposite conclusions regarding the origin of the cartilage.

Evidence against bipotentiality

In their studies on embryonic rats and in confirmation of earlier studies, Copray and colleagues claim that the condylar cartilage appears *before* the bone of the condylar process during embryonic development. This does not necessarily mean that the cartilage and bone do not arise from the same cell population, but it does mean that the cartilage would not be secondary; secondary

soft diets, scurvy, hormonal imbalance, and so forth. Others compare condylar cartilages with other cartilage types for similarities/differences in histological organization, nature of the ECMs, modes of growth, growth potential, and utility as transplant tissue.[16] Growth potential and response to *scurvy* can be taken as two examples.

(i) A series of studies on condylar cartilage development and function in rats led to the conclusion that it serves as a growth cartilage during embryonic and early postnatal life and as an articular cartilage postnatally, i.e. condylar cartilages are functional cartilage from early in development.

(ii) In guinea pigs with or recovering from scurvy, or on a scorbutic diet (Box 27.1), the patterns of circulation and mineralization differ between condylar, tibial epiphyseal and tibial articular cartilages. Condylar cartilage doubles in width after 26 days; the tibial cartilages show much less change in growth. Recovery is even more rapid; condylar cartilage is of normal width after only three days of treatment with ascorbic acid.[17]

cartilage arises from periosteal cells on membrane bone *after* initiation of osteogenesis (see below).[18] More significantly for discussions of bipotentiality, when Duterloo and Jansen (1969) examined embryos of Wistar albino rats of 14–21 days of gestation, they concluded that the condylar process has a *separate origin* from the bone of the mandible. This would make the condylar cartilage a sesamoid – a sesamoid may remain cartilaginous or be replaced by bone (Chapter 9) – removing any suggestion of bipotentiality. They provided three lines of evidence:

- the condylar blastema develops in cartilage, while the dentary blastema is osteogenic;
- the condylar cartilage – and so the condylar process – secondarily fuses with the dentary bone between the 17th and 18th day of gestation; and
- the dentary ossifies early and intramembranously, the condylar process later and endochondrally (Fig. 12.4).

The third point is not conclusive evidence for separate origins. The first and second points are.

Duterloo and Jansen (1969) cite earlier studies and other lines of evidence supporting the distinction between the two skeletogenic processes chondrogenesis and osteogenesis, including the formation of ectopic cartilages in the posterior region of the mandibular process in embryonic rats exposed to teratogens. This would seem to support bipotentiality, but the cartilage lacks any connection with the bone of the dentary. Secondly, a new mandibular condyle forms in pigs after the original condyle has been removed surgically (condylectomy; see below). This is not a regenerated condyle but a *neomorph*, a new and different cartilage that *arises apart from the bone*, only fusing with the bone during later stages of repair.

Response to condylectomy is similar in rats. The major changes observed in a study by Jolly (1961) were the formation of a *new* articular process on the dentary, followed by establishment of a new joint. Cartilage, chondroid bone and immature spongy bone all formed as mineralized tissues. Surprisingly, these were not only deposited as part of the new joint but in at least five other locations – around the periphery of the dentary, on the cut surface of the dentary, at the severed ends of the lateral pterygoid muscle, on the surface of the articular fossa, and on the lateral wall of the cranium opposite the new articular process. Even though the first three sites do not form part of the new joint, bone growth at these sites contributes to formation of the new articular process.

Evidence supporting bipotentiality

So far, I have talked only about rats and pigs, taxa in which it is clear that the *condylar cartilage has a separate origin from the bone of the condylar process of the dentary*. How did we come to believe that the condylar cartilage is a secondary cartilage that arises from periosteal cells? Because it is and it does. Not in rats and pigs, but in mice and men.

The condylar cartilage is a wonderful example of a homologous structure with different developmental origins in different taxa. This may not be important for those of you whose approach is strictly developmental, but for those of us wearing an evolutionary developmental biology hat, it is an excellent instance of the hierarchical approach to homology now taken by many biologists, in which we recognize that homologous structures may have divergent developmental origins.[19] Study of condylar cartilages also helps answer the important question of how we know when we can extrapolate results from 'model organisms' to humans. Basing reparative techniques on knowledge of how the cartilage arises in rats may (should, must?) produce a different outcome from an approach based on how the cartilage develops in mice. The types of evidence gathered from mouse and human development follow.

From histological analyses of human mandibular development it is apparent that condylar cartilage arises *after* intramembranous ossification has commenced – in embryos of 48 to 50 mm crown–rump length (Symons, 1951). In itself, this observation does not make the cartilage secondary; the cartilage has to be shown to arise from periosteal cells. Several lines of evidence indicate this to be so.

Histological studies of skeletogenic processes give this impression; progenitor cells at the distal tip of the condylar process have been described as differentiating into chondroblasts toward the cartilage, and as osteoblasts toward the bone of the process.[20]

Condylar cartilage in foetal calves contains collagen types I and II. This is an interesting class of evidence, the dogma being that cartilage contains type II collagen and bone type I. Articular and fibrocartilages contain both types, the presence of type I being attributed to the functional requirement of a cartilage at a joint, or in a tendon or ligament. Questions of a periosteal origin do not arise in these cases. However, the presumption for the condylar cartilage is that type I collagen demonstrates an affinity with bone, therefore a periosteal origin. An element of circularity imbues this reasoning. I say this, even though I have used it on several occasions. Condylar cartilage from six-month-old calves has some antigenic determinants in common with nasal and epiphyseal cartilages, but also has at least one determinant that is distinctive. A developmental study using immunological markers would be a profitable way of determining the timing of the specification of such cartilage types.[21]

Another approach to bipotentiality is physiological and involves analyzing how function regulates cell proliferation and differentiation. Under conditions of normal physiological activity the progenitor cells of the condylar periosteum continue to proliferate and differentiate as chondroblasts, adding to the bulk of the cartilage.[22] I used a similar approach to investigate secondary cartilage in avian embryos. The topic is revisited in Chapter 33 under the guise of whether cartilages have inherent growth potential or are only adaptive.

Differentiation of periosteal progenitor cells into chondroblasts requires a stimulus associated with normal joint function, shown to be a mechanical stimulation). In the absence of this stimulus, progenitors differentiate along an alternative, osteogenic pathway. Strongly in support of bipotentiality is the finding that, if the normal mechanical environment of the TMJ is disturbed by resecting the external pterygoid muscle, the progenitor cells *differentiate into osteoblasts instead of chondroblasts and deposit membrane bone not secondary cartilage.* Transplanting the TMJ intracerebrally has the same effect. Similar conclusions are drawn from a third experimental approach: transplanting the mandibular condyle from four-day-old rats into the distal metacarpal epiphysis of seven-day-old rats. Both cartilage and bone of the condyle continue to develop in this new functional environment, *provided that* articular function is maintained. Indeed, the mechanical environment can be recreated *in vitro*. While in my laboratory, Tuomo Kantomaa developed a device in

which entire TMJs can be cultured, articulating function maintained, and the joint subjected to movement.[23]

The mechanical environment can be altered in other ways; for example, by feeding rats a soft diet, or clipping the incisors to reduce forces associated with mastication (Box 12.2).[24]

Read the last sentence again. The study was done in *rats*, in which the condylar cartilage arises *outside* the periosteum as a sesamoid. Nevertheless, as the cartilage develops, it becomes responsive to mechanical forces, as does periosteally derived cartilage in mice. Divergence of developmental origin does not dictate (or even indicate) divergence of later function or structure. Similarly, the secondary cartilage of the intermaxillary suture in Sprague-Dawley rats – which is periosteal in origin – fails to form if the maxillary molar teeth are removed; the joint becomes a syndesmosis (Forbes and Al-Bareedi, 1986).

What about condylectomy, which in rats elicits formation of a new articulating cartilage, a new joint, and deposition

---

Box 12.2 Effects of consistency of diet and forces of mastication

Manipulating diet by providing softer or harder diets than animals are accustomed (adapted) to, is a palatable way of analyzing bone adaptation(s) at all levels from organ → tissue → cells → genes, and there are many studies in the dental literature relating to the effects of diet on the structure of the jaws and/or skull. Alice, in Lewis Carroll's *Alice's Adventures in Wonderland*, knew this, as we see in the poem she recited to the caterpillar:

'You are old,' said the youth, 'and your jaws are too weak
For anything tougher than suet;
Yet you finished the goose, with the bones and the beak –
Pray, how did you manage to do it?'
'In my youth,' said his father, 'I took to the law,
And argued each case with my wife;
And the muscular strength, which it gave to my jaw,
Has lasted the rest of my life.'

At the level of individual skeletal elements – bone as an *organ* – reanalysis of the altas, axis, scapulae, humeri and femora of C57BL/Gr mice, from the Hans Grüneberg collection housed at the Hubrecht Laboratory in the Netherlands, led David Johnson and his colleagues (1990) to the conclusion that diet has minimal effects on bone shape. Similarly, from a thorough analysis of a developmental series of the jaws and skulls of 342 African apes of five age classes, Taylor (2002) concluded that the evident differences in morphology (especially evident between pygmy chimpanzees or bonobos, *Pan paniscus*, and the common chimpanzee, *Pan troglodytes*) could not be accounted for by diet alone. Allometric constraints and forces associated with the development of the teeth are important shape regulators.[a]

At the level of bone as a *tissue*, cortical bone structure of the dentary of rhesus monkeys, *Macaca mulatta*, is influenced by diet through changes in bone strain. Soft diet is associated with low remodeling, hard diet with higher rates of remodeling, more Haversian canals and a deeper mandible.[b]

Examples of how harder diets enhance proliferation and/or chondrogenesis are discussed in this chapter.

Mastication influences prenatal skeletal development. Haversian systems are already present in bone of the mandibular condyle at the fifth month of human foetal development, indicative of remodeling associated with foetal suckling and swallowing, which represent the onset of masticatory function. The permanent teeth of guinea pigs, *Cavia porcellus*, erupt and begin to be worn down in response to prenatal jaw movements *in utero*.

Ecological forces also can drive mandibular morphology over evolutionary time. Woodland deer mice, *Peromyscus maniculatus*, have longer mandibles with deeper rami than do *Peromyscus* that inhabit grasslands, although prehistorically (1100–1400 AD) individuals from both habitats had a similar range of morphology.

A recent study links Igf-1 with altered masticatory loading on the nasopremaxillary bone in mice, mechanical forces altering *Igf-1* levels.

The origin and evolution of browridges in primates have been attributed to influences of mastication. However, Hylander and his colleagues convincingly demonstrated from an analysis of patterns of bone strain that both macaque, *Macaca fascicularis*, and baboon, *Papio anubis*, browridges are much too robust to be 'structural adaptations to counter intense masticatory forces [and that] macaque and baboon browridges can be considerably reduced in size and still maintain these required structural characteristics.'[c]

[a] The other two African apes studied by Taylor (2002) were the eastern mountain gorilla, *Gorilla gorilla beringei*, and the western lowland gorilla, *Gorilla gorilla gorilla*. Also see Zollikofer and Ponce de León (2004) for an evaluation of allometry, heterochrony and heterotopy in changes in the cranium early in hominid evolution.
[b] See Bouvier and Hylander (1981), Hylander (1981) and Hylander and Johnson (1989) for the macaque studies; see Fleagle (1999) for an overview of studies on rhesus monkeys; and Jaworski (1992) for Haversian systems and Haversian bone.
[c] See Goret-Nicaise and Dhem (1984) for bone resorption in human foetuses, Teaford and Walker (1983) for prenatal guinea pigs, Holbrook (1982) for deer mice, Tokimasa *et al.* (2000) for Igf-1, and Hylander *et al.* (1991a,b) and Lieberman (2000) for browridges. The quotation is from Hylander *et al.* (1991a, p. 1).

of multiple skeletal tissues in the new joint, in the dentary and in the lateral pterygoid muscle? Removing the murine condyle elicits formation of a new functional condyle, not regeneration of the old one. An initial blood clot is replaced by fibrous connective tissue, then cartilage, and finally the cartilage is replaced by bone through endochondral ossification.[25]

Another approach is to examine chondrogenic and osteogenic markers. Shibata *et al.* (1997b) localized collagen type II and type X, and alkaline phosphatase in mouse condylar cartilages. Expression of alkaline phosphatase in the prechondrogenic condensation at day 14 was taken as evidence for the periosteal origin of the cartilage (Box 5.2). In what appears to be a thorough analysis, Fukada *et al.* (1999) reviewed the earlier literature and undertook a careful study in foetal mouse using *in situ* hybridization. Their major findings are that at:

- 14.5 days of gestation, aggrecan mRNA and type I collagen mRNA (but not type II collagen mRNA) are expressed in a cellular aggregation that is continuous with the periosteum of the dentary, reminiscent of the set-aside cells described in Chapter 10;
- 15.0 days, the first cartilage appears with mRNA for collagen types I, II and X (i.e. already with features of hypertrophic chondrocytes) and continued synthesis of aggrecan mRNA; while at
- 16.0 days, cells hypertrophy with down-regulation of type I collagen, which is now only expressed in the periosteum.

Fukada and colleagues concluded that the condylar cartilage arises from the mandibular periosteum, i.e. it is a genuine secondary cartilage arising from bipotential cells.

Expression of aggrecan mRNA in the periosteum at a putative site of chondrogenesis is interesting in light of the seemingly unrelated but relevant investigation of whether a transient cartilaginous phase is present in intramembranous osteogenesis. Nah *et al.* (2000) used *in situ* hybridization of mRNA in the frontal bone of chick embryos of 12, 15 and 19 days of incubation to show that the frontal expresses mRNA for aggrecan and mRNAs for collagen types I, IIA and XI, an association that is characteristic of *prechondrogenic* condensations, and that mRNAs for aggrecan and *type II collagen* are expressed by alkaline phosphatase-positive (bipotential) cells in osteogenic condensations, periostea, the osteogenic front and osteoid.[26]

All or some?

Are *all the cells* of mouse condylar processes bipotential for osteogenesis and chondrogenesis, or does the population consist of two progenitor cells, one osteogenic (OPC) and one chondrogenic (CPC)? Stutzmann and Petrovic (1975a,b) argue for two cell types within the progenitor pool.

One line – their skeletoblast – is said to have a spontaneous tendency to form osteoblasts. *In vivo*, however, this tendency is suppressed and these cells form chondroblasts; they are bipotential. The second cell line is said to form only chondroblasts. This interpretation follows from their studies on the *chalone*-like effects of chondroblasts within the condyle, by which the presence of chondroblasts above a threshold level inhibits further chondrogenesis, even by cells with intrinsic chondrogenic potential.

Two different forms of bipotentiality are expressed later in differentiation or development. One requires dedifferentiation to progenitor cell status and redifferentiation as another cell type (Chapters 13 and 14). The other is transformation of one cell type to another, perhaps without dedifferentiation. It has been suggested that condylar cartilage hypertrophic chondrocytes survive matrix mineralization to serve as osteoprogenitor cells (Fig. 5.9; Silbermann and Frommer, 1972a,b,c, 1974; Chapter 13).

SECONDARY CARTILAGE ON AVIAN MEMBRANE BONES

Secondary cartilages, similar in histological appearance to – although probably not homologous with (Hall, 2000a) – the condylar cartilage of the mammalian dentary, develop on various membrane bones of the craniofacial skeleton in avian embryos. Murray (1963) summarized the histogenesis of the major bones that possess secondary cartilage – the quadratojugal, surangular, pterygoid, squamosal and palatine. These cartilages arise *after* initiation of intramembranous ossification. In the quadratojugal (QJ), the first of the craniofacial bones to ossify at H.H. 30.5, secondary cartilage arises some four days later, at H.H. 37 (Fig. 12.5; Table 12.3).[27]

These secondary cartilages develop from periosteal progenitor cells, which also produce bone, i.e. they are bipotential as osteochondroprogenitor cells. As concluded from several lines of evidence, secondary chondrogenesis depends upon mechanical stimulation, in the absence of which progenitor cells differentiate as osteoblasts:

- secondary cartilage forms at joints or beneath muscle or ligament insertions/attachments, locations that point toward movement as a factor in their formation;
- secondary cartilage does not form, and the progenitor cells differentiate into osteoblasts, when the QJ is removed from embryos younger than early H.H. 36 (10 days of incubation) and grafted to the CAM (Box 12.3) or maintained in organ culture (Fig. 12.5);
- similar results are obtained when embryonic chicks are paralyzed with curare or decamethonium;
- when exposed to intermittent pressure and tension *in vitro*, progenitor cells of the QJs of embryonic chicks

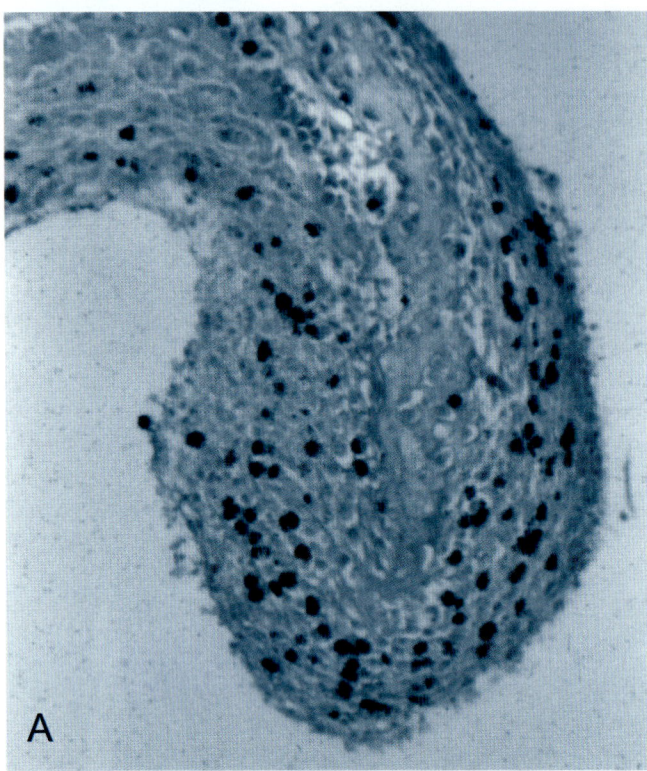

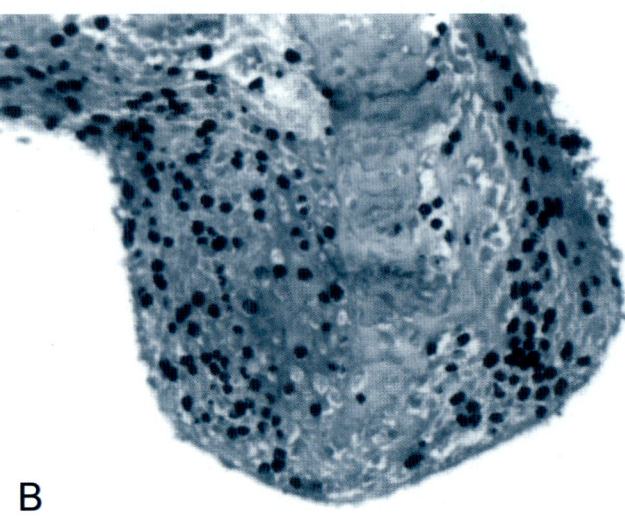

differentiate into chondroblasts. If exposed to constant pressure *in vitro*, or if allowed to differentiate within immobilized embryos, these cells differentiate as osteoblasts. Evidently, immobilization and constant pressure are recognized as the same stimulus (or lack of stimulus).

Grafting or organ culturing these progenitor cells, or paralyzing embryos, does not suppress chondrogenesis if the cells are taken from, or the drugs injected into, embryos older than H.H. 36. Either the progenitor cells are specified at that age, or they are only able to express their chondrogenic bias by that age.[28]

The initial response of these progenitor cells to intermittent movement is selective proliferation at the precise time – during the 10th day of incubation – when they switch from osteogenesis to chondrogenesis. In the absence of embryonic movement – if paralyzed, or maintained in culture – this proliferative response fails to take place and the cells continue to differentiate as osteoblasts (Fig. 12.5; Hall, 1979; Buxton *et al.*, 2003). Therefore, cessation of cell division is associated with the switch to chondrogenesis on avian membrane bones. As with condylar cartilage in mice, proliferation and differentiation are both influenced by mechanical factors. Periosteal progenitor cells selectively divide and accumulate at joint surfaces within mobile embryos. Paralysis removes some 60–75 per cent of cells from the progenitor cell pool during the 10th day, depriving the membrane bones of their ability to produce secondary cartilage (Fig. 12.5).[29]

Interestingly, chondrogenesis is initiated when QJ from paralyzed embryos are maintained as submerged cultures (Thorogood, 1979), i.e. cultured on the bottom of the Petrie dish, a condition that promotes proliferation and outgrowth of cells from the explant. Recall that Willmer based his four major cell types on results from explant culture (Chapter 10).

Figure 12.5 Autoradiographs of the posterior 'hook' of two quadratojugal bones from embryonic chicks, pulse-labeled with ^{3}H-thymidine to identify cells synthesizing DNA. (A) This bone from an 11-day-old embryo was pulse-labeled for one hour, developed and counterstained. The periosteal cells – from which secondary cartilage is developing – show a high proportion of labeled cells. Bone (centre) contains virtually no labeled cells. (B) This bone from an 11.5-day-old embryo was labeled for six hours. Again periosteal cells are heavily labeled, bone unlabeled. Secondary cartilage has begun to differentiate (right hand side). Notice that the mature chondrocytes continue to synthesize DNA, in contrast to the osteogenic cells, which cease DNA synthesis upon differentiating as osteoblasts/cytes.

Table 12.3 A summary of Hamburger–Hamilton (H.H.) stages of development of the embryonic chick over the first 12 days of incubation, and the major events of skeletal development at those stages[a]

H.H. stage	Hours or days of incubation	Developmental event
1	5–6 hr	Embryonic shield
2	6–7 hr	Initial (short) primitive streak
3	12–13 hr	Intermediate primitive streak
4	18 hr	Primitive streak with primitive groove
5	20	Initiation of notochord development
6	24	Neurulation, head fold
7	26 hr	First pair of somites appear; mesenchyme of limb fields chondrifies if grafted ectopically; antero-posterior limb axis established
8	28 hr	Four pairs of somites; neural folds appearing anteriorly; neural crest cells migrating from neural tube (H.H. 8.5); blood islands appear
9	31 hr	Seven pairs of somites; heart and eye formation; left and right heart primordia begin to fuse
10	35 hr	Ten pairs of somites and brain vesicles present; earliest stage when flank mesenchyme can be induced to form limb; dorso-ventral limb axis established; hearts loops and begins to contract; first hemoglobin synthesis
11	42 hr	13 pairs of somites with expression of anchorin C-2; dorso-ventral axis of limb buds fixed; neural tube begins to close; head fold of amnion appears
12	46 hr	16 pairs of somites; cranial flexure; sclerotomal cells migrating away from anterior somites; wing bud mesenchyme can now induce an AER from flank epidermis; *Fgf-8* is expressed in second arch epithelium, *Bmp-7* in branchial clefts and arches (H.H. 12 onwards)
13	48 hr	19 pairs of somites; head turned to the left
14	50 hr	22 pairs of somites; initiation of visceral (pharyngeal) arch formation; neural crest cells populate the optic cup; widespread distribution of type II collagen in basement membranes
15	53 hr	25–26 pairs of somites; neural crest cells in mandibular and maxillary arches
16	55 hr	26–28 pairs of somites; initiation of wing buds as outgrowths adjacent to somites 15–19; proximo-distal limb axis established progressively between H.H. 16 and H.H. 28
17	60 hr	30 pairs of somites; initiation of leg buds as outgrowths adjacent to somites 26–32. AER active until H.H. 19. Wing bud mesenchyme can no longer induce an AER from flank epidermis and flank epidermis has lost its ability to respond
18	66 hr	33–34 pairs of somites; first sign of embryonic movement; limb bud and periocular mesenchyme will form cartilage *in vitro*; *Bmp-4* in the distal tips of the branchial arches (H.H. 18 onwards); allantois appears as the first extraembryonic membrane
19	3 days	Aortic arches form; AER loses inductive ability; periocular mesenchyme chondrifies independently of the pigmented retinal epithelium
20	3 days	40–42 pairs of somites; initiation of eye pigmentation; embryo fully enclosed in amnion
21	3.5 days	44 pairs of somites
22	3.5 days	Somites extend caudally to the tip of the tail; synthesis of insulin initiated; amniotic contractions begin
23	3.75–4 days	Initial ECM around sclerotomal (vertebral) cells; first appearance of myosin in limb buds; first nerve fibres in hind limb buds; membrane bone primordia of upper and lower jaws independent of epithelial signals; initial innervation of hind limb buds
24	4 days	Prechondrogenic cells in limb buds contain 20 copies of type II procollagen mRNA and stabilized for chondrogenesis; 1500–2000 cells die within the posterior necrotic zone; chorion and allantois fuse to form the chorioallantoic membrane
25	4.5 days	Elbow and knee joints distinct; first extracellular matrix in limb bud mesenchyme; cell death maximal in the opaque patch
26	5 days	Digits demarked; first embryonic movement; Meckel's cartilage begins to chondrify; first appearance of muscle in limb buds; heart now four-chambered; first red blood cells appear; corticosteroid synthesis begins
27	5.25 days	Chondrogenesis begins in the limb buds; cartilaginous models of limb skeleton present; perichondrial cells express diffuse RARα and RARγ; the fibula loses its distal epiphysis; rhythmic contraction of amnion
28	5.75 days	Three digits distinct on wing, four on legs; beak visible
29	6–6.5 days	AER has lost the ability to initiate mesenchyme outgrowth; tibial perichondrium now a periosteum; rudiment of fifth toe appears
30	6.5–7 days	Egg tooth forming; feather germs appearing; testosterone synthesis and sexual differentiation begins; first interdigital cell death in hind limbs (continues until H.H. 35); two scleral papillae present
31	7–7.5 days	First chondroblasts in anterior vertebrae; first bone in lower and upper jaws; limb chondrocytes contain 2000 copies of type II procollagen mRNA; first interdigital cell death in wing; six scleral papillae present
32	7.5 days	6–8 scleral papillae present; fifth toe becomes rudimentary
33	7.5–8 days	13 scleral papillae present; scleral cartilage differentiating; mineralization of long bone initiated
34	8 days	Chondroblasts in vertebral primordia
35	8.5–9 days	Vertebral chondroblasts begin to deposit ECM; interdigital cell death in hind limbs complete
36	10 days	Periosteal cells on membrane bones of upper and lower jaws determined for secondary chondrogenesis; thyroid gland secretes thyroxine
37	11 days	Secondary present on membrane bones of upper and lower jaws; chondrocytes in developing vertebrae
38	12 days	Ossification of scleral ossicles and Ca^{++} absorption from shell begin

[a] The H.H. stages are based on Hamburger and Hamilton (1951). The staging table is also reproduced in Hamilton (1965), Freeman and Vince (1974) and Stern and Holland (1993). The details on skeletal development are taken from my text.
AER, apical ectodermal ridge; ECM, extracellular matrix.

Box 12.3 CAM-grafting

Grafting tissues onto the chorioallantoic membrane (the CAM) of chick embryos, a technique known as *CAM-grafting* – which is how I refer to the technique throughout – provides a wonderful way to 'culture' tissues or organs in a vascular environment (Fig. 12.6). In skilled hands, tissues can be grafted into embryos as young as six or seven days of incubation (9 or 10 is more usual) and left in place until the CAM begins to break down immediately before hatching, effectively at 18 or 19 days of incubation, allowing a maximal graft life of 13 days, 10 days being more usual.[a]

I determined (Hall, 1978d) that grafts could be recovered and re-grafted to another or a series of other host embryos, extending graft life to 30 days or more (Figs 12.6 and 12.7). Tissues can be grafted onto the CAMs of embryos established in 'shell-less culture,' in which the entire contents of the egg are emptied into a dish and the embryo allowed to develop almost to hatching. This approach has the advantage that the graft can be followed and photographed in real time, and/or treated with hormones, growth factors, drugs, implanted with beads, and so forth (Box 25.3; Fig. 12.3).

You might think that tissues from other vertebrates would be rejected if CAM-grafted, and that only avian or perhaps only chick tissues can be grafted. Fortunately, rejection mechanisms only arise after hatching so that mammalian, reptilian, even fish tissues can be CAM-grafted (Fig. 12.8). Grafting tissues from poikilotherms required some background research to determine the optimal conditions, especially optimal temperature for survival of host and graft

(Miyake and Hall, 1994). Chick embryos can be incubated at temperatures as low as 22–24°C and survive. We have taken advantage of this to graft tissues from teleost fish.

The history of grafting embryonic rudiments onto the CAM goes back 80 years. Pioneering grafts were undertaken by my Ph.D. supervisor P. D. F. Murray, in collaboration with Julian Huxley. They grafted embryonic chick limb buds to the CAM to test theories of mosaic development. The two competing theories of how embryos or embryonic regions were organized were:

(i) mosaic development, in which each part is specified early in development, often in the egg during oogenesis; and
(ii) regulative development, in which embryos could compensate (regulate) for parts that were lost or removed (Box 35.1). Mosaic embryos could not regulate.

Huxley and Murray (1924) wanted to know whether limb buds were mosaic, i.e. whether the various parts of the limb skeleton were specified early in limb-bud development (or even earlier, before limb buds arise), rather like the pattern of a mosaic floor in which each piece has a specific place, any loss of pieces disrupting the pattern or even preventing a pattern from forming. Huxley and Murray recorded the reactions of the CAM to grafting, reactions that included thickening of the CAM as the epithelium became stratified and keratinized with epithelial ingrowth and formation of deep keratinized pearls.

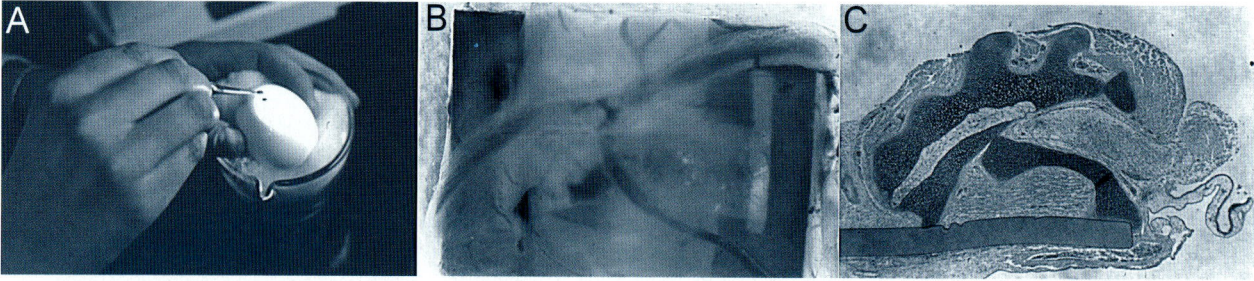

Figure 12.6 Some of the steps in CAM-grafting – grafting tissues to the chorioallantoic membrane of a chick embryo – including (A) placing a graft onto the CAM through a window in the shell; (B) a live graft upon recovery, to show the black Millipore filter substrate and the high degree of vascularization of the graft, which is enclosed in an epithelial vesicle; and (C) a histological section of vertebral cartilage that developed in a graft. The degree to which the CAM has encapsulated both graft and filter (below) is readily apparent.

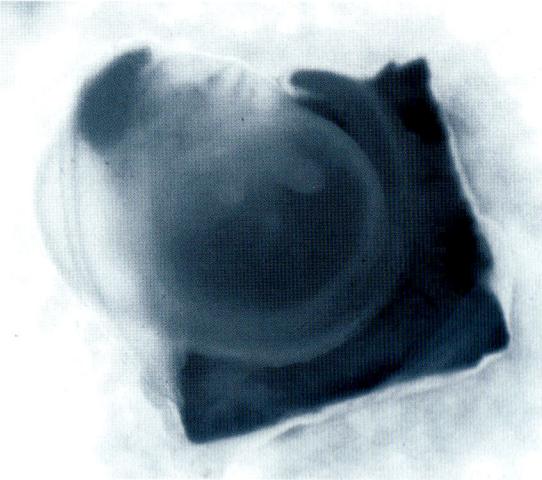

Figure 12.7 A live CAM-graft upon recovery. The black Millipore filter substrate stabilizes the tissues and facilitates finding the graft which is enclosed in an epithelial vesicle.

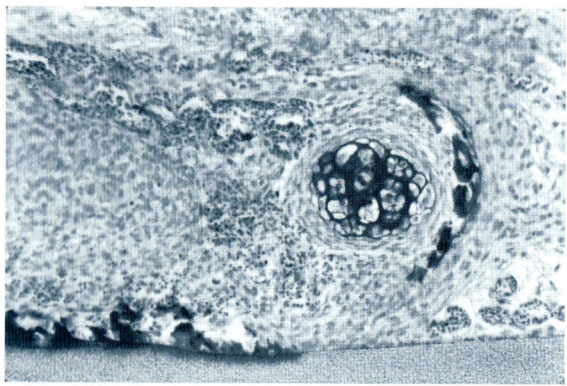

Figure 12.8 A histological section of Meckel's cartilage and dentary bone from the mandibular arch of an 11-day-old mouse embryo maintained as a CAM-graft.

Box 12.3 (*Continued*)

Below is a highly selective sample (arranged chronologically) of tissues grafted and hypotheses tested using CAM-grafting, many of which are discussed in greater detail elsewhere in the text.

Grafting avian tissues

Classic studies grafting chick limb buds to analyze skeletal development (Murray and Huxley, 1925; Murray, 1928; Hunt, 1932; Murray and Selby, 1933; Bradley, 1970; and see Fig. 12.9).

Classic studies grafting chick somites, scleral ossicles or tibiae (Hoadley, 1925; Wedlock and McCallion, 1969; O'Hare, 1972a–c; Dorey and Sorgente, 1977).

Examination of the survival of embryonic bone, and use of the results to propose that trabecular architecture is determined by blood vessels rather than mechanical factors (Hancox, 1947, 1948).

Development of a method to infuse the chorioallantoic circulation and so apply a continuous flow of drugs, hormones, etc. (Drachman and Coulombre, 1962a).

Examining nerve–muscle interactions (Bonner, 1975).

Grafting perichondrium-free quail tibial epiphyses to study reformation of the periosteum (Kahn and Simmons, 1977).

Grafting fractured embryonic quail long bones to demonstrate that osteoblasts arise from local (graft) cells, but osteoclasts from host cells (Simmons and Kahn, 1979, and see Krukowski *et al.*, 1983 for a review).

Grafting mineralized or demineralized bone, hydroxyapatite or eggshell and using TEM to follow multinucleated cells with ruffled borders that surrounded the mineralized bone within three days, indicative of ectopic and early differentiation of osteoclasts (Krukowski and Kahn, 1982).

Using the CAM as a site to bioassay angiogenesis, finding that the type of vascular response was not typical of that elicited by tumour angiogenesis factors, and that the role of chondrocyte hypertrophy in allowing vascular invasion could be tested (Fig. 12.10; Jakob *et al.*, 1978; Kuettner *et al.*, 1983; Wilting *et al.*, 1991).

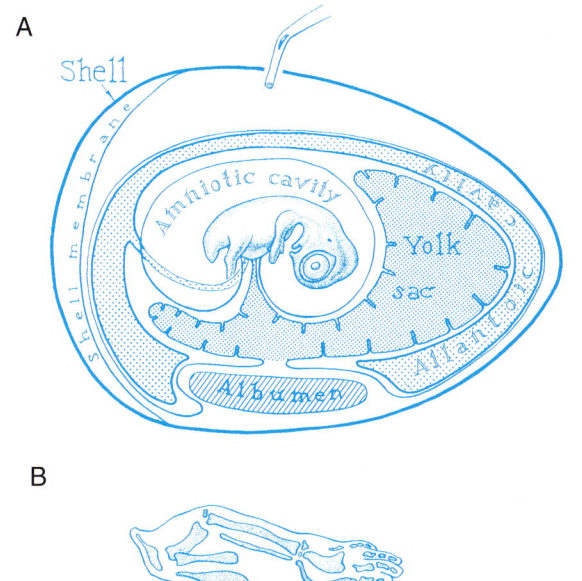

Figure 12.9 CAM-grafting. (A) A cut-away view of a chick embryo of 10 days of incubation showing the extraembryonic membranes and a hole in the shell through which a limb bud from a 3-day-old embryo is being transferred. (B) Ten and a half days later, limb skeletal elements have differentiated. Modified from Hamburger (1960) illustrating the experiments of Hunt (1932).

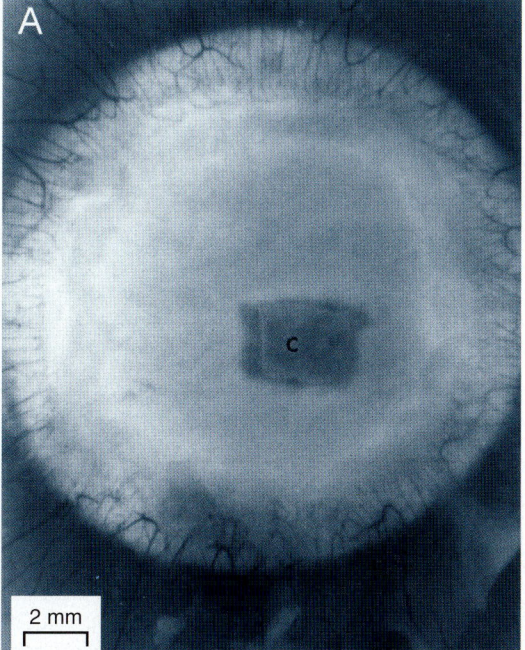
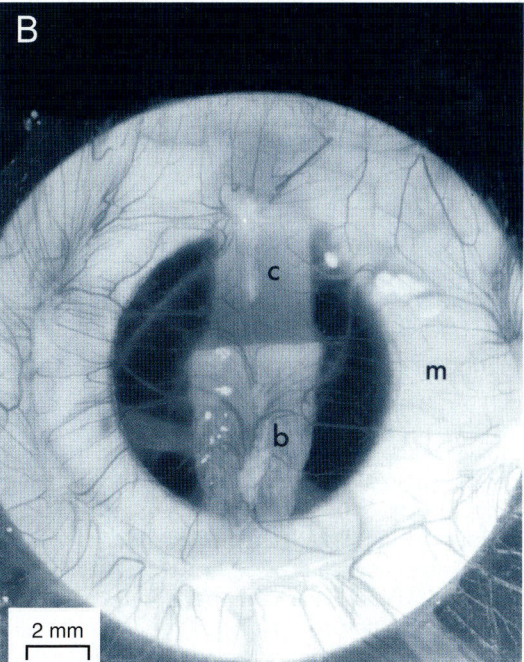

Figure 12.10 Invasion of cartilage and bone by blood vessels as demonstrated by grafting to the CAM of host chick embryos. (A) As evidenced by the exclusion of blood vessels from the graft, unmineralized cartilage (c) is not invaded by blood vessels. (B) Extracting the matrix with 1.0 M GuHCl before grafting allows both unmineralized cartilage (c) and mineralized bone (b) to be invaded by blood vessels. m is the Millipore filter on which the tissues were supported. Adapted from Hall (1978a).

Box 12.3 (*Continued*)

Grafting limb buds to examine joint development in the absence of movement (Mitrovic, 1982).

As a site to analyze the epithelial–mesenchymal interactions that initiate membrane bone development (Tyler, 1978, 1983; Hall, 1978b,d, 2000c,d) and many more, some of which are discussed in Chapters 18 and 21).

As a site to grow 15-day-old embryonic chick tibiae into which electrodes were implanted to apply direct or alternating current. A DC current of 10 μA stimulated osteoclast production at the cathode and osteoblast and periosteal formation at the anode (Noda and Sato, 1985b).

As a site to test the toxicity of such materials as titanium used in implants (Monro *et al.*, 1986).

Grafting tissues from other vertebrates

Grafting mandibles, cartilage or bone from embryonic rodents, lizards, green turtles, and frog tadpoles (Nicholas and Rudnick, 1933; Stephenson and Tomkins, 1964; Tenenbaum *et al.*, 1976; Hall, 1978d, 1980c; Fig. 12.8).

Grafting mammalian tooth buds (chapters in Dahlberg, 1971).

Grafting embryonic or larval fish tissues (Miyake and Hall, 1994).

[a] See Bradley (1970), Hall (1978d, 2000d), and Hall and Miyake (1994) for discussion of the technique of chorioallantoic grafting and its usefulness in such studies.

NOTES

1. See Heller *et al.* (1950) for spindle cells, Kember (1960) for mesenchymal cells, Pritchard (1956a,b, 1969), Tonna (1961), Owen (1963), Owen and MacPherson (1963) and Scott (1967) for preosteoblasts, and Young (1962) and Hall (1970a, 1971a, 1982b) for osteoprogenitor cells.

2. Among the first evidence for initiation of polysaccharide synthesis on rough endoplasmic reticulum, and completion of chain elongation on smooth endoplasmic reticulum, was the study by Horwitz and Dorfman (1968) using microsome fractions from epiphyseal cartilage of chick embryos.

3. For the histochemical features of progenitor cells, in what are now classical studies, see Horowitz (1942), Bevelander and Johnson (1950), Pritchard (1952), Henrichsen (1956a), Fullmer (1965), Hay (1965), Milaire (1965, 1967), Hinchliffe and Ede (1967), Searls (1967), Hall (1968a) and Scott and Glimcher (1971). For ultrastructural features, see Cameron *et al.* (1964, 1967), Scott (1965, 1969), Aho (1966), Silberberg *et al.* (1966), Anderson (1967), Hall and Shorey (1968), Luk *et al.* (1974a,b) and Thorogood and Craig Gray (1975). For functional categorization of progenitor cells, see Searls (1965a), Bingham *et al.* (1969), Pawelek (1969), Coffin and Hall (1974), Rasmussen and Bordier (1974) and Thorogood and Hall (1976, 1977).

4. See Kember (1960, 1971, 1973), Walker and Kember (1972), Young (1962, 1963, 1964), Dixon (1970), Owen (1963, 1970, 1971), and Owen and MacPherson (1963) for population kinetics of progenitor cells. Kember (1971) is a comprehensive review of the earlier work.

5. See Kimmel and Jee (1980a,b) and Table 12.2 for this cell-cycle data.

6. I reviewed much of the evidence that had accumulated up to c. 1970 for the origin of precursor cells during endochondral ossification, repair of fractures of long bones, and formation of secondary cartilage on membrane bones (Hall, 1970a). See Hall and Jacobson (1975), Fang and Hall (1997) and Buxton *et al.* (2003) for conditions that favour fibroblastic or chondroblastic differentiation. See Precious and Hall (1992) for an overview of cartilage formation on membrane bones in clinical situations. One of the first reports of repair of the rat mandible was by Sarnat and Schour (1944), who observed bone formation in the pulp cavity. For repair of membrane bone without chondrogenesis – in this case salamander lower jaws – see Hall and Hanken (1985a).

7. See Scott and Hightower (1991) and C. K. Scott *et al.* (1994) for these studies.

8. See Hall (1968a,b), Coffin and Hall (1974) and Thorogood and Hall (1976, 1977) for approaches using histochemistry and enzyme profiles.

9. See Coffin and Hall (1974) and Thorogood and Hall (1976, 1977) for LDH isozymes and LDH/MDH ratios, Faulhauber *et al.* (1986) for weak detection of subunit A (M) of LDH (LDH-5) in chick cartilage, and Granström (1986) and Granström and Magnusson (1986) for the predominance of LDH-3–5 – and the absence of LDH-1 and LDH-2 – during rat facial development. See Brighton *et al.* (1983) for lack of the glycerol-phosphate shuttle in growth-plate cartilage and consequent lactate accumulation, and Hall (1970b, 1973b) for collagen:acid mucopolysaccharide ratios. Interestingly, during *in-vitro* healing of wounds to the flexor tendon, lactate up-regulates expression of Tgfβ (Yalamanchi *et al.*, 2004).

10. See Bélanger (1954), Amprino (1955a–c, 1956), Johnston and Comar (1957), Lash *et al.* (1960), Holtzer (1964a), Searls (1965a,b), Holtzer and Abbott (1968), Royal and Goetinck (1977) and the review by Hall (1977a) for the use of synthesis and accumulation of GAG to identify chondroblasts.

11. See Miller and Matukas (1969), Miller (1972, 1973, 1976) and the review by Trelstad (1973) for the primary structure of collagens, von der Mark and Bornstein (1973), Fessler *et al.* (1973), Uitto *et al.* (1977) and von der Mark (1980) for characterization of procollagens, and Benveniste *et al.* (1973) for cell-free translation. See Wright and Leblond (1981) for the immunohistochemical method, and von der Mark (1980) and Mayne and von der Mark (1983) for reviews of cartilage collagens.

12. There is pretranslational coordination of synthesis of pro α1(I) and pro α2(I) chains in embryonic chick bone (Vuust *et al.*, 1983).

13. See Devlin *et al.* (1988) and Kosher *et al.* (1986) for level of regulation, Duchene *et al.* (1982) and Elima and Vuorio (1989) for chondrocyte dedifferentiation, Fernandez *et al.* (1985) for methylation, Kravis and Upholt (1985) for mRNA copy number, Thompson *et al.* (1985) for agarose culture of chondrocytes, and Ng *et al.* (1993) and Sandell *et al.* (1991) for alternatively spliced type II procollagen mRNA.

14. See Nah *et al.* (2001) and Zhu *et al.* (2001) for type IIA procollagen expression, and see Walsh and Lotz (2004) and Walsh *et al.* (2004) for approaches to analysis of intervertebral discs *in vivo*.

15. See Ohyama *et al.* (1997) and Salas-Vidal *et al.* (2001) for tibial and digital studies. TUNEL (<u>t</u>erminal deoxynucleotidyl transferase biotin-d<u>U</u>TP <u>n</u>ick <u>e</u>nd <u>l</u>abeling) is used in the TUNEL assay to visualize DNA cleaved by endonucleases into broken strands and indicative of apoptosis.

16. A sample of studies includes histochemical analyses of condylar cartilage in rats and mice (Greenspan and Blackwood, 1966; Hall, 1968b), overviews of the histogenesis of condylar secondary cartilage (Hall, 1970a; Durkin, 1972; Silbermann and Frommer, 1972a–c, 1973a,b; Durkin *et al.*, 1973; Beresford, 1975b; Livne *et al.*, 1990; Vinkka-Puhakka, 1991a), an overview of 216 references on the development, growth and function of this cartilage by Koski (1975), and an analysis of all the secondary cartilages in the rat by Vinkka (1982). The texts by Enlow (1975), Moore and Lavelle (1974), Moore (1981) and Ranly (1988), along with Dhem and Goret-Nicaise (1979), provide excellent reviews of the role of condylar cartilage in mandibular growth and morphogenesis (see Chapter 13).

17. See Copray *et al.* (1985a–d, 1988) for the growth studies, Irving and Durkin (1965) and Durkin *et al.* (1969a,b) for the studies involving scurvy, and Hall (1981b) for modulation of chondrogenesis by ascorbic acid. Rat condylar and epiphyseal cartilages also respond differently to thyroxine (Becks *et al.*, 1946).

18. See Copray *et al.* (1988) for this study, and see Duterloo and Jansen (1970) as an example of earlier studies. Copray (1984) brings together six important studies on regulation of the growth of condylar cartilage. Copray (1984) brings together six important studies on regulation of the growth of condylar cartilage, See Cunat *et al.* (1956) for the development of the squamoso-mandibular joint in embryonic rats.

19. For an introduction to the complex issue of homology, especially as related to development, see Hall (1994a, 1995c, 1998a, 2003a).

20. See Baume (1961, 1962a,b) and Symons (1952).

21. See Hirschmann and Shuttleworth (1976), Silbermann and von der Mark (1990), Silbermann *et al.* (1990), Pirttiniemi *et al.* (1996), Tuominen *et al.* (1996), Kantomaa and Pirttiniemi (1998) and Pietilä *et al.* (1999) for type I collagen in condylar cartilage, and Brigham *et al.* (1977) for the antigenic determinants.

22. See Charlier *et al.* (1969a), Meikle (1973a,b), Petrovic *et al.* (1973), Petrovic (1974) and Hinton (1985, 1987, 1988a) for continued development under physiological conditions.

23. See Stutzmann and Petrovic (1974) for resection of the muscle, Meikle (1973a,b) for intracerebral transplantation, and Engelsma *et al.* (1980) for grafting into the metacarpal. See Kantomaa and Hall (1988a,b, 1991) for the organ culture method and for results on cAMP and Ca^{++} metabolism.

24. See Hinton (1988a,b) and Box 12.2 for soft diet and incisor clipping. Many other studies document the same phenomenon and, in addition, analyze changes in the 3D organization of the jaws and/or skull following changes to the diet.

25. See Cohn (1964) and Poswillo (1972) for condylectomy in mice, and compare with Jolly (1961) for condylectomy in rats.

26. Aggrecan core protein is expressed in embryonic membrane bone in wild-type but not in *nanomelic* chick embryos, suggesting that *nanomelia*, which is thought of as a 'cartilage mutation,' acts directly on membrane bone (see Chapters 23 and 27).

27. Throughout the text, the stages of development of the embryonic chick are given according to Hamburger and Hamilton (1951) and are listed as H.H. Table 12.3 shows the times of incubation corresponding to each H.H. stage, along with the major events in skeletal development at each stage.

28. For secondary cartilage and secondary chondrogenesis, see Murray and Smiles (1965) and Hall (1967a, 1968c, 1970a,b, 1972b, 1975c, 1978b, 1979, 1980a–c, 1983b, 1984a, 1986a, 1987a,b, 1991b, 1994b, 2004a), Hall and Shorey (1968), Thorogood and Hall (1976, 1977), Tyler and Hall (1977), Hall and Van Exan (1982), Hall *et al.* (1983), Van Exan and Hall (1983, 1984), Hall and MacSween (1984), Hall and Coffin-Collins (1990), Pinto and Hall (1991), Dunlop and Hall (1995), Fang and Hall (1995, 1996, 1997, 1999), MacDonald and Hall (2001) and Buxton *et al.* (2003).

29. See Hall (1970a, 1978b,c, 2004a), Fang and Hall (1999) and Buxton *et al.* (2003) for discussion of the timing of the determinative events in these cells.

Dedifferentiation Provides Progenitor Cells for Jaws and Long Bones

Chapter 13

Hypertrophic chondrocytes can become osteoprogenitor cells.

A further means by which the skeleton may obtain progenitor cells is through the *dedifferentiation of already differentiated cells*.

In Chapter 10, I introduced the concept of *dedifferentiation, whereby a cell loses its specialized morphology, function and biochemistry, initiates division and reverts to a less differentiated cell, which can in turn redifferentiate*. Three well-studied examples provide clear and unequivocal evidence of dedifferentiation to progenitor cells. Each illustrates disparate ways in which dedifferentiation can be initiated and controlled. The first and second, discussed in this chapter, raise the important issue of how we distinguish dedifferentiation from transformation. The three are:

- dedifferentiation of hypertrophic chondrocytes (HCCs) to osteoprogenitor cells (OPCs) during mineralization of *condylar cartilage on the mammalian dentary*;
- survival of HCCs in the vascularized and mineralized portion of the growth plates of (some?) *mammalian long bones* and their dedifferentiation (transformation?) to osteogenic cells; and
- dedifferentiation of myoblasts and/or chondrocytes to progenitor cells to form the blastema from which an amputated *urodele limb regenerates* (see Chapter 14).

Because the condylar cartilage and periosteal progenitor cells are fresh in our minds from the previous chapter, I begin with the joint of which the condylar cartilage forms an integral part and which forms the articulation between the jaw and skull in mammals.

CONDYLAR CARTILAGE OF THE MAMMALIAN TEMPOROMANDIBULAR JOINT

The temporomandibular joint

The jaw articulation in mammals is comprised of an articulation between the temporal portion of the squamosal bone (temporo) and the condylar process of the dentary of the lower jaw (mandibular) hence *temporomandibular joint* (TMJ). Because so much of our facial pain comes from degenerative changes in this region, the TMJ is a favourite joint for study, especially in research-oriented dental schools and anatomy departments; English and Stohler (2001) edited a special issue of the journal *Cells, Tissues and Organs* devoted to TMJ research into the 21st century. The TMJ itself is discussed in Chapter 33. Here my interest is with the specialized *condylar cartilage* that was introduced in Chapter 12, and that caps the dentary component of the joint, the condylar process.

Development of the TMJ has been studied, especially histologically, in numerous mammals, including rodents, sheep, humans, guinea pigs and rabbits. In humans, and presumably in other mammals, initial ossification occurs around the alveolar nerve, an association that has been used to establish eight stages of ossification of the human TMJ. Robinson and Poswillo (1994) examined development on the TMJ in the cotton-eared marmoset, *Callithrix jacchus*, and found it similar to that in humans with a 30-day delay reflecting the differing lengths of the gestation periods.[1]

Hypertrophic chondrocytes survive

Only a portion of the condyle of the mammalian dentary forms by endochondral ossification. Even then, it is a *modified form of endochondral ossification* because the:

- cartilage arises secondarily in the periosteum of the condylar process, or as a sesamoid beside the condylar process (see Chapter 12); and
- HCCs survive to contribute to osteogenesis.

If chondrocytes survive, Durkin (1972) argued, the process is not really true endochondral ossification. Bone is not replacing cartilage; chondrocytes transform into osteoprogenitors to live on in a new guise. The same developmental processes are seen in antlers (Chapter 8), in long bones (this chapter) and in penile cartilages (Chapter 25). When cartilage transforms into bone, the process is *metaplasia* and occurs at the tissue level. When HCCs become OPCs, the process is *dedifferentiation* and occurs at the cellular level.

Michael Silbermann and Jack Frommer in Israel, and Murray Meikle in the UK, contributed much to our understanding of what happens in condylar cartilage of the TMJ.[2] Silbermann and Frommer (1972a–c, 1974) provided ultrastructural and isotopic evidence that many HCCs of the condyle of the dentaries of newborn mice survive matrix mineralization (Fig. 5.9). Many membrane-bound vesicles appear within chondrocytes at the premineralization zone, and the chondrocytes take up ^{3}H-proline, usually an indication of osteogenic, not chondrogenic cells.

Using transmission electron microscopy, Yoshioka and Yagi (1988) analyzed mandibles from rats between three and seven weeks after birth. Terminal HCCs in each cell column are not fully surrounded by extracellular matrix (ECM), lie close to capillaries, and are released near developing bone, although some are released into the marrow where they *could* transform to osteoprogenitors.

Meikle observed similar membrane-bound vesicles in the condyles of one-week-old rats, showed that they contain acid phosphatase and aryl sulfatase, and concluded that these vesicles are lysosomes, a conclusion confirmed by Silbermann and Frommer. If they were lysosomes – bags of enzymes used by cells to commit suicide – it was argued, the cells would not survive, even though at the stage when the vesicles are present the cells appear active and are continuing to synthesize glycosaminoglycans (GAGs).[3]

Another analysis of rat condylar cartilage by Heeley *et al.* (1983) visualized ^{3}H-thymidine in chondrocytes *and chondroclasts* within five minutes of administration of the isotope, with *two thirds* of the chondroclasts labeled a mere 10 minutes after isotope administration (Table 13.1). Given such rapid uptake into chondroclasts, Heeley and colleagues concluded that HCCs do not transform into chondroclasts. It appears that, even in mammals, not all cartilages are resorbed in the same way or by the same cellular mechanisms. Two types and mechanisms of cartilage resorption have been identified: erosion by cells brought in via capillary invasion in primary cartilages, and erosion by chondroclasts in secondary cartilages such as the mandibular condyle.

Table 13.1 Percentage of chondroclasts and chondroclast nuclei labeled with ^{3}H-thymidine in condylar cartilage of rats[a]

Time after injection of ^{3}H-thymidine	Chondroclasts (%)	Chondroclast nuclei (%)
10 minutes	65	33
One hour	96	76
24 hours	55	50
Five days	75	46

[a] Based on data from Heeley *et al.* (1983).

Murine condylar secondary cartilage goes through three phases of maturation (and senescence) *in vivo*. Between six and eight weeks after birth is a phase of cartilage growth and endochondral ossification. Continuing to six months of age is the phase of mineralization and hyaline cartilage formation to form a surface articular fibrocartilage. Cartilage degeneration sets in beyond six months of age. At the end of somatic growth of the crab-eating monkey, *Macaca fascicularis*, the hyaline condylar cartilage transforms into an articular cartilage. Chondrocytes cease enlarging, do not die, and the ECM no longer excludes blood vessels. With vascular invasion comes mineralization.[4]

Hypertrophic chondrocytes transform to osteoprogenitor cells

When Michael Silbermann and his colleagues maintained mandibular condyles from 19- or 20-day-old mouse foetuses in organ culture, no chondroblasts were present after five days; all had matured to hypertrophic chondrocytes. Under the influence of these HCCs, chondroprogenitor cells transformed to osteoblasts and deposited osteoid, characterized by collagen type I and mineralized matrix (Fig. 5.9). In an earlier study they had interpreted this tissue as chondroid bone, characterized by osteonectin and bone sialoprotein.[5]

Silbermann's group then developed a culture system for *progenitor cells* of the four-day-old mouse condylar cartilage and utilized it to great advantage. In this system:

- chondroblasts differentiated within two days;
- HCCs differentiated and ECM was deposited within four days;
- by six days the ECM had mineralized; and
- by nine days bone with mineralized ECM was deposited.

The spatial associations between HCCs and adjacent bone were maintained *in vitro* (Weiss *et al.*, 1986, 1988).

Shibata and colleagues carried out several studies that inform the fate of chondrocytes from rat mandibular condylar cartilage, including:

- using histology to identify large cells (>20 μm diameter), some dividing, identified as surviving and proliferating HCCs; but this evidence is circumstantial;

- an ultrastructural study of preosteoblasts in the primary spongiosa;
- a comparison of mitotic chondroblasts in condylar and tibial growth plate cartilages;
- a basic study of mouse cartilage from 14.5 days (with no evidence of a separate blastema), including possible survival of HCCs;
- an ultrastructural study of cartilage resorption in mouse condyle at 16 days; and
- localization of collagen types II and X and alkaline phosphatase in the cartilages, including alkaline phosphatase at day 14 in the cartilage condensation, and evidence for a periosteal origin of secondary cartilage.[6]

Mandibular condyles from 60-day-old human foetuses also have been maintained in organ culture. By 12 days the culture is entirely composed of HCCs and preosteoblasts producing type I collagen (Ben-Ami *et al.*, 1993).

MECKEL'S CARTILAGE

As the only cartilaginous element that has ever existed in vertebrate lower jaws, Meckel's cartilage has had a long time to evolve, although Meckelian evolution is constrained in various ways. The fact that it is the primary skeletal support for the jaws is a major constraint; lower jaws are always initiated around a single rod of cartilage (Figs 3.5 and 3.6). Outside the mammals, Meckel's rarely ossifies, remaining as a permanent cartilaginous rod in most taxa.

The fate of Meckel's cartilage varies enormously among the vertebrates. In teleost fishes Meckel's persists as a complete cartilaginous rod. In birds most of the cartilage persists, only the posterior portion undergoing endochondral ossification to form the retroarticular bone (Fig. 3.5). Lack of ossification in other taxa can be correlated with the inability of most Meckelian chondrocytes to hypertrophy, express type X collagen, or be invaded by blood vessels. Meckelian chondrocytes from chick embryos fail to hypertrophy when cultured with thyroxine. Unlike the primary cartilages of long bones, Meckelian perichondria only transform into a periosteum in some groups. Any bone that forms is usually *around Meckel's* rather than on or in Meckel's. Unmineralized cartilage in rats is resorbed by macrophages and fibroblasts in processes that are not associated with ossification, vascularization or mineralization.[7]

Mammalian Meckel's

The major exceptions to these statements are the mammals in which the fate of Meckel's cartilage is much more complex (Fig. 13.1).

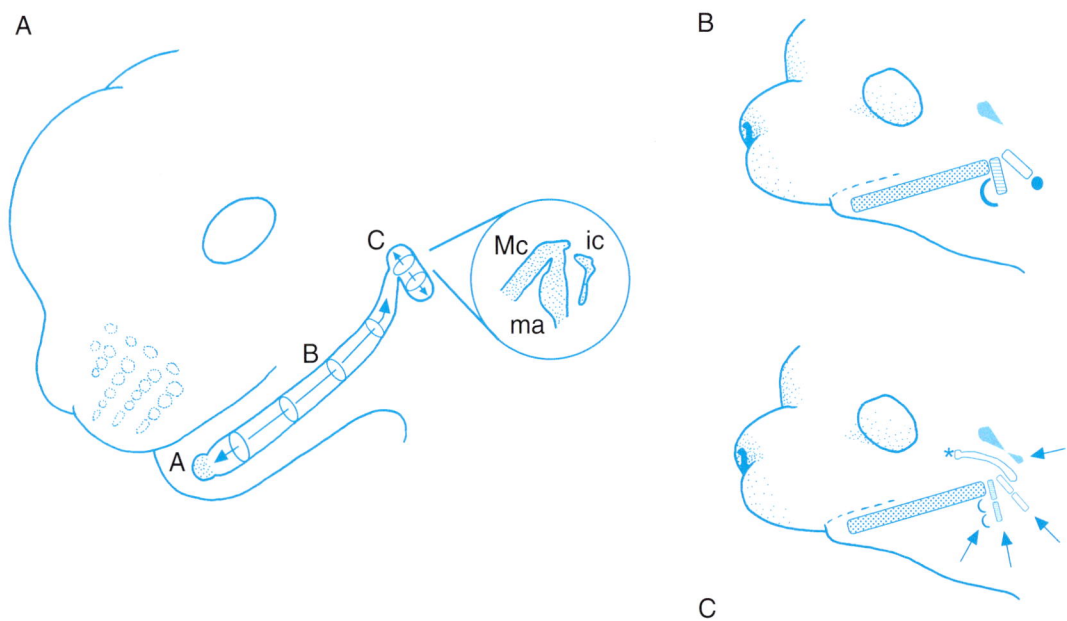

Figure 13.1 Developmental origin of the middle ear ossicles and the effect of *Hoxa-2* knockout on their development. (A) A diagrammatic representation of the components of the first arch prechondrogenic condensation in a C57BL/6 mouse embryo to demonstrate that several skeletal elements can arise from a single condensation and persist for varying periods of time. A rostral component – A in A – forms the symphyseal cartilage. The bulk of the condensation – B in A – gives rise to Meckel's cartilage. A proximal component – C in A – straddles the boundary between the mandibular and hyoid arches. Part of this component produces the proximal portion of Meckel's cartilage (Mc), the other component forms the malleus (ma). The incus (ic) arises from a separate condensation in the second arch. In B and C, the lower jaw and middle ear skeletal elements of a wild-type mouse (B) are compared with those in a mouse in which the gene *Hoxa-2* was knocked out (C). Arrows in C indicate elements that are duplicated in the knockout mice; compare C with B. The * in C identifies an ectopic cartilage that exists only as a condensation in the wild-type (not shown). The squamosal bone (top arrow in C) and all three middle ear ossicles in B and C – incus (unshaded), malleus (filled) and tympanum (black semi-circles) – are duplicated. B and C based on data in Rijli *et al.* (1994). Drawn by Tom Miyake.

Box 13.1 Middle ear ossicles

'Natural selection certainly eliminates the unfit and establishes the fit, but in my opinion it has nothing whatever to do with the creation of the fit.'
(Robert Broom, 1950, p. 7)

The poster child for evolution

The evolution of the middle ossicles – malleus, incus and stapes – has long been viewed as one of the most remarkable examples of evolutionary transformation of developmental and adult structures. In birds and reptiles, the lower jaws articulate with the skull via the quadrate and articular; see Figs 9.3 and 13.2–13.5. As mammals arose in the Late Triassic, the quadrate/articular ceased to function as the jaw joint, freeing these bones for loss or other functions. The rest of the story is nothing short of amazing:

- a series of cartilages from the lower jaws of reptiles becoming middle ear ossicles in mammals is remarkable enough;
- identifying them as essentially the same elements – i.e. as homologues – is one of the major accomplishments of comparative vertebrate anatomy;
- discovering a series of fossils that show the transitions in the sequence from reptiles → mammal-like reptiles (as the intermediate forms were long called) → to mammals is even more remarkable;
- having what is essentially the evolutionary sequence preserved in the development of each generation of marsupials tops off this extraordinary story.[a]

As in every family history, renegades occasionally appear, one of which may be what has been interpreted as an ossified Meckel's cartilage in two Early Cretaceous mammals, *Repenomamus* and *Gobiconodon*, from China, which Y. Wang *et al.* (2001) discuss in relation to current ideas on the origin of the middle ear ossicles.

Three remarkable features are seen during mandibular and middle ear ontogeny of the gray short-tailed opossum, *Monodelphis domestica*, in which:

(i) neonates have neither a typical mammalian nor a typical reptilian jaw articulation;
(ii) the first (mandibular) arch is on a faster developmental schedule than the second (hyoid) arch; and
(iii) the phylogenetically older skeletal elements develop earlier than do the phylogenetically younger elements.[b]

Embryological evidence

The middle ear ossicles lie at the junction of the first and second branchial arches (Figs 13.1 and 13.5). As a consequence, this region is subject to a disproportionately large number of developmental malformations; for example, malleus and incus are reduced in *Ap-2* mutant mice (see Box 16.4). An additional level of complexity arises from the report that in humans the anterior process of the malleus (*os goniale*) develops by intramembranous ossification independent of Meckel's cartilage, indicating a dual developmental origin for the malleus.[c]

The junction (bridge) between first and second branchial arches also is an evolutionarily important region. Malformations – especially the formation of extra elements – in this region after knocking out Hox genes in mice are often regarded as resurrected reptilian bones (Chapter 44). In *Hoxa-2* knockouts, Meckel's cartilage and the middle ear ossicles are duplicated and the second arch elements are absent (actually redirected; they form more anteriorly in *Hoxa-2* knockouts; Fig. 13.1).

Drew Noden of Cornell University obtained a duplicated set of lower jaw elements in the second arch after transplanting lower jaw-destined neural-crest cells into a more posterior region of the neural tube, from which they migrated into the second arch, where hyoid skeleton would normally develop. Noden was well aware of the evolutionary implications. In a letter he penned to the late Peter Thorogood on April 15, 1982, to accompany photographs of the results, Noden noted: 'It is important to state here that 2nd

arch crest cells normally form the retroarticular process of Meckel's and the caudal part of the angular bone [in avian embryos]; these structures form in the mesenchymal 'bridge' connecting 1st and 2nd arch crest mesenchyme. I'm sure you and Brian will recognize the evolutionary implications of this *vis-à-vis* the ear ossicles.'[d]

Retinoic acid

Other factors also influence middle ear ossicle development. Retinoic acid (RA) affects murine middle ear development in a stage-dependent manner, inhibiting formation and/or migration of neural-crest cells but not their later specification.

The tympanic ring – the embryonic precursor of the tympanic bone, which supports the medial end of the external auditory meatus to provide attachment for the tympanic membrane – is lost in RA-treated embryos, but duplicated after *Hoxa-2* knockout. Meckel's cartilage and the ossicles also are duplicated following *Hoxa-2* knockout, essentially because the second branchial arch fails to form and this second set of first arch elements forms more anteriorly than the normal position of the first arch; there is a homeotic transformation of second arch to duplicated first arch.

Changes in the palate and skull with allelic disruption of *Hoxa-1* and/or *Hoxa-2* show evidence for compensation as defects associated with *Hoxa-1* restore normal palates in *Hoxa-2* mutants. *Hoxa-1* disrupts rhombomere (r) organization in the mouse hindbrain, while *Hoxb-1* influences the fate of cells that emerge from r4. Double mutants show defects either not seen in single mutants, or expressed in mild form, including loss of r4 and r5, and loss of the second branchial arches and associated tissues leading to craniofacial defects.[e]

These responses to RA or Hox gene knockout are superimposed on a more fundamental subdivision of Meckel's cartilage in *placental* mammals, in which the distal extremity forms a permanent symphyseal cartilage between the two dentary bones; the major portion of the cartilage transforms into the mento-Meckelian ligament, while the proximal portion – which lies within the second arch – transforms into the middle ear ossicles (Figs 13.1 and 13.5). Differences identified by the fate of regions of the cartilage are reflected in the behaviours of chondrocytes isolated from each region and grafted to an intraocular site, as each region exercises what was regarded as an autonomous developmental programme. Chondrocytes from the distal (symphyseal) regions hypertrophy and undergo endochondral ossification, although some of the chondrocytes transform into osteoblasts. Only chondrocytes from this region synthesize and deposit type X collagen.[f] Each region exercises an autonomous programme.

[a] See Gregory (1913) for a critique of the (then) recent work, especially the reptilian/mammalian lower jaw and the origin of middle ear ossicles. For excellent reviews of the evolution of the mammalian middle ear, see Hopson (1966), Thomson (1966), Manley (1972), Allin (1975), Lombard and Bolt (1979), Maier (1990) and Mallo (1998, 2001). Hopson (1966) emphasized the pre-adaptation of the reptilian articular and quadrate for transformation into middle ear ossicles and described the Late Triassic tritylodontid therapsid *Bienotherium*, whose ear region has a closer resemblance to the mammalian ear than does any other therapsid.
[b] See Mallo and Gridley (1996), Mallo (1997, 1998) and Gendron-Maguire *et al.* (1994) for RA and *Hoxa-2*. The external auditory meatus mediates tympanic membrane development by producing signals (*Prx-1*?) that both enhance (*Sox-9*) and inhibit chondrogenesis (Mallo *et al.*, 2000). The foetal human middle ear is filled with mesenchyme that is resorbed and replaced by ingrowing epithelium (Rauchfuss, 1989).
[c] For morphogenesis and malformations of the ear and ear ossicles, see Van de Water *et al.* (1980), and the papers in Gorlin (1980), and Mallo (1997, 1998). For development in humans, see Whyte *et al.* (2002). For the *os goniale*, see Rodriguez-Valvquez *et al.* (1991).
[d] I thank Drew Noden of Cornell University for the quote from his letter to the late Professor Peter Thorogood.
[e] See Gendron-Maguire *et al.* (1994), Barrow and Capecchi (1999) and Rossel and Capecchi (1999) for the studies with *Hoxa-1*, *Hoxa-2* and *Hoxb-1*. Formation of the tympanic ring is controlled in part by *Gsc*, *Prx-1* and *Bapx-1* (Mallo *et al.*, 2000; Tucker *et al.*, 2004).
[f] See Richman and Diewert (1988), Chung *et al.* (1995) and Chung and Nishimura (1999) for such studies.

Development and growth of Meckel's cartilage and the mandible are both more complex than usually appreciated. Patterns of human mandibular growth and growth rates change considerably between eight and 14 weeks of embryonic life, eight weeks being when chondrocyte hypertrophy begins.[8] In *general* terms in placental mammals, the distal tip – the portion closest to where the two lower jaws articulate – becomes a symphyseal cartilage; the mid-portion transforms to the sphenomandibular (tympanomandibular) ligament (Box 13.1) while the posterior portion gives rise to the malleus and incus, two of the three middle ear ossicles (Box 13.1).

Illustrative of the fates of these different regions of mammalian Meckel's cartilages is an examination of what happens when different regions of the cartilage are taken from 16- or 17-day-old rats and transplanted:

- chondrocytes from the distal region hypertrophy, synthesize type X collagen and undergo endochondral ossification. ^{3}H-thymidine labeling demonstrates that some transform into osteoblasts;
- chondrocytes from the middle region form a fibrous tissue; while
- chondrocytes from the proximal region differentiate into two cartilages: the malleus and incus.[9]

Transformation of chondrocytes to form the ligament is initiated by tension and mediated by Egf (Boxes 3.2 and 13.2). Exogenous Egf accelerates loss of Meckel's cartilage *in vivo*, while *in vitro* Egf enhances DNA synthesis and cell division, inhibits chondrogenic differentiation, and transforms chondroblasts to fibroblasts.[10] Uniquely, and I use this word advisedly, the chondrocytes themselves contribute to the resorption of Meckel's cartilage in foetal mice. Chondrocytes survive removal of their ECM, are released and fuse with existing chondroclasts.[11]

Ishizeki and colleagues analyzed in some detail the transdifferentiation of mouse Meckel's cartilage into osteoblastic cells, either *in vitro*, or when grafted. Murine Meckelian chondrocytes and early HCCs contain metalloproteinase-1 (MMP-1), which they secrete into the pericellular matrix, and which is involved in resorption (Box 13.3). The mid-region of Meckel's from 18-day-old embryos (the hypertrophic region) undergoes endochondral ossification with survival of HCCs (which produce Bmp), accompanied by the appearance of many small osteogenic cells around the calcified matrix. These cells, which produce type I collagen, osteocalcin and alkaline phosphatase, are presumed to have arisen from HCCs. Transformation of cells into fibroblasts at the posterior extremity occurs without apoptosis.[12]

Other exceptions also exist. In fact, so few taxa have been studied that it is premature to draw too many generalizations concerning the fate of Meckel's cartilage. For example, Haines (1937) described changes at the posterior end of Meckel's cartilage and associated bones in fishes as involving formation of an 'ossification

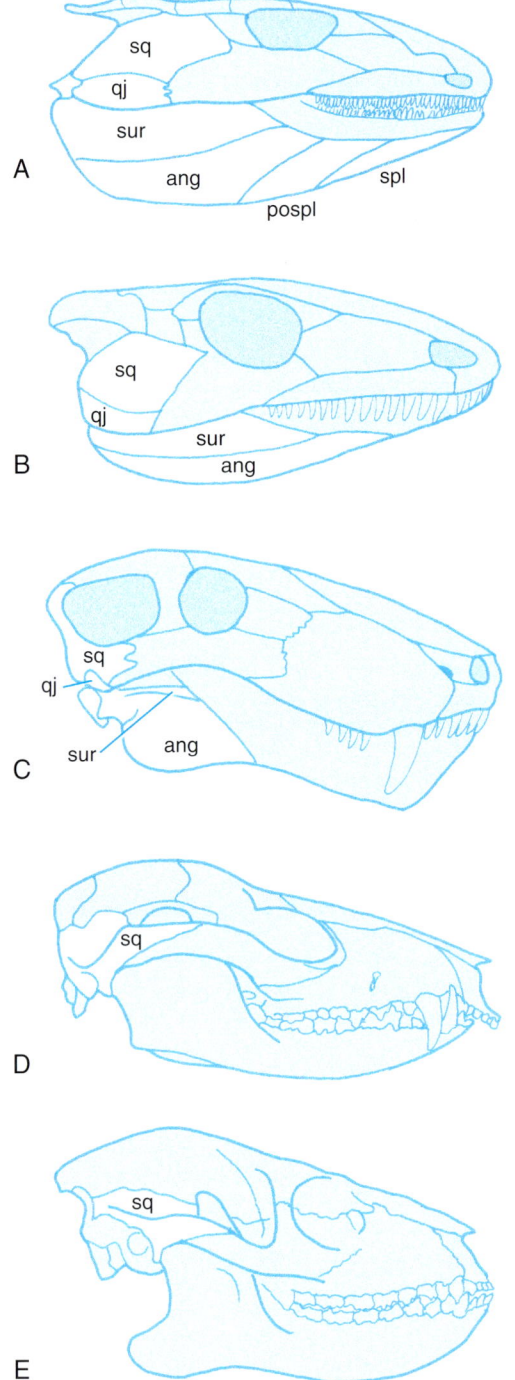

Figure 13.2 Changes in the skull associated with the evolution of the mammalian jaw joint. Bones of the lower jaw, the quadratojugal (qj) and the squamosal (sq) are unshaded. (A) An early labyrinthodont (*Palaeoherpeton*) from the Lower Carboniferous with multiple bones in the lower jaw and articulation involving the surangular, quadratojugal and squamosal. (B) *Seymouria*, a basal (cotylosaurian) reptile from the Permian–Carboniferous with loss of anterior elements of the lower jaw. (C) A mammal-like (therapsid) reptile, *Scymnognathus*, from the Permian, with reduced surangular and involvement of the angular in the jaw articulation. (D) *Eodelphis*, an early marsupial from the Upper Cretaceous, with a single bone (the dentary) in the lower jaw and involvement of the squamosal in the jaw articulation. (E) An early Eocene primate, *Notharctus*, with a single bone (the dentary) in the lower jaw articulating with the squamosal. Ang, angular; pospl, postsplenial; spl, splenial; sur, surangular. Substantially modified from Gregory (1929).

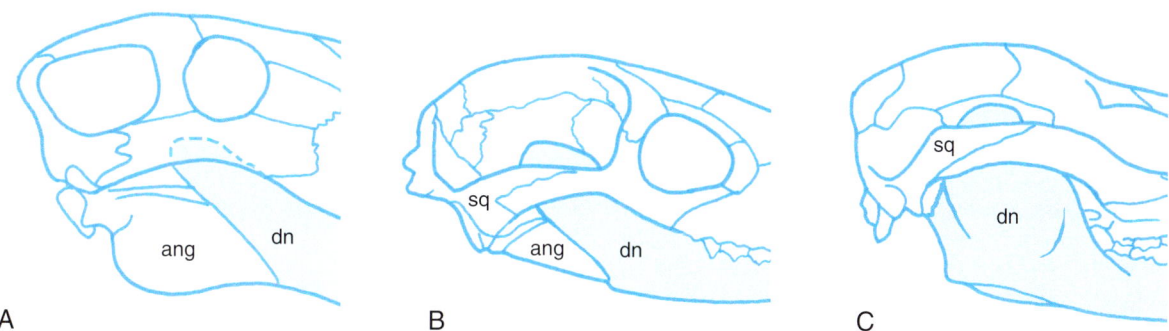

Figure 13.3 Changes in the dentary and angular bones associated with development of the mammalian jaw articulation. Elaboration and upward growth of the dentary (dn) and reduction of the angular (ang) are shown in early and late mammal-like reptiles (therapsids) *Scymnognathus* (A) and *Ictidopsis* (B), and in *Thylacinus*, an early marsupial (C). sq, squamosal. Substantially modified from Gregory (1929).

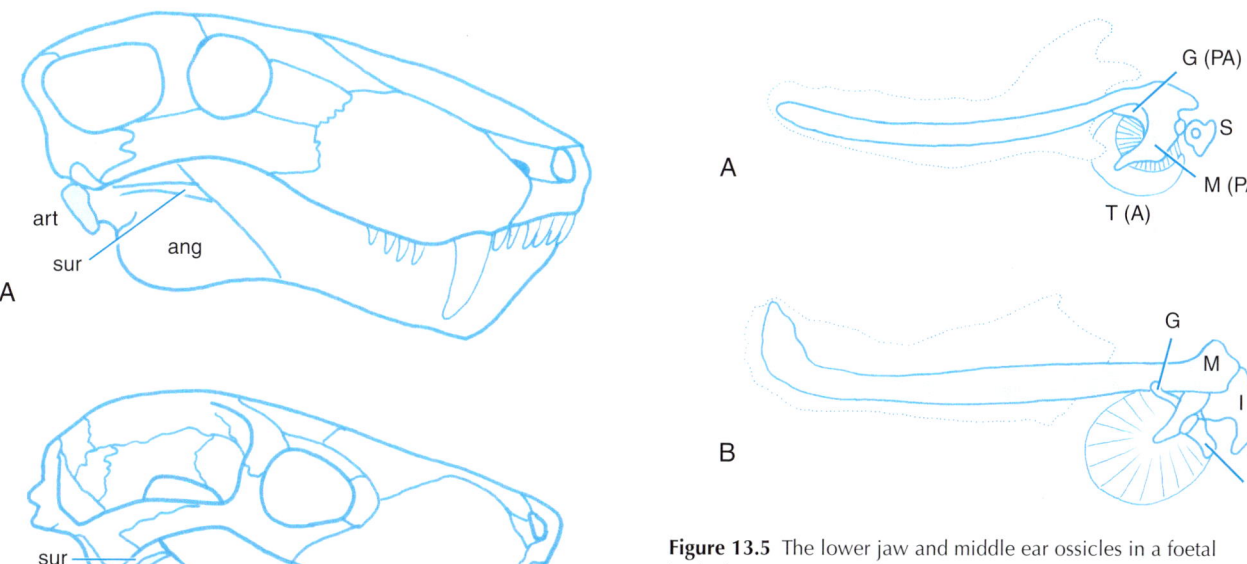

Figure 13.4 The jaw articulation in *Scymnognathus* (A), an early, and *Ictidopsis* (B), a late mammal-like reptile (therapsid) showing changes in the positions and sizes of the angular (ang), articular (art) and surangular (sur). Modified from Gregory (1929).

Figure 13.5 The lower jaw and middle ear ossicles in a foetal hedgehog, *Erinaceus europaeus* (A) and a 43-mm-long (8.5-week-old) human foetus (B). The dentary is shown in outline, Meckel's cartilage (M) is shaded. G(PA), the goniale (prearticular); I, incus; M (A), the malleus (articular); S, stapes; T (A) tympanic (angular). Names in brackets are the equivalent bones in the mammal-like reptiles. Modified from Gregory (1929).

Box 13.2 Epidermal growth factor (Egf) and its interactions with Tgfβ

In the early 1960s the first growth factor discovered was named *epidermal growth factor* (EGF) because its action was thought to be specific to epithelial cells. Similarly, fibroblast growth factor (FGF) was thought to act specifically on fibroblastic (mesenchymal) cells. Despite its name and acronym Egf also acts on mesenchymal cells, for example, promoting proliferation of osteoprogenitor cells from foetal calvariae, stimulating growth of long bones and calvariae, and suppressing proline incorporation into collagen. Subsequently, EGF was shown to influence morphogenesis of cartilages.

As we now know, these 'growth' factors are even more important as differentiation factors during embryonic development. The growth factor family has grown from two individual growth factors (Egf and Fgf) to multiple members of growth factor superfamilies.[a]

I^{125}-labeled Egf and electron microscopical autoradiography showed that osteogenic precursors possess Egf receptors, but osteoclasts and osteoblasts do not. Isolated rat calvarial osteoblasts, however, respond to low concentrations of Egf (10^{-12} to 2×10^{-8} M) by proliferating, a result which implies that osteoblasts do have receptors for and are targets of Egf. Receptors on chick epiphyseal growth-plate chondrocytes allow a dose-dependent proliferative response to Egf, while 0.4 ng Egf/ml suppresses the hydroxyproline and alkaline phosphatase content of osteoblastic cells cloned from mouse calvariae. Collagen synthesis is decreased and non-banded fibres form if the same cells are maintained in concentrations of 2–50 ng/ml. Type III collagen also appears in the cells, taken as indicative of remodeling.[b]

Action on skeletogenic cells

Adamson *et al.* (1981) localized Egf receptors in embryonic mouse tissues, with the greatest numbers present at 14 days of gestation.

Egf and Tgfβ

Tgfβ (transforming growth factor-β) is found in chondrocytes and osteoblasts of both endochondral and membrane bones (Box 15.1).

Box 13.2 *(continued)*

Bone is one of the most abundant sources of Tgfβ, and there are estimated to be 5800 Tgfβ receptors on each foetal bovine bone cell. Tgfβ also is found in skin and hair in places where epithelial–mesenchymal interactions occur.[c]

Now we have a superfamily of transforming growth factors of the β type, a superfamily that includes such important developmental regulators as the bone morphogenetic proteins (Bmps). Specific actions of individual family members are introduced in Box 15.1 and discussed as appropriate in the text. Here, I provide an overview of the early studies demonstrating how Tgfβ and Egf interact.

Osteogenesis

Tgfβ modulates proliferation of osteoblastic cells by affecting Egf receptor levels. Synthesis of Tgfβ is required for bone formation to begin, its removal for bone to mineralize. So we find that murine calvariae contain multiple forms of latent Tgfβ, the latent form requiring an acidic environment for release and so functioning as a protease.[d]

Egf has a biphasic effect on formation of bone nodules from rat calvarial cells *in vitro*. Nodule formation is reduced following continuous exposure to 10^{-12}–10^{-7} M Egf (or to Tgfβ) for 21 days or following four hours exposure on day 11. Exposure to Egf for four hours on the first or second day of culture increases nodule number, which suggested to Antosz *et al.* (1987, 1989) that Egf acts directly on osteoblasts or osteoprogenitor cells. Tgfβ inhibits both nodule formation and proliferation of chondroprogenitor cells.

Tgfβ is mitogenic for osteoblasts and enhances synthesis of collagen mRNA at concentrations of 2–4 ng/ml. This effect, first demonstrated in foetal calvarial cells by Centrella and colleagues in 1986, was shown to occur *in vivo* by Noda and Camilliere in 1989. In the latter study, 12 daily injections of Tgfβ onto the parietal bones of neonatal rats increased bone thickness by twofold as woven bone was deposited and mineralized. Platelet-derived growth factor (Pdgf), Fgf and Egf had no effect, although Pdgf, Tgfβ and Egf all stimulate proliferation of human or rat bone cells *in vitro*.[e]

Tgfβ has a 10-fold greater effect on osteoprogenitors than on other cell types. At concentrations of 1–30 ng/ml, Tgfβ increases matrix formation by some 25–40 per cent in 21-day-old foetal rat half calvariae maintained *in vitro*, 30 ng/ml being mitogenic. Centrella's group undertook a thorough investigation of interactions between Tgfβ and other growth factors, and investigated other growth factors such as Pdgf, which enhances proliferation and bone matrix deposition.[f]

Tgfβ antagonizes the action of Egf on mouse osteoblastic MC3T3-E1 cells, Tgfβ itself not being mitogenic for these cells in monolayer culture conditions. Egf stimulates proliferation and suppresses collagen synthesis; 95 per cent of the collagen synthesized in the presence of Egf is type I. Action of Tgfβ on MC3T3-E1 cells depends on their maturation status; 1 ng/ml decreases DNA synthesis during the first three days but not between three and five days in culture, after which Tgfβ increases synthesis of collagen and GAGs.[g]

Tgfβ up-regulates expression of such osteoblastic genes as collagen, osteopontin and osteonectin (which are found in both dividing chondroblasts and HCCs of long bones, in calvariae, and in epithelia in mouse, human and avian embryos; Fig. 1.6) and down-regulates osteocalcin; Tgfβ itself is regulated by Fgf.[h]

Chondrogenesis

The differential response to Tgfβ of rabbit articular chondrocytes is modulated by serum factors, most likely Egf. Thus, 0.01–10 ng/ml Tgfβ + 2% foetal calf serum (FCS) decreases cell number and proliferation while the same doses of Tgfβ with 10% FCS leads to a transient increase in proliferation. Egf + 2% FCS mimics Tgfβ + 10% FCS, implicating Egf as the serum factor that enhances the proliferative effect of Tgfβ. Rabbit mandibular condylar chondrocytes respond to Egf with enhanced proliferation, irrespective of the zone from which the chondrocytes are obtained.[i]

[a] See Canalis and Raisz (1979) and Raisz *et al.* (1980) for early studies of the action of Egf on osteogenic cells, Pratt (1987) for a review of early studies, Coffin-Collins and Hall (1989) and Hall and Coffin-Collins (1990) for regulation of mitotic activity of prechondrogenic mesenchyme by Egf, and Slavkin *et al.* (1992) and Shum *et al.* (1993) for Egf and morphogenesis of mouse Meckel's cartilages. *In vitro*, murine Meckelian chondrocytes respond to Egf with enhanced DNA synthesis and cell division, inhibition of chondrogenic differentiation, and transformation to fibroblasts. See Isheseki *et al.* (2001) for the role of Egf.
[b] See Martineau-Doizé *et al.* (1988) for the labeling study, Ng *et al.* (1983) for the study with isolated osteoblasts, Halevy *et al.* (1991) for receptors on chondrocytes, and Kumegawa *et al.* (1983) for action on collagen synthesis. Two decades on we have identified a substantial similarity between the Egf receptors in *Drosophila* and vertebrates (Mann and Casares, 2002).
[c] See Robey *et al.* (1987), Carrington *et al.* (1988) and Pelton *et al.* (1989) for these benchmark studies on the distribution of Tgfβ. See Serra and Chang (2003) for Tgfβ in human skeletal development and skeletal disorders.
[d] See Uneno *et al.* (1989) and Bonewald and Dallas (1994) for Tgfβ and initiation of osteogenesis, although see also Breen *et al.* (1994) for 0.1 ng Tgfβ/ml inhibiting osteoblast differentiation *in vitro*. See Bonewald *et al.* (1991) for multiple latent forms, and Pfeilschifter *et al.* (1990a) for Tgfβ as a protease.
[e] See Centrella *et al.* (1986, 1987a,b, 1988, 1989a) for Tgfβ as mitogenic, and Piché and Graves (1989), Hock *et al.* (1990) and Hock and Canalis (1994) for stimulation of mitotic activity by Pdgf, Tgfß and Egf. One has to be careful when interpreting the effects of Pdgf, for its effects vary with the stage of differentiation of the osteoblastic cells; 50 ng/ml Pdgf increases DNA levels in cells from mouse trabecular bone four times when the cells are in early passage, but only 1.2 times in late-passage cells (Abdennagy *et al.*, 1992). See Hoch and Soriano (2003) for the two mouse ligands, *Pdgf-a* and *Pdgf-b*, Pdgf receptors α and β, and the multiple effects of Pdgf on migration, differentiation and cell function.
[f] See Centrella *et al.* (1987a, 1989a,c) and Hock *et al.* (1990) for interactions between growth factors.
[g] See Hata *et al.* (1984), Elford *et al.* (1987) and Katagiri *et al.* (1990a) for the studies with MC3T3-E1 cells.
[h] See Nomura *et al.* (1988), Metsäranta *et al.* (1989), Oshima *et al.* (1989) and Pacifici *et al.* (1990) for up-regulation of osteoblastic genes, and Noda and Rodan (1989a,b) and Noda and Vogel (1989) for regulation of *Tgfβ*. Also see McKee *et al.* (1990a,b, 1992) for osteocalcin and osteopontin during endochondral ossification of chick tibia.
[i] See Vivien *et al.* (1991) for the serum factor, and Tsubai *et al.* (2000) for mandibular chondrocytes.

mixte,' a tissue that appears to be chondroid bone. Available comparative studies or detailed studies on a single species include comparison of Meckel's cartilage development in different strains of mice and in sheep, *Ovis aries*, and cats, *Felis domesticus*, and contain discussions of condensation and the timing and duration of condensation and chondrogenesis. Thomson published three papers on the development of Meckel's cartilage in *Xenopus laevis*. His analyses include changes that occur during metamorphosis (increase in cell number and matrix accumulation), initial ossification, effects of thyroid hormone and correlation to feeding habits of the tadpoles.[13]

As discussed in Chapter 18, Meckel's cartilage differentiates from ectomesenchyme in response to epithelial signals. Regulation of *Msx-1* by *Bmp-2* or *Bmp-4* is an important genetic pathway in these interactions

Box 13.3 Metalloproteinases

Bmp-1 was always the Bmp that did not fit. Although it falls within the Bmp family, its structure is less highly conserved than that of other members of the family, and unlike all other early members of the family (Bmp-2, Bmp-4; Box 12.1), Bmp-1 does not induce skeletal tissues. Bmp-1's outlier position is because Bmp-1 is a type 1 procollagen C-proteinase (a metalloproteinase), a class of molecules that processes procollagens to fibrillar collagens in the extracellular matrix.[a]

Bmp-1 has the specific role of cleaving latent Bmps so that they can be activated and function as chondro- or osteoinductors. Bmp-1, then, has the important role of linking growth factors, ECM and cellular activity. There are four mammalian Bmp-1/tolloid-related metalloproteinases. A novel member, tolloid-like 2, has differential distribution and enzymatic activity in the limb skeleton.[b] ADAM 12, which is both a metalloproteinase and a disintegrin, is involved with the syndecans in cell–cell and cell–matrix adhesion.[b]

Matrix metalloproteinases – a family that includes the collagenases as a major subtype – function in different skeletal tissues and in half a dozen or more processes affecting skeletogenesis:

- Murine Meckelian chondrocytes and early HCCs contain metalloproteinase-1 (MMP-1), which they secrete into the pericellular matrix and which is involved in matrix resorption (Ishizeki and Nawa, 2000).
- Matrix vesicles from rat costochondral cartilage cells are enriched in metalloproteinases that degrade proteoglycans (see Table 22.5).
- Matrix metalloproteinases have a role in various processes associated with mandibular cartilage, bone and epithelia, including cell migration and cell survival (Werb and Chin, 1998).
- Matrix metalloproteinase-9 (MMP-9) is required for invasion of mineralized hypertrophic cartilage, solubilization of unmineralized

cartilage and vascular invasion of cartilage during fracture repair (Engsig et al., 2000; Colnot et al., 2003). Callus cartilage persists in Mmp-9[-/-] mice to the point that union is delayed or fails to occur.

- MMP-9-deficient mice show delayed osteoclast recruitment. Using foetal mouse metatarsals, Blavier and Delaissé (1995) followed TRAP-positive mononuclear cells from the periosteum through the osteoid and into the marrow.[c] Inhibitors of matrix metalloproteinases block this migration, demonstrating their role in localizing osteoclast precursors.

Between 12 and 14 days of mouse embryogenesis both matrix metalloproteins and their inhibitors (TIMP) are temporally regulated. Both are important for skeletal morphogenesis; culturing murine mandibular mesenchyme with inhibitors results in abnormal cartilage morphogenesis.[d]

Endopeptidase-24.11 (E-24.11) and endopeptidase 24.18 (E-24.18) are metalloendopeptidases that cleave peptides, including some growth factors. Distribution of both enzymes and their mRNA in mouse embryogenesis shows that E-24.11 is first seen in the gut at 10 days of gestation, in the notochord and otocyst at 12 days, and by 14 days in perichondria and the palate (Spencer-Dene et al., 1994).

[a] See Ballock et al. (1993), Kessler et al. (1995), S.-W. Li et al. (1996) and Olsen (1996) for metalloproteinases, and Rapraeger (2000) for ADAM 12.
[b] See Li et al. (1996) and Sarras (1996) for the function of Bmp-1, and Scott et al. (1999) for tolloid-like 2. Disintegrins, peptides with some 70 amino acids, inhibit platelet aggregation and so prevent blood coagulation. See Kirn-Safran et al. (2004) for syndecan-3 and the other heparan sulphate proteoglycans.
[c] See Chapter 15 for TRAP-staining.
[d] See Morris-Wiman et al. (1999) for temporal regulation, and Chin and Werb (1997) for inhibitor studies.

(Box 13.4); ectopic application of Bmp-2 or Bmp-4 alters expression of Msx-1 and extends expression of Fgf-4 in the distal epithelium, leading to formation of a bifurcated Meckel's cartilage.[14]

Prx-1, Prx-2

Prx-1 (MHox) and Prx-2, paired-related homeobox genes homologous to the Drosophila Aristaless gene (Prx-1 binds to a creatine kinase enhancer), function cooperatively to maintain cell fate in craniofacial mesenchyme. Rostral mandible development is inhibited in double mutants, mandibular incisors remain at the bud stage, Meckel's cartilage fails to develop, and cells of the hyoid arch move into the first branchial arch. In chick embryos, Prx-1 is expressed in facial mesenchyme, down-regulated with chondrogenesis, and stays unexpressed in perichondria and in undifferentiated facial and limb mesenchyme. Removal of the epithelium over the somites suppresses expression of Prx-1, suggesting epithelial control of expression (Kuratani et al., 1994).

Loss-of-function mutation of Prx-1 indicates its role in early skeletogenic events in multiple sites, acting at the condensation stage (Martin et al., 1995).

Skull, limb and vertebral defects flow from loss of function of Prx-1 but not from loss of Prx-2. Novel craniofacial

and limb anomalies do appear in double knockout (Prx-1[-/-]/ Prx-2[-/-]) mice, including either a single lower incisor or absence of the lower incisor, a consequence of local reduction in proliferation in the medial portion of the mandibular arch and substantial reduction in Shh in the mandibular epithelium opposite the site of affected mesenchyme. Prx is upstream of Shh; administering jervine, an inhibitor of the hedgehog pathway, to wild-type mice partly phenocopies the double knockout with loss of the mandibular incisors. Study of the Prx-1/Prx-2 double mutants shows that both are upstream of dHand, eHand and Msx-1; expression of Msx-2 is independent of Prx pathways.[15] (Msx-1 is expressed in the progress zone of murine limb buds, expression depending on signals from the AER. Fgf-4 can substitute for the AER signals; Bmp-4 and Fgf-2 maintain Msx-1 expression [Y. Wang and Sassoon, 1995].)

Epithelial–mesenchymal interactions in avian skeletogenesis (Fig. 13.8) are mediated, in part, by Prx-1 and Prx-2. The two Prx genes are involved in epithelial–mesenchymal interactions in the inner ear, lower jaw, and in interactions between perichondrium and chondrocytes. Both genes are expressed in lateral plate mesoderm and limb mesenchyme. With single knockouts, limbs develop normally. Double knockouts lead to severe defects as early as the stage when the apical ectodermal ridge (AER) is active.

Box 13.4 *Msx* genes

Msx and *Distalless* (*Dlx*) multigene families are expressed in overlapping but distinct domains in many tissues, especially those in which epithelial–mesenchymal interactions or cell death are important developmental processes, including teeth, branchial arches (Fig. 13.6) and the limb skeleton.[a] *Msx* genes, which are a class of homeobox genes, are transcriptional repressors. *Dlx* genes are transcriptional activators in cell proliferation, differentiation, patterning and morphogenesis (Bendall and Abate-Shen, 2000 for review). Functions of *Msx* genes are discussed in particular sections. Here I provide an overview and flavour of their roles, especially where epithelial–mesenchymal signaling is involved.

Msx-1

A gene given the name *Hox 7.1* was found by Mackenzie *et al.* (1991a,b) to be expressed in various embryonic structures such as neural crest-derived mesenchyme, neuroepithelium, Rathke's pouch and the nasal pits, and at epithelial–mesenchymal junctions. The patterns of expression correlated with embryonic regions known to be defective after treatment with retinoic acid. We now know *Hox 7.1* as *Msx-1*.

Limb development
Msx-1 is expressed in the progress zone of mouse limb buds, expression depending on signals from the AER. Fgf-4 can substitute for the AER signals, expression of *Msx-1* being maintained by Bmp-4 and Ffg-2 (Y. Wang and Sassoon, 1995).

Extra digits can be induced from interdigital areas of limb buds of H.H. 29 chick embryos without any modification of *Msx* expression, which led Ros *et al.* (1994) to conclude that *Msx* maintains interdigital mesenchyme in an undifferentiated state and does not provide specific signals that control cell death in this region.

Yokouchi *et al.* (1991a) isolated *Msx-1* from domestic fowl and demonstrated expression in the lateral plate mesoderm from which limb mesenchyme arises. As limb development proceeds, expression is reduced in posterior limb mesenchyme and in subsequent prechondrogenic mesenchyme. *Msx-1* then increases dramatically in the AER and in the mesenchyme of the posterior limb border. Exposure to retinoic acid decreases expression in both mesenchyme and epithelium.

Craniofacial development
Msx-1 is expressed within growth centres in mouse embryonic mandibular mesenchyme. *Msx-2* and *Fgf-8* (Fig. 13.7; Box 14.4) are expressed in the adjacent mandibular epithelium, a spatial pattern consistent with epithelial–mesenchymal signaling. Expression is associated with programmed cell death, epithelial *Msx-2* up-regulating mRNA for mesenchymal *Msx-1*.[b]

Continuing with craniofacial gene activity, the molar teeth and the alveolar processes of mouse molar teeth fail to form in *Msx-1*-mutant mice, which also have cleft palates (Fig. 25.3). Administering Bmp-4 under the control of the *Msx-1* promoter restores wild-type levels of *Dlx-2 and* allows molar teeth and alveolar processes to form, the conclusion being that Bmp-4 can bypass *Msx-1* to restore alveolar processes and molar tooth development.[c]

Msx-2

Msx-2 plays important roles in chondrogenesis and osteogenesis, keeping sutures patent and contributing to helping to establish the balance between survival and apoptosis in skeletogenic cell populations.

In the early 1990s, a cDNA for Quox-7 (quail Hox gene 7) was cloned from Japanese quail bones shown to be expressed in facial mesenchyme, superficial ectoderm and the neural tube. Within the

mandible, expression was confined to the mid-ventral region in preosteogenic mesenchyme and required the presence of the mandibular epithelium for its expression, implicating Quox-7 in the osteogenic epithelial–mesenchymal interaction. Other studies showed that a gene named Quox-1, a homologue of *Drosophila Antennapedia*, was expressed in early somitic mesoderm, myotome and vertebral perichondria, the central nervous system, endothelial epithelium, spinal ganglia and placodal ectoderm of Japanese quail.

A third gene, Quox-8, described as a member of the Hox-7/msh family, was isolated from Japanese quail and shown to be expressed in the roof plate of the neural tube where it played a role in the normal dorso-ventral orientation of the neural tube. Quox-8 also is expressed in the dorsal mesenchyme that forms the spinous processes of the vertebrae; spinal processes failed to form when Quox-8 was down-regulated.[d]

All three genes are now known to be *msx-2*.

Craniofacial development
Msx-2 represses chondrogenic differentiation in migrating neural crest cells, probably by repressing *Sox-9* in a subpopulation of migrating cells (Takahashi *et al.*, 2001).

Sox-9 is a member of the Sox family of high-mobility group domain proteins that regulate specification of cell fate by pairing with specific partner factors that allow regulation of distinct gene sets. *Sox-9*, which is expressed in condensing prechondrogenic mesenchyme, activates the enhancer of the target gene Col2a1 which encodes type II collagen, expression of *Sox-9* and Col2a1 being parallel. Chimeric mice generated using *Sox-9*[$^{-/-}$] embryonic stem cells demonstrate that *Sox-9*[$^{-/-}$] cells are excluded from all cartilages, and that cartilage fails to form in teratomas derived from such embryonic stem cells from these chimaeras.[e]

Msx-2, which is expressed in proliferating calvarial osteoblasts, stimulates calvarial osteoblast differentiation. Mice that are transgenic for *Msx-2* show multiple craniofacial anomalies, *Msx-2* setting the balance between cell survival and apoptosis.[f]

Jabs *et al.* (1993) identified a mutation in the homeodomain of *Msx-2* in a human family with autosomal dominant craniosynostosis. *Msx-2* localizes to the same region of chromosome 5 as craniosynostosis and *Msx-2* transcripts are expressed in calvarial sutures (see Chapter 34 for craniosynostosis).

... and Bmp-4
The paired-related homeobox genes, *Prx-1* (M-Hox) and *Prx-2*, are expressed in lateral plate mesoderm and limb mesenchyme. Double knockouts lead to severe defects as early as the AER, involving absence of expression of *Msx-1* and *Msx-2*, suggesting regulation via *Bmp* (Lu *et al.*, 1999b). Grafting BMP-4-producing cells into paraxial mesoderm up-regulates *Msx-1* and *Msx-2*, resulting in production of ectopic cartilage in the pectoral girdle that develops.

Continuing with regulation of *Msx* genes by Bmp-4, Semba *et al.* (2000) showed that Bmp-4 induces ectopic cartilage in the proximal but not in the rostral mesenchyme of mouse embryonic mandibular arches. Concomitant expression of *Sox-9* and *Msx-2* (especially rostrally) accompanies chondrogenesis, suggesting that *Msx-2* normally inhibits chondrogenesis. Over-expression of *Msx-2* in proximal mandibular mesenchyme decreases ectopic and endogenous chondrogenesis, allowing the conclusion that *Bmp-4* and *Sox-9* positively regulate chondrogenesis, and that *Msx-2* negatively regulates chondrogenesis. See Box 12.1 for further information on Bmp-4.

Msx-2 is not expressed in the reduced and short-lived AERs found at early developmental stages of *wingless* or limbless mutant chick embryos (Dealy and Kosher, 1996; Fig. 36.2). Exogenous Igf-1 in limb buds of *wingless* or *limbless* mutants promotes development of

Box 13.4 (*Continued*)

a thick AER whose cells express *Msx-2* and promote limb bud outgrowth; an *Igf-1* → *Msx-2* pathway maintains the AER.

During limb development in mice *Msx-1* is expressed more distally than *Msx-2*. Foetal and newborn mice can regenerate the tips of their digits – foetuses only within the *Msx-1* domain, neonates within both domains (Reginelli *et al.*, 1995).

[a] Bone induced in response to Bmp expresses nine *Hox* and two *Msx* genes (Iimura *et al.*, 1994). Ekker *et al.* (1997) identified five *Msx* genes in the zebrafish (*Msx-A – Msx-E*), all of which are expressed in the visceral arches and head cartilages.
[b] See Mina *et al.* (1995) and McGonnell *et al.* (1998) for the primary studies, and see MacDonald and Hall (2001) and Francis-West *et al.* (2003) for overviews. See McGonnell *et al.* (1998) for fate maps of the facial primordia in chick embryos.
[c] See Satokata and Maas (1994), Richman and Mitchell (1996) and Zhao *et al.* (2000) for *Msx-1* and mandibular tooth and alveolar bone development.

[d] See Y. Takahashi and Le Douarin (1990) and Y. Takahashi *et al.* (1991) for Quox-7, Xue *et al.* (1991, 1993) for Quox-1, and Y. Takahashi *et al.* (1992) for Quox-8, all of which are msx-2.
[e] See Kamachi *et al.* (2000) and Kulyk *et al.* (2000) for *Sox-9* and its expression in prechondrogenic mesenchyme, Ng *et al.* (1997) for regulation of Col2a1, and Bi *et al.* (1999) for the chimaeric mice. Akiyama *et al.* (2002) showed that *Sox-9* acts at both precondensation and condensation stages of chondrogenesis. Neither cartilage nor bone forms, apoptosis is increased, and *Sox-5, Sox-6* and *Runx-2* are not expressed after inactivation of *Sox-9* before condensation in limb-bud mesenchymal cells. Severe chondrodysplasia develops after inactivation of *Sox-9* after condensation, a syndrome equivalent to that seen in *Sox-5⁻/Sox-6⁻* double knockout embryos: mesenchyme remains condensed; chondrogenesis is not initiated; chondroblast proliferation is inhibited severely and joints are defective.
[f] See Ryou *et al.* (1997), Dodig *et al.* (1999) and Winograd *et al.* (1997) for these three studies.

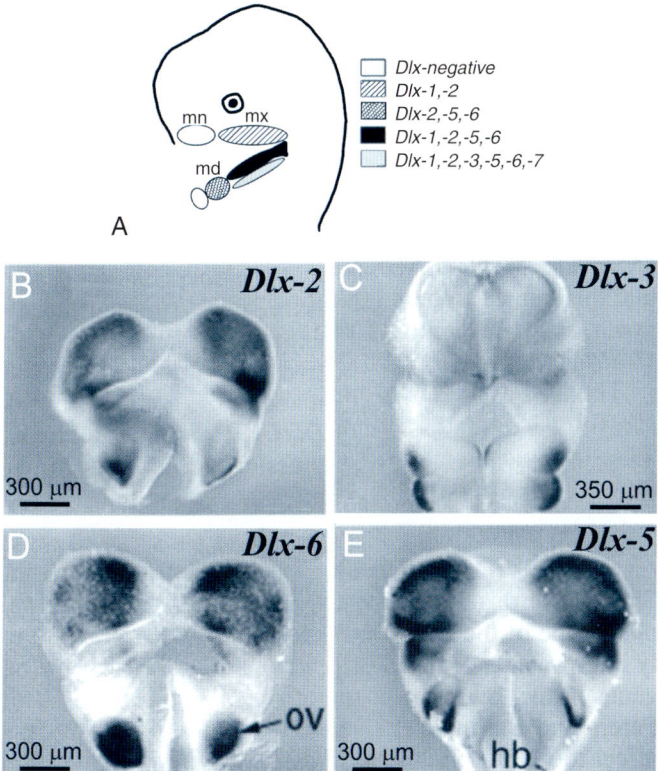

Figure 13.6 Expression of *Dlx* genes in the mandibular arch of embryonic mice of 10.5 days of gestation. (A) A diagrammatic lateral view to summarize expression of the genes *Dlx 1–7* in developing median nasal (mn) and maxillary (mx) processes, and in the mandibular arch (md). (B) In this frontal (oral) view of the mandibular (above) and second branchial arches (below), *Dlx-2* is expressed in the left and right mandibular arches and along one surface of the second arch. (C) In this frontal view of the developing brain (above) and arches (below), *Dlx-3* is expressed along the lateral borders of the first and second arches and in the antero-medial portion of the first (mandibular) arch. (D) *Dlx-6* is expressed in the mandibular arch (above), weakly in the maxillary arch, and strongly in the otic vesicle (OV). (E) *Dlx-5* shows extension expression in both mandibular and maxillary arches. hb, hindbrain. Adapted from Zhao *et al.* (2000).

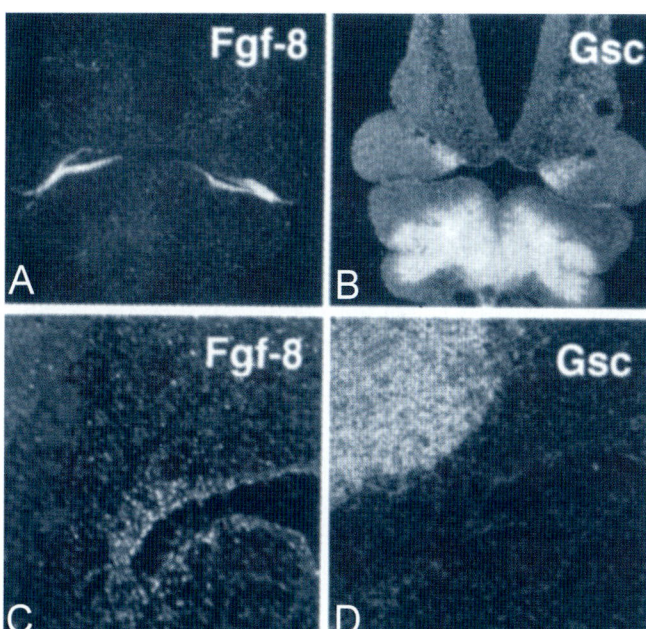

Figure 13.7 Expression of *Fgf-8* and *Goosecoid* (*Gsc*) in the branchial arches of 10.5-day-old embryonic mice. (A, B) Frontal sections showing *Fgf-8* (white) in the oral epithelium bordering the mouth cavity (A) and *Gsc* restricted to the caudal mesenchyme of the first arch (the uppermost arch in B) but expressed more extensively in the second branchial arch. (C, D) Longitudinal sections of the pharyngeal pouch, showing expression of *Fgf-8* in the pharyngeal endoderm (C) and *Gsc* in the mesenchyme (D). Compare the patterns of expression in C and D for the separation of expression of *Fgf* and *Gsc*. Modified from A. S. Tucker *et al.* (1998a).

Absence of expression of *Msx-1* and *Msx-2* in the knockout mice suggests that *Prx* genes act via *Bmp* (Lu *et al.*, 1999b).

Both genes also play a role in mediating interactions between perichondrial cells and adjacent chondrocytes, providing a mechanistic connection between condensation and perichondrial development. The presence of

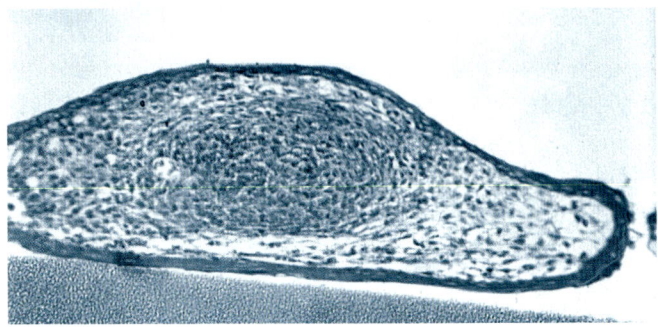

Figure 13.8 Epithelial–mesenchymal interaction in organ culture. Embryonic chick mandibular mesenchyme (H.H. 18) was combined with mandibular epithelium (H.H. 26), placed on a Millipore filter (shown below) and grown in organ culture. The epithelium has surrounded the mesenchyme and continued to differentiate. A precartilaginous condensation has developed within the mesenchyme.

Hox gene-binding sites in the upstream regulatory element of the N-CAM promoter makes N-CAM a potential downstream target for *Hox* genes. *Prx-1* is a potential upstream regulator of CAMs and therefore of condensation initiation.[16]

Alx-3

In murine embryos, *Alx-3* – a mouse paired *Aristaless*-like homeobox gene – is expressed at the condensation stage in ectomesenchyme and lateral plate mesoderm. Expression is in domains that overlap with *Prx-2* and *Cart-1*, a chondrogenic marker. *Alx-4* is found in murine mesenchymal condensations of the skull, hair, teeth and mammary glands. Null mutants have only preaxial polydactyly (preaxial referring to an additional digit(s) on the anterior [preaxial] surface of a limb; Figs 35.8 and 36.6) indicating either that *Alx-4* plays a minor role in skeletogenesis or that its function(s) overlap with other genes.[17]

Ptx-1

Pituitary homeobox-1 (*Ptx-1* or *Pitx-1*), a homeobox transcription factor related to *bicoid* in *Drosophila*, is expressed from the onset of pituitary development. *Ptx-1* – which maps in humans to the region of the chromosome associated with Treacher Collins syndrome, a major and not uncommon craniofacial syndrome – is expressed in the first arch, posterior mesoderm, anterior ectoderm and stomodeum of mouse embryos. Mice in which *Ptx-1* has been knocked out lack the proximal mandibular bones, and it appears that Meckel's cartilage is replaced by membrane bones. Other features are loss of the ilium and knee cartilage of the hind limb, and shortened long bones, the tibia and fibula being of equal lengths. The affected bones are all dorsal (Lanctôt et al., 1999).

Rieger syndrome, which results from a loss of function mutation of *Ptx-1*, has as features defective development of both mandibular and maxillary arches, arrested tooth

development, regression of the stomodeum and abnormal heart development. *Fgf-8* is not expressed, but *Bmp-4* is expressed in the branchial arch ectoderm, supporting *Ptx-1* regulating *Fgf-8* and restricting expression of Bmp-4 (Lu et al., 1999c). The role of Fgf-8 in mandibular development is discussed in Box 14.4 and depicted in Figure 13.7.

DEDIFFERENTIATION DURING ENDOCHONDRAL BONE FORMATION

The classic textbook description of the replacement of primary cartilage by bone in mammalian long bones runs as follows:

- chondrocytes in columns hypertrophy and mineralize the ECM;
- the transverse lamellae (lamellae that run across the growth plate) are dissolved while longitudinal lamellae (that run along the growth plate) remain intact;
- removal of ECM releases HCCs (usually only the terminal ones), which die;
- blood vessels invade the lamellae; and
- blood-borne cells move in, line up on the now mineralized cartilaginous trabeculae, become osteoblasts, and deposit osteoid.[18]

This description summarizes almost 80 years of research which began with Fell's (1925) light microscopical analysis, later reinforced by electron microscope studies of the epiphyseal growth plate of mammalian long bones (Fig. 13.9).[19]

As research continued, some studies began to cast doubt on whether HCCs die after they are released from their lacunae. Some survive to become OPCs, perhaps analogous to the dedifferentiation of chondrocytes during blastema formation in amphibian limb regeneration. The suggestion that chondrocytes may be precursors of OPCs is not a new one. Over a century ago, Van der Stricht (1890) and Brachet (1893) proposed that chondrocytes become osteoblasts, either directly by metaplasia, or indirectly by dedifferentiation to immature connective tissue cells and subsequent redifferentiation.[20]

When considering their fate, we have to consider HCCs in two phases of their development. One is the phase of chondrocyte hypertrophy when the cells are progressing *through the columns* in a growth plate. These maturing HCCs lie in that part of the matrix not invaded by blood vessels. Some remain to form islands of cartilage within the bone matrix (Fig. 13.9).

The second phase is represented by the most mature HCCs *at the ends* of the chondrocyte columns. These terminally hypertrophic cells *could* be released from their lacunae. I say *could* because different growth plates behave differently, even in the one individual, let alone trying to make comparisons across taxa. Zonation occurs even within a single growth plate; for example, the

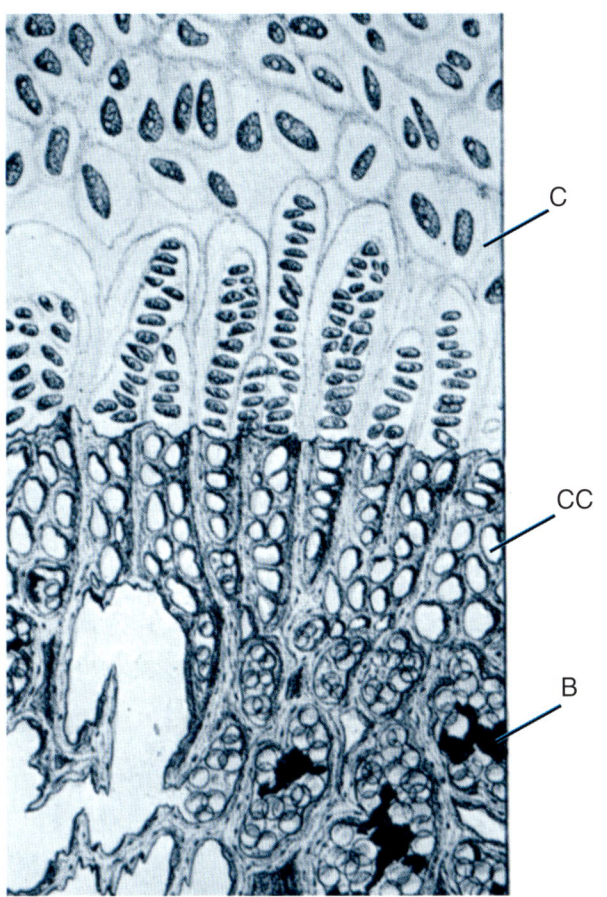

Figure 13.9 An early depiction of the growth plate. The proliferative (C), calcified (CC) and bony (B) layers of a mammalian growth plate, originally published as Figure d by Hassall (1849) in the first English language histology text. From Harris (1933).

differences in proliferative zones between lateral and central regions in the growth plates of 15-day-old rats discussed by Miralles-Flores and Deldago-Baeza (1990).

The two patterns are (i) death or (ii) survival of the last HCCs in each column. The potential for dedifferentiation/transformation exists only in those growth plates where the cells survive. Because cell behaviour/fate differs from site to site – I have already discussed the TMJ and Meckel's cartilage as two examples – I consider ribs, long bones and interpubic joints separately below.

Rodent ribs

Although it was 75 years ago that Dawson and Spark examined costal cartilages in albino rats and argued that freed HCCs transform into fibroblasts, the first modern attack on the belief that endochondral ossification prescribes a death sentence for HCCs came from studies by Marijke Holtrop in the mid-1960s.[21]

Mice

In her first study, Holtrop (1966) found that if fragments of mouse ribs – including the portion of the chondrocyte cell columns in which ossification had not commenced – were transplanted or cultured on solid media, the chondrocytes became hypertrophic and bone formed around them. The bone arose from progenitor cells within the perichondrium, which had transformed into a periosteum. Ribs deprived of their perichondria to exclude perichondrial cells as a source of progenitor cells were labeled with ^{3}H-thymidine and transplanted back into donor mice. Twelve days later, few labeled chondrocytes were found. However, labeled periosteal cells, osteoblasts and osteocytes were seen in the bone that had by then formed.

Holtrop concluded that chondrocytes were the precursors of the bone cells, probably having dedifferentiated to progenitor cells equivalent to those of the normal perichondrium–periosteum. She followed up these studies with ultrastructural investigations of the ribs of neonatal mice aged between three and 14 days. As chondrocytes matured as they passed down the cell column on the way to becoming hypertrophic, the amount of endoplasmic reticulum increased, many free ribosomes appeared, and electron-dense granules appeared within the mitochondria, granules shown with electron-microprobe analysis to be loaded with Ca^{++} and P^+. No sign of degeneration or lysis of cells was seen.[22]

Rats

Holtrop's studies were with *neonatal mice*. Using ribs from *21-day-old rats*, Brighton et al. (1973) found ultrastructural evidence for rapid chondrocyte degeneration affecting only the last four cells in each of the columns (Fig. 13.10). As Holtrop had described for mouse ribs, they confirmed an increase in cytoplasmic constituents as chondrocytes matured and moved down the cell columns. Endoplasmic reticulum, mitochondria, lysosomes and Golgi bodies all increase, while lipid vacuoles and multivesicular bodies decrease. In the *penultimate and last cells* in each column, however, the nuclear and cell membranes fragment and the cytoplasmic contents are eliminated. Brighton and colleagues claimed that Holtrop did not describe these cells because they were lost when the cell columns were dissected away from the bony metaphyses. However, Holtrop's technique was anything but sloppy. She recommended and used a two-step procedure that was the gold standard at the time and remains so: two hours at 0°C in a mixture of 2.5% glutaraldehyde:1% OsO_4 (1:2) in 0.1 M cacodylate buffer at pH 7.4, followed by post-fixation in 0.25% uranyl acetate in 0.1 M sodium acetate (pH 6.3) for two hours at 0°C.

That said, rat ribs are not mouse ribs. Perhaps degeneration had not set in at the younger ages studied by Holtrop. Furthermore, and to complicate interpretations, even the last cells in each column can survive under *in vitro* conditions in which oxygen tension is high (Sledge and Dingle, 1965).

To add a further wrinkle, high O_2 is also known to induce resorption of cartilage, although the wrinkle can

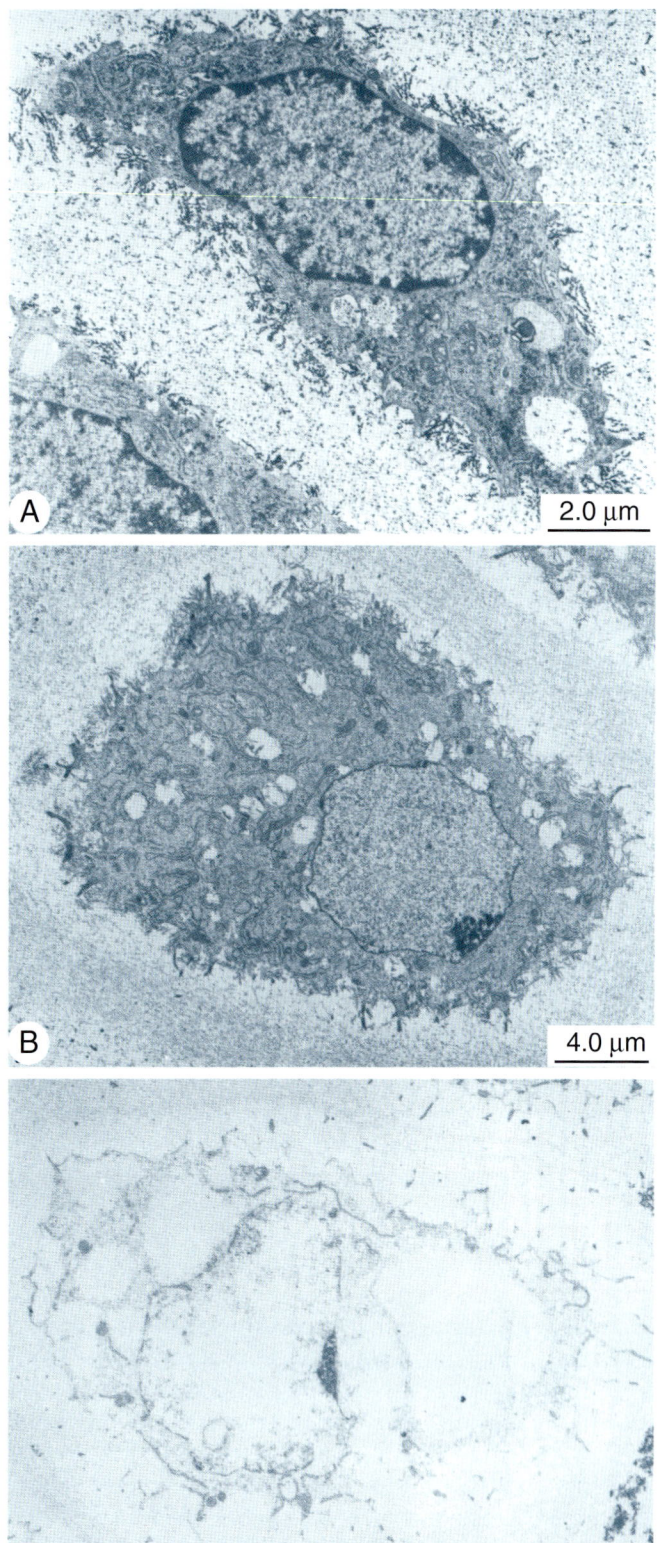

Figure 13.10 Chondrocyte degeneration in the costochondral junction of Sprague-Dawley rats as evidenced through electron microscopical analysis. (A) An active chondrocyte from the top of the cell column. (B) A chondrocyte from the hypertrophic cell zone with nucleus intact and cytoplasmic contents that signify an active cell. (C) An HCC from the base of the cell column. Nuclear and cell membranes are fragmented and most cytoplasmic constituents have been lost. Modified from Brighton *et al.* (1973).

be ironed out when we consider that resorption would facilitate release of terminal HCCs.

At the same time, Shimomura *et al.* (1973) used autoradiography to examine the costochondral junctions of five-week-old rats, and concluded that preosteoblasts at the ossification front are derived from cells of the adjacent connective tissue, primarily through modulation of cells in Ranvier's groove – which contains cells that can become osteo- or chondroblasts – and *perhaps* from HCCs. In 1975, they followed up this study with an experimental analysis and strong presumptive evidence that HCCs *induce* local cells to become osteogenic, but *do not themselves* modulate to osteoblasts. The approach was a novel one.

Chondroblasts from the resting zone and chondrocytes from the growth (hypertrophic) zones of rat ribs were placed in Millipore chambers and implanted into the abdominal cavities of host rats. Provided that host cells invaded the chambers, bone *was* produced in those chambers containing HCCs but *not* in those containing non-hypertrophic or resting chondroblasts. The conclusion: HCCs cannot modulate to osteoblasts but can induce local progenitor cells (inducible OPCs?) to become osteogenic.[23]

In studies in which unmineralized, mineralized or demineralized cartilages were implanted intramuscularly into mice, only the mineralized cartilage (which contains vital HCCs) induced new bone. In a further study from the same group, transplanted *mouse epiphyseal* chondrocytes formed cartilage that underwent endochondral ossification. Transplanted *rib* chondrocytes formed cartilage, some of which transformed to bone as chondrocytes transformed to osteocytes *without* first undergoing hypertrophy, and as the cartilaginous ECM transformed to a bony ECM.[24]

Appendicular long bones

Less ambiguous evidence from organ culture studies demonstrates specificity in the fate of HCCs from embryonic long bones of different vertebrates. HCCs from embryonic mouse tibiae survive, those from chick tibiae cytolyse and die. Diseases such as rickets influence the fate of HCCs (Fig. 3.13). HCCs from metaphyses and epiphyses of rachitic rats resume DNA synthesis – because they have dedifferentiated? – and produce osteoblasts in lacunae previously occupied by chondrocytes.[25]

Enzyme activity

The demonstration that oxidative enzyme activity is high in HCCs from rabbit femoral epiphyses was used to conclude that HCCs are active physiologically; specific activities of the major enzymes found in embryonic chick cartilage are listed in Table 13.2. Of course they are, but such an analysis does not take into account developmental changes reflected in regional differences along the

Table 13.2 Specific activities of enzymes from chick embryonic cartilage expressed as units per milligram of soluble protein or DNA[a]

Enzyme	Specific activity	
	Units $\times$ 10^3/mg protein	Units/mg DNA
β-galactosidase	0.9	0.03
β-glucuronidase	8.5	0.30
Acid deoxyribonuclease	1.3	0.04
Acid phosphatase	17.2	0.61
Acid pyrophosphatase	9.3	0.34
Acid ribonuclease	4.9	0.17
Alkaline phosphatase	21.3	0.77
Alkaline pyrophosphatase	21.5	0.77
Aryl sulfatase	4.3	0.15
Cathepsin	3.1	0.11
Cytochrome oxidase	30.7	1.10
Deoxyribonucleic acid	27.9	1.0
Esterase	12.8	0.46
Glucose-6-phosphatase	3.7	0.13
Hexosamine	86.4	3.11
Lysozyme	6.9	0.25
N-acetyl-β-D-galactosaminidase	2.4	0.24
N-acetyl-β-D-glucosaminidase	6.6	0.24
NADPH cytochrome c reductase	1.1	0.04
Ribonucleic acid	34.6	1.24
Succinate cytochrome c reductase	1.2	0.04

[a] Based on data from Arsenis et al. (1971).

cartilage columns. By way of summary, bone and cartilage contain:

- high levels of reduced diphosphopyridine nucleotide (DPNH) diaphorase, and succinic, lactic and malic dehydrogenases;
- moderate levels of isocitric dehydrogenase and glucose-6-phosphate dehydrogenase;
- little glutamic dehydrogenase; and
- no β-hydroxybutyric dehydrogenase or 6-phosphoglucuonic dehydrogenase.

Oxidative enzymes (dehydrogenases) are also found in osteoclasts, ameloblasts, odontoblasts, osteocytes, dental pulp and healing fractures. Because the same phosphohydrolytic enzymes are present in chondrocytes *and* osteoblasts, Arsenis and Huang (1971) argued that chondrocytes could transform into osteocytes. As the list above shows, similarity of enzyme profiles is not presumptive evidence for cell transformation.[26]

Evidence from [3]*H-thymidine-labeling and other approaches*

Bentley and Greer (1970) followed up the approach used by Holtrop to investigate ossification in transplanted ribs when, after labeling with ^{3}H-thymidine for three hours, they transplanted perichondrium-free rabbit metatarsal epiphyses intramuscularly. Although bone developed at the tips of the cell columns, the osteoblasts were unlabeled, even though adjacent HCCs were labeled. Bentley and Greer concluded that HCCs were *not* the source of the osteoblasts, reconciling their data with Holtrop's by invoking species specificity.

Some interesting differences in ossification of metaphyseal versus articular epiphyseal surfaces in guinea pig and mouse humeri, tibiae and femora have been described. On the metaphyseal side, where there is only slight mineralization, many chondrocytes survive. Filaments accumulate within surviving chondrocytes as they come to resemble osteoblasts ultrastructurally, changes suggestive of a metaplastic transformation similar to those discussed earlier and in transformation of secondary cartilage into bone in birds.[27]

HCCs contain type I collagen at the stage when endochondral osteoid is being deposited onto them. The ECM of HCCs in human growth plates contains type I collagen, but Yasui and colleagues could not determine whether it was a degradation product, or collagen deposited by transforming chondrocytes. The ability of mature HCCs to synthesize collagen types I and II is consistent with, but again does not prove, an osteogenic role.[28]

HCCs have been described in open lacunae in contact with primary spongiosa in the distal epiphyses of mouse femora. Ultrastructural analysis reveals substantial cell death where lacunae are open. Cells survive longer where transverse lamellae are intact but those cells also eventually die. Hanaoka stresses that cells transitional between chondrocytes and osteoblasts are *not* seen. With the S phase of the OPC cycle lasting eight hours, and terminal lacunae in cell columns opened up every four hours, the potential for transfer of label to OPCs from labeled degenerating chondrocytes is substantial.[29]

Hunziker et al. (1984) developed a technique of high-pressure freezing and freeze substitution coupled with low-temperature embedding, which they used to examine proximal growth plates from rat tibiae. They saw no degenerating cells, concluding that cartilage mineralization is controlled by chondrocytes alone.

Experiments comparing results using chondrocytes that hypertrophy and chondrocytes that don't added useful data. Xenografts of rabbit growth-plate chondrocytes – which *do* hypertrophy – form cartilaginous nodules that mineralize and induce bone formation when transplanted into nude mice. Articular chondrocytes – which *don't* hypertrophy – do not.[30]

Using the sex chromatin body to identify graft from host cells, Miki and Yamamuro (1987) transplanted growth plates intramuscularly between male and female rabbits, and determined that most, but not all, HCCs die.

An investigation of developing long bones from 17-day-old foetal rats, using a cDNA to osteopontin as an osteoblast marker, provided evidence for transformation of HCCs to osteoblast precursors. Osteopontin is found in those HCCs whose morphology is intermediate (transitional?) between chondrocytes and osteoblasts.

Vital dyes have also been used as tracers in studies seeking the fate of terminal HCCs. Farnum and Wilsman (1987) used fluorescein diacetate, Nomarski optics and transmission electron microscopy to follow the fate of cells in growth-plate cartilages, confirming that the terminal cells in the columns are metabolically active (Fig. 13.11). A subsequent study, using time-lapse cinemicroscopy and interference contrast microscopy, enabled

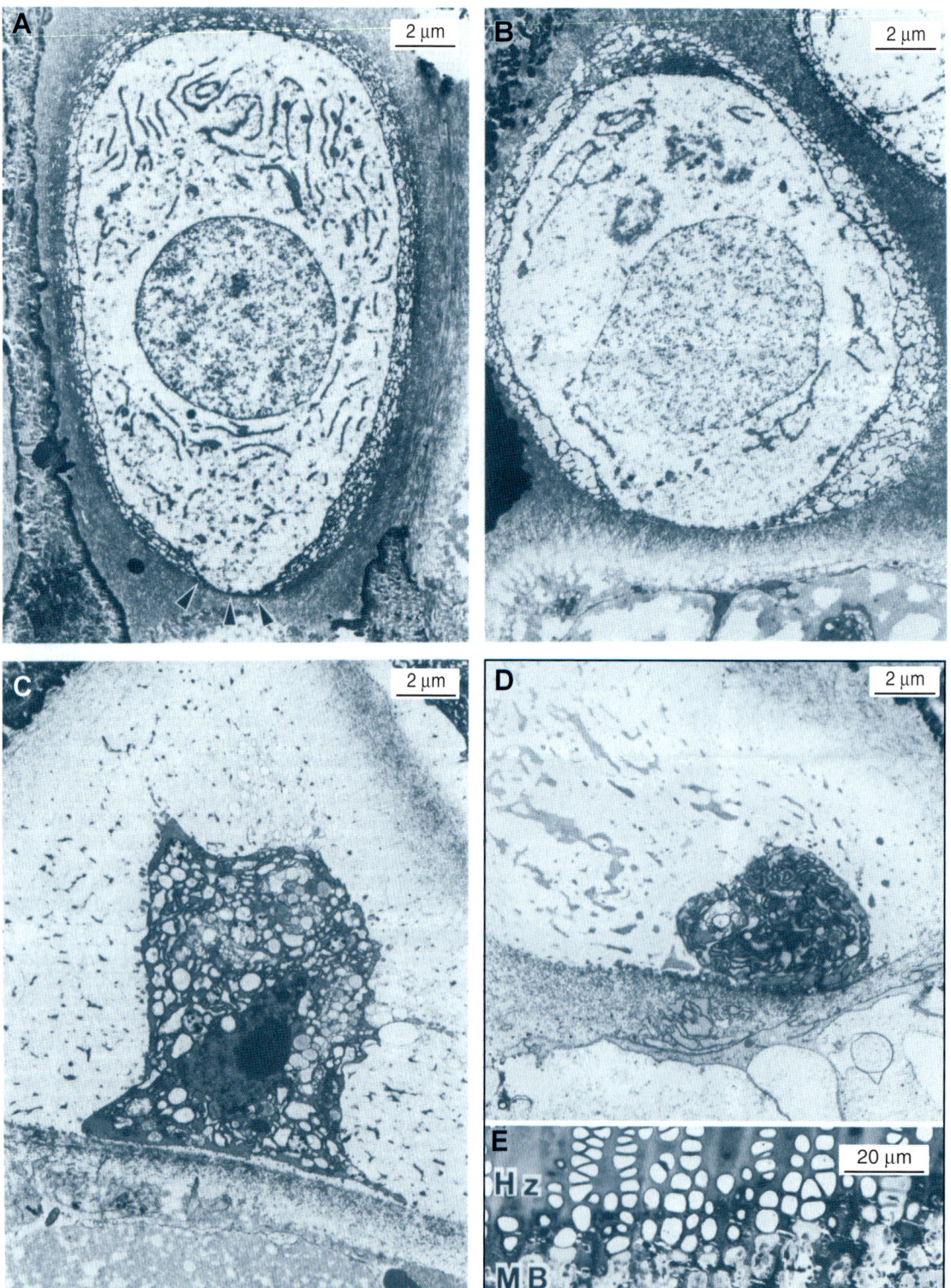

Figure 13.11 Some growth-plate chondrocytes die, as demonstrated in these transmission electron micrographs of terminal HCCs in the growth plate of four-month-old Yucatan swine. (A, B) Degenerating HCCs show variable degrees of attachment of cell membranes to the last transverse septum (arrowheads in A). (C, D) Note the condensed nuclei and detachment of the membrane from the last transverse septum. (E) The transition from the zone of chondrocyte hypertrophy (Hz) to bone in the growth plate (MB) below, and the many degenerating HCCs. Modified from Farnum and Wilsman (1987).

Farnum *et al.* (1990) to follow *live cells* in situ *in rat growth plates and confirm that the last chondrocytes in the columns are alive. As far as I am concerned this study is definitive.*

All the evidence – morphological, functional, biochemical, histochemical, molecular, labeling and cinematographic – supports the conclusion that the terminal chondrocyte survives dissolution of its ECM to transform to an OPC.

Murine interpubic joints

Murine interpubic joints show that the 'same' cells in different parts of the skeleton are not really the same. As Willmer suggested, there are lineages of osteoblasts. We know this because of a series of innovative studies begun by Ed. Crelin of Yale University 50 years ago. Here I discuss only the studies relating to dedifferentiation of chondrocytes. The specific response of cells of these joints to hormones, discussed in Chapter 23, demonstrates the non-equivalence of apparently similar skeletal cells as clearly as does the fate of HCCs in long bones and ribs.

A cartilaginous symphysis developed between the distal tips of right and left pubic bones after Crelin and Koch (1965) cultured mesenchymal joint primordia from 13-day-old mouse foetuses. Endochondral ossification was initiated within the symphyseal cartilage. It appeared to them that during ossification, *chondrocytes were released from their matrices to become chondroclasts or osteoblasts*, two very different cell fates.

Two years later, using autoradiographic evidence, Crelin and Koch confirmed the impression gained from histological analysis. Pubic joints at the mesenchymal stage were labeled for two hours with ^{3}H-thymidine and placed into organ culture. Immediately prior to endochondral ossification – which begins after five or six days *in vitro* – only HCCs were labeled. With the appearance of bone in the cultures, however, labeled osteoblasts and osteocytes appeared. As cartilaginous matrix was removed, the cytoplasm of the HCCs became basophilic. Some cells divided – something which, at the time, mammalian HCCs were not thought capable of doing. Initial osteogenesis occurred within cells with large nuclei and basophilic cytoplasm, situated in clumps within lacunae. The cytological appearance and isotopic evidence was of HCCs modulating to preosteoblasts and thence to osteoblasts.

NOTES

1. See J. H. Scott (1951) for development of the temporo-mandibular joint in sheep, and Keith (1982) and Bach-Petersen *et al.* (1994) for the association of osteogenesis with innervation and the eight stages. For development of the human TMJ, see Baume (1962a–c), Levy (1964), Sarnat (1964), Yuodelis (1966) and Shipman *et al.* (1986). For postnatal development, see D. M. Wright and Moffett (1974). See du Brul (1964), Frommer (1964), Levy (1964), Yuodeles (1966), Keith (1982), Luder and Schroeder (1992), Robinson *et al.* (1994) and Matsuda *et al.* (1997) for the development of the TMJ in other species. See Prummel (1987) for an atlas for identification of foetal skeletal elements in cattle, horses, sheep and pigs. Apoptosis plays a role in the development of the rat TMJ, facilitating adaptation of the joint to changing mechanical stresses in late embryonic development (17–20 days) and postnatally (Matsuda *et al.*, 1997). For a classic treatment of the evolution of the TMJ, see du Brul (1964).

2. See Toriumi *et al.* (1991), Hollinger and Chaudhari (1992) and Precious and Hall (1992) for discussions of the role of such materials as *rHBMP-2* (*BMP-3*) in the regeneration of the mandibular or craniofacial skeleton. English and Stohler (2001) edited a special issue of the journal *Cells, Tissues and Organs* devoted to TMJ research into the 21st century. See Beresford (1970, 1975a, 1981) for penile cartilages.

3. See Meikle (1975a) for membrane-bound vesicles, and Silbermann and Frommer (1973a,b) for the additional studies.

4. See Livne *et al.* (1990) for the mouse and Luder and Schroeder (1992) for the monkey transformations.

5. See Silbermann *et al.* (1983) for the organ culture, and Silbermann *et al.* (1978a,b) for chondroid bone. Clonal cell lines established from mandibular condyles of newborn mice do not differentiate into either cartilage or bone but do respond to retinoic acid or to vitamin D_3 with enhanced alkaline phosphatase (Bhalerao *et al.*, 1995).

6. See Shibata *et al.* (1991, 1993a,b, 1996, 1997a,b) for these studies.

7. See Kavumpurath and Hall (1990) for the chick study, and Muhlhauser (1986) for resorption of Meckel's cartilage in rats. See Thomson (1986, 1987, 1989) for the development of Meckel's cartilage in *Xenopus laevis*, including metamorphic changes, initial ossification, effects of thyroid hormone and correlation to feeding habits of the tadpoles.

8. See Goret-Nicaise and Pilet (1983) and Bareggi *et al.* (1995) for these studies.

9. See Richman and Diewert (1988), Chung *et al.* (1995) and Chung and Nishimura (1999) for the studies on which this summary is based.

10. See Richany *et al.* (1956), Burch (1966), Savostin-Asling and Asling (1973) and Asling (1976) for the transformation to ligament, and Ishizeki *et al.* (2001) for the role of Egf. Egf also influences morphogenesis of mouse Meckel's cartilage *in vitro* (Coffin-Collins and Hall, 1989; Slavkin *et al.*, 1992; Shum *et al.*, 1993; Ishizeki *et al.*, 2001). See Rodriguez-Valquez *et al.* (1997a) for the development of human Meckel's cartilage at the symphysis. The myelohyoid bridge, which is an inherited deficiency in Meckel's cartilage, involves conversion of cartilage to ligament (Ossenberg, 1974).

11. See Melcher (1972) and Savostin-Asling and Asling (1975, especially Figure 14 in the latter) for this chondrocytic origin of chondroclasts.

12. See Ishizeki and Nawa (2000) for MMP-1, Ishizeki *et al.* (1990, 1993, 1996a–c, 1997, 1998, 1999) for the transformation to osteogenic cells, and Harada and Ishizeki (1998) for transformation to fibroblasts.

13. See Rojas *et al.* (1996) for sheep and cat, and Thomson (1986, 1987, 1989) and Hall (2003b) for *Xenopus*. See Hall (1987c), Hall and Miyake (1992, 1995, 2000) and Miyake *et al.*

(1996a,b, 1997a,b) for condensation and first-arch skeletogenesis in inbred mouse strains, especially C57BL/6.

14. See Hall (1980c), Kollar and Mina (1991), Schneider *et al.* (1999) and MacDonald and Hall (2001) for epithelial–mesenchymal interactions in mouse Meckel's cartilage, and Barlow and Francis-West (1997) and Bogardi *et al.* (2000) for production of a bifurcated cartilage. Rodriguez-Valquez *et al.* (1997b) reported a 'duplicated' Meckel's cartilage on the right-hand side of a human foetus of 57-mm crown–rump length, although this may be a bifurcation of a single Meckel's cartilage rather a duplicated (i.e. a second) Meckel's cartilage; duplication of Meckel's cartilage has never been reported.

15. See Martin *et al.* (1995) for loss-of-function, Lu *et al.* (1999a) for absence of Meckel's cartilage, ten Berge *et al.* (1998, 2001) for the double knockout, and Brickell (1995) for an overview of *Prx-1* function.

16. See Hirsch *et al.* (1991), Brickell (1995), Martin *et al.* (1995), ten Berge *et al.* (1998, 2001) and Lu *et al.* (1999a) for *Prx-1*.

17. See ten Berge *et al.* (1998b) for *Alx-3*, and Hudson *et al.* (1998) for null mutants, and Grzeschik (2002) and Leroi (2003) for some of the molecular complexity underlying polydactyly in humans.

18. It is not known whether blood-borne cells are determined as OPCs while circulating in the blood stream, at the invasion front, or only after settling on the remnants of the cartilaginous trabeculae. The latter is probably the case.

19. For ultrastructural studies, see Scott and Pearse (1956), Trueta and Little (1960), Anderson and Parker (1966), Schenk *et al.* (1967, 1968), Engfeldt (1969) and Thyberg and Friberg (1971). The reviews by Ham (1974), Serafini-Fracassini and Smith (1974) and Ham and Cormack (1979) provide excellent accounts of these studies.

20. Crelin and Koch (1967) provide an excellent account of 'turn of the 20th century' views on whether chondrocytes survive the mineralization and removal of their ECMs.

21. See Holtrop (1966, 1967b, 1971, 1972a,b) for these studies on the fate of HCCs.

22. See Sutfin *et al.* (1971) for the electron-microprobe analysis.

23. In an earlier study, Goldhaber (1961) had shown that bone inside chambers constructed from Millipore filters produces a diffusible osteogenic signal that elicits osteogenesis from cells outside the chamber. In studies in which unmineralized, mineralized or demineralized cartilages were implanted intramuscularly into mice, Ksiazek (1983) and Ksiazek and Moskalewski (1983) showed that only mineralized cartilage could induce new bone.

24. See Ksiazek (1983) and Ksiazek and Moskalewski (1983) for cartilage grafts, and Moskalewski and Malejczyk (1989) for chondrocyte grafts.

25. See Cooper (1965) for the differences between mouse and chick tibiae, Dodds and Cameron (1939), Minkin and Lippiello (1969a), Fig. 3.13 and Box 7.1 for rachitic hyperchondrocytes, and Hall (1972b) and the section on chondroid bone in Chapter 5 for other examples of metaplasia of cartilage matrix.

26. See Kuhlman and McNamee (1970) for physiological activity of HCCs, and Balogh *et al.* (1961), Balogh (1964) and Balogh and Hajek (1965) for the common enzymatic profiles of bone and cartilage.

27. See Kalayjian and Cooper (1972) for the guinea-pig study, and Haines and Mohiuddin (1968) and Hall (1972b) for metaplasia in mammals and birds. Galectin-3, a β-galactoside-binding protein found in all chondroblast/cyte zones in murine long bones, regulates the survival/death of HCCs. Galectin-3-null mice display many dead terminal HCCs (Colnot *et al.*, 2001).

28. See von der Mark and von der Mark (1977a) and Yasui *et al.* (1984) for type I collagen in HCCs.

29. See by Knese and Knoop (1961a,b) and Hanaoka (1976) for these studies.

30. See G. C. Wright Jr. *et al.* (1985) for grafting into nude mice, which themselves show alterations in postnatal skeletal development; six weeks postnatally the bones are less dense on X-ray and have thinner proximal tibial growth plates than do wild type (Smetana and Holub, 1990).

Dedifferentiation and Urodele Amphibian Limb Regeneration

Chapter 14

I was envious of such an elegant study, despite the fact that frogs are not urodeles, Xenopus is not Ambystoma, and larvae are not adults.

Regeneration encompasses the three basic developmental processes in a wonderfully coordinated and orchestrated way:

(i) differentiation, with complete replacement of all lost tissues – nerves, muscles, skin, skeleton, connective and vascular tissues;
(ii) morphogenesis, with reorganization of the regenerated tissues and organs into a morphological normal and complete limb; and
(iii) growth of the regenerated limb back to very nearly the size of the original limb, an amazing series of biological processes.

Speed of regeneration is related to the size of the urodele, smaller newts regenerating faster than larger, whether larger and smaller species, larger and smaller individuals within separate populations, or larger and smaller individuals in a single population. For example, the difference in the time to regenerate a 2.5-mm-long limb ranges from 30 to 38 days in different-sized red spotted newts, *Notophthalmus viridescens* (Table 14.1). Factors other than size also come into play – rates of regeneration diverge more for hind limbs and in males than for forelimbs and in females (Table 14.1). Provision of blood vessels to pre-blastema and blastema stages in *N. viridescens* depends on reinnervation of the blastema (Rageh *et al.*, 2002) and so is slower in larger animals.

DEDIFFERENTIATION

Amputation of the limbs of adult urodele amphibians sets in motion a series of regenerative processes that replace the portion of the limb removed, *and only the portion removed*. Dedifferentiation of cells exposed at the amputation surface to form a mass or *blastema* from which regeneration ensues is a vital component. Without dedifferentiation, regeneration cannot commence.

The suggestion that tissues of the regenerate are derived from dedifferentiation of local cells at the wound surface goes back to the work of Butler (1933), who postulated dedifferentiation as an alternative to the prevailing theories of origin from red blood cells, epidermal cells, or a population of stem cells mobilized to the site of regeneration. Planarians and *Hydra* do regenerate from stem cells. Normally scattered throughout the body, they are mobilized to the wound surface to form a blastema, which is the aggregation of cells from which the regenerate arises. Regeneration is profound in such animals; entire lost portions of an animal can regenerate. Indeed, in planarians and *Hydra*, regeneration is asexual reproduction, and asexual reproduction is regeneration.

Table 14.1 Mean limb regeneration times for red spotted newts, *Notophthalmus viridescens*, of two body sizes[a]

Body size (g)	Type of limb	Sex	Number of limbs amputated	Days to attain 2.5 mm length ±SE[b]
1.0–2.3	Fore	Female	15	30.3 ± 1.0
		Male	18	30.3 ± 0.7
	Hind	Female	15	29.7 ± 0.7
		Male	19	30.4 ± 1.0
3.4–4.4	Fore	Female	17	30.0 ± 1.1
		Male	18	32.2 ± 1.3
	Hind	Female	14	32.2 ± 0.7
		Male	16	37.7 ± 1.4

[a] Based on data from Pritchett and Dent (1972).
[b] SE, standard error.

Box 14.1 Are blastemal cells embryonic, totipotent or pluripotent stem cells?

> While dedifferentiated blastemal cells might appear 'embryonic,' they do not appear to be able to function as embryonic, pluripotent cells. This finding springs from a study in which nuclei from blastemal cells on amputated hind limbs of the South African clawed frog *Xenopus laevis* were transplanted into enucleated *Xenopus* eggs (Burgess, 1974).
>
> It is known that embryonic development can be initiated and a new organism formed when a single totipotent nucleus is transplanted into an enucleated frog egg, but Burgess found that blastemal cell nuclei – she was obliged to inject many whole cells rather than a single nucleus, a criticism of the technique – would not support embryonic development beyond the cleavage stage. Cleavage in eggs of the South African clawed frog, *X. laevis*, is under the control of cytoplasmic factors deposited into the egg during oogenesis; enucleated eggs will cleave but develop no further, which means that the zygotic nucleus is not necessary for this early phase of embryonic development. Burgess concluded that blastemal cells are not totipotent and should not be equated with blastomeres of cleavage-stage embryos. Dedifferentiation is not the same as becoming embryonic.[a]
>
> [a] See Gurdon (1970), King (1966) and Gilbert (2003) for nuclear transplantation; Maclean and Hall (1987), Moody (1998) and Gilbert (2003) for dedifferentiation; and Tinsley and Kobel (1996) for the biology of *Xenopus*.

Regeneration of urodele limbs, although dramatic, is a much more local phenomenon.

Morphological dedifferentiation

Dedifferentiated blastemal cells lack the specialized morphological features associated with functional myo-, chondro- or osteoblasts. Dedifferentiation erases these features to the extent that dedifferentiated cells appear the same morphologically. You cannot tell which were previously myoblasts, chondroblasts, fibroblasts, etc. All take on the same 'embryonic' appearance as they dedifferentiate to form a single common blastema (Box 14.1). Blastemal cells have the morphological features described for skeletogenic progenitor cells in Chapter 12 – they are rounded, have a high nucleo-cytoplasmic ratio, prominent nucleoli, and many ribosomes. As discussed below, blastemal cells also show functional and molecular dedifferentiation.

Functional dedifferentiation

In addition to dedifferentiating morphologically, blastemal cells dedifferentiate functionally; they cease synthesizing such specialized products as myosin that were characteristic of their prior differentiated state (see Box 5.1). In the Mexican axolotl, *Ambystoma mexicanum*, and in the eastern (Long Island) newt, *Notophthalmus* (*Triturus*) *viridescens*[1] – two species on which much research has been undertaken – blastemal cells that were myoblasts no longer synthesize myosin, blastemal cells that were chondrocytes no longer synthesize sulphated glycosaminoglycans (GAGs), type II collagen is neither synthesized by dedifferentiating chondrocytes nor by dedifferentiated blastemal cells, and so forth.[2]

Hyaluronan

[35]S-labeling of epiphyseal chondrocytes to track synthesis of sulphated GAGs allows chondrocytes to be followed into the blastema. The level of incorporation declines as dedifferentiation proceeds until it is no different from incorporation in neighbouring fibroblasts. Interestingly, although sulphated GAGs are not synthesized by dedifferentiated chondrocytes, synthesis of hyaluronan *is* associated with formation of the blastema, providing additional support for the role of hyaluronan in cell proliferation/migration preceding differentiation. Hyaluronan performs a similar function – promoting cell accumulation – in two quite different situations in adult organisms: blastema formation in urodele limb regeneration, and around an implanted matrix in ectopic osteogenesis in mammals.[3]

Toole and Gross (1971) studied blastemal formation and subsequent regeneration following limb amputation of adult red efts, *Diemictylus viridescens*. The blastema arises by dedifferentiation of and reinitiation of division by cells at the wound surface. Hyaluronan accumulates during dedifferentiation of stump tissues and blastema formation. As regeneration ensues, hyaluronan is removed by hyaluronidase. Decline in hyaluronan parallels a rise in chondroitin sulphate as cartilage differentiates in the regenerating limb.

Formation of a blastema depends on innervation of the wound epithelium that develops over the stump. In the eastern newt the cartilaginous phase of regeneration is prolonged way beyond that of any other urodele. The regenerating skeleton is still cartilaginous 270 days after amputation and, essentially, is still regenerating. This prolongation of the cartilaginous phase recapitulates normal development. Replacement of cartilage by bone, as for example, in carpal ossification, is generally slow in urodeles. Interestingly, jaw regeneration in the eastern newt is independent of innervation.[4]

Hyaluronan levels are low and no hyaluronidase appears if the limb is denervated *before* amputation. Absence of hyaluronidase synthesis correlates with lack of differentiating chondroblasts, but *not* because of any dependence of hyaluronidase synthesis on the presence of nerves. This was shown by denervating the limb *after* blastema formation, for cartilage differentiates in these animals in association with synthesis of hyaluronidase. A similar dependence of regeneration on the presence of hyaluronan and its removal by hyaluronidase occurs in sabellid annelids, implying that this dependence is an ancient one, either because of a single evolutionary origin or through convergence.[5]

The same pattern of hyaluronan synthesis and subsequent removal is demonstrated in a completely different skeletal situation, ectopic skeletogenesis in adult rats (Iwata and Urist, 1973). Allogeneic demineralized cortical bone matrix implanted intramuscularly into rats is invaded by *circulating* progenitor cells that modulate to *osteoclasts* that resorb the matrix. Implants are invaded by *local* progenitor cells that accumulate within pores in the matrix. These host cells synthesize and accumulate

alkaline phosphatase (six days after implantation), form cartilage (seven days), woven bone (10 days) and marrow (12 days), times that depend upon the age of the donor providing the matrix and the host into which the matrix is implanted (Boxes 14.2 and 14.3). Hyaluronan is synthesized at peak levels two days after implantation as host cells invade the matrix. As cytodifferentiation ensues, hyaluronan declines, either because synthesis stops or removal by hyaluronidase sets in.

Box 14.2 Demineralized bone matrix and cartilage induction

In the 1960s, the late Marshall Urist began what would become a pioneering series of studies in which he demonstrated that demineralized bone matrix implanted into the muscles of small mammals induced *local* host cells to differentiate into chondroblasts. The chondroblasts matured and deposited a cartilage matrix, which mineralized and was then replaced by bone complete with marrow.

As an example of these studies, Iwata and Urist (1973) implanted allogeneic demineralized cortical bone matrix intramuscularly into rats. The matrix was first invaded by circulating progenitor cells, which modulated to osteoclasts and began to resorb the matrix. The implant was then invaded by surrounding progenitor cells, which accumulated within pores in the matrix; collagenous bone matrix (but not tail tendon collagen) was shown to be mitogenic (Rath and Reddi, 1979). These host cells synthesized and accumulated alkaline phosphatase by six days after implantation, formed cartilage by seven days and woven bone by 10 days. Marrow appeared by 12 days.

Because the end result is the deposition of an ossicle of bone, this process is usually referred to as bone induction. It is important to remember, however, that almost invariably the tissue that is evoked is cartilage. Replacement of that cartilage by bone is a subsequent step that involves a different set of cells (presumed to be osteoprogenitor cells) that are transported to the matrix via the vascular system. Cartilage differentiates if muscle is cultured with demineralized bone matrix, but the cartilage in neither resorbed nor replaced by bone. Cartilage arises from calvarial bone *in vitro* under the influence of bone matrix gelatin, but only with the use of specific culture media.[a]

Not all connective tissues have the same potential to initiate ectopic bone formation in response to bone morphogenetic protein. Muscle and fascia do, but spleen, liver and kidney actually suppress bone formation unless living fascia is included with the demineralized bone graft (Chalmers *et al.*, 1975). The age of the donor providing the matrix and the age of the host receiving the implant both influence the amount of cartilage or bone that forms. Somatostatin (0.25 μg/day) given to rats implanted with demineralized bone matrix inhibits proliferation and differentiation of bone and cartilage progenitor cells within the implanted matrix (Weiss *et al.*, 1981).

To a degree, the type of skeletogenic tissue produced can be modulated; articular chondrocytes placed on gel foam or on demineralized bone matrix after some time in culture produce chondromyxoid or chondroid.[b]

Induced ectopic bone shows interesting differences to bone within the skeleton, some of which are discussed in other parts of the text. Implanted matrix provides a less effective substrate for skeletogenic progenitor cells; the concentration of fibroblast-colony-forming units (CFU-F) in the marrow of induced bone is less than that found in endogenous bone (18 to 25 per cent of endogenous bone in mice, 26 per cent in rats), although the concentration of haematopoietic stem cells is similar (Reddi and Wlodarski, 1986).

Demineralized allogeneic bone matrix from vitamin-D-deficient rats implanted into control rats is less effective as a bone inducer than matrix from control animals – fewer osteoblasts are induced, perhaps because the vitamin-D-deficient matrix lacks a mitogen present in normal matrix. Similarly, demineralized bone matrix implanted into vitamin-D-deficient rats induces normal amounts of cartilage, but less bone is produced and the bone is abnormally mineralized unless vitamin D is added.[c]

The three essential elements for bone grafts are a source of osteoprogenitor cells, an osteoinductive matrix and osteoinductive growth factors (Lane *et al.*, 1999). The carrier is more than an inert vehicle. Bone can be induced directly by manipulating the matrix in which the active agent is implanted (see below).

The clinical relevance of demineralized bone matrix was evident from the outset. Powdered demineralized bone transplanted into mandibular defects in rats induces osteogenesis and promotes repair with little resorption over the ensuing six months (Fig. 14.1). Bmp was shown to be an effective means of repairing quite extensive (14–20 mm) craniotomy defects in rhesus monkeys, *Macaca speciosa*, cartilage and bone arising in cells from both dura and cranial periosteum, and of initiating repair of trephine defects in sheep skulls.[d]

Bone is normally irradiated when stored in bone banks, the radiation dose being of the order of 2.5 mrads. Irradiation affects the osteoinductive potential of demineralized bone matrix, irradiation with 3–5 mrads enhancing osteoinductive ability (Wientroub and Reddi, 1988).

Bmp

A protein, *bone morphogenetic protein* by Urist, was isolated and shown to be a 163-amino-acid 18-kDa protein (see Bessho *et al.*, 1989). These studies paved the way for the discovery that Bmp was the active molecule involved in the induction; Bmp was purified, cloned and its mRNA localized in bone in the late 1980s. Bmp has since been shown to be one of a large factor of Bmps with a plethora of functions in embryonic development (Boxes 12.1, 28.1 and 28.2).[e]

As examples: Bmp-3 (Bmp-2B) at 10 ng/ml enhances chondrogenesis from chick limb-bud mesenchymal cells, an enhancement that is inhibited by Tgfβ-1 and -2. Bmp-2 (Bmp-2A) reverses Pdgf-enhanced chondrogenesis in the same cells (Chen *et al.*, 1991, 1992).

Bmp-2 mRNA is distributed in murine embryonic limbs, heart, whiskers, hair, teeth and craniofacial mesenchyme, in patterns that differ from those of Tgfβ1–3 (Lyons *et al.*, 1990).

Bmp-3 stimulates the expression of Tgfβ-1 mRNA by monocytes (Cunningham *et al.*, 1992).

When Aspenberg and Lohmander (1989) filled the marrow cavity of demineralized bone with 75 ng Fgf-2 and implanted it into rats, they found a 25 per cent increase in the amount of mineralized tissue that formed. They argued for interaction between Bmp and Fgf-2 increasing the number of chondroblasts and promoting capillary invasion of the matrix.

Box 14.2 (*Continued*)

Carriers and bone or cartilage

Kuboki *et al.* (1995) found that manipulating the carrier determined whether chondrogenesis or intramembranous ossification

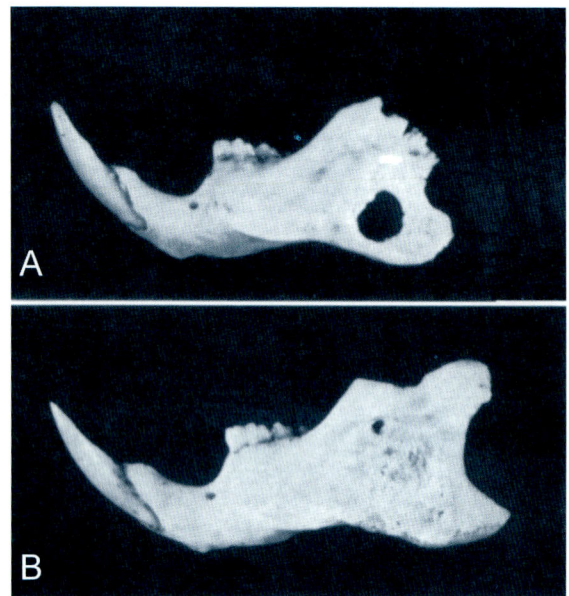

Figure 14.1 Healing defects in rat dentary bones. (A) The 'mandible' (dentary bone + teeth) of a control, two-week-old rat after a 4-mm-diameter circular defect was made in the posterior ramus. Note minimal indication of repair. (B) A similar defect in another two-week-old rat filled with freeze-dried bone powder (derived from long bones) has been filled with new bone. Modified from Kaban and Glowacki (1981).

occurred in subcutaneous sites. Bmp in a fibrous glass membrane carrier evoked bone differentiation, Bmp in a hydroxyapatite carrier evoked cartilage. The differences were attributed to the geometry of the carrier matrices. Different carrier composites for Bmp also were shown by Missana *et al.* (1994) to evoke either endochondral or membrane bone when implanted subcutaneously into rats.[f]

Using a bovine fibrous collagen membrane as a substrate, Sasano *et al.* (1993b) claimed that Bmps induce membranous bone directly and that others missed the intramembranous ossification because cartilage forms at the same time. Careful study of the time course of skeletogenesis is important so as not to miss a transitory cartilaginous phase. A ceramic containing marrow and implanted subcutaneously is said to induce bone without any cartilage precursor (Yoshikawa *et al.*, 1992), but it is not clear that they examined the samples early enough to see a cartilaginous phase.

[a] See Urist (1965), Urist and Dowell (1968), and Urist *et al.* (1969, 1983, 1984, 1985) for studies over the first 20 years, Terashima and Urist (1975) for cartilage from calvariae, Mulliken *et al.* (1984) and Kawamura and Urist (1988) for excellent reviews of these early studies, and Urist (1991) for what he then saw as the emerging concepts. Katthagen (1987) reviews studies enhancing bone regeneration with natural and artificial bone substitutes, including comment that demineralized bone matrix was first used in 1889.
[b] See Green (1971), Green and Dickens (1972) and Green and Ferguson (1975).
[c] See Turner *et al.* (1988) and Van der Steenhoven *et al.* (1988) for vitamin-D deficiency.
[d] See Kaban and Glowacki (1981, 1984), D. Ferguson *et al.* (1987) and Lindholm *et al.* (1988) for this sample of clinical studies.
[e] See Bessho *et al.* (1989) for bone morphogenetic protein, Wozney *et al.* (1988) and Rosen *et al.* (1989) for the isolation and characterization of the first Bmp, and Rosen and Thies (1992), Reddi (1992, 1994), Carrington (1994) and Hogan (1996a,b) for the actions of Bmps in bone.
[f] Cultured chick tibial osteogenic cells transferred to athymic mice produce both membrane and endochondral bone, and both chondrogenic and osteogenic cell lineages can be isolated (Nakahara *et al.*, 1990a,b, 1991a, 1992; Nakase *et al.*, 1993). For recent models of the molecular basis of the switch between chondro- and osteogenesis, see Fang and Hall (1997, 1999) and Buxton *et al.* (2003).

BLASTEMA FORMATION

In a pioneering study, Hay and Fischman (1961) used ³H-thymidine autoradiography to trace the origin of the blastema in *Notophthalmus viridescens*. Adult newts during the first 28 days of limb regeneration were injected intraperitoneally with ³H-thymidine and fixed on the same day, which revealed that DNA synthesis was initiated four or five days after amputation in the cells – myoblasts, periosteal cells, connective tissue and nerve sheaths – located within 1 mm of the amputation surface. In a second series, isotope was administered at five, 10 or 15 days after amputation and limbs fixed daily. The dedifferentiated cells incorporated label but the epithelium did not, which allowed Hay and Fischman to conclude that the dedifferentiated cells were the source of the blastema. Labeling the limb before amputation resulted in incorporation only into the epidermis.

Both cytologically and functionally, cells of regeneration blastemata become less specialized as they divide and accumulate. Consequently (and for technical reasons), several problems plagued analysis of the subsequent

fate of blastemal cells. Since dedifferentiation involves the loss of specialized morphological and functional characteristics, it is difficult to follow dedifferentiating cells for any length of time. And because the dedifferentiated products of the several mesodermal tissues of the stump – muscle, cartilage, bone and connective tissue – are morphologically equivalent, it is not possible to examine a blastema and decide which dedifferentiated cells were previously chondroblasts, which osteoblasts, which myoblasts, and so on. Consequently, it is not possible to determine whether all of the blastemal cells that form, say, chondroblasts in the regenerate, were chondroblasts before they dedifferentiated. Some may have been myoblasts or fibroblasts. Such uncertainty goes to the heart of the stability of dedifferentiated cells and whether this stability is equally shared by different cell types.

Blastemal cells are not the regenerative equivalent of pluripotent embryonic cells (Box 14.1). Nor do blastemal cells revert to the developmental program displayed by limb bud cells; monoclonal antibodies generated against blastemal cells from regenerating limbs of the Spanish

Box 14.3 The effect of age on matrix-induced bone formation

A number of studies have addressed the question – or what are really two sides of a single question – of whether the osteoinductive ability of bone matrix changes with the age of the donor providing the matrix, and/or whether the ability of the host to respond changes with age. The answer to both sides is yes. *Ability to induce and ability to respond both decline with age.*

Several laboratories in the early 1980s demonstrated that the ability of implanted bone matrix to induce bone declines with host age.[a] When Irving and colleagues (1981) implanted demineralized bone power from six-month-old rats into six-week-, six-month- or two-year-old rats, the older the host, the longer the time it took to elicit osteogenesis; membrane bone appeared after 14 days in six-week-old rats but not until 23 days in two-year-old rats. The age of the donor providing the bone powder also influences ectopic osteogenesis; matrix from six-week- or six-month-old rats implanted into two-year-old rats induces bone, while matrix from two-year-old rats transplanted into six-week-old rats does not (pers. comm. Irving, 1981). The ability to induce bone in rats is lost between six months and two years.

When Chang and colleagues took bone matrix from one-, three-, 10- and 16-month-old rats, they found that the induced bone mineralizes more quickly if the implants come from younger animals – differences could be detected between matrices from one- and three-month-old animals.

According to Nimni *et al.* (1988), who discuss some of the literature on the effects of age, calvarial cells from mouse embryos embedded in matrix cylinders evoke more ectopic bone when transplanted into senescent (26-month-old) mice than into younger ones.

Demineralized bone powder from middle-aged humans is a better osteoinducer than powder from individuals who are pre- or immediately post-puberty. Human rib periosteal cells produce cartilage and bone when transplanted *in vivo* if derived from younger but not from older individuals.[b]

What is true for bone matrix also is true for chondro-inductive extracellular matrices (ECMs). Matrigel, a basement-membrane-derived ECM, induces chondrogenesis when implanted. Matrigel obtained from older animals is less effective than Matrigel from younger, perhaps because of inhibitors in the older material. In addition to containing such ECM products as type IV collagen and fibronectin, Matrigel also contains multiple growth factors, including Egf, Igf-1, Fgf-2, Pdgf and Tgfβ. A neutralizing antibody against Tgfβ-1 blocks the action of Matrigel on cellular activity. I am not aware of this being investigated in Matrigel derived from animals of different ages.[c]

[a] See Irving *et al.* (1981), Syftestad and Urist (1982), Chang *et al.* (1985), Reddi (1985) and Urist (1991).
[b] See Jergesen *et al.* (1991) and Nakahara *et al.* (1991b).
[c] See Bradham *et al.* (1995) for Matrigel-induced cartilage, Vukicevic *et al.* (1992) for the composition of Matrigel, and Ayad *et al.* (1994) for ECM products.

ribbed newt, *Pleurodeles waltl*, do not react with limb bud mesenchyme (Fekete and Brockes, 1988).

Aneurogenic limbs

We do have to be careful with such evidence, however. The antibodies generated do not react with blastemal cells until axons from regenerating nerves reach the wound surface. Nor do they react with blastemata from what are called 'aneurogenic limbs,' limbs that develop in animals from which the nerves that would innervate the limbs are removed during early development. Such limb buds and limbs never 'see' nerves during their development.[6] Three sets of data are important:

(i) formation of a blastema is dependent upon nerves innervating the stump;
(ii) embryonic limbs do not require innervation to develop; and
(iii) aneurogenic limbs that have never been exposed to nerves can regenerate in the absence of innervation.

So you see why the antibody evidence has to be taken with **caution**. An antibody that is not expressed in the blastema until after innervation would not (might not?) be expected to cross-react with limb-bud cells before they too were innervated.[7]

More than one cell fate

Some blastemal cells can exercise more than one potential fate, which means that they can form a different cell type from that expressed before dedifferentiation. This generalization is based on several lines of evidence, some of which come from classic studies on urodele limb regeneration, differentiation, modulation and metaplasia, as outlined below.

- Beginning in the 1920s, researchers removed the skeleton – usually the humerus – from urodele forelimbs, allowed the wound to heal, and amputated the limb through the upper arm *where the humerus would have been*. Regeneration ensued and, in many cases, the regenerates developed a complete limb skeleton, *including the humerus*. As these skeletal elements could not have arisen from skeletal cells at the stump, they must have arisen from blastemal cells whose source was muscle or connective tissue; blastemal cells are of local origin, except for osteoclasts, whose precursors arrive via the vascular system to erode the skeleton at the stump and release cells to dedifferentiate.[8]
- Irradiating a urodele limb with X-rays prevents regeneration. Grafting a muscle into the limb after irradiation allows regeneration to ensue. The resulting limb contains muscle, but also connective and skeletal tissues. Implanting cartilage after irradiating a limb, however, results in a regenerate with cartilage but no muscle.[9]

These studies suggest that dedifferentiated chondrocytes are stable – once a cartilage cell always a cartilage cell – and that dedifferentiated muscle cells are not.

This dichotomy between the fates of dedifferentiated myoblasts and dedifferentiated chondroblasts is worth exploring a little further. As it turns out, grafting cartilage and grafting muscle into a limb are two very different things, so different that the dichotomy is both more apparent and more real than it seems.

MYOBLAST AND CHONDROBLAST FATES

Oberpriller (1967) approached this problem in what may seem a strange way. She amputated the intestines and tails of red efts, *Diemictylus viridescens*, to generate blastemata. Needless to say, cartilage is not found in the intestines. Nor does cartilage form in the tails of these newts; the notochord extends into the tail, but caudal somatic mesoderm does not chondrify. When ³H-thymidine-labeled blastemal cells were taken from these intestinal and tail blastemata and grafted into unlabeled limb blastemata, labeled chondrocytes were found in cartilages of the regenerated limbs. After allowing for re-utilization of label, Oberpriller concluded that dedifferentiated cells from intestinal and tail blastemata have the potential to form chondrocytes. Such a result supports the concept that dedifferentiated cells are bipotential.

For his Ph.D. research, Trygve Steen undertook a series of experiments that I have admired ever since his first paper appeared in 1968, perhaps because I was envious of such an elegant study, having completed my own thesis in the same year, and wanting to pursue postdoctoral studies on regeneration.

Triploidy provides a marker to follow cells after they are grafted into a diploid host. Steene (1968, 1970) implanted cartilage or muscle from triploid individuals injected with ³H-thymidine so that they were double-labeled (triploid+isotope), into the limbs of diploid, unlabeled individuals whose limbs were then amputated.

Only chondrocytes were labeled in the regenerate after cartilage had been grafted into the limbs; the chondrocytic cell state is stable through dedifferentiation, blastemal formation and redifferentiation. The source of the cartilage is not critical; appendicular, coracoid or Meckel's cartilages all provide cells for the blastema and for the regenerate. Namenworth (1974) later showed that dedifferentiated chondrocytes from *Ambystoma mexicanum* produce perichondrial cells, joint connective tissue, fibroblasts of connective tissue and muscle, *but not myoblasts*.[10]

Chondrocytes, myoblasts and fibroblasts of regenerates with the triploid nucleolar and ³H-thymidine markers were found in the regenerates after Steen grafted labeled muscle into the limbs. The conclusion: muscle cells are not stable and can alter their fate during or after dedifferentiation. But there is a major caveat, of which Steen was aware. While grafted cartilages are comprised of a homogeneous population of cells, muscles are not. As a tissue, muscle includes myoblasts, fibroblasts, neural cells, satellite cells,[11] and so forth.

In dealing with this problem, Steen (1973) provided perhaps the most convincing evidence for the ability of the progeny of dedifferentiated *myoblasts* to redifferentiate as chondrocytes, fibroblasts or myoblasts. This is where I admire Steen's originality and tenacity. He cultured single myoblasts or fibroblasts as clones, and allowed the clones to multiply *in vitro* until a sufficiently large population of daughter cells was available to transplant to the amputated hind limbs of larval South African clawed frogs, *Xenopus laevis*. The cultured and grafted cells were both triploid and ³H-thymidine labeled, a 'double jeopardy' biasing the outcome against any false positive results. Regenerates contained labeled fibroblasts (in connective tissue and muscle) as well as labeled myoblasts and chondrocytes. The conclusion that chondrocytes can arise from dedifferentiated myoblasts in hind limb regeneration of larval *X. laevis* seems inescapable.

However – and there always seems to be a 'however' in regeneration experiments – to obtain his clones, Steen had to switch to *Xenopus*, transplant the clones into *Xenopus* not *Ambystoma* hind limbs, and into larvae not adults. Frogs are not urodeles, *Xenopus* is not *Ambystoma*, and larvae are not adults. Frogs have much more limited regenerative potential than urodeles; even implanting additional nerves produces little additional outgrowth during regeneration of adult bullfrogs, *Rana catesbiana*. *Xenopus* displays limited dedifferentiation, minimal numbers of fibroblastic cells accumulate following amputation of the forelimbs, and regeneration is limited to replacing connective tissue and cartilage.[12]

It is perhaps not surprising, therefore, to find subsequent studies coming up with contradictory findings. Studies on *Ambystoma mexicanum* from Hugh Wallace's group contradict those of Trygve Steen and Mark Namenwirth. To destroy host cells, *Ambystoma* were exposed to 2000 rads of X-irradiation before grafting vital cartilage into the limbs, which were then amputated. Reasonable instances of regeneration ensued in which connective tissue and cartilage developed. They argued that the implanted cartilages consisted of uniform populations of chondrogenic cells so that the fibroblasts and myoblasts must have arisen from dedifferentiated chondrocytes. Even more convincing evidence of regenerates containing substantial numbers of donor cells is the destruction of regenerated limbs in allograft reactions some 16 weeks after grafting.[13]

Nathanson and his colleagues undertook a series of studies demonstrating that skeletal muscle from 19-day-old foetal rats and cloned myoblasts from 11-day-old chick embryos formed cartilage but not bone via metaplasia. Typical cartilage proteoglycans are produced and synthesis of actin and myosin continues as muscles cultured on bone matrix transform to cartilage. Ultrastructural analysis showed transformation of myoblasts to fibroblasts and then to chondroblasts, eliminating satellite cells as the progenitors of the chondroblasts.[14]

Cells from muscle, fibroblasts and the connective tissue capsules of thyroid and lung all produced cartilage when cultured. Rat myogenic cells also form cartilage in response to bone morphogenetic protein (Bmp). A line (L6) of myoblastic cells has also been shown to form cartilage when cultured on demineralized bone matrix (Box 14.2).[15]

FACTORS CONTROLLING DEDIFFERENTIATION

These labeling studies show that the proportion of cells that redifferentiate into different cell types is small, probably less than 10 per cent. Although redifferentiation into a different cell type does not make a major contribution to the blastema, it is of interest to understand how redifferentiation is controlled, and whether it could be taken advantage of. An important factor is *innervation* of the amputated stump.

Innervation

The wound epithelium that grows over the amputated surface directs cell dedifferentiation, but can do so only after it has been reinnervated. Nerves must be present until a blastema forms but can be removed without affecting regeneration once the blastema has formed. The implication is that nerves or a neuronal or neurotrophic factor(s) play an important role in 'permitting' dedifferentiation. Isolating and implanting ganglia provides evidence for the necessity of ganglionic neurons if a sufficient mass of cells is to accumulate to initiate chondrogenesis, implicating neural–mesenchymal signaling in blastema formation.[16]

Limb regeneration can be initiated if denervated forelimb stumps of larval *Ambystoma* are reinnervated. Before reinnervation, blastemal cells neither synthesize DNA nor divide. Two days after reinnervation blastemal cells have been rescued from G1 arrest, synthesize DNA and have begun to divide (Table 14.2). Neurons in the stump are reported to contain Fgf-1. The extent of the wound epithelium can be increased by implanting a bead soaked in Fgf-1 into innervated limbs, but not if implanted into a denervated limb. So Fgf-1 plays a role, but is not the elusive trophic nerve factor.[17]

Axolotl Fgf-8, which is 84–86 per cent similar to Fgf-8 from *Xenopus*, chick and mouse Fgf-8 (Box 14.4), is expressed in the epithelial cap, subjacent mesenchyme and deeper proximal mesenchyme of blastemata. The blastemal expression patterns differ from those described in chick and mouse. Nevertheless, Han et al. (2001) regard the apical cap of the blastema as equivalent to an apical ectodermal ridge (AER).

Aneurogenic limbs

As is well known in the amphibian regeneration world, urodeles whose limbs develop in the absence of nerves – because the region of the spinal cord that supplies the brachial nerves was removed before limbs buds arise – nevertheless and paradoxically regenerate amputated limbs. Such animals are known as *aneurogenic*. In a now classic study, Steen and Thornton (1963) exchanged the skin on limbs of aneurogenic *Ambystoma* with skin from individuals whose limbs were innervated and demonstrated epithelial–blastema interactions in early regeneration.[18]

Mullen et al. (1996) isolated a cDNA for a *distal-less* (*Dlx-3*) gene from regeneration blastemata of the axolotl, *A. mexicanum*, and demonstrated that *Dlx-3*:

- is expressed in the epidermis of developing limb buds in a proximo-distal gradient,
- is expressed in the apical ectoderm cap at early stages of limb regeneration,
- has maximal expression at the early nerve-dependent phase of limb regeneration,
- is rapidly down-regulated in denervated limbs, and
- is maintained in denervated limbs implanted with beads soaked in Fgf-2,

concluding that *Dlx-3* is a potential candidate for the neurotrophic factor required during limb regeneration.

Fgf-4, -8 and -10 are all found at similar levels in innervated *and* aneurogenic regenerating limbs in *Ambystoma*, suggesting that Fgf is unlikely to be the neurotrophic factor. Further, denervation prevents accumulation of Fgf-8 and -10, indicating that both genes are downstream of the nerve-dependence factor (Christensen et al., 2001; see Box 14.4).

Proliferation

Proliferation accompanies or perhaps directs dedifferentiation. This aspect of dedifferentiation is also under

Table 14.2 Mean mitotic and labeling indices of control and reinnervated regenerating limb stumps of *Ambystoma* larvae[a]

Day postamputation	Mean mitotic index	Mean labeling index
	Control regenerating limb stumps	
2	0.0 ± 0.0	4.5 ± 2.9
3	0.0 ± 0.0	19.8 ± 4.0
4	1.3 ± 0.8	35.1
5	1.6 ± 0.8	43.6 ± 7.1
6	1.8	50.6
7	0.8	41.8
8	2.4	48.4
9	2.8 ± 0.3	44.3
	Reinnervated limb stumps	
10	0.1 ± 0.0	7.0 ± 2.2
10.5	0.1 ± 0.1	7.3
11	0.2 ± 0.1	6.1 ± 0.8
11.5	0.0 ± 0.0	10.1 ± 2.6
12	0.4	17.9
12.5	0.4	15.2
13	0.3 ± 0.2	19.9 ± 1.3
13.5	1.2 ± 0.9	21.9 ± 9.1
14	1.4 ± 0.5	31.8 ± 5.5

[a] Based on data from Olsen and Tassava (1984).

Box 14.4 Fgf-4, Fgf-8 and Fgf-10

Fgf-4

In situ expression during murine embryogenesis shows that *Fgf-4* is found in the myotome, the branchial arches, the apical epithelial ridge (AER) of limb buds, and in tooth primordia. Fgf-4 has the opposite effect on mouse limb development to Bmp-2, which is found in the AER of mouse limb buds. Fgf-4 increases division in mesenchyme and therefore mimics the function of the AER; Bmp-2 decreases limb growth, *Fgf-4* regulating *Evx-1* expression via modulation of *Bmp-2*, *Fgf-4* being regulated by *Shh* from the AER. Furthermore, Fgf-4 can replace the function of the AER, *Fgf-4* and retinoic acid regulating *Shh* in limb-bud mesenchyme through positive feedbacks.[a]

Fgf-4 mutants are lethal. A conditional non-lethal mutant was constructed to study the effects of *Fgf-4* later in development. Limbs in these mutants develop normally, with normal patterns of expression of *Shh*, *Bmp-2*, *Fgf-8* and *Fgf-10*. This was surprising, as *Fgf-4* is known to regulate *Shh*. Moon *et al.* (2000), who constructed the conditional mutant, concluded that the *Fgf-4-Shh* feedback loop may not be as well substantiated as previously thought, although the possibility that Fgf-4 is being compensated for or bypassed in these mutants should be considered; we know that redundancy exists in many gene families. Such redundancy of *Fgf* gene function occurs during induction of the otic placode, in which mice with mutant alleles of *Fgf-3* and *Fgf-10* demonstrate a range of intermediate phenotypes, in inductive interactions to which *Fgf-10* also contributes.[b]

Msx-1 is expressed in the progress zone of mouse limb buds, expression depending on signals from the AER. Fgf-4 can substitute for a signal that is normally produced by the AER of mouse limb buds and that up-regulates *Msx-1* expression in the undifferentiated cells at the distal tip of the developing bud. *Fgf-2* and *Bmp-4* then maintain *Msx-1* expression initiated by *Fgf-4*. Formation of an ectopic digit was elicited from the distal mesenchyme of chick wings buds after an Fgf-4-soaked bead had been implanted. The additional Fgf-4 maintained the AER for longer than usual, implicating Fgf-4 in maintenance of the ridge.[c]

A conceptually similar situation is seen in mandibular arches in which *Msx-1* is regulated by Bmp-2 and Bmp-4. If Bmp-2 or Bmp-4 is introduced after *Msx-1* is activated, expression of *Fgf-4* extends along the distal mandibular epithelium where it signals to chondrogenic mesenchyme, resulting in formation of a bifurcated Meckel's cartilage.[d] Implanting exogenous Fgf-4 also speeds the closure of sutures in the skull (Chapter 34).

Fgf-8

As is the case with all Fgf family members, Fgf-8 plays multiple roles, regulation of *Msx-1* in several developing systems being perhaps the most prominent. Fgf-8:

- regulates mesenchymal outgrowth and induces the mesenchymal genes *Msx-1*, *Fgf-10*, *Hoxd-13* and *Bmp-4* in developing genital tubercles;
- is expressed in second arch and mandibular epithelium (Fig. 13.7) adjacent to sites of *Msx-1* expression in growth centres within the mesenchyme;

- *Bmp-4* limits the anterior expression of *Fgf-8* in the ventral ectoderm of the second arch, while ectopic Bmp-4 decreases mandibular arch development;
- up-regulates *teneurin-2* in limb-bud mesenchyme;
- up-regulates *Mtsh-1*, the mouse homologue of *Drosophila tsh*, which is expressed in the branchial arches and limb mesenchyme, implicating both genes in outgrowth of processes/buds of neural-crest and mesodermal origin;
- is expressed in the wound epithelium and blastema of regenerating axolotl limbs; and
- plays important roles in limb development, one dependent on *Shh*, one independent.[e]

Fgf-8 determines the rostro-caudal polarity of the first branchial arch in murine embryos – the rostral mandible being odontogenic and the caudal mandible skeletogenic – Fgf-8 establishing polarity via the *rostral epithelium*. Fgf-8 and *endothelin-1* maintain the polarity and regulate mesenchymal competence to respond (Fig. 13.7).

Fgf-8 has *dual functions*, being required for survival and patterning of the proximal mesenchyme of the first branchial arch; Trumpp *et al.* (1999) used Cre/loxP technology to inactivate *Fgf-8* in first-arch ectoderm. Mutants lacked most first-arch structures except the most distal, such as the lower incisor. They demonstrated a large Fgf-8-dependent proximal region and a smaller Fgf-8-independent distal region of the first branchial arch (Fig. 13.7). Note that these mutants are equivalent to the features seen in first-arch syndrome in humans, suggesting a role for Fgf-8 in this syndrome. Furthermore, there is conserved *Fgf-8*-regulated gene expression in limb and facial mesenchyme, *Clim-2* being regulated by epithelial *Fgf-8* in both (Tucker *et al.*, 1998b).

Fgf-4, *-8* and *-10* are all expressed in limb regeneration blastema in the axolotl. Denervation prevents accumulation of Fgf-8 and Fgf-10, indicating that both genes are downstream of the nerve-dependence factor (Christensen *et al.*, 2001).

Implanting Fgf-7 and Fgf-10 into the flank or dorsal midline of chick embryos induces an ectopic AER at those sites. These ectopic ridges are functional; they induce mesenchymal cells to accumulate (Yonei-Tamura *et al.*, 1999).

---

[a] See Niswander and Martin (1992) for patterns of expression of *Fgf-4*, Niswander and Martin (1993a,b) and Niswander *et al.* (1994) for the studies with Fgf-4 in the limb, and Martin (1998) for an overview.
[b] See Wright and Mansour (2003) and D. Liu *et al.* (2003) for *Fgf-3* and *Fgf-10*. Fgf-8, along with members of the Distalless family, is involved in otic placode induction in zebrafish (Solomon *et al.*, 2004).
[c] See Y. Wang and Sassoon (1995) for the AER and Nikbakht and McLachlan (1999) for the ectopic digits.
[d] See Barlow and Francis-West (1997), Bogardi *et al.* (2000), Francis-West *et al.* (2003), and Richman and Lee (2003).
[e] See Tucker *et al.* (2001) for *teneurin-2*, Han *et al.* (2001) for axolotl regeneration, Haraguchi *et al.* (2000) and Chapter 25 for genital tubercles, Tissier-Seta *et al.* (1995) and Long *et al.* (2001a) for *Mtsh-1*, McGonnell *et al.* (1998), Barlow *et al.* (1999) and Shigetani *et al.* (2000) for the second arch, Chapter 38, Martin (2001) and Chrisman *et al.* (2004) for *Fgf-8* and limb development and for a review of *Fgf-8*, *Fgf-10* and *Wnt* genes in wing formation. Also see Yang (2003) for Wnts and wing development.

---

epithelial control; chondrocytes continue to divide as they dedifferentiate, dividing at least five times in the Mexican axolotl, *A. mexicanum*, in response to interaction with the epidermis overlying the developing blastema. The parallel with condensation formation in response to epithelial–mesenchymal signaling (Chapter 18) is obvious.[19]

Not the stump

The pathway of redifferentiation – whether a cell redifferentiates into the same or another cell type – is independent of influences from the stump; blastemata grown in isolation from the limbs undergo normal differentiation.

Blastemata, like limb buds, function as autonomous morphogenetic fields. In exploring similarities between limb regeneration and embryonic limb development, and extending the positional information model proposed by Lewis Wolpert for patterning the limb skeleton during development (Chapter 38), Summerbell *et al.* (1973) argued that the fate of individual blastemal cells depends upon their position relative to a labile 'progress zone' at the tip of the blastema [20]

Electrical signals?

An outward current of between 20 and 100 μm/cm^2 flows from intact limbs of *N. viridescens*. A skin graft decreases current flow and regeneration.[21]

Urodele limb regeneration is responsive to the bioelectrical environment. Denervation of the limb, which prevents regeneration, is followed by an increase in electrical current flow through the amputated limb, driven by Na$^+$ flux across the skin. Blocking Na$^+$ flux – by applying 0.5 mM amiloride to the stump of amputated urodele limbs – blocks regeneration, leading to the conclusion that nerves are the target for the electrical current.[22] [Thinking of the links between mechanical forces, bioelectricity and cell membrane shifts (Chapter 30), Quinn and Rodan (1981) showed that intermittent compression enhances ornithine decarboxylase and Na$^+$, K$^+$ ATPase in redistribution of monovalent cations across the membranes of osteoblastoma cells.]

Hox genes

Twenty homeobox-containing genes have been isolated from the regenerating limbs of the axolotl, over half of which are in the *Hoxa* group. A closer examination of *Hoxa-9* and *Hoxa-13* reveals that, although expressed in a proximo-distal pattern in developing axolotl limbs (as they are in avian and mammalian limb buds), expression is overlapping and distal in the regeneration blastema, implying *proximo-distal patterning in limb development but disto-proximal patterning in regeneration*. Both Hox genes are expressed at a time consistent with a role in dedifferentiation of cells to form the blastema; both are re-expressed later in blastemal development. Application of vitamin A suppresses expression of *Hoxa-13* leading to a more proximal specification of blastemal cells. *Hoxa-13*⁻ mice fail to form condensations in the autopod and so fail to form digits, carpals or tarsals, a deficit that is correlated with decrease of EphA-7. Blocking EphA-7 produces a similar inability to condense.[23]

Simon and Tabin (1993) isolated *Hoxd-10* (*Hox4.5*) and *Hoxc-10* (*Hox3.6*) from *Ambystoma*. Both increase in mesenchymal cells of the blastema but not in the regenerating epithelium. *Hoxd-10*, which is not expressed during limb development, is expressed in blastemata of fore- and hind limbs, while *Hoxc-10*, which is expressed in hind-limb and

tail development, is expressed in hind-limb blastemata only. More recent studies from David Gardiner's group in Irvine, California, identified two transcripts of *Hoxc-10*. Both the short and the long transcript are expressed at high levels at the tip of developing tails and hind limbs. Additionally, the short transcript is expressed at low levels in developing forelimbs. *Both* transcripts and *Hoxb-13* are expressed in regenerating fore- and hind limbs and in the spinal cord during tail regeneration. The long transcript of *Hoxc-10*, which is not active during forelimb development, is up-regulated during forelimb regeneration, making it 'the first truly 'regeneration-specific' gene transcript identified to date' (Carlson *et al.*, 2001, p. 396).

In another recent study, Roy and Gardiner (2002) took advantage of the fact that cyclopamine specifically inhibits *Shh* to show that loss of digits occurs during regeneration of *A. mexicanum* limbs if *Shh* is inhibited, a finding of some interest as the results are similar to those seen when *Shh* is knocked out in mice, showing a conservation of this patterning mechanism between amphibians and mammals.

Simon *et al.* (1997) isolated a novel family of *T-box* transcription factors from developing and regenerating amphibian forelimb, but not hind-limb, and forelimb-bud blastemata. *Brachyury* is the primary T-box gene; *Tbx-4* and *Tbx-5* are also members (Box 35.2, Fig. 35. 4 and Chapter 43).

FgfR-1 and FgfR-2

Poulin *et al.* (1993) cloned *FgfR-1* and *FgfR-2* from the eastern newt, *Notophthalmus viridescens*, and examined their differential expression during limb regeneration (Fig. 14.2). FgfR-2 is found in the stump epithelium and periostea of the exposed bones at the preblastema stage, and in the epithelium and blastema at the blastema stage. *FgfR-1* is not expressed at this stage. At the early to mid-blastemal stages, FgfR-2 declines in condensing mesenchyme, cartilage and perichondria, and is not expressed in the epithelium; *FgfR-1* is expressed in the blastemal cells but not in the epithelium, indicating a reciprocity of expression related to epithelial–stump interactions and blastemal cell differentiation.

Radical fringe

A family of vertebrate homologues of *Drosophila fringe* has been identified and named *lunatic fringe, manic fringe* and *radical fringe*, which may say more about geneticists than genes. The newt, *N. viridescens*, homologue of *radical fringe* was isolated, cloned and sequenced by screening a forelimb blastema cDNA library. *nrFng* is expressed in the primary limb field and then intensely in developing limb buds, expression diminishing as digits develop. During forelimb regeneration, *NrFng* is up-regulated during blastema formation and down-regulated as skeletal regeneration ensues.[24]

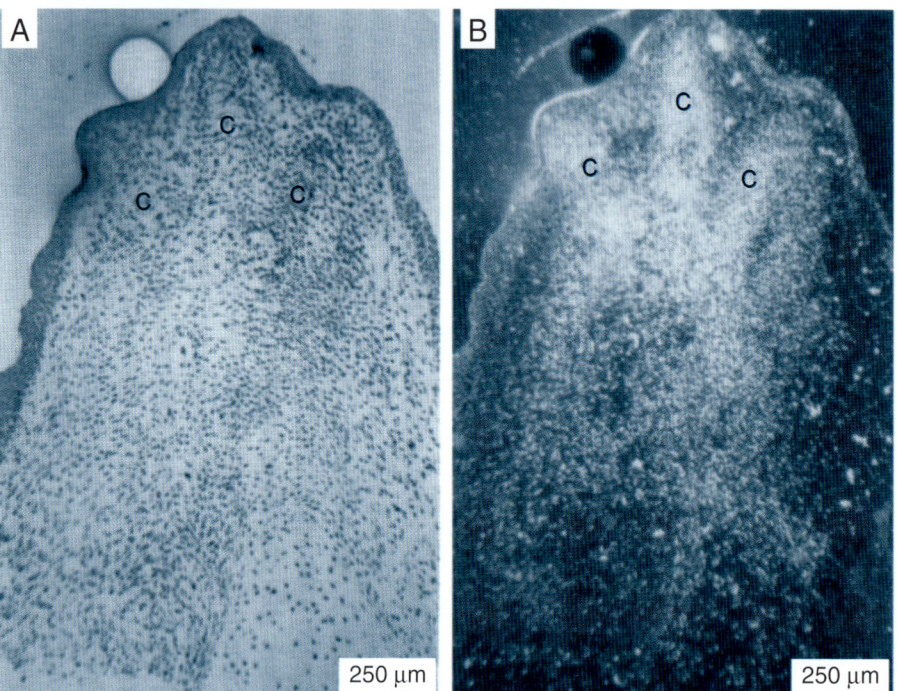

Figure 14.2 *FgfR-2* and limb regeneration in the red-spotted newt, *Notophthalmus viridescens*. Light-field (A) and dark-field (B) images of a longitudinal section of an early digit stage of regeneration showing enhanced expression of *FgfR-2* in the condensations for the digits (C). Modified from Poulin *et al.* (1993).

WHY CAN'T FROGS REGENERATE?

Many anurans (frogs and toads) cannot regenerate limbs or even digits; tail regeneration is considered in Chapter 43. Although not every species from every family has been studied, a sufficiently representative sample of species has been tested to establish a broad pattern (Table 14.3).

Regeneration has not been observed in adults of the anuran families Bufonidae (*Bufo*), Hylidae (*Hyla*, *Pseudacris*) and Ranidae (*Rana*). This may, in part, reflect how little knowledge we have of these frogs; the first postcranial ossification sequence for a hylid frog, *Hyla lanciformis*, was described only in 1988 by De Sá.

Hypermorphic regeneration – the production of a spike with little morphology – occurs in pipid frogs (*Xenopus*, *Hymenochirus*) and discoglossids (*Bombina*, *Discoglossus*), but not in the midwife toad, *Alytes obstetricans*. An interesting exception is the Kenyan reed frog, *Hyperolius viridiflavus ferniquei*, in which newly metamorphosed individuals can regenerate the digits and digit pads.

The assumption in the past has been that spike regenerates are inferior. Tassava (2004) specifically tested this using forelimb spike regenerates in *Xenopus laevis*. Regenerates from the humerus were short and infrequent, in sharp contrast to regenerates through the radius/ulna, where regeneration occurred in 100 per cent of individuals, and regenerates were used for feeding and in amplexus, grew in equilibrium with general body growth, developed nuptial pads when in males and sometimes moulted, indicating responsiveness to the hormonal milieu.

Urodeles within several families, however, do exhibit full regeneration. Unlike anurans, where all genera of a family show similar patterns – no regeneration or hypermorphic regeneration – in urodele genera with more than one species we often find that individuals of one species can regenerate but that individuals of another cannot (Table 14.3).

Conforming to the requirement of innervation for regeneration in urodeles, *Xenopus* has more and larger nerve fibres/unit area in its limbs than does the wood frog, *Rana sylvatica*, a species than does not regenerate (Table 14.3), taken to mean that *Xenopus* shows some regenerative ability because of nerve factors. However, drawing the obvious conclusion – that *Xenopus* has sufficient nerve fibres to allow it to initiate regeneration – would turn out to be an error; nerve-independent limb regeneration in *X. laevis* tadpoles is made possible by a mitogenic factor in early limb tissues, possibly Fgf-2. Implanting spinal ganglia into *Xenopus* forelimb tissue maintained *in vitro* influences cartilage regeneration; proliferation, appositional and interstitial growth are all stimulated.[25]

Despite such suggestive evidence, when Scadding (1982) investigated whether the quantity of innervation was the limiting factor for ability to regenerate limbs in five species of urodeles and eight species of anurans, he could find no correlation with innervation in the urodeles (Table 14.3). Typical of his study, '*Ambystoma maculatum* which regenerates frequently (89 percent of …cases…) [had] the lowest level of innervation observed, lower even than most of the nonregenerating species' (p. 83). Within the anurans examined, however, four species that do regenerate had greater innervation than four that do not, indicating that innervation may be limiting for anurans but not for urodeles (Tables 14.3 and 14.4). Nor did

Table 14.3 Occurrence of limb regeneration and amount of nerve fibres at the amputation site in adult amphibians[a]

Family	Species	Degree of limb regeneration	Nerve quantity (% cross-section area)[b]	
			Radius–ulna	Humerus
		Urodeles		
Ambystomatidae	*Ambystoma tigrinum* (tiger salamander)	Variable–normal, heteromorphic, or absent	–	–
	Ambystoma maculatum (spotted salamander)	Usually normal; occasionally heteromorphic or absent	0.24 ± 0.1	0.30
	Ambystoma laterale (blue-spotted salamander)	Normal	0.59 ± 0.20	–
	Ambystoma jeffersonianum (Jefferson salamander)	Normal	–	–
	Ambystoma opacum (marbled salamander)	Usually normal; occasionally heteromorphic	–	–
	Ambystoma mexicanum (Mexican axolotl)	Normal	–	–
Salamandridae	*Cynops pyrrhogaster* (Japanese fire belly newt)	Normal	–	–
	Triturus cristatus (great crested newt)	Normal	–	–
	Triturus helveticus (palmate newt)	Normal	–	–
	Triturus vulgaris (*taeniatus*) (smooth newt)	Normal	–	–
	Triturus alpestris (Alpine newt)	Normal	–	–
	Salamandra salamandra (fire salamander)	Heteromorphic	–	–
	Notophthalmus viridescens (eastern newt)	Normal	0.78 ± 0.28	1.68 ± 0.46
	Taricha torosa (California newt)	Normal	–	–
Plethodontidae	*Plethodon cinereus* (red-backed salamander)	Normal	–	–
	Plethodon dorsalis (zigzag salamander)	Normal	–	–
	Plethodon glutinosis (slimy salamander)	Normal	0.58 ± 0.18	0.58 ±0.30
	Desmognathus ochrophaeus (mountain dusky salamander)	Normal	–	–
	Desmognathus fuscus (northern dusky salamander)	Normal	–	–
	Eurycea bislineata (northern two-lined salamander)	Normal	–	–
Proteidae	*Necturus maculosus* (mudpuppy)	Absent	–	–
Sirenidae	*Siren intermedia* (lesser siren)	Absent or heteromorphic	–	–
Amphiumidae	*Amphiuma tridactylum* (three-toed amphiuma)	Heteromorphic or absent	0.68 ± 0.43	0.76 ± 0.33
	Amphiuma means (two-toed amphiuma)	Heteromorphic or absent	–	–
		Anurans		
Discoglossidae	*Alytes obstetricans* (midwife toad)	Absent	–	–
	Bombina orientalis (fire-bellied toad)	Heteromorphic	–	–
	Bombina variegata (yellow-bellied toad)	Heteromorphic	–	–
	Discoglossus pictus (spotted tree frog)	Heteromorphic	–	–
Pipidae	*Xenopus laevis* (South African clawed frog)	Heteromorphic	0.87 ± 0.19	–
	Xenopus muelleri (Muller's clawed frog)	Heteromorphic	–	–
	Hymenochirus boettgeri (South African dwarf frog)	Heteromorphic	–	–
Bufonidae	*Bufo quercicus* (oak toad)	Absent	–	–
	Bufo americanus (American toad)	Absent	0.24 ± 0.07	0.25 ± 0.04
	Bufo marinus (giant toad)	Absent	0.27 ± 0.12	0.25 ± 0.02
	Bufo regularis (Egyptian toad)	Absent	–	–
	Bufo andersonii (marbled balloon frog)	Absent	–	–
Hylidae	*Pseudacris triseriata* (Western chorus frog)	Absent or heteromorphic	0.59 ± 0.31	1.13 ± 0.73
	Pseudacris clarki (spotted chorus frog)	Heteromorphic	–	–
	Hyla crucifer (spring peepers)	Absent or heteromorphic	0.58 ± 0.26	0.97 ± 0.80
	Hyla squirella (squirrel tree frog)	Absent or heteromorphic	–	–
	Hyla versicolor (grey tree frog)	Absent	0.42 ± 0.20	–
	Hyla cinerea (green tree frog)	Absent	–	–
	Hyla septentrionalis (Cuban tree frog)	Absent	–	–
	Acris crepitans (northern cricket frog)	Absent or heteromorphic	–	–
Microhylidae	*Gastrophryne carolinensis* (eastern narrow-mouth toad)	Heteromorphic	0.61 ± 0.21	–
Pelobatidae	*Scaphiopus holbrooki* (eastern spadefoot toads)	Absent	–	–
	Scaphiopus bombifrons (plains spadefoot toad)	Absent	–	–
Hyperolidae	*Hyperolius virididlavus ferniquei* (Kenyan reed frog)	Heteromorphic	–	–

Table 14.3 (*Continued*)

Family	Species	Degree of limb regeneration	Nerve quantity (% cross-section area)[b]	
			Radius–ulna	Humerus
Ranidae	*Rana temporaria* (European common frog)	Absent	–	–
	Rana clamitans (bronze frog)	Absent	–	–
	Rana sylvatica (wood frog)	Absent	–	–
	Rana palustris (pickerel frog)	Absent	–	–
	Rana catesbeiana (bullfrog)	Absent	–	–
	Rana pipiens (northern leopard frog)	Absent; rarely heteromorphic	0.35 ± 0.16	0.64 ± 0.21
	Rana esculenta (edible frog)	Absent	–	–
	Rana ridibunda ridibunda (lake frog)	Heteromorphic	–	–
	Rana cyanophlyctis (skipping frog)	Heteromorphic or absent	–	–
	Rana tigerina (Indian frog)	Absent	–	–

[a] Based on data from Scadding (1981, 1982).
[b] Total cross-sectional area of nerves/cross-sectional area of limb $\times$ 100. Each datum in the table gives the mean $\pm$ standard deviation.

Table 14.4 Quantity of nerve compared to frequency of regeneration in several species of amphibian[a]

Species	Urodela or Anura	Nerve quantity (%)	
		Radius–ulna	Humerus
Usually regenerating normally			
Notophthalmus viridescens (eastern newt)	Urodela	0.78 ± 0.28	1.68 ± 0.46
Plethodon glutinosus (slimy salamander)	Urodela	0.58 ± 0.18	0.58 ± 0.30
Ambystoma laterale (blue-spotted salamander)	Urodela	0.59 ± 0.20	–
Frequently regenerating normally			
Ambystoma maculatum (spotted salamander)	Urodela	0.24 ± 0.10	0.30
Usually regenerating heteromorphically			
Xenopus laevis (South African clawed frog)	Anura	0.87 ± 0.19	–
Gastrophryne carolinensis (eastern narrow-mouth toad)	Anura	0.61 ± 0.21	–
Frequently regenerating heteromorphically			
Amphiuma tridactylum (three-toed amphiuma)	Urodela	0.68 ± 0.43	0.76 ± 0.33
Pseudacris trisriata (western chorus frog)	Anura	0.59 ± 0.31	1.13 ± 0.73
Occasionally regenerating heteromorphically			
Hyla crucifer (spring peepers)	Anura	0.58 ± 0.26	0.97 ± 0.80
Non-regenerating			
Hyla versicolor (grey tree frog)	Anura	0.42 ± 0.20	–
Bufo marinus (giant toad)	Anura	0.27 ± 0.12	0.35 ± 0.02
Bufo americanus (American toad)	Anura	0.24 ± 0.07	0.25 ± 0.04
Rana pipiens (northern leopard frog)	Anura	0.35 ± 0.16	0.64 ± 0.21

[a] Based on data from Scadding (1982).

Scadding find any correlation between amount of innervation and regeneration in individual *X. laevis* or *A. maculatum*, though he did for the three-toed amphiuma, *A. tridactylum*.

Augmenting regeneration

Regrowth of anuran limbs can be evoked if the nerve supply is augmented with *electrical stimulation*. Thus, in the leopard frog, *Rana pipiens*, and the bullfrog, *R. catesbiana*, 5–10 mm cones of tissues containing cartilage and muscle (some with digits) form following application of very low (0.3 V) currents. Administering 2500 units of nerve growth factor to *X. laevis* froglets promotes regeneration of cartilage (complete with perichondrium), connective tissue and epidermis. The cartilage mineralizes, but no bone, muscle or nervous tissue form. At earlier stages in ontogeny, *Xenopus* limb regeneration is dependent on Fgf-10, which is expressed in blastema but not in limbs at stages of the life cycle during which limb regeneration cannot occur.[26]

Regeneration of appendages is normally limited in frogs because they do not have a functional 'wound' epidermis capable of sustaining or initiating dedifferentiation or of maintaining cells in the cell cycle. Application of *vitamin A* to limb stumps of *R. temporaria* results in serial reduplication along the proximo-distal axis (as seen in urodeles) but in mirror-image duplication along the antero-posterior axis (as seen in birds). Vitamin A applied ectopically to the stump enhances forelimb regeneration in juvenile leopard frogs, leading to production of larger regenerates with better morphology.[27]

FINGERTIPS OF MICE, MONKEYS AND MEN

Supernumerary digits were high on the list of the many things that fascinated Charles Darwin; 'The most interesting point with respect to supernumerary digits is their occasional regrowth after amputation' (Darwin, 1861, p. 14). Here, Darwin is referring to the surgical removal of an extra digit and its subsequent partial replacement; see the introduction to Part V. He was wondering whether this was a case of one aberrant individual or an ability possessed by all children.

We can regenerate our fingertips, provided that amputation occurs early in life (Douglas, 1972; Illingworth, 1974). This is a fortunate occurrence; fingertip amputations are among the most common traumas seen in emergency rooms in the developed world. One of the major differences between urodele limb regeneration and regeneration of the fingertips of mice, monkeys and adult humans is the complete lack of dedifferentiation to form a blastema in mammalian fingertip regeneration. Regeneration of bone is periosteal.

My developmental biology class is always surprised when I tell them that young children can regenerate their fingertips, complete with fingernail, provided that a physician does not impede regeneration by suturing the wound closed. Closure prevents the epithelial–stump tissue interactions required to produce a blastema from which the tip can regenerate, favouring soft connective tissue repair and scar tissue formation. The inability of older individuals to regenerate relates to the inability of the wound epithelium to initiate or maintain dedifferentiation, or to promote proliferation of dedifferentiated cells (Tassava and Olsen, 1982). It is clear from Darwin's account that the rudiments of this story could have been told almost 140 years ago.

Mice can also regenerate the tips of their toes, including the entire nail bed and fingernail, *provided that* the toe is amputated distal to the last interphalangeal joint. The periosteum is the source of the cells for the regenerate. In mice, *bone does not contribute to a regenerate if the nail is missing*, which suggests that interactions between the regenerating nail bed and periosteal osteogenic cells are required.

Msx-1 and *Msx*-2 are expressed during regeneration, *Msx-1* more distally than *Msx-2*. Foetal mice only regenerate the tips of the digits from *within the* Msx-1 *domain*, late foetal and newborn mice regenerate within both *Msx* domains; we have yet to determine the role these homeobox genes play in this temporal restriction.[28] These and related studies can be summarized as follows.

- Bone healing complicates and to some extent prevents a regenerative response.
- Periosteal cells produce cartilage and bone, endosteal cells only produce bone.
- Dead bone persists and marrow cavities are sealed off.
- Bone regrowth is equivalent in adults and neonates.
- Amputation of the distal 40 per cent of the terminal phalanx is followed by complete regeneration, including a new nail, within five weeks; amputation of the proximal 20 per cent of the distal phalanx is followed by minimal bone regeneration and no nail regrowth, while amputation of the intermediate 40 per cent of the distal phalanx is followed by abnormal regrowth,
- The nail organ plays an essential role in bone regrowth.
- Msx-1 and Bmp-4 interact to regulate *in vitro* regeneration of digits in foetal mice.
- Fgf-2 stimulates bone regeneration after amputation (as assessed *in vitro*).
- Finally, rodent nail organs transplanted into amputations of more proximal phalanges initiate repair with substantial bone formation, formation of a nail bed and development of a keratinized nail.[29]

Comparison with urodele limb regeneration

It was inevitable that regeneration of mouse digits and newt limbs would be compared, especially to understand why murine digit regeneration is so limited spatially and temporally. Neufeld (1985) investigated whether processes initiated to repair the fractured bone and the inability to develop cartilage distal to the amputation plane are impediments to digit regeneration, concluding that they are not.

Provided that the cut is no more than 1 mm distal to the nail bed, rhesus monkeys, *Macaca mulatta*, replace amputated fingertips. Singer et al. (1987) compared this digit regeneration with urodele limb regeneration, demonstrating that digit regeneration takes place without blastema formation but with periosteal bone 'regeneration.' Osteogenesis is always direct (intramembranous) without any evidence of cartilage, in clear contrast to urodele limb regeneration.

Is this regeneration if the reaction to wounding is limited to periosteal proliferation and when no blastema forms? In that a complex part of the body is replaced with amazing fidelity, it is appropriate to regard it as regeneration. Monkey fingertips do not grow back to the length of the original and so regeneration is not as 'perfect' as in urodeles, but regeneration it is. The alternative is to speak of regrowth, the term Darwin used and a term used in some literature and followed above. But when regrowth replaces and extends a structure to the extent that replacing a fingertip does, the process is more than regrowth. Given the lack of dedifferentiation, the amount and quality of the 'replacement of the lost part' (= regeneration) is remarkable.

NOTES

1. The eastern newt, *Notophthalmus viridescens*, was formerly known as *Triturus viridescens* (Wake, 1976).

2. See DeHaan (1956) and Laufer (1959) for lack of synthesis of myosin, and Linsenmayer and Smith (1976) for lack of type II collagen synthesis.

3. See Mattson and Foret (1973) and McHenry et al. (1974) for decline in ^{35}S incorporation, and Toole and Gross (1971) and Chapter 14 for synthesis of hyaluronan in the blastema.

4. See Libbin et al. (1988, 1989) and Libbin (1992) for prolongation of the cartilaginous phase in development and regeneration and see Finch (1969) for innervation.

5. See G. N. Smith et al. (1975) for hyaluronidase synthesis after denervation and Fitzharris (1976) for annelids.

6. See Steen and Thornton (1963), Goss (1969a), G. N. Smith et al. (1975), Olsen and Tassava (1984) and Tassava and Olsen-Winner (2003) for aneurogenic limbs, produced by removal of the developing neural tube from embryos before limb buds arise.

7. The extent to and ways, if any, by which limb regeneration repeats limb development have been discussed for decades; see Goss (1969a, 1980b), Wallace (1981) and Ferretti and Géraudie (1998).

8. For studies involving removal of the humerus and regeneration, see Weiss (1925), Thornton (1938) and Goss (1958). Fishman and Hay (1962) demonstrated the migration of osteoclast precursors into the stump via the blood vascular system in their ^{3}H-thymidine-labeling studies.

9. See Butler (1933) and Thornton (1942) for muscle grafts, and Eggert (1966) for cartilage grafts into irradiated limbs.

10. One has always to beware of species differences and therefore of extrapolating results even to closely related species. Using a different species – *Ambystoma maculatum* – Foret (1970) found that Meckel's cartilage was relatively inert to dissolution when grafted into the limb and so contributed few cells to the regenerate. Although *Ambystoma mexicanum* is not an ideal species for genetic studies, some 30 mutants are known (Armstrong, 1985), some of which might provide useful markers.

11. Satellite cells are the reserve cells in muscle, their function being to replace lost myoblasts. Satellite cells have been isolated from chick muscle and from pluripotential stem cells from connective tissues that produce muscle, fibroblasts, fat, cartilage and bone (Young et al., 1992, 1993). Cultured satellite cells from muscles of adult mice differentiate into myoblasts, as expected, but can transform into osteoblasts and adipocytes when exposed to signals such as Bmp (Asakura et al., 2001).

12. See Tomlinson et al. (1985) for the studies with *Rana catesbiana*, Korneluk and Liversage (1984) for limited regeneration in *Xenopus*, and Tinsley and Kobel (1996) for *Xenopus* biology. Quite a lot of cartilage forms in the meager regenerates that develop when hind limbs of *Bufo viridis* tadpoles are amputated. Some of this cartilage is inferred to have originated from cells of muscle or connective-tissue origin (Michael and Niazi, 1972).

13. See Wallace et al. (1974) and Maden and Wallace (1975) for these studies, and see Dinsmore (1991) and Ferretti and Géraudie (1998) for overviews of historical and recent work, respectively.

14. See Nathanson et al. (1978) and Nathanson (1979) for metaplasia, Nathanson (1983a,b, 1985, 1986), Nathanson et al. (1986) and Box 5.1 for continued production of actin and myosin, and Nathanson and Hay (1980) for the ultrastructural analysis.

15. See Thompson et al. (1985) and Bettex-Galland and Wiesmann (1987).

16. See Globus and Vethamany-Globus (1977) for the ganglionic implants, and Brockes (1991) for an overview. See Rose (1948), Steen and Thornton (1963) and Scadding (1981, 1982, 1983) for analyses and reviews of early studies on the role of nerves in amphibian limb regeneration. Interestingly, during development the branchial arches attract and promote the growth of motor axons, in part through the synthesis and release of hepatocyte growth factor (Caton et al., 2000).

17. See Olsen and Tassava (1984) for the data on DNA synthesis and proliferation, and Dungan et al. (2002) for Fgf-1.

18. There is also a body of data on supplementing urodele limb regeneration using the ependymal lining from the spinal cord; see Goss (1969a) for a discussion. Grafting the ependymal lining of the spinal cord into the limbs of the desert (Yucca) night lizard, *Xantusia vigilis*, induces regeneration in 82 per cent of the implanted limbs (Bryant and Wozny, 1974).

19. See Steen (1968) for chondrocyte proliferation in *Ambystoma (Siredon) mexicanum* and see Mescher (1970) for the epithelial requirement.

20. For independence of the blastema from the stump, see Hay and Fischman (1961), Stocum (1968, 1975a,b), Goss (1969a) and Stocum and Dearlove (1972). Lovejoy et al. (2000) reviewed mammalian skeletal morphology – especially the limb – from the developmental perspective, with an emphasis on fields. See Chapter 38 for a discussion of the progress zone.

21. See Altizer et al. (2002) for current flow and see Connolly (1981) and Hall (1983f) for reviews.

22. See Borgens et al. (1979a,b) and Borgens (1982a).

23. See Gardiner et al (1995) and Gardiner and Bryant (1996) for the *Hoxa* group and Stadler et al. (2001) for the mouse study.

24. See Cadinouche et al. (1999) for *radical fringe*. Moran et al. (1999) questioned improper insertion of the selection cassette in this study. See the reply by Zhang and Gridley in *Nature* (1999) **399**, p. 742.

25. See Richards et al. (1975) for the Kenyan reed frog, Rzehak and Singer (1966) for the *Xenopus–Rana* comparison, Filoni et al. (1999) for the mitotic factor, Cannata et al. (2001) for Fgf-2, and Tsilfidis and Liversage (1991), who implanted the spinal ganglia.

26. See Bodemer (1964) and S. D. Smith (1967) for electrical stimulation, Robinson and Allenby (1974) for nerve growth factor (Ngf) and Yokoyama et al. (1998, 2000) for Fgf-10. Ngf also accelerates initiation of fin regeneration in goldfish, *Carassius auratus* (Weis, 1972).

27. See Tassava and Olsen (1982) for the wound epidermis and Maden (1983) and Cecil and Tassava (1986) for vitamin A.

28. See Borgens (1982b) for the periosteal origin of the regenerate, Zhao and Neufeld (1995) for interaction with the nail bed, and Reginelli et al. (1995) for *Msx-1* and *Msx-2*. The best recent reviews of digit regeneration are Neufeld (1992) and T. L. Muller et al. (1999), especially regeneration in young children and experimental work on mice. Also see the other chapters in Volume 5: *Fracture Repair and Regeneration* of Hall (1990–1994) for repair and regeneration.

29. See Neufeld (1985, 1992), Neufeld and Zhao (1995), Zhao and Neufeld (1995, 1996), Hall (1998b) and Mohammed et al. (1999) for the studies on which this summary is based.

Cells to Make and Cells to Break

The seemingly rock-like structure of bone constantly changes with the ebb and flow of deposition and resorption.

Another important activity accompanies regeneration. Although I have introduced this activity in relation to acellular bone, bone remodeling and transformation of hypertrophic chondrocytes to osteoprogenitor cells, I said nothing about it in the last chapter. That is because in studying the regeneration of skeletal elements we concentrate on processes such as dedifferentiation, reinitiation of cell division and the role of innervation and hormones that produce a blastema of cells. However, skeletal cells must be released from their extracellular matrices before they can dedifferentiate. The process that accomplishes this is resorption. The cells that carry out resorption form a class of *mineraloclasts*, a class that includes *osteoclasts, chondroclasts, dentinoclasts, cementoclasts* and *enameloclasts*, all of which are identified by virtue of their proximity to and active resorption of bone, cartilage, dentine, cementum and enamel.

CLASTS AND BLASTS

Are there five separate lineages of 'clasts,' or can any clast resorb any mineralized extracellular matrix? When bone is implanted into sharks it is resorbed, presumably by cells that otherwise would have been called into play to resorb cartilage. The implication is that chondro- and osteoclasts are (i) closely related (ii) the same cell type, or (iii) arise from common stem cells in response to functional demand.

Whichever one or more than one of these is the case, can the reverse occur; can osteoclasts resorb cartilage? An easy question but hard to answer. If we only recognize a clast when it is doing its job, we cannot tell whether it might be capable of resorbing another matrix. The shark study would seem to stand as an exception but, even there, we know that bone is present as part of the normal skeleton of cartilaginous fishes (Chapter 3). Perhaps the cells that resorbed the implanted bone were a separate lineage of cells that normally would have resorbed bone of the shark endoskeleton.

So, you see that we have some intriguing issues to explore. Resorption has been introduced but mostly *en passant*. Because I think it is important to provide an outline of resorption as a process, the discussion below is more or less in the reverse order in which knowledge was acquired. Resorption of bone is used as the model. I begin with resorption and *coupling, the remarkable finding that many of the factors that trigger resorption do so by acting on the bone-forming cells, osteoblasts.* Many paradoxical earlier studies take on an element of rationality when we realize that osteoblasic and osteoclastic activities are coupled.

Then I discuss how osteoclasts and their precursors can be identified. Only then do I discuss the search for osteoclast progenitors, whether osteoclast and osteoblast precursors are the same (they are not), finishing up with a discussion of the relationships between osteo- and chondroclasts, and a brief comment on resorption by the synovial cells that line joint spaces.

RESORPTION

On the one hand, resorption of bone in teleost fishes is very different from resorption in mammals. On the other hand, it is very similar. The major difference is that *mononucleated osteoclasts* are important resorptive cells in fish; in mammals, resorption is considered the provenance of osteoclasts, which by definition are *multinucleated*.[1] Although a mononucleated osteoclast sounds like an oxymoron, such a designation focuses our attention on identifying precisely which cells can resorb bone and which cells cannot. Given the complicated life history of an osteoclast, knowing the stage at which it can resorb has important implications for treatment of many metabolic bone diseases.

Osteoclasts depend on the proximity of bone matrix to acquire their characteristic morphology and to express their physiological activity. You would think, therefore, that there would be no incentive to attempt to isolate osteoclasts. Nevertheless, methods for isolating populations of osteoclasts were developed in the early 1970s. Further development of these procedures to the point where osteoclasts can be cultured with bone chips, with or without the addition of *osteoblastic* cells, has revealed how osteoclasts resorb and, perhaps as importantly, revealed a coupling between osteoclasts and osteoblasts to effect that resorption.[2]

A cohort of osteoblasts can produce a 7-μm-thick lamina of bone in a week. Humans replace trabecular bone every three to four years, and replace cortical bone once a decade. Several hundred milligrams of Ca^{++} are moved into or out of the human skeleton every day. Osteoclasts typically contain 3–5 or 12–15 nuclei depending on species. Both mammalian and teleost osteoclasts contain a vacuolar proton pump (V-ATPase) that functions in resorption. Mammalian osteoclasts express the B2 isoform of vacuolar H^+-ATPase both intracellularly and on the plasma membrane. Vacuolar ATPase is a homologue of mitochondrial or bacterial ATO-synthase. Expression of the a3 isoform of vacuolar ATPase is greater in osteoclasts with 10 or more nuclei than in osteoclasts with five or fewer nuclei, supporting the intuitive notion that larger osteoclasts are more effective in bone resorption than are smaller osteoclasts. We do not know how old osteoclasts are removed. Indeed, because osteoclasts grow by acquiring new cells, osteoclasts may be immortal.[3]

As discussed below, osteoclasts only function when in contact with bone or a mineralized surface. Consequently, it took 60 years to develop the approaches that opened up analysis of bone resorption *in vitro*. Tim Chambers and colleagues used scanning electron microscopy (SEM) to demonstrate that osteoclasts resorb slices of bone *in vitro*, and that monoclonal antibodies raised against osteoclasts inhibit bone resorption *in vitro*. Calcitonin and cytochalasin-B inhibit resorption *in vitro* as they do *in vivo*, suggesting that the *in vitro* conditions allow physiological resorption to occur.[4]

Osteoclasts are amazingly catholic in their tastes. Cultured avian and mammalian osteoclasts resorb all substrates tested, including avian and molluscan shell, calcite and aragonite, without species or substrate specificity. Osteoclasts resorb dentine when the two are co-cultured. These are important clues to the issue of clast specificity. Interestingly, sites of resorption *in vitro* are not centres

that attract osteoblasts. On the basis of SEM it was argued that neither osteoblasts nor osteocytes in rat parietal bones transforms into osteoclasts. This is intriguing, for we now know that osteoclasts are activated by (coupled to) osteoblasts; see the following section.[5]

Given the physiological conditions available *in vitro*, rates of migration and resorption were determined for rabbit osteoclasts cultured on slices of bone. Under these conditions, osteoclasts move along the bone surface at 105 ± 10 μm/hour, and resorb bone at a rate of 390 ± 109 μm³/hour. As might be expected, the range of movements, reflecting differential activity of individual osteoclasts, is large: 30–248 μm/hour for movement, and 43–1225 μm³/hour for resorption (Fig. 15.1). In part, these data reflect the fact that an osteoclast can resorb at different rates at different times and, in part, that the same osteoclast can resorb two lacunae at different rates simultaneously.

COUPLING BONE RESORPTION TO BONE FORMATION

As intimated above, bone formation and bone resorption are related. They are especially related when formation/resorption leads neither to loss nor gain of bone but rather to *remodeling*, the homeostatic situation in which removal and addition of bone are in balance. We now know that the relationship between resorption and deposition known as *coupling* exists because activation of osteoclast function – and often recruitment of osteoclast precursors as well – is mediated by osteoclasts; see Blair *et al.* (2002) for an overview of likely mechanisms.

It seems paradoxical that the cells that *form bone* are the ones that receive, translate and mediate the signals required to activate the cells that *resorb bone*. Were bone an inert tissue no coupling would be required, but remodeling requires coupling. Coupling could have evolved in other ways but appears not to have, although our knowledge outside the traditional model mammals is woefully inadequate. Hypervitaminosis D and hyperthyroidism in mammals essentially reflect *uncoupling* in which there is an excess of resorption over deposition (Box 27.1). Many other metabolic bone diseases reflect uncoupling at one stage or another of the remodeling process. Indeed, most modern therapeutic approaches to bone diseases, especially treatments that enhance bone formation, have as their target slowing bone resorption (Rodan and Martin, 2000).

COUPLING OSTEOBLASTS AND OSTEOCLASTS

An early study that hinted at coupling between bone formation and resorption was undertaken by Howard *et al.* (1980) using *in vitro* culture in serum-free medium with

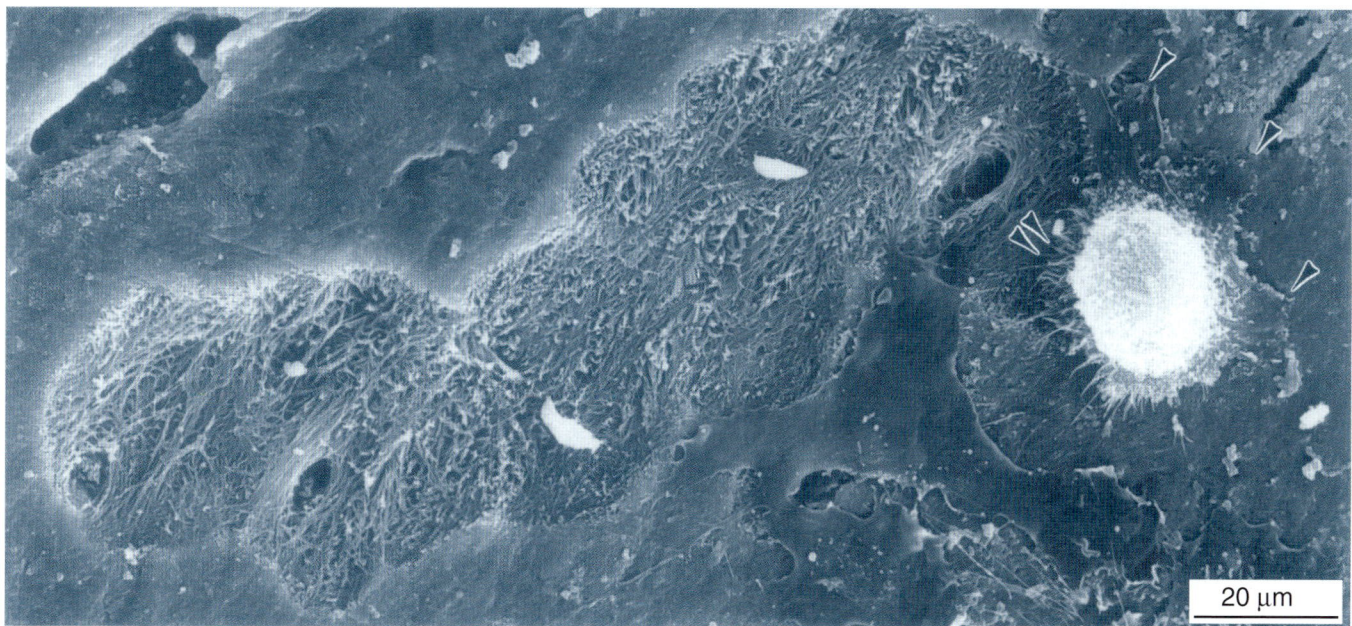

Figure 15.1 'Seeing' bone resorption in a slice of canine femur. A scanning electron micrograph of a resorption 'trail' (the roughened surface) extending from left to right, 'carved out' from this slice of cat femur in 48 hours by the osteoclast at the right (arrowheads). Modified from Kanehisa and Heersche (1988).

addition of 10^{-12} M parathyroid hormone (PTH). Bones were labeled *in vivo* with ^{3}H-proline to label hydroxyproline, and release of labeled hydroxyproline was used as an indicator of resorption. Bone *formation continued as a consequence of enhanced resorption induced by PTH*. From this and similar studies it became clear that *osteoblasts* are targets for signals (PTH, prostaglandins, 1,25-dihydroxyvitamin D_3) that effectively *remove osteoblasts* from the bone surface to expose the surface so that osteoclasts can move in, occupy the space previously occupied by osteoblasts and begin to resorb.

Rodan and Martin (1981) raised the possibility that *products of hormonally activated osteoblasts might activate osteoclasts*. Evidence supporting this notion began to come in, as outlined below.

- Wong (1982, 1984) showed that murine calvarial osteoclasts fail to show any PTH stimulation of hyaluronan synthesis unless they are co-cultured with osteoblasts or osteoblast-conditioned medium,[6] implying the presence of soluble osteoblast factors in osteoclast regulation.
- Rat osteosarcoma cells migrate in response to signals released from resorbing bone or type I collagen, demonstrating the coupling of formation to resorption (Mundy *et al.*, 1982).
- Parfitt (1982, 1984, 1990) emphasized the importance of coupling for normal skeletal development and the consequences of defective coupling for disease states such as osteoporosis (Box 8.2), proposing a five-stage remodeling cycle of quiescence, activation, resorption, reversal and formation.

- Jaworski (1984a,b) examined coupling of formation and resorption in lamellar bone, while Jilka (1986) demonstrated that osteoblasts must be present before endosteal osteoclasts can resorb bone in response to PTH; osteoclasts alone increase their levels of acid phosphatase when exposed to PTH but do not initiate resorption.
- Oursler and Osdoby (1988) showed that differentiation of *osteoclasts* from marrow in response to periosteum-free chick calvaria is unaffected by PTH.

The implication, subsequently verified by identification of PTH receptors on *osteoblasts*, is that the resorptive action elicited by PTH comes about because *osteoblasts* respond to this hormone. The conclusion from these and similar studies is clear – osteoblasts evoke osteoclast differentiation.

Are earlier stages of osteoblast and/or osteoclast cell lineages also coupled? The following evidence suggests that the answer is yes: monocytes interact with osteoblasts and so transform into osteoclasts; mononuclear cells from bone marrow induce osteoclast formation when co-cultured with calvariae; periosteum-free chick calvariae evoke osteoclasts from marrow cells (PTH has no effect); osteoclasts are obtained when cells from neonatal bone marrow are cultured on a layer of osteoblasts derived from rat tibia. However, using an *in vitro* model, S. J. Jones *et al.* (1994) found that sites of resorption do not attract osteoblasts. Such lack of chemoattraction presumably reflects the fact that osteoclast precursors are circulating in the blood stream and located on bone surfaces, and so do not have to be recruited from a distance, only (and it is a big only) held over short distances.[7]

SOME MOLECULAR PLAYERS

Bone remodeling is regulated by two factors; *osteoprotegerin* and the surface ligand *Rank* to which it binds, both of which are activated by Tgfβ-1. Inhibiting Tgfβ-receptor signaling in *osteoblasts* leads to decreased bone remodeling and increased trabecular bone mass (Box 15.1).

Transgenic mice with a cytoplasmic truncated type II Tgfβ receptor from the osteocalcin promoter show increased bone mass because of decreased osteoclastic resorption.[8]

An interesting recent study by Itonaga *et al.* (2004) shows that, in the presence of macrophage colony-stimulating factor but in the absence of Rank ligand,

---

Box 15.1 The transforming growth factor-β (Tgfβ) superfamily and its receptors

The transforming growth factor-betas (Tgfβ) are a family of 25-kDa peptides, many in latent form with proteins in serum and platelets, or in bone as large glycoprotein complexes. Tgfβ-1, β-2 and β-3 are all found in bone during mouse and avian development. Tgfβ-2, β-3 and β-4 (but not Tgfβ-1) are found in embryonic chick sternal chondrocytes and muscle cells. If we were to summarize the diverse action of these four growth factors in one sentence, we would say that Tgfβs enhance chondroblast and osteoblast proliferation and differentiation, and enhance the deposition of cartilaginous and osseous extracellular matrices.[a]

While it is invidious to select a handful of studies from the many discussed in the text, the following provide a flavour of the importance to skeletal development of the Tgfβ superfamily. Glance at the index and you will find many, many more.

Tgfβ-1

Tgfβ-1 is expressed in the perichondria of foetal mouse long bones where it acts to decrease proliferation and chondrocyte hypertrophy in endochondral ossification, providing a feedback system (Alvarez *et al.*, 2001).

In common with most Tgfβs, Tgfβ-1 interacts with other growth factors. For instance, Tgfβ-1 and Fgf-2 have additive effects on differentiation, promoting chondrogenic differentiation from micromass cultures of chick distal limb mesenchyme. However, their morphogenetic influences differ: Tgfβ-1 alone results in the chondrocytes forming sheets of cartilage, Fgf-2 produces nodules (Schofield and Wolpert, 1990).

Differential transcription and translation of Tgfβ isoforms by human articular chondrocytes is modulated by several growth factors. Articular chondrocytes can synthesize and release three isoforms, β-1, β-2 and β-3. Fgf-2 enhances release of Tgfβ but has no effect on mRNA levels. Platelet-derived growth factor (Pdgf) increases release of Tgfβ-3 and the synthesis of *Tgfβ-3* mRNA; Tgfβ-1 and β-2 are self (auto) regulatory. Tgfβ-3 increases levels of *Tgfβ-1* mRNA but not *Tgfβ-2* mRNA. Fetuin glycoprotein binds to Tgfβs and Bmps to block their osteogenic activity.[b]

Human osteoblasts from normal volunteers form condensations and go on to express osteogenic markers (alkaline phosphatase, type I collagen, osteonectin) in 3D culture when in the presence of Tgfβ-1 in serum-free medium (Kale *et al.*, 2000).

Human recombinant Tgfβ-1 is available. When applied in single or multiple injections over 3–12 days into the periosteum of the parietal bones of neonatal rats, RhTGFβ-1 evokes intramembranous ossification. Repeated application of 200 ng/day increases the thickness of the bone; a single 200 ng injection increases the size of the periosteal osteoprogenitor layer. Chondrogenesis is not seen in neonates but is seen after Tgfβ-1 is injected into adult cranial bones, demonstrating the importance of the state of responsiveness of cells – even cells in the same skeletal element at different times – and demonstrating that cartilage can arise from the periostea of membrane bones (Chapter 12). Tgfβ-1 at high concentrations (10 pM or greater) inhibits osteogenesis.[c]

Chondrogenesis is promoted during cartilage repair by low concentrations of Tgfβ-1 in the extracellular matrix or by higher concentrations delivered in artificial lipid droplets (liposomes; Table 3.2). Other members of the Tgfβ superfamily such as Bmp-2 and Bmp-13 share this biological role and have fewer side effects than Tgfβ. Members of other growth factor families (interleukin-1, Egf, Tgf-α and tenascin-C) do not share the role (Table 3.2).

Vitamin D-deficient rat bone has less Tgfβ than normal bone; administration of 1, 25 vitamin D₃ increases Tgfβ by 100 per cent (Finkelman *et al.*, 1991).

Tgfβ-2

Tgfβ-2 enhances osteogenesis in neonatal (three-day-old) rabbit tibiae following three or five injections of 20 ng Tgfβ-2. Even under these *in-vivo* conditions, action of Tgfβ-2 is dose-dependent: three injections enhance cells synthesizing collagen type I mRNA, five injections enhance chondrocytes synthesizing type II collagen mRNA. Injecting Tgfβ-2 into adults promotes only type I collagen mRNA and then only in fibrous tissue (Critchlow *et al.*, 1995), demonstrating the context and temporal dependence seen with Tgfβ-1.

Tgfβ-2 knockout mice show defects in many neural crest-derived tissues (Fig. 25.3), defects that share no phenotypic overlap with *Tgfβ-1* or *Tgfβ-3* knockout mice (Sanford *et al.*, 1997).

Tgfβ-2 enhances osteogenesis of a bovine bone fraction that contains Bmp-2 and Bmp-3 (Bentz *et al.*, 1991).

Tgfβ-2 mediates the effects of Indian hedgehog (*IHH*) on hypertrophy and on expression of PTHrP. Chondrocytes secrete *IHH* and PTHrP as they mature and hypertrophy (Fig. 15.2). PTHrP acts downstream of *Tgfβ-2*: IHH → *Tgfβ-2* → PTHrP. IHH acts via the transmembrane protein Smoothened (SMO) and cyclin-D1 to regulate chondroblast proliferation.[d]

Tgfβ-3

Tgfβ-3 is expressed in many tissues and organs, including perichondria, bone, intervertebral discs, the pleurae of the lungs, heart, palate, amnion and the central nervous system. *Tgfβ-3* mRNA and protein do not always co-localize, suggesting both local (autocrine) and more distant (paracrine) regulation of mRNA and protein, and that Tgfβ-3 does not always act where it is produced. Tgfβ-3 is required for secondary palate fusion; mice lacking this growth factor have cleft palates but no other craniofacial anomalies.[e]

Tgfβ receptors

Iseki *et al.* (1995) localized Tgfβ type I and type II receptors during mouse development, TgfβR-II preferentially in preosteogenic mesenchyme and type I in the neural tube. Both receptors are expressed in teeth, bone, Meckel's cartilage and in the

Box 15.1 (*Continued*)

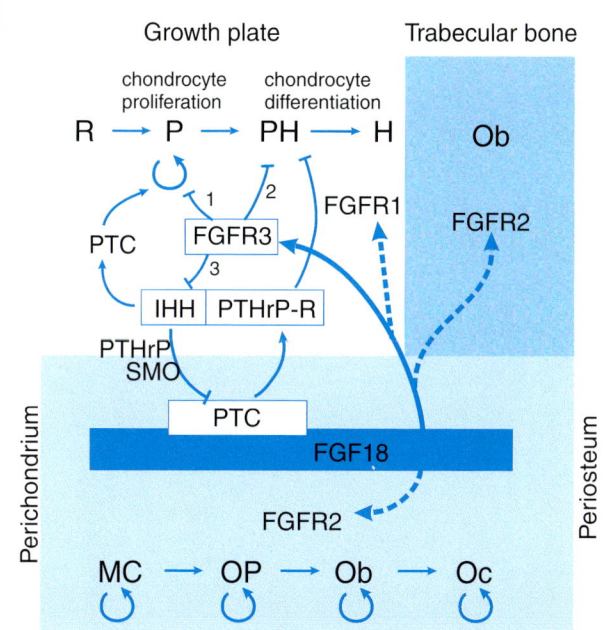

Growth plate   Trabecular bone

submandibular gland, indicating roles in both epithelial and mesenchymal morphogenesis.

Expression in mouse skeleton of a truncated kinase-defective Tgfβ type II receptor promotes terminal differentiation of chondrocytes and osteoarthritis resulting in progressive skeletal degeneration in which articular cartilage is replaced by hypertrophic cartilage with cartilage metaplasia at joints, the loss of response to Tgfβ promoting terminal differentiation.[f]

Inhibition of Tgfβ receptor signaling in osteoblasts leads to decreased bone remodeling and increased trabecular bone mass (Filvaroff et al., 1999).

[a] See Schmid et al. (1991), Jakowlew et al. (1991) and Thorp et al. (1992) for differential distribution. See Kingsley (1994a) for a review of the Tgfβ superfamily, new members and Bmps as known in the mid-1990s, and Jennings and Mohan (1990) and Ignotz (1991) for overviews of the actions of Tgfβ and for latent Tgfβ.
[b] See Villiger and Lorz (1992) for differential transcription, and Binkert et al. (1999) for fetuin.
[c] See Tanaka et al. (1993) and Binkert et al. (1999).
[d] See Alvarez et al. (2002) and F. Long et al. (2001b) for these studies.
[e] See Pelton et al. (1990a,b, 1991) for these classic studies, and Proetzel et al. (1995) for cleft palate. See Tsukahara and Hall (1994) for transmembrane signaling, and Kolodziejczyk and Hall (1996) for Tgfβ superfamily signaling.
[f] See Serra et al. (1997) for the type II receptor, Serra and Chang (2003) for an overview, and Sporn and Roberts (1990, 1992) for excellent reviews of Tgfβ and Tgfβ receptors I-III, III being beta-glycan.

Figure 15.2 A model of the role of Fgf-18 in controlling the growth of mammalian growth plates through *Patched* (PTC), Smoothened (SMO), PTHrP-R and FGF receptors 1 to 3. The model includes: (1,2) regulation of the transition from reserve cells (R) to proliferating and prehypertrophic chondroblasts (P, PH) via FgfR-3; (3) negative feedback to Indian hedgehog (IHH) via FGFR3; regulation of chondrocyte hypertrophy (H) by FGF-18 acting through FGFR-1, and of osteoblast differentiation (Ob) via FGFR-2, shown at the bottom as the transformation of mesenchymal cells (MC) → osteoprogenitor cells (OP) → osteoblasts (Ob) → osteocytes (Oc), with MC, OP and Ob shown as renewing cell populations (circular arrows). Adapted from Z. Liu *et al.* (2002).

tumour necrosis factor-α or interleukins, TGFβ-1 can induce human monocytes and RAW 264.7 cells to resorb bone using TRAP (see below). Neither osteoprotegerin nor antibodies against TNF-α or TNF-α receptors inhibited the TGFβ-1 effect.

Other factors also come into play: *osteocalcin* plays a role in recruiting osteoclasts or osteoclast precursors to bone for resorption. Implanting bone depleted of osteocalcin results in only 60 per cent of the resorption seen in control implanted bones, with only 35 per cent the number of mononuclear cells (osteoclast precursors) surrounding the implant. Osteoclast recruitment and subsequent resorption is enhanced when agarose beads loaded with Fgf-2 combined with bone fragments are grafted to the chorioallantoic membrane (CAM), implicating Fgf-2 in recruiting osteoclasts and resorption. Rat mandibular osteoblasts are more sensitive to resorptive agents such as prostaglandins than are osteoblasts from long bones.[9]

Coupling allows coordinated and appropriate reactions to external signals other than hormones; foetal mouse calvariae respond to intermittent compression with increased bone formation, decreased bone resorption and a net 16 per cent increase in bone mineral over a five-day period, a response comparable to that shown by growth-plate chondrocytes. Live (but not dead) cartilage recruits osteoblasts and osteoclasts, indicating an active role for (hypertrophic) chondrocytes in resorption and initial endochondral ossification.[10]

WHEN COUPLING GOES AWRY

What of coupling when ossification or osteogenesis goes awry? *Osteoporosis*, a well-known situation of *loss of bone mass* because of uncoupling between deposition and resorption, is discussed in Box 8.2. *Osteopetrosis*, in which *excess bone accumulates* because of uncoupling (Fig 15.3), is discussed later in this chapter.

Bone remodeling and osteoclast function is normal in parathyroidectomized suckling rats (the animals are

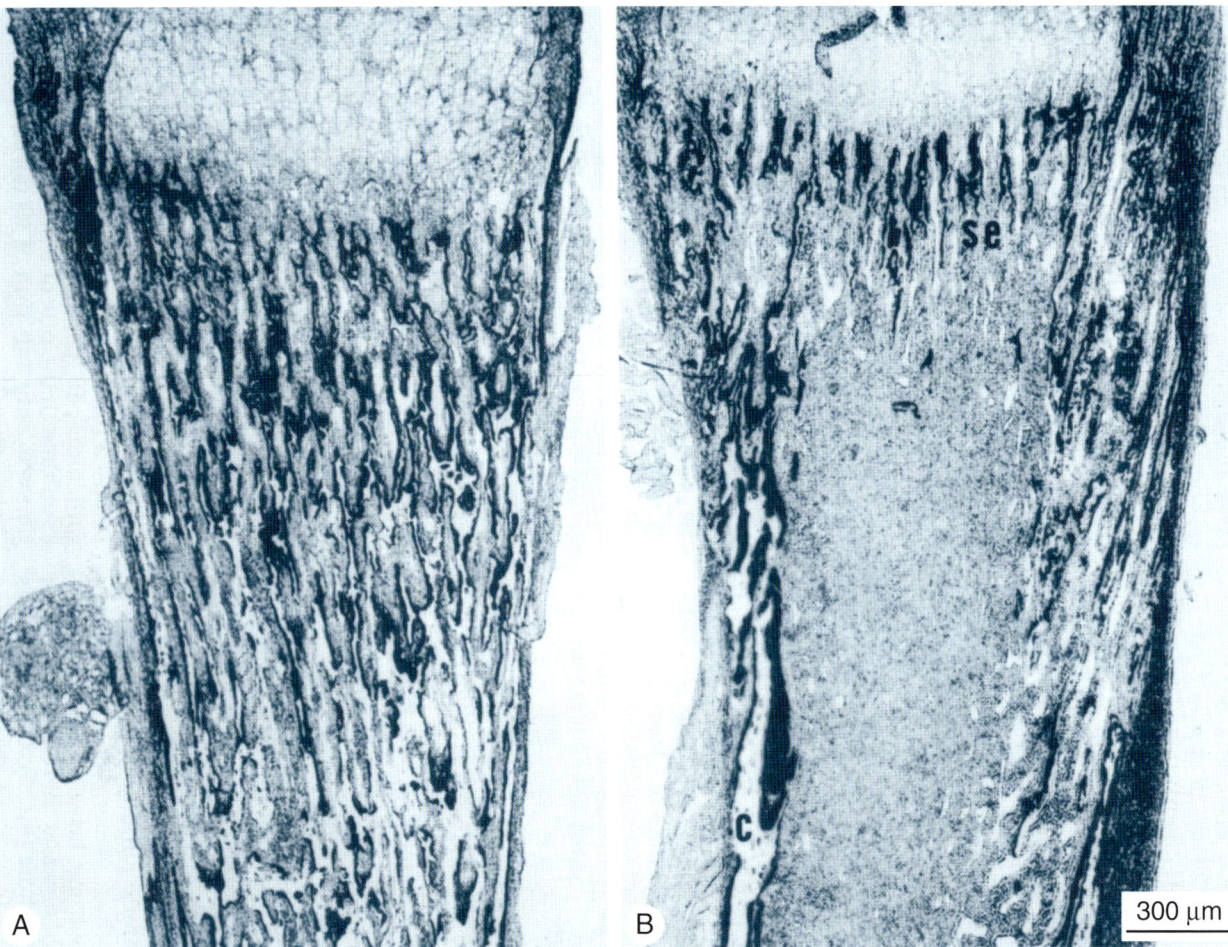

Figure 15.3 Osteopetrosis in mice. Longitudinal sections (growth plates at the top) through the proximal third of the tibiae from a 14-day-old micropthalmic (*mi/+*) mouse (A) and a wild-type (+/+) littermate (B). Both are autoradiographs taken six hours after administration of ³H-proline. The prominent marrow cavity in the wild type (B) is filled with trabecular bone in the mutant (A). Cortical bone is prominent in the wild type (B) but poorly developed in the mutant (A). Modified from Marks and Walker (1969).

hypocalcemic but skeletal development is normal) but not in adults, which led researchers to argue that a factor(s) other than PTH is controlling osteoclast recruitment in the foetal and neonatal stages. Vitamin B6 deficiency in rats uncouples the normal interaction between osteoblasts and osteoclasts and so leads to abnormalities in repair.[11]

TRAP-STAINING FOR OSTEOCLASTS

For over 20 years, *tartrate-resistant acid phosphatase* (*TRAP*)-*staining* has been an important and reliable method for identifying osteoclasts and their precursors. Conversely, absence of TRAP-staining – as for example associated with disrupted endochondral ossification and mild osteopetrosis – is an indication of deficient osteoclast activity and bone resorption. Kawata and colleagues use TRAP-staining to considerable advantage in their studies on *op/op* osteopetrotic (*op/op*) mice.[12]

Mammalian osteoclasts

A number of important findings have resulted from the application of TRAP-staining. One is that resorption in mammals is not carried out only by multinucleated osteoclasts, as shown in the following examples.

- The condylar cartilage in mice is resorbed by mononuclear and multinucleated cells, with TRAP-positive cells present from 16 days of gestation onwards. Ovariectomy of eight-week-old mice results in high levels of TRAP-positive cells in the mandibular condylar cartilage four weeks later.[13]
- Using foetal mouse metatarsals, Blavier and Delaissé (1995) followed TRAP-positive mononuclear cells from the periosteum through the osteoid and into the marrow.
- TRAP-staining facilitated the demonstration of connexin-43-mediated connections between rat osteoclasts cultured on bone slices, and between their mononucleated precursors. The gap junction inhibitor heptanal inhibits the number and activity of osteoclasts.

The proportion of *mononucleated* TRAP-positive cells also declines in the presence of heptanal, which may indicate a role for gap junctions in osteoclast fusion (Ilvesaro *et al.*, 2000).

Teleost osteoclasts

Bone resorption in *teleost fishes* is carried out by TRAP-positive *mono- and multinucleated cells* that use the same enzymatic pathways as do mammalian osteoclasts. Studies by Eckhard Witten and his colleagues described *mononucleated osteoclasts* from the Nile tilapia, *Oreochromis niloticus* (which has acellular bone). The mononucleated cells lack ruffled borders, are TRAP-positive, endosteal in location (osteoblasts are periosteal) and form aggregates that permit the intense resorption characteristic of bone in actively growing fish.[14]

In another approach to the study of teleost bone, mineral-containing bone particles transplanted intramuscularly into rainbow trout, *Oncorhynchus mykiss*, evoke formation of TRAP-positive bone-resorbing cells that 'attack' the implanted bone. TRAP is also used as a marker for scale resorption and effects of hormones such as oestradiol-17β on scale metabolism; oestradiol-17β induces osteoclastic activity.[15]

NITRIC OXIDE – IT'S A GAS

The suggestion that a gas, nitric oxide (NO), might influence cellular activity, even be produced by cells *in vivo*, has gone from being laughable, even ridiculed, to the establishment in 1997 of a journal (*Nitric Oxide*) to accommodate the growing number of findings concerning the roles of NO in cellular physiology.

Osteoclasts regulate their resorption, in part, by responsiveness to NO. Chicken osteoclasts cultivated on slices of bovine bone release NO and NO synthase. *In vivo*, osteoclasts produce NO synthase inhibitors. Osteoclasts cultured on bone slices enhance their synthesis of NO and reduce resorptive activity within 36 hours in response to nitroprusside, as evidenced by a decline in the number and area of bony pits. Expose the cells to inhibitors of NO and these parameters of bone resorption increase. NO production by osteocytes increases as one of the responses to mechanical load. The increased NO induces retraction of osteoclasts from bone surfaces and so reduces resorption.[16]

Human articular chondrocytes maintained *in vitro* synthesize NO synthase when exposed to interleukin-1β (IL-1β), tumour necrosis factor or endotoxin; the response to IL-1β is both concentration- and time-dependent, with a six-hour time lag. While NO synthase is not present in unfractured rat femora, it is synthesized by cells in the callus, and healing is suppressed if NO is suppressed. Maximal expression occurs 15 days into the healing process. Furthermore, application of exogenous NO to the fracture site results in a 30 per cent increase in the cross-sectional area of the callus 17 days post-fracture. Clearly, NO augments the repair process. NO activates guanylate cyclase and the second messenger cyclic guanosine monophosphate (cGMP).[17]

PROGENITOR CELLS FOR OSTEOBLASTS AND OSTEOCLASTS

How do osteoclasts arise? With this background on resorption, on coupling between osteoblasts and osteoclasts, and with an appreciation that mononucleated cells can resorb, we are ready to explore how osteoclasts arise. Whether a single or two separate populations of progenitor cells serve as precursors for osteoblasts *and* osteoclasts, and indeed, whether osteoblasts *arise from* osteoclasts were controversial but important issues for many decades.

My first foray into the osteoclast field (Hall, 1975b) was prompted by what I thought was a wrong-headed treatment of the origins of osteoclasts in *The Physiological and Cellular Basis of Metabolic Bone Disease* by Rasmussen and Bordier (1974). Almost three decades later, I have come back to osteoclasts through collaboration with Eckhard Witten from Germany on the formation of the kype in migrating male salmon.[18]

Because collagen fibres are not found within vacuoles in osteoclasts, Heersche (1978) reasoned that the later phases of resorption are extracellular and so proposed a *biphasic* mechanism of osteoclastic bone resorption involving *two cell types*:

- osteoclasts that degrade non-collagenous proteins, and
- mononuclear fibroblast-like cells – probably monocyte-derived – that phagocytose collagen fibres.

Direct evidence for the resorption of devitalized bone matrix by monocytes following stimulation with Pdgf was obtained a few years later.[19]

Shared properties such as hormone-binding sites and enzymatic pathways suggest affinities between various types of 'clasts' – osteo-, chondro-, odontoclasts – to be expected of a family of cells, but also suggest affinities to osteoprogenitor cells. Osteoprogenitor cells and osteoclasts possess calcitonin-binding sites, while feline osteoclasts *in vivo* respond to PTH by increasing nuclei number, presumably reflecting fusion with additional precursors. A histochemical method for making imprints of osteoclasts shows that osteoclasts from femora of 18-week-old kittens contain NADP, NADPH and glucose-6-phosphate, and succinic, malic, lactic and β-hydroxybutyrate dehydrogenases, enzymes also found in osteoblasts. Dentine is removed or resorbed by odontoclasts. Like osteoclasts, mammalian odontoclasts are multinucleated, the average number of nuclei/human odontoclasts being

7.8, five per cent of odontoclasts having 15 or more nuclei. In common with osteoclasts, mammalian odontoclasts contain β-hydroxybutyrate dehydrogenase.[20]

While osteoblasts can be distinguished from multi-nucleated osteoclasts, their precursors cannot readily be distinguished from one another. At most sites within the skeleton, deposition and resorption of bone occur side by side or sequentially (Fig. 2.3). In others, such as the femora of growing rabbits, bone deposition by osteoblasts and osteocytes occurs only on periosteal surfaces, bone resorption by osteoclasts only on endosteal surfaces. Similarly, cells in foetal rat bones respond differentially to prostaglandin E-2 (PGE-2) depending on whether they are periosteal – in which case the response is resorption *and* bone formation – or endosteal, in which case the response is resorption only (see Box 15.2). Periosteal progenitor cells are osteoblast precursors; endosteal cells are osteoclast precursors. Differential sensitivity of response is another complication; osteoclasts respond to concentrations of PGE-1 that are 200 times lower than those required to activate osteoblasts. Studies such as these, and her own investigations, led Maureen Owen to postulate separate precursors for osteoblasts and osteoclasts.[21]

Because of the evident clinical relevance, osteoclasts associated with the alveolar bone at the base of teeth have attracted attention. Alveolar bone (Chapter 25) is remodeled during normal physiological drifting of teeth in their sockets, and remodeled even more aggressively and rapidly when orthodontic appliances are applied. Bone is deposited on the leading edge of a tooth as osteoprogenitor cells respond to mechanical stresses arising from tooth movement. Bone is resorbed on the trailing surface as progenitor cells differentiate into osteo*clasts*, giving the appearance of differential distribution of progenitor cells for osteoblasts and osteoclasts. However, if the direction of tooth drift is reversed – for example, by inserting an orthodontic appliance – the pattern of progenitor cell differentiation also reverses. It *appears* that the same progenitor cells can oscillate between osteoblast and osteoclast differentiation; however, two intermingled subpopulations of progenitor cells could surround each tooth socket. A similar possibility arises in oestrogen-stimulated resorption of the pelvic girdle in pregnant mice (as discussed in Chapter 25).[22]

A number of workers claimed they could distinguish preosteoblasts from preosteoclasts on ultrastructural grounds. Young and Kember interpreted their studies

---

Box 15.2 Prostaglandins

Prostaglandins – there are around 20 – are modified fatty acids (and thus lipids) with major roles in reproduction and inflammation. Aspirin reduces inflammation by inhibiting synthesis of prostaglandins. Named for their chemical similarity, A1–A3 being related chemically to one another, B1 related to B2, and so forth, the usual acronym for a prostaglandin is PG followed by the specific prostaglandin, thus PGE-2 for prostaglandin E-2.

Although often spoken of as hormones or as having a hormone-like action, hormones act at a distance from their source of production, while prostaglandins act locally. Furthermore, most hormones have similar actions in different cells, thyroxine and amphibian metamorphosis being a major exception – thyroxine initiates tissue-specific programmes of gene expression during metamorphosis in anurans.[a] Prostaglandins often have different actions in different cells, which makes sorting out their functions difficult. To take three examples:

- rat mandibular osteoblasts are more sensitive to resorptive agents such as prostaglandins than are osteoblasts from long bones;
- depending on the bone surface, bone-lining cells in foetal rats respond differentially to PGE-2: periosteal osteoblasts respond with resorption and bone formation, endosteal cells with resorption only. Periosteal osteoblasts, which are targets for prostaglandins, are removed from bone surfaces, exposing those surfaces to osteoclastic resorption;
- osteoclasts respond to concentrations of PGE-1 that are 200 times lower than those to which osteoblasts can respond.[b]

Monocytes produce PGE-2, which enhances the release of mineral from bone matrix by increasing osteoclast numbers, the extent of their ruffled borders, and Ca[45] release. PGE-2 also is synthesized in chondrogenic condensations and as chondrocytes differentiate, functioning as a cAMP-mediated signal. Examination of prostaglandin concentrations in skeletal tissues during endochondral ossification shows high levels of PGE-2 in cartilage and bone.[c]

Mechanical loading

Finally, synthesis of prostaglandins is enhanced with mechanical loading.

The canalicular cell processes that connect osteocytes are an important physiological transport system. Mechanical loading induces fluid flow through canaliculi of human and murine calvarial and iliac crest osteoblasts, and stimulates the production of prostaglandins and the synthesis of major enzymes for prostaglandin production, indicating similar mechano-transduction pathways in the two species and bone types (Joldersma et al., 2000).

Mechanical stimulation is modulated by PGE-2 and cAMP to couple bone formation and resorption. There is a response within 20 minutes of altering the mechanical conditions *in vitro*. PGE-2 mimics application of force by elevating cAMP levels, indicating PGE-2 as the signal (Somjen et al., 1980). So, we see that prostaglandins are major players in transducing mechanical signals.

---

[a] *Xenopus* displays sexual dimorphism in the larynx, both myogenesis and chondrogenesis being dependent on androgens (Sassoon et al., 1986). Androgen-dependent changes in the larynx at metamorphosis are controlled by thyroid hormone (Robertson and Kelley, 1996). See D. D. Brown et al. (1996) for up-regulation of four transcription factors, fibronectin, integrin, four proteinases and deiodinase, in association with what they call the 'thyroid hormone-induced tail resorption program' at the outset of metamorphosis in *X. laevis*. See Tinsley and Kobel (1996) for the biology of *Xenopus*, and Shi (2000) for amphibian metamorphosis.

[b] See Hirata et al. (1983) for sensitivity and Wong and Kocour (1983) for concentration effects.

[c] See Dominguez and Mundy (1980) and Rifkin et al. (1980a) for monocytes, Gay and Kosher (1984, 1985) for chondrogenesis, and Wientroub et al. (1983) for endochondral ossification.

with pulse labeling of ^{3}H-thymidine as indicating that the *same* progenitor cell could become an osteoblast or an osteoclast. My conclusion at the close of these studies (Hall, 1975b) was that osteoclast progenitor cells are blood-borne and extraskeletal, and not sessile, intraskeletal cells, and that these migratory progenitors are a *separate* population from those osteogenic progenitors that differentiate into osteoblasts.[23]

Despite the ingenious isotopic studies of Young, Kember and others, the number of types of progenitors within the skeleton remained difficult to resolve. Ultrastructure could not unequivocally discriminate progenitor cells on morphological grounds, and the labeling studies were compromised by possible reutilization of label, especially in a system where cells fuse to generate multinucleated cells. Several groups turned to a different method – the distinctive nuclear marker in the cells of Japanese quail (Fig. 15.4) – and, *perforce*, to different model systems.

Japanese quail–domestic fowl chimaeras

Among the first studies to take advantage of quail–chick chimaeras was the grafting of limb buds, tibiae or femora from Japanese quail or mouse embryos onto the CAMs of domestic fowl. In all three approaches, progenitors from the host embryo migrated to the graft, invaded it and formed osteoclasts. Some osteoclasts contained *both* graft and host nuclei, confirming that osteoclasts incorporate new cells as they differentiate and grow via cell fusion. In contrast, the osteoblasts in these grafts are all of graft, i.e. local, origin.

A conceptually similar approach was used when quail long bones or fractured long bones were grafted onto the CAM of chick embryos (see Box 12.3 for the applications of the CAM-grafting technique). Osteoclasts of host (chick) origin form in the grafted quail bones; the osteoclasts arise from haematopoietic cells. This is so unless the CAM-grafts fail to vascularize, in which case osteoclasts do form, but they are of graft origin; perhaps, it was thought, having arisen from hypertrophic chondrocytes in the grafts. A fortuitous bonus from these studies was that the grafts were taken from seven-day-old quail embryos, which is well before osteoclasts arise in the primordia of long bones. Onset of what is precocious osteoclast differentiation in the grafts indicates that osteoclast precursors are present and can respond to signals that mediate osteoclast differentiation. These conclusions were substantiated when it was shown that host (chick) cells respond to mineralized bone grafted to the CAM by forming osteoclasts (complete with ruffled borders) within three days.[24]

These studies utilizing CAM-grafting, along with studies using *gray-lethal* and osteopetrotic mice given transfusions of stem cells (see below), support the conclusion that during normal skeletal development osteoclasts arise from migratory progenitor cells, and osteoblasts from a *separate* population of local progenitor cells.

OSTEOPETROSIS AND OSTEOCLAST ORIGINS

Osteopetrosis is characterized by the accumulation of excess bone resulting from a combination of *increased*

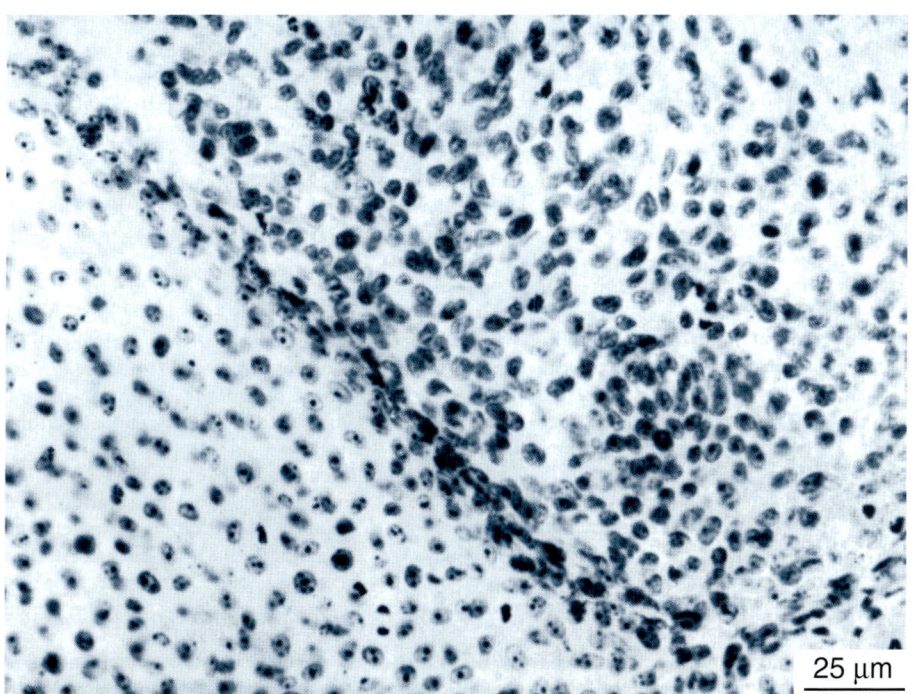

Figure 15.4 Cells of the Japanese quail, *Coturnix coturnix japonica*, possess a nuclear heterochromatin marker. Japanese quail wing-bud mesenchyme (H.H. stage 22) was recombined with chick wing-bud mesenchyme (H.H. stage 14) and cultured for seven days. Quail mesenchyme has formed cartilage, distinguishable by the heterochromatin marker in the quail cells (left). The sharp boundary from lower right to upper left demarks quail from chick cells. From Gumpel-Pinot (1982).

25 μm

bone formation and *reduced* bone resorption, tilting the normal equilibrium between deposition and resorption to the point where marrow cavities can become filled with bone and obliterated, and teeth fail to erupt. Osteopetrosis is an ancient pathology of bone, as evidenced by its presence in many different vertebrates, and by the similarity to avian osteopetrosis of a periosteal bone lesion in a long bone from a dinosaur that lived 70 mya (Fig. 15.3).[25]

Genetic models based on mutations that produce osteopetrosis are available and have provided much of our knowledge of the disease and what it can tell us about osteoclast origins. For example, incorporation into stem cells in chimaeric mice of a null mutation of the proto-oncogene *c-fos* has pleiotropic effects, including decreased body weight in all chimaeric animals, and osteopetrosis in 40 per cent (see Box 15.3). The latter animals display all the classic features of mutation-based osteopetrosis, including ossification of marrow cavities and absence of tooth eruption.[26]

The late Sandy Marks Jr – whose sudden death in 2002 robbed skeletal biology of one of its most ingenious and persistent researchers – studied osteopetrotic mutant mice for over 30 years. His studies, especially of the *incisor absent* (*ia*) osteopetrotic strain of rats, are a model for how to proceed from disease state, to cellular, to molecular control. I present his studies below more or less in the order they were published, to show the development of this knowledge base. Marks and his colleagues demonstrated that:

- the increased rates of bone matrix formation and hypocalcemia in osteopetrotic mice correlate with hyperplasia of the parafollicular cells of the thyroid gland, the source of an osteoblast-stimulating factor, possibly thyrocalcitonin (Marks, 1969; Marks and Walker, 1969);
- porcine thyrocalcitonin does not affect bone formation in rodents (Marks, 1972);
- osteopetrosis in *ia* rats is due to reduced bone resorption because of reduced osteoclast function (Marks, 1973);
- there is a discrepancy between measurements of bone resorption *in vivo* and *in vitro*, the rate of resorption in newborn osteopetrotic rats being 35 per cent *less* than control but the rate measured *in vitro* being 25 per cent *higher* than control, the difference arising from nonequivalence in release of acid phosphatase *in vitro* and *in vivo* (Marks, 1974);
- *ia* rat bone arises because of a cellular, not a humoral imbalance; grafting osteopetrotic bone in the presence or absence of serum or hormones shows that osteopetrotic osteoclasts cannot respond to normal hormonal stimuli (Nyberg and Marks, 1975);
- osteopetrosis in *ia* rats can be 'cured' by injecting spleen cells from a normal littermate, the spleen cells

providing a haematogenous source of osteoclasts (Marks, 1976);
- there is a direct relationship between lymphocytes, monocytes and osteoclasts, shown by restoring bone resorption in *ia* rats – osteoclasts with ruffled borders appear and resorb bone – by injecting spleen or thymus lymphocytes and monocytes from normal littermates, the injected cells either activating osteoclasts in the *ia* rats or producing osteoclasts directly (Marks, 1978a; Marks and Schneider, 1978);
- lymphoid or stem cells from liver, thymus or bone marrow can effect the cure, while transformation of cells in the osteoclast lineage is an early marker for the cure (Marks, 1978b, 1983; Marks and Schneider, 1982);
- reduced bone remodeling in osteopetrotic rats compromises fracture repair (Marks and Schmidt, 1978);
- tooth eruption depends on alveolar bone resorption and not on growth of the root or crown or epithelial attachment;[27]
- osteoclasts have a haematogenous origin, shown by treating microphthalmic (*mi*) mice – which are osteopetrotic – with spleen cells from beige mice, whose cells contain a giant lysosomal granule used as a cytoplasmic marker to trace osteoclast precursors (Marks and Walker, 1981; Marks and Seifert, 1985);
- reduced bone formation in another osteopetrotic mouse (*op*), which cannot be cured by transplanting spleen or marrow cells from normal littermates, is caused by deficient osteoclast precursors – which have abnormal lysosomal and vacuolar systems – and consequent compensatory hypertrophy of the ruffled borders of existing osteoclasts (Marks, 1982; Marks et al., 1984; Marks and Popoff, 1989), implying a feedback between mature or maturing osteoclasts and osteoclast precursors;[28]
- *osteoblast* abnormalities are associated with three osteopetrotic rat mutations (Shalhoub et al., 1991);
- a fourth mutation in rats, *toothless* (*tl*), which has reduced cell numbers and collagen type III in the premaxillary–maxillary suture – with consequent stunted craniofacial growth – can be rescued by treatment with colony-stimulating factor-1 (Marks et al., 1999);
- the *tl* mutation in rats is associated with progressive development of chondrodystrophy between three and six weeks after birth, a defect characterized by disorganized chondrocytes in the growth plate, failure of cartilage mineralization, and failure of cartilaginous extension into the metaphysis; mineralization of bone is normal. At two weeks after birth, collagen type X is not expressed in the few hypertrophic chondrocytes that do develop, implicating the *tl* locus in regulation of collagen X expression/function (Marks et al., 2000);
- the local bone environment in osteopetrotic rodents cannot support osteoclast migration, localization or differentiation (Marks, 1984); and that

Box 15.3 Proto-oncogenes

Proto-oncogenes are eukaryotic gene copies of tumour-inducing retroviral oncogene sequences that are involved in many developmental processes but are not themselves tumorigenic. A number of proto-oncogenes play important roles in skeletogenesis.[a]

c-fos

c-fos is involved in skeletogenesis through physiological roles in developing cartilages, bones and teeth. In osteogenesis, c-fos:

- is expressed in murine perichondria, membrane and endochondral bone, and dentine;
- when down-regulated interferes with normal bone development in transgenic mice;
- is co-regulated with the growth-factor-inducible gene Egr-1 in membrane, alveolar, periosteal and endochondral bone, and in cartilage, muscle and teeth of foetal mice;
- mediates the mitogenic effect of Tgfβ on osteoblastic cells and osteoblast maturation of osteoblasts; and
- is expressed in osteoblasts between days 10 and 28 in healing rat tibiae.[b]

c-fos also affects chondrogenesis and may play a role in modulating the transformation of certain chondrogenic cells to osteoblasts and/or of transforming extracellular matrices. Four examples follow.

(i) Expression of c-fos precedes osteogenesis of mouse mandibular condylar chondrocytes in vitro. During replacement of condylar cartilage by bone, hypertrophic chondrocytes display features of osteoblasts at the TEM level and deposit osteoid. c-fos is expressed in hypertrophic chondrocytes before they initiate DNA synthesis and transform (Closs et al., 1990).

(ii) When a null mutation of c-fos was incorporated into stem cells of chimaeric mice, pleiotropic effects ensued, including decreased body weight in all animals. Forty per cent of the mice had osteopetrosis, a disorder that includes ossification within the marrow cavities of the long bones. The teeth failed to erupt, and chondrocytes were trapped in a bone matrix characterized by features of cartilaginous matrix such as strong alcian blue staining.

(iii) Chimaeric mice generated using embryonic stem cells that express c-fos were found to produce cartilage tumours, c-fos having transformed the chondrogenic precursors and the level of c-fos being highest in the tumours. Established cell lines from these chimaeric mice express c-fos and type II collagen and produce cartilage tumours (sometimes with bone) when transplanted in vivo.[c]

(iv) Normally the expression of collagenase-3 (MMP-13) in mice is restricted to hypertrophic chondrocytes. However, osteoblasts in c-fos-induced osteosarcomas secrete MMP-13 (Tuckermann et al., 2001).

Parallel changes in two proto-oncogenes, c-fos, c-jun, and a proline-rich transcript of the brain (prtb, a serum-responsive gene) are associated with osteoblast adhesion in vitro. Expression of all three increases 30 minutes after adhesion of MC3T3-E1 cells, and is independent of proliferation during the first days in culture, of synthesis and deposition of extracellular matrix between seven and 14 days in culture, and of mineralization between 14 and 31 days (Sommerfeldt et al., 2002).

c-myc

A second proto-oncogene, c-myc, is found in all chondrocytes of chick and rat long bones. The most intense expression occurs in both dividing and hypertrophic chondrocytes (see below). Expression in proliferating chondroblasts correlates with BrdU-labeling, indicating a role in proliferation. Expression in hypertrophic cells indicates a second role in terminal differentiation of chondrocytes, a role supported by the finding that over-expression of c-myc in quail tibial chondroblasts impairs hypertrophy and mineralization. c-myc initiates dedifferentiation of terminal hypertrophic chondrocytes, which cease synthesizing type X collagen, have low levels of alkaline phosphatase, reinitiate cell division, and begin to deposit type II collagen.[d] In the cephalic portion of the embryonic chick sterna, however, c-myc is found in dividing but not in post-mitotic cells, a pattern that correlates with lack of hypertrophy in these chondrocytes.

Given the role that c-myc plays in proliferation and dedifferentiation it is not surprising to find that expression is enhanced sevenfold in epithelial and mesenchymal tissues during forelimb regeneration in young Xenopus laevis. Furthermore, levels of c-myc in non-proliferating chondrocytes are depressed in tibiae from dyschondroplastic chicks, further suggesting that c-myc plays a role in the transit of cells from the chondroprogenitor to the proliferative pools. Tgfβ also is depressed in these cells. A regulatory role for vitamin D_3 is indicated by the finding that expression of c-myc returns to normal after adding 1, 25 vitamin D_3 to the diet of dyschondroplastic chicks.[e]

c-fms

c-fms, the receptor protein of the CSF-1 (colony stimulating factor-1) proto-oncogene, is expressed on multinucleated cells around erupting molar teeth in rats. Kawakami et al. (1991) believe that it enhances recruitment of osteoclasts to the resorption surface of erupting teeth.

[a] See Maclean and Hall (1987), Grigoriadis et al. (1995), and Gilbert (2003) for oncogenes and proto-oncogenes.
[b] See Dony and Gruss (1987) and Caubet and Bernaudin (1988) for expression patterns in mice, Rüther et al. (1987) for the transgenic study, McMahon et al. (1990) for Egr-1, Machwate et al. (1995a,b) for Tgfβ, Ohta et al. (1991) for healing tibiae, and Grigoriadis et al. (1995) for a review of c-fos and bone development, including transgenic and knockout approaches.
[c] See R. S. Johnson et al. (1992) and Z.-Q. Wang et al. (1991) for the two studies with chimaeric mice. Having heart and lung defects as well as abnormal limb buds and central and peripheral nervous systems, homozygous n-myc⁻ mice die between 10 and 12 days of gestation. No distal limb elements form from cultured limb buds, a result of disrupted tissue interactions (Sawai et al., 1993).
[d] See Farquharson et al. (1992) for expression in long bone chondrocytes and BrdU-labeling, and Quarto et al. (1992b) for over-expression.
[e] See Iwamoto et al. (1993) for expression in sternal chondrocytes, Géraudie et al. (1990) for limb regeneration in Xenopus, and Loveridge et al. (1993) for the study with dyschondroplastic (avian tibial dyschondroplasia) chicks.

- mouse osteoclasts have a lifespan of six weeks unless they fuse with osteoclast precursors, in which case lifespan is prolonged potentially indefinitely with subsequent fusions (Marks and Seifert, 1985).

Other studies on osteopetrotic mice, such as those by Nisbet and colleagues on microphthalmic (mi) mice (Fig. 15.3), demonstrate that grafting normal marrow into osteopetrotic animals results in formation of normal bone without osteopetrosis, while injecting marrow (but not thymus) stimulates resorption in the Fatty/ORL-*op* rat. *Microphthalmic* mice are characterized by an early fusion defect in osteoclast precursors and in macrophage formation. Consequently, 'inefficient' mononuclear cells – cells that contain acid phosphatase but lack ruffled borders – resorb bone, but only weakly. Loss of the CIC-7 chloride channel protein, which operates in the ruffled border of osteoclasts in wild-type individuals, is associated with onset of osteopetrosis in mice.[29]

Parallel studies from Kawata and colleagues in Japan were undertaken with *op/op* osteopetrotic mice, which they refer to as 'toothless.' The osteoclast deficiency has much more widespread effects than those on subperiosteal ossification. Generalized craniofacial growth, especially mandibular growth and growth of the masticatory regions, is disrupted. The condylar head is small and composed of hypertrophic cartilage rather than bone, reflecting failure of initiation of endochondral replacement, interpreted by Kawata and colleagues as resulting from decreased numbers of osteoclasts in conjunction with insufficient mechanical stress from mastication. Because teeth fail to erupt and the functional stresses associated with feeding are reduced, changes are seen in the nasopremaxillary suture, while the nasal bones and premaxilla are deformed, emphasizing the important role of the root of the incisor for growth of the upper jaw. Lack of bone remodeling is associated with microdontia; tooth eruption either fails or is drastically reduced, and no TRAP-positive cells are found on the bone surfaces, leading to absence of bone remodeling and ankylosis of dentine and alveolar bone.[30]

OSTEOCLAST–PHAGOCYTE–MACROPHAGE OR OSTEOCLAST–MONOCYTE LINEAGES?

Substantial differences of opinion or interpretation have been expressed on the topic of whether osteoclasts can or always arise from phagocytes, and/or from macrophages or monocytes. The easy way out is to say that such differences of opinion/interpretation reflect differences between species, or different ways of gathering evidence, both of which may be true. Several lines of evidence target phagocytes/macrophages and exclude monocytes as osteoclast precursors. Some of the evidence is direct, some more inferential and some by exclusion. The types of studies undertaken are presented and discussed below.

Phagocyte/macrophage origin

Macrophages can be induced to synthesize and secrete collagenase, the presumption being that such a cell could be an osteoclastic precursor, but it is just as likely that such a cell can resorb the collagenous portion of bone matrix itself, especially if the matrix is osteoid or not

mineralized heavily. Bones produce a factor that is chemotactic for macrophages, and macrophages can erode cultured mouse calvariae, reinforcing the notion that macrophages play a direct role in resorption, whether or not they function as osteoclast precursors.[31]

PTH and calcitonin affect osteoclast function. Using the argument that related cells share similar molecules, the localization of receptors for calcitonin and PTH on mouse mononuclear *phagocytes* was thought consistent with phagocytes being migratory osteoclast precursors. Using the same reasoning, the lack of Fc receptors on osteoclasts – or on osteoblasts, for that matter – but their presence on macrophages and on the cell layer immediately adjacent to osteoclasts would argue against macrophages or cells on bone surfaces being osteoclast precursors.[32]

At about this time (the early 1980s), Tim Chambers undertook an extensive series of studies culturing mammalian (usually rabbit) cells in search of osteoclast precursors. Among his major findings are:

- osteoclasts arise by fusion of mononuclear phagocytes;
- some factors, such as osteoclast-activating factor (OAF), act on osteoclasts to stimulate bone resorption directly;
- other factors, such as PTH, activate osteoclasts or osteoclast precursors via action on osteoblastic cells, i.e. *osteoblasts regulate osteoclastic activity*; while
- calcitonin (10–50 pg/ml) suppresses osteoclastic activity by binding to a trypsin-sensitive calcitonin receptor, so that activity of cell projections (lamellipodial) ceases; osteoblasts can overcome this inhibition to release osteoclasts from calcitonin-induced quiescence.[33]

Not only did these studies bring other cells into play, they changed the ground rules. Osteoblasts regulate mammalian osteoclastic activity *in vitro* as they do *in vivo*. Any search for osteoclastic precursors that sidelines osteoblasts risks a yellow card.

Emphasis switched to more realistic approaches in which as much as possible of the normal, complex, three-dimensional interaction and architecture associated with resorption was retained. So, a study on resorption of bone implanted into calvarial defects was used to argue for a macrophage origin. Co-culture of periosteal-free embryonic bone with phagocytes demonstrated the production of osteoclasts from proliferating mononuclear phagocytes and rendered monocytes or macrophages less likely or even unlikely precursors. Osteoclasts were shown to resorb dentine when the two were co-cultured.[34]

Isolation of F4/80, a macrophage-specific antigen in the centre of mouse haematopoietic islands, cells lining periostea, connective tissue and synovial membranes, but not in fibroblasts, chondroblasts, osteoblasts, osteocytes or osteoclasts, further distanced the possibility that macrophages are osteoclast precursors. In fact, macrophages release a 43 000 kDa protein that *stimulates osteoblast growth.* Applying this protein increases the DNA content – a surrogate for cell number – of rodent calvarial osteoblasts and chondroblasts, but not fibroblasts; i.e. the specificity is to skeletal cells.[35]

Finally, isolation of 11 monoclonal antibodies, seven of which distinguish osteoclasts from macrophages, strongly supports the argument that osteoclasts arise from a separate cell lineage (Horton *et al.*, 1985a,b). Does that cell lineage consist only of mononuclear phagocytes, or can monocytes form osteoclasts?

Interleukins

Interleukins, of which there are 18 – interleukin-1 (IL-1) through interleukin-18 (IL-18) – are molecules of the immune system which direct other immune cells to divide and differentiate, i.e. interleukins are cytokines. Most are synthesized by leukocytes, some by polymorphonuclear phagocytes. All act on leukocytes, each interleukin activating a subset of cells with the appropriate receptor. Perhaps surprisingly, a number of interleukins activate skeletal cells or macrophages, and so stand in line to play a role in osteoclast function if not origin.

IL-1

A role for IL-1 in skeletal cells is explained in part by the range of cell types that synthesize this interleukin, including fibroblastic, endothelial and B cells, as well as macrophages, the latter relating to bone resorption. Many of the processes influenced or regulated by IL-1 – inflammatory responses, production of prostaglandins, release of corticosteroids – directly relate to various skeletal functions, notably fracture repair and resorption. IL-1 enhances production of IL-2.

Osteogenesis

More ectopic bone is induced in response to Bmps implanted into mice given IL-1 than in controls (Fujimori *et al.*, 1992), presumably because IL-1 suppresses the immune system, implicating IL-1 in both osteogenesis and arthritis.

IL-1 is also involved in some situations where mechanical load is translated into a cellular response, although the direction of the IL-1 response is not always the same. Compressive force inhibits IL-1β as it promotes expression of the transcriptional factor *Sox-9*, type II collagen and aggrecan in cultured limb-bud mesenchyme.[36]

Sutural cartilage ossifies early and secondary cartilage develops in the intermaxillary suture of rats in which the buccal muscles have been displaced (Fig. 15.5). Type I collagen-rich fibrous tissue develops in rat midpalatal sutures after four days of applied expansive force, after which mineralized membrane bone is deposited. The interpretation is that tension modulates the transformation of sutural secondary cartilage to fibrous tissue and then to bone under the influence of IL-1; Saitho *et al.*

(2000) showed that compressive force promotes chondrogenesis and hypertrophy (type X synthesis) in midpalatal sutural cartilage in four-week-old rats. Compressive force promotes expression of *Sox-9*, type II collagen and aggrecan (and inhibits Il-1-β) in cultured limb-bud mesenchyme from 10-day-old mouse embryos, promoting chondrogenesis, *Sox-9* activating type II collagen expression.[37]

Extensive ossification is initiated when sheep mandibles are repositioned surgically (as would be done in advancing or retracting the jaws in humans to correct malocclusion), osteogenesis being accompanied by the synthesis of much IL-1, Tgfβ and Fgf-2 (Farhadieh *et al.*, 1999).

Fujimura *et al.* (2002) investigated the effects of various concentrations of human recombinant FGF-2 (RhFGF-2) – between 2 μg and 400 ng – three weeks after bone had been induced in the muscles of rats via 2 μg of *RhBMP-2* in a type I collagen carrier being implanted into the muscles. Radiopacity (a measure of mineralized bone), the size of the bone in the muscles and alkaline phosphatase levels (an index of amounts of mature bone) all are greater after higher than lower doses of FGF-2, with strong indications of a biphasic response to FGF-2 (Table 15.1).

Chondrogenesis

Human femoral cells *in vitro* respond to IL-1 and IL-2 with increased proliferation and protein synthesis, indicative of the cytokine function of these interleukins. Human articular chondrocytes *in vitro* respond to IL-1β with synthesis of NO synthase. The response is not pharmacological; it tracks IL-1β concentration.[38]

Synthesis of collagen types IX and II is coordinately regulated in avian and human chondrocytes, in part by pretranslational regulation by IL-1. The synovium releases IL-1, and the IL-1 receptor on rabbit articular chondrocytes is down-regulated by Tgfβ.[39]

IL-6

A major function of IL-6 is inducing B cells to differentiate into antibody-forming plasma cells. Murine osteoblasts

Table 15.1 Response of bone induced in rat muscle to various levels of RhFGF-2[a]

FGF-2 concentration[b]	Radiopaque area (mm²)	Bone area (%)	Alkaline phosphatase (IU/mg protein)
0	2.8	15.2	0.1
2 μg	3.5	20	0.1
10 μg	1.7	7.6	0.75
50 μg	0	0	0.02
16 ng	5.4	30.4	6.5
80 ng	7.7	28.2	705
400 ng	9.1	32.5	17

[a] Based on data in Fujimura *et al.* (2002).
[b] Note that concentrations range from two micrograms (μg) to 400 nanograms (ng).

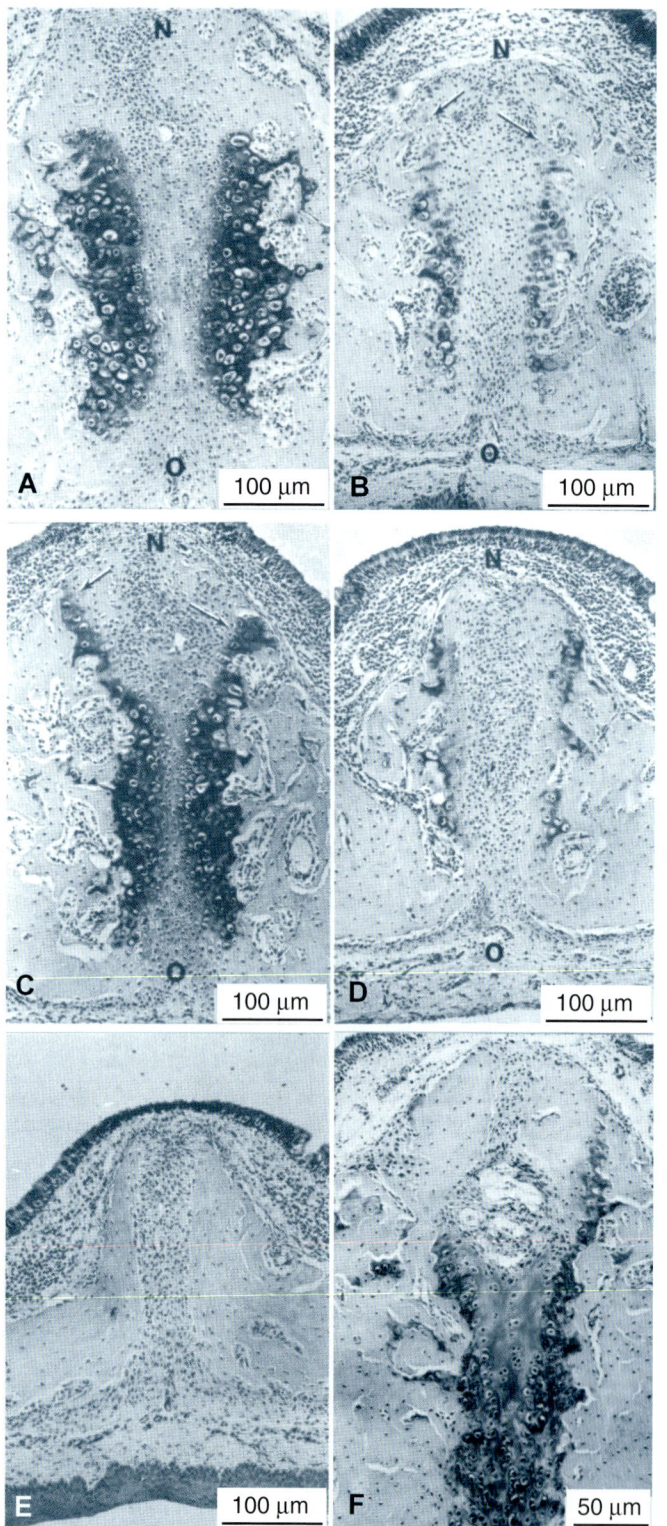

produce IL-6 in response to PTH (Rouleau *et al.*, 1988; Feyen *et al.*, 1989).

IL-10

As its major functions IL-10 activates B cells (as does IL-6) and stimulates macrophages to produce IL-1, IL-6 and tumour necrosis factor-α (Tnf-α). Thus, IL-6 is a key player in macrophage activation. Within the skeleton, IL-10 inhibits the Tgfβ synthesis required for murine bone marrow cells to become committed as osteogenic (Van Vlasselaer *et al.*, 1994).

Evidence against monocytes

Data from osteopetrotic mutants were discussed earlier in this chapter. Three other approaches used to investigate the role of monocytes were as follows.

- Tibial endostea isolated from 19-day-old chick embryos and cultured for 10 days became populated with multinucleated cells that arise from monocytes, but lack the typical morphology of osteoclasts. As they also lack acid phosphates, they seem unlikely to be resorptive cells.
- From *in vitro* studies with haematopoietic cells, it was argued that monocytes are not osteoclast precursors.
- Thirdly, receptors for 1, 25-dihydroxyvitamin D_3 cannot be detected on osteoclasts from calcium-deficient chickens, although they can be detected on monocytes, further distancing the two cell types from one another.[40]

Evidence for monocytes

Kahn *et al.* (1978) provided *direct* evidence for resorption of bone by transforming monocytes. Of course, this is not the same as showing that monocytes are osteoclast precursors, or that monocytes and osteoclasts have the same precursor, but such studies prompted searches for precursors in the monocyte lineage.

Monocytes resorb bone by *stimulating osteoclastic* activity. Feline bone cultured with monocytes or monocyte-conditioned medium releases 80 per cent more mineral than does control bone. The mechanism – monocyte production of prostaglandins – can be reversed by such inhibitors of osteoclast function as

Figure 15.5 The response of rat intermaxillary sutures to muscle displacement as depicted in frontal histological sections. (A, C, F) are controls. (B, D, E) are rats in which the buccal musculature was displaced at 30 days. (A) Suture from a control 35-day-old rat to show extensive mineralized cartilage stained with alcian blue. (B) Suture from a 35-day-old rat in which the buccal musculature had been displaced five days earlier, showing the substantial decrease in the amount of cartilage present (cf. A). (C) Suture from a control 45-day-old rat to show extension of cartilage (cf. A) over that present at 35 days (A), especially toward the nasal (N) side of the suture. (D) Suture from a 45-day-old rat in which the buccal musculature had been displaced 15 days earlier. Cartilage is reduced and ossification advanced over that seen in the control (C). (E) Suture from a 70-day-old rat in which the buccal musculature had been displaced 40 days earlier. Cartilage is absent entirely and the suture is open. (F) Suture from a control 100-day-old rat to show the further development of cartilage (centre). Modified from Ghafari (1984).

6×10^{-3} M phosphate, 10^{-6} M cortisol, or calcitonin at 50 mg.[41] Prostaglandin E_2 increases osteoclast numbers, the extent of ruffled borders and the amount of Ca^{45} released from cultured foetal long bones, all of which are indicative of actions on cell differentiation and function. Monocytes stimulated with Pdgf resorb devitalized bone matrix.[42]

These findings argue for a commonality of function between monocytes and osteoclasts. They do not speak to origins. Such approaches and the ability to separate cell populations on the basis of their differential sensitivity to PTH or to calcitonin in vitro are among methods of choice for identifying lineages of osteogenic precursor cells.[43]

Injecting leukocytes from female rats into males and using the presence of the Barr body (a male sex chromatin marker) in osteoclast nuclei provided a strong argument for monocytic origin for osteoclasts. Tissue-culture approaches also allow osteoclast precursors to be identified: mononucleated cells in marrow cultures give rise to multinucleated cells with ruffled borders; cultures of monocyte-rich blood cells resorb dentine. Radiation chimaeras with Beige mouse – in which giant lysosomes in monocytes and osteoclasts serve as a marker – have been used to derive osteoclasts from haematopoietic stem cells.[44] Adding ^{3}H-thymidine-labeled monocytes to osteoclast cultures shows that some 40 per cent of the osteoclasts incorporate labeled monocytes.[45]

CHONDROCLASTS AND OSTEOCLASTS

I have already discussed the finding that osteoclasts can resorb dentine in vitro. Does this mean that osteoclasts and odontoclasts are equivalent (identical?) cells? What of other 'clasts' such as chondroclasts or cementoclasts?

Progenitor cells for osteoclasts are not progenitor cells for osteoblasts. Chondroclasts and osteoclasts have more in common with respect to lineage relationships than do osteoclasts and osteoblasts. Chondroclasts and osteoclasts are two mineraloclasts of the amoebocyte series. The fact that macrophages can be induced to synthesize and secrete collagenase (discussed above) is consistent with Willmer's thesis of amoebocytes as a related series (Chapter 10). Dentine is resorbed by dentinoclasts, cementum by cementoclasts.[46]

What functional or lineage relationships do macrophages and clasts share? An ability to degrade other mineralized matrices is one, shared either because they are equivalent cells, or even if they are not, because they produce similar degradative enzymes. While there is no direct evidence that osteoclast collagenase can degrade type II cartilage collagen, collagenase from a single cell type or tissue can degrade more than one type of collagen.

Human gingival and rabbit leukocyte collagenase cleaves skin and bone type I collagen but not type II collagen, although Clostridium collagenase can degrade both collagen types. Such evidence has been used to argue that mechanisms of resorption of cartilage and bone collagen differ: rabbit corneal collagenase cleaves rat and lamprey skin (type I) collagen and calf articular cartilage collagen; collagenase from human skin degrades lathyritic type II cartilage collagen, although more slowly than type I collagen is degraded; collagenase from cultured synovial cells from patients with rheumatoid arthritis degrades skin and articular cartilage collagen.

Finally, cartilage itself produces catabolic factors that degrade the collagenous and mucopolysaccharide components of cartilage matrix. However, the specificity of the actions of cartilage and bone-derived collagenases remains incompletely resolved.[47]

SYNOVIAL CELLS

The heterogeneous nature of synovium, the finding that synovium only produces collagenase during arthritis, and the presence of collagen types I and II in articular cartilage, all make unraveling data from synovia less than straightforward.

Synovial linings of joints and synoviocytes have interesting properties with respect to degradation of articular cartilage. Synovial cells produce a small peptide (catabolin) and act as 'clasts.' A low-molecular-weight catabolin from synovium and connective tissue degrades cartilage in vitro and plays a role in resorption in arthritis. We now know catabolin as IL-1, discussed above.[48]

Normal human synovium releases a factor that stimulates the production of prostaglandin E and plasminogen activator by human articular chondrocytes. Synovial cells and phagocytes release a neutral proteinase. Synovial proteinase releases chondroitin sulphate and hydroxyproline from cartilage; phagocyte proteinase releases only chondroitin sulphate.[49]

Synovial fluid evokes chondrogenesis from perichondria in vivo and in vitro to the same extent as exposure to growth factors such as Egf or Pdgf (Skoog et al., 1990). Labeled chondrocytes were seen, but not after injection of anti-muscle globulin or saline. These authors propose that chondromucoprotein is released by articular chondrocytes early in their degeneration, but is recognized as foreign, initiating antibody formation and further cycles of degeneration.

NOTES

1. Mononucleated osteoclasts are known from mammals (Kaye, 1984; Chambers, 1985; Parfitt, 1988). Kaye (1984) showed that 80 per cent of all osteoclasts in 'healthy'

patients were mononucleated, but that the percentage of multinucleated osteoclasts increases under conditions of excess bone resorptions, as, for example, in renal failure. See Hattersley and Chambers (1989) for how easy it is to overlook mononucleated osteoclasts, and see the chapters in *Volume 2: The Osteoclast* of Hall (1990–1994) for overviews of osteoclast origins, structure and function.

2. See Walker (1972) and Nelson and Bauer (1977) for early studies on osteoclast isolation. Of the many reviews on osteoclast origins and function, I recommend Volume 1: *The Osteoclast* of Hall (1990–1994), Pierce et al. (1991) for osteoclast structure and function, Blair (1998) and Gunther and Schinke (2000) for osteoclasts, especially Rank, Rank ligand, and osteoprotegerin, Karsenty (1998, 2000, 2003), Chambers (2000), the papers in Cardew and Goode (2001), Karsenty and Wagner (2002), Boyle et al. (2003), Horton (2003) and Nakashima and de Crombrugghe (2003) for overviews of osteoblast and osteoclast differentiation, and Teitelbaum (2000) for regulation of osteoclast function, especially in relation to osteoporosis.

3. See Witten et al. (1999b) for V-ATPase in osteoclasts of the Nile tipalia *Oreochromis niloticus*, Lee et al. (1996) for the b2 isoform and Manolson et al. (2003) for the a3 isoform, two of the four mammalian isoforms of the 100-kDa V-ATPase. Paget's disease (Box 11.2), in which bone resorption is excessive, is associated with an increase in osteoclast size.

4. See Chambers et al. (1984a,b, 1986), for a summary of their studies, and Marotti (1990) for contributions of SEM to our understanding of bone structure.

5. See Jones et al. (1984) for the different matrices resorbed, Boyde et al. (1984) for osteoclast–dentine co-cultures, S. J. Jones et al. (1994) for lack of attraction, Jones and Boyde (1977b) for the SEM study, and H. Zhou et al. (1993) for an SEM analysis of the osteoclast–bone interface *in vivo*.

6. Conditioned medium is culture medium in which a tissue or organ has been placed so that any diffusible products can flow into and condition the medium, which is then used to test whether such products have been released.

7. See Osdoby et al. (1986), Ko and Bernard (1981), Oursler and Osdoby (1988) and Marshall et al. (1986) for these four studies. Attraction systems are known; Devlin (2000) showed that collagen III fibres in periodontal ligaments are used as guidance systems by OPCs as they move to a repair site, repair in this case being by rapid healing *without a prior resorption phase*, production of procollagenase 3 at the site of new bone formation facilitating the rapid repair.

8. See Filvaroff and Derynck (1998) and Filvaroff et al. (1999) for these studies, and Box 15.1 for Tgfβ receptors.

9. See Glowacki and Lian (1987) for osteocalcin, Collin-Osbody et al. (2002) for Fgf-2 and Hirata et al. (1983) for differential sensitivity. Osteoid can be resorbed by mononuclear cells even in the presence of inhibition of osteoclastic resorption (Rifkin and Heijl, 1980; Rifkin et al., 1980b).

10. See Klein-Nulend et al. (1987a) for differences between calvariae and long bones, and Van de Wijngaert et al. (1988) for recruitment of bone cells by cartilage.

11. See Krukowski and Kahn (1980a,b) for parathyroidectomy and Dodds et al. (1986) for vitamin B6 deficiency.

12. See Minkin (1982) for the origin of the TRAP method, Witten et al. (1999a) for the utility of TRAP-staining, and Hayman et al. (1996) for absence of TRAP-staining in osteopetrosis.

Although Bianco et al. (1988) reported TRAP-reactivity in rat osteoblasts and osteocytes and cautioned that TRAP-reactivity is not osteoclast specific, subsequent workers have found a high degree of specificity.

13. See Lewinson and Krogan (1995) and Fujita et al. (1998) for resorption by mononucleated cells and ovariectomy.

14. See Witten (1997), Witten and Villwock (1997) and Witten et al. (1999a) for TRAP-reactivity in teleost osteoclasts.

15. See Takagi and Kaneko (1995) for bone implants, and Persson et al. (1995) and Box 2.3 for scale resorption.

16. See Kasten et al. (1994) and Burger et al. (2003) for these studies on nitric oxide and osteoclast function.

17. See Palmer et al. (1993) for human articular chondrocytes and Diwan et al. (2000) for the study of fracture repair. See Chapter 30 for a discussion of cGMP and cAMP, both of which are responsive to and act as second messengers in mediating mechanical influences.

18. For overviews of the history of studies on osteoclast origins, see Hancox (1949, 1956, 1965, 1972a,b), Hall (1975b, 1990–1994), Hanaoka (1977, 1980), Chambers (1980, 1985), Bonucci (1981), Witten (1997) and Witten et al. (1999a,b, 2000). For discussions of whether osteoclasts arise by fusion of mononucleated cells, see Owen (1970, 1971), Rasmussen and Bordier (1974), Hall (1975b) and Ko and Bernard (1981). Hanaoka (1980), Yabe and Hanaoka (1985) and Hanaoka et al. (1989) argued for local, non-haematopoietic (perivascular) rather than haematopoietic cells as osteoclast precursors. For the nature of the salmon kype, see Witten and Hall (2001, 2002, 2003).

19. See Key et al. (1983) for resorption by monocytes. For the ultrastructural characteristics of osteoclasts see Cameron (1972), Hancox (1972a,b), Göthlin and Ericsson (1976) and Holtrop and King (1977).

20. See Warshawsky et al. (1980) for calcitonin-binding sites, and Addison (1978a–b, 1979, 1980) for the enzyme analyses.

21. See Nefussi and Baron (1985) for periosteal vs. endosteal response in rabbit femora, Wong and Kocour (1983) for response to PGE-1, Owen (1963, 1970, 1971) for separate precursors, and Enlow (1962a) for a good overview of the postnatal remodeling of bone.

22. See Baron (1972, 1973) and Markostamou and Baron (1973) for bone deposition/resorption with physiological drift, and Baron (1975) for reversal of the pattern.

23. See Scott (1965, 1969), Göthlin (1973), Luk et al. (1974a) and Rifkin et al. (1980c) for criteria to separate preosteoblasts from preosteoclasts, and Young (1962, 1963, 1964) and Kember (1960) for studies with ³H-thymidine.

24. See Simmons and Kahn (1979), Kahn et al. (1981) and Yabe and Hanaoka (1985) for grafts of long bones and fractured long bones, Krukowski and Kahn (1982) for rapid formation of osteoclasts and Krukowski et al. (1989) for a review of these cell lineage studies.

25. For mutant mice with osteopetrosis, see Barnicot (1941), Barnes et al. (1975), Walker (1975a,b), Loutit and Sansom (1976), Marks (1976; 1978a,b, 1984), Wiktor-Jedrzejczak et al. (1982), Schneider and Byrnes (1983), Schneider (1985), Schneider et al. (1986) and Marks et al. (2000). See Campbell (1966) for the osteopetrotic dinosaur.

26. See R. S. Johnson et al. (1992) and Z.-Q. Wang et al. (1992) for the *c-fos* null mutation, and Box 15.3 for proto-oncogenes.

27. Teeth fail to erupt in *ia* rats because of lack of bone resorption. In dogs the dental follicle attracts mononuclear cells that transform into osteoclasts to effect local resorption; teeth can be replaced with replicas that still erupt but removal of the follicle prevents eruption (Cahill and Marks, 1980; Marks, 1981; Marks *et al.*, 1983; Marks and Cahill, 1984, 1986; Gorski *et al.*, 1988).

28. *op/op* osteopetrotic mice have few macrophages and decreased numbers of monocytes; haematopoietic stromal fibroblasts release too little colony-stimulating factor, leading to reduced numbers of macrophages and monocytes (Wiktor-Jedrzejczak *et al.*, 1982). Intramembranous ossification is also disrupted in *op/op* mice – the trabecular but not the subperiosteal portion of the parietal develops normally, i.e. subperiosteal ossification is eliminated (Niida *et al.*, 1994a).

29. See Nisbet *et al.* (1982, 1983) and Thesingh and Scherft (1985) for *mi* mice and see Kornak *et al.* (2001) for loss of the chloride channel. Osteoclasts contain carbonic anhydrase on Golgi bodies, in vesicles and in membranes, and in the ruffled border, as the enzyme is released into the resorbing zone (R. E. Anderson *et al.*, 1982).

30. See Kawata *et al.* (1997, 1999b,c) and Kawasoko *et al.* (2000) for *op/op* mice. Tooth extraction influences jaw growth, either through a feedback loop or indirectly through mechanical stresses. Extracting the incisors from rabbit mandibles slows anterior growth of the mandible; extracting the incisors from the maxilla slows anterior growth of the maxilla. Extracting the molars has little effect on mandibular or maxillary growth (Ranta *et al.*, 1973).

31. See Wahl *et al.* (1975) for macrophages and Minkin *et al.* (1981) for the chemotactic factor.

32. See Minkin *et al.* (1977), Hogg *et al.* (1980) and Jones *et al.* (1981) for these studies.

33. See Chambers (1980, 1982, 1985, 2000), Chambers and Magnus (1982) and Chambers and Moore (1983) for these culture studies.

34. See Holtrop *et al.* (1982) for macrophage origin, Burger *et al.* (1982, 1984) for phagocytes, and Boyde *et al.* (1984) and S. J. Jones *et al.* (1994) for dentine and osteoblasts.

35. See Hume *et al.* (1983, 1984) for F4/80, Rifas *et al.* (1984) and Peck *et al.* (1985) for the macrophage protein.

36. Survival and maintenance as chondrogenic of a line of Syrian hamster embryonic chondrocyte-like cells is regulated by IL-1α and insulin-like growth factor-1 (Igf-1), mediated by *Sox-9*, establishing a growth factor–transcription factor link (Kolettas *et al.*, 2001). Chondrogenesis from limb-bud and cranial neural-crest cells is also dependent upon Sox-9 (Akiyama *et al.*, 2002; Mori-Akiyama *et al.*, 2003).

37. See Ghafari (1984) for musculature displacement, and I. Takahashi *et al.* (1996a, 1998) for limb mesenchyme and modulation of secondary cartilage.

38. See Wergedal *et al.* (1990) and Palmer *et al.* (1993) for these studies. IL-2 (T-cell growth factor), which is produced by large granular lymphocytes and stimulated helper T cells, activates macrophages.

39. See Yasui *et al.* (1986) and Goldring *et al.* (1988) for collagen synthesis, and Harvey *et al.* (1991) for the synovium.

40. See Osdoby *et al.* (1982, 1986) for chick tibiae, Minkin and Shapiro (1986) for the in-vitro study, and Merke *et al.* (1986) for the 1, 25-vitamin D_3 receptors. A vitamin D receptor cloned from the sea lamprey, *Petromyzon marinus*, binds 1,25-dihydroxyvitamin D_3 in mammalian COS-7 cells, but does not activate vitamin D-responsive elements in the rat osteocalcin gene (Whitfield *et al.*, 2003), perhaps because osteocalcin is involved with mineralization and lamprey cartilage is unmineralized.

41. Using foetal rat calvariae, Chyun *et al.* (1984) demonstrated that cortisol inhibits *bone formation*, in part by inhibiting proliferation of periosteal cells. See Peck (1984) for effects of glucocorticoids on bone-cell metabolism. Osteoclast activity is inhibited directly by hydrocortisone, as demonstrated in cat bone-marrow cultures (Suda *et al.*, 1983).

42. See Dominguez and Mundy (1980) for monocytes stimulating osteoclastic resorption, Rifkin *et al.* (1980a) for the studies with prostaglandin E_2, and Key *et al.* (1983) for those with Pdgf.

43. See Wong and Cohn (1974, 1975), Luben *et al.* (1977), Cohn and Wong (1979) and Chapter 24 for differential sensitivity.

44. See Allen *et al.* (1981, 1984) for marrow cultures, N. Y. Ali *et al.* (1984) for monocytes on dentine, and Ash *et al.* (1980, 1981) for Beige mice.

45. See Stanka *et al.* (1981) and Stanka and Bargsten (1983) for the Barr body studies, Allen *et al.* (1981, 1984) for marrow cultures, N. Y. Ali *et al.* (1984) for monocytes on dentine, Ash *et al.* (1980, 1981) for Beige mice, and Zambonin Zallone and Teti (1985).

46. See Knese (1972) for mineraloclasts, Willmer (1970) and Chapter 10 for amoebocytes, Takuma (1962), Schenk *et al.* (1967) and Savostin-Asling and Asling (1973) for similarities between chondroclasts and osteoclasts, and Wahl *et al.* (1975) for collagenase synthesis by macrophages. See Woessner (1973), Bilezikian *et al.* (1996) and Davoli *et al.* (2001) for mammalian collagenases.

47. See Robertson and Miller (1972) for the conclusions that flow from the studies with *Clostridium* collagenases, Davison and Berman (1973) and Wooley *et al.* (1973) for rabbit corneal and human skin collagenases, Leibovich and Weiss (1973) and Harris and McCroskery (1974) for collagenase from patients with rheumatoid arthritis, and Firestein (2003) for current concepts concerning rheumatoid arthritis.

48. See Saklatvala and Dingle (1980), Dingle (1981) and Saklatvala *et al.* (1983) for initial studies on catabolin (IL-1), Janis *et al.* (1967) and Estabrooks and Schiff (1972) for heterogeneity of synovia, and Dingle (1984) for the role of cellular interactions in joint erosion.

49. See Meats *et al.* (1980) for the stimulation study and C. H. Evans *et al.* (1981) for the neutral proteinase.

Part VI

Embryonic Origins

The stem and progenitor cells discussed in Parts IV and V that produce the vertebrate skeletons – for there is more than one – have two distinct embryological origins in two of the four germ layers, arising from mesoderm and from neural crest. Elaboration of our knowledge of the embryological origins of cartilage and bone is the topic of Chapters 16 and 17.

Skeletal Origins: Somitic Mesoderm

Chapter 16

Arise, subdivide, migrate, arrive and settle down.

Vertebrate skeletons – for there is more than one – have two distinct embryological origins from two of the four germ layers, mesoderm and neural crest.[1] Both these germ layers produce skeletogenic mesenchyme from which cartilage and bone arise: *mesoderm for endoskeleton* primarily in the trunk – limbs, vertebrae and ribs – but with contributions to the base of the skull; and *neural crest for exoskeleton and endoskeleton* of the head and branchial region (Table 1.1). The neural crest also produces odontogenic (tooth-forming) mesenchyme from which dentine and other mesenchymal tissues such as alveolar bone and the periodontal ligament arise.

Because the embryological origins of skeletal tissues have been reviewed on numerous occasions, and because my emphasis throughout the book is on how skeletal tissues develop once skeletogenic mesenchyme is established, I do not treat embryological origins exhaustively, even though I am sorely tempted to; the topic is a fascinating one. I discuss the embryological origin of the vertebral and rib mesenchyme from axial (somitic) mesoderm in this chapter, and the origin of craniofacial skeletogenic mesenchyme from the neural crest in the following chapter.[2] Skeletogenic cells do not arise in the embryo in the position where skeletal tissues will form. Cell migration is most extensive from the neural crest, but the cells that will form the vertebrae migrate a substantial distance from the somites to surround the axial notochord and spinal cord. Similarly, cells that form the ribs migrate some distance from the somites. And, while the skeletogenic cells that give rise to the limb skeletons arise *in situ* from lateral-plate mesoderm, the cells that produce the appendicular muscles arise from somitic mesoderm and migrate into the developing limb buds after the buds have begun to grow. Because migration is ubiquitous, in both chapters I discuss migration of skeletogenic mesenchyme to sites of skeletogenesis.

SOMITIC MESODERM AND THE ORIGIN OF THE VERTEBRAE

Chondrification of somitic mesoderm to produce the primordia of the vertebrae has been a favourite topic of investigation for 80 years for those interested in understanding the mechanisms underlying development and differentiation. Somitic chondrogenesis involves cellular origins, proliferation, migration, cytodifferentiation, synthesis of matrix products, tissue interactions and tissue morphogenesis, all of which are central issues in developmental biology. The topic is a large one, and my treatment will, perforce, consist of an overview of the highlights. I only briefly discuss somitogenesis – how the somites themselves form – so as to concentrate on those later stages and processes involved in vertebral development. The approach is to discuss the progression of experimental studies from the initial grafting of somites to the chorioallantoic membrane (CAM) in the 1920s, through extirpation experiments initiated in the 1940s, organ, tissue and cell culture in the 1950s, the search for 'pure' inducers to the refocusing of attention on the environment and extracellular matrix (ECM) products which dominated the 1970s, and the discovery of the molecular basis of somitic chondrogenesis from 1990 on.[3]

Just as the brain induces skeletogenic mesenchyme to produce the skull as its container (Chapter 21) so the spinal cord and notochord induce their skeletal container, the vertebral column.

PARAXIAL MESODERM → SOMITES

In tracing the origins of the vertebrae, it is appropriate to begin with the development, segmentation and subsequent subdivision of *paraxial mesoderm* into presomitic mesoderm, pairs of somites, and finally into sclerotome, dermotome and myotome. Figure 16.1 illustrates the topographical relationship between sclerotomal cells as they migrate away from the dermotome and myotome (*dermomyotome*[4], which produce connective tissue and muscle) and the spinal cord and notochord as they appear in an embryonic chick incubated for about 50 hours. Williams (1910) provided the first comprehensive account of the formation and segmentation of somites, in this case in chick embryos.[5]

As the neural plate elevates at Hamburger–Hamilton stage (H.H.) 8/9, the somitic mesoderm separates from presumptive neural tube and notochord and moves mediad. Separation is facilitated by accumulation of ECM produced and deposited by the notochord and the ventral portion of the neural tube; see Chapter 42 for the production of ECM by the notochord. By H.H. 12, bundles of fibrils connect the somatic mesoderm with notochord and ventral neural tube, both of which – *but not the dorsal neural tube* – subsequently evoke chondrogenesis from sclerotomal mesenchyme. *Shh* from the ventral neural tube and notochord promotes survival of myogenic and chondrogenic lineages in chick somites (Teillet *et al.*, 1998). Somatic mesoderm then segments into pairs of somites (see Box 16.1).

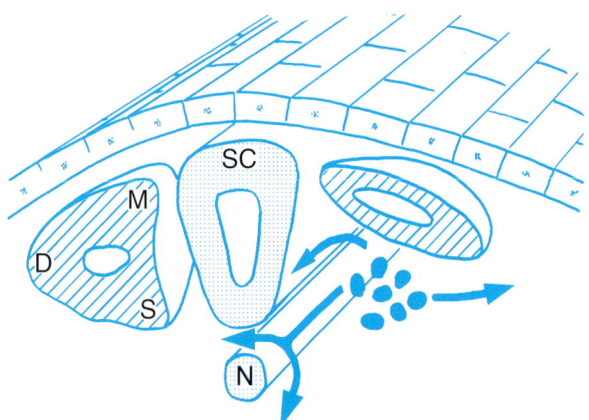

Figure 16.1 Somites and sclerotome. A diagrammatic cross section of the axis of the trunk of a chick embryo, showing the spinal cord (SC) and notochord (N). On the left is a somite as it appears in an embryo of H.H. 10 (35 hours of incubation). Future dermotome (D), myotome (M) and sclerotome (S) are shown. The situation at H.H. 14 (50 hours of incubation) is shown on the right. Sclerotomal cells (blue) are migrating from the somite to surround the notochord and the ventral portion of the spinal cord (left arrow). Other cells are migrating to the future limb buds where they will maintain the AER and produce the muscles of the limb. Modified from Hall (1977).

Box 16.1 Somitic segmentation prepattern

At least four major theories have been presented to explain *segmentation* in chick embryos and, by extension – or inference – in other vertebrates. Segmentation has been said to result from:

(i) a 'somite organizing centre' adjacent to the primitive streak;
(ii) normal movements associated with regression of the primitive streak;
(iii) influences from the notochord; and/or
(iv) influences from the neural tube.

The weight of early evidence favoured a combination of (ii), (iii) and (iv).[a]

Experimental studies support the thesis that the neural plate imposes a 'segmentation prepattern' onto the presomitic mesoderm, a prepattern that is released during subsequent migration of somatic mesoderm – a combination of (ii) and (iv) above. Part of the epiblast at H.H. 4–6 was rotated 180° to place presumptive neural-plate over presumptive lateral-plate mesoderm (not a normal source of somites). Induction of segmentally arranged 'somites' within the lateral-plate mesoderm is consistent with transfer of a prepattern (iv above).

The second test took advantage of the knowledge that somites do not form simultaneously, but in a rostro-caudal sequence, segmentation occurring just behind the advancing notochord as more and more posterior mesoderm segments progressively. An incision made alongside the primitive streak substitutes for this movement, and evokes somites simultaneously along the axis. These studies support the segmentation prepattern being released during the regressive movements of the primitive streak (ii above) as the notochord shears the mesoderm into lateral halves.[b]

The molecular basis for segmentation of paraxial mesoderm into somites has become more fully understood since Olivier Pourquié demonstrated a molecular clock based on the gene *Hairy*. mRNA for *chairy*, the avian homologue of *Drosophila hairy*, is expressed in cyclic waves every 90 minutes in the paraxial mesoderm, each wave setting out the position of one somite pair. This is an autonomous property of the paraxial mesoderm, and illustrates conservation of molecular mechanisms for patterning in vertebrates and invertebrates such as *Drosophila*.[c]

[a] See Bellairs (1971) and Deuchar (1975) for reviews of the early evidence. See Chapter 41 for the role played by the segmentally arranged spinal ganglia in somite segmentation.
[b] See Lipton and Jacobson (1974b) and Packard and Jacobson (1976) for these studies.
[c] See Palmeirim *et al.* (1997), Maroto *et al.* (2001) and Jouve *et al.* (2002) for the somite segmentation clock.

SCLEROTOME FORMATION AND MIGRATION

Each somite subdivides into three major components:

- the dorso-medial wall forms the *myotome*, from which appendicular muscle develops;
- the dorso-lateral edge forms the *dermotome*, the source of connective tissue of the dermis;
- the ventro-medial edge forms the *sclerotome*, from which vertebrae, intervertebral discs and a portion of the ribs develop (Fig. 16.1).

The sclerotome breaks up into mesenchymal cells, which take up positions between the notochord and ventral portion of the spinal cord (Fig. 16.1). A ventro-lateral migration between the myotomes delineates the *costal processes*, the forerunners of the *ribs* (see Box 16.2). Another population of cells migrates to the lateral-plate mesoderm to provide the cells that produce the appendicular muscles of the limbs. Separation and onset of migration do not occur simultaneously in all somites.

Initiation of sclerotomal mesenchymal cell separation and migration begins at H.H. 11 (40 to 45 hours; Fig. 16.2) in the chick, and during the fourth week in human embryos. In chick embryos at H.H. 17, the cranial somites are organizing into dermomyotome and sclerotome,

whilst the caudal somites are still segmenting; by H.H. 18, the more cranial sclerotomal cells are migrating around the notochord and spinal cord.[6]

As will be discussed in Chapter 42, the notochord and the ventral portion of the spinal cord produce ECM *before* sclerotomal cells begin to migrate around them. Migrating cells move into an ECM that is rich in collagen fibres and glycosaminoglycans (GAGs). Indeed, some of the preexisting ECM components may be integrated into the matrix that later surrounds the sclerotomal cells.

By H.H. 23, ECM is found between those sclerotomal cells that lie adjacent to the notochord but not around those more distant from the notochord. There is then a gradual spreading of ECM peripherally. By H.H. 23–29,

---

Box 16.2 The origin of the ribs

The statement in the text, concerning cells that migrate between the myotomes to positions where they form the costal processes of the ribs, implies that at least a portion of the ribs is somatic in origin. However, the origin of the ribs has been much debated.[a]

The principal debate concerns a basic question: are ribs derived from somatic or somitic mesoderm? A second question concerns relationships between the ribs and the sternum, especially, although not exclusively, in turtles. The debate over the primary issue seems to have been halted, at least temporarily, in favour of a somitic origin. Ribs fail to develop in chick embryos after lateral migration of somitic mesoderm is blocked with a physical barrier or the cells are removed by irradiation. Transplanting tissues between chick and quail embryos to form tissue chimaeras further confirms a somitic origin.[b]

More recently, a sclerotomal origin of the entire rib was confirmed following grafting of single sclerotome units between chick and Japanese quail. These same workers confirmed resegmentation in so far as the ribs are concerned; the sclerotome of one somite contributes the caudal portion of one rib and the cranial portion of the next most caudal rib (Huang *et al.*, 2000a). These experiments were undertaken in response to a study by Kato and Aoyama (1998), in which the distal portion of the rib was shown to arise from dermomyotome, which may have arisen from disruption of the epithelium overlying the dermomyotome; Huang and colleagues showed that removing the epithelium does lead to malformed ribs because the sclerotome–dermomyotome interactions are disrupted, not because the dermomyotome contributes cells to the ribs. In part, this difference may reflect the fact that ribs in chick embryos have a dual origin within the somitic mesoderm, the proximal portion of each rib arising from medial somitic mesoderm and the distal portion from lateral somitic mesoderm (Olivera-Martinez *et al.*, 2000).

Ossification of the ribs is often delayed. The first bone of human ribs does not appear until between age 20 and 25 years. Ossification is subperiosteal and chondrocytes express type I collagen, two patterns that differ from epiphyseal and sternal cartilages. Changes in the cartilaginous matrix are more rapid in males than in females (Claassen *et al.*, 1995).

Genes and ribs

The composite origin of mammalian ribs makes analysis of their development in gene knockout experiments of interest. With respect to mesodermal patterning:

- A mouse mutant known as *rib vertebrae* (*rv*) produces numerous vertebral defects because somitic patterning is defective in at least two ways: presomitic mesoderm is elongated in extent, and subsequently the antero-posterior polarization of the somites is disrupted. *Rib vertebrae*, which is a mutation of *Tbx-6*, interacts with *Delta-like-1* (*Dll-1*), which produces a Notch receptor, producing even more severe defects.
- *Zic-3*, which encodes a zinc-finger transcription factor, is X-linked in mice. If mutated, *Zic-3* disturbs laterality, producing anomalies of ribs and vertebrae. *Zic-3* acts through *nodal*, expression of which is initiated but not maintained in the usual asymmetrical pattern.[c]

In terms of embryological origins:

- Davies *et al.* (1998) identified six splice forms of the α1(XI) collagen gene in foetal rat cartilage. One form is expressed only in the dorsal portion of rib primordia, which is the only portion that ossifies, suggesting a role for type XI in endochondral ossification.
- The opposite ends of mouse ribs are affected by *Hoxb-8* and *Hoxc-8*, while
- *Pax-1–Pax-9* double mutants (*Pax-1⁻/Pax-9⁻*) lack the proximal portion of the ribs.[d]

In terms of homeotic transformations:

- Knocking out *Hoxa-4* results in a homeotic transformation of the seventh cervical vertebra, which develops ribs, a condition typical of a thoracic vertebra. The sternum fails to form in these animals. Over-expressing *Hoxa-4* reduces the rib anlage.
- Over-expressing a human *HOXC¹-6* (*Hox 3.3*) transgene in mice leads to ectopic expression posteriorly along the developing body axis, formation of extra ribs in the lumbar region, the posterior ribs taking on the shape of more anterior ribs, and additional ribs forming on the sternum.[e]

[a] Most anuran amphibians are characterized by loss of the ribs, although some species possess rudimentary ribs, especially around the sacral region (Blanco and Sanchiz, 2000).
[b] See Arey (1974) for migration of precursors of the ribs, Gladstone and Wakeley (1932) for early literature on the relationship between the sternum and ribs, Sweeney and Watterson (1969) and Pinot (1969b) for the barrier and irradiation study, and Chevallier (1975) for the chick–quail chimaeras.
[c] See Nacke *et al.* (2000) for the mutant rib vertebrae, P. H. White *et al.* (2003) for *Tbx-6*, and Purandare *et al.* (2002) for *Zic-3*. *Tbx-6*, which is expressed in paraxial mesoderm, is down-regulated as somites develop and essential for both specification and patterning of the posterior somites in mouse embryos.
[d] See Pollock *et al.* (1995) for *Hoxb-8* and *Hoxc-8* and Peters *et al.* (1999) for *Pax-1⁻/Pax-9⁻*.
[e] See Horan *et al.* (1994) for the *Hoxa-4* knockout, and Jegalian and de Robertis (1992) for *Hoxc-6*.

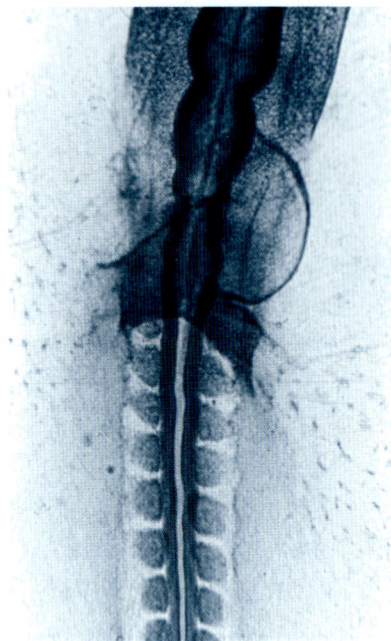

Figure 16.2 A stained whole mount of a chick embryo of H.H. 11 showing the stage of development of the heart (looping to the right) and somites, seven pairs being visible caudal to the vitelline veins, which extend out into blood islands.

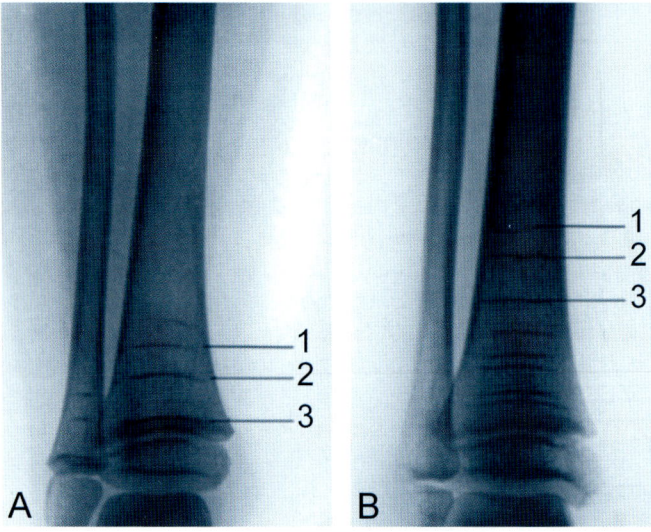

Figure 16.3 Arrested growth of bone. (A,B) Radiographs of the right ankle joint of the same girl aged five (A) and 8 (B) years showing lines of arrested growth (G1–3) corresponding to three periods of illness – measles (G1) and bronco-pneumonia (G2, G3) – between ages three and four. These and subsequent lines corresponding to subsequent illnesses (B) remain visible long into the growth period. Compare with the arrest lines associated with normal development shown in Fig. 5.9. Modified from Harris (1933).

the perinotochordal sclerotomal cells consist of a mixture of mesenchymal cells (prechondroblasts) and occasional fibroblasts. Fibroblasts predominate in the future intervertebral areas. Chondroblasts are present by H.H. 34, chondrocytes by stage 37. During H.H. 35–42, chondrocytes complete depositing ECM and begin to undergo necrosis as a prelude to endochondral ossification. Considerable osteoid is deposited onto *unmineralized* cartilage matrix – in contrast to long bones in which osteoid is deposited onto *mineralized* cartilage matrix. An organic layer – the lamina limitans – separates osteoid from mineralized bone and marks the position of arrested mineralization (Figs 5.9, 16.3 and 16.4).[7]

Hyaluronan isolated from axial tissues (notochord, spinal cord, somites) of embryonic chicks is at peak levels between H.H. 21 and 25, with more hyaluronan than chondroitin sulphate (CS) present until H.H. 25. This pattern – *high levels of hyaluronan early, declining levels later in development* – is seen in all the embryonic situations analyzed. Kvist and Finnegan suggested that initial predominance of hyaluronan over CS is associated with aggregation of sclerotomal cells around the embryonic axis. A decline in hyaluronan as vertebral chondrogenesis ensues suggested to the authors that high initial concentrations and declining levels both play developmental roles. Subsequent studies on various developing organs have proven them right; developing hearts contain non-sulphated CS, as well as CS and hyaluronan, the mix of mucopolysaccharides changing as heart muscle differentiates.[8]

RESEGMENTATION

Segmentation of somites into sclerotome, dermotome and myotome (or dermomyotome) raises the fascinating topic of *resegmentation*. The final position taken up by sclerotomal cells and where vertebrae develop does *not* correspond to the original position of the somites. Rather it is 'intersegmental' with respect to that position. The proposal that the fully formed sclerotome corresponds to the combined caudal half of one somite and the cranial half of the next most caudal somite was first presented in a unified way by Robert Remak in 1855. Such a rearrangement is achieved through the cephalad migration of cells from the cephalic portion of one sclerotome and the caudad migration of cells from the caudal half of the next somite. Each half of a sclerotome is known as a *sclerotomite*.[9]

Each sclerotomite may provide only one lineage of cells to each vertebra. This concept of a *clonal origin* of the somitic derivatives raises the issue of whether we can determine how many cells it takes to make a somite. In one approach, embryos derived from fusing blastomeres from two mouse strains were used to follow strain-specific characteristics of the vertebrae. These tetraparental mice indicate that each of the two caudal and two cranial sclerotomites may provide one cell lineage to each vertebra. Consequently, the 30 vertebrae of a mouse could derive from just 120 cell lineages or clones. Deriving this number is controversial, in part because such analyses cannot precisely pinpoint when each cell line is determined;

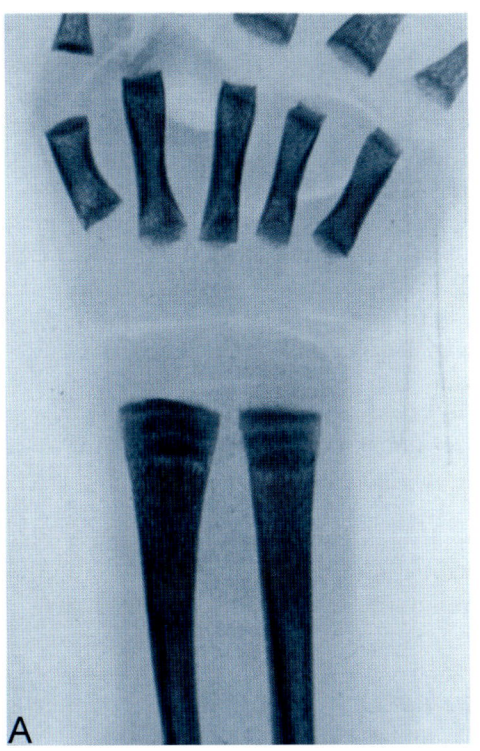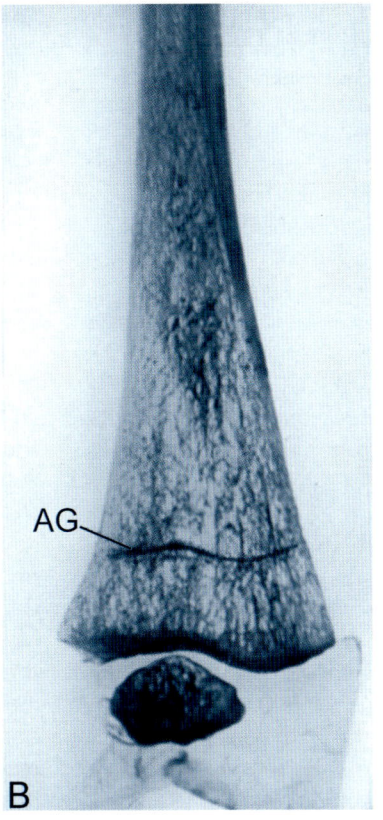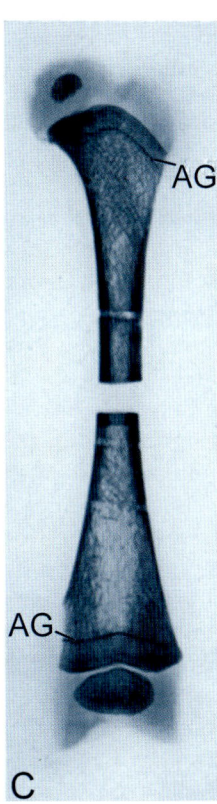

Figure 16.4 Depictions of lines of arrested growth in a stillborn and a three-month-old infant. (A) A radiograph of the wrist showing three zones of arrested growth (the pale zones) in the radius and ulna of a stillborn infant. (B) The distal femur with lines of arrested growth (AG) in a stillborn infant. (C) Distal and proximal portions of the femur of a three-month-old infant, showing two lines of arrested growth (AG) laid down at birth. From Harris (1933).

O'Higgins *et al.* (1986) repeated this study using Fourier analysis, concluding that the clonal theory is too simplistic.[10]

Mesenchymal cells situated between developing vertebrae surround the notochord to form the *intervertebral discs*. This mesenchyme comes from the dense caudal half of each sclerotome, the cells migrating cephalad to form the intervertebral disc, which comes to lie between two adjacent vertebrae and opposite the myotome in the position occupied by the original somites, an arrangement that allows muscles to insert across the joints between vertebrae and permits lateral flexure of the vertebral column. The notochord within each disc forms the *nucleus pulposus*. Should a remnant of the notochord persist within the vertebral column, it can become tumorous as a *chordoma*.

Primordia of the intervertebral discs in rats appear as condensations at 14 days of gestation, weakly positive for collagen types I and II, and with chondroitin-6- and dermatan sulphates. By 16 days the inner annulus contains cartilage (with type II collagen, chondroitin-4-sulphate and later keratan sulphate), while the outer annulus is fibroblastic with metaplasia to cartilage in adults. Again, in adults, fibrocartilage may form in the nucleus pulposus, being derived from the cartilaginous end plate and/or inner annulus. Ultrastructural analysis of human intervertebral disc, especially cells of the nucleus pulposus, from

26 weeks of foetal life to 91 years of age, demonstrates chondrocyte-like cells at all ages. Formation of the nucleus pulposus (and of the extracellular sheath of the notochord) is regulated by *Sox-5* and *Sox-6*.[11]

Cranial somites from chick embryos transplanted into the trunk produce vertebrae typical of trunk somites, including formation of intervertebral discs but without the costal processes and without the ability to support survival of dorsal root ganglia, both properties of trunk vertebrae. Surprisingly, there is no change in the pattern of expression of *Hox* genes in these transplanted somites; they do not switch to a pattern of expression appropriate to the trunk. Other genes must specify position-related morphological features of the vertebrae (Kant and Goldstein, 1999).

A variety of classes of evidence demonstrate resegmentation, notably:

- analysis of the contribution of individual labeled cells to vertebral development in chick embryos;[12]
- retroviral-mediated gene transfer to label individual somites of embryonic chicks to show that label is found in the caudal half of one somite and the rostal half of the adjacent somite;
- the demonstration that the caudal half of each somite forms the vertebral pedicle, the rostral half the fibrocartilage of each intervertebral disc; and

- using quail–chick chimaeras to show that the sclerotome of one somite contributes the caudal portion of one rib and the cranial portion of the next most caudal rib.[13]

Despite the nice story such studies tell about chick embryos, resegmentation has not been uniformly accepted in the century and a half since Remak's clear exposition. Whether individual researchers accept resegmentation depends in no small part on the group of vertebrates they study and/or the literature they read.[14] A sample follows.

Timing and sequence of vertebral development in human embryos were used by O'Rahilly and Meyer (1979) to be cautious about resegmentation but to conclude from their close developmental study that the cranial sclerotomite contributes to the centrum and the caudal sclerotomite to the intervertebral disc, neural and costal processes. On the other hand, resegmentation was described in human foetuses at the fifth and sixth weeks of gestation using the intersegmental artery as a landmark, and the fact of resegmentation was used to *divide human congenital vertebral anomalies into two types: failure of somites to form, or failure of resegmentation*, at the sixth week. In a third study, resegmentation was rejected on the basis of a study of mouse thoracic vertebrae using autoradiography. As part of an extensive analysis of the relationship between myotomes and the axial skeleton, Lauder (1980) noted that resegmentation redistributes large volumes of sclerotomal cells, but drew attention to the lack of any causal link between sclerotomal resegmentation and intersegmental vertebrae in adults.[15]

SOMITIC CONTRIBUTION TO LIMB BUDS

Chondrogenesis of limb-bud mesenchyme from embryonic chicks, which has been studied intensively and extensively for over 50 years, illustrates how prevailing theory influences the experiments we do and how we interpret them. Until the late 1970s, it was assumed that myogenesis and chondrogenesis were alternate fates of a *single* population of mesenchymal cells, a population derived from lateral-plate (*somatic*) mesoderm. Experiments were designed and interpreted against this notion. In the late 1970s however, evidence began to emerge that challenged and then overturned this view. Myogenic cells were shown to arise *outside* the limb buds from axial (*somatic*) mesoderm and migrate into the limb buds.

Extensions from the ventral wall of the somites penetrate developing chick limb buds and stimulate *proliferation* of limb-bud mesenchyme, i.e. *somitic* mesoderm invades the *somatic* mesoderm of the limb bud. Somitic cells also penetrate the developing limb buds of mouse embryos. Similar cells penetrate the limb buds of limbless tetrapods but fail to survive (Chapter 37). Furthermore, as demonstrated in chicks and mice, including somitic mesoderm with limb-bud mesenchyme *in vitro* enhances the chondrogenesis of limb mesenchyme.[16]

Formation of muscle

Until the late 1970s, there was no clear evidence that somitic cells *formed* any of the tissues of vertebrate limbs. Two studies began to change this perception.

Isotopically labeled chondrocytes were found in the humerus after ^{3}H-thymidine-labeled somites were co-cultured with unlabeled limb buds for nine days, suggesting that proximal skeletal elements in mouse forelimbs are of mixed somitic and somatic origin; labeled chondrocytes were not found in more distal skeletal elements. Of course, these results could be an artifact of the co-culture approach; somatic mesoderm itself is chondrogenic.

Destroying somitic mesoderm *in ovo* resulted in limbs that lacked extrinsic musculature. Chevallier and his colleagues thought this could be because the somites contribute myogenic cells to the limb buds, or because somites contribute a factor that maintains limb-bud mesenchyme. To test the former, they grafted somitic mesoderm from Japanese quail embryos into chick limb buds. Myoblasts of quail (donor, somatic) origin developed in the chick limb buds, demonstrating that somitic cells indeed do form muscle in the limbs.[17] Subsequently, again using quail–chick chimaeras, *all myogenic cells of wing and hind limb buds were shown to be somitic in origin, and limb cartilage somatic*. Trunk and appendicular muscles are somitic, whereas fibroblasts of connective tissues and tendons are somatopleural (somatic). The wing bud is not a homogeneous population of pluripotential mesodermal cells.[18]

We now also know, in part from studies with limbless mutants discussed in Chapter 36, and confirming Chevallier's second interpretation, that somites provide the earliest signal capable of initiating a limb bud (Chapter 35); lack of somitic cells deprives limb buds of the ability to maintain their apical epithelial ridges (AERs, Chapter 37), resulting in various degrees of limb reduction or complete limb loss in mutants and in limbless tetrapods. Newman (1977) provided experimental evidence that the cells at the distal tip of chick limb buds at H.H. 25 can form cartilage but not muscle. This could be interpreted as absence of somitic cells at the distal tip (*á la* Christ or Chevallier), stabilization of cells for chondrogenesis after H.H. 24 (*á la* Searls), or continued lability of distal mesenchymal cells.

Innervation and myogenesis

Several lines of evidence, outlined below, indicate that innervation is necessary for myogenesis to be initiated.

- Nerve fibres first enter hind-limb buds at H.H. 23, which is also when myosin is first detected. Such evidence is circumstantial but consistent with a potential role for nerves.
- Paralysis of chick embryos (Fig. 16.5) prevents myogenic differentiation, providing indirect evidence.

Figure 16.5 A 12-day-old chick embryo paralyzed by injection of 1.0 mg decamethonium at six days of incubation. The accumulation of fluid (oedema) in the body is evident, as is the underdevelopment of the lower jaw and digits fixed in position.

- CAM-grafted limb-bud mesenchyme from embryos of H.H. 22–26 only forms myogenic clones if nerves are included – more direct evidence of a role for innervation.

Genes involved in axon guidance and path finding also play a role. The three genes in the *Slit* family code for large ECM-secreted and membrane-associated glycoproteins. Interactions with their receptors from the *robo* gene family are involved in path finding by axons. *Slit-2* is expressed in peripheral (myogenic) and *Slit-3* in central chondrogenic mesenchyme in chick limb buds. *Slit-3* promotes chondrogenic cell proliferation, while *Slit-2* promotes myogenesis, peripheral mesenchyme and joint formation. Both genes are expressed in interdigital mesenchyme and in the inner cellular periosteum of long bones. (Later in skeletal development innervation sustains levels of proliferation in periosteal preosteoblasts.)[19]

Horder (1978) examined innervation in the context of the minimal number of 'rules' required to pattern developing limbs, concluding that what he termed *functional adaptability* and *morphogenetic opportunism* were the only rules required. Consequently, Horder emphasized the role of nerves, including the timing of limb-bud innervation in limb patterning, concluding that:

- contact guidance controls nerve outgrowth from the spinal cord and is sufficient to account for nerve–muscle interactions within the limb; especially as
- any nerve can innervate any muscle within the limb; while
- the pattern of skeletal elements (the condensations) sets the patterns of the muscles and nerves.

Signals to initiate a limb bud

Scatter factor/hepatocyte growth factor (SF/HGF), a large polypeptide growth and motility factor, has multiple functions, one of which is to regulate the emergence of myogenic cells from the dermomyotome and their migration into flank mesoderm where limbs will arise. It now appears that SF/HGF has further roles as limbs develop, essentially mediating patterning of what had been thought to be independent developmental units, namely limb skeleton and limb muscle. SF/HGF is maintained by a specialized epithelial ridge on the limb bud, the AER. Removing the posterior portion of a limb bud (the zone of polarizing activity; see Chapter 38), results in both enhanced and ectopic expression of SF/HGF in the posterior limb-bud mesenchyme. *Bmp-2* inhibits *SF/Hgf* and exogenous Bmp-2 functioning, as does removal of the posterior mesenchyme. In addition to acting on migrating undifferentiated myoblastic cells SF/Hgf maintains the motility of these muscle-forming cells within the limb bud, thereby keeping them in an undifferentiated state (Scaal *et al.*, 1999).

A COMMENT ON PECTORAL GIRDLES

Development evolves, unless it is constrained.

To illustrate how complex and difficult it is to unravel the mesenchymal origins of single skeletal systems, I comment briefly on the origin of the elements of the pectoral girdle in fowl. Although a single skeletal system, the pectoral girdle is comprised of a number of individual skeletal elements whose embryological origins and modes of ossification differ.

I thought the skull was complex until Matt Vickaryous began study pectoral girdles in my laboratory. Considering how difficult it has been to resolve origins of the parts of this skeletal system in the most-studied vertebrate, the embryonic chick, and that pectoral girdles exhibit substantial convergent and divergent evolution both within and between groups, we must be cautious in extrapolating the mesenchymal origins in chick to other vertebrates. Indeed, there is a rich and fertile field to be ploughed for those who are not faint hearted. In part the complexity of the components of the pectoral girdle reflects a diverse evolutionary history, as might be expected of a skeletal system with a number of elements which have different embryological origins and which can evolve with some degree of independence from one another.

Emerson (1988a,b) used morphological variation of the pectoral girdles and postcranial skeleton in frogs to investigate whether *convergence or morphological constraint* provided the best explanation for the similar morphologies seen in different and often distantly related taxa. Her approach was to investigate whether decreasing morphological diversity correlated with a decrease in the number of independent elements in the pectoral girdle. She found this to be the case for the epicoracoid cartilages, concluding that convergence and constraint both played their parts, with constraint playing the dominant role. Furthermore, Emerson saw *similar historical patterns of morphological change even in groups that are not closely related*. In part, this reflects the constraint caused by the shared embryological origin of the skeletal elements. However, the degree of sharing is now known to be less than Emerson thought.

A long-accepted study by Chevallier (1977) demonstrated that the *scapula arises from somitic mesoderm*, but that coracoid, sternum[20] and clavicle of the pectoral girdle and all elements of the pelvic girdle arise from somatopleural mesoderm (see below for the clavicle). However, more recent analysis by Huang *et al.* (2000b) demonstrates that the chick scapula is of *dual origin and segmentally organized*, the chief findings being that:

- the scapula develops in a rostral to caudal direction;
- the head and neck are derived from lateral-plate mesoderm;
- the blade of the scapula is somatic, being derived from dermomyotome – a region normally associated with producing muscle and connective tissue – leading to the conclusion that the blade is an ossifying muscle insertion and not a primary skeletal element;
- the head and neck are homologous to the true coracoid of 'higher' vertebrates;
- there is an anlage of the coracoid in chicks but it is not seen because it lacks any ossification centre; and
- the segmental organization of the scapula, including muscle insertions, parallels that described by Köntges and Lumsden (1996) for segmental skeletal elements derived from the neural crest.

At the molecular level pectoral girdle development is controlled by *Pax-1* and *Bmp-4*, both of which are expressed in equivalent antero-posterior domains during shoulder girdle and limb-bud development in chick embryos. Exogenous Bmp-2 and Bmp-4 down-regulate *Pax-1*, leading to shoulder girdle defects. Bmp-2 increases *Sox-9* expression, enhancing ectopic chondrogenesis, while noggin, an inhibitor of *Bmp-2*, suppresses *Sox-9*. Similarly, misexpressing *Sox-9 in vivo* initiates ectopic chondrogenesis. Misexpression in the dermomyotome initiates synthesis of type II collagen and *Pax-1* by dermomyotomal cells, which switch to a chondrogenic fate.[21]

Development of the blade of the scapula, which arises from the dermomyotome (see above), and the rostral ribs, is under the control of carboxypeptidase Z (CPZ), a secreted zinc-dependent enzyme that is found in the somites in early embryos and then restricted to the sclerotome. CPZ enhances Wnt-dependent expression of the homeobox gene *Cdx-1*. Mutant CPZ enhances ectopic expression of Pax-3 preceding loss of the scapular blade and rostral ribs (Moeller *et al.*, 2003).

The muscle-specific basic helix–loop–helix transcription factors *myogenin* and *Mrf-4* mediate those girdle changes that are dependent on muscle action. Homozogotes for either transcription factor display primary muscle defects with secondary defects in the thoracic skeleton, including sternal defects and fusions between ribs. The two factors act synergistically, skeletal defects being more severe in compound mutants, providing an elegant example of skeletal development dependent upon muscle development.[22]

THE CLAVICLE: EVEN MORE SURPRISING

In *Archaeopteryx* …'the two collar-bones were jointed to make a merrythought, instead of remaining separate as in reptiles…'[23]

The clavicle presents an even odder element and potentially even more complexity than does the scapula. By rights, I should discuss the clavicle in the next chapter. Why? Because the clavicles are neural crest in origin. The remainder of the endoskeleton (Table 1.1), including the other elements of the pectoral girdle, are mesodermal. Clavicles are intriguing for several other reasons, given below.

- In most birds and mammals the clavicles are the only dermal skeletal elements in the trunk.
- An evolutionary remnant of the ancient dermal skeleton incorporated and integrated so completely into the endoskeleton is unusual (Table 1.1). The clavicle is the only membrane bone associated with the pectoral girdle in birds and mammals; the dermal bones of the lower jaws are nowhere near as integrated with Meckel's cartilage; they lie outside the perichondrium of Meckel's.
- As dermal bones, clavicles develop intramembranously. This is a 'straightforward' direct ossification that can become complicated in several ways, including (i) from the presence of secondary cartilage (often transitory), (ii) the consequent endochondral ossification of a portion of the clavicle, or (iii) from fusion of the intramembranous clavicle with an endochondral element, giving the clavicle two centres of ossification, only one of which is direct.

- From the standpoint of comparative vertebrate biology, the clavicles have been lost in many lineages, providing much greater rotational mobility of the shoulders than is possible with the ventral bridge or strut they create.
- Finally, and from the evolutionary standpoint, a long-standing interest is whether the *furculae* (wishbones, merrythoughts) of birds – and from more recent analyses, of theropod dinosaurs – are homologues of the clavicles of other tetrapods.

An overview of some of these developmental, comparative and evolutionary issues follows.[24]

Humans

Studies on the embryology of the human clavicle carried out in the 1950s and 1960s emphasized that, along with the dentary, the clavicle is the first bone to ossify – as is also true for birds – and that it ossifies from two centres. Koch described clavicular primordia in foetuses of 40–68 cm length but *thought they were preformed in cartilage*, a combination of perichondrial and endochondral ossification beginning early so that the cartilage was transitory and could not mature. In a histochemical examination of the developing human clavicle (in which earlier literature is reviewed and analyzed), Helge Andersen discussed initial ossification as intramembranous, but described a *chondral phase* (Table 1.1), including perichondral and endochondral ossification, and the formation of fibrocartilage at the articular ends.[25]

The tissue-separating potential of various growth cartilages has been investigated by transplanting cartilages across the interparietal suture of rats. Under such conditions the medial clavicular cartilage, costochondral junction, proximal tibial epiphysis, basicranial synchondrosis, and mandibular condyle all have growth-promoting abilities. Basicranial synchondroses adapt to the new environment and express an intrinsic tissue-separating ability when transplanted into sutures of rat skulls.[26]

As discussed in Chapter 33, advantage has been taken of the presence of secondary cartilage on human clavicles to use them as grafts to replace mandibular condylar cartilage in situations where the condylar cartilage is damaged or malformed, essentially replacing the temporomandibular joint (TMJ) with a sternoclavicular joint. The rationale is the similarity of the two secondary cartilages and their organization into similar layers; Borstlap *et al.* (1990) compared the effectiveness of mandibular (chin) and rib grafts in repairing alveolar cleft defects, finding the mandibular grafts preferable. This has been an effective strategy; the growth capacity of costochondral grafts correlates with the amount of cartilage in the graft, although in a recent analysis Ellis *et al.* (2002) found that including too much cartilage in the graft can be detrimental to repair; costochondral cartilage is so effective that it leads to overgrowth.

A number of human syndromes are associated with congenital absence (agenesis) or underdevelopment (hypoplasia) of the clavicles, the most understood being *cleidocranial dysplasia*, for which a mouse mutant model is available (see below).[27]

Other mammals

As in humans, ossification of the clavicles begins early in rodents, earlier than other mammals such as hamsters, being the first bone to ossify in mice. Vital staining of rat clavicles with alizarin red (Box 34.1), coupled with ^{3}H-thymidine-labeling and periosteal sectioning (Chapter 31), shows that the lateral end of each clavicle is made up of the equivalent of an articular cartilage, i.e. a cartilage with little proliferation or proliferative potential. The medial end of the clavicle, also cartilaginous, consists of actively dividing cells and is a growth centre, serving clavicular growth as condylar cartilage serves dentary growth. Hence the use of clavicles as grafts to replace damaged human TMJs.[28]

Tran and Hall (1989) examined development of the clavicles and associated secondary cartilage in embryonic ICR Swiss-Webster mice and resolved part of the confusion over whether secondary chondrogenesis is a component of clavicular development. It is, but it is easily missed. Secondary cartilage is transitory, in this strain appearing at 16 days of gestation (Theiler stage 24), and having been replaced endochondrally two days later.

Others have reported secondary cartilage on the clavicles of placental mammals; in their study of the function of *Pax-1* in the development of the pectoral girdle, Timmons *et al.* (1994) figure secondary cartilage in the clavicles. Marsupial clavicles, on the other hand, lack secondary cartilage.

Secondary cartilage was described on both ends of the clavicle in the greater white-toothed shrew, *Crocidura russula*, and other placental mammals. Großman *et al.* (2002), who described secondary cartilage in the clavicles of the greater white-toothed shrew, the pygmy white-toothed shrew, *Suncus etruscus*, and the Syrian hamster, *Mesocricetus auratus*, consider clavicular secondary cartilage as a derived condition, which accords with my view on the lack of homology of avian and mammalian secondary cartilages (Hall, 2000a, 2003a).

A mutation in *runx-2* (formerly *Cbfa-1*, a member of the *runt* family of transcription factors; Box 20.4) produces *cleidocranial dysplasia* in mice. The clavicles are hypoplastic or absent, and supernumerary teeth form. Two different mutations in the mouse cartilage-specific type II collagen gene promoter Col2a1 cause changes in dentary and clavicle secondary cartilages in mice; maturation of the hypertrophic zones is altered, although the effects are less severe than those seen in primary cartilages (Box 16.3).[29]

Two reports of *Ap-2* null mutant mice were reviewed by Morriss-Kay (1996). Although neural crest appears normal at

Box 16.3 Col2a1, the type II collagen gene promoter

Targeted inactivation of Col2a1 does not affect periosteal or intramembranous ossification. It does, however, inhibit endochondral ossification in mouse embryos. The long bones consist of disorganized chondrocytes lacking extracellular matrix fibrils, epiphyseal growth plates fail to form and endochondral ossification is not initiated, although crania and ribs develop normally. Craniofacial and otic capsule abnormalities also are found in transgenic mice with a Col2a1 mutation, including short mandibles, cleft palate, misshapen otic capsules, reduction of the cartilaginous skull base, defective TMJs, and early mandibular osteoarthritis, the cartilages being unable to resist mechanical forces associated with growth. The mutation is lethal. The human homologue of the Col2a1 promoter (COL2A1) can *restore normal bone development* in Col2a1-null mice; endochondral ossification is initiated and the embryos survive.[a]

During mouse development, Col2a1 is regulated by the DNA-binding agent *Sox-9*. Mutation of *Sox-9* causes skeletal dysplasia, while a glycine–cystine mutation in Col2a1 results in disorganized synchondroses. Two different mutations in Col2a1 affect the secondary cartilages of murine dentaries and clavicle. Maturation of the hypertrophic zones is altered, although not as severely as in the primary cartilages of the long bones.[b]

A new form of *Sox-5* (*LSox-5*), *Sox-6* and *Sox-9* respectively is expressed in sites of chondrogenesis in mice and each activates the chondrocyte marker Col2a1 in 10T1/2 and murine MC615 cells by binding to a 48-base-pair chondrocyte-specific enhancer. Lefebvre *et al.* (1998) referred to these as CSEPs: *chondrocyte-specific enhancer-binding proteins*. *Sox-9* binds to DNA, activates transcription, is co-expressed with type II collagen and regulates Col2a1 during mouse chondrogenesis. *Sox-6* is a downstream regulator of BMP-2-induced chondrogenesis of C3H10T1/2 cells. Finally, there is a complex functional interplay among *Sox-8*, *-9* and *-10* in responding to signals from *Bmp* in the specification of cells as chondrogenic, and an equally complex interplay between *Sox-6*, *Sox-9* and *LSox-5* in interacting with *Bmps* to initiate chondrogenesis.[c]

T. D. Grant *et al.* (2000) developed an elegant method in which a Col2a1–green fluorescent protein reporter construct is used with confocal imaging of thick slices or whole embryos to follow the development of cartilage and chondrogenic lineages, the intensity of the fluorescence paralleling type II collagen synthesis.

[a] See S.-W. Li *et al.* (1995), Rintala *et al.* (1997) and Maddox *et al.* (1998) for targeted inactivation, Rani *et al.* (1999) for rescue, and Horton (2003) for an up-to-date evaluation of targeting the murine genome in order to understand skeletogenesis.
[b] See Rintala *et al.* (1993) and Ng *et al.* (1997) for *Sox-9*, and Rintala *et al.* (1996) for the effect on secondary cartilages.
[c] See Ng *et al.* (1997) for Col2a1, Fernández-Lloris *et al.* (2003) for the study with C3H10T1/2 cells, and Chimay-Monroy *et al.* (2003) for interaction among Sox genes and with Bmps. *Sox-5* and *Sox-6* also up-regulate *Bmp-6* in growth-plate chondrocytes (Smits *et al.*, 2004).

Box 16.4 *Ap-2*-mutant mice

Homozygote mice lacking the transcription factor *Ap-2* are anencephalic, and show many craniofacial defects. Many head and trunk bones are deformed or absent because of deficient neural crest migration. Malleus, incus and clavicles fail to form; the mandibles form but are small. *Ap-2* is expressed in ectoderm and migrating cranial neural-crest cells. Although the neural crest normally arises at 9.5 days of gestation, by 11.5 days the facial processes fail to separate in homozygotes. Increased apoptosis in proximal arch mesenchyme between 9 and 9.5 days of gestation, and development of the maxillary and mandibular arches as a single element, are consequences of neural crest and head ectoderm problems. Recent studies show that a *cis*-element in the fifth intron regulates the action of human AP-2α.[a]

From an analysis of chimaeric mice with *Ap-2+* and *Ap-2−* cells, it is now known that AP-2 is independently regulated in the neural tube, body wall and craniofacial regions, and that it plays a role in limb duplication as well as a previously unreported role in eye development (Nottoli *et al.*, 1998). Given the organs affected, you can see that the *Ap-2* phenotype overlaps with that produced by retinoic acid.

Chicken *Ap-2* is 94 per cent similar to human and mouse *Ap-2*. In chick embryos, *Ap-2* is expressed in migrating neural-crest cells and in facial mesenchyme, especially in the frontonasal and lateronasal processes, and less so in the mandibular and maxillary arches. Implanting retinoic acid decreases levels of Ap-2 and initiates cell death. Removal of the AER also reduces Ap-2, perhaps because Ap-2 determines components of basement membranes that may be involved in epithelial–mesenchymal interactions (Shen *et al.*, 1997).

[a] See J. Zhang *et al.* (1996) and Schorle *et al.* (1996), a discussion of these two papers by Morriss-Kay (1996), Zhang and Williams (2003) for *cis*-regulation, and Brewer *et al.* (2004) for multiple neural-crest defects. The related transcription factor, *Ap-1*, is required for maintenance of the notochord and formation of the intervertebral discs (Behrens *et al.*, 2003). Ten papers from a colloquium organized by The Biochemical Society in Cardiff, 16–18 July 2002 and published in *Biochem. Soc. Trans.* (2002, **30**, 829–878) provide a valuable and up-to-date overview of the biology of intervertebral discs.

have undergone extreme reduction during whale evolution. Indeed, whales are well on the way to losing their sterna as well.[30]

It may seem odd to study the development of the clavicles in animals that lack them or in which they are reduced, but like limbless tetrapods whose embryos retain limb buds (Chapter 37), most clavicle-less mammals develop clavicular rudiments, as is clear from a developmental study of cats and sheep by Rojas and Montenegor (1995). As you might expect, ossification of the clavicle is delayed relative to humans. No cartilaginous phase has been reported in either species. Cat clavicle primordia have an osteogenically active periosteum and progress further developmentally than do the primordia in sheep; the periostea of the vestigial clavicles in sheep consist of no more than a single layer of 'epithelial' cells and contain many osteoclasts. These differences between cats and sheep correlate with the absence of clavicles in sheep and the presence of vestigial clavicles in cats.

Klima (1990), who examined 64 embryos of toothed (odontocete) and baleen (mysticete) whales, found that

9.5 days, the facial processes fail to separate at 11.5 days. The mandibles are small, and the clavicles, malleus and incus all fail to develop. Because Ap-2 determines components of the basement membrane, it may play a role in epithelial–mesenchymal interactions (Fig. 10.7; Box 16.4).

Mammals that lack clavicles

Adult sheep, *Ovis aries*, lack clavicles. Clavicles are rudimentary and non-functional in cats, *Felis domesticus*, and the clavicle exists as a temporary rudiment in odontocetes, the greatest reduction being in the sperm whale, *Physeter macrocephalus*. No rudiment has been described in any mysticete whale. The rudiments in the sperm whale consist of dermal bone with no sign of cartilage or any endoskeletal component.[31]

Birds

Perhaps the most extensive study on avian clavicles (furculae) is an analysis of their variation and/or reduction in parrots (family Psittaciformes). A morphological and phylogenetic analysis of 53 species from eight avian orders demonstrates that clavicle shape is correlated with flight needs.[32] Flightless birds, such as ostriches, lack functioning clavicles but retain vestigeal elements, as Geoffroy St. Hilaire knew from the collections he made in Egypt with Napoleon Bonaparte between 1798 and 1801. In a presentation to the *Institut d'Egypt* in 1798, Geoffroy noted: 'These rudiments of the furcula have not been suppressed, because nature never advances by rapid leaps, and she always leaves the vestiges of an organ even when it is entirely superfluous, if that organ has played an important role in other species of the same family' (cited from Appel, 1987, p. 74).

A transitory secondary cartilage occurs on the clavicles of chick embryos. The clavicular mesenchyme is very sensitive to movement in that a threshold much lower than required by other secondary cartilages is sufficient to evoke clavicular secondary chondrogenesis. I found that the contraction of the amnion in embryos paralyzed with neuromuscular blocking agents was sufficient to elicit secondary cartilage from the clavicle (Hall, 1986a).

A transitory cartilaginous phase has been demonstrated during development of the clavicles in the Japanese quail, *Coturnix coturnix japonica*. The cartilage is invaded by osteoblasts and calcifies in a process described by Lansdowne (1968) as *accelerated endochondral ossification*, which means, of course, that he did not consider the clavicles as membrane bones developing by intramembranous ossification. I must say that in his figures (e.g. his Plate II (a), in which cartilage is identified) it looks very much to me like the cellular bone one sees in avian embryos. Because of Lansdowne's unexpected finding, early clavicle development in Japanese quail was reconsidered by Russell and Joffe (1985), who described mineralization at 9.5 days (H.H. 36) but failed to find any evidence of cartilage, either primary or secondary.

Wishbone or clavicles

Another unresolved issue is whether the paired, ventral thoracic elements that we call clavicles in birds (or dinosaurs) are the same bones that we identify as clavicles in other taxa. Known as the *furcula* in live animals and museum specimens and as the *wishbone* in cooked birds, a wishbone being 'a forked bone between the neck and breast of a cooked bird: when broken between two people the longer portion entitles the holder to make a wish' (OED). The term 'wishbone' was coined in the U.S. around 1850. The older English term was '*merrythought*' – the one left with the largest piece would marry first. Gavin de Beer, in a BBC Science Survey Broadcast on 13 January 1955, referred to *Archaeopteryx* (Fig. 32.2) in which 'the two collar-bones were jointed to make a merrythought, instead of remaining separate as in reptiles,' as evidence for the avian affinities of *Archaeopteryx* (de Beer, 1962, p. 68). The merrythought could be either:

- a single element that arises from fusion of the two clavicles;
- another element(s) of the ancestral pectoral girdle; or
- a neomorph, a new bone that occupies the same position as the clavicles occupy in other groups.

Bryant and Russell (1993) considered whether the furcula is a neomorph or modified clavicles inherited from avian dinosaur ancestors, concluding that the furcula is a *neomorph*; see Chapter 44 for further discussion of neomorphs. In large part, they came to this conclusion because evidence for the presence of clavicles in dinosaurs was at that time still equivocal. However, more recent discoveries have identified a furcula in tyrannosaurid theropods and other dinosaurs, including *Velociraptor* and the early Jurassic theropod dinosaur, *Syntarsus*, the oldest known dinosaurian clavicle. These recent findings make homology of avian and dinosaur furculae more likely than Bryant and Russell thought.[33]

NOTES

1. See Hall and Hörstadius (1988) and Hall (1999a) for detailed discussions of the neural crest, and Hall (1997c, 1999a, 2000a,b, 2005a) and Stone and Hall (2004) for the neural crest as the fourth germ layer.
2. For reviews of research on somites and somitic chondrogenesis, see Hall (1977a, 1986b), *Verbout (1985)*, Balling *et al.* (1996), Christ *et al.* (1998), Tajbakhsh and Spörle (1998), Brand-Saberi and Christ (2000), Huang *et al.* (2000a), Monsoro-Burq and Le Douarin (2000), Stockdale *et al.* (2000), Fleming *et al.* (2001, 2004) and Hofmann (2003).
3. To obtain a feeling for how our approaches to somitic chondrogenesis and vertebral development has progressed over the past 45 years and more, see Holtzer (1959, 1961, 1968), Lash (1963a, 1968a,c), Hall (1977a), Strudel (1967), Holtzer and Abbott (1968), Holtzer and Mayne (1973), Levitt and Dorfman (1974), Verbout (1985), Lash and Ostrovsky (1986), Balling *et al.* (1996) and Brand-Saberi and Christ (2000).
4. Since the work of Langman and Nelson (1968) there has been evidence that cells from the 'dermotome' produce myoblasts *and* fibroblasts. Consequently, the myotome and dermotome are usually collectively termed the *dermomyotome*.
5. For early stages of formation of preaxial mesoderm and somitogenesis see Lipton and Jacobson (1972a, especially Figs 3–9), Hall (1977a, 1978a) and Bellairs *et al.* (1978).

Lipton and Jacobson (1974), Bellairs and Portch (1977), the papers in Ede et al. (1977a), Bellairs (1979), Brand-Saberi and Christ (2000) and Fleming et al. (2001) provide extensive reviews and/or elegant accounts of early studies on somitogenesis, the latter including reproduction of figures from classic papers from the 19th and 20th centuries.

6. See Chevallier (1978), Arey (1974) and Minor (1963) for the timing of sclerotome migration, Dockter (2000) and Stockdale et al. (2000) for overviews of sclerotome induction and differentiation, Dockter and Ordahl (2000) for determination of the dorsoventral axis of the somitic mesoderm as determined by dorsoventral somite rotation and use of dorsal and ventral gene markers to show that sclerotome (unlike myotome) remains phenotypically plastic during early somitogenesis, and Hirsinger et al. (2000) for an overview of somite patterning.

7. See Olson and Low (1971) for these major stages, and Scherft (1978) for osteoid mineralization. Vertebral development between H.H. 35 and 42 (8.5–16 days) is described at the ultrastructural level by Crissman and Low (1974).

8. See Kvist and Finnegan (1970a,b), Manasek (1970) and Manasek et al. (1973) for early studies on hyaluronan and heart development.

9. See Hall (1977a, 1998a, 2000c), Tanaka and Uhthoff (1981), Bagnall et al. (1988), Ewan and Everett (1992) and Carlson (1996) for basic discussions of resegmentation, and see Piiper (1928) for the term sclerotomite. Brand-Saberi and Christ (2000) reviewed resegmentation in the context of a review of cell lineages derived from somites. Burke (2000) discusses resegmentation in a review of Hox genes and global patterning of somitic mesoderm, including a discussion of the origin of the scapula. Resegmentation applies to amniotes. In the zebrafish, Danio rerio, the sclerotome arises in the ventrolateral portion of each somite, each somite half contributing to more than one vertebra (leaky resegmentation), while in Atlantic salmon, Salmo salar, the centra arise within the perinotochordal sheath from a specialized population of chordoblasts (Stickney et al., 2000; Morin-Kensicki et al., 2002; Grotmol et al., 2003).

10. See Gearhart and Mintz (1972), Moore and Mintz (1972), and the reviews by Mintz (1971, 1972) for the original clonal theory of vertebrae, and Abbott et al. (1972) for a critical comment on the experimental method. See O'Higgins et al. (1997) for similar analyses on different classes of vertebrae in BALB/c and CBA mice and in humans.

11. See Rufai et al. (1995) and Trout et al. (1982) for the rat and human studies respectively, and see Smits and Lefebvre (2003) for Sox-5 and Sox-6, both of which up-regulate Bmp-6 and down-regulate Ihh, FgfR-3, and Runx-2 in growth-plate chondrocytes (Smits et al., 2004). See note e in Box 13.4 for effects on condensation of double knockout of Sox-5 and Sox-6. See Biochem. Soc. Trans. (2000) **30**, pp. 829–878, for 10 papers on the intervertebral disc and nucleus pulposus.

12. One has to be mindful of replacement (regulation; Box 35.1) in such studies. Somites with normal rostral–caudal polarity regenerate following removal of single somites from two-day-old chick embryos, provided that a gap remains after somite removal. Use of DiI labeling showed that the new somites arise from the adjacent somite and intermediate mesoderm. Lack of HNK labeling shows no contribution from NCC (Liu and Bagnall, 1995).

13. See Bagnall et al. (1988) for labeled single cells, Ewan and Everett (1992) for gene transfer, Goldstein and Kalcheim (1992) for pedicle and intervertebral disc, and Huang et al. (2000a) for ribs. Huang et al. (1996) also used chick–quail chimaeras to provide evidence for resegmentation in relation to the vertebrae.

14. Verbout (1976), who rejects resegmentation (Neugliederung) for all tetrapods, provides an extensive review of the history.

15. See O'Rahilly and Meyer (1979) for human embryos, Tanaka and Uhthoff (1981) for the types of anomalies, and Dalgleish (1985) for the autoradiographic study.

16. See Pinot (1969a, 1970), Kieny (1971) and Kieny et al. (1972) for the in-vitro studies with chick limb buds, and Houben (1976), Milaire (1976, 1977). Agnish and Kochhar (1977) and Milaire and Mulnard (1984) for mice.

17. See Agnish and Kochhar (1977) for the co-culture, and Chevallier et al. (1977) for destruction of somatic mesoderm.

18. See Christ et al. (1977, 1979), Chevallier et al. (1977), B. Beresford et al. (1978) and Chevallier (1979). Somitic myogenic cells are determined as myogenic from H.H. 19 onwards, while lateral-plate mesenchyme is determined as chondrogenic from H.H. 20 on (Wachtler et al., 1981, 1982). Of course, this experiment does not speak to the full developmental potentials of wing mesenchymal cells, only to the potential expressed under the experimental conditions.

19. See Daneo and Filogamo (1973) and Fouvet (1973) for timing of innervation and myosin synthesis, Bonner (1975) for the CAM-grafts, Ahmed (1966), Sullivan (1967) and Wu (1994, 1996) for innervation, Holmes and Niswander (2001) for Slit-2 and Slit-3, Wu (1994, 1996) and Chiego and Singh (1981) for nerves and periosteal proliferation, and Herskovits et al. (1991) for an overview of the innervation of bone.

20. Following Fell's pioneering (1939) studies on domestic fowl and budgerigar sterna, Chen (1952a,b, 1953) used the mouse sternum as a model system to explore the morphogenesis of skeletal elements in vitro. The sternum fails to develop if Hoxa-4 is knocked out in mice; over-expression of Hoxa-4 reduces the anlage for the ribs (Horan et al., 1994). Over-expressing human Hox 3.3 (HOXc-6) in mice leads to formation of additional ribs on the sternum and in the lumbar region (Jegalian and de Robertis, 1992).

21. See Hofmann et al. (1998) for regulation of Pax-1 by Bmps, and Healy et al. (1999) for expression of Pax-1 in the dermomyotome.

22. See Vivian et al. (2000) for the study with myogenin and Mrf-4. Mrf-4, also known as Myf-6 and herculin, is one of the four members of the MyoD gene family.

23. de Beer (1962, p. 68), recalling a BBC Science Survey Broadcast of 13 January 1955.

24. See Gardner, E. (1968), Hall (1986a, 2001a) and Bryant and Russell (1993) for other reviews.

25. See Gardner and Gray (1953), Gardner (1968) and O'Rahilly and Gardener (1972) for onset of ossification of the clavicle and for development of the pectoral girdle, Koch (1960) for the chondral phase, Andersen (1963) for the histochemical analysis and review, and B. D. Hall (1982) and Oppenheim et al. (1990) for fractures. A survey conducted in 1990 documented 2.7 clavicle fractures/1000 live births in humans, mostly associated with large birth weight.

26. See Kylämarkula and Rönning (1979, 1983), Rönning and Kylämarkula (1982), Rönning and Kantomaa (1988), Rönning et al. (1991a,b), Rönning (1995) and Peltomäki et al. (1997)

for studies on tissue separating ability of various cartilages. For a radiographic study of the development of human parietal bone and interparietal suture using 15 specimens to show the primary ossification centre, see Silau et al. (1995). Lieberman et al. (2000) used extensive measurements of the cranial base in primates to demonstrate correlation with brain volume and stress the importance of developmental interactions between basicranium, brain and skull. See the chapters in Minugh-Purvis and McNamara (2002) for developmental change and human evolution.

27. See Ellis and Carlson (1986), Hennig et al. (1992), Precious and Hall (1992), Hall (2001a) and Ellis et al. (2002) for clinical studies, and see Hall (1986a, 2001a) and Peltomäki (1992, 1993) for growth capacity. Yih et al. (1992) grafted autogenous auricular cartilage into the human TMJ and biopsied ankylosed joints. The cartilage cells remained viable, proliferated and deposited ECM; they were encased in fibrous capsules.

28. See M. L. Johnson (1933) and Kanazawa and Mochizuki (1974) for clavicle development in albino mice and hamsters. The clavicle may be the first bone to appear in most mammals; it is the first to appear in eight species analyzed by Sánchez-Villagra (2002).

29. See Mundlos et al. (1997) for runx-2, Rintala et al. (1996) for the collagen mutation, and Zelzer and Olsen (2003) for genetic defects including cleidocranial dysplasia.

30. See Klima (1985) and Großman et al. (2002) for analyses of the development of shoulder girdles and sterna in mammals.

31. See Hall (1984b, 1995d, 2002c, 2003a) and Bejder and Hall (2002) for further discussions of such rudiments.

32. See Glenny (1954, 1959) and Glenny and Friedmann (1954) for parrots, and Hui (2002) for the correlation with flight.

33. See Makovicky and Currie (1998), Norell et al. (1997) and Tykoski et al. (2002) for the dinosaur studies, including cladograms showing the taxa with separate clavicles or with clavicles fused as a furcula.

Skeletal Origins: Neural Crest

Before we begin, one note on terminology. To reflect its ectodermal origin, neural crest-derived mesenchyme sometimes is known as *ectomesenchyme*. I use ectomesenchyme when I want to emphasize embryological origin. Otherwise I use mesenchyme.[1]

DIFFERENT MESENCHYMES, SAME TISSUES

From time to time the notion crops up in the literature that neural crest- and mesodermally derived skeletal cells or tissues – chondroblasts/osteoblasts, cartilage/bone – *should differ* in some way, *because they originate from different germ layers*. Little solid evidence exists for this notion, although examples of two of the types of evidence proffered are worth mentioning. One relates to embryonic origin *per se*, the other to evidence from a syndrome.

Alveolar bone, which arises from the dental papilla in association with dentine, and which is of neural-crest origin, is physiologically distinct from other bone. As evidenced by loss of teeth during pregnancy in women whose diet is deficient in calcium, alveolar bone is much more susceptible to resorption than is bone at other sites. Is this because of some property of alveolar osteoblast related to embryonic origin? I take up alveolar bone again in Chapter 25.

Another related line of evidence is found in the case history of a human skeletal syndrome, a woman with *human familial osteodysplasia*, also known as Anderson syndrome, an autosomal recessive condition seen most frequently after consanguineous marriages, in this case in one female and three of her four children. It is intriguing

that in this syndrome cartilage develops normally and *only bone is deficient*. All neural crest-derived bones of the jaws, skull and clavicles, and all mesodermal bones (ribs, pelvis, limbs), are affected.

The afflicted woman in this case had severely weakened bones; she suffered a spontaneous fracture of the jaw during a tooth extraction when 13 and a spontaneous green-stick fracture of the jaw while eating a hamburger at age 22. A further fracture in the same year followed a blow to her face. She lost her adult teeth, which spontaneously loosened to the point that she could extract them manually.

The authors of this study found it surprising, as do I, that alveolar bone was histologically normal. I would seek an explanation in the fact that alveolar bone arises from the same population of cells that produces dentine, cementum, pulp and the periodontal ligament (Chapter 24), that this population is part of the exoskeleton, and that the exoskeleton has been separate from the endoskeleton for over 550 million years (see Chapter 6). Although neural crest in origin, this multipotential cell lineage of cells is separate from the lineage(s) that forms other neural crest-derived bone, the cell lineages also having been separate for over half a billion years. I would predict that the mutation was not operative in this cell lineage.[2]

A different but intriguing perspective on the issue of similarity or differences between mesodermal and neural crest-derived mesenchyme comes from the *production of a mesodermal skeletal element from NCC* under experimental conditions. Taking advantage of the knowledge that the lateral brain is mesodermal in birds but of neural-crest origin in mammals, Schneider (1999) transplanted neural-crest cells (NCC) into the appropriate mesodermal territory in chick embryos and showed that lateral braincase elements formed from the transplanted NCC.

This elegant example of regulation (Box 35.1) shows the similar differentiative capabilities of mesodermally and neural crest-derived mesenchyme.

NEURAL CREST AS A SOURCE OF SKELETAL CELLS

The neural crest is a topographic term for the ridge of cells at the junction between presumptive epidermal and neural ectoderms during early neurulation of vertebrate embryos (Fig. 17.1). The neural plate transforms into a neural tube, either by invagination of the neural folds (most fish and all tetrapods) or by cavitation of a solid rod of neural ectoderm (a neural keel) as seen in some teleost fish and in lampreys.

NCC, now in the apices of the neural folds or above the neural keel, migrate away from the neural ectoderm to many areas of the body where they differentiate into an amazing diversity of cell and tissue types.[3] Spinal ganglia, the sympathetic nervous system, the adrenal medulla (chromaffin cells), dental papillae, odontoblasts and dentine of teeth, melanocytes, connective tissue cells, chondroblasts and osteoblasts of the craniofacial skeleton – and the clavicles (Chapter 16), cardiac skeleton (see below) and skeletal tissues of antlers and horns (Chapters 7 and 8) – all arise from descendants of these wandering cells. What is the evidence for the skeletogenic potential of ectomesenchyme?[4]

EVIDENCE OF SKELETOGENIC POTENTIAL

Until perhaps the early 1980s, interest in the neural crest was motivated primarily by the suspicion that many craniofacial and dental anomalies and malformations originate in defective migration, differentiation and/or growth of NCC. Harrison (1984), who analyzed 54 patients with osseous and fibro-osseous conditions affecting craniofacial bones, concluded that each represented a separate entity and developmental defect. To cite but one of many hundreds of studies, administering retinoic acid to pregnant rats leads to death of preotic neural crest and subsequent deficiencies in the first and second branchial arches, leading to ear and jaw defects that mimic (phenocopy) the human Treacher Collins syndrome (mandibulofacial dysostosis).[5]

Evidence that NCC produce cartilage, bone and dentine comes from several types of study, including:

- selectively ablating the neural crest from neurula or earlier-aged embryos, transplanting neural crest into ectopic sites, replacing neural crest with other tissues, or exchanging neural crests between different species;
- nuclear, cellular, isotopic or genetic markers, either alone, or in conjunction with ablation and/or transplantation; and
- cultivating NCC *in vitro*.

An overview of the first and second classes of evidence follows.

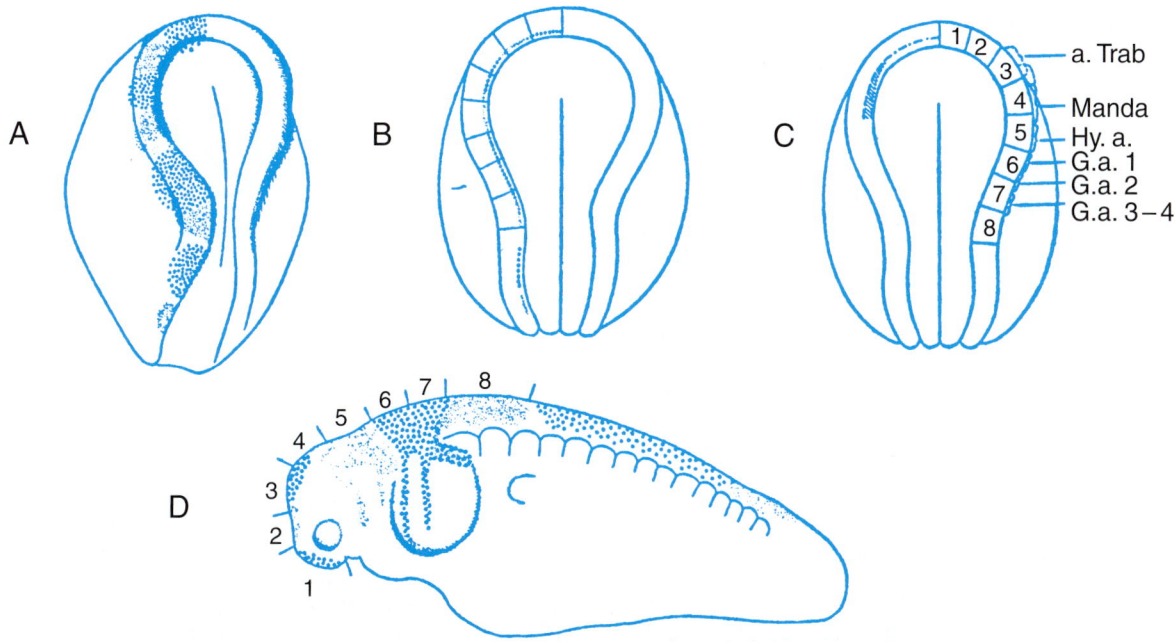

Figure 17.1 The neural folds of these neurula-stage embryos of the Mexican axolotl, *Ambystoma mexicanum* were labeled with vital dyes by Hörstadius and Sellman to map the chondrogenic neural crest. (A–C) dorsal views of neurulae with the eight regions of the crest shown in C. (D) is a lateral view of a later embryo showing the same eight regions and NCC migrating away from them. Regions 1–7 represent cranial neural crest, region 8 represents trunk neural crest. The chondrogenic crest lies within regions 3–7, with trabecular cartilage arising from region 3, mandibular arch cartilage from region 4, and hyoid and branchial arch skeleton from regions 5–7. Labels in C are: a. Trab, the anterior trabeculae – the most anterior elements of the neural crest-derived skeleton; Manda and Hy. a., the regions of the crest from which the mandibular and hyoid arch skeletons arise; G.a.1, G.a. 2, G.a. 3–4, the regions of the crest that provide the skeleton to gill arches 1–4. See text for details. Modified from Hörstadius (1950).

Ablation and transplantation experiments

Some **caution** must be exercised when interpreting deletion and transplantation experiments. NCC from one side can migrate across the midline to the other side, and NCC rostral or caudal to the ablation site can migrate into the gap. As a result, unilateral extirpation may not yield true results. Furthermore, non-neural-crest cells may, in the absence of NCC, *regulate* to form tissues normally produced only by NCC. *Regulation* is the ability of undifferentiated embryonic cells or parts of early embryos such as blastomeres to compensate for lost or deleted cells or parts by differentiating along a pathway they would normally not follow (Box 35.1). Regulation is an entirely different process to that of regeneration, which occurs either by dedifferentiation and redifferentiation of cells to form the regenerate, or from a permanent population of multipotential stem or reserve cells (Box 35.1).[6]

Ross Harrison, a pioneering American experimental embryologist and long-time Editor-in-Chief of the *Journal of Experimental Zoology*, conducted experiments that are especially elegant and pertinent.

The tiger salamander, *Ambystoma tigrinum*, uses rapid tongue projection in prey capture. Consequently, the morphogenesis of the skull is modified and the hyobranchial reorganized in comparison with other salamanders. Taking advantage of the differential growth rates of the larval organs of the spotted salamander, *Ambystoma punctatum*, and the tiger salamander, Harrison exchanged branchial neural crest – which was suspected of producing the visceral arch skeleton – between neurula-stage embryos of the two species. Development of visceral skeletons from the grafts confirmed the neural-crest origin of these skeletal elements. *Furthermore, the cartilages were always of sizes typical of the donor species*, which demonstrated to Harrison that basic size was intrinsically determined (Fig. 17.2).[7] This classic study and the figure must have been cited and used hundreds of times over the last 70 years.[8]

Similar studies – which produced similar results – were designed by Baltzer (1952) to create chimaeric teeth or visceral arches by transplanting tissues between *Triton* and the fire toad, *Bombinator*. Similarly, mouse ribs transplanted intramuscularly are not affected by the age of the host, instead growing in relation to the age of the donor, growth rates being intrinsic (Holtrop, 1967a).

Harrison's study involved only larval cartilage. What about the bony skeleton of adult amphibians? Hörstadius (1950) devoted only one paragraph of his classic monograph to the origin of membrane bones from neural crest. Even as late as 1970, Weston was not able in his review to marshal much evidence for the neural-crest origin of cranial bones. In subsequent transplant experiments with amphibians, however, craniofacial *cartilage, bone and*

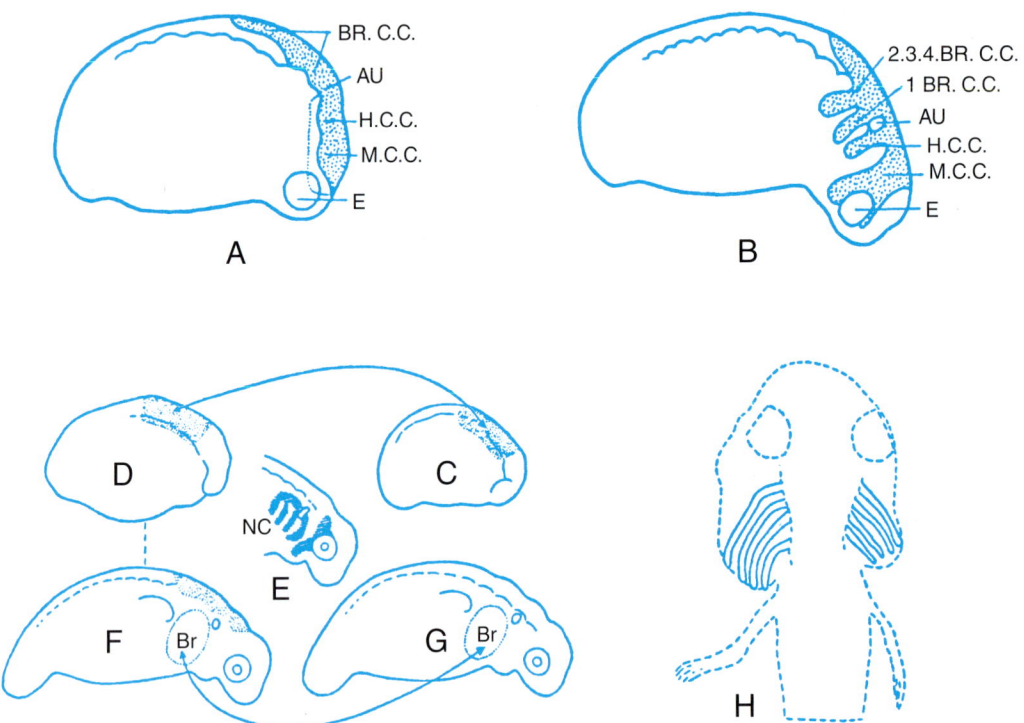

Figure 17.2 (A, B) A fate map of the cranial neural crest for the spotted salamander, *Ambystoma punctatum*, produced by Stone (1926) as seen on the right-hand side of early (A) and late (B) neurula-stage embryos. The paths of migration of NCC are also shown in B. E and AU identify the eye and auditory vesicle. M.C.C., H.C.C., and BR. C.C. identify the origin of mesenchyme for the mandibular and hyoid arch and branchial arches 1–4, respectively. (The cell populations for arch 1 and arches 2–4 are depicted in B.) (C–H) depict Ross Harrison's experiment transplanting neural crest destined for the branchial arches between neurulae of different species of *Ambystoma*, experiments that demonstrated that size of the branchial arch skeleton is intrinsic to NCC and species specific (H). See text and Harrison (1935a) for details. Modified from Hörstadius (1950).

dentine all have been shown to be derived from the mesenchyme of the neural crest, a conclusion echoed by later studies with other jawed vertebrates (teleost, avian and mammalian embryos), and one that applies to the craniofacial cartilages of jawless vertebrates, the lampreys.[9] This does not mean that amphibian adult bony cranial skeletons have been mapped out. On the contrary, knowledge is fragmentary, and only very recently have labeling methods begun to make some inroads into the origin of the dermocranium.[10]

Hammond and Yntema (1964) deleted portions of the neural tubes from chick embryos with six to 12 pairs of somites (H.H. 9 and 10+), the stage when NCC are migrating away from the neural tube. The hyoid skeleton failed to form following removal of the population of cells that normally migrate from the postotic neural crest. Depleting preotic neural crest resulted in deficiencies in Meckel's cartilage of the mandibular arch. This extirpation study suggests that *regionalized populations of NCC* produce the cartilaginous skeleton, more rostral (preotic) NCC forming more rostral cartilages (Meckel's) than more caudal (postotic) NCC, which formed more caudal (hyoid arch) cartilages. Hammond and Yntema examined these embryos at too early a developmental stage to detect whether cranial bones were crest derived. Labelling or marking NCC is a way to circumvent this problem.[11]

Because markers were unavailable at the time, we used ablation of small (200–250 μm cranio-caudal extent) pieces of neural crest from embryos to map the chondrogenic neural crest of a teleost fish, the Japanese medaka, *Oryzias latipes*, and of an agnathan, the sea lamprey, *Petromyzon marinus*, finding a pattern that was highly conserved with that found in amphibians (Langille and Hall, 1986, 1987, 1988b,c).

Marker experiments

³H-thymidine

J. A. (Jim) Weston from The University of Oregon, M. C. (Mac) Johnston from the University of North Carolina at Chapel Hill, and Pierre Chibon at Université Scientifique et Médicale de Grenoble, France, pioneered the use of isotopically labeled transplants to follow the fate of the NCC in avian and amphibian embryos. Jim Weston and Mac Johnston were both Ph.D. students at the time. Learning of each other's project, Mac (who was trained as a dentist) took the head, Jim the trunk.

In Johnston's studies with chick embryos, neural crests from embryos incubated for 30 hours (H.H. 8–9) were replaced with similar-sized pieces from embryos previously labeled with ³H-thymidine. The cartilages and connective tissues that developed in the heads and visceral arches were labeled, indicating their derivation from NCC. Weston followed the pathways of migrating NCC to detect whether cells could differentiate before they left the crest, while they were migrating, or only after

they reached their final sites. Deficiencies did not occur when neural crests from older labeled embryos were transplanted into younger unlabeled embryos, indicating that NCC are specified for chondrogenesis before migration begins.[12]

³H-thymidine labeling was also used to identify the *unique pathways of migration taken by cells from each region of the avian neural crest*, and to demonstrate that the pathways are not fixed before migration begins; pathways of migration are normal even when regions are exchanged (Noden, 1975). As introduced below and discussed more fully in Chapter 18, interactions with epithelia before, during or after migration are of prime importance for the subsequent differentiation of skeletal cells and, because of reciprocal interaction, for the continued growth of the epithelia.

Xenopus laevis–Xenopus borealis chimaeras

Sadaghiani and Thiébaud (1987) used interspecific transplantation between *Xenopus laevis* and *X. borealis* coupled with scanning electron microscopy to demonstrate the contribution of NCC to the mesenchyme of the dorsal fin, and to map and confirm three waves of cranial neural-crest migration in *X. laevis*:

- a mandibular crest of cells that migrates into the *mandibular arches* and gives rise to Meckel's cartilage, and the quadrate, ethmoid and trabecular cartilages;
- a hyoid crest that migrates into the *hyoid arch* and gives rise to the ceratohyal; and
- a branchial crest that migrates into the *branchial arches* and gives rise to the gill cartilages (Fig. 17.3).

Quail–chick chimaeras

Transplanting NCC that possess a species-specific nuclear marker has been most elegantly and profitably used by Nicole Le Douarin and her colleagues in Nogent-sur-Mer in France, and by Drew Noden at Cornell University in Ithaca, NY. The technique uses embryos of the Japanese quail, *Coturnix coturnix japonica*, whose cells contain nuclei with large accumulations of heterochromatin, which can be distinguished from the nuclei in cells of the common fowl, *Gallus domesticus*, in which heterochromatin in dispersed uniformly (Fig. 15.4). Quail–chick chimaeras were first used by Amprino *et al.* (1968) in an analysis of limb-bud development.[13]

By transplanting ³H-thymidine-labeled and/or quail neural crest to great effect in a decade-long series of studies that began in 1978, Noden demonstrated that:

- the NCC from each cranial area have a *unique pattern of migration* that is not irreversibly fixed before migration is initiated – regions can be exchanged and the migratory patterns remain normal; and that

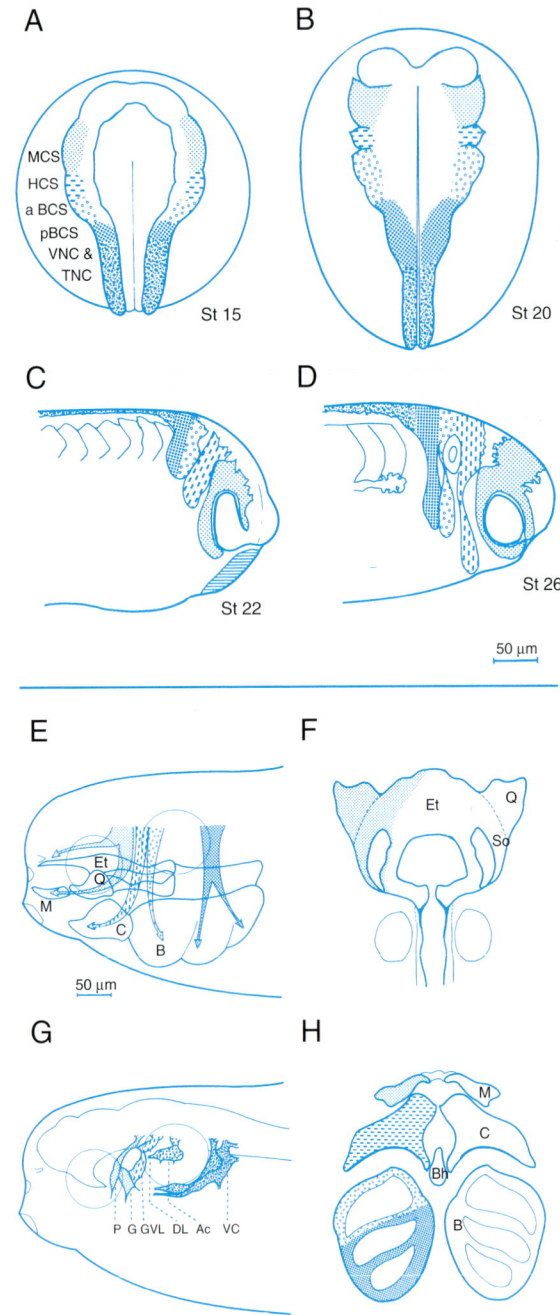

Figure 17.3 Migration of NCC and their contribution to the skeleton and cranial ganglia in the South African clawed frog, *Xenopus laevis*. The three major migration pathways of NCC in embryos of stages 15, 20, 22 and 26 are shown in dorsal (A, B) and lateral (C, D) views. (E) Migration of discrete populations into the maxillary, mandibular (M) and pharyngeal arches. (F, H) Major elements of the tadpole skull that arise from the streams of NCC shown in E; including Meckel's cartilage (M), and the ceratohyal (C), ethmoid–trabecular (Et), quadrate (Q) and subocular (So) cartilages. (G) The cranial ganglia that arise from NCC. Modified from Sadaghiani and Thiébaud (1987).

- the cranial skeleton develops normally following an exchange of regions of cranial neural crest – e.g. quail forebrain neural crest transplanted in place of chick hindbrain, and that mesencephalic (midbrain level) neural crest contributes to the bones of the skull.

Furthermore, Noden's studies:

- uncovered some of the environmental interactions directing the migration of cranial NCC, including similarities of the periocular and visceral arch environments that promote chondrogenesis and the neural-crest origin of periocular mesenchyme;
- revealed early specification of patterns of cell migration and that the *broad pattern of the craniofacial skeleton is established before NCC initiate their migration*. First arch crest transplanted to second or third arch positions in the neural tube migrates to the hyoid arch but produces mandibular (first arch) skeletal structures, including an ectopic beak (Box 17.1) and, astonishingly, elicits mandibular arch muscles from what should have been hyoid arch muscles;
- showed that trunk myotome grafted into cranial mesenchyme adjacent to the neural tube produces ectopic cartilages and ectopic membrane bones, and surprisingly that the *ectopic cartilages – which are mesodermal in origin – fail to fuse with host, neural crest-derived cartilage*, a finding that relates to the potential for interaction between neural crest- and mesodermally derived mesenchyme discussed earlier in this chapter; and
- finally, analyzed the implications of developmental processes affecting NCC for the development of craniofacial morphology.[14]

Within chick heads, only the occipital bones and bones of the otic capsule are entirely of mesodermal origin. Other bones, such as the frontal, are formed from mesodermal and ectomesenchymal cells (Couly *et al.*, 1993, 1998).

The ability of the neural crest to form cranial skeletal elements in chick embryos extends from the mid-prosencephalon caudally to the level of the fifth pair of somites, a region known as the cranial neural crest (Figs 17.1, 17.3 and 17.6). Cartilages of neural-crest origin *are* found more caudally – secondary cartilage on clavicles (Chapter 16), and cardiac cartilages (Box 17.2) – but these cartilages arise from *cranial* NCC that migrate caudally. Trunk NCC, defined as crest caudal to somite five (Figs 17.1 and 17.6), do not form cartilage.[15]

Genetic markers for murine neural crest

Being intrinsically more difficult to investigate, analysis of the neural crest in mammalian embryos has lagged behind the study of amphibians and birds. Pictet *et al.* (1976) described a technique for 'isolating' neural crest from mouse embryos, which necessitates removing the entire ectoderm and organ culturing the remaining cell layers. Although primarily used to show that insulin-producing cells are of neural-crest origin, this method might also be used profitably for the study of skeletal elements and their origins.

In an important extension of our ability to follow NCC in ontogeny, a two-component genetic marker system

Box 17.1 Bird beaks

Beaks (Fig. 7.7) form in response to epithelial–mesenchymal interaction(s). Histological studies on chick beak by Kingsbury *et al.* (1953) showed a thickening of the epidermis adjacent to proliferating mesenchyme at three days of incubation, reminiscent of that described in the mandible by Jacobson and Fell (1941).

Hayashi (1965) used a series of tissue recombinations between duck (Fig. 17.4) and chick (see Figure 17.5 for a mouse–chick recombination) to demonstrate interaction during beak development and that mesenchyme controls epithelial differentiation:

- duck oral mesenchyme plus chick oral epithelium → a duck beak complete with tooth ridge; while
- chick oral mesenchyme plus duck oral epithelium → a chick beak but lacking a tooth ridge.

Tonegawa (1973) conducted a similar series of tissue recombinations using beak and other skin derivatives to demonstrate that 'beak' epithelium from six-day-old embryos would not form a beak unless combined with mesenchyme, and that the inductive ability of beak mesenchyme persists to beyond hatching.

The epithelium of the frontonasal process in chick embryos contains a boundary region marked by *Shh/Fgf-8* expression on either side of the boundary. This region specifies the dorsoventral axis of the upper beak, and elicits ectopic upper and lower beaks when transplanted into other sites within craniofacial mesenchyme. We know that the source of that mesenchyme is the neural crest, and that beaks are species specific (Figs 7.7 and 17.4). NCC from Japanese quail grafted into duck embryos produce quail beaks and *vice versa*, and they do so by initiating patterns of gene expression typical of the donor providing the neural crest, and by modifying patterns of expression of genes such as *Shh* and *Pax-6* in host tissues to conform to the patterns typical of the donor tissues.[a]

Studies on the chick mutant *cleft primary palate* (*cpp*), in which the upper beak is shortened but the lower beak normal – a phenotype similar to that seen in paralyzed embryos – show that the growth defect resides in the epithelial covering of the frontonasal process. *Fgf-8* fails to down-regulate and so remains active in the epithelium for at least two days after it is down-regulated in wild-type embryos. Increased levels of cell death in upper beaks were also found at later stages of development in mutant embryos. Cell death in the lower beaks was normal.[b]

[a] See Hu *et al.* (2003) for the epithelial transplants, Schneider and Helms (2003) and Tucker and Lumsden (2004) for duck–quail chimaeras, and Chrisman *et al.* (2004) for an *Fgf-8/Shh* boundary in limb development and how ethanol exposure *in utero* disrupts this boundary and produces pre- or postaxial digital defects, depending on the concentration of ethanol given.
[b] See Murray and Drachman (1969) for paralyzed embryos, and MacDonald *et al.* (2004) for the studies with the *cpp* mutant.

Figure 17.4 The pink-eared duck, *Malarcorhynchus membranaceus*, photographed by the author in Armidale, NSW, Australia.

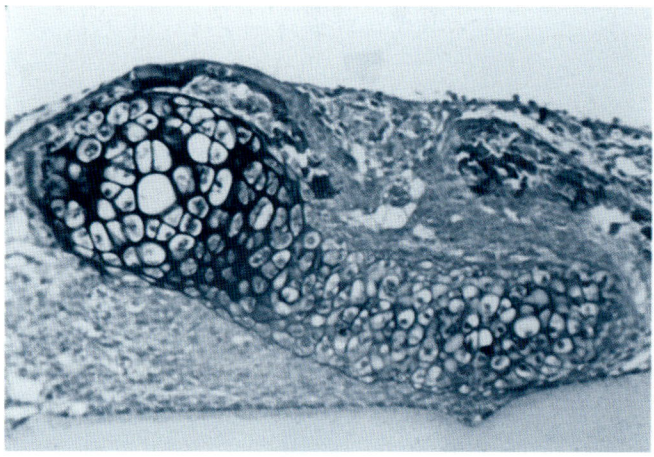

Figure 17.5 Cartilage and bone have differentiated in this CAM-graft of mandibular mesenchyme from a 13-day-old mouse embryo, recombined with mandibular epithelium from a chick embryo of H.H. 22.

(*Wnt-1-Cre/R26R*) was developed and used to follow cranial NCC in mouse embryos from 9.5 days of gestation to as late as six weeks after birth. Genetically marked cells were seen in Meckel's cartilage, the dentary of the mandible, the articular disc of the temporomandibular joint, branchial arch nerve ganglia, and in all tissues that arise from the dental papilla – mesenchyme, odontoblasts, dentine, pulp, cementum and periodontal ligament.[16]

More recently, Morriss-Kay and her colleagues used the *Wnt-1-Cre/R26R* marker to demonstrate that the frontoparietal and parietal–parietal sutures in mouse embryos arise at a boundary between neural crest and mesoderm. Cells of the frontal bone arise from neural crest, those of the parietal from head mesoderm, a tongue of neural crest-derived mesenchyme lying between the parietal bones (Fig. 17.10). Further, while mesodermally derived bones such as the parietal must interact with the neural crest-derived meninges for osteogenesis to be initiated, neural crest-derived bones such as the frontal bones require no such interaction (see Chapters 21 and 34).[17]

Surrounding tissues also influence the development of neural crest-derived skeletal elements. Knocking out the homeobox gene *Hoxa-2* leads to the second arch transforming to a first arch, *Hoxa-2* affecting branchial arch fate by altering tissue around post-migratory NCC rather than from direct action in neural crest–mesenchyme. There is also an interesting association between mandibular skeletogenesis and sites of nerves in normal and abnormal development. During normal development

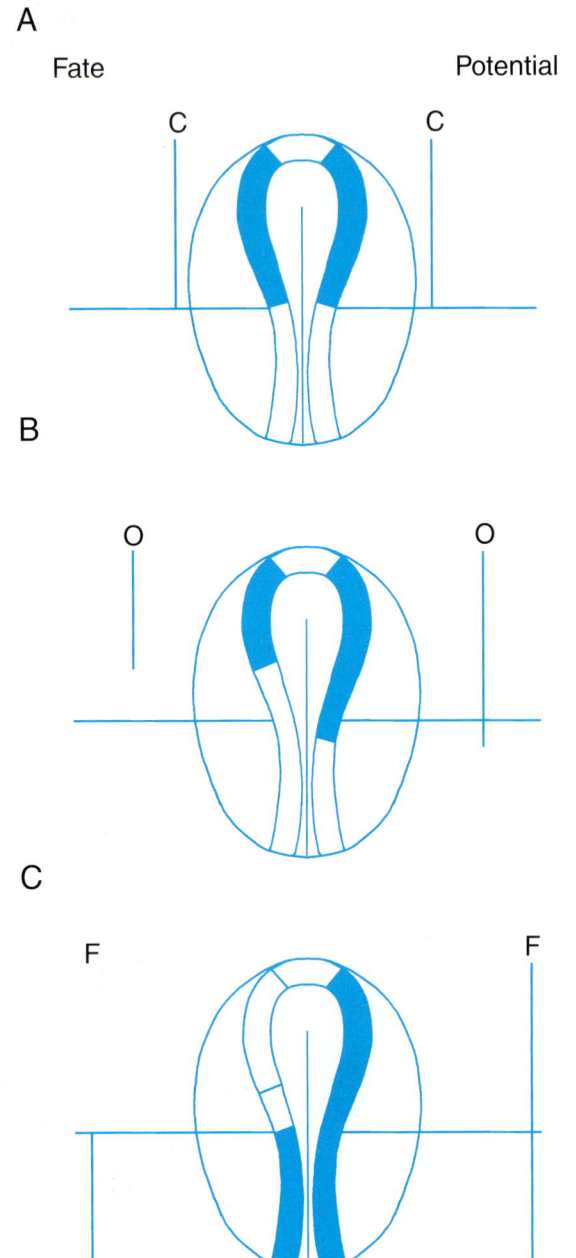

Figure 17.6 Fate maps of the formation of cartilage, teeth and median-fin mesenchyme from cranial and/or trunk NCC (CNC, TNC) from the Mexican axolotl, *Ambystoma mexicanum*. The maps are shown as dorsal views of neurala-stage embryos (anterior to the top), showing the neural folds along the centre of each neurula, with a horizontal line denoting the traditional boundary of cranial and trunk neural crest (as based on chondrogenic ability of NCC; see text for details). On the left of each neurula is the fate map (Fate) as expressed during normal development. On the right is the potential (Potential) to form the three tissues after neural crest from each level has been isolated and cultured with pharyngeal endoderm. (A) Potential to form cartilage (C) coincides with the region from which cartilage normally forms (Fate), confirming the traditional boundary between CNC and TNC *for chondrogenesis*. (B) Teeth (O, odontogenic neural crest) normally form from a more restricted region of the CNC (shown on the left) but can form from a more extensive region that extends posterior to the usually recognized CNC/TNC boundary; see Figure 18.4 for teeth formed from TNC. (C) Dorsal-fin mesenchyme (F) normally forms from TNC but can form from NCC at any level. See text and Graveson *et al.* (1997) for details.

initial mandibular ossification is around the alveolar nerve. Indeed, appearance of ossification correlates with the location of nerves in developing mandibles, maxillae, palatines and nasopalatines during human development. Goodday and Precious (1988) described a patient with cleft-lip and palate, congenital rubella syndrome, and a duplicated mental nerve, the nerve that innervates the gums, chin and eyelids. An ossification centre developed around each nerve resulting in abnormal osteogenesis.[18]

Information from mutants

Examining *mutants* has always been a means of assessing normal developmental processes, as will be seen on numerous occasions throughout the book. Loop-tail (*Lp*) mutant mice examined at 10–12 days of gestation show that the hindbrain and spinal cord have failed to close. Consequently, NCC cannot migrate and so remain in the open neural folds. Development of normal centres for the basicranium and periotic bones is taken as evidence that these are not of neural-crest origin. This is an opportunistic and piecemeal approach to mapping crest origins. An alternative approach to identifying embryonic origins is exemplified by a study in which the Bmp-antagonist noggin from *Xenopus laevis* was used to block expression of *Bmp-2* and *Bmp-4* in specific regions of the cranial neural crest in transgenic mice. Consequences included depletion of cranial NCC from those regions, underdevelopment of the branchial arches and the failure of skeletal and neural elements to form in the arches. Bmp-2 was also shown to be required for NCC migration.[19]

The mouse craniofacial mutation First Arch (*Far*) affects the anterior portion of the first branchial arch to produce defects in rhombomeres 2 and 3. *Far* maps to a Hox-4 cluster on chromosome 2, suggesting that it is a gain-of-function *Hox-4* mutant, although its action may be more complex for *Far* is partially dominant in the ICR/Bc mouse strain but recessive in Balb/c mice. Neural crest appears normal in *Far* mutant embryos but a portion of the maxillary nerve is missing. Formation of ectopic cartilage in association with nerve bifurcation was taken as indicating posteriorization of the anterior portion of the first arch, the cartilage perhaps representing the atavistic formation of the epibranchial of 'lower' vertebrates.[20]

Just as examining mutants informs normal development (Leroi, 2003; Hall, 2004c), examining ectopic expression of a Hox gene gives us some insights into the mode of action of *Far*. Ectopic expression of *Hoxa-7* (Hox-1.1) in transgenic mice is lethal and produces major craniofacial abnormalities equivalent to those seen after foetal exposure to retinoic acid or *Far* mutant (Balling *et al.*, 1989). Mice transgenic for *Hoxa-7* also show variation in cervical vertebrae, with formation of an extra vertebra, interpreted by Kessel *et al.* (1990) as a pro-atlas.

Box 17.2 Cardiac cartilages and bones

Cartilages and bones in the heart may seem an unlikely inclusion in a chapter devoted to the neural crest origin of skeletal tissues, but these elements – where investigated, and not many have been investigated experimentally – also arise from neural crest-derived mesenchyme, as do the valves and septa of the heart. So well recognized is this contribution from NCC that we now speak of the most caudal region of the cranial crest, the region from which NCC emigrate to seed the developing heart, as the *cardiac neural crest*.[a]

Constitutive cardiac cartilages

Bones and/or cartilages are surprisingly common elements of vertebrate hearts. As far as I can tell, these are normal, non-pathological, and not associated with infection, parasites or trauma. Indeed, in some species, *all individuals* have a cardiac cartilage. In such taxa, cardiac cartilage is a normal element of the endoskeleton, neither a sesamoid nor ectopic.

To my knowledge, skeletal elements have been reported in the hearts of four of the five classes of vertebrates, amphibians being the one exception. Cardiac skeletal elements may only be occasional in fish; they have been reported in the wall of the bulbus arteriosus in eight specimens from four species of teleost fishes.[b] Found in reptiles – turtles, alligators, crocodiles, 11 of 42 species of snakes studied, but not in the three species of lizards examined – the appearance of cardiac cartilage in reptiles shows no obvious correlation with size, taxon or habitat. Although there are reports from rodents, cattle and rabbits, their frequency in mammals seems to be low: 33 cases/1000 rats, 15 cases/1000 mice. Ossified cartilages complete with marrow and located in the aortic ring are more common in rabbits, a group that has an amazing proclivity to form skeletal tissue ectopically.[c]

Some of our best understanding of cardiac cartilage development comes from the recent developmental study of a turtle, the Spanish terrapin, *Mauremys leprosa*, by David López and his colleagues. The major constitutive cartilage arises from a mesenchymal condensation (see Chapter 19 for Condensations) that extends along the aorticopulmonary septum and the primordium of the pars fibrosa of the ventricular horizontal septum. The initial condensation expresses neither smooth α-muscle actin nor type II collagen. Type II collagen and chondrogenesis appear and begin in the centre of the condensation, spreading peripherally to form a hyaline cartilage that extends along the proximal part of the aorticopulmonary septum and the pars fibrosa of the horizontal septum, lacks a perichondrium and shown no signs of chondrocyte hypertrophy, mineralization or

ossification. A second hyaline cartilage, which arises between three and 18 months after birth, extends along the sinus wall of the right semilunar valve of the right aorta to penetrate the fibrous cushion that supports a portion of the valve (López *et al.*, 2003).

Cartilage is a constitutive feature of the proximal aorta, pulmonary trunk and semilunar valves of chick embryos from H.H. 37 onwards. The neural-crest origin of these cardiac cartilages has been demonstrated in chick and quail.[d] At the very least, such cartilage expands the list of neural crest-derived cartilages and demonstrates that cranial (cardiac) crest can contribute skeletal elements to the trunk.

Forty of 351 (11.4 per cent) of the pulmonary valves of Syrian (golden) hamsters, *Mesocricetus auratus*, exhibit cartilage along the fibrous attachment of the valves (Fig. 17.7). Type II collagen is a typical feature of these cartilages. The percentage may be too low to regard these as constitutive, making hamsters an interesting species in which to investigate the conditions under which cardiac cartilages arise. López and colleagues argued that these cartilages are not mechanically induced, and speculated that they are neural crest in origin.[e]

There may still be room for mechanical factors, however. In a fascinating study that does not appear to have been followed up, Kemp (1953) found that tadpoles of the northern leopard frog, *Rana pipiens*, survive for five or six days after the heart is removed! Such 'heartless tadpoles' develop small heads (microcephaly) and small eyes. Similar experiments on tadpoles of the spotted salamander, *Ambystoma punctatum*, produced similar results, with accompanying atrophy of the external gills and underdevelopment of the limb buds. Kemp and Quinn (1954) then published an interesting study in which cartilage development in larval *A. punctatum* was arrested following removal of the heart. Limb, skull and vertebral cartilages differentiated more slowly than normal because of slowed cell division. The gills gradually regressed and there was only slight further development of the limbs.

Ectopic cardiac cartilages and bones

Skeletal elements also form ectopically in the hearts of various mammals in response to a variety of abnormal circumstances or interventions:

- associated with congenital heart abnormalities in humans;
- in cardiac valves implanted into sheep (cartilage in 12/120 implants in place for longer than 13 weeks, one with bone);

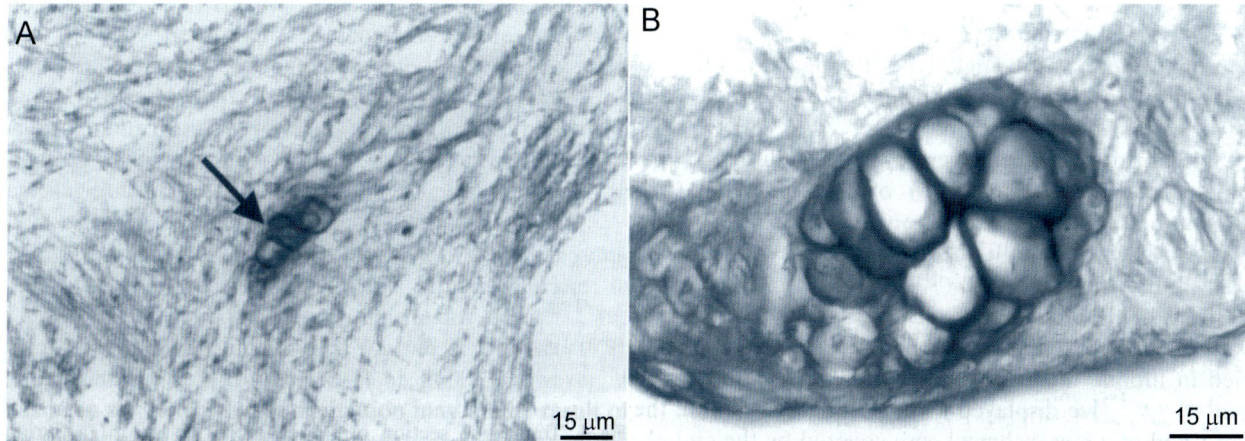

Figure 17.7 Ectopic cartilages in pulmonary valves of Syrian hamsters (*Mesocricetus auratus*). (A) Cartilage (arrow) in a three-day-old individual visualized with an antibody against type II collagen. (B) Cartilage in a five-month-old animal visualized with an antibody against type II collagen. Adapted from D. López *et al.* (2001).

Box 17.2 (*Continued*)

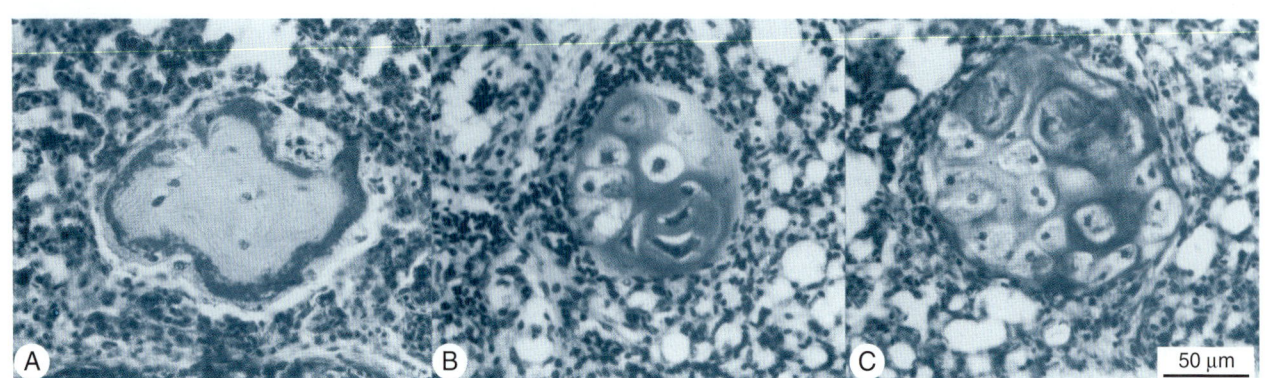

Figure 17.8 Examples of ectopic (pulmonary) cartilage and bone in domestic fowl. (A) Sparsely cellular ectopic bone, growing by appositional growth, and surrounded by lung parenchyma. (B, C) Ectopic cartilage nodules, with sparse extracellular matrix, show interstitial growth. Note the absence of any indication of an inflammatory response in A–C. Modified from Wight and Duff (1985).

- following heart–lung and double lung transplants in humans, in almost all of whom mineralization or ossification occurs (some of this cartilage may have arisen from bronchial cartilages; see below);
- in dogs given heart prostheses in which 11/15 formed cartilage and 5/15 bone; and
- in rats given auxiliary heart grafts in which cartilage and bone formed from connective tissue cells of the endocardium of the graft.[f]

Cartilages that form in the lungs, on the other hand, are unlikely to be neural crest in origin; NCC make no contribution to the lungs. Ectopic pulmonary cartilage and bone have been reported in newly hatched chicks, the assumption being that misplaced mesenchymal cells or chondrogenic precursors produced these ectopic skeletal tissues (Fig. 17.8). Irradiating guinea pig lungs results in formation of lamellar bone with haematopoietic tissue (Fig. 17.9).[g]

Carrageenan, a polysaccharide produced by red algae, consists of alternating 3-linked-β-D-galactopyranose and 4-linked-α-D-galactopyranose units. Cartilage can be induced in the walls of chicken aortae following injection of this substance, the polysaccharide presumably acting as a polysaccharide precursor. Intravenous injection of papain into rabbits also induces aortic cartilage, which is replaced by bone with marrow.[h]

[a] See Hall (1999a), Olson and Hall (2000) and Hutson and Kirby (2003) for the cardiac neural crest.

[b] The four species were a pilchard (sardine), *Sardinia pilchardus* (two specimens), the Atlantic horse mackerel, *Trachurus trachurus* (four specimens), the rainbow wrasse, *Coris julis* (one specimen), and Thor's scaldfish, *Arnoglossus thori* (one specimen).

[c] See Torres (1917), Hueper (1939, 1945), Kelsall and Visci (1970), McConnell (1970), Yamada *et al.* (1977) and Young (1994) for cardiac skeletal elements in reptiles and mammals.

[d] See López *et al.* (2000) for chick literature, and extrapolation of their results to cardiac cartilages in fish.

[e] See López *et al.* (2001, 2004) for cardiac cartilages in *Mesocricetus auratus*. All Syrian hamsters kept as pets or used in research are descended from a single litter dug from a burrow in Syria in 1930. They are the most common laboratory research mammal after mice and rats, over a million being used in the U.S. each year.

[f] See Zimmermann *et al.* (1979) for congenital anomalies, Arbustini *et al.* (1983) for valve implants in sheep, Kadowacki *et al.* (1987) for prostheses in dogs, Dittmer and Goss (1974) and Yousen *et al.* (1990) for cartilage formation following heart–lung and double lung transplants (some of which may arise from bronchial cartilages), and Dittmer *et al.* (1974) for heart grafts in rats.

[g] See Wight and Duff (1985) for the chick study, and Knowles (1984) for the irradiation study. For the presence of cartilage (so-called visceral cartilage) in respiratory tract and uterus, see Roth and Taylor (1966) and Reid (1976).

[h] See McCandless *et al.* (1963) for the carrageenan study, and Tsaltas (1962) for the papain study.

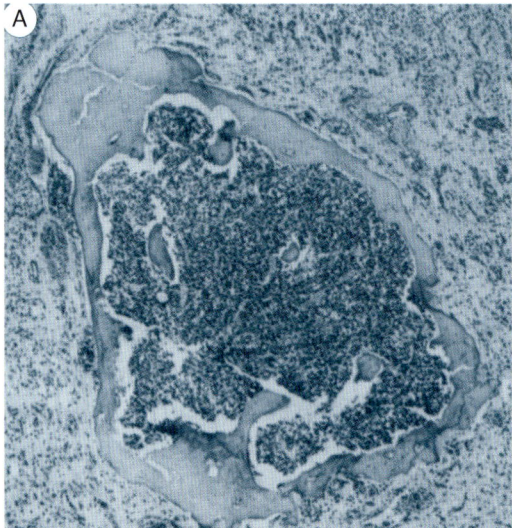

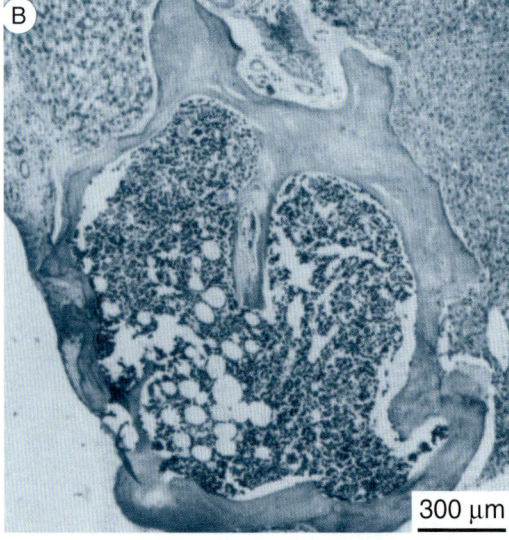

Figure 17.9 Ectopic bone – complete with marrow cavity containing haemopoietic tissue – in the lung (A) and liver (B) of guinea pigs irradiated 52 (A) or 87 (B) days before. Modified from Knowles (1984).

REGIONALIZATION OF THE CRANIAL NEURAL CREST

Largely on the basis of an extensive series of experiments undertaken with Sven Sellman, published in 1946, Sven Hörstadius (1950) discussed the evidence for a quantitative difference between the NCC that form Meckel's cartilage of the lower jaw and those that form the cartilages of the gill arches. Hörstadius distinguished *eight regions of cranial neural crest* in *Ambystoma* (Fig. 17.1):

- (i) and (ii) the most anterior neural crest – often referred to as the transverse neural crest because the anterior neural folds are oriented transverse to the embryonic axis – do not form cartilage;[21]
- (iii) the next most caudal region, forms the trabecular cartilages for the base of the skull but not visceral arch cartilages;
- (iv) a region that forms mandibular arch skeleton (Meckel's cartilage) but not trabecular or visceral arch skeleton;
- (v) to (viii) regions from which hyoid and gill skeletons arise; and finally, and most caudally
- (viii) trunk neural crest from which skeletal tissues do not arise.[22]

Le Lièvre (1974) and Le Lièvre and Le Douarin (1975) provided evidence for similar regionalization within the avian neural crest. From these pioneering labeling and transplantation studies we learn that particular regions of the cranial neural crest are competent to form particular skeletal elements. There is an even greater restriction of competence between cranial and trunk crest. Neither cartilage nor bone originates from trunk crest (region viii above). Thus, although contact with endoderm of the developing pharynx (amphibians) or cranial ectoderm (chick) is necessary before neural crest-derived cartilages will differentiate, important aspects of the regional nature of the cartilaginous skeleton are laid down within the neural tube. This is as true for jawless vertebrates (lampreys) as it is for jawed vertebrates (gnathostomes; Langille and Hall, 1993b).[23]

Regionalization of the cranial neural crest is also evidenced by the neural crest contributions to the connective tissue sheaths of the cranial muscles, as demonstrated by very fine-grained labeling, extirpation and mapping of the neural crest in chick and fire-bellied toad, *Bombina orientalis*, embryos. This regionalization reflects a prepatterning of premigratory cranial neural crest in the neural tube, reflecting the axial *Hox* code.[24]

THE VENTRAL NEURAL TUBE

By virtue of its mode of origin from the dorsal surface of the neural tube and as a consequence of earlier patterning from the notochord, which delimits the developing

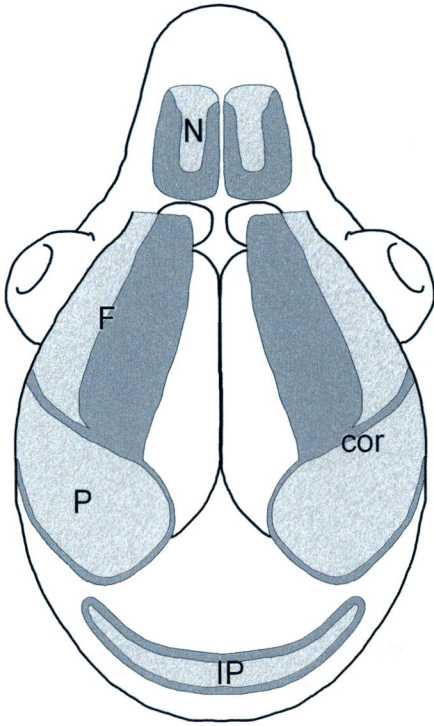

Figure 17.10 Patterns of Fgf receptor expression in the developing mouse skull, as seen in a dorsal view of the skull of a 16.5-day-old embryo. FgfR-2 (dark shading) is expressed in proliferating periosteal cells, FgfR-1 (light shading) in differentiating cells. cor, coronal suture; F, frontal bone; IP, interparietal bone; N, nasal bone; P, parietal bone. Adapted from Morriss-Kay (2001) from data in Iseki *et al.* (1997, 1999).

neural tube into dorsal and ventral regions, NCC are derivatives of the dorsal neural tube. It comes as a surprise, then, to find a body of research from Sohal's laboratory *claiming to demonstrate that some 'neural crest skeletal derivates' arise from the ventral neural tube in chick embryos*. As far as one can tell – and the evidence appears convincing to me – some chondrocytes within Meckel's cartilage and the quadrate arise from the ventral neural tube. A less direct but perhaps more important role of ventral neural tube cells is that *Shh* produced by the ventral neural tube and notochord promotes the survival of myogenic and chondrogenic lineages in chick somites.[25]

Evidence for Sohal's unexpected finding comes from experiments in which cells of the ventral neural tube were tagged with a viral LacZ marker and injected into the lumen of the rostral hindbrain of chick embryos. The injections were made *two days after CNC cell migration* to avoid any confusion with cells migrating from the dorsal neural tube. The injected ventral cells migrated from the lumen to appear in the perichondria and chondrocytes of Meckel's cartilage and the quadrate. Unlike NCCs, these cells were HNK-1 negative.[26] Consequently, they were readily distinguishable from the HNK-1-positive neural crest-derived cells that comprise the majority of the chondrocytes in the cartilage.

According to other studies by Sohal and colleagues, ventral neural tube cells are even more versatile. Labeling with the lipophilic tracer DiI (1,1-Didodecyl-3,3,3′,3′-tetramethyl indocarbocyanine) and expression of the homeobox gene *Islet-1* were used to show that cells from the ventral neural tube in duck embryos contribute to the trigeminal ganglion. Neuronal cells were shown to emigrate from the site of attachment of the trigeminal nerve to the neural tube, and migrate into the trigeminal ganglion and then into the mesenchyme of the first arch. Cells derived from the ventral hindbrain of two-day-old embryos were found in craniofacial muscles at seven days, and to migrate from the site of attachment of the vagus nerve, through the gut, into the smooth muscle of the stomach and intestine, and form hepatocytes in the liver.[27]

MIGRATION OF NCC: THE ROLE OF THE ECM

A direct way to determine whether, at what age, and under what conditions NCC form skeletal or connective tissues is to culture the neural crest, in isolation, in combination with other cells, or in the presence of growth factors. It was difficult to culture ectomesenchyme once the cells had left the neural tube and rapidly spread among the mesodermally derived mesenchyme, until Cohen and Konigsberg (1975) devised a culture technique that allows NCC to grow out from the neural tube *in vitro* under conditions whereby clonal cultures could be established (Fig. 22.3). The migratory behaviour of isolated neural crests can be modified by varying the nature of the substrate on which the NCC are cultured.[28]

The extracellular environment into which cells from the neural crest migrate *in vivo* may have important influences on their subsequent differentiation. In the Spanish ribbed toad, *Pleurodeles waltl*, this matrix initially consists of collagen and hyaluronan (Box 4.2). As development ensues, hyaluronan is removed by hyaluronidase and chondroitin-4- and chondroitin-6-sulphates begin to accumulate. In chick embryos, cranial NCC migrate into a cell-free, hyaluronan-rich space between epidermal ectoderm and mesoderm. Some matricial material may be

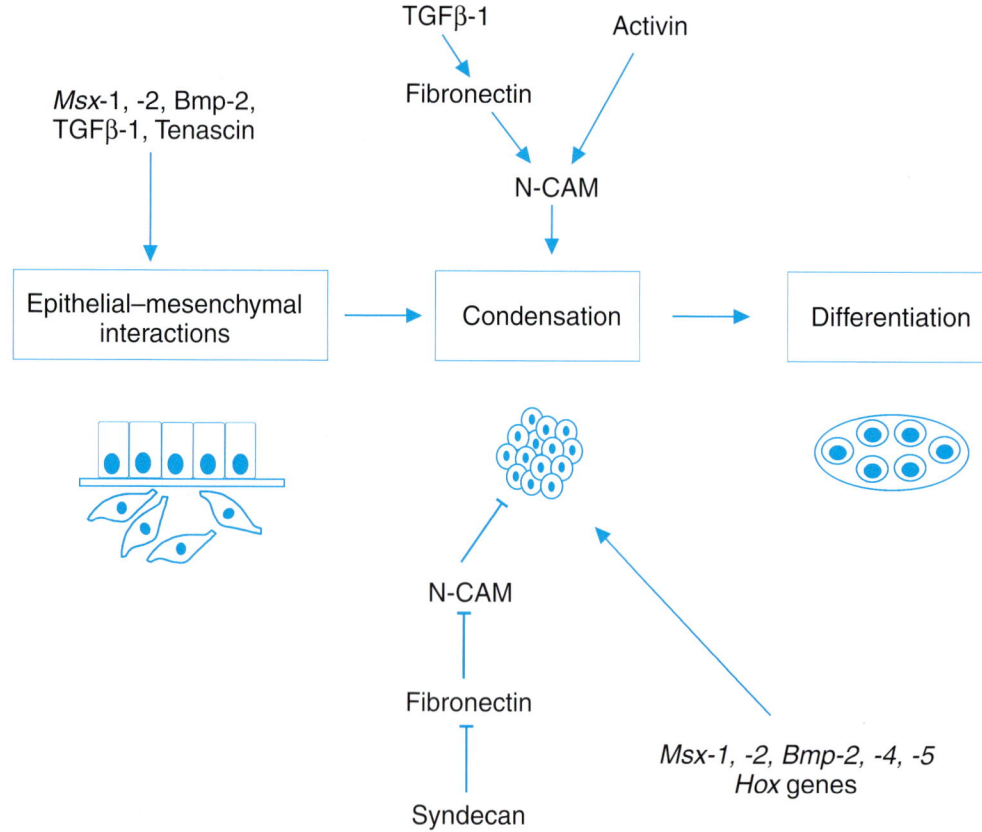

Figure 17.11 The three major phases of chondrogenesis. Epithelial–mesenchymal interactions initiate formation of a condensation, maintained by positive regulation via N-CAM, because signaling pathways for differentiation are not yet active. Transition from condensation to differentiation involves suppressing further condensation, as syndecan blocks N-CAM, and up-regulation of genes (*Msx-1, Msx-2, Bmp* and *Hox* genes) that initiate differentiation, some of the same genes having initiated the epithelial–mesenchymal interactions; see Table 19.1. Modified from Hall and Miyake (1992).

synthesized by NCC themselves; avian NCC synthesize hyaluronan and CS when cultured alone.[29]

Interactions with epithelia before, during or after migration of NCC are of prime importance for the subsequent differentiation of skeletal cells and, because of reciprocal interaction, for the continued growth of the epithelia. Migration gets the cells to the correct position. Epithelial–mesenchymal interactions allow the cells to condense and differentiate (Fig. 17.11), two critical developmental processes discussed in the next two chapters.

NOTES

1. See Hall (1999a) for the origins of terms ectomesenchyme and mesenchyme.
2. See L. G. Anderson *et al.* (1972) and Buchignani *et al.* (1972) for *familial osteodysplasia*. The latter authors were less certain of a primary neural crest involvement than were Anderson and colleagues, thinking that a separate, but related mutation may be involved, although the genetic basis remains unknown.
3. In the platyfish, *Xiphophorus maculatus* and the swordtail *X. helleri*, NCC accumulate above the neural tube during neural keel formation before migrating; the cells arise from the dorsal midline rather than from the neural keel (Sadaghiani and Vielkind, 1989), i.e. a neural ridge in the sense proposed by Balfour; see Hall (1997c) for a discussion.
4. The available literature up to around 1948, most of which was obtained from experimentation on amphibian embryos, is summarized and reviewed in depth by Hörstadius (1950) and to a lesser extent by de Beer (1958) and Holtfreter (1968). More recent monographs are those by Maderson (1985), Hall and Hörstadius (1988), Hall (1999a), Le Douarin and Kalcheim (1999) and Olsson and Jacobson (2000). See Monsoro-Burq *et al.* (2003) for a recent study on the induction of neural crest in *Xenopus*.
5. See Poswillo (1975a,b, 1976a,b) for the study cited. For migration of NCC in relation to abnormal development, see Fitch (1957), Ossenberg (1974), Bergsma (1975), Johnston (1975, 1977), Morriss and Thorogood (1978), Opitz *et al.* (1988), Osumi-Yamashita and Eto (1990), Morriss-Kay (1993) and Osumi-Yamashita *et al.* (1997).
6. See Weston (1970) for migration of cells into gaps created by surgical deletion of NCC. See McLean and Hall (1987) and Gilbert (2003) for regulation in general, and Vaglia and Hall (1999) for regulation of NCCs.
7. An essentially similar result is seen in an entirely different experiment: cross-nursing embryos of mice from small and large strains. Postnatal growth of the entire body and craniofacial growth adjusted for body size both are lower in small-bodied mice nursed by large-bodied mice and *vice versa* (Sai Htay *et al.*, 1997). Without intrinsic growth controls, we would expect larger nursing mothers to influence growth of smaller-bodied mice.
8. See Reilly and Lauder (1990) for *A. tigrinum* craniofacial skeletal development, Harrison (1935a,b) for analysis of the chimaeras, Hörstadius (1950) for a detailed discussion of the experiments, and see the papers in Dixon and Sarnat (1982, 1985) and Dixon *et al.* (1991), in Vols 6 and 7 of Hall (1990–1994) and Hall (1991c) for evaluations of factors that influence bone growth.

9. See Cassin and Capuron (1972, 1977, 1979) and Cassin (1977) for the amphibian studies, and Hall (1995a, 1999a, 2003b) for analysis of other studies. See Figure 3.7 for the developing chondrocranium of the Mexican axolotl, *Ambystoma mexicanum*, and see Sokol (1981) for a basic analysis of larval tadpole chondrocranial development, with especial emphasis on the parsley frog or mud-diver, *Pelodytes punctatus*.
10. de Beer (1947) provided suggestive evidence for *Ambystoma*. Gross and Hanken (2004) are developing fluorescent dextran labeling, with which they have shown that the nasals, frontoparietals, parasphenoids and squamosal in *Xenopus* arise from neural crest-derived cells.
11. Interestingly, Hammond and Yntema published an abstract in 1953 with essentially these results. They were extirpating neural crest from chick embryos as early as 1949, when they raised the possibility that the first NCC to migrate from the neural folds had different differentiation potentials than cells that migrated later. I do not know what caused the delay in publishing the full paper. I assume it was not because of the difficulty some working a few decades earlier had in accepting a neural-crest origin of skeletal tissue themselves, in convincing others, or in accepting a result that went so obviously against the entrenched germ-layer theory that skeletal tissues came from mesoderm; see my discussion in Hall (1999a, p. 10).
12. See Weston (1963, 1970), Johnston (1964, 1966, 1975), Weston and Butler (1966), Chibon (1967) and Johnston and Listgarten (1972) for these pioneering studies.
13. For reviews of the technique, see Le Douarin (1973, 1974) and Noden (1984a). See Graham and Meier (1975) for a staging table for the Japanese quail that includes stages, weights, and lengths at particular times of development, and see Hamburger and Hamilton (1951) and Table 12.3 for a staging table for chick embryos. For the neural-crest origin of skeletal tissues see Le Douarin (1974, 1975), Le Lièvre (1971a,b, 1974, 1978), Le Lièvre and Le Douarin (1975) and Couly *et al.* (1993). See Hall (1999a), Le Douarin and Kalcheim (1999) and Trainor *et al.* (2003) for recent reviews.
14. See Noden (1975, 1978ab, 1983, 1984a, 1986a,b, 1987, 1991a,b, 1992). Johnston *et al.* (1979) showed that the scleral cartilage of the chick eye is of neural-crest origin. The muscles are not of neural-crest origin (Noden, 1986b), but their connective tissue sheaths are, as demonstrated for chick (Köntges and Lumsden, 1996) and for the Oriental fire-bellied toad, *Bombina orientalis* (Olsson *et al.*, 2001).
15. See Hall (1986a, 2001a) and Tran and Hall (1989) for cartilage on avian and mammalian clavicles.
16. See Chai *et al.* (2000) and Jiang *et al.* (2000), who also determined the fate of cells from the cardiac neural crest, including contributions to the mesenchyme of arches three, four and six. See Hutson and Kirby (2003) for an overview of the cardiac neural crest.
17. See Jiang *et al.* (2002) for this study, Morriss-Kay (2001) for an excellent overview of the embryological and evolutionary origins of the mammalian skull, Kuhn and Zeller (1987) for an overview of morphogenesis of mammalian skulls, and Wilkie and Morriss-Kay (2001) and Wilkie *et al.* (2001) for overviews of the genetics of craniofacial development and malformations in mammals.
18. See Grammatopoulos *et al.* (2000) for the *Hoxa-2* knockout, Keith (1982) and Kjaer (1990) for the nerve–ossification association, and see Kjaer (1995) for an overview of the interrelationships between human facial, cranial and brain development in

both normal and pathological situations, the latter including cleft lip and palate, holoprosencephaly and anencephaly. Kjaer and colleagues have contributed valuable details of staging of human prenatal skeletal development; Kjaer (1989) established seven stages of prenatal development of the human maxilla, Kjaer et al. (1993) four stages for the ossification sequence of the occipital bone and the vertebrae, and Kjaer (1997) that, in human embryos, the course of Meckel's cartilage changes at the time of secondary palate formation. See Radlanski et al. (2003) for reconstructions of the mandible from human embryos of 12–117 mm crown–rump lengths, including a discussion of ossification and the mental nerve.

19. See Wilson and Wyatt (1995) for the Lp mutant and Kanzler et al. (2000) for the noggin-Bmp study

20. See Harris and Juriloff (1989), Juriloff and Harris (1991) and Juriloff et al. (1992) for the Far mutant.

21. Lack of chondrogenesis from the most rostral (transverse) neural folds was confirmed by Seufert and Hall (1990) in an analysis of tissue interactions in Xenopus head cartilages.

22. The notion of a non-skeletogenic trunk neural crest goes back 70 years, when Detwiler and van Dyke (1934) extirpated trunk NCC from Ambystoma embryos, although using Ap-2 (Box 21.4) as a marker, Epperlein et al. (2000) claimed that trunk neural crest formed cartilage when transplanted into the heads of axolotl embryos. Testing the chondrogenic potential of axolotl trunk neural crest by replacing cranial with trunk neural crest in vivo is not an effective method; we found that transplanted trunk NCC fail to migrate from the cranial neural tube (Graveson et al., 1995). The patterns of migration of axolotl NCC have been described by Cerny et al. (2004). The pivotal study with chick embryos is that the ability of trunk neural crest to produce connective tissue and dermis is restricted to crest no further caudal than the fifth pair of somites (Nakamura and Ayer-Le Lièvre, 1982). For recent studies/reviews see Smith and Hall (1990, 1993), Hall (1999a), Mayor et al. (1999) and Abzhanov et al. (2003).

23. It has been suggested that the pattern of tooth replacement in rainbow trout (Salmo gairdneri) is related to the order of migration of NCC and the local competence of the oral ectoderm (Berkovitz and Moore, 1974).

24. See Köntges and Lumsden (1996) and Olsson et al. (2001) for chick and Bombina, and see Cerny et al. (2004) for the Mexican axolotl, Ambystoma mexicanum. Olsson and Hanken (1996) and Falck et al. (2002) mapped the skeletogenic cranial neural crest in Bombina and in A. mexicanum, placing their analyses into a phylogenetic context. See Krumlauf (1993) for the axial Hox code, and see Hunt et al. (1991a–d, 1992, 1995) and Hunt and Krumlauf (1991, 1992) for the Hox code in the branchial region and its restoration after deletion of hindbrain neural crest in chicks. See Vielle-Grosjean et al. (1997) for the branchial arch Hox gene code in human embryos.

25. See Sohal et al. (1999b) for the ventral neural tube origin of chondrocytes, and Teillet et al. (1998) for Shh derived from the neural tube.

26. HNK-1 is a cell-surface carbohydrate epitope recognized by the monoclonal antibody HNK-1 (CD-57 in immunology), which was raised against a cell membrane fraction from a human T-cell line. N-CAM also expresses the HNK-1 epitope. HNK-1 is expressed on migrating neural crest-derived mesenchymal cells; antibody injection disrupts migration. The epitope is thought to function in cell-to-cell and cell–matrix interactions.

27. See Sohal et al. (1996, 1998a) for contribution to the trigeminal ganglion, and Sohal et al. (1998b, 1999a) for contributions to muscle and liver. Using a GFP probe electroporated into the ventral neural tube, Yaneza et al. (2002) failed to duplicate Sohal's results of ventral migration from mid- and hindbrain.

28. See Maxwell (1976) for a classic study on the influence of substrate, and see Hall and Ekanayake (1991, 1994a,b) for growth factors that influence skeletogenesis from NCC.

29. See Corsin (1974, 1977) for Pleurodeles, Pratt et al. (1975) and Fisher and Solursh (1977) for cranial NCC migration in vivo, and Greenberg and Pratt (1977) and Manasek and Cohen (1977) for production of hyaluronan. See Manasek (1975) Slavkin and Greulich (1975), Hay (1981), Maclean and Hall (1987), P. L. Jones et al. (1993), Comper (1996) and Gilbert (2003) for comprehensive reviews of the function of the ECM; Maclean and Hall (1987) and Behonick and Werb (2003) for relationships between chondrocytes and their peri- and extracellular matrices; and see Ayad et al. (1994) for an excellent source book for some 30 extracellular matrix products, including the 16 types of collagen.

Epithelial–Mesenchymal Interactions

Chapter 18

Extrapolating from a single trout: ideas are easy but data are hard.

Whether skeletal mesenchyme is neural crest- or meso-dermally derived, and whether limb, vertebral, craniofacial or visceral skeleton, differentiation is initiated as a result of interactions with embryonic epithelia. Such interactions, known as *epithelial–mesenchymal interactions* (Figs 13.8 and 18.1), may take place:

- *before onset of mesenchymal cell migration*, e.g. neural-crest cells (NCC) and cranial ectoderm adjacent to the neural tube (below); limb-bud mesoderm with flank ectoderm (Chapter 35);
- *during migration*, e.g. pharyngeal endoderm and mesenchyme destined to form the branchial arch skeleton in amphibians (Fig. 18.2, and see below); or
- *after migration*, when skeletogenic mesenchyme is at its final location, e.g. mandibular or maxillary arch ectoderm and mesenchyme destined to form mandibular and maxillary arch bones in birds (below); sclerotomal mesenchyme destined to form vertebral cartilage and notochord and/or ventral spinal cord in all vertebrates (Chapter 41).

These epithelial–mesenchymal interactions serve at least four functions (Fig. 18.3). They:

- *localize* skeletogenic mesenchyme within embryos;
- provide the *signals* that initiate condensation of skeletogenic mesenchyme;
- *permit* the *differentiation* of cartilage, bone (or dentine or enamel) to begin; and
- *set the fundamental number of progenitor cells* for the skeletal element.

A list of some interactions involved in skeletogenesis is provided in Table 18.1. Indeed, epithelial–mesenchymal interactions serve these four functions in the development of almost all the cells and tissues of vertebrate embryos: heart, kidneys, glands, liver, lungs, alimentary canal, skin and skin derivatives such as hair, feathers and scales.

I also introduce the fact that such interactions are reciprocal; signaling from mesenchyme or ectomesenchyme maintains epithelial proliferation and so maintains a positive feedback loop for production of skeletogenic signals from the epithelium. Mesenchymal cells lose their dependence on epithelial signals and epithelial cells become insensitive to mesenchymal signaling at the same time during development.[1]

In this chapter, I document the existence and importance of these interactions by outlining their role in initiating chondrogenesis and osteogenesis of the neural crest-derived mesenchyme that forms the mandibular skeleton in tetrapods and fish. To summarize a considerable body of evidence, *all* craniofacial membrane bones and cartilages that have been examined require an epithelial interaction before their development can be initiated (Table 18.2; Fig. 21.2).

I also address whether the lateral line or neuromasts may play a similar role in fish larvae or in urodeles, and how studies in which different germ layers have been combined and grafted shed light on whether cell–cell interactions occur in teratomas. Roles for similar interactions in skull, limb and vertebral development are discussed in Chapters 21, 35 and 41.

URODELE AMPHIBIANS: CHONDROGENESIS

Given that the nature of the environment encountered during migration of NCC is diverse and changes with time, we can ask whether cranial NCC are competent to differentiate into skeletal tissues prior to neurulation – when

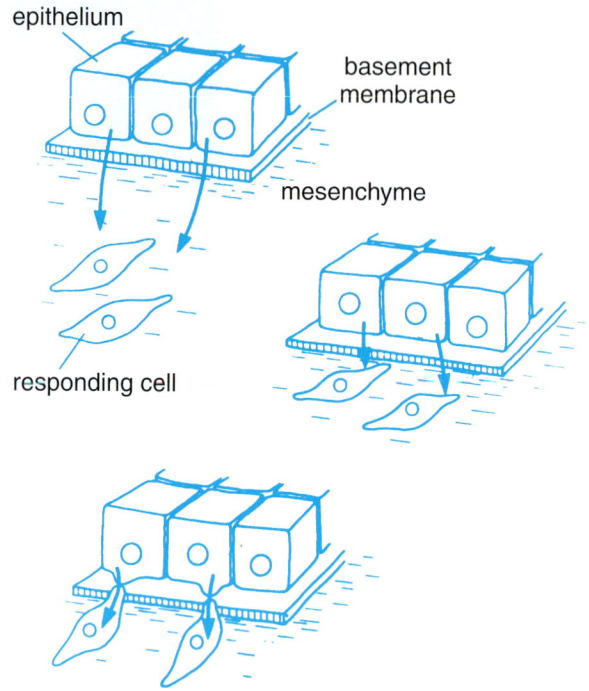

Figure 18.1 Three ways in which epithelium signals to mesenchyme are shown diagrammatically as three epithelial cells on their basement membrane and adjacent mesenchymal cells. Interaction may be (i) by epithelial release of a diffusible molecule (top); (ii) by interaction of mesenchymal cells with a product deposited by the epithelium into the basement membrane (middle), a mode requiring close interaction between mesenchymal cells and the basement membrane; or (iii) by direct cell-to-cell communication between epithelial and mesenchymal cells, a mode requiring close interaction between the two cell types, local dissolution of basement membrane, and junctional connections between the two cell types.

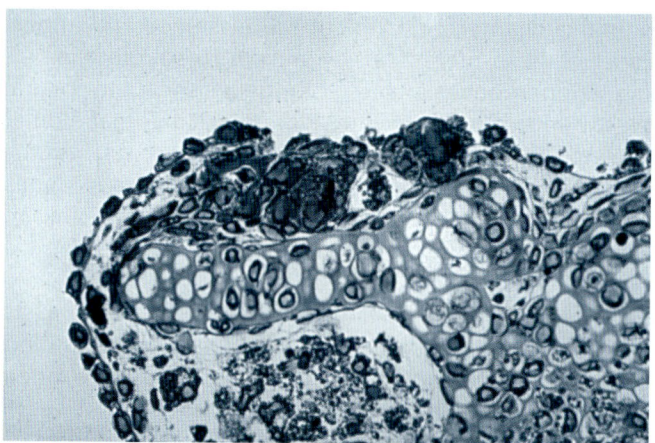

Figure 18.2 Teeth and cartilage differentiate when cranial neural crest from a Mexican axolotl neurula is combined with pharyngeal endoderm and maintained in organ culture. Also see Figs 17.6 and 18.4.

still in the neural plate or neural folds – or whether they acquire competence by association with mesoderm mesenchyme, epithelia, or extracellular matrix (ECM) products during or after their migration. An association between ectodermal or endodermal thickenings and

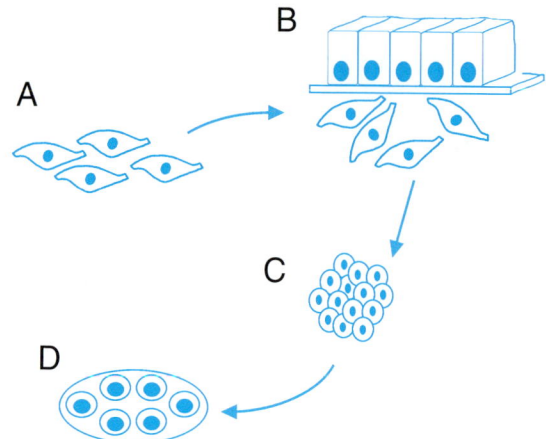

Figure 18.3 Four phases of cartilage development. (A) Origination of chondrogenic mesenchyme. (B) Epithelial–mesenchymal interaction. (C) Condensation. (D) Differentiation. Condensation and differentiation are both multistep processes.

condensations of ectomesenchymal cells has been commented upon several times, and is suggestive of tissue–tissue interactions.[2]

The early experiment that opened up this avenue of research was the demonstration of the differentiation of cartilage from embryonic amphibian NCC of the Alpine newt, *Triturus alpestris*, co-cultured with pharyngeal endoderm. No cartilage differentiated from NCC cultured alone.

Epperlein and colleagues demonstrated that for cartilage to differentiate, direct cell-to-cell contact between NCC and pharyngeal endoderm is necessary; that NCC does not preferentially migrate toward the endoderm – i.e. there is no chemoattraction – and that pharyngeal endoderm evoked mesenchyme as well as cartilage from the NCC. The action of the pharyngeal endoderm was specific; somatic and otic mesoderm, from which vertebral and otic capsule cartilages arise, respectively, chondrify in response to notochord and nasal epithelium but not in response to pharyngeal endoderm.[3]

Pharyngeal endoderm is also sufficient to evoke chondrogenesis from neural crest in the Mexican axolotl, *Ambystoma mexicanum* (Figs 18.2 and 18.4). However, we have to be careful of species specificity and/or species differences for, in another urodele, the Spanish ribbed toad, *Pleurodeles waltl*, Corsin demonstrated that contact with dorsal *mesoderm* also evokes cartilage from NCC. Corsin tested for effects of hyaluronan (30 μg/ml) or testicular hyaluronidase (2 μg/ml) on isolated neural crest to see whether these ECM products would initiate chondrogenesis, but they did not.[4]

AVIAN MANDIBULAR SKELETON: CHONDROGENESIS AND OSTEOGENESIS

An epithelial–mesenchymal interaction is also required for NCC of chick embryos to chondrify but, in this case,

Table 18.1 Osteogenic epithelial–mesenchymal interactions in embryonic development and regeneration[a]

Organ	Inductive epithelium	Osteogenic mesenchymal/ectomesenchymal cells
Scleral ossicles (chick eye)	Scleral epithelium	Scleral ectomesenchyme
Ear capsular bone	Otic placode	Otic capsule ectomesenchyme
Mandible	Mandibular epithelium	Mandibular ectomesenchyme
Maxilla	Maxillary epithelium	Maxillary ectomesenchyme
Branchial arches	Pharyngeal endoderm	Branchial arch ectomesenchyme
Limb bones	AER (limb bud)	Limb-bud mesenchyme
Tooth	Oral epithelium	Odontogenic ectomesenchyme of jaws
Regeneration (urodele limb)	Regenerating epidermis	Blastema derived by dedifferentiation of limb cells
Regeneration (deer antler)	Antler bud epidermis	Frontal bone pedicle mesenchyme

[a]Based on Anderson (1990).

Table 18.2 Tissue interactions that evoke chondrogenesis or osteogenesis from craniofacial mesenchyme in embryonic chicks[a]

Tissue	Inducing tissue(s)	H.H. stages at which induction takes place
Meckel's cartilage	Cranial ectoderm	9–10
Frontal	Prosencephalon, mesencephalon	13–15
Occipital	Rhombencephalon, neural tube	13–15
Parietal	Mesencephalon, rhombencephalon	13–15
Squamosal	Mesencephalon	13–17
Basisphenoid	Rhombencephalon, notochord	13–17
Parasphenoid	Notochord	13–17
Scleral cartilage	Pigmented retinal epithelium	14–18
Maxillary membrane bone	Mandibular epithelium	18–22
Mandibular membrane bones	Mandibular epithelium	18–24
Otic capsule cartilage	Otic vesicle	18–27
Palatal bones	Palatal epithelium	?–25
Frontal	Cranial epithelium	22–30
Scleral bones	Scleral papillae	30–36

[a] Based on studies summarized in Hall (1987d). Also see Fig. 21.2.

premigratory NCC interact with cranial epithelium adjacent to the neural tube:

- premigratory NCC cultured in isolation fail to chondrify;
- premigratory NCC taken from embryos of H.H. 9 differentiate into chondroblasts in response to H.H.10 cranial ectoderm adjacent to the neural tube.

Thus, to initiate chondrogenesis in chick embryos, premigratory NCC interact with cranial ectoderm, while in anuran and urodele amphibians, postmigratory NCC interact with pharyngeal endoderm. Osteogenesis in chick embryos is initiated following a later interaction with mandibular epithelium, an interaction that only takes place once the mesenchyme is within the mandibular arches.[5]

Exposing chick premigratory cranial NCCs to Fgf-2 or -4 *in vitro* shows that low concentrations (0.1–1 ng/ml Fgf-2) promote proliferation, while higher concentrations (10 ng/ml) promote differentiation of both cartilage and bone. With prolonged culture *both* intramembranous and endochondral ossification ensue. Thorogood and his colleagues concluded that the tissue interactions may be mediated by Fgf bound to heparan sulfate proteoglycan, which is interesting when we consider that different epithelia elicit cartilage and bone at different times during development of cranial NCC – premigration for chondrogenesis and postmigration for osteogenesis.[6]

The mandibular skeleton consists of Meckel's cartilage as a central element surrounded by membrane bones. The classic study on the development of the mandibular skeleton for any vertebrate is the organ-culture investigation of chick-embryo mandibular arches undertaken by Jacobson and Fell (1941). The authors described three centres (Condensations; Chapter 19) within each half of the mandibular arch (each mandible): *an osteogenic, a chondrogenic and a myogenic centre.*

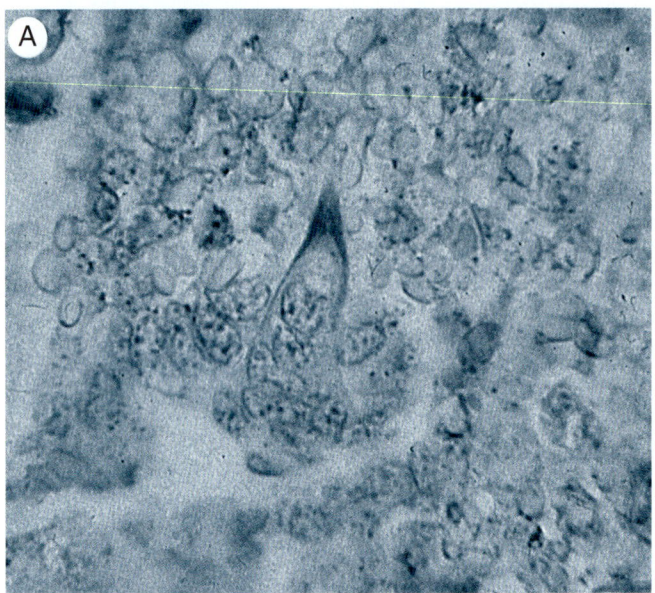

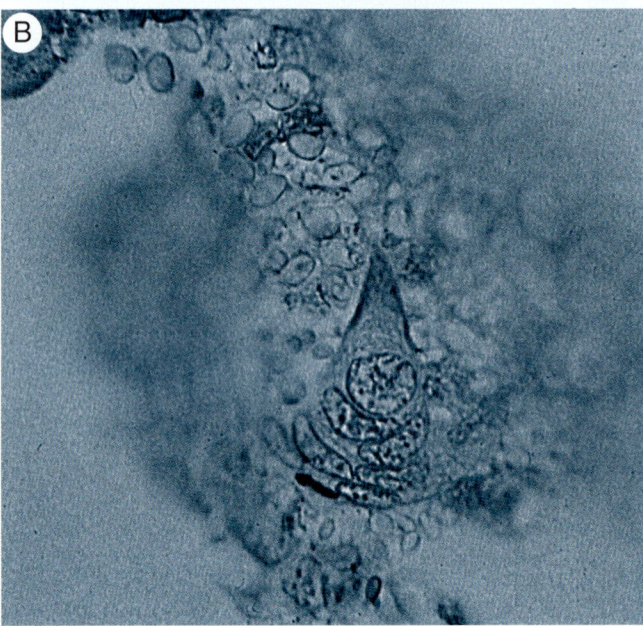

Figure 18.4 Histological sections of two teeth (A,B) formed when rostral trunk neural crest (TNC) from neurula-stage embryos of the Mexican axolotl, *Ambystoma mexicanum*, is recombined with pharyngeal endoderm and maintained as an organ culture. See Fig. 17.6 for details of the fate map of odontogenic neural crest and see Fig. 1.10 for naturally occurring teeth.

Jacobson and Fell thought that the skeletogenic centres arose within the mandible. Each centre was found close by a transitory thickening of the buccal epithelium. When maintained in isolation *in vitro*, each centre formed only one differentiated end product – bone, cartilage or muscle – and so each was interpreted as a separate lineage of committed cells. All our basic ideas about mandibular development arise from this pioneering study. We now know, however, that skeletogenic cells develop from ectomesenchyme that has migrated into the mandibular arches, and that myogenic cells are derived from local head mesoderm.

The cartilaginous and bony mandibular skeletons, along with membrane bones of the upper jaw and skull, are derived from midbrain level (mesencephalic) neural crest. In chick embryos, NCC emerge from the mesencephalon at the five-somite stage (H.H. 8.5). The last cells leave when embryos have 10 pairs of somites (H.H. 10). NCC first reach the regions of the embryo that will become the mandibular arches at H.H. 15 (Le Lièvre and Le Douarin, 1975). I say 'regions that will become the mandibular arches' because no arches or other facial processes – maxillary, nasal, frontal – exist until ectomesenchymal cells create them. Indeed – and as discussed below – crest-derived mesenchyme promotes proliferation of arch epithelium, which, in turn, promotes mesenchymal survival and outgrowth of the facial processes.[7]

Initially, the mandibular skeleton of avian embryos – indeed of all vertebrate embryos – consists of Meckel's cartilage, which begins to chondrify at five days of incubation (H.H. 26). An endochondral bone, the articular, develops in the retroarticular process of Meckel's cartilage at 14 days of incubation (H.H. 40). The remainder of Meckel's cartilage in avian embryos persists as a rod of cartilage that becomes surrounded by six membrane bones, whose ossification commences at seven days of incubation (H.H. 31). If I may classify by exclusion (invertebrates being a fine precedent), in all non-mammalian vertebrates, multiple membrane bones develop around Meckel's cartilage, the most phylogenetically conserved of which are the dentary, angular, surangular and splenial. The majority if not all of these bones persist into adult life, either separately or in various stages of fusion. In mammals – and, in various intermediate stages, in mammal-like reptiles – portions of what would be Meckel's cartilage in other vertebrates transform into the three middle ear ossicles, the malleus, incus and stapes (Box 13.3), leaving the dentary as the single bone of mammalian lower jaws (Figs 13.2–13.5).

Intact mandibular arches are comprised of ectomesenchyme (from which skeletal tissues form), a mesodermally derived mesenchymal core (from which muscles arise) and an epithelial covering. Interaction of these mesenchymal cells with pharyngeal endoderm is required to initiate Meckel's cartilage in anuran and urodele amphibians (Fig. 18.2). In the late 1970s in my laboratory we set out to ask whether this is also so for other vertebrates. Mary Tyler – the first of a wonderful series of postdoctoral fellows I have been privileged to have in my laboratory – showed that the mesenchyme that will form the membrane bones that invest Meckel's cartilage needs to interact with mandibular epithelium to initiate osteogenesis, but that neither Meckel's cartilage nor the articular depend on interactions with mandibular arch epithelium to be able to differentiate (Table 18.3).

In the initial studies, and as a control for the trypsin and pancreatin enzymatic digestion used to facilitate separating epithelium from mesenchyme (below), Mary Tyler took intact mandibular arches before either cartilage or bone had differentiated, established them in organ

Table 18.3 Results of organ culture of mandibular mesenchyme after enzymatic removal of the mandibular epithelium[a]

H.H. stage at which mandibular epithelium was removed	Per cent of cultures producing membrane bone[b]
16–18	0 (0/17)
20	0 (0/15)
21–22	0 (0/31)
23	0 (0/26)
24–25	100 (71/71)

[a] Isolated mandibular mesenchyme only forms membrane bones if epithelium has been present until H.H. 23. Based on Tyler and Hall (1997) and Hall (1978b).
[b] Cartilage forms in all of these cultures. Its formation is independent of epithelial influences. The timing of membrane bone formation is similar when mandibular mesenchyme is grafted to CAM rather than organ-cultured.

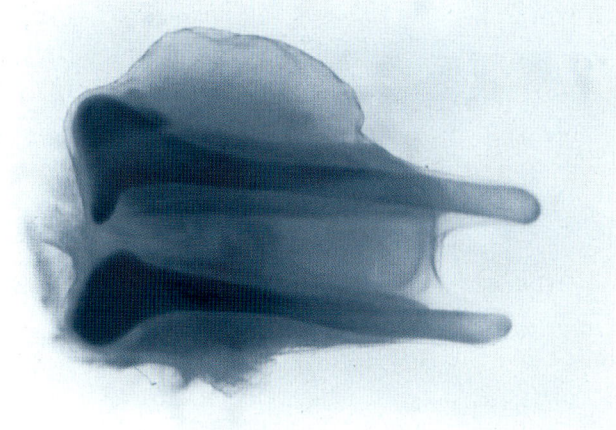

Figure 18.6 Normal morphogenesis and growth of Meckel's cartilage and adjacent membrane bones has taken place in this graft of mandibular mesenchyme from an H.H. 22 embryonic chick combined with maxillary arch epithelium and maintained as a CAM-graft.

Figure 18.5 The left and right mandibular arches from an H.H. 22 chick embryo placed on a circle of black Millipore filter in preparation for CAM-grafting.

culture, or grafted them to the CAM (see Box 12.3). These control cultures and grafts were established to ensure that cartilage and bone would differentiate normally under these artificial conditions, which indeed they do (Figs 18.5 and 18.6).

Isolated mesenchyme – chondrogenesis

Experimental cultures or grafts consisted of mandibular mesenchyme and ectomesenchyme from which the mandibular epithelial 'jacket' had been removed. I refer to such tissue as *isolated mesenchyme*, even though it consists of mesenchyme and ectomesenchyme, the mesenchyme forming a central core of cells from which myogenesis occurs. Anything else would be cumbersome and involve referring to tissue as isolated ectomesenchyme, isolated somitic mesenchyme, isolated somatic mesenchyme, isolated lateral-plate mesenchyme, and so forth. So, bear in mind that the term 'isolated mesenchyme' is shorthand and does not reveal the source of the mesenchyme. The context should.

All NCC have reached the presumptive mandibular arches by H.H. 15. Removing the mandibular epithelium

at H.H. 16 (52–56 hours of incubation) has no influence on the differentiation of Meckel's cartilage or the endochondral articular bone. We concluded that, in order for Meckel's cartilage to form, either the ectomesenchyme did not require interaction with epithelium – it certainly does not require interaction with mandibular epithelium beyond H.H. 16 – or, in line with the results from the culture of amphibian neural crest discussed in Chapter 17, there had been a prior interaction, perhaps with pharyngeal endoderm during migration before H.H. 15. I showed that Meckel's cartilage in chick embryos depends for its differentiation on a much earlier interaction of NCC with cranial ectoderm, which takes place as NCC migrate away from the neural tube. The articular, being an endochondral element with substantial subperiosteal bone, depends for its initiation on hypertrophy of the chondrocytes of the retroarticular process and transformation of the perichondrium to a periosteum. No other parts of Meckel's cartilage hypertrophy in birds.

Isolated mesenchyme – osteogenesis

Absence of mandibular epithelium did influence *osteogenesis*. *In ovo*, osteogenesis of these mandibular bones begins at seven days of incubation. No bone forms from isolated mesenchyme if mandibular epithelium is removed before H.H. 24 (4.5 days of incubation). Osteogenesis proceeds normally, however, if the mandibular epithelium is removed after H.H. 24, i.e. mandibular epithelium had to be present until H.H. 23 (four days of incubation) for osteogenesis to be initiated at H.H. 31 (Table 18.3; and see Figs 13.8, 18.7 and 18.8).

Ruling out any role for Meckel's cartilage

Because cartilage always develops in all-age cultures of mandibular mesenchyme, it is possible that Meckel's

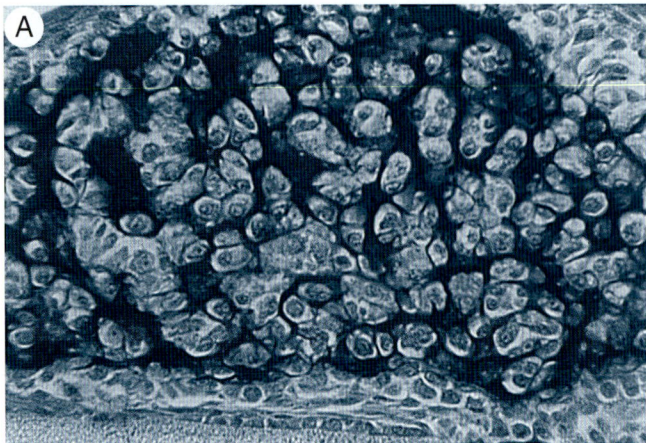

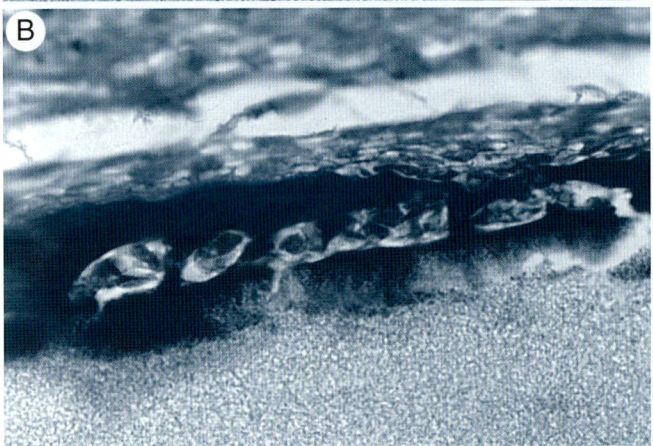

Figure 18.7 Two examples of bone development in culture. (A) Bone that developed from embryonic chick maxillary mesenchyme (H.H. 30). (B) Bone that differentiated from embryonic mouse mandibular mesenchyme (10 days of gestation). Matrix products have been deposited into the Millipore filter substrate (below).

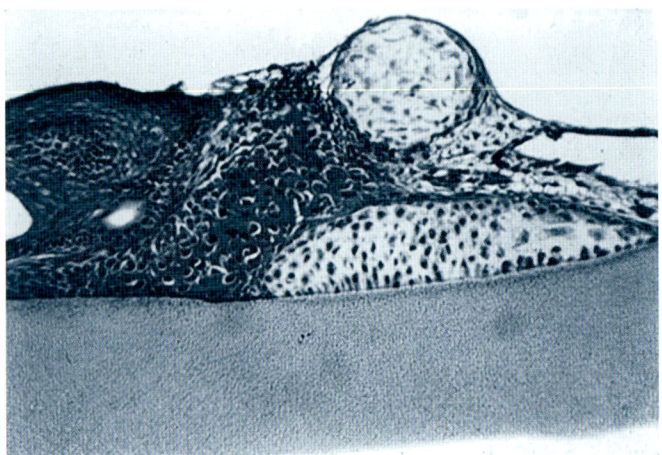

Figure 18.8 Meckel's cartilage and membrane bone formed from H.H. 22 embryonic chick mandibular mesenchyme combined with wing-bud epithelium and CAM-grafted.

cartilage itself, and not mandibular epithelium, influences mandibular ectomesenchyme to ossify. Indeed, often in the dental literature, it has been maintained – for example, by Frommer and Margolies (1971) from studies with mouse embryos – that proximity of the membrane bones (or the single dentary bone in mammals) to Meckel's cartilage is presumptive evidence for an inductive interaction between the two skeletal elements. However, osteogenesis can be initiated from mouse mandibular mesenchyme in the absence of Meckel's cartilage (Hall, 1980c; MacDonald and Hall, 2001).

Further, in our cultures and grafts of chick tissues, we often found that bone forms some distance from Meckel's cartilage. Other membrane bones, such as the maxilla or quadratojugal of the upper jaws, develop in isolation from cartilage *in ovo*, and, as discussed below, upper jaw mesenchyme must interact with maxillary epithelium before it can ossify.[8]

As discussed later in this chapter, epithelial–mesenchymal interaction is reciprocal; normal differentiation of mandibular or maxillary epithelium requires interaction with ectomesenchyme or mesenchyme, although mesenchyme of the chorioallantoic membrane (CAM) can

provide partial support for continued epithelial differentiation. The molecular basis of such reciprocal signaling is becoming apparent.

Molecular mechanisms

A series of studies by Mina Mina from the University of Connecticut, conducted over the past decade and a half, have revealed some of the molecular signals involved in these interactions in both chick and mouse mandibular development. These studies are outlined below.

- An examination of the temporal and spatial patterns of expression of mRNA for type I and type II collagen and core protein found low levels for type II collagen throughout the mandible at H.H. 15, increasing at H.H. 25 when mRNA for core protein first appeared. mRNA for type I collagen was also first expressed at H.H. 15, increasing strongly at H.H. 28–29.
- An examination of stage-specific chondrogenic potential beginning at H.H. 16 (a stage from which chondrogenesis can be obtained), found mRNA for type II collagen increasing fivefold immediately before chondrogenesis, and epithelial inhibition of chondrogenesis.
- Earlier studies demonstrating that the epithelium promotes proliferation and suppresses chondrogenesis were confirmed.
- *Msx-1* was demonstrated in growth centres in the distal mesial mesenchyme, *Msx-2* in adjacent distal mesial epithelium, and associated with programmed cell death, with epithelial up-regulation of mRNA for *Msx-1*.
- Spatial and temporal regulation of mandibular development by Fgfs and Bmps was demonstrated.[9]

Regulation of *Msx-1* by *Bmp-2* or *Bmp-4* is an important genetic pathway in these interactions (Box 13.4). Ectopic application of Bmp-2 or Bmp-4 alters expression of *Msx-1* and extends expression of *Fgf-4* in the distal epithelium, leading to formation of a bifurcated Meckel's cartilage.[10]

In embryonic chick branchial arches, *Fgf-8* is expressed in second-arch epithelium, *Bmp-7* in branchial clefts and arches from H.H. 12 on, and *Bmp-4* in the distal tips of the arches from H.H. 18 onwards. Fgf-8 in mandibular arch primordia defines the maxillo-mandibular region through epithelial–mesenchymal interactions that activate *Hox* genes. Bmp-4 in ventral ectoderm limits the anterior expression of *Fgf-8* while ectopic Bmp-4 decreases mandibular arch development. Chick mandibular *epithelial Bmp-7* increases proliferation, cell death, *Msx-1*, *Msx-2* and *Bmp-4* in lateral mandibular mesenchyme. Exogenous Bmp-7 initiates the development of ectopic mandibular and maxillary elements. Egf enhances proliferation but not gene expression, implicating *Bmp-7* as an independent part of the signaling system in chick mandibular mesenchyme (see Box 15.1).[11]

OSTEOGENESIS IN AVIAN MAXILLARY ARCH SKELETON

The membrane bones of vertebrate lower jaws develop in close association with the perichondrial surface of Meckel's cartilage. Other membrane bones at other sites do not develop in proximity to the primary cartilaginous skeleton. In pioneering discussions, Marshall Urist (1962, 1970) maintained that such bones are induced to form by interactions between potentially osteogenic mesenchyme and fibrous connective tissues, although he had no experimental evidence to support this contention.[12]

Two membrane bones – the *quadratojugal* (QJ) and the *jugal* – differentiate from mesenchyme in the suborbital region of the face of embryonic chicks at seven days of incubation, following interaction with maxillary epithelium. Neither develops in association with the primary cartilaginous skeleton. Consequently, we can study their development without any of the confounding effects of chondrogenesis in adjacent mesenchyme, as occurs in mandibular arch mesenchyme.

The QJ and jugal are derived from mesencephalic NCC. For osteogenesis to be initiated in chick embryos during the seventh day of incubation, mandibular mesenchyme must interact with mandibular epithelium until H.H. 23. Similarly, osteogenic QJ mesenchyme requires the maxillary epithelium to be present until H.H. 23 for intramembranous ossification to be initiated at H.H. 31 (Table 18.4). The similarity in timing is because preosteogenic mesenchyme for both arches accumulates at the base of the mandibular and maxillary processes as a single condensation where the interaction with epithelia takes place (Dunlop and Hall, 1995). Development of the bones of the palate in chick embryos also requires an epithelial–mesenchymal interaction.[13]

Barx-1, a mouse homeodomain transcription, factor is expressed in the mesenchyme of the first and second

Table 18.4 Results of organ culture of whole trypsin–pancreatin-treated upper beak (quadratojugal) region of the embryonic chick and of isolated upper beak mesenchyme[a]

H.H. stage of tissue cultured	Per cent of cultures producing membrane bone after seven days *in vitro*
Intact, enzyme-treated upper beaks	
17–18	100 (8/8)
23	100 (8/8)
26	100 (6/6)
29	100 (7/7)
Isolated upper beak mesenchyme	
17–19	0 (0/8)
21	0 (0/6)
23	0 (0/6)
26	100 (5/5)
29	100 (7/7)
32	100 (8/8)

[a] The enzymatic treatment does not prevent the intact tissues from forming membrane bone. The isolated mesenchyme forms membrane bone only if epithelium has been present until H.H. 23.

branchial arches from 10.5 days of gestation on. In chick maxillary primordia, *Barx-1* has an expression pattern that is complementary to *Msx-1*. Epithelial signals are required to up-regulate *Barx-1* though *Fgf-8* can substitute for epithelial signaling. *Bmp-4* down-regulates *Barx-1* and antagonizes *Fgf-8*, establishing a feedback loop: epithelial *Fgf-8* up-regulates *Barx-1*, and *Bmp-4* down-regulates *Barx-1* by inhibiting *Fgf-8*.[14]

A patterning role for *Barx-1* is reinforced from studies on mouse tooth development, in which *Barx-1* determines a molariform tooth type. *Barx-1* is lost exclusively from the molar teeth at 16.5 days of gestation. *Barx-1* in tooth primordia is inhibited by Bmp-4, restricting *Barx-1* expression to proximal premolar mesenchyme at 10 days of gestation. Inhibiting Bmp-4 with noggin elicits ectopic expression of *Barx-1* in distal incisor mesenchyme and transformation of the incisor to a molariform tooth (Tucker *et al.*, 1998c).

MAMMALIAN MANDIBULAR SKELETON

Information on epithelial involvement during the differentiation of mammalian (mostly mouse) mandibular ectomesenchyme is also available. Osteogenesis is initiated in cultured mandibular mesenchyme from mouse embryos of 13–16 days of gestation. However, since these are rather late stages of skeletogenesis when compared with chick mandibular development, cell-to-cell interactions may already have taken place. For instance, mandibular mesenchyme isolated from embryos of 12 days of gestation is able to chondrify (osteogenesis was not mentioned in this study), suggesting that if an epithelial–mesenchymal interaction is required for Meckelian chondrogenesis, it has taken place by 12 days. A thickening of the oral epithelium associated with osteogenesis in the adjacent ectomesenchyme was observed in CAM-grafts of foetal rat mandibular processes, reminiscent of the observation made on chick mandibles in culture by Fell and Jacobson 30 years before.[15]

Such experiments and observations certainly suggest the possibility of interaction between the epithelium and the mandibular ectomesenchyme in mammals, but they do not prove it. Subsequent studies demonstrated a requirement for epithelium and the timing of the epithelial–mesenchymal interaction of dentary induction in mouse embryos (Fig. 17.5).[16] Roles played in murine mandibular development by members of other families of molecules are being uncovered. Many play similar roles in chondrogenesis, osteogenesis and tooth formation, indicating substantial evolutionary conservation of what we might call 'differentiation signal pathways'. I discuss two here, Endothelin-1 and members of the distal-less (Dlx) gene family. Others are introduced in the following section on teleost fish jaw development.

Endothelin-1 (Et-1)

Endothelin-1 – a vasoactive peptide in vascular endothelial cells that regulates blood pressure – and its receptors play important roles in craniofacial development.

At 9.5 days of gestation, receptors are localized in osteogenic mesenchyme while endothelin-1 is localized in mandibular epithelium and deep mesenchyme.[17] T. Thomas et al. (1998) identified a signal cascade from endothelin-1 → dHAND → Msx-1 that regulates development of branchial arch mesenchyme (see Box 13.4). dHand, a helix–loop–helix protein, is down-regulated in endothelin-1 mutant (Et-1⁻) mice in which the first and second branchial arches are smaller than wild type (the mice have small mandibles), and the third and fourth arches fail to form. Expression in limb buds is normal, the role of endothelin-1 being specific to the craniofacial skeleton.

An inappropriate environment for early specification or emigration of NCC is created within the neural tube of Et-1⁻/⁻ mutant mice, which die just after birth. When examined at 9.5 days of gestation, homozygote mutants are found to lack any neural tissue at what would be the normal midbrain–hindbrain boundary, a defect equivalent to that seen in Wnt-1⁻/⁻ embryos, and one which disrupts the population of NCC normally arising from that region (Wurst et al., 1994).

Neural crest-derived cells are present in the arches that do form, but they fail to express Msx-1. The proposed cascade involves:

- epithelial triggering of endothelin-1; which
- regulates mesenchymal dHand; which in turn
- regulates mesenchymal Msx-1 in the distal branchial arches.

In limb development, dHAND regulates the SHH pathway to establish the anterior–posterior limb axis

(Chapter 38). Clouthier et al. (2000) identified the endothelin-A receptor in neural crest cells destined for the branchial arches and determined that endothelin-1 acts after migration of NCC to up-regulate a number of transcription factors (Dlx-2, Dlx-3, dHand, eHand) but not Prx-1, Hoxa-2 or cellular retinoic acid-binding protein (CRABP)-1. With further differentiation, members of the Dlx family play critical roles in mandibular development.

The Dlx gene family and craniofacial development

Distalless (Dlx) and Msx multigene gene families of transcriptional activators and repressors, respectively, are expressed in overlapping but distinct domains during division, differentiation, patterning and morphogenesis of various tissues such as the branchial arches (Fig. 13.6) and craniofacial skeleton in which epithelial–mesenchymal interactions or cell death occur. Some functions of Msx genes are discussed in Box 13.4. A similar outline for some of the six Dlx genes is provided here (Fig. 18.9).[18]

Dlx-1 and Dlx-2 are essential for development of the proximal regions of the first and second arches in mice. Null mutation of Dlx-2 results in failure of forebrain development and changes in the proximal portions of the first and second arch skeletons (Qiu et al., 1995, 1997).

Z. Zhao et al. (2000) examined the six Dlx genes in developing murine dentition and mandibular processes, finding that at 10.5 days of gestation all six genes are expressed in mandibular mesenchyme with Dlx-3 expression in mesial epithelium (Fig. 13.6). Dlx-2 is expressed in the proximal mesenchyme and distal epithelium of the first branchial arch. Loss of Dlx-1 or Dlx-2 results in loss of the upper molars but does not affect the lower molars, probably because of functional redundancy with Dlx-5. Thomas et al. (2000) also asked whether Dlx-1 or Dlx-2 pattern skeletal mesenchyme in the first arch. Mesenchymal and epithelial expression of Dlx-2 in the first arch is differentially regulated:

- Bmp-4 is co-expressed with Dlx-2 in distal epithelium, where it regulates expression of epithelial Dlx-2;
- Fgf-8 is expressed in proximal mesenchyme (Fig. 13.7), up-regulates mesenchymal expression of Dlx-2 and down-regulates epithelial Dlx-2.

Dlx-5 and Dlx-6 are expressed in condensations for membrane bones and in periostea around cartilage models in endochondral ossification (Fig. 18.10). Dlx-5 shows stage-specific expression in mouse calvarial osteoblasts, repressing osteocalcin with onset of mineralization. Dlx-5 is inducible by Bmp-4 in mouse embryos, in fractures, and in MC3T3-E1 cells (in which osteoblast markers are up-regulated and deposition of ECM is enhanced), demonstrating the sequence Bmp-4 → Dlx-5 → osteogenesis.[19]

Two studies of Dlx-5 knockouts confirm this role in mice and provide information on craniofacial defects, delayed

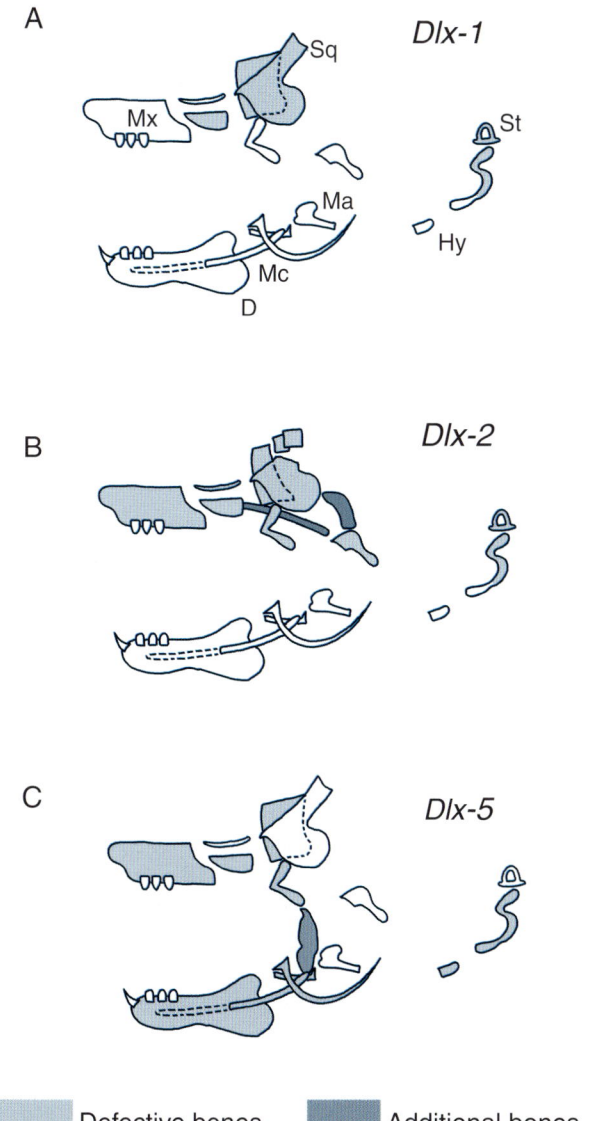

Figure 18.9 Craniofacial skeletal abnormalities in mice in which one of the genes *Dlx-1*, *-2* or *-5* has been knocked out, depicted as lateral views of the skeleton. Changes consist of shape changes (shown as defective bones, light shading) and the formation of additional bones (darker shading). (A) Morphogenesis of the squamosal (Sq) and stapes (St) is abnormal when *Dlx-1* is knocked out. (B) Skull morphogenesis is even more abnormal and additional bones develop when *Dlx-2* is knocked out. (C) Abnormalities in the lower jaw appear when *Dlx-5* is knocked out. See the text for further details. D, dentary; Hy, hyoid arch skeleton; Ma, malleus; Mc, Meckel's cartilage; Mx, maxilla. Modified from Merlo *et al.* (2000).

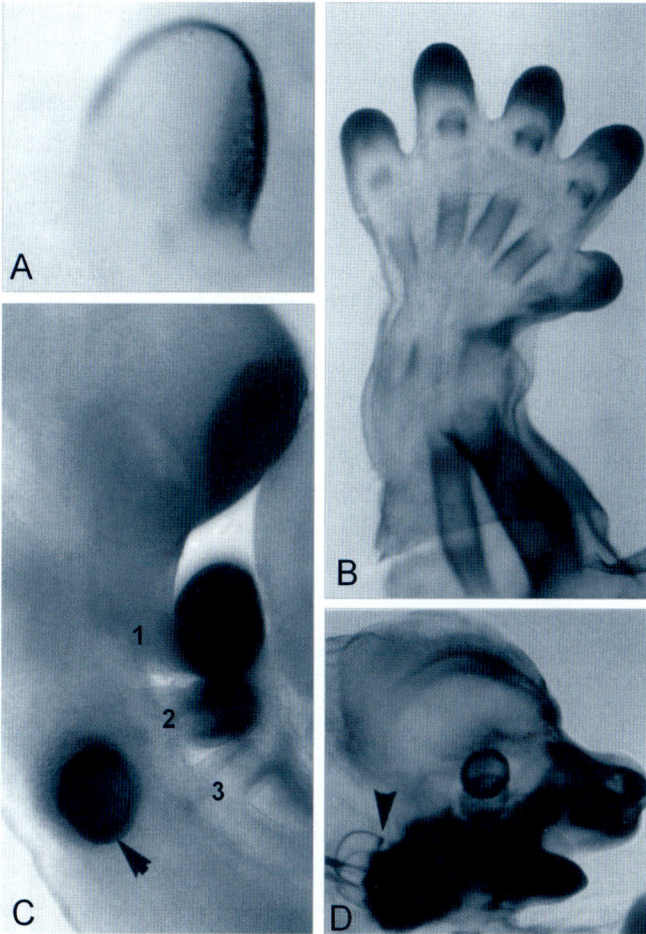

Figure 18.10 Expression of *Dlx-5* in murine limb and craniofacial development. (A) Expression in the AER and distal mesenchyme of a limb bud at 10.5 days of gestation. (B) Expression in distal digital mesenchyme and in the perichondria of more proximal skeletal elements in the forelimb at 14.5 days of gestation. (C) Expression in pharyngeal arches one and two not in three and in the otic vesicle (arrow) at 9.5 days of gestation. (D) Strong expression in maxillary and mandibular arches, and in the vestibular organ (arrowhead). Modified from Merlo *et al.* (2000).

skull ossification, hypomineralization and ectopic bone in the mandible. *Dlx-5* is expressed in the branchial arches, otic and olfactory placodes and in all bones. Homozygous *Dlx-5*$^{-/-}$ embryos die at or soon after birth with anomalies involving all branchial arches (especially the proximal portion of the mandibular arches), sense organs (ears, nose), and delayed and hypomineralization of the skull bones. *Dlx-5* and *Dlx-6* are both expressed in the distal but not in the proximal regions of the embryonic

arches (Fig. 13.6). Knocking out both genes results in a homeotic transformation of lower to upper jaws. Depew and colleagues provided a model for the role of nested domains of *Dlx* genes and the organization of proximo-distal elements, *Dlx-1* and *-2* specifying proximal, *Dlx-1*, *-2*, *-5* and *-6* specifying distal elements.[20]

Differential signaling by mandibular and maxillary ectoderm has been revealed in mouse embryos, primarily analyzed in terms of epithelial or mesenchymal control of tooth form but applicable to control of arch mesenchyme. Beginning with the knowledge that homeobox gene expression correlates with molar or incisor development and that ectoderm induces homeobox gene expression, C. A. Ferguson *et al.* (2000) showed that all arch mesenchyme can respond to ectodermal Fgf-8 in embryos younger that 10 days of gestation. By 11 days mesenchymal expression is independent of the epithelium but mandibular and maxillary mesenchyme respond differentially, mandibular and maxillary epithelium inducing *Dlx-5* and *Dlx-2* in mandibular mesenchyme, and inducing only *Dlx-2* in maxillary mesenchyme, with neither epithelium inducing *Dlx-5* in maxillary mesenchyme. Intrinsic differences between mandibular and maxillary mesenchyme are therefore established by 11 days of gestation, which is when inductive interactions with mandibular epithelium take place (Hall, 1980a; MacDonald and Hall, 2001).

Further to the specificity of action of *Dlx*, only some craniofacial cartilages develop abnormally when zebrafish are treated with retinoic acid, abnormalities associated with loss of expression of *Dlx* (which in zebrafish is normally expressed in hindbrain, NCC and visceral arches). The midbrain, which lacks *Dlx*, is less affected by retinoic acid treatment (Ellies *et al.*, 1997).

A note on limb development: *Dlx-5* is expressed in chick-wing and leg-bud epithelia. Implanting beads soaked in Fgf-2 into the flank of chick embryos induces *Dlx-5* within 12 hours. *Dlx-5* is only transiently expressed in the epithelium of limb buds from *limbless* embryos, suggesting a requirement for *Dlx-5* to maintain the AER (Ferrari *et al.*, 1999). See Figs 3.5 and 3.6 for beads implanted into the mandibular arches of chick embryos.

TELEOST MANDIBULAR ARCH SKELETON

As we saw for urodele amphibians, interaction with pharyngeal endoderm plays an important role in patterning branchial arch cartilages in the zebrafish. For fish the evidence is not from experimental manipulation but from a mutation in the zebrafish, *Danio rerio*, the gene *van gogh* (*vgo*). Mutant embryos lack the entire pharyngeal region. The defect occurs in the late stages of migration of NCC, although segmentation of the hindbrain is normal. Initial emigration from the neural tube is via the normal streams but the streams fuse peripherally. Because branchial arch segmentation requires signals from endoderm and from neural crest, arch segmentation fails to occur and, as a consequence, segmental cartilages fail to develop.[21]

Our knowledge of the transcription and growth factors involved in jaw development in teleosts outstrips our understanding of the cellular interactions involved. Nonetheless, an interesting story is emerging, one that indicates conservation of signaling systems between teleosts and tetrapods.

Fgf

Fgf is expressed in the surface ectoderm, lateral-line blastemata and otic vesicle of embryos of the Japanese flounder, *Paralichthys olivaceus*. Blocking Fgf *in vitro* impairs chondrogenesis (Suzuki and Kurokawa, 1996).

In a recent study on mechanisms of cartilage formation in the zebrafish, David *et al.* (2002) showed a requirement for endoderm mediated via Fgf-3. Even more recently, Walshe and Mason (2003) showed that an *Fgf* signal is

required for six hours after initiation of migration of NCC for neurocranial and pharyngeal cartilages to develop in zebrafish, and that *Fgf-3* and *Fgf-8 together* comprise the *Fgf*-signaling system:

- inhibiting *Fgf-3* results in complete absence of all cartilages that normally arise from pharyngeal arches three to six;
- inhibiting Fgf-8 has minimal and mild effects on chondrogenesis; and
- inhibiting both Fgf-3 and Fgf-8 results in complete loss of all pharyngeal cartilages and loss of almost all the cartilages of the neurocranium.

Hoxd-4 and retinoic acid

Retinoic acid is an important *morphogen* for limb and craniofacial development (Box 40.2). Retinoic acid also lies upstream of Hox genes in teleost fishes.

Limb development

A retinoic acid–*Hoxd-11–d-13* connection has been established in limb development, *Hoxd-11, d-12* and *d-13* being the genes involved. Retinoic acid is required for *Shh, Hoxd-12, Hoxd-13* and *Fgf-4* signaling during rat limb-bud growth, but not for initiation of the limb buds; application of retinoic acid at the 45-somite stage suppresses *Hoxd-13* and *Fgf-8* associated with a flattening of the AER and suppression of limb-bud growth. Applying retinoic acid to the anterior limb-bud mesenchyme of chick wing buds in physiological or pharmacological doses (100 or 333 μg/ml, respectively) elicits expression of *Shh* and *Hoxd-11* and induces additional digits from tissue distal to the implanted retinoic acid.[22]

Craniofacial development

Disrupting *Hoxa-1* (Hox 1.6) results in defects in the normal rostral domain of expression of the gene. *Hoxa-1* has its normal centre of expression in rhombomeres 4–7 (r4–r7) at 7.5–8.5 days of gestation, which is before migration of NCC. Homozygotes for *Hoxa-1* show delayed closure of the neural tube, lack cranial nerves and ganglia, and show inner ear malformations. Skeletal malformations are seen in elements that arise from paraxial mesoderm, while the pharyngeal arches are normal (Lufkin *et al.*, 1991).

Plant *et al.* (2000) implanted retinoic acid-soaked beads into the midbrain–hindbrain junction in chick embryos of H.H. 9, i.e. before emigration of cranial NCC. Within one day, expression boundaries of *Hoxa-1* and *Hoxb-1* were shifted anteriorly to the level of the mesencephalon; *Msx-2* was slightly down-regulated in the hindbrain but the onset of expression in the facial processes was normal. Two ectopic cartilages formed: (i) a sheet of cartilage ventral and lateral to the quadrate; and (ii) an accessory rod of cartilage from the side of Meckel's cartilage, morphologically reminiscent of the retroarticular process in the first arch domain (Meckel's is in the second arch). Initially, the quadrate was often displaced laterally and fused to the retroarticular process. Labeling cranial NCC with DiI demonstrated that a subpopulation of cells from rhombomere four (r4) of the hindbrain that normally migrate into the second arch, migrate into the wrong arch, maintain their morphogenetic identity and form second arch structures in the first arch, reminiscent of the study by Noden (1983) using chick embryos.

Fish

Hoxd-4 was cloned from the flounder, and expression and response to retinoic acid examined. *Hoxd-4* is expressed in the brain from rhombomere 7 rostrally into the spinal cord, in pharyngeal arches 2–5 and in pharyngeal cartilages. In the presence of exogenous retinoic acid the anterior border of expression moves further anteriorly, as it does in tetrapods. *Hoxb-5* is expressed in gill arch five and in the spinal cord, and may play a role in specification of arch five.

When administered later in embryonic development retinoic acid depresses *shh* and *Hoxd-4* expression in the pharyngeal arches and induces skeletal malformations. At the chondrogenic condensation stage *shh* is expressed in pharyngeal endoderm, and mandibular, hyoid and gill primordia, expression subsequently expanding to the posterior endoderm of each arch. Retinoic acid depresses *shh* and *Hoxd-4*, changing that result in malformed cartilages with their growth shifted posteriorly. Retinoids are required to maintain pharyngeal endodermal expression of a large battery of genes, including *Hoxa-1*, *Hoxb-1*, *Pax-1*, *Pax-9*, *Fgf-3* and *Fgf-8*.[23]

Jaw and pectoral fin malformations are induced in flounder following exposure to retinoic acid, exposure of shield-stage embryos for one hour suppressing development of Meckel's cartilage (Suzuki *et al.*, 2000).

Endothelin-1 (Et-1)

Lower jaws and visceral arches in zebrafish (and, we assume, in other teleosts such as the ale wife; Fig. 3.8), are specified by *endothelin-1* (*Et-1*), which is encoded by the *gene sucker* (*suc*), which is expressed in central arch mesoderm and arch epithelium but not in ectomesenchyme. It appeared to Miller *et al.* (2000) that *Et-1* expression creates an environment in which neural crest-derived skeletogenic mesenchyme is specified.

Mutants

Numerous mutations affecting the branchial and mandibular arches in zebrafish have been isolated. Neuhass *et al.* (1996) identified 48 mutations in 34 loci affecting craniofacial development. For instance, *Chinless*

disrupts skeletal fate and interactions between neural crest and muscle. *Chinless* embryos lack cartilages in all seven branchial arches and so lack Meckel's cartilage. Secondarily *chinless* mutants lack mesodermal muscle, although, interestingly, both cartilage and muscle precursors are present, giving us a clue to the time of action of the mutation.[24]

LATERAL LINE, NEUROMASTS AND DERMAL BONE

The lateral-line system is an extensive network of superficial sensory nerves in teleosts, lampreys, elasmobranchs and larval amphibians. Multiple lateral lines on the head tend to follow the sutures of dermal bones, or *vice versa*. Usually only a single lateral line is found on the trunk. In teleost fishes this single line is associated with specialized lateral-line scales. Lateral-line nerves are associated with specialized sensory receptor organs or *neuromasts* that detect mechanical, electrical and chemical signals. Neuromasts are often embedded in a bone, variously known as a *canal bone* (because it forms a canal for the lateral-line nerve) or a *lateral-line bone*. Partly because of the association of cranial lateral lines with bones of the skull, lateral-line nerves and/or neuromasts have been implicated as inducers of dermal bone (Table 1.1).[25]

Hope from a single trout

The classic experimental studies establishing a relationship between neuromasts, the lateral line and dermal-bone development in a teleost fish, the rainbow trout, *Oncorhynchus mykiss*, were undertaken over 65 years ago by Moy-Thomas (1938, 1941). Excising the lateral-line system had no effect on the development of the frontal bones, but the bony 'gutter' (the canal bone) associated with the frontal bones failed to form. Moy-Thomas suggested that if the lateral line or neuromasts were developmentally coupled in this 'higher' teleost then neuromasts and dermal bones should also be associated in such 'primitive' fishes as the bowfin, *Amia calva*. This turns out to be one of those situations where the idea is easy but the data hard.

The vertebrate palaeontologist T. S. Westoll followed up this suggestion and described a topographical association between dermal bones and the lateral-line system in *Amia* and in fossils of other 'primitive' fishes (Westoll, 1941).

In *The Development of the Vertebrate Skull*, his masterful summary and analysis of skull development and structure published in 1937, Gavin de Beer also took up the suggestion from Moy-Thomas and discussed the relationship between dermal bones and lateral-line canals. Some dermal bones in fishes – nasals in flatfish; frontal,[26] intertemporal and postparietal in *Amia* – develop in close association with neuromasts of the lateral-line system.

De Beer thought this association was a secondary one, in part because many dermal bones in fishes have no such association, in part showing the importance he placed on homology. Because homologous bones in other vertebrates arise in the absence of lateral-line organs, de Beer thought this must be true for all vertebrates. Later, de Beer came to accept that homologous structures – bones in this case – could arise by different developmental processes – presence or absence (loss) of induction by neuromasts – and that a causal developmental relationship in one group could be broken in another. Such a situation is inevitable with the bones in question; most tetrapods have neither lateral lines nor neuromasts.[27]

Although the study by Moy-Thomas is a classic – in part because it has never been repeated – it is seriously flawed. Moy-Thomas removed the lateral-line primordium from one side of a *single* rainbow trout embryo. Apart from the minimal sample size, a further objection is that the *dermal bones of many teleosts are known to develop and regenerate independently of neuromasts.* This does not mean that the frontal bone of rainbow trout or its associated lateral-line bone might not be one of the bones in one of the species with a neuromast connection. *Amia*, however, would be a better choice of experimental animal, but has to my knowledge not been explored experimentally.[28]

Heterochronic changes may also come into play (Chapter 44); paedomorphic skeletal features in the Oriental fire-bellied toad, *Bombina orientalis*, include retention of lateral lines in adulthood, decreased cranial ossification, juvenilized teeth and a reduced middle ear.[29]

In an extensive analysis of induction of the dermal skeleton in salmon (*Salmo*), DeVillers (1947, 1965) obtained descriptive and experimental support for neuromast induction of dermal bone, his chief findings being as follows.

- Osteogenic cells remain in close association with neuromasts as the neuromasts differentiate.
- Neuromasts act as aggregation centres for osteogenic cells that form the canal bones, defined by DeVillers and herein as bones developing in association with the lateral-line canal.
- De Villers thought that canal bones have a *dual composition*, each bone consisting of a laterosensory or tubular component and a basal or membranous component. Furthermore, he thought that both components were induced by neuromasts in salmon but that only the tubular component was induced in cyprinid fishes.[30]

In one of the few experimental studies to address these issues, Merrilees (1975) took semicircular canals (which are specialized neuromasts) from goldfish, *Carassius auratus*, bisected them and transplanted each into the pocket left after a scale was removed (see Box 2.3 for teleost scales). Normally scales regenerate, but regeneration was inhibited in the presence of the semicircular canals, as was any development of cartilage or bone. Merrilees described a specialized cord of epithelial cells in the lateral-line canals that inhibits osteogenesis and so controls cavitation of the tubular component of the canal bone described above. Little work has been done since, although in the first detailed analysis of scale development, Sire and Arnulf (1990) discuss the role of epithelial–mesenchymal interactions and the lateral-line system.[31]

Modern molecular and genetic studies on the induction of lateral-line bones are minimal. *Fgf* is expressed in surface ectoderm and in the lateral-line blastemata in embryos of the Japanese flounder *Paralichthys olivaceus*. Blocking *Fgf in vitro* impairs *chondrogenesis* (Suzuki and Kurokawa, 1996). Similarly, no one has investigated whether similar interactions control the development of sensory canal cartilages in elasmobranchs, although the issue has been discussed.[32]

Blind cave fish, *Astyanax mexicanus*, have more extensive development of other sense organs such as lateral lines than do sighted members of the species. Among skeletal changes in the blind fish are additional suborbital bones whose number correlates with increased numbers of cranial neuromasts, suggestive of an inductive interaction.[33]

TERATOMAS

Cartilage and bone, along with other tissues such as teeth that require epithelial–mesenchymal interactions to develop *in situ* (Figs 6.4, 12.8 and 18.3), are commonly found in teratomas. Given the normal strict requirement for epithelial–mesenchymal interactions, and the proximity of cartilage and bones to epithelia within teratomas, we have to wonder whether such interactions take place in teratomas, a possibility entertained by a number of us. The ease with which particular tissue types arise in teratomas is a clue to the stringency of the cell-to-cell interactions required for their formation; cartilage, bone, muscle, neural tissue and skin form often, lung and liver only rarely.[34]

Germ-layer combinations

Differentiation in some teratomas is restricted to neuroectodermal tissues, in others to specific germ layers or to germ-layer combinations that facilitate cell interactions.[35]

The contribution of specific germ layers to the formation of cartilage and bone in teratomas has been addressed most thoroughly by Levak-Svajger and Svajger. They used enzymes to separate rat embryonic shields into germ layers, which they then established as renal grafts or organ cultures. Single germ layers produced teratomas. Cartilage and bone formed within the teratomas only when *ectoderm and mesoderm*, or *endoderm and mesoderm* were grafted together: Respiratory epithelium + mesoderm → cartilage; oesophageal, stomach or intestinal epithelium + mesoderm → muscle. These were among the first studies to suggest inductive interactions between germ layers in mammalian embryos. The finding that *ectoderm* from head-fold-stage embryos produced bone and cartilage prompted the conclusions that presumptive 'ectoderm' contains other presumptive germ layers, a conclusion based on the assumption that mesoderm was expected to have produced the skeletal tissues, or that NCC within the presumptive ectoderm produced the cartilage and bone.[36]

Although formation of cartilage or bone in the stomach is rare, Ohtsuki *et al.* (1987) reported a case of metaplastic bone formation in a stomach polyp following transformation of the stomach epithelium to a cancerous state, and suggested that the altered epithelium induced the bone. Kumasa *et al.* (1990) examined ectopic bone formation in nine tumours, which they interpreted as having arisen by metaplastic transformation of the stroma under the inductive influences of epithelia within the tumors. Wight and Duff (1985) assumed that the ectopic pulmonary cartilage and bone they found in newly hatched chicks arose from misplaced germ cells (Fig. 17.8).

At the molecular level we have interesting data on the role of *Sox-9* in chondrogenesis within teratomas. Chimaeric mice generated using $Sox\text{-}9^{-/-}$ embryonic stem cells exclude any $Sox\text{-}9^{-/-}$ cells from the cartilages that form, and, furthermore, cartilage fails to form in teratomas derived from embryonic stem cells from these chimaeras (Bi *et al.*, 1999, and see Box 13.4).

MESENCHYME SIGNALS TO EPITHELIUM

Just as mesenchymal cells require epithelial signals, so epithelia require mesenchymal signals.

Mesenchyme maintains epithelial cell proliferation and, at least in avian embryos, prevents keratinization, which appears to be the default differentiation state for embryonic epithelia deprived of underlying mesenchyme. Both ectomesenchyme and mesenchyme derived from mesoderm play this role.[37]

Epithelial tissues or organs such as enamel, hair, feathers or mucous glands use mesenchymal signals to localize and form epithelial condensations (tooth buds, feather germs) and initiate differentiation. One example is mesenchymal evocation of teeth, gills, balancers and the dorsal median fin from the epidermal ectoderm of amphibian embryos. In this second role, mesenchyme provides signals that initiate the development of a number of organs that have both an epithelial and a mesenchymal component. In such organs, these interactions are reciprocal. Indeed, I think that the dual origin of these organs from epithelium and mesenchyme is an important clue to the requirement of reciprocal interaction between mesenchyme and epithelium. Examples include:

- teeth, which are comprised of epithelial enamel and crest-derived dentine, pulp, fibroblasts and bone of attachment;

- kidneys, whose tubules (nephrons) have both an epithelial and a mesenchymal component;
- gills, comprised of crest-derived cartilages, mesodermally derived blood vessels, and an endodermally derived absorptive/respiratory epithelium;
- dorsal fins, comprised of ectomesenchymal soft tissue with an epithelial covering and (neural crest-derived?) skeletal elements in some groups (Chapter 40); and
- balancers.[38]

Interestingly, but perhaps not surprisingly considering their location, ventral fins were regarded as being neither induced by nor to contain crest-derived mesenchyme. Hörstadius (1950) also discussed the evidence that NCC may play permissive roles in the differentiation of the cartilaginous otic and auditory capsules, although these cartilage types are of mesodermal origin.

Specification of gill number illustrates reciprocity rather nicely. Tadpoles of the red toad, *Bufo careens*, have three pairs of gills, tadpoles of the Egyptian toad, *B. regularis*, have two pairs, a *variation in number that is specified by the endodermal component* of the tissue interactions. Red toad tadpoles (normally three pairs) produce two pairs when their endoderm is replaced by endoderm from the Egyptian toad. Conversely, Egyptian toad tadpoles (normally two pairs) produce three pairs when their endoderm is replaced by endoderm from the red toad. The responses are plastic.

SPECIFICITY OF EPITHELIAL–MESENCHYMAL INTERACTIONS

The differentiation displayed by an epithelium depends upon the type of mesenchyme with which it comes into contact, reflecting different properties of different mesenchymal populations, which in turn reflect different lineages of fibroblastic cells within those populations (Chapter 11). A classic study is that by McLoughlin (1961), who combined epidermis from the hind-limb buds of five-day-old chick embryos with mesenchyme from various embryonic sites, and allowed the mesenchymal–epithelial combinations to differentiate *in vitro*. McLoughlin took advantage of the fact that epithelium from limb buds deposits keratin (keratinizes) when cultured alone but not when co-cultured with limb-bud mesenchyme or with fibroblasts from the embryonic heart.

Limb-bud epithelium in contact with mesenchyme from the embryonic gizzard does not keratinize. It becomes mucus-secreting and ciliated. Gizzard mesenchyme can initiate and maintain the mucus phenotype, heart fibroblasts cannot. Mucus secretion is initiated but not maintained in limb-bud epithelium co-cultured with mesenchyme from the embryonic proventriculus; after seven days in culture the limb-bud epithelium reverts to keratinization.[39] Mechanocytes and/or their extracellular products from these various mesenchymes are quite obviously separate lineages with respect to their ability to direct differentiation of limb-bud epithelium along disparate pathways. Of course, these mesenchymes themselves were already quite highly differentiated at the ages used in this experiment.

NOTES

1. See Hall (1982c–e, 1983b–e, 1984a, 1989, 1991a,b,c, 1994b, 1999b) for epithelial–mesenchymal interactions, growth factors and skeletogenesis in normal development; Hall (1994c) for such interactions during skeletogenesis in tumors.
2. See Jacobson and Fell (1941), de Beer (1947), Hörstadius (1950), Kingsbury et al. (1953), Holtfreter (1968), Tonegawa (1973) and Hall and Hörstadius (1988) for spatial associations of epithelial thickenings and aggregations of mesenchyme.
3. Association with endodermal epithelia may be relevant to recent discussions of patterning of NCC in development and evolution and to the evolutionary origin of teeth (Piotrowski and Nüsslein-Volhard, 2000; Wendling et al., 2000; Graham, 2001; Graham and Smith, 2001; Chambers and McGonnell, 2002; David et al., 2002; Matt et al., 2003; Ruhin et al., 2003). See Drews et al. (1972), Epperlein (1974) and Epperlein and Lehmann (1975) for the studies on *Triturus* using pharyngeal endoderm.
4. See Graveson and Armstrong (1987) for chondrogenesis in the axolotl, Graveson et al. (1997) for evocation of teeth from trunk neural crest, and Corsin (1975a,b) for the studies on *Pleurodeles*.
5. Foregut endoderm in chick embryos does play a role in patterning the pharyngeal arches and the hyoid cartilages (Ruhin et al., 2003).
6. See Hall and Tremaine (1979) and Bee and Thorogood (1980) for these initial studies with chick embryos, and see Thorogood et al. (1998) and Sarkar et al. (2001) for the studies with Fgf. Premigratory NCC from chick embryos also produces cartilage in respond to H.H. 24 maxillary arch epithelium (which normally induces maxillary mesenchyme) and to H.H. 24 pigmented retinal epithelium (which normally induces scleral cartilage, also a neural-crest derivative).
7. Hall (1982b) and Langille (1994b) describe the distribution of osteogenic and chondrogenic cells in mandibular arches of chick embryos.
8. See Tyler and Hall (1977), Minkoff and Kuntz (1978), Coffin-Collins and Hall (1989) and Hall and Coffin-Collins (1990) for mesenchymal influences on epithelial proliferation, and Hall (1978b, 1980c) for heterotopic interactions.
9. See Mina et al. (1991a, 1994) for the mRNA distributions, Mina et al. (1994) for proliferation – confirming the earlier studies by Coffin-Collins and Hall (1989) and Hall and Coffin-Collins (1990) – Mina et al. (1995) for *Msx-1* and *-2*, and Mina et al. (2002) for Fgfs and Bmps. Also see McGonnell et al. (1998) for *Msx-1* in expanding facial mesenchyme and Fgf-8 in adjacent ectoderm (Fig. 13.7).
10. See Hall (1980c), Kollar and Mina (1991) and MacDonald and Hall (2001) for epithelial–mesenchymal interactions in mouse Meckel's cartilage, and Barlow and Francis-West (1997) and Bogardi et al. (2000) for production of a bifurcated cartilage.
11. See Wall and Hogan (1995) for the expression data and ectopic elements, Y.-H. Wang et al. (1999) for Egf, and Shigetani et al. (2000) for Fgf-8.

12. Marshall Urist was an orthopedic surgeon who maintained an enormously active and productive programme of basic research and clinical practice during his career. His pioneering studies on the osteoinductive role of demineralized bone matrix – see Urist (1965, 1991), Urist et al. (1985, 1997), Johnson and Urist (1998) and Box 12.1 – led to the discovery of Bmps as skeletal inducers. The two papers cited in the text (Urist, 1962, 1970) show the breadth of his appreciation of skeletal biology. His text *Bone. Fundamentals of the Physiology of Skeletal Tissues* (McLean and Urist, 1968), was a mainstay for many of us for many years. Marshall Urist and Richard Goss in the USA and Leonard Bélanger in Canada were the three individuals I approached as postdoctoral supervisors in the late 1960s. The Vietnam War made funding for an Australian to travel to the US difficult. The Canadian Medical Research Council postdoctoral fellowship I was offered to work with Bélanger came one week after I had accepted the faculty position at Dalhousie University (and would have paid $500 more than my 'Dal.' salary).
13. See Le Lièvre (1974) and Le Lièvre and Le Douarin (1975) for mapping the upper jaw elements to the neural crest, and Tyler and Koch (1977a,b), Hall (1978b, 1980a–c, 1981a–c, 1983b–e) and Tyler (1978) for epithelial requirements.
14. See Tissier-Seta et al. (1995) and Barlow et al. (1999) for *Barx-1* in mice and chick.
15. See Kollar and Baird (1969) for 13–16 days, Svajger and Levak-Svajger (1971) for 12 days, and Tenenbaum et al. (1976) for the epithelial thickenings. See Box 12.3 for additional studies in which mammalian tissues were CAM-grafted.
16. See Hall (1980b,c), Miyake et al. (1996a,b, 1997) and MacDonald and Hall (2001) for studies with mouse embryos.
17. See Barni et al. (1995), Richman and Mitchell (1996) and Clouthier et al. (2000) for *endothelin-1* expression and role in craniofacial development, and Richman and Mitchell (1996) for a review of knockout mice.
18. See Bendall and Abate-Shen (2000) for a review. Merlo et al. (2000) review the roles of *Dlx-1, Dlx-2, Dlx-3* and *Dlx-5* in the craniofacial skeleton and in osteogenesis.
19. See Ryou et al. (1997) and Ducy et al. (2000) for expression, and Miyama et al. (1999) for the link to osteogenesis via Bmp-4.
20. See Depew et al. (1999) and Acampora et al. (1999) for *Dlx-5* knockout, Depew et al. (2002) for mice lacking *Dlx-5* and *Dlx-6*, and Bendall et al. (2003) for the role of *Dlx-5* in endochondral ossification in chicken and mouse embryos, including its role in condensation of prechondrogenic mesenchyme.
21. See Piotrowski and Nüsslein-Volhard (2000), who also provide data on *one-eye, pinhead* and *Casanova*. All three of these mutants lack endoderm; arch segmentation also requires normal endoderm. Kimmel et al. (1998) investigated the shaping of the simple columnar pharyngeal cartilages in zebrafish and mutations that deform the stacking of chondroblasts into columns.
22. See Power et al. (1999) and Helms et al. (1994) for these two studies with limb buds. Earlier, Morgan et al. (1992) had shown that retroviral vectors containing *Hoxd-11* (*Hox-4.6*) expand the *Hoxd-11* domain in chick limb buds more anteriorly and initiate posterior limb skeletal patterns in the anterior limb bud in a homeotic transformation.

23. See Suzuki *et al.* (1998, 1999a,b) for administration of retinoic acid, and Wendling *et al.* (2000) for the genes regulated by retinoic acid.

24. See Piotrowski *et al.* (1996) and Schilling *et al.* (1996b) for craniofacial mutants, and Schilling *et al.* (1996a) for *chinless*. A fish gene named *chinless* causes us to ask what is a chin and begs Stephen J. Gould's argument that chins are spandrels.

25. See Blaxter (1987), Webb and Noden (1993), Schlosser (2002a,b), and Webb and Shirey (2003) for overviews of the lateral-line system. See Hanken and Hall (1993) for relationship between lateral lines and bone induction. See Lannoo (1987a,b, 1988) and Lannoo and Smith (1989) for neuromasts in anuran and urodele amphibians with comments on caecilians and the possible relationship of neuromasts to bone induction.

26. The late Angus Bellairs, an eminent British herpetologist, wrote a novel – *The Isle of Sea Lizards* – in which we find the following: ' "You know that tome by Gavin de Beer, the late Sir Gavin, I should say?" She shook her head. "Well, nobody reads it these days, but it's a fascinating book if that's the way your mind works. But dry, one has to admit. I read it in parallel with *Tess* [of the D'Urbervilles by Thomas Hardy], one on each arm of my chair. When *Tess* made me too weepy, I turned to the timeless serenity of the frontal bones, from fish to man" ' (Bellairs, 1989, p. 93).

27. See de Beer (1937, reissued 1985), especially pp. 6, 489–490 and 508, agenda items ii.6 (p. 513) and iii.10–12 (p. 514), and the preface to the reissue by Hall and Hanken (1985b). Webb (1989a–c) investigated the distribution of neuromasts and the lateral-line system in teleost fishes, drawing attention to the relationship between neuromasts and dermal bone. See Webb and Noden (1993), Northcutt (1996) and Schlosser (2002a,b) for excellent reviews of placodes, neuromasts and lateral-line development, and see S. C. Smith *et al.* (1994) for developmental and evolutionary links between placodal ectoderm and NCC. See Smirnov (1994) for *Bombina*.

28. See Westoll (1941) and Pinganaud-Perrin (1973) for independence of teleost dermal bones from neuromasts, and (Jollie, 1984e) for head development in *Amia calva*, including relation of lateral-line canals to bones. Development and function of the frontal bones in the cichlid, *Astatotilapia elegans*, were studied by Verraes and Ismail (1980).

29. See Hall and Hanken (1993) and Maglia and Pugener (1998) for skeletal development in *Bombina orientalis*.

30. Jollie (1981, 1984a–e) and Meinke (1982a) examined times of first ossification and the topographical relationship between lateral lines and osteogenesis in salmon (especially coho, *Oncorhynchus kisutch*, and sockeye, *O. nerka*) and in *Lepisosteus* and *Polypterus*. See Meinke (1986) for the dermal skeleton in lungfish, including neuromasts and the timing of tissue interactions. See the chapters and bibliography of papers on lungfishes in Bemis *et al.* (1987), and Bartsch (1994) for the development of the cranium in *Neoceratodus*.

31. For other discussions of the possibility of such interactions, see Patterson (1977), Schaeffer (1977), Graham-Smith (1978) and Northcutt and Gans (1983).

32. See Patterson (1977) for a general discussion, and see Holmgren (1940, 1941, 1942, 1943) and Stensio (1947), especially for sharks and rays, including skull and lateral-line development and neural-crest origins.

33. See Yamamoto *et al.* (2003) for the skeletal changes in *Astyanax*, and see Strickler *et al.* (2002) and Jeffery *et al.* (2003) for an overview of how the eyes are lost in this blind cavefish.

34. See Damjanov and Solter (1974), Skreb and Svajger (1975), D. Bennett *et al.* (1977), O'Hare (1978), Hall (1987d, 1994c) and Maclean and Hall (1987) for overviews of the developmental aspects of teratomas. See Damjanov and Solter (1974), Bennett *et al.* (1977) and Hall (1987d, 1994) for the possibility of epithelial–mesenchymal interactions within teratomas.

35. See Damjanov *et al.* (1973) and Illmensee and Stevens (1979) for neuroectodermal components of teratomas.

36. See Levak-Svajger and Svajger (1971, 1974, 1979), Skreb and Svajger (1973, 1975), Svajger and Levak-Svajger (1974, 1975, 1976) and Skreb *et al.* (1976).

37. See Hörstadius (1950), Holtfreter (1968), Johnston (1975), Hall and Hörstadius (1988), Coffin-Collins and Hall (1989) and Hall and Coffin-Collins (1990) for ectomesenchymal control of epithelial proliferation, and McLoughlin (1961) for similar control by mesenchyme in the trunk.

38. See Crawford and Wake (1998) for an analysis of balancers in salamanders, which exist in approximately half of the species in three of the 10 families. Balancers contain type II collagen and may function as a non-bony skeleton.

39. Evidence for similar interactions in limb morphogenesis will be considered in Chapter 35.

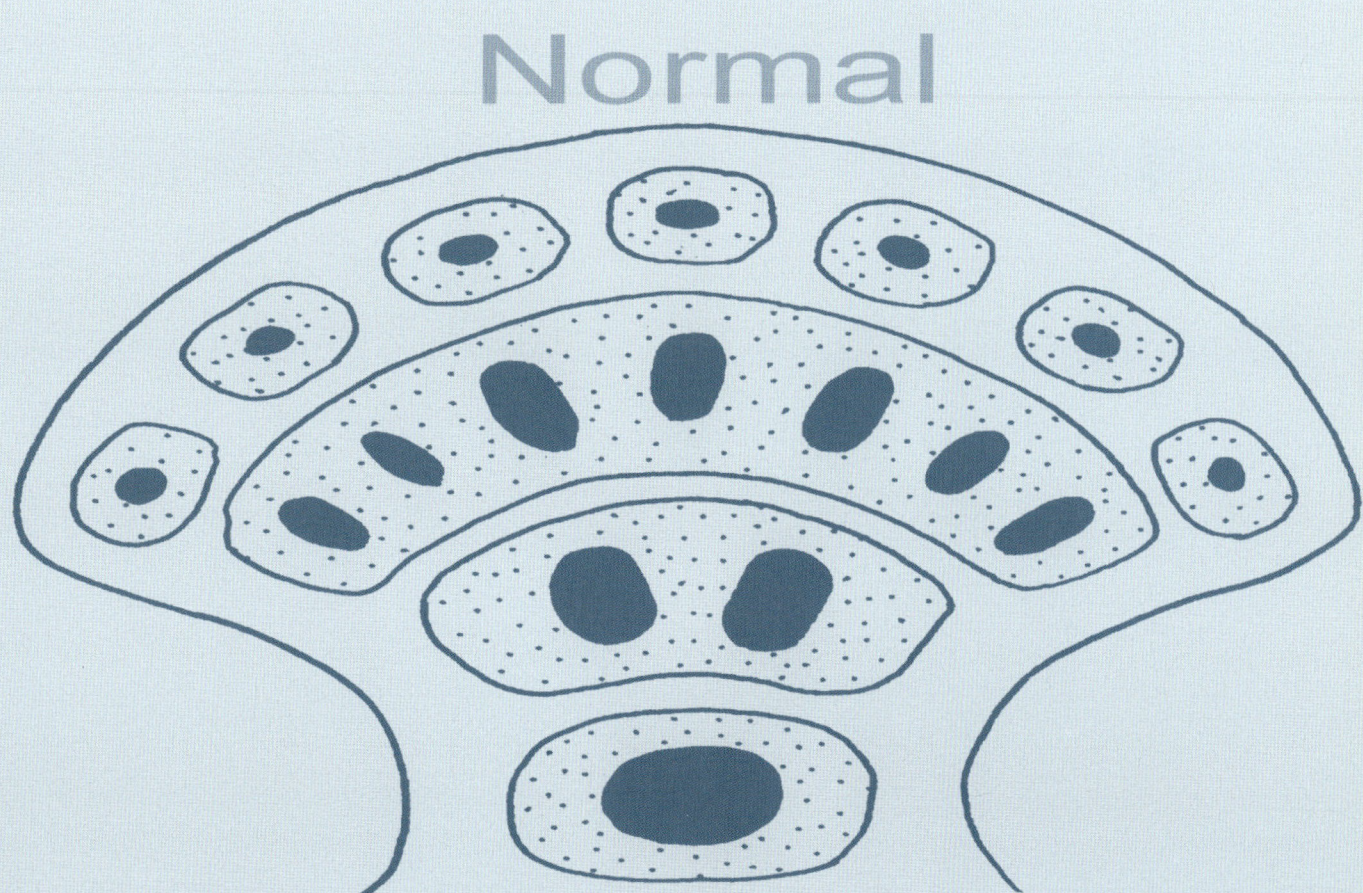

Part VII

Getting Started

Grüneberg accumulated the initial evidence – especially from mutant mouse embryos – that the membranous skeleton is just *as important and identifiable* as are the cartilaginous and osseous skeletons.

Condensations, which comprise the membranous skeleton, are the pivotal stage in the development of skeletal and other mesenchymal tissues, when a previously dispersed population of cells gather together to differentiate into a single cell/tissue type such as cartilage, bone, muscle, tendon, kidney or lung. This is the earliest stage during organ formation when tissue-specific genes are up-regulated.

Condensations are the fundamental cellular–morphogenetic units of morphological change in organogenesis during vertebrate evolution.

Normal

talpid

The Membranous Skeleton: Condensations

The membranous skeleton is as important and identifiable as are the cartilaginous and bony skeletons.

As discussed in Chapter 18, embryonic skeletal rudiments differentiate following specific interactions between mesenchyme and an epithelium. Many of these mesenchymal cells migrate to their final site from the somites to make vertebrae and ribs, or from the neural crest to make the craniofacial skeleton and teeth. Other mesenchymal populations – those that make fin and limb skeletons – arise where the skeletal elements will form, although myogenic cells migrate to limb territories from the somites (Chapter 35). At these early stages of embryogenesis, mesenchyme is uniform and apparently homogeneous. You cannot tell from histological sections which cells will form cartilage or bone and which will form connective tissue around the skeleton (Figs 10.4, 14.2, 19.1 and 19.2). Slightly later in development, each centre can be seen as an aggregation or *condensation* of cells.

Condensation, the pivotal stage in the development of skeletal and other mesenchymal tissues when a previously dispersed population of cells gather together to differentiate into a single cell/tissue type such as cartilage, bone, muscle, tendon, kidney or lung, is the earliest stage during organ formation when tissue-specific genes are up-regulated. The immediate consequence of an epithelial–mesenchymal interaction, condensation is the second of the three major processes of skeletogenesis, the three – epithelial–mesenchymal interaction, condensation and differentiation – forming a hierarchy of processes initiating differentiation and growth of skeletogenic cells.

Development of the skeleton is a stepwise set of processes, each step depending on the step before, but each involving different cellular processes – migration, adhesion, proliferation, growth – and each subject to different genetic control. Condensation is the aggregation of cells that facilitates selective regulation of genes specific for either chondro- or osteogenesis. Condensation itself is a multistep process, involving initiation, establishment of boundary conditions, cell adhesion, proliferation, growth and cessation of growth. Condensations must attain a critical size and cells must interact within a condensation for the condensation phase to cease and differentiation to be initiated; it really is 'all for one and one for all'.

Chapters 19 and 20 present a synopsis of our current understanding of the cellular and molecular bases of how condensations are identified, initiated and grow, how their boundaries and sizes are set, how condensation ceases and overt differentiation begins. Extracellular matrix (ECM) molecules, cell-surface receptors, and cell-adhesion molecules, such as fibronectin, tenascin, syndecan and N-CAM, initiate formation of condensation and set condensation boundaries. *Hox* genes (*Hoxd-11–d-13*) and other transcription factors (*Cfkh-1, Mfh-1, Runx-2*), modulate proliferation of cells within condensations. Cell adhesion is ensured through *Hoxa-2, Hoxd-13* and the cell-adhesion molecules N-CAM and N-cadherin (Fig. 19.3). Subsequent growth of condensations is regulated by Bmps activating *Pax-2, Hoxa-2* and *Hoxd-11*. Growth of a condensation ceases when noggin inhibits *Bmp* signaling, setting the stage for transition to the next stage of skeletal development, which is overt cell differentiation.

THE MEMBRANOUS SKELETON

In her pioneering studies on skeletogenesis, Honor Fell described centres of osteogenesis where mesenchymal cells appeared more tightly clumped. Similar centres are seen at the initiation of hairs, feathers, ligaments and tendons. *Condensations constitute a skeletal system even though synthesis of products characteristic of*

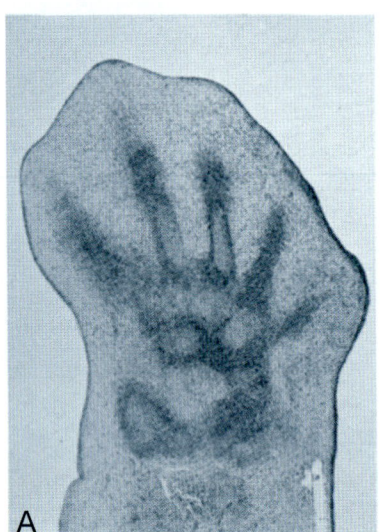
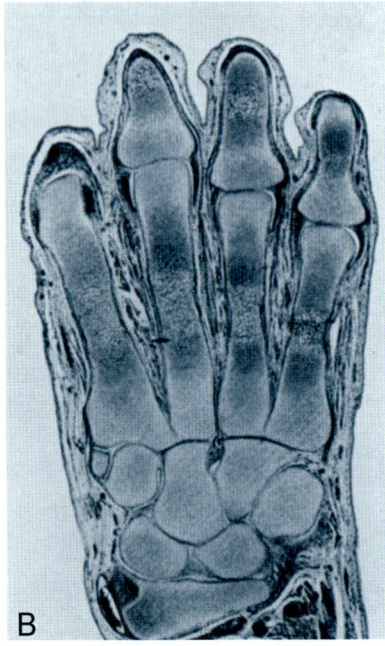
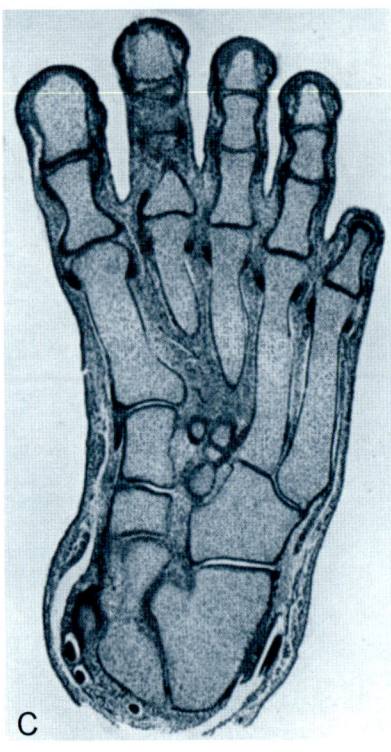

Figure 19.1 Horizontal histological sections of developing human hands and feet. (A) The hand plate at five weeks' gestation showing prechondrogenic condensations of the five digits and of wrist elements. (B) The hand at nine weeks' gestation showing chondrocyte hypertrophy and mineralization in the centre of the phalanges, and early subperiosteal osteogenesis adjacent to the hypertrophic chondrocyte zone. (C) The foot at nine weeks' gestation is at an earlier stage of chondrogenesis than are the hands (cf. B), with initial chondrocyte hypertrophy only in the proximal phalanges. From Harris (1933).

differentiating skeletal cells might not be apparent. Grüneberg (1963) referred to condensations collectively as the '*membranous skeleton*' to highlight their existence and equal status with the *cartilaginous and osseous skeletons*. Grüneberg accumulated evidence – especially from mutant mouse embryos – demonstrating that the membranous skeleton is just *as important and identifiable as are the cartilaginous and osseous skeletons*.

From the point of view of developmental mechanisms, *the membranous skeleton is of prime importance*. Differentiation cannot be initiated without it, and many skeletal mutants have their primary action at this stage. Chondrogenesis may be delayed or even not initiated if condensations fail to attain a critical size, resulting in morphological defects in the skeleton of older embryos and adults. Indeed, condensation specifies the basic shape and size of skeletal elements, mechanical forces and ECM constituents also playing important roles. As laid out in a recent overview of limb skeletal patterning: 'The formation of the condensations that prefigure the skeletal elements is arguably the most critical event in skeletal patterning, so it is surprising how little is known about what determines their size, shape and number.'[1]

Of course, the role of condensations in skeletal initiation, development and growth, is not as clear-cut as the last statement would lead us to believe. Some fuzziness

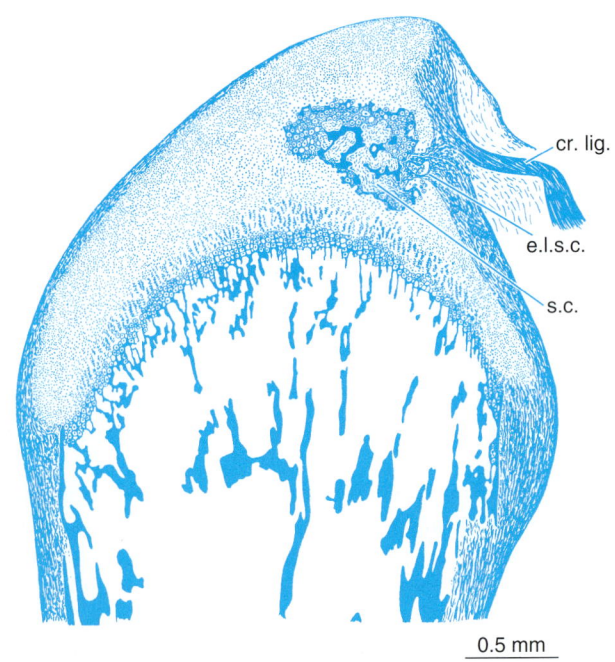

0.5 mm

Figure 19.2 An early stage in the development of a secondary centre of ossification (s.c.) in the distal epiphysis of the femur of Azara's opossum, *Didelphis azarae*. The secondary centre arises from the perichondrium (e.l.s.c.) rather than within the epiphysis as in placental mammals. cr. lig., the cruciate ligament. From Haines (1941).

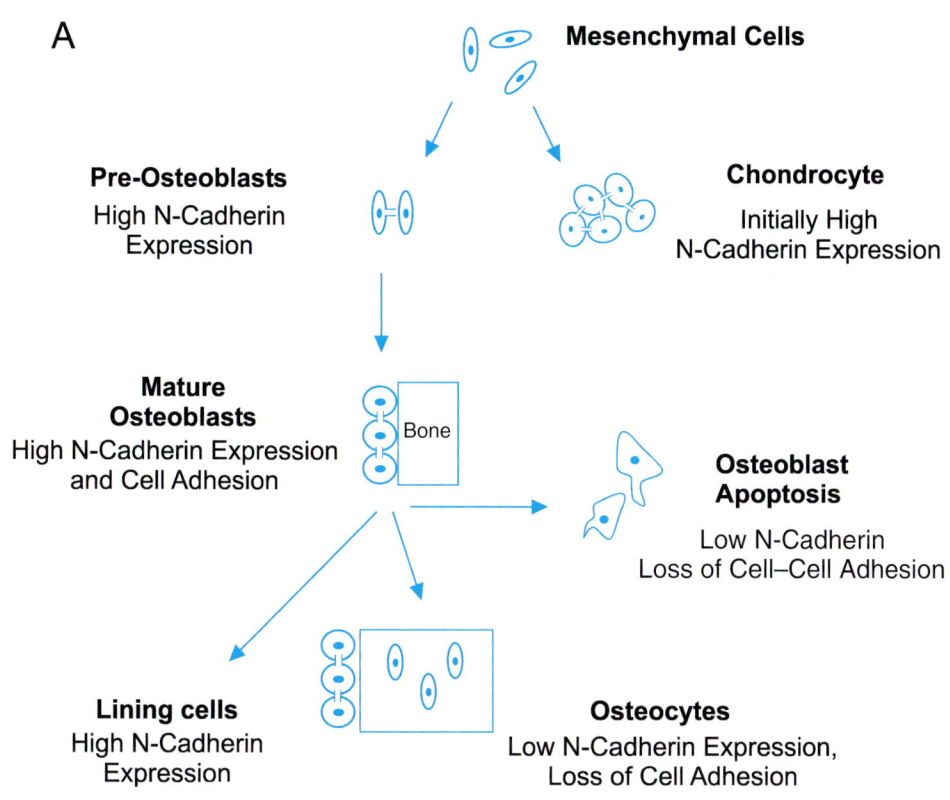

A

Mesenchymal Cells

Pre-Osteoblasts
High N-Cadherin
Expression

Chondrocyte
Initially High
N-Cadherin Expression

Mature Osteoblasts
High N-Cadherin Expression
and Cell Adhesion

Bone

Osteoblast Apoptosis
Low N-Cadherin
Loss of Cell–Cell Adhesion

Lining cells
High N-Cadherin
Expression

Osteocytes
Low N-Cadherin Expression,
Loss of Cell Adhesion

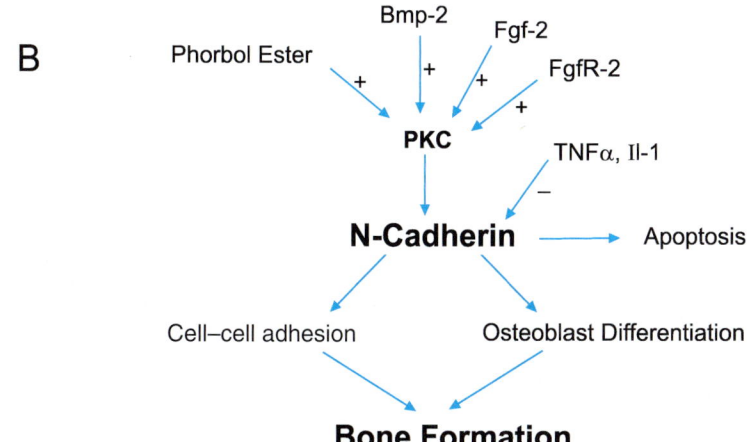

B

Phorbol Ester

Bmp-2

Fgf-2

FgfR-2

+ + +

+

PKC

TNFα, Il-1

−

N-Cadherin ⟶ Apoptosis

Cell–cell adhesion

Osteoblast Differentiation

Bone Formation

Figure 19.3 N-cadherin and skeletogenic differentiation. (A) A proposal for the roles played by high or low levels of N-cadherin and cell adhesion in the differentiation of mesenchymal cells to osteoblasts or chondrocytes. (B) N-cadherin in osteoblasts is controlled through up-regulation of protein kinase-C (PKC) by local growth factors such as Bmp-2, Fgf-2, FgfR-2 and phorbol ester. Down-regulation of N-cadherin by TNFα and interleukin-1 (Il-1) is associated with loss of cell-to-cell adhesion and apoptosis. Based on the scheme proposed by Marie (2002).

reflects our present state of knowledge, some may be more fundamental. A mutation that results in congenital hydrocephalus illustrates both aspects.

CONGENITAL HYDROCEPHALUS (ch)

Congenital hydrocephalus (*ch*), which occurs with a frequency of two per 1000 live births in humans, is characterized by skeletal defects in cells arising from the cranial neural crest, the most striking of which is *total absence or severe reduction of the skull bones*. The genetic basis of *ch* is homozygosity for a nonsense mutation in a gene,

Mf1, encoding a winged helix/forkhead transcription factor. One complication with interpreting the phenotype as entirely the result of the mutated gene is that some of the skeletal changes are secondary – a consequence of pressure exerted by excess fluid in the brain – and can be compensated for by surgical intervention. Thirty-seven hydrocephalic children aged seven to 18 years, each of whom received a shunt to drain excess fluid from the brain, showed some recovery as evidenced by thickened calvariae and increased neurocranial ossification.[2]

Congenital hydrocephalus is also seen in mice, in which some cartilages such as the trachea are missing, not because condensations fail to form but because the condensations that do form are smaller than normal. Some membrane bones, including the vomer, are absent from *ch* mice, even though condensations for the vomers form, indeed are *larger* than the equivalent condensation in wild-type embryos. To further complicate interpretation of the basis for the *ch* mutant phenotype, cartilage and membrane bone develop in *single condensations* in which *either* cartilage *or* membrane bone but *not both* develop in wild-type embryos. Similarly (paradoxically), in *chondrodystrophic* (*cho*) mice, in which the long bones are shorter than normal by late foetal life, condensations for skeletal elements of the limbs are *normal in size*. As a result of analyzing such mutants, we recognize the importance of the environment created within and around condensations, in addition to size and time of appearance of a condensation, as critical factors in initiation of skeletogenesis.[3]

Teratogens and drugs also disturb the environment in which condensations develop, or to which condensations respond. As one example, administering 6-mercaptopurine to pregnant rats massively reduced the size of condensations in the litter. During the mid 1970s, when these studies were undertaken, little was known about the factors operating within condensations, the detailed nature of local environmental influence, or how they act or interact. Attempts to understand factors controlling determination and differentiation of skeletal cells, or to extrapolate results obtained *in vitro* to the situation *in vivo*, meet with difficulty. Happily, three decades later, that has changed. We now know much about the importance of condensations, especially from studies on craniofacial and limb skeletons.[4]

CHARACTERIZING CONDENSATIONS

Fell (1925) described how all the mesenchymal cells of chick limb buds initially have a uniform cytological appearance. Subsequently, a cluster of closely packed, rounded mesenchymal cells with large nuclei forms in the centre of the limb bud preceding the appearance of ECM at H.H. 25.[5]

An extensive literature describing the histochemical characteristics of skeletal condensations accumulated through the 1950s, '60s and '70s, particularly for mammalian limb buds, as a result of studies by Jean Milaire in Brussels (Chapter 37). Studies by Drews and Drews (1972, 1973) draw our attention to the distribution of cholinesterase in skeletal condensations and the possible role of cholinesterase in initiation of ECM synthesis. This evidence may be quite important in light of the possible role of neurotransmitters as inductive stimuli (Box 19.1).[6]

Re-analysis of the histology of limb buds accompanied the burst of experimental studies on limb morphogenesis in the 1950s. With both light and electron microscopy the existence of condensations preceding chondrogenesis in chick embryos was confirmed and equivalent condensations described in developing rodent limb buds (see

below). Ultrastructural studies by Robert Searls and colleagues in 1972 showed that these cells have a high nucleo-cytoplasmic ratio, large nucleoli, a poorly developed endoplasmic reticulum, numerous free ribosomes, and small mitochondria. As discussed in Chapter 12, these are the general characteristics of osteo- and chondroprogenitor cells in a variety of situations.[7]

Condensations may be recognized either at the cellular level – through the closer packing density that distinguishes condensed from uncondensed cells (Fig. 13.8) – or at the molecular level. As shown in Fig. 19.4, prechondrogenic condensations have cell-surface molecules that bind peanut agglutinin lectin (PNA), allowing condensations to be visualized. Affinity for PNA can thus be utilized as a technique to isolate and characterize prechondrogenic cells. Condensations also have elevated levels of such

Box 19.1 Serotonin

The neurotransmitter *serotonin* has been implicated in all developmental stages from blastula (regulation of cleavage), through morphogenetic movements of the primitive streak in gastrulation, to neurulation, migration of NCC (NCC have the 5-HT receptor to which serotonin binds) and branchial arch formation, to differentiation of craniofacial mesenchyme. Serotonin appears to be involved in at least one early event in skeletogenesis, namely stimulating differentiation of dental mesenchyme (Moiseiwitsch, 2000).

Although serotonin plays a role in NCC migration, serotonin inhibits migration of mandibular mesenchyme to the extent that craniofacial malformations can result. Serotonin is expressed transiently in embryonic mouse mandibular epithelium at 10–11 days of gestation, i.e. when epithelial–mesenchymal interactions take place. 5-HT is taken up by mandibular epithelium and tooth germs, and promotes tooth formation. Exposure to serotonin uptake inhibitors decreases proliferation and increases cell death in mesenchyme five to six cell layers away from the epithelium, but increases proliferation in the subepithelial mesenchyme; a 45-kDa serotonin-binding protein is localized in craniofacial mesenchyme, especially adjacent to sites of epithelial uptake.[a]

Noting that serotonin receptor 2B (5-HT$_{2B}$) and enzymes regulating retinoic acid are coordinately expressed at sites of epithelial–mesenchymal interaction, and that both pathways have been implicated in the regulation of chondrogenesis, Bhasin *et al.* (2004) showed that chondrogenesis in micromass cultures of murine hind-limb-bud mesenchyme was enhanced by 5-HT – via activation of the p-42 map kinase pathway – but inhibited by retinoic acid, which inhibited p-38 map kinase signaling.

A second neurotransmitter, *cholinesterase*, is found in skeletal condensations, where it has been proposed to play a role in condensation formation and synthesis of ECM products. There may be a connection to the proposed role of neurotransmitters as inductive agents, although this idea has not been tested at all extensively.[b]

[a] See Moiseiwitsch and Lauder (1995) and Moiseiwitsch (2000) for a review of serotonin and other neurotransmitters in craniofacial development and for possible modes of action, and Moiseiwitsch and Lauder (1996) and Shuey *et al.* (1992, 1993) for epithelial and mesenchymal responses.
[b] See Drews and Drews (1972, 1973) for cholinesterase distribution, and McMahon (1974), Kebabian *et al.* (1975), and Landauer (1976) for neurotransmitters and induction.

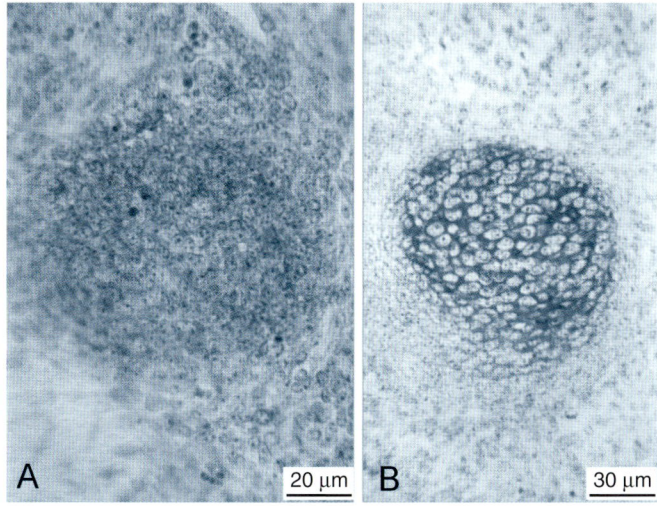

Figure 19.4 Condensation and chondrogenesis. (A) The condensation for a hyoid cartilage from a mouse embryo of 12.5 days of gestation (Theiler stage 21) visualized with peanut agglutinin lectin (PNA) and peroxidase. (B) The hyoid cartilage at 13.5 days of gestation (Theiler stage 22) visualized with alcian blue. Note the lack of ECM and the definite boundary at the condensation stage (A), and the evident ECM once chondrogenesis commences (B). Modified from Hall and Miyake (1992).

ECM or cell-surface molecules as hyaladherins, versican, tenascin, syndecan, N-CAM, and heparan-sulphate and chondroitin-sulphate proteoglycans.

Other condensation markers are being discovered. Thrombospondin-4, one of five related glycoproteins, is expressed transiently in mesenchyme associated with both chondrogenic and osteogenic condensations (see Box 9.2 and Table 19.1).[8]

Frzb-1 (*Drosophila frizzled* family) is expressed in ventral then central-core mesenchyme of chick limb buds at condensation and before the expression of markers such as aggrecan (Box 4.3) both *in ovo* and in micromass culture. *Frzb-1*, which encodes a *Wnt* receptor, is a condensation marker, but it is also expressed in the neural tube and at sites of epithelial–mesenchymal interactions. Misexpressing *Frz* genes delays chondrocyte maturation, leading to severe shortening of skeletal elements.[9]

Bagpipe homeobox gene-1 (*Bapx-1*, *Nkx3.2*), a member of the *Nkx* family of homeobox-containing transcription factors, is an early prechondrogenic marker, being expressed before the condensation stage in mouse embryos. *Pax-1* and *Pax-9* interact with the *Bapx-1*

Table 19.1 The major classes of genes and gene products associated with skeletogenic condensations along with their functions and stages of action[a]

Gene/gene product	Stage	Function
		Growth factors
Bmps	Growth, transition to differentiation	Regulate *Hox* genes (*Hoxa-2*, *Hoxd-11*, *Pax-2*) in response to *Shh*; regulate *Msx-1*, *Msx-2*
Fgf-2	Initiation, proliferation, growth	Regulates N-CAM
Tgfβ	Initiation	Regulates fibronectin
		Cell-surface, cell-adhesion and ECM molecules
Fibronectin	Initiation, proliferation	An extracellular glycoprotein regulated by *Tgfβ*; regulates N-CAM
N-cadherin	Adhesion	A cell adhesion molecule
N-CAM	Initiation, adhesion	A cell adhesion molecule regulated by fibronectin, *Prx-1*, *Prx-2* and *Fgf*
Noggin	Slows or stops growth	A secreted protein that binds to and inactivates *Bmp-2*, *Bmp-4*, *Bmp-7*
Syndecan	Sets boundary	A receptor that binds to tenascin; binds to fibronectin to inactivate N-CAM
Tenascin	Stops condensation growth, sets boundary	An extracellular glycoprotein that binds to syndecan
		Hox genes
Hoxa-2	Sets boundary, growth, prevents differentiation	Up-regulated by *Bmps*, down-regulates *runx-2*
Hoxa-13	Adhesion	Alters adhesive properties
Hoxd-11	Proliferation, growth	Up-regulated by *Bmp*
Hoxd-11-13	Transition to differentiation	Transcriptional activation
		Transcription factors
Cfkh-1	Initiation, proliferation	A chicken forkhead-Helix transcription factor that regulatess *Tgfβ* and interacts with *Smad* transcription factors
Mfh-1	Proliferation	Mesenchymal transcription factor
osf-2	Switches cells into the osteoblastic pathway	Transcriptional activating protein regulated by *Bmp-7* and vitamin D_3
Pax-1, *Pax-9*	Differentiation, growth	Encode nuclear transcription factors, regulated by *Bmp-7*
Prx-1, *Prx-2*	Initiation	Upstream regulation of N-CAM
Runx-2 (*cbfa-1*)	Differentiation of chondroblasts	Transcriptional activating protein inhibited by *Hoxa-2*
Scleraxis	Proliferation	A basic helix–loop–helix protein
Sox-9	Proliferation	Regulates Col2a1

[a] Also see Figs 17.11 and 20.2.

promoter to regulate *Bapx-1* in the sclerotome for cartilage initiation. *Bapx-1* null mice are missing vertebrae and mesodermal skull bones because of a basic failure of cartilage formation at this early condensation–precondensation stage. In zebrafish, *Bapx-1* specifies the lower jaw joint within the primordium of the first arch.[10]

HOW CONDENSATIONS ARISE

By definition, the process of condensation requires an increase in cell numbers within the condensation and/or a decrease in the number of cells around it.[11] A variety of cellular and developmental process could and do produce such an outcome:

(i)   increased proliferation within, and/or decreased proliferation outside the condensation;
(ii) shorter cell cycle time within, and/or longer cell cycle time in cells outside;
(iii) a greater fraction of the cell population dividing within, and/or a smaller fraction dividing outside;
(iv) decreased cell death within, and/or increased cell death outside; and/or
(v)   migration of cells toward a centre or focus, and/or failure of cells to move away from a centre.

These are not mutually exclusive mechanisms. The first three all relate to proliferation but can be accelerated or slowed independently or in a coordinated manner; see Atchley and Hall (1991) for a model. (iii) and (iv) could reflect subpopulations of cells, while (v) could coexist with (i) to (iv), and so forth. Mechanisms associated with secretion and/or accumulation of ECM are less important processes *initiating* a condensation – a characteristic of condensations is close contact between the cells and sparse or absent ECM – but become important after cells are recruited and accumulate, and as the condensation grows and switches from proliferation/accumulation to differentiation (see Figs 17.11 and 18.3). I use studies on avian limb buds to ask which one or more of these five mechanisms operates to initiate the chondrogenic condensations of the future limb skeleton.

Altered mitotic activity

As shown by Janners and Searls (1970), and as discussed below, limb chondrogenic condensations do not arise because of a localized increase in mitotic activity within prechondrogenic mesenchyme, although differential rates of cell division *do play a role* in other skeletal rudiments. Jacobson and Fell (1941) presented cytological evidence – but not quantitative data – that cell division within preskeletal condensations of the mandibular arch is more rapid than in the surrounding mesenchyme. Again in chick embryos, epithelial conjunctival papillae stimulate the rate of cell division in presumptive scleral

mesenchyme to produce the osteogenic condensations in which scleral ossicles form (Chapter 21). While there is no evidence for a local increase in mitotic activity associated with condensation in embryonic chick wing buds, cell density does increase, and must do so via another mechanism.[12]

Changing cell density

I use the ultrastructural analysis of wing buds from embryos of H.H. 17–27 (2–5.5 days of incubation), published by Searls and colleagues in 1972, to set the stage for a discussion of changes in cell density during condensation and early chondrogenesis in wing development. Three stages were identified on the basis of changing cell-to-cell contacts and cell associations, as outlined below.

(i)   The first, which is seen during the 24 hours that elapse between H.H. 17 and 23, is characterized by lack of regional specialization of cell types across or along wing buds, and no alteration in organelle type or configuration. Cells are in contact over broad areas of up to 4 μm of their membranes.
(ii) During the second stage, between H.H. 23 and 25 – a period of some 12–18 hours – the number of broad cell contacts declines, filopodia appear, and many cell-to-cell contacts are established via filopodial projections. Although cells in the centre of the bud are beginning to differentiate, no ECM is deposited. Searls and colleagues emphasized that the changing nature of cell contacts – from broad areas of cell membranes to filopodia – reflect neither production of ECM nor decrease in cell density. Rather, altered cell contacts reflect changes in cell membrane properties. Another sign of membrane changes is alteration in the ability of these cells to recognize cells at earlier or later stages of differentiation. This property of *selective affinity* also changes between H.H. 23 and 25 (Chapter 22).
(iii) From H.H. 25 onward, increasing numbers of cell-to-cell contacts are made by filopodial contacts less than 1 μm in diameter. Changes in cell shape or arrangement are not seen until H.H. 27, by which stage differentiation is underway and ECM is being synthesized and deposited.

Alterations in the nature of cell-to-cell contacts during initial chondrogenesis are not accompanied by – and so presumably not caused by – alterations in cell density. Nor was condensation of chondrogenic mesenchymal cells reported on the basis of ultrastructural analyses of leg and wing buds by Gould *et al.* (1972). Summerbell and Wolpert (1972) recorded that cell density increased by a factor of 1.5 between H.H. 19 and 23, and by a factor of 2.0 between H.H. 19 and 25.[13]

If a condensation represents an area of cells that are more closely packed *and* contain more cell contacts then chondrogenic mesenchyme meets this minimal definition. Nevertheless, further evidence for or against condensation was sought. Thorogood and Hinchliffe (1975) documented a considerable increase in cell packing within hind-limb-bud prechondrogenic mesenchyme – from 12 to 21 cells/1000 μm^2 between H.H. 22 and 26 (1.5 days) – accompanied by increasingly broad cell-to-cell contacts.

A conflict exists between these studies on wing and hind-limb buds. In discussing the differences, Thorogood and Hinchliffe emphasized difference in timing between wing and hind-limb-bud development, and the transitory nature of condensations. Perhaps more importantly, when they employed the fixation protocol used by Gould and colleagues, the specimen shrank considerably, to the point that chondrogenic condensations could not be observed electron microscopically. Thorogood and Hinchliffe therefore concluded that condensations are present in both sets of limb buds. My re-analysis of the data on cell density (Hall, 1978a) supports that conclusion, and provides consensus regarding increase in cell density within prechondrogenic mesenchyme after H.H. 22 but not before, and that cell density at any age is greater in prechondrogenic than premyogenic mesenchyme.

Aggregation and/or failure to disperse

Evidence from *in-vitro* studies on the amphibian neural crest carried out by Epperlein and Lehmann (1975) shows that cells can fail to move away from a centre and form a chondrogenic condensation. After initial contact of cultured neural folds with pharyngeal endoderm, NCC proliferate and increase their adhesiveness to one another, thereby preventing migration and allowing a condensation to accumulate.[14]

Of the five bases for condensation formation outlined earlier, (v) – lack of movement of cells away from a centre or aggregation of cells toward a centre (two mechanisms that need not be mutually exclusive) – are the cellular processes responsible for producing chondrogenic condensations in two disparate skeletal systems: developing avian limb buds and amphibian limb regeneration.[15]

Limb buds and limb regeneration

When H.H. 21 limb-bud cells are cultured at a sufficiently high density, chondrogenesis is always preceded by condensation, initiated by a few cells, and grows by addition of cells at the periphery (Fig. 19.5). Without condensation, chondrogenesis does not occur; dispersal of prechondrogenic cells favours fibroblastic not chondrogenic differentiation. Chondrogenesis *has* been initiated without condensation, but the cells had to be tricked into behaving as if they were in a condensation. This can be done by culturing cells in a collagen gel or in agar so that secreted matrix products accumulate around them as they would *in vivo*. One of the consequences of the evolution of multicellularity is that cells become dependent on their neighbours, and your neighbours have to be cells of the same type – not just from the same species or individual, but of the same differentiative type and stage of differentiation (Chapter 22).[16]

In what remains one of the nicest comparative analyses, Gould and his colleagues examined wing buds at H.H. 19–28 and a regeneration blastema of amputated limbs of adult European crested newts, *Triturus cristatus*. Cells condensed within the regeneration blastema, but whether in the same way as during embryonic development is unclear, just as it is not clear how much of regeneration is a recapitulation of embryonic development. Figures 1–5 in their 1974 paper provide a clear visualization of the formation and growth of a condensation during digit development.

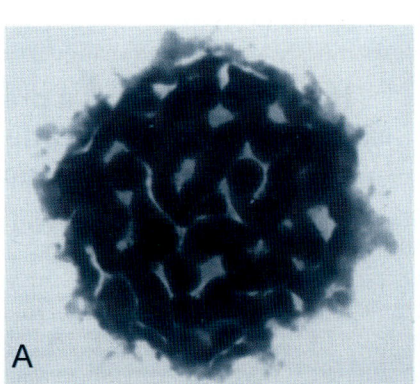 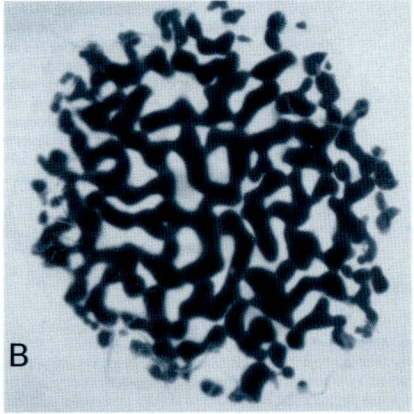 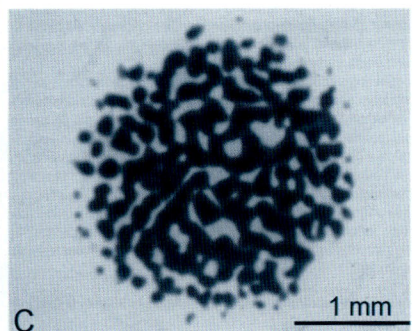

Figure 19.5 Chondrogenesis in micromass culture. Mandibular mesenchyme removed from a chick embryo of H.H. stage 21 (3.5 days of incubation), dissociated, established in micromass culture (10 μl drops of 2×10^7 cells/ml) and cultured for six days, forms cartilage nodules, here visualized by alcian blue staining. (A) Extensive chondrogenesis with large nodules in control medium (Ham's F12 : BGJ$_b$ [3:1] + 10% fetal calf serum). More but smaller nodules form with the addition of ascorbic acid at 150 μg/ml (B) or ascorbic acid + 10^{-7} M dexamethasone (C). Modified from Ekanayake and Hall (1994b).

Molecular control

By 1992, it was evident that Tgfβ and fibronectin are involved in condensation formation. Tgfβ regulates fibronectin, which in turn regulates the cell-adhesion molecule, N-CAM (Table 19.1; Figs 17.11 and 20.2). By 1995, we and others could model condensation as resulting from epithelial–mesenchymal interactions controlled by Tgfβ, Bmp-2, *Msx-1* and tenascin. Bmp receptors (Box 39.2) have now been characterized and localized to condensing mesenchyme in chick limbs and in condensations in mouse embryos.[17]

In the early 1990s when N-CAM was shown to be expressed in osteoblasts coincident with type I collagen and alkaline phosphatase but after fibronectin and tenascin, it was suggested that N-CAM might play a role switching cells from division (condensation) to differentiation, which indeed N-CAM does; tenascin is expressed in limb condensations after N-CAM, both are expressed in fracture callus, while we demonstrated N-CAM as a mediator of secondary chondrogenesis from periosteal cells in response to embryonic movement (Fang and Hall, 1995, 1996, 1997, 1999).[18]

Widelitz *et al.* (1993) used Fab' fragments of an antibody against N-CAM to inhibit aggregation of chick chondrogenic mesenchyme. Overexpression of N-CAM increased the number of cartilage nodules that formed. *Talpid* chick mutants have a larger than normal number of condensations (Fig. 19.6), with enhanced N-CAM levels in those condensations (Chapter 34). Widelitz and colleagues also showed that *Hox* genes must act on condensations to alter N-CAM, and that activin up-regulates N-CAM and increases condensation size.

The activin–N-CAM link is important for condensation. There is a fivefold increase in chondrogenesis in micromass cultures of limb-bud mesenchyme exposed to activin. The size of condensations increases without any concomitant increase in cell number, which means that activin recruits cells into condensations rather than promoting proliferation of cells already within condensations.

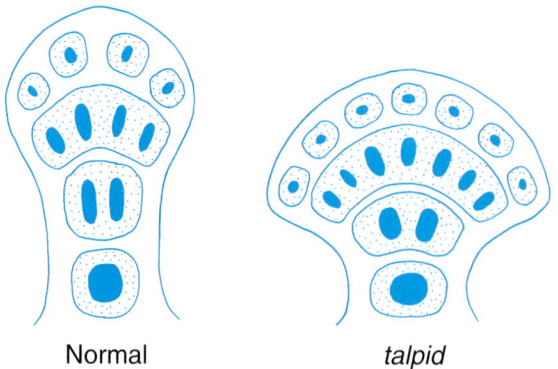

Figure 19.6 Diagrammatic representations of a normal (wild-type) and a *talpid*³ wing bud to show the progressive formation of additional condensations from proximal to distal within the *talpid*³ bud. Modified from Ede and Agerbak (1968).

Activin also enhances Bmp-induced ectopic bone formation.[19]

Levels of intracellular cAMP increase when prechondrogenic cells condense. The increase is associated with phosphorylation of p35, the nuclear substrate for cAMP-dependent protein kinase (PKA). Indeed, within the chondrogenic pathway, PKA is found only in the nuclei of condensing cells, not in chondrocyte nuclei (Zhang *et al.*, 1996). The cell-to-cell interactions concomitant with condensation – and which elevate cAMP levels – may in turn up-regulate chondrogenic genes, providing a mechanism initiating condensation and skeletal development.

Pax-1 and *Pax-9*, genes that share the paired-type DNA-binding homeodomain and encode nuclear transcription factors, have their strongest expression at the condensation stage, consistent with a role in condensation (Table 19.1; Fig. 20.2). *Pax* genes are important regulators of epithelial–mesenchymal interactions in many tissues and organs, including the skeleton, kidneys, sense organs (eyes, ears, nose), limb muscle and brain. *Pax-2*, which is regulated by *Bmp-7*, is also an important regulator of condensation size.[20]

Epithelial–mesenchymal interactions in avian skeletogenesis are mediated, in part, by *Prx-1* (formerly *MHox*) and *Prx-2*, two members of a subclass of the paired-class of homeobox genes related to *Drosophila aristaless*. Both genes also play a role in mediating interactions between perichondrial cells and adjacent chondrocytes, providing a mechanistic connection between condensation and perichondrial development. The presence of *Hox*-gene-binding sites in the upstream regulatory element of the N-CAM promoter makes N-CAM a potential downstream target for *Hox* genes. *Prx-1* is a potential upstream regulator of CAMs and therefore of condensation initiation.[21]

In mouse embryos, *Alx-3* – a mouse paired (*aristaless*)-like homeobox gene – is expressed at the condensation stage in ectomesenchyme and lateral-plate mesoderm. Expression is in domains that overlap with *Prx-2* and *Cart-1*, a chondrogenic marker. *Alx-4* is found in mouse mesenchymal condensations of the skull, hair, teeth and mammary glands. Null mutants have only preaxial polydactyly – preaxial referring to an additional digit(s) on the anterior (preaxial) surface of a limb – indicating either that *Alx-4* plays a minor role in skeletogenesis or that its function(s) overlap with those of other genes.[22]

Fgfs may have a greater role in condensation than has been appreciated so far. Homologues of human fibroblast growth factor (FGF) receptors -1, -3 and -4 are found in high levels in cranial, branchial arch, limb and axial skeletogenic mesenchyme in the European salamander, *Pleurodeles waltl* (Launay *et al.*, 1994). Such high levels are indicative of a role for FGFs in condensation initiation or maintenance.

N-cadherin (*Cadherin-2*) is an active player in condensation, chondrogenesis and osteogenesis, including terminal differentiation of growth-plate chondrocytes, apoptosis, and interacting with hormones and growth factors.

N-cadherin and N-CAM are involved in cell–cell interactions at condensation in limb chondrogenesis, N-cadherin being present at high levels in the distal mesenchyme of the chick limb buds, both being expressed at condensation in cultured H.H. 28–30 chick tibial chondrocytes. Both are lost with hypertrophy but reappear with osteogenesis. Tavella *et al.* (1994) concluded that N-cadherin initiates and N-CAM stabilizes condensations. Tgfβ-1 increases N-cadherin, N-CAM, fibronectin and tenascin at the condensation stage of chondrogenesis in mouse limb mesenchyme.[23]

Anti-N-cadherin antibodies block cell sorting as if the AER had been removed or the limb buds exposed to retinoic acid, implicating N-cadherin in organizing mesenchyme along the proximo-distal axis. Condensation formation for zebrafish fish pectoral fins is disrupted when the function of N-cadherin is disrupted using morpholinos; N-cadherin normally is expressed in fin condensations.[24]

After DeLise and Tuan (2002a,b) transfected chick limbbud mesenchymal cells with constructs of wild-type N-cadherin or one of two deletion mutants, one deleted in the intra- and one in the extracellular domain, those with the wild-type construct differentiated normally. Those with the deletion mutation showed decreased condensation formation, the deletion in the extracellular domain having the greatest effect. Transfection of cells with either deletion construct inhibited subsequent differentiation, a period of prolonged cell adhesion associated with defective condensation being the likely mechanism.[25]

ESTABLISHING BOUNDARIES

The association between Tgfβ and tenascin in epithelial–mesenchymal interactions documented in Hall and Miyake (1995) has been further confirmed experimentally using mouse-limb mesenchyme in culture. Tgfβ-1 up-regulates a number of molecules associated with prechondrogenic condensations, including tenascin, fibronectin, N-CAM and N-cadherin. N-cadherin is associated with adhesion of cells in condensations, N-CAM with stabilization or maintenance of condensations. Eliminating N-CAM does not affect condensation initiation.[26]

Syndecan and tenascin

Syndecans, of which there are four, are transmembrane heparan-sulphate proteoglycans that act as receptors for components of the ECM and as co-receptors for growth factors and signaling molecules. Indeed, signaling between cells and ECM is an important component of morphogenesis, the basic unit being the cell plus its pericellular, and for some cells its extracellular, matrix.

Our understanding of the roles of tenascin and syndecan has been enhanced by the demonstration that early condensations of embryonic chick-limb cartilages are surrounded by a strongly syndecan-3-positive cell layer.

Syndecan plays an important role in establishing the boundary conditions that set the limits to condensation size. Syndecan-3 has also been demonstrated in other areas of epithelial–mesenchymal interactions in association with the formation of condensations, including those involved in the development of the lens, otic placode, sclerotome and feathers. Syndecan-3 is also found in areas outside the developing skeleton where basic patterning occurs, namely in the floor plate of the ventral neural tube and in distal-limb mesenchyme.[27]

Syndecans mediate cell–cell adhesion, cell signaling via heparan-sulphate-binding growth factors, and cell–ECM adhesion via adam 12, which is a disintegrin and metalloproteinase, implicating syndecans in the regulation of integrin activity, and raising the *interesting possibility of involvement of syndecans in integrin-mediated transduction of mechanical stimuli.*[28]

Syndecan-3 also mediates cell–ECM and cell–cell interactions during chondrogenesis (perhaps also osteogenesis) including joint formation and proliferation of epiphyseal chondrocytes. Syndecan is expressed in:

- distal-limb mesenchyme, where it may function in patterning and outgrowth of the mesenchyme;
- condensing mesenchyme;
- perichondria of developing cartilages;
- future joint primordia;
- periostea along the diaphyses of long bone rudiments; and
- is lost from mesenchyme with chondro- and myogenesis, but remains in the epithelium and in peripheral and distal mesenchyme.[29]

Syndecan expression is up-regulated in the presence of epithelia, suggesting potential involvement in epithelial–mesenchymal interactions (Solursh *et al.*, 1990; Dunlop and Hall, 1995).

Both syndecan-3 and tenascin-C are expressed at the boundaries of condensations and then in the diaphyseal perichondrium and periosteum, and in periosteal bone and in the epiphyseal periosteum, in which tenascin-C declines (Koyama *et al.*, 1995, 1996b). Tenascin is expressed by osteoblastic cell lines from humans, rats and chicks, especially in aggregates (Mackie and Tucker, 1992). Vakeva *et al.* (1990) compared the distribution patterns of tenascin (and alkaline phosphatase) in rat and mouse embryos.

Syndecan and tenascin also set *boundary conditions* in perichondria and periostea. Thus, tenascin and syndecan are deployed in those situations where a chondrogenic or osteogenic population must be set off from surrounding non-skeletogenic mesenchyme. Tenascin-C is also associated with the ectomesenchymal condensations of the mandibular skeleton in chick embryos, both the prechondrogenic condensation for Meckel's cartilage and the preosteogenic condensations for the membrane bones of the mandible. Thus, tenascin-C and syndecan-3 are key players in the initiation of ectomesenchymal skeletogenic condensations and in condensations derived from mesoderm (axial and appendicular skeleton).[30]

Although involved in condensation, expression of *tenascin may not be essential for condensation or for skeletal development*, both of which are normal in mice after the single copy of the tenascin gene is knocked out (Saga *et al.*, 1992). I say 'may not be essential,' because at this time the possibility that another gene(s) compensates for the absence of tenascin cannot be ruled out; because of partial functional redundancy and/or overlapping gene functions, knockout experiments are not as clear-cut as originally thought.

Fgfs

Fgfs are another class of molecules that mediate condensation in various organ systems, often acting at more than one stage in a cascade of interactions. Bmps and Fgfs are involved in another situation where skeletogenic cells are maintained at a boundary with non-skeletogenic cells, viz. in developing sutures. Expansion of the osteogenic front is associated with intense expression of *Bmp-2* and *Bmp-4* and up-regulation of *Msx-1* and *Msx-2*. *Msx-1* is also found in growing facial mesenchyme in which condensations will form (Box 13.4). Implanting beads loaded with anti-Fgf-2 and anti-Fgf-4 into skull bones grafted to the CAM evokes a three- to fourfold increase in bone over untreated controls. Closure of sutures is speeded up when beads soaked in Fgf-4 are implanted into the suture front. Therefore, as in condensation, maintenance of the population of preosteogenic cells at suture margins is mediated by signals involving Fgf, Bmp and *Msx*.[31]

Wnt-7a

Wnt-7a transfected into chondrocytes *in vivo* or *in vitro* blocks chondrogenesis. *Wnt-7a* functions as a mitogen at the condensation phase; blocking the gene at condensation blocks chondroblast differentiation early.[32]

Stott *et al.* (1999) described five stages through which chick limb-bud mesenchyme progresses *in vitro* – dissociated, small aggregates, cell clusters, precartilage condensation, nodule formation – and confirmed that *Wnt-7a* blocks the transition from condensation → nodule formation. *Wnt-7a*-treated cultures retain N-CAM, N-cadherin, integrin β$_1$, fibronectin and tenascin-C. Application of *Wnt-7a* and *Bmp-2* suppresses cartilage differentiation, while BMP-2 alone enhances differentiation. Stott and colleagues concluded that *Wnt-7a* and *Bmp-2* antagonize one another to set the boundaries of chondrogenic mesenchyme, and control chondroblast control shape via local concentrations and interactions. Tufan and Tuan (2001) demonstrated that N-cadherin is modulated by *Wnt-7a* and -*5a* in high-density micromass cultures of chick limb-bud mesenchymal cells.

NOTES

1. See Fell (1925), Grüneberg (1963), Jollie (1971) and the literature discussed by Hall and Miyake (1992, 1995, 2000) and Hall (2003d) for the uniform mesenchyme that precedes aggregation of preskeletogenic centres. See p. 319 in Mariani and Martin (2003) for the quotation. See Grüneberg (1963) for his analysis of skeletal mutants and development of the concept of the membranous skeleton; Kochhar (1973) for the effect of condensation size on skeletogenesis in mice; Oster *et al.* (1983, 1985, 1988), Oster and Murray (1989), Atchley and Hall (1991) and Hall and Miyake (2000) for more recent overviews of condensations and the membranous skeleton; and Pavlov *et al.* (2003) for a detailed study of the ontogeny of the murine nasal cartilage (strain unspecified) between 12.5 and 18.5 days of gestation.

2. See Hong *et al.* (1999) for the genetic basis, Kantomaa *et al.* (1987) for morphology of the cranial base in untreated patients with hydrocephaly, Huggare *et al.* (1986) for responses to shunts, and Jeffery and Spoor (2004) for normal patterns of ossification of the human cranial base.

3. See Grüneberg and des Wickramaratne (1974) for the initial discussion on effects of condensation size, and Moore and Lavelle (1974), Johnson (1986), Frenz *et al.* (1989), Atchley and Hall (1991), Hall and Miyake (1992, 1995, 2000), Gehris *et al.* (1997), Fang and Hall (1999), Stott *et al.* (1999), Pizette and Niswander (2000) and Hall (2003d) for subsequent discussion. See Seegmiller *et al.* (1971) for *cho*. Other examples of chondrodystrophic mutants are discussed in Chapter 27.

4. See Merker *et al.* (1975) for the study with 6-mercaptopurine. For comments on the state of knowledge in the early 1970s, see Willmer (1970), Hall (1971a) and Saxén (1976).

5. See Fell (1925) and Fell and Canti (1934) for the pioneering studies, and Murray (1936), Hinchliffe and Johnson (1980), Hall and Miyake (1992, 1995, 2000) and Wezeman (1998) for reviews.

6. See Milaire (1965, 1967, 1974, 1978, 1983) and Milaire and Rooze (1982) for mammalian limb buds. For comprehensive reviews of the histochemical literature, see Pritchard (1952), Cabrini (1961), Fullmer (1965) and Kobayashi (1971). See McMahon (1974), Kebabian *et al.* (1975) and Landauer (1976) for neurotransmitters as inductive signals.

7. See Saunders (1948), Godman and Porter (1960), Jurand (1965) and Searls *et al.* (1972) for the ultrastructural studies, and Hall and Shorey (1968) for characteristics of progenitor cells.

8. See Miyake *et al.* (1997a), Zschäbitz *et al.* (1995a,b) and Zschäbitz (1998) for cell-surface markers, Aulthouse and Solursh (1987), Alber *et al.* (1994) and Stringa and Tuan (1996) for PNA, and Tucker *et al.* (1995) for osteogenic condensations. *Distribution of glycoconjugates at the condensation stage of various cartilages in various mammal species was assessed by Zschäbitz* et al. *(1995a,b) and Zschäbitz (1998).* Sasano *et al.* (1992) could not detect PNA staining in the nasal septal or Meckel's cartilages in Swiss Webster mice, because it is either absent or masked; see Miyake *et al.* (1996b) for visualization in C57BL/6 inbred mice.

9. See Wada *et al.* (1999) and Ladher *et al.* (2000) for *Frzb-1* in chick embryos, and Orsulic and Peifer (1996) for misexpression.

10. See Triboli and Lufkin (1999), Rodrigo *et al.* (2003) and Tucker *et al.* (2004) for *Bapx-1* in mice, and C. T. Miller *et al.* (2003) for zebrafish.

11. See Fyfe and Hall (1983) and Hall and Miyake (1992, 1995, 2000) for how condensations form.

12. Nor does differential division appear to be a factor in the formation of the dermal lamina or the enamel organ in mice (Osman and Ruch, 1975). See Janners and Searls (1970) and Searls (1973a,b) for lack of increased mitotic activity and increase in cell packing.

13. In a companion study on myogenesis in wing buds, Hilfer *et al.* (1973) observed a *broadening* of cell-to-cell contacts in presumptive muscle-forming peripheral wing mesenchyme, in contrast to the narrowing of cell contacts in the more central prospective chondrogenic mesenchyme, giving the appearance that presumptive myogenic cells were more compact (condensed) than presumptive chondrogenic cells.

14. Holtfreter (1968) had previously proposed that this inductive message was spread by diffusion rather than by proliferation of induced cells.

15. See Gould *et al.* (1972, 1974) for lack of movement away from, and Ede and Agerback (1968), Thorogood and Hinchliffe (1975) and Ede (1980) for aggregation of cells to a centre.

16. See Ede and Flint (1972) for condensation and chondrogenesis; Grobstein and Holtzer (1955) and Searls (1968) for exceptions; and Niven (1933), Weiss and Amprino (1940), Abbott and Holtzer (1966) and Fell (1969, 1976) for cell dispersal and fibrogenesis.

17. See Hall and Miyake (1992, 1995), Raff (1996), Hall (1998a) and Peters (1999) for the growth factors and tenascin, Zou *et al.* (1997) for Bmp receptors, and Ayad *et al.* (1994) for fibronectin.

18. See Y.-S. Lee and Chuong (1992) and Chuong *et al.* (1993), and see Fang and Hall (1999) and Crossin and Krushel (2000) for reviews of cellular signaling by N-CAM.

19. See Jiang *et al.* (1993) for activin and condensation and Ogawa *et al.* (1992) for activin and ectopic bone formation. Activin has now been found to elicit chondrogenesis from ectoderm of *Xenopus* embryos (Furue *et al.*, 2002), presumably by activating neural-crest cells (NCC), for which see Seufert and Hall (1990). See Hall and Tremaine (1979) and Bee and Thorogood (1980) for chondrogensis from premigratory NCC of chick embryos.

20. See Dahl *et al.* (1997) and Le Clair *et al.* (1999) for *Pax-1* and *Pax-9*.

21. See Hirsch *et al.* (1991), Brickell (1995), Martin *et al.* (1995), ten Berge *et al.* (1998a, 2001) and Lu *et al.* (1999a,b) for *Prx-1*.

22. See ten Berge *et al.* (1998b) for Alx-3, and Hudson *et al.* (1998) for null mutants.

23. See Oberlender and Tuan (1994) for expression, and Chimal-Monroy and Diez de León (1999) for Tgfβ effects.

24. See Yajima *et al.* (1999, 2002) for the blocking study; Chimal-Monroy and Diez de León (1999), Vleminckx and Kemler (1999) and Delise and Tuan (2002a,b) for the role of cadherins in cell adhesion and signaling and for N-cadherin in cartilage; and see Q. Liu *et al.* (2003) for N-cadherin in pectoral fin development.

25. See Marie (2002) for an excellent review of the role of N-cadherin in bone formation (Fig. 19.3), including expression profiles in different osteogenic cells, the use of neural antibodies and antisense probes, cleavage of N-cadherin in apoptosis and possible role in terminal growth-plate chondrocytes, and interactions with hormones and growth factors. Delise *et al.* (2000) review cellular interactions and signaling in cartilage development using limb chondrogenesis as the model and with emphasis on condensation.

26. See Tavella *et al.* (1994) and Chimal-Monroy and Diaz de León (1999).

27. See Gould *et al.* (1995) and Koyama *et al.* (1995) for syndecan-3, and see Kirn-Safran *et al.* (2004) for syndecan-3 and the other heparan-sulphate proteoglycans.

28. See Maclean and Hall (1987), Ettinger and Doljanski (1992), Flaumenhaft and Rifkin (1992) and Jones *et al.* (1993) for signaling between cells and their matrices. See Rapraeger (2000) for an overview of syndecan signaling pathways, and see Shimo *et al.* (2004) for evidence that Indian hedgehog signaling involves syndecan-3.

29. See Solursh *et al.* (1990), Gould *et al.* (1992, 1995) and reviews by Bernfield *et al.* (1993) and Kosher (1998).

30. See Koyama *et al.* (1996b) for boundaries, and Gluhak *et al.* (1996) for tenascin in craniofacial mesenchyme.

31. See Perantoni *et al.* (1995) for Fgf-2 and kidney induction, McGonnell *et al.* (1998) for facial mesenchyme, and Kim *et al.* (1998) and Thorogood *et al.* (1998) for the studies with Fgfs.

32. See Rudnicki and Brown (1997) and Yang (2003) for overviews of the role of *Wnt-7a*.

From Condensation to Differentiation

Condensations are the primary resource from which the skeleton is built and through which the skeleton is modified in development and evolution.

Atchley and Hall (1991) identified condensations as the fundamental cellular units of morphological change in organogenesis during vertebrate evolution.[1] As discussed in Chapters 19 and 20, each phase of the multiphase process that is condensation involves different cellular processes and separate genetic control as a previously dispersed population of mesenchymal cell forms aggregates and accumulates to the point where a skeletal element (or elements) begins to form. I say *elements* because patterning within condensations can be quite complex with more than one bone or cartilage arising from a single condensation. For example:

- two cartilages of the lower jaw (Meckel's and the symphyseal cartilage) and two of the bones of the middle ear (the incus and the malleus; Fig. 13.5) in mouse embryos arise from a single condensation (Fig. 13.1);
- the seven bones of the lower beak of chick embryos arise from a single condensation; while
- three bones of the zebrafish head arise from a single condensation.

Condensations can also be temporally complex; the subunits of a condensation may persist for different lengths of time before making the transition to overt cell differentiation. For example, elastic cartilage has a prolonged condensation phase before any matrix is deposited, and, even then, the specialized elastic fibres are deposited weeks after initial matrix deposition (Chapter 3).[2]

CONDENSATION GROWTH

Gould and colleagues (1972) proposed that condensation growth is generated by a mechanism involving synthesis, secretion and accumulation of extracellular matrix (ECM), especially at the centre of the cell mass. Matrix accumulation separates the central cells, enlarges the cell mass, and orients the cells perpendicular to the long axis of the condensation. Cells at the edge of the condensation produce little ECM, flatten circumferentially, and form a superficial layer that becomes the perichondrium (Figs 10.4, 13.8 and 19.4).

Is there sufficient spatial restriction at the edge of a condensation to flatten the peripheral cells? Once chondroblasts begin to secrete appreciable amounts of ECM, the presence of such a hydrated matrix would separate cells and so expand a growing condensation. However, and as noted in the previous chapter, during developmental stages when the condensation is *initiated* (H.H. 20–22), little ECM is synthesized. Furthermore, presumptive chondro- and myogenic cells of the limb bud synthesize glycosaminoglycans (GAGs) at the same rate, but only chondrogenic condensations form a flattened peripheral layer.[3]

In light of studies by Vinson and Seyer (1974) on initiation of feather condensations in chick embryos, perhaps differential synthesis of collagen should be examined. Synthesis of a tissue-specific collagen is enhanced during formation of feather condensations. Applying collagenase, or a lathyrogen,[4] prevents condensation, implicating collagen synthesis in condensation. Feathers also fail to form in *scaleless* mutant fowl. Examination of early-stage mutant embryos shows that condensations fail to form, illustrating the lessons that can be learned from mutants. The same is true for chondrogenic condensations.

Lessons from mutants

talpid³

Using wild type and the *talpid³* polydactylous mutant (for details on *talpid*, see Box 20.1 and Chapters 20 and 39), Donald Ede and his students, initially in Edinburgh, then at Glasgow (UK), led a major assault on mechanisms initiating condensations in wing buds. Their approach was twofold: compare wild-type and mutant embryos, and use cell culture as a surrogate for condensation. Arguing that reaggregation of dissociated cells in culture is the *in-vitro* equivalent of *in-vivo* condensation, and using knowledge reported in a 1964 study that condensations are defective in *talpid³* embryos, Ede and Agerback (1968)

dissociated cells from wing buds of H.H. 24 and 26 embryos and observed their behaviour during reaggregation *in vitro*. Cells from wild-type embryos produce fewer and larger aggregates than cells from *talpid³* embryos.[5]

A word of **caution**: as a method, dissociating cells may have unrecognized consequences. In some studies, dissociated and reaggregated limb-bud cells fail to differentiate as chondroblasts, or to deposit cartilage-type proteoglycans. In other studies, cells that are first reaggregated into pellets form cartilage nodules containing cartilage-type proteoglycan. Perhaps even more disconcerting, 30 or more digits develop when *Xenopus laevis* limb-bud mesenchyme is dissociated and grafted; mesenchyme grafted without first dissociating the cells forms a normal numbers of digits. Fgf-8 is distributed in patches within the distal dissociated mesenchyme, *Shh* more broadly across the mesenchyme, both patterns being unusual.[6] One can imagine both molecules providing multiple signaling centres for chondrogenesis.

Ede and colleagues proposed that cells behave during condensation as they do during reaggregation, aggregating to a central focus. Their analogy was aggregation in the slime mould, *Dictyosteleum*, in which a colony (an aggregate or condensation) arises by the 'huddling' of peripheral cells towards a centre. In chondrogenesis, *in-vitro* aggregation increases the size of the aggregate by adding cells at the periphery. Similar aggregation of cells toward a centre is seen during development of the horny process on the posterior ray of the anal fin of female Japanese medaka, *Oryzias latipes* (Fig. 20.1), and in the fin spines of sharks.[7]

Cell death occurs between chondrogenic nodules that form *in vitro* from chick limb-bud mesenchyme, especially if the mesenchyme is obtained from embryos of H.H. 25–26. Bmp-2, Bmp-4 and Bmp-7 are also localized around the nodules, corresponding in position to the foci of cell death. Coupling this finding with Ede's demonstration that cartilage cells can recruit neighbours, Omi *et al.* (2000) proposed that cartilage cells induce cell death via cell migration and production of Bmps.

Box 20.1 *Talpid* mutants in domestic fowl

Because so much has been learnt from analysis of *talpid* (*ta*) mutants in domestic fowl I gather some basic information about their development here.

Talpid, a polydactylous lethal mutant of domestic fowl, has been isolated on at least three occasions. All three mutants show extreme polydactyly – eight to 10 digits in a foreshortened, broad limb bud reminiscent of the forelimbs of the mole, *Talpa*, after which the mutants are named; see Fig. 19.6 and Chapter 34 for details.[a] All three mutations are homozygous recessive and lethal.

Talpid¹, now extinct, dies between five and 13 days of incubation.

Talpid², studied extensively by Abbott and her colleagues in California, dies at around 13 days of incubation.

Talpid³, studied by Ede and Hinchliffe and their colleagues in the UK, dies between five and six days of incubation.

Talpid² and *talpid³* both exhibit abnormalities of face and beak, severe polydactyly, shortened vertebral columns, shortened proximal long bones of the appendicular skeleton, fusion of lateral elements such as radius with ulna, and fusion of metacarpals.

Limb buds of *talpid²* mutants are symmetrical, with nine or 10 digits joined by connective tissue and arranged on a broad and short limb and fused proximal skeletal elements. It appears that the antero-posterior axis is never established in *talpid²* limb buds.

The major cellular defect contributing to the characteristic phenotype is the mesenchymal cells being much more adhesive than in wild type; the cells consequently cannot separate from one another. The limb buds of *talpid³* embryos have an abnormally large AER, lack the zones of programmed cell death that narrow the size of the limb bud and separate radius and ulna during development in wild-type embryos, and have defective condensations for limb skeletal elements (Fig. 19.6; Chapter 34).

[a] See Abbott (1967) for details on *talpid* mutants. See Lessa and Stein (1992) and Vizcaino and Milne (2002) for morphological constraints on the skeleton of digging mammals: five species of pocket gophers, in which we see contrasting constraints on the muscles of the jaws and forelimbs; 14 species of armadillos, in which biomechanical properties of forelimbs but not hind limbs correlate with digging. See Leroi (2003) and Hall (2004c) for the insights gathered from studying mutations that affect limb development.

Figure 20.1 A female Japanese medaka, *Oryzias latipes*, in the foreground, is releasing eggs, while a male lurks in the background.

The distance between foci could determine whether or not adjacent foci fuse. The more distal in the limb bud, the greater the distance between centres, hence the greater the separation of skeletal elements. When cells are more adhesive – as limb-bud cells from *talpid³* embryos are – while cell motility should be reduced, condensations form less easily and fuse more readily, contributing to the polydactyly typical of *talpid* embryos.

bp^H

Any factors that reduce adhesion of cells to one another or to their substrate will facilitate formation of condensations and initiation of chondrogenesis (Watanabe and Okada, 1975). The studies by William Elmer and his colleagues from Emory (Georgia, USA) on *brachypod* (bp^H) mutant mice provided additional confirmation of an adhesion-based model. Mesenchymal cells from limb buds of bp^H embryos are more adhesive than wild type. Consequently, the rate of decline in the number of single cells in rotation culture is greater in bp^H, chondrogenesis is more limited, and the amount of ECM produced less than that produced in wild type (see Chapter 22).[8]

ADHERE, PROLIFERATE AND GROW

Gap junctions

Gap junctions joining adjacent cells are an important means by which cells communicate and exchange signals and metabolites. Gap junctions form between epithelial cells from the outset; the cells are always in apposition. Mesenchymal cells must first attain a sufficient density (proximity to be more precise) before gap junctional connections can form. Therefore, the timing of the appearance of gap junctions enables mesenchymal cells to communicate with one another in ways they could not before the gap junctions were established, and provides the physical evidence that such communication could now be taking place. Formation of gap junctional connections has been studied in limb bud and craniofacial mesenchyme; see Box 31.1 for gap junctional connections between osteogenic cells.

Limb-bud mesenchyme

Assembly and disassembly of gap junctions is associated with the formation of condensations of limb skeletal elements in mice. Such connections would serve at least two functions: allowing communication between cells (molecules of <1200–1500 kDa can travel through such junctions) and helping maintain the cells in aggregation.[9]

Gap junctions also have a role in patterning chick limb buds, since communication via gap junctions allows signals from the posterior region to be transferred to more anterior regions of a limb bud. Transfer of the dye Lucifer yellow between cells in micromass cultures of limb mesenchyme shows that formation of gap junctions is selective; gap junctions form between chondrogenic but not between connective tissue cells. Expression of the homeobox gene *Msx-1* is required to maintain gap junctions between limb mesenchymal cells. The gap junctional protein connexin-43 is expressed in the specialized epithelial ridge and adjacent mesenchyme of chick limb buds. Antisense probes against connexin-43 disrupt limb patterning, following which split or truncated limbs develop, providing additional evidence of the importance of gap junctions.[10]

Craniofacial mesenchyme

Gap junctions in embryonic chick facial primordia (H.H. 22–28) are distributed uniformly in epithelium and mesenchyme, but in unique patterns in the nasal placodes and retina. Connexin-43, which plays an important role in facial morphogenesis in chick embryos, is found throughout mesenchyme at the condensation stage before initiation of mandibular osteogenesis. Antisense probes to connexin-43 decrease connexin-43 levels, decrease proliferation of facial processes and *Msx-1* levels, and produce facial defects. Connexin-43 is thus upstream of *Msx-1* in an important developmental pathway in limb-bud and craniofacial mesenchyme.[11]

Connexin-43 is utilized in gap junctions when rat nasal septal cartilage cells dedifferentiate and redifferentiate in culture. Labeling with connexin 43 or injection of Lucifer yellow shows *clusters of some 20 connected neighbouring cells, essentially making a mini-condensation.* The essential role for gap junctions was shown when the connexin-43 inhibitor, glycyrrhetinic acid, blocked cell-to-cell coupling and blocked chondrogenesis.[12]

Finally, elaborate cell processes emanating from fibrocartilaginous cells in tendons and ligaments form a three-dimensional (almost canalicular) network through the ECM.[13] These cell processes are connected by gap junctions, indicative of extensive cell-to-cell communication and a potential load-sensing system in these fibrocartilages.

Transcription Factors and Hox genes

Several transcription factors that play pivotal roles in condensation formation or in the transition from condensation to overt differentiation have been identified over the past several years.

Cfkh-1 – a chicken forkhead (winged)-Helix transcription factor expressed in condensations of mesodermal (ribs, vertebrae, limbs) and neural crest (branchial arch) origin – is down-regulated as chondrocytes or osteoblasts differentiate. The most likely function of *Cfkh-1 is in promoting proliferation at the condensation stage* (Table 19.1; Fig. 20.2). *Cfkh*-1 is thought to act by mediating the

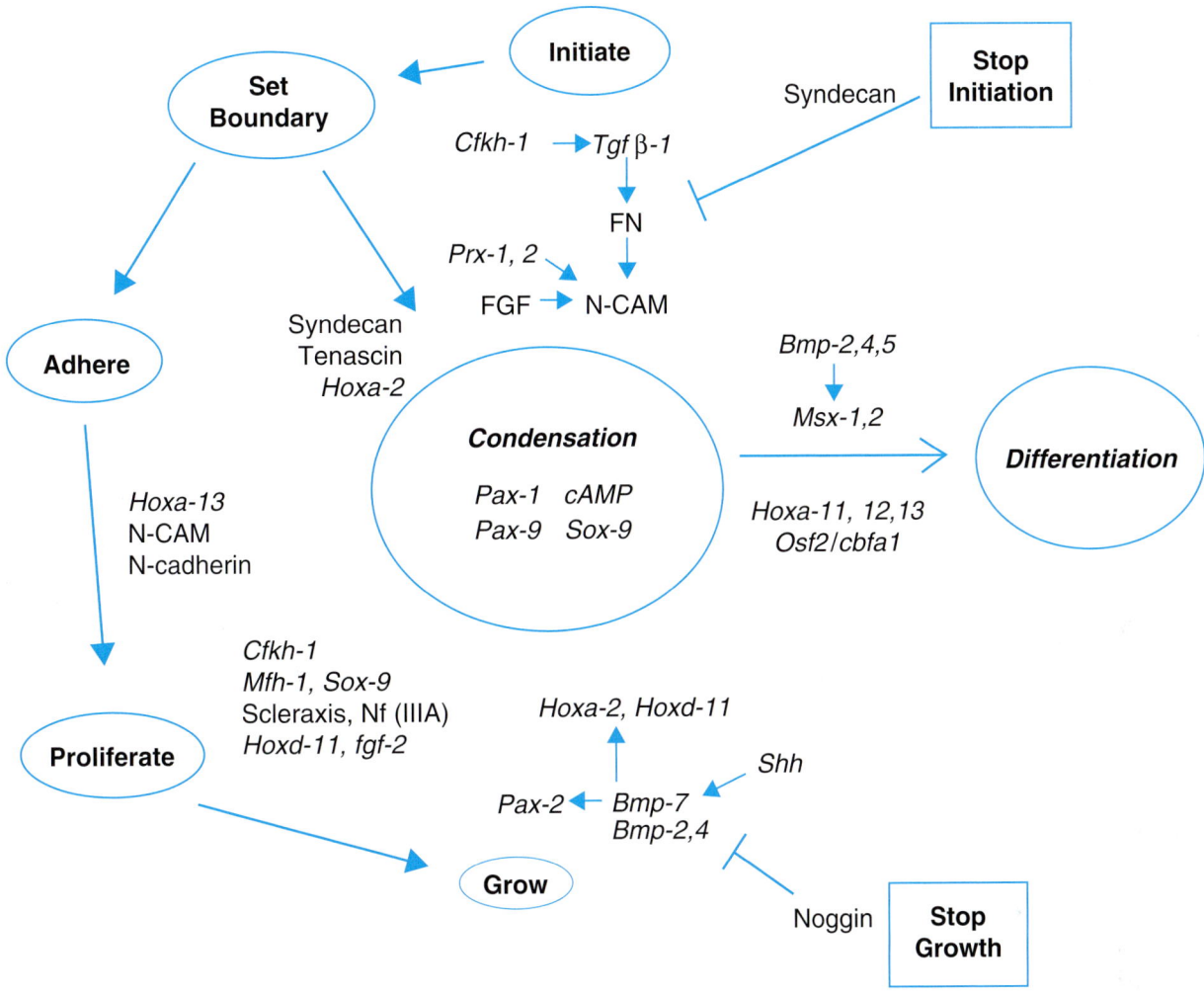

Figure 20.2 A summary of the major genes and gene products involved in prechondrogenic condensation formation, and in the transition from condensation to overt chondrogenic differentiation. Elevated levels of cAMP, *Pax-1*, *Pax-9* and *Sox-9* characterize prechondrogenic condensations. The major genes and proteins associated with five stages in condensation are shown, the five being Initiate, Set Boundary, Adhere, Proliferate, and Grow. Two pathways that stop condensation initiation (Stop Initiation) and condensation growth (Stop Growth) are shown. Cessation of condensation leads to differentiation, which involves both up-regulation of genes to initiate differentiation and down-regulation of genes to terminate condensation. See Table 19.1 and the text for details. Modified from Hall and Miyake (2000).

effects of Tgfβ on transcription of its target genes (fibronectin, for example; Box 20.2), by interacting with Smad proteins using both combinatorial and heteromeric receptor interactions.[14]

Mesenchyme forkhead 1 (*Mfh-1*) is a transcription factor expressed in sclerotomal and craniofacial mesenchyme and in differentiating cartilage of mouse embryos. *Mfh-1*-deficient mice display skeletal defects consistent with a possible role for *Mfh-1* in condensation proliferation. A further transcription factor, *Sox-9*, is expressed in murine chondrogenic condensations, where it regulates the colα21 gene. As with *Mfh-1*, deficiencies in *Sox-9* lead to skeletal defects. Scleraxis, a basic helix–loop–helix protein expressed at maximal levels in axial, appendicular and craniofacial chondrogenic condensations, is down-regulated with chondrogenic differentiation (see Chapter 19).[15] Clearly, further research will be

required to understand the relative roles of these various transcription factors.

Hoxa-2 is specifically excluded from chondrogenic condensations in the second branchial arch of mouse embryos, a pattern of expression that is consistent with *Hoxa-2* playing a role in establishing the boundary and/or size of the condensation. *Hoxa-13* alters the adhesive properties of prechondrogenic cells. Therefore, *Hox* genes have to be considered as modulators of cell adhesion. Infection of limb mesenchyme with a *Hoxa-13* construct and subsequent culture of these cells is followed by aggregation of the *Hoxa-13*-expressing cells into multiple clusters. Because limb mesenchyme contains cells from both somitic and lateral-plate mesoderm that give rise to myogenic, chondrogenic and fibroblastic cell lineages, similar experiments need to be undertaken using cells from individual condensations.[16]

Box 20.2 Fibronectin: some background

In the early 1980s, Weiss and Reddi demonstrated that *fibronectin*, a large extracellular dimeric glycoprotein, was present in cartilage and bone matrix and differentiating chondrocytes; fibronectin, procollagen fibres, hyaluronan and chondroitin sulphate form pericellular fibres within the chondrocyte pericellular matrix, fibronectin also binding to type II collagen. Treatment with hyaluronidase was required to visualize the fibronectin, which otherwise is masked by GAGs in the matrix. They also showed that fibronectin inhibited prechondroblast proliferation by 60 per cent and inhibited chondrogenesis in 43 per cent of their cultures. Fibronectin is one of the first molecules synthesized by chondrocytes after their ECM has been removed and the cells established in culture. As discussed in the text of Chapter 20, we now know that fibronectin plays an important role in the condensation phase of chondrogenesis; at least 10 cell-surface molecules and six sites on fibronectin facilitate binding of cells to fibronectin.[a]

The precondensation stage of chondrogenesis is characterized by deposition of fibronectin and type I collagen, the condensation stage by fibronectin and type II collagen. As chondrocytes differentiate, levels of fibronectin and of type I collagen drop. No fibronectin remains in mature cartilage. Direct evidence for a role for fibronectin in condensation was obtained in studies of H.H. 22–23 wing mesenchyme. Antibodies against the amino-terminal heparin-binding domain of fibronectin (but not antibodies against other fibronectin domains) decrease condensation formation, supporting interaction between fibronectin and heparin as a requirement for condensation.[b]

Fibronectin is associated with early stages of limb development: fibronectin is found in the basement membrane adjacent to the AER and in the condensing core (chondrogenic) mesenchyme between H.H. 19 and 25. Fibronectin persists in cartilage until H.H. 27. After predigestion with hyaluronidase, Melnick et al. (1981) first detected fibronectin at H.H. 25 in chick limb mesenchyme. Several matrix molecules, including fibronectin, are sensitive to

TgfB-1 when cells are at the condensation stage. TgfB-1 increases fibronectin, tenascin, N-cadherin and N-CAM in mouse limb mesenchyme.[c]

When chick limb mesenchyme is cultured at high density, fibronectin is found in the fibroblasts at day one and in chondroblasts and chondrocytes as differentiation ensues, but is lost with elaboration of the first ECM. Differentiation will not progress when fibronectin levels are high. Chondrocytes cultured in the presence of 15–150 µg of fibronectin/ml of medium were found to assume a fibroblastic morphology and to produce collagen typical of fibroblasts, findings that led to the suggestion that reduction of fibronectin levels is associated with chondrogenic differentiation. Also the degree of GAG synthesis by chondrocytes was found to be higher in the absence than in the presence of fibronectin, rabbit articular chondrocytes being especially sensitive.[d]

Although there is a family of fibronectins, all are alternatively spliced products of a single fibronectin gene. Chick limb-bud prechondrogenic mesenchyme contains the B^+A^+ or *mesenchymal isoform*, chondroblasts and chondrocytes the B^+A^- or cartilage isoform, plasma the B^-A^- isoform. Denise White and colleagues (2003) showed that, although attachment to different isoforms is similar, limb-bud mesenchymal cells spread 40 per cent less on the mesenchymal than on either the cartilage or plasma isoforms, and that the mesenchymal isoform promotes condensation.

[a] See Weiss and Reddi (1980, 1981a,b) for the early study, Hedman et al. (1982) and Knudson and Toole (1985) for the pericellular fibres, and Argraves and Gehlsen (1991) for binding activity.
[b] See Dessau et al. (1980) for early synthesis *in vitro*, and Frenz et al. (1989) and Widelitz et al. (1993) for the antibody studies.
[c] See Kosher et al. (1982) for fibronectin in limb development, and Chimal-Monroy and Diez de León (1999) for response to TgfB-1.
[d] See Glant et al. (1985) for the high-density cell culture, and Pennypacker et al. (1979), West et al. (1979) and Pennypacker (1981) for these studies establishing the inhibitory action of fibronectin on chondrogenesis.

POSITION AND SHAPE

Homeobox genes are now known to play greater roles in epithelial–mesenchymal interactions and condensation formation than previously suspected. Indeed, *Hox* genes are involved in determination of such fundamental attributes of condensations as timing and position and how condensations acquire their fundamental shapes.

5' *Hoxd* genes (*Hoxd 9–13*), which are known to be involved in skeletal patterning, have now been shown to function at the mesenchymal condensation stage of skeletogenesis. Their patterns of expression in chick-limb mesenchyme maintained *in vitro* are complex, *Hoxd-9* and *Hoxd-13* exhibiting one pattern, *Hoxd-10–d-12* another. Early in culture, *Hoxd-9* and *d-13* are expressed only in some cells, while *Hoxd-10*, *Hoxd-11* and *Hoxd-12* are expressed in all cells. As chondrogenesis ensues, *Hoxd-9* and *d-13* are only expressed in chondrogenic cells, *Hoxd-10–d-12* in both chondrogenic and non-chondrogenic cells. Furthermore, at least in chick limb buds, cartilage can function as a signaling centre, inducing expression of *Hoxd-12*, *Hoxd-13* and *Shh*, which are in turn associated with induction of supernumerary cartilages.[17]

Doublefoot (*Dbf*) limb mutants in mice have severe polydactyly with six to eight digits/limb. Expression of *Shh* is normal but *Ihh* is expressed ectopically in the distal limb mesenchyme, where it activates *Fgf-4*, *Hoxd-13* and *Bmp-2*; in wild-type limb buds, *Shh* acts via BMP in regulating antero-posterior polarity in chick limb bud. From analysis of the ectopic expression of *Ihh* in *doublefoot*, other mutants such as *extra toes* and *talpid*, and the presence of a zone of polarizing activity (ZPA) activity in craniofacial primordia, Schneider et al. (1999) presented a cogent set of arguments for the conservation of molecular signaling between limb and craniofacial development, especially the roles of *Shh* and retinoic acid in patterning.[18]

Misexpressing *Hoxd-13* and *Hoxd-11* in chick limb buds results in skeletal defects that can be traced to two stages of chondrogenesis: *Hoxd-11* acts at both the condensation stage and during later chondrogenesis; *Hoxd-13* acts only at the later stage (Goff and Tabin, 1997).

Double knockout of *Hoxa-11* and *Hoxd-11* results in skeletal defects traceable to the condensation stage and illustrates the functional cooperation between paralogous *Hox* genes. There is also cooperation between non-paralogous *Hox* genes; *Hoxd-11* and *Hoxa-10* exhibit functional cooperation, double mutants having more extreme phenotypes that single mutants. Differential expression of *Hoxa-11* at the condensation stage in *Xenopus* fore- and hind limbs forms the basis of the finding that the *proximal* tarsal elements are patterned by genes that pattern the lower not the upper limb. Previously, all the tarsals were presumed to be under the control of more distally expressed Hox genes.

ESTABLISHING CONDENSATION SIZE

Individual condensations have predictable sizes. The size of a condensation and therefore, by inference, the number of cells within it, affects whether skeletogenesis will be initiated. Reduce a condensation below a critical threshold and skeletogenesis may not begin. Increase the size of a condensation and an overly large skeletal element can form. This statement is supported by studies on mutant embryos in which missing, small or abnormally shaped skeletal elements can be traced to deficiencies in the membranous skeleton. *Talpid³* chick embryos and *Brachypod* (*bp^H*), *Congenital hydrocephalus* (*ch*), *Short ear* and *Phocomelia* (*Pc*) mice are well-studied mutations with initial action at the condensation stage but which act through different genetic pathways. *Talpid³* reflects overexpression of N-CAM; *Brachypod* is a frameshift mutation in *growth and differentiation factor-5* (*gdf-5*), while *Short ear* is a mutation in *Bmp-5* (Box 22.1). The number of cells in a condensation is therefore a realistic reflection of the growth potential of individual skeletal elements, as Grüneberg concluded 40 years ago.

The initial action of mutations at the condensation stage underscores condensation as the first major stage of selective gene activation in skeletogenesis, or indeed in odontogenesis, myogenesis or ligamentogenesis. Genes specific to the differentiation pathway are upregulated at condensation, genes not specific to the pathway are not.

Bmps

Condensation size is regulated through signaling pathways involving *Bmp-2* and *Bmp-4* (Table 19.1; Fig. 20.2). Over-expression of both of these growth factors in chick embryos is followed by dramatic increases in both size and shape of skeletal elements. Enhanced recruitment of mesenchymal precursors to cartilage condensations – and to perichondria, which contribute cells to anlage *in vivo* or to nodules *in vitro* in the same way that condensations grow by accretion – are likely mechanisms of the Bmp effect; *Bmp-2* is expressed in mesenchyme surrounding condensations and so is critically located to modulate expansion of condensation size. Recruitment of cells to condensations by Bmp is also consistent with the known action of Bmp in inducing ectopic bone in adults (Box 14.2). Implanting Bmp outside the skeleton elicits a cascade of events – condensation, overt differentiation of cartilage, vascular invasion, and replacement of cartilage by bone. Calcifying vascular cells obtained from aortic cell cultures also condense before forming nodules that mineralize in the presence of cAMP – in what Tintut *et al.* (1998) described as an osteoblast-like differentiation – further reinforcing condensation as an essential initial step in ectopic mineralization, as it is in normal and ectopic skeletogenesis.[19]

Fibronectin

Fibronectin is critical for early embryogenesis. Homozygote mutant embryos lacking fibronectin implant normally into the uterus and begin gastrulation, but fail to form notochords or somites and have deformed hearts and circulation (George *et al.*, 1993).

An alternatively spliced exon (IIIa), one of the functional portions of fibronectin involved in condensation, is downregulated immediately after the condensation phase during limb chondrogenesis (Gehris *et al.*, 1997). Exposing micromass cultures of limb mesenchyme to an antibody against the exon disrupts condensation and inhibits subsequent chondrogenesis. Similarly, injection of an antibody into limb buds *in ovo* results in embryos with smaller limbs and fewer limb elements.

Surprisingly, levels of fibronectin and fibronectin mRNA are higher in prechondrogenic condensations from chick leg buds than in wing-bud condensations. Condensations that develop when mesenchyme from the two limb types is maintained *in vitro* differ in morphology and in distribution of fibronectin. Wing condensations are broad and flat with much diffusely organized fibronectin; leg condensations are compact and spherical and connected by fibronectin-rich fibres. Wing-bud condensations are more sensitive to Tgfβ, which increases fibronectin levels and in turn increases condensation size. Treating cultured wing-bud mesenchyme with an antibody against the amino-terminal heparin-binding domain of fibronectin inhibits condensation formation. Similar treatment of leg-bud mesenchyme has no effect on condensation development (Downie and Newman, 1995). Comparative analysis of wing- and leg-bud condensations is therefore a promising system in which to analyze the basis for differences between condensations and how condensation size is set. For the molecular basis of specification of limb or fin type, see Chapters 38–40.

Hyaluronan

Studies initiated in the 1970s on the synthesis, distribution and degradation of hyaluronan (Box 4.2) demonstrated that this component of the ECM plays a vital role in the initiation of chondrogenic condensations in a number of skeletal systems. Bryan Toole and his colleagues investigated several organ systems, all of which involve two phases – hyaluronan synthesis as cells accumulate, followed by removal of hyaluronan during cell differentiation. The avian cornea was one of the organ systems Toole explored in depth.

Hyaluronan is the major GAG of the cornea at H.H. 24 and 25, which is when ectomesenchymal cells invade the cornea and accumulate. Levels of hyaluronan decline while chondroitin sulphate (CS) and hyaluronidase rise as corneal differentiation progresses. Toole and Trelstad (1971) suggested that hyaluronan affects the behaviour of invading mesenchymal cells by providing a substrate for migration. Similar relationships and an association with migration have been explored in developing limb buds and somitic mesoderm.[20] This range of studies indicates roles for hyaluronan in the formation of prechondrogenic condensations and for removal of hyaluronan (or the presence of hyaluronidase) for chondrogenesis to be initiated. Several authors have sought more direct experimental verification of these presumptive roles.

Migration of cells into the ectopic site of skeletogenesis parallels cellular events described in condensation and chondrogenic differentiation, and in the earlier migration, accumulation and differentiation of cranial NCC. As assessed by autoradiographic and biochemical analyses, the cell-free space between ectoderm and mesoderm into which cranial NCC migrate in chick embryos contains much hyaluronan, a temporal and spatial distribution that is consistent with hyaluronan playing a role in cell migration similar to that discussed above. A role for hyaluronan in suppressing the differentiation phase is consistent with addition of hyaluronan (30 μg/ml) to culture media in which amphibian neural crest and pharyngeal endoderm are maintained. The exogenous hyaluronan prevents chondrogenesis, although chondrogenesis is initiated under the same culture conditions in the absence of hyaluronan.[21]

Additional support for a relationship between cell aggregation and synthesis of hyaluronan comes from the association between production of proteoglycans and cell density. Low-density cultures of amniotic cells from 10-day-old embryonic chicks produce hyaluronan almost exclusively. Higher density cultures produce hyaluronan and CS in almost equal amounts; as cell density increases, synthesis of hyaluronan declines and synthesis of CS rises. Adding as little as 1 ng/ml hyaluronan to limb-bud mesenchymal cells blocks colony formation by up to 90 per cent; inhibiting cell aggregation into colonies prevents initiation of chondrogenesis.[22]

Several established non-skeletal cell lines – some fibroblastic, some epithelioid and some lymphoblastic – produce a factor that is inactivated by hyaluronidase, induces cell aggregation, and can be mimicked by addition of hyaluronan. There is a positive correlation between the concentration of added hyaluronan and the rate of proliferation of fibroblasts from embryonic chicks maintained in vitro.[23] Hyaluronan, therefore, favours cell proliferation in connective and skeletal tissues.

Although we have far to go before we understand all the cellular aspects involved in the initiation of condensations preceding chondrogenesis or osteogenesis, these studies on the roles of hyaluronan and hyaluronidase indicate convincingly that the microenvironment surrounding the cells plays a key role.

Extrinsic control

As well as being set by such intrinsic mechanisms as recruitment of Bmp-2-positive cells to a condensation, the size of a condensation can also be set extrinsically. The condensations for avian dorsal root ganglia (DRG) in brachial segments 14 and 15 are more than 80 per cent larger than the condensations for DRG in cervical segments 5 and 6. The increased cell number in brachial DRG is not because of colonization of more NCC to brachial ganglia (i.e. is not intrinsic to the condensations), but is imposed extrinsically by specific influences from the sclerotomal mesoderm in the regions in which the DRG develop (Goldstein et al., 1995). How such extrinsic influences as mechanical and endocrine control, which are so important to subsequent skeletal development, influence the size of skeletal condensations is worthy of further study.

FROM CONDENSATION TO OVERT DIFFERENTIATION

Condensation is required for limb-bud mesenchyme to initiate chondrogenesis. In what can now be seen as a classic study, Umansky (1966) obtained both cell clusters and cells that formed monolayers when he cultured isolated mesenchymal cells from limb buds of 11-day-old foetal mice, 11 days being the stage immediately preceding condensation. Chondrogenesis was initiated in the clusters, myogenesis in the monolayers. However, chondrogenesis was only initiated if the initial cell density exceeded 5000 cells/mm². Umansky concluded that limb mesenchymal cells could produce more than one differentiated cell type depending upon the conditions influencing cell-to-cell contacts.[24]

Umansky's experiments are elegant and his interpretation the appropriate one given the received wisdom of the day, which was a limb bud as a homogeneous population of cells. We now know that limb mesenchyme is not a homogeneous population of cells (Chapter 35). Nevertheless, his data remain sound. It is our interpretation that has changed.[25]

Recent studies affirm the importance of a critical condensation size if cells are to progress from condensation to initiation of chondrogenic or osteogenic differentiation. For example, Msx-1 decreases the size of the alveolar units (tooth plus alveolar bone) in murine mandibles (Fig. 25.3; Richman and Mitchell, 1996). Mice in which Hoxa-2, Dlx-2, Prx-1, Otx, or the retinoic acid receptor have been knocked out display ectopic or apparently duplicated skeletal elements (Box 20.3).[26] Two examples

Box 20.3 Retinoic acid receptors

The effects of retinoic acid are mediated via three nuclear retinoic acid receptors (RAR-α, RAR-β and RAR-γ) and by cellular retinol-binding proteins (CRBPs) and retinoic acid-binding proteins (CRABPs), which modify accessibility of retinoic acid to its three receptors.

Mice

During mouse development, retinoic acid receptors are found on both epithelial and mesenchymal cells. RAR-γ is expressed at eight days of gestation in presomitic mesoderm, from 9.5 to 11.5 days in frontonasal processes, pharyngeal arches and limb mesoderm sclerotome, at 12.5 days in all precartilage tissue, and at 13.5 days in *all cartilages*, keratinized epithelium, teeth and whiskers. Double mutants of RARs in mice reveal multiple abnormalities at different stages of embryogenesis, including abnormal development of the neck, trunk and abdomen. BMS453, a synthetic compound that functions as an RAR-β agonist, elicits fusion or hypoplasia of the first and second branchial arches by activating RAR heterodimers. RAR-α and RAR-γ agonists fail to elicit arch defects.[a]

Chicks

Limb-bud and craniofacial mesenchyme of chick embryos express three forms of RAR-β of 3.2, 3.4 and 4.6 kb, respectively. Developing forebrain and frontal nasal processes are coordinated in their development through retinoid signaling (Schneider *et al.*, 2001). Disrupting signaling so that RAR cannot bind results in failure of the upper beak to develop. Retinoic signaling is important for migration of NCC, but the upper beak loss is not because neural crest precursors fail to migrate to the region. Migration is normal but *Fgf-8* and *Shh* are disrupted; the phenotype can be rescued with exogenous retinoic acid, *Fgf-8* or *Shh*.

Mutants

Inactivating the mouse homeobox gene *Cdx1* results in anterior homeotic transformations of the vertebrae by causing a posterior shift in the mesodermal expression domains of *Hox* genes (Subramanian *et al.*, 1995).

Using mouse embryos, Allan *et al.* (2001) examined the role played by interactions between RAR-γ and *Cdx-1* in patterning

of the vertebrae by generating a complete allelic series of *Cdx-1*/RAR-γ embryos and examining their response to exogenous retinoic acid. A member of the caudal-type homeobox gene family (which specify antero-posterior position in *Drosophila*), *Cdx-1* produces a transcription factor expressed in the endoderm. Synergy was found between the alleles in these compound mutants. The full effect of retinoic acid required *Cdx-1* expression, indicating that in patterning the vertebrae, retinoic acid acts upstream of *Cdx-1*. However, depending on your interpretation, redundancy or fail-safe mechanisms are in place; RAR-γ and *Cdx-1* can operate in parallel.

Chondrogenesis

Chondrocyte maturation and endochondral ossification within chick limb buds requires retinoid signaling. Chondrocytes and especially perichondral cells at H.H. 27 express diffuse RAR-α and RAR-γ. Between eight and 10 days RAR-γ is enhanced in hypertrophic chondrocytes. Implanting beads with an RAR antagonist results in short and bent humeri although the bony collar (which is intramembranous) develops normally. Overexpression of RAR-α results in skeletal defects because cells remain prechondrogenic and fail to differentiate, even in the presence of Bmps, indicating that RAR-α and Bmps operate separately to regulate cartilage differentiation.[b]

Cellular retinoic acid-binding protein (CRABP)

In chick embryos, cellular CRABP is expressed in those tissues that are malformed after treating embryos with retinoic acid, viz. the central nervous system, craniofacial and limb-bud mesenchyme, peripheral ganglia, and the visceral arches. CRABP also is expressed in preosteoblastic cells six to eight days into healing of rat tooth-extraction sockets. Some of these CRABP-positive cells also express alkaline phosphatase or osteopontin.[c]

[a] See Noji *et al.* (1989) and Ruberte *et al.* (1990, 1992) for RARs in mouse embryos, Mendelsohn *et al.* (1994) and Lohnes *et al.* (1994) for the double receptor mutants, and Matt *et al.* (2003) for the RAR agonist study.
[b] See Koyama *et al.* (1999) for the antagonist, and A. D. Weston *et al.* (2000) for overexpression.
[c] See Vaessen *et al.* (1990) for embryonic distribution of CRABP in mouse embryos, and Shyng *et al.* (1999) for expression in preosteoblasts.

---

will be used to illustrate this approach. Both involve targeted deletion of *Hoxa-2*.

Gendron-Maguire *et al.* (1994) deleted exon 1, the first 32 base pairs of exon 2, the splice acceptor site, and all introns from *Hoxa-2*. Rijli *et al.* (1993) deleted exon 1, the first 72 base pairs of exon 2, and the translation initiation site. Both studies demonstrated abnormal development of the malleus and incus of the middle ear and what appeared to be mirror-image duplications of the malleus and incus (Box 13.3), and the tympanic bone of the skull. An ectopic squamosal bone also formed. Stapes and stylohyal cartilages were missing in both sets of embryo but an ectopic cartilage formed (Figs 13.1 and 18.9).

Lack of *Hoxa-2* results in absence of second arch elements and their transformation into first arch structures, which has been interpreted as an atavistic reptilian palatoquadrate. In *Hoxa-2*⁻/⁻ mice, the second arch skeleton

is replaced by first arch skeletal elements. *Hoxa-2* is normally excluded from chondrogenic condensations in the second arch, but in the absence of *Hoxa-2*, there is ectopic activation of cartilage within the arch, with misexpression of *Sox-9*, which is shifted into the normal *Hoxa-2* domain. *Sox-9* is therefore equivalent to the *Hoxa-2* mutant phenotype, with *Hoxa-2* acting upstream of *Sox-9*. Misexpressing *Hoxa-2* is also accompanied by inhibition of membrane bone formation.

Overexpressing *Hoxa-2* in the first arch of chick embryos transforms the first into a second arch. Knocking out *Hoxa-2* in the second arch transforms it into a first arch. *Hoxa-2* selects arch fate by altering expression in tissue around NCC and so represents a late patterning event. Exposing *Xenopus* embryos to exogenous *Hoxa-2* after migration of the neural crest induces a mirror-image homeotic transformation of the lower jaw to a hyoid arch, a response which is the reverse of a *Hoxa-2* knockout, indicating that *Hoxa-2* plays a fundamental role in the specification of the hyoid arch.[27]

In a thoughtful comparative analysis of the skeletal phenotypes of such mice, Smith and Schneider (1998) detected a common underlying theme. An excess of mesenchymal cells accumulates at sites that normally possess very small condensations that do not differentiate. Because of this accumulation of additional mesenchyme, chondrogenesis is initiated, resulting in the formation of ectopic cartilages, as shown in Figs 13.1 and 18.9. The site of this condensation – between the mouse alisphenoid and incus (epipterygoid and quadrate in reptiles) – and its differentiation in knockout mice allow a cartilage to form that bears a superficial resemblance to the reptilian palatoquadrate. Indeed, ectopic 'quadrate-like' and 'pterygoquadrate-like' bones were interpreted as atavistic reversals to a reptilian condition (Rijli *et al.*, 1993). Smith and Schneider (1998) however, dispute this interpretation, seeing no need to invoke atavistic interpretations when a more mechanistic explanation is sufficient: the differentiation of cartilage or bone in centres of condensation that are normally too small to initiate skeletogenesis. While neomorphic in the sense that the cartilages do not normally form, the elements are not atavistic; the rudiment is normally present in all individuals although unexpressed. Neomorphs and atavisms are discussed in Chapter 44.[28]

Miyake *et al.* (1996) had emphasized the significance of the origin of these skeletal elements from NCC from more than one region of the hindbrain, and that mesenchyme from both mandibular and hyoid arches contributes to the skeletal condensations of the auditory region. Smith and Schneider (1998) also propose disruption of normal migration of NCC as the most likely mechanism responsible for increased condensation size. Their **cautionary** emphasis – that genes supply products for differentiative and morphogenetic pathways rather than specifying phenotypes – is a very important one. In order to elucidate the effects on the development of condensations of gene knockout experiments, developmental series of embryos must be evaluated; analysis of the phenotype at a single end-point is not sufficient to establish underlying mechanisms.

The molecular cascades

Progressing from condensation to overt differentiation of cells identifiable as chondroblasts or osteoblasts requires down-regulation of N-CAM and genes controlling proliferation and up-regulation of genes associated with differentiation. Thus, progression to the differentiation phase is achieved both through signals that *stop condensation growth* and so *favour differentiation indirectly* (syndecan binding to fibronectin to down-regulate N-CAM), and *signals that initiate differentiation directly* through such pathways as *Bmp-2, Bmp-4, Bmp-5* and activation of homeobox genes such as *Msx-1* and *Msx-2* (Box 13.4, Table 19.1; Fig. 20.2). A subsequent transition in endochondral ossification – from proliferating chondroblasts to chondrocytes and hypertrophic chondrocytes – is regulated via Indian hedgehog (*Ihh*), which induces parathyroid (PTH) and PTH-related peptide receptor. PTHrP in turn acts via negative feedback to regulate *Ihh* and so control the size of developing growth plates (Fig. 15.2; Box 31.2).[29]

Bmps

Bmp-2 and *Bmp-5*, which are expressed in condensations *in vivo*, are expressed in C3H10T1/2 cells as they enter chondrogenic differentiation (see Box 10.2). Not all cell lines are similar, however. Exposure to Bmp-2 or Bmp-4 allows ATDC5 cells to progress to chondroblast differentiation, bypassing a condensation phase.[30]

Bmp-5 is expressed at the condensation stage of skeletogenesis and plays a role in induction of the condensation. The *short ear* mouse results from a nonsense mutation of *Bmp-5* that results in altered size and shape of the condensation. *This was the first mutation of Bmp to be identified.* Some bones are lost from *short ear* mice, which also have reduced ability to repair fractures.[31]

Bmp favors recruitment of cells into condensations, a recruitment that is antagonized by the secreted protein Noggin. Bmp-Noggin feedback is therefore an important means of regulating condensation location, size and duration (Table 19.1; Fig. 20.2). Noggin has been demonstrated in condensations and immature chondroblasts in the mammalian skull and to interact antagonistically with Bmps during limb development.[32] Condensations can form in the absence of Noggin, but they rapidly become hyperplastic. Subsequently, although cartilage maturation is normal, joints between adjacent skeletal elements fail to form.

Bmps do not induce fibronectin and act primarily *after* condensation initiation. Tgfβ regulates fibronectin and so both Tgfβ and fibronectin are involved in condensation *initiation*.[33]

Tenascin and N-CAM

Tenascin and its cell-surface receptor syndecan play important roles in epithelial–mesenchymal interactions leading to condensation (up-regulation of tenascin) and in the transition from condensation to overt differentiation. Binding of syndecan to fibronectin blocks N-CAM in the late condensation stage (see Chapter 20). Consequently, further condensation growth is blocked and differentiation is facilitated (Hall and Miyake, 1995). N-CAM, which plays an important role in mediating cell-to-cell adhesion regulating condensation formation (N-CAM contains type I collagen and heparan-sulphate-proteoglycan-binding domains), is expressed in periostea in association with osteogenesis but is down-regulated when periosteal cells switch to chondrogenesis, as occurs when secondary cartilage differentiates in membrane bone periostea. Curiously, healthy fertile mice are born after knocking out the N-CAM gene. Given all that we know about N-CAM's role in condensation formation and the need to block N-CAM to halt condensation growth, other factors must normally act with N-CAM.[34]

Runx-2

The transcriptional activator *Runx-2* or *cbfa-1* (Box 20.4) is expressed in prechondrogenic and preosteogenic condensations but then strictly only in the osteogenic cell lineage, being down-regulated in chondrogenic lineages.[35] The central function of *Runx-2* in osteoblast differentiation

was confirmed by the demonstration that osteogenic genes are expressed in non-osteogenic cells in which *Runx-2* has been expressed. Regulated by *Bmp-7* and vitamin D$_3$, *Runx-2* is a strong candidate as a switching gene directing cells into the osteogenic pathway. *Runx-2* is inhibited by *Hoxa-2*; *Runx-2* is expressed in the second arch of *Hoxa-2*$^-$ embryos (Box 20.4).[36]

Runx-2 is only expressed in cells in which proliferation has been arrested. p-107 and p-130 are both expressed in cells that have exited the cell cycle. The retinoblastoma (pRB) family proteins p-107 and p-130 regulate the cell cycle by regulating cyclin-dependent kinases and cyclins. Skeletal development is defective in p-107/p-130 knockout mice. P-107 is expressed in prechondrogenic condensations and is required for cells to withdraw from the cell

---

Box 20.4 The skeleton gene?

One of the most exciting discoveries in skeletal biology in the past decade is a gene which, when knocked out, prevents all bone from forming. It seems not unreasonable to regard such a gene as a major upstream signaling gene for osteogenesis, a 'skeleton gene,' perhaps 'the skeleton gene' as far as osteogenesis is concerned.[a] The gene, a specific *cis*-acting element in the osteocalcin promoter that acts as a transcriptional activator of osteoblast differentiation, goes by various acronyms:

- *Osf-2* – osteoblast-stimulating factor-2;
- *Cbfa-1* – core-binding factor a1;
- *Runx-2*, because the gene is now known to be a member of the *runt*-domain family of core-binding transcription factors; *run* from the *Drosophila runt* gene and *x* to denote the mammalian gene; hence *Runx*, the name now most commonly in use.[b]

Runx-2

Ducy et al. (1997) cloned the cDNA for the Runx-2 protein. It is first expressed in embryonic mesenchyme, then in prechondrogenic and preosteogenic condensations, and finally strictly in osteoblastic lineages. Expressing the gene in non-osteoblasts such as MC3T3-E1 cells up-regulates osteogenic genes. *Runx-2* is regulated by *Bmp-7* and by vitamin D$_3$. Targeted disruption results in complete lack of formation of all membrane and endochondral bones in homozygotes (*Runx-2*$^-$/*Runx-2*$^-$) because of arrest of osteoblast maturation. A few alkaline phosphatase-positive osteoblasts develop, but no post-condensation osteoblasts develop.

Runx-2$^-$/*Runx-2*$^-$ homozygotes do not express osteopontin, bone sialoprotein or collagenase-3, but surprisingly express lots of Runx-2 protein in chondrocytes, especially hypertrophic chondrocytes. PTH, PTHrP, *Ihh*, type X collagen and *Bmp-6* are not expressed in the hypertrophic chondrocytes, indicating a block of chondrogenesis at the pre-hypertrophic stage (Fig. 15.2; Box 31.2).[c] Given this ongoing expression it is not surprising to find that *Runx-2* promotes chondrogenesis *in vitro*. Over-expressing *Runx-2* leads to dwarfism via an increase in endochondral ossification because of precocious chondrocyte differentiation. Over-expression of the dominant negative form of *Runx-2* also results in dwarfism but from decreased maturation of chondrocytes and delayed endochondral ossification. In both cases, joints fail to develop (Ueta et al., 2001).

Runx-2-deficient mice lack all membrane and endochondral bones (and so do not express osteopontin, bone sialoprotein or collagenase 3) but surprisingly express lots of *Runx-2* in chondrocytes,

especially in hypertrophic chondrocytes. PTH, PTHrP, *Ihh*, type X collagen and *Bmp-6* are not expressed in the hypertrophic chondrocytes, indicating a block of chondrogenesis at the pre-hypertrophic stage.[d]

A mutation in *Runx-2* produces *cleidocranial dysplasia* in mice. In heterozygotes, the clavicles are underdeveloped or absent and supernumerary teeth form. Homozygotes lack bone or osteoblasts. *Runx-2* is up-regulated in the second arch of *Hoxa-2* mutants, suggesting that *Runx-2* is inhibited by *Hoxa-2*, i.e. that *Hoxa-2* inhibits osteogenesis by down-regulating *Runx-2*.[e]

Osterix

A second important transcription factor, *Osterix* (*Osx*), acts downstream of *Runx-2*. Like *Runx-2*-knockout mice, *Osx*-null embryos lack bone formation. Rounded cells with chondrogenic properties (expression of *Sox-9*, *Sox-5*, *Ihh* and *Col2a1*) develop in the mesenchyme where osteoblasts normally form in both endochondral and intramembranous ossification. Although further research is needed, it appears that *Runx-2* regulates transformation of progenitor cells to preosteoblasts and that *Osx* is required for the transformation of preosteoblasts to osteoblasts, and may act to switch cells from chondrogenesis into osteogenesis.[f]

[a] Perhaps the most unexpected recent discovery in 'skeletal biology' was that the 'fat proteins' (leptins) which control fat deposition through a central signaling system in the hypothalamus also control bone mass through the same central system (Ducy et al., 2000; Amling et al., 2000; Harada and Rodan, 2003).
[b] For overviews of skeletal cell differentiation, especially gene cascades, including the roles of *Runx-2* (*Osf-2*/*Cbfa-1*) in osteoblast differentiation and *Sox-9* in chondrogenesis, see Ducy et al. (1997, 1998, 2000), Efstratiadis (1998), Karsenty (1998, 2000, 2003), Ducy (2000), Gunther and Schinke (2000), the papers in Cardew and Goode (2001), Karsenty and Wagner (2002) and Nakashima and de Crombrugghe (2003).
[c] See Komori et al. (1997), Otto et al. (1997) and Rodan and Harada (1997) for these studies, and see Ferguson et al. (1998), Komori and Kishimoto (1998), Ducy et al. (2000), Cohen (2000b,c), Stricker et al. (2002) and Nakashima and de Crombrugghe (2003) for *Runx-2* in skeletal development and repair.
[d] See Inada et al. (1999) for Runx-2 protein and for *Runx-2* (*Cbfa-1*)-deficient mice, Box 31.2 and Ueta et al. (2001) for over-expression, and Volk and LeBoy (1999) for a review of how chondrocyte hypertrophy is regulated by *Runx-2*.
[e] See Mundlos et al. (1997) Otto et al. (1997) and Zelzer and Olsen (2003) for *cleidocranial dysplasia*, and Kanzler et al. (1998) for *Hoxa-2* mutants.
[f] See Nakashima et al. (2002) and Nakashima and de Crombrugghe (2003) for *Osterix*.

cycle and enter differentiation, but is not required for differentiation itself (Rossi et al., 2002).

Thus, a variety of pathways are utilized as cells make the transition from condensation to overt differentiation. Each stage is discrete and each stage is controlled by a specific cascade(s) of factors. The major genes, gene knockouts and mutations for each phase are now known (Table 19.1; Fig. 20.2). Challenges for the future are to fully understand the cascades of genes regulating each phase of condensation formation, and to identify functionally redundancy between genes acting at different phases to understand how condensation ceases and the differentiation phase begins.

NOTES

1. See Fell (1925) and Hall (1985a) for a commentary on her work, Grüneberg (1963), Atchley and Hall (1991), Smith and Hall (1990, 1993), Ettinger and Doljanski (1992) and Newman and Tomasek (1996).

2. See Hall and Miyake (1995), Miyake et al. (1997a), Dunlop and Hall (1995) and Schilling and Kimmel (1997) for the three examples, and Hall and Miyake (1995) and Miyake et al. (1997a,b) for levels of complexity.

3. Goel and Jurand (1975) questioned whether cells could flatten at the periphery of a condensation. See Searls (1965a,b) for synthesis of GAG by presumptive cells. Goel (1970) published some of the earliest studies on the synthesis and organization of ECM materials by chondrocytes.

4. A lathyrogen is a compound that interferes with collagen synthesis. In osteolathyrism, lysyl oxidase is inhibited and collagen is released in a soluble form that cannot aggregate into fibrils/fibres (Barrow et al., 1974). See Goetinck and Sekellick (1970) for the scaleless mutant.

5. See Ede and Kelly (1964a,b) for defective condensation in talpid³ embryos, and Agerback (1968), Ede (1971, 1976), Ede and Flint (1972), Ede et al. (1977b) and Ede (1980, 1983) for these classic approaches to condensation in vitro. Recall that wild type refers to embryos that are wild type for the particular gene for which a mutation has been studied.

6. See Karasawa et al. (1979) for the caution concerning reaggregation, and Yokoyama et al. (1998) for the study with Xenopus.

7. See Uwa (1969, 1974) and Uwa and Nagata (1976) for Japanese medaka fins, and Maisey (1979) for shark fins.

8. See Elmer and Selleck (1975), Duke and Elmer (1977, 1978, 1979), Elmer (1977), Shambaugh and Elmer (1980), Elmer (1983) and Elmer et al. (1988) for studies on reduced adhesiveness of mesenchymal cells from bp^H embryos.

9. See Zimmermann et al. (1982), Zimmerman (1984) for the initial studies, and Hall and Miyake (1992, 1995, 2000) for overviews.

10. See Allen et al. (1990) for posterior limb mesenchyme signaling, Coelho and Kosher (1991a,b) for transfer of Lucifer yellow, Ferrari et al. (1994) for the role of Msx-1, Green et al. (1994) and Makarenkova and Patel (1999) for regulation of connexin-43 in limb buds.

11. See Minkoff et al. (1991, 1994) for distribution of gap junctions and connexin-43, and McGonnell et al. (2001) for the role of Msx-1.

12. See Loty et al. (2002) for this study. Glycyrrhetinic acid, with the glycoside glucuronic acid (1:2), forms the steroid-like glycoside glycyrrhizin, found in and extracted from licorice root, Glycyrrhiza glabra, a perennial (and widely cultivated) herb native to southern Europe, Asia and the Mediterranean, and used medicinally since at least the third century BC in Egypt; King Tutankhamen was buried with a supply of licorice.

13. See Benjamin and Ralphs (1997a,b, 1998) for the network.

14. See Buchberger et al. (1998) for Cfkh-1, Labbé et al. (1998) for the relationship with Tgfβ, and Derynck and Zhang (2003) for TGFβ interaction with Smads. Smad-2 homozygote mutant mice fail to form mesoderm. Heterozygotes lack mandibles and eyes (Nomura and Li, 1998). Bmp regulates Egf during mouse mandibular chondrogenesis via the Smad-1 pathway (Nonaka et al., 1999). The Tob family of proteins has anti-proliferative properties and they act as negative regulators of Bmp/Smad in osteoblasts. Deletion of Tob is followed by increased bone mass and osteoblast numbers; Bmp-2-induced osteogenesis is enhanced in such animals (Yoshida et al., 2000).

15. See Iida et al. (1997) for Mfh-1, Ng et al. (1997) for Sox-9, and Cserjesi et al. (1995) and Chapter 9 for scleraxis.

16. See Kanzler et al. (1998) for Hoxa-2, and Yogouchi et al. (1995) and Newman (1996) for Hoxa-13.

17. See Jung and Tsonis (1998) for Hoxd-10 to Hoxd-12, and Koyama et al. (1996a) for cartilage as a signaling centre.

18. See Yang et al. (1997, 1998) for the Dbf mutant, and see Blanc et al. (2002) for extensive expression of Shh within anterior limb-bud mesenchyme of Hemimelia extra toes (Hx) mutant mouse embryos.

19. See Langille (1994a) for recruitment, Duprez et al. (1996) and Urist et al. (1997, 1998) for Bmp-2 inside and outside the skeleton, and Tintut et al. (1998) for vascular cells.

20. See Toole (1972, 1973a,b) for the data on limb buds and somites; and see Toole (1990, 2001) and Spicer and Tien (2004) for reviews of hyaluronan.

21. See Pratt et al. (1975) for hyaluronan in the ECM, and Corsin (1975b) for addition of hyaluronan to NCC/endoderm cultures.

22. See Mayne et al. (1971) for the amniotic cells, and Toole et al. (1972a) for exogenous hyaluronan to limb mesenchyme.

23. See Pessac and Defendi (1972) for production of hyaluronan or hyaluronan-like molecule, and Moscatelli and Rubin (1975) for the correlation with proliferation.

24. Also see Ahrens et al. (1977a, 1979) and Solursh et al. (1978) for early studies on condensation in limb-bud mesenchyme. Dissociation and reaggregation of mesenchyme from H.H. 19 chick limb buds favours chondrogenesis over myogenesis (Moscona, 1961; Medoff, 1967).

25. A parallel and favourite example of mine is the experiment – reproduced in many laboratories – demonstrating that one out of 100 nuclei from tadpole intestinal cells allows complete development of another tadpole when transplanted into an enucleated egg. Interpretations of these results are: one per cent of intestinal cell nuclei are totipotent; all intestinal cell nuclei are totipotent; no intestinal cell nuclei are totipotent. I shall be very pleased to provide a free copy of this book to the first person who tells me (bkh@dal.ca) how three such divergent interpretations can follow from this experiment. Because of his extensive knowledge, SFG is disqualified from this competition.

26. Retinoic acid inhibits the migration of cranial NCC in cultured mouse embryos; the cells fail to leave the

neuroepithelium or stay close by the neural tube (Pratt et al., 1987). According to a recent study by Wendling et al. (2000), retinoids that lead to ectopic migration of NCC normally destined for the third and fourth arches do not inhibit migration directly but indirectly via inhibition of the pharyngeal endoderm.

27. See Grammatopoulos et al. (2000) and Pasqualetti et al. (2000) for the chick and Xenopus studies.

28. Newman and Tomasek (1996) and Hall (1998a, 2003a) discuss the physical/proximate and evolutionary/genetic mechanisms in skeletal evolution.

29. See Vortkamp et al. (1996), Smith and Schneider (1998) and Minina et al. (2001).

30. See Zhang et al. (1996) and Atkinson et al. (1997) for CEH10T1/2 cells, and Shukunami et al. (1998) and Wahl et al. (2004) for ATDC5 cells.

31. See Kingsley (1994b), Kingsley et al. (1992) and J. A. King et al. (1994, 1996) for Bmp-5 and the Short ear mutation, and see Vortkamp (1997) for review of Bmp-5, joint formation and mutants.

32. See Brunet et al. (1998) and Pizette and Niswander (1999) for noggin, whose action on the skeleton is complex. Mice which lack noggin form skeletal condensations, but the condensations are hyperplastic. Cartilage maturation proceeds, skeletal defects are mild, but joints fail to form (Brunet et al., 1998). Double knockout of Noggin and Chordin produces much more severe defects, ranging from mild truncation of mandibular development to agnathia. Indeed some embryos develop normal mandibles (Stottman et al., 2001).

33. See Roark and Greer (1994) and Merino et al. (1998).

34. See Fang and Hall (1995, 1997, 1999) for secondary chondrogenesis, and Cremer et al. (1994) for co-factors.

35. See Ducy et al. (1997, 1998) and Nakashima and de Crombrugghe (2003) for cbfa-1 (runx-2).

36. See Komori et al. (1997) and Kanzler et al. (1998) for Hoxa-2 and Runx-2 (cbfa-1).

Skulls, Eyes and Ears: Condensations and Tissue Interactions

Chapter 21

In cranial development, the contents induce the container.

The epithelial–mesenchymal interactions that elicit mesenchymal condensations set the stage for the differentiation of chondroblasts and osteoblasts to form cartilages and bones. In Chapter 18, I used skeletal development in the jaws to illustrate how epithelial–mesenchymal interactions set that stage. In the present chapter, I examine four additional skeletal systems in which these processes are especially well understood. Two – *the skull and scleral ossicles* – are based on intramembranous ossification, two – *the cartilages of the eye and ear* – on chondrogenesis.

The bony skull is familiar to every individual who has visited a museum, found a dried skeleton on a beach or in a field, or engaged in Halloween 'trick or treating.' The scleral ossicles may be less familiar. They form a ring of overlapping flat circular bones protecting the eyes of many vertebrates (Fig. 21.1).

A number of cartilages comprise the skeleton of the eye. I discuss both the cartilaginous *optic* and *otic capsule* that supports the eye and middle ears, the tympanic cartilage that supports the tympanic membrane, and a ring of cartilage known as *scleral cartilage* that supports the sclera of the eye. The middle and inner ear is surrounded and protected by an *otic capsule*. The elastic cartilage of the external ears (the pinnae) of mammals was discussed in Chapter 3. Depending on taxon, *scleral cartilages may be permanent, replaced by bone, or in some taxa transformed into bone.* Some taxa have a scleral cartilage and scleral ossicles to support their eyes, others only one of the two. Other taxa have additional dermal ossifications as sesamoids associated with the scleral ossicles or with the orbital bones.

Each of these skeletal systems arises following interactions with one or more epithelia, the:

- bony skull with the developing brain,
- cartilaginous base of the skull with the notochord,[1]
- otic and optic capsules with epithelial otic and optic vesicles,
- scleral cartilage with pigmented retinal epithelium (PRE), and
- scleral ossicles with the scleral epithelium.

I begin with the skull.

THE BONY SKULL

The membrane bones of the skull – cranium, dermocranium, dermatocranium, calvaria or bony skull – so named to distinguish it from the neurocranium, the initially cartilaginous base of the skull, surround and protect the brain. The mechanical role of bone and bones is often equated with the need for mechanical factors to initiate bone formation. However, while mechanical stresses associated with skull expansion play a major role in patterning skull shape and directing skull growth (Box 21.1 and Chapter 34), mechanical forces are not required to initiation cranial osteogenesis. Mechanical forces do affect the symmetry of the skull. Effects range from: (i) minor deviations from the perfect left–right symmetry apparently desirable in mates (such random deviations are termed fluctuating asymmetry; Box 8.1) through (ii) culturally desirable traits in head shape acquired through head binding, (iii) deviations associated with mutational, drug- or alcohol-induced syndromes, and finally (iv) asymmetries that characterize

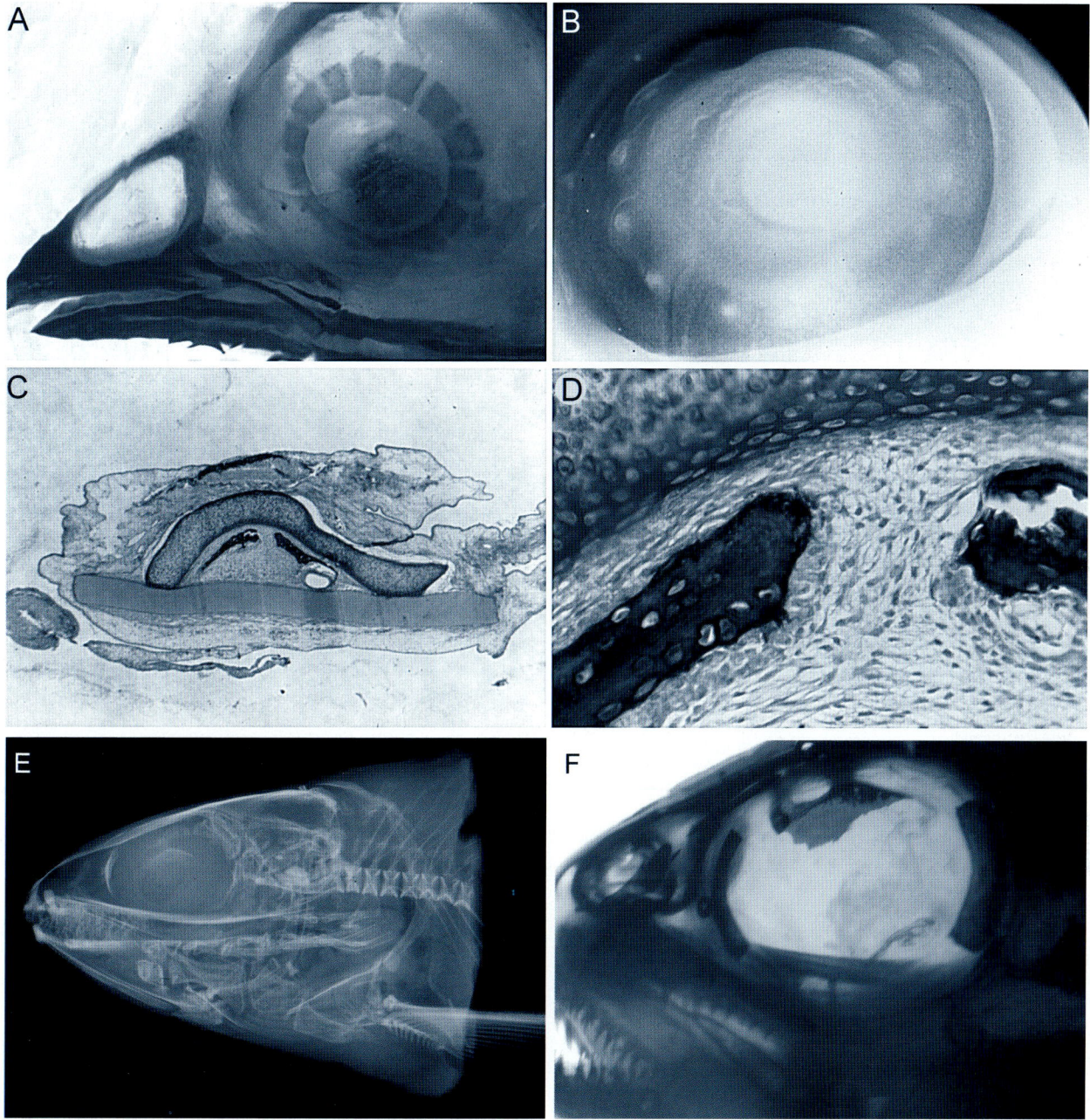

Figure 21.1 Development of scleral ossicles in chick (A–D) and salmon (E,F) embryos. (A) The ring of scleral ossicles supporting the left eye in an alizarin-red-stained chick embryo. (B) The scleral papillae that arise before and induce ossicle formation, seen in an unfixed embryo. (C, D) Low and high magnification views of a CAM-graft of scleral mesenchyme (H.H. 36) combined with mandibular epithelium (H.H. 22). Scleral cartilage and two scleral ossicles are shown in (C) and in greater detail in (D). Salmon (*Salmo salar*) only have two scleral ossicles, one anterior (toward the snout) and one posterior, as seen in the radiograph (E) and in an alizarin-red-stained preparation (F) of young salmon (parr). Images E and F provided by J. Andrew Gillis and T. Franz-Odendaal, respectively.

individual taxa – for example, both eyes on one side of the head in flatfishes (Box 21.2), crossed beaks in many birds such as Tengmalm's owl, *Aegolius funereus*, or the single spiral tooth (tusk) in narwhals, *Monodon monoceros*.

The roofing bones of the skull – paired frontals, nasals and parietals – develop from ectomesenchyme whose skeletogenic potential is activated by interactions with the developing brain. As elegantly demonstrated by Schowing using chick embryos and Petricioni with anurans, these

membrane bones are site-specific centres of skull development and growth.[2]

Avian skull development

The basic technique used by Schowing and colleagues was to remove selected areas of the developing brain (or notochord) from chick embryos at 35–40 hours' incubation (H.H. 10+ to 11) and look for deletions or deficiencies in

Box 21.1 Kinesis

Bird skulls exhibit several interesting adaptations to the mechanical forces that impact on them. In a woodpecker, the brain has a sling to protect it while drilling into trees. A more general adaptation is to make the skull moveable (*kinesis*), which is common, and to evolve a secondary jaw joint, which is uncommon but fascinating (see Chapter 44).

Birds skulls are kinetic, the two *forms of kinesis* being:

- *prokinesis*, in which the whole of the upper jaw moves on the brain case; and
- *rhynchokinesis*, in which only the anterior portion of the upper jaw moves.

Two *methods of kinesis* accompany these two modes:

- *uncoupled kinesis*, in which the two jaws are independent and there is no postorbital ligament; and
- *coupled kinesis*, in which the jaws are either coupled by a postorbital or lacrymo-mandibular ligament, or interlocked via a hinge between the quadrate and the articular.

Bock (1959, 1964, 1966, 1974) described both forms and methods in the context of the evolution of cranial kinesis and approaches to functional analysis of bill shape.

the crania at 15 days of incubation, the assumption being that a missing portion of the cranium would indicate dependence of that region on influences from the portion of the brain removed. Their map of the interactions (which are listed in Table 18.2) is shown in Fig. 21.2. As an example, differentiation of the frontal bones depends upon influences from both prosencephalon and mesencephalon (forebrain and midbrain). If both brain regions are excised, the skull develops, but without frontal bones; other elements of the skull are normal.

As demonstrated using a different approach, frontal bones fail to form after the developing brain is rotated 180° to place the hindbrain into the position normally occupied by the forebrain, a result indicative of the site specificity of these interactions. CAM-grafting presumptive ectomesenchyme for cranial bones from embryos of various ages demonstrates that these interactions occur between 48 and 60 hours of incubation, H.H. 13 to 16+. By way of summary we can assert that *the brain induces the cranial vault* and *the notochord induces the cartilaginous floor of the skull* (Fig. 21.2). Schowing neatly encapsulated these results as 'in cranial development, the contents induce the container.'

Box 21.2 Eye migration in flatfish

Transformation of the skeleton in 'flatfish' – members of the family Pleuronectiformes – fascinates all who encounter it.[a] Individuals *transform* from bilaterally symmetrical larvae to highly asymmetrical adults with both eyes on one side of the head. I use the term *transformation* of the skeleton because the eye that migrates does so after skeletal formation has been initiated – indeed, in some species after the skeleton has mineralized. In the Dover sole, *Solea solea*, cartilage reduction begins at day 14, metamorphosis at day 18, the left eye finally moving to the median crest of the head at day 20 (Wagemans and Vandewalle, 1999).

This is not an embryonic process but rather part of the metamorphosis these larvae undergo.[a] The eye and associated neural and soft tissues move through hard skeletal tissue that must be resorbed to allow the eye to 'pass.' Some species are preadapted for the eye migration. In some species, for example, the infraorbital series of bones associated with the orbit of the eye only develops on the side over which the eye migrates, facilitating movement (Burgin, 1986). The parasphenoid may be the only ossified bone present when migration begins. In other areas of the skull, cartilages have to make way for bones, for example when movements of head tissues place a dermal bone over the position previously occupied by a cartilaginous bone. In evaluating the literature on these changes, Brewster (1987) discussed five separate developmental processes involved in migration of the eye and the resulting cranial asymmetry:

- anterior and posterior portions of the cranium rotate in opposite directions;
- a portion of the cranium is resorbed, allowing space for the eye to move;
- the migrating eye creates its own passage through the head;
- a ligament that later ossifies is responsible for 'pulling' the eye into its new position; and finally,
- the eye migrates between the skull and the dorsal fin.

Not all five processes occur in all species. Processes in operation in the starry flounder, *Platichthys stellatus*, include migration of the eye through incompletely ossified bone, rotation of the neurocranium (the brain and orbits) but not the rest of the cranial skeleton

or head, and some bone resorption. In this species migration does not involve 'tunneling' through ossified bone. Cranial cartilages and bone in the Japanese flounder, *Paralichthys olivaceus*, are asymmetrical *before* metamorphosis begins, developing with a twist that presages the direction of eye migration. The pseudomesial bar, a bone found only in flounders (and only on the blind side) develops in association with thickened skin and retrorbital vesicles – which may be part of the lymphatic system, and which are more prominent on the blind side – both of which may play mechanical roles in eye migration.[b]

There is still much to be learned about flatfish comparative skeletal biology. Even less is known about the molecular (or even cellular) bases of such changes, although Fgfs may be good candidates. A 22.5-kDa *Fgf* (which may be *Fgf-4*) is expressed in the surface ectoderm, in primordia of the lateral lines and otic vesicles, and in prechondroblasts in embryos of the Japanese flounder, *Paralichthys olivaceus*. Blocking this *Fgf* with antiserum *in vitro* impairs chondrogenesis (Suzuki and Kurokawa, 1996).

The expression of Fgf in primordia of the lateral line is interesting for another reason. A number of dermal bones in teleosts – including the nasal bones in flatfish – develop in close association with neuromasts of the lateral-line system. As many dermal bones in fishes have no such association, and because the homologous bones in tetrapods arise in the absence of lateral-line organs, de Beer (1937) believed this association in flatfish to be secondary (evolutionarily derived). But the developmental bases for skeletogenesis can evolve and have evolved. As Jim Hanken and I discussed in a preface to the reissue of *The Development of the Vertebrate Skull*, there is considerable evidence for an inductive link between lateral lines and/or the neuromasts derived from them and ossification in teleosts and chondrification in elasmobranchs.[c]

[a] See Hall and Wake (1999), Shi (2000) and Hall *et al.* (2004) for overviews of the processes associated with larval development and metamorphosis.
[b] See Policansky (1982) for an overview of mechanisms of eye migration and see Okada *et al.* (2001, 2003) for the studies with the Japanese flounder.
[c] See Hall and Hanken (1985b) for the preface to de Beer's *The Development of the Vertebrate Skull*, and see Chapter 18 herein for a discussion of epithelial–mesenchymal interactions.

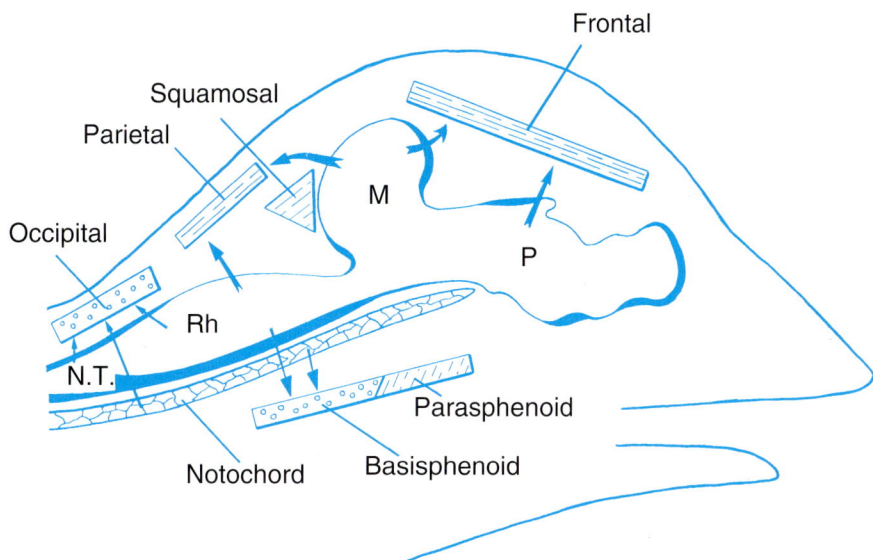

Figure 21.2 Interactions between developing brain/notochord and cranial mesenchyme are depicted in this diagram of the head of an early chick embryo. Arrows indicate epithelial–mesenchymal interactions required to initiate chondrogenesis or osteogenesis. Endochondral bones are filled with small circles, intramembranous bones are cross-hatched. M, mesencephalon; N.T., neural tube; P, prosencephalon; Rh, rhombencephalon. Also see Table 18.2. Modified from Hall (1978a) after Schowing (1968c).

I should comment briefly on terminology. I have been discussing these roles of the brain and notochord in terms previously used for epithelial–mesenchymal interactions. The notochord is epithelial at these stages of embryonic development, surrounded by a pericellular matrix deposited by notochordal cells, the nature of which is discussed in some detail in Chapter 42. The brain is also surrounded by a matrix that contains growth factors and peptides, to which mesenchymal cells respond.

But the relationship is more than spatial. Spinal cord, which induces somitic mesoderm to chondrify (Chapter 42), will not substitute for brain, although quail brain can substitute for chick to produce a normal, albeit small skull – quail being smaller than chicks and basic size being intrinsic (Chapter 17). Using an ingenious, if indirect approach that demonstrated this role of the brain, Schowing's group removed the embryonic brain, treated it with actinomycin D to suppress cell proliferation, and reimplanted it. Such embryos develop small brains and cyclopia, and the skull fails to form.[3]

Although interactions underlying cranial vault development were initially understood in any depth only for chick embryos, more recent studies provide details on mammalian skull development.

Mammalian skull development

By implanting a porous (pore diameter = 300 μm) filter between the periosteum and the calvarial bone surface in rats, Shimizu et al. (2001) showed that bone would only develop where cellular contact was made through the holes, supporting a bone matrix–periosteal cell interaction in calvarial osteogenesis.

Mammalian skulls arise immediately over the dura mater of the developing skull, the dura mater being the tough connective tissue membrane that envelops brain and spinal cord. The dura is closely associated with the periostea of the skull bones, perhaps to the point where the two fuse; the dura cannot be dissected away from either brain or skull with ease. In Wistar rat embryos, the dura develops one day before the appearance of bone, a timing that is consistent with dura–ectomesenchymal interaction in bone development (Angelov, 1989). Signals emanate from the dura during embryonic life to maintain mouse skull sutures open before birth. Does the dura induce bone? Several observations and experiments suggest that it should.

Bone forms preferentially on the dura side of healing defects in rat calvariae. If defects are large (8-mm diameter), healing is incomplete and areas of cartilage form. Chondrocytes from the hypertrophic or proliferative zones of 16-day foetal mouse metatarsals co-cultured with brain tissue differentiate into osteoblasts that deposit osteoid and then a mineralized matrix. Cultured hypertrophic chondrocytes from metatarsals of 17-day-old foetal mice transform into stromal cells, but if co-cultured with cerebrum transform into osteoblasts that synthesize type I collagen and osteocalcin, indicating a brain-controlled transdifferentiation.[4]

The dura itself has skeletal-inducing properties and, depending on the region from which the dura is taken, can induce bone or cartilage. Dura underlying the squamosal bone induces osteogenesis; dura underlying the sutures induces chondro- and osteogenesis. This is not a reflection of different properties of sutural and non-sutural bone for, in further analyses, skull bone was rotated so that it overlay the posterior frontal dura, the repositioned dura and bone cultured, and the effect shown to be specific to the dura.[5] The posterior frontal dura continues to act after birth, evoking sutural fusion (Chapter 34).

THE CARTILAGINOUS SKULL

The *cartilaginous skull* or *neurocranium* is built on the basis of three sets of cartilaginous capsules that surround the three major sense organs (eyes, ears and nose), a cartilaginous base to the skull formed from trabecular cartilages, and cartilaginous joints at the base of the skull (synchondroses) connecting the skull with the vertebral columns. An important component in localizing chondrogenic cells to positions where the major sense capsules can form is the distribution of type II collagen within epithelia with which migrating neural crest cells (NCC) interact and which will then provide the signals to initiate chondrogenic differentiation.

Type II collagen

Thorogood *et al.* (1986) demonstrated transient expression of collagen type II in the basement membranes of developing brain and notochord at sites where future epithelial–mesenchymal interactions would initiate chondrogenesis for the base of the chondrocranium and the sense capsules. They called the model that arose from these studies the 'flypaper model' to reflect their observation that these basement-membrane sites captured only some cells and not others (Fig. 10.7).

Subsequently, it was shown that mesenchymal cells with the collagen type II binding protein, anchorin C-2, a member of the annexin family, attach at sites of collagen type II. With this in mind, the more recent finding by Takahashi *et al.* (2001), that *Msx-2* represses chondrogenic differentiation in migrating NCC because it represses expression of *Sox-9* in a subpopulation of cells, is of interest. Is this also the population of cells that expresses anchorin C-2? A role for anchorin C-2 is also suggested by the finding by Hofmann *et al.* (1992) that anchorin C-2 is expressed in somites at H.H. 11 (Fig. 16.2), in the anterior sclerotome at H.H. 24, and then in the epithelia of the otic vesicle and mandibular mesenchyme at H.H. 20.

However, Kosher and Solursh (1989) demonstrated a widespread occurrence of type II collagen in chick embryos of H.H. 14–19 – in the basement membranes of the neural tube, notochord, otic vesicle, surface epidermis, gut endoderm, mesonephros, splanchnic mesoderm and heart. In a similar widespread pattern, expression of type II collagen in *Xenopus laevis* begins at stage 21 and is intense at stages 33–36, with expression in the ventrolateral brain, sensory vesicles and notochord *after* migration of NCC, and so too late to serve a flypaper function. Type II is found surrounding the notochord and otic vesicle

in zebrafish, *Danio rerio*, embryos. Finally, type II collagen is expressed transiently during mouse development; on the surfaces of otic and optic vesicles, ventral brain, pharyngeal endoderm and branchial arch ectoderm. Immunogold-staining showed that the type II lies near the basal lamina on 10–15 nm fibrils.[6] In its original version, the flypaper model comes unstuck.

Interestingly, type II collagen is deposited subepithelially during chick limb-bud development, initially beneath the epithelium at the chondrogenic condensation stage (H.H. 25), first proximally and spreading distally. This type II appears unconnected to events of chondrogenesis and is regulated independently of proteoglycan core protein (Mallein-Gerin, 1990). Of course, type II collagen is not type II collagen, is not type II collagen, and so forth; α-II collagen mRNA is alternately spliced in cartilage in human embryos, as well as in AER, tooth and sites of remodeling (Lui *et al.*, 1995).

Otic, optic and nasal capsules

Vertebrate eyes, and inner and middle ears, are supported and protected by cartilaginous capsules, optic and otic. The optic cartilages arise from neural crest, the otic capsules from head mesoderm, although there is some evidence that ectomesenchyme provides a permissive environment for chondrogenesis of otic mesenchyme.[7]

The otic vesicle

In amphibian embryos, chondrogenesis to form the otic capsule follows interaction between cranial mesoderm and the epithelial otic vesicle (otocyst), depicted in Fig. 21.3. In a study now 80 years old, Luther (1924) found that removing the otic vesicles from embryos of the edible frog, *Rana esculenta*, prevents otic cartilage from forming but has no effect on development of the columella auris; unlike the cartilaginous otic capsule, the columella forms without influence from the otic vesicle (Reagan, 1917). The nasal capsule also depends on the nasal sac for its formation, as demonstrated by Schmalhausen (1939) for urodeles. Following Burr (1916), who removed the nasal pits from *Ambystoma* embryos, Reiss (1998) extirpated the nasal placode from ranid frog embryos, arguing for homology of skeletal elements because of the common response of anuran and urodele embryos to nasal placode extirpation.

Other studies showed that cartilage can be differentiated from local mesenchyme if an otic vesicle is transplanted elsewhere in the head – otic vesicle acts as an inducer, and other head mesenchyme can respond – but that cartilage fails to form after transplanting the otic vesicle into trunk mesenchyme. Luther concluded that otic vesicles could not induce trunk mesenchyme, but it may be that trunk mesenchyme cannot respond (Fig. 21.3). Transplanting the otic vesicles elsewhere in spotted salamander, *Ambystoma punctatum*, embryos elicits cartilage

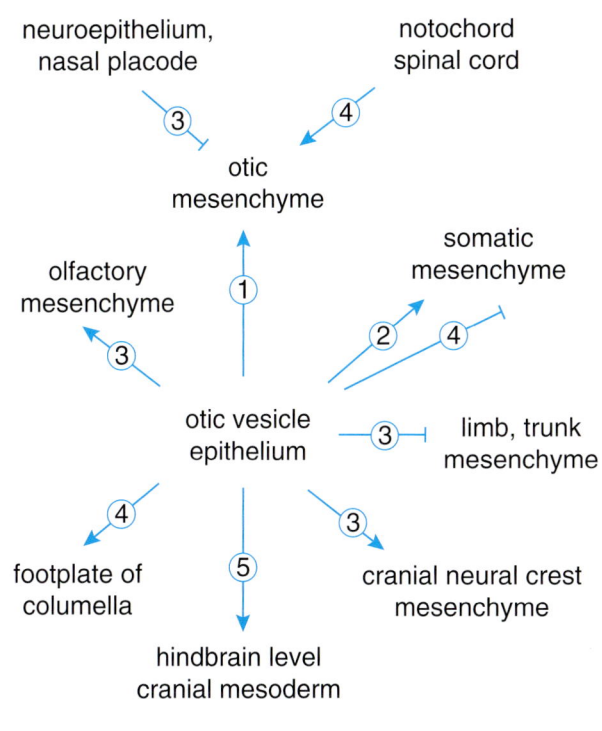

1. trout, frogs, urodeles, turtle, chick, mouse, rabbit
2. amphibian, mouse   3. amphibian
4. chick   5. frogs, urodeles, chick

Figure 21.3 A comparative analysis of the ability (———▶) of otic vesicle epithelium from various vertebrates to elicit chondrogenesis from different embryonic mesenchymal tissues, along with inability (———⊣) of amphibian and chick otic vesicle to elicit chondrogenesis from limb/trunk somatic mesenchyme. Whether neuroepithelium, nasal placode, notochord and/or spinal cord elicit chondrogenesis from otic mesenchyme also is shown. See text and Hall (1991b) for details.

from the ectomesenchyme of the head and gills *and from* limb-bud and pronephric mesenchyme, both of which are of mesodermal origin. Did Luther fail to graft the vesicle in association with the appropriate mesenchyme, or do anurans and urodeles differ?

We now know that in vertebrates from teleosts to mammals (trout, turtles, birds, mice, humans) determination of the position within the head where the otic capsules will form, and where otic capsular chondrogenesis will occur, depends upon an interaction(s) between otic mesenchyme and the otic vesicle (otocyst).[8]

Interesting patterns of specificity of interactions within individuals and between species have emerged (Figs 21.3 and 21.4). For example, chick otic epithelium will not induce amphibian ectomesenchyme or somitic mesoderm to chondrify (Hörstadius, 1950). *Triturus* otic mesenchyme chondrifies in response to nasal epithelium but not in response to pharyngeal endoderm.[9]

Interestingly, synthesis and deposition of type II collagen precedes chondrogenesis of mouse otic capsules by some two days, a delay that is unusual. Normally synthesis

of type II collagen marks the onset of chondrogenic differentiation.[10]

The role of the otic vesicle in murine otic capsule formation was studied in some depth during the 1980s and early '90s at the Albert Einstein College of Medicine in the Bronx (NY) by Tom Van de Water and his colleagues Joseph McPhee and Dorothy Frenz.

In the initial study at an international conference on prenatal craniofacial development, Van de Water (1980) showed that reducing the amount of cephalic mesenchyme in primordia maintained *in vitro* resulted in abnormalities of the sensory elements of the inner ear, indicating a morphogenetic interaction between mesenchyme and labyrinthine sensory structures.[11] Two reviews published in the same year emphasized the role of epithelial–mesenchymal interactions in otic capsule formation throughout the vertebrates, provided preliminary data using mouse embryos and examined deficiencies of the inner, middle and external ears within a developmental framework (Van de Water *et al.*, 1980a,b). Attention then was directed to chondrogenesis of the otic capsule as follows.

● Determination of five basic developmental stages and the profile of collagen and GAG changes – using ^{3}H-proline- and ^{35}S-labeled products – during otic capsule formation *in vivo* and using an *in-vitro* model system, and a proposal that continuous reciprocal interactions between epithelium and otic mesenchyme were required (McPhee and Van de Water, 1982, 1985).

● Demonstration that continuous exposure to the L-proline analogue L-azetidine-2-carboxylic acid (LACA) disrupted otic capsule induction (Van de Water and Galinovic-Schwartz, 1986).

● Determination that the otocyst induces capsule formation between 12 and 13 days of gestation, that the otocyst alone from embryos younger than 12 days of gestation is insufficient to induce cartilage, that otic mesenchyme from 13-day-old embryos chondrifies but displays abnormal cartilage morphogenesis: i.e. that the epithelium is involved in initiating both differentiation and morphogenesis (McPhee and Van de Water, 1986).

● Demonstration that collagen synthesis in otic-capsule mesenchyme increases by 50 per cent between 12 and 13 days of gestation, and that appearance of type II collagen precedes overt chondrogenesis by some 48 hours (D'Amico-Martel *et al.*, 1987).

● Determination that, as with limb-bud and mandibular epithelia,[12] otic-capsule epithelium first enhances and then inhibits chondrogenesis, and that the Tgfβ-1 and Fgf-2 – both of which occur in the epithelium – are sufficient to elicit chondrogenesis (Frenz and Van de Water, 1991; Frenz *et al.*, 1992, 1994).

Morphogenesis

Once formed, the otic capsule influences the morphogenesis of other regions of the skull. Initiation of chondrogenesis

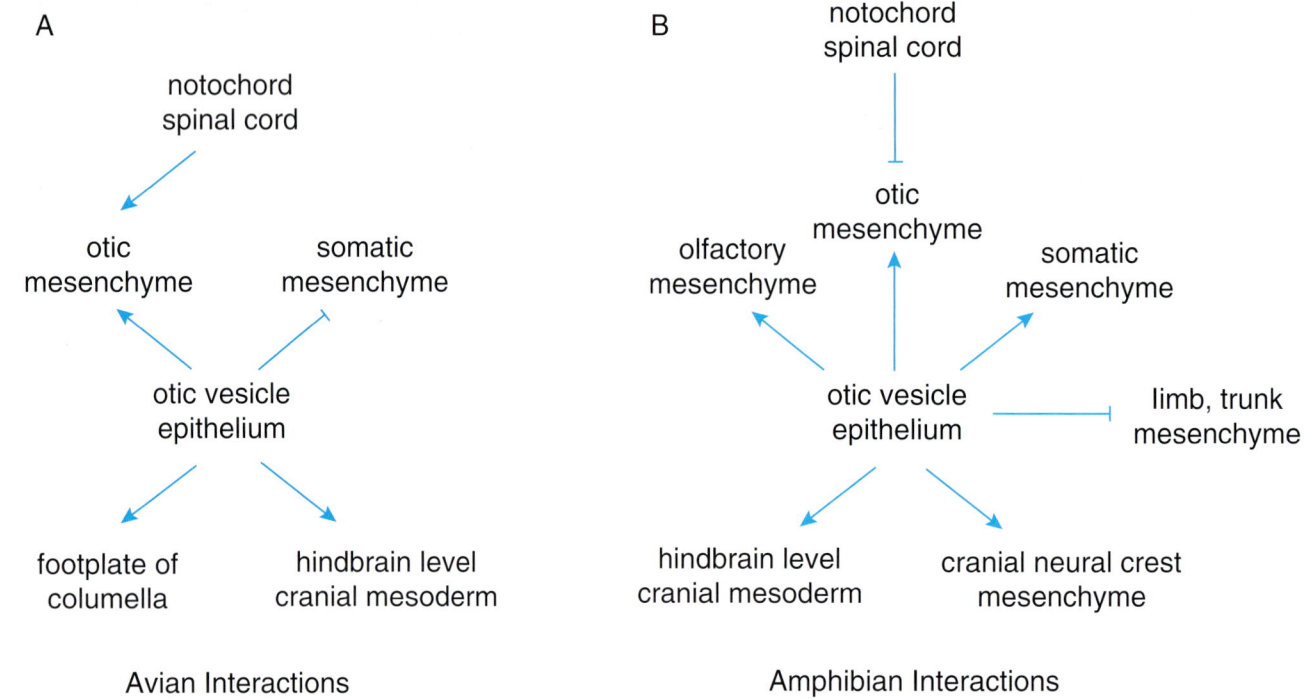

Figure 21.4 A summary of the ability of the epithelial otic vesicle of avian (A) and amphibian (B) embryos to induce otic mesenchyme to chondrify, along with the ability (⟶) or inability (⊣) to evoke chondrogenesis from other embryonic mesenchymal tissues. Note that avian but not amphibian notochord elicits cartilage from otic mesenchyme.

in response to the otic vesicle is one level of interaction – the level that initiates differentiation – in a hierarchy of interactions. Interaction of otic capsule with other components of the skull is a second level of interaction – as Leibel (1976) demonstrated by removing the otic capsule or altering the volume of the capsular cartilage – an example of the epigenetic/extrinsic factors that regulate skull morphogenesis.[13]

TYMPANIC CARTILAGES

The tympanic cartilage supports the tympanic membrane of the middle ear. A series of studies involving transplanting tissues associated with the tympanic cartilage and tympanic membrane in the northern leopard frog, *Rana pipiens*, the pickerel frog, *R. palustris*, and the bullfrog, *R. catesbiana* demonstrated a series of inductive interactions, as follows.

- Epidermis from anywhere on the embryonic body can form a tympanic membrane if transplanted to the normal site of tympanic membrane formation.
- Removal of the annular tympanic cartilage results in failure of the tympanic membrane to form, implicating the cartilage in induction of the membrane.
- Both quadrate and suprascapular (to a lesser degree) can induce a tympanic membrane in the pickerel frog, *Rana palustris*, but only if in direct contact with the skin,

while non-living annular tympanic cartilage can induce a tympanic membrane.

- Removal of the cartilage after the tympanic membrane has begun to form results in arrested membrane development, indicating that the cartilage is involved in continued development (and maintenance) of the membrane.
- Transplanting the annular tympanic cartilage to an ectopic site is followed by induction of a tympanic membrane from the skin at that site, the degree of development of the membrane depending on the stage of development of the cartilage when transplanted.
- Following application of thyroid hormone, the metapterygoid region of the quadrate transforms to an annular tympanic cartilage, and a tympanic membrane is induced in the overlying skin by that cartilage.
- Remove the columella and the tympanic membrane fails to thin or to form the elastic fibres normally characteristic of the membrane, implicating the columella in tympanic membrane differentiation.[14]

Development of the tympanum and the role of the columella in inducing otic capsular chondrocyte regression, dedifferentiation and redifferentiation to the annular ligament have been studied in birds. The external auditory meatus mediates tympanic membrane development by producing signals that enhance and inhibit chondrogenesis. In murine embryos, *Gsc* and *Prx-1* are required for formation of the tympanic ring, while the external auditory meatus produces a signal – possibly *Sox-9* – that opposes tympanic ring formation and enhances chondrogenesis, and a second signal – possibly *Prx-1* – that inhibits chondrogenesis.[15]

SCLERAL CARTILAGE

Heterogeneity

The optic cups of all vertebrate eyes are enveloped by mesenchyme of neural-crest origin, forming the sclera. In snakes and mammals, the sclera is collagenous. In fish, amphibians, lizards and birds, however, *scleral cartilages* develop within the mesenchyme to support the eye, especially during accommodation. In birds, scleral cartilages consist of thin, curved plates of cartilage whose characteristic morphology is moulded by the mechanisms of the growing eye and can be modified by tension applied *in vitro* (Fig. 21.5).

Other cartilages also respond to tension. Chick femoral epiphyseal chondrocytes increase proliferation, synthesis of acid mucopolysaccharides and cAMP levels in response to tension applied *in vitro*. We cannot, however, generalize the mechanism underlying this response to all cartilages. Indeed, response may vary from place to place within a single cartilage. One example – heterogeneity in the kinetics and biochemistry of epiphyseal cartilages, even early in their development – is shown by which epiphyseal cartilage-cell zones from the long bones of chicks, mice and rabbits show the highest labeling index following administration of ³H-thymidine (Table 21.1).[16]

Chondrogenic mesenchyme

Neural crest cells (NCC) populate the optic cup of chick embryos by H.H. 14 (50 hours of incubation) although scleral chondrogenesis does not commence until the

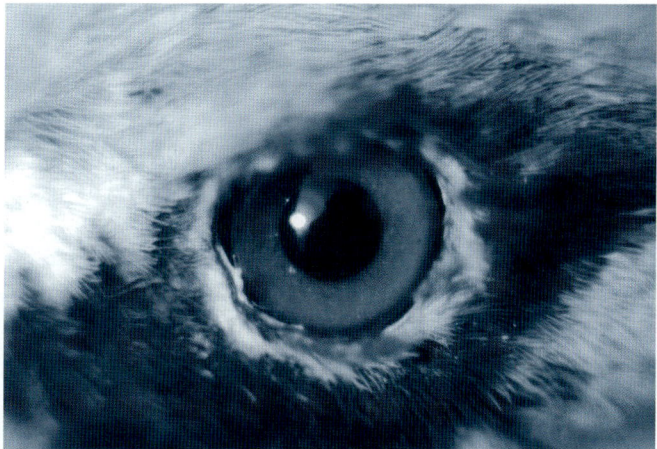

Figure 21.5 The eye of a pink-eared duck, *Malarcorhynchus membranaceus*, photographed by the author (who is reflected in the pupil) in Armidale, NSW, Australia.

Table 21.1 ³H-thymidine labeling indices of epiphyseal regions from hind-limb long bones of chick, mouse and rabbit embryos at equivalent developmental stages[a]

Species	Age (days[b])	Zone I[c]	Zone II[d]	Zone III[e]
Chick	3–4		Even distribution	
	6	Second highest	Highest	Lowest
Mouse	13	High	High	Low
	17	–	High	–
Rabbit	28	Relatively high	Relatively high	–

[a] Based on data from Diao *et al.* (1989).
[b] Chick: days of incubation; mouse and rabbit: days of gestation.
[c] Zone I: immediately beneath the joint surface.
[d] Zone II: the metaphyseal–chondroepiphyseal region.
[e] Zone III: the proximal metaphysis.

Table 21.2 Per cent of grafts of periocular mesenchyme that form scleral cartilage as a function of the age of the donor embryos[a]

Donor age (H.H. stage)	Per cent of grafts with cartilage[b]
17	0
18	0
19	14
20	38
23	82

[a] Adapted from data in Newsome (1972).
[b] Periocular mesenchyme was grafted to the CAM. Specification of mesenchyme for chondrogenesis is initiated between H.H. stages 18 and 19.

seventh or eighth day of incubation. Type II collagen can be detected by immunofluorescence from H.H. 30 (6.5–7 days of incubation) on. Because periocular ectomesenchyme only chondrifies *in vitro* if taken from embryos older than four days of incubation, Weiss and Amprino (1940) thought that specification for chondrogenesis was established by the fourth day. However, isolated periocular mesenchyme from some but not all embryos as young as 3–3.5 days of incubation (H.H. 19) forms cartilage when grafted to the CAM. These experiments, summarized in Table 21.2, indicate that chondrogenic potential can be elicited after H.H. 18. Because NCC reach the optic cups at H.H. 14, specification occurs within the optic cup.[17]

Pigmented retinal epithelium (PRE)

Scleral cartilages develop adjacent to and often in close contact with the pigmented retinal epithelium (PRE; Fig. 21.6) Given what we have seen of epithelial–mesenchymal interactions, we can ask whether PRE is required for scleral mesenchyme to chondrify. Several lines of evidence using chick embryos indicate that the ability of periocular mesenchyme to chondrify is established by H.H. 19 and dependent upon interaction with the PRE (Fig. 21.6).[18]

The action of the pigmented retinal epithelium on scleral ectomesenchyme is mediated by products within an ECM. PRE deposits a collagenous, acellular matrix closely associated with the basement membrane. Cloned PRE cells deposit similar extracellular materials onto filters on which they are cultured. Chondrogenesis is initiated after the epithelial cells are removed and replaced with periocular mesenchyme. Neither type I nor type II collagen can substitute for the PRE extracellular material. Chondrogenesis is initiated in periocular mesenchyme cultured transfilter to PRE when the filter pore size is 0.8 μm, but not when it is 0.2 μm. implicating either epithelial–mesenchymal contact or a matrix product in the interaction. Ectopic cartilage forms when PRE is grafted into head mesenchyme, indicating that PRE can evoke chondrogenesis from non-scleral mesenchyme, but in this case still cranial ectomesenchyme (Chapter 21).[19]

Morphogenesis

The morphology of the sheet of cartilage supporting the eye is readily distinguishable from the block-like vertebral cartilages that arise from somitic mesoderm or the elongate, rod-like cartilages that form from limb-bud mesenchyme. How are such fundamental shapes determined? In part, we recognize these shapes because of the position cartilages occupy *in vivo*. But shape is retained when cartilages are placed *in vitro*. When discussing the neural-crest origin of Meckel's and branchial-arch cartilages (Chapter 17), we saw that morphogenetic specificity can be intrinsic to mesenchyme arising from particular rostrocaudal regions of the cranial neural crest. Is this also true for scleral cartilages? Two types of experiment reveal the basic form of scleral cartilages as intrinsic to scleral mesenchyme, as follows.

(i) Prechondrogenic scleral mesenchyme grafted to the CAM or organ-cultured forms sheets of cartilage, not rods or nodules.
(ii) In what is now a classic study, chondrocytes disassociated from scleral cartilage, resuspended, and maintained *in vitro*, reform sheets of cartilage. *Once differentiated, morphogenetic specificity is maintained.*[20]

In confirmatory studies by Archer and colleagues (1985), cultured scleral chondrocytes formed sheets and cultured limb chondrocytes nodules, *unless* interactions between chondroblasts were limited, by maintaining the cells at a low cell density, or co-culturing them with non-chondrogenic cells, when nodules rather than sheets formed. Table 21.3 shows the results obtained with

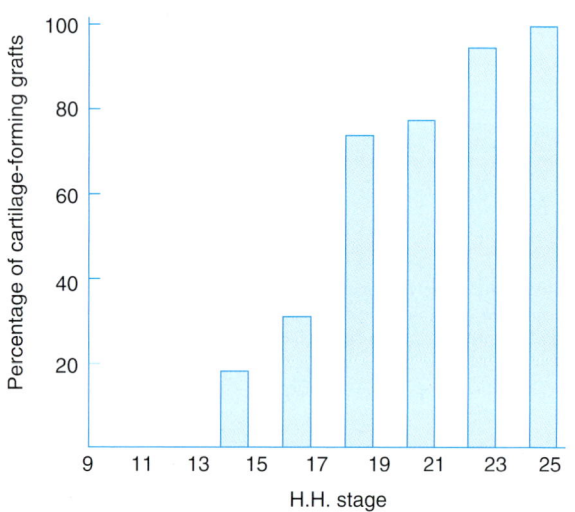

Figure 21.6 Scleral cartilage is induced by pigmented retinal epithelium (PRE). The histogram shows the percentage of grafts of scleral mesenchyme that form cartilage when combined with PRE from embryos of H.H. stages 9 to 25.

Table 21.3 Morphogenesis of embryonic scleral or limb cartilages maintained *in vitro* at varying densities or cell-type combinations[a,b]

Culture combinations	Cartilage morphology
High-density scleral mesenchyme	Sheet (homogeneous)
Medium-density scleral mesenchyme	Nodules separated by fibroblast cells
High-density limb mesenchyme	Nodules as concentric aggregations
Medium-density limb mesenchyme	Little chondrogenesis
High-density limb mesenchyme:heart cells (3:1)	Little chondrogenesis
High-density scleral mesenchyme:heart cells (3:1)	Little chondrogenesis
High-density limb mesenchyme:quail scleral mesenchyme (1:1)	Nodules as concentric aggregations
High-density epiphyseal chondrocytes	Sheet (homogeneous)
Medium-density epiphyseal chondrocytes	Nodules separated by fibroblast cells
High-density scleral chondrocytes	Sheet (homogeneous)
Medium-density scleral chondrocytes	Nodules separated by fibroblast cells
High-density epiphyseal:high density scleral chondrocytes	Sheet (homogeneous)
High-density epiphyseal chondrocytes:heart cells (3:1)	Nodules separated by fibroblast cells
High-density scleral chondrocytes:heart cells (3:1)	Nodules separated by fibroblast cells

[a] Based on data from Archer *et al.* (1985).
[b] All cells were from chick embryos derived unless otherwise stated.

embryonic scleral or limb mesenchyme, or chondroblasts maintained *in vitro* at various cell densities, in some cases in combination with heart cells.

Localization of scleral cartilages to the developing optic cup results from specific interactions between extracellular products of the PRE and component periocular mesenchyme. *Morphogenesis* of scleral cartilages results from interactions among scleral chondroblasts, *independent* of the PRE that elicited differentiation.

SCLERAL OSSICLES

Staying within the eye, I turn to the localization, differentiation and morphogenesis of scleral bones. As we saw with chondrogenesis to form scleral cartilage, we will see that initiation of osteogenesis to form scleral ossicles involves precise epithelial–mesenchymal interactions but that morphogenesis is under subsequent, independent, mesenchymal control.

Scleral ossicles present most commonly as a pair (teleost fishes) or ring (birds) of trapezoidal, overlapping bony plates surrounding the margins of the cornea in many non-mammalian vertebrates (Fig 21.1). These ossicles – which clearly consist of membrane bone in birds, are cartilaginous in elasmobranchs, and have been reported to be endochondral in turtles – are distinct from scleral cartilages, although it has been argued that scleral ossicles in teleost fishes arise by endochondral replacement of scleral cartilage rather than separate intramembranous ossification adjacent to the cartilage as in birds. If true, *we would not consider teleost scleral ossicles as homologues of tetrapod scleral ossicles.*[21]

It has been suggested that, in reptiles, scleral ossicles form by direct intramembranous or indirect endochondral ossification depending upon the taxon studied. If so, this would provide an interesting example of how evolutionary modification of developmental process can produce homologous elements. However, no comprehensive comparative developmental studies have been undertaken, and it is easy to confuse early unmineralized ossicles with cartilage, especially in whole-mount preparations, in which osteoid takes up alcian or methylene blue. That said, there may be unrecognized variation and variability to be uncovered; Watanabe *et al.* (1994) demonstrated transient expression of type II collagen during scleral ossicle development in chick embryos at H.H. 37–39. No other chondrogenic markers were positive. If you regard expression of type II collagen as synonymous with cartilage, these chick scleral ossicles go through a cryptic cartilaginous phase. Or, perhaps, type II expression signifies the deposition of an intermediate tissue such as chondroid. Or, more likely, initial intramembranous ossification is associated with deposition of type II collagen.

One further complication: sclera of the common potoo, *Nictibius griseus*, and 10 species of owl contain an additional sesamoid bone that forms in cartilage, lies on the trajectory of the tendon of the pyramidal muscle and plays a role in contracting the nictitating membrane.[22]

Ossicle number

Ossicle number varies *within species*, often for reasons that are neither obvious nor intuitive.

Most species of sceloporine iguanid lizards[23] have 14 ossicles per eye. Fourteen is also the modal number in domestic fowl and the minimal number required to encircle the eye, according to de Queiroz (1982).

Scleral number is canalized at 14 in six species of birds examined (Table 21.4). In larger birds such as turkeys and crows, 95 per cent or more individuals have 14 ossicles per eye. In smaller birds such as quail, the percentage can be as low as 67 per cent, with almost 30 per cent of individuals having between 15 and 18 ossicles per eye (Table 21.4). In 44 species of pelicans, ossicle number ranges between 10 and 15, providing sufficient variation to make these ossicles a taxonomically useful character. Although most fowl have 14 scleral ossicles, the greatest range in ossicles number is seen in the domestic (and highly inbred) chicken, where as few as 11 and as many as 17 ossicles have been recorded, the tendency being toward extra rather than fewer ossicles (Table 21.5).[24]

The basis for variation in ossicle number has been investigated experimentally. Using a neat experimental technique – intubation of the eye of developing chick embryos to drain away fluid to slow growth of the eye – Coulombre and Coulombre (1975) showed that the number

Table 21.4 Percentage distribution of scleral ossicles in turkeys, chicken, quail, Cornish hens and crows[a]

Number of ossicles	Turkey	Chicken	Quail	Cornish hen	Crow
11	0	0.13	0	0	0
12	0.04	0.58	0.06	0	2.7
13	2.34	4.05	3.72	1.79	0
14	96.7	87.78	67.37	92.26	94.6
15	0.92	7.13	21.28	5.95	2.7
16	0	0.3	6.12	0	0
17	0	0.03	1.32	0	0
18	0	0	0.13	0	0

[a] Based on data from Canavese *et al.* (1986).

Table 21.5 Percentage distribution of scleral ossicles in chick embryos after 17 days of incubation[a]

Number of ossicles	Percentage
13	<1
14	80
15	19
16	<1

[a] Based on a sample of 612 embryos reported by Coulombre and Coulombre (1973).

of ossicles that form, their distribution in the sclera, and their pattern of overlap all depend on the growth rate of the optic cup and eye between four and eight days of incubation, and (in part) on the growth of the ossicles themselves between 12 and 14 days of incubation. The slower the growth of the eye, the fewer the ossicles that form – evidently a problem in the minimal spacing between the primordia of the epithelial papillae that induce ossicle formation (see below).

Scleral papillae

Scleral ossicles do not self-differentiate but are induced by specializations of the scleral epithelium known as *scleral papillae*. Studied most intensively in chick embryos, perhaps one of the first reports was from my Ph.D. supervisor, P. D. F. Murray, who described ossicle development between eight and 12 days of incubation using histology.[25]

In chick embryos, ossicles are preceded by a series of thickenings of the scleral epithelium. Each thickening becomes a papilla, the number and position of papillae corresponding exactly to the number and future positions of the ossicles. Papillae do not arise simultaneously, although all appear during the eighth day of incubation. Similarly, but on an even more exaggerated time scale, scleral ossicles arise sequentially and in a strict order that is conserved from individual to individual. One to two ossicles are present at seven days, 11 by 7.5 days and 14 (the modal number) a day later.

Ingrowth of an epithelial placode into the scleral mesenchyme is followed by withdrawal and outpushing of the epithelium, leaving what have been interpreted as epithelially derived collagen and granular material within the ectomesenchyme. Extensive cell death occurs within the epithelium, followed by disappearance of the papillae during the 12th day. Morphogenesis of the ossicles may also be influenced by cell death, which destroys cells within the osteogenic centres (Hale, 1956b). Intramembranous ossification begins in the adjacent ectomesenchyme in centres isolated from those on either side, each underlying an epithelial papilla. During days 13 and 14, adjacent centres gradually overlap, producing the imbricated bony ring (Fig. 21.1).

Correspondence between number and location of epithelial papillae and number and location of subjacent ossicles – even when ossicle number deviates from normal because of mutation or after experimental manipulation – coupled with correlation between timing of papilla degeneration and onset of osteogenesis, is consistent with the possibility of interaction between epithelial papillae and adjacent mesenchyme.

Between seven and 10 days of incubation, cell density within the mesenchymal primordium underlying each papilla increases almost twofold – from 4×10^5 to 14×10^5 cells/mm^3 – while cell density between primordia (in mesenchyme beneath non-papilla epithelium) increases to only 9×10^5 cells/mm^3 (Hale, 1956a). The implication is that proximity of mesenchyme to a papilla results in increased cellularity, either because the papilla stimulates mesenchymal cell division as Hale thought, or because papillae influence the migration of mesenchyme to a position beneath each papilla, each ossicle arising through the aggregation of mesenchymal cells toward a centre, as Putchkov (1964) thought, and as Ede demonstrated in limb development (Chapter 20).

David MacGregor Fyfe, a Nova Scotian of obvious Scottish descent, now a dentist in the Falkland Islands and whose father was the eminent anatomist Forest Fyfe, restudied scleral papilla development and analyzed how condensations arise beneath papillae using ^{3}H-thymidine autoradiography (Fyfe and Hall 1981, 1983).

An epithelial–mesenchymal interaction

Although not tested by removing papillae and examining changes in mesenchymal cell density, it looks as if the stimulus to form preosteogenic condensations comes from adjacent scleral papillae via an epithelial–mesenchymal interaction, a possibility that became more likely when administering hydrocortisone to chick embryos was found to reduce the number of scleral papillae, the number lost depending on time and number of treatments during days six to eight of incubation (Table 21.6). Failure of a papilla to form correlated with failure of the adjacent ossicle to form.

In 1962, Chris Coulombre and his colleagues provided the first direct experimental evidence for dependence of ossicle formation on interactions with the adjacent scleral papilla. Each ossicle in a chick eye is numbered using a convention based on when and where the ossicle arises. I use the same convention for the papillae adjacent to each ossicle. They removed papilla number 12 by microdissection at seven, eight or nine days of incubation, and examined the embryos later in development to determine whether ossicle number 12 had developed (Table 21.7). The results are interesting.

- If papilla #12 is removed at *seven days* of incubation, 70 per cent of the embryos lack ossicle #12. The other 30 per cent have a smaller ossicle #12 than controls.

Table 21.6 Average number of conjunctival papillae in chick embryos after injections of hydrocortisone on different days of incubation[a]

Day(s) injected with hydrocortisone	Average number of papillae
Embryos recovered at 9 days	
6	10
7	7
6, 7	1
Embryos recovered at 11 days	
6	7.5
7	9
6, 7	1
7, 8	4.5

[a] Based on data from L. G. Johnson (1973).

- If papilla #12 is removed at *nine days*, 40 per cent of the embryos lack ossicle #12. The other 60 per cent have a smaller ossicle #12 than controls.
- In contrast, if removal of papilla #12 is delayed until *12 days of incubation*, all embryos develop ossicle #12, 90 per cent of which are normal in size (Table 21.7).

Coulombre and his colleagues concluded that papilla influence development of the ossicles between seven and nine days of incubation. By CAM-grafting periocular mesenchyme free of scleral epithelium they showed that the interaction occurs during the eighth day of incubation (Table 21.8).

Hall (1981c) examined the specificity of the epithelial–mesenchymal interactions required to initiate scleral ossicles in chick embryos by combining scleral mesenchyme with mandibular epithelium and *vice versa*. Scleral mesenchyme was shown to respond to mandibular epithelium by initiating intramembranous osteogenesis in the form of bony ossicles typical of scleral ossicles. Mandibular mesenchyme responded to scleral epithelium by forming bony rods typical of mandibular membrane bones, i.e. morphogenetic specificity is a property of the mesenchyme. The same holds for scleral and Meckelian cartilages; morphogenesis is mesenchymal (Hall, 1989).

Scaleless mutant fowl

As with so many other developmental processes, mutations provide natural experiments that aid our understanding of scleral ossicle development. In this case, the mutant is *scaleless* in fowl. As the name implies, *scaleless* mutants lack the scales normally found on the legs and feet, but the mutation affects other epithelial derivatives as well. Feathers, scales, spurs, footpads and *scleral papillae* are among the epithelial structures altered or missing in *scaleless* embryos.

Interestingly for our purposes, *scaleless* inhibits the development of some but not all scleral papillae. Three papillae form but only one persists beyond the ninth day of incubation, and it is not always the same one in different individuals. Consequently, only a single condensation and a single scleral ossicle develop. The ossicle that forms is always adjacent to the one surviving papilla, although which papilla survives varies from embryo to embryo. Reciprocal grafts of periocular mesenchyme or scleral epithelium between *scaleless* and wild-type embryos establish that the genetic defect is initially epithelial, preventing the scleral epithelium from signalling.[26]

A role for tenascin?

Tenascin was introduced in Box 9.3 in relation to Egf-repeats, in Chapter 9 in relation to the composition of tendon fibrocartilage, where it plays a mechanical role, in Chapters 19 and 20 in the context of its important role with syndecan, the tenascin receptor, in condensation, and in this chapter in relation to Tgfβ and epithelial–mesenchymal interactions. Because tenascin has emerged on so many occasions, and will emerge again, an overview of tenascin and its actions is provided in Box 21.3.

Murray described fibres that appear as scleral papillae degenerate. He thought they were collagenous strands running from each papilla to each site of ossicle formation. David Fyfe and colleagues described columns of tenascin within the mesenchyme beneath each papilla but not in mesenchyme between papillae. Tenascin was described as disappearing as bone formed; tenascin was expressed as a halo surrounding each plate of osteoid. These are in all probability the fibres Murray described 45 years earlier.[27]

The distribution of tenascin in chick embryos coincides with the migration pathways of NCC, tenascin serving as a template for migration. Tamara Franz-Odendaal in my laboratory is testing whether tenascin acts as a pathway for movement of scleral mesenchymal cells (Franz-Odendaal and Hall, 2004). The hypothesis is that mesenchymal cells receive epithelial signals when close to the epithelium, as a consequence of which they migrate to the site of ossicle formation some 200–300 μm beneath the scleral epithelium. We know from transfilter tissue recombinations using Nucleopore filters of various porosities (0.03–0.8 μm) and thicknesses (5 and 10 μm) that direct cell-to-cell contact between scleral epithelium and mesenchyme is not required (Pinto and Hall, 1991). Furthermore, using dialysis membranes which excluded molecules of three size ranges – 2000–3500, 6000–8000 or 12 000–14 000 Da – we demonstrated that a diffusible

Table 21.7 Effects on formation of scleral ossicle number 12 of removing papilla number 12 from chick embryos at 7, 9, or 12 days of incubation[a]

Day of incubation papilla removed	% Ossicles missing	% Ossicles reduced in size	% Ossicles of normal size
7	70	30	0
9	40	60	0
12	0	9	91

[a] Based on data from Coulombre *et al.* (1962).

Table 21.8 Percentage of grafts of periocular mesenchyme that form scleral ossicles as a function of embryonic age[a]

Age of donor embryos		Percentage of grafts with bone[b]
Days	H.H. stage	
7	31	0
8	34	7
9	35	50
10	36	100

[a] Determination for osteogenesis occurs during the eighth day of incubation. Adapted from data in Coulombre *et al.* (1962).
[b] All grafts examined 7–10 days after initial grafting to the CAM.

Box 21.3 Tenascin

A six-armed (hexabrachion) ECM protein with 13 Egf repeats (Box 9.3), *tenascin*, binds to chondroitin-sulphate proteoglycans but not to fibronectin. The Egf repeats reflect a common structural motif in proteins of the ECM, the *repeats serving related functions in apparently unrelated proteins* (Table 21.9).[a]

Tenascin, which is found in chondrogenic condensations, perichondria, osteogenic cells and differentiated cartilage and bone, plays a key role in mesenchymal cell condensation (Chapter 19). Tenascin also is found in embryonic epithelia, including the neural epithelium as NCC emerge, and in the extracellular space and anterior sclerotome through which NCC migrate (Chapter 17). Mesenchyme cultured with tenascin produces enhanced amounts of cartilage, presumably because tenascin allows cell rounding and promotes condensation.[b]

Tenascin is inducible by *Tgfβ* and is mitogenic for several cell lines. The mitogenic action of tenascin is in part direct and mediated via the Egf repeats, and in part indirect by enhancing the mitogenic activity of Egf itself (Box 9.3, Table 21.9). Both *Fgf-2* and *Tgfβ-1* induce tenascin. *Fgf-2* is more potent, and functions either alone or with *Tgfβ* to induce the three isoforms of tenascin that arise from alternative splicing of the fibronectin type III domain.

The highest similarity (homology if you will) of the Egf repeats in tenascin is with the cluster of fibronectin type III repeats in the class III region of the human major histocompatability complex. It has been suggested that the mitogenic activity of the Egf domains in ECM proteins reflects common ancestral genes present in various classes of molecules in metazoans, including ECM proteins such as tenascin, cell-adhesion molecules (CAMs), immunoglobulins, growth factors and a variety of receptors.[c]

Ancient origins of these genes is reinforced by the study by Nicole King and her colleagues in which they demonstrated the presence of 'metazoan' cell signaling and cell adhesion molecules in a distinctly non-metazoan group, the choanoflagellates (*Monosiga brevicollis* and a *Proterospongia* sp.), which are unicellular (*Monosiga*) or form small colonies of undifferentiated cells (*Proterospongia* ATCC50818), and which possess cahderins, tyrosine kinases, typosine phosphoproteins and a G-protein-coupled receptor. The basis for cell–cell interaction was established before the origin of the metazoans.[d]

An example of an immunoglobulin is the multifunctional molecule neuroleukin (NLK), a neuronal growth factor, cytokine and lymphokine produced by lectin-stimulated T cells in association with immunoglobulin secretion. NLK levels increase 3.5-fold in MC3T3-E1 cells concomitant with deposition of ECM and matrix mineralization. High levels of neuroleukin are found in osteoblasts and superficial chondrocytes during bone development, in fibroblasts, in proliferating chondroblasts and osteoblasts during fracture repair, but not in terminally differentiated skeletogenic cells, i.e. hypertrophic chondrocytes or osteocytes (Zhi *et al.*, 2001).

The important role played by tenascin and its interactions with syndecan (the cell-surface receptor for tenascin) and with ECM components such as N-CAM in the formation of *skeletogenic condensation* is discussed in detail in Chapters 19 and 20. An overview of the nature of the evidence includes that:

- during chick development, tenascin is found in condensing mesenchyme of cartilages and membrane bones, and in perichondria;
- splice variants of tenascin mRNA and thrombospondin-2 mRNA are localized in prechondrogenic condensation in chick embryos;
- tenascin is expressed in the condensation for the dentaries of developing mice at 14 days of gestation, fibronectin in bone matrix at 15 days;
- tenascin co-localizes in human and rat prechondrogenic condensations with versican (PG-M), a hyaluronan-binding chondroitin-sulphate proteoglycan;
- versican also is expressed in condensing limb mesenchyme, in the perinotochordal sheath, and in the neuroepithelial basement membrane in chick embryos; and
- versican inhibits chondrogenesis at the post-condensation stage.[e]

In an immunolocalization of collagen and tenascin in intramembranous bone of normal and dysplastic human mandibles, Carter *et al.* (1991) localized tenascin on type II collagen fibres which they thought provide a preliminary framework for modeling and alignment of bone. High levels of tenascin were deposited around chondrocytes and fibroblasts of the tendon, and around collagen fibres within the body of the tendon.

Tenascin, along with collagen XII, is *modulated by mechanical influences* during synovial joint formation. In paralyzed chick embryos (Fig. 16.5), tenascin and collagen types I and XII decrease

Table 21.9 Binding domains and their distribution within ECM proteins[a]

Domain	Protein
Fibronectin type I (4C[b])	Fibronectin, tissue plasminogen activator, factor XIIIa
Fibronectin type II (4C)	Fibronectin, factor XIIIa, seminal plasma protein, mannose-6-phosphate receptor, gelatinase
Fibronectin type III	Fibronectin, tenascin, collagen VI, integrin β4, collagen XII and XIV, LI, contactin, fasicilin II, neurglian, N-CAM, twitchin, undulin, titin
Epidermal growth factor-like (6C)	Agrin, tenascin, thrombospondin, nidogen/entactin, cartilage matrix protein, fibulin, versican, many other proteins
Epidermal growth factor-like (8C)	Laminin, agrin, perlecan
Cysteine-rich (8C)	Integrin β4, other β subunits
Inhibitor (6C)	Collagen VI, agrin, kazal family of protease inhibitors
Von Willebrands factor type A	Collagen VI, von Willebrand factor, cartilage matrix protein, undulin
Fibrinogen β, γ	Tenascin, fibrinogen
Properdin (6C)	Thrombospondin, properdin
Sex hormone-binding globulin	Laminin A chain, sex hormone-binding globulin
Heptad repeat	Laminin, tenascin, thrombospondin, fibrinogen, many cytoskeletal and other fibrous proteins, transcription factors, Jun/Fos
GXY repeat	Collagen VI, all collagens, proteins with collagenous domains
Transmembrane	Integrin β4, integrin α and β subunits, integral membrane proteins

[a] Based on data from Engel (1991).

[b] C, number of cysteine residues in small (5–10 kDa) cysteine-rich domains.

Box 21.3 (*Continued*)

at sites where joints form in mobile embryos. Tenascin is found in high amounts at the attachment site of the quadriceps muscle to bone in rats; immobilization in a cast for three weeks leads to removal of almost all the tenascin (Chapter 9).[f]

Tenascin may play a role in *maintaining articular chondrocytes* in their specialized differentiated state and/or preventing endochondral ossification of articular cartilage. Some articular chondrocytes from pig ankle joints synthesize tenascin, while meniscal cartilage contains thick collagen fibres, elastin and fibronectin deposited by fibrochondrocytes.[g]

[a] Although tenascin does not bind to fibronectin it decreases integrin-mediated attachment of fibroblasts to fibronectin, laminin and GRDS-peptide (Chiquet-Ehrismann et al., 1988).
[b] See Mackie et al. (1987), Pearson et al. (1988), Erickson and Bourdon (1989), End et al. (1992), Riou et al. (1992) and Ayad et al. (1994) for tenascin in epithelia, and Mackie et al. (1987) for mesenchyme cultured on tenascin.
[c] See Matsumoto et al. (1992) for the nature of the Egf repeats in tenascin, and Levesque et al. (1991) for retention of ancient mitogenic genes.
[d] See King et al. (2003) for the evolutionary scenario, and see Tsukahara and Hall (1994) and Kolodziejczyk and Hall (1996) for G-protein signaling.
[e] See Gluhak et al. (1996) and Sasano et al. (2000) for tenascin in condensations, R. P. Tucker (1993) for splice variants, Perides et al. (1993), Yamagata et al. (1993) for the relation to versican, and Ayad et al. (1994) for versican. The G3 domain of versican (Table 21.9) is an equally effective

inhibitor of chondrogenesis but not if the two Egf-like repeats are deleted (Box 9.3). Conversely. the G3 domain of aggrecan – which lacks Egf-like repeats – does not inhibit chondrogenesis (Y. Zhang et al., 1998). See Chuong et al. (1993), Crossin and Krushel (2000) and Jones and Jones (2000) for reviews of the tenascin family of ECM glycoproteins, cellular signaling by N-CAM and tenascin, and the multiple roles of tenascin; Y.-S. Lee and Chuong (1992), Mackie and Tucker (1992) and Dunlop and Hall (1995) for osteoblasts; Sato et al. (1999) and Sasano et al. (2000) for expression in the dentary and condylar cartilage; Saga et al. (1992) for tenascin knockout; and Mackie and Tucker (1999) for detection of abnormalities. Mizoguchi et al. (1990, 1992) identified chondroid bone as a component of the mandibular condyle in rats. They comment on the presence of tenascin and fibronectin in mandibular condylar and intermaxillary sutural secondary cartilages, but its absence from the tibial growth plate and nasal septal, sphenoccipital and intermaxillary sutural cartilages. Vakeva et al. (1990) compared the distribution patterns of tenascin and alkaline phosphatase in rat and mouse embryos.
[f] See Mikic et al. (2000a) for immobilization, and Järvinen et al. (1999) for the immobilization study. Tucker et al. (1994) identified a novel tenascin in tendons, ligaments and mesenchyme at sites of epithelial–mesenchymal interactions in chick embryos.
[g] See Miller and McDevitt (1988) and McDevitt and Webber (1990) for porcine articular cartilage, and see Pacifici (1995) for review of tenascin in developing articular cartilage and discussion of whether tenascin prevents endochondral ossification.

signal between 3500 and 6000 Da was involved. When combined with studies using Millipore filters of 150 or 300 μm thickness, we showed that the epithelial component acted over distances of 150–300 μm (Pinto and Hall, 1991).

NOTES

1. The majority of the notochord interacts with somitic mesoderm to induce cartilage from which the vertebrae develop (see Chapter 41).
2. See Schowing (1961, 1968a–c) and Benoit and Schowing (1970) for chicks, and Petricini (1964) for anurans. de Beer (1937) presents some fragmentary evidence for other species. See Murray (1936) and the update in Hall (1985a) for mechanical factors and the initiation of intramembranous osteogenesis. In studies conducted in the early 1990s, Lengelé et al. (1990, 1996a,b) claimed to have found a cartilaginous plate underlying the cranial vault in chick embryos. I thought from their published figures that this tissue was, at least in part, secondary cartilage. They regard it as chondroid, used chimaeras between chick and Japanese quail to map secondary cartilage and chondroid, and claim that chondroid appears at nine days of incubation and woven bone at 12. From our histological studies, the first bone appears much earlier.
3. See Schowing and Robadey (1971), Robadey and Schowing (1972) and Diethelm and Schowing (1973, 1974) for transplants of quail brains to chick embryos, and Schowing (1974) for spinal cord not substituting for brain.
4. See Schmitz et al. (1990) for calvarial repair, and Thesingh et al. (1991) and Groot et al. (1994) for co-culture with brain.
5. See Bradley et al. (1997) and Yu et al. (1997) for regional properties of the dura, and Rice (1999) for a review and for basic studies of molecular mechanisms in calvarial bone and suture development, including discussion of Fgf, FgfR, *Hox*, *Msx*, *Tgfβ* and dura signals. See Coumoul and Deng (2003)

for an overview of Fgf receptors in both normal mammalian development and in congenital diseases.
6. See Seufert et al. (1994) for *Xenopus*, Yan et al. (1995) for *Danio*, and Wood et al. (1991) for mice.
7. See Hörstadius (1950) and Holtfreter (1968) on the point of permissive environments.
8. Epithelial–mesenchymal interaction has been demonstrated in trout (Benoit, 1960c), turtles (Toerien, 1965), amphibians (Hall, 1991b), birds (Reagan, 1915, 1917; Benoit, 1960a,b, 1963, 1965; Benoit and Schowing, 1970; Simons, 1974, 1979, and Milaire, 1974, for a discussion) and in mice and men (Goedbloed, 1964; Pugin, 1972; McPhee and Van de Water, 1986).
9. See Drews et al. (1972), Epperlein (1974) and Epperlein and Lehmann (1975) for the studies on *Triturus*, and see Hall (1991b) for an overview of studies involving the otic vesicle as a cartilage inducer.
10. See D'Amico-Martel et al. (1987) for type II collagen deposition, and McPhee and Van de Water (1985) for the GAG of the mouse otic capsule ECM.
11. In a later study, Zhou and Van de Water (1987) quantified the effect of target tissues on the differentiation and survival of statoacoustic ganglia maintained in organ culture. Ganglia maintained alone had a basal level of 18.5 per cent neurons. Co-culture with otic sensory epithelium or rhombencephalon increased that percentage to 97 and 87, respectively.
12. See Coffin-Collins and Hall (1989) and Solursh and Reiter (1988) for mandibular (Meckelian) and limb-bud chondrogenesis, respectively.
13. See Lewis (1907), Kaan (1930, 1938), Ichikawa (1936) and Hall (1991b) for inductive activity of the otic vesicle, Leibel (1976) for morphogenetic influences of the otic capsule, and Thorogood (1983) for a review that compares and contrasts the morphogenesis of cartilage in birds and amphibians.
14. See Helff (1928, 1931, 1934, 1937, 1940) for these studies and see Hall (1987d, 1999a, 2003b) for more detailed discussions of them.

15. See Jaskoll and Maderson (1978) for birds, and see Mallo *et al.* (2000) and Mallo (2001, 2003) for molecular signals involved in murine embryos.

16. See Weiss and Amprino (1940), Coulombre (1965) and Johnston *et al.* (1979) for the response of scleral cartilage to mechanical stress *in vivo* and *in vitro*, De Witt *et al.* (1984) for chick femoral chondrocytes, and Diao *et al.* (1989) for specificity of kinetics early in development.

17. See Stewart and McCallion (1975) for when NCC reach the optic cup, von der Mark *et al.* (1977b) and von der Mark (1980) for initiation of type II collagen synthesis, and Newsome (1972) for timing of specification for chondrogenesis.

18. See Reinbold (1968), Newsome (1972) and Stewart and McCallion (1975) for these three studies.

19. See Reinbold (1968) and Stewart and McCallion (1975) for grafts into cephalic mesenchyme, Newsome and Kenyon (1973) for the ECM deposited by PRE, Newsome (1976) for chondrogenesis in response to matricial products, and L. Smith and Thorogood (1983) for the transfilter study. Watanabe *et al.* (1992) cultured scleral fibroblasts in soft agar and obtained chondrogenic differentiation.

20. See Newsome (1972) and Smith and Thorogood (1983) for the studies with scleral mesenchyme, Weiss and Amprino (1958) for the resuspension studies, and Hall (1981c, 1989) for further studies with scleral and mandibular mesenchyme.

21. See Walls (1942) for the skeletons of vertebrate eyes, Nakamura and Yamaguchi (1991) for scleral ossicles in 21 species of teleost fishes, and see Canavese *et al.* (1986), Canavese (1987) and Warheit *et al.* (1989) birds.

22. See Bohúrquez-Mahecha *et al.* (1998).

23. Scleral ossicles are well developed in crocodiles. In a phylogenetic tree published in association with the description of *Calsoyasuchus valliceps*, an Early Jurassic crocodile, Tykoski *et al.* (2002) illustrate ossicles in the oldest known crocodile, *Euparkeria*, from the Triassic.

24. See Canavese *et al.* (1986) and Canavese (1987) for ossicle number in birds, and Warheit *et al.* (1989) for the data on pelicans.

25. For histological development of scleral papillae see Murray (1941, 1943), O'Rahilly (1962), Coulombre *et al.* (1962), van de Kamp (1968) and Ambrosi *et al.* (1975).

26. See Palmoski and Goetinck (1970) and Goetinck and Sekellick (1970) for these pioneering studies on the *scaleless* mutant, and Abbott (1975) for a discussion.

27. See Murray (1941, 1943) and Fyfe *et al.* (1988) for these fibres.

Part VIII

Similarity and Diversity

We have little systematic information on the basis for the heterogeneity of cartilage and bones as tissue, or of chondrocytes and osteocytes as differentiated cells. Does heterogeneity originate in the initial determination or setting-aside of precursor cells for chondrogenesis/osteogenesis, or is it an aspect of later stages of chondrocyte/osteocyte differentiation? As two instances:

Is hypertrophic cartilage and/or are HCCs specified as hypertrophic at the outset or is it/are they initially specified as cartilage/chondrocytes, only acquiring specificity as hypertrophic later in chondrogenesis?

Is membrane and/or are the osteocytes in membrane bone specified as intramembranous at the outset or is it/are they initially specified as bone/osteocytes, only acquiring specificity as membranous later in chondrogenesis?

Similar questions could be asked with respect to fish vs. mammalian cartilage/chondrocytes or reptilian vs. amphibian bone/osteocytes.

Chondrocyte Diversity

Chapter 22

A chondrocyte is not a chondrocyte is not a chondrocyte.

Cartilages vary in origin, biochemistry, structure, physiological and developmental roles, and in their final fates. In discussing cartilage, chondrocytes and chondrogenesis in earlier chapters, I gave some indication that not all cartilages are the same. Cartilages differ histologically – hyaline, articular and elastic (Chapter 3). Synthesis and accumulation of extracellular matrix (ECM) components differ with cartilage type. Other fundamental differences that allow us to classify cartilages into types include cartilages that:

- precede osteogenesis and have a temporary existence – *primary cartilage for endochondral bones,* and those that succeed osteogenesis and have a more permanent existence – *secondary cartilage on membrane bones*;
- undergo *chondrocyte hypertrophy,* and those that do not;
- are *avascular* – the common situation – or highly vascular, as in antlerogenesis;
- are based on *type II collagen* (vertebrates) and those that are not (invertebrates, lampreys, hagfishes); and
- *grade into* connective tissue, chondroid, or chondroid bone (Chapter 5).

We have little systematic information on the basis for this heterogeneity of cartilage as a tissue, or of chondrocytes as differentiated cells. Does heterogeneity originate in the initial determination or setting-aside of precursor cells for chondrogenesis, or is it part of later stages of chondrocyte differentiation? We do not know whether cell lines become determined initially to form 'cartilage' or to form particular types of cartilage, although the information available points towards some specification of heterogeneity early in development. I present a number of approaches and types of evidence to illustrate the diversity of cartilage and chondrocytes. The approaches I have selected are: segregation from precursors; the nature of the perichondrium; morphogenetic specificity; embryological origin; vascularity; hypertrophy; and ability to induce osteogenesis.

SEGREGATION FROM PRECURSORS

One approach to identify distinct phases of chondrogenesis that may be regulated independently is to investigate the ability of chondrogenic cells at different stages of their differentiation to segregate from one another or from other cell types. Table 22.1 contains a summary of experiments from the laboratories of Ed Zwilling, Robert Searls and Donald Ede, addressed when (and how?) myogenic and chondrogenic cells acquire the ability to segregate, segregation representing the stage in differentiation when the two cell types no longer recognize one another as similar.

In Searls' 1971 study, embryos of one developmental age were labeled *in ovo* with ³H-thymidine, mesenchymal cells from their limb buds were mixed in various proportions (4:1 to 1:4) with unlabeled cells from embryos of a different developmental stage, and the cell mixtures were cultured on nutrient agar or grafted to the CAMs of host embryos. Terminally differentiated chondrocytes would be expected to segregate from terminally differentiated myoblasts, and indeed they do (Table 22.1).

Mesenchyme from 'young' limb buds (H.H. 20–22) segregates from differentiated myoblasts but not from chondrocytes until the latter attain later stages of differentiation (Table 22.1), the interpretation being that future chondroblasts segregate from other cell types earlier than they segregate from later stages of their own cell type, in this case, chondrocytes. Consequently, younger limb mesenchyme segregates from dermomyotome from embryos of the same stage, while even younger presumptive limb mesoderm segregates from flank mesoderm from embryos of the same age (Table 22.1). The conclusion: mesenchymal cells of early limb buds recognize cartilage but not muscle as a similar cell type.

Searls (1971, 1972) investigated further whether terminally differentiated chondrocytes segregate from their precursors, found no segregation until after H.H. 27

Table 22.1 Summary of the results of studies on the segregation of chondrogenic or prechondrogenic cells from myogenic or premyogenic cells[a]

Source of chondrogenic cells	Source of myogenic cells	Segregation	Reference
H.H. 31 limb chondrocytes	H.H. 38 limb myoblasts	+	Searls (1971)
H.H. 31 limb chondrocytes	H.H. 25 ventricular myoblasts	+	Searls (1971)
H.H. 27 limb chondrocytes	H.H. 24 limb myoblasts	+	Zwilling (1968)
H.H. 20–22 limb mesenchyme	H.H. 38 limb myoblasts	+	Searls (1971)
H.H. 20–22 limb mesenchyme	H.H. 25 ventricular myoblasts	+	Searls (1971)
H.H. 18 or 19 limb mesenchyme	H.H. 18 or 19 dermamyotome	+	Zwilling (1968)
H.H. 14, 15, or 18 limb mesenchyme	H.H. 14, 15, or 18 flank mesenchyme	+	Zwilling (1968)
H.H. 26 chondrogenic limb mesenchyme	H.H. 26 myogenic limb mesenchyme	−	Ede and Flint (1972)[b]

[a] Results are expressed as +, segregation or −, lack of segregation.
[b] These cultures were in liquid medium under conditions or agitation. All others were on solid media or as grafts to the CAM.

Table 22.2 Summary of the results of studies on the segregation of chondrocytes from their precursor cells within the limb bud[a]

Age and source of cells		Segregation	Reference
H.H. 20 or 21 mesenchyme	with H.H. 24 mesenchyme	−	Searls (1972)
H.H. 20–22 mesenchyme	with H.H. 25 mesenchyme	−	Searls (1972)
H.H. 20 mesenchyme	with H.H. 26 mesenchyme	−	Searls (1972)
H.H. 20–22 mesenchyme	with H.H. 27 mesenchyme	−	Searls (1972)
H.H. 24 mesenchyme	with H.H. 27 mesenchyme	+	Zwilling (1972)
H.H. 20–22, 24, or 25 mesenchyme	with H.H. 31 chondrocytes	+	Searls (1971, 1972)
H.H. 26 mesenchyme	with H.H. 31 chondrocytes	+[b] −[c]	Searls (1972)

[a] Results are expressed as +, segregation or −, lack of segregation.
[b] Cultured on an agar substrate.
[c] Grafted to the CAM.

(Table 22.2) and concluded that the ability of chondrocytes to segregate from their precursors was associated with late events in differentiation, later than stabilization of prechondrogenic cells for chondrogenesis – which is H.H. 24/25 – and after initiation of deposition of ECM, which begins at H.H. 27.

Cell-surface and cell-adhesion molecules are obvious candidates to mediate segregation. Some are discussed in the context of the condensation of prechondrogenic cells in Chapters 19 and 20. Another candidate is Ephrin-A2, a ligand that is anchored to glycosylphosphatidylinositol (GPI) cell-surface proteins. Ephrin-A2 is uniformly distributed through early limb-bud mesenchyme in chick embryos, then within proximal and mid (but not distal) mesenchyme, the expression pattern being regulated by *Fgf-8*. Over-expressing *ephrin-A2* initiates patterning malformation of the autopod *in ovo*, disrupts nodule formation *in vitro*, but does not inhibit chondrogenesis; all results consist with a role in regulating position-specific cell affinity of prechondrogenic cells (Wada *et al.*, 2003).

PERICHONDRIA

Somewhat surprisingly, little attention has been directed to detailed descriptions of perichondria or perichondrial formation. We know rather more about periostea, perhaps because the bi-layered periosteum – with its outer fibrous non-osteogenic fibroblastic covering and inner cellular (cambial) osteogenic layer – is intrinsically more interesting and, being permanent, of more importance. One aspect of the perichondrium – its transformation to a periosteum under the influence of HCCs – has intrigued me for decades. Of course, chondrogenesis need not always be associated with or require a perichondrium:

- Chondrogenesis begins from a condensation *in vivo* before a perichondrium is present.
- Chondrogenesis *in vitro* virtually always occurs in the absence of a perichondrium.
- Ectopic cartilage induced in soft tissue rarely develops a perichondrium.
- Cartilage that forms during regeneration of holes in rabbit ears arises in part from a blastema and in part from the perichondrium of the auricular cartilage at the edges of the hole.
- During regeneration of the caudal fin of the glass knifefish, *Eigenmannia virescens*, cartilage develops at the end of the notochord and is replaced by perichondrial ossification without any involvement of a perichondrium, the cartilage arising from the notochord.[1]

The impression one receives is that formation of a perichondrium and chondrogenesis are two quite distinct processes.

In the absence of the perichondrium, chick tibiotarsi cultured *in vitro* show enhanced development of the hypertrophic zone as cells exit from the cell cycle and differentiate (Long and Linsenmayer, 1998). Tgfβ-1 in perichondria decreases proliferation and chondrocyte hypertrophy in endochondral ossification, providing a feedback system (Fig. 15.2).

Chondrocyte maturation and endochondral ossification of chick long bones require retinoid signaling. Chondrocytes but especially perichondrial cells at H.H. 27 express diffuse RARα and RARγ.

In an analysis of the ultrastructure of perichondria of embryonic chick metatarsals, tibiotarsals and sterna, Bairati and colleagues identified a temporary *syncytial* stage when cells are prechondroblastic and mitotically active. Fgf-5 may maintain the fibrous perichondrium; ectopic expression of human *FGF-5* in chick hind limbs stimulates development of the fibrous perichondrium through enhanced proliferation of a subpopulation of connective tissue cells. Fgf-5 plays a similar role in enhancing proliferation of the fibroblasts that produce the connective tissue sheaths of the muscles.[2]

Interestingly, two types of cartilage arise after perichondria are grafted into Sprague-Dawley rats. One has its origin in pericytes, the other in fibroblast-like cells in the connective tissue at the graft site (Diaz-Flores *et al.*, 1991).

Another way to understand how a perichondrium forms is in situations where a perichondrium reforms after the perichondrium has been removed. Kavumpurath and Hall (1989) found that Meckel's cartilage can reform a perichondrium from superficial chondrocytes after the perichondrium has been removed.[3]

MORPHOGENETIC SPECIFICITY

Interesting patterning differences appear when different chondrogenic cells are cultured.

Mesenchymal cells from limb buds and from the periocular (scleral) region of early chick embryos were cultured before differentiation of chondroblasts had begun. The patterns of reaggregation of the differentiated chondrocytes *in vitro* were distinctive, specific to the origin of the mesenchyme, and to the cartilages that arose from that mesenchyme. Mesenchyme from limb buds formed nodules, scleral mesenchyme forms sheets of cartilage.[4] Further studies confirmed that scleral chondrocytes form sheets while limb chondrocytes form nodules (Fig. 22.1), *unless* interaction between chondroblasts is low, either because of culture at low cell density or because of co-culture with non-chondrogenic cells.

Altering cell shape has a direct effect on glycosaminoglycan (GAG) synthesis. Rabbit costal chondrocytes transform from polygonal to rounded and increase GAG synthesis in the presence of cytochalasin, but flatten in the presence of colchicine. Cytochalasin-induced change in cell shape of condylar articular cartilage in long-term organ culture is associated with a change in the ratio of types I and II collagen produced (Pirttiniemi and Kantomaa, 1998).

Stage of differentiation of chondrogenic mesenchyme also affects the type of morphogenesis displayed: H.H. 23/24 limb mesenchyme in micromass culture forms nodules interspersed with connective tissue, while H.H. 25 mesenchyme forms sheets of cartilage without connective tissue. Even *prechondrogenic cells* can sort out. Thus, in chimaeric aggregates of distal and proximal embryonic chick limb mesenchyme, or of distal mesenchyme with tendon fibroblasts in micromass culture, the cells sort out to form aggregates according to cell type. A threshold aggregate size (condensation; Figs 13.8 and 19.4) is required for chondrogenesis to begin, as is interaction between similar cell types. Quail limb-bud mesenchyme cultured as monolayers, or on a reciprocal shaking plate to minimize cell-to-cell interactions, differentiates into fibroblasts; the same mesenchyme differentiates as cartilage when cultured as a pellet, or on a gyratory shaker to

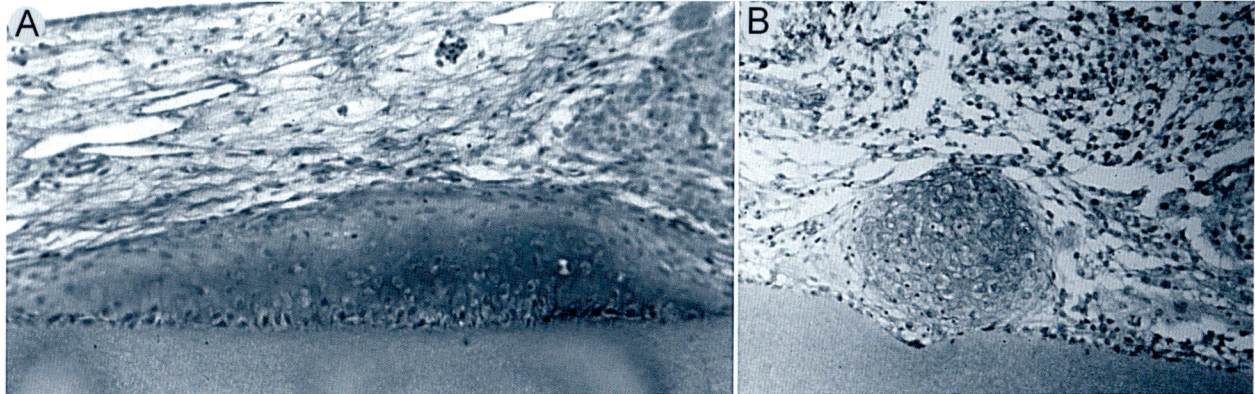

Figure 22.1 Morphogenesis of neural crest-derived cartilage is independent of the type of epithelium used to initiate chondrogenesis. (A) Nodular scleral cartilage formed when embryonic chick scleral mesenchyme was combined with maxillary arch epithelium and CAM-grafted. Absence of a perichondrium means that the chondrocytes have spread along the Millipore filter substrate (seen at the bottom). (B) A transverse section through a rod of cartilage formed when mandibular mesenchyme (future Meckel's cartilage) was combined with maxillary epithelium and CAM-grafted. Note that the rod of cartilage is surrounded by a perichondrium and so lies apart from the Millipore filter substrate.

maximize cell-to-cell interactions. The same is true for bone. Rat bone-marrow stromal cells produce fewer mineralized bone nodules when skin or periodontal ligament fibroblasts are mixed with the cultures. A <1 kDa fraction of fibroblast-conditioned medium has a similar effect on an inhibition that is blocked by indomethacin, implicating a soluble factor such as a prostaglandin.[5]

A further example of cartilage diversity, discussed in more detail at the end of this chapter, is revealed in *Brachypod* (*bp*^H*) mice, in which the fibulae fail to undergo the chondrocyte hypertrophy typical of fibulae from wild-type littermates. This defect can be traced back to condensation of limb mesenchyme at 10 days of gestation (Theiler stage 16) when *bp*^H* cells cannot provide the interactions required to initiate chondrogenesis. By Theiler stage 20 (12 days of gestation) maturation of *Brachypod* limb mesenchyme has accelerated; distal cells are spontaneously chondrogenic, while proximal cells require stimulation from cAMP to differentiate, indicative of some catch-up to wild type. These patterns of morphogenesis, and the ability to undergo hypertrophy, are specified in the *precursors* of the cells that exhibit the behaviours.[6]

I presented evidence above that chondrocytes acquire the ability to segregate from their own precursors as they differentiate. Similar studies show that chondrocytes from one portion of the skeleton – such as a limb – recognize and reaggregate with chondrocytes from another portion of the skeleton – e.g. a vertebra – but that *precursors* of the two sets of chondrocytes segregate from one another (Table 22.3). Heterogeneity, which is present early in the life of the cell lines, is lost with differentiation, perhaps concomitant with deposition of the ECM, which isolates cells from one another, or reflecting cell membrane changes.

It turns out that, *provided that chondrocytes have the same embryological origin*, they do not segregate from

chondrocytes from other sites or species (Table 22.3).[7] The story is different when embryonic origins differ. By embryonic origins I mean neural crest or mesoderm, a topic that is discussed more fully in Part VI.

CARTILAGES OF DIFFERENT EMBRYOLOGICAL ORIGINS

Cartilages show distinctive morphogenetic but not metabolic or differentiative properties that reflect their embryological origins in neural crest or mesoderm. The differentiative behaviour *in vitro* of chondrocytes from Meckel's cartilage – which are of neural-crest origin, and tibial or femoral chondrocytes – which are mesodermal in origin, provides a nice example.[8]

One technique used to effect in several studies involves fusion or lack of fusion between cartilaginous rods of neural-crest or mesodermal origin. Chiakulas (1957) excised cartilage from larval spotted salamanders, *Ambystoma maculatum*, and grafted it into the tail fin in contact with a second cartilaginous rod. The two cartilages fused when both partners had the same embryological origin; femur against femur, humerus against humerus, Meckel's with Meckel's. Fusion was not seen when the embryological origins differed, for example, femur against Meckel's cartilage. Chiakulas took this as evidence of heterogeneity based on embryological origin, but heterogeneity based on ability to hypertrophy or undergo endochondral ossification should not be ruled out.

We obtained similar evidence when neural-crest and mesodermally derived cartilages were cultured in association. Not only did avian Meckel's cartilage not fuse when cultured in contact with tibial cartilage; the two cartilages were segregated from one another by connective tissue. Autoradiographic analysis in which one cartilage was prelabeled with ³H-thymidine established the demarcation between adjacent cartilaginous rods with some precision. Similarly, when trunk myotomes are grafted into the cranial regions of chick embryos, ectopic cartilage develops but fails to fuse with the neural crest-derived cartilage in the head.[9]

In a different approach, humeral or Meckelian cartilages were transplanted into the forearms of *A. maculatum* and limb and graft amputated, and regeneration of host limbs was examined. Some fusion of Meckel's with the host humerus was observed. Such *in-vivo* studies, of course, are complicated by the presence of other host cells, differential vascularity, graft survival, and so forth. Cultivation *in vitro*, as in our study, overcomes such problems.

The Meckelian symphysis in hamster lower jaws consists of a fibrocartilage that forms by fusion of Meckel's cartilage with two neural crest-derived secondary cartilages that arise within the joint. This symphysis would provide a good *in-vivo* model for testing cartilage fusion or lack of fusion after transplanting non-neural-crest-derived cartilage into the joint.[10]

Table 22.3 Summary of the results of studies on the segregation of avian limb-bud chondrocytes or their precursors from avian vertebral chondrocytes or their precursors, or from mouse limb-bud mesenchyme

Age of limb-bud cells	Age of avian vertebral or mouse limb-bud cells	Segregation[a]	Reference
H.H. 34 chondrocytes	H.H. 34 vertebral chondrocytes	–	Zwilling (1968)
H.H. 22 mesenchyme	14-day mouse mesenchyme	–	Levak-Svajger and Moscona (1964)
H.H. 22 mesenchyme	12-day mouse mesenchyme	–	Moscona (1957)
H.H. 18 mesenchyme	H.H. 18 sclerotomal mesenchyme	+	Zwilling (1968)
H.H. 18 mesenchyme	H.H. 13 somatic mesenchyme	+	Zwilling (1968)

[a] Results are expressed as +, segregation or –, lack of segregation.

Silberzahn (1968) cultured Meckelian and femoral cartilages from chick and mouse embryos in various combinations for 14 days. Incidences of fusion were: femur against femur 77 per cent, Meckel's against Meckel's 56 per cent, and Meckel's against femur 37 per cent. Fusion involved elaboration of ECM, proliferation, and chondrification of perichondrial cells at the contact points. In many instances, a common perichondrium formed. While fusion of homologous cartilages of either embryological origin began by the third day, fusion of heterologous cartilages was delayed and did not commence until the sixth day. 'Chondrification of perichondrial cells at the contact points' or formation of 'a common perichondrium,' raises the two issues of just what the perichondrium is and how it forms, a topic that is taken up in the next section.

A common cartilage forms – with intermingling of chondrocytes from both sources – when Meckel's and femur are dissociated into single cell suspensions, reaggregated and cultured. Segregation appears to be a property of the ECM. Any conflict between these studies and those of Chiakulas and Foret could reflect differences in class, species, age or experimental approach (culture/graft). Such 'traditional' ways of explaining differences between different studies merely indicates that we don't understand the basis for segregation of cells of the 'same' phenotypes but with differing developmental histories.

CHONDROCYTE HYPERTROPHY

Hypertrophy is the normal fate of those temporary chondrocytes that are removed and replaced during endochondral ossification. Whether chondrocytes hypertrophy, synthesize type X collagen or mineralize their ECM sets such chondrocytes and cartilages apart as a distinctive group (Box 4.3).

Non-hypertrophic chondrocytes (HCCs) within an unmineralized matrix resist vascular invasion (Chapter 23); the ECM of HCCs is readily invaded and mineralized. Why the difference?

Attainment of chondrocyte hypertrophy is usually accompanied by mineralization of the ECM. Ability to resist vascular invasion is lost coincident with mineralization. Breakdown of cartilaginous ECM begins with vascular invasion. Mineralization without hypertrophy is rare, but can occur if ingrowth of blood vessels is initiated experimentally. The ability of vascular buds to penetrate the perichondrium may be related to chondrocyte death, an event that accompanies mineralization of the ECM in some but not all situations.[11]

Synthesis of type X collagen is intimately connected with vascular invasion. This is so both in cartilages of long bones after chondrocytes hypertrophy and in the (cephalic) portion of the avian sternum where chondrocytes hypertrophy and are replaced by bone. Regulation of Type X synthesis in this portion of the chick sternum is

transcriptional, mRNA only being found in the cephalic half. *In ovo*, sternal chondrocytes undergo apoptosis after terminally differentiating and before resorption and vascular invasion associated with endochondral ossification.[12]

TYPE X COLLAGEN

The role of type X collagen appears in many places through the text; type X is regarded as synonymous with chondrocyte hypertrophy and is an important regulator of whether cartilage can be invaded by blood vessels. But synthesis of type X collagen and hypertrophy do not always go hand in hand (Fig. 22.2). Consequently, and to provide a more inclusive look at the role(s) of type X collagen, I have gathered a variety of types of evidence into this section. Inevitably, some elements of the discussion occur elsewhere.

Discovery and regulation of synthesis

In a study published in 1976, Benya and colleagues used cyanogen bromide digestion to identify a previously unknown collagen type produced by rabbit articular chondrocytes in monolayer culture. By 1979, Burgeson and Hollister had identified six types of collagen in human hyaline cartilage, bringing to nine the number of structural genes for collagen molecules. By the mid-1980s, synthesis of type X was known to be associated with chondrocyte hypertrophy *in vivo* and *in vitro*.[13]

As visualized in proximal tibial growth plates from seven-day-old chickens, type X collagen is associated with type II collagen fibres but is neither found in the mineralization centre nor in matrix vesicles, the implication being that type X does not provide a nucleation site for matrix mineralization.

Synthesis of type X is most prominently associated with *vascular invasion* of the hypertrophic cartilage of long bones and with the cephalic but not the caudal portion of the avian sternum, the cephalic being the only portion that hypertrophies and is replaced by bone. Regulation of synthesis of type X in the cephalic portion is transcriptional; mRNA is found only in the cephalic half. Type X protein is present in the caudal half of the sternum, which can be induced to hypertrophy if cultured in collagen or agarose. Sternal chondrocytes show reduced hypertrophy and deposit type X collagen under conditions of calcium deficiency, deposition.

Van de Wijngaert *et al.* (1988) used periosteal-free, live or dead metatarsals from 17-day-old murine embryos to show that live but not dead cartilage recruited osteoblasts and osteoclasts, indicating an active role for (hypertrophic) chondrocytes in resorption and initial endochondral ossification.

Just as hypertrophy can be inhibited, so hypertrophy can be induced *in vitro*: foetal bovine serum, Igf-1, Fgf-2 or Pdgf induces the synthesis of type X collagen by, and

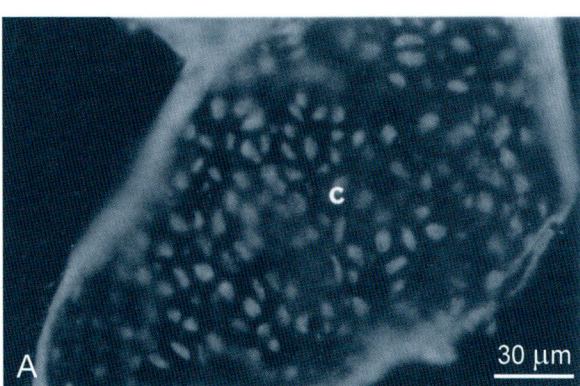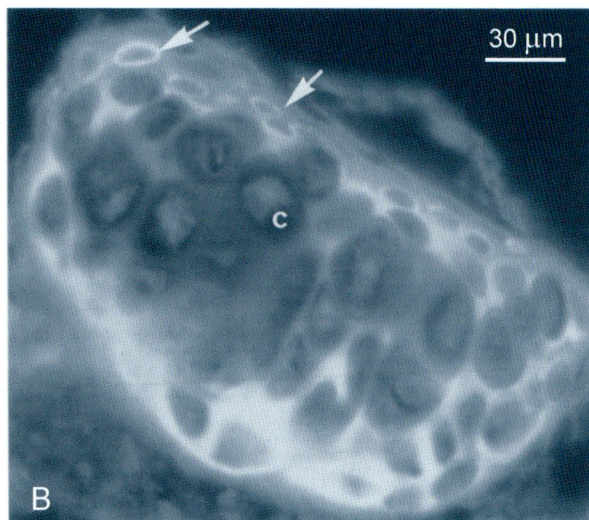

Figure 22.2 Chondrogenesis, synthesis of type X collagen and chondrocyte hypertrophy in micromass culture. Mandibular mesenchyme removed from a chick embryo of H.H. stage 21 (3.5 days of incubation), dissociated and established in micromass culture (10 μl drops of 2×10^7 cells/ml) undergoes chondrogenesis with synthesis of type X collagen in a culture medium of Ham's F12:BGJ$_b$ [3:1] + 10% foetal calf serum + ascorbic acid (150 μg/ml). Type X collagen visualized with fluorescent label. (A) An eight-day culture with chondrogenesis (c) but no type X collagen. (B) After 10 days *in vitro* type X collagen is seen in cells at the periphery of the culture (arrows), irrespective of cell size. Modified from Ekanayake and Hall (1994b).

hypertrophy of, chick sternal chondrocytes. Constitutive expression of the proto-oncogene *v-myc in quail* tibial chondroblasts impairs hypertrophy and mineralization; the cells fail to synthesize type X collagen, have low levels of alkaline phosphatase, initiate cell division and deposit type II collagen.[14]

In studies on collagen types IX and X in developing chick tibiotarsi, Linsenmeyer *et al.* (1991, found that the onset of mRNA for type X collagen correlates with the appearance of the first protein – regulation is not transcriptional – but that type IX collagen is found in the ECM of HCCs *without* concomitant expression of mRNA in the oldest hypertrophic cells. The highest levels of mRNA for type X are found in HCCs, suggesting that these cells are maintained and not undergoing apoptosis.

As determined by transfecting the chick collagen X promoter into chondrogenic cells, multiple negative elements in the gene inhibit type X gene expression in non-HCCs. In contrast, type X can be activated in normally non-chondrogenic cells; chondrocytes – identified by expression of collagen IIB and type X – can be generated from calvarial cell populations from 12-day-old mouse embryos (when preosteoblasts are at the condensation stage), but not from 17-day-old embryos when the cells are osteoblastic.[15]

It has been known for some time that *chondrocyte shape* modulates phenotypic expression; e.g. poly 2-hydroxyethyl-methacrylate (poly Hema), which modifies the shape and substrate adhesion of human chondrocytes, also modifies proliferation and matrix synthesis. Similarly, synthesis of type X collagen by HCCs is modulated by cell shape: type X mRNA increases sixfold in monolayer compared with suspension culture, conditions that are conducive to different cell shapes.[16]

Epiphycan (proteoglycan-LB, PG-Lb), a little understood proteoglycan, named epiphycan because of its predominant expression in epiphyseal cartilage, appears after the initial expression of type II collagen. Present in the hypertrophic cartilage of the growth plate, epiphycan is excluded from areas that contain type X collagen, unless type X is inactivated, in which case epiphycan is expressed throughout the entire growth plate. Epiphycan is then a potentially useful marker for intermediate stages of chondrogenesis (J. Johnson *et al.*, 1999).

Retinoic acid acts as a transient up-regulator of type X collagen in cultured chick vertebral chondrocytes, mRNAs for type X increasing threefold, for fibronectin fivefold, and for type I several fold. mRNAs for proteoglycan core protein and for type II collagen, on the other hand, *decrease* several fold.[17]

Retrovirus-mediated over-expression of *Shh* in chick limb-bud mesenchymal micromass cultures results in the production of novel tight nodules of cells that are positive for alkaline phosphatase and type X collagen and deposit some ECM, implicating *Shh* in chondrocyte hypertrophy and/or type X collagen synthesis (Stott and Chuong, 1997).

Syndromes and mutations

Several syndromes and/or mutations illuminate the role of type X collagen.

Growth-plate compression, decreased deposition of trabecular bone and altered haematopoiesis (*haematopoietic aplasia*) all are seen in collagen X null mice, leading to what Gress and Jacenko (2000) called a *variable skeleto-haematopoietic phenotype*. This study indicates, perhaps for the first time, that hypertrophic cartilage and endochondral ossification contribute to providing the proper marrow microenvironment for blood cells to form. In a subsequent study, Calvi and colleagues (2003) used murine osteoblast-specific activated Pth/PTHrP receptors to show that osteoblasts regulate haematopoietic stem cells through activation of the Notch ligand jagged 1.

Spondylometaphyseal chondrodysplasia (*smc*) in mice is an unusual syndrome in which cartilage columns fail to form and development of secondary centres of ossification is delayed. The primary defect is a mutation in the collagen type X gene, resulting in compressed hypertrophic cells and growth plates, and reduced ossification.[18]

The *toothless* (*tl*) mutation in rats is associated with progressive development of chondrodystrophy between three and six weeks after birth. Collagen type X is not expressed in the few HCCs that develop two weeks after birth, implicating the *tl* locus in regulating collagen X expression and/or function (Marks *et al.*, 2000). The chondrodystrophy that develops is characterized by disorganized chondrocytes in growth plates, failure of cartilage mineralization, and failure of cartilaginous extension into the metaphysis. Mineralization of the bone that forms appears normal.

Runx-2 (*Cbfa-1*)-deficient mice lack all membrane and endochondral bones. Cartilage develops, but type X collagen is not expressed in the HCCs (nor are PTH, PTHrP, Ihh or Bmp-6) indicating a block of chondrogenesis at the pre-hypertrophic stage.[19]

Chick tibial dyschondroplasia is an unusual limb mutation. Tibiae of mutant chicks have *two growth plates*, with avascular unmineralized cartilage in the proximal metaphysis of the tibiotarsus. Type X collagen is expressed in the cartilage, but there is no mRNA for type II or type X, indicating a late disruption in chondrogenesis (Chen *et al.*, 1993).

Type X does not always indicate hypertrophy

Expression of type X collagen is generally considered synonymous with chondrocyte hypertrophy and *vice versa.* Although we are aware that notochord also synthesizes and deposits type X – only when notochordal cells are undergoing hypertrophy (Chapter 42) – in studies where notochord is not present and so could not be confused with cartilage, type X is regarded as a marker for chondrocyte hypertrophy. Similarly, the knowledge that type II collagen is expressed in epithelial basement membranes is not a barrier to using type II to identify cartilage, provided that no epithelia are present.

However, as intimated above, the association between type X collagen and hypertrophy is not always straightforward.

- Adult chick tibial articular chondrocytes and sternal chondrocytes *in vitro* synthesize type X collagen independently of the size of the chondrocytes, implying that hypertrophy and type X synthesis are not co-regulated.

- Synthesis of type X mRNA by embryonic chick sternal chondrocytes exposed to 100 or 10–35 nM retinoic acid is independent of cell size.
- In human foetal cartilages, type X is expressed in a narrow band within the hypertrophic cartilage, in mineralized cartilage and in remnants of spongy bone. Hypertrophy precedes type X expression, which in turn precedes mineralization.[20]

The complexity of the relationship(s) between type X and chondrocyte maturation was revealed using a chondrogenic cell line (N1511) derived from rib cartilage of a p53 null mouse and exposing the cells to two different culture media (Kamiya *et al.*, 2002):[21]

(i) Bmp-2 (50 ng/ml) and bovine insulin (1×10^{-6} M);
(ii) PTH (1×10^{-7} M) and dexamethasone (1×10^{-6} M).

The cells proliferated and differentiated in both media. Once confluence was attained, however, differentiation diverged:

- No cartilage nodules developed in medium (i). The differentiated cells were positive for alizarin red and alkaline phosphatase, indicative of an *osteogenic phenotype*. However, collagen types II, IX and X were deposited in a temporal sequence – first type II, then type IX, finally type X – *more indicative of HCCs* than of osteoblasts.
- Cells in medium (ii) formed alcian blue-positive nodules that were negative for alizarin red and alkaline phosphatase, indicative of a *chondrogenic phenotype*. Collagen type II appeared slowly followed by type IX. Type X collagen was not seen, indicative of a *chondrogenic pathway limited to pre-HCCs*.

We obtained chondrogenesis in clonal cultures of mandibular mesenchyme derived from H.H. 17 (60 hours' incubation) chick embryos (Fig. 22.3). Hypertrophy was not required for type X collagen to be expressed (Fig. 22.2).[22]

I continue this discussion of type X collagen and hypertrophy by considering some of the situations under which chondrocyte hypertrophy has been elicited to see if we can discover the *range of controls regulating hypertrophy.*

Regulation of chondrocyte hypertrophy

We have a reasonable idea of the factors required to initiate, promote and enhance chondrocyte hypertrophy. As most of this information comes from work *in vitro*, we have to be a little careful when extrapolating back to organisms. That said, we can generalize and say that *thyroxine promotes the hypertrophy of mammalian chondrocytes* although the response to thyroxine varies with cartilage type; rat condylar and epiphyseal cartilages respond differently while, in our hands, thyroxine failed to promote hypertrophy of avian Meckelian chondrocytes.

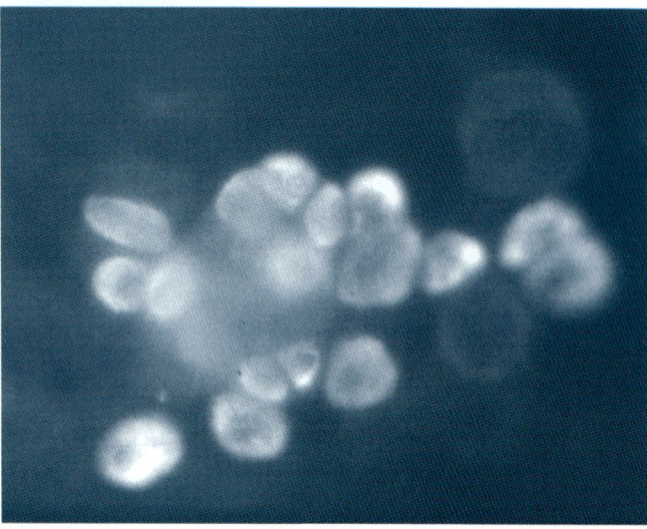

Figure 22.3 A cluster of chondrocytes that differentiated from a single mandibular mesenchymal cell, obtained from the mandibular arch of a chick embryo (H.H. 17), established in clonal cell culture and labeled with an antibody to type II collagen.

Table 22.4 Effect of ascorbic acid ± β-glycerophosphate on mineralization and collagen synthesis of vertebral chondrocytes from 12-day-old chick embryos maintained *in vitro*[a]

	Control	Ascorbic acid	Ascorbic acid + βGPO$_4$[b]	NEM[c] + ascorbic acid	NEM + ascorbic acid + βGPO$_4$
Mineralization	−	+	++	+++	++++
% Collagen synthesis	34 ± 8	48 ± 12	53 ± 14	58 ± 14	62 ± 15

[a] Based on data from Gerstenfeld and Landis (1991).
[b] βGPO$_4$, β-glycerophosphate.
[c] NEM, Nutrient-enriched medium.

Full expression of chondrocyte hypertrophy, including deposition and mineralization of ECM, can be elicited from bovine foetal growth-plate chondrocytes cultured in serum-free conditions in the presence of thyroid hormone (T3 or T4), the most immature cells in the growth plate being targets of these two hormones.[23]

If exposed to ascorbic acid or to combinations of ascorbic acid and 20 mM sodium-β-glycerophosphate, vertebral chondrocytes from 12-day-old chick embryos hypertrophy in culture. In the transformation, collagen synthesis is effectively replaced by proteoglycan synthesis (Table 22.4).

Hypertrophy of some cells and clonal lines derived from them occurs in serum-free medium, supplemented with thyroxine (T3) dexamethasone and insulin. Some clones deposit type X and type II collagens, others only type X. Ascorbic acid is required – as a co-factor for general collagen synthesis or with a more specific role? – for HCCs to complete their maturation.[24]

Igf-1, the main serum factor regulating chondroblast proliferation, binds to proliferating rat rib growth-plate chondrocytes at twice the level that it binds to resting chondroblasts or to hypertrophic chondroblasts (Makower *et al.*, 1989a,b).

Tgfβ

Gelb *et al.* (1990) measured production of Tgfβ-1 by chick growth-plate chondrocytes maintained in short-term monolayer culture. HCCs produced 4.5 ng/10^6 cells, small chondroblasts 2.3 ng/10^6 cells. Exposure to Fgf-2 led to a sixfold increase in Tgfβ-1 production.

Tgfβ regulates chondrocyte growth, in part, through inhibition of chondrocyte hypertrophy. Several types of evidence support this connection.

Despite our understanding of collagen type X and signaling between PTHrP and Ihh (Chapter 23, Fig. 15.2, Box 31.2), chondrocyte hypertrophy remains incompletely explained. Yang *et al.* (2001) showed that the *Tgfβ/Smad-3* pathway was down-regulated with hypertrophy in mice. Disruption of exon 8 was used to create a *Smad-3* mutant (*Smad-3*$^{ex8/ex8}$), expression of which was followed by development of a degenerative joint disease phenotypically similar to osteoarthritis in humans. The features included loss of articular cartilage, development of large osteophytes, decreased levels of proteoglycan and increase in type X collagen in chondroblasts at the synovial surface. Interestingly, epiphyseal growth-plate chondrocytes exhibited enhanced terminal differentiation. In follow-up *in-vitro* studies, Tgfβ-1 was shown to decrease chondrocyte differentiation in metatarsals from wild-type embryos and to diminish chondrocyte differentiation in metatarsals from mutant embryos. Yang and colleagues concluded that the *Tgfβ/Smad-3* pathway regulates articular cartilage formation.

Bmps

Bmp-7 (osteogenic protein-1) mRNA is expressed in human and murine hypertrophic cartilage (and in perichondria and periostea) and in the growth-plate cartilage and bone of chick embryos, especially in the HCCs near metaphyseal blood vessels. Introduction of dominant or negative receptors of Bmps to sternal chondrocytes shows how these growth factors control proliferation and hypertrophy.[25]

Bmp-6 (*Vgr-1*) is expressed in murine HCCs. Chinese hamster ovarian (CHO) cells that express Vgr-1 induce tumours that contain cartilage and bone and that undergo endochondral ossification. *Bmp-6* also is found in chick sternal chondrocytes as they differentiate. Growth factor expression precedes expression of type X collagen. Both sternal chondrocyte differentiation and synthesis of type X collagen are increased following treatment with 10^{-7} M PTHrP. As *Bmp-6* enhances Ihh mRNA, it has been proposed that *Bmp-6* has a direct action on PTHrP while Ihh has an indirect action.[26]

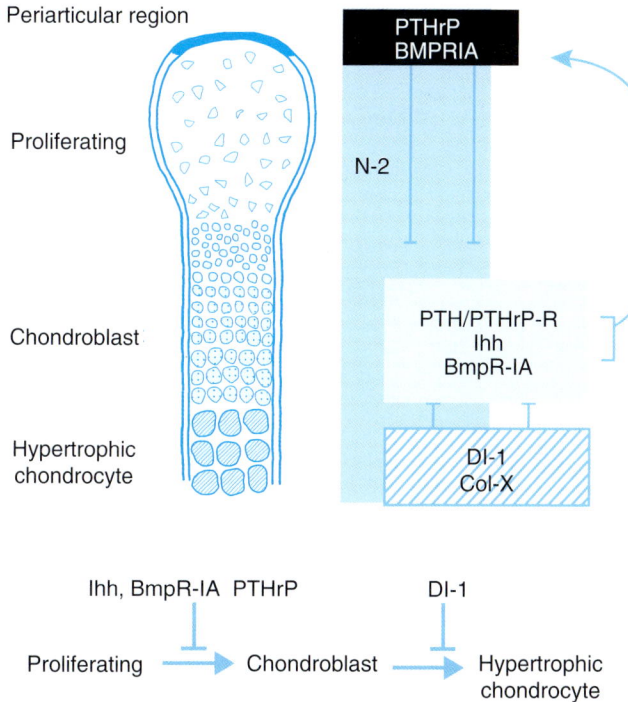

Figure 22.4 Molecular control of the transition from chondrocyte to hypertrophic chondrocyte (HCC) as seen in the long bones of birds. The major cellular zones in the cartilage model are shown at the left. Molecules expressed in those zones are shown on the right. As summarized at the bottom of the figure, *Ihh* and BMPR1A in the chondroblast zone dampen proliferation by feedback to PTHrP and BMPR1A in the proliferative zone. *Delta-1* (*DI-1*) in the HCC zone negatively regulates transition from chondroblast to HCC by signaling to the *DI-1* receptor N-2, which is expressed throughout the cartilage column. Based on Crowe *et al.* (1999).

Delta-1 is expressed in HCCs during murine limb development. Misexpression of *Delta-1* blocks hypertrophy, *Delta-1* being a negative regulator of hypertrophy (Fig. 22.4).

Type X and mineralization

Birds

Long-bone development in chick embryos is characterized by resorption of cartilage before mineralization; a novel phosphatase is up-regulated fivefold in HCCs from chick growth plates where it is found in concentrations 100 times higher than in non-chondrogenic tissues. Resorption is carried out by mononucleated phagocytes. Chondrocytes that are released after their matrix is removed transform to osteoblasts and deposit osteoid in empty chondrocyte lacunae. HCCs reinitiate cell division, transform into osteoblasts and deposit bone matrix, which they mineralize. Division is asymmetric, one cell remaining viable and one undergoing apoptosis.[27]

If cultured without the perichondrium, chick tibiotarsi show enhanced development of the hypertrophic zone as cells exit from the cell cycle and differentiate (Long and Linsenmayer, 1998).

Frogs

During long-bone development in the North American bullfrog, *Rana catesbeiana*, cartilage cells are not aligned in columns, and hypertrophy is neither associated with cartilage mineralization nor endochondral ossification, unlike in birds and mammals (Felisbino and Carvalho, 1999). Indeed, there is little endochondral ossification in anuran long bones; increase in the length of long bones is by intramembranous ossification, the role of the cartilage being to increase width only (Hall, 2003b).

Rickets

Deficiency of vitamin D = *rickets* (inability to mineralize because HCCs fail to mature; Fig. 3.13) leading to osteomalacia in adults. Collagen synthesis and calcium nucleation are both enhanced when rachitic bone is exposed to vitamin D (Cousins and DeLuca, 1972; see Stern, 1980, for a review of vitamin D and bone).

MATRIX VESICLES

What controls ECM mineralization? Much attention has focused on matrix vesicles.

Mineralization is initiated within membrane-bound ECM vesicles. Consequently, matrix vesicles are found in mineralizing cartilage, bone, dentine, enamel, antlers, aorta, membrane and endochondral bone, intraskeletal and ectopic skeletal elements, and in all vertebrates from fish to mammals.

Matrix vesicles arise by budding from plasma membranes; they contain calcium, phosphate, lipids, GAGs, alkaline phosphatase, pyrophosphorylase, ATPase and 5'-AMPase. Matrix vesicles from rat costochondral cartilage cells are enriched in metalloproteinases that degrade proteoglycans (Table 22.5 and see Box 13.3).[28]

Yousef Ali and his colleagues isolated matrix vesicles in 1970. When incubated with ATP under alkaline conditions, they accumulate calcium ions. Matrix vesicles isolated from foetal bovine cartilage or artificially reconstituted vesicles mineralize *in vitro*, with active involvement of ATP and phospholipids. Mineralization of matrix deposited by clonal osteogenic cell lines is initially on vesicles and only then on collagen fibres. Similarly, the first mineral crystals in rachitic rat growth-plate chondrocytes are deposited on vesicles without any involvement of collagen (Fig. 3.13).[29]

How matrix vesicles develop from chondrocyte microvilli has been investigated using growth-plate chondrocytes *in vitro*. Cytochalasin inhibits the microfilament assembly required to release matrix vesicles, phalloidin inhibits disassembly – and so inhibits vesicle release – while colchicine (which inhibits microtubule assembly) results in cell fragmentation. Chemical analyses show that matrix vesicles and microvilli are equivalent cellular structures (Table 22.6). Mature chondrocytes, as from the tibiae of

Table 22.5 Active and total amounts of metalloproteinase in matrix vesicles and plasma membranes of growth- and resting-zone chondrocytes from rat costochondral cartilages[a]

Organelle	Acid metalloproteinase[b]				Neutral metalloproteinase[b]			
	Growth-zone chondrocytes		Resting-zone chondrocytes		Growth-zone chondrocytes		Resting-zone chondrocytes	
	Active[c]	Total[d]	Active	Total	Active	Total	Active	Total
Matrix vesicle	1.5	1.6	0.25	0.5	2.25	2.5	1.0	1.5
Plasma membrane	0.5	0.6	0.4	0.75	1.0	1.25	0.2	1.0

[a] Based on data from Dean et al. (1992).
[b] Units are enzyme units/mg protein.
[c] Content of active enzymes found in each membrane fraction (%).
[d] Content of total (active + latent) enzyme found in each fraction (%).

Table 22.6 Comparison between % phospholipid composition of matrix vesicles and microvilli[a]

Lipid	Microvilli fraction (%)	Matrix vesicle	
		Wuthier (1975)	Peress et al. (1974)
Sphingomyelin	13.2 ± 3.4	13.3 ± 2.5	14.9 ± 2.4
Phosphatidylcholine	26.5 ± 1.5	36.8 ± 3.0	33.8 ± 0.3
Lysophosphatidylcholine	3.8 ± 1.4	3.3 ± 1.3	–
Phosphatidylethanolamine	17.3 ± 0.7	13.1 ± 1.9	20.9 ± 1.5
Lysophosphatidylethanolamine	4.8 ± 2.4	7.8 ± 1.2	–
Phosphatidylserine	14.6 ± 0.2	12.8 ± 0.4	13.2 ± 0.8
Phosphatidylinositol	9.7 ± 1.5	4.6 ± 0.4	8.3 ± 0.8
Phosphatidic acid	2.1 ± 1.9	0.4 ± 0.2	2.6 ± 0.9
Diphosphatidylglycerol and phosphatidylglycerol	7.8 ± 0.9	2.8 ± 0.9	2.9 ± 0.9

[a] Based on data summarized in Wuthier (1987).

chick embryos of H.H. 28–30, or dedifferentiated chondrocytes, hypertrophy in suspension culture in the presence of ascorbic acid and sodium-β-glycerophosphate, synthesize type X collagen and alkaline phosphatase, release matrix vesicles, and deposit and mineralize an ECM.[30]

A gradual shift from *intramitochondrial calcium* to *intravesicular calcium* occurs from the top to the bottom of the cell columns in long-bone growth plates (Fig. 3.13). Matrix vesicles are not, however, mitochondrial in origin; they contain no respiratory enzymes. In a freeze-fracture study of avian epiphyseal cartilages, matrix vesicles were demonstrated in the lacunae of HCCs and then in the ECM. Matrix vesicles of 0.02–0.22 μm diameter are found within 3 μm of the mineralization front in rat tibial bone remodeling following injury. As chondrocytes mature and die they secrete empty vesicles that accumulate Ca^{++} and PO_4^{---}. Similarly, fibrocartilage of the rat *os penis* only displays matrix vesicles when degenerating during puberty.[31]

Two words of **caution**: one on the uniformity of matrix vesicles, the other on their existence. Not all ECM

vesicles are the same, nor need all be associated with mineralization. Vesicles in the cartilage of rat femora and tibiae contain two peptides not found in bone; auricular cartilage contains matrix vesicles whose function relates to secretion of elastic fibres not mineralization.[32]

The following three studies either failed to find matrix vesicles, found them but with no association with mineralization, or found them in a different location within the growth plate than discussed above.

A study using electron optical and analytical methods failed to find a mineral phase in matrix vesicles from rat growth plates. Rather, dispersed particles were designated as the initial solid phase of calcium phosphate deposition in the ECM. Despite using high-resolution electron spectroscopic imaging, matrix vesicles were not found with initial ossification of the *perichondrial ring* of mouse femora.[33] Vesicles were found in the epiphyseal growth plate of male rats (England Wright Y strain), but the highest density was in the resting zone, the lowest density in the zone of mineralization. The argument that most vesicles arise from degenerating chondrocytes does not, of itself, disqualify matrix vesicles from a role in mineralization.[34]

HYPERTROPHIC CHONDROCYTES AND SUBPERIOSTEAL OSSIFICATION

Several centuries ago, prevailing opinion was that bone growth was entirely appositional, and from the periosteum: 'as an exogenous stem grows from the inner layer of the bark' (Duhamel de Monceau, 1742). Not until 1912 did MacEwen show that bone formed from origins other than the periosteum, i.e. endochondrally. Although its life span is short, hypertrophic cartilage plays an important role in eliciting osteogenesis from the adjacent perichondrium, converting it into a periosteum. This role is the prerogative of HCCs, and so beautifully illustrates cartilage/chondrocyte specificity.[35]

The possibility that liberated HCCs provide osteoprogenitor cells during *endochondral* ossification has already been discussed (Chapter 13). *Subperiosteal* ossification is intramembranous. Indeed, membrane bone makes up a much greater proportion of the bulk of so-called endochondral bones than is usually thought. Virtually all of the bone deposited during embryonic life in avian long bones is subperiosteal and not endochondral. In a condition such as a hypertrophic form of chondrodystrophy, metatropic dwarfism, in humans and *talpid* mutants in fowl (Box 20.1), distinctiveness of the types of osteogenesis is readily seen; endochondral and perichondrial ossification are uncoupled and there is no primary spongiosa.[36]

Cartilages such as Meckel's and those of the ear and trachea, in which chondrocytes do not normally undergo hypertrophy and where ECM remains unmineralized, may remain intact throughout life. Subperiosteal ossification is not seen within the perichondria of such cartilages, with exceptions such as Addison's disease, above. When subperiosteal ossification is seen, it is found adjacent to intact HCCs, initially as a collar of bone surrounding the diaphyses of embryonic long bones or adjacent to HCCs in ectopic sites (Fig. 19.1).[37]

Growth of endochondral bone depends on maintaining the growth of a primary cartilaginous model. Since this model requires functional stimuli such as mechanical stress, continued deposition of endochondral bone depends secondarily upon such biomechanical stimuli. Deposition of membrane bones, on the other hand, is far less dependent on mechanical factors, as illustrated by the following examples.

- One year after rabbit ilia or frontoparietal bones are grafted subcutaneously or subperiosteally, the endochondral bone of the ilia has being resorbed and reduced to 25 per cent of its original volume, but the membrane bone of the frontoparietals has increased in volume.

- Patterns of bone deposition in rat calvariae are similar *in vitro* to those seen *in vivo*.
- Intramembranous ossification occurs in fixed (immobilized) fractures, while endochondral ossification occurs in free fractures.[38]

Initiation of subperiosteal ossification involves converting the perichondrium to a periosteum. In long-bone rudiments this conversion first occurs adjacent to the zone of chondrocyte hypertrophy in the shaft or diaphysis. An association of chondrocyte hypertrophy and initiation of subperiosteal osteogenesis has been demonstrated experimentally. Examples of four lines of evidence will be followed by a discussion of a mutation – bp^H – in mice in which hypertrophy and, consequently, ossification, is delayed.

(i) When bladder mucosa is transplanted in contact with *hyaline* costal cartilage in dogs, the 'hyaline' chondrocytes hypertrophy and ossification is initiated.

(ii) The experiments of Shimomura *et al.* were mentioned (Chapter 13). Cartilage within diffusion chambers implanted into mammals induces osteogenic activity from host cells, *provided that* the implanted cartilage cells are hypertrophic.[39]

(iii) Retarding chondrocyte hypertrophy *in vitro* suppresses subperiosteal osteogenesis, discussed in Chapter 30 in the context of the *Creeper* (*cp*) mutant in domestic fowl.

(iv) Mareel (1967) cultured intact HCC zones in contact with the perichondria from nine-day-old embryonic chick tibiae and obtained rapid build-up of what he called 'contact cells', which he thought were young osteoblasts. No osteoid or bone was deposited, however. The tibiae were probably too old – the tibial perichondrium becomes a periosteum at H.H. 29 (6–6.5 days) – and the culture period too short to obtain osteogenesis. In my laboratory, Peter Scott-Savage repeated and extended Mareel's experiments without obtaining initiation of osteoblasts. His experiments involved isolated perichondria from six-day-old embryos co-cultured with HCCs to determine whether this association is involved in the initiation of subperiosteal osteogenesis.

Failure of chondrocyte hypertrophy is seen in the *Brachypod* mutant in mice, which warrants further discussion, as more detailed information is available on the origins of the defect.

Brachypod (*bp^H*) in mice

The autosomal recessive mutant *Brachypod* has a mode of action that makes it valuable for studying relationships between chondrocyte hypertrophy and subsequent periosteal osteogenesis. Tibiae and femora of *Brachypod* mice are similar to wild type, but fibular chondrocytes fail to hypertrophy, and do not initiate osteogenesis until three weeks after birth.[40]

Early changes

Brachypod acts on hind-limb mesenchymal condensations at 12 days of gestation or earlier; condensations for hind-limb elements are reduced at that age.

Formation of cartilage nodules from mesenchyme cultured from 12-day-old *Brachypod* embryos is delayed and nodule size reduced in comparison to wild type. Cell division and DNA synthesis are prolonged in the mutant, in contrast to the decline with chondrogenesis that occurs in wild-type cells. Cells with a fibroblastic morphology are much more evident, and confluence is not attained.

These lines of evidence were all taken to indicate a cell membrane dysfunction that retards condensation of cells. Subsequently, changes in the plasma membranes, and decreased galactosyltransferase activity at the condensation stage, were found. Intensity of proximo-distal defects increases when hind-limb rudiments from 12-day-old *bp^H* embryos are cultured for six days. Further analysis revealed deficiency in proteoglycans and thickened collagen fibres in the ECM.[41]

Confirmation of the absence of the cell membrane changes that occur in wild-type embryos between 11 and 12 days of gestation comes from demonstrations that: (i) agglutination to concanavalin A or to wheat-germ agglutinin seen in wild-type cells does not occur in mutant cells, and (ii) that the rate of decline of single cells in rotation culture is reduced with only half the *bp^H* cells entering chondrogenesis. Clearly, an early defect in cell membrane properties affects cell-to-cell contact and the ability of *bp^H* cells to accumulate and chondrify.[42]

Although a 76 000 kDa protein that inhibits growth of tissues from wild-type limb buds is present in 13-day-old *Brachypod* embryos (Pleskova *et al.*, 1974), production of this inhibitor and failure of cells to pack into condensations have not been correlated experimentally. However, failure of the cells to chondrify normally *in vitro* and the delay *in vivo* clearly are consequences of an earlier event.

Recent studies on *brachypodism* have shown it to be a defect in *Gdf-5*, a member of the Tgβ-superfamily (Box 22.1). *Gdf-5* is encoded by the *brachypodism* locus. The genetic defect is seen at the condensation stage, confirming the studies at the cellular level. Involvement of *Gdf-5* is interesting as it was the first growth factor shown to link chondrogenesis and joint formation. *Gdf-5* is expressed in transverse bands across developing limb anlage at positions where joints will develop; *Gdf-5* is both necessary and sufficient for cartilage development and to restore joint development in *Brachypod* mutants (Box 22.1).[43]

Fibulae

The fibulae of *bp^H* mutants contain chondrocytes that do not hypertrophy. The possibility that this failure stems

from alterations in cell behaviour at the precondensation phase has important implications for studies on the timing of the specificity of cartilage types; specification of the ability to undergo hypertrophy may occur early in development, not during subsequent differentiation of chondroblasts or chondrocytes.

Collagen produced by cultured *bp^H* cells is type II. As a consequence of dissociation of the normal tight correlation between collagen synthesis and degradation, levels of collagen and GAGs increase in *Brachypod* fibulae when compared with wild-type fibulae (see Hall, 1973b). A general slowdown in protein synthesis reduces collagen synthesis to half the normal rate. Collagen degradation also slows, the *net* result being an increase in collagen content.

Hydroxylation of proline is retarded in mutant fibulae, while collagen synthesis – as assessed by incorporation of ^{3}H-proline – is independent of ascorbic acid stimulation, which is unusual. A proline:hydroxyproline ratio of 28:1 versus 1.4:1 in wild type results. Rate of GAG synthesis is normal, but, as with collagen, degradation is slowed, so that GAGs accumulate. Whether such alterations in the products of differentiated chondroblasts are consequences of the initial defect at 12 days of gestation, or represent an independent later action(s) of the mutation, is not known.[44]

Chondrocytes from *Brachypod* fibulae neither hypertrophy nor trigger subperiosteal osteogenesis. Because alkaline phosphatase is associated with chondrocyte hypertrophy preceding osteogenesis in various species, Krotoski and Elmer (1973) examined levels of alkaline phosphatase between 14 and 17 days of gestation in tibiae, femora and fibulae from *Brachypod* and wild-type mice. Alkaline phosphatase accumulates slowly in wild-type tibiae and femora between 14 and 16 days, more rapidly between 16 and 17 days (Table 22.7). Both phases of this accumulation are depressed in *Brachypod* tibiae and femora. Normally, the level of alkaline phosphatase is low in wild-type fibulae at 17 days; no alkaline phosphatase is found in *Brachypod* fibulae at this age (Table 22.7), and none is present one day after birth.[45]

Osteogenesis begins belatedly in the fibulae of *bp^H* mice two weeks after birth. Whether its initiation is accompanied or preceded by chondrocyte hypertrophy and/or accumulation of alkaline phosphatase, and whether both subperiosteal and endochondral ossification begin at the same time have not been analyzed.

Obviously, *Brachypod* is a useful mutant in which to examine dependence of subperiosteal osteogenesis on chondrocyte hypertrophy and timing of the onset of the ability of chondrocytes to hypertrophy, both of which are important aspects of establishing cartilage diversity.

A role for Wnts

An analysis of chondrogenesis in chick limb-bud development goes some distance toward revealing the role of

Box 22.1 Growth and differentiation factors

The growth and differentiation factors (Gdfs) are members of the Bmp family and the Tgfβ superfamily of growth factors. Myostatin is Gdf-8. Three of the 11 Gdfs identified so far (*Gdf-5, Gdf-10* and *Gdf-11*) are involved in skeletal development; neither Gdf-7 (Bmp-12) nor Gdf-6 (Bmp-13) elicits skeletogenic markers from MC615 cells, although both Bmp-2 and Bmp-4 do (Valcourt *et al.*, 1999). A mutation of *Gdf-5* (*brachypodism, bp*) and production of its product Gdf-5 (cartilage-derived morphogenetic protein-1, Cdmp-1) are responsible for *brachypodism* (Chapter 22).

Gdf-5

Gdf-5, expressed in neurons and bones, is involved in chondrogenesis, joint positioning and joint morphogenesis, *Gdf-5* and *Wnt-14* being the major determinants of the positions of joint anlage.

Wnt-5 is expressed in the pre-joint region of chick long bones. Misexpression in adjacent sites initiates the first steps in joint formation and suppresses formation of adjacent joints, i.e. *Wnt-14* spaces joints along a skeletal system.

Gdf-5, the first growth factor shown to link chondrogenesis and joint formation, is expressed in transverse bands across developing limb anlage at positions where joints will develop. Overexpressing *Gdf-5* in chick-limb mesenchyme increases the size of chondrogenic condensations and of the skeletal elements that develop from them in what appears to be a two-step mechanism. The second step has recently been shown by Coleman and Tuan (2003) to include regulation (and perhaps induction) of chondrocyte hypertrophy. Sites of future joints act as signaling centres for *Gdf-5*, as they do for *wnt-14*.[a]

Gdf-5 is regulated by *Bmps* which act with hedgehog genes to control epiphyseal differentiation and growth. The Bmp antagonist gremlin, which antagonizes *Bmp-2, -4* and *-7*, helps maintain the AER – and so maintains bud outgrowth and proximo-distal specification – and restricts cell death, confining cartilage to the central mesenchyme.[b]

A model for foetal bone repair which involved a 'pinch' fracture of the murine cartilaginous ulna at 14.5 days of gestation was developed by Stone (2000) using Muneoka's technique of exteriorizing the embryo. Union was rapid – within two days – but, surprisingly, did not involve changes in *Gdf-5, Bmp-2* or *Bmp-4*.

In zebrafish median fins studied 6–45 days post-fertilization, Crotwell *et al.* (2001) found initial *Gdf-5* expression in the mesenchyme between condensations of future fin rays and then distalward expansion. No expression was seen in stages older than 7.5-mm length, at which stage the hypurals were mature.

... and BmpR1B

Bmp receptors play multiple roles in skeletogenesis including roles in digit morphogenesis, which in chick embryos is regulated by *Fgfs, Tgfβs, Gdf-5* and noggin acting through Bmp signaling.

Mutants lacking the Bmp receptor BmpRIB fail to form digits. BmpRIB regulates chondrogenesis through *Gdf-5*-dependent and *Gdf-5*-independent processes (Baur *et al.*, 2000), providing a nice demonstration of how a single receptor can be triggered by different ligands and how a single ligand can bind to different receptors.

Generating *BMPR1B/Gdf-5* double mutants shows *that Gdf-5 is a ligand for BMPR1B*. Generating *BMPR1B/Bmp-7* double mutants shows that receptor and ligand have overlapping pathways. In the absence of *BMPR1B*, *BMP-7* plays an essential role in limb development (Yi *et al.*, 2000).

Brachypodism
Recent studies on *brachypodism* show this mutant to be a frame-shift mutation in *Gdf-5*, which is encoded by the *brachypodism* locus. The genetic defect is seen at the condensation stage, confirming the earlier studies at the cellular level (Chapter 22 text). *Gdf-5* is both necessary and sufficient for cartilage development and to restore joint development in *Brachypod* mutants.[c]

Gdf-10 and Gdf-11

Although *Gdf-10* is expressed in skeletal tissues, its function is not yet known.

Gdf-11 (*Bmp-11*), expressed in somites, maxillary, mandibular and hyoid arches, and limb- and tail-bud mesenchyme, is a negative regulator of chondrogenesis and myogenesis in developing chick limbs, and plays a role in specification of vertebral identity.[d]

Gdf-11 is found in sub-ridge mesenchyme of limb buds, but not in chondrogenic condensations, although it is expressed between condensations. Implanting beads loaded with Gdf-11 truncates both limb skeleton and muscle. *Gdf-11* activates *Hoxd-11* and *Hoxd-13* but not *Hoxa-11, Hoxa-13* or *Msx*, indicating control of late distal expansion of *Hoxd* genes in limb mesenchyme. *Gdf-11* also induces its own antagonist, *follistatin*, implying negative feedback (Gamer *et al.*, 2001).

Additional thoracic and lumbar vertebrae form, the tail is reduced or absent, and the hind limbs are located more posteriorly when *Gdf-11* is disrupted in mice. Although initial somite numbers are normal, the number of thoracic vertebrae increases from the normal 13 to 17 or 18. Cervical vertebrae are unaffected. Gdf-11 provides a posteriorizing signal, acting upstream of *Hox* genes and via at least four *Hox* genes – *Hoxc-6, Hoxc-8, Hoxc-10* and *Hoxc-11* (McPherron *et al.*, 1999).

[a] See Hartmann and Tabin (2001) for *wnt-14*, Francis-West *et al.* (1999a) for *Gdf-5*, and Vortkamp (1997) and Bostrom *et al.* (1999) for reviews of *Gdf-5, Bmp-5* (the mutant *Short ear* is a mutation in *Bmp-5*) and *Bmp-7* involved in condensation and joint formation, and in wild-type and mutant embryos.
[b] See Merino *et al.* (1998, 1999a,b) and Vogt and Duboule (1999) for these regulatory pathways.
[c] See Storm *et al.* (1994), Storm and Kingsley (1996, 1999) and Kingsley (2001) for Gdf-5. For nomenclatures of growth factors in bone, see Veis (1987).
[d] See J. A. King *et al.* (1996), Gad and Tam (1999) and McPherron *et al.* (1999) for Gdf-11.

Wnt-5a, Wnt-5b and *Wnt-4* in regulating chondrogenesis. All three genes are expressed in chondrogenic tissues: *Wnt-5a* in the perichondrium, *Wnt-5b* in a subpopulation of pre-HCCs and in the fibrous layer of the perichondrium, and *Wnt-4* in chondrocytes of the joint. *Wnt-5a* and *Wnt-4* have opposing effects on chondrocyte differentiation, effects that are mediated through different signaling pathways:

- misexpressing *Wnt-5a* delays chondrocyte maturation and delays the onset of subperiosteal bone formation around the shaft;
- misexpressing *Wnt-4* accelerates chondrocyte maturation and periosteal bone formation.[46]

Table 22.7 Activity of alkaline Phosphatase within femora, tibiae, and fibulae of wild-type and *Brachypod* (*bp^H*) mice[a,b]

		Gestation age (days)			
		14	15	16	17
Femur	Wild type	0.1	0.25	0.45	1.4
	bp^H	0.1	0.20	0.25	0.5
	%[c]	100	80	55	36
Tibia	Wild type	0.1	0.35	0.6	1.7
	bp^H	0.1	0.15	0.2	0.5
	%	100	43	33	29
Fibula	Wild type	[d]	[d]	[d]	0.4
	bp^H	[d]	[d]	[d]	0
	%				0

[a] Activity is expressed as μmoles phenol liberated/10 min/rudiment.

[b] Based on data in Krotoski and Elmer (1973).

[c] %, *brachypod*/normal × 100.

[d] Alkaline phosphatase does not appear within the fibula until the 17th day of gestation.

NOTES

1. See Williams-Boyce and Daniel (1986) for rabbit ear cartilage regeneration, and Kirschbaum and Meunier (1981, 1988) for caudal fin regeneration.
2. See Bairati et al. (1996) for the syncytial stage, and Clase et al. (2000) for *Fgf-5*.
3. O'Driscoll (1999) provides a good review for similar studies on periostea, including the use of periosteal grafts to repair articular cartilage, in which the perichondrium is minimal.
4. See Weiss and Amprino (1940) and Weiss and Moscona (1958) for the classic studies, and Archer et al. (1985) for the follow-up study. Similarly, frontonasal mesenchyme forms sheets of cartilage while mandibular mesenchyme forms nodules (Wedden et al., 1986). Langille (1994b) showed that cultures from lateral mandibular mesenchyme form sheets of cartilage.
5. See Takigawa et al. (1984a) for the studies with colchicine and cytochalasin, Cottrill et al. (1987) for the chimaeric cultures, Pirttiniemi and Kantomaa (1998) for condylar cartilage, Gay and Kosher (1984) for stage specificity, and Matsutani and Kuroda (1980) for the threshold studies. See Ogiso et al. (1991) for the study with bone cells.
6. See Elmer and Selleck (1975) and Owens and Solursh (1982) for *bp^H*, and Archer et al. (1984) for a discussion of early specification of hypertrophy.
7. See Moscona (1957, 1965), Levak-Svajger and Moscona (1964) and Zwilling (1968) for these studies. Vogel and Kelley (1977) provided information on the types of GAGs found on the surfaces of human limb-bud cells. See Roth (1973), Edelman (1976, 1986, 1988) and Chapters 19 and 20 for the cell-surface and cell-adhesion molecules involved.
8. See Levenson (1969, 1970) for a summary of some of the differences.
9. See Fyfe and Hall (1979) for the co-cultures, and Noden (1986b) for the grafts.
10. See Foret (1970) for the transplant study, and Trevisan and Scapino (1976a) for the hamster Meckelian symphysis.
11. See Alcock (1972), Stockwell (1979) Russell et al. (1986), Boskey et al. (2001) and Sasagawa (2002) for cartilage mineralization,

and Riede et al. (1971) for initiation of vascular invasion. Kaminski et al. (1977) found that viable chondrocytes can inhibit lymphocyte-induced formation of blood vessels *in vitro*. Dead chondrocytes cannot.

12. See Gibson and Flint (1985), Gibson et al. (1986), Solursh et al. (1986), LuValle et al. (1989) and Iyama et al. (1991) for type X synthesis and vascular invasion, and see Gibson et al. (1985) for sternal chondrocyte apoptosis. Type X collagen is found in the caudal half of the avian sternum, which can be induced to hypertrophy if cultured in collagen or agarose (Chapter 23).
13. See Kielty et al. (1985), Schmid and Linsenmayer (1985a,b), Castagnola et al. (1986, 1987) and A. R. Poole and Pidoux (1989) for type X and hypertrophy.
14. See A. R. Poole and Pidoux (1989) for type X and II collagen association, Gibson and Flint (1985), Gibson et al. (1986), Solursh et al. (1986), Bruckner et al. (1989), LuValle et al. (1989), Bohme et al. (1992) and Reginato et al. (1993) for type X in hypertrophic cartilage of long bones and sterna, and Quarto et al. (1992b) for the *v-myc* study. *Osteonectin* functions as a secreted heat-shock protein, being found in dedifferentiated and hypertrophic chondrocytes *in vitro* (Nori et al., 1992).
15. See Lu Valle et al. (1993) for the type X collagen promoter, and Toma et al. (1997b) for type X induction in calvarial cells. Saitho et al. (2000) showed that compressive force promotes chondrogenesis, hypertrophy and the synthesis of type X collagen in midpalatal sutural cartilage in growing (four-week-old) rats.
16. See Glowacki et al. (1983) for the poly (Hema) study, and Adams et al. (1989) for modification of type X synthesis by cell shape.
17. See Oettinger and Pacifici (1990) and Pacifici et al. (1991b). Sternal chondrocytes from 19-day-old embryos exposed to 100^{-35} or 10^{-35} nM retinoic acid synthesize type X mRNA, without any change in type II collagen or core protein mRNAs.
18. See Gress and Jacenko (2000) for haematopoietic aplasia, and Thurston et al. (1985b) and Jacenko et al. (1993) for spondylometaphyseal chondrodysplasia. Wallis (1993) discusses mutations of collagen type X and their effects on endochondral bone formation, Gorlin et al. (2001) and Cohen (2002) review chondrodysplasias with short limbs, including metaphyseal chondrodysplasia.
19. See Inada et al. (1999) for *Runx-2*, and Volk and LeBoy (1999) for a review of how chondrocyte hypertrophy is regulated.
20. See Pacifici et al. (1991a,b) for the chick studies, and Kirsch and von der Mark (1992) for the foetal human study.
21. p53 is a protein involved in the transcription of DNA, especially during cell proliferation, functioning to suppress abnormal proliferation and in programmed cell death. When mutation, p53 can no longer block and may even enhance abnormal proliferation. It is estimated that as many as half of human cancers contain a mutated *p53* gene.
22. See Kavumpurath and Hall (1990) and Ekanayake and Hall (1994a,b) for studies with Meckel's cartilage and mandibular mesenchyme. For a discussion and literature review of whether chondrocyte hypertrophy is required for subperiosteal osteogenesis, see Scott-Savage and Hall (1980) and Amprino (1985).
23. See Becks et al. (1946), Kavumpurath and Hall (1990) and Alini et al. (1996b) for the rat, avian and bovine studies.

24. See Quarto et al. (1990, 1992a). Clonal efficiency was low and varied with how the medium was conditioned: 13% in medium conditioned by dedifferentiated chondrocytes, 1.4% in medium conditioned with HCCs, and zero with unconditioned media. See Cancedda et al. (1993, 1995, 2000) for overviews of their studies including the cell-culture model, fate of hypertrophic cells, bone-marrow stromal cells and condensation.

25. See Houston et al. (1994) and Helder et al. (1995) for expression of Bmp-7, and see Enomoto-Iwamoto et al. (1998) for the receptor study, a tree of the relationship among bone morphogenetic proteins, and a discussion of Bmps and chondrocyte hypertrophy.

26. See Gitelman et al. (1994) for the study with CHO cells, and Grimsrud et al. (1999) for PTHrP.

27. See Houston et al. (1999) for the phosphatase, and Roach (1992b), Roach and Shearer, 1989 and Roach et al. (1995) for resorption. Cultured chick long bones (14-day-old embryonic femora) show virtually no resorption in long-term culture, and show mineralization outside the cartilage and bone in response to addition of β-glycerophosphate (Roach, 1990, 1992a).

28. For classic studies on the isolation of matrix vesicles, see Anderson (1969, 1985), Ali et al. (1970), Anderson et al. (1970), Anderson and Reynolds (1973), Katchburian (1973), Thyberg et al. (1973), Landis et al. (1977a,b), Dickson (1981, 1982) and Katchburian and Severs (1983). See the papers in Ascenzi et al. (1981) for an overview of the studies from the Third International Conference devoted to matrix vesicles, and Christoffersen and Landis (1991) for a review.

29. See Ali and Evans (1973b), Hsu and Anderson (1978) and Hsu et al. (1978) for ATP-driven Ca^{++} accumulation; Sudo et al. (1983) and Wu et al. (1989) for the associations between collagen and matrix vesicles; and Väänänen et al. (1983) for involvement of matrix vesicles in rickets.

30. See Hale and Wuthier (1987) and Table 22.5 for matrix vesicle development, and Tacchetti et al. (1987, 1989) for mineralization in vitro.

31. See Brighton and Hunt (1974) and Anderson et al. (1975) for the transition from mitochondrial to vesicle calcium, Person et al. (1977) for absence of respiratory enzymes, Borg et al. (1981) for the chick study, Sela et al. (1987) for vesicles in injured bone, and Izumi et al. (2000) for the os penis.

32. See Gabrovska and Vancov (1982) for vesicles and elastic fibres, Muhlrad et al. (1983) for cartilage vesicles, and Newbrey and Banks (1983) and Chapter 8 for another role for vesicles in antler formation. Ushiki (2002) has provided a valuable comparitive analysis of the organization of elastic, reticular and collagen fibers.

33. The perichondrial ring surrounds the distal epiphysis. Arsenault and Ottensmeyer (1984) used high-resolution electron spectroscopic imaging of the perichondrial ring of mouse femora to demonstrate that it is a a site of intramembranous ossification. Excision of the perichondrial ring from rat radii – which involves removing the fibrous cover and a piece of the epiphysis – elicits enlargement of the growth plate, proliferation of trabeculae and bending of the bone rudiment (Rodriguez et al., 1985).

34. See Landis and Glimcher (1982), Arsenault and Ottensmeyer (1984) and Reinholt et al. (1982) for these three studies.

35. See Eyre-Brook (1984) for an overview of the periosteum back to MacEwen (1912), and see Keith (1918b) for much of the earlier literature, including the contributions of Ollier and Macewen. See Fell and Robinson (1929, 1930), Henrichsen (1958b) and Dickson (1978, 1982) for relationships between chondrocyte hypertrophy and initiation of periosteal osteogenesis.

36. See Rinaldi et al. (1974) for subperiosteal osteogenesis in birds, and Boden et al. (1987) for metatropic dwarfism.

37. See Fell (1925, 1928) for the classic study in situ, and Copher (1935), Lacroix (1951) and Shimomura et al. (1975) for subperiosteal osteogenesis in ectopic sites.

38. See Lewis and Irving (1970) and Jarry and Uhthoff (1971). Intramembranous ossification also is more sensitive than endochondral to disruption of collagen biosynthesis (Chapter 2 and see Diegelmann and Peterkofsky, 1972, and Klein and Zika, 1976).

39. Cooper (1965), who found that chondrocytes undergoing hypertrophy would act inductively on somite mesoderm to evoke chondrogenesis, presents a good discussion of the literature on the association between chondrocyte hypertrophy and osteogenesis.

40. See Konyukhov and Ginter (1966) and Grüneberg and Lee (1973) for basic information on Brachypod.

41. See Grüneberg and Lee (1973) and Elmer and Selleck (1975) for reduced condensations at 12 days, Elmer (1977) and Elmer et al. (1988) for cell membrane dysfunction, and Kwasigroch et al. (1992) for the culture approach.

42. See Hewitt and Elmer (1976, 1978) and Duke and Elmer (1977) for the agglutination and cell culture studies.

43. See Storm et al. (1994), Storm and Kingsley (1996, 1999) and Kingsley (2001) for Gdf-5. For nomenclatures of growth factors in bone, see Veis (1987) and Khan et al. (2000).

44. See Elmer and Selleck (1975) and Rhodes and Elmer (1975) for collagen.

45. See Fell and Robinson (1929, 1930), Henrichsen (1958b), Dickson (1978, 1982), Scott-Savage and Hall (1979) and Ekanayake and Hall (1994b) for relationships between chondrocyte hypertrophy and the initiation of periosteal osteogenesis.

46. See Hartmann and Tabin (2000) for the study, Spitz and Duboule (2001) for an analysis of its significance, and see Yang (2003) for an overview of the roles of Wnts in limb development.

Cartilage Diversity

Chondrocytes that have proven very amenable for the analysis of chondrogenesis are those from the developing sterna of avian embryos. In part this is because the sternum consists of two components: a cephalic portion in which chondrocytes synthesize type X collagen, hypertrophy and are replaced by bone, and a second portion in which the cells remain as small-celled chondrocytes. Consequently, we can use one portion of the sternum as an internal control for the other in studies aimed at understanding the basis for cartilage diversity. I begin this chapter with a discussion of sternal chondrocytes.

Vascularization and mineralization of cartilage matrix go hand in hand. Conversely, is unmineralized cartilage not invaded by blood vessels because of a distinctive property of unmineralized cartilage, or is it fortuitous? A similar question could be asked about tumours, for cartilage tumours are uncommon. Discussion of tumour invasion and cartilage vascularity – the two go hand in hand – follows the discussion of sternal cartilage and leads into a discussion of a very specialized transformation, that of cartilage to ligament. This transformation is restricted to female mammals and reflects a specific response of the lineage of chondrocytes found in the pubic symphysis.

STERNAL CHONDROCYTES

A flavour of what sternal chondrocytes have taught us about chondrocyte diversity is presented below. The majority of these topics are treated in greater detail later in the book.

Synthesis of collagen and glycosaminoglycan (GAG)

Using an autoradiographic analysis of double-labeled sternal chondrocytes from 13-day-old embryos, Smith (1972) showed that all chondrocytes incorporate both ^{3}H-proline and ^{35}S-sulphate over the same time period. Short pulses were used to overcome the problems of re-utilization or exchange of label. Visualization of the chondrocytes was such that Smith could state that the same cells were synthesizing collagen and GAGs at the same time.

In common with most (all?) chondrocytes, and using feedback control, sternal chondrocytes synthesize GAGs in proportion to the rate of extracellular matrix (ECM) accumulation. A non-dialyzable, trypsin- and heat-sensitive molecule of molecular weight 30 000–150 000 that stimulates syntheses of chondroitin sulphate and collagen is liberated into the medium. Proteoglycan monomers from sternal cartilage also evoke chondrogenesis from somitic mesoderm implicating ECM components in inductive interaction.[1]

Differential expression of type II collagen

Mesenchymal cells of H.H. 24 limb buds contain only 20 copies of the type II procollagen mRNA, a number that rises to 2000 copies when chondrocytes are fully differentiated at H.H. 31 (7.5 days). Sternal chondrocytes from 17-day-embryos, on the other hand, contain 10 000 copies, indicative of differential expression in the two cartilages. The chondroblast Col2a1 gene for the pro α1(II) poylpeptide is under-methylated when compared with fibroblasts or red blood cells. Perhaps this explains why the degree of methylation does not change when chondroblasts dedifferentiate.[2]

Differential synthesis and organization of collagen types

Some collagens are ubiquitous, collagen type VI being perhaps the prime example (Box 3.1). The types of collagens synthesized by chondrocytes vary depending on the source of the cells. Chondrocytes from rabbit sterna tend to produce a predominantly type II collagen, articular chondrocytes a mix of types I and II. Embryonic chick vertebral chondrocytes synthesize type X collagen and type I procollagen; sternal chondrocytes synthesize type IX and X collagen, but neither type I procollagen nor type X collagen.

The organization of collagen types also varies from cartilage to cartilage. Chick sternal chondrocytes incorporate collagen types II, IX and XI into single fibrils. Type IX collagen stabilizes what would otherwise be thin collagen fibrils; some one fifth of the collagen synthesized by sternal chondrocytes is type IX. If cultured in a collagen gel, sternal chondrocytes can synthesize additional collagen types, indicating considerable lability (Box 23.1).[3]

Box 23.1 Collagen types IX, XI and XII

Type IX

Collagen types IX, XII, XIV and XVI form a subfamily characterized by involvement with formation of type II collagen molecules, association with the surface of type II collagen fibres, and roles in linking type II collagen with other components of the ECM. Of the three collagenous domains in type IX, one (Col-1) is shared between collagen types IX, XI and XII. Four non-collagenous domains (designated NC1-4) complete the molecule. Synthesis of types IX and II collagen are coordinately regulated, in part by pretranslational regulation by interleukin-1.[a]

Type IX is found in cartilage, intervertebral discs, and the vitreous humour of the eye, but not in periostea or membrane bone. Type IX is highly conserved; four monoclonal antibodies against the low-molecular-weight fragment of bovine type IX cross-react with human, rat and chick. Chick sternal chondrocytes incorporate collagen types II, IX and XI into single fibrils, the type IX being covalently cross-linked to type II, stabilizing what otherwise would be thin collagen fibrils.[b]

In studies on collagen types IX and X in developing chick tibiotarsi, Linsenmeyer et al. (1991) used in-situ hybridization and immunostaining to demonstrate that type IX is found in the matrix of hypertrophic chondrocytes (HCCs) without concomitant expression of mRNA in the oldest hypertrophic cells; onset of mRNA for type X collagen correlates with appearance of the first protein, suggesting that regulation is not transcriptional.

Type IX and other 'cartilage-specific collagens' (types II, V, XI) are found at the ligament–bone interface in bovine medial collateral and anterior cruciate ligaments.[c]

Developmental regulation of alternate splicing of the $\alpha 1$(IX) collagen chain in the cartilage of foetal rat long bones has been shown. Tissue-specific alternate exon splicing creates two peptides, P6b and P8. P6b is expressed in the periphery of the cartilage under the perichondrium in the diaphysis from 14 days of gestation on. P8 is expressed in prechondrogenic mesenchyme and in chondrocytes up to but excluding the hypertrophic stage. Expression remains at the periarticular surfaces where articular cartilage forms (Morris et al., 2000).

During healing of alveolar bone following tooth extraction, an $\alpha 1$(II) mRNA that differs from that found in hyaline cartilage is expressed. $\alpha 1$(IX) mRNA also is found in cells within the bone matrix, which indicated to Tink et al. (1993) that osteoblasts synthesize type II and IX collagens, a promoter switch in the type IX collagen being associated with the switch of progenitor cells to cartilage or bone.

Type XI

Type XI collagen is regarded as a minor collagen, in part because its distribution is restricted to cartilage and related tissues (cartilage of the knee joint and chondrosarcomas; Fig. 23.1) and in part because it does not have fibrils of its own, only in conjunction with collagen types II and IX; chick sternal chondrocytes incorporate collagen types II, IX and XI into single fibrils, type IX collagen being covalently cross-linked with type II, stabilizing what otherwise would be thin collagen fibrils.

As demonstrated from an analysis of mice with chondrodysplasia (cho), a fibrillar collagen type XI gene, ColXIa1, is essential for skeletal morphogenesis, playing a role in organization of the ECM, cho resulting from a cysteine deletion in $\alpha 1$(XI). Davies et al. (1998) identified six splice forms of the $\alpha 1$(XI) collagen gene in foetal rat cartilage, one form being expressed only in the dorsal portion of the ribs, which is the portion of the rib primordium that ossifies, suggesting a role for type XI in endochondral ossification. Differential expression of cis-elements of the $\alpha 2$(XI) collagen gene occurs in different mouse cartilages.[d]

Type XII

Type XII collagen shares the features described under type IX, being a member of the same subfamily. Its distribution is primarily in tissues rich in type I collagen, but it also is found associated with type II collagen in cartilages. Type XII collagen is found in developing murine connective tissue, tendon, membrane bone, fibrocartilage and perichondria (Oh et al., 1993).

Using a polyclonal antibody to type XII collagen, Gregory et al. (2001) showed that type XII is expressed in the joint interzone of rat forelimbs at 16 days of gestation and in the future articular cartilage at 18 days. In juvenile rats, articular cartilage and both longitudinal and transverse septa of hypertrophic growth-plate cartilages (Fig 4.4) express type XII. Expression is weak in epiphyseal cartilages. No type XII is found in secondary ossification centres.

Immobilizing chick embryos with paralyzing drugs allowed Mikic et al. (2000a) to show that type XII collagen and tenascin are mechanically modulated during synovial joint formation. Both decrease at the sites of future joints in cells that initiate chondrogenesis in these paralyzed embryos. Consistent with a mechanical role, type XII is found in developing murine tendon as well as in connective tissue, membrane bone, fibrocartilage and perichondria.[e]

[a] See Yasui et al. (1986), Goldring et al. (1988) and Ayad et al. (1994) for collagen type IX and its allies.
[b] See Gibson and Flint (1985), Kielty et al. (1985), Gibson et al. (1986), Solursh et al. (1986) and Gerstenfeld et al. (1989) for different collagens produced by different chondrocytes. See Gibson et al. (1982, 1983), Van der Rest and Mayne (1988) and Mendler et al. (1989) for the stabilizing properties of type IX collagen. See Müller-Glauser et al. (1986) for distribution, and Ye et al. (1991) for conservation of type IX. Aspects of selective assembly and remodeling of collagens II and IX, based on chondrocyte cultures derived from foetal bovine growth plates, are elaborated in Mwale et al. (2000).
[c] See Visconti et al. (1996) for types IX. For an overview of collagen in tendon, ligament and bone healing, see S. H. Liu et al. (1995).
[d] See Y. Li et al. (1995) for ColX1a1 and chondrodysplasia, Tsumaki et al. (1996) for cis regulation, Furuto (1991) for chondrosarcomas, and Ayad et al. (1994) for collagen type XI. Dozens of human chondrodysplasias are known. A fifth of 25 chondrodysplasias with short limbs reviewed by Cohen (2002) involved mutations in collagen genes. See H. B. Sarnat and Flores-Sarnat (2004) for an approach to identification and classification of malformations based on patterns of gene expression during development.
[e] See Keene et al. (1991), Oh et al. (1993), Ayad et al. (1994) and Benjamin and Ralphs (1998) for collagen type XII. For an overview of collagen in tendon, ligament and bone healing, see S. H. Liu et al. (1995).

Type X collagen and hypertrophy

Given the appropriate chondrocyte population – and this is very much a property of only some chondrocytes – hypertrophy can be inhibited or induced *in vitro*. Foetal bovine serum, insulin-like growth factor-1, and fibroblast or platelet-derived growth factors induce the synthesis of type X collagen and hypertrophy of chick sternal chondrocytes. As determined by transfecting the chick collagen X promoter into chondrocytes, multiple negative elements in the gene inhibit gene expression in non-hypertrophic chondrocytes (non-HCCs).

In ovo, type X synthesis in chondrocytes within the cephalic (hypertrophic) portion of the sternum is associated with vascular invasion, as it is in long bones. Again, *in ovo*, sternal chondrocytes undergo apoptosis after terminally differentiating, but before the matrix resorption and vascular invasion associated with endochondral ossification.[4]

Sternal and tibial articular chondrocytes from chicks maintained *in vitro* synthesize type X collagen *independently of the size of the chondrocytes*, implying that hypertrophy and type X synthesis are not co-regulated. Sternal chondrocytes from 19-day-old embryos exposed to 100^{-35} or 10^{-35} nM retinoic acid synthesize type X mRNA, without any change in type II collagen or core protein mRNAs, again independent of cell size (Pacifici *et al.*, 1991a,b).

Bmp-6 is expressed in chick sternal chondrocytes as they differentiate and before the appearance of type X collagen. Both Bmp-6 and type X collagen are increased by treating sternal chondrocytes with 10^{-7} M 1,25 $(OH)_2$-D_3 (Box 31.2), suggesting that sternal chondrocyte proliferation/differentiation is regulated by a pathway(s) involving PTHrP. *Bmp-6* also is expressed in mouse HCCs, while CHO cells that express Bmp-6 induce tumours that contain cartilage and bone and that undergo endochondral ossification (Gitelman *et al.*, 1994). As *Bmp-6* enhances *Ihh* mRNA, it has been proposed that *Bmp-6* has a direct action on PTHrP while Ihh has an indirect action (Grimsrud *et al.*, 1999).[5]

Fibronectin

A few general comments on the important matrix glycoprotein fibronectin may be found in Box 20.2. Sternal chondrocytes cultured at high density in the presence of vitamin A bind almost three times more fibronectin to their cell surfaces than do chondrocytes cultured in the absence of vitamin A, the vitamin affecting binding rather than synthesis of fibronectin.

Alternative splicing of fibronectin is temporally and spatially regulated in sternal chondrocytes. The region encoded by the alternatively spliced exon IIIa is required for condensation of chondrogenic mesenchyme:

- levels of exon IIIa decline immediately after condensation;
- micromass cultures exposed to an antibody against exon IIIa show disrupted condensation and inhibition of chondrogenesis; and

- injection of antibody into limb buds results in smaller limbs and loss of skeletal elements.

Fibronectin also modified the type of collagen synthesized by sternal chondrocytes in collagen gels, two disulphide-linked molecules being produced in the presence of fibronectin. Similarly, sternal chondrocytes from 13-day-old embryos retain their differentiated state and continue to deposit ECM when cultured in collagen gels.[6]

Nanomelia

Mutants are enormously informative about the normal (wild-type) structure and function of organs, tissues and cells. Mutants that affect the skeleton have provided some of the most important information concerning skeletogenesis. Synthesis of chondroitin sulphate (CS) in chondrocytes from sternal and limb cartilage of chick embryos with *nanomelia* (*nm*) is depressed 90 per cent below wild-type level; synthesis in the skin is normal. The level of synthesis in sterna *in ovo* is no greater than that found in the H.H. 18 normal limb-bud mesenchyme, pointing to a block in the enhancement of synthesis in the mutant. That block is because nanomelic chondrocytes do not synthesize any proteoglycan core protein. Cultured wild-type sternal, vertebral and epiphyseal chondrocytes synthesize two major peaks of proteochondroitin sulphate. The major peak, which is missing from the medium in which *nm* sterna are cultured and severely reduced in the chondrocytes, may represent a cartilage-specific proteoglycan.[7]

TUMOUR INVASION

As a tissue, cartilage is rarely subject to tumours. When it is, the tumours do not invade cartilage *per se*, but rather encapsulate nodules of cartilages or infiltrate the cartilage via connective tissue canals.

Conversely, tumour invasion can be regulated by cartilage-derived anti-invasion factors. Thus, bovine metacarpophalangeal articular cartilage is resistant to invasion by osteosarcoma cells, even after freeze thawing to devitalize the cartilage. However, extracting cartilage with 1.3 M guanidium chloride ($GuCl_2$) to remove GAGs from the ECM (Fig 12.10) allows the cartilage to be invaded by some, but not all, osteosarcoma cell lines. Invasion can be inhibited by adding a cartilage-derived anti-invasion factor. Interestingly, and as an example of abnormally differentiated cells possessing some features of normal cells, collagen type XI is found in the articular cartilages of knee joints and in chondrosarcomas (Fig. 23.1).[8]

When human osteosarcoma or metastatic mammary carcinoma cells (Box 23.2) are co-cultured with human growth plates, the tumour cells invade the bone but not any unmineralized cartilage. Clonal cell lines derived from mouse osteosarcomas retain their tumorigenicity for

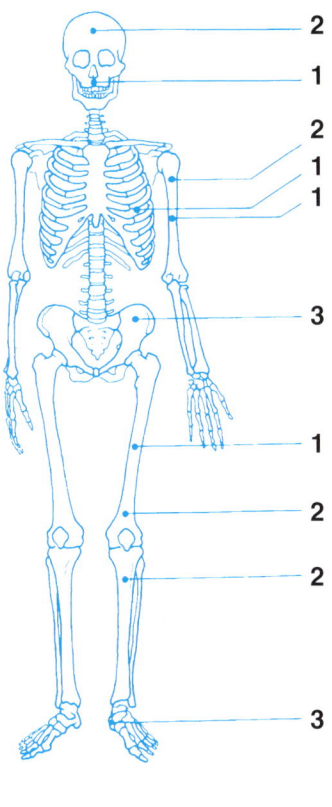

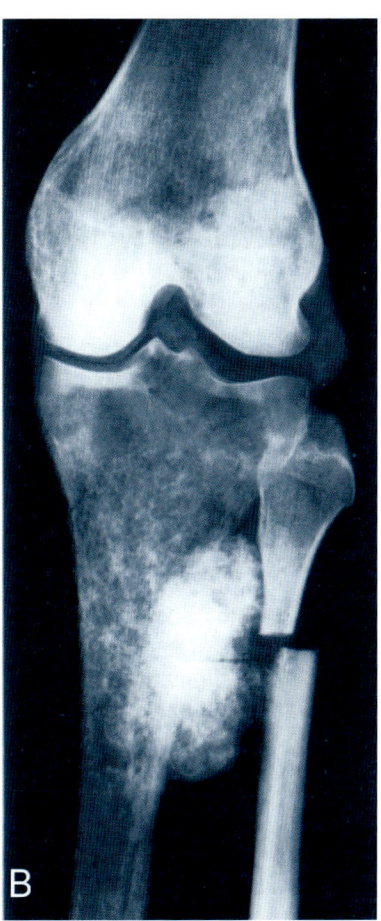

Figure 23.1 Mesenchymal chondrosarcomas in humans. (A) The distribution of 18 mesenchymal chondrosarcomas in the skeleton. One extraosseous tumor was located in the soft tissue of the upper arm. (B) A radiograph of a mesenchymal chondrosarcoma in the proximal tibia showing both loss of bone and a mineralized soft-tissue tumour (white mass adjacent to the fractured fibula). Modified from Dabska and Huvos (1983).

Box 23.2 Metastases, breast and prostate cancer

Eighty per cent of women with advanced breast cancer have bony metastases outside the breast. All cell lines tested from such patients can *resorb* bone in the absence of the usual stimulation required to activate osteoclasts. Breast carcinomas also synthesize soluble factors that enhance osteoblast differentiation.

Prostatic cancers also metastasize. Bone releases factors that:

- are chemotactic for prostatic carcinoma cells;
- facilitate cell attachment; and
- facilitate release of growth factors by carcinoma cells, factors that can initiate osteogenesis.

Prostatic cells, in turn, release osteoblastic and osteolytic factors that regulate bone remodeling and the release of growth factors from bone matrix, all of which indicates a complex interplay or cross-talk between the two cell types. Prostate glands (along with uteri) release mitogens with selective actions on osteoblasts. One possible mechanism for the spread of osteoblastic metastases from the prostate is the mitogenic response of osteoblasts to prostate-specific antigen (PSA), which activates latent Tgfβ and is associated with proteolytic modulation of cell adhesion receptors.[a]

[a] See Eilon and Mundy (1978) for resorption of bone, Evans *et al.* (1991) for factors enhancing osteoblasts, Killian *et al.* (1993) and Keller and Brown (2004) for prostate studies, and Koutsilieris *et al.* (1986) and Koutsilieris (1989) for mitogens.

many years in culture, and secrete a protein that induces bone when the cells are implanted *in vivo*. Although the osteoinductive factor is lost with prolonged culture, it is regained after implantation.[9]

VASCULARITY

Vascularity has long been known to be important in skeletogenesis:

- vascularity promotes osteogenesis, avascularity favours chondrogenesis;
- vascularity enhances higher oxygen tensions, and so indirectly favours osteogenesis over chondrogenesis;
- formation of a haematoma is the first step in repair of a fractured bone, both to fill the gap, and to provide a source of cells and growth factors; while
- patterns of vascularity vary by cartilage type and with disease state.[6]

Healing of fractures of foetal rabbit long bones *in utero* involves less inflammation and haematoma formation than seen in repair of fractures in newborn bunnies, with more periosteal and endosteal proliferation, a larger callus and earlier chondrogenesis (Ris and Wray, 1972). Even without a fracture, subperiosteal bone can form in haematomas in the human orbit (Sabet *et al.*, 2001).

As an example of *change with cartilage type*, at the end of somatic growth in the long-tailed or crab-eating macaque, *Macaca fascicularis*, the hyaline cartilage of the mandibular condyle – which is not invaded by blood vessels – transforms into an articular cartilage, which is invaded by vessels. Chondrocytes cease enlarging, do not die, and the ECM no longer excludes blood vessels. With vascular invasion comes mineralization. A similar set of changes accompanies hypertrophy and replacement of cartilage by bone during endochondral ossification.[11]

Cartilage *develops* quite happily from mesenchyme in the absence of a vascular supply – as, for example, in organ culture – and can be maintained as a differentiated tissue without vascularization, obtaining metabolites by diffusion through the ECM. During early development, and at those sites where it is retained as a permanent tissue, cartilage is avascular and unmineralized. Cartilage *may* be invaded by vascular channel as a prelude to endochondral ossification. In such situations the cartilaginous matrix mineralizes and chondrocytes degenerate or transform to osteoprogenitor cells (Chapter 12).

As an example of a *disease state*, mandibular condylar, tibial epiphyseal and articular cartilages of guinea pigs, *Cavia porcellus*, show different changes in vascularity – and different patterns of mineralization – in response to scurvy. Condylar cartilage doubles in width after 26 days on a scorbutic diet, but can return to normal width after three days of ascorbic acid treatment. Tibial cartilage is much less responsive.[12]

RESISTING VASCULAR INVASION

It became evident in the 1970s that:

- grafted cartilage actively resists invasion by blood vessels;
- an active component confers this ability (and the ability to resist invasion by tumours?); and
- the agent could be isolated from cartilage.

It did not escape the attention of those engaged in these studies that such a factor might be useful for containing the growth of a variety of tumours by inhibiting their vascularization, although few believed that such a mechanism would work, or be sufficiently selective to be effective. Understanding the mode of action of such a factor also should provide a better understanding of how some cartilages remain as permanent unmineralized tissues. Interestingly, cartilage develops in the avascular cores of limb buds. Hyaluronan accumulates under the AER and in the distal sub-ridge mesenchyme where it aids in blocking differentiation. Epithelia with high levels of hyaluronan (amniotic, dorsal and limb-bud epidermis) enhance avascularization of mesenchyme; epithelia with low levels (pigmented retinal, lens, corneal and yolk-sac epithelia) promote vascularization.[13]

One body of data, derived from the grafting of cartilage from one species to another, indicates that *intact cartilage is immunologically privileged*; homografts of mammalian cartilage survive longer than any grafted tissue other than cornea. GAGs of the ECM prevent exposure of chondrocytes to products from other tissues or to antisera. Conversely, chondrocytes separated from their ECMs and grafted, sensitize the host, and elicit an immunological response. Antibodies against cartilage GAGs can suppress chondrocyte hypertrophy and reduce replacement of cartilage by bone, a finding consistent with the concept that cartilage must be intact to resist replacement by bone. Type II collagen from articular cartilage has antigenic specificity associated with the glycine–proline–alanine sequences of one fraction. This antigenicity may be masked by GAGs of the ECM and so not elicit a response.[14]

Evidence for a more active resistance to vascular invasion came from a study in which various tissues were tested for their resistance to invasion when grafted on Millipore filters to the CAM of host embryonic chicks (Box 12.3). Tissues normally are vascularized rapidly by invasion of blood vessels from the chorionic mesenchyme. Live and devitalized hyaline cartilage resists vascular invasion, provided that it is not mineralized, and provided that the ECM is intact (Fig. 12.10). Mineralized cartilage and bone are invaded rapidly by blood vessels (Fig. 12.10). Resistance to vascular invasion is reduced when unmineralized cartilage is extracted with 1 M $GuHCl_2$ (see above) and grafted (Fig. 12.10).[15]

Extraction with guanidium chloride was used to isolate a multiprotein fraction from calf scapular cartilage that inhibits vascular invasion of tumours induced in rabbit corneas. These tumours are normally vascularized at a maximum rate of capillary growth of 600 μm/day. In the presence of the extract, capillary growth is reduced to 200 μm/day, after which vessel regression sets in.

The inhibitory factor has protease activity. An inhibitor fraction isolated from bovine nasal septum, characterized by lack of collagen and less than 10 per cent GAGs, may also be a protease inhibitor; human growth-plate epiphyseal cartilage also produces a neutral protease (first isolated in 1982) that degrades proteoglycan. This fraction inhibits growth of endothelial cells *in vitro* – the possibility of cartilage actively retarding blood vessel growth is real – and is mimicked by known protease inhibitors. Inhibition of endothelial cell growth is dose dependent (Table 23.1), suppressing the action of collagenase from human skin and osteosarcomas; cells derived from human osteosarcomas secrete collagenase *in vitro* if challenged with heparin unless a less than 50 000-Da cartilage extract is added. The anti-invasion factor in bovine cartilaginous matrix resides in the ECM, and is blocked by anti-Tgfβ antibodies.

Table 23.1 Dose-dependent inhibition of growth of endothelial cells *in vitro* in the presence of extracts isolated from bovine nasal septal cartilage[a]

	Concentration of cartilage extract added (μg/ml)	Number of endothelial cells/dish[b]
Crude extract	0	700 000
	20	510 000
	100	285 000
	500	52 000
<50 000 MW fraction	5	300 000
	20	260 000
	100	65 000
	500	22 000
>50 000 MW fraction	500	710 000[c]

[a] Based on data from Eisenstein *et al.* (1975), who should be consulted for the method of preparation and purification of the extract.
[b] Cell number counted four days after seeding 40 000 cells/dish.
[c] This fraction was inactive.

Interestingly, cultured endothelial cells release a mitogen with potent action on proliferation but not on collagen synthesis of osteoblasts from one-day-old rats; it neither acts on fibroblasts nor on articular chondrocytes. Osteogenesis is enhanced (induced) when foetal calvarial osteoblasts are co-cultured in agarose with endothelial cells, perhaps because of mitogens released by the endothelial cells. A recent study (Shin *et al.*, 2004) shows that endothelial cells secrete Bmp and that medium conditioned by endothelial cells elicits osteogenesis from marrow stromal cells.[16]

One can readily see how the inhibition of collagenase activity could enable cartilage to resist vascular invasion, maintain its differentiated and unmineralized state, and resist replacement by bone. It would be necessary to inactivate or remove the inhibitor before cartilage could either produce its own collagenase – none has been found – or be invaded.[17]

Ossification, with onset of ageing or in pathologies such as Addison's disease, of tracheal, epiglottal and elastic cartilages that normally do not ossify, indicates that resistance to vascular invasion is lost. Acquisition of this ability would be a welcome replacement for the distress of prolonged corticosteroid treatment individuals with Addison's disease must suffer.[18]

Vascular invasion of long-bone cartilage is mediated by degradation of cartilaginous ECM. The enzyme gelatinase B is found in the periphery and immediately outside cartilages in associated with type II collagen fibres. Gelatinase B and collagenase-3 lyse collagen fibrils, and proteinases cleave the core protein of aggrecan to facilitate vascular invasion and replacement of cartilage by bone (see Box 4.2).[19]

In contrast to vascular inhibitors, a low-molecular-weight, angiogenic procollagenase *activator* isolated from epiphyseal growth-plate cartilage, and from mineralized but not from unmineralized chondrocytes in cultures, was claimed to be the first report of an angiogenic compound produced by chondrocytes in culture. Others claimed that, when they isolated a component from chick scapular chondrocytes, they were the first to show *in-vitro* production of a vascular inhibitor by cartilage. The first chemo-attractant for angiogenesis was isolated from HCCs by Alini and colleagues in 1996.[20]

INHIBITORS OF ANGIOGENESIS AND VASCULAR INVASION

The molecular nature of these vascular inhibitors is slowly being revealed.

Chondromodulin-I, a 25-kDa glycoprotein that functions as an endothelial cell-growth inhibitor, is found in cartilage and chondrogenic precursors in chick embryos, inhibits vascular invasion of cartilage, and therefore decreases endochondral bone formation. Murine *calmodulin* has been cloned, is expressed with type II collagen during chondrogenesis, but is not found in the hypertrophic zone. Similarly, chondromodulin-1 and type II collagen can be induced in a line of embryonic carcinoma cells (ATDC5) in which expression of chondromodulin-1 decreases as type X collagen appears with hypertrophy of the cells.[21]

Vascular endothelial growth factor (Vegf)

A major difference between chondrogenic and osteogenic condensations is that the former are not vascularized but the latter are; vasculature must regress from the condensed mesenchyme in mouse limb buds before chondrogenesis can begin (Figs 23.2 and 23.3). Vascular endothelial growth factor (Vegf-A), a growth factor that stimulates proliferation of endothelial cells, enables vascular invasion, and consequently influences various stages of skeletogenesis stages from initial condensation to bone remodeling, is a major player.

Vegf-A mRNA is expressed in sclerotomal and limb-bud mesenchyme at 10.5 days, declines during chondrogenesis, and is up-regulated again during chondrocyte hypertrophy preceding vascular invasion; chondrocytes create the environment for vascular invasion and subsequent endochondral ossification. Vegf further enhances endochondral ossification by acting as a chemoattractant for osteoclasts and hence for invasion of the hypertrophic cartilage of growth plates. Finally, Vegf influences skeletal patterning, exogenous Vegf increasing vascularization and decreasing digit formation (Fig. 23.2).[22]

Inactivation of *Vegf-A* is lethal in most heterozygote mice. The few foetuses that survive to 17.5 days have abnormal endochondral bone formation. *Vegf*'s role in osteoblast activity is further revealed by the skeletal deficiencies in *Vegf*[120/120] double homozygous mice, in which recruitment of blood vessels into the perichondria and invasion of vessels into the primary ossification centre of the long

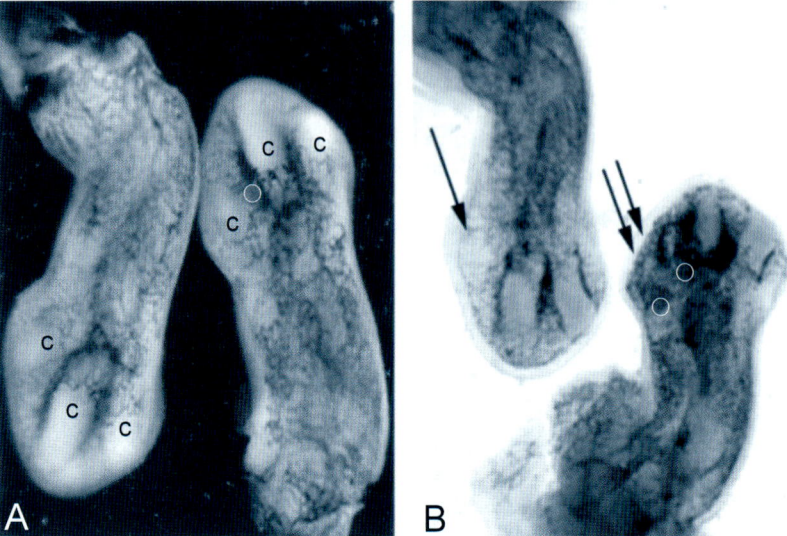

Figure 23.2 Vascular endothelial growth factor (Vegf) and vascular development during chick wing development. (A) On the left is a control, on the right a wing bud into which an unloaded bead (circle) was implanted at H.H. 26. One day later the embryos were perfused with India ink (black). Note the absence of blood vessels from the condensations (c) for the digits. (B) Developing wing buds from a single embryo. On the left is the control with blood vessels excluded from the condensations (arrow). On the right is the contralateral wing into which two beads (circles) loaded with 1 mg/ml Vegf were implanted at H.H. 26. The ectopic Vegf initiated additional vascularization (arrows). Modified from Yin and Pacifici (2001).

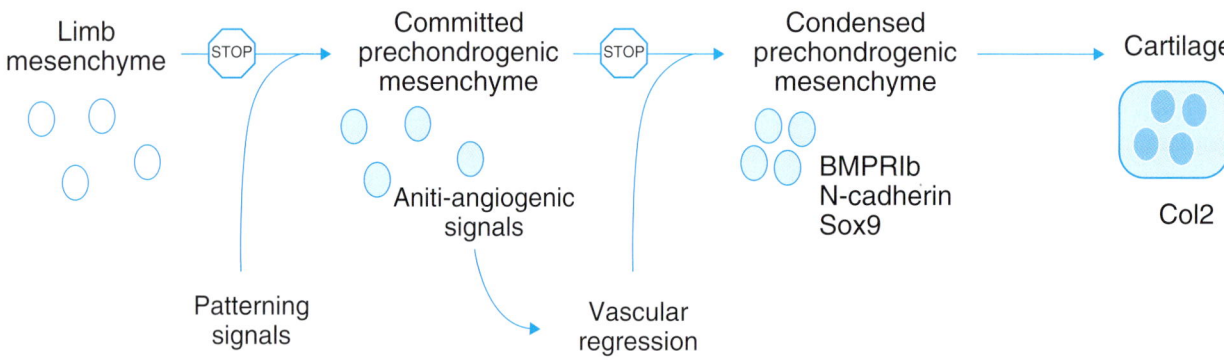

Figure 23.3 A model of the role of vascularization in chondrogenesis of wing-bud mesenchyme. Commitment of mesenchymal cells as prechondrogenic is followed by synthesis of anti-angiogenic signals that limit vascular invasion, facilitating condensation, synthesis of BMPR1b, N-cadherin and Sox–9 and differentiation of cartilage containing collagen type II (Col2). The stop signs indicate positions in the pathway where lack of signaling either inhibits chondrogenesis or results in abnormal chondrogenesis. Based on Yin and Pacifici (2001).

bones are both delayed, indicating that Vegf is involved in early and late stages of cartilage vascularization.[23]

Recent study demonstrates that expression of *Vegf* is decreased in the expanded growth plates associated with impaired endochondral ossification in *CTgf*-deficient mice (Ivkovic *et al.*, 2003). Connective tissue growth factor (CTgf) is a secreted protein with several domains that facilitate interactions with other growth factors. CTgf has its highest expression in the developing vascular system and chondrocytes, regulates cartilage remodeling, mediates the deposition of collagen in wound healing, and enhances neovascularization *in vitro*.

PTH-PTHrP

Vascular invasion of primary cartilage also is regulated, in part, via the PTH–PTHrP pathway; mice transgenic for the hormone or the hormone-related protein show decreased

vascular invasion following a delay of chondrocyte hypertrophy. The hypertrophic phase is prolonged and mineralization delayed. Fgf also plays a role; injecting Fgf-2 into proximal tibial growth plates of six-week-old New Zealand white rabbits promotes vascular invasion and ossification, chondrocytes being targets for Fgf-2 (Box 23.3).[24]

INTERPUBIC JOINTS AND THE TRANSFORMATION OF CARTILAGE TO LIGAMENT

Publications appeared during the latter two thirds of the 18th century whose purpose was to demonstrate that differences between the human sexes were not merely skin-deep, but extended to internal differences evident in the skeleton. The ramifications of these publications go well beyond the scope of this book, but they do shed

Box 23.3 Fgf-2 and a few of its relatives

As discussed in Box 3.2, the first growth factor, Egf, was discovered in the early 1960s. Fibroblast growth factor (Fgf), the next growth factor to be discovered, is now known to be one of 18 members of a fibroblast growth family with important roles in embryonic development, including skeletal development.[a]

More recently, Fgf receptors have been shown to have major roles in skeletogenesis, especially in maintaining the sutures in growing mammalian skulls. Receptor action is primarily discussed in Box 31.2 and Chapter 34. The Fgf with the greatest influence on skeletal development is Fgf-2 (bFGF). Discussion of the skeletogenic roles of Fgf-2 abounds in the text. A telegraphic summary of those roles is that Fgf-2 is produced by skeletal cells and can initiate or enhance both osteogenesis and chondrogenesis.

Osteogenesis

Cultured bovine bone cells synthesize and secrete Fgf-2 and store it in bone matrix. Fgf-2 is mitogenic for osteogenic cells, its mitogenic action being potentiated by Tgfβ (which is not itself a mitogen for osteogenic cells). Fgf-1 enhances proliferation of osteogenic cells under the control of Tgfβ but is nowhere near as effective a mitogen as Fgf-2.

Synergistic interaction of Fgf-2 with other growth factors in bone matrix has been assessed using mouse calvarial cells. Fgf-2, Tgfβ-1 and Pdgf each increase DNA synthesis and decrease alkaline phosphates activity, while Igf-2 has little effect on alkaline phosphatase. In combination at low concentrations, Fgf-2, Tgfβ-1 and Igf-2 act synergistically to modulate the proliferation of osteogenic cells (Kasperk et al., 1990).

Fgf-2 enhances osteogenesis from bone marrow cells: 0.3 ng/ml increases proliferation fourfold (the half maximum concentration is a low 15 pg/ml), increases alkaline phosphatase 3.5-fold, and after 21 days of exposure results in six times more bone at 3 ng/ml.[b]

Proliferation of osteoblast and osteoclast precursors is controlled in part by Fgf-2 acting through hepatocyte growth factor/scatter factor (HGF/SC), which up-regulates expression of the Fgf-2 receptor. Injecting Fgf-2 directly into the proximal tibial growth plates of six-week-old rabbits promotes vascular invasion and ossification, chondrocytes being targets for Fgf-2.[c]

Chondrogenesis

Colony formation by rabbit chondrocytes in vitro is enhanced three- to fourfold by Tgfβ in the presence of Fgf-2, while addition of Tgfβ decreases the concentration of Fgf-2 required to enhance chondrogenesis by 40–100-fold. Tgfβ and Fgf combine to stimulate the proliferation of rabbit growth-plate chondrocytes maintained in vitro (Kato et al., 1988).

Tgfβ increases the responsiveness of chondrocytes to Fgf: Tgfβ at 1 ng/ml and Fgf-2 at 10 ng/ml act synergistically to regulate DNA sequences to inhibit the synthesis of collagen type II by chondrocytes. Fgf-2 enhances proliferation and synthesis of ECM by chondrocytes but inhibits the differentiation of HCCs. Growth-plate chondrocytes bind I[125]-labeled Fgf-2, HCCs do not; they lack the Fgf-2 receptor that mediates differentiation of earlier stage cells.[d]

Fgf-2 modulates growth of the condylar secondary cartilage of mouse mandibles: Fgf-2 is expressed in the prechondrogenic condensation, perichondrium and proliferative cell zone. Culturing condyles in rhFgf-2 (25–1000 ng/ml) reduces proliferation and the size of the HCC zone. At the molecular level mRNA for type X collagen and aggrecan are reduced, while Ihh, PTHrP, BMP-4 and Runx-2 are all down-regulated (Ogawa et al., 2003).

Limb buds

Riley et al. (1993) provided the first evidence of involvement of Fgf-2 in limb development; ectopic expression of Fgf-2 in a retroviral vector duplicated the anterior skeleton (consisting of the proximal humerus, distal radius and digits I and II) in 90 per cent of the individuals. The more posteriorly in the limb bud the Fgf-2 was placed, the fewer were the duplications. Implanting Fgf-2-soaked beads produces an additional digit II. See Chapter 33 for the role of Fgf-2 in establishing antero-posterior polarity in limb buds, and see Fig. 12.2 for beads implanted into chick embryos.

Other family members

Additional Fgf family members are being discovered.

Embryonic chick-limb mesenchyme cultured in the absence of epithelium forms cartilage in thin sheets but rounded cartilaginous nodules if epithelium is present. This epithelial morphogenetic influence is mediated by Fgf-2 and Fgf-8, either of which can substitute for the epithelial signal to activate FgfR-2. Fgf-2 and -8 are both epithelially localized in limb buds. Fgf-7, a mesenchymal Fgf, cannot substitute for the epithelium (Moftah et al., 2002). Actions on skeletogenesis of Fgf-4, -8 and -10 are discussed in Box 14.4.

Two further members, Fgf-12 and Fgf-13 – also known as fibroblast growth factor homologous factors -1 and -2 (Fhf-1 and Fhf-2) – are expressed in limb-bud mesenchyme, Fgf-12 in the zone of polarizing activity within the posterior mesenchyme (for which, see Chapter 38), and Fgf-13 distally and anterior in the progress zone (for which, see Chapter 35). Consistent with a role in early specification of limb buds, neither Fgf-12 nor Fgf-13 are expressed in limb buds in wingless or limbless mutant embryos. Consistent with a role in specification of distal skeletal elements, expression of Fgf-13 is expanded in talpid² embryos. Fgf-13 up-regulates Hoxd-13, Hoxd-14, Fgf-4 and Bmp-2 in patterning the antero-posterior limb axis (Munoz-Sanjuan et al., 1999).

Fgf-17 and Bmp-2 are expressed in the specialized ventral epithelial ridge (VER) on the tail buds of mouse embryos (Chapter 43). Fgf-18 plays an important role in regulating growth plates, while in zebrafish Fgf-24 (a member of the Fgf-8, -17, -18 family) activates Fgf-10 in lateral-plate mesoderm and is required for Tbx-5-positive cells to migrate to the fin bud.[e]

[a] For a major review of Fgfs as multifunctional signaling factors see Szebenyi and Fallon (1999).
[b] See Globus et al. (1988, 1989) for Fgf-2 as a mitogen, and Noff et al. (1989) for the bone marrow study.
[c] See Blanquaert et al. (1999) and Baron et al. (1994). See Riancho and Mundy (1995) for a review of interactions of growth factors with systemic hormones, vitamin D, calcitonin and corticosteroids in bone.
[d] See Iwamoto et al. (1989, 1991) and Horton et al. (1989).
[e] See Goldman et al. (2000) for the VER, and Fischer et al. (2003) for Fgf-24.

light on how visualization of the skeleton is used to serve a multitude of ends,[25] sexual dimorphism of the interpubic joints being one.

Sexual dimorphism of the mammalian pelvic girdle is evident in the size and shape of the ilium, ischium and pubic bones (which, collectively and when fused, are known as the innominate bone),[26] and in the nature of the tissues within the pubic symphysis, the latter being the juxtaposition of left and right pubic bones to form the interpubic joint of the pelvic girdle. The bones of the pelvic girdle display one of three fates that allow three types of mammalian pubic symphyses to be recognized. They:

- fuse at the pubic symphysis, forming a *synostosis* (some rodents);
- are connected by an interpubic cartilage forming a *synchondrosis* (some rodents, carnivores, primates and the Artiodactyla); or
- are connected by a ligament forming a *syndesmosis* (some rodents and bats).[27]

Between them, synostosis, synchondrosis and syndesmosis encompass the three types of joints found between skeletal elements. Sutures, as seen between skull bones, are normally not mobile, and so are not joints (Figs 15.5 and 17.10). By examining the functional significance of the mandibular symphysis in 22 species from six orders of mammals, a sequence from ligamentous symphysis → cartilage → mineralized cartilage and finally → bony symphysis can be recognized (Beecher, 1979). The sequence parallels toughness of the diet, the presumption being that the evolutionary response to stress imposed by the diet favoured change in composition of the symphyseal tissues (Table 23.2 and see Box 12.2). So, functionally equivalent levels of stress and strain are maintained across the mandibular symphysis during ontogeny of two species of macaques, the crab-eating macaque, *Macaca fascicularis*, and the pig-tailed macaque, *M. nemestrina* (Vinyard and Ravosa, 1998).

In what follows, I concentrate on those vertebrates in which the interpubic joint is a synchondrosis, within which several types of joint may be distinguished on the basis of the histological changes that occur within the interpubic cartilage during pregnancy.

Cartilage → ligament

The behaviour and fate of the interpubic cartilage varies from taxon to taxon. In primates, there is some *softening* of the cartilage so that by parturition the joint is flexible.

Table 23.2 Structural differences between mandibular symphyses of 22 species of mammals compared with their diets[a]

Symphyseal structure	Ligament orientation	Species	Diet
Ligamentous with fibrocartilage attachment	Cruciate	*Didelphis azarae* (Azara's opposum)	Omnivorous
Ligamentous with fibrocartilage attachment	Transverse	*Ochotona princepes* (pika)	Leaves
Ligamentous	Cruciate	*Sciurus carolinesis* (northern gray squirrel)	
Ligamentous with fibrocartilage attachment	Transverse and cruciate	*Rattus norvegicus* (Norway rat)	
Ligamentous with fibrocartilage attachment	Cruciate	*Microcebus murinus* (mouse lemur)	Fruit
Ligamentous with fibrocartilage attachment	Cruciate	*Galago demidovii* (dwarf galago)	Insect
Interdigitation of bone		*Lemur fulvus* (brown lemur)	Leaves
Interdigitation of bone		*Propithecus verreauxi* (Verreaux's sifaka)	Leaves
Interdigitation of bone		*Hapalemur griseus* (gray gentle lemur)	Leaves
Interdigitation of bone	Transverse and cruciate	*Sylvilagus brasiliensis* (South American rabbit)	Leaves
Interdigitation of bone	Irregular orientation	*Herpestes auropunctatus* (mongoose)	Leaves
Interdigitation of bone		*Pteronotus psilotis* (naked-backed bat)	
		Galago crassicaudatus (thick-tailed galago)	Fruit
Ligamentous	Cruciate	*Galago senegalensis* (lesser bush *baby*)	Insects
		Perodicticus potto (potto)	Fruit
		Lemur macaco (black lemur)	Leaves
		Tarsius bancanus (Horsfield's tarsier)	Insects
		Tupaia glis (tree shrew)	Insects
		Molossus coibensis (little mastiff bat)	
		Desmodus rotundus (common vampire bat)	
	Transverse	*Phyllostomus hastatus* (carnivorous bat)	Carnivore

[a] Based on data from Beecher (1979).

Exceptions are species such as African lorises, in which the foetus is small even at birth.[28]

The interpubic cartilage in rodents and guinea pigs is *converted* into a flaccid interpubic *ligament*, accompanied by some resorption of the medial edges of the pelvic bones during pregnancy. Extensive resorption of the medial edges accompanies formation of the interpubic ligament in mice; collagenase is only produced by symphyseal tissues immediately before, during and immediately after parturition. Modulation of the cartilage to ligament in guinea pigs is controlled by *relaxin*, a peptide hormone secreted by the corpus luteum during pregnancy, and that also may modulate fluid balance within the body.[29]

Sex-based differences in resorption of the parietal bones in mice have also been reported. Following observations published in 1983 that the outer surfaces of parietal bones are flat with few osteoclasts, but the inner endocranial surfaces are rough with many osteoclasts, Abe *et al.* (1984) used the amount of rough surface to determine that after puberty the rough area occupied 40 per cent of the endocranial surface in females compared with 10 per cent in males, a difference attributed to hormonal influences.

Removing chondroitin sulphate unmasks the fibrous component of articular cartilage, leading to the production of an avascular connective tissue on the human ulna during postnatal growth (Haines, 1976), but this is not quite the same as transformation to a ligament. Cartilage can form in ligaments; however, as discussed in Chapter 12, I know of several other instances of transformation of cartilage to ligamentous tissue during development. All involve Meckel's cartilage.

- A portion of Meckel's cartilage in mammals transforms to the sphenomandibular (tympanomandibular) ligament. Tension, rather than hormonal stimulation, is the initiating factor (Chapter 13).
- Mouse Meckelian chondrocytes *in vitro* respond to Egf with enhanced DNA synthesis and cell division, inhibition of chondrogenic differentiation, and transformation to fibroblasts.
- Formation of the myelohyoid bridge, which is an inherited deficiency in Meckel's cartilage, involves conversion of cartilage to ligament.[30]

Mediation by oestrogen and relaxin

Transformation of cartilage to ligament is an example of a sexually dimorphic skeletal character and hormonal response. In this case, the transformation normally only occurs in females.

Different patterns of modeling of the bone of the mandibular condyle in male *and* female mice are also known to occur in response to oestrogen and androgen, as demonstrated using ovariectomy and orchiectomy (Fujita *et al.*, 2000).

A further fascinating example of a sexually dimorphic skeletal change has been described in the mammal with the most derived life style and reproductive behaviour, the naked mole rat, *Heterocephalus glaber*, of Kenya, Ethiopia and Somalia. In this species, as in such social insects as ants, one female is the reproductive female (queen) for the underground colony, other individuals (male and female) aiding her in foraging for food, raising the young and defending the colony. Average litter size is 12, but as many as 27 pups have been recorded in a single litter, and a single queen can produce as many as five litters a year. The lumbar vertebrae in breeding females are 50 per cent longer than the lumbar vertebrae in virgin and non-breeding females, a difference that O'Riain *et al.* (2000) think is probably caused by prostaglandins associated with the later stages of pregnancy, or by oxytocin and prolactin levels associated with birth and lactation.

Gardner (1936) first confirmed experimentally that formation of the interpubic ligament and resorption of the medial edges of the pelvic girdle in female mice are controlled by oestrogen. Resorption of the bone is discussed fully in Chapter 15. Both transformations occur in males if the testes are removed – to remove testosterone – and if they are given oestrogen. Gardner thought these differential responses were attained at sexual maturity, but they occur earlier; oestrogen produces a specific hydration and swelling of interpubic cartilage when administered to virgin female mice. If oestrogen and relaxin are both administered, the cartilage swells and is converted into an interpubic ligament. The same changes can be observed after administering hormone to gonadectomized males or females, even when the pelves are transplanted to an ectopic site. Such hormone-induced changes are rapid; dissolution of cartilage is apparent after one day's treatment.[31]

The bulk of the experimental work on the hormonal control of the sexual dimorphism in mice, was undertaken by E. S. Crelin and his colleagues at Yale University. The predetermined, agonal bony pelvic type in mice is female. Sexual dimorphism of the pelvis first becomes evident between 17 and 20 days after birth.

If mice of either sex are gonadectomized at birth, the female type of pelvis is present at sexual maturity in all cases. Administration of testosterone to gonadectomized mice produces a male-type bony pelvis. Muscle function does not influence the type of pelvis that develops (Crelin, 1960), but testosterone transforms the agonal (female) type into a male pelvis. Implanting immature pelvis joints to ectopic sites (usually into the pectoral musculature), or cultivating embryonic pelvic primordia for a time before reimplanting them to the pelves of five-day-old females, shows that the unique hormonal responses of the interpubic cartilage are *acquired* by precursor cells of the cartilage during development.[32]

Chronology of the histogenesis of the female interpubic joint is shown in Table 23.3. One week after birth, cartilage is present at the medial edges of the pelvic bones.

Table 23.3 The chronology and histogenesis of the interpubic joints in female mice[a]

Age (weeks after birth)	Histology
One	Ventral portion of the pelvis is made of a solid, inflexible bar of small-celled hyaline cartilage and connects the pubic and ischial bones; a cartilage lamina is present on the cranial half of the bar
Two	Slight growth of interpubic bar; cartilage lamina develops into fibrocartilage
Three	Inferior rami of pubic bone extends caudally on each side of interpubic bar; fibrocartilaginous lamina extends almost to caudal border of interpubic cartilage bar
Four	Inferior rami of pubic bones forms lateral walls of the pubic synthesis; intermediate fibrocartilage lamina is broad and extends completely through the interpubic tissue
Five	Little change in the pelvis from four weeks of age
Six	The symphysis resembles that of a mature virgin female, differing from that of the male with wider, thicker bones forming the lateral walls of the symphysis

[a] Based on data from Crelin and Levin (1955).

After three weeks, an immature pubic symphysis is present, and by four weeks fibrocartilage appears within the interpubic cartilage. By six weeks postnatally, ligament formation and sexual dimorphism of the pubic symphysis are complete (Crelin and Levin, 1955).

Conversion of symphyseal cartilage into an interpubic ligament involves release of enzymes from chondrocytes and *liberation* of chondrocytes from ECMs. After modulating to a fibroblastic state these cells deposit collagen, which they align across the interpubic joint, the collagen in the cartilage having been arranged randomly. Mitosis is reinitiated within the chondrocytes as they transform to fibroblasts; the role played by hormones in this part of the process has not been determined. Other possible fates of these chondrocytes were discussed in Chapter 13. The parallel to the behaviour and fate of HCCs released from their matrices (discussed above) is striking.

Formation of the interpubic ligament is reversible. If females with an interpubic ligament are ovariectomized, the ligament reverts to cartilage, indicating that maintenance and initiation depend upon a continued supply of oestrogen and/or relaxin. Perhaps reversal is possible because the ligament retains some features of cartilage – it is flaccid, hydrated, and contains considerable quantities of GAGs.[33]

The response of interpubic cartilage to oestrogen and relaxin cannot be initiated simply by placing the tissues in direct contact with the hormones, whether by implanting pellets, intrasymphyseal injection, or culturing them in the presence of hormone. The longer the ovary is present before puberty, the greater the enhancement of ligament formation. Accordingly, the longer the testis is present, the greater the inhibition of ligament formation.

Conditioning by growth hormone and thyroxine also is necessary; growth hormone is required for oestrogen to act on the symphyseal cartilage and pubic bones in female mice. Calcitonin inhibits this resorption, although the precise roles of these other hormones with regard to the interpubic cartilage have not been determined.[34]

In addition, interpubic cartilages may in part serve a mechanical function. If the symphysis in nulliparous females is severed, the pubic bones separate, and the cartilage disappears. The ligament produced after an interpubic joint is grafted back to a pelvic site is longer than that produced when the joint is transplanted back to an ectopic site, indicating a mechanical involvement in the determination of ligament length (Crelin, 1960, 1963). However, since paralysis of pelvic musculature *in situ* does not affect the length of the interpubic ligament, muscle action may not play a vital role in normal development.

Several examples cited in this and earlier chapters illustrate some of the conditions under which the chondrocyte phenotype can be altered. Such alterations can lead to modulation of the phenotype, modulation of function, and/or dedifferentiation and subsequent redifferentiation, alterations that are evident following injury, hormonal or vitamin deficiency or excess, and diseases or syndromes, the latter often being mutation based. Examples in each of these categories for how chondrocyte stability is maintained and/or can be disrupted, are taken up in Chapters 26 and 27. Before that, and because cartilage and bone, and chondrocytes and osteocytes, are so interrelated, I use Chapters 24 and 25 to discuss osteoblasts/cytes and bone diversity to complement the discussion of cartilage and chondrocyte diversity in Chapters 22 and 23, beginning with a consideration of whether osteocytes continue forever or whether old osteocytes just fade away. We will discover that some, like the good ageing actors they are, take on new roles that better suit their mature status.

NOTES

1. See Solursh and Meier (1973, 1974) and Solursh et al. (1973) for this factor, and Vasan and Miller (1985) for the monomers. Meier and Hay (1974a,b) and Solursh and Karp (1975) contain excellent discussions of the possible mechanisms of such feedback for chondrocytes and corneal epithelium.
2. See Kravis and Upholt (1985) for mRNA copy number, and Fernandez et al. (1985) for methylation.
3. See Schindler et al. (1975, Gibson and Flint (1985), Kielty et al. (1985), Gibson et al. (1986), Solursh et al. (1986), Gerstenfeld et al. (1989) and Iyama et al. (1991) for differential synthesis of collagen types, and Gibson et al. (1982, 1983) and Mendler et al. (1989) for type IX and collagen fibril formation.
4. See Bruckner et al. (1989) and Bohme et al. (1992) for induced hypertrophy, Lu Valle et al. (1993) for the promoter study, Gibson and Flint (1985), Gibson et al. (1986), Solursh et al. (1986), Lu Valle et al. (1989) and Iyama et al. (1991) for vascular

invasion, and Gibson *et al.* (1985) for sternal chondrocyte apoptosis.

5. Bmp-6 protein also is found in various epithelial tissues in mice beyond 9.5 days of gestation (Wall *et al.*, 1993).

6. See Hassell *et al.* (1979) for the response to vitamin A, ffrench-Constant and Hynes (1989) and Gehris *et al.* (1997) for alternative splicing, and Gibson *et al.* (1983) and Kimura *et al.* (1984b) for culture in collagen gels.

7. See Mathews (1967b) and Fraser and Goetinck (1971) for reduced CS synthesis, Argraves *et al.* (1981), Goetinck *et al.* (1981) and Kosher *et al.* (1986b) for lack of proteoglycan core protein, and Palmoski and Goetinck (1972), Pennypacker and Goetinck (1976), Okayama *et al.* (1976) and Wiebkin and Muir (1977) for the missing proteoglycan.

8. See Kuettner and Pauli (1983) for the pattern of invasion, Pauli *et al.* (1981a,b) for addition of the anti-invasion factor, and Furuto *et al.* (1991) for collagen type XI.

9. See Kuettner *et al.* (1977b, 1978) and Kuettner and Pauli (1983) for the cancer cells, and Amitani *et al.* (1974) and Amitani and Nakata (1975a,b, 1977) for the clonal cell lines.

10. See Trueta and Little (1960), Trueta (1968), Brooks (1971), Little (1973) and Singh *et al.* (1991) for the vascularity of bone, and Schenk *et al.* (1968) and Kuettner and Pauli (1983) for the vascularity of cartilage.

11. See Luder and Schroeder (1992).

12. See Irving and Durkin (1965) and Durkin *et al.* (1969a,b) for these guinea pig studies. For the prenatal and postnatal ossification of fore and hind limbs, tarsus, pelvic girdle, axial skeleton and skull in guinea pigs, see Rajtova (1966, 1967a,b, 1968a,b, 1969a,b).

13. See Kosher and Savage (1981) and Kosher *et al.* (1981) for hyaluronan distribution, and Feinberg and Beebe (1983) for epithelia that promote or inhibit vascularization.

14. See Heyner (1969), Dingle *et al.* (1975) and Fell (1975) for the role of the ECM, Hadházy *et al.* (1972), Elves (1974, 1976, 1978, 1983) and Langer and Gross (1974) for effects of removal of ECM, Jain and Sabet (1974) for the antibody studies, and Faulk *et al.* (1975), Hahn *et al.* (1975) and Furthmayr and Timpl (1976) for antigenicity of type II collagen.

15. See Eisenstein *et al.* (1973) and Dorey and Sorgente (1977) for CAM grafts and resistance to vascular invasion, and Sorgente *et al.* (1975) for extraction of the ECM.

16. See Brem and Folkman (1975) and Langer *et al.* (1976), and see Folkman (1976) for a review of these initial studies. See Eisenstein *et al.* (1975, 1976) and Kuettner *et al.* (1976a, 1977a) for isolation of the proteinaceous factor, Kuettner *et al.* (1976b, 1977b, 1978) for the factor from bovine nasal septum, and Anderson and Sajdera (1971) for a technique using guanidium chloride or 1.9 M $CaCl_2$ to remove GAG from nasal cartilage. See Ehrlich *et al.* (1982) for the protease from human growth plate, Wezeman and Childs (1982) for localization in the ECM, and Pepper *et al.* (1991) for the anti-Tgfβ study. Receptors for members of the Tgfβ family have been cloned, R1, RII and β-glycan being the most widespread (Massagué, 1992 and Box 15.1). See

Guenther *et al.* (1986) for the endothelial cell mitogen, and Villanueva and Nimni (1990) for action on calvarial cells.

17. See Ehrlich *et al.* (1977) and Wolley *et al.* (1977).

18. See Beneke *et al.* (1966), Andrew (1971) and Siebenmann (1977) for Addison's disease. Interestingly, more ectopic bone can be induced in response to implanted Bmp in mice with collagen-induced arthritis than in controls (Fujimori *et al.*, 1992).

19. See E. R. Lee *et al.* (1999) for gelatinase B and Davoli *et al.* (2001) for collagenase-3.

20. See R. A. Brown *et al.* (1987) for the procollagenase activator, and Moses *et. al.* (1992) for the isolation from scapular chondrocytes.

21. See Dietz *et al.* (1999) and Shukunami *et al.* (1999) for chondromodulin-1. Exposing ATDC5 cells to Bmp-2 takes the cells to the chondroblast stage without condensation (Shukunami *et al.*, 1999).

22. See Haigh *et al.* (2000) for the pattern of expression of *Vegf-A*, Engsis *et al.* (2000) for osteoclast attraction, and Yin and Pacifici (2001) for digit formation in mouse embryos. See Patton and Kaufman (1995) for the ossification sequence of long bones in laboratory mouse embryos from 15 to 19 days of gestation, and at one, seven and 14 days postnatally.

23. See Haigh *et al.* (2000) for inactivation, Zelzer *et al.* (2002) for the double homozygotes, and Zelzer and Olsen (2003) for genetic approaches to skeletal development.

24. See Schipani *et al.* (1997) and Kronenberg (2003) for PTH–PTHrP, and Baron *et al.* (1994) for Fgf-2.

25. See Schiebinger (1986, 2003), Stolberg (2003) and Laqueur (2003) for the widespread influence of these depictions.

26. Innominate bone is an unfortunate term. This is not a single bone but rather the fusion product of ilium, ischium and pubic bones. I use the more neutral term of pelvic girdle.

27. See Ruth (1932) for the classification and a more recent confirmation of interpubic joint types by Ortega *et al.* (2003).

28. See Gingerich (1972) and Rawlins (1975) for the typical pattern and Leutenegger (1973) for exceptions.

29. See Ruth (1935, 1936a,b), Crelin and Brightman (1957) and Bernstein and Crelin (1967) for the conversion to a ligament, Wahl (1971) for collagenase production, and Talmage (1947) for an early study on the role of relaxin.

30. See Richany *et al.* (1956), Burch (1966), Savostin-Asling and Asling (1973) and Asling (1976) for the role of tension in the transformation of Meckel's, Isheseki *et al.* (2001) for the role of Egf, and Ossenberg (1974) for the myelohyoid bridge.

31. See Crelin (1954a,b, 1959) and Crelin and Levin (1955) for the studies with oestrogen and relaxin, gonadectomy and transplantation.

32. See Crelin and Haines (1955), Crelin (1959, 1960, 1963, 1969) and Crelin and Koch (1965).

33. See K. Hall (1947), Crelin (1957, 1969) and Linck *et al.* (1975).

34. See Crelin and Haines (1955) and Harkey and Crelin (1963a,b) for exposure to hormones, Horn (1958), Steinetz and Beach (1963) and Steinetz *et al.* (1965) for growth hormone, and Steinetz *et al.* (1973) for calcitonin.

Osteoblast and Osteocyte Diversity

Chapter 24

Bone cells without bone and bones without cells.

Osteoblasts and osteocytes are distinguished by their characteristic morphology, by the synthesis of type I collagen and other specialized 'bone' proteins such as osteocalcin, osteonectin and osteopontin (Boxes 24.1–24.3), and by the deposition of these products into an extracellular matrix (ECM) that then mineralizes.

Osteoblasts are active cells, depositing 470 μm^3 of matrix/osteoblast/day into rat parietal bones. Osteocytes initiate, control and survive mineralization of their ECM, connections between osteocytes and with other osteoblastic cells enabling them to survive and remain active, in contrast to terminally differentiated chondrocytes (Chapter 22).[1]

So, what happens to old osteocytes?

OSTEOCYTIC OSTEOLYSIS

Largely from the studies undertaken in the 1960s and '70s by Leonard Bélanger and S. S. Jande at the University of Ottawa, we know that, in addition to their long-recognized role in bone *deposition*, osteocytes can resorb bone, a process known as *osteocytic osteolysis*. Individual osteocytes go through a cycle of deposition, *osteolysis* and senescence. During osteolysis, mature osteocytes lying some distance from the bone surface become responsive to stimulation by parathyroid hormone (PTH), produce lysosomal enzymes, can and do resorb the surrounding matrix, senesce and die. Whether osteocyte senescence is a form of programmed cell death (apoptosis) is unknown.[2]

Administering PTH to chicks *decreases* the total numbers of osteocytes but *increases* the number involved in resorption (Table 24.1). Since osteo*clasts* also resorb bone, there has been considerable discussion concerning the role of hormones in activating precursor cells and maintaining or redirecting the physiological activity of osteocytes and osteoclasts.[3]

Many cancers metastasize (Box 23.2). Those that metastasize to the skeleton normally induce osteolysis and bone *resorption*, although when prostate cancers metastasize to the skeleton they primarily evoke osteoblastic activity, with only limited osteolysis, and that associated with remodeling. Bone releases factors that are chemotactic for prostatic carcinoma cells, facilitate their attachment, and regulate the release of growth factors that initiate osteogenesis. The prostatic cells, in turn, release osteoblastic and osteolytic factors that regulate bone remodeling and release of growth factors from bone matrix, all of which indicates a complex cross-talk between the two cell types.[4]

I should note that not all workers see osteolysis as a primary function of osteocytes and, indeed, osteolysis is associated primarily with mature or even ageing osteocytes. Cullinane (2001) regards the primary role of mammalian osteocytes as mechanoreception, not osteolysis, citing such statistics as osteocyte density being inversely related to body mass ($R^2 = 0.86$) and proportional to metabolic rate. Such an analysis does not, however, separate out mature or ageing osteocytes from resting osteocytes. Osteocytic osteolysis is an active mode of resorption in hibernating mammals, and perhaps also in snakes (Box 24.4).

Not all osteocytes and osteoclasts respond similarly to hormonal treatment; gender- and region-specific responses exist. Two particularly nice examples illustrating differential responses are oestrogen-triggered deposition of medullary bones in female birds during egg laying, and resorption of the pelvic bones during pregnancy/birth in some mammals.

Box 24.1 Osteocalcin

Osteocalcin or bone gla protein is a 5800-MW, vitamin K-dependent, γ-carboxyglutamic acid-containing, Ca^{++}-binding protein. The seventh most abundant protein in human bone, osteocalcin is the most abundant non-collagenous protein in bone, accounting for 10–20 per cent of the non-collagenous proteins. Osteocalcin binds to hydroxyapatite. With a negatively charged surface, each osteocalcin molecule can coordinate five calcium ions in a 3D-organization that is complementary to the calcium ions in the hydroxyapatite crystal lattice. Osteocalcin recruits osteoclasts or osteoclast precursors to bone for resorption; implanting bone depleted of osteocalcin results in only 60 per cent of the resorption seen after implanting control bone, with only 35 per cent the number of mononuclear cells (osteoclast precursors) surrounding the implant.[a]

The gene for osteocalcin was cloned from rat and mouse osteosarcoma cell lines in 1986. Osteocalcin plays an important role in regulating chondrocyte mineralization, a secondary role in endochondral ossification, and a less critical tertiary role in intramembranous ossification. Although absent from osteoid, osteocalcin:

- can be purified from rat bone;
- can be detected using a radioimmunoassay with the first perichondral mineralization in chick long bones at seven or eight days of incubation;
- appears – as determined by immunohistochemistry – at the mineralization front but not in mineralized cartilage of radii or mandibular condylar cartilage from 20-day-old rat foetuses; and
- coincides with cartilage mineralization in the bone that is induced following subcutaneous implantation of bone matrix.[b]

Consequently, osteocalcin, along with *osteopontin* (Boxes 1.5 and 24.2), is nowadays used as a *marker* for chondrocyte hypertrophy and bone mineralization (Fig. 1.6) and to localize cartilaginous ECM, even when a tissue cannot clearly be identified as cartilage using other criteria, as happens, for example, with chondroid bone of the rat glenoid fossa, which contains type I and type II collagens and osteocalcin, and thus has features of both hypertrophic chondrocytes (HCCs) and osteoblasts/cytes.[c]

In perhaps the first *in-situ* demonstration in adult bone – in this instance, in rats – mRNA for osteocalcin was demonstrated in periosteal and endosteal osteoblasts (and mRNA for osteopontin in osteocytes and HCCs). Levels of osteocalcin, calcium and hydroxyproline all declined in vertebrae and humeri from growing rats examined after one week in space on board NASA Space Lab 3.[d]

Osteocalcin is responsive to vitamin D. The response of rat osteoblastic cells is pleiotropic, the extent of the response depending on basal levels of gene expression, duration of exposure, and competency of the bone matrix. Three examples of the value of osteocalcin as a marker and for unraveling regulatory pathways in bone are:

- Increased levels of mRNAs for osteocalcin, alkaline phosphatase and *Bmp-2* signal the promotion of osteogenesis in foetal rat calvarial cells exposed to recombinant *Bmp-2*.
- Synthesis of osteocalcin, increased alkaline phosphatase, and development of a cAMP response to PTH, all signify osteogenic differentiation from a bone-marrow stromal cell line (W-20-17) exposed to RhBmp-2, but there is no effect on proliferation.
- Murine metatarsal rudiments or chondrocytes cultured with retinoic acid display retinoic acid-induced differentiation of

mineralized bone, deposition of collagen type II, and increased synthesis of osteocalcin and *Runx-2* (*Cbfa-1*), forming a regulatory cascade.[e]

Inhibition of osteocalcin by the organic pesticide warfarin enhances mineralization of hypertrophic but not immature chondrocytes *in vitro*. Viral transfection with osteocalcin decreases mineralization, blocks chondrocyte maturation, and inhibits endochondral ossification *in ovo*. The consequent blocking of subperiosteal intramembranous ossification is a secondary response to the lack of inductive interaction between HCCs and the perichondrium. This interaction is required to transform the perichondrium into a periosteum, and to direct progenitor cells into the osteogenic lineage, allowing initiation of intramembranous ossification around the shafts of long bones (Chapter 22).

The osteocalcin gene is transcriptionally regulated by *Tgfβ*. Levels of osteocalcin decrease two- to threefold, and osteocalcin mRNA decreases threefold, when cells from the osteoblastic cell line ROS 17/2.8 are cultured with Tgfβ. A specific *cis*-acting element in the osteocalcin promoter (*Osf2/cbfa-1*, now *Runx-2*) acts as a transcriptional activator of osteoblast differentiation. Ducy and colleagues cloned the cDNA for the promoter, which is expressed in embryonic mesenchyme and then strictly in osteoblastic lineages. Expression in non-osteoblasts such as MC3T3-E1 cells up-regulates osteogenic genes. *Osf2* is regulated by *Bmp-7* and by vitamin D_3. Filvaroff *et al.* (1999) used a cytoplasmic truncated type II Tgfβ receptor from the osteocalcin promoter to generate transgenic mice in which bone mass increased because of decreased osteoclastic resorption.[f]

Osteocalcin is up-regulated by *Fgf-2*. Cultured bovine bone cells synthesize Fgf-2, secrete it, and store it in the bone matrix (Box 23.3). Proliferation of bone cells is regulated by Fgf-2, which increases osteocalcin and cell division in a dose-dependent manner. Osteocalcin also is repressed by a member of the *Distal-less* gene family, *Dlx-5*, which shows stage-specific expression in mouse calvarial osteoblasts, repressing osteocalcin as the matrix mineralizes.[g]

[a] See Lian and Gundberg (1988) for osteocalcin levels in humans, Hoang *et al.* (2003) for binding to hydroxyapatite, Glowacki and Lian (1987) for bone implant studies, and An and Draughn (2000) for the mechanical properties of bone implants.

[b] See Celeste *et al.* (1986) for cloning of osteocalcin (bone Gla protein), Yagami *et al.* (1999) for its role, and T. A. Owens *et al.* (1991) for regulation. See Hauschka and Reddi (1980), Otawara *et al.* (1980) and Price *et al.* (1980a,b, 1983) and Hauschka *et al.* for osteocalcin in rat bone (1983).

[c] See Tuominen *et al.* (1996) and Mizoguchi *et al.* (1997a) for osteocalcin as a marker for chondroid bone, McKee *et al.* (1990a,b, 1992) and Lian *et al.* (1993) for osteocalcin as a marker for hypertrophy, Hauschka (1986) and Hauschka *et al.* (1989) for reviews, Price (1988) for the role of vitamin K-dependent proteins in bone metabolism, and the chapters in Volume 4: *Bone Metabolism and Mineralization* of Hall (1990–1994) for overviews.

[d] See Ikeda *et al.* (1992) for the *in-situ* demonstration, and Patterson-Buckendahl *et al.* (1987) for the NASA study.

[e] See Thies *et al.* (1992) and Harris *et al.* (1995) for the studies with *Bmp-2*, and Jiménez *et al.* (2001) for the study with retinoic acid.

[f] See Noda (1989) and Efstratiadis (1998) for *Tgfβ* regulation, and Ducy *et al.* (1997), who cloned the osteocalcin promoter.

[g] See Ryou *et al.* (1997) for *Dlx-5*, and Globus *et al.* (1988, 1989) for Fgf-2. Proliferation also is stimulated by αFgf but to a lesser extent. Tgfβ does not directly affect osteoblastic proliferation but does potentiate the mitogenic actions of Fgf-1 and -2.

Box 24.2 Osteopontin

Osteopontin is found in mouse endochondral and membrane bone, osteoid, preosteoblasts, osteoblasts and osteocytes (Fig. 1.6). Chicken osteopontin has been cloned as a 66-kDa bone phosphoprotein, the avian homologue of mammalian osteopontin expressed in resting and hypertrophic chondrocytes (HCCs).[a]

Although *Tgfβ* down-regulates the expression of osteocalcin, *Tgfβ* up-regulates osteopontin, osteonectin and collagen in dividing and hypertrophic chondroblasts of long bones, calvarial bones of the skull, and embryonic epithelia in murine, human and avian embryos.[b]

Cell survival and cell migration are enhanced, and mineralization inhibited by osteopontin. In cell culture, foetal rat calvarial cells synthesize osteopontin in association with the deposition of a mineralized ECM. Levels of osteopontin are highest closest to mineralized matrix and at interfaces between mineralized and unmineralized matrices. Osteopontin mRNA is sensitive to mechanical stimuli; isolated osteoblasts respond to strain with enhanced synthesis, and osteopontin is expressed in cranial sutures where expression parallels up-regulation of *FgfR-1* and decline in *FgfR-2*.[c]

[a] For osteopontin in mammalian skeletal tissues, see Mark *et al.* (1987a, 1988), Yoon *et al.* (1987), Nomura *et al.* (1988), Franzen *et al.* (1989) and Chen *et al.* (1993). For chicken osteopontin, see Castagnola *et al.* (1991) and M. A. Moore *et al.* (1991).
[b] See Nomura *et al.* (1988), Metsäranta *et al.* (1989), Oshima *et al.* (1989) and Pacifici *et al.* (1990) for up-regulation of various genes by *Tgfβ*. See Noda and Rodan (1989a,b) and Noda and Vogel (1989) for Fgf and Tgfβ, and McKee *et al.* (1990a,b, 1992) for osteopontin and osteocalcin during endochondral ossification of chick tibia.
[c] See Nagata *et al.* (1991), T. A. Owens *et al.* (1991), McKee and Nanci (1996), Toma *et al.* (1997a) and Yagami *et al.* (1999) for osteopontin action on cells and ECMs, and see Iseki *et al.* (1997, 1999) and Yu *et al.* (2003) for response to mechanical factors.

Box 24.3 Osteonectin

Some 10 per cent of the protein in bone is made up of the 32 000-MW ECM protein *osteonectin* (or SPARC; *secreted protein, acidic, cysteine-rich*). Appearing in bone matrix with mineralization, osteonectin links collagen to hydroxyapatite, serves as a nucleus for mineralization, and regulates the formation and growth of hydroxyapatite crystals. Osteonectin also functions as a secreted heat-shock protein (Nori *et al.*, 1992). Cataracts and lens anomalies are consequences of osteonectin deficiency in mice.

Patterns of expression of osteonectin mRNA, while similar in rats and mice, are not universal: osteonectin is found in lamprey but not mammalian livers. Invertebrates are reported to lack osteonectin (Ringuette *et al.*, 1991), although Alison Cole in my laboratory finds that a vertebrate antibody reacts with the ECM of sabellid polychaete and hemichordate skeletons.[a]

Within skeletal tissues osteonectin is expressed in osteoprogenitor cells, osteoblasts, young (but not old) osteocytes, hypertrophic chondrocytes (HCCs) of mineralizing cartilage, fracture callus and ectopic ossification, as seen in *myositis ossificans*. In human and avian embryos calvariae, growth plates, skin and perivascular cells contain the highest levels of osteonectin mRNA.[b]

In chicken tibial growth cartilage osteonectin is found in the cells of all zones but only in the matrix in the mineralizing zone. Similarly, osteonectin is expressed in the hypertrophic chondroblasts, mineralized matrix and perichondrium of the mandibular condylar cartilages. Chondroid also contains osteonectin (Chapter 5). Osteonectin mRNA is up-regulated by dexamethasone, retinoic acid and vitamin D_3.[c]

[a] See Ringuette *et al.* (1991) for absence of osteonectin and Cole and Hall (2004b) for its possible presence.
[b] See Termine *et al.* (1981), Stenner *et al.* (1986), Jundt *et al.* (1987), Nomura *et al.* (1988) and Mothe and Brown (2001) for expression data, and Metsäranta *et al.* (1989), Oshima *et al.* (1989) and Mundlos *et al.* (1992) for the highest mRNA levels.
[c] See Pacifici *et al.* (1990) and Copray *et al.* (1989) for tibial and condylar cartilages, and Leboy *et al.* (1991) for mRNA regulation.

Table 24.1 Composition of osteocytes in the diaphyses of four-week-old chicks treated with 125 grams of parathyroid hormone (PTH) daily for 2, 24, 48 and 74 hours[a]

Time exposed to PTH (hours)	% Formative phase osteocytes/ population	% Resorptive phase osteocytes/ population	% Degenerative phase osteocytes/ population	% Empty lacunae/ population
Control	33 ± 2.7	59 ± 2.1	4 ± 1.5	4 ± 2.1
2	24 ± 2.5	64 ± 1.9	10 ± 1.8	2 ± 1.7
24	20 ± 2.1	76 ± 4.5	3 ± 1.1	1 ± 0.9
48	11.5 ± 2.5	46.5 ± 2.5	31 ± 1.7	11 ± 3.2
74	11 ± 2.2	55.5 ± 4.1	32.5 ± 4.1	1 ± 1.5

[a] Based on data from Jande (1972).

INITIATING OSTEOGENESIS *IN VITRO* FROM EMBRYONIC MESENCHYME

Many studies indicate that osteogenesis can be initiated in *organ culture* from preosteogenic embryonic mesenchyme that retains the three-dimensional organization it had within the embryo (Figs 18.7 and 18.8; Chapter 18).[5] Surprisingly few studies have been published on the ability of *isolated preosteogenic mesenchymal cells* to *initiate* osteogenesis *in vitro* in situations other than those in which cartilage is replaced by bone, although Fitton

Jackson (1965) obtained osteogenesis from cultured embryonic chick fibroblasts.

The best study on the initiation of osteogenesis by isolated cells *in vitro* was published by Marvaso and Bernard (1977). Cells from the calvarial mesenchyme of 14-day-old foetal mice – i.e. before onset of osteogenesis – were dissociated using collagenase following the method of Peck *et al.* (1964). The isolated cells were cultured in BGJ + 20 per cent foetal calf serum and 5 per cent beef embryo extract. Over the first week in culture, the cells differentiated into osteoblasts and synthesized and deposited collagen into an ECM that mineralized as the cells formed nodules of woven bone. The quality of the bone produced in such a short time and under such avascular conditions is impressive.[6]

OSTEOGENIC CELLS *IN VITRO*

The matrix surrounding osteocytes plays a critical role in modifying their response to environmental perturbations.

Box 24.4 Hibernation

A surprising number of mammals and other tetrapods hibernate; that is, go into a prolonged period of physical inactivity and reduced metabolism, usually annually. Hibernation normally takes place in winter, when resources are scarce and the weather is less conducive to being out and about. Consequently, most hibernators inhabit areas of extreme climate. Depending on hemisphere, hibernation will be limited to individuals at the far northerly or southerly limits of the geographical range of their species.[a] Groups with hibernating members include ground squirrels (gophers), groundhogs, bears, bats, frogs and snakes.

As most will know, hibernators put on fat before hibernation, fat they consume during the winter hibernation. Less well known is that hibernation also affects skeletal metabolism. One of the most interesting adaptations of the skeleton to hibernation is *osteocytic osteolysis*, the resorption of bone by osteocytes. Osteocytic osteolysis has been described for three hibernators but may be more common. Some of the few studies are detailed below.

During hibernation, and in response to cold-stress or prolonged inactivity outside the winter season, golden hamsters, *Mesocricetus auratus*, exhibit increased osteocytic osteolysis. Hibernating thirteen-lined (Franklin) ground squirrels, *Citellus tridecemlineatus*, show a higher level of mineral removal (osteocytic osteolysis) from alveolar bone than do non-hibernating individuals, the Ca^{++} released being used in metabolism. Vacuolated osteocytes and what was described as 'foamy bone' attest to the resorptive activity of the osteocytes. There is an interesting parallel here with loss of teeth in women whose pregnancy coincides with calcium deprivation. Indeed, the old wives' tale that women lose one tooth per pregnancy is not without foundation, as Ca^{++} is mobilized from the metabolically active alveolar bone at the base of the teeth (see Chapters 25 and 30 for alveolar bone).[b]

When the little brown bat, *Myotis lucifugus*, is in hibernation, the marrow cavities of its long bones fill with lipid, although there are no signs of haematopoietic or osteoclastic cells that could be resorbing bone, and the bone surfaces are covered with quiescent bone-lining cells. No new bone forms during hibernation, all new bone having been deposited in the summer. Following arousal from hibernation, osteoclasts are reactivated, bone surfaces become repopulated with bone cells, and the lipid of the marrow cavity is replaced by haematopoietic cells. However, *pregnant females lose bone in the summer; active males do not*. As there is no increase in osteoclast numbers during lactation, and as resorption cavities form deep within the bone, osteocytic osteolysis is the implied mechanism of bone resorption.[c]

Snakes also hibernate. Vertebral cortical bone in the southern European asp, *Vipera aspis*, exhibits more perilacunar haloes in the winter (57 per cent) than in the summer (22 per cent), taken as indicative of seasonal demineralization associated with winter hibernation. Removal of bone is selective; Ca^{++} is mobilized from vertebrae but not from ribs, mandibles, or skull.[d]

[a] If hibernation is spending the winter in a dormant state (as defined by the OED), then many humans, especially those north of the 49th parallel, hibernate.
[b] See Steinberg et al. (1981) for the golden hamster, Haller and Zimny (1978) for the ground squirrel, and see the chapters in Volume 4: *Bone Metabolism and Mineralization* of Hall (1990–1994) for overviews of bone metabolism.
[c] See Doty and Nunez (1985) and Kwiecinski et al. (1987) for studies on the little brown bat.
[d] See Alcobendas and Castanet (1985) and Alcobendas and Baud (1988) for skeletal changes in the asp.

Box 24.5 Growth hormone

Growth hormone (GH) is just that: a hormone that controls growth. Beyond that, the story becomes complicated. Different hormones interact. GH is no exception. Some hormones act directly via receptors, others via an intermediary. GH does both (see Box 27.5). Cells are responsive to hormones at different phases of their differentiation. Cells that respond to GH are no exception.

Growth hormone can stimulate longitudinal bone growth directly as demonstrated by local administration of growth hormone directly into the growth plates of proximal tibiae in hypophysectomized rats (Isaksson et al., 1982). Growth hormone usually acts through the liver intermediary somatomedin (see text).

At least in foetal mouse calvariae, growth hormone is mitogenic for osteoblasts but not for osteoprogenitor cells. Growth hormone receptors have been isolated from a clonal osteoblast-like cell line (UMR106.06) and mediate the mitogenic response of these cells to growth hormone.[a]

Administering growth hormone is an effective treatment in individuals of short stature and can augment growth in individuals of 'normal' height. Van Vliet et al. (1983) took 15 'normal' individuals between ages 4.3 and 15.5 years, all of short stature (i.e. none carried a known genetic defect affecting height) and gave them pituitary growth hormone (0.1 μg/kg body weight) three times a week for six months. Growth was enhanced, especially in the youngest children.

Differences in growth rates and final size between inbred strains of mice may have genetic or hormonal causes. In one experiment, transgenic mice were engineered to begin to express the gene for sheep growth hormone 21 days after birth, chosen because it marks the end of weaning and the onset of the normal growth spurt. The transgenic mice showed increased long-bone and mandibular growth, especially at sites of muscle attachment. Differences *earlier* in development are more likely to be genetic in origin; the biggest differences in mandibular development between C3H and C57 inbred strains of mice occur at 15 days after birth.[b]

Growth hormone influences bone repair. Administration of GH (0–10 mg/kg body weight) influences the mechanical properties of repairing rat tibial fractures in a dose-dependent manner. Growth hormone administered daily for the first 40 days of repair in old rats enhanced mechanical properties at 80 days (assessed using a three-point-bending test) and, interestingly, increased mechanical properties in the contralateral control.[c]

[a] See Slootweg et al. (1988) for the mouse calvariae, and R. Barnard et al. (1991) for the UMR106.06 cell line.
[b] See Vogl et al. (1993, 1994) and Shea et al. (1990) for the transgenic data.
[c] See Bak et al. (1990) and Bak and Andreassen (1991) for GH and strength of repair.

Consequently, it is difficult to study the physiology of osteocytes embedded in a heavily mineralized matrix. This led to attempts to obtain populations of osteoblastic and osteocytic cells *in vitro*. Osteocytes do not survive for long once released from their ECMs, and any information gained *in vitro* must be applied cautiously to understanding their physiological activity *in vivo*.

As one example, *in-vitro* and *in-vivo* studies with growth hormone give quite disparate results (Box 24.5). *In vivo*, growth hormone stimulates the liver to synthesize and/or release somatomedin, which mediates growth hormone effects on the skeleton, a step and effect that should not occur *in vitro* (see Box 31.3). Furthermore, responsiveness and response both change with age; increased sensitivity to somatomedin during puberty contributes to the adolescent growth spurt. Even accounting for such differences may not provide a sufficient explanation of the differences observed. Further hidden complexities exist. For example, we would think that somatomedin cannot be involved when growth hormone stimulates the proliferation of rabbit ear and rat rib growth-plate chondrocytes *in vitro*. This conclusion would stand unless the chondrocytes themselves produced somatomedin, something that was not even considered. It turns out that embryonic cartilage does secrete a somatomedin-like peptide that stimulates cartilage growth *in vitro*. The peptide stimulates growth in a dose-dependent manner; this is blocked by anti-somatomedins, and reversed by insulin.[7]

Much of the response of individual osteoblasts and osteocytes *in vivo* is initiated because they are in contact and communicate through cell-to-cell junctions (Box 24.6). These connections are broken when osteogenic cells are isolated in preparation for cultivation *in vitro*, which is yet another factor distancing *in-vitro* from *in-vivo* responses. Populations of osteoblasts and osteocytes isolated *in vitro* may not and perhaps cannot show the physiological responses seen *in vivo*, even though they may be terminally differentiated and metabolically active. Nevertheless, and with all these qualifications, *in-vitro* studies have provided valuable information on the responses of osteogenic cells. One approach has been to isolate populations of osteogenic cells *in vitro* with as much of their *in-vivo* integrity and three-dimensional organization as possible. Folded periostea do this very well indeed.

Folded periostea

Pieter Nijweide from Amsterdam developed an *in-vitro* model for osteogenesis when he devised a means of culturing folded periostea from embryonic chick calvaria and used it to study osteoid formation, analyze regulation of calcium transport and assess effects of hormones and metabolic inhibitors. Using this system, Van der Plas and Nijweide (1988) identified cell compartments in which osteoblasts regulate the proliferation of periosteal fibroblasts and so regulate osteoprogenitor cell production. This same system transmits stimulatory signals from PTH to periosteal fibroblasts.[8]

Howard (Howie) Tenenbaum in Toronto demonstrated that folded periostea from calvariae of 17-day-old chick embryos deposit osteoid after six days in culture on plasma clots and mineralize the osteoid, provided that 5 or 10 mM β-glycerophosphate is added. Dependent on time of application, β-glycerophosphate affects phosphate accumulation in chick periosteal cultures, leading to a 50 per cent reduction in type I collagen.[9]

Box 24.6 Bone: A network that functions as a syncytium

A network: 'anything reticulated or decussated, at equal distances, with interstices between the intersections.'[a]

A *syncytium* is a multinucleated mass of cytoplasm undivided by cell membranes, i.e. not divided into cells. Functionally, bone is a syncytium whose cells, from osteoprogenitor cells to mature osteocytes, are connected in a vast network, able to communicate and transfer information because of cell-to-cell connections.[b]

Connections between these skeletal cells are dynamic, not static, as has been demonstrated in various ways; Menton *et al.* (1984) used SEM to follow loss of cell processes during the transformation of osteoblasts to osteocytes in rat femoral endosteal bone (Fig. 1.3). As discussed in Box 31.1, the connections consist of gap junctions through which information can flow.

The canalicular cell processes that connect osteocytes are an important physiological transport system. Loading induces fluid flow through canaliculi of mouse calvarial and human iliac crest osteoblasts, and stimulates production of prostaglandin and synthesis of major enzymes in prostaglandin production, indicating similar mechanotransduction pathways in the two species and bone types. Formation of canalicular cell processes is blocked by antibodies against laminin or a 32/67-kDa laminin receptor with discrete domains within the laminin molecule (but not the RGD-peptide) regulating cell process formation.[c]

Knowing the syncytial nature of bone explains what were, when they appeared, intriguing and convincing data, but data that were hard to explain. For example, wounding one area of a bone not only initiates a compensatory response at the wound site – which is to be expected – but elicits responses *some distance* from the wound, and for a substantial time after the initial stimulus.[d] The explanation could involve release of a stimulus – implanting an intramedullary nail in *one* tibia of Sprague-Dawley rats initiates osteogenic responses in *both tibiae*, presumably because of release of circulating factors – but release of stimulus does not explain communication of that stimulus.

[a] Cited from de Beer (1962), p. 134.
[b] See Furseth (1973) and Weinger and Holtrop (1974) for pioneering studies on connections between osteoblasts, and the review of osteocyte function by Aarden *et al.* (1994).
[c] See Joldersma *et al.* (2000) for loading, and Vukicevic *et al.* (1990b) for the role of laminin.
[d] See Melcher and Accursi (1972), Pritchard (1972b) and Hyldebrandt *et al.* (1974) for compensatory response within a bone, and Einhorn *et al.* (1990) for response of the contralateral tibia.

PGE-2 and cAMP modulate mechanical stimulation and so couple formation and resorption, as Somjen *et al.* (1980) demonstrated when they created a device to apply force to tissue-culture dishes in which bone fragments are cultured. Levels of PGE-2 and cAMP rise within 20 minutes of force application. Adding PGE-2 mimics force application in elevating cAMP levels, indicative of stress-induced bone remodeling mediated by PGE-2-mediated changes in cAMP levels. Furthermore, osteoclasts respond to PGE-1 concentrations that are 200 times lower than those required to stimulate osteoblasts. Production of PGE-2 is therefore one of the pathways coupling bone formation and bone resorption, increasing cAMP and activating *Igf-1* in foetal rat osteoblasts.[10]

Foetal rat calvarial cells synthesize osteopontin and bone sialoprotein and deposit mineralized matrix. Both molecules bind to mineral. Bone sialoprotein also binds to collagen. *In vitro*, osteopontin levels are only high in osteoblasts close to the bone matrix. *In vivo*, osteopontin is expressed at mineralized tissue interfaces in bones and teeth such as at cement lines, and in the lamina limitans. Osteopontin mRNA in chick calvarial osteoblasts is sensitive to mechanical stimuli; nine hours after exposing osteoblasts to dynamic strain at 0.25 Hz for two hours, mRNA levels increase fourfold. Existing transcription factors require integrity of microfilaments and are activated via protein kinase A and tyrosine kinase but not via protein kinase C.[11]

Establishing isolated osteoblasts and initiating osteogenesis *in vitro*

Two main types of isolated osteogenic cells have been studied to probe two different aspects of osteogenesis:

- osteoblasts isolated from *bone after osteogenesis has begun in vivo* to study *ongoing osteogenesis*; and
- mesenchymal cells isolated *before initiation of in-vivo osteogenesis* to determine whether osteogenesis *can be initiated in vitro*.

In either case, the source of the cells may be embryonic, neonatal or adult, from intraskeletal or ectopic sites, intramembranous or endochondral bone, or normal or neoplastic tissues.

In a pioneering study, Peck *et al.* (1964) used buffered collagenase to isolate osteoblasts from rat *calvariae* and established them in culture. The cells were still viable after seven days, although they appeared fibroblastic. Was this because the osteoblasts had dedifferentiated or because no osseous matrix had been deposited? In 1972, Bard *et al.* reported isolation of viable osteoblasts from the femoral heads of various adult animals, providing a model for isolating osteoblastic cells from an *endochondral* source.

Peck's method, or modifications of it, provided the basis for much subsequent work. Peck and colleagues demonstrated that the osteoblastic cells synthesize collagen, respond to ascorbic acid with enhanced collagen synthesis, react to hormones such as hydrocortisone and PTH, react to adenosine by synthesis of adenylate cyclase, and proliferate in serum-free medium. The last finding was pivotal for it opened up the potential for a wide range of biochemical and metabolic studies on isolated bone cells in controlled chemical environments.[12]

Osteoblasts isolated from calvariae of newborn rats synthesize lipid *in vitro*. Data on oxygen consumption and lactate production over short-term (2–4-hr) incubation periods were obtained. Calcium uptake of 17 mmol/kg wet weight was observed and shown to be independent of parathyroid stimulation but increased by the ionophore A23187.[13]

These and other observations were from relatively short-term cultures. *Ongoing osteogenesis was not reported.* It is worth emphasizing this point. Incubation in a chemically defined medium for two or three hours is *not* cell culture. Maintain such cells for several more hours *in vitro* and the responses disappear. Cell culture implies and requires long-term maintenance *in vitro*. And not merely maintenance. The cells have to proliferate and maintain their differentiated state, or if undifferentiated at the outset, initiate and maintain a differentiated state. Otherwise, all we see is the short-term continuation of processes that were ongoing *in vivo* at the time the cells were isolated.

Calvarial osteoblasts in vitro

Differentiated osteogenic cells can be maintained *in vitro*, respond in physiological fashion to exogenous hormone stimulation, and – perhaps of greatest importance – deposit and mineralize an extracellular osseous matrix.

The first convincing evidence for continued osteogenesis of an isolated population of osteoblastic cells was provided in 1974 (Fig. 24.1). Using trypsin and EDTA, Binderman and colleagues isolated cells from calvariae and calvarial periostea from 16- to 20-day-old foetal rats. In some isolates, the periosteum was traumatized to initiate cell division to augment the population of cells obtained. Earlier, when seeking periostea to use as grafts, it had been shown that adult periostea with limited osteogenic potential *in vivo* could be activated if the periostea were scraped or peeled before being grafted; after 24 hours a burst of proliferation occurs. (There also is a tendency for such periostea to form cartilage nodules, which attests to the bipotentiality of calvarial periosteal cells and shows how altered circumstances can initiate the normally dormant chondrogenic potential.)[14]

The cells cultured in Binderman's laboratory were maintained for up to eight weeks in BGJ medium plus 10 per cent foetal calf serum in an atmosphere of 10% CO_2 in air. The cells retained the general morphological appearance of osteoblasts (Fig. 24.2), continued to produce and deposit a collagenous ECM containing collagen with a banding pattern typical of bone collagen, mineralized the ECM, and responded to calcitonin stimulation with increased calcium uptake (Fig. 24.1). Similar cells, examined using scanning electron microscopy, have a distinctive morphology and distinctive cell-to-cell contacts, and can be separated from fibroblasts by their ability to bind, but not to be lysed by, anti-rat tail collagen serum.[15]

A word of **caution**: The features displayed by these cells are all features of active osteoblastic cells. In *aggregate*, they demonstrate that the cells continue to function as osteoblasts/osteocytes for a prolonged time in culture.

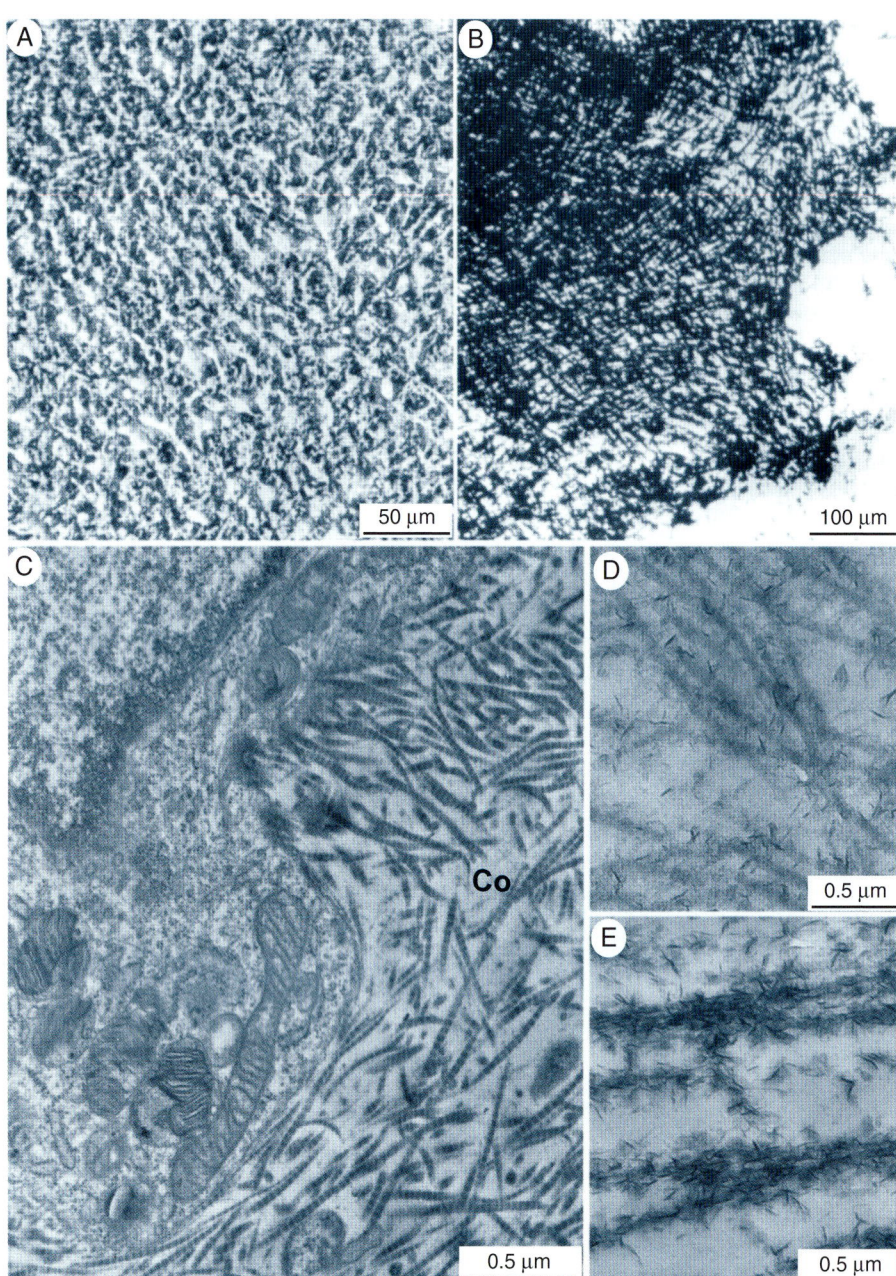

Figure 24.1 Progressive deposition of matrix by rat osteoblasts maintained *in vitro*. (A) Closely packed cells and ECM after 30 days in culture. Phase contrast microscopy. (B) Deposition of mineral after eight weeks in culture; Kossa stain. (C–E) Transmission electron micrographs after three (C), six (D) and eight (E) weeks in culture to show extensive deposition of fibrous collagen (Co) in C, and mineralized collagen fibres in D and E. Modified from Binderman *et al.* (1974).

Using *any single feature* alone to identify these cells as osteogenic might be problematic. Morphology is notoriously plastic. Other cell types can deposit and mineralize an ECM or respond to a hormonal stimulation.

For example, response to calcitonin and/or PTH as a marker for differentiated osteogenic cells should be used with care. High-density (5×10^6/ml) cultures of limb-bud mesenchymal cells respond to PTH after only one or two days and chondrify within a few days. After about 13 days they deposit a mineralized tissue with features of an osteogenic matrix rather than mineralized cartilage. However, cultures established at initial densities of 2 or 2.5×10^6 cells/ml form fibroblasts that subsequently mineralize and deposit a bone-specific alkaline phosphatase. In these cultures, PTH responsiveness is associated with a particular type of ECM rather than identifying cells as osteoblastic.[16]

Isolating subpopulations of calvarial osteogenic cells

In an important refinement of the methods used to isolate osteoblastic cells, EDTA, trypsin and collagenase, in three sequential extractions of mouse calvariae, were used to isolate *separate* populations of osteogenic cells on the basis of their closeness to the bone surface.[17]

- The *first* population extracted consists of surface and superficial periosteal cells, remains fibroblastic *in vitro*, is unresponsive to PTH or calcitonin, and consists largely of osteoprogenitor cells.

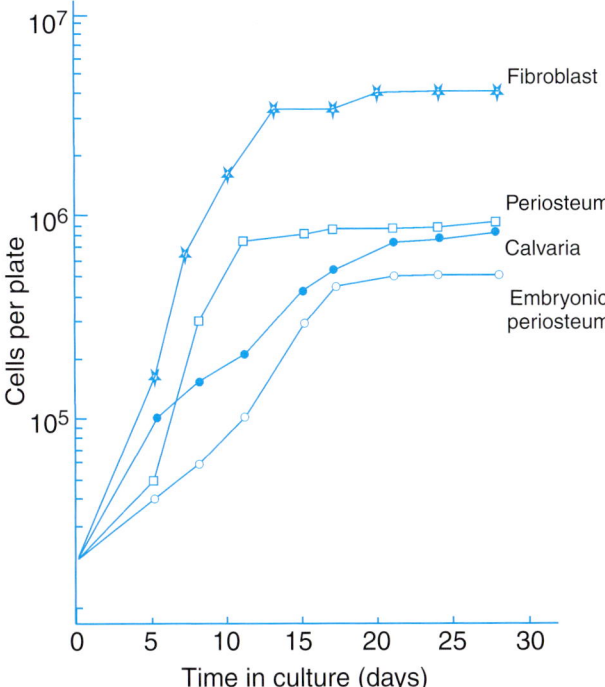

Figure 24.2 Growth rates of four cell types isolated from rats, established in culture at 2×10^4 cells/plate in BGJ medium + 10% foetal calf serum, and maintained for 30 days. Adult fibroblasts and embryonic periosteal cells show the highest and lowest growth rates, respectively. From data in Binderman *et al.* (1974).

- The *third* population, which consists of stellate cells with many connecting cell processes, shows the greatest response to hormone stimulation, as measured by production of 3′,5′-cAMP. Responses to PTH and calcitonin occur in parallel: both peak in cells cultured for one week, and both decline together. These cells are primarily osteocytes and mature osteoblasts.
- The *second* population is a mix of the types found in populations one and three.

To test whether the *same cells* respond to PTH and calcitonin, gentler enzymatic treatment was used to further subdivide the *third* population into three subpopulations. One was responsive only to PTH, one to PTH and calcitonin, the third unresponsive to either hormone. Evidently, different cells, or different stages of the same cells, are present, at least as assessed by hormonal responsiveness.[18]

Rao *et al.* (1977) used a slightly different enzymatic digestion – collagenase, elastase and DNAase, but no trypsin or EDTA – to isolate *five populations of rat calvarial osteogenic cells*, which retained the ability to respond to PTH and to prostaglandin E_1 (PGE-1), but not to calcitonin after being subcultured several times and stored for a prolonged time at $-80°C$.

Foetal rat calvarial cells maintained *in vitro* were shown by Bellows and Heersche (2001) to consist overwhelmingly (92 per cent) of committed osteogenic cells, with a small percentage (3%) of adipocyte progenitors responsive to 1,25-dihydroxyvitamin D_3 and dexamethasone. The remaining 5 per cent are common osteogenic and adipogenic stem cells.

The morphological arrangement of calvarial osteoblasts *in vitro* also is ordered by PTH: osteoblasts are migratory and respond to PTH by migrating along rat parietal bones previously denuded of osteoblasts (Fig. 15.1). The importance of substrate in allowing osteogenic cells to express their potential can be demonstrated using calvarial cells isolated from 19- to 21-day-old foetal mice. Cells cultured apart from bone form fibroblasts. Cells cultured with bone form osteoblasts and osteoclasts. Culturing rat calvarial osteoblasts on type I collagen modulates various aspects of cell growth, adhesion and matrix mineralization of the ECM.[19]

Differentiation of osteoprogenitor cells from foetal rat calvariae can be suppressed reversibly by PTH at concentrations of 2×10^{-5} to 2×10^{-2} IU/ml, the PTH acting at the late preosteoblast stage. Interestingly, continuous exposure to PTH results in formation of osseous nodules, but single two-day pulses of PTH do not (Bellows *et al.*, 1990).

Chondrogenesis from rodent and avian osteogenic cells

Toma *et al.* (1997b) used calvarial cell populations from 12- and 17-day-old chick embryos – the former when the cells are at the condensation stage, the latter at the osteoblastic stage – to show that chondrocytes (identified by expression of collagen IIB and collagen type X), can be generated from the 12-day but not from 17-day specimens. In cultures from the younger embryos maintained under conditions that promote chondrogenesis, the number of chondrogenic cells increases from 15 to 50 per cent while the number of osteogenic cells remains constant at around 34–40 per cent, because of selective proliferation of the chondrogenic cells, a mechanism already seen in other situations of bipotentiality (Chapter 12). Toma and colleagues concluded that there is negative selection against progressive growth of chondrogenic cells between 12 and 17 days (i.e. between condensation and osteoblastic differentiation), unless the cells are provided with a permissive or instructive environment for chondrogenesis. They wondered further whether the progressive loss of chondrogenic potential parallels the progressive loss of the ability of membrane bones to repair.

Rat calvarial cells exposed to Bmp-7 (OP-1) for 6 days at concentrations of 4–100 ng/ml produce cartilage. Delaying the introduction of Bmp-7 until the cultures have been established for a week initiates osteoblastic differentiation, an action that is blocked by Tgfβ (Asahina *et al.*, 1993).

Chick calvarial cells that have been sequentially digested and cultured produce chondroblasts and osteoblasts. They synthesize collagen type I, osteocalcin and alkaline phosphatase (whether Na, β-glycerophosphate is present or not), mineralize an ECM in the presence of Na, β-glycerophosphate, and deposit a typical osteogenic matrix. Like mouse calvarial cells, chick cells chondrify if cultured with dexamethasone, producing a core of osteoid surrounded by cartilage and new osteoid. Fibronectin accumulates and is required for initial attachment of preosteoblasts and osteoblasts to the ECM. *In vivo*, fibronectin first appears at 16–17 days of gestation, at the condensation stage.[20]

Stein *et al.* (1989) used foetal rat and chick calvarial osteoblast cultures to demonstrate how the onset and progression of osteoblastic differentiation is functionally related to proliferation. T. A. Owen *et al.* (1990) and Stein *et al.* (1990) proposed a three-step model – the three steps being cell division, deposition of ECM and mineralization – with two restriction points requiring activation of a new suite of genes to proceed to the next step. Lian and Stein (1991) used a two-threshold model for osteoblast proliferation and differentiation, using antibodies to stages of osteoblast differentiation as markers.[21] Their approach using calvarial osteoblasts is congruent with our approach using craniofacial chondrogenesis (Hall and Miyake, 1995, 2000).

Clonal cultures

In the early 1980s, several groups dissociated and cloned foetal rat and mouse calvaria cells. Jane Aubin's group used enzymatic isolation of cells, while Ecarot-Charrier used a non-enzymatic approach, allowing cells to migrate from calvariae onto glass fragments. In both situations osteogenic cells synthesize alkaline phosphatase and differentiate into osteoblasts, with accompanying synthesis of osteonectin and collagen type I but not type II. Preosteoblasts cannot deposit a mineralized matrix until they synthesize alkaline phosphatase (Weinreb *et al.*, 1990). These cloned cells, and a human osteoblastic cell line derived from periostea, deposit a mineralized matrix, provided that a source of organic phosphate such as Na, β-glycerophosphate is included in the medium.[22]

Seven clonal lines from rat calvarial cells were isolated by Guenther *et al.* (1989) and shown to display some osteoblastic markers. Some lines displayed all the markers tested, suggesting that the clones represented a differentiation series, some further down the differentiation pathway than others. These clonal lines produce mineralized matrix when transplanted *in vivo* in diffusion chambers. They also produce cartilage.

A sub-clone of the foetal rat calvarial cell line RCJ 3.1 (designated RCJ 3.1C5) initiates chondrogenesis, especially in the presence of dexamethasone. Four different phenotypes – muscle, adipocyte, cartilage and bone – can be obtained from this one sub-clone; 20 per cent as monopotential cells, 10 per cent as bipotential and 1 per cent as tripotential. One tripotential and three bipotential cell lines were established:

- adipocyte and chondrocyte,
- adipocyte and myocyte,
- chondrocyte and myocyte, and
- chondrocyte, myocyte and adipocyte as the tripotential lineage.

Exposing these cells lines to retinoic acid moves them away from chondrogenesis and into the alternate pathway. Chondrogenesis is inhibited if the cells have not yet differentiated, while cartilage that has already formed is not maintained, because synthesis of GAGs is inhibited and degradation enhanced.

In studies from another laboratory, five osteoblastic clones isolated from neonatal rat calvariae express myogenic and adipocytic potential. Bmp-2 both stimulates osteoblastic and inhibits myogenic potential, and converts myoblastic to osteoblastic cells (Box 12.1).[23]

The isolation of adipocyte clones with chondrogenic or myogenic potential is interesting in light of the distribution of clonal lines with adipocyte potential within organisms, as detailed below.

- Marrow adipocytes have osteogenic but not chondrogenic potential and can be modulated between osteogenesis and adipogenesis by adding dexamethasone, foetal calf serum or vitamin D_3.
- Marrow from adult mice also contains progenitor cells that can form chondrocytes, adipocytes, osteoblasts and stromal cells. These marrow stromal cells respond to culture on type I collagen by producing mineralized nodules. Culture in collagen gels enhances the action of Bmp-2 in promoting osteogenesis.
- Adipocytes can also enhance osteogenesis from other cell types; conditioned medium from adipocytes enhances proliferation, collagen synthesis (although not the synthesis of non-collagenous proteins) and the deposition of alkaline phosphatase by osteoblasts.[24]

NOTES

1. See Jones (1974) for the rate of matrix deposition, and see the chapters in Volume 1: *The Osteoblast and Osteocyte* of Hall (1990–1994) for the structure and function of osteoblasts and osteocytes.
2. For osteocytic osteolysis, see Bélanger *et al.* (1966), Bélanger (1969), Jande and Bélanger (1969, 1971, 1973), Jande (1971), Aaron (1973), Aaron and Pautard (1973), Schulz *et al.* (1974), Yeager *et al.* (1975) and Kwiecinski *et al.* (1987). Interestingly, osteocytes do not resorb bone in lactating rats (Mercer and Crenshaw, 1985).
3. See Jande (1972) for the chick study, and McGuire and Marks (1974), Rasmussen and Bordier (1974) and Hall (1975b) for overviews of hormonal action on osteocytes and osteoclasts. Eriksen *et al.* (1988) first demonstrated oestrogen receptors on human osteoblast-like cells.
4. See Keller *et al.* (2001) and Keller and Brown (2004) for details and the growth factors involved.
5. See Glasstone (1968, 1971) and Hall (1971) for osteogenesis from intact mesenchyme *in vitro*.

6. See Bernard and Pease (1969), Boyde and Hobdell (1969), Bernard (1978), Boivin *et al.* (1983), Grynpas *et al.* (1989) and Thompson *et al.* (1989) for ultrastructural studies of initial intramembranous ossification of mammalian or avian calvariae and mandibles *in vivo*, *in vitro* and after transplantation. Basic early studies on the ultrastructure structure of developing bone in chick embryos are those by Fitton-Jackson (1957) and Hancox and Boothroyd (1965). A study by Boivin *et al.* (1983) contains the first TEM study of sutural cartilage in neonatal mouse calvariae. The cartilage is described as unmineralized, less hypertrophic than other cartilages, and with a fine fibrillar ECM. Weiner and Traub (1992) provide an excellent overview of bone structure over the angstrom to micron scale.

7. See McConaghey and Sledge (1970), Daughaday (1971), Urist (1972), Sledge (1973), Ash and Francis (1975) and Ashton and Matheson (1979) for how responses vary *in vivo* and *in vitro*, and see Madsen *et al.* (1983a) and Burch *et al.* (1986) for the somatomedin-like peptide.

8. See Nijweide (1975), Nijweide and van der Plas (1979) and Nijweide *et al.* (1981, 1982a,b, 1986) for this methodology.

9. See Tenenbaum and Heersche (1982, 1986) and Tenenbaum *et al.* (1986) for osteoid deposition, Tenenbaum *et al.* (1992a) for phosphate accumulation, Tenenbaum *et al.* (1989) for the regulatory effects of phosphates on bone metabolism *in vitro*, and Tenenbaum *et al.* (1992b) for effects of bisphosphonates and inorganic pyrophosphate.

10. See Wong and Kocour (1983) and McCarthy *et al.* (1991) for osteoblast–osteoclast responses and coupling.

11. See Nagata *et al.* (1991) and McKee and Nanci (1996) for the studies *in vitro* and *in vivo*, and see Toma *et al.* (1997a) for the mRNA study. Using inhibitors of the mitogen-activated protein kinase pathway, Bobick *et al.* (2004) revealed that an extracellular signal-regulated kinase, mitogen-activated protein kinase pathway inhibited chondrogenesis in micromass cultures of chick limb-bud mesenchymal cells.

12. See Yagiela and Woodbury (1977) for modifications of the method to isolate osteoblasts, and Peck *et al.* (1967a,b, 1974), Burks and Peck (1978) and Peck (1984) for characterization of these cells *in vitro*.

13. See Schuster *et al.* (1975) for *de novo* synthesis of lipids, D. M. Smith *et al.* (1973) for O_2 consumption and lactate production, and Dziak and Brandt (1974a,b) and Dziak and Stern (1976) for Ca^{++} uptake.

14. See Binderman *et al.* (1974) for the method, and Piatier-Piketty and Zucman (1971) for grafting periostea.

15. See Boyde *et al.* (1976) for the SEM and Duksin *et al.* (1975) for response to the anti-collagen serum.

16. See Osdoby and Caplan (1976) and Oakes *et al.* (1977) for the high-density cell cultures, and Osdoby and Caplan (1980, 1981) for effects of cell density.

17. See Wong and Cohn (1974, 1975), Luben *et al.* (1977), Wong *et al.* (1977) and Cohn and Wong (1979) for development, refinement and exploitation of this method.

18. Later studies showed that isolated osteoblasts/osteoblast-like cells, respond to PTH and to 1,25-vitamin D_3 (Rosen and Clark, 1982).

19. See Jones and Boyde (1976a, 1977a) for response to PTH, Altman *et al.* (1978) for the importance of substrate, and Lynch *et al.* (1995) for culture on type I collagen.

20. See Gerstenfeld *et al.* (1987, 1988) and Berry and Shuttleworth (1989) for isolation of embryonic chick calvarial cells, Heersche *et al.* (1984) for deposition of osteoid, Grynpas *et al.* (1989) for maturation of the initial apatite crystals, and Winnard *et al.* (1995) and Cowles *et al.* (1998) for fibronectin *in vitro* and *in vivo*. Layers of necrotic cells accumulate on top of the osteoid in cultured *whole calvariae* from 21-day-old foetuses, the assumption being that these cells release large calcium stores used by osteoblasts near the mineralization front (Zimmermann, 1992).

21. Reviews of the molecules involved in these transitions were provided by Stein and Lian (1993a,b) and Stein *et al.* (1996).

22. See Kadis *et al.* (1980), Aubin *et al.* (1982), Ecarot-Charier *et al.* (1983, 1988), Sudo *et al.* (1983), Bellows *et al.* (1986), Bhargava *et al.* (1988), Benayahu *et al.* (1989) and Binderman *et al.* (1989) for rodent calvarial cell clones, and Koshihara *et al.* (1987) for human.

23. See Yamaguchi and Kahn (1991), Yamaguchi *et al.* (1991) and Yamaguchi (1995) for chondrogenesis, Grigoriadis *et al.* (1988, 1989, 1990) for isolation of the five cell lines, and Lau *et al.* (1993) for a retinoic acid study.

24. See Bennett *et al.* (1991) and Beresford *et al.* (1992) for osteogenic potential, Hasegawa *et al.* (1994), Hashimoto *et al.* (1995) and Dennis *et al.* (1999) for marrow from adult mice, and Benayahu *et al.* (1993) for adipocyte enhancement of osteogenesis from other cells.

Bone Diversity

The phalli, which were hung around the necks of the Roman ladies, or worn in their hair, might have effect in producing a greater proportion of male children.[1]

Much of the interest in fluoride is a consequence of the addition of fluoride to drinking water in many parts of the world. We know that adding fluoride at 100 ppm to the drinking water provided to rats increases bone surface density and increases the amount of active bone surface on induced ectopic bone.

The action(s) of fluoride on bone cells has been studied for some decades, often with seemingly contradictory results. The contradictions are real. Individual populations of osteogenic cells or seemingly equivalent populations from different species give different results, even when the same methods of analysis are used. More worrying, for it makes understanding mechanisms of action so difficult, dramatically different results often are obtained from studies in vivo and in vitro. Consequently, responsiveness of osteogenic cells to sodium fluoride (NaF) provides a link between the osteoblast diversity discussed in the last chapter and the diversity of bone as a tissue or as an organ discussed in this chapter. I then move on to discuss several examples of bones with distinctive physiological capabilities or responsiveness to hormonal stimulation – alveolar bone; penile and clitoral bones; and the medullary bone in laying hens and resorption of the pubic joint, both of which are hormonally mediated.

HETEROGENEITY OF RESPONSE TO SODIUM FLUORIDE

As a general statement, NaF enhances proliferation of pre-osteoblasts, enhances initial differentiation of osteogenic cells and bone mineralization, and increases the mechanical strength of bone, but not all responses are elicited from the same cell population (Fig. 25.1).[2]

- As measured by alkaline phosphatase production, fluoride increases osteoblastic proliferation and differentiation in rats in vivo, but not when the same osteoblasts are cultured with fluoride.
- Implanting fluoride encased in polymer into rabbit vertebrae, femora or tibiae increases the number of vertebral and tibial trabeculae but not the number of femoral trabeculae.
- Administering fluoride (5–25 mg/kg body weight) to mice 20 days after periosteal or ectopic bone had been induced by human tumour KC cells (20 days is when proliferation of induced bone is greatest) has no impact on bone, a finding that led the investigators to conclude that fluoride is not mitogenic for human osteoblasts.

Mode of application must be controlled carefully. Topical application of a 2 per cent acid solution of NaF in 0.1 M H_3PO_4 for 20 minutes to standard 2.8-mm drill holes in rat calvariae accelerates osteogenesis and repair. The action of drilling the holes, however, accelerates remodeling and vascular changes around the holes, as shown in beagle dogs of various ages. Possible toxicological effects of NaF itself must also be borne in mind.[3]

Enhanced proliferation and osteogenesis

Lundy and colleagues developed an avian model using 14-day-old chicks to study the action of fluoride. Adding fluoride to the drinking water increased bone formation (assessed in the tibiae) and bone density but had no effect on osteoclast numbers. They found a strong positive correlation between serum fluoride and tibial alkaline

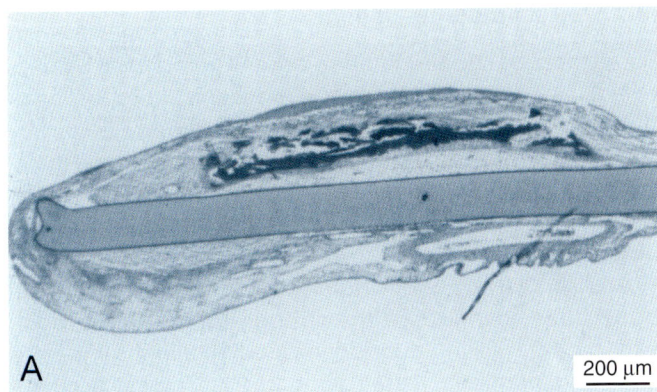

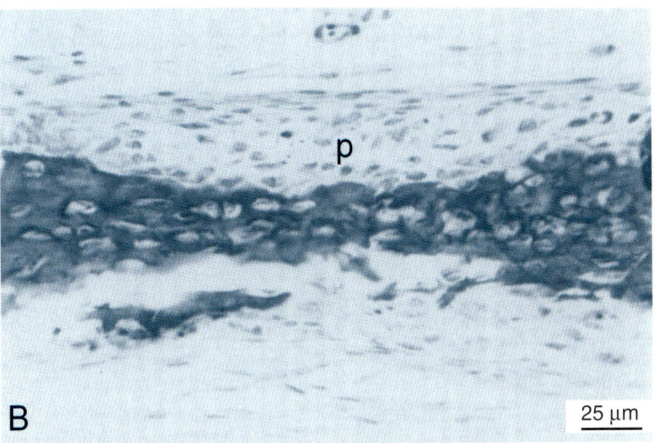

Figure 25.1 Intramembranous osteogenesis initiated in embryonic chick mandibular mesenchyme (H.H. stage 22) cultured for 10 days in the presence of 10^{-5} M NaF and then grafted to the CAM of a host embryo. (A) A low magnification micrograph of the bone (black) on the Millipore filter substrate and surrounded by chorioallantoic tissue. (B) A higher magnification of the bone to show its cellular nature and the multicellular periosteum (p). From Hall (1987a).

phosphatase ($r^2 = 0.88$) and ash content ($r = 0.93$). On the other hand, and reflecting specific skeletal responses, the same group of investigators determined that osteoblasts from chick *calvariae* contain fluoride-sensitive acid phosphatase activity, 100 μM Fl being sufficient to inhibit osteoblastic acid phosphatase (Lundy *et al.*, 1986, 1988, 1992).

Alteration of G-protein-dependent tyrosine phosphorylation via inhibition of tyrosine phosphatase or activation of tyrosine kinase, depending on the particular situation, is a potential mechanism of action. At concentrations that are mitogenic for osteogenic cells, fluoride increases the steady-state phosphotyrosyl phosphorylation level of cellular proteins in human bone cells. Such a proliferative role would be consistent with the role tyrosine phosphatase plays in the proliferation → differentiation transition at two stages of chondrogenesis.

The γ (-) splice variant of cytosolic tyrosine phosphatase is expressed at two borders during mouse mandibular chondrogenesis: at the border of condensations and at the border between chondrocytes and mature chondrocytes. Loss-of-function experiments

in vitro using antisense oligodeoxynucleotides result in abnormal patterns of Meckel's cartilage and increased amounts of cartilage because of enhanced proliferation at the border regions, the γ isoform being involved in regulation of proliferation and transition between chondrogenic stages (Augustine *et al.*, 2000).

Farley and Baylink (1986) used alkaline phosphatase: hydroxyproline ratios as a marker for osteogenesis *in vitro* in situations where the effects on bone of agents such as parathyroid hormone and NaF were being tested. Our knowledge of the mode of action of NaF has been expanded substantially in the hands of this group, who found NaF to be one of the most potent stimulators of bone formation *because of its direct action on bone cells*. Chick calvarial cells respond to NaF with enhanced proliferation, a response that is dependent on the phosphate concentration in the medium and mediated by increasing the activity of endogenous mitogens released by local resorption. Indeed, Na-dependent phosphate transport is selectively stimulated by fluoride in osteoblastic UMR-106 cells as a result of new protein synthesis.[4]

Human and chick osteoblasts respond to fluoride by increased proliferation and synthesis of alkaline phosphatase; human skin cells do not respond, although a human osteosarcoma cell line (SAOS-2) does respond by increased proliferation (Fig. 25.1). A concentration of >250 μmol NaF/l elicits these changes, which are inhibited by verapamil, a calcium channel blocker.[5]

Rat marrow stromal cells respond to 10^{-12}–10^{-9} M fluoride by enhanced formation of colony-forming units (CFUs) but to 10^{-7} or 10^{-6} M fluoride by decreased CFU formation, indicating a concentration-dependent proliferative response; the osteoid deposited by rat osteoblasts in culture increases in response to fluoride. The osteogenic competence of marrow stromal cells displays a circadian rhythm; the ability to form fibroblast CFUs is maximal in the late dark and early light phases (Simmons *et al.*, 1974, 1984).

Such concentration effects cannot be translated directly to other species. Human stromal osteoblastic precursors are more sensitive to 10^{-5} M NaF than are mature osteoblasts, in part explaining the action of NaF on osteoblast precursor proliferation. Others, however, failed to find a stimulation of proliferation or protein synthesis in human osteoblasts exposed to NaF. For example, inconsistent results were obtained when first-passage human osteoblastic cells were cultured in the presence or absence of serum and 10^{-6}–10^{-3} M NaF[6]; 10^{-5} M may be a critical concentration.[6]

Interaction with hormonal action

Fluoride stimulates bone formation *and* bone resorption in rats from which both thyroid and parathyroid glands have been removed, and stimulates bone cell number in ovariectomized rats. A synthetic testosterone, norethindrome, which acts through a different mechanism than

NaF, nevertheless allows NaF to act, as assessed by effects on proliferation and alkaline phosphatase synthesis by human Te 85 osteosarcoma cells. Dexamethasone also enhances fluoride effects on these cells.[7]

Osteoporosis

Eleven osteoporotic patients responded positively to prolonged fluoride therapy with increased bone formation (and with a decrease in those for whom treatment was stopped). A comparison of the efficacy of intermittent slow-release fluoride (25 mg twice daily for three months) on vertebral bone mass and on inhibition of fractures with or without 1,25 vitamin D_3 (2 µg/day for two weeks), in 24 osteoporotic patients, found fluoride more efficacious. An assessment of osteoblastic cells from 12 osteoporotic patients found enhanced proliferation in response to fluoride, indicative of fluoride acting by enhancing proliferation not differentiation. Bone mass increases, but the strength of the bone produced is suboptimal; see Box 8.2 for further details on osteoporosis.[8]

Chondrogenesis

Administration of excess fluoride (10 mg/kg body weight/day) to rabbits for up to 10 months results in cartilage forming within cancellous bone of the iliac crest, vertebral body and median spine, either because osteoblasts/cytes dedifferentiate to progenitor cells, or from pluripotential mesenchymal cells resident within the bone. Cartilage regeneration in the ears of rabbits and pikas (Box 25.1) is greater in males, reduced by cortisone, which stops cartilage proliferation, but accelerated by sodium fluoride.[9] Little else has been reported on effects on chondrogenesis, perhaps because, as in the first study, prolonged exposure is required to elicit any response, or because rabbits show a response not shared by other mammals.

Mineralization

A conclusion that does seem to have generality is that bone formed in the presence of fluoride is denser and more heavily mineralized than what we might term 'non-fluoridated bone,' as demonstrated in Wistar albino rats. Studies using chick long bones and calvariae illustrate this point. Fluoride *enhances* mineralization of intact embryonic chick tibiae maintained *in vitro*. At concentrations of 0–20 µm, fluoride increases alkaline phosphatase and proliferation and *suppresses* mineralization in chick periostea maintained *in vitro*. Subsequently, two phases of mineralization of osteoid deposited by chick calvarial osteoblasts *in vitro* were identified, each differentially affected by fluoride: (i) an alkaline-phosphatase-*dependent* phase associated with initiation of mineralization and inhibited by fluoride, and (ii) an alkaline-phosphatase-*independent* phase associated with progression of mineralization and stimulated by fluoride.[10]

Mechanical properties of bone

Four doses of sodium fluoride administered over 80 days to adult rats did not impair the mechanical properties of the limb skeletons, even though half the skeleton was turned over through remodeling during that time. Again, different mammals respond differently; fluoride administered to rabbits increases serum Igf-1, bone turnover and bone mass but not bone strength.[11]

Adding NaF to 5-mm holes in rabbit femoral cortical bone does not increase the strength of the repair bone over defects lacking fluoride. Indeed, healing is impaired and unmineralized osteogenic mesenchyme accumulates. On the other hand, when a biodegradable system is used to deliver known amounts of sodium fluoride around rods inserted in the medullary cavities of rabbit femora, bone strength and bone growth increase locally around the application site.[12]

ALVEOLAR BONE OF MAMMALIAN TEETH

Teeth are not in my remit, which is to say that I have elected not to discuss development of teeth *per se*, treating them only in so far as they inform discussion of skeletal tissues. That said, an important type of bone, alveolar bone, develops at the base of teeth to anchor them into the jaws, although this anchor can drift or be dragged by the tide of tooth movement. *Alveolar bone* nicely illustrates the diversity of bone as a tissue.

Origin

Alveolar bone is of interest developmentally because its osteoblasts arise from the same population of cells which produce the odontoblasts that deposit dentine, demonstrated for mouse alveolar bone using ^{3}H-thymidine labeling and grafting. Alveolar bone will not form unless dental follicle mesenchyme is present. If teeth are absent, so too is alveolar bone; see Box 25.2.[13]

Structures and tissues for tooth attachment – alveolar bone, osteodentine and cementum – depend upon extrinsic factors for their morphogenesis, although this may be less true for anurans than for mammals; teeth of the northern leopard frog, *Rana pipiens*, transplanted with or without the bone of attachment – the frog's equivalent of alveolar bone – grow to normal size and retain their normal morphology, even if transplantation leads to ankylosis.[14]

Physiology and circadian rhythms

Alveolar bone is among the most physiologically active bone in mammals, perhaps the bone that can be resorbed the fastest. Indeed, it has been argued that alveolar bone *is* the most metabolically active bone.[15]

The periosteum of alveolar bone and the adjacent cementum in young mice display a *circadian rhythm* of

Box 25.1 Holes in the ears of rabbits and mice

As few mammals can heal defects in their external ears, we can use ear tags to keep track of experimental animals in the laboratory. Rabbits and pikas are exceptions, pikas being tailless, hamster-size[a] relatives of rabbits (order Lagomorpha) from the mountains of western North America and Asia. Another exception is a single strain of mice. Even quite large holes punched out of rabbit and pika ears heal, healing involving regeneration of new cartilage across the defect.

Rabbits

Healing of large holes in the external ears (pinnae) of rabbits, especially in males, includes regeneration of new cartilage across the defect. Cortisone reduces regeneration by inhibiting proliferation of cartilage at the edges of the hole; sodium fluoride speeds up regeneration by promoting proliferation. In a comprehensive analysis of regeneration of holes punched in rabbits' ears it was shown that:

- 75 per cent of males and 20 per cent of females regenerate;
- 75 per cent of pregnant females but only 25 per cent of females in oestrus or lactating regenerate;
- 70 per cent or more of ovariectomized females given testosterone regenerate;
- successive rounds of regeneration following successive wounding result in thicker cartilage and an increase in the regeneration rate of 30 to almost 70 per cent; while
- regeneration of proximal ear tissue is more frequent (78 per cent) than that of distal tissue (12 per cent), perhaps because of differential thickness of the ear cartilage.[b]

In a follow-up study comparing ear-tissue regeneration in mammals in general, Williams-Boyce and Daniel (1986) showed that cartilage in regenerating rabbit ears arises from both the blastema and from preexisting perichondrial cells. Regeneration of a wounded ear depends entirely on cartilage being present; remove the cartilage and regeneration is not initiated.

Regeneration of cartilage also depends upon interactions between the cartilage that remains and the ear epithelium; replace ear epithelium with epithelium from the belly and the wound regenerates but cartilage does not. Transplant rabbit-ear skin onto the tail and amputate the tail after the transplant has healed in place, and ectopic cartilage overlain by normal tail skin develops. Furthermore, tongues of epidermis grow down around the edges of the wound during the first one to two weeks following wounding. Such down-growth fails to occur if cartilage is removed before the wound is made, and does not occur in species that cannot regenerate wounds in the ear. Thus, the sequence is cartilage → epidermis → epidermal down-growth → interaction with mesenchyme → regeneration.[c]

Irradiating ear *cartilage* with 3000 rads prevents chondrogenesis but not epidermal down-growth; irradiating the *skin* has no effect on either process. Morphogenetic specificity is maintained during regeneration; a single cartilage regenerates after a second cartilage has been grafted into the ear, allowed to heal, and a full-thickness hole punched through both cartilages (Grimes, 1974a,b).

Just as amphibian limb regeneration is associated with local bio-electrical activity, so too is rabbit ear regeneration; the surface potential of the skin, the sequence of events, and the level of electro-negativity (-20 mV negative potential) associated with growth all resemble conditions seen in amphibian limb regeneration (Chang and Snellen, 1982).

Essentially no molecular studies have been undertaken, although we know that *Bmp-2* is expressed in condensing mesenchyme during rabbit ear regeneration (Urist et al., 1997).

MRL mice

One strain of mice (MRL/MpJ) also can regenerate holes punched or made in the ears. MRL mice, which are homozogous for the *lpr* (lymphoproliferation) gene, are used as a model for rheumatoid arthritis. Large and docile, these mice die between 18 and 23 weeks, usually with chronic glomerulonephritis. Regeneration involves tissue remodeling, formation of a blastema, deposition of ECM, initiation of growth, and formation of new hair follicles and sebaceous glands. Keratin mRNAs and molecules that suppress keratinocyte growth are up-regulated within two days of wounding. A thickened epidermis and considerable epithelial down-growth, reminiscent of that seen in the early stages of regeneration of the ears of rabbits and antlers (Chapter 8) is seen from day five onwards.[d]

Interestingly, from about 10 days, tenascin is seen in the blastema of MRL mice but not in ear wounds in strains that cannot regenerate. Even more interestingly, mice of other strains develop normally when tenascin is knocked out, with normal fibronectin, laminin, collagen and proteoglycan (Saga et al., 1992). Whether MRL mice could regenerate ear cartilage in the absence of tenascin has not been tested.

---

[a] For non-North American readers, hamster-sized will be about as useful as the Canadian unit of measure, the 'Prince Edward Island', where x is said to be y times the size of Prince Edward Island. The North American pika, *Ochotona rufescens*, has a body weight of 150–200 grams, a crown–rump length of 140 mm, and stands about 40 cm tall when reared up on its hind legs.
[b] See Joseph and Tydd (1973a,b) for rabbit-ear cartilage regeneration, and Williams-Boyce and Daniel (1980) for the comprehensive analysis.
[c] See Goss and Grimes (1972, 1974) and Grimes and Kulis (1979) for epithelium–cartilage interactions, and see Joseph and Tydd (1973a,b) for rabbit-ear cartilage regeneration.
[d] See Clark et al. (1998) and Heber-Katz (1999) for ear-cartilage regeneration in MRL/MpJ mice, and see Firestein (2003) for current approaches to rheumatoid arthritis.

---

resorption with a single peak at 6.00 p.m. in the outer fibrous layer of the periosteum and two peaks in the inner periosteum and cementum (Tonna et al., 1987). Cycles with a longer periodicity than 24 hours are also known. Alveolar bone displays a six-day cycle of formation and resorption consisting of resorption for 1.5 days, a reversal phase for 3.5 days, and deposition for one day (Table 25.1). Human alveolar bone displays *seasonal rates* of turnover of formation and resorption (Table 25.2).

The periodontal ligament (Fig. 1.11) displays a diurnal periodicity when at rest – at rest meaning unstimulated by mechanical or other factors. Minimal numbers of mitoses occur at 9.00 p.m., maximal numbers at 9.00 a.m. and

3.00 p.m. Robert (1975) interpreted these results as demonstrating three populations within the periodontal ligament – one blocked at G1 and one at G2 of the cell cycle, and a third partly differentiated population.

Turnover of alveolar bone in male rats correlates with sites of fast and slow growth and with age. Rodent alveolar osteoclasts respond differentially to calcium or phosphorus deprivation depending on their location – whether endosteal or along the surface of the tooth socket. Osteoclasts along the socket are more numerous and have more nuclei than endosteal osteoclasts (Table 25.3). Calcium deprivation increases endosteal osteoclast numbers 5.5-fold and nuclear number 6.5-fold but has *no*

Box 25.2 The diastema and toothlessness

A diagnostic feature of rodents as a group is reduction in dentition. Rodents have only incisors and molars, not the full range of incisors, canines, premolars and molars that characterize our own dentition (Fig. 25.2). Because rodent teeth reside in the same positions in the mouth as they do in mammals with a full dentition, a large tooth-free (edentulous) space known as the *diastema* lies between incisors and molars (Fig. 5.8). Tooth formation, and therefore formation of alveolar bone (because it arises from the same population of cells as does dentine), fails to occur in this gap.[a]

In a classic experiment, Ed Kollar (1972) recombined epithelium and mesenchyme from diastemal and tooth-forming regions of

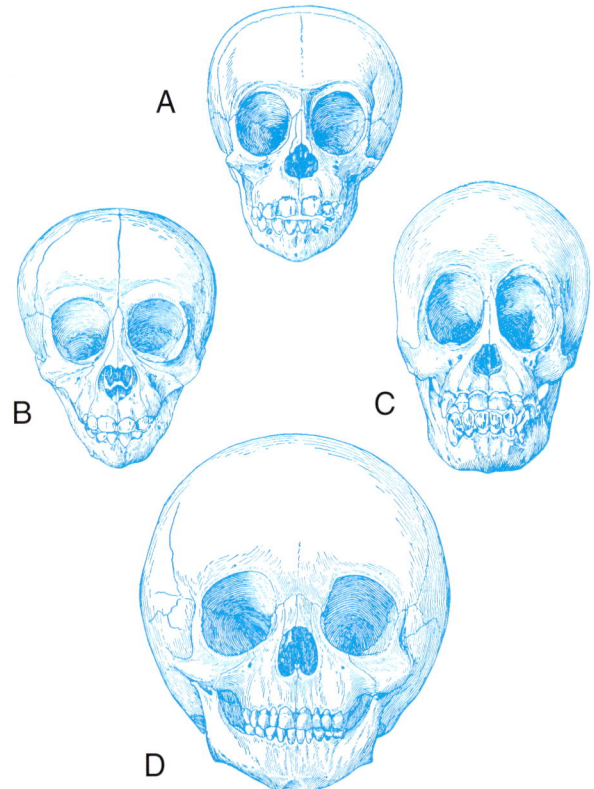

Figure 25.2 Primate skulls. Skulls of juvenile chimpanzee, *Pan troglodytes* (A), gorilla, *Gorilla gorilla* (B), orangutan, *Pongo pygmaeus* (C), and human, *Homo sapiens* (D) showing the difference in proportions. Modified from Gregory (1929).

mouse embryos, and demonstrated that the inability of the diastemal region to make teeth results from a primary deficiency in the mesenchyme, not in the oral epithelium. *Teeth fail to develop* if epithelium that would normally form an enamel organ is recombined with diastemal mesenchyme. *Teeth develop* when diastemal epithelium is recombined with odontogenic mesenchyme. These teeth are complete in two senses: (i) they contain enamel, meaning that diastemal epithelium responded to the mesenchyme to initiate ameloblast differentiation; and (ii) dentine forms, meaning that the diastemal epithelium signaled to the mesenchyme to initiate odontoblast differentiation and formation of dentine.

A comparative investigation of the molecular basis of mouse and vole diastemata revealed that three tooth germs do form in the diastema of the upper jaws in mouse embryos, but that they arrest at the early bud stage. In voles (Fig. 5.8), one large tooth germ forms in the upper diastema, but it arrests at the late bud stage. Expression patterns of *all* the major genes involved in odontogenesis are abnormal in these diastemata (Keränen *et al*., 1999).

The range of genes examined included *Bmp-2* and *-4*, *Fgf-4* and *-8*, *Lef-1*, *Msx-1*, *Msx-2*, *Pitx-2*, *Pax-9* and *Shh*. From other studies with these genes we know that teeth fail to develop when only one gene is inactive.

- *Msx-1*-deficient mice cannot form teeth and have deficient alveolar bone and cleft palate (Fig. 25.3).
- *Pax-9*-deficient mice also are missing all their teeth. Development is initiated but arrests at the bud stage, as occurs in the diastemata of wild-type mice. *Pax-9* mutants also have supernumerary preaxial digits, fail to develop derivatives of the pharyngeal arches such as the thymus, pituitary and ultimobranchial glands, and lack the angular and coronoid processes of the dentary.

Mutation in the gene for the transmembrane protein ectodysplasin (EDA), a protein with affinities to the tumour necrosis gene family, elicits ectodermal dysplasia in humans, in which teeth and hair fail to form. The ectodysplasin-A receptor is required for scale development in the Japanese medaka, *Oryzias latipes*, a finding that led Sharpe (2001) to reexamine the homology of hairs, teeth and scales.[b]

Other genes influence tooth development but are not required for teeth to form; *Gli-2*-mutant mice show severe skeletal abnormalities in neural crest-derived craniofacial and somitic skeleton, including tooth defects (Mo *et al*., 1997).

[a] Suggestions that innervation or some property of nerves specifies where teeth will form along the oral epithelium or dental lamina arose in part from observations that nerve fibres enter the region before tooth buds arise but do not enter the diastemal region (Pearson, 1977).
[b] See Satokata and Maas (1994) for *Msx-1*, Peters *et al*. (1998) for *Pax-9*.

Table 25.1 Duration of phases of bone remodeling in alveolar bone of rats[a]

Phase	Formation phase (σF)	Resorption phase (σR)	Reversal phase (σRev)	Resting phase (σO)	Total remodeling cycle (σBMU)
Time (days)	1	1.5	3.5	5	11

[a] Based on data from Vignery and Baron (1980).

effect on osteoclasts lining the tooth socket (Table 25.4). Consequently, bone is preferentially removed from the endosteal surface. Phosphorus deprivation causes a doubling of endosteal osteoclast numbers but decreases

nuclear numbers to a quarter of control levels. As with calcium deprivation, endosteal osteoclasts are unaffected (Table 25.4).[16]

PENILE AND CLITORAL CARTILAGES AND BONES

Human males are a major exception to the general pattern of male mammals possessing a skeletal element – a penile bone, *os penis* or baculum – to support and stiffen the penis. A parsimonious interpretation of humans as baculum-less is loss of an ancestral character. A recent

Table 25.2 Seasonal rate of turnover in human alveolar bone of two age groups[a]

Season	Bone formation (average)				Bone resorption (average)			
	[45]Ca uptake (cpm)		Alkaline phenylphosphatase (activity/gram bone)		β-glucuronidase (activity/mg protein)		Acid phenylphosphatase (activity/gram bone)	
	11–13 years	20–25 years	11–13 years	20–25 years	11–13 years	20–25 years	11–13 years	20–25 years
Spring[b]	12 496	3239	164	35.6	37.63	11.22	15.54	6.59
Fall[c]	9910	2574	134	30.2	23.76	8.68	12.75	5.05

[a] Based on data from Stutzmann et al. (1981) and Stutzmann and Petrovic (1989).
[b] April 1 to July 1.
[c] October 15 to January 15.

Table 25.3 Comparison of mesial and distal alveolar bone resorption and formation in adult and young rats[a]

Location	Bone formation (average)						Bone resorption (average)					
	Double-labeled surface (%)		Mineral apposition rate (μm/day)		Bone formation rate (square μm/day)		Osteoclast (number/mm)		Osteoclast surface (%)		Osteoblast surface (%)	
	Adult[b]	Young[c]	Adult	Young	Adult	Young	Adult	Young	Adult	Young	Adult	Young
Distal	18	20	2.5	2.8	1	3	5	2.5	14	12	50	33
Mesial	24	55	2.3	5.3	0.5	14	1	0.8	2	5	65	40

[a] Based on data in King et al. (1995). % is per cent of bone surface occupied by osteoclasts, osteoblasts or both (double-labeled).
[b] 89–94 days old.
[c] 30–35 days old.

Table 25.4 Number of osteoclasts and osteoclast nuclei elicited in rat endosteal and incisor pocket bone in response to calcium or phosphorus deprivation[a]

Group	Number of osteoclasts/section		Number of nuclei/section	
	Endosteum	Socket	Endosteum	Socket
Control	17.6 ± 3.0	34.2 ± 1.4	30.2 ± 5.3	51.6 ± 3.1
Ca[++] deprivation	99.0 ± 13.8[b]	32.6 ± 3.1[d]	197.8 ± 34.0[c]	56.2 ± 5.0[d]
P[+] deprivation	35.8 ± 2.2[c]	37.5 ± 5.7[d]	7.2 ± 14.0[c]	60.0 ± 5.5[d]

[a] Based on data from Liu and Baylink (1984).
Comparisons with control: [b] $P < 0.001$; [c] $P < 0.002$; [d] not statistically significant.

paper in the *American Journal of Medical Genetics* by a developmental biologist and a student of biblical litera- ture notes that mineralization of ossification of the penis in human males has been reported in association with congenital anomalies (the developmental biology part) and wonders whether the Hebrew noun *tzela*, translated as rib, should really have been translated to indicate a structural supporting beam or baculum, i.e. Adam's rib was really Adam's baculum (Gilbert and Zevit, 2000).

Penile bones develop in genital tubercles and have a surprisingly complex histogenesis. Their responsiveness to hormonal control is not unexpected. What is surprising is to find that they share many genes and gene networks with developing limbs and faces; the major phenotypic features of a rare human syndrome, Aarskog syndrome, (faciogenital dysplasia) include stunted growth, unusually

broad facial features, short broad hands and feet, and genital abnormalities. *Goosecoid* (*gsc*) may be a shared regulatory gene; inactivating *gsc* results in neonatal death of mouse embryos preceded by abnormalities of the genitalia, mandibles, tongue, nasal cavity, trachea, ear, and shoulder and hip joints. Coronoid and angular processes on the dentary bones are reduced in size or fail to form (Fig. 25.3), the size of the bony processes correlat- ing with the proportion of Gsc-null cells.[17]

Goosecoid cells act autonomously in craniofacial mes- enchyme. Thus chimeric embryos composed of *gsc*- expressing and *gsc*-null cells have nasal capsule and mandibular defects equivalent to those found in *gsc* mutant embryos, the decrease in size of condylar and angular processes correlating with the proportion of *gsc*-null cells in those elements. Double knock out of *Gsc*

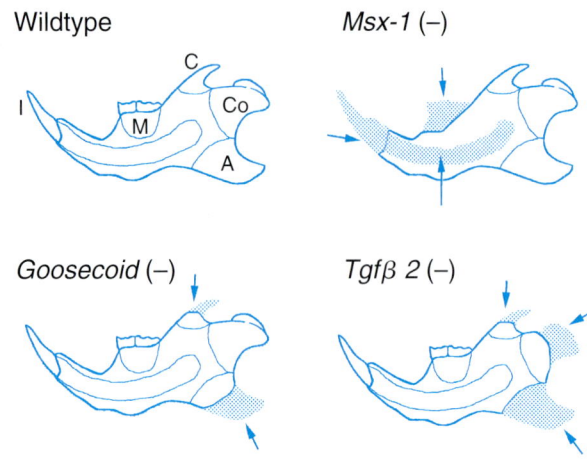

Wildtype   Msx-1 (–)

Goosecoid (–)   Tgfβ 2 (–)

Figure 25.3 Effects of gene knockout on the murine dentary. The wild type shows the morphological/morphogenetic components of the dentary. These consist of two alveolar units associated with incisor (I) and molar (M) teeth, and associated alveolar bone (note the extensive root of the incisor extending into the body of the dentary), and the coronoid (C), condylar (Co) and angular (A) processes. Different genes act on different morphogenetic units. The units affected are stippled and identified by arrows. Knocking out *Msx-1* results in alveolar units failing to form. Knocking out *Goosecoid* results in smaller coronoid and angular processes but has no effect on the condylar or alveolar units. Knocking out *Tgfβ-2* results in decreased growth of all three processes (especially the angular process) but has no effect on the alveolar units. See text and Hall (2003d) for details.

and *Msx-1* produces defects that are additive over single knock outs and that affect only first arch derivatives (Fig. 25.3). The patterning of the distal part of the malleus was shown to depend on the tympanic membrane which requires both genes (Kuratani *et al.*, 1999).

Os penis

Formation and transformational growth of rat penile bones is understood in some detail, in part because skeletogenesis in rat penile bones is predictable from individual to individual. Not so for other mammals. In the bobcat, *Felis rufus*, cartilage is only present in some juvenile specimens and the timing of baculum chondrogenesis varies considerably from individual to individual. Proximal and distal growth cartilages and endochondral ossification have been identified, and the skeletal tissues shown to develop in a proximo-distal sequence, just as in the limb skeleton (Chapter 38). The proximal growth cartilage is comparable to the secondary cartilage of foetal mandibular condyles. Chondrocytes from the distal segment remain viable and are liberated from their lacunae as the cartilage mineralizes.[18]

Os clitoridis

Given penile bones, obvious questions are whether female mammals have a clitoral bone (*os clitoridis*), and whether

such bones (if they exist) are hormonally induced and, if so, whether by hormones similar to those that elicit penile bones in males. The answers are yes, yes and maybe.

Depending on the genetic strain, female *rats* either lack an *os clitoridis* or possess a membrane bone component equivalent to the proximal segment of the male *os penis*. An *os clitoridis* can be induced in female rats within two days of birth by injecting testosterone. The skeletal element that develops parallels the penile bone in basic organization in that it is jointed, with a distal cartilage that ossifies and a proximal membrane bone. With prolonged androgen treatment, growth of these induced bones can be promoted and the distal element extended.[19]

Female *mice* have an *os clitoridis* as a membrane bone only but develop a condensation for the cartilage. So, mice have a clitoral bone. A more fully developed – which, I am afraid, means more male-like – *os clitoridis* can be induced with testosterone. A potential role for epithelial–mesenchymal interactions in regulating the growth of these skeletal tissues was suggested when skeletal elements were found to be larger and developed more cartilage when genital mesenchyme was cultured or grafted with an epithelium. Otherwise, chondrogenic tissue in females arrests at the condensation stage unless testosterone is given.[20]

Hormonal control

Androgens bind to genital tubercle mesenchyme not to the epithelium. The condensation phase is *androgen-independent* in male and female mice. Only undifferentiated mesenchyme forms when genital tubercles are excised from castrated rats, unless the rats are pretreated with androgens, in which case the full range of skeletal tissues develops. Differentiation of cartilage and bone is therefore dependent on androgens in males, and the clitoral bone that can be induced in females requires androgen administration.[21]

Murakami and colleagues investigated aspects of androgen dependence of rat penile bones, which they described as composed of two segments: a proximal segment of Haversian bone with a proximal hyaline growth cartilage, and a distal segment of fibrocartilage that ossifies after puberty (which occurs at about six weeks of age). Penile primordia cultured at the blastemal stage form only the fibrocartilage of the distal segment, and then only if testosterone is included in the medium. Otherwise, only fibroblastic tissue develops. A collagenous matrix is deposited in prepubertal rats (birth to four weeks); chondrocyte maturation and mineralization are initiated during puberty, chondrocyte degeneration and accumulation of matrix vesicles after puberty. Chondrocytes fail to mature, matrix vesicles fail to form and cartilage fails to degenerate in castrated rats. A cartilaginous callus forms if rat penile bones are fractured. Speed of fracture repair and speed of ossification depend on circulating hormones.[22]

Digits and penile bones

A fascinating correlation exists between digit and genital bud development (Fig. 19.1).

When posterior *Hox* genes are knocked out, both digits and the genital bud are lost. In mice, digit number, digit size and baculum size are regulated by dose-dependent production of the posterior *Hox* genes, *Hoxd-11, Hoxd-12* and *Hoxd-13* (Zákány *et al.*, 1997). In one sense the relationship is straightforward; progressive reduction of the doses of Hox genes progressively reduces digit number to zero (adactyly). But the relationship is more complex than it appears; digit number increases (polydactyly) during the dose-related transition to adactyly.

A surprising degree of similarity in genetic regulation underlies formation of the genital tubercle and limb bud. Grafting mouse polarizing zones or the anterior portion of embryonic primitive streaks into limb buds of *talpid*[3] chick mutants up-regulates expression of *Hoxd-4* and evokes posterior skeletal elements. Grafting genital tubercles into *talpid*[3] limb buds also evokes posterior skeletal elements but does not alter *Hox-4* expression.

Fgf-8 (but not *Fgf-10*) is expressed in the epithelium of the murine genital tubercle, regulates mesenchymal outgrowth and induces the mesenchymal genes *Msx-1, Fgf-10, Hoxd-13* and *Bmp-4*. Although not involved in genital tubercle formation, *Fgf-10* does play a role in the development of the glans penis and glans clitoris.[23]

Mutations in *Hoxd-12* and *Hoxd-13* are associated with the formation of extra phalanges (*polyphalangy*; Box 36.2) within digits, providing a further demonstration of the genetic basis of bone diversity.

Hoxd-12, Hoxd-13 AND POLYPHALANGY

Hox genes are involved in specification of the number of phalanges that form in a digit. Targeted disruption of murine *Hoxd-13* results in loss of the metacarpals and phalanges. In a double knockout, targeted disruption of murine *Hoxd-13* and *Hoxd-12* delays ossification and results in formation of an ectopic postaxial digit VI.

In humans, a mutation in *Hoxd-13* causes several fingers to fuse into a single large digit (*synpolydactyly*) – the metacarpal/metatarsal is transformed into a short carpal/tarsal. The mutation is revealing because it seems to indicate that a long 'long bone' can be readily converted into a short 'long bone.'[24]

Extra toes (*xt*), a syndrome based on a recessive mutation in mice, identified after insertional mutagenesis in transgenic mice, disrupts development of the anterior portion of the forelimbs, resulting in alterations in digit I (the thumb) and sometimes resulting in development of an extra phalanx (polyphalangy) in digit II. The mutation (*add*, anterior digit-pattern deformity) maps to the centromere of chromosome 13 where the gene *extra toes*

resides. From analysis of double mutants of *add/xt*, Pohl *et al.* (1990) showed that *add* is an allele of *xt*. Expression of the zinc finger gene *Gli-3*, which is normally expressed in limb development, is reduced by 50 per cent in heterozygote and to zero in homozygous mutants (Schimmang *et al.*, 1992).

Dactylaplasia in mice, an autosomal dominant mutation in which the phalangeal bones are missing from the middle digit (digit III), often only on one limb, is of interest because loss of digit III is regarded as a difficult task evolutionarily, complete loss of digit III being a 'forbidden morphology' on the basis of theories of limb development and morphogenesis. Targeted disruption of *Hoxd-12* and *Hoxd-13* leads either to minor defects in the autopodium (*Hoxd-12*) or to more extensive defects such as absent metacarpus and phalanges with disruption of *Hoxd-13*. Disruption of both leads to new defects, such as delayed ossification and an ectopic postaxial digit VI, not seen when either is disrupted alone. Within the limbs, two enhancers of *Hoxd* compete for interaction with the same promoter. Consequently, the position of the *Hox* gene relative to the enhancers could determine expression patterns (Hérault *et al.*, 1999). The regulatory elements near *Hoxd-12* are conserved in tetrapods but not in fish, mutations affecting the zeugopod and trunk skeleton.[25]

OESTROGEN-STIMULATED DEPOSITION OF MEDULLARY BONE IN LAYING HENS

The general – usual, majority, expected or anticipated – action of oestrogen on skeletal tissues is *stimulation* of bone formation. Examples abound:

- oestrogen stimulates the differentiation and activity of osteoblasts in intramembranous bones;
- the long bones of female rats are stronger at the end of pregnancy – during which oestrogen levels peak – than in virgin animals; and
- birds when producing and laying eggs accumulate bone within the medullary cavities of their long bones.

One of the most studied actions of oestrogen is how it triggers the formation of medullary bone in laying hens (domestic fowl) and Japanese quail (Fig. 6.2). Oestrogen stimulates osteoprogenitor cells lining bone surfaces to divide and modulate into osteoblasts, and stimulates bone deposition by osteocytes; PTH stimulates osteoclastic resorption. If oestrogen is given to Japanese quail hens, bone-lining cells are stimulated to proliferate and produce osteoblasts. Oestrogen has no direct effect on bone resorption in males.

Medullary bone serves as a reserve of ions to mineralize the egg shell; a hen's egg contains some 120 mg of calcium (Box 25.3). Maintaining laying hens on a hypocalcemic diet results in 20 per cent of the osteocytes displaying *osteoclastic* activity as they functionally replace

Box 25.3 Shell-less culture

Shell-less culture of chick embryos is a technique that has been used in a number of ways to study skeletogenesis and the significance of the shell for embryonic development. Essentially, the technique involves emptying the contents of a fertilized egg – embryo and all – into a container, sometimes with the addition of egg albumen or culture media, and allowing the embryo to develop in the container but outside the shell. Growth is similar to *in ovo* control embryos up to 11 days of incubation (De Gennaro *et al.*, 1980).

Various techniques of shell-less culture have been developed, going back 25 years to when Dunn *et al.* (1981) sought the best containers and conditions for survival and growth, beginning with embryos of different ages. In our hands, the yolk sac of some 90 per cent of the eggs breaks either on emptying the contents or within the first 24 hours of culture. After that inauspicious beginning survival rates are high. Shell-less culture can provide a convenient alternative to CAM-grafting, in that the grafts can be visualized as they become vascularized and develop.[a]

There is a caveat, however. The shell provides some 120 mg of calcium to the developing embryo, and shell-less culture deprives the embryo of this source of calcium. As a consequence, development of endochondral bone is retarded. Islands of cartilage, preceded by peanut agglutin lectin (PNA)-positive cells in the sub-cambial layer of the calvarial bones, appear in the bones of the skull. PNA-negative periosteal cells remain fibroblastic and do not form cartilage, observations that confirm the existence of a chondrogenic subpopulation in chick calvariae.[b]

This drawback can be put to advantage; shell-less culture provides a model system for studying the effects of calcium deficiency. Olena Jacenko and Rocky Tuan published a series of studies on the mechanism by which calcium deficiency elicits expression of a cartilage-like phenotype in chick embryo calvariae. Characterized by type II collagen and deposition of GAGs, this phenotype appears in undermineralized parts of the calvaria. The mineral deficit promotes cell-to-cell association and chondrogenesis, while the presence of mineralized matrix suppresses chondrogenesis. It also seems that situations allowing interactions between cells suppress chondrogenesis and allow osteogenesis. Type X collagen is synthesized by calvarial cells unless calcium is added to counteract the calcium deficiency. Interestingly, Bmp-2 enhances chondrogenesis from murine calvariae or cultures from mouse calvariae, but only from the portions of the calvaria that normally form cartilage. Bmp-2 enhances chondrogenesis from chondrogenic cells but appears not to bring non-chondrogenic cells into the chondrogenic family.[c]

A further link between early calvarial bone and chondrogenesis is the finding that embryonic chick calvariae contain proteoglycans that immunologically cross-react with proteoglycans in chick cartilage, although they are smaller molecules than found in cartilage. Calvarial proteoglycans are comprised of 67–88 per cent chondroitin sulphate, 5–18 per cent heparan sulphate and 1–2.5 per cent keratin sulphate (Sugahara *et al.*, 1981).

[a] See Jaskoll and Melnick (1982) and Tuan *et al.* (1991) for long-term shell-less culture.

[b] See Watanabe and Imura (1983) for how avian embryos make use of their shells, Tuan and Lynch (1983) for the Ca++ provided by the shell, and Stringa and Tuan (1996) for the PNA-positive cells.

[c] See Jacenko and Tuan (1986, 1995), Jacenko *et al.* (1995) and Wong and Tuan (1995) for these studies, and Komaki *et al.* (1996) for Bmp-2.

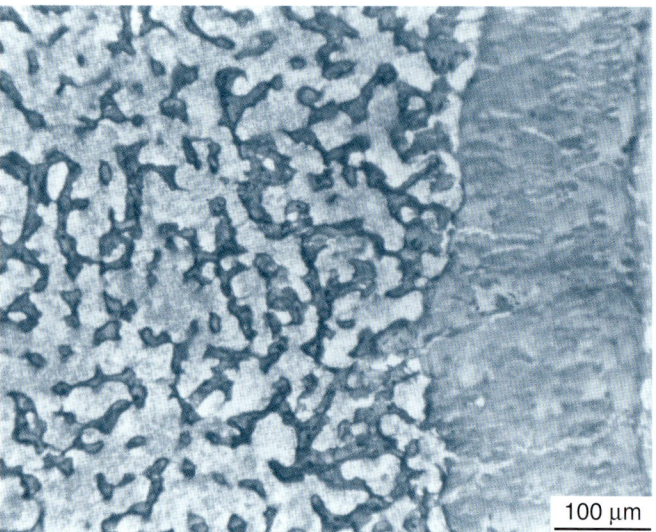

Figure 25.4 Undemineralized frozen section of the femur of a female Japanese quail, *Coturnix coturnix japonica*, showing cortical bone (right) and islets of medullary bone filling the medullary cavity (left). Modified from Van de Velde *et al.* (1985). Also see Fig. 6.2.

osteoclasts. The osteoclasts break down into mononuclear and polynuclear cell fragments.[26] Studies by Van de Velde *et al.* (1985), using whole-body technetium[99] complexed with stannium and methylene-diphosphonate, demonstrate separate regulation of matrix formation and mineralization of *medullary bone* (Figs 6.2 and 25.4).[27]

OESTROGEN-STIMULATED RESORPTION OF PELVIC BONES IN MICE

Another unusual action of oestrogen is stimulation of *resorption* of the medial edges of the fused bones of the pelvic girdle during pregnancy in some mammals. Indeed, the pelvic bones of mice are resorbed completely and so disappear entirely in response to exogenous oestrogen. The mechanism of resorption involves oestrogen-stimulated modulation of osteoprogenitor cells to *osteoclasts*.[28] The uniqueness of this physiological response was demonstrated convincingly in three studies performed by Ed Crelin and his colleagues, outlined below.

(i) Pieces of rib were transplanted in contact with the pelvic bones and pubic symphysis of castrated male mice, which were then treated with oestrogen and relaxin. The median edges of the bones were resorbed, and an interpubic ligament formed in the symphysis. Ribs grafted as controls remain intact.

(ii) Gonads were removed from three-week-old male and female mice. Two weeks later, capsules containing oestrogen were implanted, either in contact with the pelvic symphysis or under the patellar ligament in contact with the patella. Thirty days later, extensive amounts of medullary bone had formed in the patella, femur and tibia, sufficient to almost completely fill the medullary cavities.

(iii) In the third study, pieces of pelvic bone or rib were placed into defects in the tibiae of host mice and oestrogen was placed into the back musculature of the hosts. Six weeks later, the grafted pelvic bones were undergoing resorption, while rib grafts and host tibiae (controls) were depositing new bone.[29]

In a study of 14 strains of mice, Uesugi et al. (1991) showed further that the shape of the pelvic bones is determined by sex hormones before the mice are two months old. Possibilities for studying the differential behaviour of osteoblastic and osteoclastic cell populations and their precursors under the conditions they used are considerable.

So, just as cartilage is not cartilage is not cartilage, we see that bone is not bone is not bone. It also appears that a chondrocyte is not a chondrocyte is not a chondrocyte, and than an osteoblast is not an osteoblast is not an osteoblast. Families of osteoblastic and chondrogenic cells exist, a concept that takes us back to the concepts elaborated by Neville Willmer and discussed in Chapter 10.

I now turn, in Part IX, to a discussion of how cartilages are maintained under normal (Chapter 26) and abnormal (achondroplasia, Chapter 27) conditions, how proliferation can be reinitiated in the non-dividing chondrocytes of articular cartilage (Chapter 28), and whether growth-plate cartilages can be repaired or regenerated (Chapter 29).

NOTES

1. Erasmus Darwin (1794), p. 524.
2. See Mohn and Kragstrup (1991) for addition to drinking water, Cheng and Bader (1990) and Chavassieux et al. (1993a,b) for the studies with rats, Guise et al. (1992) for trabecular bone in rabbits, and Thomas et al. (1996) for phosphotyrosyl phosphorylation levels. Gruber and Baylink (1991) reviewed the effects of fluoride on bone, especially interaction with other mitogens and the unequal response of different skeletons to NaF. See Caverzasio et al. (1998) for mechanisms of action and an overview of the actions of NaF in vivo and in vitro.
3. See Biller et al. (1977) for the rabbit study, P. K. Wlodarski et al. (1995) for the rat study, Shih and Norrdin (1985) for the study with beagle dogs, and Boivin and Meunier (1990) for an overview of toxicological and therapeutic studies.
4. See Farley et al. (1983, 1987, 1988a,b), and Selz et al. (1991) for the study with UMR-106 cells. Type I collagen modulates signal transduction in UMR-106 and rat calvarial cells by altering PKC activity, modulating Ca^{++} concentration and cAMP signaling (Green et al., 1995).
5. See Wergedal et al. (1988) and Farley et al. (1991, 1993) for enhanced proliferation, and Khokher and Dandona (1990) for the blocking study.
6. See Kassen et al. (1994) for the stromal osteoblast precursors, and Kopp and Robey (1990) for the first-passage cells.
7. See Liu and Baylink (1977) for thyroparathyroidectomized, and Modrowski et al. (1992) for ovariectomized rats, and see Takada et al. (1995, 1996) for action on Te 85 cells.

8. See Lundy et al. (1989) for the study with 11 patients, Pak et al. (1989) for the comparison with vitamin D_3, and Marie et al. (1992) for the study with 12 patients.
9. See Joseph and Tydd (1973a,b) for rabbit-ear cartilage regeneration.
10. See Grynpas et al. (1986) for fluoridated and non-fluoridated bone; Hicks and Ramp (1975) for enhancement and Tenenbaum et al. (1992c) for suppression of mineralization; Bellows et al. (1993) for osteoid mineralization; and the chapters in Volume 4: Bone Metabolism and Mineralization of Hall (1990–1994) for overviews of mineralization.
11. See Einhorn et al. (1992) for the rat, and Turner et al. (1997) for the rabbit study.
12. See McCormack et al. (1993) and Tencer et al. (1989) for these two studies.
13. See Palmer and Lumsden (1987) and Osborn and Price (1988) for experimental evidence. Braut et al. (2003) used a Col1a1-2.3-GFP transgenic marker to show that cells from dental pulp can produce osteocytes.
14. See Ten Cate et al. (1974), Ten Cate (1975), Ten Cate and Mills (1972) and Freeman et al. (1975) for morphogenesis, and Howes (1977a,b) for studies with teeth in Rana.
15. See Vignery and Baron (1980), Atkinson et al. (1977) and Stutzmann and Petrovic (1989) for metabolic studies on alveolar bone.
16. See Stutzmann et al. (1981) for seasonal activity in human alveolar bone, and Vignery and Baron (1989), Liu and Baylink (1984) and King et al. (1995) for the rodent studies. Similarly, osteoclast numbers and bone resorption both increase in teleosts on phosphorus-deficient diets (Roy et al., 2004).
17. See Gilbert and Zevit (2001) for baculum deficiency; Gorski et al. (2000) for Aarskog syndrome; G. Yamada et al. (1995), Zhu et al. (1998) and Rivera-Pérez et al. (1999) for goosecoid; and Kondo et al. (1997) and Burke and Rosa Molinar (2002) for genes shared between faces, limbs and genitalia.
18. See Vilmann and Vilmann (1979, 1983), Vilmann (1982b), Murakami and Mizuno (1984, 1986), Rasmussen et al. (1986), and Yamamoto (1989) for the rat os penis, and Tumlinson and McDaniel (1984) for bobcats.
19. See Glücksmann and Cherry (1972) for induction of a more complete os clitoridis.
20. See Murakami and Mizuno (1984, 1986) for histogenesis of the os clitoridis (and os penis) in rats, Glücksmann et al. (1976) for the os clitoridis in mice, Murakami (1987a) and Kurzroch et al. (1999) for tissue interactions, and Glücksmann and Cherry (1972) for induction by testosterone. Both membrane bone and cartilage develop when genital tubercles are grown beneath the renal capsule of adult rats.
21. See Murakami (1986, 1987a,b) for these androgen studies.
22. See Murakami et al. (1994) and Izumi et al. (2000) for these studies. Williams-Ashman and Reddi (1991) review the role of androgens, tissue interactions, and growth factors in differentiation of the os penis. See Beresford (1972) and Jeffery (1974) for fracture repair.
23. See Kondo et al. (1997) for the Hox gene knockout, and Izpisúa-Belmonte et al. (1992b) and Haraguchi et al. (2000) for the roles of Fgf-8 and -10.
24. See Muragaki et al. (1996) for synpolydactyly, and see Zákány and Duboule (1996) for triple knockout of Hoxd-13, Hoxd-12 and Hoxd-11 in mice, which results in small digit primordia, and small and disorganized cartilages, a phenotype that is similar to synpolydactyly in humans. More recent

analyses have revealed that *Hoxa-11* and *Hoxd-11* act at several steps during chondrogenesis, notably smaller pre-chondrogenic condensations and failure of radius and ulna to form normal growth plates (Boulet and Capecchi, 2004).

25. See Dollé *et al.* (1993) and Davis and Capecchi (1996) for *Hoxd-13* and *Hoxd-12* knockout, Chai (1981) for *Dactylaplasia*, and Hérault *et al.* (1998, 1999) and Kmita *et al.* (2002) for *Hoxd* enhancers.

26. See R. T. Turner *et al.* (1992) for intramembranous ossification, Currey (1973) for oestrogen during pregnancy, and Zambonin Zallone and Teti (1981) and Zambonin Zallone *et al.* (1983) for the studies with hens on a hypocalcemic diet.

27. See Gardner and Pfeiffer (1938), Haines (1957), Bonucci and Gherardi (1975), Zambonin Zallone and Teti (1981), Hunter and Schrarer (1983), Schrarer and Hunter (1985), Bowman and Miller (1986), Miller and Bowman (1981), and Miller *et al.* (1984) for oestrogen and medullary bone, and Nutik and Creuss (1974) and Liskova (1976) for oestrogen effects in males. Nijweide *et al.* (1985) isolated five monoclonal antibodies against quail medullary bone osteoclasts, facilitating their identification. Nijweide and Mulder (1986) also generated a monoclonal antibody – OB 7.3 – against osteocytes and used it to detect osteocytes in cultured chick calvarial cells.

28. See Gardner (1936) for complete resorption, and Corwin and Morehead (1971) for an overview of the mechanism of action on OPCs.

29. See Crelin (1954c) for the rib transplants, Crelin and Haines (1955) for gonadectomy and oestrogen replacement, and Pinnell and Crelin (1963) for grafting to the ectopic sites.

Part IX

Maintaining Cartilage in Good Times and Bad

Chondrogenesis can be initiated without condensation, but the cells have to be tricked into behaving as if they were in a condensation.

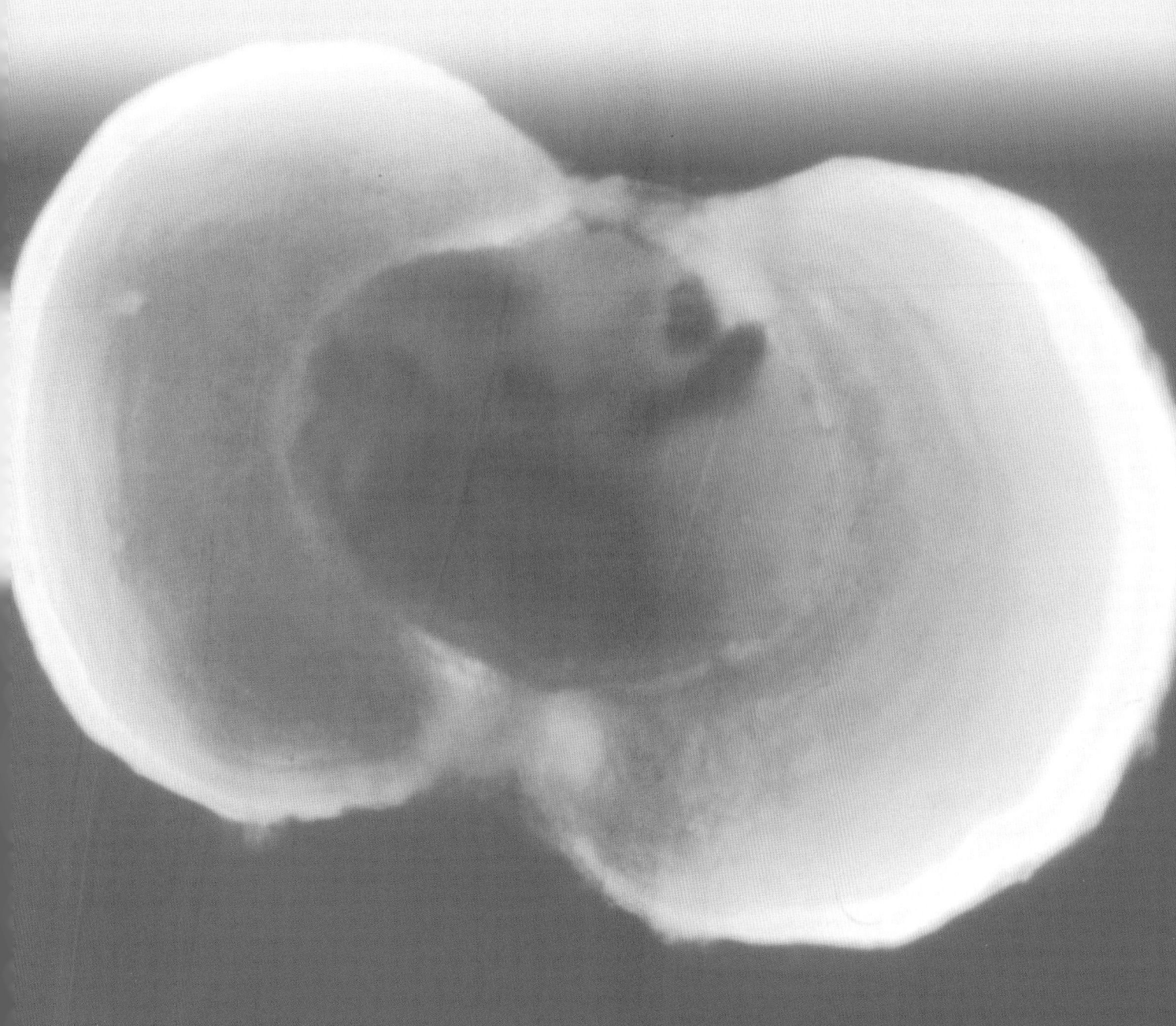

Maintaining Differentiated Chondrocytes

Chapter 26

Chondrocyte hypertrophy can be induced prematurely.

A vast literature exists on factors that enhance or retard expression of the differentiated state of chondroblasts and osteoblasts *in vitro*. The majority of the studies involve populations of chondrocytes that were already terminally differentiated when placed into culture. As a result, these studies provide valuable information on the *continued expression* of the differentiated state and on the ability of differentiated cells to modulate their differentiated state. They do not, however, aid our understanding of the *acquisition* of the differentiated state or the *process* of differentiation from precursor cells.

We know considerably more about factors involved in maintaining differentiated chondroblasts/cytes than osteoblasts/cytes, in part because in most skeletal sites the chondrocytes are not permanent. They degenerate and are replaced by osteoblasts and bone. The permanent mineralized matrix surrounding osteocytes makes study of their activity difficult *in vitro*; successful approaches were discussed in Chapter 24. I spend some time in this chapter discussing how the chondrocyte phenotype is maintained. Failure to maintain individual chondrocyte populations can result in modulation of chondrocyte or cartilage metabolism, function and/or phenotype, any of which can lead to dedifferentiation and subsequent redifferentiation. In this chapter I concentrate on *in-vitro* studies that demonstrate how the chondrocyte phenotype is maintained. In Chapter 27, I turn to achondroplasias (chondrodystrophies) as examples of how failure to maintain the chondrocytic phenotype can result in clinical syndromes.

DIFFERENTIATED CHONDROCYTES

Morphology and specialized function are the means by which we distinguish chondroblasts and chondrocytes from the cells of connective and skeletal tissues. The distinction between chondroblast and chondrocyte is often arbitrary; a chondrocyte is more mature than a chondroblast. Distinguishing chondrocytes from hypertrophic chondrocytes (HCCs) is more straightforward, facilitated as it is by the substantial change in morphology, synthesis of new products and altered developmental function that accompany hypertrophy (Chapter 13).

The characteristic surface projections of a rounded chondrocyte resting within its lacuna (Fig. 1.7),[1] synthesis of chondroitin sulphate (CS) and type II collagen and their deposition into an extracellular matrix (ECM), and the ability to yield chondrogenic clones (Chapter 22), all identify chondrogenic cells. Understanding how precursor cells differentiate into chondroblasts requires knowledge of the synthesis and deposition of the products of the ECM (Chapter 22). Hyaluronan must be removed before differentiation can be initiated (Chapter 20), synthesis of CS enhanced, and synthesis of type II collagen initiated. Maintenance of that differentiated state requires continued synthesis and deposition of the specialized ECM. Thus, we might expect differentiated chondrocytes to be sensitive to their extracellular environment and use it to monitor – and thereby regulate – their synthetic activity. Indeed, without their ECM, chondrocytes dedifferentiate.

In Chapter 20, I discussed hyaluronan as a component of the ECM. Here, I concentrate on CS and collagen, especially as regards their synthesis by populations of chondrocytes and by the same chondrocyte, regulation of rates of synthesis, positive and negative feedback to chondrocytes, their deposition, and interactions between them in the ECM.

SYNTHESIS AND DEPOSITION OF CARTILAGINOUS EXTRACELLULAR MATRIX

Initiation of chondrogenesis, indeed initiation of the differentiation of any cell type, is usually identified by onset of synthesis of the specialized molecules associated with the particular differentiated cell type. Chondroitin sulphate and type II collagen serve this role for chondroblast differentiation. I discuss them here with respect to chondrogenesis of limb mesenchyme and in Chapter 42 in the context of the formation of vertebrae from sclerotomal mesenchyme.

Synthesis of chondroitin sulphate

Detecting ^{35}S incorporation autoradiographically has long been the method of choice to quantify synthesis and deposition of CS.

Accumulation of CS is gradual. Low levels are present at the sites of future limb buds in chick embryos as early as H.H. 15, although ^{35}S is still taken up uniformly by flank and limb-bud mesenchyme at H.H. 16 and 17. By H.H. 22, ^{35}S is distributed uniformly throughout limb-bud mesenchyme in cells synthesizing chondroitin-4- and chondroitin-6-sulphates. Robert Searls, who did some of the first studies in the mid-1960s, concluded that *all* mesenchymal cells of early limb buds synthesize CS. This uniform distribution of ^{35}S uptake changes between H.H. 22 and 26, increasing in prechondrogenic mesenchyme and decreasing within premyogenic mesenchyme.[2]

The pattern of accumulation of UDP glucose dehydrogenase and of UDP-*N*-acetylglucosamine-4-epimerase – two key enzymes in the synthesis of CS (Fig. 26.1) – is of steady increase in whole limb buds in parallel with the accumulation of ^{35}S in the central mesenchyme until chondrogenesis commences, when there is a sharp increase in accumulation. Judith Medoff interpreted this pattern as *amplification of preexisting enzymatic activity in prechondrogenic but not in myogenic mesenchyme.*[3] So we see that synthesis of CS is neither an exclusive property of, nor diagnostic for, chondrogenic cells. The property that *is* diagnostic for chondrogenic cells is the ability to *accumulate* CS into an ECM and to synthesize a cartilage-specific CS or proteochondroitin sulphate.[4]

If, among other things, becoming chondrogenic involves amplification of the synthesis of CS and its deposition into an ECM, then knowledge of the control of these processes should illuminate the mechanism by which chondroprogenitor cells differentiate into chondroblasts and chondrocytes, and perhaps even how bipotential progenitor cells initiate chondrogenesis.

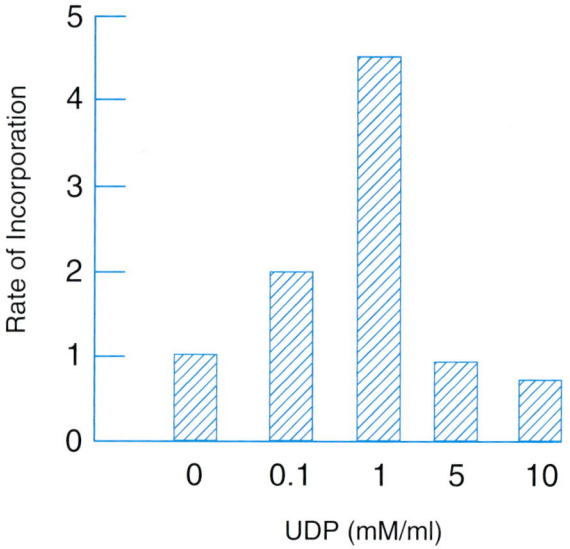

Figure 26.1 Proteoglycan synthesis is regulated by uridine diphosphate (UDP). The histogram shows the effect of adding UDP (0.1–10 mM/ml culture medium) on the incorporation of [35]SO$_4$ into slices of rabbit articular cartilage *in vitro*. Incorporation of [35]SO$_4$ in the absence of UDP is shown as a rate of 1. From data in Ehrlich *et al.* (1974). Reproduced from Hall (1978a).

The synthetic pathway for CS is multistep, involving complex and multilayered control, with several steps in the metabolic pathway where synthesis and hence accumulation are controlled. The enzymatic machinery for CS synthesis is established and functioning at a low level in many, if not most, embryonic cells and tissues from as early as the blastodisc stage, and certainly in presumptive myo- and chondrogenic mesenchyme of limb buds. Given the low level of synthesis in many cells and up-regulation with chondrogenesis, controls act on rate of synthesis, accumulation, and/or degradation and removal, not on initiation of synthesis. Suggested control steps are one or more than one of:

(i) transformation of *N*-acetylglucosamine-1-phosphate to UDP-*N*-acetylglucosamine by UTP and UDP-*N*-acetylglucosamine pyrophosphorylase (Lash, 1968a,b);

(ii) transformation of UDP-*N*-acetylglucosamine to UDP-*N*-acetylgalactosamine by UDP-*N*-acetylglucosamine-4-epimerase (Medoff, 1967; Marzullo and Desiderio, 1972; Manasek, 1973; and Ross, 1976);

(iii) transformation of glucosamine to glucosamine-6-phosphate by hexokinase (Winterburn and Phelps, 1970; Kim and Conrad, 1974);

(iv) UDP-xylose or xylosyltransferase (Balduini *et al.*, 1973; Caplan and Stoolmiller, 1973; DeLuca *et al.*, 1975 and Schwartz and Dorfman, 1975); and/or

(v) mRNA for the core protein, which accumulates with condensation concomitant with accumulation of mRNA for type II collagen (Herington *et al.*, 1972; Schwartz and Dorfman, 1975; Ho *et al.*, 1977; Lash and Vasan, 1977; Kosher *et al.*, 1986b and Mallein-Gerin *et al.*, 1988).[5]

Synthesis of type II collagen

The same situation does not pertain to the synthesis of type II collagen, for, unlike CS, type II collagen is not synthesized ubiquitously. Consequently, onset of type II collagen is a sensitive marker for differentiation of precursor cells as chondroblasts, and may coincide with irreversible commitment to chondrogenesis. Up-regulation of type II collagen mRNA certainly coincides with the differentiation of condensed mesenchymal cells as prechondroblasts and chondroblasts. The appearance of type II collagen *protein* in limb-bud mesenchyme *coincides* with the differentiation of precursor cells into chondroblasts.[6]

SYNTHESIS OF COLLAGEN AND CHONDROITIN SULPHATE BY THE SAME CHONDROCYTE

Prockop *et al.* (1964) set out to determine whether collagen and CS were synthesized by the same or different chondrocytes. Uptake of ^{3}H- and ^{14}C-labeled proline into hydroxyproline confirmed that isolated vertebral chondrocytes from 10-day-old chick embryos synthesize collagen *in vitro*. Double labeling with $^{35}SO_4$ and ^{3}H-proline visualized incorporation into glycosaminoglycans (GAGs) and collagen, respectively. Synthesis of both products stopped under conditions – growth on fibrin clots – that favour (permit, direct?) chondrocyte dedifferentiation. Prockop and colleagues argued that since dedifferentiated chondrocytes lose the ability to synthesize both products simultaneously their synthesis might be closely interrelated. However, Bhatnagar and Prockop (1966) went on to show that GAGs and collagen could be inhibited independently. Using an autoradiographic analysis of double-labeled sternal chondrocytes from 13-day-old chick embryos, Smith (1972) showed that all the chondrocytes incorporate both labels over the same time period. Smith used short pulses to overcome re-utilization or exchange of label. Visualization of the chondrocytes was such that Smith could state that the same cells were synthesizing collagen and GAGs at the same time.[7]

As seen in earlier discussions, the type(s) of collagens synthesized by chondrocytes vary: vertebral chondrocytes from chick embryos synthesize type X collagen and type I procollagen; sternal chondrocytes synthesize type IX and X collagen, 20 per cent of sternal cartilage collagen being type IX.

Collagen types II and IX are co-distributed in the pericellular and extracellular matrices of avian chondrocytes. Chick sternal chondrocytes incorporate collagen types II, IX and XI into single fibrils; type IX collagen is covalently cross-linked to type II, stabilizing what otherwise would be thin collagen fibrils.[8]

Collagen gel culture

Similarly, sternal chondrocytes from 13-day-old embryos retain their differentiated state and continue to deposit ECM when cultured in collagen gels. They can synthesize additional collagen types if cultured in collagen gels, fibronectin modifying the type of collagen synthesized, two disulphide-linked molecules being produced in the presence of fibronectin. Under calcium deficiency sternal chondrocytes show reduced hypertrophy and type X collagen deposition. *In ovo*, they undergo apoptosis after terminally differentiating before the resorption and vascular invasion associated with endochondral ossification.

H.H. 22/23 limb-bud chondrocytes fail to differentiate if cultured as monolayers in collagen gels, but do differentiate if cultured as aggregates, the enclosed environment provided by collagen or agarose gels being vital for differentiation.[9]

FEEDBACK CONTROL OF THE SYNTHESIS OF GLYCOSAMINOGLYCANS

Reports began to accumulate in the late 1970s that the amount and nature of the GAGs in chondrocyte ECMs regulated the rate of synthesis of GAGs by those same chondrocytes. Whether that feedback is positive or negative depends on whether the three-dimensional architecture of cartilage is retained – as it is in whole skeletal rudiments, organ or micromass culture – or whether the chondrocytes are liberated from their matrices and established in cell culture.

Evidence from organ culture

When testicular hyaluronidase was used to selectively remove a specific GAG (hyaluronan) from the ECM, organ-cultured cartilaginous tibiae initiated a compensatory increase in hyaluronan synthesis. Collagen synthesis was unaffected. Recovery of GAG synthesis depends on the dose of hyaluronidase used, indicating regulation via the level of matrix GAG.

This stimulation of GAG synthesis by *negative feedback* contrasts with stimulation by *positive feedback* in cultured *isolated* chondrocytes. Production of GAGs by somites from early chick embryos is increased two- to threefold by the addition of GAGs derived from embryonic cartilage. The magnitude of the stimulation indicates that either chondroblast precursors or chondroblasts in the early stages of differentiation – when little matrix is present – are especially sensitive to regulation of GAG synthesis by end-product stimulation.[10]

Stimulation may also come from other intermediates in the CS synthetic pathway. When Kim and Conrad (1974) cultured vertebral chondrocytes from eight-day-old mouse embryos with exogenous D-glucosamine, GAG synthesis increased as the concentration of added D-glucosamine increased up to 2 mM, but was suppressed at 4 and 25 mM. Glucosamine, UDP-N-acetylglucosamine, and UDP – at concentrations of 1 mM/ml – stimulate GAG synthesis by rabbit articular cartilage *in vitro*.[11] UDP was most effective, the chondrocytes exhibiting differential responses depending on the concentration of added UDP (Fig. 26.1). It appears, therefore, that chondrocytes can respond differentially depending on the local concentration of individual ECM components. Exogenous glucose does not exert a regulatory role.[12]

The mechanism whereby organ-cultured cartilage senses and regulates synthesis of GAGs is not fully understood. As noted above there are be several points of regulation within the metabolic pathway. Even increasing medium Ca^{++} concentration from 0.5 to 1 mM can stimulate GAG synthesis. Other matrix products may also be regulated by end-product stimulation. For example, the concentration of lysozyme within the ECM regulates the rate at which lysozyme is synthesized.[13]

Evidence from chondrocyte cell cultures

Organ-cultured cartilage responds to enzymatic depletion of ECM by initiating GAG synthesis. Cultured chondrocytes respond to hyaluronidase, chondroitinase, or trypsin with decreased incorporation of ^{35}S. The reason for the differential response of rudiments and isolated cells is not clear.[14] Nevertheless, *chondrocytes are sensitive to the amount and quality of their ECMs, and respond by regulating the rate of GAG synthesis, thereby aiding maintenance of their differentiated phenotype.*

Nevo and Dorfman (1972) obtained dose-dependent enhancement of GAG synthesis over a 2.5-hour period when they maintained epiphyseal chondrocytes in suspension culture and added chondromucoprotein and other polyanionic GAGs (Table 26.1). Synthesis of collagen and

Table 26.1 Results of stimulation of the synthesis of glycosaminoglycans by cultured chondrocytes as a function of the amount of exogenous chondromucoprotein added to the medium[a]

Chondromucoprotein added (µg/ml)	GAG synthesis (cpm ^{35}S)
0	5790
20	8207
200	22896
2000	26521

[a] Based on data in Nevo and Dorfman (1972).

total protein was unchanged indicating specificity of the response to GAG synthesis. Enzyme induction was ruled out, as xylosyltransferase and N-acetylgalactosaminyl-transferase activities were not enhanced.

Initiation of core protein synthesis does not appear to be the mechanism either, although protein synthesis is required; puromycin inhibits synthesis of CS. By adding CS or heparin to monolayered chondrocytes, Schwartz and Dorfman (1975) stimulated synthesis of CS. When β-D-xyloside and CS were added together, stimulation of synthesis exceeded that with either agent alone. Hyaluronan was subsequently shown to interact with GAGs at the chondrocyte surface to regulate GAG synthesis.[15]

INTERACTIONS BETWEEN GLYCOSAMINOGLYCANS AND COLLAGENS WITHIN THE EXTRACELLULAR MATRIX

Interaction between GAGs and collagen fibril/fibres of the ECM are complex but important; the integrity of matrix and cartilaginous rudiments/tissues depends on them.

Complex formation occurs between collagen and CS, and between collagen and hyaluronan. These complexes are highly ordered and provide the biochemical basis for the integrity of cartilage, indeed of the early embryo. Table 26.2 summarizes differences in the composition of several types of bovine cartilage and bone.[16] According to Serafini-Fracassini and Smith (1974): 'In cartilage, the main function of collagen is probably that it determines the ordering within the matrix of other physiologically important components, such as proteoglycan macromolecules.'

Synthesis of collagen and chondroitin sulphate are regulated independently

The rate of CS synthesis is unaffected when collagen synthesis is inhibited with α,α-dipyridyl. When CS synthesis is inhibited with 6-diazo-5-oxonorleucine (DON, a glutamine analogue), collagen synthesis is unaffected. While CS can be synthesized in the absence of collagen, collagen is required if the polysaccharide is to be deposited into an insoluble matrix.[17]

Table 26.2 The composition (%) of extracellular matrix products in bovine cartilages and bones[a]

ECM Component	Cartilage (%)			Bone (%)	
	Articular	Epiphyseal	Primary	Secondary spongiosa	Cortex spongiosa
Collagen	63.7	56.2	75.8	79.5	79.2
Chondroitin sulphate	25.3	25.3	1.9	1.1	0.8
Keratan sulphate	3.7	4.4	3.3	0.7	0.9
Sialic acid	0.5	0.7	0.4	0.35	0.25

[a] Based on data from Campo and Tourtellote (1967).

Box 26.1 Diabetes

Surprisingly – to me, at least – demineralized bone matrix from rats made diabetic after administration of the anti-cancer agent streptozocin is a *better* inducer of bone that bone from non-diabetic rats. I say surprisingly because healing fractures in diabetic rats have only half the collagen and 40 per cent of the DNA of control fractures during the initial phases of repair (4–11 days post-fracture), while tensile strength and stiffness are 30–50 per cent lower two weeks post-fracture. Chondrocyte maturation (hypertrophy) in fractures in diabetic rats is delayed, with reduced deposition of type X collagen around those chondrocytes that do hypertrophy. Given the 16 million Americans with diabetes, it is important to learn more of skeletal changes with diabetes. Interesting site-specific effects also are seen in diabetic rats; mandibular bone is more sensitive than the limb skeleton to teratogenesis when maintained *in vitro*. D-glucose, β-hydroxybutyric acid, glucose and growth factors such as Igf also have differential effects on mandibular and limb pre-chondrocytes from diabetic animals.[a]

In the offspring of rats made diabetic by administering vitamin E to the mother, Meckel's cartilage is malformed, thyroid and thymus are small, ears are low set, the parathyroid gland fails to develop, and cardiac anomalies form – all similar to di George syndrome and indicative of effects of neural crest-derived structures (Siman *et al.*, 2000).

[a] See Landesman and Reddi (1985) for streptozocin-induced diabetes. Streptozocin, isolated from *Streptomyces achromogenes*, induced diabetes mellitus in rats following an intravenous dose of 50 mg/kg body weight. The LD$_{50}$ i.v. is 138 mg/kg body weight. See Macey *et al.* (1989) and Gooch *et al.* (2000) for the fracture data, and Styrud and Eriksson (1990, 1991) and Unger and Eriksson (1992) for the mandibular and limb studies.

Continuous RNA synthesis is required to maintain synthesis of CS but not of collagen, indicating that control of CS is transcriptional, whereas control of collagen synthesis is translational or post-translational. BrdU inhibits CS synthesis but not synthesis of collagen.[18]

Synthesis of collagen was unaffected when Dondi and Muir (1976) used the GAG initiator 4-methylumbelliferyl-β-D-xyloside to inhibit synthesis in pig laryngeal cartilage. Insulin stimulates synthesis of GAGs, has little effect on collagen synthesis and inhibits cartilage differentiation, *preventing* attainment of chondrocyte hypertrophy (see Box 26.1).[19]

Hypertrophy

Some aspects of chondrocyte hypertrophy were discussed in Chapter 23, especially synthesis of type X collagen, loss of resistance to vascular invasion, matrix mineralization, and the ability of HCCs to induce subperiosteal bone formation.

Hypertrophy is normally an autonomous process, the primary mechanism being cytoplasmic and nuclear swelling. Chondrocytes from the flattened cell zone of primary cartilages of long bones hypertrophy 'on time' when maintained *in vitro*, as do pre-HCCs from the avian columella. If ascorbic acid and Na, β-glycerophosphate are added to the medium, mature chondrocytes in suspension culture – for example, from the tibiae of chick embryos of H.H. 28–30 – and dedifferentiated chondrocytes both hypertrophy, synthesize type X collagen and alkaline phosphatase, release matrix vesicles, and deposit and mineralize an ECM.[20]

Hypertrophy can, however, be induced prematurely. Foetal bovine serum, Igf-1, Fgf and Pdgf all induce synthesis of type X collagen by, and hypertrophy of, chick sternal chondrocytes. Synergistic interactions are involved; two different isoforms of Pdgf (a and b) with or without Igf-1 have differential effects on proliferation and proteoglycan synthesis by rat rib growth-plate chondrocytes. Stimulation of embryonic chick pelvic cartilage with tri-iodothyroxine over three days in organ culture induces alkaline phosphatase-positive HCCs. Cartilage dry weight increases by 77 per cent, length by 35 per cent, protein by 67 per cent, but DNA synthesis by only 2 per cent, indicating that the transformation neither involves proliferation (nor dedifferentiation).[21]

THE INTERACTIVE EXTRACELLULAR MATRIX

This sampling of experimental approaches shows that collagen and GAGs are synthesized under independent control. Also, *the two classes of molecule need not interact with one another for either to be transported through or secreted from the cell.* However, certain interactions between collagen and CS within the ECM are necessary before either can be deposited into an insoluble complex. Consequently, concentrations of collagen and GAGs within ECMs of cartilages are closely correlated, even after a variety of disparate treatments (Hall, 1973b; also see Fig. 4.11).[22]

Glycoproteins and GAGs, such as CS and dermatan sulphate, interact with tropocollagen to aid collagen fibrillogenesis. Ultrastructural evidence demonstrates spacing of GAGs along collagen fibres in bovine and rabbit articular, epiphyseal and condylar secondary cartilages (Fig. 26.2).[23] Eisenstein *et al.* (1971) used ruthenium red, protamine and lysozyme to distinguish three proteoglycans in the epiphyseal growth cartilage of dogs and mice, each having different associations with collagen fibres.

Collagen – and hence its triple helical structure – is thermally stabilized by interaction with chondroitin-6-sulphate; adding C-6-S to collagen in multiples of 5.5 disaccharide residues/100 amino acids raises the 'melting' temperature of calf skin collagen from 38 to 46°C, stabilizing collagen fibres. The more GAGs bound within the ECM, the more able cartilage is to resist stresses; osteoarthritic human articular cartilage contains considerably more type I collagen and so is less able to bind GAGs and therefore more liable to destruction from abrasion.[24]

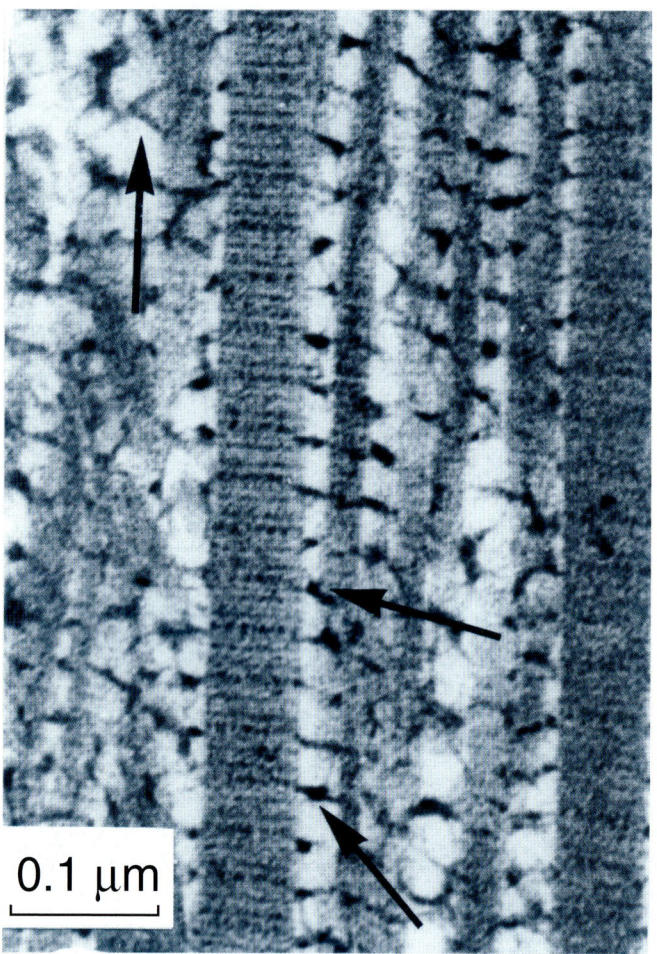

Figure 26.2 Proteoglycan (arrows) forms regularly repeated attachments to collagen fibres, as shown in this transmission electron micrograph of the ECM of rabbit articular cartilage stained with 0.1% safranin O and counterstained with uranyl acetate and lead citrate. From Hall (1978a).

Type II collagen renatures slowly and is particularly able to bind to GAGs, maintaining integrity of the ECM. Although other areas of collagen chains are highly conservative, type II collagen contains four or five times more hydroxylysine and glycosidically linked GAGs than does type I collagen. Hydroxylysine provides binding sites for hexose sugars. Any reduction in glysosylation or hydroxylation disrupts the integrity of cartilage matrix; thyroxine, corticosteroids, growth hormone, and β-aminopropionitrile all disrupt cartilage by reducing the glycosylation of type II collagen, a characteristic thought to be responsible for the characteristic morphological appearance of cartilage collagen fibres.[25]

Disruption in interactions between and among matrix molecules is seen in osteoarthritis and other degenerative joint diseases. Disruption of collagen–GAG interactions in osteoarthritic patients leads to deposition of swollen and disorganized collagen fibres, and GAGs that are less firmly bound to the ECM, both changes correlating with later alterations in the structure of the weight-bearing portions of the skeleton.[26]

Just as they provide valuable insights into other aspects of chondrogenesis, so mutants contribute to our understanding of interactions between matricial products. Naturally occurring *achondroplastic mutants* and the changes produced in chemically induced *micromelia* (literally, shortened limbs) indicate that impairment of interaction between GAGs and collagen in the ECM leads to instability of the differentiated chondrocyte phenotype and abnormal chondrogenesis. This is the topic of the next chapter.

NOTES

1. Lawton *et al.* (1995) claimed to have identified an interlacunar network in chick digital and rat femoral growth plates, which they analyzed using a method to prevent tissue shrinkage.
2. See Searls (1965a,b) and Medoff (1967) for the studies on chondroitin sulphate, and Pinot (1969a, 1970) for the *in-vitro* study.
3. See Medoff (1967), Zwilling (1968) and Medoff and Zwilling (1972) on amplification.
4. See Daniel (1976), Holtzer and Mayne (1973), Kosher and Searls (1973), Mattson and Foret (1973) and Ahrens *et al.* (1977b) for synthesis not being a diagnostic feature of chondrogenesis, Abrahamsohn *et al.* (1975) for accumulation, and Goetinck *et al.* (1974), Pennypacker and Goetinck (1976), Ho *et al.* (1977) and Wiebkin and Muir (1977), Vynios and Tsiganos (1990), Hall and Newman (1991), Ayad *et al.* (1994), P. G. Scott *et al.* (1995), Selleck (2000) and Knudson and Knudson (2001) for cartilage-specific proteoglycan.
5. See Delbrück (1970) and Abrahamsohn *et al.* (1975) for low-level synthesis in many tissues. Thorpe and Dorfman (1967), Kleine (1972), Serafini-Fracassini and Smith (1974), Lamberg and Stoolmiller (1974), Levitt and Dorfman (1973, 1974) and Levitt *et al.* (1974) review possible control mechanisms as they relate to the initiation of chondrogenesis.
6. See Linsenmayer *et al.* (1973a), von der Mark and von der Mark (1977b), H. von der Mark *et al.* (1976) and von der Mark (1980) for initial synthesis of type II collagen, and Toole *et al.* (1972b), Linsenmayer *et al.* (1973a,b), Linsenmayer (1974), H. von der Mark *et al.* (1976), Oohira *et al.* (1974) and von der Mark and von der Mark (1977b) for distribution at later stages.
7. Nimni (1973), Bornstein (1974), Jacoby and Jason (1975b), Ross (1976), Ovadia *et al.* (1980) and Vasan *et al.* (1986a,b) contain excellent reviews on the routes by which GAGs and collagen are synthesized.
8. See Gibson and Flint (1985), Kielty *et al.* (1985), Gibson *et al.* (1986), Solursh *et al.* (1986) and Gerstenfeld *et al.* (1989) for different collagens produced by different chondrocytes. See Gibson *et al.* (1982, 1983), Van der Rest and Mayne (1988) and Mendler *et al.* (1989) for stabilizing properties of type IX collagen. Further aspects of selective assembly and remodeling of collagens II and IX, based on chondrocyte cultures derived from foetal bovine growth plates, are elaborated in Mwale *et al.* (2000).
9. See Gibson *et al.* (1983, 1986, 1995) and Kimura *et al.* (1984a,b) for sternal and limb-bud chondrocytes, and see Reginato *et al.* (1993) for calcium deficiency.

10. See Fitton Jackson (1970, 1976) for early studies on matrix depletion, Hardingham *et al.* (1972) for the type of GAG synthesized, Nevo and Dorfman (1972) for studies with isolated chondrocytes, and Kosher *et al.* (1973) and Takeichi (1973) for end-product regulation.
11. There are two proliferative zones in rabbit articular cartilages, one superficial and responsible for the interstitial growth of the epiphysis, and one deeper (Pribylova and Hert, 1971).
12. See Ehrlich *et al.* (1974) for addition of glucosamine and UDP, and Delcher *et al.* (1973) for the lack of effect of glucose.
13. See Shulman and Opler (1974) for effects of Ca^{++} concentration, and Kuettner *et al.* (1972) for lysozyme regulation.
14. See Huang (1974, 1977) and Solursh and Karp (1975) for ^{35}S incorporation.
15. See Turner and Burger (1973) for cell-membrane receptors, Wiebkin and Muir (1975) for hyaluronan, and Solursh and Meier (1973, 1974) and Solursh *et al.* (1973) for stimulation of CS and collagen. Meier and Hay (1974a,b) and Solursh and Karp (1975) provide excellent discussions of the possible mechanisms of these stimulations for chondrocytes and corneal epithelia. See Muir (1995) for a basic review of chondrocytes and their ECM, including aggregan, the function of hyaluronan, and mutations.
16. See Bazin *et al.* (1962), Mathews (1965), Wiebkin and Muir (1973) and Campo (1974) for collagen–GAG complexes. See A. R. Poole *et al.* (1982b, 1984), Hunziker *et al.* (1984), Hunziker and Schenk (1987), C. A. Poole (1997) and Mwale *et al.* (2000) for the structural organization of proteoglycan in cartilage.
17. See Bhatnagar and Prockop (1966) and Bhatnagar and Rapaka (1971) for inhibition with α,α-dipyridyl and DON, and Lavietes (1971) for deposition into an insoluble matrix.
18. See Solursh and Meier (1972) for continued RNA synthesis, Nimni (1973) for control of synthesis, Levitt and Dorfman (1972) for inhibition by BrdU, and Agnish and Kochhar (1976) for the mouse limb-bud study. BrdU also has pharmacological effects; exposing post-implantation mouse embryos to BrdU and culturing their limb buds irreversibly inhibits limb differentiation (*ibid.*)
19. See Hajek and Solursh (1975) and Maor *et al.* (1993a) for GAG and collagen, and Chen (1954) for insulin and hypertrophy.
20. See Archer *et al.* (1984), Buckwalter *et al.* (1986) and Eavey *et al.* (1984) for hypertrophy 'on time,' and Tacchetti *et al.* (1987, 1989) for chick tibiae.
21. See Bruckner *et al.* (1989) and Bohme *et al.* (1992) for the studies with chick sternal cartilage, Wroblewski and Edwall (1992) for Pdgf, Hoch and Soriano (2003) for an overview of *Pdgf-a* and *-b* and the two Pdgf receptors, and Burch and Lebovitz (1982) for chick pelvic cartilages. *Pdgf-a* also stimulates chondrogenesis in micromass cultures of somitic cells, reciprocal signaling between the myotomal and sclerotomal compartments of the somites involving *Pdgf-a* and *Myf-5* (Tallquist *et al.*, 2000). The pioneering study on the direct effects of thyroxine on embryonic bones *in vitro* was by Fell and Mellanby (1955), in which cartilage maturation in young bones was enhanced, but growth of older bones retarded.
22. See Nimni (1973) and Ross (1975) for transport and secretion. The chemistry of these interactions is reviewed by Steven (1972), Walton (1974), and Scott (1975).
23. See Smith and Serafini-Fracassini (1967) and Shepard and Mitchell (1977) for bovine and rabbit articular cartilage, Eisenstein *et al.* (1971), Campo and Phillips (1973) and Thyberg (1977) for epiphyseal cartilage, and Silbermann and Frommer (1974) for condylar secondary cartilage.
24. See Toole and Lowther (1968a,b), Toole (1969), Öbrink and Wasterson (1971), Lowther and Natarajan (1972) and Neméth-Csóka (1974) for collagen fibrillogenesis, and Gelman and Blackwell (1973) for thermal stability. See Nimni and Deshmukh (1973) for osteoarthritic cartilage.
25. See Igarashi *et al.* (1973) for speed of renaturation of type II collagen, Nimni and Deshmukh (1973) and Butler *et al.* (1974) for chemical composition, Blumenkrantz and Prockop (1970), Stark *et al.* (1972) and Bruns *et al.* (1973) for disruption of glycosylation, Lee-Owen and Anderson (1975) for GAG content and resistance to stress, and Hunziker and Schenk (1987) for the structural organization of proteoglycan in cartilage. Lane and Weiss (1975) reviewed the organization of collagen within articular cartilage. See Hall (1972c,d) for β-aminopropionitrile-induced inhibition of chondrogenesis and skeletal defects in chick embryos.
26. See Ridler (1974) for collagen fibres, Brandt (1974) for GAGs, and Pugh *et al.* (1974) for loss of weight bearing. For reviews of the biology of osteoarthritis, see Bollet (1969), Jaffe (1972) and Brandt (2001).

Maintenance Awry – Achondroplasia

We study rare events to understand common occurrences; the storm of the century to understand how weather works; a world war to understand normal human behaviour; mutants to understand the processes of normal growth and development.

(Hall, 2004c, p. 356)

The importance of interaction between extracellular collagen and glycosaminoglycans (GAGs) in maintaining the differentiated state and chondrocyte integrity is illustrated by mutations underlying achondroplasias, in each of which the abnormality involves some aspect of collagen and/or GAG synthesis, transport, deposition or interaction. While the genetic, cellular and/or metabolic bases of these changes differ from mutant to mutant, the gross morphological manifestation is often similar. The range of defects subsumed under the rubric of achondroplasia is great. As one example, *Langer-Saldino achondrogenesis* is a human condition in which type II collagen is not expressed in cartilage at all (Eyre *et al.*, 1986).

Achondroplasia as a class of syndromes reflects a defect(s) in the conversion of cartilage into bone. The result is *asymmetrical dwarfism*, by which is meant that the legs are disproportionately short in relation to the rest of the body. Achondroplasia may be inherited as an autosomal recessive – as in 'bulldog' calves of the Dexter breed of cattle, in dachshunds, and in mice – or as an autosomal dominant, as in birds and humans. Although histological changes usually accompany achondroplasia, only in some types are they diagnostic. Some show no histological abnormality at all. Chondrocyte necrosis is a feature of many of the genetic achondroplasias in humans, mice, rabbits and the crooked-neck dwarf chick, of achondroplasia induced by insulin or 6-aminonicotinamide-, and of thallium-induced micromelia. Necrosis may be reduced or reversed with ascorbic acid or sodium ascorbate, suggesting impairment in collagen metabolism as also occurs in vitamin C deficiency (Box 27.1).[1]

GENETIC DISORDERS OF COLLAGEN METABOLISM

As has been known for at least two decades, defective collagen metabolism can impair skeletal development, structure and/or function. Many of these defects have a genetic basis; two involving type X collagen are summarized below. Such syndromes as Ehlers-Danlos, Marfan, cutis laxa, osteochondrodysplasia, osteogenesis imperfecta and achondroplasia (our topic) can be traced to defective collagen synthesis, processing, post-translational modification, secretion, assembly or degradation.[2]

Spondylometaphyseal chondrodysplasia (smc) in mice is an unusual syndrome in which cartilage columns fail to form and the development of secondary centres is delayed. The primary defect is a mutation in the collagen type X gene resulting in compressed hypertrophic chondrocytes (HCCs) and growth plates and reduced ossification.[3]

The *toothless* (tl) mutation in rats is associated with progressive development of chondrodystrophy three to six weeks after birth. Characteristic features are disorganized growth-plate chondrocytes, failure of cartilage mineralization, and failure of cartilaginous extension into the metaphysis. Bone mineralization is normal. Two weeks after birth, collagen type X is not expressed in the few HCCs that do develop, implicating the *tl* locus in regulation of collagen X expression/function (Marks *et al.*, 2000).

The literature on achondroplasia is enormous, as is true for all topics thus far. Almost 35 years ago, Walter Landauer

Box 27.1 Vitamins A, C and D

Vitamin A

Vitamin A, a fat-soluble vitamin, affects skeletogenesis when present in excess (*hypervitaminosis-A*) or in deficiency (*hypovitaminosis-A*).

Hypervitaminosis-A can be induced in humans or rodents by administering 25 000–50 000 IU vitamin A/day over one to three weeks. Changes include dissolution of cartilage matrix, formation of thick cellular periostea, and increased osteoclastic bone resorption. As demonstrated in one of the early autoradiographic studies, hypervitaminosis-A in rats is associated with the production of large amounts of metachromatic cancellous periosteal bone. In hypervitaminosis A and D bone deposition is so accelerated that the bones become fragile and liable to spontaneous fracture.[a]

Excess vitamin A has dramatic influences on craniofacial development, largely because it substantially reduces the number of neural-crest cells (NCC) arising and/or migrating from the neural tube. As a consequence, facial processes may be reduced or absent and skeletal elements such as Meckel's or mandibular condylar cartilages fail to form. Oddly, ectopic cartilage forms in the zygomatic arches of mouse foetuses whose mothers were given 10 000 IU vitamin A on day 8.7 of gestation.[b]

The primary action of hypovitaminosis-A on the skeleton is to facilitate deposition of new bone and the formation of islands of vascularized connective tissue. Night blindness, dryness and thickening of the sclera (xerophthalmia) and formation of keratinized epithelia also characterize hypovitaminosis. Maternal deficiency in vitamin A decreases growth of the craniofacial region in foetal rats but *catch-up growth* restores growth by 21 days after birth.[c] Always we should have in the backs of our minds that embryos can compensate, sometimes in surprising and complete ways, for insults hurled at them by vagaries of nutrition or experimentation.

Vitamin A also is an important signaling molecule in development, perhaps shown most dramatically by the fact that applying vitamin A onto the wound stump can transform the regenerating tail buds of some anuran tadpoles so that multiple limbs form from the tail stumps (Box 43.2).

Vitamin C

Vitamin C (ascorbic acid) is an important additive to culture media to maximize synthesis of hydroxyproline and deposition of collagen,

150 μg/ml being optimal (Fig. 4.2). Vitamin C also can modulate cell differentiation and matrix synthesis; chick hypertrophic chondrocytes transform into osteoblast-like cells and deposit a collagenous, mineralized bony ECM when challenged with vitamin C (Gentili *et al.*, 1993). In excess, vitamin C increases sulphate incorporation into GAGs and inhibits mineral deposition into the ECM. Vitamin C deficiency produces scurvy.

Vitamin D

Vitamin D plays an important role in the mineralization of cartilage and bone (Box 1.1). Vitamin D deficiency can result in rickets in children and in osteomalacia in adults. Excess vitamin D (hypervitaminosis-D) and hyperthyroidism in mammals uncouple bone formation from resorption, two processes normally tightly correlated (Chapter 15).

Implanting demineralized bone matrix into soft tissue or muscle is an effective way to induce ectopic bone formation. Demineralized allogeneic bone matrix from vitamin D-deficient rats, implanted into rats with normal vitamin D levels, is a less effective bone inducer than matrix from control animals, primarily because fewer osteoblasts are induced – perhaps, it was thought, because the vitamin D-deficient matrix lacks a mitogen present in normal matrix. We now know that vitamin D-deficient rat bone has less Tgfβ than normal bone and that administering 1,25 vitamin D_3 to vitamin D-deficient rats increases Tgfβ by 100 per cent. Similarly (or conversely?), demineralized bone matrix from normal rats implanted *into vitamin D-deficient rats* induces normal amounts of cartilage; however, less bone is produced than in control rats, and the bone is abnormally mineralized unless vitamin D is added.[d]

[a] See Bélanger and Clark (1967) for the autoradiographic study, and Mundy and Raisz (1982) for an overview of disorders of bone resorption, including hypervitaminosis.
[b] See Hall (1999a) for an overview of vitamin A and migration of NCC, and see Kay (1987) for the mouse study.
[c] See Barnicot and Datta (1972) and Franquin and Baume (1981) for hypovitaminosis-A. See Johnston and Bronsky (1995) and Sperber (2001) for thorough overviews of mammalian prenatal craniofacial development.
[d] See Turner *et al.* (1988) for lack of a mitogen, and Finkelman *et al.* (1991) for that mitogen being Tgfβ. See Van der Steenhoven *et al.* (1988) for demineralized matrix from and into vitamin D-deficient rats.

---

indexed 1755 references (Landauer, 1969b). Here, I discuss only a few types, concentrating on those for which some knowledge of the mechanism of action of the mutated gene is known.[4]

CARTILAGE ANOMALY (Can) IN MICE

David Johnson and his colleagues in Leeds investigated the recessive mutant *Can*, which arose spontaneously in their mouse colony in 1965. *Can* displays all the classic features of achondroplasia; growth is retarded, and the skull, axial, caudal and appendicular skeletons are all disproportionately shortened. The homozygote (*Can/Can*) is lethal at about 10 days after birth. The primary defect lies in cartilage.[5]

The first sign of biochemical alterations in these skeletons – which occurs before mutant embryos can be identified morphologically – is reduced protein synthesis at 17 days of gestation. Protein synthesis rises to above-normal levels after birth, when the defect is first manifest morphologically. While the biochemical changes are complex, cartilage ultrastructure and collagen synthesis are normal. Cartilaginous extracellular matrix (ECM) is diminished, as assessed by light and electron microscopy and incorporation of ^{14}C-glucose. Although the levels of UDPG dehydrogenase and UDP-*N*-acetylglucosamine-4-epimerase are reduced, glucose-6-phosphate dehydrogenase is not – a pattern of change consistent with a specific reduction in GAG synthesis. Oxidative phosphorylation is normal (cf. the *Ac* rabbit below). *Can* therefore involves decreased protein synthesis and a mechanism that results in reduced deposition of ECM by chondrocytes.

ACHONDROPLASIA (Ac) IN RABBITS

In this autosomal recessive achondroplastic mutant, synthesis of cartilaginous ECM is reduced and there is much cell death and necrosis, especially in the centre of the cartilaginous primordia, an observation that prompted the suggestion that isolation from vascularity is involved. Incorporation of glucose and galactose is increased in comparison with wild-type littermates, indicating defective glucose metabolism. Oxidative phosphorylation in liver mitochondria also is defective.[6] Other than these data, we have no information about how the defective ECM is produced in this mutant.

ACHONDROPLASIA (Cn) IN MICE

As in human achondroplasia, Cn in mice affects growth in early postnatal life. In fact, the growth rate is *only* below normal during the first three weeks of life. The mutation primarily inhibits chondrocyte maturation and growth. Shortening of cartilage columns, decreased deposition of GAGs into ECM, accumulation of glycogen within chondrocytes, and consequent delays in ossification, characterize Cn mutant embryos.[7]

Such mutants which inhibit intrinsic regulation of skeletal growth are notoriously difficult to understand. Homozygote (Cn/Cn) mice have smaller zones in long-bone growth plates, smaller cells and reduced proliferation. Histochemical and ultrastructural criteria were used to claim that the difference in differentiation between Cn and wild-type cartilage is premature ageing of the chondrocytes (see Box 27.2). However, the defect can be traced back to the condensation stage. Many HCCs survive the delayed matrix mineralization. Subperiosteal osteogenesis is unaffected, probably because – as shown by reciprocal transplantation of cartilage between wild-type and Cn mice – the gene acts directly on chondrocytes.[8] Some of the biochemical information on mutants conflicts. In separate studies, levels of hydroxyproline and sialic acid were reported to increase and to lie within normal limits. Perhaps such heterogeneity in a syndrome affecting growth is to be expected.

Interstitial growth of the spheno-occipital, midsphenoidal and nasal cartilages is inhibited. However, mandibular condylar cartilage, which has *appositional* growth, grows normally, as it does in human achondroplasia.[9]

The major gene responsible for achondroplasia is now known and well studied. Achondroplasia results from mutation in an FGF receptor gene, FGFR-3. Indeed, at least three skeletal dysplasias with clinically related symptoms – achondroplasia, hypochondroplasia and thanatophoric (death-bringing) dysplasia – result from mutations in the FGFR-3 gene, a single mutation in the single kinase domain of FGFR-3 in type II thanatophoric dysplasia, and mutations in either the intra- or extracellular domains in type I thanatophoric dysplasia.[10]

FgfR-3

Mutations in FgfR-3 are associated with and may in a sense be said to cause skeletal overgrowth (and deafness) in mice. FgfR-3 is a negative regulator of bone growth. Disrupting this receptor in mice leads to bone dysplasia and prolonged endochondral growth, demonstrating that FgfR-3 limits endochondral ossification. Following this thinking, human achondroplasia would represent a gain-of-function mutation in FGFR-3.[11]

Y. Wang et al. (1999) induced a point mutation in the FGFR-3 receptor that results in achondroplasia. The human mutation inserted into the mouse genome also produces achondroplasia. Delezoide et al. (1998) mapped the temporal and spatial patterns of expression of FGFR-1, R-2 and R-3 during human development. Within developing limb buds:

- FGFR-1 and R-2 but not FGFR-3 are expressed in epithelia and mesenchyme;
- FGFR-1 and R-2 are expressed in perichondria and in periostea; while
- FGFR-3 is expressed in mature chondrocytes, indicating a role for FGFR-3 in skeletal growth.

In the membrane bones of the skull all three receptors are expressed in osteoprogenitor cells and osteoblasts, perhaps the first record of all three receptors co-expressed in a single tissue. All three play critical roles in skull development, especially regulating osteogenesis at sutural margins; mutations in these receptors initiate fusion of sutures (craniosynostosis) in humans (Chapter 34).

As achondroplasia results from a mutation in FGFR-3, we would expect FGFR-3 to influence proliferation of chondroblasts *in vitro*. It does not, although it does enhance proliferation *in vivo*. FGFR-1 does provide a proliferative signal for chondroblast proliferation *in vitro*. Using transgenic mice, Q. Wang et al. (2001) demonstrated that this action of FgfR-3 results from a unique property of growth-plate chondrocytes to respond to FgfR-3.

Fgf-18 is expressed in perichondria. *Fgf-18* null mice have a growth-plate phenotype almost indistinguishable from that seen in *FgfR-3* null mice, but with the addition of ossification defects at sites of *FgfR-2* expression. Exogenous Fgf-18 or FgfR-3 leads to increased proliferation and enhancement of the HCC zone, increased proliferation and differentiation of chondrocytes, and expression of *Ihh*, implicating Fgf-18 as a ligand for FgfR-3. *Fgf-18* null mice also display decreased ossification accompanied by depression of osteogenic markers, a decrease not seen in *FgfR-3⁻* mice, indicating that *Fgf-18* is signaling via another Fgf receptor to control osteoblastic differentiation, i.e. Fgf-18 signals to multiple FgfRs controlling chondrogenesis and osteogenesis at the growth plates, coordinating the two processes (Fig. 15.2).

Box 27.2 Ageing of cartilage

Just as bone ages, cartilage also shows age-related changes. For example, ageing of rat costal and tracheal chondrocytes over the first two years of life is evident from the decrease in the number of chondrocytes, increase in glycogen and lipid contents, decrease in CS – but increase in keratan sulphate – the increasing thickness of collagen fibres, and increased mineralization (Bonucci *et al.* 1974). Some signs of ageing – such as changes in composition of the matrix – reflect *physiological deterioration*. Others reflect greater deviations from what we consider the normal state of cartilage or cartilages. As illustrated by three examples below, both types usually are cartilage-specific.

(i) Regions of articular cartilage in rabbit knee joints show signs of ageing – accumulation of fibrillar collagen – from as early as late in gestation.

(ii) Age-related and regional changes in chondroitin-4-sulphate, chondroitin-6-sulphate and keratan sulphate occur in the mandibular condylar cartilage of rats aged four to 62 weeks.

(iii) In four- to 18-month-old rabbits, 13–24 per cent of auricular chondrocytes are binuclear, while the vast majority of articular chondrocytes have a single nucleus. An interesting finding: unlike articular chondrocytes, ageing of auricular chondrocytes *in vivo* is *not* because they exhaust their potential to proliferate. However, auricular chondrocytes from young (one-week-old) and old (30-month-old) rabbits behave differently *in vitro*, suggesting an age programme for synthesis and turnover of such products as collagen, proteoglycan and elastin (Table 27.1). Relative synthesis of proteoglycans is 32 per cent higher in auricular cartilage from two-week-old rabbits than from one-week-old, but 50 per cent lower in cartilage from 30-month-old rabbits. Even more dramatic changes are seen in the synthesis of collagen and elastin.[a]

Chondrocyte density in human articular (medial femoral condyle) and costal cartilages shows a different pattern of decline with age (Fig. 27.1). Cell density declines between birth and 30 years in both cartilages. Although there is no further decline in the articular cartilage between 31 and 89 years, cell density in the costal cartilage decreases by 25 per cent over the same five decades (Stockwell, 1967). Such site-specific differences must be accounted for in analyses such as those that claim a relationship between cartilage cellularity and individual lifespan, one example of which is the inverse relationship between cellularity of mammalian femoral articular cartilage and species lifespan (Table 27.2). By examining the number of population doublings – which correlate with species life span – Evans and

Georgescu investigated senescence *in vitro* of cells derived from articular cartilages of rabbits, dogs and humans. Cell doublings decline dramatically – threefold or more – between neonates and 3.5-year-old rabbits (Table 27.3). Differentiative behavior of the cells

Table 27.2 Articular chondrocyte cell doublings *in vitro* in relation to maximum lifespan of species[a]

Species	Maximum lifespan (years)	Cell doublings
Rabbit	6	8–10
Dog (breed not specified)	18	19
Domestic fowl	30	30
Human	100	35–40

[a] Data from Evans and Georgescu (1983) for adult organisms.

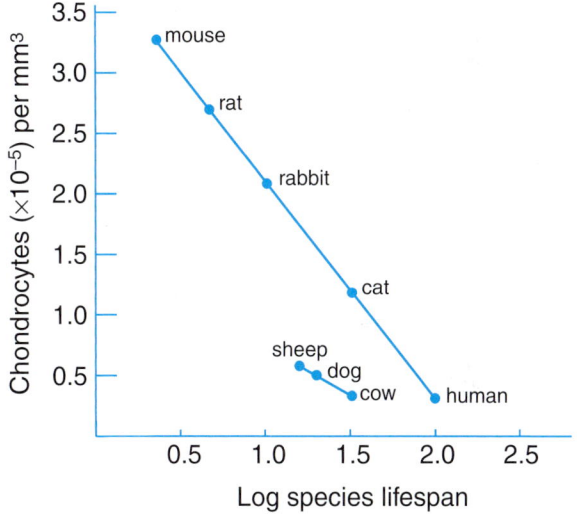

Figure 27.1 Chondrocytes and ageing. Chondrocyte density in femoral articular cartilages of eight mammals shows a positive correlation with lifespan (depicted as $\log_{10}$ lifespan in years), perhaps, in part, because of the very low proliferation in articular cartilage. From C. H. Evans (1983).

Table 27.1 Amount and relative activity of proteoglycan, collagen and elastin in auricular chondrocytes of rabbits of different ages[a]

Rabbit age	Proteoglycan		Collagen		Elastin	
	Total (%)[b]	Relative activity (%)[c]	Total (%)	Relative activity (%)	Total (%)	Relative activity (%)
2 weeks	358	132	332	240	89	65
4 weeks	264	137	155	142	76	22
8 weeks	312	166	355	122	32	12
5 months	454	157	344	100	10	2
12 months	115	52	163	78	–	–
30 months	158	50	184	56	1	–

[a] Based on data from Madsen *et al.* (1983b). Levels in one-week-old rabbits are taken as the baseline.
[b] Total amount of macromolecule (cpm) expressed as percentage of total amount of macromolecule found in auricular chondrocytes of one-week-old rabbits.
[c] Relative activity of macromolecule (cpm/µg) expressed as percentage of total amount of macromolecule found in auricular chondrocytes of one-week-old rabbits.

Box 27.2 (*Continued*)

Table 27.3 Rabbit articular chondrocyte cell doublings *in vitro* in relation to age[a]

Age	Cell doublings
4 months	11–12
6 months	8–10
3.5 years	3–4

[a] Data from Evans and Georgescu (1983). Age represents age of the rabbits from which chondrocytes were obtained.

also differs; rabbit articular chondrocytes always transform into fibroblasts, dog chondrocytes sometimes do, while human chondrocytes never do.

Ageing is not confined to permanent cartilages or to cartilage that is replaced endochondrally; fibrocartilages that form within or in association with the Achilles tendon in rabbits show signs of ageing after birth (Rufai *et al.*, 1992).

[a] See Bland and Ashhurst (1996) for fibrillar collagen, I. Takahashi *et al.* (1996b) for condylar cartilage, Lipman *et al.* (1984) for binucleate chondrocytes, and Moskalewski *et al.* (1980) and Madsen *et al.* (1983b) for rabbit auricular chondrocytes. See Ralphs and Benjamin (1994) for an overview of the joint capsule in ageing, and Fig. 3.4 for the joint.

CHONDRODYSPLASIA (*Cho*) IN MICE

Chondrodysplasia (*Cho*), a recessive lethal mutant in the C57 mouse strain, is characterized by shortened but wide long bones, a shortened face, cleft palate and collapsed tracheae. The diminished ECM contains large 600–2000 Å diameter banded collagen fibres. Similar thick collagen fibres are seen in the cloverleaf skull syndrome, which is a chondrodystrophy of the skull base in humans. Increase in Golgi body size, especially the number of Golgi vacuoles, is characteristic.[12]

Seegmiller and his group showed that chondroblasts fail to align in columns within the long bones. Instead they spread laterally, contributing to expanded appositional growth of the epiphyses. They proposed that GAGs were not released from the chondrocytes, but were retained within Golgi vacuoles. In 1998 they compared three genetically distinct chondrodystrophies (*Cho, Cmd* and *Dmm*), each of which displays distinctive ultrastructural derangement.

Cartilage matrix deficient (*Cmd*) mutant mice contain two chondroitin sulphate (CS) proteoglycan core proteins that are expressed differentially. The link proteins appear to be normal but the proteoglycans fail to aggregate. Muscles and connective tissues of *Cmd* mutants are also weakened because of deficiencies in CS proteoglycans. Expression of fibronectin mRNA is enhanced four to eight times in *Cmd* chondrocytes. Administering proteoglycan corrects the proteoglycan and the collagen deficiencies in *Cmd/Cmd* chondrocytes.

In metaphyseal dysostosis and pseudoachondroplastic dwarfism – two chondrodystrophies affecting humans – there is retention of material within distended cisternae of the endoplasmic reticulum: granular (GAGs?) in the former, and lamellar (protein–glycoprotein?) in the latter.[13]

Reduced GAG content of the ECM of *Cho* would enhance polymerization of tropocollagen into collagen fibrils. A matrix with abnormally large collagen fibres but reduced GAG levels would be produced and, as a result, would be weakened, affecting alignment of chondroblasts and causing collapse of the tracheal cartilage, resulting in asphyxiation. Kochhar *et al.* (1976) applied the glutamine analogue 6-diazo-5-oxonorleucine to mouse limb buds

in vitro and observed formation of a dense meshwork of unusually wide collagen fibres. Collagen synthesis was unaffected. Thus, reduced deposition of GAG into the ECM of cartilage can affect collagen fibrillogenesis and, consequently, cartilage integrity. In nanomelic chicks (see below), in which synthesis and deposition of GAGs are retarded severely, however, collagen fibrillogenesis is normal.[14]

Stephens and Seegmiller (1976) provided further biochemical data on rib and limb cartilages from *Cho* mutants, indicating normal levels of incorporation of $Na_2^{35}SO_4$-labeled glucosamine and glucose, total uronic acid, and protein, and that GAGs of normal molecular weight were synthesized. Less GAG is deposited into the ECM – more ^{35}S is washed out of *Cho* cartilage than from wild type – but the collagen content also is below normal. A similar failure of deposition of GAGs into the ECM is seen in a human achondroplastic mutant in which GAGs are structurally normal but increased linkage to proteoglycan produces more aggregates with higher viscosity, i.e. the defect is in processing. Whether defective collagen fibrillogenesis in *Cho* results from a primary defect in the deposition of GAGs into the ECM or a primary defect in collagen itself remained to be determined until it was demonstrated that *Cho* results from a cysteine deletion in $\alpha1(XI)$, the gene for type XI collagen.[15]

Sprouty

Overexpression of *sprouty* genes, which are downstream of *Fgf* and induced by *Fgf* signaling, can cause chondrodysplasia.

Drosophila sprouty encodes an antagonist of *Fgf* and *Fgf* signaling. Regulation of *Fgf* by *sprouty* is conserved between *Drosophila* and chick:

- *sprouty* is expressed in *Fgf*-signaling centres;
- gain- or loss-of-function experiments show that *Fgf* induces *sprouty*;
- over-expression of *sprouty* in wing buds reduces early limb-bud growth (equivalent to the role that Fgf from the AER plays; Chapter 38), and dramatically reduces the lengths of skeletal elements later in development (Minowada *et al.*, 1999).

Long-bone growth is reduced so dramatically because of inhibition of chondrocyte differentiation. Given that a mutation in *FGFR-3* is critical to induction of chondrodysplasia in humans, *sprouty* also is most likely involved in chondrodysplasia.

BRACHYMORPHIC (*Bm*) MICE

Another mutant producing disproportionate dwarfism in mice is *Brachymorphic* (*Bm/Bm*). Although the basic genetics was described by Lane and Dickie in 1968, little is known of the mechanism of action of this mutant. The collagen is normal type II collagen, but CS is undersulphated and fails to aggregate into proteo-CS granules. As fibroblasts show similar defects, this may not be as specifically a skeletal mutant as those discussed above. A similar failure to aggregate is seen in nanomelic chicks, although the mechanism is quite different (see below). Connective and epithelial tissues in *Bm/Bm* mice show reduced carbohydrate synthesis.[16]

Wikström *et al.* (1984a) coupled immunohistochemistry with TEM to analyze changes in the epiphyseal growth zones of long bones from *Bm/Bm* mice. Their findings, at variance with earlier studies, show cessation of progression from the proliferative to the hypertrophic zones but normal production of collagens II and V (Figs 4.8 and 4.9). In a companion study published in the same year, evidence of abnormal distribution of matrix vesicles and increase in vesicle size associated with cell degeneration led to the conclusion that matrix vesicles represent cell debris and therefore cell death.

STUMPY (*Stm*) MICE

Another chondrodystrophic mutant in mice is *Stumpy* (*Stm*). Although the rate of chondroblast proliferation is normal, chondrocytes fail to separate from one another, HCCs are small, and endochondral ossification disorganized, perhaps indicating a cell membrane dysfunction, as seen in *Brachypod* (*Bp^H*) mice (Chapter 22).[17]

Johnson *et al.* (1991) undertook a morphometric analysis of shape changes in the scapulae and basioccipitals of three chondrodystrophic mouse mutants (*Achondroplasia, Brachymorphic, Stumpy*) (Fig. 4.9) and of grey lethal osteopetrotic mouse mutants.

NANOMELIA (*nm*) IN DOMESTIC FOWL

Nanomelia (*nm*) is a recessive lethal chondrodystrophic mutant in domestic fowl that causes extreme (homozygote) or moderate (heterozygote) shortening (*micromelia*) of the legs.

Synthesis of CS in nanomelic limb and sternal cartilage is 90 per cent below that in wild-type individuals, but synthesis in the skin is normal; *nm* does not involve a systemic alteration in synthesis of CS but a tissue-specific effect. The level of synthesis is similar to that found in H.H. 18 wild-type limb-bud mesenchyme, pointing toward a block in enhancement of synthesis in the mutant. An explanation for the drastically reduced proteoglycan levels is that nanomelic chondrocytes are completely deficient in proteoglycan core protein, the gene for which is transcribed at condensation in wild-type embryos. Aggrecan core protein is expressed in embryonic membrane bone in wild-type but not in nanomelic embryos, suggesting that the mutation also acts directly on membrane bone.[18]

Cultured wild-type sternal, vertebral and epiphyseal chondrocytes synthesize two major peaks of proteo-CS. The major peak is missing completely from media in which *nm* sterna are cultured, is reduced severely in chondrocytes, and may represent a cartilage-specific proteoglycan. The number of GAG granules in the ECM, and the amount of matrix itself, are severely reduced below wild-type levels.[19]

Xyloside stimulates synthesis of CS in *nm* chondrocytes, indicating that glycosyltransferases for CS are present and functional, as they are in another mutant in fowl – *micromelia-Abbott* – in which the proteoglycan backbone is abnormal and xylosylated protein reduced. GAGs produced in *nm* cannot aggregate with hyaluronan, although the reduced GAG content has minimal effect on collagen fibrillogenesis in *nm* chondrocytes. Collagen synthesis and the ultrastructure of the collagen fibres also appear normal.[20]

A considerable number of genes are now known to modify GAG, especially cell-surface proteoglycans (Table 27.4). The *nm* gene suppresses – or does not allow – the augmented synthesis of proteo-CS characteristic of chondrocytes in wild-type embryos. Failure to aggregate with hyaluronan results in absence of matrix granules. Although chondrocytes from nanomelic chicks lack almost all cartilage-specific GAGs and have lower concentrations of all GAGs than chondrocytes from wild type, their response to pressure – as measured by changes in cAMP – is the same. This indicated to Bourret *et al.* (1979) that cartilage-specific GAGs played no role in the chondrocyte's response to mechanical stress.

A phenotypically similar mutation in mice, *disproportionate micromelia* (*Dmm*), an incomplete dominant mutation introduced above, also is characterized by deposition of abnormal ECM in which proteoglycan is reduced by 70 per cent and total collagen by 30 per cent, with type II collagen especially affected (Brown *et al.*, 1981; Berggren *et al.*, 1997).

A further and quite unusual mutation in domestic fowl is *tibial dyschondroplasia*, in which *tibiae have two growth plates* and avascular, unmineralized cartilage with apoptotic chondrocytes in the proximal metaphysis of the tibiotarsus. Apoptosis arrests differentiation early in hypertrophy. While type X collagen is found in the cartilage, no mRNA for type II or type X is found, indicating a late disruption in chondrogenesis.[21]

Table 27.4 Genes required for biosynthesis of GAG modifications of cell-surface proteoglycans and the phenotype associated with mutation of that gene in *Drosophila*, mice or humans[a]

Gene	Protein activity	Associated phenotype
Glypican 3 (GPC-3), human	Glypican-linked heparan sulphate modifies proteoglycan	SGBD[b], overgrowth, skeletal abnormalities, Wilms' tumour
Division abnormally delayed (dally), Drosophila	Glypican-linked heparan sulphate modifies proteoglycan	Defects in Wingless and Dpp[c] signaling
Gpc-3, mouse	Glypican-linked heparan sulphate modifies proteoglycan	Overgrowth, dysplastic kidney
Sugarless, Drosophila	Required for synthesis of glucuronate-containing polymers	Defects in Wingless, FgfR[d], Dpp signaling
Sulphateless, Drosophila	Heparan sulphate-modifying enzyme	Defects in Wingless, FgfR, Hedgehog signaling
N-deacetylase N-sulfotransferase (NDST-2), mouse	Heparan sulphate-modifying enzyme	Loss of mast cell heparan, defects in secretory granules
Exostoses-1 (EXT-1), human	Encodes heparan-sulphate co-polymerase	Multiple exostoses, tumours
EXT-2, human	Encodes heparan-sulphate co-polymerase	Multiple exostoses
tout-velu, Drosophila	Encodes heparan-sulphate co-polymerase	Selective defect in Hedgehog signaling and distribution
Heparan 2-O-sulfotransferase, mouse	Heparan sulphate-modifying enzyme	Absence of kidneys, eye and skeletal abnormalities
pipe, Drosophila	Heparan sulphate-modifying enzyme	Defects in dorsal–ventral patterning, controls generation of ventralizing signals

[a] Based on data from Johnston and Bronsky (1995).
[b] SGBD, Simpson-Golabi-Benmel dysmorphia.
[c] Dpp, decapentaplegic.
[d] FgfR, fibroblast growth-factor receptor.

INDUCED MICROMELIA

Micromelia can be induced in chick embryos following administration of a variety of teratogens, including insulin and 6-aminonicotinamide (6-AN). *In ovo*, 6-AN is converted to 6-aminonicotinamide adenine dinucleotide (6-AN-NAD), which displaces endogenous NAD, reduces oxidative phosphorylation and impairs the dehydrogenase-linked reactions involved in utilization of glucose. Co-treatment *in vivo* or *in vitro* with NAD, ascorbic acid, sodium or calcium ascorbate prevents the teratological action of 6-AN.[22]

In the presence of 6-AN, clonally cultured myogenic cells synthesize collagen and GAGs. Sulphation of CS is reduced, although incorporation of glucosamine into GAGs and proline into collagen is normal. The number of chondrocyte Golgi vesicles is reduced greatly. This, along with some reduction in the rough endoplasmic reticulum, indicates reduced sites of synthesis and slowed transport of collagen and GAGs. Deposition of ECM is inhibited, especially in the chondrogenic core of the limb bud.[23]

Insulin-induced micromelia in long bones of chick embryos, demonstrated almost half a century ago, is associated with abnormal bending of the tibia and development of zones of cell death that spread throughout the knee joint. Both changes are induced after insulin is administered *in ovo*, but not when long bones are cultured in the presence of insulin, i.e. the action is not direct. Subsequently it has been shown that insulin also acts to inhibit chondrocyte hypertrophy (Box 26.1).[24]

Induced micromelia and achondroplastic mutants demonstrate that any impairment in interactions between ECM GAGs and collagen destabilizes the chondrocyte phenotype, leading to abnormal chondrogenesis.

METABOLIC REGULATION AND STABILITY OF DIFFERENTIATION

The introduction of 400–600 μg of the nicotinic acid antagonist, 3-acetylpyridine (methyl pyridyl ketone), into chick eggs incubated for at least 96 hours produces severe muscular hypoplasia later in development because of direct action on myoblasts; proliferation and myoblast fusion into myotubes are both slowed when myogenic cells are exposed to 3-acetylpyridine *in vitro*. Co-treatment with nicotinamide *in ovo* or *in vitro* counteracts the metabolic effects of 3-acetylpyridine.[25]

Taking advantage of the fact that mesenchymal cells from limb buds are nor irreversibly committed to chondrogenesis until H.H. 25, Caplan and associates asked: 'What is the fate of cells that 3-acetylpyridine prevents from becoming myogenic? Do they die, become fibroblastic or become chondrogenic?' High-density cell cultures (5–30 × 10^6 cells/plate) of trypsinized limb buds from embryos of H.H. 20–32 were assessed for the production of metachromatic ECM (Table 27.5; Caplan *et al.*, 1968; Caplan, 1970). At 30 × 10^6 cells/plate and with limb buds from embryos younger than H.H. 26, 100 per cent of the areas of the plates chondrified within seven days. Nodules of

Table 27.5 Per cent cartilage expression in cultures of H.H. 24 chick limb-bud mesenchymal cells in relation to initial cell count and presence or absence of 3-acetylpyridine[a,b]

Initial cell count	Cartilage expression (%)	
	Control medium	Addition of 3-acetylpyridine
10×10^6	18	80
12×10^6	30	100
15×10^6	50	100
20×10^6	80	100
30×10^6	100	100

[a] Based on data from Caplan (1970).
[b] 3-Acetylpyridine (1300–1500 μg/ml) was introduced on day two; trials lasted seven days.

chondrocytes invested with perichondrial-like membranes were interspersed with flatter zones of chondrocytes. Mesenchymal cells from limb buds of H.H. 26 or older did not chondrify. Addition of 3-acetylpyridine increased the number of limb-bud cells producing metachromatic ECM such that all cells chondrified at densities as low as 12×10^6 cells/plate (Table 27.5).

Thus, under these culture conditions, 3-acetylpyridine stimulated chondrogenesis in cells already able to undergo chondrogenesis – those from embryos younger than H.H. 26 – and initiated chondrogenesis in cells from older embryos that otherwise would not chondrify. Caplan speculated that 3-acetylpyridine acts as an exogenous derepressor or inducer.

The action of 3-acetylpyridine is prevented by addition of equivalent amounts of nicotinamide because, it was argued, 3-acetylpyridine depletes endogenous levels of nicotinamide and NAD. 3-Acetylpyridine does increase incorporation of nicotinamide into mesenchymal cells from H.H. 24 embryos cultured at moderately high densities (12.5×10^6 cells/plate); total pools of nicotinamide and NAD are reduced, as is synthesis of RNA, protein and GAGs. Levels of three key enzymes involved in the synthesis of CS – UDP-glucose pyrophosphorylase, xylosyltransferase and N-acetylgalactosaminotransferase – all increase sharply, perhaps reflecting the increasing number of cells involved in chondrogenesis.[26]

There is a general increase in total NAD with limb-bud development *in ovo*, while *in vitro* high levels of NAD correlate with myogenesis, low levels with chondrogenesis. Similar trends are seen in 6-AN-induced micromelia. Rosenberg and Caplan (1974) proposed that high levels of nicotinamide inhibit chondrogenesis while low levels favour chondrogenesis and inhibit myogenesis. Further, they proposed that the level of nicotinamide is influenced by exogenous 3-acetylpyridine, which in turn influences the internal pool of NAD.

Although cells are not permeable to NAD, they are permeable to nicotinamide. Elevated levels of extracellular nicotinamide would lead to elevated intracellular levels of NAD. Indications are that NAD acts via conversion to poly(ADP-ribose), which is bound to histone and to

non-histone chromosomal proteins. Synthesis of poly (ADP-ribose) parallels differentiation of mesenchymal cells into chondrocytes *in vitro* and is suppressed by 3-acetylpyridine and by nicotinamide. The suggestion is that alteration of the internal pool size of NAD may influence transcription by altering levels of the NAD-dependent ADP-ribosylation of chromatin, thereby enhancing or initiating chondrogenesis. Moderately repetitive cartilage-specific sequences of DNA are amplified and modified during differentiation of chondrocytes *in vitro*. Conversely, transcriptional diversity decreases as myoblasts differentiate.[27]

So, we see that metabolic status can play a role in regulating pathways of differentiation.

NOTES

1. See Rimoin *et al.* (1970, 1976) and Johnson (1986) for fundamental features of achondroplasia, Shepard *et al.* (1969), Nardi *et al.* (1974), Allenspach and Babiarz (1975) and Silberberg *et al.* (1976) for achondroplasia in mammals and fowl, Zwilling (1959), Seegmiller *et al.* (1972a), Hall (1972a,e, 1977a), Daimon (1973a,b) and Hinchliffe (1974) for experimentally induced micromelia, and Barnes *et al.* (1969), Hall (1972e) and Overman *et al.* (1976) for ascorbic acid.
2. See Hollister *et al.* (1982) and Gorlin *et al.* (2001) for reviews of human collagen-based syndromes, Papadatos and Bartocas (1982) for skeletal dysplasias, and Cohen (2002) for chondrodysplasias, including osteogenesis imperfecta and achondroplasia.
3. See Thurston *et al.* (1985b) for the description of the syndrome, and Jacenko *et al.* (1993) for the type X defect.
4. For reviews of achondroplasia, see Grüneberg (1963), McKusick (1972) and Rimoin (1975). See Hollister *et al.* (1982) for such disorders as Ehlers-Danlos, Marfan, cutis laxa, osteochondrodysplasia, osteogenesis imperfecta and achondroplasia, Bauze *et al.* (1975) for osteogenesis imperfecta, and see Papadatos and Bartocas (1982) for skeletal dysplasias.
5. See Johnson and Wise (1971) and Johnson (1974, 1986) for *cartilage anomaly* in mice, and Johnson and Hunt (1974) for the biochemical changes.
6. See Shepard *et al.* (1969), Shepard (1971) and Bargman *et al.* (1972) for achondroplasia in rabbits, and Bruce (1941) for the time and order of appearance of ossification centres in rabbit skulls.
7. See Lane and Dickie (1968), Kalter (1980) and Johnson (1986) for the genetics and growth characteristics of this achondroplastic recessive mutant in the mouse.
8. See Silberberg and Lesker (1975) and Sannasgala and Johnson (1990) for reduction of the growth plates, Bonnucci *et al.* (1976, 1977) and Silberberg *et al.* (1976) for premature ageing of chondrocytes, Thurston *et al.* (1985a) for onset at condensation, and Konyukhov and Paschin (1967) for the transplant study.
9. See Shepard *et al.* (1969), Jolly and Moore (1975), Brewer *et al.* (1977), Hall (1983a) and Chapter 3 (Cartilage growth).
10. See Silberberg and Lesker (1975) for increase of hydroxyproline, and Kleinman *et al.* (1977) for normal levels. See Papadatos and Bartocas (1982), Gorlin *et al.* (2001) and Cohen (2002) for skeletal dysplasias, and Webster and Donoghue (1997), Cohen (1998), Wilkie and Morriss-Kay

(2001), Cohen (2002) and Coumoul and Deng (2003) for syndromes, dysplasias and craniosynostosis resulting from mutations in Fgf receptor genes. As assessed in 13 patients with thanatophoric dysplasia, the growth plate and periosteum are replaced by an abnormal mesenchyme that differentiates into abnormal bone (Ornoy et al., 1985).

11. See Colvin et al. (1996) for mutations in *FgfR-3*, and Dean et al. (1996) for achondroplasia as a gain-of-function mutation.

12. See Seegmiller et al. (1971, 1972b, 1988), Seegmiller and Fraser (1977) and Johnson (1986) for characteristic features, including collagen and Golgi body, and Bonnucci and Nardi (1972) for the cloverleaf skull syndrome.

13. See Kimata et al. (1981, 1983, 1984) and Brennan et al. (1983) for CS proteoglycans, Cooper and Ponsetti (1973) and Cooper et al. (1973) for metaphyseal dysostosis and pseudoachondroplastic dwarfism (*dmm*), and see Takeda et al. (1986) for administration of proteoglycans to *cmd/cmd* chondrocytes. For studies on the organization of proteoglycan monomer, link protein and collagen in cartilage and for the persistence of proteoglycan and link protein during endochondral ossification until secondary spongiosa form, see A. R. Poole et al. (1980, 1982a,b,c) and Hunziker and Schenk (1987).

14. See Pennypacker and Goetinck (1976), Bourret et al. (1979), and Sawyer and Goetinck (1981) for nanomelic chick embryos.

15. See Pedrini-Mille and Pedrini (1971, 1982) for the human achondroplasia, and Li et al. (1995) for the collagen type XI defect.

16. See Orkin et al. (1976, 1977) and Pennypacker et al. (1981) for data on collagen and CS, and Yamada et al. (1984) for carbohydrate synthesis.

17. See Johnson (1977, 1978) and Thurston et al. (1983) for *stumpy*.

18. See Landauer (1965) for the nanomelic phenotype, Mathews (1967b) and Fraser and Goetinck (1971) for levels of CS synthesis, and Argraves et al. (1981), Goetinck et al. (1981), Kosher et al. (1986b) and M. Wong et al. (1992) for deficiency in core protein.

19. See Palmoski and Goetinck (1972), Okayama et al. (1976), Pennypacker and Goetinck (1976) and Wiebkin and Muir (1977) for these studies.

20. See Pennypacker and Goetinck (1976) for the study with xyloside, Quintner and Goetinck (1998) for *micromelia-Abbott*, and Goetinck and Pennypacker (1977) for inability to interact with hyaluronan.

21. See Praul et al. (1997) for apoptosis, and Chen et al. (1993) for collagen.

22. See Landauer (1957) for 6-AN administration, Caplan (1972c) and Seegmiller and Runner (1974) for glucose metabolism, and Caplan (1972f) and Overman et al. (1972, 1976) for the prevention studies.

23. See Schubert and Lacorbiere (1976) for the clonal cultures, Overman et al. (1972), Seegmiller et al. (1972) and Seegmiller and Runner (1974) for Golgi, CS and collagen, and Caplan (1972c,d) and Seegmiller (1977) for the studies on matrix deposition.

24. See Zwilling (1959) for insulin-induced micromelia, and J. M. Chen (1954) for insulin and hypertrophy.

25. See Landauer (1957) and Caplan (1971, 1972a) for the *in-ovo* and *in-vitro* studies.

26. See Caplan (1970, 1972b) and Caplan and Stoolmiller (1973) for these studies.

27. See Caplan and Rosenberg (1975) and Caplan (1977) for poly(ADP-ribose), Holliday and Pugh (1975), and Strom and Dorfman (1976a,b) and Ordahl and Caplan (1976) for moderately repetitive sequences.

Restarting Mammalian Articular Chondrocytes

Mammalian articular chondrocytes, normally in a post-mitotic state, can respond to degenerative changes by reinitiating DNA synthesis and proliferating. While this represents a loss of the differentiated state of *individual* chondrocytes, it is a means of maintaining the differentiated state of the cell *population* and the cartilaginous *tissue*, and consequently of repairing defects within that cartilage.

A profitable way to study factors that maintain the differentiated chondrocyte state is to perturb chondrocytes in cell culture. Consequently, there is a large volume of literature on cultivating avian chondrocytes *in vitro*. Coon (1966) established that differentiated avian chondrocytes could produce clones that were stable for many generations. Although phenotypic expression is lost with morphological dedifferentiation, it can be reexpressed if the chondrocytes are transferred to a permissive medium. As discussed in the previous chapter, depletion of extracellular matrix (ECM) initiates a compensating synthetic response from chondrocytes. Altering the ECM or environmental conditions also can result in modulation of the phenotype, enabling chondrocytes to express characteristics normally associated with other differentiated cell types.[1]

Development of methods to clone chondrogenic cells facilitated analysis of cell differentiation enormously. Culture conditions are critical to enable a cell to express its determined state, and to maintain that differentiated state. For example, clonally cultured chondrocytes (Fig. 10.8) switch from the synthesis of type II to synthesis of type I collagen under conditions where the chondrogenic phenotype is unstable. Furthermore, clonally derived myoblast cultures can be induced to cease production of the cell products characteristic of myoblasts, initiate synthesis of collagen and glycosaminoglycans (GAGs), and transform to cells that are indistinguishable from chondroblasts. Flower (1972) used equilibrium density gradient centrifugation to isolate and clone mesenchymal cells from limb buds of chick embryos of H.H. stage 18–28. Separate subpopulations yielded colonies of chondrocytes, myoblasts or fibroblasts.[2]

During the 1970s, attention was directed toward investigating the properties and responses of chondrocytes isolated from *mammalian* articular cartilage and maintained *in vitro*. The impetus for such studies came from the need to understand the metabolism and maintenance of articular cartilage, so that degenerative bone diseases and osteoarthroses involving articular cartilage might be treated. Of particular significance are differences between the behaviour *in vitro* of mammalian articular chondrocytes and avian (usually limb) chondrocytes, differences that could make transference of information from avian studies to mammals hazardous. Because of these differences, because these studies are not as well known to developmental biologists as are the avian studies, and because I discuss re-initiation of cell division in mammalian articular cartilage in the context of cartilage maintenance and the reparative response to degenerative bone diseases, I introduce the behaviour of mammalian articular chondrocytes *in vitro*. Previously thought unable to divide, the proliferative ability of articular chondrocytes when released from their matrices has been established.[3]

MAMMALIAN ARTICULAR CHONDROCYTES *IN VITRO*

Mammalian articular cartilage and articular chondrocytes already have been used to illustrate a number of topics: that chondrocytes contain cilia (Fig. 1.5); the ultrastructure of a human articular chondrocyte (Fig. 1.7); the distribution of collagen types I and II in rat articular cartilage (Fig. 1.8);

the junction between articular cartilage and subchondral bone (Fig. 3.11); and the correlation of chondrocyte density with lifespan in eight mammals (Fig. 27.1).

Articular chondrocytes from adult humans were first cultured in 1967 by Manning and Bonner, who used collagenase digestion and mechanical disruption to isolate chondrocytes from individuals up to 84 years old. Although the cells appeared fibroblastic in monolayer culture, if pelleted by centrifugation prior to being cultured they continued to produce ECM, a parallel to the situation described in 1966 for avian vertebral chondrocytes. Collagenase digestion of femoral articular cartilage from mature female rabbits releases chondrocytes that – when cultured for 14 days in serum-supplemented Ham's F12 medium – formed colonies containing metachromatic extracellular matrix.[4]

Synthesis of GAGs by mammalian articular chondrocytes continues *in vitro* even under monolayer conditions that are unfavourable for GAG synthesis by avian chondrocytes; you cannot extrapolate cell behaviour *in vitro* from birds to mammals. The activities of glucose-metabolizing enzymes are typical of fully functional chondrocytes. In cultured chondrocytes, lactate/malate dehydrogenase (LDH/MDH) ratios are greater than those in fibroblasts, the levels of both dehydrogenases increasing with increasing oxygen tension. Mammalian articular chondrocytes produce A and B arylsulphatases and, *in vitro*, modify their rates of synthesis in response to hormones and ascorbic acid, indicating that they maintain *in vitro* the physiological activity and specific responses of articular chondrocytes *in vivo*.[5]

A role for oxygen

Ever since the classic studies by C. A. L. Bassett and Carl Brighton, numerous studies have shown that hypoxia favours initiation and maintenance of chondrogenesis, cartilage development being maximal at around 21 per cent O_2, and osteogenesis at around 5 per cent.[6] One contrary view was put forward by Marcus (1973), who found that hypoxic conditions (7 per cent O_2) did not promote chondrogenesis from monolayered rabbit articular chondrocytes, and so questioned the general role of hypoxia in chondrogenesis. Although a lone voice, Marcus did point out the alarming fact that plastic culture dishes – on which most cells were and are cultured – were found by Chapman and colleagues to contain considerable diffusible oxygen, which diffused into the culture media at rates ranging from 10^{-1} to 10^{-2} μl/min over the first three hours to 10^{-2}–10^{-4} μl/min after 24 hours, depending on the composition of the dishes – polycarbonate, polystyrene and Perspex were tested – giving O_2 concentrations of the order of 10 μm up to 750 μm from the bottom of the dishes.

Oxygen consumption of articular cartilage *in vivo* is low (6 μl O_2/g dry wt/hr) in comparison to epiphyseal cartilage, in which oxygen consumption ranges from 500 to 900 μl O_2/g dry wt/hr. In part, this may reflect the adaptation of chondrocytes to differing circulatory systems found in different cartilages. Such regional specificity notwithstanding, when maintained in organ culture, human articular chondrocytes, like avian chondrocytes, respond to low (20 per cent) oxygen tension by increasing GAG synthesis and respond to higher (50 per cent) tensions by decreased synthesis or resorption. Furthermore, production of GAGs can continue at a high rate for prolonged periods of time.[7]

Responsiveness to environmental signals

Cultured articular chondrocytes respond to environmental perturbations by modifying their synthetic activity. Under clonal conditions, differentiated hyaline chondrocytes can be maintained through many cell generations. The type of ECM produced – indeed, the type of cartilaginous tissue produced – can be modified by varying the substrate on which the cells are cultured, but (see above) substrates may have undesirable and unknown effects. If the cells are placed on gel foam or on demineralized bone matrix after some time in culture, they produce chondromyxoid or chondroid.[8]

Under other conditions – monolayer or prolonged culture, or treatment with liver lysosomes, vitamin A, cAMP, $CaCl_2$, parathyroid hormone or calcitonin – synthesis of type II collagen ceases and synthesis of type I is initiated, *a switch that is reversible, and so a modulation*. Non-articular chondrocytes – e.g. from rabbit sternum – tend to produce predominantly type II collagen, while articular chondrocytes produce a mix of types I and II. As articular cartilage from osteoarthritic patients synthesizes type I collagen, exploration of these shifts may reveal relationships to the osteoarthritic state.[9]

As it happens, not all mammalian articular chondrocytes behave similarly *in vitro*. Human and rabbit articular chondrocytes are quite different. Growth of human cells is poorer, and they produce a wider range of GAGs. In this study by Srivastava *et al.* (1974b), however, the samples were obtained from a variety of human sarcomas, so one might expect there to be abnormalities, and/or the chondrocytes to exhibit poor growth.

As one example of a metabolic anomaly and variability between tumours, S-100 protein, an acidic nerve protein found in the brain, is found in chondrocytes and chondrogenic tumours – chondromas, chondroblastomas, amesenchymal chondrosarcomas and osteosarcomas – but not in osteogenic tumours such as osteomas, osteoblastomas, giant cell tumours or Ewing's sarcoma. As a further complication, a human salivary acinar cell line exposed to 1,25 vitamin D_3 produces type II collagen, S-100 protein and proteoglycans, all typical of chondroblasts. This effect is reversible upon removal of the 1,25 vitamin D_3. In an analogous situation, chondroitin sulphate (CS), identical to bovine nasal cartilage CS proteoglycan, is found in cerebellar glial cells unless they transform into oligodendrocytes, while a transition between astrocytes and chondrocytes (accumulation of glycogen and GAGs) was documented by Kepes et al. (1984) in a study of gliomas from four patients[10]

Webber et al. (1977) examined the behaviour, in monolayer culture, of articular chondrocytes from eight mammalian species. Cell proliferation, synthesis of matrix products and responsiveness to exogenous vitamin C are all species-related (Table 28.1).[11]

Bovine articular chondrocytes respond to Bmp-4 and Tgfβ-1 by increasing incorporation of ^{35}S into the ECM and enhanced synthesis of type I collagen and core pro-teoglycan (Table 28.2), the chondrocytes having high-affinity binding sites of the order of 6000 receptors/cell. Indeed, Fgf-2, Tgfβ and Bmp all bind to proteoglycans before binding to their receptors, nine different proteins binding to Tgfβ (Box 15.1). Bovine osteogenin and RhBMP2B are equipotential in their ability to maintain proteoglycans in bovine articular cartilage explants, 'both' growth factors increasing proteoglycan synthesis and reducing degradation to levels equivalent to those elicited by Igf or serum. I say 'both' because we now know

Table 28.2 Effect of Bmp-4 and Tgfβ-1 on synthesis of type II collagen and proteoglycan in bovine articular cartilage after 72 hours *in vitro*[a]

Treatment	Fold increase as compared to controls	
	Type II collagen	Proteoglycans
Control	1.0	1.0
Bmp-4 (30 ng/ml)	5.6	2.4
Tgfβ-1 (10 ng/ml)	3.2	1.7

[a] Based on data from Luyten et al. (1994).

that osteogenin and rhBMP2B are one and the same molecule, namely BMP-3 (Box 28.1).[12] Osteogenin protein-1 or OP-1 is now known as Bmp-7 (Box 28.2).

So we see that differentiated chondrocytes are sensitive to their extracellular environment and maintain their terminally differentiated state by initiating appropriate responses to changes in that environment.

Derangement in ECM can lead to abnormal expression of the differentiated state when chondrocytes are maintained *in vitro*, or in such *in-vivo* situations as achondroplasia (Chapter 27). Abnormal expression is stable and maintained in both situations. Chondrocytes may respond to altered environments by altering the nature of the products produced, released and/or deposited. Synthesis of abnormal GAGs may be initiated, or collagen synthesis switch from type II to type I. *In vitro*, such changes are often accompanied by phenotypic 'dedifferentiation' so that chondrocytes appear more like fibroblasts. Indeed, such shifts may represent a modulation to the fibroblastic state. When these changes occur *in vivo*, foci of abnormally differentiating cells may appear among normal cells. Such foci are typical of the histopathological changes in articular cartilage during the early stages of osteoarthritis and degenerative joint diseases.

MECHANISMS OF ARTICULAR CARTILAGE REPAIR

Hypertrophic chondrocytes (HCCs) are maintained as long as their mineralized ECM is intact (Chapter 12). Once mineralized matrix is removed, they resume DNA synthesis and divide or die, depending on the cartilage(s). Releasing chondrocytes from their ECMs, therefore, does not necessarily mean death; HCCs liberated *in vivo* from the cartilaginous models of mammalian ribs and from the condylar process of mammalian dentaries can dedifferentiate to osteoprogenitor cells (Chapters 12 and 13).

Table 28.1 Response of articular chondrocytes from eight mammalian species to different media ± vitamin C as measured by doubling time and production of ECM[a]

Species	Age (months)	Medium	Doubling time (h)[b]	Matrix production
European rabbit, *Oryctolagus cuniculus*	Three	DMEM[c]	14.6	–
		DMEM	16.0	–
		DMEM + Vit-C	–	–
		DMEM	17.2	–
		DMEM + Vit-C	–	–
		DMEM	–	–
Domestic cat, *Felis domesticus*	Six–nine	DMEM	14.2[d]	Cells did not survive
		DMEM	14.2[d]	
		DMEM + Vit-C	–	
Virginia possum, *Didelphis virginianus*	Young	DMEM	>108.0[d]	–
	Mature	DMEM + Vit-C	>108.0[d]	–
	Young	F12[e]	29.2[d]	–
	Mature	F12 + Vit-C	30.4[d]	–
Woodchuck, *Marmota monax*	Mature	DMEM	44.6	–
		DMEM	>72.0[d]	–
		F12	>72.0[d]	–
		F12 + Vit-C	50.0[d]	–
Domestic dog, *Canis familiaris*	Five	DMEM	17.8	–
Dall's sheep, *Ovis dalli*	Nine	DMEM	22.6	Little extracellular deposits
Rhesus monkey, *Macaca mulatta*	Several	DMEM	39.6	–
Cebus monkey, *Cebus apella*	Several	DMEM	33.4	–

[a] Based on data from Webber et al. (1977).
[b] Based on the accumulation of DNA in culture flasks at intervals of 8–24 hours beginning one day after inoculation.
[c] DMEM: Dulbecco-Vogt minimal essential medium.
[d] Based on stained cell count instead of DNA.
[e] F12: Ham's F12 medium.

Dividing again *in vitro*

Parallel events occur *in vitro*. Releasing chondrocytes from their ECM and culturing them under low cell density allows

Box 28.1 Bmp-3 (osteogenin, Bmp2B, rhBmp2B)

In the late 1980s a new bone inductive protein was isolated from the ECM of rat using dissociative extraction and partial purification. It was named *osteogenin*. A year later osteogenin was purified from bovine bone and a partial amino acid sequence obtained of this 30–40 kDa protein, which we now know to be a member of the Bmp family of growth factors. Described in the literature under various names – osteogenin, Bmp2B and human recombinant BMP2B (rhBMP2B) – osteogenin was the third Bmp isolated, and so is Bmp-3.[a]

Terminology

Bmp-3, a heparin-binding growth factor, binds avidly to type IV collagen and less strongly to collagen types I and IX. The binding specificity ensures that Bmp-3 localizes to sites such as basement membranes. Type IV collagen is a normal constituent of basement membranes, being especially high in those associated with the roof of the developing mouth cavity. Tgfβ-1 also binds to collagen type IV. Neither *Tgfβ-1* nor *Bmp-3* are inhibited by basement membrane products such as laminin or fibronectin.

Cellular binding proteins for recombinant human BMP-3 were identified and characterized on MC3T3-E1 and NIH 3T3 fibroblastic cells. The receptors have a high specificity; they do not cross-react with Pdgf, Igf, Fgf-2, Egf or with Tgfβ; Tgfβ-2 was found to enhance osteogenesis of a bovine bone fraction that contained both Bmp-2 and Bmp-3, while osteogenin (then thought to be a separate molecule) stimulates expression of *Tgfβ-1* mRNA by monocytes.[b]

Vukicevik and his colleagues carried out a series of studies that provide much of our knowledge of localization and function of Bmp-3. They:

- showed that osteogenin stimulates the expression of osteogenic and chondrogenic phenotypes *in vitro*, decreases proliferation and increases alkaline phosphatase and collagen synthesis by periosteal cells, increases the cAMP response to PTH in osteoblasts, and increases proteoglycan synthesis by chondrocytes;
- purified and sequenced *Bmp-3*, and localized it autoradiographically in 11–20-day-old rats in cartilage, bone, periostea and perichondria;
- found that Bmp-3 binds with high affinity to type IV collagen in basement membrane; and
- used an antibody to localize Bmp-3 to perichondria, periostea, osteoblasts, hypertrophic cartilage, bone matrix and basement membranes in 5–14-week-old human foetuses, and identified the lungs and kidneys as major sources of synthesis.[c]

Osteogenin functions at stages of skeletogenesis when the balance between proliferation and differentiation is altered.

Osteoprogenitor cells

Bmp-3 inhibits proliferation and enhances differentiation of human marrow osteoprogenitor cells (assessed by up-regulation of type I collagen, cAMP, alkaline phosphatase and osteocalcin), being active at 1–10 ng/ml (Amedee *et al.*, 1994). These results suggest that Bmp-3 may switch osteoprogenitor cells between proliferation and differentiation pathways.

Chondrogenesis

Bmp-3 stimulates chondrogenesis from chick limb-bud cells *in vitro*; instead of isolated nodules forming with intervening connective tissue, the whole culture chondrifies. Although Bmp-3 had no action on proliferation (Carrington *et al.*, 1991) it does promote re-expression of the cartilage phenotype by dedifferentiated rabbit articular chondrocytes (Harrison *et al.*, 1991).[d]

Induced osteogenesis

Bmp-3 induces bone formation in athymic rats and in baboons, composites of Bmp-3 and porous hydroxyapatite working especially well in baboons. Of a number of matrices tested as carriers for Bmp-3 in promoting bone induction collagen was found to be the best. Sterilization by irradiation of insoluble collagenous bone matrix with one or three million Rads had no effect on the utility of the matrix as a functional carrier for Bmp-3. Perhaps surprisingly, the geometry of the porous hydroxyapatite delivery system is critical; Bmp-3 implanted in discs of matrix is inductive, Bmp-3 on granular matrix is not. Optimal porosity is 300–400 μm.[e]

Fracture repair

Collagenous bone matrix with or without Bmp-3 enhances chondro- and osteogenesis in 8-mm defects in rat calvariae. Pdgf administered with Bmp-3 to rat craniotomy defects depressed Bmp-3-induced bone formation and enhanced soft tissue repair (Marden *et al.*, 1993a,b).

[a] See Sampath *et al.* (1987) and Katz and Reddi (1988) for isolation of Bmp-3 from teeth, and Luyten *et al.* (1989) from bovine bone. See Sampath *et al.* (1990) and Reddi *et al.* (1989) for bone induction.
[b] See Paralkar *et al.* (1990, 1992) for binding to type IV collagen, Xu *et al.* (1990b) for type IV in basement membranes, Paralkar *et al.* (1991a,b) for Tgfβ-1 and the receptors, Bentz *et al.* (1991) for Tgfβ-2, and Cunningham *et al.* (1992) for the monocyte study.
[c] See Vukicevic *et al.* (1989, 1990a, 1993, 1994a,b) for these studies.
[d] See Carrington *et al.* (1991) and Harrison *et al.* (1991) for these two studies.
[e] See Ripamonti *et al.* (1991, 1992a,c, 1993) for the rat and baboon studies, Ma *et al.* (1990) and Katz *et al.* (1990) for collagen as carrier, Ripamonti *et al.* (1992b) for carrier geometry, Tsuruga *et al.* (1997) for optimal porosity, and Ripamonti and Reddi (1992, 1997) for reviews of the role of Bmp-3 in craniofacial and periodontal bone repair. Synthetic sponge implanted into pig skin or adipose tissue is invaded by cells that produce ectopic bone within 9–13 weeks, provided that the pores in the sponge are around 40 μm in diameter (Winter and Simpson, 1969; Winter, 1971).

the cells to resume DNA synthesis and proliferate. When 'apparently' terminally differentiated HCCs from mouse Meckel's cartilages or chick tibiae are exposed *in vitro* to hydrocortisone or to complement-sufficient antiserum, respectively, DNA synthesis is reinitiated. Response *in vitro*, however, can vary from that elicited *in vivo*. Cortisone inhibits chondrocyte growth *in ovo* – chondrocytes are smaller with diminished amounts of ECM – but at physiological concentrations *in vitro*, cortisone reduces synthesis of CS by embryonic chondrocytes.[13]

Many cells lose the ability to divide upon terminal differentiation, although division and differentiation are not necessarily mutually exclusive. For many years it was thought that mammalian articular chondrocytes lose the ability to divide when the organism reaches adulthood. Consequently, no attention was paid to the possible

Box 28.2 Bmp-7 (osteogenin protein-1, OP-1)

Bmp-7 (osteogenin protein-1, OP-1) is another member of the Bmp family active in skeletogenesis. *Bmp-7* is expressed in branchial clefts and arches from H.H. 12 on, with exogenous Bmp-7 initiating ectopic mandibular and maxillary elements (Chapter 21).[a]

Lens induction

Bmp-7 is expressed in mouse embryonic head ectoderm coincident with induction of the lens placode. The association is more than spatial for *Bmp-7* antagonists inhibit lens induction at the placode stage. *Pax-6* is lost from the placode, indicating that *Bmp-7* regulates *Pax-6*.

The *Small eye* (*Sey/Sey*) mutation in mice, in which the lens and nasal placodes fail to form, palatal closure is delayed, extra upper incisors form, and which develops a median cartilage between the maxillae, is a mutation of *Pax-6*. From studies on induction of the lens placode, it was established that *Bmp-7* regulates *Pax-6*. *Bmp-7*-deficient mice die soon after birth. The metanephrogenic kidney fails to develop, the lens of the eye is not induced and there are patterning defects of the ribs, skull and hind limbs.[b]

Limb-bud development

mRNAs for *Bmp-7* and *Bmp-2* are co-localized in the murine ZPA, AER and notochord. Heterodimers have the same functions as Bmp-2 and Bmp-7, indicating cooperative mediation of tissue interactions (Lyons et al., 1995).[c]

Bmps are negative regulators of the AER in chick limb buds. By delivering the BMP antagonist noggin in a retroviral vector to chick limb-bud cells, cartilage can be completely suppressed. Bmp is required for cells to condense – in the presence of noggin, cells fail to condense and accumulate as undifferentiated mesenchyme – and for cells to differentiate as chondrocytes – noggin blocks differentiation and leads to up-regulation of joint markers such as Gdf-5, indicating that cells are not switching into chondrogenesis. Muscle and tendon morphogenesis is also altered.

Implanting beads soaked in Bmp-7 or Bmp-2 into the interdigital regions of avian limb buds affects mesenchymal cell death but not epithelial cell death or cartilage differentiation (Macias et al., 1997).

Endochondral ossification

Bmp-7 is expressed in growth-plate cartilages and bone of chick and human embryos, especially in the HCCs near metaphyseal blood vessels, perichondria and periostea.

Bmp-7 regulates *runx-2*. Targeted disruption of *runx-2* results in complete loss of bone formation in homozygotes because of arrest of maturation of osteoblasts. Although a few alkaline phosphatase-positive osteoblasts develop, no post-condensation osteoblasts form. Acting via vitamin D_3, Bmp-7 also enhances the proliferation of cells derived from an osteosarcoma.[d]

Induced skeletogenesis

Bmp-7 is a potent inducer of several skeletal tissues, inducing new bone formation *in vivo* with similar specific activity to bovine osteogenic protein (Sampath et al., 1992).[e] *Cementogenesis* in the baboon, *Papio ursinus*, is promoted following single applications of 100 or 500 µg/g bovine collagenous matrix (Ripamonti et al., 1996). Implanting Bmp-7 in a collagen carrier enhances the repair of large defects in the ulnar diaphyses of rabbits, dogs and primates, as outlined below.

- Application of Bmp-7 at concentrations of between 6.25 and 400 µg to 1.5-cm defects in the ulna of adult rabbits results in complete healing in eight weeks. The new bone has mechanical properties comparable to intact bone. Torsional strength is 95 per cent, angular deformation to failure 92 per cent, and energy absorption to failure 94 per cent of control values.
- Application of 1200 µg to 2.5-cm defects in the ulna of adult male dogs elicits complete healing by eight weeks, at which time torsional strength is 72 per cent, angular deformation to failure 92 per cent, and energy absorption to failure 67 per cent of control values.
- Applying 1000 µg to 2.0-cm defects in the ulna or 250–2000 µg to defects in the tibiae of adult African green monkeys, *Cercopithecus aethiops*, elicited complete healing by eight weeks in five of six ulnae and in four of five tibiae. Bone formed in the other individuals but did not lead to healing (Cook et al., 1994a,b, 1995).

Skeletogenesis from cell lines *in vitro*

Calf periosteal cells exposed to Bmp-7 show increased proliferation and enhanced osteogenesis and chondrogenesis (Brüber et al., 2001).

Bmp-7 induces chondroblastic, osteoblastic and/or adipocyte differentiation from cloned murine cells, the cell type that is elicited *depending on the cell line*. ATDC5 embryonic carcinoma cells produce cartilage, MC3T3-E1 calvarial cells produce osteoblasts and C3H10T1/2 cells produce adipocytes in response to low concentrations of Bmp-7. C3H10T1/2 cells produce cartilage in response to high concentrations (500 ng/ml) of Bmp-7. A concentration of 40 ng/ml selects and amplifies rat bone marrow cells to initiate chondro- and osteogenesis.[f]

Delaying the time of application also can affect the differentiative outcome. Rat calvarial OPCs exposed to 4–100 ng/ml of Bmp-7 for 6 days produce cartilage. Delaying addition of Bmp-7 until day 7 results in osteoblastic differentiation, an action that is blocked by Tgfβ, although Tgfβ-1 at 0.3–1 ng/ml enhances chondrogenesis and inhibits osteogenesis from cultured periosteal cells derived from one-day-old chick tibiae.[g]

[a] See Bostrom et al. (1999) for a review including Bmp-7 and other Bmps and growth-derived factors (Gdfs), Khan et al. (2000) for Bmps, Tgfβ, Pdgf, Igf and Fgf, and Wall and Hogan (1995) for expression data and ectopic elements.

[b] See Kaufman et al. (1995), Luo et al. (1995), Quinn et al. (1997), and Wawersik et al. (1999) for *Bmp-7* and lens induction.

[c] Dimers of *Xenopus* Bmp-4 and Bmp-7 implanted into rats induce membrane bone and up-regulate alkaline phosphatase in primary cultures of rat marrow (Aono et al., 1995).

[d] See Houston et al. (1994) and Helder et al. (1995) for expression in growth-plate chondrocytes, Komori et al. (1997), Otto et al. (1997), Rodan and Harada (1997) and Stricker et al. (2002) for *Bmp-7* and *runx-2*, and see Knutsen et al. (1993) for *Bmp-7* and osteosarcomas.

[e] Two members of the *Tgfβ* superfamily in *Drosophila*, decapentaplegic protein and 60A, induce cartilage, bone and marrow in mammals (Sampath et al., 1993). The *Drosophila* 60A gene encodes a product that is more similar to Bmps 5–7 than to decapentaplegic protein (Wharton et al., 1991).

[f] See Asahina et al. (1996) and Andrades et al. (2001) for these *in-vitro* studies.

[g] See Asahina et al. (1993) and Iwasaki et al. (1993) for these studies.

regenerative or reparative abilities of articular cartilage as a means of overcoming degenerative conditions. Theories of 'wear and tear' predominated, with the implication that wear could not be reversed by 'repair and restore.' Interesting as an exception are articular chondrocytes of the pubic symphyseal cartilage of adult mice, in which mitosis is frequently observed.[14]

Reinitiation of DNA synthesis and mitosis within chondrocytes is a compensatory mechanism that maintains the population of differentiated chondrocytes in the face of environmental influences seeking to destroy it. Mammalian articular chondrocytes provide a useful example. In the early stages of osteoarthritis or degenerative joint disease a *proportion of the chondrocytes resume DNA synthesis and perhaps even divide*, thereby maintaining the population of differentiated chondrocytes, at least for a short time.

I emphasize *perhaps* because many authors assume that evidence of DNA synthesis – incorporation of ^{3}H-thymidine into DNA – is evidence that the cell is about to enter mitosis. DNA synthesis need not, however, be followed by initiation of mitosis, as has been demonstrated in periosteal osteogenesis, in fibroblasts of the periodontal ligament, and in the blastema of amputated amphibian appendages (Box 28.3).[15]

Dividing again *in vivo*

Proliferation *in vitro* shows that the ability of chondrocytes from adult mammalian articular cartilage to reenter mitosis is not irreversibly lost with terminal differentiation. Building on this finding, Cedric Mankin carried out a series of studies on the proliferative abilities of articular cartilage *in vivo*. Using ^{3}H-thymidine autoradiography in analyses of the distal femoral articular cartilage of immature rabbits, Mankin showed that a small percentage of chondrocytes near the articular surface incorporate label. Trauma to these cartilages increased the number of labeled cells, but these additional cells made little contribution to repair; synthesis of DNA should not be equated with cell division.[16] Is mitosis blocked, perhaps because of a long cell cycle?

DNA synthesis vs. division

[3]H-thymidine also is incorporated into chondrocytes from osteoarthritic patients. Some labeled cells are present in nests, presumptive evidence for localization of daughter cells after division. Quantitative data show that total DNA and RNA decrease to some 50 per cent of the levels in articular cartilage from non-osteoarthritic patients, representing *cell loss*. Synthesis of DNA/µg DNA, however, *increases* in osteoarthritic cartilage (Table 28.3). This is unexpected. Normally, synthesis of DNA and RNA, and total DNA and RNA, *decreases* with age in healthy articular cartilage. The conclusion from these studies is that the remaining cells are metabolically active.[17]

Box 28.3 Uncoupling DNA synthesis from mitosis

One must be aware that incorporation of ^{3}H-thymidine and mitotic activity need not show a one-to-one correspondence.

On the basis of grain counts of labeled periodontal ligament cells, some osteogenic precursor cells labeled with ^{3}H-thymidine had not divided 72 hours after labeling. These cells could have a long S phase, or be precursors of cells whose activity is not required during that period. The same grain-counting procedure indicates that only 37 per cent of fibroblasts in orthodontically stimulated rat periodontal ligaments (PDLs) had divided 27 hours later. Roberts and his colleagues made valuable use of the kinetics of cell proliferation of both resting and orthodontically stimulated rat PDLs *in vivo*, following pulse labeling with ^{3}H-thymidine, to demonstrate:

- two waves of mitosis in the PDL;
- diurnal periodicity in PDLs at rest (at rest meaning unstimulated by mechanical or other factors), with minimal mitoses at 9.00 p.m, and two maxima (one at 9.00 a.m. and one at 3.00 p.m.), which were taken as demonstrating three populations – one blocked at G1, one at G2, and a partly differentiated population;
- differentiation of preosteoblasts of the PDL into osteoblasts *without* initiation of proliferation;
- nuclear size as a cell kinetic marker, fibroblast precursors having a nuclear volume of <80 µm^3, preosteoblasts a nuclear volume of >170 µm^3; and
- four size classes of nuclei and five cell compartments: (i) self-perpetuating precursors, (ii) committed osteoprogenitor cells, (iii) preosteoblasts in G1 of the cell cycle, (iv) preosteoblasts in G2 of the cell cycle, and (v) osteoblasts; transformation of precursors to osteoblasts take some 60 hours, with stages up to G1 preosteoblasts being influenced by mechanical stimuli.[a]

The periodontal ligament behaves similarly when maintained *in vitro*, i.e. it is probably not the orthodontic treatment that blocks cell division *in vivo*. *In vitro*, rat PDL cells transform into OPCs and deposit mineralized bone. Uncoupling DNA synthesis from cell division also is seen during regeneration of newt limbs after denervation. In normal regenerates, DNA synthesis and proliferation increase in parallel as the blastema forms (Chapter 14). DNA synthesis is initiated in denervated animals – the ^{3}H-thymidine labeling index rises to 54 per cent of control value – but proliferation is not – no mitoses are seen, the cells remaining blocked at G2 of the cell cycle.[b]

[a] See Owen and MacPherson (1963) for the periosteal study, and Roberts and Jee (1974) for studies with the PDL.
[b] See Melcher and Turnbull (1974, 1976) and Mukai *et al.* (1993) for the PDL *in vivo*, and see Tassava *et al.* (1974), Mescher and Tassava (1975), Ferretti and J. Géraudie (1998) and Chapter 14 for these aspects of amphibian limb regeneration.

This same group of investigators devised an experimental model to elicit osteoarthritis in rabbits. Excising or severing the ligaments of the knee joint initiates what becomes prolonged joint degeneration. ^{3}H-thymidine-labeled chondrocytes are seen as early as five days after the trauma. Labeled chondrocytes are also seen within tibial and femoral condyles, indicating that uninjured nearby tissues respond to the trauma. Total immobilization of the joint is not a sufficient stimulus to initiate mitosis. Following total

Table 28.3 Total DNA and rates of DNA synthesis in articular cartilage from healthy (control) and osteoarthritic individuals[a]

	Total DNA (μg DNA/ mg wet wt)	Rates of DNA synthesis	
		(cpm DNA/ mg wet wt)	(cpm DNA/ μg DNA)
Control	2.2	49.7	24.1
Osteoarthritic	1.0	33.8	44.0
Osteoarthritic/control	−54%	−32%	+82%

[a] Based on data in Telhag (1976).

immobilization the cells die and the cartilage degenerates. Slight movements may induce some chondrocytes to divide, but the numbers are negligible and must be accumulated with colchicine arrest to be seen.[18]

Osteotomy and trauma

Removing wedges or small blocks of bone (*osteotomy*) to facilitate repair, taking out tumours and excising dead bone all fall within the everyday life of orthopaedic surgeons. Repair is enhanced by a variety of factors, listed below, that stabilize the osteotomy site mechanically, enhance proliferation or differentiation of cells at the site and encourage other cells to migrate into the site.

- In osteotomy defects in 20 dogs, *stable* mechanical conditions are followed by the deposition of woven bone, *instability* by the deposition of cartilage followed by endochondral ossification, and *maximum instability* by fibrocartilage.
- Osteotomy and/or damage to articular cartilage reduces mitotic activity substantially; in one study only one in 15 knee joints had dividing chondrocytes.
- In a male rabbit ulnar osteotomy model, rhBMP-2 (BMP-3) implanted in a collagen sponge increases the rate of bone repair by a third.
- Collagen barrier membranes have also been used to guide tissue regeneration in zygomatic arch osteotomies.[19]

Articular chondrocytes can respond to other stimuli by initiating mitotic activity. Local freezing of rabbit femoral articular cartilage produces degenerative changes accompanied by nests of new chondrocytes, especially at the margins of the defect. Similar nests are seen at the chondrosynovial junction, where fibrous tissue transforms into chondroid tissue after intra-articular injection of anti-chondromucoprotein immunoglobulin. Implanting active electrodes also produces an electrochemical environment favouring regeneration of fibrocartilage, especially from marginal chondrocytes. It may well be that if a greater proportion of chondrocytes could be brought into the proliferative and metabolically active pool, larger defects could be repaired, and/or repair made more permanent.[20]

Transcription factor

The first transcription factor involved in the differentiation of articular chondrocytes has been cloned and analyzed. Named C-1-1, it is a variant of the *ets* transcription factor *ch*-ERG. C-1-1 is expressed in articular chondrocytes, *ch*-ERG in HCCs. Expressing C-1-1 in chondrocytes *in vitro* maintains them in a stable immature state, prevents hypertrophy, and up-regulates tenascin-C, a matrix glycoprotein characteristic of articular chondrocytes. *ch*-ERG, on the other hand, enhances chondrocyte maturation *in vitro* (Iwamato et al., 2000).

NOTES

1. Excellent reviews and discussions of the early literature on avian chondrocytes *in vitro* may be found in Lasher (1971), Green (1971), Searls (1973a,b), Sokoloff et al. (1973), Levitt and Dorfman (1974), Levitt et al. (1974), and Wigley (1975).
2. For studies on the behaviour and ultrastructure of clonally cultured chondrocytes, see Fig. 10.8, Chacko et al. (1969), Eguchi and Okada (1971), Flower (1972), and Mayne et al. (1976). See Flower (1972) for equilibrium density gradient centrifugation.
3. Sokoloff (1974) offers a useful review of studies with avian and mammalian articular chondrocytes in the context of osteoarthritis.
4. See Abbott and Holtzer (1966) for avian vertebral chondrocytes, Ham and Sattler (1968) and Ham et al. (1970) for rabbit articular chondrocytes.
5. See Sokoloff et al. (1970) and Srivastava et al. (1974a) for continued GAG synthesis, Marcus and Srivastava (1973) for glucose-metabolizing enzymes and LDH/MDH ratios, Coffin and Hall (1974) and Thorogood and Hall (1976, 1977) for LDH/MDH ratios to identify chondrogenic cell types, and Jones and Addison (1975) and Schwartz and Adamy (1976) for the work on the arylsulphatases.
6. There is a vast literature on effects of oxygen tension on skeletogenesis *in vitro*, to which the following citations provide an entrée; Bassett and Herrmann (1961), Bassett (1962), Brighton et al. (1969), Hall (1969, 1970a), Pawelek (1969), Brighton and Schaffzin (1970), Brighton and Heppenstall (1971), Hadházy et al. (1974), Brighton and Kregs (1972a,b), Nevo et al. (1972), Brighton et al. (1974), Heppenstall et al. (1975) and Heppenstall (1980). There also is a literature on the effect of hypoxia on vertebral development in mice and rats *in vivo* (Hunter and Clegg, 1973a,b), and a smaller body of literature on oxygen gradients (Malda et al., 2003).
7. See Serafini-Fracassini and Smith (1974) for oxygen consumption, Durkin et al. (1969a,b) for differential vascularity of different cartilage types, Lane and Brighton (1974), Jacoby and Jayson (1975a,b) and Lemperg et al. (1975) for GAG synthesis, and Sledge and Dingle (1965) for resorption in response to oxygen levels.
8. See Green and Dickens (1972) for prolonged culture, and Green (1971) and Green and Ferguson (1975) for the transformations.
9. See Layman et al. (1972), Sokoloff et al. (1973), Deshmukh (1974), Schindler et al. (1975), Gay et al. (1976b), Deshmukh

and Kline (1976), Benya *et al.* (1977), Desmukh and Sawyer (1977), Deshmukh *et al.* (1977) and Norby *et al.* (1977) for modulation of collagen synthesis, and Nimni and Deshmukh (1973), Gay *et al.* (1976b) and Müller and Kühn (1977) for synthesis of type I collagen in articular cartilage from patients with osteoarthritis.

10. Schindler *et al.* (1975) analyzed the sternal chondrocytes. See Fell *et al.* (1968) and Melcher (1971a,b) for mouse Meckelian and chick tibial chondrocytes, Azuma *et al.* (1989) for the acinar cell line, Stefansson *et al.* (1982), Y. Nakamura *et al.* (1983) and Landry *et al.* (1990) for S-100 protein, and Gallo *et al.* (1987) for the glial cells.

11. See Bourne (1972) for a review of the actions of vitamin C on the skeleton, and see Ornoy and Zusman (1983), Bilezikian *et al.* (1996) and Boskey *et al.* (2001) for the action of vitamins on cartilage.

12. See Luyten *et al.* (1992, 1994) and Box 12.1 for response to Bmp-4 and Tgfβ-1, and Massagué (1991, 1992) for binding to proteoglycans.

13. See Holtzer *et al.* (1960) and Abbott and Holtzer (1966, 1968) for culture under low cell densities, Fell *et al.* (1968) and Melcher (1971a,b) for mouse Meckelian and avian tibial chondrocytes, and Badran and Provenza (1969), Barrett *et al.* (1966) and Hall (1972c) for response to cortisone *in vivo* and *in vitro*.

14. See Elliott (1936), Crelin (1957) and Barnett *et al.* (1961) for non-proliferation of articular chondrocytes, Crelin (1957), Crelin and Koch (1965, 1967) and Chapter 23 for pubic symphyseal cartilage, and Ali *et al.* (1974) for 'wear and tear' theories. Hamrick (1999) revisited the chondral modeling theory proposed by Frost, i.e. the theory that articular cartilage growth is regulated mechanically through processes affecting both cell–cell and cell–ECM interactions, and so maintains the congruence of the elements at joints.

15. See Cahn and Lasher (1967) and Cameron and Jeter (1971) for thymidine uptake and cell division.

16. See Mankin (1962a,b, 1964, 1973) for these pioneering studies.

17. See Hulth *et al.* (1972) for ^{3}H-thymidine incorporation, Telhag (1976) for the DNA and RNA data, Telhag and Havdrup (1975) for the normal decrease in RNA and DNA with age, and Mankin (1973) and Mayor and Moskowitz (1974) for the metabolic activity of such cells.

18. See Telhag (1972, 1973) and Havdrup *et al.* (1975) for the rabbit model, and Crelin and Southwick (1960, 1964) for the study involving colchicine arrest.

19. See Schatzker *et al.* (1989) for the dog study, Johnell and Telhag (1977) for articular cartilage, Bouxsein *et al.* (2001) for *rhBMP-2*, and Mundell *et al.* (1993) for the collagen membranes. As a basic (now classic) orthopaedics text, see Albright and Brand (1979). See An and Friedman (1998) and Bronner and Worrell (1999) for more recent overviews.

20. See Simon *et al.* (1976) for local freezing, Eguro and Goldner (1974) for the chondrosynovial junction, Baker *et al.* (1974a,b) for the electrode implantation, and Bertram (2001) for a biomechanical role of subarticular trabecular bone in attenuating the loading of articular cartilage.

Repair of Fractures and Regeneration of Growth Plates

The optimal outcome for repair of a fractured long bone is rapid union of the two ends of the bone with restoration of normal morphology, stability, strength and function.

Every day thousands of individuals fracture a bone and the process of reparative cartilage formation is set in motion (Figs 23.1 and 29.1). Inhabitants of the United States of America fracture over six million bones a year. Investigation into the mechanisms underlying the repair of fractured bones goes back to the 17th century if not earlier. By the 1830s, the four fundamental steps in repair of fractures had been discovered.[1]

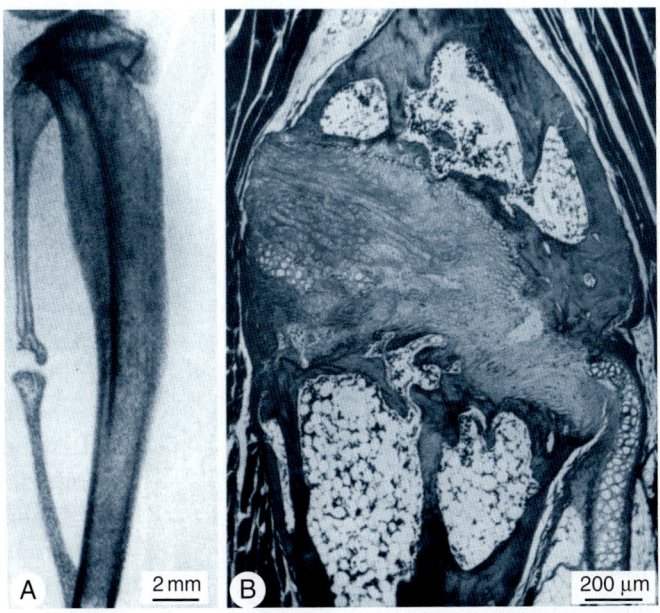

Figure 29.1 Fracture repair. (A) A radiograph of a repairing fracture of the fibula. The tibia (right) is intact. (B) Extensive development of cartilage in the fracture gap is shown in A. Modified from Altmann (1964).

A BRIEF HISTORY OF FRACTURE REPAIR

'Modern' orthopaedic practices have ancient origins.[2]

Evidence from mummies from Fifth Dynasty Egypt shows that fractured limb bones were splinted to immobilize healing fractures as long as 4600 years ago. By 1572 sufficient knowledge had accumulated that Ambroise Paré could produce a treatise on fracture healing. Heyde (1684) carried out what appears to be the first recorded experimental investigation into fracture repair, in this case on frogs. Observing that a callus had formed between the broken ends of the bone, he concluded that the callus went on to form the bone by which the fracture healed, and proposed mineralization of the blood clot that formed in the fracture gap as the mechanism, a perfectly logical conclusion, as the one invariably followed the other. Treatises on bone – especially on orthopaedics and fracture healing – were in wide circulation from at least the mid-1700s, Andry (1741) and Nesbitt (1736) being two notable examples.

A century and a half later, John Hunter, a pupil of Albrech von Haller – both of whom supported Heyde's interpretation – carried out a detailed analysis of repair of fractured long bones in chickens.[3] Hunter (1835) defined four basic processes in a repairing fracture, phases that he thought applied across all tetrapods. We recognize the same four today:

(i) formation of a vascular haematoma, from which Hunter believed the bone was deposited, as had Heyde and Haller;

(ii) formation of an early soft callus and a later hard, mineralized callus;

(iii) tissue transformation or metaplasia as the blood clot or cartilage of the soft callus is replaced by another mineralized tissue, bone; and

(iv) bone remodeling or turnover, as the new bone adapts to local conditions of existence.

The ability of bone to repair – and the subsequent use of bone transplants to stimulate or augment repair – generated much debate about the source of the cells that carry out the repair. Two broad views were held: (i) that the ability to produce new bone is limited to osteoblasts or osteogenic cells, a view maintained by Ollier in the 19th century and Axhausen and Lexer in the early 20th; and (ii) that less fully differentiated cells, undifferentiated cells, or stem cells can be induced to ossify in fractures or when transplanted.[4]

After J. Müller (1836) demonstrated that undifferentiated cells – embryonic connective tissue cells – could deposit bone during normal skeletogenesis, cell types as diverse as chondroblasts and white blood cells were proposed as osteoblast progenitors for both normal and reparative osteogenesis.[5]

The first steps in the repair of a fractured mammalian long bone are formation of a blood clot and callus. Differentiation of cartilage in the callus is a consequence of *periosteal progenitor cells* differentiating into chondroblasts. Had there not been a fracture the periosteal cells would have continued to form osteoblasts.

Such transformations occur following fractures of membrane bones, which perhaps is more striking. Periosteal cells on endochondral bones arise from what was the perichondrium of the cartilage model, now transformed into a periosteum. Osteoprogenitor cells in membrane bone periostea have no such history, and may not previously have been exposed to a stimulus evoking chondrogenic potential.[6]

Standardizing the fracture

A number of different methods are available to produce reproducible fractures in long bones of small mammals with minimal variation from animal to animal. Two examples for producing a standard transverse 'closed' fracture (closed meaning that no surgery is required) of the long bones of small mammals (rodents, rabbits) with minimal displacement at the fracture site are: a small pneumatic punch press driven by compressed air (see Figure 1 in

Jackson *et al.*, 1970), and a small brass guillotine, mounted on a wooden base and operated by a lever arm (see Figures 1A and 1B in Bourque *et al.*, 1992).

As a model for nonunion, Neto and Volpon (1984) resected three *mm* of bone along with one *cm* of the periosteum of dog long bones. Pseudarthroses with synovial cavities and fibrocartilage formed in some but not all of the dogs so treated.

Ashhurst *et al.* (1982, 1986) developed a method for producing standardized fractures in rabbit tibiae, involving two means of stabilizing the fracture. One uses steel rods to produce rigidity, which favours primary healing and Haversian remodeling. The other uses plastic to produce unstable fractures, secondary healing and cartilage formation.

T. N. Gardner *et al.* (2000) determined the mechanical function of each region of the callus in diaphyseal tibial fractures in 20-year-old human males at four stages in healing, and were able to correlate timing of callus growth and maturation with the stimulus provided by the original loss of stability caused by the fracture.

We are largish mammals. This will sound surprising to those accustomed to working with larger mammals, but in some (usually small) taxa whether the fracture site is mobile or not may be of little consequence. The same biochemical profile is seen in the calluses of immobilized and non-immobilized fractured rat tibiae (Table 29.1). The biochemical profile of the callus produced five days after closure of fractures of rat tibiae includes cells that express mRNA for both collagen types I and II (Sandberg *et al.*, 1989).

Motion

Perhaps not surprisingly, and in confirmation of many previous studies, Rooij *et al.* (2001) found that cartilage fails to form in mechanically unloaded fractured long bones, and that cartilage disappears when loading is removed.

Application of intermittent stress is the basic rationale for 'walking casts' rather than immobilization in bed as therapy for those who are unfortunate enough to fracture a long bone. Scar tissue develops if fractured rabbit femora are immobilized completely. If a normal range of

Table 29.1 The profiles of hexosamine, uronic acid and sialic acid (μg/mg) in callus tissue from immobilized and non-immobilized repairing rat tibiae[a]

	Days after fracture	Hexosamine	Uronic acid	Sialic acid
One day	Non-immobilized	19.2 ± 4.2	28.1 ± 1.2	6.1 ± 0.4
	Immobilized	21.0 ± 2.2	30.5 ± 9.9	6.1 ± 0.9
Three days	Non-immobilized	27.1 ± 2.3	25.0 ± 2.4	6.5 ± 0.5
	Immobilized	29.9 ± 2.5	20.1 ± 11.1	6.5 ± 0.5
Seven days	Non-immobilized	8.2 ± 0.7	14.8 ± 1.3	5.4 ± 0.2
	Immobilized	5.9 ± 0.3	9.3 ± 2.1	4.7 ± 0.5

[a] Based on data in Lane *et al.* (1982).

movement is allowed, a mixture of fibrous tissue and fibrocartilage develops and healing takes about six months. Placing the fracture in continuous motion favours rapid differentiation of progenitor cells into chondroblasts. Defects 203 mm in diameter created in long bones, plugged with screws and subjected to alternating motion, are either filled with cartilage or characterized by cycles of deposition and resorption of bone as cells modulate their differentiative activity.[7]

Intermittency of the signal is important if progenitors are to differentiate into chondroblasts. Mechanical stability modifies the GAGs and collagens produced during repair: the proportion of sulphated GAGs increases with increasing mechanical stability, while the mix of collagen types deposited – types I to V, and/or IX – is modulated in response to mechanical conditions.[8]

Intermittent stresses favour chondrogenesis in the repair of fractures of endochondral and of membrane bones in birds and mammals; membrane bones in amphibians and reptiles do not produce chondroblasts in repair. Clinically, the concept and implementation of *continuous passive motion* was found to maximize initiation of chondrogenesis and therefore initiation of repair (Box 29.1 and see Chapter 32).

Molecular studies reinforce and extend these findings. *Ihh*, *Gli-3* and *Bmp-6* appear earlier in mobile than in stabilized tibial fractures in mice. As *Bmp-6* enhances synthesis of *Ihh* mRNA, it has been proposed that it has a direct action on PTHrP, *Ihh* acting indirectly (see Box 31.2). Involvement of other growth factors is outlined below. *Bmp-6* is expressed in mouse hypertrophic chondrocytes (HCCs). CHO cells that express *Bmp-6* induce tumours that contain cartilage and bone and that undergo endochondral ossification. Bmp-6 also is found in chick sternal chondrocytes as they differentiate, its expression preceding type X collagen. Expression of both molecules is increased by treatment with 10^{-7} M PTHrP.[9]

Non-unions and persistent non-unions

The optimal outcome for repair of a fractured bone is rapid union of the two ends of the bone with restoration of normal morphology and function. Alas, not all unions behave so nicely. Repair may be delayed or incomplete. Between 300 000 and 600 000 of the six million bones Americans fracture each year either fail to heal or exhibit delayed healing. The reparative tissue may not bond strongly to the existing bone. The united ends of the bone may be misaligned. In extreme situations, a fracture may fail to heal, leaving a *non-union*. If repair of human long bones is delayed more than six to eight months we speak of *persistent non-union*. The distinction between *delayed union* and *non-union* first becomes apparent in the cellular response at the fracture site. In delayed unions the block to timely repair is cessation of the periosteal response as cells are unable to produce a callus at the expected rate. Non-unions reflect cessation of periosteal and endosteal responses.[10]

Box 29.1 Continuous passive motion (CPM)

After more than a century of experimentation and trial and error, it now appears that *continuous passive motion* (CMP), a concept pioneered by the distinguished Canadian orthopaedic surgeon Robert B. Salter, provides the optimal biomechanical conditions for fracture healing or when bone grafts must be used in mechanically unstable situations. The 'walking cast,' which places a fracture under continuous passive motion, is an obvious and highly desirable outcome of such studies. Salter's pioneering studies some 25–30 years ago demonstrated that in the repair of fractured long bones:

- *complete immobilization* of rabbit long bone joints elicits development of fibrous scar tissue;
- *normal motion* initiates repair with a combination of fibrous tissue and fibrocartilage in 20 per cent of rabbits after six months; and
- *continuous passive motion* permits healing with hyaline cartilage in 80 per cent of rabbits in only four weeks.[a]

These are impressive figures, but, as outlined below, their demonstration of the utility of continuous passive motion did not stop here.

In a study in which intra-articular periosteal autografts were used to repair articular cartilage, chondrogenesis was only initiated in eight per cent of the immobilized joints but in 60 per cent of those joints exposed to continuous passive motion.

Hyaline cartilage formed in one tenth of both immobilized joints and joints given intermittent motion, but in 70 per cent of those joints exposed to continuous passive motion during repair of major osteochondral defects in rabbit femoral condyles following autogenous periosteal grafts. Furthermore, newly developed cartilage was much more fully bound to the bone in animals that had undergone CMP.

Similarly, when CMP is used, much more type II collagen is deposited after resurfacing full-thickness patellar defects with autogenous tibial periosteal grafts (93 per cent of the collagen deposited being type II) than following immobilization (two per cent) or intermittent motion (47 per cent). Deposition of hexosamine and chondroitin and keratan sulphates (markers of deposition of cartilaginous ECM) also is much greater following CMP.[b]

Finally, when Schatzker et al. (1989) created osteotomy defects in 20 dogs, they found that stable mechanical conditions result in deposition of woven bone, instability results in deposition of cartilage followed by endochondral ossification, and maximal instability elicits fibrocartilage.

At about the same time (the late 1980s) Elizabet Burger and her colleagues from Amsterdam developed a way to apply intermittent pressure to foetal mouse long bones or calvariae in organ culture. Chief among their findings are that:

- calvariae respond to intermittent compression with increased bone formation, decreased bone resorption and a net 16 per cent increase in bone mineral, a response that is comparable to that shown by growth-plate chondrocytes;
- long bones respond by increased ^{35}S incorporation into ECM; and
- mineralization of growth-plate cartilage is enhanced by application of intermittent or continuous compressive force *in vitro*.[c]

[a] See Salter (1975), Salter et al. (1980), O'Driscoll et al. (1986) and O'Driscoll and Salter (1986).
[b] See O'Driscoll and Salter (1984) for intra-articular grafts, O'Driscoll and Salter (1986) and O'Driscoll (1999) for osteochondral defects, and O'Driscoll et al. (1986) for the collagen type II and ECM data.
[c] See Klein-Nulend et al. (1987a,b), Nulend et al. (1985) and Burger et al. (1991) for these three studies. Klein-Nulend et al. (1993, 1995) also demonstrated that Tgfβ and soluble bone factors are released by mouse calvariae and periosteal cells loaded in intermittent hydrostatic compression. See also Burger et al. (1991), Wright et al. (1992) and Joldersma et al. (2000) for additional studies.

Box 29.2 Coal trimmers and shoemakers and pseudarthroses

Coal trimmers worked in gangs in the holds of steam ships into which coal was being loaded for a voyage. Regarded as the hardest and dirtiest of all jobs in the coal industry, the coal trimmer's job was to spread the coal evenly throughout the hold so that the ship would be balanced and 'in trim.' A vessel out of trim was in danger of capsizing. The 562-ton steam ship, *Steam Governor General*, which sailed from Melbourne to Sydney 150 years ago (14 November 1854),[a] had six coal trimmers out of a crew of 38 to trim a ship carrying 12 saloon passengers, 14 intermediate passengers and 53 passengers (including seven children) in steerage. The term coal trimmer was also used for those who positioned barges under a coal tipple and used ropes or winches to tip the coal into the barge.

Sir William Arbuthnot Lane (1856–1943), a surgeon at Guy's Hospital and the Great Ormond Street Hospital for Sick Children in London, described the skeletal changes in a coal trimmer in considerable detail. The most striking alteration was the formation of a joint between the fourth and fifth lumbar vertebrae, a subdivision of the neural arch of the fourth vertebra, changes that Lane attributed to the forces imposed on the spine when rotated around the vertical axis as coal was thrown with force and over a wide compass.

Lane invented a number of procedures using internal fixation to align the ends of a broken bone, progressing from silver wire to steel screws and plates (Lane, 1894, 1907), the screws and plates still being known as *Lane's plates and screws*. His technique of designing each instrument with a long handle to keep the surgeon's hands as far from the wound as possible demonstrated that fractures of long bones could be treated in an open operation. The programmes of health care in Britain today are the direct descendants of Lane's reforms in health education.[b]

Shoemakers need no introduction. In another publication (1888), Lane described the skeleton of a 73-year-old man whose body was sent to Guy's Hospital for dissection. Lane emphasized a buttress of bone (an exostosis; also see Fig. 9.1) extending dorsally from the atlas on one side and articulating with the occipital bone by a pseudarthrosis. Lane attributed these changes to the stresses associated with a life spent mending shoes: 'Having concluded, from a careful examination of the changes which the body presented, that the man had been a shoemaker, I wrote to the medical officer of the infirmary in which he died for any information he could give me, and he kindly informed me that the man was entered on the books as a shoemaker.' Arthur Conan Doyle owed much to Lane for the skills of close observation and reasoning with which Doyle imbued Sherlock Holmes.[c]

[a] The 14th of November 1854 – the day of the great hurricane that battered the English after the disastrous Charge of the Light Brigade (25 October) and the bloodiest engagement of the Crimean War, the Battle of Inkerman (5 November). Also the day (at 5.00 pm) of the death, in agony, from an abscess on the kidney, of Edward Forbes (1815–1854), zoologist, palaeontologist, marine biologist and one of the first biogeographers. Forbes abandoned art – insufficient talent, and medicine – insufficient interest, for the study of marine animals, especially their distribution, producing one of the first general studies of oceanography. Edwin Lankester named his three sons Edwin Ray after the great botanist Edwin Ray, Edward Forbes after the zoologist, and Alfred Owen, known as Owen, after Richard Owen. Thomas Henry Huxley, who replaced Forbes at the Royal School of Mines in 1854, ranked Forbes with Richard Owen as: 'the … leader of Zoological Science in this country [with] more influence by his personal weight and example upon the rising generation of scientific naturalists than Owen will have if he write from now till Doomsday' [letter from Huxley to William Sharp MacLeay, 9 November 1851, cited from L. Huxley (1900), Vol. 1, p. 94]. A founder of the 'Red Lions' dining club that meet at annual meetings of the British Association for the Advancement of Science from 1839 onwards, Forbes was responsible for many of the verses that accompanied the consumption of quantities of inexpensive beef washed down with copious amounts of beer with much roaring and flourishing of coat tails at the annual dinner. A sample of Forbes' verse – **caution**, it may not appeal to all tastes – written in 1847, follows.

Song of the Dodo [pronounced *doo-doo*]
> Do-Do! although we can't see him,
> His picture is hung in the British Museum:
> For the creature itself, we may judge what a loss it is,
> When its claws and its bill are such great curiosities.
> Do-do! Do-do!
> Ornithologists all have been puzzled by you.

Do-do! Monsieur de Blainville,[1]
Who hits very hard all the nails on his anvil,
Maintains that the bird was a vulture rapacious,
> And neither a wader, nor else gallinaceous;
> A do-do! a do-do,
> And not a cock-a-doodle-doo!

… and so on. Cited from Wilson and Geikie (1861), p. 421, which contains other samples of Forbes' verse and song.

[1] Henri Blainville was Cuvier's deputy and his successor to the Chair of Comparative Anatomy at the Natural History Museum in Paris.

[b] We also have the Murphy-Lane bone skid, a 30-cm-long instrument used in operations involving the head of the femur.

[c] The quotation is from Lane (1888), p. 593. The comment concerning Conan Doyle is from Tanner (1946), p. 68. The papers by Lane on the shoemaker and coal trimmer, along with others on similar topics, are included in his book, *The Operative Treatment of Chronic Intestinal Stasis* (1915). Lane is perhaps best known to medical students through 'Lane's kinks', obstructive twists of the ileum, but medical students also had to learn Lane's bands (Lane's bridle) – adhesions between loops of the ileum, Lane's operation – the drastic procedure of removing the entire colon to treat constipation, and Lane's syndrome (Arbuthnot Lane syndrome; Lane's disease) – contraction of the colon, obstruction of the rectum and inability to contract the muscles of the pelvis – all in all, rather an unhappy epitaph.

Exposing a non-union to mechanical stress can elicit formation of a false joint (pseudarthrosis; Box 29.2 and Fig. 29.2). Understanding why some fractures fail to unite is therefore of considerable interest. Facilitating healing would save the lives of as many as 600 000 individuals a year in the United States alone, prevent the protracted pain and discomfort that accompany non-unions, and save the health-care system untold numbers of dollars.

Are the reasons for a non-union local or external to the fracture site? Both, most likely. As measured by ATP in rabbits, levels of metabolism are higher in non-unions than in unions. One week after a fracture, those that are healing have a haematoma and granulation tissue, non-unions a haematoma only. At eight weeks, bone has been deposited in healing fractures with growth factors (Tgfβ-1, Pdgf, Fgf-2, Bmp-2 and Bmp-4) at the fracture site. Non-unions show only fibrous tissue and none of the growth factors, a dramatic difference.[11]

How mechanical stimuli are transduced during skeletogenesis has been a perennial problem, with similar factors playing different roles at different stages. Take Tgfβ-1. Using Chinese mountain goats, Yeung *et al.* (2002) showed that expression of *Tgfβ-1* was enhanced in osteoblasts and osteocytes during the distraction phase of long-bone healing, but that levels dropped after the distraction phase. We made the same observations in healing fractures of mouse long bones (Bourque *et al.*, 1992, 1993a,c).

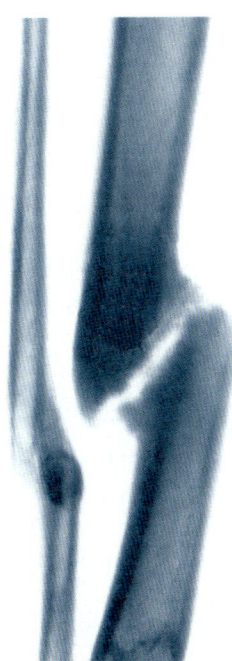

Figure 29.2 Pseudarthrosis. An X-ray of a pseudarthrosis in a human tibia (right) and partial healing of the fibula (left). Modified from Altmann (1964).

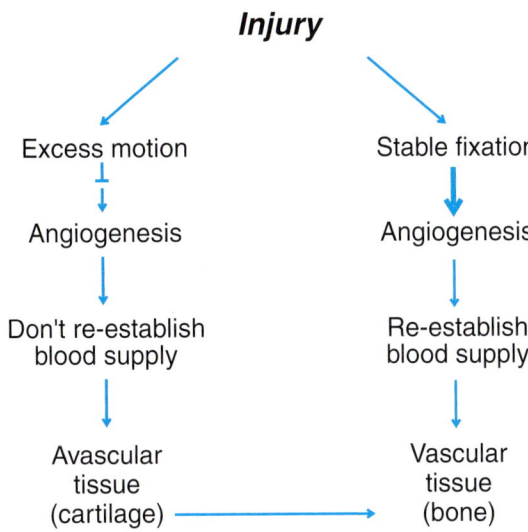

Figure 29.3 The model proposed by Ferguson et al. (1999) for the role of movement at a fracture site in a long bone. As shown on the right, stability promotes angiogenesis, reestablishment of the blood supply and development of a vascular tissue, bone. As shown on the left, excess motion reduces angiogenesis and vascularization, leading to the development of an avascular tissue (cartilage), which is subsequently replaced by bone if the vascular system is reestablished.

If cells at the fracture site function, but only slowly, then approaches to enhance their activity could (should?) speed up or initiate repair. Two approaches are to implant growth factors known to enhance chondro- and osteogenesis, or to apply an external agent that will alter the mechanical conditions at the fracture site.

Growth factors and fracture repair

Much information is available on the role of growth factors in the various stages of repair of fractured long bones in several animal models.

Ferguson et al. (1999) specifically investigated whether repair of murine tibial fractures repeats embryonic cartilage and bone development by examining *Ihh*, *runx-2*, matrix metalloproteinase 3, collagen type II, *gli-1* and *osteocalcin*. As previous workers have done, they placed some emphasis on the role of differential angiogenesis in establishing sites of chondrogenesis – avascular sites, or osteogenesis – vascular sites (Fig. 29.3). This may be of clinical relevance, as non-unions in humans (Box 32.1) are categorized on the basis of presence or absence of a vascular supply.[12]

J. R. Lieberman et al. (2002) reviewed the role of growth factors in bone repair, with special emphasis on Tgfβ, Bmps, Fgfs, Igf and Pdgf and their clinical use in enhancing repair (especially of persistent non-unions), enhancing spinal fusion, and in treatment of pseudarthroses of the spine (Box 29.2). Rudert (2002) examined the adequacy of the rabbit as a model experimental animal for evaluating osteochondral defects, especially those involving defects of articular cartilage.

Studies undertaken by William Bourque in my laboratory established a reproducible model for fracturing mouse long bones, and a methodology to analyze growth factors involved in repair in mouse histological sections (Bourque et al., 1992, 1993a,c). Some growth factors that especially influence critical stages of repair are highlighted as follows.

- *Tumour necrosis factor-α* inhibits cartilage formation during repair of rib fractures in rats, a fibrous callus forming instead. Periosteal osteogenesis is not affected (Hashimoto et al., 1989).
- Injecting Fgf-1 during the first nine days of repair in rats stimulates enlargement of cartilage while inhibiting mRNA for procollagen II and proteoglycan core protein. Injection of Fgf in a sodium hyaluronan gel enhances callus formation, chondrogenesis and subsequent ossification, resulting in reparative bone that is twice as strong as bone formed in the absence of exogenous Fgf.[13]
- *Insulin-like growth factors* I and II are expressed in osteoblasts and non-HCCs in human callus (Andrew et al., 1993b); Igf-1 increases with bone lengthening in rabbits (Schumacher et al., 1996).

Tgfβ is an important player in fracture repair. It is expressed at the onset of repair and in cartilage and bone during repair, the association being with proliferation and differentiation; levels are lowest in the haematoma and the HCCs that form. Subperiosteal injection of Tgfβ-1 or -2 enhances chondrogenesis and osteogenesis (endochondral and intramembranous) of rat femora, the ratio of cartilage to bone varying with the dose administered.

Furthermore, injecting Tgfβ-1 enhances local synthesis of Tgfβ-1, indicating self- or autoregulation. Similarly, applying human TGFβ-1 and demineralized bone paste into cranial periosteal defects in rabbits accelerates repair. Injecting Tgfβ-1 or -2 over the frontal or parietal bones in mice (five daily injections for a total of 1 μg/day) results in heavier bones, enhanced periosteal proliferation and increased vascularization of the bone.[14]

Bmps

Analysis of Bmp-2 and repair of dog mandibles and sheep femora, along with studies with C3H10T1/2 cells and Bmp-2 at concentrations of 0–1000 ng/ml, 10 ng being sufficient to evoke bone, began to appear in the early 1990s. Low concentrations evoke fat and higher concentrations differentiation of cartilage and bone, the bone formed being stable when Bmp-2 is withdrawn from the cultures.[15]

Implanting a growth factor requires finding the appropriate matrix in which to imbed the factor. Healing can be initiated in dogs by treating non-unions with Bmp immobilized in polyactic acid; implanting the Bmp in demineralized bone matrix does not. Species specificity also is important; bovine Bmp is not active in dogs. Allogenic bone infused with cortical bone Bmp and non-collagenous proteins improves healing of femoral non-unions in humans, with increases in length of individual femora of 1.5–5 cm, making this a potential substitute for autogenous (self-produced) bone.[16]

Jump-starting repair

In a rat mid-femoral defect model, Cullinane *et al.* (2002) found that constant fixation leads to union and bony repair, while daily bending leads to non-union, prominent cartilage in the callus site, and indications that a pseudarthrosis (Fig. 29.2) is forming. In the same rat mid-femoral defect model, control fractures with constant (rigid) fixation showed full or partial bony union. Application of a pulsed electromagnetic field (PEMF) has been found to be an effective way to 'jump-start' repair of persistent non-unions or pseudarthroses that can form following prolonged misalignment and abnormal mechanical stresses (Boxes 29.2 and 32.1) although Cullinane and colleagues found that application of 1 Hz for 10 minutes a day did not initiate repair.

REGENERATION OF GROWTH PLATES IN RATS, OPOSSUMS AND MEN

Growth plates can repair minor damage, provided that the damage is confined to the articular surface and does not penetrate the subchondral bone (Fig. 13.9). Billions of dollars are spent annually replacing joints (usually hips) because of more extensive damage than we can repair

ourselves. The therapeutic value and cost–benefits of growth-plate regeneration would be enormous. What are the prospects for regenerating a growth plate?[17]

An important early study that is not as well known as it should be is that by Selye (1934), who amputated the legs of 12–15-day-old rats above the level of the femoral epiphyseal cartilage. Within days proliferating cells and osteoblasts appeared at the stump. Selye described these 'osteoblasts' transforming into chondroblasts by 11 days and the production of islets of cartilage from which the epiphyseal cartilage regenerated. Growth in length of the femur then resumed. Selye reported that no regeneration was initiated in the amputated limbs of rats older than five months.

Amputation of the forelimbs of rats above the elbow initiates partial regeneration. Cartilage and bone form if muscle is sutured over the stump. Open amputation[18] of the hind limbs of 10- to 12-day-old rats through the femur or tibio-fibular was the model system used by Libbin and Weinstein (1986). The responses of the two bones differed, which is in keeping with the pattern for digital regeneration, namely that distal elements (the tibio-fibular) can regenerate but a proximal element (the femur) cannot.

Cartilage developed in 30 per cent of the femoral amputations, 60 per cent showed only soft-tissue repair, but two individuals (12 per cent of the sample) regenerated incomplete growth plates. However, well-organized growth plates regenerated in half the tibio-fibular fractures and a complete growth plate regenerated in one. Regeneration was rapid, with growth plates replaced within two weeks. Libbin and Weinstein (1987) determined that the regenerated cartilage was periosteal in origin and began to form within two days of the operation.[19]

A flurry of research in the 1970s sought to augment repair/regeneration with electrical current. Becker (1972b) applied electrical currents of 3–6 nA (nano amperes) to the amputated limbs of 15 rats and obtained a high percentage of regeneration and formation of epiphyseal growth centres, but the flow of such studies has slowed.[20]

In the early 1980s, a number of groups turned to the Virginia opossum, *Didelphis virginiana*, as an experimental model, primarily because the limb skeleton is at a relatively immature (embryonic) stage when the pups are born; the gestation period is only 12.5 days, which compares with 19 days for mice and 22 for rats. At birth Virginia opossum embryos are developmentally equivalent to rat embryos at 12 days of gestation and human embryos in the second month (Figs 19.1 and 19.2).

Hind limbs were amputated either through the ankle (tarsals) or through the digits (phalanges) four days after birth. The results, although initially interpreted as regeneration, are more consistent with further embryonic development and growth (or regulation; see Box 35.1) rather than regeneration. Much the same type of 'repair'

Table 29.2 An annotated list of features of marsupial skeletal development and skeletal tissues

Articular cartilage development in *the short-tailed possum, Monodelphis,* including the role of appositional growth (Archer et al., 1994)

Carpal development in the eastern native cat (quoll), *Dasyurus viverrinus* (Sanchez-Villagro and Dottling, 2003)

Cartilage canals lacking in long bones of immature specimens of four species of marsupials (a wallaby, ring-tail possum koala and antechinus; Thorp, 1990; Thorp and Dixon, 1991)

Central nervous system development is retarded relative to skeletal and muscle development – based on comparison of five placental and four marsupial species (K. K. Smith, 1997)

Comparative development rates in *Monodelphis* and in the Virginia opossum, *Didelphis* (K. K. Smith and van Nievelt, 1997)

Cranial osteogenesis of the nectar-feeding honey possum, *Tarsipes rostratus* (Rosenberg and Richardson, 1995)

Cranial osteogenesis in *Monodelphis domestica* and in the tammar wallaby, *Macropus eugenii,* in which dermal ossification starts before endochondral, and facial skeleton develops faster than neurocranial (Clark and Smith, 1993).

Craniofacial development comparing marsupials with placentals; emphasis on sequence changes and statistical methods (Nunn and Smith, 1998)

Embryology of marsupials (Klima and Bangma, 1987)

Embryonic growth rates of marsupials and comparison with monotremes (R. W. Rose, 1989)

Epiphyses in marsupials (*Didelphis, Macropus, Dendrolagus*) lack cartilage canals and have secondary centres of ossification that ossify directly from the perichondrium (Haines, 1941, 1942b, 1969; Thorp, 1990)

Evolutionary analysis of osteological characters in marsupials (Szalay, 1994)

Mandibular arch structures in *Monodelphis domestica* on a faster developmental schedule than those of the hyoid arch (Filan, 1991)

Middle ear ossicle development in *Monodelphis domestica,* the neonate having neither a mammalian nor a reptilian jaw articulation (Filan, 1991)

Osteogenesis in the bandicoot, *Isodon macrourus,* and the possum *Trichosurus vulpecula* (Gemmell et al., 1988)

Scaling of limb bones during ontogeny of *Monodelphis domestica* and *Didelphis virginiana* (Maunz and German, 1997)

Scapula development for the comparative analysis, which includes, embryos, neonates and postnatal specimens of the gray short-tailed possum (*Monodelphis domestica*), the four-eyed opossum (*Philander opossum*), the bare-tailed woolly opossum (*Caluromys philander*) and the red-cheeked dunnart (*Sminthopsis virginae*) (Sánchez-Villagra and Maier, 2003)

Viscerocranial growth in opossum *Monodelphis domestica* (Maunz and German, 1996) where growth is faster than neurocranial and for longer. Also includes data on sexual dimorphism

is seen when chick radii are fractured at 6.5–7 days of incubation, when they are still cartilaginous models. No callus cartilage forms but bony healing occurs in 85 per cent of the embryos (Table 29.2).[21]

What of human growth plates?[22]

NOTES

1. From the enormous literature on fracture of long bones, the following are classic studies: Ham (1930), Pritchard and Ruzicka (1950), Murray (1954), Tonna and Cronkite (1962), Pritchard (1965, 1969, 1972b, 1974), Hall (1970a), Ham and Harris (1971), Bassett (1972a), Becker (1972a), Tonna and Pentel (1972), Kernek and Wray (1973), White (1975), Lane (1987), McGibbon (1978), Simmons (1980, 1985) and Frost (1989a,b).

2. See Trueta (1968), Hall (1992) and Einhorn (1998) for the history of studies on bone repair, and see Hulth (1980, 1981, 1989), Nunamaker (1998) and the chapters in Volume 5: *Fracture Repair and Regeneration* of Hall (1990–1994) for reviews of mammalian and avian models for fracture repair.

3. As discussed by Keith (1917), Heyde's view of the mechanism was shared by Albrecht von Haller 80 years later (Haller, 1763), but denied by Duhamel de Monceau.

4. See Ollier (1867), Axhausen (1907, 1909) and Lexer (1908) for osteoblasts as precursors of repair. For repair from other cell types, see Barth (1895), Curtis (1893), Axhausen (1907), Phemister (1914, 1951), Levander (1934, 1964), Lacroix (1947, 1951, 1953), Urist and MacLean (1952) and Urist (1953, 1980).

5. For chondroblasts as precursors, see Kölliker (1853), Müller (1858), Gegenbauer (1867), Van der Stricht (1890) and Brachet (1893). For white blood cells, see Wolff (1879) and Schaffer (1888), connective tissue cells, Müller (1836) and Renault (1886), and for cells lining vascular canals, see Minot (1898).

6. For chondrogenesis during repair of fractured membrane bones, see Pritchard (1946), Girgis and Pritchard (1958), Craft et al. (1974), James et al. (1974), Hall and Jacobson (1975) and Precious and Hall (1994). For overviews of fracture repair see McKibbin (1978), Heppenstall (1980), Simmons (1980, 1985), Hulty (1980), Lane (1987), Frost (1989a,b) and Volume 5: *Fracture Repair and Regeneration* of Hall (1990–1994).

7. See Schenk (1973) and Uhthoff and Germain (1977) for the defect studies

8. See Page and Ashhurst (1987) and Page et al. (1986). See Mwale et al. (2000) for selective assembly and remodeling of collagens II and IX in association with maturation of the ECM and of the chondrocyte.

9. See Le et al. (2001) for the fracture study, Gitelman et al. (1994) for CHO cells, and Grimsrud et al. (1999) for sternal chondrocytes.

10. See Marsh (1998) for delayed vs. non-unions, and Delaporte (1983) for a history of the respective roles of periostea and endostea in osteogenesis.

11. See Brownlow and Simpson (2000) for the ATP-based study, and Brownlow et al. (2001) for the growth factors. ATP may be more than a convenient measure of metabolism in chondrogenesis. The G-protein-coupled ECM receptor, cP22-/1, a receptor for extracellular ATP, is found in chick-limb mesenchyme, is unregulated with differentiation, and inhibits chondrogenesis in micromass cultures (Meyer et al., 2001).

12. See Ferguson et al. (1998) for an overview of fracture repair and normal skeletal development, especially the roles of *Shh, Sox-9* and *runx-2,* Glowacki (1998) for the role of growth factors in the angiogenesis phase of fracture repair, Bostrom et al. (1999) for an overview covering Bmp-7, Gdf-5 and other growth factors, and Bourque et al. (1993c) and Khan et al. (2000) for Bmps, Tgfβ, Pdgf, Igf and Fgfs.

13. See Cuevas et al. (1988), Jingushi et al. (1990), Andreshak et al. (1997) and Radmosky et al. (1998) for Fgf in repair in various mammalian species.

14. See Joyce et al. (1990a) and Andrew et al. (1993a) for the initial expression studies, Joyce et al. (1990b) for autoregulation, and Kibblewhite et al. (1993) and Mackie and Trechsel (1990)

for the studies with calvariae; as this reaction could not be inhibited by indomethacin, prostaglandins are not involved.

15. See Wang *et al.* (1992, 1993) and Wang (1993) for these early studies with Bmp-2. Wozney (1992) discusses the activity of BMP-2 and studies on rat and sheep femora and dog mandibles; see Wozney *et al.* (1993) for Bmp expression in cartilage and bone development.

16. See Heckman *et al.* (1991) for the matrix study, Johnson and Urist (1998) for allogenic bone, and J. R. Lieberman *et al.* (2002) for the role of growth factors in initiating repair of persistent non-unions.

17. Libbin (1992) reviewed the prospects, including past research and much of the primary literature on growth-plate regeneration.

18. *Open amputation* requires invasive surgery to expose and fracture or amputate a growth plate. A *closed fracture* or 'closed' amputation is made by breaking the limb bone without surgical intervention. See Bourque *et al.* (1992) for the device we designed and constructed to apply reproducible closed fractures to mouse long bones.

19. See Person *et al.* (1979) and Person (1983b) for these regeneration studies.

20. See Becker and Spadaro (1972), Becker (1978), Borgens (1982a) and Hall (1983f) for reviews of the literature on bioelectricity.

21. See Mizell (1968), Mizell and Isaacs (1980), Fleming and Tassava (1981) and Neufeld (1992) for studies on Virginia opossums, and McCullagh *et al.* (1990) for the embryonic chick study. See Table 29.2 for an annotated bibliography of some studies on skeletal tissues and systems in marsupials.

22. See Goss (1980b) for an early but still pertinent summary of the prospects for regeneration in humans.

Part X

Growing Together

Growth: permanent increase in size

Morphogenesis: attainment of form or change in form (shape)

At the very root of growth as a developmental process are mechanisms that determine how many cells are available for an individual skeletal element. Consequently, and paradoxically, growth is best analyzed before the growth phase.

'To the elbow I go, from the knee I flee,' is an aphorism that medical students learn as a way to remember that the human humerus grows more rapidly from its proximal than from its distal growth plate, and that this pattern of growth is reversed for the femur. The aphorism assumes differential growth rates in different growth plates. The basis for such differences is a topic discussed in the chapters of this part, as we consider the events and processes associated with two bones growing together, and why, when they approximate one another, some bones form joints while others fuse.

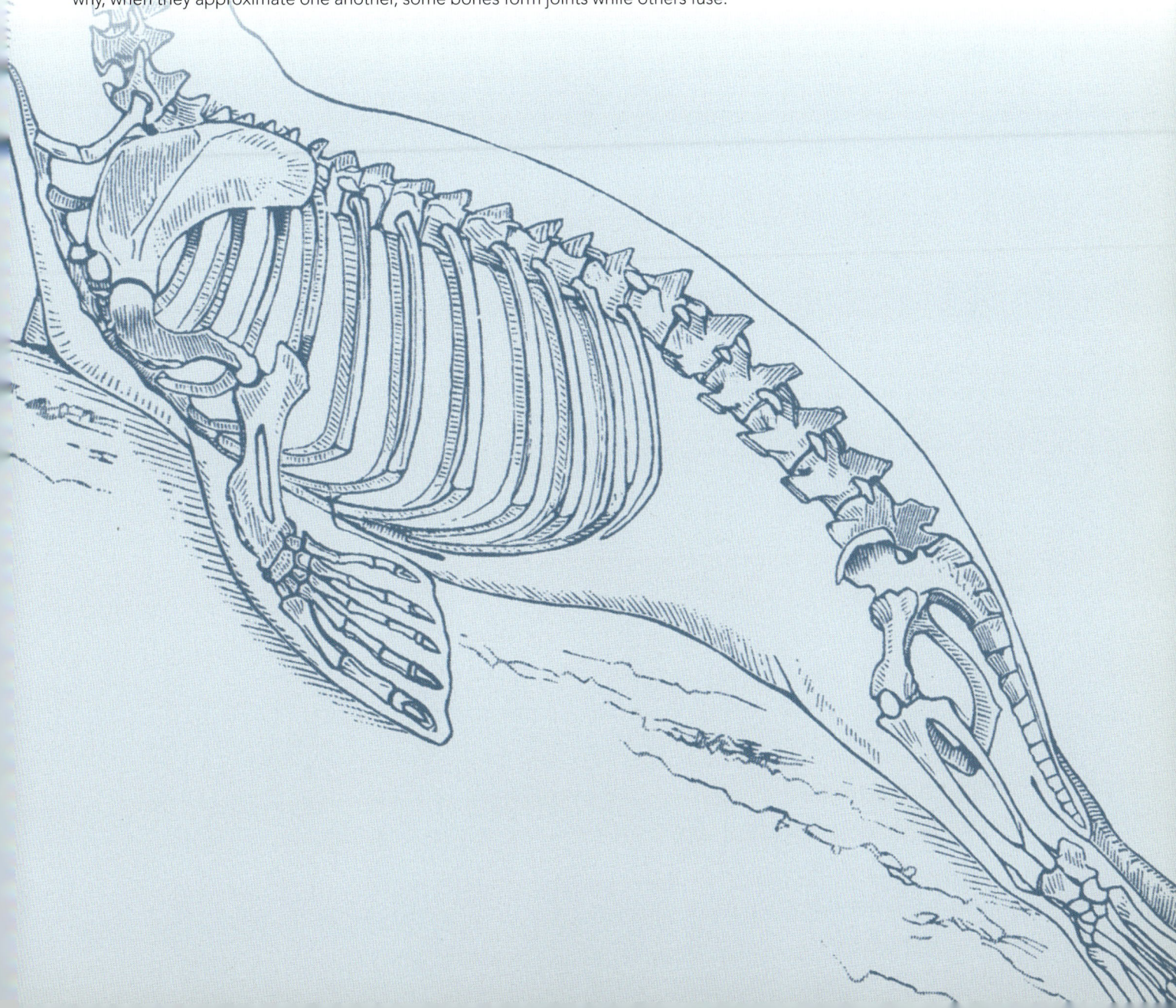

Initiating Skeletal Growth

Chapter 30

To the elbow I go, from the knee I flee.

Morphogenesis and growth, both of which are directional (multidirectional), four-dimensional processes, can be analyzed with reference to axes and polarity. D'Arcy Thompson's (1917) analysis of polarization and mainte-nance of form is a classic example of this approach. Thompson analyzed form as adaptation to the environ-ment, but was unaware of (or chose not to consider), the relationship between ontogenetic adaptation and the inheritance of form and pattern. He modeled change in form as transformation from one adult form to another, whereas new adult forms arise evolutionarily by change in embryo form, as indeed they do during development.

WHAT IS GROWTH?

I am attracted to the definition of growth proposed by Melvin Moss: 'In the broadest sense, growth may be defined as any temporal change in any parameter that is measurable' (Moss, 1972b). Unfortunately, this definition applies also to morphogenesis, and I find it important to distinguish growth from morphogenesis, growth being permanent increase in size and morphogenesis the attainment of form or change in form (shape). In a sense, much of the discussion of the preceding chapters informs skeletal growth and morphogenesis. In this chapter, I con-centrate not on the ways of measuring rates or amounts of growth, but rather on the processes that initiate and regulate growth as a developmental process.

Growth is a composite set of processes, consisting as it does of increase in cell populations and regulation of the physiological activity of those populations. The produc-tion, differentiation and maintenance of scleroblasts, along with deposition of extracellular matrices (ECMs)

(Chapters 2–5), provide major components of skeletal growth, although the mechanisms controlling these processes differ from those that operate during cell prolif-eration.[1] At the very basis of growth as a developmental process are mechanisms that determine how many cells will be available for an individual skeletal element. This topic begins our discussion of growth.

NUMBERS OF STEM CELLS

Proliferation of progenitor cells and their accumulation into condensations represent the earliest growth of the skeleton, growth that is largely determined by the num-ber of stem cells and their rates of proliferation, accumu-lation and/or loss. While we know that condensation size influences the size of a skeletal element (Chapters 19 and 20), little information is available on the number of stem cells required to produce a condensation or a skeletal element. Exceptions are the studies of Beatrice Mintz and her colleagues who created 'tetraparental' mice (cellular genetic mosaics) by fusing blastocyst cells of two strains (C3H and C57BL/6) having morphologically distinct verte-bral characteristics. Individual vertebrae in these chi-maeras also are cellular genetic mosaics, being C3H on one side and C57BL/6 on the other, demonstrating that the left and right halves of each vertebra are of independ-ent origin (Chapter 41). The interpretation given to this finding is that vertebrae are clonally derived, each lateral sclerotomite providing one cell line to each vertebra.[2]

The presence within a single somite of both forms of the dimeric enzyme glucosephosphate isomerase, one dimer from each parent, was used by Gearhart and Mintz (1972) as evidence that individual somites are not derived from a single clone. Resegmentation leads us to the same conclusion (Chapter 16). On this basis, each vertebra would be derived from four clones of cells. Mintz and her colleagues also used their data to argue that vertebral cell lines are established after five days of gestation – when a sufficient number of cells are first present to form all the necessary clones for the vertebral column – and before seven days, which is when somites first appear. A further finding was that the paired bones of the skull – notably the palatines, pterygoids and occipital condyles – occasionally show evidence of mosaic organization, suggesting that they too originate from at least two cell lines.

Growth of the components of each vertebra could be regulated by, or vertebral defects could arise from:

(i) eliminating a clonal line or portion of a clonal line, for example, by apoptosis, resulting in a smaller than normal mesenchymal condensation;

(ii) excess proliferation or duplication of a clonal cell line or part thereof, resulting in a larger than normal mesenchymal condensation; or

(iii) disruption of a whole body or skeletal element gradient as seen, for example, when temperature perturbs vertebral development and changes vertebral number (Chapter 43).

The operation of any of these mechanisms makes it necessary to look for control of growth processes during early embryonic development, not during the 'growth period.'[3]

CELL MOVEMENT AND CELL VIABILITY

Maintaining that cell movement within limb buds is an important component of limb outgrowth, Donald Ede and his colleagues in Glasgow provided ultrastructural evidence of the attachment of mesenchymal cells to, and movement toward, the limb-bud epithelium. Ede's studies on cell adhesion in the *talpid³* mutant are important in this regard (Chapter 39). Culturing limb buds in the presence of vitamin A or applying Vitamin A to pregnant mice, decreases cell motility and so decreases limb size. Vitamin A also initiates apoptosis via caspase-3. Similarly, vitamin A blocks migration and proliferation of neural-crest cells and so inhibits the development of the upper and lower jaws (Box 27.1). Any reduction in cell motility or decrease in surface adhesiveness will tend to increase the generation time for cell division and will appear as a lower labeling index after ^{3}H-thymidine.[4]

Epithelia and Fgf/FgfR-2

From the data discussed in Parts VI and VII we know that proliferation and viability of skeletogenic mesenchymal cells depend on interactions with epithelia.

Within the developing head, facial epithelia are required for mesenchymal viability. This requirement is stage dependent and cell-to-cell contact dependent. Consequently, growth of facial mesenchyme is promoted by influences from facial epithelia. While epithelia from different facial processes in chick embryos are interchangeable in this role, interestingly, frontonasal epithelium promotes the outgrowth of limb buds but mandibular and maxillary epithelia do not.

As can be demonstrated using micromass culture, differential growth of facial primordia is regulated by growth factors. Fgf-2 stimulates growth of embryonic chick frontonasal but not maxillary or mandibular processes, and enhances chondrogenesis from frontonasal but not the other processes. Fgf can *replace* epithelium to promote outgrowth of facial processes.[5]

Epithelia and mesenchyme signal through FgfR-1 and FgfR-2; epithelial FgfR-1 maintains FgfR-2 expression in chick facial mesenchyme. In the absence of the epithelium, *FgfR-2* and type II collagen are down-regulated (Matovinovic and Richman, 1997). Both frontonasal and maxillary processes contain high levels of Fgf-2, but FgfR-2 is high in the frontonasal and lower in the maxillary process, correlating with differential growth of the two facial processes.

Similarly in limb buds, the epithelial *b* variant of *FgfR-2* is active in limb epithelium and in the AER where it plays a role in limb outgrowth. Chimaeric mouse embryos containing a high proportion of embryonic stem cells with a lack of function mutation of *FgfR-2* fail to form limbs. Mutant cells are predominently found in limb epithelium (not the AER) in the subjacent mesenchyme where *Shh* and *Msx-1* are down-regulated. Epithelial expression of *Engrailed-1* and *Wnt-7a* is discontinuous, reduced and ectopic, implying that both are under the control of *FgfR-2*.[6]

METABOLIC REGULATION

We know that use/disuse, amount and quality of diet, nutrition, and metabolic rate all have important influences on skeletal growth in neonates, juveniles and adults. Metabolic regulation is important for initiation and initial growth of skeletal elements. The microenvironment immediately surrounding skeletal cells – local shifts in pH, temperature, pO_2 and pCO_2 tensions, osmolarity, metabolic inhibitors – modify the rates of physiological processes and so alter growth rates.

Little is known about this microenvironment *in vivo*, so it is difficult to determine accurately the relative roles played by each of the factors *in vivo*. Alterations in *pericellular* ECMs are more amenable to analysis and have yielded valuable information on the control of cartilage growth by *positive feedback from matrix products* to chondroblastic and chondrocytic biosynthetic activity. As I discussed when evaluating achondroplasia in Chapter 27, growth is retarded by any disruption in feedback or interactions between ECM products.

To turn to mutants again, the *creeper* (*cp*) mutant in domestic fowl provides a nice example of how metabolic inhibition can severely retard growth and morphogenesis.

Creeper (*cp*) fowl

Creeper (*cp*), a dominant mutation on the C chromosome of the common fowl, is homozygous lethal. Homozygous embryos, which die at the end of the third day, have retarded heads, limbs and extraembryonic vasculature. *Creeper* is genetically linked with *rose comb*, a useful linkage, as *creeper* embryos are often difficult to distinguish from wild-type embryos at early ages. Heterozygotes are viable but phocomelic; proximal limb elements are reduced so that the digits emerge much closer to the body than is normal. Thalidomide taken during sensitive phases of pregnancy can result in a phocomelic phenotype (Chapter 38).[7]

Tibia/fibula

Long bones, especially those of the hind limbs, and especially the tibiae, are severely shortened. The tibiae are severely bent, although the fibulae may be larger than normal. Dissimilarity in the action of *cp* on tibia and fibula is thought to reflect a faster growth rate of the tibia between seven and nine days of incubation in heterozygotes and that *cp* acts by retarding the growth rates. Initially, the prechondrogenic condensations for tibia and fibula are of similar size, differential growth of the tibia subsequently outstripping that of the fibula. This is true even when the two skeletal elements are maintained *in vitro*, suggesting an intrinsic (preprogrammed) pattern of growth that is disrupted in the mutant. Interestingly, maintaining tissues from *cp* embryos *in vitro* revealed that only the heart showed decreased growth, implying that growth retardation in other tissues/organs is secondary.[8]

For a fibula (or any skeletal element) to develop at all, a critical volume of mesenchyme must be present in the early limb bud. Studies published in the 1950s and '60s were interpreted as indicating 'competition' for mesenchyme between presumptive tibia and presumptive fibula. Rotating fibular mesenchyme through 90 degrees and reimplanting it into a limb bud changed the direction of fibular growth, allowing it to grow to the same length as the tibia. Madeleine Kieny obtained experimental data that she interpreted as evidence for a competitive interaction for cells between fibular and tibial primordia, the 'stronger' tibial rudiment taking cells from the 'weaker' fibular rudiment. Hampé subsequently published a fascinating set of experiments on fibular growth that have been taken as having evolutionary implications (Box 30.1).[9]

During development in wild-type embryos, the fibula loses its distal epiphysis by H.H. 27 or 28, a normal part of skeletal development in most birds. Archer *et al.* (1983) argue that the loss of the epiphysis rather than competition

Box 30.1 Hampé's experiments

Armand Hampé published some now-classic experiments, in which he manipulated the developing limb buds of chick embryos and obtained what were interpreted as reversals of skeletal and muscle patterns to those more typical of reptiles.

This seeming ability to create atavisms experimentally is intriguing. Gerd Müller repeated Hampé's experiments by implanting a foil barrier between tibial and fibular rudiments in the hind-limb bud between H.H. 22 and 24. Patterning of the *muscles* changed to patterns typical of reptilian muscle, especially *M. flexor perforans*, *M. popliteus*, *M. fibularis*, and *M. brevis*. Rather than evoking atavisms as if that term described a mechanism (which it does not), Müller invoked a process-oriented interpretation, heterochrony (Chapter 44).

Subsequent experiments in which mesenchyme was excised from limb buds showed that the fibula was especially affected, its condensation being differentially sensitive to reduction; a finding that illustrates the importance of cell number and condensation size for chondrogenesis to be initiated.

Müller and Streicher (1989) used a similar experimental approach in their examination of the ontogeny of the *syndesmosis tibiofibularis* in bird hind limbs. This element appears initially as a cartilage – essentially a sesamoid – and is incorporated subsequently into the tibiofibularis as a bony crest. A similar crest is found in theropod dinosaurs, but not in other tetrapods. Because the cartilaginous precursor is mechanically induced, it fails to form in paralyzed embryos. One can imagine a scenario in which altered movement evoked the cartilage in theropod dinosaurs.

[a] See Hampé (1958, 1959, 1960) for the early studies, and Müller (1985, 1986, 1989, 1991), Streicher (1991) and Streicher and Müller (1992) for the foil barrier and extirpation studies.

between condensations is a more likely explanation for the slowed growth of the fibula and therefore for differences between fibula and tibia – *differential growth, not competition* – although as they also showed, the diameter of the tibia is already greater than that of the fibula at the onset of chondrogenesis, which could result from competition.

Growth retardation

All who have studied *creeper* embryos attribute the slowed long-bone growth to the failure of tibial chondrocytes to hypertrophy on schedule, which in turn slows subperiosteal osteogenesis.[10]

The theory that growth retardation can lead to specific skeletal defects was tested 70 years ago by Fell and Landauer (1935), who cultured limb buds from wild-type chick embryos of 3.5–5 days of incubation in a medium that would restrict their growth. The medium was modified from the one they had developed for organ culture of skeletal elements by diluting the embryo extract. Limb-bud growth was retarded, as now we would perhaps expect, but those were the pioneering days of organ culture, immune to revisionist interpretations.

Of greater interest, although the smaller limb buds produced chondrocytes, they did not hypertrophy. Consequently, the perichondrium failed to transform to a periosteum and so subperiosteal osteogenesis was not initiated. This finding forms the basis of the concept that chondrocyte hypertrophy plays a role in transforming the adjacent perichondrium to a periosteum and initiating subperiosteal osteogenesis (Chapter 22). Mandibular membrane bones cultured on the same growth-restricting medium ossified normally. Fell and Landauer extrapolated these findings from organ culture with wild-type elements to argue that retardation of chondrocyte maturation is sufficient to explain defects seen in creeper mutants in ovo.[11]

This study stimulated research on the cp mutant. Indeed, more studies using in vitro cultivation or transplantation may have been initiated to investigate the mode of action of cp than for any other skeletal mutant. Despite this, the action(s) of cp, like the thalidomide syndrome (Chapter 38), is not fully understood; both are fundamentally defects in growth, and little is known about how the many processes involved in the control of growth are regulated and coordinated. Morphogenesis of cp limb buds remains abnormal after transplantation into wild-type hosts, indicating that cp acts directly on the skeleton.[12]

There is, however, a systemic effect of the creeper gene on the whole embryo that potentiates the action of other agents with influence over skeletal development and growth. As two examples: creeper embryos show earlier and more pronounced rickets in response to vitamin D than do wild-type embryos; and retardation of tibial growth in response to administration of sex hormones (testosterone, oestradiol, dehydroepiandrosterone) is more pronounced in creeper than in wild-type embryos.[13]

A growth inhibitor

Creeper embryos produce a factor that inhibits growth but not morphogenesis of tibiae from wild-type embryos cultured with it. In addition to providing presumptive evidence for a growth inhibitor, this is a nice illustration of how morphogenesis and growth can be controlled independently. Tibial growth is retarded by some 12 per cent when cultured in a medium containing embryo extract from creeper embryos rather than embryo extract from wild-type embryos. Total protein is reduced because of deficient leucine metabolism. Co-culture of wild-type and creeper tibiae retards growth of wild-type tibiae in comparison to their growth when co-cultured with a second wild-type tibia. The tibiae of creeper mutants seem, therefore, to be one source of the growth inhibitor, releasing it in vitro, and perhaps synthesizing it as well.[14]

Tibiae from wild-type embryos show an accelerated accumulation of protein and rate of growth between eight and nine days of incubation. Both parameters are slowed in tibiae from creeper embryos. Protein/μg DNA and hydroxyproline/μg protein decrease at the beginning of the eighth day in creeper tibiae relative to wild-type embryos, but levels of hydroxyproline/μg DNA are within wild-type levels, suggesting that non-collagenous protein is reduced in creeper.[15]

MECHANICAL STIMULATION AND CHONDROBLAST DIFFERENTIATION/GROWTH

Different types of mechanical stress, as listed below, affect skeletal tissues and initiation of skeletal growth differently.

- Cyclic motion and associated shear promotes mitosis.
- High tensile stress promotes the differentiation of fibrous tissue.
- Tensile strain accompanied by hydrostatic compressive stress promotes fibrocartilage.
- Hydrostatic compressive stress promotes chondrogenesis.
- Intermittent loads of low stress or low tensile strain promote intramembranous ossification.
- Intermittent hydrostatic compressive stress inhibits endochondral ossification.[16]

Here are two examples of chondrogenesis and intermittent load. In the first, chondrogenic cells from limb buds of H.H. 23 or 24 chick embryos are cultured in agarose and exposed to static compressive loading (a constant 4.5-kPa stress) or cyclic compressive loading (9 kPa peak at 0.3 Hz). Cyclic loading doubles the number of cartilage nodules that form (doubling the number of chondrogenic cells) and increases incorporation of ^{35}S. Static loading has little effect on either parameter (Table 30.1). In the second, chick epiphyseal chondrocytes cultured under high cell density produce more proteoglycan and more aggregated proteoglycan when exposed to intermittent compression than when not. On the other hand, cultured embryonic chick tibiae respond to 20 minutes of continuous exposure to 0.4 Hz by maintaining levels of alkaline phosphates (levels decline in controls) and with enhanced synthesis of type I collagen.[17]

For progenitor cells to respond to mechanical stimuli by initiating chondrogenesis rather than osteogenesis, the mechanical stimulus must be intermittent (Box 29.1). Several classes of independent evidence support this

Table 30.1 Average number of cartilage nodules formed after two hours of static or cyclic loading of chondrogenic cultures from wing and limb buds of chick embryos[a]

Area (μm^2× 10^3)	Nodules/10 mm^2		
	Control	Static loading (constant 4.5 kPa)	Cyclic loading (0.25–9 kPa at 0.3 Hz)
0–2	7.5	9	19
2–4	7	6	22
4–6	7	6	16.5
6–8	7	8	13
8–10	3	3	6
10 +	11	5.5	8

[a] Based on data from Elder et al. (2000).

statement, as follows.

- The number of chondrogenic cells entering the S phase of mitosis is regulated by intermittent compressive forces.
- Progenitor cells within mesenchyme take up lipid and modulate to fat cells (adipocytes) in the absence of mechanical stress, but synthesize sulphated GAGs and differentiate into chondroblasts in response to high rates of change in compression.
- Sensitivity to intermittent compression induces cardiac cartilage formation (Box 17.2).
- Bipotential progenitor cells on avian membrane bones switch to secondary chondrogenesis in response to intermittent mechanical forces. I explored this class of evidence in Chapter 12.[18]

MECHANICAL STIMULI AND METABOLIC ACTIVITY

If progenitor cells differentiate into chondroblasts instead of osteoblasts or fibroblasts in response to intermittent mechanical stimulation, we might expect particular cellular metabolic activities to be especially sensitive to changes in the mechanical environment surrounding progenitor cells. Several types of response have been documented.

(i) Hyaluronan, a molecule already shown to play a role in condensation (Box 4.1, Chapter 20), is extremely sensitive to mechanical stresses. Indeed, seeking any correlation between cellular constituents and biomechanical data from the same cells reveals that GAGs are often the only component whose levels correlate with the mechanical data.

(ii) Synthesis of hyaluronan, collagen, chondroitin-6-sulphate and DNA all are enhanced when smooth muscle cells are stretched on membranes, particularly when stretching is cyclical.

(iii) Rat calvarial osteoblasts respond to intermittent or constant stretching with enhanced cell division and enhanced synthesis of non-collagenous proteins.

(iv) Osteoblast proliferation, behaviour and collagen synthesis all respond to mechanical stresses,

(v) Avian calvarial osteoblasts respond to cyclic tension *in vitro* by increasing DNA synthesis and enhancing mitosis.

(vi) Osteoblasts exhibit behavioural changes by orienting at 90° to the imposed strain.

(vii) Subjecting cranial bones from embryonic chicks to high levels of compression *in vitro* slows conversion of procollagen into collagen, while

(viii) Filling root canals stimulates collagen synthesis in the alveolar bone that anchors the teeth to the jaw.[19]

Transduction

Obviously, levels of major ECM components of cartilage and bone vary in accordance with the mechanical environment, reflecting the responses of cells to alterations in that environment. How do skeletal cells sense those changes? Bioelectrical changes are one way by which cells could transduce mechanical stimuli into signals they can recognize. Other possibilities include *indirect* sensing by changes in blood vessels or nerves resulting from the liberation of metabolites, or *direct* sensing by membrane components of skeletal cells – membrane receptors act as binding sites for ions, hormones and macromolecules.

The periostea of mammalian long bones are innervated by postganglionic sympathetic neurons. Adrenergic (norepinephrine-based) innervation in neonatal rat calvariae is restricted to the bone and not found in the sutures. Adrenergic neurons arise from the neural crest and their differentiation is enhanced in culture media conditioned by neural tubes, but not in media conditioned by notochord or somites. Given that the sutures of mouse skull are neural crest in origin (Chapter 17), one wonders whether sutures might inhibit adrenergic neuronal differentiation.[20]

With respect to periosteal–nerve interactions, Asmus *et al.* (2000) examined the ability of periostea to change the transmitter properties of the sympathetic neurons that innervate the periosteum. Working from knowledge that sweat glands modulate neurons from noradrenergic to cholinergic (acetylcholine-mediated) and peptidergic, they showed that rat sternal periostea could modify neurons from cholinergic, which is their state when they reach the periosteum, to acetylcholine-secreting neurons after contact with the periosteum. As a final proof of this role, periostea transplanted to the skin induce skin neurons to switch from noradrenergic to cholinergic and peptidergic.

Sensory and autonomic innervation augments osteoblastic activity; reduced incorporation of ³H-proline into hydroxyproline for collagen synthesis is seen in mandibular and femoral diaphyseal osteoblasts following resection of the inferior alveolar nerve or chemically induced sympathectomy. As is so often the case, alveolar bone responds differently; surgical sympathectomy in rats induces resorption at the base of the incisors within a day. Do cell also have receptors/mechanisms to allow them to respond directly to changes in the mechanical characteristics of the environment?[21]

It has been suggested that *integrins*, which play an important cell-signaling role in condensation (Fig. 5.1), also act as mechanochemical transducers and transducers of mechanical signals. As assessed using wing-bud cells an association between focal adhesion kinase and fibronectin is required for precartilage condensation, condensation being enhanced via β-integrin-mediated interaction with fibronectin via tyrosine phosphorylation of focal adhesion kinase (Bang *et al.*, 2000).

The proposal for integrins and mechanical transduction arises, in part, from the theory of *tensional integrity vs. compressional continuity* used in architecture. Application of the principle of tensional integrity goes back to the Romans, the Roman arch being held in place by gravity pulling the arch tensionally downwards.

R. Buckminster Fuller pulled exactly the opposite way when he argued, not for structural weight providing the compressive continuity, but for a structure being pulled outwards (and thus supported) by tensional forces inherent in and restrained by the structure. Ingber extends this theory to the cytoskeleton in his theory of *cellular tensegrity*. Although little empirical information is available, it is known that integrin β1 promotes the formation of focal adhesions in mediating changes in the ECM to the cell surface and then to cytoskeletal mechanotransduction. Integrin subtypes change as osteoblasts differentiate but- whether such changes drive or react to changing mechanotransduction was unclear until Ichiro Takahashi and his colleagues (2003) demonstrated that both β1 integrin and focal adhesion kinase are up-regulated in experimentally expanded (stretched) midpalatal sutures of rats.[22]

Membrane potential

Depending on the amount of deformation to which their membranes are exposed, the electrical potential of fibroblast cell membranes varies from −8 to −17 mV (Bard and Wright, 1974).

From the information on transmembrane potential we can conclude that osteoblasts possess metabolic pumps, the resting potential of osteoblasts cell membranes being quite similar in different cell populations *in vitro*: −20.3 ± 3.8 mV, with a range of −11 to −30 mV for isolated murine calvarial osteoblasts *in vitro* vs. −16.9 ± 0.64 mV for osteoblasts from cortical endosteal rabbit long bones. The resting potential of osteoblasts from rabbit parietal bones *in situ* appears to be much lower − −3.93 mV.[23] Surface osteoblasts sense strain via *electric coupling* between adjacent cells; membrane polarization being responsive to hormones, PTH eliciting depolarization and calcitonin hyperpolarization.

If mechanoreceptors exist − and there is evidence that they do − then there must be a mechanism(s) for transducing the mechanical stimulus into a biochemical signal(s) to which cells can respond. Transduction via cAMP − activation of adenylate cyclase and alteration in intracellular levels of cyclic AMP − is an attractive possibility.

SKELETAL RESPONSES MEDIATED BY cAMP

A molecule with an ubiquitous distribution, adenosine 3′, 5′-monophosphate or cyclic AMP (cAMP) mediates external signals received via a cell-surface mechanism, thereby regulating intracellular metabolic activity in a great diversity of cell types and over a wide range of metabolic activities, including but by no means limited to:

- induction of flagellar protein and sugar-digesting enzymes in bacteria, with starvation as the external signal;
- functioning as the attractant (acrasin) in the aggregation of slime moulds, again with starvation as the stimulus;

- aggregation of polymorphonuclear leukocytes;
- catalyzing phosphorylation of glycogen to glucose 1-phosphate; and
- mediating synaptic transmission, activation of melanophores, the immune response, neural induction, cell division, assembly of microtubules and microfilaments, and release and activation of steroid hormones.

A stimulus perceived by a cell activates the synthesis of membrane-bound adenylate cyclase and catalyzes the transformation of ATP into cAMP, which activates inactive enzymes and protein kinases to stimulate metabolic processes. In the presence of phosphodiesterase, cAMP is transformed into the inactive 5′-AMP (adenylic acid). Therefore, levels of cAMP are *increased* via activation of adenylate cyclase or inhibition of phosphodiesterase, and *decreased* by inhibition of adenylate cyclase or activation of phosphodiesterase. Levels of cAMP-phosphodiesterase are elevated in limb mesenchyme in the mouse mutant *Hemimelia-extra toes* (Hm^x), an elevation that could enhance mitosis and initiate the abnormal outgrowth.[24]

Matrix synthesis and condensation

Alterations in cAMP levels can affect the synthesis of ECM products characteristic of skeletal cells. Dibutyryl 3′, 5′-cAMP modulates transformation of hamster ovarian cells into fibroblasts and induces synthesis of collagen within those transformed cells, increasing rates of synthesis and secretion of GAGs by transformed fibroblasts, and blocking hyaluronan-induced inhibition of chondrogenesis by somitic cells maintained *in vitro*.[25]

cAMP increases cell-to-cell communication as cells condense, stimulating chondrogenesis of limb mesenchymal cells in a density-dependent mechanism, *provided that* the cells are cultured at high densities (high being 2.5–5.0 × 10⁴ cells/10 μl; low 1–2 × 10⁵ cells/10 μl). Importantly, limb mesenchymal cells can chondrify without condensation if cAMP is added to the medium.[26]

cAMP can regulate collagen synthesis. In the presence of dibutyryl cAMP or 1.8 mM $CaCl_2$, rabbit articular chondrocytes in suspension culture can be induced to switch from synthesis of type II to synthesis of type I collagen. Direct application of dibutyryl cAMP to clonal myogenic cells depresses myogenesis and stimulates the synthesis of collagen and GAGs.[27]

However, chondrocytes from *nanomelic chicks* − which lack almost all cartilage-specific GAG (Chapter 23) − respond to pressure by altering cAMP in the same way that wild-type chondrocytes respond, which suggested to Bourret *et al.* (1979) that cartilage-specific GAGs play no role in the chondrocytic response to mechanical stress.

Hormones

Hormones such as thyrocalcitonin, thyroxine, and growth and parathyroid hormones elevate intracellular levels of cAMP in the entire skeleton and in isolated populations of osteoblasts, although the mechanisms vary. Johanne Heersche and his co-workers showed that elevated levels of cAMP in rat calvariae exposed to calcitonin are due to activation of adenylate cyclase. cAMP is elevated following treatment of calvariae with dibutyryl 3', 5'-cAMP because of inhibition of phosphodiesterase.[28]

I have not discussed the possible role of *cyclic GMP*, which is antagonistic to cAMP in hormone induction. Levels of cGMP in skeletal tissues follow the pattern of cAMP. Facial bones respond to PTH by a rapid elevation of the levels of *both* cyclic nucleotides; over a one-hour period, cGMP increases by 750 per cent, cAMP by 150 per cent. In mechanically compressed tibiae, cGMP decreases in the HCC zone, but increases in the proliferative and resting chondroblast zones, paralleling changes in cAMP.[29]

Teeth and alveolar bone

Davidovitch and Shanfeld assayed levels of cAMP in tibiae from cats and chicks – in which levels range from 0.2 to 0.4 pmol cAMP/mg wet wt – and in alveolar and mandibular basal bone of cats, in which levels are somewhat lower at 0.1–0.3 pmol/mg wet wt. Orthodontic tipping of cat canines, by applying an initial force of 100 g by elastic and maintaining the elastic in position for seven or 15 days, increases levels of cAMP by 50–130 per cent. When 80 g of initial force is applied with a coiled spring, an initial (24-hour) decrease in cAMP levels is followed by a prolonged period (28 days) when cAMP levels are some 50 per cent above the levels in untipped canines. When cAMP is localized intracellularly using an immunohistochemical reaction, however, only a few cells have elevated levels.[30]

So, we know that cAMP is stimulated at pressure sites when cat canines are tipped with orthodontic force. We do not know (i) the precise cellular response that causes this elevation, (ii) which cells produce the response, (iii) how the mechanical stimulus is perceived by these cells, or (iv) whether alteration in cAMP levels is causally related to the remodeling that follows orthodontic tipping. One possibility is that stress applied to alveolar bone alters the electrical environment, which in turn regulates the differentiation of osteogenic progenitor cells. Certainly, application of electric current as low as $15 \pm 2\ \mu A$ to feline alveolar bone increases cAMP and cGMP levels and the numbers of cAMP- and cGMP-positive cells.[31]

Electrical stimulation

Only sparse information is available on how cAMP levels change in response to electrical stimulation. Norton *et al.* (1976) exposed embryonic tibiae to electrical stimulation and found a 20 per cent increase in cAMP, *provided that* the long axes of the tibiae were parallel to the applied electrical field. They speculated that receptors on the cells or within the ECM were oriented along the long axis

of the bones, which would provide an elegant mechanism to direct the response along the growing axis.

cAMP AND PRECHONDROBLAST PROLIFERATION

Prechondroblasts can translate pressure changes into altered levels of cAMP. Three situations investigated are long-bone development, limb regeneration and condylar cartilage.

Long bones

Gideon Rodan and colleagues adapted a tuberculin syringe to deliver a known compressive force to embryonic tibiae *in vitro* and used it to evaluate the effects of compression on glucose metabolism and ^{3}H-thymidine incorporation into DNA. Subsequently, they used the device to show that whole tibiae, slices of tibial epiphyses, and isolated epiphyseal cells from 16-day-old chick embryos all respond to a pressure of 60 g/cm^2 by accumulating or losing cAMP and cGMP. A preferential decrease in cAMP in the proliferative zone matches the decrease in whole tibiae, meaning that *the change in prechondroblasts is sufficient to explain the decrease in the entire skeletal element.*

Prechondroblasts from the proliferative zones and hypertrophic chondrocytes (HCCs) from embryonic chicks tibiae were then exposed to the same pressure for 15 minutes. cAMP levels were 20 per cent lower in the pressure-treated proliferating cells than in non-pressure-treated, although levels of HCCs were unaffected (Table 30.2). This response to short exposure to pressure on proliferating chondroblasts was mimicked by the calcium ionophore A23187 *in the absence* of pressure, and the effects of pressure on cAMP levels abolished in the presence of the chelating agent EGTA (Table 30.2). Bourett and Rodan concluded that an increase in Ca^{++} concentration mediates a decrease in cAMP levels within proliferating cells.[32]

Table 30.2 A summary of the levels of cAMP in proliferating prechondroblasts from tibiae of 16-day-old chick embryos after exposure to pressure (60 g/cm), the calcium ionophore A23187, and/or the chelating agent EGTA[a,b]

Treatment	cAMP[c]	Per cent of control
Krebs–Ringer–glucose (control)	4.08 ± 0.37	
Pressure	3.21 ± 0.36	−21
A23187	3.47 ± 0.47	−15
Pressure + A23187	3.48 ± 0.42	−15
EGTA	6.93 ± 0.31	+70
Pressure + EGTA	6.79 ± 0.11	+66

[a] EGTA, ethylenebis(oxyethylenenitrilo)tetraacetic acid.
[b] Data adapted from Bourret and Rodan (1976a,b).
[c] pmol cAMP/10^6 cells, mean ± SEM, based on 6–12 replicates.

However, pressure increased calcium levels within HCCs without altering cAMP. This is because the adenylate cyclase within HCCs is not sensitive to changing calcium levels; calcium inhibits adenylate cyclase in plasma membrane preparations derived from proliferating chondroblasts, but not in membrane preparations from HCCs (Rodan et al., 1977). Therefore:

- short, 15-minute exposure to physiological pressure (60 g/cm^2) enhances Ca^{++} uptake into chondroblasts and chondrocytes;
- proliferating chondroblasts respond to increased Ca^{++} levels by a decrease in adenylate cyclase and lowering cAMP levels; and
- the effects of pressure are localized on immature cells by the loss of the ability to respond to heightened Ca^{++} as chondroblasts differentiate.

An intriguing recent suggestion raises *Indian hedgehog* (*Ihh*) in transduction of mechanical stimuli that influence chondroblast proliferation. *Ihh* is known to play a role in regulating proliferation (Fig. 15.2). Q.-q. Wu and colleagues (2001) have shown that cyclic mechanical stress induces expression of *Ihh* by chondroblasts, and the expression is abolished by gadolinium, an inhibitor of stretch-activated channels. Block *Ihh* during mechanical loading of chick sternal chondrocytes and the stimulatory effect of loading on proliferation is abolished. *Bmp-2* and *Bmp-4* are also up-regulated in response to mechanical loading, an up-regulation that is mediated by *Ihh*.

Limb regeneration

Levels of cAMP are high in blastemata during the first 10 days of regeneration of amputated amphibian limbs, possibly aiding the accumulation of immature blastemal cells, perhaps by enhancing their proliferation. cGMP has been implicated in regulating proliferation during amphibian limb regeneration and cAMP in proliferation of chondrogenic cells.[33]

Condylar cartilage

When occlusal changes are created in rat mandibular condyles, cells in the intermediate but not in the hypertrophic layer show reduced levels of cAMP, but enhanced levels of cGMP associated with proliferation.

Continuous compression applied to condylar cartilage from four-day-old rats maintained *in vitro* reduces the levels of both acid and alkaline phosphatase, increases proliferation, and decreases ^{35}S incorporation into acid mucopolysaccharides. *Intermittent compression* increases alkaline phosphatase in the hypertrophic zone, decreases proliferation, and increases ^{35}S incorporation, inhibiting condylar cartilage growth beyond that seen with continuous compression (Table 30.3). Proliferation is inhibited with

Table 30.3 Effects of continuous and intermittent compression on behavior of mandibular condylar cartilage isolated from four-day-old rats and maintained *in vitro*[a]

Parameter measured	Continuous compression	Intermittent compression
Alkaline phosphatase activity	Decreased in hypertrophic and erosive zones	Increased in hypertrophic zones
Acid phosphatase activity	Decreased	Unaltered
Prechondroblast proliferation	Increased	Decreased
Incorporation of ^{35}S-sulphate	Decreased	Increased

[a] Based on data from Copray et al. (1985a,b).

intermittent compression. At least some of these changes are mediated by cyclic nucleotides:

- an increase in intracellular cAMP is associated with reduced proliferation and enhanced hypertrophy, i.e. with *enhanced differentiation*; while
- an increase in intracellular cGMP is associated with increased proliferation without affecting hypertrophy, i.e. *enhancing proliferation* without action on differentiation.[34]

Alteration in intracellular cAMP is then an attractive means of allowing proliferating chondroprogenitor cells to respond to mechanical changes in their environment and transform to chondroblasts to initiate cartilage growth.

NOTES

1. See Cook (1974), Bryant and Simpson (1984), Atchley and Hall (1991) and Hall (1991c) for control of skeletal growth. These processes are sometimes spoken of as *epigenetic*, meaning factors that influence genetic activity. The science of epigenetics is concerned with the causal analysis of development and, in particular, with the mechanisms by which genes express their phenotypic effects (Hall, 1983g, 1984a–c, 1998a). Murray and Selby (1930), Storey (1972), Moss (1972b,c), the volume edited by Moffett (1972) and Bryant and Simpson (1984), Hall (1985a), Atchley and Hall (1991), Storey et al. (1992) and Herring (1993, 1994) provide excellent discussions of the relative contributions of intrinsic and extrinsic factors to the growth of cartilage and bone and facial development. For the importance of an epigenetic approach to the evolution of the skeleton see Hall (1983g, 1984d,e, 1987a, 1990b,c, 1991d, 1995a, 1998a).

2. See Mintz (1971, 1972), Gearhart and Mintz (1972) and Moore and Mintz (1972) for blastocyst fusion as an approach to skeletal cell lineage. Severe criticism of our ability to determine the time of clonal initiation using this technique was

raised by Lewis *et al.* (1972) and McLaren (1972). Cartilage and bone also develop when blastocysts are cultivated *in vitro* or ova are grafted into the testes (Stevens, 1968; Hogan and Tilly, 1977).

3. See Grüneberg (1954), Grüneberg and des Wickramaratne (1974) and Cooke (1975) for the three possible mechanisms. See Laird (1966) and Saxén (1976) for development of a similar concept in the context of comparative growth analysis and teratogenesis, and see Atchley and Hall (1991) for a recent model.

4. See Ede and Agerback (1968), Ede and Law (1969), Ede *et al.* (1974) and Wilby and Ede (1975) for cell movement in chick limb buds, Ali-Khan and Hales (2003) for apoptosis, Kwasigroch and Kochhar (1975) for the mouse studies, and Riley (1974) for the link with generation time.

5. See Minkoff and Kuntz (1977, 1978), Minkoff (1980, 1984) and Minkoff and Martin (1984) for the growth of craniofacial processes, Richman and Tickle (1989, 1992) for the role of facial epithelia, Richman and Crosby (1990), Richman *et al.* (1997) and Richman and Lee (2003) for Fgf-2, and Hall (1980a), Coffin-Collins and Hall (1989), Saber *et al.* (1989) and Hall and Coffin-Collins (1990) for epithelia and mesenchymal viability.

6. See Gorivodsky and Lonai (2003) for the chimaeras. ES cells that form embryoid bodies can redifferentiate as chondrocytes that hypertrophy, mineralize and are then associated with an osteoblast-like phenotype, a redifferentiation that is blocked by Tgfβ-3. Clonal cultures established from these chondrocytes express an adipogenic potential (Hegert *et al.*, 2002).

7. See Landauer and Dunn (1930), Landauer (1932a,b, 1934), David (1936), Lerner (1936), Hamilton (1941), Rudnick (1945b), Kieny and Abbott (1962), Elmer (1968a,b), Kuroda and Shibuya (1969) and Dinner (1970) for the *Creeper* phenotype. See Landauer (1932b) for the link to *rose comb*.

8. See Hicks and Hinchliffe (1979) for the organ culture of the tibia and fibula from six-day-old embryos in various combinations. I have a typescript of the paper presented by Martin Hicks and Richard Hinchliffe in 1979, and Hinchliffe and Johnson (1983, pp. 276–280 and Figures 7–10 therein) discuss this study. See David (1936) for the culture of other tissues.

9. See Wolff and Hampé (1954) and Hampé (1958, 1960) for the rotation experiments, and Kieny (1957) for competition.

10. See Landauer and Dunn (1930), Landauer (1932a), Lerner (1936), Cock (1966) and Perlov (1968) for growth retardation, and Perlov (1968) for tibial hypertrophy.

11. See Van Limborgh (1970, 1972, 1982) for useful discussions of the differing controls operating in the growth of cartilage and bone. The *Cp* strain that arose in Japan is said to show praecocious differentiation of chondrocytes coupled with decreased rates of proliferation of precursor cells (Kuroda and Shibuya, 1969; Shibuya *et al.*, 1972; Shibuya and Kuroda, 1973). However, I am unable to completely follow their evidence.

12. Hamburger and Waugh (1940) showed that, when grown as transplants, limb buds from *Cp* embryos were less retarded than limb buds from normal embryos. Accordingly, they cautioned against using transplantation as the means to analyze growth and its control. Similar **cautions** were voiced by Hall (1978d) in the context of chorioallantoic grafting;

both techniques can impose space limitations that impede full growth potential.

13. See Hamburger (1941) and Rudnick (1945b) for delayed morphogenesis, Wallace *et al.* (1969) for the systemic effect, Landauer (1934, 1969a) for induced rickets, and Scopelliti (1975) for response to sex hormones. Discussions of the potentiation of agents influencing skeletal growth may be found in Hall (1972a,e, 1977b).

14. See Wolff and Kieny (1957, 1963), Kieny and Abbott (1962) and Elmer (1968a,b) for these metabolic studies.

15. See Elmer (1968b), Perlov (1968) and Dinner (1970) for these studies, and Loewenthal (1957) for some histochemical data on young embryos.

16. See Carter *et al.* (1998a). Dennis Carter has promoted the role of intermittent stress in skeletal differentiation, assigning a role to mechanical forces early in ontogeny (Carter, 1987; Carter *et al.*, 1987, 1988, 1991, 1996; Carter and Wong, 1988a,b; and Carter and Orr, 1992). Carter also argues that if mechanical loading is important in bone development, it must also play an important role in bone evolution. See Wright *et al.* (1992) for the response of chondrocyte cell membranes to hydrostatic pressure.

17. See Elder *et al.* (2000) and Van Kampen *et al.* (1985) for the two studies with intermittent load, and Zaman *et al.* (1992) for the study with chick tibiae. Burger *et al.* (1991) provides a good review of mechanical stimulation modulating osteogenesis from foetal bones *in vitro*, especially onset of movement and its relation to mineralization *in vivo* and *in vitro*.

18. See Veldhuijzen *et al.* (1979) for proliferation and Rodbard (1970) for modulation to chondroblasts.

19. See Åkeson *et al.* (1974) and Woo *et al.* (1975) for data on levels of GAGs and mechanical stimulation, Leung *et al.* (1976, 1977) for culture of smooth muscle cells, Hasegawa *et al.* (1985) for the response of calvarial osteoblasts, Buckley *et al.* (1988) for the data from avian osteoblasts, Ehrlich and Bornstein (1972) for procollagen, and Espie (1975) for responses to root canals. See Cowin *et al.* (1992a) for an excellent analysis of mechanosensory system in bone.

20. See Hohman *et al.* (1986) and Alberius and Skagerberg (1990) for long bones and calvariae, Howard and Bronner-Fraser (1985) for the culture study, and Herskovits *et al.* (1991) for innervation of bone.

21. See Singh *et al.* (1981, 1982) for the ^{3}H-proline study, Sandhu *et al.* (1990) for the incisor study, Herskovits *et al.* (1991) for an overview of innervation, and Roth (1973), Edelman (1976), Salomon and Pratt (1976), Hogg *et al.* (1980), Jones *et al.* (1981), Maclean and Hall (1987), Knudson and Knudson (1991), Matzner *et al.* (1992) and Killian *et al.* (1993) for membrane receptors.

22. See Ingber (1991, 1993) for cellular tensegrity, N. Wang *et al.* (1993) for promotion of focal adhesions, and Bennett *et al.* (2001) for changing subtypes.

23. See Schusterman *et al.* (1974) and Chow *et al.* (1984) for the resting potential data, Harrigan and Hamilton (1993) for strain detection, and Jeansonne *et al.* (1978) for polarization.

24. See Knudsen *et al.* (1985) for *Hemimelia-extra toes*, now known to be a defect in a *cis*-regulatory element that controls the express of *Shh* (Sagai *et al.*, 2004) and that is conserved in tetrapods and in teleosts.

25. See Hsie *et al.* (1971) and Goggins *et al.* (1972) for the fibroblast data, and Toole (1973a) for somatic chondrogenesis.

26. See Hattori and Ede (1985) and Rodgers *et al.* (1989).

27. See Deskmukh and Sawyer (1977) and Schubert and Lacorbiere (1976) for the studies with articular chondrocytes and myogenic cells.

28. See Toole (1973a), Rodan and Rodan (1974) and Davidovitch *et al.* (1977b) for hormonal elevation of cAMP, and Heersche *et al.* (1971, 1974) and Heersche (1972) for the mechanism.

29. See Davidovitch *et al.* (1977a,b) for the cGMP in facial bones, and Rodan *et al.* (1975b) for the data on tibiae. As discussed in Chapter 15, cGMP mediates the action of nitric oxide in enhancing fracture repair.

30. See Davidovitch and Shanfeld (1975), Davidovitch *et al.* (1976a, 1978a,b) and Shanfield *et al.* (1975) for levels of cAMP in bone, Davidovitch (1973) and Davidovitch and Shanfield (1975) for the approaches using orthodontic appliances, and Davidovitch *et al.* (1976a,b, 1978b) and Gustafson *et al.* (1977) for the response being restricted to only a few cells.

31. See De Angelis (1970) and Davidovitch *et al.* (1980) for these studies on alveolar bone.

32. See Rodan *et al.* (1975a,b) for the method and application to tibiae, and Bourett and Rodan (1976a,b) for the subsequent studies.

33. See Jabaily *et al.* (1975) and Liversage *et al.* (1977) for regeneration, and Burger *et al.* (1972) and Rodan *et al.* (1975, 1978) for cAMP and proliferation. See McMahon (1974), Whitfield *et al.* (1973), Rebhun (1977) and Rasmussen and Goodman (1977) for overviews of the interplay between Ca^{++} and cAMP in regulating cell proliferation and differentiation. Liversage *et al.* (1977) documented the possible involvement of cGMP in proliferation during amphibian limb regeneration.

34. See Ehrlich *et al.* (1980) for occlusal changes, and Copray (1984), Copray and Jansen (1985), Copray *et al.* (1985a–c) and Kantomaa and Rönning (1992) for continuous and intermittent compression.

Form, Polarity and Long-Bone Growth

We stimulate our bones with every breath, bite or step we take. Respiration and locomotion produce cycles of bone deformation as tension and compression are applied alternately to vertebrae, ribs and long bones.

Although environmental factors play a role in morphogenesis, the embryonic rudiments of bones and cartilages exhibit many of their characteristic morphological features even when allowed to develop in such unnatural environments as organ culture or when implanted subcutaneously, intramuscularly or intracerebrally or grafted onto the chorioallantoic membrane (CAM) of a host embryo. Morphogenesis has major intrinsic components.[1]

FUNDAMENTAL FORM

Attainment of the *fundamental form of an endochondral bone* – presence and position of condyles, articular surfaces, tuberosities and grooves – is independent of functional demand and biomechanical factors; the basic form is outlined in the three-dimensional organization of the cartilaginous model that precedes bone formation. Congenital bowing of human long bones and experimental production of bowing in rats are both associated with hypertrophy of the periosteum on the concave (bowed) surface and manifest earlier in the cartilaginous model.

The *basic form of subperiosteal intramembranous bone* is laid down by osteogenic precursors in the periosteum, although initiation of their osteogenic activity depends upon stimuli from the adjacent cartilage (Chapter 22). Both aspects are illustrated nicely when tibial periosteum is grafted over defects in the tracheal cartilage of rabbits. The transplanted periosteum continues to produce bone, but osteogenesis is intramembranous and the bone forms as flat sheets.[2]

Development and maintenance of *minor morphogenetic features* of the skeleton – e.g. ridges for attachment of muscles and ligaments – depend much more upon the functional demand to which the skeleton is subjected.[3]

The blueprint for skeletal form contained within the genes is neither expressed rigidly nor without reference to context. The skeleton of a particular individual or a particular skeletal element can adapt to its owner's way of life, including life *in utero*. Although the four young in each litter of the armadillo result from the fission of a single zygote, their features are not identical. In other species in which litter size varies from female to female or from pregnancy to pregnancy, maturity of the skeleton at particular developmental ages is greater when litter size is small, and reduced when litter size increases and body size is reduced (Table 39.1).

The uterine environment modifies form, function and growth. For example, mice transferred to the uteri of females from different strains develop differences in the size of the craniofacial region, including strain-specific effects on mandibular and incisor growth. The implication is of strain-specific uterine effects, but one has to be careful with such an interpretation. There is the equivalent of a placebo effect in which a beneficial effect is gained no matter what the pill contains. The act of transferring an embryo to another uterus reduces mandibular growth *no matter what the genotype of the host*.[4]

The uterine environment certainly is no protection against many environmental influences. Drugs such as thalidomide that cross the placenta are one class of influences (see Chapter 38 for thalidomide). Even sound can influence osteogenesis *in utero*. High sound levels inhibit osteogenesis, as demonstrated by a study in which pregnant rats were exposed to 74–94 decibels – the noise pollution level for humans – for 10 per cent of each

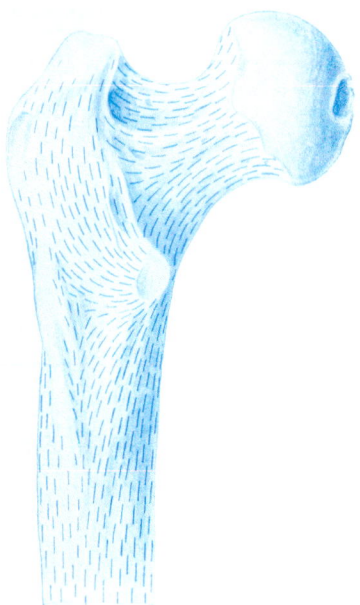

Figure 31.1 The arrangement of the trabeculae in the proximal region of the human femur. Also see Figure 2.6 from Murray (1936).

24-hour period delivered from 'a wide variety of electric horns, gongs, and alarm bells ... along with six 10-inch speakers . . .'[5]

Structural evidence within trabecular bone of the axes along which bone growth and form were moulded has long been available (Fig. 31.1). Orientation of collagen fibres (Wolff's Law), and permanent electrical polarization of the skeleton provide additional evidence of the directional nature of these processes. Effects of gravity on the skeleton are discussed in the following chapter.[6]

POLARITY

Despite evidence that morphogenesis and growth are polarized processes, little attention has been paid to analyzing how polarity and axial symmetry are expressed during differentiation. Tantalizing information is available, however, from dental laminae and tooth development.

The dental laminae of larval salamanders are *unpolarized*, as evidenced by the finding that tooth regeneration can occur in either direction, from anterior to posterior or from posterior to anterior. Adult dental laminae are polarized. Regeneration still occurs in adults, but only from posterior to anterior. The hormonal milieu associated with metamorphosis plays a role in this restriction of polarity. But so does the associated bone; the dental lamina in *Ambystoma mexicanum* is only active when in the presence of bone, otherwise it remains encased in oral epithelium.[7]

Goss and Stagg (1958) demonstrated that the lower jaws of adult newts can regenerate following complete amputation of the jaw, the regenerating tissues arising from a blastema rather than from metaplasia of existing differentiated cells; see Hay (1970) for metaplasia during limb regeneration.

When Graver (1978) re-amputated jaws of the eastern newt, *Notophthalmus viridescens*, that were already one-third of the way through regeneration, he found that the bone of the mandible and the dental lamina could both regenerate more than once, and that regrowth was more rapid in the re-regenerates. To Graver, this demonstrated that recently differentiated cells in a regenerating structure can dedifferentiate to form a blastema more readily than cells that had differentiated longer ago. This important observation suggests a possible experimental approach to the question of how the stability of differentiation and morphogenesis are maintained.[8]

Molars are reduced in size in the mouse X-linked mutant, *Tabby*, and there is abnormal proliferation of the dental lamina anterior to the first molar. Positioning and formation of the other molar teeth appears to be determined by influences from the first molar tooth germ. Intraocular grafting of the first molar germ initiates the formation of three molar teeth, but only one tooth develops if molar two or molar three is grafted alone. In insightful recent analyses, Renata Peterková and her colleagues in Prague identified five distinct morphotypes within the dentition of *Tabby* mutants. Each arises from a defect in subdivision associated with the mesio-distal axis of the dental epithelium, establishing boundaries between tooth primordia different from those established in wild-type embryos. Indeed, a single *Tabby* molar tooth may arise from what would be two molar primordia in wild-type embryos.[9]

Reflecting a different form of polarity, odontoblasts in some rodents deposit dentine on the mesial face of the incisor but a vascularized dentine (vasodentine) on the lingual face. Moss-Salentijn and Moss (1975) interpreted this finding as a consequence of, or response to, continuous growth of the rodent incisor.

Polarized cells

Odontoblasts and ameloblasts are obviously polarized cells. In another sense of polarization, we saw that acellular bone is secreted in a polarized way by osteoblasts in teleost fish (Figs 2.9 and 2.10). In yet another sense, scanning electron microscopy has been used to assess orientation and polarization of osteoblasts within membrane bone, orientation of osteoid fibres in relation to osteoblast orientation, and how parathyroid hormone (PTH) can modulate osteoblast orientation.[10]

Microfilaments and/or microtubules are obvious contenders as active agents in the polarization of skeletal cells; the high degree of polarization of odontoblasts and ameloblasts is dependent upon intact, polymerized microtubules and microfilaments. Directional transport of metabolites from the surface to the depth of bones is facilitated by 240 Å diameter microtubules *within* osteoblasts and osteocytes, 50–70 Å diameter microfilaments *within* osteocyte processes, and tight junctions that couple cells to one another within the bone (Box 31.1).[11]

Box 31.1 Gap junctions connect bone cells

Bone as a network of connected cells is a concept discussed in Box 24.6. Evidence that cell-to-cell communication between bone cells is via *gap junctions* comes from:

- lanthanum colloid, which shows that the junctions among osteoblasts and those between osteoblasts and osteocytes in rat tibiae and femora are gap junctions that allow molecules of some 1200 MW to pass;
- dye injection and measurement of electrical conductance to demonstrate osteoblast-to-osteoblast communication within rat calvariae;
- visualization of junctional connections between osteocytic processes in mouse mandibular bone, emphasizing that the bone cells effectively form a syncytium.[a]

Inactive bone surfaces – surfaces in which active formation or resorption of bone is not taking place – are lined by a continuous 'membrane' of bone-lining cells connected by gap junctions, part of the marrow stromal system. Essentially these consist of determined osteoprogenitor cells (DOPC) that can be induced to divide and differentiate, and that control the inductive environment for haematopoiesis. They may also propagate activation signals for remodeling.[b]

Using antibodies against gap junction proteins, S. J. Jones *et al.* (1993) investigated the incidence and size of gap junctions in rat calvariae, describing junctions between preosteoblasts, osteoblasts and osteocytes. Palumbo *et al.* (1990a,b) investigated cytoplasmic processes between osteocytes, identified gap and adherens junctions between preosteoblasts, osteoblasts and osteocytes, and created informative 3D reconstructions (Fig. 2.1).

Gap junctions between osteoblast-like cells in rat calvariae are connected by connexins. If Lucifer yellow is injected, it labels up to 30 surrounding cells, demonstrating the extent of connectivity (Schirrmacher *et al.*, 1992). *Osteoclasts* also are connected by connexin-43, as verified by immunohistochemical observation of rat osteoclasts cultured on bone slices and by the finding that the gap junction inhibitor heptanal inhibits the number and activity of osteoclasts. The proportion of mononucleated TRAP-positive cells also declines in the presence of heptanal, which may indicate a role for gap junctions in controlling fusion of osteoclast precursors (Ilvesaro *et al.*, 2000).

In an elegant analysis of junctional connections between skeletal and skeletogenic cells in the femoral metaphyses of C57BL/6 mice Yamazaki and Eyden (1995) demonstrated gap junctions among osteocytes, among osteoblasts and among marrow reticular cells, as well as between osteocytes and osteoblasts and between osteoblasts and marrow reticular cells, demonstrating an intimate connection network between those cells that will *become* osteogenic (marrow reticular cells) and those that *are*.

Finally, the importance of gap-junctional connection is seen when disrupting such junctions selectively down-regulates specific genes. Lecanda *et al.* (2000) over-expressed *connexin-43* (*Cx-43*), the major gap-junction protein in osteoblasts, and found that cell-to-cell communication is reduced and osteoblast-specific genes are down-regulated. They also investigated a *Cx43*-null mouse in which both endochondral and intramembranous ossification in the skull are reduced, leaving open foramina. Mandible size was reduced with smaller incisors. Ossification of the clavicles, ribs, vertebrae and limbs also is reduced embryonically, though the axial and appendicular skeletons are normal at birth. They interpreted their results as indicating a great effect on neural crest-derived skeletal tissues.

[a] See Doty (1981) for lanthanum colloid, Jeansonne *et al.* (1979) for electrical conductance, and Johnson and Highison (1983) for the essentially syncytial nature of bone cells.
[b] See Miller and Jee (1987, 1992) and Miller *et al.* (1989) for bone-lining cells, and see Boivin *et al.* (1990) for an ultrastructural analysis.

A number of studies utilized orientation of Golgi bodies to show that chondrogenic and prechondrogenic cells are polarized but that polarization can change during morphogenesis and in mutations such as *talpid* that affect skeletogenesis (Chapter 39).[12]

LONG-BONE GROWTH

Bone appears rigid and has a mineralized extracellular matrix (ECM) that prevents interstitial growth. The rock-like nature of bone was evident to early workers, primarily surgeons, who dealt with bone on a daily basis and could see that bone grew. But how to measure bone growth? An experimental method was required, especially to show not only that bone grows but to determine *where* growth occurs along the length of a bone.

Stephen Hales is usually credited with reporting the first experiments in his treatise, *Vegetable Staticks*, published in 1727. Hales reported an experiment in which holes were bored into the shafts of the long bones of growing chickens to act as markers. Distances between the holes and from the holes to the ends of the bones were recorded immediately after the holes were made and two months later. The distance between the holes remained constant, while the distance from the holes to the ends of the bones increased. Clearly, the bones grew from their ends and not from the middle of the shaft.[13]

Others confirmed the lack of interstitial growth of long bones, notably Jean Duhamel (1742), who used silver stylets to investigate long-bone growth in pigeons and dogs, and John Hunter (1771), who used lead shot to study tibial growth in pigs and tarso-metatarsal growth in chickens. Hunter bored a hole at each end of a tibia and precisely measured the space between the holes. They were two inches apart. He then placed a lead shot in each hole and allowed the pigs to complete their growth. At the end of the growth period the space between the two lead shots still measured two inches, confirming Hales' conclusion that long bones grow from their ends. Julius Wolff (1885) carried out a conceptually similar study of craniofacial growth by implanting metal markers into rabbit nasal and frontal bones.

Subsequently, especially after the discovery and utilization of X-rays and radiographs late in the 19th century (Fig. 29.1) metallic pins or rods were inserted into holes drilled in long bones and the distances measured on radiographs with increased precision. Although coming almost 200 years after Hales' experiments, these 'high-tech' studies added little to our basic understanding of how bone growth occurs, merely confirming the veracity of the conclusions of Hales, Duhamel and Hunter and extending their observations to other species.[14]

Radiographs provided the ability to 'see' the structure of bone and how it changes over time (Figs 16.3, 16.4, 31.2 and 31.4), although, as demonstrated in Fig. 31.3, talented anatomists were depicting the skeleton in great deal before the advent of radiographs. Publication in

1931 of a method for cephalometric radiography enabled serial studies of bone growth to be performed in the same individual. The first human studies, published six and 10 years later, included a longitudinal study covering three months to eight years of age. Björk subsequently (1955) described a method for combining cephalometric radiography with metal implants (shades of Stephen Hales), a method still in use today.[15]

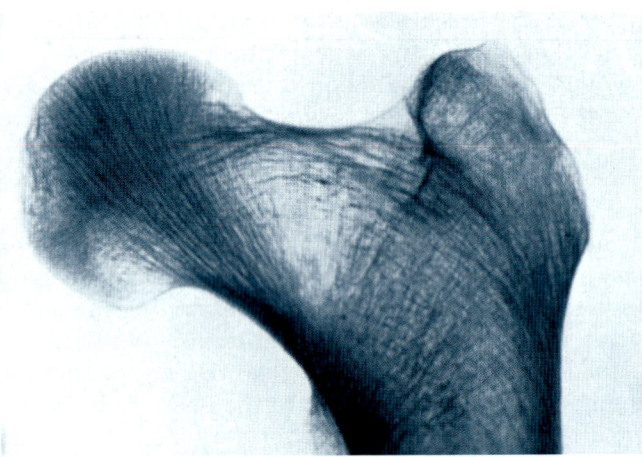

Figure 31.2 Loss of bone with age. A radiograph of the proximal femur of a 70-year-old woman showing loss of trabecular in the centre ('neck') of the femoral head. From Harris (1933).

Growth plates

From the 1830s on, the important contribution made by the *growth plate* to bone structure, histological organization and growth was recognized by most prominent and influential anatomists (Fig. 13.9). The most important advance was William Sharpey's theory that cartilage of the long-bone model is replaced by the bone of the mature individual (Sharpey, 1848).[16]

Now we know that the longitudinal growth of long bones is due to proliferation of cells in the growth plate, their transformation into chondroblasts,[17] chondrocytes and hypertrophic chondrocytes (HCCs), and the concomitant deposition and mineralization of ECM (Fig. 31.5). It is hypothesized that growth in length of mammalian long bones ceases when the supply of growth-plate progenitor cells is exhausted, although hormonal changes also may be involved. The supply of progenitor cells varies from bone to bone and, indeed, from end to end of single bones; see Growth at opposite ends, below.[18]

In introducing his discussion of the mode of growth of bones, Brash (1934) took five propositions, listed below, as given. They serve as a summary of the understanding of mechanisms of bone growth in the early 1930s and, indeed, of our current understanding.

- The diaphyses of long bones grow in length by cartilage replacement at their ends only.

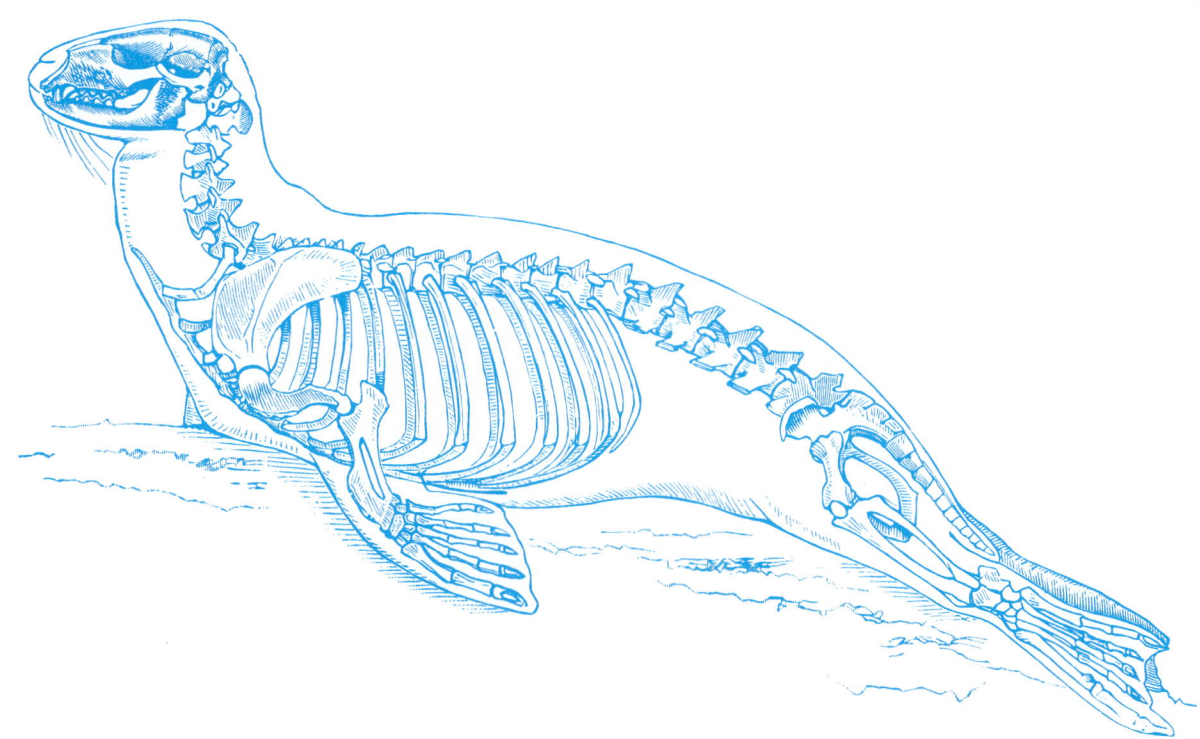

Figure 31.3 The skeleton of an adult seal, probably a harbour seal, *Phoca vitulina*. The skeletal elements of the flippers are especially well shown. From Romanes (1901).

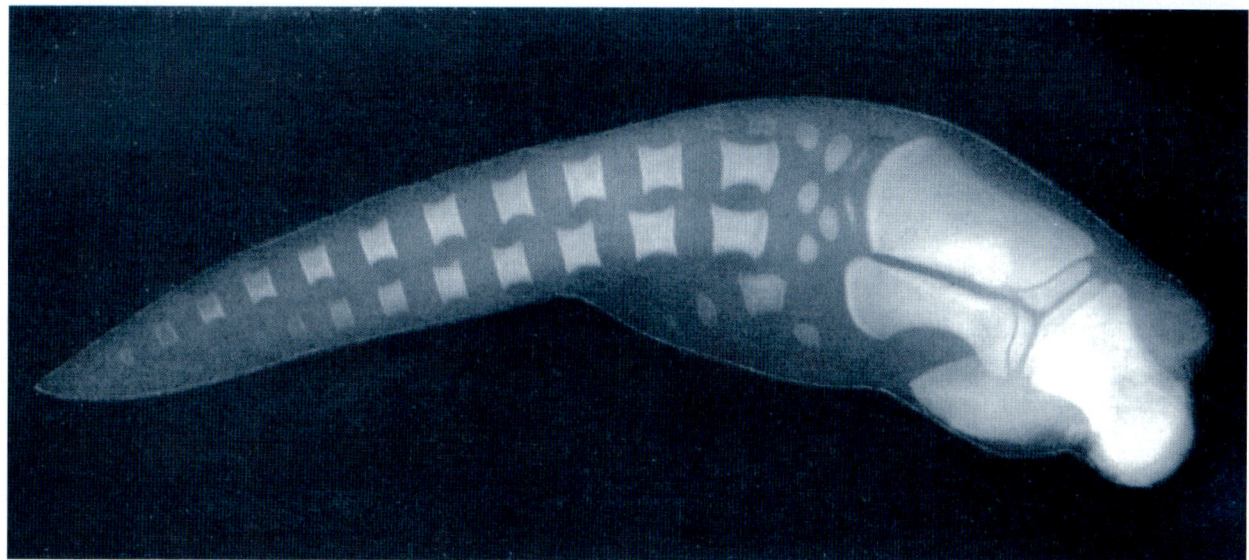

Figure 31.4 A radiograph of the left flipper of a yearling pilot whale, *Globicephala melaena*. Distal is to the left. From proximal to distal the humerus, radius and ulna, carpals, metacarpals and phalanges can be seen. Digits I to V (top to bottom) have, respectively, two, 11, eight, two and zero phalanges, digits II and III therefore showing hyperphalangy. From Felts and Spurrell (1966).

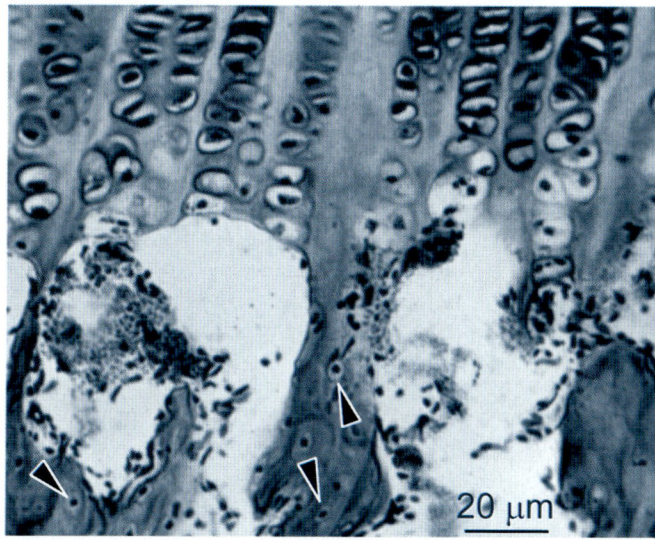

Figure 31.5 A histological section of the proximal end a human tibia, interpreted as showing chondrocytes being incorporated into bony trabecular as osteoblasts (arrowheads), with other osteoblasts forming on the surfaces of mineralized cartilage. Modified from Harris (1933).

- Long bones grow in width by surface accretion through intramembranous (periosteal) ossification.
- Long bones maintain their shape by modeling absorption and deposition principally towards their ends.
- Interstitial expansion of bone does not occur.
- No feature of a bone, however permanent it may seem, can be assumed to be 'fixed' for the purpose of comparison of different growth stages (Brash, 1934, pp. 315–316).

Tissues adjacent to the skeleton – its functional matrix – impose limits on or enhance skeletal growth depending on circumstances (Fig. 31.6). These adjacent tissues include tendons, the fibrous periosteum, muscles, and vascular and nerve supplies. They can:

- control *primary skeletal growth* – as when tension in the periosteum of a long bone changes (Fig.31.7);
- initiate *compensatory skeletal growth* – as, for example, when the brain undergoes expansive growth; or
- themselves *adapt secondarily to bone growth* – as when tendons adapt to bone shortening.

Growth-plate cartilage imposes a polarity on development of the adjacent bone, as demonstrated beautifully by Abad *et al.* (1999), who removed, inverted and reinserted the distal ulnar growth plate in four-week-old rabbits. Over a three-week period the *cartilaginous growth plate and the adjacent bone maintained their polarity* as hypertrophic cells deposited metaphyseal bone, a telling example of the primacy of cartilage and of coupling between growth plate and adjacent bone in mammalian growth plates, a coupling that is believed to be maintained through PTHrP (Fig. 15.2; Box 31.2).

Growth-plate dynamics

Alongside Brash's five propositions it has long been established that increase in the length of long bones results from two cellular events within the growth plate cartilage – proliferation and hypertrophy – along with the deposition of ECM by chondroblasts and chondrocytes.[19]

New cells, bigger cells and matrix

Long-bone growth can be analyzed by following cells and ECM region by region through growth plates as the

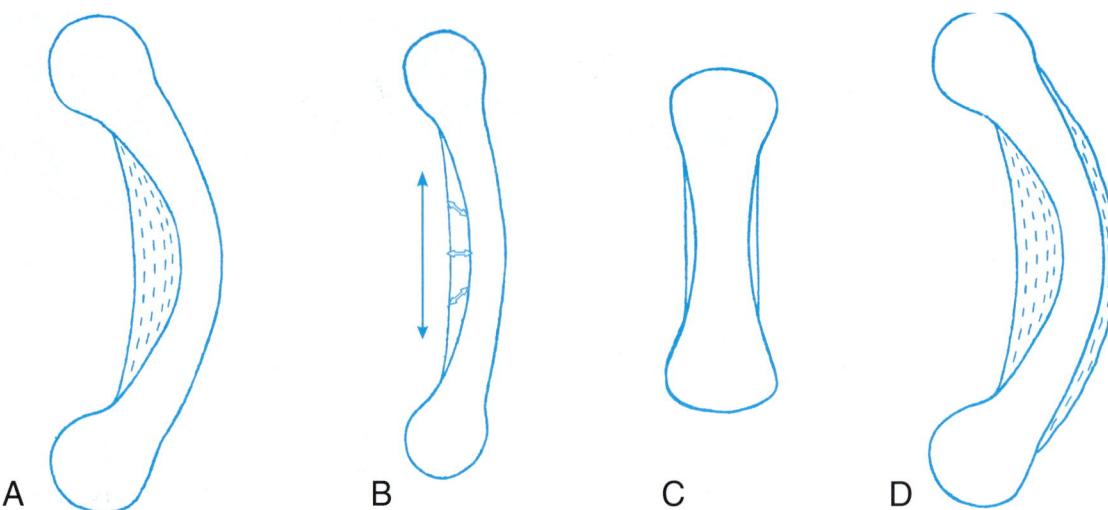

Figure 31.6 Bone formation in a long bone showing response to pressure and or tension. (A) Arrangement of trabeculae (hatched) if orientated along lines of pressure. (B) The direction of tension required to produce the trabecular architecture in A is illustrated by the arrows. (C) The periosteum as seen in normal development *in vivo*. (D) Formation of excess periosteal bone on the pressure side of a bent long bone. Also see Fig. 31.7. Adapted from Murray (1936).

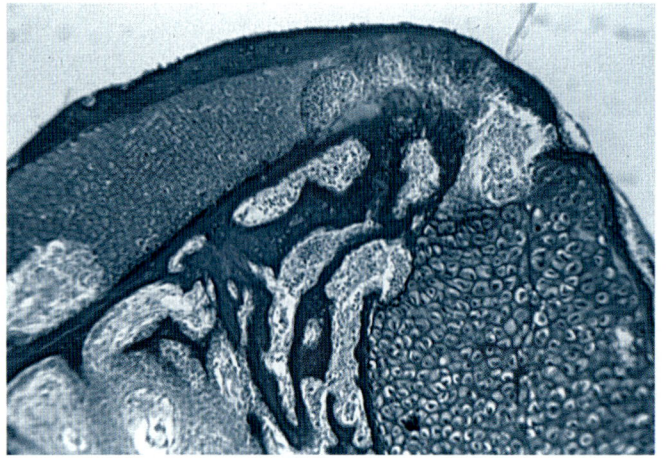

Figure 31.7 Mechanical factors influence bone growth. This tibia from a seven-day-old chick embryo grafted onto the CAM of a host embryo for five days became bent, creating tension on the periosteum on the concave face, which promoted excess osteogenesis on that side. Also see Fig. 31.6.

chondrogenic differentiation cascade is converted into longitudinal growth. Results obtained depend, in part, on how rapidly the animal (and therefore its bones) are growing. As determined in four different growth plates in 28-day-old rats, at least eight variables relating to chondrocyte function control the differential growth of growth plates. For all four growth plates the number of new chondrocytes produced per day balances the number lost per day. Proliferation, however, is a minor contributor to long-bone growth. The relative contribution of different processes to growth of the proximal tibial growth plate (Figs 4.8 and 4.9), is proliferation 9 per cent, deposition of ECM 32 per cent, and chondrocyte hypertrophy 59 per cent, i.e. 90 per cent of bone growth is attributable to increase in cell size and deposition of ECM.

Differences in inherent growth potential between growth plates *are* attributable to the numbers of cells produced. The proximal tibial growth plate produces four times more chondrocytes a day than does the proximal growth plate of the radius – 16 400 vs. 3700 respectively – providing a partial explanation of the different intrinsic growth rates of these two long bones. Of course, the situation is not static; different values would be obtained if data were collected at different times.[20]

Cell proliferation

Neville Kember, to whom we owe much of the data and theory on long-bone growth in mammals, argued that 40–50 divisions of the starting stem-cell population are sufficient to produce a rat tibia, bone length being determined by the product of the number of stem cells and how many HCCs are produced. Consequently, in a study of long bones of Yucatan swine, the number of HCCs was found to be inversely proportional to growth of the growth plate, while in a study of various rabbit epiphyseal cartilage plates, Kember demonstrated that *each growth plate shares a constant rate of cell division, the only difference between growth plates being the number of starting cells*.

In their comparative analysis of the growth of mammalian and avian growth plates, Starck and Ricklefs (1998) also concluded that the growth rates of avian long bones vary according to the number of cells in the flattened (proliferative) cell zone. Consistent with these findings is an examination of long-bone growth in six species of herons and egrets from the family Ardeidae, in which the number of dividing cells rather than the rate of cell division, shows the best correlation with tarso-metatarsal growth (Table 31.1).[21]

Box 31.2 PTHrP–*Ihh* and cartilage growth

PTHrP

Parathyroid hormone-related protein (PTHrP), a polypeptide of the terminus of parathyroid hormone, was described in the late 1980s as modulating the effect of Tgfβ on osteoblast division and collagen synthesis. Tgfβ has PTHrP-dependent and PTHrP-independent effects on endochondral ossification, as detected in mouse metatarsals *in vitro*. Tgfβ decreases proliferation, hypertrophy and mineralization and increases PTHrP, while PTHrP decreases hypertrophy and mineralization without affecting proliferation. Tgfβ is therefore upstream of PTHrP in regulating chondrocyte hypertrophy (Serra *et al.*, 1999).

In conjunction with *Indian hedgehog* (*Ihh*), PTHrP is now known to be a major regulator of the growth of many if not all cartilages; *Ihh*, *FgfR-3* and *Bmp-4* all are expressed in growth plates (Fig. 15.2). Within the craniofacial skeleton, PTHrP and its type I receptor have been localized to the anterior portion of mouse Meckel's cartilage and to mandibular condylar cartilage in a distribution that correlates with cartilage growth. PTHrP expression in the condylar cartilage occurs in some chondrocytes at 15 days of gestation, in the flattened and hypertrophic cell layers at 16 days, and in the flattened cells and in some hypertrophic cells at 17 days.[a]

PTHrP⁻ mice have defects in craniofacial and limb skeletons. The effects are cartilage specific:

- the anterior cartilaginous cranial base is normal;
- chondrocyte differentiation and replacement by bone is accelerated in the posterior cranial base and cranial base synchondroses;
- mandibular condylar cartilage is reduced in size but displays normal zonation;
- proliferating chondroblasts are reduced and chondrocyte differentiation is accelerated in the growth plates of the long bones; and finally,
- ectopic bone forms in the medial portion of Meckel's cartilage, which is the portion that normally transforms into a ligament.[b]

PTHrP is widely thought to slow cartilage maturation. This is true for growth-plate cartilages, but is by no means a universal action of this protein, as is clear from the above results. In terms of growth-plate cartilages, PTH⁻ or PTHrP⁻ null mice show delayed mineralization, delay of chondrocyte differentiation to hypertrophy, a prolongation of the hypertrophic phase, decreased vascular invasion and consequent reduced long-bone growth.[c]

In another example of specificity of action, Shibata *et al.* (2000) examined PTHrP-deficient mice and found that the mandibular ramus is bent, a prominent bony crest develops, the masseter muscle is attached to the crest by a tendon and bone forms along the tendon. The excessive number of 2B muscle fibres in the masseter indicates premature maturation.

PTHrP–*Ihh*

Ihh is expressed in prehypertrophic growth-plate chondrocytes, where it regulates the rate of transformation of chondrocytes to hypertrophic chondrocytes (HCCs) by inducing PTHrP in the periarticular perichondrium. Feedback from PTHrP to *Ihh* regulates hypertrophy by repression of PTHrP in a feedback loop (Fig. 15.2).

In addition to regulating proliferation and differentiation of cartilage, *Ihh* is required during osteogenesis, a conclusion based on a study of *Ihh*-null mutant mice, in which chondrocyte proliferation is reduced, chondroblasts mature in incorrect positions, and no osteoblasts form in endochondral bone. Cartilage mineralizes normally although no bony collar forms. As discussed in Chapter 22, subperiosteal bone formation requires hypertrophy of the underlying chondrocytes, a system in which *Ihh* operates by PTHrP-*dependent* and -*independent* pathways.[d]

More recent analysis by Karp *et al.* (2000) adds significantly to our understanding of the relationship between *Ihh*, PTHrP and the PTHrP receptor PTHrPR, as follows.

- Deleting *Ihh* from mice results in short-limbed dwarfism, in which chondrocytes show reduced proliferation and accelerated hypertrophy, results that parallel those seen in *PTHrP* or *PTHrPR* mutants.
- Activating *Ihh* up-regulates *PTHrP* and decreases hypertrophy in wild-type but not in *PTHrP*⁻ mutants, suggesting that *Ihh* acts via PTHrP.
- Double mutants (*Ihh*⁻/*PTHrP*⁻) show the same defects as *Ihh* mutants, indicating that *Ihh* is required for PTHrP function.
- Double mutants (*PTHrP*⁻/*PTHrPR*⁻) with the PTHrP receptor under the collagen II promoter (designated PTHrPR*) rescue *PTHrP*⁻/⁻ embryos.

PTHrPR* is expressed in those chondrocytes that are *Ihh*⁻/⁻, stops premature hypertrophy, has no action on proliferation and does not prevent dwarfism, indicating that the mechanisms regulating hypertrophy and proliferation differ. *Ihh regulates PTHrP to prevent hypertrophy. Ihh is required for proliferation but acts independently of PTHrP.*

Kobayashi *et al.* (2002) outlined a multi-step scenario, given below, involved in regulating mammalian growth-plate chondrocytes by PTHrP and *Ihh*, concluding that *Ihh* promotes chondrocyte differentiation via two pathways, one independent of PTHrP and one acting with PTHrP.

(i) *Ihh* is localized to pre-HCCs, where it enhances expression of PTHrP and so delays chondrocyte differentiation via the PTH/PTHrP, which they designate as PPR.

(ii) Chondrocyte-specific ablation of PPR or reduction in PPR expression leads to premature chondrocyte differentiation and an increase in *Ihh* in early-stage chondrocytes.

(iii) Disruption of PPR in one-third of the chondrocytes in a growth plate leads – paradoxically – to *enhanced* expression of *Ihh* because of ectopic and early differentiation of PPR⁻ HCCs.

Tgfβ-2 mediates the effects of *Ihh* on chondrocyte hypertrophy and on expression of PTHrP. Chondrocytes secrete *Ihh* and PTHrP as they mature and hypertrophy, and PTHrP acts downstream of *Tgfβ-2*, which means that *Ihh* activates *Tgfβ*, which activates PTHrP (Alvarez *et al.*, 2002). Inhibitors of cellular retinoic acid-binding protein (CRABP; Box 20.3) are localized in growth plates where they have a major impact, suppressing proliferation and delaying hypertrophy. The gene for the *Ihh* receptor *patched* (Fig. 15.2) also is down-regulated.

FgfR2–PTHrP

The Fgf receptor FgfR2 comes in two splice variants. IIIb is expressed in epithelia, IIIc in mesenchyme condensations and in the periosteal collar during osteogenesis of membrane and endochondral bones (Fig. 15.2). Disrupting synthesis of IIIc delays the onset of ossification, shifting the balance of cell behaviour toward differentiation, and resulting in a loss of growth, dwarfism and fusion of cranial bones. These disruptions are associated with reduction or down-regulation of *Spp-1* and *Runx-2* in osteoblasts and down-regulation of *Ihh* and PTHrP in chondrocytes, indicating autonomous action in both skeletogenic cell lineages (Eswarakumar *et al.*, 2002).

Box 31.2 (*Continued*)

Proliferation of the secondary cartilage of the rat mandibular condyle can be regulated in explant cultures exposed to IgfR-1, FgfR-1 and FgfR-3 – which primarily influence chondroblasts and HCCs, and by FgfR-2 – which primarily influences articular and prechondroblastic cells. Igf-1 has a greater influence on proliferation than does Fgf-2, but the threshold concentration for Fgf-2 is lower than that for Igf-1. Further, during the transition from prepuberty to puberty (31–42 days), Igf-1, FgfR-2 and FgfR-3 increase while Fgf-2 decreases (Fuentes *et al.*, 2002).

[a] See Centrella *et al.* (1989b) for modulation of Tgfβ actions, Naski *et al.* (1998) for Ihh and FgfR-3, and Serra *et al.* (1999) for PTHrP-dependent and -independent actions. Although few comparative studies have been undertaken, PTHrP is expressed in the dermal denticles (and in the gills, spinal cord, pituitary and rectal gland) of sharks and rays, and in lamprey kidney, skin and muscle (Trivett *et al.*, 1999). PTHrP is expressed in chick sternal chondrocytes as they differentiate, preceding type X collagen.

Levels of PTHrP and collagen X are increased by treatment with 10^{-7} M PTHrP (Grimsrud *et al.*, 1999).
[b] See Yamazaki *et al.* (1997, 1999) for the distribution of PTHrP in mouse cartilages. Using serum-free organ culture, Shum *et al.* (2003) demonstrated that Bmp-4 is required to maintain the growth of murine cranial base cartilage, exogenous Bmp-4 enhancing cartilage differentiation and evoking ectopic hypertrophy, the latter accompanied by increased expression of PTHrP receptor.
[c] See Lanske *et al.* (1996), Schipani *et al.* (1997), Ishii-Suzuki *et al.* (1999), Kronenberg and Chung (2001) and Kronenberg (2003) for PTH and PTHrP.
[d] See Vortkamp *et al.* (1996) and Kronenberg (2003) for Ihh expression, Wallis (1996) for an overview of how negative feedback between Ihh and PTHrP regulates chondrocyte hypertrophy, St-Jacques *et al.* (1999) for the link to osteogenesis and the likelihood that Ihh operates through PTHrP-dependent and PTHrP-independent pathways, and Minina *et al.* (2001) for modulation of Ihh by Bmp-2.

Table 31.1 Comparison in mean number of proliferating cells, rate of cell division, and longitudinal growth rate in the tarsometatarsus of herons and egrets (family Ardeidae)[a]

Species	Mean number of proliferating endochondral cells	Mean rate of endochondral cell division (cells/day)	Mean longitudinal growth rate (mm/day)
Nycticorax nycticorax (black-crowned night heron)	809 ± 313	0.61 ± 0.24	–
Bubulcus ibis (cattle egret)	681 ± 93	0.23 ± 0.06	1.69
Ardeola ralloides (squacco heron)	385 ± 137	0.76 ± 0.30	2.79
Ixobrychus minutus (little bittern)	234	1.27	3.11
Egretta garzetta (little egret)	798 ± 257	0.43 ± 0.21	3.78
Ardea purpurea (purple heron)	702 ± 142	0.55 ± 0.18	4.19
Ardea cinerea (grey heron)	1027 ± 246	0.37 ± 0.03	4.23

[a] Based on data from Cubo *et al.* (2000).

Birds and mammals

Caution: We cannot compare avian and mammalian long bones in quite the cavalier way I have just done. Modes of osteogenesis differ significantly. Much of the development of the long bones in avian embryos is subperiosteal (Fig. 1.12), endochondral ossification not beginning until after hatching, although even here patterns vary between *altricial* and *precocial* birds.

Much of the cartilage in long-bone primordia in embryonic chicks is replaced by marrow, not by bone. Haines and Mohuiddin, 1960, provided some evidence that human epiphyseal plates incorporate cartilage derived from precursors within the marrow to replenish exhausted reserves of undifferentiated cells. If this occurs, the contribution to growth is probably small, but could perhaps be enhanced experimentally. There is some indication that proteoglycans are retained during the mineralization of avian but not mammalian epiphyseal growth plates, at least not in those of the most commonly studied mammals. When chick growth plates develop in the absence of functional skeletal stimuli, proliferation is reduced by between 27 and 55 per cent but hypertrophy is unaffected, not the pattern described for mammals above.[22]

Clones and timing

Computer-aided three-dimensional reconstruction of chondrocyte columns in rabbit growth plates by Letty Moss-Salentijn confirmed and extended Kember's studies. Her simulation of clonal chondrocyte (*chondrone*) behaviour in rabbit growth plates shows that the model of best fit is one with variable numbers of clones and variable interclonal length. Growth in length ceases once the growth plate has used up its supply of precursor cells. Consequently, animals with long-lived proliferating stem-cell populations in their growth plates could potentially grow throughout life, as indeed is seen in amphibians, many reptiles, and some fish. Such animals show *indeterminate growth*. Those with a finite number of proliferating cells show *determinate growth*.[23]

Duration of the cell cycle influences gene expression in developing limbs, as growth factors modify patterning via the cell cycle. Growth rate is not a wholly intrinsic mechanism, however. The rates of growth of distal and proximal epiphyses can be modified experimentally; control is more subtle than mere exhaustion of a proliferative cell pool.

Regulation of growth in growth plates is not a constant but changes with age, especially postnatally. Furthermore, regional differences in growth rate within a single growth plate can be enormous; the growth rate of the distal femoral growth plate of six-week-old New Zealand white rabbits is as much as 700 μm/day, but regional variation within the growth plate is of the order of 300 μm/day (Lerner and Kuhn, 1997).

In order to study chondrocytic dynamics in relationship to longitudinal growth, Hunziker and his colleagues used a combination of a fluorochrome and ³H-thymidine to label cells in rat proximal tibial growth plates and showed that:

- changes in cell shape, not proliferation, regulate growth between 21 and 35 days postnatally;
- changes in cell shape and proliferation regulate growth between 35 and 80 days, one chondrocyte being lost every four hours but compensated for by high replacement; and
- deposition of ECM plays a minor role in growth.[24]

Hormonal involvement

Some of the dynamics of the cell populations within growth plates, including the balance between proliferation and hypertrophy, is hormonally regulated. Injecting Igf-1 or growth hormone into hypophysectomized rats drastically reduces the cycle times of reserve cells, division rates of proliferating cells and the duration of the hypertrophic phase. For example, cycle time of prechondroblasts is reduced from 50 to 15 days after injection of Igf-1 (Table 31.2).

Thyroxine regulates morphogenesis of rat growth-plate chondrocytes *in vitro*; adding thyroxine increases alkaline phosphatase and type X collagen and the cells organize into columns as they do *in vivo*. To investigate whether thyroid hormone action is direct or indirect, Ballock *et al.* (1999) examined the expression of thyroid hormone receptors on rat growth-plate cartilage, finding three of the four isoforms of the receptor but only TRα-1 and TRβ-1 expressed as protein, implicating post-translational regulation of thyroid hormone receptors on growth-plate chondrocytes. Epiphyseal chondrocytes respond to thyroid hormone by decreased DNA synthesis and enhanced terminal differentiation via cyclin-dependent kinase inhibitors.[25]

Growth at opposite ends

As was known almost 150 years ago, rates of increase in growth vary between different long bones and between the distal and proximal extremities of individual bones. In the study on New Zealand white rabbits introduced above, anywhere between 45 and 70 per cent of femoral growth resulted from growth of the distal growth plate, the precise contribution varying with the age of the rabbits. In humans, three-quarters of the growth of the humerus takes place at the proximal growth plate, while three quarters of the growth of the radius occurs at the distal growth plate. 'To the elbow I go, from the knee I flee,' was something every medical student learnt about the patterns of growth of the upper arm and leg bones.

Left–right asymmetry also occurs. A radiographic study of ossification in human long bones between 8 and 26 weeks, including comparison of left and right elements, showed that growth of the left humerus, tibia and fibula dominates the right, while growth of the right femur dominates the left. Gender also comes into play; after 21 weeks of gestation, female foetuses are in advance of male (Bagnall *et al.*, 1982). It was generally thought that proliferation of precursor cells was controlled intrinsically perhaps by specification of the length of the cell cycle early in foetal life, an interesting idea that is only partly true.

Mechanical factors also play a role, although not an exclusive role, in controlling these growth rates. Stapling *one* growth plate, by stapling the epiphysis to the metaphysis, or irradiating a growth plate prevents growth at *that* extremity. However, the *opposite* growth plate is stimulated to increase cell proliferation and enhance its growth rate in compensation. Independence of the growth plate also is amply illustrated by its ability to reform (regenerate) after removing the distal end of a long bone, as demonstrated for rat femora and as discussed in the following section. In one of the nicest studies of this phenomenon, Farnum *et al.* (2000) stapled the proximal tibial growth plates of four-week-old rats and used tetracycline and BrdU-labeling to show that changes in *both* proliferative and hypertrophic cells contribute to the reduced growth seen three days later. Volume changes during hypertrophy were affected by mechanical factors but the limit of ability to respond was reached quickly; data obtained three and six days after labeling were similar.[26]

Diurnal and circadian rhythms

Many of life's activities are rhythmic, including reproduction, skeletal development and growth.

Comparing mice mated in the morning with those mated overnight shows that embryos from morning matings exhibit a more rapid rate of digit formation than do embryos from females mated in the evening (Ishikawa *et al.*, 1992). Failure to take account of rhythms when

Table 31.2 Effects of insulin-like growth factor-1 (Igf-1) or growth hormone on the duration (in days) of growth-plate cells in hypophysectomized rats[a]

Cell population	Control	Igf-1	Growth hormone
Prechondrogenic cell cycle time	50	15	8
Chondroblast cycle time	11	4.5	3
Hypertrophic phase	6	4	2.8

[a] Data from Hunziker *et al.* (1994). Also see Hunziker (1994).

designing experiments can compromise the data collected. Samples taken at different times of the day can be confounded by natural daily variation. For example, peak acid phosphatase levels occur around 10 p.m. in humans. Sample at another time and you miss the peak. Sample at different times on different days and the data have little meaning.[27]

Osteogenesis in humans also shows endogenous biorhythms. A biorhythm in bone growth in rabbits has a periodicity of circa two days whether the bones are endochondral or intramembranous, weight or non-weight bearing. Growth of bones following fracture (maximal ossification and remodeling of the callus) follows a diurnal rhythm, growth being greatest if the fracture occurs at midday. Hypophysectomy abolishes this rhythm. Alveolar bone and the fibroblasts of the periodontal ligament display circadian rhythms of proliferation and a longer periodicity of formation and resorption (Chapter 25). Diseases also have peaks during the day, e.g. aspects of osteoarthritis at around 6–8 p.m.[28]

Mammalian growth plates exhibit daily growth rhythms.

Using fluorochrome labeling, Stevenson et al. (1990) detected a peak in proliferation at 6 a.m. in chondrocytes of the growth plates of Wistar rats; the daily growth rate is 284 μm/day. Simmons (1962) detected a diurnal periodicity in epiphyseal cartilages after he used 0.1 mg of colchicine per gram body weight to arrest mitosis. Two decades later, Simmons et al. (1983b) documented a circadian rhythm in various aspects of rat epiphyseal cartilage, including mineralization, alkaline phosphatase levels (especially levels of matrix vesicle alkaline phosphatase), the calcium-binding activity of cell membranes, and enzyme activity. They also found that appositional (subperiosteal) growth in width and linear (endochondral) bone growth are out of phase diurnally:

- epiphyseal cartilage and metaphyses are most active during the day;
- articular cartilage and diaphyses are most active during the first four hours of dark;
- lysozymal enzymes and alkaline phosphatase show two peaks in epiphyseal cartilage, one at midday and one at midnight; and
- lysozymal enzymes and alkaline phosphatase show one peak in articular cartilage, at midnight.

Rhythms are under hormonal control

Circadian influences are damped in the growth plate of the long bones of Yucatan swine, perhaps because the HCCs undergo condensation and death at the chondroosseous junction. In three-week-old Sprague-Dawley rats, the chondrocytes in longitudinal columns of the growth plate are more synchronized than those chondrocytes that lie between columns, the cell cycle being some 24 to 28 hours and the cells spending four days in the proliferative and two days in the hypertrophic zone.[29]

Light acting through the *pineal* is one mechanism underlying rhythms of cartilage growth, as demonstrated in epiphyseal cartilage growth in female rats under constant light or dark, or alternating dark and light, and with and without pinealectomy. Cartilage growth is diminished under constant dark or light, and growth retardation persists in pinealectomized rats maintained under constant light or dark but not in those maintained under alternating light and dark (Nir et al., 1972).

Immobilization abolishes rhythmic skeletal growth patterns. Adrenalectomy abolishes the rhythm of DNA synthesis in epiphyseal cartilage and bone, while parathyroidectomy abolishes rhythms in the synthesis of collagenous and non-collagenous proteins (Russell et al., 1984).

Igf-1 and growth hormone may also regulate growth-plate rhythms (Box 31.3). In a study of hypophysectomized rats, a hypertrophic phase (in days) was measured in growth-plate chondrocytes at four and three days following injection of Igf-1 and growth hormone, respectively. This compared with a six-day hypertrophic phase in uninjected animals.[30]

The osteogenic competence of marrow stromal cells also displays a circadian rhythm: the ability to form fibroblast colony-forming units (FCFUs) is maximal in the late dark and early light phases. Allograft-induced osteoinduction in rats also follows a circadian rhythm, as seen in a study in which allografts were implanted at four-hourly intervals during the day. Maximal host response – measured by numbers of osteoclasts and inflammatory cells – occurs between midday and mid-afternoon (Table 31.3). *Implanted dead bone elicits the same response at any time of day.*[31]

A role for the periosteum in regulation of the growth plate?

A variety of studies of interactions between the cartilaginous growth plate and the periosteum suggest that *cell proliferation within the growth plate is damped by tension exerted by the periosteum.* Periosteal strength varies along the length of bovine long bones, being lowest in the diaphysis and increasing toward the metaphysis (Table 31.4). The suggestion from this study is that periostea can exert sufficient pressure to influence proliferation within the growth plate. Surprisingly, little information is available on the mechanical properties of periostea. I discuss two examples below, one from pigs and one from quail.

Periosteal migration with bone growth controls orientation of the masseter muscle in miniature pigs, *Sus scrofa*, as assessed following the implantation of titanium granules. A further study using pigs examined both the stiffness and the strength of periostea from the body of the dentary, zygomatic arch and metacarpals, and properties associated with fracture of the bones under constant tension. Periostea from bones at a ligament or muscle interface (zygomatic

Box 31.3 Insulin-like growth factor-1 (Igf-1)

Insulin-like growth factor-1 (Igf-1 or somatomedin-C) and Igf-2 mediate the actions of growth hormone *in vivo*, growth hormone being converted to Igf-1 and Igf-2 in the liver. Consequently, Igf-1 has wide-ranging actions in promoting the survival and proliferation of all normal and some cancerous cells, and in mediating growth hormone effects on metabolic activity, including growth. Circulating in the blood stream, Egf-1 tracks changes both in growth hormone in normal functioning and when in excess or deficiency. Igf-1 is a mitogen for many cells in culture, its mitogenic action being mediated through the IGF-1 receptor, receptor tyrosine kinase, activation of the ras and Map kinase cascades, and subsequent modification of transcription.

As the mediator of growth hormone action, Igf-1 is required for normal embryonic growth; heterozygote mice in which *Igf-1* has been deleted are 10–20 per cent smaller than wild type, while homozygotes are less than 60 per cent smaller and die perinatally with underdeveloped muscles and lungs (Powell-Braxton *et al.*, 1993).

In mammalian long bones, however, growth hormone acts directly on chondrocytes of the growth plate. In turn, the chondrocytes synthesize Igf-1, which acts in an autocrine manner within the growth plate. Circulating Igf-1 does not play a primary role in regulating growth-plate growth.

Igf-1 and -2 are found in the perichondria and proliferating chondrocytes – but not in the osteocytes – of the long bones of growing rats between birth and adulthood. In chick embryos Igf-1 peptides are at a low concentration in prechondrogenic mesenchyme, absent from condensing prechondrogenic mesenchyme and from osteogenic mesenchyme, but present in developing limb cartilage and bone, associations more reflective of action on differentiation than on proliferation.[a]

In contrast, Igf-1 and -2 and Igf-binding protein-2 are expressed in the epithelium and mesenchyme of rat limb buds, where they facilitate proliferation of the AER. Exogenous Igf-1 applied to limb buds of *wingless* or *limbless* chick mutants maintains an active AER, up-regulates *Msx-2* and promotes limb-bud outgrowth.[b]

Chondrogenesis

Igf-1 is the main serum factor regulating chondroblast proliferation, binding to proliferating chondrocytes of rat ribs at twice the level that it binds to resting chondroblasts or to hypertrophic chondroblasts. Both growth hormone and Igf-1 stimulate cell function in distinct zones of the rat epiphyseal growth plate, low concentrations enhancing mitosis in the proliferative zone. Igf-1 (50 μg) and TGFβ-1 (10 μg) also enhance proliferation during the early phases of repair of long bones without compromising callus composition or biomechanical properties.[c]

Igf-1 enhances the proliferation and synthesis of proteoglycan core protein by progenitor cells from mouse mandibular condylar cartilage. Blocking Igf-1 leads to a 90 per cent decrease·in ³H-thymidine incorporation into chondroprogenitor and chondroblastic cells (Maor *et al.*, 1993b,c).

Outside the tetrapods both Igfs stimulate ³⁵S incorporation into cultured cartilage from the Japanese eel, *Anguilla japonica*. In the first *in-vitro* study of the hormonal regulation of elasmobranch cartilage – vertebral cartilage from the clear-nose skate, *Raja eglanteria* – Igf-1 and corticosterone were shown to have the same effects on glycosaminoglycan synthesis as they do in tetrapods.[d]

Osteogenesis

Igf-1 and -2 are stored in bone matrix and increase DNA synthesis and collagen type I production when applied to cultured bone cells. Igf-1 and Igf-2 receptors on foetal rat bone cells were characterized by Centrella *et al.* (1990).

Igf-1 is mitogenic for human osteosarcoma cells (MG-63) and for rat osteoblasts, activating the glucose-6-phosphate dehydrogenase pathway. In induced bone formation in rats, *Igf-1* is expressed early during the proliferative phase, *Igf-2* during the later mineralization phase.[e]

Igf-2 and Tgfβ both increase total protein, total collagen and type I procollagen mRNA in human osteoblast-like cells *in vitro*, each via a distinct mechanism (Strong *et al.*, 1991).

[a] See Beck *et al.* (1988) and Ralphs *et al.* (1990) for rat and chick, and see Dupont and Holzenberger (2003) for an overview of insulin-like growth factors.
[b] See Streck *et al.* (1992) and Dealy and Kosher (1996) for rat and chick mutants.
[c] See Makower *et al.* (1989a,b) and Oberbauer and Peng (1995) for growth-plate chondroblasts/cytes, and Wildermann *et al.* (2003) and Schmidmaier *et al.* (2004) for fracture repair.
[d] See Duan and Hirano (1990) and Gelsleichter and Musick (1999) for eels and skates.
[e] See Farquharson *et al.* (1993) for osteosarcoma cells, and Prisell *et al.* (1993) for induced bone.

Table 31.3 Circadian rhythm of induced bone formation elicited by allogenic rat cortical bone three weeks after transplantation[a]

Light level	Clock hour transplant occurred	Per cent of allograft surface covered in osteoclasts
Light	0800	8
	1200	10
	1600	13
Dark	2000	2
	2400	1
	0400	–
Light	0800	7

[a] Based on data from Simmons *et al.* (1974).

Table 31.4 Periosteal stength along the length of calf tiabiae[a]

Location along the tibia	Periosteum strength (kg/cm)
Proximal metaphysis	30
	15
Diaphysis	5
Distal metaphysis	4.5
	13

[a] Based on data from Sebek *et al.* (1972).

arch) had a higher degree of stiffness (91.7 MPa) than the metacarpals (84.7 MPa) whose periosteal surfaces interface with connective tissue. Periosteal strength was also higher; 12.3 and 11.3 MPa, respectively. Peak load for the zygomatic arch was seven orders of magnitude higher than for the ramus of the dentary, and three orders of magnitude higher than for metacarpals (see Table 31.5 for these and additional data).[32]

In order to ascertain whether periostea have the strength to resist growth forces from the growth plate,

Table 31.5 Properties of periostea from three sites in the pig skeleton[a]

Parameter	Ramus of dentary	Zygomatic arch	Metacarpal
Thickness (mm)	0.3	1.2	0.5
Width (mm)	10.5	10.7	10.0
Cross-section (mm²)	3.2	12.7	5.0
Stiffness (MPa)	63.0	91.7	84.7
Peak load (kg)	2.4	16.4	5.5
Peak stress (MPa)	8.2	12.3	11.3
Peak strain (%)	15.6	17.7	17.9

[a] See Popowicz et al. (2002) for details and statistics.

Bertram et al. (1991) determined the breaking strength of periostea from avian long bones. This was a challenging study involving two approaches. In one, measurements were made on the small long bones of Japanese quail, Coturnix coturnix japonica, after the periosteum around the diaphysis had been sectioned – a technique known as periosteal sectioning (see following section). The second involved measuring the tensile properties of isolated periostea. Periosteal sectioning had no significant effect on the growth of tibiotarsi or radii. In-vivo periosteal stress was calculated as 0.85 ± 0.45 MPa, which was estimated to result in a stress on the growth plate of $<10^4$ Pa, a level too low for the periosteum to influence longitudinal bone growth.

Other experiments using periosteal sectioning (but without measuring the properties of the periosteum) have suggested a mechanical role for periostea in long bone growth.

Periosteal sectioning

Chickens. Severing the periosteum of chicken radii across the long axis of the bone stimulates growth in length. Splitting the periosteum longitudinally has little effect on bone growth. Crilly (1972) proposed that the periosteum maintains the growth plate under tension. Proliferating cells of the growth plate are stimulated when periosteal tension is released, growth of the growth plate being opposed by resistance of the periosteum and thereby regulated. Others confirmed the results obtained from chickens using a similar in-situ approach with rodents and rabbits.[33]

Rodents. A study in which rat tibial periostea were sectioned transversely or longitudinally, with or without removal of the periosteum (periosteal stripping), produced the following results, expressed as increase (↑) in length of the tibiae:

- transverse sectioning ↑ by 0.41 mm;
- longitudinal sectioning ↑ by 0.03 mm;
- transverse sectioning + periosteal stripping ↑ 0.69 mm;
- longitudinal sectioning + periosteal stripping ↑ by 0.45 mm.

These results demonstrate that longitudinal sectioning has little effect (a 30 μm increase) in comparison with transverse sectioning (a 410 μm increase). Add periosteal sectioning and the increase jumps to 690 μm. Subsequently, Taylor and his colleagues showed that the response involved increased proliferation and increased chondrocyte hypertrophy.[34]

Tibial growth is stimulated when the periostea of rat tibiae are sectioned and the bones transplanted subcutaneously. Sectioning the periosteum at the proximal end produces a 30 per cent increase in length from growth at distal and proximal growth plates. Sectioning the periosteum at the distal end elicited a 16 per cent increase in length, with enhanced growth only from the proximal growth plate, growth distally being within control values (Harkness and Trotter, 1978). These results are similar to those obtained by Crilly.

On the other hand, McLain and Vig (1983) found that transverse sectioning of the periostea of rat femora resulted in only a slight decrease in length and minimal alteration in shape. They concluded that the periosteum restrains epiphyseal growth where bone is normally under tension but not where bone is under compression.

Rabbits. An examination of the growth of rabbit femora after longitudinal or circumferential sectioning ± periosteal stripping (Kuijpers-Jagtman et al., 1988) showed that the length of the femur increases after longitudinal sectioning, growth of the distal epiphysis is accelerated after a combination of periosteal sectioning and stripping, while circumferential sectioning of the periosteum enhances growth in length from the distal growth plate.

Feedback control

Moss (1972b) saw such feedback as a cycle. As a growth plate elongates, the fibrous periosteum adapts to the tension placed upon it by initiating its own compensatory growth, decreasing tension on the growth plate and slowing proliferation of cartilage precursors, allowing the cycle to be repeated.

Such a feedback loop has been implicated in subperiosteal osteogenesis of chick tibiae in a study that spanned the age range from four-day-old embryos to 32-week-old chicks. As the periosteum lengthens, local mechanical stimuli enhance deposition of subperiosteal bone, enabling the shaft to grow in width. Of course, the causal sequence may be the reverse: this is a cycle, deposition of subperiosteal bone causing the periosteum to lengthen. In either case there is periosteal feedback. The issue is whether periosteal lengthening initiates the sequence. The experiments reported above and clinical experience suggest that it can and does; stripping periostea in a clinical study to take advantage of the feedback loop between periosteum and cartilaginous growth plate stimulated bone growth in children with poliomyelitis.[35]

Epiphyses grow in width from the zone where the periosteum attaches to the epiphysis, the so-called Ranvier's groove (zone of Ranvier, perichondrial or periosteal ring), a site of fibrogenesis, chondrogenesis and osteogenesis. Interstitial growth within epiphyses also contributes to their growth in width.[36]

In conclusion, the proliferative activity of precursors within mammalian growth plates is not intrinsically fixed but can respond to functional demands. Nevertheless, growth-plate prechondrogenic cells are distinguished by their ability to continue to divide *in the absence of functional demands*. Consequently, long bones continue to grow in the absence of functional demands. *Rate of division, not ability to divide, is controlled extrinsically.*

This is not true for other types of cartilage, such as condylar cartilages. Growth of the condylar cartilage is discussed in Chapter 33, but as the following studies utilized periosteal sectioning I include them here.

Circumferential periosteal sectioning of rat mandibular rami initiates an increase in vertical growth and a decrease in length because of a redirection in growth potential. Three days after removing the periosteum from the condylar process of 21-day-old rats, the *mitotic index* in the condylar cartilage *drops by 20 per cent*, indicating to Koski and his colleagues that periosteal tension plays a part in maintaining growth of the condylar process. They also came to the same conclusion using an entirely different technique – injection of a lathyritic agent to disrupt collagen, which reduces tension in the mandibular condylar periosteum, following which cell proliferation falls.[37]

NOTES

1. See Murray (1926, 1928), Felts (1961), Mawdsley and Ainsworth-Harrison (1963) and Hall (1985a) for how fundamental form is set.
2. See Howell (1917), Duerden (1920), Glücksmann (1938, 1939, 1942), Monson and Felts (1961), Chalmers and Ray (1962), Grüneberg (1963), Hall (1971a, 1975a, 1982c,d, 1985a, 1988, 1989, 1990c, 1998b), Y. Yasuda (1973) and Howes (1977a,b) for the fundamental form of endochondral bones. See Ritsila and Alhopuro (1973) for transplanted periostea, and Nogami et al. (1986) for bowing of long bones.
3. See Murray (1936), Chalmers (1965), Drachman and Sokoloff (1966), Murray and Drachman (1969) and Hall (1971a, 1975a, 1985a, 1997a), Moore and Lavelle (1974) and Herring (1993, 1994) for minor features of skeletal form.
4. See Nonaka et al. (1993) and Sasaki et al. (1994a,b, 1995) for uterine transplants.
5. This analysis, which Geber (1973) based on cleared and alizarin red-stained embryos, would bear repeating with more fine-grained analysis and examination of any associated hormonal changes or indirect effects related to sound-induced stress. The novelist Margaret Atwood is well aware of the medical uses of ultrasound, as we see in a feature Crake installed into the children he designed in *Oryx and Crake*. 'Once he'd [Crake] discovered that the cat family purred at the same frequency as the ultrasound used on bone fractures and skin lesions and were thus equipped with their own self-healing mechanism, he'd turned himself inside out in the attempt to install that feature' (Atwood, 2003, p. 191).
6. See Murray (1936), Krompecher (1937), Enlow (1968a, 1973), Lanyon (1974), Moore and Lavelle (1974), Starck (1975, 1979) and Herring (1993, 1994) for treatments of the classic literature on morphogenesis. See Athenstaedt (1969, 1970, 1974) for permanent polarization of the skeleton.
7. See Graver (1973, 1978) for directionality, and Clemen (1988) for the transplantation study. Plethodontid salamanders within the genus *Bilitoglossa* have either two or three dental laminae in the upper jaws associated with production of teeth on the premaxillary and maxillary bones and with the production of specialized teeth in males during the breeding season (Ehmcke et al., 2004).
8. See Maclean and Hall (1987) and Moody (1998) for the issues involved in maintaining the differentiated and morphogenetically stable states of cells. See Volumes 8 and 9 of Hall (1990–1994) for differentiation and morphogenesis of bone.
9. See Sofaer (1975) for the grafting studied with *Tabby*, Lumsden and Osborn (1976) for the influence of the first molar, and Peterková et al. (2002a,b) for subdivision of the tooth primordia in *Tabby* in comparison with wild type. The human equivalent of *Tabby* is *anhidrotic ectodermal dysplasia*, a syndrome in which teeth, hair and sweat glands all fail to form.
10. See Jones and Boyde (1976a,b) and Jones et al. (1975).
11. See Holtrop and Weinger (1972), Stanka (1975) and see Karcher-Djuricic et al. (1975) for the dental cells.
12. See Ede et al. (1977b), Holmes and Trelstad (1977), Trelstad (1977) and Ede and Wilby (1981) for studies using Golgi body orientation.
13. Hales' *Vegetable Staticks* (1727) and *Statistical Essays* (1738), the latter of which contains a reprint of the 1727 work, are both available in reprinted versions. See Keith (1919) for an excellent presentation and evaluation of these early studies.
14. Dubreuil (1913a–c), Haas (1926), Gatewood and Mullen (1927), Troitzky (1932) and Bisgard and Bisgard (1935) used radiography to obtain a time series of epiphyseal long-bone growth in rabbits, goats and dogs.
15. See Broadbent (1931) and Hofrath (1931) for the methodology, and Broadbent (1937) and Brodie (1941) for the first human studies.
16. See Miescher (1836), Todd and Bowman (1845), Sharpey (1848), Leidy (1849), Tomes and De Morgan (1853) and Müller (1858) for important studies from this period. See Brighton (1978) for the structure and function of growth plates as known 130–140 years later.
17. Transitions between stages of cartilage cell differentiation are not easy to identity (Chapters 1 and 22). As in discussions of other skeletal regions throughout the text, I use the terms chondroblasts, chondrocytes and hypertrophic chondrocytes for three stages of chondrogenesis. Cornelia (Nelly) Farnum pointed out to me that most 'growth-plate aficionados' use the term proliferating chondrocytes for the cells I call chondroblasts, pointing out that all cells of the growth plate fulfil the definition of chondrocytes, as all have the ability to make type II collagen. For ease of comparison with other systems – see Figs 5.9, 15.2 and 22.4 – I retain chondroblasts for those cells that lie between the actively proliferating layer of cells not yet producing ECM, and

those cells in which proliferation has slowed and matrix synthesis and deposition have been enhanced.

18. See Hinchliffe and Johnson (1983), Ham and Cormack (1979), Farnum (1994) and Farnum and Wilsman (1998) for differential growth at either end of a long bone.

19. From the vast literature available, see Hales (1727), Bhaskar et al. (1954), Kember (1960, 1971, 1978, 1979, 1983, 1985), Kember and Walker (1971), Walker and Kember (1971), Thorngren and Hansson (1973a,b), Silbermann and Kadar (1977), Brighton (1978), Kember and Kirkwood (1987), Farnum and Wilsman (1989, 1993), Kember and Kirkwood (1991), Pines and Hurwitz (1991), Kirkwood and Kember (1993) and Farnum (1994).

20. See Farnum (1994) and Farnum and Wilsman (1998, 2001) for differential growth rates of mammalian long bones.

21. See Kember (1978, 1979) for rat tibiae, Farnum and Wilsman (1989) for swine, Kirkwood and Kember (1993) for comparative analysis of avian and mammalian growth plates, and Cubo et al. (2000) for the study on herons and egrets. See Pines and Hurwitz (1991), Farnum (1994) and Starck and Ricklefs (1998) for reviews of the role of the growth plate in longitudinal bone growth in birds. For those interested in mechanisms of heterochrony, Cubo et al. (2000) used their data (Table 31.1) to argue for a non-heterochronic basis for morphological heterochrony.

22. See Davis et al. (1982) for the proteoglycans, Starck (1993, 1994, 1996) and Starck and Ricklefs (1998) for altricial vs. precocial birds, and Germiller and Goldstein (1997) for the functional analysis. For a review of the biology and biochemistry of mammalian growth plates, see Howell and Dean (1992).

23. See Moss-Salentijn et al. (1987, 1991) for the three-dimensional reconstructions and computer simulation, and Goss (1974) and Maclean and Hall (1987) for determinate and indeterminate growth.

24. See Dixon (1970) and Moss-Salentijn (1974) for intrinsic control, Ohsugi et al. (1997) for length of the cell cycle and gene expression, and Hunziker et al. (1987) and Hunziker and Schenk (1989) for rat tibial growth plates.

25. See Ballock and Reddi (1994) for the in-vitro study, and Ballock et al. (2000) for the kinase inhibition.

26. See Humphry (1861) for the earliest study, Hall-Craggs (1969) and Dawson and Kember (1974) for stapling or irradiation, and Selye (1934) for growth-plate regeneration.

27. See Ishikawa et al. (1992) for morning and overnight mating, Lampl et al. (1992) for human foetal growth, and Simmons (1992) for an overview of circadian aspects of bone biology.

28. See Swinson et al. (1975) for osteogenesis in humans, Burns (2000) for an overview, Tam et al. (1974) for the rabbit study, and Simmons and Cohen (1980) for fracture repair.

29. See Farnum (1994) and Farnum and Wilsman (1989, 1993) for overviews.

30. See Nir et al. (1972) for the pineal, and Hunziker et al. (1994) for Igf-1 and growth hormone. Using foetal rat metatarsals from 18- to 19-day-old foetal rats, Scheven and Hamilton (1991) showed that growth hormone effects on growth are mediated via Igf-1. See Dupont and Holxenberger (2003) for the actions of Igfs during development.

31. See Simmons et al. (1974, 1984).

32. See Herring et al. (1993) and Popowics et al. (2002) for these two studies.

33. See Hert (1969), Kéry (1972) and Bex and Maltha (1976) for studies similar to that published by Crilly (1972).

34. See Warrell and Taylor (1979) and Taylor et al. (1987). Sectioning the periosteum and scoring the underlying cortical bone is a means of enhancing the number of osteoprogenitor cells obtained from the periosteum (Simon et al., 2003).

35. See Lutfi (1974) for the chick study, and Jenkins et al. (1975) for periosteal stripping and poliomyelitis.

36. See Badi (1972a, 1978), Shimomura et al. (1973), Ogden et al. (1975) and Shapiro et al. (1977) for growth in width at Ranvier's groove, and Hert (1972) for interstitial epiphyseal growth.

37. See McLain et al. (1982) for sectioning the mandibular ramus, Koski and Rönning (1982) and Koski et al. (1985) for the condylar process, and Nakamura and Koski (1983) for the lathyritic study.

Long Bone Growth: A Case of Crying Wolff?

Chapter 32

The trajectory theory presents idealized structures formed in response to pure forces by an isolated bone whose only function is to resist those mechanical forces. Theoretical approaches which treat bones as idealized, isolated units ... simply fall short of reality. A particular bone's response to altered mechanical stress might be compromised by the simultaneous response of the attached muscles or connective tissue ... by altered blood flow, by associated mineral requirements, etc ... Murray foresaw this interactive view and elegantly expressed it in his concluding chapter (1936, pp. 177–180): 'Every bony structure is a compromise and no compromise is perfect ... It [the trajectorial theory] requires, not rejection, but dilution' (p. 179). (Hall, 1985a, pp. xxvi–xxvii)

WOLFF, VON MEYER OR ROUX

The reality of history is not always reflected in modern-day accounts of how the architecture of bone reflects the forces imposed upon it. Wolff (1870, 1892) is credited with originating the *trajectory theory* relating bone structure to the mechanical forces imposed upon it, a theory now known as *Wolff's Law*. But Julius Ward (1838) had already compared the architecture of the proximal portion of the femur with a bracket, and it was von Meyer not Wolff, who first proposed the theory in 1867 (Fig. 2.7). Wolff was to von Meyer's trajectory theory as Huxley was to Darwin's theory of evolution – advocate, defender and popularizer – although Lee and Taylor (1999) argue that Wolff misinterpreted the mechanical data and rejected any role for bone resorption in determination of bone structure (Fig. 32.1). Recognition of the strength of bone and of correlations between strength and the size and shape of bones is, of course, even older than Ward's comparison of cranes and femora. Galileo (1638) was well aware of the additional strength that came from bones as tubes rather than as boxes.

Following Ward's observation that the architecture of the proximal portion of the human femur resembles a bracket (Fig. 31.1) studies of the mechanical properties of intact bone were carried out in the mid-1800s by Wertheim (1847), von Meyer (1867) and Rauber (1876). When Meyer compared the trabecular structure within bone with the stress to which the bone had been subjected, he found a correlation: the orientation of the trabeculae follows lines of stress (Figs 2.6 and 32.1). Meyer was aided in this analysis by the Swiss engineer Culmann (1866) but Wolff compared the architecture of the femur to the design used in constructing the Fairbairn steam crane.[1]

Wolff explicitly related bone structure to bone function in terms of the forces and loads imposed on living active bone. Development and dissemination of Wolff's Law owes much to Roux (1885) and Koch (1917). In fact, it has been argued that Roux was the first to accurately describe the adaptation of bone to altered load, and that consequently *'Wolff's law' should really be 'Roux's law.'*[2]

Two 20th century restatements of Wolff's Law follow.

- 'Wheresoever stresses of pressure and tension are caused in a bone ... formation of bone takes place' (Jansen, 1920, p. 5).
- 'The form of a bone being given, the bone elements place or displace themselves in the direction of functional pressure and increase or decrease their mass to reflect the amount of functional pressure' (Bassett, 1968, p. 260).

But trabecular structure of cancellous bone is not merely a set of actualized trajectories and the trajectory theory has been modified in various ways, mostly to account for factors other than mechanical history that leave their

Figure 32.1 Julius Wolff (1870) compared the stress trajectories in the Culmann crane (left) with the pattern of trabeculae in the proximal end of the human femur (right). See text for details. Modified from Wolff (1870) and Lee and Taylor (1999).

mark on bone structure. As one example, microcracks of the order of 50 ± 10 μm in length are an important stimulus for remodeling of loaded bone (Lee *et al.*, 2002).

Results from studies on CAM-grafted embryonic bones were used to argue that blood vessels and vascular influences, rather than mechanical factors, determine trabecular architecture (Box 12.3). From studies on patterns of force transmission in cat skulls, it was proposed that stresses were best visualized in *functional units* within the skull rather than in individual elements, one element compensating for forces on another. Venous stasis, which is known to accelerate both fracture repair and periosteal osteogenesis, has no effect on longitudinal growth of long bones in rabbits, other than a decline one or two days after surgery.[3]

I wrote the epigraph to this chapter in an overview introducing a reprinting of *Bones* by P. D. F. Murray. As Murray himself saw it, 'every bony structure is a compromise and no compromise is perfect … [the trajectorial theory] requires, not rejection, but dilution' (Murray, 1936, p. 179).

Four detailed papers by Howard Frost (1990a–d) set out a major review of skeletal adaptation to mechanical usage and discuss modeling and remodeling of bone, cartilage and fibrous tissue. The most insightful recent analysis of Wolff's Law also appeared in the early 1990s (Bertram and Swartz, 1991). A major concern and source of confusion concerning the differing responses of bone to what appear to be similar mechanical conditions is the use of different models and the underlying assumption that what is true for one bone/situation will be true for all – the universal and over-riding primacy of Wolff's Law. John Bertram is especially concerned with differences between: load applied during bone development (when

the effect is primarily on differentiation and morphogenesis rather than on growth), load applied during growth, and load applied to adult, non-growing animals/bones. John considers that changes in adults – especially during fracture repair – are more properly attributed to inflammatory mediated processes than to Wolff's law.[4]

RESPONSE TO PRESSURE

There was much debate in the latter part of the 19th century over whether *pressure* stimulates or retards bone growth (Fig. 31.5). Because of the responsiveness of epiphyseal cartilage to pressure, Wolff maintained that growth was stimulated by pressure. Hueter and Volkmann took the opposite view. From both medico-surgical experience and anthropological studies, it was clear that pressure could initiate bone formation in situations where no cartilage was present, as in the production of Wormian bones in skulls deformed by binding.[5]

More recent studies from three groups take some of the strain out of the system by resolving the apparent paradox. Hert and Liskova, Denis Carter and his colleagues, and Lanyon and Rubin all provide *compelling theoretical and experimental verification that the effects of pressure on bone growth vary with the magnitude of the pressure applied.* Pressure beyond normal physiological limits inhibits bone growth. Relaxation of pressure stimulates growth. Perhaps more importantly, bone is – and bones are – especially adapted to respond to *intermittent loading*. Rapid transformation from quiescence to bone formation and increased synthetic activity can be recorded in osteocytes in adult periostea following a single brief period of bone loading.[6]

CONTINUOUS OR INTERMITTENT MECHANICAL STIMULI

We stimulate our bones with every breath, bite or step we take. Respiration and locomotion produce cycles of bone deformation as tension and compression are applied alternately to vertebrae, ribs and long bones.[7]

In their studies on the responses of rabbit tibiae to mechanical stress, J. Hert and his colleagues emphasized the importance of *intermittent loading* if skeletal cells are to be activated. Because mechanical loading of the skeleton is normally intermittent, Hert and Zalud (1971) argued that piezoelectric models based on differential responses of cells to pressure or tension are inadequate. Instead, they proposed that skeletal cells sense local changes in electrical potential and ionic fluxes. A contrary view came from C. A. L. Bassett's laboratory. Using *in-vitro* studies, Bassett *et al.* (1964) demonstrated that mechanical deformation of the apatite–collagen piezoelectrical junctions in bone matrix produces an electrical signal, which in turn elicits a cellular response, resulting in synthesis and oriented deposition of newly formed collagen fibrils.[8] See Box 29.1 for a discussion of the efficacy of intermittent mechanical stimulation in promoting chondrogenesis in fracture repair, and Box 32.1 for the use of pulsed electromagnetic fields (PEMF) in initiating repair in persistent non-unions.

Box 32.1 Pulsed electromagnetic fields (PEMF)

A substantial although not entirely concordant body of basic and clinical studies documents enhanced fracture repair following application of pulsed electromagnetic fields (PEMF), especially to initiate repair of persistent non-unions and the false joints (pseudarthroses, Fig. 29.2) that can form following prolonged misalignment and abnormal mechanical stresses (Box 29.2). The pioneering studies were initiated by the late Cal Bassett, then at Columbia University in New York.[a]

It is unusual for clinical studies with humans to demonstrate more consistent enhanced healing than is found in studies with other animal models. For example, healing of fibula fractures in dogs was enhanced, and the mechanical quality of the bone improved, 28 days after application of 65-Hz pulses of PEMF. But when 15-Hz pulses were applied, there appeared to be no effect when the bones were examined two or six months later, nor did Law *et al.* (1985) see any improvement in sheep tibiae.

In part this is because the wave form of PEMF is important: four different wave forms stimulate healing of fractures of the rodent radius (including increasing the tensile strength of the callus by up to 30 per cent), three other wave forms having no effect.[b]

After Grace *et al.* (1998) applied 72-Hz square wave PEMF to an osteochondral defect in the patello-femoral groove of rats they observed enhanced early vascular invasion, early initiation of chondrogenesis, early initiation of osteogenesis and advanced restoration of normal trabecular architecture, concluding that *PEMF advances the early phases of repair*. However, they counsel against prolonged use – chondrogenesis can be prolonged to the stage where development of normal bone architecture is delayed – too much of a good thing too early.

McLeod and Rubin (1989) determined that frequency bands of <75 Hz were optimal for bone adaptation, implying that bones are extremely sensitive to low-frequency electric fields. For comparison, the refresh rate of a good monitor operates at around 70–75 Hz/second and neural oscillators in the brain at 35–75 Hz.

In an early clinical trial, Bassett reported the successful repair of 96 per cent of human tibial non-unions. In studies in other animals, pulsed electromagnetic fields applied to rats improved the effects of disuse osteoporosis and prevented congenital pseudarthroses; a pulse train of PEMF increased cellular calcium and promoted mineralization while a single pulse decreased cellular calcium but promoted bone formation, a desirable outcome in patients with osteoporosis or avascular necrosis.[c]

Initially, many were skeptical about the efficacy of PEMF. Barker and Lunt (1983) cautioned about the total absence of controlled trials and the lack of experimental verification of effects caused by magnetic fields (see below). Smith and Nagel (1983) applied PEMF to rabbit long bones for two hours/day for eight weeks and found no change in bone growth but did find a 22 per cent increase in articular cartilage glycosaminoglycans (GAGs). Of course, there was no fracture or non-union in these animals. These authors were unduly pessimistic.

By 1982 Bassett could report that 5000 patients had been treated with PEMF, that PEMF was beginning to be used in dentistry, and that evidence had accumulated to show that PEMF provided a signal for chondrogenesis and for vascular invasion of cartilage, although not for osteogenesis *per se*. Studies with isolated cell populations bear this out: tibial chondroblasts from seven-day-old chick embryos react to continuous exposure to pulsed magnetic fields for seven days *in vitro* with reduced collagen synthesis, a less marked reduction in both acid mucopolysaccharide production and proliferation, but with enhanced mineralization. Rabbit costal chondrocytes respond to PEMF by enhanced responsiveness to parathyroid hormone and enhanced differentiation, while 16-day-old chick embryo tibial and sternal chondrocytes (and skin fibroblasts) respond to PEMF with enhanced synthesis of GAGs.[d]

Ongoing investigations in both experimental and clinical situations continue to explore the efficacy of PEMF in such situations as: the healing of the flexor tendon in chickens, in which tension decreases and tendinous adhesions increase; a double-blind clinical trial in which 27 individuals with osteoarthritis of the knee were treated with PEMF over a month, with between 20 and 60 improvement in the parameters measured; and a study in which $TGF\beta$-1 increased in cells from non-unions exposed to PEMF, while cell number, alkaline phosphatase activity, collagen synthesis, prostaglandin E-2 and osteocalcin production were unaffected.[e]

[a] See Bassett *et al.* (1974a,b, 1977, 1978, 1979), Fitton-Jackson and Bassett (1980) and Goodman *et al.* (1983). For a review of electro-stimulation and bone-fracture repair, see Behari (1991). In what was perhaps their first study, Bassett and Herrmann (1968) showed that fibroblasts are sensitive to electrostatic fields, synthesis of collagen increasing by as much as 100 per cent.
[b] See Bassett *et al.* (1974a,b) and Miller *et al.* (1984) for 65- and 15-Hz pulses, and Christel *et al.* (1980) for wave form.
[c] See Bassett *et al.* (1978) for the early trial, L. S. Bassett *et al.* (1979) and C. A. L. Bassett *et al.* (1981a,b) for disuse osteoporosis and congenital pseudarthroses, and Goodman *et al.* (1983) for mineralization.
[d] See Archer and Ratcliffe (1983) for tibial chondrocytes, Hiraki *et al.* (1987) for costal chondrocytes, and Norton *et al.* (1988) for enhanced GAG synthesis.
[e] See Robotti *et al.* (1999) for chick flexor tendons, Trock *et al.* (1993) for the study of osteoarthritis, and Guerkov *et al.* (2001) for the study with non-union cells.

SCALING AND VARIATION: WHEN WOLFF MEETS THE DWARFS

Are rats 'merely' scaled-up mice and are dwarf (miniaturized) taxa 'merely' scaled down versions of their larger relatives, or does taxonomic diversification accompanied by size increase involve modification of the processes of development and growth? It has, for example, been argued that all members of the genus *Australopithecus* are variations of the same animal over a range of sizes, but here we are dealing with a single taxon. Similarly, an examination of the scaling of skeletal growth in six specimens of *Archaeopteryx lithographica*, all of which are sub-adult (Fig. 32.2), led to the twofold conclusions that these represent a single growth series of a single species.[9]

Going in the other direction, an allometric analysis of skull development in the common hippopotamus, *Hippopotamus amphibius*, and the pygmy hippopotamus, *Hexaprotodon liberiensis*, shows that the dwarf species is not simply a scaled-down version of its larger relative (Weston, 2003).

Regional variation in bone growth is related to local parameters of mechanical stress and strain, as demonstrated, for example, in the distal growth plates of rabbit femora, in which stress patterns correlate with growth rates, high compressive stress correlating with lower growth. These associations, however, explain only 15 per cent of the variation in growth rate. Responses to strenuous exercise also result from local loading effects, as determined in rat tibiae and metatarsals, just as total body size and regional variation within individual skeletal elements influence response to mechanical loading.[10] van der Meulen and Carter (1995) modelled intrinsic long-bone growth with extrinsic adaptive modeling to simulate allometric morphological relationships over the size range from mice to mammals. In this model, scaling is adaptive, not intrinsic.

There are situations where Wolff's law seems not to apply, especially in such small mammals as bats and shrews, where terrestrial locomotion is limited. For instance, no correlation between mechanical forces and bone structure was found in a sample of 94 femora from the short-tailed shrew, *Blarina brevicauda*, the little brown bat, *Myotis lucifugus*, and another bat, the eastern pipistrelle, *Pipistrellus subflavus*. All are small – the average weights of these animals are 8.6, 6.4 and 6.3 grams, respectively, very low for mammals; squirrels weigh around 900–1200 grams. As David Dawson concluded: 'There is no evidence that bones in these diminutive mammals respond to mechanical forces, and the applicability of Wolff's Law is not indicated … it is hypothesized that intrinsic tissue strength is sufficient to resist mechanical deformation, and femoral anatomy in these species is dictated by genetic and inherent physiologic conditions' (Dawson, 1980, p. 1).[11]

GRAVITY

Many have investigated the skeletons of numerous species under varied conditions of micro- or zero gravity, including space flight. If one can generalize from the data obtained, it would appear that bone formation is inhibited during space flight – there is complete cessation of periosteal bone formation in rats, for example – but that resorption is unchanged. *Increased gravity* increases resorption of mouse calvarial bone, and more Ca^{++} is released (Gazit, 1980).

In a study on the effects of 18.5 days of space flight on trabecular bone in rats (Table 32.1), fat mass increased while mineralized tissue mass declined. Osteoblast numbers declined but osteoclast numbers remained unaffected, confirming the effect on formation, not resorption, and demonstrating an uncoupling of the normal tight association between formation and resorption, a coupling that is influenced by mechanical conditions. By 29 days

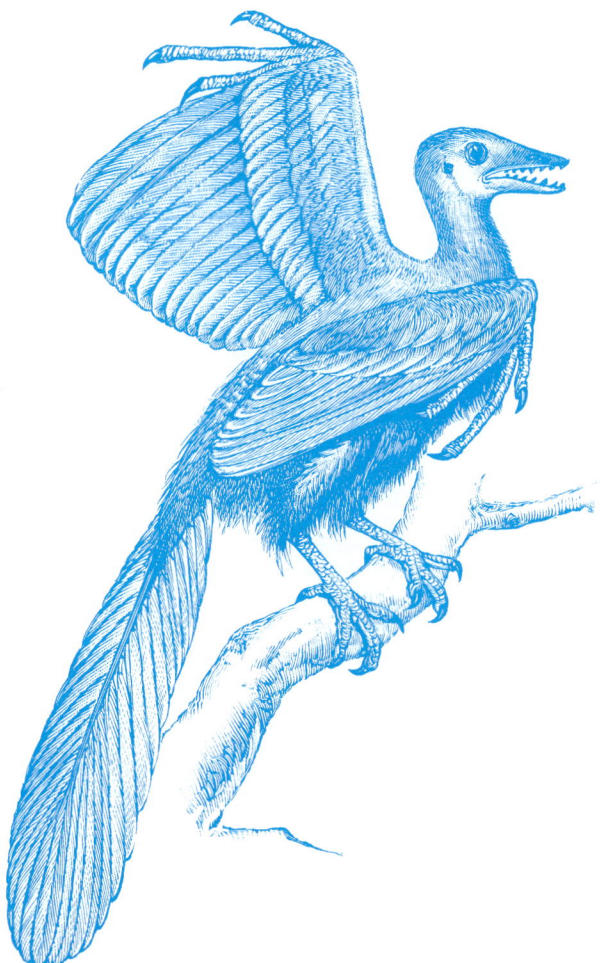

Figure 32.2 *Archaeopteryx lithographica*, approximately one-third natural size. From Romanes (1901).

Table 32.1 Effects of spaceflight on periosteal formation, periosteal apposition, and length of arrest lines in rat tibiae[a]

Rats	Mean periosteal formation rate (10^{-3} mm^3/day)		Mean periosteal apposition rate (10^{-3} mm^3/day)		Arrest line length (mm)
	Bone	Matrix	Bone	Matrix	
Flight period					
Flight	9.4 ± 2.8	7.2 ± 2.8	1.3 ± 0.4	1.0 ± 0.4	
Flight control[b]	15.8 ± 1.5	13.8 ± 1.4	2.2 ± 0.2	1.9 ± 0.2	
Vivarium control[c]	16.0 ± 1.4	14.0 ± 1.4	1.8 ± 0.2	2.3 ± 0.2	
Post-flight period					
Flight	17.1 ± 2.2	17.8 ± 2.1	2.2 ± 0.2	2.3 ± 0.2	5.3 ± 0.6
Flight control					2.1 ± 0.6
Vivarium control	11.3 ± 1.4	11.2 ± 0.4	1.4 ± 1.4	1.4 ± 0.2	1.5 ± 0.7

[a] Based on data from Morey and Baylink (1978).
[b] Controls were housed in an identical, ground-based spacecraft for the duration of the experiment.
[c] Controls were housed in standard cages and animal quarters.

post-flight most parameters had returned to normal. Although mineral metabolism is disrupted in zero gravity, there is a six- to nine-month musculoskeletal 'safety period' for humans in space. Any bone or mineral loss during this period is reversed upon returning to earth's gravity.[12] A sample of studies follows, mostly representative of results obtained on Kosmos or NASA flights.

(i) Rats maintained in prolonged hypogravity in the 1975 US–USSR Kosmos Space Laboratory suffered substantial bone loss (Asling, 1977).

(ii) Growing rats given tetracycline to label bone growth, and then flown on board Kosmos 782 and Kosmos 936 for 19 days, showed defects at the periosteal surfaces of the tibial diaphyses, consisting of a 3 μm arrest line, hypomineralized matrix and abnormal collagen orientation (Turner *et al.*, 1985).

(iii) Five days of weightlessness on Kosmos 1514 increased osteoclast numbers in pregnant rats and so decreased bone mass via excess resorption (Vico *et al.*, 1987).

(iv) Kosmos 2044 involved a 13.8-day flight, and a 10-hour recovery that proved to be sufficient time to replenish preosteoblasts in the rat maxillary periodontal ligament lost during the flight (Garetto *et al.*, 1992).

(v) Teeth are not immune to the effects of gravity. Rat incisor dentine showed a 10–15 per cent increase in Ca^{++}, a 20–30 per cent increase in PO_4^{3-} and abnormal distribution of GAGs after 18.5 days on board Kosmos 1129 (Rosenberg *et al.*, 1984).

(vi) Growing rat bones were examined in NASA Space Lab 3 after one week in space. Levels of osteocalcin, calcium and hydroxyproline had all declined in vertebrae and humeri. A one-week space flight reduced the growth rates of humeri but not tibiae of growing rats, but resulted in tibiae that were weaker, as assessed by a three-point bending test (Patterson-Buckendahl *et al.*, 1987; S. R. Shaw *et al.*, 1988).[13]

Both weight-bearing and non-weight-bearing elements of the skeleton respond to weightlessness with decreased histogenesis, as demonstrated in a comparative study of periodontal ligament and tibial metaphysis in Sprague-Dawley rats in simulated weightlessness and in the non-weight-bearing portions of the rat mandible. Scleral ossicles, which are non-weight-bearing, developed normally in quail embryos flown on the Mir US–Russian joint project.[14]

Developing skeletons also are responsive to increased gravity. Mouse limb buds exposed *in vitro* to 2.6 *g* in a centrifuge show a proximo-distal gradient of sensitivity, the proximal mesenchyme being less sensitive. Indeed, proximal elements present at the beginning of the experiment disappeared, while the differentiation of distal elements was accelerated. Pre-metatarsal elements at an early developmental stage produced cartilage rods when flown in the space shuttle, suggesting normal differentiation.[15]

Studies on humans in space are relevant to the skeletal loss consequent to prolonged bed rest, menopause (one per cent and more loss of skeletal mass/year) or in middle to old age (during which 15 per cent of skeletal mass is lost).[16] During the manned space flights, astronauts lost 3–15 per cent of their heel-bone density (Gemini) and 0.2 per cent of body calcium (Apollo) (Goode and Rambaut, 1985). In a review of the consequences of space flight published in 1986, Wheldon concluded that bone loss in space parallels bone loss during bed rest or in individuals with polio, and is of the order of 0.4 per cent of body calcium/month. Bone density does not return to normal if the duration of the space flight is prolonged, presumably because loss isolates bone spicules or trabeculae to the point of no recovery. Other approaches may enable us to address these issues; rats maintained under conditions of 2.76 or 4.15 *g* for prolonged periods (810 days in this case) show effects similar to those seen after immobilization (Table 32.2).

Table 32.2 Rats maintained under increased gravity conditions (2.76 or 4.15 *g*) for 810 days show an effect similar to that seen after immobilization[a]

Parameters	Gravity of 2.76 *g*	Gravity of 4.15 *g*
Body weight	19.3	29.1
Femur length	6.6	9
Cross-section of femur	15.5	19.1

[a] Results shown as % of control values.

TRANSDUCTION OF MECHANICAL STIMULI

Despite the studies discussed above and others, we do not fully understand how the skeleton responds to mechanical stimulation, although cAMP as a mediator of prechondroblast proliferation to conform with changing mechanical condition represents an important feedback control system (Chapter 31).

Muscles can and do insert onto resorptive surfaces on bone so that mechanical action can stimulate resorption and deposition directly (Figs 35.2 and 35.4). One potential mechanism receiving wide support in the 1970s was translation of mechanical stimuli into bioelectrical activity to which skeletal cells could respond. I discuss this approach and then turn to three other approaches.[17]

There is no doubt but that electrical currents in the μamp range influence the proliferation, differentiation and activity of skeletal cells, although a mechanistic relationship between mechanical stimulation and bioelectrical potential is elusive. Empirical evidence for the efficacy of exposure to electric current/fields includes:

- enhanced chondrogenesis and acceleration of the growth of epiphyseal cartilage and bone following direct application of an electrical field;
- acceleration of the proliferation of induced cartilage following application of a 5 μA current to muscle cultured with Bmp – the cartilage having been induced by Bmp;
- promotion of precocious initiation of mineralization by hypertrophic chondrocytes following 10 μA DC current applied *in vitro* to femora from nine-day-old chick embryos, at which age the cartilage is unmineralized;
- enhancement of osteogenesis in chick tibiae and mouse calvariae exposed *in vitro* to as little as 10^{-5} V/m for 30 minutes/day;
- promotion of osteogenesis in foetal rat long bones in response to direct or pulsating current *in vitro*; and
- alteration of oxygen tension in the immediate vicinity of skeletal cells.[18]

Bone is deposited in regions of electro-negativity. Healing fractures respond to low-voltage electric fields with enhanced repair. For example, new bone is deposited around an active electrode implanted in the medullary canal of rabbits, but not around an inactive electrode. Indeed, bone disappears around the inactive electrode but can be regained if current is allowed to flow. The new bone forms from polymorphic perivascular cells associated with invading blood vessels. Mineral is differentially deposited around the electrodes – mineralizing collagen around both active and inactive electrodes, and degenerating cells around the inactive electrode (Brighton and Hunt, 1986).

In studies testing skeletal responsiveness to electrical stimuli, as in all experimental studies, adequate controls must be used. Inserting the wire into medullary canals or marrow *in vivo* can enhance osteogenesis, even with no current flowing through the wire. The effect is not simply the result of trauma – inserting and removing a wire does not enhance osteogenesis – indicating an effect of the wire itself. Furthermore, the authors of these studies found that electrical and mechanical stimuli were additive. Not all mechanical stimuli are perceived by cells as electrical change.[19]

NOTES

1. The Fairbairn steam crane built in 1875 for the dockyard at Bristol, designed by the Scottish engineer William Fairbairn (1789–1874) to lift 35 tons from a height of 115 feet, was in continuous operation at the dockyards until 1973. Now refurbished, the crane can be visited as part of the Bristol dockyard restoration. Fairbairn operated a shipyard in London where the first iron-hulled steamship, the *Lord Dundas*, was built. He conceived using tubular steel for construction when assisting Robert Louis Stephenson in the construction of the Conway and the Menai Strait bridges in North Wales.

2. Wolff's classic 1892 study was reprinted in translation by Maquet and Furlong (1986). See Lee and Taylor (1999) for Roux's Law. Benninghoff (1925, 1930) performed some of the more detailed analyses on trabecular fibre structure (Fig. 2.6); Murray (1936) and Altmann (1964) provide thorough evaluation of these studies. Glücksmann (1938) used *in-vitro* cultivation of skeletal elements to demonstrate the influence of pressure on structural orientation, while Murray and Selby (1930), Murray (1936) and Altmann (1964) provide what are still among the best accounts of the early literature on the response of the skeleton to mechanical forces. For later studies, see Pauwels (1973), and see Oxnard (1991) for an insightful analysis of the three-dimensional architecture – the morphanalysis – of bone.

3. See Hancox (1947) for the CAM-grafts, Buckland-Wright (1978) for the functional unit approach, and Hansson *et al.* (1975) for venous stasis. Under conditions of prolonged venous status, as when the inferior vena cava is ligated, osteocytes dedifferentiate to chondrocytes as areas of bone cells transform from acidophilic to basophilic and are replaced by an island of cartilage cells (Abdalla and Harrison, 1966).

4. See also Bertram (2001) for an analysis of how the trabecular bone immediately beneath the articular cartilage of mammalian long bones may attenuate loading of both bone and cartilage. I am grateful to John Bertram for his analyses and his

comments on this chapter. John considers that growth 'appears to have a strong element of determination (however that is mediated), whereas the modifications of mature bone appear to have some stringent limitations' (pers. comm.).

5. See Wolff (1892), Hueter (1862, 1863) and Volkmann (1862) for pressure and epiphyseal cartilages, and Dorsey (1897) for Wormian bones.

6. See Hert (1964a–c, 1969), Hert and Liskova (1964), Lanyon and Rubin (1984), Carter (1987), Carter *et al.* (1987), Rubin and Lanyon (1987), Carter and Wong (1988), Rubin *et al.* (1989), Wong and Carter (1990), Carter *et al.* (1991, 1996) and Carter and Orr (1992). See Pead *et al.* (1988a,b) for rapid response to brief loading.

7. See Lanyon (1972), Lanyon *et al.* (1975) and Piekarski and Munro (1977) for data.

8. See Hert *et al.* (1969, 1971a,b, 1972a,b), Liskova and Hert (1971) and Chamay and Tschantz (1972). For the contrary view, see Bassett and Becker (1962), Becker *et al.* (1964), Bassett (1968), Bassett and Pawluk (1972) and Becker (1978). For mechanical loading of tibial periostea stimulating rapid changes in periosteal gene expression, see Raab-Cullen *et al.* (1994).

9. See Nonaka *et al.* (1993) and Sasaki *et al.* (1994a,b, 1995) for uterine transplants, Pilbeam and Gould (1974) for australopithecines, and Houck *et al.* (1990) for *Archaeopteryx*.

10. See Lerner *et al.* (1998) and Li *et al.* (1991) for these rat and rabbit studies, and see Felts and Spurell (1965, 1966) for analyses of long-bone structure in cetaceans.

11. The selective pressure for evolution of the distinctive interdigital membrane in bat wings seems to have been elongation of the metacarpal bone (Kovtun, 1985). See Adams (1998) for the developmental and functional integration of the elements of bat wings.

12. See Morey and Baylink (1978) for the general statement, Whedon *et al.* (1976) for mineral metabolism, Jee *et al.* (1983) for trabecular bone, and Whedon and Heaney (1993) for an overview of the effects of weightlessness, paralysis and inactivity on bone growth.

13. John Bertram commented on a 'rat-in-space' study in which there was a decrease in the deposition of periosteal bone in the dentary (which is not a weight-bearing bone) and that the decrease occurred in areas not associated with muscle attachment as well as where muscles attach, the implication being that some effects of zero gravity may be systemic (Bertram, pers. comm.).

14. See Fielder *et al.* (1986), Simmons *et al.* (1983a) and Barrett *et al.* (2000) for these three studies.

15. See Duke (1983) for excess gravity, Klement and Spooner (1994) for the space shuttle, and Klement *et al.* (2004) for subsequent studies using the NASA rotating wall vessel which simulates microgravity by randomizing the direction of the gravitational forces on cells.

16. This is because immobilization, more or less whatever the cause, leads to hypoplasia of bone. As one example, four weeks after unilateral resection of the sciatic nerve to the rat femur, periosteal (cortical) bone and ash content are both reduced, the latter indicative of reduced mineral content (Pennock *et al.*, 1972).

17. See Brash (1924, 1934) and Hoyte and Enlow (1966) for stimulation of resorption and deposition, and see the reviews by Bassett (1972b), Marino and Becker (1977), Spadaro (1977), Hall (1983f), and chapters in the volumes edited by Liboff and Rinaldi (1974), Attinger and Parakkal (1977), Brighton *et al.* (1979) and Connolly (1981).

18. See Watson *et al.* (1975), Brighton *et al.* (1976) and Norton *et al.* (1977) for cartilage, Norton and Moore (1972), Spadaro (1982) and Noda and Satoh (1985b) for bone, Nogami *et al.* (1982) for proliferation of induced cartilage, Noda and Sato (1985a) for precocious mineralization, Fitzsimmons *et al.* (1986) for tibiae and calvariae *in vitro*, Theharne *et al.* (1980) for direct or pulsating current, and Brighton and Friedenberg (1974) and Brighton *et al.* (1975) for enhanced local O_2 tension. Electrical stimulation enhances bone and muscle mass in paraplegics (Pacy *et al.*, 1988).

19. See Spadaro *et al.* (1986) and Schaberg *et al.* (1985) for these studies, and see the volumes edited by Brighton *et al.* (1979), and Connolly (1981) and Hall (1983f) for the electrical properties of bone and cartilage as seen in both experimental and clinical studies. Becker (1972b) applied electrical currents of 3–6 nA to the amputated limbs of 15 rats and obtained a high percentage of regeneration and formation of epiphyseal growth centres.

Part XI

Staying Apart

The condylar process is not a primary growth centre for the mandible. It is a centre of adaptive and compensatory growth.

Part of the difference between the growth of synchondroseal and condylar cartilages is that growth of condylar secondary cartilage is appositional, while growth of synchondroseal cartilage is interstitial. Consequently, growth of the jaws responds to functional demand, as Lewis Carroll tells us in *Alice's Adventures in wonderland*.

"You are old," said the youth, "and your jaws are too weak
For anything tougher than suet;
Yet you finished the goose, with the bones and the beak —
Pray, how did you manage to do it?"
"In my youth, said his father, "I took to the law
And argued each case with my wife;
And the muscular strength, which it gave to my jaw,
Has lasted the rest of my life."

The Temporomandibular Joint and Synchondroses

Chapter 33

Passive boundaries or active growth centres?

Initiation of the formation of a joint between the elements of a long bone does not depend upon functional demand or extrinsic stimulation. Maintaining the joint does.

Hamburger (1929) demonstrated this in a classic study in which development of the limbs of embryos proceeded normally after the developing lumbosacral spinal cord had been extirpated. Articular cavities formed, although they were smaller than normal. In a similar vein, the mesenchyme of a presumptive joint region placed *in vitro* differentiates into a morphologically normal joint with articulating surfaces.[1] Innervation *is* important in maintaining fine-scale features and responses of joints. Sensory neurons are found in cartilages at articular surfaces in various joints – hyaline cartilage in long bones and vertebrae, secondary cartilage in temporomandibular joints (TMJs), fibrocartilage of the intervertebral discs and menisci – indicating a role in local sensing (Schwab and Funk, 1998). Prolonged cultivation *in vitro*, transplantation to an ectopic site or *in-ovo* paralysis (Fig. 16.5), all lead to fusion of tissues across preexisting joints, and fusions of joints with other connective or skeletal tissues.[2]

Such studies also reveal how sensitive extracellular matrix (ECM) molecules are to mechanical factors. Keratan and chondroitin-6-sulphates are expressed differentially in the flattened cells of future joint regions. Syndecan-3, tenascin and collagens I and XII are expressed at sites of future joints. Hyaluronan is required for joint cavitation. Immobilizing chick embryos using paralyzing drugs allowed Mikic *et al.* (2000a) to demonstrate that tenascin and collagen types I and XII are mechanically modulated during synovial joint formation; all three molecules decrease at what would have been sites of joint formation as cells in those regions switch to chondrogenesis.[3]

Ruano-Gil *et al.* (1985) took the opposite approach to paralysis. Larger than normal joints form when reserpine is injected into chick embryos at H.H. 25/26 to *enhance* embryonic motility; reserpine acts by depleting the supply of dopamine, norepinephrine and serotonin in nerve terminals. Such studies established that the *maintenance* of functioning long-bone joints requires extrinsic mechanical factors. Some of these factors can come from unexpected places. Unilateral amputation of a rat hind paw initiates remodeling of the articular cartilage in *both hip joints*, a bilateral response to altered biomechanics in both limbs (Threlkeld and Smith, 1988).

In contrast, although the TMJ can form in the absence of extrinsic stimuli (Glasstone, 1971), maintaining proliferating progenitor cells in the condylar process requires mechanical stimuli normally associated with joint function. The condylar process is not a primary growth centre for the mandible. It is a centre of adaptive and compensatory growth. I discuss the evidence supporting this claim before turning to a discussion of two other types of joint, cranial synchondroses and cranial sutures.

THE MAMMALIAN TEMPOROMANDIBULAR JOINT (TMJ)

The contribution made by the condylar process of the mammalian dentary to the morphogenesis and growth of the human face has been a controversial topic for many years. The issue is not only academic; a great deal of our TMJ pain comes from imbalances in the cartilage of the condylar process. Simply put, the topic can be reduced to one question: is the condylar cartilage a primary growth centre for the mandible, or is its growth compensatory and adaptive?[4] Although the question is simple, the answer is not. Age and species differences both confound interpretation.

Mechanical factors

The condylar process

A sizeable body of evidence indicates that both the presence and the continued growth of a condylar process depend on biomechanical stimuli provided by functioning musculature. In particular, maintenance of a pool of proliferating precursor cells on the apex of the condylar process depends upon local mechanical stimulation, primarily from forces generated by the lateral pterygoid muscle, which inserts onto the condylar process. Mechanoreceptors within the capsule of the temporomandibular joint play a role in coordinating the activity of these muscle fibres (Abe et al., 1973). Preferential differentiation of chondroblasts over osteoblasts in the condylar process also requires movement and associated mechanical forces (Chapter 12).

Similarly, continued deposition of bone on the condylar process depends on local mechanical stimuli; condylar growth is inhibited in human congenital muscular dystrophy because of reduced muscle function. This also is true for growth of the other two posterior processes of the mammalian dentary, the angular and coronoid processes, on which cartilage is more transitory. Although we think of muscular dystrophy in terms of the limb skeleton, in two murine muscular dystrophic mutants – dy/dy and dy^{25}/dy^{25} – the dentary is affected more than the long bones, with the condylar, coronoid and angular processes being most affected. Size rather than shape changes are seen in the long bones of these mutants and have been analyzed in the context of heterochrony and ontogenetic allometry (Boxes 6.1 and 33.1).[5]

Box 33.1 Allometry

Proportional changes in size and/or shape (allometry) of skeletal rudiments can persist for long time periods, as illustrated by the following four examples.

Horses

Aiming to resolve contradictions in previous studies, Radinsky (1983, 1984) carried out an interesting study on allometric changes associated with evolution of the skull in some 25 species of horses. The results demonstrate that allometric reorganization – in this case an increasing attachment of the masseter and internal pterygoid muscles associated with reorganization of skull proportions – took place some 15–25 mya, with little evidence of such change before or since.

Such a pattern is indicative of a proportional change in growth that was effective and so stabilized over evolutionary time. Ossification of the postorbital ligament to produce a postorbital bone that would resist compression, a second change in horse lineages, also occurred between 15 and 25 mya. The most likely triggers of these changes were ecological – the development of grassland and associated altered diet and feeding strategies; the horse lineage underwent a rapid decline in body size before becoming extinct in the Pleistocene of Alaska, a shift in size that has been correlated by Guthrie (2003) with climatic or vegetational change.

Mice

Corner and Shea (1995) used finite-element scaling to examine growth allometry of the mandibles of giant transgenic mice, giant because they were expressing additional growth hormone. The study, which was also a test of the model proposed by Atchley and Hall (1991), compared growth using 18 landmarks on the mandibles. Growth trajectories were similar to those in the control mice, proportional differences being attributed to ontogenetic scaling (see Chapter 32). Also confirming the model of independent units within the mandible proposed by Atchley and Hall, Daegling (1996) found that growth and morphology of the mandibles of the chimpanzee, Pan troglodytes, and the mountain gorilla, Gorilla gorilla, are a composite of independent growth events; also see Lieberman et al. (2004) in the context of the Atchley and Hall (1991) model, and tinkering with basic developmental processes.

Fish

'Needlefish' is the common name of some 50 species of teleost fishes with long, toothed jaws, belonging to the family Belonidae.

'Halfbeaks,' the common name for 70 or so species in the family Hemiramphidae, also are named for their unusual jaws, the upper jaw being short and triangular, the lower jaw long and beak-like. Needlefish go through a developmental 'halfbeak' stage in which their jaws resemble those of adult halfbeaks. Two possible interpretations have been suggested for this situation: (i) either halfbeaks are developmentally arrested (paedomorphic) or else (ii) needlefishes evolved from a halfbeak-like ancestor, repeating their phylogenetic history during their development. Recently, however, Lovejoy (2000) used a phylogenetic approach based on mitochondrial DNA to show that halfbeaks are not paedomorphic needlefish, though the developmental aspects remain in limbo.

Ontogeny of the rostrum and mandible in the swordfish, Xiphias gladius, begins with equal growth of maxilla and mandible, but negative allometry of mandibular growth in larvae produces the jaw asymmetry characteristic of adults (McGowan, 1988).

Birds

An impressive example of acceleration of skeletal growth is seen in the relative sizes and proportions of the hind limbs (positive allometry) of one of the rarest seabirds, Xantus's murrelet, Synthliboramphus hypoleucus, which breeds only in the Channel Islands off the Californian coast and in northwest Baja. Chicks hatch at a large body size with hind limbs that are 98 per cent of the size of those of the adult. At hatching, the metatarsals are 2.27 cm long, those of adults being only 2.31 cm long (Eppley, 1984). Females lay a clutch of two eggs whose weight equals 45 per cent of that of the female. After an incubation of 34 days, the extremely precocious young (at 30 grams, they are already 15 per cent of adult weight) leave the nest after two days and swim to forage at sea (Murray et al., 1983).

Fossils

Ontogenetic, including allometric, analyses can be undertaken on fossils if a sufficiently good growth series is available that can be used as an ontogenetic series. Partial growth series of the Carboniferous tetrapod, Greererpeton burkemorani, allowed Godfrey (1989) to analyse ontogenetic changes in skull development over a 4.4-fold range of skull lengths. Allometric changes were limited to the orbits, skull width in restricted regions, brain case–skull length relationships and the stapedial foot plate.

Although most studies have used rodent or monkey mandibles, the mandibular condyles of other mammals show adaptive responses to mandibular displacement. Ma *et al.* (2002) used an intraoral appliance to displace the mandibles of castrated four-month-old merino sheep, and used four fluorochromes to assess how the bone responded. Their data primarily show increased osteoblastic activity leading to adaptation to the changing functional demands. The subchondral and central regions of the condylar cancellous bone responded differently, reflecting the fact that ECM composition varies from region to region.

There is an interesting exception: in the Syrian (golden) hamster, *Mesocricetus auratus*, mandibular secondary cartilage can be initiated *in vitro* in the absence of muscle function. Secondary cartilage also is found in the auditory bulla of the hamster between 5 and 15 days postnatally where it is associated with translational growth.[6]

The angular process

Although it does not form a joint, the angular process, which consists partially of membrane bone and partially of secondary cartilage, is dependent upon local muscle action for its growth. Duterloo and Wolters (1971) observed that the secondary cartilage of the angular process of rats is not maintained beyond 17 days after birth. They removed the condyle and transplanted the angular process into the position normally occupied by the condyle. A new condyle and temporomandibular joint formed, and the secondary cartilage of the transplanted angular process persisted well beyond 17 days in its new functional environment.[7]

The medial pterygoid muscle inserts onto the angular process, maintaining it as a stable skeletal site. During the first six weeks of postnatal life in rats, growth of the angular process is rapid as the number of muscle fibres in the medial pterygoid muscle doubles (Table 33.1). Approximately half the angular process is resorbed if either the medial pterygoid or masseter muscle is extirpated. The angular process is resorbed completely if both muscles are extirpated. Similarly, the angular process fails to form if there is congenital absence of these muscles, and becomes unusually large if the muscles are

Table 33.1 Mean number of muscle fibres in the medial pterygoid muscle during growth in rats[a]

Gender	Age	Number of muscles tested	Mean number of fibres/muscle
Male	Newborn	7	8 270
	One week	7	14 100
	Six weeks	7	17 540
	Adult	9	15 560
Female	Six weeks	6	15 100
	Adult	7	12 960

[a] Based on data from Rayne and Crawford (1975).

congenitally hypertrophic. Thus, for both initiation and maintenance the angular process depends upon local muscle action, adjusting the balance between bone deposition and resorption (bone growth or bone reduction) in response to local mechanical conditions. Similar reductions in the size of the coronoid process flow from removal of the masseter or temporal muscles. Consistent with this observation, the anlage for the temporal muscle appears before the coronoid process in development.[8]

Depending on the biomechanical circumstances impinging on cells on these processes, bone may replace the lost cartilage. Hyperactivity of the rat lateral pterygoid muscle, induced by electrical stimulation, leads to loss of cartilage in areas of the condyle loaded in tension within two days and replacement of the cartilage by membrane bone by seven days (I. Takahashi *et al.*, 1995).

Diet

Another experimental approach used to analyze maintenance and continued growth of the condylar process is modification of the diet of laboratory animals (Box 12.2).

Providing a softer than normal diet to rodents or removing their incisors reduces the muscular activity associated with mastication, alters the shape of the mandibular ramus, and slows growth of the condylar process by slowing chondroblast proliferation and synthesis of ECM. The ECM of rat mandibular condylar cartilages can be changed in as little as a week following a change in dietary consistency. Osteogenesis also is altered in rats fed a soft diet; the level of alkaline phosphatase in the condylar cartilage decreases, as does the thickness of the alkaline phosphatase-positive cell layer.[9]

Similarly, a series of reproducible changes follows *increased* force of the muscles of mastication. These changes encompass (i) alteration in the transverse and vertical dimensions of the face, (ii) increased jaw loading, which increases sutural growth and bone apposition, leading to (iii) increased transverse growth of the maxilla, followed by (iv) anterior rotation of the mandible and development of prominent angular, condylar and coronoid processes on the dentary.[10]

Chondrogenesis stops about 10 days after the condylar secondary cartilage is transplanted intracerebrally, a finding that was taken to indicate that this cartilage lacks independent growth potential. Similarly, if cultured under conditions that allow epiphyseal or Meckelian cartilage to divide and grow, condylar secondary cartilage survives but neither divides nor grows.[11]

In an interesting study with postnatal mice using a combination of radiography, cephalometry and histochemistry, Nakata (1981) found that the growth of cranial bones corresponds to the *attachment sites* of masticatory muscles, not necessarily to the growth or development of individual muscles. Similarly, there is a correlation between cross-sectional area of the masseter muscle and medial pterygoid and skull shape in humans, but not between temporalis or lateral pterygoid and skull shape or facial dimensions, each muscle playing a distinctive role in shaping the skeleton. The myosin isoforms found in masseter muscles from adult humans are the same as those found in the early stages of trunk and limb muscles, indicating a differential regulation of myosin differentiation in different muscles, a finding relevant to understanding the functioning of individual regions through ontogeny.[12]

The obverse of this story is how bone robusticity reflects bone function. The mandibular condyle of the miniature pig, Sus scrofa, has far more robust and dense trabeculae than those in other parts of the skeleton, indicating regionalized loading on the TMJ as mandibular muscles exert site-specific compression on the joint. Compare the compression exerted on the temporal fossa with that exerted on the medial zygomatic arch or mandibular angle (Table 33.2, and see Teng and Herring, 1995, 1998). Deposition of modified proteins or proteoglycans is another means of local adaptation. S. Roth et al. (1997) demonstrated that the chondroitin-sulphate proteoglycans found in pig mandibular condylar cartilage – especially in the articular zone – are especially adapted for growth, both in composition and in type: less keratan sulphate than aggregan in femoral cartilage, and the presence of versican.

Other functional approaches

Charlier and Petrovic developed a model for a most promising control system for investigating condylar secondary cartilage growth in response to functional demand from the lateral pterygoid muscle, i.e. growth of a skeletal component via the functional matrix.

Charlier and Petrovic placed a chin cup onto the heads of rats so that the lower incisor was projected in front of the upper incisor for four weeks. Within a week of employing the chin cup to reduce functional stimulus on the joint, proliferation slowed and the thickness of the prechondroblast and chondroblast zones decreased. This explains the clinical value of chin-cup therapy; continuous proliferation of prechondroblasts on the condylar process depends on mechanical stimuli.[13]

Petrovic et al. (1973) sought to relate these findings to muscular activity in vivo. Hyperpropulsion increases, and the chin cup decreases, the number of sarcomeres in the lateral pterygoid muscle and the amount of proliferation in the prechondroblastic zone. They proposed that contraction of the lateral pterygoid muscle controls the rate of proliferation of prechondroblasts, providing the means by which other stimuli such as hormones and tongue movement are translated to the prechondroblastic zone. Condylar cartilage growth was viewed as entirely adaptive and responsive to muscular control, and the model extended to the coronoid and angular cartilages and their growth control via the temporal and masseter muscles.

This model of condylar cartilage growth was tested further by resecting the lateral pterygoid muscle of young growing rats and following the proliferation of prechondroblasts autoradiographically. Proliferation slowed and, as noted in Chapter 12, the prechondroblasts differentiated into osteoblasts, not chondrocytes.[14]

However, when the lateral pterygoid muscle is sectioned on one side in growing (four- to six-week old) rats, and oxytetracycline and alizarin are used as markers for bone growth/displacement, no statistically significant inhibition of condylar growth is found, although growth of the posterior region of the condyle is diminished. After

Table 33.2 Measurements of compression (kPa) exerted by the mandibular muscles on the temporal fossa, medial zygomatic arch and mandibular angle of miniature pigs, Sus scrofa[a]

Age (weeks)	Body weight (kg)	Temporal fossa (kPa)[b]	Medial zygomatic arch (kPa)[c]	Mandibular angle (kPa)[d]
12.5	14.0	164.8 ± 2.0	43.2 ± 2.7	48.5 ± 2.5
13.5	19.9	73.7 ± 2.5	42.1 ± 2.4	–
14.0	17.9	49.6 ± 0.8	20.4 ± 0.5	–
14.0	17.7	68.4 ± 5.1	20.4 ± 1.2	46.9 ± 3.2
14.5	17.0	118.9 ± 11.3	24.5 ± 2.9	–
17.0	17.6	47.5 ± 3.6	16.7 ± 2.9	–
17.5	19.9	218.9	86.3 ± 2.1	–
18.0	23.5	116.1 ± 6.8	50.1 ± 3.2	–
18.0	20.4	207.7 ± 20.3	66.8 ± 0.9	–
18.0	15.6	65.1 ± 3.1	30.8 ± 1.2	37.8 ± 0.1
20.0	17.6	82.8 ± 4.5	47.5 ± 2.9	–
21.0	22.7	118.8 ± 2.7	17.7 ± 0.7	–
21.0	22.7	170.1 ± 0.5	22.8 ± 0.5	100.1 ± 0.5
24.0	31.7	116.5 ± 5.3	36.8 ± 0.9	–
24.0	31.2	53.5 ± 0.9	32.0 ± 0.4	–

[a] Based on data from Teng and Herring (1998).
[b] Temporal fossa pressure is caused by contraction of the temporalis muscle, [c] medial zygomatic arch pressure by contraction of the zygomatico-mandibularis muscle, and [d] mandibular angle pressure by the masseter muscle.

the lateral pterygoid muscle was removed unilaterally or bilaterally from 10-week-old rats in a second study, a complex pattern of resorption of the anterior border and growth of the posterior border was seen, without overall growth retardation of the mandibular condyle.[15]

Also deserving of discussion are studies by Murray Meikle (1973a,b), whose experiments led him to conclude that proliferation of prechondroblasts is *independent* of functional demand. Meikle transplanted the mandibular joint intracerebrally. *Three weeks later* he introduced ^{3}H-thymidine to label any dividing cells. Because cell proliferation was still active at this time, Meikle concluded that proliferation must be independent of function.[16]

It is difficult to reconcile this study with Petrovic's. Comparative data on cell proliferation *in situ* and in the transplants might indicate that cell proliferation within the transplants was indeed slowed. Data on transplantation or culture of whole TMJs versus isolated condylar cartilage, coupled with quantitative analysis of mitotic activity, and close comparison with controls, might help to resolve this rather important question. Meikle did find, as had Petrovic's group, that prechondroblasts in the transplants differentiated into osteoblasts rather than into chondrocytes.

In an extensive series of studies, Tuomo Kantomaa and his colleagues in Oulu, Finland and Canada documented adaptive changes in the condylar cartilage (listed below), concluding that the mandible has to be moved gradually for the condylar cartilage to adapt to pressure changes.

- Fixation of the maxilla by gluing the frontonasal, frontal-premaxillary and frontal-maxillary sutures, after which the length of the mandible is reduced but height advanced as the condylar cartilage thins and the mandible is carried forward passively (Kantomaa and Rönning, 1985, 1992).[17]
- Resection of the bony process carrying the articulation for the glenoid fossa, after which the prechondroblast layer increases and metachromasia of existing ECM decreases (Kantomaa, 1987a).
- Creation of an artificial cranial synostosis, following which prechondroblasts increase and chondrogenesis is disturbed, with areas of acellular necrosis (Kantomaa, 1987b).
- Development of an organ culture system that allowed the mouse TMJ to be manipulated *in vitro* and which provides essentially normal articulating function (Kantomaa and Hall, 1988a,b).
- Use of the organ culture system to demonstrate the importance of cAMP and Ca^{++} in mandibular condylar growth and adaptation (Kantomaa and Hall, 1991), to maintain the TMJ and demonstrate that electrical stimulation of the masseter muscles maintains condylar cartilage in long-term organ culture (Pirttiniemi and Kantomaa, 1996).
- Demonstration of increased proliferation in prechondroblasts immediately posterior to the articular surface

following relocation of the glenoid fossa (Pirttiniemi *et al.*, 1993).
- Simulation of the application of an orthodontic appliance, which leads to increased proliferation and enhanced deposition of ECM in the postero-superior portion of the joint (Kantomaa *et al.*, 1994a).
- Feeding a soft diet or cutting the incisors, both of which lead to regional differences in the rate of cartilage differentiation, including decrease in deposition of type II collagen and decrease in chondroblast number by one third (Kantomaa *et al.*, 1992, 1994b; Pirttiniemi *et al.*, 1996).
- Demonstrating that type II collagen in rat glenoid fossae is regulated by tension and compression (Tuominen *et al.*, 1996).[18]
- Demonstrating that posterior displacement of the glenoid fossa leads to reduced collagen and proteoglycan deposition (Kantomaa and Pirttiniemi, 1998).
- Alteration in proteoglycan content – especially aggregating proteoglycan – after altering masticatory function by grinding the molars of rabbits on the right-hand side at 10 days postnatally and twice a week thereafter (Poikela *et al.*, 2000).[19]
- Use of one- to 28-day-old rats to show that type II collagen is only distributed in the upper areas of the condylar cartilage compared with epiphyseal cartilage where type II is found throughout (Visnapuu *et al.*, 2000).

Studies are beginning to correlate such changes with patterns of gene expression. A nice example, using 28-day-old Sprague-Dawley rats, repositioned the mandibles so that forces acted asymmetrically on left and right condyles. Chondroblast proliferation and condylar cartilage thickness increased on the more loaded side. Using the semi-quantitative reverse-transcription polymerase chain reaction, expression of *Igf-1* and *Fgf-2* increased three- and twofold, respectively, expression of *FgfR-1* and *FgfR-2* decreased by 80 and 60 per cent, respectively, expression of *IgfR-1* decreased by 70 per cent, and FgfR-1 increased 2.5-fold (Fuentes *et al.*, 2003a,b).

From the condylar cartilage of the mammalian TMJ, in which growth is adaptive, I turn to a discussion of whether other joints contribute to skeletal growth.

CRANIAL SYNCHONDROSES

Synchondroses are those joints whose bony surfaces are united by an intervening cartilage. Organizationally, they resemble two epiphyseal growth plates back to back, but with a common central zone of resting cells. Because of this arrangement, synchondroses are exquisitely adapted to resist pressure and might be expected to regulate their growth in response to pressure.[20]

Of several synchondroses in the crania of developing mammals, the midsphenoidal, septoethmoidal and sphenooccipital are present at birth in humans. The *sphenooccipital synchondrosis*, which lies at the base of the skull, has received much attention. Since it is the last synchondrosis to fuse, it has been argued that its role in cranial growth is proportionately greater than the roles of the other two. The veracity of this proposition was demonstrated dramatically over 40 years ago, when du Brul and Laskin (1961) found that removing the sphenooccipital synchondrosis modified all the synchondroses of newborn rats by producing primate-like features normally associated with the adoption of upright posture. Nevertheless, at least three answers to the question 'Do synchondroses direct growth of the cranium or do they respond adaptively to cranial growth?' have been provided by different workers, at different times, using different experimental techniques. The three possible answers are that synchondroses are pacemakers, have inherent growth potential, or are adaptive.

As pacemakers

The oldest view, which goes back at least to Scott (1954), is that the cartilaginous synchondroses – cartilage in general, e.g. nasal and condylar cartilage – serve as growth centres (*pacemakers*) for the growth of the bony skull, equivalent in their growth potential to epiphyseal growth cartilages. In incisive reviews of dermocranial and chondrocranial growth, Van Limborgh developed *an extreme version of this position in which bone growth is primarily adaptive and cartilage growth the expression of inherent growth potential* (Fig. 33.1). Growth of sutures and synchondroses provides strong support for his view. Growth of condylar cartilages does not; cartilage is not cartilage is not cartilage.[21]

Experimental evidence for the pacemaker concept comes from the inherent growth potential demonstrated by synchondroses when isolated *in vitro* or as grafts. For example, chondrocranial chondrocytes isolated from 19- or 20-day-old rats continue to grow in serum-free medium, an ability not shared with other cartilages and

indicative of inherent growth potential. In an autoradiographic study, progenitor cells were followed from resting to proliferative zones. The presence of labeled hypertrophic chondrocytes showed that the cells had matured in the absence of function, although some might debate whether completion of maturation can be equated with intrinsic growth potential.[22]

Clinical experience and histological examination of human septal cartilages is consistent with the nasal septum having higher growth potential than the adjacent regions. Lines of evidence provided to me by a highly experienced maxillofacial surgeon, David Precious, Dean of the Dental Faculty at Dalhousie University, are that: (i) the nasal septum is the most anterior element in the anterior portion of the face; (ii) histological examination is consistent with septal cartilage growing at a faster rate than the surrounding connective tissue; (iii) the face becomes asymmetrical in clinical situations where the nasal septum is bent to one side because muscles are only attached to that side; and (iv) facial symmetry is restored when the nasal septum and the nasolabial muscles are repositioned to the midline (Fig. 33.2; David Precious, pers. comm.).

Basicranial synchondroses adapt when transplanted into sutures in rat skulls. These synchondroses have tissue-separating ability, as do epiphyseal growth plates, and mandibular condylar and clavicular cartilage. Part of the difference between the growth of synchondroseal and condylar cartilages is that growth of condylar secondary cartilage is appositional, while growth of synchondroseal cartilage is interstitial.[23]

Chondrocytes from rabbit sphenooccipital synchondroses show less responsiveness to bovine parathyroid hormone (PTH) than do chondrocytes from costal or nasal septal cartilages. Nevertheless, they show considerably more response than do mandibular condylar chondrocytes, which fail to respond unless at high cell densities (Table 33.3; and see Box 33.2).

As might be expected from their different patterns of development, growth and function, the amounts and

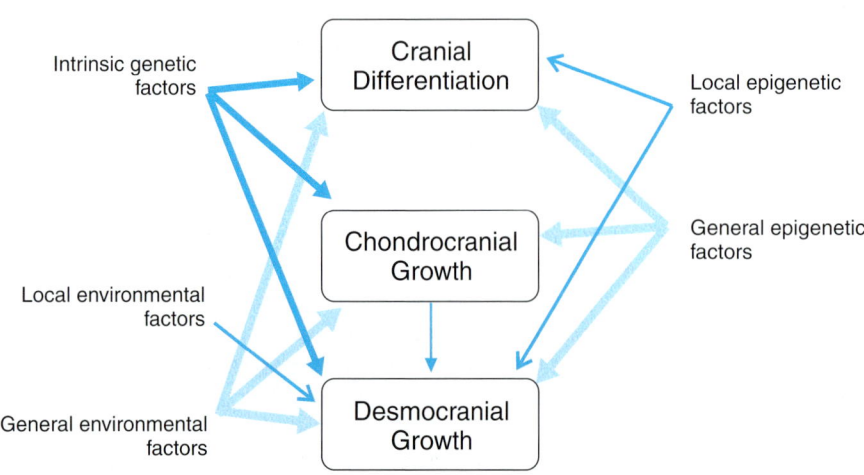

Figure 33.1 Van Limborgh's view of the roles of genetic, epigenetic and environmental factors in the differentiation and growth of cranial cartilages and bones. All factors influence chondrogenesis and osteogenesis with the exception of local environmental factors (e.g. muscle action), which only influence osteogenesis in the desmo (dermal) cranium. Based on Dullemeijer (1985) after van Limborgh (1970).

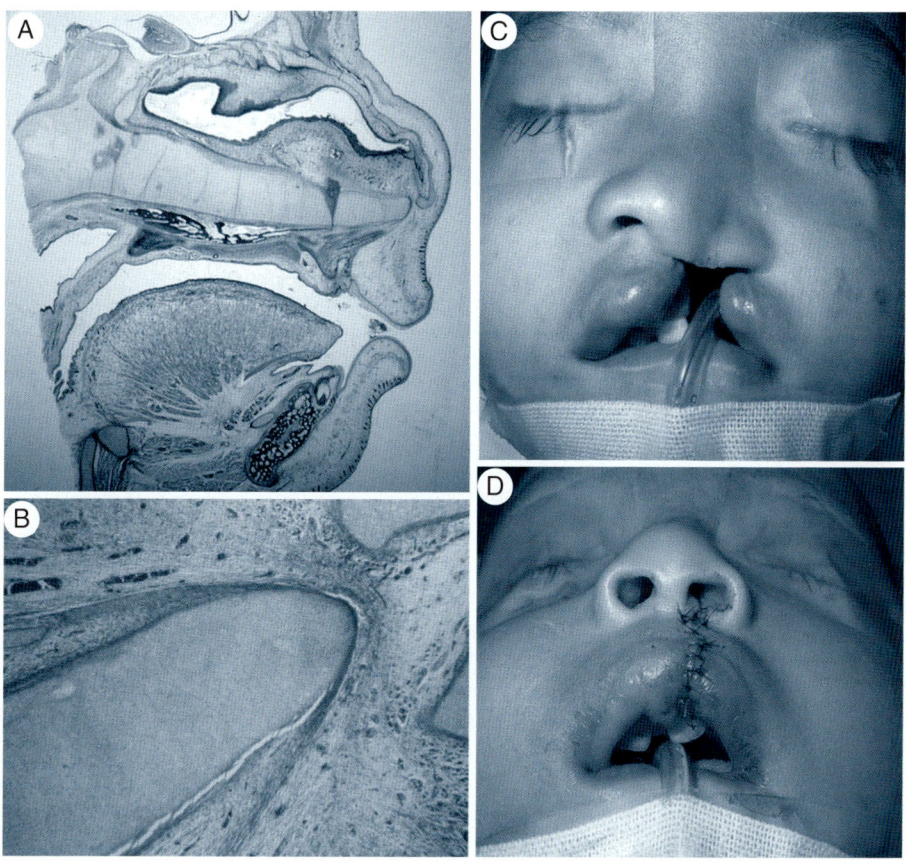

Figure 33.2 The nasal septal cartilage in human embryos as seen in longitudinal histological sections at 16 (A) and 24 (B) weeks of gestation is the most anterior element of the face (A); it is seen to grow rapidly (B) and asymmetrically (C) in a situation (unrepaired cleft lip and palate) where muscles are attached only to one side, a pattern of growth that can be corrected surgically (D). Figures provided by David Precious.

Table 33.3 Differential responses of chondrocytes to PTH as measured by percentage increase in GAG synthesis over control[a]

Chondrocyte type	% increase
Mandibular condylar	137
Nasal septal	138
Sphenooccipital synchondrosis	133
Costal growth cartilage	162

[a] Based on data from Takigawa *et al.* (1984a). 10^{-7} M bovine PTH was administered for 24 hours after chondrocytes had been in culture for 48 hours.

properties of the collagens differ among different cartilages. Pietilä *et al.* (1999) examined this in 25- and 35-day-old rabbits, using mandibular condylar, costal and nasal cartilages (Table 33.4). The content of proteoglycans and proteglycan aggregate formation are greatest in nasal, lower in costal and very low in condylar cartilages, a trend Pietilä and colleagues correlate with independent growth potential; the higher the growth potential, the lower the proteglycan content. However, other correlations are feasible. Nasal cartilage has the highest collagen content. Collagen counterbalances osmosis changes. And so increased collagen correlates with nasal cartilaginous function.

In yet another approach, the sphenooccipital synchondrosis is not obliterated – although periosteal ossification is modified – after the neck muscles of rats are partly detached, a result indicating independence from the altered mechanical conditions. Harkness and Trotter (1980, 1982) saw a growth spurt when they transplanted rat cranial bases into older animals, concluding that intrinsic genetic factors regulate cranial base growth, although they did report more resorption in transplanted cranial bases.

Limited growth potential

An alternate view, compatible with some of the studies discussed above, is that synchondroseal growth potential is not equivalent to that of epiphyseal growth plates, rather that synchondroses have *limited independent growth potential* – less than epiphyses, but more than condylar secondary cartilage. In this view, synchondroses are not primary growth centres pushing cranial bones apart. Their growth is largely adaptive. What is the evidence?

If synchondroseal cartilage from the presphenoid–basisphenoid synchondrosis of five- to seven-day-old rats is transplanted *intracerebrally* for between 30 and 60 days, growth potential is found to be intermediate between that shown by condylar and epiphyseal cartilages under the same conditions. When transplanted *subcutaneously*, synchondroses survive but fail to grow (Koski and Rönning, 1969, 1970, 1992), although Harkness (1976) reported that some growth does occur at these sites, albeit less than normally seen *in vivo*. Such transplant studies

Box 33.2 Nasal septal cartilage as pacemaker?

It has been a controversial question for three quarters of a century or more: Are mammalian nasal septal cartilages centres of primary skeletal growth? There are two aspects of this question, aspects that apply to any cartilage thought to be a primary growth centre. Does the cartilage display an inherent potential for growth and, if it does, does the resultant growth influence the growth of other skeletal elements, regions or functional units, i.e. is the cartilage a pacemaker? The jury is still out on both questions for nasal septal cartilage.

When Verwoerd et al. (1980) partially resected rabbit nasal septal cartilage they found that the basic growth of the upper jaws and nose region *continued*, suggesting no major influence on growth of adjacent regions of the craniofacial skeleton. In an *in-vitro* study, however, Copray (1986) found that the growth of rat nasal septal cartilage was comparable to that of epiphyseal cartilages, which are growth cartilages.

In human infancy, septal cartilage exhibits regional patterns of growth (Fig. 33.2). Using patients with congenital labiomaxillary clefts (Fig. 33.2), as well as three cases of physical injury to the face, it has been shown that nasal septal cartilage has a *direct* effect on premaxillary growth but only an *indirect* effect on maxillary growth.[a]

Although we do not understand fully what makes one cartilage a pacemaker and another not (see Chapter 28), two molecular features of nasal septal cartilage are worth mentioning:

At the critical condensation stage, Sasano et al. (1992) claimed they could not detect peanut agglutinin lectin staining in either nasal septal or Meckel's cartilages in Swiss Webster mice. If absent and not masked, initiation of these two cartilages may differ from the other craniofacial cartilage (and Meckel's is a primary growth cartilage for the mandible).

Rat mandibular condylar and intermaxillary sutural secondary cartilages contain collagen types I and II, fibronectin and tenascin. Tibial growth plate, *nasal septal* and sphenooccipital cartilages all lack type I collagen.[b] I interpret this pattern as a consequence of the periosteal origin of the condylar and secondary cartilages, the other cartilages being primary. Perhaps it also is telling us something about growth potential.

[a] See Vetter et al. (1983, 1984) for patterns of regional growth, Delaire and Precious (1986) and Precious et al. (1988) for clinical studies, Precious and Delaire (1987) and Precious and Hall (1992) for overviews of balanced facial growth in humans, Radlanski et al. (2003) for 3D reconstructions of patterns of apposition and resorption in the embryonic human mandible (12–117-mm crown–rump length), and the papers in Glowacki (2004) for the repair and regeneration of oral tissues.
[b] See Mizoguchi et al. (1990, 1992) for these immunohistochemical studies.

Table 33.4 Relative levels of collagen and chondroitin sulphate (CS) in costal, nasal and condylar cartilages from 25- and 35-day-old rabbits[a]

	Nasal	Costal	Condylar
	25-day-old rabbits		
Collagen[b]	102.5	207	166
CS[c]	343	408	320
	35-day-old rabbits		
Collagen	176	210	173
CS	603	543	331

[a] Based on data from Pietilä et al. (1999).
[b] nM hydroxyproline/mg dry weight guanidium chloride extract.
[c] μg/mg dry weight guanidium chloride extract.

primary growth centre) with the septoethmoidal joint (which was thought to be partially adaptive and partially growth plate-like). These criteria, coupled with the differential distribution of alkaline and acid phosphatases, were then used to distinguish synchondroseal from condylar and epiphyseal cartilage.[24]

Another approach was to administer repeated doses of papain to growing rats to disrupt deposition of the cartilaginous ECM. The consequent reduced growth of the entire skull was attributed to secondary consequences of the impaired growth of the synchondroseal cartilages (Kvinnsland, 1974).

An autosomal semidominant mutation, *Brachyrrhine* (*Br*), arose in a hybrid mouse strain being used to study the effects of neutron radiation on chromosome structure. *Br* mice are retrognathic ('pugnose'), i.e. the midface is severely foreshortened with respect to the skull and jaws. The primary defect is in the sphenoethmoidal joint (the nasal capsule being virtually unaffected), providing a strong indication that the sphenoethmoid contributes in a substantial way to craniofacial growth, as indeed does the sphenooccipital synchondrosis. A recent study by McBratney et al. (2003) shows that *Br/Br* mice lack the presphenoid bone, and display frontonasal dysplasia, median facial clefts and a bifurcated cranium. Chondrocytes from the sphenoethmoidal region of *Br* mice respond much more strongly to nerve growth factor (1 ng/ml) than do chondrocytes from wild type.

As adaptive

A third view posits the growth of synchondroseal cartilages as entirely adaptive, occurring only in response to the functional matrices in which they develop. Response to the mass of neural tissue present, and to its growth, is seen as especially important – but only *as a response, not as a driving force*. This view was promoted most strongly by Melvin Moss in the context of functional matrices – skeletal elements must be considered in context – and by Jan van Limborgh in the context of the intrinsic growth properties of craniofacial cartilage (above). In this approach, the origin, growth and maintenance of all

support the concept that synchondroses have some inherent potential for growth in the absence of functional stimuli.

But is the amount of growth expressed after transplantation or upon cultivation *in vitro* the best criterion for evaluating the inherent growth potential of cartilage?

In an attempt to differentiate adaptive from intrinsic growth-plate-like patterns of growth, Durkin and colleagues established a set of histological criteria to monitor growth. The criteria included arrangement of cells, patterns of mineralization, mechanisms of erosion, and presence or absence of subchondral bone. These criteria were used to compare the septopresphenoidal joint (a skeletal tissues is secondary and compensatory. While it does appear that some cartilaginous growth is adaptive, it is equally clear that it is not entirely adaptive.[25]

NOTES

1. See Fell and Canti (1934), Glasstone (1968, 1971, 1973), Hall (1968d) and Tyler and Hall (1977) for initiation of joint formation *in vitro*. See Barnett (1954) and Barnett *et al.* (1961), Sokoloff (1978, 1980), Carter and Wong (1988b) and Archer *et al.* (1994) for the structure and development of joints between long bones. For the nature of the osteochondral junction of mammalian joints, including electron microprobe analyses for Ca^{++}, S^- and Fe, see Hough *et al.* (1974).

2. See Pellegrini (1934), Lelkes (1958), Barnett *et al.* (1961), Drachman and Coulombre (1962b), Drachman (1964), Chalmers (1965), Murray and Stiles (1965), Drachman and Sokoloff (1966), Sullivan (1966, 1974, 1976), Hall (1967b, 1968d,e, 1972b, 2004a), Murray and Drachman (1969), Mitrovic (1971, 1972, 1974, 1982), Persson (1983, 1995), Hall and Herring (1990), Hosseini and Hogg (1991a,b), Hogg and Hosseini (1992), Bertram *et al.* (1997), Wu *et al.* (2001) and Buxton *et al.* (2003) for a sampling of such studies.

3. See Craig *et al.* (1987) and Sorell and Caterson (1989) for keratan and chondroitin-6-sulphates, Solursh *et al.* (1990), Gould *et al.* (1992, 1995), Bernfield *et al.* (1993) and Kosher (1998) for syndecan, and Archer *et al.* (1994) and Spicer and Tien (2004) for hyaluronan. Interestingly, levels of keratan and chondroitin-6-sulphates rise in articular cartilage with age.

4. For reviews of how the TMJ functions, see Moss (1962c, 1968d, 1981), Moffett (1972), Moore and Lavelle (1974), Enlow (1975), Dhem and Goret-Nicaise (1979), Kantomaa and Rönning (1992) Sperber (2001) and two edited issues of the journal, *Cells, Tissues and Organs* (English and Stohler, 2001 and English, 2003).

5. See Kreiberg *et al.* (1978) and Ghafari and Cowin (1989) for congenital muscular dystrophy in humans, Vilmann *et al.* (1985) for the dentary in muscular dystrophic mice, Lammers *et al.* (1998) for the two muscular dystrophic strains, and Lightfoot and German (1998) for effects of muscular dystrophy on craniofacial growth in two strains of mice. In artificial monozygotic twin mice, generated by dividing the morula at the 16-cell stage and reimplantation, an analysis of postnatal skeletal growth indicates greater genetic control over craniofacial than over somatic growth (Watanabe *et al.*, 1998b).

6. See Vinkka-Puhakka and Thesleff (1993) for the *in-vitro* study and Vinkka-Puhakka (1991b) for the bulla.

7. For longer-range effects of mandibular condylectomy, see Sarnat and Muchnic (1971). Poswillo (1972) and Soni and Malloy (1976) used three fluorochromes to monitor diminished condylar growth after condylectomy of guinea pigs. See Sarnat (1968, 1986, 1992, 2001) and Sarnat and Selman (1982) for reviews of the use of markers to study mandibular and craniofacial growth. See Vilmann (1982a) for angular secondary cartilage in the rat, Vinkka (1982) for all the secondary cartilages in the rat, and Vinkka-Puhakka (1991a) for a summary of secondary cartilages in rat and hamster.

8. See Pratt (1943), Washburn (1947), Moore (1969), Soni and Mallory (1974) and Rayne and Crawford (1975) for these studies on muscle extirpation, and Spyroloulos (1977) for timing.

9. Moore (1965), Simon (1977), Lavelle (1983), Hinton (1988a), Kiliaridis (1995) and Kiliaridis *et al.* (1996) are typical of the many studies dealing with soft diet or reduction of the incisors. See Hinton (1993) and Bouvier (1987) for ECM changes. Working with rats Tuomo Kantomaa (pers. comm.) finds that a soft diet enhances osteogenesis within the mandibular condyle, accompanied by both more rapid hypertrophy of chondrocytes and replacement of cartilage by bone.

10. See Kiliaridis (1995) for this study. Breakdown of matrix components such as collagen can also play an important role in the metabolic response of condylar cartilage to loading (Pirttiniemi *et al.*, 2004).

11. See Koski and Mäkinen (1963), Rönning (1966) and Duterloo and Wolters (1971) for transplantation, and Charlier and Petrovic (1967) for the culture study.

12. See Nakata (1981) for the combined study, Hillen (1984, 1986) for the human data and Soussi-Yanicostas *et al.* (1990) for the myosin isoforms.

13. See Charlier *et al.* (1969a,b) for the orthodontic device and Graber (1977) for chin-cup therapy. Petrovic (1970, 1972) reviews the early work of Charlier *et al.* on this model; Petrovic *et al.* (1990) provide a more recent update of the cellular basis of the efficacy of functional appliances. Despite this elegant theory, we have only incomplete understanding of the links between regulatory factors and the growth of condylar cartilage.

14. See Stutzman and Petrovic (1974) for this study, and Petrovic (1974) for a review of this model and for a cybernetic model for growth of the entire mandible.

15. See Awn *et al.* (1987) and Goret-Nicaise *et al.* (1983) for these studies.

16. In an earlier study using the same approach with epiphyseal cartilages from the third metacarpals of rats transplanted intracerebrally and ^3H-thymidine-labeled, Meikle (1975b) found that (i) division of these epiphyseal chondrocytes slowed, (ii) the diameter of the cartilages did not increase as it normally would have, while (iii) perichondrial cells transformed into osteoblasts instead of chondroblasts.

17. From an analysis of 45 human patients in whom fixation devices were implanted in genioplasty, Precious *et al.* 1992) determined that fixation in areas of bone resorption was best, having earlier shown that rigid fixation should be avoided (Precious *et al*, 1990).

18. Condylar and glenoid fossa remodeling is a major contributor to the mandibular prognathism that follows use of the Herbst appliance (Ruf and Pancherz, 1998).

19. Proteoglycans, especially small proteoglycans, differ between peripheral and deeper regions of the fibrocartilaginous bovine TMJ disc (P. G. Scott *et al.*, 1995). The articular disc of the rat TMJ is equivalent to cartilage in terms of collagen and GAG content (Carvalho *et al.*, 1993).

20. Several excellent reviews on various aspects of synchondroses and their role in skull growth include Felts (1961), Enlow (1968b, 1975), van Limborgh (1972, 1982), Moore and Lavelle (1974), Koski (1975, 1985) and Moss (1975).

21. See Petrovic and Charlier (1967), Baume (1968, 1970), Servoss (1973), Kylämarkula and Rönning (1979) and Rifas *et al.* (1982) for synchondrosis as pacemakers. See Fig. 33.1 and van Limborgh (1970, 1972, 1982) for cartilage growth as intrinsic. Dullemeijer (1985) analyzed the significance of van Limborgh's approach, an analysis that claims – wrongly I believe – that van Limborgh was not interested in function,

causal relationships between form and function, or mechanisms. van Limborgh may not have performed the experiments but he certainly provided the theoretical underpinning for causal analyses.

22. See Rifas et al. (1982) for the organ-culture study, and Servoss (1973) for the autoradiographic study.

23. See Jolly and Moore (1975), Kvinnsland and Kvinnsland (1975), Brewer et al. (1977), Kylämarkula and Rönning (1979), Rönning and Kylämarkula (1982) and Rönning et al. (1991a,b). Adams and Harkness (1972) described the cartilage of the sphenooccipital synchondrosis in the cynomolgus monkey, Macaqueirus.

24. See Durkin et al. (1973) for the histological criteria, and Brown and Heeley (1974) and Kenrad and Vilmann (1977) for the tests.

25. See Moss (1962c, 1981), Moss et al. (1972) and especially Moss (1997a–d) for a detailed overview of his approach, and see Carlson (1999) for an evaluation of it.

Sutures and Craniosynostosis

Sutural growth may differ in rate, duration and with time. Sutures are not merely reactive, and certainly not mere spandrels[1]

Sutures, as seen between bones of the skull, are joints in which one bone articulates with another through intervening fibrous connective tissue. Sutures may link two membrane bones – as in the skull – or they may link a membrane and an endochondral bone, as in the sphenofrontal suture in humans. Sutures link bone of neural crest origin and they link neural-crest and mesodermally derived bones. Because of the connective tissue, sutures are especially well adapted to resist *tension*; although the basic *structure* of the coronal and sagittal sutures in rabbits is determined, environmental releasers are required to elicit that basic form (Oudhof, 1982).[2]

SUTURAL GROWTH AS SECONDARY AND ADAPTIVE

An early view of mammalian (mostly human) skull growth contended that osteogenesis at sutural margins makes no contribution to cranial growth and that, when viewed as a single entity, the growth rate of the skull is entirely determined (i) by growth of cranial cartilages and (ii) by secondary and adaptive sutural growth. However, deposition and resorption of bone do occur at sutural margins, as is demonstrated using alizarin red and fluorescent calcium-binding markers (see Box 34.1). Furthermore, growth on the opposite sides of a single suture may differ in rate, duration, direction and with time. Sutures are neither merely reactive nor spandrels.[3]

The growth potential of sutures can be assessed in several ways. One is to examine the properties of sutural fibroblasts, which respond to alterations in their mechanical environment by modulating collagen synthesis. In four studies, Murray Meikle exposed fibroblasts from rabbit cranial sutures to tension and compared their responses with that of control fibroblasts. Those fibroblasts under tension synthesized two or three times more protein and twice as much collagen within six hours, three times as much DNA after 48 hours, began to synthesize type III collagen, and showed increased levels of metalloproteinase inhibitors and enzymes associated with hydrolysis.[4]

Does deposition and resorption of bone and connective tissue at sutures provide a primary growth force for the dermocranium, or is such growth secondary and adaptive? Experimental excision, transplantation and fusion of cranial sutures indicate that *sutural growth is secondary and adaptive*. Results from a sample of approaches underpinning this conclusion follow.[5]

Closure of the intermaxillary and transverse palatine sutures in rabbits and humans is not associated with cartilage formation, altered vasculature or intramembranous ossification. As determined in Wistar rats, the timing of the closure of the interfrontal suture is adaptive not intrinsic; sutures from three-day-old rats transplanted into 10-week old rats delay their closure, closing between four and five weeks after birth rather than one week after birth.[6]

Moss, who showed that bone is deposited over the original suture line after sutures are extirpated from rat crania, argued that the position of the suture is not fixed intrinsically. This may be true once the suture has arisen. The finding that calvarial bones can be regenerated, complete with sutures in their correct positions, in young humans who have undergone craniectomy to treat sutural fusion (craniosynostosis), suggests and perhaps even indicates that sutural position is predetermined in this situation.

Box 34.1 Madder, alizarin, and early studies on bone growth

Osteomancy, the art of prophecy by examining a bone of a sacrificed animal, was widely used in the mid-19th century in Sindh, in the valley of the Indus, the bone being a scapula, the sacrificial animal a sheep: 'The osteomancers divide the bone into twelve areas, or "houses," each answering a different question about the future. If in the first "house" the bone was clear and smooth, the omen was propitious and the consulter would prove to be a good man. If, in the second "house," which pertained to the herds, the bone was clear and clean, the herds would thrive, but if there were *layers of red and white streaks*, robbers must be expected.'[a]

Nowadays we have many ways to label and follow the growth of bone; tetracycline (Figs 5.7 and 34.1), fluorochromes, lead acetate and isotopes perhaps being the most useful.

The technique of staining bone, however, is almost 300 years old, owing its origin to the London surgeon John Belchier, who made his discovery at a dinner table in 1736. The pork bones on his plate were stained red, the meat was not. Upon inquiry of his host, Belchier found that the pigs had been fed madder, a dye obtained from the roots of madder (*Rubia tinctorum* L.), a herbaceous climbing plant. Belchier saw the use to which madder could be put and fed madder to pigs and birds to study the growth of their bones. As a result he discovered that blood circulates through the compact bone of the entire shafts of long bones. The active agent in madder is *alizarin* (see below).[b]

Madder is a member of the family Rubiaceae.[c] Several early Greek writers (Discorides, Hippocrates, Herodotus) and Pliny refer to madder as a cloth dye, a property well known in ancient Egypt, Persia and India; the shrouds wrapped around Egyptian mummies were stained with madder. Madder was also used medicinally. In his 1597 *Herball*, Gerard notes that the madder plant was cultivated in gardens, although no modern studies have revealed any medicinal properties. As can be seen from the epigraph, osteomancers may have had the secret of madder staining.

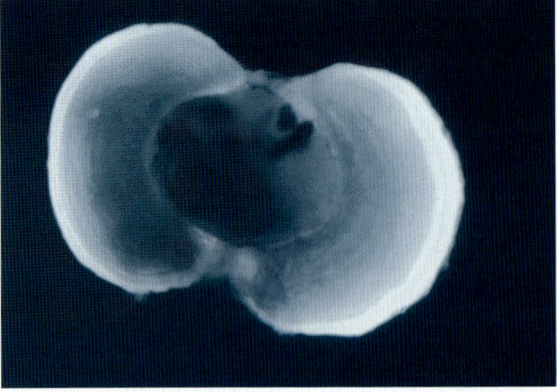

Figure 34.1 A cross-section through an ectopic ossicle of bone and chondroid bone, induced in a yellow perch, *Perca flavescens*, in response to infection with metacercaria of the trematode parasite, *Apophallus brevis*. Three intraperitoneal injections of tetracycline reveal deposition of lamellar bone peripherally (white rings). Photographed under fluorescent light. Also see Figs 5.6 and 5.7 and see Taylor *et al.* (1994) for details.

How does bone grow?

The 1740s were exciting times for those fascinated by how bones grow. Following on Belchier's heels, Duhamel discovered that madder dyed only those parts of a bone deposited when madder was part of the diet. Adding madder to or withdrawing madder from grain fed to chickens, turkeys and pigeons enabled Duhamel to establish the basic precepts of long bone growth, as follows.

- Bones grow from their ends.
- Growth in width is by addition of new bone at the periphery.
- The periosteum is the source of the red (madder)-coloured bone.

Duhamel also found that bone 'turned over.' The red colour left the bones a month after he stopped feeding madder to the animals. Madder is taken up in bone laid down at sites of fracture repair and appears to interfere with development of a fracture callus.[d]

Duhamel (1742) also used experimental approaches to investigate how bones grow. When he placed a circle of silver wire subperiosteally against the cortex, with subsequent growth the wire became imbedded in the shaft of the bone. Duhamel concluded that bone grew by apposition, as do trees, depending on the periosteum 'as an exogenous stem grows from the inner layer of the bark.' Indeed, he argued that appositional growth is the *only* method of bone growth, a view supported into the 19th century by Syme (1840), Flourens (1847) and Ollier (1867).[b]

These pioneering studies provided the foundation for the justly famous work of John Hunter, for whom a major concern was how the proportions of bone are maintained, and how the marrow cavity can increase in diameter during bone growth. Like Duhamel, Hunter recognized that the red and white bands in animals fed madder corresponded to the times when madder was added to or withheld from the diet. Hunter recognized *deposition* and *resorption* as two fundamentally different processes governing bone growth. Deposition at the periphery – in the periosteum – increases the width of a long bone; resorption on the inner, marrow surface allows the marrow cavity to expand with increasing bone width, a process now known as *metaphyseal bone remodeling*.[e]

Hunter recognized that the seemingly rock-like structure of bone was constantly changing with the ebb and flow of deposition and resorption. The concept of cycles of resorption and deposition was extended by Tomes and Tomes and De Morgan, who determined that the Haversian systems of cortical bone develop through deposition of bony layers on the inside of cavities created by resorption. Haversian canals were described by Havers in the late 17th century and in some detail by Howship, 120 years later.[f]

Hunter (1771) was one of the first to apply quantitative methods to bone growth using madder feeding. By aligning human mandibles at their symphyses and along their lower borders, he showed that between five years of age and adulthood:

- growth of the body of the dentary takes place posteriorly;
- growth of the condylar and coronoid processes occurs superiorly; and
- the body of the dentary grows in width through resorption anteriorly and deposition posteriorly (Fig. 34.2).

Almost a century later, using wires wrapped around or implanted into the body of pig dentaries, Humphry (1861, 1863, 1864) confirmed Hunter's observations. Humphry attributed the different growth rates between bone and periosteum to the periosteum being firmly attached to the epiphysis, but loosely connected to the metaphysis and therefore growing more slowly than the bone, sliding over the bony shaft as it elongated.

Box 34.1 *(Continued)*

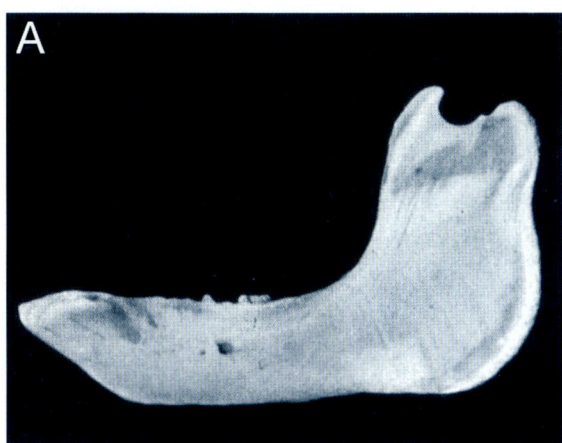

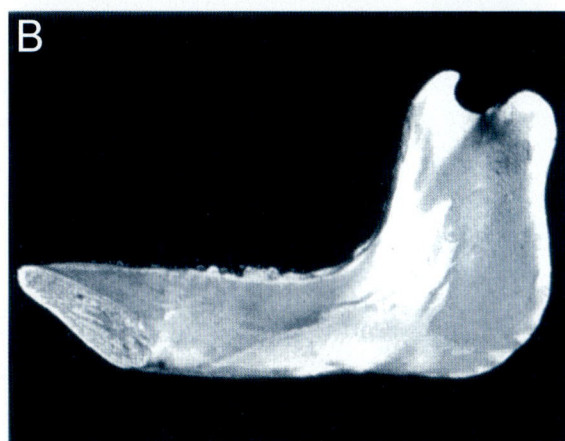

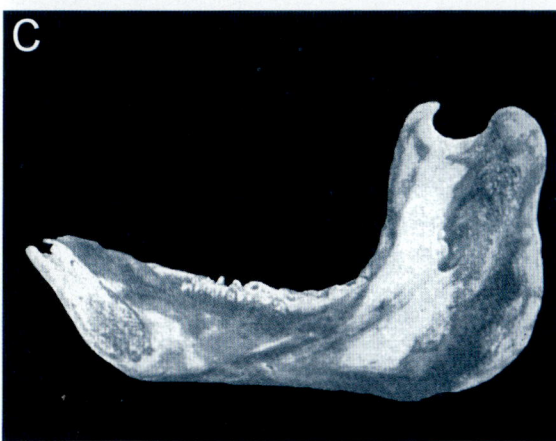

Figure 34.2 Madder feeding reveals bone growth. Addition and withdrawal of madder from the diets of pigs demonstrates bone growth, as illustrated in a classic study by Brash (1934). (A, B) Lateral (A) and medial (B) surfaces of the left dentary bone of a 20-week-old pig. The pale areas (especially evident in B) indicate areas of new bone deposited during the previous 19 days when madder was withdrawn from the diet. (C) The medial surface of the right dentary bone of a 40-week-old pig. New bone (white) was deposited during the previous 54 days when madder was withdrawn from the diet.

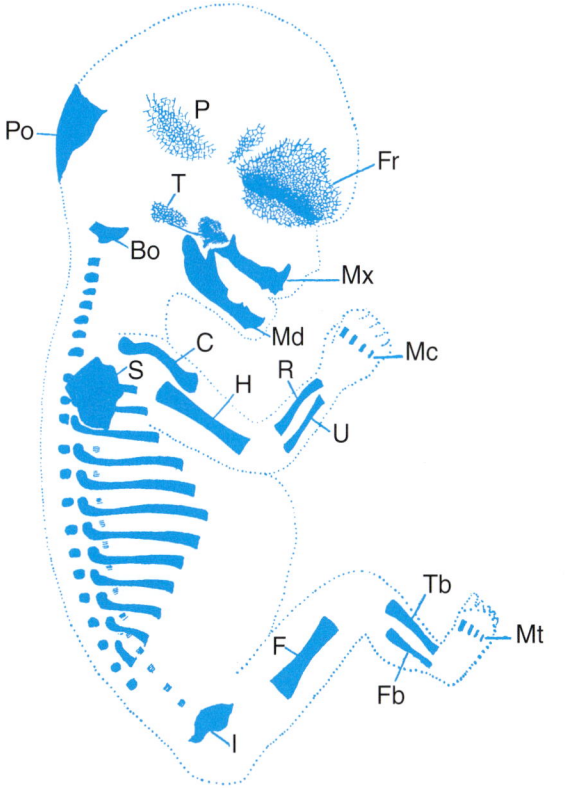

Figure 34.3 A drawing of an alizarin red-stained human embryo showing the stage of ossification attained at 13 weeks' gestation. Within the craniofacial skeleton, ossification is ongoing in the basioccipital (Bo), frontal (Fr), mandible (dentary bone, Md); maxilla (Mx), parietal (P), postoccipital (Po) and squamous portion of the temporal bone (T). Scapula (S), clavicle (C) ischium (I) and ribs are well ossified. Proximal limb elements – humerus, femur (H, F), radius/ulna (R, U), tibia/fibula (Tb, Fb), are well ossified, and ossification has begun in the metacarpals (Mc) and metatarsals (Mt). From Le Gros Clark (1958).

Alizarin

The active agent in madder is *alizarin* (1,2-dihydroxy-9, 10-anthracenedione; 1,2-dihydroxyanthraquinone; $C_{14}H_8O_4$, MW, 240.2). Alizarin was synthesized in 1870 and adopted as a marker for studies on bone growth. Alizarin has the major advantage over madder feeding that a single injection is sufficient to label bone, and can be detected within one hour of intraperitoneal injection of a 2% solution (Fig. 34.3). Nevertheless, madder was used well into the 20th century; the autochrome in Brash (1934) of tibiae, skulls and jaws of madder-fed pigs (Figs 34.2 and 34.4) illustrates brilliantly how the technique can be employed to identify regions of deposition and resorption during bone growth, as analyzed in detail for the pig skull.[g]

Alizarin red also is the bone-staining agent in the most common method used to clear and selectively stain the bony skeleton of entire organisms (Figs 5.6, 5.7, 9.2, 10.3, 21.1, 34.3 and 37.2). Alizarin is used either alone or in combination with a stain such as alcian blue or methylene blue to stain cartilage in a double-staining method of great sensitivity (Figs 3.8, 34.5 and 34.6); but see the **cautions** and comment below.[h]

Box 34.1 *(Continued)*

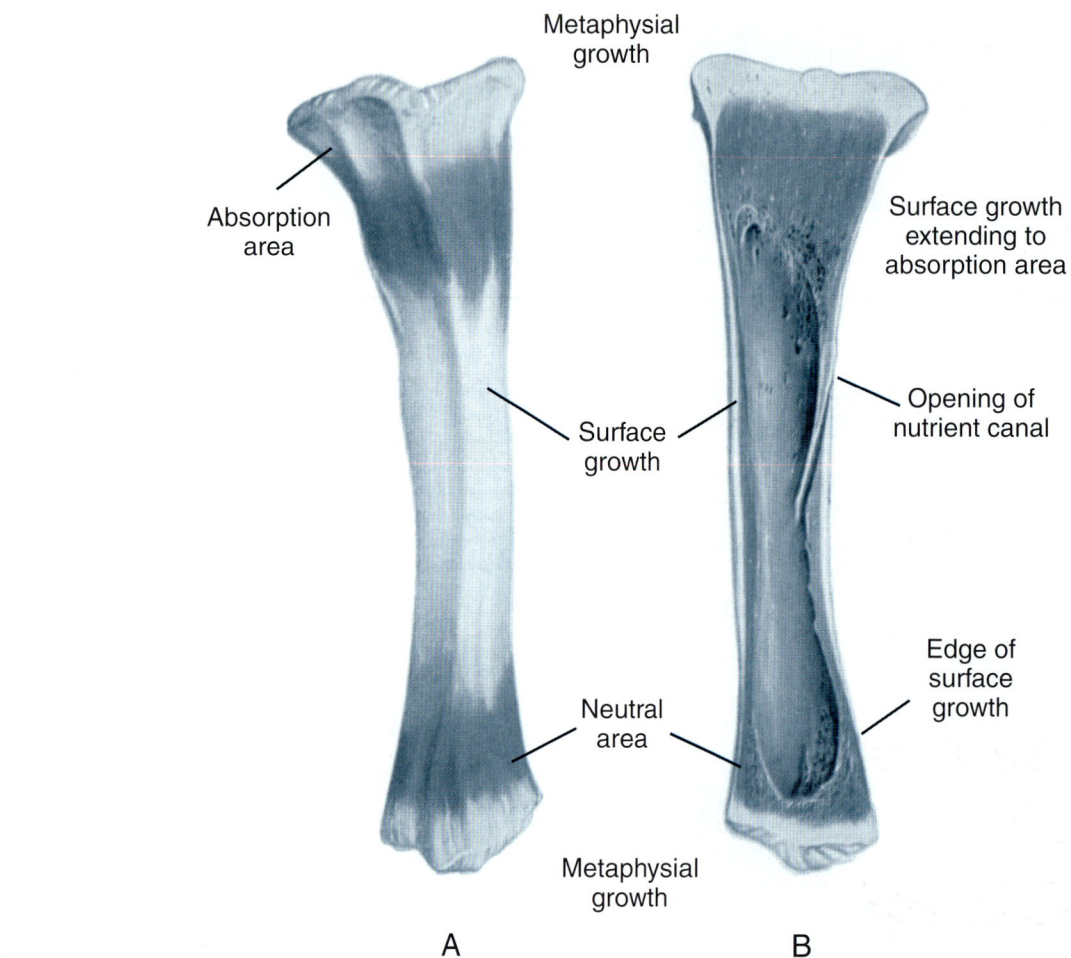

Metaphysial
growth

Absorption
area

Surface growth
extending to
absorption area

Opening of
nutrient canal

Surface
growth

Edge of
surface
growth

Neutral
area

Metaphysial
growth

A   B

Figure 34.4 The tibia from a 9-month (40 week)-old pig fed madder for the first eight months of life. (A) A surface view. (B) A longitudinal section. New bone added during the ninth month – when madder was omitted from the diet – is white and found subperiosteally and at the metaphyses. From Le Gros Clark (1958).

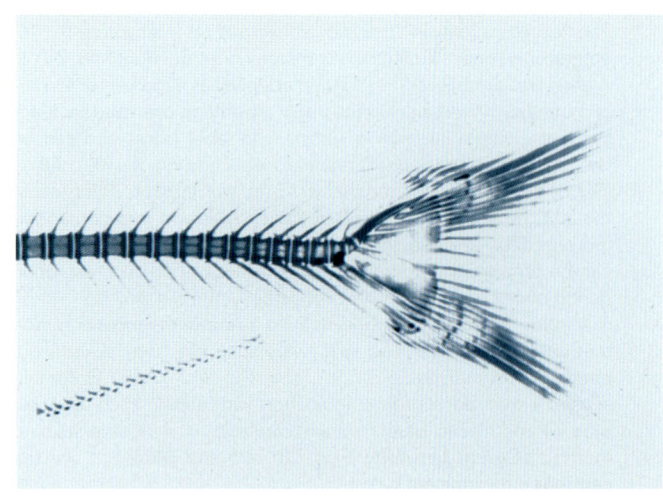

Figure 34.5 A specimen of the alewife, *Alosa pseudoharengus*, cleared and stained with alcian blue and alizarin red reveals the fine details of the skeleton in exquisite detail, including rays of the caudal fin, vertebrae and the cartilaginous endoskeletal elements of the ventral and caudal fins. Photographed by Tom Miyake. Other common names for this herring are bigeye, branch, freshwater, gray, river and white herring; gaspareau; grayback; kyack/kiack; and sawbelly.

Box 34.1 *(Continued)*

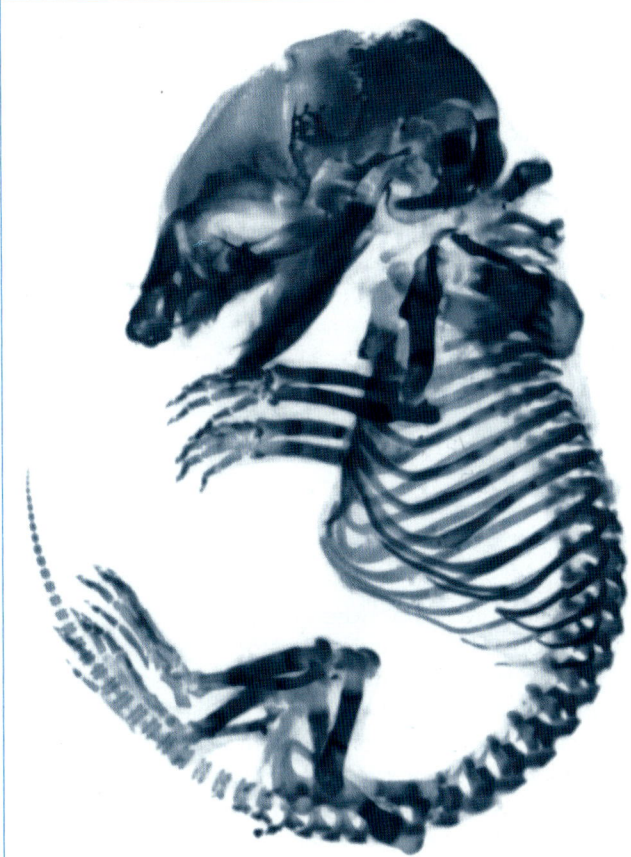

Figure 34.6 This 17-day-old (Theiler stage 25) mouse embryo has been cleared and stained with alcian blue and alizarin red to visualize the cartilages and bones of the skeleton.

[a] Boorstin (1983), p. 16, emphasis mine.

[b] See Dickson (1984) for methods for the study of bone and other mineralized tissues, and see Belchier (1736a,b). The first to record bone staining by madder root may have been Lemnius (1567) in his *Occulta Naturae Miracula*; see Cameron (1930). Volumes 6 and 7 of Hall (1990–1994) contain evaluations of bone growth from many perspectives.

[c] Other economically important plants in the family Rubiaceae produce coffee (*Coffea arabica*), the anti-malarial alkaloid quinine (the fever-bark tree, *Cinchona officinalis*), the emetic and purgative ipecacuahna (the South American shrubby plant, *Cephaelis ipecacuanha*), and the dye and tanning agent gambier (cat's claw, *Uncaria tomentosa*).

[d] Duhamel (1739, 1740, 1742, 1743); see Keith (1919) for an excellent discussion. Madder also stains keratinized structures such as the claws and beaks of birds and the claws of foetal rats (Muther, 1988a), although the latter observation evoked a contrary response and rebuttal (Runner, 1988; Muther, 1988b). Jennifer Legere in my laboratory has also found that alizarin stains the spicular skeleton of larval echinoderms (pers. comm.).

[e] Hunter (1771), and see Hunter (1835) for a compilation of his works. Delaporte (1983) is a good review of views on osteogenesis in the 18th century, especially bone growth and remodeling, madder feeding, and the roles of periostea and endostea.

[f] See Tomes (1839), Tomes and De Morgan (1853) and Jaworski (1992) for the Haversian system, and Havers (1691/2) cited in Howship (1815) for Haversian canals.

[g] See Hoyte (1960) for the alizarin method, Fieser (1930) for the chemistry of alizarin, Flourens (1845, 1847), Gottlieb (1914), Schour (1936) and Schour *et al.* (1941) for early use of alizarin, and Hoyte (1960), Puchtler *et al.* (1969) and Hall (1992) for histories of the use of alizarin. See Macklin (1917), Proell (1926) and Brash (1924, 1926, 1934) for some studies from the first quarter of the 20th century using madder.

[h] T. Yamada (1991) and Yamada *et al.* (1993) developed a method to selectively stain rat foetuses previously prepared as whole mounts with alizarin red S, showing that bromophenol blue could be used to stain cartilages even after specimens had been in glycerol for 10 years. Two important **cautions**: Because it contains a large amount of mucopolysaccharides, newly deposited osteoid stains with alcian or methylene blue. Such elements can be and have been confused with or identified as cartilages.

Skull growth continues after sutures are removed, indicating that sutures do not generate the forces required to separate cranial bones one from another. On the other hand, isolated sutures lack inherent growth potential: the midpalatal sutures of rapidly growing five-day-old rats can be cultured for 18 days without observing any growth; growth and morphology of extirpated or repositioned sutures depends upon the functional characteristics of the new site.[7]

Another approach is to use a mechanical device to expand sutures (a vice in reverse) to see whether a response can be elicited. Proliferation is confined to flattened cells at the sutural margin of rat midpalatal sutures after sutural expansion (Kobayashi *et al.*, 1999). These cells, not the non-dividing stretched cells in the middle of the suture, express osteogenic markers such as alkaline phosphatase and osteocalcin.

Fusion of sutures disturbs the growth of other components of the skull. Premature fusion (synostosis) is thus to be avoided, although there are instances when premature fusion evolved as a mechanism for strengthening the face during rapid growth.[8]

Working with the functional matrix

Melvin Moss discussed the adaptiveness of sutural growth in the context of the skull and its associated structures as a *functional matrix*, emphasizing that we must analyze growth of the *functional unit* to understand growth of the individual components. For example, the calvaria is displaced outwardly as the brain expands, subjecting sutures to tension. Sutures respond adaptively by depositing bone to counteract the tension. This adaptive response to tension can be used experimentally to initiate bone deposition to heal bone defects, as follows.

- A screw device used to place the palatomaxillary suture under tension induces sufficient bone deposition on the palatine bones that they grow to approximate one another and can obliterate a cleft palate.
- Facial sutures have been placed under tension and the resorption initiated used to reposition the maxilla.
- Applying tension for appropriate periods of time can establish cycles of deposition and resorption.[9]

SUTURAL CARTILAGE

Two complications not often considered in analysis of sutural growth are the presence of cartilage within sutures and the contribution that cartilage growth makes to sutural growth.

Secondary cartilage is a normal component of many sutures, and can develop in unexpected places when sutures are repositioned or in such clinical situations as cleft lip and cleft palate. A remnant of primary cartilage also may persist in sutures or during sutural development, as for example in a 2.7-month-old human embryo (75-mm crown–rump length) as a transient cartilage in the sphenofrontal suture, which forms the boundary between neuro- and chondrocranium and unites the endochondral lesser wing (*ala orbitalis*) of the sphenoid bone with the intramembranous portion of the orbital part of the frontal bone (Captier *et al.*, 2003).

Antibodies against constituents normally associated with cartilage – chondrocalcin, collagen type II, decorin, biglycan, proteoglycans – react with the coronal suture in rats, a reactivity that is suggestive of cartilage in the suture. Similarly, cartilage antibodies react with repairing rat parietal bones, but no cartilage forms during the repair process. The potential for chondrogenesis is present in rabbit coronal sutures, in which direct current promotes cartilage formation.[10] The contribution made by such cartilages to sutural growth in growing mammals may explain why 'apparently' similar sutures exhibit a diversity of responses when subjected to identical treatments. Following are several instances.

Displacing the buccal musculature in Sprague-Dawley rats is followed by earlier ossification of sutural cartilage and the development of new secondary cartilage in the intermaxillary suture (Fig. 15.5).

Fibrous tissue rich in type I collagen develops in rat midpalatal sutures after four days of applied expansive force, after which mineralized membrane bone is deposited. In this case, tension appears to modulate the differentiation of preexisting secondary cartilage in the suture through fibrous tissue to bone, interleukin-1 (IL-1) being the modulator.[11]

Cartilage has been described *underlying* the coronal suture in mice of the C57/BL6 inbred strain, with the suggestion that it may mediate suture–brain interactions and maintain the coronal suture as a growth centre. Something similar may occur in fish. An unexpected finding in a study of the development of the occipital portion of the skull in the Japanese musk shrew, *Suncus murinus*, was the presence of membrane bone arising from a periosteal band in the perichondrium of the supraoccipital cartilage, i.e. arising as subperiosteal bone. The membranous portion of the squamous part of the occipital in humans and mice ossifies in two parts from a supraoccipital (which is both cartilage and bone) and an interparietal.[12]

The coronal sutures of human infants three and seven months old, and of rats, were described by Markens

(1975), who included a discussion of ectopic cartilage in the analysis. Markens and Taverne (1978) followed this up with a study of the development of cartilage in rat coronal sutures after sutures from 19-day-old embryos had been transplanted into the dura mater of adults for between one and six days. Ectopic cartilage formed in all transplants. The cartilage was always on the cerebral side, always arose from sutural connective tissues, and was presumed to have been induced by the dura. The cartilage disappeared by 21 days, perhaps because mechanical conditions were inappropriate to maintain it. This study highlights the role of the dura mater in sutural biology, a topic introduced in Chapter 21 and elaborated below.

THE DURA

The dura (dura mater) is a tough dense connective tissue membrane enveloping the brain and spinal cord. In the head, the dura lies between the brain and the skull but is continuous with the periosteum of the skull.

Signals from the osteogenic front of each cranial bone predominate after birth. Tyler (1983) investigated the induction of frontal bones in embryonic chicks using a tissue-interaction approach and grafting to the CAM. Cranial epithelium is required for frontal bone induction, the epithelial–mesenchymal interaction taking place between H.H. 22 and 30 (H.H. 30 being immediately before osteogenesis begins).

Our knowledge of dura function comes primarily from an elegant series of studies undertaken by Lynne Opperman and colleagues over the past decade. We now know that:

- the dura functions to maintain patent sutures but does not initiate suture formation;
- the dura plays a role in epithelial–mesenchymal interactions but not a mechanical role, as determined from an analysis of the interparietal suture in rats from 19-day-old foetuses to adult;
- in the absence of the periosteum, coronal sutures transplanted to the parietal bone are not obliterated by bone; sutures fuse if the dura is not present;
- the dura signal can cross a 0.4-μm filter, and so is diffusible;
- the dura secretes a heparin-binding component that prevents sutural closure (based on studies *in vitro*); and
- the dura regulates sutural cell proliferation and collagen production.[13]

From further studies in her laboratory, we know that members of the Tgfβ superfamily mediate interactions between dura and skull. *Tgfβ-1*, *-2* and *-3* are all expressed in the dura. Tgfβ-1 and -2 have distinct patterns of sutural expression that correlate with obliteration. Tgfβ-3 is found in unossified regions of sutures.

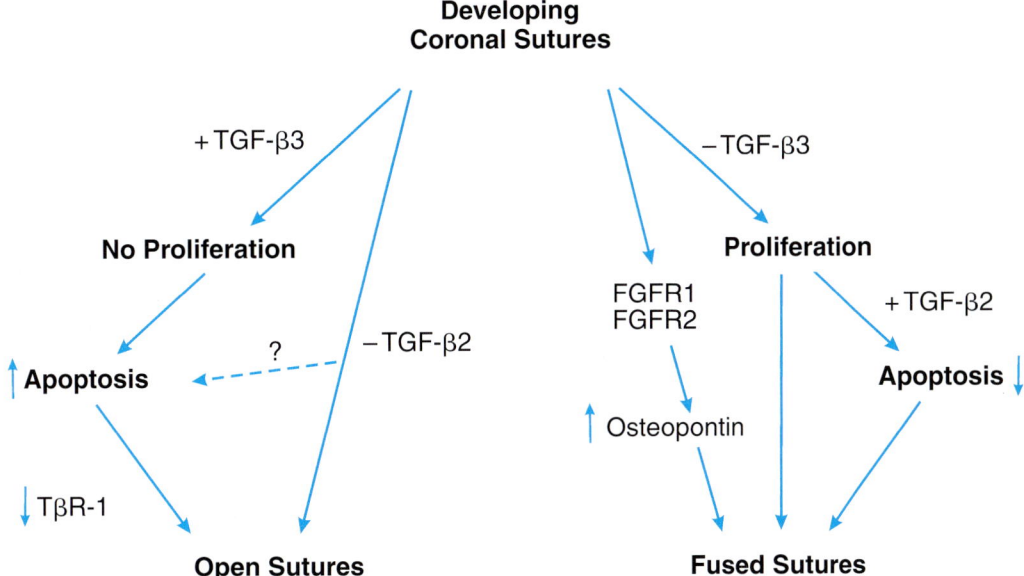

Figure 34.7 A proposal for the roles of Tgfβs and Fgf receptors in suture development, shown as controlled by the presence (+) or absence (−) of Tgf-β3. TGF-β3 inhibits proliferation, enhances cell death (apoptosis), which in association with down-regulation of TGFβreceptor-1 (TβR-1) maintains sutures open. TGF-β2 also may play a role. In the absence of TGF-β3 (and the presence of TGFβb-2, Msx, Fgf-4 and Bmp-4): (i) proliferation is enhanced and sutural fusion facilitated while (ii) TGFβ-2 inhibits apoptosis, further enhancing sutural fusion. Osteopontin is up-regulated by Fgf receptors 1 and 2 (FGFR1, FGFR2), reinforcing sutural fusion. From a scheme presented by Opperman et al. (2000).

Removing Tgfβ-3 from the coronal suture induces sutural fusion. Removing Tgfβ-3 from calvarial cultures prevents maintenance of the suture, i.e. Tgfβ-3 maintains sutures patent. Antibodies that block Tgfβ-3 invoke premature closure of rat calvarial sutures maintained *in vitro*. Premature closure is preceded by enhanced proliferation, equivalent to that seen after removing the dura. Addition of Tgfβ-3 to calvaria cultured without the dura stops closure and decreases proliferation. Addition of Tgfβ-2 has the opposite effect, leading to suture closure and enhanced proliferation. Tgfβ action is therefore linked to cell number and cell division (Fig. 34.7).[14]

Other growth factors are also involved:

- *Msx-1* and *Msx-2* are expressed in sutural mesenchyme and in the dura. *Bmp-2* and *-4* are expressed intensely in the osteogenic front (Box 12.1), and Fgf-9 in calvarial mesenchyme, dura and overlying skin.[15]
- Implanting beads soaked in Fgf-4 into the sutural front in organ-cultured calvariae leads to premature sutural closure, an experimental result that may parallel the situation seen in FgfR mutants with craniosynostosis (see below).
- Injecting Fgf-2 or -4 near the coronal suture increases apoptosis, with ectopic expression of type I collagen and enhanced mineralization to form the equivalent of a synostotic suture.
- Implanting Bmp-4 expands tissue volume but has no effect on sutural closure. Implanting Bmp is an effective means of repairing quite extensive (14–20 mm) craniotomy defects in stump-tailed macaques, *Macaca speciosa*, bone *and* cartilage arising from cranial periostea *and* from cells in the dura.[16]

Information also can be gleaned from studies on the healing of skull wounds in adult mammals (Box 34.2), which demonstrate that the dura–mesenchyme relationship is activated during bone repair.

Craniotomy in neonatal rabbits illustrates the importance of the integrity of the dura and periosteal envelope if sutures are to redevelop during repair. In skull regeneration in dogs and rabbits, new bone only forms if there is continuity of the dura and contact between periosteum and dura. This may be because the dura also plays a mechanical role – for example providing appropriate tension – or a more direct and inductive role. Placing Millipore filter barriers between dura and overlying mesenchyme produces results consistent with an inductive role for the dura in regenerating skull vaults in rabbits. Periostea contain osteogenic cells but they have to be induced by the dura.[17]

CRANIOSYNOSTOSIS

Sutural fusion is normal in many situations but can occur as an abnormality.

Craniosynostosis is exactly what the term implies: fusion or synostosis of cranial (skull) bones. The syndromic implications of the term are that sutural fusion is premature. Craniosynostosis, which has a prevalence of between 30 and 50 individuals per 100 000 live human births, has a basis in mutations of *Fgf* or Fgf receptor genes but, as might be expected if fusion of skull bones occurs during the growth period (which it inevitably does), expansive growth of the brain influences bone growth and morphogenesis secondarily.

Box 34.2 Trephination

The skull, followed by the temperament; the two hardest parts of the body[a]

A *trepan* is a cylindrical saw used for millennia to remove portions of the skull. A *trephine*, a more modern version of a trepan, has a central pin to guide the surgeon's hand; hence trephination. The deliberate creation of wounds in bone, and especially trephination of the skull, was practised in western Europe, Peru, Bolivia and Mexico in Neolithic time; a skull from 5100 BC shows signs of stone-age cranial surgery. Trephination continued in Mexico until as recently as AD 700. An interesting application of digital imaging has been to provide evidence of trephination in North America, to argue that Neanderthals were cannibals, and to visualize the results of pre-Columbian dentistry.[b]

Trephination continues to be used both clinically and in experimental studies on wound repair. For example, implanting Bmp is an effective way to initiate repair of trephine holes in sheep skulls and of repairing 14–20 mm craniotomy defects in stump-tailed macaques, *Macaca speciosa*. So too is implanting octacalcium phosphate, which evokes bone from osteoprogenitor cells on calvarial surfaces in rats. Twenty-four weeks later the defects are completely filled in with bone, while control defects are merely rimmed with bone.[c]

[a] J. M. Coetzee, *Disgrace*, 1999, p. 20.
[b] See Bishop (1961) and Wilkinson (1975) for trephination, Gilbert and Richards (2000) for the use of digital imaging, and Alt *et al*. (1997) for stone-age cranial surgery. Excellent source books for the history of studies on bone, especially in relation to orthopaedic surgery, are Bick (1968) and Trueta (1968).
[c] See Lindholm *et al*. (1988) and D. Ferguson *et al*. (1987) for the sheep and monkey studies, and Kamakura *et al*. (1999) and Sasano *et al*. (1999) for the octacalcium phosphate. New agents effective in promoting bone healing are always being sought and tested, a recent example being a derivative of dextran (a polysaccharide composed of glucose monomers) given the trade name RGTA™, that protects heparin-binding growth factors released during repair. When tested in 3-mm diameter defects in rat calvariae a single dose promoted bone formation without inflammation, and promoted more peripheral resorption after seven days than in controls (Lafont *et al*., 1998; Colombier *et al*., 1999).

The reverse also holds true. Changes associated with craniosynostosis have been found in the human central nervous system. Considering what we know of relationships between the brain, dura and skull discussed above and in Chapter 21, some compensatory influences back to the brain are not unexpected. Others changes are predicted consequences of sutural fusion. For example, quantitative analysis shows that abnormal locations of the ossification centre within the fronto-parietal suture in craniosynostotic rabbits are secondary, not primary. Adaptive growth of the cranial base has been described at 50 and 100 days after premature synostosis was induced in five-day-old rabbits.[18] Other changes are a more direct result of altered gene action following mutations in *Fgfs*, Fgf receptors, *Msx-1* or *Msx-2*.

Msx-2

Jabs *et al*. (1993) identified a mutation in the homeodomain of *MSX-2* in a family with autosomal dominant craniosynostosis. *MSX-2* localizes to the same region of chromosome 5 as craniosynostosis, and *MSX-2* transcripts are expressed in the sagittal suture; *Msx-2* transcripts are more widespread in murine sutures. We know that mice with a gain-of-function mutation of *Msx-2* show premature suture closure and craniosynostosis, and deposit ectopic bones, all of which are consistent with the features of craniosynostoses in humans, rodents and rabbits. *Msx-2* gene dosage influences the number of proliferating osteogenic cells in growth centres of the mouse skull, determined by using a segment of the *Msx-2* promoter to direct a reporter gene expression to a subset of sutural cells. Overexpression of *Msx-2* enhances parietal bone growth. In craniosynostosis, osteogenic cell differentiation is inhibited and proliferating osteoprogenitor cells enhanced.[19]

Fgf receptors

Fgf receptors were introduced in Chapter 14 in the context of urodele limb regeneration and in Chapter 27 in relation to achondroplasia, both of which exemplify the roles of FGF receptor expression in skeletal growth, especially sutural growth and sutural fusion.

Sutural growth

The recruitment of osteoblasts during mammalian skull development is regulated by intricate cross-talk between *Fgf-2*, *FgfR-1* and *FgfR-2*. Fgf-2 protein, which has a high affinity for FgfR-2, is locally abundant. Immunohistochemical detection shows it to be present at low levels in FgfR-2 expression domains and at high levels in differentiated areas. Implanting Fgf-2-soaked beads onto the foetal coronal suture by *ex utero* surgery results in ectopic osteopontin expression, encircled by FgfR-2 expression. When Fgf-2 is low, osteoprogenitor cells divide and express *FgfR-2*. The growth factor Fgf-2 then increases as progenitor cells differentiate into preosteoblasts, FgfR-2 declines, *FgfR-1* is up-regulated and osteopontin is switched on. Once osteogenic differentiation is completed, FgfR-1 declines. Inactivation of *FgfR-3* in mice results in dwarfism and reduced bone density because osteoprogenitor proliferation and functioning of mature osteoblasts is disrupted, although osteoblast differentiation is normal.[20]

Mansukhani *et al*. (2000) used primary cultures of mouse osteoblasts and an osteoblastic cell line (OB1) to explore Fgf-2 and FgfR-2 signaling in osteogenesis, finding a dual effect of Fgf:

- Fgf activates endogenous Fgfs to produce a complex that activates MAP kinase;
- immature osteoblasts respond to Fgf with increased proliferation; while
- differentiating osteoblasts respond to Fgf with apoptosis not proliferation, expression of alkaline phosphatase and mineralization both being blocked.

Sutural fusion

Much also is known about Fgf receptors (FgfR) and the fusion or synostosis of skull bones (craniosynostosis), of which the following is a woefully inadequate summary. Muenke *et al.* (1998) reviews FgfR-related skeletal disorders, which include mutations in all three receptors, as follows.

- A unique point mutation in *FGFR-3* defines a new *human craniosynostosis syndrome* identified in 61 individuals from 20 families.
- At least three types of *dwarfism* result from mutations of the *FGFR-3* gene, chondroblast proliferation and differentiation being affected in all three (Table 34.1).
- Crouzon-like syndrome in mice and other craniosynostoses result from an insertional mutation of the *FGF-3/FGF-4* locus. Centres of bone formation at sutures are displaced, including those arising following mutations of *FGFR* or *TWIST* (see Fig. 34.8).

Table 34.1 Dwarfism syndromes caused by mutations of *FGFR-3* and the mutations associated with each syndrome[a]

Syndrome	Associated mutation
Thanatophoric dysplasia type I	Arg248Cus, Ser249Cys, Gly370Cys, Ser371Cys, Tyr373Cys, Stop807Gly, Stop807Arg, Stop807Cys
Thanatophoric dysplasia type II	Lys650Glu
Achondroplasia	Gly346Glu, Gly375Cys, Gly380Arg
Hypochondroplasia	Asn540Lys

[a] Based on data from Webster and Donoghue (1997).

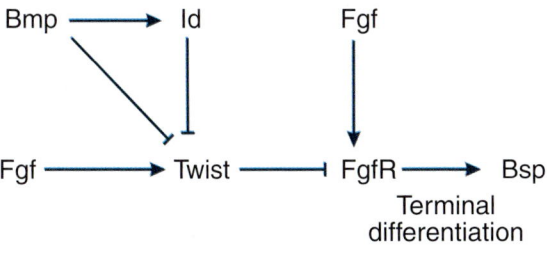

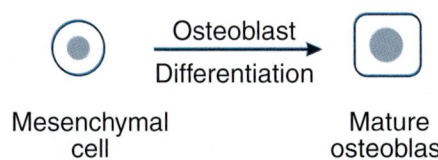

Figure 34.8 BMP-2 and osteoblast differentiation in calvarial bones. Bmp-2 and Inhibitor of differentiation (Id) inhibit Twist and FGFR, keeping cells within the mesenchymal cell pool. In up-regulating Twist, Fgf antagonizes Bmp-2, removing inhibition of Fgf receptors (FgfR). Coupled with direct binding to FgfR, FGF facilitates terminal differentiation, as evidenced by the synthesis of bone sialoprotein (Bsp). Based on the model proposed by Rice *et al.* (2000).

- In individuals with Apert or Pfeiffer syndrome FGFR-1 and FGFR-2 isoforms are differentially activated by several possible mutations. Accumulation of fibronectin, collagen and GAGs, and an altered response to FGF-2, can be detected in periosteal fibroblasts from both Crouzon and Apert patients expressing a certain subset of mutations.

Introducing an *FGFR-2* carrying either the *C342Y* (*Crouzon syndrome*) or the *S252W* (*Apert syndrome*) mutation into the OB-1 cell line inhibits differentiation and has a dramatic effect on apoptosis. Overexpressing *Fgf-2* in transgenic mice also dramatically increases apoptosis in calvariae.[21] So, *FGF can trigger proliferation or cell death depending on the state of differentiation of the osteoblastic cells*.

Fgf-2 and the basic helix–loop–helix factor Twist are in the same pathway, acting at early and late stages of osteogenesis (Fig. 34.8). Homozygous Twist null mice die before birth with neural tube, craniofacial and limb defects. Heterozygotes have a duplicated digit I in the hind limbs and defects in skull bones. TWIST was the first helix–loop–helix protein shown to produce a human syndrome when in mutated form. The syndrome is Saethre-Chotzen syndrome, characterized by craniosynostosis. In their study of skull and sutural development, Rice *et al.* (2002, 2003) showed that:

- expression of *Fgf-2* is lower in sutures;
- Twist is expressed in preosteoblasts in patterns that overlap with Fgf, expression decreasing as osteoblasts differentiate; and
- implanting Fgf-2-soaked beads up-regulates Twist and down-regulates bone sialoprotein (an osteoblast marker; Fig. 34.8).[22]

Twist binds to the *periostin* promoter in undifferentiated preosteoblasts and promotes *periostin* gene expression *in vitro*. Periostin (osteoblast-specific factor-2), localized to periostea and periodontal ligaments, is a 90-kDa protein deposited by osteoblasts and involved in cell adhesion. Expression is enhanced by Tgfβ and both Tgfβ-1 and Tgfβ-3 are modulated in mouse sutures subjected to intrauterine constraint. As demonstrated using MC3T3-E1 cells periostin plays a role in recruiting osteoblasts and attaching them to the periosteum. In mouse mammary epithelial cells, periostin is regulated by *wnt-3*.[23]

NOTES

1. In architecture, spandrel refers to the triangular space resulting from where the outer curve of an arch contacts the wall and ceiling. Spandrels are not part of the design but rather inevitable consequences of arches contacting horizontal or vertical structures. Spandrels were brought into the biological literature by Gould and Lewontin (1979) in what is usually referred to as the 'Spandrels of San Marco' paper, which had the twofold aim of sounding the death

knell for adaptationism and bringing developmental analyses back into evolutionary studies. See Hall (2005b) for an analysis of the impact of the paper, especially on developmental biologists studying skeletal systems.

2. The basic structure of sutures is well documented by Pritchard et al. (1956), Moss (1957a), Koskinen et al. (1976), Decker and Hall (1985) and in the chapters in Cohen (1986). See Cole et al. (2003) for sutures in vertebrates and invertebrates. See Johansen and Hall (1982) for morphogenesis of the coronal suture from 15 to 71 days in mice. This study places special emphasis on how overlapping osteogenic fronts fix the position of the suture, and is an excellent review of the literature. Wagemans et al. (1988) is an excellent review of the response of sutures to mechanical forces. See Jiang et al. (2002) for sutures linking neural-crest and mesodermal bones in mouse skulls.

3. See Brash (1934), Scott (1954) and Hoyte (1966) for this view. See Isotupa et al. (1965), Hoyte (1971) and Cleall (1974) for the labeling studies, and Dixon and Hoyte (1963) for a comparison of alizarin red and autoradiographic techniques for measuring bone growth. See de Angelis (1969), Moore and Lavelle (1974) and Smith and McKeown (1974) for differential growth on opposite sides of a suture. See Note 1 for spandrels.

4. See Meikle et al. (1979, 1980, 1982, 1984). Meikle et al. (1979) developed a method for maintaining rabbit cranial sutures in vitro.

5. For detailed discussions of various aspects of the literature on skull growth, see Hoyte (1966), van Limborgh (1970, 1982), Moss (1972b,c), Moss et al. (1972), Persson (1973, 1995), Storey (1973), Moore and Lavelle (1974), Enlow (1975), Persson et al. (1978) and Volume 6: Bone Growth – A in Hall (1990–1994).

6. See Persson et al. (1978) and Oudhof and Markens (1982) for these two studies.

7. See Moss (1954, 1957a,b, 1959) for extirpation of cranial sutures, Drake et al. (1993) for the craniectomy study, Stutzmann and Petrovic (1970) for culture of midpalatal sutures, and Smith and McKeown (1974) and Dixon and Sarnat (1982, 1985) for extirpation and repositioning sutures. I thank Lynne Opperman for bringing the craniectomy study to my attention.

8. See Ritsila et al. (1973) and Alhopuro (1978) for growth disturbances following sutural fusion, and Herring (1974, 2000) and Rafferty and Herring (1999) for situations in which premature fusion is advantageous.

9. See Moss (1972b) and Moss et al. (1972) for the functional matrix, Latham et al. (1973) for the screw device, Elder and Tuenge (1974) for maxillary repositioning, and Hinrichsen and Storey (1968), Murray and Cleall (1971) and Ten Cate et al. (1977) for cycles of deposition and resorption.

10. See Alberius and Johnell (1990, 1991) for the antibody studies, and Hathaway (1997) for cartilage induction in rabbits. See Pritchard et al. (1956), Moss (1958), Hall (1970a), Stutzmann and Petrovic (1970), Aaron (1973) and Boivin et al. (1983) for cartilage as a normal component of sutures, Smith and McKeown (1974) for its development in repositioned sutures, and Curtin and Pruzansky (1971), Pruzansky (1971) and Friede (1973) for association with cleft lip and palate.

11. See Ghafari (1984) for musculature displacement, I. Takahashi et al. (1996a) for modulation of secondary cartilage, and see Moore and Lavelle (1974) for additional examples and for an excellent analysis of the literature.

12. See Iseki et al. (1999) for cartilage under the coronal suture, Niida et al. (1994b) for the Japanese musk shrew, and Srivastava (1992) for ossification of the human supraoccipital, a mode that may be common to many mammals. See Silau et al. (1995) for a radiographic study of the development of the human parietal bone and interparietal suture using 15 specimens to show the primary ossification centre.

13. See Opperman et al. (1993, 1994, 1995, 1996, 1998) for these studies, which are based on analysis of parietal suture in rats from 19 days of foetal life to adult, and on culturing calvaria with or without the dura.

14. See Opperman et al. (1997, 1999, 2000) for the Tgfβ studies, and Opperman (2000) for an overview of these and other studies. TGFβ-1, -2 and -3 (and Msx-2) also are expressed in the frontonasal suture (a facial suture) of Sprague-Dawley rats but in a different temporal sequence to that seen in cranial sutures (Adab et al., 2002).

15. Msx-1 (Hox-7.1) is expressed in developing murine skull bones from 10.5 days onwards. Msx-1 is widely distributed in various embryonic structures such as neural crest-derived mesenchyme, neuroepithelium, Rathke's pouch, the nasal pits, and at epithelial–mesenchymal junctions, all sites that correlate with regions affected by embryonic exposure to excess retinoic acid (Mackenzie et al., 1991a,b).

16. See Kim et al. (1998) for the expression data, Mathijssen et al. (2000) for Fgf injection, and D. Ferguson et al. (1987) for the study with the rhesus monkeys. See Rice (1999) for molecular mechanisms in calvarial bone and suture development, including discussion of Fgf, FgfR, Hox, Msx, Tgfβ and dura signals.

17. See Kantorova (1972, 1981a, 1983) and Mossaz and Kokich (1981) for dura continuity and the filter barrier experiments, and Mabbutt and Kokich (1979) for continuity and suture formation. See Melcher (1969) for the role of the periosteum in repair of wounds to parietal bones in rats.

18. See K. Aldridge et al. (2002) for secondary changes in the brain, Dechant et al. (1999) for ossification centres in craniosynostotic rabbits, and Kantomaa et al. (1991) for adaptive changes in the cranial base. See Mooney et al. (1996a,b) for the basic genetics of an autosomal dominant craniosynostosis (coronal sutural synostosis) that appeared in a rabbit colony. Herring (2000) is an excellent review of sutures and craniosynostosis, especially in a phylogenetic context, discussing sutures and craniosynostosis in each of the vertebrate groups. The volumes edited by Cohen (1986) and Cohen and Maclean (2000) contain valuable reviews of craniosynostosis, including general chapters on sutural development and growth. For a study of suture development and sagittal synostosis, see Richtsmeier et al. (1998).

19. See Y. H. Liu et al. (1995, 1996, 1999) for the transgenic studies with Msx-2, and Cohen (2000a) for a summary of human craniofacial disorders caused by mutations of MSX-1 or MSX-2, especially hypodontia, Boston-type craniosynostosis and formation of parietal foramina. Ignelz et al. (1995) survey the utility of transgenic mice for craniofacial development, including Msx-2 and premature suture closure.

20. See Iseki et al. (1997, 1999) for the sutural study and Yu et al. (2003) for FgfR-2 inactivation.

21. See Muenke et al. (1997) for the point mutation, Webster and Donoghue (1997) for dwarfisms, Carlton et al. (1998)

for the Crouzon-like syndrome, Mathijssen *et al.* (1999) for displacement of ossification centres, and Britto *et al.* (2001) and Carinci *et al.* (2000) for Apert and Pfeiffer syndromes.

22. See Iseki *et al.* (1997) for lower levels of FGF-2 protein in the sutures.

23. See Kirschner *et al.* (2003) for intrauterine constraint and Horiuchi *et al.* (1999) for the study with MC3T3-E1 cells. Periostin, which is up-regulated by Tgfβ and Bmp-2, is also expressed in developing teeth, initially – 9.5 days of gestation – in dental epithelium and then in the dental mesenchyme (Kruzynska-Frejtag *et al.*, 2004). A second basic helix–loop–helix protein, DERMO-1, which has considerable homology to TWIST, is expressed in human osteoblasts and has been cloned by Lee *et al.* (2000). Osteogenic cell lines over- or underexpressing *DERMO-1* show that it keeps cells as preosteoblasts, stopping their differentiation, and so has to be down-regulated for differentiation of preosteoblasts → osteoblasts.

Part XII

Limb Buds

Limbs are paired sets of appendages with digits
Tetrapods are vertebrates with limbs
Snakes are limbless
Snakes are tetrapods?

The variety of interactions underlying apparently similar skeletal tissues makes the developing skeleton a fascinating and profitable organ system to study. Perhaps the best-studied interactions responsible for establishing skeletal systems are (i) those between presumptive limb mesoderm/mesenchyme and the apical ectodermal (epithelial) ridge (AER), and (ii) those between presumptive vertebral mesenchyme and the epithelial notochord and spinal cord. In both cases, we have learned most from study of and experimentation on embryonic chicks. Limb and vertebral development are the topics of the three parts that follow.

The Limb Field and the AER

The idea of morphogenetic fields has pervaded embryology so thoroughly that it cannot be shaken off, any more than a religious upbringing.[1]

Because limb buds are used to illustrate aspects of cell condensation (Chapter 19), chondrogenesis (Chapters 22 and 26), polarity, cell death, morphogenesis and growth (Chapter 38 and 39), I provide a summary of the major morphogenetic and differentiative events of limb-bud development in embryonic chicks up to H.H. 30 (6.5–7 days of incubation) (Table 35.1), in the hope that such a ready reference will be useful.

Mapping the presumptive areas of wing buds by implanting carbon particles and following them during development allowed Saunders (1948) to demonstrate that wing-bud mesenchyme is derived from *somatic mesoderm of the lateral plate and not from somitic mesoderm* (Chapter 16).

THE MESODERMAL LIMB FIELD

Demonstrating as they did that *limb* mesoderm is the site of the *limb field* the experiments performed by Ross Harrison (1918) on limb development in embryos of *Ambystoma* are classics in experimental embryology. A field (*morphogenetic field*) is an area of the early embryo specified to form an organ system. As noted in the epigraph the field concept, although difficult, has persisted. We remain 'held down still by the gossamer threads of the past, like a giant in a fairy tale, disabled by magic.'[2]

Evidence for fields arose initially from experimental studies such as those by Harrison, from which it became clear that regions of early embryos were specified to make individual organ systems. Transplant the field elsewhere on the embryo and the organ forms ectopically in the new site. Just as, importantly, cells in the original site cannot compensate for the removal of the field – they are not specified to make the particular organ system (Figs 35.1 and 35.2). So we can speak of the heart field, limb field, eye field and so on. Cells outside these fields do not have the ability to make heart, limb or eye. You can think of each of the imaginal discs in *Drosophila* as a developmental field specified to produce a particular part of the adult. In important recent analyses, the mutation *obake* (*obk*) was shown to result in duplication/multiplication of the antenna morphogenetic field in *Drosophila* and to be influenced by other mutations affecting antennal imaginal discs and by environmental factors such as larval crowding which suppresses the affect of the mutant.[3] Morphogenetic fields are alive and well as evolvable units of morphology.

An important property of developmental fields is their ability to regulate, where *regulation* is the ability of entire early embryos or embryonic fields to compensate for loss in such a way that part of the field can produce the entire structure (see Box 35.1). As an instance, a partial limb field can produce a whole limb but cells outside the field cannot contribute to limb formation.

When prospective limb-bud mesoderm is grafted under flank ectoderm *away from where limbs would form*, the mesoderm grows, flank ectoderm is drawn into limb development and a supernumerary limb grows out from the side of the host (Figs 35.1 and 35.3). No limb develops when mesoderm from *outside* the prospective limb field is grafted under ectoderm within the limb field. A limb fails to develop if prospective limb mesoderm is removed from neurula-stage embryos but the ectoderm left intact. Specificity of mesoderm as the initial 'limb inductor' and evocation of ectoderm in response to limb mesoderm are

Table 35.1 A summary of the major events in the development of the wing bud in the embryonic chick

H.H. stage[a]	Length of H.H. stage (hours)	Developmental event	References
7	3	Presumptive limb mesenchyme can chondrify ectopically	13, 15
10	7	Earliest point at which flank mesenchyme can become limb	3
10–11	10	A-P axis already fixed; dorsoventral axis fixed at this point	2
12	3	Earliest point at which wing mesenchyme will induce formation of an AER[b] from flank epidermis	17
15	2–3	Earliest age for autonomous differentiation	13
17	14	Limb buds present (wing bud adjacent to somites 15–19; leg bud adjacent to somites 26–32); AER present; wing mesenchyme loses ability to induce AER in the flank and flank epithelium loses ability to respond to limb mesenchyme	10, 16
18	6	Limb mesenchyme forms cartilage *in vitro*; humerus specified	20, 25
20	4	Vascular pattern initiated; ulna and radius specified	1, 25
21	8	Specification of wrist starts (lasts until H.H. 24)	25
22	6	S^{35} uniformly distributed over mesenchyme; sinusoids demark myo- from chondrogenic areas; proximal, central mesenchyme condensing; PNZ[c] irreversibly determined; mitotic index starts to decline, especially centrally	27
23	12	Myosin first appears; first innervation of hind limb bud	6, 11
24	6	Y-shaped condensation and opaque patch appear; vascular pattern well established; cell packing in prechondrogenic mesenchyme increased 60% over that at H.H. 22; proximo-distal gradient in cell division present; first necrotic cells in PNZ; ability to form cartilage stabilized between H.H. 24 and 25	8, 23, 26
25	6	ECM of cartilage appears; proximal mesenchyme will no longer support an AER (i.e. morphogenetic properties lost), nor will it produce distal structures; innervation reaches stylopod	5, 6, 18
26	12	Cell density in chondrogenic mesenchyme increased 10% over that at H.H. 24; ability of cells to segregate appears; dissociated chondrogenic mesenchyme produces condensations; digit III and metacarpals specified (H.H. 25–26); muscle appears	4, 7, 21, 22, 25, 26
27	12	Cartilaginous models of limb skeleton present; proximal phalanx specified; innervation reached zeugopod (H.H. 27–29)	5, 6, 25
28	6	Opaque patch disappeared; distal phalanx specified; innervation reached autopod	6, 25, 26
29	12	AER loses inductive ability	14
30	6	Interdigital cell death (H.H. 30 in hind limb, H.H. 31 in wing); definitive nerve pattern	16

[a] For hours of incubation corresponding to these H.H. stages, see Table 12.3.
[b] AER, apical epithelial ridge.
[c] PNZ, posterior necrotic zone.
References: (1) Caplan and Koutroupas (1973); (2) Chaube (1959); (3) Dhouailly and Kieny (1972); (4) Ede ànd Flint (1972); (5) Finch and Zwilling (1971); (6) Fouvet (1970); (7) Hilfer *et al.* (1973); (8) Hornbruch and Wolpert (1970); (9) Janners and Searls (1970); (10) Kieny (1960); (11) Medoff and Zwilling (1972); (12) Mottet and Hammer (1972); (13) Pinot (1970); (14) Rubin and Saunders (1972); (15) Rudnick (1945); (16) Saunders and Fallon (1967); (17) Saunders and Reuss (1974); (18) Saunders *et al.* (1959); (19) Searls (1965a); (20) Searls (1968); (21) Searls (1971); (22) Searls (1972); (23) Searls and Janners (1969); (24) Searls *et al.* (1972); (25) Summerbell (1974b); (26) Thorogood and Hinchliffe (1975); (27) Zwilling (1966).

both strongly suggested in this pioneering study, which was one of the founding studies of the concept of developmental fields.

ECTODERMAL RESPONSIVENESS

The position along the embryonic axis occupied by the limbs is set by the location of the mesodermal limb field and by the ability of flank ectoderm to respond to signals from the limb field.

Collagen accumulates beneath the flank ectoderm but not beneath prospective limb ectoderm of embryonic chicks between H.H. 12 and 17 (A. A. Smith *et al.*, 1975).

Deposition of this fibrous collagenous *barrier* partly explains why flank ectoderm loses its ability to support limb development at and beyond H.H. 17. Limbs fail to form at all in some mutant chick embryos because such a barrier accumulates beneath the apical epithelial ridge (AER), which is the specialized ectodermal thickening that develops a little later in limb development. As discussed in the context of wingless mutants in Chapter 37, the first definitive evidence for a *reciprocal* interaction between mesoderm and ectoderm came from Zwilling's experiments on the chick *wingless* mutant.

Experiments similar to Zwilling's were performed using limb and non-limb sites in wild-type embryos. The site selected was a region of the flank between wing and hind

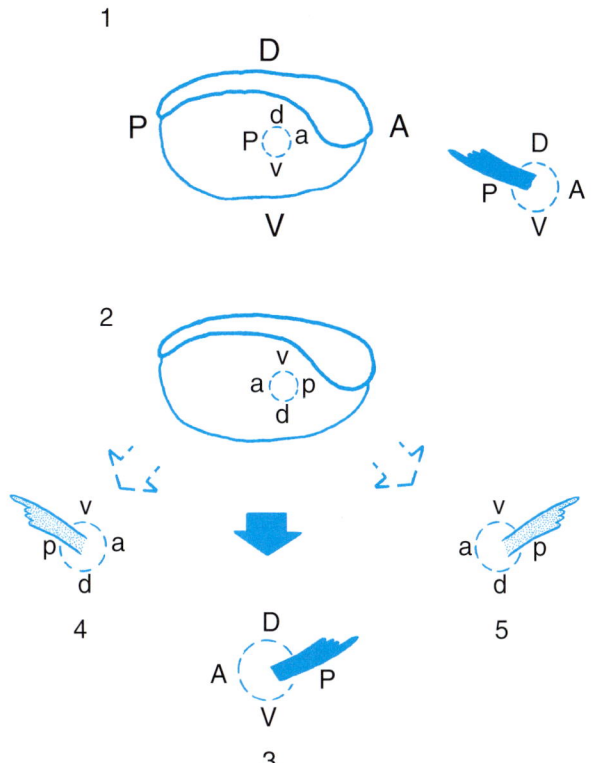

Figure 35.1 Limb fields, axes and axis determination, illustrated using amphibian embryos. (1) An outline of a neurula (anterior to the right) to show the forelimb bud field (circle) and the four planes: anterior (a), dorsal (d), posterior (p) and ventral (v). Note that the planes and axes (A-P, D-V) of the limb field correspond to those of the embryo (A, D, P, V). The orientation of the posteriorly directed forelimb is shown on the right arising from the limb field. (2) The limb field has been removed, rotated 180° and reimplanted so that limb field and embryonic axes no longer correspond. (3) The A-P but not the D-V axis is reversed after the rotation shown in (2), indicating that the A-P but not the D-V axis had been determined before field rotation, i.e. early in development. (4) The result expected in the experiment in (2) if the D-V but not the A-P axis had already been determined. (5) The result expected in the experiment in (2) if both D-V and A-P axes had been determined before field rotation. Also see Figs 35.2 and 35.3. From Hall (1978a).

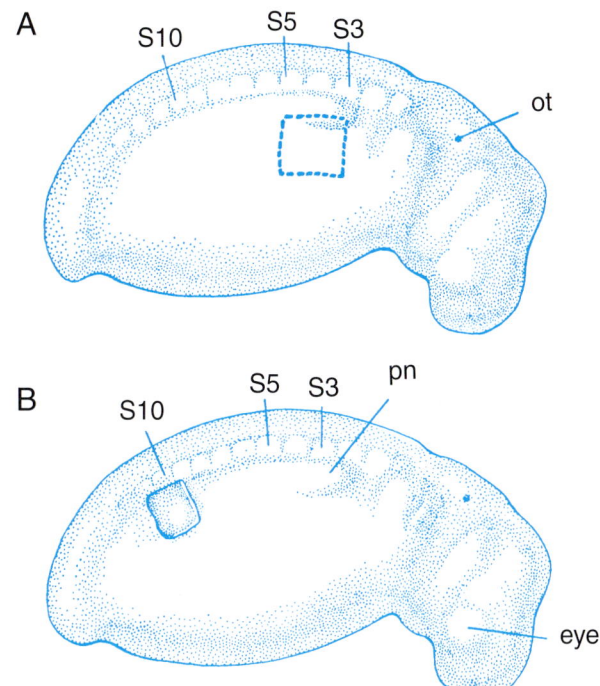

Figure 35.2 Views of the right side of neurula-stage embryos (anterior to the right) of the Mexican axolotl showing the position of the forelimb field – outlined in the square in (A) – and its transplantation to a more posterior region, shown as the square in (B). ot, otocyst; S3, S5, S10, somites three, five and 10; pn, the position of the pronephros (future kidney). Modified from Hamburger (1960).

limb buds, a region that contains mesenchyme and epithelium but does not produce limbs. Flank ectoderm from embryos of H.H. 10–17 can participate in limb-bud formation if brought into contact with limb-field mesoderm from embryos of H.H. 13–17 (Fig. 35.1). Grafting pre-limb-bud mesoderm to the flank elicited an AER from flank epithelium. Formation of a supernumerary wing followed.

Timing of this interaction provides a clue to the normal time course of the interaction of limb mesenchyme and epithelium. Wing mesenchyme loses the ability to induce an AER in flank ectoderm at H.H. 17. Reciprocally, flank ectoderm loses the ability to respond at H.H. 17 (see Table 35.1), this being the stage with the first external sign of limb buds. Presumptive limb-area mesoderm from embryos of H.H. 11 (Fig. 16.2) or younger will not elicit a ridge from flank ectoderm. Although mesoderm from embryos of H.H. 12–17 will act, mesoderm from older embryos will not. Localizing the time of interaction

between H.H. 12 and 17 ensures that once a limb bud forms, additional limbs are unlikely to be produced.[4]

[3]H-thymidine labeling, the ability to identify nuclei in Japanese quail cells, and the type of epidermal differentiation seen in grafts between chick and Japanese quail all indicate that flank ectoderm contributes to the supernumerary limb. Limb-bud mesenchyme from embryos of H.H. 19 and 20 has also been grafted beneath tail-bud epithelium of H.H. 21 and 22 embryos, where it induces tail epithelium to form an AER, followed by limb outgrowth and digit formation. Specificity to initiate limb development resides with mesoderm of the limb field. Mesoderm also specifies the type of limb that will form – forelimb (wing/arm/flipper) or hind limb (leg).[5]

MESODERM SPECIFIES FORE- vs. HIND LIMB

For at least half a century, experimental embryologists have wanted to know how fore- and hind limbs are specified. As discussed above, the limb field is a property of the mesoderm. Whether a limb will be fore or hind also is a mesodermal property, a conclusion based on swapping wing- and hind-limb-bud epithelium and mesenchyme as follows.

Box 35.1 Regulation

Regulation is the ability of entire early embryos or embryonic fields to compensate for loss in such a way that part of the field can produce the entire structure. Examples include the ability of each of the two blastomeres of a two-cell-stage embryo to produce an entire embryo after being separated, of a bisected limb field to produce an entire limb, or of the left half of the heart field to produce a complete heart. Regulation also occurs on a smaller scale; removing a single somite from a two-day-old chick embryo elicits regulation from the adjacent somite and from intermediate mesoderm, the new somite fitting into the rostro-caudal patterning sequence. Removing a small section of neural crest from zebrafish embryos elicits regulation from the adjacent neural crest.[a]

Regulative ability is lost as soon as regions within fields are specified. Regulation is the embryonic equivalent of regeneration, the difference being – and it is an important difference – that regeneration is the replacement of differentiated cells in structures that have already formed, while regulation is the replacement of cells that have yet to differentiate. Limb fields show the property of regulation. In this case, half a limb field can produce a complete limb bud. Regulative ability also exists during the transition from limb field to limb bud. In this case the limb bud can compensate for a cell population that is lost. Once regulative ability is lost, the cells in question can be considered as specified and their fate set.

A pivotal study and an elegant demonstration of regulation is that published by Searls and Janners in 1969. When prechondrogenic mesenchyme from chick embryos younger than H.H. 24 is grafted to the premyogenic area of a limb bud of a similarly aged embryo, almost all the cells regulate to form myoblasts. If, however, prechondrogenic mesenchyme is taken from limb buds of embryos older than H.H. 25, the majority of the cells produce cartilage, readily identified as nodules of ectopic cartilage within the muscle of the host limb. The ability of central cells to chondrify stabilizes between H.H. 24 and 25 or, to put it another way, the ability to regulate is lost once chondrogenesis has been specified. Several other pieces of evidence, outlined below, lead us to this same conclusion.

Zwilling (1966) cultured or CAM-grafted premyogenic, prechondrogenic and what he called intermediate mesenchyme (at the boundary of premyogenic and prechondrogenic regions) limb buds of embryos of H.H. 22–24. Mesenchyme from all three sites chondrified in similar proportions – 67 per cent of the 'myogenic,' 64 per cent of the 'chondrogenic,' and 71 per cent of 'intermediate' mesenchyme. Cells whose progeny would have differentiated into myoblasts differentiated as chondroblasts.

Stark and Searls (1974) excised various prospective long-bone areas from wing buds to determine whether adjacent premyogenic mesenchyme would regulate and replace them (Table 35.2). They excised:

(i) mesenchyme that normally forms the humerus, radius *and* ulna, representing some 60–70 per cent of limb mesenchyme;
(ii) prospective radius and ulna mesenchyme, representing some 20–40 per cent of limb mesenchyme; and
(iii) humeral, radial *or* ulnar mesenchyme independently.

Procedures (ii) and (iii) gave similar results, and are grouped together in Table 35.3. The greater the amount of mesenchyme removed and the older the embryo from which mesenchyme is removed, the greater the incidence of abnormal limbs and the lower the regulative ability. Regulative ability is lost at H.H. 24, the stage at which prechondrogenic mesenchyme becomes stabilized for chondrogenesis (see above). Finally, regulation occurs irrespective of the position along the proximo-distal axis from which the mesenchyme is excised.

In a second experiment Stark and Searls rotated the prospective elbow region through 180° and reimplanted it into the wing bud. Epithelium adjacent to the graft was rotated with the mesenchyme or left in place. Normal development of the joint after rotation would

Table 35.2 Regulative ability of limb mesenchyme as assessed by the ability to replace excised prospective humeral, radial or ulna mesenchyme and produce a morphologically normal limb[a,b]

Embryonic age at excision (H.H. stage)	Per cent (N) of normal limbs six or seven days post-excision	
	Prospective humerus, radius and ulna all excised	Prospective ulna/radius or humerus, radius, or ulna excised
19	67 (8/12)	
20	60 (3/5)	100 (9)
21	37 (3/8)	100 (18)
22	33 (4/12)	77 (27/35)
23	10 (1/10)	33 (11/33)
24	0 (0/2)	5 (2/40)

[a] Regulative ability decreases with embryonic age and is lost after H.H. 24.
[b] Based on data in Stark and Searls (1974).

Table 35.3 Regulative ability of limb mesenchyme as assessed by the ability to compensate for rotation of the prospective elbow region – with or without rotation of adjacent ectoderm – and produce a normal limb bud[a,b]

Embryo age at rotation (H.H. stage)	Prospective elbow mesenchyme and ectoderm rotated		Prospective elbow mesenchyme rotated		Dorsal ectoderm rotated
	Per cent of specimens with				
	normal joint	ectopic cartilage	normal joint	ectopic cartilage	normal joint
20	25	50	–	–	–
21	17	67	100	0	60
22	5	80	76	23	36
23	0	100	0	100	19
24	0	100	0	100	–

[a] Presence of nodules of ectopic cartilage at the normal site of the joint indicates inability to regulate. Regulative ability is lost after H.H. 22.
[b] Results are expressed as per cent of specimens with normal joints or with nodules of ectopic cartilage. Based on data in Stark and Searls (1974) and in Searls (1976).

Box 35.1 (*Continued*)

signify regulation, while formation of nodules of ectopic cartilage at the joint site would demonstrate failure of regulation. As shown in Table 35.4, provided that the epithelium retains its original orientation, regulation occurs until H.H. 22; rotation of dorsal epithelium of the joint region alone is sufficient to prevent regulation. Regulation at this stage of development begins to appear as if it is not entirely independent of influences from adjacent tissues.

Other workers obtained similar results with slightly different stages for onset or offset of regulation:

- for removal of 90 per cent of the wing mesoderm until H.H. 19;
- until H.H. 23 after limb mesenchyme from embryos of Japanese quail was grafted into various levels of embryonic chick limb buds;
- until H.H. 22 after slices along the proximo-distal axis of the chick wing bud were removed; and
- within the common condensation for the tibia and fibula.[b]

In summary, regulation is possible before H.H. 22. Between H.H. 22 and 24, cells (regions?) stabilize and are less able to regulate. Regulative ability is lost after H.H. 24 (Table 35.3). Loss of regulative ability in such experiments is interpreted as loss of the ability of mesenchymal cells to change fate. Alternatively, it could be that cells are less capable of filling the larger wounds required to excise

mesenchyme in older limb buds, or less able to migrate across the limb bud to fill the gap. This seems unlikely; Barasa (1962, 1964) found that regulation is more complete in larger than in smaller wounds.

Mouse embryos can regulate for loss of limb-bud mesenchyme. The experimental approach was to remove forelimb buds from 11.5-day-old embryos and maintain the whole embryos *in vitro* (well, whole embryos minus the forelimb buds). Within 24 hours, 90 per cent (24/27) of the embryos produced bud-like outgrowths, a third of which formed AERs. In a second approach, a block of mesoderm two to three somites wide was excised from the forelimb bud region of 10-day-old mouse embryos and the embryos were cultured for six to 24 hours. Two thirds of these embryos restored normal morphology and formed AERs as assessed using SEM.[c]

[a] See Vaglia and Hall (1999) for a discussion of regulation following removal of regions of neural crest, and Liu and Bagnall (1995) for regulation following somite removal.
[b] See Barasa (1962, 1964) and Searls (1976) for regulation after 90 per cent removal, Kieny and Pautou (1976) for the quail–chick chimaeras, Summerbell (1977b, 1981) for removal of P-D slices of wing buds, and Kieny (1967) for regulation within the condensation.
[c] See K. K. H. Lee (1992) and Lee and Chan (1991) for these experimental studies.

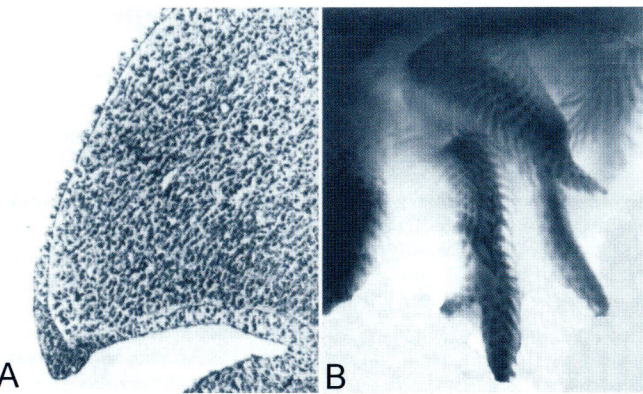

Figure 35.3 Wing-bud mesenchyme can evoke an AER from flank ectoderm, resulting in supernumerary wing formation. (A) An outgrowth, complete with AER formed after wing-bud mesenchyme (H.H. 14) was grafted beneath flank ectoderm of an H.H. 13 host embryo. (B) A supernumerary wing formed after wing-bud mesenchyme from H.H. 13 was grafted beneath flank ectoderm of an H.H. 14 embryo. Also see Fig. 35.1. Modified from Saunders and Reuss (1974).

- When epithelium from a hind-limb bud is grafted onto *wing-bud mesenchyme* the distal structures (digits) that develop are *wing* digits.
- When epithelium from a wing bud is grafted onto *hind-limb-bud mesenchyme* leg skeletal elements develop.
- Conversely, grafting *hind-limb-bud mesenchyme* in the place of wing-bud mesoderm produces digits ('toes') typical of the *leg*, while
- *wing mesenchyme* grafted onto hind-limb bud produces digits typical of wing at the end of the other leg skeletal elements.

From such studies, we see that the distal structures are always determined by the type of limb bud supplying the

mesenchyme. When the exchange is between different species – e.g. between domestic fowl and duck – the results are even more striking, as species-specific characteristics reinforce the evidence of mesenchymal specification of limb type.[6] Of course, to say that mesoderm specifies limb type tells us nothing of the molecular basis for that specification, for which see Box 35.2.

Once limb epithelium is specified to form an AER, reciprocal interactions between mesenchyme and AER determine the size, outgrowth and differentiation of the limb buds and the resulting limb skeleton. The function of the AER in specification of the proximo-distal sequence of skeletal elements is discussed in Chapter 39. In the present chapter I evaluate evidence for interactions between the AER and underlying mesenchyme under three headings:

- the role of the limb-bud epithelium;
- induction of the AER by limb mesenchyme; and
- the maintenance of the AER by limb mesenchyme.

ROLES FOR THE ECTODERM ASSOCIATED WITH THE LIMB FIELD

By grafting regions of the blastoderm into the body cavities of host embryos and following their fate, Rudnick (1945a) mapped the presumptive wing territory in chick embryos of H.H. 6 and the leg territory in embryos of H.H. 8.

Early limb buds of embryonic chicks – indeed the limb buds of any tetrapod – consist of a core of mesenchyme (the mesoblast) underlying a thin cap of cuboidal epithelium. Late in H.H. 17, the epithelial cells at the apex of the limb bud become columnar. By H.H. 19, more rapid development post-axially produces an asymmetrical bud with a nipple-like ridge of ectoderm distally, *the apical epithelial ridge*

Box 35.2 Fore or aft

Fore–hind limb, arm/wing–leg, pectoral fin–pelvic fin

Until recently, palaeontologists and evolutionary biologists (often one and the same) gave little consideration to the possibility that fore- and hind limbs (or pectoral and pelvic fins) might have evolved independently as anterior and posterior paired appendages. Recently, a non-simultaneous origin appears more likely, primarily because we now know something of how anterior and posterior paired appendages are specified in fishes and tetrapods.

As discussed in the text, specification of limb type is a property of limb mesoderm, not epithelium. Traditionally, limb type was assessed by examining the skeletal elements that form after experimental manipulation, e.g. combine wing-bud mesenchyme with leg epithelium and ask whether the skeleton that forms is wing or leg (it is wing). Three further aspects, listed below, are addressed here.

- How is the position along the flank where paired appendages will develop specified?
- What is the molecular basis of specification of limb or fin type: fore or hind?
- Can differences associated with limb type be detected before the skeletal elements form?

All three are important questions. The first and second are easier in the sense that the genes that specify location and limb type have been identified. That done, we tend to think that we 'know' how limb type is specified although, of course, much remains to be uncovered. The third aspect seeks a mechanism or mechanisms: what are the downstream genetic and cellular activities that allow wing or leg to arise from seemingly similar mesenchyme[a]?

Positioning paired appendages

Tabin and Laufer (1993), recognizing the serial homology of fore- and hind limbs, argued that *Hox* genes originated in the hind limbs with a homeotic transformation transferring them to the fore limbs. From the elegant study by Burke *et al.* (1995) we know that a *Hox* gene code expressed in paraxial and associated mesoderm specifies vertebral type into three broad classes, cervical, thoracic and caudal. Further, we know that both fore- and hind limb buds arise adjacent to the anterior expression boundary of *Hoxc-5*, even if that boundary is expressed adjacent to different somites along the body axis in different species, which it is; the forelimb boundary is seven somites more posterior in the chick when compared with the mouse. In an important study, Nelson *et al.* (1996) cloned 23 genes and identified three phases of development, each with different sets of genes in both fore- and hind limb buds. The *Hox* pattern arises in response to *Shh*.

More recently, Gaunt (2000) reviewed evolutionary shifts in vertebral structures in relation to *Hox* gene expression, concentrating on the positions of the neck/thorax (cervical/thoracic vertebrae) and forelimbs. Part of such specificity resides in *cis*-regulatory elements in Hox genes, as demonstrated for *Hoxc-8*.

***Hoxc-8 cis*-regulation**
Belting *et al.* (1998) examined the evolution of *cis*-regulatory elements of *Hoxc-8* in chick and mouse embryos in relation to the divergent axial morphology of birds and mice. Although *Hoxc-8* is expressed in the mid-thoracic (brachial) mesoderm and neural tube in both species, activation is delayed in the chick so that expression is more posterior and over a smaller area of mesoderm which, they argue, explains the shorter thorax in chicks as compared with mice.

Baleen whales lack four base pairs in element C of *Hoxc-8*. When expressed in transgenic mice lack of this *cis*-acting element directs gene expression in the more posterior neural tube but not in the posterior mesoderm (where *Hoxc-8* is not normally expressed). Shashikant *et al.* (1998) correlated his loss with specific traits in whales. *Hoxc-8* is expressed in prechondrogenic limb condensations, over-expression leading to the accumulation of dividing chondrocytes. The severity of the over-expression is gene dosage dependent and specific to limb condensations, cranial elements being unaffected. Yueh *et al.* (1998) argue that *Hoxc-8* controls the transition from dividing cells to differentiating cells. Tsumaki *et al.* (1996) identified separable *cis*-regulatory elements controlling tissue- and site-specific α2(XI) collagen gene expression in embryonic mouse cartilage. Weatherbee and Carroll (1999) explained regulation of target regions by selector genes in sub-regions of developing limbs through field-specific expression of *cis*-regulatory elements.

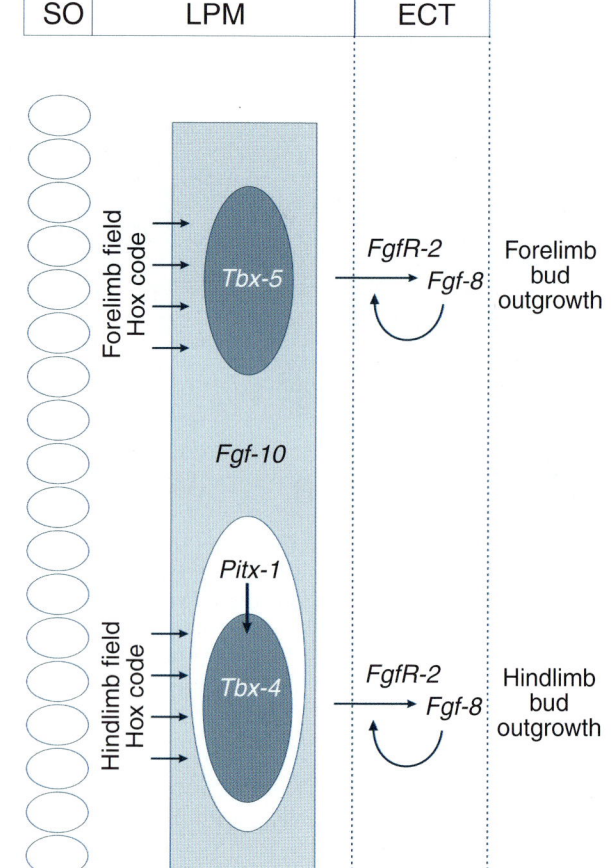

Figure 35.4 A model for how the position of the fore- and hindlimbs is specified along the embryonic axis and for how outgrowth of the limb buds is initiated. The right-hand side of an embryos is shown as viewed from the dorsal surface (anterior to the top), with the embryonic axis shown as somites (SO), future limb and flank lateral-plate mesoderm (LPM) and the overlying ectoderm (ECT). The axial Hox gene code expressed in LPM activates *Tbx-5* or *Tbx-4* to specify fore- and hind-limb-bud fields. Outgrowth is controlled via *FgfR-2* and *Fgf-8*. Expression of *Tbx-5* and *Tbx-4* is maintained by Fgf-10 within the LPM. Additionally *Tbx-4* is maintained by *Pitx-1*. Based on Ruvinsky *et al.* (2000).

Box 35.2 (*Continued*)

Fore or aft?

Specification of fore- vs. hind limbs and of pectoral vs. pelvic fins is controlled by members of the T-box gene family. *Brachyury*, perhaps the best-known and most well-studied family member, is expressed in axial mesoderm, lateral-plate mesoderm and limb-bud mesenchyme adjacent to the AER, and may play a role in maintaining the AER (C. Liu *et al.*, 2003).

Tbx-5 is expressed in restricted zones along the right and left flanks of teleost and tetrapod embryos before fin or limb buds arise (Fig. 35.4). *Tbx-5* is expressed in the wing and flank but not in the hind limb buds of chick embryos, while *Tbx-4* is expressed in leg but not wing buds. Implanting Fgf-2 into chick embryo flanks (i.e. in the area between where the wings and the legs arise) up-regulates either *Tbx-4* or *Tbx-5* (depending on position along the flank) and induces a supernumerary wing or leg, again depending on position along the flank. Limbs that are part wing and part leg do not occur. The response is all or none and mediated by members of the *Wnt* and *Fgf* families. Tbx-5 controls wingness; Tbx-4 controls legness.[b]

Pitx-1, which encodes a transcription factor in hind limb but not forelimb buds, is upstream of *Tbx-4* and involved in specification of hind-limb identity. Misexpressing *Pitx-1* in wing buds elicits more distal expression of *Tbx-4*, *Hoxc-10* and *Hoxc-11*, resulting in the development of hind-limb characters in wing buds, including hind-limb muscle. Recent analysis demonstrates that *Pitx-1* and *Pitx-2* are both required for hind-limb development, both being expressed in the mesodermal hind-limb field, *Pitx-1* also being expressed throughout hind-limb development.[c]

One paradox is that synthesis of *Tbx-5* and *Tbx-4* is initiated at H.H. 13, but limb type is specified as early as H.H. 9 (see text, Chapter 35). Saito *et al.* (2002) speculate that the midline tissue medial to lateral-plate mesoderm provides inhibitory signals that regulate *Tbx* expression, and thus plays an early role in specification of limb type. Signals from midline tissues can transform a potential leg site into wing, the wing field being specified earlier than the leg field.

Knocking out *Tbx-5* in mice or zebrafish prevents forelimb-bud/pectoral fin-bud development but has no effect on hind limb-bud/pelvic fin-bud development. The knockout phenotype in mice is essentially that seen in *brachyury* mutant embryos. The knockout phenotype in zebrafish is essentially that seen in *spadetail* mutant embryos, *spadetail* being a mutation of *Tbx-5*. Both mutations inhibit migration of the lateral-plate mesoderm, which normally migrates to the flank to form fin or limb buds.

Recently it has been shown that Holt-Oram syndrome in humans, which is characterized by upper arm and heart defects, results from a haploinsufficiency of *TBX-5*. In chick and mouse, *Tbx-5* is required both for specification of limb buds as fore rather than hind early in development *and* for limb growth later in development (Rallis *et al.*, 2003).

Tbx-6 (mice) and *Tbx-16* (zebrafish) are expressed in restricted zones along the right and left flanks before limb or fin buds arise. Knocking out *Tbx-6* in mice or *Tbx-16* in zebrafish results in lack of hind limb-bud/pelvic fin-bud development but has no effect on forelimb-bud or pectoral fin-bud development.[d]

Early differences

Differences between wing- and leg-bud mesenchyme have been identified as early as the condensation stage. One difference is morphogenetic. Leg and wing mesenchyme from H.H. 24 embryos – which means that the leg mesenchyme is a little younger – both chondrify when placed in culture, but leg mesenchyme forms nodules of cartilage while wing mesenchyme forms sheets (Downie and Newman, 1994).

The differences are apparent as early as condensation and relate in part to fibronectin; prechondrogenic mesenchyme from hind limb buds has higher levels of fibronectin mRNA and fibronectin than does mesenchyme from wing buds. Wing condensations are broad and flat with much diffusely organized fibronectin; leg condensations are compact and spherical and connected by fibronectin-rich fibres.

Leg- and wing-bud mesenchyme also respond differentially to foetal bovine serum, Tgfβ-1 and retinoic acid. Wing-bud condensations are more sensitive to Tgfβ, which up-regulates fibronectin levels (see Chapter 20), and in turn increases condensation size. Treating cultured wing-bud mesenchyme with an antibody against the amino-terminal heparin-binding domain of fibronectin inhibits condensation formation in wing- but not in leg-bud mesenchyme. Comparative analyses of wing- and leg-bud condensations would be a profitable way to analyze how differences between fore- and hind-limb mesenchyme are established.[e]

Misexpressing *Hoxa-13* leads to reduction of the zeugopod and arrest of cartilage growth in chick limb buds. A homeotic transformation of the long bones of the zeugopod into distal carpals/tarsals then occurs. Yokouchi *et al.* (1995) concluded that Hoxa-13 plays an essential role in switching long bones to short bones by altering cell-adhesion properties. Newman (1996) related this fascinating study to his own extensive studies on the differential role of cell adhesion molecules such as N-CAM in wing- and leg-bud chondrogenesis.

Tgfβ-2 also plays a role in sorting out prechondrogenic cells and in governing their differentiation. Exogenous Tgfβ-2 enhances chondrogenesis from mouse limb buds and promotes the production of *Tgfβ-2* mRNA in a positive feedback loop. Beads soaked in Tgfβ-2 suppress chondrogenesis, suggesting lateral inhibition within aggregations. Tgfβ-2 is chemotactic for proximal and distal limb mesenchyme and promotes the expression of N-cadherin, which plays a central role in condensation.[f]

[a] The same issue arises when we ask how proximal and distal skeletal elements are specified from apparently similar chondrocytes, proximal chondrocytes making a humerus, distal ones a digit.
[b] See Isaac *et al.* (1998), Rodriguez-Esteban *et al.* (1999), Takeuchi *et al.* (2003) and Yang (2003).
[c] See Logan and Tabin (1999) for *Pitx-1*, Niswander (1999) and Kawakami *et al.* (2003) for overviews, and Marcil *et al.* (2003) for *Pitx-1* and *Pitx-2* and hind-limb development.
[d] See Ruvinsky *et al.* (1998). *Tbx-16* from zebrafish is an orthologue of chicken *Tbx-6*. Zebrafish and mouse *Tbx-6* are not orthologues but distantly related paralogues (Ruvinsky *et al.*, 1998).
[e] See Leonard *et al.* (1991) for Tgfβ as a stimulator of fibronectin gene expression and initiation of condensation, and Downie and Newman (1995) for differential responses of wing and leg mesenchyme. See Newman and Cooper (1990), Newman (1992), Newman and Tomasek (1996) for morphogenetic mechanisms operating at condensation, and Newman (1996) for the importance of cell interactions and adhesivity in condensation and for differences between fore- and hind limbs.
[f] See Oberlender and Tuan (1994) and Miura and Shiota (2000) for these studies with TGFβ-2.

(AER, but see Box 35.3 for terminology). As development proceeds, this asymmetry becomes more pronounced as the ridge directs limb bud growth posteriorly.

It has been known for 125 years that a portion of the ectodermal covering of the developing limb bud thickens into a ridge. It was not until 1948, however, that experimental evidence indicated that this ectodermal ridge directs developmental events in the underlying mesoderm.[7] In that year, the talented American experimental embryologist John Saunders surgically removed the ridge – which he named the apical ectodermal ridge (see Box 35.3) – from wing buds of embryonic chicks. Subsequent outgrowth of wing mesenchyme ceased. The partial wing that developed lacked some or all of such distal skeletal elements as radius, ulna and/or digits. Proximal limb elements – humerus, pectoral girdle – were normal in morphology and size; overall *shortening of the limb resulted from absence of the distal skeletal elements, not from a global decrease in limb length*.

Limb-bud growth

The initiation of growth within a limb bud is a complex series of processes that includes initiating cell division, altering rates of division over time, establishing gradients in cell division across or along limb buds, and, as limbs develop, altering such cell-surface-related properties as attachment and cell movement.

Cell proliferation

In Chapters 19 and 20, I discussed rates of proliferation of progenitor cells and the accumulation of precursor cells into condensations. Here, I concentrate on *initiation of*

Box 35.3   Mesoderm is not synonymous with mesenchyme, nor ectoderm with epithelium

It is most unfortunate that Saunders named the AER an *ectodermal* and not an *epithelial* ridge. The stage in development when the AER appears is way beyond any stage when ectoderm is still present, ectoderm being the name of a germ layer, epithelium (epithelia) being a term for the type of cellular organization in which cells exist as a layer(s) of connected cells on a basement membrane. Similarly, limb *mesenchyme* is often, perhaps usually, referred to as limb *mesoderm*. However, the embryonic stages when limb buds arise are well beyond the germ-layer stage, and mesoderm is a germ layer.

I will try to speak of limb-bud mesenchyme and limb-bud epithelium rather than mesoderm and ectoderm, and encourage you to begin to do the same.

Mesoderm, ectoderm and endoderm are three germ layers; epithelia and mesenchyme are two types of cellular organization. Mesoderm ≠ mesenchyme, nor does ectoderm ≠ epithelium.

Most embryologists of the 1940s probably thought that epithelia only arose from ectoderm or endoderm, and so the terms ectoderm and epithelium were often used interchangeably. However, epithelia can arise from mesoderm; mesodermal somites are initially epithelial. Indeed, all cells of metazoans are organized either as an epithelium or as mesenchyme, for these are the only two types of cellular organization. This is true even when the mesenchymal matrix may be liquid (as in blood) or when it is solid (as in bone) or whether the epithelium is a single layer or multiple layers of cells. I doubt that many will begin to refer to the apical ectoderm ridge as the *apical epithelial ridge*, but I do hope that we begin to subconsciously replace epithelial for ectodermal when we speak of the AER, which is an ectodermally derived epithelial ridge. Similarly, limb-bud mesoderm or limb mesoderm is mesenchyme derived from mesoderm.

---

cell proliferation as a factor in growth. The outgrowth of avian embryonic limb buds received the most attention and provides perhaps the best-understood system for the analysis of *gradients of cell proliferation* and their role in outgrowth of structures such as limb buds.[8]

Since the carbon-marking experiments by Saunders (1948), we have known that limb buds grow from their tips, i.e. growth is apical. Continued bud outgrowth depends on the presence of the AER, but removing the AER does not affect mitotic activity in flank mesoderm (Janners and Searls, 1971).

Limb-bud growth consists of more than the production of more cells. Outgrowth involves the number of cells, their proliferation, position, movement, size, shape, packing density and constraint from the limb epithelium. In a brave attempt to discriminate between these factors and to determine the least number of components necessary to produce a limb bud, Ede and Law (1969) formulated a computer simulation of limb-bud outgrowth. A combination of cell proliferation in a proximo-distal gradient with distal movement of the cells produced form and growth patterns remarkably similar to those seen in normal limb buds. The computer simulation required *a gradient of cell division* within limb buds; Amprino (1965) had produced the first evidence of such a gradient *in ovo*.

Mitotic activity is relatively constant in wing buds between H.H. 16 and early stage 22, when mitotic activity declines to a lower but again relatively constant rate. The greatest decline is in proximal chondrogenic mesenchyme, where the labeling index drops by 75 per cent between H.H. 19 and 24. Stark and Searls (1973) concluded that cell proliferation without cell migration was sufficient to account for limb-bud outgrowth.[9]

Suppressing the flank

Since limb buds protrude from the flank, we assume that the growth mechanisms lie completely within limb-bud cells, i.e. limb buds grow out and, indeed, limb-bud cells do divide at a higher rate than flank cells (see below). However, initial limb-bud outgrowth results from *suppressing mitotic activity in the flank* between H.H. 16 – when the limb bud appears – and H.H. 20. During this period the mitotic rate in flank mesenchyme *declines* by 25 per cent while the mitotic rate of limb-bud mesenchyme remains constant (Searls and Janners, 1971). What appears to be *outgrowth* of limb mesenchyme is really *regression* of flank mesenchyme.

So, initial 'outgrowth' of the limb bud is more apparent than real; the flank recedes away from the limb buds because mitotic activity in flank mesenchyme declines. This also is true in other regions, such as the facial processes from which the jaws and face arise. Differential rates of *decline* in cell division and/or the appearance of more slowly cycling subpopulations are responsible for the growth and morphogenesis of the frontonasal and maxillary processes in embryonic chicks. The pattern is one of proliferation declining proximally within each facial process but remaining high distally or at boundaries.

Mitotic rate in limb mesenchyme

Hornbruch and Wolpert (1970) counted mitotic activity in wing buds from embryos between H.H. 18 and 30. Throughout all stages, mitotic activity was constant within the epithelium at around two per cent. Mitotic activity in the limb mesenchyme declined with developmental stage – from 12 per cent at H.H. 18 to 2 per cent at H.H. 30. No proximo-distal gradient was observed until H.H. 24, by which stage the commitment of mesenchymal cells as chondrogenic had been stabilized, and at which stage the mitotic index in distal mesenchyme exceeds that in proximal mesenchyme.

Although they found no statistically significant gradient before H.H. 24, there was a difference in mitotic activity at H.H. 20 of 6.5 per cent distally to 9.5 per cent proximally.

This became apparent when Lewis (1975) reworked their data, which I present in another form in Figure 32 in Hall (1978a), using as a common base the length of the limb bud, with the distal tip as zero and the proximal base of the limb bud *at that* H.H. stage as 10; the per cent mitotic index is then plotted as a per cent of the rate at the distal tip at that H.H. (rate at distal tip as 100 per cent). These plots show no gradient at H.H. 18, a slight decline in distal-proximal rate during H.H. 19/20, with a rise at H.H. 21/22 midway along the limb bud, steepening considerably between H.H. 21 and 23, the latter stage corresponding to condensation of proximal skeletal elements. A sharp decline in mitotic activity at H.H. 24 and establishment of a distal-proximal gradient at stage 25 complete the pattern.

A proximo-distal decline in agglutination of limb mesenchymal cells to the plant lectin concanavalin between H.H. 19 and 26 parallels establishment of the mitotic gradient, stabilization of cells for chondrogenesis and a decline in morphogenetic potential. The possibility that cell-surface changes and cell-to-cell adhesion play a role in establishing the mitotic gradient should be considered.[10]

More detailed information shows a gradient in mitotic rate *within* the chondrogenic mesenchyme during H.H. 24–27: highest distally, a sharp drop in the distal one-third of the limb bud, and a slower decline proximally. A much more gradual gradient is found in myogenic mesenchyme, suggesting that mitotic activity may be differentially controlled in chondro- and myogenic mesenchymal cells. Blocking mitosis at 6.5 days does block mitosis in myogenic but not chondrogenic mesenchyme. Similar studies on earlier embryos might: provide valuable information on the establishment of regional differences in mitotic activity across limb buds; show that the differences reflect central (chondrogenic) mesenchyme as more homogeneous than peripheral (myogenic, fibroblastic) mesenchyme; or else show that myogenic and chondrogenic cells are at different stages of cell proliferation.[11] The AER does not maintain the high mitotic activity in distal mesenchyme, although influences from the epithelium as a resistant layer have not been investigated adequately.

Since time spent within the distal tip of the limb bud (the progress zone) determines positional morphology (Chapter 38), any change in mitotic rate should have pronounced effects on the length of particular segments of the skeleton. Between four and 10 days of incubation, each mesenchymal cell divides three times. Given the number of starting cells, rate of proliferation, and rate of accretion of extracellular matrix (ECM), growth rates and final size should be predictable with considerable accuracy, especially as specification of the length of the humerus, ulna and digits between H.H. 22 and 36 (4–10 days of incubation) is accurate to ± 4–5 per cent. Such are the bounds within which wild-type development must remain.[12]

Proximo-distal patterning of the limb skeleton

Saunders noted that the older the embryo from which the AER was removed the less marked were the deficiencies in wing development. In particular, Saunders observed that increasingly distal cartilages were present when the ridge was removed from progressively older embryos, which he interpreted as the AER controlling skeletal development by specifying a proximo-distal sequence of limb elements and limb-bud outgrowth. Early in development, mesenchyme adjacent to the AER would be specified for proximal skeletal elements such as the humerus. Progressively later in development, mesenchyme near the ridge (distal mesenchyme) would be specified for increasingly distal elements.

Limb cartilages *are* patterned according to a proximo-distal sequence.

Rowe and Fallon (1982) demonstrated proximo-distal specification of hind-limb skeletal elements, the most proximal element – the humerus – being specified earliest and the most distal elements – the phalanges of the digits – being specified last. Furthermore, and surprisingly, it takes much longer to specify a short complex region with multiple skeletal elements than to specify a longer region with a single element. Specification of the humerus takes some 12 hours; elements of wrist or ankle – the carpal and tarsal regions – take 24 hours. The ridge is active from H.H. 17 to 19 (see Table 35.1). It follows that the form of the mature limb skeleton can be mapped out in the limb bud as a *fate map*. Indeed, as far back as the early 1920s Murray and Huxley had used CAM-grafting to test whether the limb bud is a mosaic, i.e. whether all the parts are prefigured in the earliest limb buds (Box 12.3).

Is the maintenance of an AER an intrinsic property of limb-bud epithelium or does it depend on influences from the adjacent flank epithelium or from limb mesenchyme?

MESENCHYMAL FACTORS MAINTAIN THE AER

As will be discussed in the context of limbless mutants (Chapter 37), much of the evidence for a mesenchymal factor that maintains the AER – an *apical epidermal maintenance factor* (*AEMF*) – comes from studies on *wingless* and *polydactylous* mutants in domestic fowl. A succinct summary of *wingless* is that the AER in *wingless* embryos regresses because AEMF is lacking; *wingless* mesenchyme grafted to wild-type limb epithelium is followed by regression of the AER and a wingless phenotype.[13]

AEMF

AEMF is distributed within the postaxial region of each limb bud close to an area with high levels of cell death known as the posterior necrotic zone (PNZ; see below). The initial evidence for this conclusion was experimental, showing that you don't have to isolate a factor to know that one must exist.

Placing a filter across the long (proximo-distal) axis of a limb bud divides it into pre- and post-axial regions and is followed by regression of the AER (Saunders and Gasseling, 1963). If AEMF is only found post-axially, then 180° rotation of the distal half of the limb bud would place AEMF into the anterior (preaxial) face of the limb bud, where it should maintain a second AER, resulting in limb duplication. Saunders and Gasseling (1968) showed that this is just what happens.

Polarity of the AER is maintained by a gradient of AEMF. MacCabe and Parker (1975) found that the AER flattens – is not maintained as a specialized ridge – if preaxial (anterior) wing mesenchyme is cultured alone, with flank mesenchyme or with anterior limb mesenchyme. However, the AER is maintained with no indication of cell death when *anterior and posterior* (*preaxial and postaxial*) *mesenchymes are co-cultured*. When cultured with mesenchyme from the mid-region of the limb bud – i.e. mesenchyme in a lower part of the proposed gradient of AEMF – some loss of AER and some cell death ensues, interpreted as indicating lower levels of AEMF in that mesenchyme.

Consistent with the presence of a diffusible molecular factor, a cell-free extract of post-axial – but not preaxial – limb buds has AEMF activity, shown to result from two diffusible components of high (>300 000-kDa) molecular weight. Conversely, mitotic inhibitors administered *in ovo* diminish the rate of accumulation of mesenchymal cells, a diminution that in turn leads to premature loss of the AER and loss of skeletal elements.[14]

Mesenchymal cells derived from limb buds produce AEMF when maintained in monolayer culture. If covered with a sheet of limb epithelium containing an AER, the monolayered mesenchymal cells grow out and pile up to produce a limb-like bud with a normal AER that is maintained morphologically and functionally (Fig. 35.5).

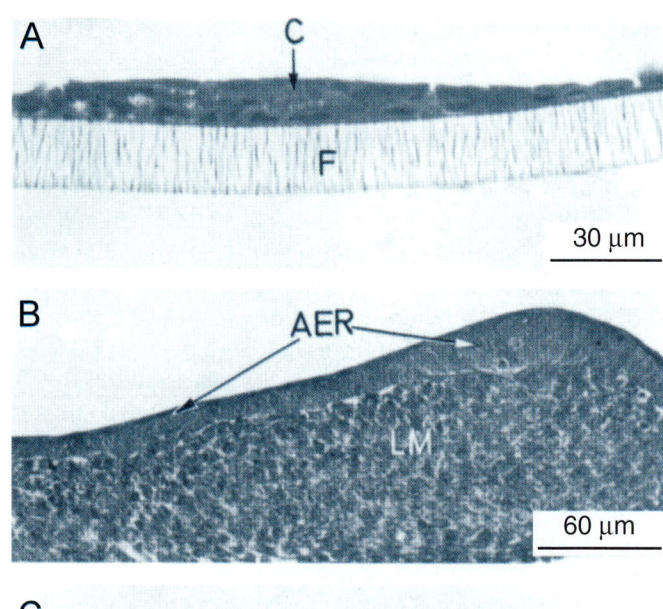

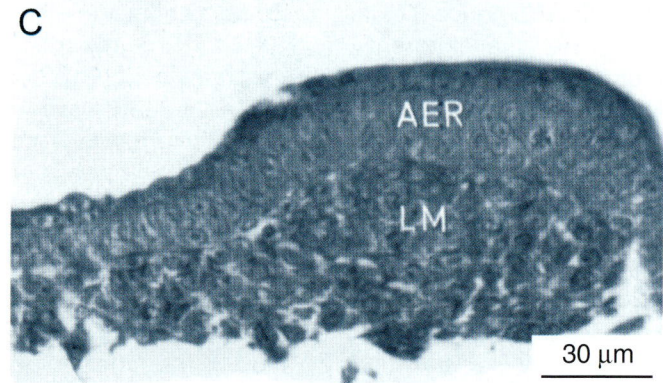

Figure 35.5 Embryonic chick-limb development mimicked *in vitro*. (A) A monolayered culture of mesenchyme cells from limb buds of embryos of H.H. 19/20 differentiates cartilage nodules (C) after 48 hours' culture. F, filter substrate. (B, C) Substantial mesenchyme accumulates when monolayered limb mesenchyme (LM) is cultured in contact with an AER. Modified from Globus and Vethamany-Globus (1976).

Limb-bud epithelium lacking the AER (i.e. non-AER epithelium) does not support outgrowth under the same culture conditions (Globus and Vethamany-Globus, 1976). Whether non-limb mesenchymal cells would have maintained the AER was not tested, but would not be expected.

The chick mutant *diplopodia*$_4$ is polydactylous with extra digits preaxially (Figs 35.6 and 35.7). Additional AEMF is produced in anterior or preaxial limb mesenchyme in *diplopodia*$_4$ mutants, resulting in preaxial thickening of the ectoderm and preaxial polydactyly. The same argument is used for the mechanism underlying the polydactylous mutants *talpid*^{2} (Box 20.1), and for similar mutants in murine embryos (Fig. 35.8).[15]

The PNZ

The *posterior necrotic zone* (*PNZ*) was described in wing buds in 1968.

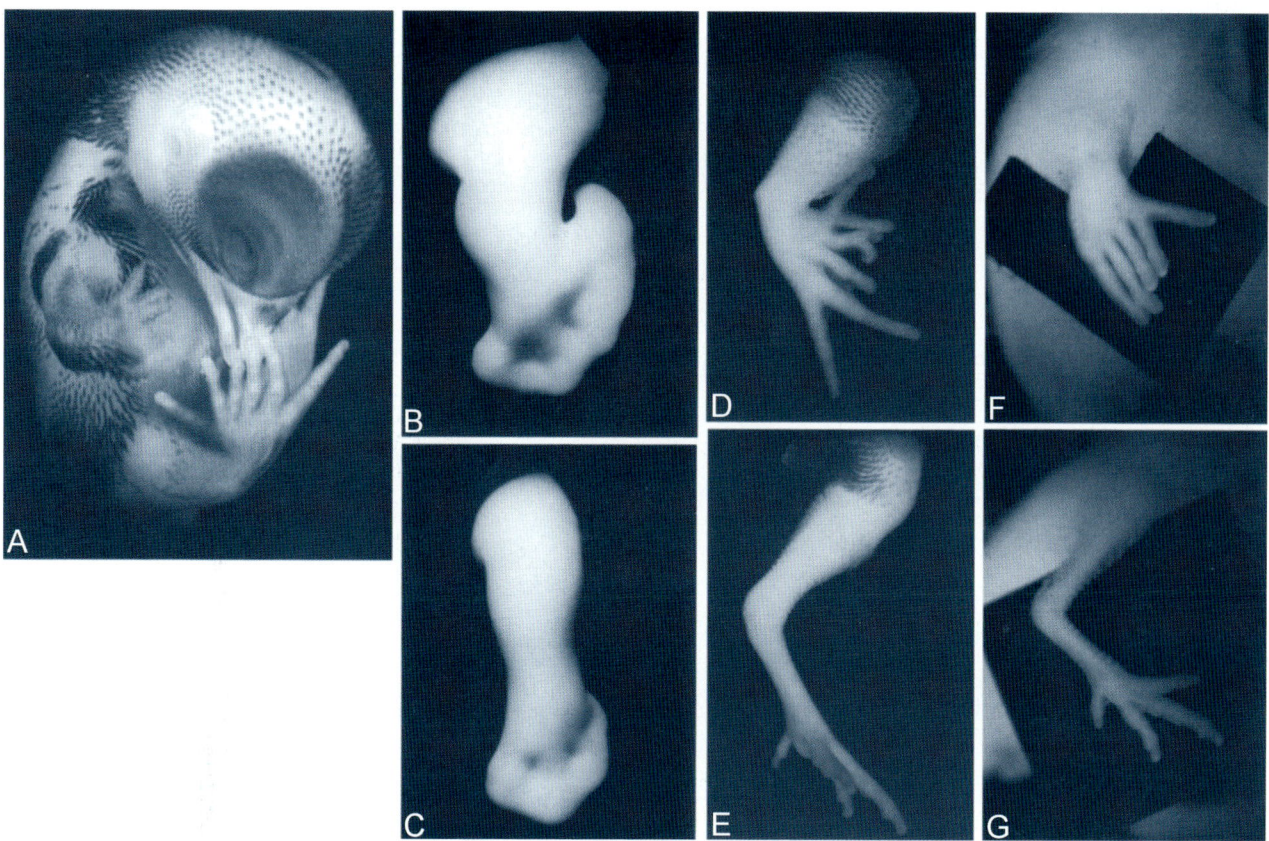

Figure 35.6 Hind-limb development in *diplopodia₄*, a polydactylous mutant of domestic fowl. (A) A *diplopodia₄* embryo at 11 days of incubation to show the extra digits. (B, C) The hind limbs of *diplopodia₄* (B) and wild-type (C) embryos at seven days of incubation. Note the broad, spatulate foot-plate in the mutant. (D, E) The hind limbs of *diplopodia₄* (B) and wild-type (C) embryo at 11 days of incubation. Note the extra and bifurcated digits in the mutant. The skeletons of these two hind limbs are shown in C and D in Fig. 35.7. (F) A polydactylous hind limb developed when limb-bud mesenchyme from a *diplopodia₄* embryo of H.H. 20 was combined with limb-bud epithelium from an H.H. 21 wild-type embryo. (G) A normal hind limb developed when limb-bud mesenchyme from a wild-type embryo of H.H. 20 was combined with limb-bud epithelium from an H.H. 21 *diplopodia₄* mutant embryo. The results shown in F and G demonstrate that the limb-bud epithelium not the mesenchyme is defective in *diplopodia₄* embryos. Modified from MacCabe *et al.* (1975).

If the PNZ – or the equivalent region from hind-limb buds – is grafted below the AER, the AER regresses and disappears as expected. However, surprising at the time, perhaps even now, a *new AER* forms immediately preaxial to the graft and a limb grows out from this preaxial location. Paradoxically, although the PNZ does not produce AEMF, it seems capable of inducing AEMF from mesenchymal cells immediately preaxial to it. Perhaps this explains the normal post-axial location of the AEMF and asymmetry of the AER, whose position is partially governed by the location of the PNZ. Since the AER is normally thicker at the posterior margin, this zone may play a role during normal limb development, a suggestion consistent with the observation that grafting a portion of the PNZ beneath the epithelium results in preaxial thickening of the AER.[16]

All well and good, but except for changes in cell death, surprisingly, removing the PNZ does not interfere with normal limb development. Therefore, if the PNZ does influence the production of AEMF *in ovo*, its action must be rapid, early in development and/or based on a long-lasting message(s) passed to adjacent mesenchyme.

In summary, limb mesoderm:

- specifies limb type as wing or leg;
- responds to the AER by proliferation and outgrowth;
- maintains or supports formation or regeneration of an AER;
- transmits AEMF largely in a proximo-distal direction;
- forms the skeleton in a proximo-distal sequence (see below).

SPECIFICITY OF LIMB-BUD EPITHELIUM

Limb mesenchyme can maintain a second ridge in limb epithelium (see above and Zwilling, 1956a). Can limb mesenchyme *elicit* a ridge from non-limb epithelium?

Zwilling (1964) tested this possibility by grafting flank epithelium to a limb bud from which the epithelium had been removed after the normal stage of initiation of the AER. The limb failed to develop further. Therefore, either the ability to initiate an AER is time specific or flank

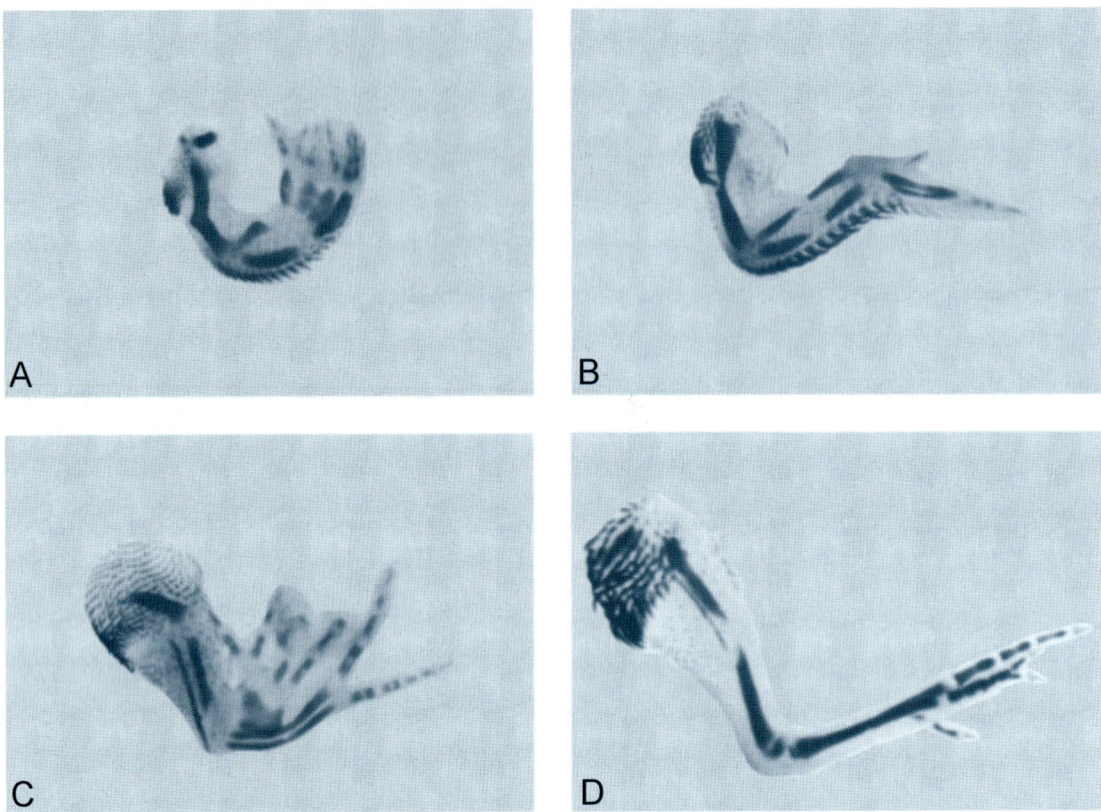

Figure 35.7 Limb development in *diplopodia₄*, a polydactylous mutant of domestic fowl. (A, B) The skeleton of the wings of *diplopodia₄* (A) and wild-type (B) chick embryos at 11 days of incubation to show the polydactyly in the mutant. (B, C) The skeleton of the hind limbs of a *diplopodia₄* (C) and a wild-type (D) chick embryo at 11 days of incubation to show the polydactyly in the mutant. C and D are the same hind limbs shown in D and E in Fig. 35.6. Modified from MacCabe *et al.* (1975).

ectoderm is non-responsive. Could younger mesoderm induce flank ectoderm to produce a ridge?

Zwilling (1964) dissociated limb mesenchyme into single cells, reaggregated them into a pellet, wrapped them in limb or flank epithelium and grafted the package to the CAM. The mesenchyme only formed cartilage when associated with limb epithelium. Therefore, limb epithelium is both site and time specific. Other experiments involving the response of flank ectoderm to limb mesoderm were discussed under 'Ectodermal responsiveness' above, including evidence for the time of onset of this specificity. Once flank ectoderm of a future limb field has been in contact with presumptive limb mesoderm for a time it 'becomes' limb ectoderm; ectoderm not having this contact loses the ability to respond to limb mesoderm and can no longer become limb ectoderm. Timing of positioning of limb buds along the axis is thereby specified (see Table 35.1).

There is, however, an earlier action of limb-bud epithelium that occurs immediately before and as the AER is forming. This is a requirement on an epithelial signal for limb-bud mesenchyme to chondrify. This is not the patterning interaction that occurs later in development, but rather a differentiation signal akin to those discussed

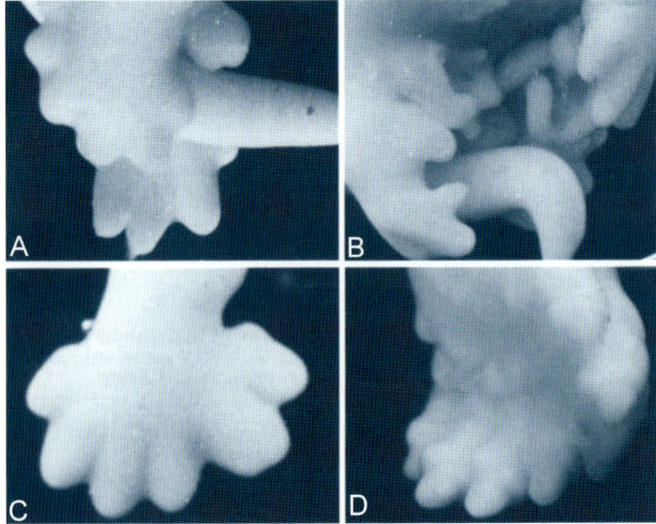

Figure 35.8 Polydactylous hind-limb buds of mutant mouse embryos. (A) Normal hind-limb buds on either side of the tail of a wild-type (+/+) embryo of 15 days of gestation. (B) Additional hind-limb buds in an embryo of 15 days of gestation that is heterozygous (*Ds*/+) for the mutation *disorganization* (*Ds*). (C) Preaxial polydactyly (right) in a hind-limb bud from a 15-day-old *Ds*/+ embryo. (D) A hind-limb bud from an 18-day-old *Ds*/+ embryo with more than double the normal number of digits. Modified from Ede (1980).

for the induction of ectopic bone (Chapter 12), jaws (Chapter 18) and skull bones and cartilages (Chapter 21). Gumpel-Pinot (1980) established a series of transfilter tissue interactions involving limb epithelium and mesenchyme to investigate whether initial chondrogenesis from limb mesenchyme required an epithelial signal. She investigated the time immediately before and after initial appearance of the AER, i.e. H.H. 14–18. As you can see from the results in Table 35.4 an epithelial interaction is required until H.H. 18 if mesenchyme is to chondrify.

Using SEM analysis of the tissue recombinations, Gumpel-Pinot demonstrated mesenchymal cell processes penetrating the pore of the Nucleopore filters used to separate epithelium from mesenchyme. The percentage of cultures forming cartilage increased with the porosity of the filters over the range 0.2–0.8 μm porosity. Subsequently, Gumpel-Pinot (1981, 1982) showed that proximity but not direct contact of epithelium and mesenchyme was required for the interaction to occur. Mesenchymal cell processes crossed the Nucleopore filter within 15 minutes and deposited an ECM onto the filter (Table 35.5). She noted that the epithelial–mesenchymal interaction preceded condensation of limb mesenchyme; we now know that such interactions are required for condensation (Fig. 18.3, and see Chapter 18).

Around the same time it was shown that avian limb-bud epithelium establishes a *peripheral non-chondrogenic area of avascular mesenchyme*, characterized by flattened fibroblastic cells through a diffusible factor that crosses 25-μm but not 150-μm Millipore filters. Limb-bud epithelium inhibits chondrogenesis from limb mesenchyme in collagen gel cultures via a diffusible factor that can travel up to 200 μm through the gel or that acts in gels preconditioned with ectoderm. Consistent with Gumpel-Pinot's demonstration of the requirement of pre-limb-bud mesenchyme for epithelial signals to become chondrogenic, limb-bud territory epithelium from H.H. 15 stimulates chondrogenesis, while limb-bud epithelium of H.H. 23/24 inhibits chondrogenesis.[17]

SPECIFICITY OF DISTAL LIMB MESENCHYME

The mesenchyme that responds to an AER *in ovo* is the most distal mesenchyme immediately subjacent to the AER. Can more proximal limb-bud mesenchyme or non-limb-bud mesenchyme respond to the AER?

In the mid- to late-1950s, Saunders and his colleagues transplanted proximal prospective *hind-limb* (thigh) mesenchyme adjacent to the AER of chick *wing buds* at various times between H.H. 18 and 27. Their question was whether transplanted proximal mesenchyme would produce proximal or distal limb structures – or no structures at all – and whether those structures would be typical of the hind limb (the source of the mesenchyme) or of the wing (the source of the AER and the donor site). Proximal mesenchyme *did* respond, provided that it was taken from embryos younger than H.H. 24 and provided that it was in contact with the AER. The transplanted 'proximal' mesenchyme forms distal skeletal elements, but those are distal *hind-limb* structures, not wing (Table 35.6). Proximal mesenchyme responds by producing skeletal elements appropriate to its new position relative to the AER, but retains its limb-type specificity by producing toes rather than wing digits. Leaving a barrier of distal mesenchyme between the AER and the transplant prevents the transplanted mesenchyme from responding. The influence of the AER is local.[18]

Amprino and Bonetti (1964) also grafted proximal mesenchyme beneath the most distal limb mesenchyme in embryos of H.H. 25 and 26. Even though they left

Table 35.4 Incidence (% of cultures) of chondrogenesis from limb mesenchyme maintained in the presence or absence of limb-bud epithelium[a]

H.H. stage of limb tissues	Mesenchyme cultured transfilter to epithelium	Mesenchyme cultured alone
14	60	0
15–16	58	6
17	81	57
18	100	100

[a] Based on data in Gumpel-Pinot (1980).

Table 35.5 Activity of mesenchymal cells from chick wing buds as they travel across a Nucleopore filter with pore sizes of 0.6–0.8 μm[a]

Elapsed time[b]	Mesenchymal cell activity
15 min	Appearance of cell processes on the other surface of the filter, some projecting into the filter
30 min	More cell processes and filopodia (some as small as 0.1–0.2 μm in diameter) cross the filter
1–4 hours	Filter surface covered with processes; maximum coverage between 2 and 4 hours

[a] Based on data from Gumpel-Pinot (1981).
[b] Time after epithelium and mesenchyme established transfilter.

Table 35.6 Results of typical experiments grafting proximal hind limb-bud mesenchyme subjacent to the AER of the wing bud[a]

Donor age (H.H. stage)	Per cent (N) of grafts producing distal foot structures
17–19	86 (74/86)
20–22	62 (24/39)
23–25[b]	29 (15/51)
26–27[b]	0 (0/33)

[a] Based on data in Saunders *et al.* (1959).
[b] The ability of proximal limb mesoderm to produce distal skeletal structures is lost between H.H. stages 25 and 26.

15–20 rows of mesenchymal cells at the tip, they still obtained distal structures from the transplant, and so concluded that the proximal mesenchyme did *not* have to be in contact with the AER. At such a late developmental stage, however, the graft may have received the epithelial message secondarily from already-determined distal mesenchyme with which it had come into contact before being transplanted. Indeed, the experiment does not necessarily rule out epithelial involvement at earlier stages.

THE TEMPORAL COMPONENT

Rubin and Saunders (1972) examined temporal aspects of interaction within early limb buds by combining mesenchyme from younger (H.H. 18–20) with epithelium from older (H.H. 23–25) embryos, and mesenchyme from older with epithelium from younger embryos. Regardless of age, epithelium elicited complete limbs from limb mesenchyme when tissue recombinants were grown as flank grafts. Epithelium from the older embryos was just as capable of eliciting limb development as epithelium from younger embryos. Rubin and Saunders proposed that signals from the epithelium – presumably from the AER – are constant over time, and that any proximo-distal sequence within the limb bud must be a property of limb mesenchyme or earlier limb-field mesoderm. A decade later it was confirmed that *flank* epithelium of embryos of H.H. 15 and 20 could respond to ridge induction. The AER finally loses its ability to induce limb outgrowth during H.H. 29 (Table 35.7). Interestingly, the loss of inductive ability is neither accelerated by combining limb epithelium with mesenchyme from older embryos nor slowed down by combining epithelium with mesenchyme from younger embryos.[19]

A MECHANICAL ROLE FOR THE EPITHELIUM?

Amprino and his colleagues produced evidence that they thought contradicted what, by the mid-1960s, was known as the *Saunders-Zwilling model* of epithelial–mesenchymal interaction. Their view did not gain wide support, although one of their studies supporting a biomechanical role for limb epithelium did gain some acceptance.

Using wing and hind-limb buds from embryos between H.H. 18 and 25 (3–5 days of incubation), Amprino and Ambrosi (1973) placed chips of dyed agar immediately beneath the AER or at various depths within the distal mesenchyme. By monitoring subsequent limb outgrowth *in ovo* they observed a proximo-distal *sliding of the entire epithelial cover, moving the AER progressively more and more distally*. 'Sliding' ceased if the epithelium was pinned dorso-ventrally, growth of the limb bud stopped, and distal deficiencies arose. Such results had been seen before, but after removing or destroying the AER. The AER was not injured in the study; rather, movement of the entire limb epithelium was stopped.

Amprino and Ambrosi proposed that the entire limb epithelium – a coherent sheet of epithelium, which we could regard as a compartment (Box 35.4) – plays a role in modeling mesenchymal outgrowth. They did not say – indeed, could not resolve from their experimental design – whether epithelial movement governs mesenchymal growth or mesenchymal growth governs epithelial movement.[20]

The notion that epithelial sliding largely maintains the AER is in line with the low mitotic activity in the AER.

Table 35.7 Results of experiments to test the time dependency of the ability of the AER to induce limb-bud mesenchyme to grow and produce the skeleton of the limb[a,b]

Age of embryo providing Limb-bud ectoderm (H.H. stage)	Per cent (N) of grafts forming wings
26 to early 29	78 (36/46)
29	18 (4/22)
30	0 (0/9)

[a] Ectoderm from the limb buds of embryos of various ages was recombined with limb-bud mesoderm from embryos of H.H. 18–20 and maintained as a flank graft. Based on data in Rubin and Saunders (1972).
[b] This ability to induce limb-bud mesoderm is lost between H.H. 29 and 30.

Box 35.4 Compartments

Compartments are important features of the development of some invertebrate embryos (*Drosophila* being the most studied), and have been invoked in stochastic models of cell differentiation (MacLean and Hall, 1987).

The proposal by Amprino and Ambrosi (1973) that the epithelium of the limb functions as a coherent sheet in directing mesenchymal cell movement essentially treats the epithelium as a compartment.

Altabef *et al.* (1997) identified dorsal and ventral ectodermal compartments associated with developing chick limb buds, claiming them as the first non-neural compartments in vertebrates (but see text for ectomeres). No equivalent mesodermal compartments were found, indicating primacy of limb ectoderm. The future AER was fate-mapped as scattered over the dorsal and ventral ectoderm.

On the basis of labeling with DiI or Lucifer yellow and expression patterns of genes such as *Shh*, *Wnt* and *Bmp*, the sclerotome has also been postulated as consisting of compartments.[a]

Nowicki and Burke (2000) transplanted segmental plate in chick embryos and identified two compartments: (i) a dorsal compartment of *somitic mesoderm* that retains its *Hox* gene code when transplanted along the axis, and (ii) a ventral compartment of *somitic and lateral plate mesoderm* that adapts to the Hox code appropriate to the level to which it is transplanted. Here compartmentalization is associated with (based on?) independent paraxial and lateral-plate mesoderm Hox codes.

[a] See Bagnell (1992), Bagnell *et al.* (1992) and Brand-Saberi *et al.* (1996) for the DiI and Lucifer yellow labeling, Christ *et al.* (1998) for the gene expression patterns, and Brand-Saberi and Christ (2000) for a review of the development skel and evolution of somitic cell lineages/compartments.

Mitosis is slight at H.H. 17 when the ridge arises and ceases by H.H. 21. Epithelial cells adjacent to the ridge do divide and could provide a mechanism to cause, secondarily, piling-up of cells into a ridge.[21]

The experimental analysis by Amprino and Ambrosi shows that mesenchymal outgrowth is greater distally than proximally, and that during growth, superficial blood vessels maintain a constant distance from the epithelium as a consequence of ongoing vascular reconstruction. They considered the possibility that pressure exerted by the vascular tissue might be important for mesenchymal growth. As Amprino (1974) summarized the situation: the pressure of the vascularized mesoderm is greater along the proximo-distal axis than along other axes, exerting a stress on the ectoderm that is balanced by division and sliding of ectoderm, i.e. control is initially mesodermal. Later in limb-bud development, regression (or exclusion) of blood vessels from the core mesenchyme precedes chondrogenesis (Hallmann et al., 1987).

The proposal that mesenchymal growth might be greater distally raises the question of the mechanism of that growth. Amprino found no evidence for cell migration – which would bring additional cells into the limb bud – or cellular hypertrophy, the enlargement of cells seen as chondrocytes mature. Unless pressure from the superficial blood vessels is sufficient in itself – as Amprino thought it might be – he was left with the possibility of a proximo-distal gradient in cell division within the mesenchyme. After reviewing the available data, Amprino and colleagues emphasized the lack of evidence for any significant gradient but concluded that *some slight gradient must be operative*. Interestingly, and as discussed above, a proliferation gradient must be incorporated into computer models if they are to simulate limb outgrowth with any accuracy.

NOTES

1. Wallace (1981), p. 224.
2. Trollope (1999), p. 157.
3. See Dworkin et al. (2001) and Atallah et al. (2004) for the studies on *Drosophila* morphogenetic fields, and see Hall and Wake (1999) and Hall (1998a, 2004a) for how larval density can affect developmental processes in individual larvae.
4. See Kieny (1960), Reuss and Saunders (1965) and Saunders and Reuss (1974) for the experiments inducing ectopic AERs and limbs. See Carrington and Fallon (1984a,b, 1986, 1988) and Wilson and Hinchliffe (1985) for the timing of the responsiveness of flank ectoderm and for recombinations between *wingless* and wild-type embryos. See Abbott (1975) and Hall (1983d, 2000d) for the utility of the tissue-recombination approach for studies within and between wild type or mutant individuals of the same species, and between species or classes (mouse–chick, reptile–chick).
5. See Dhouailly and Kieny (1972) and Searls and Zwilling (1964) for these two studies. Normally, tail development is

initiated by the tail equivalent of an AER, a ventral ectodermal (epithelial) ridge (VER); see Chapter 43.

6. See Cairns and Saunders (1954), Zwilling (1955) and Saunders et al. (1957, 1958) for specification of limb type.
7. See Fallon (2002) for an interview with John Saunders, including how he, Saunders, came to investigate the AER. Kelley (1973) has traced identification of the ridge back to studies by Kölliker in 1879 and by Balfour in 1885. Kelley (1975) and Kieny and Fallon (1976) published an informative paper on the ultrastructure of the AER of human limb buds.
8. Some general reviews in which cell proliferation is considered are Faber (1971), Zwilling (1972), Searls (1973a,b) and Hinchliffe and Johnson (1980, 1983).
9. See Janners and Searls (1970) and Searls and Janners (1971) for mitotic rate between H.H. 16 and 22.
10. See Lewis (1975) for the data upon which this discussion is based, including data for the level of mitotic activity at each stage, and see Paulsen and Finch (1977) for the lectin data.
11. See Ede et al. (1977b) for the gradient, and Kieny (1975) for blocking mitosis. This gradient is reversed in *talpid³*.
12. See Amprino (1965), Hornbruch and Wolpert (1970), Amprino and Ambrosi (1973) and Summerbell (1974b, 1977a) for possible physical constraints of the ectoderm. See Summerbell and Wolpert (1973) for the precision of skeletal growth.
13. See Zwilling (1956b, 1974) and Zwilling and Hansborough (1956) for the original experiments, and Chapter 36 for a discussion of them.
14. See Calandra and MacCabe (1978) and MacCabe et al. (1977) for the cell-free extract and molecular weight analyses, and Kieny (1975) and W. J. Scott et al. (1977) for the mitotic inhibitors.
15. See MacCabe and Abbott (1974) and MacCabe et al. (1975) for the mouse mutants.
16. Gasseling and Saunders (1964) first described the PNZ in wing buds – see the summary by Saunders and Gasseling (1968) – and did the transplant studies, which were confirmed by Summerbell (1974a).
17. See Solursh (1984) for the non-chondrogenic zone, and Solursh et al. (1981, 1984) and Solursh and Reiter (1988) for epithelial inhibition/stimulation of chondrogenesis.
18. See Saunders et al. (1955, 1957, 1959) for transplantation of proximal mesenchyme to distal sites in the limb bud, Amprino (1964, 1968) for transplantation to the more proximal sites, and Amprino (1984) for an evaluation of these studies.
19. See Carrington and Fallon (1984b, 1986, 1988) for the later studies, and see Fallon (2002) for an interview with John Saunders.
20. See Amprino (1965, 1975a, 1977b, 1978) for his data and views against the generally accepted model of limb outgrowth. Coffin-Collins and Hall (1989) and Hall and Coffin-Collins (1990) demonstrated regulation of mitotic activity of craniofacial prechondrogenic epithelia and mesenchyme by Egf.
21. See Amprino (1974) for the mitotic data, and Errick and Saunders (1976) and Saunders et al. (1976) for a contrary view. Mitotic rates are similarly low in mandibular-arch epithelia (Hall and Coffin-Collins, 1990).

Adding or Deleting an AER

Paradoxically, wingless (wl) mutant chick embryos form wing buds with an AER.

Surgical manipulation of avian embryos has taught us a great deal about the function of the AER (apical epithelial ridge), and Saunders' observations on the AER's essential role, and the consequences of removing it, have been confirmed and expanded considerably.

AER REGENERATION

Any ability of an AER to regenerate, or of flank or limb-bud epithelium to regulate to replace a lost AER, would greatly complicate the studies I have been discussing. Apropos, and given the appropriate circumstance, avian AERs *can* regenerate following extirpation of the AER *in ovo*, provided that the cut edges of the epithelium can make contact.

Searls and Zwilling (1964), who first made this observation, isolated limb-bud epithelium from embryos of H.H. 19 and 20 and cultured the epithelium with agar, somitic mesoderm or limb-bud mesenchyme. In all cases, the AER degenerated within 7–10 hours. A new ridge formed when limb mesenchyme was used as the substrate. When these cultures were grafted to the CAM to provide a vascularized environment, the new ridge supported limb outgrowth and skeletal development. Even after 10 hours contact with agar – by which time the AER has regressed – development could be restarted (continued) by replacing the agar with limb-bud mesenchyme.

An elegant set of experiments demonstrating reformation of the AER undertaken by Errick and Saunders (1974, 1976) involved placing hind limb-bud mesenchyme from embryos of H.H. 21–23 (when an AER is present) into jackets of wing- or hind limb-bud epithelium which had been turned inside out, and then grafting these tissue recombinants into somites *in ovo*. Normal ectodermal polarity and histology were restored rapidly. Growth of the limb buds resumed within 24 hours, but it was three days before an AER reformed, implying that the epithelium continued to act on the limb mesenchyme even though an AER could not be seen.

Epithelia from other embryonic regions do not substitute for limb-bud epithelium. Dissociated AERs grafted with hind limb-bud mesenchyme from H.H. 19 embryos also direct limb development. Active epithelium, not organization of that epithelium into an AER, is required for epithelial–mesenchymal signaling. This is an important caveat to bear in mind when looking for an AER on limb buds of other tetrapods, as we shall do in the next chapter.

This **cautionary** note applies to the interpretation Amprino and Aiello-Malmsberg (1971) gave to a study in which the developmental role of limb epithelium was questioned. Some 75 per cent of the distal and central mesenchyme was removed from hind limb buds of H.H. 23–25 embryos. The two peripheral, proximal pieces were joined together. Provided that epithelium grew over the cut surface, some digits developed from this proximal mesenchyme, but an AER was never seen. They concluded that the mesenchyme provided all the necessary stimulation, although it may be that, even though it had not formed a ridge, the epithelium was still acting on the mesenchyme to direct limb outgrowth and digit formation.

Given that the AER has a controlling role in limb development, variations in the size, shape and location of an AER on an early limb bud should be reflected in deviations from normal during limb development. In this chapter, I consider three situations, listed below, and discuss whether such limb buds provide evidence in support of a directive role for the AER. In each situation, information may be obtained from *mutant embryos* and from *experimental manipulation* of limb buds.

- *Limb buds lacking an AER.* In nature, such limb buds are found in limbless tetrapods and skeletal mutants (Chapter 37) and, as shown by Saunders and discussed below, they can be produced surgically.
- *Limb buds with an additional or enlarged AER.* Again, mutants exist and such limb buds can be produced experimentally.
- *Limb buds with a narrow or subdivided AER.*[1]

EXPERIMENTAL REMOVAL OF THE AER

In what is a classic study, Denis Summerbell established the timing of the specification of skeletal elements in the wing by removing the AER from embryos of different ages (see Table 36.1 and Fig. 36.1). He proposed that the proximo-distal sequence of skeletal specification is continuous and graded, rather than stepwise. If specification were stepwise, large elements such as those of the forearm would be specified at one time. As Fig. 36.1 indicates, however, the elements of the forearm are determined progressively over a 12-hour period between H.H. 18 and 21, while the much shorter – but developmentally more complex – wrist takes 24 hours (H.H. 21–24) to be specified.

Not all workers accepted that skeletal deletions following removal of the AER provided evidence of epithelial control. Rudolfo Amprino from Bari in Italy was a particularly effective critic of extirpation experiments, suggesting that damage to mesenchyme adjacent to the AER must be considered. He may have had a point. For example, more deficiencies occur if the entire epithelium of the limb bud is removed than if only the AER is removed. Amprino and colleagues assumed that mesenchymal damage and *non-ridge* epithelium both play a role, which indeed they do; if only lateral epithelium is removed and the AER left intact, deficiencies still arise. Furthermore, some outgrowth of the mesenchymal core of early limb rudiments – *anlage* (primordium, rudiment, character), to use a German term popular in embryology – occurs

before the AER develops. Either the pre-AER epithelium is not essential for this early outgrowth – as Amprino (1965) thought – or flank epithelium can direct outgrowth before it acquires its distinctive ridge-like morphology (see Chapter 35).[2]

Does the AER govern limb-bud outgrowth by maintaining a high rate of mitosis in subridge mesenchyme? No, it does not. Although the labeling index of subridge mesenchyme at H.H. 23 and 24 is slightly higher than the index in more proximal mesenchyme, removing the AER at H.H. 19 does not affect the labeling index (Manners and Seals, 1971).

A wave of *cell death* does engulf subridge mesenchyme after the AER is removed; this was part of Amprino's legitimate concern over damage to distal mesenchyme. Perhaps the AER functions to slow cell death in distal mesenchyme and so maintain a larger pool of cells than in the surrounding mesenchyme. Such a mechanism would lead to outgrowth of the limb bud. One of my favourite experiments provides evidence on this point. It is a favourite because of its elegance and simplicity, and because it showed us the

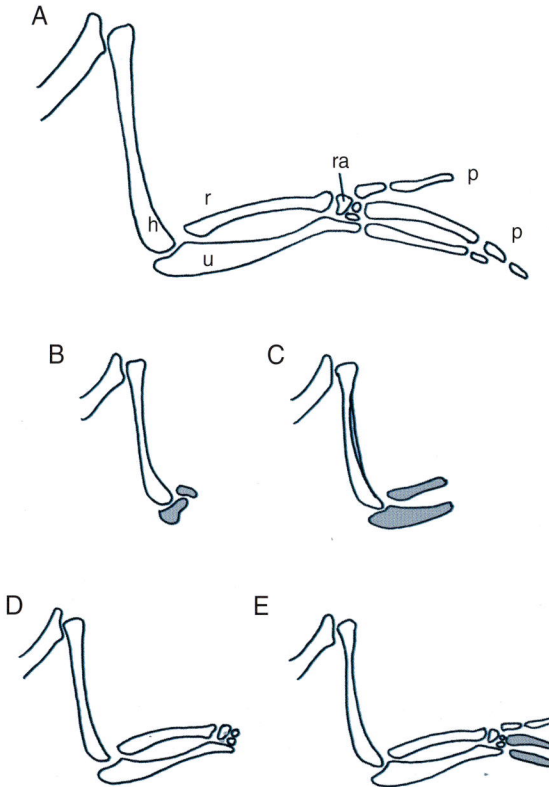

Figure 36.1 The later the AER is removed from a limb bud, the greater the skeletal deficiencies. (A) The wing skeleton as it appears normally and as it appears if the AER is removed after H.H. 28. h, humerus; p, phalanges; r, radius; ra, radiale; u, ulna. (B)–(E) show the extent of development of the wing skeleton after removal of the AER at successively later stages of embryonic development, viz. H.H. 18 (B), 19/20 (C), 21–25 (D) and H.H. 26 (E). Distal deficient elements are shaded. So, in B only the proximal portions of the radius and ulna have formed. In E digit I is complete, but digits II and III have distal deficiencies. Based on data from Summerbell (1974b). Modified from Hall (1978a).

Table 36.1 The timing of specification of the skeletal elements in the wing buds of embryonic chicks[a]

H.H. stage	Skeletal element specified
Before stage 18	Shoulder girdle, humerus
Stages 19–20	Ulna, radius
Stages 21–24	Wrist
Stages 25–26	Metacarpals
Stage 27	Proximal phalanx of digit III
Stage 28	Distal phalanx of digit III

[a] Based on the latest H.H. stage at which deletion of the AER results in deficiency of that element, from data in Summerbell (1974b).

path to follow in the search for the biochemical/molecular basis of epithelial–mesenchymal interactions, wherever and whenever they occur. AERs were maintained in *vitro* in small wire baskets suspended *over* limb mesenchyme but *not in contact* with the mesenchyme (Table 36.2). Suspending AERs over the cultures prevents the cell death that occurs when limb mesenchyme is cultured alone, providing presumptive evidence for the transmission of a diffusible stimulus from epithelium to mesenchyme.[3]

On balance, the evidence is that ablation of the AER prevents further development or outgrowth of the mesenchymal core of a limb bud. This conclusion supports the thesis that interaction between AER and mesenchyme is essential to initiate and maintain limb buds.

Several other lines of evidence indicate that the AER controls limb development. One of the most fascinating is the work initiated by Edward (Ed) Zwilling (1949) on *wingless* mutant chicks, which together with Saunders' studies led to the Saunders–Zwilling model of limb development.

FAILURE TO MAINTAIN AN AER: *WINGLESS* (*wl*) MUTANTS

Wingless (*wl*) mutants are exactly what their name implies, adult fowl without wings. Their legs are perfectly normal (Fig. 36.2). When faced with an adult lacking such a major structure as the wings, our automatic reaction is to assume that the mutant lacks wings because it cannot form them. Indeed, we extend this thinking to the loss of any structure. After all, what mechanism could lead to a major part of the phenotype failing to develop, other than loss of the developmental capability to make that structure? What does it profit an animal to lose its wings but keep its buds?

Table 36.2 Amount (%) of cell death in limb-bud mesenchyme from chicks of H.H. Stages 17–23 after 10 hours culture with ectoderm or the AER[a]

Test	Per cent of cell debris			
	None	Little	Some	Much
Central distal mesenchyme – no ectoderm	17	15	59	8
Central distal mesenchyme – no ectoderm	40	27	28	5
Distal mesenchyme with AER	70	18	11	2
Distal mesenchyme with ectoderm but no AER	4	20	48	28
Proximal mesenchyme with ectoderm, no AER	–	33	33	33
Mesenchyme at bottom, six AERs suspended	75	12	12	–

[a] Based on data from Cairns (1975).

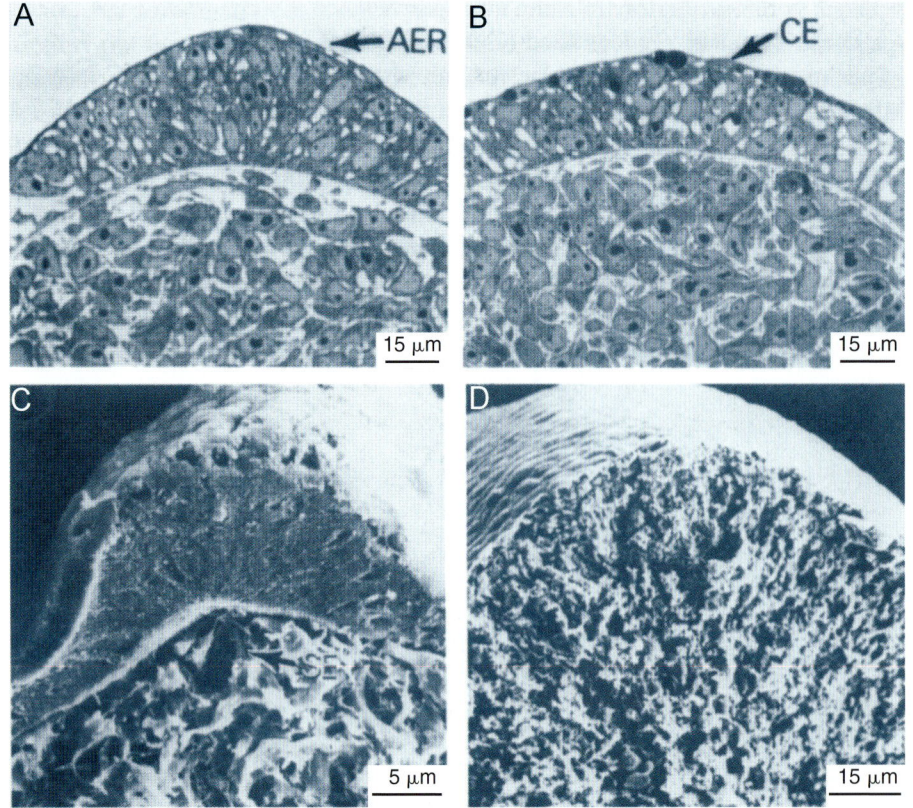

Figure 36.2 Wing-bud development in *wingless* mutant chick embryos. (A) The wing bud from a wild-type embryo of H.H. 20 (three days of incubation) with prominent AER. (B) The wing bud from a *wingless* mutant embryo of H.H. 20 with compact epithelium (CE) where the AER is found in wild-type embryos; compare with A. A and B are 0.5 μm sections embedded in Epon. (C) A scanning electron micrograph of a wing bud from a wild-type embryo of H.H. 25 (4.5 days of incubation) showing the prominent AER and subepithelial space (SE) between the AER and the most distal mesenchyme. (D) The wing bud from a *wingless* embryo at H.H. 25 lacks any evidence of maintenance of an AER; compare with C. Modified from Sawyer (1982).

While it seems paradoxical that wing buds should develop in wingless individuals, this is the normal situation for such mutants, and not only for limbs. The tails of mice develop under the influence of a ventral epithelial ridge (VER). Embryos carrying the *vestigial tail* (*vt*) mutation form tail buds but no VER, a lack that leads to tail-bud regression and tailless adults (Chapter 43). In an analogous manner, embryos of limbless snakes produce rudimentary hind limb (but not forelimb) buds, embryonic whales produce hind limb buds but no hind limbs (Chapter 44), and hairless mice begin their lives with a fine coat of hair, rather like men who go bald. There is a selective advantage to maintaining the ability to form these organ primordia (Chapter 44).[4]

Paradoxically, *wingless* embryos initiate the development of wing buds *complete with an AER*. Rather than being maintained, however, the AER regresses and the wing buds survive only to the third day of incubation (Fig. 36.2). Gap junctions are not maintained between the epithelial cells, and gaps appear in the basement membrane as it breaks down and the AER starts to regress. Consequently, wing buds fail to show any further growth or development and are overtaken by the growing body wall. Although *wingless* embryos fail to form any skeletal elements in their wing buds – the buds regress well before the skeleton begins to form in wild type – 40 per cent of the wing musculature develops. The primary muscles subdivide normally but later muscles fail to differentiate. Interestingly, one muscle, the *supracoracoideus*, is missing, but there is a space in the wing where the muscle should be. It appears that connective tissue determines the space normally occupied by muscle.[5]

Mutual interaction

The first definitive evidence for a *mutual interaction* between potential limb mesenchyme and epithelium came from experiments conducted by Ed Zwilling in the mid-1950s on *wingless*. By combining mesenchyme and epithelium from limb buds of wild-type and *wingless* embryos, Zwilling demonstrated that the primary defect in *wingless* mutants is epithelial, not mesenchymal. Wings do not form when mesenchyme from wild-type embryos is grafted beneath the limb-bud epithelium of *wingless* embryos. Zwilling also grafted mesenchyme from *wingless* embryos beneath limb-bud epithelium of wild-type embryos *after* the AER had formed. Over the next three days the existing AER flattened and was not maintained. Subsequently, distal limb structures failed to from. Zwilling concluded that information was flowing from mesenchyme to epithelium and *vice versa*, and hypothesized that the mesenchymal factor discussed above maintains the epithelial ridge.

Indeed, much of the evidence for a mesenchymal factor able to maintain the AER – the apical ectodermal maintenance factor – comes from studies on wingless or polydactylous mutant fowl. Recall Zwilling's classic study with *wingless* mutants.

Presumptive evidence for an association between programmed cell death, loss of the AER, and consequent loss of the wings is seen in another chick mutant and in the rudimentary limb buds of limbless tetrapods (Chapter 37). In *wingless* (*ws*) – a different mutation from *wingless* (*wl*) – precocious cell death in the posterior margin of the limb bud begins at H.H. 19 and extends well beyond the posterior margin during H.H. 20–23. Removal of this additional mesenchyme may rob the limb bud of apical ectodermal maintenance factor (AEMF) and so contribute to the lack of an AER.[6]

The molecular bases of these interactions are also being revealed. *Msx-2* is not expressed in the reduced AERs found at early developmental stages of *wingless* embryos (Dealy and Kosher, 1996). Adding Igf-1, however, promotes development of a thick AER, whose cells express *Msx-2* and promote limb-bud outgrowth. This also is true for the *limbless* (*ll*) mutant, which lacks both fore- and hind limbs. Therefore, an *Igf-1* → *Msx-2* pathway maintains the AER.

Tgf-α is expressed in mesenchyme *and* in the epithelia of wild-type limb buds. *Tgf-α* is expressed in the mesenchyme but not in the epithelia of limb buds from *wingless* and *limbless* embryos. Expression of *Tgf-α* and *Egf* in the AER and subridge mesenchyme of wild-type limb buds, and of *Egf* in ventral but not in dorsal limb-bud epithelium, suggested a potential basis for epithelial–mesenchymal signaling. Indeed, exogenous Tgf-α and Egf increase outgrowth of limb buds in *wingless* and *limbless* mutants *in the absence* of an AER. Therefore, a *Tgf-α* → *Egf* pathway also maintains the AER.[7]

Dlx-5, which is expressed in wild-type limb buds (Fig. 18.10), is expressed only transiently in the ectoderm of *limbless* limb buds, a down-regulation that may explain, in part, the lack of maintenance of the AER (Ferrari *et al.*, 1999). Therefore, *Dlx-5* also maintains the AER, and we begin to see a molecular basis linking all three pathways.

The two primary Fgf receptors, FgfR-1 and FgfR-2, are differentially expressed in epithelium and mesenchyme in mouse developing limb buds, FgfR-1 in limb and somitic mesenchyme, FgfR-2 in the epithelium (K. G. Peters *et al.*, 1992). Epithelium and mesenchyme both signal through these Fgf receptors.

Differential splicing of the products of the *Fgfr-2* gene produces two transmembrane tyrosine kinase receptor proteins. Deleting one (FgfR-2[IIIb]) results in severe limb defects and less severe defects in teeth and skull (DeMoerlooze *et al.*, 2000). The IIIb splice variant helps maintain but not initiate the AER in mouse limb buds; could this be the magic AEMF? This conclusion is based on analyses of *FgfR-2-IIIb* mice, which lack teeth, limbs, hair and many glands. Expression of *Fgf-8* and *-10*, *Bmp-4* and *Msx-1* is normal, expression of *Shh* and *Fgf-4* are not, indicating that the receptor acts upstream of *Shh* and *Fgf-4*. The mechanisms of cellular action are through apoptosis of ectodermal and mesenchymal cells at 10–10.5 days of gestation (Revest *et al.*, 2001).

Two further members, *Fgf-12* and *Fgf-13* – *fibroblast growth factor homologous factors -1 and -2* (*Fhf-1*, *Fhf-2*) – are expressed in limb-bud mesenchyme, *Fgf-12* in the posterior mesenchyme – the zone of polarizing activity (Chapter 38), *Fgf-13* distally and anterior – the progress zone (Chapter 35). Consistent with a role in early specification of limb buds, neither *Fgf-12* nor *Fgf-13* is expressed in limb buds in *wingless* or *limbless* mutant embryos. Consistent with a role in specification of distal skeletal elements, expression of *Fgf-13* is expanded in *talpid*² embryos. *Fgf-13* up-regulates *Hoxd-13*, *Hoxd-14*, *Fgf-4* and *Bmp-2* in patterning the A-P limb axis (Munoz-Sanjuan *et al.*, 1999).

EXPERIMENTAL ADDITION OF AN AER

Zwilling (1956a) also provided the first evidence that when an extra AER is grafted to a limb bud, both transplant and host AER are maintained, provided that they are sufficiently far apart, as when the graft is to the anterior surface of the limb bud. A duplicated limb results with one limb axis beneath each AER. The younger the host embryo when the AER is grafted, the more complete the duplication. Conversely, the older the host, the less complete the duplication. Therefore, the presence of one AER does not inhibit maintenance of another. Mesenchyme can respond to the second AER by morphogenetic duplication of a limb axis and duplication of a limb skeleton.

MUTANTS WITH DUPLICATED LIMBS

The converse of wingless mutant are mutants with duplicated limbs. Complete duplication of the limbs in nature is unusual. However, *polydactyly* – a limb normal in all respects other than having additional digits – is common. (Polyphalangy – additional phalanges in the digits, which may or may not be associated with polydactyly – is discussed in Chapter 25 and in Box 36.2.) The founder of population genetics, Sewall Wright, undertook a substantial genetic analysis of polydactyly in guinea pigs, placing emphasis on the contribution of genetic *and* non-genetic factors (such as season), and how threshold effects determine whether additional digits will form (Wright, 1934b,c).

Another word of **caution**: polydactyly refers to having additional digits in excess of the normal mammalian number of five. However, if the normal number of digits for a species is four, as it is for the guinea pigs studied by Wright, then having five digits is polydactylous *for that species*. Secondly, identifying an additional digit(s) says nothing of the mechanism by which the extra digit(s) arise. The differences are important. Two mechanisms are possible:

(i) *subdivision of an original digital primordial*, so that there might be, for example, six digits, the additional digit being a duplicated digit V; or

(ii) *specification of an additional digit*, so that there might be, using the same example, six digits, the additional digit being a digit VI.

For developmental and/or evolutionary studies, the differences in mechanisms are crucial, especially if we attempt to homologize individual digits (Figs 36.3 and 36.4). In mechanism (i) the additional digit arises by branching or bifurcation of the primordium of digit V and is a homologue of digit V. In mechanism (ii) the additional digit is novel and so not homologous to any of the five digits (Fig. 31.3). See Boxes 36.1 and 36.2 for an example of how knowing your digits helps to unravel phylogenetic relationships.

Application of cytosine arabinofuranoside, a cell-proliferation inhibitor, to embryos of the green lizard, *Lacerta viridis*, at the early limb-bud stage arrests digital development, leading to loss of digits. When digit number is reduced from five to four, it is always digit I that is lost. When digit number is reduced to three, either digits I and V or digits I and II are lost. The basis for such specific loss of digits is discussed in Box 36.1.[8]

Embryos of wingless mutants and limbless tetrapods initiate limb buds that then regress (this chapter and Chapter 37). We might expect the reverse pattern in polydactylous limbs, in that the life or activity of a limb bud might be prolonged or enhanced, and indeed both occur. Prolongation or enhancement can come about by two means: increasing the size of an existing AER or duplicating the AER.

An enlarged AER

Commonly, embryos carrying mutations for polydactyly have limb buds with AERs that are as much as 50 per cent larger than the AERs in wild-type embryos, suggesting that the increased size of the AER may be causally

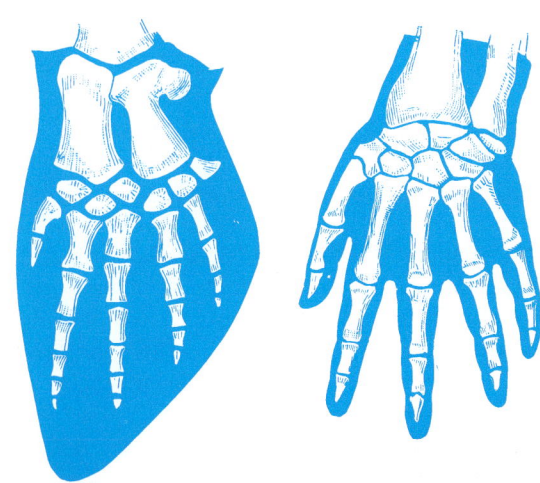

Figure 36.3 The 'paddle' of a whale flipper (left) compared with a human hand to show the homology between the two. From Romanes (1901).

A

B

C

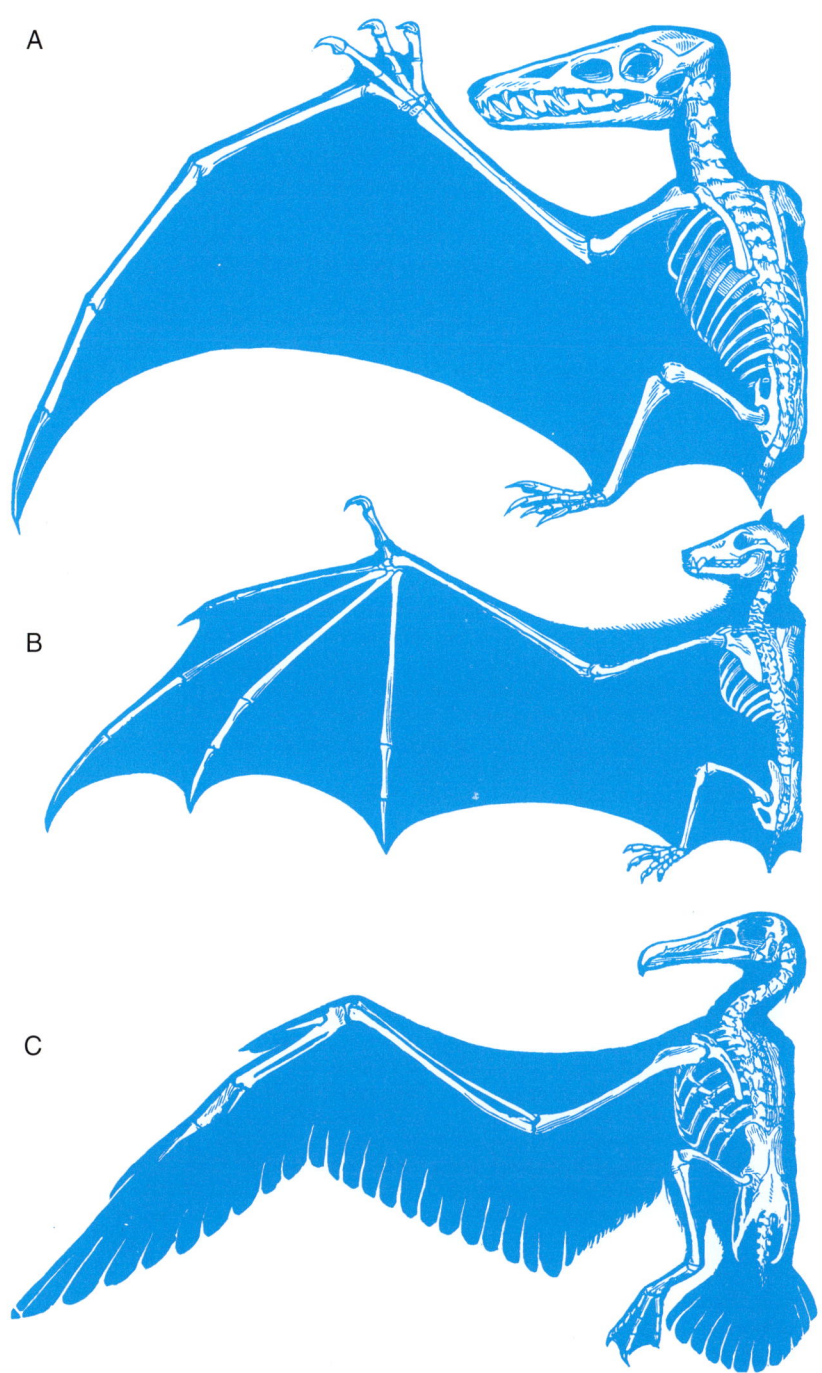

Figure 36.4 Homology of 'wing' skeletons as seen when comparing (A) a pterodactyl, *Pterodactylus*, (B) a bat, and (C) a bird, probably a great black-backed gull, *Larus marinus*. From Romanes (1901).

related to formation of the additional digits. At least four polydactylous mutants in domestic fowl fit this description: *dominant polydactyly, duplicate, talpid²* and *talpid³*. So, too does *preaxial polydactyly* in humans, preaxial here referring to an additional digit(s) on the anterior (preaxial) surface of the limb (Figs 19.1, 19.6 and 36.6). Zwilling and Hansborough used *dominant poly-dactyly* to provide some of the most convincing evidence for a directive role for the AER in limb development.[9]

Wild-type limb buds and 'polydactylous limb buds'[10] are indistinguishable until the third day of incubation,

when the mutant AER enlarges. If one-third of the AER of polydactylous limb buds is surgically removed on the third day, a normal limb develops, providing clear evidence that the enlarged AER plays a role in the mutation. Indeed, the form of a limb can be manipulated by appropriate recombination of wild-type and mutant tissues; when mesenchyme from a polydactylous limb bud is grafted beneath the epithelium of a wild-type limb bud, the wild-type AER thickens preaxially. As in the mutant, extra distal limb elements develop (Fig. 35.6).

Box 36.1 Dinosaur digits and bird origins

You would think that counting digits would be dead easy. After all, one of the first things we learn as children is to count to five using our fingers, and that we have five fingers. Counting digits when reconstructing the origins of birds turns out to be rather more difficult (Fig. 36.4), especially when considering a dinosaur origin for birds. The problem of the homology of the digits of birds and dinosaurs has divided embryologists from palaeontologists for at least two decades.

Hinchliffe (1985, 2002) reviewed the embryological approach to what he termed the 'One, two, three' or 'Two, three, four' problem. The issue is whether the digits of birds and dinosaurs are digits I, II and III or II, III and IV. Palaeontologists regard theropod dinosaur digits as I, II and III. Embryologists see bird digits as II, III and IV, a difference that may not seem great but that is sufficient for some to exclude dinosaurs as bird ancestors.[a] How did we get to this impasse?

In large part we assign digit identity on the basis of connections between digits and the wrist (carpal) or ankle (tarsal) elements. Avian embryos have five precartilaginous carpals. Two, the radiale and ulnare, do not continue to develop and are lost, leaving three carpals. Zones of cell death on both the anterior and the posterior faces of limb bud remove mesenchyme that would otherwise form digits I and V. So, embryonically, birds have the primordia for five digits but lose two during development. The issue remains, which two?

Avian embryos have three distal carpals and, in the same row, a median cartilage known as the *pisiform cartilage*. The position of the pisiform allows the three carpals and associated digits to be identified as II, III and IV. Hinchliffe (1977b) was the first to demonstrate that the classically recognized patterns of condensations in the carpus of the chick wing are not what they had been purported to be on the basis of analysis of developed skeletons. There are not primitively five condensations but four – ulnare, carpal 3, pisiform and radiale.

Analysis by Kundrát et al. (2002) claims evidence for a pentadactyl pattern early in chick and ostrich limb development. The pattern consists of five avascular zones for metacarpals. Three metacarpals develop, the other two – the first and fifth – being evident as transient vestiges. These are nice solid data, but metacarpals and digits need not evolve in parallel.[b] Tim Fedak and I discuss this issue in relation to polyphalangy (Box 36.2) and hyperphalangy in whales where a similar conceptual problem exists: how do we tell which digit(s) has been added in polydactyly if the numbers of carpals or metacarpals do not coincide with digit number? The question is not rhetorical; see Fedak and Hall (2004).

In an experimental study, Nikbakht and McLachlan (1999) claim to have restored digit V in the chick wing by implanting a bead soaked in Fgf-4 into the distal portion of the wing bud at H.H. 25–26, stages when the AER is coming to the end of its action. The additional Fgf-4 maintained the AER for longer than normal, a prolongation that was followed by the formation of an extra digit, which they interpret as digit V and not as a carpal.[c]

A conceptually based solution to the 1, 2, 3 – 2, 3, 4 dilemma, proposed by Wagner and Gauthier (1999), is that the digital condensations (C) in avian limb buds are C-II, C-III and C-IV. These three condensations would be expected to produce the three digits II, III and IV. During later development, according to their solution, a 'frame-shift' occurs such that condensations C-II to C-IV form digits

I, II and III. This may seem like sleight of hand; it certainly generated discussion and comment. Their subsequent analyses using peanut agglutinin lectin (PNA), a much more sensitive marker for condensation – see Dunlop and Hall, 1995, whose protocol they followed – demonstrated five digital condensations, those for digits II, III and IV developing into digits (Larsson and Wagner, 2002).

From their analysis of the effect of Shh and BMP-2 on limb-bud development, in which they demonstrated that *Shh is required to initiate and to maintain BMP-2 expression during digit specification*, Drossopoulou et al. (2000) and Sanz-Ezquerro and Tickle (2000) developed a new model for digit patterning in which long-range signaling from Shh sets the competence of the digital mesenchyme, and, in effect, controls digit number. Subsequent short-range signaling from Shh induces BMP expression to specify digit identity.

Galis (2001a) explicitly addressed the digit identity problem addressed implicitly by Drossopoulou and colleagues in terms of a homeotic change in digit identity, as proposed by Wagner and Gauthier. In a further analysis, Galis et al. (2001) also address the 'why five fingers?' problem, attributing this stability to the fact that the limb buds develop at the phylotypic stage – the conserved stage shared by embryos of taxa within specific phyla – when there are many epithelial–mesenchymal interactions, which would be linked (mechanistically at least) via pleiotropy.[d]

As our understanding of how digit number is specified increases, molecular approaches may help to resolve the dilemma. Litingtung et al. (2002) found that Shh and Gli-3 are dispensable in formation of the chick limb skeleton, but do play a role in regulating digit number and digit identity, Gli-3 negatively regulating Shh expression. $Shh^{-/-}/Gli\text{-}3^{-/-}$ double mutant mice develop additional digits – they are polydactylous – but the digits lack wild type digit identity.[e] They concluded that the pentadactyl pattern acts as a constraint on the potential for polydactyly.

[a] See the chapters in Hecht et al. (1985) for overviews of the debate until 1985, and Hecht and Hecht (1994) for a later update. See Hinchliffe (1985, 1997, 2002) for the embryological stance over the same time period. See Thulborn and Hamley (1982) and Thulborn (1985) for the origin of birds from theropods by neoteny. Differences in the patterns of ossification of the tibia, astralagus and calcaneum in theropods and birds, along with serrated theropod but unserrated avian teeth, have been raised as further difficulties in deriving birds from theropods (Martin et al., 1980).
[b] The situation is even more complicated in avian hind limbs. Two patterns of tarsal development are found in birds: fusion of the pretibial bone with the calcaneum in carinates, and fusion of the ascending process with the astragalus in ostriches, rheas and emu (ratites; McGowan, 1984, 1985).
[c] Earlier, Vogel and Tickle (1993) had shown that Fgf-4 maintains polarizing activity of cells in the posterior portion of the chick limb bud, both in ovo and in vitro. They removed the posterior AER, following which polarizing activity declined in posterior mesoderm. Implanting an Fgf-4-soaked bead reversed the effect of removing the AER, while limb-bud epithelium (or Fgf-4 in the absence of epithelium) maintains polarizing activity in cultures. For reviews of limb development, especially the role of Fgf in inducing the limb, see Tickle (1996, 2000).
[d] See Alberch (1985) and Hall (1996, 1997b,c, 1998a, 2002d) for constraints on body plans and the phylotypic stage.
[e] Digit number and digit size in mice, as is the baculum, are regulated by a dose-dependent mechanism of posterior Hox genes, Hoxd-11, d-12 and d-13 (Zákány et al., 1997; Boulet and Capecchi, 2004). Interaction of Hoxd-12 with Gli-3 contributes to patterning the digits by enabling Gli-3 to activate rather than suppress target genes of Shh (Y. Chen et al., 2004).

Duplicating the AER

Alternatively, embryos may have a second AER on the limb bud, as in the mutation *eudiplopodia* in fowl. Mutants with an enlarged AER have their basis in a primary mesodermal/mesenchymal defect; those with an extra ridge reflect a primary ectodermal or epithelial defect. Far fewer mutations act by disrupting the ectoderm – *eudiplopodia* is the only one in domestic

Box 36.2 Polyphalangy and extra joints

Formation of additional bony elements (phalanges) beyond the normal number (*polyphalangy*) in the digits of a taxon requires either (i) subdivision of existing digital primordia and/or (ii) provision and differentiation of additional distal mesenchyme in which additional digits can form. In either situation, extra joints must be specified.[a] Three types of data inform this question for we see polyphalangy:

  (i) as an evolved feature or variant in individual taxa;
 (ii) resulting from mutation; or
(iii) appearing during regeneration or after experimental manipulation.

The role of *Hoxd-12* and *Hoxd-13* in specification of phalange number was discussed in Chapter 25.

Polyphalangy as normal phenotype

The digits of many cetaceans are naturally polydactylous, a function of the large size of these mammoths and of the evolution of their forelimbs as flippers used in locomotion through water; a long blade-like flipper with many phalanges and soft-tissue fusion between the digits makes for a most effective paddle. Sedmera *et al.* (1997b) examined developing flippers in spotted dolphins, *Stenella attenuata* (Fig. 36.5), finding that the phalangeal number increases as development progresses until the final phalangeal formula of 3,7,7,5,3 is reached. One specimen had transitory cartilaginous rudiments for two carpal elements, even though adults lack these elements.[b]

Miniaturization can be thought of as the opposite trend to that displayed during cetacean evolution (Chapter 44). The most extensive data available come from an examination of the effects of miniaturization on the skeletons of 129 species of frog from 12 families (Yeh, 2002). The frogs were assessed for presence or absence of individual bones and for changes in skeletal morphogenesis and growth. The most recurrent changes were the loss of phalanges and bones of the skulls. The most likely mechanism is the truncation of development (pedomorphosis) shown by these frogs. Bones that are most frequently lost are those that develop late. With pedomorphosis, development would be over before the 'standard' number of phalanges could be laid down. Function may also constrain phalangeal number.

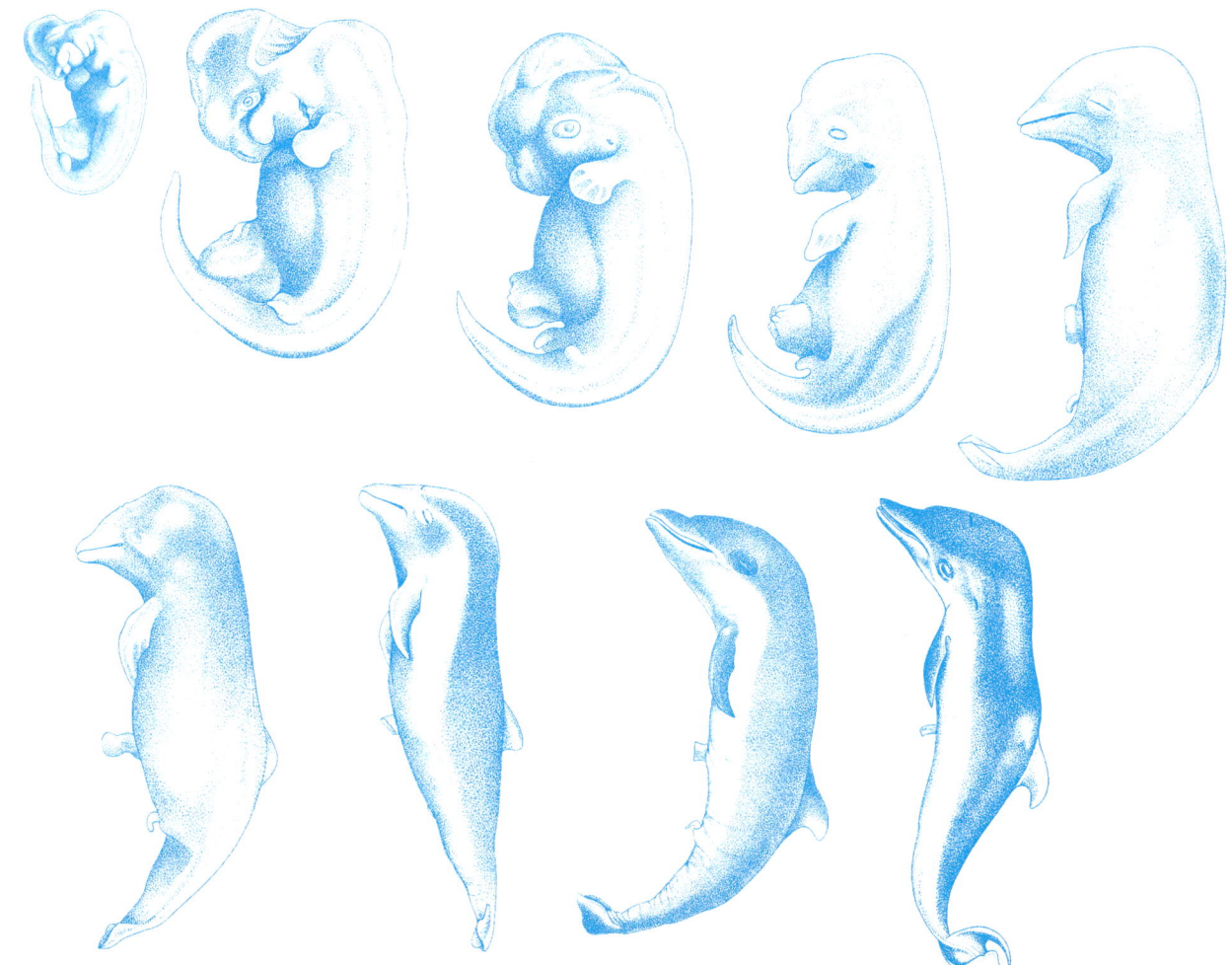

Figure 36.5 Ontogeny of the pan-tropical spotted dolphin, *Stenella attenuata*, to show the progressive development of body form, the flippers, and origin and loss of the hind limb buds between 24 days of gestation (top left) and 205 days (bottom right). The total gestation period is estimated as between 179 and 190 days. In sequence the embryos from top left to bottom right represent 24, 30, 38, 46, 57, 70, 82, 104 and 205 days of gestation. Hind limb buds are evident in the first three stages. Modified from Sterba *et al.* (2000).

Box 36.2 (*Continued*)

Polyphalangy as variant phenotype

A study by Fabrezi and Barg (2001) of 24 species from five anuran families revealed that the pattern of fusion between distal but not proximal phalanges is characteristic of some orders (variation being intraordinal) but that different developmental patterns can result in convergent carpal morphology. In the phalangeal diaphyses of the European common frog, *Rana temporaria*, layers of bone – having survived subsequent endosteal resorption – can be seen up to two years after metamorphosis is completed, and be used to determine age.[c]

Using radiographs of 2550 people as the data-set, Le Minor (1995) investigated the incidence of two or three additional toes (bi- and triphalangy). Triphalangy of toes 2–5 (digits II–V to use the more formal notation, where the big toe is digit I) and biphalangy of the fifth toe (digit V) accounted for 97.5 per cent of the cases (Table 36.3). Biphalangy of toes 4, 3 and 2 in that order accounted for 2.5, 0.2 and 0.12 per cent of the patterns, suggesting a pattern of addition that is more common in postaxial (digit IV) than in preaxial digits (digit II). The primary mechanism was failure of the distal biphalangeal joint to form.

Polyphalangy and regeneration

The ability of tetrapods to regenerate digits is discussed in Chapter 14.

Digits can regenerate in mice but the ability is limited spatially; mice can regenerate the tips of the toes on their forelimbs provided that the amputation is distal and not proximal to the last interphalangeal joint. Msx-1 and Bmp-4 are key players in regeneration of digits of mid-foetal-stage (14.5-day) embryos.[d]

Table 36.3 Occurrence of bi- and triphalangy of human toes[a]

Bi- or triphalangy	Digit(s) involved[b]	Percentage
Triphalangy	II–V	56.5
Biphalangy	V	41.0
Biphalangy	IV	2.5
Biphalangy	III	0.2
Biphalangy	II	0.1

[a] Data from Le Minor (1995) from 2550 individuals.
[b] Digit I is the big toe.

Sesamoids and phalanges

A survey by Russell and Bauer (1988) found extra cartilages (*paraphalanges*) alongside the phalanges where they supported the footpad in 57 species belonging to 16 genera of gecko lizards. No mere nubbins, these broad flexible cartilages extend across most of the portion of the digits that contacts the surface, and can be classified into at least seven patterns suggesting multiple evolutionary origins (Fig. 9.2). In five out of six species examined from the Madagascan genus *Uroplatus* – typified by the leaf-tailed gecko, *U. fimbriatus* – and in the Namibian web-footed gecko, *Palmatogecko rangei*, the most proximal paraphalanges and associated muscles appeared to be specializations associated with digging and climbing life styles.

Fibrocartilaginous sesamoids that develop in the capsule of the proximal interphalangeal joint of human fingers and toes can show a high degree of differentiation, prominent GAGs, and much type II collagen.[e]

Experimental manipulation

As discussed in Chapter 35, once the limb skeleton is specified, limb-bud epithelium suppresses further chondrogenesis. This is especially true of epithelium on the dorsal and ventral surfaces of limb buds and of the interdigital epithelium. Removing dorsal epithelium from limb buds up to H.H. 31 elicits ectopic nodules of cartilage or digits containing two phalanges (Hurlé and Gañan, 1986, 1987).

[a] See Henrikson and Cohen (1965) for a light and transmission electron microscopical analysis of developing interphalangeal joints in chick embryos, especially separation of the adjacent phalanges and histogenesis of the synovial area. Macias *et al.* (1993) demonstrated that retinoic acid, Tgfβ-1, Tgfβ-2 and the kinase-inhibitor staurosporin are all required during development of interphalangeal joints in chick embryos. See Joyce and Cohen (1970) for the interphalangeal joint of the leopard frog, *Rana pipiens*, which is a symphysis with fibrocartilage.
[b] See Bejder and Hall (2002) for evolutionary and developmental gain and loss of limbs in cetaceans, Richardson and Oelschläager (2002), Fedak and Hall (2004) and Richardson *et al.* (2004) for hyperphalangy in cetaceans, and Padian (1992) for phalangeal formulae.
[c] See Fabrezi and Barg (2001) for phalangeal fusion and Guyetant *et al.* (1984) for ossification.
[d] See Becker and Spadaro (1972), Becker (1978) and Borgens (1982a,b) for data and reviews of studies on digit regeneration, and see Han *et al.* (2003) for the roles of Msx-1 and Bmp-4.
[e] See Lewis *et al.* (1998) and Milz *et al.* (1998) for the interphalangeal joint, and Ralphs and Benjamin (1994) for an overview of the joint capsule in normal development, ageing and disease states.

---

fowl – than by disrupting mesoderm, perhaps signifying that 'ectodermal' mutants are more often lethal.[11]

Fraser and Abbott (1971a,b) provided compelling evidence for the role of the AER in *eudiplopodia*, in which distal elements such as digits are duplicated on the dorsal face of the limb bud. These additional skeletal elements arise in response to a second AER, which develops one day later than, and independent of, the original AER. An extended ability of *eudiplopodia* ectoderm to form a ridge until H.H. 23 – wild-type limb bud ectoderm loses this ability at H.H. 18 – is the primary action of the gene. Ectoderm from these mutants can respond to mesenchyme from wild-type, *talpid*[2], or Japanese quail

embryos by producing a secondary AER and distal skeletal duplication.

In *normal* limb-bud development, i.e. in embryos that are wild type for mutations affecting limb development, *Hox-7* and *Hox-8* are expressed in early lateral flank mesenchyme and then only in subridge mesenchyme. Grafting an additional AER to a limb bud induces new sites of expression, i.e. the AER controls expression of these Hox genes in limb mesenchyme. Consistent with a role for *Hox-7* and *Hox-8* in ridge formation, *eudiplopodia* limb mesenchyme has two sets of expression; limb buds in *limbless* embryos show no expression (Robert *et al.*, 1991).

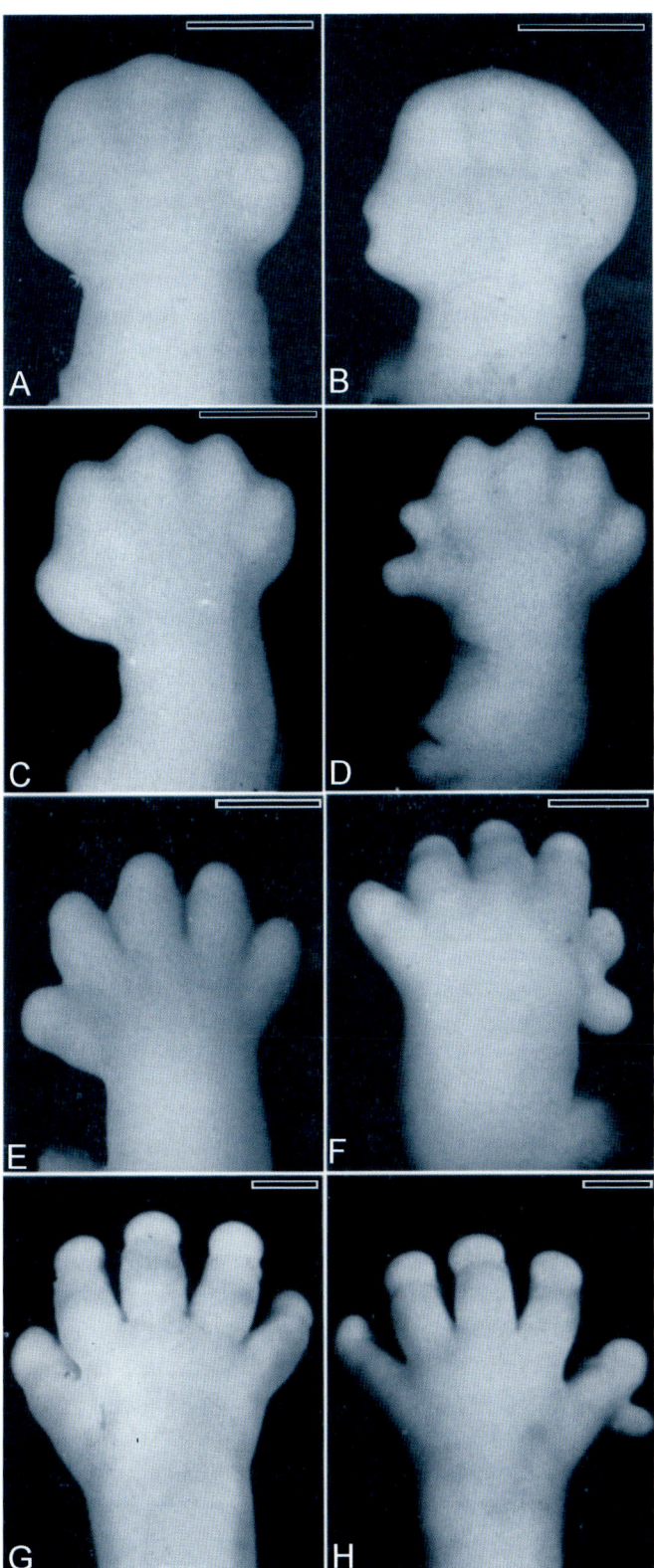

Figure 36.6 Human digit development (left) and preaxial polydactyly (right) as seen in dorsal views of forelimb buds at Carnegie stages 18–23. (A, B) Normal (A) limb bud and preaxial extra digit primordium (B), at stage 18 (44 days post-ovulation). (C, D) Normal limb bud (C) and preaxial extra digit primordium (D) at stage 19 (48 days post-ovulation). (E, G) Hand plates at stages 20 and 23 (50 and 56 days, respectively). (F, H) Polydactylous hand plates at stages 20 and 23, respectively. Modified from Yasuda (1975).

But what about mutations that reduce or subdivide the AER?

Narrow or subdivided AERs

We know of no mutations that narrow or subdivide the AER. Cleft hand deformity in humans might arise because of subdivision of the AER (Yasuda, 1975), although the evidence is not compelling.

AERs grafted side by side onto a limb bud fuse into a giant ridge. A single limb axis with much enlarged skeletal elements results. AERs cultured in contact also fuse. Thus, unless they are far apart, AERs appear not to be able to retain their independence. Another possibility is that subdivided AERs arise but fuse secondarily before the limbs develop; see the section on regeneration of AERs at the beginning of this chapter.

NOTES

1. Hampé (1959), Barasa (1959, 1960), Summerbell (1974b) and Rowe and Fallon (1982) were among those to verify the directive role of the AER. See Abbott (1967, 1975), Kalter (1980), Goetinck (1983) and Johnson (1986) for mutants affecting limb-bud development.
2. See Amprino (1965, 1975a,b) and Amprino and Amprino-Bonetti (1967a,b) for these studies, and see Amprino (1982) for a complete list of his publications and an evaluation by G. Marotti of this work on the occasion of Amprino's 70th birthday.
3. See Barasa (1960) and Janners and Searls (1971) for the wave of cell death after AER removal, Cairns (1975) for the diffusible substance from the AER, and Kerr et al. (1972), Bowen and Lockshin (1981), Maclean and Hall (1987), Hirata and Hall (2000) and Gilbert (2003) for studies or analyses of studies on the significance of cell death in normal and abnormal development.
4. See Abbott (1967, 1975) for general reviews of avian skeletal mutants, Grüneberg (1956), Grüneberg and Wickramaratne (1974), Hall (2000c) and Chapter 43 for the VER, and see Bejder and Hall (2002) and Hall (1984b, 2003a) for retention of organ rudiments when the organ is lost.
5. See Zwilling (1949, 1956b, 1974) and Zwilling and Hansborough (1956) for the classic studies, Sawyer (1982) for basic descriptions of the *wingless* mutant, and Lanser and Fallon (1987) for the musculature.
6. See Hinchliffe and Ede (1973) and Hinchliffe (1977a) for *wingless* (*ws*).
7. See Dealy et al. (1998) for *Tgf-α* expression. Like Egf, *Tgf-α* inhibits chondrogenesis from rat craniofacial mesenchyme, an inhibition that can be reversed by adding Egf receptor antibody (Huang et al., 1996). *Tgf-α* is first detected in mouse craniofacial mesenchyme at day 16. Egf also maintains proliferation of mandibular epithelium at stages when epithelial–mesenchymal interactions are maintaining epithelial proliferation and mesenchymal viability (Coffin-Collins and Hall, 1989; Hall and Coffin-Collins, 1990).
8. See Raynaud (1981, 1986) for the inhibitor study, and Raynaud and Clergue-Gazeau (1986) for digit loss.

9. See Zwilling and Hansborough (1956) for *dominant polydactyly* and *duplicate*, Goetinck (1966, 1983), Ede and Agerback (1968), Ede (1980), Dvorak and Fallon (1991) and Caruccio *et al.* (1999) for *talpid²* and *talpid³*, and Yasuda (1975) for human *preaxial polydactyly*. See Prentiss (1903), Grüneberg (1963) and Johnson (1986) for reviews of polydactyly in humans and domestic animals.

10. I use this shorthand – 'polydactylous limb buds' – and similar phrases such as 'eudiplopodia limb buds' for limb buds from embryos carrying mutation x, where x is *dominant polydactyly, eudiplopodia* or another mutation.

11. See Goetinck (1964) and Fraser and Abbott (1971a,b) for *eudiplopodia*, and Goetinck (1983) for the ectodermal or mesodermal primary action of polydactylous mutants.

AERs in Limbed and Limbless Tetrapods

Bipes canaliculatus or *axolotl con dos patitas* – the salamander with two little feet – is not a salamander but a 'legless' Baja worm lizard with big feet.

The obviously significant function served by the avian apical epithelial ridge (AER) prompted embryologists and then developmental biologists to ask whether such structures and/or epithelial–mesenchymal interaction are a universal property of tetrapod limb buds. As we have seen, absence of a ridge does not necessarily mean absence of epithelial–mesenchymal interactions. Here, I briefly evaluate the other three classes of tetrapods: amphibians, reptiles and mammals.

AERs ACROSS THE TETRAPODS

We can generalize in concluding that localization and development of tetrapod limbs require interaction between presumptive limb mesenchyme and a physiologically – and, in some cases, cytologically – specialized epithelium. Urodele amphibians may be an exception, although, even here, **caution** must be exercised. Experiments on urodele embryos may not have been performed at the appropriate developmental stages. Even in avian embryos, where the AER is such a prominent structural feature, dissociated limb-bud epithelium signals to limb mesenchyme without re-forming an AER.

Amphibians

Anurans

The limb buds of that most popular of model organisms, the South African clawed toad, *Xenopus laevis*, possess an AER. Limb buds of the Puerto Rican coqui, *Eleutherodactylous coqui*, now the model organism for frogs without tadpoles (direct-developers) and in which limbs develop precociously during metamorphosis, do not possess an AER (Box 10.1) and the limb buds continue to develop and grow after the distal limb-bud epithelium is removed. A zone of polarizing activity (ZPA) is present – it induces additional digits when transplanted to the anterior face of a limb bud – but loses its activity much earlier in limb-bud development than is typical for birds or mammals.[1] Which, if either, represents the 'typical' anuran pattern?

An early study by Tschumi (1957) showed that development of *Xenopus* limb buds continues after surgical removal of the ridge but that the resulting limbs lack digits. If the ridge is rotated by 180° and then replaced, digits do form but their orientation is abnormal. Either the *Xenopus* ridge serves a different function than the AER in birds – in which the AER controls outgrowth and skeletal patterning – or the ridge in *Xenopus* was not removed early enough to reveal its full role. It appears that the ridge in *Xenopus* (and limb-bud epithelium in coqui; see Box 10.1) specifies only the digits, not more proximal skeletal elements, although a complicating factor is that the AER regenerates. In fact, regeneration is so rapid that the limb epithelium has to be removed repeatedly in experiments where absence of the epithelium is being tested (Tschumi, 1957). We may not have uncovered the full role of the AER in *Xenopus* limb development.

Ultrastructural analysis of limb-bud development in *Xenopus* indicates that presumptive limb mesenchymal cells arrive at the epithelium at the site of the future limb and accumulate within a complex ECM directly beneath the epidermis. This ECM is then partially degraded and rearranged, allowing close contact between epithelial and mesenchymal cells, following which specialized junctions form between mesenchymal cells. A similar alteration of morphogenetic behaviour of mesenchymal cells occurs when *Xenopus* limb buds are cultured on isolated basement lamellae. What appears to be an equivalent disorganization of subepidermal basement membrane and interaction between ectoderm and the neural folds may also be a necessary condition for the induction of supernumerary limbs of the common toads in the genus *Bufo*.[2]

Urodeles

Although Lauthier could find no enzymatic specificity in the epithelium of the limb buds of the European salamander, *Pleurodeles waltl*, penetration of filopodia from the mesenchyme into the basal layer of the multi-layered epithelium is consistent with interaction. Neither this species nor the newts *Notophthalmus cristatus* and *N. vulgaris* have a morphologically specialized AER. Indeed, replacing *Pleurodeles* limb epithelium with tadpole dorsal or ventral fin epithelium has no adverse effect on proximo-distal skeletal development. The *impression* is either that the limb epithelium has a limited role to play, or that its role is permissive and can be duplicated by other epithelia. In either case, the situation is very different from anurans in which the ridge appears to control digit formation. The situation may be even more different from birds in which the AER controls outgrowth and patterning. Furthermore, separation of digital primordia is based on differential growth in urodeles but on apoptosis in all other tetrapods, further distancing the mechanisms of limb development in urodeles.[3]

Reptiles

Limbed reptiles such as lizards have an AER on their limb buds (Milaire, 1957). *Limbless reptiles*, including snakes and legless lizards, develop hind limb buds with an active AER, at least during the early developmental stage of limb development. The AER is not maintained and regresses. With AER regression, the limb bud also regresses. See the section on limbless tetrapods below for further discussion.

Mammals

Mice

Almost single-handedly over the 35 years between 1956 and 1991, Jean Milaire of Brussels University carried out studies on mouse limb buds that provided our basic information on mammalian limb development. Other studies at that time dealt primarily with limb mutants; the studies by Hans Grüneberg on mouse mutants spanned almost the same period, 1954 to 1974. Both workers published monographic treatments of their work, Grüneberg in 1963, Milaire in 1965. Milaire gave us patterns of normal development, Grüneberg the membranous skeleton (Chapter 19) and deviations from normal in mutants affecting the murine skeleton.[4]

Milaire took a morphological, cytochemical and histochemical approach in his search for the metabolic and biochemical bases of mouse limb development. His research led to the discovery of a functional AER in mouse limb buds (and a VER on the tail, Chapter 43). The ridge arose at a stage consistent with the possibility of interaction with limb mesenchyme. An important stimulus to further study of limb-bud development in mammalian (especially primate) embryos in the early 1960s was the discovery of the devastating effects on limb development of the administration of thalidomide to pregnant women; see below.

Excise the ridge from the limb buds of mouse embryos younger than 10.5 or 11 days of gestation and the limb cartilages fail to form. Removing the ridge at or later than 11 days of gestation – a developmental time coincident with the precartilaginous stage of limb development – has no effect on chondrogenesis. As we know that filopodia from mesenchymal cells contact the basement membrane underlying distal limb epithelium, interaction may require close contact between the two cell layers.

A requirement for interaction with epithelium to initiate chondrogenesis may not be specific to limb epithelium or to the AER; mesenchyme from limb buds from 10-day-old embryos will grow and chondrify when in contact with spinal cord. Furthermore, non-ridge limb epithelium may direct proximal skeletal differentiation, with the AER governing distal organization, which is quite a different situation from that seen in chick limb buds. Experimental verification of this divergence comes from recent studies on *Engrailed-1* (see below).[5]

In-vitro analyses have been used to study chondrogenesis from mouse limb mesenchyme. Owens and Solursh (1981) determined the basic timing of chondrification of forelimb bud mesenchyme isolated from embryos of Theiler stages 15–21 (9.5–13 days of gestation) maintained *in vitro*. Mesenchyme from embryos as young as stage 16 forms cartilage but the amount declines with embryo age, indicative of some change in subpopulations of cells. Dependence on limb-bud epithelium is therefore over by Theiler stage 16 or 10 days of gestation.

As in embryonic chicks, mouse flank ectoderm is competent to form an AER in response to Fgf-4, while flank mesoderm (extending as far anteriorly as into the neck) is competent to produce a polarizing signal associated with onset of *Shh* or induction of *Shh*. Further, Tanaka *et al.* (2000) showed that chick tail mesenchyme can contribute to digits, i.e. be drawn into the limb field.

In contrast to what we see in chick limb buds, mouse limb buds have two lineage boundaries in future limb-bud epithelium. One is a transient boundary at the limit of expression of *Engrailed-1* (*En-1*), a position that is equivalent to the dorso-ventral midline of the AER. The other is equivalent to the dorsal margin of the AER. R. A. Kimmel *et al.* (2000) showed that only some of the epithelial cells that express AER-specific genes contribute to the AER, *En-1* playing the role of setting the AER boundary.

En-1 provides the signaling to position and maintain *both* lineage boundaries; misexpressing *En-1* in the dorsal region of the AER shifts the AER dorsally, while high levels of *En-1* prevent formation of an AER. Homozygote and hemizygote embryos expressing an *Msx-2/En-1* transgene have truncated fore- and hind limbs. Homozygotes truncate the forelimbs at the proximal stylopod and the hind limbs at the proximal femur, while hemizygotes lose digits IV and V from the forelimb and sometimes lose the entire autopod from the hind limbs. *En-1*$^{-/-}$ mutant mice die just after birth, development of the digits of the forelimbs having been disrupted (Wurst et al., 1994).

Chimaeras

Similarities in the early stages of limb development among tetrapods mean that tissues from different organs or taxa can be combined. We saw examples of recombination between mutant and wild-type epithelia/mesenchyme when discussing mutants affecting chick limb development in Chapter 36. Similar recombinations using avian and mammalian tissues create bird–mammal limb chimaeras (and can be used to evoke murine skeletogenesis using chick epithelia or *vice versa* (Figs 17.5 and 37.1).

Chick–mouse chimaeras were created by grafting distal mesenchyme from limb buds from younger and older embryonic mice beneath the epithelium of limb buds of chick embryos of H.H. 18–20 (Table 37.1). The younger mouse epithelium, which according to Cairns (1965) was equivalent in development to limb buds in chick embryos of H.H. 18–24 (a rather wide range) allowed development of mouse-like terminal digits on the chick limbs. (In comparison, growth slows and digits fail to form when mesenchyme from older mouse embryos is grafted.

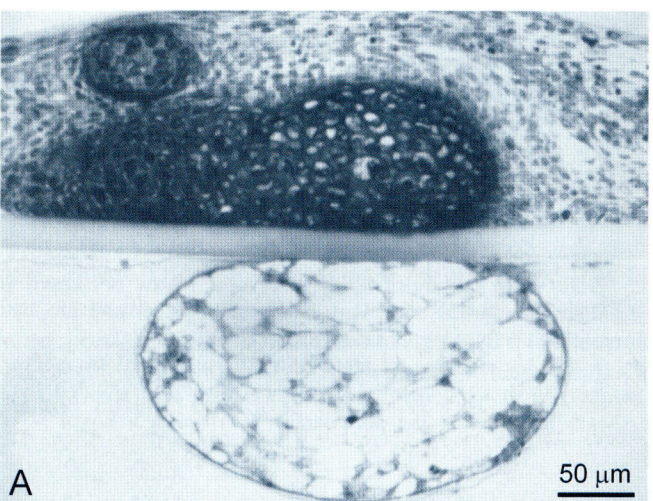

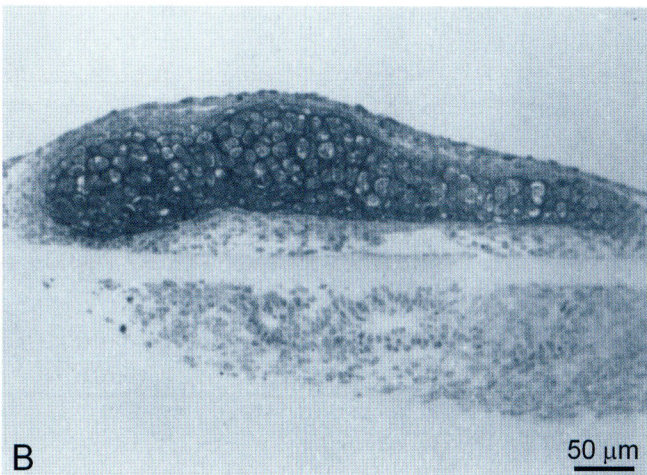

Figure 37.1 Notochord and spinal cord induce somitic mesenchyme to chondrify. (A) Cartilage (above) has developed in somites from a mouse embryo cultured on the opposite side of a filter on which H.H. 33 chick notochord was placed. (B) Cartilage (above) has developed in somites from a mouse embryo cultured on the opposite side of a filter on which ventral spinal cord from a nine-day-old mouse embryos was placed. Adapted from Cooper (1965).

Table 37.1 Results of transplanting tissues from mouse limb-buds onto chick wing buds of H.H. stages 18–20[a]

Transplant	Recognizable toe	Mouse cells in apical region	Mouse cells present but only proximally	Mouse cells present but not in apical region or wing amputated[b]
Mouse limb-bud apical zone transplanted to dorsal surface of chick wing bud	4/4	–	–	–
Mouse limb-bud mesoderm	–[c]	11/12	1/12	–
inserted under stretched AER of chick wing bud	–[d]	5/6	1/6	–
Older limb mesoderm or maxillary process	–[e]	1/7	–	6/7
mesenchyme inserted under stretched AER of chick wing bud	–[d]	1	–	2

[a] Based on data from Cairns (1965).
[b] Wing buds were short and truncated, as if the AER was removed.
[c] Chicks were fixed up to 54 hours after transplant.
[d] Chicks were fixed after skeletal morphogenesis.
[e] Chicks were fixed up to 8 hours after transplant.

Maxillary mesenchyme also fails to respond under similar conditions.)[6] Chick and mouse mesenchymal cells did not intermingle. Either this confirms lack of large-scale migration of mesenchymal cells within the limb bud or reflects incompatibility between mouse and chick cells. Cairns concluded that, in order to chondrify, mouse limb-bud mesenchyme has to interact with limb-bud epithelium, and that the interaction is time dependent.

Chick tail-bud mesenchyme can contribute to digits if transplanted under the AER in the posterior face of a limb bud at H.H. 20, the tail mesenchyme forming additional phalanges on existing digits but not additional digits (Tanaka *et al.*, 2000).

Humans

An AER appears on the limb buds of human embryos at around five weeks of gestation (Streeter stages 15 and 16). The active AER of human limb buds can be distinguished from adjacent ectoderm on ultrastructural grounds. Gap junctions and bundles of microfilaments characterize the ultrastructure of the inductively active AERs on avian and mammalian (including primate) limb buds. By the seventh week (Streeter stages 17 and 18) of human development, the AER has thinned in areas where it is underlain by necrotic mesenchyme.[7]

LIMBLESS TETRAPODS

Tetrapods are vertebrates with four limbs with digits, arranged in two sets, one anterior and one posterior. *Limbless tetrapods* – if that is not an oxymoron – are tetrapods without limbs. We know they are tetrapods because they nest phylogenetically within Tetrapoda and are derived from ancestors with limbs, i.e. from tetrapods.

Many tetrapod taxa in several major groups – amphibians (Caecilia), reptiles (Ophidia, Amphisbaenia) and mammals (Cetacea) – lack one or both sets of limbs as adults, and/or have only rudimentary elements of limbs or girdles. I exclude amphibian tadpoles from this list because frogs have limbs. I also exclude flightless birds because all have wings, although the wings may be much reduced. There are wingless avian mutants (Chapter 36) but no wingless avian taxa. No birds or mammals (other than cetaceans) have lost their hind limbs.

Evolutionary patterns

Limb loss evolved many times within tetrapods. Caecilians, snakes, amphisbaenians, legless lizards and whales are major examples discussed below and in Chapter 44. Limblessness is always secondary – a reduction from an ancestral limbed condition or a rudimentary form of an ancestral condition. The bases of such evolutionary changes are discussed in the last chapter.[8]

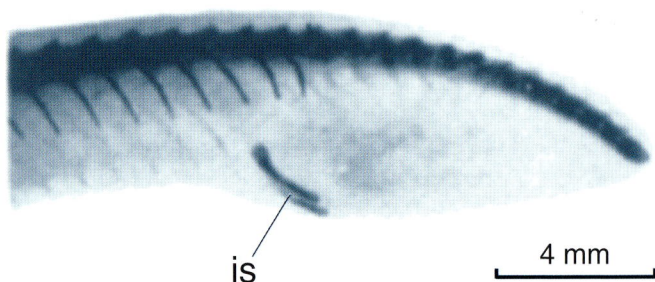

Figure 37.2 An alizarin-red-stained preparation of the posterior half of the body of the worm snake, *Typhlops jamaicensis*, showing vertebral ribs and the well-ossified ischium (is), which lack connection with the vertebral column. Ilium and pubis, which are variably present, are usually cartilaginous. Modified from Evans (1955).

Limb reduction may be *complete*, by which I mean that no limb or girdle elements are present in adults, or *partial*, in which one or more rudimentary skeletal elements is present, the assumption being that such elements are rudiments of the more complete limbs/girdles possessed by ancestral taxa. An alternative but less parsimonious explanation is that the limb skeleton was lost completely in the ancestors and that new elements (neomorphs) arose in the descendant.[9]

Complete loss is seen in the forelimbs of snakes, the hind limbs of whales and both sets of limbs in caecilians.

One example of partial loss, the South African skink, *Scelotes brevipes*, has an ossified femur, while individuals within a sister species, *S. gronopii*, have ossified but reduced femora, tibiae and fibulae. Neither skink has any forelimb elements. In describing the osteology of the worm snake, *Typhlops jamaicensis*, Howard Evans of Cornell University, one of the most knowledgeable comparative anatomists on the planet, noted that an ossified ischium along with cartilaginous rudiments of the ilium and pubis lie as rudiments free of the vertebral column (Fig. 37.2).

Haasiophis terrasanctus is a limbed fossil snake from the Middle Cretaceous (95 mya) of what is now the Middle East. *Haasiophis* is terrestrial, allied to pythons and boas, and advanced, i.e. it is not an early or primitive limbed snake (or, in cladistic analyses, *Haasiophis* nests among modern snakes, sharing a close affinity with pythons and boas). Presence of legs in such an advanced (deeply nested) snake raises a number of issues, including the possibility that there has been more than one origin of hind limbs in tetrapods, and/or that the limbs in *Haasiophis* may have evolved from rudimentary limb buds (see Fig. 37.5) or rudimentary limbs, as seen in python (see Fig. 37.6).[10]

Gaining limbs back

Reversals from the limbless condition in living taxa, albeit rare, are known.

We have to be careful when we speak of reversals. The reappearance of limbs in a taxon whose ancestors were limbless certainly is a reversal in the sense that the adult features were lost and regained. However, given the presence of limb buds in limbless tetrapods, the reappearance is not a reversal but the extended development of embryonic rudiments that are present in the ancestors. What could be regarded as the beginnings of such a reappearance are the occasional atavistic elements found in isolated individuals within a population (Chapter 44).

Identifying a phylogenetic reversal – i.e. a return to an earlier condition – requires a well-resolved phylogeny and expert knowledge of the group. Crumly (1990) described three species of the Baja worm lizards, *Bipes*, with legs. *Bipes* inhabits Baja California and parts of western Mexico, where the locals call them *axolotl con dos patitas* – the salamander with two little feet.

Salamanders they are not. The three species (*Bipes canaliculatus*, *B. biporus*, *B. tridactylus*) are in reality legless burrowing 'worm lizards' (amphisbaenians) but they are legless worm lizards with legs, in the same way that snakes are tetrapods (limbed vertebrates) without limbs. *Nor are the feet little.* These are not mere rudiments or vestiges but powerful front legs with claws used for digging and tunneling.

All 130 species of amphisbaenians burrow, using their bullet-shaped heads to advance through soft soil. With regard to the reappearance of such features, it is interesting to note that Papenfuss (1982) reported digit reduction in individuals of *B. canaliculatus* from Petacalco, Mexico. The interpretation that taxa in the group went from limbed to limbless and back to limbed can only be made in the context of a robust phylogeny. *Bipes* is now known to be the most basal of the amphisbaenians. Its basal position could imply that it lay on a lineage from whose more derived members limbs would be lost, but whose basal members had not yet lost their limbs, rather like a legged snake or whale with hind limbs.[11]

Ecological correlates of limblessness

Pretty much wherever it is found within the tetrapods, limblessness is correlated with one or more of:

- body elongation, often as a result of increase in vertebral number;
- reduction in body diameter and mass;
- a sheltering or burrowing mode of life (a category that includes some birds but excludes many snakes); and/or
- locomotion by lateral undulation, again with the exception of all birds and with exceptions among snakes.

Although first-rate functional morphologists like Carl Gans and Russell Lande have examined these points and other anatomical and ecological modifications of limblessness, the selective pressures that lead to limb reduction remain in some doubt.[12]

For example, body elongation and increase in vertebral number don't always go hand in hand. Increase in vertebral number has a more complex explanation than elongation of the body, as seen in body elongation by increase in the length of trunk vertebrae in the salamander *Lineatriton*, and from an examination of the correlation between numbers of body and tail vertebrae in male and female snakes, some lineages showing a negative correlation, some a positive correlation, some no correlation at all.[13]

Reduction that is less extreme than limblessness can be and perhaps usually is driven by functional changes. Limb reduction in the Italian three-toed skink, *Chalcides chalcides*, is associated with (driven by?) a shift from limb to trunk locomotion. The remaining digits are II, III and IV. Digit I is missing and digit V is rudimentary (Greer *et al.*, 1998).

Wiens and Slingluff (2001) studied 27 species of anguid lizards to understand correlations between increased body length and limb reduction and between limb reduction and digital reduction, both of which they found. They did not, however, find any evidence of a strict sequence of increased body length → limb reduction → digit reduction. Nor was a burrowing habit a necessary precondition for limb reduction. The authors disagree with the Hox-driven model of limb reduction in snakes proposed by Cohn and Tickle.[14]

In perhaps the best overview available for a single genus of reptiles until the work of Michael Shapiro this century, Presch (1975) evaluated the distribution of limb reduction in the worm lizard *Bachia*, a genus of some 15 species from Costa Rica, Panama, northern Colombia, Venezuela, the Guianas and the Amazon basin of Brazil, Ecuador, Peru, and Bolivia. Limb reduction in *Bachia* correlates with increasing body length. The longer the body the greater the limb reduction (Fig. 37.3). Reduction of the hind limbs has progressed furthest; no species shows any external sign of hind limbs. Forelimbs are reduced to some two or three mm in length and are visible externally (Fig. 37.4). Presch regards the loss as explained by paedomorphosis, and compared the pattern with that seen in other genera with limb reduction.

In a more recent study, Shapiro (2002) detailed limb reduction (including variation and/or reduction in digits) in the three species of the scincid lizard *Hemiergis* from Western Australia, in which anywhere from two to five digits may be present. The two-toed earless skink, *H. quadrilineata*, has two digits on fore- and hind limbs, the lowlands earless skink, *H. peronii*, has three on the fore- and four on the hind limbs (three specimens had an atavistic digit V), while the earless skink, *H. initialis*, has five digits on each limb. Digit number was not a simple consequence of changing limb length, although both parameters correlated with habitat and burrowing life styles. Neither truncation of development nor heterochrony explained the observed patterns. Subsequent analysis showed a correlation of reduced

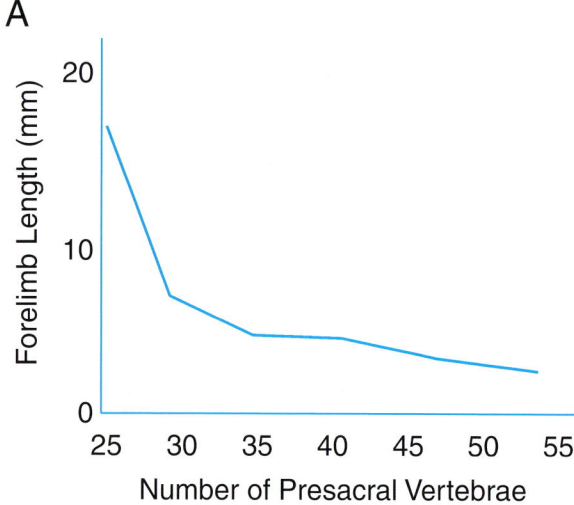

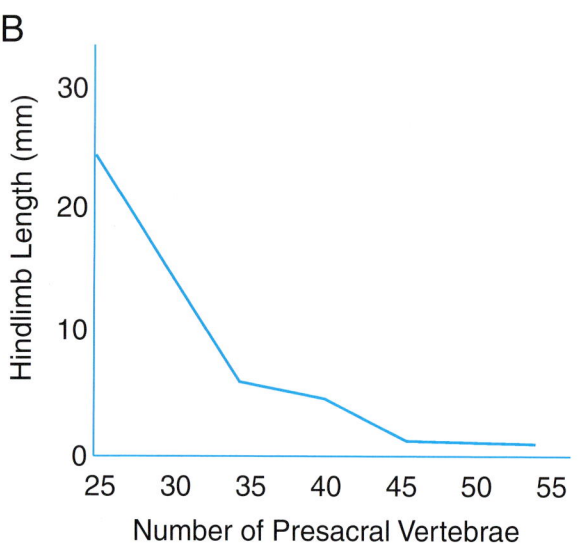

Figure 37.3 Limb reduction and body elongation are correlated as shown in these plots of forelimb (A) and hind-limb (B) length against a measure of body length – the number of presacral vertebrae – for 13 species of the South American worm lizard, genus *Bachia*. From data in Presch (1975).

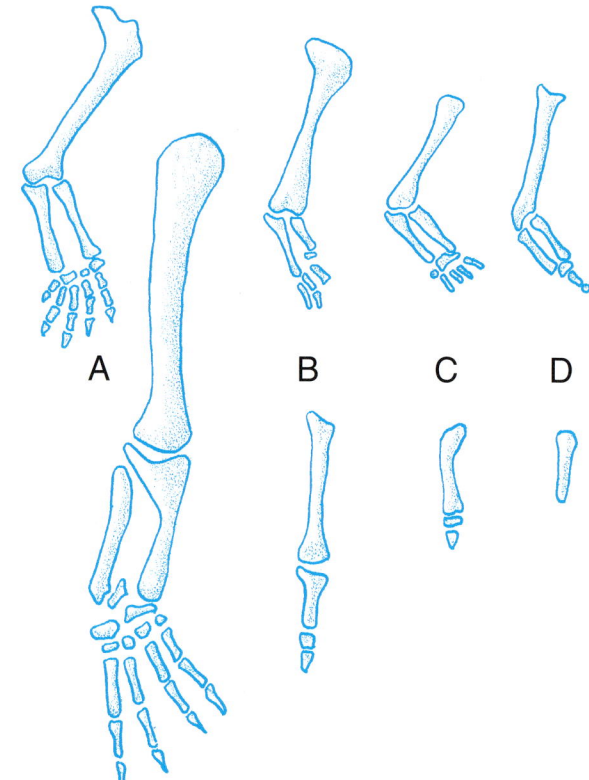

Figure 37.4 Progressive reduction in fore- and hind limbs of South American worm lizards in the genus *Bachia*. Forelimbs above, hind limbs below. (A) *Bachia heterotopa* has unreduced fore- and hind limbs. (B) *Bachia dorbignyi* shows loss of digits in the forelimb and reduction to a single digit and loss of the fibula in the hind limb. (C) *Bachia monodactylus* has even more reduced hind limbs. (D) *Bachia trisanale* has only a single digit in the forelimb and only a reduced (rudimentary) femur in the hind limb. Modified from Presch (1975).

expression of *Shh* with decreased cell proliferation and decreased digit number in the limb buds and limbs, indicating that *Shh* may have played an important role in digit loss in the three closely related species. Expression of *Msx* and *Dlx* in the limb buds of the three species was unchanged, ruling out a role for either of these genes in digit loss (Shapiro *et al.*, 2003).

Some of this complexity in limb morphology also is reflected in vertebral response to tail amputation, which we expect to involve addition of new vertebrae from a blastema at the amputation site. Vaglia *et al.* (1997) demonstrated that tail regeneration in the larval four-toed salamander, *Hemidactylium scutatum*, is *associated with increase in the length of tail vertebrae anterior to the injury site*, raising whether we can speak of regeneration

when the lost tissue is effectively not replaced but compensated for.

The developmental basis of limblessness in snakes and legless lizards

Most – but certainly not all – limbless tetrapods possess limb buds early in embryonic development and so resemble wingless and limbless mutants. Limblessness represents *arrested limb development* and subsequent *regression of the limb bud, not regression of a fully formed limb or failure of limb buds to develop at all.* As such, limblessness provides an excellent signpost to the need for continued interactions between epithelium and mesenchyme for normal and complete limb development. Indeed, as you will see from the studies below, limblessness results from the inability to maintain these interactions.

Publications on limb reduction in reptiles go back to at least the early 19th century, to studies by such founders of vertebrate morphology as Cuvier (1825) and Gegenbaur (1864–1865). Sewertzoff (1931) may have been the first to seek an embryological explanation for evolutionary limb reduction in reptiles, concluding that limb loss is progressive; reduction in the overall size of the limb is followed by loss of the last elements to form (distal elements) and then loss of progressively more proximal elements. Given how limbs develop (Box 37.1), loss of limb elements could not be otherwise without major disruption.[15]

Development of limb buds in snakes and legless lizards has been studied most extensively by Albert Raynaud and his colleagues in France. Their publications – which span 1972–1993 – provide detailed analyses of the slow worm, *Anguis fragilis*, and its close relative the legless lizard or glass snake, *Ophisaurus apodus*; the green lizard, *Lacerta viridis*; the reticulated python, *Python reticulates*, and its relative, *Tropidonotus tessellata*; and two South African skinks, *Scelotes brevipes* and *S. gronopi*. One of the nice features of these studies is the comparison that Raynaud and his colleagues were able to make of patterns of limb reduction between species.[16]

In *Anguis fragilis* embryos, limb buds arise and an AER forms. Both this ridge and that of *Lacerta viridis* show morphological specialization and higher levels of protein and RNA than in the adjacent epithelium. Some of the fine scanning electron micrographs obtained by Raynaud et al. (1974b) are shown in Fig. 37.5.[17]

Inability to maintain an AER

In limbed and legless lizards, as in other tetrapods, four anterior somites send ventral extensions into the lateral mesoderm (somatopleure) at the level of the future fore-limb buds, and three somites send processes into the mesoderm of the hind limb buds. Regression of limb buds begins at this developmental stage in legless lizards. Cellular degeneration begins in the distal tips of the somitic extensions and is followed by degeneration of the AER – evidenced by the accumulation of substantial numbers of lysosomes – and, finally, degeneration of limb-bud mesenchyme. Raynaud suggested that regression is initiated because too few somitic processes penetrate the somatopleure – because somatopleural mesenchyme lacks a factor(s) required to maintain the processes? In subsequent studies, Raynaud and Kan showed that degeneration of the AER triggers decreased DNA synthesis and decreased proliferation in the adjacent mesenchyme.[18]

An experimental analysis using the European pond turtle, *Emys orbicularis*, provided direct evidence of the necessity of somitic extensions to stimulate proliferation of somatopleural mesenchyme and allow limb development to continue. Removing somite pairs 6–13 from embryos with 20–33 somite pairs prevented any further limb-bud growth. Inserting a mechanical barrier into the path of the somitic extensions also prevented further development (Vasse, 1974, 1977). *Thus, somitic extensions activate the proliferation of limb-bud mesenchyme.*

In the European glass lizard, *Ophisaurus apodus*, a species from the same group (the family Anguidae) as *Anguis fragilis*, three somites send extensions into the future hind limb buds. An AER forms, and the developing limb buds become vascularized and innervated. The AER, somitic extensions and mesenchyme then all necrose, reducing the limb buds to a rudiment in which a rudimentary skeletal rod develops. Similar patterns of somite extensions, formation of an AER and regression are seen in the reticulated python, *Python reticulates*, in skinks, and in the Greek tortoise, *Testudo graeca*.[19] The characteristic pattern in all limbless lizards and snakes so far studied is initiation of hind limb-bud development and subsequent regression.

Experimentation on chick embryos supports a role for somitic cells in the maintenance of limb development. The normal role somites play in tetrapod limbs is disrupted in limbless tetrapods. In chick embryos, somitic cells stimulate limb mesenchyme to proliferate and penetrate the somatopleure, where they contribute myoblasts and fibroblasts to the limb buds (Chapter 16).[20] Coordinated positioning of limb field and 'limb-level' somites is a necessary precondition for subsequent interactions between mesenchyme and epithelium to form the AER.

Box 37.1 Development of reptilian limb skeletons

The following list provides an entrée into the literature on reptilian limb development.

Alligator mississippiensis; development of limb skeleton (Müller and Alberch, 1990; Rieppel, 1993d); growth dynamics of femur (A. H. Lee, 2004).

Blanus cinereus (Iberian worm lizard); vestigial pelvic appendages (Renous et al., 1991).

Calotes versicolor (the crested tree lizard); histogenesis and metaplastic mineralization of humeri and femora, especially condensation, perichondrial and periosteal formation, and sesamoids (Mathur and Goel, 1976; Mathur, 1979).

Chamaeleo hoehnelii (the high casqued chameleon); includes discussion of derived autopodial characters in chameleons (Rieppel, 1993b).

Chelydra serpentina (snapping turtle); development of carpus and tarsus (Burke and Alberch, 1985).

Chrysemys picta (painted turtle); development of carpus and tarsus (Burke and Alberch, 1985).

Cyrtodactylus pubisculcus (gecko); especially homology of the astragalus (Rieppel, 1992a). Epiphyses of lizards (Haines, 1941).

Geckoes; extra cartilages beside the phalanges in 57 species of 16 genera (Russell and Bauer, 1988).

Gehyra oceanica (the big tree gecko); patterns of ossification in the limb skeleton, especially the independence of chondrogenesis and (Rieppel, 1994).

Lacerta vivipara (common lizard); patterns of ossification (Rieppel, 1992b).

Lepidodactylus lugubris (the mourning gecko); patterns of ossification in the limb skeleton, especially the independence of chondro- and osteogenesis (Rieppel, 1994).

Limblessness and the origin of snakes (Rieppel, 1988).

Sphenodon (the tuatara); epiphyses and primitive secondary centres of ossification (Haines, 1939).

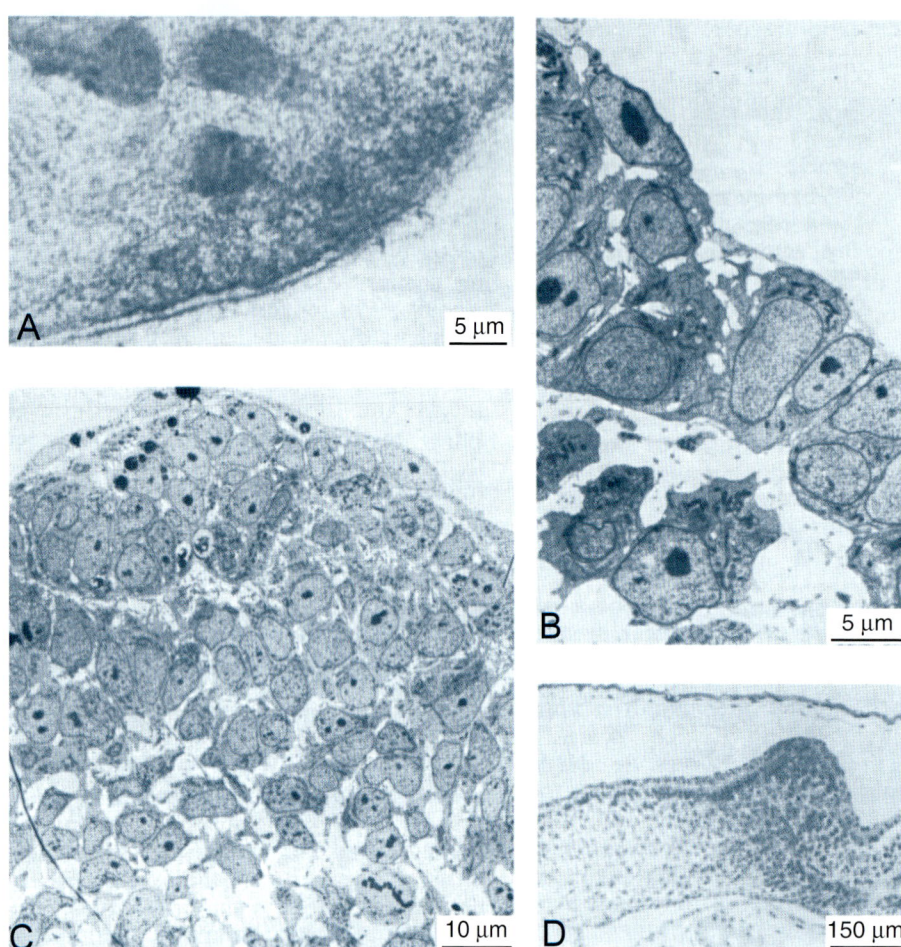

Figure 37.5 Development of limb buds in the slow worm, *Anguis fragilis*, and the green lizard, *Lacerta viridis*. (A) A cell at the apical surface of a limb bud 24 hours before degeneration of the apical ridge shows some initial disruption of the epithelium. (B) A lower-power electron micrograph of the apical ectodermal ridge, again before degeneration. (C) A low-power electron micrograph of the apex of the forelimb bud with evidence of degeneration in the apical ectoderm (black granules, upper left). (D) A limb bud at the onset of degeneration of the AER. Modified from Raynaud *et al.* (1974a).

Another **cautionary** note: We have to be careful when comparing somites across even closely related taxa. There are two more somites in the parachordal plate in *Anguis* than in *Lacerta*. Consequently, the vertebrae in the two genera are not homologous, the third cervical vertebra in *Anguis* corresponding to the fifth cervical vertebra in *Lacerta*.[21]

Molecular mechanisms

Where the genetic basis has been analyzed, limblessness has been found to be polygenic, involving many genes with pleiotropic effects. Consequently, we might expect genes that function within limb buds to be retained if they also function in other organs. Such other organs could be retained or vestigial appendages – flippers in whales (Fig. 31.3), the femur in pythons (Fig. 37.6) – or other organ systems such as teeth, jaws, the heart or genitalia.

Retaining limb buds and major genes involved in skeletogenesis creates the potential for partial or even complete reappearance of limbs as seen in *Bipes* (above). Such reappearance is seen in the atavistic skeletal elements of hind-limb rudiments in whales and snakes (Chapter 44).[22]

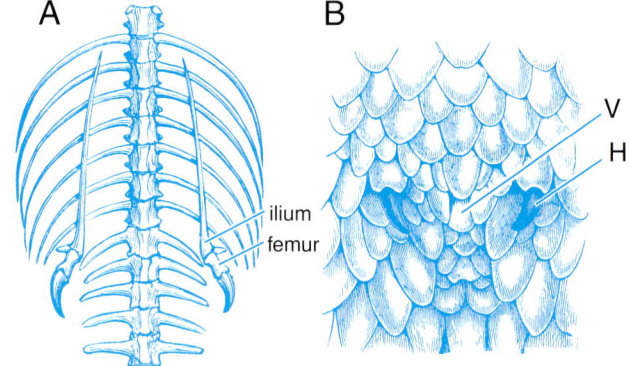

Figure 37.6 Vestigial hind-limb and pelvic-girdle elements in a python. (A) Portion of the skeleton, showing the ilium (a remnant of the pelvic girdle), femur and the spur associated with each femora. (B) The spurs of the vestigial hind-limb skeleton (H) on either side of the vent (V) as seen on the ventral surface of a python. From Romanes (1901).

NOTES

1. See Richardson *et al.* (1998) and Hanken *et al.* (2001) for detailed analyses of limb-bud development in coqui, including data on absence of an AER, activity of the ZPA, and *Dlx* and *Shh* expression in distal epithelium and the ZPA, respectively.

2. See Tarin and Sturdee (1971, 1974) and Kelley and Bluemink (1974) for the ultrastructural studies, Overton (1977) for culture on basement lamellae, and Balinsky (1974) for induction of supernumerary limbs.

3. See Lauthier (1974, 1977, 1978) for *Pleurodeles* (de Ricqlès, 1964, 1965 for the long bones), Sturdee and Connock (1975) for *Notophthalmus*, and Hinchliffe (2000) and Hall (2003b) for digit formation in urodele limb buds.

4. See, for example, Milaire (1956, 1963) and his monographic volume published in 1965, and Grüneberg (1956), Grüneberg and Wickramaratne (1974) and Grüneberg's monographic volume published in 1963.

5. See Milaire and Mulnard (1968, 1984) for excision of the AER, Milaire (1965) for recombinations with non-ridge epithelium, and Houben (1976) for epithelial–mesenchymal contact via filopodia.

6. Of course, many factors may have contributed to the negative result (lack of chondrogenesis). For instance, the maxillary mesenchyme may have been too old or perhaps limb epithelium promotes chondrogenesis but not osteogenesis (maxillary mesenchyme forms intramembranous bone). Mandibular mesenchyme would have been a better population to graft.

7. See O'Rahilly *et al.* (1956) for timing of human limb-bud development, Kelley (1975), Kelley and Fallon (1976), Fallon and Kelley (1977) and Kaprio (1977 for ultrastructure and gap junctions, and Swinyard (1969) and Bergsma and Lenz (1977) for reviews of human limb-bud development.

8. See Rieppel (1988) for a discussion of limb reduction in the context of the origin of snakes, Bejder and Hall (2002) for the loss of limbs in snakes and whales, and Hall (2003a) for rudimentation within the broader context of rudiments, vestiges, homology and homoplasy.

9. See Chapter 44, Hall (1984b, 1998a, 2003a) and Bejder and Hall (2002) for further discussions of these alternatives.

10. See Greene and Cundall (2000), Tchernov *et al.* (2000) and Rieppel *et al.* (2003) for descriptions and discussion of *Haasiophis terrasanctus*. Rieppel and Kearney (2001) review the origins of the 2700 extant snake species.

11. See Kearney (2003) for an overview of the appendicular skeleton in amphisbaenians. Earlier reports that either the first or the fourth/fifth digits were reduced in *B.canaliculatus* depending on location were not confirmed by Papenfuss (1982), who found that the fifth digit is always reduced, with digits I–IV bearing well-developed claws.

12. See Gans (1974, 1975), Lande (1978), Greer (1987), Hall (2002), Shapiro (2002) and Caldwell (2003) for overviews of the ecological and evolutionary aspects of limblessness.

13. See Parra-Olea and Wake (2001) for *Lineatriton* and Shine (2000) for the snakes.

14. Also see Graham and McGonnell (1999) for a commentary on Hox model, and see Bejder and Hall (2002) for further evaluation of Hox-driven models associated with appendage loss, in this case, in cetaceans. Another possible genomic change is loss of a limb-specific *cis*-acting regulator of *Shh*. This element undergoes base substitutions in the mouse mutant *Hemimelic-extra toes* (*Hx*) – which is characterized by preaxial polydactyly – and is absent from such limbless tetrapods as the Japanese four-lined rat snake, *Elaphe quadrivirgata*, the Japanese rat snake, *Elaphe climacophora*, and from a limbless newt, the rubber eel, *Typhlonectes*, a caecilian (Sagai *et al.*, 2004).

15. See Presch (1975) and Shapiro (2002) for overview of limb loss in the lizard genera *Bachia* and *Hemiergis*, respectively, and see Bejder and Hall (2002) for hind-limb loss in whales.

16. See the papers discussed in Raynaud *et al.* (1989), Raynaud (1990) and Raynaud *et multi al.* For overviews, see Raynaud (1977, 1990), Cohn and Tickle (1999), Bejder and Hall (2000) and Shapiro (2002).

17. See Raynaud *et al.* (1974a, 1975, 1979), Raynaud and Adrian (1975a,b) and Raynaud and Brabet (1979) for the presence and nature of the AER, and Raynaud *et al.* (1974b) for scanning electron microscopy of the AER.

18. See Raynaud (1972) for the proposal that too few somitic cells enter the limb buds, Raynaud *et al.* (1973a,b, 1974a) for cellular degeneration, Raynaud (1977, 1990) for reviews of somitic deficiency and premature loss of the AER, and Raynaud and Kan (1992) for data on DNA synthesis and cell proliferation.

19. See M-Z Rahmani (1974) for the studies on the glass snake, *Ophisaurus apodus*, and Raynaud (1974), Raynaud *et al.* (1974c), Pieau and Raynaud (1976) and Raynaud and van den Elzen (1976) for the studies on the python and turtles.

20. See Pinot (1970) and Kieny (1971) and Chapter 16 for experimental studies using chick embryos.

21. See Raynaud *et al.* (1990) for the studies with *Anguis* and *Lacerta*, and see Box 35.2 for a discussion of how fore- and hind limb buds are positioned with respect to somite number along the embryonic axis.

22. See Rieppel (1988) for the origin of snakes, including a discussion of limb reduction, and see Nopsca (1923) and Lande (1978) for atavisms in reptiles in the context of discussions of the reversibility of evolution and the evolution of limblessness.

Part XIII

Limbs and Limb Skeletons

Limbs as appendages and the appendicular skeletons they contain are patterned in time and space. Ultimately, all factors contributing to morphogenesis do so by regulating one or more of a finite and surprisingly small number of processes: cell proliferation, differentiation, selective activation or inhibition of osteo- and chondroprogenitor cells, synthesis of cellular products and the deposition/removal of those products from extracellular matrices. The fact that all these processes are multidimensional presents a major challenge to the analysis of morphogenesis. So too does the fact that these same processes underlie growth.

Axes and Polarity

The blueprint for skeletal form contained within the genes is expressed neither rigidly nor without reference to context.

ESTABLISHING AXES AND POLARITY

A substantial amount of evidence exists on how the major axes and polarity of developing limbs buds are determined and expressed (Figs. 35.1 and 35.2). Limb buds, the limbs that develop from them and limb skeletons all share three major axes of symmetry: *antero-posterior* (*A-P*), *dorso-ventral* (*D-V*) and *proximo-distal* (*P-D*).

The *antero-posterior* (A-P) axis corresponds to the primary A-P embryonic axis. Hold your arms out at right angles to your body as if being 'wanded' in an airport security line. Flare your digits. Your thumb should now be pointing toward your head, the most anterior position of the A-P axis. Your little finger (pinkie, digit V) is pointing posteriorly. The A-P axis runs through the digits, digit I being the most anterior, digit V the most posterior.[1]

The *dorso-ventral* (D-V) axis fixes the dorsal and ventral aspects of limb buds and the skeletal and muscular tissues that develop. Like the A-P axis, the D-V axis is fixed early in development, long before any localization or outgrowth of limb mesenchyme; i.e. before H.H. 8 (ca. 28 hours of incubation) for the A-P axis and before H.H. 11 (ca. 42 hours of incubation; Fig. 16.2) for the D-V axis.

The A-P axis is established by properties within limb mesenchyme, the D-V axis by properties within the epithelium of the limb bud; when pieces of limb mesenchyme are wrapped in epithelial jackets and grafted into the somites, or if limb-bud epithelium is rotated, the skeletal structures that form have an A-P axis corresponding to the orientation of the mesenchyme and a D-V axis corresponding to the orientation of the epithelium (see below).[2]

The third axis, *proximo-distal* (P-D), corresponds to the direction of limb outgrowth away from the body, where distal is furthest from the body, and proximal closest to the body. The AER plays a major role in determining the P-D axis; remove the ridge and any further distal limb-bud outgrowth ceases, as does the development of more distal limb structures (Chapter 35). Unlike the A-P and D-V axes, the P-D axis is determined progressively and *after* development of the limb bud begins, in chick embryos between H.H. 16 and 28.

THE A-P AXIS AND THE ZPA

Initial studies on these axes formed part of the classic study on amphibians carried out by Ross Harrison in 1918. Rotating a limb field of the urodele *Ambystoma* at the tail-bud stage – so that the A-P and D-V axes of the field no longer correspond to those of the body – resulted in formation of a limb with its A-P axis reversed *with respect to the body* but with a normal D-V axis (Fig. 35.2). The A-P axis had already been fixed, the D-V axis had not, and so conformed to the D-V axis of the host. Similar experiments with avian embryos give similar results. The A-P axis is determined by properties within the limb mesoderm, the D-V axis by properties within the ectoderm.[3] Information on A-P polarity lies in the '*zone of polarizing activity*,' affectionately known as the ZPA, and which occupies the proximal and posterior region of a limb bud, overlapping with the zone of cell death known as the posterior necrotic zone or PNZ.[4]

The A-P axis is determined rapidly.

When an area of proximal and posterior bud mesenchyme corresponding to the location of the PNZ of the wing (or the equivalent but non-necrotic area of the leg) is grafted beneath the epithelium anteriorly in a second bud, an accessory limb bud develops immediately *posterior* to the graft. The posterior face of this new limb bud always faces the grafted mesenchyme. Other areas of limb mesenchyme or dissociated limb mesenchyme lacking the ZPA produce a distorted limb bud without recognizable A-P polarity, whereas dissociated ZPA implanted anteriorly elicits a normal limb.

Similarly, a 180° rotation of the distal tip of a wing bud results in duplication of the wing *only* if the posterior zone is included. The ZPA functions only for a short time; removing it between H.H. 15 and 24 has no effect on limb development, although this may be because the ZPA was not removed completely. Excising the posterior portion of a limb bud – which removes the entire ZPA – results in the development of a single element anteriorly, taken by Hinchliffe and Gumpel-Pinot (1981) as evidence for action of the ZPA *in ovo*.[5]

Tickle *et al.* (1975) published a set of elegant experiments to test whether a gradient of morphogenetic information emanates from the ZPA. Digits arise from mesenchyme that can be traced back and fate-mapped to the posterior half of the wing bud adjacent to somites 18 and 19. The ZPA lies immediately posterior to this zone, at the level of somites 19 and 20.

If the ZPA represents the posterior portion of a gradient diffusing anteriorly, then grafting a ZPA into the anterior face of a limb-bud should induce supernumerary digits, with the posterior digit located closest to the ZPA. If grafted sufficiently posterior in the limb bud, only posterior supernumerary digits should form. Both hypotheses were confirmed when ZPAs from H.H. 18–20 embryos were grafted. Tickle and co-workers calculated that the gradient covered 500–1000 μm, and formed within approximately 10 hours. As cells leave the distal portion of the limb bud (the zone that specifies the P-D character of the limb; Chapter 38), their position within the A-P gradient determines whether they form an anterior or a posterior skeletal element (Wolpert, 1971), a prediction confirmed by Summerbell (1974a); cells from the distal tip can alter their A-P character when grafted anteriorly into the limb bud and proximal cells can be distalized.

In this, as in so many other situations, mutants test and confirm theory. Limb buds of *talpid*[2] mutants (Box 20.1) are symmetrical, with nine or 10 syndactylous digits arranged on a broad, short limb. The symmetry is expressed as symmetrical digits and in fusion of adjacent proximal elements, such as fusion of adjacent metacarpals and of radius and ulna (Abbott, 1967). Symmetry is evident as early as H.H. 18 in wing buds and H.H. 20 in leg buds, which suggests that the A-P limb axis is probably never established. ZPAs are normal in *talpid*[2] embryos; a ZPA from a *talpid*[2] embryo can induce duplicated limbs with normal A-P polarity from wild-type mesenchyme.

Analysis of the *hands off* mutant in the zebrafish, *Danio rerio*, in which small fin buds form later than usual, lack A-P patterning and have poor expression of *Tbx-5*, led to

identification of a *hands off* locus as encoded by the basic helix–loop–helix transcription factor *hand-2*, which is a *dHand* null mutation. *Hand-2* embryos die perinatally with a range of craniofacial defects, cleft palate and underdeveloped cartilage. In zebrafish, *Danio rerio*, *hand-2* under the control of *endothelin-1* (*sucker, edn-1*) specifies the cartilages of the lower jaw.[6]

A role for Fgf-2

An early indication that the ZPA in embryonic chick wing buds functions through an Fgf-like molecule came when Aono and Ede (1988) dissected limb buds from H.H. 22/23 embryos, divided each bud into four areas, and dissociated and cultured the cells separately. While Egf and insulin promoted growth of *all* regions of the limb bud, a gradient of responsiveness to the ZPA was found to be associated with an Fgf-like molecule. Subsequently, Fallon *et al.* (1994) described expression of *Fgf-2* in chick limb buds and demonstrated that Fgf-2 could replace the AER and direct limb outgrowth (see Box 23.3 for actions of Fgf-2 and see below for Fgf-4).

Fgf-2 also affects growth and morphogenesis. Implanting Fgf-2 into the posterior (but not the anterior or mid) aspect of a limb bud results in decreased skeletal length and loss of digits. Fgf-2 also induces digit regeneration in 85 per cent of chick limb buds into which Fgf-2-soaked beads are implanted.[7]

Implanting beads soaked in Fgf-2 into limb buds maintains AER-dependent expression of *Shh* and ZPA signaling. Interestingly, *cartilages* from early chick limb buds also possess polarizing activity. Grafted cartilage acts like a ZPA in that it induces *Hoxd-12* and *Hoxd-13* and supernumerary cartilages that in turn can function as ZPAs. The common link appears to be secretion of a member of the hedgehog family of genes; ZPA expresses *Shh*, cartilage expresses *Ihh*. This polarizing activity of limb cartilages is acquired first posteriorly, then distally and anteriorly as *Shh* spreads with cartilage differentiation (Koyama *et al.*, 1996a).

Slug, a zinc finger gene, is expressed in association with ZPA, the progress zone and interdigital areas in developing chick limb buds. *Slug* is maintained by AER signals that are independent of *Shh* (Ros *et al.*, 1997b).

dHand and *Shh*

The helix–loop–helix protein dHand plays an important role in establishing both the A-P axis and the ZPA in chick limb and in fin development, regulating *Shh* expression in both fins and limbs (Fig. 38.1). As noted above, fins fail to develop in the *hands off* mutant in zebrafish, which is a dHand mutant. A regulatory sequence has now been established leading from retinoic acid → overexpression of *Hoxb-8* → dHand → *Shh*.[8] As discussed in Chapter 18, dHand also is central to a signal cascade (*endothelin-1* → dHand → *Msx-1*) that regulates development of branchial arch mesenchyme.

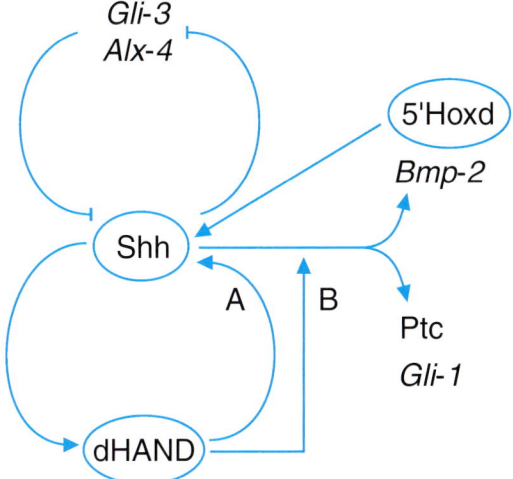

Figure 38.1 A model for the role of *dHand* in patterning the antero-posterior (A-P) axis of chick limb buds in which 5′-Hoxd positively regulates *Shh*, while *Gli-3* and *Alx-4* negatively regulate *dHand* via *Shh*. Positive feedback from *dHand* to *Shh* patterns *Shh* expression in the posterior distal limb bud (feedback loop A). Positive feedback to *Ptc/Gli-1* and *Bmp-2* (feedback loop B) bypasses *Shh* to control A-P polarity of the limb bud. Based on the model proposed by Fernandez-Teran *et al.* (2000).

dHand is expressed in the lateral plate before limb-bud emergence, after which it is restricted to the posterior half of the limb bud. Expression of dHand precedes *Shh* and dHand is upstream of *Shh*. Homozygous mutants lack *Shh* expression in limb buds, which are underdeveloped. Ectopic expression in the anterior face of a limb leads to ectopic expression of *Shh* and mirror-image duplication of limb elements associated with mirror-image duplication of dHand along the A-P axis.

Overexpression of dHand leads to preaxial polydactyly; misexpression leads to ectopic anterior activation of genes *Gli* and *Patched* in the *Shh* pathway, and of such posterior genes as 5′*Hoxd* and *Bmp-2* (Fig. 38.1). Patched-1 (Ptc-1), a binding protein for hedgehog genes, alters *Shh* in the anterior limb bud, resulting in polydactyly. Rescue of *Ptc-1* homozygotes results in loss of anterior digits and an increase in overall embryo size. *Patched* also is expressed with *Shh*, *Bmp-2* and *Fgf-8* in mandibular epithelium, *Shh* (and therefore *Patched*) being required for development of the avian frontonasal and maxillary processes. The dorsoventral axis of the upper beak is established by an *Fgf-8/Shh* boundary in the frontonasal process. Transplanting this boundary duplicates upper and lower beaks ectopically.[9]

Interestingly, perhaps surprisingly, *hypodactyly* – which results from a mutation in *Hoxa-13*, and in which digit number is reduced – and which is a common evolutionary change, is a far less common mutation than polydactyly, and entails far less modification of the genes controlling digit formation. A murine hypodactylous mutant shows normal patterns of expression of *Shh*, *Fgf-4* and *Hoxa-13*. The assumption is that the Hox proteins must be abnormal

(Robertson *et al.*, 1997),[10] but hypodactyly could be a simple consequence of retardation of growth of the limb buds, shortened development time or delayed proliferation; alteration in patterning genes may not and probably need not be involved.

A recently described limb mutant in chick embryos, *oligozeugodactyly* (*ozd*), which affects wings and legs, is characterized by hypodactyly – absence of all digits except digit-I in the legs – and lack of the posterior elements (the zeugopod; ulna and fibula) in wings and legs. This limb mesenchymal defect, which is associated with lack of expression of *Shh*, is presumed to result from a mutation of the regulatory region that drives expression of *Shh* in the polarizing region. Indeed, *Shh* is expressed normally throughout mutant embryos, with the exception of the developing limb buds, and application of *Shh* anteriorly is associated with mirror-image duplications.[11]

Wnts and Fgf

Several *Wnt* genes interact with Fgfs, particularly Fgf-8 and Fgf-10, in orchestrating the function of the AER. From the studies by Kawakami *et al.* (2001) we learn that:

- Fgf-8 and -10 initiate the formation of an AER;
- Fgf-10 is induced by *Wnt-2b*, which can evoke extra forelimbs;
- Fgf-10 is induced by *Wnt-8c*, which can evoke extra hind limbs; and
- Fgf-10 signals back to Fgf-8 in limb-bud epithelium via *Wnt-3a*.

The hypothesis is that *Fgf-8* in the intermediate mesoderm induces *Wnt-2b*, which in turn induces *Tbx-5* (Box 35.2) in the lateral plate (pre-limb mesoderm), *Tbx-5* activating *Fgf-10* which signals to the limb epithelium (Fig. 35.4). In zebrafish, *Fgf-24* (which is in the Fgf-8, -17, -18 superfamily), functions downstream of *Tbx-5* to activate *Fgf-10* in lateral-plate mesoderm, and is required for *Tbx-5*-positive cells to be able to migrate into the fin bud (Fischer *et al.*, 2003).

Wnt-5a is expressed in a caudal gradient in mouse gastrulae. Consistent with this caudal expression is a role in limb development; loss-of-function mutants cannot extend the A-P axis, one consequence of which is stunted proximal skeletal structures and an inability to form digits. *Wnt-5a* also acts to reduce proliferation of cells in the progress zone. As limb-bud development proceeds, *Wnt-5a* (and *Wnt-7a*) modulate the expression of N-cadherin.[12]

Dominant hemimelia (*Dh*) is a mutation that uncouples epithelial–mesenchymal interactions and disrupts formation of anterior mesenchyme in mouse hind limbs (Lettice *et al.*, 1999). *Fgf-8*, which is normally expressed throughout the AER, is not expressed in the anterior domain of the AER in *Dh* embryos. The regulatory loop between *Fgf-8* in the AER and *Fgf-10* in the underlying mesenchyme is uncoupled in *Dh* mutants, leading to selective loss of anterior mesenchyme and consequent severe preaxial defects. Disruption of the *Fgf-8/Shh* feedback loop between AER and ZPA following exposure to ethanol *in utero* produces pre- and postaxial digital anomalies in a dose-dependent way.

ZPAs abound

The hind limb buds of *Xenopus laevis* possess a polarizing zone similar to that in chick embryos in its location and ability to induce a polarized supernumerary limb after the tip of the limb is rotated 180° or transplanted into the anterior region of a *Xenopus* limb bud. Transplanting *Xenopus* cells into avian limb buds does not produce duplication; however, the equivalent zone from limb buds of pigs, ferrets, mice, partridge, guinea fowl, turtles, hamsters and humans does induce duplications when implanted into embryonic chick limb bud. The existence of polarizing zones is widespread.[13]

D-V POLARITY

Considerably less research has been undertaken on D-V polarity although an early study reported evidence for the role of dorsal and ventral limb epithelium in stabilizing limb pattern. Authors of a later study identified dorsal and ventral *ectodermal compartments* associated with developing chick limb buds, claiming them to be the first non-neural compartments discovered in vertebrates, although craniofacial ectodermal compartments (*ectomeres*) were discovered earlier (see Box 35.4). The future AER could be fate-mapped as scattered cells over dorsal and ventral ectoderm. The fact that no equivalent mesodermal compartments were found could be taken as indicative of the primary role exercised by limb ectoderm in positioning the AER.[14]

Dorsal and ventral epithelia of embryonic chick wing buds play important roles in patterning muscle, skeleton and joints. When Akita (1996) reversed the D-V polarity of wing buds, skeletal and muscular tissue patterning were reversed, *provided that* the epithelium was switched early enough (H.H. 17), with dorsal limb epithelium modifying the behaviour of ventral epithelium.

P-D POLARITY AND THE PROGRESS ZONE

Skeletal elements of embryonic limb buds are laid down in a proximo-distal sequence, proximal elements first. The older the limb bud, the more restricted the remaining undifferentiated mesenchyme in its ability to form proximal structures. The role of the AER in this phenomenon has already been discussed (Chapter 35).

A region extending approximately 300–500 μm from the tip of the limb buds in chick embryos contains undifferentiated proliferating mesenchymal cells that retain lability to form P-D skeletal elements. Temporal changes in the AER do not control the P-D sequence; P-D patterning is normal after an AER from an 'old' embryo is placed onto limb mesoderm from a 'young' embryo, and *vice versa*. The AER may, however, help maintain lability of the distal mesenchyme.[15]

A *progress zone*, through which mesenchymal cells pass and within which the proximo-distal level of organization is specified, was proposed in 1973 by Summerbell and colleagues. Immediately behind the tip of the limb bud is a labile zone in which differentiation does not take place. The length of time spent by cells in this labile zone specifies whether *those cells* will form proximal or distal skeletal elements. Importantly, this model for polarity of the P-D limb axis *links pattern formation and limb-bud growth*. Slowing down or speeding up limb-bud growth should alter the pattern of skeletal elements as does stopping or extending limb-bud growth. This model also is *time graded*; the P-D sequence is established according to the amount of time spent within the progress zone, rather than by the AER or proximal tissues.

Summerbell and colleagues tested this proposal by repeating Saunders' classic (1948) experiment – removing the AER from limb buds of various developmental ages. They found, as Saunders had, that the older the embryo when the ridge is removed, the more distal the structures affected, a result they interpreted as halting change in positional value within the progress zone. Cells in the progress zone produce a skeletal element at the level specified when the ridge is removed. Until at least H.H. 19, both Japanese quail and chick limb buds can shift their proximo-distal 'values' either more distally or more proximally. Once differentiation commences, polarity of outgrowth and morphogenesis of cartilaginous elements are maintained via differential synthesis and deposition of ECM.

Tabin (1998) applied the progress zone model to explain how thalidomide inhibits limb morphogenesis, specifically how cessation of limb-bud growth results in specific loss of proximal elements. Tabin argued that blocking proliferation stops cells from leaving the progress zone, with the consequence that any limb mesenchyme that forms is specified only for distal skeletal elements. Neubert *et al.* (1999) claim that Tabin's model cannot explain the specific effects of thalidomide, including species specificity (below), and propose instead a model based on inhibiting mesenchymal cell migration before AER formation.

Extension to amphibian limb regeneration

The year after it was proposed, the progress zone concept was extended to amphibian limb regeneration, especially to the reinitiation during regeneration of distal transformation and growth, which are the chief processes regulated by the progress zone in limb development. The proposal is that a progress zone is set up when the blastema is established. Spatial information in the progress zone of a blastema corresponds to the level at which the limb is amputated. Amputation through the humerus establishes a progress zone 'for' humerus, amputation through the wrist a zone 'for' wrist, and so forth. As the blastema grows and regeneration ensues, the amount of time cells spend in the progress zone will impose more and more distal values onto blastemal cells, as occurs in limb development. One way of specifying such increasingly distal properties is through centres of mitotic activity; neurons can specify such centres in vivo.[16]

This model was not embraced by all those investigating amphibian limb regeneration. From a ^{3}H-thymidine-labeling study, Maden (1976) maintained that neither a progress zone nor centres of proliferation exist within blastemata; level-specific, not growth-related properties specify proximal or distal differentiation. Subsequently, Bryant and her co-workers established an influential model – the complete circle rule – that does depend on level-specific properties: the shortest intercalation to complete a segment, and completion of a circle of polar coordinates, determine polarity of the regenerating structures.[17]

Much more recently, the progress-zone model has come under review with respect to its operating in developing limb buds. Vargesson et al. (1997) obtained a detailed fate map of chick limb buds in which they mapped the majority of the skeleton to the posterior half of the limb bud and the digits to subapical mesoderm. Despite their contention, this fate map is not necessarily at variance with the progress-zone model. Indeed, it does not speak of the presence of a progress zone. Subsequent studies mapping even earlier stages of limb-bud development do.

Dudley et al. (2002) re-examined how early in development the skeletal elements of the chick limb bud are specified, finding what they interpreted as early specification as distinct domains which then expand through proliferation and differentiation. Such early specification, of course, is not consistent with the progress-zone model.

In a separate study, Sun et al. (2002) inactivated either Fgf-4 or Fgf-8 in the AER of mouse limb buds and demonstrated that both regulate number and survival of mesenchymal cells some distance from the AER. Double knockouts resulted in failure of limb buds to form. The mechanism of action of these two Fgfs is not consistent with a progress-zone model, double knockout giving results consistent with both progress-zone and early-specification models.

Richardson and colleagues (2004) used an evolutionary approach to argue that repeated skeletal elements within one region of a limb bud – additional phalanges in hyperphalangy in dolphins and porpoises; repeated units in the pectoral fins of the African lungfish, Protopterus aethiopicus – are not consistent with the progress-zone model but consistent with a two-step model of proximo-distal patterning. So, the progress zone and branching/segmentation are both under attack.[18]

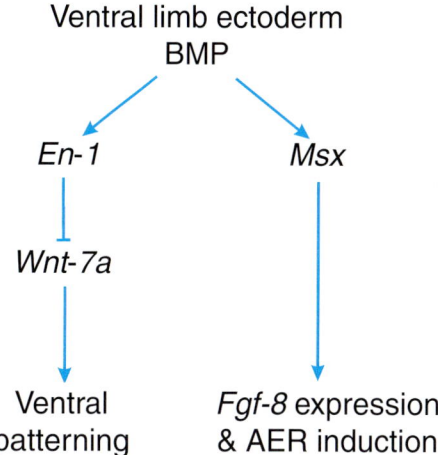

Figure 38.2 A model for the role of Bmp signaling in chick limb development. BMP in the ventral limb ectoderm controls the patterning of ventral ectoderm via Engrailed-1 (En-1) and Wnt-7a, and induction of the AER by up-regulating Fgf-8 expression via Msx. Modified from Pizette et al. (2001).

CONNECTING D-V AND P-D POLARITY

Pizette et al. (2001) identified an important molecular link between specification of the D-V and P-D axes of the limb skeleton in chick embryos. The D-V axis is set by action of Engrailed 1 (En-1) in ventral ectoderm. En-1 suppresses dorsalizing factors in ventral limb mesenchyme.[19]

The link to the P-D axis is through Bmp, which acts in both D-V and P-D axes and plays two roles: (i) Bmp within the ventral limb epithelium up-regulates En-1 leading to expression (specification?) of the D-V axis, while (ii) Bmp acts through Msx during induction of the AER for the P-D axis. There is thus a bifurcation in the downstream action of Bmp. Up-regulation of En-1 in ventral limb mesenchyme by Bmp inhibits Wnt-7a in ventral patterning. Up-regulation of Msx by Bmp regulates expression of Fgf-8 in the induction of the AER (Fig. 38.2 and Box 13.4).

Knocking out BmpR-1 demonstrates that this Bmp receptor is also involved in linking D-V and P-D (AER) polarity. AER development is disrupted in BmpR mutants and En-1 is not expressed in non-ridge ectoderm, resulting in dorsal transformation of ventral limb structures (Ahn et al., 2001).

THALIDOMIDE AND LIMB DEFECTS

Thalidomide (DL-α-phthalimidoglutaridime), synthesized in 1956 in the laboratories of Chemie Grunenthal Gmbh, in Stolberg, Germany, has a structure that resembles glutamine and glutamic acid, both of which are important in glycogen metabolism. When administered in trials to humans in a single oral therapeutic dose of 100–300 mg, thalidomide is an effective sedative, particularly for women in the early stages of pregnancy. No undesirable side effects were reported. Consequently, thalidomide was licensed for use as a drug in 1957. Two years later thalidomide was a component in some 50 commercial preparations.

During 1960–61 reports began to appear that prolonged use of thalidomide-containing drugs was associated with symptoms of peripheral neuritis – tingling sensations in the extremities indicating nerve dysfunction. Accordingly, prolonged use of the drugs was not recommended. To the credit of the USA, these neuropathologic symptoms were sufficient to prevent the drug from being licensed for commercial distribution in that country.

The first hint of the disastrous situation that was to develop appeared in 1960. W. Lenz, a German pediatrician, began to notice an increase in a virtually unknown human syndrome. The defects varied from minor malformations of the toes and fingers, through *meromelia* – shortened limbs – to *phocomelia* – a flipper-like hand or foot joined directly to the pectoral or pelvic girdle – or, in the most extreme cases, *amelia*, complete absence of one or other set of limbs. Deficiencies of the ears, heart and intestines often accompanied these skeletal defects, as did early onset of arthritis. Lenz found 20 cases in a town with a population of only 20 000. Investigating the clinical history of his patients, Lenz suspected a correlation between prescription of thalidomide as a sedative during early pregnancy and the syndrome. He reported this correlation to a meeting of pediatricians in Dusseldorf on 18 November 1961. A similar, independent finding was reported in Australia by William McBride in 1961.[20]

In comparison to previous years, during 1959 and 1960 there was a 1000 per cent increase in limb malformations in West Germany; 6000 cases were reported between January 1960 and October 1962. Epidemiological evidence for thalidomide as the cause of these defects, although indirect, was overwhelming. One study involved 1624 females who had given birth to infants with limb defects in North Rhine and Westphalia between January 1960 and the end of March 1962. A staggering 90 per cent of those who took thalidomide in the first month of pregnancy produced infants with limb defects. The frequency of malformations in Germany and Japan paralleled the sale of thalidomide with an eight- or nine-month lag, and the epidemic ceased eight months after thalidomide was withdrawn from the market. In all, 7000 infants were malformed by thalidomide in West Germany alone.

Nevertheless, a proposal that a sedative administrated to the mother could cause such a profound and deforming effect on her foetus was not accepted readily. This was entirely understandable. At the time only two syndromes or *embryopathies* attributable to environmental agents were known: that produced by exposure to X-rays, and that produced following maternal infection with rubella or German measles. No syndromes following ingestion of chemicals were known or even suspected – the placenta was regarded as a 100 per cent effective barrier, isolating foetus from mother and environment.

The correlation reported independently by Lenz and McBride focused attention on possible maternal–foetal interactions. It was, however, not until early 1962, when Somers reported limb deformities in rabbits given thalidomide (150 mg/g body weight) that thalidomide was withdrawn from the market. Later in 1962 it was shown that the 'thalidomide syndrome' could be produced in rodents. *Some* non-human primates also responded to thalidomide with amelia, phocomelia or spontaneous abortion. Differences in the responses of different primates were attributed to differences in placental structure. For a drug with a half-life of two or three hours and inactive metabolites, transit time across the placenta is critical. The placenta as a porous barrier and the advantages/disadvantages of tests using animal models have been recognized ever since.[21]

Time of action

Before pursuing theories of the mode of action of thalidomide, I should consider its time of action on embryonic development. Clues to the mechanisms of action of a teratogen are often found in the developmental events occurring at that *critical time* or critical period (Box 38.1).

The major organ system affected by thalidomide is the limbs (Fig. 19.1). During human development, the legs lag some two to three days behind the arms in initiation and development. Arm buds appear at the 26th day of gestation. Proximal limb mesenchyme condenses by the 37th day; chondrogenesis begins on the 41st day. The most severe deformities of arms or legs occur when thalidomide is administered between the 25th and 30th or 30th and 34th days of gestation, ages that correspond, respectively, to early stages of arm- and leg-bud development. Administering thalidomide after the 36th day, even as late as the 50th day, still produces deficiencies, but these are less severe than when thalidomide is administered earlier, normally being confined to absence of digits.

Mode of action

Despite the malformations, the reputation of the pharmaceutical industry – which was at stake – and the intrinsic scientific interest in understanding the first known drug-induced embryonic syndrome, we still do not understand thalidomide's mode of action. Thalidomide is unstable in neutral or slightly alkaline solutions. In the body thalidomide produces many metabolites, none of which is teratogenic. Presumably the intact molecule is active, but *[14]C-labeled thalidomide binds directly to tissues, but not to tissues of the limb buds* (Box 38.2). Perhaps thalidomide acts in combination with another compound natural to the body. Such synergisms are well known, but aside from an indication that a particular vitamin deficiency (absence of pantothenic acid) intensifies the action of thalidomide in rats, we have no real clues to any synergism involving thalidomide.

Box 38.1 Critical periods

The concept of critical or sensitive periods is an important one, especially when assessing the effects of exogenous agents such as drugs. Indeed, knowledge of critical periods underpins the field of teratology.[a]

Stockard (1921) introduced the term and concept of 'critical period' in a study of interactions among organs as they arise and develop in normal, twin and malformed embryos.[a] Stockard used the term for a time of heightened cell division during which developing systems are more sensitive. Hamilton (1952) identified three critical periods in avian development in relation to temperature, humidity, gravity and exposure to chemicals. The three are:

- *three to four days of incubation*, when many organ systems are being initiated and cell division is high;
- *13–14 days of incubation*, when major endocrine systems come into play and when proliferation is again high; and
- *19–20 days of incubation* when the embryo is preparing to hatch.

A number of critical periods have been identified during the ontogeny of teleost fishes (Browman, 1989), including:

- embryological, coinciding with periods of embryonic induction;
- neurobiological, when the neural circuits are plastic;
- ethological, when imprinting occurs and behavior can be modified; and
- a 'fisheries science critical period' – perhaps better called a larval period – an ecological concept relating to the time of massive larval mortalities.

At least some of these critical periods are under ecological control. For example: fish not exposed to food stimuli during the neurobiological critical period do not establish the neural networks required to forage for food; skeletal changes associated with the onset of feeding and respiration in Atlantic cod, *Gadus morhua*, are tightly integrated (Hunt von Herbing et al., 1996a–c); life history changes are saltational, with individual stages triggered internally (hormonally) and externally (by temperature, density, day length, predation) and so forth.[b]

[a] See J. P. Scott (1986) for a review of critical periods in human development, and see Sucheson (1993) for the effects of drugs on bone growth. Merker (1977) reviewed the concept of the critical period as it pertains to limb development, Hall and Miyake (1997) in the context of how embryos tell time, and heterochrony, Stoleson and Beissinger (1995) in relation to hatching of birds. Hall (1985b) used thallium-induced embryopathy to assess critical periods for skeletal development in chick embryos.
[b] See Hunt von Herbing et al. (1996a–c) for Atlantic cod, Balon (1985, 1989, 2004) for saltation, and Hall and Wake (1999), Urho (2002) and Hall et al. (2004) for overviews in the context of larval biology.

Box 38.2 The mesonephros and chondrogenesis

The mesonephros is the name given to the rudiment of the developing kidney in tetrapods. Mesonephric tissues metabolize glucosamine, a finding that is relevant to the longstanding issue of whether mesonephric cells can chondrify, and/or whether they influence chondrogenesis in adjacent limb or somitic mesenchyme. Strudel and Pinot (1965) thought the cartilage that formed when they cultured mesonephros from embryos of H.H. 16–22 had differentiated from contaminating somitic cells. These were their findings:

- mesonephros alone – no cartilage;
- somitic mesoderm alone – 25 per cent formed cartilage;
- mesonephros plus adjacent mesenchyme – one of 38 cultures (2.5 per cent) formed cartilage;
- mesonephros plus somitic mesoderm – 25 per cent formed cartilage; and
- mesonephros plus limb buds – 100 per cent formed cartilage (as did limb buds cultured alone).

The discovery by Stirpe et al. (1990) that the mesonephros of embryonic chicks synthesizes the link protein for proteoglycan (although neither core protein nor type II collagen) made the issue even more intriguing. Indeed, mesonephros from 5–8.5-week-old human embryos exposed to thalidomide *in vitro* synthesizes a cartilaginous matrix. The proposal from these findings was that mesonephric mesenchyme stimulates limb mesenchyme to chondrify, a stimulation that is inhibited by thalidomide. The long-recognized connection between mesonephric mesenchyme and chondrogenesis exists in part because the mesonephros produces a cartilage growth promoter; ablating the mesonephros decreases proliferation of limb mesenchyme, and so decreases chondrogenesis.[a]

[a] See Lash and Saxén (1972), Lash et al. (1974) and Geduspan and Solursh (1992) for cartilage promotion, and see the discussion on thalidomide, the mesonephros and limb abnormalities in this chapter. Earlier, Stirpe and Goetinck (1989) provided some of the first information on the time of appearance of link and core proteins during chick wing bud development, finding link protein and core protein at H.H. 25.

Several lines of evidence suggest that thalidomide does *not* act directly on the cells of developing limb buds.

Local haemorrhaging has been correlated with limb and facial defects induced by thalidomide in a variety of species including monkeys, implicating the vascular system. Anti-angiogenic properties of thalidomide are now being taken advantage of and used in the treatment of some cancer patients.

McCredie, who in collaboration with McBride did much work on affected children, favoured the theory that thalidomide injures sensory peripheral nerves, lack of normal nerve function resulting in defective limb development. However, substantial evidence indicates that initiation of limb development *precedes* innervation of the limb buds; spinal nerves enter embryonic chick limb buds at H.H. 24. Furthermore, limb buds produce normal skeletal elements when they develop in the absence of nerves, as, for example, in organ culture or as CAM grafts. Innervation does affect later aspects of skeletal growth and morphogenesis – resection of the brachial plexus in

25-day-old rats results in smaller deltoid tuberosities, and shorter, narrower humeri with more growth distally than proximally at 53 days[22] – but these are variations in morphogenesis and growth, not initiation of skeletogenesis.

Tissue culture provides some of the best evidence that thalidomide does not act directly on limb buds. Culturing limb buds from human embryos of five to eight weeks of gestation with thalidomide does not impair their ability to form cartilage. Cartilage formation is inhibited, however, when limb buds are cultured with developing kidney tubules. In such co-cultures, radioactively labeled thalidomide binds to kidney cells, not to limb cells. Interestingly, it is known from other studies that the developing kidney has a positive effect on chondrogenesis (see Box 38.2). Nevertheless, almost 45 years after the thalidomide syndrome first made its appearance, we remain largely ignorant as to how its devastating effects come about.

NOTES

1. We have common names for the five fingers on our hands. From digit I to digit V these are the thumb, index finger, pointer, ring finger and pinkie. My niece, Karen Binstadt, recently asked me (on behalf of her two-year-old daughter Carly) whether there are equivalent names for toes. My immediate response was 'big toe, little toe and the toes in between.' I then thought that 'ring-toe' might be a suitable name, but we could not decide which toe rings are most often worn on, although four of the women in my laboratory thought the second toe (digit II) the most common 'ring toe.' I asked the members of my laboratory and Charles Oxnard (University of Western Australia) but no one knew of any common names for toes. Can any of the readers of this endnote help me get to the bottom of this question?

2. See Errick and Saunders (1974), MacCabe et al. (1974) and Pautou (1977) for the experimental studies.

3. See Harrison (1918) and Swett (1937) for the amphibian studies, and Chaube (1959) for chick.

4. The ZPA as a distinct region was discovered by Gasseling and Saunders (1964) and named ZPA by Balcuns et al. (1970). Most studies have been undertaken with chick wing buds, although Hinchliffe and Sansom (1985) mapped the ZAP in chick leg buds. For reviews of early studies on the role of the ZPA in A-P orientation see Saunders and Gasseling (1968) and Saunders (1977). For a more recent overview, see Tickle (2002).

5. See Amprino and Camosso (1959, 1963) for rapid determination of the A-P axis, MacCabe et al. (1973), Crosby and Fallon (1975) and MacCabe and Parker (1975, 1976a) for grafting beneath the ectoderm, and Fallon and Crosby (1975a,b) for the rotation studies. The single element obtained by Hinchliffe and Gumpel-Pinot (1981) is a phenotype similar to *Shh*[-/-]-mouse limbs (C. Tickle, pers. comm.).

6. See Yelon et al. (2000) for the *hands off* mutant, Yanagisawa et al. (2003) for *Hand-2* mutants, and C. T. Miller et al. (2003) for zebrafish. *Edn-1* mutant zebrafish have abnormal dermal bones including fused bones in the mandibles and absence of the opercle bone from the hyoid arch (Kimmel et al., 2003).

7. See S. Li et al. (1996) and Taylor et al. (1994) for these two studies.

8. See Charité et al. (2000), Fernandez-Teran et al. (2000) and McFadden et al. (2002) for the role of dHand. Cohn (2000) reviews these two papers and a third on the role of dHand in fin and limb development, the essential role being to regulate *Shh* in both types of limbs, perhaps mediated by protein–protein interactions independent of any direct binding of *dHand* to DNA (McFadden et al., 2002). For the role of *Hoxb-8* in positioning the ZPA to establish A-P polarity in chick fore- but not hind limbs, see Stratford et al. (1997). For unique roles for Hoxb-8 and Hoxd-8 in patterning hind limbs and thoracic and sacral vertebrae, see van den Akker et al. (2001).

9. See Milenkovic et al. (1999) and Helms et al. (1997) for polydactyly and mandibular epithelium, and Hu and Helms (1999) and Hu et al. (2003) for facial process growth.

10. See Robertson et al. (1997) for the normal expression patterns. Mortlock et al. (1996) described a 50-base-pair deletion in the first exon of the *hypodactyly* (*Hd*) allele in mice and emphasized the similar but more severe defects in *Hd/Hd* mice when compared with *Hoxd-13*-deficient mice.

11. See Ros et al. (2003) and Maas and Fallon (2004) for *oligozeugodactyly*. *eHand*, which is not expressed in *talpid*[2] mutant limb buds, is expressed normally in *oligozeugodactyly* (Fernanez-Teran et al., 2003).

12. See T. P. Yamaguchi et al. (1999) for *Wnt-5a* and the progress zone, and see Tufan and Tuan (2001) and Yang (2003) for *Wnt-5a, Wnt-7a* and N-cadherin.

13. See MacCabe and Parker (1976b), Tickle et al. (1976), Cameron and Fallon (1977a) and Fallon and Crosby (1977) for these studies, Honig (1984) for posterior signaling regions in turtle and alligator limb buds, and A. H. Lee (2004) for the dynamics of long-bone growth in *Alligator mississippiensis*.

14. See Stark and Searls (1974) for the early study, and Altabef et al. (1997) for compartments.

15. See Rubin and Saunders (1972) for the AER transplants, Finch and Zwilling (1971), Stark and Searls (1973) and Summerbell and Lewis (1975) for the labile progress zone, and Rubin and Saunders (1972) for the AER.

16. See A. R. Smith et al. (1974) and Smith and Crawley (1977) for this extension to limb regeneration, and Globus and Vethamany-Globus (1977) for neuronal specification of mitotic centres, recalling the role of neurotrophic factors in blastema formation.

17. See Bryant (1976, 1977), French et al. (1976), Bryant et al. (1977) and Glass (1977) for the complete circle rule, and Slack and Savage (1978) for evidence against it.

18. See Duboule (2002) for a review of the 2002 studies, and see comments and rebuttals by John Saunders (2002), who discovered the role of the AER (Saunders, 1948), and by Lewis Wolpert (2002), who with Denis Summerbell and Julian Lewis proposed the progress-zone model (Summerbell et al., 1973).

19. As will be discussed in connection with the zones of cell death found where limb bud meets flank, *radical fringe*, the gene that sets the position of the AER along the limb-bud epithelium, is co-expressed with *engrailed*.

20. See the report in *The Lancet* (1962) 1, 45.

21. Manouvrier-Hanu *et al.* (1999) provided a useful analysis of human limb anomalies on various bases, including clinical, clinical/embryological and clinical/genetic. They recognize six categories – reduction anomalies; hypoplasia; brachydactyly; syndactyly; polydactyly; and abnormal patterning – and note that human limb malformations occur in 1/1000 neonates. See Dufresne and Richtsmeier (1995) for an ordering of deformation, malformation, disruption, dysplasia and syndrome as a parallel classification based on response to surgery. A poor response indicates a growth disorder, a good response reflects a change where growth is not affected.

22. See Flint (1984) for the data on chick and rodent limb buds, and Dysart *et al.* (1989) for brachial plexus resection.

Patterning Limb Skeletons

The shapes, sizes and external forms of bones, cartilages and teeth are the final expressions of interaction between genotype and environment that constitute morphogenesis.

MORPHOGENESIS AND GROWTH

Morphogenesis – the shaping of form – should not be equated with growth, which is permanent increase in size (Chapter 31; Table 39.1). Although these two developmental processes can occur together, form can change without concomitant change in size, and *vice versa*. Unraveling the intricacies of morphogenesis is a formidable problem. Ultimately, all factors contributing to morphogenesis do so by regulating one or more of a finite and surprisingly small number of processes: cell proliferation, differentiation, selective activation or inhibition of osteo- and chondroprogenitor cells, synthesis of cellular products, and their deposition/removal from an extracellular matrix (ECM). The fact that all these processes are multidimensional presents a major challenge to the analysis of morphogenesis. So too does the fact that these same processes underlie growth.[1]

Differential deposition of ECM by central and peripheral cells within a mesenchymal condensation is one mechanism whereby the shape of a cartilaginous skeletal element is moulded. I discuss two further mechanisms of morphogenesis. One, *programmed cell death*, moulds the basic morphology of limb buds, limbs, parts of appendages such as hands and feet, and limb skeletons. For example, the toes of the forelimbs (but not the hind limbs) of the mouse mutant *Fused toes* (*Ft*) are fused (Fig. 39.1). Programmed cell death is enhanced in forelimbs and not in hind limbs, consistent with the mutation selectively enhancing cell death in the distal portion of the forelimbs during digit development. The thymus also is underdeveloped in *Ft* mutants; enhanced cell death occurs selectively in the developing thymus gland (Van der Hoeven *et al.*, 1994).

The second, *differential adhesion of cell membranes*, is expressed in exaggerated form in *talpid²* and *talpid³* mutants.

Table 39.1 Mean skeletal maturity scores of several strains of 12-day-old mice reared at 32°C as a function of litter size[a]

Mouse strain	Litters of four offspring				Litters of seven offspring			
	Thorax	Forelimbs	Hind limbs	Tail	Thorax	Forelimbs	Hind limbs	Tail
BALB	10.9	18.3	25.7	30.3	–	–	–	–
F₁ (BALB (M) × C57 (F))	10.1	18.3	24.1	30.3	–	–	–	–
F₁ (C57 (M) × BALB (F))	12.9	21.4	28.5	40.9	9.1	17.3	23.0	28.3

[a] Scores were determined by assigning a number corresponding to skeletal maturity to each element at 12 days of gestation and adding the numbers together. High scores = high skeletal maturity. Based on data from Garrard *et al.* (1974).

A

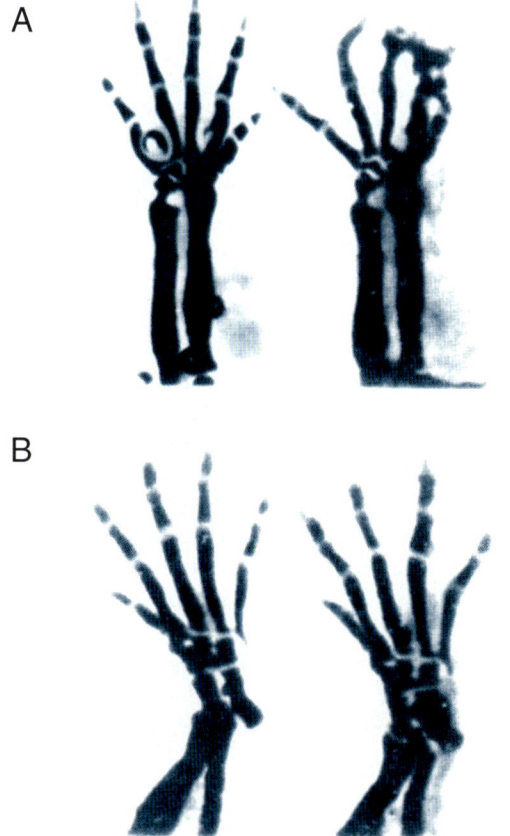

B

Figure 39.1 The mouse mutant *Fused toes* (*Ft*). A comparison of stained preparations of fore- (A) and hind limbs (B) from 15.5-day-old littermates of wild-type (left) and *Ft* (right) mutant embryos.
(A) Forelimbs to show fusion of distal phalanges between digits I and IV in an *Ft*/+ embryo (right). Compare with the wild-type embryo on the left. (B) There is slight lateral expansion of the distal phalange on digit II of the hind limb from a heterozygous *Ft*/+ embryo on the right but no phalangeal fusion. Modified from van der Hoeven *et al.* (1994).

PROGRAMMED CELL DEATH (APOPTOSIS)

Recognition of cell death or 'amplification of the currently mysterious control of catabolism, which ultimately crosses the threshold of irreversibility'[2] is as old as the cell theory (Box 39.1). Programmed cell death (apoptosis) in limb development is:

- a way of fashioning the entire limb and moulding individual skeletal elements;
- metabolically controlled – in chick limb buds, time-to-die is slowed at 30°C and postponed indefinitely at 20°C (Fallon and Saunders, 1968);
- not suicide – neither phagocytic nor lysosomal enzymes are produced;
- when precocious, a means of preventing continued development of limb buds in limbless tetrapods (Chapter 37).

Developing limb buds also affect cell death in nearby tissues. The ventral horns of the grey matter in the spinal

Box 39.1 Cell death and the cell theory

We tend to think that recognition of programmed cell death – now more generally known as *apoptosis* – is very recent and, indeed, it was only a little over 30 years ago that Kerr *et al.* (1972) drew the distinction between necrosis and apoptosis. Cell death, however, was identified soon after Schleiden and Schwan established the cell theory in the 1830s.

An analysis published in 1996 listed and reviewed over 100 papers on cell death published in the 19th century, the term for apoptosis then being *chromatolysis*. Those papers with particular reference to the skeleton described cell death in the cartilage and notochord of tadpoles of the midwife toad, *Alytes obstetricans*, during metamorphosis (Vogt, 1842) and the death of chondrocytes during endochondral ossification in the metatarsal of a newborn dog, a phenomenon which Stieda (1872) may have been the first to report.[a]

[a] See Clarke and Clarke (1996) for 19th century studies, especially p. 96 for chromatolysis and Table 2 for papers from the late 19th century on cell death in cartilage of mammals (L. Stieda), birds (Z. Strelzoff) and amphibians (N. Kastschenk). It is fortunate that the priority rules for systematic nomenclature and species names do not hold for cell biology terminology.

cord are enlarged where they lie adjacent to the limb buds, in part because limb buds slow the rate of neuroblast death (Deuchar, 1975, and see Box 38.2).

Programmed cell death characterizes various aspects of normal limb development, including:

- where limb buds emerge from the flank (*posterior and anterior necrotic zones; PNZ, ANZ*);
- between the presumptive digital primordia (*interdigital zones* of cell death);
- in the *opaque patch* in the centre of the wing bud, in which cell death separates tibia from fibula and radius from ulna; and
- in zones of mesenchyme where *joints* will form.

Cell death also has been analyzed in the ectomesenchyme from which scleral ossicles develop (Chapter 21) and at sutures between skull bones (Chapter 34).[3]

Posterior and anterior necrotic zones (PNZ, ANZ)

Localized areas of cell death on the posterior face of the wing buds (the PNZ) and on the anterior (preaxial) border of hind limb buds (the ANZ) were described in 1962. Cell death is extensive and rapid; some 2500 cells die within the PNZ, 1500–2000 of them within eight hours during H.H. 24, which is when necrotic cells first appear and the limb is sculpted into a paddle-shape.[4]

From as early as H.H. 17, these cells are programmed to die at H.H. 24, but this programme can be altered until H.H. 22, in which case the cells survive (Saunders and Fallon, 1967). The 'death clock' can be altered by grafting the cells to other areas of the limb bud or into non-limb mesenchyme *where they survive to differentiate into chondroblasts*. Cells from outside the PNZ do not show programmed cell death if grafted into the PNZ region. Evidently, cells of the PNZ are a subpopulation of limb mesenchymal cells set aside for this particular developmental role.

Attainment of irreversible commitment for cell death at H.H. 22 is accompanied by a sharp decline of DNA synthesis, a finding that prompted the proposal that cell death results from increased cross-linking of DNA. Necrotic cells appear all over embryos injected with chemicals to increase cross-linking, However, DNA in the PNZ is no more highly cross-linked than is the DNA in other cells.[5]

Four different approaches provided evidence for biochemical degeneration *leading up* to programmed cell death in the PNZ and ANZ:

- intracellular protein and ribosomal crystals appear at H.H. 22–24;
- protein synthesis decreases significantly from H.H. 22 onward (when compared with synthesis in cells at the distal tip of the limb bud);
- DNA and RNA synthesis decline between H.H. 21 and 22, approximately six hours before the decline in protein synthesis; and
- acid phosphatase accumulates within Golgi bodies of the dying cells.[6]

Does this programmed cell death help sculpt the bases of limb buds as they *grow out* from the body? Recall that *recession of the flank* from the body is as prominent a growth mechanism as is outgrowth of the limb bud at these stages (Chapter 35). Like many a cherished hypothesis, this one became somewhat tarnished when confronted with initial experimental data: limb morphogenesis is normal after the zone of cell death, the PNZ, is removed *in ovo*.[7] Presumptive evidence was obtained, however, from two domestic fowl mutants.

In the wing buds of *wingless* (*ws*) embryos precocious cell death begins in the ANZ at H.H. 19. By H.H. 20–23, cell death has extended well beyond the normal bounds of the ANZ. Removing this additional mesenchyme may rob the limb bud of apical ectodermal maintenance factor (AEMF), and so contribute to loss of the AER. *Talpid*[3] embryos, which have a longer AER than wild-type embryos, lack ANZ and PNZ, a finding that prompted Hinchliffe and Ede (1967) to propose that these zones of cell death in wild type act as 'end-stops' for the AER, limiting its extent. They felt that altered cell-to-cell adhesiveness was involved in cell death in these necrotic zones.[8]

We now know that the gene involved in regulating epithelial cell activity to position the AER on the limb bud is *radical fringe*, a homologue of the *Drosophila* gene *fringe*, and one of a family of genes including *lunatic fringe, manic fringe* and *radical fringe*: the 'Adams Family' of the gene world? *Radical fringe* is expressed at the dorso-ventral boundary of developing limb buds in an expression domain that correlates with expression of *engrailed-1* (*En-1*). *Radical fringe* therefore positions the AER; *En-1* represses *radical fringe* in the ventral limb-bud epithelium, the ridge forming at the boundary between expression/non-expression of *radical fringe*.[9]

Interdigital cell death

Areas of necrotic cells appear at the distal tips of the limb buds of birds, lizards, turtles, rodents and humans but not in urodele amphibians. Several lines of evidence show that interdigital necrosis sculpts the interdigital area, removing cells entirely or reducing the area to a thin membrane depending on limb and species.[10] The pattern of webbing between the digits of duck feet is related to the amount of interdigital cell death; necrotic areas are prominent where no web forms – as between the first and second toes – but are reduced between other digits where webs do form. In analogous situations, interdigital necrosis is not seen between the digits in mice – which have connective tissue connections (a web, syndactyly) between the digits – in the syndactylous limb buds of *talpid*[3] embryos, or after syndactyly is induced with Janus Green in embryonic chicks.[11] Interdigital cell death occurs between H.H. 30 and 35 in chick embryos, although biochemical degeneration may set in earlier.[12] Phagocytes engulf the dead cells.

Some studies in the 1960s and early 1970s suggested that the AER suppresses cell death in adjacent mesenchyme at the time the digits are specified.[13]

Although it was then not possible to selectively test and compare the activity of epithelium overlying digital and interdigital mesenchyme, such experiments have since been performed as part of an important programme initiated by Juan Hurlé in Spain. Hurlé and his colleagues compared the epithelial changes associated with interdigital cell death in chick embryos – in which death removes interdigital tissue completely – with duck embryos – in which a web of tissue remains, except between digits I and II (see above). As these studies progressed, they revealed the cellular and molecular bases of interdigital patterning by cell death.

- In chick but not duck embryos, epithelial cells transform from polygonal to rounded with epithelial and interdigital (mesenchymal) cell death (Hurlé and Colvee, 1982).
- Cell death is followed by regression of blood vessels in chick but not duck limb buds (Hurlé et al., 1985).
- In chicks and ducks, changes at the epithelial–mesenchymal junction lead to breaks in the basal lamina, deposition of collagen at the interface, and detachment of epithelial cells into the amniotic sac (Hurlé and Fernandez-Teran, 1983, 1984).
- Removing the AER or dorsal epithelium from chick limb buds up to H.H. 31 elicits ectopic nodules of cartilage – or even extra digits containing two phalanges, but lacking

tendons – indicating that the epithelium normally inhibits chondrogenesis (Hurlé and Gañan, 1986, 1987), just as mandibular epithelium inhibits Meckelian chondrogenesis (Coffin-Collins and Hall, 1989).

- Removing marginal epithelium from the third interdigital region at H.H. 27–30 is followed by the appearance of prechondrogenic mesenchyme, which condenses. Because some condensations disintegrate under local inhibitory influences from the remaining epithelium, fewer cartilages than condensations form (Hurlé et al., 1989).

- Transplanting a zone of polarizing activity (ZPA) into interdigital epithelium does not alter the patterns of the cartilages that form (Macias and Gañan, 1991).

- Removing epithelium from the interdigital area of duck hind limb buds elicits ectopic cartilages, the interdigital mesenchyme having a high chondrogenic potential (Macias et al., 1992).

- Local microinjection of retinoic acid, the kinase inhibitor staurosporin, or Tgfβ-1 or -2 shows that the interphalangeal joint will not develop unless all four are present (Macias et al., 1993).

- Extra digits can be induced from interdigitial areas of limb buds of H.H. 29 without any modification of the normal patterns of Msx or Hoxd expression, Msx maintaining the interdigital mesenchyme as undifferentiated rather than providing specific signals that control cell death in this region (Ros et al., 1994).

- Local administration of Fgf-2 or Fgf-4 into the interdigital area of chick hind-limb buds inhibits cell death and promotes increased proliferation and formation of webs between the digits (Macias et al., 1996).

- Implanting Bmp-2- and Bmp-7-soaked beads affects mesenchymal cell death but neither epithelial cell death nor cartilage differentiation, both growth factors regulating the amount and distribution of prechondrogenic mesenchyme (Macias et al., 1997);[14] and

- Cells that would normally die interdigitally can be diverted to form a new digit, with analysis of expression patterns of Msx, Hox and Bmp (Ros et al., 1997a).

A role for BmpR-1

As discussed in Box 39.2, BMP receptors play multiple roles in skeletogenesis including roles in digit morphogenesis, which in chick embryos is regulated by Fgfs, Tgfβs, Gdf-5 and noggin acting through BMP signaling. Bmps are regulated by different antagonists in complementary ways. The evidence and the sequence appear to be as follows.

- Tgfβ-2 and noggin are expressed in chondrogenic condensations.
- Application of exogenous Tgfβ, noggin or Fgf alters levels of BmpR-1B.

- Fgfs reduce BmpR-1B and maintain the progress zone.
- Tgfβ increases BmpR-1B and leads to formation of ectopic cartilage.
- Therefore, a Bmp–noggin loop maintains the size and shape of the skeletal elements.

Gdf-5 also is regulated by Bmps acting with hedgehog genes to control epiphyseal differentiation and growth. The Bmp antagonist gremlin, which antagonizes Bmp-2, -4 and -7, helps maintain the AER – thus maintaining bud outgrowth and proximo-distal specification – and restricts cell death, confining cartilage to the central mesenchyme.[15]

Following the pioneering studies on cell death in rat and mouse limb development by Milaire and Rooze (1983), timing of the interdigital cell death during various phases of the development of mouse limb buds has been determined by K. K. H. Lee and colleagues. Chondrogenic capability of interdigital cells declines as cell death ensues. Retinoic acid induces cell death, but medium conditioned by epithelia from limb buds does not. The timing of interdigital cell death and loss of chondrogenic potential in mouse limb buds is:

- 14.5 days: cell death in the AER; interdigital mesenchyme can form cartilage if transplanted under the kidney capsule;
- 15.5 days: cell death in AER and mesenchyme; less cartilage produced in grafts; and
- 16.5 days: interdigital cell death and inability of interdigital mesenchyme to form cartilage.

Lee's group then identified growth-arrest-specific-2 (gas 2) gene and peptide in 11.5–14.5-day-old embryos in the interdigital mesenchyme where it may regulate apoptosis, acting as a substrate for caspase enzymes; the peptide is part of the microfilament network. Other members of this gene family are expressed in growth-arrested fibroblasts where they regulate proliferation and apoptosis.[16]

Bmp-4 initiates praecocious cell death in the interdigital mesenchyme of 12.5-day-old mouse embryos maintained ex utero. In vitro, Bmp-4 promotes chondrogenesis not cell death, evidently because of the presence of a digit primordium. Bmp-4 does induce apoptosis if a digit primordium is included in the cultures. Indeed, it now appears that whether digits are anterior or posterior in character is determined by Bmp localized within the interdigital mesenchyme, for more posterior interdigital mesenchyme or a higher dose of Bmp specifies more posterior digits. Fedak and Hall (2004) modeled this in the context of the development and evolution of extra phalanges (hyperphalangy) in whales.[17] We saw the developmental prerequisites for hyperphalangy as:

- lack of interdigital cell death – a requirement to produce a flipper-like limb – and
- maintenance of a secondary AER – to initiate digit elongation and the specification of additional joint.

Box 39.2 Bmp receptors

Receptors for Bmps were not discovered until the mid-1990s, when the 532-amino-acid rat Bmp type IA receptor was cloned from a dental pulp library (and its expression followed in bone induced by Bmp) and type I receptors for osteogenic protein-1 and BMP-4 were identified from rat and human cell cultures.[a]

Tissue localization of the rat type 1 receptor for Bmp-2 and Bmp-4 has been compared with spatial distribution of the Bmp ligand. Receptor expression is widespread at 18 days of gestation – cartilage, bone, hair, whiskers, teeth – and more widespread than either Bmp-2 or Bmp-4. The type I receptor, which can be regarded as ubiquitously distributed in early mouse embryos, is essential for mouse embryogenesis: BmpR-1 mutant embryos die at 9.5 days, having failed to gastrulate, to form the majority of the mesoderm, and with teratomas in all three germ layers,[b] Bmp-2 and Bmp-4 being central to gastrulation and germ layer formation.

Given these broad patterns of expression and central roles, you might expect Bmp receptors to be expressed from early in skeletogenesis, which they are: the two Bmp receptors (BmpR1A and BmpR1B) are co-expressed in condensing chondrogenic mesenchyme of developing chick limb buds and in condensing mesenchyme in 13.5-day-old rat embryos. Both receptors are up-regulated three days after bone fracture.[c]

Receptor mutations also are known (Baur et al., 2000). A recessive mutation (BmpR1B or Alk-6) results in brachydactyly. BmpR1B is present in two major classes of transcripts representing two promoters, one distal controlling limb skeletogenesis, the other proximal involved in neural development. Mutants lacking BmpR1B fail to form digits, the receptor regulating chondrogenesis through Gdf-5-dependent and -independent processes. This study emphasizes how a single receptor can be triggered by different ligands and how a single ligand can bind to different receptors.

Activation of BmpR1A inhibits chondrogenesis in chick embryos in a manner equivalent to misexpressing Ihh. The proximo-distal boundary of expression of Ihh in chondrocytes is reflected in changes in Bmps in the adjacent perichondrium, misexpression of Ihh increasing Bmp-2 and Bmp-4 and altering noggin and chordin domains. Four 70-amino-acid cystein-rich domains allow chordin to bind to and inactivate Bmp-4. Chordin-like protein (CHL) is expressed in dermotome, limb buds and chondrocytes and also acts to regulate Bmp. Changes in Bmp-5 and Bmp-7 in the perichondrium alter differentiation in adjacent chondrocytes.[d]

In a second series of studies, injection of the gene for a dominant negative type I Bmp receptor in a viral vector into chick limb buds blocks Bmp; cell death is suppressed and interdigital webs develop. The digits and tails are truncated in 10 per cent of these embryos, and scales transform to feathers. BmpR1B is expressed even earlier in limb mesenchyme at condensation, with initial chondrogenesis, and at sites of apoptosis. BmpR1A is expressed in prehypertrophic cells, both receptors being regulated by Ihh.[e]

Targeted disruption of BmpR1B produces limb defects in mice. Although initial digit formation is normal, the digital mesenchyme fails to proliferate or to differentiate. BmpR1B/Gdf-5 double mutants reveal that Gdf-5 is a ligand for BmpR1B. BmpR1B/Bmp-7 double mutants indicate that receptor and ligand have overlapping pathways and that in the absence of BmpR1B, Bmp-7 plays an essential role in limb development (Yi et al., 2000).

Knowledge of the roles of type 1A, 1B and type II Bmp receptors has been advanced by an analysis by Ashique et al. (2002) of their expression and the consequences of misexpression in chick embryos. This study concentrated on roles of the receptors in intramembranous ossification, chondrogenesis and feather formation. In summary, in chick embryos:

- BmpR1A is expressed in cartilage condensations, but down-regulated as differentiation ensues;
- BmpR1B is expressed in all cartilages;
- BmpR-II is expressed in low levels in the nasal cartilage, but in higher levels in other cartilages;
- both type I receptors are expressed in maxillary membrane bones;
- misexpression of BmpR1B results in smaller cartilages and bones; while
- misexpression of BmpR1B has no effect on skeletogenesis, despite its normal expression in condensations.

[a] See Takeda et al. (1994) for rat BmpR1A, ten Dijke et al. (1994) for type I receptors, and Kingsley (1994a) and Rosen et al. (1996) for Bmp receptor signaling in bone formation.
[b] See Ikeda et al. (1996) and Mishina et al. (1995) for these studies on rat and mouse Bmp receptors.
[c] See Kawakami et al. (1996) for receptor expression and Ishidou et al. (1995) for fracture repair.
[d] See Pathi et al. (1999) for BmpRIA, Larraín et al. (2000) for chordin and Bmp-4 inactivation, and Nakayama et al. (2001) for chordin-like protein.
[e] See Zou and Niswander (1996) and Zou et al. (1997).

Changes at the molecular level include antagonism of Bmp to limit cell death and prolonged expression of Fgf to form an elongate limb bud. Both these molecular changes allow continued expression of ck-erg – and therefore Wnt-14 and Gdf-5 – to specify and maintain additional joints (see Figure 7 in Fedak and Hall, 2004).

Stone (2000) described fracture repair in mouse foetuses that had undergone a 'pinch' fracture of the ulna at 14.5 days of gestation and been kept ex utero. Union was rapid – within two days – as might be expected at such a stage. Surprisingly, repair did not involve any changes in Bmp-2 or -4, or Gdf-5.

The opaque patch

The opaque patch is the name for an area of cell death seen as less translucent mesenchyme in the centre of the wing buds of embryonic chicks between H.H. 24 and 25; cell death commences at H.H. 23 and is maximal at H.H. 24/25.

The loss of the cells in this patch through cell death divides the single mesenchymal condensation for ulna and radius into separate condensations. Before H.H. 23, cells in the opaque patch incorporate ^{35}S into glycosaminoglycans (GAGs) but GAG synthesis is suppressed when cell death is maximal. The opaque patch is absent or reduced in talpid[3] mutants, in which radius and ulna fail to separate; cells within the patch continue to synthesize GAGs and go on to chondrify, i.e. as in interdigital cell death, chondrogenesis is expressed when cell death is suppressed.[18]

CELL ADHESION AND MORPHOGENESIS: *TALPID (ta)* MUTANT FOWL

The particular, often highly specialized, even novel morphology of a skeletal element is generated from the accumulated actions of cell division, aggregation and/or migration, differentiation, and deposition of ECM. Consequently, any alteration in cell-to-cell adhesion early in development will affect any or all of these developmental processes. See Chapter 22 for a discussion of how altered cell adhesion during mesenchymal condensation elicits abnormal limb morphogenesis in *Brachypod* mice.

Increased cell-to-cell adhesion has also been proposed as the mechanism underlying the generation of the abnormal morphology of the limbs in *talpid* chick mutants. Isolation of three *talpid* mutants and features of their developing limb buds are outlined in Box 20.1. Despite similarities in morphology, each arose independently and so I discuss each mutant separately.

Talpid²

The AER is normal in mutant embryos until H.H. 22, when it begins to enlarge praecociously. By H.H. 25 the AER is 50 per cent larger than the ridge of similarly aged wild-type embryos. AEMF activity, which is distributed more extensively within *talpid* mesenchyme than within wild type, allows the larger-than-normal ectodermal ridge to be maintained. Non-histone chromatin protein of 125 000 kDa, present in 10^6 copies/nucleus in precartilaginous mesenchyme, is lost with cartilage differentiation in wild-type embryos. In comparison loss is reduced in cartilage from the *talpid²* embryo, indicating the immature differentiation status of mutant cartilage.[19]

The primary defect lies in the mesenchyme, as convincingly shown by recombining *talpid²* and wild-type limb-bud mesenchyme and epithelium (Table 39.2). Epithelium from *talpid²* limb buds responds to wild-type limb mesenchyme by forming a normal limb. The ZPA is distributed normally, although *talpid²* limb buds lack antero-posterior polarity (Box 20.1).[20]

In wild-type limb buds, *Hoxd-11* (GHox-4.6) is expressed in the posterior mesenchyme and in the skeleton, and *Msx-2* (GHox-8) in anterior mesenchyme but not in the skeleton. In both *talpid²* and *diplopodia-5* mutants, ectopic *Hoxd-11* is expressed in anterior limb-bud mesenchyme, which is the mesenchyme from which additional digits arise to generate the polydactylous phenotype. Expression of *Msx-2* is normal in the AER,

Table 39.2 Results of recombinations between *talpid²*, *scaleless* and wild-type limb-bud mesenchyme and epithelium to show that *talpid²* is a mesenchymal deficiency[a,b]

Source of ectoderm	Source of mesoderm	Results
talpid²	++	85% ++
++[c]	++	82% ++
Scaleless	*talpid²*	87% *talpid²*
Scaleless	++	71% ++

[a] Recombinants were grown as flank grafts within normal hosts.
[b] Based on data in Goetinck and Abbott (1964).
[c] ++, Normal (wild-type) embryos.

demonstrating independence of mesenchymal *Msx-2* expression from expression in the AER, and a role for *Msx-2* in digit specification.

Gli-3, a zinc finger gene, is antagonized by *Shh* from the posterior margin of the limb bud. Consequently, a gradient of *Gli-3* repression exists across the limb bud from posterior (where *Shh* is highest) to anterior (where *Shh* is lowest). *Gli-3* is perturbed in *talpid²*, providing a potential basis for the lack of A-P symmetry in the extra digits, an interpretation enhanced by the finding that expression of *Gli-3* is reduced by half in heterozygote and to zero in homozygous *extra toes* (*xt*) mouse mutants.[21]

The pattern of expression of *Shh* is unchanged in *talpid²* limb buds when compared with wild type (Fig. 39.2). However, expression of *Bmp-2*, *Fgf-4*, *Hoxd-13* and *patched* extends more anteriorly in the limbs of *talpid²* than of wild-type embryos.

Talpid³

We know far more about *talpid³* embryos in which prechordal mesenchyme fails to separate and bilateral symmetry of the head is not established. As in *talpid²* the basic skeletal defect resides in the segregation of prechondrogenic limb mesenchyme into condensations that are laid down in the correct proximo-distal sequence, but within which antero-posterior polarity is disturbed (Fig 19.6). Consequently, some elements fuse and others are symmetrically duplicated. Chondrogenesis is abnormal, and osteogenesis is not initiated in long-bone primordia. *Intramembranous ossification, however, is normal.*[22]

The AER is abnormally elongated, especially at H.H. 27 and 28, while the junction between ridge and subjacent mesenchyme is unusually wide, leaving a gap that could be sufficient to prevent AER–mesenchymal interaction. Areas of programmed cell death seen in wild-type limb buds are absent. As in *talpid²*, reduced cell death provides additional cells for the broad spatulate limbs of *talpid³* embryos.[23]

Donald Ede and his colleagues explored the thesis that cell adhesion and movement are disturbed in *talpid³* and responsible for the abnormal morphogenesis. Limb buds from H.H. 24 and 26 wild-type and *talpid³* embryos were dissociated into single cells and patterns of reaggregation in rotation culture observed. Rate and duration of cell movement in *talpid³* cells increase the probability of cell-to-cell contact *in vitro*. Viscometric analysis confirmed that *talpid³* cells are more adhesive than cells from wild-type embryos. Because they are both more adhesive and less mobile, *talpid³* cells aggregate more rapidly and form smaller aggregates than do wild-type cells. *In vivo*, this behaviour is expressed in the failure of condensations to separate from one another and fusion of adjacent skeletal elements.[24]

Niederman and Armstrong (1972) reaggregated *talpid²* cells with wild-type limb mesenchymal cells, expecting to

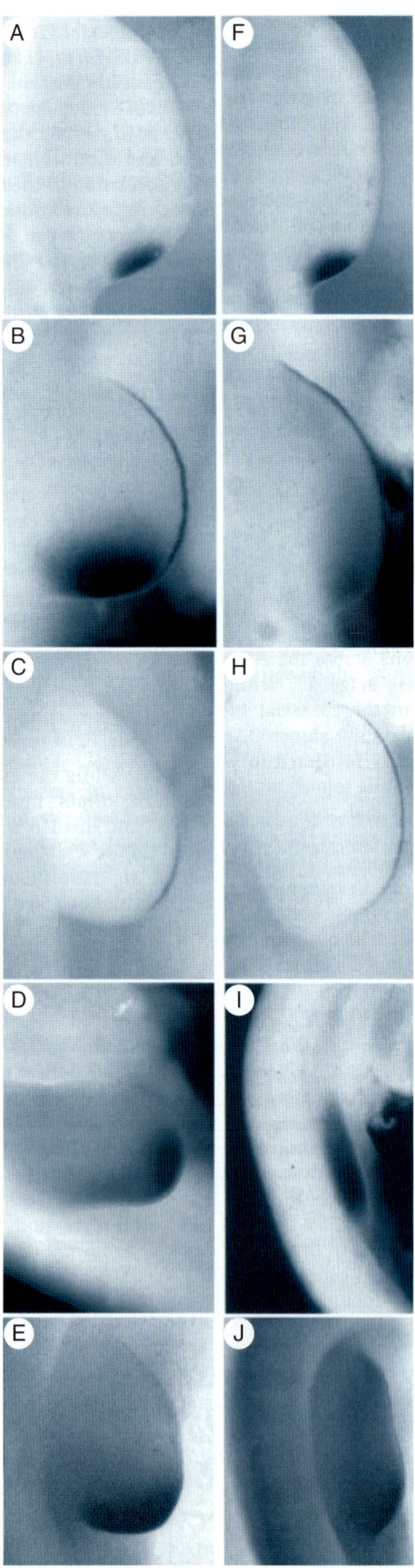

see sorting out if *talpid²* cells also are more adhesive than wild type. No sorting out was obtained, although the methodology used by Ede's group with *talpid³* has not been applied to *talpid²*.

Distal movement of mesenchymal cells was thought to be a normal component of limb outgrowth. According to this thesis, altered properties of *talpid³* cells should slow limb-bud outgrowth; a computer simulation of limb outgrowth generates a *talpid³*-like limb pattern when distal movement of cells is removed from the input. Subsequently, Ede and his colleagues found that the gradient in mitotic activity in *talpid³* limb buds is reversed: it is distal → proximal, not proximal → distal (as it is in wild-type limb buds); the computer simulation incorporated a proximo-distal gradient. Lee and Ede (1989) transplanted wild-type or *talpid³* myogenic cells into quail limb buds. Seventy per cent (10/14) of the control (wild-type) transplants migrated extensively in a proximo-distal direction into the limb bud. In *talpid³* transplants only 18 per cent showed moderate migration and that was all distalward.[25]

Hox-4 is both expressed early and misexpressed in limb buds of *talpid³* embryos. In wild type expression is confined to the posterior of the limb bud, but expression is throughout the limb bud in mutants. The polarizing zone from mouse limb buds or the mouse anterior primitive streak (which possesses polarizing activity) elicits *Hox-4* expression and formation of posterior skeletal elements after being grafted into chick limb buds, giving the impression that up-regulation of *Hox-4* and formation of additional skeletal elements are causally related. However, grafts of neural tube or genital tubercle induce less posterior skeleton and do not alter *Hox-4* expression. In part, proliferation is inhibited in *talpid³* because of defective distribution of *Shh* and *Bmps*. Similarly, injecting anti-*Shh* antibody into cranial mesenchyme adjacent to the hindbrain of wild-type chick embryos results in decreased head size and loss of branchial arch structures.[26]

So, *talpid³* limb buds are distinguished by increased cell adhesion, decreased cell movement, decreased or lack of cell death, decreased rates of cell proliferation, and reversal of the wild-type gradient of cell division. Defective mesenchymal condensations produced as a consequence of these cell behaviors have their later expression in abnormal chondrogenesis and defective limb morphogenesis.

Figure 39.2 Patterns of expression of *Shh, Bmp-2, Fgf-4, Hoxd-13* and *patched* in limb buds of wild-type (A–E) and *talpid²* (F–J) chick embryos. Expression of *Shh* is unchanged in *talpid²* (F) when compared with wild type (A). Expression of *Bmp-2* (G) and *Fgf-4* (H) extends more anteriorly in the mutant than in wild type (c.f. G with B and H with C). Expression of *Hoxd-13* and *patched* also extends more anteriorly in mutant limb buds; compare (I) with (D) for *Hoxd-13* and (J) with (E) for *patched* in *talpid²* and wild type, respectively. Modified from Caruccio *et al.* (1999).

NOTES

1. See Oppenheimer (1974) for an excellent statement of this problem.
2. See Lockshin and Beaulaton (1975).
3. For overviews of cell death associated with skeletal development see Hale (1956b), Saunders (1966), Saunders and Fallon (1967), Whitten (1969), Mitrovic (1971, 1977), Lockshin and Beaulaton (1975), Ten Cate et al. (1977), Hinchliffe (1981, 1982), Hirata and Hall (2000), and the chapters in Ede et al. (1977a) and Bowen and Lockshin (1981). For patterns of cell death during development of embryonic chicks from gastrulation through organogenesis, see Hirata and Hall (2000).
4. See Saunders et al. (1962) and Saunders (1966) for the initial studies and Zuzarte-Luís and Hurlé (2002) for a brief history and overview.
5. See Saunders and Fallon (1967) for decline in DNA, and Webster and Gross (1970) for cross-linking.
6. See Mottet and Hammar (1972) and Pollak and Fallon (1974) for the protein data, Pollak and Fallon (1976) for DNA and RNA synthesis, and Hurlé and Hinchliffe (1978) for acid phosphatase.
7. See Saunders (1966) and Saunders and Gasseling (1968).
8. See Hinchliffe and Ede (1973) and Hinchliffe (1977a) for the *wingless* embryos. Ede and Agerback (1968) also supported altered adhesiveness.
9. See Laufer et al. (1997) and Rodriguez-Esteban et al. (1997) for *radical fringe*, and Zeller and Duboule (1997) for a review of dorso-ventral limb polarity and the role of *radical fringe*.
10. See Saunders and Fallon (1967) for birds, Fallon and Cameron (1977) for reptiles, Milaire (1963, 1965), Menkes et al. (1965), Ballard and Holt (1968) and Kelley (1973) for mammals, and Cameron and Fallon (1977b) for absence of interdigital cell death in urodele amphibians.
11. See Pautou (1974) for webbing in duck feet, Milaire (1965) and Hinchliffe and Thorogood (1974) for mice and the *talpid*[3] mutants, and Saunders and Fallon (1967) for induced cell death.
12. In a histochemical analysis, Hammar and Mottet (1971) found that succinate dehydrogenase was absent from the interdigital area at H.H. 26 and 27. Fallon et al. (1974) used biochemical methods of assay and histochemistry on frozen sections to repeat Hammar and Mottet's study and found succinate dehydrogenase in the interdigital areas before necrosis set in. However, Fallon's group also obtained a reaction in the whole limb in the absence of the enzyme substrate.
13. See Barasa (1960, 1964), Janners and Searls (1971), Ede and Flint (1972) and Cairns (1975) for the early studies.
14. Both Bmp-2 and Bmp-7 play roles in later limb development through involvement in specification of the location of joints (Macias et al., 1997). See Fig. 12.2 for beads implanted into chick embryos.
15. See Merino et al. (1998, 1999a,b) and Vogt and Duboule (1999) for these regulatory pathways.
16. See Lee et al. (1993, 1994, 1999) for these studies and for an introduction to the literature on *gas-2*.
17. See Tang et al. (2000) for *ex utero*, Dan and Fallon (2000) for how the concentration of Bmp patterns the digits, Dawson (2003) for hyperphalangy in digits II and III in the harbour porpoise, *Phocoena phocoena*, Gol'din (2004) for continuation of flipper skeletal growth for four to eight years and allometry of the first phalanges in the harbour porpoise, Richardson and Oelschläger (2002) for hyperphalangy in the spotted dolphin, *Stenella attenuata*, and see Crumly and Sánchez-Villagra (2004) for what may be the minimal possible 'hyperphalangy,' the multiple and independent evolutionary increase from one to two phalanges in digit V in land tortoises.
18. See Dawd and Hinchliffe (1971) for GAG synthesis before H.H. 23, and Hinchliffe and Thorogood (1974) for the *talpid*[3] mutant.
19. See Perle and Newman (1980) and Perle et al. (1982) for the chromatin study.
20. See Abbott et al. (1960) for the *talpid*[2] phenotype, Goetinck and Abbott (1964) and MacCabe and Abbott (1974)) for praecocious enlargement of the AER and distribution of AEMF, and MacCabe and Abbott (1974) for the PNZ.
21. See Rogina et al. (1992) and Coelho et al. (1992) for *Hoxd-11* and *Msx-2*, B. Wang et al. (2000) for *talpid*[2], and Schimmang et al. (1992) for *extra-toes*.
22. See Ede and Kelley (1964a) and Ede (1971, 1980) for generic defects in *talpid*[2] embryos, and Ede and Kelley (1964b) and Hinchliffe and Ede (1967, 1968) for defective condensation and skeletogenesis.
23. See Hinchliffe and Ede (1967) for the abnormally long AER, Ede et al. (1974) and Ede (1980) for separation of the AER from distal mesenchyme, and Hinchliffe and Ede (1967), Dawd and Hinchliffe (1971), Hinchliffe and Thorogood (1974) for programmed cell death.
24. See Ede and Agerback (1968), Ede and Law (1969), Ede (1971) and Ede and Flint (1972, 1975a,b).
25. See Ede and Agerback (1968) for distal movement, Ede and Law (1969) for the computer model, and Ede et al. (1975) for reversal of the gradient.
26. See Izpisúa-Belmonte et al. (1992b) for *Hox-4*, Francis-West et al. (1995) for *Shh* and *Bmp*, Ahlgren and Bronner-Fraser (1999) for the anti-*Shh* antibody, and see Britto et al. (2000), Francis-West et al. (2003) and Richman and Lee (2003) for overviews of Shh in both limb and craniofacial mesenchyme.

Before Limbs There Were Fins

Which fish are the only extant vertebrates with three functional sets of paired appendages?[1]

The term 'fin' can be misleading. Bony and cartilaginous fishes have two types of fins: (i) *paired pectoral and pelvic fins* that project laterally from the body (Figs 40.1 and 40.2); and (ii) *unpaired median fins* along the dorsal, ventral and posterior surfaces of the body (Figs 40.3–40.5). A further complication is that not all fish have all fins. The pelvic fins can be lost or turned into a sucker on the underside of the body (Figs 40.6–40.8). Any one of the three median fins (*dorsal, caudal and anal*) can be duplicated, fused or lost. I discuss median fins, especially the dorsal, before moving on to paired fins.

DORSAL MEDIAN UNPAIRED FINS

Amphibian tadpoles, fish larvae and many adult fishes have one or more median unpaired fins that run along the dorsal midline and are known as *dorsal (median, unpaired) fins*. These should not be confused with the paired pectoral and pelvic fins, although, like paired fins, the dorsal fins of fishes (but not amphibians) contain skeletal elements as *fin rays (lepidotrichia)* (Figs 5.6 and 40.1–40.4).[2]

Teleost fish

The three median unpaired fins of teleost fishes – a dorsal, caudal (tail) and anal (ventral) fin – can be identified even when they occupy most unusual positions on the body (Box 40.1), perhaps the most extreme example being the long-spine razorfish or shrimpfish, *Aeoliscus strigatus*, from tropical northern Queensland and the Indo-Pacific. The highly compressed bodies of these 10- to 14-cm-long fish have all three median fins on the ventral surface (Fig. 40.9), a most unusual arrangement until you realize that the razorfish swims with its body vertical, head down and with the dorsal surface facing the direction in which it is swimming.

The distribution of fins within some of the more well-known groups of teleost fishes is shown in Box 40.1. Those taxa with dorsal fins often also have one or more *ventral (median, unpaired) fins* on the ventral body surface. Depending on clade or taxon, dorsal and ventral fins may be separate or continuous. The ventral fin in teleost fishes is usually referred to as the anal fin (also see Box 43.1). At the level of developmental mechanisms and evolutionary transformations, we would like to know whether (i) median ventral unpaired, (ii) median dorsal unpaired and (iii) lateral paired fins share any history, genes, development, structure or function.[3]

Life style

Location, size and shape of the median fins are closely associated with body shape, life style and food-gathering strategies, as shown below.

- *Roving predators* such as minnows (Cyprindae) have a streamlined body form with the fins more or less evenly distributed over the body to provide stability and manoeuverability.
- *Lie-in-waiting predators* such as pike (Esocidae) and barracudas (Belonidae) have the dorsal and anal fins positioned far back on the body and opposed to each other to provide greater thrust when lunging at passing prey.

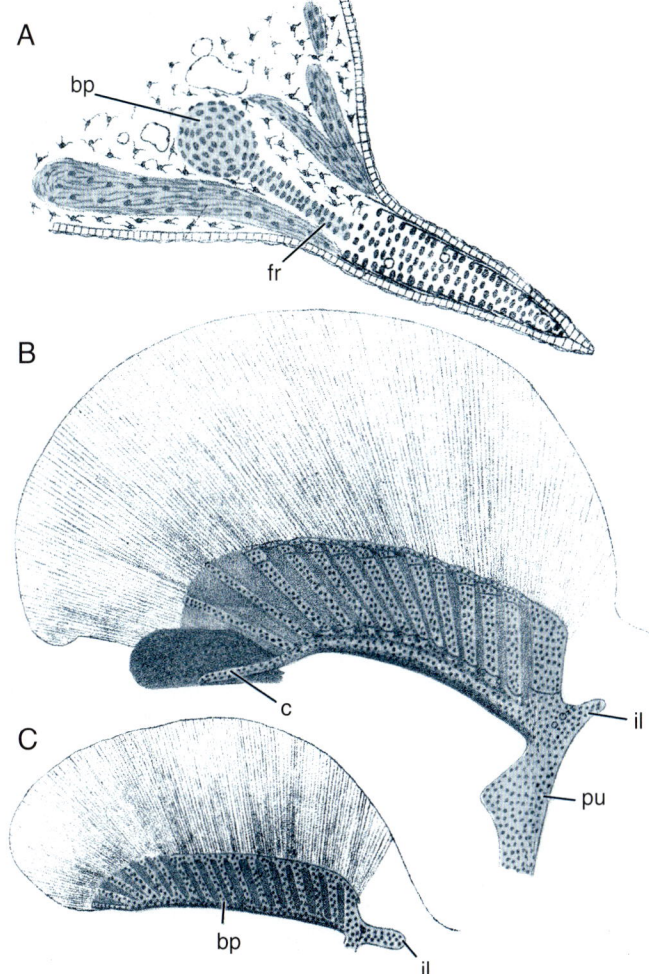

Figure 40.1 Balfour's depiction of the development of the skeleton of the pelvic fins of the 'dogfish,' *Scyllium*. (A) A transverse section of an embryonic pelvic fin, showing the cartilaginous basipterygium (bp) and an initial fin ray (fr). (B) A longitudinal view of the pelvic fin from a male embryo with a well-developed blade of fin rays and endoskeletal supporting basipterygium. A posterior process of the basipterygium continues as the clasper (c). Iliac (il) and pubic (pu) sections of the pelvic girdle are labeled. (C) A longitudinal view of a pelvic fin from an older female embryo with basipterygium (bp) and the iliac process (il) of the pelvic girdle. (B) and (C) are both drawn at the same magnification. Compare with the pectoral fins in Fig. 40.2. Modified from Balfour (1882).

- *Surface-oriented predators* such as the killifish (*Fundulus*) or flying fishes (Exocoetidae) are dorso-ventrally flattened and have the dorsal fin far posterior on the body.
- In *eel-like fishes* the dorsal fin frequently runs along most of the length of the body and may merge with the caudal fin (Fig. 40.9).

Developmental origins

Little is known about the developmental origins of dorsal fins: a study in the early 1950s showed that the dorsal, unpaired fin of amphibians is induced by neural crest and contains mesenchyme of neural-crest origin. Almost four decades later a study using interspecific transplantation between the South African and the Kenyan clawed frogs, *Xenopus laevis* and *X. borealis*, confirmed a neural-crest cell contribution to the mesenchyme of the dorsal fin.[4]

M. M. Smith *et al.* (1994) used DiI labeling of trunk neural crest or somitic mesoderm of zebrafish embryos to follow trunk neural-crest cells (NCC) into the caudal fin mesenchyme. The embryos could not be followed for long enough to determine whether the bony fin rays arose from these NCC. The presumption is that they do.

Interestingly, but perhaps not surprising considering their location, tadpole ventral fins are not induced by neural crest but do contain neural crest-derived mesenchyme. We do not know about teleosts. We do know from a study of median fin development in the zebrafish, *Danio rerio*, that between six and 45 days after fertilization, *Gdf-5* is expressed in mesenchyme *between* the condensations that will form the endoskeletal elements of the dorsal, caudal and anal fins, expanding distally as development proceeds, being expressed later as the radials segment. The similarity to the role of Gdf-5 in the segmentation of cartilages in the limbs buds is considerable.[5]

Implanting Fgf-7 and Fgf-10 into the flank or dorsal midline of chick embryos induces an ectopic apical epithelial ridge (AER) at those sites. These ectopic ridges are functional; they induce mesenchymal cells to accumulate. Yonei-Tamura *et al.* (1999) put an evolutionary spin on their study in discussing whether the ectopic dorsal AER is a vestige of the dorsal fin. Replacing limb-bud epithelium of embryos of the sharp-ribbed newt, *Pleurodeles waltl*, with tadpole dorsal or ventral fin epithelium has no adverse effect on proximo-distal skeletal development (Lauthier, 1978), although given that limb-bud epithelium in urodeles may only pattern the digits and not the more proximal skeleton (Chapter 37), I don't know quite what to make of this result. It certainly is worth exploring.

Evolutionary origins

Interest in the evolutionary origins and relationships of dorsal fins goes back at least to the fin-fold theory in which paired fins are seen as the remnants of a once continuous fin fold that circled the body of early fish.[6]

Ahlberg (1992) argued that the posterior dorsal and anal median fins of coelacanths are equivalent to paired fins, the median fins having secondarily acquired their skeletal pattern from the paired fins. More recently, and more generally, Minelli (2000) argued that arthropod *and* vertebrate appendages are duplications of the body axis (minus the endodermal component). While Ahlberg sees the transformation from paired to medial fins as homeotic, Minelli sees appendages *and* body axis as homoplastic.

Figure 40.2 Balfour's depiction of the development of the skeleton of the pectoral fins of the 'dogfish,' *Scyllium*. (Top) A transverse section of an embryonic pelvic fin showing the cartilaginous basipterygium (bp) and fin ray (fr). (Left) A longitudinal section of an embryonic pectoral fin showing the cartilaginous basipterygium (bp), fin rays (fr) and pectoral girdle (pg). (Right) A longitudinal view of the pectoral fin and portion of the pectoral girdle of a more advanced embryo. Compare with the pelvic fins in Fig. 40.1. Modified from Balfour (1882).

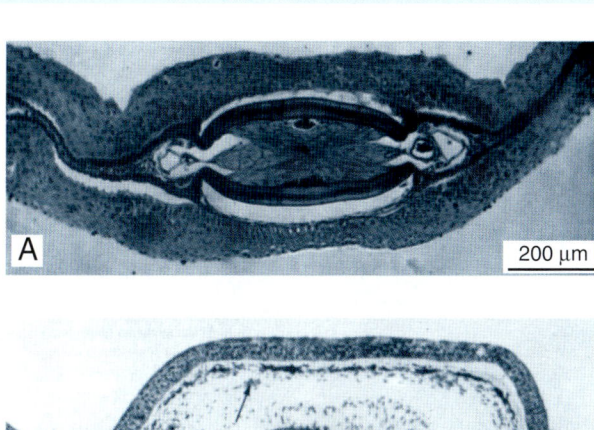

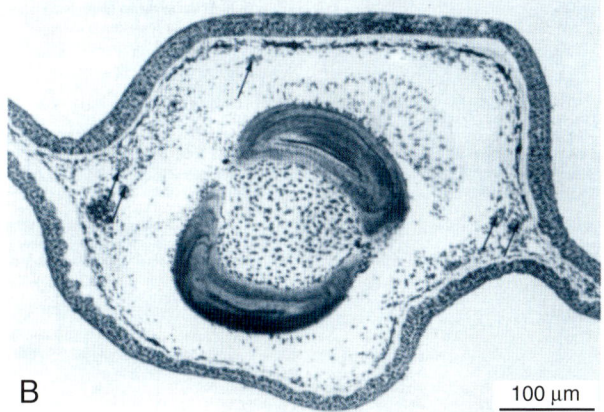

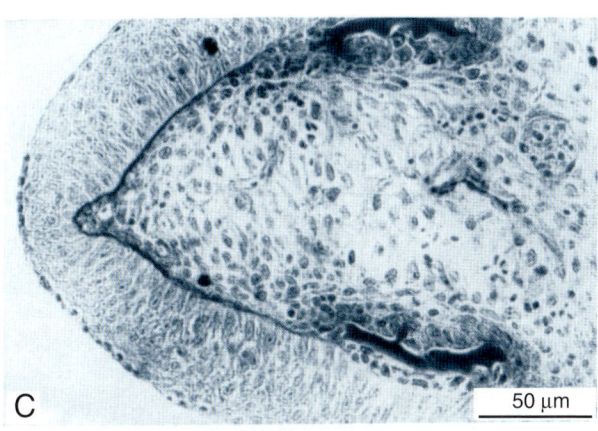

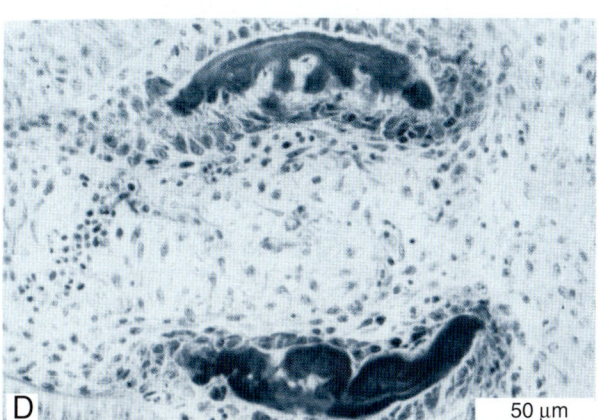

Figure 40.3 Bony fin rays in teleosts. Cross sections through the tail fin of the blue gourami, *Trichogaster sumatranus* (A), and through the pectoral fin of the European scaleless blenny (shanny), *Blennius pholis* (B), showing the bony plates of double rays. (C, D) Cross sections through two regions of a regenerating tail fin in the goldfish, *Carassius auratus*, to show ossification of the fin rays. (A), (C), (D) modified from Haas (1962), (B) from Géraudie and Singer (1977).

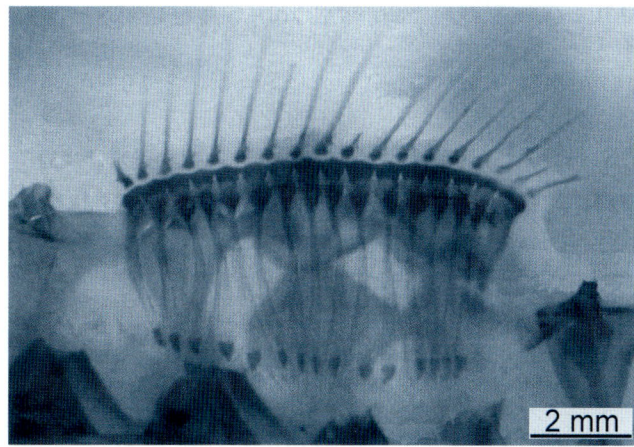

Figure 40.4 The skeleton of the dorsal fin of the northern sea horse, *Hippocampus hudsonius,* showing the bony fin rays above and the proximal radials below. Adapted from Ashley-Ross (2002).

Figure 40.5 A 230-kg Queensland grouper, *Epinephelus lanceolatus,* with anteriorly placed pectoral fins and sturdy dorsal and ventral fins. This species is known to grow to 2.7 m in length and attain weights of 400 kg. Photographed by the author.

Figure 40.6 The sand fish (Jamaica sand fish), *Awaous tajasica,* has a prominent, circular ventral sucker (disc) that lies ventrally between the pectoral fins; also see Fig. 40.7. Note the absence of pelvic fins. Image courtesy of Patricia Avendaño.

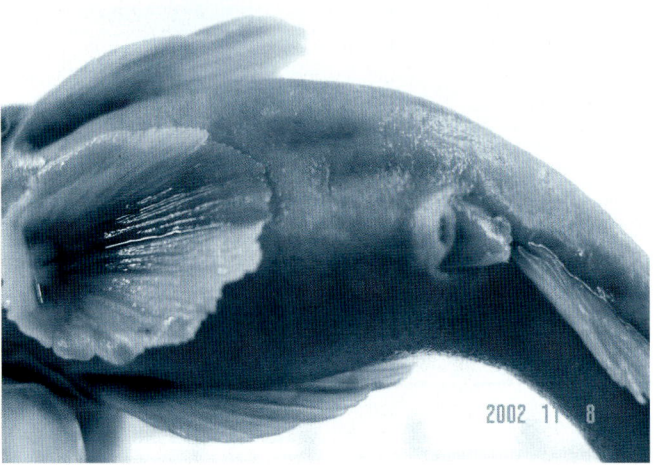

Figure 40.7 The sucker of the sand fish, *Awaous tajasica*, to show the relationship to the pectoral fins; also see Fig. 40.6. Image courtesy of Patricia Avendaño.

Figure 40.8 The Atlantic spiny lumpsucker, *Eumicrotremus spinosus*, to show the large, fleshy sucker on the ventral surface. Anterior to the right. Image courtesy of Patricia Avendaño.

Box 40.1 Fins that come and fins that go

Most teleosts have a dorsal fin, presence of an anal fin being more variable. A fish without an anal (caudal) fin is technically *tailless*. Surprisingly, tailless fish are known (Box 43.1).

Members of some families, such as the Scombridae (mackerels, tunas, bonitos), possess two dorsal fins. A few others such as the Gadidae (cod, haddock) have three dorsal spines and, in the case of the Gadidae, two anal fins. Phylogenetically, the dorsal fin began as a stiff spine; members of some families have a combination of spines and soft-rayed fins. A survey of families of fish with *additional* dorsal, anal and/or ventral (median) fins or that *lack* one or more of the median fins follows.

Dorsal fin

Two dorsal fins – mackerels, tunas, bonitos (Scombridae)
Two dorsal fins with or without spines; anal (caudal) fin absent – dogfish, sharks (Squaliformes)
Two dorsal fins, anal fin absent – rays (Rajiformes)
One or two dorsal fins – lampreys (Petromyzontidae), longnose chimaeras (Rhinochimaeridae)
One dorsal fin preceded by a venomous spine – shortnose chimaeras or ratfishes (Chimaeridae)
No dorsal fin[a]; anal fin extremely long (from near pectoral fin origin to near tip of body), absent or greatly reduced; eel-like body form – South American knifefishes (Gymnotiformes)[b]
No dorsal fin – hagfishes (Myxinidae)[c]

Anal fin

No anal fin – stingrays (Dasyatidae), river stingrays (Potamotrygonidae), devil rays (Mobulidae), and eagle and manta rays (Myliobatididae)[d]
Anal fin variably present – sharks (Selachimorpha)
No anal fin and no pelvic fins – knifefishes (Gymnarchidae)
Anal fin absent or formed by a few rays of a pseudocaudal fin derived from posteriorly migrated dorsal and anal fin rays – molas (family Molidae in the order Tetraodontiformes)

Base of the anal fin long and merged with what remains of the caudal fin; caudal fin skeleton reduced or absent; tail easily regenerated when lost – spiny eels (Notacanthiformes)

Dorsal, anal and pectoral fins

Dorsal, anal and pectoral fins may be absent in adults of some species of pipefishes and sea horses (family Sygnathidae in the order Syngnathiformes)

[a] Members of one family of knifefishes, the ghost knifefishes (family Apteronotidae), do have have a distinct caudal fin.
[b] Species such as the banded knifefish, *Gymnotus carapo* (Gymnotiformes), lack a dorsal fin but possess a long anal fin and a deep but thin body that allows undulatory locomotion. The dorsal section of the body is sufficiently thin that it acts like a dorsal fin.
[c] Hags do possess folds of skin, which are similar to the fin fold seen in teleost fishes during early life and which function as dorsal and anal fins, respectively.
[d] Manta rays, members of the family Myliobatididae, are the only extant vertebrates with three functional sets of paired appendages; the pectoral fins are structurally and functionally subdivided. The species are the giant manta or Pacific manta ray, *Manta birostris*, and the smoothtail mobula, *Moula thurstoni*. The three pairs of appendages are: (i) the cephalic pair, which project forward around the head, assist in feeding and are an anterior subdivision of the pectoral fins; (ii) the greatly expanded pectoral fins; and (iii) the pelvic fins, including the paired claspers which could be said to represent a fourth pair of appendages in males.
Little is know about the skeletons of these appendages in adults. Nothing is known of their development. Left and right cephalic fins are connected basally but separate from the pectoral fins. We do know that the skeletal elements of the cephalic fin are derived from radials of the pectoral fin and from the anterior dorsal-most region of the neurocranium extending over the nasal capsules. The scapulocoracoid, which supports the pectoral fins, is extended anteriorly and the muscles of the cephalic fins derive from the anterior-most muscle fibres of the pectoral fins (Nishida, 1990). See the text, Holst and Bone (1993) and Lucifora and Vassallo (2002) for modification of radials in the pelvic girdles of those 'legged,' 'bipedal' or 'walking' skates that display pelvic fin walking.

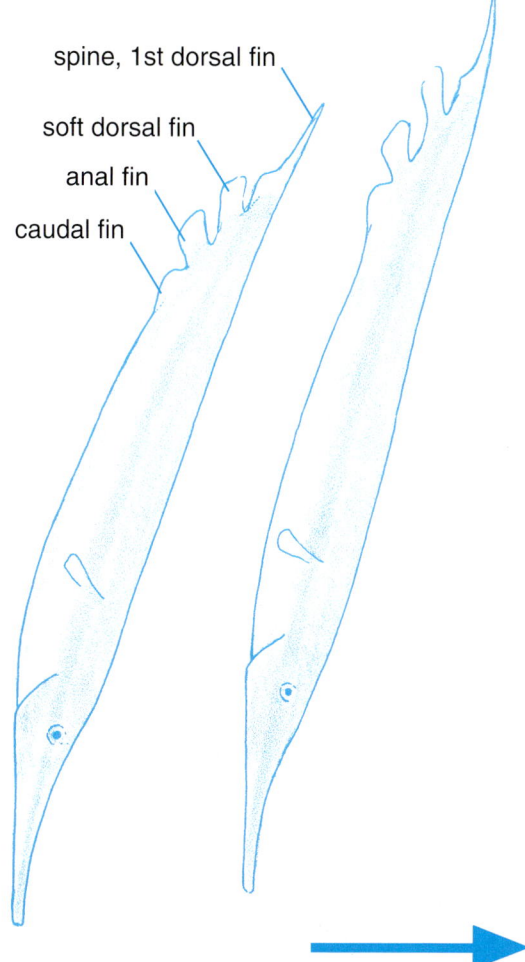

spine, 1st dorsal fin

soft dorsal fin

anal fin

caudal fin

Figure 40.9 Dorsal, anal and caudal fins of the long-spine razorfish (shrimpfish), *Aeoliscus strigatus*; all lie on the ventral surface, an arrangement that only makes sense when we realize that these slender, flattened fish swim (as shown) with the body vertical in the water column and the dorsal surface (to the right) facing the direction of movement, shown by the arrow.

PAIRED FINS

Fins minus fin rays plus digits = limbs

Fins and limbs on the one hand and fin buds and limb buds on the other have much in common.

(1a) Fin buds are homologues of tetrapod limb buds as developmental precursors of paired appendages.

(1b) Fins and limbs are homologous as paired appendages.

(2a) The T-box gene, *Tbx-5*, specifies the location of the forelimbs along the antero-posterior tetrapod body axis (Fig. 35.4).

(2b) In teleost fish, *Tbx-5* is expressed before fin buds appear in embryogenesis and specifies the anterior location of the pectoral fins – if the demonstration by Han *et al.* (2002) with zebrafish is typical of other teleosts.

(3a) Knocking out *Tbx-5* in mice prevents forelimb but not hind limb development.

(3b) Knocking out *Tbx-5* in zebrafish prevents pectoral but not pelvic fin development.

(4a) The *Tbx-5* mouse phenotype is equivalent to that seen in *Brachyury* mutant embryos. *Brachyury* is an important T-box gene (Box 35.2).

(4b) The zebrafish *Tbx-5*-knockout phenotype is equivalent morphologically to that seen in *spadetail*, a mutation in *Tbx-5*. Both result from defective migration of the lateral-plate mesoderm that normally moves into flank to create the fin primordium.

Further similarities (homologies) will emerge as I review fin buds, fin skeletons, the action of RA on fins and limbs, and fin regeneration. The primary basis for these homologies is discussed at the end of the chapter: limbs evolved from fins or, to put it another way, fins minus fin rays + digits = limbs.

In a quite remarkable instance of convergence between fins and limbs, the pelvic fins of the Rio skate, *Rioraja agassizi*, the small nose sand skate, *Sympterygia bonapartii*, and skates within the genus *Psammobatis* (the smallthorn sand skate, *P. rudis*, blotched sand skate, *P. bergi* and shortfin sand skate, *P. normani*) are modified for walking along the bottom. Each pelvic fin consists of a posterior fin-like lobe and an anterior leg-like lobe, the latter supported by radials that are functionally equivalent to femur, tibia and digits, with a broad condyle on the pelvic girdle accommodating a wide range of movements.[7]

Fin buds and fin folds

Fin buds of bony and cartilaginous fishes (teleosts and elasmobranchs) possess a ridge-like structure, often referred to as a *pseudoapical ridge*. In teleosts this ridge transforms into an extensive fold, the *fin fold*, at the apex of the fin bud. Fin folds are considerably larger than the AER of a limb bud but function like an AER. As in AER–mesenchyme interactions, ridge and fin fold interact with fin-bud mesenchyme; centres of proliferation within pectoral fin mesenchyme are at their height when the ridge is present. Molecular signaling also is similar.

The gene *dackel* (*dak*) acts in the epithelium of the zebrafish pectoral fin to maintain fin-fold signaling. *Dak* acts with *Shh* to activate *Fgf-4* and *-8*, but can act independently of *Shh* to maintain the morphology of the fin fold. Furthermore, *en-1* in ventral non-ridge epithelium needs the signal from the ridge.[8]

Shh is expressed in the posterior zone of zebrafish fin buds in a location equivalent to the ZPA of amniote limb buds. *Shh* is required to establish some aspects of A-P polarity, to activate posterior *Hoxd* genes, for development of the fin fold, growth of the fin bud and formation of the fin endoskeleton. As seen in *Sonic you*, a mutation that disrupts *Shh*, A-P polarity *can* be established, albeit transiently, in the zebrafish pectoral fin bud in the absence of *Shh* (C. J. Neumann et al., 1999).

Ruvinsky et al. (2002) isolated seven new *T-Box* genes from amphioxus, each of which is equivalent to two or three vertebrate *T-Box* genes. This finding supports the single (or perhaps double) whole genome duplication at the origin of the vertebrates. Previously, Ruvinsky and Gibson-Brown (2002) had reviewed tetrapod limb evolution in relation to serial homology, tinkering, the two models proposed (a gill-arch or a lateral fin-fold origin) and the role of T-box genes (Fig. 35.4).

Examination of the genomic organization and expression of 32 *Hox* genes in zebrafish revealed: several genes not found in tetrapods; the possibility of an extra *Hox* cluster; and that anterior *Hox* genes are compacted over a shorter antero-posterior region than in tetrapods. Sordino et al. (1995), who cloned *Hoxd* and *Hoxa* from zebrafish (*Danio rerio*), showed that *Hoxd-11, -12* and *-13* are not expressed in the anterior half of fin buds but *are* expressed in the anterior half of limb buds. This expression pattern, coupled with what we know of the function of these *Hox* genes in distal limb development, leads to the conclusion that the entire distal portion of a fin bud is equivalent to the posterior portion of a limb bud. Within limb buds, two enhancers of *HoxD* compete for interaction with the same promoter. Consequently, the position of the *Hox* gene relative to the enhancers could determine expression patterns. The regulatory elements near *Hoxd-12* are conserved in tetrapods but not in fish.[9]

Fin skeletons

Fins consist of basal cartilaginous elements (Figs 40.1, 40.2 and 40.4) that are homologous to proximal tetrapod limb elements (humerus/femur, radius/ulna, tibia/fibula). As in the limb skeleton, the basal fin cartilages are derived from lateral-plate mesoderm. Unlike limbs, where these cartilages form elongate long bones (digits being small in comparison), basal fin cartilages are quite short in most taxa. The bulk of the fin skeleton consists of the fin rays (*lepidotrichia*), which are thought to be of neural-crest origin; no skeletal element in any tetrapod limb has a neural-crest origin. Presence of lepidotrichia in fins and their absence in limbs, and presence of digits in limbs and their absence in fins, are the two major features distinguishing fins from limbs.

Examination of the mineral phase of the bony fin rays in the trout, *Oncorhynchus mykiss*, demonstrates needle-like crystals composed of hydroxyapatite. Kemp and Park (1970) described fusion of collagen fibres as the basis for how the bony and soft rays (lepidotrichia and actinotrichia) form during regeneration of the tail fins in tilapia, *Tilapia mossambica*, a process they describe as the equivalent of mineralizing dermal bone. Once formed, the collagen is not inert. Regeneration of the tail fin in the goldfish, *Carassius auratus*, is accompanied by turnover of collagen and elastoidin within actinotrichia as blastemal cells and

basal laminae control the number and distribution of the new actinotrichia that form. Fin rays grow as successive segments are added distally to existing rays. In zebrafish, initiation of these segments, addition of new segments, and control of segment length are regulated independently during growth of the caudal fin – which is isometric with growth of the body – as demonstrated by analysis of two mutants, *short fin* (*sof*), in which segment number and segment size are both reduced, and *long fin* (*lof*), which appears to allow additional segmentation of fin rays. Patterning of the fin rays also is under the influence of *Shh*, *Patched-1* and *Bmp-2*.[10]

The fin skeleton has been studied extensively and intensively in a comparatively small number of taxa; of course, the same can be said of tetrapod limbs. Haas (1962) carried out pioneering studies on fin-ray development (Fig. 40.3). Lanzing (1976) examined the fine structure of the fins and fin rays in tilapia, *Tilapia mossambica*. Also studied in some detail are the:

- development of the appendicular support for the anal fin (including sexual dimorphism) in the western mosquito fish, *Gambusia affinis* (Rosa-Molinar et al., 1994);
- fin skeleton of Siamese fighting fish, *Betta splendens*, including intraspecific variation and interspecific comparisons (Mabee and Trendler, 1996);
- development of dorsal, anal and paired fins in dentex, *Dentex dentex* (Koumoundouros et al., 2001); and
- ontogeny of the skeleton in the anemone fish, *Amphiprion melanopus* (Green and McCormick, 2001), which has the advantages of rapid development, early mineralization of the jaws, and mineralization of the fins at metamorphosis.[11]

Table 40.1 shows the order of onset of chondrification and ossification of the head, vertebral and fin skeletons of the Red Sea bream, *Pagrus major*, but bear in mind that such sequences vary from taxon to taxon, even within closely related species (Tables 5.7–5.9).

Jacqueline Géraudie from Paris has carried out and continues to carry out a major series of studies on pelvic and pectoral fin development and regeneration in numerous species of teleost fishes, including salmon and trout (*Salmo gairdneri, S. trutta fario*), the killifish (*Fundulus heteroclitus*), zebrafish (*Danio rerio*), the Senegal bichir, *Polypterus senegalus*, and the reedfish (snake fish, ropefish), *Calamoichthys calabaricus*.[12]

Her transmission and scanning electron microscopical studies of pelvic fin development in trout and salmon included details on the epithelial–mesenchymal interface and innervation of the fin bud. In another study, Géraudie used ^{3}H-thymidine autoradiography to show that cells of the pseudoapical ridge of the pelvic fin buds of the trout are mitotically active. Nerves are present early – earlier than at equivalent stages of limb buds – and so could play a role in fin development, as Géraudie went on to show (see below). An analysis of trout pelvic fins showed that bony fin rays mineralize within basal

Table 40.1 Order of onset of chondrification and ossification of the skeleton of the Red Sea bream, *Pagrus major*[a]

Length (mm)	Head skeleton	Vertebral column	Fins
2.80	–	–	Cleithrum (ossification) Coracoid-scapula (chondrification)
2.90	Neurocranium, branchial arches and hyoid arch (chondrification)	–	–
3.5	–	–	Pectoral fin-supports (chondrification)
4.5	–	–	Caudal fin (chondrification)
4.85	–	Neural arches (chondrification)	–
5.20	–	Haemal arches (chondrification)	–
5.30	–	–	Caudal fin (ossification)
5.90	Neurocranoum and hyoid arch (ossification)	–	–
6.00	Branchial arches (ossification)	Neural arches and vertebrae (ossification)	Anal fin-supports (chondrification) Pectoral fin-supports (ossification)
6.5	–	–	Pelvic fin-supports (chondrification)
7.0	–	–	Dorsal fin-supports (chondrification)
7.05	–	Haemal arches (ossification)	–
8.0	–	Plural ribs (chondrification)	Scapula and coracoid (ossification)
8.50	–	Plural ribs (ossification)	Dorsal, anal and pelvic fin-supports (ossification)
9.0	–	Dorsal ribs (ossification)	–

[a]Based on data from Matsuoka (1985, 1987).

lamellar collagen, while actinotrichia in trout and the coelacanth, *Latimeria chalumnae*, are composed of elastoidin.[13]

You will recall from Chapter 14 that urodele limbs require a minimal threshold of nerve fibres if regeneration is to ensue. Géraudie demonstrated the requirement for a minimal density of nerve fibres for *Fundulus* pectoral fins to regenerate. Species that cannot regenerate their pectoral fins have a nerve fibre density of around 16 fibres/fin stump, which compares with 25–35 fibres/fin stump in species that can regenerate. The synthesis and accumulation of protein, and synthesis of DNA during regeneration, are both nerve-dependent.[14]

TEM and SEM analyses of the pectoral fin dermoskeletons of *Polypterus senegalus* and *Calamoichthys calabaricus* showed that the fin rays consist of ganoine on cellular bone without dentine. Two types of ganoine are known, enamel and enameloid-like. The specialized camptotrichia of the fin skeleton were shown to be composed of cellular bone in the Queensland lungfish, *Neoceratodus forsteri*, and of acellular bone in the African lungfish, *Protopterus annectens*. Superficial layers are mineralized, deeper layers are not.[15]

Meinke *et al.* (1979) used X-ray diffraction to study the structure of the mineralized scales and bone in the bichir, *Polypterus senegalus*. The mineral is hydroxyapatite and whitlockite, the structure of the dermal skeleton resembling those of Palaeozoic and Mesozoic actinopterygian fishes; whitlockite also is deposited in arthritic human cartilage (Ali, 1985). In her developmental study, Meinke (1982a) identified vascular and lamellar bone, and dentine and enameloid in the dermal skeleton, bone and dentine arising from *the same population of mesenchymal cells*, with bone developing first and dentine fusing to it. Meinke (1982b) also studied tooth microstructure in *Polypterus*, especially with regard to timing of deposition of enamel, enameloid and dentine.

Retinoic acid

Retinoic acid (RA) is an important *morphogen* for skeletal tissues in vertebrates, a morphogen being a molecule whose primary or exclusive role is in patterning elements rather than triggering cell differentiation. Applying exogenous noggin and RA transforms the identity of facial processes in chick embryos. Applying both molecules on beads to the maxillary field of embryos of H.H. 15 results in duplication of the frontonasal processes (but not the maxillary processes) to produce duplicated arches containing cartilage but not maxillary bone. The *legless* insertional mutation produces limb and craniofacial deformities in homozygous mice, including absence of the AER and increased mesodermal and ectodermal cell death in embryos that are especially sensitive to RA.[16]

The important role that RA plays in patterning and morphogenesis during development, in the growth and regeneration of tetrapod limbs (Box 40.2), and in the craniofacial skeleton did not arise with the tetrapods. RA plays similar roles in craniofacial development in lampreys and in teleost fins. Treating Japanese lampreys, *Entosphenus japonicus*, with all-*trans* RA leads to truncation of the rostral neural tube and to loss of the pharynx and branchial arches, but has no effect on the myotomes, indicating separate somitomeric (vertebral) and branchiomeric (branchial arches) programs and providing insights into the interface between head and trunk (Kuratani *et al.*, 1998, 2001, 2002).

RA is required as early in zebrafish embryonic development as the pre-segmentation stage and serves multiple roles thereafter, as outlined below.

- Exposing *Fundulus* to 5×10^{-7} to 5×10^{-4} molar RA for two hours during gastrulation leads to cell death, reduced growth, duplication of pectoral fins and deletion of craniofacial cartilages.[17]

Box 40.2 Retinoic acid as a morphogen

With a flurry of activity beginning in the mid-1980s it appeared that the limb morphogen had been discovered. The molecule that caused all the excitement, including the declarative commentary 'We have a morphogen!' which accompanied the report by Thaller and Eichelle (1987) in the journal *Nature*, was retinoic acid (RA), which:

- is distributed in a gradient across limb-bud mesenchyme, with a source in the region of the zone of polarizing activity (ZPA), declining to an anterior sink;
- acts directly on limb mesenchyme – a nuclear RA receptor (RAR) is localized in limb mesenchyme (Box 20.3);
- acts on as few as 100 cells to induce mirror-image duplication;
- functions in the limb in a position- and dose-dependent manner, with action that is irreversible after 12–18 hours' exposure;
- evokes normal A-P patterning when implanted preaxially into wing or hind limb buds (i.e. in the absence of the normal ZPA); and
- correlates with the length of the AER and with the number of digits specified.[a]

To investigate how limb mesenchyme is reprogrammed by retinoids, Wilde *et al.* (1987) implanted RA-soaked beads into the future limb-bud regions of chick embryos at H.H. 10, a treatment that led to limb duplication or limb loss (Table 3.3). To characterize the effective time of action of the RA, beads were removed after various times *in ovo*, demonstrating that early limb buds require longer exposure to RA to produce duplications than do limb buds of older embryos, while embryos of all stages respond to similar time periods of exposure by reducing limb buds.

RA in proximal mesenchyme plays a role in patterning the proximal limb elements through its regulation of *Fgf*, which in turns regulates the Hox-DNA-binding cofactors *Meis-1* and *Meis-2* (Mercader *et al.*, 2000).

The *legless* insertional mutation produces limb and craniofacial mutations in homozygous mice (less severe on a BALB/c background), including absence of the AER and increased mesodermal and ectodermal cell death in embryos that are especially sensitive to RA.[b]

RA also patterns regenerating limbs – implanting antagonists of RA receptors on silastic blocks modifies the pattern of limb regeneration in the axolotl, *Ambystoma mexicanum* (del Rincón and Scadding, 2002) – and can homeotically transform tail regenerates into limb regenerates (Box 43.2).

[a] See Eichele (1986, 1989) and Thaller and Eichelle (1987) for the studies, Slack (1987) for 'We have a morphogen!', Lee and Tickle (1985, 1990) and Brickell and Tickle (1989) for reviews of RA as a limb morphogen, Underhill and Weston (1998) and Underhill *et al.* (2001) for reviews of retinoids in skeletal development, including their role in prechondrogenic condensations, and Tabata and Takei (2004) for an overview of morphogens.

[b] See Singh *et al.* (1991) for the *legless* mutation. The importance of genetic background in modifying the phenotypic expression of gene action, although often forgotten, is important. The gene *Tcof-1* encodes the nucleolar protein Treacle. Mutations in *Tcof-1* elicit severe Treacher Collins syndrome, delayed ossification of long bones, fusion between ribs, abnormalities of the digits and embryo lethality in a mixed CBA/Ca and 129 genetic background, but normal and viable embryos form when the gene is in a BALB/c and 129 background (Dixon and Dixon, 2004).

- The retinaldehyde dehydrogenase-2 gene (*RA-2*), which is required for synthesis of RA, is expressed in mesoderm, branchial arches and fin buds.
- The zebrafish mutation *no-fin* (*Nof*) maps to the chromosome location of *RA-2*; mutant embryos lack pectoral fin buds and gill cartilages but have normal pelvic fin development.
- Mutant embryonic mice that lack *RA-2* secrete no RA except in the retina and at 9.5–10 days of gestation, i.e. before forming limb buds. Maternal RA can prolong the embryonic life of these mutants, allowing hind limb buds but not forelimb buds to form. The area where the forelimbs should form expresses either no or little *Shh*, and diminished amounts of Fgf-4 are expressed ectopically.[18]

The zebrafish RA receptor ZfRAR-γ, a homologue of the RAR-γ of mice and humans, was cloned in the early 1990s. Expression is high in blastemal cells at the distal ends of the rays in regenerating caudal fins (White *et al.*, 1994).

...Regeneration

RA appears to play interesting roles in regeneration.

In tetrapods and zebrafish, extra RA produces an interesting morphogenetic difference. In tetrapods RA results in duplication of the *proximo-distal limb axis*. In zebrafish the *dorso-ventral axis* is duplicated during caudal fin regeneration in wild-type and *long-fin* mutant zebrafish (Géraudie *et al.*, 1995).

Jaw and pectoral fin malformations are induced in flounder following exposure to RA or to agents such as α, α'-dipyridyl or L-azetidine-2-carboxylic acid (LACA) that block collagen synthesis. Exposure of very early (embryonic shield) stage embryos to RA for one hour completely suppresses development of Meckel's cartilage, exposure to collagen-blocking agents resulting in abnormal morphogenesis or growth of pharyngeal and pectoral fin elements (Suzuki *et al.*, 2000).

An RA-Shh link

Shh and Bmp-3 are both involved in regulating proliferation and differentiation of the cells that deposit new bone in regenerating zebrafish caudal fins (Quint *et al.*, 2002).

By regulating *shh* expression, the helix–loop–helix protein dHAND plays an important role in establishing the A-P axis and the ZPA in teleost fin development, the essential role of dHand being to regulate *Shh*. The regulatory sequence is RA → overexpression of *Hoxb-8* → dHand → *Shh*. dHand also is central to the signal cascade [endothelin-1 (*en-1*) → dHand → *Msx-1*] that regulates development of branchial arch mesenchyme. Fins fail to develop in the *hands off* zebrafish mutant, a mutation of dHand.

Treating zebrafish embryos with RA results in ectopic expression of *Shh* in the pectoral find buds and abnormal fin morphogenesis, a genetic mechanism that is conserved between fish fins and tetrapod limbs. Disrupting

RA signaling induces duplication of pectoral fins in almost 100 per cent of zebrafish or killifish, three or four duplicated fins often developing. Exposing mouse embryos to ethanol also produces limb defects – especially affecting postaxial digits – by disrupting the *Shh–Fgf-8* feedback loop.[19]

FIN REGENERATION

Pioneering studies on fin regeneration in the killifish, *Fundulus*, were undertaken by Goss and Stagg (1957), who showed that a blastema forms from connective tissue and osteoblasts of the stump (indeed, regeneration does not proceed unless part of the ray remains to provide these osteoblasts), and that the blastema is closely associated with the epidermis, regeneration returning the size of the fin and fins rays to that of the original (Fig. 40.3).[20]

Such studies were few and far between until recently. Fin regeneration is now a popular system for investigation, in part because the intrinsic beauty of fish fins attracts investigators, in part because the fin is often treated as a two-dimensional organ system and so analyzed more readily than structures such as regenerating amphibian limbs. Interesting differences and similarities between fin and limb regeneration also attract investigators, not the least because they make fin regeneration a fundable enterprise.

Marí-Beffa and colleagues (1996, 1999) investigated the regeneration of *individual fin rays* and *entire tail fins* in teleost fishes, including zebrafish and goldfish. Four epidermal cell types, scleroblasts, fibroblasts and epithelial–mesenchymal interactions all contribute to blastema formation. Cell–ECM interactions in tail-fin regeneration, evidence of the production of two blastemata, interactions between such lepidotrichial matrix components as collagen and GAGs, and discussions of the relationships (if any) between fin regeneration and the fin-fold theory are contained in studies by Santamaria and Becerra (1991) and Santamaria *et al.* (1992).

Information on the roles of growth factors, *Wnt* genes, cell interactions and regeneration fields also has been obtained. Nerve growth factor accelerates the initiation of fin regeneration in the goldfish, *Carassius auratus* (Weis, 1972), raising issues of whether nerves are involved in fin regeneration, as they are in fin development and limb regeneration. Fgf and FgfR-1 play important roles in zebrafish pectoral fin regeneration, as shown below.

- Using a novel technique of gene electroporation developed for zebrafish embryos, Tawk *et al.* (2002) showed that fin regeneration can be disrupted after disrupting FgfR-1.
- *FgfR*-1 is expressed in mesenchyme underneath the wound epidermis at the blastema stage and in distal blastemal mesenchyme during regeneration. *Inhibiting FgfR-1 after fin amputation blocks blastema formation* by blocking proliferation of mesenchyme and the onset of *Msx* gene activity.

- A zebrafish *Fgf*, designated *zfgf*, is expressed in the epidermis of regenerating fins during outgrowth. Inhibiting *zfgf* during this stage prevents further outgrowth and down-regulates *Msx* and epidermal *Shh* (Poss *et al.*, 2000a).
- *msx-C* is differentially expressed in swords – the extensions of the ventral fin rays of the caudal fin in male swordtail fishes of the genus *Xiphophorus* – being up-regulated in the caudal fins of developing and regenerating fins in males but not in females (Zauner *et al.*, 2003).

Such strong similarities between fin regeneration and limb and craniofacial development and regeneration, suture development and maintenance, and craniosynostosis further enhance the fascination of fins.

Lef-1, a transcription factor for members of the *Wnt* gene family, is expressed in the epidermis of regenerating zebrafish fins near scleroblasts and sites of active bone formation, but is not expressed in epithelia adjacent to mature bone or during blastema formation. A similar pattern is seen with *Shh*, indicating a role for *Lef-1* and *Shh* in scleroblast alignment. Subsequent studies identified two cell populations (two compartments?) within the blastemata of regenerating zebrafish fins, with the proximal population dividing 50 times faster than the distal population. The absence of a gradient between the two populations suggests that they may indeed represent separate compartments.[21]

Misof and Wagner (1992) used the eastern Atlantic peacock blenny, *Salaria pavo*, to examine pectoral fin regeneration. They identified a dorsal and a ventral hook field as two regeneration fields, as well as separate mechanisms for morphogenesis and maintenance of the regenerating/regenerated fins. Additionally, their examination of the regenerative ability of pectoral fins in 14 species from six families provides some of the first inklings of the ecological correlates of fin regeneration; regenerative ability is impaired in bottom-dwelling species.

FINS → SUCKERS

Lumpfishes and snailfishes of the family Cyclopteridae have modified their fin rays to form a ventral, anterior sucker (or disc) that develops before hatching (Figs 40.6 to 40.8). In their study of the mechanisms of sucker function in the lumpsucker, *Cyclopterus lumpus*, Davenport and Thorsteinsson (1990) describe how the elasticity of the cartilaginous sucker skeleton antagonizes a horizontal muscle pair between the sucker and the hyoid arch. They also found that suckers are larger in males than in females on a per unit weight basis.

FINS → LIMBS[22]

Much interest has been expressed and a sizeable literature accumulated regarding the transformation of fish fins to limbs during the origin of the tetrapods. Some brief insights into the skeletal changes involved are provided in this section. Several excellent books treat the topic in much greater depth and breadth, Jenny Clack's (2002) *Gaining Ground: The Origin and Evolution of Tetrapods*, being perhaps the most recent.

Classic contributions on the gradual transition from the fins of Palaeozoic fishes to the limbs of the earliest tetrapods by such eminent Swedish palaeontologists as Nils Holmgren and Eric Jarvik were based on analyses of the structures of *adult* limbs and their transformation. Later, palaeontologists examined functional and structural perspectives.

For example, from a functional analysis of the fin of the angler fish, *Lophius piscatorius* (which shows convergent features with tetrapod limbs), Edwards (1989) argued that limbs (i.e. digits) evolved for aquatic, not terrestrial locomotion. Although convergence of structure often implies that equivalent forces (function, environmental change, selection) drove evolution of the structures, it need not. An anticipated intermediate stage between fins and limbs might be a fish with fin rays functioning as rudimentary digits – if fin rays became digits, which they do not, or a fish with fin rays and rudimentary digits – if digits replace fin rays, which they do.

Daeschler and Shubin (1998) described what looked like a classic missing link (Box 44.2) – a fossil fish with radials encased in fin rays – and discussed it in the context of digits having evolved in an aquatic environment. From phylogenetic analyses it now appears that this taxon is off the main line leading to tetrapods, in some ways a far more interesting finding as it shows that more than one group was 'experimenting' with digits in the Palaeozoic.[23]

Over the past decade or so palaeontologists have increasingly incorporated data from developmental and molecular biology into their analyses, some of which are discussed below. Developmental biologists have also contributed much to our understanding of how the transition is likely to have occurred. For example, based on knowledge acquired from a close analysis of the *talpid³* mutant and its effects on chick limb development, Ede (1977) proposed a cellular basis for the transformation of crossopterygian fins to tetrapod limbs, involving modulation of A-P gradients in cell division and establishment of differential cell movement.

Maderson (1967) introduced the concept of differential rates of differentiation resulting from a time lag in induction, an idea developed more fully by Thorogood (1991), who laid out a model for the transition from fin fold to limb ridge that relied on heterochrony. Prolonging the activity of the fin fold during development would increase the portion of the limb bud derived from mesoderm and allow it to be specified for cartilaginous elements that would form more distally in the fin/limb bud. (A concomitant change would be suppression of neural-crest cell migration to reduce the mesenchymal cell populations from which fin rays are

thought to arise.) Such a mechanism could lead to the production of additional cartilages, but without additional changes the cartilages would resemble those already present more proximally. A molecular understanding of how the new distal elements could acquire new features – develop as digits – came a few years later with our understanding of fin and limb skeletal patterning by *Hox* genes.[24]

Changes in the musculature occurred in parallel with the skeletal changes as fin muscles were replaced by limb muscles. An analysis of the developmental origin of the fin musculature in zebrafish and dogfish demonstrated that zebrafish appendicular muscles arise from migratory mesenchymal precursors that are *morphologically and molecularly equivalent to those used by tetrapods*. Appendicular muscles in dogfish (Figs 40.1 and 40.2), on the other hand, arise as epithelial extensions of the somites. What we think of as the basic 'tetrapod' pattern of limb muscles predates the tetrapods and arose before the radiation of the sarcopterygian fishes. Evolution of even such an apparently dramatic transformation as fin → limb works by tinkering with existing cells and developmental processes.[25]

FROM MANY TO FEWER DIGITS

Five or 10 as the canonical number?

For the 150 years since Richard Owen (Fig. 6.3) produced his monographic analysis *On The Nature of Limbs* (Owen, 1849)[26] we have known that tetrapod limbs are built on a primitive *pentadactyl* plan, five digits being the basal (archetypical, primitive, original, fundamental, conserved) digit *Bauplan*.[27]

A major reason for believing in such an archetype was the underlying presumption of the homology of adult limb structure reflecting (indeed, being caused by) an underlying homology of limb development. At the beginning of this chapter I separated out homology of fin and limb buds as developmental processes from homology of fins and limbs as appendages. I did so quite deliberately, for – to reiterate a point made several times in other contexts in the book – homology of process or origin is no guarantee of homology of the structures those origins or processes produce.

Around the mid-1980s it became clear that no archetypal pattern underlies all tetrapod limbs. Adult tetrapods with less than five digits do not all have embryonic primordia for five digits; no evolutionarily primitive plan is conserved in limb development (Box 36.1).

Then in the early 1990s came the remarkable discovery that the earliest known tetrapods had 10 or more digits on each limb. Even more remarkable, this evidence had been embedded quite literally in the rock surrounding one of the prime specimens described by Erik Jarvik, who, seeing five digits exposed, had no reason to remove additional matrix to look for more; five was the canonical number.

Ten digits on a *talpid* mutant limb bud is regarded as polydactyly, exceeding by five the standard number of digits (setting aside the fact that the standard number of digits in birds is not five). So these early tetrapods were polydactylous with respect to modern tetrapods. Or should we say that subsequent tetrapods all show reduction from the origin condition of decadactyly?[28]

Evidence for 'polydactyly' comes from three early and *still aquatic* tetrapods from the Upper Devonian: *Acanthostega* with eight digits in the forelimbs, *Ichthyostega* with seven digits in the hind limbs, and *Tulerpeton* with six digits in the hind limbs. So, harking back to earlier in the chapter, digits did not evolve on land.[29]

Although *Acanthostega*, *Ichthyostega* and *Tulerpeton* are early tetrapods they are not basal tetrapods; many of their features are already more derived than those found in fishes of the time. This gap in the tetrapod fossil record came to be known as 'Romer's Gap.'[30] In 2002, Jenny Clack (2002b) filled the gap with her description of *Pederpes finneyae*, from the Early Carboniferous, 348–344 mya. *Pederpes* is an early tetrapod with five digits on the hind limbs but with polydactylous fore limbs. Consequently, we are all going to have to revisit pentadactyly/polydactyly and the origin and evolution of tetrapod digits. The shift from five to 10 (or 10+) is not a difficult evolutionary step; subdivision of existing digital primordia will do it. The broader limb that comes with ten digits would be advantageous in an aquatic environment. The really interesting question takes us back to the pentadactyl limb. Why five and not 10 at the outset?

NOTES

1. See the section on variation at the beginning of Chapter 44.
2. Koumoundouros *et al.* (2001) evaluate the development of dorsal, anal and paired fins in the popular Mediterranean and tropical food fish *Dentex dentex*, the common dentex.
3. See Mabee *et al.* (2002) for a phylogeny of median fin evolution in fishes and for discussions of some of the theoretical issues raised by similarities and differences among and between dorsal, caudal and anal fins in fishes. The sarcopterygian fish, *Eusthenopteron foordi*, is perhaps one of the best preserved of the Devonian fishes. A complete brain case was described and shown to be tetrapod-like but with an intracranial joint, changes in brain case and limbs being tightly integrated (Ahlberg and Milner, 1994; Ahlberg *et al.*, 1996). Cote *et al.* (1992) undertook a careful analysis of vertebral development of 27 specimens and, surprisingly, found that ossification began caudally and in the second dorsal fin. Ossification of vertebrae from posterior to anterior (i.e. from tail to trunk) was argued to be unique and, given the systematic position of *Eusthenopteron*, has implications for our views of vertebral evolution. Long (1990), however, considered that the cranial characters were paedomorphic and the limb characters peromorphic, and so proposed dissociation in timing of these two skeletal systems (dissociated heterochrony) in the transition from fish to tetrapods.

4. See Bodenstein (1952) and Sadaghiani and Thiébaud (1987) for these two studies, and see Hall (1999a) for a discussion.
5. See Hall (1999a) and Tucker and Slack (2004) for ventral fins and trunk neural crest and see Crotwell *et al.* (2001) for Gdf-5.
6. See Goodrich (1958), Jarvik (1965a, 1980), Maderson (1967), Hall (1991d), Santamaria and Becerra (1991), Thorogood (1991), Santamaria *et al.* (1992), Bowler (1996) and Bemis and Grande (1999) for analyses of the fin-fold theory.
7. See Lucifora and Vassallo (2002) for this study and see Holst and Bone (1993 for further documentation of 'bipedalism' in skates and rays.
8. See Bouvet (1974, 1976, 1978) for fin folds and proliferation, Wood (1982), Hall (1991d) and Thorogood (1991) for transformation of the pseudoapical ridge to a fin fold, and Grandel *et al.* (2000) for the genetic analysis.
9. See Prince *et al.* (1998a, 1999b) for the studies on *Hox* genes, and Hérault *et al.* (1998, 1999) for the enhancers.
10. See Landis and Géraudie (1990) for the mineral study, Marí-Beffa *et al.* (1989) for collagen turnover, Laforest *et al.* (1998) for growth, and Iovine and Johnson (2000) for the two mutants, *sof* and *lof*.
11. See J. R. Dunn (1984) for basic developmental osteology, including brief discussions of the major skeletal regions, sequences of ossification – the skeleton associated with feeding and respiration invariably develops first – patterns of chondrification, and scale development. Cleared and stained skeletal preparations are often used as samples to provide metric data such as the length of an element or of the entire organism (Moser, 1984). In a study of 611 fixed individual *Tilapia mossambica* ranging in length from 6.0 to 61.5 mm, Mabee *et al.* (1998) demonstrated shrinkage of 3–6 per cent, especially in the larger specimens.
12. The pectoral girdle has been less well studied in teleost fishes, cellular or molecular analyses being minimal. Perhaps the most useful developmental studies are those on *Acipenser*, *Amia*, *Esox*, *Lepisosteus* and *Polypterus* by Jollie (1975, 1980, 1984c–e) and on *Danio rerio* by Cubbage and Mabee (1996).
13. See Géraudie (1978a,b, 1985) for the ultrastructure of pelvic fin development, Géraudie (1980) for mitotic activity, and Géraudie and Landis (1982) and Géraudie and Meunier (1980, 1982) for collagen and elastoidin.
14. See Géraudie and Singer (1977, 1979, 1985) for studies on pectoral fin regeneration.
15. See Géraudie and Meunier (1982, 1984), Géraudie (1988) and Richter and Smith (1995) for ganoine in various recent and fossil actinopterygians and acanthodians.
16. See S.-H. Lee *et al.* (2002) for the chick study, and Singh *et al.* (1991) for *legless*.
17. See Hirata and Hall (2000) for a discussion of apoptosis in relation to pre- and post-gastrula development.
18. See Grandel *et al.* (2002) for *RA-2*, Vandersea *et al.* (1998b) for *no-fin*, and Niederreither *et al.* (2002) for the mouse study.
19. See Vandersea *et al.* (1998a) for RA in zebrafish embryos, and Chrisman *et al.* (2004) for ethanol and mouse embryos.
20. Chapter 8 in Goss (1969) provides an excellent overview of early work on scale and fin regeneration. Liversage (1973) showed that regeneration of the pectoral fins of killifish, *Fundulus heteroclitus*, continues after hypophysectomy.
21. See Poss *et al.* (2000b) for *Lef-1* and *Shh*, and see Nederbragt *et al.* (2002) and Nechiporuk and Keating (2002) for the two cell populations.

22. A forthcoming edited volume, *Fins into Limbs* (Hall, 2005c) examines historical, evolutionary, developmental and life history aspects of fins, limbs and the transition of fins → limbs.

23. See Holmgren (1933) and Jarvik (1959, 1965a,b, 1980) for the classic studies from adult structure, Thomson (1993) for an overview of the origin of tetrapods from osteolepiform fishes (which were polydactylous, had partial gills and inhabited mid-Devonian coastal wetlands), Vorobyeva and Hinchliffe (1996) for an overview of developmental perspective and the fossil evidence, Sordino and Duboule (1996) for a review that evaluates the heterochrony model, and Capdevila and Izpisúa-Belmonte (2000) for a review of the evolutionary transition from fin → limb, including molecular control of limb development. Laurin *et al.* (2000) review early tetrapod evolution, especially how tetrapods are defined, whether digits arose from radials of sarcopterygian fishes and how to integrate *Hox* data with the fossil record.

24. Maderson (1967) and Hall (1975a, 1991d, 1998a) discuss the lateral fin-fold theory. Ruvinsky and Gibson-Brown (2002) review tetrapod limb evolution in relation to serial homology, tinkering, gill-arch or lateral fin-fold origins, and the role of *T-box* genes.

25. See Neyt *et al.* (2000) for this study, and see the discussion by Galis (2001b), who emphasizes mosaic patterns and intermediate patterns in amphibians, reptiles and fish.

26. Owen's 1849 discourse *On The Nature of Limbs* is being reprinted, with accompanying essays, and under the editorship of Ron Amundson, by The University of Chicago Press. I anticipate publication in 2005.

27. Carl Gegenbaur is equally renowned for his development of skeletal archetypes, e.g. the *archipterygium* as the archetypal limb, and the shark chondrocranium as the archetypal vertebrate skull. See Nyhart (1995), Bowler (1996), Hall (1997b,c, 1999b), Coates (2003) and Kuratani (2003) for discussions, and see the special issue of *Theory in Biosciences* edited by Ho[eβet]feld *et al.* (2003) for an evaluation of Gegenbaur and evolutionary morphology on the 100th anniversary of his death. Thorogood and Hanken (1992) discuss four papers on how to build the basic plan of the skull.

28. See Hinchliffe and Johnson (1980) for an excellent overview of the pentadactyl limb, Hinchliffe and Hecht (1984) and Hinchliffe (1989, 2002) for the initial embryological studies that led to the shift away from the concept of pentadactyly as primitive, and see Coates (2003) for an evaluation of the work of Jarvik and Gegenbaur.

29. See Coates and Clack (1990, 1995) for the three taxa and Shubin (1991, 1995), Coates (1994), Shubin *et al.* (1997) and Coates and Cohn (1998, 1999) for reviews that include discussion of the independent evolution of pectoral and pelvic limbs, a topic I have glossed over. See Ahlberg (1991) for a discussion of *Panderichthys*, especially the presence of eight digits. Laurin (1998) questions the concept that pentadactyly arose twice, considering *Tulerpeton*, a stem tetrapod, and illustrating how important it is to distinguish stem from crown taxa and to have a rigorous phylogeny when assessing origins. A remarkable example of convergence is represented in the recent description of a polydactylous marine *reptile* from the Triassic period in which the preaxial polydactyly closely resembles that of the Late Devonian basal tetrapods (Wu *et al.*, 2003).

30. The term Romer's gap was coined by Coates and Clack (1995). See Carroll (2002) for an evaluation of the gap and the significance of *Pederpes finneyae*.

Part XIV

Backbones and Tails

The question of whether vertebral development is initiated because of self-differentiation of mesenchymal cells or as a consequence of interactions with one or more tissues (notochord, spinal cord) adjacent to the body axis preoccupied many developmental biologists for the first two-thirds of the 20th century. The quest was so long-lived because both are true.

With respect to the notochord, and in the context of an evaluation of segmentation, it was noted that 'Given that the lower chordates do not possess a sclerotome but display segmentation of the trunk, and that in many vertebrate classes the notochord displays obvious metamerism, the foregoing discussion suggests that the notochord is the archetypal segmented structure in the trunk of vertebrates'

(Claudio Stern, 1990, p. 111).

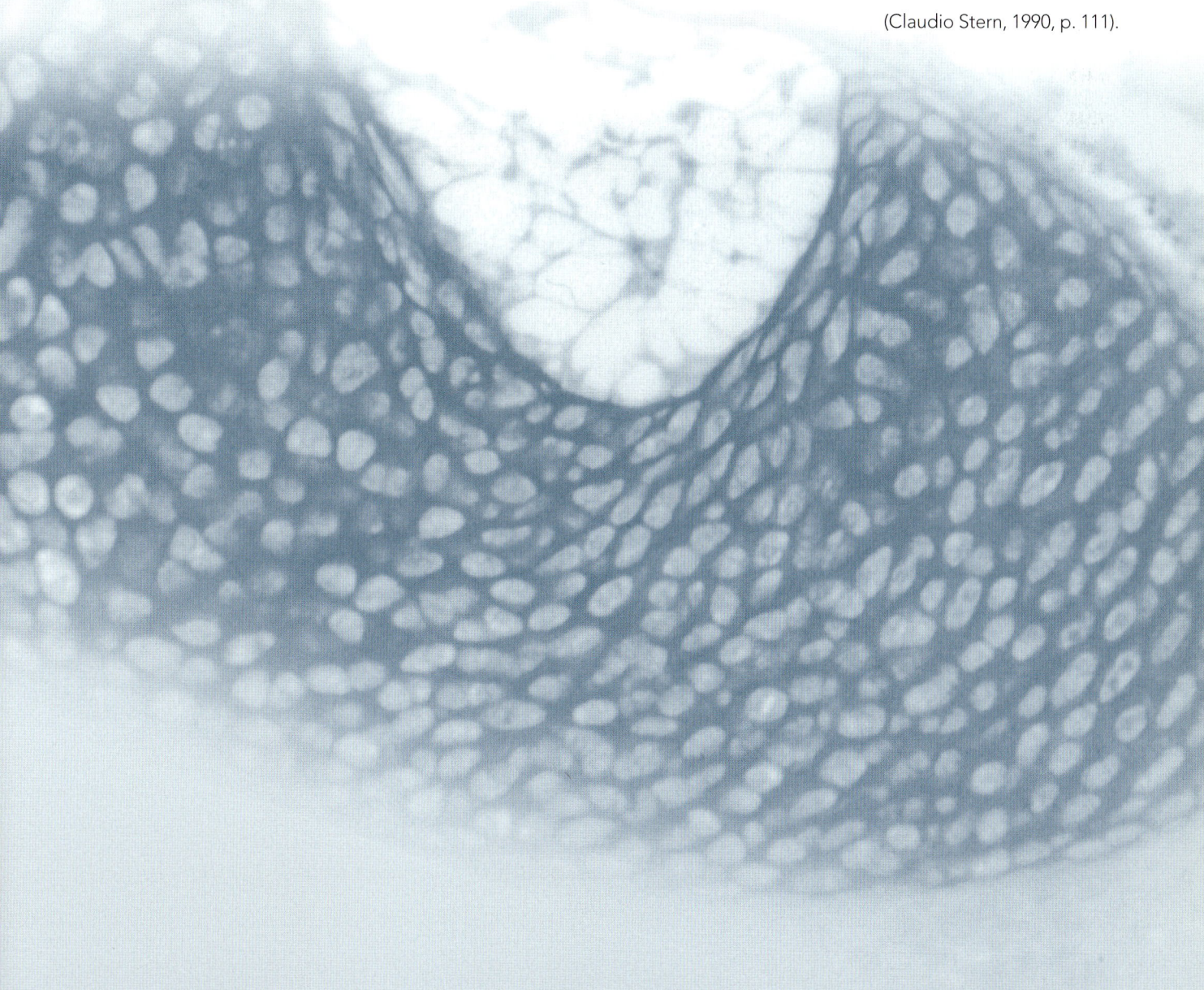

Vertebral Chondrogenesis: Spontaneous or Not?

Chapter 41

Priming the pump or building the pump?

Whether vertebrae are initially cartilaginous and only later replaced by bone – which is the situation for the majority of vertebrates – or whether they arise by intramembranous ossification as in some teleost fishes, their embryological origin is the same. All vertebrae form from sclerotomal mesenchyme that delaminates from the epithelial somites to migrate around the notochord and sides of the spinal cord (Chapter 16).

The question of whether vertebral development is initiated because of self-differentiation of mesenchymal cells or as a consequence of interactions with one or more tissues (notochord, spinal cord; Fig. 41.1) adjacent to the body axis, preoccupied many developmental biologists for the first two-thirds of the 20th century. As it turned out, the quest was so long-lived *because self-differentiation and induction both occur and because mechanisms of vertebral differentiation (although not morphogenesis) vary among different groups of vertebrates.*

These 20th century studies on vertebral development coincided with our quest to understand the broader question of how chondrogenesis is initiated and maintained, a quest that is ongoing. Consequently, an examination of how vertebrae develop reveals much about chondrogenesis in general. The studies also coincided with decades when embryonic induction was under intense investigation. Interpretations of experiments designed to understand vertebral development had implications way beyond vertebrae or chondrogenesis.

Finally, and more specifically in regard to vertebral chondrogenesis itself, many tissues and organs develop in proximity to somitic mesoderm and the primary body axis. Any one or combination of these tissues could potentially influence sclerotomal development or vertebral differentiation, morphogenesis and/or growth. Some

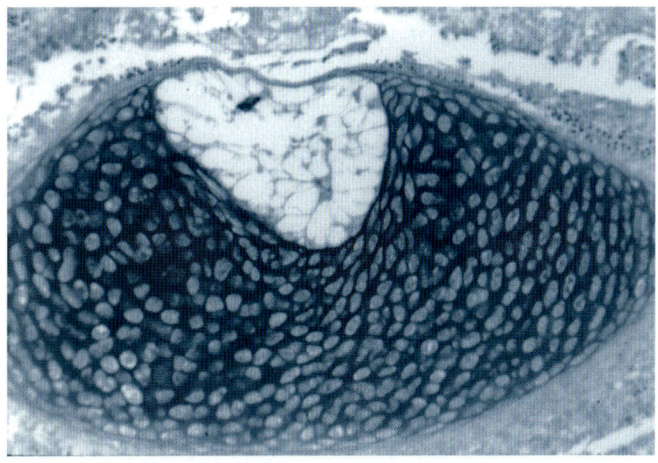

Figure 41.1 A histological section of vertebral cartilage and notochord differentiated from axial tissues from a chick embryo grafted to the CAM of a host embryo.

of these tissues are epithelial – notochord, spinal cord, epidermal ectoderm and endoderm – and so would be involved in epithelial–mesenchymal interactions, as set out in Part VI (Figs 18.3 and 41.1). Others, while not epithelial, have interesting segmental distributions suggestive of potential roles in patterning segmentally arranged vertebrae; recall segmentation and resegmentation of somitic mesoderm (Chapter 16). These tissues include spinal ganglia, spinal nerves, and populations of neural-crest cells throughout the vertebrates, as well as lateral-line nerves in fishes and amphibians.

SELF-DIFFERENTIATION OR INDUCTION?

The first experiments on the differentiation of somitic mesoderm were carried out by Hoadley (1925), who utilized the then newly developed technique of grafting tissues to the chorioallantoic membrane (CAM) of embryonic chick hosts, a technique that dominated studies until the early 1940s (See Box 12.3). Cartilage was shown to develop from somites or from sclerotomal mesenchyme grafted *without* notochord or spinal cord, although some authors maintained that these tissues enhanced ongoing chondrogenesis. Pioneering studies grafting somitic mesoderm from avian embryos to the CAM showed that more cartilage forms in more grafts when spinal cord or notochord is included (an example from my laboratory is shown in Fig. 41.1), cartilage forms from somites isolated from embryos as young as H.H. 10 and from as few as one somite, and that paraxial unsegmented mesoderm forms an abundance of cartilage when CAM-grafted.[1]

The study by Murray and Selby (1933), which predated later interest in the inductive role of the spinal cord, is, in retrospect, illuminating on this point. Somites (probably along with overlying ectoderm) were taken from embryos of H.H. 12 and 13 and CAM-grafted. Five to eight per cent of somites grafted singly or as strips of three formed cartilaginous nodules, bone and muscle. Chondrogenesis from single somites increased from 5 to 60 per cent when spinal cord was included.[2] As 30 per cent of grafts of unsegmental paraxial mesoderm from younger (H.H. 10) embryos also produced cartilage, Murray and Smiles concluded that paraxial mesoderm has a *chondrogenic bias before segmenting into somites*. For the same reason, they did not think the low incidence of cartilage obtained in somites grafted with spinal cord resulted from influences from contaminating spinal cord or notochord. They thought it reflected *inherent chondrogenic ability*. It was many years before this conclusion was rediscovered and confirmed.

Williams (1942) also grafted avian somites with or without notochord. Although the frequency of cartilage was greatest when notochord was present, Williams concluded that the notochord serves a mechanical role and has no inductive properties. Twelve years later Watterson and colleagues published a now classic study that used a variety of extirpation, implantation and grafting techniques to address the following four questions in vertebral chondrogenesis.

● What are the relative roles of neural tube and notochord?
● If induction occurs, how much association between inducer and sclerotomal mesenchyme is required?
● Can any portion of a somite become sclerotome?
● Can various regions of the neural tube evoke chondrogenesis?

Simultaneous removal of notochord and neural tube from embryos of H.H. 11–16 prevented chondrogenesis. Medial sclerotome formed vertebral centra and intervertebral discs, while the lateral sclerotome formed neural arches and the proximal portions of the ribs. Removing notochord alone eliminated centra. Removing neural tube alone eliminated neural arches. Williams concluded that notochord and spinal cord induce the cartilage that surrounds them, and can function independently of one another.

No cartilage formed when Watterson and colleagues CAM-grafted somites from embryos of H.H. 12–16 embryos, as Murray and Selby had shown in 1933. Somites from slightly older embryos (H.H. 18/19) did chondrify, indicating that induction takes place during the last half of the third day of incubation. Demonstrating conservation of signaling between birds and mammals, Pugin (1973) showed that spinal cord from mouse embryos can replace spinal cord from chick embryos to elicit chondrogenesis.

MORPHOGENESIS

Hörstadius (1944) developed procedures to remove notochord or spinal cord from early embryos of the spotted salamander *Ambystoma punctatum*. Vertebral cartilages differentiated but morphogenesis and regional organization were severely disrupted. Similarly, when Kitchin (1949) removed the notochord from slightly later (neural plate-stage) embryos of *A. mexicanum*, vertebral cartilages developed but as a fused rod without neural arches. These pioneering studies indicated that, at least in these two species of *Ambystoma*, notochord *was essential for morphogenesis but not for initiation of differentiation* of vertebral cartilages.

Each vertebra consists of a dorsal component, the *neural arch*, which develops adjacent to the spinal cord and spinal ganglia, and a ventral component, the *centrum*, which develops from mesenchyme surrounding the notochord (Fig. 41.2). We might expect, therefore, that if adjacent tissues influence vertebral development, the spinal cord would influence neural-arch development and the notochord centrum development, which is precisely what has been found to be the case.

When Howard Holtzer inverted and reimplanted the spinal cord in embryonic urodeles, the axial skeleton was inverted dorso-ventrally. Not only is regionalization of the vertebral column dependent on the notochord (as Hörstadius and Kitchin had shown) but the spinal cord – and evidently only the ventral spinal cord – plays a role.

In further careful morphogenetic studies, Holtzer found that manipulating the size of the neural tube affected the size of the vertebral column. Using embryos from three salamanders – *Ambystoma punctatum*, *A. tigrinum* and *Triturus torosus* – Holtzer found that removing the notochord resulted in formation of a massive cartilaginous rod ventral to the spinal cord. Placing the spinal cord dorsal to the somites resulted in vertebral arches developing around the spinal cord. Holtzer concluded that the role of the notochord was secondary to that of the spinal cord. Subsequent studies substantiated this conclusion; in *Ambystoma* and *Triturus* the notochord influences segmentation of vertebral cartilage *after* the neural tube has induced the somites to chondrify.[3]

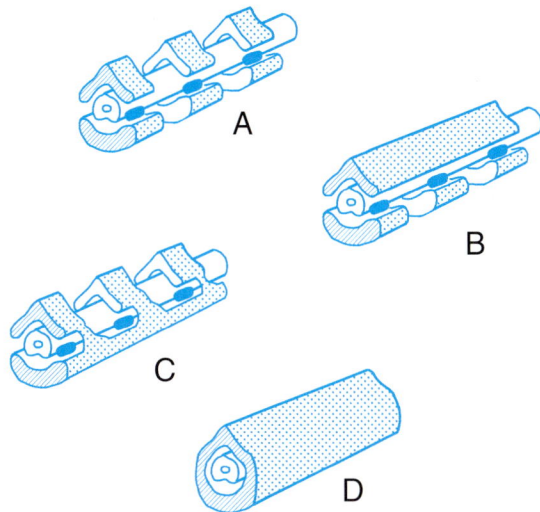

Figure 41.2 Spinal ganglia and notochord influence vertebral morphogenesis. (A) The axial structures of a chick embryo showing the segmental arrangement of the spinal ganglia (black), neural arches and centra of the developing vertebrae (stippled), the neural arches lying above and the centra below the spinal cord (white) and notochord. (B) Excising the spinal ganglia results in development of a continuous rod of neural arches but does not influence morphogenesis of the centra. (C) Excising the notochord results in development of a continuous rod of centra but does not influence morphogenesis of the neural arches. (D) Excising both spinal ganglia and the notochord results in the development of a continuous tube of cartilage around the spinal cord. Modified from Hall (1978a).

Similarly in chick embryos, morphogenesis is controlled by notochord and spinal cord (and by spinal ganglia). For some two and a half decades, Georges Strudel in Paris was one of the most active workers using chick embryos. Initially he removed the neural tube from early embryos and found that neural arches failed to develop. Removing the notochord resulted in vertebrae that lacked the centrum but had normal neural arches *and ribs* (Fig. 41.2). Strudel postulated that notochord induces differentiation of the centra and the neural tube induces neural arches. He extended this study with embryos of H.H. 10–17 to confirm that removing the neural tube results in an abnormally segmented cartilaginous sheath around the notochord, while removing the notochord results in an unsegmented cartilaginous 'gutter' ventral to the spinal cord that is associated with normally segmented neural arches (Fig. 41.2). It appeared that in chick embryos notochord acting alone influences segmentation, while notochord in concert with spinal cord influences differentiation (Strudel, 1953a, 1955). Both conclusions are in line with results from urodeles.

Spinal ganglia and vertebral morphogenesis

The cartilaginous rudiments of the vertebral column are not a continuous cylinder around the notochord and spinal cord but are interrupted by segmentally arranged neural crest-derived spinal ganglia. By deflecting migrating sclerotomal mesenchyme, the spinal ganglia play a role in positioning vertebral *anlage*, an idea that has its origin in an analysis of sectioned human embryos published by Feller and Sternberg (1934). Spinal ganglia develop from neural-crest cells (NCC) that migrate to the median face of each somite *before* the segmentally organized vertebrae develop, indeed before sclerotome migration. Consequently, spinal ganglia could serve as barriers deflecting migrating sclerotomal cells cephalad and caudad to bring about resegmentation (Chapter 16).

Removing NCC from early amphibian or avian embryos results in absence of spinal ganglia or production of small ganglia. The neural arches develop as an unsegmented, continuous cartilaginous 'roof' over the spinal cord, and the centra develops normally (Fig. 41.2). Similarly, somites grafted to the CAM only form segmented neural arches if spinal ganglia are grafted along with them. Removing the notochord gives the reverse result, which is formation of normally segmented neural arches and an unsegmented cartilaginous 'gutter' or trough (Fig. 41.2).[4]

One implication of these experiments was that *both segmented spinal ganglia and unsegmented notochord impose segmentation onto vertebrae*, a suggestion that requires that 'unsegmented notochord' really is segmentally organized, and/or that it provides influences that are interpreted as segmental by sclerotome mesenchyme.

On the basis of an analysis of paraxial mesodermal development Claudio Stern argued that 'the notochord is the archetypal segmented structure in the trunk of vertebrates' and, on the basis of the type of evidence outlined above, went on to conclude that 'even in those animals in which the notochord is not overtly segmented, this structure may imprint segmented information onto the adjacent paraxial mesoderm.' And, indeed, in developing Atlantic salmon, *Salmo salar*, a perinotochordal structure consisting of *chordablast* cells forms on the notochord sheath, preceding development of *chordacentra* as an acellular mineralized zone within the sheath of the notochord, which, combined with sclerotomal bone, forms the vertebrae; i.e. in this species, the vertebrae have a dual origin (dual segmentation): neural and haemal arches from the sclerotome, centra from chordacentra.[5]

Direct evidence for notochordal segmentation now exists. Vicky Prince's group found regional, albeit transient, expression of *Hox* genes along the zebrafish notochord, indicative of cryptic 'segmentation' (Prince *et al.*, 1999b). *Hoxb-1*, *Hoxb-5*, *Hoxc-6* and *Hoxc-8* are the four genes expressed in an antero-posterior sequence, the most 3′ genes being expressed most anteriorly. The anterior expression boundaries of the four genes are:

- *Hoxb-1* at the boundary between somites one and two;
- *Hoxb-5* at the boundary between somites four and five;
- *Hoxc-6* at the boundary between somites five and six; and
- *Hoxc-8* at the boundary between somites six and seven.

So, spinal ganglia and notochord influence the morphogenesis of dorsal and ventral components of vertebrae, respectively. What about the prior event of initiation of differentiation? Do notochord and spinal cord prime the pump or tell mesenchyme how to build the pump? From the mid-1950s on, analysis of chondrogenesis *in vitro* increasingly became the approach of choice to answer this question. Grobstein and Holtzer (1955) used somites from foetal mice to carry out the first *in-vitro* analysis combining somitic mesoderm with spinal cord. The result: cartilage forms *only* when the spinal cord is included. Vertebral chondrogenesis was regarded as the rigidly controlled 'switching-on' of a new developmental pathway within sclerotomal mesenchyme.

CHONDROGENESIS *IN VITRO*

Clifford Grobstein and his students were the first to publish attempts to culture somites, using somites 5–12 from foetal mice on the ninth day of gestation (Theiler stage 14). Cartilage developed provided that the spinal cord was included. [A word of **caution**: You cannot use differentiation or lack of differentiation *in vitro* to indicate the full potential of a population of cells; cartilage, bone, fat, haematopoietic and lymphoid tissues all differentiate in somites implanted into the anterior chamber of the eye.] Only the ventral half of the spinal cord evoked chondrogenesis, and then only for a limited time; spinal cord from 12-day-old embryos was active, spinal cord from 15-day-old embryos was not.[6]

Such interactions are neither species nor even class specific: ventral spinal cord from chick embryos will induce chondrogenesis from somitic mesoderm from foetal mice; spinal cord from mouse embryos induces somatic chondrogenesis after transplantation into chick embryos of H.H. 12–14 (Fig. 37.1).[7]

Grobstein's pioneering studies were soon adapted for studies using intact or dissociated somites from chick embryos, which chondrify, again provided that spinal cord or notochord are included.[8]

SPONTANEOUS CHONDROGENESIS?

This phase of research, which established that somites from chick embryos younger than H.H. 19 would not chondrify *in vitro* unless associated with spinal cord or notochord, was followed in the 1960s by a series of studies that contradicted earlier studies on dependence on axial organs by showing that isolated somites free of spinal cord or notochord *could* form cartilage – so-called *spontaneous cartilage formation* – if cultured in an appropriate medium. The concept of *environmental enhancement of latent potential* within somites began to appear in the literature. New observations *in vitro* brought new interpretations, with attention focusing on *somites as fully*

primed responding tissues rather than on the inducer as a provider of developmental information. Concepts of induction changed from 'provision of instructive information' to 'permissive induction,' inductors moved from magic molecules to environmental switches, and induction moved from imposing new potentials onto naive responding tissues to enhancing preexisting potential.

Avery *et al.* (1956) showed that somites could form cartilage *in vitro* and on the CAM *without* an inductor (Table 41.1). Murray and Selby had made a similar observation 23 years earlier but, as with so many developmental phenomena, their finding had to be rediscovered again and again. Notochord and spinal cord appeared equally effective; either both are equivalent inducers or each conditions the somites or somitic environment to the same extent. Doubt was shed on the concept that somites from embryos younger than H.H. 18 can only form cartilage when in contact with an inducer, and on the concept that induction at that stage confers specificity on the somites.

Others capitalized on *in-vitro* analysis. A variety of perturbations of culture media or environment affect chondrogenesis from somites cultured *without* notochord or spinal cord. For instance, somites from embryos as young as H.H. 12 chondrify if cultured in agar and embryo extract. Without supplementing the medium, somites disperse and become fibroblastic.[9] Just as Murray and Selby (1933) discovered and others confirmed when CAM-grafting, somite mass influences the extent of chondrogenesis *in vitro*. Somites from embryos as young as H.H. 13/14 chondrify if cultured as clusters or rows of somites. Such 'mass' effects are now well known as problems associated with culturing small amounts of embryonic mesenchyme.[10]

Table 41.1 Chondrogenesis from somites from chick embryos with or without notochord or spinal cord[a]

Tissue	Method of cultivation	H.H. stage of donor	Incidence of chondrogenesis (%)
Somites alone	Tissue culture	14	0
		16	0
		18	11
		20	100
		24	100
	CAM-graft	14	0
		16	17
		18	86
		20	100
Somites + notochord	Tissue culture	14	0
		16	82
	CAM-graft	14	76
Somites + spinal cord	Tissue culture	14	0
		16	87
		20	100
	CAM-graft	14	78

[a] Based on data from Avery *et al.* (1956).

Environmental influences

Ellison *et al.* (1969) provided a detailed study of *environmental* effects on chondrogenesis, including information on the stage of development from which chondrogenesis can be initiated as a function of the number of somites cultured. The younger the embryonic stage, the more somites required to allow chondrogenesis (Table 41.2). Subsequently, Ellison and Lash (1971) showed that somites from H.H. 17 embryos cultured in the presence of foetal calf serum produced as much cartilage, and produced it as early, as somites cultured with the inducer. Therefore, spinal cord and notochord are not unique in their actions, or in the degree of their actions on somitic mesoderm.[11]

Only 10 hours of contact with notochord or spinal cord was required to evoke chondrogenesis from somitic mesoderm, whereas continuous exposure to enriched medium is necessary to initiate spontaneous chondrogenesis. Jay Lash concluded *that all somites cultured from embryos of any age chondrify if appropriate conditioned media are used*, although he recognized that chondrogenic potential within somites from different craniocaudal levels is not equivalent (Lash, 1967).[12] I explore this point a little further.

In 1910, Williams had recognized that somites arise and chondrify in a cranio-caudal sequence that reflects the cranio-caudal developmental gradient of the entire embryo. Craniad somites, therefore, should be more advanced. It was therefore surprising when Lash observed that 80 per cent of posterior somites but only 20 per cent of anterior somites from H.H. 16 embryos chondrified *in vitro* in the absence of the notochord, and that cranial somites from H.H. 17 embryos only chondrified when notochord was present. Lash suggested two possible explanations: either the more cranial somites had insufficient contact with the notochord *in ovo*, or their histoarchitecture did not permit survival *in vitro*. The normal *in-vivo* cranio-caudal gradient is expressed when somites are grafted to the CAM. For every stage and position studied, clusters of four somites produced fewer cartilage nodules than did the four somites immediately cranial to them.[13]

The finding that *somites can induce themselves via positive feedback from products within their extracellular matrices* led Robert Kosher and his group to propose that 'self-induction' was the basis for spontaneous chondrogenesis. This was a radical shift in position from earlier studies of the complete dependence of somitic mesoderm on information received from inducers.[14]

Other environmental components were shown to promote chondrogenesis: increasing the K^+ concentration in the medium from 2.7 to 4.9 mM increases the amount of glycosaminoglycan (GAG) synthesized by somites during the first 24 hours in culture, but decreases by 20 per cent the number of cultures that form cartilage. Addition of GAG to the medium, however, increases the amount of cartilage produced, shown later to reflect positive feedback from matrix products to synthesis, secretion and deposition (Chapter 26).[15]

Cell division and cell death

Cell proliferation and cell death also come into play. As emphasized by Howard Holtzer, initial interaction between spinal cord or notochord and sclerotomal mesenchyme (which occurs around H.H. 12, some 30–40 hours before chondrogenesis is initiated) *is with precursors of the cells that chondrify. The cells that chondrify are the progeny of the cells exposed to the inducer.*

Rate of cell division and therefore number of cells produced is higher in somites cultured with notochord than in somites cultured alone, suggesting that notochord enhances sclerotomal cell proliferation. But notochord also could minimize cell loss, as the AER does for limb mesenchyme (Chapter 35). According to Gordon and Lash (1974), precartilage cells *in vitro* are more susceptible to cell death than other somitic cells, notochord minimizes cell death, and so a critical mass of cells – a condensation – accumulates. Cell death is seen only infrequently in somites *in vivo*, and then only at some distance from the notochord.[16]

Abbott *et al.* (1972) exposed 10 of the caudal somites from embryos of H.H. 17 and 18 to 10 µg/ml BrdU for three-day periods and then cultured dissociated somitic cells with or without notochord for a further 10–25 days (Table 41.3). Exposing somitic cells to notochord for the three-day periods between day zero and day three, or day one and day four, prevented chondrogenesis. Exposure between days two and five reduced chondrogenesis to one-third of that seen in control cultures. Exposure between three and six days has no effect, the interpretation being that blocking cells during the *proliferative* phase has the greatest effect on *chondrogenic differentiation*.[17]

Table 41.2 Percentage of somites from chick embryos producing cartilage when cultured on different media[a]

| H.H. Stage | Medium | | | | |
	Horse serum (HS)	HS + egg[b]	Foetal calf serum (FCS)	FCS + egg	FCS + liquid paraffin layer
9	–	–	0	0	38
10	–	–	0	0	20
11	–	–	3	5	–
12	–	–	19	31	–
13	–	–	20	53	–
14	–	0	69	86	–
15	0	17	92	100	87
16	13	33	100	100	–
17	0	47	89	100	–

[a] Eight of the 10 most caudal somites were used per experiment. Based on data from Ellison et al. (1969).
[b] Beaten contents of unincubated hen's egg.

Table 41.3 Effects of various times of exposure to BrdU on percentage cartilage development in somite cultured with or without notochord[a]

Days in 10 mg/ml BrdU[b]	Somites 1 notochord[c]	Somites alone[c]
0–3	0 (48)	0 (12)
1–4	0 (12)	(0)
2–5	33 (3/9)	(0)
3–6	100 (5/5)	(0)
Controls	100 (44)	100 (32)

[a] Based on data from Abbott et al. (1972) using somites and notochord from embryos of H.H. 17–18.
[b] BrdU, 5-bromo-2'-deoxyuridine.
[c] % (N) of cartilage-positive cultures.

NOTES

1. See Hoadley (1925), Murray and Selby (1933) and Williams (1942) for these CAM-grafting studies, see Hall (1977a) for an overview of studies on somitic chondrogenesis using CAM-grafting, and see Box 12.3 for CAM-grafting.

2. In light of future studies on the importance of a critical mass of tissue for differentiation to be initiated, it is of interest that single somites produce cartilage.

3. See Holtzer (1951) for dorso-ventral rotation and reimplantation of the spinal cord, Holtzer (1952a,b) for manipulation of the size of the neural tube, Holtzer (1952b) for his conclusions, and Holtzer and Detwiler (1953), Detwiler and Holtzer (1956) and Holtfreter (1968) for the further studies. For an overview of the structure of anuran notochords and their role in induction, see Malacinski and Youn (1982). The only study using anurans was a study of embryos of the northern leopard frog, *Rana pipiens*, which showed that transplanting tail somites into the abdomens of neurulae allowed muscle to develop but did not support chondrogenesis (Smithberg, 1964).

4. See Detwiler (1934, 1937), Detwiler and Van Dyke (1934) and Strudel (1953a, 1967) for excision of the neural-crest primordia of the spinal ganglia, and Williams (1942) for the CAM-grafting. See Holtzer (1952a), Strudel (1967) and Balinsky (1975) for early ideas on the role played by spinal ganglia, and see O'Rahilly and Müller (2003) for relationships of vertebrae to spinal ganglia in human development.

5. See Stern (1990, p. 111) for the quotations and see Grotmol et al. (2003) for the notochord in Atlantic salmon.

6. This time sequence is reversed with respect to the ability of spinal cord to evoke differentiation of kidney tubules from intermediate mesoderm, spinal cord from older embryos being more active than spinal cord from younger embryos (Grobstein and Holtzer, 1955; Flower and Grobstein, 1967), reflecting both the later development of kidneys during ontogeny (Box 38.2), and the ability of an organ such as the spinal cord to play multiple roles during development.

7. See Grobstein and Parker (1954) and Grobstein and Holtzer (1955) for the pioneering studies, and Grobstein (1955), Cooper (1965) and Pugin (1973) for the mouse–chick interactions.

8. See Lash et al. (1957), Lash (1963a), and Holtzer (1964b) for these pioneering studies with avian somites *in vitro*, and Stockdale et al. (1961) for chondrogenesis from dissociated somitic cells.

9. For these studies on perturbing culture conditions, see Strudel (1962, 1963), Lash (1964), Lash et al. (1964), Cooper (1965) and Zilliken (1967).

10. See Strudel (1962, 1963), Holtzer (1964a) and Thorp and Dorfman (1967) for cultures, and O'Hare (1972a) for CAM-grafts.

11. Serum concentration also affects differentiation of H.H. 23/24 chick limb-bud mesenchyme. Fibroblasts form in 10% foetal calf serum, chondroblasts in 0.1% serum, either by selection of divergent cell populations or through commitment of bipotential cells (Hattori and Ide, 1984).

12. Periosteum and perichondrium produce soluble factors to enhance and slow cartilage growth, respectively, during endochondral ossification of chick limbs. Medium conditioned with articular or hypertrophic cartilage, or with perichondria, releases the most potent enhancers (Di Nino et al., 2001).

13. See Gordon and Lash (1974) for the two explanations and O'Hare (1972a) for the data from CAM-grafting. Later stages may be out of phase with the early cranio-caudal gradient; the sequence of ossification in human vertebral neural arches does not follow a strict cranio-caudal sequence (Ford et al., 1982).

14. See Holtzer (1964a) and Lash (1968a) for time of exposure to axial tissues or conditioned media, Gordon and Lash (1974) and Lash (1967) for the cranio-caudal gradient, and Kosher (1976, 1978), Kosher and Savage (1979) and Kosher et al. (1979) for self-induction.

15. See Lash et al. (1973) for K^+ concentration, and Kosher et al. (1973) for the effects of exogenous GAGs.

16. See Holtzer (1964a) and Holtzer and Mayne (1973) for the importance of cell division, and Minor (1973) and Hirata and Hall (2000) for cell death *in ovo*.

17. BrdU is now used to detect proliferation using a monoclonal antibody, for example to follow proliferation in growth-plate chondrocytes *in vivo* (Apte, 1988).

The Search for the Magic Bullet

Chapter 42

The message is 'speed up, not start up.'

Although neither notochord nor spinal cord provides information to sclerotomal mesenchyme that is essential for chondrogenesis to commence *in vitro*, both tissues provide important and necessary information *in vivo*. While the inducer can be 'bypassed' or 'mimicked' *in vitro* by conditioned medium, embryos cannot compensate for the absence of spinal cord or notochord. Rather, vertebral differentiation and/or morphogenesis will be disrupted. Consequently, I think we are justified in continuing to refer to spinal cord and notochord as inducers of sclerotomal chondrogenesis and to their interaction with sclerotome as inductive. In this Chapter I explore what we know of the inducers and of their mechanisms of action on vertebral chondrogenesis and morphogenesis.

INTEGRITY OF NOTOCHORD/SPINAL CORD AND VERTEBRAL MORPHOGENESIS

As discussed at the beginning of Chapter 41, intact spinal cord, notochord and spinal ganglia influence vertebral morphogenesis. But must notochord and spinal cord be intact to act? What type of registration, if any, exists between the axial tissues and vertebral segmentation?

Normal neural arches develop in the presence of disorganized neural tissue in the killifish, *Fundulus heteroclitus*. In urodele amphibians and birds, on the other hand, abnormalities of the spinal cord lead to vertebral anomalies; administering nicotine sulphide to two-day-old chick embryos results in twisting of the notochord in the neck within hours and vertebral deformities within days. An interesting difference in the integrity of cartilage appears to depend on whether induction is by spinal cord or notochord. Cartilage induced by spinal cord dissipates between the 9th and 16th days in culture. Notochord-induced cartilage persists until at least the 25th day. Lash extrapolated this result to suggest that the spinal cord helps to remodel the vertebral column as it grows *in vivo*, thereby preventing compression of the spinal cord, an interesting idea that should be explored further, especially in relation to vertebral anomalies in fish raised under aquaculture conditions.[1]

Fish skeletal defects

Laboratory rearing, like domestication (Box 7.2), influences variation; there is greater variation in all parts of the vertebral column in laboratory-reared than in wild-caught Red Sea bream, *Pagrus major* (Matsuoka, 1982). Defects represent variation outside that expected for the species. Skeletal defects in hatchery-reared fish are sufficiently common to be an accepted cost of aquaculture and a major concern for aquaculturalists. Depending on species and hatchery conditions, as many as 50 per cent of the fish will have vertebral defects. In one study, 44 per cent of hatchery-reared Senegal sole, *Solea senegalensis*, showed vertebral abnormalities, 28 per cent of which affected the tail and vertebral column. Although clues to the aetiology of some defects can be obtained from teratological studies – wavy notochords and bent gill arches result when retinoic acid or collagen maturation is blocked in flounder embryos (Suzuki *et al.*, 2001) – determining the origins of vertebral defects has been an intractable problem.

From studies on Atlantic salmon, *Salmo salar*, cultured in Tasmania, Australia, we know that gender has no influence but that chromosome number (ploidy) significantly affects the incidence of deformities. For example, only

1 per cent of *diploid* freshwater smolts have lower jaw deformities. By contrast, the incidence of deformities in *triploid* fry, freshwater and seawater smolts is 2, 7 and 14 per cent, respectively. Similarly, primary gill filaments fail to form in up to 60 per cent of triploid individuals but only in 4 per cent of diploids. Triploidy provides the clue. The underlying mechanism remains elusive. Triploid rainbow trout, *Oncorhynchus mykiss*, were found to develop an additional caudal vertebra than found in diploid individuals, No other skeletal changes were found.[2]

Development of 'short-tails' is a common problem in farmed Atlantic salmon. These are fish in which the vertebral column is compressed along the A-P axis to such an extent that body length is reduced and body depth so enlarged that the body form hardly resembles that of a salmon (see Figure 1 in Witten *et al.*, 2005). The defect at the cellular level involves the metaplasia of intervertebral tissues to cartilage, essentially replacing the notochord with cartilage (Witten *et al.*, 2005).

FOR HOW LONG ARE NOTOCHORD AND SPINAL CORD ACTIVE?

In many inductive interactions the inducing cells lose their ability to induce at the same time as the responding cells lose their ability to respond, a mutually convenient temporal association that assures that induction will not be prolonged. As the spinal cord and notochord appear to play specific roles in inducing somitic cartilage, it is of interest to know whether they lose their inductive ability. I can summarize by stating that, *as late as has been tested, notochords retain inductive activity but spinal cords do not*. The evidence comes from combining notochord or spinal cord from older embryos or neonates with somites from younger embryos to determine whether cartilage forms. Some typical studies are summarized below and in Tables 36.1–36.3.

Cooper (1965) used chick–chick and chick–mouse chimaeras to show that notochord from embryos as old as H.H. 37 (11.5 days) can induce chick or mouse somites to form cartilage (Fig. 37.1 and Table 42.1). Notochords from older chick embryos were not tested but notochord from 19-day-old mice remains inductively active (Fig. 37.1). The spinal cord does not retain inductive ability for the same period, which may reflect, in part, earlier maturation of spinal cord in comparison with notochord. Spinal cords from mouse embryos older than nine days of gestation no longer induce (Tables 42.1 and 42.2). We found that spinal cord from chick embryos as old as H.H. 44 (18 days of incubation) allow H.H. 17 somites to chondrify *in vitro* under conditions wherein somites alone fail to chondrify (Tremaine and Hall, 1979).

Ability or inability to induce correlates with state of differentiation. Notochord (and cartilage; Chapter 22) only induces when the cells are undergoing hypertrophy or vacuolation (Table 42.3).[3]

Table 42.1 Inductive ability of notochord and spinal cord from chick and mice on somites of 9-day-old mice and H.H. 15, 17 or 18 chick embryos[a]

H.H. stage or age in days of notochord/ spinal cord donor	Cartilage induced in transfilter recombinations (%)	Cartilage induced in direct recombination (%)
Chick–chick chimaeras – notochord		
Stage 18	100	–
Stages 27–37.5	100	100
Chick–mouse chimaeras – notochord		
Stages 21–24	100	100
Stages 12–15	83	–
Chick–mouse chimaeras – spinal cord		
5.5 days[b]	100	–
7.5 days[b]	100	–
5.5–6.5 days[c]	0	–
Mouse–mouse chimaeras – notochord		
9 days	83	–
Mouse–mouse chimaeras – spinal cord		
9 days[b]	87.5	100

[a] Based on data from Cooper (1965).
[b] Ventral half of spinal cord.
[c] Dorsal half of spinal cord.

Table 42.2 Relative amount of cartilage in somites cultured with spinal cord from mice of different ages[a]

Age of spinal cord (days)	Relative % of cartilage/culture
9	100
11	80
12 and 13	25

[a] Based on data from Grobstein and Holtzer (1955).

CAN DERMOMYOTOME OR LATERAL-PLATE MESODERM CHONDRIFY?

During embryonic development the sclerotome chondrifies, the dermotome produces connective tissue, and the myotome forms muscle (Fig. 16.1). Can dermotome or myotome form cartilage if suitably challenged? An unstated assumption underlying such a question is that distance from the normal inducers is all that prevents dermotome and myotome from chondrifying. Three major types of experiments have been designed to answer this question.

Although muscle development is delayed, cartilage differentiation proceeds normally after 75 per cent of the medial aspect of the somites is removed from tail bud-stage urodele embryos. Apparently, areas of the somites normally destined for connective tissue and muscle regulate (Box 35.1) and chondrify.

Implanting spinal cord or notochord into the *dermomyotome* of embryonic chicks elicits a secondary (ectopic) neural arch as chondrogenic differentiation and morphogenesis are initiated in non-sclerotomal mesenchyme.

Table 42.3 Cartilage-inducing activity of embryonic chick and mouse notochord at four developmental stages[a]

Stage of developing notochord	H.H. stage/age of embryo	Number of cultures tested in notochord stage	Cartilage-inducing activity
1 Aggregating chordamesoblast cells	Chick H.H. 12–15	2	None
	Mice 9 days	3	None
2 Intercellular vacuolation	Chick H.H. 18	20	Cartilage induced
	Chick H.H. 21–24	18	Cartilage induced
3 Cellular hypertrophy	Chick H.H. 27–37.5	77	Cartilage induced
4 Non-hypertrophied state	Chick H.H. 37.5 vertebral canal	2	None

[a] Based on data from Cooper (1965). Notochord and somites were either recombined directly or transfilter and maintained *in vitro*.

Table 42.4 Chondrification in lateral-plate mesoderm from chick embryos of H.H. 9–12 grafted to the CAM[a]

Grafted tissue	Chondrification (%)
Lateral-plate mesoderm	4
Lateral-plate mesoderm + ectoderm and endoderm	37
Lateral-plate mesoderm + spinal cord	42

[a] Based on data from O'Hare (1972c).

Cartilage forms in CAM-grafted *lateral-plate meso-derm* adjacent to the last four pairs of somites from embryos of H.H. 9–12, provided that ectoderm/endoderm or spinal cord is included in the grafts (Table 42.4). This result removes an interpretation applied to some earlier experiments – that transplanted spinal cord or notochord did not induce lateral mesoderm, but attracted sclerotomal cells into the lateral mesoderm where they formed the cartilage.[4]

THE SEARCH FOR THE MAGIC BULLET

The search for a pure inducer that pervaded so many areas of embryology and developmental biology throughout the 20th century also directed, channeled, perhaps even hamstrung, students of somite chondrogenesis. For a brief period, beginning with a study by George Strudel in 1953, it looked as if progress was being made, but no significant advances were made in the two or three decades that followed.[5]

Strudel found that a saline extract of spinal cords or notochords from H.H. 18–21 chick embryos could induce somitic chondrogenesis *in vitro*.[6] A decade later, he extended this finding back to earlier stages, concluding:

…even very young somite cells are genetically determined cells and all that they need to undergo or accomplish their phenotypic differentiation is a microenvironment favouring chondrogenesis. This does not mean that the inducing action of the spinal cord and the notochord is dispensable. It may be that the spinal cord and/or the notochord exercise their inducing effect very early. (pers. comm.)

In 1957, Jay Lash and his colleagues introduced an ingenious experimental procedure adopted by a number of us. Embryonic spinal cords were placed transfilter to somites *in vitro* for 10 hours, during which time the spinal cord deposited extracellular material onto and into the filter. The spinal cords were removed and somites then placed onto the matrix products deposited into the filter and allowed to develop. Three days later, cartilage differentiated. 'Control' somites cultured on filters without ECM products formed no cartilage. Lash and colleagues concluded that only a short period of contact between somite and inducing spinal cord is needed to transfer a chemical signal between spinal cord and somites (Lash *et al.*, 1957).

A role for ectoderm?

In studies carried out in the 1920s to 1940s ectoderm and possibly also subjacent endoderm were often included with grafted somitic mesoderm, either deliberately or inadvertently (see Chapter 41). Potential problems associated with ectodermal contamination were taken up as researchers asked whether ectoderm adjacent to the somites plays any role in cartilage induction. The results are contradictory and the final story is not in even now. We can say that *epithelia exert no direct inductive action in vivo*; chondrogenesis is not initiated when notochord or spinal cord are extirpated but dorsal epidermis is left intact. If epithelia play a role, it is secondary to and dependent on notochord and/or spinal cord. A sample of the results and conclusions include the following.

- In somites from embryos of H.H. 10–13 grafted with or without ectoderm or endoderm, cartilage forms only when an epithelium is present. Seventy-five per cent form cartilage when ecto- and endoderm are present, 57 per cent when only ectoderm, and 14 per cent when only endoderm is present. Seno and Büyüközer (1958) concluded that epithelia, especially ectoderm, provide mechanical support for sclerotomal mesenchyme, thus preventing cells from dispersing. They ruled out a more physiological role.

- Stockdale *et al.* (1961) wrapped ectoderm around pellets of somitic cells equivalent in size to two to four somites and cultured them. Cartilage failed to develop, a result that does not necessarily rule out a role for ectoderm acting on intact somites *in ovo*. Similarly, no cartilage formed in any of 90 cultures of strips or clusters of somite wrapped in ectoderm or endoderm and either cultured or grafted into the coelomic cavities of host embryos (Lash, 1963b).

- Among the tissues Holtzer (1964b) tested for inductive ability, epidermis (source and age not specified) would not induce H.H. 18 somites to chondrify *in vitro*.

The British came to the rescue with an extensive (and definitive?) series of experiments by M. J. O'Hare, published in three parts in 1972. O'Hare grafted the last four somites from embryos of H.H. 9–12 after associating them with one of a variety of epithelia. As you can see from Table 42.5, somites grafted alone fail to chondrify, but a variety of epithelia permit chondrogenesis in anywhere from 10 to 30 per cent of the grafts.

O'Hare also showed that the ability to augment chondrogenesis was not restricted to epithelia adjacent to somitic mesoderm but also was present in limb bud and trunk epithelia from older embryos (Table 42.5). He explained these temporal differences as based on a correlation with presence of a basement membrane and associated extracellular matrix (ECM) in 'active' epithelia. Such temporal differences argue against a strict mechanical effect, as does the ability of lateral-plate mesoderm to chondrify when associated with ectoderm (Table 42.5).

The tentative conclusion from these studies is that epithelia allow somites to chondrify in the absence of notochord or spinal cord, if somites are CAM-grafted but not if cultured. Conclusions relating to any role for epithelia *in vivo* must be confined to the statement that epithelia cannot act in the absence of spinal cord or notochord.

Cartilage cells as cartilage inducers

When considering skull and vertebral column development we can draw the generalization that the contents induce

the container – brain and notochord induce skull, notochord and spinal cord induce vertebral column. The inducer is a very different tissue than the tissue induced. It may come as a surprise then, if I tell you that the container can induce more container. *Chondroblasts* and *chondrocytes* from a variety of skeletal sites – including ribs, long bones, vertebrae and trachea – can induce sclerotomal mesenchyme to chondrify. But the cartilage cells have to be at a particular stage in their differentiation.

Cooper (1965) established this relationship by culturing mouse or chick somites transfilter to cartilage (Table 42.6). Only chondrocytes that were enlarging or becoming hypertrophic elicited cartilage. Thus, with mouse tissues, induced cartilage was found adjacent to flattened or enlarging cells, but not adjacent to fully hypertrophic chondrocytes (Table 42.6). Small-celled chondroblasts fail to become flattened chondroblasts *in vitro* and fail to induce. Murine cartilage remains terminally hypertrophic *in vitro* and fails to induce, while chick chondrocytes regress from their hypertrophic state, cytolyse and fail to induce. This association of onset of inductive ability with onset of chondrocyte hypertrophy and concomitant loss of inductive ability with attainment of hypertrophy is analogous to the attainment and loss of inductive ability by the notochord as its cells hypertrophy, discussed in Chapter 41. Are similar factors involved?[7]

It is curious indeed that the tissue to be induced (cartilage) has the ability to induce cartilage. It was therefore exciting to find that the products deposited by chondroblasts into their ECM are strikingly similar to those produced and deposited by inductively active cartilage, spinal cord and notochord. Do contents induce container (vertebral column) using the same molecules that will come to characterize the cartilaginous container? To begin to answer this question, I review briefly the nature of the ECMs produced by chondroblasts/cytes before

Table 42.5 Chondrogenesis in the last four somites from chick embryos of H.H. 9–12 grafted in contact with different epithelia[a]

Treatment	Chondrogenesis % (N)
Isolated somites	0 (0/89)
Somites + ectoderm and endoderm	21 (17/81)
Somites + ectoderm and endoderm after trypsinization	10 (5/48)
Somites + trunk epithelium from two-day-old embryos	0 (0/21)
Somites + trunk epithelium from three-day-old embryos	30 (10/33)
Somites + trunk epithelium from four-day-old embryos	23 (7/30)
Somites + limb-bud epithelium from four-day-old embryos	24 (7/29)

[a] Based on data from O'Hare (1972b).

Table 42.6 Incidence of cartilage in somites from mouse and chick embryos exposed to chondroblasts of different differentiation stages[a]

Stage of cartilage differentiation[b]	Incidence of cartilage % (N)
Mouse	
Small-celled chondroblasts	0 (0/51)
Flattened and enlarging chondrocytes	91 (51/56)
Enlarging or terminal chondrocytes	0 (0/12)
Chick	
Small-celled chondroblasts	0 (0/17)
Flattened chondrocytes	93 (27/28 mouse somites)
	88 (15/17 chick somites)
Enlarging chondrocytes	21 (3/14)
Terminal chondrocytes	95 (20/21)

[a] Based on data from Hall (1977a). Mouse embryos were nine days of gestation, chick embryos were H.H. 12–15 and 17/18. Somites were cultured transfilter to chondroblasts.
[b] Murine terminal chondrocytes are hypertrophic; chick terminal chondrocytes have lost their hypertrophic state.

discussing the matrices produced by notochord and spinal cord.

CHONDROCYTE EXTRACELLULAR MATRIX

The production of glycosaminoglycans (GAGs), especially chondroitin sulphate (CS), is characteristic of functioning chondroblasts and chondrocytes. Onset of such synthesis, as detected by ^{35}S autoradiography, was used for a long time to detect chondrogenic cells. ^{35}S is taken up by pre-chondrogenic somitic mesoderm of chick and mammalian embryos, with maximal uptake in mesenchyme immediately adjacent to the embryonic axis. Lash and colleagues sought to determine whether GAG accumulation preceded histological differentiation of cartilage, but could not detect incorporation of ^{35}S chondroitin sulphate earlier than they could detect metachromatic ECM.[8]

Somites explanted *in vitro* without spinal cord or notochord neither chondrify nor do they produce collagen or CS. Somites from embryos of H.H. 16 and 17 can be cultured for eight days without observing any activity of the enzymes ATP-sulphurylase or APS-kinase – both of which are involved in sulphation of chondroitin – unless spinal cord or notochord is included. With refinements in technique it became possible to detect CS before ECM could be detected histochemically. For example, H.H. 11 somites (Fig. 16.2) cultured with notochord or spinal cord produce what were then called chondroitin sulphates A and C (now chondroitin-4-sulphate and chondroitin-6-sulphate).[9]

With the development of thin-layer chromatographic techniques to isolate precursors of CS, it became apparent that many early embryonic tissues – somites, epidermis, spinal cord, notochord, extraembryonic membranes and mesonephros (see Box 38.2 for the latter) – metabolize glucosamine to form UDP-N-acetylgalactosamine (UDP-NaGal), but in much smaller amounts than cartilage (Fig. 26.1). Two conclusions were reached: (i) many tissues can synthesize CS but not all tissues can accumulate it, and (ii) isolated somites can produce CS but need a stimulus to augment and stabilize the rate of deposition. Exposing somites to spinal cords or notochord does not confer upon them the ability to synthesize CS but rather enhances the rate at which already functioning synthetic pathways operate. The 'instruction' from the 'inducer' is speed-up, not start-up.

Which extracellular molecules are being deposited by notochord and spinal cord when these two tissues interact with sclerotomal mesenchyme?

NOTOCHORD AND SPINAL CORD EXTRACELLULAR MATRICES

The evident role played by the notochord and spinal cord in evoking chondrogenesis *in vivo*, and the accumulating evidence that normal constituents of cartilage ECM such as CS, collagen and hyaluronan can influence the rate of synthesis of cartilage ECM through feedback inhibition and stimulation (Hall, 1973a; and see Chapter 26), led to a search for ECM around the notochord and spinal cord, a search that demonstrated that notochord and the ventral portion of the spinal cord in chick embryos as young as H.H. 10 synthesize and accumulate GAGs into an ECM. Table 42.7 presents a summary of the major events in the formation of notochordal and spinal cordal ECMs and the changes occurring in sclerotomal mesenchyme at the same stages.

GLYCOSAMINOGLYCANS

The first studies utilized autoradiographic techniques to show that GAGs can be visualized around the notochord when chondrogenesis is commencing within adjacent sclerotome. The impetus came when Johnson and Comar

Table 42.7 Chronology of vertebral chondrogenesis in the embryonic chick[a]

H.H. stage	Age	Notochord, spinal cord	Sclerotome
3$^+$	12 hours	^{35}S in presumptive notochord	
10	36 hours	Collagen, GAGs as ECM	Intact mass of ovoid cells
11	42 hours	Basement membrane present	Nest of cells breaking up
12	47 hours		Mesenchymal cells beginning to migrate
13	50 hours	Youngest age shown to produce collagen when isolated *in vitro*	Sclerotomites starting to form
17	60 hours	Considerable hyaluronan around notochord. Treatment with collagenase or hyaluronidase prevents induction	
18	3 days	Notochord vacuolated, now induces 100% of cultured somites to chondrify	Sclerotomal cells finished migration
23	4 days		ECM around cells closes to notochord – now prechondroblasts
27–30	5–7 days	More CS than hyaluronan	ECM spreading into peripheral cells
31–36	7–10 days	Ventral spinal cord loses inductive ability	Chondroblasts and chondrocytes
36–39	10–13 days	Notochord loses inductive ability (H.H. 37.5)	Cell death in area nearest notochord

[a] Based on data from Hall (1977a).

(1957) injected ^{35}S into the albumen of eggs and detected uptake into notochord and primitive streak in embryos as young as H.H. 3+, *long before somite chondrogenesis*. Notochord and spinal cord from embryos of H.H. 11 and older are surrounded by sulphated GAGs. [35]S was localized over the spinal cord, notochord and immediately adjacent ECM when these regions were cultured, but Lash and his colleagues could not judge where this material was synthesized.[10]

Application of electron microscopy and histochemistry demonstrated sulphated GAGs (hyaluronan and CS) adjacent to the notochord at H.H. 16 and 17, respectively. By H.H. 17, there was 2.5 times more hyaluronate than CS. By H.H. 28 the ratio had reversed. Therefore, it was suggested that hyaluronan plays a role in sclerotome aggregation. In 1972, my fellow Australian Bryan Toole began what became a career-long study of hyaluronan and its relationship with CS (see Chapters 14 and 20). The ratio of hyaluronan to CS is high at H.H. 23 but low thereafter, the decline correlating with increasing levels of hyaluronidase (Table 42.8) and with formation of metachromatic ECM in perinotochordal mesenchyme; removal of hyaluronan accompanies formation of ECM.[11]

Further application of electron microscopy laid the basis for our understanding of the organization of the ECM, demonstrating:

- amorphous material (GAGs) on microfibrils around the notochords of two- to four-day-old embryonic chicks;
- the importance of the metachromatic ECM between sclerotome and spinal cord and around the notochord, with the interpretation that these matrix products progressively *become* the ECM of perinotochordal cartilage (Box 4.4); and
- the accumulation of 200–400-Å GAG granules along with 150-Å unbanded collagen fibrils around the notochord and the *ventral* portion of the spinal cord at H.H. 10; recall that ventral but not dorsal spinal cord is inductively active. By H.H. 17, this ECM was prominent and a basement membrane surrounded notochord and spinal cord, a matrix maturation preceding the arrival of migrating sclerotomal mesenchyme around the notochord at H.H. 18.[12]

Both ECM and basement membrane are lost following trypsinization to isolate somites. The finding that products of the ECM *reform* when notochord or spinal cord is maintained *in vitro* provided evidence for the production of the ECM *by these tissues*. Analysis of enzymatic digests of products from cultures of isolated notochords and neural tubes reveals mostly chondroitin and heparan sulphates, including shared CS–protein complexes.[13]

In a series of insightful papers, Nagaswamisri Vasan showed that perinotochordal ECM from chick embryos contains small proteoglycans and cartilage-type proteoglycans, and that only the large aggregated forms are required for notochord to induce somitic cartilage. Unstimulated somites – i.e. somites not exposed to notochord – do not contain any link protein for proteoglycan. Adding notochord to somite cultures activates the synthesis of link protein and stabilizes the ECM. On the other hand neither hyaluronan nor hyaluronidase were affected by adding notochord to somite cultures. Proteoglycan monomers from sternal cartilage also evoke chondrogenesis from somitic mesoderm, further implicating ECM components in the inductive interaction.[14]

Collagen was also being studied. Indeed, these studies were among the first to demonstrate a family of collagen molecules and tissue-specific collagen types.

Collagens

Until the late 1950s, and in large part as a consequence of the methodologies available, collagen had been demonstrated only in mesenchymal tissues. Consequently, collagen was regarded as a mesenchymal protein. From the early 1960s onwards, there was a suggestion that epithelial structures such as notochord and spinal cord might have the ability to synthesize and export collagen. The initial findings came from ultrastructural studies, as TEM became an important tool revealing cellular organization.

The ECM of the notochord in embryonic chicks and mice was described as containing perinotochordal fibrils, some of which were thought to be derivatives of sclerotomal cells, some from the notochord sheath. Then came a series of studies by Low and colleagues describing microfilaments on the notochord and between notochord and spinal cord at H.H. 11. Amorphous material – interpreted as GAG – is attached to these microfibrils, which are of two types: 150–200-Å diameter, unbanded and beaded fibrils close to the notochord and sensitive to removal by collagenase, and 100-Å tubular banded fibrils some distance from the notochord, sensitive to removal by hyaluronidase and amylase but not by collagenase.[15]

Georges Strudel's group described collagen fibrils in the ECMs of spinal cord in chick embryos. Because these fibrils were sensitive to collagenase and coincided in position with localized uptake of ^{3}H-proline as measured autoradiographically, Strudel suggested that they are incorporated into the ECM of differentiating vertebral cartilage.[16] Direct evidence that collagen around the spinal cord is *produced by the spinal cord* was obtained by Cohen and Hay (1971), who showed that: (i) ventral

Table 42.8 Ratios of hyaluronan to CS in somites isolated from chick embryos and *maintained in vitro*[a]

H.H. stage	Ratio of hyaluronan to CS
23	1.92
26	0.60
28	0.38
30	0.16

[a] Based on data from Toole (1972). Somite numbers 20–32 were isolated from embryos of H.H. 23–30.

spinal cord from two-day-old chick embryos synthesizes collagen when cultured alone; (ii) the collagen is deposited as fibrils into the ECM; and (iii) the deposited collagen is associated with the basement membrane.

Similarly, direct evidence that the collagen around the notochord is *produced by the notochord* was obtained by isolating notochords from two-day-old chick embryos using trypsinization to remove ECM and basement lamina and culturing the 'naked' notochords. After two or three days ECM microfibrils accumulated next to a reconstituted basal lamina. Culture for longer times in the presence of 100 µg/ml ascorbic acid (a co-factor for collagen synthesis) demonstrated that the fibrils – some as wide as 1500 Å and laid down in sheets – were cross-striated with an axial periodicity of 510 Å, within the range for collagen.

With advances in knowledge of collagen types, these perinotochordal and perispinal cordal fibrils were shown to be type II (cartilage-type) collagen as is the collagen synthesized *in vitro* by spinal cords from two-day-old chick embryos. Subsequently, antibodies against type II collagen were shown to bind to embryonic notochord. A decade later, notochords from chick embryos were shown to secrete type IX collagen just prior to vertebral chondrogenesis.[17]

The developmental significance of the discovery that notochord and spinal cord, which promote chondrogenesis in sclerotomal cells, produce the same collagen type (type II) as the tissue they induce is the possibility that type II collagen may play an inductive role in chondrogenesis.

The evolutionary significance is that notochord is phylogenetically older than cartilage; notochord may have acquired the ability to synthesize and deposit type II collagen before cartilage did. Rather than saying that notochord possesses cartilage-type collagen, we should be saying that cartilage possesses notochord-type collagen.

FUNCTION OF NOTOCHORD AND SPINAL CORD MATRIX PRODUCTS

An obvious question that follows from these studies is whether collagen or GAGs produced by notochord and/or spinal cord play any role in initiating chondrogenesis from sclerotomal mesenchyme; the desire to answer this question motivated the search for matricial materials in the first place. Familiar names are associated with these studies – Strudel, O'Hare, Lash and Kosher – as two experimental approaches were initiated in the early 1970s:

- removal of the ECM from notochord or spinal cord following digestion with enzymes, and testing for retention of inductive activities; and
- adding matrix products, individually or in combinations, to the media in which otherwise unstimulated somites are cultured, and looking for chondrogenesis.

Differentiation of sclerotomal cartilage was inhibited when Strudel cultured *axial rudiments* from chick embryos in the presence of 20 µg collagenase to remove collagen, or 10 µg testicular hyaluronidase to remove GAGs. Transferring these rudiments to a control medium allowed chondrogenesis to commence. Of course, enzymatic treatment of *entire* axial rudiments removes collagen and GAG from somites as well as from the other tissues. Inhibiting chondrogenesis could result from the inability of the sclerotomal cells to secrete ECM, rather than an inability of the notochord or spinal cord to induce. This difficulty in interpretation applies to other studies in which the proline analogue L-azetidine-carboxylic acid (LACA) was injected *in ovo* or added to culture media containing somites and axial organs, and in which secretion of ECM was retarded and myotomal differentiation favoured at the expense of chondrogenesis.[18]

O'Hare (1972c) surmounted such difficulties by impregnating Millipore filters with collagenase and hyaluronidase, placing embryonic somites and spinal cord onto the filters, and CAM-grafting them. Without enzyme pretreatment of the filters, 52 per cent of the grafts produced cartilage. Only 14 per cent produced cartilage on filters impregnated with enzymes. O'Hare attributed the lack of chondrogenesis to loss of the spinal-cord basement membrane and/or ECM rather than to a direct effect of the enzymes on the somites.

Another approach was to treat isolated notochord with chondroitinase or testicular hyaluronidase, combine the treated notochords with intact somites, and maintain them *in vitro*. Cartilage did not form unless some ECM remained on the notochord. Trypsinizing notochords prevents them from inducing, but after a time *in vitro*, perinotochordal materials reform – including a new basement membrane, reappearance of microfibrils and uptake of ³H-proline – and the notochords regain their ability to induce.[19]

Experiments in which matrix products were added to somites in Robert Kosher's laboratory did not produce results allowing such clear-cut interpretations. Chondroitin-4- and chondroitin-6-sulphates extracted from vertebrae were added to the media in which somites were cultured. Although the percentage of cultures forming cartilage was the same in treated and control media, accumulation of GAG dropped off after two days in control medium but was maintained at the high initial rate in treated media. Similar results for maintenance of collagen synthesis were obtained when procollagen or collagen was added to the medium used to maintain somites; synthesis of type II collagen is maintained at a high level, indicating positive feedback from product to synthesis of product.[20]

KEY ROLES FOR *Pax-1* AND *Pax-9*

Pax-1, a member of the paired box of homeobox genes, is expressed in all somites early in development, but is down-regulated in the five most caudal somites. Expression also is seen in the perichordal zone (Table 1.1) of the vertebral column of embryonic but not adult mice. It appears that Pax-1 plays two roles. Transient expression in condensing mesenchyme *before* the onset of chondrogenesis suggests an early role in sclerotomal differentiation. The second role is in the differentiation of intervertebral discs.[21]

Pax-9 plays an even more ancient role, a role that was uncovered from studies on the Japanese lamprey, *Entosphenus* (*Lampetra*) *japonicus*, in which *Pax-9* is expressed in the endodermal pharyngeal pouches, mesenchyme of the velum (but not the velar muscle), hyoid arch and nasohypophyseal plate, *but not in somitic mesoderm*; involvement of *Pax-9* in somitic chondrogenesis is a feature of jawed not jawless vertebrates. Ogasawara et al. (2000) used these expression patterns to argue that *Pax-9* tracks neural crest-derived mesenchyme in the velum (which first arises from a premandibular segment) and that, as a consequence, the agnathan velum is homologous with the gnathostome jaw. If correct, this homology has important consequences for our views on the origin of gnathostome jaws. Reinforcing this possibility, *Pax-9*-deficient mice lack the angular and coronoid processes of the dentary, have supernumerary preaxial digits, are missing all teeth, which are arrested at the bud stage, and fail to develop derivatives of the pharyngeal arches such as the thymus, pituitary and ultimobranchial glands (Peters et al., 1998).

Pax-1 and *Pax-9* – a paired box-containing gene closely related to *Pax-1* – act synergistically in vertebral development. *Pax-9* is expressed in the vertebral column, pharyngeal pouches, tail, head and limbs; the mutant *Danforth's short tail* involves loss of caudal expression of *Pax-9* and therefore loss of the notochord inducer. *Pax-1* is involved in ventralizing the sclerotome; loss of *Pax-1* is associated with sclerotomal and vertebral anomalies.

Similarly, Hox group 3 paralogous genes (*Hoxa-3*, *Hoxb-3* and *Hoxd-3*) act synergistically to alter neural tissues, and neural-crest and somitic mesenchyme. Surprisingly, the identity of specific Hox genes is less critical than the number of genes expressed in a particular region or tissue. An especially nice example is that redundancy among *Hox-10* and *Hox-11* paralogues is so great that transformatioin of vertebral identity from sacral → lumbar or lumbar → thoracic does not take place in mice that are mutant for five of the six alleles, all six having to be knocked out to effect the transformation.[22]

Defects seen in *Pax-1* mutant mice are not seen in *Pax-9* mutants. Double mutants (*Pax-1⁻*/*Pax-9⁻*) lack vertebral bodies, intervertebral discs and the proximal portion of the ribs (see Box 16.2 for genes that influence rib development). Neural arches are normal. The chondrocyte lineage is induced to produce a loose mesenchyme rather than segmental elements. *Sox-9* and collagen type II are down-regulated in this loose mesenchyme, which shows a low rate of proliferation but is not maintained,

undergoing apoptosis. Misexpressing *Sox-9* in the dermomyotome leads to production of type II collagen and *Pax-1* by dermomyotomal cells, which switch to a chondrogenic fate.[23]

In a substantive study of mouse sclerotome development, Furumoto et al. (1999) demonstrated that *Pax-1* and *Mfh-1* – both of which depend on *Shh* from the notochord for their expression – act synergistically to control vertebral column development. Double mutants (*Pax-1⁻*/*Mfh-1⁻*) lack the dorso-medial elements of the vertebrae and consequently fail to form either vertebral bodies *or* intervertebral discs. Both genes regulate somitogenesis at the proliferation (condensation) stage.[24]

Shh-null mice have normal molecular markers in the three portions of the somites. Marcelle et al. (1999) therefore separated the paraxial mesoderm from the axial structures and examined the role of *Shh* in expression of *Pax-1*, *Myo-D* and *Pax-3*. *Pax-1* is rescued by *Shh*, *Myo-D* is maintained but not induced by *Shh*, while *Pax-3* is expressed independently of *Shh*. Thus, *Shh* is a potent mitogen for somitic cells. Murtaugh et al. (1999) showed that *Shh* has several actions on somitic precursor cells in chick embryos. Presomitic mesodermal cells respond to *Shh* by initiating chondrogenesis, the later stages of their response depending on Bmp signaling. *Shh* enhances the response of somitic cells to Bmps, suggesting at least two phases. Finally, *Shh* enhances precursor cells competent to respond to Bmp by initiating chondrogenesis.

Msx-2 and notochordal influences are differentially expressed in dorsal and neural-arch development in chick embryos. Bmps are also involved; there is parallel expression of *Bmp-4* and *Msx-1* and *Msx-2* in the lateral neural plate and then in the dorsal neural tube and midline ectoderm in chick embryos. These are locations from which signaling could be transferred to the sclerotome. Grafting Bmp-4 or Bmp-2 into the neural tube up-regulates *Msx-1* and *Msx-2* in the adjacent mesenchyme, resulting in production of ectopic cartilage in the pectoral girdle that develops. Grafting Bmp-4 or Bmp-2 lateral to the neural tube down-regulates *Pax-1* and *Pax-3*. Subsequently, *Shh* was shown to antagonize *Bmp-4* and *Msx* during vertebral development.[25]

The paired type homeodomain transcription factor *Uncx4.1*, which acts upstream of *Pax-9*, is required before neural arches can form from lateral sclerotomal cells. Mice homozygous for a targeted mutation in *Uncx4.1* die perinatally with severe malformations of the axial skeleton – all lateral sclerotomal derivatives (neural arches) fail to form because chondrogenesis is inhibited. The defect is early – anlagen form but fail to condense – further reinforcing the roles of these genes at the condensation stage of vertebral chondrogenesis.[26]

By using morpholino antisense oligonucleotides against the transcription factor *twist*, Yasutake et al. (2004) showed that *twist* is required for neural arches to form in embryos of the Japanese medaka, *Oryzias latipes*, because of involvement of *twist* after sclerotomal-cell migration.

CONCLUSIONS

The major events occurring in sclerotome, ventral spinal cord and notochord during chick development discussed in Chapters 41 and 42 are summarized in Table 42.7 and below.

Sclerotomal mesenchyme has an inherent potential for chondrogenesis. Nevertheless, expression of chondrogenic potential is exquisitely sensitive to the environment. Depending on the nature of supplements added *in vitro*, sclerotomal mesenchyme from embryonic chicks as young as H.H. 10 (36 hours of incubation) can chondrify. Grafting somites to a vascularized environment such as the CAM permits somites from even younger embryos to chondrify, presumably not because of vascularization *per se*, but because of exposure to molecules in the circulation.

The presence of the synthetic machinery for, and the synthesis of CS – attributes previously thought to apply only to induced mesoderm – are shared with many nonchondrogenic tissues. What is special about sclerotomal mesenchyme and its interaction with notochord and spinal cord is the ability to augment the rate of synthesis of CS and other GAGs, and to deposit these products into an ECM. Control is not at the level of presence or absence of the synthetic pathways, but in regulating those pathways.

Notochord and spinal cord provide the major *in-vivo* environmental factors that augment somitic chondrogenesis. Notochord and spinal cord are also the most potent agents active *in vitro*. Whilst other factors can substitute *in vitro*, their absence is not compensated for *in vivo*. Collagen and CS enhance and maintain chondrogenesis by positive feedback to matrix products, chondrogenesis having been initiated at the condensation stage through the action of such regulatory genes as *Pax-1* and *Pax-9*.

Normal morphogenesis of the neural arches and centra also depends on influences from spinal cord, notochord and spinal ganglia. Ablating these tissues provides the evidence for this conclusion. The underlying mechanisms await resolution, although *Msx* and *Bmps* are possible players. If epidermal ectoderm plays a role *in vivo*, it is subordinate to and dependent upon notochord and/or spinal cord.

NOTES

1. See Watterson (1952) for the studies on *Fundulus*, Detwiler and Holtzer (1956) and Strudel and Gateau (1971) for abnormalities in urodele spinal cord and chick notochord, Lash (1968b) for maintenance of cartilage by notochordal influences, and Roy et al. (2004) and Witten et al. (2005) for vertebral anomalies in aquacultured fish.

2. See Sadler et al. (2001) for triploid Atlantic salmon, and Kacem et al. (2004) for triploid rainbow trout.

3. See Cooper (1965) for the correlation with hypertrophy, and Bancroft and Bellairs (1976) for associated ultrastructural features.

4. See Holtzer and Detwiler (1953) for the urodele study, Strudel (1953b) and Watterson et al. (1954) for secondary neural arches, and O'Hare (1972c) for the CAM-grafts. Culturing myotome plus notochord and spinal cord from chick embryos of H.H. 18 did not result in chondrogenesis (Cooper, 1965). It may be, however, that by H.H. 18, the myotome is fixed with respect to ability to form muscle and cannot switch.

5. See Lash (1963a) for a review of the first decade of the search (1953–1963), Hall (1977a) for the research to 1977, and Monsoro-Burq and Le Douarin (2000) and Stockdale et al. (2000) for more recent approaches.

6. See Strudel (1953b, 1962, 1962).

7. Epithelial otic vesicles from mouse and chick embryos are amongst the other tissues that have been tested for an ability to induce sclerotomal mesenchyme and found not to induce (Grobstein and Holtzer, 1955; Strudel, 1955, 1962; Benoit, 1960b; O'Hare, 1972b). See Benoit (1955, 1960a,c, 1964) and Hall (1991b) for the ability of the otic capsule to form in ectopic sites.

8. See Amprino (1955a) and Johnson and Comar (1957) for ^{35}S uptake into somatic mesoderm, and Lash et al. (1960) for uptake and CS. Lash (1963a) showed that CS is sensitive to RNA'ase digestion.

9. See Glick et al. (1964) for production of ATP-sulphurylase and APS-kinase, Franco-Browder et al. (1963) for production of chondroitin-4-sulphate and chondroitin-6-sulphate, and Okayama et al. (1976) for a discussion of early determination for chondrogenesis.

10. See Franco-Browder et al. (1963) for the studies with H.H. 11 embryos, and Lash (1963) and Lash et al. (1964) for ^{35}S uptake.

11. See O'Connell and Low (1970) for ultrastructure, Kvist and Finnegan (1970a,b) for histochemistry, Toole (1972) for HA:CS, and Goldberg and Toole (1984) for hyaluronan as part of the pericellular coat.

12. See Frederickson and Low (1971), Strudel (1971) and Corsin (1974) for the amorphous material and importance of the ECM, and Minor (1973) and Bancroft and Bellairs (1976) for the GAG granules and collagen fibrils.

13. See Hay and Meier (1974) for separation of chondroitin and heparin sulphates, and Mathews (1971, 1975) for the similarity of CS–protein complexes.

14. See Vasan (1981, 1983, 1987) for proteoglycan accumulation, Vasan et al. (1986a,b) for core protein, hyaluronan and hyaluronidase, and Vasan and Miller (1985) for sternal cartilage monomers.

15. See Jurand (1962, 1974), O'Connell and Low (1970), Frederickson and Low (1971), Minor (1973) and Bancroft and Bellairs (1976) for perinotochordal fibrils.

16. See Bazin and Strudel (1972, 1973) and Ruggeri (1972) for independent autoradiographic studies.

17. See Carlson et al. (1974) and Carlson and Upson (1974). Hydroxylation of proline and lysine is inhibited in ascorbic acid (vitamin C) deficiency. Consequently, biosynthesis of collagen and elastin is impaired (Barnes et al., 1969; Wu et al., 1989). See Linsenmayer et al. (1973a) and Trelstad et al. (1973) for notochord and spinal-cord type II collagen, von der Mark et al. (1976) and von der Mark (1980) for the work with type II antibodies, and Hayashi et al. (1992) for notochordal secretion of type X collagen.

18. See Strudel (1972, 1973a–c) for the enzyme studies and Strudel (1975a–c) for the studies with LACA (which inhibits

excretion of collagen and consequently inhibits growth and delays chondrogenesis, for which see Aydelotte and Kochhar, 1972 and Hall, 1978c).

19. See Kosher and Lash (1975) and Lauscher and Carlson (1975) for studies with chondroitinase and hyaluronidase. Mice immunized with chondroitinase ABC, proteoglycan and adjuvant develop polyarthritis, an ankylosing spondylitis (Glant *et al.*, 1987).

20. See Kosher *et al.* (1973) for the studies with CSs, and Kosher and Church (1975) and Kosher and Savage (1979) for the studies with procollagen and collagen.

21. See Deutsch *et al.* (1988) for *Pax-1* expression, Wallin *et al.* (1994) for early and late roles in chondrogenesis, and Wilting *et al.* (1995) for caudal expression. Wilting and colleagues also used the boundary of expression to determine the head/trunk (cervical/occipital) boundary in Japanese quail and mice.

22. See Manley and Capecchi (1997) for the concept of redundancy among *Hox* genes and see Wellik and Capecchi (2003) for the *Hox-10*/*Hox-11* example.

23. See Peters *et al.* (1998, 1999) for *Pax-1* and *Pax-9*, Neubüser *et al.* (1995) for the distribution of *Pax-9* and *Danforth's short tail*, and Balling *et al.* (1996) for *Pax-1* knockout. *Pax-1* and *Pax-9* are also expressed in complex, non-overlapping patterns in chick limb mesenchyme, especially at the condensation stage, when expression is strongest (Le Clair *et al.*, 1999). See Healy *et al.* (1999) for expression of *Pax-1* in the dermomyotome.

24. The winged helix transcription factor, *MFH-1*, is expressed in mouse cranial neural crest and head mesoderm and then in cartilage. *MFH-1⁻* mice die as embryos or at birth (Iida *et al.*, 1997).

25. See Monsoro-Burq *et al.* (1994, 1996), Liem *et al.* (1995), Pourquié *et al.* (1996), Watanabe and Le Douarin (1996) and Watanabe *et al.* (1998a). Tgfβ-1 enhances chondrogenesis from chick sclerotomal micromass cultures (Sanders *et al.*, 1993), probably because of a generic proliferative effect.

26. See Leitges *et al.* (2000) and Mansouri *et al.* (2000) for the studies with *Uncx4.1*.

Tail Buds, Tails and Taillessness

Chapter 43

Thereby hangs a tail.

Many tetrapods and almost all fish (Box 43.1) have tails, we humans being in the minority. The smallest terrestrial tailed tetrapods are 15 species of lungless (plethodontid) salamanders in the genus *Thorius*. With a snout–vent length (SVL) of 13 mm, the pygmy salamander, *Thorius pennatulus*, is rated the smallest. although, curiously, this standard measure of length (SVL) specifically and deliberately excludes the tail. Determining which animal has the longest tail depends very much on how you define a tail. The standard definition (by exclusion) is any (segmented) part of the body posterior to the vent (anus). Indeed, we often speak of the *post-anal tail* to distinguish the 'true' tail from other posterior portions of the body. We do so because the body merges imperceptibly into the tail in animals such as snakes and crocs, at least as viewed from the outside. Internally, there may be specialized caudal vertebrae (or even no vertebrae at all, as in anuran tadpole tails), distinctiveness in the notochord, or other features to tell where tail starts and body ends.[1]

EMBRYOLOGICAL ORIGIN

We know most about the early origins of tail-bud mesenchyme from studies with the South African clawed frog, *Xenopus laevis*, but given the animal's bizarre and derived early development, this pattern may not translate to other tetrapods.[2]

The tail bud arises following interaction between two neural-plate territories (fields) and a posterior mesodermal territory. Two lines of evidence underpin this conclusion, as follows.

(i) Tails fail to form in gastrulae made to develop abnormally by evaginating rather than invaginating their mesoderm, a process known as exogastrulation. Mesodermal and neural-plate fields cannot interact correctly in these exogastrulae, resulting in inability to initiate a tail.

(ii) Transplanting these territories elsewhere on the embryo induces extra tails to grow (Tucker and Slack, 1995a,b). Tail ectoderm is more extensive than mesoderm. Interestingly, much of the tail in *Xenopus* tadpoles forms from caudal-ward displacement of axial tissues, not from the tail bud.

Beck and Slack (2002) demonstrated a requirement for the transmembrane signaling protein Notch in *Xenopus* tail development; ectopic expression leads to development of an ectopic tail; inhibition of *notch* signaling stops tail development.

Delta proteins are ligands for Notch. Several members of the delta family (*Dl-1* to *Dl-4*) have been characterized from mice. Three mutations of *Dl-3* are associated with axial defects in humans with *spondylocostal dysplasia*. Loss-of-function of *Dl-3* in mice also results in axial defects, the result of delayed somite formation and abnormal antero-posterior somite polarity (Dunwoodie et al., 2002).

The questions as to how tail somites develop, whether they segment or resegment, and whether tail somites chondrify are largely unexplored and likely to vary from taxon to taxon. Unknown to many who work with limbs or fins, tails develop from tail buds under the influence of an epithelial ridge, the ventral epithelial ridge or VER.

THE VENTRAL EPITHELIAL RIDGE (VER)

Knowledge of epithelial–mesenchymal interaction in tail-bud development goes back at least 35 years to

Box 43.1 Tailless fish

If the caudal fin in teleost fish is equated with the tail, then a number of interesting fishes lack tails (Figs 34.5 and 43.1).

Tailless fish are not closely related, which suggests that loss of the caudal fin occurred multiple times. All tailless fish do have elongate bodies. Indeed many have *eel* as part of their common names – gulper eels, swamp eels – although not all are true eels. The driving forces to tail-fin reduction appear to have been extreme elongation of the body under a variety of ecological conditions and the changed mode of locomotion, steering and manoeuvrability associated with the long body rather than the tail.

One example, the umbrella-mouth gulper eel, *Eurypharynx pelecanoides* – a true eel – inhabits temperate and tropical waters around the world, living at depths ranging from 1500 to 2750 metres. Highly derived, the gulper eel is essentially a 0.6-metre, weakly muscularized body surmounted by a small head with huge but weak jaws. Median fins form a flattened continuous fold around the entire body. A plankton feeder, this gulper swims with its mouth open to engulf whatever it encounters. The huge mouth has been likened to that of a pelican, but unlike the pelican (whose beak holds more than its belly can), the eel's stomach is as long as the body and the jaws are capable of being 'unhinged' to engulf a large volume of prey.

A second example, the Asian swamp eel or rice eel, *Monopterus albus*, is not in the eel family. Its elongate body, which is surprisingly snake-like, may be as long as a metre. The similarity to snakes extends to their ability to breathe air and move across land, and even further to the lack of any paired fins.

Figure 43.1 A butterfish (*Stromateus* sp.) with pigmented dorsal and ventral fins and distinctive caudal fin. Photographed by the author.

Rodolfo Amprino and his colleagues, who implanted tail buds with or without tail-bud epithelium from H.H. 16–23 chick embryos into the flanks of host embryos. Intact tail buds continued to develop because, according to Amprino *et al.* (1969), tail-bud mesenchyme grafted without the epithelium degenerates because the necessary epithelial–mesenchymal interaction cannot take place.

Can tail-bud epithelium respond to *non*-tail-bud mesenchyme? This question was tested by grafting *limb*-bud mesenchyme from embryos of H.H. 19 and 20 beneath *tail*-bud epithelium of H.H. 21/22 embryos. The limb-bud mesenchyme induced tail epithelium to form a ridge, but was it an AER or a VER? As the ridge maintains outgrowth of the limb mesenchyme and as digits form, the ridge is functioning as an AER. Further studies demonstrated that when suitably grafted, chick *tail* mesenchyme can be drawn into the *limb* field and contribute to digit formation.[3]

Subsequent studies confirmed the existence of such interactions, identifying a specialized region of tail-bud epithelium originating from the cloacal membrane of mouse embryos at 9–9.5 days of gestation. Grüneberg (1956) named this the *ventral ectodermal ridge* (VER).[4] Gajovic and Kostovic-Knezevic (1995) used transmission (TEM) and scanning (SEM) electron microscopy to describe the VER of rat embryos between 10.5 and 14 days of gestation. I described the VER on mouse embryos, demonstrated maximal extent at 10.5 days of gestation and an epithelial–mesenchymal interaction, and showed that growth of tail-bud mesenchyme stops if

the VER is removed (Hall, 2000c). No equivalent ridge has been described for *Xenopus* or for any other amphibian.

Goldman *et al.* (2000) examined the function and fate of the VER in mouse embryos, demonstrating that fewer somites are specified when the VER is ablated, the middle and posterior portions of the VER being the most active, as is the case for an AER. *Fgf-17* and *Bmp-2* are expressed in the VER, suggesting that *Bmp-2* functions to up-regulate *noggin* in tail mesenchyme.

Looking for clues in amphioxus, *Branchiostoma*, Schubert *et al.* (2001) cloned three *Wnt* genes from amphioxus (*AmphiWnt-3*, *-5* and *-6*) associated with the blastopore and tail-bud development, as the homologous genes are in vertebrates. Amphioxus, however, adds new somites to the tail throughout development; vertebrates lay down all their somites early in development. These workers suggested that the emergence of epithelial–mesenchymal interactions at the origin of the vertebrates allowed for the evolution of presomitic mesoderm. I took a similar position with regard to elaboration of the neural crest (Hall, 1999a).

A unique transcription factor, *Tcf-1*, a nuclear effector of *Wnt* signaling, is expressed in developing ribs, thoracic vertebrae and craniofacial region, while a unique lymphoid-enhancer factor, *Lef-1* (also a transcription factor for *Wnt*), is expressed in tail vertebrae. In mouse embryos, mRNA for both transcription factors is expressed in limb buds, neural crest, pharyngeal arches and nasal processes at 10.5 days of gestation (Oosterwegel *et al.*, 1993).

Gdf-11 (*Bmp-11*) is expressed in somites and tail-bud mesenchyme, where it plays a role in specification of vertebral identity. When *Gdf-11* is disrupted in mice, the tail is reduced or absent, additional thoracic and lumbar vertebrae form at the expense of the tail, and the hind limbs are located more posteriorly. Although somite numbers are normal early in development, the number of thoracic vertebrae increases from 13 to 17 or 18. As cervical vertebrae are unaffected we can conclude that *Gdf-11* provides a posteriorizing signal, acting upstream of and via at least four *Hox* genes – *Hoxc-6*, *Hoxc-8*, *Hoxc-10*, *Hoxc-11*.[5]

The human *hand–foot–genital syndrome* is associated with a mutation in *HOXA-13*. Overexpression of the human gene in the caudal endoderm of chick embryos produces caudal malformations and tailless embryos, interpreted by de Santa Barbara and Roberts (2001) as indicating a role for the epithelium in tail development. In mice, on the other hand, knocking out *Hoxb-13* causes *overgrowth* of structures derived from the tail bud, implying repression of tail-bud growth in wild type (Economides *et al.*, 2003). We still have much to learn about the genetic control of tail development.

Tbx GENES

In 1997, a novel family of T-box transcription factors was isolated from developing amphibian fore- but not hind limbs, and from regenerating forelimb-bud blastemata. T-box genes play important roles in patterning and specification of major embryonic regions (fields), including wing, limb and tail territories.

Brachyury, the quintessential T-box gene, is an important signaling gene in tail development. Axial truncation in mice – primarily from tail loss – can be induced with retinoic acid, *Brachyury* mediating both the loss and subsequent down-regulation of all mesoderm markers. The tail defects in *Brachyury* mice can be rescued in a dose-dependent way with single copy transgenic implants. Retinoic acid can also elicit a homeotic transformation of regenerating tail-bud blastemata in some urodeles, such that limbs regenerate from the tail buds (Box 43.2).[6]

Another member of the T-box family, *Tbx-5*, is best known because it specifies the location of tetrapod forelimbs along the body axis (Fig. 35.4; Box 35.2). *Tbx-5* has an even more ancient role; it specifies the anterior location of the pectoral fins in teleost fishes, or at least it does in the zebrafish, and so we extrapolate to other teleosts. Knocking out *Tbx-5* prevents pectoral fin buds from developing but has no effect on pelvic-fin formation.

Zebrafish *Tbx-16*, which is expressed in paraxial mesoderm and tail buds, specifies both location and development of pelvic fins. The knockout phenotype in zebrafish is equivalent to the phenotype of the *spadetail* mutant in zebrafish and *Brachyury* in mice, indicating conservation of function of the gene in these two groups (Ruvinsky *et al.*, 1998).

TAIL GROWTH

Genes or environment

An issue that emerges whenever and wherever growth is analyzed is how much growth is under genetic control and how much is reactive to environmental conditions, a version of the nature–nurture conundrum that plagues our understanding of intelligence, behaviour and emotions.

Removing tail vertebral primordia from seven- or eight-day-old mice and transplanting them under the kidney capsules of host mice for six months shows that *growth of tail vertebrae correlates with the age of the host*, presumably because of hormonal influences. Kidney capsules are not a normal functional site for tail vertebrae so one has to be aware of the possibility of fusion (ankylosis) of transplanted tail vertebrae. When tail vertebrae from younger (four-day-old) rats are transplanted, the connective tissue between the vertebrae undergoes metaplasia to a chondroid tissue. As pressure compresses the annulus fibrosus the chondroid bone is replaced by metaplastic bone. Feik and Storey (1982) likened these changes to the late stages of *ankylosing spondylitis*.

Furthermore, basic morphogenetic features associated with mouse vertebrae from different axial regions (cervical, thoracic) are established before somites form, as determined by Sofaer (1985) from an analysis of six mutants and the effects of starvation.[7] Tail vertebrae also remodel when transplanted into non-functional sites, further illustrating their inherent capacity to form bones of particular shapes, although remodeling is modulated by mechanical stresses.

We know that individual genes have major effects on growth, either because they produce hormones such as growth hormone that have systemic effects, or because they direct growth of a particular cell type, tissue, organ or organ system. Stunted limb growth (achondroplasia) was discussed as one example in Chapter 27. Mutations that affect tail growth are discussed above. These are genes of large effect. But growth, as with any quantitative or continuous character, is influenced by many genes, each thought to have a small effect. Some are genes that were isolated because of their role in other systems; *myocyte stress-1* (*Ms-1*), a gene involved in the response of striated and cardiac muscle to pressure, is expressed in osteoblasts, osteoclasts, chondrocytes and chondroclasts within sutures in the mandibles, alveolar bone, and intramembranous and endochondral bone in growing mice. Study of these genes as quantitative trait loci (QTL), the provenance of quantitative genetics, is providing important insights into the genetic control of skeletal growth and development (Box 43.3).[8]

Temperature

Temperature can influence tail growth both during development and postnatally. The effects of short-term

Box 43.2 Homeotic transformation and anuran tail regeneration

Most studies on regeneration in anurans have concentrated on limb regeneration, but anuran tadpole tails and larval urodele tails also can regenerate. As might be expected from all that has gone before in our discussion, mechanisms of regeneration vary widely.[a] Following are two examples of urodele tail regeneration at quite different levels in the hierarchy of processes controlling regeneration.

Vaglia et al. (1997) demonstrated that larval tail regeneration in the four-toed salamander, *Hemidactylium scutatum*, is associated with an *increase in the lengths of the tail vertebrae anterior to the injury site*, illustrating that regeneration can affect existing structures, and act via more than one mechanism.

Secondly, some of the differences between limb and tail regeneration – and between fore- and hind limb regeneration – reside in patterning roles of specific *Hox* genes. *Hox 4.5*, which is *not* expressed during newt *limb development*, is expressed in the blastemata of *both* fore- and hind limbs following amputation. *Hox 3.6*, which *is* expressed in developing hind limb buds and tails, is found in the hind- but not forelimb blastemata following amputation (Simon and Tabin, 1993).

Retinoids influence morphogenetic phases of regeneration and regulate *Hox* genes. When applied to the amputation stumps of hind limbs of tadpoles from some anuran species, retinoids reprogramme the tissues to regenerate a hind limb or multiple hind limbs rather than a single tail.

Sixty-five per cent of regenerates are normal when tadpoles of the black-spined toad, *Bufo melanostictus*, are maintained in vitamin A (15 IU retinol palmitate/ml) for three days before the tails are amputated. However, no normal regenerates form if tadpoles are maintained in vitamin A for three days before and three days after amputation, indicating a clear differential effect depending on stage of regeneration. To the authors of these studies, this meant that dedifferentiation was enhanced, the blastemal cells becoming even more like embryonic limb-bud cells than is the case in normal regeneration. This is an interesting idea that needs testing and which, to a certain degree, has now been tested, for regenerating tails of *Xenopus laevis* tadpoles have a pattern of gene expression typical of the distal portion of developing tadpole tails rather than of the embryonic tail bud. In *Xenopus* at least, tail regeneration is just that – regeneration of a larval tail – and not the formation of an embryonic tail.[b]

Considerable information is now available on the axial *Hox* gene code and how modification of that code (in part using retinoic acid, RA) leads to an *anterior* shift in *Hox* gene expression and homeotic transformation of vertebrae to a more *posterior* vertebral type. Functional relationships exist between the *Hox* code and the segmentation of somitic mesoderm that precedes vertebral development, and respecification by vitamin A of the *Hox* code in migrating sclerotomal cells.[c]

The first vitamin A-mediated homeotic transformation in vertebrates was reported in 1992 when Mohanty-Hejmandi and colleagues obtained limb regeneration from tail blastema after applying vitamin A to tail amputation sites in the marbled balloon frog, *Uperdon systoma*. Only a few species show this remarkable response and ability.

Retinoids induce a homeotic transformation of tail buds to limbs in the European common frog, *Rana temporaria*, the response being correlated with concentration, time and stage of administration. As many as nine supernumerary limbs develop on a single tail stump, usually in pairs, always hind limb in morphology, and with normal proximo-distal and antero-posterior polarity (Maden, 1993).

Gerd Müller and his colleagues took analysis of transformation in *R. temporaria* considerably further when they identified a homeotic duplication of the whole posterior body segment, including vertebrae, pelvic girdle and limbs, following amputation through the tails of tadpoles which were then exposed to RA (10 IU/ml) for three days. Their conclusion was that 'the order of magnitude of this experimentally induced homeotic transformation is equivalent to major homeotic mutations in insects.' Mice also duplicate the limbs and lower body following RA treatment at 4.5–5.5 days of gestation, which is just before and during gastrulation.[d]

[a] See Goss (1969) and Ferreti and Géraudie (1998) for an overview of studies on tadpole tail regeneration.
[b] See Niazi and Alam (1984) and Niazi and Ratnasamy (1984) for the studies on the black-spined (common Asiatic) toad, and see Sugiura et al. (2004) for regeneration of *Xenopus* tadpole tails.
[c] See Kessel (1991, 1992) and Kessell and Gruss (1990, 1991) for the *Hox* gene code and vertebral transformation. Johnson and O'Higgins (1996) undertook a morphometric analysis of shape changes in cervical vertebrae in specimens transformed by modification of *Hox* gene patterns of expression in the study by Kessel and Gruss (1991).
[d] See Müller et al. (1996) for *R. temporaria* (the quotation is from p. 347), and Rutledge et al. (1994) for the mouse study.

temperature shocks at specific times during tail development are a nice example of how a particular environmental factor influences growth and how the effects of that influence can persist throughout life.

Exposing white mice in the early stages of pregnancy to hypothermia affects the number and morphology of the vertebrae that form, vertebral number decreasing with hypothermia. Reducing the temperature at which the female is housed for 24 hours when embryos are at 7.5 days of gestation (when somitogenesis is commencing) or between 8.5 days and 12.5 days (the latter being when the last tail somites form) produces differential effects on vertebral development (see Table 43.1). Exposing *postnatal* mice to short-term temperature changes affects tail growth without altering vertebral number. Two days at 33° C or 38° C causes a rapid

increase in tail growth in 31-day-old mice. The effect is transitory, lasting only 12 hours, but is enough to result in longer tails than in control animals (Al-Hilli and Wright, 1983).

Surprisingly, even genetically homogeneous inbred strains of mice can respond differently *as adults* to the same selective pressures. Mice respond to selection for increased tail length either by adding on an additional caudal vertebra or by decreasing the number and increasing the size of the remaining vertebrae (Rutledge et al., 1974).

These are two very different developmental mechanisms, one requiring *additional segmentation of somitic mesoderm or subdivision of an existing somite* to provide the additional vertebra, the other requiring *decreased segmentation of somitic mesoderm and compensatory*

Box 43.3 Quantitative trait loci (QTLs)

The combination of QTL analysis with identification of genes of major or minor effect is the most exciting genetic approach to skeletal growth and development available today.

QTLs are essentially the gene loci in a particular region of a chromosome. A QTL analysis determines the combined effects those loci have on a quantitative character such as growth, size or shape. Therefore, a QTL analysis reveals the cumulative action of an unknown number of genes in that region of the chromosome. The assumption underlying such analyses is that QTLs represent many genes, each of small effect, some acting positively on the feature, some negatively. Genes with a QTL may interact with one another or influence similar features by pleiotropy. Analysis of QTLs and skeletal growth shows that this may not be so. While QTL analysis reveals a genetic contribution to growth it masks what may be important actions of few genes. Unfortunately, research fields rarely collide. Using quantitative polymerase chain reaction (PCR) analysis, Yamaza *et al.* (2001) detected 15 genes in mouse mandible that are active in tooth formation between 10.5 and 12 days of gestation. Many of these genes also are active in chondro- and osteogenesis, but teeth were the focus of the analysis. Few would think to carry out a QTL analysis in parallel. Most quantitative geneticists are not interested in identifying the genes at a quantitative trait locus; their definition of a gene is a QTL. Most molecular geneticists focus on single genes of known effects, their definition of a gene being a sequence of DNA with a particular function. Happily, as you will see below, the two fields of quantitative and molecular genetics are intersecting in a few laboratories.

Mice

In a quantitative genetic analysis of six lines of rats selected for post-weaning weight gain, Cheverud *et al.* (1983) identified four QTLs (regions of chromosomes, sets of genes) with effects on tail growth and body weight, namely:

- genes that act *throughout ontogeny* and account for most of the genetic variance associated with tail length or body weight;
- genes with a *small effect early in ontogeny and a larger effect later in ontogeny*, which account for some genetic variance;
- genes with a *large effect early in ontogeny and a small effect later in ontogeny*, which account for some of the genetic variance; and
- genes *that act at only one or two stages, and have only minor effects* on tail length or body weight.

Surprisingly, some QTLs have large effects, accounting for as much as a quarter of the variation in the character being measured. This occurs either because these loci (which may be quite large) contain many genes of small effect, or because they contain one or more genes of major effect, a distinction not always appreciated by devotees of QTL analysis.

Other skeletal regions are moulded by the pleiotropic effects of individual genes. In 1997, Cheverud and colleagues identified 27 QTLs that affect different functional regions of the mouse mandible. Half influence the muscles of the ascending ramus, 27 per cent the alveolar processes associated with the teeth and 23 per cent the whole mandible. A follow-up study using geometric morphometrics confirmed the two modular regions of the dentary (alveolar and ascending ramus), identified 33 QTLs and showed that the QTLs contribute to each module, i.e. discrete sets of QTLs are not required to established discrete morphological units.[a]

This QTL analysis, as with other studies Jim Cheverud has carried out on mouse mandibles, confirms the modular composition of the dentary proposed by Atchley and Hall (1991). This is so, despite the fact that our study used an entirely different class of evidence – condensations as morphogenetic modules and cellular properties such as rate of proliferation, proportion of the populating dividing, amount of cell death, and so forth. A reanalysis of Cheverud and Colleagues' 1997 data confirmed that the ramus of the mandible is modular, taking their lead from the 'imitatory epigenotype concept' proposed to show that functional units of the phenotype should be genetically modular.[b]

The applications of QTL analysis are numerous. Take the following situation. Mice were introduced to the Isle of Man (UK) in April 1982. Within 18 months – during which quite a number of generations would have been produced – heritable changes in skeletal characters could already be measured; there was a six to eight per cent increase in size as well as shape changes in the coronoid and angular processes of the dentary. Here is a system ripe for QTL analysis. Or take the selection experiment for increased tail length in mice, in which replicates from a genetically homogeneous population responded to the same selection pressure by increasing tail length to the same extent, but via two vastly different developmental mechanisms – addition of an extra vertebra or increasing the size of existing vertebrae. Again, the time is ripe for a QTL analysis to identify the complexity underlying the genetic changes involved.[c]

Cichlids

Applying QTL analysis to Lake Malawi cichlid fishes showed that hybrids differ from both their parents in morphometric differences associated with biting and sucking. This study was followed up with further QTL analyses on the feeding apparatus to demonstrate directional selection. Several of the QTL were identified as segregating to where *Bmp-4* is localized on the chromosome, implicating *Bmp-4* in these QTL effects. The mouse QTLs also may contain one or more Bmp genes.[d]

Sticklebacks

In what will prove to be an exciting project, David Kingsley's laboratory developed a genome-wide linkage map for the threespine stickleback, *Gasterosteus aculeatus*, which they used to study the development of body armour (dermal skeleton) and feeding morphology in benthic (deep) and limnetic (lake-dwelling) pairs of sympatric species in six lakes in British Columbia, Canada. Benthic species have reduced body armour, increased body depth and decreased numbers of gill rakers, in contrast to limnetic species with their increased body armour, more slender bodies and increased numbers of gill rakers.

Their initial finding (Peichel *et al.*, 2001) is that independent chromosome regions control plate number, spine length and gill raker number. No major QTL contributes significantly to the number of long gill rakers, leading to the conclusion that *many genes, each of small effect, control the number of rakers*, à la the murine mandible as modeled by Bailey (1986). Two QTLs contribute significantly to the number of short gill rakers, explaining two-thirds of the variance in number. QTLs belonging to several linkage groups also influence the length of dorsal spines one and two, the pelvic spines, and the number of lateral plates, explaining 17–20 per cent of the variance in these characters.

Threespine sticklebacks are covered with dermal plates. Plate number is polymorphic, the number of plates varying between populations in a predictable manner. Different variants – we are dealing with a single species – are known as *lateral-plate morphs* and result from genetic polymorphism for plate number in combination with paedomorphosis, which influences the duration of

Box 43.3 (*Continued*)

ontogeny when the plates are laid down. Using sticklebacks from an insular lake in the Queen Charlotte Islands, British Columbia, individual fish (especially large adults) with high *or* low plate numbers were shown to have increased asymmetry of plate numbers, a phenomenon these researchers linked to parasite burden. Recent QTL analysis shows that a single major locus contributes to most of the variation in both number and patterning of lateral plates. Additional QTL analysis would allow genetic and environmental influences to be partitioned and/or integrated further.[e]

Different populations of threespine sticklebacks also show various stages of reduction of their pelvic girdles. Again, the mechanism is paedomorphosis, as truncation of development removes individual elements one at a time, or removes entire pelvic girdles. Pelvic reduction, as analyzed in populations in 179 lakes, requires the *combination* of absence of predatory pike from the lake *and* low Ca^{++} concentrations in the water, indicating how unrelated environmental variables can interact and influence a part of the skeletal system.[f] This system is ripe for a QTL analysis.

The combination of QTL analysis and identification of genes of major or minor effect is the most exciting genetic approach to skeletal growth and development available today.

Add in environmental and life history components and the prospects for understanding skeletal development in the real world looms on the horizon.

These are exciting times to be a skeletal biologist.

[a] See Cheverud *et al.* (1997), Ehrich *et al.* (2003), Cheverud (2004) and Klingenberg *et al.* (2003, 2004) for these QTL analyses on mouse mandibles.
[b] See Mezey *et al* (2000) for the reanalysis and Riedl (1978) for the 'imitatory epigenotype concept.'
[c] See Scriven and Bauchau (1992) for the study on the Isle of Man and Rutledge *et al.* (1974) for selection on tail length.
[d] See Albertson and Kocher (2001) and Albertson *et al.* (2000) for these studies on Lake Malawi cichlids.
[e] See Bell (1981) and Reimchen and Nosil (2001) for these studies on lateral-plate morphs, and see Bell *et al.* (2004) for the rapid evolution of lateral-plate morphs in Loberg Lake, Alaska.
[f] See Bell (1987, 1988) and Bell *et al.* (1993) for changes in the pelvic girdles.

Table 43.1 Effects of hypothermia during pregnancy on the vertebrae of the offspring in mice[a]

Treatment	Number of females	Number of young	Number of vertebral anomalies[b]
Temperature reduced to 20°C at 7.5 days of gestation	737	253	131 (52%)
Temperature reduced to 25°C at 8.5 days of gestation	216	281	50 (18%)
Control maintained at 36°C	42	175	9 (5%)

[a] Based on data in Lecyk (1965).
[b] Anomalies included changes in the number of vertebrae of a particular region. Teratological changes comprised fusion of the centra and arches of two or more neighbouring vertebrae, split-out centra, and wedge-shaped half-vertebrae.

growth of the remaining vertebrae. This would be an ideal system to serve as the basis of a longitudinal study, especially now that we have the imaging techniques required to follow development/growth in real time.

I explore temperature effects on vertebral development in a little more detail in the following section, discussing effects on cervical, thoracic, lumbar and caudal vertebrae, before moving on to discuss taillessness.

TEMPERATURE-INDUCED CHANGE IN VERTEBRAL NUMBER: MERISTIC VARIATION

In part because they are easy to count, but also because they respond so beautifully to perturbation, influences of temperature on vertebral number have received much attention. Frogs, fish, birds and mammals all respond to exposure to an unusual temperature or heat shock with disruptions in the mechanisms that generate vertebrae. Often the effect is specific to the time during development

(or adult life in some cases) when the temperature changes. We can consider such times as *critical periods for vertebral development, morphogenesis or growth* (Box 38.1).

Temperature is a major environmental variable affecting skeletal morphology and the numbers of repeated skeletal elements that form. The human skeleton has adapted to cold climates with reduced head size (brachycephalization). *Meristic variation* – alteration in number of serially repeated units such as vertebrae – is affected by temperature and is seen in all vertebrates including, perhaps surprisingly, homoeotherms. We often think, mistakenly, that having a constant body temperature insulates homoeotherms from temperature variation. Fevers, walking to the laboratory at –20°C, hibernation (Box 24.4), and much more, demonstrate reactivity to temperature change.

Natural variation and adaptive value

Responsiveness to environmental factors such as temperature in part reflects the natural variation in vertebral number resulting from lability in development. For example, in a single population of the spotted salamander, *Ambystoma maculatum*, Worthington (1974) reports a modal vertebral number of 14, with one-third of the population having 13, 13.5, 14.5 or 15 vertebrae. Eighty-five per cent of the individuals also had limb anomalies such as syndactyly, oligodactyly and polydactyly. Worthington thought temperature was the most likely factor leading to variability and the anomalies. In addition to this type of variation *within* a population, there also is variation *between* populations, as seen in spotted salamanders from five locations in Maryland or Mississippi (Table 43.2).

Analyses such as Worthington's lead one to ask whether variation in vertebral number has adaptive value. It appears that it has. Predation on sticklebacks by sunfish

Table 43.2 Vertebral numbers (%) in various North American populations of the spotted salamander, *Ambystoma maculatum*[a]

Number of trunk vertebrae	Largo, MD	Howard Co., MD	Washington Co., MD	Mt Olive, MS	Hattiesburg, MS	Total mean
13	0	0	0	15	2.5	2
13.5	7	2	0	0	0	2
14	80	92	76	60	70	77
14.5	3	4	9	5	5	5
15	10	2	13[b]	20	22.5	14

[a] Based on data from Worthington (1974)
[b] One individual contained 10 trunk vertebrae.

Table 43.3 Effect of temperature on number of vertebrae in chick embryos[a]

Temperature (°C)	Number of vertebrae
34	45
36	44.85
37.5	45
39	45.5
41	48

[a] Based on data in Lindsey and Moddie (1967). Embryos were maintained at the temperatures indicated from onset of incubation and were examined at 15 days (34°C), 13–15 days (36°C), 12–13 days (37.5°C), 10–12 days (39°C) or six days (41°C).

is 1.3–1.7 times higher on individuals with 31 vertebrae than on individuals with 32 vertebrae.

Parental reproductive history can also affect meristic variation, as detected in the cyprinodontid mangrove killifish, *Rivulus marmoratus*. Embryos that begin to develop soon after the eggs are laid have fewer vertebrae than embryos in which development is delayed. The same is true for embryos whose parents and eggs were maintained at 30°C rather than 25°C. Predation correlates with length and vertebral number, burst speed correlates with vertebral phenotype, while selection (via predation) correlates with ratio of abdominal to tail vertebrae.[9]

The effects of temperature may be subtle (the number of costal grooves in the northwestern salamander, *Ambystoma gracile*, increases with increasing temperature, reflecting minor changes in morphogenesis and growth) or more generalized (vertebral number is reduced at higher temperature, reflecting more fundamental changes in mesodermal patterning and/or segmentation).[10] I survey, briefly, experiments and results on induced meristic variation in the various vertebrate classes.

Studies with teleost fish

Increased temperatures reduce vertebral number in the cyprinid neotropical killifish, *Rivulus marmoratus*. As one might anticipate, genetic background and temperature interact as explanations for meristic variation, as in the Japanese medaka, *Oryzias latipes* (Fig. 20.1). Exposure to increased temperature need not be prolonged; a single break in temperature during development is sufficient to increase vertebral number in rainbow trout.[11]

Temperature also affects rate of mineralization of the vertebral and fin skeletons (Fig. 43.2).

Another factor or series of factors causing meristic variation is rearing fish in the laboratory. Laboratory-reared Red Sea bream, *Pagrus major*, have greater variability in vertebral number than do wild-caught individuals (Matsuoka, 1982).

Studies with avian embryos

Chick embryos respond to heat shock with somitic and skeletal anomalies in one or two adjacent segments

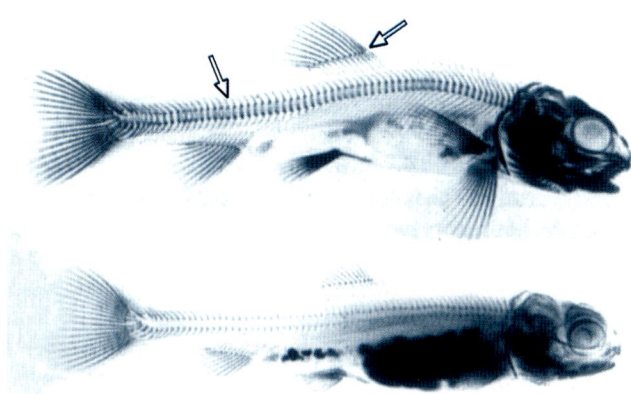

Figure 43.2 Temperature influences mineralization of fish bone. Mineralization of the teleost skeleton is influenced by temperature, as shown in these skeletal preparations of Sunapee charr, *Salvelinus alpinus oquassa*, incubated under warm (top) or cold (bottom) conditions. Note, for example, the extent of mineralization of the vertebral column and the skeleton of the dorsal fins (arrows) in the two specimens. Modified from Balon (2002).

blocks, separated by blocks of six or seven unaffected somites (Primmett *et al.*, 1988). Exposure to a range of temperatures (34–41°C) induces increasing numbers of vertebrae; 45 at 34°C, 48 at 41°C (Table 43.3).

Studies with mammals

Exposing adult mice to short-term temperature changes affects tail growth without altering vertebral number. As an example, in one study, albino mice were weaned at 23 days after birth, transferred to a cold room (8°C) at 25 days for a week, and then either left in the cold room or transferred to 33°C for two days at 31 days after birth. Exposure to the higher temperature caused a rapid increase in tail growth that lasted for some 12 hours, followed by elevated growth for the next day and a half (when measurements stopped), resulting in tail hypertrophy (Al-Hilli and Wright, 1983).

Exposing white mice in the early stages of pregnancy to hypothermia affects vertebral number and morphology, vertebral number decreasing with hypothermia

(Table 43.1). Temperature was reduced to 20°C or 25°C for 24 hours when embryos were at 7.5 days of gestation (when somitogenesis would be commencing), at 8.5 days, and through to 12.5 days, when the last somites would be forming. Vertebral anomalies reported (mostly associated with exposure to 20°C) included six instead of seven cervical vertebrae, 11 or 12 instead of 13 thoracic vertebrae, four or five instead of six lumbar vertebrae, and vertebral fusions.

We know the molecular if not the cellular basis for these changes. Exposing pregnant mice to a temperature shock of 42°C or 43°C induces stage-specific homeotic vertebral transformation in the foetuses. Exposure when embryos are at:

- 7.5 days of gestation affects cervical vertebrae (C1–C7), each of which is transformed to a more cranial morphology as expression of *Hoxa-5, Hoxa-7, Hoxc-8* and *Hoxc-9* shifts cranially by one somite;
- 8.5 days affects more caudal vertebrae (T6–S1), each of which is transformed to a more cranial morphology by one or two somites as expression of *Hoxc-8* and *Hoxc-9* shifts cranially;
- 9.5 days affects trunk and lumbar vertebrae (T13–L4), which are transformed caudally (Li and Shiota, 1999).

At the cellular level, the changes were mediated via heat-induced suppression of proliferation.

Studies with amphibian embryos

In a comprehensive analysis, the effects of five different temperatures, three levels of salinity and three light regimes on vertebral number in the spotted salamander, *Ambystoma maculatum*, were examined (Table 43.4). Trunk vertebral counts varied between 13 and 15 – the modal normally is four in the trunk – being highest at the lowest temperature, increasing with increasing salinity and unaffected by photoperiod, i.e. slow development is associated with high vertebral number, faster development with low number.

Of course, pathologies can be induced by temperature; tadpoles of the Iberian green frog, *Rana perezi,*

raised on one of 10 diets at one of two temperatures under hatchery conditions show skeletal defects, including failure of ossification, as a consequence of altered collagen metabolism (Martinez *et al.*, 1992).

Temperature plus…

Clearly, developing and growing skeletal tissues are sensitive to temperature changes; even a pulsed temperature change can alter vertebral number. Although temperature is a major environmental variable affecting skeletal morphology and number of skeletal elements, bear in mind that isolating temperature from other environmental parameters such as salinity or diurnal changes is not always easy; the number of trunk vertebrae in species of slender salamanders in the *Batrachoseps* increases or decreases geographically with temperature but the correlation is not sufficient to explain natural variation in trunk vertebral number (Jockusch, 1997).

TAILLESSNESS

Humans are tailless, although as embryos we each have a tail bud, which in some individuals can develop into a rudimentary tail. Like limb buds in limbless tetrapods, tail buds in tailless tetrapods and larval tails are removed by cell death and subsequent phagocytosis by macrophages. Terminal mesoderm fails to form somites because this mesenchyme is a site of cell death rather than cell proliferation, although caudal tail mesoderm can be partially rescued *in vitro*. Similarly, cell death removes the tail buds from the tailless mouse mutants *Brachyury* and *Rumpless*.[12]

Other mouse mutants reveal other mechanisms for ceasing to elaborate a tail. Tailless *T/btm*, in which half the offspring lack tails, is first evident phenotypically as a branched notochord and duplicated neural tube fused to – or more likely, not separated from – tail mesenchyme. *T/btm* mutant mice also lack sacral and caudal vertebrae. This mutant suggests that the mechanism described

Table 43.4 Effects of temperature, salinity and photoperiod on vertebral number in the spotted salamander, *Ambystoma maculatum*[a]

Number of vertebrae	Percentage of Individuals									
	Temperature (°C)				Salinity (%)			Photoperiod (hours light)		
	5	10	15	20	1.5	3	6	11	12	14
13	0	5	0	7	5	0	0	2	0	3
13.5	5	4	4	7	4	1	0	0	4	6
14	10	89	87	81	89	65	29	98	87	79
14.5	10	1	4	0	1	17	14	0	4	3
15	74	1	6	4	1	17	57	0	6	9

[a] Based on data from Peabody and Brodie (1975). Embryos were exposed to the conditions from the one- to four-cell stages onwards.

above for tail-bud initiation in *Xenopus* – at the junction of neural and mesodermal fields – might operate in mice.

Two further mouse mutants illustrate how different defects in the VER can disrupt tail development. The *repeated epilation* mutant has a VER that is unusually thickened. Tail-bud growth slows, resulting in a shortened tail. Embryos carrying the *vestigial tail* (*vt*) mutation form tail buds but have only a rudimentary VER, a lack that leads to tail bud regression and tailless adults.[13]

AND THEREBY HANGS A TAIL

From our knowledge of limb regeneration in urodele amphibians we do not expect tissues proximal to the amputation site to show any changes other than dedifferentiation at the amputation surface. But tails are not limbs and tail regeneration is very different from limb regeneration. Indeed, the tails of most tailed tetrapods do not regenerate, if by regenerate we mean replacing with considerable fidelity the structure that has been lost. As one of the few examples in the literature, regeneration of the larval tail in the four-toed salamander, *Hemidactylium scutatum*, is associated with increase in the length of tail vertebrae anterior to the injury site (Vaglia *et al.*, 1997).

A fascinating aspect of amphibian tadpole tail regeneration is the ability of tail-bud blastemata to undergo homeotic transformation in response to vitamin A, to the extent that *limbs regenerate from the tail stump* in what, in at least one species, is associated with duplication of the posterior portion of the body (Box 43.2). Tenascin increases in blastemata and in muscle condensations during tail regeneration of the Spanish ribbed newt, *Pleurodeles waltl* (Arsanto *et al.*, 1990).

Fish tails

Most students of regeneration in fishes use one or other of the paired fins as their study site (Chapter 40), although a few have investigated regeneration of the caudal skeleton and/or tail.

Kirschbaum and Meunier (1981, 1988) studied tail regeneration in the glass knifefish, *Eigenmannia virescens*, during which cartilage develops at the end of the notochord. The absence of a perichondrium is consistent with the cartilage arising from the existing perinotochordal (perhaps even notochordal) cells. The chondrocytes do not hypertrophy nor does the matrix mineralize. Rather, the matrix on the ventral side of the cartilage is removed by 'chondroclasts' and replaced by perichondral ossification (see Box 4.4), resulting in a regenerated tailbone with 2–3 mm of cartilage at the tip.

Regeneration of the tail fin in various teleosts is accompanied by turnover of collagen within both the actinotrichia blastemal cells and the basal laminae from the adjacent epithelium which control the number and distribution of any new actinotrichia that form.

Fin rays grow in the regenerate as they grew in the embryo – by successive distal addition of new segments to existing rays. The role of cell–matrix interactions in producing blastemata and in interactions between such lepidotrichial matrix components as collagen and glycosaminoglycans, and relationships (if any) of fin regeneration to the fin-fold theory, have received some attention.[14]

LIZARDS' TAILS: AUTOTOMY

Amputating the tails, limbs or digits of most lizards initiates development of a blastema but little if any subsequent growth, most blastemal cells only forming a cartilaginous cap on the end of the amputated bone (Bellairs and Bryant, 1968).

Tail regeneration *is* more complete in lizards in which growth of the regenerate may be fast or slow: *2.6 mm/day* over five weeks in the Madeira wall lizard, *Lacerta dugesii*, producing a 90-mm-long regenerated tail, compared with *5 µm/day* over 14 weeks in the slow worm, *Anguis fragilis*, to produce a 5-mm-long regenerate.

Embryos of some lizards can regenerate their tails while still in the amniotic cavity, as occurs after amnioallantoic constriction bands 'amputate' the tails. In other species, such as the common lizard, *Lacerta vivipara*, regenerative ability is not present in young animals.[15]

Regeneration of tail and tail skeleton in some lizards, for example in the Mediterranean gecko, *Hemidactylus turcicus*, occurs only from a specific position along the tail known as an *autotomy plane*. In insects such as the German cockroach, *Blatella germanica*, leg regeneration only occurs *at the next larval stage* if the leg is removed before a critical period during the preceding larval stage. Otherwise, there is a delay until the next moult before the leg will regenerate. *Hemidactylus* and other geckos also display a latent period: a quarter of a vertebra is resorbed after autotomy before regeneration of an unsegmented cartilaginous tube can ensue. Autotomy, autotomy planes and all that flow from them would make an intriguing Ph.D. project.[16]

And thereby ends my developmental tale.

NOTES

1. *The Guinness Book of Records* recently changed its method for determining the world's largest kite to include the tail. Size was previously determined from the area of the 'head' or lifting part of the kite. A New Zealand kite, *Megaray*, with a lifting area of 635 m^2, has been trumped by a kite from China with a tiny head but enormously long tail.
2. See Duellman and Trueb (1986) for an overview of amphibians and amphibian relationships, and see the chapters in Tinsley and Kobel (1996) for the biology of *Xenopus*.
3. See Dhouailly and Kieny (1972) and Searls and Zwilling (1964) for these two studies. The proto-oncogene *int-2*

(*Fgf-3*) plays a role in tail and inner ear development; mice homozygous for *int-2* show defects in both systems resulting from defective expression in the posterior primitive streak and hindbrain/otic vesicle. Penetrance is 100 per cent in the tail, but more variable in the inner ear, indicative of stochastic expression (Mansour *et al.*, 1993).

4. As with the AER (see Box 35.3), it is more correct to refer to the VER as an epithelial not an ectodermal ridge. See Hall (2000c) for an overview of past literature on the VER and for demonstration of its presence in mouse embryos, and see Hall (1997c), Handrigan (2003) and C. Liu *et al.* (2004) for how tail development (secondary neurulation) differs from head development (primary neurulation), and whether secondary neurulation is a primitive vertebrate character (Hall) or derived (Handrigan).

5. *Gdf-11* also is expressed in maxillary, mandibular and hyoid-arch mesenchyme (J. A. King *et al.*, 1996; Gad and Tam, 1999; McPherron *et al.*, 1999). Galis (1999) argues that the number of cervical vertebrae in mammals is fixed at seven because changes in Hox gene expression that would affect vertebral number are associated with neurological problems and early cancer and so are heavily selected against.

6. See Simon *et al.* (1997) for isolation of T-box genes, J. Smith (1999) for a review, Iulianella *et al.* (1999) for the retinoic acid study, and Stott *et al.* (1993) for transgenic restoration of tail defects.

7. See Noel and Wright (1972) and Noel (1973) for transplantation for six months, Sofaer (1985) for the mutants and starvation, and Storey and Feik (1982, 1985), Feik and Storey (1983), Storey *et al.* (1992) and Feik (1993) for remodeling and response to mechanical influences.

8. See Bailey (1986) for many genes of small effect, Orestes-Cardoso *et al.* (2001) for *Ms-1*, and Atchley and Hall (1991) for a model derived from quantitative genetics.

9. See Swain and Lindsey (1984, 1986a,b) for predation on sticklebacks and temperature effects on *Rivulus*, and Swain (1992a,b) for selection and predation.

10. See Beals *et al.* (1983) for the human skeleton and Lindsey (1966) for the salamander data.

11. See Lindsey and Harrington (1972) for *Rivulus*, Ali and Lindsey (1974) for *Oryzias*, and Lindsey *et al.* (1984) for rainbow trout.

12. See Mills and Bellairs (1989) and Mills *et al.* (1990) for tail segmentation in avian embryos (which is inhibited by a high level of integrin receptors), and see Fallon and Simanal (1978), Johnson (1986) and Fujimoto *et al.* (1994) for mouse mutants.

13. See Salzgeber and Guénet (1984) for *repeated epilation*, Grüneberg (1956) and Grüneberg and Wickramaratne (1974) for *vestigial tail*, and Hall (2000c) for an overview of the VER, relevant literature, and a discussion of secondary neurulation – the process by which the tail bud arises not from differentiated germ layers but from a posterior caudal field of cells that does not segregate into germ layers (Hall, 1997c, 1999a).

14. See Marí-Beffa *et al.* (1989) for collagen, and Santamaría and Becerra (1991) and Santamaría *et al.* (1992) for cell–matrix interactions.

15. See Bellairs and Bryant (1968) for blastema formation, and Bryant and Bellairs (1967a, 1970) for speed of regeneration or lack of regeneration in young animals. See Maderson and Licht (1968) and Maderson and Salthe (1971) for variability associated with tail regeneration in an iguanid lizard, the green anole, *Anolis carolinensis*. Susan Bryant and Paul Maderson were both Ph.D. students of Angus Bellairs.

16. See Werner (1967) for the study with *Hemidactylus*, and Arnold (1984) for the evolutionary significance of tail shedding and the autotomy plane in lizards and their allies.

Part XV

Evolutionary Skeletal Biology

Variation, the stuff and very essence of evolution, is forgotten, ignored or unknown in the vast majority of laboratory-based studies of skeletal development.

Even structures that have been lost from a lineage often leave a vestige, as Geoffroy St. Hilaire knew from the collections he made in Egypt with Napoleon Bonaparte: 'These rudiments of the furcula have not been suppressed, because nature never advances by rapid leaps, and she always leaves the vestiges of an organ even when it is entirely superfluous, if that organ has played an important role in other species of the same family.'

(Appel, 1987, p. 74)

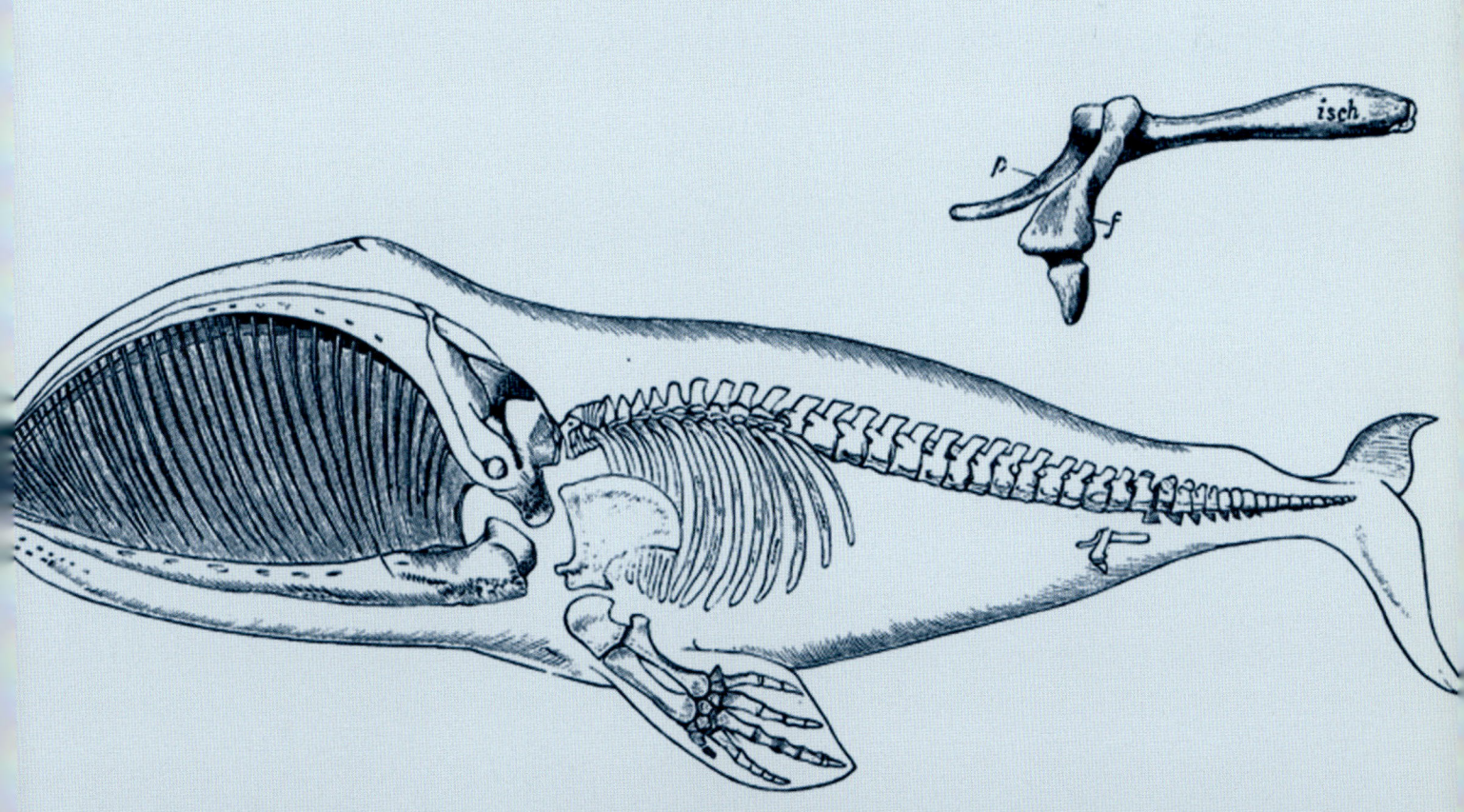

Evolutionary Experimentation Revisited

VARIATION

'It should be remembered that systematists are far from pleased at finding variability in important characters, and that there are not many men who will labouriously examine internal and important organs, and compare them in many specimens of the same species' (Darwin, 1959, p. 45). Darwin's concern is as true today as it was 145 years ago. Variation, the stuff and very essence of evolution, is forgotten, ignored or unknown in the vast majority of laboratory-based studies of skeletal development. In part this is because, although enormously important, the analysis of variation and of variability is anything but straightforward. It is also because developmental biology is based on constancy; for example, we stage 20 chick embryos that have been incubated for 48 hours and assume that all will be identical in size and state of development. We experiment with three-day-old mice and tacitly assume that all are the same. We separate the sexes – although with mice we don't even do this until they are mature at, say, six weeks of age – but otherwise we assume constancy. We could not devise experiments without these underlying assumptions. But, and it is a very big BUT indeed, without variation we would have nothing to work on or to experiment with. All organisms on the planet would be the same. I do not know what they would look like, but without variation and its handmaiden natural selection, all would be identical. Not necessarily clones, but invariant.

Some aspects of variation are alluded to in the text, often in the context of a perturbation of development such as a natural or induced mutation. Variation is more evident in mutants, as:

- mutation removes some of the buffering associated with the wild-type genotype;
- mutations affect mean phenotypes but also affect phenotypic variances; and
- selection on phenotypes associated with mutations is one way to introduce variation into a population.

Other instances of variation are presented more explicitly – fluctuating asymmetry, variation in growth, effects of scale, and diurnal or circadian rhythms. Here I provide only some examples of the types of variation of which we need to be aware.

Variation of individual elements

Variation may be as simple as two character states of a skeletal element; the variants coded by systematists in keys to organisms or by phylogeneticists in constructing phylogenetic trees. Bock's (1960a) study of variation in the palatine process of the premaxilla in passerine birds is a fine example of the value of a close analysis of a single character across a group. The palatine process is either fused or unfused and so variation of the pattern is limited to the two conditions. Variation in the developmental processes that bring about the fusion may be more extensive – they certainly are more profound – but we should not confuse variation in pattern with variation of process.

Variability, the ability or propensity to vary, is yet another level of analysis, as recent analyses of canalization and variability in hominoid postcranial skeletons illustrate (Wahner *et al.*, 1997; Young, 2003).

Prompted by an earlier study by Larson and De Sá (1998), in which previously unreported variability was used to argue for parallel evolution of the postcranial skeletons of living apes, Young applied principal components and cluster analyses in a cladistic analysis to a large set of published and unpublished data obtained from 10 anthropologists, finding that his analysis was much more discriminatory than Larson's. Young was able to distinguish functional changes associated with suspensory and quadrupedal locomotion, hominoids and *Ateles* from the other primates, and distinguish among the apes.

Primate evolution during a relatively short span of evolutionary time (the Caenozoic era) was characterized by considerable morphological change based on patterns that recur over and over; what Chiu and Hamrick termed 'variational tendencies.' Benedikt Hallgrímsson and I chose *Variation: A Hierarchical Examination of a Central Concept in Biology*, as the title for an edited work on the importance of these issues in developmental and evolutionary biology (Hallgrímsson and Hall, 2005).

The palatine process evolved repeatedly among passerine birds. We know this because it appears and disappears on cladograms representing the phylogenetic history of the group. Lots of interesting issues and questions present themselves. The phylogenetic tree(s) has to be extremely good (robust) to allow interpretations of repeated evolution. Is the pattern of fusion or lack of fusion always the same with repeated evolution? If not, what aspects of development and/or function have changed to modify the character? Are the 'forces' that evoke the repeated evolution of a character always the same? Those forces could be any combination of changed environment/habitus, natural selection or internal constraints.

A nice example of constraint on skeletal development is the conservation of skeletal characters in the painted (yellow-blotched) ensatina, *Ensatina eschscholtzii*, a lungless (plethodontid) salamander that exists as a polymorphic ring species in the Central Valley of California. Proteins and colour patterns vary markedly around the range, to the extent that they are used to distinguish seven subspecies. Limb and tail skeletons are highly conserved (invariant), the small amount of variation that does exist – loss of phalanges, mineralization of tarsals, absence of ribs from caudal trunk vertebrae, variation in the number of tail vertebrae between 17 and 46 (the latter in part due to sexual dimorphism) – showing no correlation with geographical location.[2]

When a skeletal element (or an entire skeletal organ such as a limb) is lost, must a remnant (rudiment, vestige) remain in embryonic development for the character to reappear, or can characters reappear *de novo* as neomorphs? These questions/issues, which appear throughout the book, sometimes with answers, sometimes not, are taken up later in this chapter.

A second example of single character analysis is that undertaken by Cracraft (1968) on the lacrimal–ectethmoid complex in 2700 specimens representing most families of birds. Variation is considerable because the biological roles for this complex are many, including bracing the palatines during feeding, bracing the jugal bar to prevent breakage during kinesis, as a 'stop' to prevent abnormal retraction of or to protect the eye, and as an attachment site for connective tissue associated with the eye. Evolutionary alteration of different combinations of these features generates different variants.

A third example, discussed in Chapter 16, is a series of studies from the 1950s on variation of the clavicles and dorsal coracoid arteries during avian evolution, including a detailed analysis of clavicle variation and reduction in parrots, in which 19 of 67 genera (221 species and subspecies) show anywhere from some to advanced reduction of the clavicles. Glenny identified four classes of clavicles: (1) complete and united; (2) reduced to a ligament at the sternal end but with an extensive bony element; (3) where the greater part of the clavicle is reduced to a ligament but a bony element is still present; and (4) complete reduction (transformation) to a ligament.[3]

Is it significant that these three examples are all from birds? Don't the skeletons of other vertebrates vary? You decide, now that you are approaching the end of the book.

Variation that tests a hypothesis

Tague (2002) examined variation in the hands of three primates, each of which has a rudimentary digit(s). The three taxa were the black-handed spider monkey, *Ateles geoffroyi*, the Abyssinian black-and-white colobus monkey, *Colobus guereza*, and the potto, *Perodicticus potto*. Metapodials (metacarpals in the wrist, metatarsals in the ankle) are the bones that link the phalanges of the digits to the more proximal bones of the limb. Tague proposed that metapodial variation should be greater in animals with vestigial digits; stabilizing selection would be relaxed. The second metacarpal is rudimentary in the potto, associated with a rudimentary index finger. In two of the species (colobus and spider monkeys) the first metacarpal is variable to the point of being vestigial, and is associated with a rudimentary thumb. But the metacarpals associated with the other fully formed and functional digits also exhibit variation.

Cullinane (2000) investigated whether vertebral or limb skeletons show greater variation by comparing their variability in 15 species of mammals (11 orders, 29 skeletal elements). The limb skeleton was less variable than the vertebral column, distal limb elements more variable than proximal ones. The greatest variability was in the fifth digit (digit V) on fore- and hind limbs and the associated metacarpal V. Digit V (the little finger, toe) is the last to develop. Does this alone explain its greater variability? Is there less selective pressure to minimize variation of digit V? It certainly plays a less important role in locomotion than does digit I.[4]

Milne and O'Higgins examined interspecific variation in size and shape of the skulls of seven species of kangaroos and wallabies in the genus *Macropus*. From geometric morphometrics, and by testing their data against a consensus phylogeny of the taxa, they were able to tease out influences of diet and function from phylogenetic influences. Allometric scaling of the skulls – essentially imposing a phylogenetic and ontogenetic constraint – coupled with the ability to respond to changing climate and diet – essentially phylogenetic and ontogenetic plasticity – obscures what would otherwise be relatively straightforward influences of phylogeny on morphology. Jukka Jernvall and his colleagues, who came to essentially similar conclusions from their analyses of tooth morphology in ungulates of the Cenozoic, went on to demonstrate that variability in cusp patterns is most often seen in the shortest cups which are also the last to appear in ontogeny, arguing that small changes in development produce large changes in both size and number of tooth cusps.[5]

Pattern variation

Hanken (1983b) reported a high incidence of skeletal variation in the carpals and tarsals between individuals in a single population of the red-backed salamander, *Plethodon cinereus*, at the extreme edge of its range in Nova Scotia. Of the nine carpal and five tarsal patterns, three of the carpal and two of the tarsal patterns had not been reported in any lungless salamander. A follow-up study described a low frequency of variation from populations in three other locations, making local environmental or habitat-based explanations of the variation unlikely.

The variants are not abnormalities caused by disease or parasitism nor do they result from regeneration. The natural fusion of tarsal elements is lateral (across the ankles); fusion in regenerated limbs is proximo-distal (up and down the limb) and occurs 10 times more frequently. Patterns found in the wild reflect variation in limb development. The regeneration patterns are the mechanical consequences of regenerating a limb in a laboratory (or terrestrial) environment where the limb has to be used in locomotion while regenerating. Variant or abnormal patterns can arise during regeneration. Limb abnormalities resulting from abnormal regeneration account for some four per cent of individuals with supernumerary or bifurcated limbs, and for 3.7 per cent of those with atrophic limbs or missing digits in 'natural populations' (a pond in the grounds of a London college of higher education) of smooth newts, *Triturus vulgaris*.[6]

Variation often is geographical. A study of natural variation in the appendicular skeleton of the Italian crested (Vienna) newt, *Triturus carnifex*, from northern and central Italian populations shows that forelimb variation is present in 36 per cent of the northern but only in 13 per cent of the southern population, while variation in the hind limbs is present in 55 per cent of the northern and 23 per cent of the southern population. Variation in the north is two-thirds or more greater than in the south, with 30 per cent less variation in forelimbs than in hind limbs in both locations (Pacces Zaffaroni *et al.*, 1996).

Gollmann (1991) reported rare bilateral basipodial patterns in the two Australian frogs in the genus *Geocrinia*, the southern smooth froglet, *Geocrinia laevis*, and the Victoria smooth froglet, *G. Victoriana*, and their hybrids. Patterns such as fusion of the first and second vertebrae and one of the patterns of carpal arrangement are typical of patterns seen more commonly in other genera. Similarly, among 500 individuals of the rough-skinned newt, *Taricha granulosa*, from a single population studied by Shubin *et al.* (1995), 70 per cent had the standard number of tarsal elements while 30 per cent had variants of that pattern. Some of the variants were ancestral, i.e. patterns characteristic of the majority of individuals in an ancestral lineage. Others represented derived patterns; variants in few individuals represented patterns possessed by most individuals in a more derived taxon.

ADAPTIVE VALUE

The existence and persistence of variants such as those described above causes us to ask about their adaptive value. Does the persistence of a variant in a minority of individuals necessarily mean that it is as functional (adaptive if you will) as the majority pattern? Can such a variant persist if it is not adaptive? Arguments from developmental constraint would allow us to answer yes, and this may be a sufficient explanation in many cases. In other cases, the adaptive value of variation is evident.

Variation in vertebral number can have adaptive value (Chapter 43). Swain and Lindsey (1984) showed that predation on sticklebacks by sunfish was up to 1.7 times higher on individual sticklebacks with 31 vertebrae than on those with 32 vertebrae. Interestingly, parental reproductive history also affects such meristic variation, as detected in the neotropical killifish, the mangrove rivulus, *Rivulus marmoratus*. Embryos produced soon after eggs are laid have fewer vertebrae than those without such delays. Similarly, embryos from eggs and females maintained at 30°C have fewer vertebrae than do those from eggs and females maintained at 25°C. The adaptive value is complex. There are trade-offs. From life history and ecological studies we know that predation correlates with stickleback body length and vertebral number – selection via predation correlates with the ratio of abdominal to tail vertebrae – but that stickleback burst speed correlates with vertebral phenotype. Maintaining variation is the fastest way to respond to altered selection pressure.[7]

Variation in vertebral number is much greater in laboratory-reared Red Sea bream, *Pagrus majo*, than in fish caught from the wild (Matsuoka, 1982). On the other hand, sometimes variation, including variation in skeletal structure, looks like random noise. Enormous variation is found in the skeleton of the sturgeon, *Acipenser*, including large numbers of anamestic (Wormian) bones which are small, 'randomly placed' bones that fill in spaces between other larger bones in the skull; they are spandrels. So many anamestic bones develop with so much variability that it is impossible to name (homologize) many of the skull bones in sturgeons (Hilton and Bemis, 1999). I admit this is an extreme example, but it is the end of a continuum, not an isolated case.

One end of the continuum is that variation is bounded and can be used systematically as, for example, cranial variation in Siamese fighting fish, *Betta splendens*. Somewhere along the continuum lies intraspecific variation, as in the number of ceratobranchials in the four-toed salamander, *Hemidactylium scutatum*, in which the fourth ceratobranchial is present in 92 per cent of a sample of 49 larvae. Interesting, the skeleton is more highly conserved than other organ systems in the polymorphic ring species of the plethodontid Monterey ensatina, *Ensatina eschscholtzii*.[8]

METAMORPHOSIS

Any animal with metamorphosis in its life history will exhibit more variation in ossification than direct-developing species. This is especially true for groups such as urodele amphibians in which paedomorphosis is common, as verified in many studies, for example in smooth and alpine newts, *Triturus vulgaris* and *T. alpestris*. Skull development in two species of salamanders in the genus *Desmognathus* (*D. quadramaculatus* and *D. aeneus*) at extremes of the life history strategies of indirect (biphasic) and direct development, shows ontogenetic repatterning in the direct developer *D. aeneus* evidenced in loss of such ancestral larval structures as the palatopterygoid and alterations in timing of appearance of elements as in praecocious development of the maxilla. The skull of the neotenic alpine newt, *Triturus alpestris*, shows the same abbreviated development as seen in Tertiary salamanders.[9]

Reilly (1986) investigated the ontogeny of cranial ossification in the eastern newt, *Notophthalmus viridescens*, with especial emphasis on metamorphosis and neoteny. Neoteny does not involve the skull, and so is not global. Both completely metamorphosing and partial neotenic populations are known. In comparing *Ambystoma talpoideum* with *N. viridescens*, Reilly (1987) compared the completion of development of the hyobranchial apparatus in *Ambystoma* before metamorphosis (because metamorphosis is delayed), with the limited development in *Notophthalmus*, in which metamorphosis is incomplete. These are two strategies for neoteny. *En passant*, Reilly and Lauder (1988) used the identification of an atavistic epibranchial in *Notophthalmus* to argue for the homology of the amphibian epibranchials with those elements labeled epibranchials in other vertebrates.[10]

Variation associated with speeding up or slowing down metamorphosis can be considerable, but also may be subtle. *Pyxicephalus adspersus*, an African frog that avoids desiccation in the dry season by creating a 'cocoon' from multiple layers of skin, develops rapidly, metamorphosis being completed in 17 days at 29°C. Despite this exceptional speed of development and that ossification is initiated early, the stage of skeletal development reached at metamorphosis is the same as seen in much more slowly developing frogs (Haas, 1999).

Wake *et al.* (1983) examined the ossification sequence in the direct developing plethodontid salamander, *Aneides lugubris*, the species with the most prolonged ontogeny of any plethodontid; ossification of the long bones continues throughout life, indeed is never completed. Wake and Shubin (1998) examined limb development in the Pacific giant salamander, *Dicamptodon tenebrosus*, a stream developer with praecocious limb development, including formation of a complete set of digits at hatching. Despite its large size and modified ecology (stream-dwelling), *Dicamptodon* conserves the basic salamander pattern of preaxial rather than postaxial dominance during limb development and development of digits from an isolated digital arch rather than from basal (postaxial) elements.

On the other hand, some taxa, including those with metamorphosis in the life cycle, have skeletons that are highly conserved. All groups of amphibians (anurans, urodeles, caecilians) have highly conserved skeletal development, unless trends such as miniaturization intervene (see below). This can be seen by comparing the remarkably similar sequence of ossification in the Palaeozoic temnospondyl, *Apateon*, to that of the extant Asiatic mountain salamander *Ranodon sibiricus* (Table 44.1).

One final aspect of variation is dissociation in the timing of development of different skeletal systems. When Dunlap and Sanchez (1996) examined temporal dissociation between the cranial and limb skeletons in the common European toad, *Bufo bufo*, they found as few as 10 or as many as 19 ossified limb bones when the cranium first ossifies, skull development being delayed relative to limb development.

MINIATURIZATION

Transplant elephants to island habitats and over a surprisingly small amount of time (number of generations) their body size reduces dramatically. *Miniaturization*, the evolutionary process of reduction in overall body size, also has profound effects on the skeleton.

The smallest terrestrial tailed tetrapods are 15 species of plethodontid salamanders of the genus *Thorius* (Chapter 43). The smallest species of *Thorius* is the pygmy (minute) salamander, *T. pennatulus*, the adults of which can have a snout–vent length of only 13 mm. The cartilaginous capsules

Table 44.1 Similarity in sequence of skull ossification in the extant Kazakhstan Mountain salamander *Ranodon sibiricus* and the Paleozoic temnospondyl (amphibian) *Apateon*[a]

Growth stage (1 is the youngest)	*Ranodon*	*Apateon*
1	Pterygoid (palatine process)	Pterygoid (palatine process)
2	Premaxilla	Premaxilla
	Vomer	Vomer
		Prearticular
		Pterygoid (quadrate process)
		Maxilla
3	Frontal	Frontal
	Squamosal	Squamosal
	Parietal	Parietal
	Prearticular	
	Pterygoid (quadrate process)	
		Ectopterygoid
		Supratemporal
4	Posterior parietal primordium	Postparietal
5	Nasal	Nasal
	Maxilla	
6	Prefrontal	Prefrontal
	Lachrymal	Lachrymal
		Postfrontal
		Postorbital
		Jugal
		Tabular
7	Septomaxilla	Septomaxilla

[a] Based on data in Schoch (2002); also see Schoch and Carroll (2003).

Table 44.2 A generalized sequence of ossification of the craniofacial skeleton in anuran amphibians[a]

Skeletal element	Bone type and function
Premetamorphosis	
Parasphenoid	Dermal, flooring
Frontoparietal	Dermal, roofing
Exoccipital	Endochondral, braincase
Early–mid-metamorphosis	
Prootic	Endochondral, braincase
Septomaxilla	Dermal, internal support
Maxilla and premaxilla	Dermal, upper jaw
Nasal	Dermal, roofing
Mid–late-metamorphosis	
Dentary and angulosplenial	Dermal, mandible
Squamosal	Dermal, suspensory
Pterygoid	Dermal, brace (jaw to braincase)
Vomer (+)	Dermal, flooring and support
Late–post-metamorphosis	
Mentomeckelian	Endochondral, lower jaw
Palatine (+)	Dermal, brace (jaw to braincase)
Quadratojugal (+)	Dermal, brace (jaw to quadrate)
Columella (+)	Dermal, sound transmission
Sphenethmoid	Endochondral, braincase

[a] Based on data from Trueb and Alberch (1985).

around the sense organs appear to be all that exists in their skulls, so reduced or absent are other skeletal elements. Bones that remain are heavily mineralized (hypermineralization) to compensate for what would otherwise be a weakened skeleton.[11]

Trueb (1985) and Trueb and Alberch (1983) examined miniaturization of anuran skulls demonstrating loss of elements in small individuals *and in species with individuals of average size*, in both cases the loss reflecting attainment of sexual maturity at an early age (paedomorphosis) and consequently less time during ontogeny for elements to form; ossification of skull and vertebrae is later in the plains spadefoot toad, *Spea bombifrons* than in most other anurans. The greatest variation is seen in the bones of the jaws and jaw suspension that develop during metamorphosis; Table 44.2 shows a generalized scheme (there is much variation) of when skeletal elements form in relation to metamorphosis. An analysis of five species of ranid frogs shows that the major changes in the order of appearance of elements occur during and especially after metamorphosis (Table 44.3).[12]

As another example of conservation, Perotti (2001) examined the sequence of ossification in the nest-building leptodactyline anuran, *Leptodactylus chaquensis*, finding that it is typical of anurans, with postcranial ossification and ossification of exoccipital, parasphenoid and frontoparietal at premetamorphosis (stages 33–36), major modification of the chondrocranium later in metamorphosis (stages 40 and 41) and further ossifications at the metamorphic climax.

Yeh (2002) examined the effects of miniaturization of body size on the skeletons of 129 species of frogs from 12 families, providing the most extensive database available. Frogs were assessed for presence or absence of individual bones and for changes in skeletal morphogenesis and growth. Bones were lost from the skull (and from the digits) and changes were seen in skull shape. The changes were interpreted as resulting from functional constraints and paedomorphosis.

Miniaturization has also been described in lizards, especially burrowing lizards. As you might expect, the skull, which bears the brunt of the forces experienced during burrowing, bears the brunt of the changes. Rieppel (1984a,b) examined the functional and evolutionary implications of miniaturization of lizard skulls, emphasizing how much we need to know about skull development if we are to understand the functional and evolutionary changes.

Although many fish are very small, few studies have specifically addressed the effects of miniaturization on the skeleton, one exception being studies on the development of the cartilaginous skull (the chondrocranium) of the dwarf sea horse, *Hippocampus* and the pipefish, *Nerophis*.[13]

Miniaturization, as is true of so many changes in shape and size during development and evolution, is brought about by changes in the timing of developmental processes, which is the evolutionary/developmental mechanism of *heterochrony*.

Table 44.3 Sequence of ossification of cranial bones of five species of frogs in the genus *Rana*[a]

	R. aurora	R. cascade	R. pipiens	R. pretiosa	R. temporaria
Pre-metamorphosis	Parasphenoid **Exoccipital** Frontoparietal **Prootic**	Parasphenoid **Exoccipital** Frontoparietal **Prootic** Premaxilla	Parasphenoid **Exoccipital** Frontoparietal **Prootic**	Parasphenoid **Exoccipital** Frontoparietal **Prootic** Premaxilla Septomaxilla	Parasphenoid Frontoparietal **Exoccipital** Premaxilla Septomaxilla
Metamorphosis	Premaxilla Septomaxilla Maxilla Nasal Angulosplenial Dentary Squamosal Pterygoid Vomer **Mentomeckelian** Quadratojugal Palatine	Septomaxilla Maxilla Nasal Angulosplenial Dentary Squamosal Pterygoid Quadratojugal Palatine Vomer	Premaxilla Septomaxilla Maxilla Nasal Angulosplenial Dentary Squamosal	Maxilla Nasal Dentary Squamosal Angulosplenial Vomer Palatine Pterygoid **Mentomeckelian** Quadratojugal	Maxilla **Prootic** Angulosplenial Dentary Quadratojugal Pterygoid Nasal **Mentomeckelian**
Post-metamorphosis	**Columella** **Sphenethmoid**	**Mentomeckelian** **Columella** **Sphenethmoid**	Pterygoid Vomer **Mentomeckelian** Quadratojugal Palatine **Columella**	Columella Sphenethmoid	Vomer Palatine

[a] Based on data in Trueb (1985). Endochondral bones are shown in bold face. The first bone in each column is the first to appear, the last in each column the last to appear. Order within each column represents the sequence of appearance and whether particular bones appear in pre-metamorphosis, during metamorphosis or post-metamorphosis. Alignment between columns is not significant. The common names of the five frogs are: Californian red-frog (*R. aurora*), cascades frog (*R. cascadae*), northern leopard frog (*R. pipiens*), Oregon spotted frog (*R. pretiosa*), and European common frog (*R. temporaria*).

HETEROCHRONY

As discussed in Box 6.1, heterochrony has had a long history, culminating in a concept that is today used in two quite different ways: comparison with an ancestral condition or comparison within an individual (Box 6.1). Alteration in the timing of developmental, cellular or genetic pathways influences skeletogenesis within and between generations. Below I outline some examples of both usages to describe/understand various aspects of skeletal change. Some relate to changes in size and shape (heterochrony à la Gould), others to more novel aspects of skeletal change, a topic that is taken up in the following section under the heading of neomorphs.

Process heterochrony

An important but virtually unknown and woefully underutilized cellular model for heterochrony and allometry was proposed by Katz (1980). Katz took the standard allometric formula $y = bx^{\alpha}$ agreed to by Huxley and Teissier in 1935 and applied it at the cellular level, equating α with the relative frequency of the rate of division between any two parts and b with the relative number of proliferation centres of the two parts. Atchley and Hall (1991) essentially extend this in their quantitative genetics approach to morphological change.[14]

Placing his emphasis on developmental processes, Müller (1991) recognized two mechanisms of heterochronic changes affecting the limb skeleton. One is heterochrony of primary patterning mechanisms, by which novel or unanticipated changes can occur. The second is changes in secondary patterning mechanisms, in sizes and shapes. I take a similar approach and place much emphasis on developmental mechanisms underlying heterochrony and how those can change.[15] Cubo (2000) has the same notion in mind when discussing what he terms 'process heterochronies' affecting endochondral ossification, processes that included onset, rate and offset of major events such as condensation, the number of cells set aside in a growth plate (see Chapter 30), initial differentiation, the time at which cell proliferation is maximal, and so forth, and which demonstrate the efficacy of the model Bill Atchley and I developed.

Heterochrony may reflect a simple shift in onset of growth, rate of growth, and/or when growth stops. Much of the heterochrony literature consists of such studies. For example, Fiorello and German (1997) examined 15 skeletal parameters of craniofacial growth in giant, standard and dwarf breeds of rabbits. The main difference between them is the length of time spent growing at the maximal rate. This is extremely useful information but is really only step one; many of us want to know which underlying developmental process(es) has/have changed.

Miriam Zelditch used a process orientation when she evaluated models of developmental integration of the skull in laboratory rats, the cotton rat *Sigmodon fulviventer*, and in eight other *Sigmodon* species. The more tightly integrated the functional units of the skull, the less likely it will be that simple heterochronic shifts will be effective. She and her colleagues specifically test whether such integration is because of shared bases of induction, differential growth, tissue of origin, and/or developmental or functional integration. Influences over the duration of ontogeny change; in neonates it makes a difference whether tissues arise in mesoderm or neural crest but this is not true in older individuals. This integrative approach of developmentally individualizing parts that change with region and over time allows heterochrony to be examined at the mechanistic level, and shows which aspects of the skeleton are more or less constrained in their ability to change, and whether such constraint is constant over ontogeny.[16]

Nasal bones fail to form in white-lined bats of the genus *Chiroderma* because of delayed induction, providing a nice example of how heterochrony can prevent an element from forming. Similarly, altered development time allows some individuals in southern populations of the Olympic salamander, *Rhyacotriton olympicus*, to form nasal bones.[17]

Smirnov (1991) thought that the heterochronies he documented during development of the middle ear in frogs were not related to function directly but via paedomorphosis, possibly because there are non-tympanic routes for hearing in frogs.

Smirnov (1992) reared the common Eurasian spadefoot toad, *Pelobates fuscus*, in the laboratory over several years, prolonging the time at which they metamorphosed, and thus prolonging the tadpole (larval) period; these animals became sexually mature as juveniles. Variation in duration of the tadpole stage influenced adult cranial structure. With slowed development, bones that formed early experienced a longer developmental period and became hypermorphic. Bones that formed close to metamorphosis were unchanged. An examination of old individuals of three *Pelobates* species revealed between two and four extra dermal bones (all postorbitals), demonstrating that paedomorphosis had not robbed these species of the requisite developmental programme, 'merely' not allowed enough time for it to be initiated early in development (Smirnov, 1995).

Coupling and uncoupling dermal and endochondral ossification

Interesting patterns emerge when heterochrony is studied by comparative timing of the development of different organ systems or parts of the skeleton within individuals. I am not thinking here of the separation of normally coupled events under experimental conditions (although such experiments can be illuminating) but of uncoupling in nature.

Several studies demonstrate uncoupling of the timing of onset of dermal and endochondral ossification:

- The approach Jim Hanken and I took – implanting pellets loaded with thyroxine into specific regions of the skulls of tadpoles of the Oriental fire-bellied toad, *Bombina orientalis* – demonstrated that chondro- and osteogenesis could be uncoupled during subsequent metamorphosis.[18]

- Myobatrachine frogs are a group of 22 genera endemic to Australia and New Guinea. Most are small (20–40 mm snout–vent length), although the gastric-brooding frog, *Rheobatrachus*, which as the name suggests broods developing embryos in its stomach, reaches 80 mm. Uncoupling often results from praecocious maturation (paedomorphosis) and so is seen in some species but not in others. The more paedomorphic the species the greater the number of uncoupled features. For instance, of three genera of Australian frogs, *Pseudophryne* has two uncoupled features, *Crinia* four, and *Uperoleia* eleven (Davies, 1989).

- Strauss (1990, 1991) identified sequence changes in cranial ossification in five species of live-bearing top-minnows (poeciliid fishes). Through mapping developmental characters to a robust phylogenetic tree, Strauss showed decoupling of larval and adult growth patterns, the patterns among larvae being more similar than those among adults.

Primates

Heterochrony has been much invoked in anthropological research, not always with uniformity of approach or results and certainly not with any uniformity of conclusions. For example, Brian Shea found that extension of a common growth trend in animals of different sizes explains much skull growth in African apes. This may not be unexpected when one considers the variation in life history among primates and the role of function in the development of craniofacial form. Length of gestation, weight, size at first litter, age at weaning, age at sexual maturity, age at first breeding, longevity and length of the oestrous cycle all vary, with some 85 per cent of the variation at the subfamily level. Most of these variables show a high correlation with body size.[19]

On the other hand, analysis of the growth of the cranial base in hominoids (*Homo sapiens*, *Gorilla gorilla*, the chimpanzee, *Pan troglodytes*, and the orangutan, *Pongo pygmaeus*) indicated to Dean and Wood (1984) that a relatively simple modification of timing or patterning of growth is *insufficient* to account for the phylogenetic changes seen in these four taxa. The cranial base is not a simple single functional unit and has influences as far forward as the anterior portion of the skull. Dainton and Macho (1999) examined heterochrony in skeletal and dental development in *Gorilla*, *Homo* and *Pan*, tried to distinguish genetic from epigenetic control, and treated issues such as whether size or age is the better baseline against which to make comparisons (Fig. 25.2), and the importance of condensations and early growth trajectories that cannot be overcome by selection later in life (Box 44.1).[20]

I should really place these primates into an appropriate phylogenetic context, for without such a context we cannot determine the direction of evolutionary change. Dainton and Macho's comparison of three species (*Gorilla*, *Homo*, *Pan*; Fig. 25.2) is no accident; at least three species (and a robust phylogeny) are required to draw sensible conclusions about direction of change. Agreeing on a robust phylogeny is a problem in primatology and an even greater problem in palaeoanthropology; analysis of hominoid and hominid systematics evokes polarized positions, even though our knowledge of hominid phylogeny is strong. So we find one group speaking of the desire to make revolutionary finds having 'wreaked havoc with recognition of early human ancestors,' of the history of palaeoanthropology as 'one of repeated misidentification of fossil ancestors and of occasional fraud,' and of hominoid and hominid taxonomy as 'based largely on preevolutionary notions and on misinterpretations.'[21]

Christine Tardieu from France uses heterochrony effectively in her studies on the evolution of hominid ontogeny, examining ontogeny of the femora in humans and great apes to show that the first hominids had a short developmental period, with infantile and adolescent growth spurts as seen in modern great apes. Used in this way, these analyses invoke process heterochrony, but Tardieu uses femora of *Australopithecus afarensis* as a phylogenetic baseline, and so also is able to evaluate de Beerian *and* Gouldian heterochrony, which is an ideal situation (Tardieu, 1997, 1998).

NEOMORPHS

I use the term *neomorph* for a cartilage or bone that appears in a species, taxon, clade or lineage when sister species/taxa, related clades or ancestors in the lineage do not possess the feature. Neomorphs appear to appear out of nowhere, *de novo*, but are found in most if not all individuals of a species (Hall, 2003a). You can think of a neomorph as the obverse of an *atavism*, an atavism being the reappearance of an ancestral character in an individual in a descendant species when the *immediate ancestors* of that species lacked the character. Individual extant horses with three-toed hooves are one example. The three-toed atavistic condition reminds us of ancestral horse species in which all individuals possessed three toes (Hall, 1984b, 1998a).

Neomorphic elements (neomorphs) can be used as valuable natural experiments through which to explore the basis for the evolution of new skeletal elements. Let's look at several examples.

The preglossale of the common pigeon

A medial bone is found at the tip of the tongue in the common pigeon, *Passer domesticus*. This bone – *the preglossale* – appears in newly hatched chicks as a weakly ossified endochondral bone and has no ancestral rudiments in any other birds.

Box 44.1 Functional units and anthropology

Much of our understanding of our primate relatives and ancestry comes from analysis of bones or teeth, often only minute pieces of bone or a single tooth. Unraveling the processes by which the skeleton evolved within the primates requires a sophisticated appreciation and analysis of functional units and functional integration, Sadly, all too seldom is this approach brought to primatology.

Functional integration

Dan Lieberman is using developmental, functional, adaptationist, morphological and phylogenetic approaches to assess the processes underlying human evolution. For example, using 33 commonly employed cranial, dental and mandibular characters, Lieberman found that many are not homologous from taxon to taxon, are evolutionarily derived (homoplastic), and/or are not found in all taxa.[a] Despite these limitations – and Lieberman (1995) thinks we should focus on the integration of functional morphology rather than individual characters – signal does emerge from the noise; the data support a recent African origin for humans.

Subsequently, Lieberman and his colleagues (2001) examined whether the articular surface of a long bone (especially articular surface area) is as good a predictor of body mass, joint function and other somatic parameters as it has been proposed to be in palaeoanthropology. They found that, although developmentally constrained, articular surface area *is* correlated with locomotion at the species level and with body mass at the individual level.

Although located at the base of the skull, the joint at the cranial base plays an important role in craniofacial growth. In a further analysis, Lieberman et al. (2002b, 2004) showed that the base of the skull is an important driver of the growth of the front of the skull and the face. This informed analysis of primate cranial bases demonstrates integration of cranial base and brain, and integration of cranial base and face, two interrelationships that might not have been predicted before the analysis – I was surprised – but which inform our understanding.

Genus *Homo*

Another ongoing problem is how to define the genus *Homo* in the fossil record. One approach is to seek features only possessed by *Homo*, in the way that the supraorbital ridge was used as diagnostic for Neanderthals (Broom, 1950). However, identifying a genus on the basis of a single feature is dodgy at best, sacrilege at worst, as Richard Owen discovered to his cost, and at the hands of Thomas Huxley, when he (Owen) attempted to separate humans from the apes on the basis of the presence of the hippocampus minor in the brains of the former but not the latter. *Vanity Fair* had enormous fun with such luminaries (Fig. 6.3).

Lieberman et al. (2002) addressed this conundrum by searching for unique shared derived structural and functional cranial units (*autapomorphies*). They found two: facial retraction and the degree of globularity of the neurocranium, the second measured from the angle of the cranial base, the width/length ratio of the cranial fossae and facial length. Developmental and evolutionary changes were then investigated.

In a conceptually similar approach, Ackerman and Krovitz (2002) demonstrated common patterns of facial ontogeny in hominids, which they summarize as early prenatal divergence in all lineages followed by parallel postnatal development when Neanderthals are compared with humans. In their analysis of cranial anatomy Zollikofer and Ponce de León (2004) concluded that modification in but a few parameters early in ontogeny explains many of the changes during early hominid evolution. As discussed in this chapter, Tardieu (1997, 1998) used heterochrony to identify more global patterns in the evolution of hominid ontogeny than can be identified from single characters. The integrated functional system she used was the ontogeny of the femora in humans and great apes. Her conclusions: the first hominids had a short developmental period with infantile and adolescent growth spurts, as seen in modern great apes. Apply the appropriate methods and much is there to be discovered.

[a] For an introduction to the complex issue of the relationship between homology and homoplasty, see Wake (1991, 1994, 1996, 1999), Hall (1994a, 1995c, 1998a, 2003a) and Hall and Olson (2003).

The existence of a bone such as the preglossale raises numerous questions. Is it of neural crest or mesodermal in origin? Does it arise following a tissue interaction(s) or does it self-differentiate? If induced, do nerves or muscles provide the necessary signals for its development? Bock and Morony (1978), who described the bone, claim that it 'is associated with, and indeed is dependent, upon the presence of the *M. hg. Anterior* [anterior hypoglossal muscle],' but have no experimental evidence to support this claim. So, what is the function of the preglossale? And why have other birds not developed an equivalent bone?

Digits

A second example that is much more widespread and understood is the array of *digits* on tetrapod limbs. As discussed in Chapter 40, we know a lot about the origin of limbs from fins. Why are digits regarded as neomorphs but the other elements of the limb skeletons not? Quite simply, because we can trace the history of the other elements (humerus/femur, radius–ulna/tibia–fibula) in the fin skeleton, in which homologous elements occur.

Transformation of fin to limb (Chapter 40) involved retention of these proximal endoskeletal elements, loss of the bony fin rays, and the elaboration of distal mesenchyme as digits; fins minus fin rays plus digits equals limbs. We cannot trace a similar history for the digits which appear *de novo*.[22]

Secondary jaw articulations

The beaks of many birds contain a secondary articulation where the median process of the mandible abuts the lateral or medial process of the basitemporal plate to form a secondary brace for the mandible (Bock, 1959, 1960b). As a consequence of this median brace the normal quadrate joint is underdeveloped. Secondary articulations may be:

- a simple syndesmosis as in the plover, *Charadrius*, consisting of fibres and ligaments without an articular cavity or cartilage;
- a true diarthrosis, as in the black skimmer, *Rynchops niger*, with an articular cavity, dense fibrocartilaginous pads and a 'synovial' membrane lining the joint cavity.

Bock and Wahlert (1965) used such examples to develop an influential model of the adaptation and evolution of complex morphologies and functional units (see Box 44.1). The degree of development of a secondary articulation correlates with the strength of the forces acting on the depressed mandible. Thus the skimmer, which flies with its knife-like lower beak just beneath the water's surface, imposing massive forces on the jaws, has the most highly developed secondary articulation.

In another variant, sesamoids develop as integrated parts of the articulation between the quadrate and the mandible. Thus in a New Zealand bird, the North Island kokako, *Callaeas cinerea*, two sesamoids occur as ossifications within the internal jugomandibular ligament, one between quadrate and mandible with cartilaginous pads facing each (Burton, 1973), i.e. the sesamoid is an integrated component of the joint (Fig. 9.3).

Intra-ramal mandibular joints in pelicans and their allies (Pelecaniformes), which facilitate swallowing large prey, are synovial and develop by independent and internal ossification of Meckel's cartilage, not as secondary cartilages or from bones such as the splenial or angular in the lower jaws. Zusi and Warheit (1992) discuss the presence of such a joint in loons, grebes, albatrosses, penguins, herons, storks and New World vultures on the basis of examination of skeletons, and in the great frigatebird, *Fregata minor*, and in the brown booby, *Sula leucogaster*, on the basis of histological analysis. Similar joints are found in the 'western bird' *Hesperornis*, and in the 'fish bird' *Ichthyornis*, two toothed-birds of the Cretaceous period.

A Boid intramaxillary joint

The maxilla is the single bone of the tetrapod upper jaw. One species of snake, however, has a joint within the maxilla, an *intramaxillary joint*. The species is an endangered boa, the Round Island Boa, *Casarea dussumieri*, a 1.5-metre nocturnal snake that feeds on lizards and is found only in the tropical rain forests on Round Island off Africa. Although endangered and suffering from loss of habitat, strict control of access of humans to the island and removal of the rabbit population from the island has allowed some recovery in numbers.

The joint has a functional advantage; it enhances elevation of the anterior maxillary teeth to capture larger prey items (Cundall and Irish, 1989). The older idea that such a novel structure could only have arisen by snakes with fractured maxillae being selected for was never satisfactory. The origin of a second centre of ossification for the maxilla is much more plausible and, indeed, a snake with two centres is known. Cranial development in the eastern or black rat snake, *Elaphe obsoleta* (Ophidia, Colubridae), includes two centres of ossification for the maxilla and, as a second unusual feature, a laterosphenoid that is partly intramembranous and partly endochondral (Haluska and Alberch, 1983).

Regenerated joints

A fourth example involves an experimentally produced neomorph; atavisms also can be created experimentally.

The experimental procedure was surgical removal of the condyle (condylectomy) of the dentary bone from pig temporomandibular joints (TMJs). In some situations/species, the condylar cartilage regenerates, provided that a remnant remains. In this study, *a new mandibular condyle developed*, not from the condylar process of the dentary bone but *from a new cartilage that arose independently of the bone, essentially a sesamoid* (Chapter 9), only fusing with the bone during later stages of repair.[23]

Wishbones

A fifth example, mentioned here but discussed in detail in Chapter 16, is whether the wishbone in birds represents the clavicles of other tetrapods, another element of the pectoral girdle, or a neomorph – a new bone, which happens to be located where clavicles are found in other groups.

Limb rudiments in whales

The sixth example is the presence of rudiments of skeletal elements in some whales at the position where hind limbs are found on other tetrapods. Rudimentary limb bones are retained in some modern baleen (mysticete) whales but not in any of the toothed (odontocete) whales.

In his analysis of rudimentation of the hind limbs in the sperm whale, *Physeter macrocephalus*, Deimer (1977) described and illustrated rudimentary femora. The humpback whale, *Megaptera nodosa*, has a cartilaginous femur, the blue whale, *Balaenoptera* (*Sibbaldus*) *musculus*, a bony femur, connected to the pelvic girdle by a ligament. The least reduced limb is possessed by the bowhead whale, *Balaena mysticetus*, which has a bony femur some 10–22 cm long connected to the pelvic girdle by a synovial joint (Fig. 44.1). Depending on the phylogenetic history of whales, such element(s) could be *vestiges* or *rudiments* of a limb element (assuming that the limb skeleton was incompletely lost in ancestors) or a *neomorph* (if the limb skeleton was lost completely in the ancestors and a new element evolved in a similar position). As discussed in the next section, from time to time *atavisms* also appear (Hall, 2003a).

Sperm whales, *Physeter catodon*, have isolated bones in the position where the hind limbs would have been in their limbed ancestors. These bones lack any connection to the vertebral column, are activated by abdominal rather than appendicular muscles, and are sexually dimorphic, those in males being twice the size of those in females.[24]

A developmental analysis is an obvious way to determine whether hind-limb elements are vestiges or neomorphs, although such series for cetaceans are few and far between. That said, Stêrba *et al.* (2000) produced a monographic treatment on staging of dolphin embryos and foetuses that includes data on the initial development

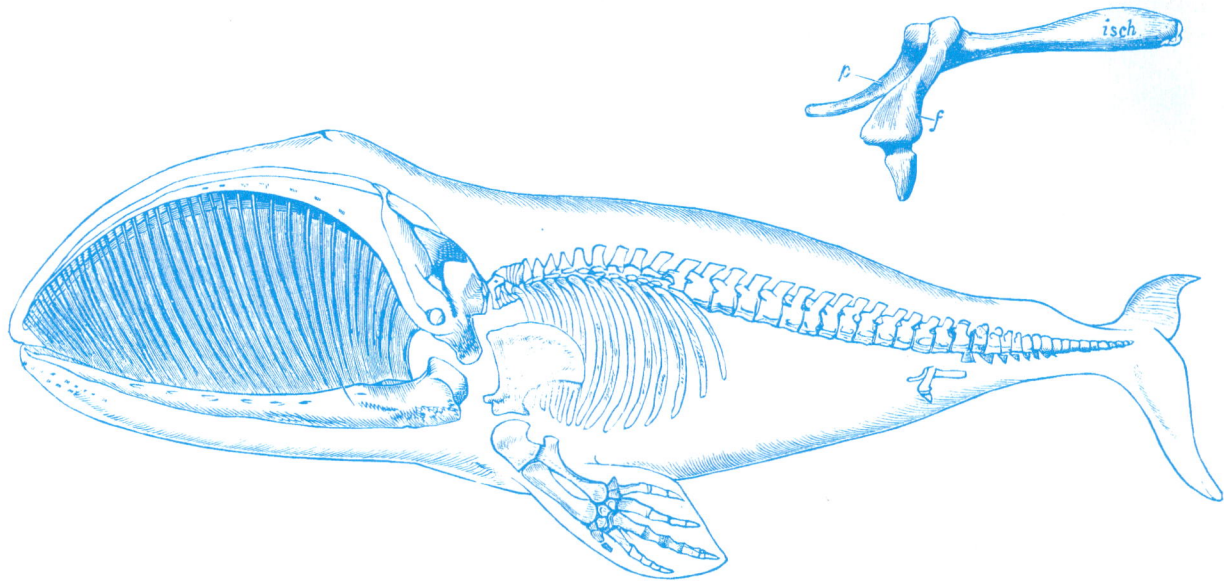

Figure 44.1 The skeleton of a bowhead (Greenland) whale, *Balaena mysticetus*. Note the baleen and the skeletal elements of the flipper. The rudimentary pelvic girdle and hind limb are shown in position and in higher magnification. f, femur, isch, ischium; p, pubis. From Romanes (1901).

and regression of the pelvic limb buds, which never exceed 0.8 cm in length – at stages 3 and 4, when the hand plate is present in the forelimbs – and which are reduced to 0.2–0.3 cm by stage 6, which is when ossification begins in the flippers (Figs 31.4 and 44.2).[25]

Turtle shells

The seventh example is familiar to all of us. One of the most fascinating skeletal features of turtles is their shell or, more correctly, their *shells*, for turtles have a dorsal and a ventral shell, the *carapace* and *plastron*, respectively.

The fossil record of turtles is very poor. In part, this is because when turtles first appeared as fossils they already had fully developed shells. There are no missing links between turtles and 'non-turtles,' which is how Annie Burke likes to subdivide the tetrapods (Box 44.2). How did the shell evolve, and how does it develop?

Development

Developmentally, we know much more about the carapace than the plastron. The carapace fascinates developmental biologists because it is the only vertebrate skeletal element that develops *inside out* with respect to the ribs. Indeed, portions of the ribs and parts of the vertebrae fuse with dermal elements to create the carapace. The interest of the nascent evolutionary–developmental biology community was captured by several independent studies that appeared 15 years ago.

- The fundamental elements of carapacial development were described, including condensation of mesenchyme and the presence of what Burke called a *carapacial ridge*, a mound-like outgrowth with an epithelial cap situated in the position on the flank where carapace devel-

opment starts. She homologized the carapacial ridge with the AER of limb buds, hypothesizing that epithelial–mesenchymal interactions control carapace development as they control limb development. Despite subsequent work, we still do not know if this is so.
- Embryonic development of the marine leatherback turtle, *Dermochelys coriacea*, was described and divided into 22 stages from the appearance of the first somites to hatching, providing a staging table that facilitated experimental work. A description of the development of a carapacial ridge was included.
- Growth of the carapace of the red-eared terrapin, *Chrysemys scripta*, was described using Fourier analysis.[26]

Turtle shells are being revisited by developmental biologists interested in the origin of novelties. Development of the shell (which can contain as many as 50 dermal bones) has been described in two species, the red-eared slider turtle, *Trachemys scripta*, and the common snapping turtle, *Chelydra serpentina*. Dramatic hypertrophy of the dermis and a carapacial ridge that appears to 'trap' the ribs was described (Gilbert et al., 2001). Subsequent study of *Fgf-8* and *Fgf-10* in *T. scripta* showed that *Fgf-10* is expressed in mesenchyme associated with the carapacial ridge. *Fgf-8* was expressed in the AER of the developing limbs but not in association with the carapacial ridge (Loredo et al., 2001). Perhaps it is not surprising that the shell and limb buds do not share the same genes. The shell as an evolutionary novelty would be expected to have evolved because of novel use of existing gene networks.

Evolutionary history

An important evolutionary element was added in 1991 with the description of a new specimen and new genus

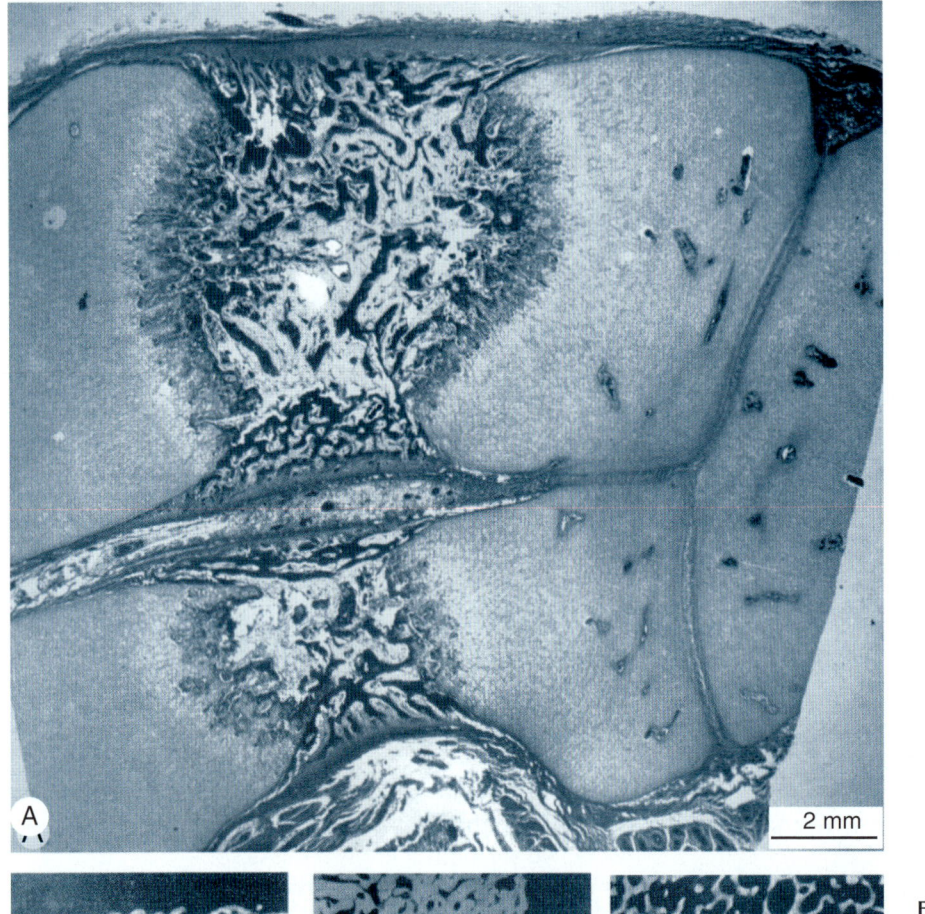

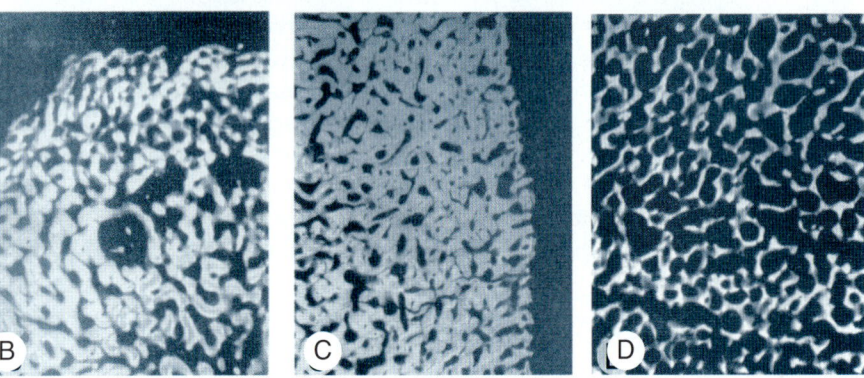

Figure 44.2 Long-bone development in whale flippers. (A) A longitudinal histological section through the radius (above) and ulna (below) of a foetal (165-mm snout–tail notch length) pilot whale, *Globicephala melaena*, showing the vascularized epiphyses, absence of longitudinal columns within the growth plates, and ongoing endochondral ossification. (B–D). Microradiographs of different regions of a radius of a late juvenile beluga (*Delphinapterus leucas*) showing local variation in trabecular structure and density. From Felts and Spurrell (1966).

from the Upper Permian–Lower Triassic, named *Owenetta* after Richard Owen ('Old Bones,' Fig. 6.3). *Owenetta* is in the procolophonids, recognized by some as a sister group to the Captorhinidae, the family to which turtles are traditionally assigned. The relationships of turtles, however, are much debated.[27]

Pareiasaurs, a group of Upper Permian anapsids, regarded by some as the nearest relatives of turtles, possess many turtle traits. Although pareiasaurs lack shells they do have isolated osteoderms, thought by Lee to be precursors of the shell. A complete mitochondrial genome sequence (16 787 bp) of the East African side-necked turtle, *Pelusios subniger*, has been used to show that turtles are a sister group of alligators/birds.[28]

Moss (1969) who compared dermal ossification among reptiles, demonstrated that most are tendinous

ossification rather than periosteal, producing a range of structures from mineralized tendons to bone. Further analysis of how reptiles are capable of such substantial dermal and metaplastic ossification may well contribute to understanding the origin of the shells in turtles. I say this, in part, because the formation of *osteoderms*, literally bones in the skin, is common in extant reptiles (Fig. 44.4): in skinks, especially in the broad-headed skink, *Eumeces laticeps*, in which the osteoderms form an elaborate ornamentation that increases with the age of the individual; and in the slow worm, *Anguis fragilis*, in which osteoderms have been shown to be comprised of two layers.[29]

Geckos have supraorbital ossifications associated with the dermis and epidermal scales as osteoderms and parafrontal bones, the latter lacking the association that

Box 44.2 Missing links

The paradoxical platypus

Several classic (archetypical) missing links are known, *Pithecanthropus erectus* perhaps being the most famous as well as the only species named before it was discovered. Another is *Paranthropus robustus*, described by Robert Broom in the August 20, 1938, issue of the *Illustrated London News* under the headline 'The Missing Link No Longer Missing' (Broom, 1950). Other organisms sought as missing links turned out on discovery not to be links at all, e.g. the platypus, *Ornithorhynchus anatinus* (Fig. 44.3), thought to link reptiles and mammals, and the Senegal bichir, *Polypterus senegalus*, thought to link fishes and amphibians.[a]

In Box 13.1, I discuss taxa that link and show the transitions in the evolution of the middle ear ossicles. Although we no longer rely on missing links to determine transitions between taxa, the absence of intermediate forms can be a real nuisance, as I noted when discussing the origin of the turtle shell; there are no missing links between turtles and 'non-turtles.'

In the introduction to *Kant and the Platypus*, Umberto Eco recounts a dislike for the platypus as one of the reasons given by Jorges Luis Borges for never having been to Australia: '…the platypus, which is a horrible animal, made from the pieces of other animals.' Eco takes a more enlightened and evolutionarily appropriate view: '…the platypus is not horrible, but prodigious and providential…given the platypus's very early appearance in the development of the species, [in the sense of natural kinds] I insinuate that it was not made from the pieces of other animals, but that other animals were made from pieces of the platypus' (Eco, 1997, p. 6).

As I have tried to show in this book, evolution has tinkered with skeletal tissues as much as she has tinkered with the animals that contain them.

[a] See Shipman (2001) for an engaging account of the discovery of *P. erectus*, and see Hall (1999b, 2001d) for platypus and bichir as missing links and for the extraordinary lengths to which late 19th century naturalists went to obtain embryos of such species.

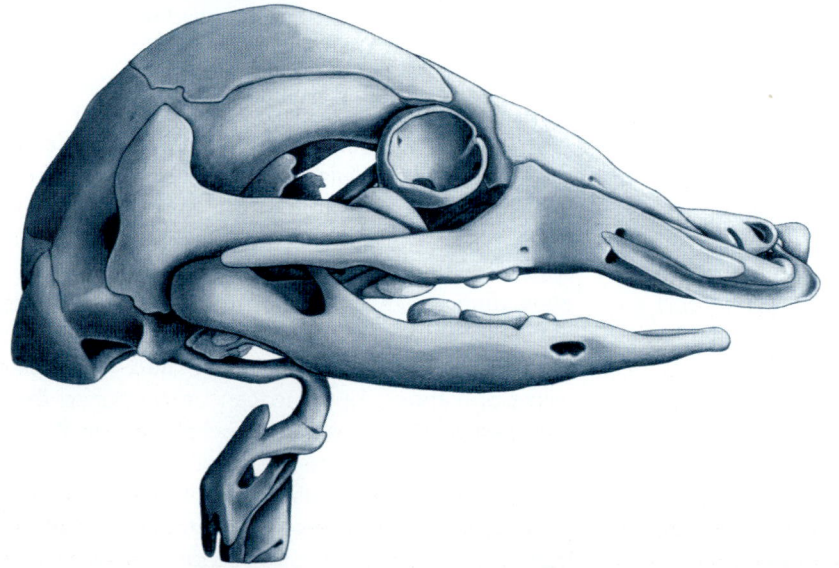

Figure 44.3 A nestling platypus (*Ornithorhynchus anatinus*) embryo, 180 mm long, showing the stage of development of the embryo (above) and of the cranial skeleton (below) at the same stage (stage 39). Modified from Zeller (1988).

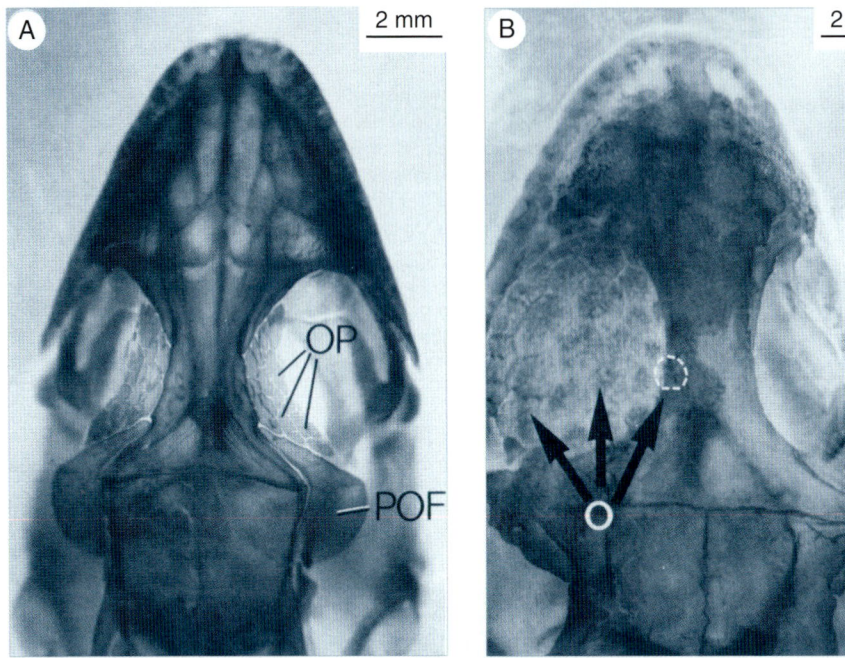

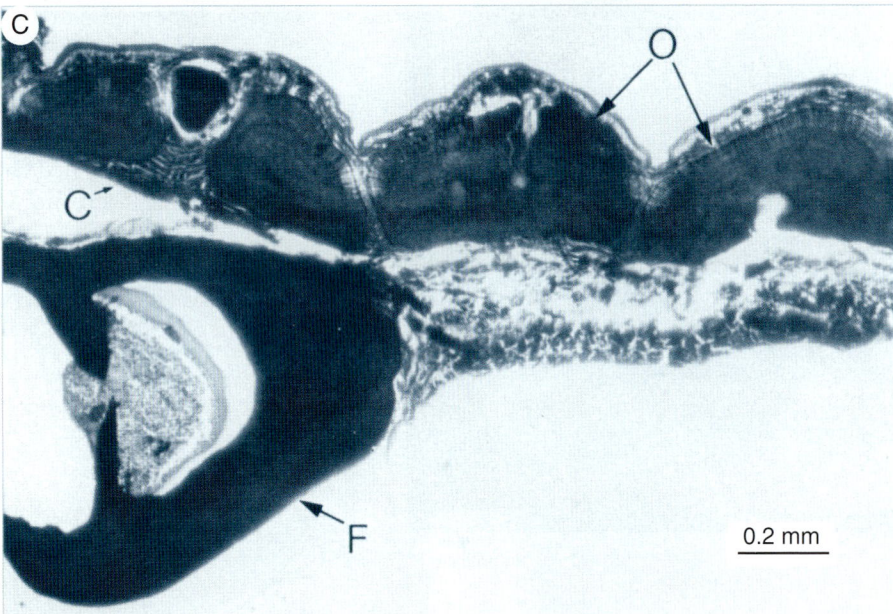

Figure 44.4 Osteoderms and/or expanded dermal bones associated with the orbits in three species of gecko. (A) The Cayman gecko, *Aristelliger praesignis*, has parafrontal bones (OP) as supraorbital bones (not osteoderms) over the midportion of the orbits. The postfrontal bones (POF) are also enlarged. (B) Supraorbital osteoderms (O, arrows), present in the helmeted gecko, *Geckonia chazaliae*, differ from the parafrontal bones in the Cayman gecko (A) in being continuous with osteoderms (one is circled) over the rest of the skull and in being separate from epidermal scales. The supraorbital osteoderms have been removed from the right side of this specimen. (C) is a histological section of the Moorish gecko, *Tarentola mauritanica*, showing osteoderms (O) linked to the epidermis, connected basally by collagen bands (C), and clearly separate from and lying above the frontal bone (F). Modified from Bauer and Russell (1989).

osteoderms have with epidermal scales (Fig. 44.4). Comparative study of the osteoderms in the Moorish gecko, *Tarentola mauritanica*, and the neglected gecko, *T. neglecta*, demonstrated metaplastic ossification, as in the slow worm: two layers in each osteoderm, a basal mineralized layer with abundant collagen, and an outer mineralized layer with minimal collagen and mineral granules.[30]

ATAVISMS

Structures or behaviours that appear to be throwbacks or reversions to an ancestral condition are always fascinating.

An atavism is the occasional reappearance of an ancestral character in an individual in a descendant species when the immediate ancestors of that species lacked the character. We use the term 'atavism' or 'atavistic' in both the specialized and arcane literature of the sciences and in every day speech and writing; 'his behaviour was atavistic;' 'he showed atavistic tendencies,' and so forth.[31]

Atavisms occur naturally in isolated individuals within wild populations, but also can be produced experimentally. Partial hind limbs in some whales or a snake are examples of natural atavisms (see below). Eliciting teeth from avian tissues when birds have not possessed teeth since they separated from reptiles is an example of an experimental atavism. Implicit in recognizing such features as atavisms is (a) that we can recognize a structure in isolated individuals within a recent population or species as equivalent to (homologous with)[32] an ancestral character, and (b) that some vestige of the developmental programme for the structure has remained in the individuals within the recent population.

Given these two requirements, it is perhaps not surprising that skeletal atavisms can be recognized most readily in limbless tetrapods – they retain the developmental programme in limb buds that are initiated and then normally regress (Chapter 37). A mutation, parasitic infection or environmental agent that permits the limb buds of such limbless tetrapods to persist and to continue to develop could allow sufficient time for a skeletal element(s) to form. Having lost their limbs 50 mya but retaining limb buds (and sometimes vestigial limb elements; see Fig. 37.6), whales make an excellent case study.

Limb skeletal elements in whales

Whales evolved from limbed terrestrial mammals. Some of the earliest whale fossils, such as *Pakicetus inachus* from the Early Eocene of Pakistan, had morphological features indicating that they were partly terrestrial and partly aquatic, mosaic evolution resulting in different organs being modified at different times and rates and under different selective pressures. For example, *Pakicetus* had ears that were still adapted for hearing on land but limb structures indicating that they spent part of their lives in water. Another Eocene species (46 mya) from what is now Pakistan had a reduced femur indicative of further adaptation toward an aquatic life.[33]

Sedmera *et al.* (1997a) examined hind-limb development in embryos of the spotted dolphin, *Stenella attenuata*, between 10 and 30 cm crown–rump length. They described limb buds with AERs, chondrogenic condensations and documented nerves growing into the limb buds. As the limb buds deteriorated, some mesenchymal cells died, others being incorporated into the body wall (Fig. 36.5). Using X-ray analysis, DiGiancamillo *et al.* (1998) described postnatal ossification of the hind-limb skeleton of the striped dolphin, *Stenella coeruleoalba* (Fig. 36.5).

As discussed above, some whales contain what may either be remnants of the hind limbs or neomorphic elements in the position where their four-footed ancestors had hind limbs. Atavisms also arise, one in every 5000 individual sperm whales (Hall, 1984b).

Minke whales grow to some eight or nine metres in length and six to seven tonnes body weight. The 1978–79 Japanese whaling season in the Antarctic yielded 72 pelvic bones from the southern minke whale, *Balaenoptera bonaerensis* (Omura, 1980). For a species that has lost its hind limbs, the number of individuals with a skeletal element in the hind-limb region is enormous; 39 per cent of the males and 36 per cent of the females have an ossified femoral rudiment. This hardly ranks as the occasional reappearance of an ancestral character. Rather it appears that all the more distal skeletal elements of the hind limbs have been lost completely but that the proximal femur is still undergoing evolutionary reduction.

Mammalian teeth

Rodents have a reduced dentition consisting of an incisor and molar teeth in the lower jaw (Figs 5.8 and 14.1). From experiments involving the recombination of buccal epithelium and mesenchyme between and across toothed and toothless regions it has long been known that the buccal epithelium, not the mesenchyme, of the intervening toothless region – the diastema – is deficient. Recent studies have identified vestigial tooth primordia in the diastema that regress because of epithelial apoptosis, and that within the functional murine dentition a single tooth can be formed from several adjacent primordia. Such plasticity and retention of primordia provide preconditions (latent homologues) that facilitate the evolution of an atavism.[34]

An oft-cited example of an atavism in mammals is the gradual reappearance during the Pleistocene of the talonid and metaconid cusps of the first molar (M1) and the reappearance of a second molar (M2) in northern lynx, *Felis lynx*, whose ancestors lost both the cusps and M2 in the Miocene (Kurtén, 1963). As with hind-limb elements in whales, developmental explanations for the reappearance of these dental features are not difficult to imagine. In describing vestigial teeth in rabbits, rats and mice, Moss-Salentijn (1978) used such developmental explanations as retardation of development, disruption of epithelial–mesenchymal interactions, and the activity of particular cell populations to explain the evolutionary loss of teeth. Remove the inhibition, as in limbless tetrapods, and tooth evolution can change.

Teleosts and taxic atavisms

You will have noted a curious aspect of the lynx tooth atavism example. All the members of the species have the features, but an atavism was described as the occasional reappearance of a feature in isolated individuals. The apparent conundrum reflects the thinking that an atavism may be the first stage in the reappearance of a feature; we saw this when discussing limbs in *Bipes* in Chapter 37.

Taxa that show much reversal of evolution – by which I mean multiple gain and loss of features – are known, especially among teleost fishes. Stiassny (1992) discussed such gains and losses as *taxic* or *phyletic atavisms* to emphasize that the developmental basis for the characters had been retained and could be reinvoked and inactivated multiple times. Of course, interpretation of features as taxic atavisms is only as good as the robustness of the phylogeny and the knowledge of the worker.

More traditional atavisms also have been reported from teleosts, one example being atavistic teeth on the vomer in one specimen of a member of the perch family, the western Atlantic black drum, *Pogonias cromis*. The primitive condition for bony fishes is to have teeth on the palatine. Vomerine teeth were lost some 65–70 mya.

Late-developing bones in anurans

Some skeletal elements that develop late in the lives of the South African clawed frog, *Xenopus laevis*, and the Oriental fire-bellied toad, *Bombina orientalis*, are regarded by Smirnov (1994, 1997) as atavisms.

Smirnov examined *Xenopus* that were between 2.5 and 12 years old and found skeletal elements in the skull that are not present in younger individuals: palatine, preorbital processes of the maxillary, lateral wings of the parasphenoids, and paired vomers (a single vomer is present in younger individuals).[35] Formation of these elements can be explained without consideration of atavisms. However, two extra dermal ossifications also appear as elements that Smirnov thought were homologous with the postrostrals of Devonian crossopterygian fishes. If this interpretation is correct and given the absence of postrostrals in the immediate ancestors of *Xenopus*, these elements would be atavisms, unless we regard the crossopterygian fishes to be so distant from Xenopus that we regard the elements as neomorphs. Which you choose rather depends on how far back you think conservation of developmental information goes; the Devonian is a stretch.

Similarly, in *Bombina* additional ossifications appear as occasional bones between the nasal and frontoparietal (equivalent in position to a crossopterygian element), between the frontoparietals, and on the dorsal connection of the auditory capsules (the *tectum synoticum*) behind the frontoparietals.[36]

NEOMORPH OR ATAVISM?

Identifying an atavistic character can be tricky. For example, there is intraspecific variation in ceratobranchial number in the four-toed plethodontid salamander, *Hemidactylium scutatum*, a fourth ceratobranchial being found in 92 per cent of larvae. C. S. Rose (1995d) could not determine whether it was atavistic or primitive. Stiassny (1992) spoke of phyletic atavism to describe patterns in teleost fish evolution in which characters were lost and gained repeatedly. The fourth ceratobranchial does not fit this category, nor is it a neomorph.

The insight gained from the the 500 rough-skinned newts, *Taricha granulosa*, was discussed under 'Adaptive value' earlier in the chapter. Detailed understanding of the phylogeny enabled some patterns of arrangement of the tarsal bones to be identified as atavistic while others were recognized as rare variants of patterns that are the norm in more derived taxa. The latter are neomorphs in *T. granulosa*.

Experimental production of an atavism sometimes can provide useful information. For example, treating pregnant rats with aspirin (500 mg) when the embryos were at day 10 of gestation elicits formation of a novel accessory bone between the nasal and frontal bones in one fifth of the offspring. Although previously unreported in rats this bone is found in mice and guinea pigs (Mitala *et al.*, 1984). Similar studies in those species, combined with an analysis of the relationships between the three groups, would be one way of investigating loss and reappearance.

Skeletal elements that have proven difficult to interpret – where is the fossil record when you need it? – are those that appear following gene knockout in mice. Whether we regard these as neomorphs, atavisms, or neither, depends on whether we seek an evolutionary or a more proximate developmental explanation, and on the level of analysis.

Ectopic expression of *Hoxd-4* (*Hox-4.2*) in mice so that it is expressed more rostrally in the mesoderm leads to homeotic transformation of occipital bones into cervical vertebrae. Lufkin *et al.* (1992) discuss whether these are atavisms.

Other transformations are homeotic but not atavistic, in that they result in the transformation of a skeletal element into a more anterior (rostral) or posterior (caudal) member of a series that is normally present.

- A homeotic transformation of cervical vertebrae – the formation of ribs on the 7th vertebra – follows knockout of *Hoxa-4*.
- Mice mutant for *Hoxb-4* show a partial homeotic transformation of the first cervical vertebra (the atlas) to an axis (the second cervical vertebra).
- Compound mutants, involving double knockout of pair-wise combinations of the paralogous genes *Hoxa-4*, *Hoxb-4* and *Hoxd-4*, lead to complete transformation of the atlas to the axis with a dose-dependent increase in the number of vertebrae transformed. Tissue-specific enhancers between *Hoxb-4* and *Hoxb-5* operate in limb and nervous system as partial explanation for such selectivity, sharing and competitive interactions of *Hoxb* genes.
- Placing *Hoxa-5* into mice with a different genetic background alters transcription and elicits homeotic transformations of the third cervical and second thoracic vertebrae.
- Targeted disruption of *Hoxd-3* in mice is followed by anterior transformation of the atlas and axis such that the anterior half of the atlas becomes an extension of the basioccipital, the lateral half becoming an exoccipital and the axis taking on an atlas-like appearance. It is especially interesting that different parts of a single vertebra can take on new appearances, reflecting the embryological origins of each vertebra from separate populations of cells (Chapter 16).
- Targeted disruption of *Hoxa-11* leads to homeotic repatterning of the sacrum.
- Double knockout of *Hoxa-11* (*Hoxa-11$^{-/-}$*) transforms the 13th thoracic vertebra into a first lumbar vertebra as the sacral region is transformed into lumbar (Small and Potter, 1993).

- Knocking out *Hox-10* genes in mice results in the formation of lumbar vertebrae not sacral vertebrae, but all posterior vertebrae develop ribs, while knocking out *Hox-11* genes results in sacral vertebrae not forming but assuming a lumbar identity.[37]

These homeotic transformations indicate the role these genes play in specification of the elements of the axial skeleton that normally form, and point to a mechanism whereby an ectopic, neomorphic and/or atavistic element could form. In *Hoxa-2* knockout mice, the additional element develops where a rudimentary condensation (but not a skeletal element) forms in wild-type mice (Fig. 13.1). Only with the gene knockout does the condensation attain a size sufficient to initiate chondrogenesis. Because this element is not normally present, it is a neomorph. Because the rudiment (condensation) is normally present, the element is not an atavism; see Chapter 20 for details, including the position of the element itself, which I have deliberately not stated here, hoping to whet your appetite.

Another example comes from mouse embryos in which both retinoic acid receptors are knocked out. Such double mutants die *in utero* or just after birth. Because of the role retinoic acid plays in craniofacial development, the mutants display severe craniofacial abnormalities. The frontonasal processes from which the mid-face arises are absent entirely and as much as two-thirds of the pharyngeal arch skeleton is affected. You would expect skeletal elements to be missing, which they are, but extra (supernumerary) cranial skeletal elements also form. As these reside in positions where no skeletal element normally forms, but rather in positions occupied by skeletal elements in ancestral lineages, they have been interpreted as atavisms.[38]

A closer study of development is needed, however, to determine whether condensations for such elements are normally present. If condensations are present, then the skeletal elements that form under experimental conditions would be rudiments, not atavisms, illustrating the insights to be gained from combined developmental and evolutionary approaches to the skeleton, which seems an appropriate note upon which to finish a book on developmental and evolutionary skeletal biology.

NOTES

1. Charles Kingsley (1863), *The Water Babies*, p. 21.
2. See Frolich (1991) for this study. Mean number of caudal vertebrae is 26 for females and 34 for males. Also see D. B. Wake (1997) for incipient speciation within the *Ensatina* complex, and Wake and Jockusch (2000) for detecting species boundaries and incipient speciation in plethodontid salamanders.
3. See Glenny (1954), Glenny and Friedmann (1954) and Glenny (1959) for reduction of the clavicles in parrots. Class 3 clavicles are found as variants in the varied lorikeet, *Psitteuteles versicolor*, and in *Opopsitta* sp., and as the major pattern in the lovebirds (*Agapornis* spp.), parakeets and rosellas (*Cyanoramphus, Platycercus*), budgerigars (*Melopsittacus* spp.), pygmy and grass parrots (*Micropsitta, Neophema, Psephotus*), grass parakeets (*Northiella* spp.) and the red-capped parrot (*Purpureicephalus spurius*).
4. Also see Hallgrímsson *et al.* (2002, 2003) for variability in distal versus proximal elements of the limb skeleton, and see Hallgrímsson and Hall (2005) for further developmental examples.
5. See Hunter and Jernvall (1995), Jernvall (1996) and Jernvall *et al.* (1996) for the initial studies on tooth evolution, and Jernvall (2000) for the developmental study. Also see Hall (1983g, 1984c, 1990a, 1995a,c, 1998a, 2003c, 2004b, 2005a) for how small developmental changes elicit large morphological changes in evolution.
6. See Hanken and Dinsmore (1986) for the follow-up study, Dinsmore and Hanken (1986) for regeneration, and Griffiths (1981) for *Triturus*. Hanken (1982) documented patterns of carpal fusion in the minute plethodontid salamander, *Thorius*.
7. See Swain and Lindsey (1986a,b) and Swain (1992a,b) for these two studies with *Rivulus marmoratus*.
8. See Mabee and Trendler (1996) for *Betta*, C. S. Rose (1995d) for *Hemidactylium*, and Frolich (1991) for *Ensatina*.
9. See Roček (1996) and Djorovic and Kalezic (2000) for *Triturus*, S. B. Marks (2000) for *Desmognathus*, Heatwole and Carroll (2000) and Heatwole and Davies (2003) for overviews of amphibian skeletal development and evolution, and Reiss (2002) for an initial analysis of the phylogeny of amphibian metamorphosis using characters and transformation sequences associated with the skull.
10. Haas (1997) analyzed the larval hyobranchial skeleton in 11 species of discoglossid frogs.
11. See Hanken (1983a, 1984, 1985) for *Thorius*, Hanken (1982) for hypermineralization, Hanken and Wake (1998) for carpal/tarsal patterns in nine species of *Thorius*, and Hanken (1993) for bone growth and miniaturization. The smallest tetrapods are tailless frogs of the genus *Brachycephalus* from Brazil, with a snout–vent length of some 9 mm. With such miniaturization comes loss of digits. These species have three digits on their hind limbs and only two on the forelimbs.
12. See Wiens (1989) for the plains spadefoot toad. Trueb (1996) examined constraints on skeletal systems in pipid frogs, a highly derived group, with *Xenopus laevis* as perhaps the most bizarre (Trueb and Hanken, 1992), while Trueb *et al.* (2000) described skeletal development in the direct-developing Surinam toad, *Pipa pipa*, which has a hyperossified skull with major skeletal changes occurring after the metamorphic climax. In another direct-developer, the Puerto Rican coqui, *Eleutherodactylus coqui*, most cranial cartilages typical of larvae in indirect-developers do not form, and in many regions the adult morphology or a mid-metamorphic morphology is present from the start (Hanken *et al.*, 1992). See Hall (2003b) for an overview of skeletal development in amphibians.
13. See Kadam (1958, 1961) for studies on the chondrocranium and see Azzarello (1989a,b) for the developmental osteology of two species of sea horses.
14. Both Julian Huxley in England and Georges Teissier in France had introduced terms and mathematical expressions for changes in the relative dimensions of parts of the body (or of plants) as size increased during growth. In 1935 Teissier

and Huxley agreed on the term 'allometry' to replace Huxley's term *heterogony* and Teissier's term *disharmony*, and agreed on $y = bx^\alpha$ as the form in which the relationship would be expressed. In 1936 they published their agreement in two papers, one in French and one in English (Huxley and Teissier, 1936a,b). See Gayon (2000) for an analysis of these issues and for a conceptual history of allometry (Box 33.1), whose origins are much earlier.

15. See Hall (1984c, 1990a, 2000a, 2001c, 2002b), especially Hall (1984c, 2003c,d), for the basic developmental mechanisms that underlie heterochrony.

16. See Zelditch (1987, 1988), Zelditch and Carmichael (1989a,b) and Zelditch et al. (1990, 1992, 1993) for these studies. The integrated approach is especially well developed in Zelditch et al. (1992, 1993).

17. See Straney (1984) for bats and Wake (1980) for salamanders.

18. See Hanken and Hall (1984, 1988a,b) and Hanken et al. (1989) for the studies implanting thyroxine, Figs 10.3 and 10.4 and Kemp and Hoyt (1969a,b) for thyroxine and skeletogenesis in the northern leopard frog, *Rana pipiens*, Trueb (1996) for an overview of constraint and flexibility in skeletal development in pipid frogs, Maglia et al. (2001) for an analysis of conserved and variable skeletal features in the larvae of 22 taxa of frogs representing eight families, and Rose (2004) for thyroid hormone-mediated development.

19. See Shea (1983a,c, 1985, 2002) for samples of his extensive allometric analyses of craniofacial growth in African apes and on paedomorphosis and neoteny in the bonobo, *Pan paniscus*, Kean and Houghton (1987) and Harvey and Clutton-Brock (1985) for the life-history parameters (including the role of function in development of human craniofacial form), and the chapters in Jungers (1985), Fleagle (1999), Hallgrímsson (1999) and Minugh-Purvis and McNamara (2002) for overviews of allometry, size and scaling in primates.

20. See Hall and Miyake (1997) for a discussion of the best baseline to use for comparisons. Aliverti et al. (1979) suggested an *index of foetal ossification* as a guide to assessing the effects of teratogens and as an index of delayed development. The index, which involves counting the number of ossification centres in the sternum, metacarpals, metatarsals, proximal phalanges, and cervical and caudal vertebrae, was developed using rat embryos between 19 and 21 days of gestation.

21. See Sarmiento et al. (2002, p. 50) for the quotations, made in the context of a morphology-based analysis of hominoid and hominid systematics.

22. See Shubin (1991, 1995), Sordino et al. (1995), Shubin et al. (1997), Carroll (1997) and Hall (1998a, 2005c) for fins into limbs.

23. See Hall (1984b, 1998a, 2002c, 2003a) for experimental creation of atavisms, and see Duterloo and Jansen (1969) for the TMJ study.

24. See Bejder and Hall (2002) for elements in sperm whales. Renous et al. (1991) described vestigial pelvic appendages in the Iberian worm lizard, *Blanus cinereus*.

25. See Slijper (1979) for an overview of whale biology. Pelvic girdle reduction was already well underway in *Chrysocetus bealyorum*, a member of a newly discovered genus of early archaeocytes (early whales) from the Middle-to-Late Eocene of South Carolina. Reduction had progressed sufficiently that the modified pelvic girdles would not have supported the body on land (Uhen and Gingerich, 2001). Thewissen

and Williams (2002) review the early radiation of cetaceans, species by species; also see Thewissen et al. (2001).

26. See Burke (1989) and Renous et al. (1989) for *Dermochelys*, Letsrel et al. (1989) for the Fourier analysis, and Ewert (1985) for an overview of turtle embryology.

27. Reisz and Laurin (1991) and Rieppel and de Braga (1996), who described *Owenetta*, regard turtles as diapsids, not primitive reptiles. Rieppel and Reisz (1999) reviewed the origin and early evolution of turtles, demonstrating the high homoplasty index that emerges – which they consider evidence of convergence – when one attempts to relate turtles to marine diapsids; Rieppel (2001) reviews turtles as hopeful monsters, emphasizing how novelties need not appear gradually, but rather accumulate through stepwise changes.

28. See M. S. Y. Lee (1993, 1996) for pareiasaurs and the origin of the turtle shell, and Zardoya and Meyer (1998) for the mitochondrial genome study.

29. See Oliver (1951) for skinks, and Zylberberg and Castanet (1985) for the slow worm.

30. See Bauer and Russell (1989), Levrat-Calviac (1986) and Levrat-Calviac and Zylberberg (1986) for osteoderms in geckos.

31. See Hall (1984b, 1998a, 2002c) for overviews of atavisms.

32. See Hall (1994a, 1995c, 1998a, 2003a) for the vexing issue of determining homology.

33. See Gingerich et al (1983, 1984), Thewissen and Hussain (1993), Thewissen et al. (1994) and Thewissen and Fish (1997) for the discovery and descriptions of individual fossils on the evolutionary line that led to whales. For reviews and analyses of the origin of whales see Wyss (1990), O'Leary and Geisler (1999), Luo (2000), Clack (2002a) and Bejder and Hall (2002).

34. See Kollar (1972, 1975) for tissue recombination studies, Stone and Hall (2004) and Hall (2005a) for latent homology, and Viriot et al. (2000), Keränen et al. (1999) and Peterková et al. (2002a,b) for vestigial primordia.

35. The vomers and vomeronasal complex (Jacobson's organ) is just that, complex. Wible and Bhatnagar (1996) used the variable presence of the vomeronasal complex in bats (it is absent from cetaceans and sirenians) to argue for multiple evolution of the complex.

36. See Hanken and Hall (1984, 1988a) for variation and timing of the ossification sequence in *B. orientalis*, the first three bones to form being the exoccipital, parasphenoid and frontoparietal. Osteoderms also are quite common in anurans, developing late in postmetamorphic life and not being present in juveniles. Anuran osteoderms may be vascularized or not. They are not associated with enamel or dentin and so are not equivalent to the dermal scales in caecilians (Ruibal and Shoemaker, 1984).

37. See Horan et al. (1994, 1995) for *Hoxa-4*, *Hoxb-4* and *Hoxd-4* knockout, Ramirez-Solis et al. (1993) for *Hoxb-4*, J. Sharpe et al. (1998) for tissue-specific enhancers, Aubin et al. (1998) for the studies with *Hoxa-5*, Condie and Capecchi (1993) for *Hoxd-3*, Davis and Capecchi (1994) for *Hoxa-11*, Small and Potter (1993) for the double knockout, Wellik and Capecchi (2003) for knockout of *Hox-10* and *Hox-11* genes, and Van den Akker et al. (2001) for redundancy and unique roles of *Hoxb-8*, *c-8* and *d-8* in patterning hind-limb and trunk axial skeletons.

38. See Lohnes et al. (1994) and Mendelsohn et al. (1994) for the receptor knockout studies.

References

A

Aarden, E. M., Burger, E. H., and Nijweide, P. J. (1994). Function of osteocytes in bone. *J. Cell Biochem.* **55**, 287–299.

Aaron, J. E. (1973). Osteocyte types in developing mouse calvarium. *Calcif. Tissue Res.* **12**, 259–279.

Aaron, J. E., and Pautard, F. G. E. (1973). A cell cycle in bone mineralization. In *The Cell Cycle in Development and Differentiation* (M. Balls and F. S. Billett, eds), pp. 325–330. Cambridge University Press, Cambridge.

Abad, V., Uyeda, J. A., Temple, H. T., de Luca, F., and Baron, J. (1999). Determinants of spatial polarity in the growth plate. *Endocrinology* **140**, 958–962.

Abbott, J., and Holtzer, H. (1966). The loss of phenotypic traits by differentiated cells. III. The reversible behaviour of chondrocytes in primary culture. *J. Cell Biol.* **28**, 473–487.

Abbott, J., and Holtzer, H. (1968). The loss of phenotypic traits by differentiated cells. V. The effect of 5-bromodeoxyuridine on cloned chondrocytes. *Proc. Natl Acad. Sci. U.S.A.* **59**, 1144–1151.

Abbott, J., Mayne, R., and Holtzer, H. (1972). Inhibition of cartilage development in organ cultures of chick somites by the thymidine analog, 5-bromo-2'-deoxyuridine. *Devel. Biol.* **28**, 430–441.

Abbott, U. K. (1967). Avian developmental genetics. In *Methods in Developmental Biology* (F. H. Wilt and N. K. Wessells, eds), pp. 13–52. Crowell-Collier, New York, NY.

Abbott, U. K. (1975). Genetic approaches to studies of tissue interactions. *Genetics Lectures*, Volume IV, pp. 69–84. Oregon State University Press, Corvalis, OR.

Abbott, U. K., Taylor, L. W., and Ablanalp, H. (1960). Studies with *talpid²*, an embryonic lethal in the fowl. *J. Hered.* **51**, 195–202.

Abdalla, A. B. E., and Harrison, R. G. (1966). Observations on the reaction of tubular bone to venous stasis. *J. Anat.* **100**, 627–638.

Abdalla, O. (1979). Ossification and mineralization in the tendons of the chicken (*Gallus domesticus*). *J. Anat.* **129**, 351–360.

Abdennagy, B., Hott, M., and Marie, P. J. (1992). Effects of platelet-derived growth factor on human and mouse osteoblastic cells isolated from the trabecular bone surface. *Cell Biol. Intern. Rep.* **16**, 235–247.

Abdin, M., and Friedenstein, A. Y. (1972). Electron microscopic study of bone induction by the transitional epithelium of the bladder in guinea pigs. *Clin. Orthop. Rel. Res.* **82**, 182–194.

Abe, K., Kanno, T., Kitao, K., and Schneider, G. B. (1984). Sex differences in bone resorption: A scanning electron microscopic study of mouse parietal bones. *Archs Histol. Japon.* **47**, 429–440.

Abe, K., Kanno, T., and Schneider, G. B. (1983). Surface structures and osteoclasts of mouse parietal bones: A light and scanning electron microscopic study. *Archs Histol. Japon.* **46**, 663–676.

Abe, K., Takata, M., and Kawamura, Y. (1973). A study on inhibition of masseteric a-motor fibre discharges by mechanical stimulation of the temporomandibular joint in the cat. *Archs Oral Biol.* **18**, 301–304.

Able, K. W., Markle, D. F., and Fahay, M. P. (1984). Cyclopteridae: Development. In *Ontogeny and Systematics of Fishes* (H. G. Moser, ed.), pp. 428–437, Special Publication Number 1, Amer. Soc. Ichthyologists and Herpetologists, Lawrence, KS.

Abrahamsohn, P. A., Lash, J. W., Kosher, R. A., and Minor, R. R. (1975). The ubiquitous occurrence of chondroitin sulfates in chick embryos. *J. Exp. Zool.* **194**, 511–518.

Abzhanov, A., Tzahor, E., Lassar, A. B., and Tabin, C. J. (2003). Dissimilar regulation of cell differentiation in mesencephalic (cranial) and sacral (trunk) neural crest cells in vitro. *Development* **130**, 4567–4579.

Acampora, D., Merlo, G. R., Paleari, L., Zerega, B., Postiglione, M. P., Mantero, S., Bober, E., Barbieri, O., Simeone, A., and Levi, G. (1999). Craniofacial, vestibular and bone defects in mice lacking the *Distal-less-related gene Dlx5*. *Development* **126**, 3795–3809.

Ackermann, R. R., and Krovitz, G. E. (2002). Common patterns of facial ontogeny in the hominoid lineage. *Anat. Rec.* **269**, 142–147.

Adab, K., Sayne. J. R., Carlson, D. S., and Opperman, L. A. (2002). Tfg-β1, Tfg-β2, Tfg-β3 and Msx2 expression is elevated during frontonasal suture morphogenesis and during active postnatal facial growth. *J. Orthod. Craniofacial Res.* **5**, 227–237.

Adams, A. E. (1924). An experimental study of the development of the mouth in the amphibian embryo. *J. Exp. Zool.* **40**, 311–380.

Adams, D., and Harkness, E. M. (1972). Histological and radiographic study of the spheno-occipital synchondrosis in cynomolgus monkeys, *Macaque irus. Anat. Rec.* **172**, 127–136.

Adams, E. (1978). Invertebrate collagens. *Science* **202**, 591–598.

Adams, J. C., and Tucker, R. P. (2000). The thrombospondin type 1 repeat (TSR) superfamily: diverse proteins with related roles in neuronal development. *Devel. Dynam.* **218**, 280–299.

Adams, R. A. (1992a). Stages of development and sequence of bone formation in the little brown bat, *Myotis lucifugus. J. Mammal.* **73**, 160–167.

Adams, R. A. (1992b). Comparative skeletogenesis of the forearm of the little brown bat (*Myotis lucifugus*) and the Norway rat (*Rattus norvegicus*). *J. Morphol.* **214**, 251–260.

Adams, R. A. (1998). Evolutionary implications of developmental and functional integration in bat wings. *J. Zool.* **246**, 165–174.

Adams, S. L., Boettiger, D., Focht, R. J., Holtzer, H., and Pacifici, M. (1982). Regulation of the synthesis of extracellular matrix components in chondroblasts transformed by a temperature-sensitive mutant of *Rous sarcoma* virus. *Cell* **30**, 373–384.

Adams, S. L., Pacifici, M., Boettiger, D., and Pallante, K. M. (1987). Modulation of fibronectin gene expression in chondrocytes by viral transformation and substrate attachment. *J. Cell Biol.* **105**, 483–488.

Adams, S. L., Pallante, K. M., and Pacifici, M. (1989). Effects of cell shape on the type X collagen gene expression in hypertrophic chondrocytes. *Conn. Tissue Res.* **20**, 223–232.

Adamson, E. D., Deller, M. J., and Warshaw, J. B. (1981). Functional EGF receptors are present on mouse embryo tissues. *Nature* **291**, 656–659.

Addison, W. C. (1978a). Enzyme histochemical properties of kitten osteoclasts in bone imprint preparations. *Histochem. J.* **10**, 645–656.

Addison, W. C. (1978b). β-hydroxybutyrate dehydrogenase activity in human and kitten odontoclasts and kitten osteoclasts. *Histochem. J.* **10**, 731–737.

Addison, W. C. (1978c). The distribution of nuclei in human odontoclasts in whole cell preparations. *Archs Oral Biol.* **23**, 1167–1171.

Addison, W. C. (1979). The distribution of nuclei in imprints of feline osteoclasts. *J. Anat.* **129**, 63–68.

Addison, W. C. (1980). The effect of parathyroid hormone on the numbers of nuclei in feline osteoclasts in vivo. *J. Anat.* **130**, 479–486.

Adelstein, R. S., and Conti, M. A. (1971). The characterization of contractile proteins from platelets and fibroblasts. *Cold Spring Harbor Symp. Quant. Biol.* **37**, 599–606.

Adelstein, R. S., Conti, M. A., Johnson, G. S., Pastan, I., and Pollard, T. D. (1972). Isolation and characterization of myosin from cloned mouse fibroblasts. *Proc. Natl Acad. Sci. U.S.A.* **69**, 3693–3697.

Adelstein, R. S., Pollard, T. D., and Kuehl, W. M. (1971). Isolation and characterization of myosin and two myosin fragments from human blood platelets. *Proc. Natl Acad. Sci. U.S.A.* **68**, 2703–2707.

Adriaens, D., and Verraes, W. (1998). Ontogeny of the osteocranium in the African catfish, *Clarias gariepinus* Burchell (1822) (Siluriformes, Clariidae). Ossification sequence as a response to functional demands. *J. Morphol.* **235**, 183–237.

Adriaens, D., Verraes, W., and Taverne, L. (1997). The cranial lateral-line system in *Clarias gariepinus* (Burchell (1822) (Siluroidei: Clariidae): Morphology and development of canal related bones. *Eur. J. Morphol.* **35**, 181–208.

Agnish, N. D., and Kochhar, D. M. (1976). Direct exposure of post-implantation mouse embryos to 5-bromodeoxyuridine *in vitro* and its effects on subsequent chondrogenesis in the limbs. *J. Embryol. Exp. Morphol.* **36**, 623–638.

Agnish, N. D., and Kochhar, D. M. (1977). The role of somites in the growth and early development of mouse limb buds. *Devel. Biol.* **56**, 174–183.

Agrawal, P., Atre, P. R., and Kulkarni, D. S. (1984). The role of cartilage canals in the ossification of the talus. *Acta Anat.* **119**, 238–240.

Agrawal, P., Atre, P. R., Kulkarni, D. S., and Arole, V. V. (1983). Differential phases of vascularization in temporary cartilages of human fetal limbs. *J. Anat. Soc. India* **32**, 119–123.

Ahlberg, P. E. (1991). Tetrapod or near-tetrapod fossils from the Upper Devonian of Scotland. *Nature* **354**, 298–301.

Ahlberg, P. E. (1992). Coelacanth fins and evolution. *Nature* **358**, 459.

Ahlberg, P. E., Clack, J. A., and Luksevics, E. (1996). Rapid brain case evolution between *Panderichthys* and the earliest tetrapods. *Nature* **381**, 61–64.

Ahlberg, P. E., and Milner, A. R. (1994). The origin and early diversification of tetrapods. *Nature* **368**, 507–514.

Ahlgren, S. C., and Bronner-Fraser, M. (1999). Inhibition of sonic hedgehog *in vivo* results in craniofacial neural crest cell death. *Curr. Biol.* **9**, 1304–1314.

Ahmed, Y. Y. (1966). The effect of a muscle relaxant on the growth and differentiation of skeletal muscles in the chick embryo. *Anat. Rec.* **155**, 133–138.

Ahn, D.-G., Kourakia, M. J., Rohde, L. A., Silver, L. M., and Ho, R. K. (2002). T-box gene *tbx5* is essential for formation of the pectoral limb bud. *Nature* **417**, 754–758.

Ahn, K., Mishina, Y., Hanks, M. C., Behringer, R. R., and Crenshaw, E. B. III. (2001). BMPR-IA signaling is required for

the formation of the apical ectodermal ridge and dorsal–ventral patterning of the limb. *Development* **128**, 4449–4461.

Aho, A. J. (1966). Electron microscopic and histological observations on fracture repair in young and old rats. *Acta Pathol. Microbiol. Scand.* Suppl. **184**, 1–95.

Ahrens, P. B., Solursh, M., and Reiter, R. S. (1977a). Stage-related capacity for limb chondrogenesis in cell culture. *Devel. Biol.* **60**, 69–82.

Ahrens, P. B., Solursh, M., and Meier, S. (1977b). The synthesis and localization of glycosaminoglycans in striated muscle differentiating in cell culture. *J. Exp. Zool.* **202**, 375–388.

Ahrens, P. B., Solursh, M., Reiter, R. S., and Singley, C. T. (1979). Position-related capacity for differentiation of limb mesenchyme in cell culture. *Devel. Biol.* **69**, 436–450.

Åkeson, W. H., Woo, S. L. Y., Amiel, D., and Matthews, J. V. (1974). Biomechanical and biochemical changes in the periarticular connective tissue during contracture development in the immobilized rabbit knee. *Connect. Tissue Res.* **2**, 315–324.

Akimenko, M.-A., and Ekker, M. (1995). Anterior duplication of the *Sonic hedgehog* expression pattern in the pectoral fin buds of zebrafish treated with retinoic acid. *Devel. Biol.* **170**, 243–247.

Akita, K. (1996). The effect of the ectoderm on the dorsoventral pattern of epidermis, muscles and joints in the developing chick leg: a new model. *Anat. Embryol.* **193**, 377–386.

Akiyama, H., Chaboissier, M.-C., Martin, J. F., Schedl, A., and de Crombrugghe, B. (2002). The transcription factor Sox9 has essential roles in successive steps of the chondrocyte differentiation pathway and is required for expression of *Sox5* and *Sox6*. *Genes & Devel.* **16**, 2813–2828.

Alber, R., Sporus, O., Weikert, T., Willbold, E., and Layer, P. G. (1994). Cholinesterases and peanut agglutinin binding related to cell proliferation and axonal growth in embryonic chick limbs. *Anat. Embryol.* **190**, 429–438.

Alberch, P. (1985). Developmental constraints: Why St. Bernards often have an extra digit and poodles never do. *Amer. Nat.* **126**, 430–433.

Alberch, P., and Gale, E. A. (1983). Size dependence during the development of the amphibian foot. Colchicine-induced digital loss and reduction. *J. Embryol. Exp. Morphol.* **76**, 177–197.

Alberch, P., and Gale, E. A. (1985). A developmental analysis of an evolutionary trend: Digital reduction in amphibians. *Evolution* **39**, 8–23.

Alberch, P., and Gale, E. A. (1986). Pathways of cytodifferentiation during the metamorphosis of the epibranchial cartilage in the salamander *Eurycea bislineata*. *Devel. Biol.* **117**, 233–244.

Alberch, P., Lewbart, G. A., and Gale, E. A. (1985). The fate of larval chondrocytes during the metamorphosis of the epibranchial in the salamander, *Eurycea bislineata*. *J. Embryol. Exp. Morphol.* **88**, 71–84.

Alberius, P., and Johnell, O. (1990). Immunohistochemical assessment of cranial suture development in rats. *J. Anat.* **173**, 61–68.

Alberius, P., and Johnell, O. (1991). Repair of intramembranous bone fractures and effects in rats. Immunolocalization of bone and cartilage proteins and proteoglycans. *J. Cranio-Max.-Fac. Surg.* **19**, 15–20.

Alberius, P., and Skagerberg, G. (1990). Adrenergic innervation of the calvarium of the neonatal rat. Its relationship to the sagittal suture and developing parietal bones. *Anat. Embryol.* **182**, 493–498.

Albertson, R. C., and Kocher, T. D. (2001). Assessing morphological differences in an adaptive trait: a landmark-based morphometric approach. *J. Exp. Zool.* **289**, 385–403.

Albertson, R. C., Streelman, J. T., and Kocher, T. D. (2000). The beak of the fish: genetic basis of adaptive shape differences among Lake Malawi cichlid fishes. Assessing morphological differences in an adaptive trait: a landmark-based morphometric approach. *J. Exp. Zool.* **289**, 385–403.

Albright, J. A., and Brand, R. A. (eds) (1979). *The Scientific Basis of Orthopaedics*. Appleton-Century-Crofts, New York, NY.

Alcock, N. W. (1972). Calcification of cartilage. *Clin. Orthop. Rel. Res.* **86**, 287–311.

Alcobendas, M., and Baud, C.-A. (1988). Halos périlacunaires de déminéralisation de l'os cortical chez *Vipera aspis* (L) (Reptilia, Ophidia) dans diverses conditions physiologiques. *CR. Acad. Sci. Paris, Ser. 3*, **307**, 177–182.

Alcobendas, M., and Castanet, J. (1985). Variation du degré de minéralisation osseuse au cours du cycle annuel chez *Vipera aspis* (L). Ophidia, Viperidae. *CR. Acad. Sci. Paris, Ser. 4*, **301**, 187–190.

Aldridge, K., Marsh, J. L., Govier, D., and Richtsmeier, J. T. (2002). Central nervous system phenotypes in craniosynostosis. *J. Anat.* **201**, 31–39.

Aldridge, R. J., Briggs, D. E. G., Sansom, I. J., and Smith, M. P. (1994). The latest vertebrates are the earliest. *Geology Today* **10**, 141–145.

Aldridge, R. J., Briggs, D. E. G., Smith, M. P., Clarkson, E. N. K., and Clark, N. D. L. (1993). The anatomy of conodonts. *Phil. Trans. R. Soc. B.* **340**, 405–421.

Aldridge, R. J., and Donoghue, P. C. J. (1998). Conodonts: A sister group to hagfishes? In *The Biology of Hagfishes* (J. M. Jørgensen, J. P. Lomholt, R. E. Weber, and H. Malte, eds), pp. 15–31. Chapman & Hall, London.

Aldridge, R. J., and Purnell, M. A. (1996). The conodont controversies. *Trends Ecol. Evol.* **11**, 463–468.

Al-Hilli, F., and Wright, E. A. (1983). The short term effects of a supra-lethal dose of irradiation and changes in the environmental temperature on the growth of tail bones of the mouse. *Brit. J. Exp. Pathol.* **64**, 684–692.

Alhopuro, S. (1978). Premature fusion of facial sutures with free periosteal grafts. *Scand. J. Plast. Reconst. Surg.* Suppl. 17, 1–68.

Ali, M. Y., and Lindsey, C. C. (1974). Heritable and temperature-induced meristic variation in the medaka, *Oryzias latipes*. *Can. J. Zool.* **52**, 959–976.

Ali, N.-N., Jones, S. J., and Boyde, A. (1984). Monocyte-enriched cells on calcified tissues. *Anat. Embryol.* **170**, 169–175.

Ali, S. Y. (1983). Calcification of cartilage. In *Cartilage, Volume 1: Structure, Function and Biochemistry* (B. K. Hall, ed.), pp. 343–378. Academic Press, New York, NY.

Ali, S. Y. (1985). Apatite-type crystal deposition in arthritic cartilage. *SEM* **1985**, 1555–1566.

Ali, S. Y. (1987). Mechanism of calcification in cartilage and bone. *Bone* **4**, P8–P21.

Ali, S. Y., Elves, M. W., and Leaback, D. H. (eds) (1974). *Normal and Osteoarthritic Articular Cartilage*. Institute of Orthopaedics, University of London, London.

Ali, S. Y., and Evans, L. (1973a). Enzymic degradation of cartilage in osteoarthritis. *Fed. Proc., Fed. Amer. Soc. Exp. Biol.* **32**, 1494–1498.

Ali, S. Y., and Evans, L. (1973b). The uptake of [^{45}Ca] calcium ions by matrix vesicles isolated from calcifying cartilage. *Biochem. J.* **134**, 647–650.

Ali, S. Y., Sajdera, S. W., and Anderson, H. C. (1970). Isolation and characterization of calcifying matrix vesicles from epiphyseal cartilage. *Proc. Natl Acad. Sci. U.S.A.* **67**, 1513–1520.

Alini, M., Kofsky, Y., Wu. W., Pidoux, I., and Poole, A. R. (1996). In serum-free culture thyroid hormone can induce full expression of chondrocyte hypertrophy leading to matrix calcification. *J. Bone Min. Res.* **11**, 105–113.

Alini, M., Marriott, A., Chen, T., Abe, S., and Poole, A. R. (1996a). A novel angiogenic molecule produced at the time of chondrocyte hypertrophy during endochondral bone formation. *Devel. Biol.* **176**, 124–132.

Ali-Khan, S. E., and Hales, B. F. (2003). Caspase-3 mediates retinoid-induced apoptosis in the organogenesis-stage mouse limb. *Birth Defects Res. (Part A)* **67**, 848–860.

Aliverti, V., Bonanomi, L., Giavini, E., Leone, V. G., and Mariani, L. (1979). The extent of fetal ossification as an index of delayed development in teratogenic studies on the rat. *Teratology* **20**, 237–242.

Allan, D., Houle, M., Bouchard, N., Meyer, B. I., Gruss, P., and Lohnes, D. (2001). Rar$_\gamma$ and CDX1 interaction in vertebral patterning. *Devel. Biol.* **240**, 46–60.

Allen, F., Tickle, C., and Warner, A. (1990). The role of gap junctions in patterning of the chick limb bud. *Development* **108**, 623–634.

Allen, T. D., Testa, N.-G., Suda, T., Schor, S. L., Onions, D., Jarrett, O., and Boyde, A. (1981). The production of putative osteoclasts in tissue culture – ultrastructure, formation and behavior. *SEM* **1981**, 347–354.

Allen, T. D., Testa, N.-G., Suda, T., Schor, S. L., Onions, D., Jarrett, O., and Boyde, A. (1984). The production of putative osteoclasts in tissue culture – ultrastructure, formation and behavior. In *Scanning Electron Microscopy of Cells in Culture* (P. B. Bell, ed.), pp. 275–282. SEM Inc., Chicago, IL.

Allenspach, A. L., and Barbiarz, B. S. (1975). Intramitochondrial binding of ruthenium red in degenerating chondroblasts. *J: Ultrastruct. Res.* **51**, 348–353.

Allin, E. F. (1975). Evolution of the mammalian middle ear. *J. Morphol.* **147**, 403–438.

Alluchon-Gérard, M.-J. (1982). Influence of thyroidectomy and PTU treatment on cartilage ultrastructure in the embryo and very young dogfish (*Scyllium canicula*, Chondrichthyes). *Arch. Anat. Microsc. Morphol. Exp.* **71**, 51–70.

Almedia, M. J., Milet, C., Peduzzi, J., Pereira, L., Haigle, J., Barthélemy, M., and Lopez, E. (2000). Effect of water-soluble matrix fraction extracted from the nacre of *Pinctada maxima* on the alkaline phosphatase activity of cultured fibroblasts. *J. Exp. Zool. (Mol. Devel. Evol.)* **288**, 327–334.

Alt, K. W., Jeunnesse, C., Buitrago-Tellez, C. H., Wächter, R., Boes, E., and Pichler, S. L. (1997). Evidence for stone age cranial surgery. *Nature* **387**, 360.

Altabef, M., Clarke, J. D. W., and Tickle, C. (1997). Dorsoventral ectodermal compartments and origin of apical ectodermal ridge in developing chick limb. *Development* **124**, 4547–4556.

Altizer, A. M., Stewart, S. G., Albertson, B. K., and Borgens, R. B. (2002). Skin flaps inhibit both the current of injury at the amputation surface and regeneration of that limb in newts. *J. Exp. Zool.* **293**, 467–477.

Altman, A. J., Bandelin, J. G., Dominguez, J. H., and Mundy, G. R. (1978). Differentiation of isolated calvarial cells into a mature heterogeneous bone cell population in culture. *Metab. Bone Dis. Rel. Res.* **1**, 75–79.

Altmann, K. (1964). *Zur Kausalen Histogenese des Knorpels. W. Roux's Theorie und die Experimentelle Wirklichkeit.* Springer-Verlag, Berlin.

Alvarez, J., Horton, J., Sohn, P., and Serra, R. (2001). The perichondrium plays an important role in mediating the effects of TGFβ-1 on endochondral bone formation. *Devel. Dynam.* **221**, 311–321.

Alvarez, J., Sohn, P., Zeng, X., Doetschman, T., Robbins, D. J., and Serra, R. (2002). TGFβ2 mediates the effects of hedgehog on hypertrophic differentiation and PTHrP expression. *Development* **129**, 1913–1924.

Amar, S., Sires, B., Clohisy, J., and Veis, A. (1991). The isolation and partial characterization of a rat incisor dentin matrix polypeptide with *in vitro* chondrogenic activity. *J. Biol. Chem.* b266, 8609–8618.

Ambrosi, G., Camosso, M. E., and Roncali, L. (1973). Morphological data regarding conjunctival papillae and scleral ossicles in chick embryos. *Boll. Soc. Ital. Biol. Sper.* **49**, 135–140.

Amedee, J., Bareille, R., Rouais, F., Cunningham, N., Reddi, H., and Harmand, M.-F. (1994). Osteogenin (bone morphogenetic protein-3) inhibits proliferation and stimulates differentiation of osteoprogenitors in human bone marrow. *Differentiation* **58**, 157–164.

Amitani, K., and Nakata, Y. (1975a). Establishment and alkaline phosphatase activity of clonal cell lines of murine osteosarcomas – preliminary study. *Clin. Orthop. Rel. Res.* **113**, 164–167.

Amitani, K., and Nakata, Y. (1975b). Studies on a factor responsible for new bone formation from osteosarcoma in mice. *Calcif. Tissue Res.* **17**, 139–150.

Amitani, K., and Nakata, Y. (1977). Characteristics of osteosarcoma cells in culture. *Clin. Orthop. Rel. Res.* **122**, 315–324.

Amitani, K., Nakata, Y., and Stevens, J. (1974). Bone induction by lyophilized osteosarcoma in mice. *Calcif. Tissue Res.* **16**, 305–314.

Amling, M., Takeda, S., and Karsenty, G. (2000). A neuro (endo)crine regulation of bone remodeling. *BioEssays* **22**, 970–975.

Amprino, R. (1955a). Autoradiographic research on the S^{35}-sulphate metabolism in cartilage and bone differentiation and growth. *Acta Anat.* **24**, 121–163.

Amprino, R. (1955b). Distribution of S^{35}-sodium sulfate in early chick embryos. *Experientia* **11**, 19–21.

Amprino, R. (1955c). On the incorporation of radiosulfate in the cartilage. *Experientia* **11**, 65–67.

Amprino, R. (1956). Uptake of S^{35} in the differentiation and growth of cartilages and bones. *Bone Struct. Metab., Ciba Found. Symp.* 1955, pp. 89–100.

Amprino, R. (1964). Relations morphogénétiques entre l'ectoderme et le mésoderm dans le développement des Membres du poulet. *Arch. Biol.* **75** (Suppl), 1047–1080.

Amprino, R. (1965). Aspects of limb morphogenesis in the chicken. In *Organogenesis* (R. L. DeHaan and H. Ursprung, eds), pp. 255–282. Holt, New York, NY.

Amprino, R. (1968). Developmental changes of the hand skeleton induced by grafting leg mesoderm to the wing bud in the chicken. *W. Roux. Archiv. EntwicklMech. Org.* **161**, 1–22.

Amprino, R. (1974). Cell density as a factor of negative control of tissue proliferation in the early development of the chick embryo limb. *Rech. Biol. Contemp.*, pp. 245–250.

Amprino, R. (1975a). Developmental ability of the apical mesoderm of the chick embryo limb bud in sectors deprived of the ectodermal ridge. *Nova Acta Leopold.* [N. S.] **41**, 235–270.

Amprino, R. (1975b). Observations sur les relations ecto-mésodermiques dans la morphogenèse du bourgeon des membres. *Arch. Anat., Histol., Embryol.* **58**, 29–40.

Amprino, R. (1976). On the topography of the presumptive skeletal mesenchyme in the early wing bud of chick embryos. *Arch. Biol.* **87**, 1–41.

Amprino, R. (1977a). Further observations on the site of bone prospective areas in the chick embryo wing bud. *Acta Anat.* **98**, 295–312.

Amprino, R. (1977b). Experimental study on the morphogenetic relationships between ectoderm and mesoderm in the developing chick embryo limb bud. *Acta Embryol. Exp.* **1**, 51–70.

Amprino, R. (1978). Relationship between ectoderm and skeletal morphogenesis in the chick embryo limb bud. *Anat. Embryol.* **153**, 305–320.

Amprino, R. (1982). A tribute to honor his 70th birthday, by G. A. Marotti. *Calcif. Tissue Int.* **34**, 515–518.

Amprino, R. (1984). The development of the vertebrate limb. *Clin. Orthop. Rel. Res.* **188**, 263–284.

Amprino, R. (1985). The influence of stress and strain in the early development of shaft bones. An experimental study of the chick embryo tibia. *Anat. Embryol.* **172**, 49–60.

Amprino, R., and Aiello-Malmberg, P. (1971). Digit formation from the prospective stylozeugopod of the hind-limb in the chick. *Acta Anat.* **80**, 183–203.

Amprino, R., and Aiello-Malmberg, P. (1973). Observations on the avascular cultures of the chick embryo limb bud. *Acta Embryol. Exp.* **15**, 3–28.

Amprino, R., and Ambrosi, A. (1973). Experimental analysis of chick embryo limb bud growth. *Arch. Biol.* **84**, 35–86.

Amprino, R., and Amprino-Bonetti, D. (1967a). Relations morphogénétiques entre ecto-et mésoderme au cours du développement du bourgeon des membres du poulet. *Bull. Assoc. Anat.* **52**, 134–140.

Amprino, R., and Amprino-Bonetti, D. (1967b). Experimental observations in the development of ectoderm-free mesoderm of the limb bud in chick embryos. *Nature* **214**, 826–827.

Amprino, R., Amprino-Bonetti, D., and Ambrosi, G. (1968). Resultants de transplantations hétérotopiques de la queue chez l'embryon de poulet. *Bull. L'Assoc. Anat.* **53**, 409–414.

Amprino, R., Amprino-Bonetti, D., and Ambrosi, G. (1969). Observations on the developmental relations between ectoderm and mesoderm of the chick embryo tail. *Acta Anat. Suppl.* **56**, 1–26.

Amprino, R., and Bonetti, D. (1964). Effect of the implantation site on the development of grafted limb bud mesoderm in chick embryos. *Nature* **204**, 298.

Amprino, R., and Camosso, M. E. (1959). Observations sur les duplications experimentales de la partie distale de l'ebauche de l'aile chez l'embryon de poulet. *Arch. Anat. Microsc. Morphol. Exp.* **48**, 261–306.

Amprino, R., and Camosso, M. E. (1963). Effects of exchanging the AP-reoriented apex between wing & hind-limb buds. *Acta Embryol. Morphol. Exp.* **6**, 241–259.

Amsel, S., and Dell, E. S. (1971). The radiosensitivity of the bone-forming process of heterotopically-grafted rat bone marrow. *Int. J. Radiat. Biol.* **20**, 119–127.

Amsel, S., and Dell, E. S. (1972). Bone formation by hemopoietic tissue: Separation of preosteoblast from hemopoietic stem cell function in the rat. *Blood* **39**, 267–273.

Amsel, S., Maniatis, A., Tavassoli, M., and Crosby, W. H. (1969). The significance of intramedullary cancellous bone formation in the repair of bone marrow tissue. *Anat. Rec.* **164**, 101–112.

Amtmann, E., and Oyama, J. (1973). Changes in functional construction of bone in rats under conditions of simulated increased gravity. *Z. Anat. Entwicklungs.* **139**, 307–318.

An, Y. H. (ed.) (2003). *Orthopaedic Issues in Osteoporosis*. CRC Press, Boca Raton, FL.

An, Y. H., and Draughn, R. A. (eds) (2000). *Mechanical Testing of Bone and the Bone-Implant Interface*. CRC Press, Boca Raton, FL.

An, Y. H., and Friedman, R. J. (eds) (1998). *Animal Models in Orthopaedic Research*. CRC Press, Boca Raton, FL.

Andersen, H. (1963). Histochemistry and development of the human shoulder and acromioclavicular joints with particular reference to the early development of the clavicle. *Acta Anat.* **55**, 124–165.

Anderson, C., and Danylchuk, K. D. (1978). Bone-remodeling rates of the Beagle: A comparison between different sites on the same rib. *Amer. J. Vet. Res.* **39**, 1763–1765.

Anderson, C. E., and Parker, J. (1966). Invasion and resorption in endochondral ossification: An electron microscopic study. *J. Bone Joint Surg. Amer. Vol.* **48**, 899–914.

Anderson, H. C. (1967). Electron microscopic studies of induced cartilage development and calcification. *J. Cell Biol.* **35**, 81–102.

Anderson, H. C. (1969). Vesicles associated with calcification in the matrix of epiphyseal cartilage. *J. Cell Biol.* **41**, 59–72.

Anderson, H. C. (1976a). Osteogenetic epithelial–mesenchymal cell interactions. *Clin. Orthop. Rel. Res.* **119**, 211–224.

Anderson, H. C. (1976b). Glycoproteins of the connective tissue matrix. *Int. Rev. Connect. Tissue Res.* **7**, 251–322.

Anderson, H. C. (1985). Matrix vesicle calcification: Review and update. In *Bone and Mineral Research/3* (W. A. Peck, ed.), pp. 109–149. Elsevier Science Publishers B. V., Amsterdam.

Anderson, H. C. (1990). The role of cells versus matrix in bone induction. *Conn. Tissue Res.* **24**, 3–12.

Anderson, H. C., Cecil, R., and Sajdera, S. W. (1975). Calcification of rachitic rat cartilage *in vitro* by extracellular matrix vesicles. *Amer. J. Pathol.* **79**, 237–254.

Anderson, H. C., and Coulter, P. R. (1967). Bone formation induced in mouse thigh by cultured human cells. *J. Cell Biol.* **30**, 165–177.

Anderson, H. C., Hodges, P. T., Aguilera, X. M., Missana, L., and Moylan, P. E. (2000). Bone morphogenetic protein (BMP) localization in developing human and rat growth plate, metaphysis, epiphysis, and articular cartilage. *J. Histochem. Cytochem.* **48**, 1493–1502.

Anderson, H. C., Marker, P. C., and Fogh, J. (1964). Formation of tumors containing bone after intramuscular injection of transformed human amniotic cells (FL) into cortisone-treated mice. *Amer. J. Pathol.* **54**, 507–519.

Anderson, H. C., Matsuzawa, T., Sajdera, S. W., and Ali, S. Y. (1970). Membranous particles in calcifying cartilage matrix. *Trans. NY. Acad. Sci.* [2] **32**, 619–630.

Anderson, H. C., and Reynolds, J. J. (1973). Pyrophosphate stimulation of calcium uptake into cultured embryonic bones. Fine structure of matrix vesicles and their role in calcification. *Devel. Biol.* **34**, 211–227.

Anderson, H. C., and Sajdera, S. W. (1971). The fine structure of bovine nasal cartilage. Extraction as a technique to study proteoglycans and collagen in cartilage matrix. *J. Cell Biol.* **49**, 650–663.

Anderson, J. C. (1982). Bovine flexor tendon contains aggregating proteoglycan similar to that of cartilage. *Biochem. Biophys. Res. Commun.* **107**, 1390–1394.

Anderson, L. G., Cook, A. J., Coccaro, P. J., Coro, C. J., and Bosma, J. F. (1972). Familial Osteodysplasia. *J. Amer. Med. Assoc.* **220**, 1687–1693.

Anderson, R. E., Schrarer, H., and Gay, C. V. (1982). Ultrastructural immunocytochemical localization of carbonic anhydrase in normal and calcitonin-treated chick osteoclasts. *Anat. Rec.* **204**, 9–20.

Andrades, J. A., Santamaría, J. A., Nimni, M. E., and Becerra, J. (2001). Selection and amplification of a bone marrow cell population and its induction to the chondro-osteogenic lineage by rhOP-1: an *in vitro* and *in vivo* study. *Int. J. Devel. Biol.* **45**, 689–693.

Andreshak, J. L., Rabin, S. I., Parwardhan, A. G., and Wezeman, F. H. (1997). Tibial segmental defect repair – chondrogenesis and biomechanical strength modulated by basic fibroblast growth factor. *Anat. Rec.* **248**, 198–204.

Andreucci, R. D., and Blumen, G. (1971). Radioautographic study of *Spheroides testudineus* denticles (checkered puffer). *Acta Anat.* **79**, 76–83.

Andrew, J. G., Hoyland, J., Andrew, S. M., Freemont, A. J., and Marsh, D. (1993a). Demonstration of TGF-β mRNA by *in situ* hybridization in normal human fracture healing. *Calcif. Tissue Int.* **52**, 74–78.

Andrew, J. G., Hoyland, J., Freemont, A. J., and Marsh, D. (1993b). Insulin-like growth factor gene expression in human fracture callus. *Calcif. Tissue Int.* **53**, 97–102.

Andrew, R. C. (1921). A remarkable case of external hind limb in a humpback whale. *Amer. Mus. Novitates* **9**, 1–6.

Andrew, W. (1971). *The Anatomy of Aging in Man and Animals*. Heinemann, London.

Andry, N. (1741). *L'Orthopédie, or l'art de Prévenir et Corriger dans le Enfants les Defformites du Corp*. 2 volumes. Alix, Paris.

Angelov, D. N. (1989). Morphogenesis of rat cranial meninges. A light-microscopic and electron microscopic study. *Cell Tissue Res.* **257**, 207–216.

Anker, G. Ch. (1986). The morphology of joints and ligaments in the head of a generalized *Haplochromis* species: *H. elegans* Trewavas 1933 (Teleostei, Cichlidae). I. *Neth. J. Zool.* **36**, 498–530.

Anker, G. Ch. (1987). The morphology of joints and ligaments in the head of a generalized *Haplochromis* species: *H. elegans* Trewavas 1933 (Teleostei, Cichlidae). II. The jaw apparatus. *Neth. J. Zool.* **37**, 394–427.

Anker, G. Ch. (1989). The morphology of joints and ligaments in the head of a generalized *Haplochromis* species: *H. elegans* Trewavas 1933 (Teleostei, Cichlidae). III. *Neth. J. Zool.* **39**, 1–40.

Antosz, M. E., Bellows, C. G., and Aubin, J. E. (1987). Biphasic effects of epidermal growth factor on bone nodule formation by isolated rat calvarial cells *in vitro*. *J. Bone Min. Res.* **2**, 385–394.

Antosz, M. E., Bellows, C. G., and Aubin, J. E. (1989). Effects of transforming growth factor β and epidermal growth factor on cell proliferation and the formation of bony nodules in isolated fetal rat calvarial cells. *J. Cell Physiol.* **140**, 386–395.

Aono, A., Hazama, M., Notoya, K, Taketomi, S., Yamasaki, H., Tsukuda, R., Sasaki, S., and Fujisawa, Y. (1995). Potent ectopic bone-inducing activity of bone morphogenetic protein-4/7 heterodimer. *Biochem. Biophys. Res. Commun.* **210**, 670–677.

Aono, H., and Ide, H. (1988). A gradient of responsiveness to the growth-promoting activity of ZPA (zone of polarizing activity) in the chick limb bud. *Devel. Biol.* **128**, 136–141.

Appel, T. A. (1987). *The Cuvier-Geoffroy Debate. French Biology in the Decades before Darwin.* Oxford University Press, New York, NY.

Applegate, S. P. (1967). A survey of shark hard parts. In *Sharks, Skates and Rays* (P. W. Gilbert, R. F. Matheson, and D. P. Rall, eds), pp. 37–67. Johns Hopkins University Press, Baltimore, MD.

Apte, S. S. (1988). Application of monoclonal antibody to bromodeoxyuridine to detect chondrocyte proliferation in growth plate cartilage *in vivo*. *Med. Sci. Res.* **16**, 405–406.

Arbustini, E., Jones, M., and Ferrans, V. J. (1983). Formation of cartilage in bioprosthetic cardiac valves implanted in sheep – A morphologic study. *Amer. J. Cardiology* **52**, 632–656.

Archer, C. W., Cottrill, C. P., and Rooney, P. (1984). Cellular aspects of cartilage differentiation and morphogenesis. In *Matrices and Differentiation* (R. B. Kemp and J. R. Hinchliffe, eds), pp. 409–426. Alan R. Liss, New York, NY.

Archer, C. W., Hornbruch, A., and Wolpert, L. (1983). Growth and morphogenesis of the fibula in the chick embryo. *J. Embryol. Exp. Morphol.* **75**, 101–116.

Archer, C. W., Morrison, H., and Pitsillides, A. A. (1994). Cellular aspects of the development of diarthrodial joints and articular cartilage. *J. Anat.* **184**, 447–456.

Archer, C. W., and Ratcliffe, N. A. (1983). The effects of pulsed magnetic fields on chick embryo cartilaginous skeletal rudiments *in vitro*. *J. Exp. Zool.* **225**, 243–256.

Archer, C. W., Rooney, P., and Cottrill, C. P. (1985). Cartilage morphogenesis *in vitro*. *J. Embryol. Exp. Morphol.* **90**, 33–48.

Arey, L. B. (1974). *Developmental Anatomy: A Textbook and Laboratory Manual of Embryology.* 7th Edn. Saunders, Philadelphia, PA.

Argraves, W. S., and Gehlsen, K. R. (1991). Cellular interactions with fibronectin as a model for redundant binding of cells to other extracellular matrix proteins. *In Vivo* **5**, 489–492.

Argraves, W. S., McKeown-Longo, P. J., and Goetinck, P. F. (1981). Absence of proteoglycan core protein in the cartilages of mutant *nanomelia*. *FEBS Letts* **131**, 265–268.

Armstrong, J. B. (1985). The axolotl mutants. *Devel. Genet.* **6**, 1–26.

Armstrong, L. A., Wright, G. M., and Youson, J. H. (1987). Transformation of mucocartilage to a definitive cartilage during metamorphosis in the sea lamprey, *Petromyzon marinus*. *J. Morphol.* **194**, 1–22.

Armstrong, W. G., and Halstead Tarlo, L. B. (1966). Amino-acid components in fossil calcified tissues. *Nature* **210**, 481–482.

Armstrong, W. G., Halstead, L. B., Reed, F. B., and Wood, L. (1983). Fossil proteins in vertebrate calcified tissues. *Phil. Trans R. Soc. Lond. B.* **301**, 302–343.

Arnold, E. N. (1984). Evolutionary aspects of tail shedding in lizards and their relatives. *J. Nat. Hist.* **19**, 127–169.

Arratia, G. (1991). The caudal skeleton of Jurassic teleosts; a phylogenetic analysis. In *Early Vertebrates and Related Problems in Evolutionary Biology* (M.-M. Chang, L. Hai, and Z. Guo-Rui, eds), pp. 249–340. Science Press, Beijing.

Arratia, G., and Schultze, H.-P. (1990). The urohyal: Development and homology within osteichthyans. *J. Morphol.* **203**, 247–282.

Arratia, G., and Schultze, H.-P. (1991). Palatoquadrate and its ossifications: Development and homology within osteichthyans. *J. Morphol.* **208**, 1–81.

Arratia, G., and Schultze, H.-P. (1992). Reevaluation of the caudal skeleton of certain Actinopterygian fishes. III. Salmonidae. Homologization of caudal skeletal structures. *J. Morphol.* **214**, 187–249.

Arrow, G. H. (1951). *Horned Beetles.* Dr. W. Junk, The Hague.

Arsanto, J.-P., Diano, M., Thouverny, Y., Thiery, J.-P., and Levi, G. (1990). Patterns of tenascin expression during tail regeneration of the amphibian urodele *Pleurodeles waltl*. *Development* **109**, 177–188.

Arsenis, C., Eisenstein, R., Soble, L. W., and Kuettner, K. E. (1971). Enzyme activities in chick embryonic cartilage. Their subcellular distribution in isolated chondrocytes. *J. Cell Biol.* **49**, 459–467.

Arsenis, C., and Huang, S.-M. (1977). Distribution and immunologic cross reactivity of a phosphohydrolytic activity of calcifying cartilage and metaphyseal bone. *Differentiation* **8**, 183–190.

Arsenault, A. L., and Ottensmeyer, F. P. (1984). Visualization of early intramembranous ossification by electron microscopic and spectroscopic imaging. *J. Cell Biol.* **98**, 911–921.

Arvy, L. (1976). Some critical remarks on the subject of the cetacean 'girdles.' In *Investigations on Cetacea* (G. Pilleri, ed.), Volume 1, pp. 179–186. Brain Anatomy Institute, Berne.

Arvy, L. (1979). The abdominal bones of cetaceans. In *Investigations on Cetacea* (G. Pilleri, ed.), Volume 10, pp. 215–227. Brain Anatomy Institute, Berne.

Arzate, H., Chimal-Monroy, J., Hernandez-Lagunas, L., and Diaz de Leon, L. (1996). Human cementum protein extract promotes chondrogenesis and mineralization in mesenchymal cells. *J. Periodont. Res.* **31**, 144–148.

Asahina, I., Sampath, T. K., Nishimura, I., and Hauschka, P. V. (1993). Human osteogenin protein-1 induces both chondroblastic and osteoblastic differentiation of osteoprogenitor cells derived from newborn rat calvaria. *J. Cell Biol.* **123**, 921–933.

Asahina, I., Sampath, T. K., and Hauschka, P. V. (1996). Human osteogenin protein-1 induces chondroblastic, osteoblastic, and/or adipocytic differentiation of clonal murine target cells. *Exp. Cell Res.* **222**, 38–47.

Asakura, A., Komaki, M., and Rudnicki, M. A. (2001). Muscle satellite cells are multipotential stem cells that exhibit myogenic, osteogenic, and adipogenic differentiation. *Differentiation* **68**, 245–253.

Ascenzi, A., Bonucci, E., and Bocciarelli, D. S. (1965). An electron microscopy study of osteon calcification. *J. Ultrastruct. Res.* **12**, 287–303.

Ascenzi, A., Bonucci, E., and Bocciarelli, D. S. (1967). An electron microscopy study of primary periosteal bone. *J. Ultrastruct. Res.* **18**, 605–618.

Ascenzi, A., Bonucci, E., and de Bernard, B. (eds) (1981). *Matrix Vesicles. Proceedings of the Third International Conference on Matrix Vesicles.* Wichtig Editore SRL, Milan.

Ash, P., and Francis, M. J. O. (1975). Response of isolated rabbit articular and epiphyseal chondrocytes to rat liver somatomedin. *J. Endocrinol.* **66**, 71–78.

Ash, P., Loutit, J. F., and Townsend, K. M. S. (1980). Osteoclasts derived from hematopoietic stem cells. *Nature* **283**, 669–670.

Ash, P., Loutit, J. F., and Townsend, K. M. S. (1981). Osteoclasts derive from hematopoietic stem cells according to marker, giant lysosomes of Beige mice. *Clin. Orthop. Rel. Res.* **155**, 249–258.

Ashhurst, D. E. (1986). The influence of mechanical conditions on the healing of experimental fractures in the rabbit: A microscopical study. *Phil. Trans R. Soc. London B.* **313**, 271–302.

Ashhurst, D. E. (2004). The cartilaginous skeleton of an elasmobranch fish does not heal. *Matrix Biol.* **23**, 15–22.

Ashhurst, D. E., Hogg, J., and Perren, S. M. (1982). A method for making reproducible experimental fractures of the rabbit tibia. *Injury* **14**, 236–242.

Ashique, A. M., Fu, K., and Richman, J. M. (2002). Signaling via type IA and type IB bone morphogenetic protein receptors (BMPR) regulates intramembranous bone formation, chondrogenesis and feather formation in the chicken embryo. *Int. J. Devel. Biol.* **46**, 243–253.

Ashley, D. J. B. (1970). Bone metaplasia in trachea and bronchi. *J. Pathol.* **102**, 186–188.

Ashley-Ross, M. A. (2002). Mechanical properties of the dorsal fin muscle of seahorse (*Hippocampus*) and pipefish (*Syngnathus*). *J. Exp. Zool.* **293**, 561–577.

Ashton, B. A., Allen, T. D., Howlett, C. R., Eagleson, C. C., Hattori, A., and Owen, M. (1980). Formation of bone and cartilage by marrow stromal cells in diffusion chambers *in vivo*. *Clin. Orthop. Rel. Res.* **151**, 294–307.

Ashton, B. A., Eagleson, C. C., Bab, I., and Owen, M. (1984). Distribution of fibroblast colony-forming cells in rabbit bone marrow and assay of their osteogenic potential by an *in vivo* diffusion chambers method. *Calcif. Tissue Res.* **36**, 83–86.

Ashton, I. K., and Matheson, J. A. (1979). Change in response with age of human articular cartilage to plasma somatomedin activity. *Calcif. Tissue Int.* **29**, 89–94.

Asling, C. W. (1976). Fibrous metaplasia in hyaline cartilage. *J. Dent. Res.* **55**, B111.

Asmus, S. E., Parsons, S., and Landis, S. C. (2000). Developmental changes in the transmitter properties of sympathetic neurons that innervate the periosteum. *J. Neurosci.* **20**, 1495–1504.

Aspenberg, P., and Lohmander, L. S. (1989). Fibroblast growth factor stimulates bone formation – bone induction studied in rats. *Acta Orthop. Scand.* **60**, 473–476.

Atallah, J., Dworkin, I., Cheung, U., Greene, A., Ing, B., Leung, L., and Larsen, E. (2004). The environmental and genetic regulation of *obake* expressivity: morphogenetic fields as evolvable systems. *Evol. & Devel.* **6**, 114–122.

Atchley, W. R., and Hall, B. K. (1991). A model for development and evolution of complex morphological structures. *Biol. Rev. Camb. Philos. Soc.* **66**, 101–157.

Athenstaedt, H. (1969). Permanent electric polarization and pyroelectric behaviour of the vertebrate skeleton. IV. The cranial bones of man. *Z. Zellforsch. Mikrosk. Anat.* **97**, 537–548.

Athenstaedt, H. (1970). Permanent longitudinal electric polarization and pyroelectric behaviour of collagenous structures and nervous tissue in man and other vertebrates, *Nature* **228**, 830–834.

Athenstaedt, H. (1974). Pyroelectric and piezoelectric properties of vertebrates. *Ann. NY. Acad. Sci.* **238**, 68–94.

Atkinson, B. L., Fantle, K. S., Benedict, J. J., Huffer, W. E., and Gutierrez-Hartmann, A. (1997). Combination of osteoinductive bone proteins differentiates mesenchymal C3H10T1/2 cells specifically to the cartilage lineage. *J. Cell Biochem.* **65**, 325–339.

Atkinson, P. J., and Hallsworth, A. S. (1983). The changing pore structure of aging human mandibular bone. *Gerodontology* **2**, 57–66.

Atkinson, P. J., Powell, K., and Woodhead, C. (1977). Cortical structure of the pig mandible after the insertion of metallic implants into alveolar bone. *Archs Oral Biol.* **22**, 383–392.

Attinger, E., and Parakkal, P. (1977). *The Relevance of Biomedical Engineering to Dentistry*. DHEW Publ. (NIH) 77–1198. DHEW, Washington, DC.

Atwood, M. (2003). *Oryx and Crake*. Seal Books, Canada.

Atzenkern, J. (1923). Zur Entwicklung der Os Cornu der Carvicornier. *Anat. Anz.* **57**, 125–130.

Aubin, D. J. St., Stinson, R. H., and Geraci, J. R. (1984). Aspects of the structure and composition of baleen, and some effects of exposure to petroleum hydrocarbons. *Can. J. Zool.* **62**, 193–198.

Aubin, J. E., Heersche, J. N. M., Merrilees, M. J., and Sodek, J. (1982). Isolation of bone cell clones with differences in growth, hormone responses, and extracellular matrix production. *J. Cell Biol.* **92**, 452–461.

Aubin, J., Lemieux, M., Tremblay, M., Behringer, R. R., and Jeannotte, L. (1998). Transcriptional interferences at the *Hoxa4/Hoxa5* locus: Importance of correct *Hoxa5* expression for the proper specification of the axial skeleton. *Devel. Dynam.* **212**, 141–156.

Augustine, K. S., Rossi, R. M., Silbiger, S. M., Bucay, N., Duryer, D., Marshall, W. S., and Medlock, E. S. (2000). Evidence that the protein tyrosine phosphatase (PC12, Br7, SI) γ (-) isoform modulates chondrogenic patterning and growth. *Int. J. Devel. Biol.* **44**, 361–371.

Aulthouse, A. L., and Solursh, M. (1987). The detection of a precartilage, blastema-specific marker. *Devel. Biol.* **120**, 377–384.

Avery, G., Chow, M., and Holtzer, H. (1956). An experimental analysis of the development of the spinal column. V. Reactivity of chick somites. *J. Exp. Zool.* **132**, 409–426.

Awn, M., Goret-Nicaise, M., and Dhem, A. (1987). Unilateral section of the lateral pterygoid muscle in the growing rat does not alter condylar growth. *Eur. J. Orthod.* **9**, 122–128.

Axhausen, G. (1907). Histologische Untersuchungen bei Knochentransplantationen am Menschen. *Deutsch Z. Chir.* **91**, 388.

Axhausen, G. (1909). Die histologischen Grundlagen der freien Osteoplastik auf Grund von Tierversuchen. *Langenbecks Arch. Klin. Chir.* **88**, 23–145.

Ayad, S., Boot-Handford, R. P., Humphries, M. J., Kadler, K. E., and Shuttleworth, C. A. (1994). *The Extracellular Matrix FactsBook*. Academic Press, London.

Aydelotte, M. B., and Kochhar, D. M. (1972). Development of mouse limb buds in organ culture: Chondrogenesis in the presence of a proline analog, L-azetidine-2-carboxylic acid. *Devel. Biol.* **28**, 191–201.

Ayers, D. C., Pellegri, V. D., and Evarts, C. M. (1991). Prevention of heterotopic ossification in high-rink patients by radiation therapy. *Clin. Orthop. Rel. Res.* **263**, 87–93.

Aziz, A., and Dhem, A. (1987). Actin and myosin like structure induced in the growth cartilage by acute ischemia. *Calcif. Tissue Int.* **40**, 16–20.

Azuma, M., Kawamata, H., Kasai, Y., Yanagawa, T., Yoshida, H., and Sato, M. (1989). Induction of cells with a chondrocyte-like phenotype by treatment with 1–25-dihydroxyvitamin-D3 in a human salivary acinar cell line. *Cancer Res.* **49**, 5435–5442.

Azzarello, M. Y. (1989a). A comparative study of the developmental osteology of *Sungnathus scovelli* and *Hippocampus zosterae* (Pisces: Syngnathidae) and its phylogenetic implications. 90 pp. Ph.D. Thesis. University of South Florida, St. Petersburg, Fl.

Azzarello, M. Y. (1989b). The pterygoid series in *Hippocampus zosterae* and *Sygnathus scovelli* (Pisces, Syngnathidae). *Copeia* **1989**, 621–628.

B

Bab, I., Ashton, B. A., Gazit, D., Marx, G., Williamson, M. C., and Owen, M. E. (1986). Kinetics and differentiation of marrow stromal cells in diffusion chambers *in vivo*. *J. Cell Sci.* **84**, 139–151.

Bab, I., Ashton, B. A., Syftestad, G. T., and Owen, M. E. (1984a). Assessment of an *in vivo* diffusion chamber method as a quantitative assay for osteogenesis. *Calcif. Tissue Int.* **36**, 77–82.

Bab, I., Howlett, C. R., Ashton, B. A., and Owen, M. E. (1984b). Ultrastructure of bone and cartilage formed *in vivo* in diffusion chambers. *Clin. Orthop. Rel. Res.* **187**, 243–254.

Bachiller, D., Kingensmith, J., Shneyder, N., Tran, U., Anderson, R., Rossant, J., and De Robertis, E. M. (2003). The role of chordin/Bmp signals in mammalian pharyngeal development and DiGeorge syndrome. *Development* **130**, 3567–3578.

Bächner, D., Ahrens, M., Schröder, D., Hoffmann, A., Lauber, J., Betat, N., Steinert, P., Flohe, L., and Gross, G. (1998). BMP-2 downstream targets in mesenchymal development identified by subtractive cloning from recombinant mesenchymal progenitors (C3H10T1/2). *Devel. Dynam.* **213**, 398–411.

Bach-Petersen, S., Kjaer, I., and Fischer-Hansen, B. (1994). Prenatal development of the human osseous temporomandibular region. *J. Craniofac. Genet. Devel. Biol.* **14**, 135–143.

Bada, J. L., Wang, X. S., and Hamilton, H. (1999). Preservation of key biomolecules in the fossil record: Current knowledge and future challenges. *Phil. Trans R. Soc. Lond. B.* **354**, 77–87.

Badi, M. H. (1972a). Ossification in the fibrous growth plate at the proximal end of the tibia in the rat. *J. Anat.* **111**, 201–210.

Badi, M. H. (1972b). Calcification and ossification of fibrocartilage in the attachment of the patellar ligament in the rat. *J. Anat.* **112**, 415–422.

Badi, M. H. (1978). The fibrous growth plate of the rat tibia: Tritiated thymidine autoradiographic study. *J. Anat.* **126**, 547–554.

Badran, A., and Provenza, D. V. (1969). Studies of the growth-inhibitory action of cortisone on chick embryos. *Teratology* **2**, 221–234.

Bagnall, K. M. (1992). The migration and distribution of somite cells after labelling with the carbocyanine dye, DiI: The relationship of the distribution to segmentation in the vertebrate body. *Anat. Embryol.* **185**, 317–324.

Bagnall, K. M., Harris, P. F., and Jones, P. R. M. (1982). A radiographic study of the longitudinal growth of primary ossification centers in limb long bones of the human fetus. *Anat. Rec.* **203**, 293–299.

Bagnall, K. M., Higgins, S. J., and Sanders, E. J. (1988). The contribution made by a single somite to the vertebral column: Experimental evidence in support of resegmentation using the chick–quail chimaera model. *Development* **103**, 69–85.

Bagnall, K. M., Sanders, E. J., and Berdan, R. C. (1992). Communication compartments in the axial mesoderm of the chick embryo. *Anat. Embryol.* **186**, 195–204.

Bahrami, S., Stratmann, U., Wiesmann, H. P., Mokrys, K., Bruckner, P., and Szuwart, T. (2000). Periosteally derived osteoblast-like cells differentiate into chondrocytes in suspension culture in agarose. *Anat. Rec.* **259**, 124–130.

Bailey, D. (1986). Genes that affect morphogenesis of the murine mandible, recombinant inbred strain analysis. *J. Hered.* **77**, 17–25.

Bairati, A., de Biasi, S., Cheli, F., and Oggioni, A. (1987). The head cartilage of cephalopods. I. Architecture and ultrastructure of the extracellular matrix. *Tissue Cell* **19**, 673–686.

Bairati, A., Cheli, F., Oggioni, A., and Vitellaro-Zuccarello, L. (1989). The head cartilage of cephalopods. II. Ultrastructure of isolated native collagen fibrils and of polymeric aggregates obtained *in vitro*: comparison with the cartilage of mammals. *J. Ultra. Mol. Struct. Res.* **102**, 132–138.

Bairati, A., Comazzi, M., and Gioria, M. (1995). A comparative microscopic and ultrastructural study of perichondrial tissue in cartilages of *Octopus vulgaris*. *Tissue Cell* **27**, 515–523.

Bairati, A., Comazzi, M., and Gioria, M. (1996). An ultrastructural study of the perichondrium in cartilages of the chick embryo. *Anat. Embryol.* **194**, 155–167.

Bairati, A., Comazzi, M., Gioria, M., and Rigo, C. (1998). The ultrastructure of chondrocytes in the cartilage of *Sepia officinalis* and *Octopus vulgaris* (Mollusca, Cephalopoda). *Tissue Cell* **30**, 340–351.

Bairati, A., Comazzi, M., Gioria, M., Hartmann, D. J., Leone, F., and Rigo, C. (1999). Immunological study of collagens of the extracellular matrix in cartilage of *Sepia officinalis*. *Eur. J. Histochem.* **43**, 211–225.

Bak, B., and Andreassen, T. T. (1989). The effect of aging on fracture healing in the rat. *Calcif. Tissue Int.* **45**, 292–297.

Bak, B., and Andreassen, T. T. (1991). The effect of growth hormone on fracture healing in old rats. *Bone* **12**, 151–154.

Bak, B., Jørgensen, P. H., and Andreassen, T. T. (1990). Dose response of growth hormone on fracture healing in the rat. *Acta Orthop. Scand.* **61**, 54–57.

Baker, B. E., Becker, R. O., and Spadaro, J. (1974a). A study of electrochemical enhancement of articular cartilage repair. *Clin. Orthop. Rel. Res.* **102**, 251–267.

Baker, B. E., Spadaro, J., Marino, A. W., and Becker, R. O. (1974b). Electrical stimulation of articular cartilage regeneration. *Ann. NY. Acad. Sci.* **238**, 491–499.

Baksi, S. N., and Newbrey, J. W. (1989). Bone metabolism during antler growth in female reindeer. *Calcif. Tissue Int.* **45**, 314–317.

Balcuns, A., Gasseling, M. T., and Saunders, J. W., Jr. (1970). Spatio-temporal distribution of a zone that controls antero-posterior polarity in the limb bud of the chick and other bird embryos. *Amer. Zool.* **10**, 323.

Balduini, C., Brovelli, A., deLuca, G., Galligani, L., and Castellani, A. A. (1973). Uridine diphosphate glucose dehydrogenase from cornea and epiphyseal-plate cartilage. *Biochem. J.* **133**, 243–254.

Balfour, F. M. (1881). On the development of the skeleton of the paired fins of Elasmobranchii, considered in relation to its bearing on the nature of the limbs of the Vertebrata. *Proc. Zool. Soc. Lond.*, **1881**, 656–671, Plates IV and V. (Reprinted in *Studies from the Morphological Laboratory in the University of Cambridge*. Part II. **1882**, 51–68.)

Balinsky, B. I. (1956). Discussion. *Cold Spring Harb. Symp. Quant. Biol.* **21**, 354.

Balinsky, B. I. (1974). Supernumerary limb induction in the anura. *J. Exp. Zool.* **188**, 195–202.

Balinsky, B. I. (1975). *An Introduction to Embryology*. 4th Edn. W. B. Saunders, Philadelphia, PA.

Ball, J., Cook, T. A., Lynch, P. G., and Tumperley, W. R. (1975). Mixed mesenchymal differentiation in meningiomas. *J. Pathol.* **116**, 253–258.

Ballantyne, F. M. (1930). Notes on the development of *Callichthys littoralis*. *Trans. R. Soc. Edin.* **lvi**, 437–466.

Ballard, K. J., and Holt, S. J. (1968). Cytological and cytochemical studies on cell death and digestion in the foetal rat foot: The role of macrophages and hydrolytic enzymes. *J. Cell Sci.* **3**, 245–262.

Balling, R., Mutter, G., Gruss, P., and Kessel, M. (1989). Craniofacial abnormalities induced by ectopic expression of the homeobox gene Hox-1.1 in transgenic mice. *Cell* **58**, 337–347.

Balling, R., Neubüser, A., and Christ, B. (1996). Pax genes and sclerotome development. *Sem. Cell Devel. Biol.* **1**, 129–136.

Ballock, R. T., Heydemann, A., Wakefield, L. M., Flanders, K. C., Roberts, A. B., and Sporn, M. B. (1993). TGF-β1 prevents hypertrophy of epiphyseal chondrocytes: Regulation of gene expression for cartilage matrix proteins and metalloproteases. *Devel. Biol.* **158**, 414–429.

Ballock, R. T., Mita, B. C., Zhou, X. L., Chen, D. H. C., and Mink, L. M. (1999). Expression of thyroid hormone receptor isoforms in rat growth plate cartilage *in vivo*. *J. Bone Min. Res.* **14**, 1550–1556.

Ballock, R. T., and Reddi, A. H. (1994). Thyroxine is the serum factor that regulates morphogenesis of columnar cartilage from isolated chondrocytes in chemically defined medium. *J. Cell Biol.* **136**, 1311–1318.

Ballock, R. T., Zhou, X. L., Mink, L. M., Chen, D. H. C., Mita, B. C., and Stewart, M. C. (2000). Expression of cyclin-dependent kinase inhibitors in epiphyseal chondrocytes induced to terminally differentiate with thyroid hormone. *Endocrinology* **141**, 4552–4557.

Balogh, K. (1964). Further observations on oxidative enzyme activity in decalcified bone and teeth. *J. Histochem. Cytochem.* **12**, 485–486.

Balogh, K., Dudley, R. H., and Cohen, R. B. (1961). Oxidative enzyme activity in skeletal cartilage and bone. *Lab. Invest.* **10**, 839–845.

Balogh, K., and Hajek, J. V. (1965). Oxidative enzymes in intermediary metabolism in healing bone fractures. *Amer. J. Anat.* **116**, 429–448.

Balon, E. K. (ed.). (1985). *Early Life Histories of Fishes: New Developmental, Ecological and Evolutionary Perspectives.* Dr. W. Junk Publishers, Dordrecht.

Balon, E. K. (1989). The epigenetic mechanisms of bifurcation and alternative life-history styles. In *Alternate Life-History Styles of Animals* (M. N. Bruton, ed.), pp. 467–501. Kluwer Academic Publishers, Dordrecht.

Balon, E. K. (2002). Epigenetic processes, when *nature non facit saltum* becomes a myth, and alternative ontogenies a mechanism of evolution. *Envtl. Biol. Fishes* **65**, 1–35.

Balon, E. K. (2004). Alternative ontogenies and evolution: A farewell to Gradualism. In *Environment, Development and Evolution: Toward a Synthesis* (B. K. Hall, R. D. Pearson, and G. B. Müller, eds), pp. 37–66. The MIT Press, Cambridge, MA.

Baltzer, F. von. (1952). Experimentelle Beiträge zur der Homologie. *Experientia* **8**, 285–297.

Bancroft, M., and Bellairs, R. (1976). The development of the notochord in the chick embryo, studied by scanning and transmission electron microscopy. *J. Embryol. Exp. Morphol.* **35**, 383–401.

Bang, G. (1972). Induction of heterotopic bone formation by demineralized dentin in guinea pigs: Antigenicity of the dentin matrix. *J. Oral Pathol.* **1**, 172–185.

Bang, G. (1973a). Induction of heterotopic bone formation by demineralized dentin in rats and guinea pigs. *Scand. J. Dent. Res.* **81**, 230–239.

Bang, G. (1973b). Induction of heterotopic bone formation by demineralized dentin: An experimental model in guinea pigs. *Scand. J. Dent. Res.* **81**, 240–250.

Bang, G. (1973c). Induction of heterotopic bone formation by demineralized dentin in guinea pigs: Relationship to time. *Acta Pathol. Microbiol. Scand.* (A) Suppl. **236**, 60–70.

Bang, G., and Johannessen, J. V. (1972a). The effect of proteolytic enzymes on the induction of heterotopic bone formation by demineralized dentin in guinea pigs. *J. Oral Pathol.* **1**, 221–230.

Bang, G., and Johannessen, J. V. (1972b). The effect of physical treatment on the induction of heterotopic bone formation by demineralized dentin in guinea pigs. *J. Oral Pathol.* **1**, 231–243.

Bang, O. S., Kim, E. J., Chung, J. G., Lee, S. R., Park, T. K., and Kang, S. S. (2000). Association of focal adhesion kinase with fibronectin and paxillin is required for precartilage condensation of chick mesenchymal cells. *Biochem. Biophys. Res. Commun.* **278**, 522–529.

Banks, W. J. (1974). The ossification process of the developing antler in the white-tailed deer (*Odocoileus virginianus*). *Calcif. Tissue Res.* **14**, 257–274.

Banks, W. J., Epling, G. P., Kainer, R. A., and Davis, R. W. (1968a). Antler growth and osteoporosis. I. Morphological and morphometric changes in the costal compacta during the antler growth cycle. *Anat. Rec.* **162**, 387–397.

Banks, W. J., Epling, G. P., Kainer, R. A., and Davis, R. W. (1968b). Antler growth and osteoporosis. II Gravimetric and chemical changes in the costal compacta during the antler growth cycle. *Anat. Rec.* **162**, 399–405.

Barasa, A. (1959). Conséquences de l'ablation de l'éplaississement épidermique apical de l'ebauche de l'aile du poulet (1). *Extrait CR. Assoc. Anat.* **46**, 78–81.

Barasa, A. (1960). Consequenze dell'albazione della cresta ectodermica apicale sullo sviluppo dell'abbozzo dell'ala nell'embrione di pollo. *Riv. Biol.* **52**, 257–292.

Barasa, A. (1962). Reglazione di deficenze nell'abbozzo dell'ala dell'embrione di pollo. *Atti Soc. Ital. Sci. Vet.* **16**, 416–420.

Barasa, A. (1964). On the regulative capacity of the chick embryo limb bud. *Experientia* **20**, 1–3.

Baratz, R. S. (1974). Fish tooth enameloid – an unusual mineralized system. *J. Dent. Res.* **53**, 77.

Bard, D. R., Dickens, M. J., Smith, A. U., and Zarek, J. M. (1972). Isolation of living cells from mature mammalian bone. *Nature* **236**, 314–315.

Bard, J., and Wright, M. O. (1974). The membrane potentials of fibroblasts in different environments. *J. Cell. Physiol.* **84**, 141–146.

Bareggi, R., Grill, V., Zweyer, M., Sandrucci, M. A., Narducci, P., and Forabosco, A. (1994). The growth of long bones in human embryological and fetal upper limbs and its relationship to other developmental patterns. *Anat. Embryol.* **189**, 19–24.

Bareggi, R., Sandrucci, M. A., Baldini, G., Grill, V., Zweyer, M., and Narducci, P. (1995). Mandibular growth rates in human fetal development. *Archs Oral Biol.* **40**, 119–125.

Bargman, G. J., Mackler, B., and Shepard, T. H. (1972). Studies of oxidative energy deficiency. I. Achondroplasia in the rabbit. *Arch. Biochem. Biophys.* **150**, 137–146.

Barker, A. T., and Lunt, M. J. (1983). The effects of pulsed magnetic fields of the type used in the stimulation of bone fracture repair. *Clin. Phys. Physiol. Meas.* **4**, 1–27.

Barker, D., Wright, E., Nguyen, K., Cannon, L., Fain, P., Goldgar, D., Bishop, D. T., Carey, J., Baty, B., Kivlin, J., Willard, H., Waye, J. S., Greig, G., Leinwand, L., Nakamura, Y., O'Connell, P., Leppert, M., Lalouel, J.-M., White, R., and Skolnick, M. (1987).

Gene for von Recklinghausen neurofibromatosis is in the pericentromeric region of chromosome 17. *Science* **236**, 1100–1102.

Barlow, A. J., Bogardi, J.-P., Ladher, R., and Francis-West, P. H. (1999). Expression of chick *Barx-1* and its differential regulation by FGF8 and BMP signaling in the maxillary primordium. *Devel. Dynam.* **214**, 291–302.

Barlow, A. J., and Francis-West, P. H. (1997). Ectopic application of recombinant BMP-2 and BMP-4 can change patterning of developing chick facial primordia. *Development* **124**, 391–398.

Barnard, R., Ng, K. W., Martin, T. J., and Waters, M. J. (1991). Growth hormone (GH) receptors in clonal osteoblast-like cells mediate a mitogenic response to GH. *Endocrinology* **128**, 1459–1464.

Barnes, D. W. H., Loutit, J. F., and Sansom, J. M. (1975). Histocompatible cells for the resolution of osteopetrosis in microphthalmic mice. *Proc. R. Soc. London, Ser. B.* **188**, 505–515.

Barnes, M. J., Constable, B. J., and Kodicek, E. (1969). Studies *in vivo* on the biosynthesis of collagen and elastin in ascorbic acid-deficient guinea pigs. *Biochem. J.* **113**, 387–398.

Barnett, C. H. (1954). The structure and functions of fibrocartilages within vertebrate joints. *J. Anat.* **8**, 363–368.

Barnett, C. H., Davies, D. V., and MacConaill, M. A. (1961). *Synovial Joints. Their Structure and Mechanics.* C. C. Thomas, Springfield, IL.

Barnett, C. H., and Lewis, O. J. (1958). The evolution of some traction epiphyses in birds and mammals. *J. Anat.* **92**, 593–601.

Barni, T., Maggi, M., Fantoni, G., Serio, M., Tollaro, I., Gloria, L., and Vannelli, G. B. (1995). Identification and localization of endothelin-1 and its receptors in human fetal jaws. *Devel. Biol.* **169**, 373–377.

Barnicott, N. A. (1941). Studies on the factors involved in bone resorption. I. The effect of subcutaneous transplantation of bones of the grey-lethal house mouse into normal hosts and or normal bones into grey-lethal hosts. *Amer. J. Anat.* **68**, 497–531.

Barnicott, N. A., and Datta, S. P. (1972). Vitamin A and bone. In *Biochemistry and Physiology of Bone* (G. H. Bourne, ed.), 2nd Edn, Volume 2, pp. 197–230. Academic Press, New York, NY.

Baron, J., Klein, K. O., Yanovski, J. A., Novosad, J. A., Bacher, J. D., Bolander, M. E., and Cutler, G. B., Jr. (1994). Induction of growth plate cartilage ossification by basic fibroblast growth factor. *Endocrinology* **135**, 2790–2793.

Baron, R. (1972). Contribution à l'etude de l'os avléolaire structure, formation et remaniement de la lame cribiforme. Thèse Doctoral Troisième Cycle. *Science Odontol.*, pp. 1–239. Université de Paris.

Baron, R. (1973). Remaniement de l'os alvéolaire et des fibres desmondantales au cours de la migration physioloque. *J. Biol. Buccale* **1**, 151–170.

Barratt, M. E. J. (1973). The role of soft connective tissue in the response of pig articular cartilage in organ culture to excess of retinol. *J. Cell Sci.* **13**, 205–220.

Barrett, A. J., Sledge, C. B., and Dingle, J. T. (1966). Effect of cortisol on the synthesis of chondroitin sulphate by embryonic cartilage. *Nature* **211**, 83–84.

Barrett, J. E., Wells, D. C., Paulsen, A. Q., and Conrad, G. W. (2000). Embryonic quail eye development in microgravity. *J. Appl. Physiol.* **88**, 1614–1622.

Barretto, C., Albrecht, R. M., Bjorling, D. E., Horner, J. R., and Wilsman, N. J. (1993). Evidence of the growth plate and the growth of long bones in juvenile dinosaurs. *Science* **262**, 2020–2023.

Barrow, J. R., and Capecchi, M. R. (1999). Compensatory defects associated with mutations in *Hoxa1* restore normal palatogenesis to *Hoxa2* mutants. *Development* **126**, 5011–5026.

Barrow, M. V., Simpson, C. F., and Miller, E. J. (1974). Lathyrism: A review. *Q. Rev. Biol.* **49**, 101–128.

Barth, A. (1895). Histologische Untersuchungen über Knochenimplantationen. *Beitrag Path. Anat.* **17**, 65–142.

Bartholinus, C. (1676). Diaphragmatis Structura Nova. *Acta Med. Phil. Hafn.* **4**, 14.

Bartsch, P. (1994). Development of the cranium of *Neoceratodus forsteri* with a discussion of the suspensorium and the opercular apparatus in Dipnoi. *Zoomorphol.* **114**, 1–31.

Basle, M., Minard, M. F., and Rebel, A. (1978). Structure et ultrastructure des ostéoblastes et du tissu ostéoide dans la maladie osseuse de Paget. *Pathol. Biol.* **26**, 475–479.

Bassett, C. A. L. (1962). Current concepts of bone formation. *J. Bone Joint Surg. Am. Vol.* **44**, 1217–1244.

Bassett, C. A. L. (1964). Environmental and cellular factors regulating osteogenesis. In *Bone Biodynamics* (H. M. Frost, ed.), pp. 233–244. Little, Brown & Co., Boston, MA.

Bassett, C. A. L. (1968). Biologic significance of piezoelectricity. *Calcif. Tissue Res.* **1**, 252–272.

Bassett, C. A. L. (1972a). Clinical implications of cell function in bone grafting. *Clin. Orthop. Rel. Res.* **87**, 49–59.

Bassett, C. A. L. (1972b). A biophysical approach to craniofacial morphogenesis. *Acta Morphol. Neerl. Scand.* **10**, 71–86.

Bassett, C. A. L. (1982). Pulsing electromagnetic fields: A new method to modify cell behavior in calcified and non-calcified tissues. *Calcif. Tissue Int.* **34**, 1–8.

Bassett, C. A. L., and Becker, R. O. (1962). Generation of electrical potentials by bone in response to mechanical stress. *Science* **137**, 1063.

Bassett, C. A. L., Caulo, N., and Kort, J. (1981a). Congenital pseudarthroses of the tibia: treatment with pulsing electromagnetic fields. *Clin. Orthop. Rel. Res.* **154**, 136–149.

Bassett, C. A. L., and Herrmann, L. (1961). Influence of oxygen concentration and mechanical factors on differentiation of connective tissue *in vitro*. *Nature* **190**, 460–461.

Bassett, C. A. L., and Herrmann, L. (1968). The effect of electrostatic fields on macromolecular synthesis by fibroblasts *in vitro*. *J. Cell Biol.* **39**, 9A.

Bassett, C. A. L., Mitchell, S. N., and Gaston, S. R. (1981b). Treatment of ununited tibial diaphyseal fractures with pulsing electromagnetic fields. *J. Bone Joint Surg. Am. Vol.* **63**, 511–523.

Bassett, C. A. L., Mitchell, S. N., Norton, L., and Pilla, A. A. (1978). Repair of non-unions by pulsing electromagnetic fields. *Acta Orthop. Belg.* **44**, 706–724.

Bassett, C. A. L., Mitchell, S. N., Norton, L., Caulo, N., and Gaston, S. R. (1979). Electromagnetic repairs of non-unions. In *Electrical Properties of Bone and Cartilage: Experimental Effects and Clinical Applications* (C. T. Brighton, J. Black and S. R. Pollack, eds), pp. 605–630. Grune & Stratton, New York, NY.

Bassett, C. A. L., and Pawluk, R. J. (1972). Electrical behavior of cartilage during loading. *Science* **178**, 982–983.

Bassett, C. A. L., Pawluk, R. J., and Becker, R. O. (1964). Effects of electric currents on bone *in vivo*. *Nature* **204**, 653–654.

Bassett, C. A. L., Pawluk, R. J., and Pilla, A. A. (1974a). Augmentation of bone repair by inductively-coupled electromagnetic fields. *Science* **184**, 575–577.

Bassett, C. A. L., Pawluk, R. J, and Pilla, A. A. (1974b). Acceleration of fracture repair by electromagnetic fields. A surgically noninvasive method. *Ann. N. Y. Acad. Sci.* **238**, 242–262.

Bassett, C. A. L., Pilla, A. A., and Pawluk, R. J. (1977). A non-operative salvage of surgically resistant pseudoarthroses and non-unions by pulsing electromagnetic fields. *Clin. Orthop. Rel. Res.* **124**, 128–143.

Bassett, C. A. L., and Ruedi, T. P. (1966). Transformation of fibrous tissue to bone *in vivo*. *Nature* **209**, 988–989.

Bassett, L. S., Tzitzikalakis, G., Pawluk, R. J., and Bassett, C. A. L. (1979). Prevention of disuse osteoporosis in the rat by means of pulsing electromagnetic fields. In *Electrical Properties of Bone and Cartilage: Experimental Effects and Clinical Applications* (C. T. Brighton, J. Black, and S. R. Pollack, eds), pp. 311–331. Grune & Stratton, New York, NY.

Bassini, E. (1872). Sul proceso Istologico di riassorbimento del tessuto oseo. Nota Preventica.

Bauer, A. M., and Russell, A. P. (1989). Supraorbital ossifications in geckos (Reptilia: Gekkonidae). *Can. J. Zool.* **67**, 678–684.

Baume, L. J. (1961). Response of condylar growth cartilage to induced stresses. *Science* **134**, 53–54.

Baume, L. J. (1962a). Embryogenesis of the human temporomandibular joint. *Science* **138**, 904–905.

Baume, L. J. (1962b). The prenatal and postnatal development of the human temporomandibular joint. *Trans. Eur. Orthod. Soc.* pp. 1–11.

Baume, L. J. (1962c). Ontogenesis of the human temporomandibular joint. I. Development of the condyles. *J. Dent. Res.* **41**, 1327–1339.

Baume, L. J. (1968). Patterns of cephalofacial growth and development. *Int. Dent. J.* **18**, 489–513.

Baume, L. J. (1970). Differential response of condylar, epiphyseal, synchondrotic, and articular cartilages in the rat to varying levels of vitamin A. *Amer. J. Orthod.* **58**, 537–551.

Baur, S. T., Mai, J. J., and Dymecki, S. M. (2000). Combinatorial signaling through BMP receptor IB and GDF5: Shaping of the distal mouse limb and the genetics of distal limb diversity. *Development* **127**, 605–619.

Bauze, R. J., Smith, R., and Francis, M. J. O. (1975). A new look at osteogenesis imperfecta. *J. Bone Joint Surg. Br. Vol.* **57**, 2–12.

Baylink, D. J., Wergedal, J., and Thompson, E. (1972). Loss of protein polysaccharides at sites where bone mineralization is initiated. *J. Histochem. Cytochem.* **20**, 279–292.

Bazin, S., Girard, H., and Delaunay, A. (1962). Effects biochimiques exerces par la papaine sur des ebauches cartilagineuses d'embryos de poulet. *Ann. Inst. Pasteur Paris* **102**, 346–352.

Bazin, S., and Strudel, G. (1972). Mise en évidence de collagène dans le matériel extracellulaire des organes axiaux de jeunes embryons de poulet. *C.R. Acad. Sci. (Paris)* **275**, 1167–1170.

Bazin, S., and Strudel, G. (1973). Biosynthesis of collagen in the axial organs of young chick embryos. In *Biology of Fibroblasts* (E. Kulonen and J. Pikkarairen, eds), pp. 411–416. Academic Press, New York, NY.

Beals, K. L., Smith, C. L., and Dodd, S. M. (1983). Climate and the evolution of brachycephalization. *Amer. J. Phys. Anthrop.* **62**, 425–438.

Beck, C. W., and Slack, J. M. W. (2002). Notch is required for outgrowth of the *Xenopus* tail bud. *Int. J. Devel. Biol.* **46**, 255–258.

Beck, F., Samani, N. J., Byrne, S., Morgan, K., Gebhard, R., and Brammar, W. J. (1988). Histochemical localization of IGF-I and IGF-II mRNA in the rat between birth and adulthood. *Development* **104**, 29–39.

Becker, R. O. (1972a). Augmentation of regenerative healing in man. A possible alternative to prosthetic implantation. *Clin. Orthop. Rel. Res.* **83**, 255–262.

Becker, R. O. (1972b). Stimulation of partial limb regeneration in rats. *Nature* **235**, 109–111.

Becker, R. O. (1978). Electrical osteogenesis – pro and con. *Calcif. Tissue Res.* **26**, 93–98.

Becker, R. O., Bassett, C. A. L., and Bachman, C. H. (1964). Bioelectrical factors controlling bone structure. In *Bone Biodynamics* (H. M. Frost, ed.), pp. 209–232. Little, Brown & Co., Boston, MA.

Becker, R. O., and Spadaro, J. A. (1972). Electrical stimulation of partial limb regeneration in mammals. *Bull. N. Y. Acad. Sci.* **48**, 627–641.

Beckham, C., Dimond, R., and Greenlee, T. K. (1977). The role of movement in the development of a digital flexor tendon. *Amer. J. Anat.* **150**, 443–460.

Becks, H., Collins, D. A., Simpson, M. E., and Evans, H. M. (1946). Growth and transformation of the mandibular joint in the rat. III. The effect of growth hormone and thyroxine injections in hypophysectomized female rats. *Amer. J. Orthod.* **32**, 447–451.

Bee, J., and Thorogood, P. V. (1980). The role of tissue interactions in the skeletogenic differentiation of avian neural crest cells. *Devel. Biol.* **78**, 47–62.

Beecher, R. M. (1979). Functional significance of the mandibular symphysis. *J. Morphol.* **159**, 117–130.

Beertsen, W., Everts, V., and van den Hooff, A. (1974a). Fine structure of fibroblasts in the periodontal ligament of the rat incisor and their possible role in tooth eruption. *Archs Oral. Biol.* **19**, 1087–1098.

Beertsen, W., Everts, V., and van den Hooff, A. (1974b). Fins structure and possible function of cells containing leptomeric organelles in the periodontal ligament of the rat incisor. *Archs Oral. Biol.* **19**, 1099–1100.

Beertsen, W., and van den Bos, T. (1992). Alkaline phosphatase induces the mineralization of sheets of collagen implanted subcutaneously in the rat. *J. Clin. Invest.* **89**, 1974–1980.

Behari, J. (1991). Electrostimulation and bone fracture healing. *Crit. Rev. Biomed. Eng.* **18**, 235–255.

Behonick, D. J., and Werb, Z. (2003). A bit of give and take: the relationship between the extracellular matrix and the developing chondrocyte. *Mech. Devel.* **120**, 1327–1336.

Behrens, A., Haigh, J., Metchta-Grigoriou, F., Nagy, A., and Yaniv, M. (2003). Impaired intervertebral disc formation in the absence of *Jun. Development* **130**, 103–109.

Bejder, L., and Hall, B. K. (2002). Limbs in whales and limblessness in other vertebrates: mechanisms of evolutionary and developmental transformation and loss. *Evol. & Devel.* **4**, 445–458.

Bélanger, L. F. (1954). Autoradiographic visualization of the entry and transit of S^{35} in cartilage, bone, and dentine of young rats and the effect of hyaluronidase *in vitro*. *Can. J. Biochem. Physiol.* **32**, 161–169.

Bélanger, L. F. (1969). Osteocytic osteolysis. *Calcif. Tissue Res.* **4**, 1–12.

Bélanger, L. F., and Clark, I. (1967). Alpharadiographic and histological observations on the skeletal effects of hypervitaminosis A and D in the rat. *Anat. Rec.* **158**, 443–452.

Bélanger, L. F., Semba, T., Tolnai, S., Copp, D. H., Krook, L., and Gries, C. (1966). The two faces of resorption. *Calcif. Tissue Proc. Europe Symp.* **3**, 1–10.

Belchier, J. B. (1736a). An account of the bone of animals being changed to a red color by aliment only. *Phil. Trans. R. Soc. London* **39**, 286–288.

Belchier, J. B. (1736b). II. A further account of the bones of animals being made red by aliment only. *Phil. Trans. R. Soc. London* **39**, 299–303.

Bell, M. A. (1981). Lateral plate morphology and ontogeny of the complete plate morph of threespine sticklebacks (*Gasterosteus aculeatus*). *Evolution* **35**, 67–74.

Bell, M. A. (1987). Interacting evolutionary constraints in pelvic reduction of threespine sticklebacks, *Gasterosteus aculeatus* (Pisces, Gasterosteidae). *Biol. J. Linn. Soc.* **31**, 347–382.

Bell, M. A. (1988). Stickleback fishes: Bridging the gap between population biology and paleobiology. *Trends Ecol. Evol.* **3**, 320–325.

Bell, M. A. (1993). Evolution of pelvic reduction in threespine stickleback fish: A test of competing hypotheses. *Evolution* **47**, 906–914.

Bell, M. A., Aguirre, W. E., and Buck, N. J. (2004). Twelve years of contemporary armor evolution in a threespine stickleback population. *Evolution* **58**, 814–824.

Bell, M. A., Ortí, G., Walker, J. A., and Koenings, J. P. (1993) Evolution of pelvic reduction in threespine stickleback fish: A test of competing hypotheses. *Evolution* **47**, 906–914.

Bellairs, A. d'A. (Angus). (1989). *The Isle of Sea Lizards.* Durrell Institute of Conservation and Ecology, University of Kent, Canterbury, on the occasion of the First World Congress of Herpetology, September 1989.

Bellairs, A. d'A., and Bryant, S. V. (1968). Effects of amputation of limbs and digits of lacertid lizards. *Anat. Rec.* **161**, 489–496.

Bellairs, A. d'A., and Gans, C. (1983). A reinterpretation of the Amphisbaenian orbitosphenoid. *Nature* **302**, 243–244.

Bellairs, A. d'A., and Jenkin, C. R. (1960). The skeleton of birds. In *Biology and Comparative Physiology of Birds* (A. J. Marshall, ed.), Volume 1, pp. 241–300. Academic Press, New York, NY.

Bellairs, A. d'A., and Kamal, A. M. (1981). The chondrocranium and the development of the skull in recent reptiles. In *Biology of the Reptilia. Volume 11, Morphology F* (C. Gans and T. S. Parsons, eds), pp. 1–264. Academic Press, New York, NY.

Bellairs, R. (1971). *Developmental Processes in Higher Vertebrates.* Logos Press, London.

Bellairs, R. (1979). The mechanism of somite segmentation in the chick embryo. *J. Embryol. Exp. Morphol.* **51**, 227–243.

Bellairs, R., Curtis, A. S. G., and Sanders, E. J. (1978). Cell adhesiveness and embryonic differentiation. *J. Embryol. Exp. Morphol.* **46**, 207–213.

Bellairs, R., and Portch, P. A. (1977). Somite formation in the chick embryo. In *Vertebrate Limb and Somite Morphogenesis* (D. A. Ede, J. R. Hinchliffe, and M. Balls, eds), pp. 449–462. Cambridge University Press, Cambridge.

Bellows, C. G., Aubin, J. E., and Heersche, J. N. (1993). Differential effects of fluoride during initiation and progression of mineralization of osteoid nodules formed *in vitro*. *J. Bone Min.* **8**, 1357–1363.

Bellows, C. G., Aubin, J. E., Heersche, J. N. M., and Antosz, M. E. (1986). Mineralized bone nodules formed *in vitro* from enzymatically released rat calvarial cell populations. *Calcif. Tissue Int.* **38**, 143–154.

Bellows, C. G., and Heersche, J. N. M. (2001). The frequency of common progenitors for adipocytes and osteoblasts and of committed and restricted adipocyte and osteoblast

progenitors in fetal rat calvaria cell populations. *J. Bone Min. Res.* **16**, 1983–1993.

Bellows, C. G., Heersche, J. N. M., and Aubin, J. E. (1989). Effects of dexamethasone on expression and maintenance of cartilage in serum-containing cultures of calvarial cells. *Cell Tissue Res.* **256**, 145–152.

Bellows, C. G., Ishida, H., Aubin, J. E., and Heersche, J. N. (1990). Parathyroid hormone reversibly suppresses the differentiation of osteoprogenitor cells into functional osteoblasts. *Endocrinology* **127**, 3111–3116.

Bellows, C. G., Pei, W., Jia, Y., and Heersche, J. N. (2003). Proliferation, differentiation and self-renewal of osteoprogenitors in vertebral cell populations from aged and young female rats. *Mech. Ageing Devel.* **18**, 747–757.

Belting, H.-G., Shashikant, C. S., and Ruddle, F. H. (1998). Modification of expression and *cis*-regulation of *Hoxc8* in the evolution of diverged axial morphology. *Proc. Natl Acad. Sci. U.S.A.* **95**, 2355–2360.

Bemis, W. E., Burggren, W. W., and Kemp, N. E. (eds) (1987). *The Biology and Evolution of Lungfishes.* Alan R. Liss, Inc., New York, NY.

Bemis, W. E., and Grande, L. (1999). Development of the median fins of the North American paddlefish (*Polyodon spathula*) and a reevaluation of the lateral fin-fold theory. In *Mesozoic Fishes: Systematics and Fossil Record* (G. Arratia and H.-P. Schultze, eds), pp. 41–68. Verlag Dr. Friedrich Pfeil, Munich.

Ben-Ami, Y., von der Mark, K., Franzen, A., de Bernard, B., Lunazzi, G. C., and Silbermann, M. (1993). Transformation of fetal secondary cartilage into embryonic bone in organ cultures of human mandibular condyles. *Cell Tissue Res.* **271**, 317–322.

Benayahu, D., Kletter, Y., Zipori, D., and Wientroub, S. (1989). Bone marrow-derived stromal cell line expressing osteoblastic phenotype *in vitro* and osteogenic capacity *in vivo*. *J. Cell Physiol.* **140**, 1–7.

Benayahu, D., Zipori, D., and Wientroub, S. (1993). Marrow adipocytes regulate growth and differentiation of osteoblasts. *Biochem. Biophys. Res. Commun.* **197**, 1245–1252.

Bendall, A. J., and Abate-Shen, C. (2000). Roles for Msx and Dlx homeoproteins in vertebrate development. *Gene* **247**, 17–31.

Bendall, A. J., Hu, G., Levi, G., and Abate-Shen, C. (2003). Dlx5 regulates chondrocyte differentiation at multiple stages. *Int. J. Devel. Biol.* **47**, 335–344.

Beneke, G., Endres, O., Becker, H., and Kulka, R. (1966). Über Wachstum und Degeneration des Trachealknorpels. *Virchows Arch. Pathol. Anat. Physiol.* **341**, 365–380.

Benjamin, M. (1986). The oral sucker of *Gyrinocheilus aymonieri* (Teleostei: Cypriniformes). *J. Zool. London Ser. B.* **1**, 211–254.

Benjamin, M. (1988). Mucochondroid (mucous connective) tissues in the heads of teleosts. *Anat. Embryol.* **178**, 461–474.

Benjamin, M. (1989a). Hyaline-cell cartilage (chondroid) in the heads of teleosts. *Anat. Embryol.* **179**, 285–303.

Benjamin, M. (1989b). The development of hyaline-cell cartilage in the head of the black molly, '*Poecilia sphenops.*' Evidence for secondary cartilage in a teleost. *J. Anat.* **164**, 145–154.

Benjamin, M. (1990). The cranial cartilages of teleosts and their classification. *J. Anat.* **169**, 153–172.

Benjamin, M., Qin, S., and Ralphs, J. R. (1995). Fibrocartilage associated with human tendons and their pulleys. *J. Anat.* **187**, 625–633.

Benjamin, M., and Ralphs, J. R. (1991). Extracellular matrix of connective tissues in the heads of teleosts. *J. Anat.* **179**, 137–148.

Benjamin, M., and Ralphs, J. R. (1995). Fibrocartilage associated with human tendons and their pulleys. *J. Anat.* **187**, 625–633.

Benjamin, M., and Ralphs, J. R. (1997a). Tendons and ligaments – an overview. *Histol. Histopathol.* **12**, 1135–1144.

Benjamin, M., and Ralphs, J. R. (1997b). Cells, tissues, and structures of the musculoskeletal system. *Curr. Opin. Orthopaed.* **8**(6), 29–33.

Benjamin, M., and Ralphs, J. R. (1998). Fibrocartilage in tendons and ligaments – an adaptation to compressive load. *J. Anat.* **193**, 481–494.

Benjamin, M., Ralphs, J. R., and Eberewariye, O. S. (1992). Cartilage and related tissues in the trunk and fins of teleosts. *J. Anat.* **181**, 113–118.

Benjamin, M., Rufai, A., and Ralphs, J. R. (2000). The mechanism of formation of bony spurs (enthesophytes) in the Achilles tendon. *Arthritis Rheum.* **43**, 576–583.

Benjamin, M., and Sandhu, J. S. (1990). The structure and ultrastructure of the rostral cartilage in the spiny eel, *Macrognathus siamensis* (Teleostei: Mastacembeloidei). *J. Anat.* **169**, 37–47.

Bennett, A. F., and Dalzell, B. (1973). Dinosaur physiology: A critique. *Evolution* **27**, 170–174.

Bennett, D., Artzt, K., Magnuson, T., and Spiegelman, M. (1977). Developmental interactions studied with experimental teratomas derived from mutants at the *T/t* locus in the mouse. In *Cell Interactions in Differentiation* (M. Karkinen-Jääskeläinen, L. Saxén and L. Weiss, eds), pp. 359–398. Academic Press, London.

Bennett, J. H., Carter, D. H., Alavi, A. L., Beresford, J. N., and Walsh, S. (2001). Patterns of integrin expression in a human mandibular explant model of osteoblast differentiation. *Archs Oral Biol.* **46**, 229–238.

Bennett, J. H., Joyner, C. J., Triffitt, J. T., and Owen, M. E. (1991). Adipocytic cells cultured from marrow have osteogenic potential. *J. Cell Sci.* **99**, 131–139.

Bennett, V. D., and Adams, S. L. (1990). Identification of a cartilage-specific promoter within intron 2 of the chick $\alpha 2$(I) collagen gene. *J. Biol. Chem.* **265**, 2223–2230.

Benninghoff, A. (1925). Spaltlininien am Knochen, eine Methode zur Ermittlung der Architektur platter Knochen. *Verh. Anat. Ges. Vers.* **34**, 189.

Benninghoff, A. (1930). Über Leitsysteme der Knochencompakta. Studien zur Architektur der Knochen. 3. *Morphol. Jb.* **65**, 1–44.

Benoit, J. A. A. (1960a). Induction de cartilage *in vitro* par l'extrait d'otocystes d'embryons de poulet. *J. Embryol. Exp. Morphol.* **8**, 33–38.

Benoit, J. A. A. (1960b). L'otocyste exerce-t-il une action inductrice sur le mésenchyme somatique chez l'embryon de poulet? *J. Embryol. Exp. Morphol.* **8**, 39–46.

Benoit, J. A. A. (1960c). Etude expérimentale des facteurs de l'induction du cartilage otique chez les embryons de poulet et de truite. *Ann. Sci. Nat. Zool.* **12**, Ser. 2, 323–385.

Benoit, J. A. A. (1963). Chronologie de l'induction du cartilage otique chez l'embryon de poulet. *Arch. Anat. Microsc. Morphol. Exp.* **52**, 573–590.

Benoit, J. A. A. (1965). De l'excision de l'otocyste chez l'embryon de poulet et des conséquences sur la morphogenèse de la capsule otique cartilagineuse. *CR. Soc. Biol. Paris* **149**, 998–1000.

Benoit, J. A. A., and Schowing, J. (1970). Morphogenesis of the neurocranium. In *Tissue Interactions during Organogenesis* (E. Wolff, ed.), pp. 105–130. Gordon & Breach, New York, NY.

Bentley, G., and Greer, R. B. (1970). The fate of chondrocytes in endochondral ossification in the rabbit. *J. Bone Joint Surg. Br. Vol.* **52**, 571–577.

Benton, M. J. (1987). Conodonts classified at last. *Nature* **325**, 482–483.

Bentz, H., Thompson, A. Y., Armstrong, R., Chang, R. J., Piez, K. A., and Rosen, D. M. (1991). Transforming growth factor-β2 enhances the osteoinductive activity of a bovine bone-derived fraction containing bone morphogenetic protein-2 and protein-3. *Matrix* **11**, 269–275.

Benveniste, K., Wilczek, J., and Stern, R. (1973). Translation of collagen mRNA from chick embryo calvaria in a cell-free system derived from Krebs II ascites cells. *Nature* **246**, 303–305.

Benya, P. D., Harvey, W., Cheung, H., and Nimni, M. (1976). The CNBr peptide pattern of an isolated a-chain indicates a new kind of collagen. *J. Dent. Res.* **55** (Special issue B), 78.

Benya, P. D., Padilla, S. R., and Nimni, M. E. (1977). The progeny of rabbit articular chondrocytes synthesize collagen types I and III and type I trimer but not type II. Verification by cyanogen bromide peptide analysis. *Biochemistry* **16**, 865–872.

Beresford, B., Le Lievre, C., and Rathbone, M. P. (1978). Chimaera studies of the origin and formation of the pectoral musculature of the avian embryo. *J. Exp. Zool.* **205**, 321–326.

Beresford, J. N. (1989). Osteogenic stem cells and the stromal system of bone and marrow. *Clin. Orthop. Rel. Res.* **240**, 270–280.

Beresford, J. N., Bennett, J. H., Devlin, C., Leboy, P. S., and Owen, M. E. (1992). Evidence for an inverse relationship between the differentiation of adipocytic and osteogenic cells in rat marrow stromal cell cultures. *J. Cell Sci.* **102**, 341–351.

Beresford, W. A. (1970). Healing of the experimentally fractured os priapi of the rat. *Acta Orthop. Scand.* **41**, 134–149.

Beresford, W. A. (1972). The influence of castration on fracture repair in the penile bone of the rat. *J. Anat.* **112**, 19–26.

Beresford, W. A. (1975a). Growth cartilages of the mandibular condyle and penile bones: How alike? *J. Dent. Res.* **54**, 417.

Beresford, W. A. (1975b). Schemes of zonation in the mandibular condyle. *Amer. J. Orthod.* **68**, 189–195.

Beresford, W. A. (1981). *Chondroid Bone, Secondary Cartilage and Metaplasia.* Urban and Schwarzenberg, Munich.

Beresford, W. A. (1983). Ectopic cartilage, neoplasia and metaplasia. In *Cartilage, Volume 3. Biomedical Aspects* (B. K. Hall, ed.), pp. 1–48. Academic Press, New York, NY.

Beresford, W. A., and Clayton, S. P. (1977). Intracerebral transplantation of the genital tubercle in the rat: The fate of the penile bone and cartilages. *J. Anat.* **123**, 297–312.

Beresford, W. A., and Hancox, N. M. (1965). The influence of snips of bladder dispersed in polyurethane sponge or deproteinized bone on the repair of skull defects in guinea pigs and rats. *J. Anat.* **99**, 407–408.

Beresford, W. A., and Hancox, N. M. (1967). Urinary bladder mucosa and bone regeneration in guinea-pig and rat. *Acta Anat.* **66**, 78–117.

Berggren, D., Frenz, D. A., Galinovic-Schwartz, V., and Van de Water, T. R. (1997). Fine structure of extracellular matrix and basal laminae in 2 types of abnormal collagen production – L-proline-analog-treated otocyst cultures and disproportionate micromelia (*Dmm/Dmm*) mutants. *Hearing Res.* **107**, 125–135.

Bergot, C. (1975). Morphogenèse et structure des dents d'un téléostéen (*Salmo fario* L). *J. Biol. Buccale* **3**, 301–324.

Bergsma, D. (ed.) (1975). *Morphogenesis and Malformation of Face and Brain*, Birth Defects Orig. Artic. Ser., Volume 11, No. 7. Alan R. Liss, Inc., New York, NY.

Bergsma, D., and Lenz, W. (eds) (1977). *Morphogenesis and Malformation of the Limb*, Birth Defects Orig. Artic. Ser. 13, Volume 11, No. 1. Alan R. Liss, Inc., New York, NY.

Berkovitz, B. K. B. (1978). Tooth ontogeny in the upper jaw and tongue of the rainbow trout (*Salmo gairdneri*). *J. Biol. Buccale* **6**, 205–215.

Berkovitz, B. K. B., and Moore, M. H. (1974). A longitudinal study of replacement patterns of teeth on the lower jaw and tongue in the rainbow trout *Salmo gairdneri*. *Archs Oral Biol.* **19**, 1111–1120.

Bernard, G. W. (1978). Ultrastructural localization of alkaline phosphatase in initial; intramembranous osteogenesis. *Clin. Orthop. Rel. Res.* **135**, 218–225.

Bernard, G. W., and Dodson, S. A. (1992). Differences in osteogenic and hemopoietic potential between young and aged bone marrow stem cells in vitro. In *Biological Mechanisms of Tooth Movement and Craniofacial Adaptation* (Z. Davidovitch, ed.), Ohio State University, Birmingham, Alabama, 17–28.

Bernard, G. W., and Pease, D. C. (1969). An electron microscopic study of initial intramembranous osteogenesis. *Amer. J. Anat.* **125**, 271–290.

Bernfield, M., Hinkes, M. T., and Gallo, R. L. (1993). Developmental expression of the syndecans: Possible function and regulation. *Development* Suppl. 205–212.

Bernstein, P., and Crelin, E. S. (1967). Bony pelvic sexual dimorphism in the rat. *Anat. Rec.* **157**, 517–526.

Berrill, N. J. (1955). *The Origin of the Vertebrates*. Oxford University Press, Oxford.

Berry, L., and Shuttleworth, C. A. (1989). Expression of the chondrogenic phenotype by mineralizing cultures of embryonic chick calvarial bone cells. *Bone Min.* **7**, 31–46.

Bert, P. (1865). Sur la greffe animale. *C.R. Acad. Sci. Paris* **51**, 587.

Bertram, J. E. A. (2001). What does trabecular bone do (and why should it follow Wolff's law)? *Osteologie* **10**, 17–27.

Bertram, J. E. A., and Gosline, J. M. (1986). Fracture toughness design in horse hoof keratin. *J. Exp. Biol.* **125**, 29–47.

Bertram, J. E. A., Greenberg, L., Miyake, T., and Hall, B. K. (1997). Paralysis and long bone growth in the chick: growth trajectories of the pelvic limb. *Growth, Devel., Aging* **61**, 51–60.

Bertram, J. E. A., Hall, B. K., and Demont, M. E. (1991). Does diaphyseal periosteum restrict linear growth of long bones? In *Fundamentals of Bone Growth* (A. D. Dixon, B. G. Sarnat, and D. A. N. Hoyte, eds), pp. 363–374. CRC Press, Boca Raton, FL.

Bertram, J. E. A., and Swartz, S. M. (1991). The 'Law of bone transformation': A case of crying Wolff? *Biol. Rev. Camb. Phil. Soc.* **66**, 245–273.

Bessho, K., Tagawa, T., and Murata, M. (1989). Purification of bone morphogenetic protein derived from bovine bone matrix. *Biochem. Biophys. Res. Commun.* **165**, 595–601.

Bettex-Galland, M., and Wiesmann, U. (1987). Differentiation of L6 myoblastic cells into chondrocytes. *Experientia* **43**, 610–611.

Berzin, A. A. (1972). *The Sperm Whale*. Keter Press, Jerusalem.

Bevelander, G., and Johnson, P. L. (1950). A histochemical study of the development of membrane bones. *Anat. Rec.* **108**, 1–21.

Bex, J. H. M., and Maltha, J. C. (1976). Effects of periosteal cutting on the growth of long bones. *J. Dent. Res.* **55**, D163.

Beyer, P. E., and Chernoff, N. (1986). The induction of supernumerary ribs in rodents: Role of the maternal stress. *Teratol. Carcinog. Mutag.* **6**, 419–430.

Bhalerao, J., Bogers, J., van Marck, E., and Merregaert, J. (1995). Establishment and characterization of two clonal cell lines derived from murine mandibular condyles. *Tissue & Cell* **27**, 369–382.

Bhargava, U., Bar-Lev, M., Bellows, C. G., and Aubin, J. E. (1988). Ultrastructural analysis of bone nodules formed in vitro by isolated fetal rat calvarial cells. *Bone* **9**, 155–163.

Bhasin, N., Kernick, E., Luo, X., Seidel, H. E., Weiss, E. R., and Lauder, J. M. (2004). Differential regulation of chondrogenic differentiation by the serotonin$_{2B}$ receptor and retinoic acid in the embryonic mouse hindlimb. *Devel. Dyn.* **230**, 201–209.

Bhaskar, S. N., Weinmann, J. P., and Schour, I. (1954). The growth rate of the tibias of the ia rat from 17 days insemination age to 30 days after birth. *Anat. Rec.* **119**, 231–245.

Bhatnagar, R. S., and Prockop, D. J. (1966). Dissociation of the synthesis of sulphated mucopolysaccharides and the synthesis of collagen in embryonic cartilage. *Biochim. Biophys. Acta* **130**, 382–392.

Bhatnagar, R. S., and Rapaka, S. S. R. (1971). Cellular regulation of collagen biosynthesis. *Nature, New Biol.* **234**, 92–93.

Bhatti, H. K. (1938). The integument and dermal skeleton of Siluroidea. *Trans. Zool. Soc. London* **24**, 1–82.

Bi, W., Deng, J. M., Zhang, Z., Behringer, R. R., and de Crombrugghe, B. (1999). Sox9 is required for cartilage formation. *Nature Genet.* **22**, 85–89.

Bianco, P., Ballanti, P., and Bonucci, E. (1988). Tartrate-resistant acid phosphatase activity in rat osteoblasts and osteocytes. *Calcif. Tissue Int.* **43**, 167–171.

Bick, E. M. (1968). *Source Book of Orthopaedics*. Hafner, New York, NY.

Bienkowski, R. S., Baum, B. J., and Crystal, R. G. (1978). Fibroblasts degrade newly synthesized collagen within the cell before secretion. *Nature* **276**, 413–416.

Bierly, J. A., Sottosanti, J. S., Costley, J. M., and Cherrick, H. M. (1975). An evaluation of the osteogenic potential of marrow. *J. Periodontol.* **46**, 277–283.

Biggers, J. D. (1960). The growth of embryonic chick tibiotarsi on a chemically-defined medium, with appendix on the preparation of the medium. *J. Exp. Zool.* **144**, 233–252.

Biggers, J. D. (1965). Cartilage and bone. In *Cells and Tissues in Cultures. Methods, Biology and Physiology* (E. N. Willmer, ed.), Volume 2, pp. 198–260. Academic Press, New York, NY.

Biggers, J. D., Gwatkin, R. B. L., and Heyner, S. (1961). Growth of embryonic avian and mammalian tibiae on a relatively simple chemically defined medium. *Exp. Cell Res.* **25**, 41–58.

Biggers, J. D., and Heyner, S. (1963). Growth of embryonic chick and rat long bones in vitro on natural and chemically defined media. *J. Exp. Zool.* **152**, 41–56.

Biggers, J. D., Webb, M., Parker, R. C., and Healy, G. M. (1957). Cultivation of embryonic chick bones on chemically defined media. *Nature* **180**, 825–828.

Bilezikian, J. P., Raisz, L. G., and Rodan, G. A. (eds) (1996). *Principles of Bone Biology*. Academic Press, San Diego, CA.

Biller, T., Yosipovitch, Z., and Gedalia, I. (1977). Effect of topical application of fluoride on the healing rate of experimental calvarial defects in rats. *J. Dent. Res.* **56**, 53–56.

Billo, R., and Wake, M. H. (1987). Tentacle development in *Dermophis mexicanus* (Amphibia, Gymnophiona) with an hypothesis of tentacle origin. *J. Morphol.* **192**, 101–111.

Biltz, R. M., and Pellegrino, E. O. (1969). The chemical anatomy of bone. I. A comparative study of bone composition in sixteen vertebrates. *J. Bone Joint Surg. Am. Vol.* **51**, 456–466.

Binderman, I., Berger, E., Fine, N., Shimshoni, Z., Harell, A., and Somjen, D. (1989). Calvaria derived osteogenic cells: Phenotypic expression in culture. *Conn. Tissue Res.* **20**, 41–47.

Binderman, I., Duksin, D., Harell, A., Katzir (Katchalski), E., and Sachs, L. (1974). Formation of bone tissue in culture from isolated bone cells. *J. Cell Biol.* **61**, 427–439.

Bingham, P. J., Barzell, I. A., and Owen, M. J. (1969). The effect of parathyroid extract on cellular activity and plasma calcium levels *in vivo*. *J. Endocrinol.* **45**, 387–400.

Binkert, C., Demetriou, M., Sukhu, B., Szweras, M., Tenenbaum, H. C., and Dennis, J. W. (1999). Regulation of osteogenesis by fetuin. *J. Biol. Chem.* **274**, 28514–28520.

Bischoff, R. (1971). Acid mucopolysaccharide synthesis by chick amnion cell cultures inhibited by 5-bromodeoxyuridine. *Exp. Cell Res.* **66**, 224–236.

Bisgard, J. D., and Bisgard, M. E. (1935). Longitudinal growth of long bones. *Arch. Surg.* **31**, 568–578.

Bishop, W. J. (1961). *The Early History of Surgery*. Robert Hale Ltd., London.

Björk, A. (1955). Facial growth in man studied with the aid of metallic implants. *Acta Odontol. Scand.* **13**, 9–34.

Blair, H. C. (1998). How the osteoclast degrades bone. *BioEssays* **20**, 837–846.

Blair, H. C., Zaidi, M., and Schlesinger, P. H. (2002). Mechanisms balancing skeletal matrix synthesis and degradation. *Biochem. J.* **364**, 329–341.

Blanc, I., Bach, A., and Robert, B. (2002). Unusual pattern of *Sonic hedgehog* expression in the polydactylous mouse mutant Hemimelic extra-toes. *Int. J. Devel. Biol.* **46**, 969–974.

Blanc, M. (1944). L'ossifications des arcs branchiaux chez les poissons téléostéens. *Bull. Soc. Zool. France* **69**, 226–230.

Blanc, M. (1953). Contributions à l'etude de l'ostéogénèse chez les poissons téléostéens. *Mem. Mus. Natl Hist. Nat. Ser. A* **7**, 1–146.

Blanc, Y.-S., and Ashhurst, D. E. (1996). Development and ageing of the articular cartilage of the rabbit knee joint: distribution of the fibrillar collagens. *Anat. Embryol.* **194**, 607–619.

Blanco, C., López, D., DeAndres, A. V., Schib, J. L., Gallego, A., Duran, A. C., and Sans-Coma, V. (2001). Cartilage in the *bulbus arteriosus* of teleostean fishes. *Neth. J. Zool.* **51**, 361–370.

Blanco, M. J., Misof, B. Y., and Wagner, G. P. (1998). Heterochronic differences of *Hoxa-11* expression in *Xenopus* fore-limb and hind-limb development – Evidence for lower limb patterning of the Anuran ankle bones. *Devel. Genes Evol.* **208**, 175–187.

Blanco, M. J., and Sanchiz, B. (2000). Evolutionary mechanisms of rib loss in anurans: a comparative developmental approach. *J. Morphol.* **244**, 57–67.

Bland, Y. S., and Asshhurst, D. E. (1996). Development and ageing of the articular cartilage of the rabbit knee joint: distribution of the fibrillar collagens. *Anat. Embryol.* **194**, 607–619.

Blanquaert, F., Delany, A. M., and Canalis, E. (1999). Fibroblast growth factor-2 induces hepatocyte growth factor/scatter factor expression in osteoblasts. *Endocrinology* **140**, 1069–1074.

Blavier, L., and Delaissé, J. M. (1995). Matrix metalloproteinases are obligatory for the migration of preosteoclasts to the developing marrow cavity of primitive long bones. *J. Cell Sci.* **108**, 3649–3659.

Blaxter, J. H. S. (1987). Structure and development of the lateral line. *Biol. Rev. Camb. Philos. Soc.* **62**, 471–514.

Bledsoe, A. H., Raikow, R. J., and Glasgow, A. G. (1993). Evolution and functional significance of tendon ossification in woodcreepers (Aves: Passeriformes: Dendrocolaptinae). *J. Morphol.* **215**, 289–300.

Blumenkrantz, N., and Prockop, D. J. (1970). Variations in the glycosylation of the collagen synthesized by chick embryo cartilage: Effects of development and several hormones. *Biochim. Biophys. Acta* **208**, 461–466.

Blumer, M. J. F., Fritsch, H., Pfaller, K., and Brenner, E., (2004). Cartilage canals in the chicken embryo: ultrastructure and function. *Anat. Embryol.* **207**, 453–462.

Bobick, B. E., and Kulyk, W. M. (2004). The MEK-ERK signaling pathway is a negative regulator of cartilage-specific gene expression in embryonic limb mesenchyme. *J. Biol. Chem.* **279**, 4588–4595.

Bock, W. J. (1960a). The palatine process of the premaxilla of the Passeres. A study of the variation, function, evolution and taxonomic value of a single character throughout an avian order. *Bull. Mus. Comp. Zool. Harvard Univ.* **122**, 361–488.

Bock, W. J. (1959). Preadaptation and multiple evolutionary pathways. *Evolution* **13**, 194–211.

Bock, W. J. (1960b). Secondary articulations of the avian mandible. *The Auk* **77**, 19–55.

Bock, W. J. (1964). Kinetics of the avian skull. *J. Morphol.* **114**, 1–42.

Bock, W. J. (1966). An approach to the functional analysis of bill shape. *Auk* **83**, 10–51.

Bock, W. J. (1974). The avian skeletomuscular system. In *Avian Biology* (D. S. Farner, J. R. King, and K. C. Parkes, eds), Volume 4, pp. 120–259. Academic Press, New York, NY.

Bock, W. J., and Morony, J. (1978). The preglossale in *Passer* (Aves: Passeriformes) – a skeletal neomorph. *J. Morphol.* **155**, 99–110.

Bock, W. J., and Wahlert, G. von (1965). Adaptation and the form–function complex. *Evolution* **19**, 169–299.

Bodemer, C. W. (1964). Evocation of regrowth phenomena in anural limbs by electrical stimulation of the nerve supply. *Anat. Rec.* **148**, 441–458.

Boden, S. D., Kaplan, F. S., Fallon, M. D., Ruddy, R., Belik, J., Anday, E., Zackai, E., and Ellis, J. (1987). Metatropic dwarfism – uncoupling of endochondral and perichondral growth. *J. Bone Joint Surg. Am. Vol.* **69**, 174–184.

Bodenstein, D. (1952). Studies on the development of the dorsal fin in amphibia. *J. Exp. Zool.* **120**, 213–245.

Boer, H. H. de (1988). *Vascularized Fibular Transfer*. Thesis, Drukkerij Groen B. V., Leiden.

Bogardi, J.-P., Barlow, A. J., and Francis-West, P. (2000). The role of FGF-8 and BMP-4 in the outgrowth and patterning of the chick embryonic maxillary primordium. In *Regulatory Processes in Development* (L. Olsson and C.-O. Jacobson, eds), pp. 173–177. Portland Press, London.

Bohme, K., Conscience-Egli, M., Tschan, T., Winterhalter, K. H., and Bruckner, P. (1992). Induction of proliferation or hypertrophy of chondrocytes in serum-free culture: The role of insulin-like growth factor-1, insulin or thyroxine. *J. Cell Biol.* **116**, 1035–1042.

Bohúrquez-Mahecha, G. A., and Aparecida de Oliveira, C. (1998). An additional bone in the sclera of the eyes of owls and the common potoo (*Nictibius griseus*) and its role in the contraction of the nictitating membrane. *Acta Anat.* **163**, 201–211.

Boivin, G., Anthoine-Terrier, C., and Obrant, K. J. (1990). Transmission electron microscopy of bone tissue – A review. *Acta Orthop. Scand.* **61**, 170–180.

Boivin, G., and Meunier, P. J. (1990). Fluoride and bone: Toxicological and therapeutic aspects. In *The Metabolic and*

Molecular Basis of Acquired Disease (R. D. Cohen, B. Lewis, K. G. M. M. Alberti, and A. M. Denman, eds), pp. 1803–1823. Baillière Tindall, London.

Boivin, G., Morel, G., Meunier, P. J., and Dubois, P. M. (1983). Ultrastructural aspects after cryoultramicrotomy of bone tissue and sutural cartilage in neonatal mice calvaria. *Biol. Cell* **49**, 227–230.

Bollet, A. J. (1969). An essay on the biology of osteoarthritis. *Arthritis Res.* **12**, 152–163.

Bona, C. A., Stânescu, V., Dumitrescu, M. S., and Ionescu, V. (1967). Histochemical and cytoenzymological studies in Myositis Ossificans. *Acta Histochem.* **27**, 207–224.

Bone, Q. (1972). *The Origin of Chordates.* Oxford Biology Reader, #18. Oxford University Press, Oxford.

Bonewald, L. F., and Dallas, S. L. (1994). Role of active and latent transforming growth factor β in bone formation. *J. Cell. Biochem.* **55**, 350–357.

Bonewald, L. F., Wakefield, L., Oreffo, R. O. C., Escobedo, A., Twardzik, D. R., and Mundy, G. R. (1991). Latent forms of transforming growth factor-β (TGF-β) derived from bone cultures: Identification of a naturally occurring 100-kDa complex with similarity to recombinant latent TGFβ. *Mol. Endocrinol.* **5**, 741–751.

Bonner, J. J. (1974). Method for evaluating intrauterine versus genetic influences on craniofacial anomalies. *J. Dent. Res.* **53**, 1313–1316.

Bonner, P. H. (1975). Clonal analysis of vertebrate myogenesis. V. Nerve–muscle interaction in chick limb bud chorio-allantoic membrane grafts. *Devel. Biol.* **47**, 222–227.

Bonucci, E. (1981). New knowledge on the origin, function and fate of osteoclasts. *Clin. Orthop. Rel. Res.* **158**, 252–269.

Bonucci, E., Cuicchio, M., and Dearden, L. C. (1974). Investigations of ageing in costal and tracheal cartilage in rats. *Z. Zellforsch.* **147**, 505–528.

Bonucci, E., Delmarco, A., Nicoletti, B., Petrinelli, P., and Pozzi, L. (1976). Histological and histochemical investigations of achondroplasic mice – possible model of human achondroplasia. *Growth* **40**, 241–252.

Bonucci, E., and Gherardi, G. (1975). Histochemical and electron microscopic investigations on medullary bone. *Cell Tissue Res.* **163**, 81–98.

Bonucci, E., Gherardi, G., Delmarco, A., Nicoletti, B., and Petrinelli, P. (1971). An electron microscope investigation of cartilage and bone in achondroplastic (cn/cn) mice. *J. Submicrosc. Cytol.* **9**, 299–306.

Bonucci, E., and Nardi, F. (1972). The cloverleaf skull syndrome: Histological, histochemical and ultrastructural findings. *Virchows Arch. A Pathol. Anat.* **357**, 199–212.

Boorstin, D. J. (1983). *The Discoverers.* Vintage Books, New York, NY.

Boot-Handford, R. P., and Tuckwell, D. S. (2003). Fibrillar collagen: the key to vertebrate evolution? *BioEssays* **25**, 142–151.

Bordat, C. (1987). Ultrastructural study of the vertebrae of the selachian *Scyliorhinus canicula* L. *Can. J. Zool.* **65**, 1435–1444.

Bordier, P., Hioco, D., Rouquier, M., Hepner, G. W., and Thompson, G. R. (1969). Effects of intravenous vitamin D on bone and phosphate metabolism in osteomalacia. *Calcif. Tissue Res.* **4**, 78–83.

Borg, T. K., Runyan, R., and Wuthier, R. E. (1981). A freeze-fracture study of avian epiphyseal cartilage differentiation. *Anat. Rec.* **199**, 449–457.

Borgens, R. B. (1982a). What is the role of naturally produced electric current in vertebrate regeneration and healing? *Int. Rev. Cytol.* **76**, 245–298.

Borgens, R. B. (1982b). Mice regrow the tips of their foretoes. *Science* **217**, 747–750.

Borgens, R. B., Vanable, J. W., Jr., and Jaffe, L. F. (1979a). Bioelectricity and regeneration. *BioScience* **29**, 468–474.

Borgens, R. B., Vanable, J. W., Jr., and Jaffe, L. F. (1979b). Reduction of sodium dependent stump currents disturbs urodele limb regeneration. *J. Exp. Zool.* **209**, 377–386.

Borkhvardt, V. G. (1983). The development of the cranio-vertebral joint in salamanders. *Vest. Leningrad. Univ.* (*Biol.*) **1983**, 26–33 (in Russian).

Bornstein, P. (1974). The biosynthesis of collagen. *Annu. Rev. Biochem.* **43**, 567–604.

Borstlap, W. A., Neidbuch, K. L., Freihofe, H. P., and Kulipers, A. M. (1990). Early secondary bone grafting of alveolar cleft defects: A comparison between chin and rib grafts. *J. Cranio-Max.-Fac. Surg.* **18**, 201–205.

Bos, P. K., van Osch, G. J. V. M., Frenz, D. A., Verhaar, J. A. N., and Verwoerd-Verhoef, H. L. (2001). Growth factor expression in cartilage wound healing: temporal and spatial immunolocalization in a rabbit auricular cartilage wound model. *Osteoarth. & Cart.* **9**, 382–389.

Boskey, A. L., Blank, R. D., and Doty, S. B. (2001). Vitamin C-sulfate inhibits mineralization in chondrocyte cultures: a caveat. *Matrix Biol.* **20**, 99–106.

Bostrom, M. P. G., Saleh, K. J., and Einhorn, T. A. (1999). Osteoinductive growth factors in preclinical fracture and long bone defect models. *Orthop. Clin. N. Amer.* **30**, 647–658.

Boughton, D. A., Collette, B. B., and McCune, A. R. (1991). Heterochrony in jaw morphology of needlefishes (Teleostei: Belonidae). *Syst. Zool.* **40**, 329–354.

Boulet, A. M., and Capecchi, M. R. (2004). Multiple roles of Hoxa11 and Hoxd11 in the formation of the mammalian forelimb zeugopod. *Development* **131**, 299–309.

Bourne, G. H. (ed.) (1971–1976). *The Biochemistry and Physiology of Bone*, 2nd Edn, Volumes 1–4. Academic Press, New York, NY.

Bourne, G. H. (ed.) (1972). Vitamin C and bone. In *The Biochemistry and Physiology of Bone*, 2nd Edn, Volume 2, pp. 231–281. Academic Press, New York, NY.

Bourque, W. T., Gross, M., and Hall, B. K. (1992). A reproducible method for producing and quantifying the stages of fracture repair. *Lab. Animal Sci.* **42**, 369–374.

Bourque, W. T., Gross, M., and Hall, B. K. (1993a). A histological processing technique which preserves the integrity of calcified tissues (bone, enamel), yolky amphibian embryos and growth factor antigens in skeletal tissues. *J. Histochem. Cytochem.* **41**, 1429–1434.

Bourque, W. T., Gross, M., and Hall, B. K. (1993b). The presence of PDGF, aFGF, IGF-I, and TGF-β during four stages of fracture repair. *Int. J. Devel. Biol.* **37**, 573–579.

Bourret, L. A., Goetinck, P. F., Hintz, R., and Rodan, G. A. (1979). Cyclic 3′,5′-AMP changes in chondrocytes of the proteoglycan deficient chick embryonic mutant, nanomelia. *FEBS Letts* **108**, 353–355.

Bourret, L. A., and Rodan, G. A. (1976b). The role of calcium in the inhibition of cAMP accumulation in epiphyseal cartilage cells exposed to physiological pressure. *J. Cell. Physiol.* **88**, 353–362.

Bourret, L. A., and Rodan, G. A. (1976b). Inhibition of cAMP accumulation in epiphyseal cartilage cells exposed to physiological pressure. *Calcif. Tissue Res.* **21**, Suppl., 431–436.

Bouvet, J. (1974). Différenciation et ultrastructure du squelette distal de la nageoire pectorale chez la truite indigène (*Salmo trutta* fario L). I. Différenciation et ultrastructure des actinotriches. *Arch. Anat. Microsc. Morphol. Exp.* **63**, 79–86.

Bouvet, J. (1976). Enveloping layer and periderm of the trout embryo (*Salmo trutta fario* L). *Cell Tissue Res.* **179**, 367–382.

Bouvet, J. (1978). Cell proliferation and morphogenesis of apical ectodermal ridge in pectoral fin bud of trout embryos (*Salmo-trutta-fario* L). *W. Roux. Arch. Entwicklungs. Org.* **185**, 137–154.

Bouvier, M. (1987). Variation in alkaline phosphatase activity with changing load on the mandibular condylar cartilage in the rat. *Archs Oral Biol.* **32**, 671–677.

Bouvier, M., Couble, M. L., Hartmann, D. J., and Magloire, H. (1991). Isolation and characterization of rat alveolar bone cells. *Cell. Mol. Biol.* **37**, 509–517.

Bouvier, M., and Hylander, W. L. (1981). Effects of bone strain on cortical bone structure in Macaques (*Macaca mulatta*). *J. Morphol.* **167**, 1–12.

Bouxsein, M. L., Turek, T. J., Blake, C. A., D'Augusta, D., Li, X., Stevens, M., Seehermah, H. J., and Wozney, J. M. (2001). Recombinant human bone morphogenetic protein-2 accelerates healing in a rabbit ulnar osteotomy model. *J. Bone Joint Surg.* **83A**, 1219–1230.

Bowen, I. D., and Lockshin, R. A. (eds) (1981). *Cell Death in Biology and Pathology.* Chapman & Hall, London.

Bowler, P. J. (1996) *Life's Splendid Drama. Evolutionary Biology and the Reconstruction of Life's Ancestry 1860–1940.* The University of Chicago Press, Chicago, IL.

Bowman, B. M., and Miller, S. C. (1986). The proliferation and differentiation of the bone-lining cells in estrogen-induced osteogenesis. *Bone* **7**, 351–357.

Boyde, A., Ali, N. N., and Jones, S. J. (1984). Resorption of dentine by isolated osteoclasts *in vitro*. *Br. Dent. J.* **156**, 216–220.

Boyde, A., and Hobdell, M. H. (1969). Scanning electron microscopy of primary membrane bone. *Zellforschung Mikr. Anat.* **99**, 98–108.

Boyde, A., and Jones, S. J. (1983). Scanning electron microscopy of cartilage. In *Cartilage, Volume 1, Structure, Function and Biochemistry* (B. K. Hall, ed.), pp. 105–148. Academic Press, New York, NY.

Boyde, A., Jones, S. J., Binderman, I., and Harell, A. (1976). Scanning electron microscopy of bone cells in culture. *Cell Tissue Res.* **166**, 65–70.

Boyde, A., and Shapiro, I. M. (1986). Mineralization pattern and morphology of the normal and rachitic chick growth cartilage. *Bone* **7**, 161–162.

Boyde, A., and Shapiro, I. M. (1987). Morphological observations concerning the pattern of mineralization of the normal and the rachitic chick growth cartilage. *Anat. Embryol.* **175**, 457–466.

Boyle, W. J., Simonet, W. S., and Lacey, D. L. (2003). Osteoclast differentiation and activation. *Nature* **423**, 337–342.

Brachet, A. (1893). Étude sur la resorption des cartilages et la développement des os longs chez les oiseaux. *Int. Monatsschr. Anat. Physiol.* **10**, 391–417.

Bradamante, Z., and Hall, B. K. (1980). The role of epithelial collagen and proteoglycan in the initiation of osteogenesis by avian neural crest cells. *Anat. Rec.* **197**, 305–315.

Bradamante, Z., Kostovic-Knezevic, L., Levak-Svajger, B., and Svajger, A. (1991). Differentiation of the secondary elastic cartilage in the external ear of the rat. *Int. J. Devel. Biol.* **35**, 311–320.

Bradamante, Z., Kostovic-Knezevic, L., and Svajger, A. (1975). Light and electron-microscopic observations on the presence of pre-elastic (oxytalan) fibres around the mature cartilage in the external ear of the rat. *Experientia* **31**, 979–980.

Bradamante, Z., and Svajger, A. (1981). Scanning electron microscopy of elastic cartilage in the rat external ear. *Folia Anat. Jugosl.* **11**, 25–34.

Bradham, D. M., Passaniti, A., and Horton, W. E., Jr. (1995). Mesenchymal cell chondrogenesis is stimulated by basement membrane matrix and inhibited by age-associated factors. *Matrix Biol.* **14**, 561–571.

Bradley, J. P., Levine, J. P., McCarthy, J. G., and Longaker, M. T. (1997). Studies in cranial suture biology – regional dura mater determines in vitro cranial suture fusion. *Plast. Reconstruct. Surg.* **100**, 1091–1099.

Bradley, S. J. (1970). An analysis of self-differentiation of chick limb buds in chorio-allantoic grafts. *J. Anat.* **107**, 479–490.

Brand, R. A. (1979). Fracture healing. In *The Scientific Basis of Orthopaedics* (J. A. Albright and R. A. Brand, eds), pp. 289–312. Appleton-Century-Crofts, New York, NY.

Brand-Saberi, B., and Christ, B. (2000). Evolution and development of distinct cell lineages derived from somites. *Curr. Top. Devel. Biol.* **48**, 1–42.

Brand-Saberi, B., Wilting, J., Ebensperger, C., and Christ, B. (1996). The formation of somite compartments in the avian embryo. *Int. J. Devel. Biol.* **40**, 411–420.

Brandt, K. D. (1974). Enhanced extractability of articular cartilage proteoglycans in osteoarthritis. *Biochem. J.* **143**, 475–478.

Brandt, K. D. (2001). *An Atlas of Osteoarthritis.* CRC Press, Boca Raton, FL.

Brash, J. C. (1924). *The Growth of the Jaws and Palate. Five Lectures on The Growth of the Jaws, Normal and Abnormal, in Health and Disease.* Dental Board of the United Kingdom, London.

Brash J. C. (1926). The growth of the alveolar bone. *Trans. Brit. Soc. Study Orthodontics,* **1926**, 43–96.

Brash, J. C. (1934). Some problems in the growth and developmental mechanics of bone. *Edin. Med. J.* **41**, 305–319, 363–386.

Braut, A., Kollar, E. J., and Mina, M. (2003). Analysis of the odontogenic and osteogenic potentials of dental pulp *in vivo* using a Col1a1–2.3-GFP transgene. *Int. J. Devel. Biol.* **47**, 281–291.

Brazelton, T. R., Rossi, F. M. V., Keshet, G. I., and Blau, H. M. (2000). From marrow to brain: expression of neuronal phenotypes in adult mice. *Science* **290**, 1775–1779.

Breen, E. C., Ignotz, R. A., McCabe, L., Stein, J. L., Stein, G. S., and Lian, J. B. (1994). TGFβ alters growth and differentiation related gene expression in proliferating osteoblasts *in vitro*, preventing development of the mature bone phenotype. *J. Cell. Physiol.* **160**, 323–335.

Brehe, J. E., and Fleming, W. R. (1976). Calcium mobilization from acellular bone and effects of hypophysectomy on calcium metabolism in *Fundulus kansae*. *J. Comp. Physiol. B.* **110**, 159–170.

Brem, H., and Folkman, J. (1975). Inhibition of tumor angiogenesis mediated by cartilage. *J. Exp. Med.* **141**, 427–439.

Brennan, M. J., Oldberg, Å., Ruoslahti, E., Brown, K., and Schwartz, N. (1983). Immunological evidence for two distinct chondroitin sulfate proteoglycan core proteins: Differential expression in cartilage matrix deficient mice. *Devel. Biol.* **98**, 139–147.

Brewer, A. K., Johnson, D. R., and Moore, W. J. (1977). Further studies on skull growth in achondroplastic (*cn*) mice. *J. Embryol. Exp. Morphol.* **39**, 59–70.

Brewer, S., Feng, W. G., Huang, J., Sullivan, S., and Williams, T. (2004). Wnt1-Cre-mediated deletion of AP-2 alpha causes multiple neural crest-related defects. *Devel. Biol.* **267**, 135–152.

Brewster, B. (1987). Eye migration and cranial development during flatfish metamorphosis: A reappraisal (Teleostei: Pleuronectiformes). *J. Fish. Biol.* **31**, 805–833.

Brickell, P. M. (1995). *MHox* and vertebrate skeletogenesis: The long and the short of it. *BioEssays* **17**, 750–753.

Brickell, P. M., and Tickle, C. (1989). Morphogens in chick limb development. *BioEssays* **11**, 145–149.

Bridges, J. B. (1959). Experimental heterotopic ossification. *Int. Rev. Cytol.* **8**, 253–278.

Bridges, J. B., and Pritchard, J. J. (1958). Bone and cartilage induction in the rabbit. *J. Anat.* **92**, 28–38.

Brigham, G., Scaletta, L., Johnson, L., Jr., and Occhino, J. (1977). Antigenic differences among condylar, epiphyseal and nasal septal cartilages. In *The Biology of Occlusal Development* (J. A. M. McNamara, Jr., ed.), pp. 313–331. University of Michigan, Ann Arbor.

Brighton, C. T. (1978). Structure and function of growth plate. *Clin. Orthop. Rel. Res.* **136**, 22–32.

Brighton, C. T., Adler, S., Black, J., Itada, N., and Friedenberg, Z. B. (1975). Cathodic oxygen consumption and electrically induced osteogenesis. *Clin. Orthop. Rel. Res.* **107**, 277–282.

Brighton, C. T., Black, J., and Pollack, S. R. (eds) (1979). *Electrical Properties of Bone and Cartilage. Experimental Effects and Clinical Applications.* Grune & Stratton, New York, NY.

Brighton, C. T., Cronkey, J. E., and Osterman, A. L. (1976). *In vitro* epiphyseal-plate growth in various constant electrical fields. *J. Bone Joint Surg. Am. Vol.* **58**, 971–977.

Brighton, C. T., and Friedenberg, Z. B. (1974). Electrical stimulation and oxygen tension. *Ann. NY. Acad. Sci.* **238**, 314–320.

Brighton, C. T., and Heppenstall, R. B. (1971). Oxygen tension in zones of the epiphyseal plate, the metaphysis, and diaphysis. An *in vitro* and *in vivo* study in rats and rabbits. *J. Bone Joint Surg. Am. Vol.* **53**, 719–728.

Brighton, C. T., and Hunt, R. M. (1974). Mitochondrial calcium and its role in calcification. Histochemical localization of calcium in electron micrographs of the epiphyseal growth plate with K-pyroantiminate. *Clin. Orthop. Rel. Res.* **100**, 406–416.

Brighton, C. T., and Hunt, R. M. (1986). Ultrastructure of electrically induced osteogenesis in the rabbit medullary canal. *J. Orthop. Rel. Res.* **4**, 27–36.

Brighton, C. T., and Krebs, A. G. (1972a). Oxygen tension of healing fractures in the rabbit. *J. Bone Joint Surg. Am. Vol.* **54**, 323–332.

Brighton, C. T., and Krebs, A. G. (1972b). Oxygen tension of nonunion of fractured femurs in the rabbit. *Surg. Gynecol. Obstet.* **135**, 379–385.

Brighton, C. T., Lackman, R. D., and Cuckler, J. M. (1983). Absence of the glycerol phosphate shuttle in the various zones of the growth plate. *J. Bone Joint Surg. Am. Vol.* **65**, 663–666.

Brighton, C. T., Lane, J. M., and Koh, J. K. (1974). *In vitro* rabbit articular cartilage organ model. II. ^{35}S incorporation in various oxygen tensions. *Arthritis Res.* **17**, 245–252.

Brighton, C. T., Ray, R. D., Soble, L. W., and Kuettner, K. E. (1969). *In vitro* epiphyseal-plate growth in various oxygen tensions. *J. Bone Joint Surg. Am. Vol.* **51**, 1383–1396.

Brighton, C. T., and Schaffzin, E. A. (1970). Comparison of the effects of excess vitamin A and high oxygen tension on *in vitro* epiphyseal plate growth. 1. Morphological response. *Calcif. Tissue Res.* **6**, 151–161.

Brighton, C. T., Sugioka, Y., and Hunt, R. M. (1973). Cytoplasmic structures of epiphyseal plate chondrocytes. Quantitative evaluation using electron micrographs of rat costochondral junctions with special reference to the fate of hypertrophic cells. *J. Bone Joint Surg. Am. Vol.* **55**, 771–784.

Britto, J. A., Moore, R. L., Evans, R. D., Hayward, R. D., and Jones, B. M. (2001). Negative autoregulation of fibroblast growth factor receptor 2 expression characterizing cranial development in cases of Apert (P253R mutation) and Pfeiffer (C278F mutation) syndromes and suggesting a basis for differences in their cranial phenotypes. *J. Neurosurg.* **95**, 660–673.

Britto, J. M., Tannahill, D., and Keynes, R. J. (2000). Life, death and sonic hedgehog. *BioEssays* **22**, 499–502.

Broadbent, B. H. (1931). A new x-ray technique and its application to orthodontia. *Angle Orthodontist* **1**, 45–66.

Broadbent, B. H. (1937). The face of the normal child. *Angle Orthodontist* **7**, 209–233.

Brochu, C. A. (1996). Closure of neurocentral sutures during crocodilian ontogeny: Implications for maturity assessment in fossil archosaurs. *J. Vert. Paleontol.* **16**, 49–57.

Brochu, C. A. (2003). Osteology of Tyrannosaurus rex: insights from a nearly complete skeleton and high-resolution computed tomographic analysis of the skull. *J. Vert. Paleontol.* **22** (Suppl.), 1–138 (Mem. Soc. Vert. Paleont, #7).

Brockes, J. P. (1991). Some current problems in amphibian limb regeneration. *Phil. Trans R. Soc. B.* **331**, 287–290.

Brodie, A. G. (1941). On the growth pattern of the human head from the third month to the eighth year of life. *Amer. J. Anat.* **68**, 209–262.

Brodie, P. F. (1969). Mandibular layering in *Delphinapterus leucas* and age determination. *Nature* **221**, 956–958.

Bromage, T. G. (1987). The scanning electron microscopy replica technique and recent applications to the study of fossil bone. *Scan. Microsc.* **1**, 607–614.

Bronner, F., and Worrell, R. V. (1999). *Orthopaedics. Principles of Basic and Clinical Science.* CRC Press, Boca Raton, FL.

Brooker, A. F., Bowerman, J. W., and Robinson, R. A. (1973). Ectopic ossification following total hip replacement. Incidence and a method of classification. *J. Bone Joint Surg. Am. Vol.* **55**, 1629–1632.

Brooks, M. (1971). *The Blood Supply of Bone. An Approach to Bone Biology.* Butterworths, London.

Broom, R. (1950). *Finding the Missing Link. An Account of Recent Discoveries Throwing New Light on the Origin of Man.* Watts & Co., London.

Brothers, E. B. (1984). Otolith studies. In *Ontogeny and Systematics of Fishes* (H. G. Moser, ed.), pp. 50–57. Special Publication Number 1, American Society of Ichthyologists and Herpetologists, Lawrence, KS.

Browman, H. I. (1989). Embryology, ethology and ecology of odtogenetic critical periods in fish. *Brain Behav. Evol.* **34**, 5–12.

Brown, D., Wagner, D., Li, X-q., Richardson, J. A., and Olson, E. N. (1999). Dual role of the basic helix–loop–helix transcription factor scleraxis in mesoderm formation and chondrogenesis during mouse embryogenesis. *Development* **126**, 4317–4329.

Brown, D. D., Wang, Z., Furlow, J. D., Kanamori, A., Schwartzman, R. A., Remo, B. F., and Pinder, A. (1996). The thyroid hormone-induced tail resorption program during *Xenopus laevis* metamorphosis. *Proc. Natl Acad. Sci. U.S.A.* **93**, 1924–1929.

Brown, K. S., Cranley, R. E., Greene, R., and Kleinman, H. K., and Pennypacker, J. P. (1981). Disproportionate micromelia

(*Dmm*): An incomplete dominant mouse dwarfism with abnormal cartilage matrix. *J. Embryol. Exp. Morphol.* **62**, 165–182.

Brown, N. A., Hoyle, C. I., McCarthy, A., and Wolpert, L. (1989). The development of asymmetry: the sidedness of drug-induced limb abnormalities is reversed in *situs inversus* mice. *Development* **107**, 637–642.

Brown, R. A., Taylor, C., McLaughlin, B., McFarland, C. D., Weiss, J. B., and Ali, S. Y. (1987). Epiphyseal growth plate cartilage and chondrocytes in mineralizing cultures produce a low molecular mass angiogenic procollagenase activator. *Bone Min.* **3**, 143–158.

Brown, R. D., Chao, C. C., and Faulkner, L. W. (1983a). Hormone levels and antler development in white-tailed and Sika deer. *Comp. Biochem. Physiol. A Comp. Physiol.* **75**, 385–390.

Brown, R. D., Chao, C. C., and Faulkner, L. W. (1983b). The endocrine control of the initiation and growth of antlers in white-tailed deer. *Acta Endocrinol.* **103**, 138–144.

Brown, T. A., and Heeley, J. D. (1974). Maturational changes in the cartilaginous nasal septum of the rat. *J. Dent. Res.* **53**, 246.

Brownell, A. G., Gerth, N., and Finerman, G. A. M. (1988). Isolation, partial purification and *in vitro* characterization of osteogenic inhibitory protein. *Conn. Tissue Res.* **17**, 261–276.

Brownlow, H. C., Reed, A., and Simpson, A. H. R. W. (2001). Growth factor expression during the development of atrophic non-union injury. *Int. J. Care Injured* **32**, 519–524.

Brownlow, H. C., and Simpson, A. H. R. W. (2000). Metabolic activity of a new atrophic nonunion model in rabbits. *J. Orthoped. Res.* **18**, 438–442.

Bruce, J. A. (1941). Time and order of appearance of ossification centres and their development in the skull of the rabbit. *Amer. J. Anat.* **68**, 41–67.

Bruckner, P., Hörler, I., Mendler, M., Houze, Y., Winterhalter, K. H., Eich-Bender, S. G., and Spycher, M. A. (1989). Induction and prevention of chondrocyte hypertrophy in culture. *J. Cell Biol.* **109**, 2537–2545.

Bruder, S. P., Fink, D. J., and Caplan, A. I. (1994). Mesenchymal stem cells in bone development, bone repair, and skeletal regeneration therapy. *J. Cell. Biochem.* **56**, 283–294.

Brunet, L. J., McMahon, J. A., McMahon, A. P., and Harland, R. M. (1998). Noggin, cartilage morphogenesis and joint formation in the mammalian skeleton. *Science* **280**, 1455–1457.

Bruns, R. R., Trelstad, R. L., and Gross, J. (1973). Cartilage collagen: A staggered substructure in reconstituted fibrils. *Science* **181**, 269–271.

Bryant, H. N., and Russell, A. P. (1993). The occurrence of clavicles within Dinosauria: Implications for the homology of the avian furcula and the utility of negative evidence. *J. Vert. Paleont.* **13**, 171–184.

Bryant, P. J., Bryant, S. V., and French, V. (1977). Biological regeneration and pattern formation. *Sci. Amer.* **237**, 66–81.

Bryant, P. J., and Simpson, P. (1984). Intrinsic and extrinsic control of growth in developing organs. *Q. Rev. Biol.* **59**, 387–415.

Bryant, S. V. (1976). Regenerative failure of double half limbs of *Notophthalmus viridescens*. *Nature* **263**, 676–679.

Bryant, S. V. (1977). Pattern regulation in amphibian limbs. In *Vertebrate Limb and Somite Morphogenesis* (D. A. Ede, J. R. Hinchliffe, and M. Balls, eds), pp. 311–318. Cambridge University Press, Cambridge.

Bryant, S. V., and Bellairs, A. d'A. (1967a). Tail regeneration in the lizards *Anguis fragilis* and *Lacerta dugesii*. *J. Linn. Soc. London* **46**, 297–305.

Bryant, S. V., and Bellairs, A. d'A. (1967b). Amnio-allantoic constriction bands in lizard embryos and their effect on tail regeneration. *J. Zool. London* **152**, 155–162.

Bryant, S. V., and Bellairs, A. d'A. (1967). Development of regenerative ability in the lizard, *Lacerta vivipara*. *Amer. Zool.* **10**, 167–173.

Bryant, S. V., and Wozny, K. J. (1974). Stimulation of limb regeneration in the lizard *Xantusia vigilis* by means of ependymal implants. *J. Exp. Zool.* **189**, 339–352.

Bryden, M. M., and Felts, W. J. L. (1974). Quantitative anatomical observations on the skeletal and muscular systems of four species of Antarctic seals. *J. Anat.* **118**, 589–600.

Bubenik, G. A., Brown, G. M., Bubenik, A. B., and Grota, L. J. (1974). Immunolocalization localization of testosterone in the growing antler of the white-tailed deer (*Odocoileus virginianus*). *Calcif. Tissue Res.* **14**, 121–130.

Bubenik, G. A., and Bubenik, A. B. (eds) (1990). *Horns, Pronghorns and Antlers. Evolution, Morphology, Physiology and Social Significance*. Springer-Verlag, New York, NY.

Bubenik, G. A., Bubenik, A. B., Brown, G. M., and Wilson, D. A. (1975). The role of sex hormones in the growth of antler bone tissue. I. Endocrine and metabolic effects of antiandrogen therapy. *J. Exp. Zool.* **194**, 349–358.

Bubenik, G. A., Bubenik, A. B., Stevens, E. D., and Binnington, A. G. (1982). The effect of neurogenic stimulation on the development and growth of bony tissues. *J. Exp. Zool.* **219**, 205–216.

Bubenik, G. A., Reyes, E., Schams, D., Lobos, A., Bartos, L., and Koerner, F. (2002). Effect of antiandrogen cyproterone acetate on the development of the antler cycle in Southern pudu (*Puda puda*). *J. Exp. Zool.* **292**, 393–401.

Buchberger, A., Schwarzer, M., Brand, T., Pabst, O., Seidl, K., and Arnold, H.-H. (1998). Chicken winged-helix transcription factor CFKH-1 prefigures axial and appendicular skeletal structures during chicken embryogenesis. *Devel. Dynam.* **212**, 94–101.

Buchignani, J. S., Cook, A. J., and Anderson, L. G. (1972). Roentgenographic findings in familial osteodysplasia. *Amer. J. Roentgenol.* **116**, 602–608.

Buckland-Wright, J. C. (1978). Bone structure and the patterns of force transmission in the cat skull (*Felis catus*). *J. Morphol.* **155**, 35–62.

Buckley, M. J., Banes, A. J., Levin, L. G., Sumpio, B. E., Sato, M., Jordan, R., Gilbert, J., Link, G. W., and Tran Son Tay, R. (1988). Osteoblasts increase their rate of division and align in response to cyclic, mechanical tension *in vitro*. *Bone Min.* **4**, 225–236.

Buckwater, J. A., Mower, D., Ungar, R., Schaeffer, J., and Ginsberg, B. (1986). Morphometric analysis of chondrocyte hypertrophy. *J. Bone Joint Surg. Am. Vol.* **68**, 243–255.

Budenz, R. W., and Bernard, G. W. (1980). Osteogenesis and leukopoiesis within diffusion-chamber implants of isolated bone marrow subpopulations. *Amer. J. Anat.* **159**, 455–474.

Burch, J. G. (1966). The cranial attachment of the spheno-mandibular (tympanomandibular) ligament. *Anat. Rec.* **156**, 433–438.

Burch, W. M., and Lebovitz, H. E. (1982). Triiodothyroxine stimulation of *in vitro* growth and maturation of embryonic chick cartilage. *Endocrinology* **111**, 462–468.

Burch, W. M., Weir, S., and Van Wyk, J. J. (1986). Embryonic chick cartilage produces its own somatomedin-like peptide to stimulate cartilage growth *in vitro*. *Endocrinology* **119**, 1370–1376.

Burger, E. H., Boonekamp, P. M., and Nijweide, P. J. (1986). Osteoblast and osteoclast precursors in primary cultures of calvarial bone cells. *Anat. Rec.* **214**, 32–40.

Burger, E. H., Klein-Nulend, J., and Veldhuijzen, J. P. (1991). Modulation of osteogenesis in fetal bone rudiments by mechanical stress *in vitro*. *J. Biomech.* **24** (Suppl. 1), 101–109.

Burger, E. H., van der Meer, J. W. M., van de Gevel, J. S., Gribnau, J. C., Thesingh, C. W., and van Furth, R. (1982). *In vitro* formation of osteoclasts from long-term cultures of bone marrow mononuclear phagocytes. *J. Exp. Med.* **156**, 1604–1614.

Burger, E. H., van der Meer, J. W. M., and Nijweide, P. J. (1984). Osteoclast formation from mononuclear phagocytes: Role of bone-forming cells. *J. Cell Biol.* **99**, 1901–1906.

Burger, M. M., Bombik, B. M., Breckingridge, B. M., and Sheppard, J. R. (1972). Growth control and cyclic alterations of cyclic AMP in the cell cycle. *Nature, New Biol.* **239**, 161–163.

Burgeson, R. E., and Hollister, D. W. (1979). Collagen heterogeneity in human cartilage: Identification of several new collagen chains. *Biochem. Biophys. Res. Commun.* **87**, 1124–1131.

Burgess, A. M. C. (1974). Genome control and the genetic potentialities of the nuclei of dedifferentiated regeneration blastema cells. In *Neoplasia and Cell Differentiation* (G. V. Sherbet, ed.), pp. 106–152. Karger, Basel.

Burgin, T. (1986). *Beiträge zur Kopfanatomie der Plattfische (Teleostei, Pleuronectiformes). Eine beschreibend-morphologische Studie, unter Einbeziehung funktioneller Gesichtspunkte.* Ph.D. Thesis, University of Basel, Basel, Switzerland.

Buring, K. (1974). Studies on bone induction. *Akad. Avh., Malmo*, pp. 1–20.

Buring, K. (1975). On the origin of cells in heterotopic bone formation. *Clin. Orthop. Rel. Res.* **110**, 293–302.

Burke, A. C. (1989). Development of the turtle carapace: Implications for the evolution of a novel bauplan. *J. Morphol.* **199**, 363–378.

Burke, A. C. (2000). *Hox* genes and the global patterning of the somitic mesoderm. *Curr. Top. Devel. Biol.* **47**, 155–180.

Burke, A. C., and Alberch, P. (1985). The development and homology of the chelonian carpus and tarsus. *J. Morphol.* **186**, 119–131.

Burke, A. C., and Rosa Molinar, E. (2002). Starting from fins: Parallelism in the evolution of fins and genitalia. Part two: Fins to limbs. *Evol. & Devel.* **4**, 375–377.

Burke, A. C., Nelson, C. E., Morgan, B. A., and Tabin, C. (1995). *Hox* genes and the evolution of vertebrate axial morphology. *Development* **121**, 333–346.

Burke, A. C., and Nowicki, J. L. (2003). A new view of patterning domains in the vertebrate mesoderm. *Developmental Cell* **4**, 159–165.

Burkhart, J. M., and Jowsey, J. (1967). Parathyroid and thyroid hormones in the development of immobilization osteoporosis. *Endocrinology* **81**, 1053–1062.

Burks, J. K., and Peck, W. A. (1978). Bone cells: A serum-free medium supports proliferation in primary culture. *Science* **199**, 542–544.

Burns, E. R. (2000). Biological time and *in vivo* research: A field guide to pitfalls. *Anat. Rec.* (*New Anat.*) **261**, 141–152.

Burr, H. S. (1916). The effects of the removal of the nasal pits in *Ambystoma* embryos. *J. Exp. Zool.* **20**, 27–57.

Burrow, C. J. (2003). Comment on 'Separate evolutionary origins of teeth from evidence in fossil jawed vertebrates' *Science* **300**, 1661.

Burstone, M. S. (1959). Histochemical demonstration of acid phosphatase activity in osteoclasts. *J. Histochem. Cytochem.* **7**, 39–41.

Burton, P. J. K. (1973). Structure of the depressor mandibulae muscle in the kokako *Callaeas cinerea*. *Ibis* **115**, 138–139.

Butler, E. G. (1933). The effects of x-radiation on the regeneration of the forelimbs of *Ambystoma* larvae. *J. Exp. Zool.* **65**, 271–315.

Butler, W. T., Mikulski, A., and Urist, M. R. (1977). Noncollagenous proteins of a rat dentin matrix possessing bone morphogenetic activity. *J. Dent. Res.* **56**, 228–232.

Butler, W. T., Miller, E. J., Finch, J. E., Jr., and Inagami, T. (1974). Homologous regions of collagen. *Biochem. Biophys. Res. Commun.* **57**, a1(I) and a1(II) chains: Apparent clustering of variable and invariant amino acid residues. 190–195.

Buxton, P. G., Hall, B. K., Archer, C. W., and Francis-West, P. (2003). Secondary chondrocyte-derived Ihh stimulates proliferation of periosteal cells during chick cranial development. *Development* **130**, 4729–4739.

C

Cabrini, R. L. (1961). Histochemistry of ossification. *Int. Rev. Cytol.* **11**, 283–306.

Cadinouche, M. Z. A., Liversage, R. A., Muller, W., and Tsilfidis, C. (1999). Molecular cloning of the *Notophthalmus viridescens Radical Fringe* cDNA and characterization of its expression during forelimb development and adult forelimb regeneration. *Devel. Dynam.* **214**, 259–268.

Cahill, D. R., and Marks, S. C., Jr. (1980). Tooth eruption: Evidence for the central role of the dental follicle. *J. Oral Pathol.* **9**, 189–200.

Cahn, R. D., and Lasher, R. (1967). Simultaneous synthesis of DNA and specialized cellular products by differentiated cartilage cells *in vitro*. *Proc. Natl Acad. Sci. U.S.A.* **58**, 1131–1138.

Cairns, J. M. (1965). Development of grafts from mouse embryos to the wing bud of the chick embryo. *Devel. Biol.* **12**, 36–52.

Cairns, J. M. (1975). The function of the ectodermal apical ridge and distinctive characteristics of adjacent distal mesoderm in the avian wing bud. *J. Embryol. Exp. Morphol.* **34**, 155–169.

Cairns, J. M., and Saunders, J. W., Jr. (1954). The influence of embryonic mesoderm on the regional specification of epidermal derivatives in the chick. *J. Exp. Zool.* **127**, 221–248.

Calandra, A. J., and MacCabe, J. A. (1978). The in vitro maintenance of the limb-bud apical ridge by cell-free preparations. *Devel. Biol.* **62**, 258–269.

Caldwell, M. W. (1996). Ontogeny and phylogeny of the mesopodial skeleton in mosasauroid reptiles. *Zool. J. Linn. Soc. Lond.* **116**, 407–436.

Caldwell, M. W. (1997a). Limb ossification patterns of the ichthyosaur *Stenopterygius*, and a discussion of the proximal tarsal row of ichthyosaurs and other neodiapsid reptiles. *Zool. J. Linn. Soc. Lond.* **120**, 1–25.

Caldwell, M. W. (1997b). Modified perichondral ossification patterns and the evolution of paddle-like limbs in fossil marine reptiles. *J. Vert. Paleont.* **17**, 534–547.

Caldwell, M. W. (1997c). Limb osteology and ossification patterns in *Cryptoclidus* (Reptilia: Plesiosauroidea) with a review of sauropterygian limbs. *J. Vert. Paleont.* **17**, 295–307.

Caldwell, M. W. (2003). 'Without a leg to stand on': On the evolution and development of axial elongation and limblessness in tetrapods. *J. Earth Sci.* **40**, 573–588.

Caldwell, M. W., and Lee, M. S. Y. (1997). A snake with legs from the marine Cretaceous of the Middle East. *Nature* **386**, 705–709.

Calhoun, N. R., Smith, C., and Becker, K. L. (1974). The role of zinc in bone metabolism. *Clin. Orthop. Rel. Res.* **103**, 212–234.

Callery, E. M., and Elinson, R. P. (2000). Thyroid hormone-dependent metamorphosis in a direct developing frog. *Proc. Natl Acad. Sci. U.S.A.* **97**, 2615–2620.

Callery, E. M., Fang, H., and Elinson, R. P. (2001). Frogs without polliwogs: evolution of anuran direct development. *BioEssays* **23**, 233–241.

Callis, P. D. (1982). Bone development following transplants of urinary bladder wall – a quantitative histological and ultra-structural study. *J. Anat.* **135**, 53–64.

Calvi, L. M., Adams, G. B., Welbrecht, K. W., Webber, J. M., Olson, D. P., Knight, M. C., martin, R. P., Schipani, E., Divieti, P., Brinhurst, F. R., Milner, L. A., Kronenberg, H. M., and Scadden, D. T. (2003). Osteoblastic cells regulate the haematopoietic stem cell niche. *Nature* **425**, 841–846.

Cameron, D. A. (1972). The ultrastructure of bone. In *The Biochemistry and Physiology of Bone* (G. H. Bourne, ed.), 2nd Edn, Volume 1, pp. 191–236. Academic Press, New York, NY.

Cameron, D. A., Paschall, H. A., and Robinson, R. A. (1964). The ultrastructure of bone cells. In *Bone Biodynamics* (H. M. Frost, ed.), pp. 91–104. Little, Brown, Boston, MA.

Cameron, D. A., Paschall, H. A., and Robinson, R. A. (1967). Changes in the fine structure of bone cells after the administration of parathyroid extract. *J. Cell Biol.* **33**, 1–14.

Cameron, G. R. (1930). The staining of calcium. *J. Pathol. Bact.* **33**, 929–955.

Cameron, I. L., and Jeter, J. R., Jr. (1971). Relationship between cell proliferation and cytodifferentiation in embryonic chick tissues. In *Developmental Aspects of the Cell Cycle* (I. L. Cameron, G. M. Padilla, and A. M. Zimmerman, eds), pp. 191–222. Academic Press, New York, NY.

Cameron, J. A., and Fallon, J. F. (1977a). Evidence for polarizing zone in the limb bud of *Xenopus laevis. Devel. Biol.* **55**, 320–330.

Cameron, J. A., and Fallon, J. F. (1977b). The absence of cell death during development of free digits in amphibians. *Devel. Biol.* **55**, 331–337.

Campana, S. E. (1983). Feeding periodicity and the production of daily growth increments in otoliths of steelhead trout (*Salmo gairdneri*) and starry flounder (*Platichthys stellatus*). *Can. J. Zool.* **61**, 1591–1597.

Campana, S. E. (2004). *Photographic Atlas of Fish Otoliths of the Northwest Atlantic Ocean.* NRC Press, Ottawa.

Campbell, C. J. (1972). Homotransplantation of a half or whole joint. *Clin. Orthop. Rel. Res.* **87**, 146–155.

Campbell, J. G. (1966). A dinosaur bone lesion resembling avian osteopetrosis and some remarks on the mode of development of the lesions. *J. Royal Microsc. Soc.* **85**, 163–174.

Campo, R. D. (1974). Soluble and resistant proteoglycans in epiphyseal plate cartilage. *Calcif. Tissue Res.* **14**, 105–119.

Campo, R. D., and Phillips, S. J. (1973). Electron microscopic visualization of proteoglycans and collagen in bovine costal cartilage. *Calcif. Tissue Res.* **13**, 83–92.

Campo, R. D., and Tourtellote, C. D. (1967). The composition of bovine cartilage and bone. *Biochim. Biophys. Acta* **141**, 614–624.

Canalis, E., and Raisz, L. G. (1979). Effect of epidermal growth factor on bone formation *in vitro. Endocrinology* **104**, 862–869.

Canavase, B. (1987). Variabilita e corrispondenza numerica degli ossicini sclerali negli occhi di tacchino, pollo & quaglia. *Summa* **4**, 27–31.

Canavese, B., Durio, P., Bellardi, S., and Porporato, P. (1986). Caratteristiche morfologiche degli ossicini sclerali e dell'os opticus in alcune specie di uccelli domestici e selvatici. *Ann. Fac. Med. Vet. Torino* **31**, 1–23.

Cancedda, F. D., Gentili, C., Manduca, P., and Cancedda, R. (1993). Hypertrophic chondrocytes undergo further differentiation in culture. *J. Cell Biol.* **117**, 427–435.

Cancedda, R., Cancedda, F. D., and Castagnola, P. (1995). Chondrocyte differentiation. *Int. Rev. Cytol.* **159**, 265–358.

Cancedda, R., Castagnola, P., Descalzi Cancedda, F., Dozin, B., and Quarto, R. (2000). Developmental control of chondrogenesis and osteogenesis. *Int. J. Devel. Biol.* **44**, 707–714.

Cannata, S. M., Bagni, C., Bernardini, S., Christen, B., and Filoni, S. (2001). Nerve-independence of limb regeneration in larval *Xenopus laevis* is correlated to the level of fgf-2 mRNA expression in limb tissues. *Devel. Biol.* **231**, 436–446.

Capdevila, J., and Izpisúa-Belmonte, J. C. V. (2000). Perspectives on the evolutionary origin of tetrapod limbs. *J. Exp. Zool. (Mol. Devel. Evol.)*, **288**, 287–303.

Caplan, A. I. (1970). Effects of the nicotinamide-sensitive teratogen 3-acetylpyridine on chick limb cells in culture. *Exp. Cell Res.* **62**, 341–355.

Caplan, A. I. (1971). The teratogenic action of the nicotinamide analogs 3-acetylpyridine and 6-aminonicotinamide on developing chick embryos. *J. Exp. Zool.* **178**, 351–358.

Caplan, A. I. (1972a). Effects of the nicotinamide sensitive teratogen 6-aminonicotinamide on chick limb cells in culture. *Exp. Cell Res.* **70**, 185–195.

Caplan, A. I. (1972b). The effects of the nicotinamide-sensitive teratogen 3-acetylpyridine on chick limb mesodermal cells in culture: Biochemical parameters. *J. Exp. Zool.* **180**, 351–362.

Caplan, A. I. (1972c). The site and sequence of action of 6-aminonicotinamide in causing bone malformations of embryonic chick limb and its relationship to normal development. *Devel. Biol.* **28**, 71–83.

Caplan, A. I. (1972d). Comparison of the capacity of nicotinamide and nicotinic acid to relieve the effects of muscle and cartilage teratogens in developing chick embryos. *Devel. Biol.* **28**, 344–351.

Caplan, A. I. (1977). Muscle, cartilage and bone development and differentiation from chick limb mesenchymal cells. In *Vertebrate Limb and Somite Morphogenesis* (D. A. Ede, J. R. Hinchliffe, and M. Balls, eds), pp. 199–214. Cambridge University Press, Cambridge.

Caplan, A. I., and Koutroupas, S. (1973). The control of muscle and cartilage development in the chick limb: The role of differential vascularization. *J. Embryol. Exp. Morphol.* **29**, 571–584.

Caplan, A. I., and Rosenberg, M. J. (1975). Interrelationships between poly-(adenosine diphospho ribose) synthesis, intracellular NAD levels and muscle or cartilage differentiation from embryonic chick limb mesodermal cells. *Proc. Natl Acad. Sci. U.S.A.* **72**, 1852–1857.

Caplan, A. I., and Stoolmiller, A. C. (1973). Control of chondrogenic expression in mesodermal cells of embryonic chick limb. *Proc. Natl Acad. Sci. U.S.A.* **70**, 1713–1717.

Caplan, A. I., Zwilling, E., and Kaplan, N. O. (1968). 3-acetylpyridine: Effects *in vitro* related to teratogenic activity in chicken embryos. *Science* **160**, 1009–1010.

Captier, G., Cristol, R., Montoya, P., Prudhomme, M., and Godlewski, G. (2003). Prenatal chondrogenesis and morphogenesis of the sphenofrontal suture in humans. *Cells Tissues Organs* **175**, 98–104.

Cardew, G., Goode, J. (eds); *and* Hall, B. K. (Chair) (2001). *The Molecular Basis of Skeletogenesis*. Novartis Foundation Symposium 232, John Wiley & Sons, Chichester.

Carinci, P., Becchetti, E., and Bodo, M. (2000). Role of the extracellular matrix and growth factors in skull morphogenesis and in the pathogenesis of craniosynostosis. *Int. J. Devel. Biol.* **44**, 715–723.

Carlberg, A. L., Pucci, B., Rallapalli, R., Tuan, R. S., and Hall, D. J. (2000). Efficient chondrogenic differentiation of mesenchymal cells in micromass culture by retroviral transfer of BMP-2. *Differentiation* **67**, 128–138.

Carlson, B. M. (1996). *Patten's Foundations of Embryology*. Sixth Edn. McGraw Hill, New York, NY.

Carlson, D. S. (ed.) (1990). *Craniofacial Growth Theory and Orthodontic Treatment*. Monograph 23, Craniofacial Growth Series, Center for Human Growth and Development, University of Michigan, Ann Arbor, MI.

Carlson, D. S. (1999). Growth modification: from molecules to mandibles. In *Growth Modification: What Works, What Doesn't, and Why* (J. A. McNamara, Jr., ed.), pp. 17–61, Craniofacial Growth Series 35, Center for Human Growth and Development. Ann Arbor, The University of Michigan.

Carlson, E. C., and Upson, R. H. (1974). 'Native' striated collagen produced by chick notochordal epithelial cells *in vitro*. *Amer. J. Anat.* **141**, 441–446.

Carlson, E. C., Upson, R. H., and Evans, D. K. (1974). The production of extracellular connective tissue fibrils by chick notochordal epithelium *in vitro*. *Anat. Rec.* **179**, 361–374.

Carlson, M. R. J., Komine, Y., Bryant, S. V., and Gardiner, D. M. (2001). Expression of Hoxb13 and Hoxc10 in developing and regenerating axolotl limbs and tails. *Devel. Biol.* **229**, 396–406.

Carlton, M. B. L., Colledge, W. H., and Evans, M. J. (1998). Crouzon-like craniofacial dysmorphology in the mouse is caused by an insertional mutation at the FGF3/FGF4 locus. *Devel. Dynam.* **212**, 242–249.

Carriker, M. R., Person, P., Libben, R. M., and van Zandt, D. (1972). Regeneration of the proboscis of muricid gastropods after amputation, with emphasis on the radula and cartilages. *Biol. Bull.* **143**, 317–331.

Carrington, J. L. (1994). Bone morphogenetic proteins and the induction of embryonic and adult bone. In *Bone, Volume 8 Mechanisms of Bone Development and Growth* (B. K. Hall, ed.), pp. 85–108. CRC Press, Boca Raton, FL.

Carrington, J. L., Chen, P., Yanagishita, M., and Reddi, A. H. (1991). Osteogenin (bone morphogenetic protein-3) stimulates cartilage formation by chick limb bud cells *in vitro*. *Devel. Biol.* **146**, 406–415.

Carrington, J. L., and Fallon, J. F. (1984a). Evidence that the ectoderm is the affected germ layer in the *wingless* mutant chick embryo. *J. Exp. Zool.* **232**, 297–308.

Carrington, J. L., and Fallon, J. F. (1984b). The stages of flank ectoderm capable of responding to ridge induction in the chick embryo. *J. Embryol. Exp. Morphol.* **84**, 19–34.

Carrington, J. L., and Fallon, J. F. (1986). Experimental manipulation leading to induction of dorsal ectodermal ridges on normal limb buds results in a phenocopy of the *Eudilopodia* chick mutant. *Devel. Biol.* **116**, 130–137.

Carrington, J. L., and Fallon, J. F. (1988). Initial limb budding is independent of apical ectodermal ridge activity; evidence from a limbless mutant. *Development* **104**, 361–367.

Carrington, J. L., Roberts, A. B., Flanders, K. C., Roche, N. S., and Reddi, A. H. (1988). Accumulation, localization and compartmentation of transforming growth factor β during endochondral bone development. *J. Cell Biol.* **107**, 1969–1975.

Carroll, R. L. (1977). The origin of lizards. *Linn. Soc. Symp. Ser.* **#4**, 359–396.

Carroll, R. L. (1986). Developmental processes and the origin of lepospondyls. In *Studies in Herpetology* (Z. Roček, ed.), pp. 45–48. Charles University Press, Prague.

Carroll, R. L. (1989). Developmental aspects of lepospondyl vertebrae in Paleozoic tetrapods. *Hist. Biol.* **3**, 1–25.

Carroll, R. L. (1997) *Patterns and Processes of Vertebrate Evolution*. Cambridge University Press, Cambridge.

Carroll, R. L. (2002). Early land vertebrates. *Nature* **418**, 35–36.

Carroll, R. L., Kuntz, A., and Albright, K. (1999). Vertebral development and amphibian evolution. *Evol. & Devel.* **1**, 36–48.

Carter, D. H., Sloan, P., and Aaron, J. E. (1991). Immunolocalization of collagen types I and III, tenascin and fibronectin in intramembranous bone. *J. Histochem. Cytochem.* **39**, 599–606.

Carter, D. R. (1987). Mechanical loading history and skeletal biology. *J. Biomech.* **20**, 1095–1109.

Carter, D. R., Beaupre, G. S., Giori, N. J., and Helms, J. A. (1998a). Mechanobiology of skeletal regeneration. *Clin. Orthop. Rel. Res.* **355**, S41–S55.

Carter, D. R., Blenman, P. R., and Beaupré, G. S. (1988). Correlations between mechanical stress history and tissue differentiation in initial fracture healing. *J. Orthop. Res.* **6**, 736–748.

Carter, D. R., Mikic, B., and Padian, K. (1998b). Epigenetic mechanical factors in the evolution of long bone epiphyses. *Zool. J. Linn. Soc.* **123**, 163–178.

Carter, D. R., and Orr, T. E. (1992). Skeletal development and bone functional adaptation. *J. Bone Min. Res.* **7**, S389–S395.

Carter, D. R., Orr, T. E., Fyhrie, D. P., and Schurman, D. J. (1987). Influences of mechanical stress on prenatal and postnatal skeletal development. *Clin. Orthop. Rel. Res.* **219**, 237–250.

Carter, D. R., van der Meulen, M. C. H., and Beaupre, G. S. (1996). Mechanical factors in bone growth and development. *Bone* **18**, S5–S10.

Carter, D. R., and Wong, M. (1988a). Mechanical stresses and endochondral ossification in the chondroepiphysis. *J. Orthop. Res.* **6**, 148–154.

Carter, D. R., and Wong, M. (1988b). The role of mechanical loading histories in the development of diarthrodial joints. *J. Orthop. Res.* **6**, 804–816.

Carter, D. R., Wong, M., and Orr, T. E. (1991). Musculoskeletal ontogeny, phylogeny and functional adaptation. *J. Biomech.* **24**, Suppl. 1, 3–16.

Carter, J. G. (ed.) (1990). *Skeletal Biomineralization: Patterns, Processes and Evolutionary Trends. Volumes 1 and 2*. Van Nostrand Reinhold, New York, NY.

Caruccio, N. C., Martinez-Lopez, A., Harris, M., Dvorak, L., Bitgood, J., Simandl, B. K., and Fallon, J. F. (1999). Constitutive activation of sonic hedgehog signaling in the chicken mutant *talpid*[2]: Shh-independent outgrowth and polarizing activity. *Devel. Biol.* **212**, 137–149.

Carvalho, R. S., Yen, E. H. K., and Suga, D. M. (1993). The effect of growth on collagen and glycosaminoglycans in the articular disc of the rat temporomandibular joint. *Archs Oral Biol.* **38**, 457–466.

Casey, J., and Lawson, R. (1977). Amphibians with scales: the structure of the scale in the caecilian: *Hypogeophis rostratus.* *Brit. J. Herpet.* **5**, 831–833.

Casser-Bette, M., Murray, A. B., Closs, E. I., Erfle, V., and Schmidt, J. (1990). Bone formation by osteoblast-like cells in a three-dimensional cell culture. *Calcif. Tissue Int.* **46**, 46–56.

Cassin, C. (1977). Crête neurale et capacité morphogénétique du stomodeum chez *Pleurodeles waltlii* (Amphibien urodèle). *Rev. Stomato-Odontol.* **118**, 149–162.

Cassin, C., and Capuron, A. (1972). Obtention d'ouvertures buccales et de bouches complètes des implantation dans le blastocoèle, de tissues embryonnaires de *Pleurodeles waltlii* Michah (Amphibien Urodèle). *C.R. Hebd. Sciences Acad. Sci.* **275**, 2953–2956.

Cassin, C., and Capuron, A. (1977). Evolution de la capacité morphogénétique de la région stomodéale chez l'embryon de *Pleurodeles waltlii* Michah (Amphibien Urodéle). Etude par transplantation intrablastocélienne et par culture *in vitro*. *Wilhelm Roux Arch. Entwicklungsmech. Org.* **181**, 107–112.

Cassin, C., and Capuron, A. (1979). Buccal organogenesis in *Pleurodeles waltl* Michah (Urodele, Amphibian). Study by intrablastocoelic transplantation and *in vitro* culture. *J. Biol. Buccale* **7**, 61–76.

Castagnola, P., Bet, P., Quarto, R., Gennari, M., and Cancedda, R. (1991). cDNA cloning and gene expression of chicken osteopontin – expression of osteopontin messenger RNA in chondrocytes is enhanced by trypsin treatment of cells. *J. Biol. Chem.* **266**, 9944–9949.

Castagnola, P., Moro, G., Descalzi-Cancedda, F., and Cancedda, R. (1986). Type X collagen synthesis during *in vitro* development of chick embryo tibial chondrocytes. *J. Cell Biol.* **102**, 2310–2317.

Castagnola, P., Torella, G., and Cancedda, R. (1987). Type X collagen synthesis by cultured chondrocytes derived from the permanent cartilaginous region of chick embryo sternum. *Devel. Biol.* **123**, 332–337.

Castanet, J., and Cheylan, M. (1979). Les marques de croissance des os et des écailles comme indicateur de l'age chez *Testudo hermanni* et *Testudo graeca* (Reptilia, Chelonia, Testudinidae). *Can. J. Zool.* **57**, 1649–1665.

Castanet, J., Francillon-Vieillot, H., Meunier, F. J., and de Ricqlès, A. (1993). Bone and individual aging. In *Bone, Volume 7: Bone Growth – B* (B. K. Hall, ed.), pp. 245–283. CRC Press, Boca Raton, FL.

Castro, R. F., Jackson, K. A., Goodell, M. A., Robertson, C. S., Liu, H., and Shine, H. D. (2002). Failure of bone marrow cells to transdifferentiate into neural cells *in vivo*. *Science* **297**, 1299.

Caton, A., Hacker, A., Naeem, A., Livet, J., Maina, F., Bladt, F., Klein, R., Birchmeier, C., and Guthire, S. (2000). The branchial arches and HGF are growth-promoting and chemoattractant for cranial motor axons. *Development* **127**, 1751–1766.

Caubet, J.-F., and Bernaudin, J.-F. (1988). Expression of the *c-fos* oncogene in bone, cartilage and tooth forming tissues during mouse development. *Biol. Cell* **64**, 101–104.

Caverzasio, J., Palmer, G., and Bonjour, J. P. (1998). Fluoride – Mode of action. *Bone* **22**, 585–589.

Cecil, M. L., and Tassava, R. A. (1986). Vitamin A enhances fore-limb regeneration in juvenile leopard frogs *Rana pipiens.* *J. Exp. Zool.* **237**, 57–61.

Celeste, A. J., Rosen, V., Buecker, J. L., Kriz, R., Wang, E. A., and Wozney, J. M. (1986). Isolation of the human gene for bone GLA protein utilizing mouse and rat cDNA clones. *EMBO J.* **5**, 1885–1890.

Centrella, M., Canalis, E., McCarthy, T. L., Stewart, A. F., Orloff, J. J., and Insogna, K. L. (1989b). Parathyroid hormone-related protein modulates the effect of transforming growth factor-β on deoxyribonucleic acid and collagen synthesis in fetal rat bone cells. *Endocrinology* **125**, 199–208.

Centrella, M., Massagué, J., and Canalis, E. (1986). Human platelet-derived transforming growth factor-β stimulates parameters of bone growth in fetal rat calvariae. *Endocrinology* **119**, 2306–2312.

Centrella, M., McCarthy, T. L., and Canalis, E. (1987a). Mitogenesis in fetal rat bone cells simultaneously exposed to type β transforming growth factor and other growth regulators. *FASEB J.* **1**, 312–317.

Centrella, M., McCarthy, T. L., and Canalis, E. (1987b). Transforming growth factor β is a bifunctional regulator of replication and collagen synthesis in osteoblast-enriched cell cultures from fetal rat bone. *J. Biol. Chem.* **262**, 2869–2874.

Centrella, M., McCarthy, T. L., and Canalis, E. (1988). Skeletal tissue and transforming growth factor β. *FASEB J.* **2**, 3066–3073.

Centrella, M., McCarthy, T. L., and Canalis, E. (1989a). Effects of transforming growth factors on bone cells. *Conn. Tissue Int.* **20**, 267–275.

Centrella, M., McCarthy, T. L., and Canalis, E. (1989c). Platelet-derived growth factor enhances deoxyribonucleic acid and collagen synthesis in osteoblast-enriched cultures from fetal rat parietal bone. *Endocrinology* **125**, 13–19.

Centrella, M., McCarthy, T. L., and Canalis, E. (1990). Receptors for insulin-like growth factors-I and -II in osteoblast-enriched cultures from fetal rat bone. *Endocrinology* **126**, 39–44.

Cerny, R., Meulemans, D., Berger, J., Wilsch-Bräuninger, M., Kurth, T., Bronner-Fraser, M., and Epperlein, H.-H. (2004). Combined intrinsic and extrinsic influences pattern cranial neural crest migration and pharyngeal arch morphogenesis in axolotl. *Devel. Biol.* **266**, 252–269.

Chacko, S., Abbott, J., Holtzer, S., and Holtzer, H. (1969). The loss of phenotypic traits by differentiated cells. VI. Behavior of the progeny of a single chondrocyte. *J. Exp. Med.* **130**, 417–442.

Chai, C. K. (1981). Dactylaplasia in mice. A two-locus model for developmental anomalies. *J. Hered.* **72**, 234–237.

Chai, Y., Jiang, X., Ito, Y., Bringas, P., Jr., Hun, J., Rowitch, D. H., Soriano, P., McMahon, A. P., and Sucov, H. M. (2000). Fate of the mammalian cranial neural crest during tooth and mandibular morphogenesis. *Development* **127**, 1671–1679.

Chalmers, J. (1965). A study of some of the factors controlling growth of transplanted skeletal tissues (fetal mouse). *Proc. Eur. Calcif. Tissues Congr. Colloq. Unit. Liège, 1964,* 2nd Volume **31**, pp. 177–184.

Chalmers, J., Gray, D. H., and Rush, J. (1975). Observations on the induction of bone in soft tissues. *J. Bone Joint Surg. Br. Vol.* **57**, 36–45.

Chalmers, J., and Ray, R. D. (1962). Transplantation immunity in bone homografting. *J. Bone Joint Surg. Br. Vol.* **44**, 149–164.

Chamay, A., and Tschantz, P. (1972). Mechanical influence in bone remodeling. Experimental research on Wolff's law. *J. Biomech.* **5**, 173–180.

Chambers, D., and McGonnell, I. M. (2002). Neural crest: facing the facts of head development. *Trends Genet.* **18**, 381–384.

Chambers, T. J. (1980). The cellular basis of bone resorption. *Clin. Orthop. Rel. Res.* **151**, 283–293.

Chambers, T. J. (1983). Osteoblasts release osteoclasts from calcitonin-induced quiescence. *J. Cell Sci.* **57**, 247–260.

Chambers, T. J. (1985). The pathobiology of the osteoclast. *J. Clin. Pathol.* **38**, 241–252.

Chambers, T. J. (2000). Regulation of the differentiation and function of osteoclasts. *J. Pathol.* **192**, 4–13.

Chambers, T. J., Fuller, K., Darby, J. A., Pringle, J. A. S., and Horton, M. A. (1986). Monoclonal antibodies against osteoclasts inhibit bone resorption *in vitro*. *Bone Min.* **1**, 127–136.

Chambers, T. J., and Magnus, C. J. (1982). Calcitonin alters behaviour of isolated osteoclasts. *J. Pathol.* **136**, 27–40.

Chambers, T. J., and Moore, A. (1983). The sensitivity of isolated osteoclasts to morphological transformation by calcitonin. *J. Clin. Endocrinol. Metabol.* **57**, 819–824.

Chambers, T. J., Revell, P. A., Fuller, K., and Athanson, N. A. (1984a). Resorption of bone by isolated rabbit osteoclasts. *J. Cell Sci.* **66**, 383–400.

Chambers, T. J., Thomson, B. M., and Fuller, K. (1984b). Effect of substrate composition on bone resorption by rabbit osteoclasts. *J. Cell Sci.* **70**, 61–72.

Chandraraj, S., and Briggs, C. A. (1988). Role of cartilage canals in osteogenesis and growth of the vertebral centra. *J. Anat.* **158**, 121–136.

Chang, C., Gendler, E., Nishimoto, S. K., Stryker, W. F., and Nimni, M. E. (1985). Age dependence of bone matrix induced bone formation. In *Current Advances in Skeletogenesis* (A. Ornoy, A. Harell, and J. Sela, eds), pp. 27–32. Excerpta Medica, Amsterdam.

Chang, K. S., and Snellen, J. W. (1982). Bioelectric activity in the rabbit ear regeneration. *J. Exp. Zool.* **221**, 193–204.

Chao, C. C., and Brown, R. D. (1984). Measurement of parathyroid hormone in white-tailed deer. *Ann. Rep. Caesar Kleberg Wildlife Res. Inst.*, June 1984, 34–35.

Chapman, J. D., Sturrock, J., Boad, J. W., and Crookall, J. O. (1970). Factors affecting the oxygen tension around cells growing in plastic Petri dishes. *Int. J. Radiat. Biol.* **17**, 305–328.

Chappard, D., Alexandre, C., Chol, R., and Riffat, G. (1983). Les canaux intrachondraux: Histogénèse, anatomie et histophysiologie de la vascularisation cartilagineuse du foetus humain. *Lyon Med.* **149**, 417–428.

Chappard, D., Alexandre, C., Chol, R., and Riffat, G. (1986). Uncalcified cartilage resorption in human fetal cartilage canals. *Tissue Cell* **18**, 701–708.

Charité, J., McFadden, D. G., and Olson, E. N. (2000). The bHLH transcription factor dHAND controls *Sonic hedgehog* expression and establishment of the zone of polarizing activity during limb development. *Development* **127**, 2461–2470.

Charlier, J.-P., and Petrovic, A. (1967). Recherches sur la mandibule de rat en culture d'organes: Le cartilage condylien a-t-ill un potentiel de croissance independant? *Orthod. Fr.* **38**, 1–11.

Charlier, J.-P., Petrovic, A., and Herrmann-Stutzmann, J. (1969a). Effects of mandibular hyperpropulsion on the prechondroblastic zone of young rat condyle. *Amer. J. Orthod.* **55**, 71–74.

Charlier, J.-P., Petrovic, A., and Linck, G. (1969b). La fronde mentonnière et son action sur la croissance mandibulaire. Recherches expérimentales chez le rat. *Orthod. Fr.* **40**, 99–113.

Chaube, S. (1959). On axiation and symmetry in transplanted wing of the chick. *J. Exp. Zool.* **140**, 29–78.

Chavassieux, P., Boivin, G., Serre, C. M., and Meunier, P. J. (1993a). Fluoride increases rat osteoblast function and population after *in vivo* administration but not after *in vitro* exposure. *Bone* **14**, 721–725.

Chavassieux, P., Chenu, C., Valentin-Opran, A., Delmas, P. D., Boivin, G., Chapuy, M. C., and Meunier, P. J. (1993b). *In vitro* exposure to sodium fluoride does not modify activity or proliferation of human osteoblastic cells in primary culture. *J. Bone Min. Res.* **8**, 37–44.

Chen, J., Singh, K., Mukherjee, B. B., and Sodek, J. (1993). Developmental expression of osteopontin (OPN) mRNA in rat tissues: Evidence for a role for OPN in bone formation and resorption. *Matrix* **13**, 113–123.

Chen, J. M. (1952a). Studies on the morphogenesis of the mouse sternum. I. Normal embryonic development. *J. Anat.* **86**, 373–386.

Chen, J. M. (1952b). Studies on the morphogenesis of the mouse sternum. II. Experiments on the origin of the sternum and its capacity for self-differentiation *in vitro*. *J. Anat.* **86**, 387–401.

Chen, J. M. (1953). Studies on the morphogenesis of the mouse sternum. III. Experiments on the closure and segmentation of the sternal bands. *J. Anat.* **87**, 130–149.

Chen. J. M. (1954). The effect of insulin on embryonic limb bones cultivated *in vitro*. *J. Physiol.* **125**, 148–162.

Chen, J.-Y., Huang, D.-Y., and Li, C.-W. (1999). An early Cambrian craniate-like chordate. *Nature* **402**, 518–522.

Chen, P., Carrington, J. L., Hammonds, R. G., and Reddi, A. H. (1991). Stimulation of chondrogenesis in limb bud mesoderm cells by recombinant human bone morphogenetic protein-2B (BMP-2B) and modulation by transforming growth factor-β1 and factor-β2. *Exp. Cell Res.* **195**, 509–515.

Chen, P., Carrington, J. L., Paralkar, V. M., Pierce, G. F., and Reddi, A. H. (1992). Chick limb bud mesodermal cell chondrogenesis: Inhibition by isoforms of platelet-derived growth factor and reversal by recombinant bone morphogenetic protein. *Exp. Cell Res.* **200**, 110–117.

Chen, Q., Gibney, E. P., Leach, R. M., and Linsenmayer, T. F. (1993). Chicken tibial dyschondroplasia: A limb mutant with two growth plates and possible defects of collagen crosslinking. *Devel. Dynam.* **196**, 54–61.

Chen, Q., Johnson, D. M., Haudenschild, D. R., and Goetinck, P. F. (1995). Progression and recapitulation of the chondrocyte differentiation program: Cartilage matrix protein is a marker for cartilage maturation. *Devel. Biol.* **172**, 293–306.

Chen, Y., Knezevic, V., Ervin, V., Hutson, R., Ward, Y., and Mackem, S. (2004). Direct interaction with Hoxd proteins reverses Gli3-repressor function to pronote digit formation downstream of Shh. *Development* **131**, 2339–2347.

Cheng, P. T., and Bader, S. M. (1990). Effects of fluoride on rat cancellous bone. *Bone & Min.* **11**, 153–162.

Cheselden, W. (1733). *Osteographia, or the Anatomy of the Bones*. London.

Chevallier, A. (1975). Rôle du mésoderme somitic dans le développement de la cage thoracique de l'embryon d'oiseau. 1. Origine du segment sternal et méchanismes de la différenciation des côtes. *J. Embryol. Exp. Morphol.* **33**, 291–333.

Chevallier, A. (1977). Origine des ceintures scapulaires et pelviennes chez l'embryon d'oiseau. *J. Embryol. Exp. Morphol.* **42**, 275–295.

Chevallier, A. (1978). Migration of somitic cells into the somatopleural mesoderm of the limb anlage. *W. Roux. Archiv. Entwicklungs. Mech.* **184**, 57–74.

Chevallier, A. (1979). Role of the somitic mesoderm in the development of the thorax in bird embryos. II. Origin of thoracic and appendicular musculature. *J. Embryol. Exp. Morphol.* **49**, 73–88.

Chevallier, A., Kieny, M., and Mauger, A. (1977). Limb-somite relationship: Origin of the limb musculature. *J. Embryol. Exp. Morphol.* **41**, 245–258.

Cheverud, J. M. (2004). Modular pleiotropic effects of quantitative trait loci on morphological traits. In *Modularity in Development and Evolution* (G. Schlosser and G. P. Wagner, eds), pp. 132–153. The University of Chicago Press, Chicago, IL.

Cheverud, J. M., and Richtsmeier, J. T. (1986). Finite-element scaling applied to sexual dimorphism in Rhesus macaque (*Macaca mulatta*) facial growth. *Syst. Zool.* **35**, 381–399.

Cheverud, J. M., Routman, E. J., and Irschick, D. J. (1997). Pleiotropic effects of individual gene loci on mandibular morphology. *Evolution* **51**, 2006–2016.

Cheverud, J. M., Rutledge, J. J., and Atchley, W. R. (1983). Quantitative genetics of development: genetic correlations among age-specific trait values and the evolution of ontogeny. *Evolution* **37**, 895–905.

Chiakulas, J. J. (1957). The specificity and differential fusion of cartilage derived from mesoderm and mesectoderm. *J. Exp. Zool.* **136**, 287–300.

Chiappe, L. M., Salgado, L., and Coria, R. A. (2001). Embryonic skulls of titanosaur sauropod dinosaurs. *Science* **293**, 2444–2446.

Chibon, P. (1967). Marquage nucléaire par la thymidine tritée des dérivés de la crête neurale chez l'amphibien Urodèle *Pleurodeles waltlii* Michah. *J. Embryol. Exp. Morphol.* **18**, 343–358.

Chiego, D. J., and Singh, I. J. (1981). Evaluation of the effects of sensory denervation on osteoblasts by ^{3}H-proline autoradiography. *Cell Tissue Res.* **217**, 569–576.

Chimal-Monroy, J., and Diez de León, L. (1999). Expression of N-cadherin, N-CAM, fibronectin and tenascin is stimulated by TGF-β1, β2, β3 and β5 during the formation of precartilage condensations. *Int. J. Devel. Biol.* **43**, 59–67.

Chimal-Monroy, J., Rodriguez-Leon, J., Montero, J. A., Gañan, Y., Macias, D., Merino, R., and Hurlé, J. M. (2003). Analysis of the molecular cascade for mesodermal limb chondrogenesis: *Sox* genes and BMP signaling. *Devel. Biol.* **257**, 292–301.

Chin, J. R., and Werb, Z. (1997). Matrix metalloproteinases regulate morphogenesis, migration and remodeling of epithelium, tongue, skeletal muscle and cartilage in the mandibular arch. *Development* **124**, 1519–1530.

Chiquet-Ehrismann, R., Kalla, P., Pearson, C. A., Beck, K., and Chiquet, M. (1988). Tenascin interferes with fibronectin action. *Cell* **53**, 383–390.

Chiu, C.-H., and Hamrick, M. W. (2002). Evolution and development of the primate limb skeleton. *Evol. Anthropol.* **11**, 94–107.

Cho, M.-I., and Garant, P. R. (1984). Formation of multinucleated fibroblasts in the periodontal ligaments of old mice. *Anat. Rec.* **208**, 185–196.

Chong, D. A., Evans, C. A., and Heeley, J. D. (1982). Morphology and maturation of the periosteum of the rat mandible. *Archs Oral Biol.* **27**, 777–785.

Chow, S. Y., Chow, Y. C., Jee, W. S. S., and Woodbury, D. M. (1984). Electrophysiological properties of osteoblast-like cells from the cortical endosteal surface of rabbit long bones. *Calcif. Tissue Res.* **36**, 401–408.

Chrisman, K., Kenney, R., Comin, J., Thal, T., Suchocki, L., Yueh, Y. G., and Gardner, D. P. (2004). Gestational ethanol exposure disrupts the expression of FGF8 and sonic hedgehog during limb patterning. *Birth defects Res. (Part A)* **70**, 163–171.

Christ, B., Jacob, H. J., and Jacob, M. (1977). Experimental analysis of the origin of the wing musculature in avian embryos. *Anat. Embryol.* **150**, 171–186.

Christ, B., Jacob, H. J., and Jacob, M. (1979). Differentiating abilities of avian somatopleural mesoderm. *Experientia* **35**, 1376–1378.

Christ, B., Schmidt, C., Huang, R., Wilting, J., and Brand-Saberi, B. (1998). Segmentation of the vertebrate body. *Anat. Embryol.* **197**, 1–8.

Christel, P., Cerf, G., and Pilla, A. A. (1980). Modulation of rat radial osteotomy repair using electromagnetic current induction. In *Mechanisms of Growth Control* (R. O. Becker, ed.), pp. 237–250. C. C. Thomas, Springfield, IL.

Christensen, R. N., Weinstein, M., and Tassave, R. A. (2001). Fibroblast growth factors in regenerating limbs of *Ambystoma*: Cloning and semi-quantitative RT-PCR expression studies. *J. Exp. Zool.* **290**, 529–540.

Christoffersen, J., and Landis, W. J. (1991). A contribution with review to the description of mineralization of bone and other calcified tissues *in vivo*. *Anat. Rec.* **230**, 435–450.

Chung, K. S., and Nishimura, I. (1999). Maintenance of regional histodifferentiation patterns and a spatially restricted expression of type X collagen in rat Meckel's cartilage explants in vitro. *Archs Oral Biol.* **44**, 489–497.

Chung, K. S., Park, H. H., Ting, K., Takita, H., Apte, S. S., Kuboki, Y., and Nishimura, I. (1995). Modulated expression of type X collagen in the Meckel's cartilage with different developmental fates. *Devel. Biol.* **170**, 387–396.

Chung, U.-I., Lanske, B., Lee, K., Li, E., and Kronenberg, H. (1998). The parathyroid hormone/parathyroid hormone-related peptide receptor coordinates endochondral bone development by directly controlling chondrocyte differentiation. *Proc. Natl Acad. Sci. U.S.A.* **95**, 13030–13035.

Chuong, C.-M., Widelitz, R. B., Jiang, T.-X., Abbott, U. K., Lee, Y.-S., and Chen, H.-M. (1993). Roles of adhesion molecules NCAM and tenascin in limb skeletogenesis: Analysis with antibody perturbation, exogenous gene expression, *talpid²* mutants and activin stimulation. In *Limb Development and Regeneration* (J. Fallon, ed.), pp. 465–474. Wiley-Liss, New York, NY.

Church, L. E., and Johnson, L. C. (1964). Growth of long bones in the chick. Rates of growth in length, and diameter of the humerus, tibia, metatarsus. *Amer. J. Anat.* **114**, 521–538.

Chyun, Y. S., Kream, B. E., and Raisz, L. G. (1984). Cortisol decreases bone formation by inhibiting periosteal cell proliferation. *Endocrinology* **114**, 477–480.

Ciochon, R. L., Nisbett, R. A., and Corruccini, R. S. (1997). Dietary consistency and craniofacial development related to masticatory function in minipigs. *J. Craniofac. Genet. Devel. Biol.* **17**, 96–102.

Cione, A. L., and Torno, A. E. (1987). Atavistic vomerine teeth in a specimen of *Pogonias cromis* (Linnaeus, 1776) (Teleostei, Perciformes). *Copeia* **1987**, 1057–1059.

Claassen, H., Kampen, W. U., and Kirsch, T. (1995). Localization of type I and II collagen during development of human first rib cartilage. *Anat. Embryol.* **192**, 329–334.

Claassen, H., and Kirsch, T. (1994). Temporal and spatial localization of type I and II collagens in human thyroid cartilage. *Anat. Embryol.* **189**, 237–242.

Claassen, H., Kirsch, T., and Simons, G. (1996). Cartilage canals in human thyroid cartilage characterized by immunolocalization of collagen types I, II, pro-III, IV and X. *Anat. Embryol.* **194**, 147–153.

Clack, J. A. (2002a). *Gaining Ground. The Origin and Evolution of Tetrapods.* Indiana University Press, Bloomington, IN.

Clack, J. A. (2002b). An early tetrapod from 'Romer's Gap.' *Nature* **418**, 72–76.

Clark, C. T., and Smith, K. K. (1993). Cranial osteogenesis in *Monodelphis domestica* (Didelphidae) and *Macropus eugenii* (Macropodidae). *J. Morphol.* **215**, 119–149.

Clark, L. D., Clark, R. K., and Heber-Katz, E. (1998). A new murine model for mammalian wound repair and regeneration. *Clin. Immunol. Immunopathol.* **88**, 35–45.

Clarke, P. G. H., and Clarke, S. (1996). Nineteenth century research on naturally occurring cell death and related phenomena. *Anat. Embryol.* **193**, 81–99.

Clase, K. L., Mitchell, P. J., Ward, P. J., Dorman, C. M., Johnson, S. E., and Hannon, K. (2000). FGF5 stimulates expansion of connective tissue fibroblasts and inhibits skeletal muscle development in the limb. *Devel. Dynam.* **219**, 368–380.

Cleall, J. F. (1974). Growth of the palate and maxillary dental arch. *J. Dent. Res.* **53**, 1226–1234.

Clemen, G. (1988). Experimental analysis of the capacity of dental lamina in *Ambystoma mexicanum* Shaw. *Archs Oral Biol.* **99**, 111–132.

Clemen, G., Wanningger, A.-C., and Greven, H. (1997). The development of the dentigerous bones and teeth in the hemiramphid fish *Dermogenys pusillus* (Atheriniformes, Teleostei). *Ann. Anat.* **179**, 165–174.

Clement, J. G. (1992). Re-examination of the fine structure of endoskeletal mineralization in chondrichthyans: Implications for growth, ageing and calcium homeostasis. *Aust. J. Mar. Freshwater Res.* **43**, 157–181.

Closs, E. I., Murray, A. B., Schmidt, J., Schön, A., Erfle, V., and Strauss, P. G. (1990). c-fos expression precedes osteogenic differentiation of cartilage cells *in vitro*. *J Cell Biol.* **111**, 1313–1323.

Clouthier, D. E., Williams, S. C., Yanagisawa, H., Wieduwilt, M., Richardson, J. A., and Yanagisawa, M. (2000). Signaling pathways crucial for craniofacial development revealed by endothelin-A receptor-deficient mice. *Devel. Biol.* **217**, 10–24.

Coates, M. I. (1994). The origin of vertebrate limbs. *Development* Suppl. 169–180.

Coates, M. I. (2003). The evolution of paired fins. *Theory Biosci.* **122**, 266–287.

Coates, M. I., and Clack, J. A. (1990). Polydactyly in the earliest known tetrapod limbs. *Nature* **347**, 66–69.

Coates, M. I., and Clack, J. A. (1995). Romers gap: Tetrapod origins and terrestriality. *Bull. Mus. Natl. d'Hist. Nat.* **17**, 373–388.

Coates, M. I., and Cohn, M. J. (1998). Fins, limbs, and tails: Outgrowths and axial patterning in vertebrate evolution. *BioEssays* **20**, 371–381.

Coates, M. I., and Cohn, M. J. (1999). Vertebrate axial and appendicular patterning: The early development of paired appendages. *Amer. Zool.* **39**, 676–685.

Cock, A. G. (1966). Genetical aspects of metrical growth and form in animals. *Q. Rev. Biol.* **41**, 131–190.

Coelho, C. N. D., and Kosher, R. A. (1991a). Gap junctional communication during limb cartilage differentiation. *Devel. Biol.* **144**, 47–53.

Coelho, C. N. D., and Kosher, R. A. (1991b). A gradient of gap junctional communication along the anterior–posterior axis of the developing chick limb bud. *Devel. Biol.* **148**, 529–535.

Coelho, C. N. D., Upholt, W. B., and Kosher, R. A. (1992). Role of the chick homeobox-containing genes *GHox-4.6* and *GHox-8* in the specification of positional identities during the development of normal and polydactylous chick limb buds. *Development* **115**, 629–637.

Coetzee, J. M. (1999). *Disgrace.* Penguin Books, New York, NY.

Coffin, P. A., and Hall, B. K. (1974). Isozymes of lactate dehydrogenase (LDH) in skeletal tissues of the embryonic and newly hatched chick. *J. Embryol. Exp. Morphol.* **31**, 169–181.

Coffin-Collins, P. A., and Hall, B. K. (1989). Chondrogenesis of mandibular mesenchyme from the embryonic chick is inhibited by mandibular epithelium and by epidermal growth factor. *Int. J. Devel. Biol.* **33**, 297–311.

Cogliano, A., Mock, D., Birek, C., Pawson, A., and Tenenbaum, H. C. (1987). *In vitro* transformation of osteoblasts: Putative formation of osteosarcoma *in vitro*. *Bone* **8**, 299–304.

Cohen, A. M., and Hay, E. D. (1971). Secretion of collagen by embryonic neuroepithelium at the time of spinal cord–somite interaction. *Devel. Biol.* **26**, 578–605.

Cohen, A. M., and Konigsberg, I. R. (1975). A clonal approach to the problem of neural crest determination. *Devel. Biol.* **46**, 262–280.

Cohen, M. M., Jr. (ed.) (1986). *Craniosynostosis. Diagnosis, Evaluation and Management.* Raven Press, New York, NY.

Cohen, M. M., Jr. (1998). Achondroplasia, hypochondroplasia and thanatophoric dysplasia: clinically related skeletal dysplasias that are also related at the molecular level. *Int. J. Oral Maxillofac. Surg.* **27**, 451–455.

Cohen, M. M., Jr. (2000a). Craniofacial disorders caused by mutations in homeobox genes *MSX1* and *MSX2*. *J. Craniofac. Genet. Devel. Biol.* **20**, 19–25.

Cohen, M. M., Jr. (2000b). Merging the old skeletal biology with the new. I. Intramembranous ossification, endochondral ossification, ectopic bone, secondary cartilage and pathologic considerations. *J. Craniofac. Genet. Devel. Biol.* **20**, 84–93.

Cohen, M. M. Jr. (2000c). Merging the old skeletal biology with the new. II. Molecular aspects of bone formation and bone growth. *J. Craniofac. Genet. Devel. Biol.* **20**, 94–106.

Cohen, M. M., Jr. (2001). RUNX genes, neoplasia, and cleidocranial dysplasia. *Amer. J. Med. Genet.* **104**, 185–188.

Cohen, M. M., Jr. (2002). Some chondrodysplasias with short limbs: molecular perspectives. *Amer. J. Med. Genet.* **112**, 304–313.

Cohen, M. M., Jr. (2003). TGFβ/Smad signaling system and its pathologic correlates. *Amer. J. Med. Genet.* **11A**, 1–10.

Cohen, M. M., Jr., and MacLean, R. E. (eds) (2000). *Craniosynostosis: Diagnosis, Evaluation, and Management.* Oxford University Press, New York, NY.

Cohn, D. V., and Wong, G. L. (1979). Isolated bone cells. In *Skeletal Research: An Experimental Approach* (D. J. Simmons and A. S. Kunin, eds), Volume 1, pp. 3–20. Academic Press, New York, NY.

Cohn, M. J. (2000). Giving limbs a hand. *Nature* **406**, 953–954.

Cohn, M. J. (2002). Lamprey *Hox* genes and the origin of jaws. *Nature* **416**, 386–387.

Cohn, M. J., Lovejoy, C. O., Wolpert, L., and Coates, M. I. (2002). Branching, segmentation and the metapterygial axis: pattern versus process in the vertebrate limb. *BioEssays* **24**, 460–465.

Cohn, M. J., and Tickle, C. (1999). Developmental basis of limblessness and axial patterning in snakes. *Nature* **399**, 474–479.

Cohn, S. A. (1964). Regeneration of the mandibular condyle in the mouse after condylectomy. *Anat. Rec.* **148**, 272.

Coiter, V. (1559). *Tractatus anatomicus & medicus, De Ossibus Infantis., cognoscendis, conservandis, & curandis.* Groningen.

Cole, A. A., and Cole, M. B., Jr. (1989). Are perivascular cells in cartilage canals chondrocytes? *J. Anat.* **165**, 1–8.

Cole, A. A., and Wezeman, F. H. (1985). Perivascular cells in cartilage canals of the developing mouse epiphysis. *Amer. J. Anat.* **174**, 119–129.

Cole, A, Fedak, T. J., Hall, B. K., Olson, W., and Vickaryous, M. (2003). Sutures joining ontogeny and fossils. *Palaeont. Assoc. Newsletter* **52**, 29–32.

Cole, A. G., and Hall, B. K. (2002). Non-vertebrate cartilage: a phylogenetic analysis. *Amer. Zool.* **41**, 1413–1414.

Cole, A. G., and Hall, B. K. (2003). Constructing cartilage: How do cephalopods do it? *Integ. Comp. Biol.* **42**, 1210.

Cole, A. G., and Hall, B. K. (2004a). Cartilage is a metazoan tissue; integrating data from non-vertebrate sources. *Acta Zoologica* **85**, 65–80.

Cole, A. G., and Hall, B. K. (2004b). The nature and significance of invertebrate cartilages revisited: Distribution and histology of cartilage and cartilage-like tissues within the Metazoa. *Zoology* **107**, 261–274.

Coleman, C. M., and Tuan, R. S. (2003). Growth/differentiation factor 5 enhances chondrocyte maturation. *Devel. Dynam.* **228**, 208–216.

Collin-Osdoby, P., Rothe, L., Bekker, S., Anderson, F., Huang, Y., and Osdoby, P. (2002). Basic fibroblast growth factor stimulates osteoclast recruitment, development, and bone pit resorption in association with angiogenesis *in vivo* on the chick chorioallantoic membrane and activates isolated avian osteoclast resorption *in vitro*. *J. Bone Min. Res.* **17**, 1859–1871.

Colnot, C., Sidhu, S. S., Balmain, N., and Poirier, F. (2001). Uncoupling of chondrocyte death and vascular invasion in mouse galectin 3 null mutant bones. *Devel. Biol.* **229**, 203–214.

Colnot, C., Thompson, Z., Miclau, T., Werb, Z., and Helms, J. A. (2003). Altered fracture repair in the absence of MMP9. *Development* **130**, 4123–4133.

Colognato, H., and Yurchenco, P. D. (2000). Form and function: The laminin family of heterodimers. *Devel. Dynam.* **218**, 213–234.

Colombier, M. L., Lafont, J., Blanquaert, F., Caruelle, J. P., Barritault, D., and Saffar, J. L. (1999). A single low dose of RGTA®, a new healing agent, hastens wound maturation and enhances bone deposition in rat craniotomy defects. *Cells Tissue Org.* **164**, 131–140.

Coltman, D. W., O'Donoghue, P., Jorgerson, J. T., Hogg, J. T., Strobeck, C., and Festa-Blanchet, M. (2003). Undesirable evolutionary consequences of trophy hunting. *Nature* **426**, 655–658.

Colvin, J. S., Bohne, B. A., Harding, G. W., McEwen, D. G., and Ornitz, D. M. (1996). Skeletal overgrowth and deafness in mice lacking fibroblast growth factor receptor 3. *Nature Genet.* **12**, 390–397.

Comper, W. D. (ed.) (1996). *Extracellular Matrix*. 2 Vols. Harwood Academic Publishers, Amsterdam.

Condie, B. G., and Capecchi, M. R. (1993). Mice homozygous for a targeted disruption of Hoxd-3 (Hox-4.1) exhibit anterior transformations of the first and second vertebrae, the atlas and axis. *Development* **119**, 579–595.

Connolly, J. F. (ed.) (1981). *Clinical Applications of Bioelectrical Effects*. *Clin. Orthop. Rel. Res.* **161**, 1–162.

Connor, J. M. (1983). *Soft Tissue Ossification*. Springer-Verlag, Berlin.

Conrad, G. W. (1970). Collagen and mucopolysaccharide synthesis in mass cultures and clones of chick corneal fibroblasts *in vitro*. *Devel. Biol.* **21**, 611–635.

Conrad, G. W., Hart, G. W., and Chen, Y. (1977). Differences *in vitro* between fibroblast-like cells from cornea, heart, and skin of embryonic chicks. *J. Cell Sci.* **26**, 119–138.

Conway Morris, S. (1994). Why molecular biology needs palaeontology. *Development* **1994** Suppl., 1–13.

Conway Morris, S. (1998). Palaeontology: grasping the opportunity in the science of the twenty-first century. *Geobios*, **30**, 895–904.

Cook, P. R. (1974). On the inheritance of differentiated traits. *Biol. Rev. Cambridge Philos. Soc.* **49**, 51–84.

Cook, S. D., Baffes, G. C., Wolfe, M. W., Sampath, T. K., and Rueger, D. C. (1994a). Recombinant human bone morphogenetic protein-7 induces healing in a canine long-bone segmental defect model. *Clin. Orthop.* **301**, 302–312.

Cook, S. D., Baffes, G. C., Wolfe, M. W., Sampath, T. K., Rueger, D. C., and Whitecloud, T. S. II. (1994b). The effect of recombinant human osteogenic protein-1 on healing of large segmental bone defects. *J. Bone Joint Surg.* **76A**, 827–838.

Cook, S. D., Wolfe, M. W., Salkeld, S. L., and Rueger, D. C. (1995). Effect of recombinant human osteogenic protein-1 on healing of segmental defects in non-human primates. bone defects. *J. Bone Joint Surg.* **77A**, 734–750.

Cook, T. A. (1914). *The Curves of Life, being an Account of Spiral Formations, and their Application to growth in Nature, to Science, and to Art; with special reference to The Manuscripts of Leonardo Da Vinci*. Constable and Co., London. [Reprinted 1979 by Dover Publications]

Cooke, J. (1975). Control of somite number during morphogenesis of a vertebrate, *Xenopus laevis. Nature* **254**, 196–199.

Coon, H. G. (1966). Clonal stability and phenotypic expression of chick cartilage cells *in vitro*. *Proc. Natl Acad. Sci. U.S.A.* **55**, 66–73.

Cooper, G. W. (1965). Induction of somite chondrogenesis by cartilage and notochord: A correlation between inductive activity and specific stages of cytodifferentiation. *Devel. Biol.* **12**, 185–212.

Cooper, R. R., and Ponseti, I. V. (1973). Metaphyseal dysostosis: Description of an ultrastructural defect in the epiphyseal plate chondrocytes. *J. Bone Joint Surg. Am. Vol.* **55**, 485–495.

Cooper, R. R., Ponseti, I. V., and Maynard, J. A. (1973). Pseudoachondroplastic dwarfism: A rough-surfaced endoplasmic reticulum storage disorder. *J. Bone Joint Surg. Am. Vol.* **55**, 475–484.

Copher, G. H. (1935). Influence of urinary bladder transplants on hyaline cartilage. *Ann. Surg.* **102**, 927–940.

Copping, R. R. (1978). Microscopic age changes in the frontal bone of the domestic rabbit. *J. Morphol.* **155**, 123–130.

Copray, J. C. V. M. (1984). *Growth Regulation of Mandibular Condylar Cartilage* in vitro. Ph.D. Dissertation, Groningen University, Krips Repro Meppel, Groningen, the Netherlands.

Copray, J. C. V. M. (1986). Growth of the nasal septal cartilage of the rat *in vitro*. *J. Anat.* **144**, 99–112.

Copray, J. C. V. M., Dibbets, J. M. N., and Kantomaa, T. (1988). The role of condylar cartilage in the development of the temporomandibular joint. *Angle Orthod.* **58**, 369–380.

Copray, J. C. V. M., and Jansen, H. W. B. (1985). Cyclic nucleotides and growth retardation of the mandibular condylar cartilage of the rat *in vitro*. *Archs Oral Biol.* **30**, 749–752.

Copray, J. C. V. M., Jansen, H. W. B., and Duterloo, H. S. (1985a). Effect of compressive forces on phosphatase activity in mandibular condylar cartilage of the rat *in vitro*. *J. Anat.* **140**, 479–490.

Copray, J. C. V. M., Jansen, H. W. B., and Duterloo, H. S. (1985b). Effects of compressive forces on proliferation and matrix synthesis in mandibular condylar cartilage of the rat *in vitro*. *Archs Oral Biol.* **30**, 299–304.

Copray, J. C. V. M., Jansen, H. W. B., and Duterloo, H. S. (1985c). An *in vitro* system for studying the effect of variable compressive forces on the mandibular condylar cartilage of the rat *in vitro*. *Archs Oral Biol.* **30**, 305–311.

Copray, J. C. V. M., Johnson, P. M., Decker, J. D., and Hall, S. H. (1989). Presence of osteonectin/SPARC in mandibular condylar cartilage of the rat. *J. Anat.* **162**, 43–51.

Corner, B. D., and Shea, B. T. (1995). Growth allometry of the mandibles of giant transgenic mice: An analysis based on the finite-element scaling method. *J. Craniofac. Genet. Devel. Biol.* **15**, 125–139.

Corsin, J. (1966). The development of the osteocranium of *Pleurodeles waltlii* Michahelles. *J. Morphol.* **119**, 209–216.

Corsin, J. (1974). Matériel extracellulaire et chondrogenèse chez les amphibiens. *Arch. Anat. Microsc. Morphol. Exp.* **63**, 231–238.

Corsin, J. (1975a). Différenciation *in vitro* de cartilage à partir des crêtes neurales céphaliques chez *Pleurodeles waltlii* Michah. *J. Embryol. Exp. Morphol.* **33**, 335–342.

Corsin, J. (1975b). Influence du hyaluronate et de la hyaluronidase sur la chondrogenèse cephalique chez les amphibiens. *Acta Embryol. Exp.* **1**, 15–22.

Corsin, J. (1977). Le matériel extracellulaire au cours du développement du chondrocrâne des amphibiens: Mise en place et constitution. *J. Embryol. Exp. Morphol.* **38**, 139–150.

Corwin, J. A., and Morehead, J. R. (1971). The origin of osteoclasts in estrogen-stimulated bone resorption of the pubic symphysis of the mouse. *Anat. Rec.* **171**, 509–516.

Cossu, G., Warren, L., Boettiger, D., Holtzer, H., and Pacifici, M. (1982). Similar glycopeptides in normal chondroblasts and in *Rous sarcoma* virus-transformed fibroblasts. *J. Biol. Chem.* **257**, 4463–4468.

Cote, S., Carroll, R., Clouthier, R., and Bar-Sagi, L. (2002). Vertebral development in the Devonian sarcopterygian fish *Eusthenopteron foordi* and the polarity of vertebral evolution in non-amniote tetrapods. *J. Vert. Paleontol.* **22**, 487–502.

Cotta-Pereira, G., Del-Caro, L. M., and Montes, G. S. (1984). Distribution of elastic system fibers in hyaline and fibrous cartilages of the rat. *Acta Anat.* **119**, 80–85.

Cottrill, C. P., Archer, C. W., and Wolpert, L. (1987). Cell sorting and chondrogenic aggregate formation in micromass culture. *Devel. Biol.* **122**, 503–515.

Coulombre, A. J. (1965). The eye. In *Organogenesis* (R. L. DeHaan and H. Ursprung, eds), pp. 219–251. Holt, New York, NY.

Coulombre, A. J., and Coulombre, J. L. (1973). The skeleton of the eye. II. Overlap of the scleral ossicles of the domestic fowl. *Devel. Biol.* **33**, 257–267.

Coulombre, A. J., Coulombre, J. L., and Mehta, H. (1962). The skeleton of the eye. I. Conjunctival papillae and scleral ossicles. *Devel. Biol.* **5**, 382–401.

Coulombre, J. L., and Coulombre, A. J. (1975). Corneal development. V. Treatment of five-day-old embryos of domestic fowl with 6-diazo-5-oxo-L-norleucine (DON). *Devel. Biol.* **45**, 291–303.

Couly, G. F., Coltey, P. M., and Le Douarin, N. M. (1993). The triple origin of skull in higher vertebrates. A study in quail-chick chimeras. *Development* **117**, 409–429.

Couly, G. F., Grapin-Botton, A., Coltey, P., Ruhin, B., and Le Douarin, N. M. (1998). Determination of the identity of the derivatives of the cephalic neural crest: Incompatibility between *Hox* gene expression and lower jaw development. *Development* **125**, 3445–3459.

Coumoul, X., and Deng, C.-X. (2003). Roles of FGF receptors in mammalian development and congenital diseases. *Birth Defects Res. (Part C)* **69**, 286–304.

Courtland, H.-W., Wright, G. M., Root, R. G., and DeMont, M. E. (2003). Comparative equilibrium mechanical properties of bovine and lamprey cartilaginous tissues. *J. Exp. Biol.* **206**, 1397–1408.

Cousins, R. J., and DeLuca, H. F. (1972). Vitamin D and bone. In *The Biochemistry and Physiology of Bone* (G. H. Bourne, ed.), Volume 2, pp. 282–336. Academic Press, New York, NY.

Cowden, R. R. (1967). A histochemical study of chondroid tissue in *Limulus* and *Octopus*. *Histochemie* **9**, 149–163.

Cowden, R. R., and Fitzharris, T. P. (1975). The histochemistry and structure of tentacle cartilage tissues in the marine polychaete, *Sabella melanostigma*. *Histochemistry* **43**, 1–10.

Cowin, S. C., Moss-Salentijn, L., and Moss, M. L. (1992a). Candidates for the mechanosensory system in bone. *J. Biomech. Engin.* **113**, 191–197.

Cowin, S. C., Sadegh, A. M., and Luo, G. M. (1992). An evolutionary Wolff law for trabecular architecture. *J. Biomech. Engin.* **114**, 129–136.

Cowles, E. A., DeRome, M. E., Pastizzo, G., Brailey, L. L., and Gronowicz, G. A. (1998). Mineralization and the expression of matrix proteins during *in vivo* bone development. *Calcif. Tissue Int.* **62**, 74–82.

Cowles, R. B. (1947). Comments by readers. *Science* **105**, 282.

Cracraft, J. (1968). The lacrimal–ectethmoid bone complex in birds: A single character analysis. *Amer. Midl. Naturalist* **80**, 316–359.

Craft, P. D., Mani, M. M., Pazel, J., and Masters, F. W. (1974). Experimental study of healing in fractures of membranous bone. *Plastic Reconstruct. Surg.* **53**, 321–325.

Craig, F. M., Bentley, G., and Archer, C. W. (1987). The spatial and temporal pattern of collagens I and II and keratan sulphate in the developing chick metatarso-phalangeal joint. *Development* **99**, 383–391.

Crawford, A. J., and Wake, D. B. (1998). Phylogenetic and evolutionary perspectives on an enigmatic organ: The balancer of larval caudate amphibians. *Zoology* **101**, 107–123.

Crelin, E. S. (1954a). The effect of androgen, estrogen and relaxin on intact and transplanted pelves in mice. *Amer. J. Anat.* **95**, 47–73.

Crelin, E. S. (1954b). Prevention of innominate bone separation during pregnancy in the mouse. *Proc. Soc. Exp. Biol. Med.* **86**, 22–24.

Crelin, E. S. (1954c). The effects of estrogen and relaxin on the symphysis pubis and transplanted ribs in mice. *Anat. Rec.* **120**, 23–32.

Crelin, E. S. (1957). Mitosis in adult cartilage. *Science* **125**, 650.

Crelin, E. S. (1960). The development of bony pelvic sexual dimorphism in mice. *Ann. NY. Acad. Sci.* **84**, 479–512.

Crelin, E. S. (1963). The development and hormonal response of the autotransplanted interpubic joint in mice. *Anat. Rec.* **146**, 149–163.

Crelin, E. S. (1969). The development of the bony pelvis and its changes during pregnancy and parturition. *Trans NY. Acad. Sci.* [2] **31**, 1049–1058.

Crelin, E. S., and Brightman, M. W. (1957). The pelvis of the rat: Its response to estrogen and relaxin. *Anat. Rec.* **128**, 467–484.

Crelin, E. S., and Haines, A. L. (1955). The effect of locally applied estrogen on the pubic symphysis and knee joint in castrated mice. *Endocrinology* **56**, 461–470.

Crelin, E. S., and Koch, W. E. (1965). Development of the mouse pubic joint *in vivo* following initial differentiation *in vitro*. *Anat. Rec.* **153**, 161–172.

Crelin, E. S., and Koch, W. E. (1967). An autoradiographic study of chondrocyte transformation into chondroclasts and osteocytes during bone formation *in vitro*. *Anat. Rec.* **158**, 473–483.

Crelin, E. S., and Levin, J. (1955). The prepuberal pubic symphysis and uterus in the mouse; their response to estrogen and relaxin. *Endocrinology* **57**, 730–747.

Crelin, E. S., and Southwick, W. D. (1960). Mitosis of chondrocytes induced in the knee joint articular cartilage of adult rabbits. *Yale J. Biol. Med.* **33**, 243–244.

Crelin, E. S., and Southwick, W. D. (1964). Changes induced by sustained pressure in the knee-joint articular cartilage of adult rabbits. *Anat. Rec.* **149**, 113–134.

Cremer, H., Lange, R., Christoph, A., Plomann, M., Vopper, G., Roes, J., Brown, R., Baldwin, S., Kraemer, P., Scheff, S., Barthels, D., Rajewsky, K., and Wille, W. (1994). Inactivation of the N-CAM gene in mice results in size reduction of the olfactory bulb and deficits in spatial learning. *Nature* **367**, 455–459.

Crilly, R. G. (1972). Longitudinal overgrowth of chicken radii. *J. Anat.* **112**, 11–18.

Crissman, R. S., and Low, F. N. (1974). A study of fine structural changes in the cartilage-to-bone transition within the developing chick vertebra. *Amer. J. Anat.* **140**, 451–470.

Critchlow, M. A., Bland, Y. S., and Ashhurst, D. E. (1995). The expression of collagen mRNAs in normally developing neonatal long bones and after treatment of neonatal and adult rabbit tibiae with transforming growth factor-β2. *Histochem. J.* **27**, 505–515.

Crosby, G. M., and Fallon, J. F. (1975). Inhibitory effect on limb morphogenesis by cells of the polarizing zone coaggregated with pre- or postaxial wing bud mesoderm. *Devel. Biol.* **46**, 28–39.

Crossin, K. L., and Krushel, L. A. (2000). Cellular signaling by neural crest adhesion molecules of the immunoglobulin superfamily. *Devel. Dynam.* **218**, 260–279.

Crotwell, P. L., Clark, T. G., and Mabee, P. M. (2001). GDF5 is expressed in the developing skeleton of median fins of late-stage zebrafish, *Danio rerio*. *Devel. Genes Evol.* **211**, 555–558.

Crowe, R., Zikherman, J., and Niswander, L. (1999). Delta-1 negatively regulates the transition from prehypertrophic to hypertrophic chondrocytes during cartilage formation. *Development* **126**, 987–998.

Cruess, R. L. (ed.) (1982). *The Musculoskeletal System. Embryology, Biochemistry and Physiology*. Churchill Livingstone, New York, NY.

Crumly, C. R. (1990). The case of the two-legged 'lizard.' *Environment West* **21**, 20–24.

Crumly, C. R., and Sánchez-Villagra, M. R. (2004). Patterns of variation in the phalangeal forlumae of land tortoises (Testudinidae): Developmental constraint, size, and phylogenetic history. *J. Exp. Zool. (Mol. Devel. Evol.)* **302B**, 134–146.

Cserjesi, P., Brown, D., Ligon, K. L., Lyons, G. E., Copeland, N. G., Gilbert, D. J., Jenkins, N. A., and Olson, E. N. (1995). Scleraxis: A basic helix–loop–helix protein that prefigures skeletal formation during mouse embryogenesis. *Development* **121**, 1099–1110.

Cubbage, C. C., and Mabee, P. M. (1996). Development of the cranium and paired fins in the zebrafish *Danio rerio* (Ostariophysi, Cyprinidae). *J. Morphol.* **229**, 121–160.

Cubo, J. (2000). Process heterochronies in endochondral ossification. *J. Theor. Biol.* **205**, 343–353.

Cubo, J., and Casinos, A. (2000). Mechanical properties and chemical composition of avian long bones. *Eur. J. Morphol.* **38**, 112–121.

Cubo, J., Fouces, V., Gonzáles-Martín, M., Pedrocchi, V., and Ruiz, X. (2000). Nonheterochronic developmental changes underlie morphological heterochrony in the evolution of the Ardeidae. *J. Evol. Biol.* **13**, 269–276.

Cuevas, P., Burgos, J., and Baird, A. (1988). Basic fibroblast growth factor (FGF) promotes cartilage repair *in vivo*. *Biochem. Biophys. Res. Commun.* **156**, 611–618.

Cullinane, D. M. (2000). Axial *versus* appendicular: Constraint *versus* selection. *Amer. Zool.* **40**, 136–145.

Cullinane, D. M. (2001). The role of osteocytes in bone regulation: mineral homeostasis versus mechanoreception. 31st Inter. Sun. Valley Hard Tissue Workshop, 6–10 August 2001, 2 pp.

Cullinane, D. M., Frederick, A., Eisenberg, S. R., Pacicca, D., Elman, M. V., Lee, C., Salisbury, K., Gerstenfeld, L. C., and Einhorn, T. A. (2002). Induction of a neoarthrosis by precisely controlled motion in an experimental mid-femoral defect. *J. Orthop. Res.* **20**, 579–586.

Culmann, C. (1866). *Die Graphische Statik*, Zurich.

Cunat, J. J., Bhaskar, S. N., and Weinmann, J. P. (1956). Development of the squamoso-mandibular articulation in the rat. *J. Dent. Res.* **35**, 533–546.

Cundall, D., and Irish, F. J. (1989). The function of the intramaxillary joint in the Round Island Boa, *Casarea dussumieri*. *J. Zool. London* **217**, 569–598.

Cunningham, N. S., Paralkar, V., and Reddi, A. H. (1992). Osteogenic and recombinant bone morphogenetic protein 2B are chemotactic for human monocytes and stimulate transforming growth factor β1 mRNA expression. *Proc. Natl Acad. Sci. U.S.A.* **89**, 11740–11744.

Currey, J. D. (1962). The histology of the bone of a prosauropod dinosaur. *Palaeontology* **5**, 238–246.

Currey, J. D. (1970). The mechanical properties of bone. *Clin. Orthop. Rel. Res.* **73**, 210–231.

Currey, J. D. (1973). Interactions between age, pregnancy and lactation, and some mechanical properties of the femora of rats. *Calcif. Tissue Res.* **13**, 99–112.

Currey, J. D. (1982). Osteons in biomechanical literature. *J. Biomech.* **15**, 717–718.

Currey, J. D. (1984a). *The Mechanical Adaptations of Bones*. Princeton University Press, Princeton, NJ.

Currey, J. D. (1984b). Comparative mechanical properties and histology of bone. *Amer. Zool.* **24**, 5–12.

Curtin, J. W., and Pruzansky, S. (1971). The stapling operation for complete bilateral cleft lip and palate. *World Congr. Plast. Reconstr. Surg. 5th, 1970*, pp. 181–184.

Curtis, B. F. (1893). Cases of bone implantation and transplantation for cyst of tibia, osteomyelitic cavities and ununited fractures. *Amer. J. Med. Sci.* **106**, 30–37.

Cutright, D. E. (1972). Osseous and chondromatous metaplasia caused by dentures. *Oral Surg. Oral Med. Oral Pathol.* **34**, 625–633.

Cuvier, G. (1825). *Recherches sur les ossemens fossiles, Volume 5*. Third Edn. Paris.

D

Dabska, M., and Huvos, A. G. (1983). Mesenchymal chondrosarcoma in the young. A clinicopathologic study of 19 patients with explanation of histogenesis. *Virchows Arch. A. Path. Anat. Histopathol.* **399**, 89–104.

Dacke, C. G. (1979). *Calcium Regulation in sub-Mammalian Vertebrates.* Academic Press, New York, NY.

Daegling, D. J. (1996). Growth in the mandible of African Apes. *J. Human Evol.* **30**, 315–341.

Daeschler, E. B., and Shubin, N. (1998). Fish with fingers? *Nature* **391**, 133.

Daget, J. (1964). Le crâne des Teleostéens. *Mém. Mus. Natl Hist. Nat. Ser. a, Zool.* **31**, 163–340.

Dahl, E., Koseki, H., and Balling, R. (1997). *Pax* genes and organogenesis. *BioEssays* **19**, 755–765.

Dahlberg, A. A. (ed.) (1971). *Dental Morphology and Evolution.* The University of Chicago Press, Chicago, IL.

Dahn, R. D., and Fallon, J. F. (2000). Interdigital regulation of digit identity and homeotic transformation by modulated BMP signaling. *Science* **289**, 438–441.

Daimon, T. (1973a). Effects of large dose of insulin on the chondrogenesis of the tibiotarsus in developing chick embryos. 1. A light microscopic study. *Acta Histochem. Cytochem.* **6**, 280–293.

Daimon, T. (1973b). Effects of large dose of insulin on the chondrogenesis of the tibiotarsus in developing chick embryos. 2. Ultrastructural and histochemical studies. *Acta Histochem. Cytochem.* **6**, 294–311.

Dainton, M., and Macho, G. A. (1999). Heterochrony: Somatic, skeletal and dental development in *Gorilla, Homo,* and *Pan.* In *Human Growth in the Past: Studies from Bones and Teeth* (R. D. Hoppa and C. M. Fitzgerald, eds), pp. 32–64. Cambridge University Press, Cambridge.

Dale, L., and Jones, C. M. (1999). BMP signalling in early *Xenopus* development. *BioEssays* **21**, 751–760.

Dalgleish, A. E. (1985). A study of the development of thoracic vertebrae in the mouse assisted by autoradiography. *Acta Anat.* **122**, 91–98.

D'Amico-Martel, A., Van de Water, T. R., Wootton, J. A. M., and Minor, R. R. (1987). Changes in the types of collagen synthesized during chondrogenesis of the mouse otic capsule. *Devel. Biol.* **120**, 542–555.

Damjanov, I., and Solter, D. (1974). Experimental teratoma. *Curr. Topics Pathol.* **59**, 69–130.

Damjanov, I., Solter, D., and Serman, D. (1973). Teratocarcinoma with the capacity for differentiation restricted to neuroectodermal tissue. *Virchows Arch. Cell Pathol.* **13**, 179–195.

Damjanov, I., and Urbanke, A. (1969). Heterotopic ossification in implantation metastasis from a carcinoma of the bladder. *J. Urol.* **101**, 863–865.

Daneo, L. S., and Filogamo, G. (1973). Ultrastructure of early neuro-muscular contacts in the chick embryo. *J. Submicrosc. Cytol.* **5**, 219–225.

Daniel, J. C. (1976). Changes in type of collagen synthesized by chick fibroblasts *in vitro* in the presence of 5-bromodeoxyuridine. *Cell Differ.* **5**, 247–254.

Daniel, J. C., Kosher, R. A., Lash, J. W., and Hertz, J. (1973). The synthesis of matrix components by chondrocytes *in vitro* in the presence of 5-bromodeoxyuridine. *Cell Differ.* **2**, 285–298.

Daniels, K., and Solursh, M. (1991). Modulation of chondrogenesis by the cytoskeleton and extracellular matrix. *J. Cell Sci.* **100**, 249–254.

Daniels, S., Ellis, S., and Carlson, D. S. (1987). Histologic analysis of costochondral and sternoclavicular grafts in the TMJ of the juvenile monkey. *J. Oral Maxillofac. Surg.* **45**, 672–682.

Danke, J., Miyake, T., Powers, T., Schein, J., Shin, H., Bosdet, I., Erdmann, M., Caldwell, R., and Amemiya, C. (2004). Genome resource for the Indonesian coelacanth, *Latimeria menadoensis. J. Exp. Zool.* **301A**, 228–234.

Darwin, C. R. (1859). *On the Origin of Species by Means of Natural Selection.* John Murray, London.

Darwin, C. R. (1861). *The Variation of Animals and Plants under Domestication.* John Murray, London.

Darwin, E. (1794). *Zoonomia; or The Laws of Organic Life,* Volume 1. J. Johnson, London.

Daughaday, W. H. (1971). Sulfation factor regulation of skeletal growth: A stable mechanism dependent on intermittent growth hormone secretion. *Amer. J. Med.* **50**, 277–280.

Davenport, J., and Thorsteinsson, V. (1990). Sucker action in the lumpsucker *Cyclopterus lumpus* L. *Sarsia* **75**, 33–42.

David, N. B., Saint-Etienne, L., Tsang, M., Schilling, T. F., and Rosa, F. M. (2002). Requirement for endoderm and GFG3 in ventral head skeleton formation. *Development* **129**, 4457–4468.

David, P. R. (1936). Studies on the creeper fowl. X. A study of the mode of action of a lethal factor by explantation methods. *Archs. EntwMech.* **135**, 521–551.

Davidovitch, Z. (1973). The production of 3′,5′ cAMP by orthodontic forces. *Amer. J. Orthod.* **64**, 314–315.

Davidovitch, Z., Montgomery, P. C., Eckerdal, O., and Gustafson, G. T. (1976a). Demonstration of cyclic AMP in bone cells by immuno-histochemical methods. *Calcif. Tissue Res.* **19**, 305–316.

Davidovitch, Z., Montgomery, P. C., Eckerdal, O., and Gustafson, G. T. (1976b). Cellular localization of cyclic AMP in periodontal tissues during experimental tooth movement in cats. *Calcif. Tissue Res.* **19**, 317–330.

Davidovitch, Z., Montgomery, P. C., and Shanfeld, J. L. (1977a). Cellular localization and concentration of bone cyclic AMP nucleotides in response to acute PTE administration. *Calcif. Tissue Res.* **24**, 81–91.

Davidovitch, Z., Montgomery, P. C., and Shanfeld, J. L. (1977b). Guanosine 3′,5′-monophosphate in bone: Microscopic visualization by an immuno-histochemical technique. *Calcif. Tissue Res.* **24**, 73–79.

Davidovitch, Z., Montgomery, P. C., Yost, R. W., and Shanfeld, J. L. (1978a). Immunohistochemical localization of cyclic nucleotides in the periodontium: Mechanically-stressed cells *in vivo. Anat. Rec.* **192**, 351–362.

Davidovitch, Z., Montgomery, P. C., Yost, R. W., and Shanfeld, J. L. (1978b). Immunohistochemical localization of cyclic nucleotides in mineralized tissues: Mechanically-stressed osteoblasts *in vivo. Anat. Rec.* **192**, 363–374.

Davidovitch, Z., and Shanfeld, J. L. (1975). Cyclic AMP levels in alveolar bone of orthodontically-treated cats. *Archs Oral Biol.* **20**, 567–574.

Davidovitch, Z., Steigman, S., Finkelson, M. D., Yost, R. W., Montgomery, P. C., Shanfeld, J. L., and Korostoff, E. (1980). Immunohistochemical evidence that electric currents increase periosteal cell cyclic nucleotide levels in feline alveolar bone *in vivo. Archs Oral Biol.* **25**, 321–327.

Davies, G. B. M., Oxford, J. T., Hausafus, L. C., Smoody, B. F., and Morris, N. P. (1998). Temporal and spatial expression of alternative splice forms of the α1(XI) collagen gene in fetal rat cartilage. *Devel. Dynam.* **213**, 12–26.

Davies, M. (1989). Ontogeny of bone and the role of hete-rochrony in the Myobatrachine genera *Uperoleia*, *Crinia* and *Pseudophryne* (Anura: Leptodactylidae: Myobatrachinae). *J. Morphol.* **200**, 269–300.

Davis, A. P., and Capecchi, M. R. (1994). Axial homeosis and appendicular skeleton defects in mice with a targeted disruption of hoxd-11. *Development* **120**, 2187–2198.

Davis, A. P., and Capecchi, M. R. (1996). A mutational analysis of the 5′ HoxD genes: Dissection of genetic interactions during limb development in the mouse. *Development* **122**, 1175–1185.

Davis, A. P., Witte, D. P., Hsleh-li, H. M., Potter, S. S., and Capecchi, M. R. (1995). Absence of radius and ulna in mice lacking *hoxa-11 and hoxd-11*. *Nature* **375**, 791–795.

Davis, W. L., Jones, R. G., Knight, J. P., and Hagler, H. K. (1982). Cartilage calcification: An ultrastructural, histochemical and analytical X-ray microprobe study of the zone of calcification in the normal avian epiphyseal growth plate. *J. Histochem. Cytochem.* **30**, 221–234.

Davison, P. F., and Berman, M. (1973). Corneal collagenase: Specific cleavage of types $(\alpha 1)_2\alpha 2$ and $(\alpha 1)_3$ collagens. *Connect. Tissue Res.* **2**, 57–64.

Davoli, M. A., Lamplugh, L., Lee, E. R., Beauchemin, A., Chan, K., Mordier, S., Mort, J. S., Murphy, G., Docherty, A. J. P., Leblond, C. P., and Lee, E. K. (2001). Enzymes active in the areas undergoing cartilage resorption during the development of the secondary ossification center in the tibiae of rats ages 0–21 days: II. Two proteinases, gelatinase B and collagenase-3, are implicated in the lysis of collagen fibrils. *Devel. Dynam.* **222**, 71–88.

Dawd, D. S., and Hinchliffe, J. R. (1971). Cell death in the 'opaque patch' in the central mesenchyme of the developing chick limb: A cytological, cytochemical and electron microscopic analysis. *J. Embryol. Exp. Morphol.* **26**, 401–424.

Dawson, A., and Kember, N. F. (1974). Compensatory growth in the rat tibia. *Cell Tissue Kinet.* **7**, 285–292.

Dawson, A. B., and Spark, C. (1928). The fibrous transformation and architecture of the costal cartilage of the albino rat. *Amer. J. Anat.* **42**, 109–138.

Dawson, D. L. (1980). Functional interpretations of the radiographic anatomy of the femora of *Myotis lucifugus*, *Pipistrellus subflavus* and *Blarina brevicauda*. *Amer. J. Anat.* **157**, 1–15.

Dawson, S. D. (2003). Pattern of ossification in the manus of the harbor porpoise (Phocoena phocoena): Hyperphalangy and delta-shaped bones. *J. Morphol.* **258**, 200–206.

Dealy, C. N., and Kosher, R. A. (1996). IGF-1, insulin and FGFs induce outgrowth of the limb buds of amelic mutant chick embryos. *Development* **122**, 1323–1330.

Dealy, C. N., Scranton, V., and Cheng, H. C. (1998). Roles of transforming growth factor alpha and epidermal growth factor in chick limb development. *Devel. Biol.* **202**, 43–55.

Dean, D. D., Schwartz, Z., Muniz, O. E., Gomez, R., Howell, D. S., and Boyan, B. D. (1992). Matrix vesicles are enriched in metalloproteinases that degrade proteoglycan. *Calcif. Tissue Int.* **50**, 342–349.

Dean, M. C., and Wood, B. A. (1984). Phylogeny, neoteny and growth of the cranial base in Hominoids. *Folia Primatol.* **43**, 157–180.

deAngelis, V. (1969). Autoradiographic investigation of calvarial growth in the rat. *Amer. J. Anat.* **123**, 359–368.

deAngelis, V. (1970). Observations on the response of alveolar bone to orthodontic force. *Amer. J. Orthod.* **58**, 284–294.

de Beer, G. R. (1937). *Development of the Vertebrate Skull.* Oxford University Press, Oxford.

de Beer, G. R. (1947). The differentiation of neural crest cells into visceral cartilages and odontoblasts in *Amblystoma*, and a re-examination of the germ-layer theory. *Proc. R. Soc. London, Ser. B.* **134**, 377–398.

de Beer, G. R. (1958). *Embryos and Ancestors*, 3rd Edn. Oxford University Press, Oxford.

de Beer, G. R. (1962). *Reflections of a Darwinian.* Thomas Nelson and Sons Ltd., London.

de Bernard, B., Biano, P., Bonucci, E., Costantini, M., Lunazzi, G. C., Martinuzzi, P., Modricky, C., Moro, L., Panfili, E., Pollesello, P., Stagni, N., and Vittur, F. (1986). Biochemical and immunohistochemical evidence that in cartilage an alkaline phosphatase is a Ca^{2+} binding glycoprotein. *J. Cell Biol.* **103**, 1615–1623.

de Bernard, B., Stagni, N., Colautti, I., Vittur, F., and Bonucci, E. (1977). Glycosaminoglycans and endochondral ossification. *Clin. Orthop. Rel. Res.* **126**, 285–291.

de Boer, H. H. (1988). The history of bone grafts. *Clin. Orthop. Rel. Res.* **226**, 292–297.

de Buffrènil, V., and Buffetaut, E. (1981). Skeletal growth lines in an Eocene crocodilian skull from Wyoming as an indicator of ontogenic age and paleoclimatic conditions. *J. Vert. Paleont.* **1**, 57–66.

de Buffrènil, V., Farlow, J. O., and de Ricqlès (1986). Growth and function of *Stegosaurus* plates: evidence from bone histology. *Paleobiology* **12**, 459–473.

de Buffrènil, V., and Mazin, J.-M. (1990). Bone histology of the ichthysaurs: Comparative data and functional interpretations. *Paleobiology* **16**, 435–447.

de Buffrènil, V., and Schoevaert, D. (1988). On how the periosteal bone of the Delphinid humerus becomes cancellous: Ontogeny of a histological specialization. *J. Morphol.* **198**, 149–164.

de Buffrènil, V., de Ricqlès, A. J., Ray, C. E., and Domming, D. P. (1990). Bone histology of the ribs of the Archaeocetes (Mammalia: Cetacea). *J. Vert. Paleont.* **10**, 455–466.

Dechant, J. J., Mooney, M. P., Cooper, G. M., Smith, T. D., Burrows, A. M., Losken, H. W., Mathijssen, I. M. J., and Siehel, M. I. (1999). Positional changes of the frontoparietal ossification centers in perinatal craniosynostotic rabbits. *J. Craniofac. Genet. Devel. Biol.* **19**, 64–74.

Decker, J. D., and Hall, S. H. (1985). Light and electron microscopy of the new born sagittal suture. *Anat. Rec.* **212**, 81–89.

Decker, J. D., Marshall, J. J., and Herring, S. W. (1996). Differential cell replication within the periosteum of the pig mandibular ramus. *Acta Anat.* **157**, 144–150.

Deflitch, C. J., and Stryker, J. A. (1993). Postoperative hip irradiation in prevention of heterotopic ossification – cases of treatment failure. *Radiology* **188**, 265–270.

De Gennaro, L. D., Packard, D. S., Jr., Stach, R. W., and Wagner, B. J. (1980). Growth and differentiation of chicken embryos in simplified shell-less cultures under ordinary conditions of incubation. *Growth* **64**, 343–354.

DeHaan, R. L. (1956). The serological determination of developing muscle protein in the regenerating limb of *Amblystoma mexicanum*. *J. Exp. Zool.* **133**, 73–85.

Deimer, P. (1977). Der rudimentäre hintere Extremitätengürtel der Pottwals (*Physeter macrocephalus* Linnaeus, 1758), seine Variabilität und Wachstumsallometrie. *Z. SäugetierKunde* **42**, 88–101.

de Jonge-Strobel, G. E. F., Veldhuijzen, J. P., Vermeiden, J. W. P., van de Wijngaert, F. P., and Prahl-Andersen, B. (1987). Development and use of shell-less quail chorio-allantoic-membrane cultures to study developing skeletal tissues: A qualitative study. *Experientia* **43**, 199–200.

Delaire, J., and Precious, D. (1986). Influence of the nasal septum on maxillonasal growth in patients with congenital labiomaxillary cleft. *Cleft Palate J.* **23**, 270–277.

Delaporte, F. (1983). Theories of osteogenesis in the eighteenth century. *J. Hist. Biol.* **16**, 343–360.

Delbrück, A. (1970). Enzyme activity determinations in bone and cartilage. *Enzymol. Biol. Clin.* **11**, 130–153.

Delcher, H. K., Eisenbarth, G. S., and Lebovitz, H. E. (1973). Fatty acid inhibition of sulfation factor-stimulated $^{35}SO_4$ incorporation into embryonic chicken cartilage. *J. Biol. Chem.* **248**, 1901–1905.

Deleskamp, G. (1906). Das Verhalten der Knochenarterien bei Knochenerkrankungen und Frakturen. *Fortschr. Röntgenst.* **10**, 219.

Delezoide, A. L., Benoist-Lasselin, C., Legeai-Mallet, L., Le Merrer, M., Munnich, A., Vekemans, M., and Bonaventure, J. (1998). Spatiotemporal expression of Fgfr-1, Fgfr-2 and Fgfr-3 genes during human embryo-fetal ossification. *Mech. Devel.* **77**, 19–30.

Delgado-Baeza, E., Nieto-Chagvaceda, A., Miralles-Flores, C., and Santos-Alvarez, I. (1992). Cartilage canal growth: Experimental approaches in the rat tibia. *Acta Anat.* **145**, 143–148.

Delides, G. S. (1972). Bone and cartilage in malignant tumours of the urinary bladder. *Br. J. Urol.* **44**, 571–581.

Delise, A. M., Fischer, L., and Tuan, R. S. (2000). Cellular interactions and signaling in cartilage development. *Osteoarthritis Cart.* **8**, 309–334.

Delise, A. M., and Tuan, R. S. (2002a). Alterations in the spatiotemporal expression pattern and function of N-cadherin inhibit cellular condensation and chondrogenesis of limb mesenchymal cells *in vitro*. *J. Cell. Biochem.* **87**, 342–359.

Delise, A. M., and Tuan, R. S. (2002b). Analysis of N-cadherin function in limb mesenchymal chondrogenesis in vitro. *Devel. Dynam.* **225**, 195–204.

Dell'Orbo, C., Gioglio, L., and Quacci, D. (1992). Morphology of epiphyseal apparatus of a ranid frog (*Rana esculenta*). *Histol. Histopathol.* **7**, 267–273.

del Rincón, S. V., and Scadding, S. R. (2002). Retinoid antagonists inhibit normal patterning during limb regeneration in the axolotl, *Ambystoma mexicanum*. *J. Exp. Zool.* **292**, 435–443.

DeLuca, G., Speziale, P., Balduini, C., and Castellani, A. A. (1975). Uridine diphosphate glucose 4′-epimerase from cornea and epiphyseal-plate chondrocytes. *Connect. Tissue Res.* **3**, 39–48.

de Margeri, E. (2002). Laminar bone as an adaptation to torsional loads in flapping flight. *J. Anat.* **201**, 521–526.

DeMoerlooze, L., Spencer-Dene, B., Revest, J.-M., Hajihosseini, M., Rosewell, I., and Dickson, C. (2000). An important role for the IIIB isoform of fibroblast growth factor receptor 2 (FGF-R2) in mesenchymal–epithelial signaling during mouse organogenesis. *Development* **127**, 483–490.

Denker, A. E., Haas, A. R., Nicoll, S. B., and Tuan, R. S. (1999). Chondrogenic differentiation of murine C3H101T1/2 multipotential mesenchymal cells: I. Stimulation by bone morphogenetic protein-2 in high density micromass cultures. *Differentiation* **64**, 67–76.

Deng, C., Wynshaw-Boris, A., Zhou, F., Kuo, A., and Leder, P. (1996). Fibroblast growth factor receptor 3 is a negative regulator of bone growth. *Cell* **84**, 911–921.

Denison, R. H. (1963). The early history of the vertebrate calcified skeleton. *Clin. Orthop. Rel. Res.* **31**, 141–152.

Denker, A. E., Nicoll, S. B., and Tuan, R. S. (1995). Formation of cartilage-like spheroids by micromass cultures of murine C3H10T1/2 cells upon treatment with transforming growth factor-β1. *Differentiation* **59**, 25–34.

Dennis, J. E., Merriam, A., Awadallah, A., Yoo, J. U., Johnstone, B., and Caplan, A. I. (1999). A quadripotential mesenchymal progenitor cell isolated from marrow of an adult mouse. *J. Bone Min. Res.* **14**, 700–709.

Depew, M. J., Liu, J. K., Long, J. E., Presley, R., Meneses, J. J., Pedersen, R. A., and Rubenstein, J. L. R. (1999). *Dlx5* regulates regional development of the branchial arches and sensory capsules. *Development* **126**, 3831–3846.

Depew, M. J., Lufkin, T., and Rubenstein, J. L. R. (2002). Specification of jaw subdivisions by *Dlx* genes. *Science* **298**, 381–385.

Deporter, D. A., and Ten Cate, A. R. (1973). Fine structural localization of acid and alkaline phosphatase in collagen-containing vesicles of fibroblasts. *J. Anat.* **114**, 457–461.

de Queiroz, K. (1982). The scleral ossicles of Sceloporine iguanids: A reexamination with comments on their phylogenetic significance. *Herpetologia* **38**, 302–311.

de Ricqlès, A. J. (1964). La formation des os longs des membres de *Pleurodeles waltlii* (Michahelles). *Bull. Soc. Zool. France* **89**, 797–808.

de Ricqlès, A. J. (1965). La formation des os longs des membres de *Pleurodeles waltlii* (Michahelles). *Bull. Soc. Zool. France* **90**, 267–286.

de Ricqlès, A. J. (1973). *Recherches Paléohistologiques sur les Os Longs des Tétrapods*. Thèse de Doctoral d'Etat, University de Paris VII.

de Ricqlès, A. J. (1974). Evolution of endothermy: Histological evidence. *Evol. Theory* **1**, 51–80.

de Ricqlès, A. J. (1975). Quelques remarques paléo-histologiques sur le problème de la néotenie chez les stégocephales. *Colloq. Int. C.N.R.S. Paris* **218**, 351–363.

de Ricqlès, A. J. (1976). On bone histology of fossil and living reptiles, with comments on its functional and evolutionary significance. In *Morphology and Biology of Reptiles* (A. d'A. Bellairs and C. V. Cox, eds), pp. 123–150. Academic Press, London.

de Ricqlès, A. J. (1989). Les mécanismes hétèrochroniques dans le retour des tétrapods au milieu aquatique. *Geobios, Mém. Spéciale* **12**, 337–348.

de Ricqlès, A. J. (2000). L'origine dinosaurienne des oiseaux et de l'endothermie avienne: Les arguments histogiques. *Ann. Biol.* **39**, 69–100.

de Ricqlès, A. J., Mateus, O., Antunes, M. T., and Taquet, P. (2001). Histomorphogenesis of embryos of Upper Jurassic theropods from Lourinhâ (Portugal). *C. R. Acad. Sci. Paris, Earth & Planet. Sci.*, **332**, 647–656.

de Ricqlès, A. J., Meunier, F. J., Castanet, J., and Francillon-Vieillot, H. (1991). Comparative microstructure of bone. In *Bone, Volume 3: Bone Matrix and Bone Specific Products* (B. K. Hall, ed.), pp. 1–78. CRC Press, Boca Raton, FL.

de Ricqlès, A. J., Padian, K., Horner, J. R., and Francillon-Vieillot, H. (2000). Palaeohistology of the bones of pterosaurs (Reptilia: Archosauria): anatomy, ontogeny, and biomechanical implications. *Zool. J. Linn. Soc. London* **129**, 349–385.

Derynck, R., and Zhang, Y. (2003). Smad-dependent and Samd-independent pathways in TGF-β family signalling. *Nature* **425**, 577–584.

De Sá, R. O. (1988). Chondrocranium and ossification sequence of *Hyla lanciformis*. *J. Morphol.* **195**, 345–356.

De Sá, R. O., and Hill, S. (1998). Chondrocranial anatomy and skeletogenesis in *Dendrobates auratus*. *J. Herpetol.* **32**, 205–210.

de Santa Barbara, P., and Roberts, D. J. (2001). Tail gut endoderm and gut/genitourinary/tail development: a new tissue-specific role for *Hoxa13*. *Development* **129**, 551–561.

Deshmukh, K. (1974). Synthesis of tissue nonspecific collagen by bovine articular cartilage as a result of aging *in vitro*. *Proc. Soc. Exp. Biol. Med.* **147**, 726–729.

Deshmukh, K., and Kline, W. G. (1976). Characterization of collagen and its precursors synthesized by rabbit-articular-cartilage cells in various culture systems. *Eur. J. Biochem.* **69**, 117–124.

Deshmukh, K., Kline, K. D., and Sawyer, B. D. (1977). Effects of calcitonin and parathyroid hormone on the metabolism of chondrocytes in culture. *Biochim. Biophys. Acta* **499**, 28–35.

Deshmukh, K., and Sawyer, B. D. (1977). Synthesis of collagen by chondrocytes in suspension culture: Modulation by calcium, 3′,5′-cyclic AMP, and prostaglandins. *Proc. Natl Acad. Sci. U.S.A.* **74**, 3864–3868.

Desmond, A. J. (1976). *The Hot-Blooded Dinosaurs: A Revolution in Palaeontology.* Blond & Briggs, London.

Dessau, W., Sasse, J., Timpl, R., Jilek, F., and von der Mark, K. (1978). Synthesis and extracellular deposition of fibronectin in chondrocyte cultures. Response to the removal of extracellular cartilage matrix. *J. Cell Biol.* **79**, 342–355.

Dessau, W., von der Mark, H., von der Mark, K., and Fischer, S. (1980). Changes in the patterns of collagens and fibronectin during limb-bud chondrogenesis. *J. Embryol. Exp. Morphol.* **57**, 51–60.

Detwiler, S. R. (1934). An experimental study of spinal nerve segmentation in *Amblystoma* with reference to the plurisegmental contribution to the brachial plexus. *J. Exp. Zool.* **67**, 393–443.

Detwiler, S. R. (1937). Observations upon the migration of neural crest cells, and upon the development of the spinal ganglia and vertebral arches in *Amblystoma*. *Amer. J. Anat.* **61**, 63–94.

Detwiler, S. R., and Holtzer, H. (1956). The developmental dependence of the vertebral column upon the spinal cord in the urodeles. *J. Exp. Zool.* **132**, 299–310.

Detwiler, S. R., and van Dyke, R. H. (1934). The development and functions of deafferented fore limbs in *Amblystoma*. *J. Exp. Zool.* **68**, 321–346.

Deuchar, E. M. (1975). *Cellular Interactions in Animal Development.* Chapman & Hall, London.

Deutsch, U., Dressler, G. R., and Gruss, P. (1988). *Pax 1*, a member of a paired box homologous murine gene family, is expressed in segmented structures during development. *Cell* **53**, 617–625.

de Vernejoul, M. C., Bielakoff, J., Herve, M., Gueris, J., Hott, M., Modrowski, D., Kuntz, D., Miravet, L., and Ryckewaert, A. (1983). Evidence for defective osteoblastic function: A role for alcohol and tobacco consumption in osteoporosis in middle-aged men. *Clin. Orthop. Rel. Res.* **179**, 107–115.

DeVillers, C. (1947). Recherches sur la crâne dermique des téléostéens. *Ann. Paléont.* **33**, 1–94.

DeVillers, C. (1965). The role of morphogenesis in the origin of higher levels of organization. *Syst. Zool.* **14**, 259–271.

Devlin, C. J., Brickell, P. M., Taylor, E. R., Hornbruch, A., Craig, R. K., and Wolpert, L. (1988). *In situ* hybridization reveals differential spatial distribution of mRNAs for type I and type II collagen in the chick limb bud. *Development* **102**, 111–118.

Devlin, H. (2000). Early bone healing events following rat molar tooth extraction. *Cells Tissues Organs* **167**, 33–37.

DeWitt, M. T., Handley, C. J., Oakes, B. W., and Lowther, D. A. (1984). *In vitro* response of chondrocytes to mechanical loading. The effect of short term mechanical tension. *Conn. Tissue Res.* **12**, 98–110.

DeWitt, N. (2003). Nature Insight: Bone and cartilage. *Nature* **423**, 315.

Dhem, A., and Goret-Nicaise, M. (1979). Rôle du cartilage condylien dans la croissance mandibulaire. *Arch. Anat. Histol. Embryol.* **62**, 95–102.

Dhem, A., Goret-Nicaise, M., Dambrain, R., Nyssen-Behets, C., Lengele, B., and Manzanares, M. C. (1989). Skeletal growth and chondroid tissue. *Arch. Ital. Anat. Embryol.* **94**, 237–241.

Dhordain, P., Dewitte, F., Desbiens, X., Stehelin, D., and Duterque-Coquillaud, M. (1995). Mesodermal expression of the chicken *erg* gene associated with precartilaginous condensation and cartilage differentiation. *Mech. Devel.* **50**, 17–28.

Dhouailly, D., and Kieny, M. (1972). The capacity of the flank somatic mesoderm of early bird embryos to participate in limb development. *Devel. Biol.* **28**, 162–175.

Diao, E., Zaleske, D. J., Avella, D., Trahan, C., Armstrong, A., Ehrlich, M. G., and Mankin, H. J. (1989). Kinetic and biochemical heterogeneity in vertebrate chondroepiphyseal regions during development. *J. Orthop. Res.* **1**, 501–510.

Dias, E. V., and Richter, M. (2002). On the squamation of *Australerpeton cosgriffi* Barberena, a temnospondyl amphibian from the Upper Permian of Brazil. *Anais Acad. Bras. Ciências* **74**, 477–490.

Diaz-Flores, L., Gutierrez, R., Gonzales, P., and Varela, H. (1991). Inducible perivascular cells contribute to the neochondrogenesis in grafted perichondrium. *Anat. Rec.* **229**, 1–8.

Dickson, G. R. (1978). Ultrastructural localization of alkaline phosphatase in the hypertrophic chondrocyte of the frog. *Histochemistry* **57**, 343–347.

Dickson, G. R. (1981). The role of matrix vesicles in the calcification of amphibian growth cartilage. In *Matrix Vesicles. Proceedings of the Third International Conference on Matrix Vesicles*, pp. 191–196. Wichtig Editore, Milan.

Dickson, G. R. (1982). Ultrastructure of growth cartilage in the proximal femur of the frog, *Rana temporaria*. *J. Anat.* **135**, 549–564.

Dickson, G. R. (1984). *Methods of Calcified Tissue Preparation.* Elsevier, Amsterdam.

Diegelmann, R. F., and Peterkofsky, B. (1972). Collagen biosynthesis during connective tissue development in chick embryo. *Devel. Biol.* **28**, 443–453.

Diekwisch, T. G. H. (2001). The developmental biology of cementum. *Int. J. Devel. Biol.* **45**, 695–706.

Diekwisch, T. G. H. (ed.). (2002). Development and histochemistry of vertebrate teeth. *Microsc. Res. Tech.* **59**, 339–459.

Diethelm, F., and Schowing, J. (1973). Action de l'actinomycine D sur l'encéphale embryonnaire de poulet. des conséquences sur le développement cephalique. *Bull. Assoc. Anat.* **57**, 1–8.

Diethelm, F., and Schowing, J. (1974). Action directe de l'actinomycine D sur l'encéphale embryonnaire du poulet. Obtention expérimentale de la cyclopie. *Wilhelm Roux Arch. Entwicklungsmech. Org.* **175**, 163–172.

Dietz, V. H., Ziegelmeier, G., Bittner, K., Bruckner, P., and Balling, R. (1999). Spatio-temporal distribution of chondromodulin-I mRNA in the chicken embryo: Expression during cartilage development and formation of the heart and eye. *Devel. Dynam.* **216**, 233–243.

DiGiancamillo, M., Rattegni, G., Podestà, M., Cagnolaro, L., Cozzi, N., and Leonardi, L. (1998). Postnatal ossification of the thoracic limb in striped dolphins (*Stenella coeruleoalba*) (Meyen, 1833) from the Mediterranean sea. *Can. J. Zool.* **76**, 1286–1293.

Dingerkus, G., Seret, B., and Guilbert, E. (1991). Multiple prismatic calcium phosphate layers in the jaws of present-day sharks (Chondrichthyes, Selachii). *Experientia* **47**, 38–39.

Dingerkus, G., and Uhler, L. D. (1977). Enzyme clearing of alcian blue stained whole vertebrates for demonstration of cartilage. *Stain Technol.* **52**, 229–232.

Dingle, J. T. (1981). Catabolin – A cartilage catabolic factor from synovium. *Clin. Orthop. Rel. Res.* **156**, 219–231.

Dingle, J. T. (1984). The role of cellular interactions in joint erosions. *Clin. Orthop. Rel. Res.* **182**, 24–30.

Dingle, J. T., Horsfield, P., Fell, H. B., and Barratt, M. E. J. (1975). Breakdown of proteoglycan and collagen induced in pig articular cartilage in organ culture. *Ann. Rheum. Dis.* **34**, 303–311.

Dinner, B. J. (1970). A comparison of selected biochemical components of normal and creeper chick embryo tibiae. M.Sc. Thesis, Dept. of Orthodontics, University of Nebraska, Lincoln, NE.

Di Nino, D. L., Long, F., and Linsenmayer, T. F. (2001). Regulation of endochondral cartilage growth in the developing avian limb: cooperative involvement of perichondrium and periosteum. *Devel. Biol.* **240**, 433–442.

Dinsmore, C. E. (ed.) (1991). *A History of Regeneration Research. Milestones in the Evolution of a Science.* Cambridge University Press, Cambridge.

Dinsmore, C. E., and Hanken, J. (1986). Native variant limb skeletal patterns in the red-backed salamander, *Plethodon cinereus*, are not regenerated. *J. Morphol.* **190**, 191–200.

Diogo, R., Oliveira, C., and Chardon, M. (2001). On the homologies of the skeletal components of catfish (Teleostei: Siluriformes) suspensorium. *Belg. J. Zool.* **131**, 93–109.

Dittmer, J. E., and Goss, R. J. (1974). Size changes of auxillary heart grafts in rats. *Cardiology* **58**, 355–363.

Dittmer, J. E., Goss, R. J., and Dinsmore, C. E. (1974). The growth of infant hearts grafted to young and adult rats. *Amer. J. Anat.* **141**, 155–160.

Diwan, A. D., Wang, M. X., Jang, D., Zhu, W., and Murrell, G. A. C. (2000). Nitric oxide modulates fracture healing. *J. Bone Min. Res.* **15**, 342–351.

Dixon, A. D., and Hoyte, D. A. N. (1963). A comparison of autoradiographic and alizarin techniques in the study of bone growth. *Anat. Rec.* **145**, 101–113.

Dixon, A. D., and Sarnat, B. G. (eds) (1982). *Factors and Mechanisms Influencing Bone Growth.* Alan R. Liss Inc., New York, NY.

Dixon, A. D., and Sarnat, B. G. (eds) (1985). *Normal and Abnormal Bone Growth. Basic and Clinical Research.* Alan R. Liss Inc., New York, NY.

Dixon, A. D., Sarnat, B. G., and Hoyte, D. A. N. (eds) (1991). *Fundamentals of Bone Growth: Methodology and Applications.* CRC Press, Boca Raton, FL.

Dixon, B. (1970). Regional variation in the cycle time of cells in epiphyseal cartilage. *Rev. Eur. Etud. Clin. Biol.* **15**, 875–878.

Dixon, J., and Dixon, M. J. (2004). Genetic background has a major effect on the penetrance and severity of craniofacial defects in mice heterozygous for the gene encoding the nucleolar protein Treacle. *Devel. Dynam.* **229**, 907–914.

Djorivic, A., and Kalezic, M. L. (2000). Paedogenesis in European newts (*Triturus*: Salamandridae): Cranial morphology during ontogeny. *J. Morphol.* **243**, 127–139.

Dockter, J. L. (2000). Sclerotome induction and differentiation. *Curr. Top. Devel. Biol.* **48**, 77–127.

Dockter, J. L., and Ordahl, C. P. (2000). Dorsoventral axis determination in the somite: a re-examination. *Development* **127**, 2201–2206.

Dodig, M., Tadic, T., Kronenberg, M. S., Dacic, S., Liu, Y.-H., Maxson, R., Rowe, D. W., and Lichtler, A. C. (1999). Ectopic *Msx2* overexpression inhibits and *Msx2* antisense stimulates calvarial osteoblast differentiation. *Devel. Biol.* **209**, 298–307.

Dodds, G. S., and Cameron, H. C. (1939). Studies on experimental rickets in rats. III. Behavior and fate of the cartilage remnants in the rachitic metaphysis. *Amer. J. Pathol.* **15**, 723–740.

Dodds, R. A., Catterall, A., Bitensky, L., and Chayen, L. (1986). Abnormalities in fracture healing induced by vitamin B6-deficiency in rats. *Bone* **1**, 489–496.

Doi, M., Nagano, A., and Nakamura, Y. (2004). Molecular cloning and characterization of a novel gene, EMILIN-5, and its possible involvement in skeletal development. *Biochim. Biophys. Res. Co.* **313**, 888–893.

Dollé, P., Dierich, A., LeMeur, M., Schimmang, T., Schuhbaur, B., Chambon, P., and Duboule, D. (1993). Disruption of the Hoxd-13 gene induces localised heterochrony leading to mice with neotenic limbs. *Cell* **75**, 431–441.

Dominguez, J. H., and Mundy, G. R. (1980). Monocytes mediate osteoclastic bone resorption by prostaglandin production. *Calcif. Tissue Int.* **31**, 29–34.

Dondi, P., and Muir, H. (1976). Collagen synthesis and deposition in cartilage during disrupted proteoglycan production. *Biochem. J.* **160**, 117–120.

Donoghue, P. C. J. (1998). Growth and patterning in the conodont skeleton. *Phil. Trans R. Soc. B.* **353**, 633–666.

Donoghue, P. C. J. (2001). Microstructural variation in conodont enamel is a functional adaptation. *Proc. R. Soc. Lond. B.* **268**, 1691–1698.

Donoghue, P. C. J. (2002). Evolution of development of the vertebrate dermal and oral skeletons: unraveling concepts, regulatory theories, and homologies. *Paleobiology* **28**, 474–507.

Donoghue, P. C. J., and Aldridge, R. J. (2001). Origin of a mineralized skeleton. In *Early Vertebrate Evolution: Palaeontology, Phylogeny, Genetics and Evolution* (P. E. Ahlberg, ed.), pp. 85–105. Taylor Francis, London.

Donoghue, P. C. J., and Chauffe, K. M. (1998). *Conchodontus, Mitrellataxis* and *Fungulodus*: Conodonts, fish, or both? *Lethaia* **31**, 283–292.

Donoghue, P. C. J., Forey, P. L., and Aldridge, R. J. (2000). Conodont affinity and chordate phylogeny. *Biol. Rev. Camb. Philos. Soc.* **75**, 191–251.

Donoghue, P. C. J., and Purnell, M. A. (1999a). Growth, function and the conodont fossil record. *Geology* **27**, 251–254.

Donoghue, P. C. J., Purnell, M. A., and Aldridge, R. J. (1998). Conodont anatomy, chordate phylogeny and vertebrate classification. *Lethaia* **31**, 211–219.

Donoghue, P. C. J., and Purnell, M. A. (1999b). Mammal-like occlusion in conodonts. *Paleobiology* **25**, 58–74.

Donoghue, P. C. J., and Sansom, I. J. (2002). Origin and early evolution of vertebrate skeletonization. *Micr. Res. Tech.* **59**, 352–372.

Dony, C., and Gruss, P. (1987). Proto-oncogene *c-fos* expression in growth regions of fetal bone and mesodermal web tissue. *Nature* **328**, 711–714.

Dorey, C. K., and Sorgente, N. (1977). Bone induction without vascular contribution and biochemical determinants of vascularization. *J. Dent. Res.* **56**, A81.

Dorsey, G. A. (1897). Wormian bones in artificially deformed Kwakintl crania. *Amer. Anthrop.* **10**, 169–173.

Doty, S. B. (1981). Morphological evidence of gap junctions between bone cells. *Calcif. Tissue Int.* **33**, 509–512.

Doty, S. B., and Nunez, E. A. (1985). Activation of osteoclasts and the repopulation of bone surfaces following hibernation in the bat, *Myotis lucifugus. Anat. Rec.* **213**, 481–495.

Douglas, B. S. (1972). Conservative management of guillotine amputation of the finger in children. *Aust. Paediatr.* **8**, 86–89.

Downie, S. A., and Newman, S. A. (1994). Morphogenetic differences between fore and hind limb precartilage mesenchyme: Relation to mechanisms of skeletal pattern formation. *Devel. Biol.* **162**, 195–208.

Downie, S. A., and Newman, S. A. (1995). Different roles for fibronectin in the generation of fore and hind limb precartilage condensations. *Devel. Biol.* **172**, 519–530.

Drachman, D. B. (1964). Atrophy of skeletal muscle in chick embryos treated with botulinum toxin. *Science* **145**, 719–721.

Drachman, D. B., and Coulombre, A. J. (1962a). Method for continuous infusion of fluids into the chorioallantoic circulation of the chick embryo. *Science* **138**, 144–145.

Drachman, D. B., and Coulombre, A. J. (1962b). Experimental clubfoot and arthrogryposis multiplex congenita. *Lancet* **2**, 523–526.

Drachman, D. B., and Sokoloff, L. (1966). The role of movement in embryonic joint development. *Devel. Biol.* **4**, 401–420.

Drake, D. B., Persing, J. A., Berman, D. E., and Ogle, R. C. (1993). Calvarial deformity regeneration following subtotal craniectomy for craniosynostosis: a case report and theoretical implications. *J. Craniofac. Surg.* **4**, 85–89.

Drew, I. M., Perkins, D., Jr., and Daly, P. (1971). Prehistoric domestication of animals: Effects on bone structure. *Science* **171**, 280–282.

Drews, Ulrich, and Drews, Ute (1972). Cholinesterase in development of chick limb. I. Phases of cholinesterase activity in early bud and in demarcating cartilage and muscle anlagen. *Wilhelm Roux Arch. Entwicklungsmech. Org.* **169**, 70–86.

Drews, Ulrich, and Drews, Ute (1973). Cholinesterase in development of chick limb. II. Enzyme activity and locomotory behavior of the presumptive cartilage cells. *Wilhelm Roux Arch. Entwicklungsmech. Org.* **173**, 208–227.

Drews, Ulrich, Kocher-Becker, U., and Drews, Ute (1972). The induction of visceral cartilage from cranial neural crest by pharyngeal endoderm in hanging drop cultures and the locomotory behaviour of the neural crest cells during cartilage differentiation. *Wilhelm Roux Arch. Entwicklungsmech. Org.* **171**, 17–37.

Drossopoulou, G., Lewis, K. E., Sanz-Ezquerro, J. J., Nikbakht, N., McMahon, A. P., Hofmann, C., and Tickle, C. (2000). A model for anteroposterior patterning of the vertebrate limb based on sequential long- and short-range Shh signalling and Bmp signalling. *Development* **127**, 1337–1348.

Duan, C., and Hirano, T. (1990). Stimulation of ^{35}S-sulfate uptake by mammalian insulin-like growth factors I and II in cultured cartilages of the Japanese eel, *Anguilla japonica. J. Exp. Zool.* **256**, 347–350.

Duboule, D. (ed.) (1994). *Guidebook to the Homeobox Genes.* Oxford University Press, Oxford.

Duboule, D. (2002). Making progress with limb models. *Nature* **418**, 492–493.

Dubreuil, G. (1913a). La croissance des os des mammifères. I. Méthode de précision pour la mesure de la croissance des os. *C.R. Séanc. Soc. Biol.* **74**, 756–758.

Dubreuil, G. (1913b). La croissance des os des mammifères. II. Croissance au niveau du cartilage de conjugaison. *C.R. Séanc. Soc. Biol.* **74**, 888–890.

Dubreuil, G. (1913c). La croissance des os des mammifères. III. L'accroissement interstitiel n'existe pas dans les os longs. *C.R. Séanc. Soc. Biol.* **74**, 935–937.

du Brul, E. L. (1964). Evolution of the temporomandibular joint. In *The Temporomandibular Joint* (B. G. Sarnat, ed.), 2nd Edn, pp. 3–27. C. C. Thomas, Springfield, IL.

du Brul, E. L., and Laskin, D. M. (1961). Preadaptive potentialities of the mammalian skull: An experiment in growth and form. *Amer. J. Anat.* **109**, 117–132.

Ducy, P. (2000). Cbfa1: a molecular switch in osteoblast biology. *Devel. Dynam.* **219**, 461–471.

Ducy, P., and Karsenty, G. (1998). Genetic control of cell differentiation in the skeleton. *Curr. Opin. Cell Biol.* **10**, 614–619.

Ducy, P., Schinke, T., and Karsenty, G. (2000). The osteoblast: Sophisticated fibroblast under central surveillance. *Science* **289**, 1501–1504.

Ducy, P., Zhang, R., Geoffroy, V., Ridall, A. L., and Karsenty, G. (1997). Osf2/cbfa1: A transcriptional activator of osteoblast differentiation. *Cell* **89**, 747–754.

Dudley, A. T., Ros, M. A., and Tabin, C. J. (2002). A re-examination of proximodistal patterning during vertebrate limb development. *Nature* **418**, 539–544.

Dudley, A. T., and Tabin, C. J. (2000). Constructive antagonism in limb development. *Curr. Opin. Genet. Devel.* **10**, 387–392.

Duellman, W. E., and Trueb, L. (1986). *Biology of Amphibians.* McGraw-Hill, New York, NY.

Duerden, J. E. (1920). The inheritance of callosities in the ostrich. *Amer. Nat.* **54**, 289–312.

Dufresne, C., and Richtsmeier, J. T. (1995). Interaction of craniofacial dysmorphology, growth and prediction of surgical outcome. *J. Craniofac. Surg.* **6**, 270–281.

Duhene, M., Sobel, M. E., and Müller, P. K. (1982). Levels of collagen mRNA in dedifferentiating chondrocytes. *Exp. Cell Res.* **142**, 317–324.

Duhamel de Monceau, H. (1739). Sur une racine qui a la faculté de teindre en rouge les os des animaux vivants. *Mém. Acad. Roy. Sci. Paris* **52**, 1–13

Duhamel de Monceau, H. (1740). Observations and experiments with madder root. *Phil. Trans R. Soc. London* **8**, 420–434.

Duhamel de Monceau, H. (1742). Sur le Développement et la crue des os des Animaux. *Hist. Mem. Acad. Inscriptions Belles Letters* **2**, 481.

Duhamel de Monceau, H. (1743a). IVᵉ Mémoire sur les os. *Mém. Acad. R Sci.* **56**, 87–111.

Duhamel de Monceau, H. (1743b). Vᵉ Mémoire sur les os. *Mém. Acad. R. Sci.* **56**, 111–146.

Duke, J., and Elmer, W. A. (1977). Effect of the brachypod mutation on cell adhesion and chondrogenesis in aggregates of mouse limb mesenchyme. *J. Embryol. Exp. Morphol.* **42**, 209–217.

Duke, J., and Elmer, W. A. (1978). Cell adhesion and chondrogenesis in brachypod mouse limb mesenchyme: Fragment fusion studies. *J. Embryol. Exp. Morphol.* **48**, 161–168.

Duke, J., and Elmer, W. A. (1979). Effect of the brachypod mutation on early stages of chondrogenesis in mouse embryonic hind limbs: An ultrastructural analysis. *Teratology* **19**, 367–376.

Duke, J. C. (1983). Suppression of morphogenesis in embryonic mouse limbs exposed *in vitro* to excess gravity. *Teratology* **27**, 427–436.

Duksin, D., Maoz, A., and Fuchs, S. (1975). Differential cytotoxic activity of anticollagen serum on rat osteoblasts and fibroblasts in tissue culture. *Cell* **5**, 83–86.

Dullemeijer, P. (1985). The significance of van Limborgh's approach to craniofacial biology. *Acta Morphol. Neerl.-Scand.* **23**, 317–324.

Dungan, K. M., Wei, T. Y., Nace, J. D., Poulin, M. L., Chiu, I.-M., Lang, J. C., and Tassava, R. A. (2002). Expression and biological effect of urodele fibroblast growth factor 1: Relationship to limb regeneration. *J. Exp. Zool.* **292**, 540–544.

Dunlap, K. D., and Sanchiz, B. (1996). Temporal dissociation between the development of the cranial and appendicular skeletons in *Bufo bufo* (Amphibia: Bufonidae). *J. Herpetol.* **30**, 506–513.

Dunlop, L.-L. T., and Hall, B. K. (1995). Relationships between cellular condensation, preosteoblast formation and epithelial-mesenchymal interactions in initiation of osteogenesis. *Int. J. Devel. Biol.* **39**, 357–371.

Dunn, B. E., Fitzharris, T. P., and Barnett, B. D. (1981). Effects of varying chamber construction and embryo pre-incubation age in survival and growth of chick embryos in shell-less culture. *Anat. Rec.* **199**, 33–43.

Dunn, J. R. (1984). Developmental osteology. In *Ontogeny and Systematics of Fishes* (H. G. Moser, ed.), pp. 48–50, Special Publication Number 1, Amer. Soc. Ichthyologists and Herpetologists, Lawrence, KS.

Dunwoodie, S. L., Clements, M., Sparrow, D. B., Sa, X., Conclon, R. A., and Beddington, R. S. P. (2002). Axial skeletal defects caused by mutation in the spondylocostal dysplasia/pudgy gene *Dll3* are associated with disruption of the segmentation clock within the presomitic mesoderm. *Development* **129**, 1795–1806.

Dupont, J., and Holzenberger, M. (2003). Biology of insulin-like growth factors in development. *Birth Defects Res. (Part C)*, **69**, 257–271.

Duprez, D., Bell, E. J. de H., Richardson, M. K., Archer, C. W., Wolpert, L., Brickell, P. M., and Francis-West, P. H. (1996). Overexpression of BMP-2 and BMP-4 alters the size and shape of developing skeletal elements in the chick limb. *Mech. Devel.* **57**, 145–157.

Durkin, J. F. (1972). Secondary cartilage: A misnomer? *Amer. J. Orthod.* **62**, 15–41.

Durkin, J. F., Heely, J., and Irving, J. T. (1973). The cartilage of the mandibular condyle. *Oral Sci. Rev.* **2**, 29–99.

Durkin, J. F., Irving, J. T., and Heely, J. (1969a). A comparison of the circulatory and calcification patterns in the mandibular condyle of the guinea pig with those found in the tibial epiphyseal and articular cartilages. *Archs Oral Biol.* **14**, 1365–1372.

Durkin, J. F., Irving, J. T., and Heely, J. (1969b). A comparison of circulatory and calcification changes induced in the mandibular condyle, tibial epiphyseal and articular cartilages of the guinea pig by the onset and healing of scurvy. *Archs Oral Biol.* **14**, 1373–1382.

Duterloo, H. S., and Jansen, H. W. B. (1969). Chondrogenesis and osteogenesis in the mandibular condylar blastema. *Report Cong. Eur. Orthodontic Soc.* 109–118.

Duterloo, H. S., and Wolters, J. M. (1971). Experiments on the significance of articular function as a stimulating chondrogenic factor for the growth of secondary cartilages of the rat mandible. In *Transactions of the European Orthodontic Society*, pp. 103–116.

Duverney, G. J. (1700). Sur la structure intime et le sentiment de laxmoelle des os. *Bull. Mem. Soc. Sci. Paris* **8**.

Dvorak, L., and Fallon, J. F. (1991). *Talpid*² mutant chick limb has anteroposterior polarity and altered patterns of programmed cell death. *Anat. Rec.* **231**, 251–260.

Dworkin, I., Tanda, S., and Larsen, E. (2001). Are entrenched characters developmentally constrained? Creating biramous limbs in an insect. *Evol. & Devel.* **3**, 424–431.

Dysart, P. S., Harkness, E. M., and Herbison, G. P. (1989). Growth of the humerus after denervation. An experimental study in the rat. *J. Anat.* **167**, 147–159.

Dziak, R., and Brand, J. S. (1974a). Calcium transport in isolated bone cells. I. Bone cell isolation procedures. *J. Cell. Physiol.* **84**, 75–84.

Dziak, R., and Brand, J. S. (1974b). Calcium transport in isolated bone cells. II. Calcium transport studies. *J. Cell. Physiol.* **84**, 85–96.

Dziak, R., and Stern, P. H. (1976). Responses of fetal rat bone cells and bone organ cultures to the ionophore A 23187. *Calcif. Tissue Res.* **22**, 137–147.

Dziedzic-Goclawska, A., Emerich, J., Grzesik, W., Stachowicz, W., Michalik, J., and Ostrowski, K. (1988). Differences in the kinetics of the mineralization process in endochondral and intramembranous osteogenesis in human fetal development. *J. Bone Min. Res.* **3**, 533–540.

Dzik, J. (1986). Chordate affinities of the conodonts. In *Problematic Fossil Taxa* (A. Hoffman and M. H. Nitecki, eds), pp. 240–254. Oxford University Press, New York, NY.

Dzik, J. (2000). The origin of the mineral skeleton in chordates. *Evol. Biol.* **31**, 105–154.

E

Eastoe, J. E. (1970). The place of cartilage and bone among vertebrate mineralized tissues. *Calcif. Tissue Res.* **4** (Suppl.), 24–27.

Eavey, R. D., Schmid, T. M., and Linsenmayer, T. F. (1988). Intrinsic and extrinsic controls of the hypertrophic program of chondrocytes in the avian columella. *Devel. Biol.* **126**, 57–62.

Ecarot-Charrier, B., Glorieux, F. H., van der Rest, M., and Pereira, G. (1983). Osteoblasts isolated from mouse calvaria initiate matrix mineralization in culture. *J. Cell Biol.* **96**, 639–643.

Ecarot-Charrier, B., Shephard, N., Charette, G., Grynpas, M., and Glorieux, F. H. (1988). Mineralization in osteoblast cultures: A light and electron microscopic study. *Bone* **9**, 147–154.

Eco, U. (1984). *The Name of the Rose*. Translated by William Weaver. Warner Books, New York, NY.

Eco, U. (1997). *Kant and the Platypus: Essays on Language and Cognition*. Harcourt, Brace & Company, New York, NY.

Economides, K. D., Zeltser, L., and Capecchi, M. R. (2003). *Hoxb13* mutations cause overgrowth of caudal spinal cord and tail vertebrae. *Devel. Biol.* **256**, 317–330.

Ede, D. A. (1971). Control of form and pattern in the vertebrate limb. *Symp. Soc. Exp. Biol.* **25**, 235–254.

Ede, D. A. (1976). Cell interactions in vertebrate limb development. In *Cell Surface in Animal Embryogenesis* (G. Poste and G. L. Nicolson, eds), pp. 495–544. Elsevier/North Holland, Amsterdam.

Ede, D. A. (1977). Relations between ontogeny and phylogeny of the vertebrate limbs suggested by studies on the *talpid³* mutant of the chick. *Coll. Int. CNRS Paris* **266**, 410–411.

Ede, D. A. (1980). Role of the ectoderm in limb development of normal and mutant mouse (disorganized, pupoid foetus) and fowl (*talpid³*) embryos. In *Teratology of the Limbs* (H. J. Merker, H. Nav, and D. Neubert, eds), pp. 53–66. Walter de Gruyter & Co., Berlin.

Ede, D. A. (1983). Cellular condensation and chondrogenesis. In *Cartilage, Volume 2. Development, Differentiation and Growth* (B. K. Hall, ed.), pp. 143–186. Academic Press, New York, NY.

Ede, D. A., and Agerback, G. S. (1968). Cell adhesion and movement in relation to the developing limb pattern in normal and *talpid³* mutant chick embryos. *J. Embryol. Exp. Morphol.* **20**, 81–100.

Ede, D. A., Bellairs, R., and Bancroft, M. (1974). Scanning electron microscope study of early limb-bud in normal and *talpid³* mutant chick embryos. *J. Embryol. Exp. Morphol.* **31**, 761–786.

Ede, D. A., and Flint, O. P. (1972). Patterns of cell division, cell death and chondrogenesis in cultured aggregates of normal and *talpid³* mutant chick limb mesenchyme cells. *J. Embryol. Exp. Morphol.* **27**, 245–260. ·

Ede, D. A., and Flint, O. P. (1975a). Intercellular adhesion and formation of aggregates in normal and *talpid³* mutant chick limb mesenchyme. *J. Cell Sci.* **18**, 97–112.

Ede, D. A., and Flint, O. P. (1975b). Cell movement and adhesion in the developing chick wing bud: Studies on cultured mesenchyme cells from normal and *talpid³* mutant embryos. *J. Cell Sci.* **18**, 301–314.

Ede, D. A., Flint, O. P., and Teague, P. (1975). Cell proliferation in the developing wing-bud of normal and *talpid³* mutant chick embryos. *J. Embryol. Exp. Morphol.* **34**, 589–607.

Ede, D. A., Flint, O. P., Wilby, O. K., and Colquhoun, P. (1977b). The development of precartilage condensations in limb-bud mesenchyme *in vivo* and *in vitro*. In *Vertebrate Limb and Somite Morphogenesis* (D. A. Ede, J. R. Hinchliffe, and M. Balls, eds), pp. 161–180. Cambridge University Press, Cambridge.

Ede, D. A., Hinchliffe, J. R., and Balls, M. (eds) (1977a). *Vertebrate Limb and Somite Morphogenesis.* Third Symp., Brit. Soc. Devel. Biol. Cambridge University Press, Cambridge.

Ede, D. A., and Kelly, W. A. (1964a). Developmental abnormalities in the head region of the *talpid³* mutant of the fowl. *J. Embryol. Exp. Morphol.* **12**, 161–182.

Ede, D. A., and Kelly, W. A. (1964b). Developmental abnormalities in the trunk and limbs of the *talpid³* mutant of the fowl. *J. Embryol. Exp. Morphol.* **12**, 339–356.

Ede, D. A., and Law, J. T. (1969). Computer simulation of vertebrate limb morphogenesis. *Nature* **221**, 244–248.

Ede, D. A., and Wilby, O. K. (1981). Golgi orientation and cell behaviour in the developing pattern of chondrogenic condensations in chick limb-bud mesenchyme. *Histochem. J.* **13**, 615–630.

Edelman, G. M. (1976). Surface modulation in cell recognition and cell growth. *Science* **192**, 218–226.

Edelman, G. M. (1986). Cell adhesion molecules in the regulation of animal form and tissue pattern. *Annu. Rev. Cell Biol.* **2**, 81–116.

Edelman, G. M. (1988). *Topobiology. An Introduction to Molecular Embryology.* Basic Books, New York, NY.

Edom-Vovard, F., and Duprez, D. (2004). Signals regulating tendon formation during chick embryonic development. *Devel. Dyn.* **229**, 449–457.

Edwards, J. L. (1989). Two perspectives on the evolution of the tetrapod limb. *Amer. Zool.* **29**, 235–254.

Efstratiadis, A. (1998). Genetics of mouse growth. *Int. J. Devel. Biol.* **42**, 955–976.

Egerbacher, M., Krestan, R., and Böck, P. (1995). Morphology, histochemistry and differentiation of the cat's epiglottic cartilage: A supporting organ composed of elastic cartilage, fibrous cartilage, myxoid tissue, and fat tissue. *Anat. Rec.* **242**, 471–482.

Eggert, R. C. (1966). The response of x-irradiated limbs of adult urodeles to autografts of normal cartilage. *J. Exp. Zool.* **161**, 369–390.

Eguchi, G., and Okada, T. S. (1971). Ultrastructure of the differentiated cell colony derived from a singly isolated chondrocyte in *in vitro* culture. *Devel. Growth Differ.* **12**, 297–312.

Eguro, H, and Goldner, J. L. (1974). Antigenic properties of chondromucoprotein and inducibility of experimental arthritis by antichondromucoprotein immune globulin. *J. Bone Joint Surg. Am. Vol.* **56**, 129–141.

Ehmcke, J., Wistuba, J., and Clemen, G. (2004). Separated dental laminae are present in the upper jaw of Mesoamerican lungless salamanders (Amphibia, Plethodontidae). *Ann. Anat.* **186**, 45–53.

Ehrich, T. H., Vaughn, T. T., Koreishi, S. F., Linsey, R. B., Pletscher, L. S., and Cheverud, J. M. (2003). Pleiotropic effects on mandibular morphology I. Developmental morphological integration and differential dominance. *J. Exp. Zool. (Mol. Dev. Evol.)* **296B**, 58–79.

Ehrlich, J., Yaffe, A., Shanfeld, J. L., Montgomery, P. C., and Davidovitch, Z. (1980). Immunohistochemical localization and distribution of cyclic nucleotides in the rat mandibular condyle in response to an induced occlusal change. *Archs Oral Biol.* **25**, 545–552.

Ehrlich, M. G., Armstrong, A. L., Neuman, R. G., Davis, M. W., and Mankin, H. J. (1982). Patterns of proteoglycan degradation by a neutral protease from human growth-plate epiphyseal cartilage. *J. Bone Joint Surg. Am. Vol.* **64**, 1350–1354.

Ehrlich, M. G., Mankin, H. J., Treadwell, B. V., and Jones, H. (1974). Uridine phosphate (UDP) stimulation of protein polysaccharide production. A preliminary report. *J. Bone Joint Surg. Am. Vol.* **56**, 1239–1245.

Ehrlich, M. G., Mankin, H. J., Jones, H., Wright, R., Crispen, C., and Vigliani, G. (1977). Collagenase and collagenase inhibitors in osteoarthritic and normal human cartilage. *J. Clin. Invest.* **59**, 226–233.

Ehrlich, P. H., and Bornstein, P. (1972). Microtubules in transcellular movement of procollagen. *Nature New Biol.* **238**, 257–260.

Eiben, B., Scharla, S., Fischer, K., and Schmidtgayk, H. (1984). Seasonal variations of serum 1,25-dihydroxyvitamin-D3 and alkaline phosphatase in relation to the antler formation in the Fallow Deer (*Dama dama* L). *Acta Endocrinol.* **107**, 141–144.

Eichele, G. (1986). Retinoids induce duplications in developing vertebrate limbs. *BioScience* **36**, 534–540.

Eichele, G. (1989). Retinoic acid induces a pattern of digits in anterior half wing buds that lack the zone of polarizing activity. *Development* **107**, 863–868.

Eilberg, R. G., Person, P., and Zuckerberg, D. (1974). Mineralization of invertebrate cartilage. *J. Dental Res.* **53**, 246 (abstract 767).

Eilberg, R. G., Zuckerberg, D., and Person, P. (1975). Mineralization of invertebrate cartilage. *Calcif. Tissue Res.* **19**, 85–90.

Eilon, G., and Mundy, G. R. (1978). Direct resorption of bone by human breast cancer cells *in vitro*. *Nature* **276**, 726–728.

Einhorn, T. A. (1998). The cell and molecular biology of fracture healing. *Clin. Orthop. Rel. Res.* **355**, S7–S21.

Einhorn, T. A., Simon, G., Devlin, V. J., Warman, J., Sidhu, S. P. S., and Vigorita, V. J. (1990). The osteogenic response to distant skeletal injury. *J. Bone Joint Surg. Am. Vol.* **72**, 1374–1378.

Einhorn, T. A., Wakley, G. K., Linkhart, S., Rush, E. B., Maloney, S., Faierman, E., and Baylink, D. J. (1992). Incorporation of sodium fluoride into cortical bone does not impair the mechanical properties of the appendicular skeleton in rats. *Calcif. Tissue Int.* **51**, 127–131.

Eisenstein, R., Kuettner, K. E., Neapolitan, C., Soble, L. W., and Sorgente, N. (1975). The resistance of certain tissues to invasion. III. Cartilage extracts inhibit the growth of fibroblasts and endothelial cells in culture. *Amer. J. Pathol.* **81**, 337–348.

Eisenstein, R., Kuettner, K. E., Soble, L. W., and Sorgente, N. (1976). Tissue inhibitors and cell growth regulators. In *Protides of the Biological Fluids* (H. Peeters, ed.), pp. 217–229. Pergamon Press, Oxford.

Eisenstein, R., Sorgente, N., and Kuettner, K. E. (1971). Organization of extracellular matrix in epiphyseal growth plate. *Amer. J. Pathol.* **65**, 515–534.

Eisenstein, R., Sorgente, N., Soble, L. W., Miller, A. W., and Kuettner, K. E. (1973). The resistance of certain tissues to invasion: Penetrability of explanted tissues by vascularized mesenchyme. *Amer. J. Pathol.* **73**, 765–774.

Ekanayake, S., and Hall, B. K. (1987). The development of acellularity of the vertebral bone of the Japanese medaka, *Oryzias latipes* (Teleostei, Cyprinidontidae). *J. Morphol.* **193**, 253–261.

Ekanayake, S., and Hall, B. K. (1988). Ultrastructure of the osteogenesis of acellular vertebral bone in the Japanese medaka, *Oryzias latipes* (Teleostei, Cyprinidontidae). *Amer. J. Anat.* **182**, 241–249.

Ekanayake, S., and Hall, B. K. (1991). Development of the notochord in the Japanese medaka, *Oryzias latipes* (Teleostei, Cyprinidontidae) with special reference to desmosomal connections and functional integration with adjacent tissues. *Can. J. Zool.* **69**, 1171–1177.

Ekanayake, S., and Hall, B. K. (1994a). Formation of cartilaginous nodules and heterogeneity in clones of HH 17 mandibular ectomesenchyme from the embryonic chick. *Acta Anat.* **151**, 171–179.

Ekanayake, S., and Hall, B. K. (1994b). Hypertrophy is not a prerequisite for type X collagen expression or mineralization of chondrocytes derived from cultured chick mandibular ectomesenchyme. *Int. J. Devel. Biol.* **38**, 683–694.

Ekanayake, S., and Hall, B. K. (1997). The *in vivo* and *in vitro* effects of bone morphogenetic protein-2 on the development of the chick mandible. *Int. J. Devel. Biol.* **41**, 67–81.

Eke, N., and Warrington, A. J. (1981). Ossification in abdominal scars. *J. Roy. Soc. Med.* **74**, 653–655.

Ekker, M., Akimenko, M.-A., Allende, M. L., Smith, R., Drouin, G., Langille, R. M., Weinberg, E. S., and Westerfield, M. (1997). Relationships among Msx gene structure and function in zebrafish and other vertebrates. *Mol. Biol. Evol.* **14**, 1008–1022.

Elder, J. R., and Tuenge, R. (1974). Cephalometric and histologic changes produced by extraoral high pull traction to the maxilla in *Macaca mulatta*. *Amer. J. Orthod.* **66**, 599–617.

Elder, S. H., Kimura, J. H., Soslowsky, L. J., Lavagnino, M., and Goldstein, S. A. (2000). Effect of compressive loading on chondrocyte differentiation in agarose cultures of chick limb-bud cells. *J. Orthop. Res.* **18**, 78–86.

Elford, P. R., Guenther, H. L., Felix, R., Cecchini, M. G., and Fleisch, H. (1987). Transforming growth factor-β reduces the phenotypic expression of osteoblastic MC3T3-E1 cells in monolayer culture. *Bone* **8**, 259–262.

Elima, K. (1993). Osteoinductive proteins. *Ann. Med.* **25**, 395–402.

Elima, K., and Vuorio, E. (1989). Expression of messenger RNAs for collagens and other matrix components in dedifferentiating and redifferentiating human chondrocytes in culture. *FEBS Letts* **258**, 195–198.

Ellies, D. L., Langille, R. M., Martin, C. C., Akimenko, M.-A., and Ekker, M. (1997). Specific craniofacial cartilage dysmorphogenesis coincides with a loss of *dlx* gene expression in retinoic acid-treated zebrafish embryos. *Mech. Devel.* **61**, 23–36.

Elliott, H. C. (1936). Studies on articular cartilages. 1. Growth mechanisms. *Amer. J. Anat.* **58**, 127–145.

Ellis, E. III., and Carlson, D. S. (1986). Histologic comparisons of the costochondral, sternoclavicular and temporomandibular joints during growth in *Macaca mulatta*. *J. Oral Maxillofac. Surg.* **44**, 312–321.

Ellis, E. III., Schneiderman, E. D., and Carlson, D. S. (2002). *J. Oral Maxillofac. Surg.* **60**, 1461–1470.

Ellison, M. L., Ambrose, E. J., and Easty, G. C. (1969). Myogenesis in chick embryo somites *in vitro*. *J. Embryol. Exp. Morphol.* **21**, 331–340.

Ellison, M. L., and Lash, J. W. (1971). Environmental enhancement of *in vitro* chondrogenesis. *Devel. Biol.* **26**, 486–496.

Elmer, W. A. (1968a). Experimental analysis of the creeper condition in chickens. Effect of embryo extract on elongation, protein content, and incorporation of amino acids by cartilaginous tibiotarsi. *Devel. Biol.* **18**, 76–92.

Elmer, W. A. (1968b). *In vitro* and *in situ* analysis of the inhibitory effect of creeper tissues. *J. Exp. Zool.* **169**, 381–390.

Elmer, W. A. (1977). Morphological and biochemical modifications of cartilage differentiation in the brachypod and other micromelic mouse embryos. *Colloq. Int. C.N.R.S. Paris* **266**, 235–241.

Elmer, W. A. (1983). Growth Factors and Cartilage. In *Cartilage, Volume 2. Development, Differentiation, and Growth* (B. K. Hall, ed.), pp. 369–400. Academic Press, New York, NY.

Elmer, W. A., Pennypacker, M. F., Knudsen, T. B., and Kwasigroch, T. E. (1988). Alterations in cell surface galactosyltransferase activity during limb chondrogenesis in *Brachypod* mutant mouse embryos. *Teratology* **38**, 475–484.

Ellmer, W. A., and Selleck, D. K. (1975). *In vitro* chondrogenesis of limb mesoderm from normal and brachypod mouse embryos. *J. Embryol. Exp. Morphol.* **33**, 371–386.

Elves, M. W. (1974). A study of the transplantation antigens on chondrocytes from articular cartilage. *J. Bone Joint Surg. Br. Vol.* **56**, 178–185.

Elves, M. W. (1976). Newer knowledge of immunology of bone and cartilage. *Clin. Orthop. Rel. Res.* **120**, 232–259.

Elves, M. W. (1978). The immunobiology of joints. In *The Joints and Synovial Fluid* (L. Sokoloff, ed.), Volume 1, pp. 332–406. Academic Press, New York, NY.

Elves, M. W. (1983). Immunology of cartilage. In *Cartilage, Volume 3, Biomedical Aspects* (B. K. Hall, ed.), pp. 229–266. Academic Press, New York, NY.

Elves, M. W., and Pratt, L. M. (1975). The pattern of new bone formation in isografts of bone. *Acta Orthop. Scand.* **46**, 549–560.

Emerson, S. B. (1988a). Convergence and morphological constraint in frogs: Variance in postcranial morphology. *Fieldiana Zool.* n.s. **43**, 1–19.

Emerson, S. B. (1988b). Testing for historical patterns of change: A case study with frog pectoral girdles. *Paleobiology* **14**, 174–186.

Emlen, D. J. (2000). Integrating development with evolution: A case study with beetle horns. *BioScience* **50**, 403–418.

Emlen, D. J., and Nijhout, H. F. (1999). Hormonal control of male horn length dimorphism in the dung beetle *Ontophagus taurus* (Coleoptera: Scarabaeidae). *J. Insect Physiol.* **45**, 45–53.

Emlen, D. J., and Nijhout, H. F. (2000). The development and evolution of exaggerated morphologies in insects. *Annu. Rev. Entomol.* **45**, 661–708.

End, P., Panayotou, G., Entwistle, A., Waterfield, M. D., and Chiquet, M. (1992). Tenascin – A modulator of cell growth. *Eur. J. Biochem.* **209**, 1041–1051.

Engel, J. (1991). Common structural motifs in proteins of the extracellular matrix. *Curr. Opin. Cell Biol.* **3**, 779–785.

Engelsma, S. O., Jansen, H. W. B., and Duterloo, H. S. (1980). An *in vivo* transplantation study of growth of the mandibular condyle in a functional position in the rat. *Archs Oral Biol.* **25**, 305–311.

Engfeldt, B. (1969). Studies on the epiphyseal growth zone. 3. Electron microscopic studies on the normal epiphyseal growth zone. *Acta Pathol. Microbiol. Scand.* **75**, 201–219.

English, A. W. (ed.) (2003). Joint and muscle dysfunction of the temporomandibular joint. *Cells Tissues Organs* **174**(1–2), 1–96.

English, A. W., and Stohler, C. S. (eds) (2001). *Moving Temporomandibular Joint Research into the 21st Century. Cells Tissues Organs* **169**(3), 181–324.

Engsis, M. T., Chen, Q.-J., Vu, T. H., Pedersen, A.-C., Therkidsen, B., Lund, L. R., Henriksen, K., Lenhard, T., Foged, N. T., Werb, Z., and Delaissé, J.-M. (2000). Matrix metalloproteinase 9 and vascular endothelial growth factor are essential for osteoclast recruitment into developing long bones. *J. Cell Biol.* **151**, 879–889.

Enlow, D. H. (1962a). A study of the postnatal growth and remodeling of bone. *Amer. J. Anat.* **110**, 79–102.

Enlow, D. H. (1962b). Functions of the Haversian system. *Amer. J. Anat.* **110**, 269–306.

Enlow, D. H. (1966a). Osteocyte necrosis in normal bone. *J. Dent. Res.* **45**, 213.

Enlow, D. H. (1966b). An evaluation of the use of bone histology in forensic medicine and anthropology. In *Studies on the Anatomy and Function of Bone and Joints* (F. G. Evans, ed.), pp. 93–112. Springer-Verlag, Berlin.

Enlow, D. H. (1968a). Wolff's law and the factor of architectonic circumstance. *Amer. J. Orthod.* **54**, 803–822.

Enlow, D. H. (1968b). *The Human Face. An Account of the Postnatal Growth and Development of the Craniofacial Skeleton.* Harper (Hoeber), New York, NY.

Enlow, D. H. (1969). The bone of reptiles. In *Biology of the Reptilia* (C. Gans, ed.), Volume 1, pp. 45–80. Academic Press, New York, NY.

Enlow, D. H. (1973). Growth and the problem of the local control mechanism. *Amer. J. Anat.* **136**, 403–406.

Enlow, D. H. (1975). *Handbook of Facial Growth.* Saunders, Philadelphia, PA.

Enlow, D. H., and Brown, S. O. (1956). A comparative histological study of fossil and recent bone. Part I. *Texas J. Sci.* **8**, 405–443.

Enlow, D. H., and Brown, S. O. (1957). A comparative histological study of fossil and recent bone. Part II. *Texas J. Sci.* **9**, 186–214.

Enlow, D. H., and Brown, S. O. (1958). A comparative histological study of fossil and recent bone. Part III. *Texas J. Sci.* **10**, 187–230.

Enomoto-Iwamoto, M., Iwamoto, M., Mukudai, Y., Kawakami, Y., Nohno, T., Higuchi, Y., Takemoto, S., Ohuchi, H., Noji, S., and Kurisu, K. (1998). Bone morphogenetic protein signaling is required for maintenance of differentiated phenotype, control of proliferation, and hypertrophy in chondrocytes. *J. Cell Biol.* **140**, 409–418.

Epperlein, H. H. (1974). The ectomesenchymal–endodermal interaction-system [EEIS] of *Triturus alpestris* in tissue culture. 1. Observations on attachment, migration, and differentiation of neural crest cells. *Differentiation* **2**, 151–168.

Epperlein, H. H., and Lehmann, R. (1975). The ectomesenchymal–endodermal interaction-system [EEIS] of *Triturus alpestris* in tissue culture. 2. Observations on differentiation of visceral cartilage. *Differentiation* **4**, 159–174.

Epperlein, H. H., Meulemans, D., Bronner Fraser, M., Steinbeisser, H., and Selleck, M. A. J. (2000). Analysis of cranial neural crest migratory pathways in axolotl using cell markers and transplantation. *Development* **127**, 2751–2761.

Eppley, Z. A. (1984). Development of thermoregulatory abilities in Xanthus murrelet chicks *Synthliboramphus hypoleucus*. *Physiol. Zool.* **57**, 307–317.

Erickson, G. M. (1996a). Incremental lines of von Ebner in dinosaurs and the assessment of tooth replacement rates using growth line counts. *Proc. Natl Acad. Sci. U.S.A.* **93**, 14623–14627

Erickson, G. M. (1996b). Toothlessness in American alligators, *Alligator mississippiensis. Copeia* **1996**, 739–743.

Erickson, H. P., and Bourdon, M. A. (1989). Tenascin: An extracellular matrix protein prominent in specialized embryonic tissues and tumors. *Annu. Rev. Cell Biol.* **5**, 71–92.

Eriksen, E. F., Covard, D. S., Berg, N. J., Graham, M. L., Mann, K. G., Spelsberg, T. C., and Riggs, B. L. (1988). Evidence of estrogen receptors in normal human osteoblast-like cells. *Science* **241**, 84–86.

Errick, J. E., and Saunders, J. W., Jr. (1974). Effect of an 'inside-out' limb-bud ectoderm on development of the avian limb. *Devel. Biol.* **41**, 338–351.

Errick, J. E., and Saunders, J. W., Jr. (1976). Limb outgrowth in the chick embryo induced by dissociated and reaggregated cells of the apical ectodermal ridge. *Devel. Biol.* **50**, 26–34.

Espie, J. H. (1975). Autoradiographic study with tritiated glycine on the effect of pulpectomy followed by root canal filling on the collagen synthesis of the periodontal ligament and alveolar bone. *J. Biol. Buccale* **3**, 157–166.

Estabrooks, L., and Schiff, R. (1972). Esterase isozymes from rabbit synovial fluids. Normal and artificial joints. *J. Histochem. Cytochem.* **20**, 211–219.

Eswarakumar, V. P., Monsonego-Ornan, E., Pines, M., Antonopoulou, I., Morriss-Kay, G. M., and Lonai, P. (2002). The *IIIc* alternative of *Fgfr 2* is a positive regulator of bone formation. *Development* **129**, 3783–3793.

Ettinger, L., and Doljanski, F. (1992). On the generation of form by the continuous interactions between cells and their extracellular matrix. *Biol. Rev. Camb. Philos. Soc.* **67**, 459–489.

Evanko, S. P., and Vogel, K. G. (1993). Proteoglycan synthesis in fetal tendon is differentially regulated by cyclic compression *in vitro. Archs Biochem. Biophys.* **307**, 153–164.

Evans, C. E., Ward, C., and Braidman, I. P. (1991). Breast carcinomas synthesize factors which influence osteoblast-like cells independently of osteoclasts *in vitro. J. Endocrinol.* **128**, R5–R8.

Evans, C. H. (1983). An inverse relationship between mammalian lifespan and cartilage cellularity. *Exp. Gerontol.* **18**, 137–138.

Evans, C. H., and Georgescu, H. I. (1983). Observations on the senescence of cells derived from articular cartilage. *Mech. Ageing Devel.* **22**, 179–191.

Evans, C. H., Mears, D. C., and Cosgrove, J. L. (1981). Release of neutral proteinase from mononuclear phagocytes and synovial cells in response to cartilage wear particles *in vitro*. *Biochim. Biophys. Acta* **677**, 287–294.

Evans, E. J., Benjamin, M., and Pemberton, D. J. (1990) Fibrocartilage in the attachment zones of the quadriceps tendon and patellar ligament of man. *J. Anat.* **171**, 155–162.

Evans, F. G. (1976). Mechanical properties and histology of cortical bone from younger and older men. *Anat. Rec.* **185**, 1–12.

Evans, H. E. (1948). Clearing and staining small vertebrates, *in toto*, for demonstrating ossification. *Turtox News* **26**, 42–47.

Evans, H. E. (1955). The osteology of a worm snake, *Typhlops jamaicensis* (Shaw). *Anat. Rec.* **122**, 381–396.

Evans, S. E. (1983). Mandibular fracture and inferred behavior in a fossil reptile. *Copeia* **1983**, 845–847.

Ewan, K. B. R., and Everett, A. W. (1992). Evidence for resegmentation in the formation of the vertebral column using the novel approach of retroviral-mediated gene transfer. *Exp. Cell Res.* **198**, 315–320.

Ewert, M. A. (1985). Embryology of turtles. In *The Biology of the Reptilia, Volume 14 Development A* (C. Gans, F. Billett, and P. F. A. Maderson, eds), pp. 75–267, John Wiley & Sons, New York, NY.

Exposito, J. Y., and Garrone, R. (1990). Characterization of a fibrillar collagen gene in sponges reveals the early evolutionary appearance of two collagen gene families. *Proc. Natl Acad. Sci. U.S.A.* **87**, 6669–6673.

Eyre, D. R., Upton, M. P., Shapiro, F. D., Wilkinson, R. H., and Vawter, G. F. (1986). Nonexpression of cartilage type II collagen in a case of Langer-Saldino achondrogenesis. *Amer. J. Human Genet.* **39**, 52–67.

Eyre-Brook, A. L. (1984). The periosteum: Its function reassessed. *Clin. Orthop. Rel. Res.* **189**, 300–307.

F

Faber, J. (1971). Vertebrate limb ontogeny and limb regeneration: Morphogenetic parallels. *Adv. Morphogen.* **9**, 127–148.

Fabrezi, M., and Barg, M. (2001). Patterns of carpal development among anuran amphibians. *J. Morphol.* **249**, 210–220.

Falck, P., Hanken, J. and Olsson, L. (2002). Cranial neural crest emergence and migration in the Mexican axolotl (*Ambystoma mexicanum*). *Zoology,* **105**, 195–202.

Falkner, F., and Tanner, J. M. (eds) (1986). *Human Growth. A Comprehensive Treatise*. 2nd Edn. *Volume 1. Developmental Biology, Prenatal Growth*. Plenum Press, New York, NY.

Fallon, J. F. (2002). How serendipity shaped a life. An interview with John W. Saunders, Jr. *Int. J. Devel. Biol.* **46**, 853–861.

Fallon, J. F., Brucker, R. F., and Harris, C. M. (1974). A re-examination of succinic dehydrogenase activity and its association with cell death in the interdigit of the chick foot. *J. Cell Sci.* **15**, 17–29.

Fallon, J. F., and Cameron, J. A. (1977). Interdigital cell death during limb development of the turtle and lizard with an interpretation of evolutionary significance. *J. Embryol. Exp. Morphol.* **40**, 285–289.

Fallon, J. F., and Caplan, A. I. (eds) (1983). *Limb Development and Regeneration. Part A*. Alan R. Liss Inc., New York, NY.

Fallon, J. F., and Crosby, G. M. (1975a). The relationship of the zone of polarizing activity to supernumerary limb formation (twinning) in the chick wing bud. *Devel. Biol.* **42**, 24–34.

Fallon, J. F., and Crosby, G. M. (1975b). Normal development of the chick wing following removal of the polarizing zone. *J. Exp. Zool.* **193**, 449–456.

Fallon, J. F., and Crosby, G. M. (1977). Polarizing zone activity in limb buds of amniotes. In *Vertebrate Limb and Somite Morphogenesis* (D. A. Ede, J. R. Hinchliffe, and M. Balls, eds), pp. 55–70. Cambridge University Press, Cambridge.

Fallon, J. F., and Kelley, R. O. (1977). Ultrastructural analysis of the apical ectodermal ridge during vertebrate limb morphogenesis. II. Gap junctions as distinctive ridge structures common to birds and mammals. *J. Embryol. Exp. Morphol.* **41**, 223–232.

Fallon, J. F., López, A., Ros, M. A., Savage, M. P., Olwin, B. B., and Simandl, B. K. (1994). FGF-2: Apical ectodermal ridge growth signal for chick limb development. *Science* **264**, 104–107.

Fallon, J. F., and Saunders, J. W., Jr. (1968). *In vitro* analysis of the control of cell death in a zone of prospective necrosis from the chick wing bud. *Devel. Biol.* **18**, 553–570.

Fallon, J. F., and Simanal, B. K. (1978). Evidence of a role for cell death in the disappearance of the embryonic human tail. *Amer. J. Anat.* **152**, 111–130.

Falshaw, R., Hubl, U., Ofman, D., Slim, G. C., Amjad Tariq, M., Watt, D. K., and Yorke, S. C. (2000). Comparison of the glycosaminoglycans isolated from the skin and head cartilage of Gould's arrow squid (*Nototodarus gouldi*). *Carbohydrate Polymers* **41**, 357–364.

Fang, C., Carlson, C. S., Leslie, M. P., Tulli, H., Stolerman, E., Perris, R., Ni, L. G., and Di Cesare, P. E. (2000). Molecular cloning, sequencing, and tissue and developmental expression of mouse cartilage oligomeric matrix protein (COMP). *J. Orthop. Res.* **18**, 593–603.

Fang, J., and Hall, B. K. (1995). Differential expression of neural cell adhesion molecule (N-CAM) during osteogenesis and secondary chondrogenesis in the embryonic chick. *Int. J. Devel. Biol.* **39**, 519–528.

Fang, J., and Hall, B. K. (1996). *In vitro* differentiation potential of the periosteal cells from a membrane bone, the quadratojugal of the embryonic chick. *Devel. Biol.* **180**, 701–712.

Fang, J., and Hall, B. K. (1997). Chondrogenic cell differentiation from membrane bone periostea. *Anat. Embryol.* **196**, 349–362.

Fang, J., and Hall, B. K. (1999). N-CAM is not required for initiation of secondary chondrogenesis: The role of N-CAM in skeletal condensation and differentiation. *Int. J. Devel. Biol.* **43**, 335–342.

Farhadieh, R. D., Dickinson, R., Yu, Y., Fianoitsis, M. P., and Walsh, W. R. (1999). The role of transforming growth factor-beta, insulin-like growth factor I, and basic fibroblast growth factor in distraction osteogenesis of the mandible. *J. Craniofacial. Surg.* **10**, 80–86.

Farley, J. R., and Baylink, D. J. (1986). Skeletal alkaline phosphatase activity as a bone formation index *in vitro*. *Metabolism Clin. & Exper.* **35**, 563–571.

Farley, J. R., Hall, S. L., Herring, S., and Tanner, M. A. (1993). Fluoride increases net Ca^{45} uptake by SAOS-2 cells – the effect is phosphate dependent. *Calcif. Tissue Res.* **53**, 187–192.

Farley, J. R., Hall, S. L., Herring, S., Tarbaux, N. M., Matsuyama, T., and Wergedal, J. E. (1991). Skeletal alkaline-phosphatase-specific activity is an index of the osteoblastic phenotype in subpopulations of the human osteosarcoma cell line SAOS-2. *Metabolism* **40**, 664–671.

Farley, J. R., Tarbaux, N., Hall, S., and Baylink, D. J. (1988b). Evidence that fluoride-stimulated [H-3]-thymidine incorporation in embryonic chick calvarial cell cultures is dependent on the presence of a bone cell mitogen, sensitive to changes in the phosphate concentration, and modulated by systemic skeletal effectors. *Metabolism* **37**, 988–995.

Farley, J. R., Tarbaux, N., Murphy, L. A., Matsuda, T., and Baylink, D. J. (1987). *In vitro* evidence that bone formation may be coupled to resorption by release of mitogen(s) from resorbing bone. *Metabolism* **36**, 314–321.

Farley, J. R., Tarbaux, N., Vermeiden, J. P. W., and Baylink, D. J. (1988a). *In vitro* evidence that local and systemic effectors can regulate ³[H]-thymidine incorporation in chick calvarial cell cultures and modulate the stimulatory action(s) of embryonic chick bone extract. *Calcif. Tissue Int.* **42**, 23–33.

Farley, J. R., Wergedal, J. E., and Baylink, D. J. (1983). Fluoride directly stimulates proliferation and alkaline phosphatase activity of bone-forming cells. *Science* **222**, 330–332.

Farnum, C. E. (1994). Differential growth rates of long bones. In *Bone, Volume 8: Mechanisms of Bone Development and Growth* (B. K. Hall, ed.), pp. 193–222. CRC Press, Boca Raton, FL.

Farnum, C. E., Nixon, A., Lee, A. O., Kwan, D. T., Belanger, L., and Wilsman, N. J. (2000). Quantitative three-dimensional analysis of chondrocytic kinetic responses to short-term stapling of the rat proximal tibial growth plate. *Cells Tissues Organs* **167**, 247–258.

Farnum, C. E., Turgai, J., and Wilsman, N. J. (1990). Visualization of living terminal hypertrophic chondrocytes of growth plate cartilage *in situ* by differential interference contrast microscopy and time lapse cinematography. *J. Orthop. Res.* **8**, 750–763.

Farnum, C. E., and Wilsman, N. J. (1987). Morphologic stages of the terminal hypertrophic chondrocyte of growth plate cartilage. *Anat. Rec.* **219**, 221–232.

Farnum, C. E., and Wilsman, N. J. (1989). Condensation of hypertrophic chondrocytes at the chondro-osseous junction of growth plate cartilage in Yucatan swine: Relationship to long bone growth. *Amer. J. Anat.* **186**, 346–358.

Farnum, C. E., and Wilsman, N. J. (1993). Determination of proliferative characteristics of growth plate chondrocytes by labeling with bromodeoxyuridine. *Calcif. Tissue Int.* **52**, 110–119.

Farnum, C. E., and Wilsman, N. J. (1998). Growth plate cellular function. In *Skeletal Growth and Development: Clinical Issues and Basic Science Advances* (The Symposium Series), American Academy of Orthopaedic Surgeons (J. A. Buckwalter, S. B. Trippel, M. G. Ehrlich, and L. J. Sandell, eds), pp. 203–223. NIH and Orthopaedic Research and Education Foundation, Washington.

Farnum, C. E., and Wilsman, N. J. (2001). Converting a differentiation cascade into longitudinal growth: stereology and analysis of transgenic animals as tools for understanding growth plate function. *Curr. Opin. Orthopaedics* **12**, 428–433.

Farquharson, C., Hesketh, J. E., and Loveridge, N. (1992). The proto-oncogene c-myc is involved in cell differentiation as well as cell proliferation: Studies on growth plate chondrocytes *in situ*. *J. Cell Physiol.* **152**, 135–144.

Farquharson, C., Milne, J., and Loveridge, N. (1993). Mitogenic action of insulin-like growth factor-1 on human osteosarcoma MG-63 cells and rat osteoblasts maintained *in situ*: The role of glucose-6-phosphate dehydrogenase. *Bone Min.* **22**, 105–115.

Faulhauber, I., Mohs, H., and Öchsner, I. (1986). Immunohistochemical detection of subunit A(M) of lactate dehydrogenase in tissues of the developing chick. *W. Roux. Archiv. Devel. Biol.* **195**, 409–416.

Faulk, W. P., Conochie, L. B., Temple, A., and Papamichail, M. (1975). Immunobiology of membrane-bound collagen on mouse fibroblasts. *Nature* **256**, 123–125.

Faustino, M., and Power, D. M. (2001). Osteologic development of the viscerocranial skeleton in sea bream: alternative ossification strategies in teleost fish. *J. Fish Biol.* **58**, 537–572.

Favier, B., Rijli, F. M., Fromental Ramain, C., Fraulob, V., Chambon, P., and Dollé, P. (1996). Functional cooperation between the non-paralogous genes Hoxa-10 and Hoxd-11 in the developing forelimb and axial skeleton. *Development* **122**, 449–460.

Fawcett, H. A., and Marsden, R. A. (1983). Hereditary osteoma cutis. *J. R. Soc. Med.* **76**, 697–699.

Fedak, T. J., and Hall, B. K (2004). Perspectives on hyperphalangy: patterns and processes. *J. Anat.* **204**, 151–163.

Fedak, T. J., Hall, B. K., Olson, W., Stone, J., and Vickaryous, M. (2002a). Morphology, embryos and fossils: Palaeontology and Evo-Devo. *Palaeontol. Assoc. Newsletter* **49**, 41–42.

Fedak, T. J., Hall, B. K., Olson, W., Stone J., and Vickaryous, M. (2002b). Sinking teeth into morphology through cell homology. *Palaeontol. Assoc. Newsletter* **51**, 27–30.

Federman, M., and Nichols, G., Jr. (1974). Bone cell cilia: Vestigial or functional organelles? *Calcif. Tissue Res.* **17**, 81–86.

Feik, S. A. (1993). Early periosteal changes in translation-induced bone modelling. *J. Anat.* **182**, 389–401.

Feik, S. A., and Storey, E. (1982). Joint changes in transplanted caudal vertebrae. *Pathology* **14**, 139–147.

Feik, S. A., and Storey, E. (1983). Remodelling of bone and bones: Growth of normal and transplanted caudal vertebrae. *J. Anat.* **136**, 1–14.

Feinberg, R. N., and Beebe, D. C. (1983). Hyaluronate in vasculogenesis. *Science* **220**, 1177–1179.

Fekete, D. M., and Brockes, J. P. (1988). Evidence that the nerve controls molecular identity of progenitor cells for limb regeneration. *Development* **103**, 567–573.

Felisbino, S. L., and Carvalho, H. F. (1999). The epiphyseal cartilage and growth of long bones in *Rana catesbiana*. *Tissue & Cell* **31**, 301–307.

Felisbino, S. L., and Carvalho, H. F. (2000). The osteochondral ligament: a fibrous attachment between bone and articular cartilage in *Rana catesbiana*. *Tissue & Cell* **32**, 527–536.

Felisbino, S. L., and Carvalho, H. F. (2001). Growth cartilage calcification and formation of bone trabeculae are late and dissociated events in the endochondral ossification of *Rana catesbiana*. *Cell Tissue Res.* **306**, 319–323.

Fell, H. B. (1925). The histogenesis of cartilage and bone in the long bones of the embryonic fowl. *J. Morphol. Physiol.* **40**, 417–459.

Fell, H. B. (1928). Experiments on the differentiation *in vitro* of cartilage and bone. Part 1. *Archs Exp. Zellforsch.* **7**, 390–412.

Fell, H. B. (1931). Osteogenesis *in vitro*. *Archs Exp. Zellforsch.* **11**, 245–252.

Fell, H. B. (1932). The osteogenic capacity *in vitro* of periosteum and endosteum isolated from the limb skeleton of fowl embryos and young chicks. *J. Anat.* **66**, 157–180.

Fell, H. B. (1933). Chondrogenesis in cultures of endosteum. *Proc. R. Soc. London B.* **112**, 417–427.

Fell, H. B. (1939). The origin and developmental mechanics of the avian sternum. *Phil. Trans. R. Soc. London B.* **229**, 402–464.

Fell, H. B. (1956). Skeletal development in tissue culture. In *Biochemistry and Physiology of Bone* (G. H. Bourne, ed.), pp. 401–442. Academic Press, New York, NY.

Fell, H. B. (1969). The effect of environment on skeletal tissue in culture. *Embryologia* **10**, 181–205.

Fell, H. B. (1975). The role of mucopolysaccharides in the protection of cartilage cells against immune reactions. *Philos. Trans. R. Soc. London, Ser. B.* **271**, 325–342.

Fell, H. B. (1976). The development of organ culture. In *Organ Culture in Biomedical Research* (M. Balls and M. A. Monnickendam, eds), pp. 1–13. Cambridge University Press, Cambridge.

Fell, H. B., and Canti, G. R. (1934). Experiments on the development *in vitro* of the avian knee-joint. *Proc. R. Soc. London, Ser. B.* **116**, 316–351.

Fell, H. B., Dingle, J. T., Coombs, R. R. A., and Lachmann, P. J. (1968). The reversible 'dedifferentiation' of embryonic skeletal tissues in culture in response to complement-sufficient antiserum. *Symp. Int. Soc. Cell Biol.* **7**, 49–68.

Fell, H. B., and Landauer, W. (1935). Experiments on skeletal growth and development *in vitro* in relation to the problem of avian phokomelia. *Proc. R. Soc. London, Ser. B.* **118**, 133–154.

Fell, H. B., and Mellanby, E. (1955). The biological action of thyroxine on embryonic bones grown in tissue culture. *J. Physiol.* **127**, 427–447.

Fell, H. B., and Robison, R. (1929). The growth, development and phosphatase activity of embryonic avian femora and limb-buds *in vitro*. *Biochem. J.* **23**, 767–783.

Fell, H. B., and Robison, R. (1930). The development and phosphatase activity *in vivo* and *in vitro* of the mandibular skeletal tissue of the embryonic fowl. *Biochem. J.* **24**, 1905–1921.

Feller, A., and Sternberg, H. (1934). Zur Kenntnis der Fehlbildungen der Wirbelkörper bie Spaltbildungen des Zentralnervensystems und ihre Formale Genese. *Z. Anat. Entwick.-Gesch.* **103**, 606–633.

Felts, W. J. L. (1961). *In vivo* implantation as a technique in skeletal biology. *Int. Rev. Cytol.* **12**, 243–302.

Felts, W. J. L., and Spurrell, F. A. (1965). Structural orientation and density in Cetacean humeri. *Amer. J. Anat.* **116**, 171–204.

Felts, W. J. L., and Spurrell, F. A. (1966). Some structural and developmental characteristics of cetacean (odontocete) radii. A study of adaptive osteogenesis. *Amer. J. Anat.* **118**, 103–134.

Feng, J.-Q., Chen, D., Cooney, A. J., Tsai, M.-J., Harris, M. A., Tsai, S. Y., Feng, M., Mundy, G. R., and Harris, S. E. (1995). The mouse bone morphogenetic protein-4 gene: Analysis of promoter utilization in fetal calf calvarial osteoblasts and regulation by coup-TF1 orphan receptor. *J. Biol. Chem.* **270**, 28364–28373.

Feng, J.-Q., Harris, M. A., Ghosh-Choudhury, N., Feng, M., Mundy, G. R., and Harris, S. E. (1994). Structure and sequence of mouse bone morphogenetic protein-2 gene (BMP-2): Comparison of the structures and promoter regions of BMP-2 and BMP-4 genes. *BBA Gene Struct. Exp.* **1218**, 221–224.

Fenwick, J. C. (1974). The comparative in vitro release of calcium from dried and defatted eel and rat bone fragments and its possible significance to calcium homeostasis in teleosts. *Can. J. Zool.* **52**, 755–764.

Ferrario, V. F., Sforza, C., Miani, A., Jr., and Sigurta, D. (1997). Asymmetry of normal mandibular condylar shape. *Acta Anat.* **158**, 266–273.

Fertber, D. (2000). Cholesterol drugs show promise as bone builders. *Science* **288**, 2297–2298.

Ferguson, C., Alpern, E., Miclau, T., and Helms, J. A. (1999). Does adult fracture repair recapitulate embryonic skeletal formation? *Mech. Devel.* **87**, 57–66.

Ferguson, C. A., Tucker, A. S., and Sharpe, P. T. (2000). Temporospatial cell interactions regulating mandibular and maxillary arch patterning. *Development* **127**, 403–412.

Ferguson, C. M., Miclau, T., Hu, D., Alpern, E., and Helms, J. A. (1998). Common molecular pathways in skeletal morphogenesis and repair. *Ann. N. Y. Acad. Sci.* **857**, 33–42.

Ferguson, D., Davis, W. L., Urist, M. R., Hurt, W. C., and Allen, E. P. (1987). Bovine bone morphogenetic protein (bBMP) fraction-induced repair of craniotomy defects in the Rhesus monkey (*Macaca speciosa*). *Clin. Orthop. Rel. Res.* **219**, 251–258.

Ferguson, M. W. J. (1984). Craniofacial development in *Alligator mississippiensis*. *Symp. Zool. Soc. London* #**52**, 223–274.

Fernandez, M. P., Young, M. F., and Sobel, M. E. (1985). Methylation of type II and type I collagen genes in differentiated and dedifferentiated chondrocytes. *J. Biol. Chem.* **260**, 2374–2378.

Fernández-Lloris, R., Viñals, F., López-Rovira, T., Harley, V., Bartrons, R., Rosa, J. L., and Ventura, F. (2003). Induction of the Sry-related factor SOX6 contributes to bone morphogenetic protein-2-induced chondroblastic differentiation of C3H10T1/2 cells. *Molec. Endocr.* **17**, 1332–1343.

Fernandez-Teran, M., Piedra, M. E., Ros, M. A., and Fallon, J. F. (1999). The recombinant limb as a model for the study of limb patterning, and its application to muscle development. *Cell Tissue Res.* **296**, 121–129.

Fernandez-Teran, M., Piedra, M. E., Kathiriya, I. S., Srivastava, D., Rodriguez-Rey, J. C., and Ros, M. A. (2000). Role of dHAND in the anterior–posterior polarization of the limb bud: implications for the sonic hedgehog pathway. *Development* **127**, 2133–2142.

Fernandez-Teran, M., Piedra, M. E., Rodriguez-Rey, J. C., Talamillo, A., and Ros, M. A. (2003). Expression and regulation of *eHAND* during limb development. *Devel. Dynam.* **226**, 690–701.

Ferrari, D., Harrington, A., Dealy, C. N., and Kosher, R. A. (1999). *Dlx-5* in limb initiation in the chick embryo. *Devel. Dynam.* **216**, 10–15.

Ferrari, D., Kosher, R. A., and Dealy, C. N. (1994). Limb mesenchymal cells inhibited from undergoing cartilage differentiation by a tumor promoting phorbol ester maintain expression of the homeobox-containing gene Msx1 and fail to exhibit gap junctional communication. *Biochem. Biophys. Res. Commun.* **205**, 429–434.

Ferretti, P., and Géraudie, J. (1998). *Cellular and Molecular Basis of Regeneration: From Invertebrates to Humans*. John Wiley & Sons Ltd., London.

Fessler, L. I., Burgeson, R. E., Morris, N. P., and Fessler, J. H. (1973). Collagen synthesis: A disulfide-linked collagen precursor in chick bone. *Proc. Natl Acad. Sci. U.S.A.* **70**, 2993–2996.

Feyen, J. H. M., Elford, P., Di Padova, F. E., and Trechsel, U. (1989). Interleukin-6 is produced by bone and modulated by parathyroid hormone. *J. Bone Min. Res.* **4**, 633–638.

ffrench-Constant, C., and Hynes, R. O. (1989). Alternative splicing of fibronectin is temporally and spatially regulated in the chicken embryo. *Development* **106**, 375–388.

Fielder, P. J., Morey, E. R., and Roberts, W. E. (1986). Osteoblast histogenesis in periodontal ligament and tibial metaphysis during simulated weightlessness. *Aviation Space & Envtl. Med.* **57**, 1125–1130.

Fieser, L. F. (1930). The discovery of synthetic alizarin. *J. Chem. Educ.* **7**, 2609–2633.

Fietzek, P. P., Allmann, H., Rauterberg, J., and Wachter, E. (1977). Ordering of cyanogen bromide peptides of type II collagen based on their homology to type I collagen: Preservation of sites for crosslinking formation during evolution. *Proc. Natl Acad. Sci. U.S.A.* **74**, 84–86.

Filan, S. L. (1991). Development of the middle ear region in *Monodelphis domestica* (Marsupialia, Delphidae): Marsupial solutions to an early birth. *J. Zool. London* **225**, 577–588.

Filoni, S., Bernardini, S., Cannata, S. M., and Ghittoni, R. (1999). Nerve-independence of limb regeneration in larval *Xenopus laevis* is related to the presence of mitogenic factors in early limb tissues. *J. Exp. Zool.* **284**, 188–196.

Filvaroff, E., and Derynck, R. (1998). Bone remodelling: A signalling system for osteoclast regulation. *Curr. Biol.* **8**, R679–R682.

Filvaroff, E., Erlebacher, A., Ye, J.-Q., Gitelman, S. E., Lotz, J., Heillman, M., and Derynck, R. (1999). Inhibition of TGF-β receptor signaling in osteoblasts leads to decreased bone remodeling and increased trabecular bone mass. *Development* **126**, 4267–4279.

Finch, R. A. (1969). The influence of the nerve on lower jaw regeneration in the adult newt, *Triturus viridescens. J. Morphol.* **129**, 401–414.

Finch, R. A., and Zwilling, E. (1971). Culture stability of morphogenetic properties of chick limb-bud mesoderm. *J. Exp. Zool.* **176**, 397–408.

Finerty, M. (1981). The homology of collagens. *J. Theor. Biol.* **93**, 279–301.

Fink, W. L. (1981). Ontogeny and phylogeny of tooth attachment modes in Actinopterygian fishes. *J. Morphol.* **167**, 167–184.

Fink, W. L. (1989). Ontogeny and phylogeny of shape and diet in the South American fishes called piranhas. *Geobios, Mém. spéciale* **12**, 167–172.

Finkelman, R. D., Linkhart, T. A., Mohan, S., Lau, K.-H. W., Baylink, D. J., and Bell, N. H. (1991). Vitamin D deficiency causes a selective reduction in deposition of transforming growth factor β in rat bone: Possible mechanism for impaired osteoinduction. *Proc. Natl Acad. Sci. U.S.A.* **88**, 3657–3660.

Finkle, H. I., and Goldman, R. L. (1973). Heterotopic cartilage in the thyroid. *Arch. Pathol.* **95**, 48–49.

Fiorello, C. V., and German, R. Z. (1997). Heterochrony within species: Craniofacial growth in giant, standard and dwarf rabbits. *Evolution* **51**, 250–261.

Firestein, G. S. (2003). Evolving concepts of rheumatoid arthritis. *Nature* **423**, 356–361.

Fischer, M., and Tillmann, B. (1991). Tendinous insertions in the human thyroid cartilage plate: macroscopic and histologic studies. *Anat. Embryol.* **183**, 251–257.

Fischer, S., Draper, B. W., and Neumann, C. J. (2003). The zebrafish *fgf24* mutant identifies and additional level of Fgf signaling involved in vertebrate forelimb initiation. *Development* **130**, 3515–3524.

Fischman, J. (1993). A closer look at the Dinosaur–Bird link. *Science* **262**, 1975.

Fischmann, D. A., and Hay, E. D. (1962). Origin of osteoclasts from mononuclear leucocytes in regenerating newt limbs. *Anat. Rec.* **143**, 329–338.

Fisher, M., and Solursh, M. (1977). Glycosaminoglycan localization and role in maintenance of tissue space in the early chick embryo. *J. Embryol. Exp. Morphol.* **42**, 195–207.

Fitch, N. (1957). An embryological analysis of two mutants in the house mouse both producing cleft palate. *J. Exp. Zool.* **136**, 329–362.

Fitton-Jackson, S. (1957). The fine structure of developing bone in the embryonic fowl. *Proc. R. Soc. London B.* **146**, 270–280.

Fitton-Jackson, S. (1965). Antecedent phases in matrix formation. In *Structure and Function of Connective and Skeletal Tissues* (G. R. Tristram and S. Fitton Jackson, eds), pp. 277–281. Butterworths, London.

Fitton Jackson, S. (1970). Environmental control of macromolecular synthesis in cartilage and bone: Morphogenetic response to hyaluronidase. *Proc. R. Soc. London, Ser. B.* **175**, 405–453.

Fitton Jackson, S. (1976). The maintenance of differentiation in skeletal tissues. In *Organ Culture in Biomedical Research* (M. Balls and M. A. Monnickenham, eds), pp. 165–178. Cambridge University Press, Cambridge.

Fitton Jackson, S. (1980). The response of skeletal tissues to pulsed magnetic fields. In *Use of Tissue Culture in Medical Research (II)* (R. J. Richards and K. T. Rajan, eds), pp. 21–28. Pergamon Press, London.

Fitzharris, T. P. (1976). Regeneration in sabellid annelids. *Amer. Zool.* **16**, 593–616.

Fitzsimmons, R. J., Farley, J., Adey, W. R., and Baylink, D. J. (1986). Embryonic bone matrix formation is increased after exposure to a low-amplitude capacitively coupled electric field *in vitro. Biochim. Biophys. Acta* **882**, 51–56.

Flaumenhaft, R., and Rifkin, D. B. (1992). The extracellular regulation of growth factor action. *Mol. Biol. Cell* **3**, 1057–1065.

Fleagle, J. G. (1999). *Primate Adaptation and Evolution.* 2nd Edn. Academic Press, San Diego, CA.

Fleming, A., Keynes, R. J., and Tannahill, D. (2001). The role of the notochord in vertebral column formation. *J. Anat.* **199**, 177–180.

Fleming, A., Keynes, R., and Tannahill, D. (2004). A central role for the notochord in vertebral patterning. *Development* **131**, 873–880.

Fleming, M. W., and Tassava, R. A. (1981). Preamputation and postamputation histology of the neonatal opossum hindlimb: implications for regeneration experiments. *J. Exp. Zool.* **215**, 143–150.

Flint, O. P. (1984). Interactions between differentiating nerve and limb cells *in vitro* : Implications for limb pattern formation. In *Matrices and Cell Differentiation* (R. B. Kemp and J. R. Hinchliffe, eds), pp. 399–408. Alan R. Liss Inc., New York, NY.

Flour, M.-P., Ronot, X., Vincent, F., Benoit, B., and Adolphe, M. (1992). Differential temperature sensitivity of cultured cells from cartilaginous or bone origin. *Biol. Cell* **75**, 83–87.

Flourens, J. P. M. (1845). Experiences sur la resorption et la reproduction successives des tetes des os. *Ann. Sci. Nat.* **4**, 358–363.

Flourens, J. P. M. (1847). *Théorie expérimentelle de la formation des os.* Paris.

Flourens, J. P. M. (1892). *Recherches sur le Développement des Os et des Dents.* Gide, Paris.

Flower, M. (1972). Cellular heterogeneity in the chick wing mesoblast. I. Analysis by equilibrium density gradient centrifugation. *Devel. Biol.* **28**, 583–602.

Flower, M., and Grobstein, C. (1967). Interconvertibility of induced morphogenetic responses of mouse embryonic somites to notochord and ventral spinal cord. *Devel. Biol.* **15**, 193–205.

Folkman, J. (1976). The vascularization of tumors. *Sci. Amer.* **234**, 59–73.

Fontana, F., and Rubini, M. (1990). Chromosomal evolution in Cervidae. *Bio. Systems* **24**, 157–174.

Forand, K. J., Marchinton, R. L., and Miller, K. V. (1985). Influence of dominance rank on the antler cycle of white-tailed deer. *J. Mammal.* **66**, 58–62.

Forbes, D. P., and Al-Bareedi, S. (1986). Inhibition of secondary cartilage of the intermaxillary suture in Sprague-Dawley rats following the enucleation of maxillary molars. *J. Craniofac. Genet. Devel. Biol.* **6**, 73–88.

Ford, D. M., McFadden, K. D., and Bagnall, K. M. (1982). Sequence of ossification in human vertebral neural arch centers. *Anat. Rec.* **203**, 175–178.

Foret, J. E. (1970). Regeneration of larval urodele limbs containing homplastic transplants. *J. Exp. Zool.* **175**, 297–322.

Forey, P., and Janvier, P. (1993). Agnathans and the origin of jawed vertebrates. *Nature* **361**, 129–134.

Fornasier, V. L. (1977). Osteoid: An ultrastructural study. *Human Pathol.* **8**, 243–254.

Fouvet, B. (1973). Nerve supply and morphogenesis in the leg of the chick embryo. I. Ontogeny of normal innervation. *Arch. Anat. Microsc. Morphol. Exp.* **62**, 269–280.

Francillon, H. (1974). Développement de la partie postérieure de la mandibule de *Salmo trutta fario* L (Pisces, Teleostei, Salmonidae). *Zool. Scripta* **3**, 41–51.

Francillon, H. (1977). Développement de la partie antérieure de la mandibule de *Salmo trutta fario* L (Pisces, Teleostei, Salmonidae). *Zool. Scripta* **6**, 145–251.

Francillon, H., and Castanet, J. (1985). Mise en évidence expérimentale du caractère annuel des lignes d'arrêt de croissance squellitique chez *Rana esculenta* (Amphibia: anura). *C.R. Acad. Sci. Paris* **300**, 327–332.

Francillon, H., Meunier, F., Ngo Tuan Phong, D., and de Ricqlès, A. J. (1975). Données preliminaires sur les structures histologiques du squelette de *Latimeria chalmunae*. II. Tissus osseux et cartilages. *Colloq. Int. C.N.R.S. Paris* **218**, 169–174.

Francillon-Vieillot, H., de Buffrénil, V., Castanet, J., Géraudie, J., Meunier, F. J., Sire, J.-Y., Zylberberg, L., and de Ricqlès, A. J. (1990). Microstructure and mineralization of vertebrate skeletal tissues. In *Skeletal Biomineralization; Patterns, Processes and Evolutionary Trends* (J. G. Carter, ed.), Volume 1, 471–548. Van Nostrand Reinhold, New York, NY.

Francis-West, P. H., Abdelfattah, A., Chen, P., Allen, C., Parish, J., Ladher, R., Allen, S., MacPherson, S., Luyten, F. P., and Archer, C. W. (1999a). Mechanism of GDF-5 action during skeletal development. *Development* **126**, 1305–1315.

Francis-West, P. H., Parish, J., Lee, K., and Archer, C. W. (1999b). BMP/GDF-signalling interactions during synovial joint development. *Cell Tissue Res.* **296**, 111–119.

Francis-West, P. H., Robertson, K. E., Ede, D. A., Rodriguez, C., Izpisua-Belmonte, J. C., Houston, B., Burt, D. W., Gribbin, C., Brickell, P. M., and Tickle, C. (1995). Expression of genes encoding bone morphogenetic proteins and sonic hedgehog in *talpid* (*ta³*) limb buds: Their relationships in the signalling cascade involved in limb patterning. *Devel. Dynam.* **203**, 187–197.

Francis-West, P. H., Robson, L., and Evans, D. J. R. (2003). Craniofacial development: the tissue and molecular interactions that control development of the head. *Adv. Anat. Embryol. Cell Biol.* **169**, 1–144.

Francis-West, P. H., Tatla, T., and Brickell, P. M. (1994). Expression patterns of the bone morphogenetic protein genes *Bmp-4* and *Bmp-2* in the developing chick face suggest a role

in outgrowth of the primordium. *Devel. Dynam.* **201**, 168–178.

Franco-Browder, D., de Rydt, J., and Dorfman, A. (1963). The identification of a sulfated mucopolysaccharide in chick embryos stages 11–23. *Proc. Natl Acad. Sci. U.S.A.* **49**, 643–647.

Franquin, J. C., and Baume, L. J. (1981). Effets d'une carence maternelle en vitamine A sur les structures cranio-dentaires du rat. *J. Biol. Buccale* **9**, 163–182.

Franzen, A., Oldberg, A., and Solursh, M. (1989). Possible recruitment of osteoblastic precursor cells from hypertrophic chondrocytes during initial osteogenesis in cartilaginous limbs of young rats. *Matrix* **9**, 261–265.

Franz-Odendaal, T., Cole, A. G., Fedak, T. J., Vickaryous, M., and Hall, B. K. (2003). Inside and outside skeletons. *Palaeont. Assoc. Newsletter* **54**, 17–21.

Franz-Odendaal, T. A., and Hall, B. K. (2004). The development and distribution of scleral ossicles in vertebrates: who has them and how do they form? *Integ. Comp. Biol.* **43**, 941.

Fraser, R. A., and Abbott, U. K. (1971a). Studies on limb morphogenesis. V. The expression of eudiplopodia and its experimental modification. *J. Exp. Zool.* **176**, 219–236.

Fraser, R. A., and Abbott, U. K. (1971b). Studies on limb morphogenesis. VI. Experiments with early stages of the polydactylous mutant eudiplopodia. *J. Exp. Zool.* **176**, 237–248.

Fraser, R. A., and Goetinck, P. F. (1971). Reduced synthesis of chondroitin sulfate by cartilage from the mutant nanomelia. *Biochem. Biophys. Res. Comm.* **43**, 494–503.

Frasier, M. B., Banks, W. J., and Newbrey, J. W. (1975). Characterization of developing antler cartilage matrix. I. Selected histochemical and enzymatic assessment. *Calcif. Tissue Res.* **17**, 273–288.

Frederickson, R. G., and Low, F. N. (1971). The fine structure of perichordal microfibrils in control and enzyme-treated chick embryos. *Amer. J. Anat.* **130**, 347–376.

Freeman, B. M., and Vince, M. A. (1974). *Development of the Avian Embryo. A Behavioural and Physiological Study.* Chapman & Hall, London.

Freeman, E., Ten Cate, A. R., and Dickinson, J. B. (1975). Development of a gomphosis by tooth germ implants in the parietal bone of the mouse. *Archs Oral Biol.* **20**, 139–140.

French, M. M., Smith, S. E., Akanbi, K., Sanford, T., Hecht, J., Farach-Carson, M. C., and Carson, D. D. (1999). Expression of the heparan sulfate proteoglycan perlecan, during mouse embryogenesis and perlecan chondrogenic activity in vitro. *J. Cell Biol.* **145**, 1103–1115.

French, V., Bryant, P. J., and Bryant, S. V. (1976). Pattern regulation in epimorphic fields. *Science* **193**, 969–981.

Frenz, D. A., and Van de Water, T. R. (1991). Epithelial control of periotic mesenchyme chondrogenesis. *Devel. Biol.* **144**, 38–46.

Frenz, D. A., Galinovic-Schwartz, V., Liu, W., Flanders, K. C., and Van de Water, T. R. (1992). Transforming growth factor β1 is an epithelial-derived signal peptide that influences otic capsule formation. *Devel. Biol.* **153**, 324–336.

Frenz, D. A., Jaikaria, N. S., and Newman, S. A. (1989). The mechanism of pre cartilage mesenchymal condensation: A major role for interaction of the cell surface with the amino-terminal heparin-binding domain of fibronectin. *Devel. Biol.* **136**, 97–103.

Frenz, D. A., Liu, W., Williams, J. D., Hatcher, V., Galinovic-Schwartz, V., Flanders, K. C., and Van de Water, T. R. (1994). Induction of chondrogenesis: Requirement for synergistic interaction of basic fibroblast growth factor and transforming growth factor beta. *Development* **120**, 415–424.

Friant, M. (1964). Sur l'ossification du cartilage de Meckel d'une chauve-souris le grand Murin [Chiroptera, *Myotis myotis* (Borkh.)]. *Acta Anat.* **57**, 66–71.

Friant, M. (1968). L'evolution du cartilage de Meckel du porc (*Sus scrofa* dom. Gray). *Ann. Fac. Med. Vet. Pisa. Univ. Studi. Pisa* **21**, 1–17.

Friede, H. (1973). Histology of the premaxillary–vomerine suture in a bilateral cleft case. *Cleft Palate J.* **10**, 14–22.

Friedenstein, A. Y. (1962). The humoral nature of the osteogenetic activity of transitional epithelium. *Byull. Eksp. Biol. Med.* **12**, 82–83.

Friedenstein, A. Y. (1973). Determined and inducible osteogenic precursor cells. *Hard Tissue Growth, Repair, Remineralization, Ciba Found. 1972 Symp.* No. 11, pp. 170–185.

Friedenstein, A. Y. (1976). Precursor cells of mechanocytes. *Int. Rev. Cytol.* **47**, 327–359.

Friedenstein, A. J., Chailakhyan, R. K., and Gerasimov, U. V. (1987). Bone marrow osteogenic stem cells: *in vitro* cultivation and transplantation in diffusion chambers. *Cell Tissue Kinetics* **20**, 263–272.

Friedenstein, A. Y., Chailakhjan, R. K., and Lalykina, K. S. (1970). The development of fibroblast colonies in monolayer cultures of guinea-pig bone marrow and spleen cells. *Cell Tissue Kinet.* **3**, 393–403.

Friedenstein, A. Y., Gorskaja, U. F., and Kulagina, N. N. (1976). Fibroblast precursors in normal and irradiated mouse hematopoietic organs. *Exp. Hematol.* **4**, 267–274.

Friedenstein, A. Y., and Kuralesova, A. I. (1971). Osteogenic precursor cells of bone marrow in radiation chimeras. *Transplantation* **12**, 99–108.

Friedenstein, A. Y., and Lalykina, K. S. (1970). Lymphoid cell populations are competent systems for induced osteogenesis. *Calcif. Tissue Res.* **4**, Suppl., 105–106.

Friedenstein, A. Y., and Lalykina, K. S. (1972). Thymus cells are inducible to osteogenesis. *Eur. J. Immunol.* **2**, 602–603.

Friedenstein, A. Y., and Lalykina, K. S. (1973). *Bone Induction and Osteogenic Precursor Cells.* Medical Publishing House, Moscow (in Russian).

Friedenstein, A. Y., Lalykina, K. S., and Tolmacheva, A. A. (1967). Osteogenic activity of peritoneal fluid cells induced by transitional epithelium. *Acta Anat.* **68**, 532–549.

Friedenstein, A. Y., Piatelzky-Shapiro, I. I., and Petrakova, K. V. (1966). Osteogenesis in transplants of bone marrow cells. *J. Embryol. Exp. Morphol.* **16**, 381–390.

Friedenstein, A. Y., Petrakova, K. V., Kurolesova, A. I., and Frolova, G. P. (1968). Heterotopic transplants of bone marrow. Analysis of precursor cells for osteogenic and hematopoietic tissues. *Transplantation* **6**, 230–247.

Friedman, D. L. (1976). Role of cyclic nucleotides in cell growth and differentiation. *Physiol. Rev.* **56**, 652–708.

Frolich, L. M. (1991). Osteological conservation and developmental constraint in the polymorphic 'ring species' *Ensatina eschscholtzii* (Amphibia: Plethodontidae). *Biol. J. Linn. Soc.* **43**, 81–100.

Frommer, J. (1964). Prenatal development of the mandibular joint in mice. *Anat. Rec.* **150**, 449–462.

Frommer, J., and Margolies, M. R. (1971). Contribution of Meckel's cartilage to ossification of the mandible in mice. *J. Dent. Res.* **50**, 1260–1267.

Frommer, J., and Monroe, C. W. (1966). Development and distribution of elastic fibers in the mandibular joint of the mouse. A comparison of fetal, suckling, juvenile and adult stages. *Anat. Rec.* **156**, 333–346.

Frost, H. M. (1989a). The biology of fracture healing: an overview for clinicians. Part I. *Clin. Orthop. Rel. Res.* **248**, 283–293.

Frost, H. M. (1989b). The biology of fracture healing: an overview for clinicians. Part II. *Clin. Orthop. Rel. Res.* **248**, 294–309.

Frost, H. M. (1990a). Skeletal structural adaptations to mechanical usage (SATMU): I. Redefining Wolff's law: the bone modeling problem. *Anat. Rec.* **226**, 403–413.

Frost, H. M. (1990b). Skeletal structural adaptations to mechanical usage (SATMU): 2. Redefining Wolff's law: the remodeling problem. *Anat. Rec.* **226**, 414–422.

Frost, H. M. (1990c). Skeletal structural adaptations to mechanical usage (SATMU): 3. The hyaline cartilage modeling problem. *Anat. Rec.* **226**, 423–432.

Frost, H. M. (1990d). Skeletal structural adaptations to mechanical usage (SATMU): 4. Mechanical influences on intact fibrous tissues. *Anat. Rec.* **226**, 433–439.

Frowen, P., and Benjamin, M. (1995). Variations in the quantity of uncalcified fibrocartilage a the insertions of the extrinsic calf muscles in the foot. *J. Anat.* **186**, 417–421.

Frylestam, B., and von Schantz, T. (1977). Age determination of European hares based on periosteal growth lines. *Mammal Rev.* **7**, 151–154.

Fuchs, H. (1906). Untersuchungen über die Entwicklung der Gehörknöchelchen des Squamosums und des Kiefergelenkes der Säugetiere, nebst einigen vergleichend-Anatomischen Betrachtungen über Articulare, Quadratum und Gehörknöchelchen. *Archs Anat. Entwkgscht, Anat. Abt. Suppl.* 1–90.

Fuchs, H. (1909). Über Knorpelbildung in Deckknochen nebst Untersuchungen und Betrachtungen über Gehörknöchelchen, Kiefer und Kiefergelenk der Wirbeltiere. *Archs Anat. Physiol., Anat. Suppl.* 1–256.

Fuentes, M. A., Opperman, L. A., Bellinger, L. L., Carlson, D. S., and Hinton, R. J. (2002). Regulation of cell proliferation in rat mandibular condylar cartilage in explant culture by insulin-like growth factor-1 and fibroblast growth factor-2. *Archs Oral Biol.* **47**, 643–654.

Fuentes, M. An., Opperman, L. A., Buschang, P., Bellinger, L. L., Carlson, D. S., and Hinton, R. J. (2003). Lateral functional shift of the mandible: Part I. Effects on condylar cartilage thickness and proliferation. *Am. J. Orthod. Dentofacial Orthop.* **123**, 153–159.

Fuentes, M. An., Opperman, L. A., Buschang, P., Bellinger, L. L., Carlson, D. S., and Hinton, R. J. (2003). Lateral functional shift of the mandible: Part II. Effects on gene expression in condylar cartilage. *Am. J. Orthod. Dentofacial Orthop.* **123**, 160–163.

Fuenzalida, M., Illanes, J., Lemus, R., Guerrero, A., Oyarzun, A., Acuna, O., and Lemus, D. (1999). Microscopic and histochemical study of odontoclasts in physiologic resorption of teeth of the polyphyodont lizard, *Liolaemus gravenhorsti*. *J. Morphol.* **242**, 295–309.

Fujimori, Y., Nakamura, T., Ijiri, S., Shimizu, K., and Yamamuro, T. (1992). Heterotopic bone formation induced by bone morphogenetic protein in mice with collagen-induced arthritis. *Biochem. Biophys. Res. Commun.* **186**, 1362–1367.

Fujimoto, A., Wakasugi, N., and Tomita, T. (1994). The developmental and morphological studies on the neural and skeletal abnormalities in the *T/btm* tailless mice. *Devel. Growth Differ.* **36**, 409–417.

Fujimura, K., Bessho, K., Okubo, Y., Kusumoto, K., Segami, N., and Iizuka, T. (2002). The effect of fibroblast growth factor-2 on the osteoinductive activity of recombinant human bone morphogenetic protein-2 in rat muscle. *Archs Oral Biol.* **47**, 577–584.

Fujita, T., Kawata, T., Tokimasa, C., Kaku, M., Kawasoko, S., and Tanne, K. (1998). Influences of ovariectomy and orchiectomy on the remodeling of mandibular condyle in mice. *J. Craniofac. Genet. Devel. Biol.* **18**, 164–170.

Fujita, T., Kawata, T., Tokimasa, C., and Tanne, K. (2000). Influence of oestrogen and androgen on modelling of the mandibular condyle in ovariectomized and orchiectomized growing mice. *Archs Oral Biol.* **46**, 57–65

Fukada, K., Shibata, S., Susuki, S., Ohya, K., and Kuroda, T. (1999). In situ hybridization study of types I, II, X collagens and aggrecan mRNAs in the developing condylar cartilage of fetal mouse mandible. *J. Anat.* **195**, 321–329.

Fullmer, H. M. (1965). The histochemistry of the connective tissues. *Int. Rev. Connect. Tissue Res.* **3**, 1–76.

Furseth, R. (1973). Tight junctions between osteocyte processes. *Scand. J. Dent. Res.* **81**, 339–341.

Furthmayr, H., and Timpl, R. (1976). Immunochemistry of collagens and procollagens. *Int. Rev. Connect. Tissue Res.* **7**, 61–101.

Furue, M., Myoishi, Y., Fukui, Y., Ariizumi, T., Okamoto, T., and Asashima, M. (2002). Craniofacial cartilage from undifferentiated *Xenopus* ectoderm *in vitro*. *Proc. Natl Acad. Sci. U.S.A* **99**, 15474–15479.

Furumoto, T-a., Miura, N., Akasaka, T., Mizutani-Koseki, Y., Sudo, H., Fukuda, K., Maekawa, M., Yuasa, S., Fu, Y., Moriya, H., Taniguchi, M., Imai, K., Dahl, E., Balling, R., Pavlova, M., Gossler, A., and Koseki, H. (1999). Notochord-dependent expression of MFH1 and Pax1 cooperates to maintain the proliferation of sclerotome cells during the vertebral column development. *Devel. Biol.* **210**, 15–29.

Furuto, D. K., Gay, R. E., Stewart, T. E., Miller, E. J., and Gay, S. (1991). Immunolocalization of types V and XI collagen in cartilage using monoclonal antibodies. *Matrix* **11**, 144–149.

Fyfe, D. M., Ferguson, M. W. J., and Chiquet-Ehrismann, R. (1988). Immunochemical localisation of tenascin during the development of scleral papillae and scleral ossicles in the embryonic chick. *J. Anat.* **159**, 117–127.

Fyfe, D. M., and Hall, B. K. (1979). Lack of association between avian cartilages of different embryological origins when maintained *in vitro*. *Amer. J. Anat.* **154**, 485–496.

Fyfe, D. M., and Hall, B. K. (1981). A scanning electron microscopic study of the developing epithelial scleral papillae in the eye of the embryonic chick. *J. Morphol.* **167**, 201–209.

Fyfe, D. M., and Hall, B. K. (1983). The origin of the ectomesenchymal condensations which precede the development of the bony scleral ossicles in the eyes of embryonic chicks. *J. Embryol. Exp. Morphol.* **73**, 69–86.

G

Gabbott, S. E., Aldridge, R. J., and Theron, J. N. (1995). A giant conodont with preserved muscle tissue from the Upper Ordovician of South Africa. *Nature* **374**, 800–803.

Gabrovska, M., and Vancov, V. (1982). Secretory features of chondrocytes related to morphogenesis of the elastic elements in the auricular cartilage. *Acta Morphol. Acad. Sci. Hung.* **30**, 127–134.

Gad, J. M., and Tam, P. P. L. (1999). Axis development: The mouse becomes a dachshund. *Curr. Biol.* **9**, R783–R786.

Gadow, H. (1902). The evolution of horns and antlers. *Proc. Zool. Soc. London* **1**, 206–222.

Gaillard, P. J. (1974). The influence of ascorbic acid on the effect of parathyroid extract on the histology of explanted mouse radius rudiments. I. *Proc. K. Ned. Akad. Wet. Ser. C* **77**, 101–115.

Gajovic, S., and Kostovic-Knezevic, L. (1995). Ventral ectodermal ridge and ventral ectodermal groove: Two distinct morphological features in the developing rat embryo tail. *Anat. Embryol.* **192**, 181–187.

Gakunga, P. T., Kuboki, Y., and Opperman, L. A. (2000). Hyaluronan is essential for the expansion of the cranial base growth plates. *J. Craniofac. Genet. Devel. Biol.* **20**, 53–63.

Galileo, G. (1638). Discourses and Mathematical Demonstrations concerning two new Sciences pertaining to Mechanics and Motion. Translated by H. Crew and A. de Salvio, 1933. MacMillan Co., New York, NY.

Galis, F. (1999). Why do almost all mammals have seven cervical vertebrae? Developmental constraints, *Hox* genes, and cancer. *J. Exp. Zool. (Mol. Dev. Evol.)*, **285**, 19–26

Galis, F. (2001a). Digit identity and digit number: indirect support for the descent of birds from tetrapod dinosaurs. *Trends Ecol. Evol.* **16**, 16.

Galis, F. (2001b). Evolutionary history of vertebrate appendicular muscle. *BioEssays* **23**, 383–387.

Galis, F., Van Alphen, J. J. M., and Metz, J. A. J. (2001). Why five fingers? Evolutionary constraints on digit numbers. *Trends Ecol. Evol.* **16**, 637–646.

Gallo, V., Bertolotto, A., and Levi, G. (1987). The proteoglycan chondroitin sulfate is present in a subpopulation of cultured astrocytes and in their precursors. *Devel. Biol.* **123**, 282–285.

Gamer, L. W., Cox, K. A., Small, C., and Rosen, V. (2001). *Gdf-11* is a negative regulator of chondrogenesis and myogenesis in the developing chick limb. *Devel. Biol.* **229**, 407–420.

Ganey, T. M., Ogden, J. A., and Olsen, J. (1990). Development of the Giraffe horn and its blood supply. *Anat. Rec.* **227**, 497–507.

Ganey, T. M., Ogden, J. A., Sasse, J., Neame, P. J., and Hilbelink, D. R. (1995). Basement membrane composition of cartilage canals during development and ossification of the epiphysis. *Anat. Rec.* **241**, 425–437.

Gans, C. (1974). *Biomechanics: Approach to Vertebrate Biology.* J. B. Lippincott Co., Philadelphia, PA.

Gans, C. (1975). Tetrapod limblessness: Evolution and functional corollaries. *Amer. Zool.* **15**, 455–467.

Gans, C., and Northcutt, R. G. (1983). Neural crest and the origin of vertebrates: a new head. *Science* **220**, 268–274.

Gao, J., Messner, K., Ralphs, J. R., and Benjamin, M. (1996). An immunohistochemical study of enthesis development in the medial collateral ligament of the rat knee joint. *Anat. Embryol.* **194**, 399–406.

Garant, P. A., and Cho, M-Il. (1979). Autoradiographic evidence of the coordination of the genesis of Sharpey's fibers with new bone formation in the periodontium of the mouse. *J Periodont. Res.* **14**, 107–115.

Gardiner, D. M., and Bryant, S. V. (1996). Molecular mechanisms in the control of limb regeneration: the role of homeobox genes. *Int. J. Devel. Biol.* **40**, 797–805.

Gardiner, D. M., Blumberg, B., Komine, Y., and Bryant, S. V. (1995). Regulation of *HOXA* expression in developing and regenerating axolotl limbs. *Development* **121**, 1731–1741.

Gardner, D. P., and Kappen, C. (2000). Developmental characterization and chromosomal mapping of a LacZ transgene expressed in the mouse apical ectoderm ridge. *J. Exp. Zool.* **287**, 106–111.

Gardner, E. (1950). Physiology of moveable joints. *Physiol. Rev.* **30**, 127–176.

Gardner, E. (1968). The embryology of the clavicle. *Clin. Orthop. Rel. Res.* **58**, 9–16.

Gardner, E. (1971). Osteogenesis in the human embryo and fetus. In *The Biochemistry and Physiology of Bone* (G. H. Bourne, ed.), 2nd Edn, Volume 3, pp. 77–118. Academic Press, New York, NY.

Gardner, E., and Gray, D. J. (1953). Prenatal development of the human shoulder and acromioclavicular joints. *Amer. J. Anat.* **92**, 219–276.

Gardner, E., and O'Rahilly, R. (1968). The early development of the knee joint in staged human embryos. *J. Anat.* **102**, 289–300.

Gardner, R. J. M., Yun, K., and Craw, S. M. (1988). Familial ectopic ossification. *J. Med. Genet.* **25**, 113–117.

Gardner, T. N., Stoll, T., Marks, L., Mishra, S., and Tate, M. K. (2000). The influence of mechanical stimulus on the pattern of tissue differentiation in a long bone fracture – an FEM study. *J. Biomech.* **33**, 15–425.

Gardner, W. U. (1936). Sexual dimorphism of the pelvis of the mouse and effect of estrogenic hormones upon the pelvis and upon the development of scrotal hernias. *Amer. J. Anat.* **59**, 459–483.

Gardner, W. U., and Pfeiffer, C. A. (1938). Skeletal changes in mice receiving estrogens. *Proc. Soc. Exp. Biol. Med.* **37**, 678–679.

Garetto, L. P., Morey, E. R., Durnova, G. N., Kaplansky, A. S., and Roberts, W. E. (1992). Preosteoblast production in Cosmos 2044 rats: Short-term recovery of osteogenic potential. *J. Appl. Physiol.* **73**, 14–18.

Garg, V., Yamagishi, C., Hu, T., Kathiriya, I. S., Yamagishi, H., and Srivastava, D. (2001). *Tbx1*, a Di George syndrome candidate gene, is regulated by sonic hedgehog during pharyngeal arch development. *Devel. Biol.* **235**, 62–73.

Garrard, G., Harrison, G. A., and Weiner, J. S. (1974). Genotypic differences in the ossification of 12 day old mice at 23° C and 32° C. *J. Anat.* **117**, 531–540.

Garrett, I. R., and Mundy, G. R. (2002). The role of statins as potential targets for bone formation. *Arthritis Res* **4**, 237–240.

Garrone, R. (1978) *Phylogenesis of Connective Tissues.* S. Karger, Basel.

Garrone, R. (1998). Evolution of metazoan collagens. In *Progress in Molecular and Subcellular Biology, Volume 21* (W. E. G. Muuller, ed.), pp. 119–139. Springer-Verlag, Berlin.

Garrone, R., and Pottu, J. (1973). Collagen biosynthesis in sponges: elaboration of sponging by spongocytes. *J. Submicr. Cytol.* **5**, 199–218.

Gasc, J.-P., and Renous, S. (1979). La région pelvi-cloacale de *Dibamus* (Squamata, Reptilia). Nouvelle contribution à les position systématique. *Bull. Mus. Natn. Hist. Nat. Paris 4th ser.*, No 1, 659–684.

Gasseling, M. T., and Saunders, J. W., Jr. (1964). Effect of the 'posterior necrotic zone' of the early chick wing bud on the pattern and symmetry of limb outgrowth. *Amer. Zool.* **4**, 303–304.

Gatewood, L., and Mullen, B. (1927). Experimental observations on the growth of long bones. *Arch. Surg.* **15**, 215–221.

Gaunt, S. J. (2000). Evolutionary shifts of vertebrate structures and *Hox* expression up and down the axial series of segments: a consideration of possible mechanisms. *Int. J. Devel. Biol.* **44** Sp. Issue, 109–117.

Gavaia, P. J., Dinis, M. T., and Cancela, M. L. (2002). Osteological development and abnormalities of the vertebral column and caudal skeleton in larval and juvenile stages of hatchery-reared Senegal sole (*Solea senegalensis*). *Aquaculture* **211**, 305–323.

Gay, S., Martin, G. R., Müller, P. K., Timpl, R., and Kühn, K. (1976a). Simultaneous synthesis of types I and III collagen by fibroblasts in culture. *Proc. Natl Acad. Sci. U.S.A.* **73**, 4037–4040.

Gay, S., Müller, P. K., Lemmen, C., Remberger, K., Matzen, K., and Kühn, K. (1976b). Immunohistochemical study on collagen in cartilage-bone metamorphosis and degenerative osteoarthritis. *Klin. Wochenschr.* **54**, 969–976.

Gay, S. W., and Kosher, R. A. (1984). Uniform cartilage differentiation in micromass cultures prepared from a relatively homogeneous population of chondrogenic progenitor cells of the chick limb bud: Effect of prostaglandins. *J. Exp. Zool.* **232**, 317–326.

Gay, S. W., and Kosher, R. A. (1985). Prostaglandin synthesis during the course of limb cartilage differentiation *in vitro*. *J. Embryol. Exp. Morphol.* **89**, 367–382.

Gayon, J. (2000). History of the concept of allometry. *Amer. Zool.* **40**, 748–758.

Gazit, E. (1980). Effect of increased gravity on bone resorption in the mouse calvaria explant system. *Israel J. Med. Sci.* **16**, 867–869.

Gearhart, J. D., and Mintz, B. (1972). Clonal origins of somites and their muscle derivatives: Evidence from allophenic mice. *Devel. Biol.* **29**, 27–37.

Geber, W. F. (1973). Inhibition of fetal osteogenesis by maternal noise stress. *Fed. Proc.* **32**, 2101–2104.

Geduspan, J. S., and Solursh, M. (1992). A growth-promoting influence from the mesonephros during limb outgrowth. *Devel. Biol.* **151**, 242–250.

Gegenbaur, C. (1864). Ueber die Bildung Knochenstruktur. *Jena Zeitsch. Med. Naturwiss.* **1**, 343–369.

Gegenbaur, C. (1864–1865). *Untersuchungen zur vergleichenden Anatomie der Wirbeltiere, Heft 1 & 2*, Schultergürtel. Leipzig.

Gegenbaur, C. (1867). Über die Bildung des Knockengewebes. *Jena Zeitsch. Med. Naturwiss.* **3**, 206–246.

Gehris, A. L., Stringa, E., Spina, J., Desmond, M. E., Tuan, R. S., and Bennett, V. D. (1997). The region encoded by the alternatively spliced exon IIIa in mesenchymal fibronectin appears essential for chondrogenesis at the level of cellular condensation. *Devel. Biol.* **190**, 191–205.

Geist, V. (1966a). The evolution of horn-like organs. *Behaviour* **27**, 175–214.

Geist, V. (1966b). The evolutionary significance of mountain sheep horns. *Evolution* **20**, 558–566.

Geist, V. (1986). New evidence of high frequency of antler wounding in cervids. *Can. J. Zool.* **64**, 380–384.

Geist, V. (1998) *Deer of the World. Their Evolution, Behaviour, and Ecology*. Stackpole Books, Mechanicsburg, PA.

Geist, V., and Jones, T. D. (1996). Juvenile skeletal structures and the reproductive habits of dinosaurs. *Science* **272**, 712–714.

Gelb, D. E., Rosier, R. N., and Puzas, J. E. (1990). The production of transforming growth factor-β by chick growth plate chondrocytes in short term monolayer culture. *Endocrinology* **127**, 1941–1947.

Gelman, R. A., and Blackwell, J. (1973). Interaction between collagen and chondroitin-6-sulfate. *Connect. Tissue Res.* **2**, 31–36.

Gelsleichter, J., and Musick, J. A. (1999). Effects of insulin-like growth factor-I, corticosterone, and 3, 3′, 5-triiodo-L-thyroxine on glycosaminoglycan synthesis in vertebral cartilage of the clearnose skate, *Raja eglanteria. J. Exp. Zool.* **284**, 549–556.

Gemmell, R. T., Johnston, G., and Bryden, M. M. (1988). Osteogenesis in two marsupial species, the bandicoot, *Isodon macrourus* and the possum *Trichosurus vulpecula. J. Anat.* **159**, 155–164.

Gendron-Maguire, M., Mallo, M., Zhang, M., and Gridley, T. (1994). *Hoxa-2* mutant mice exhibit homeotic transformation of skeletal elements derived from cranial neural crest. *Cell* **75**, 1317–1331.

Gentili, C., Bianco, P., Neri, M., Malpeli, M., Campanile, G., Castagnola, P., Cancedda, R., and Cancedda, F. D. (1993). Cell proliferation, extracellular matrix mineralization, and ovotransferrin transient expression during *in vitro* differentiation of chick hypertrophic chondrocytes into osteoblast-like cells. *J. Cell Biol.* **122**, 703–712.

George, E. L., Georges-Labouesse, E. N., Patel-King, R. S., Rayburn, H., and Hynes, R. O. (1993). Defects in mesoderm, neural tube and vascular development in mouse embryo lacking fibronectin. *Development* **119**, 1079–1091.

Gerard, J. (1597). *Herball, or generall Historie of Plantes.* London.

Géraudie, J. (1978a). The fine structure of the early pelvic fin bud of the trouts *Salmo gairdneri* and *S. trutta fario. Acta Zool.* (*Stockholm*) **59**, 85–96.

Géraudie, J. (1978b). Scanning electron microscope study of the developing trout pelvic fin bud. *Anat. Rec.* **191**, 391–396.

Géraudie, J. (1980). Mitotic activity in the pseudoapical ridge of the trout pelvic fin bus, *Salmo gairdneri. J. Exp. Zool.* **214**, 311–316.

Géraudie, J. (1985). Innervation of the early pelvic fin bud of the trout embryo, *Salmo gairdneri. J. Morphol.* **184**, 61–73.

Géraudie, J. (1988). Fine structural peculiarities of the pectoral fin dermoskeleton of two brachiopterygii, *Polypterus senegalus* and *Calamoichthys calabaricus* (Pisces, Osteichthyes). *Anat. Rec.* **221**, 455–468.

Géraudie, J., Hourdry, J., Vriz, S., Singer, M., and Mechali, M. (1990). Enhanced *c-myc* gene expression during forelimb regenerative outgrowth in the young *Xenopus laevis. Proc. Natl Acad. Sci. U.S.A.* **87**, 3797–3801.

Géraudie, J., and Landis, W. J. (1982). The fine structure of the developing pelvic fin dermal skeleton in the trout, *Salmo gairdneri. Amer. J. Anat.* **163**, 141–156.

Géraudie, J., and Meunier, F. J. (1980). Elastoidin actinotrichia in coelacanth fins: A comparison with teleosts. *Tissue Cell* **12**, 637–645.

Géraudie, J., and Meunier, F. J. (1982). Comparative fine structure of the osteichthyan dermotrichia. *Anat. Rec.* **202**, 325–328.

Géraudie, J., and Meunier, F. J. (1984). Structure and comparative morphology of camptotrichia of lungfish fins. *Tissue & Cell* **16**, 217–236.

Géraudie, J., Monnot, M. J., Brulfert, A., and Ferretti, P. (1995). Caudal fin regeneration in wild type and *long-fin* mutant zebrafish is affected by retinoic acid. *Int. J. Devel. Biol.* **39**, 373–381.

Géraudie, J., and Singer, M. (1977). Relation between nerve fiber number and pectoral fin regeneration in the teleost. *J. Exp. Zool.* **199**, 1–8.

Géraudie, J., and Singer, M. (1979). Nerve dependent macromolecular synthesis in the pectoral fin regenerate of the fish, *Fundulus. J. Exp. Zool.* **208**, 281–286.

Géraudie, J., and Singer, M. (1985). Necessity of an adequate nerve supply for regeneration of the amputated pectoral fin in the teleost *Fundulus. J. Exp. Zool.* **234**, 367–374.

Géraudie, J., and Singer, M. (1985). Relation between nerve fiber number and pectoral fin regeneration in the teleost. *J. Exp. Zool.* **199**, 1–8.

Germiller, J. A., and Goldstein, S. A. (1997). Structure and function of embryonic growth plate in the absence of functioning skeletal muscle. *J. Orthopaed. Res.* **15**, 362–370.

Gerstenfeld, L. C., Chipman, S. D., Glowacki, J., and Lian, J. B. (1987). Expression of differentiated function by mineralizing cultures of chicken osteoblasts. *Devel. Biol.* **122**, 49–60.

Gerstenfeld, L. C., Chipman, S. D., Kelly, C. M., Hodgens, K. J., Lee, D. D., and Landis, W. J. (1988). Collagen expression, ultrastructural assembly, and mineralization in cultures of chicken embryo osteoblasts. *J. Cell Biol.* **106**, 979–990.

Gerstenfeld, L. C., Finer, M. H., and Boedtker, H. (1989). Quantitative analysis of collagen expression in embryonic chick chondrocytes having different developmental fates. *J. Biol. Chem.* **264**, 5112–5120.

Gerstenfeld, L. C., and Landis, W. J. (1991). Gene expression and extracellular matrix ultrastructure of a mineralizing chondrocyte cell culture system. *J. Cell Biol.* **112**, 501–513.

Ghafari, J. (1984). Palatal sutural response to buccal muscular displacement in the rat. *Amer. J. Orthod.* **85**, 351–356.

Ghafari, J., and Cowin, D. H. (1989). Condylar cartilage in the muscular dystrophic mouse. *Amer. J. Orthod.* **95**, 107–114.

Ghalambor, N., Matta, J. M., and Bernstein, L. (1994). Heterotopic ossification following operative treatment of acetabular fracture – An analysis of risk factors. *Clin. Orthop. Rel. Res.* **305**, 96–105.

Ghosh-Choudhury, N., Harris, M. A., Feng, J. Q., Mundy, G. R., and Harris, S. E. (1994). Expression of the BMP 2 gene during bone cell differentiation. *Crit. Rev. Eukary. Gene Exp.* **4**, 345–355.

Ghosh-Choudhury, N., Windle, J. J., Koop, B. A., Harris, M. A., Guerrero, D. L., Wozney, J. M., Mundy, G. R., and Harris, S. E. (1996). Immortalized murine osteoblasts derived from BMP-2-T-antigen expressing transgenic mice. *Endocrinology* **137**, 331–339.

Gibson, G. J., Bearman, C. H., and Flint, M. H. (1986). The immunoperoxidase localization of type-X collagen in chick cartilage and lung. *Collagen Rel. Res.* **6**, 163–184.

Gibson, G. J., and Flint, M. H. (1985). Type X collagen synthesis by chick sternal chondrocytes and its relationship to endochondral development. *J. Cell Biol.* **101**, 277–284.

Gibson, G. J., Kielty, C. M., Garner, C., Schor, S. L., and Grant, M. E. (1983). Identification and partial characterization of three low-molecular-weight collagenous polypeptides synthesized by chondrocytes cultured within collagen gels in the absence and in the presence of fibronectin. *Biochem. J.* **211**, 417–426.

Gibson, G. J., Kohler, W. J., and Schaffler, M. B. (1995). Chondrocyte apoptosis in endochondral ossification of chick sterna. *Devel. Dynam.* **203**, 468–476.

Gibson, G. J., Schor, S. L., and Grant, M. E. (1982). Effects of matrix macromolecules on chondrocyte gene expression: Synthesis of a low molecular weight collagen species by cells cultured within collagen gels. *J. Cell Biol.* **93**, 767–774.

Gilbert, G. H., and Gorman, H. A. (1971). Transplantation of urinary bladder mucosa for osteogenic effect. *J. Amer. Vet. Med. Assoc.* **158**, 77–81.

Gilbert, L. (1997). From Joseph Banks to Joseph Maiden: Towards a scientific botanical garden. *Hist. Rec. Aust. Sci.* **11**, 283–300.

Gilbert, S. F. (2003). *Developmental Biology.* Seventh Edn. Sinauer Associates, Sunderland, MA.

Gilbert, S. F., Loredo, G. A., Brukman, A., and Burke, A. C. (2001). Morphogenesis of the turtle shell: the development of a novel structure in tetrapod evolution. *Evol. & Devel.* **3**, 47–58.

Gilbert, S. F., and Zevit, Z. (2001). Congenital human baculum deficiency: The generative bone of Genesis 2: 21–23. *Amer. J. Med. Genet.* **101**, 284–285.

Gilbert, W. H., and Richards, G. D. (2000). Digital imaging of bone and tooth modification. *Anat. Rec. (New Anat.)* **261**, 237–246.

Gimlich, R. L., and Braun, J. (1985). Improved fluorescent compounds for tracing cell lineage. *Devel. Biol.* **109**, 509–514.

Gingerich, P. D. (1972). The development of sexual dimorphism in the bony pelvis of the squirrel monkey. *Anat. Rec.* **172**, 589–596.

Gingerich, P. D., Raza, S. M., Arif, M., Anwar, M., and Zhou, X. (1994). New whale from the Eocene of Pakistan and the origin of cetacean swimming. *Nature* **368**, 844–847.

Gingerich, P. D., Wells, N. A., Russell, D. E., and Shah, S. M. I. (1983). Origin of whales in epicontinental remnant seas: New evidence from the early Eocene of Pakistan. *Science* **220**, 403–406.

Girgis, F. G., and Pritchard, J. J. (1958). Experimental production of cartilage during the repair of fractures of the skull vault in rats. *J. Bone Joint Surg. Br. Vol.* **40**, 274–281.

Gitelman, S. E., Kobrin, M. S., Ye, J.-Q., Lopez, A. R., Lee, A., and Derynck, R. (1994). Recombinant Vgr-1/BMP-6-expressing tumors induce fibrosis and endochondral bone formation *in vivo. J. Cell Biol.* **126**, 1595–1609.

Gladstone, R. J., and Wakeley, C. P. G. (1932). The morphology of the sternum and its relation to the ribs. *J. Anat.* **66**, 508–564.

Glant, T. T., Hadházy, Cs., Mikecz, K., and Sipos, A. (1985). Appearance and persistence of fibronectin in cartilage. Specific interaction of fibronectin with collagen type II. *Histochemistry* **82**, 149–158.

Glant, T. T., Mikecz, K., Arzoumanian, A., and Poole, A. R. (1987). Proteoglycan-induced arthritis in BALB/c mice. Clinical features and histopathology. *Arth. Rheum.* **30**, 201–212.

Glass, L. (1977). Patterns of supernumerary limb regeneration. *Science* **198**, 321–322.

Glasstone, S. (1968). Tissue culture of the mandible and mandibular joint of mouse embryos. *Nature* **220**, 705–706.

Glasstone, S. (1971). Differentiation of the mouse embryonic mandible and squamo-mandibular joint in organ culture. *Archs. Oral Biol.* **16**, 723–730.

Glasstone, S. (1973). The development of teeth and jaws in tissue culture. *J. Dent. Res. S. Afr.* **28**, 328–334.

Glenister, T. W. (1976). An embryological view of cartilage. *J. Anat.* **122**, 323–330.

Glenny, F. H. (1954). The clavicles and dorsal coracoid arteries as indices of phyletic relationships and levels of avian evolution. *Anat. Anz.* **101**, 95–100.

Glenny, F. H. (1959). Specific and individual variation in reduction of the clavicles in the parrots. *Ohio J. Sci.* **59**, 321–322.

Glenny, F. H., and Friedmann, H. (1954). Reduction of the clavicles in the mesoenatidae with some remarks concerning the relationship of the clavicles to flight-function in birds. *Ohio J. Sci.* **54**, 111–113.

Glick, M. C., Lash, J. W., and Madden, J. W. (1964). Enzymic activities associated with the induction of chondrogenesis *in vitro. Biochim. Biophys. Acta* **83**, 84–92.

Glimcher, M. J., Cohen-Solal, L., Kossiva, D., and de Ricqlès, A. J. (1990). Biochemical analyses of fossil enamel and dentin. *Paleobiology* **16**, 219–232.

Glimcher, M. J., Daniel, E. J., Travis, D. F., and Kahmis, S. (1965). Electron optical and X-ray diffraction studies of the organization of the inorganic crystals in embryonic bovine enamel. *J. Ultrastruct. Res.* **7**, Suppl., 1–77.

Globus, M., and Vethamany-Globus, S. (1976). An *in vitro* analogue of early chick limb bud outgrowth. *Differentiation* **6**, 91–96.

Globus, M., and Vethamany-Globus, S. (1977). Transfilter mitogenic effect of dorsal root ganglia on cultured regeneration blastemata, in the newt, *Notophthalmus viridsecens. Devel. Biol.* **56**, 316–328.

Globus, R. K., Patterson-Buckendahl, P., and Gospodarowicz, D. (1988). Regulation of bovine bone cell proliferation by fibroblast growth factor and transforming growth factor β. *Endocrinology* **123**, 98–105.

Globus, R. K., Plouet, J., and Gospodarowicz, D. (1989). Cultured bovine bone cells synthesize basic fibroblast growth factor and store it in their extracellular matrix. *Endocrinology* **124**, 1539–1547.

Glowacki, J. (1998). Angiogenesis in fracture repair. *Clin. Orthop. Rel. Res.* **355**, S82–S89.

Glowacki, J. (ed.). (2004). Repair and regeneration of oral and craniofacial tissues (Special Issue). *Cells Tissues Organs* **176** (1–3), 1–165.

Glowacki, J., Cox, K. A., O'Sullivan, J., Wilkie, D., and Deftos, L. J. (1986). Osteoclasts can be induced in fish having an acellular bony skeleton. *Proc. Natl Acad. Sci. U.S.A.* **83**, 4104–4107.

Glowacki, J., and Lian, J. B. (1987). Impaired recruitment and differentiation of osteoclast progenitors by osteocalcin-deplete bone implants. *Cell Differ.* **21**, 247–254.

Glowacki, J., Trepman, E., and Folkman, J. (1983). Cell shape and phenotypic expression in chondrocytes. *Proc. Soc. Exp. Biol. Med.* **172**, 93–98.

Glücksmann, A. (1938). Studies on bone mechanics *in vitro.* I. Influence of pressure on orientation of structure. *Anat. Rec.* **72**, 97–113.

Glücksmann, A. (1939). Studies on bone mechanics *in vitro.* II. The role of tension and pressure in chondrogenesis. *Anat. Rec.* **73**, 39–55.

Glücksmann, A. (1942). The role of mechanical stresses on bone formation *in vitro. J. Anat.* **76**, 231–236.

Glücksmann, A., and Cherry, C. P. (1972). The hormonal induction of an os clitoridis in the neonatal and adult rat. *J. Anat.* **112**, 223–231.

Glücksmann, A., Ooka-Souda, S., Miura-Yasugi, E., and Mizuno, T. (1976). The effect of neonatal treatment of male mice with antiandrogens and of females with androgens on the development of the os penis and os clitoridis. *J. Anat.* **121**, 363–370.

Gluhak, J., Mais, A., and Mina, M. (1996). Tenascin-C is associated with early stages of chondrogenesis by chick mandibular ectomesenchymal cells *in vivo* and *in vitro*. *Devel. Dynam.* **205**, 24–40.

Godfrey, S. J. (1989). Ontogenetic changes in the skull of the Carboniferous tetrapod *Greererpeton burkemorani* Romer 1969. *Phil. Trans. R. Soc. London B.* **323**, 135–153.

Godman, G. C., and Porter, K. R. (1960). Chondrogenesis, studied with the electron microscope. *J. Biophys. Biochem. Cytol.* **8**, 719–760.

Goedbloed, J. F. (1964). The early development of the middle ear and the mouth cavity. A study of the interaction of processes in the epithelium and the mesenchyme. *Archs Biol.* **75**, 207–244.

Goel, S. C. (1970). Electron microscopic studies on developing cartilage. I. The membrane system related to the synthesis of extracellular materials. *J. Embryol. Exp. Morphol.* **23**, 169–184.

Goel, S. C., and Jacob, J. (1976). Reinterpretation of the ultrastructure of cartilage matrix. *Experientia* **32**, 216–217.

Goel, S. C., and Jurand, A. (1975). Electron microscopic studies on chick limb cartilage differentiated in tissue culture. *J. Embryol. Exp. Morphol.* **34**, 327–337.

Goetinck, P. F. (1964). Studies on avian limb morphogenesis. II. Experiments with the polydactylous mutant *Eudiplopodia*. *Devel. Biol.* **10**, 71–91.

Goetinck, P. F. (1966). Genetic aspects of skin and limb development. *Curr. Top. Devel. Biol.* **1**, 253–283.

Goetinck, P. F. (1983). Mutations affecting limb cartilages. In *Cartilage, Volume 3 Biomedical Aspects* (B. K. Hall, ed.), pp. 165–190. Academic Press, New York, NY.

Goetinck, P. F., and Abbott, U. K. (1964). Studies on limb morphogenesis. I. Experiments with the polydactylous mutant, *talpid.*[2] *J. Exp. Zool.* **155**, 161–170.

Goetinck, P. F., Lever-Fischer, P. L., McKeowin-Longo, P. J., Quintner, M. I., Sawyer, L. M., Sparks, I. J., and Argraves, W. S. (1981). Chondrogenesis in normal and mutant avian embryos. In *Levels of Genetic Control in Development* (S. Subtelny and U. K. Abbott, eds), pp. 15–35. Alan R. Liss Inc., New York, NY.

Goetinck, P. F., and Pennypacker, J. P. (1977). Controls in the acquisition and maintenance of chondrogenic expression. In *Vertebrate Limb and Somite Morphogenesis* (D. A. Ede, J. R. Hinchliffe, and M. Balls, eds), pp. 139–160. Cambridge University Press, Cambridge.

Goetinck, P. F., Pennypacker, J. P., and Royal, P. D. (1974). Proteochondroitin sulfate synthesis and chondrogenic expression. *Exp. Cell Res.* **87**, 241–248.

Goetinck, P. F., and Sekellick, M. J. (1970). Early morphogenetic events in normal and mutant skin development in the chick embryo and their relationship to alkaline phosphatase activity. *Devel. Biol.* **21**, 349–363.

Goggins, J. F., and Billups, L. C. (1972). *In vitro* synthesis of acid mucopolysaccharides by human gingival fibroblasts. *J. Dent. Res.* **51**, 303–307.

Goggins, J. F., Johnson, G S., and Pastan, I. (1972). The effect of dibutyryl cyclic adenosine monophosphate on synthesis of sulfated acid mucopolysaccharides by transformed fibroblasts. *J. Biol. Chem.* **247**, 5759–5764.

Goldberg, R. L., and Toole, B. P. (1984). Pericellular coat of chick embryo chondrocytes: Structural role of hyaluronate. *J. Cell Biol.* **99**, 2114–2122.

Goldhaber, P. (1961). Osteogenic induction across Millipore filters *in vitro*. *Science* **133**, 2065–2067.

Goldhaber, P., Rabadjija, L., Beyer, W. R., and Kornhauser, A. (1973). Bone resorption in tissue culture and its relevance to periodontal disease. *J. Amer. Dent. Assoc.* **87**, 1027–1033.

Gol'din, P. E. (2004). Postnatal growth and ossification of forelimb skeleton in harbour porpoise (*Phocoena phocoena*, Linnaeus, 1758) in the Sea of Azov and the Black Sea. *Sci. Rep. Taurida Nat. Univ. Ser. Biol. Chem.* **17**, 66–81 (in Russian).

Goldman, D. C., Martin, G. R., and Tam, P. P. L. (2000). Fate and function of the ventral ectodermal ridge during mouse tail development. *Development* **127**, 2113–2123.

Goldring, M. B., Birkhead, J., Sandell, L., Kimura, T., and Krane, S. M. (1988). Interleukin-1 suppresses expression of cartilage-specific types II and IX collagens and increases types I and II collagens in human chondrocytes. *J. Clin. Invest.* **82**, 2026–2037.

Goldschmidt, T. (1996). *Darwin's Dreampond: Drama in Lake Victoria*. Translated by S. Marx-Macdonald. The MIT Press, Cambridge, MA.

Goldstein, R.S., Avivi, C., and Geffen, R. (1995). Initial axial-level-dependent differences in size of avian dorsal root ganglia are imposed by the sclerotome. *Devel. Biol.* **168**, 214–222.

Goldstein, R. S., and Kalcheim, C. (1992). Determination of epithelial half-somites in skeletal morphogenesis. *Development* **116**, 441–445.

Gollmann, G. (1991). Osteological variation in *Geocrinia laevis*, *Geocrinia victoriana* and their hybrid populations (Amphibia, Anura, Myobatrachinae). *Z. Zool. Syst. Evolut.-Forsch.* **29**, 289–303.

Gomori, G. (1943). Calcification and phosphatase. *Amer. J. Pathol.* **19**, 197–209.

Gooch, H. L., Hale, J. E., Fujioka, H., Balian, G., and Hurwitz, S. R. (2000). Alterations of cartilage and collagen expression during fracture healing in experimental diabetes. *Conn. Tissue Res.* **41**, 81–94.

Goodday, R. H. B., and Precious, D. S. (1988). Duplication of mental nerve in a patient with cleft-lip-palate and Rubella syndrome. *Oral Surg., Oral Med., Oral Pathol.* **65**, 157–160.

Goode, A. W., and Rambaut, P. C. (1985). The skeleton in space. *Nature* **317**, 204–205.

Goodman, R., Bassett, C. A. L., and Henderson, A. S. (1983). Pulsing electromagnetic fields induce cellular transcription. *Science* **220**, 1283–1285.

Goodrich, E. S. (1958). *Studies on the Structure and Development of Vertebrates*. Two vols. Dover Publications, New York, NY; Constable and Co., London.

Goodsir, J. (1845). The structure and economy of bone. In *Anatomical and Pathological Observations* (J. Goodsir and H. D. S. Goodsir, eds). Myles MacPhail, Edinburgh.

Gordon, J. S., and Lash, J. W. (1974). *In vitro* chondrogenesis and cell viability. *Devel. Biol.* **36**, 88–104.

Goret-Nicaise, M. (1981). Influence des insertions des muscles masticateurs sur la structure mandibulaire du nouveau-ne. *Bull. L'Assoc. Anat.* **65**, 287–296.

Goret-Nicaise, M. (1982). La symphyse mandibulaire du nouveau-né. Etude histologique et microradiographique. *Rev. Stomalologie Chir. Maxillo-faciale* **83**, 266–272.

Goret-Nicaise, M. (1984a). Die symphysis menti beim menschlichen feten. *Anat. Anz. Jena* **156**, 217–224.

Goret-Nicaise, M. (1984b). Identification of collagen type I and type II in chondroid tissue. *Calcif. Tissue Int.* **36**, 682–689.

Goret-Nicaise, M. (1986). *La Croissance de la Mandible Humaines conception actuelle*. Thèse d'Agrégé de L'Enseignement Supérieur Catholic University Louvain, Belgique.

Goret-Nicaise, M., Awn, M., and Dhem, A. (1983). The morphological effects on the rat mandibular condyle of section of the lateral pterygoid muscle. *Eur. J. Orthdont.* **5**, 315–321.

Goret-Nicaise, M., and Dhem, A. (1982). Presence of chondroid tissue in the symphyseal region of the growing human mandible. *Acta Anat.* **113**, 189–195.

Goret-Nicaise, M., and Dhem, A. (1984). The mandibular body of the human fetus. Histologic analysis of the basilar part. *Anat. Embryol.* **169**, 231–236.

Goret-Nicaise, M., and Dhem, A. (1985). Comparison of the calcium content of different tissues present in the human mandible. *Acta Anat.* **124**, 167–172.

Goret-Nicaise, M., Manzanares, M. C., Bulpa, P., Nolmans, E., and Dhem, A. (1988). Calcified tissues involved in the ontogenesis of the human cranial vault. *Anat. Embryol.* **178**, 399–406.

Goret-Nicaise, M., and Pilet, D. (1983). A few observations about Meckel's cartilage in the human. *Anat. Embryol.* **167**, 365–370.

Gorlin, R. J. (ed.) (1980). *Morphogenesis and Malformation of the Ear.* Birth Defects, Original Article Series **16**(4), 1980. Alan R. Liss Inc., New York, NY.

Gorlin, R. J., Cohen, M. M., Jr., and Hennekman, R. C. M. (2001). *Syndromes of the Head and Neck.* Oxford University Press, New York, NY.

Gorivodsky, M., and Lonai, P. (2003). Novel roles of *Fgfr2* in AER differentiation and positioning of the dorsoventral limb interface. *Development* **130**, 5471–5479.

Gorski, J. L., Estrada, L., Hu, C., and Liu, Z. (2000). Skeletal-specific expression of *Fgd1* during bone formation and skeletal defects in faciogenital dysplasia (FGDY; Aarskog syndrome). *Devel. Dynam.* **218**, 573–586.

Gorski, J. P. (1998). Is all bone the same? Distinctive distributions and properties of non-collagenous matrix proteins in lamellar vs. woven bone imply the existence of different underlying mechanisms. *Crit. Rev. Oral Biol. Med.* **9**, 201–223.

Gorski, J. P., Marks, S. C., Jr., Cahill, D. R., and Wise, G. E. (1988). Developmental changes in the extracellular matrix of the dental follicle during tooth eruption. *Conn. Tissue Res.* **18**, 175–190.

Goss, R. J. (1954). The role of the central cartilaginous rod in the regeneration of the catfish taste barbel. *J. Exp. Zool.* **127**, 181–199.

Goss, R. J. (1958). Skeletal regeneration in amphibians. *J. Embryol. Exp. Morphol.* **6**, 638–644.

Goss, R. J. (1961). Experimental investigations of morphogenesis in the growing antler. *J. Embryol. Exp. Morphol.* **9**, 342–354.

Goss, R. J. (1963). The deciduous nature of deer antlers. In *Mechanisms of Hard Tissue Destruction*, Publication No. 75, pp. 339–369. Amer. Assoc. Adv. Sci. Washington, DC.

Goss, R. J. (1964). The role of skin in antler regeneration. *Adv. Biol. Skin* **5**, 194–207.

Goss, R. J. (1968). Inhibition of growth and shedding of antlers by sex hormones. *Nature* **220**, 83–85.

Goss, R. J. (1969a). *Principles of Regeneration.* Academic Press, New York, NY.

Goss, R. J. (1969b). Photoperiodic control of antler cycles in deer. I. Phase shift and frequency changes. *J. Exp. Zool.* **170**, 311–324.

Goss, R. J. (1969c). Photoperiodic control of antler cycles in deer. II. Alterations in amplitude. *J. Exp. Zool.* **171**, 223–234.

Goss, R. J. (1970). Problems in antlerogenesis. *Clin. Orthop. Rel. Res.* **69**, 227–238.

Goss, R. J. (1974). Aging versus growth. *Perspect. Biol. Med.* **17**, 485–494.

Goss, R. J. (1976). Photoperiodic control of antler cycles in deer. III. Decreasing versus increasing day lengths. *J. Exp. Zool.* **197**, 307–312.

Goss, R. J. (1977). Photoperiodic control of antler cycles in deer. IV. Effects of constant light : dark ratios on circannual rhythms. *J. Exp. Zool.* **201**, 379–382.

Goss, R. J. (1980a). Photoperiodic control of antler cycles in deer. V. Reversed seasons. *J. Exp. Zool.* **211**, 101–106.

Goss, R. J. (1980b). Prospects for regeneration in man. *Clin. Orthop. Rel. Res.* **151**, 270–282.

Goss, R. J. (1983). *Deer Antlers. Regeneration, Function and Evolution.* Academic Press, New York, NY.

Goss, R. J. (1984). Photoperiodic control of antler cycles in deer. VI. Circannual rhythms on altered day lengths. *J. Exp. Zool.* **230**, 265–275.

Goss, R. J. (1987a). Induction of deer antlers by transplanted periosteum. II. Regional competence for velvet transformation in ectopic skin. *J. Exp. Zool.* **244**, 101–111.

Goss, R. J. (1987b). Photoperiodic control of antler cycles in deer. VII. Role of light vs. dark in suppression of circannual rhythms. *J. Exp. Zool.* **244**, 113–116.

Goss, R. J. (1990a). Interactions between asymmetric brow tines in caribou and reindeer antlers. *Can. J. Zool.* **68**, 1115–1119.

Goss, R. J. (1990b). Tumor-like growth of antlers in castrated fallow deer – an electron microscopic study. *Scann. Microscopy* **4**, 715–722.

Goss, R. J. (1991). Induction of deer antlers by transplanted periosteum. III. Orientation. *J. Exp. Zool.* **259**, 246–251.

Goss, R. J., and Grimes, L. N. (1972). Tissue interactions in the regeneration of rabbit ear holes. *Amer. Zool.* **12**, 151–157.

Goss, R. J., and Grimes, L. N. (1975). Epidermal downgrowths in regenerating rabbit ear holes. *J. Morphol.* **146**, 533–542.

Goss, R. J., and Powell, R. S. (1985). Induction of deer antlers by transplanted periosteum. I. Graft size and shape. *J. Exp. Zool.* **235**, 359–373.

Goss, R. J., Severinghaus, C. W., and Free, S. (1964). Tissue relationships in the development of pedicles and antlers in the Virginia deer. *J. Mammal.* **45**, 61–68.

Goss, R. J., and Stagg, M. W. (1957). The regeneration of fins and fin rays in *Fundulus heteroclitus*. *J. Exp. Zool.* **136**, 487–508.

Goss, R. J., and Stagg, M. W. (1958). Regeneration of the lower jaw in adult newts. *J. Morphol.* **102**, 289–309.

Goss, R. J., van Praagh, A., and Brewer, P. (1992). The mechanism of antler casting in the fallow deer. *J. Exp. Zool.* **264**, 429–436.

Göthlin, G. (1973). Electron microscopic observations on fracture repair in the rat. *Acta Pathol. Microbiol. Scand., Ser. A* **81**, 507–522.

Göthlin, G., and Ericsson, J. L. E. (1973). On the histogenesis of the cells in fracture callus. Electron microscopic autoradiographic observations in parabiotic rats and studies on labeled monocytes. *Virchows Arch. B Zellpathol.* **12**, 318–329.

Göthlin, G., and Ericsson, J. L. E. (1976). The osteoclast. Review of ultrastructure, origin, and structure–function relationships. *Clin. Orthop. Rel. Res.* **120**, 201–231.

Gottlieb, B. (1914). Die vitale Färbung der Kalkhaltigen Gewebe. *Anat. Anz.* **46**, 179–194.

Gould, G. M., and Pyle, W. L. (1896). *Anomalies and Curiosities of Medicine.* W. B. Saunders, Philadelphia, PA.

Gould, R. P., Day, A., and Wolpert, L. (1972). Mesenchymal condensation and cell contact in early morphogenesis of the chick limb. *Exp. Cell Res.* **72**, 325–336.

Gould, R. P., Selwood, L., Day, A., and Wolpert, L. (1974). The mechanism of cellular orientation during early cartilage formation in the chick limb and regenerating amphibian limb. *Exp. Cell Res.* **83**, 287–296.

Gould, S. E., Rhee, J. M., Tay, B. K. B., Otsuka, N. Y., and Bradford, D. S. (2000). Cellular contribution of bone graft to fusion. *J. Orthoped. Res.* **18**, 920–927.

Gould, S. E., Upholt, W. B., and Kosher, R. A. (1992). Syndecan 3: A member of the syndecan family of membrane-intercalated proteoglycans that is expressed in high amounts at the onset of chicken limb cartilage differentiation. *Proc. Natl Acad. Sci. U.S.A.* **89**, 3271–3275.

Gould, S. E., Upholt, W. B., and Kosher, R. A. (1995). Characterization of chicken syndecan 3 as a heparan sulfate proteoglycan and its expression during embryogenesis. *Devel. Biol.* **168**, 438–451.

Gould, S. J. (1977). *Ontogeny and Phylogeny.* The Belknap Press of Harvard University Press, Cambridge, MA.

Gould, S. J. (2002). Both neonate and elder: the first fossil of 1557. *Paleobiology* **28**, 1–8.

Gould, S. J., and Lewontin, R. C. (1979). The spandrels of San Marco and the panglossian paradigm: a critique of the adaptationist programme. *Proc. R. Soc. Lond. B.* **205**, 581–598.

Graber, L. W. (1977). Chin cup therapy for mandibular prognathism. *Amer. J. Orthod.* **72**, 23–41.

Grace, K. L. R., Revell, W. J., and Brookes, M. (1998). The effects of pulsed electromagnetism on fresh fracture healing: Osteochondral repair in the rat femoral groove. *Orthopedics* **21**, 297–302.

Graham, A. (2001). The development and evolution of the pharyngeal arches. *J. Anat.* **199**, 133–141.

Graham, A., and McGonnell, I. (1999). Developmental evolution: This side of paradise. *Curr. Biol.* **9**, R630–R632.

Graham, A., and Smith, A. (2001). Patterning the pharyngeal arches. *BioEssays* **23**, 54–61.

Graham, D. L., and Meier, G. W. (1975). Standards of morphological development of the quail, *Coturnix coturnix japonica* embryo. *Growth* **39**, 389–400.

Graham, E. A., Rainey, R., Kuhlman, R. E., Houghton, E. H., and Moyer, C. A. (1962). Biochemical investigations of deer antler growth. Part I. Alterations of deer blood chemistry resulting from antlerogenesis. *J. Bone Joint Surg. Am. Vol.* **44**, 482–488.

Graham-Smith, W. (1978). On the lateral lines and dermal bones in the parietal region of some Crossopterygian and Dipnoan fishes. *Phil. Trans R. Soc. Lond. B.* **212**, 41–105.

Grammatopoulos, G. A., Bell, E., Tode, L., Lumsden, A., and Tucker, A. S. (2000). Homeotic transformation of branchial arch identity after *Hoxa2* overexpression. *Development* **127**, 5355–5365.

Grande, L., and Bemis, W. E. (1991). Osteology and phylogenetic relationships of fossil and recent paddlefishes (Polyodontidae) with comments on the interrelationships of Acipenseriformes. *J. Vert. Paleont.* **11** (Suppl.), 1–121.

Grandel, H., Draper, B. W., and Schulte-Merker, S. (2000). *Dackel* acts in the ectoderm of the zebrafish pectoral fin bud to maintain AER signaling. *Development* **127**, 4169–4178.

Grandel, H., Lun, K., Rauch, G.-J., Rhinn, M., Piotrawski, T., Houart, C., Sordino, P., Kuechler, A. M., Schulte, M. S., Geisler, R., Holder, N., Wilson, S. W., and Brand, M. (2002). Retinoic acid signalling in the zebrafish embryo is necessary during pre-segmentation stages to pattern the anterior–posterior axis of the CNS and to induce a pectoral fin bud. *Development* **129**, 2851–2865.

Granström, G. (1986). Isoenzyme changes during rat facial development. *Scand. J. Dent. Res.* **94**, 1–14.

Granström, G., and Magnusson, B. C. (1986). Lactate dehydrogenase isoenzymes during facial development. *J. Anat.* **148**, 183–192.

Grant, P. G., Buschang, P. H., Drolet, D. W., and Pickerell, C. (1980). Invariance of the relative positions of structures attached to long bones during growth: Cross-sectional and longitudinal studies. *Acta Anat.* **107**, 26–34.

Grant, T. D., Cho, J., Ariail, K. S., Weksler, N. B., Smith, R. W., and Horton, W. A. (2000). *Col2-GFP* reporter marks chondrocyte lineage and chondrogenesis during mouse skeletal development. *Devel. Dynam.* **218**, 394–400.

Graver, H. B. (1973). The polarity of the dental lamina in the regenerating salamander jaw. *J. Embryol. Exp. Morphol.* **30**, 635–646.

Graver, H. B. (1978). Re-regeneration of lower jaws and the dental lamina in adult urodeles. *J. Morphol.* **157**, 269–280.

Graveson, A. C., and Armstrong, J. B. (1987). Differentiation of cartilage from cranial neural crest in the axolotl (*Ambystoma mexicanum*). *Differentiation* **35**, 16–20.

Graveson, A. C., Hall, B. K., and Armstrong, J. B. (1995). The relationship between migration and chondrogenic potential of trunk neural crest cells in *Ambystoma maculatum*. *W. Roux. Arch. Devel. Biol.* **204**, 477–483.

Graveson, A. C., Smith, M. M., and Hall, B. K. (1997). Neural crest potential for tooth development in a urodele amphibian: developmental and evolutionary significance. *Devel. Biol.* **188**, 34–42.

Green, B. S., and McCormick, M. I. (2001). Ontogeny of the digestive and feeding systems in the anemone fish *Amphiprion melanopus*. *Envtl. Biol. Fishes* **61**, 73–83.

Green, C. R., Bowles, L., Crawley, A., and Tickle, C. (1994). Expression of the connexin 43 gap junctional protein in tissues at the tip of the chick limb bud is related to the epithelial–mesenchymal interactions that mediate morphogenesis. *Devel. Biol.* **161**, 12–21.

Green. H., and Goldberg, B. (1968). Differentiation for collagen synthesis in cultured cells. In *Differentiation and Immunology* (K. B. Warren, ed.), pp. 123–134. Academic Press, New York, NY.

Green, J., Schotland, S., Stauber, D. J., Kleeman, C. R., and Clemens, T. L. (1995). Cell–matrix interaction in bone: Type I collagen modulates signal transduction in osteoblast-like cells. *Amer. J. Physiol.* **268**, C1090–C1103.

Green. M. C., and Witham, B. A. (eds) (1991). *Handbook on Genetically Standardized JAX Mice.* The Jackson Laboratory, Bar Harbor, Maine.

Green, W. T., Jr. (1971). Behavior of articular chondrocytes in cell culture. *Clin. Orthop. Rel. Res.* **75**, 248–260.

Green, W. T., Jr., and Dickens, R. (1972). Culture model for the investigation of cartilage matrix formation. *Surg. Forum* **23**, 453–455.

Green. W. T., Jr., and Ferguson, R. J. (1975). Histochemical and electron microscopic comparison of tissue produced by rabbit articular chondrocytes *in vivo* and *in vitro*. *Arthritis Rheum.* **18**, 273–280.

Greenberg, J. H., and Pratt, R. M. (1977). Glycosaminoglycan and glycoprotein synthesis by cranial neural crest cells *in vitro*. *Cell Differ.* **6**, 119–132.

Greene, H. W., and Cundall, D. (2000). Limbless tetrapods and snakes with legs. *Science* **287**, 1939–1941.

Greenspan, J. S., and Blackwood, H. J. J. (1966). Histochemical studies of chondrocyte function in the cartilage of the mandibular condyle of the rat. *J. Anat.* **100**, 615–626.

Greer, A. E. (1987). Limb reduction in the lizard genus *Lerista*. I. Variation in the number of phalanges and presacral vertebrae. *J. Herpetol.* **21**, 267–276.

Greer, A. E., Caputo, V., Lanza, B., and Palmieri, R. (1998). Observations on limb reduction in the scincid lizard genus *Chalcides*. *J. Herpetol.* **32**, 244–252.

Gregory, K. E., Keene, D. R., Tufa, S. F., Lunstroim, G. P., and Morris, N. P. (2001). Developmental distribution of collagen type XII in cartilage: association with articular cartilage and the growth plate. *J. Bone Min. Res.* **16**, 2005–2016.

Gregory, W. K. (1913). Critique of recent work on the morphology of the vertebrate skull, especially in relation to the origin of mammals. *J. Morphol.* **24**, 1–42.

Gregory, W. K. (1929). *Our Face from Fish to Man. A Portrait Gallery of our Ancient Ancestors and Kinfolk together with a Concise History of Our Best Features*. G. P. Putnam's Cons, London.

Gress, C. J., and Jacenko, O. (2000). Growth plate compression and altered hematopoiesis in collagen X null mice. *J. Cell Biol.* **149**, 983–993.

Griffiths, R. A. (1981). Physical abnormalities and accessory limb growth in the smooth newt, *Triturus vulgaris*. *Brit. J. Herpetol.* **6**, 180–182.

Grigoriadis, A. E., Aubin, J. E., and Heersche, J. N. M. (1989). Effects of dexamethasone and vitamin D3 on cartilage differentiation in a clonal chondrogenic cell population. *Endocrinology* **125**, 2103–2110.

Grigoriadis, A. E., Heersche, J. N. M., and Aubin, J. E. (1988). Differentiation of muscle, fat, cartilage and bone from progenitor cells present in a bone-derived clonal cell population: Effect of dexamethasone. *J. Cell Biol.* **106**, 2139–2152.

Grigoriadis, A. E., Heersche, J. N. M., and Aubin, J. E. (1990). Continuously growing bipotential and monopotential myogenic, adipogenic, and chondrogenic subclones isolated from the multipotential RCJ 3.1 clonal cell line. *Devel. Biol.* **142**, 313–318.

Grigoriadis, A. E., Wang, Z.-Q., and Wagner, E. F. (1995). *Fos* and bone cell development: Lessons from a nuclear oncogene. *Trends Genet.* **11**, 436–441.

Grimes, L. N. (1974a). Selective X-irradiation of the cartilage at the regeneration margin of rabbit ear holes. *J. Exp. Zool.* **190**, 237–240.

Grimes, L. N. (1974b). The effect of supernumerary cartilaginous implants upon rabbit ear regeneration. *Amer. J. Anat.* **141**, 447–452.

Grimes, L. N., and Kulis, E. J. (1979). The interaction between rabbit ear skin and rabbit tail mesodermal tissues at healing rabbit tail amputation sites. *Anat. Rec.* **193**, 554–555.

Grimsrud, C. D., Romano, P. R., D'Souza, M., Puzas, J. E., Reynolds, P. R., Rosier, R. H., and O'Keefe, R. J. (1999). BMP-6 is an autocrine stimulator of chondrocyte differentiation. *J. Bone Min. Res.* **14**, 475–482.

Gritzka, T. L., Fry, L. R., Cheesman, R. L., and LaVigne, A. (1973). Deterioration of articular cartilage caused by continuous compression in a moving rabbit joint. A light and electron microscopic study. *J. Bone Joint Surg. Am. Vol.* **55**, 1698–1720.

Groβman, M., Sánchez-Villagra, M. R., and Maier, W. (2002). On the development of the shoulder girdle in *Crocidura russula* (Soricidae) and other placental mammals: evolutionary and functional aspects. *J. Anat.* **201**, 371–381.

Grobstein, C. (1955). Inductive interaction in the development of the mouse metanephros. *J. Exp. Zool.* **130**, 319–339.

Grobstein, C., and Holtzer, H. (1955). *In vitro* studies of cartilage induction in mouse somite mesoderm. *J. Exp. Zool.* **128**, 333–357.

Grobstein, C., and Parker, G. (1954). *In vitro* induction of cartilage in mouse somite mesoderm by embryonic spinal cord. *Proc. Soc. Exp. Biol. Med.* **85**, 477–481.

Groman, D. B. (1982). Histology of the striped bass. *Amer. Fisheries Soc. Monograph #3*, 116 pp. Bethesda.

Gronowicz, G., Woodiel, F. N., McCarthy, M.-B., and Raisz, L. G. (1989). *In vitro* mineralization of fetal rat parietal bones in defined serum-free medium: Effect of glycerol phosphate. *J. Bone Min. Res.* **4**, 313–324.

Groot, C. G., Danes, J. K., Blok, J., Hoogendijk, A., and Hauschka, P. V. (1986). Light and electron microscopic demonstration of osteocalcin antigenicity in embryonic and adult rat bone. *Bone* **7**, 379–385.

Groot, C. G., Thesingh, C. W., Wassenaar, A. M., and Scherft, J. P. (1994). Osteoblasts develop from isolated fetal mouse chondrocytes when cocultured in high density with brain tissue. *In Vitro Cell Devel. Biol. Animal* **30A**, 547–554.

Gross, J. B., and Hanken, J. (2004). Use of fluorescent dextran conjugates as a long-term marker of osteogenic neural crest in frogs. *Devel. Dyn.* **230**, 100–106.

Grotmol, S., Kryvi, H., Nordvik, K., and Totland, G. K. (2003). Notochord segmentation may lay down the pathway for the development of the vertebral bodies in the Atlantic salmon. *Anat. Embryol.* **207**, 263–272.

Gruber, H. E., and Baylink, D. J. (1991). The effects of fluoride on bone. *Clin. Orthop. Rel. Res.* **267**, 264–277.

Grüber, R., Mayer, C., Bobacz, K., Kraut, M.-T., Grainger, W., Luyten, F. P., and Erlacher, L. (2001). Effects of cartilage-derived morphogenetic proteins and osteogenic protein-1 on osteochondrogenic differentiation of periosteum-derived cells. *Endocrinology* **142**, 2087–2094.

Grüneberg, H. (1954). Genetical studies on the skeleton of the mouse. XII. The development of *undulated*. *J. Genet.* **52**, 441–455.

Grüneberg, H. (1956). A ventral ectodermal ridge of the tail in mouse embryos. *Nature* **177**, 787–788.

Grüneberg, H. (1963). *The Pathology of Development. A Study of Inherited Skeletal Disorders in Animals*. Blackwell, Oxford.

Grüneberg, H., and Lee, A. J. (1973). The anatomy and development of brachypodism in the mouse. *J. Embryol. Exp. Morphol.* **30**, 119–141.

Grüneberg, H., and Wickramaratne, G. A. (1974). A re-examination of two skeletal mutants of the mouse, vestigial tail (vt) and congenital hydrocephalus (ch). *J. Embryol. Exp. Morphol.* **31**, 207–222.

Grynpas, M. D., Simmons, E. D., Pritzker, K. P. H., Hancock, R. V., and Harrison, J. E. (1986). Is fluoridated bone different from non-fluoridated bone? In *Cell Mediated Calcification and Matrix Vesicles* (S. Y. Ali, ed.), pp. 409–414. Elsevier, London.

Grynpas, M. D., Tenenbaum, H. C., and Holmyard, D. P. (1989). The emergence and maturation of the first apatite crystals in an *in vitro* bone formation system. *Conn. Tissue Res.* **21**, 227–237.

Grzeschik, K.-H. (2002). Human limb malformations; an approach to the molecular basis of development. *Int. J. Devel. Biol.* **46**, 983–991.

Guenther, H. L., Fleisch, H., and Sorgente, N. (1986). Endothelial cells in culture synthesize a potent bone cell active mitogen. *Endocrinology* **119**, 193–201.

Guenther, H. L., Hofstetter, W., Stutzer, A., Mühlbauer, R., and Fleisch, H. (1989). Evidence for heterogeneity of the osteoblastic phenotype determined with clonal rat bone cells established from transforming growth factor-β-induced cell colonies grown anchorage independently in semisolid medium. *Endocrinology* **125**, 2092–2102.

Guerkov, H. H., Lohmann, C. H., Liu, Y., Dean, D. D., Simon, B. J., Heckman, J. D., Schwartz, Z., and Boyan, B. D. (2001). Pulsed electromagnetic fields increase growth factor release by nonunion cells. *Clin. Orthop. Rel. Res.* **384**, 265–279.

Guise, J. M., McCormack, A., Anderson, P. A., and Tencer, A. F. (1992). Effect of controlled local release of sodium fluoride on trabecular bone. *J. Orthop. Res.* **10**, 588–595.

Gumpel-Pinot, M. (1980). Ectoderm and mesoderm interactions in the limb bud of the chick embryo, studied by transfilter culture; cartilage differentiation and ultrastructural observations. *J. Embryol. Exp. Morphol.* **59**, 157–173.

Gumpel-Pinot, M. (1981). Ectoderm–mesoderm interactions in relation to limb-bud chondrogenesis in the chick embryo: Transfilter cultures and ultrastructural studies. *J. Embryol. Exp. Morphol.* **65**, 73–87.

Gumpel-Pinot, M. (1982). Cartilage differentiation in the limb bud of the chick embryo: Ultrastructural observations, culture and grafting experiments. *Arch. Anat. Microsc.* **71**, 241–256.

Gunasekaran, S., Hall, G. E., and Kenny, A. D. (1986b). Vitamin D_3 and avian bone *in vitro*: Stimulation of calcium movement into Japanese quail calvaria. *Calcif. Tissue Int.* **39**, 396–403.

Gunasekaran, S., and Kenny, A. D. (1986a). Vitamin D_3 and avian bone *in vitro*: Specificity of effect on Japanese quail calvaria. *Calcif. Tissue Int.* **39**, 404–411.

Gunther, C. C., Lipscomb, H. L., and Sharp, J. G. (1980). Effects of thymectomy on bone growth in the rat. *J. Anat.* **131**, 693–704.

Gunther, T., and Schinke, T. (2000). Mouse genetics have uncovered new paradigms in bone biology. *Trends Endocrinol. Metab.* **11**, 189–193.

Guralnick, R., and Smith, K. (1999). Historical and biomechanical analysis of integration and dissociation in molluscan feeding with special emphasis on the true limpets (Patellogastropoda: Gastropoda). *J. Morphol.* **241**, 175–195.

Gurdon, J. B. (1970). Nuclear transplantation and the control of gene activity in animal development. *Proc. R. Soc. London, Ser. B.* **176**, 303–314.

Gustafson, G. T., Eckerdal, O., Leever, D. L., Shanfield, J. L., Montgomery, P. C., and Davidovitch, Z. (1977). Cyclic AMP in dental and periodontal tissues during tooth eruption in kittens. *J. Dent. Res.* **56**, 407–415.

Guthrie, R. D. (2003). Rapid body size decline in Alaskan Pleistocene horses before extinction. *Nature* **426**, 169–171.

Guyetant, R., Castanet, J., and Pinston, H. (1984). Détermination de l'âge de jeunes grenouilles, *Rana temporaria* L., par l'analyse des marques de croissance de coupes transversales d'os compact. *C.R. Soc. Biol. Paris* **178**, 271–277.

H

Haas, A. (1997). The larval hyobranchial apparatus of discoglossid frogs – its structure and bearing on the systematics of the Anura (Amphibia, Anura). *J. Zool. Syst. Evol. Res.* **35**, 179–197.

Haas, A. (1999). Larval and metamorphic skeletal development in the fast-developing frog *Pyxicephalus adspersus* (Anura, Ranidae). *Zoomorphol.* **119**, 23–35.

Haas, D. W., and Holick, M. F. (1996). Enhanced osteonectin expression in the chondroid matrix of the unloaded mandibular condyle. *Calcif. Tissue Int.* **59**, 200–206.

Haas, H. J. (1962). Studies on mechanisms of joint and bone formation in the skeletal rays of fish fins. *Devel. Biol.* **5**, 1–34.

Haas, S. L. (1926). Interstitial growth in growing long bones. *Arch. Surg. Chicago* **12**, 887–900.

Habuchi, O., Taen, Y., Sato, T., Washino, Y., and Takeuchi, Y. (1983). Glucuronic acid-containing glycopeptide from squid cartilage. *Biochim. Biophys. Acta* **760**, 318–326.

Häckel, C., Radig, K., Röse, I., and Roessner, A. (1995). The urokinase plasminogen activator (u-PA) and its inhibitor (PAI-1) in embryo-fetal bone development in the human: An immunohistochemical study. *Anat. Embryol.* **192**, 363–368.

Hadházy, Cs., Conti, G., Spreca, A., Musy, J. P., Cappeli-Gotzos, B., and Gotzos, V. (1971). Effets de différentes concentrations d'oxygène sur des fibroblasts d'embryons de poulet cultivés *in vitro*. I. Etude cytologique. *Acta Anat.* **78**, 362–382.

Hadházy, Cs., Glant, T., Mandi, B., and Miltényi, L. (1972). On the antigenicity of cartilage tissue. *Acta Biol. Acad. Sci. Hung.* **23**, 277–288.

Hadly, E. A. (2003). The interface of paleontology and mammalogy: past, present, and future. *J. Mammal.* **84**, 347–353.

Hahn, E., Timpl, R., and Miller, E. J. (1975). Demonstration of a unique antigenic specificity for the collagen α1(II) chain from cartilaginous tissue. *Immunology* **28**, 561–568.

Haigh, J. J., Gerber, H.-P., Ferrara, N., and Wagner, E. F. (2000). Conditional inactivation of VEGF-A in areas of collagen 2a1 expression results in embryonic lethality in the heterozygous state. *Development* **127**, 1445–1453.

Haines, A. L. (1957). The effect of estrogen on cartilage and bone in castrate C3H mice. *Yale J. Biol. Med.* **30**, 121–136.

Haines, R. W. (1934). Epiphyseal growth in the branchial skeleton of fishes. *Q. J. Microsc. Sci.* **77**, 77–97.

Haines, R. W. (1937). The posterior end of Meckel's cartilage and related ossifications in bony fishes. *Q. J. Microsc. Sci.* **80**, 1–38.

Haines, R. W. (1939). The structure of the epiphyses in *Sphenodon* and the primitive form of secondary centre. *J. Anat.* **74**, 80–90.

Haines, R. W. (1940). Note on the independence of sesamoid and epiphyseal centres of ossification. *J. Anat.* **75**, 101–105.

Haines, R. W. (1941). Epiphyseal structure in lizards and marsupials. *J. Anat.* **75**, 282–294.

Haines, R. W. (1942a). The tetrapod knee joint. *J. Anat.* **76**, 270–301.

Haines, R. W. (1947). The development of joints. *J. Anat.* **81**, 33–35.

Haines, R. W. (1942b). The evolution of epiphyses and of endochondral bone. *Biol. Rev. Camb. Philos. Soc.* **17**, 267–292.

Haines, R. W. (1969). Epiphyses and sesamoids. In *Biology of the Reptilia* (C. Gans, ed.), Volume 1, pp. 81–116. Academic Press, New York, NY.

Haines, R. W. (1976). Destruction of hyaline cartilage in the sigmoid notch of the human ulna. *J. Anat.* **122**, 331–334.

Haines, R. W., and Mohiuddin, A. (1960). Supplementary cartilage in human epiphyseal plates. *J. Fac. Med. Baghdad* [N.S.] **2**, 16–22.

Haines, R. W., and Mohiuddin, A. (1968). Metaplastic bone. *J. Anat.* **103**, 527–538.

Hajek, A. S., and Solursh, M. (1975). Stimulation of growth and mucopolysaccharide synthesis by insulin treatment of chick embryo chondrocytes in cell culture. *Gen. Comp. Endocrinol.* **25**, 432–446.

Hale, J. E., and Wuthier, R. E. (1987). The mechanism of matrix vesicle formation: Studies on the composition of chondrocyte microvilli and on the effects of microfilament-perturbing agents on cellular vesiculation. *J. Biol. Chem.* **262**, 1916–1925.

Hale, L. J. (1956a). Mitotic activity during the early skeletal differentiation of the scleral bones in the chick. *Q. J. Microsc. Sci.* [N.S.] **97**, 333–353.

Hale, L. J. (1956b). Mesodermal cell death during the early development of the scleral bones in the chick. *Q. J. Microsc. Sci.* [N.S.] **97**, 355–368.

Hales, S. (1727). *Vegetable Staticks*. W. & J. Innys, London.

Hales, S. (1738). *Statistical Essays: Containing Vegetable Statistics, or, an Account of some Statistical Experiments on the Sap in Vegetables*. Volume 1, 2nd Edn. Innys, Woodward and Peele, London.

Halevy, O., Schindler, D., Hurwitz, S., and Pines, M. (1991). Epidermal growth factor receptor gene expression in avian epiphyseal growth-plate cartilage cells – effect of serum, parathyroid hormone and atrial natriuretic peptide. *Mol. Cell Endocrinol.* **75**, 229–236.

Hall, B. D. (1982). Syndromes and situations associated with congenital clavicular hypoplasia or agenesis. In *Skeletal Dysplasias* (C. J. Papadatos and C. S. Bartsocas, eds), pp. 279–288. Alan R. Liss Inc., New York, NY.

Hall, B. K. (1967a). The formation of adventitious cartilage by membrane bones under the influence of mechanical stimulation applied *in vitro*. *Life Sci.* **6**, 663–667.

Hall, B. K. (1967b). The distribution and fate of adventitious cartilage in the skull of the eastern rosella, *Platycercus eximius* (Aves: Psittaciformes). *Aust. J. Zool.* **15**, 685–698.

Hall, B. K. (1968a). Histochemical aspects of the differentiation of adventitious cartilage on the membrane bones of the embryo chick. *Histochemie* **16**, 206–220.

Hall, B. K. (1968b). A histochemical study of the condylar secondary cartilage of the mouse, *Mus musculus* (Mammalia: Rodentia). *Aust. J. Zool.* **16**, 807–815.

Hall, B. K. (1968c). *In vitro* studies on the mechanical evocation of adventitious cartilage in the chick. *J. Exp. Zool.* **168**, 283–206.

Hall, B. K. (1968d). Studies on the nature and evocation of the articular cartilage on the avian pterygoid. *Aust. J. Zool.* **16**, 815–821.

Hall, B. K. (1968e). The fate of adventitious and embryonic articular cartilage in the skull of the common fowl, *Gallus domesticus* (Aves: Phasianidae). *Aust. J. Zool.* **16**, 795–806.

Hall, B. K. (1969). Hypoxia and differentiation of cartilage and bone from common germinal cells *in vitro*. *Life Sci.* **8**, 553–558.

Hall, B. K. (1970a). Cellular differentiation in skeletal tissues. *Biol. Rev. Cambridge Philos. Soc.* **45**, 455–484.

Hall, B. K. (1970b). Differentiation of cartilage and bone from common germinal cells. I. The role of acid mucopolysaccharides and collagen. *J. Exp. Zool.* **173**, 383–394.

Hall, B. K. (1971a). Histogenesis and morphogenesis of bone. *Clin. Orthop. Rel. Res.* **74**, 249–268.

Hall, B. K. (1971b). Calcification of the cartilage formed on avian membrane bones. *Clin. Orthop. Rel. Res.* **78**, 182–190.

Hall, B. K. (1972a). Thallium-induced achondroplasia in the embryonic chick. *Devel. Biol.* **28**, 47–60.

Hall, B. K. (1972b). Immobilization and cartilage transformation into bone in the embryonic chick. *Anat. Rec.* **173**, 391–404.

Hall, B. K. (1972c). Differentiation of cartilage and bone from common germinal cells. II. β-aminopropionitrile-induced inhibition of chondrogenesis. *Calcif. Tissue Res.* **8**, 276–286.

Hall, B. K. (1972d). Skeletal defects in embryonic chicks induced by administration of β-aminopropionitrile. *Teratology* **5**, 81–88.

Hall, B. K. (1972e). Achondroplasia in the embryonic chick: its potentiation by cortisone acetate and alleviation by Vitamin C. *Can. J. Zool.* **50**, 1527–1536.

Hall, B. K. (1973a). Thyroxine and the development of the tibia in the embryonic chick. *Anat. Rec.* **176**, 49–64.

Hall, B. K. (1973b). Correlations between the concentrations of acid mucopolysaccharides and collagen in the tibia of the embryonic chick. *Can. J. Zool.* **51**, 771–776.

Hall, B. K. (1975a). Evolutionary consequences of skeletal differentiation. *Amer. Zool.* **15**, 329–350.

Hall, B. K. (1975b). The origin and fate of osteoclasts. *Anat. Rec.* **183**, 1–11.

Hall, B. K. (1975c). A simple, single-injection method for inducing long-term paralysis in embryonic chicks and preliminary observations on growth of the tibia. *Anat. Rec.* **181**, 767–778.

Hall, B. K. (1977a). Chondrogenesis of the somitic mesoderm. *Adv. Anat. Embryol. Cell Biol.* **53**, Part IV, 1–50.

Hall, B. K. (1977b). Thallium-induced achondroplasia in the chicken embryo and the concept of critical periods during development. *Teratology* **15**, 1–16.

Hall, B. K. (1978a). *Developmental and Cellular Skeletal Biology*. Academic Press, New York, NY.

Hall, B. K. (1978b). Initiation of osteogenesis by mandibular mesenchyme of the embryonic chick in response to mandibular and non-mandibular epithelia. *Archs Oral Biol.* **23**, 1157–1161.

Hall, B. K. (1978c). Use of the L-proline analog, L-azetidine-2-carboxylic acid (LACA) to analyze embryonic growth and determination and expression of the chondrogenic phenotype *in vivo* and *in vitro*. *Anat. Rec.* **190**, 243–256.

Hall, B. K. (1978d). Grafting of organs and tissues to the chorioallantoic membrane of the embryonic chick. *Tissue Cult. Assoc. Manual* **4**, 881–884.

Hall, B. K. (1979). Selective proliferation and accumulation of chondroprogenitor cells as the mode of action of biomechanical factors during secondary chondrogenesis. *Teratology* **20**, 81–92.

Hall, B. K. (1980a). Viability and proliferation of epithelia and the initiation of osteogenesis within mandibular ectomesenchyme in the embryonic chick. *J. Embryol. Exp. Morphol.* **56**, 71–89.

Hall, B. K. (1980b). Chondrogenesis and osteogenesis of cranial neural crest cells. In *Current Research Trends in Prenatal Craniofacial Development* (R. M. Pratt and R. L. Christiansen, eds), pp. 47–63. Elsevier, North-Holland Inc., New York, NY.

Hall, B. K. (1980c). Tissue interactions and the initiation of osteogenesis and chondrogenesis in the neural crest-derived mandibular skeleton of the embryonic mouse as seen in isolated murine tissues and in recombinations of murine and avian tissues. *J. Embryol. Exp. Morphol.* **58**, 251–264.

Hall, B. K. (1981a). The induction of neural crest-derived cartilage and bone by embryonic epithelia: an analysis of the mode of action of an epithelial–mesenchymal interaction. *J. Embryol. Exp. Morphol.* **64**, 305–320.

Hall, B. K. (1981b). Modulation of chondrocyte activity *in vitro* in response to ascorbic acid. *Acta Anat.* **109**, 51–63.

Hall, B. K. (1981c). Specificity in the differentiation and morphogenesis of neural crest-derived scleral ossicles and of epithelial scleral papillae in the eye of the embryonic chick. *J. Embryol. Exp. Morphol.* **66**, 175–190.

Hall, B. K. (1982a). Bone in the cartilaginous fishes. *Nature* **298**, 324.

Hall, B. K. (1982b). Distribution of osteo- and chondrogenic neural crest cells and of osteogenically inductive epithelia in mandibular arches of embryonic chicks. *J. Embryol. Exp. Morphol.* **68**, 127–136.

Hall, B. K. (1982c). How is mandibular growth controlled during development and evolution? *J. Craniofac. Genet. Devel. Biol.* **2**, 45–49.

Hall, B. K. (1982d). Review: Mandibular morphogenesis and craniofacial malformations. *J. Craniofac. Genet. Devel. Biol.* **2**, 309–322.

Hall, B. K. (1982e). The Role of Tissue Interactions in the Growth of Bone. In *Factors and Mechanisms Influencing Bone Growth* (A. D. Dixon and B. G. Sarnat, eds), pp. 205–216. Alan R. Liss Inc., New York, NY.

Hall, B. K. (ed.) (1983a). *Cartilage, Volumes 1–3*. Academic Press, New York, NY and London.

Hall, B. K. (1983b). Epithelial–mesenchymal interactions in cartilage and bone development. In *Epithelial–Mesenchymal Interactions in Development* (R. H. Sawyer and J. F. Fallon, eds), pp. 189–214. Praeger Press, New York, NY.

Hall, B. K. (1983c). Cell–tissue interactions – a rationale and resume. *J. Craniofac. Genet. Devel. Biol.* **3**, 75–82.

Hall, B. K. (1983d). Embryogenesis: Cell–tissue interactions. In *Skeletal Research – An Experimental Approach* (D. J. Simmons and A. S. Kunin, eds), Volume 2, pp. 53–87. Academic Press, New York, NY.

Hall, B. K. (1983e). Tissue interactions and chondrogenesis. In *Cartilage, Volume 2. Development, Differentiation and Growth* (B. K. Hall, ed.), pp. 187–222. Academic Press, New York, NY.

Hall, B. K. (1983f). Bioelectricity and Cartilage. In *Cartilage, Volume 3. Biomedical Aspects* (B. K. Hall, ed.), pp. 309–338. Academic Press, New York, NY and London.

Hall, B. K. (1983g). Epigenetic control in development and evolution. In *Development and Evolution*, British Society for Developmental Biology Symposium 6. (B. C. Goodwin, N. J. Holder, and C. C. Wylie, eds), pp. 353–379. Cambridge University Press, Cambridge.

Hall, B. K. (1984a). Matrices control the differentiation of cartilage and bone. In *Matrices and Differentiation*, Proceedings of a joint British Society for Developmental Biology and British Society for Cell Biology Symposium, Aberystwyth, Wales, September 14–16 (R. B. Kemp and J. R. Hinchliffe, eds), pp. 147–169. Alan R. Liss Inc., New York, NY.

Hall, B. K. (1984b). Developmental mechanisms underlying the formation of atavisms. *Biol. Rev. Camb. Philos. Soc.* **59**, 89–124.

Hall, B. K. (1984c). Developmental processes underlying heterochrony as an evolutionary mechanism. *Can. J. Zool.* **62**, 1–7.

Hall, B. K. (1984d). Genetic and epigenetic control of connective tissues in the craniofacial structures. In *Hereditary and Induced Connective Tissue Disorders of Craniofacial Structures* (K. S. Brown and C. F. Salinas, eds), pp. 1–17. Alan R. Liss Inc., New York, NY.

Hall, B. K. (1984e). Developmental processes underlying the evolution of cartilage and bone. In *The Structure, Development and Evolution of Reptiles*, Symposium No. 52 of the Zoological Society of London (M. W. J. Ferguson ed.), pp. 155–176. Academic Press, London.

Hall, B. K. (1985a). Research in the Development and Structure of the Skeleton since the Publication of Bones. An introduction to *Bones: A Study of the development and Structure of the Vertebrate Skeleton* by P. D. F. Murray, originally published

1936, reissued 1985. pp. xi–xlix. Cambridge University Press, Cambridge.

Hall, B. K. (1985b). Critical periods during development as assessed by thallium-induced inhibition of growth of embryonic chick tibiae *in vitro*. *Teratology* **31**, 353–361.

Hall, B. K. (1986a). The role of movement and tissue interactions in the development and growth of bone and secondary cartilage in the clavicle of the embryonic chick. *J. Embryol. Exp. Morphol.* **93**, 133–152.

Hall, B. K. (1986b). Initiation of chondrogenesis from somitic, limb and craniofacial mesenchyme: search for a common mechanism. In *Somites in Developing Embryos* (R. Bellairs, D. A. Ede, and J. W. Lash, eds), NATO Advanced Science Institutes Series A: Life Science, pp. 247–259. Plenum Publishing Corp., New York, NY.

Hall, B. K. (1987a). Sodium fluoride as an initiator of osteogenesis from embryonic mesenchyme. *Bone* **8**, 111–116.

Hall, B. K. (1987b). Development of the mandibular skeleton in the embryonic chick as evaluated using the DNA-inhibiting agent 5-fluoro-2'-deoxyuridine (FUDR). *J. Craniofac. Genet. Devel. Biol.* **7**, 145–159.

Hall, B. K. (1987c). Earliest evidence of cartilage and bone development in embryonic life. *Clin. Orthop. & Rel. Res.* **225**, 255–272.

Hall, B. K. (1987d). Tissue interactions in the development and evolution of the vertebrate head. In *Developmental and Evolutionary Aspects of the Neural Crest* (P. F. A. Maderson, ed.), pp. 215–259. Wiley, New York, NY.

Hall, B. K. (1988). The embryonic development of bone. *Amer. Scientist* **76**, 174–181.

Hall, B. K. (1989). Morphogenesis of the skeleton: epithelial or mesenchymal control? In *Trends in Vertebrate Morphology*. Proceedings of the 2nd International Symposium on Vertebrate Morphology, Vienna, 1986 (H. Splechtna and H. Hilgers, eds), Fortschritte der Zoologie Progress in Zoology, Volume 35, pp. 198–201. Gustav Fischer Verlag, Stuttgart.

Hall, B. K. (1990a). Heterochronic changes in vertebrate development. *Sem. Devel. Biol.* **1**, 237–243.

Hall, B. K. (1990b). Evolutionary issues in craniofacial biology. Proceedings of the Symposium on Advances in Craniofacial Developmental Biology and Clinical Implications, San Francisco, CA, April, 1989. *Cleft Palate J.* **27**, 95–100.

Hall, B. K. (1990c). Genetic and epigenetic control of vertebrate development. *Netherlands J. Zool.* **40**, 362–361.

Hall, B. K. (ed.) (1990–1994). *Bone: Volumes 1–9*. CRC Press, Boca Raton, FL.

Hall, B. K. (1991a). Bone: embryonic development. In *Encyclopedia of Human Biology* (R. Dulbecco, ed.), Volume 1, pp. 781–791. Academic Press, Orlando, FL.

Hall, B. K. (1991b). Cellular interactions during cartilage and bone development. Proceedings of the NIDR-NIEH sponsored Conference on Research Advances in Prenatal Craniofacial Development. *J. Craniofac. Genet. Devel. Biol.* **11**, 238–250.

Hall, B. K. (1991c). Closing address: What is bone growth? In *Fundamentals of Bone Growth: Methodology and Applications* (A. D. Dixon, B. G. Sarnat, and D. A. N. Hoyte, eds). Proceedings of the International Conference held at the University of California Center for the Health Sciences, Los Angeles, California, January 2–5, 1990, pp. 605–612. CRC Press, Inc., Boca Raton, FL.

Hall, B. K. (1991d). The evolution of connective and skeletal tissues. In *Developmental Patterning of the Vertebrate Limb*

(J. R. Hinchliffe, J. Hurlé, and D. Summerbell, eds), NATO Advanced Science Institutes Series A: Life Science, pp. 303–311. Plenum Publishing Corp., New York, NY.

Hall, B. K. (1992). Historical overview of studies on bone growth and repair. In *Bone, Volume 6: Bone Growth – A* (B. K. Hall, ed.), pp. 1–19. CRC Press, Boca Raton, FL.

Hall, B. K. (1994a). *Homology: The Hierarchical Basis of Comparative Biology*. Academic Press, San Diego, CA (paperback issued 2001).

Hall, B. K. (1994b). Embryonic bone formation with special reference to epithelial–mesenchymal interactions and growth factors. In *Bone, Volume 8: Mechanisms of Bone Development and Growth* (B. K. Hall, ed.), pp. 137–192. CRC Press, Boca Raton, FL.

Hall, B. K. (1994c). Biology and mechanisms of tissue interactions in developing systems. In *Developmental Biology and Cancer* (G. M. Hodges and C. Rowlett, eds), pp. 161–185. CRC Press Inc., Boca Raton, FL.

Hall, B. K. (1995a). Evolutionary aspects of craniofacial structures and development. *Cleft Palate Craniofac. J.* **32**, 520–522.

Hall, B. K. (1995b). Embryogenesis. In *1996 McGraw-Hill Yearbook of Science and Technology*, pp. 100–101. McGraw-Hill Inc., New York, NY.

Hall, B. K. (1995c). Homology and embryonic development. *Evol. Biol.* **28**, 1–37.

Hall, B. K. (1995d). Atavisms and atavistic mutations: evolutionary conservation of genetic and developmental information. *Nature Genetics* **10**, 126–127.

Hall, B. K. (1995e). Evolutionary Developmental Biology. In *1996 McGraw-Hill Yearbook of Science and Technology*, pp. 110–112. McGraw-Hill Inc., New York, NY.

Hall, B. K. (1996). Baupläne, phylotypic stages and constraint: Why there are so few types of animals. *Evol. Biol.* **29**, 215–261.

Hall, B. K. (1997a). Bone: Embryonic development. In *Encyclopedia of Human Biology* (R. Dulbecco, ed.), 2nd Edn, Volume 2, pp. 105–114. Academic Press.

Hall, B. K. (1997b). Phylotypic stage or phantom: is there a highly conserved embryonic stage in vertebrates? *Trends Ecol. Evol.* **12**, 461–463.

Hall, B. K. (1997c). Germ layers and the germ-layer theory revisited: Primary and secondary germ layers, neural crest as a fourth germ layer, homology, demise of the germ-layer theory. *Evol. Biol.* **30**, 121–186.

Hall, B. K. (1998a). *Evolutionary Developmental Biology*. Kluwer Academic Publishers, Dordrecht.

Hall, B. K. (1998b). The bone. In *Cellular and Molecular Basis of Regeneration: From Invertebrates to Humans* (P. Ferretti and J. Géraudie, eds), pp. 289–307. John Wiley & Sons Ltd., London.

Hall, B. K. (1998c). Bibliography on supportive and connective tissues, *Encyclopaedia Britannica*, Britannica Center, Chicago, IL.

Hall, B. K. (1999a). *The Neural Crest in Development and Evolution*. Springer-Verlag, New York, NY.

Hall, B. K. (1999b). The paradoxical platypus. *BioScience* **49**, 211–218.

Hall, B. K. (2000a). The evolution of the neural crest in vertebrates. In *Regulatory Processes in Development*, Wenner-Gren International Series Volume 76 (L. Olsson and C.-O. Jacobson, eds), pp. 101–113. The Portland Press, London.

Hall, B. K. (2000b). The neural crest as a fourth germ layer and vertebrates as quadroblastic not triploblastic. *Evol. Devel.* **2**, 1–3.

Hall, B. K. (2000c). A role for epithelial–mesenchymal interactions in tail growth/morphogenesis and chondrogenesis in embryonic mice. *Cells Tissues Organs* **166**, 6–14.

Hall, B. K. (2000d) Epithelial–mesenchymal interactions. In *Methods in Molecular Biology, Volume 125: Developmental Biology Protocols*, Volume 3 (R. S. Tuan and C. W. Lo, eds), pp. 235–243. Humana Press Inc., Totowa, NJ.

Hall, B. K. (2001a). Development of the clavicles in birds and mammals. *J. Exp. Zool.* **289**, 153–161.

Hall, B. K. (2001b). Introduction. In *The Molecular Basis of Skeletogenesis* (G. Cardew and J. Goode, eds), Novartis Foundation Symposium 232, pp. 1–5. John Wiley & Sons, Chichester.

Hall, B. K. (2001c). *Foreword to Beyond Heterochrony: The Evolution of* Development (M. L. Zelditch, ed.), vii-ix, Wiley-Liss, New York, NY.

Hall, B. K. (2001d). John Samuel Budgett (1872–1904): In pursuit of *Polypterus. BioScience* **51**, 399–407.

Hall, B. K. (2002a). Palaeontology and evolutionary developmental biology: A science of the 19th and 21st centuries. *Palaeontology* **45**, 647–669.

Hall, B. K. (2002b). Evolutionary developmental biology: Where embryos and fossils meet. In *Human Evolution Through Developmental Change* (N. Minugh-Purves and K. J. McNamara, eds), pp. 7–27. Johns Hopkins University Press, Baltimore, MD.

Hall, B. K. (2002c). Atavisms. In *Encyclopedia of Evolution* (M. Pagel, ed. in chief), Volume 1, pp. 86–87. Oxford University Press, New York, NY.

Hall, B. K. (2002d). Phylotypic stages. In *Encyclopedia of Evolution* (M. Pagel, ed. in chief), Volume 2, 904–905. Oxford University Press, New York, NY.

Hall, B. K. (2003a). Descent with modification: the unity underlying homology and homoplasy as seen through an analysis of development and evolution. *Biol. Rev. Camb. Philos. Soc.* **78**, 409–433.

Hall, B. K. (2003b). Developmental and cellular origins of the amphibian skeleton. In *Amphibian Biology, Volume 5, Osteology* (H. Heatwole and M. Davies, eds), pp. 1551–1597, Surrey Beatty & Sons, Chipping Norton, NSW.

Hall, B. K. (2003c). Evo-devo: evolutionary developmental mechanisms. *Int. J. Devel. Biol.* **47**, 491–495.

Hall, B. K. (2003d). Unlocking the black box between genotype and phenotype: Cell condensations as morphogenetic (modular) units. *Biol. & Philos.* **18**, 219–247. (Special issue devoted to evolutionary developmental biology.)

Hall, B. K. (2004a). In search of evolutionary developmental mechanisms: The 30-year gap between 1944 and 1974. [Special Issue, Papers of the 2001 Kowalevsky Medal Winner Symposium] *J. Exp. Zool. (Mol. Dev. Evol)* **302B**, 5–18.

Hall, B. K. (2004b). Evolution as the control of development by ecology. In *Environment, Development and Evolution: Toward a Synthesis* (B. K. Hall, R. D. Pearson, and G. B. Müller, eds), pp. ix-xxiii. The MIT Press, Cambridge, MA.

Hall, B. K. (2004c). Mutatis mutandis. *Trends Ecol. Evol.* **19**, 356–357.

Hall, B. K. (2004d). Arthur W. Ham, F.R.S.C. (1902–1992). *Proc. Royal Soc. Canada.* (in press).

Hall, B. K. (2005a). A Consideration of the Neural Crest and its Skeletal Derivatives in the Context of Novelty/Innovation. *Integ. Comp. Biol.* (in press).

Hall, B. K. (2005b). 'Spandrels': metaphor for morphological residue or entrée into evolutionary developmental mechanisms?

In *The Spandrels of San Marco 25 years Later* (D. Walsh, ed.), Oxford University Press, Oxford (in press).

Hall, B. K. (ed.) (2005c). *Fins into Limbs*. The University of Chicago Press, Chicago, IL.

Hall, B. K., and Coffin-Collins, P. A. (1990). Reciprocal interactions between epithelium, mesenchyme and epidermal growth factor (EGF) in the regulation of mitotic activity of mandibular epithelium and mesenchyme in the embryonic chick. *J. Craniofac. Genet. Devel. Biol.* **10**, 241–261.

Hall, B. K., and Ekanayake, S. (1991). Effects of growth factors on the differentiation of neural crest cells and neural crest cell-derivatives. *Int. J. Devel. Biol.* **35**, 367–386.

Hall, B. K., and Hanken, J. (1985a). Repair of fractured lower jaws in the spotted salamander: Do amphibians form secondary cartilage? *J. Exp. Zool.* **233**, 359–368.

Hall, B. K., and Hanken, J. (1985b). Foreword to reissue of *The Development of the Vertebrate Skull* by Sir G. R. de Beer originally published 1937, reissued 1985, pp. vii–xxviii. University of Chicago Press, Chicago, IL.

Hall, B. K., and Hanken, J. (1993). Bibliography of skull development: The last fifty years. In *The Skull. Volume 1. Development* (J. Hanken and B. K. Hall, eds), pp. 378–577. The University of Chicago Press, Chicago, IL.

Hall, B. K., and Herring, S. W. (1990). Paralysis and growth of the musculoskeletal system in the embryonic chick. *J. Morphol.* **206**, 45–56.

Hall, B. K., and Hörstadius, S. (1988). *The Neural Crest*. Including a facsimile reprint of *The Neural Crest* by Sven Hörsdtadius. Oxford University Press, Oxford.

Hall, B. K., and Jacobson, H. N. (1975). The repair of fractured membrane bones in the newly hatched chick. *Anat. Rec.* **181**, 55–70.

Hall, B. K., and MacSween, M. C. (1984). An SEM analysis of the epithelial–mesenchymal interface in the mandible of the embryonic chick. *J. Craniofac. Genet. Devel. Biol.* **4**, 59–76.

Hall, B. K., and Miyake, T. (1992). The membranous skeleton: the role of cell condensations in vertebrate skeletogenesis. *Anat. Embryol.* **186**, 107–124.

Hall, B. K., and Miyake, T. (1995). Divide, accumulate, differentiate: Cell condensation in skeletal development revisited. *Int. J. Devel. Biol.* **39**, 881–893.

Hall, B. K., and Miyake, T. (1997). How do embryos tell time. In *Evolution through Heterochrony* (K. J. McNamara, ed.), pp. 1–20. John Wiley & Sons, Chichester.

Hall, B. K., and Miyake, T. (1999). Craniofacial development in avian and rodent embryos. In *Methods in Molecular Biology, Volume 125: Developmental Biology Protocols*, (R. S. Tuan and C. W. Lo, eds) Volume 1, pp.127–137. Humana Press Inc., Totowa, NJ.

Hall, B. K., and Miyake, T. (2000). All for one and one for all: Condensations and the initiation of skeletal development. *BioEssays* **22**, 138–147.

Hall, B. K., and Newman, S. (eds) (1991). *Cartilage: Molecular Aspects*. CRC Press, Boca Raton, FL.

Hall, B. K., and Olson, W. M. (eds) (2003). *Keywords and Concepts in Evolutionary Developmental Biology*. Harvard University Press, Cambridge, MA.

Hall, B. K., Pearson, R., and Müller, G. B. (eds) (2004). *Environment, Development, and Evolution: Toward a Synthesis*. The MIT Press, Cambridge, MA.

Hall, B. K., and Shorey, C. D. (1968). Ultrastructural aspects of cartilage and membrane bone differentiation from common germinal cells. *Aust. J. Zool.* **16**, 821–840.

Hall, B. K., and Tremaine, R. (1979). Ability of neural crest cells from the embryonic chick to differentiate into cartilage before their migration away from the neural tube. *Anat. Rec.* **194**, 469–476.

Hall, B. K., and Van Exan, R. J. (1982). Induction of bone by epithelial cell products. *J. Embryol. Exp. Morphol.* **69**, 37–46.

Hall, B. K., Van Exan, R. J., and Brunt, S. L. (1983). Retention of epithelial basal lamina on isolated mandibular mesenchyme permits the induction of bone. *J. Craniofac. Genet. Devel. Biol.* **3**, 253–267.

Hall, B. K., and Wake, M. H. (1999). *The Origin and Evolution of Larval Forms*. Academic Press, San Diego, CA.

Hall, K. (1947). The effects of pregnancy and relaxin on the histology of the pubic symphysis in the mouse. *J. Endocrinol.* **5**, 174–182.

Hall-Craggs, E. C. B. (1969). Influence of epiphyses on the regulation of bone growth. *Nature* **21**, 1245.

Haller, A. von (1763). *Experimentorum di Ossium Formatione. Opera Minora*. Volume 2. Francisci Graset, Lausanne.

Haller, A. C., and Zimny, M. L. (1978). Effects of hibernation on interradicular alveolar bone. *J. Dent. Res.* **56**, 1552–1557.

Hallgrímsson, B. (1998). Fluctuating asymmetry in the mammalian skeleton. Evolutionary and developmental implications. *Evol. Biol.* **30**, 187–251.

Hallgrímsson, B. (1999). Ontogenetic patterning of skeletal fluctuating asymmetry in Rhesus monkeys and humans: Evolutionary and developmental implication. *Int. J. Primatol.* **20**, 121–151.

Hallgrímsson, B., and Hall, B. K. (eds) (2005). *Variation: A Hierarchical Examination of a Central Concept in Biology*. Elsevier, New York, NY.

Hallgrímsson, B., and Maiorana, V. (2000). Variability and size in mammals and birds. *Biol. J. Linn. Soc. Lond.* **70**, 571–595.

Hallgrímsson, B., Miyake, T., Willmore, K., and Hall, B. K. (2003). Embryological origins of developmental stability: Size, shape, and fluctuating asymmetry in prenatal random bred mice. *J. Exp. Zool. (Mol. Devel. Evol.)* **296B**, 1–18.

Hallgrímsson, B., Willmore, K., and Hall, B. K. (2002). Canalization, developmental stability, and morphological integration in primate limbs. *Amer. J. Phys. Anthropol. (Yearbook)* **S45**, 131–158.

Hallmann, R., Feinberg, R. N., Latker, C. H., Sasse, J., and Risau, W. (1987). Regression of blood vessels precedes cartilage differentiation during chick limb development. *Differentiation* **34**, 98–105.

Halstead, L. B. (1969a). *The Pattern of Vertebrate Evolution*. Oliver & Boyd, Edinburgh.

Halstead, L. B. (1969b). The origin and early evolution of calcified tissues in the vertebrates. *Proc. Malacol. Soc. London* **38**, 552–553.

Halstead, L. B. (1969c). Calcified tissues in the earliest vertebrates. *Calcif. Tissue Res.* **3**, 107–124.

Halstead, L. B. (1973). The heterostracan fishes. *Biol. Rev. Cambridge Philos. Soc.* **48**, 279–332.

Halstead Tarlo, B. J., and Halstead Tarlo, L. B. (1965). The origin of teeth. *Discovery* **26**, 20–26.

Halstead Tarlo, L. B., and Mercer, J. R. (1966). Decalcified fossil dentine. *J. Royal Microsc. Soc.* **86**, 137–140.

Haluska, F., and Alberch, P. (1983). The cranial development of *Elaphe obsoletai* (Ophidia, Colubridae). *J. Morphol.* **178**, 37–55.

Ham, A. W. (1930). A histological study of the early phases of bone repair. *J. Bone Joint Surg. Am. Vol.* **12**, 827–844.

Ham, A. W. (1974). *Histology*, 7th Edn. J. B. Lippincott Co., Philadelphia, PA.

Ham, A. W., and Cormack, D. H. (1979). *Histophysiology of Cartilage, Bone and Joints*. J. B. Lippincott Co., Philadelphia, PA.

Ham, A. W., and Harris, W. R. (1971). Repair and transplantation of bone. In *The Biochemistry and Physiology of Bone* (G. H. Bourne, ed.), 2nd Edn, Volume 3, pp. 337–399. Academic Press, New York, NY.

Ham, R. G., Murray, L. W., and Sattler, G. L. (1970). Beneficial effects of embryo extract on cultured rabbit cartilage cells. *J. Cell. Physiol.* **75**, 353–360.

Ham, R. G., and Sattler, G. L. (1968). Clonal growth of differentiated rabbit cartilage cells. *J. Cell. Physiol.* **72**, 109–114.

Hamburger, V. (1929). Die Entwicklung experimentell erzeugter nervenloser und schwach innervierter Extremitäten von Anuren. *W. Roux. Arch. EntMech. Org.* **114**, 272–363.

Hamburger, V. (1941). Transplantation of limb primordia of homozygous and heterozygous chondrodystrophic (Creeper) chick embryos. *Physiol. Zool.* **14**, 355–364.

Hamburger, V. (1960). *A Manual of Experimental Embryology*. Revised Edn. The University of Chicago Press, Chicago, IL.

Hamburger, V., and Hamilton, H. L. (1951). A series of normal stages in development of the chick embryo. *J. Morphol.* **88**, 49–92.

Hamburger, V., and Waugh, M. (1940). The primary development of the skeleton in nerveless and poorly innervated limb transplants of the chick embryo. *Physiol. Zool.* **13**, 367–382.

Hamilton, H. L. (1952). Sensitive periods during development. *Ann. NY. Acad. Sci.*, **55**, 177–187.

Hamilton, H. L. (1965). *Lillie's Development of the Chick. An Introduction to Embryology*, Third Edn. Holt, Rinehart and Winston, New York, NY.

Hammar, S. P., and Mottet, N. K. (1971). Tetrazolium salt and electron-microscopic studies of cellular degeneration and necrosis in the interdigital areas of the developing chick limb. *J. Cell Sci.* **8**, 229–252.

Hammond, W. S., and Yntema, C. L. (1949). Depletions in the thoraco-lumbar sympathetic system following removal of neural crest in the chick. *J. Comp. Neurol.* **86**, 237–266.

Hammond, W. S., and Yntema, C. L. (1953). Deficiencies in visceral skeleton of the chick after removal of cranial neural crest. *Anat. Rec.* **115**, 393–394.

Hammond, W. S., and Yntema, C. L. (1964). Depletions of pharyngeal arch cartilages following extirpation of cranial neural crest in chick embryos. *Acta Anat.* **56**, 21–34.

Hampé, A. (1958). Le développement du péroné dans les expériences sur la régulation des déficiencies et des excédents dans la patte du poulet. *J. Embryol. Exp. Morphol.* **6**, 215–222.

Hampé, A. (1959). Contribution à l'étude du développement et de la régulation des déficiencies et des excédents dans la patte de l'embryon de poulet. *Arch. Anat. Microsc. Morphol. Exp.* **48**, 345–478.

Hampé, A. (1960). La compétition entre les éléments osseux du zygopode de poulet. *J. Embryol. Exp. Morphol.* **8**, 241–245.

Hamrick, M. W. (1999). A chondral modeling theory revisited. *J. Theoret. Biol.* **201**, 201–208.

Hamrick, M. W. (2001). Development and evolution of the mammalian limb: adaptive diversification of nails, hooves, and claws. *Evol. & Devel.* **3**, 355–363.

Han, M., An, J.-Y., and Kim, W.-S. (2001). Expression patterns of *Fgf-8* during development and limb regeneration of the axolotl. *Devel. Dynam.* **220**, 40–48.

Han M., Yang, X., Farrington, J. E., and Muneoka, K. (2003). Digit regeneration is regulated by *Msx1* and BMP4 in fetal mice. *Development* **130**, 5123–5132.

Hanaoka, H. (1976). Fate of hypertrophic chondrocytes of epiphyseal plate – electron microscopic study. *J. Bone Joint Surg. Am. Vol.* **58**, 226–229.

Hanaoka, H. (1977). On the hypothesis of modulation of osteoclasts to osteoblasts on the endosteal bone surface – a critical review. *J. Jap. Orthop. Assoc.* **51**, 613–616.

Hanaoka, H. (1980). The origin of the osteoclast. *Clin. Orthop. Rel. Res.* **145**, 252–263.

Hanaoka, H., Yabe, H., and Bun, H. (1989). The origin of the osteoclast. *Clin. Orthop. Rel. Res.* **239**, 286–298.

Hancox, N. M. (1947). The survival of transplanted embryo bone grafted to chorioallantoic membrane and subsequent osteogenesis. *J. Physiol.* **106**, 279–285.

Hancox, N. M. (1948). Osteogenesis around multiple fragments of chick embryo bone grafted to developing chick chorioallantois. *J. Physiol.* **107**, 513–517.

Hancox, N. M. (1949). The osteoclast. *Biol. Rev. Cambridge Philos. Soc.* **24**, 448–471.

Hancox, N. M. (1956). The osteoclast. In *The Biochemistry and Physiology of Bone* (G. H. Bourne, ed.), pp. 213–250. Academic Press, New York, NY.

Hancox, N. M. (1965). The osteoclast. In *Cells and Tissues in Culture. Methods, Biology and Physiology* (E. N. Willmer, ed.), Volume 2, pp. 261–272. Academic Press, New York, NY.

Hancox, N. M. (1972a). *Biology of Bone*. Cambridge University Press, Cambridge.

Hancox, N. M. (1972b). The osteoclast. In *The Biochemistry and Physiology of Bone* (G. H. Bourne, ed.), 2nd Edn, Volume 1, pp. 45–69. Academic Press, New York, NY.

Hancox, N. M., and Boothroyd, B. (1965). Electron microscopy of the early stages of osteogenesis. *Clin. Orthop. Rel. Res.* **40**, 153–161.

Hancox, N. M., and Wlodarski, K. H. (1972). The role of host site in bone induction by transplanted xenogenic epithelial cells. *Calcif. Tissue Res.* **8**, 258–261.

Handrigan, G. R. (2003). *Concordia discors*: duality in the origin of the vertebrate tail. *J. Anat.* **202**, 255–267.

Hanken, J. (1982). Appendicular skeletal morphology in minute salamanders, Genus *Thorius* (Amphibia: Plethodontidae): Growth, regulation, adult size, determination and natural variation. *J. Morphol.* **174**, 57–77.

Hanken, J. (1983a). Miniaturization and its effect on cranial morphology in Plethodontid salamanders, genus *Thorius* (Amphibia, Plethodontidae). II. The fate of the brain and sense organs and their role in skull morphogenesis and evolution. *J. Morphol.* **177**, 255–268.

Hanken, J. (1983b). High incidence of limb skeletal variants in a peripheral population of the red-backed salamander, *Plethodon cinereus* (Amphibia, Plethodontidae), from Nova Scotia. *Can. J. Zool.* **61**, 1925–1931.

Hanken, J. (1984). Miniaturization and its effect on cranial morphology in Plethodontid salamanders, genus *Thorius* (Amphibia, Plethodontidae). I. Osteological variation. *Biol. J. Linn. Soc.* **23**, 55–75.

Hanken, J. (1985). Morphological novelty in the limb skeleton accompanies miniaturization in salamanders. *Science* **229**, 871–874.

Hanken, J. (1993). Adaptation of bone growth to miniaturization. In *Bone, Volume 7: Bone Growth – B.* (B. K. Hall, ed.), pp. 105–131. CRC Press, Boca Raton, FL.

Hanken, J., Carl, T. F., Richardson, M. K., Olsson, L., Schlosser, G., Osabutey, C. K., and Klymkowsky, M. W. (2001). Limb development in the 'nonmodel' vertebrate, the direct-developing frog *Eleutherodactylus coqui*. *J. Exp. Zool. (Mol. Devel. Evol.)* **291**, 375–388.

Hanken, J., and Dinsmore, C. E. (1986). Geographic variation in the limb skeleton of the red-backed salamander, *Plethodon cinereus*. *J. Herpetol.* **20**, 97–101.

Hanken, J., and Hall, B. K. (1983). Evolution of the vertebrate skeleton. *Natural History* **92**, 28–39.

Hanken, J., and Hall, B. K. (1984). Variation and timing of the cranial ossification sequence of the Oriental fire-bellied toad, *Bombina orientalis* (Amphibia, Discoglossidae). *J. Morphol.* **182**, 245–255.

Hanken, J., and Hall, B. K. (1988a). Skull development during anuran metamorphosis. I. Early development of the first three bones to form – the exoccipital, the parasphenoid and the frontoparietal. *J. Morphol.* **195**, 247–256.

Hanken, J., and Hall, B. K. (1988b). Skull development during anuran metamorphosis. II. Role of thyroid hormone in osteogenesis. *Anat. Embryol.* **178**, 219–227.

Hanken, J., and Hall, B. K. (1993). General introduction. In *The Skull. Volume 1. Development* (J. Hanken and B. K. Hall, eds), pp. ix–xiii. University of Chicago Press, Chicago, IL.

Hanken, J., Klymkowsky, M. W., Summers, C. H., Seufert, D. W., and Ingebrigtsen, N. (1992). Cranial ontogeny in the direct-developing frog, *Eleutherodactylus coqui* (Anura: Leptodactylidae), analyzed using whole-mount immunohistochemistry. *J. Morphol.* **211**, 95–118.

Hanken, J., and Summers, C. H. (1988). Skull development during anuran metamorphosis. III. Role of thyroid hormone in chondrogenesis. *J. Exp. Zool.* **246**, 156–170.

Hanken, J., Summers, C. H., and Hall, B. K. (1989). Morphological integration in the cranium during anuran metamorphosis. *Experientia* **45**, 872–875.

Hanken, J., and Thorogood, P. (1993). Evolution and development of the vertebrate skull: The role of pattern formation. *Trends Ecol. Evol.* **8**, 9–14.

Hanken, J., and Wake, D. B. (1998). Biology of tiny animals: Systematics of the minute salamanders (*Thorius*: Plethodontidae) from Veracruz and Puebla, Mexico, with descriptions of five new species. *Copeia* **1998**, 312–345.

Hansson, L. I. (1967). Daily growth in length of diaphysis measured by oxytetracycline in rabbits normally and after medullary plugging. *Acta Orthop. Scand.* Suppl. **101**, 1–199.

Hansson, L. I., Stenström, A., and Thorngren, K.-G. (1975). Effects of venous stasis on longitudinal bone growth in the rabbit. *Acta Orthop. Scand.* **46**, 177–184.

Harada, S-i., and Rodan, G. A. (2003). Control of osteoblast function and regulation of bone mass. *Nature* **423**, 349–355.

Harada, Y., and Ishizeki, K. (1998). Evidence for transformation of chondrocytes and site-specific resorption during the degradation of Meckel's cartilage. *Anat. Embryol.* **197**, 439–450.

Haraguchi, R., Suzuki, K., Murakami, R., Sakai, M., Kamikawa, M., Kengaku, M., Sekine, K., Kawano, H., Kato, S., Ueno, N., and Yamada, G. (2000). Molecular analysis of external genitalia formation: The role of gibroblast growth factor (Fgf) genes during genital tubercle formation. *Development* **127**, 2471–2479.

Hardesty, R. A., and Marsh, J. L. (1990). Craniofacial onlay bone grafting – a prospective evaluation of graft morphology, orientation and embryonic origin. *Plast. Reconst. Surg.* **85**, 5–14.

Hardingham, T. E., Fitton Jackson, S., and Muir, H. (1972). Replacement of proteoglycans in embryonic chick cartilage in organ culture after treatment with testicular hyaluronidase. *Biochem. J.* **129**, 101–112.

Hardisty, M. W. (1979). *Biology of the Cyclostomes*. Chapman & Hall, London.

Hardisty, M. W. (1981). The skeleton. In *The Biology of Lampreys* (M. W. Hardisty and I. C. Potter, eds), Volume 3, pp. 333–376. Academic Press, London.

Har-El, R., and Tanzer, M. L. (1993). Extracellular matrix 3: Evolution of the extracellular matrix in invertebrates. *FASEB J.* **7**, 1115–1123.

Harkey, M. E., and Crelin, E. S. (1963a). *In vitro* study of pubic symphyseal cartilage from adult virgin female mice. *Anat. Rec.* **145**, 322–323.

Harkey, M. E., and Crelin, E. S. (1963b). Hormonal response of the pubic symphysis in adult mice gonadectomized at different times before puberty. *Anat. Rec.* **145**, 323.

Harkness, E. M. (1976). Growth of transplants of the rat cranial base. *J. Dent. Res.* **55**, 1134.

Harkness, E. M., and Trotter, W. D. (1978). Growth of transplants of rat humerus following circumferential periosteal division of the periosteum. *J. Anat.* **126**, 275–289.

Harkness, E. M., and Trotter, W. D. (1980). Growth spurt in rat cranial bases transplanted into adult hosts. *J. Anat.* **131**, 39–56.

Harkness, E. M., and Trotter, W. D. (1982). The influence of host age on the growth of transplanted rat cranial bones and humeri. *J. Anat.* **135**, 353–369.

Harrigan, T. P., and Hamilton, J. J. (1993). Bone strain sensation via transmembrane potential changes in surface osteoblasts: loading rate and microstructural implications. *J. Biomech.* **26**, 183–200.

Harris, E. D., Jr., and McCroskery, P. A. (1974). The influence of temperature and fibril stability on degradation of cartilage collagen by rheumatic synovial collagenase. *New Engl. J. Med.* **290**, 1–6.

Harris, H. A. (1933). *Bone growth in Health and Disease. The Biological Principles Underlying the Clinical, Radiological, and Histological Diagnosos of Perversions of growth and Disease in the Skeleton*. Oxford University Press, Oxford.

Harris, M. J., and Juriloff, D. M. (1989). Test of the isoallele hypothesis at the mouse first arch (*far*) locus. *J. Hered.* **80**, 127–131.

Harris, S. E., Feng, J.-Q., Harris, M. A., Ghosh-Choudhury, N., Dallas, M. R., Wozney, J., and Mundy, G. R. (1995). Recombinant bone morphogenetic protein 2 accelerates bone cell differentiation and stimulates BMP-2 mRNA expression and BMP-2 promoter activity in primary fetal rat calvarial osteoblast cultures. *Mol. Cell Differ.* **3**, 137–155.

Harrison, D. F. N. (1984). Osseous and fibro-osseous conditions affecting the craniofacial bones. *Ann. Oto. Rhinol. Laryngol.* **93**, 199–203.

Harrison, E. T., Luyten, F. P., and Reddi, A. H. (1991). Osteogenin promotes reexpression of cartilage phenotype by dedifferentiated articular chondrocytes in serum-free medium. *Exp. Cell Res.* **192**, 340–345.

Harrison, R. G. (1918). Experiments on the development of the forelimb in *Amblystoma*, a self-differentiating, equipotential system. *J. Exp. Zool.* **25**, 413–462.

Harrison, R. G. (1935a) Heteroplastic grafting in embryology. *The Harvey Lectures* **29**, 116–157.

Harrison, R. G. (1935b) On the origin and development of the nervous system studied by the methods of experimental embryology. The Croonian Lecture. *Proc. R. Soc. Lond. B.* **118**, 155–196.

Hartmann, C., and Tabin, C. J. (2000). Dual roles of Wnt signaling during chondrogenesis in the chicken limb. *Development* **127**, 3141–3159.

Hartmann, C., and Tabin, C. J. (2001). Wnt-14 plays a pivotal role in inducing synovial joint formation in the developing appendicular skeleton. *Cell* **104**, 341–351.

Hartwig, H. (1968). Heterotopic frontal horn formation on deer experimentally caused by shifting of the periost. *Z. Saeugetierk.* **33**, 246–248.

Hartwig, H. (1972). 'Fegeverhalten' bei einem gehörnlosen Rehbock. *Z. Jagdwiss.* **18**, 166–168.

Hartwig, H., and Schrudde, J. (1974). Experimentelle untersuchungen zur bildung der primären stirnauswüchse reim reh (*Capreolus capreolus* L.). *Z. Jagdwiss.* **20**, 1–13.

Harvey, A. K., Hrubey, P. S., and Chandrasekhar, S. (1991). Transforming growth factor-β inhibition of interleukin-I activity involves down-regulation of interleukin-I receptors on chondrocytes. *Exp. Cell Res.* **195**, 376–385.

Harvey, P. H., and Clutton-Brock, T. H. (1985). Life history variation in primates. *Evolution* **39**, 559–581.

Hascall, V. C. (1977). Interaction of cartilage proteoglycans with hyaluronic acid. *J. Supramol. Struct.* **1**, 101–120.

Hasegawa, S., Sato, S., Saito, S., Suzukim Y., and Brunette, D. M. (1985). Mechanical stretching increases the number of cultured bone cells synthesizing DNA and alters their pattern of protein synthesis. *Calcif. Tissue Int.* **37**, 431–436.

Hasegawa, T., Oguchi, H., Mizuno, M., and Kuboki, Y. (1994). The effect of the extracellular matrix on differentiation of bone marrow stromal cells to osteoblasts. *Jap. J. Oral Biol.* **36**, 383–394.

Hashimoto, J., Joshikawa, H., Takaoka, K., Shimizu, N., Masuhara, K., Tsuda, T., Miyamoto, S., and Ono, K. (1989). Inhibitory effects of tumor necrosis factor alpha on fracture healing in rats. *Bone* **10**, 453–457.

Hashimoto, Y., Shimokobe, H., Mizuno, M., and Kuboki, Y. (1995). Effect of BMP on the differentiation of bone marrow stromal cells into osteoblasts in a collagen-gel culture. *Jap. J. Oral Biol.* **37**, 130–143.

Hassell, J. R., Greenberg, J. H., and Johnston, M. C. (1977). Inhibition of cranial neural crest cell development by vitamin A in cultured chick embryo. *J. Embryol. Exp. Morphol.* **39**, 267–271.

Hassell, J. R., Pennypacker, J. P., Kleinman, H. K., Pratt, R. M., and Yamada, K. M. (1979). Enhanced cellular fibronectin accumulation in chondrocytes treated with vitamin A. *Cell* **17**, 821–826.

Hata, R.-I., Hori, N., Nagai, Y., Tanaka, S., Kondo, M., Hiramatsu, M., Utsumi, N., and Kumegawa, M. (1984). Selective inhibition of type I collagen synthesis in osteoblastic cells by epidermal growth factor. *Endocrinology* **115**, 867–876.

Hathaway, R. R. (1997). Direct current stimulation of the coronal suture in rabbits. *J. Craniofac. Surg.* **8**, 360–366.

Hattersley, G., and Chambers, T. J. (1989). Generation of osteoclastic function in mouse bone marrow cultures: multinuclearity and tartrate-resistant acid phosphatase are unreliable markers for osteoclastic differentiation. *Endocrinology* **124**, 1689–1696.

Hattori, T., and Ide, H. (1984). Limb bud chondrogenesis in cell culture with particular reference to serum concentration in the culture medium. *Exp. Cell Res.* **150**, 338–346.

Hattori, T., and Ide, H. (1985). Effects of cyclic AMP on limb bud chondrogenesis in low cell density. *Exp. Cell Res.* **157**, 371–378.

Hauschka, P. V. (1986). Osteocalcin: The vitamin K-dependent Ca^{2+}-binding protein of bone matrix. *Haemostasis* **16**, 258–272.

Hauschka, P. V., Frenkel, J., DeMuth, R., and Gundberg, C. M. (1983). Presence of osteocalcin and related higher molecular weight-γ-carboxyglutamic-acid-containing proteins in developing bone. *J. Biol. Chem.* **258**, 176–182.

Hauschka, P. V., Lian, J. B., Cole, D. E. C., and Gundberg, C. M. (1989). Osteocalcin and matrix Gla proteins: Vitamin K-dependent proteins in bone. *Physiol. Res.* **69**, 990–1047.

Hauschka, P. V., and Reddi, A. H. (1980). Correlation of the appearance of γ-carboxyglutamic acid with the onset of mineralization in developing endochondral bone. *Biochem. Biophys. Res. Commun.* **92**, 1037–1041.

Havdrup, T., Hulth, A., and Telhag, H. (1975). Scattered mitoses in mature joint cartilage in rabbits after local trauma – chalone effect. *Clin. Orthop. Rel. Res.* **113**, 246–248.

Havelka, S., Horn, V., Spohrova, D., and Valouch, P. (1984). The calcified-noncalcified cartilage interface: The tidemark. *Acta Biol. Hung.* **35**, 271–279.

Havers, C. (1691/1692). *Osteologia Nova: or Some New Observations of the Bones, and the Parts Belonging to Them.* London. (Reprinted in *Principles of Bone Remodeling* (D. H. Enlow, ed.), Charles Thomas, Springfield, IL, 1963.)

Hay, E. D. (1965). Metabolic patterns in limb development and regeneration. In *Organogenesis* (R. L. DeHaan and H. Ursprung, eds), pp. 315–336. Holt, New York, NY.

Hay, E. D. (1970). Reversibility of the differentiated state. In *Congenital Malformations* (F. C. Fraser and V. A. McKusick, eds), pp. 91–105. Excerpta Medica, Amsterdam.

Hay, E. D. (ed.) (1981). *Cell Biology of Extracellular Matrix.* Plenum Press, New York, NY.

Hay, E. D., and Dodson, J. W. (1973). Secretion of collagen by corneal epithelium. I. Morphology of the collagenous products produced by isolated epithelia grown on frozen-killed lens. *J. Cell Biol.* **57**, 190–213.

Hay, E. D., and Fischman, D. A. (1961). Origin of the blastema in regenerating limbs of the newt *Triturus viridsecens*. An autoradiographic study using tritiated thymidine to follow cell proliferation and migration. *Devel. Biol.* **3**, 26–59.

Hay, E. D., and Meier, S. (1974). Glycosaminoglycan synthesis by embryonic inductors – neural tube, notochord, and lens. *J. Cell Biol.* **62**, 889–898.

Hayashi, M., Hayashi, K., Iyama, K.-I., Trelstad, R. L., Linsenmayer, T. F., and Mayne, R. (1992). Notochord of chick embryos secretes short-form type IX collagen prior to the onset of vertebral chondrogenesis. *Devel. Dynam.* **194**, 169–176.

Hayashi, Y. (1965). Differentiation of the beak epithelium as studied by a xenoplastic induction system. *Jap. J. Exp. Morphol.* **19**, 116.

Hayman, A. R., Jones, S. J., Boyde, A., Foster, D., Colledge, W. H., Carlton, M. B., Evans, M. J., and Cox, T. M. (1996). Mice lacking tartrate-resistant acid phosphatase (Acps) have disrupted endochondral ossification and mild osteopetrosis. *Development* **122**, 3151–3162.

Healy, C., Uwanogho, D., and Sharpe, P. T. (1999). Regulation and role of Sox9 in cartilage formation. *Devel. Dynam.* **215**, 69–78.

Heatwole, H., and Carroll, R. L. (eds) (2000). *Amphibian Biology, Volume 4, Paleontology.* Surrey Beatty & Sons, Chipping Norton, NSW.

Heatwole, H., and Davies, M. (eds) (2003). *Amphibian Biology, Volume 5, Osteology.* Surrey Beatty & Sons, Chipping Norton, NSW.

Heber-Katz, E. (1999). The regenerating mouse ear. *Sem. Cell Devel. Biol.* **10**, 415–419.

Hecht, M. K., and Hecht, B. M. (1994). Conflicting developmental and paleontological data: The case of the bird manus. *Acta Paleontol. Polenica* **38**, 329–338.

Hecht, M. K., Ostrom, J. H., Viohl, G., and Wellnhofer, P. (eds) (1985). *The Beginnings of Birds*. Proceedings of the International *Archaeopteryx* Conference, Eihstätt, 1984. Fruende des Jura-Museums, Eichstätt.

Heckman, J. D., Boyan, B. D., Aufdemorte, T. B., and Abbott, J. T. (1991). The use of bone morphogenetic protein in the treatment of non-union in a canine model. *J. Bone Joint Surg. Am. Vol.* **73**, 750–764.

Hedberg, A., Messnek, K., Persliden, J., and Hilderbrand, C. (1995). Transient local presence of nerve fibers at onset of secondary ossification in the rat knee joint. *Anat. Embryol.* **192**, 247–255.

Hedman, K., Johansson, S., Vartio, T., Kjellen, L., Vaheri, A., and Hook, M. (1982). Structure of the pericellular matrix: Association of heparan and chondroitin sulfates with fibronectin– procollagen fibers. *Cell* **28**, 663–671.

Heeley, J. D., Dobeck, J. M., and Derice, R. A. (1983). [³H] thymidine uptake in cells of rat condylar cartilage. *Amer. J. Anat.* **167**, 451–462.

Heermeier, K., Strauss, P. G., Erfle, V., and Schmidt, J. (1994). Adipose differentiation of cartilage *in vitro*. *Differentiation* **56**, 46–53.

Heersche, J. N. M. (1972). Cyclic AMP as a transmitter in the endocrine regulation of calcium metabolism. *Endocrinol. Proc. Int. Congr., 4th, 1972* Int. Congr. Ser. No. 273, pp. 359–364. Excerpta Medica, Amsterdam.

Heersche, J. N. M. (1978). Mechanisms of osteoclastic bone resorption: A new hypothesis. *Calcif. Tissue Res.* **26**, 81–84.

Heersche, J. N. M., Fedak, S. A., and Aurbach, G. D. (1971). The mode of action of dibutyryl adenosine 3',5'-monophosphate on bone tissue *in vitro*. *J. Biol. Chem.* **246**, 6770–6775.

Heersche, J. N. M., Marcus, M. R., and Aurbach, G. D. (1974). Calcitonin and the formation of 3',5'-AMP in bone and kidney. *Endocrinology* **94**, 241–247.

Heersche, J. N. M., Pitaru, S., Aubin, J. E., and Tenenbaum, H. C. (1984). Corticosteroid-induced expression of cartilage phenotype in cultured membrane bone periosteum. In *Endocrine Control of Bone and Calcium Metabolism* (D. V. Cohn, T. Fujita, J. T. Potts, and R. V. Talmage, eds), pp. 147–150. Elsevier Science Publications, Amsterdam.

Hegert, C., Kramer, J., Hargus, G., Müller, J., Guan, K., Wobus, A. M., Müller, P. K., and Rohwedel, J. (2002). Differentiation plasticity of chondrocytes derived from mouse embryonic stem cells. *J. Cell Sci.* **115**, 4617–4628.

Heggie, A. D. (1977). Growth inhibition of human embryonic and fetal rat bones in organ culture by Rubella virus. *Teratology* **15**, 47–56.

Helder, M. N., Ozkaynak, E., Sampath, K. T., Luyten, F. P., Latin, V., Oppermann, H., and Vukicevic, S. (1995). Expression pattern of osteogenic protein-1 (bone morphogenetic protein-7) in human and mouse development. *J. Histochem. Cytochem.* **43**, 1035–1044.

Helff, O. M. (1928). Studies on amphibian metamorphosis. III. The influence of the annular tympanic cartilage on the formation of the tympanic membrane. *Physiol. Zool.* **1**, 463–495.

Helff, O. M. (1931). Studies on amphibian metamorphosis. VII. The influence of the columella on the formation of the lamina propria of the tympanic membrane. *J. Exp. Zool.* **59**, 179–196.

Helff, O. M. (1934). Studies on amphibian metamorphosis. XII. Potential influences of the quadrate and supra-scapular on tympanic membrane formation in the anuran. *J. Exp. Zool.* **68**, 305–318.

Helff, O. M. (1937). Studies on amphibian metamorphosis. XV. Direct tympanic membrane formation from dermal plicae integument transplanted to the ear region. *J. Exp. Biol.* **14**, 1–15.

Helff, O. M. (1940). Studies on amphibian metamorphosis. XVII. Influence of non-living annular tympanic cartilage on tympanic membrane formation. *J. Exp. Biol.* **17**, 45–60.

Heller, M., McLean, F. C., and Bloom, W. (1950). Cellular transformations in mammalian bones induced by parathyroid extract. *Amer. J. Anat.* **87**, 315–348.

Helm, J. W., and German, R. Z. (1996). The epigenetic impact of weaning on craniofacial morphology during growth. *J. Exp. Zool.* **276**, 243–253.

Helms, J. A., Kim, C. H., Hu, D., Minkoff, R., Thaller, C., and Eichele, G. (1997). *Sonic hedgehog* participates in craniofacial morphogenesis and is down-regulated by teratogenic doses of retinoic acid. *Devel. Biol.* **187**, 25–35.

Helms, J. A., Thaller, C., and Eichele, G. (1994). Relationship between retinoic acid and sonic hedgehog, two polarizing signals in the chick wing bud. *Development* **120**, 3267–3274.

Hennig, T. B., Ellis, E., and Carlson, D. S. (1992). Growth of the mandible following replacement of the mandibular condyle with the sternal end of the clavicle – an experimental investigation in *Macaca mulatta*. *J. Oral Max. Surg.* **50**, 1196–1206.

Henrichsen, E. (1956a). Alkaline phosphatase in osteoblasts and fibroblasts cultivated *in vitro*. *Exp. Cell Res.* **11**, 115–127.

Henrichsen, E. (1956b). Alkaline phosphatase and the calcification in tissue culture. *Exp. Cell Res.* **11**, 403–416.

Henrichsen, E. (1958a). Alkaline phosphatase and calcification. *Acta Orthop. Scand.* Suppl. **34**, 1–82.

Henrichsen, E. (1958a). Bone formation and calcification in cartilage. *Acta Orthop. Scand.* **27**, 173–191.

Henrikson, R. C., and Cohen, A. S. (1965). Light and electron microscopic observations on the developing chick interphalangeal joint. *J. Ultrastruct. Res.* **13**, 129–162.

Heppenstall, R. B. (1980). Fracture healing. In *Fracture Treatment and Healing* (R. B. Heppenstall, ed.), pp. 35–64. W. B. Saunders, Philadelphia, PA.

Heppenstall, R. B., Grislis, G., and Hunt, T. K. (1975). Tissue gas tensions and oxygen consumption in healing bone defects. *Clin. Orthop. Rel. Res.* **106**, 357–365.

Hérault, Y., Beckers, J., Kondo, T., Fraudeau, N., and Duboule, D. (1998). Genetic analysis of a *Hoxd-12* regulatory element reveals global versus local modes of control in the *HoxD* complex. *Development* **125**, 1669–1677.

Hérault, Y., Beckers, J., Gérard, M., and Duboule, D. (1999). *Hox* gene expression in limbs: Colinearity by opposite regulatory controls. *Devel. Biol.* **208**, 157–165.

Herington, A. C., Adamson, L. F., and Bornstein, J. (1972). Differentiation on the basis of glucose requirements between effects of somatomedin on protein synthesis and sulfate incorporation in embryonic chick cartilage. *Biochim. Biophys. Acta* **286**, 164–174.

Herold, H. Z., Hurwitz, A., Lupo, L., and Tadmor, A. (1971). Influence of cartilage extracts on the osteoblastic response in calvarial defects in rats. *Israel J. Med. Sci.* **7**, 1164–1170.

Herring, S. W. (1974). A biometric study of suture fusion and skull growth in Peccaries. *Anat. Embryol.* **146**, 167–180.

Herring, S. W. (1993). Formation of the vertebrate face: Epigenetic and functional influences. *Amer. Zool.* **33**, 472–483.

Herring, S. W. (1994). Development of functional interactions between skeletal and muscular systems. In *Bone, Volume 9: Differentiation and Morphogenesis of Bone* (B. K. Hall, ed.), pp. 165–191. CRC Press, Boca Raton, FL.

Herring, S. W. (2000). Sutures and craniosynostosis: A comparative, functional, and evolutionary perspective. In *Craniosynostosis: Diagnosis, Evaluation, and Management* (M. M. Cohen, Jr., and R. MacLean, eds), pp. 3–10. Oxford Univertsity Press, New York, NY.

Herring, S. W., Muhl, Z. F., and Obrez, A. (1993). Bone growth and periosteal migration control masseter muscle orientation in pigs (*Sus scrofa*). *Anat. Rec.* **235**, 215–222.

Herskovits, M. S., Singh, I. J., and Sandhu, H. S. (1991). Innervation of bone. In *Bone, Volume 3: Bone Matrix and Bone Specific Products* (B. K. Hall, ed.), pp. 165–185. CRC Press, Boca Raton, FL.

Hert, J. (1964a). The regulation of the longitudinal growth at long bones. *Acta Chir. Orthop. Traum. Czech.* **7**, 85–91.

Hert, J. (1964b). Regulation of the longitudinal growth of long bones. Experimental study on the significance of mechanical factors. *Plzensky Sbornik* Suppl. **12**, 5–132.

Hert, J. (1964c). Langenwachstum des radius nach freier auto-transplantation beim Kaninchen. *Anat. Anz.* **114**, 412–424.

Hert, J. (1969). Acceleration of the growth after decrease of load on epiphyseal plates by means of spring distractors. *Folia Morphol. (Prague)* **17**, 194–203.

Hert, J. (1972). Growth of the epiphyseal plate in circumference. *Acta Anat.* **82**, 420–436.

Hert, J., Koudela, K., and Hert, J. (1972a). Wachstum der Rippen und Entwicklung der form des Brustkorbs im Querschnitt. *Anat. Anz.* **131**, 105–123.

Hert, J., and Liskova, M. (1964). Proliferation der Epiphysen-Knorpelzellen nach Anderung der Belastung. *Z. Mikrosc. Anat.* **71**, 185–197.

Hert, J., Liskova, M., and Landrgot, B. (1969). Influence of the long-term continuous bending on the bone. An experimental study on the tibia of the rabbit. *Folia Morphol. (Prague)* **17**, 389–399.

Hert, J., Liskova, M., and Landa, J. (1971a). Reaction of bone to mechanical stimuli. Part 1. Continuous and intermittent loading of tibia in rabbits. *Folia Morphol. (Prague)* **19**, 290–300.

Hert, J. Pribylová, E., and Liskova, M. (1972b). Reaction of bone to mechanical stimuli. Part 3. Microstructure of compact bone of rabbit tibia after intermittent loading. *Acta Anat.* **82**, 218–230.

Hert, J., Sklenská, A., and Liskova, M. (1971b). Reaction of bone to mechanical stimuli. Part 5. Effect of intermittent stress on the rabbit tibia after resection of the peripheral nerves. *Folia Morphol. (Prague)* **19**, 378–387.

Hert, J., and Zalud, V. (1971). The reaction of bones to mechanical stimuli. Part 6. Bioelectrical theory of the functional adaptation mechanism of bones. *Acta Chir. Orthop. Traum. Cech.* **5**, 280–288.

Hewitt, A. T., and Elmer, W. A. (1976). Reactivity of normal and brachypod mouse limb mesenchymal cells with con A. *Nature* **264**, 177–178.

Hewitt, A. T., and Elmer, W. A. (1978). Developmental modulation of lectin-binding sites on the surface membranes of normal and brachypod mouse limb mesenchymal cells. *Differentiation* **10**, 31–38.

Hewitt, R. A. (1983). Teleost hyperostoses; a case of Miocene problematica from Tunisia. *Tertiary Res.* **5**, 63–70.

Heyde, A. de (1684). *Anatomia Mytuli, Subjecta Centuria Observatorium*. Janssonio Waesbergios, Amsterdam.

Heyner, S. (1969). The significance of the intercellular matrix in the survival of cartilage allografts. *Transplantation* **8**, 666–677.

Hibiya, T. (ed.) (1982). *An Atlas of Fish Histology. Normal and Pathological Features*. Gustav Fischer Verlag, Stuttgart,

Hicks, E. P., and Ramp, W. K. (1975). The effects of fluoride on the mineralization of embryonic chick tibiae in organ culture. *Calcif. Tissue Res.* **17**, 205–218.

Hicks, M. J., and Hinchliffe, J. R. (1979). Analysis of the differential growth of the tibia and fibula chondrogenic elements in the chick hind limb. Paper given at the Autumn meeting of the British Society for Developmental Biology, Durham, UK (4–7 September). Abstract 8.

Hierton, C., Blomgren, G., and Lindgren, U. (1983). Factors associated with heterotopic bone formation in cemented total hip prostheses. *Acta Orthop. Scand.* **54**, 698–702.

Hilfer, S. R., Searls, R. L., and Fonte, V. G. (1973). An ultrastructural study of early myogenesis in the chick wing bud. *Devel. Biol.* **30**, 374–391.

Hillman, J. R., Davis, R. W., and Abdelbaki, Y. Z. (1973). Cyclic bone remodeling in deer. *Calcif. Tissue Res.* **12**, 323–330.

Hilton, E. J., and Bemis, W. E. (1999). Skeletal variation in short-nose sturgeon (*Acipenser brevirostrum*) from the Connecticut River: Implications for comparative osteological studies of fossil and living fishes. In *Mesozoic Fishes 2 – Systematics and Fossil Record* (G. Arratia and H.-P. Schultze, eds), pp. 69–94. Verlag Dr. Friedrich Pfeil, Munich.

Hinchliffe, J. R. (1974). Experimental modification of patterns of cell death and chondrogenesis in insulin-induced micromelia of the developing chick limb: An autoradiographic analysis of $^{35}SO_4$ uptake into chondroitin sulfate. *Teratology* **9**, 263–298.

Hinchliffe, J. R. (1977a). The development of winglessness (ws) in the chick embryo. *Colloq. Int. C.N.R.S. Paris* **266**, 175–182.

Hinchliffe, J. R. (1977b). 'Rudimentation', reduction and specialization in the development and evolution of the bird wing. *Colloq. Int. C.N.R.S. Paris* **266**, 411–414.

Hinchliffe, J. R. (1981). Cell death in embryogenesis. In *Cell Death in Biology and Pathology* (I. D. Bowen and R. A. Lockshin, eds), pp. 38–78. Chapman & Hall, London.

Hinchliffe, J. R. (1982). Cell death in vertebrate limb morphogenesis. In *Progress in Anatomy* (R. J. Harrison and V. Navaratnam, eds), Volume 2, pp. 1–27. Cambridge University Press, Cambridge.

Hinchliffe, J. R. (1985). 'One, two, three' or 'Two, three, four': An embryologist's view of the homologies of the digits and carpus of modern birds. In *The Beginnings of Birds*. Proceedings of the International *Archaeopteryx* Conference, Eichstätt, 1984 (M. K. Hecht, J. H. Ostrom, G. Viohl, and P. Wellnhofer, eds), pp. 141–147. Freunde des Jura-Museums, Eichstätt.

Hinchliffe, J. R. (1989). Reconstructing the Archetype: Innovation and conservation in the evolution and development of the pentadactyl limb. In *Complex Organismal Functions: Integration and Evolution in Vertebrates* (D. B. Wake and G. Roth, eds), pp. 171–189. John Wiley & Sons, New York, NY.

Hinchliffe, R. (1997). The forward march of the bird-dinosaur halted? *Science* **278**, 596–597.

Hinchliffe, J. R. (2002). Developmental basis of limb evolution. *Int. J. Devel. Biol.* **46**, 835–845.

Hinchliffe, J. R., and Ede, D. A. (1967). Limb development in the polydactylous *talpid*[3] mutant of the fowl. *J. Embryol. Exp. Morphol.* **17**, 385–404.

Hinchliffe, J. R., and Ede, D. A. (1968). Abnormalities in bone and cartilage development in the *talpid*[3] mutant of the fowl. *J. Embryol. Exp. Morphol.* **19**, 327–339.

Hinchliffe, J. R., and Ede, D. A. (1973). Cell death and the development of limb form and skeletal pattern in normal and wingless (ws) chick embryos. *J. Embryol. Exp. Morphol.* **30**, 753–772.

Hinchliffe, J. R., and Gumpel-Pinot, M. (1981). Control of maintenance and anteroposterior skeletal differentiation of the anterior mesenchyme of the chick wing bud by its posterior margin (the ZPA). *J. Embryol. Exp. Morphol.* **62**, 63–82.

Hinchliffe, J. R., and Hecht, M. K. (1984). Homology of the bird wing skeleton: Embryological versus paleontological evidence. *Evol. Biol.* **18**, 21–29.

Hinchliffe, J. R., and Johnson, D. R. (1980). *The Development of the Vertebrate Limb*. Oxford University Press, Oxford.

Hinchliffe, J. R., and Johnson, D. R. (1983). Growth of cartilage. In *Cartilage, Volume 2. Development, Differentiation and Growth* (B. K. Hall, ed.), pp. 255–296. Academic Press, New York, NY.

Hinchliffe, J. R., and Sansom, A. (1985). The distribution of the polarizing zone (ZPA) in the legbud of the chick embryo. *J. Embryol. Exp. Morphol.* **86**, 169–175.

Hinchliffe, J. R., and Thorogood, P. V. (1974). Genetic inhibition of mesenchymal cell death and the development of form and skeletal pattern in the limbs of *talpid*[3] (*ta*[3]) mutant chick embryos. *J. Embryol. Exp. Morphol.* **31**, 747–760.

Hinrichsen, G. J., and Storey, E. (1968). The effect of force on bone and bones. *Angle Orthod.* **38**, 155–165.

Hinrichsen, K. (1985). The early development of morphology and pattern of the face in the human embryo. *Adv. Anat. Embryol. Cell Biol.* **98**, 1–79.

Hinton, R. J. (1985). Adaptive response of the articular eminence and mandibular fossa to altered function of the lower jaw: An overview. In *Developmental Aspects of Temporomandibular Joint Disorders* (D. S. Carlson, J. A. McNamara, and K. A. Ribbens, eds), pp. 207–234. The Center for Human Growth and Development, Ann Arbor, MI.

Hinton, R. J. (1987). Effect of condylectomy on DNA synthesis in cells of the mandibular condylar cartilage in the rat. *Archs Oral Biol.* **32**, 865–872.

Hinton, R. J. (1988a). Effect of altered masticatory function on [³H]-thymidine and [³⁵S]-sulfate incorporation in the condylar cartilage of the rat. *Acta Anat.* **131**, 136–139.

Hinton, R. J. (1988b). Response of the intermaxillary suture cartilage to alterations in masticatory function. *Anat. Rec.* **220**, 376–387.

Hinton, R. J. (1993). Effect of dietary consistency on matrix synthesis and composition in the rat condylar cartilage. *Acta Anat.* **147**, 97–104.

Hiraki, Y., Endo, N., Takigawa, M., Asada, A., Takahashi, H., and Suzuki, F. (1987). Enhanced responsiveness to parathyroid hormone and induction of functional differentiation of cultured rabbit costal chondrocytes by a pulsed electromagnetic field. *Biochim. Biophys. Acta* **931**, 94–100.

Hirata, H., Dohi, T., Terada, H., Tanmaka, S., Okamoto, H., and Tsujimoto, A. (1983). Labelled-calcium release from rat mandibles exposed to prostaglandins *in vitro*. *Archs Oral Biol.* **28**, 963–965.

Hirata, M., and Hall, B. K. (2000). An overview of temporospatial patterns of programmed cell death (apoptosis) in chick embryos during the morphogenetic period of development. *Int. J. Devel. Biol.* **44**, 757–768.

Hirsch, M.-R., Valarché, I., Deagostini-Bazin, H., Pernelle, C., Joliot, A., and Gordis, C. (1991). An upstream regulatory element of the NCAM promoter contains a binding site for homeodomain. *FEBS Lett.* **287**, 197–202.

Hirschman, A., and Dziewiatkowski, D. D. (1966). Protein-polysaccharide loss during endochondral ossification: Immunological evidence. *Science* **154**, 393–395.

Hirschmann, P. N., and Shuttleworth, C. A. (1976). The collagen composition of the mandibular joint of the foetal calf. *Archs Oral Biol.* **21**, 771–774.

Hirsinger, E., Jouve, C., Dubrulle, J., and Pourquié, O. (2000). Somite formation and patterning. *Int. Rev. Cytol.* **198**, 1–65.

Hjerpe, A., Engfeldt, B., Tsegenides, T., Antonopoulos, C. A., Vynios, D. H., and Tsiganos, C. P. (1983). Analysis of the acid polysaccharides from squid cartilage and examination of a novel polysaccharide. *Biochim. Biophys. Acta* **757**, 85–91.

Ho, P. L., Levitt, D., and Dorfman, A. (1977). A radioimmune study of the effect of bromodeoxyuridine on the synthesis of proteoglycan by differentiating limb bud cultures. *Devel. Biol.* **55**, 233–243.

Ho, T.-Y. (1967a). Relationship between imino acid content of mammalian bone collagen and body temperature as a basis for estimation of body temperature of prehistoric mammals. *Comp. Biochem. Physiol.* **22**, 113–120.

Ho, T.-Y. (1967b). Comparative biochemistry of amino acid composition of bone and dentine collagens in Pleistocene mammals. *Biochim. Biophys. Acta* **133**, 568–573.

Hoadley, L. (1925). The differentiation of isolated chick primordia in chorio-allantoic grafts. II. The effect of the presence of the spinal cord, i.e., innervation, on the differentiation of the somitic region. *J. Exp. Zool.* **42**, 143–162.

Hoang, Q. Q., Sicheri, F., Howard, A. J., and Yang, D. S. C. (2003). Bone recognition mechanism of porcine osteocalcin from crystal structure. *Nature* **425**, 977–980.

Hoch, R. V., and Soriano, P. (2003). Roles of PDGF in animal development. *Development* **130**, 4769–4784.

Hock, J. M., and Canalis, E. (1994). Platelet-derived growth factor enhances bone cell replication but not differentiated function of osteoblasts. *Endocrinology* **134**, 1423–1428.

Hock, J. M., Canalis, E., and Centrellam, M. (1990). Transforming growth factor-β stimulates bone matrix apposition and bone cell replication in cultured fetal rat calvariae. *Endocrinology* **126**, 421–426.

Hoenig, J. M., and Walsh, A. H. (1982). The occurrence of cartilage canals in shark vertebrae. *Can. J. Zool.* **60**, 483–485.

Hofmann, C. (2003). Somite and axial development in vertebrates. In *Patterning in Vertebrate Development* (C. Tickle, ed.), pp. 48–89. Oxford University Press, Oxford.

Hofmann, C., Drossopoulou, G., McMahon, A., Balling, R., and Tickle, C. (1998). Inhibitory action of BMPs on *Pax1* expression and on shoulder girdle formation during limb development. *Devel. Dynam.* **213**, 199–206.

Hofmann, C., Gropp, R., and von der Mark, K. (1992). Expression of anchorin CII, a collagen-binding protein of the annexin family, in the developing chick embryo. *Devel. Biol.* **151**, 391–400.

Hofrath, H. (1931). Die bedeutung der röntgenfern und abstandsaufnahme für die diagnostik der kieferanomalien. *Fortschr. Orthod.* **1**, 232–242.

Hogan, B. L. M., and Tilly, R. (1977). *In vitro* culture and differentiation of normal mouse blastocysts. *Nature* **265**, 626–629.

Hogan, B. L. M. (1996a). Bone morphogenetic proteins: Multifunctional regulators of vertebrate development. *Genes & Devel.* **10**, 1580–1594.

Hogan, B. L. M. (1996b). Bone morphogenetic proteins in development. *Curr. Opin. Genet. Devel.* **6**, 432–438.

Hogg, D. A., and Hosseini, A. (1992). The effects of paralysis on skeletal development in the chick. *Comp. Biochem. Physiol. A* **103**, 25–28.

Hogg, N., Shapiro, I. M., Jones, S. J., Slusarenko, M., and Boyde, A. (1980). Lack of Fc receptors on osteoclasts. *Cell Tissue Res.* **212**, 509–516.

Hohman, E. L., Elde, R. P., Rysavy, J. A., Einzig, S., and Gebhard, R. L. (1986). Innervation of periosteum and bone by sympathetic vasoactive intestinal peptide-containing nerve fibers. *Science* **232**, 868–871.

Holbrook, S. J. (1982). Ecological inferences from mandibular morphology of *Peromyscus maniculatus*. *J. Mammal.* **63**, 399–408.

Holder, N. (1977). An experimental investigation into the early development of the chick elbow joint. *J. Embryol. Exp. Morphol.* **39**, 115–127.

Holland, L. Z., and Holland, N. D. (2001). Evolution of neural crest and placodes: amphioxus as a model for the ancestral vertebrate? *J. Anat.* **199**, 85–98.

Holliday, R., and Pugh, J. E. (1975). DNA modification mechanisms and gene activity during development. *Science* **187**, 226–232.

Hollinger, J., and Chaudhari, A. (1992). Bone regeneration materials for the mandibular and craniofacial complex. *Cells & Materials* **2**, 143–151.

Hollister, D. W., Byers, P. H., and Holbrook, K. A. (1982). Genetic disorders of collagen metabolism. *Adv. Human Genet.* **12**, 1–88.

Holmbeck, K., Bianco, P., Caterina, J., Yamada, S., Kromer, M., Kuznetsov, S. A., Mankani, M., Robey, P. G., Poole, A. R., Pidoux, I., Ward, J. M., and Birkedal-Hansen, H. (1999). Mt1-MMP-deficient mice develop dwarfism, osteopenia, arthritis, and connective tissue disease due to inadequate collagen turnover. *Cell* **99**, 81–92.

Holmes, F. A., van Leeuwen, G., and Zilliken, F. (1962). Induction of cell differentiation. II. The isolation of a chondrogenic factors from embryonic chick spinal cords and notochords. *Biochim. Biophys. Acta* **56**, 320–325.

Holmes, G., and Niswander, L. (2001). Expression of *slit-2* and *slit-3* during chick development. *Devel. Dynam.* **222**, 301–307.

Holmes, L. B., and Trelstad, R. L. (1977). Patterns of cell polarity in the developing mouse limb. *Devel. Biol.* **59**, 164–173.

Holmgren. N. (1933). On the origin of the tetrapod limb. *Acta Zool.* (*Stockholm*) **14**, 185–295.

Holmgren, N. (1940). Studies on the head in fishes. Embryological, morphological, and phylogenetical researches. Part I. Development of the skull in sharks and rays. *Acta Zool.* (*Stockholm*) **21**, 51–267.

Holmgren, N. (1941). Studies on the head in fishes. Embryological, morphological, and phylogenetical researches. Part II. Comparative anatomy of the adult selachian skull, with remarks on the dorsal fin in sharks. *Acta Zool.* (*Stockholm*) **22**, 1–90.

Holmgren, N. (1942). Studies on the head in fishes. Embryological, morphological, and phylogenetical researches. Part III. The phylogeny of elasmobranch fishes. *Acta Zool.* (*Stockholm*) **23**, 129–261.

Holmgren, N. (1943). Studies on the head in fishes. Embryological, morphological, and phylogenetical researches. Part IV. General morphology of the head in fish. *Acta Zool.* (*Stockholm*) **24**, 1–188.

Holst, R. and Bone, Q. (1993). On bipedalism in skates and rays. *Phil. Trans R. Soc. Lond. B.* **339**, 105–108.

Holtfreter, J. (1968). Mesenchyme and epithelia in inductive and morphogenetic processes. In *Epithelial–Mesenchymal Interactions* (R. Fleischmajer and R. E. Billingham, eds), pp. 1–30. Williams & Wilkins, Baltimore, MD.

Holtrop, M. E. (1966). The origin of bone cells in endochondral ossification. In *Calcified Tissues 1965*, Proceedings of the Third European Symposium on Calcified Tissues held at Davos. (H. Fleisch, H. J. J. Blackwood, and M. Owen, eds), pp. 32–36. Springer-Verlag, New York, NY.

Holtrop, M. E. (1967a). Factors influencing the growth rate in endochondral ossification. *Proc. K. Ned. Akad. Wet. Ser. C* **70**, 29–38.

Holtrop, M. E. (1967b). The potency of the epiphyseal cartilage in endochondral ossification. *Proc. K. Ned. Akad. Wet. Ser. C* **70**, 21–28.

Holtrop, M. E. (1971). The ultrastructure of the hypertrophic chondrocyte. *Israel J. Med. Sci.* **7**, 473–476.

Holtrop, M. E. (1972a). The ultrastructure of the epiphyseal plate. I. The flattened chondrocyte. *Calcif. Tissue Res.* **9**, 131–139.

Holtrop, M. E. (1972b). The ultrastructure of the epiphyseal plate. II. The hypertrophic chondrocyte. *Calcif. Tissue Res.* **9**, 140–151.

Holtrop, M. E. (1990). Light and electron microscopic structure of bone-forming cells. In *Bone, Volume 1: The Osteoblast and Osteocyte* (B. K. Hall, ed.), pp. 1–39. The Telford Press, Caldwell, NJ.

Holtrop, M. E., Cox, K. A., and Glowacki, J. (1982). Cells of the mononuclear phagocytic system resorb implanted bone matrix: A histologic and ultrastructural study. *Calcif. Tissue Int.* **34**, 488–494.

Holtrop, M. E., and King, G. J. (1977). The ultrastructure of the osteoclast and its functional implications. *Clin. Orthop. Rel. Res.* **123**, 177–196.

Holtrop, M. E., and Weinger, J. M. (1971). Ultrastructural evidence for a transport system in bone. *Proc. Parathyroid Conf., 4th, 1971 Int. Congr. Ser.* No. 243, pp. 365–374.

Holtzer, H. (1951). Morphogenetic influence of the spinal cord on the axial skeleton and musculature. *Anat. Rec.* **109**, 373–374.

Holtzer, H. (1952a). An experimental analysis of the development of the spinal column. I. Response of pre-cartilage cells to size variations of the spinal cord. *J. Exp. Zool.* **121**, 121–148.

Holtzer, H. (1952b). An experimental analysis of the development of the spinal column. II. The dispensability of the notochord. *J. Exp. Zool.* **121**, 573–591.

Holtzer, H. (1959). The development of mesodermal axial structures in regeneration and embryogenesis. In *Regeneration in Vertebrates* (C. S. Thornton, ed.), pp. 15–33. The University of Chicago Press, Chicago, IL.

Holtzer, H. (1961). Aspects of chondrogenesis and myogenesis. In *Synthesis of Molecular and Cellular Structure* (D. Rudnick, ed.), pp. 335–370. Ronald Press, New York, NY.

Holtzer, H. (1964a). Control of chondrogenesis in the embryo. *Biophys. J.* **4**, 239–250.

Holzer, H. (1964b). The induction and maintenance of the vertebral cartilages. In *Second International Conference on Congenital Malformations* (E. Fishbein, ed.), pp. 233–239. International Medical Congress, New York, NY.

Holtzer, H. (1968). Induction of chondrogenesis: A concept in search of mechanisms. In *Epithelial–Mesenchymal Interactions* (R. Fleischmajer and R. E. Billingham, eds), pp. 152–164. Williams and Wilkins, Baltimore, MD.

Holtzer, H., and Abbott, J. (1968). Oscillations of the chondrogenic phenotype *in vitro*. In *The Atability of the Differentiated State* (H. Ursprung, ed.), pp. 1–16. Springer-Verlag, Berlin.

Holtzer, H., Abbott, J., Lash, J. W., and Holtzer, S. (1960). The loss of phenotypic traits by differentiated cells *in vitro*. 1. Dedifferentiation of cartilage cells. *Proc. Natl Acad. Sci. U.S.A.* **46**, 1533–1542.

Holtzer, H., and Detwiler, S. R. (1953). An experimental analysis of the development of the spinal column. III. Induction of skeletogenous cells. *J. Exp. Zool.* **123**, 335–366.

Holtzer, H., and Mayne, R. (1973). Experimental morphogenesis. The induction of somitic chondrogenesis by embryonic spinal cord and notochord. In *Pathobiology of Development – or Ontogeny Revisited* (E. V. D. Perrin and M. J. Finegold, eds), pp. 52–65. Williams & Wilkins, Baltimore, MD, MD.

Hong, H.-K., Lass, J. H., and Chakravarti, A. (1999). Pleiotropic skeletal and ocular phenotypes of the mouse mutation congenital hydrocephalus (*ch/MF1*) arise from a winged helix/forkhead transcription factor gene. *Human Mol. Genet.* **8**, 625–637.

Honig, L. S. (1984). Pattern formation during development of the amniote limb. In *The Structure, Development and Evolution of Reptiles*, Symposium No. 52 of the Zoological Society of London. (M. W. J. Ferguson, ed.), pp. 197–221. Academic Press, London.

Hoppa, R. D., and Fitzgerald, C. M. (eds) (1999). *Human Growth in the Past: Studies from Bones and Teeth.* Cambridge University Press, Cambridge.

Hopson, J. A. (1966). The origin of the mammalian middle ear. *Amer. Zool.* **6**, 437–450.

Horan, G. S. B., Ramirez-Solis, R., Featherstone, M. S., Wolgemuth, D. J., Bradley, A., and Behringer, R. R. (1995). Compound mutants for the paralogous *hoxa-4, hoxb-4,* and *hoxd-4* genes show more complete homeotic transformations and a dose-dependent increase in the number of vertebrae transformed. *Genes & Devel.* **9**, 1667–1677.

Horan, G. S. B., Wu, K., Wolgemuth, D. J., and Behringer, R. R. (1994). Homeotic transformation of cervical vertebrae in *Hoxa-4* mutant mice. *Proc. Natl Acad. Sci. U.S.A* **91**, 12644–12648.

Horder, T. J. (1978). Functional adaptability and morphogenetic opportunism, the only rules for limb development? *Zoon* **6**, 181–192.

Horiuchi, K., Amizuka, N., Takeshita, S., Takamatsu, H., Katsuura, M., Ozawa, H., Toyama, Y., Bonewald, L. F., and Kudo, A. (1999). Identification and characterization of a novel protein, periostin, with restricted expression to periosteum and peridontal ligament and increased expression by transforming growth factor β. *J. Bone Min. Res.* **14**, 1239–1249.

Horn, E. H. (1958). Effect of feeding thiouracil and/or thyroid powder upon pubic symphyseal separation in female mice. *Endocrinology* **63**, 481–486.

Hornbruch, A., and Wolpert, L. (1970). Cell division in the early growth and morphogenesis of the chick limb. *Nature* **226**, 764–766.

Horner, J. R., de Ricqlès, A., and Padian, K. (1999). Variation in dinosaur skeletochronology indicators: implications for age assessment and physiology. *Paleobiology* **25**, 295–304.

Horner, J. R., de Ricqlès, A., and Padian, K. (2000). Long bone histology of the hadrosaurid dinosaur *Maiasaura peeblesorum*: growth dynamics and physiology based on an ontogenetic series of skeletal elements. *J. Vert. Paleo.* **20**, 115–129.

Horner, J. R., Padian, K., and de Ricqlès, A. (2001). Comparative osteohistology of some embryonic and perinatal archosaurs: developmental and behavioral implications for dinosaurs. *Paleobiology* **27**, 39–58.

Horner, J. R., and Weishampel, D. B. (1988). A comparative embryological study of two ornithischian dinosaurs. *Nature* **332**, 256–257.

Horowitz, N. H. (1942). Histochemical study of phosphatase and glycogen in foetal heads. *J. Dent. Res.* **21**, 519–527.

Hörstadius, S. (1944). Ueber die Folgen von Chordaexstirpation an spaeten Gastrulae und Neurulae von *Amblystoma punctatum. Acta Zool. (Stockholm)* **25**, 75–87.

Hörstadius, S. (1950). *The Neural Crest: Its Properties and Derivatives in the Light of Experimental Research.* Oxford University Press, Oxford.

Hörstadius, S., and Sellman, S. (1946). Experimentelle untersuchungen über die Determination des Knorpeligen Kopfskelettes bei Urodelen. *Nova Acta Reg. Soc. Sci. Upsaliensis* Ser 4 **13**, 1–170.

Horton, D. R., and Samuel, J. (1978). Palaeopathology of a fossil macropod population. *Aust. J. Zool.* **26**, 279–292.

Horton, M. A., Lewis, D., McNulty, K., Pringle, J. A. S., and Chambers, T. J. (1985a). Monoclonal antibodies to osteoclastomas (giant cell bone tumors): Definition of osteoclast-specific cellular antigens. *Cancer Res.* **45**, 5663–5669.

Horton, M. A., Rimmer, E. F., Moore, A., and Chambers, T. J. (1985b). On the origin of the osteoclast: The cell surface phenotype of rodent osteoclasts. *Calcif. Tissue Int.* **37**, 46–50.

Horton, W. A. (2003). Skeletal development: insights from targeting the mouse genome. *Lancet* **362**, 560–569.

Horton, W. E., Jr., Higginbotham, J. D., and Chandrasekhar, S. (1989). Transforming growth factor-beta and fibroblast growth factor act synergistically to inhibit collagen II synthesis through a mechanism involving regulatory DNA sequences. *J. Cell Physiol.* **141**, 8–15.

Horwitz, A. L., and Dorfman, A. (1968). Subcellular sites for synthesis of chondromucoprotein of cartilage. *J. Cell Biol.* **38**, 358–368.

Hosseini, A., and Hogg, D. A. (1991a). The effects of paralysis on skeletal development in the chick embryo. I. General effects. *J. Anat.* **177**, 159–168.

Hosseini, A., and Hogg, D. A. (1991b). The effects of paralysis on skeletal development in the chick embryo. II. Effects on histogenesis of the tibia. *J. Anat.* **177**, 169–178.

Hoβfeld, U., Olsson, L., and Breidbach, O. (eds) (2003). Carl Gegenbaur and evolutionary morphology. *Theory BioSci.* **122**, 105–301.

Houben, J.-J. G. (1976). Aspects ultrastructuraux de la migration de cellulés somitiques dans les bourgeons de membres postérieurs de souris. *Arch. Biol.* **87**, 345–365.

Houck, M. A., Gauthier, J. A., and Strauss, R. E. (1990). Allometric scaling in the earliest fossil bird, *Archaeopteryx lithographica. Science* **247**, 195–198.

Houde, P. (1987). Histological evidence for the systematic position of *Hesperornis* (Odontornithes, Hesperornithiformes). *Auk* **104**, 125–129.

Houdelier, C., Guyonarc'h, C., Lumineau, S., and Richard, J.-P. (2004). Daily organization of laying in Japanese and European quail: Effect of domestication. *J. Exp. Zool.* **301A**, 186–194.

Hough, A. J., Banfield, W. G., Mottram, F. C., and Sokoloff, L. (1974). The osteochondral junction of mammalian joints: An ultrastructural and microanalytic study. *Lab. Invest.* **31**, 685–695.

Houston, B., Seawright, E., Jefferies, D., Hoogland, E., Lester, D., Whitehead, C., and Farquharson, C. (1999). Identification and cloning of a novel phosphatase expressed at high levels in differentiating growth plate chondrocytes. *Biochim. Biophys. Acta* **1448**, 500–506.

Houston, B., Thorp, B. H., and Burt, D. W. (1994). Molecular cloning and expression of bone morphogenetic protein-7 in the chick epiphyseal growth plate. *J. Molec. Evol.* **13**, 289–301.

Howard, G. A., Bottemiller, B. L., and Baylink, D. J. (1980). Evidence for the coupling of bone formation to bone resorption in vitro. *Metab. Bone Dis. Rel. Res.* **2**, 131–135.

Howard, M. J., and Bronner-Fraser, M. (1985). The influence of neural tube-derived factors in differentiation of neural crest cells *in vitro*. I. Histochemical study of the appearance of adrenergic cells. *J. NeuroSci.* **5**, 3302–3309.

Howell, A. B. (1930). *Aquatic Mammals. Their Adaptation to Life in the Water*. Charles C. C. Thomas, Springfield, IL. (Reprinted 1970 by Dover Publications Inc., New York, NY.)

Howell, D. S., and Dean, D. D. (1992). The biology, chemistry and biochemistry of the mammalian growth plate. In *Disorders of Bone and Mineral Metabolism* (F. L. Coe and M. J. Favus, eds), pp. 313–353. Raven Press, New York, NY.

Howell, J. A. (1917). An experimental study of the effects of stress and strain on bone development. *Anat. Rec.* **13**, 233–252.

Howes, G. J., and Teugels, G. G. (1989). Observations on the ontogeny and homology of the pterygoid bones in *Corydoras paleatus* and some other catfish. *J. Zool. London* **219**, 441–456.

Howes, R. I., Jr. (1977a). Tooth and root formation in transplanted fibrously attached shark teeth. *J. Dent. Res.* **56**, A70.

Howes, R. I., Jr. (1977b). Root formation in ectopically transplanted teeth of the frog, *Rana pipiens*. 1. Tooth morphogenesis. *Acta Anat.* **97**, 151–165.

Howlett, C. R. (1980). The fine structure of the proximal growth plate and metaphysis of the avian tibia: Endochondral osteogenesis. *J. Anat.* **130**, 745–768.

Howship, J. (1815). Experiments and observations in order to ascertain the means employed by the animal economy in the formation of bone. *Med. Chir. Trans.* **6**, 263–295.

Howship, J. (1817). Observations on the morbid structure of bone and an attempt at an arrangement of their diseases. *Med. Chir. Trans.* **8**, 57–107.

Hoyte, D. A. N. (1960). Alizarin as an indicator of bone growth. *J. Anat.* **94**, 432–442.

Hoyte, D. A. N. (1966). Experimental investigations of skull morphology and growth. *Int. Rev. Gen. Exp. Zool.* **2**, 345–407.

Hoyte, D. A. N. (1971). Mechanisms of growth in the cranial vault and base. *J. Dent. Res.* **50**, 1447–1461.

Hoyte, D. A. N., and Enlow, D. H. (1966). Wolff's law and the problem of muscle attachment on resorptive surfaces of bone. *J. Phys. Anthropol.* **24**, 205–214.

Hsie, A. W., Jones, C., and Puck, T. T. (1971). Further changes in differentiation state accompanying the conversion of Chinese hamster cells to fibroblastic form by dibutyryl adenosine cyclic 3′,5′-monophosphate and hormones. *Proc. Natl Acad. Sci. U.S.A.* **68**, 1648–1652.

Hsu, H. H. T., and Anderson, H. C. (1978). Calcification of isolated matrix vesicles and reconstituted vesicles from fetal bovine cartilage. *Proc. Natl Acad. Sci. U.S.A.* **75**, 3805–3808.

Hsu, H. H. T., Cecil, R. N. A., and Anderson, H. C. (1978). Role of adenosine triphosphate, phospholipid, and vesicular structure in the calcification of isolated and reconstituted matrix vesicles. *Metab. Bone Dis. Rel. Res.* **1**, 169–172.

Hsu, J. D., and Robinson, R. A. (1969). Studies on the healing of long-bone fractures in hereditary insufficient mice. *J. Surg. Res.* **9**, 535–536.

Hu, D., and Helms, J. A. (1999). The role of sonic hedgehog in normal and abnormal craniofacial morphogenesis. *Development* **126**, 4873–4884.

Hu, D., Marcucio, R. S., and Helms, J. A. (2003). A zone of frontonasal ectoderm regulates patterning and growth in the face. *Development* **130**, 1749–1758.

Huang, D. (1974). Effect of extracellular chondroitin sulphate on cultured chondrocytes. *J. Cell Biol.* **62**, 881–886.

Huang, D. (1977). Extracellular matrix–cell interactions and chondrogenesis. *Clin. Orthop. Rel. Res.* **123**, 169–176.

Huang, J. I., Zuk, P. A., Jones, N. F., Zhu, M., Lorenz, H. P., Hedrick, M. H., and Benhaim, P. (2004). Chondrogenic potential of multipotential cells from human adipose tissue. *Plast. Reconst. Surg.* **113**, 585–594.

Huang, L., Solursh, M., and Sandra, A. (1996). The role of transforming growth factor alpha in rat craniofacial development and chondrogenesis. *J. Anat.* **189**, 73–86.

Huang, L. F., Fukai, N., Selby, P. B., Olsen, B. R., and Mundlos, S. (1997). Mouse clavicular development – analysis of wild-type and cleidocranial dysplasia mutant mice. *Devel. Dynam.* **210**, 33–40.

Huang, R., Zhi, Q., Nebüser, A., Müller, T. S., Brand-Saberi, B., Christ, B., and Wilting, J. (1996). Function of somite and somitocoele cells in the formation of the vertebral motion segment in avian embryos. *Acta Anat.* **155**, 231–234.

Huang, R., Zhi, Q., Schmidt, C., Wilting, J., Brand-Saberi, B., and Christ, B. (2000a). Sclerotomal origin of the ribs. *Development* **127**, 527–532.

Huang, R., Zhi, Q., Patel, K., Wilting, J., and Christ, B. (2000b). Dual origin and segmental organisation of the avian scapula. *Development* **127**, 3789–3794.

Hubrecht, A. A. W. (1897). *The Descent of the Primates. Lectures Delivered on the Occasion of the Sequicentennial Celebration of Princeton University*. Charles Scribner's Sons, New York, NY.

Hudson, R., Taniguchi-Sidle, A., Boras, K., Wiggan, O. N., and Hanel, P. A. (1998). Alx-4, a transcriptional activator whose expression is restricted to sites of epithelial-mesenchymal interactions. *Devel. Dynam.* **213**, 159–169.

Hueper, W. C. (1939). Cartilaginous foci in the hearts of white rats and of mice. *Arch. Pathol.* **27**, 466–468.

Hueper, W. C. (1945). Ossified cartilage with myeloid fat marrow in the aortic ring of a rabbit. *Arch. Pathol.* **39**, 89–90.

Hueter, C. (1862). Anatomische Studien an den Extremitätengelenken Neugeborenen und Erwachsener. *Virchows Arch. Pathol. Anat.* **25**, 572–599.

Hueter, C. (1863). Anatomische Studien an den Extremitätengelenken Neugeborenen und Erwachsener. *Virchows Arch. Pathol. Anat.* **26**, 484–519.

Huggare, J. Å., Kantomaa, T. J., Rönning, O. V., and Serlo, W. S. (1986). Craniofacial morphology in shunt-treated hydrocephalic children. *Cleft Palate J.* **23**, 261–269.

Huggins, C. B. (1931a). The formation of bone under the influence of epithelium of the urinary tract. *Arch. Surg. (Chicago)* **22**, 377–408.

Huggins, C. B. (1931b). The phosphatase activity of transplants of the epithelium of the urinary bladder to the abdominal wall producing heterotopic ossification. *Biochem. J.* **25**, 728–732.

Huggins, C. B. (1969). Epithelial osteogenesis – A biological chain reaction. *Proc. Amer. Philos. Soc.* **113**, 458–463.

Huggins, R. A., Huggins, S. E., Hellwig, I. H., and Deutschlander, G. (1942). Ossification in the nestling house wren. *Auk* **59**, 532–543.

Huggins, C. B., McCarroll, H. R., and Blocksom, B. H., Jr. (1936). Experiments on the theory of osteogenesis. The influence of local calcium deposits on ossification; the osteogenic stimulus of epithelium. *Arch. Surg. (Chicago)* **32**, 915–931.

Huggins, C. B., and Sammett, J. F. (1933). Function of the gall bladder epithelium as an osteogenic stimulus and the physiological differentiation of connective tissue. *J. Exp. Med.* **58**, 393–400.

Huggins, C. B., and Urist, M. R. (1970). Dentin matrix transformation: Rapid induction of alkaline phosphatase and cartilage. *Science* **167**, 896–898.

Hughes, D. R., Bassett, J. R., and Moffat, L. A. (1994a). Structure and origin of the tooth pedicel (the so-called bone of attachment) and dental-ridge bone in the mandibles of the sea breams *Acanthopagrus australis*, *Pagrus auratus* and *Rhabdosargus sarba* (Sparidae, Perciformes, Teleostei). *Anat. Embryol.* **189**, 51–69.

Hughes, D. R., Bassett, J. R., and Moffat, L. A. (1994b). Histological identification of osteocytes in the allegedly acellular bone of the sea breams *Acanthopagrus australis*, *Pagrus auratus* and *Rhabdosargus sarba* (Sparidae, Perciformes, Teleostei). *Anat. Embryol.* **190**, 163–179.

Hughes, F. J., Collyer, J., Stanfield, M., and Goodman, S. A. (1995). The effect of bone morphogenetic protein-2, -4, and -6 on differentiation of rat osteoblast cells *in vitro*. *Endocrinology* **136**, 2671–2677.

Hughes, F. J., and McCulloch, C. A. G. (1991). Stimulation of the differentiation of osteogenic rat bone marrow stromal cells by osteoblast cultures. *Lab. Invest.* **64**, 617–622.

Hui, C. A. (2002). Avian furcula morphology may indicate relationships of flight requirements among birds. *J. Morphol.* **251**, 284–293.

Hulth, A. (1980). Fracture healing. A concept of competing healing factors. *Acta Orthop. Scand.* **51**, 5–8.

Hulth, A. (1981). Fracture healing – more biology than mechanics. *Clin. Orthop. Rel. Res.* **156**, 259–261.

Hulth, A. (1989). Current concepts of fracture healing. *Clin. Orthop. Rel. Res.* **249**, 265–284.

Hulth, A., Lindberg, L., and Telhag, H. (1972). Mitosis in human osteoarthritic cartilage. *Clin. Orthop. Rel. Res.* **84**, 197–199.

Hume, D. A., Loutit, J. F., and Gordon, S. (1984). The mononuclear phagocyte system of mouse defined by immunohistochemical localization of antigen F4/80: Macrophages of bone and associated connective tissue. *J. Cell Sci.* **66**, 189–194.

Hume, D. A., Robinson, A. P., MacPherson, G. G., and Gordon, S. (1983). The mononuclear phagocyte system of the mouse defined by immunohistochemical localization of antigen F4/80. *J. Exp. Med.* **158**, 1522–1536.

Humphry, G. M. (1861). Observations on the growth of the long bones and of stumps. *Med.-Chir. Soc. Trans.* **44**, 117–134.

Humphry, G. M. (1863). Results of experiments on the growth of the jaws. *Br. J. Dent. Sci.* **6**, 548–550.

Humphry, G. M. (1864). On the growth of the jaws. *Phil. Trans R. Soc. London* **11**, 1–5.

Hunt, E. A. (1932). The differentiation of chick limb buds in chorio-allantoic grafts, with special reference to the muscles. *J. Exp. Zool.* **62**, 57–91.

Hunt, P., Boncinelli, E., and Krumlauf, R. (1992). Hox genes and the development of the branchial region. In *Development of the Central Nervous System in Vertebrates* (S. C. Sharpe and A. M. Goffinet, eds), NATO ASI Series 234, pp. 49–73. Plenum Press, New York, NY.

Hunt, P., Ferretti, P., Krumlauf, R., and Thorogood, P. (1995). Restoration of normal Hox code and branchial arch morphogenesis after extensive deletion of hindbrain neural crest. *Devel. Biol.* **168**, 584–597.

Hunt, P., Gulisano, M., Cook, M., Sham, M.-H., Faiella, A., Wilkinson, D., Boncinelli, E., and Krumlauf, R. (1991a). A distinct *Hox* code for the branchial region of the vertebrate head. *Nature* **353**, 861–864.

Hunt, P., and Krumlauf, R. (1991). Deciphering the Hox code; Clues to patterning branchial regions of the head. *Cell* **66**, 1075–1078.

Hunt, P., and Krumlauf, R. (1992). *Hox* codes and positional specification in vertebrate embryonic axis. *Annu. R. Cell* **8**, 227–256.

Hunt, P., Whiting, J., Muchamore, I., Marshall, H., and Krumlauf, R. (1991b). Homeobox genes and models for patterning the hindbrain and branchial arches. *Development* **1991**, Suppl. 1, 187–196.

Hunt, P., Whiting, J., Nonchev, S., Sham, M.-H., Marshall, H., Graham, A., Cook, M., Allemann, R., Rigby, P. W. J., Gulisano, M., Faiella, A., Boncinelli, E., and Krumlauf, R. (1991c). The branchial *Hox* code and its implications for gene regulation, patterning of the nervous system and head evolution. *Development* **1991**, Suppl. 2, 63–78.

Hunt, P., Wilkinson, D., and Krumlauf, R. (1991d). Patterning the vertebrate head: Murine Hox 2 genes mark distinct subpopulations of premigratory and migrating cranial neural crest. *Development* **112**, 43–50.

Hunter, C., and Clegg, E. J. (1973a). Changes in skeletal proportions of the rat in response to hypoxic stress. *J. Anat.* **114**, 201–220.

Hunter, C., and Clegg, E. J. (1973b). The effects of hypoxia on the caudal vertebrae of growing mice and rats. *J. Anat.* **116**, 227–244.

Hunter, J. (1771). *The Natural History of the Human Teeth*. J. Johnson, London.

Hunter, J. (1835). *The Works of John Hunter, F.R.S.* (J. F. Palmer, ed.), Longman, Rees, Orme, Brown, Green and Longman, London.

Hunter, S. J., and Schraer, H. (1983). *In vitro* synthesis of proteoglycans associated with medullary bone in Japanese quail. *Archs Biochim. Biophys.* **220**, 272–279.

Hunt von Herbing, I., Boutilier, R. G., Miyake, T., and Hall, B. K. (1996a). Effects of temperature on morphological landmarks critical to growth and survival in larval Atlantic cod (*Gadus morhua*). *Marine Biol.* **124**, 593–606.

Hunt von Herbing, I., Miyake, T., Hall, B. K., and Boutilier, R. G. (1996b). The ontogeny of feeding and respiration in larval Atlantic cod, *Gadus morhua* (Teleostei; Gadiformes): (1) Morphology. *J. Morphol.* **227**, 15–35.

Hunt von Herbing, I., Miyake, T., Hall, B. K., and Boutilier, R. G. (1996c). The ontogeny of feeding and respiration in larval Atlantic cod, *Gadus morhua* (Teleostei; Gadiformes): (2) Function. *J. Morphol.* **227**, 37–50.

Hunziker, E. B. (1994). Mechanism of longitudinal bone-growth and its regulation by growth plate chondrocytes. *Micros. Res. Tech.* **28**, 505–519.

Hunziker, E. B., Driesang, I. M. K., and Morris, E. A. (2001). Chondrogenesis in cartilage repair is induced by members of the transforming growth factor-beta superfamily. *Clin. Orthop. Rel. Res.* **391S**, S171–S181.

Hunziker, E. B., Herrmann, W., Schenk, R. K., Mueller, M., and Moor, H. (1984). Cartilage ultrastructure after high pressure freezing, freeze substitution and low temperature embedding. I. Chondrocyte ultrastructure – implications for the theories of mineralization and vascular invasion. *J. Cell Biol.* **98**, 267–276.

Hunziker, E. B., and Schenk, R. K. (1987). Structural organization of proteoglycan in cartilage. In *Biology of Proteoglycans* (T. N. Wight, and R. P. Mecham, eds), pp. 155–185. Academic Press, New York, NY.

Hunziker, E. B., and Schenk, R. K. (1989). Physiological mechanisms adopted by chondrocytes in regulating longitudinal bone growth in rats. *J. Physiol.* **414**, 55–72.

Hunziker, E. B., Schenk, R. K., and Cruzorive, L. M. (1987). Quantitation of chondrocyte performance in growth-plate cartilage during longitudinal bone growth. *J. Bone Joint Surg. Am. Vol.* **69**, 162–173.

Hunziker, E. B., Wagner, J., and Zapf, J. (1994). Differential effects of insulin-like growth-factor-I and growth hormone on developmental stages of rat growth-plate chondrocytes in vivo. *J. Clin. Invest.* **93**, 1078–1086.

Hurlé, J. M., and Colvee, E. (1982). Surface changes in the embryonic interdigital epithelium during the formation of the free digits. A comparative study in the chick and duck foot. *J. Embryol. Exp. Morphol.* **69**, 251–263.

Hurlé, J. M., Colvee, E., and Fernandez-Teran, M. A. (1985). Vascular regression during the formation of the free digits in the avian limb. A comparative study in chick and duck embryos. *J. Embryol. Exp. Morphol.* **85**, 239–250.

Hurlé, J. M., and Fernandez-Teran, M. A. (1983). Fine structure of the regressing interdigital membranes during the formation of the digits of the chick embryo leg bud. *J. Embryol. Exp. Morphol.* **78**, 195–209.

Hurlé, J. M., and Fernandez-Teran, M. A. (1984). Fine structure of the interdigital membranes during the morphogenesis of the digits of the webbed foot of the duck embryo. *J. Embryol. Exp. Morphol.* **79**, 201–210.

Hurlé, J. M., and Gañan, Y. (1986). Interdigital tissue chondrogenesis induced by surgical removal of the ectoderm in the embryonic chick leg bud. *J. Embryol. Exp. Morphol.* **94**, 231–244.

Hurlé, J. M., and Gañan, Y. (1987). Formation of extra-digits induced by surgical removal of the apical ectodermal ridge of the chick embryo leg bud in the stages previous to the onset of interdigital cell death. *Anat. Embryol.* **176**, 393–399.

Hurlé, J. M., Gañan, Y., and Macias, D. (1989). Experimental analysis of the *in vivo* chondrogenic potential of the interdigital mesenchyme of the chick leg bud subjected to local ectodermal removal. *Devel. Biol.* **132**, 368–374.

Hurlé, J. M., and Hinchliffe, J. R. (1978). Cell death in the posterior necrotic zone (PNZ) of the chick wing bud: A stereoscan and ultrastructural survey of autolysis and cell fragmentation. *J. Embryol. Exp. Morphol.* **43**, 123–136.

Hurov, J. R. (1986). Soft-tissue bone interface: How do attachments of muscles, tendons, and ligaments change during growth? A light microscopic study. *J. Morphol.* **189**, 313–326.

Hutson, M. R., and Kirby, M. L. (2003). Neural crest and cardiovascular development: A 20-year perspective. *Birth Defects Res. (Part C)* **69**, 2–13.

Huxley, J. S., and Murray, P. D. F. (1924). A note on the reactions of chick chorio-allantois to grafting. *Anat. Rec.* **28**, 385–389.

Huxley, J. S., and Teissier, G. (1936a). Terminology of relative growth. *Nature* **137**, 780–781.

Huxley, J. S., and Teissier, G. (1936b). Terminologie et notation dans la description de la croissance relative. *C. R. Séances Soc. Biol. Fil.* **121**, 934–937.

Huxley, L. (1900). *Life and Letters of Thomas Henry Huxley.* 2 vols. London.

Huysseune, A. (1983). Observations on tooth development and implantation in the upper pharyngeal jaws in *Astatotilapia elegans* (Teleostei, Cichlidae). *J. Morphol.* **175**, 217–234.

Huysseune, A. (1985). The opercular cartilage in *Astatotilapia elegans. Fortsch. Zool.* **30**, 371–373.

Huysseune, A. (1986). Late skeletal development at the articulation between upper pharyngeal jaws and neurocranial base in the fish, *Astatotilapia elegans*, with the participation of a chondroid form of bone. *Amer. J. Anat.* **177**, 119–137.

Huysseune, A. (1989). Morphogenetic aspects of the pharyngeal jaws and neurocranial apophysis in postembryonic *Astatotilapia elegans* (Trewavas, 1933) (Teleostei, Cichlidae). *Med. Kon. Acad. Wetensch. Letteren Schone Kunsten Belgie* **51**, 11–35.

Huysseune, A. (2000). Skeletal Systems. In *Microscopic Functional Anatomy* (G. K. Ostrander, ed.), pp. 307–317. Academic Press, San Diego.

Huysseune, A., and Sire, J.-Y. (1990). Ultrastructural observations on chondroid bone in the teleost fish *Hemichromis bimaculatus. Tissue & Cell* **22**, 371–383.

Huysseune, A., and Sire, J.-Y. (1992a). Development of cartilage and bone tissues of the anterior part of the mandible in cichlid fishes: a light and TEM study. *Anat. Rec.* **233**, 357–375.

Huysseune, A., and Sire, J.-Y. (1992b). Bone and cartilage resorption in relation to tooth development in the anterior part of the mandible in cichlid fishes: a light and TEM study. *Anat. Rec.* **234**, 1–14.

Huysseune, A., Sire, J.-Y., and Meunier, F. J. (1994). Comparative study of lower pharyngeal jaw structure in two phenotypes of *Astatoreochromis alluaudi* (Teleostei: Cichlidae). *J. Morphol.* **221**, 25–43.

Huysseune, A., and Verraes, W. (1986). Chondroid bone on the upper pharyngeal jaws and neurocranial base in the adult fish *Astatotilapia elegans. Amer. J. Anat.* **177**, 527–535.

Huysseune, A., and Verraes, W. (1990). Carbohydrate histochemistry of mature chondroid bone in *Astatotilapia elegans* (Teleostei, Cichlidae) with a comparison to acellular bone and cartilage. *Ann. Sci. Nat. Zool.* **11**, 29–43.

Huysseune, A., Van den Berghe, W., and Verraes, W. (1986). The contribution of chondroid bone in the growth of the parasphenoid bone of a cichlid fish as studied by oblique computer-aided reconstruction. *Biol. Jb. Dodonaea* **54**, 131–141.

Hylander, W. L. (1981). Patterns of stress and strain in the macaque mandible. In *Craniofacial Biology* (D. S. Carlson, ed.), pp. 1–35. Center for Human Growth and Development, Ann Arbor, MI.

Hylander, W. L., and Johnson, K. R. (1989). The relationship between masseter force and masseter electromyogram dueing mastication in the monkey *Macaca fascicularis. Archs Oral Biol.* **34**, 713–722.

Hylander, W. L., Picq, P. G., and Johnson, K. R. (1991a). Masticatory-stress hypotheses and the supraorbital region of primates. *Amer. J. Phys. Anthropol.* **86**, 1–36.

Hylander, W. L., Picq, P. G., and Johnson, K. R. (1991b). Function of the supraorbital region of primates. *Archs Oral Biol.* **36**, 273–281.

Hyldebrandt, N., Damholt, W., and Nordentoft, E. (1974). Investigation of the cellular response to fracture assessed by autoradiography of the periosteum. *Acta Orthop. Scand.* **45**, 175–181.

I

Ichikawa, M. (1936). Experimental studies on the formation of the auditory capsule of amphibians. *Bot. Zool.* **4**, 1211–1223.

Igarashi, S., Trelstad, R. L., and Kang, A. H. (1973). Physical chemical properties of chick cartilage collagen. *Biochim. Biophys. Acta* **P24**, 514–519.

Ignelzi, M. A., Jr., Liu, Y.-H., Maxson, R. E., Jr., and Snead, M. L. (1995). Genetically engineered mice: tools to understand craniofacial development. *Crit. Rev. Oral Biol. Med.* **6**, 181–201.

Ignotz, R. A. (1991). TGF-β and extracellular matrix related influences on gene expression and phenotype. *Crit. Rev. Eukary. Gene Exp.* **1**, 75–84.

Iida, K., Koseki, H., Kakinuma, H., Kato, N., Mizutaniko-Seki, Y., Ohuchi, H., Yoshioka, H., Noji, S., Kawamura, K., Kataoka, Y., Ueno, F., Taniguchi, M., Yoshida, N., Sugiyama, T., and Miura, N. (1997). Essential roles of the winged helix transcription factor Mfh-1 in aortic-arch patterning and skeletogenesis. *Development* **124**, 4627–4638.

Iida, S., Kawasaki, T., Mizuno, M., and Kuboki, Y. (1994). Bone formation in osteoporotic rats, reduced effect of bone morphogenetic protein (BMP) ascribed to the suppressed osteoblast differentiation. *Jap. J. Oral Biol.* **36**, 249–262.

Iimura, T., Oida, S., Takeda, K., Maruoka, Y., and Sasaki, S. (1994). Changes in homeobox-containing gene expression during ectopic bone formation induced by bone morphogenetic protein. *Biochem. Biophys. Res. Commun.* **201**, 980–987.

Ikeda, T., Nomura, S., Yamaguchi, A., Suda, T., and Yoshiki, S. (1992). *In situ* hybridization of bone matrix proteins in undecalcified adult rat bone sections. *J. Histochem. Cytochem.* **40**, 1079–1088.

Ikeda, T., Takahashi, H., and Suzuki, A., Ueno, N., Yokose, S., Yamaguchi, A., and Yoshiki, S. (1996). Cloning of rat type 1 receptor cDNA for bone morphogenetic protein-2 and bone morphogenetic protein-4, and the localization compared with that of the ligands. *Devel. Dynam.* **206**, 318–329.

Ikeda, Y., Ozaki, H., and Yasuda, H. (1973). Growth of scales in goldfish. *Bull. Jap. Soc. Scient. Fish.* **39**, 25–33.

Illingworth, C. M. (1974). Trapped fingers and amputated finger tips in children. *J. Pediatr. Surg.* **9**, 853–858.

Illmensee, K., and Stevens, L. C. (1979). Teratomas and chimeras. *Sci. Amer.* **240**, 121–132.

Ilvesaro, J., Väänänen, K., and Tuukkanen, J. (2000). Bone-resorbing osteoclasts contain gap-junctional connexin-43. *J. Bone Min. Res.* **15**, 919–926.

Inada, M., Yasui, T., Nomura, S., Miyake, S., Deguchi, K., Himeno, M., Sato, M., Yamagiwa, H., Kimura, T., Yasui, N., Ochi, T., Endo, N., Kitamura, Y., Kishimoto, T., and Komori, T. (1999). Maturational disturbance of chondrocytes in *Cbfa-1*-deficient mice. *Devel. Dynam.* **214**, 279–290.

Ingber, D. E. (1991). Integrins as mechanochemical transducers. *Curr. Opin. Cell Biol.* **3**, 841–848.

Ingber, D. E. (1993). Cellular tensegrity: Defining new rules of biological design that govern the cytoskeleton. *J. Cell Sci.* **104**, 613–627.

Ingram, R. T., Bonde, S. K., Riggs, B. L., and Fitzpatrick, L. A. (1994). Effects of transforming growth factor beta (TGF-β) and 1,25 dihydroxyvitamin D_3 on the function, cytochemistry and morphology of normal human osteoblast-like cells. *Differentiation* **55**, 153–163.

Inoue, T., Deporter, D. A., and Melcher, A. H. (1986). Induction of chondrogenesis in muscle, skin, bone marrow and periodontal ligament by demineralized dentin and bone matrix *in vivo* and *in vitro*. *J. Dent. Res.* **65**, 12–22.

Inuzuka, H., Redies, C., and Takeichi, M. (1991). Differential expression of R- and N-cadherin in neural and mesodermal tissues during early chicken development. *Development* **113**, 959–967.

Ioseliani, D. G. (1972). The use of tritiated thymidine in the study of bone induction by transitional epithelium. *Clin. Orthop. Rel. Res.* **88**, 183–196.

Iovine, M. K., and Johnson, S. L. (2000). Genetic analysis of isometric growth control mechanisms in the zebrafish caudal fin. *Genetics* **155**, 1321–1329.

Irving, J. T. (1973). Theories of mineralization of bone. *Clin. Orthop. Rel. Res.* **97**, 225–236.

Irving, J. T., and Durkin, J. F. (1965). A comparison of the changes in the mandibular condyle with those in the upper tibial epiphysis during the onset and healing of scurvy. *Archs Oral Biol.* **10**, 179–185.

Irving, J. T., Le Bolt, S. A., and Schneider, E. L. (1981). Ectopic bone formation and aging. *Clin. Orthop. Rel. Res.* **154**, 249–253.

Irving, J. T., and Wuthier, R. E. (1968). Histochemistry and biochemistry of calcification with special reference to the role of lipids. *Clin. Orthop. Rel. Res.* **56**, 237–260.

Irwin, C. R., and Ferguson, M. W. J. (1986). Fracture repair of reptilian dermal bones: Can reptiles form secondary cartilage? *J. Anat.* **146**, 53–64.

Isaac, A., Rodriguez-Esteban, C., Ryan, A., Altabef, M., Tsuki, T., Patel, K., Tickle, C., and Izpisua-Belmonte, J.-C. (1998). Tbx genes and limb identity in chick embryo development. *Development* **125**, 1867–1875.

Isaksson, O. G. P., Jansson, J.-O., and Gause, I. A. M. (1982). Growth hormone stimulates longitudinal bone growth directly. *Science* **216**, 1237–1239.

Işcan, M. Y., and Kennedy, K. A. R. (1989). *Reconstructing Life from the Skeleton.* Alan R. Liss, Inc., New York, NY.

Iseki, S., Osumi-Yamashita, N., Miyazono, K., Franzen, P., Ichijo, H., Ohtani, H., Hayashi, Y., and Eto, K. (1995). Localization of transforming growth factor-β type I and type II receptors in mouse development. *Exp. Cell Res.* **219**, 339–347.

Iseki, S., Wilkie, A.-O. M., Heath, J. K., Ishimaru, T., Eto, K., and Morriss-Kay, G. M. (1997). *FgfR2* and *osteopontin* domains in the developing skull vault are mutually exclusive and can be altered by locally applied FGF2. *Development* **124**, 3375–3384.

Iseki, S., Wilkie, A.-O. M., and Morriss-Kay, G. M. (1999). *Fgfr-1* and *Fgfr2* have distinct differentiation- and proliferation related roles in the developing mouse skull vault. *Development* **126**, 5611–5620.

Ishidou, Y., Kitajima, I., Obama, H., Maruyama, I., Murata, F., Imamura, T., Yamada, N., Ten Dijke, P., Miyazono, K., and Sakou, T. (1995). Enhanced expression of type I receptors for bone morphogenetic proteins during bone formation. *J. Bone Min. Res.* **10**, 1651–1659.

Ishii, M., Suda. N., Tengan, T., Susuki, S., and Kuroda, T. (1998). Immunohistochemical findings of type-I and type-II collagen in prenatal mouse mandibular condylar cartilage compared with the tibial anlage. *Archs Oral Biol.* **43**, 545–550.

Ishii-Suzuki, M., Suda, N., Yamazaki, K., Kuroda, T., Senior, P. V., Beck, F., and Hammond, V. E. (1999). Differential responses to parathyroid hormone-related protein (PTHrP) deficiency in the various craniofacial cartilages. *Anat. Rec.* **255**, 452–457.

Ishikawa, H., Omoe, K., and Endo, A. (1992). Growth and differentiation schedule of mouse embryos obtained from delayed matings. *Teratology* **45**, 655–659.

Ishizeki, K., Chida, T., Yamamoto, H., and Nawa, T. (1996a). Light and electron microscopy of stage-specific features of the transdifferentiation of mouse Meckel's cartilage chondrocytes *in vitro*. *Acta. Anat.* **157**, 1–10.

Ishizeki, K., Fujiwara, N., Sugawara, M., and Nawa, T. (1990). Endochondral calcification by hypertrophic chondrocytes in the Meckel's cartilage grafted into the isogenic mouse spleen. *Arch. Histol. Cytol.* **53**, 187–192.

Ishizeki, K., Hiraki, Y., Kubo, M., and Nawa, T. (1997). Sequential synthesis of cartilage and bone marker proteins during trans-differentiation of mouse Meckel's cartilage chondrocytes *in vitro*. *Int. J. Devel. Biol.* **41**, 83–89.

Ishizeki, K., Kubo, M., Yamamoto, H., and Nawa, T. (1998). Immunocytochemical expression of type I and type II collagens by rat Meckel's chondrocytes in culture during phenotypic transformation. *Archs Oral Biol.* **43**, 117–126.

Ishizeki, K., Nagano, H., Fujiwara, N., and Nawa, T. (1994). Morphological changes during survival, cellular transformation, and calcification of the embryonic mouse: Meckel's cartilage transplanted into heterotopic sites. *J. Craniofac. Genet. Devel. Biol.* **14**, 33–42.

Ishizeki, K., and Nawa, T. (2000). Further evidence for secretion of matrix metalloproteinase-1 by Meckel's chondrocytes during degradation of the extracellular matrix. *Tissue & Cell* **32**, 207–215.

Ishizeki, K., Saiot, H., Shinagawa, T., Fujiwara, N., and Nawa, T. (1999). Histochemical and immunohistochemical analysis of the mechanism of calcification of Meckel's cartilage during mandible development in rodents. *J. Anat.* **194**, 265–277.

Ishizeki, K., Takigawa, M., Harada, Y., Suzuki, F., and Nawa, T. (1996b). Meckel's cartilage chondrocytes in organ culture synthesize bone-type proteins accompanying osteocytic phenotype expression. *Anat. Embryol.* **193**, 61–71.

Ishizeki, K., Takigawa, M., Nawa, T., and Susuki, F. (1996c). Mouse Meckel's cartilage chondrocytes evoke bone-like matrix and further transform into osteocyte-like cells in culture. *Anat. Rec.* **245**, 25–35.

Ishizeki, K., Takahashi, N., and Nawa, T. (2001). Formation of the sphenomandibular ligament by Meckel's cartilage in the mouse: possible involvement of epidermal growth factor as revealed by studies in vivo and in vitro. *Cell Tissue Res.* **304**, 67–80.

Ismail, M. H., Verraes, W., and Huysseune, A. (1982). Developmental aspects of the pharyngeal jaws in *Astatotilapia elegans* (Trewavas, 1933) (Teleostei, Cichlidae). *Neth. J. Zool.* **32**, 513–543.

Isogai, N., Landis, W. J., Mori, R., Gotoh, Y., Gerstenfeld, L. C., Upton, J., and Vacanti, J. P. (2000). Experimental use of fibrin glue to induce site-directed osteogenesis from cultured periosteal cells. *Plast. Reconst. Surg.* **105**, 953–963.

Isotupa, K., Koski, K., and Makinen, L. (1965). Changing architecture of growing cranial bones at sutures as revealed by vital staining with Alizarin Red S in the rabbit. *Amer. J. Phys. Anthrop.* **23**, 19–22.

Iten, L. E., and Murphy, D. J. (1980). Growth of quail wing skeletal element in a host chick wing. *Amer. Zool.* **20**, 739.

Ito, Y., Fitzsimmons, J. S., Sanyal, A., Mello, M. A., Mukherjee, N., and O'Driscoll, S. W. (2001). Localization of chondrocyte precursors in periosteum. *Osteoarth. & Cart.* **9**, 215–223.

Itonaga, I., Sabokbar, A., Sun, S. G., Kudo, O., Danks, L., Ferguson, D., Fujikawa, Y., and Athanasou, N. A. (2004). Transforming growth factor-β induces osteoclast formation in the absence of RANKL. *Bone* **34**, 57–64.

Iulionella, A., Beckett, B., Petkovich, M., and Lohnes, D. (1999). A molecular basis for retinoic acid-induced axial truncation. *Devel. Biol.* **205**, 33–48.

Iverson, J. B. (1982). Ontogenetic changes in relative skeletal mass in the painted turtle *Chrysemys picta*. *J. Herpetol.* **16**, 412–414.

Ivkovic, S., Yoon, B. S., Popoff, S. N., Safadi, F. F., Libuda, D. E., Stephenson, R. C., Daluiski, A., and Lyons, K M. (2003). Connective tissue growth factor coordinates chondrogenesis and angiogenesis during skeletal development. *Development* **130**, 2779–2791.

Iwamoto, M., Higuchi, Y., Koyama, E., Enomoto-Iwamoto, M., Kurisu, K., Yeh, H., Abrams, W. R., Rosenbloom, J., and Pacifici, M. (2000). Transcription factor ERG variants and functional diversification of chondrocytes during limb long bone development. *J. Cell Biol.* **150**, 27–39.

Iwamoto, M., Sato, K., Nakashima, K., Fuchihata, H., Suzuki, F., and Kato, Y. (1989). Regulation of colony formation of differentiated chondrocytes in soft agar by transforming growth factor-β. *Biochem. Biophys. Res. Commun.* **159**, 1006–1011.

Iwamoto, M., Shimazu, A., Nakashima, K., Suzuki, F., and Kato, Y. (1991). Reduction in basic fibroblast growth factor receptor is coupled with terminal differentiation of chondrocytes. *J. Biol. Chem.* **266**, 461–467.

Iwamoto, M., Yagami, K., Lu Valle, P., Olsen, B. R., Petrodoulos, C. J., Ewert, D. L., and Pacifici, M. (1993). Expression and role of *c-myc* in chondrocytes undergoing endochondral ossification. *J. Biol. Chem.* **268**, 9645–9652.

Iwasaki, M., Nakata, K., Nakahara, H., Nakase, T., Kimura, T., Kimata, K., Caplan, A. I., and Ono, K. (1993). Transforming growth factor-β1 stimulates chondrogenesis and inhibits osteogenesis in high density culture of periosteum-derived cells. *Endocrinology* **132**, 1603–1608.

Iwata, H., and Urist, M. R. (1973). Hyaluronic acid production and removal during bone morphogenesis in implants of bone matrix in rats. *Clin. Orthop. Rel. Res.* **90**, 236–245.

Iyama, K.-I., Ninomiya, Y., Olsen, B. R., Linsenmayer, T. F., Trelstad, R. L., and Hayashi, M. (1991). Spatiotemporal pattern of type X collagen gene expression and collagen deposition in embryonic chick vertebrae undergoing endochondral ossification. *Anat. Rec.* **229**, 462–472.

Izpisúa-Belmonte, J.-C., Brown, J. M., Crawley, A., Duboule, D., and Tickle, C. (1992b). *Hox-4* gene expression in mouse/chicken heterospecific grafts of signalling regions to limb buds reveals similarities in patterning mechanisms. *Development* **115**, 553–560.

Izpisúa-Belmonte, J.-C., Ede, D. A., Tickle, C., and Duboule, D. (1992a). The mis-expression of posterior *Hox-4* genes in talpid (ta³) mutant wings correlates with the absence of antero-posterior polarity. *Development* **114**, 959–963.

Izumi, K., Yamaoka, I., and Murakami, R. (2000). Ultrastructure of the developing fibrocartilage of the os penis of rat. *J. Morphol.* **243**, 187–191.

J

Jabaily, J., Rall, T. W., and Singer, M. (1975). Assay of cyclic 3',5'-monophosphate in the regenerating forelimb of the newt, *Triturus*. *J. Morphol.* **147**, 379–384.

Jabs, E. W., Müller, U., Li, X., Ma, L., Luo, W., Haworth, I. S., Klisak, I., Sparkes, R., Warman, M. L., Mulliken, J. B., Snead, M. L., and Maxson, R. (1993). A mutation in the homeodomain of the human Msx2 gene in a family affected with autosomal dominant craniosynostosis. *Cell* **75**, 443–450.

Jacenko, O., Lu Valle, P. A., and Olsen, B. R. (1993). Spondylometaphyseal dysplasia in mice carrying a dominant negative mutation in a matrix protein specific for cartilage-to-bone transition. *Nature* **365**, 56–61.

Jacenko, O., San Antonio, J. D., and Tuan, R. S. (1995). Chondrogenic potential of chick embryonic calvaria: II. Matrix calcium may repress cartilage differentiation. *Devel. Dynam.* **202**, 27–41.

Jacenko, O., and Tuan, R. S. (1986). Calcium deficiency induces expression of cartilage-like phenotype in chick embryonic calvaria. *Devel. Biol.* **115**, 215–232.

Jacenko, O., and Tuan, R. S. (1995). Chondrogenic potential of chick embryonic calvaria: 1. Low calcium permits cartilage differentiation. *Devel. Dynam.* **202**, 13–26.

Jackson, R. W., Reed, C. A., Israel, J. A., Abou-Keer, F. K., and Garside, H. (1970). Production of a standard experimental fracture. *Can. J. Surg.* **13**, 415–420.

Jacobson, W., and Fell, H. B. (1941). The developmental mechanisms and potency of the undifferentiated mesenchyme of the mandible. *Q. J. Microsc. Sci.* **82**, 563–586.

Jacoby, R. K., and Jayson, M. I. U. (1975a). Organ culture of adult human articular cartilage. I. The effect of hyperoxia on synthesis of glycosaminoglycans. *J. Rheumatol.* **2**, 270–279.

Jacoby, R. K., and Jayson, M. I. U. (1975b). Organ culture of adult human articular cartilage. II. The differential rate of glycosaminoglycan synthesis in layers of articular cartilage matrix. effect of hyperoxia on synthesis of glycosaminoglycans. *J. Rheumatol.* **2**, 280–286.

Jaczewski, Z., and Krzywinska, K. (1974). The induction of antler growth in a red deer male castrated before puberty by traumatization of the pedicle. *Bull. Acad. Polish Sci.* **22**, 67–72.

Jaffe, H. L. (1968). *Tumors and Tumorous Conditions of the Bones and Joints.* Lea & Febiger, Philadelphia, PA.

Jaffe, H. L. (1972). *Metabolic, Degenerative, and Inflammatory Diseases of Bones and Joints.* Lea & Febiger, Philadelphia, PA.

Jain, P., and Sabet, T. Y. (1974). Fracture healing and antibodies against cartilage proteoglycans. *J. Dent. Res.* **53**, 282.

Jakob, W., Jentzsch, K. D., Mauersberger, B., and Heder, G. (1978). Chick embryo chorioallantoic membrane as a bioassay for angiogenesis factors – reactions induced by carrier materials. *Exper. Pathol.* **15**, 241–249.

Jakowlew, S. B., Dillard, P. J., Winokur, T. S., Flanders, K. C., Sporn, M. B., and Roberts, A. B. (1991). Expression of transforming growth factor-βs 1–4 in chicken embryo chondrocytes and myocytes. *Devel. Biol.* **143**, 135–148.

James, R. B., Alexander, R. W., and Traver, J. G., Jr. (1974). Osteochondroma of the mandibular coronoid process. *Oral Surg., Oral Med. Oral Pathol.* **37**, 189–195.

Jande, S. S. (1971). Fine structural study of osteocytes and their surrounding bone matrix with respect to their age in young chicks. *J. Ultrastruct. Res.* **37**, 279–300.

Jande, S. S. (1972). Effects of parathormone on osteocytes and their surrounding bone matrix: An electron microscopic study. *Z. Zellforsch.* **130**, 463–470.

Jande, S. S., and Bélanger, L. F. (1969). Ultrastructural changes associated with osteocytic osteolysis in normal trabecular bone. *Anat. Rec.* **163**, 204.

Jande, S. S., and Bélanger, L. F. (1971). Electron microscopy of osteocytes and the pericellular matrix in rat trabecular bone. *Calcif. Tissue Res.* **6**, 280–289.

Jande, S. S., and Bélanger, L. F. (1973). The life cycle of the osteocyte. *Clin. Orthop. Rel. Res.* **94**, 281–305.

Janis, C. M. (1982). Evolution of horns in ungulates: Ecology and paleoecology. *Biol. Rev. Camb. Philos. Soc.* **57**, 261–318.

Janis, C. M. (1986). Evolution of horns and related structures in hoofed mammals. *Discovery* **19**(2), 8–17.

Janis, R., Sandson, J., Smith, C., and Hamerman, D. (1967). Synovial cell synthesis of a substance immunologically like cartilage proteinpolysaccharide. *Science* **158**, 1464–1467.

Janners, M. Y., and Searls, R. L. (1970). Changes in rate of cellular proliferation during the differentiation of cartilage and muscle in the mesenchyme of the embryonic chick wing. *Devel. Biol.* **23**, 136–155.

Janners, M. Y., and Searls, R. L. (1971). Effect of removal of the apical ectodermal ridge on the rate of cell division in the sub-ridge mesenchyme of the embryonic chick wing. *Devel. Biol.* **24**, 465–276.

Jansen, M. (1920). On bone formation: its relation to tension and pressure. University of Manchester Med. Ser. # 16, Manchester, 114 pp.

Janvier, P. (1996). *Early Vertebrates.* The Clarendon Press, Oxford.

Janvier, P. (2001). Classical French Recipe: Matelotte de Lamproie. *The Linnean* **17**, 27.

Janvier, P., and Arsenault, M. (2002). Calcification of early vertebrate cartilage. *Nature* **417**, 609.

Jarry, L., and Uhthoff, H. K. (1971). Differences in healing of metaphyseal and diaphyseal fractures. *Can. J. Surg.* **14**, 127–135.

Järvinen, T. A. H., Jozsa, L., Kannus, P., Järvinen, T. L. N., Kvist, M., Hurme, T., Isola, J., Kalimo, H., and Järvinen, M. (1999). Mechanical loading regulates tenascin-C expression in the osteotendious junction. *J. Cell Sci.* **112**, 3157–3166.

Jarvik, E. (1959). Dermal fin-rays and Holmgren's principle of delamination. *Kungl. Svensk. Vetensk. Handl.* **6**, 3–49.

Jarvik, E. (1965a). On the origin of girdles and paired fins. *Israel J. Zool.* **14**, 141–172.

Jarvik, E. (1965b). Die Raspelzunge der Cyclostomen und die pentadactyle Extremität der Tetrapoden als Bewise für monophyletische Herkunft. *Zool. Anz.* **175**, 8–143.

Jarvik, E. (1980). *Basic Structure and Evolution of Vertebrates. Volumes 1 and 2.* Academic Press, London.

Jaskoll, T. F., and Maderson, P. F. A. (1978). A histological study of the development of the avian middle ear and tympanum. *Anat. Rec.* **190**, 177–200.

Jaskoll, T., and Melnick, M. (1982). The effects of long-term fetal constraint *in vitro* on the cranial base and other skeletal components. *Amer. J. Med. Genet.* **12**, 289–300.

Javor, P. (1074). The chondrogenesis in the extremities of staged human embryos. *Z. Anat. Entwick.* **145**, 227–242.

Jaworski, Z. F. G. (ed.) (1976). *Bone Morphometry.* University of Ottawa Press, Ottawa, Ontario.

Jaworski, Z. F. G. (1984a). Coupling of bone formation to bone resorption: A broader view. *Calcif. Tissue Int.* **36**, 531–535.

Jaworski, Z. F. G. (1984b). Lamellar bone turnover system and its effector organ. *Calcif. Tissue Int.* **36**, 546–555.

Jaworski, Z. F. G. (1992). Haversian systems and Haversian bone. In *Bone, Volume 4: Bone Metabolism and Mineralization* (B. K. Hall, ed.), pp. 21–45. CRC Press, Boca Raton, FL.

Jaworski, Z. F. G., Duck, B., and Sekaly, G. (1981). Kinetics of osteoclasts and their nuclei in evolving secondary Haversian systems. *J. Anat.* **133**, 397–406.

Jaworski, Z. F. G., and Hooper, C. (1980). Study of cell kinetics within evolving secondary Haversian systems. *J. Anat.* **131**, 91–102.

Jeanloz, R. W. (1960). The nomenclature of mucopolysaccharides. *Arthritis Rheum.* **3**, 233–234.

Jeansonne, B. G., Feagin, F. F., Shoemaker, R. L., and Rehm, W. S. (1978). Transmembrane potential of osteoblasts. *J. Dent. Res.* **57**, 361–364.

Jeansonne, B. G., Feagin, F. F., McMinn, R. W., Shoemaker, R. L., and Rehm, W. S. (1979). Cell-to-cell communication of osteoblasts. *J. Dent. Res.* **58**, 1415–1423.

Jee, W. S. S., Wronski, T. J., Morey, E. R., and Kimmel, D. B. (1983). Effects of space flight on trabecular bone in rats. *Amer. J. Physiol.* **244**, R310–R314.

Jeffery, K. L. (1974). Fracture of the os penis in a dog. *J. Amer. Animal Hosp. Assoc.* **10**, 41–44.

Jeffery, N., and Spoor, F. (2004). Ossification and midline shape changes of the human fetal cranial base. *Amer. J. Phys. Anthrop.* **123**, 78–90.

Jeffery, W. R., Strickler, A. G., and Yamamoto, Y. (2003). To see or not to see: Evolution of eye degeneration in Mexican blind cavefish. *Integ. Comp. Biol.* **4**, 531–541.

Jegalian, B. G., and de Robertis, E. M. (1992). Homeotic transformations in the mouse induced by overexpression of a human Hox3.3 transgene. *Cell* **71**, 901–910.

Jenkins, D. H. R., Cheng, D. H. F., and Hodgson, A. R. (1975). Stimulation of bone growth by periosteal stripping. *J. Bone Joint Surg. Br. Vol.* **57**, 482–484.

Jenkins, F. A., Jr., and Walsh, D. M. (1993). An early Jurassic caecilian with limbs. *Nature* **365**, 246–250.

Jennings, J. C., and Mohan, S. (1990). Heterogeneity of latent transforming growth factor-β isolated from bone matrix proteins. *Endocrinology* **126**, 1014–1021.

Jergesen, H. E., Chua, J., Kao, R. T., and Kaben, L. B. (1991). Age effects on bone induction by demineralized bone powder. *Clin. Orthop. Rel. Res.* **268**, 253–259.

Jernvall, J. (2000). Linking development with generation of novelty in mammalian teeth. *Proc. Natl Acad. Sci. U.S.A.* **97**, 2641–2645.

Jernvall, J, Hunter, J. P., and Fortelius, M. (1996). Molar tooth diversity, disparity, and ecology in Cenozoic ungulate radiations. *Science* **274**, 1489–1492.

Jha, M., and Sushella, A. K. (1982). *In vivo* chondrogenesis and histochemical appearance of dermatan sulphate in rabbit cancellous bone. *Differentiation* **22**, 235–236.

Jiang, H., Knudson, C. B., and Knudson, W. (2001). Antisense inhibition of CD44 tailless splice variant in human articular chondrocytes promotes hyaluronan internalization. *Arthr. Rheum.* **44**, 2599–2610.

Jiang, T.-X., Yi, J.-R., Ying, S.-Y., and Chuong, C.-M. (1993). Activin enhances chondrogenesis of limb bud cells: Stimulation of precartilaginous mesenchymal condensations and expression of N-CAM. *Devel. Biol.* **155**, 545–557.

Jiang, X., Iseki, S., Maxson, R. E., Sucov, H. M., and Morriss-Kay, G. M. (2002). Tissue origins and interactions in the mammalian skull vault. *Devel. Biol.* **241**, 106–116.

Jiang, X., Rowitch, D. H., Soriano, P., McMahon, A. P., and Sucov, H. M. (2000). Fate of the mammalian cardiac neural crest. *Development* **127**, 1607–1616.

Jilka, R. L. (1986). Are osteoblastic cells required for the control of osteoclast activity by parathyroid hormone? *Bone & Mineral* **1**, 261–266.

Jiménez, M. J. G., Balbín, M., Alvarez, J., Komori, T., Bianco, P., Holmbeck, K., Birkedal, H. H., Lopez, J. M., and Lopez-Otin, C. (2001). A regulatory cascade involving retinoic acid, Cbfa1,

and matrix metalloproteinases is coupled to the development of a process of perichondral invasion and osteogenic differentiation during bone formation. *J. Cell Biol.* **155**, 1333–1344.

Jingushi, S., Heydemann, A., Kana, S. K., Macey, L. R., and Bolander, M. E. (1990). Acidic fibroblast growth factor (aFGF) injection stimulates cartilage enlargement and inhibits cartilage gene expression in rat fracture healing. *J. Orthop. Res.* **8**, 364–371.

Jockusch, E. L. (1997). Geographic variation and phenotypic plasticity of trunk vertebrae in slender salamanders, *Batrachoseps* (Caudata: Plethodontidae). *Evolution* **51**, 1966–1982.

Johansen, V. A., and Hall, S. H. (1982). Morphogenesis of the mouse coronal suture. *Acta Anat.* **104**, 58–67.

Johnell, O., and Hulth, A. (1980). The response of bone marrow cells, thymocytes and osteoclasts to hydrocortisone. *Brit. J. Exp. Pathol.* **61**, 411–414.

Johnell, O., and Telhag, H. (1977). Effect of osteotomy and cartilage damage on mitotic activity – experimental study in rabbits. *Acta Orthop. Scand.* **48**, 263–265.

Johnels, A. G. (1949). On the development and morphology of the skeleton of the head of *Petromyzon. Acta Zool. (Stockholm)* **29**, 139–279.

Johnson, D. R. (1974). The *in vivo* behaviour of achondroplastic cartilage from the cartilage anomaly (*can/can*) mouse. *J. Embryol. Exp. Morphol.* **31**, 313–318.

Johnson, D. R. (1977). Ultrastructural observations on *stumpy* (*stm*), a new chondrodystrophic mutant in the mouse. *J. Embryol. Exp. Morphol.* **39**, 279–284.

Johnson, D. R. (1978). The growth of femur and tibia in three genetically distinct chondrodystrophic mutants in the mouse. *J. Anat.* **125**, 267–276.

Johnson, D. R. (1980). Formation of marrow cavity and ossification in mouse limb buds grown *in vitro. J. Embryol. Exp. Morphol.* **56**, 301–307.

Johnson, D. R. (1986). *The Genetics of the Skeleton.* Clarendon Press, Oxford.

Johnson, D. R., and Hunt, R. M. (1974). Biochemical observations on the cartilage if achondroplastic (*can*) mice. *J. Embryol. Exp. Morphol.* **31**, 319–328.

Johnson, D. R., and O'Higgins, P. (1996). Is there a link between changes in the vertebral 'Hox code' and the shape of vertebrae? A quantitative study of shape change in the cervical vertebral column of mice. *J. Theor. Biol.* **183**, 89–93.

Johnson, D. R., O'Higgins, P., and McAndrew, T. J. (1990). The effect of diet on bone shape in the mouse. *J. Anat.* **172**, 213–220.

Johnson, D. R., O'Higgins, P., and McAndrew, T. J. (1991). The effects of disorders of cartilage formation and bone resorption on bone shape: A study with chondrodystrophic and osteopetrotic mouse mutants. *J. Anat.* **176**, 81–88.

Johnson, D. R., and Wise, J. M. (1971). Cartilage anomaly (*can*): A new mutant gene in the mouse. *J. Embryol. Exp. Morphol.* **25**, 21–31.

Johnson, E. E., and Urist, M. R. (1998). One-stage lengthening of femoral nonunion augmented with human bone morphogenetic protein. *Clin. Orthop. Rel. Res.* **347**, 105–116.

Johnson, F. R., and McMinn, R. M. H. (1956). Transitional epithelium and osteogenesis. *J. Anat.* **90**, 106–116.

Johnson, J., Shinomura, T., Eberspaecher, H., Pinero, G., de Crombrugghe, B., and Höök, M. (1999). Expression and localization of *Pg-Lb/epiphycan* during mouse development. *Devel. Dynam.* **216**, 499–510.

Johnson, L. G. (1973). Development of chick embryo conjunctival papillae and scleral ossicles after hydrocortisone treatment. *Devel. Biol.* **30**, 223–227.

Johnson, R. B., and Highison, G. J. (1983). A re-examination of the osteocytic network of interdental bone. *J. Submicrosc. Cytol.* **15**, 619–626.

Johnson, R. S., Spiegelman, B. M., and Papaioannou, V. (1992). Pleiotropic effects of a null mutation in the *c-fos* proto-oncogene. *Cell* **71**, 577–586.

Johnston, M. C. (1964). Facial malformation in chick embryos resulting from removal of neural crest. *J. Dent. Res.* **43**, 822.

Johnston, M. C. (1966). A radioautographic study of the migration and fate of cranial neural crest cells in the chick embryo. *Anat. Rec.* **156**, 143–156.

Johnston, M. C. (1975). The neural crest in abnormalities of the face and brain. *Birth Defects, Orig. Artic. Ser.* **11**, 1–18.

Johnston, M. C. (1977). Pathogenesis of selected craniofacial malformations. *Birth Defects, Orig. Artic. Ser.* **13**, 1–6.

Johnston, M. C., and Bronsky, P. T. (1995). Prenatal craniofacial development: New insights on normal and abnormal mechanisms. *Crit. Rev. Oral Biol. Med.* **6**, 368–422.

Johnston, M. C., and Listgarten, M. A. (1972). Observations on the migration, interaction, and early differentiation of orofacial tissues. In *Developmental Aspects of Oral Biology* (H. C. Slavkin and L. A. Bavetta, eds), pp. 55–80. Academic Press, New York, NY.

Johnston, M. C., Noden, D. M., Hazelton, R. D., Coulombre, J. L., and Coulombre, A. J. (1979). Origins of avian ocular and periocular tissues. *Exp. Eye Res.* **29**, 27–45.

Johnson, M. L. (1933). The time and order of appearance of ossification centres in the albino mouse. *Amer. J. Anat.* **52**, 241–271.

Johnson, P. A., Atkinson, P. J., and Moore, W. J. (1976). The development and structure of the chimpanzee mandible. *J. Anat.* **122**, 467–477.

Johnston, P. A. (1979). Growth rings in dinosaur teeth. *Nature* **178**, 635–636.

Johnston, P. M., and Comar, C. L. (1957). Autoradiographic studies of the utilization of S^{35}-sulfate by the chick embryo. *J. Biophys. Biochem. Cytol.* **3**, 231–238.

Johnstone, B., Hering, T. M., Caplan, A. I., Goldberg, V. M., and Yoo, J. U. (1998). In vitro chondrogenesis of bone-marrow-derived mesenchymal progenitor cells. *Exp. Cell Res.* **238**, 265–272.

Joldersma, M., Burger, E. H., Semeins, C. M., and Klein-Nulend, J. (2000). Mechanical stress induces COX-2 mRNA expression in bone cells from elderly women. *J. Biomech.* **33**, 53–61.

Jollie, M. T. (1957). The head skeleton of the chicken and remarks on the anatomy of this region in other birds. *J. Morphol.* **100**, 389–436.

Jollie, M. T. (1960). The head skeleton of the lizard. *Acta Zool. (Stockholm)* **41**, 1–64.

Jollie, M. T. (1968). The head skeleton of a new-born *Manis javanica* with comments on the ontogeny and phylogeny of the mammal head skeleton. *Acta Zool. (Stockholm)* **49**, 227–305.

Jollie, M. T. (1971). Some developmental aspects of the head skeleton of the 35–37 mm *Squalus acanthias* foetus. *J. Morphol.* **133**, 17–40.

Jollie, M. T. (1975). Development of the head skeleton and pectoral girdle in *Esox*. *J. Morphol.* **147**, 61–88.

Jollie, M. T. (1980). Development of head and pectoral girdle skeleton and scales in *Acipenser*. *Copeia* **1980**, 226–249.

Jollie, M. T. (1981). Segment theory and the homologizing of cranial bones. *Amer. Nat.* **118**, 785–802.

Jollie, M. T. (1984a). Development of the head skeleton and pectoral girdle of salmons, with a note on the scales. *Can. J. Zool.* **62**, 1757–1778.

Jollie, M. T. (1984b). The vertebrate head – Segmented or a single morphogenetic structure? *J. Vert. Paleont.* **4**, 320–329.

Jollie, M. T. (1984c). Development of the head and pectoral skeleton of *Polypterus* with a note on scales (Pisces: Actinopterygii). *J. Zool. London* **204**, 469–507.

Jollie, M. T. (1984d). Development of cranial and pectoral girdle bones of *Lepisosteus* with a note on scales. *Copeia* **1984**, 476–501.

Jollie, M. T. (1984e). Development of the head and pectoral skeleton of *Amia* with a note on the scales. *Gegenb. Morphol. Jb. Leipzig* **130**, 315–351.

Jolly, M. T. (1961). Condylectomy in the rat. An investigation of the ensuing repair processes in the region of the temporo-mandibular articulation. *Aust. Dent. J.* **6**, 243–256.

Jolly, R. J., and Moore, W. J. (1975). Skull growth in achondroplastic (*cn*) mice; A craniometric study. *J. Embryol. Exp. Morphol.* **33**, 1013–1022.

Jones, F. S., and Jones, P. L. (2000). The tenascin family of ECM glycoproteins: Structure, function, and regulation during embryonic development and tissue remodeling. *Devel. Dynam.* **218**, 235–259.

Jones, K. L., and Addison, J. (1975). Pituitary fibroblast growth factor as a stimulator of growth in cultured rabbit articular chondrocytes. *Endocrinology* **97**, 359–365.

Jones, P. L., Schmidhauser, C., and Bissell, M. J. (1993). Regulation of gene expression and cell function by extracellular matrix. *Crit. Rev. Eukary. Gene Exp.* **3**, 137–154.

Jones, S. J. (1974). Secretory territories and rate of matrix formation of osteoblasts. *Calcif. Tissue Res.* **14**, 309–316.

Jones, S. J., and Boyde, A. (1976a). Morphological changes of osteoblasts in vitro. *Cell Tissue Res.* **166**, 101–108.

Jones, S. J., and Boyde, A. (1976b). Experimental study of changes in osteoblastic shape induced by calcitonin and parathyroid extract in an organ culture system. *Cell Tissue Res.* **169**, 449–466.

Jones, S. J., and Boyde, A. (1977a). The migration of osteoblasts. *Cell Tissue Res.* **184**, 179–194.

Jones, S. J., and Boyde, A. (1977a). Some morphological observations on osteoclasts. *Cell Tissue Res.* **185**, 387–398.

Jones, S. J., Boyde, A., and Ali, N. N. (1984). The resorption of biological and non-biological substrates by cultured avian and mammalian osteoclasts. *Anat. Embryol.* **170**, 247–256.

Jones, S. J., Boyde, A., and Pawley, J. B. (1975). Osteoblasts and collagen orientation. *Cell Tissue Res.* **159**, 73–80.

Jones, S. J., Gray, C., and Boyde, A. (1994). Simulation of bone resorption-repair coupling in vitro. *Anat. Embryol.* **190**, 339–340.

Jones, S. J., Gray, C., Sakamaki, H., Arora, M., Boyde, A., Gourdie, R., and Grewen, C. (1993). The incidence and size of gap junctions between the bone cells in rat calvariae. *Anat. Embryol.* **187**, 343–352.

Jones, S. J., Hogg, N. M., Shapiro, I. M., Slusarenko, M., and Boyde, A. (1981). Cells with Fc receptors in the cell layer next to osteoblasts and osteoclasts on bone. *Metab. Bone Dis.* **2**, 357–362.

Jørgensen, J. M., Lomholt, J. P., Weber, R. E., and Malte, H. (eds) (1998). *The Biology of Hagfishes*. Chapman & Hall, London.

Joseph, J., and Tydd, M. (1973a). The effects of cortisone acetate on tissue regeneration in the rabbit's ear. *J. Anat.* **115**, 445–460.

Joseph, J., and Tydd, M. (1973b). Effect of sodium fluoride on regeneration in the rabbit ear. *Nature* **246**, 165–166.

Jotereau, F., and Le Douarin, N. M. (1978). The developmental relationship between osteocytes and osteoclasts: A study using the quail-chick nuclear marker in endochondral ossification. *Devel. Biol.* **63**, 253–265.

Jouve, C., Iimura, T., and Pourquié, O. (2002). Onset of the segmentation clock in the chick embryo: evidence for oscillations in the somite precursors in the primitive streak. *Development* **129**, 1107–1117.

Jowsey, J. (1968). Age and species differences in bone. *Cornell Vet.* **58**, 74–94.

Joyce, M. E., Jingushi, S., and Bolander, M. E. (1990a). Transforming growth factor-β in the regulation of fracture repair. *Orthop. Clinics N. Amer.* **21**, 199–209.

Joyce, M. E., Roberts, A. B., Sporn, M. B., and Bolander, M. E. (1990b). Transforming growth factor-β and the initiation of chondrogenesis and osteogenesis in the rat femur. *J. Cell Biol.* **110**, 2195–2207.

Joyce, S. J., and Cohen, A. S. (1970). The interphalangeal area of *Rana pipiens*: A light and electron microscopic study of the development of a fibrocartilaginous joint (symphysis). *J. Morphol.* **130**, 315–336.

Jundt, G., Berghäuser, K.-H., Termine, J. D., and Schulz, A. (1987). Osteonectin – A differentiation marker of bone cells. *Cell Tissue Res.* **248**, 409–415.

Jung, J.-C., and Tsonis, P. A. (1998). Role of 5′ *HoxD* genes in chondrogenesis *in vitro*. *Int. J. Devel. Biol.* **42**, 609–615.

Jungers, W. L. (ed.) (1985). *Size and Scaling in Primate Biology.* Plenum Press, New York, NY.

Junqueira, L. C. U., Toledo, O. M. S., and Montes, G. S. (1983). Histochemical and morphological studies on a new type of acellular cartilage. *Basic Appl. Histochem.* **27**, 1–8.

Jurand, A. (1962). The development of the notochord in chick embryos. *J. Embryol. Exp. Morphol.* **10**, 602–621.

Jurand, A. (1965). Ultrastructural aspects of early development of the fore-limb buds in the chick and the mouse. *Proc. R. Soc. London, Ser. B.* **162**, 387–405.

Jurand, A. (1974). Some aspects of the development of the notochord in mouse embryos. *J. Embryol. Exp. Morphol.* **32**, 1–34.

Jurié-Lekic, G., Bradamante, Z., and Svajger, A. (1982). Phenotype modification in zones of appositional growth of human elastic cartilage. *Folia Anat. Yugos.* **12**, 93–100.

Juriloff, D. M., and Harris, M. J. (1991). Mapping the mouse craniofacial mutation first arch (*Far*) to chromosome 2. *J. Hered.* **82**, 402–405.

Juriloff, D. M., Harris, M. J., Miller, J. E., Jacobson, D., and Martin, P. (1992). Is *Far* a *Hox* mutation? *J. Craniofac. Genet. Devel. Biol.* **12**, 119–129 (1991).

K

Kaan, H. W. (1930). The relation of the developing auditory vesicle to the formation of the cartilage capsule in *Ambystoma punctatum*. *J. Exp. Zool.* **55**, 263–291.

Kaan, H. W. (1938). Further studies on the auditory vesicle and cartilaginous capsule of *Ambystoma punctatum*. *J. Exp. Zool.* **78**, 159–183.

Kaban, L. B., and Glowacki, J. (1981). Induced osteogenesis in the repair of experimental defects in rats. *J. Dent. Res.* **60**, 1356–1364.

Kaban, L. B., and Glowacki, J. (1984). Augmentation of rat mandibular ridge with demineralized bone implants. *J. Dent. Res.* **63**, 998–1002.

Kacem, A., and Meunier, F. J. (2000). Mise en évidence de l'ostèolyse périostéocytaire vertébral chez le salmon Atlantique *Salmo salar* (Salmonidae, Teleostei), au cours de sa migration anadrome. *Cybium* **24**(3) Suppl. 105–112.

Kacem, A., and Meunier, F. J. (2003). Halastatic demineralization in the vertebrae of Atlantic salmon during their spawning migration. *J. Fish Biol.* **63**, 1122–1130.

Kacem, A., Meunier, F. J., Aubin, J., and Haffray, P. (2004). Caractérisation histo-morphologiques des malformations du squelette vertebral chez la truitte arc-en-ciel (*Oncorhynchus mykiss*) après différents traitements de triploïdisation. *Cybium* **28**(1) Suppl. 15–23.

Kacem, A., Meunier, F. J., and Baglinière, J. L. (1998). A quantitative study of morphological and histological changes in the skeleton of *Salmo salar* during its anadromous migration. *J. Fish Biol.* **53**, 1096–1109.

Kadam, K. M. (1958). The development of the chondrocranium of the sea horse, *Hippocampus* [Lophobranchii]. *J. Linn. Soc. Zool.* **43**, 557–573.

Kadam, K. M. (1961). The development of the skull in *Nerophis* (Lophobranchii). *Acta Zool. (Stockholm)* **42**, 257–298.

Kadis, B., Goodson, J. M., Offenbacher, S., Bruns, J. W., and Seibert, S. (1980). Characterization of osteoblast-like cells from fetal rat calvaria. *J. Dent. Res.* **59**, 2006–2013.

Kadowacki, M. H., Levett, J. M., Manjoney, D. L., Wilson, C. A., and Glagov, S. (1987). Comparative studies of prosthetic materials in the left atrium of the dog. *Virchows. Arch. A.* **411**, 173–177.

Kahn, A. J., and Simmons, D. J. (1975). Investigation of cell lineage in bone using a chimera of chick and quail embryonic tissue. *Nature* **258**, 325–327.

Kahn, A. J., and Simmons, D. J. (1977). Chondrocyte-to-osteocyte transformation in grafts of perichondrium-free epiphyseal cartilage. *Clin. Orthop. Rel. Res.* **129**, 299–304.

Kahn, A. J., Simmons, D. J., and Krukowski, M. (1981). Osteoclast precursor cells are present in the blood of preossification chick embryos. *Devel. Biol.* **84**, 230–234.

Kahn, A. J., Stewart, C. C., and Teitelbaum, S. L. (1978). Contact-mediated bone resorption by human monocytes *in vitro*. *Science* **199**, 988–990.

Kaku, M., Kawata, T., Kawasoko, S., Fujita, T., Tokimasa, C., and Tanne, K. (1999). Remodelling of the sagittal suture in osteopetrotic (*op/op*) mice associated with cranial flat bone growth. *J. Craniofac. Genet. Devel. Biol.* **19**, 109–112.

Kalayjian, D. B., and Cooper, R. R. (1972). Osteogenesis of the epiphysis: A light and electron microscopic study. *Clin. Orthop. Rel. Res.* **85**, 242–256.

Kale, S., Biermann, S., Edwards, C., Tarnowski, C., Morris, M., and Long, M. W. (2000). Three-dimensional cellular development is essential for ex vivo formation of human bone. *Nature Biotechnol.* **18**, 954–958.

Kalter, H. (1960). The teratogenic effects of hypervitaminosis A upon the face and mouth of inbred mice. *Ann. NY. Acad. Sci.* **85**, 42–55.

Kalter, H. (1980). A compendium of the genetically induced congenital malformations of the house mouse. *Teratology* **21**, 397–429.

Kamachi, Y., Uchikawa, M., and Kondoh, H. (2000). Pairing SOX off with partners in the regulation of embryonic development. *Trends Genet.* **16**, 182–187.

Kamakura, S., Sasano, Y., Homma, H., Susuki, O., Kagayama, M., and Motegi, K. (1999). Implantation of octacalcium phosphate (OCP) in rat skull defects enhances bone repair. *J. Dent. Res.* **78**, 1682–1687.

Kamalia, N., McCulloch, C. A. G., Tenenbaum, H. C., and Limeback, H. (1992). Dexamethasone recruitment of self-renewing osteoprogenitor cells in chick bone marrow stromal cell cultures. *Blood* **79**, 320–326.

Kaminski, M., Kaminski, G., Jakobisiak, M., and Brzeninski, W. (1977). Inhibition of lymphocyte-induced angiogenesis by isolated chondrocytes. *Nature* **268**, 238–240.

Kamiya, N., Jikko, A., Kimata, K., Damsky, C., Shimizu, K., and Watanabe, H. (2002). Establishment of a novel chondrocytic cell line N1511 derived from p53-null mice. *J. Bone Min. Res.* **17**, 1832–1843.

Kanan, C. V. (1962). Observations on the development of the osteocranium in *Camelus fromeoarius* (Camel). *Acta Zool. (Stockholm)* **43**, 297–310.

Kanazawa, E., and Mochizuki, K. (1974). The time and order of appearance of ossification centers in the hamster before birth. *Exp. Animals* **23**, 113–122.

Kanazawa, E., and Takano, K. (1979). Studies on the development of the Meckel's cartilage and the mandible in the hamster. *Nihon Univ. J. Oral Sci.* 5, 88–97.

Kanehisa, J., and Heersche, J. N. M. (1988). Osteoclastic bone resorption: *In vitro* analysis of the rate of resorption and migration of individual osteoclasts. *Bone* **9**, 73–79.

Kant, R., and Goldstein, R. S. (1999). Plasticity of axial identity among somites: cranial somites can generate vertebrae without expressing *Hox* genes appropriate to the trunk. *Devel. Biol.* **216**, 507–520.

Kantomaa, T. (1986a). The effect of prenatally increased oxygen tension on the development of the mandibular condyle. *Acta Odontol. Scand.* **44**, 301–305.

Kantomaa, T. (1986b). The effect of increased oxygen tension on the growth of the mandibular condyle. *Acta Odontol. Scand.* **44**, 307–312.

Kantomaa, T. (1987a). Effect of functional change in cell differentiation in the condylar cartilage. *J. Anat.* **152**, 133–244.

Kantomaa, T. (1987b). Reaction of the condylar tissues to attempts to increase mandibular growth. *Scand. J. Dent. Res.* **95**, 335–339.

Kantomaa, T., and Hall, B. K. (1988a). Mechanisms of mandibular condyle adaptation: An organ culture study in the mouse. *Acta Anat.* **132**, 114–119.

Kantomaa, T., and Hall, B. K. (1988b). Organ culture providing articulating function for the temporomandibular joint. *J. Anat.* **161**, 195–201.

Kantomaa, T., and Hall, B. K. (1991). On the importance of cAMP and Ca^{++} in mandibular condylar growth and adaptation. *Amer. J. Orthodont. Dentofac. Orthop.* **99**, 418–426.

Kantomaa, T., Huggare, J., Rönning, O., and Wendt, L. v. (1987). Cranial base morphology in untreated hydrocephalics. *Child's Nervous Syst.* **3**, 222–224.

Kantomaa, T., and Pirttiniemi, P. (1998). Changes in proteoglycan and collagen content in the mandibular condylar cartilage of the rabbit caused by an altered relationship between the condyle and glenoid fossa. *Eur. J. Orthodontics* **20**, 435–441.

Kantomaa, T., Pirttiniemi, P., and Tuominen, M. (1991). Cranial base and the growth of the cranial vault: an experimental study on the rabbit. *Proc. Finn. Dent. Soc.* **87**, 93–98.

Kantomaa, T., Pirttiniemi, P., Tuominen, M., and Poikela, A. (1994a). Glycosaminoglycan synthesis in the mandibular condyle during growth adaptation. *Acta Anat.* **151**, 88–96.

Kantomaa, T., and Rönning, O. (1985). Effect of growth of the maxilla on that of the mandible. *Europ. J. Orthodontics* **7**, 267–272.

Kantomaa, T., and Rönning, O. (1992). Growth of the mandible. In *Bone, Volume 6: Bone Growth – A.* (B. K. Hall, ed.), pp. 157–183. CRC Press, Boca Raton, FL.

Kantomaa, T., Tuominen, M., and Pirttiniemi, P. (1994b). Effect of mechanical forces on chondrocyte maturation and differentiation in the mandibular condyle of the rat. *J. Dent. Res.* **73**, 1150–1156.

Kantomaa, T., Tuominen, M., Pirttiniemi, P., and Rönning, O. (1992). Weaning and the histology of the mandibular condyle in the rat. *Acta Anat.* **144**, 311–315.

Kantorova, V. I. (1972). On the role of dura matter by the induction of regeneration of skull vault bones. *Ontogenez* **3**, 448–455 (in Russian).

Kantorova, V. I. (1975). Osteogenic role of dura matter in adult rabbits during regeneration of skull vault bones. *Ontogenez* **6**, 63–70 (in Russian).

Kantorova, V. I. (1976). Ectopic induction of bone and cartilaginous tissues in diffusion chambers in adult rabbits under the influence of crushed bone tissue. *Ontogenez* **7**, 262–270 (in Russian).

Kantorova, V. I. (1981a). Inducing ability of osteogenic cells – precursors of periosteum of the skull vault bones in adult rabbits. *Ontogenesis* **12**, 145–153.

Kantorova, V. I. (1981b). Induction of skull vault bone regeneration in adult dogs under the effect of conserved minced bone tissues. *Ontogenesis* **12**, 352–362.

Kantorova, V. I. (1983). Osteogenic possibilities of dura mater under conditions of its cultivation: *in vivo* in diffusion chambers. *DAN SSSR* **270**, 449–452.

Kanzler, B., Foreman, R. K., Labosky, P. A., and Mallo, M. (2000). BMP signaling is essential for development of skeletogenic and neurogenic cranial neural crest. *Development* **127**, 1095–1104.

Kanzler, B., Kuschert, S. J., Liu, Y.-H., and Mallo, M. (1998). *Hoxa-2* restricts the chondrogenic domain and inhibits bone formation during development of the branchial area. *Development* **125**, 2587–2597.

Kaprio, E. A. (1977). Ectodermal–mesenchymal interspace during the formation of the chick leg bud. A scanning and transmission microscopic study. *Wilhelm Roux. Arch. Entwicklungsmech. Org.* **182**, 213–226.

Karasawa, K., Kimata, K., Ito, K., Kato, Y., and Suzuki, S. (1979). Morphological and biochemical differentiation of limb bud cells cultured in chemically defined medium. *Devel. Biol.* **70**, 287–305.

Karcher-Djuricic, V., Ruch, J.-V., Stäubli, A., and Fabre, M. (1975). The role of microfilaments and microtubules in polarization of odontoblasts and ameloblasts. *Folia Anat. Jugosl.* **4**, 53–58.

Karp. S. J., Schipani, E., St.-Jacques, B., Hunzelman, J., Kronenberg, H., and McMahon, A. P. (2000). *Indian hedgehog* coordinates endochondral bone growth and morphogenesis via parathyroid hormone-related-protein-dependent and independent pathways. *Development* **127**, 543–548.

Karsenty, G. (1998). Genetics of skeletogenesis. *Devel. Genet.* **22**, 301–313.

Karsenty, G. (2000). The genetic transformation of bone biology. *Genes & Devel.* **13**, 3037–3051.

Karsenty, G. (2003). The complexities of skeletal biology. *Nature* **423**, 316–318.

Karsenty, G., and Wagner, E. F. (2002). Reaching a genetic and molecular understanding of skeletal development. *Devel. Cell* **2**, 389–406.

Kasperk, C. H., Wergedal, J. E., Mohan, S., Long, D. L., Lau, K. H. W., and Baylink, D. J. (1990). Interactions of growth factors present in bone matrix with bone cells: effects on DNA synthesis and alkaline phosphatase. *Growth Factors* **3**, 147–158.

Kasperk, C. H., Wergedal, J. E., Strong, D., Farley, J., Wangerin, K., Gropp, H., Ziegler, R., and Baylink, D. J. (1995). Human bone cell phenotypes differ depending on their skeletal site of origin. *J. Clin. Endocrinol. Metab.* **80**, 2511–2517.

Kassen, M., Mosekilde, L., and Eriksen, E. F. (1994). Effects of fluoride on human bone cells *in vitro* – differences in responsiveness between stromal osteoblast precursors and mature osteoblasts. *Eur. J. Endocrinol.* **130**, 381–386.

Kasten, T. P., Collin-Osdoby, P., Patel, N., Osdoby, P., Krukowski, M., Misko, T. P., Settle, S. L., Curie, M. G., and Nichols, G. A. (1994). Potentiation of osteoclast bone-resorption activity by inhibition of nitric oxide synthase. *Proc. Natl Acad. Sci. U.S.A.* **91**, 3569–3573.

Katagiri, T., Lee, T., Takeshima, H., Suda, T., Tanaka, H., and Omura, S. (1990a). Transforming growth factor-beta modulates proliferation and differentiation of mouse clonal osteoblastic MC 3T3-E1 cells depending on their maturation stages. *Bone Min.* **11**, 285–294.

Katagiri, T., Yamaguchi, A., Ikeda, T., Yoshiki, S., Wozney, J. M., Rosen, V., Wang, E. A., Tanaka, H., Omura, S., and Suda, T. (1990b). The non-osteogenic mouse pluripotent cell line C3H1T1/2 (C3H 10T12) is induced to differentiate into osteoblastic cells by recombinant human bone morphogenetic protein-2. *Biochem. Biophys. Res. Commun.* **172**, 295–299.

Katagiri, T., Yamaguchi, A., Komaki, M., Abe, E., Takahashi, N., Ikeda, T., Rosen, V., Wozney, J. M., Fujisawa-Sehara, A., and Suda, T. (1994). Bone morphogenetic protein-2 converts the differentiation pathway of C2C12 myoblasts into the osteoblast lineage. *J. Cell Biol.* **127**, 1755–1766.

Katchburian, E. (1973). Membrane-bound bodies as initiators of mineralization of dentine. *J. Anat.* **116**, 285–302.

Katchburian, E., and Severs, N. J. (1983). Matrix constituents of early developing bone examined by freeze fracture. *Cell Biol. Intern. Rep.* **7**, 1063–1070.

Kato, N., and Aoyama, H. (1998). Dermomyotomal origin of the ribs as revealed by extirpation and transplantation experiments in chick and quail embryos. *Development* **125**, 3437–3443.

Kato, Y., Iwamoto, M., Koike, T., Susuki, F., and Takano, Y. (1988). Terminal differentiation and calcification in rabbit chondrocyte cultures grown in centrifuge tubes: Regulation by transforming growth factor β and serum factors. *Proc. Natl Acad. Sci. U.S.A.* **85**, 9552–9556.

Katthagen, B.-D. (1987). *Bone Regeneration with Bone Substitutes. An Animal Study*. Springer-Verlag, New York, NY.

Katz, M. J. (1980). Allometry formula: A cellular model. *Growth* **44**, 89–96.

Katz, R. W., Felthousen, G. C., and Reddi, A. H. (1990). Radiation-sterilized insoluble collagenous bone matrix is a functional carrier of osteogenin for bone induction. *Calcif. Tissue Int.* **47**, 183–185.

Katz, R. W., and Reddi, A. H. (1988). Dissociative extraction and partial purification of osteogenin; a bone inductive protein, from rat tooth matrix by heparin affinity chromatography. *Biochem. Biophys. Res. Commun.* **157**, 1253–1257.

Katzman, R. L., and Jeanloz, R. G. (1969). Acid polysaccharides from invertebrate connective tisue: Phylogenetic aspects. *Science* **166**, 758–759.

Kaufman, M. H., Chang, H.-H., and Shaw, J. P. (1995). Craniofacial abnormalities in homozygous *Small eye* (*Sey/Sey*) embryos and newborn mice. *J. Anat.* **186**, 607–617.

Kaur, S., Singh, G., Stock, J. L., Shreiner, C. M., Kier, A. B., Yager, K. L., Mucenski, M. L., Scott, W. J., Jr., and Potter, S. S. (1992). Dominant mutation of the murine Hox-2.2 gene results in developmental abnormalities. *J. Exp. Zool.* **264**, 323–336.

Kavumpurath, S., and Hall, B. K. (1989). *In vitro* reformation of the perichondrium from perichondrial-free Meckel's cartilage of the embryonic chick. *J. Craniofac. Genet. Devel. Biol.* **9**, 173–184.

Kavumpurath, S., and Hall, B. K. (1990). Lack of either chondrocyte hypertrophy or osteogenesis in Meckel's cartilage of the embryonic chick exposed to epithelia and to thyroxine *in vitro*. *J. Craniofac. Genet. Devel. Biol.* **10**, 263–275.

Kawakami, M., Kuroda, S., Yamashita, K., Yoshida, C. A., Nakagawa, K., and Takeda, K. (1999). Expression of CSF-1 receptor on TRAP-positive multinuclear cells around the erupting molars in rats. *J. Craniofac. Genet. Devel. Biol.* **19**, 213–220.

Kawakami, Y., Capdevela, J., Büscher, D., Itoh, T., Rodríquez Esteban, C., and Izpisúa Belmonte, J. C. (2001). Wnt signals control FGF-dependent limb initiation and SER induction in the chick embryo. *Cell* **104**, 891–900.

Kawakami, Y., Ishikawa, T., Shimabara, M., Tanda, N., Enomoto-Iwamoto, M., Iwamoto, M., Kuwana, T., Ueki, A., Noji, S., and Nohno, T. (1996). BMP signaling during bone pattern development in the developing limb. *Development* **122**, 3557–3566.

Kawakami, Y., Tsukui, T., Ng, J. K., and Izpisúa-Belmonte, J. C. (2003). Insights into the molecular basis of vertebrate forelimb and hindlimb identity. In *Patterning in Vertebrate Development* (C. Tickle, ed.), pp. 198–213. Oxford University Press, Oxford.

Kawamura, M., and Urist, M. R. (1988). Growth factors, mitogens, cytokines, and bone morphogenetic protein in induced chondrogenesis in tissue culture. *Devel. Biol.* **130**, 435–442.

Kawasaki, K., and Weiss, K. M. (2003). Mineralized tissue and vertebrate evolution: The secretory calcium-binding phosphoprotein gene cluster. *Proc. Natl Acad. Sci. U.S.A.* **100**, 4060–4065.

Kawasoko, S., Niida, S., Kawata, T., Sugiyama, H., Kaku, M., Fujita, T., Tokimasa, C., Maeda, N., and Tanne, L. (2000). Influences of osteoclast deficiency on craniofacial growth in osteopetrotic (*op/op*) mice. *J. Craniofac. Genet. Devel. Biol.* **20**, 76–83.

Kawata, T., Niida, S., Kawasoko, S., Kaku, M., Fujita, T., Sugiyama, H., and Tanne, K. (1997). Morphology of the mandibular condyle in 'toothless' osteopetrotic (*op/op*) mice. *J. Craniofac. Genet. Devel. Biol.* **17**, 198–203.

Kawata, T., Tokimasa, C., Nowroozi, N., Fujita, T., Kaku, M., Kawasoto, S., Sugiyama, H., Ozawa, S., Zernik, J. H., and Tanne, K. (1999b). Lack of bone remodeling in osteopetrotic (*op/op*) mice associated with microdontia. *J. Craniofac. Genet. Devel. Biol.* **19**, 113–117.

Kawata, T., Tokimasa, C., Fujita, T., Ozawa, S., Sugiyama, H., and Tanne, K. (1999c). Facial skeletal growth in growing 'toothless' osteopetrotic (op/op) mice: radiographic findings. *J. Craniofac. Genet. Devel. Biol.* **19**, 221–225.

Kay, E. D. (1987). Craniofacial dysmorphogenesis following hypervitaminosis A in mice. *Teratology* **35**, 105–117.

Kay, S., and Harrison, J. M. (1969). Unusual pleiomorphic carcinoma of the pancreas featuring production of osteoid. *Cancer* **23**, 1158–1162.

Kaye, M. (1984). When is it an osteoclast? *J. Clin. Pathol.* **37**, 398–400.

Kean, M. R., and Houghton, P. (1987). The role of function in the development of human craniofacial form – a perspective. *Anat. Rec.* **218**, 107–110.

Kearney, M. (2003). Appendicular skeleton in amphisbaenians (Reptilia; Squamata). *Copeia* **2003**, 719–738.

Kebabian, J. W., Bloom, F. E., Steiner, A. L., and Greengard, P. (1975). Neurotransmitters increase cyclic nucleotides in postganglionic neurons: Immunocytochemical demonstration. *Science* **190**, 157–159.

Kedes, L., Ng, S.-Y., Lin, C.-S., Gunning, P., Eddy, R., Shows, T., and Leavitt, J. (1985). The human beta-actin multigene family. *Trans. Assoc. Amer. Phys.* **98**, 42–46.

Keene, D. R., Sakai, L. Y., and Burgeson, R. E. (1991). Human bone contains type-III collagen, type-VI collagen and fibrillin – type II collagen is present on specific fibers that may mediate attachment of tendons, ligaments and periosteum to calcified bone cortex. *J. Histochem. Cytochem.* **39**, 59–70.

Keilisborok, I. V., Raevskaya, M. V., and Friedenstein, A. Y. (1982). Properties of bone tissue induced by transitional epithelium. *Biull. Eksper. Biol. Med.* **94**, 1725–1728.

Keith, A. (1917). The foundation of our knowledge of bone growth by Duhamel and Hunter. *Brit. J. Surg.* **5**, 685–693.

Keith, A. (1918a). Researches made by Syme and Goodsir regarding the growth and repair of bones. *Brit. J. Surg.* **6**, 19–23.

Keith, A. (1918b). Researches into bone growth and bone reproduction by Ollier of Lyons and MacEwen of Glasgow. *Brit. J. Surg.* **6**, 160–165.

Keith, A. (1919). *Menders of the Maimed.* Oxford University Press, Oxford.

Keith, D. A. (1982). Development of the human temporomandibular joint. *Brit. J. Oral Surg.* **20**, 217–224.

Keith, D. A., Paz, M. A., Gallop, P. M., and Glimcher, M. J. (1977). Histological and biochemical identification and characterization of an elastin in cartilage. *J. Histochem. Cytochem.* **25**, 1154–1162.

Keller, E. T., and Brown, J. (2004). Prostate cancer bone metastases promote both osteolytic and osteoblastic activity. *J. Cell. Biochem.* **91**, 718–729.

Keller, E. T., Zhang, J., Cooper, C. R., Smith, P. C., McCauley, L. K., Pienta, K. J., and Taichman, R. S. (2001). Prostate carcinoma skeletal metastases: Cross-talk between tumor and bone. *Cancer & Metastasis Revs* **20**, 333–349.

Kelley, R. O. (1973). Fine structure of the apical rim-mesenchyme complex during limb morphogenesis in man. *J. Embryol. Exp. Morphol.* **29**, 117–131.

Kelley, R. O. (1975). Ultrastructural features of chondrogenesis in the human hand plate: a cytochemical and autoradiographic study. *J. Embryol. Exp. Morphol.* **33**, 387–401.

Kelley, R. O., and Bluemink, J. G. (1974). An ultrastructural analysis of cell and matrix differentiation during early limb development in *Xenopus laevis. Devel. Biol.* **37**, 1–17.

Kelley, R. O., and Fallon, J. F. (1976). Ultrastructural analysis of the apical ectodermal ridge during vertebrate limb morphogenesis. I. The human forelimb with special reference to gap junctions. *Devel. Biol.* **51**, 241–256.

Kelley, R. O., Goetinck, P. F., and MacCabe, J. A. (eds) (1982). *Limb Development and Regeneration. Part B.* Alan R. Liss Inc., New York, NY.

Kelsall, M. A., and Visci, M. (1970). Aortic cartilage in the heart of Syrian hamsters. *Anat. Rec.* **166**, 627–634.

Kember, N. F. (1960). Cell division in endochondral ossification. A study of cell proliferation in rat bones by the method of tritiated thymidine autoradiography. *J. Bone Joint Surg. Br. Vol.* **42**, 824–837.

Kember, N. F. (1971). Cell proliferation kinetics of bone growth: The first ten years of autoradiographic studies with tritiated thymidine. *Clin. Orthop. Rel. Res.* **76**, 213–230.

Kember, N. F. (1973). Patterns of cell division in the growth plates of the rat pelvis. *J. Anat.* **116**, 445–452.

Kember, N. F. (1978). Cell kinetics and the control of growth in long bones. *Cell Tissue Kinet.* **11**, 477–485.

Kember, N. F. (1979). Proliferation controls in a linear growth system: Theoretical studies of cell division in the cartilage growth plate. *J. Theor. Biol.* **78**, 365–374.

Kember, N. F. (1983). Cell kinetics of cartilage. In *Cartilage, Volume 1, Structure, Function and Biochemistry* (B. K. Hall, ed.), pp. 149–180. Academic Press, New York, NY.

Kember, N. F. (1985). Comparative patterns of cell division in epiphyseal cartilage plates in the rabbit. *J. Anat.* **142**, 185–190.

Kember, N. F., and Kirkwood, J. K. (1987). Cell kinetics and longitudinal bone growth in birds. *Cell Tissue Kinetics* **20**, 625–630.

Kember, N. F., and Kirkwood, J. K. (1991). Cell kinetics and the study of longitudinal bone growth: a perspective. In *Fundamentals of Bone Growth: Methodology and Applications* (A. D. Dixon, B. G. Sarnat, and D. A. N. Hoyte, eds), pp. 153–162. CRC Press, Boca Raton, FL.

Kember, N. F., and Walker, K. V. R. (1971). Control of bone growth in rats. *Nature* **229**, 428–429.

Kemp, A., and Nicoll, R. S. (1996). A histochemical analysis of biological residues in conodont elements. *Mod. Geol.* **20**, 287–302.

Kemp, N. E. (1953). Morphogenesis and metabolism of amphibian larvae after excision of heart. I. Morphogenesis of heartless tadpoles of *Rana pipiens. Anat. Rec.* **117**, 405–425.

Kemp, N. E. (1977a). Calcification of the endoskeleton in elasmobranchs. *Amer. Zool.* **17**, 932 (Abstract 402).

Kemp, N. E. (1977b). Banding pattern and fibrillogenesis of ceratotrichia in shark fins. *J. Morphol.* **154**, 187–204.

Kemp, N. E. (1979). Growth of calcified tissue in the endoskeleton of elasmobranchs. *Amer. Zool.* **19**, 977 (abstract 641).

Kemp, N. E. (1984). Organic matrices and mineral crystallites in vertebrate scales, teeth and skeletons. *Amer. Zool.* **24**, 965–976.

Kemp, N. E. (1985). Ameloblastic secretion and calcification of the enamel layer in shark teeth. *J. Morphol.* 184, 215–230.

Kemp, N. E., and Hoyt, J. A. (1965a). Influence of thyroxine on ossification of the femur in *Rana pipiens. J. Cell Biol.* **27**, 51A (abstract 98).

Kemp, N. E., and Hoyt, J. A. (1965b). Influence of thyroxine on order of ossification of bones of the skull of *Rana pipiens. Amer. Zool.* **5**, 719 (abstract 444).

Kemp, N. E., and Hoyt, J. A. (1965c). Influence of thyroxine on order of ossification of the parasphenoid bone in the skull of *Rana pipiens. Amer. Zool.* **5**, 710 (abstract 412).

Kemp, N. E., and Hoyt, J. A. (1969a). Sequence of ossification in the skeleton of growing and metamorphosing tadpoles of *Rana pipiens. J. Morphol.* **129**, 415–444.

Kemp, N. E., and Hoyt, J. A. (1969b). Ossification of the femur in thyroxine-treated tadpoles of *Rana pipiens. Devel. Biol.* **20**, 387–410.

Kemp, N. E., and Park, J. H. (1970). Regeneration of lepidotrichia and actinotrichia in the tailfin of the teleost, *Tilapia mossambica. Devel. Biol.* **22**, 321–342.

Kemp, N. E., and Quinn, B. L. (1951). Differentiation of *Amblystoma* larvae after extirpation of heart. *Anat. Rec.* **111**, 543–544.

Kemp, N. E., and Quinn, B. L. (1954). Morphogenesis and metabolism of amphibian larvae after excision of heart. II. Morphogenesis of heartless larvae of *Ambystoma punctatum. Anat. Rec.* **118**, 773–787.

Kemp, N. E., and Smith, S. K. (1976). Scanning electron microscopy of crystalline pattern in shark scales, teeth and calcified portions of skeleton. *Amer. Zool.* **16**, 183 (abstract 27).

Kemp, N. E., Smith, S. K., and Jacobs, R. A. (1975). Ultrastructure of calcified jaw tesserae in the lemon *shark Negaprion brevirostris. Amer. Zool.* **15**, 829 (abstract 828).

Kemp, N. E., Susko, M. A., and Christian, B. J. (1970). Repression of osteogenesis of the femur in parathyroid-treated tadpoles of *Rana pipiens. Amer. Zool.* **10**, 534 (abstract 340).

Kemp, N. E., and Westrin, S. K. (1979). Ultrastructure of calcified cartilage in the endoskeletal tesserae of sharks. Calcification of the endoskeleton in elasmobranchs. *J. Morphol.* **160**, 75–102.

Kemp, R. B., and Hinchliffe, J. R. (eds) (1984). *Matrices and Differentiation.* Alan R. Liss, New York, NY.

Kenrad, B., and Vilmann, H. (1977). Phosphomonesterases in growth cartilages of the rat. *Calcif. Tissue Res.* **181**, 349–359.

Kepes, J. J., Rubinstein, L. J., and Chiang, H. (1984). The role of astrocytes in the formation of cartilage in gliomas: An immunohistochemical study of 4 cases. *Amer. J. Pathol.* **117**, 471–483.

Keränen, S. V. E., Kettunen, P., Åberg, T., Thesleff, I., and Jernvall, J. (1999). Gene expression patterns associated with suppression of odontogenesis in mouse and vole diastema regions. *Devel. Genes Evol.* **209**, 495–506.

Kernek, C. B., and Wray, J. B. (1973). Cellular proliferation in the formation of fracture callus in the rat tibia. *Clin. Orthop. Rel. Res.* **91**, 197–209.

Kerr, J. F. R., Wyllie, A. H., and Currie, A. R. (1972). Apoptosis: a basic biological phenomenon with wide-ranging implications in tissue kinetics. *Brit. J. Cancer* **26**, 239–257.

Kéry, L. (1972). Effect of periosteal stripping and incision of cortical bone on the longitudinal growth of long bones: An experimental study. *Acta Chir. Acad. Sci. Hung.* **13**, 133–140.

Kessel, M. (1991). Molecular coding of axial positions by *Hox* genes. *Sem. Devel. Biol.* **2**, 367–373.

Kessel, M. (1992). Respecification of vertebral identities by retinoic acid. *Development* **115**, 487–501.

Kessel, M., Balling, R., and Gruss, P. (1990). Variations of cervical vertebrae after expression of a Hox-1.1 transgene in mice. *Cell* **61**, 301–308.

Kessel, M., and Gruss, P. (1990). Murine developmental control genes. *Science* **249**, 374–379.

Kessel, M., and Gruss, P. (1991). Homeotic transformations of murine vertebrae and concomitant alteration of *Hox* codes induced by retinoic acid. *Cell* **67**, 89–104.

Kessler, E., Takahara, K., Biniamunov, L., Brusel, M., and Greenspan, D. S. (1995). Bone morphogenetic protein-1: The type 1 procollagen C-proteinase. *Science* **271**, 360–362.

Kester, H. A., Ward-Van Oostwaard, Th. M. J., Goumans, M. J., Van Rooijen, M. A., Van der Saag, P. T., Van der Burg, B., and Mummery, C. L. (2000). Expression of TGF-β stimulated clone-22 (TSC-22) in mouse development and TGF-β signalling. *Devel. Dynam.* **218**, 563–572.

Kesteven, H. L. (1942). The ossification of the avian chondrocranium, with special reference to that of the emu. *Proc. Linn. Soc. N.S.W.* **67**, 213–237.

Kesteven, H. L. (1957a). Notes on the skull and cephalic muscles of the Amphisbaenia. *Proc. Linn. Soc. N.S.W.* **82**, 109–116.

Kesteven, H. L. (1957b). On the development of the crocodile skull. *Proc. Linn. Soc. N.S.W.* **82**, 117–124.

Key, L. L., Jr., Carnes, D. L., Jr., Weichselbaum, R., and Anast, C. S. (1983). Platelet-derived growth factor stimulates bone resorption by monocyte monolayers. *Endocrinology* **112**, 761–762.

Khan, S. N., Bostrom, M. G. P., and Lane, J. M. (2000). Bone growth factors. *Orthop. Clin. N. Amer.* **31**, 375–388.

Khokher, M. A., and Dandona, P. (1990). Fluoride stimulates [^{3}H]-thymidine incorporation and alkaline phosphatase production by human osteoblasts. *Metab. Clin. Exp.* **39**, 1118–1121.

Kibblewhite, D. J., Bruce, A. G., Strong, D. M., Ott, S. M., Purchio, A. F., and Larrabee, W. F., Jr. (1993). Transforming growth factor-β accelerates osteoinduction in a craniofacial onlay model. *Growth Factors* **9**, 185–193.

Kielty, C. M., Kwan, A. P. L., Holmes, D. F., Schor, S. L., and Grant, M. E. (1985). Type X collagen, a product of hypertrophic chondrocytes. *Biochem. J.* **227**, 545–554.

Kieny, M. (1958). Contribution a l'étude besoins nutritifs des tibias embryonnaires d'oiseaux cultivées en mileux naturels et synthétiques. *Arch. Anat. Microsc. Morphol. Exp.* **47**, 86–186.

Kieny, M. (1960). Rôle inducteur du mésodérme dans la différenciation précose du bourgeon de membre chez l'embryon de poulet. *J. Embryol. Exp. Morphol.* **8**, 457–467.

Kieny, M. (1967). Phénomènones de régulation de l'ébauche de membre chez l'embryon de poulet. *Rev. Anat. Morphol. Exper.* **39**, 1–39.

Kieny, M. (1971). Les phases d'activité morphogènes du mésoderme somatopleural pendant le développement précose du membre chez l'embryon de poulet. *Ann. Embryol. Morphog.* **4**, 281–298.

Kieny, M. (1975). Effets de la vinblastine sur la morphogenèse du pied de l'embryon de poulet. Aspects histologiques. *J. Embryol. Exp. Morphol.* **34**, 609–632.

Kieny, M. (1977). Proximo-distal pattern formation in avian limb development. In *Vertebrate Limb and Somite Morphogenesis* (D. A. Ede, J. R. Hinchliffe, and M. Balls, eds), pp. 87–104. Cambridge University Press, Cambridge.

Kieny, M., and Abbott, U. K. (1962). Contribution à l'étude de la diplodie liée au sexe et de l'achondroplasis Creeper chez l'embryon de poulet: Culture *in vitro* des ébauches cartilagineuses du tibio-tarse et du péroné. *Devel. Biol.* **4**, 473–488.

Kieny, M., Mauger, A., and Sengel, P. (1972). Early regionalization of the somitic mesoderm as studied by the development of the axial skeleton of the chick embryo. *Devel. Biol.* **28**, 142–161.

Kieny, M., and Pautou, M. P. (1976). Experimental analysis of excendentary regulation in xenoplastic quail–chick limb bud recombinants. *Wilhelm Roux Arch. Entwicklungsmech. Org.* **179**, 327–338.

Kieny, M., and Pautou, M. P. (1977). Proximo-distal pattern regulation in deficient avian limb buds. *Wilhelm Roux Arch. Entwicklungsmech. Org.* **183**, 177–192.

Kierdorf, H., Kierdorf, U., Szuwart, T., and Clemen, G. (1995). A light microscopic study of primary antler development in fallow deer (*Dama dama*). *Ann. Anat.* **177**, 525–532.

Kierdorf, H., Kierdorf, U., Szuwart, T., Gath, U., and Clemen, G. (1994). Light microscopic observations on the ossification process in the early developing pedicle of fallow deer (*Dama dama*). *Ann. Anat.* **176**, 243–249.

Kierdorf, U., Stoffels, E., Stoffels, D., Kierdorf, H., Szuwart, T., and Clemen, G. (2003). Histological studies of bone formation during pedicle restoration and early antler regeneration in roe deer and fallow deer. *Anat. Rec.* **273A**, 741–751.

Kiliaridis, S. (1995). Masticatory muscle. Influence on craniofacial growth. *Acta Odontol. Scand.* **53**, 196–202.

Kiliaridis, S., Bresin, A., Holm, J., and Strid, K. G. (1996). Effects of masticatory muscle function on bone mass in the mandible of the growing rat. *Acta Anat.* **155**, 200–205.

Killian, C. S., Corral, D. A., Kawinski, E., and Constantine, R. I. (1993). Mitogenic response of osteoblast cells to prostate-specific antigen suggests an activation of latent TGF-β and a proteolytic modulation of cell adhesion receptors. *Biochem. Biophys. Res. Commun.* **192**, 940–947.

Kiltie, R. A. (1985). Evolution and function of horns and hornlike organs in female ungulates. *Biol. J. Linn. Soc.* **24**, 299–320.

Kim, H., Lee, J.-H., and Suh, H. (2003). Interaction of mesenchymal stem cells and osteoblasts for *in vivo* osteogenesis. *Yonsei Med. J.* **44**, 187–197.

Kim, H.-J., Rice, D. P. C., Kettunen, P. J., and Thesleff, I. (1998). FGF-, BMP- and Shh-mediated signalling pathways in the regulation of cranial suture morphogenesis and calvarial bone development. *Development* **125**, 1241–1251.

Kim, H. T., Olson, W. M., and Hall, B. K. (2002). Effects of hind limb denervation on development of sesamoids in *Hymenochirus boettgeri*. *Integ. Comp. Biol.* **42**, 1256.

Kim, J. J., and Conrad, H. E. (1974). Effect of D-glucosamine concentration on the kinetics of mucopolysaccharide biosynthesis in cultured chick embryo vertebral cartilage. *J. Biol. Chem.* **249**, 3091–3097.

Kimata, K., Barrach, H.-J., Brown, K. S., and Pennypacker, J. P. (1981). Absence of proteoglycan core protein in cartilage from the *cmd/cmd* (*cartilage matrix deficiency*) mouse. *J. Biol. Chem.* **256**, 6961–6968.

Kimata, K., Brown, K. S., Shimizu, S., Murata, H., and Yamada, K. (1984). Complex carbohydrates in cartilaginous and other tissues of *cartilage matrix deficiency* (*cmd/cmd*) mice as studies by light microscopic histochemical methods. *Histochemistry* **80**, 539–546.

Kimata, K., Takeda, M., Suzuki, S., Pennypacker, J. P., Barrach, H.-J., and Brown, K. S. (1983). Presence of link protein in cartilage from *cmd/cmd* (*cartilage matrix deficiency*) mice. *Archs Biochem. Biophys.* **226**, 506–516.

Kimmel, C. B., Miller, C. T., Kruze, G., Ullmann, B., BreMiller, R. A., Larison, K. D., and Synder, H. C. (1998). The shaping of pharyngeal cartilages during early development of the zebrafish. *Devel. Biol.* **203**, 245–263.

Kimmel, C. B., Ullmann, B., Walker, M., Miller, C. T., and Crump, J. G. (2003). Endothelin 1-mediated regulation of pharyngeal bone development in zebrafish. *Development* **130**, 1339–1351.

Kimmel, D. B., and Jee, W. S. S. (1980a). Bone cell kinetics during longitudinal bone growth in the rat. *Calcif. Tissue Int.* **32**, 123–133.

Kimmel, D. B., and Jee, W. S. S. (1980b). A quantitative histological analysis of the growing long bone metaphysis. *Calcif. Tissue Int.* **32**, 113–122.

Kimmel, R. A., Turnbull, D. H., Blanquet, V., Wurst, W., Loomis, C. A., and Joyner, A. L. (2000). Two lineage boundaries coordinate vertebrate apical ectodermal ridge formation. *Genes & Devel.* **14**, 1377–1389.

Kimura, S., and Karasawa, K. (1985). Squid cartilage collagen: Isolation of type I collagen rich in carbohydrate. *Comp. Biochem. Physiol.* **81**, 361–366.

Kimura, S., and Matsui, R. (1990). Characterization of two genetically distinct type I-like collagens from hagfish (*Eptatretus burgeri*). *Comp. Biochem. Physiol. B.* **95**, 137–143.

Kimura, S., Miyauchi, Y., and Uchida, N. (1991). Scale and bone type-1 collagens of carp (*Cyprinus carpio*). *Comp. Biochem. Physiol. B.* **99**, 473–476.

Kimura, T., Yasui, N., Ohsawa, S., and Ono, K. (1984a). Chondrogenic differentiation of limb bud cells in collagen gel culture. *Biomed. Res.* **5**, 465–472.

Kimura, T., Yasui, N., Ohsawa, S., and Ono, K. (1984b). Chondrocytes embedded in collagen gels maintain cartilage phenotype during long-term culture. *Clin. Orthop. Rel. Res.* **186**, 231–239.

King, G. J., and Holtrop, M. E. (1975). Actin-like filaments in bone cells of cultured mouse calvaria as demonstrated by binding to heavy meromyosin. *J. Cell Biol.* **66**, 445, 450.

King, G. J., Latta, L., Rutenberg, J., Ossi, A., and Keeling, S. D. (1995). Alveolar bone turnover in male rats: site and age-specific changes. *Anat. Rec.* **242**, 321–328.

King, J. A., Marker, P. C., Seung Kwonjune, J., and Kingsley, D. M. (1994). BMP5, and the molecular skeletal and soft-tissue alterations in *short ear* mice. *Devel. Biol.* **166**, 112–122.

King, J. A., Storm, E. E., Marker, P. C., Sileone, R. J., and Kingsley, D. M. (1996). The role of BMPs and GDFs in development and region-specific skeletal structures. *Ann. N. Y. Acad. Sci.* **785**, 70–79.

King, M. C., and Wilson, A. C. (1975). Evolution at two levels in humans and chimpanzee. *Science* **188**, 107–116.

King, N., Hittinger, C. T., and Carroll, S. B. (2003). Evolution of key signaling and adhesion protein families predates animal origins. *Science* **301**, 361–363.

King, T. J. (1966). Nuclear transplantation in amphibia. *Methods Cell Physiol.* **2**, 1–36.

Kingsbury, J. E., Allen, C. G., and Rotheram, B. A. (1953). The histological structure of the beak in the chick. *Anat. Rec.* **116**, 95–111.

Kingsley, C. (1863). *The Water Babies. A Fairy Tale for a Land Baby*. Chapman & Hall, London.

Kingsley, D. M. (1994a). The TGF-β superfamily: new members, new receptors, and new genetic tests of function in different organisms. *Genes & Devel.* **8**, 133–146.

Kingsley, D. M. (1994b). What do BMP2 do in mammals? Clues from the mouse short-ear mutation. *Trends Genet.* **10**, 16–21.

Kingsley, D. M. (2001). Genetic control of bone and joint formation. In *The Molecular Basis of Skeletogenesis* (G. Cardew and J. Goode, eds; B. K. Hall, chair), Novartis Foundation Symposium 232, pp. 213–234. John Wiley & Sons, Chichester.

Kingsley, D. M., Bland, A. E., Grubber, J. M., Marker, P. C., Russell, L. B., Copeland, N. G., and Jenkins, N. A. (1992). The mouse *shortear* skeletal morphogenesis locus is associated with defects in a bone morphogenetic member of the TGFβ superfamily. *Cell* **71**, 399–410.

Kinoshita, A., Yamada, S., Haslam, S. M., Morris, H. R., Dell, A., and Sugahara, K. (1997). Novel tetrasaccharides isolated from squid cartilage chondroitin sulfate E contain unusual sulfated disaccharide units GlcA (3-*O*-sulfate) β1–3-GalNAc(6-*O*-sulfate) or GlcA(3-*O*-sulfate)β1–3GalNAc(4,5-*O*-disulfate). *J. Biol. Chem.* **272**, 19656–19665.

Kirkwood, J. K., and Kember, N. F. (1993). Comparative quantitative histology of mammalian growth plates. *J. Zool.* **231**, 543–562.

Kirn-Safran, C. B., Gomes, R. R., Brown, A. J., and Carson, D. D. (2004). Heparan sulfate proteoglycans: Coordinators of multiple signaling pathways during chondrogenesis. *Birth Defects Res. (Part C)*, **72**, 69–88.

Kirschbaum, F., and Meunier, F. J. (1981). Experimental regeneration of the caudal skeleton of the glass knifefish, *Eigenmannia virescens* (Rhamphicthydae, Gymnotoidei). *J. Morphol.* **168**, 121–136.

Kirschbaum, F., and Meunier, F. J. (1988). South American Gymnotiform fishes as model animals for regeneration experiments. In *Control of Cell Proliferation and Differentiation during Regeneration* (H. J. Anton, ed.), pp. 112–123. S. Karger, Basel.

Kirsch, T., and von der Mark, K. (1992). Remodelling of collagen types I, Ii and X and calcification of human fetal cartilage. *Bone Min.* **18**, 107–117.

Kirschner, R. E., Gannon, F. H., Xu, J., Wang, J., Karmacharya, J., Bartlett, S. P., and Whitaker, L. A. (2003). Craniosynostosis and altered patterns of fetal TGF-β expression induced by intrauterine constraint. *Plast. Recons. Surgery* **109**, 2338–2346.

Kitchin, I. C. (1949). The effects of notochordectomy in *Amblystoma mexicanum. J. Exp. Zool.* **112**, 393–415.

Kjaer, I. (1989). Prenatal sksletal maturation of the human maxilla. *J. Craniofac. Genet. Devel. Biol.* **9**, 257–264.

Kjaer, I. (1990). Correlated appearance of ossification and nerve tissue in human fetal jaws. *J. Craniofac. Genet. Devel. Biol.* **10**, 329–336.

Kjaer, I. (1995). Human prenatal craniofacial development related to brain development under normal and pathologic conditions. *Acta Odontol. Scand.* **53**, 133–143.

Kjaer, I. (1997). Mandibular movements during elevation and fusion of palatal shelves evaluated from the course of Meckel's cartilage. *J. Craniofac. Genet. Devel. Biol.* **17**, 80–85.

Kjaer, I., Kjaer, T. W., and Graem, N. (1993). Ossification sequence of occipital bone and vertebrae in human fetuses. *J. Craniofac. Genet. Devel. Biol.* **13**, 83–88.

Klein, I. E. (1975). The effect of thyrocalcitonin and growth hormones on bone metabolism. *J. Prosthet. Dentistry* **33**, 365–379.

Klein, L., and Zika, J. M. (1976). Comparison of whole calvarial bones and long bones during early growth in rats. II. Turnover of calcified and uncalcified collagen masses. *Calcif. Tissue Res.* **20**, 217–228.

Kleine, T. O. (1972). Structure, biosynthesis and heterogeneity of chondroitin sulfate proteins. *Naturwissenschaften* **59**, 64–71.

Kleinman, H. K., Pennypacker, J. P., and Brown, K. S. (1977). Proteoglycan and collagen of achondroplastic (*CN-CN*) neonatal mouse cartilage. *Growth* **41**, 171–178.

Klein-Nulend, J., Roelofsen, J., Sterck, J. G. H., Semeins, C. M., and Burger, E. H. (1995). Mechanical loading stimulates the release of transforming growth factor-β activity by cultured mouse calvarial and periosteal cells. *J. Cell Physiol.* **163**, 115–119.

Klein-Nulend, J., Semeins, C. M., Veldhuijzen, J. P., and Burger, E. H. (1993). Effect of mechanical stimulation on the production of soluble bone factors in cultured fetal mouse calvariae. *Cell Tissue Res.* **271**, 513–517.

Klein-Nulend, J., Veldhuijzen, J. P., de Jong, M., and Burger, E. H. (1987a). Increased bone formation and decreased bone resorption in fetal mouse calvaria as a result of intermittent compressive force *in vitro. Bone & Mineral* **2**, 442–448.

Klein-Nulend, J., Veldhuijzen, J. P., Vandesta, R. J., VanKampe, G. P., Kuijer, R., and Burger, E. H. (1987b). Influence of intermittent compressive force on proteoglycan content in calcifying growth plate cartilage *in vitro. J. Biol. Chem.* **262**, 5490–5495.

Klembara, J. (2001). Postparietal and prehatching ontogeny of the supraoccipital in *Alligator mississippiensis* (Archosauria, Crocodylia). *Anat. Embryol.* **249**, 147–153.

Klement, B. J., and Spooner, B. S. (1994). Pre-metatarsal skeletal development in tissue culture at unit- and microgravity. *J. Exp. Zool.* **269**, 230–241.

Klement, B. J., Young, Q. M., George, B. J., and Nokkaew, M. (2004). Skeletal tissue growth, differentiation and mineralization in the NASA Rotating Wall Vessel. *Bone* **34**, 487–498.

Klima, M. (1985). Development of shoulder girdle and sternum in mammals. *Fortschr. Zool.* **30**, 81–83.

Klima, M. (1990). Rudiments of the clavicle in the embryon of whales (Cetacea). *Z. SäugetierKunde* **55**, 202–212.

Klima, M., and Bangma, G. C. (1987). Unpublished drawings of marsupial embryos from the Hill collection and some problems of marsupial ontogeny. *Z. Säugetierkunde* **52**, 201–211.

Klingenberg, C. P., Mebus, K., and Auffray, J.-C. (2003). Developmental integration in a complex morphological structure: how distinct are the modules in the mouse mandible? *Evol. & Devel.* **5**, 522–531.

Klingenberg, C. P., and Nijhout, H. F. (1998). Competition among growing organs and developmental control of morphological asymmetry. *Proc. R. Soc. Lond. B.* **265**, 1135–1138.

Klingenberg, C. P., Leamy, L. J., and Cheverud, J. M. (2004). Integration and modularity of quantitative trait locus effects on geometric shape in the mouse mandible. *Genetics* **166**, 1909–1921.

Kmita, M., Fraudeau, N., Hérault, Y., and Duboule, D. (2002). Serial deletions and duplications suggest a mechanism for the colinearity of *Hoxd* genes in limbs. *Nature* **420**, 145–150.

Knese, K. H. (1972). Osteoblasten, Chondroklasten, Mineraloklasten, Kollagenoklasten. *Acta Anat.* **83**, 275–288.

Knese, K. H., and Knopp, A. M. (1961a). Über den Ort der Bildung des Mucopolysaccharisproteinkomplexes in Knorpelgewebe. Elektronen-Mikroskopische und histochemische Untersuchungen. *Z. Zellforsch. Mikrosk. Anat.* **53**, 201–258.

Knese, K. H., and Knopp, A. M. (1961b). Elektronenmikroskopische Bedbachtungen über die Zellen in der Eröffnungszone des epiphysenknorpels. *Z. Zellforsch. Mikrosk. Anat.* **54**, 1–38.

Knowles, J. F. (1984). Bone in the irradiated lung of the guinea-pig. *J. Comp. Pathol.* **94**, 529–534.

Knudsen, T. B., Elmer, W. A., and Kochhar, D. M. (1985). Elevated rates of DNA synthesis and its correlation to cAMP-phosphodiesterase activity during induction of polydactyly in mouse embryos heterozygous for *Hemimelia-extra toes* (*Hm*[x]). *Teratology* **31**, 155–166.

Knudsen, C. B., and Knudson, W. (2001). Cartilage proteoglycans. *Sem. Cell Devel. Biol.* **12**, 69–78.

Knudson, C. B., Munaim, S. I., and Toole, B. P. (1995). Ectodermal stimulation of the production of hyaluronan-dependent pericellular matrix by embryonic limb mesodermal cells. *Devel. Dynam.* **204**, 186–191.

Knudson, C. B., and Toole, B. P. (1985). Changes in the pericellular matrix during differentiation of limb bud mesoderm. *Devel. Biol.* **112**, 308–318.

Knudson, C. B., and Toole, B. P. (1987). Hyaluronate–cell interactions during differentiation of chick embryo limb mesoderm. *Devel. Biol.* **124**, 82–90.

Knudson, W., and Knudson, C. B. (1991). Assembly of a chondrocyte-like pericellular matrix on non-chondrogenic cells. Role of the cell surface hyaluronan receptors in the assembly of a pericelluar matrix. *J. Cell Sci.* **99**, 227–235.

Knutsen, R., Wergedal, J. E., Sampath, T. K., Baylink, D. J., and Mohan, S. (1993). Osteogenic protein-1 stimulates proliferation and differentiation of human bone cells *in vitro. Biochem. Biophys. Res. Commun.* **194**, 1352–1358.

Ko, J. S., and Bernard, G. W. (1981). Osteoblast formation in vitro from bone marrow mononuclear cells in osteoclast-free bone. *Amer. J. Anat.* **161**, 415–425.

Kobayashi, E. T., Hashimoto, F., Kobayashi, Y., Sakai, E., Miyazaki, Y., Kamiya, T., Kobayashi, K., Kato, Y., and Sakai, H. (1999). Force-induced rapid changes in cell fate at midpalatal suture cartilage of growing rats. *J. Dent. Res.* **78**, 1495–1504.

Kobayashi, S. (1971). Acid mucopolysaccharides in calcified tissues. *Int. Rev. Cytol.* **30**, 257–371.

Kobayashi, T., Chung, U.-i., Schipani, E., Starbuck, M., Karsenty, G., Katagiri, T., Goad, D.L., and Lanske, B. (2002). PTHrP and Indian hedgehog control differentiation of growth plate chondrocytes at multiple steps. *Development* **129**, 2977–2986.

Koch, A. R. (1960). Die Frühentwicklung der Clavicula beim Menschen. *Acta Anat.* **42**, 177–212.

Koch, J. C. (1917). Laws of bone architecture. *Amer. J. Anat.* **21**, 177.

Kochhar, D. M. (1973). Limb development in mouse embryos. I. Analysis of teratogenic effects of retinoic acid. *Teratology* **7**, 289–295.

Kochhar, D. M., Aydelotte, M. B., and Vest, T. K. (1976). Altered collagen fibrillogenesis in embryonic mouse limb cartilage deficient in matrix granules. *Exp. Cell Res.* **102**, 213–222.

Kohyama, J., Abe, H. Shimazaki, T., Koizumi, A., Nakashima, K., Gojo, S., Taga, T., Okano, H., Hata, J., and Umezawa, A. (2001). Brain from bone: efficient 'meta-differentiation' of marrow stroma-derived mature osteoblasts to neurons with Noggin or a demethylating agent. *Differentiation* **68**, 235–244.

Kolettas, E., Muri, H. I., Barrett, J. C., and Hardingham, T. E. (2001). Chondrocyte phenotype and cell survival are regulated by culture conditions and by specific cytokines through the expression of Sox-9 transcription factor. *Rheumatology* **40**, 1146–1156.

Kollar, E. J. (1972). The development of the integument: spatial, temporal, and phylogenetic factors. *Amer. Zool.* **12**, 125–135.

Kollar, E. J. (1975). Gene–environment interactions during tooth development. *Dental Clin. North Amer.* **19**, 141–150.

Kollar, E. J., and Baird, G. R. (1969). The influence of the dental papilla on the development of tooth shape in embryonic mouse tooth germs. *J. Embryol. Exp. Morphol.* **21**, 131–148.

Kollar, E. J., and Baird, G. R. (1970). Tissue interactions in embryonic mouse tooth germs. II. The inductive role of the dental papilla. *J. Embryol. Exp. Morphol.* **24**, 173–186.

Kollar, E. J., and Mina, M. (1991). Role of the early epithelium in the patterning of the teeth and Meckel's cartilage. *J. Craniofac. Genet. Devel. Biol.* **11**, 223–228.

Kölliker, A. von (1849). Allgemeine Betrachtungen über die Entstehung des Knöchernen Schädels der Wirbeltiere. *Ber. Zool. Anat. Würzburg* **2**, 35–52.

Kölliker, A. von (1853). *A Manual of Microscopic Anatomy.* Translated and edited by G. Busk and T. Huxley. New Sydenham, London.

Kölliker, A. von (1859). On the different types in the microstructure of the skeleton of osseous fishes. *Proc. R. Soc. London* **9**, 656–668.

Kölliker, A. von (1873). *Die normale Resorption des Knochengewebe und ihre Bedeutung für die Entstehung der typischen Knochenformen.* Vogel, Leipzig.

Kolodziejczyk, S., and Hall, B. K. (1996). TGF-β superfamily members and signal transduction. *Biochem. Cell Biol.* **74**, 299–314.

Komaki, M., Katagini, T., and Suda, T. (1996). Bone morphogenetic protein-2 does not alter the differentiation pathway of committed progenitors of osteoblasts and chondroblasts. *Cell Tissue Res.* **284**, 9–17.

Komori, T., and Kishimoto, T. (1998). Cbfa1 in bone development. *Curr. Opin. Genet. Devel.* **8**, 494–499.

Komori, T., Yagi, H., Nomura, S., Yamaguchi, A., Sasaki, K., Deguchi, K., Shimizu, Y., Bronson, R. T., Gao, Y.-H., Inada, M., Sato, M., Okamoto, R., Kitamura, Y., Yoshika, S., and Kishimoto, T. (1997). Targeted disruption of *cbfa1* results in a complete lack of bone formation owing to maturational arrest of osteoblasts. *Cell* **89**, 755–764.

Kondo, T., Zákány, J., Innis, J. W., and Duboule, D. (1997). Of fingers, toes and penises. *Nature* **390**, 29.

Köntges, G., and Lumsden, A. (1996). Rhombencephalic neural crest segmentation is preserved throughout craniofacial ontogeny. *Development* **122**, 3229–3242.

Konyukhov, B., and Ginter, E. (1966). A study of the action of the Brachypodism-H gene on development of the long bones of the hind limbs in the mouse. *Folia Biol. (Prague)* **12**, 199–206.

Konyukhov, B., and Paschin, Y. V. (1967). Experimental study of the achondroplasia gene effects in the mouse. *Acta Biol. Acad. Sci. Hung.* **18**, 285–294.

Koob, T. J., and Summers, A. P. (2002). Tendon – bridging the gap. *Comp. Biochem. Physiol. Part A* **133**, 905–909.

Koole, R. (1994a). Ectomesenchymal mandibular symphysis bone grafts – an improvement in alveolar cleft grafting. *Cleft Palate Craniofac. J.* **31**, 217–223.

Koole, R. (1994b). *The Bone Graft in the Alveolar Cleft.* Ph.D. Thesis, University of Utrecht, the Netherlands.

Koole, R., Bosker, H., and Van der Dussen, F. N. (1989). Late secondary autogenous bone grafting in cleft patients comparing mandibular (ectomesenchymal) and iliac crest (mesenchymal) grafts. *J. Cranio-Max.-Fac. Surg.* **17**, 28–30.

Koop, J. B., and Robey, P. G. (1990). Sodium fluoride does not increase human bone cell proliferation or protein synthesis *in vitro. Calcif. Tissue Int.* **47**, 221–229.

Kornak, V., Kasper, D., Büsh, M. R., Kaiser, E., Schweizer, M., Schulz, A., Friedrich, W., Delling, G., and Jentsch, T. J. (2001). Loss of the ClC-7 chloride channel leads to osteopetrosis in mice and man. *Cell* **104**, 205–215.

Korneluk, R. G., and Liversage, R. A. (1984). Tissue regeneration in the amputated forelimb of *Xenopus laevis* froglets. *Can. J. Zool.* **62**, 2382–2391.

Kosher, R. A. (1976). Inhibition of 'spontaneous,' notochord-induced, and collagen-induced *in vitro* somite chondrogenesis by cyclic AMP derivatives and theophylline. *Devel. Biol.* **53**, 265–276.

Kosher, R. A. (1978). Inhibition of 'spontaneous,' notochord-induced and collagen-induced *in vitro* somite chondrogenesis by the calcium ionophore, A23187. *J. Exp. Zool.* **203**, 215–222.

Kosher, R. A. (1998). Syndecan-3 in limb skeletal development. *Microscopy Res. Tech.* **43**, 123–130.

Kosher, R. A., and Church, R. L. (1975). Stimulation of *in vitro* somite chondrogenesis by procollagen and collagen. *Nature* **258**, 327–329.

Kosher, R. A., Gay, S. W., Kamanitz, J. R., Kulyk, W. M., Rodgers, B. J., Sai, S., Tanner, T., and Tanzer, M. L. (1986b). Cartilage proteoglycan core protein gene expression during limb cartilage differentiation. *Devel. Biol.* **118**, 112–117.

Kosher, R. A., Kulyk, W. M., and Gay, S. W. (1986a). Collagen gene expression during limb cartilage differentiation. *J. Cell Biol.* **102**, 1151–1156.

Kosher, R. A., and Lash, J. W. (1975). Notochordal stimulation of *in vitro* somite chondrogenesis before and after enzymatic removal of perinotochordal materials. *Devel. Biol.* **42**, 362–378.

Kosher, R. A., Lash, J. W., and Minor, R. R. (1973). Environmental enhancement of *in vitro* chondrogenesis. IV. Stimulation of somite chondrogenesis by exogenous chondromucoprotein. *Devel. Biol.* **35**, 210–220.

Kosher, R. A., and Savage, M. P. (1979). The effect of collagen on the cyclic AMP content of embryonic somites. *J. Exp. Zool.* **208**, 35–40.

Kosher, R. A., and Savage, M. P. (1981). Glycosaminoglycan synthesis by the apical ectodermal ridge of chick wing bud. *Nature* **291**, 231–232.

Kosher, R. A., Savage, M. P., and Chan, S. C. (1979). Cyclic AMP derivatives stimulate the chondrogenic differentiation of the mesoderm subjacent to the apical ectodermal ridge of the chick limb bud. *J. Exp. Zool.* **209**, 221–228.

Kosher, R. A., Savage, M. P., and Walker, K. H. (1981). A gradation of hyaluronate accumulation along the proximodistal axis of the embryonic chick limb bud. *J. Embryol. Exp. Morphol.* **63**, 85–98.

Kosher, R. A., and Searls, R. L. (1973). Sulfated mucopolysaccharide synthesis during the development of *Rana pipiens*. *Devel. Biol.* **32**, 50–68.

Kosher, R. A., and Solursh, M. (1989). Widespread distribution of type Ii collagen during embryonic chick development. *Devel. Biol.* **131**, 558–566.

Kosher, R. A., Walker, K. A., and Ledger, P. W. (1982). Temporal and spatial distribution of fibronectin during development of the embryonic chick limb bud. *Cell Differentiation* **11**, 217–288.

Koshihara, Y., Kawamura, M., Oda, H., and Higaki, S. (1987). *In vitro* calcification in human osteoblastic cell line derived from periosteum. *Biochem. Biophys. Res. Commun.* **145**, 651–657.

Koski, K. (1975). Cartilage in the face. *Birth Defects, Orig. Artic. Ser. 11*, 231–254.

Koski, K. (1985). Reflections on craniofacial growth research. *Acta Morphol. Neerl.-Scand.* **23**, 357–368.

Koski, K., and Mäkinen, L. (1963). Growth potential of transplanted components of the mandibular ramus of rats. *Fin. Tandlak. Sallsk. Forh.* **59**, 296–308.

Koski, K., and Rönning, O. (1969). Growth potential of subcutaneously transplanted cranial base synchondroses of the rat. *Acta Odontol. Scand.* **27**, 343–357.

Koski, K., and Rönning, O. (1970). Growth potential of intracerebrally transplanted cranial base synchondroses in the rat. *Archs Oral Biol.* **15**, 1107–1108.

Koski, K., and Rönning, O. (1982). Condyle neck periostomy and the mitotic activity in the condylar tissues of young rats. *Swedish Dent. J. Suppl.* **15**, 109–113.

Koski, K., Rönning, O., and Nakamura, T. (1985). Periosteal control of mandibular condyle growth. In *Normal and Abnormal Bone Growth: Basic and Clinical Research* (A. Dixon and B. G. Sarnat, eds), pp. 413–423. Alan R. Liss Inc., New York, NY.

Koskinen, E. V. S., Isotupa, K., and Koski, K. (1976). A note on craniofacial sutural growth. *Amer. J. Phys. Anthropol.* **45**, 511–516.

Koskinen, E. V. S., Ryöppy, S. A., and Lindholm, T. S. (1972). Osteoinduction and osteogenesis in implants of allogeneic bone matrix. Influence of somatotropin, thyrotropin and cortisone. *Clin. Orthop. Rel. Res.* **87**, 116–131.

Kostovic-Knezevic, L., Bradamante, Z., and Svajger, A. (1981). Ultrastructure of elastic cartilage in the rat external ear. *Cell Tissue Res.* **218**, 149–160.

Kostovic-Knezevic, L., Bradamante, Z., and Svajger, A. (1986). On the ultrastructure of the developing elastic cartilage in the rat external ear. *Anat. Embryol.* **173**, 385–391.

Koumans, J. T. M., and Sire, J.-Y. (1996). An in vitro, serum-free organ culture technique for the study of development and growth of the dermal skeleton in fish. *In Vitro Cell Devel. Biol.* **36**, 612–626.

Koumoundouros, G., Divanach, P., and Kentouri, M. (1999). Osteological development of the vertebral column of the caudal complex in *Dentex dentex*. *J. Fish. Biol.* **54**, 424–435.

Koumoundouros, G., Divanach, P., and Kentouri, M. (2000). Development of the skull in *Dentex dentex* (Osteichthyes: Sparidae). *Marine Biol.* **136**, 175–184.

Koumoundouros, G., Divanach, P., and Kentouri, M. (2001). Osteological development of *Dentex dentex* (Osteichthyes: Sparidae): dorsal, anal, paired fins and squamation. *Marine Biol.* **138**, 399–406.

Koutsilieris, M. (1989). Human uterus-derived growth substances for rat bone cells and fibroblasts. *Amer. J. Obstet. Gynecol.* **161**, 1313–1317.

Koutsilieris, M., Rabbani, S. A., and Goltzman, D. (1986). Selective osteoblast mitogens can be extracted from prostatic tissue. *Prostate* **9**, 109–116.

Kovtun, M. F. (1985). The evolutionary morphology of locomotion organ system in bats (Mammalia: Chiroptera). In *Evolution and Morphogenesis* (J. Mlikovsky and J. A. Novak, eds), pp. 589–596. Academia, Prague.

Koyama, E., Golden, E. B., Kirsch, T., Adams, S. L., Chandraratna, R. A. S., Michaille, J.-J., and Pacifici, M. (1999). Retinoid signaling is required for chondrocyte maturation and endochondral bone formation during limb skeletogenesis. *Devel. Biol.* **208**, 375–391.

Koyama, E., Leatherman, J. L., Shimazu, A., Hyun-Duck, N., and Pacifici, M. (1995). Syndecan-3, tenascin-C, and the development of cartilaginous skeletal elements and joints in chick limbs. *Devel. Dynam.* **203**, 152–162.

Koyama, E., Leatherman, J. L., Noji, S., and Pacifici, M. (1996a). Early chick limb cartilaginous elements possess polarizing activity and express *Hedgehog*-related morphogenetic factors. *Devel. Dynam.* **207**, 344–354.

Koyama, E., Shimazu, A., Leatherman, J. L., Golden, E. B., Nah, H. D., and Pacifici, M. (1996b). Expression of syndecan-3 and tenascin-C: Possible involvement in periosteum development. *J. Orthop. Res.* **14**, 403–412.

Kravis, D., and Upholt, W. B. (1985). Quantitation of type II procollagen mRNA levels during chick limb cartilage differentiation. *Devel. Biol.* **108**, 164–172.

Kreiberg, S., Leth, J. B., Moller, E., and Bjork, A. (1978). Craniofacial growth in a case of congenital muscular dystrophy. A roentgencephalometric and electromyographic investigation. *Amer. J. Orthod.* **74**, 207–215.

Krompecher, S. (1937). *Die Knochenbildung*. J. Fischer, Jena.

Kronenberg, H. M. (2003). Developmental regulation of the growth plate. *Nature* **423**, 332–336.

Kronenberg, H. M., and Chung, U.-i. (2001). The parathyroid hormone-related protein and Indian hedgehox feedback loop in the growth plate. In *The Molecular Basis of Skeletogenesis* (G. Cardew and J. Goode, eds; B. K. Hall, chair), Novartis Foundation Symposium 232, pp. 144–157. John Wiley & Sons, Chichester.

Kronmiller, J. E., Upholt, W. B., and Kollar, E. J. (1991). Expression of epidermal growth factor mRNA in the developing mouse mandibular process. *Archs Oral Biol.* **36**, 405–410.

Kronmiller, J. E., Upholt, W. B., and Kollar, E. J. (1993). Effects of retinol on the temporal expression of transforming growth factor-β mRNA in the embryonic mouse mandible. *Archs Oral Biol.* **38**, 185–188.

Krotoski, D. M., and Elmer, W. A. (1973). Alkaline phosphatase activity in fetal hind limbs of the mouse mutation brachypodism. *Teratology* **7**, 99–106.

Krukowski, M., Iler, H. D., and Kahn, A. J. (1980). Intra-allantoic implantation: An alternative to the classical chorioallantoic membrane grafting technique. *J. Exp. Zool.* **214**, 365–367.

Krukowski, M., and Kahn, A. J. (1980a). The role of parathyroid hormone in mineral homeostasis and bone modeling in suckling rat pups. *Metab. Bone Dis. Rel. Res.* **2**, 257–260.

Krukowski, M., and Kahn, A. J. (1980b). Normal osteoclast number and function in rat pups lacking parathyroid hormone. *Experientia* **36**, 871–872.

Krukowski, M., and Kahn, A. J. (1982). Inductive specificity of mineralized bone matrix in ectopic osteoclast differentiation. *Calcif. Tissue Res.* **34**, 474–479.

Krukowski, M., Simmons, D. J., and Kahn, A. J. (1983). Cell lineage studies. In *Skeletal Research: An Experimental Approach* (A. S. Kunin and D. J. Simmons, eds), Volume 2, pp. 89–120. Academic Press, New York, NY. *Calcif. Tissue Res.* **34**, 474–479.

Krumlauf, R. (1993). *Hox* genes and pattern formation in the branchial region of the vertebrate head. *Trends Genet.* **9**, 106–112.

Kruuk, L. E. B., Slate, J., Pemberton, J. M., Brotherstone, S., Guinness, F., and Clutton-Brock, T. (2002). Antler size in red deer: heritability and selection but no evolution. *Evolution* **56**, 1683–1695.

Kruzynska-Frejtag, A., Wang, J., Maeda, M., Rogers, R., Krug, E., Hoffman, S., Markwald, R. R., and Conway, S. J (2004). *Periostin* is expressed within the developing teeth at the sites of epithelial–mesenchymal interaction. *Devel. Dynam.* **229**, 857–868.

Ksiazek, T. (1983). Bone induction by calcified cartilage transplants. *Clin. Orthop. Rel. Res.* **172**, 243–250.

Ksiazek, T., and Moskalewski, S. (1983). Studies on bone formation by cartilage reconstructed by isolated epiphyseal chondrocytes, transplanted syngeneically or across known histocompatibility barriers *in vitro*. *Clin. Orthop. Rel. Res.* **172**, 233–242.

Kuboki, Y., Saito, T., Murata, M., Takita, H, Mizuno, M., Inoue, M., Nagai, N., and Poole, A. R. (1995). Two distinctive BMP carriers induce local zonal chondrogenesis and membranous ossification, respectively – geometrical factors of matrices for cell differentiation. *Conn. Tissue Res.* **32**, 219–226.

Kubota, S., Tashiro, K., and Yamada, Y. (1992). Signaling site of laminin with mitogenic activity. *J. Biol. Chem.* **267**, 4285–4288.

Kuettner, K. E., Harper, E., and Eisenstein, R. (1977a). Protease inhibitors in cartilage. *Arthritis Rheum.* **20** (Suppl.) S124–S129.

Kuettner, K. E., Hiti, J., Eisenstein, R., and Harper, E. (1976a). Collagenase inhibition by cationic proteins derived from cartilage and aorta. *Biochem. Biophys. Res. Commun.* **72**, 40–46.

Kuettner, K. E., and Pauli, B. U. (1983). Vascularity of cartilage. In *Cartilage, Volume 1. Structure, Function and Biochemistry* (B. K. Hall, ed.), pp. 281–312. Academic Press, New York, NY.

Kuettner, K. E., Pauli, B. U., and Soble, L. (1978). Morphological studies on the resistance of cartilage to invasion by osteosarcoma cells *in vivo* and *in vitro*. *Cancer Res.* **38**, 277–287.

Kuettner, K. E., Schleyerbach, R., and Hascall, V. C. (eds) (1986). *Articular Cartilage Biochemistry*. Raven Press, New York, NY.

Kuettner, K. E., Soble, L. W., Sorgente, N., and Eisenstein, R. (1976b). The possible role of protease inhibitors in cartilage metabolism. In *Protides of the Biological Fluids* (H. Peeters, ed.), pp. 221–225. Pergamon Press, Oxford.

Kuettner, K. E., Soble, L., Croxon, R. L., Marczynska, B., Hiti, J., and Harper, E. (1977b). Tumor cell collagenase and its inhibition by a cartilage derived protease inhibitor. *Science* **196**, 653–654.

Kuettner, K. E., Wezeman, F. H., Simmons, D. J., Lisk, P. Y., Croxen, R. L., Soble, L. W., and Eisenstein, R. (1972). Lysozyme in preosseous cartilage. V. The response of embryonic chick cartilage to antilysozyme antibodies in organ culture. *Lab. Invest.* **27**, 324–330.

Kugler, J. H., Tomlinson, A., Wagstaff, A., and Ward, S. M. (1979). The role of cartilage canals in the formation of secondary centres of ossification. *J. Anat.* **129**, 493–506.

Kuhlman, R. E., and McNamee, M. J. (1970). The biochemical importance of the hypertrophic cartilage cell area to endochondral bone formation. *J. Bone Joint Surg. Am. Vol.* **52**, 1025–1032.

Kuhlman, R. E., Rainey, R., and O'Neill, R. (1963). Biochemical investigations of deer antler. Part II. Quantitative microchemical changes associated with antler bone formation. *J. Bone Joint Surg. Am. Vol.* **45**, 345–350.

Kuhn, H.-J., and Zeller, U. (eds) (1987). *Morphogenesis of the Mammalian Skull*. Verlag Paul Parey, Hamburg.

Kuijpers-Jagtman, A. M., Bex, J. H. M., Maltha, J. C., and Daggers, J. G. (1988). Longitudinal growth of the rabbit femur after vascular and periosteal interference. *Anat. Anz.* **167**, 349–358.

Kulonen, E., and Pikkarairen, J. (eds) (1973). *Biology of Fibroblast*. Academic Press, New York, NY.

Kulyk, W. M., Coelho, C. N. D., and Kosher, R. A. (1991). Type-IX collagen gene expression during limb cartilage differentiation. *Matrix* **11**, 282–288.

Kulyk, W. M., Franklin, J. L., and Hoffman, L. M. (2000). SOX9 expression during chondrogenesis in micromass cultures of embryonic limb mesenchyme. *Exp. Cell Res.* **255**, 327–332.

Kumasa, S., Mori, H., Mori, M., Shibutani, T., Iwayama, Y., Tsujimura, T., Ohnishi, T., Arakaki, N., Nakata, M., and Kurisu, K. (1990). Heterotopic bone formation in tumor stromal tissue. *Acta Histochem. Cytochem.* **23**, 427–440.

Kumegawa, M., Hiramatsu, M., Hatakeyama, K., Yajima, T., Kudama, H., Osaki, T., and Kurisu, J. (1983). Effects of epidermal growth factor on osteoblastic cells. *in vitro*. *Calif. Tissue Int.* **35**, 542–548.

Kunin, A. S., and Simmons, D. J. (eds) (1983). *Skeletal Research: An Experimental Approach. Volume 2*. Academic Press, New York, NY.

Kundrát. M., Seichert, V., Russell, A. P., and Smetana, K., Jr. (2002). Pentadactyl pattern of the avian wing autopodium and pyramid reduction hypothesis. *J. Exp. Zool. (Mol. Dev. Evol.)* **294**, 152–159.

Kuralesova, A. I. (1971). Osteogenic potencies of the bone marrow of irradiated mice revealed by means of heterotopic transplantation. *Byull. Eksp. Biol. Med.* **71**, 92–95.

Kuratani, S. (1987). The development of the orbital region of *Caretta caretta* (Chelonia, Reptilia). *J. Anat.* **154**, 187–200.

Kuratani, S. (1999). Development of the chondrocranium of the loggerhead turtle, *Caretta caretta*. *Zool. Sci.* **16**, 803–818.

Kuratani, S. (2003). Evolutionary developmental biology and vertebrate head segmentation: A perspective from developmental constraint. *Theory Biosci.* **122**, 230–251.

Kuratani, S., Kuraku, S., and Murakami, Y. (2002). Lamprey as an evo-devo model: Lessons from comparative embryology and molecular phylogenetics. *Genesis* **34**, 175–183.

Kuratani, S., Martin, J. F., Wawersik, S., Lilly, B., Eichele, G., and Olson, E. N. (1994). The expression pattern of the chick homeobox gene *gMHox* suggests a role in patterning of the limbs and face and in compartmentalization of somites. *Devel. Biol.* **161**, 357–369.

Kuratani, S., Matsuo, I., and Aizawa, S. (1997). Developmental patterning and evolution of the mammalian viscerocranium – genetic insights into comparative morphology. *Devel. Dynam.* **209**, 139–155.

Kuratani, S., Nobusada, Y., Horigome, N., and Shigetani, Y. (2001). Embryology of the lamprey and evolution of the vertebrate jaw: insights from molecular and developmental perspectives. *Phil. Trans R. Soc. Lond. B.* **356**, 1615–1632.

Kuratani, S., Satokata, I., Blum, M., Komatsu, Y., Haraguchi, R., Nakamura, S., Suzuki, K., Kosai, K., Maas, R., and Yamada, G. (1999). Middle ear defects associated with the double knock out mutation of murine goosecoid and Msx1 genes. *Cell Mol. Biol.* **45**, 589–599.

Kuratani, S., Ueki, T., Hirano, S., and Aizawa, S. (1998). Rostral truncation of a cyclostome, *Lampetra japonica*, induced by All-*trans* retinoic acid defines the head/trunk interface of the vertebrate body. *Devel. Dynam.* **211**, 35–51.

Kuroda, Y., and Shibuya, T. (1969). Clonal analysis of embryonic limb cartilage cells in Creeper mutant of chicken. *Annu. Rep. Inst. Genet.* 20, 29–30.

Kurtén, B. (1963). Return of a lost structure in the evolution of the felid dentition. *Comment. Biol. Soc. Sci. Fenn.* **26**, 1–12.

Kurzroch, E. A., Baskin, L. S., Li, Y., and Cunha, G. R. (1999). Epithelial–mesenchymal interactions in development of the mouse fetal genital tubercle. *Cells Tissues & Organs* **164**, 125–130.

Kutschera, U., and Niklas, K. J. (2004). The modern theory of biological evolution: an expanded synthesis. *Naturwissenschaften* **91**, 255–276.

Kvinnsland, S. (1974). Craniofacial skeletal changes in young rats induced by prolonged papain administration. *Growth* **38**, 381–388.

Kvinnsland, S., and Kvinnsland, S. (1975). Growth in craniofacial cartilages studied by ^{3}H-thymidine incorporation. *Growth* **39**, 305–314.

Kvist, T. N., and Finnegan, C. V. (1970a). The distribution of glycosaminoglycans in the axial region of the developing chick embryo. I. Histochemical analysis. *J. Exp. Zool.* **175**, 221–240.

Kvist, T. N., and Finnegan, C. V. (1970b). The distribution of glycosaminoglycans in the axial region of the developing

chick embryo. II. Biochemical analysis. *J. Exp. Zool.* **175**, 241–257.

Kwasigroch, T. E., and Kochhar, F. M. (1975). Locomotory behavior of limb bud cells. Effect of excess vitamin A *in vivo* and *in vitro*. *Exp. Cell Res.* **95**, 269–278.

Kwasigroch, T. E., Curtis, S. K., Knudsen, T. B., Barrach, H.-J., and Elmer, W. A. (1992). Morphological analysis of abnormal digit chondrogenesis in the *Brachypod* (*bp^H*) mouse limb in organ culture. *Anat. Embryol.* **185**, 307–315.

Kwiecinski, G. G., Krook, L., and Wimsatt, W. A. (1987). Annual skeletal changes in the little brown bat, *Myotis lucifugus lucifugus*, with particular reference to pregnancy and lactation. *Amer. J. Anat.* **178**, 410–420.

Kylämarkula, S. (1988). Growth changes in the skull and upper cervical skeleton after partial detachment of neck muscles. An experimental study in the rat. *J. Anat.* **159**, 197–205.

Kylämarkula, S., and Rönning, O. (1979). Transplantation of a basicranial synchondrosis to a sutural area in the isogenic rat. *Eur. J. Orthod.* **1**, 145–153.

Kylämarkula, S., and Rönning, O. (1983). Morphogenetic potential of autogenous costal cartilage transplanted into the interparietal suture area of the rat. *J. Neurosurg.* **58**, 755–759.

L

Labbé, E., Silvestri, C., Hoodless, P. A., Wrana, J. L., and Attisano, L. (1998). Smad2 and Smad3 positively and negatively regulate TGFβ-dependent transcription through the forkhead DNA-binding protein FAST2. *Mol. Cell.* **2**, 109–120.

Laborde, C. (1988). New observations on the development of the embryonic chick femur: Cartilage calcification before resorption. *Bone & Mineral* **4**, 147–156.

Lacroix, P. (1947). Organisers and the growth of bone. *J. Bone Joint Surg. Am. Vol.* **29**, 292–296.

Lacroix, P. (1951). *The Organisation of Bones* (Trans. by S. Gilder). Churchill, London.

Lacroix, P. (1953). Sur la réparation des fractures, Les méchanismes locaux. *C.R. Soc. Intern. Chir.* 15th Congress, 553–563.

Lacy, R. C., and Horner, B. E. (1996). Effects of inbreeding on skeletal development of *Rattus villosissimus*. *J. Hered.* **87**, 277–287.

Ladher, R. K., Church, V. L., Allen, S., Robson, K., Abdelfattah, A., Brown, N. A., Hattersley, G., Rosen, V., Luyten, F. P., Dale, K., and Francis-West, P. H. (2000). Cloning and expression of the Wnt antagonists Sfrp-2 and Frzb during chick development. *Devel. Biol.* **218**, 183–198.

Laerm, J. (1976). The development, function and design of amphicoelous vertebrae in teleost fishes. *Zool. J. Linn. Soc.* **58**, 237–254.

Laerm, J. (1979). The origin and homology of the chondrostean vertebral centrum. *Can. J. Zool.* **57**, 475–485.

Lafont, J., Baroukh, B., Berdal, A., Colombier, M. L., Barritault, D., Caruelle, J. P., and Saffar, J. L. (1998). Rgta11, a new healing agent, triggers developmental events during healing of craniotomy defects in adult rats. *Growth Factors* **16**, 23–38.

Laforest, L., Brown, C. W., Poleo, G., Géraudie, J., Tada, M., Ekker, M., and Akimenko, M.-A. (1998). Involvement of the *Sonic, Hedgehog, patched 1* and *bmp2* genes in patterning of the zebrafish dermal fin rays. *Development* **125**, 4175–4184.

Lagueux, O. (2003). Geoffroy's giraffe: the hagiography of a charismatic mammal. *J. Hist. Biol.* **36**, 225–247.

Laird, A. K. (1966). Dynamics of Bone Growth. *Growth* **30**, 263–275.

Lalykina, K. S., and Friedenstein, A. Y. (1969). Induction of the bone tissue in populations of lymphoid cells in guinea pigs. *Byull. Eksp. Biol. Med.* **67**, 105–108.

Lamarck, J. B. (1809). *Zoological Philosophy* (translated by H. Elliott, 1984). The University of Chicago Press, Chicago, IL.

Lamberg, S. I., and Stoolmiller, A. C. (1974). Glycosaminoglycans. A biochemical and clinical review. *J. Invest. Dermatol.* **63**, 433–449.

Lammers, A. R., German, R. Z., and Lightfoot, P. S. (1998). The impact of muscular dystrophy on limb bone growth and scaling in mice. *Acta Anat.* **162**, 199–208.

Lampl, M., Veldhuis, J. D., and Johnson, M. L. (1992). Saltation and stasis: a model of human growth. *Science* **258**, 801–803.

Lanctôt, C., Moreau, A., Chamberland, M., Tremblay, M. L., and Drouin, J. (1999). Hindlimb patterning and mandible development require the *Ptx1* gene. *Development* **126**, 1805–1810.

Landauer, W. (1932a). Studies on the creeper fowl. III. The early development and lethal expression of homozygous creeper embryos. *J. Genet.* **25**, 367–394.

Landauer, W. (1932b). Studies on the creeper fowl. V. The linkage of the genes for creeper and single-comb. *J. Genet.* **26**, 285–290.

Landauer, W. (1934). Studies on the creeper fowl. VII. The expression of vitamin D deficiency (rickets) in creeper chicks as compared with normal chicks. *Amer. J. Anat.* **55**, 229–252.

Landauer, W. (1957). Niacin antagonists and chick development. *J. Exp. Zool.* **136**, 509–530.

Landauer, W. (1965). Nanomelia, a lethal mutation of the fowl. *J. Hered.* **56**, 131–138.

Landauer, W. (1969a). A bibliography on micromelia. In *Limb Development and Deformity. Problems of Evaluation and Rehabilitation* (C. A. Swinyard, ed.), pp. 120–135. C. C. Thomas, Springfield, IL.

Landauer, W. (1969b). Dynamic aspects of hereditary and induced limb malformations. In *Limb Development and Deformity. Problems of Evaluation and Rehabilitation* (C. A. Swinyard, ed.), pp. 540–621. C. C. Thomas, Springfield, IL.

Landauer, W. (1976). Cholinomimetic teratogens. III. Interaction with amino acids known as neurotransmitters. *Teratology* **13**, 41–46.

Landauer, W., and Dunn, L. C. (1930). Studies on the creeper fowl. I. Genetics. *J. Genet.* **23**, 397–413.

Lande, R. (1978). Evolutionary mechanisms of limb loss in tetrapods. *Evolution* **32**, 73–92.

Landesman, R., and Reddi, A. H. (1985). Induction of endochondral bone by demineralized bone matrix from diabetic rats. *Calcif. Tissue Int.* **37**, 630–634.

Landini, G. (1991). Immunohistochemical demonstration of type-II collagen in the chondroid tissue of pleomorphic adenomas of the salivary glands. *Acta Pathol. Japon* **41**, 270–276.

Landis, W. J., and Géraudie, J. (1990). Organization and development of the mineral phase during early ontogenesis of the bony fin rays of the trout *Oncorhynchus mykiss*. *Anat. Rec.* **228**, 383–391.

Landis, W. J., and Glimcher, M. J. (1982). Electron optical and analytical observations of rat growth plate cartilage prepared by ultracryomicrotomy: the failure to detect a mineral phase in matrix vesicles and the identification of heterodispersed particles as the initial solid phase of calcium phosphate deposition in the extracellular matrix. *J. Ultrast. Res.* **78**, 227–268.

Landis, W. J., Hauschka, B. T., Rogerson, C. A., and Glimcher, M. J. (1977b). Electron microscopic observations of bone tissue prepared by ultracryomicrotomy. *J. Ultrastruct. Res.* **59**, 185–206.

Landis, W. J., Paine, M. C., and Glimcher, M. J. (1977a). Electron microscopic observations of bone tissue prepared anhydrously in organic solvents. *J. Ultrastruct. Res.* **59**, 1–30.

Landry, C. F., Youson, J. H., and Brown, I. R. (1990). Expression of the Beta-S100 gene in brain and craniofacial cartilage of the embryonic rat. *Devel. Neurosci.* **12**, 225–234.

Lane, J. M. (1987). *Fracture Healing*. Churchill Livingstone, London.

Lane, J. M., and Brighton, C. T. (1974). *In vitro* rabbit articular cartilage organ model. I. Morphology and glycosaminoglycan metabolism. *Arthritis Rheum.* **17**, 235–244.

Lane, J. M., Golembieqski, G., Boskey, A. L., and Posner, A. S. (1982). Comparative biochemical studies of the callus matrix in immobilized and non-immobilized fractures. *Metab. Bone Dis. Rel. Res.* **4**, 61–68.

Lane, J. M., Romin, E., and Bostrom, M. P. G. (1999). Biosynthetic bone grafting. *Clin. Orthop. Rel. Res.* **367**, S107–S117.

Lane, J. M., and Weiss, C. (1975). Review of articular cartilage collagen research. *Arthritis Rheum.* **18**, 553–562.

Lane, P. W., and Dickie, M. M. (1968). Three recessive mutations producing disproportionate dwarfism in mice. *J. Hered.* **59**, 300–308.

Lane, W. A. (1888). The anatomy and physiology of the shoemaker. *J. Anat. Physiol.* **22**, 593–620.

Lane, W. A. (1894). A method of treating simple oblique fractures of the tibia and fibula more efficient than those in common use. *Trans Clin. Soc. Lond.* **27**, 165–175.

Lane, W. A. (1907). Clinical remarks on the operative treatment of fractures *Brit. Med. J.* **1**, 1037–1038

Langer, C. (1876). Über das Gefässsystem der Röhrenknoche, mit Beiträgen zur Kenntniss der Baues und der Entwicklung des Knochengewebes. *Dentschr. Akad. Wiss. Wien.* **36**, 1–40.

Langer, F., and Gross, A. E. (1974). Immunogenicity of allograft articular cartilage. *J. Bone Joint Surg. Am. Vol.* **56**, 297–304.

Langer, R., Brem, H., Falterman, K., Klein, M., and Folkman, J. (1976). Isolation of a cartilage factor that inhibits tumor neovascularization. *Science* **193**, 70–72.

Langille, R. M. (1993). Formation of the vertebrate face: Differentiation and development. *Amer. Zool.* **33**, 462–471.

Langille R. M. (1994a). Chondrogenic differentiation in cultures of embryonic rat mesenchyme. *Microsc. Res. Tech.* **28**, 455–469.

Langille, R. M. (1994b). *In vitro* analysis of the spatial organization of chondrogenic regions of avian mandibular mesenchyme. *Devel. Dynam.* **201**, 55–62.

Langille, R. M., and Hall, B. K. (1986). Evidence of cranial neural crest cell contribution to the skeleton of the sea lamprey, *Petromyzon marinus*. In *New Discoveries and Technologies in Developmental Biology* (H. C. Slavkin, ed.), Part B. pp. 263–266. Alan R. Liss Inc., New York, NY.

Langille, R. M., and Hall, B. K. (1987). Development of the head skeleton of the Japanese medaka, *Oryzias latipes* (Teleostei). *J. Morphol.* **193**, 135–158.

Langille, R. M., and Hall, B. K. (1988a). The organ culture and grafting of lamprey cartilage and teeth. *In Vitro Cell Devel. Biol.* **24**, 1–8.

Langille, R. M., and Hall, B. K. (1988b). Role of the neural crest in development of the trabecular and branchial arches in embryonic sea lamprey, *Petromyzon marinus* (L). *Devel. Biol.* **102**, 301–310.

Langille, R. M., and Hall, B. K. (1988c). The role of the neural crest in the development of the cartilaginous cranial and visceral skeleton of the medaka, *Oryzias latipes* (Teleostei). *Anat. Embryol.* **177**, 297–305.

Langille, R. M., and Hall, B. K. (1989a). Developmental processes, developmental sequences and early vertebrate phylogeny. *Biol. Rev. Camb. Philos. Soc.* **64**, 73–91.

Langille, R. M., and Hall, B. K. (1989b). Neural crest-derived branchial arches link lampreys and gnathostomes. In *Trends in Vertebrate Morphology*. Proceedings of the 2nd International Symposium on Vertebrate Morphology, Vienna, 1986 (H. Splechtna and H. Hilgers, eds), Fortschritte der Zoologie Progress in Zoology, Volume 35, pp. 210–212. Gustav Fischer Verlag, Stuttgart.

Langille, R. M., and Hall, B. K. (1993a). *In vitro* calcification of cartilage from the lamprey, *Petromyzon marinus* (L). *Acta Zoologica* **74**, 31–41.

Langille, R. M., and Hall, B. K. (1993b). Patterning and the neural crest. In *The Skull. Volume 1. Development* (J. Hanken and B. K. Hall, eds), pp. 77–111. The University of Chicago Press, Chicago, IL.

Langman, J., and Nelson, G. R. (1968). A radioautographic study of the development of the somite in the chick embryo. *J. Embryol. Exp. Morphol.* **19**, 217–226.

Langness, U., and Udenfriend, S. (1974) Collagen biosynthesis in nonfibroblastic cell lines. *Proc. Natl Acad. Sci. U.S.A.* **71**, 50–51.

Lankester, E. R. (1907). On the origin of the lateral horns of the giraffe in foetal life on the area of the parietal bones. *Proc. Zool. Soc. Lond.* **1907**(1), 100–115.

Lannoo, M. J. (1987a). Neuromast topography in anuran amphibians. *J. Morphol.* **191**, 115–129.

Lannoo, M. J. (1987b). Neuromast topography in urodele amphibians. *J. Morphol.* **191**, 247–263.

Lannoo, M. J. (1988). The evolution of the amphibian lateral line system and its bearing on amphibian phylogeny. *Z. Zool. Syst.* **26**, 128–134.

Lannoo, M. J., and Smith, S. C. (1989). The lateral line system. In *Developmental Biology of the Axolotl* (J. B. Armstrong and G. M Malacinski, eds), pp. 176–184. Oxford University Press, Oxford.

Lansdowne, A. B. G. (1968). The origin and early development of the clavicle in the quail (*Coturnix c. japonica*). *J. Zool. London* **156**, 307–312.

Lanser, M. E., and Fallon, J. F. (1987). Development of wing bud-derived muscles in normal and *wingless* chick embryos: A computer-assisted three-dimensional reconstruction study of muscle pattern formation in the absence of skeletal elements. *Anat. Rec.* **217**, 61–78.

Lanske, B., Karaplis, A. C., Lee, K., Luz, A., Vortkamp, A., Pirro, A., Karperien, M., Defize, L. H. K., Ho, C., Mulligan, R. C., Abou-Samra, A.-B., Jüpper, H., Segre, G. V., and Kronenberg, H. M. (1996). PTH/PTHrP receptor in early development and Indian hedgehog-regulated bone growth. *Science* **273**, 663–666.

Lanyon, L. E. (1972). The prospect of encouraging osteogenesis. *Vet. Annu.* **13**, 126–129.

Lanyon, L. E. (1974). Experimental support for the trajectorial theory of bone structure. *J. Bone Joint Surg. Br. Vol.* **56**, 160–166.

Lanyon, L. E., Hampson, W. G. J., Goodship, A. E., and Shah, J. S. (1975). Bone deformation recorded *in vivo* from strain gauges attached to the human tibial shaft. *Acta Orthop. Scand.* **46**, 256–268.

Lanyon, L. E., and Rubin, C. T. (1984). Static vs. dynamic loads as an influence on bone remodeling. *J. Biomech.* **17**, 897–906.

Lanzing, W. J. R. (1976). The fine structure of fins and finrays of *Tilapia mossambica* (Peters). *Calcif. Tissue Res.* **173**, 349–356.

Lanzing, W. J. R., and Higginbotham, D. R. (1974). Scanning microscopy of surface structure of *Tilapia mossambica* (Peters) scales. *J. Fish. Biol.* **6**, 307–310.

Lanzing, W. J. R., and Wright, R. G. (1976). The ultrastructure and calcification of the scales of *Tilapia mossambica* (Peters). *Cell Tissue Res.* **167**, 37–47.

Laqueur, T. W. (2003). Sex in the flesh. *Isis* **94**, 300–306.

Larraín, J., Bachiller, D., Lu, B., Agius, E., Piccole, S., and de Robertis, E. M. (2000). BPM-binding modules in chordin: a model for signalling regulation in the extracellular space. *Development* **127**, 821–830.

Larson, P. M., and De Sá, R. O. (1998). Chondrocranial morphology of *Leptodactylus* larvae (Leptodactylidae: Leptodactylinae): its utility in phylogenetic reconstruction. *J. Morphol.* **238**, 287–305.

Larsson, H. C. E., and Wagner, G. P. (2002). Pentadactyl ground state of the avian wing. *J. Exp. Zool. (Mol. Devel. Evol.)* **294**, 146–151.

Larsson, S.-E., and Kuettner, K. E. (1974). Microchemical studies of acid glycosaminoglycans from isolated chondrocytes in suspension. *Calcif. Tissue Res.* **14**, 49–58.

Lash, J. W. (1959). Presence of myoglobin in 'cartilage' of the marine snail, *Busycon*. *Science* **130**, 334.

Lash, J. W. (1963a). Tissue interaction and specific metabolic responses: Chondrogenesis induction and differentiation. In *Cytodifferentiation and Macromolecular Synthesis* (M. Locke, ed.), pp. 235–260. Academic Press, New York, NY.

Lash, J. W. (1963b). Studies on the ability of embryonic mesonephros explants to form cartilage. *Devel. Biol.* **6**, 219–232.

Lash, J. W. (1964). Normal embryology and teratogenesis. *Amer. J. Obstet. Gynecol.* **90**, 1193–1207.

Lash, J. W. (1967). Differential behavior of anterior and posterior embryonic chick somites *in vitro*. *J. Exp. Zool.* **165**, 47–56.

Lash, J. W. (1968a). Chondrogenesis: Genotypic and phenotypic expression. *J. Cell. Physiol.* **71** Suppl. 1, 35–46.

Lash, J. W. (1968b). Somite mesenchyme and its response to cartilage induction. In *Epithelial–Mesenchymal Interactions* (R. Fleischmajer and R. E. Billingham, eds), pp. 165–172. Williams & Wilkins, Baltimore, MD.

Lash, J. W. (1968c). Phenotypic expression and differentiation: *In vitro* chondrogenesis. In *The Stability of the Differentiated State* (H. Ursprung, ed.), Volume 1, pp. 17–24. Springer-Verlag, Berlin.

Lash, J. W., Glick, M. C., and Madden, J. W. (1964). Cartilage induction in vitro and sulfate-activating enzymes. *Natl Cancer Inst., Monogr.* **13**, 39–49.

Lash, J. W., Holtzer, S., and Holtzer, H. (1957). An experimental analysis of the development of the spinal column. VI. Aspects of cartilage induction. *Exp. Cell Res.* **13**, 292–303.

Lash, J. W., Holtzer, H., and Whitehouse, M. W. (1960). *In vitro* studies on chondrogenesis: The uptake of radioactive sulphate during cartilage induction. *Devel. Biol.* **2**, 76–89.

Lash, J. W., Holmes, F. A., and Zilliken, F. (1962). Induction of cell differentiation. 1. The *in vitro* induction of vertebral cartilage with a low molecular weight tissue component. *Biochim. Biophys. Acta* **56**, 313–319.

Lash, J. W., and Ostrovsky, D. (1986). On the formation of somites. In *Developmental Biology* (L. W. Browder, ed.), Volume 2, pp. 547–563. Plenum Publishing Co., New York, NY.

Lash, J. W., and Saxén, L. (1972). Human teratogenesis: *In vitro* studies on thalidomide-inhibited chondrogenesis. *Devel. Biol.* **28**, 61–70.

Lash, J. W., Saxén, L., and Kosher, R. A. (1974). Human chondrogenesis: Glycosaminoglycan content of embryonic human cartilage. *J. Exp. Zool.* **189**, 127–131.

Lash, J. W., and Whitehouse, M. W. (1960). An unusual polysaccharide in the chondroid tissue of the snail, *Busycon*; polyglucose sulfate. *Biochem. J.* **74**, 351–355.

Lasher, R. (1977). Studies on cellular proliferation and chondrogenesis. In *Developmental Aspects of the Cell Cycle* (I. L. Cameron, G. M. Padilla, and A. M. Zimmerman, eds), pp. 223–241. Academic Press, New York, NY.

Latham, R. A., Smiley, G. R., and Greeg, J. M. (1973). The problem of tissue deficiency in cleft palate: An experiment in mobilizing the palatine bones of cleft dogs. *Brit. J. Plast. Surg.* **26**, 252–260.

Lau, W. F., Tertinegg, I., and Heersche, J. K. M. (1993). Effects of retinoic acid on cartilage differentiation in a chondrogenic cell line. *Teratology* **47**, 555–563.

Lauder, G. V. (1980). On the relationship of the myotome to the axial skeleton in vertebrate evolution. *Paleobiology* **6**, 51–56.

Lauder, G. V. (1983). Functional design and evolution of the pharyngeal jaw apparatus in euteleostean fishes. *Zool. J. Linn. Soc.* **77**, 1–33.

Lauder, G. V., and Liem, K. F. (1989). The role of historical factors in the evolution of complex organismal functions. In *Complex Organismal Functions: Integration and Evolution in Vertebrates* (D. B. Wake and G. Roth, eds), pp. 63–78. John Wiley Sons, Chichester.

Laufer, E., Dahn, R., Orozco, O. E., Yeo, C.-Y., Pisenti, J., Henriqués, D., Abbott, U. K., Fallon, J. F., and Tabin, C. (1997). Expression of *Radical fringe* in limb-bud ectoderm regulates apical ectodermal ridge formation. *Nature* **386**, 366–373.

Laufer, H. (1959). Immunochemical studies of muscle proteins in mature and regenerating limbs of the adult newt, *Triturus viridescens*. *J. Embryol. Exp. Morphol.* **7**, 431–458.

Launay, C., Fromentoux, V., Thery, C., Shi, D. L., and Boucaut, J. C. (1994). Comparative analysis of the tissue distribution of 3 fibroblast growth-factor receptor messenger-RNAs during amphibian morphogenesis. *Differentiation* **58**, 101–111.

Laurin, M. (1998). A reevaluation of the origin of pentadactyly. *Evolution* **52**, 1476–1482.

Laurin, M., Girondot, M., and de Ricqlès, A. J. (2000). Early tetrapod evolution. *Trends Ecol. Evol.* **15**, 118–123.

Lauscher, C. K., and Carlson, E. C. (1975). The development of proline containing extracellular connective tissue fibrils by chick notochordal epithelium *in vitro*. *Anat. Rec.* **182**, 151–168.

Lauthier, M. (1974). Histochemical data on the first stages of the fore and hind limb development in *Pleurodeles waltlii* Michah. (Urodela, Amphibia). *Wilhelm Roux Arch. Entwicklungsmech. Org.* **175**, 185–198.

Lauthier, M. (1977). Etude ultrastructurale des stades précoces du développement du membre postérieur de *Pleurodeles waltlii* Michah (Amphibien, Urodèle). *J. Embryol. Exp. Morphol.* **38**, 1–18.

Lauthier, M. (1978). Study of the hind limb development in the newt *Pleurodeles waltlii* Michah (Amphibiae, Urodela) after epidermis removal of the bud. *Experientia* **34**, 790–791.

Lavelle, C. L. B. (1974). The ramus of the mandible between Anglo-Saxon and nineteenth century periods. *Acta Anat.* **89**, 80–88.

Lavelle, C. L. B. (1983). Study of mandibular shape in the mouse. *Acta Anat.* **117**, 314–320.

Lavelle, C. L. B. (1985). The use of medial axis transformation to examine evolutionary changes in mandibular shape. *Anat. Anz.* **158**, 305–314.

Lavietes, B. B. (1971). Kinetics of matrix synthesis in cartilage cell culture. *Exp. Cell Res.* **68**, 43–48.

Law, H. T., Annan, I., and McCarthy, I. D. (1985). The effect of induced electrical currents on bone after experimental osteotomy in sheep. *J. Bone Joint Surg. Br. Vol.* **67**, 463–469.

Lawson, R. (1966). The development of the centrum of *Hypogeophis rostratus* (Amphibia, Apoda) with special reference to the notochordal (intravertebral) cartilage. *J. Morphol.* **118**, 137–148.

Lawton, D. M., Oswald, W. B., and McClure, J. (1995). The biological reality of the interlacuna network in the embryonic, cartilaginous skeleton – a thiazine dye absolute ethanol LR white resin protocol for visualizing the network with minimal tissue shrinkage. *J. Microsc. Oxford* **178**, 66–85.

Layman, D. L., Sokoloff, L., and Miller, E. J. (1972). Collagen synthesis by articular chondrocytes in monolayer culture. *Exp. Cell Res.* **73**, 107–112.

Le, A.-X., Michau, T., Hu, D., and Helms, J. A. (2001). Molecular aspects of healing in stabilized and non-stabilized fractures. *J. Orthop. Res.* **19**, 78–84.

Leamy, L. (1993). Morphological integration of fluctuating asymmetry in the mouse mandible. *Genetica* **89**, 139–153.

Leamy, L. (1999). Heritability of directional and fluctuating asymmetry for mandibular characters in random-bred mice. *J. Evol. Biol.* **12**, 146–155.

Leamy, L. J., Routman, E. J., and Cheverud, J. M. (1997). A search for quantitative trait loci affecting asymmetry of mandibular characters in mice. *Evolution* **51**, 957–969.

Leboy, P. S., Beresford, J. N., Deulin, C., and Owen, M. E. (1991). Dexamethasone induction of osteoblast mRNAs in rat marrow stromal cell cultures. *J. Cell Physiol.* **146**, 370–378.

Lecanda, F., Warlow, P. M., Sheikh, S., Furlan, F., Steinberg, T.-H., and Civitelli, R. (2000). Connexin43 deficiency causes delayed ossification, craniofacial abnormalities, and osteoblast dysfunction. *J. Cell Biol.* **151**, 931–943.

LeClair, E. E., Bonfiglio, L., and Tuan, R. S. (1999). Expression of the paired-box genes *Pax-1* and *Pax-9* in limb skeleton development. *Devel. Dynam.* **214**, 101–115.

Lecyk, M. (1965). The effect of hypothermia applied in the given stages of pregnancy on the number and form of vertebrae in the offspring of white mice. *Experientia* **21**, 452–453.

Le Douarin, N. M. (1973). A biological cell labeling technique and its use in experimental embryology. *Devel. Biol.* **30**, 217–222.

Le Douarin, N. M. (1974). Cell recognition based on natural morphological nuclear markers. *Med. Biol.* **52**, 281–319.

Le Douarin, N. M. (1975). The neural crest in the neck and other parts of the body. *Birth defects, Orig. Artic. Ser.* **11**, 19–50.

Le Douarin, N. M., and Kalcheim, C. (1999). *The Neural Crest.* 2nd Edn. Cambridge University Press, Cambridge.

Le Douarin, N. M., and Teillet, M.-A. (1974). Experimental analysis of the migration and differentiation of neuroblasts of the autonomic nervous system and of neurectodermal mesenchymal derivatives, using a biological cell marking technique. *Devel. Biol.* **41**, 162–184.

Lee, A. H. (2004). Histological organization and its relationship to function in the femur of *Alligator mississippiensis*. *J. Anat.* **204**, 197–207.

Lee, A. K., and Langer, R. (1983). Shark cartilage contains inhibitors of tumor angiogenesis. *Science* **221**, 1185–1187.

Lee, A. K., Beuzekom, M. van., Glowacki, J., and Langer, R. (1984). Inhibitors, enzymes and growth factors from shark cartilage. *Comp. Biochem. Physiol. B.* **78**, 609–616.

Lee, B. S., Halliday, S., Ojikutu, B., Krits, I., and Gluck, S. L. (1996). Osteoclasts express the B2 isoform of vacuolar H$^+$-ATPase intracellularly and on their plasma membranes. *Amer. J. Physiol. Cell Physiol.* **39**, C382–C388.

Lee, E. R., Lamplugh, L., Davoli, M. A., Beauchemin, A., Chan, K., Mort, J. S., and Leblond, C. P. (2001). Enzymes active in the areas undergoing cartilage resorption during the development of the secondary ossification center in the tibiae of rats ages 0–21 days: I. Two groups of proteinases cleave the core protein of aggrecan. *Devel. Dynam.* **222**, 52–70.

Lee, E. R., Murphy, G., El-Alfy, M., Davoli, M. A., Lamplugh, L., Docherty, A. J., and Leblond, C. P. (1999). Active gelatinase B is identified by histozymography in the cartilage resorption sites of developing long bones. *Devel. Dynam.* **215**, 190–205.

Lee, J., and Tickle, C. (1985). Retinoic acid and pattern formation in the developing chick wing: SEM and quantitative studies of early effects on the apical ectodermal ridge and bud outgrowth. *J. Embryol. Exp. Morphol.* **90**, 139–169.

Lee, K. A., Pierce, R. A., Davis, E. C., Mecham, R. P., and Parks, W. C. (1994). Conversion to an elastogenic phenotype by fetal hyaline chondrocytes is accompanied by altered expression of elastin-related macromolecules. *Devel. Biol.* **163**, 241–252.

Lee, K.-H., Marden, J. J., Thompson, M. S., MacLennan, H., Kishimoto, Y., Pratt, S. J., Schulte-Merker, S., Hammerschmidt, M., Johnson, S. L., Postelthwaite, J. H., Veier, D. C., and Zon, L. I. (1998). Cloning and genetic mapping of zebrafish BMP-2. *Devel. Genetics* **23**, 97–103.

Lee, K. K. H. (1992). The regulative potential of the limb region in 11.5-day rat embryos following the amputation of the forelimb bud. *Anat. Embryol.* **186**, 67–74.

Lee, K. K. H., and Chan, W. Y. (1991). A study of the regenerative potential of partially excised mouse embryonic fore-limb buds. *Anat. Rec.* **184**, 153–157.

Lee, K. K. H., Chan, W. Y., and Sze, L. Y. (1993). Histogenetic potential of rat hind-limb interdigital tissue prior to and during the onset of programmed cell death. *Anat. Rec.* **236**, 568–572.

Lee, K. K. H., and Ede, D. A. (1989). The capacity of normal and *talpid*³ mutant fowl myogenic cells to migrate in quail limb buds. *Anat. Rec.* **179**, 395–402.

Lee, K. K. H., Li, F. C. H., Yung, W. T., Kung, J. L. S., Ng, J. L., and Cheah, K. S. E. (1994). Influence of digits, ectoderm and retinoic acid on chondrogenesis by mouse interdigital mesoderm in culture. *Devel. Dynam.* **201**, 297–309.

Lee, K. K. H., Tang, M. K., Yew, D. T. W., Chow, P. H., Yee, S. P., Schiender, C., and Brancolini, C. (1999). gas2 is a multifunctional gene involved in the regulation of apoptosis and chondrogenesis in the developing mouse limb. *Devel. Genet.* **207**, 14–25.

Lee, M. S., Loew, G., Flanagan, S., Kuchler, K., and Glackin, C. A. (2000). Human dermo-1 has attributes similar to twist in early bone development. *Bone* **27**, 591–602.

Lee, M. S. Y. (1993). The origin of the turtle body plan: bridging a famous morphological gap. *Science* **261**, 1716–1720.

Lee, M. S. Y. (1996). Correlated progression and the origin of turtles. *Nature* **379**, 812–815.

Lee, S.-H., Fu, K. K., Hui, J. N., and Richman, J. (2002). Noggin and retinoic acid transform the identify of avian facial prominences. *Nature* **414**, 909–912.

Lee, T. C., Staines, A., and Taylor, D. (2002). Bone adaptation to load: microdamage as a stimulus for bone remodelling. *J. Anat.* **201**, 437–446.

Lee, T. C., and Taylor, D. (1999). Bone remodelling: Should we cry Wolff? *Irish J. Med. Sci.* **168**, 102–105.

Lee, Y.-S., and Chuong, C.-M. (1992). Adhesion molecules in skeletogenesis: 1. Transient expression of neural cell adhesion molecules (NCAM) in osteoblasts during endochondral and intramembranous ossification. *J. Bone Min. Res.* **7**, 1435–1446.

Lee-Owen, V., and Anderson, J. C. (1975). The isolation of collagen associated proteoglycan from bovine nasal cartilage and its preferential interaction with alpha2 chains of type I collagen. *Biochem. J.* **149**, 57–64.

Leeuwenhoek, A van (1693). An extract of a letter from Mr Anthony van Leeuwenhoek, containing several observations on the texture of the bones of animals compared with that of wood: on the bark of trees: on the little scales formed on the cuticula, & etc. *Phil. Trans R. Soc.* **17**, 838.

Lefebvre, V., Garofalo, S., and de Crombrugghe, B. (1995). Type-X collagen gene expression in mouse chondrocytes immortalized by a temperature-sensitive Simian-virus–40 large tumour antigen. *J. Cell Biol.* **128**, 239–245.

Lefebvre, V., Li, P., and de Crombrugghe, B. (1998). A new long form of Sox 5 (L-Sox5), Sox6 and Sox9 are coexpressed in chondrogenesis and cooperatively activate the type II collagen gene. *EMBO J.* **17**, 5718–5733.

Le Gros Clark, W. E. (1958). *The Tissues of the Body. An Introduction to the Study of Anatomy.* The Clarendon Press, Oxford.

Leibel, W. S. (1976). The influence of the otic capsule in Ambystomid skull formation. *J. Exp. Zool.* **196**, 85–104.

Leibovich, S. J., and Weiss, J. B. (1973). Elucidation of the exact sites of cleavage of tropocollagen by rheumatoid synovial collagenase: correlation of cleavage sites with fibril structure. *Connect. Tissue Res.* **2**, 11–20.

Leidy, J. (1849). Observations on the development of bone, the structure of articular cartilage, and on the relation of the areolar tissue with muscle and tendon. *Proc. Acad. Nat. Sci. Philadelphia* **4**, 116–117.

Leitges, M., Neidhardt, L., Haenig, B., Herrmann, B. G., and Kispert, A. (2000). The paired homeobox gene *Uncx4.1* specifies pedicles, transverse processes and proximal ribs of the vertebral column. *Development* **127**, 2259–2267.

Le Lièvre, C. (1971a). Recherches sur l'origine embryologique des arcs viscéraux chez l'embryon d'Oiseau par la méthode des greffes interspécifiques entre caille et poulet. *C.R. Séanc. Soc. Biol. Paris* **165**, 195–400.

Le Lièvre, C. (1971b). Recherches sur l'origine embryologique du squelette viscéral chez l'embryon d'Oiseau. *C.R. Ass. Anat. Paris* **152**, 575–583.

Le Lièvre, C. (1974). Rôle des cellules mésectodermiques issues des crêtes neurales céphaliques dans la formation des arcs branchiaux et du squelette viscéral. *J. Embryol. Exp. Morphol.* **31**, 453–477.

Le Lièvre, C. (1978). Participation of neural crest derived cells in the genesis of the skull in birds. *J. Embryol. Exp. Morphol.* **47**, 17–37.

Le Lièvre, C., and Le Douarin, N. M. (1975). Mesenchymal derivatives of the neural crest: Analysis of chimaeric quail and chick embryos. *J. Embryol. Exp. Morphol.* **34**, 125–154.

Lelkes, G. (1958). Experiments *in vitro* on the role of movement in the development of joints. *J. Embryol. Exp. Morphol.* **6**, 183–186.

Le Minor, J. M. (1995). Biphalangeal and triphalangeal toes in the evolution of the human foot. *Acta Anat.* **154**, 236–241.

Lemperg, R. K., Bergenholtz, A., and Smith, T. W. D. (1975). Calf articular cartilage in organ culture in a chemically defined medium. 2. Concentrations of glycosaminoglycans and [^{35}S]-sulfate incorporation at different oxygen tensions. *In Vitro* **11**, 291–301.

Lengelé, B., Dhem, A., and Schowing, J. (1990). Early development of the primitive cranial vault in the chick embryo. *J. Craniofac. Genet. Devel. Biol.* **10**, 103–112.

Lengelé, B., Schowing, J., and Dhem, A. (1996a). Embryonic origin and fate of chondroid tissues and secondary cartilages in the avian skull. *Anat. Rec.* **246**, 377–393.

Lengelé, B., Schowing, J., and Dhem, A. (1996b). Chondroid tissue in the early facial morphogenesis of the chick embryo. *Anat. Embryol.* **193**, 505–513.

Lennon, D. P., Haynesworth, S. E., Arm, D. M., Baber, M. A., and Caplan, A. I. (2000). Dilution of human mesenchymal stem cells with dermal fibroblast and the effects on *in vitro* and *in vivo* osteochondrogenesis. *Devel. Dynam.* **219**, 50–62.

Lennon, D. P., Haynesworth, S. E., Young, R. G., Dennis, J. E., and Caplan, A. I. (1995). A chemically defined medium supports *in vitro* proliferation and maintains the osteochondral potential of rat marrow-derived mesenchymal stem cells. *Exp. Cell Res.* **219**, 211–222.

Leonard, C. M., Fuld, H. M., Frenz, D. A., Downie, S. A., Massagué, J., and Newman, S. A. (1991). Role of transforming growth factor-β in chondrogenic pattern formation in the embryonic limb: Stimulation of mesenchymal condensation and fibronectin gene expression by exogenous TGF-β and evidence for endogenous TGF-β-like activity. *Devel. Biol.* **145**, 99–109.

Leonard, K., and Sharaway, M. (1974). Histochemistry of myosin-like fibrils in the cells of the human dental pulp. *J. Dent. Res.* **53**, 157.

Lerner, A. L., and Kuhn, J. L. (1997). Characterization of regional and age-related variations in the growth of the rabbit distal femur. *J. Orthop. Res.* **15**, 353–361.

Lerner, A. L., Kuhn, J. L., and Hollister, S. J. (1998). Are regional variations in bone growth related to mechanical stress and strain parameters? *J. Biomech.* **31**, 327–335.

Lerner, I. M. (1936). Heterogony in the axial skeleton of the creeper fowl. *Amer. Nat.* **70**, 595–598.

Leroi, A. M. (2003). *Mutants. On Genetic Variety and the Human Body*. Viking, New York, NY.

Lessa, E. P., and Stein, B. R. (1992). Morphological constraints in the digging apparatus of pocket gophers (Mammalia: Geomyidae). *Biol. J. Linn. Soc.* **47**, 439–453.

Lessa, E. P., and Wake, M. H. (1992). Morphometric analysis of the skull of *Dermophis mexicanus* (Amphibia: Gymnophiona). *Zool. J. Linn. Soc.* **106**, 1–15.

Lestrel, P. E., Sarnat, B. G., and McNabb, E. G. (1989). Carapace growth of the turtle *Chrysemys scripta*: A longitudinal study of shape using Fourier analysis. *Anat. Anz.* **168**, 135–143.

Lettice, L., Heckster-Sørensen, J., and Hill, R. E. (1999). The dominant hemimelia mutation uncouples epithelial–mesenchymal interactions and disrupts anterior mesenchyme formation in mouse hindlimbs. *Development* **126**, 4729–4736.

Leung, D. Y. M., Glagov, S., and Mathews, M. B. (1976). Cyclic stretching stimulates synthesis of matrix components by arterial smooth muscle cells *in vitro*. *Science* **191**, 475–477.

Leung, D. Y. M., Glagov, S., and Mathews, M. B. (1977). A new *in vitro* system for studying cell response to mechanical stimulation. Different effects of cyclic stretching and agitation on smooth muscle cell biosynthesis. *Exp. Cell Res.* **109**, 285–298.

Leutenegger, W. (1973). Sexual dimorphism in the pelvis of African lorises. *Amer. J. Phys. Anthropol.* **38**, 251–254.

Levak-Svajger, B., and Moscona, A. A. (1964). Differentiation in grafts of aggregates of embryonic chick and mouse cells. *Exp. Cell Res.* **36**, 692–695.

Levak-Svajger, B., and Svajger, A. (1971). Differentiation of endodermal tissues in homografts of primitive ectoderm from two-layered rat embryonic shields. *Experientia* **27**, 683–684.

Levak-Svajger, B., and Svajger, A. (1974). Investigation on the origin of the definitive ectoderm in the rat embryo. *J. Embryol. Exp. Morphol.* **32**, 445–459.

Levak-Svajger, B., and Svajger, A. (1979). Course of development of isolated rat embryonic ectoderm as renal homografts. *Experientia* **35**, 258–260.

Levander, G. (1934). On the formation of new bone in bone transplantation. *Acta Chir. Scand.* **74**, 425–426.

Levander, G. (1964). *Induction Phenomena in Tissue Regeneration*. Williams & Wilkins, Baltimore, MD.

Levene, C. I., and Bates, C. J. (1970). Growth and macromolecular synthesis in the 3T6 mouse fibroblast. 1. General description and the role of ascorbic acid. *J. Cell Sci.* **7**, 671–682.

Levenson, G. E. (1969). The effect of ascorbic acid on monolayer cultures of three types of chondrocytes. *Exp. Cell Res.* **55**, 225–228.

Levenson, G. E. (1970). Behaviour in culture of three types of chondrocytes, and their response to ascorbic acid. *Exp. Cell Res.* **62**, 271–285.

Levesque, J.-P., Hatzfeld, A., and Hatzfeld, J. (1991). Mitogenic properties of major extracellular proteins. *Immunol. Today* **12**, 258–262.

Levi, G., Topilko, P., Schneider-Maunoury, S., Mantero, S., Cancedda, R., and Charnay, P. (1996). Defective bone formation in *Krox-20* mutant mice. *Development* **122**, 113–120.

Levitt, D., and Dorfman, A. (1972). The irreversible inhibition of differentiation of limb-bud mesenchyme by bromodeoxyuridine. *Proc. Natl Acad. Sci. U.S.A.* **69**, 1253–1257.

Levitt, D., and Dorfman, A. (1973). Control of chondrogenesis in limb-bud cell cultures by bromodeoxyuridine. *Proc. Natl Acad. Sci. U.S.A.* **70**, 2201–2205.

Levitt, D., and Dorfman, A. (1974). Concepts and mechanisms of cartilage differentiation. *Curr. Top. Devel. Biol.* **8**, 103–109.

Levitt, D., Ho, P. L., and Dorfman, A. (1974). Differentiation of cartilage. In *The Cell Surface in Development* (A. A. Moscona, ed.), pp. 101–126. John Wiley & Sons, New York, NY.

Levrat-Calviac, V. (1986). Comparative study of the osteoderms in *Tarentola mauritanica* and *Tarentola neglecta* (Gekkonidae, Squamata). *Archs Anat. Microsc. Morphol. Exp.* **75**, 29–44.

Levrat-Calviac, V., and Zylberberg, L. (1986). The structure of the osteoderm in the gekko: *Tarentola mauritanica*. *Amer. J. Anat.* **176**, 437–446.

Levy, B. M. (1964). Embryological development of the temporomandibular joint. In *The Temporomandibular Joint* (B. G. Sarnat, ed.), pp. 59–70. C. C. Thomas, Springfield, IL.

Lewinson, D., Bialik, G. M., and Hochberg, Z. (1994). Differential effects of hypothyroidism on the cartilage and the osteogenic process in the mandibular condyle: recovery by growth hormone and thyroxine. *Endocrinology* **135**, 1504–1510.

Lewinson, D., and Boskey, A. L. (1984). Calmodulin localization in bone and cartilage. *Cell Biol. Intern. Rep.* **8**, 11–18.

Lewinson, D., and Krogan, Y. (1995). Ontogenesis of chondro/osteoclasts and their precursors in the mandibular condyle of the mouse. *Bone* **17**, 293–299.

Lewinson, D., and Silbermann, M. (1986). Parathyroid hormone stimulates proliferation of chondroprogenitor cells *in vitro*. *Calcif. Tissue Int.* **38**, 155–162.

Lewinson, D., Toister, Z., and Silbermann, M. (1982). Quantitative and distributional changes in the activity of alkaline phosphatase during the maturation of cartilage. *J. Histochem. Cytochem.* **30**, 261–269.

Lewis, A. R., Ralphs, J. R., Kneafsey, B., and Benjamin, M. (1998). Distribution of collagens and glycosaminoglycans in the joint capsule of the proximal interphalangeal joint of the human finger. *Anat. Rec.* **250**, 281–291.

Lewis, E. A., and Irving, J. T. (1970). An autoradiographic investigation of bone remodelling in the rat calvarium grown in organ culture. *Archs Oral Biol.* **15**, 769–776.

Lewis, J. H., and Holder, N. (1977). The development of the tetrapod limb: Embryological mechanisms and evolutionary possibilities. In *Major Patterns in Vertebrate Evolution* (M. K. Hecht, P. C. Goody, and B. M. Hecht, eds), pp. 139–148. Plenum Press, New York, NY.

Lewis, J. H. (1975). Fate map and pattern of cell division: A calculation for the chick wing-bud. *J. Embryol. Exp. Morphol.* **33**, 419–434.

Lewis, J. H., Summerbell, D., and Wolpert, L. (1972). Chimaeras and cell lineage in development. *Nature* **239**, 276–279.

Lewis, W. H. (1907). On the origin and differentiation of the otic vesicle in amphibian embryos. *Anat. Rec.* **1**, 141–145.

Lexer, E. (1908). Ueber Gelenktransplantation. *Med. Klin.* **4**, 817–820.

Lexer, E., Kuliga, P., and Turk, W. (1904). *Untersuchungen der Knochenarterien mittelst Roentgenaufnahmen injizierter Knochen und ihre Bedeutung fuer Einzelne Pathologische Vorgaenge am Knochen System.* A. Hirschwald, Berlin.

Li, C. Y., Harris, A. J., and Suttie, J. M. (2001). Tissue interactions and antlerogenesis: New findings revealed by a xenograft approach. *J. Exp. Zool.* **290**, 18–30.

Li, C. Y., and Suttie, J. M. (1994). Light microscopic studies of pedicle and early first antler development in red deer (*Cervus elephas*). *Anat. Rec.* **239**, 198–215.

Li, C. Y., Waldrup, K. A., Corson, I. D., Littlejohn, R. P., and Suttie, J. M. (1995). Histogenesis of antlerogenic tissues cultivated in diffusion chambers in vivo in red deer (*Cervus elaphus*). *J. Exp. Zool.* **272**, 345–355.

Li, K.-C., Zernicke, R. F., Barnard, R. J., and Li, A. F.-Y. (1991). Differential response of rat limb bones to strenuous exercise. *J. Appl. Physiol.* **70**, 554–560.

Li, R. S., and Denbeste, P. K. (1993). Expression of bone protein messenger RNA at physiological fluoride concentrations in rat osteoblast culture. *Bone Min.* **22**, 187–196.

Li, S., Anderson, R., Reginelli, A. D., and Muneoka, K. (1996). FGF-2 influences cell movements and gene expression during limb development. *J. Exp. Zool.* **274**, 234–247.

Li, S.-W., Prockop, D. J., Helminen, H., Fässler, R., Lapveteláinen, T., Kiraly, K., Peltarri, A., Arokoski, J., Lui, H., Arita, M., and Khillan, J. S. (1995). Transgenic mice with targeted inactivation of the *Col2a1* gene for collagen II develop a skeleton with membrane and periosteal bone but no endochondral bone. *Genes & Devel.* **9**, 2821–2830.

Li, S.-W., Sieron, A. L., Fertala, A., Hojima, Y., Arnold, W. V., and Prockop, D. J. (1996). The c-proteinase that processes procollagens to fibrillar collagens is identical to the protein previously identified as bone morphogenetic protein-1. *Proc. Natl Acad. Sci. U.S.A.* **93**, 5127–5130.

Li, Y., Lacerda, A., Warman, M. L., Beier, D. R., Yoshioka, H., Ninomiya, Y., Oxford, J. T., Morris, N. P., Andrikopoulos, K., Ramirez, F., Wardell, B. B., Lifferth, G. D., Teuscher, C., Woodward, S. R., Taylor, B. A., Seegmiller, R. E., and Olsen, B. R. (1995). A fibrillar collagen gene, *Col11a1*, is essential for skeletal morphogenesis. *Cell* **80**, 423–430.

Li, Z.-L., and Shiota, K. (1999). Stage-specific homeotic vertebral transformations in mouse fetuses induced by maternal hyperthermia during somitogenesis. *Devel. Dynam.* **216**, 336–348.

Lian, J. B., and Gundberg, C. M. (1988). Osteocalcin: Biochemical considerations and clinical applications. *Clin Orthop. Rel. Res.* **226**, 267–291.

Lian, J. B., McKee, A. M. P., Todd, A. M., and Gerstenfeld, L. C. (1993). Induction of bone-related proteins, osteocalcin and osteopontin, and their matrix ultrastructural localization with development of chondrocyte hypertrophy *in vitro*. *J. Cell Biol.* **52**, 206–219.

Lian, J. B., and Stein, G. S. (1991). Concepts of osteoblast growth and differentiation: basis for modulation of bone cell development and tissue formation. *CRC Revs Oral Biol. Med.* **3**, 269–305.

Libbin, R. M. (1992). Prospects for regeneration of growth plates in mammals. In *Bone, Volume 5: Fracture Repair and Regeneration* (B. K. Hall, ed.), pp. 287–312. CRC Press, Boca Raton, FL.

Libbin, R. M., Mitchell, O. G., Guerra, L., and Person, P. (1989). Delayed carpal ossification in *N. viridescens* efts: Relation to the progress of meospodial completion in newt forelimb regenerates. *J. Exp. Zool.* **252**, 207–211.

Libbin, R. M., Ozer, R., Person, P., and Hirschman, A. (1976). *In vitro* accumulation of mineral components by invertebrate cartilage. *Calcif. Tissue Res.* **22**, 67–76.

Libbin, R. M., Singh, I. J., Hirschman, A., and Mitchell, O. G. (1988). A prolonged cartilaginous phase in newt forelimb skeletal regeneration. *J. Exp. Zool.* **248**, 238–242.

Libbin, R. M., and Weinstein, M. (1986). Regeneration of growth plates in the long bones of the neonatal rat hindlimb. *Amer. J. Anat.* **177**, 369–383.

Libbin, R. M., and Weinstein, M. (1987). Sequence of development in innately regenerated growth-plates cartilage in the hindlimbs of the neonatal rat. *Amer. J. Anat.* **180**, 255–265.

Liboff, A. R., and Rinaldi, R. A. (eds) (1974). Electrically mediated growth mechanisms in living systems. *Ann. NY. Acad. Science* **238**, 1–593.

Lidauer, R. M., Plenk, H., Jr., and Grundschober, F. (1985). Sternal histogenesis in blackbirds with respect to the use of wings. *Fortschr. Zool.* **30**, 85–88.

Lieberman, D. E. (1995). Testing hypotheses about recent human evolution from skulls: Integrating morphology, function, development, and phylogeny. *Curr. Anthropol.* **36**, 159–197.

Lieberman, D. E. (1996). How and why humans grow thin skulls: experimental evidence for systemic cortical robusticity. *Amer. J. Phys. Anthropol.* **101**, 217–236.

Lieberman, D. E. (1997). Making behavioral and phylogenetic inferences from hominid fossils: considering the developmental influence of mechanical forces. *Annu. Rev. Anthropol.* **26**, 185–210.

Liebermann, D. E. (2000). Ontogeny, homology, and phylogeny in the hominid craniofacial skeleton. The problem of the brow ridge. In *Development, Growth and Evolution. Implications for the Study of the Hominid Skeleton*, Linnaean Society of London Symposium, Number 20, pp. 85–122. Academic Press, San Diego.

Lieberman, D. E., Devlin, M. J., and Pearson, O. M. (2001). Articular area responses to mechanical loading: Effects of exercise, age, and skeletal location. *Amer. J. Phys. Anthrop.* **116**, 266–277.

Lieberman, D. E., Krovitz, G. E., and McBratney-Owen, B. M (2004). Testing hypotheses about tinkering in the fossil record: the case of the human skull. *J. Exp. Zool. (Mol. Devel. Evol.)* **302B**, 284–301.

Lieberman, D. E., McBratney, B. M., and Krovitz, G. E. (2002). The evolution and development of cranial form in *Homo sapiens*. *Proc. Natl Acad. Sci. U.S.A.* **99**, 1134–1139.

Lieberman, D. E., Pearson, O. M., and Mowbray, K. M. (2000a). Basicranial influence on overall cranial shape. *J. Human Evol.* **38**, 291–315.

Lieberman, D. E., Ross, C. F., and Ravosa, M. J. (2000b). The primate cranial base: ontogeny, function, and integration. *Yrbook Phys. Anthrop.* **43**, 117–169.

Lieberman, D. E., Wood, B. A., and Pilbeam, D. R. (1996). Homoplasy and early *Homo*: an analysis of the evolutionary relationships of *H. habilis sensu stricto* and *H. rudolfensis*. *J. Human Evol.* **30**, 97–120.

Lieberman, J. R., Daluiski, A., and Einhorn, T. A. (2002). The role of growth factors in the repair of bone: biology and clinical applications. *J. Bone Joint Surg.* **84A**, 1032–1044.

Lieberman, J. R., Le, L. Q., Wu, L., Finerman, G. A. M., Berk, A., Witte, D. N., and Stevenson, O. (1998). Regional gene therapy with a BMP-2-producing murine stromal cell line induces heterotopic and orthotopic bone formation in rodents. *J. Orthop. Res.* **16**, 330–339.

Liem, K. F. (1973). Evolutionary strategies and morphological innovations: Cichlid pharyngeal jaws. *Syst. Jaws* **22**, 425–441.

Liem, K. F., Tremml, G., Roelink, H., and Jessell, T. M. (1995). Dorsal differentiation of neural plate cells induced by BMP-mediated signals from epidermal ectoderm. *Cell* **82**, 969–979.

Lightfoot, P. S., and German, R. Z. (1998). The effects of muscular dystrophy on craniofacial growth in mice: a study of heterochrony and ontogenetic allometry. *J. Morphol.* **235**, 1–16.

Linck, G., Oudet, C., and Petrovic, A. (1975). Multiplication des differentes varietes cellulairès de la symphyse pubienne de souris au cours de la croissance et de la premiere gestation: Etude radioautographique a l'aide de la thymidine tritee. *Bull. Assoc. Anat.* **59**, 467–478.

Lincoln, G. A. (1973). Appearance of antler pedicles in early foetal life of red deer. *J. Embryol. Exp. Morphol.* **29**, 431–437.

Lincoln, G. A. (1992). Biology of antlers. *J. Zool. Lond.* **226**, 517–528.

Lincoln, G. A., and Fletcher, T. J. (1976). Induction of antler growth in a congenitally polled Scottish red deer stag. *J. Exp. Zool.* **195**, 247–252.

Linde, A. (ed.) (1984). *Dentin and Dentinogenesis.* Volume 1. CRC Press, Boca Raton, FL.

Lindholm, T. C., Lindholm, T. S., Alitalo, I., and Urist, M. R. (1988). Bovine bone morphogenetic protein (bBMP) induced repair of skull trephine defects in sheep. *Clin. Orthop. Rel. Res.* **227**, 265–268.

Lindsay, K. N. (1977). An autoradiographic study of cellular proliferation of the mandibular condyle after induced dental malocclusion in the mature rat. *Archs Oral Biol.* **22**, 711–714.

Lindsey, C. C. (1966). Temperature-controlled meristic variation in the salamander *Ambystoma gracile*. *Nature* **209**, 1152–1153.

Lindsey, C. C., Brett, A. M., and Swain, D. P. (1984). Responses of vertebral numbers in rainbow trout to temperature changes during development. *Can. J. Zool.* **62**, 391–396.

Lindsey, C. C., and Harrington, R. W., Jr. (1972). Extreme vertebral variation induced by temperature in a homozygous clone of a self-fertilizing cyprinodontid fish *Rivulus marmoratus*. *Can. J. Zool.* **50**, 733–744.

Lindsey, C. C., and Moddie, G. E. E. (1967). The effect of incubation temperature on vertebral count in the chicken. *Can. J. Zool.* **45**, 891–892.

Linsenmayer, T. F. (1974). Temporal and spatial transitions in collagen types during embryonic chick limb development. II. Comparison of the embryonic cartilage collagen molecule with that from adult cartilage. *Devel. Biol.* **40**, 372–377.

Linsenmeyer, T. F., Chen, Q., Gibney, E., Gordon, M. K., Marchant, J. K., Mayne, R., and Schmid, T. M. (1991). Collagen types IX and X in the developing chick tibiotarsus: analyses of mRNAs and proteins. *Development* **111**, 191–196.

Linsenmayer, T. F., and Smith, G. N., Jr. (1976). The biosynthesis of cartilage type collagen during limb regeneration in the larval salamander. *Devel. Biol.* **52**, 19–30.

Linsenmayer, T. F., Toole, B. P., and Trelstad, R. L. (1973a). Temporal and spatial transitions in collagen types during embryonic chick limb development. *Devel. Biol.* **35**, 232–239.

Linsenmayer, T. F., Trelstad, R. L., Toole, B. P., and Gross, J. (1973b). The collagen of osteogenic cartilage in the embryonic chick. *Biochem. Biophys. Res. Commun.* **52**, 870–876.

Linsenmayer, T. F., Trelstad, R. L., and Gross, J. (1973c). The collagen of chick embryonic notochord. *Biochem. Biophys. Res. Commun.* **53**, 39–45.

Lipman, J. M., Hicks, B. J., and Sokoloff, L. (1984). Rabbit chondrocytes are binucleate in auricular but not articular cartilage. *Experientia* **40**, 553–554.

Lipton, B. H., and Jacobson, A. G. (1974a). Analysis of normal somite development. *Devel. Biol.* **38**, 73–90.

Lipton, B. H., and Jacobson, A. G. (1974b). Experimental analysis of the mechanisms of somite development. *Devel. Biol.* **38**, 91–103.

Liskova, M. (1976). Influence of estrogens on bone resorption in organ culture. *Calcif. Tissue Res.* **22**, 207–218.

Liskova, M., and Hert, J. (1971). Reaction of bone to mechanical stimuli. Part 2. Periosteal and endosteal reaction of tibial diaphysis in rabbit to intermittent loading. *Folia Morphol. (Prague)* **19**, 301–317.

Listgarten, M. A. (1974). Intracellular collagen fibrils in the periodontal ligament of the mouse, rat, hamster, guinea pig and rabbit. *J. Periodont. Res.* **8**, 335–342.

Litingtung, Y., Dahn, R. D., Li, Y., Fallon, J. F., and Chiang, C. (2002). *Shh* and *Gli3* are dispensable for limb skeleton formation but regulate digit number and identity. *Nature* **418**, 979–983.

Little, K. (1973). *Bone Behaviour.* Academic Press, London.

Liu, A. L., and Bagnall, K. M. (1995). Regeneration following somite removal in chick embryos. *Anat. Embryol.* **192**, 459–469.

Liu, C., Knezevic, V., and Mackem, S. (2004). Central tail bud mesenchyme is a signaling center for tail paraxial mesoderm induction. *Devel. Dyn.* **229**, 600–606.

Liu, C., Nakamura, E., Knezevic, V., Hunter, S., Thompson, K., and Mackem, S. (2003). A role for the mesenchymal T-box gene *Brachyury* in AER formation during limb development. *Development* **130**, 1327–1337.

Liu, C.-C., and Baylink, D. J. (1977). Stimulation of bone formation and bone resorption by fluoride in thyroparathyroidectomized rats. *J. Dent. Res.* **56**, 304–311.

Liu, C.-C., and Baylink, D. J. (1984). Differential response in alveolar bone osteoclasts residing at two different bone sites. *Calcif. Tissue Int.* **36**, 182–188.

Liu, C.-C., Baylink, D. J., and Wergedal, J. (1974). Vitamin-D-enhanced osteoclastic bone resorption at vascular canals. *Endocrinology* **95**, 1011–1018.

Liu, D., Chu, H., Maves, L., Yan, Y.- L., Morcos, P. A., Postlethwait, J. H., and Westerfield, M. (2003). Fgf3 and Fgf8 dependent and independent transcription factors are required for otic placode specification. *Development* **130**, 2213–2224.

Liu, F., Malaval, L., Gupta, A. K., and Aubin, J. E. (1994). Simultaneous detection of multiple bone-related mRNAs and protein expression during osteoblast differentiation: polymerase chain reaction and immunocytochemical studies at the single cell level. *Devel. Biol.* **166**, 220–234.

Liu, Q., Kerstetter, A. E., Azodi, E., and Marrs, J. A. (2003). Cadherin-1, -2, and -11 expression and cadherin-2 function in the pectoral limb bud and fin of the developing zebrafish. *Devel. Dynam.* **228**, 734–739.

Liu, S. H., Yang, R. S., Alshaikh, R., and Lane, J. M. (1995). Collagen in tendon, ligament, and bone healing – a current review. *Clin. Orthop. Rel. Res.* **318**, 265–278.

Liu, Y. H., Kundu, R., Wu, L., Luo, W., Ignelzi, M. A., Jr., Snead, M. L., and Maxson, R. E., Jr. (1995). Premature suture closure and ectopic cranial bone in mice expressing *Msx2* transgenes in the developing skull. *Proc. Natl Acad. Sci. U.S.A.* **92**, 6137–6141.

Liu, Y. H., Ma, L., Kundu, R., Ignelzi, M. A., Jr., Sangiorgi, F., Wu, L., Luo, W., Snead, M. L., and Maxson, R. E., Jr. (1996). Function of the *Msx2* gene in the morphogenesis of the skull. *Ann. NY. Acad. Sci.* **785**, 48–58.

Liu, Y. H., Tang, Z., Kundu, R., Wu, L., Luo, W., Zhu, D., Sangiorgi, F., Snead, M. L., and Maxson, R. E., Jr. (1999). *Msx2* gene dosage influences the number of proliferative osteogenic cells in growth centers of the developing murine skull: a possible mechanism for Msx-2-mediated craniosynostosis in humans. *Devel. Biol.* **205**, 260–274.

Liu, Z., Xu, J., Colvin, J. S., and Ornitz, D. M. (2002). Coordination of chondrogenesis and osteogenesis of fibroblast growth factor 18. *Genes & Devel.* **16**, 859–869.

Liversage, R. A. (1973). Hypophysectomy and pectoral fin regeneration in adult *Fundulus heteroclitus* (Killifish). *Can. J. Zool.* **51**, 1047–1054.

Liversage, R. A., Rathbone, M. P., and McLaughlin, H. M. G. (1977). Changes in cyclic GMP levels during forelimb regeneration in adult *Notophthalmus viridescens*. *J. Exp. Zool.* **200**, 169–175.

Livne, E., Weiss, A., and Silbermann, M. (1990). Changes in growth patterns in mouse condylar cartilages associated with skeletal maturation and senescence. *Growth, Devel. Aging* **54**, 183–193.

Locke, M. (2004). Structure of long bones in mammals. *J. Morphol.* **262**, 546–565.

Lockshin, R. A., and Beaulaton, J. (1975). Programmed cell death. *Life Sci.* **15**, 1549–1565.

Loewenthal, L. A. (1957). Histological and histochemical studies on the homozygous creeper embryo. *Anat. Rec.* **128**, 201–210.

Logan, M., and Tabin, C. J. (1999). Role of Pitx1 upstream of Tbx4 in specification of hindlimb identity. *Science* **283**, 1736–1739.

Lohnes, D., Mark, M., Mendelsohn, C., Dollé, P., Direich, A., Gorry, P., Gansmuller, A., and Chambon, P. (1994). Function of the retinoic acid receptors (RARs) during development. (I). Craniofacial and skeletal abnormalities in RAR double mutants. *Development* **120**, 2723–2748.

Lombard, R. E., and Bolt, J. R. (1979). Evolution of the tetrapod ear: An analysis and reinterpretation. *Biol. J. Linn. Soc.* **11**, 19–76.

Long, F., Chung, U.-i., Ohba, S., McMahon, J., Kronenberg, H., and McMahon, A. P. (2004). Ihh signaling is directly required for the osteoblast lineage in the endochondral skeleton. *Development* **131**, 1309–1318.

Long, F., and Linsenmayer, T. F. (1998). Regulation of growth region cartilage proliferation and differentiation by perichondrium. *Development* **125**, 1067–1073.

Long, F., Schipani, E., Asahara, H., Kronenberg, H., and Montminy, M. (2001a). The CREB family of activators is required for endochondral bone development. *Development* **128**, 541–550.

Long, F., Zhang, X., Karp, S., Yang, Y., and McMahon, A. P. (2001b). Genetic manipulation of the hedgehog signaling in the endochondral skeleton reveals a direct role in he regulation of chondrocyte proliferation. *Development* **128**, 5099–5108.

Long, J. A. (1990). Heterochrony and the origin of tetrapods. *Lethaia* **23**, 157–166.

Long, Q., Park, B. K., and Ekker, M. (2001). Expression and regulation of mouse *Mtsh1* during limb and branchial arch development. *Devel. Dynam.* **222**, 308–312.

López, D., Durán, A. C., de Andrés, A. V., Guerrero, A., Blasco, M., and Sans-Coma, V. (2003). Formation of cartilage in the heart of the Spanish terrapin, *Mauremys leprosa* (Reptilia, Chelonia). *J. Morphol.* **258**, 97–105.

López, D., Duran, A. C., and Sans-Coma, V. (2000). Formation of cartilage in cardiac semilunar valves of chick and quail. *Ann. Anat.-Anat. Anz.* **182**, 349–359.

López, D., Fernandez, M. C., Duran, A. C., and Sans-Coma, A. C. (2001). Cartilage in pulmonary valves of Syrian hamsters. *Ann. Anat.-Anat. Anz.* **183**, 383–388.

Lopez, E. (1970a). L'os cellulaire d'un Poisson téléostéen *Anguilla anguilla* L. I. Étude histocytologique et histophysique. *Z. Zellforsch. Mikrosk. Anat.* **109**, 552–565.

Lopez, E. (1970b). L'os cellulaire d'un Poisson téléostéen *Anguilla anguilla* L. II. Action de l'ablation des corpuscles de Stannius. *Z. Zellforsch. Mikrosk. Anat.* **109**, 566–572.

Lopez, E., and Deville, J. (1973). Effect of prolonged administration of synthetic Salmon calcitonin (SCT) on vertebral bone morphology and on the ultimobranchial body (UB) activity of the mature female eel (*Anguilla anguilla* L.). In *Calcified Tissue* (H. Czitober and J. Eschberger, eds), pp. 169–174. Facta-Publications.

Lopez, E., and Martelly-Bagot, E. (1971). L'os cellulaire d'un Poisson téléostéen *Anguilla anguilla* L. III. Étude histologique et histophysique au cours de la maturation provoquée par injections d'extrait hypophysaire de Carpe. *Z. Zellforsch. Mikrosk. Anat.* **117**, 176–190.

Lopez, E., MacIntyre, I., Martelly, E., Lallier, F., and Vidal, B. (1980). Paradoxical effect of 1,25 dihydroxycholecalciferol on osteoblastic and osteoclastic activity in the skeleton of the eel, *Anguilla anguilla* L. *Calcif. Tissue Int.* **32**, 83–97.

Lorch, I. J. (1947). Localization of alkaline phosphatase in mammalian bone. *Q. J. Microsc. Sci.* **88**, 367–381.

Lorch, I. J. (1949a). The distribution of alkaline phosphatase in the skull of the developing trout. *Q. J. Microsc. Sci.* **90**, 183–207.

Lorch, I. J. (1949b). The distribution of alkaline phosphatase in relation to calcification in *Scyliorhinus canicula*. *Q. J. Microsc. Sci.* **90**, 381–391.

Lorch, I. J. (1949c). Alkaline phosphatase and the mechanism of ossification. *J. Bone Joint Surg. Br Vol.* **31**, 94–99.

Loredo, G. A., Brukman, A., Harris, M. P., Kable, D., Leclair, E. E., Gutman, R., Denney, E., Henkelman, E., Murray, B. P., Fallon, J. F., Tuan, R. S., and Gilbert, S. F. (2001). Development of an evolutionarily novel structure: Fibroblast growth factor expression in the carapacial ridge of turtle embryos. *J. Exp. Zool. (Mol. Devel. Evol.)*, **291**, 274–281.

Loty, S., Foll, C., Forest, N., and Sautier, J.-M. (2000). Association of enhanced expression of gap junctions with in vitro chondrogenic differentiation of rat nasal septal cartilage-released cells following their dedifferentiation and redifferentiation. *Archs Oral Biol.* **45**, 843–856.

Loudon, A. S. I., and Curlewis, J. D. (1988). Cycles of antler and testicular growth in an aseasonal tropical deer (*Axis axis*). *J. Reprod. Fertil.* **83**, 729–738.

Loutit, J. F., and Sansom, J. M. (1976). Osteopetrosis of microphthalmic mice – a defect of the hematopoietic stem cells? *Calcif. Tissue Res.* **20**, 251–259.

Lovejoy, C. O., Cohn, M. J., and White, T. D. (2000). The evolution of mammalian morphology: a developmental perspective. In *Development, Growth and Evolution. Implications for the Study of the Hominid Skeleton* (P. O'Higgins and M. J. Cohn, eds), pp. 41–55. Academic Press, San Diego.

Lovejoy, N. R. (2000). Reinterpreting recapitulation: Systematics of needlefishes and their allies (Teleostei: Beloniformes). *Evolution* **54**, 1349–1362.

Loveridge, N., Farquharson, C., Hesketh, J. E., Jakowlew, S. B., Whitehead, C. C., and Thorp, B. H. (1993). The control of chondrocyte differentiation during endochondral bone growth in vivo: changes in TGF-β and the proto-oncogene c-myc. *J. Cell Sci.* **105**, 949–956.

Løvtrop, S. (1974). *Epigenetics*. John Wiley & Sons, New York, NY.

Løvtrup, S. (1977). *The Phylogeny of Vertebrata*. John Wiley & Sons, New York, NY.

Lowther, D. A., and Natarajan, M. (1972). The influence of glycoprotein on collagen fibril formation in the presence of chondroitin sulphate proteoglycan. *Biochem. J.* **127**, 607–608.

Lu, M.-F., Cheng, H.-T., Kern, M. J., Potter, S. S., Tran, B., Diekwisch, T. G. H., and Martin, J. F. (1999a). *Prx-1* functions cooperatively with another *paired*-related homeobox gene, *Prx-2*, to maintain cell fate within the craniofacial mesenchyme. *Development* **126**, 495–504.

Lu, M.-F., Cheng, H.-T., Lacy, A. R., Kern, M. J., Argao, E. A., Potter, S. S., Olson, E. N., and Martin, J. F. (1999b). *Paired*-related homeobox genes cooperate in handplate and hindlimb zeugopod morphogenesis. *Devel. Biol.* **205**, 145–157.

Lu, M.-F., Pressman, C., Dyer, R., Johnson, R. L., and Martin, J. F. (1999c). Function of Rieger syndrome gene in left-right asymmetry and craniofacial development. *Nature* **401**, 276–278.

Luben, R. A., Wong, G. L., and Cohn, D. V. (1977). Parathormone-stimulated resorption of devitalized bone by cultured osteoclast-type bone cells. *Nature* **265**, 629–630.

Lucht, U. (1971). Acid phosphatase of osteoclasts demonstrated by electron microscopic histochemistry. *Histochemie* **28**, 103–117.

Lucht, U. (1980). Osteoclasts – ultrastructure and functions. In *The Reticuloendothelial System. Volume 1. A Comprehensive Treatise. Morphology.* (I. Carr and W. T. Daems, eds), pp. 705–734. Plenum Press, New York, NY.

Lucifora, L. O., and Vassallo, A. I. (2002). Walking in skates (Chondrichthyes, Rajidae): anatomy, behaviour and analogies to tetrapod locomotion. *Biol. J. Linn. Soc. B* **77**, 35–41.

Luder, H. U., and Schroeder, H. E. (1992). Light and electron microscopic morphology of the temporomandibular joint in growing and mature crab-eating monkeys (*Macaca fascicularis*): the condylar calcified cartilage. *Anat. Embryol.* **185**, 189–199.

Lufkin, T., Dierich, A., LeMeur, M., Mark, M., and Chambon, P. (1991). Disruption of the Hox-1.6 homeobox gene results in defects in a region corresponding to its rostral domain of expression. *Cell* **66**, 1105–1119.

Lufkin, T., Mark, M., Hart, C. P., Dollé, P., LeMeur, M., and Chambon, P. (1992). Homeotic transformation of the occipital bones of the skull by ectopic expression of a homeobox gene. *Nature* **359**, 835–841.

Lui, V. C. H., Ng, L. J., Nicholls, J., Tam, P. P. L., and Cheah, K. S. E. (1995). Tissue-specific and differential expression of alternately spliced alpha-1 (II) collagen messenger RNAs in early human embryos. *Devel. Dynam.* **203**, 198–211.

Luk, S. C., Nopagaroonsri, C., and Simon, G. T. (1974a). The ultrastructure of endosteum. A topographic study in young adult rabbits *J. Ultrastruct. Res.* **46**, 165–183.

Luk, S. C., Nopagaroonsri, C., and Simon, G. T. (1974b). The ultrastructure of cortical bone in young adult rabbits. *J. Ultrastruct. Res.* **46**, 184–203.

Lumsden, A., and Osborn, J. (1976). Development of the mouse dentition in culture. *J. Dent. Res.* **55**, D136.

Lundy, M. W., Farley, J. R., and Baylink, D. J. (1986). Characterization of a rapidly responding animal model for fluoride-stimulated bone formation. *Bone* **7**, 289–294.

Lundy, M. W., Lau, K. H. W., Blair, H. C., and Baylink, D. J. (1988). Chick osteoblasts contain fluoride-sensitive acid phosphatase activity. *J. Histochem. Cytochem.* **36**, 1175–1180.

Lundy, M. W., Russell, J. E., Avery, J., Wergedal, J. E., and Baylink, D. J. (1992). Effect of sodium fluoride on bone density in chickens. *Calcif. Tissue Int.* **50**, 420–426.

Lundy, M. W., Wergedal, J. E., Teubner, E., Burnell, J., Sherrard, D., and Baylink, D. J. (1989). The effect of prolonged fluoride therapy for osteoporosis: Bone composition and histology. *Bone* **10**, 321–327.

Lüning, C., Rass, A., Rozell, B., Wroblewski, J., and Öbrink, B. (1994). Expression of E-cadherin during craniofacial development. *J. Craniofac. Genet. Devel. Biol.* **14**, 207–216.

Luo, G., Hofmann, C., Bronckers, A. L. J. J., Sohocki, M., Bradley, A., and Karsenty, G. (1995). BMP-7 is an inducer of nephrogenesis and is also required for eye development and skeletal patterning. *Genes & Devel.* **9**, 2808–2820.

Luo, Z. (2000). In search of whale's sisters. *Nature* **404**, 235–239.

Lutfi, A. M. (1970a). Study of cell multiplication in ther cartilaginous upper end of the tibia of the domestic fowl by tritiated thymidine autoradiography. *Acta Anat.* **76**, 454–463.

Lutfi, A. M. (1970b). Mode of growth, fate and function of cartilage canals. *J. Anat.* **106**, 135–146.

Lutfi, A. M. (1971). The fate of chondrocytes during cartilage erosion in the growing tibia in the domestic fowl (*Gallus domesticus*). *Acta Anat.* **79**, 27–35.

Lutfi, A. M. (1974). The role of cartilage in long bone growth: A reappraisal. *J. Anat.* **117**, 413–418.

Luther, A. (1924). Entwicklungsmechanische Untersuchungen an labyrinth einiger Anuren. *Comment. Biol. Soc. Sci. Fenn* **2**, 1–24.

Lu Valle, P., Hayashi, M., and Olsen, B. R. (1989). Transcriptional regulation of type X collagen during chondrocyte maturation. *Devel. Biol.* **133**, 613–616.

Lu Valle, P., Iwamoto, M., Fanning, P., and Pacifici, M. (1993). Multiple negative elements in a gene that codes for an extracellular matrix protein, collagen X, restrict expression to hypertrophic chondrocytes. *J. Cell Biol.* **121**, 1173–1179.

Luyten, F. P., Chen, P., Paralkar, V., and Reddi, A. H. (1994). Recombinant bone morphogenetic protein-4, transforming growth factor-β1, and activin A enhance the cartilage phenotype of articular chondrocytes *in vitro*. *Exp. Cell Res.* **210**, 224–229.

Luyten, F. P., Cunningham, N. S., Ma, S., Muthukumaran, N., Hammonds, R. G., Nevins, W. B., Wood, W. I., and Reddi, A. H. (1989). Purification and partial amino acid sequence of osteogenin, a protein initiating bone differentiation. *J. Biol. Chem.* **264**, 13377–13380.

Luyten, F. P., Yu, Y. M., Yanagishita, M., Vukicevic, S., Hammonds, R. G., and Reddi, A. H. (1992). Natural bovine osteogenic and recombinant human bone morphogenetic protein-2B are equipotent in the maintenance of proteoglycans in bovine articular cartilage explant cultures. *J. Biol. Chem.* **267**, 3691–3695.

Lydekker, R. (1904). On the subspecies of *Giraffa camelolopardalis*. *Proc. Zool. Soc. Lond.* **1904**(1), 202–227.

Lynch, M. P., Stein, J. L., Stein, G. S., and Lian, J. B. (1995). The influence of type 1 collagen on the development and maintenance of the osteoblast phenotype in primary and passaged rat calvarial osteoblasts: modification of expression of genes supporting cell growth, adhesion, and extracellular matrix mineralization. *Exp. Cell Res.* **216**, 35–45.

Lyons, K. M., Hogan, B. L. M., and Robertson, E. J. (1995). Colocalization of BMP7 and BMP2 RNAs suggests that these factors cooperatively mediate tissue interactions during murine development. *Mech. Devel.* **50**, 71–83.

Lyons, K. M., Pelton, R. W., and Hogan, B. L. M. (1990). Organogenesis and pattern formation in the mouse: RNA distribution patterns suggest a role for bone morphogenetic protein-2A (BMP-2A). *Development* **109**, 833–844.

M

Ma, B., Sampson, W., Wilson, D., Wiebkin, O., and Fazzalari, N. (2002). A histomorphometric study of adaptive responses of cancellous bone in different regions in the sheep mandibular condyle following experimental forward mandibular displacement. *Archs Oral Biol.* **47**, 519–527.

Ma, S., Chen, G., and Reddi, A. H. (1990). Collaboration between collagenous matrix and osteogenin is required for bone induction. *Ann. NY. Acad. Sci.* **580**, 524–525.

Ma, W., and Lozanoff, S. (2002). Differential *in vitro* response to epidermal growth factor by prenatal murine cranial-base chondrocytes. *Archs Oral Biol.* **47**, 155–163.

Maas, S. A., and Fallon, J. F. (2004). Isolation of the chicken *Lmbr1* coding sequence and characterization of its role during chick limb development. *Devel. Dynam.* **229**, 520–528.

Mabbutt, L. W., and Kokich, V. G. (1979). Calvarial and sutural re-development following craniectomy in the neonatal rabbit. *J. Anat.* **129**, 413–422.

Mabee, P. M., Aldridge, E., Warren, E., and Helenurm, K. (1998). Effect of cleaving and staining on fish length. *Copeia* **1998**, 346–353.

Mabee, P. M., Crotwell, P. L., Bird, N. C., and Burke, A. C. (2002). Evolution of median fin modules in the axial skeleton of fishes. *J. Exp. Zool. (Mol. Devel. Evol.)* **294**, 77–90.

Mabee, P. M., and Trendler, T. A. (1996). Development of the cranium and paired fins in *Betta splendens* (Teleostei: Percomorpha): Intraspecific variation and interspecific comparisons. *J. Morphol.* **227**, 249–287.

McBratney, B. M., Margaryan, E., Ma, W., Urban, Z., and Lozanoff, S. (2003). Frontonasal dysplasia in 3H1 *Br/Br* mice. *Anat. Rec.* Part A 271A, 291–302.

McBurney, K. M., Kelley, F. W., Kibenge, F. S. B., and Wright, G. M. (1996). Spatial and temporal distribution of lamprin mRNA during chondrogenesis of trabecular cartilage in the sea lamprey. *Anat. Embryol.* **193**, 419–426.

McBurney, K. M., and Wright, G. M. (1996). Chondrogenesis of a non-collagen-based cartilage in the sea lamprey, *Petromyzon marinus*. *Can. J. Zool.* **74**, 2118–2130.

McCandless, E. L., Lehoczky, J. M., and Rodbard, S. (1963). Aortic cartilage produced by intramural carrageenan. *Archs Pathol.* **75**, 507–516.

McCarthy, T. L., Centrella, M., Raisz, L. G., and Canalis, E. (1991). Prostaglandin E2 stimulates insulin-like growth factor I synthesis in osteoblast-enriched cultures from fetal rat bone. *Endocrinology* **128**, 2895–2900.

McCollum, M. A., and Sharpe, P. T. (2001). Developmental genetics and early hominid craniodental evolution. *BioEssays* **23**, 481–493.

McConaghey, P., and Sledge, C. B. (1970). Production of 'sulphation factor' by the perfused liver. *Nature* **225**, 1249–1250.

McConnell, T. H. (1970). Bony and cartilaginous tumors of the heart and great vessels: Report of an osteosarcoma of the pulmonary artery. *Cancer* **25**, 611–617.

McCormack, A. P., Anderson, P. A., and Tencer, A. F. (1993). Effect of controlled local release of sodium fluoride on bone formation – filling a defect in the proximal femoral cortex. *J. Orthop. Res.* **11**, 548–555.

McCullagh, J. J., Gill, P., and Wilson, D. J. (1990). Repair of cartilaginous fractures during chick limb development. *J. Orthop. Res.* **8**, 127–131.

McCullagh, J. J., and Wilson, D. J. (1993). Antero-posterior skeletal patterning is not dependent on the continuity of the apical ectodermal ridge in the chick wing bud. *Anat. Embryol.* **188**, 371–379.

McCulloch, C. A. G., Fair, C. A., Tenenbaum, H. C., Limeback, H., and Homareau, R. (1990). Clonal distribution of osteoprogenitor cells in cultured chick periostea: functional relationship to bone formation. *Devel. Biol.* **140**, 352–361.

McCulloch, C. A. G., Tenenbaum, H. C., Fair, C. A., and Birek, C. (1989). Site-specific regulation of osteogenesis maintenance of discrete levels of phenotypic expression *in vitro*. *Anat. Rec.* **223**, 27–34.

McDevitt, C. A., and Webber, R. J. (1990). The ultrastructure and biochemistry of meniscal cartilage. *Clin. Orthop. Rel. Res.* **252**, 8–18.

McDonald, S. A., and Tuan, R. S. (1989). Expression of collagen type transcripts in chick embryonic bone detected by *in situ* cDNA-mRNA hybridization. *Devel. Biol.* **133**, 221–234.

McFadden, D. G., McAnally, J., Richardson, J. A., Charité, J., and Olson, E. N. (2002). Misexpression of dHAND induces ectopic digits in the developing limb bud in the absence of direct DNA binding. *Development* **129**, 3077–3088.

McGonnell, I. M., Clarke, J. D. W., and Tickle, C. (1998). Fate maps of the developing chick face: analysis of expansion of facial primordia and establishment of the primary palate. *Devel. Dynam.* **212**, 102–118.

McGonnell, I. M., Green, C. R., Tickle, C., and Becker, D. L. (2001). Connexin 43 gap junction protein plays an esential role in morphoge nesis of the embryonic chick face. *Devel. Dynam.* **222**, 420–438.

McGowan, C. (1984). Evolutionary relationships of ratites and carinates: Evidence from ontogeny of the tarsus. *Nature* **307**, 733–735.

McGowan, C. (1985). Tarsal development in birds: Evidence for homology with the theropod conditions. *J. Zool. London* **206**, 53–67.

McGowan, C. (1988). Differential development of the rostrum and mandible of the swordfish (*Xiphias gladius*) during ontogeny and its possible functional significance. *Can. J. Zool.* **66**, 496–503.

McGuire, J. L., and Marks, S. C., Jr. (1974). The effect of parathyroid hormone on bone cell structure and function. *Clin. Orthop. Rel. Res.* **100**, 392–405.

McHenry, F. A., Hoffman, P. N., and Salpeter, M. M. (1974). Uptake of ^{35}S-sulfate by morphologically differentiated replicating chondrocytes *in vivo*: A double isotope electron microscope autoradiographic study. *Devel. Biol.* **39**, 96–104.

McKee, M. D., Glimcher, M. J., and Nanci, A. (1990a). Developmental appearance and ultrastructural immunolocalization of a major 66 kDA phosphoprotein in embryonic and post-natal chicken bone. *Anat. Rec.* **228**, 77–92.

McKee, M. D., Glimcher, M. J., and Nanci, A. (1992). High-resolution immunolocalization of osteopontin and osteocalcin in bone and cartilage during endochondral ossification in the chicken tibia. *Anat. Rec.* **234**, 479–492.

McKee, M. D., and Nanci, A. (1996). Osteopontin at mineralized tissue interfaces in bone, teeth, and osseointegrated implants: ultrastructural distribution and implications for mineralized tissue formation, turnover, and repair. *Microsc. Res. Tech.* **33**, 141–164.

McKee, M. D., Nanci, A., Landis, W. J., Gotoh, Y., Gerstenfeld, L. C., and Glimcher, M. J. (1990b). Expression and ultrastructural immunolocalization of a major 66 kDA phosphoprotein in embryonic and post-natal chicken bone. *Anat. Rec.* **228**, 93–103.

McKeown, M. (1975). The influence of environment on the growth of the craniofacial complex – a study of domestication. *Angle Orthodontist* **45**, 137–140.

McKibbin, B. (1978). The biology of fracture healing in long bones. *J. Bone Joint Surg. Br. Vol.* **60**, 150–162.

McKusick, V. A. (1972). *Heritable Disorders of Connective Tissue.* C. V. Mosby, St. Louis, MI.

McLachlan, J. C., Bateman, M., and Wolpert, L. (1976). Effect of 3-acetylpyridine on tissue differentiation of the embryonic chick limb. *Nature* **264**, 267–269.

McLain, J. B., and Vig, P. S. (1983). Transverse periosteal sectioning and femur growth in the rat. *Anat. Rec.* **207**, 339–348.

McLain, J. B., Vig, P. S., and Hamilton, D. C. (1982). The influence of circumferential periosteal section on mandibular morphology and growth in the rat. In *Factors and Mechanisms Influencing Bone Growth* Dixer, A. D., and Sarnat B. G. pp. 581–596. Alan R. Liss, Inc., New York, NY.

McLain, K., Schreiner, C., Yager, K. L., Stock, J. L., and Potter, S. S. (1992). Ectopic expression of *Hox-2.3* induces craniofacial and skeletal malformations in transgenic mice. *Mech. Devel.* **39**, 3–16.

McLaren, A. (1972). Numerology of development. *Nature* **239**, 274–276.

McLean, F. C., and Urist, M. R. (1968). *Bone. Fundamentals of the Physiology of Skeletal Tissues*, 3rd Edn. University of Chicago Press, Chicago, IL.

McLeod, K. J., and Rubin, C. T. (1989). Frequency specific modulation of bone adaptation by induced electric fields. *Theoret. Biol.* **145**, 385–396.

McLoughlin, C. B. (1961). The importance of mesenchymal factors in the differentiation of chick epidermis. II. Modification of epidermal differentiation by contact with different types of mesenchyme. *J. Embryol. Exp. Morphol.* **9**, 385–409.

McMahon, A. P., Champion, J. E., McMahon, J. A., and Sukhatme, V. P. (1990). Developmental expression of the putative transcription factor *Egr-1* suggests that *Eghr-1* and *c-fos* are co-regulated in some tissues. *Development* **108**, 281–287.

McMahon, D. (1974). Chemical messengers in development: A hypothesis. *Science* **185**, 1012–1021.

McNamara, J. A., Jr. (1980). Functional determinants of craniofacial size and shape. *J. Orthodont.* **2**, 131–159.

McNamara, J. A., Jr., Hinton, R. J., and Hoffman, D. L. (1982). Histologic analysis of temporomandibular joint adaptation to protrusive function in young adult Rhesus monkeys (*Macaca mulatta*). *Amer. J. Orthod.* **82**, 288–298.

McPhee, J. R., and Van de Water, T. R. (1982). A biochemical profile of the ECM during the sequential stages of otic capsule formation *in vivo* and *in vitro*. In *Extracellular Matrix* (S. Hawkes and J. L. Wang, eds), pp. 289–294. Academic Press, New York, NY.

McPhee, J. R., and Van de Water, T. R. (1985). A comparison of morphological stages and sulfated glycosaminoglycan production during otic capsule formation: *In vivo* and *in vitro*. *Anat. Rec.* **213**, 566–577.

McPhee, J. R., and Van de Water, T. R. (1986). Epithelial–mesenchymal tissue interactions guiding otic capsule formation: the role of the otocyst. *J. Embryol. Exp. Morphol.* **97**, 1–24.

McPherron, A. C., Lawler, A. M., and Lee, S.-J. (1999). Regulation of anterior/posterior patterning of the axial skeleton by growth/differentiation factor 11. *Nature Genet.* **22**, 260–264.

MacCabe, J. A., and Abbott, U. K. (1974). Polarizing and maintenance activities in 2 polydactylous mutants of fowl – *diplopodia*[1] and *talpid*[2]. *J. Embryol. Exp. Morphol.* **31**, 735–746.

MacCabe, J. A., Calandra, A. J., and Parker, B. W. (1977). *In vitro* analysis of the distribution and nature of a morphogenetic factor in the developing chick wing. In *Vertebrate Limb and Somite Morphogenesis* (D. A. Ede, J. R. Hinchliffe, and M. Balls, eds), pp. 25–40. Cambridge University Press, Cambridge.

MacCabe, J. A., Errick, J. E., and Saunders, J. W., Jr. (1974). Ectodermal control of the dorsoventral axis in the leg bud of the chick embryo. *Devel. Biol.* **39**, 69–82.

MacCabe, J. A., MacCabe, A. B., Abbott, U. K., and McCarrey, J. R. (1975). Limb development in *Diplopodia₄*: A polydactylous mutation in the chicken. *J. Exp. Zool.* **191**, 383–394.

MacCabe, J. A., and Parker, B. W. (1975). The *in vitro* maintenance of the apical ectodermal ridge of the chick embryo wing bud: An assay for polarizing activity. *Devel. Biol.* **45**, 349–357.

MacCabe, J. A., and Parker, B. W. (1976a). Evidence for a gradient of a morphogenetic substance in the developing limb. *Devel. Biol.* **54**, 297–303.

MacCabe, J. A., and Parker, B. W. (1976b). Polarizing activity in the developing limb of the Syrian hamster. *J. Exp. Zool.* **195**, 311–317.

MacCabe, J. A., Saunders, J. W., Jr., and Pickett, M. (1973). The control of the anteroposterior and dorsoventral axes in embryonic chick limbs constructed of dissociated and reaggregated limb-bud ectoderm. *Devel. Biol.* **31**, 323–335.

Macey, L. R., Kana, S. M., Jingushi, S., Terek, R. M., Boirretos, J., and Bolander, M. E. (1989). Defects of early fracture-healing in experimental diabetes. *J. Bone Joint Surg. Am. Vol.* **71**, 721–733.

MacDonald, M. E., and Hall, B. K. (2001). Altered timing of the extracellular-matrix-mediated epithelial–mesenchymal interaction that initiates mandibular skeletogenesis in three inbred strains of mice: Development, heterochrony, and evolutionary change in morphology. *J. Exp. Zool.* **291**, 258–273.

MacDonald, M. E., Abbott, U. K., and Richman, J. M. (2004). Upper beak truncation in chicken embryos with the cleft primary palate mutation is due to an epithelial defect in the frontonasal mass. *Devel. Dyn.* **230**, 335–349.

MacEwen, W. (1881). Observations concerning transplantation on bone. Illustrated by a case of inter-human osseous transplantation whereby over two thirds of the shaft of the humerus was restored. *Proc. R. Soc. London* **32**, 232–247.

MacEwen, W. (1912). *The Growth of Bone. Observations on Osteogenesis. An Experimental Enquiry into the Development and Reproduction of Diaphyseal Bone.* James Maclehose & Sons, Glasgow.

Machwate, M., Jullienne, A., Moukhtar, M., Lomri, A., and Marie, P. J. (1995a). *C-fos* protooncogene is involved in the mitogenic effect of transforming growth factor-β in osteoblastic cells. *Mol. Endocrinol.* **9**, 187–198.

Machwate, M., Jullienne, A., Moukhtar, M., and Marie, P. J. (1995b). Temporal variation of c-fos proto-oncogene expression during osteoblast differentiation and osteogenesis in developing rat bone. *J. Cell Biol.* **57**, 62–70.

Macias, D., and Gañan, Y. (1991). The role of the polarizing zone in the pattern of experimental chondrogenesis in the chick embryo interdigital space. *Int. J. Devel. Biol.* **35**, 63–67.

Macias, D., Gañan, Y., and Hurlé, J. M. (1992). Interdigital chondrogenesis and extra digit formation in the duck leg bud subjected to local ectoderm removal. *Anat. Embryol.* **186**, 27–32.

Macias, D., Gañan, Y., and Hurlé, J. M. (1993). Modification of the phalangeal pattern of the digits in the chick embryo leg bud by local microinjection of RA, staurosporin and TGFβs. *Anat. Embryol.* **188**, 201–208.

Macias, D., Gañan, Y., Ros, M. A., and Hurlé, J. M. (1996). Inhibition of programmed cell death by local administration of FGF-2 and FGF-4 in the interdigital areas of the embryonic chick leg bud. *Anat. Embryol.* **193**, 533–541.

Macias, D., Gañan, Y., Sampath, T. K., Piedra, M. E., Ros, M. A., and Hurlé, J. M. (1997). Role of BMP-2 and OP-1 (BMP-7) in programmed cell death and skeletogenesis during chick limb development. *Development* **124**, 1109–1117.

Mackem, S., and Mahon, K. A. (1991). *GHox 4.7*: a chick homeobox gene expressed primarily in limb buds with limb-type differences in expression. *Development* **112**, 791–806.

MacKenzie, A., Ferguson, M. W. J., and Sharpe, P. T. (1991b). Hox-7 expression during murine craniofacial development. *Development* **113**, 601–611.

MacKenzie, A., Leeming, G. L., Jowett, A. K., Ferguson, M. W. J., and Sharpe, P. T. (1991a). The homeobox gene Hox 7.1 has specific regional and temporal expression patterns during early murine craniofacial embryogenesis, especially tooth development *in vivo* and *in vitro*. *Development* **111**, 269–285.

Mackie, E. J., Thesleff, I., and Chiquet-Ehrismann, R. (1987). Tenascin is associated with chondrogenic differentiation *in vivo* and promotes chondrogenesis *in vitro*. *J. Cell Biol.* **105**, 2569–2580.

Mackie, E. J., and Trechsel, U. (1990). Stimulation of bone formation *in vivo* by transforming growth factor-β. Remodeling of woven bone and lack of inhibition by indomethacin. *Bone* **11**, 295–300.

Mackie, E. J., Trechsel, U., and Bruns, C. (1990). Somatostatin receptors are restricted to a subpopulation of osteoblast-like cells during endochondral bone formation. *Development* **110**, 1233–1239.

Mackie, E. J., and Tucker, R. P. (1992). Tenascin in bone morphogenesis: expression by osteoblasts and cell type-specific expression of splice variants. *J. Cell Sci.* **103**, 765–771.

Mackie, E. J., and Tucker, R. P. (1999). The tenascin-C knockout revisited. *J. Cell Sci.* **112**, 3847–3853.

Mackie, E. J., Tucker, R. P., Halfter, W., Chiquet-Ehrismann, R., and Epperlein, H. H. (1988). The distribution of tenascin coincides with pathways of neural crest cell migration. *Development* **102**, 237–250.

Macklin, C. C. (1917). Notes on the preparation of bones from madder fed animals. *Anat. Rec.* **12**, 403–405.

Maclean, N., and B. K. Hall (1987). *Cell Commitment and Differentiation.* Cambridge University Press, Cambridge.

MacLeod, C. D. (2002). Possible function of the ultradense bone in the rostrum of Blainville's beaked whale (*Mesoplodon densirostris*). *Can. J. Zool.* **80**, 178–184.

Maddox, B. K., Garofalo, S., Horton, W. A., Richardson, M. D., and Trune, D. R. (1998). Craniofacial and otic capsule abnormalities in a transgenic mouse strain with a *Col2a1* mutation. *J. Craniofac. Genet. Devel. Biol.* **18**, 195–201.

Maden, M. (1976). Blastemal kinetics and pattern formation during amphibian limb regeneration. *J. Embryol. Exp. Morphol.* **36**, 561–574.

Maden, M. (1983). The effect of vitamin A on limb regeneration in *Rana temporaria*. *Devel. Biol.* **98**, 409–416.

Maden, M. (1993). The homeotic transformation of tails into limbs in *Rana temporaria* by retinoids. *Devel. Biol.* **159**, 379–391.

Maden, M. (2002). Retinoic acid and limb regeneration. *Int. J. Devel. Biol.* **46**, 883–886.

Maden, M., and Wallace, H. (1975). The origin of limb regenerates from cartilage grafts. *Acta Embryol. Exp.* **2**, 77–86.

Maderson, P. F. A. (1967). A comment on the evolutionary origin of vertebrate appendages. *Amer. Nat.* **101**, 71–78.

Maderson, P. F. A. (1975). Embryonic tissue interactions as the basis for morphological change in evolution. *Amer. Zool.* **15**, 315–328.

Maderson, P. F. A. (ed.) (1987). *Developmental and Evolutionary Aspects of the Neural Crest.* John Wiley & Sons., New York, NY.

Maderson, P. F. A., and Licht, P. (1968). Factors influencing rates of tail regeneration in the lizard *Anolis carolinensis. Experientia* **24**, 1083–1086.

Maderson, P. F. A., and Salthe, S. N. (1971). Further observations on tail regeneration in *Anolis carolinensis* (Iguanidae, Lacertilia). *J. Exp. Zool.* **177**, 185–190.

Madsen, K., Friberg, U., Roos, P., Eden, S., and Isaksson, O. (1983a). Growth hormone stimulates the proliferation of cultured chondrocytes from rabbit ear and rat rib growth cartilage. *Nature* **304**, 545–547.

Madsen, K., Moskalewski, S., von der Mark, K., and Friberg, U. (1983b). Synthesis of proteoglycans, collagen and elastin by cultures of rabbit auricular chondrocytes – relation to age of the donor. *Devel. Biol.* **96**, 63–73.

Maekawa, K., and Yamada, J. (1970). Some histochemical and fine structural aspects of growing scales of the rainbow trout. *Bull. Fac. Fish. Hokkaida Univ.* **21**, 70–78.

Maglia, A. M., and Pugener, L. A. (1998). Skeletal development and adult osteology of *Bombina orientalis* (Anura, Bombinatoridae). *Herpetologica* **54**, 344–363.

Maglia, A. M., Pugener, L. A., and Trueb, L. (2001). Comparative development of anurans: Usinf phylogeny to understand ontogeny. *Amer. Zool.* **41**, 538–551.

Maier, W. (1990). Phylogeny and ontogeny of mammalian middle ear structures. *Neth. J. Zool.* **40**, 55–74.

Maisey, J. G. (1979). Finspine morphogenesis in squalid and heterodontid sharks. *Zool. J. Linn. Soc.* **66**, 161–184.

Maisey, J. G. (1987). Notes on the structure and phylogeny of vertebrate otoliths. *Copeia* **1987**, 495–499.

Maisey, J. G. (1988). Phylogeny of early vertebrate skeletal induction and ossification patterns. *Evol. Biol.* **22**, 1–36.

Makarenkova, H., and Patel, K. (1999). Gap junction signalling mediated through connexin-43 is required for chick limb development. *Devel. Biol.* **207**, 380–392.

Makovicky, P. J., and Currie, P. J. (1998). The presence of a furcula in tyrannosaurid theropods, and its phylogenetic and functional implications. *J. Vert. Paleont.* **18**, 143–149.

Makower, A.-M., Skottner, A., and Wroblewski, J. (1989a). Effects of IGF-1, rGH, FGF, EGF and NCS on DNA-synthesis, cell proliferation and morphology of chondrocytes isolated from rat rib growth cartilage. *Cell Biol. Int. Rep.* **13**, 259–270.

Makower, A.-M., Skottner, A., and Wroblewski, J. (1989b). Binding of insulin-like growth factor (IGF-1) to primary cultures of chondrocytes from rat rib growth cartilage. *Cell Biol. Int. Rep.* **13**, 655–665.

Malacinski, G. M., and Youn, B. W. (1982). The structure of the anuran amphibian notochord and a re-evaluation of its presumed role in early development. *Differentiation* **21**, 13–21.

Malda, J., Martens, D. E., Tramper, J., van Blitterswijk, C. A., and Riesle, J. (2003). Cartilage tissue engineering: Controversy in the effect of oxygen. *Crit. Rev. Biotechnol.* **23**, 175–194.

Malemud, C. J., Killeen, W., Hering, T. M., and Purchio, A. F. (1991). Enhanced sulfated-proteoglycan core protein synthesis by incubation of rabbit chondrocytes with recombinant transforming growth factor-β. *J. Cell Physiol.* **149**, 152–159.

Mallatt, J. (1996). Ventilation and the origin of jawed vertebrates: a new mouth. *Zool. J. Linn. Soc. Lond.* **117**, 329–404.

Mallein-Gerin, F. (1990). Subepithelial type II collagen deposition during embryonic chick limb development. *Roux's Arch. Devel. Biol.* **198**, 363–369.

Mallein-Gerin, F., Kosher, R. A., Upholt, W. B., and Tanzer, M. L. (1988). Temporal and spatial analysis of cartilage proteoglycan core protein gene expression during limb development by *in situ* hybridization. *Devel. Biol.* **126**, 337–345.

Mallinger, R., and Böck, P. (1985). Differentiation of extracellular matrix in the cellular cartilage ('Zellknorpel') of the mouse pinna. *Anat. Embryol.* **172**, 69–74.

Mallo, M. (1997). Retinoic acid disturbs mouse middle ear development in a stage-dependent fashion. *Devel. Biol.* **184**, 175–186.

Mallo, M. (1998). Embryological and genetic aspects of middle ear development. *Int. J. Devel. Biol.* **42**, 11–22.

Mallo, M. (2001). Formation of the middle ear: recent progress on the developmental and molecular mechanisms. *Devel. Biol.* **231**, 410–419.

Mallo, M. (2003). Formation of the outer and middle ear, molecular mechanisms. *Curr. Top. Devel. Biol.* **57**, 85–103.

Mallo, M., and Gridley, T. (1996). Development of the mammalian ear: coordinate regulation of formation of the tympanic ring and the external acoustic meatus. *Development* **122**, 173–179.

Mallo, M., Schrewe, H., Martin, J. F., Olson, E. N., and Ohnemus, S. (2000). Assembling a functional tympanic membrane: signals from the external acoustic meatus coordinate development of the malleal manubrium. *Development* **127**, 4127–4136.

Manasek, F. J. (1970). Sulfated extracellular matrix production in the embryonic heart and adjacent tissues. *J. Exp. Zool.* **174**, 415–440.

Manasek, F. J. (1973). Some comparative aspects of cardiac and skeletal myogenesis. In *Developmental Regulation: Aspects of Cell Differentiation* (S. J. Coward, ed.), pp. 193–218. Academic Press, New York, NY.

Manasek, F. J. (1975). Extracellular matrix – dynamic component of developing embryos. *Curr. Top. Devel. Biol.* **10**, 35–102.

Manasek, F. J., and Cohen, A. M. (1977). Anionic glycopeptides and glycosaminoglycans synthesized by embryonic neural tube and neural crest. *Proc. Natl Acad. Sci. U.S.A.* **74**, 1057–1061.

Manasek, F. J., Reid, M., Vinson, W., Seyer, J., and Johnson, R. (1973). Glycosaminoglycan synthesis by the early embryonic chick heart. *Devel. Biol.* **35**, 332–348.

Mankin, H. J. (1962a). Localization of tritiated thymidine in articular cartilage of rabbits. (i). Growth in immature cartilage. *J. Bone Joint Surg. Am. Vol.* **44**, 682–688.

Mankin, H. J. (1962b). Localization of tritiated thymidine in articular cartilage of rabbits. (ii). Repair in immature cartilage. *J. Bone Joint Surg. Am. Vol.* **44**, 689–698.

Mankin, H. J. (1964). Mitosis in the articular cartilage of rabbits. *Clin. Orthop. Rel. Res.* **34**, 170–183.

Mankin, H. J. (1973). Biochemical and metabolic abnormalities in osteoarthritic human cartilage. *Fed. Proc. Fed. Amer. Soc. Exp. Biol.* **32**, 1478–1480.

Mankin, H. J., and Lippiello, L. (1969a). Nucleic acid and protein synthesis in epiphyseal plates in rachitic rats. *J. Bone Joint Surg. Am. Vol.* **51**, 862–874.

Mankin, H. J., and Lippiello, L. (1969b). The turnover of adult rabbit articular cartilage. *J. Bone Joint Surg. Am. Vol.* **51**, 1591–1600.

Manley, G. A. (1972). A review of some current concepts of the functional evolution of the ear in terrestrial vertebrates. *Evolution* **26**, 608–621.

Manley, N. R., and Capecchi, M. R. (1997). Hox group 3 paralogous genes act synergistically in the formation of somitic and neural crest-derived structures. *Devel. Biol.* **192**, 274–288.

Mann, R. S., and Casares, F. (2002). Signalling legacies. *Nature* **418**, 737–739.

Manning, W. K., and Bonner, W. M., Jr. (1967). Isolation and culture of chondrocytes from human adult articular cartilage. *Arthritis Rheum.* **10**, 235–239.

Manolagas, S. C. (2000). Birth and death of bone cells. Basic regulatory mechanisms and implications for the pathogenesis and treatment of osteoporosis. *Endrocrine Rev.* **21**, 115–137.

Manolson, M. F., Yu, H., Chen, W., Yao, Y., Li, K., Lees, R. L., and Heersche, J. N. M. (2003). The a3 isoform of the 100-kDa V-Tpase subunit is highly but differentially expressed in large (≥10 nuclei) and small (≤5 nuclei) osteoclasts. *J. Biol. Chem.* **278**, 49271–49278.

Manouvrier-Hanu, S., Holder-Espinasse, M., and Lyonnet, S. (1999). Genetics of limb anomalies in humans. *Trends Genet.* **15**, 409–417.

Mansour, S. L., Goddard, J. M., and Capecchi, M. R. (1993). Mice homozygous for a targeted disruption of the protooncogene *int-2* have developmental defects in the tail and inner ear. *Development* **117**, 13–28.

Mansouri, A., Voss, A. K., Thomas, T., Yokota, Y., and Gruss, P. (2000). Uncx4.1 is required for the formation of the pedicles and proximal ribs and acts upstream of *Pax9*. *Development* **127**, 2251–2258.

Mansukhani, A., Bellosta, P., Sahni, M., and Basilico, C. (2000). Signaling by fibrolast growth factors (FGF) and fibroblast growth factor receptor 2 (FGFR2)-activating mutations blocks mineralization and induces apoptosis in osteoblasts. *J. Cell Biol.* **149**, 1297–1308.

Manzanares, M. C., Goret-Nicaise, M., and Dhem, A. (1988). Metopic sutural closure in the human skull. *J. Anat.* **161**, 203–215.

Maor, G., Hochberg, Z., and Silbermann, M. (1993b). Insulin-like growth factor accelerates proliferation and differentiation of cartilage progenitor cells in cultures of neonatal mandibular condyle. *Acta Endocrinol.* **128**, 56–64.

Maor, G., Laron, Z., Eshet, R., and Silbermann, M. (1993c). The early postnatal development of the murine mandibular condyle is regulated by endogenous insulin-like growth factor-1. *J. Endocrinol.* **137**, 21–26.

Maor, G., Silbermann, M., von der Mark, K., Heingard, D., and Laron, Z. (1993a). Insulin enhances the growth of cartilage in organ and tissue cultures of mouse neonatal mandibular condyle. *Calcif. Tissue Int.* **52**, 291–299.

Maquet, P., and Furlong, R. (1986). *The Law of Bone Remodeling (Das Gesetz der Transformation der Knochen)*. Translation of text by J. Wolff, 1892. Springer-Verlag, Berlin.

Marcelle, C., Ahlgren, S., and Bronner-Fraser, M. (1999). *In vivo* regulation of somite differentiation and proliferation by sonic hedgehog. *Devel. Biol.* **214**, 277–287.

Marcil, A., Dumontier, E., Chamberland, M., Camper, S. A., and Douin, J. (2003). *Pitx1* and *Pitx2* are required for development of hindlimb buds. *Development* **130**, 45–55.

Marcus, R. E. (1973). The effect of low oxygen concentration on growth, glycolysis, and sulfate incorporation by articular chondrocytes in monolayer culture. *Arthritis Rheum.* **16**, 646–656.

Marcus, R. E., and Sribastava, V. M. L. (1973). Effect of low oxygen tensions on glucose-metabolising enzymes in cultured articular chondrocytes. *Proc. Soc. Exp. Biol. Med.* **143**, 488–491.

Marden, L. J., Fan, R. S. P., Pierce, G. F., Reddi, A. H., and Hollinger, J. O. (1993b). Platelet-derived growth factor inhibits bone regeneration induced by osteogenin, a bone morphogenetic protein, in rat craniotomy defects. *J. Clin. Invest.* **92**, 2897–2905.

Marden, L. J., Quigley, N. C., Reddi, A. H., and Hollinger, J. O. (1993a). Temporal changes during bone regeneration in the calvarium induced by osteogenin. *Calcif. Tissue Int.* **53**, 262–268.

Mareel, M. (1967). Recherches dur la relation inductrice entre chondrocytes et périoste dans le tibia embryonnaire du poulet. *Arch. Biol.* **78**, 145–166.

Marí-Beffa, M., Carmona, M. C., and Becerra, J. (1989). Elastoidin turn-over during tail fin regeneration in teleosts. *Anat. Embryol.* **180**, 465–470.

Marí-Beffa, M., Palmqvist, P., Marín-Girón, F., Montes, G. S., and Becerra, J. (1999). Morphometric study of the regeneration of individual rays in teleost tail fins. *J. Anat.* **195**, 393–405.

Marí-Beffa, M., Santamaría, J. A., Fernández-LLebrez, P., and Becerra, J. (1996). Histochemically defined cell states during tail fin regeneration in teleost fishes. *Differentiation* **60**, 139–149.

Mariani, F. V., and Martin, G. R. (2003). Deciphering skeletal patterning: clues from the limb. *Nature* **423**, 319–325.

Marie, P. J. (2002). Role of N-cadherin in bone formation. *J. Cell. Physiol.* **190**, 297–305.

Marie, P. J., Devernej, M. C., and Lomri, A. (1992). A stimulation of bone formation in osteoporosis patients treated with fluoride associated with increased DNA synthesis by osteoblastic cells *in vitro*. *J. Bone Min. Res.* **7**, 103–113.

Marino, A. W., and Becker, R. O. (1977). Piezoelectric effect and growth control of bone. *Clin. Orthop. Rel. Res.* **123**, 280–282.

Mark, M. P., Butler, W. T., Finkelman, R. D., and Ruch, J.-V. (1987b). Bone γ-carboxyglutamic acid-containing protein (osteocalcin) expression by osteoblasts during mandibular bone development in fetal rats – absence of correlation with the mineralization process. *Med. Sci. Res: Biochem.* **15**, 1299–1300.

Mark, M. P., Butler, W. T., Prince, C. W., Finkelman, R. D., and Ruch, J.-V. (1988). Developmental expression of 44 kDa bone phosphoprotein (osteopontin) and bone γ-carboxyglutamic acid (Gla)-containing protein (osteocalcin) in calcifying tissues of rat. *Differentiation* **37**, 123–136.

Mark, M. P., Prince, C. W., Oosawa, T., Gay, S., Bronckers, A. L. J. J., and Butler, W. T. (1987a). Immunohistochemical demonstration of a 44-KD phosphoprotein in developing rat bones. *J. Histochem. Cytochem.* **35**, 707–716.

Markens, I. S. (1975). Embryonic development of the coronal suture in man and rat. *Acta Anat.* **93**, 257–273.

Markens, I. S., and Taverne, A. A. R. (1978). Development of cartilage in transplanted future coronal sutures. *Acta Anat.* **100**, 428–434.

Markostamou, K., and Baron, R. (1973). Etude quantitative de l'ostéoclasie sur la paroi avleolaire en orthodontie expérimentale chez le rat. *Orthod. Fr.* **44**, 245–256.

Marks, S. B. (2000). Skull development in two plethodontid salamanders (genus *Desmognathus*) with different life histories. In *The Biology of Plethodontid Salamanders* (R. C. Bruce, R. G. Jaeger, and L. D. Houcj, eds), pp. 261–276. Kluwer Academic/Plenum Publishers, New York, NY.

Marks, S. C., Jr. (1969). The parafollicular cell of the thyroid gland as the source of an osteoblast-stimulating factor. Evidence from experimentally osteopetrotic mice. *J. Bone Joint Surg. Am. Vol.* **51**, 875–890.

Marks, S. C., Jr. (1972). Lack of effect of thyrocalcitonin on formation of bone matrix in mice and rats. *Hormone Metab. Res.* **4**, 296–300.

Marks, S. C., Jr. (1973). Pathogenesis of osteopetrosis in the ia rat: Reduced bone resorption due to reduced osteoclast function. *Amer. J. Anat.* **138**, 165–180.

Marks, S. C., Jr. (1974). A discrepancy between measurements of bone resorption *in vivo* and *in vitro* in newborn osteopetrotic rats. *Amer. J. Anat.* **141**, 329–340.

Marks, S. C., Jr. (1976). Osteopetrosis in the ia rat cured by spleen cells from a normal littermate. *Amer. J. Anat.* **146**, 331–338.

Marks, S. C., Jr. (1978a). Studies of the mechanism of spleen cell cure for osteopetrosis in ia rats: Appearance of osteoclasts with ruffled borders. *Amer. J. Anat.* **151**, 119–130.

Marks, S. C., Jr. (1978b). Studies on the cellular cure for osteopetrosis by transplanted cells: Specificity of the cell type in ia rats. *Amer. J. Anat.* **151**, 131–137.

Marks, S. C., Jr. (1981). Tooth eruption depends on bone resorption: Experimental evidence from osteopetrotic (ia) rats. *Metab. Bone Dis. Rel. Res.* **3**, 107–115.

Marks, S. C., Jr. (1982). Morphological evidence of reduced bone resorption in osteopetrotic (op) mice. *Amer. J. Anat.* **163**, 157–168.

Marks, S. C., Jr. (1983). The origin of osteoclasts: Evidence, clinical implications and investigative challenges of an extraskeletal source. *J. Oral Pathol.* **12**, 225–256.

Marks, S. C., Jr. (1984). Congenital osteopetrotic mutations as probes of the origin, structure and function of osteoclasts. *Clin. Orthop. Rel. Res.* **189**, 239–263.

Marks, S. C., Jr., and Cahill, D. R. (1984). Experimental study in the dog of the non-active role of the tooth in the eruptive process. *Archs Oral Biol.* **29**, 311–322.

Marks, S. C., Jr., and Cahill, D. R. (1986). Ultrastructure of alveolar bone during tooth eruption in the dog. *Amer. J. Anat.* **177**, 427–438.

Marks, S. C., Jr., and Cahill, D. R. (1987). Regional control by the dental follicle of alterations in alveolar bone metabolism during tooth eruption. *J. Oral Pathol.* **16**, 164–169.

Marks, S. C., Jr., Cahill, D. R., and Wise, G. E. (1983). The cytology of the dental follicle and adjacent alveolar bone during tooth eruption in the dog. *Amer. J. Anat.* **168**, 277–290.

Marks, S. C., Jr., Lundmark, C., Christersson, C., Wurtz, T., Odgren, P. R., Seifert, M. F., Mackay, C. A., Mason-Savas, A., and Popoff, S. N. (2000). Endochondral bone formation in toothless (osteopetrotic) rats: failure of chondrocyte patterning and type X collagen expression. *Int. J. Devel. Biol.* **44**, 309–316.

Marks, S. C., Jr., Lundmark, C., Wurtz, T., Odgren, P. R., Mackay, C. A., Mason-Savas, A., and Popoff, S. N. (1999). Facial development and type III collagen RNA expression: concurrent repression in the osteopetrotic (*Toothless, tl*) rat and rescue after treatment with colony-stimulating factor-1. *Devel. Dynam.* **215**, 117–125.

Marks, S. C., Jr., and Popoff, S. N. (1989). Osteoclast biology in the osteopetrotic (op) rat. *Amer. J. Anat.* **186**, 325–334.

Marks, S. C., Jr., and Schmidt, C. J. (1978). Bone remodeling as an expression of altered phenotype – studies of fracture healing in untreated and cured osteopetrotic rats. *Clin. Orthop. Rel. Res.* **137**, 259–264.

Marks, S. C., Jr., and Schneider, G. B. (1978). Evidence for a relationship between lymphoid cells and osteoclasts: Bone resorption restored in ia (osteopetrotic) rats by lymphocytes, monocytes and macrophages from a normal littermate. *Amer. J. Anat.* **152**, 331–342.

Marks, S. C., Jr., and Schneider, G. B. (1982). Transformation of osteoclast phenotype in *ia* rats cured of congenital osteopetrosis. *J. Morphol.* **174**, 141–147.

Marks, S. C., Jr., and Seifert, M. F. (1985). The lifespan of osteoclasts: Experimental studies using the giant granule cytoplasmic marker characteristic of Beige mice. *Bone* **6**, 451–455.

Marks, S. C., Jr., Seifert, M. F., and McGuire, J. L. (1984). Congenitally osteopetrotic (*op/op*) mice are not cured by transplants of spleen or bone marrow cells from normal littermates. *Metab. Bone Dis. Rel. Res.* **5**, 183–186.

Marks, S. C., Jr., and Walker, D. G. (1969). The role of the parafollicular cells of the thyroid gland in the pathogenesis of congenital osteopetrosis in mice. *Amer. J. Anat.* **126**, 299–314.

Marks, S. C., Jr., and Walker, D. G. (1981). The hematogenous origin of osteoclasts: Experimental evidence from osteopetrotic (microphthalmic) mice treated with spleen cells from beige mouse donors. *Amer. J. Anat.* **161**, 1–10.

Maroto, M., and Pourquié, O. (2001). A molecular clock involved in somitogenesis. *Curr. Topics Devel. Biol.* **51**, 221–248.

Marotti, G. (1990). The original contributions of the scanning electron microscope to the knowledge of bone structure. In *Ultrastructure of Skeletal Tissues* (E. Bonucci and P. M. Motta, eds), pp. 19–39. Kluwer Academic Publishers, Dordrecht.

Marsh, D. (1998). Concepts of fracture union, delayed union, and nonunion. *Clin. Orthop. Rel. Res.* **355**, S32–S30.

Marshall, M. J., Nisbet, N. W., and Green, P. M. (1986). Evidence for osteoclast production in mixed bone cell culture. *Calcif. Tissue Int.* **38**, 268–274.

Martin, G. R. (1998). The roles of FGFs in the early development of vertebrate limbs. *Genes & Devel.* **12**, 1571–1586.

Martin, G. R. (2001). Making a vertebrate limb: new players enter from the wings. *BioEssays* **23**, 865–868.

Martin, J. F., Bradley, A., and Olson, E. N. (1995). The *paired*-like homeo box gene *MHox* is required for early events of skeletogenesis in multiple lineages. *Genes & Devel.* **9**, 1237–1249.

Martin, L. D., Stewart, J. D., and Whetstone, K. N. (1980). The origin of birds: Structure of the tarsus and teeth. *The Auk* **97**, 86–93.

Martineau-Doizé, B., Lai, W. H., Warshawsky, H., and Bergeron, J. J. M. (1988). *In vivo* demonstration of cell types in bone that harbor epidermal growth factor receptors. *Endocrinology* **123**, 841–858.

Martinez, I., Alvarez, R., Herraes, I., and Herraes, P. (1992). Skeletal malformations in hatchery reared *Rana pereri* tadpoles. *Anat. Rec.* **233**, 314–320.

Marvaso, V., and Bernard, G. W. (1977). Initial intramembranous osteogenesis *in vitro*. *Amer. J. Anat.* **149**, 453–468.

Marx, J. L. (1980). Osteoporosis: New help for thinning bones. *Science* **207**, 628–630.

Marx, R. E., and Kline, S. N. (1983). Principles and methods of osseous reconstruction. *Int. Adv. Surg. Oncol.* **6**, 167–228.

Marzullo, G., and Desiderio, E. (1972). Tissue distribution of some cartilage enzymes. *Devel. Biol.* **27**, 13–19.

Marzullo, G., and Lash, J. W. (1967a). Separation of glycosaminoglycans on thin layers of silica gel. *Anal. Biochem.* **18**, 575–578.

Marzulo, G., and Lash, J. W. (1967b). Acquisition of the chondrocytic phenotype. *Exp. Biol. Med.* **1**, 213–219.

Massagué, J. (1991). A helping hand from proteoglycans. *Curr. Biol.* **1**, 117–119.

Massagué, J. (1992). Receptors for the TGF-β family. *Cell* **69**, 1067–1070.

Masters, P. M., and Zimmerman, M. R. (1978). Age determination of an Alaskan mummy: Morphological and biochemical correlation. *Science* **201**, 810–811.

Matalon, R., and Dorfman, A. (1966). Hurler's syndrome: biosynthesis of acid mucopolysaccharides in tissue culture. *Proc. Natl Acad. Sci. U.S.A.* **56**, 1310–1316.

Mathews, M. B. (1965). The interaction of collagen and acid mucopolysaccharides. A model for connective tissue. *Biochem. J.* **96**, 710–716.

Mathews, M. B. (1967a). Macromolecular evolution of connective tissue. *Biol. Rev. Cambridge Philos. Soc.* **42**, 499–551.

Mathews, M. B. (1967b). Chondroitin sulfate and collagen in inherited skeletal defects of chickens. *Nature* **213**, 1255–1256.

Mathews, M. B. (1968). Molecular evolution of connective tissue. In *Biology of the Mouth* (P. Person, ed.), Publ. No. 89, pp. 199–236. Amer. Assoc. Adv. Sci., Washington, DC.

Mathews, M. B. (1971). Comparative biochemistry of chondroitin sulphate proteins of cartilage and notochord. *Biochem. J.* **125**, 37–46.

Mathews, M. B. (ed.) (1975). *Connective Tissue Macromolecular Structure and Evolution*, Molecular Biology, Biochemistry and Biophysics, Volume 19, pp. 1–318. Springer-Verlag, Berlin.

Mathews, M. B., Duh, J., and Person, P. (1962). Acid mucopolysaccharides of invertebrate cartilage. *Nature* **193**, 378–379.

Mathijssen, I. M. J., Van Leeuwen, J. P. T. M., and Vermeij-Keers, C. (2000). Simultaneous induction of apoptosis, collagen type I expression and mineralization in the developing coronal suture following FGF4 and FGF2 application. *J. Craniofac. Genet. Devel. Biol.* **20**, 127–136.

Mathijssen, I. M. J., Van Splunder, J., Vermeij-Keers, C., Pieterman, H., deJong, T. H. R., Mooney, M. P., and Vaandrager, J. M. (1999). Tracing craniosynostosis to its developmental stage through bone center displacement. *J. Craniofac. Genet. Devel. Biol.* **19**, 57–63.

Mathur, J. K. (1979). Histogenesis of cartilage and bone in humerus and femur of lizard *Calotes versicolor*. *Indian J. Exp. Biol.* **17**, 533–537.

Mathur, J. K., and Goel, S. C. (1976). Pattern of chondrogenesis and calcification in the developing limb of the lizard *Calotes versicolor*. *J. Morphol.* **149**, 401–420.

Matovinovic, E., and Richman, J. M. (1997). Epithelium is required for maintaining FGFR-2 expression levels in facial mesenchyme of the developing chick embryo. *Devel. Dynam.* **210**, 407–416.

Matsuda, S., Mishima, K., Yoshimura, Y., Hatta, T., and Otani, H. (1997). Apoptosis in the development of the temporomandibular joint. *Anat. Embryol.* **196**, 383–391.

Matsui, R., Ishida, M., and Kimura, S. (1990). Characterization of two genetically distinct type I-like collagens from lamprey (*Entosphenus japonicus*). *Comp. Biochem. Physiol. B* **95**, 669–675.

Matsumoto, K., Matsunaga, S., Imamura, T., Ishidou, Y., Yoshida, H., and Sakou, T. (1994). Expression and distribution of transforming growth factor-β and decorin during fracture repair. *In Vivo* **8**, 215–220.

Matsumoto, K.-I., Ari, M., Ishihara, N., Ando, Inoko, H., and Ikemura, T. (1992). Cluster of fibronectin type III repeats found in the human major histocompability complex class III region shows the highest homology with the repeats in the extracellular matrix protein, tenascin. *Genomics* **12**, 485–491.

Matsuoka, M. (1982). Development of vertebral column and caudal skeleton of the Red Sea Bream, *Pagrus major*. *Jap. J. Ichthyol.* **29**, 285–294.

Matsuoka, M. (1985). Osteological development in the Red Sea Bream, *Pagrus major*. *Jap. J. Ichthyol.* **32**, 35–51.

Matsuoka, M. (1987). Development of the skeletal tissues and skeletal muscles in the Red Sea Bream. *Bull. Seikai Regional Fish. Res. Lab.* **65**, 1–114.

Matsuoka, M., and Iwai, T. (1983). Adipose fin cartilage found in some teleostean fishes. *Jap. J. Ichthyol.* **30**, 37–46.

Matsutani, E., and Kuroda, Y. (1980). Effect of cell association on *in vitro* chondrogenesis of mesenchyme cells from quail limb buds. *Cell Struct. Funct.* **5**, 239–246.

Matt, N., Ghyselinck, N. B., Wendling, O., Chambon, P., and Mark, M. (2003). Retinoic acid-induced developmental defects are mediated by RARβ/RXR heterodimers in the pharyngeal endoderm. *Development* **130**, 2083–2093.

Matthew, W. D. (1925). The value of paleontology. *Nat. Hist.* 25, 166–168.

Mattson, P., and Foret, J. E. (1973). Autoradiographic analyses of ^{35}S-sulfate uptake in regenerating limbs of larval *Ambystoma*. *Wilhelm Roux Arch. Entwicklungsmech. Org.* **173**, 169–182.

Matzner, U., Figura, K., and Pohlmann, R. (1992). Expression of the two mannose 6-phosphate receptors is spatially and temporally different during mouse embryogenesis. *Development* **114**, 965–972.

Mauer, P. H., and Hudack, S. S. (1952). Isolation of hyaluronic acid from callus tissue during early healing. *Arch. Biochem.* **38**, 49–53.

Maunz, M., and German, R. Z. (1996). Craniofacial heterochrony and sexual dimorphism in the short tailed opossum (*Monodelphis domestica*). *J. Mammal.* **77**, 992–1005.

Maunz, M., and German, R. Z. (1997). Ontogeny and limb bone scaling in two new world marsupials, *Monodelphis domestica* and *Didelphis virginiana*. *J. Morphol.* **231**, 117–130.

Mawdsley, R., and Ainsworth-Harrison, G. (1963). Environmental factors determine the growth and development of whole bone transplants. *J. Embryol. Exp. Morphol.* **11**, 537–547.

Maxwell, G. D. (1976). Substrate dependence of cell migration from explanted neural tubes *in vitro*. *Cell Tissue Res.* **172**, 325–330.

Maxwell, W. A., Spicer, S. S., Miller, R. L., Halushica, P. V., Westphal, M. C., and Setser, M. E. (1977). Histochemical and ultrastructural studies in fibrodysplasia ossificans progressiva (Myositis Ossificans Progressiva). *Amer. J. Pathol.* **87**, 483–498.

Mayne, R., Abbott, J., and Schiltz, J. (1972). Studies concerning the divergence of myoblast and fibroblast precursor cells during myogenesis *in vitro*. *J. Cell Biol.* **55**, 168a

Mayne, R., Abbott, J., and Schiltz, J. (1973). Requirement for cell proliferation for the effects of 5-bromo-2'-deoxyuridine on cultures of chick chondrocytes. *Exp. Cell Res.* **77**, 255–263.

Mayne, R., Sanger, J. W., and Holtzer, H. (1971). Inhibition of mucopolysaccharide synthesis by 5-bromodeoxyuridine in cultures of chick amnion cells. *Devel. Biol.* **25**, 547–567.

Mayne, R., Vail, M. S., Mayne, P. M., and Miller, E. J. (1976). Changes in type of collagen synthesized as clones of chick chondrocytes grow and eventually lose division capacity. *Proc. Natl Acad. Sci. U.S.A.* **73**, 1674–1678.

Mayne, R., and von der Mark, K. (1983). Collagens of cartilage. In *Cartilage, Volume 1. Structure, Function and Biochemistry* (B. K. Hall, ed.), pp. 181–214. Academic Press, New York, NY.

Mayor, M. B., and Moskowitz, R. W. (1974). Metabolic studies in experimentally-induced degenerative joint disease in the rabbit. *J. Rheumatol.* **1**, 17–23.

Mayor, R., Young, R., and Vargas, A. (1999). Development of neural crest in *Xenopus. Curr. Top. Devel. Biol.* **43**, 85–113.

Meats, J. E., McGuire, M. B., and Russell, R. G. G. (1980). Human synovium releases a factor which stimulates chondrocyte production of PGE and plasminogen activator. *Nature* **286**, 891–892.

Medoff, J. (1967). Enzymatic events during cartilage differentiation in the chick embryonic limb bud. *Devel. Biol.* **16**, 118–143.

Medoff, J., and Zwilling, E. (1972). Appearance of myosin in the chick limb bud. *Devel. Biol.* **28**, 138–141.

Meekeren, J. van. (1668). *Heel-en Geneeskonstige Aanmerkingen.* Commelijn, Amsterdam.

Meier, S., and Hay, E. D. (1973). Synthesis of sulfated glycosaminoglycans by embryonic corneal epithelium. *Devel. Biol.* **35**, 318–331.

Meier, S., and Hay, E. D. (1974a). Stimulation of extracellular matrix synthesis in the developing cornea by glycosaminoglycans. *Proc. Natl Acad. Sci. U.S.A.* **71**, 2310–2313.

Meier, S., and Hay, E. D. (1974b). Control of corneal differentiation by extracellular materials. Collagen as a promoter and stabilizer of epithelial stroma production. *Devel. Biol.* **38**, 249–270.

Meijer, R., and Walia, I. S. (1983). Preserved cartilage to fill facial bone defects. In *Biomaterials in Reconstructive Surgery* (L. R. Rubin, ed.), pp. 509–515. C. V. Mosby Co., St. Louis, MO.

Meikle, M. C. (1973a). The role of the condyle in the postnatal growth of the mandible. *Amer. J. Orthod.* **64**, 50–62.

Meikle, M. C. (1973b). *In vivo* transplantation of the mandibular joint of the rat: An autoradiographic investigation into cellular changes at the condyle. *Archs Oral Biol.* **18**, 1011–1020.

Meikle, M. C. (1975a). The distribution and function of lysosomes in condylar cartilage. *J. Anat.* **119**, 85–96.

Meikle, M. C. (1975b). The influence of function on chondrogenesis at the epiphyseal cartilage of a growing long bone. *Anat. Rec.* **182**, 387–400.

Meikle, M. C., Heath, J. K., Hembrey, R. M., and Reynolds, J. J. (1982). Rabbit cranial suture fibroblasts under tension express a different collagen phenotype. *Archs Oral Biol.* **27**, 609–614.

Meikle, M. C., Heath, J. K., and Reynolds, J. J. (1984). The use of *in vitro* models for investigating the response of fibrous joints to tensile mechanical stress. *Amer. J. Orthodont.* **85**, 141–153.

Meikle, M. C., Reynolds, J. J., Sellers, A., and Dingle, J. T. (1979). Rabbit cranial sutures *in vitro*: A new experimental model for studying the response of fibrous joints to mechanical stress. *Calcif. Tissue Intern.* **28**, 137–144.

Meikle, M. C., Sellers, A., and Reynolds, J. J. (1980). Effect of tensile mechanical stress on the synthesis of metalloproteinases by rabbit coronal sutures *in vitro. Calcif. Tissue Intern.* **30**, 77–82.

Meinke, D. K. (1982a). A light and scanning electron microscope study of microstructure, growth and development of the dermal skeleton of *Polypterus* (Pisces: Actinopterygii). *J. Zool. London* **197**, 355–382.

Meinke, D. K. (1982b). A histological and histochemical study of developing teeth in *Polypterus* (Pisces: Actinopterygii). *Archs Oral Biol.* **27**, 197–206.

Meinke, D. K. (1984). A review of cosmine: Its structure, development, and relationship to other forms of the dermal skeleton in osteichthyans. *J. Vert. Paleontol.* **4**, 457–470.

Meinke, D. K. (1986). Morphology and evolution of the dermal skeleton in lungfishes. *J. Morphol. Suppl.* **1**, 133–149.

Meinke, D. K., Skinner, H. C. W., and Thomson, K. S. (1979). X-ray diffraction of the calcified tissues of *Polypterus. Calcif. Tissue Intern.* **28**, 37–42.

Meinke, D. K., and Thomson, K. S. (1983). The distribution and significance of enamel and enameloid in the dermal skeleton of osteolepiform rhipidistian fishes. *Paleobiology* **9**, 138–149.

Melcher, A. H. (1969). Role of the periosteum in repair of wounds of the parietal bone of the rat. *Archs Oral Biol.* **14**, 1101–1109.

Melcher, A. H. (1971a). *In vitro* effect of oxygen, hydrocortisone and triiodothyronine on cells of Meckel's cartilage. *Isr. J. Med. Sci.* **7**, 374–376.

Melcher, A. H. (1971b). Behaviour of cells of condylar cartilage of foetal mouse mandible maintained *in vitro. Archs Oral Biol.* **16**, 1379–1391.

Melcher, A. H. (1972). Role of chondrocytes and hydrocortisone in resorption of proximal fragment of Meckel's cartilage: An *in vitro* and *in vivo* study. *Anat. Rec.* **172**, 21–36.

Melcher, A. H., and Accursi, G. E. (1972). Transmission of an 'osteogenic message' through intact bone after wounding. *Anat. Rec.* **173**, 265–276.

Melcher, A. H., McCulloch, C. A. G., Cheong, T., Nemeth, E., and Shiga, A. (1987). Cells from bone synthesize cementum-like and bone-like tissue *in vitro* and may migrate into periodontal ligament *in vivo. J. Periodont. Res.* **22**, 246–247.

Melcher, A. H., and Turnbull, R. S. (1974). Effect of hydrocortisone on DNA synthesis in periodontal ligament *in vitro. J. Dent. Res.* **53**, 223.

Melcher, A. H., and Turnbull, R. S. (1976). Organ culture in studies on the periodontium. In *Organ Culture in Biomedical Research* (M. Balls and M. A. Monnickendam, eds), pp. 149–164. Cambridge University Press, Cambridge.

Melnick, M., Jaskoll, T., Brownell, A. G., MacDougall, M., Bessem, C., and Slavkin, H. C. (1981). Spatiotemporal patterns of fibronectin distributed during embryonic development. 1. Chick limbs. *J. Embryol. Exp. Morphol.* **63**, 193–206.

Mendelsohn, C., Lohnes, D., Décimo, D., Lufkin, T., LeMeur, M., Chambon, P., and Mark, M. (1994). Function of the retinoic acid receptors (RARs) during development. (II). Multiple abnormalities at various stages of organogenesis in RAR double mutants. Craniofacial and skeletal abnormalities in RAR double mutants. *Development* **120**, 2749–2771.

Mendler, M., Eich-Bender, S. G., Vaughan, L., Winterhalter, K. H., and Bruckner, P. (1989). Cartilage contains mixed fibrils of collagen types II, IX, and XI. *J. Cell Biol.* **108**, 191–198.

Menkes, B., Deleanu, M., and Ilies, A. (1965). Comparative study of some areas of physiological necrosis at the embryo of man, some laboratory mammalians and fowl. *Rev. Roum. Embryol. Cytol., Ser. Embryol.* **2**, 162–171.

Menton, D. N., Simmons, D. J., Orr, B. Y., and Plurad, S. B. (1982). A cellular investment of bone marrow. *Anat. Rec.* **203**, 157–164.

Menton, D. N., Simmons, D. J., Chang, S.-L., and Orr, B. Y. (1984). From bone lining cell to osteocyte – an SEM study. *Anat. Rec.* **209**, 29–30.

Mercader, N., Leonardo, E., Piedra, M. E., Martinez A. C., Ros, M. A., and Torres, M. (2000). Opposing RA and FGF signals control proximodistal vertebrate limb development through regulation of Meis genes. *Development* **127**, 3961–3970.

Mercer, R. R., and Crenshaw, M. A. (1985). The role of osteocytes in bone resorption during lactation: Morphometric observations. *Bone* **6**, 269–274.

Merchant, T. E., Nguyen, L., Nguyen, D., Wu, S., Hudson, M. M., and Kaste, S. C. (2004). Differential attenuation of clavicle growth after asymmetric mantle radiotherapy. *Int. J. Radiation Oncology Biol. Phys.* **59**, 556–561.

Meredith Smith, M., and Miles, A. E. W. (1971).The ultrastructure of odontogenesis in larval and adult urodeles; differentiation of the dental epithelial cells. *Z. Zellforsch. Mikrosc. Anat.* **121**, 470–498.

Meredith Smith, M., Hobdell, M. H., and Miller, W. A. (1972). The structure of the scales of *Latimeria chalumnae*. *J. Zool. (London)* **167**, 501–509.

Merino, R., Gañan, Y., Macias, D., Economides, A. N., Sampath, K. T., and Hurlé, J. M. (1998). Morphogenesis of digits in the avian limb is controlled by FGFs, TGFβs, and noggin through BMP signaling. *Devel. Biol.* **200**, 35–45.

Merino, R., Macias, D., Gañan, Y., Economides, A. N., Wang, X., Wu, Q., Stahl, N., Sampath, K. T., Varons, P., and Hurlé, J. M. (1999a). Expression and function of *Gdf-5* during digit skeletogenesis in the embryonic chick leg bud. *Devel. Biol.* **206**, 33–45.

Merino, R., Rodriguez-Leon, J., Macias, D., Gañan, Y., Economides, A. N., and Hurlé, J. M. (1999b). The BMP antagonist Gremlin regulates outgrowth, chondrogenesis and programmed cell death in the developing limb. *Development* **126**, 5515–5522.

Merke, J., Klaus, G., Hugel, U., Waldherr, R., and Ritz, E. (1986). No 1,25-dihydroxyvitamin-D3 receptors on osteoclasts of calcium-deficient chicken despite demonstrable receptors on circulating monocytes. *J. Clin. Invest.* **77**, 312–314.

Merker, H. J. (1977). Considerations on the problem of criétical period during the development of limb skeleton. *Birth Defects, Orig. Art. Ser.* **13**(1), 179–202.

Merker, H. J., Pospisil, M., and Mewes, P. (1975). Cytotoxic effects of 6-mercaptopurine on the limb-bud blastemal cells of rat embryos. *Teratology* **11**, 199–218.

Merlo, G. R., Zerega, B., Paleari, L., Trombino, S., Mantero, S., and Levi, G. (2000). Multiple functions of *Dlx* genes. *Int. J. Devel. Biol.* **44**, 619–626.

Merrilees, M. J. (1975). Tissue interaction: Morphogenesis of the lateral-line system and labyrinth of vertebrates. *J. Exp. Zool.* **192**, 113–118.

Merrilees, M. J., and Flint, M. H. (1980). Ultrastructural study of tension and pressure zones in a rabbit flexor tendon. *Amer. J. Anat.* **157**, 87–106.

Mescher, A. L. (1976). Effects on adult newt limb regeneration of partial and complete skin flaps over the amputation surface. *J. Exp. Zool.* **195**, 117–128.

Mescher, A. L., and Tassava, R. A. (1975). Denervation effects on DNA replication and mitosis during the initiation of limb regeneration in adult newts. *Devel. Biol.* **44**, 187–197.

Metsäranta, M., Young, M. F., Sandberg, M., Termine, J., and Vuorio, E. (1989). Localization of osteonectin expression in human fetal skeletal tissues by *in situ* hybridization. *Calcif. Tissue Int.* **45**, 146–152.

Meunier, F.-J. (1989). The acellularisation process in Osteichthyan bone. *Fortsch. Zool.* **35**, 443–446.

Meunier, F.-J., and Boivin, G. (1974). Divers aspects de la fixation du chlorhydrate de tétracycline sur les tissus squelettiques de quelques téléostéens. *Bull. Soc. Zool. Fr.* **99**, 495–504.

Meunier, F.-J., and Desse, G. (1986). Les hyperostoses chez les téléosténs: Description, histologie et problèmes étiologiques. *Ichthyophysiologia Acta* **10**, 130–142.

Meunier, F.-J., François, Y., and Castanet, J. (1978). Étude histologique et microradiographique des écailles de quelques Actinopterigiens primitifs actuels. *Bull. Soc. Zool. France* **103**, 309–318.

Meunier, F. J., and Huysseune, A. (1992). The concept of bone tissue in osteichthyes. *Neth. J. Zool.* **42**, 445–458.

Meunier, F.-J., Pascal, M., and Loubens, G. (1979). Comparison de methodes squelettochronologiques et considerations fonctionelles sur le tissue osseux, acellulaire d'un osteichthyen du lagon neo-Caledonien, *Lethrinus nebulosus* (Forskal, 1775). *Aquaculture* **17**, 137–157.

Meyer, D. B., and O'Rahilly, R. (1976). The onset of ossification in the human calcaneus. *Anat. Embryol.* **150**, 19–33.

Meyer, G. (1867). Die Architektur des Spongiosa. *Arch. Anat. Physiol.* **34**, 615–628.

Meyer, M. P. (2001). The extracellular ATP receptor cP2-/1, inhibits cartilage formation in micromass cultures of chick limb mesenchyme. *Devel. Dynam.* **222**, 494–505.

Mezey, E., Chandross, K. J., Harta, G., Maki, R. A., and McKercher, S. R. (2000). Turning blood into brain: cells bearing neuronal antigens generated *in vivo* from bone marrow. *Science* **290**, 1779–1782.

Mezey, J.-G., Cheverud, J. M., and Wagner, G. P. (2000). Is the genotype–phenotype map modular? A statistical approach using mouse quantitative trait loci data. *Genetics* **156**, 305–311.

Michael, M. I., and Niazi, I. A. (1972). Hind limb regeneration in tadpoles of *Bufo viridis viridis* Laurenti, and cartilage formation from cells of non-chondrogenic origin in the thigh. *Acta Embryol. Exp.* **14**, 349–363.

Meischer, F. (1836). *De Inhlammatione ossium eorumque anatome generali.* Dissertation. G. Eichler, Berlin.

Miki, T., and Yamamuro, T. (1987). The fate of hypertrophic chondrocytes in growth plates transplanted intramuscularly in the rabbit. *Clin. Orthop. Rel. Res.* **218**, 276–282.

Mikic, B., Johnson, T. L., Chhabra, A. B., Schalet, B. J., Wong, M., and Hunziker, E. B. (2000b). Differential effects of embryonic immobilization on the development of fibrocartilaginous skeletal elements. *J. Rehab. Res. Devel.* **37**, 127–133.

Mikic, B., Wong, M., Chiquet, M., and Hunziker, E. B. (2000a). Mechanical modulation of tenascin-C and collagen-XII expression during avian synovial joint formation. *J. Orthop. Res.* **18**, 406–415.

Milaire, J. (1956). Contribution à l'étude morphologique et cytochimique des bourgeons de membres chez le rat. *Arch. Biol.* **67**, 297–394.

Milaire, J. (1957). Contribution à la connaissance morphologique et cytochimique des bourgeons de membres chez quelques reptiles. *Arch. Biol.* **68**, 429–514.

Milaire, J. (1963). Morphological and cytochemical study of development of the limbs of the mouse and mole. *Arch. Biol.* **74**, 129–317.

Milaire, J. (1965). Aspects of limb morphogenesis in mammals. In *Organogenesis* (R. L. DeHaan and H. Ursprung, eds), pp. 283–300. Holt, New York, NY.

Milaire, J. (1967). Histochemical observations on the developing foot of normal, oligosyndactylous (os/+) and syndactylous (sm/sm) mouse embryos. *Arch. Biol.* **78**, 223–288.

Milaire, J. (1974). Histochemical aspects of organogenesis in vertebrates. *Handb. Histochemie* **8**, Part 3, Suppl., 1–135.

Milaire, J. (1976). Contribution cellulaires des somites à la genèse des bourgeons de membres postereurs chez la souris. *Arch. Biol.* **87**, 315–343.

Milaire, J. (1977). Histochemical expression of morphogenetic gradients during limb morphogenesis (with particular reference to mammalian embryos). *Birth Defects: Orig. Art. Ser.* **13** (1), 37–68.

Milaire, J. (1978). Étude morphologique, histochemique et autoradiographique du développement du squelette des membres chez l'embryon de Souris. I. Membres antérieurs. *Arch. Biol. (Bruxelles)* **89**, 169–216.

Milaire, J. (1983). Patterns of dephosphorylating activities in the mesoderm of developing mouse limb buds. I. 5′ nucleotidase, non specific ATP-phosphorydrolase and alkaline phosphatase in normal forelimb buds. *Arch. Biol. (Bruxelles)* **94**, 301–344

Milaire, J., and Mulnard, J. (1968). Le rôle de l'épiblasts dans la chondrogénèse du bourgeon de membre chez la souris. *J. Embryol. Exp. Morphol.* **20**, 215–236.

Milaire, J., and Mulnard, J. (1984). Histogenesis in 1-day mouse embryo limb buds explanted in organ culture. *J. Exp. Zool.* **232**, 359–377.

Milaire, J., and Rooze, M. (1982). Étude morphologique, histochemique et autoradiographique du développement du squelette des membres chez l'embryon de Souris. II. Membres postérieurs. *Arch. Biol. (Bruxelles)* **93**, 311–342.

Milaire, J., and Rooze, M. (1983). Hereditary and induced modifications of the normal mecrotic patterns in the developing limb buds of the rat and mouse: facts and hypotheses. *Arch. Biol. (Bruxelles)* **94**, 459–490.

Milenkovic, L., Goodrich, L. V., Higgins, K. M., and Scott, M. P. (1999). Mouse *patched 1* controls body size determination and limb patterning. *Development* **126**, 4431–4440.

Miles, A. E. W. (ed.) (1967). *Structural and Chemical Organization of Teeth*. Volumes 1 and 2. Academic Press, New York, NY.

Miles, A. E. W., and Dawson, J. A. (1962). Elastic fibres in the articular fibrous tissue of some joints. *Archs Oral Biol.* **7**, 249–252.

Miller, C. T., Schilling, T. F., Lee, K.-H., Parker, J., and Kimmel, C. B. (2000). *Sucker* encodes a zebrafish endothelin-1 required for ventral pharyngeal arch development. *Development* **127**, 3815–3828.

Miller, C. T., Yelon, D., Stainier, Y. R. S., and Kimmel, C. B. (2003). Two *endothelin 1* effectors, *hand2* and *bapx1*, pattern ventral pharyngeal cartilage and the jaw joint. *Development* **130**, 1353–1365.

Miller, E. J. (1972). The biochemical characterization of various collagens. In *Developmental Aspects of Oral Biology* (H. C. Slavkin and L. A. Bavetta, eds), pp. 275–291. Academic Press, New York, NY.

Miller, E. J. (1973). A review of biochemical studies on the genetically distinct collagens of the skeletal system. *Clin. Orthop. Rel. Res.* **92**, 260–280.

Miller, E. J. (1976). Biochemical characteristics and biological significance of the genetically distinct collagens. *Mol. Cell. Biochem.* **13**, 165–192.

Miller, E. J. (1985). Recent information on the chemistry of the collagens. In *The Chemistry and Biology of Mineralized Tissues* (W. T. Butler, ed.), pp. 80–93. Ebsco-Medica Inc., Birmingham, AL.

Miller, E. J., and Mathews, M. B. (1974). Characterization of notochord collagen as a cartilage-type collagen. *Biochem. Biophys. Res. Commun.* **60**, 424–430.

Miller, E. J., and Matukas, V. J. (1969). Chick cartilage collagen: A new type of α1 chain not present in bone or skin of the species. *Proc. Natl Acad. Sci. U.S.A.* **64**, 1264–1268.

Miller, G. J., Burchardt, H., Enneking, W. F., and Tylkowski, C. M. (1984). Electromagnetic stimulation of canine bone grafts. *J. Bone Joint Surg. Am. Vol.* **66**, 693–698.

Miller, R. F., Cloutier, R., and Turner, S. (2003). The oldest articulated chondrichthyan from the Early Devonian period. *Nature* **425**, 501–504.

Miller, R. R., and McDevitt, C. A. (1988). Thrombospondin is present in articular cartilage and is synthesized by articular chondrocytes. *Biochem. Biophys. Res. Commun.* **153**, 708–714.

Miller, S. C., and Bowman, B. M. (1981). Medullary bone osteogenesis following estrogen administration to mature male Japanese quail. *Devel. Biol.* **87**, 52–63.

Miller, S. C., Bowman, B. M., and Myers, R. L. (1984). Morphological and ultrastructural aspects of the activation of avian medullary bone osteoclasts by parathyroid hormone. *Anat. Rec.* **208**, 223–231.

Miller, S. C., and Jee, W. S. S. (1987). The bone lining cells: A distinct phenotype? *Calcif. Tissue Int.* **41**, 1–5.

Miller, S. C., and Jee, W. S. S. (1992). Bone lining cells: In *Bone, Volume 4: Bone Metabolism and Mineralization* (B. K. Hall, ed.), pp. 1–19. CRC Press, Boca Raton, FL.

Miller, S. C., Saintgeorges, L., Bowman, B. M., and Jee, W. S. S. (1989). Bone lining cells – structure and function – review. *Scann. Microscopy* **3**, 953–962.

Miller, W. A. (1979). Observations on the structure of mineralized tissues of the Coelacanth, including the scales and their associated odontodes. *Occ. Pap. Cal. Acad. Sci.* #**134**, 68–78.

Mills, B. G. (1991). Bone resorbing cells and human clinical conditions. In *Bone, Volume 2: The Osteoclast* (B. K. Hall, ed.), pp. 175–252. CRC Press, Boca Raton, FL.

Mills, B. G., Holst, P. A., Stabile, E. K., Adams, J. S., Rude, R. K., Fernie, B. F., and Singer, F. R. (1985). A viral antigen-bearing cell line derived from culture of Paget's bone cells. *Bone* **6**, 257–268.

Mills, B. G., Singer, F. R., Weiner, L. P., and Holst, P. A. (1979). Long-term culture of cells from bone affected by Paget's disease. *Calcif. Tissue Intern.* **29**, 79–87.

Mills, B. G., Singer, F. R., Weiner, L. P., and Holst, P. A. (1980). Cell cultures from bone affected by Paget's disease. *Arth. Rheum.* **23**, 1115–1120.

Mills, B. G., Singer, F. R., Weiner, L. P., and Holst, P. A. (1981). Immunohistological demonstration of respiratory syncytial virus antigens in Paget disease of bone. *Proc. Natl Acad. Sci. U.S.A.* **78**, 1209–1213.

Mills, C. L., Ariyo, O., Yamada, K. M., Lash, J. W., and Bellairs, R. (1990). Evidence for the involvement of receptors for fibronectin in the production of chick tail segmentation. *Anat. Embryol.* **182**, 425–434.

Mills, C. L., and Bellairs, R. (1989). Mitosis and cell death in the tail of the chick embryo. *Anat. Embryol.* **180**, 301–308.

Milne, N., and O'Higgins, P. (2002). Inter-specific variation in *Macropus* crania: form, function and phylogeny. *J. Zool. Lond.* **256**, 523–535.

Milz, S., McNeilly, C., Putz, R., Ralphs, J. R., and Benjamin, M. (1998). Fibrocartilage in the extensor tendons of the interphalangeal joints of human toes. *Anat. Rec.* **252**, 264–270.

Mina, M., Gluhak, J., Upholt, W. B., Kollar, E. J., and Rogers, B. (1995). Experimental analysis of *Msx-1* and *Msx-2* gene expression during chick mandibular morphogenesis. *Devel. Dynam.* **202**, 195–214.

Mina, M., Kollar, E. J., Bishop, J. A., and Rohrbach, D. H. (1990). Interaction between the neural crest and extracellular matrix proteins in craniofacial skeletogenesis. *Crit. Rev. Oral Biol. Med.* **1**, 79–87.

Mina, M., Kollar, E. J., and Upholt, W. B. (1991b). Temporal and spatial expression of genes for cartilage extracellular matrix proteins during avian mandibular arch development. *Differentiation* **48**, 17–24.

Mina, M., Upholt, W. B., and Kollar, E. J. (1991a). Stage-related chondrogenic potential of avian mandibular ectomesenchymal cells. *Differentiation* **48**, 9–16.

Mina, M., Upholt, W. B., and Kollar, E. J. (1994). Enhancement of avian mandibular chondrogenesis *in vitro* in the absence of epithelium. *Archs Oral Biol.* **39**, 551–562.

Mina, M., Wang, Y.-H., Ivanisevic, A.-M., Upholt, W. B., and Rodgers, B. (2002). Region- and stage-specific effects of FGFs and BMPs in chick mandibular morphogenesis. *Devel. Dynam.* **223**, 333–352.

Minelli, A. (2000). Limbs and tail as evolutionary diverging duplicates of the main body axis. *Evol. & Devel.* **2**, 157–165.

Miner, R. W. (ed.) (1950). The ground substance of the mesenchyme and hyaluronidase. *Ann. NY. Acad. Sci.* **52**, 943–1196.

Minina, E., Wenzel, H. M., Kreschel, C., Karp, S., Gaffield, W., Mcmahon, A. P., and Vortkamp, A. (2001). BMP and Ihh/PTHrP signaling interact to coordinate chondrocyte proliferation and differentiation. *Development* **128**, 4553–4534.

Minkin, C. (1982). Bone acid phosphatase: Tartrate-resistant acid phosphatase as a marker of osteoclast function. *Calcif. Tissue Int.* **34**, 285–290.

Minkin, C., Blackman, L., Newbrey, J., Pokress, S., Posek, R., and Walling, M. (1977). Effects of parathyroid hormone and calcitonin n adenylate cyclase in murine mononuclear phagocytes. *Biochem. Biophys. Res. Commun.* **76**, 875–881.

Minkin, C., Posen, R., and Newbrey, J. (1981). Mononuclear phagocytes and bone resorption: Identification and preliminary characterization of a bone-derived macrophage chemotactic factor. *Metab. Bone Dis.* **2**, 363–369.

Minkin, C., and Shapiro, I. M. (1986). Osteoclasts, mononuclear phagocytes and physiological bone resorption. *Calcif. Tissue Int.* **39**, 357–359.

Minkoff, R. (1980). Regional variation of cell proliferation within the facial processes of the chick embryo: A study of the role of 'merging' during development. *J. Embryol. Exp. Morphol.* **57**, 37–49.

Minkoff, R. (1984). Cell cycle analysis of facial mesenchyme in the chick embryo. I. Labelled mitoses and continuous labelling studies. *J. Embryol. Exp. Morphol.* **81**, 49–59.

Minkoff, R., and Kuntz, A. J. (1977). Cell proliferation during morphogenetic change: Analysis of frontonasal morphogenesis in the chick embryo employing DNA labeling indices. *J. Embryol. Exp. Morphol.* **40**, 101–113.

Minkoff, R., and Kuntz, A. J. (1978). Cell proliferation and cell density of mesenchyme in the maxillary process and adjacent regions during facial development in the chick embryo. *J. Embryol. Exp. Morphol.* **46**, 65–74.

Minkoff, R., and Martin, R. E. (1984). Cell cycle analysis of facial mesenchyme in the chick embryo. II Label dilution studies and developmental fate of slow cycling cells. Labelled mitoses and continuous labelling studies. *J. Embryol. Exp. Morphol.* **81**, 61–73.

Minkoff, R., Parker, S. B., and Hertzberg, E. L. (1991). Analysis of distribution patterns of gap junctions during development of embryonic chick facial primordia and brain. *Development* **111**, 509–522.

Minkoff, R., Rundos, V. R., Parker, S. B., Hertzberg, E. L., Laing, J.-G., and Beyer, E. C. (1994). Gap junction proteins exhibit early and specific expression during intramembranous bone formation in the developing chick mandible. *Anat. Embryol.* **190**, 231–241.

Minor, R. R. (1973). Somite chondrogenesis. A structural analysis. *J. Cell Biol.* **56**, 27–50.

Minot, C. S. (1898). On a hitherto unrecognized form of blood circulation without capillaries in the organs of vertebrates. *J. Boston Soc. Med. Sci.* **4**, 133–134.

Minowada, G., Jarvis, L. A., Chi, C. L., Neubüser, A., Sun, X., Hacohen, N., Krasnow, M. A., and Martin, G. R. (1999). Vertebrate *Sprouty* genes are induced by FGF signaling and can cause chondrodysplasia when overexpressed. *Development* **126**, 4465–4475.

Mintz, B. (1971). Allophenic mice of multi-embryo origin. In *Methods in Mammalian Embryology* (J. Daniel, ed.), pp. 186–214. Freeman, San Francisco, CA.

Mintz, B. (1972). Clonal units of gene control in mammalian differentiation, In *Cell Differentiation* (R. Harris, P. Allin, and D. Viza, eds), pp. 267–271. Munksgaard, Copenhagen.

Minugh-Purvis, N., and McNamara, K. J. (eds). (2002). *Human Evolution Through Developmental Change.* Johns Hopkins University Press, Baltimore, MD.

Miralles-Flores, C., and Delgado-Baeza, E. (1990). Histomorphometric differences between the lateral region and central region of the growth plate in fifteen-day-old rats. *Acta Anat.* **139**, 209–213.

Mishina, Y., Suzuki, A., Ueno, N., and Behringer, R. R. (1995). *Bmpr* encodes a type I bone morphogenetic protein receptor that is essential for gastrulation during mouse embryogenesis. *Genes & Devel.* **9**, 3027–3037.

Misof, B. Y., and Wagner, G. P. (1992). Regeneration in *Salaria pavo* (Blenniidae, Teleostei). Histogenesis of the regenerating pectoral fin suggests different mechanisms for morphogenesis and structural maintenance. *Anat. Embryol.* **186**, 153–165.

Missana, L., Nagai, N., and Kuboki, Y. (1994). Comparative histological studies of bone and cartilage formations induced by various BMP-carrier composites. *Jap. J. Oral Biol.* **36**, 9–19.

Mitala, J. J., Boardman, J. P., Carrano, R. A., and Iuliucci, J. D. (1984). Novel accessory skull bone in fetal rats after exposure to aspirin. *Teratology* **30**, 95–98.

Mitchell, G. J. (1980). *The Pronghorn Antelope of Alberta.* Publication of the Alberta Department of Lands and Forests, Fish and Wildlife Division.

Mitrovic, D. (1971). La nécrose physiologique dans le mésenchyme articulaire des embryons de rat et de poulet. *C.R. Hebd. Seances Acad. Sci.* **273**, 642–645.

Mitrovic, D. (1972). Régression des fentes articulaires normalement constituées chez l'embryon de poulet paralysée. *C.R. Hebd. Seances Acad. Sci.* **274**, 288–291.

Mitrovic, D. (1974). Développement du mésenchyme articulaire dans les greffons de bourgeons de pattes chez l'embryon de poulet. *C.R. Hebd. Seances Acad. Sci.* **278**, 1629–1632.

Mitrovic, D. (1977). Development of the metatarso-phalangeal joint of the chick embryo: Morphological, ultrastructural and histochemical studies. *Amer. J. Anat.* **150**, 333–348.

Mitrovic, D. (1982). Development of the articular cavity in paralyzed chick embryos and in chick embryo limb buds cultured on chorioallantoic membranes. *Acta Anat.* **113**, 313–324.

Miura, T., and Shiota, K. (2000). TGFβ2 acts as an 'Activator' molecule in reaction-diffusion model and is involved in cell sorting phenomenon in mouse limb micromass culture. *Devel. Dynam.* **217**, 241–249.

Miyake, T., Cameron, A. M., and Hall, B. K. (1996a). Detailed staging of inbred C57BL/6 mice between Theiler's (1972) stages 18 and 21 (11–13 days of gestation) based on craniofacial development. *J. Craniofac. Genet. Devel. Biol.* **16**, 1–31.

Miyake, T., Cameron, A. M., and Hall, B. K. (1996b). Stage-specific onset of condensation and matrix deposition for Meckel's and other first arch cartilages in inbred C57BL/6 mice. *J. Craniofac. Genet. Devel. Biol.* **16**, 32–47.

Miyake, T., Cameron, A. M., and Hall, B. K. (1997a). Stage-specific expression patterns of alkaline phosphatase during development of the first arch skeleton in inbred C57BL/6 mouse embryos. *J. Anat.* **190**, 239–260.

Miyake, T., Cameron, A. C., and Hall, B. K. (1997b). Variability of embryonic development among three inbred strains of mice. *Growth, Devel. Aging* **61**, 141–155.

Miyake, T., and Hall, B. K. (1994). Development of *in vitro* organ culture techniques for differentiation and growth of cartilages and bones from teleost fish and comparisons with *in vivo* skeletal development. *J. Exp. Zool.* **268**, 22–43.

Miyake, T., and McEachran, J. D. (1991). The morphology and evolution of the ventral gill arch skeleton in batoid fishes (Chondrichthyes: Batoidea). *Zool. J. Linn. Soc.* **102**, 75–100.

Miyake, T., McEachran, J. D., Walton, P. J., and Hall, B. K. (1992). Development and morphology of rostral cartilages in batoid fishes (Chondrichthyes: Batoidea), with comments on homology within vertebrates. *Biol. J. Linn. Soc.* **46**, 259–298.

Miyake, T., and Uyeno, T. (1987). The urodermals in lanternfish Family Myctophidae (Pisces: Myctophiformes). *Copeia* **1987**, 176–181.

Miyake, T., Vaglia, J. L., Taylor, L. H., and Hall, B. K. (1999). Development of dermal denticles in skates (Chondrichthyes, Batoidea): Patterning and cellular differentiation. *J. Morphol.* **241**, 61–81.

Miyama, K., Yamada, G., Yamamoto, T. S., Takagi, C., Miyado, K., Sakai, M., Ueno, N., and Shibuya, H. (1999). A BMP-inducible gene, *Dlx5*, regulates osteoblast differentiation and mesoderm induction. *Devel. Biol.* **208**, 123–133.

Mizell, M. (1968). Limb regeneration: Induction in the newborn opossum. *Science* **161**, 283–286.

Mizoguchi, I., Nakamura, M., Takahashi, I., Kagayama, M., and Mitrani, H. (1990). An immunohistochemical study of localization of type I and type II collagens in mandibular condylar cartilage compared with tibial growth plate. *Histochemistry* **93**, 593–600.

Mizoguchi, I., Nakamura, M., Takahashi, I., Kagayama, M., and Mitrani, H. (1992). A comparison of the immunohistochemical localization of type I and type II collagens in craniofacial cartilages of the rat. *Acta Anat.* **144**, 59–64.

Mizoguchi, I., Nakamura, M., Takahashi, I., Sasano, Y., Kagayama, M., and Mitrani, H. (1993). Presence of chondroid bone on rat mandibular condylar cartilage. An immunohistochemical study. *Acta Anat.* **187**, 9–15.

Mizoguchi, I., Takahashi, I., Nakamura, M., Sasano, Y., Sato, S., Kagayama, M., and Mitrani, H. (1996). An immunohistochemical study of regional differences in the distribution of type I and type II collagens in rat mandibular condylar cartilage. *Archs Oral Biol.* **41**, 863–869.

Mizoguchi, I., Takahashi, I., Sasano, Y., Kagayama, M., Kuboki, Y., and Mitrani, H. (1997a). Localization of types I, II and X collagen and osteocalcin in intramembranous, endochondral and chondroid bone of rats. *Anat. Embryol.* **196**, 291–297.

Mizoguchi, I., Takahashi, I., Sasano, Y., Kagayama, M., and Mitrani, H. (1997b). Localization of type I, type II and type III collagen and glycosaminoglycans in the mandibular condyle of growing monkeys – an immunohistochemical study. *Anat. Embryol.* **195**, 127–135.

Mo, R., Freer, A.M., Zinyk, D. L., Crackower, M. A., Michaud, J., Heng, H. H.-Q., Chik, K. W., Shi, X.-M., Tsui, L.-C., Cheng, S. H., Joyner, A. L., and Hui, C.-c. (1997). Specific and redundant function of *Gli2* and *Gli3* zinc finger genes in skeletal patterning and development. *Development* **124**, 113–123.

Modell, W. (1969). Horns and antlers. *Sci. Amer.* **220**, 114–122.

Modell, W., and Noback, C. V. (1931). Histogenesis of bone in the growing antler of the cervidae. *Amer. J. Anat.* **49**, 65–95.

Modrowski, D., and Marie, P. J. (1993). Cells isolated from the endosteal bone surface of adult rats express differentiated osteoblastic characteristics *in vitro*. *Cell Tissue Res.* **271**, 499–505.

Modrowski, D., Miravet, L., Feuga, M., Bannie, F., and Marie, P. J. (1992). Effect of fluoride on bone and bone cells in ovarectomized rats. *J. Bone Min.* **7**, 961–969.

Moeller, C., Swindell, E. C., Kispert, A., and Eichele, A. (2003). Carboxypeptidase Z (CPZ) modulates Wnt signaling and regulates the development of skeletal elements in the chicken. *Development* **130**, 5103–5111.

Moen, C. (1982). Orthopaedic aspects of progeria. *J. Bone Joint Surg. Am. Vol.* **64**, 542–546.

Moffett, B. C., Jr. (1965). The morphogenesis of joints. In *Organogenesis* (R. L. De Haan and H. Ursprung, eds), pp. 301–313. Holt, Rinehart & Winston, New York, NY.

Moffett, B. C., Jr. (ed.) (1972). *Mechanisms and Regulation of Craniofacial Morphogenesis*. Swets & Zeitlander, Amsterdam.

Moftah, M. Z., Downie, S. A., Bronstein, N. B., Mezentseva, N., Pu, J., Maher, P. A., and Newman, S. A. (2002). Ectodermal FGFs induce perinodular inhibition of limb chondrogenesis *in vitro* and *in vivo* via FGF Receptor 2. *Devel. Biol.* **249**, 270–282.

Mohammad, K. S., Day, F. A., and Neufeld, D. A. (1999). Bone growth is induced by nail transplantation in amputated proximal phalanges. *Calcif. Tissue Int.* **65**, 408–410.

Mohanty-Hejmandi, P., Dutta, S. K., and Mahapatra, P. (1992). Limbs generated at site of tail amputation in marbled balloon frog after vitamin A treatment. *Nature* **355**, 352–353.

Mohr, H., and Kragstrup, J. (1991). A histomorphometric analysis of the effects of fluoride on experimental ectopic bone formation in the rat. *J. Dent. Res.* **70**, 957–960.

Moiseiwitsch, J. R. D. (2000). The role of serotonin and neurotransmitters in craniofacial development. *Crit. Rev. Oral Biol. Med.* **11**, 230–239.

Moiseiwitsch, J. R. D., and Lauder, P. M. (1995). Serotonin regulates mouse cranial neural crest migration. *Proc. Natl Acad. Sci. U.S.A.* **92**, 7182–7186.

Moiseiwitsch, J. R. D., and Lauder, P. M. (1996). Stimulation of murine tooth development in organotypic culture by the neurotransmitter serotonin. *Archs Oral Biol.* **41**, 161–165.

Moiseiwitsch, J. R. D., and Lauder, P. M. (1997). Regulation of gene expression in cultured embryonic mouse mandibular mesenchyme by serotonin antagonists. *Anat. Embryol.* **195**, 71–78.

Moiseiwitsch, J. R. D., Raymond, J. R., Tamir, H., and Lauder, P. M. (1998). Regulation by serotonin of tooth-germ morphogenesis and gene expression in mouse mandibular explant cultures. *Archs Oral Biol.* **43**, 789–800.

Monnot, M.-J., Babin, P. K., Poleo, G., Andre, M., Laforest, L., Ballagny, C., and Akimenko, M.-A. (1999). Epidermal expression of apolipoprotein E gene during fin and scale development and fin regeneration in zebrafish. *Devel. Dynam.* **214**, 207–215.

Monro, P. P., Knight, D. P., Pringle, W. S., Fyfe, D. M., and Shearer, J. R. (1986). The use of the chorioallantoic membrane (CAM) of the embryonic chick for the direct assessment of implant toxicity. *Alternatives to Lab. Animals* **13**, 261–266.

Monson, J. W., and Felts, W. J. L. (1961). Transplantation studies of factors in skeletal organogenesis. II. The response of the immature mouse humerus to longitudinal compressive forces. *Amer. J. Phys. Anthropol.* **19**, 63–77.

Monsoro-Burq, A.-H., Bontoux, M., Teillet, M.-A., and Le Douarin, N. M. (1994). Heterogeneity in the development of the vertebra. *Proc. Natl Acad. Sci. U.S.A.* **91**, 10435–10439.

Monsoro-Burq, A.-H., Duprez, D., Watanabe, Y., Bontoux, M., Vincent, C., Brickell, P., and Le Douarin, N. M. (1996). The role of bone morphogenetic protein in vertebral development. *Development* **122**, 3607–3616.

Monsoro-Burq, A.-H., Fletcher, R. B., and Harland, R. M. (2003). Neural crest induction by paraxial mesoderm in *Xenopus* embryos requires FGF signals. *Development* **130**, 3111–3124.

Monsoro-Burq, A.-H., and Le Douarin, N. M. (2000). Duality of molecular signaling involved in vertebral chondrogenesis. *Curr. Top. Devel. Biol.* **48**, 43–75.

Monteiro, L. R., and Abe, A. S. (1999). Functional and historical determinants of shape in the scapula of Xenarthran mammals: Evolution of a complex morphological structure. *J. Morphol.* **241**, 251–263.

Moody, S. (1998). *Cell Lineage and Fate Determination.* Academic Press, San Diego, CA.

Moon, A. M., Bowler, A. M., and Capecchi, M. R. (2000). Normal limb development in conditional mutants of *fgf4. Development* **127**, 989–996.

Mooney, M. P., Aston, C. E., Siegel, M. I., Losken, H. W., Smith, T. D., Burrows, A. M., Wenger, S. H., Caruso, K., Siegel, B., and Ferrell, R. E. (1996a). Craniosynostosis with autosomal dominant transmission in New Zealand white rabbits. *J. Craniofac. Genet. Devel. Biol.* **16**, 52–63.

Mooney, M. P., Aston, C. E., Siegel, M. I., Losken, W. L., Smith, T. D., Burrows, A. M., Wenger, S. L., Caruso, K., Siegel, B., and Ferrell, R. E. (1996b). Coronal suture pathology with synostotic progression in rabbits with congenital craniosynostosis. *Cleft Pal. Craniofac. J.* **33**, 369–378.

Moore, M. A., Gotoh, Y., Rafidi, K., and Gerstenfeld, L. C. (1991). Characterization of a cDNA for chicken osteopontin: expression during bone development, osteoblast differentiation and tissue distribution. *Biochemistry* **30**, 2501–2508.

Moore, W. J. (1965). Masticatory function and skull growth. *J. Zool.* **146**, 123–131.

Moore, W. J. (1969). Muscular function and skull growth in the laboratory rat (*Rattus norvegicus*). *J. Zool.* **152**, 287–296.

Moore, W. J. (1981). *The Mammalian Skull.* Cambridge University Press, Cambridge.

Moore, W. J., and Lavelle, C. L. B. (1974). *Growth of the Facial Skeleton in the Hominoidea.* Academic Press, New York, NY.

Moore, W. J., and Mintz, B. (1972). Clonal model of vertebral column and skull development derived from genetically mosaic skeletons of allophenic mice. *Devel. Biol.* **27**, 55–70.

Moran, J. L., Levorse, J. M., and Vogt, T. F. (1999). Limbs move beyond the radical fringe. *Nature* **399**, 742–743.

Morey, E. R., and Baylink, D. J. (1978). Inhibition of bone formation during space flight. *Science* **201**, 1138–1141.

Morgan, B. A., Izpisúa-Belmonte, J.-C., Duboule, D., and Tabin, C. J. (1992). Targeted misexpression of *Hox-4.6* in the avian limb bud causes apparent homeotic transformations. *Nature* **358**, 236–239.

Mori, I., Kodaka, T., Sano, T., Yamagishi, N., Asari, M., and Naito, Y. (2003). Comparative histology of the laminar bone between young calves and foals. *Cells Tissues Organs* **175**, 43–50.

Mori-Akiyama, Y., Akiyama, H., Rowitch, D. H., and de Crombrugghe, B. (2003). Sox9 is required for determination of the chondrogenic cell lineage in the cranial neural crest. *Proc. Natl Acad. Sci. U.S.A.* **100**, 9360–9365.

Morin-Kensicki, E. M., Melancon, E., and Eisen, J. S. (2002). Segmental relationship between somites and vertebral column in zebrafish. *Development* **129**, 3851–3860.

Morris, N. P., Oxford, J. T., Davies, G. B. M., Smoody, B. F., and Keene, D. R. (2000). Developmentally regulated alternative splicing of the a1(IX) collagen chain: Spatial and temporal segregation of isoforms in the cartilage of fetal rat long bones. *J. Histochem. Cytochem.* **48**, 725–741.

Morris, P. J. (1993). The developmental role of the extracellular matrix suggests a monophyletic origin of the Kingdom Animalia. *Evolution* **47**, 152–165.

Morrison, A. (1873). Bone absorption by means of giant cells. *Edin. Med. J.* **19**, 305.

Morrison, C. M. (1993). Histology of the Atlantic cod, *Gadus morhua*: An atlas. Part Four. Eleutheroembryo and larva. Canadian Special Publications of Fisheries and Aquatic Sciences 119. NRC, Canada.

Morrison, C. M., Miyake, T., and Wright, J. R., Jr. (2001). Histological study of the development of the embryo and early larva of *Oreochromis niloticus* (Pisces: Cichlidae). *J. Morphol.* **247**, 172–195.

Morrison, S. L., Campbell, C. K., and Wright, G. M. (2000). Chondrogenesis of the branchial skeleton in embryonic sea lamprey, *Petromyzon marinus. Anat. Rec.* **260**, 252–267.

Morriss, G. M. and Thorogood, P. V. (1978). An approach to cranial neural crest cell migration and differentiation in mammalian embryos. In *Development of Mammals* (M. Johnson, ed.), Volume 3., pp. 363–412. North Holland, Amsterdam.

Morriss-Kay, G. M. (1993). Retinoic acid and craniofacial development: molecules and morphogenesis. *BioEssays* **15**, 9–15.

Morriss-Kay, G. M. (1996). Craniofacial defects in AP-2 null mutant mice. *BioEssays* **18**, 785–788.

Morriss-Kay, G. M. (2001). Derivation of the mammalian skull vault. *J. Anat.* **199**, 143–155.

Morris-Wiman, J., Du, Y., and Brinkley, L. (1999). Occurrence and temporal variation in matrix metalloproteinases and their inhibitors during murine secondary palatal morphogenesis. *J. Craniofac. Genet. Devel. Biol.* **19**, 201–212.

Mortlock, D. P., Post, L. C., and Innis, J. W. (1996). The molecular basis of hypodactyly (Hd): a deletion in Hoxa 13 leads to arrest of digital arch formation. *Nat. Genet.* **13**, 284–289.

Moscatelli, D., and Rubin, H. (1975). Increased hyaluronic acid production on stimulation of DNA synthesis in chick embryonic fibroblasts. *Nature* **254**, 184–194.

Moscona, A. A. (1957). The development *in vitro* of chimeric aggregations of dissociated embryonic chick and mouse cells. *Proc. Natl Acad. Sci. U.S.A.* **43**, 184–194.

Moscona, A. A. (1961). Rotation-mediated histogenetic aggregations of dissociated cells. A quantifiable approach to cell interactions *in vitro. Exp. Cell Res.* **22**, 455–475.

Moscona, A. A. (1965). Recombination of dissociated cells and the development of cell aggregates. In *Cells and Tissues in*

Culture. Methods, Biology and Physiology (E. N. Willmer, ed.), Vol. 1, pp. 489–531. Academic Press, New York, NY.

Moser, H. G. (ed.) (1984). *Ontogeny and Systematics of Fishes.* International Symposium dedicated to the memory of Elbert Halvor Ahlstrom, August 15–18, 1983, La Jolla, California. Special Publication Number 1, American Society of Ichthyologists and Herpetologists. Lawrence, KS.

Moser, H. G., and Ahlstrom, E. H. (1970). Development of lanternfishes (Family Myctophidae) in the California Current. Part 1. Species with narrow-eyed larvae. *Bull. L. A. County Mus. Nat. Hist. Sci.* **#7**, 1–145.

Moses, M. A., Sudhalter, J., and Langer, R. (1992). Isolation and characterization of an inhibitor of neovascularization from scapular chondrocytes. *J. Cell Biol.* **119**, 475–482.

Moskalewski, S. (1976). Elastic fiber formation in monolayer and organ cultures of chondrocytes isolated from auricular cartilage. *Amer. J. Anat.* **146**, 443–448.

Moskalewski, S., Golaszewska, A., and Ksiazek, T. (1980). *In situ* aging of auricular chondrocytes is not due to the exhaustion of their replicative potential. *Experientia* **36**, 1294.

Moskalewski, S., Hye, A., Osiecka, A., and Malejczyk, J. (1990). Comparison of bone formed by transplants of isolated scapular and vertebral osteoblasts. *Fol. Hist. Cytobiol.* **28**, 35–42.

Moskalewski, S., and Malejczyk, J. (1989). Bone formation following intrarenal transplantation of isolated murine chondrocytes: Chondrocyte-bone cell transformation? *Development* **107**, 473–380.

Moskalewski, S., Malejczyk, J., and Osiecka, A. (1986). Structural differences between bone formed intramuscularly following the transplantation of isolated calvarial bone cells or chondrocytes. *Anat. Embryol.* **175**, 271–277.

Moskalewski, S., Osiecka, A., and Malejczyk, J. (1988). Comparison of bone formed intramuscularly after transplantation of scapular and calvarial osteoblasts. *Bone* **9**, 101–106.

Moss, M. L. (1954). Growth of the calvaria in the rat; the determination of osseous morphology. *Amer. J. Anat.* **94**, 333–362.

Moss, M. L. (1957a). Experimental alteration of sutural area morphology. *Anat. Rec.* **127**, 569–590.

Moss, M. L. (1957b). Premature synostosis of the frontal suture in the cleft palate skull. *Plast. Reconst. Surg.* **20**, 199–205.

Moss, M. L. (1958). Fusion of the frontal suture in the rat. *Amer. J. Anat.* **102**, 141–166.

Moss, M. L. (1959). The pathogenesis of premature cranial synostosis in man. *Acta Anat.* **37**, 351–370.

Moss, M. L. (1961a). Studies on the acellular bone of teleost fish. I. Morphological and systematic variation. *Acta Anat.* **46**, 343–362.

Moss, M. L. (1961b). Osteogenesis of acellular teleost bone. *Amer. J. Anat.* **108**, 99–110.

Moss, M. L. (1962a). Studies on the acellular bone of teleost fish. II. Response to fracture under normal and acalcemic conditions. *Acta Anat.* **48**, 46–60.

Moss, M. L. (1962b). Studies on the acellular bone of teleost fish. III. Intraskeletal heterografts in the rat. *Acta Anat.* **49**, 266–280.

Moss, M. L. (1962c). The functional matrix. In *Vistas in Orthodontics* (B. S. Kraus and R. A. Riedel, eds), pp. 85–98. Lea & Febiger, Philadelphia, PA.

Moss, M. L. (1963). The biology of acellular teleost bone. *Ann. NY. Acad. Sci.* **109**, 337–350.

Moss, M. L. (1964a). The phylogeny of mineralized tissues. *Int. Rev. Gen. Exp. Zool.* **1**, 297–331.

Moss, M. L. (1964b). Development of cellular dentin and lepidosteal tubules in the bowfin, *Amia calva. Acta Anat.* **58**, 333–354.

Moss, M. L. (1965). Studies on the acellular bone of teleost fish. V. Histology and mineral homeostasis of fresh-water species. *Acta Anat.* **60**, 262–276.

Moss, M. L. (1968a). Bone, dentin, and enamel and the evolution of vertebrates. In *Biology of the Mouth* (P. Person, ed.), Publ. No. 89, pp. 37–65. Amer. Assoc. Adv. Sci., Washington, DC.

Moss, M. L. (1968b). Comparative anatomy of vertebrate dermal bone and teeth. I. The epidermal-co-participation hypothesis. *Acta Anat.* **71**, 178–208.

Moss, M. L. (1968c). The origin of vertebrate calcified tissues. In *Current Problems in Lower Vertebrate Phylogeny* (T. Ørvig, ed.), pp. 359–371. Wiley (Interscience), New York, NY.

Moss, M. L. (1968d). Functional cranial analysis of mammalian mandibular ramal morphology. *Acta Anat.* **71**, 423–447.

Moss, M. L. (1969). Comparative histology of dermal sclerifications in reptiles. *Acta Anat.* **73**, 510–533.

Moss, M. L. (1970). Enamel and bone in shark teeth: with a note on fibrous enamel in fishes. *Acta Anat.* **77**, 161–187.

Moss, M. L. (1972a). The vertebrate dermis and the integumental skeleton. *Amer. Zool.* **12**, 27–34.

Moss, M. L. (1972b). The regulation of skeletal growth. In *Regulation of Organ and Tissue Growth* (R. J. Goss, ed.), pp. 127–142. Academic Press, New York, NY.

Moss, M. L. (1972c). New research objectives in craniofacial morphogenesis. *Acta Morphol. Neerl. Scand.* **10**, 103–110.

Moss, M. L. (1975). New studies of cranial growth. *Birth Defects, Orig. Artic. Ser.* **11**, 283–295.

Moss, M. L. (1977). Skeletal tissues in sharks. *Amer. Zool.* **17**, 335–342.

Moss, M. L. (1981). Genetics, epigenetics and caudation. *Amer. J. Orthod.* **80**, 366–377.

Moss, M. L. (1997a). The functional matrix hypothesis revisited. 1. The role of mechanotransduction. *Amer. J. Orthod. Dentofac. Orthop.* **112**, 8–11.

Moss, M. L. (1997b). The functional matrix hypothesis revisited. 2. The role of an osseous connected cellular network. *Amer. J. Orthod. Dentofac. Orthop.* **112**, 221–226.

Moss, M. L. (1997c). The functional matrix hypothesis revisited. 3. The genomic thesis. *Amer. J. Orthod. Dentofac. Orthop.* **112**, 338–342.

Moss, M. L. (1997d). The functional matrix hypothesis revisited. 4. The epigenetic antithesis and the resolving synthesis. *Amer. J. Orthod. Dentofac. Orthop.* **112**, 410–417.

Moss, M. L., Jones, S. J., and Piez, K. A. (1964). Calcified ectodermal collagens of shark tooth enamel and teleost scale. *Science* **145**, 940–942.

Moss, M. L., Meehan, M. A., and Salentijn, L. (1972). Transformative and translative growth processes in neurocranial development of the rat. *Acta Anat.* **81**, 161–182.

Moss, M. L., and Moss-Salentijn, L. (1983). Vertebrate cartilages. In *Cartilage, Volume 12, Structure, Function and Biochemistry* (B. K. Hall, ed.), pp. 1–30. Academic Press, New York, NY.

Mossaz, C. F., and Kokich, V. G. (1981). Redevelopment of the calvaria partial craniectomy in growing rabbits: The effects of altering dural continuity. *Acta Anat.* **109**, 321–331.

Moss-Salentijn, L. (1974). Studies on long bone growth. I. Determination of differential elongation in paired growth plates of the rat. *Acta Anat.* **90**, 145–160.

Moss-Salentijn, L. (1975). Cartilage canals in the human spheno-occipital synchondrosis during fetal life. *Acta Anat.* **92**, 595–606.

Moss-Salentijn, L. (1978). Vestigial teeth in the rabbit, rat and mouse; their relationship to the problem of lacteal dentitions. In *Development, Function and Evolution of Teeth* (P. M. Butler and K. A. Joysey, eds), pp. 13–29. Academic Press, London.

Moss-Salentijn, L., Kember, N. F., Shinozuka, M., Wu, W., and Bose, A. (1991). Computer simulations of chondrocytic clone behaviour in rabbit growth plates. *J. Anat.* **175**, 7–17.

Moss-Salentijn, L., and Moss, M. L. (1975). Studies on dentin. 2. Transient vasodentin in the incisor teeth of a rodent (*Perognathus longimembris*). *Acta Anat.* **91**, 386–404.

Moss-Salentijn, L., Moss, M. L., Masanobu, S., and Skalak, R. (1987). Morphological analysis and computer-aided three dimensional reconstruction of chondrocytic columns in rabbit growth plates. *J. Anat.* **151**, 157–168.

Mothe, A. J., and Brown, I. R. (2001). Differential mRNA expression of the related extracellular matrix glycoproteins SC1 and SPARC in the rat embryonic nervous system and skeletal structures. *Brain Res.* **892**, 27–41.

Mottershead, S. (1988). Sesamoid bones and cartilages: An enquiry into their function. *Clin. Anat.* **1**, 59–62.

Mottet, N. K., and Hammar, S. P. (1972). Ribosome crystals in necrotizing cells from the posterior necrotic zone of the developing chick limb. *J. Cell Sci.* **11**, 403–411.

Mow, V. C., Roth, V., and Armstrong, C. G. (1980). Biomechanics of Joint Cartilage. In *Basic Biomechanics of the Locomotor System* (M. Nordin and V. Frankel, eds), pp. 61–86. Lea & Febiger, Philadelphia, PA.

Moy-Thomas, J. A. (1938). The problem of the evolution of the dermal bones in fishes. In *Evolution. Essays on Aspects of Evolutionary Biology* (G. R. de Beer, ed.), pp. 305–319. Clarendon Press, Oxford.

Moy-Thomas, J. A. (1941). Development of the frontal bones of the rainbow trout. *Nature* **147**, 681–682.

Muenke, M., Francomano, C. A., Cohen, M. M. Jr., and Wang, Jabs, E. (1998). Fibroblast growth factor receptor-related skeletal disorders. In *Principles of Molecular Medicine* (J. L. Jameson, ed.), pp. 1029–1038. Humana Press Inc., Totowa, NJ.

Muenke, M., Gripp, K. W., McDonald-McGinn, D. M., Gadenz, K., Whitaker, L. A., Bartlett, S. P., Markowitz, R. I., Robin, N. H., Nwokoro, N., Mulvihill, J. J., Losken, H. W., Mulliken, J. B., Guttmacher, A. E., Wilroy, R. S., Clarke, L. A., Hollway, G., Adès, L. C., Haan, E. A., Mulley, J. C., Cohen, M. M., Jr., Bellus, G. A., Francomano, C. A., Moloney, D. M., Wall, S. A., Wilkie, A. O. M., and Zackai, E. H. (1997). A unique point mutation in the fibroblast growth factor receptor 3 gene (FGFR3) defines a new craniosynostosis syndrome. *Amer. J. Human Genet.* **60**, 555–564.

Muhlhauser, J. (1986). Resorption of the unmineralized proximal part of Meckel's cartilage in the rat – a light and electron microscopic study. *J. Submicrosc. Cytol.* **18**, 717–724.

Muhlrad, A., Setton, A., Sela, J., Bab, I., and Deutsch, D. (1983). Biochemical characterization of matrix vesicles from bone and cartilage. *Metab. Bone Dis. Rel. Res.* **5**, 93–99.

Muir, H. (1995). The chondrocyte, architect of cartilage. Biomechanics, structure, function and molecular biology of cartilage matrix molecules. *BioEssays* **17**, 1039–1048.

Muirden, K. D., Deutschmann, P., and Philipps, M. (1974). Articular cartilage in rheumatoid arthritis: Ultrastructure and enzymology. *J. Rheumatol.* **1**, 24–33.

Mukai, M., Yoshimine, Y., Akamine, A., and Maeda, K. (1993). Bone-like nodules formed *in vitro* by rat periodontal ligament cells. *Cell Tissue Res.* **271**, 453–460.

Mulder, E. W. A. (2001). Co-ossified vertebrae of mososaurs and cetaceans: implications for the mode of locomotion of extinct marine reptiles. *Paleobiology* **27**, 724–734.

Mullen, L. M., Bryant, S. V., Torok, M. A., Blumberg, B., and Gardiner, D. M. (1996). Nerve dependency of regeneration: The role of *Distal-less* and FGF signaling in amphibian limb regeneration. *Development* **122**, 3487–3497.

Müller, F., and O'Rahilly, R. (1980). The human chondrocranium at the end of the embryonic period proper with particular reference to the nervous system. *Amer. J. Anat.* **159**, 33–58.

Müller, G. B. (1985). Experimental reestablishment of ancestral patterns in the chick limb. In *Evolution and Morphogenesis* (J. Mlikovsky and V. J. A. Novak, eds), pp. 439–446. Academia, Prague.

Müller, G. B. (1986). Effects of skeletal change on muscle pattern formation. In *Development and Regeneration of Skeletal Muscles* (B. Christ and R. Cihak, eds), pp. 91–108. Karger, Berlin.

Müller, G. B. (1989). Ancestral patterns in bird limb development: A new look at Hampé's experiment. *J. Evol. Biol.* **2**, 31–48.

Müller, G. B. (1991). Evolutionary transformation of limb pattern: heterochrony and secondary fusion. In *Developmental Patterning of the Vertebrate Limb* (J. R. Hinchliffe, J. M. Hurlé., and D. Summerbell, eds), pp. 395–405. Plenum Press, New York, NY.

Müller, G. B., and Alberch, P. (1990). Ontogeny of the limb skeleton in *Alligator mississippiensis*: developmental invariance and change in the evolution of Archosaur limbs. *J. Morphol.* **203**, 151–164.

Müller, G. B., and Streicher, J. (1989). Ontogeny of the syndesmosis tibiofibularis and the evolution of the bird hindlimb: A caenogenetic feature triggers phenotypic novelty. *Anat. Embryol.* **179**, 327–339.

Müller, G. B., Streicher, J., and Müller, R. J. (1996). Homeotic duplication of the pelvic body segment in regenerating tadpole tails induced by retinoic acid. *Devel. Genes Evol.* **206**, 344–348.

Müller, G. B., and Wagner, G. P. (1991). Novelty in evolution: restructuring the concept. *Annu. Rev. Ecol. Syst.* **22**, 229–256.

Müller, H. (1858). Über die Entwickelung der Knochensunstanz nebst Bemerkungen über den Bau rachitiachen knochen. *Wschr. Wiss. Zool.* **9**, 147–233.

Müller, P. K., and Kühn, K. (1977). Kollagenbiosynthese als ein Beispiel für die Regulation der Genexpression in mesenchymalen Zellen. *Arzneim. Forsch.* **27**, 199–202.

Muller, T. L., Ngo-Muller, V., Reginelli, A., Taylor, G., Anderson, R., and Muneoka, K. (1999). Regeneration in higher vertebrates: Limb buds and digit tips. *Sem. Devel. Biol.* **10**, 405–413.

Müller, W., Fricke, H., Halliday, A. N., McCulloch, M. T., and Wartho, J.-A. (2003). Origin and migration of the Alpine Iceman. *Science* **302**, 862–866.

Müller-Glauser, W., Humbel, B., Glatt, M., Sträuli, P., Winterhalter, K. H., and Bruckner, P. (1986). On the role of type IX collagen in the extracellular matrix of cartilage: Type IX collagen is localized to intersections of collagen fibrils. *J. Cell Biol.* **102**, 1931–1939.

Mulliken, J. B., Kaban, L. B., and Glowacki, J. (1984). Induced osteogenesis – the biological principle and clinical applications. *J. Surg. Res.* **37**, 487–496.

Mundell, R. D., Mooney, M. P., Siegel, M. I., and Losken, A. (1993). Osseous guided tissue regeneration using a collagen barrier membrane. *J. Oral Maxillofac. Surg.* **51**, 1004–1012.

Mundlos, S., Otto, F., Mundlos, C., Milliken, J. B., Aylsworth, A. S., Albright, S., Linghout, D., Cole, W. G., Henn, W., Knoll, J. H. M., Owen, M. J., Zabel, B. U., Mertelsmann, R., and Olsen, B. R. (1997). Mutations involving the transcription factor CBFA1 cause cleidocranial dysplasia. *Cell* **98**, 73–779.

Mundlos, S., Schwahn, B., Reichert, T., and Zabel, B. (1992). Distribution of osteonectin mRNA and protein during human embryonic and fetal development. *J. Histochem. Cytochem.* **40**, 283–291.

Mundy, G. R., Atman, A. J., Gonder, M. D., and Bandelin, J. G. (1977). Direct resorption of bone by human monocytes. *Science* **196**, 1109–1111.

Mundy, G. R, Garrett, R., Harris, S., Chan, J., Chen, D., Rossini, G., Boyce, B., Zhao, M., and Gutierrez, G. (1999). Stimulation of bone formation *in vitro* and in rodents by statins. *Science* **286**, 1946–1949.

Mundy, G. R., and Raisz, L. G. (1981). Disorders of bone resorption. In *Disorders of Mineral Metabolism. Volume 3, Pathophysiology of Calcium, Phosphorus and Magnesium* (F. Bronner and J. W. Coburn, eds), pp. 1–66. Academic Press, New York, NY.

Mundy, G. R., Rodan, S. B., Majeska, R. J., DeMartino, S., Trimmier, C., Martin, T. J., and Rodan, G. A. (1982). Unidirectional migration of osteosarcoma cells with osteoblast characteristics in response to products of bone resorption. *Calcif. Tissue Int.* **34**, 542–546,

Mundy, G. R., and Roodman, G. D. (1987). Osteoclast ontogeny and function. In *Bone and Mineral Research/5* (W. A. Peck, ed.), pp. 209–280. Elsevier Science Publishers BV, Amsterdam.

Munoz-Sanjuan, I., Simandl, B. K., Fallon, J. F., and Nathans, J. (1999). Expression of chicken fibroblast growth factor homologous factor (FHF)-1 and of differentially spliced isoforms of FHF-2 during development and involvement of FHF-2 in chicken limb development. *Development* **126**, 409–421.

Muragaki, Y., Mundlos, S., Upton, J., and Olsen, B. R. (1996). Altered growth and branching patterns in synpolydactyly caused by mutations in HoxD13. *Science* **272**, 548–551.

Murakami, R. (1986). Development of the os penis in genital tubercles cultured beneath the renal capsule of adult rats. *J. Anat.* **149**, 11–20.

Murakami, R. (1987a). Autoradiographic studies of the localization of androgen-binding cells in the genital tubercles of fetal rats. *J. Anat.* **151**, 209–219.

Murakami, R. (1987b). A histological study of the development of the penis of wild-type and androgen-insensitive mice. *J. Anat.* **153**, 223–231.

Murakami, R., Miyake, K., and Yamaoka, I. (1994). Androgen-induced differentiation of the fibrocartilage of os penis cultured *in vitro. Zool. Sci.* **11**, 847–853.

Murakami, R., and Mizuno, T. (1984). Histogenesis of the os penis and os clitoridis in rats. *Devel. Growth Differ.* **26**, 419–426.

Murakami, R., and Mizuno, T. (1986). Proximal-distal sequence of development of the skeletal tissues in the penis of rat and the inductive effect of epithelium. *J. Embryol. Exp. Morphol.* **92**, 133–143.

Murie, J. (1872). On the horns, viscera, and muscles of the giraffe. *Ann. Mag. Nat. Hist.* 4th Ser. **9**, 177–185,

Murray, B. M., and Wilson, D. J. (1994). A scanning electron microscopic study of the normal development of the chick wing from stages 19 to 36. *Anat. Embryol.* **189**, 147–155.

Murray, J. Mc.G., and Cleall, J. F. (1971). Early tissue response to rapid maxillary expansion in the midpalatal suture of the Rhesus monkey. *J. Dent. Res.* **50**, 1654–1660.

Murray, K. G., Winnett-Murray, K., Eppley, Z. A., Hunt, G. L., Jr., and Schwartz, D. B. (1983). Breeding biology of the Xanthus' murrelet. *Condor* **85**, 12–21.

Murray, P. D. F. (1926). An experimental study of the development of the limbs of the chick. *Proc. Linn. Soc. N.S.W.* **51**, 187–263.

Murray, P. D. F. (1928). Chorio-allantoic grafts of fragments of the two-day chick, with special reference to the development of the limbs, intestine and skin. *Aust. J. Exp. Biol. Med. Sci.* **5**, 237–256.

Murray, P. D. F. (1936). *Bones. A Study of the Development and Structure of the Vertebrate Skeleton.* Cambridge University Press, Cambridge. (Reprinted 1985 with an introduction by B. K. Hall.)

Murray, P. D. F. (1941). Epidermal papillae and dermal bones of the chick skeleton. *Nature* **148**, 471.

Murray, P. D. F. (1943). The development of the conjunctival papillar and of the scleral bones in the chick embryo. *J. Anat.* **77**, 225–240.

Murray, P. D. F. (1954). The fusion of parallel long bones and the formation of secondary cartilage. *Aust. J. Zool.* **2**, 364–380.

Murray, P. D. F. (1963). Adventitious (secondary) cartilage in the chick embryo, and the development of certain bones and articulations in the chick skull. *Aust. J. Zool.* **11**, 368–430.

Murray, P. D. F., and Drachman, D. B. (1969). The role of movement in the development of joints and related structures: The head and neck in the chick embryo. *J. Embryol. Exp. Morphol.* **22**, 349–371.

Murray, P. D. F., and Huxley, J. S. (1925). Self-differentiation in the grafted limb bud of the chick. *J. Anat.* **59**, 379–384.

Murray, P. D. F., and Selby, D. (1930). Intrinsic and extrinsic factors in the primary development of the skeleton. *W. Roux. Arch. EntwickMech.* **122**, 629–662.

Murray, P. D. F., and Selby, D. S. (1933). Chorio-allantoic grafts of single somites and of the unsegmented paraxial region of the two-day chick embryo. *J. Anat.* **67**, 563–572.

Murray, P. D. F., and Smiles, M. (1965). Factors in the evocation of adventitious (secondary) cartilage in the chick embryo. *Aust. J. Zool.* **13**, 351–381.

Murtaugh, L. C., Chyung, J. H., and Lassar, A. B. (1999). Sonic hedgehog promotes somitic chondrogenesis by altering the cellular response to BMP signaling. *Genes & Devel.* **13**, 225–237.

Muschler, G. F., and Midura, R. J. (2002). Connective tissue progenitors: practical concepts for clinical applications. *Clin. Orthop. Rel. Res.* **395**, 66–80.

Muther, T. F. (1988a). Caffeine and reduction of fetal ossification in the rat: fact or artifact? *Teratology* **37**, 239–247.

Muther, T. F. (1988b). Response to 'Caffeine, an exquisitely specific inhibitor of osteogenic differentiation.' *Teratology* **38**, 605–606.

Mwale, F., Billinghurst, C., Wu, W., Alini, M., Webber, C., Reiner, A., Ionescu, M., Poole, J., and Poole, A. R. (2000). Selective asembly and remodelling of collagens II and IX associated with expression of the chondrocyte hypertrophic phenotype. *Devel. Dynam.* **218**, 648–662.

M-Z Rahmani, T. (1974). Morphogenesis of the rudimentary hind-limb of the glass snake (*Ophisaurus apodus* Pallas). *J. Embryol. Exp. Morphol.* **32**, 431–443.

N

Nacke, S., Schäfer, R., Hrabé De Angelis, M., and Mundlos, S. (2000). Mouse mutant 'rib-vertebrae' (*rv*): a defect in somite polarity. *Devel. Dynam.* **219**, 192–200.

Nagamoto, N., Iyama, K.-I., Kitaoka, M., Ninomiya, Y., Yoshioka, H., Mizuta, H., and Tagagi, K. (1993). Rapid expression of collagen type X gene of non-hypertrophic chondrocytes in the grafted chick periosteum demonstrated by *in situ* hybridization. *J. Histochem. Cytochem.* **41**, 679–684.

Nagata, T., Bellows, C. G., Kasugai, S., Butler, W. T., and Sodek, J. (1991). Biosynthesis of bone proteins (Spp-1 (secreted phosphoprotein-1, osteopontin), BSP (bone sialoprotein) and SPARC (osteonectin) in association with mineralized-tissue formation by fetal-rat calvarial cells in culture. *Biochem. J.* **274**, 513–520.

Nah, H. D., Pacifici, M., Gerstenfeld, L. C., Adams, S. L., and Kirsch, T. (2000). Transient chondrogenic phase in the intramembranous pathway during normal skeletal development. *J. Bone Min. Res.* **15**, 522–533.

Nah, H. D., Swoboda, B., Birk, D. E., and Kirsch, T. (2001). Type IIA procollagen: expression in developing chicken limb cartilage and human osteoarthritic articular cartilage. *Devel. Dynam.* **220**, 307–322.

Nakahara, H., Bruder, S. P., Goldberg, V. M., and Caplan, A. I. (1990a). *In vivo* osteochondrogenic potential of cultured cells derived from the periosteum. *Clin. Orthop. Rel. Res.* **259**, 223–232.

Nakahara, H., Dennis, J. E., Bruder, S. P., Haynesworth, S. E., Lennon, D. P., and Caplan, A. I. (1991a). *In vitro* differentiation of bone and hypertrophic cartilage from periosteal-derived cells. *Exp. Cell Res.* **195**, 492–503.

Nakahara, H., Goldberg, V. M., and Caplan, A. I. (1991b). Culture expanded human periosteal-derived cells exhibit osteochondral potential *in vivo*. *J. Orthop. Res.* **9**, 465–476.

Nakahara, H., Goldberg, V. M., and Caplan, A. I. (1992). Culture expanded periosteal-derived cells exhibit osteochondral potential in porous calcium phosphate ceramics *in vivo*. *Clin. Orthop. Rel. Res.* **276**, 291–298.

Nakahara, H., Watanabe, H., Sugrue, S. P., Olsen, B. R., and Caplan, A. I. (1990b). Temporal and spatial distribution of type XII collagen in high cell density culture of periosteal-derived cells. *Devel. Biol.* **142**, 481–485.

Nakajima, T. (1984). Larval vs. adult pharyngeal dentition in some Japanese cyprinid fishes. *J. Dent. Res.* **63**, 1140–1146.

Nakajima, T. (1987). Development of pharyngeal dentition in the Cobitid Fishes, *Misgurnus anguillicaudatus* and *Cobitis biwae* with a consideration of evolution of Cypriniform dentitions. *Copeia* **1987**, 208–213.

Nakajima, T., and Yue, P. (1989). Development of pharyngeal teeth in the Big Head, *Aristichthye nobilis* (Cyprinidae). *Jap. J. Ichthyol.* **36**, 42–47.

Nakamoto, T., Porter, J. R., and Winkler, M. M. (1983). The effect of prenatal protein-energy malnutrition on the development of mandibles and long bones in newborn rats. *Brit. J. Nutrition* **50**, 75–80.

Nakamoto, T., and Shaye, R. (1984). Effects of caffeine on the growth of mandible and long bone in protein-energy malnourished newborn rats. *Proc. Soc. Exp. Biol. Med.* **177**, 55–61.

Nakamura, H., and Ayer-Le Lièvre, C. S. (1982). Mesectodermal capabilities of the trunk neural crest of birds. *J. Embryol. Exp. Morphol.* **70**, 1–18.

Nakamura, K., and Yamaguchi, H. (1991). Distribution of scleral ossicles in teleost fishes. *Mem. Fac. Fish. Kagoshima Univ.* **40**, 1–20.

Nakamura, T., Hanada, K., Tamura, M., Shibanushi, T., Nigi, H., Tagawa, M., Fukumoto, S., and Matsumoto, T. (1995). Stimulation of endosteal bone formation by systemic injections of recombinant basic fibroblast growth factor in rats. *Endocrinology* **136**, 1276–1284.

Nakamura, T., and Koski, K. (1983). Periosteal control of mitotic activity in the mandibular condyle of the rat. *IRCS-Biol.* **11**, 307–308.

Nakamura, Y., Becker, L. E., and Marks, A. (1983). S-100 protein in tumors of cartilage and bone: An immunohistochemical study. *Cancer* **52**, 1820–1824.

Nakase, T., Nakahara, H., Iwasaki, M., Kimura, T., Kimata, K., Watanabe, K., Caplan, A. I., and Ono, K. (1993). Clonal analysis for developmental potential of chick periosteum-derived cells: Agar gel culture system. *Biochem. Biophys. Res. Commun.* **195**, 1422–1428.

Nakashima, K., and de Crombrugghe, B. (2003). Transcriptional mechanisms in osteoblast differentiation and bone formation. *Trends Genet.* **19**, 458–466.

Nakashima, K., Zhou, X., Kunkel, G., Zhang, Z., Deng, J. M., Behringer, R. R., and de Crombrugghe, B. (2002). The novel zinc finger-containing transcription factor osterix is required for osteoblast differentiation and bone formation. *Cell* 108, 17–29.

Nakata, S. (1981). Relationship between the development and growth of cranial bones and masticatory muscles in postnatal mice. *J. Dent. Res.* **60**, 1440–1450.

Nakayama, N., Han, C-y. E., Scully, S., Nishinakamura, R., He, C., Zeni, L., Yamane, H., Chang, D, Yu, D., Yokota, T., and Wen, D. (2001). A novel chordin-like protein inhibitor for bone morphogenetic proteins expressed preferentially in mesenchymal cell lineages. *Devel. Biol.* **232**, 372–387.

Namenwirth, M. (1974). The inheritance of cell differentiation during limb regeneration in the axolotl. *Devel. Biol.* **41**, 42–56.

Narbaitz, R. (1992). Effects of vitamins A, C, D, and K on bone growth, mineralization, and resorption. In *Bone, Volume 4: Bone Metabolism and Mineralization* (B. K. Hall, ed.), pp. 141–169. CRC Press, Boca Raton, FL.

Nardi, F., Gerlini, G., and Bonucci, E. (1974). Achondrogenesis: Report on a case, with particular reference to ultrastructure and histochemistry. *Virchows Arch. Path. Anat., A* **363**, 311–322.

Naski, M. C., Colvin, J. S., Coffin, J. D., and Ornitz, D. M. (1998). Repression of hedgehog signaling and BMP4 expression in growth plate cartilage by fibroblast growth factor receptor 3. *Development* **125**, 4977–4988.

Natanson, L. J., and Cailliet, G. M. (1990). Vertebral growth zone deposition in Pacific Angel sharks. *Copeia* **1990**, 1133–1145.

Nathanson, M. A. (1979). Skeletal muscle metaplasia – Formation of cartilage by differentiated skeletal muscle. In *Muscle Regeneration* (A. Mauro, ed.), pp. 83–90. Raven Press, New York, NY.

Nathanson, M. A. (1983a). Proteoglycan synthesis by skeletal muscle undergoing bone matrix-directed transformations into cartilage *in vitro*. *J. Biol. Chem.* **258**, 10325–10334.

Nathanson, M. A. (1983b). Analysis of cartilage differentiation from skeletal muscle grown on bone matrix. III. Environmental regulation of glycosaminoglycan and proteoglycan synthesis. *Devel. Biol.* **96**, 46–62.

Nathanson, M. A. (1985). Bone matrix-directed chondrogenesis of muscle in vitro. Clin. Orthop. Rel. Res. **200**, 142–158.

Nathanson, M. A. (1986). Transdifferentiation of skeletal muscle into cartilage: Transformation or differentiation? Curr. Top. Devel. Biol. **20**, 39–62.

Nathanson, M. A., Bush, E. W., and Venderburg, C. (1986). Transcriptional-translational regulation of muscle-specific protein synthesis and its relationship to chondrogenic stimuli. J. Biol. Chem. **261**, 1477–1486.

Nathanson, M. A., and Hay, E. D. (1980). Analysis of cartilage differentiation from skeletal muscle grown on bone matrix. I. Ultrastructural aspects. Devel. Biol. **78**, 301–331.

Nathanson, M. A., Hilfer, S. R., and Searls, R. L. (1978). Formation of cartilage by nonchondrogenic cell types. Devel. Biol. **64**, 99–117.

Nechiporuk, A., and Keating, M. T. (2002). A proliferation gradient between proximal and msxb-expressing distal blastema directs zebrafish fin regeneration. Development **127**, 2607–2617.

Nederbragt, A. J., van Loon, A. E., and Dictus, W. J. A. G. (2002). Hedgehog crosses the snail's midline. Nature **417**, 811–812.

Nefussi, J.-R., and Baron, R. (1985). PGE2 stimulates bone resorption and formation of bone in vitro: Differential responses of the periosteum and the endosteum in fetal rat long bone cultures. Anat. Rec. **211**, 9–16.

Negulesco, J. A. (1971). Bone fracture repair in normal and hypophysectomized chickens. Anat. Rec. **169**, 386–397.

Nelson, C. E., Morgan, B. A., Burke, A. C., Laufer, E., Di Mambro, E., Murtaugh, L. C., Gonzales, E., Tessarollo, L., Parada, L. F., and Tabin, C. (1996). Analysis of Hox gene expression in the chick limb bud. Development **122**, 1449–1466.

Nelson, R. L., and Bauer, G. E. (1977). Isolation of osteoclasts by velocity sedimentation at unit gravity. Calcif. Tissue Res. **22**, 303–314.

Nelson, S. W. (1975). Developmental potentialities of marrow tissue culture cells surviving programmed freezing. J. Dent. Res. **54**, 265–268.

Nemeschkal, H. L. (1999). Morphometric correlation patterns of adult birds (Fringillidae: Passeriformes and Columbiformes) mirrow the expression of developmental control genes. Evolution **53**, 899–918.

Neméth-Csóka, M. (1974). The effect of acid mucopolysaccharides on the activation energy of collagen fibril-formation. Exp. Pathol. **9**, 256–262.

Nesbitt, R. (1736). Human Osteogeny Explained in Two Lectures. London.

Nesslinger, C. L. (1956). Ossification centres and skeletal development in the postnatal Virginia opossum. J. Mammal. **37**, 382–394.

Neto, F. L. D. S., and Volpon, J. B. (1984). Experimental nonunion in digs. Clin. Orthop. Rel. Res. **187**, 260–271.

Neubert, R., Merker, H.-J., and Neubert, D. (1999). Developmental model for thalidomide action. Nature **400**, 419–420.

Neubüser, A., Koseki, H., and Balling, R. (1995). Characterization and developmental expression of Pax9, a paired-box-containing gene related to Pax1. Devel. Biol. **170**, 701–706.

Neufeld, D. A. (1985). Bone healing after amputation of mouse digits and newt limbs: Implication for induced regeneration in mammals. Anat. Rec. **211**, 156–165.

Neufeld, D. A. (1992). Digital regeneration in mammals. In Bone, Volume 5: Fracture Repair and Regeneration (B. K. Hall, ed.), pp. 313–338. CRC Press, Boca Raton, FL.

Neufeld, D. A., and Zhao, W. G. (1995). Bone regrowth after digit tip amputation in mice is equivalent in adults and neonates. Wound Repair. Regen. **3**, 461–466.

Neuhass, S. C. F., Solnica-Krezel, L., Schier, A. F., Zwartkruis, F., Stemple, D. L., Malicki, J., Abdelilah, S., Stainier, D. Y. R., and Driever, W. (1996). Mutations affecting craniofacial development in zebrafish. Development **123**, 357–367.

Neuhof, H. (1917). Fascia transplantation into visceral defects. Surg. Gynecol. Obstet. **24**, 383–387.

Neuman, W. F. (1969). The milieu interieur of bone. Claude Bernard revisited. Fed. Proc., Fed. Amer. Soc. Exp. Biol. **28**, 1846–1850.

Neumann, C. J., Grandel, H., Gaffield, W., Schultze-Merker, S., and Nusslein-Volhard, C. (1999). Transient establishment of antero-posterior polarity in the zebrafish pectoral fin bud in the absence of sonic hedgehog activity. Development **126**, 4817–4826.

Neumann, K., Moegelin, A., Temminghoff, N., Radlanski, R. J., Langford, A., Unger, M., Langer, R., and Bier, J. (1997). 3D-computed tomography: a new method for the evaluation of fetal cranial morphology. J. Craniofac. Genet. Devel. Biol. **17**, 9–22.

Nevo, Z., and Dorfman, A. (1972). Stimulation of chondromuco-protein synthesis in chondrocytes by extracellular chondro-mucoprotein. Proc. Natl Acad. Sci. U.S.A. **69**, 2069–2072.

Nevo, Z., Horwitz, A. L., and Dorfman, A. (1972). Synthesis of chondromucoprotein by chondrocytes in suspension culture. Devel. Biol. **28**, 219–228.

Newbrey, J. W., and Banks, W. J. (1975). Characterization of developing antler cartilage matrix. II. An ultrastructural study. Calcif. Tissue Res. **17**, 289–302.

Newbrey, J. W., and Banks, W. J. (1983). Ultrastructural changes associated with the mineralization of deer antler cartilage. Amer. J. Anat. **166**, 1–18.

Newman, S. A. (1977). Lineage and pattern in the developing wing bud. In Vertebrate Limb and Somite Morphogenesis (D. A. Ede, J. R. Hinchliffe, and M. Balls, eds), pp. 181–198. Cambridge University Press, Cambridge.

Newman, S. A. (1992). Generic physical mechanisms of morpho-genesis and pattern formation as determinants in the evolu-tion of multicellular organization. In Principles of Organization in Organisms (J. Mittenthal and A. Baskin, eds), pp. 241–267. Addison-Wesley, New York, NY.

Newman, S. A. (1996). Sticky fingers – Hox genes and cell adhe-sion in vertebrate limb development. BioEssays **18**, 171–174.

Newman, S. A., and Comper, W. D. (1990). 'Generic' physical mechanisms of morphogenesis and pattern formation. Development **110**, 1–18.

Newman, S. A., and Tomasek, J. J. (1996). Morphogenesis of con-nective tissues. In Extracellular Matrix, Volume 2, Molecular Components and Interactions (W. D. Comper, ed.), pp. 335–369. Harwood Academic Publishers, Amsterdam.

Newsome, D. A. (1972). Cartilage induction by retinal pigmented epithelium of chick embryos. Devel. Biol. **27**, 575–579.

Newsome, D. A. (1976). In vitro stimulation of cartilage in embry-onic chick neural crest cells by products of retinal pigmented epithelium. Devel. Biol. **49**, 496–507.

Newsome, D. A., and Kenyon, K. R. (1973). Collagen production in vitro by the retinal pigmented epithelium of the chick embryo. Devel. Biol. **32**, 387–400.

Neyt, C., Jagla, K., Thisse, C., Thisse, B., Haines, L., and Currie, P. D. (2000). Evolutionary origins of vertebrate appendicular muscle. Nature **408**, 82–86.

Ng, K. W., Partridge, N. C., Niall, M., and Martin, T. J. (1983). Stimulation of DNA synthesis by epidermal growth factor in osteoblast-like cells. *Calcif. Tissue Int.* **35**, 624–628.

Ng, L. G., Tam, P. P. L., and Cheah, K. S. E. (1993). Preferential expression of alternatively spliced mRNAs encoding type II procollagen with a cysteine-rich amino-propeptide in differentiating cartilage and nonchondrogenic tissues during early mouse development. *Devel. Biol.* **159**, 403–417.

Ng, L.-J., Wheatley, S., Muscat, G. E. O., Conway-Campbell, J., Bowles, J., Wright, E., Bell, D. M., Tam, P. L. L., Cheah, K. S. E., and Koopman, P. (1997). Sox9 binds DNA, activates transcription, and coexpresses with type II collagen during chondrogenesis in the mouse. *Devel. Biol.* **183**, 108–121.

Niazi, I. A., and Alam, S. (1984). Regeneration of whole limbs in toad tadpoles treated with retinol palmitate after the wound-healing stage. *J. Exp. Zool.* **230**, 501–505.

Niazi, I. A., and Ratnasamy, C. S. (1984). Regeneration of whole limbs from shank stumps in toad tadpoles treated with vitamin A. *W. Roux. Arch. Entwickl.* **193**, 111–116.

Nicholas, J. S., and Rudnick, D. (1933). The development of embryonic rat tissues upon the chick chorioallantois. *J. Exp. Zool.* **66**, 193–262.

Nichols, S., Gelsleichter, J., Manire, C. A., and Cailliet, G. M. (2003). Calcitonin-like immunoreactivity in serum and tissue in the bonnethead shark, *Sphyrna tiburo*. *J. Exp. Zool.* **298A**, 160–161.

Niederman, R., and Armstrong, P. B. (1972). Is abnormal limb bud morphology in the mutant *talpid*² chick embryo a result of altered intercellular adhesion? Studies employing cell sorting and fragment fusion. *J. Exp. Zool.* **181**, 17–32.

Niederreither, K., Vermot, J., Schuhbaur, B., Chambon, P., and Dollé, P. (2002). Embryonic retinoic acid synthesis is required for forelimb growth and anteroposterior patterning in the mouse. *Development* **129**, 3563–3574.

Nifuji, A., Kellermann, O., and Noda, M. (2004). Noggin inhibits chondrogenic but not osteogenic differentiation in mesodermal stem cells line C1 and skeletal cells. *Endocrinology* **145**, 3434–3442.

Nigrelli, R. F., and Gordon, M. (1946). Spontaneous neoplasms in fishes. I. Osteochondroma in the jewelfish, *Hemichromis bimaculatus*. *Zoologica* **31**, 89–92.

Niida, S., Okada, N., Wakisaka, H., Miyata, K., and Maeda, N. (1994b). Occipital roof development in the Japanese musk shrew, *Suncus murinus*. *J. Anat.* **185**, 433–437.

Niida, S., Wakisaka, H., Kanno, E., and Yamasaki, A. (1994a). Cranial flat bone formation in the osteopetrotic (Op/Op) mouse. *Biomed. Res.* **15**, 37–44.

Nijhout, H. F., and Emlen, D. J. (1998). Competition among body parts in the development and evolution of insect morphology. *Proc. Natl Acad. Sci. U.S.A.* **95**, 3685–3689.

Nijweide, P. J. (1975). Embryonic chicken periosteum in tissue culture, osteoid formation and calcium uptake. *Proc. Kon. Ned. Akad. Wed. Ser. C. Biol. Med. Sci.* **78**, 410–417.

Nijweide, P. J., Burger, E. H., and Feyen, J. H. M. (1986). Cells of bone: Proliferation, differentiation and hormonal regulation. *Physiol. Rev.* **66**, 855–886.

Nijweide, P. J., Burger, E. H., Hekkelman, J. W., Herrmann-Erlee, M. P. M., and Gaillard, P. J. (1982b). Regulatory mechanisms in the development of bone and cartilage: The use of tissue culture techniques in the study of the development of embryonic bone and cartilage: A perspective. In *Factors and Mechanisms Influencing Bone Growth* (A. D. Dixon and B. G. Sarnat, eds), pp. 457–480. Alan R. Liss Inc., New York, NY.

Nijweide, P. J., and Mulder, R. J. P. (1986). Identification of osteocytes in osteoblast-like cultures using a monoclonal antibody specifically directed against osteocytes. *Histochemistry* **84**, 342–347.

Nijweide, P. J., and Van der Plas, A. (1979). Regulation of calcium transport in isolated periosteal cells, effects of hormones and metabolic inhibitors. *Calcif. Tissue Int.* **29**, 155–161.

Nijweide, P. J., Van der Plas, A., and Scherft, J. P. (1981). Biochemical and histological studies on various bone cell preparations. *Calcif. Tissue Int.* **33**, 529–540.

Nijweide, P. J., Van-Iperin-Van Gent, A. S., Kawilarang-de Haas, E. W. M., Van der Plas, A., and Wassenaar, A. M. (1982a). Bone formation and calcification by isolated osteoblastlike cells. *J. Cell Biol.* **93**, 318–323.

Nijweide, P. J., Vrijheid-Lammers, T., Mulder, R. J. P., and Blok, J. (1985). Cell surface antigens on osteoclasts and related cells in the quail studied with monoclonal antibodies. *Histochemistry* **83**, 315–324.

Nikbakht, N., and McLachlan, J. C. (1999). Restoring avian wing digits. *Proc. R. Soc. Lond. B* **266**, 1101–1104.

Nilsen, R. (1977). Electron microscopy of induced heterotopic bone formation in guinea pigs. *Archs Oral Biol.* **22**, 485–494.

Nilsen, R. (1980a). Microfilaments in cells associated with induced heterotopic bone formation in guinea pigs. An immunofluorescence and ultrastructural study. *Acta Pathol. Microbiol. Scand. A* **88**, 129–134.

Nilsen, R. (1980b). Electronmicroscopic study of mineralization in induced heterotopic bone formation in guinea pigs. *Scand. J. Dent. Res.* **88**, 340–347.

Nilsen, R., and Magnusson, B. C. (1979). Enzyme histochemistry of induced heterotopic bone formation in guinea pigs. *Archs Oral Biol.* **24**, 833–841.

Nilsen-Hamilton, M. (ed.) (1990). *Growth Factors and Development*. Current Topics in Developmental Biology, Volume 24. Academic Press, San Diego, CA.

Nimni, M. E. (1973). Metabolic pathways and control mechanisms involved in the biosynthesis and turnover of collagen in normal and pathological connective tissues. *J. Oral Pathol.* **2**, 175–202.

Nimni, M. E., Bernick, S., Ertl, D., Nishimoto, S. K., Paule, W., Strates, B. S., and Villanueva, J. (1988). Ectopic bone formation is enhanced in senescent animals implanted with embryonic cells. *Clin. Orthop. Rel. Res.* **234**, 255–266.

Nimni, M. E., and Deshmukh, K. (1973). Differences in collagen metabolism between normal and osteoarthritic human articular cartilage. *Science* **181**, 751–752.

Nir, I., Shani, J., Locker, D., and Sulman, F. G. (1972). Effect of light and pinealectomy on body weight and tibia cartilage of female rats. *Life Sci.* **11**, 41–49.

Nisbet, N. W., Menage, J., and Loutit, J. F. (1982). Osteogenesis in osteopetrotic mice. *Calcif. Tissue Int.* **34**, 37–42.

Nisbet, N. W., Waldron, S. F., and Marshall, M. J. (1983). Failure of thymic grafts to stimulate resorption of bone in the Fatty/ORL-*op* rat. *Calcif. Tissue Int.* **35**, 122–125.

Nishida, K. (1990). Phylogeny of the suborder Myliobatidoidei. *Mem. Fac. Fish. Hokkaido Univ.* **37**, 1–108.

Nishimatsu, S.-i., Susuki, A., Shoda, A., Murakami, K., and Ueno, N. (1992). Genes for bone morphogenetic proteins are differentially transcribed in early amphibian embryos. *Biochem. Biophys. Res. Commun.* **186**, 1487–1495.

Nishimatsu, S.-i., Takebayashi, K., Susuki, A., Murakami, K., and Ueno, N. (1993). Immunodetection of *Xenopus* morphogenetic protein-4 in early embryos. *Growth Factors* **8**, 173–176.

Nishimoto, S. K., Chang, C-H., Gendler, E., Stryker, W. F., and Nimni, M. E. (1985). The effect of aging on bone formation in rats: Biochemical and histological evidence for decreased bone formation capacity. *Calcif. Tissue Int.* **37**, 617–624.

Nishiyama, A., Dahlin, K. J., and Stallcup, W. B. (1991). The expression of NG₂ proteoglycan in the developing rat limb. *Development* **111**, 933–944.

Niswander, L. (1997). Limb mutants: What can they tell us about normal limb development? *Curr. Opin. Genet. Devel.* **7**, 530–536.

Niswander, L. (1999). Legs to wings and back again. *Nature* **398**, 751–752.

Niswander, L., Jeffrey, S., Martin, G. R., and Tickle, C. (1994). A positive feedback loop coordinates growth and patterning in the vertebrate limb. *Nature* **371**, 609–612.

Niswander, L., and Martin, G. R. (1992). FGF-4 expression during gastrulation, myogenesis, limb and tooth development in the mouse. *Development* **114**, 755–768.

Niswander, L., and Martin, G. R. (1993a). FGF-4 and BMP-2 have opposite effects on limb growth. *Nature* **361**, 68–71.

Niswander, L., and Martin, G. R. (1993b). FGF-4 regulates expression of *Evx-1* in the developing mouse limb. *Development* **119**, 287–294.

Niven, J. S. F. (1933). The development *in vivo* and *in vitro* of the avian patella. *Wilhelm Roux Arch. Entwicklungsmech. Org.* **128**, 480–501.

Nixon, J. E. (1983). Avascular necrosis of bone: A review. *J. Roy. Soc. Med.* **76**, 681–692.

Noda, M. (1989). Transcriptional regulation of osteocalcin production by transforming growth factor-β in rat osteoblast-like cells. *Endocrinology* **124**, 612–617.

Noda, M., and Camilliere, J. J. (1989). *In vivo* stimulation of bone formation by transforming growth factor-β. *Endocrinology* **124**, 2991–2994.

Noda, M., and Rodan, G. A. (1989a). Type beta transforming growth factor regulates expression of genes encoding bone matrix proteins. *Conn. Tissue Res.* **21**, 71–75.

Noda, M., and Rodan, G. A. (1989b). Type β transforming growth factor regulates expression of genes encoding bone matrix proteins. *Conn. Tissue Res.* **21**, 402–408.

Noda, M., and Sato, A. (1985a). Calcification of cartilaginous matrix in culture by constant direct-current stimulation. *Clin. Orthop. Rel. Res.* **193**, 281–287.

Noda, M., and Sato, A. (1985b). Appearance of osteoclasts and osteoblasts in electrically stimulated bones cultured on chorioallantoic membranes. *Clin. Orthop. Rel. Res.* **193**, 288–298.

Noda, M., and Vogel, R. (1989). Fibroblast growth factor enhances type β1 transforming growth factor gene expression in osteoblast-like cells. *J. Cell Biol.* **109**, 2529–2535.

Nodder, S., and Martin, P. (1997). Wound healing in embryos: a review. *Anat. Embryol.* **195**, 215–228.

Noden, D. M. (1975). An analysis of the migratory behavior of avian cephalic neural crest cells. *Devel. Biol.* **42**, 106–130.

Noden, D. M. (1978a). The control of avian cephalic neural crest cytodifferentiation. I. Skeletal and connective tissues. *Devel. Biol.* **67**, 296–312.

Noden, D. M. (1978b). Interactions directing the migration and cytodifferentiation of avian neural crest cells. In *The Specificity of Embryological Interactions* (D. Garrod, ed.), pp. 3–50. Chapman & Hall, London.

Noden, D. M. (1983). The role of the neural crest in patterning of avian cranial skeletal, connective and muscle tissues. *Devel. Biol.* **96**, 144–165.

Noden, D. M. (1984a). The use of chimeras in analysis of craniofacial development. In *Chimeras in Developmental Biology* (N. M. Le Douarin and A. McLaren, eds), pp. 241–280. Academic Press, London.

Noden, D. M. (1984b). Craniofacial development: New views on old problems. *Anat. Rec.* **208**, 1–13.

Noden, D. M. (1986a). Origins and patterning of craniofacial mesenchymal tissues. *J. Craniofac. Genet. Devel. Biol. Suppl.* **2**, 15–32.

Noden, D. M. (1986b). Patterning of avian craniofacial muscles. *Devel. Biol.* **116**, 347–356.

Noden, D. M. (1987). Interactions between cephalic neural crest and mesodermal populations. In *Developmental and Evolutionary Aspects of the Neural Crest* (P. F. A. Maderson, ed.), pp. 89–119. John Wiley & Sons, New York, NY.

Noden, D. M. (1991a). Vertebrate craniofacial development: The relation between ontogenetic process and morphological outcome. *Brain, Behav. Evol.* **38**, 190–225.

Noden, D. M. (1991b). Cell movements and control of patterned tissue assembly during craniofacial development. *J. Craniofac. Genet. Devel. Biol.* **11**, 192–213.

Noden, D. M. (1992). Vertebrate craniofacial development: Novel approaches and new dilemmas. *Curr. Opin. Genet. Devel.* **2**, 576–581.

Noel, J. F. (1973). The control of growth in transplanted mammalian cartilage. *J. Embryol. Exp. Morphol.* **29**, 53–64.

Noel, J. F., and Wright, E. A. (1972). The growth of transplanted mouse vertebrae; Effects of transplantation under the renal capsule, and the relationship between the rate of growth of the transplant and the age of the host. *J. Embryol. Exp. Morphol.* **28**, 633–645.

Noff, D., Pitaru, S., and Savion, N. (1989). Basic fibroblast growth factor enhances the capacity of bone marrow cells to form bone-like nodules *in vitro*. *FEBS Letters* **250**, 619–621.

Nogami, H., Aoki, H., Okagawa, T., and Mimatsu, K. (1982). Effects of electric current on chondrogenesis *in vitro*. *Clin. Orthop. Rel. Res.* **163**, 243–247.

Nogami, H., Oohira, A., Murotanagi, M., and Mizutani, A. (1986). Congenital bowing of limb bones: Clinical and experimental studies. *Teratology* **33**, 1–7.

Noji, S., Yamaai, T., Koyama, E., Nohno, T., and Taniguchi, S. (1989). Spatial and temporal expression pattern of retinoic acid receptor genes during mouse bone development. *FEBS Letters* **257**, 93–96.

Nomura, M., and Li, E. (1998). Smad2 role in mesoderm formation, left-right patterning and craniofacial development. *Nature* **393**, 786–790.

Nomura, S., Wills, A. J., Edwards, D. R., Heath, J. K., and Hogan, B. L. M. (1988). Developmental expression of 2ar (osteopontin) and SPARC (osteonectin) RNA as revealed by *in situ* hybridization. *J. Cell Biol.* **106**, 441–450.

Nonaka, K., Sasaki, Y., Yanagita, K.-i., Matsaumoto, T., Watanabe, Y., and Nakata, M. (1993). II. Intrauterine effect of dam on prenatal development of craniofacial complex of mouse embryo. *J. Craniofac. Genet. Devel. Biol.* **13**, 206–212.

Nonaka, K., Shum, L., Takahashi, I., Takahashi, K., Ikura, T., Dashner, R., Nuckolls, G. H., and Alavkin, H. C. (1999). Convergence of the BMP and EGF-signaling pathways on Smad1 in the regulation of chondrogenesis. *Int. J. Devel. Biol.* **43**, 795–807.

Nopsca, F. Baron (1923). Reversible and irreversible evolution; a study based on reptiles. *Proc. Zool. Soc. London* **68**, 1045–1059.

Norby, D. P., Malemud, C. J., and Sokoloff, L. (1977). Differences in the collagen types synthesized by lapine articular chondrocytes in spinner and monolayer culture. *Arthritis Rheum.* **20**, 709–716.

Nordeide, J. T., Holm, J. C., Ottera, H., Blom, G., and Borge, A. (1992). The use of oxytetracycline as a marker for juvenile cod (*Gadus morhua* L). *J. Fish. Biol.* **41**, 21–30.

Norell, M. A., Makovicky, P., and Clark, J. M. (1997). A *Velociraptor* wishbone. *Nature* **389**, 447.

Nori, M., Descalzi-Cancedda, F., and Cancedda, R. (1992). Heat-shock response in cultured chick embryo chondrocytes. Osteonectin is a secreted heat-shock protein. *Eur. J. Biochem.* **205**, 569–574.

Norman, J. R. (1926). The development of the chondrocranium of the eel (*Anguis vulgaris*), with observations on the comparative morphology and development of the chondrocranium in bony fishes. *Phil. Trans. R. Soc. London* **214**, 369–464.

Northcutt, R. G. (1996). The origin of craniates – neural crest, neurogenic placodes, and homeobox genes. *Israel J. Zool.* **42**, S273–S313.

Northcutt, R. G., and Gans, C. (1983). The genesis of neural crest and epidermal placodes: A reinterpretation of vertebrate origins. *Q. Rev. Biol.* **58**, 1–28.

Norton, L. A., Bourret, L. A., and Rodan, G. A. (1977). Electric field enhancement of thymidine incorporation in chondrocytes. *J. Dent. Res.* **56**, A105.

Norton, L. A., and Moore, R. R. (1972). Bone growth in organ culture modified by an electric field. *J. Dent. Res.* **51**, 1492–1499.

Norton, L. A., Rodan, G. A., and Bourret, L. A. (1976). Cyclic AMP fluctuation in bones grown in electric fields. *J. Dent. Res.* **55**, B215.

Norton, L. A., Witt, D. W., and Rovetti, L. A. (1988). Pulsed electromagnetic fields alter phenotypic expression in chondroblasts in tissue culture. *J. Orthop. Res.* **6**, 685–689.

Nottoli, T., Hagopian-Donaldson, S., Zhang, J., Perkins, A., and Williams, T. (1998). AP-2-null cells disrupt morphogenesis of the eye, face, and limbs in chimeric mice. *Proc. Natl Acad. Sci. U.S.A.* **95**, 13714–13719.

Novak, V. (1964). Vascularization of cartilage. *Plzen. Lek. Sb., Suppl.* **12**, 133–213.

Nowicki, J. L., and Burke, A. (2000). *Hox* genes and morphological identity: axial versus lateral patterning in the vertebrate mesoderm. *Development* **127**, 4265–4275.

Nozwaw-Inoue, K., Ajima, H., Takagi, R., and Maeda, T. (1999). Immunocytochemical demonstration of laminin in the synovial lining layer of the rat temporomandibular joint. *Archs Oral Biol.* **44**, 531–534.

Nulend, J. K., Veldhuijzen, J. P., and Burger, E. H. (1985). Calcification of growth plate cartilage *in vitro*: Effects of intermittent versus continuous compressive force. *Bone* **6**, 479–480.

Nunamaker, D. M. (1998). Experimental models of fracture repair. *Clin. Orthop. Rel. Res.* **355**, S56–S65.

Nunn, C. L., and Smith, K. K. (1998). Statistical analyses of developmental sequences – the craniofacial region in marsupials and placental mammals. *Amer. Nat.* **152**, 82–101.

Nutik, G., and Creuss, R. L. (1974). Estrogen receptors in bone. An evaluation of the uptake of estrogen into bone cells. *Proc. Soc. Exp. Biol. Med.* **146**, 265–268.

Nyberg, L. M., and Marks, S. C., Jr. (1975). Organ culture of osteopetrotic (ia) rat bone; Evidence that the defect is cellular. *Amer. J. Anat.* **144**, 373–378.

Nyhart, L. K. (1995). *Biology Takes Form. Animal Morphology and the German Universities, 1800–1900.* The University of Chicago Press, Chicago, IL.

O

Oakes, B. W., Handley, C. J., Lisner, F., and Lowther, D. A. (1977). An ultrastructural and biochemical study of high density primary cultures of embryonic chick chondrocytes. *J. Embryol. Exp. Morphol.* **38**, 239–263.

Oberbauer, A. M., and Peng, R. (1995). Growth hormone and IGF-1 stimulate cell function in distinct zones of the rat epiphyseal growth plate. *Conn. Tissue Res.* **31**, 189–195.

Oberlender, S. A., and Tuan, R. S. (1994). Expression and functional involvement on n-cadherin in embryonic limb chondrogenesis. *Development* **120**, 177–187.

Oberpriller, J. (1967). A radioautographic analysis of the potency of blastemal cells in the adult newt, *Diemictylus viridescens*. *Growth* **31**, 251–296.

Öbrink, B., and Wasterson, A. (1971). Nature of the interaction of chondroitin 4-sulphate and chondroitin sulphate-proteoglycan with collagen. *Biochem. J.* **121**, 227–234.

O'Connell, J. J., and Low, F. N. (1970). A histochemical and fine structural study of early extracellular connective tissue in the chick embryo. *Anat. Rec.* **167**, 425–438.

O'Dell, N. L., Sharawy, M., Pennington, C. B., and Marlow, R. K. (1989). Distribution of putative elastic fibers in rabbit temporomandibular joint tissues. *Acta Anat.* **135**, 239–244.

Odense, P. H., and Logan, V. H. (1974). Marking Atlantic salmon (*Salmo salar*) with oxytetracycline. *J. Fish. Res. Bd. Canada* **31**, 348–350.

O'Driscoll, S. W. (1999) Articular cartilage regeneration using periosteum. *Clin. Orthop. Rel. Res.* **367**, 186–203.

O'Driscoll, S. W., Fitzsimmons, J. S., and Commisso, C. N. (1997). Role of oxygen tension during cartilage formation by periosteum. *J. Orthop. Res.* **15**, 682–687.

O'Driscoll, S. W., and Fitzsimmons, J. S. (2001). The role of periosteum in cartilage repair. *Clin. Orthop. Rel. Res.* **391S**, S190–S207.

O'Driscoll, S. W., Keeley, F. W., and Salter, R. B. (1986). The chondrogenic potential of free autogenous periosteal grafts for biological resurfacing of major full-thickness defects in joint surfaces under the influence of continuous passive motion. *J. Bone Joint Surg. Am. Vol.* **68**, 1017–1035.

O'Driscoll, S. W., and Salter, R. B. (1984). The induction of neochondrogenesis in free intra-articular periosteal autografts under the influence of continuous passive motion. *J. Bone Joint Surg. Am. Vol.* **66**, 1248–1257.

O'Driscoll, S. W., and Salter, R. B. (1986). The repair of major osteochondral defects in joint surfaces by neochondrogenesis with autografts osteoperiosteal grafts stimulated by continuous passive motion – an experimental investigation in the rabbit. *Clin. Orthop. Rel. Res.* **208**, 131–140.

Oettinger, H. F., and Pacifici, M. (1990) Type-X collagen gene expression is transiently up-regulated by retinoic acid treatment in chick chondrocyte cultures. *Exp. Cell Res.* **191**, 292–298.

Officer, R. A., Clement, J. G., and Rowler, D. K. (1995). Vertebral deformities in a school shark, *Galeorhinus galeus* – circumstantial evidence for endoskeletal resorption. *J. Fish. Biol.* **46**, 85–98.

O'Gara, B. W., and Matson, G. (1975). Growth and casting of horns by pronghorns and exfoliation of horns by bovids. *J. Mammal.* **56**, 829–846.

O'Gara, B. W., Moy, R. F., and Bear, G. D. (1971). The annual testicular cycle of horn casting in the pronghorn(*Antilocapra americana*). *J. Mammal.* **52**, 537–544.

Ogasawara, M., Shigetani, Y., Hirano, S., Satoh, N., and Kuratani, S. (2000). *Pax1/Pax9*-related genes in an agnathan vertebrate, *Lampetra japonica*: expression pattern of *LjPax9* implies sequential evolutionary events toward the gnathostome body plan. *Devel. Biol.* **223**, 399–410.

Ogata, T., Wozney, J. M., Benezra, R., and Noda, M. (1993). Bone morphogenetic protein 2 transiently enhances expression of a gene, Id (inhibitor of differentiation) encoding a helix-loop-helix molecule in osteoblast-like cells. *Proc. Natl Acad. Sci. U.S.A.* **90**, 9219–9222.

Ogata, T., Wozney, J. M., Rodan, G. A., and Noda, M. (1994). Bone morphogenetic protein-2 (BMP-2) acts both synergistically with and antagonistically against retinoic acid in regulating expression of phenotypic genes in osteoblast-like cells. *Endocrinol. J.* **2**, 237–240.

Ogawa, T., Shimokawa, H., Fukada, K., Suzuki, S., Shibata, S., Ohya, K., and Kuroda, T. (2003). Localization and inhibitory effect of basic fibroblast growth factor on chondrogenesis in cultures mouse mandibular condyle. *J. Bone Min. Metab.* **21**, 145–153.

Ogawa, Y., Schmidt, D. K., Nathan, R. M., Armstrong, R. M., Miller, K. L., Sawamura, S. J., Ziman, J. M., Erickson, K. L., De Leon, E. R., Rosen, D. M., Seyedin, S. M., Glaser, C. B., Chang, R.-J., Corrigan, A. Z., and Vale, W. (1992). Bovine bone activin enhances bone morphogenetic protein-induced ectopic bone formation. *J. Biol. Chem.* **267**, 14233–14237.

Ogden, J. A., Hempton, R. F., and Southwick, W. O. (1975). Development of the tibial tuberosity. *Anat. Rec.* **182**, 431–446.

Ogiso, B., Hughes, F. J., Melcher, A. H., and McCulloch, C. A. G. (1991). Fibroblasts inhibit mineralized bone nodule formation by rat bone marrow stromal cells *in vitro*. *J. Cell Physiol.* **146**, 442–450.

Oh, S. P., Griffith, C. M., Hay, E. D., and Olsen, B. R. (1993). Tissue-specific expression of type XII collagen during mouse embryonic development. *Devel. Dynam.* **196**, 37–46.

O'Hare, M. J. (1972a). Differentiation of chick embryo somites in chorioallantoic culture. *J. Embryol. Exp. Morphol.* **27**, 215–228.

O'Hare, M. J. (1972b). Chondrogenesis in chick embryo somites grafted with adjacent and heterologous tissues. *J. Embryol. Exp. Morphol.* **27**, 229–234.

O'Hare, M. J. (1972c). Aspects of spinal cord induction of chondrogenesis in chick embryo somites. *J. Embryol. Exp. Morphol.* **27**, 235–243.

O'Hare, M. J. (1978). Teratomas, neoplasia and differentiation: A biological overview. I. The natural history of teratomas. *Invest. Cell Pathol.* **1**, 39–64.

O'Higgins, P., Johnson, D. R., and McAndrew, T. J. (1986). The clonal model of vertebral column development: A reinvestigation of vertebral shape using Fourier analysis. *J. Embryol. Exp. Morphol.* **96**,171–182.

O'Higgins, P., Milne, N., Johnson, D. R., Runnion, C. K., and Oxnard, C. E. (1997). Adaptation in the vertebral column: a comparative study of patterns of metameric variation in mice and men. *J. Anat.* **190**, 105–113

Ohsugi, K., Gardiner, D. M., and Bryant, S. V. (1997). Cell cycle length affects gene expression and pattern formation in limbs. *Devel. Biol.* **189**, 13–21.

Ohta, S., Yamamuro, T., Lee, K., Okumura, H., Kasai, R., Hiraki, Y., Ikeda, T., Iwasaki, R., Kikuchi, H., Konishi. J., and Shigeno, C. (1991). Fracture healing induces expression of the proto-oncogene *c-fos in vivo*. Possible involvement of the *fos* protein in osteoblastic differentiation. *FEBS Lett.* **284**, 42–45.

Ohta, Y., and Matsunaga, H. (1974). Bone lesions in divers. *J. Bone Joint Surg. Br. Vol.* **56**, 3–16.

Ohtsuki, Y., Danbara, Y., Takeda, I., Takahashi, K., Hayashi, K., Sonobe, H., Yoshino, T., and Akagi, T. (1987). Metaplastic bone formation in a hyperplastic polyp of the stomach: A case report. *Acta Med. Okayama* **41**, 43–46.

Ohyama, K., Chung, C.-H., Chen, E., Gibson, C. W., Misof, K., Fratzl, P., and Shapiro, I. M. (1997). p53 influences mice skeletal development. *J. Craniofac. Genet. Devel. Biol.* **17**, 161–171.

Okada, E., and Ichikawa, M. (1956). Isolationsversuche zur Analyse der Knorpelbildung aus Neuralleistenzellen beim Anurenkeim. *Mem. Coll. Sci. Kyoto Ser. B* **23**, 27–36.

Okada, N., Takagi, Y., Eeikai, T., Tanaka, M., and Tagawa, M. (2001). Asymmetrical development of bones and soft tissues during eye migration in metamorphosing Japanese flounder, *Paralichthys olivaceus*. *Cell Tissue Res.* **304**, 59–66.

Okada, N., Takagi, Y., Tanaka, M., and Tagawa, M. (2003). Fine structure of soft and hard tissues involved in eye migration in metamorphosing Japanese flounder (*Paralichthys olivaceus*). *Anat. Rec. Part A* **273A**, 663–668.

Okayama, M., Pacifici, M., and Holtzer, H. (1976). Differences among sulfated proteoglycans synthesized in nonchondrogenic cells, presumptive chondroblasts, and chondroblasts. *Proc. Natl Acad. Sci. U.S.A.* **73**, 3224–3228.

Okihano, H., and Shimomura, Y. (1992). Osteogenic activity of growth cartilage examined by implanting decalcified or devitalized ribs and costal cartilage zones, and living growth cartilage cells. *Bone* **13**, 387–393.

Okihano, H., and Shimomura, Y. (1993). A monoclonal antibody distinguishes growth cartilage from other types of cartilage: a new probe for osteogenic cartilage. *Histochem. J.* **25**, 166–171.

O'Leary, M., and Geisler, J. H. (1999). The position of Cetacea within mammalia: phylogenetic analysis of morphological data from extinct and extant taxa. *Syst. Biol.* **48**, 455–490.

Oliver, J. A. (1951). Ontogenetic changes in osteodermal ornamentation in skinks. *Copeia* **1951**, 127–130.

Olivera-Martinez, I., Coltey, M., Dhouailly, D., and Pourquié, O. (2000). Mediolateral somitic origin of ribs and dermis determined by quail-chick chimeras. *Development* **127**, 4611–4617.

Ollier, L. (1867). *Traité Expérimental et Clinique de la Régénération des Os et de la Production Artificielle du Tissue Osseux*. 2 Volumes. Victor Mason & Fils, Paris.

Olsen, B. R. (1996). Collagen it takes and bone it makes. *Curr. Biol.* **6**, 645–647.

Olsen, B. R., Reginato, A. M., and Wang, W. F. (2000). Bone development. *Annu. Rev. Cell Biol. Devel.* **16**, 191–220.

Olsen, C. L., and Tassava, R. A. (1984). Cell cycle and histological effects of reinnervation in denervated forelimb stumps of larval *Ambystome*. *J. Exp. Zool.* **229**, 247–258.

Olson, M. D., and Low, F. N. (1971). The fine structure of developing cartilage in the chick embryo. *Amer. J. Anat.* **131**, 197–216.

Olson, W. M. (2000). Phylogeny, ontogeny, and function: extraskeletal bones in the tendons and joints of *Hymenochirus boettgeri* (Amphibia: Anura: Pipidae). *Zoology* **103**, 15–24.

Olson, W. M., and Hall, B. K. (2000). Heart morphogenesis and neural crest cell migration in cardia bifida chick embryos. *Amer. Zool.* **40**, 1158.

Olsson, L., Falck, P., Lopez, K., Cobb, J., and Hanken, J. (2001). Cranial neural crest cells contribute to connective tissue in cranial muscles in the anuran amphibian, *Bombina orientalis*. *Devel. Biol.* **237**, 354–367.

Olsson, L., and Hanken, J. (1996). Cranial neural-crest migration and chondrogenic fate in the Oriental Fire-Bellied toad, *Bombina orientalis*: Defining the ancestral pattern of head development in anuran amphibians. *J. Morphol.* **229**, 105–120.

Olsson, L., and Jacobson, C.-O. (eds) (2000) *Regulatory Processes in Development*, Wenner-Gren International Series Volume 76. The Portland Press, London.

Olsson, L., Moury, J. D., Carl, T. F., Hastad, O., and Hanken, J. (2001). Cranial neural crest-cell migration in the direct-developing frog, *Eleutherodactylous coqui*; molecular heterogeneity within and among migratory streams. *Zoology* **105**, 3–13.

Omi, M., Sato-Maeda, M., and Ide, H. (2000). Role of chondrogenic tissue in programmed cell death and *BMP* expression in chick limb buds. *Int. J. Devel. Biol.* **44**, 381–388.

Omura, H. (1980). Morphological study of pelvic bones of the minke whale from the Antarctic. *Sci. Rep. Whales Res. Inst.* **32**, 25–37.

Onda, H., Goldhamer, D. J., and Tassava, R. A. (1990). An extracellular matrix molecule of newt and axolotl regenerating limb blastemas and embryonic limb buds: immunological relationships of MT1 antigen with tenascin. *Development* **108**, 657–668.

Oohira, A., Kimata, K., Susuki, S., Takata, K., and Susuki, I. (1974). A correlation between synthetic activities for matrix macromolecules and specific stages of cytodifferentiation in developing cartilage. *J. Biol. Chem.* **249**, 1637–1645.

Oosterwegel, M., Van de Wetering, M., Timmerman, J., Kruisbeek, A., Destree, O., Meijilink, F., and Clever, H. (1993). Differential expression of the HMG box factors *TCF-1* and *LEF-1* during murine embryogenesis. *Development* **118**, 439–448.

Opitz, J. M., Gorlin, R. J., Reynolds, J. F., and Spano, L. M. (eds) (1988). *Neural Crest and Craniofacial Disorders. Genetic Aspects*. Alan R. Liss, New York, NY.

Oppenheim, W. L., Davis, A., Growdon, W. A., Dorey, F. J., and Davlin, L. B. (1990). Clavicle fractures of the newborn. *Clin. Orthop. Rel. Res.* **250**, 176–180.

Oppenheimer, J. M. (1974). Asymmetry revisited. *Amer. Zool.* **14**, 867–879.

Opperman, L. A. (2000). Cranial sutures as intramembranous bone growth sites. *Devel. Dynam.* **219**, 472–485.

Opperman, L. A., Adab, K., and Gakunga, P. T. (2000). Transforming growth factor-β2 and TGFβ-3 regulate fetal rat cranial suture morphogenesis by regulating rates of cell proliferatyion and apoptosis. *Devel. Dynam.* **219**, 237–247.

Opperman, L. A., Chhabra, A., Cho, R. W., and Ogle, R. C. (1999). Cranial suture obliteration is induced by removal of transforming growth factor (TGF)-β3 activity and prevented by removal of TGF-β2 activity from fetal rat calvaria *in vitro*. *J. Craniofac. Genet. Devel. Biol.* **18**, 164–173.

Opperman, L. A., Chhabra, A., Nolen, A. A., Bao, Y., and Ogle, R. C. (1998). Dura mater maintains rat cranial sutures *in vitro* by regulating suture cell proliferation and collagen production. *J. Craniofac. Genet. Devel. Biol.* **18**, 150–158.

Opperman, L. A., Nolen, A. A., and Ogle, R. C. (1997). TGF-β1, TGF-β2, and TGF-β3 exhibit distinct patterns of expression during cranial suture formation and obliteration *in vivo* and *in vitro*. *J. Bone Min. Res.* **12**, 301–310.

Opperman, L. A., Passarelli, R. W., Morgan, E. P., Reintjes, M., and Ogle, R. C. (1995). Cranial sutures require tissue interactions with dura mater to resist osseous obliteration *in vitro*. *J. Bone Min. Res.* **10**, 1978–1987.

Opperman, L. A., Passarelli, R. W., Nolen, A. A., Gampper, T. J., Lin, K. Y. K., and Ogle, R. C. (1996). Dura mater secretes soluble heparin-binding factors required for cranial suture morphogenesis. *In Vitro Cell Devel. Biol.* **32**, 627–632.

Opperman, L. A., Persing, J. A., Sheen, R., and Ogle, R. C. (1994). In the absence of periosteum, transplanted fetal and neonatal rat coronal sutures resist osseous obliteration. *J. Craniofac. Surg.* **5**, 327–332.

Opperman, L. A., Sweeney, T. M., Redman, J., Persing, J. A., and Ogle, R. C. (1993). Tissue interactions with underlying dura mater inhibit osseous obliteration of developing cranial sutures. *Devel. Dynam.* **198**, 312–322.

O'Rahilly, R. (1962). The development of the sclera and the choroid in staged chick embryos. *Acta Anat.* **48**, 335–346.

O'Rahilly, R., and Gardener, E. (1956). The development of the knee joint in the chick and its correlation with embryonic staging. *J. Morphol.* **98**, 49–88.

O'Rahilly, R., and Gardener, E. (1972). The initial appearance of ossification in stages human embryos. *Amer. J. Anat.* **134**, 291–308.

O'Rahilly, R., Gardner, E., and Gray, D. J. (1956). The ectodermal thickening and ridge in the limbs of staged human embryos. *J. Embryol. Exp. Morphol.* **4**, 254–264.

O'Rahilly, R., and Meyer, D. B. (1979). The timing and sequence of events in the development of the human vertebral column during the embryonic period proper. *Anat. Embryol.* **157**, 167–176.

O'Rahilly, R., and Müller, F. (2003). Somites, spinal ganglia, and centra. *Cells Tissues Organs* **173**, 75–92.

Ordahl, C. P., and Caplan, A. I. (1976). Transcriptional diversity in myogenesis. *Devel. Biol.* **54**, 61–72.

Orestes-Cardoso, S. M., Nefussi, J. R., Hotton, D., Mesbah, M., Do Socorro Orestes-Cardosa, M., Robert, B., and Berdal, A. (2001). Postnatal MS1 expression pattern in craniofacial, axial, and appendicular skeleton of transgenic mice from the first week until the second year. *Devel. Dynam.* **221**, 1–13.

O'Riain, M. J., Jarvis, J. U. M., Alexander, R., Buffenstein, R., and Peeters, C. (2000). Morphological castes in a vertebrate. *Proc. Natl Acad. Sci. U.S.A.* **97**, 13194–13197.

Orkin, R. W., Pollard, T. D., and Hay, E. D. (1973). SDS gel analysis of muscle proteins in embryonic cells. *Devel. Biol.* **35**, 388–394.

Orkin, R. W., Pratt, R. M., and Martin, G. R. (1976). Undersulfated chondroitin sulfate in the cartilage matrix of brachymorphic mice. *Devel. Biol.* **50**, 82–94.

Orkin, R. W., Williams, B. R., Cranley, R. E., Poppke, D. C., and Brown, K. S. (1977). Defects in the cartilaginous growth plates of brachymorphic mice. *J. Cell Biol.* **73**, 287–299.

Ornitz, D. M. (2001). Regulation of chondrocyte growth and differentiation by fibroblast growth factor receptor 3. In *The Molecular Basis of Skeletogenesis* (G. Cardew and J. Goode, eds; B. K. Hall, chair), Novartis Foundation Symposium 232, pp. 63–80. John Wiley & Sons, Chichester.

Ornitz, D. M., and Marie, P. J. (2002). FGF signaling pathways in endochondral and intramambranous bone development and human genetic disease. *Genes & Devel.* **16**, 1446–1465.

Ornoy, A., Adomian, G. E., Eteson, D. J., Burgeson, R. E., and Rimoin, D. L. (1985). A role of mesenchyme-like tissue in the pathogenesis of thanatophoric dysplasia. *Amer. J. Med. Genet.* **21**, 613–630.

Ornoy, A., and Zusman, I. (1983). Vitamins and Cartilage. In *Cartilage, Volume 2, Development, Differentiation and Growth* (B. K. Hall, ed.), pp. 297–326. Academic Press, New York, NY.

Orsulic, S., and Peifer, M. (1996). Cell-cell signalling: wingless lands at last. *Curr. Biol.* **6**, 1363–1367.

Ortega, H. H., Munoz-de-Toro, M., Luque, E. H., and Montes, G. S. (2003). Morphological characteristics of the interpubic joint (*Symphysis pubica*) of rats, guinea pigs and mice in different physiological situations. *Cells Tissue Organs* **173**, 105–114.

Ortner, D. J., and Putschar, W. G. J. (1982). *Identification of Pathological Conditions in Human Skeletal Remains.* Smithsonian Contributions to Anthropology #28, 479 pp. Smithsonian Institution Press/US Government Printing Office, Washington DC.

Ørvig, T. (1951). Histologic studies of Placoderms and fossil Elasmobranchs. I. The endoskeleton with remarks on the hard tissues of lower vertebrates in general. *Ark. Zool.* **2**, 323–456.

Ørvig, T. (1957). Palaeohistological notes. I. On the structure of the bone tissue in the scales of certain Palaeonisciformes. *Ark. Zool.* **10**, 481–490.

Ørvig, T. (1965). Palaeohistological notes. 2. Certain comments on the phyletic significance of acellular bone tissue in early vertebrates. *Ark. Zool.* **16**, 551–556.

Ørvig, T. (1967). Phylogeny of tooth tissues: Evolution of some calcified tissues in early vertebrates. In *Structural and Chemical Organization of Teeth* (A. E. W. Miles, ed.), Volume 1, pp. 45–110. Academic Press, New York, NY.

Ørvig, T. (1968). *Current Problems of Lower Vertebrate Phylogeny.* Nobel Symp. No. 4. Almqvist & Wiksell, Stockholm.

Ørvig, T. (1977). A survey of odontodes ('dermal teeth') from developmental, structural, functional, and phyletic points of view. In *Problems in Vertebrate Evolution* (S. M. Andrews, R. S. Miles, and A. D. Walker, eds), pp. 53–75. Academic Press, London.

Ørvig, T. (1989). Histologic studies of ostracoderms, placoderms and fossil elasmobranchs. 6. Hard tissue of Ordovician vertebrates. *Zool. Scripta* **18**, 427–446.

Orzel, J. A., and Rudd, T. G. (1985). Heterotopic bone formation: Clinical, laboratory, and imaging correlations. *J. Nucl. Med.* **26**, 125–132.

Osborn, H. F. (1929). The Titanotheres of ancient Wyoming, Dakota and Nebraska. US *Geological Survey Monograph* **55**, two volumes.

Osborn, J. W. (ed.) (1981). *Dental Anatomy and Embryology.* Blackwell Scientific Publications, Oxford.

Osborn, J. W. (1984). From reptile to mammal: Evolutionary considerations of the dentition with emphasis on tooth attachment. *Symp. Zool. Soc. Lond.* #52, 549–574.

Osborn, J. W., and Price, D. G. (1988). An autoradiographic study of periodontal development in the mouse. *J. Dent. Res.* **67**, 455–461.

Osdoby, P., and Caplan, A. I. (1976). The possible differentiation of osteogenic elements *in vitro* from chick limb mesodermal cells. *Devel. Biol.* **52**, 283–299.

Osdoby, P., and Caplan, A. I. (1980). A scanning electron microscopic investigation of *in vitro* osteogenesis. *Calcif. Tissue Int.* **30**, 43–50.

Osdoby, P., and Caplan, A. I. (1981). Characterization of a bone-specific alkaline phosphatase in chick limb mesenchymal cell cultures. *Devel. Biol.* **86**, 136–146.

Osdoby, P., Martini, M. C., and Caplan, A. I. (1982). Isolated osteoclasts and their presumed progenitor cells, the monocyte, in culture. *J. Exp. Zool.* **224**, 331–344.

Osdoby, P., Oursler, M. J., and Anderson, F. (1986). The osteoblast and osteoclast cytodifferentiation. In *Progress in Developmental Biology, Part B* (H. C. Slavkin, ed.), pp. 409–414. Alan R. Liss Inc., New York, NY.

Oshima, O., Leboy, P. S., McDonald, S. A., Tuan, R. S., and Shapiro, I. M. (1989). Developmental expression of genes in chick growth cartilage detected by *in situ* hybridization. *Calcif. Tissue Int.* **45**, 182–192.

Osman, A., and Ruch, J. V. (1975). Topographical distribution of mitosis in odontogenic fields of the lower jaw in mice embryos. *J. Biol. Buccale* **3**, 117–132.

Ossenberg, N. S. (1974). The myohyoid bridge: An anomalous derivative of Meckel's cartilage. *J. Dent. Res.* **53**, 77–82.

Oster, G. F., and Murray, J. D. (1989). Pattern formation models and developmental constraints. *J. Exp. Zool.* **251**, 186–202.

Oster, G. F., Murray, J. D., and Harris, A. K. (1983). Mechanical aspects of mesenchymal morphogenesis. *J. Embryol. Exp. Morphol.* **78**, 83–125.

Oster, G. F., Murray, J. D., and Maini, P. K. (1985). A model for chondrogenic condensations in the developing limb: The role of extracellular matrix and cell tractions. *J. Embryol. Exp. Morphol.* **89**, 93–112.

Oster, G. F., Shubin, N. H., Murray, J. D., and Alberch, P. (1988). Evolution and morphogenetic rules: The shape of the vertebrate limb in ontogeny and phylogeny. *Evolution* **42**, 862–884.

Ostrowski, K., and Wlodarski, K. H. (1971). Induction of heterotopic bone formation. In *The Biochemistry and Physiology of Bone* (G. H. Bourne, ed.), 2nd Edn, Volume 3, pp. 299–337. Academic Press, New York, NY.

Osumi-Yamashita, N., and Eto, K. (1990). Review: Mammalian cranial neural crest cells and facial development. *Devel. Growth Differ.* **32**, 451–460.

Osumi-Yamashita, N., Ninomiya, Y., and Eto, K. (1997). Mammalian craniofacial embryology *in vitro.* *Int. J. Devel. Biol.* **41**, 187–194.

Otawara, Y., Hosoya, N., Kasai, H., Okuyama, N., and Moriuchi, S. (1980). Purification and characterization of calcium-binding protein containing γ-carboxyglutamic acid from rat bone. *J. Nutrit. Sci. Vitaminol.* **26**, 209–220.

Otto, F., Thornell, A. P., Crompton, T., Denzel, A., Gilmour, K. C., Rosewell, I. R., Stamp, G. W. H., Beddington, R. S. P., Mundlos, S., and Olsen, B. R. (1997). *Cbfa1*, a candidate gene for cleidocranial dysplasia syndrome, is essential for osteoblast differentiation and bone development. *Cell* **89**, 765–771.

Ouchi, K., Yamada, J., and Kosaka, S. (1972). On the resorption of scales and associated cells in precocious male parr of the Masu salmon (*Oncorhynchus masou*). *Bull. Jap. Soc. Scient. Fish.* **38**, 423–438.

Oudhof, H. A. J. (1982). Sutural growth. *Acta Anat.* **112**, 58–68.

Oudhof, H. A. J., and Markens, I. S. (1982). Transplantation of the interfrontal suture in the Wistar rat. *Acta Anat.* **113**, 39–46.

Oursler, M. J., and Osdoby, P. (1988). Osteoclast development in marrow cultured in calvaria-conditioned media. *Devel. Biol.* **127**, 170–178.

Ovadia, M., Parker, C. H., and Lash, J. W. (1980). Changing patterns of proteoglycan synthesis during chondrogenic differentiation. *J. Embryol. Exp. Morphol.* **56**, 59–70.

Overman, D. O., Seegmiller, R. E., and Runner, M. N. (1976). Coenzyme competition and precursor specificity during teratogenesis induced by 6-aminonicotinamide. *Devel. Biol.* **28**, 573–582.

Overman, D. O., Graham, M. N., and Roy, W. (1976). Ascorbate inhibition of 6-aminonicotinamide teratogenesis in chicken embryos. *Teratology* **13**, 85–94.

Overton, J. (1977). Response of epithelial and mesenchymal cells to culture on basement lamella observed by scanning microscopy. *Exp. Cell Res.* **105**, 313–324.

Owen, M. J. (1963). Cell population kinetics of an osteogenic tissue. I. *J. Cell Biol.* **19**, 19–32.

Owen, M. J. (1970). The origin of bone cells. *Int. Rev. Cytol.* **28**, 213–238.

Owen, M. J. (1971). Cellular dynamics of bone. In *The Biochemistry and Physiology of Bone* (G. H. Bourne, ed.), 2nd Edn, Volume 3, pp. 271–298. Academic Press, New York, NY.

Owen, M. J. (1985). Lineage of osteogenic cells and their relationship to the stromal system. In *Bone and Mineral Research/3* (W. A. Peck, ed.), pp. 1–25. Elsevier Science Publishers, BV, Amsterdam.

Owen, M. J. (1988). Marrow stromal stem cells. *J. Cell Sci.* (Suppl.), **10**, 63–76.

Owen, M. J., Cavé, J., and Joyner, C. J. (1987). Clonal analysis *in vitro* of osteogenic differentiation of marrow CFU-F. *J. Cell Sci.* **87**, 731–738.

Owen, M. J., and MacPherson, S. (1963). Cell population kinetics of an osteogenic tissue. II. *J. Cell Biol.* **19**, 33–44.

Owen, R. (1841). Notes on the anatomy of the Nubian giraffe. *Trans. Zool. Soc. Lond.* **2**, 217–248.

Owen, R. (1849). *On The Nature of Limbs. A Discourse Delivered on Friday, February 9, at an Evening Meeting of the Royal Institution of Great Britain.* John Van Voorst, London.

Owen, T. A., Aronow, M. S., Barone, L. M., Bettencourt, D., Stein, G. S., and Lian, J. B. (1991). Pleiotropic effects of vitamin D on osteoblast gene expression are related to the proliferative and differentiated state of the bone cell phenotype: Dependency upon basal levels of gene expression, duration of exposure and bone matrix competency in normal rat osteoblast cultures. *Endocrinology* **128**, 1496–1504.

Owen, T. A., Aronow, M. S., Shalhoub, V., Barone, L. M., Wilming, L., Tassinari, M. S., Kennedy, M. B., Pockwinse, S., Lian, J. B., and Stein, G. S. (1990). Progressive development of the rat osteoblast phenotype *in vitro*: Reciprocal relationships in expression of genes associated with osteoblast proliferation and differentiation during formation of the bone extracellular matrix. *J. Cell Physiol.* **143**, 420–430.

Owens, E. M., and Solursh, M. (1981). *In vitro* histogenic capacities of limb mesenchyme from various stage mouse embryos. *Devel. Biol.* **88**, 297–311.

Owens, E. M., and Solursh, M. (1982). Cell-cell interactions by mouse limb cells during *in vitro Devel. Biol.* **91**, 376–388.

Owens, E. M., and Solursh, M. (1983). Accelerated maturation of limb mesenchyme by the Brachypod[H] mouse mutation. *Differentiation* **24**, 145–148.

Oxnard, C. E. (1991). Morphanalysis of bone: a perspective from the outside. In *Fundamentals of Bone Growth: Methodology and Applications* (A. D. Dixon, B. G. Sarnat and D. A. N. Hoyte, eds), pp. 455–469. CRC Press, Boca Raton, FL.

P

Pacces Zaffaroni, N., Arias, E., Lombardi, S., and Zavanella, T. (1996). Natural variation in the appendicular skeleton of *Triturus carnifex* (Amphibia: Salamandridae). *J. Morphol.* **230**, 167–175.

Pacifici, M. (1995). Tenascin-C and the development of articular cartilage. *Matrix Biol.* **14**, 689–698.

Pacifici, M., Cossu, G., Molinard, M., and Tato, F. (1980). Vitamin A inhibits chondrogenesis but not myogenesis. *Exp. Cell Res.* **129**, 469–474.

Pacifici, M., Fellini, S. A., Holtzer, H., and de Luca, S. (1981). Changes in the sulfated proteoglycans synthesized by 'aging' chondrocytes. I. Dispersed cultured chondrocytes and *in vivo* cartilages. *J. Biol. Chem.* **256**, 1029–1037.

Pacifici, M., Golden, E. B., Adams, S. L., and Shapiro, I. M. (1991a). Cell hypertrophy and type-X collagen synthesis in cultured articular chondrocytes. *Exp. Cell Res.* **192**, 266–270.

Pacifici, M., Golden, E. B., Iwamoto, M., and Adams, S. L. (1991b). Retinoic acid treatment induces type-X collagen gene expression in cultured chick chondrocytes. *Exp. Cell Res.* **195**, 38–46.

Pacifici, M., Oshima, O., Fisher, L. W., Young, M. F., Shapiro, I. M., and Leboy, P. S. (1990). Changes in osteonectin distribution and levels are associated with mineralization of the chicken tibial growth cartilage. *Calcif. Tissue Int.* **47**, 51–61.

Packard, D. S., Jr., and Jacobson, A. G. (1976). The influence of axial structures on chick somite formation. *Devel. Biol.* **53**, 36–48.

Packer, C. (1983). Sexual dimorphism: The horns of African antelopes. *Science* **221**, 1191–1193.

Pacy, P. J., Hesp, R., Halliday, D. A., Katz, D., Cameron, G., and Reeve, J. (1988). Muscle and bone in paraplegic patients and the effect of functional electrical stimulation. *Clin. Sci.* **75**, 481–487.

Padian, K. (1983). A functional analysis of flying and walking in pterosaurs. *Paleobiology* **9**, 218–239.

Padian, K. (1992). A proposal to standardize tetrapod phalangeal formula designations. *J. Vert. Paleont.* **12**, 260–262.

Padian, K., and Rayner, J. M. V. (1993). The wings of pterosaurs. *Amer. J. Sci.* **293A**, 91–166.

Page, M., and Ashhurst, D. E. (1987). The effects of mechanical stability on the macromolecules of the connective tissue matrices produced during fracture healing. II. The glycosaminoglycans. *Histochem. J.* **19**, 39–61.

Page, M., Hogg, J., and Ashhurst, D. E. (1986). The effects of mechanical stability on the macromolecules of the connective tissue matrices produced during fracture healing. I. The collagens. *Histochem. J.* **18**, 251–265.

Pak, C. Y. C., Sakhaee, K., Zerwekh, J. E., Parcel, C., Peterson, R., and Johnson, K. (1989). Safe and effective treatment of osteoporosis with intermittent slow release sodium fluoride: Augmentation of vertebral bone mass and inhibition of fractures. *J. Clin. Endocrinol. Metabol.* **68**, 150–159.

Palfrey, A. J., and Davies, D. V. (1966). The fine structure of chondrocytes. *J. Anat.* **100**, 213–226.

Palmeirim, I., Henrique, D., Ish-Horowicz, D., and Pourquié, O. (1997). Avian *hairy* gene expression identifies a molecular clock linked to vertebrate segmentation and somitogenesis. *Cell* **91**, 639–648.

Palmer, R. M., and Lumsden, A. G. S. (1987). Development of periodontal ligament and alveolar bone in homografted recombinations of enamel organs and papillary, pulpal and follicular mesenchyme in the mouse. *Archs Oral Biol.* **32**, 281–289.

Palmer, R. M. J., Hickery, M. S., Charles, I. G., Moncada, S., and Bayliss, M. T. (1993). Induction of nitric oxide synthase in human chondrocytes. *Biochem. Biophys. Res. Commun.* **193**, 398–405.

Palmoski, M., and Goetinck, P. F. (1970). An analysis of the development of conjunctival papillae and scleral ossicles in the eye of the scaleless mutant. *J. Exp. Zool.* **174**, 157–164.

Palmoski, M., and Goetinck, P. F. (1972). Synthesis of proteochondroitin sulfate by normal, nanomelic and 5-bromodeoxyuridine-treated chondrocytes in cell culture. *Proc. Natl Acad. Sci. U.S.A.* **69**, 3385–3388.

Palumbo, C. (1986). A three-dimensional ultrastructural study of osteoid-osteocytes in the tibia of chick embryos. *Cell Tissue Res.* **246**, 125–131.

Palumbo, C., Palazzini, S., and Marotti, G. (1990a). Morphological study of intercellular junctions during osteocyte differentiation. *Bone* **11**, 401–406.

Palumbo, C., Palazzini, S., Zappe, D., and Marotti, G. (1990b). Osteocyte differentiation in the tibia of newborn rabbits: an ultrastructural study of the formation of cytoplasmic processes. *Acta Anat.* **137**, 350–358.

Panayotou, G., End, P., Aumailley, M, Timpl, R., and Engel, J. (1989). Domains of laminin with growth-factor activity. *Cell* **56**, 93–101.

Papadatos, C. J., and Bartsocas, C. S. (eds) (1982). *Skeletal Dysplasias.* Alan R. Liss Inc., New York, NY.

Papenfuss, T. J. (1982). The ecology and systematics of the Amphisbaenian genus *Bipes. Occas. Pap. Calif. Acad. Sci.* **#136**, 1–42.

Paralkar, V. M., Hammonds, R. G., and Reddi, A. H. (1991b). Identification and characterization of cellular binding proteins (receptors) for recombinant human bone morphogenetic protein 2B, an initiator of bone differentiation cascade. *Proc. Natl Acad. Sci. U.S.A.* **88**, 3397–3401.

Paralkar, V. M., Nandedkar, A. K. N., Pointers, R. H., Kleinman, H. K., and Reddi, A. H. (1990). Interaction of osteogenin, a heparin binding bone morphogenetic protein, with type IV collagen. *J. Biol. Chem.* **265**, 17281–17284.

Paralkar, V. M., Vukicevic, S., and Reddi, A. H. (1991). Transforming growth factor β type 1 binds to collagen IV of basement membrane matrix: implications for development. *Devel. Biol.* **143**, 303–308.

Paralkar, V. M., Weeks, B. S., Yu, Y. M., Kleinman, H. K., and Reddi, A. H. (1992). Recombinant human bone morphogenetic protein 2B stimulates PC12 cell differentiation: potentiation and binding to type IV collagen. *J. Cell Biol.* **119**, 1721–1728.

Paré, A. (1572). *Fractures du Coll du Femur.* Cinc Livres de Chirurgiè, Paris.

Parenti, L. R. (1986). The phylogenetic significance of bone types in euteleost fishes. *Zool. J. Linn. Soc. London* **87**, 37–51.

Parfitt, A. M. (1982). The coupling of bone formation to bone resorption: A clinical analysis of the concept and of its relevance to the pathogenesis of osteoporosis. *Metab. Bone Dis. Rel. Res.* **4**, 1–6.

Parfitt, A. M. (1984). The cellular basis of bone remodelling: The quantum concept reexamined in light of recent advances in the cell biology of bone. *Calcif. Tissue Int.* **36**, S37–S45.

Parfitt, A. M. (1988). Bone remodelling: Relationship to amount and structure of bone, and the pathogenesis and prevention

of fractures. In *Osteoporosis: Etiology, Diagnosis and Management* (B. L. Riggs and L. J. Melton, eds), pp. 45–93. Raven Press, New York, NY.

Parfitt, A. M. (1990). Bone forming cells in clinical conditions. In *Bone, Volume 1: The Osteoblast and Osteocyte* (B. K. Hall, ed.), pp. 351–429. The Telford Press, Caldwell, NJ.

Parker, W. K. (1873). On the structure and development of the skull of the salmon, *Salmo salar*, Linn. *Phil. Trans. R. Soc. Lond.* **163**, 95–145.

Parra-Olea, G., and Wake, D. B. (2001). Extreme morphological and ecological homoplasy in tropical salamanders. *Proc. Natl Acad. Sci. U.S.A.* **98**, 7888–7891.

Parsons, F. G., and Keith, A. (1897). The presence of sesamoid bodies in either head of the gastrocnemius, and in the tendon of the peroneus longus. *J. Anat. Physiol.* **32**, 182–186.

Pasqualetti, M., Ori, M., Nardi, I., and Rjli, F. M. (2000). Ectopic *Hoxa2* induction after neural crest migration results in homeosis of jaw elements in *Xenopus. Development* **127**, 5367–5378.

Pathi, S., Rutenberg, J. B., Johnson, R. L., and Vortkamp, A. (1999). Interaction of Ihh and BMP-Noggin signaling during cartilage differentiation. *Devel. Biol.* **209**, 239–253.

Patterson, C. (1977). Cartilage bones, dermal bones and membrane bones, or the exoskeleton versus the endoskeleton. In *Problems in Vertebrate Evolution* (S. M. Andrews, R. S. Miles, and A. D. Walker, eds), Linnaean Soc. Symposium Series No. 4, pp. 77–122. Academic Press, London.

Patterson-Buckendahl, P., Arnaud, S. B., Mechanic, G. L., Martin, R. B., Grindeland, R. E., and Cann, C. E. (1987). Fragility and composition of growing rat bone after one week in space flight. *Amer. J. Physiol.* **252**, R240–R246.

Patton, J. T., and Kaufman, M. H. (1995). The timing of ossification of the limb bones, and growth rates of various long bones of the fore and hind limbs of the prenatal and early postnatal laboratory mouse. *J. Anat.* **186**, 175–185.

Pauli, B. U., Memoli, V. A., and Kuettner, K. E. (1981a). Regulation of tumor invasion by cartilage-derived anti-invasion factor *in vitro. J. Natl Cancer Inst.* **67**, 65–74.

Pauli, B. U., Memoli, V. A., and Kuettner, K. E. (1981b). *In vitro* determination of tumor invasiveness using extracted hyaline cartilage. *Cancer Res.* **41**, 2084–2091.

Paulsen, D. F., and Finch, R. A. (1977). Age and region dependent concanavalin A reactivity of chick wing-bud mesoderm cells. *Nature* **268**, 639–641.

Pautou, M.-P. (1974). Evolution comparée de la nécrose morphogène interdigitale dans le pied de l'embryon de poulet et de canard. *CR. Hebd. Seances Acad. Sci.* **278**, 2209–2212.

Pautou, M.-P. (1977). Establissement de l'axe dorso-ventral dans le pied de l'embryon de poulet. *J. Embryol. Exp. Morphol.* **42**, 177–194.

Pauwels, F. (1973). Kurzer Überblick über die mechanische Beanspruchung des Knochens und ihre Bedeutung für die Funktionells Anspassung. *Z. Orthop.* **111**, 681–704.

Pauwels, F. (1974). Über die Bedeutung des markhöhle für mechanische Beanspruchung des Röhrenknochens. *Z. Anat. Entwick.* **145**, 81–86.

Pavasant, P., Shizari, T., and Underhill, C. B. (1996). Hyaluronan contributes to the enlargement of hypertrophic lacunae in the growth plate. *J. Cell Sci.* **109**, 327–334.

Pavlov, M. I., Sautier, J.-M., Obouef, M., Asselin, A., and Berdal, A. (2003). Chondrogenic differentiation during midfacial development in the mouse: in vivo and in vitro studies. *Biol. Cell* **95**, 75–86.

Pawelek, J. M. (1969). Effects of thyroxine and low oxygen tension on chondrogenic expression in cell culture. *Devel. Biol.* **19**, 52–72.

Pawlicki, R. (1974). An electron microscopic study of the structure of the wall of the bone canaliculus with particular consideration of the place of its branching. *Z. Mikrosk. Anat. Forsch. Leipzig* **88**, 537–544.

Pawlicki, R. (1975a). Studies of the fossil dinosaur bone in the scanning electron microscope. *Z. Mikrosk. Anat. Forsch. Leipzig.* **89**, 393–398.

Pawlicki, R. (1975b). Bone canaliculus endings in the area of the osteocyte lacuna. Electron-microscopic studies. *Acta Anat.* **91**, 292–304.

Pawlicki, R. (1977a). Topochemical localization of lipids in dinosaur bone by means of Sudan B black. *Acta Histochem.* **59**, 40–46.

Pawlicki, R. (1977b). Histochemical reactions for mucopolysaccharides in dinosaur bone: Studies on epon-embedded and methacrylate-embedded semithin sections as well as on isolated osteocytes and ground sections of bone. *Acta Histochem.* **58**, 75–78.

Pawlicki, R. (1978). Morphological differentiation of the fossil dinosaur bone cells: light, transmission electron-, and scanning electron-microscopic studies. *Acta Anat.* **100**, 411–418.

Pawlicki, R. (1983). Metabolic pathways of the fossil dinosaur bones. I. Vascular communication system. *Folia Histochem. Cytochem.* **21**, 253–262.

Pawlicki, R. (1984a). Metabolic pathways of the fossil dinosaur bones. Part III. Intermediary and other osteocytes in the system of metabolic pathways of dinosaur bone. *Folia Histochem. Cytochem.* **22**, 91–98.

Pawlicki, R. (1984b). Metabolic pathways of the fossil dinosaur bones. Part IV. Modes of linkage between osteocytes and a variety of nexuses of osteocytic processes. Intermediary and other osteocytes in the system of metabolic pathways of dinosaur bone. *Folia Histochem. Cytochem.* **22**, 99–104.

Pawlicki, R. (1985). Metabolic pathways of the fossil dinosaur bones. 4. Morphological differentiation of osteocyte lacunae and bone canaliculi and their significance in the system of extracellular communication. *Folia Histochem. Cytochem.* **23**, 165–174.

Peabody, R. B., and Brodie, E. D., Jr. (1975). Effect of temperature, salinity and photoperiod on the number of trunk vertebrate in *Ambystoma maculatum. Copeia* **1975**, 741–746.

Pead, M. J., Skerry, T. M., and Lanyon, L. E. (1988a). Direct transformation from quiescence to bone formation in the adult periosteum following a single brief period of bone loading. *J. Bone Min. Res.* **3**, 647–656.

Pead, M. J., Suswillo, R., Skerry, T. M., Vedi, S., and Lanyon, L. E. (1988b). Increased ^{3}H-uridine levels in osteocytes following a single period of dynamic bone loading *in vivo. Calcif. Tissue int.* **43**, 92–96.

Pearson, A. A. (1977). The early innervation of the developing deciduous teeth. *J. Anat.* **123**, 563–578.

Pearson, C. A., Pearson, D., Shibahara, S., Hofsteenge, J., and Chiquet-Ehrismann, R. (1988). Tenascin: cDNA cloning and induction by TGF-β. *EMBO J.* **7**, 2677–2981.

Pearson, O. P. (1954). Habits of a lizard. *Copeia* **1954**, 111–116.

Peck, W. A. (1984). The effect of glucocorticoids on bone cell metabolism and function. *Adv. Exp. Med. Biol.* **171**, 111–119.

Peck, W. A., Birge, S. J., and Brandt, J. (1967a). Collagen synthesis by isolated bone cells: Stimulation by ascorbic acid *in vitro. Biochim. Biophys. Acta* **142**, 512–525.

Peck, W. A., Birge, S. J., and Fedak, S. A. (1964). Bone cells: Biochemical and biological studies after enzymatic isolation. *Science* **146**, 1476–1477.

Peck, W. A., Brandt, J., and Miller, I. D. A. (1967b). *Proc. Natl Acad. Sci. U.S.A.* **57**, 1599–1606.

Peck, W. A., Carpenter, A. J., and Messinger, K. (1974). Cyclic 3′,5′-adenosine monophosphate in isolated bone cells. II. Responses to adenosine and parathyroid hormone. *Endocrinology* **94**, 148–154.

Peck, W. A., Rifas, L., and Shen, V. (1985). Macrophages release a peptide stimulator of osteoblast growth. *Ann. Biol. Clinique* **43**, 751–754.

Pedrini-Mille, A., and Pedrini, V. (1971). Studies of human iliac crest cartilage. III. Proteinpolysaccharides in human achondroplasia. *Calcif. Tissue Res.* **8**, 106–113.

Pedrini-Mille, A., and Pedrini, V. (1982). Proteoglycans and glycosaminoglycans in human achondroplastic cartilage. Studies of human iliac crest cartilage. *J. Bone Joint Surg. Am. Vol.* **64**, 39–46.

Pehrson, T. (1945). Some problems concerning the development of the skull of turtles. *Acta Zool.* (*Stockholm*) **26**, 157–181.

Peichel, C. L., Nereng, K. S., Ohgi, K. A., Cole, B. L. E., Colosimo, P. F., Buerkle, C. A., Schluter, D., and Kingsley, D. M. (2001). The genetic architecture of divergence between three-spine stickleback species. *Nature* **414**, 901–905.

Peignoux-Deville, J., Baud, C. A., Allier, F., and Vidal, B. (1985). Perichondral ossification of vertebral arches from dogfish to man. *Fortschr. Zool.* **30**, 65–68.

Peignoux-Deville, J., Bordat, C., and Vidal, B. (1989). Demonstration of bone resorbing cells in elasmobranchs: Comparison with osteoclasts. *Tissue & Cell* **21**, 925–934.

Peignoux-Deville, J., and Janvier, P. (1984). L'os du requin ou la biologie au rendez-vous de la palêontologie. *La Recherche* **15**, 1140–1142.

Peignoux-Deville, J., Lallier, F., and Vidal, B. (1981). Evidence for the presence of osseous tissue in dogfish vertebra. *C.R. Acad. Sci. Paris* **292**, 73–78.

Peignoux-Deville, J., Lallier, F., and Vidal, B. (1982). Evidence for the presence of osseous tissue in dogfish vertebra. *Cell Tissue Res.* **222**, 605–614.

Pellegrini, O. (1934). Lo sviluppo di abbozzi di articolazioni impiantati nell membrana corioallantoidee. *Atti. Soc. Med.-Chir. Padova Fac. Med. Chir. Univ. Padova* **11**, 927–941.

Peltomäki, T. (1992). Growth of a costochondral graft in the rat temporomandibular joint. *J. Oral Maxillofac. Surg.* **50**, 851–857.

Peltomäki, T. (1993). Growth of the costochondral junction and its potential applicability for the reconstruction of the mandibular condyle. *Ann. Univ. Turk., Ser. D. Med.-Odontol.* **108**, 1–77.

Peltomäki, T., Kylamarkula, S., Vinkkapu-Hakka, H., Rintala, M., Kantomaa, T., and Rönning, O. (1997). Tissue-separating capacity of growth cartilages. *Eur. J. Orthodont.* **19**, 473–481.

Pelton, R. W., Dickinson, M. E., Moses, H. L., and Hogan, B. L. M. (1990). *In situ* hybridization analysis of TGFβ3 RNA expression during mouse development: comparative studies with TGFβ1 and 2. *Development* **110**, 609–620.

Pelton, R. W., Hogan, B. L. M., Miller, D. A., and Moses, H. L. L. (1991a). Differential expression of genes encoding TGFβ1, β2, and β3 during murine palate formation. *Devel. Biol.* **141**, 456–460.

Pelton, R. W., Nomura, S., Moses, H. L., and Hogan, B. L. M. (1989). Expression of transforming growth factor β2 RNA during murine embryogenesis. *Development* **106**, 759–768.

Pelton, R. W., Saxena, B., Jones, M., Moses, H. L., and Gold, L. I. (1991b). *Immunohistochemical* localization of TGFβ1, TGFβ2, and TGFβ3 in the mouse embryo: expression patterns suggest multiple roles during embryonic development. *J. Cell Biol.* **115**, 1091–1105.

Pennock, J. M., Kalu, D. N., Clark, M. B., Foster, G. V., and Doyle, F. H. (1972). Hypoplasia of bone induced by immobilization. *Brit. J. Radiol.* **45**, 641–646.

Pennypacker, J. P. (1981). Modulation of chondrogenic expression in cell culture by fibronectin. *Vision Res.* **21**, 65–70.

Pennypacker, J. P., and Goetinck, P. F. (1976). Biochemical and ultrastructural studies on collagen and proteochondroitin sulfate in normal and nanomelic cartilage. *Devel. Biol.* **50**, 35–47.

Pennypacker, J. P., Hassell, J. R., Yamada, K. M., and Pratt, R. M. (1979). The influence of an adhesive cell surface protein on chondrogenic expression *in vitro*. *Exp. Cell Res.* **121**, 411–415.

Pennypacker, J. P., Kimata, K., and brown, K. S. (1981). Brachymorphic mice (*bm/bm*): A generalized biochemical defect expressed primarily in cartilage. *Devel. Biol.* **81**, 280–287.

Penttinen, R., Rantanen, J., and Kulonen, E. (1971). Effect of reduced air pressure on fracture healing. A biochemical study on rats *in vivo*. *Israel J. Med. Sci.* **7**, 444–446.

Penttinen, R., Rantanen, J., and Kulonen, E. (1972). Fracture healing at reduced atmospheric: A biochemical study with rats *in vivo*. *Acta Chirurg. Scand.* **138**, 147–151.

Pepper, M. S., Montesano, R., Vassalli, J. D., and Orci, L. (1991). Chondrocytes inhibit endothelial sprout formation *in vitro* – evidence for involvement of a transforming growth factor-β. *J. Cell Physiol.* **146**, 170–179.

Perantoni, A. O., Dove, L.F., and Karavanova, I. (1995). Basic fibroblast growth factor can mediate the early inductive events in renal development. *Proc. Natl Acad. Sci. U.S.A.* **92**, 4696–4700.

Perides, G., Erickson, H. P., Rahentulla, F., and Bignami, A. (1993). Colocalization of tenascin with versican, a hyaluronate-binding chondroitin sulfate proteoglycan. *Anat. Embryol.* **188**, 467–479.

Perle, M. A., Leonard, C. M., and Newman, S. A. (1982). Developmentally regulated nonhistone proteins: Evidence for deoxyribonucleic acid binding role and localization near deoxyribonuclease I sensitive domains of precartilage cell chromatin. *Biochemistry* **21**, 2379–2386.

Perle, M. A., and Newman, S. A. (1980). *Talpid*[2] mutant of the chicken with perturbed cartilage development has an altered precartilage-specific chromatin protein. *Proc. Natl Acad. Sci. U.S.A.* **77**, 4828–4830.

Perlov, F. A. (1968). Growth mechanisms in normal and chondrodysplastic (creeper) tibias using a genetic marker for skeletal type determination. M.Sc. in Dentistry Thesis, University of Nebraska, Lincoln, Nebraska.

Perotti, M. G. (2001). Skeletal development of *Leptodactylus chaquensis* (Anura, Leptodactylidae). *Herpetologica* **57**, 318–335.

Person, P. (1960). Some observations on the evolution of oral tissues. *Ann. NY. Acad. Sci.* **85**, 9–16.

Person, P. (1969). Cartilaginous dermal scales in cephalopods. *Science* **164**, 1404–1405.

Person, P. (1983a). Invertebrate cartilages. In *Cartilage, Volume 1. Structure, Function and Biochemistry* (B. K. Hall, ed.), pp. 31–58. Academic Press, New York, NY.

Person, P. (1983b). Mammalian limb regeneration. In *Nerve, Tissue and Organ Regneration: Research Perspectives* (F. J. Seil, ed.). Academic Press, New York, NY.

Person, P., and Fine, A. S. (1959). Cytochrome oxidase and succinoxidase activity of *Limulus* gill cartilage. *Arch. Biochem. Biophys.* **84**, 123–133.

Person, P., Lash, J. W., and Fine, A. S. (1959). On the presence of myoglobin and cytochrome oxidase in the cartilaginous odontophore of the marine snail, *Busycon*. *Biol. Bull.* **117**, 504–510.

Person, P., Libbin, R. M., Shah, D., and Papierman, S. (1979). Partial regeneration of the above-elbow amputated rat forelimb. I. Innate responses. *J. Morphol.* **159**, 427–438.

Person, P., and Mathews, M. B. (1967). Endoskeletal cartilage in a marine polychaete, *Eudistylia polymorpha*. *Biol. Bull.* **132**, 244–252.

Person, P., Papierman, S., Zipper, H., and Libbin, R. M. (1977). Absence of mitochondrial terminal respiratory enzymes in cartilage matrix vesicles. *Calcif. Tissue Res.* **24**, 37–39.

Person, P., and Philpott, D. E. (1963). Invertebrate cartilages. *Ann. NY. Acad. Sci.* **109**, 113–116.

Person, P., and Philpott, D. E. (1969a). The nature and significance of invertebrate cartilages. *Biol. Rev. Cambridge Philos. Soc.* **44**, 1–16.

Person, P., and Philpott, D. E. (1969b). The biology of cartilage. I. Invertebrate cartilages: *Limulus* gill cartilage. *J. Morphol.* **128**, 67–94.

Persson, M. (1973). Structure and growth of facial sutures. Histologic, microangiographic and autoradiographic studies in rats and a histologic study in man. *Odontol. Revy* **24**, Suppl. 26, 1–146.

Persson, M. (1983). The role of movements in the development of sutural and diarthrodial joints tested by long-term paralysis of chick embryos. *J. Anat.* **137**, 591–599.

Persson, M. (1995). The role of sutures in normal and abnmormal craniofacial growth. *Acta Odontol. Scand.* **53**, 152–161.

Persson, M., Magnusson, B. C., and Thilander, B. (1978). Sutural closure in rabbit and man: A morphological and histochemical study. *J. Anat.* **125**, 313–322.

Persson, P. (1997). *Calcium Regulation during Sexual Maturation of Female Salmonids: Estradiol-17S and Calcified Tissues.* Ph.D. Thesis. Dept. Zoophysiology, Göteborg University, Göteborg, Sweden.

Persson, P., Björnsson, B. Th., and Takagi, Y. (1995). Tartrate resistant acid phosphatase as a marker for scale resorption in rainbow trout, *Oncorhynchus mykiss*: effects of estradiol-17β treatment and refeeding. *Fish Physiol. Biochem.* **14**, 329–339.

Persson, P., Johannsson, S. H., Takagi, Y., and Björnsson, B. Th. (1997). Estradiol-17β and nutritional status affect calcium balance, scale and bone resorption, and bone formation in rainbow trout, *Oncorhynchus mykiss*. *J. Comp. Physiol. B* **167**, 468–473.

Persson, P., Shrimpton, J. M., McCormick, S. D., and Björnsson, B. Th. (2000). The presence of high-affinity, low-capacity estradiol-17β binding in rainbow trout scale indicates a possible endocrine route for the regulation of scale resorption. *Gen. Comp. Endocrinol.* **120**, 35–43.

Persson, P., Sundell, K., and Björnsson, B. Th. (1994). Estradiol-17β-induced calcium uptake and resorption in juvenile rainbow trout, *Oncorhynchus mykiss*. *Fish Physiol. Biochem.* **13**, 379–386.

Persson, P., Sundell, K., Björnsson, B. Th., and Lundqvist, H. (1998). Calcium metabolism and osmoregulation during sexual maturation of river running Atlantic salmon. *J. Fish. Biol.* **52**, 334–349.

Pertoldi, C., Loeschchke, V., Braun, A., Madsen, A. B., and Randi, E. (2000). Craniometrical variability and developmental stability. Two useful tools for assessing the population viability of Eurasian otter (*Lutra lutra*) populations in Europe. *Biol. J. Linn. Soc.* **70**, 309–323.

Pessac, B., and Defendi, V. (1972). Cell aggregation: Role of acid mucopolysaccharides. *Science* **175**, 898–900.

Peterková, R., Kristenová, P., Lesot, H., Lisi, S., Vonesch, J.-L., Gendrault, J.-L., and Peterka, M. (2002a). Different morpho-types of the tabbly (EDA) dentition in the mouse mandible result from a defect in the mesio-distal segmentation of dental epithelium. *Orthod. Craniofac. Res.* **5**, 215–226.

Peterková, R., Peterka, M., Viriot, L., and Lesot, H.(2002b). Development of the vestigial tooth primordia as part of mouse odontogenesis. *Conn. Tissue Res.* **43**, 120–128.

Peters, H., and Balling, R. (1999). Teeth: Where and how to make them. *Trends Genet.* **15**, 59–65.

Peters, H., Neubüser, A., Kratochwil, K., and Balling, R. (1998). *Pax9*-deficient mice lack pharyngeal pouch derivatives and teeth and exhibit craniofacial and limb abnormalities. *Genes & Devel.* **12**, 2735–2747.

Peters, H., Wilm, B., Sakai, N., Imai, K., Maas, R., and Balling, R. (1999). Pax1 and Pax9 synergistically regulate vertebral column development. *Development* **126**, 5399–5408.

Peters, K. G., Werner, S., Chen, G., and Williams, L. T. (1992). Two FGF receptor genes are differentially expressed in epithelial and mesenchymal tissues during limb formation and organo-genesis in the mouse. *Development* **114**, 233–243.

Petite, H., Viateau, V., Bensaid, W., Meunier, A., de Pollak, C., Bourguignon, M., Oudina, K., Sedel, L., and Guillemin, G. (2000). Tissue-engineered bone regeneration. *Nature Biotechnol.* **18**, 959–963.

Petrakova, K. V., and Friedenstein, A. Y. (1965). Resorption of the transitional epithelium homografts inducing osteogenesis in the surrounding connective tissue. *Byull. Eksp. Biol. Med.* **2**, 98–101.

Petricioni, V. (1964). Entwicklungsphysiologische Untersuchungen über die Induzierbarkeit von Skelettelementen des Anuren-schädels durch flüssigen Organextrakt. *W. Roux. Arch. EntwickMech.* **155**, 358–390.

Petrovic, A. (1970). Recherches sur les méchanismes histophysi-ologiques de la croissance osseuse craniofaciale. *Annee Biol.* **9**, 304–311.

Petrovic, A. (1972). Mechanisms and regulation of mandibular condylar growth. *Acta Morphol. Neerl. Scand.* **10**, 25–34.

Petrovic, A. (1974). Control of postnatal growth of secondary cartilages of the mandible by mechanisms regulating occlu-sion. Cybernetic model. *Trans. Eur. Orthod. Soc.* pp. 1–7.

Petrovic, A., and Charlier, J.-P. (1967). La synchondrose sphénooccipitale de jeune rat en culture d'organes: Mise en évidence d'un potentiel de croissance indépendant. *C.R. Hebd. Seances Acad. Sci.* **265**, 1511–1513.

Petrovic, A., Oudet, C., and Gasson, N. (1973). Effets des appareils de propulsion et de rétropulsion mandibulaire sur le numbre des sarcomères en série du muscle ptérygoidien externe et sur les croissance du cartilage condylien du jeune rat. *Orthod. Fr.* **44**, 191–210.

Petrovic, A., Stutzmann, J. J., and Lavergne, J. M. (1990). Mechanisms of craniofacial growth and modus operandi of functional appliances: a cell level and cybernetic approach to orthodontic decision making. In *Craniofacial Growth Theory and Orthodontic Treatment* (D. S. Carlson, ed.), Monograph 23,

Craniofacial Growth Series, pp. 13–74. Center for Human Growth and Development, University of Michigan, Ann Arbor, MI.

Pfeilschifter, J., Bonewald, L., and Mundy, G. R. (1990a). Characterization of the latent transforming growth factor β complex in bone. *J. Bone Min. Res.* **5**, 49–58.

Pfeilschifter, J., Oechsner, M., Naumann, A., Grunwald, R. G. K., Minne, H. W., and Ziegler, R. (1990b). Stimulation of bone matrix apposition *in vitro* by local growth factors: a comparison between insulin-like growth factor 1, platelet-derived growth factor, and transforming growth factor β. *Endocrinology* **127**, 69–75.

Pfeilschifter, J., Wolf, O., Naumann, A., Minne, H. W., Mundy, G. R., and Ziegler, R. (1990c). Chemotactic response of osteoblast-like cells to transforming growth factor β. *J. Bone Min. Res.* **5**, 825–830.

Phelps, A. M. (1891). Transplantation of tissue from lower animals to man. *Med. Rec.* **39**, 221–225.

Phemister, D. B. (1914). The fate of transplanted bone and regenerative power of its various constituents. *Surg. Gynecol. Obstet.* **19**, 303–333.

Phemister, D. B. (1951). Biologic principles in the healing of frac-tures and their bearing on treatment. *Ann. Surg.* **133**, 433–436.

Phillips, C., Shapiro, P. A., and Luschei, E. S. (1982). Morphologic alterations in *Macaca mulatta* following destruction of the motor nucleus of the trigeminal nerve. *Amer. J. Orthod.* **81**, 292–298.

Philpott, D. E., and Person, P. (1966). Intracellular aggregates and granules of *Limulus* gill cartilage. In *Electron Microscopy* (R. Uyeda, ed.), Volume 2, pp. 565–566, Maruzen, Tokyo.

Philpott, D. E., and Person, P. (1970). The biology of cartilage. II. Invertebrate cartilages: Squid head cartilages. *J. Morphol.* **131**, 417–430.

Piatier-Piketty, D., and Zucman, J. (1971). The cellular response of the periosteum to trauma. An autoradiographic study of periosteal autografts. *Israel J. Med. Sci.* **1**, 447–448.

Piché, J. E., and Graves, D. T. (1989). Study of the growth factor requirements of human bone-derived cells: A comparison with human fibroblasts. *Bone* **10**, 131–138.

Pictet, R. L., Rall, L. B., Phelps, P., and Rutter, W. J. (1976). The neural crest and the origin of the insulin-producing and other gastrointestinal hormone-producing cells. *Science* **191**, 191–192.

Pieau, C., and Raynaud, A. (1976). Dégénérescence cellulaire dans la crête apicale de l'ébauche du membre de la Tortue mauresque (*Testudo graeca* L, Chelonien). *CR. Hebd. Seances Acad. Sci.* **282**, 1797–1800.

Piekarski, K., and Munro, M. (1977). Transport mechanisms operating between blood supply and osteocytes in long bone. *Nature* **269**, 80–82.

Pierce, A. M., Lindskog, S., and Hammarstrom, L. (1991). Osteoclasts: Structure and function. *Elect. Microsc. Rev.* **4**, 1–45.

Pietilä, K., Kantomaa, T., Pirttiniemi, P., and Poikela, A. (1999). Comparison of amounts and properties of collagen and pro-teoglycans in condylar, costal and nasal cartilages. *Cells Tissues & Organs* **164**, 30–36.

Pietsch, T. W. (1984). Enlarged cartilages in the protrusible upper jaws of teleost fishes: phylogenetic and functional implica-tions. *Copeia* **1984**, 1011–1015.

Piiper, J. (1928). On the evolution of the vertebral column in birds, illustrated by its development in *Larus* and *Struthio*. *Phil. Trans. R. Soc. London, Ser. B* **216**, 285–351.

Pike, A. W., and Burt, M. D. B. (1983). The tissue response of Yellow Perch, *Perca flavescens* Mitchill to infections with the metacercarial cyst of *Apophallus brevis* Ransom, 1920. *Parasitology* **87**, 393–404.

Pilbeam, D., and Gould, S. J. (1974). Size and scaling in human evolution. *Science* **186**, 892–901.

Pilloni, A., and Bernard, G. W. (1998). The effect of hyaluronan on mouse intramembranous osteogenesis *in vitro*. *Cell Tissue Res.* **294**, 323–333.

Pines, M., and Hurwitz, S. (1991). The role of the growth plate in longitudinal bone growth. *Poultry Sci.* **70**, 1806–1814.

Pinganaud-Perrin, G. (1973). Conséquences de l'ablation de l'os frontal sur la form des os du toit crânien de la truite(*Salmo irideus* Gib, Pisces-Teleostei). *C.R. Acad. Sci. Paris* **276**, 2809–2811.

Pinnell, S. R., and Crelin, E. S. (1963). Fate of pubic bone auto-transplanted to the tibia in estrogen-treated adult female mice. *Anat. Rec.* **145**, 345.

Pinot, M. (1969a). *In vitro* culture of the primordium or of the presumptive territory of the wing of the chick embryo: Differentiation of the cartilage. *Arch. Anat. Microsc. Morphol. Exp.* **58**, 123–144.

Pinot, M. (1969b). Etude expérimentale de la morphogenèse de la cage thoracique chez l'embryon de poulet. Méchanisme et origine du matériel. *J. Embryol. Exp. Morphol.* **21**, 149–164.

Pinot, M. (1970). Le rôle du mésoderme somitique dans la morphogenèse précose des membres de l'embryon de poulet. *J. Embryol. Exp. Morphol.* **23**, 109–151.

Pinto, C. B., and Hall, B. K. (1991). Toward an understanding of the epithelial requirement for osteogenesis in scleral mesenchyme of the embryonic chick. *J. Exp. Zool.* **259**, 92–108.

Piotrowski, T., and Nüsslein-Volhard, C. (2000). The endoderm plays an important role in patterning the segmental pharyngeal region in zebrafish (*Danio rerio*). *Devel. Biol.* **225**, 339–356.

Piotrowski, T., Schilling, T. F., Brand, M., Jiang, Y.-J., Heisenberg, C.-P., Beuchle, D., Grandel, H., van Eeden, F. J. M., Furutani-Seiki, M., Granato, M., Haffter, P., Hammerschmidt, M., Kane, D. A., Kelsh, R. N., Mullins, M. C., Odenthal, J., Warga, R. M., and Nüsslein-Volhard, C. (1996). Jaw and branchial arch mutants in zebrafish. II: Anterior arches and cartilage differentiation. *Development* **123**, 345–356.

Pirttiniemi, P., and Kantomaa, T. (1996). Electrical stimulation of masseter muscles maintains condylar cartilage in long-term organ culture. *J. Dent. Res.* **75**, 1365–1371.

Pirttiniemi, P., and Kantomaa, T. (1998). Effect of cytochalasin D on articular cartilage cell phenotype and shape in long-term organ culture. *Eur. J. Orthod.* **20**, 491–499.

Pirttiniemi, P., Kantomaa, T., Salo, L., and Tuominen, M. (1996). Effect of reduced articular function in deposition of type I and type II collagens in the mandibular condylar cartilage of the rat. *Archs Oral Biol.* **41**, 127–131.

Pirttiniemi, P., Kantomaa, T., Sorsa, T. (2004). Effects of decreased loading on the metabolic activity of the mandibular condylar cartilage in the rat. *Eur. J. Orthod.* **26**, 1–5.

Pirttiniemi, P., Kantomaa, T., and Tuominen, M. (1993). Increased condylar growth after experimental relocation of the glenoid fossa. *J. Dent. Res.* **72**, 1356–1359.

Pisano, M. M., and Greene, R. M. (1986). Hormone and growth factor involvement in craniofacial development. *IRCS Med. Sci.* **14**, 635–640.

Pizette, S., Abate-Shen, C., and Niswander, L. (2001). BMP controls proximodistal outgrowth, via induction of the apical ectodermal ridge, and dorsoventral patterning in the vertebrate limb. *Development* **128**, 4463–4474.

Pizette, S., and Niswander, L. (1999). BMPs negatively regulate structure and function of the limb apical ectodermal ridge. *Development* **126**, 883–894.

Pizette, S., and Niswander, L. (2000). BMPs are required at two steps of limb chondrogenesis: Formation of prechondrogenic condensations and their differentiation into chondrocytes. *Devel. Biol.* **219**, 237–249.

Plant, M. R., MacDonald, M. E., Grad, L. I., Ritchie, S. J., and Richman, J. M. (2000). Locally released retinoic acid repatterns the first branchial arch cartilages *in vivo*. *Devel. Biol.* **222**, 12–26.

Pleskova, M. V., Rodionov, V. M., Bugrilova, R. S., and Konyukhov, B. (1974). The partial purification of growth-inhibiting factor of the brachypodism-H mouse embryo. *Devel. Biol.* **37**, 417–421.

Plessow, S., Koster, M., and Knochel, W. (1991). cDNA sequence of *Xenopus laevis* bone morphogenetic protein-2 (BMP-2). *Biochem. Biophys. Acta* **1089**, 280–281.

Pockwinse, S. M., Stein, J. L., Lian, J. B., and Stein, G. S. (1995). Developmental stage-specific cellular responses to vitamin D and glucocorticoids during differentiation of the osteoblast phenotype: interrelationships of morphology and gene expression by *in situ* hybridization. *Exp. Cell Res.* **216**, 244–260.

Pohl, T. M., Mattei, M.-G., and Rüther, U. (1990). Evidence for allelism of the recessive insertional mutation *add* and the dominant mouse mutation *extra toes* (*Xt*). *Development* **110**, 1153–1157.

Poikela, A., Kantomaa, T., Pirttiniemi, P., Tuukkanen, J., and Pietilä, K. (2000). Unilateral masticatory function changes the proteoglycan content of mandibular condylar cartilage in rabbit. *Cells Tissues Organs* **167**, 49–57.

Policansky, D. (1982). The asymmetry of flounders. *Sci. Amer.* **246**, 116–122.

Pollak, R. D., and Fallon, J. F. (1974). Autoradiographic analysis of macromolecular synthesis in prospectively necrotic cells of the chick limb bud. *Exp. Cell Res.* **86**, 9–14.

Pollak, R. D., and Fallon, J. F. (1976). Autoradiographic analysis of macromolecular synthesis in prospectively necrotic cells of the chick limb bud. II. Nucleic acids. *Exp. Cell Res.* **100**, 15–22.

Pollock, R. A., Sreenath, T., Ngo, L., and Bieberich, C. J. (1995). Gain of function mutations for paralogous *Hox* genes: Implications for the evolution of *Hox* gene function. *Proc. Natl Acad. Sci. U.S.A.* **92**, 4492–4496.

Ponder, W. F., and Lindberg, D. R. (1997). Towards a phylogeny of gastropod molluscs – an analysis using morphological characters. *Zool. J. Linn. Soc. London* **119**, 83–265.

Poole, A. R., Matsui, Y., Hinek, A., and Lee, E. R. (1989). Cartilage macromolecules and the calcification of cartilage matrix. *Anat. Rec.* **224**, 167–179.

Poole, A. R., and Pidoux, I. (1989). Immunoelectron microscopic studies of type X collagen in endochondral ossification. *J. Cell Biol.* **109**, 2547–2554.

Poole, A. R., Pidoux, I., Reiner, A., Choi, H., and Rosenberg, L. C. (1984). Association of an extracellular protein (chondrocalcin) with the calcification of cartilage in endochondral bone formation. *J. Cell Biol.* **98**, 54–65.

Poole, A. R., Pidoux, I., Reiner, A., and Rosenberg, L. C. (1982b). An immunoelectron microscope study of the organization of proteoglycan monomer, link protein and collagen in the matrix of articular cartilage. *J. Cell Biol.* **93**, 921–937.

Poole, A. R., Pidoux, I., Reiner, A., Tang, L.-H., Choi, H., and Rosenberg, L. C. (1980). Localization of proteoglycan monomer and link protein in the matrix of bovine articular cartilage: An immunohistochemical study. *J. Histochem. Cytochem.* **28**, 621–635.

Poole, A. R., Pidoux, I., and Rosenberg, L. C. (1982a). Role of proteoglycans in endochondral ossification: Immunofluorescent localization of link protein and proteoglycan monomer in bovine fetal epiphyseal growth plate. *J. Cell Biol.* **92**, 249–260.

Poole, A. R., Reddi, A. H., and Rosenberg, L. C. (1982c). Persistence of cartilage proteoglycan and link protein during matrix-induced endochondral bone development: an immunofluorescent study. *Devel. Biol.* **89**, 532–539.

Poole, C. A. (1997). Articular cartilage chondrons: Form, function and failure. *J. Anat.* **191**, 1–13.

Poole, C. A., Ayad, S., and Schofield, J. R. (1988a). Chondrons from articular cartilage I. Immunolocalization of type VI collagen in the pericellular capsule of isolated canine tibial chondrons. *J. Cell Sci.* **90**, 635–643.

Poole, C. A., Flint, M. H., and Beaumont, B. W. (1984). Morphological and functional interrelationships of articular cartilage matrices. *J. Anat.* **138**, 113–138.

Poole, C. A., Flint, M. H., and Beaumont, B. W. (1987). Chondrons in cartilage: Ultrastructural analysis of the pericellular microenvironment in adult human articular cartilages. *J. Orthop. Res.* **5**, 509–522.

Poole, C. A., Flint, M. H., and Beaumont, B. W. (1988b). Chondrons extracted from canine tibial cartilage: Preliminary report on their isolation and structure. *J. Orthop. Res.* **6**, 408–419.

Poole, C. A., Glant, T. T., and Schofield, J. R. (1991a). Chondrons from articular cartilage. (IV) Immunolocalization of proteoglycan epitopes in isolated canine tibial chondrons. *J. Histochem. Cytochem.* **39**, 1175–1187.

Poole, C. A., Matsuoka, A., and Schofield, J. R. (1991b). Chondrons from articular cartilage. (III) Morphologic changes in the cellular microenvironment of chondrons isolated from osteoarthritic cartilage. *Arths. Rheum.* **34**, 22–35.

Popowics, T. E., Zhu, Z., and Herring, S. W. (2002). Mechanical properties of the periosteum in the pig, *Sus scrofa. Archs Oral Biol.* **47**, 733–741.

Portal, A. (1770). *Histoire de l'Anatomie et de la Chirurgie.* Didot, Paris.

Poss, K. D., Shen, J., and Keating, M. T. (2000a). Induction of *Lef1* during zebrafish fin regeneration. *Devel. Dynam.* **219**, 282–286.

Poss, K. D., Shen, J., Nechiporuk, A., McMahon, G., Thisse, B., Thisse, C., and Keating, M. T. (2000b). Roles for Fgf signaling during zebrafish fin regeneration. *Devel. Biol.* **222**, 347–358.

Poswillo, D. (1972). The late effects of mandibular condylectomy. *Oral Surg.* **33**, 500–511.

Poswillo, D. (1975a). Hemorrhage in development of the face. *Brit. Med. Bull.* **31**, 101–106.

Poswillo, D. (1975b). The pathogenesis of the treacher Collins syndrome (mandibulofacial dysostosis). *Brit J. Oral. Surg.* **13**, 1–26.

Poswillo, D. (1976a). Mechanisms and pathogenesis of malformation. *Brit. Med. Bull.* **32**, 59–164.

Poswillo, D. (1976b). Causal mechanisms of craniofacial deformity. *Brit. Med. Bull.* **32**, 159–64.

Potthoff, T., Kelley, S., and Collins, L. A. (1988). Osteological development of the red snapper, *Lutjanus campechanus* (Lutjanidae). *Bull. Mar. Sci.* **43**, 1–40.

Potts, J. D., and Carrington, J. L. (1993). Selective expression of the chicken platelet-derived growth factor alpha (PDGFα) receptor during limb bud development. *Devel. Dynam.* **198**, 14–21.

Poulin, M. L., Patrie, K. M., Botelho, M. J., Tassava, R. A., and Chiu, I. M. (1993). Heterogeneity in the expresion of fibroblast growth factor receptors during limb regeneration in newts (*Notophthalmus viridescens*). *Development* **119**, 353–361.

Pourquié, O. (2003). Vertebrate somitogenesis: a novel paradigm for animal segmentation? *Int. J. Devel. Biol.* **47**, 497–603.

Pourquié, O., Fan, C.-M., Coltey, M., Hirsinger, E., Watanabe, Y., Bréant, C., Francis-West, P., Brickell, P., Tessier-Lavigne, M., and Le Douarin, N. M. (1996). Lateral and axial signals involved in avian somite patterning: a role for BMP4. *Cell* **84**, 461–471.

Powell-Braxton, L., Hollingshead, P., Warburton, C., Dowd, M., Pitts-Meek, S., Salton, D., Gillett, N., and Stewart, T. A. (1993). IGF-1 is required for normal embryonic growth in mice. *Genes & Devel.* **7**, 2609–2617.

Power, S. C., Lancman, J., and Smith, S. M. (1999). Retinoic acid is essential for Shh/Hodx signaling during rat limb outgrowth but not for limb initiation. *Devel. Dynam.* **216**, 469–480.

Pratt, L. W. (1943). Experimental masseterectomy in the laboratory rat. *J. Mammal.* **24**, 204–211.

Pratt, R. M. (1987). Role of epidermal growth factor in embryonic development. *Curr. Top. Devel. Biol.* **22**, 175–193.

Pratt, R. M., Goulding, E. H., and Abbott, B. D. (1987). Retinoic acid inhibits migration of cranial neural crest cells in the cultured mouse embryos. *J. Craniofac. Genet. Devel. Biol.* **1**, 205–218.

Pratt, R. M., Larsen, M. A., and Johnston, M. C. (1975). Migration of cranial neural crest cells in a cell-free hyaluronate-rich matrix. *Devel. Biol.* **44**, 298–305.

Praul, C. A., Gay, C. V., and Leach, R. M., Jr. (1997). Chondrocytes of the tibial dyschondroplastic lesion are apoptotic. *Int. J. Devel. Biol.* **41**, 621–626.

Precious, D. S., Armstrong, J. E., and Morais, D. (1992). Anatomic placement of fixation devices in genioplasty. *Oral Surg., Oral Med., Oral Pathol.* **73**, 2–8.

Precious, D. S., and Delaire, J. (1987). Balanced facial growth: A schematic interpretation. *Oral Surg., Oral Med., Oral Pathol.* **63**, 637–644.

Precious, D. S., Delaire, J., and Hoffman, C. D. (1988). The effect of nasomaxillary injury on future facial growth. Balanced facial growth: A schematic interpretation. *Oral Surg., Oral Med., Oral Pathol.* **66**, 525–530.

Precious, D. S., and Hall, B. K. (1992). Growth and development of the maxillofacial region. In *Principles of Oral and Maxillofacial Surgery* (L. J. Peterson, ed.), Volume 3, pp. 1211–1236. J. B. Lippincott Co., Philadelphia, PA.

Precious, D. S., and Hall, B. K. (1994). Repair of fractured membrane bones. In *Bone, Volume 9: Differentiation and Morphogenesis of Bone* (B. K. Hall, ed.), pp. 145–163. CRC Press, Boca Raton, FL.

Precious, D. S., Morais, D., Armstrong, J. E. (1990). L'intérêt d'éviter l'utilisation de la fixation rigide lors de la génioplastie fonctionnelle. *Rev. Stomatol. Chir. Maxillofac.* **91**, 349–356.

Prentiss, C. W. (1903). Polydactylism in man and domestic animals, with especial reference to digital variations in swine. *Bull. Mus. Comp. Zool. Harvard* **40**, 245–303.

Presch, W. (1975). The evolution of limb skeleton in the teiid lizard genus *Bachia. Bull. So. Cal. Acad. Sci.* **74(3)**, 113–121.

Presley, R. (1983). A shaky foundation in the structure of the skull? *Nature* **302**, 210–211.

Pribylova, E., and Hert, J. (1971). Proliferation zones in articular cartilage of young rabbits. *Folia Morphol.* **19**, 233–241.

Price, P. A. (1988) Role of vitamin-K-dependent proteins in bone metabolism. *Annu. Rev. Nutrit.* **8**, 565–583.

Price, P. A., Lothringer, J. W., and Nishimoto, S. K. (1980a). Absence of the vitamin K-dependent bone protein in fetal rat mineral. Evidence of another γ-carboxyglutamic acid-containing component in bone. *J. Biol. Chem.* **255**, 2938–2942.

Price, P. A., Parthemore, J. G., Deftos, L. J., and Nishimoto, S. K. (1980b). New biochemical marker for bone metabolism – measurement by radioimmunoassay of bone GLA protein in the plasma of normal subjects and patients with bone disease. *J. Clin. Invest.* **66**, 878–883.

Price, P. A., Urist, M. A., and Otawara, Y. (1983). Matrix GLA protein, a new γ-carboxyglutamic acid-containing protein which is associated with the organic matrix of bone. *Biochem. Biophys. Res. Commun.* **117**, 765–771.

Primmett, D. R. N., Stern, C. D., and Keynes, R. J. (1988). Heat shock causes repeated segmental anomalies in the chick embryo. *Development* **104**, 331–339.

Prince, V. E., Joly, L., Ekker, M., and Ho, R. K. (1998a). Zebrafish *hox* genes: genomic organization and modified colinear expression patterns in the trunk. *Development* **125**, 407–420.

Prince, V. E., Prince, A. L., and Ho, R. K. (1998b). *Hox* gene expression reveals regionalization along the anteroposterior axis of the zebrafish notochord. *Devel. Genes Evol.* **208**, 517–522.

Prisell, P. T., Edwall, D., Lindblad, J. B., Levinovitz, A., and Norstedt, G. (1993). Expression of insulin-like growth factors during bone induction in rat. *Calcif. Tissue Int.* **53**, 201–205.

Pritchard, J. J. (1946). Repair of fracture of the parietal bone in rats. *J. Anat.* **80**, 55–60.

Pritchard, J. J. (1952). A cytological and histochemical study of bone and cartilage formation in the rat. *J. Anat.* **86**, 259–277.

Pritchard, J. J. (1956a). General anatomy and histology of bone. In *The Biochemistry and Physiology of Bone* (G. H. Bourne, ed.), 1st Edn, pp. 1–26. Academic Press, New York, NY.

Pritchard, J. J. (1956b). The osteoblast. In *The Biochemistry and Physiology of Bone* (G. H. Bourne, ed.), 1st Edn, pp. 179–212. Academic Press, New York, NY.

Pritchard, J. J. (1962). Heterotopic ossification and bone induction. *Anat. Anz.* **109**, 662–669.

Pritchard, J. J. (1965). Comparison of tendon and bone repair. In *Studied in Physiology* (D. R. Curtis and A. K. McIntyre, eds), pp. 232–238. Springer-Verlag, Berlin.

Pritchard, J. J. (1969). Bone. In *Tissue Repair* (R. H. McMinn, ed.), pp. 148–168. Academic Press, New York, NY.

Pritchard, J. J. (1972a). General histology of bone. In *The Biochemistry and Physiology of Bone* (G. H. Bourne, ed.), 2nd Edn, Volume 1, pp. 1–20. Academic Press, New York, NY.

Pritchard, J. J. (1972b). The control or trigger mechanism induced by mechanical forces which causes responses of mesenchymal cells in general and bone apposition and resorption in particular. *Acta Morphol. Neerl. Scand.* **10**, 63–69.

Pritchard, J. J. (1974). Growth and differentiation of bone and connective tissue. In *Differentiation and Growth of Cells in Vertebrate Tissues* (G. Goldspink, ed.), pp. 101–128. Chapman & Hall, London.

Pritchard, J. J., and Ruzicka, A. J. (1950). Comparison of fracture repair in the frog, lizard and rat. *J. Anat.* **84**, 236–261.

Pritchard, J. J., Scott, J. H., and Girgis, F. G. (1956). The structure and development of cranial and facial sutures. *J. Anat.* **90**, 73–86.

Pritchett, W. H., and Dent, J. N. (1972). The role of size in the rate of limb regeneration in the adult newt. *Growth* **36**, 275–290.

Prockop, D. J., Pettengill, O., and Holtzer, H. (1964). Incorporation of sulfate and the synthesis of collagen by cultures of embryonic chondrocytes. *Biochim. Biophys. Acta* **83**, 189–196.

Proell, F. (1926). Beiträge zur Vitalen Knochenfärbung. *Z. Zellforsch.* **3**, 461–471.

Proetzel, G., Pawlowski, S. A., Wiles, M. V., Yin, M., Boivin, G. P., Howles, P. N., Ding, J., Ferguson, M. W. J., and Doetschman, T. (1995). Transforming growth factor-β3 is required for secondary palate fusion. *Nature Genet.* **11**, 409–414.

Prostak, K., Seifert, P., and Skobe, Z. (1990). The effects of colchicine on the ultrastructure of odontogenic cells in the common skate. *Raja erinanae. Amer. J. Anat.* **189**, 77–91.

Prothero, R. S., and Schoch, R. M. (1994). *Major Features of Vertebrate Evolution*. Short Courses in Paleontology Number 7. University of Tennessee Knoxville for the Paleontology Society.

Prothero, R. S., and Schoch, R. M. (2002). *Horns, Tusks, and Flippers: The Evolution of Hoofed Mammals*. Johns Hopkins University Press, Baltimore, MD and London.

Prudden, J. F., Nishihara, G., and Baker, L. (1957). The acceleration of wound healing with cartilage. *Surg. Gynecol. Obstet.* **105**, 283–286.

Prummel, W. (1987). Atlas for identification of foetal skeletal elements of cattle, horse, sheep and pig. *Archaeozoologia* **1**, 23–30.

Pruzansky, S. (1971). The growth of the premaxillary-vomerine complex in complete bilateral cleft lip and palate. *Tandlaegebladet* **75**, 1157–1169.

Puchtler, H., Meldan, S. N., and Terry, M. S. (1969). On the history and mechanisms of alizarin and alizarin red S stains for calcium. *J Histochem. Cytochem.* **17**, 110–124.

Pugh, J. W., Radin, E. L., and Rose, R. M. (1974). Quantitative studies of human subchondral cancellous bone. Its relationship to the state of its overlying cartilage. *J. Bone Joint Surg. Am. Vol.* **56**, 313–321.

Pugin, E. M. (1972). Induction de cartilage, après excision de la cupule otique chez l'embryon de poulet, par des greffons d'organes embryonnaires de Souris. *C.R. Hebd. Seances Acad. Sci.* **275**, 2543–2546.

Pugin, E. M. (1973). Sur le comportement des troncons du tube neural et du la corde d'embryon de souris greffés à la place des organes homologues chez l'embryon de poulet. *CR. Acad. Sci. (D) Paris* **276**, 3477–3480.

Purandare, S. M., Ware, S. M., Kwan, K. M., Gebbia, M., Bassi, M. T., Deng, J. M., Vogel, H., Behringer, R. R., Belmont, J. W., and Casey, B. (2002). A complex syndrome of left-right axis, central nervous system and axial skeleton defects in *Zic3* mutant mice. *Development* **129**, 2293–2302.

Purnell, M. A. (1999). Conodonts: Functional analysis of disarticulated skeletal features. In *Functional Morphology of the Invertebrate Skeleton* (E. Savazzi, ed.). pp. 129–146. John Wiley & Sons, New York, NY.

Purnell, M. A., and Donoghue, P. C. J. (1997). Architecture and functional morphology of the skeletal apparatus of ozarkodinid conodonts. *Phil. Trans R. Soc. Lond. B.* **352**, 1545–1564.

Purnell, M. A., and von Bitter, P. H. (1992). Blade-shaped conodont elements functioned as cutting teeth. *Nature* **359**, 629–631.

Putchkov, V. F. (1964). Mechanism by which the scleral papillae develop in the embryonic chick eye. *Arkh. Anat. Gistol. Embriol.* **46**, 16–24.

Putman, R. J., Sullivan, M. S., and Langbein, J. (2000). Fluctuating asymmetry in antlers of fallow deer (*Dama dama*): The relative roles of environmental stress and sexual selection. *Biol. J. Linn. Soc.* **70**, 27–36.

Q

Qiu, M., Bulfone, A., Ghattas, I., Meneses, J. J., Christensen, L., Sharpe, P. T., Presley, R., Pedersen, R. A., and Rubenstein, J. L. R. (1997). Role of the Dlx homeobox genes in proximodistal patterning of the branchial arches: mutations of *Dlx-1*, *Dlx-2*, and *Dlx-1* and *-2* alter morphogenesis of proximal skeletal and soft tissue structures derived from the first and second arches. *Devel. Biol.* **185**, 165–184.

Qiu, M., Bulfone, A., Martinez, S., Meneses, J. J., Shimamura, K., Pedersen, R. A., and Rubenstein, J. L. R. (1995). Null mutation of *Dlx-2* results in abnormal morphogenesis of proximal first and second branchial arch derivatives and abnormal differentiation in the forebrain. *Genes & Devel.* **9**, 2523–2538.

Quarto, R., Campanile, G., Cancedda, R., and Dozin, B. (1992a). Thyroid hormone, insulin, and glucocorticoids are sufficient to support chondrocyte differentiation to hypertrophy: a serum-free analysis. *J. Cell Biol.* **119**, 989–995.

Quarto, R., Dozin, B., Bonaldo, P., Cancedda, R., and Colombatti, A. (1993). Type VI collagen expression is upregulated in the early events of chondrocyte differentiation. *Development* **117**, 245–251.

Quarto, R., Dozin, R., Tacchetti, C., Campanile, G., Malfatto, C., and Cancedda, R. (1990). *In vitro* development of hypertrophic chondrocytes starting from selected clones of dedifferentiated cells. *J. Cell Biol.* **110**, 1379–1386.

Quarto, R., Dozin, B., Tacchetti, C., Robino, G., Zenke, M., Campanile, G., and Cancedda, R. (1992b). Constitutive c-myc expression impairs hypertrophy and calcification in cartilage. *Devel. Biol.* **149**, 168–176.

Quekett, J. (1846). On the intimate structure of bone. *Trans. Microsc. Soc. London* **2**, 46–58.

Quilhac, A., and Sire, J.-Y. (1998). Restoration of the sub-epidermal tissues and scale regeneration after wounding a cichlid fish, *Hemicromis bimaculatus. J. Exp. Zool.* **281**, 305–327.

Quinn, J. C., West, J. D., and Kaufman, M. H. (1997). Genetic background effects on dental and other craniofacial abnormalities in homozygous small eye (Pax6[Sey]/Pax6[Sey]) mice. *Anat. Embryol.* **196**, 311–321.

Quinn, R. S., and Rodan, G. A. (1981). Enhancement of ornithine decarboxylase and Na[+], K[+] ATPase in osteoblastoma cells by intermittent compression. *Biochem. Biophys. Res. Commun.* **100**, 1696–1702.

Quint, E., Smith, A., Avaron, F., Laforest, L., Miles, J., Gaffield, W., and Akimenko, M-A. (2002). Bone patterning is altered in the regenerating zebrafish caudal fin after ectopic expression of *sonic hedgehog* and *bmp2b* or exposure to cyclopamine. *Proc. Natl Acad. Sci. U.S.A.* **99**, 8713–8718.

Quintner, M. I., and Goetinck, P. F. (1981). A biochemical analysis of cartilage proteoglycan in the avian mutant *micromelia-Abbott. Devel. Genet.* **2**, 35–48.

R

Raab-Cullen, D. M., Thiede, M. A., Petersen, D. N., Kimmel, D. B., and Recker, R. R. (1994). Mechanical loading stimulates rapid changes in periosteal gene expression. *Calcif. Tissue Int.* **55**, 473–478.

Rabinowitz, J. L., Tavares, C. J., Lipson, R., and Person, P. (1976). Lipid components and *in vitro* mineralization of some invertebrate cartilages. *Biol. Bull.* **150**, 69–79.

Rabinowitz, T., Syftestad, G. T., and Caplan, A. I. (1990). Chondrogenic stimulation of embryonic chick limb mesenchyme cells by factors in bovine and human dentine extracts. *Archs Oral Biol.* **35**, 49–55.

Radin, E. L., and Lanyon, L. E. (1982). The effects of ageing on the skeleton. In *Lectures on Gerontology, Volume 1: On Biology of Ageing, Part B* (A Viidik, ed.), pp. 379–406. Academic Press, London.

Radinsky, L. (1983). Allometry and reorganization in horse skull proportions. *Science* **221**, 1189–1191.

Radinsky, L. (1984). Ontogeny and phylogeny in horse skull evolution. *Evolution* **38**, 1–15.

Radlansky, R. J., Renz, H., and Klarkowski, M. C. (2003). Prenatal development of the human mandible. 3D reconstructions. Morphometry and bone remodelling patterns, sizes 12–117 mm CRL. *Anat. Embryol.* **210**, 221–232.

Radmosky, M. L., Thompson, A. Y., Spiro, R. L., and Poser, J. W. (1998). Potential role of fibroblast growth factor in enhancement of fracture healing. *Clin. Orthop. Rel. Res.* **355**, S283–S293.

Raff, M. (1996). Size control: the regulation of cell numbers in animal development. *Cell* **86**, 173–175.

Rafferty, K. L., and Herring, S. W. (1999). Craniofacial sutures: Morphology, growth and *in vivo* masticatory strains. *J. Morphol.* **242**, 167–179.

Rageh, M. A. E., Mendenhall, L., Moussad, E. E. A., Abbey, S. E., Mescher, A. L., and Tassava, R. A. (2002). Vasculature in pre-blastema and nerve-dependent blastema stages of regenerating forelimbs of the adult newt, *Notophthalmus viridescens. J. Exp. Zool.* **292**, 255–266.

Rahr, H. (1981). Ultrastructure of gill bars of *Branchiostoma lanceolatum* with special reference to gill skeleton and blood vessels (Cephalochordata). *Zoomorphology* **99**, 167–189.

Raisz, L. G., Simmons, H. A., Sandberg, A. L., and Canalis, E. (1980). Direct stimulation of bone resorption by epidermal growth factor. *Endocrinology* **107**, 270–273.

Rajtova, V. (1966). Skeletogeny in the guinea pig. I. Prenatal and postnatal ossification of the forelimb skeleton. *Folia Morphol.* **14**, 99–106.

Rajtova, V. (1967a). Skeletogeny in the guinea pig. II. The morphogenesis of the carpus in the guinea-pig (*Cavia porcellus*). *Folia Morphol.* **15**, 132–139.

Rajtova, V. (1967b). Skeletogeny in the guinea pig. III. Prenatal and postnatal ossification of the skeleton of the pelvic limb. *Folia Morphol.* **15**, 258–267.

Rajtova, V. (1968a). Skeletogeny in the guinea pig. 1V. Morphogenesis of the tarsus in the guinea pig (*Cavia porcellus*). *Folia Morphol.* **16**, 162–170.

Rajtova, V. (1968b). Skeletogeny in the guinea pig. V. Prenatal and postnatal ossification of the axial skeleton in the *Cavia porcellus. Folia Morphol.* **16**, 233–242.

Rajtova, V. (1969a). The development of the skeleton in the guinea pig. VI. Prenatal and postnatal ossification of the bones of the neurocranium in the guinea pig. *Folia Morphol.* **17**, 48–55.

Rajtova, V. (1969b). The development of the skeleton in the guinea pig. VI. Prenatal and postnatal ossification of the bones of the splanchnocranium in the guinea pig (*Cavia porcellus* L). *Folia Morphol.* **17**, 56–65.

Rajtova, V. (1971). Les transformations du cartilage de Meckelet l'ossification dela mandible chez *Cavia porcellu*s L. Part I. *Anat. Anz.* **128**, 392–401.

Ralis, Z. A., and Watkins, G. (1992). Modified tetrachrome method for osteoid and defectively mineralized bone in paraffin sections. *Biotech. Hist.* **67**, 339–345.

Rallis, C., Bruneau, B. G., Del Buono, J., Seidman, C. E., Seidman, J. G., Nissim, S., Tabin, C. J., and Logan, M. P. O. (2003). *Tbx5* is required for forelimb bud formation and continued outrowth. *Development* **130**, 2741–2751.

Ralphs, J. R., and Benjamin, M. (1994). The joint capsule: structure, composition, aging and disease. *J. Anat.* **184**, 503–509.

Ralphs, J. R., Tyers, R. N. S., and Benjamin, M. (1992). Development of functionally distinct fibrocartilages at two sites in the quadriceps tendon of the rat: the suprapatella and the attachment of the patella. *Anat. Embryol.* **185**, 181–187.

Ralphs, J. R., Wylie, L., and Hill, D. J. (1990). Distribution of insulin-like growth factor peptides in the developing chick embryo. *Development* **109**, 51–58.

Rama, S., and Chandrakasan, G. (1984). Distribution of different molecular species of collagen in the vertebral cartilage of shark (*Carcharhinus [Carcharius] acutus*). *Conn. Tissue Res.* **12**, 111–118.

Ramirez-Solis, R., Zheng, H., Whiting, J., Krumlauf, R., and Bradley, A. (1993). *Hoxb-4* (Hox-2.6) mutant mice show homeotic transformation of a cervical vertebra and defects in the closure of the sternal rudiment. *Cell* **73**, 279–294.

Ramp, W. K. (1975). Cellular control of calcium movements in bone. Interrelationships of the bone membrane, parathyroid hormone and alkaline phosphatase. *Clin. Orthop. Rel. Res.* **106**, 311–322.

Rani, P. U., Stringa, E., Dharmavaram, R., Chatterjee, D., Tuan, R. S., and Khillan, J. S. (1999). Restoration of normal bone development by human homologue of collagen type II (Col2a1) gene in Col2a1 null mice. *Devel. Dynam.* **214**, 26–33.

Ranly, D. M. (1988). *A Synopsis of Craniofacial Growth.* 2nd Edn. Appleton & Lange, Norwal, Conn.

Ranta, R., Alhopuro, S., and Ritsula, V. (1973). The effect of tooth extractions on the growth of the jaws in rabbits. *Proc. Fin. Dent. Soc.* **69**, 116–119.

Rao, L. G., Ng, B., Brunette, D. M., and Heersche, J. N. M. (1977). Parathyroid hormone- and prostaglandin E_1-response in a selected population of bone cells after repeated subculture and storage at $-80°C$. *Endocrinology* **100**, 1233–1241.

Rapraeger, A. C. (2000). Syndecan-regulkated receptor signaling. *J. Cell Biol.* **149**, 995–997.

Rasmussen, H., and Bordier, P. (1974). *The Physiological and Cellular Basis of Metabolic Bone Disease.* Williams & Wilkins, Baltimore, MD, MD.

Rasmussen, H., and Goodman, D. B. P. (1977). Relationships between calcium and cyclic nucleotides in cell activation. *Physiol. Rev.* **57**, 421–509.

Rasmussen, K. K., Vilmann, H., and Juhl, M. (1986). Os penis of the rat. V. The distal cartilage process. *Acta Anat* **125**, 208–212.

Rath, N. C., and Reddi, A. H. (1979). Collagenous bone matrix is a local mitogen. *Nature* **278**, 855–857.

Rauber, A. A. (1876). *Elasticität und Festigheit der Knochen.* Engelmann, Leipzig.

Rauchfuss, A. (1989). Pneumatization and mesenchyme in the human middle ear. *Acta Anat.* **136**, 285–290.

Rawlins, R. G. (1975). Age changes in the pubic symphysis of *Macaca mulatta. Amer. J. Phys. Anthropol.* **42**, 477–488.

Raynaud, A. (1972). Morphogenèse des membres rudimentaires chez les reptiles: Un probléme d'embryologie et d'évolution. *Bull. Soc. Zool. Fr.* **97**, 469–485.

Raynaud, A. (1974). Stades précoces du développement de la région cloacale et des appendices postérieurs chez l'embryon de *Python reticulatus* (Schneider, 1801). *Bull. Mus. Natl Hist. Nat.* **225**, 705–720.

Raynaud, A. (1977). Somites and early morphogenesis in reptile limbs. In *Vertebrate Limb and Somite Morphogenesis* (D. A. Ede, J. R. Hinchliffe, and M. Balls, eds), pp. 373–386. Cambridge University Press, Cambridge.

Raynaud, A. (1981). La réduction du nombre des doigts aux mains et aux pieds, chez les embryons de lézard vert (*Lacerta viridis* Laud), sous l'effet de la cytosine-arabinofuranoside. *C.R. Acad. Sci. Paris* **293**, 383–388.

Raynaud, A. (1986). Modifications précoces de l'ontogenèse des membres d'embryos de *Lacerta viridis* (Laur.) sous l'effet de la cytosine-arabinofuranoside comparaison avec l'ontogenèse des membres de reptiles serpentiformes. *C.R. Acad. Sci. Paris* **303**, 37–42.

Raynaud, A. (1987). Modalités ontogènétiques et la réduction digitale dans la patte des embryons de seps tridactyle (*Chalcides chalcides*, L.). *C.R. Acad. Sci. Paris* **304**, 359–364.

Raynaud, A. (1990). Developmental mechanisms involved in the embryonic reduction of limbs in reptiles. *Int. J. Devel. Biol.* **34**, 233–243.

Raynaud, A., and Adrian, M. (1975a). Mise en évidence, au moyen de la microscopie électronique de la pénétration des cellules somitiques dans le mesoblaste de l'ébauche des membres des embryons de reptiles (*Anguis fragilis, Lacerta viridis*). *Arch. Anat. Microsc. Morphol. Exp.* **64**, 287–316.

Raynaud, A., and Adrian, M. (1975b). Caractéristiques ultrastructurales des divers constituants des ébauches des membres, chez les embryons d'Orvet (*Anguis fragilis* L) et de lézard (*Lacerta viridis* Laur). *C.R. Hebd. Seances Acad. Sci.* **280**, 2591–2594.

Raynaud, A., Adrian, M., and Kouprach, S. (1973a). Étude au microscope électronique de la dégénérescence de la crête apicale épiblastique des ébauches de membres de l'embryon d'Orvet (*Anguis fragilis* L) *C.R. Acad. Sci. Paris* **277**, 1503–1505.

Raynaud, A., Adrian, M., and Kouprach, S. (1973b). Étude ultrastructurale du mésoblasts de l'ébauches des membres de l'embryon d'Orvet (*Anguis fragilis* L) et cours de la période de règression de la crête apicale. *C.R. Acad. Sci. Paris* **277**, 1671–1673.

Raynaud, A., Adrian, M., and Kouprach, S. (1974a). Étude, au microscope électronique des ébauches des membres de l'orvet (*Anguis fragilis* L) et du lézard vert (*Lacerta viridis* Laur). *Ann. Embryol. Morphol.* **7**, 243–264.

Raynaud, A., and Brabet, J. (1979). On the ultrastructure of the apical crest of the limb bud of *Anguis fragilis* and *Lacerta viridis*. *C.R. Acad. Sci. Paris* **288**, 1675–1678.

Raynaud, A., Brabet, J., and Adrian, M. (1979). Comparative ultrastructural study of the apical ridge of the limb buds of embryos of the slow-worm (*Anguis fragilis* L) and embryos of the green lizard (*Lacerta viridis* Laur). *Arch. d'Anat. Micros. Morphol. Exp.* **68**, 301–331.

Raynaud, A., and Clergue-Gazeau, M. (1986). Identifications des doigts reduits ou manquants dans les pattes des embryos de lezard vert (*Lacerta viridis*, L.aur.) traites par la cytosine-arabi-nofuranoside. Comparison avec les reductions digitales naturelles des especes de reptiles serpentiformes. *Arch. Biol. Bruxelles* **97**, 279–299.

Raynaud, A., Gasc, J. P., Vassé, J., Renous, S., and Pieau, C. (1974c). Relations entre les somites et les ébauches des membres antérieurs chez les jeunes embryons de '*Scelotes brevides*' (Hewitt). *Bull. Soc. Zool. Fr.* **99**, 165–173.

Raynaud, A., Jeanny, J.-C., and Gontcharoff, M. (1975). Données cytophotométriques sur les teneurs en protéines et en ARN, des noyaux de la crête apicale et de l'epiblaste de la jeune ébauche du membre du Lézard vert (*Lacerta viridis* Laur) et de l'Orvet (*Anguis fragilis* L). *C.R. Hebd. Seances Acad. Sci.* **280**, 2693–2696.

Raynaud, A., Jeanny, J.-C., and Gontcharoff, M. (1977). Histochemical and cytophotometric data on the apical crest, mesoblastic cells and the somitic processes of the limb buds of reptile embryos (*Lacerta viridis* Laur. and *Anguis fragilis* L). *Arch. d'Anat. Micros. Morphol. Exp.* **66**, 73–96.

Raynaud, A., and Kan, P. (1992). DNA synthesis decline involved in the developmental arrest of the limb bud in the embryo of the slow worm, *Anguis gragilis* (L.). *Int. J. Devel. Biol.* **36**, 303–310.

Raynaud, A., Okuzumi, H., and Kouprach, S. (1974b). Morphologie externe des stades précoces du développement des ébauches des membres du Lézard vert '*Lacerta viridis* Laur' et de l'Orvet '*Anguis fragilis* L', etudiee au moyen de la microscopie électronique à balayage. *Bull. Soc. Zool. Fr.* **99**, 149–153.

Raynaud, A., Perret, J.-L., Bons, J., Clergue-Gazeau, M. (1989). Pattern of digital reduction in some African Scincidae (Reptilia). *Rev. Suisse Zool.* **96**, 779–802.

Raynaud, A., Renous, S., Gasc, J.-P., Clergue-Gazeau, M. (1990). Contribution a la recherche des homologies dans la region cervicale de l'embryon de lezard vert (*Lacerta viridis*, Laur) et de l'embryon d'orvet (*Anguis fragilis*, L.). *Amphib.-Reptil.* **11**, 339–350.

Raynaud, A., and van den Elzen, P. (1976). La rudimentation des membres chez les embryons de *Scelotes gronovii* (Daudin), reptile scincidé Sud-Africain. *Arch. Anat. Microsc. Morphol. Exp.* **65**, 17–36.

Raynaud, A., and van den Elzen, P. (1978). Structure histologique aux stades avancés du développement et chex l'adulte, des membres postérieurs rudimentaires et deux scincidés Sid-Africains (*Scelotes brevipes* et *Scelotes gronovii*) et leur utilisation pour la locomotion chez *Scelotes gronovii*. *Bull. Soc. Hist. Nat. Toulouse* **114**, 360–372.

Rayne, J., and Crawford, G. N. C. (1975). Increase in fibre numbers of the rat pterygoid muscles during postnatal growth. *J. Anat.* **119**, 347–357.

Reagan, F. P. (1915). A genetic interpretation of the stapes, based on a study of avian embryos in which the development of the cartilaginous otic capsules has been experimentally inhibited. *Anat. Rec.* **9**, 114–115.

Reagan, F. P. (1917). The role of the auditory sensory epithelium in the formation of the stapedial plate. *J. Exp. Zool.* **23**, 85–108.

Rebhun, L. I. (1977). Cyclic nucleotides, calcium and cell division. *Int. Rev. Cytol.* **39**, 1–54.

Recklinghausen, F. von (1910). *Untersuchungen über Rachitis und Osteomalacie*. Gustav Fisher, Jena.

Reddi, A. H. (1985). Age-dependent decline in extracellular matrix-induced local bone differentiation. *Israel J. Med. Sci.* **21**, 312–313.

Reddi, A. H. (1992). Regulation of cartilage and bone differentiation by bone morphogenetic proteins. *Curr. Opin. Cell Biol.* **4**, 850–855.

Reddi, A. H. (1994). Bone and cartilage differentiation. *Curr. Opin. Genet. Devel.* **4**, 737–744.

Reddi, A. H., Muthukumaran, N., Ma, S., Carrington, J. L., Luyten, F. P., Paralkar, V. M., and Cunningham, N. S. (1989). Initiation of bone development by osteogenin and promotion by growth factors. *Conn. Tissue Res.* **20**, 303–312.

Reddi, A. H., and Wlodarski, K. H. (1986). Precursors of the fibroblast colony forming units (CFU-F) in heterotopically induced bone marrow of rats and mice. *Bull. Polish Acad. Sci. Biol. Sci.* **34**, 23–27.

Reddy, N. P., and Joshi, A. M. (1987). A stochastic compartmental model of bone cells. *Int. J. Biomed. Computing* **21**, 163–174.

Redi, L. (1976). Visceral cartilage. *J. Anat.* **122**, 349–356.

Redini, F., Galera, P., Mauviel, A., Loyau, G., and Pujol, J.-P. (1988). Transforming growth factor β stimulates collagen and glycosaminoglycan biosynthesis in cultured rabbit articular chondrocytes. *FEBS Letters* **234**, 172–176.

Redler, I. (1974). A scanning electron microscopic study of human normal and osteoarthritic cartilage. *Clin. Orthop. Rel. Res.* **103**, 262–268.

Reed, C. G., and Cloney, R. A. (1977). Brachiopod tentacles: ultrastructure and functional significance of the connective tissue and myoepithelial cells in *Terebratalia*. *Cell Tissue Res.* **185**, 17–42.

Reginato, A. M., Tuan, R. S., Ono, T., Jimenez, S. A., and Jacenko, O. (1993). Effects of calcium deficiency on chondrocyte hypertrophy and type X collagen expression in chick embryonic sternum. *Devel. Dynam.* **198**, 284–295.

Reginelli, A. D., Wang, Y.-Q., Sassoon, D., and Muneoka, K. (1995). Distal tip regeneration correlated with regions of *Msx1* (*Hox 7*) expression in fetal and newborn mice. *Development* **121**, 1065–1076.

Reid, R. E. H. (1981). Lamellar-zonal bone with zones and annuli in the pelvis of a sauropod dinosaur. *Nature* **292**, 49–51.

Reid, R. E. H. (1984). The histology of dinosaurian bone, and its possible bearing on dinosaurian physiology. *Symp. Zool. Soc. London # **52**, 629–664.

Reif, W.-E. (1978). Types of morphogenesis of the dermal skeleton in fossil sharks. *Paläont. Z.* **52**, 110–128.

Reif, W.-E. (1980a). Development of dentition and dermal skeleton in embryonic *Scyliorhinus canicula*. *J. Morphol.* **166**, 275–288.

Reif, W.-E. (1980b). A model of morphogenetic processes in the dermal skeleton of elasmobranchs. *N. Jb. Geol. Paläont. Abh.* **159**, 339–359.

Reif, W.-E. (1982). Evolution of dermal skeleton and dentition in vertebrates: The odontode regulation theory. *Evol. Biol.* **15**, 287–368.

Reif, W. E., and Richter, M. (2001). Revisiting the lepidomorial and the odontode regulation theories of dermoskeletal morphogenesis. *Neues Jahrb. Geol. Palaont.-Abt.* **219**, 285–304.

Reilly, S. M. (1986). Ontogeny of cranial ossification in the Eastern Newt, *Notophthalmus viridsecens* (Caudata: Salamandridae), and its relationship to metamorphosis and neoteny. *J. Morphol.* **188**, 315–326.

Reilly, S. M. (1987). Ontogeny of the hypobranchial apparatus in the salamanders *Ambystoma talpoideum* (Ambystomatidae) and *Notophthalmus viridescens* (Salamandridae); The ecological morphology of two neotenic strategies. *J. Morphol.* **191**, 205–214.

Reilly, S. M., and Lauder, G. V. (1988). Atavisms and the homology of hyobranchial elements in lower vertebrates. *J. Morphol.* **195**, 237–246.

Reilly, S. M., and Lauder, G. V. (1990). Metamorphosis of cranial design in tiger salamanders (*Ambystoma tigrinum*) – A morphometric analysis of ontogenetic change. *J. Morphol.* **204**, 121–137.

Reimchen, T. E., and Nosil, P. (2001). Lateral plate asymmetry, diet and parasitism in threespine sticklebacks. *J. Evol. Biol.* **14**, 632–645.

Reinbold, R. (1968). Rôle du tapetum dans la différenciation de la sclérotique chez l'embryon de poulet. *J. Embryol. Exp. Morphol.* **19**, 43–47.

Reinholt, F. P., Engfeldt, B., Hjerpe, A., and Jansson, K. (1982). Stereological studies on the epiphyseal growth plate with special reference to the distribution of matrix vesicles. *J. Ultrastruct. Res.* **80**, 270–279.

Reiss, J. O. (1997). Early development of chondrocranium in the tailed frog *Ascaphus truei* (Amphibia: Anura): Implications for anuran palatoquadrate homologies. *J. Morphol.* **231**, 63–100.

Reiss, J. O. (1998). Anuran postnasal wall homology: an experimental extirpation study. *J. Morphol.* **238**, 343–353.

Reiss, J. O. (2002). The phylogeny of amphibian metamorphosis. *Zoology* **105**, 85–96.

Reisz, R. R., and Laurin, M. (1991). *Owenetta* and the origin of turtles. *Nature* **349**, 324–326.

Reisz, R. R., and Smith, M. M. (2001). Lungfish dental pattern conserved for 360 Mya. *Nature* **411**, 548.

Remak, R. (1855). Untersuchungen über die Entwicklung der Wirbelthiere. G. Reimer, Berlin.

Renault, J. (1886). Note sur le malle osseuse et le disposition anatomique en raport avec ses proprietes osteogeniques comunes. *Gaz. Med. Paris* **57**, 15–18.

Renous, S., Exbrayat, J. M., and Estabel, J. (1997). Recherche d'indices de membres chez les Amphibiens Gymnophiones. *Ann. Sci. Nat. Zool.* **18**, 11–26.

Renous, S., Gasc, J.-P., and Raynaud, A. (1991). Comments on the pelvic appendicular vestiges in an amphisbaenian: *Blanus cinereus* (Reptilia: Squamata). *J. Morphol.* **209**, 23–28.

Renous, S., Rimblot-Baly, F., Fretey, J., and Pieau, C. (1989). Embryonic development characteristics of the leatherback *Dermochelys coriacea* (Vandelli, 1761). *Ann. Sci. Nat. Zool. Biol. Anim.* **10**, 97–229.

Rensberger, J. M., and Watabe, M. (2000). Fine structure of bone in dinosaurs, birds and mammals. *Nature* **406**, 619–622.

Révillion-Carette, F., Desbiens, X., Meunier, L., and Bart. A. (1986). Chondrogenesis in mouse limb buds *in vitro*: effects of dibutyryl cyclic AMP treatment. *Differentiation* **33**, 121–129.

Reuss, C., and Saunders, J. W., Jr. (1965). Inductive and axial properties of prospective limb mesoderm in the early chick embryo. *Amer. Zool.* **5**, 214.

Revest, J.-M., Spencer, D. B., Kerr, K., DeMoerlooze, L., Rosewell, I., and Dickson, C. (2001). Fibroblast growth factor receptor 2-IIIB acts upstream of *Shh* and *Fgf4* and is required for limb bud maintenance, but not for the induction of *Fgf8*, *Fgf10*, or *BMP4*. *Devel. Biol.* **231**, 47–62.

Rhodes, R. K., and Elmer, W. A. (1975). Aberrant metabolism of matrix components in neonatal fibular cartilage of brachypod (*bp^H*). *Devel. Biol.* **46**, 14–27.

Rhodin, A. G. J., Ogden, J. A., and Conlogue, G. J. (1981). Chondro-osseous morphology of *Dermochelys coriacea*, a marine reptile with mammalian skeletal features. *Nature* **290**, 244–246.

Riancho, J. A., and Mundy, G. R. (1995). The role of cytokines and growth factors as mediators of the effects of systemic hormones at the bone local level. *Crit. Rev. Eukary. Gene Exp.* **5**, 193–217.

Riccardi, V. M., and Eichner, J. E. (1986). *Neurofibromatosis: Phenotype, Natural History, and Pathogenesis.* Johns Hopkins University Press, Baltimore, MD.

Riccardi, V. M., Mulvihill, J. J., and Wade, W. M. (eds) (1981). *Neurofibromatosis (von Recklinghausen Disease).* Advances in Neurobiology Volume 20, 282 pp. Raven Press, New York, NY.

Rice, D. (1999). *Molecular Mechanisms in Calvarial Bone and Suture Development.* Ph.D. Dissertationes Biocentri Viikki Universitatis Helsingiensis, Helsinki, Finland.

Rice, D. P. C., Åberg, T., Chan, Y.-S., Tang, Z., Kettunen, P. J., Pakarinen, L., Maxson, R. E., Jr., and Thesleff, I. (2000). Integration of FGF and TWIST in calvarial bone and suture development. *Development* **127**, 1845–1855.

Rice, D. P. C., Rice, R., and Thesleff, I. (2003). Molecular mechanisms in calvarial bone and suture development, and their relation to craniosynostosis. *Eur. J. Orthodont.* **25**, 139–148.

Richany, S. F., Bast, T. H., and Anson, B. J. (1956). The development of the first branchial arch in man, and the fate of Meckel's cartilage. *Q. Bull. Northwest. Univ. Med. School.* **30**, 331–355.

Richards, C. M., Carlson, B. M., and Rogers, S. L. (1975). Regeneration of digits and forelimbs in the Kenyan reed frog, *Hyperolius viridiflavus ferniquei. J. Morphol.* **146**, 431–446.

Richardson, M. K., Carl, T. F., Hanken, J., Elinson, R. P., Cope, C., and Bagley, P. (1998). Limb development and evolution: a frog embryo with no apical ectodermal ridge (AER). *J. Anat.* **192**, 379–390.

Richardson, M. K., Jeffery, J. E., and Tabin, C. J. (2004). Proximodistal patterning of the limb: Insights from evolutionary morphology. *Evol. & Devel.* **6**, 1–5.

Richardson, M. K., and Oelschläger, H. A. (2002). Time, pattern, and heterochrony: a study of hyperphalangy in the dolphin embryo flipper. *Evol. & Devel.* **4**, 435–444.

Richman, J. M., and Crosby, Z. (1990). Differential growth of facial primordia in chick embryos: responses of facial mesenchyme to basic fibroblast growth factor (bFGF) and serum in micromass culture. *Development* **109**, 341–348.

Richman, J. M., and Diewert, V. M. (1988). The fate of Meckel's cartilage chondrocytes in ocular culture. *Devel. Biol.* **129**, 48–60.

Richman, J. M., Herbert, M., Matovinovic, E., and Walin, J. (1997). Effect of fibroblast growth factors on outgrowth of facial mesenchyme. *Devel. Biol.* **189**, 135–147.

Richman, J. M., and Lee, S.-H. (2003). About face: signals and genes controlling jaw patterning and identity in vertebrates. *BioEssays* **25**, 554–568.

Richman, J. M., and Mitchell, P. J. (1996). Craniofacial development: Knockout mice take one on the chin. *Curr. Biol.* **6**, 364–367.

Richman, J. M., and Tickle, C. (1989). Epithelia are interchangeable between facial primordia of chick embryos and morphogenesis is controlled by the mesenchyme. *Devel. Biol.* **136**, 201–210.

Richman, J. M., and Tickle, C. (1992). Epithelial-mesenchymal interactions in the outgrowth of limb buds and facial primordia in chick embryos. *Devel. Biol.* **154**, 299–308.

Richter, M., and Smith, M. M. (1995). A microstructural study of the ganoine tissue of selected lower vertebrates. *Zool. J. Linn. Soc.* **114**, 173–212.

Richtsmeier, J. T., and Cheverud, J. M. (1986). Finite element scaling analysis of human craniofacial growth. *J. Craniofac. Genet. Devel. Biol.* **6**, 289–324.

Richtsmeier, J. T., Cole, T. M. III., Krovitz, G., Valeri, C. J., and Lele, S. (1998). Preoperative morphology and development in sagittal synostosis. *J. Craniofac. Genet. Devel. Biol.* **18**, 64–78.

Richtsmeier, J. T., Corner, B. D., Grausz, H. M., Cheverud, J. M., and Danahey, S. E. (1993). The role of postnatal growth patterns in the production of facial morphology. *Syst. Biol.* **42**, 307–330.

Richtsmeier, J. T., and Lele, S. (1993). A coordinate-free approach to the analysis of growth patterns: models and theoretical considerations. *Biol. Rev. Camb. Philos. Soc.* **68**, 381–411.

Rickard, D. J., Sullivan, T. A., Shenker, B. J., Leboy, P. S., and Kazhdan, I. (1994). Induction of rapid osteoblast differentiation in rat bone marrow stromal cell cultures by dexamethasone and BMP-2. *Devel. Biol.* **161**, 218–228.

Riede, U. N., Zinkernagel, R., Remagen, W., and Villiger, W. (1971). Zellen und Matrix im Blasenknorpel einer durch D-Penicillamin veränderten Ratten tibiaepiphysenfuge. *Beitr. Pathol.* **143**, 271–282.

Riedl, R. (1978). *Order in Living Organisms.* John Wiley & Sons, New York, NY.

Riegler, H. F., and Harris, C. M. (1976). Heterotopic bone formation after total hip arthroplasty. *Clin. Orthop. Rel. Res.* **117**, 209–216.

Rieppel, O. (1984a). Miniaturization of the lizard skull: It's functional and evolutionary implications. *Symp. Zool. Soc. London* **# 52**, 503–520.

Rieppel, O. (1984b). The upper temporal arcade of lizards: An ontogenetic problem. *Rev. Suisse Zool.* **91**, 475–482.

Rieppel, O. (1988). A review of the origin of snakes. *Evol. Biol.* **22**, 37–130.

Rieppel, O. (1992a). Studies on skeletal formation in reptiles. I. The postembryonic development of the skeleton in *Cyrtodactylus pubisculus* (Reptilia: Gekkonidae). *J. Zool.* **227**, 87–100.

Rieppel, O. (1992b). Studies on skeletal formation in reptiles. III. Patterns of ossification in the skeleton of *Lacerta vivipara* Jacquin (Reptilia, Squamata). *Fieldiana, Zool.* ns #68 (1437), 1–25.

Rieppel, O. (1993a). Studies on skeletal formation in reptiles. Patterns of ossification in the skeleton of *Chelydra serpentina* (Reptilia, Testudines). *J. Zool.* **231**, 487–509.

Rieppel, O. (1993b). Studies on skeletal formation in reptiles. II. *Chamaeleo hoehnelii* (Squamata: Chamaeleoninae), with comments on the homology of carpal and tarsal bones. *Herpetologica* **49**, 66–78.

Rieppel, O. (1993c). Studies on skeletal formation in reptiles. IV. The homology of the reptilian (amniote) astragalus revisited. *J. Vert. Paleont.* **13**, 31–47.

Rieppel, O. (1993d). Studies on skeletal formation in reptiles. V. Patterns of ossification in the skeleton of *Alligator mississippiensis* Daudin (Reptilia, Crocodylia). *Zool. J. Linn. Soc.* **109**, 301–325.

Rieppel, O. (1994). Studies on skeletal formation in reptiles. I. Patterns of ossification in the limb skeleton of *Gehyra oceanica* (Lesson) and *Lepidodactylus lugubris* (Dumeril & Bibron). *Ann. Sci. Nat. Zool. Biol. Anim.* **15**, 83–91.

Rieppel, O. (2001). Turtles as hopeful monsters. *BioEssays* **23**, 987–991.

Rieppel, O., and de Braga, M. (1996). Turtles as diapsid reptiles. *Nature* **384**, 453–455.

Rieppel, O., and Crumly, C. (1997). Paedomorphosis and skull structure in Malagasy chamaeleons (Reptilia: Chamaeleoninae). *J. Zool. Lond.* **243**, 351–380.

Rieppel, O., and Kearney, M. (2001). The origin of snakes: limits of a scientific debate. *Biologist* **48**, 110–114.

Rieppel, O., and Reisz, R. R. (1999). The origin and early evolution of turtles. *Annu. Rev. Ecol. Syst.* **30**, 1–22.

Rieppel, O., and Zaher, H. (2000). The intramandibular joint in squamates, and the phylogenetic relationships of the fossil snake *Pachyrhachis problematicus* Haas. *Fieldiana Geol.* ns #43, **1507**, 1–69.

Rieppel, O., Zaher, H., Tchernov, E., and Polcyn, M. J. (2003). The anatomy and relationships of *Haasiophis terrasanctus*, a fossil snake with well-developed hind limbs from the Mid-Cretaceous of the Middle East. *J. Paleont.* **77**, 536–558.

Rifas, L., Shen, V., Mitchell, K., and Peck, W. A. (1984). Macrophage-derived growth factor for osteoblast-like cells and chondrocytes. *Proc. Natl Acad. Sci. U.S.A.* **81**, 4558–4562.

Rifas, L., Uitto, J., Memoli, V. A., Kuettner, K. E., Henry, R. W., and Peck, W. A. (1982). Selective emergence of differentiated chondrocytes during serum-free culture of cells derived from fetal rat calvaria. *J. Cell Biol.* **92**, 493–504.

Rifkin, B. R., Baker, R. L., and Coleman, S. J. (1980a). Effects of prostaglandin E2 on macrophages and osteoclasts in cultured fetal long bones. *Cell Tissue Res.* **207**, 341–346.

Rifkin, B. R., Baker, R. L., and Coleman, S. J. (1980b). Osteoid resorption by mononuclear cells *in vitro*. *Cell Tissue Res.* **210**, 493–500.

Rifkin, B. R., Brand, J. S., Cushing, J. E., Coleman, S. J., and Sanavi, F. (1980c). Fine structure of fetal rat calvarium; provisional identification of preosteoclasts. *Calcif. Tissue Int.* **31**, 21–28.

Rifkin, B. R., and Heijl, L. (1980). The occurrence of mononuclear cells at sites of osteoclastic bone resorption in experimental periodontitis. *J. Periodont.* **50**, 636–640.

Rigo, C., and Bairati, A. (1998). Use of rotary shadowing electron microscopy to investigate the collagen fibrils in the extracellular matrix in cuttle-fish (*Sepia officinalis*) and chick cartilage. *Tissue & Cell* **30**, 112–117.

Rijli, F. M., and Chambon, P. (1997). Genetic interactions of *Hox* genes in limb development: learning from compound mutants. *Curr. Opin. Genet. Devel.* **7**, 481–487.

Rijli, F. M., Mark, M., Lakkaraju, S., Dierich, A., Dollé, P., and Chambon, P. (1993). A homeotic transformation is generated in the rostral branchial region of the head by disruption of *Hoxa-2*, which acts as a selector gene. *Cell* **75**, 1333–1349.

Riley, B. B., Savage, M. P., Simandi, B. K., Olwin, B. B., and Fallon, J. F. (1993). Retroviral expression of FGF-2 (bFGF) affects patterning in chick limb bud. *Development* **118**, 95–104.

Riley, P. A. (1974). The effect on cell proliferation of reduced substrate adhesiveness. *Cell Differ.* **3**, 233–238.

Rimoin, D. L. (1975). The chondrodystrophies. *Adv. Hum. Genet.* **5**, 1–118.

Rimoin, D. L., Hughes, G. N., Kaufman, R. L., Rosenthal, R. E., McAlister, W. H., and Silberberg, R. (1970). Endochondral ossification in achondroplastic dwarfism. *New Engl. J. Med.* **283**, 728–735.

Rimoin, D. L., Silberberg, R., and Hollister, D. W. (1976). Chondro-osseous pathology in chondrodystrophies. *Clin. Orthop. Rel. Res.* **114**, 137–152.

Rinaldi, L. (1972). Morfogenesi della regione prossimale del tarso-metatarso nell'embrione di Pollo. *Ist. Lombardo (Rend. Sc.)* **106**, 83–91.

Rinaldi, L., and Caronna, E. W. (1971). L'ossificazione embrionale del Pollo (*Gallus domesticus* L.). *Arch. Ital. Anat. Embriol.* **76**, 201–237.

Rinaldi, L., Caronna, E. W., and Cinquetti, R. (1974). Processi di ossificazione nelle ossa lunghe del Pollo. *Rend., 1st. Lomb. Accad. Sci. Lett. B* **108**, 52–62.

Ringe, J.-D., and Steinhagen-Thiessen, E. (1985). Prevention of physiological age-dependent bone atrophy by controlled exercise in mice. *Age* **8**, 44–47.

Ringuette, M., Damjanovski, S., and Wheeler, D. (1991). Expression of SPARC/osteonectin in tissues of bony and cartilaginous vertebrates. *Biochem. Cell Biol.* **69**, 245–250.

Rintala, M., Metsäranta, M., Garofalo, S., de Crombrugghe, B., Vvorio, E., Rönning, O. (1993). Abnormal craniofacial morphology and cartilage structure in transgenic mice harboring a Gly-Cys mutation in the cartilage-specific type II collagen gene. *J. Craniofac. Genet. Devel. Biol.* **13**, 137–146.

Rintala, M., Metsäranta, M., Säämänen, A.-M., Vvorio, E., Rönning, O. (1997). Abnormal craniofacial growth and early mandibular osteoarthritis in mice harboring a mutant type II collagen transgene. *J. Anat.* **190**, 201–208.

Rintala, M., Metsäranta, M., Vvorio, E., Rönning, O. (1996). Abnormalities in secondary cartilages in four lines of transgenic mice harboring two different types of mutations in the cartilage-specific type II collagen gene. *J. Craniofac. Genet. Devel. Biol.* **16**, 148–155.

Riou, J.-F., Umbhauer, M., Shi, D. L., and Boucart, J.-C. (1992). Tenascin: a potential modulator of cell-extracellular matrix interactions during vertebrate embryogenesis. *Biol. Cell* **75**, 1–9.

Ripamonti, U., Heliotis, M., Rueger, D. C., and Sampath, T. K. (1996). Induction of cementogenesis by recombinant human osteogenic protein-1 (HOP-1/BMP-7) in the baboon (*Papio ursinus*). *Archs Oral Biol.* **41**, 121–126.

Ripamonti, U., Ma, S., Cunningham, N. S., Yeates, L., and Reddi, A. H. (1992c). Initiation of bone regeneration in adult baboons by osteogenin, a bone morphogenetic protein. *Matrix* **12**, 369–380.

Ripamonti, U., Ma, S., and Reddi, A. H. (1992a). Induction of bone in composites of osteogenin and porous hydroxyapatite in baboons. *Plast. Reconst. Surg.* **89**, 731–739.

Ripamonti, U., Ma, S., and Reddi, A. H. (1992b). The critical role of geometry of porous hydroxyapatite delivery system in induction of bone by osteogenin, a bone morphogenetic protein. *Matrix* **12**, 202–212.

Ripamonti, U., Magan, A., Ma, S., Van den Heever, B., Moehlt, T., and Reddi, A. H. (1991). Xenogeneic osteogenin, a bone morphogenetic protein, and demineralized bone matrices, including human, induce bone differentiation in athymic rats and baboons. *Matrix* **11**, 404–411.

Ripamonti, U., and Reddi, A. H. (1992). Growth and morphogenetic factors in bone induction: role of osteogenin and related bone morphogenetic proteins in craniofacial and periodontal bone repair. *Crit. Rev. Oral Biol. Med.* **3**, 1–14.

Ripamonti, U., and Reddi, A. H. (1997). Tissue engineering, morphogenesis, and regeneration of the periodontal tissues by bone morphogenetic proteins. *Crit. Rev. Oral Biol. Med.* **8**, 154–163.

Ripamonti, U., Yeates, L., and van den Heever, B. (1993). Initiation of heterotopic ossification in primates after chromatographic adsorption of osteogenin, a bone morphogenetic protein, onto porous hydroxyapatite. *Biochem. Biophys. Res. Commun.* **193**, 509–517.

Ris, P. M., and Wray, J. B. (1972). A histological study of fracture healing within the uterus of the rabbit. *Clin. Orthop. Rel. Res.* **87**, 318–321.

Ritsila, V., and Alhopuro, S. (1973). Reconstruction of experimental tracheal cartilage defects with free periosteum. A preliminary report. *Scand. J. Plast. Reconstr. Surg.* **7**, 116–119.

Ritsila, V., Alhopuro, S., and Ranta, R. (1973). The role of the zygomatic arch in the growth of the skull in rabbits. *Proc. Finn. Dent. Soc.* **69**, 164–165.

Rivera-Pérez, J. A., Wakamiya, M., and Behringer, R. R. (1999). *Goosecoid* acts cell autonomously in mesenchyme-derived tissues during craniofacial development. *Development* **126**, 3811–3821.

Roach, H. I. (1990). Long-term organ culture of embryonic chick femora: a system for investigating bone and cartilage formation at an intermediate level of organization. *J. Bone Min. Res.* **5**, 85–100.

Roach, H. I. (1992a). Induction of normal and dystrophic mineralization by glycerophosphates in long-term bone organ culture. *Calcif. Tissue Int.* **50**, 553–563.

Roach, H. I. (1992b). Trans-differentiation of hypertrophic chondrocytes into cells capable of producing a mineralized bone matrix. *Bone Min.* **19**, 1–20.

Roach, H. I., Erenpreisa, J., and Aigner, T. (1995). Osteogenic differentiation of hypertrophic chondrocytes involves asymmetric cell divisions and apoptosis. *J. Cell Biol.* **131**, 483–494.

Roach, H. I., and Shearer, J. R. (1989). Cartilage resorption and endochondral bone formation during the development of long bones in chick embryos. *Bone & Min.* **6**, 289–310.

Roark, E. F., and Greer, K. (1994). Transforming growth factor-β and bone morphogenetic protein-2 act by distinct mechanisms to promote chick limb cartilage differentiation *in vitro*. *Devel. Dynam.* **200**, 103–116.

Robadey, M., and Schowing, J. (1972). Trapianto di territori encefalici embrionali di quaglia (*C. coturnix japonica*) su embrione di pollo (*G. gallus*) allo stesso stadio di sviluppo. *Boll. Zool. Ital., Atti 16th Conv. Triesti*, **39**, 656

Robert, B., Lyons, G., Simandl, B. K., Kuroiwa, A., and Buckingham, M. (1991). The apical ectodermal ridge regulates Hox-7 and Hox-8 gene expression in developing chick limb buds. *Genes & Devel.* **5**, 2363–2374.

Roberts, G. J., and Blackwood, H. J. J. (1983). Growth of the cartilages of the mid-line cranial base: A radioautographic and histological study. *J. Anat.* **136**, 307–320.

Roberts, G. J., and Blackwood, H. J. J. (1984). Growth of the cartilages of the mid-line cranial base: An autoradiographic study using tritium labelled thymidine. *J. Anat.* **138**, 525–535.

Roberts, W. E. (1975). Cell kinetic nature and diurnal periodicity of the rat periodontal ligament. *Archs Oral Biol.* **20**, 465–471.

Roberts, W. E., and Chase, D. C. (1981). Kinetics of cell proliferation and migration associated with orthodontically-induced osteogenesis. *J. Dent. Res.* **60**, 174–181.

Roberts, W. E., Chase, D. C., and Jee, W. S. S. (1974). Counts of labelled mitosis in the orthodontically-stimulated periodontal ligament in the rat. *Archs Oral Biol.* **19**, 665–670.

Roberts, W. E., and Jee, W. S. S. (1974). Cell kinetics of ortho- dontically-stimulated and non-stimulated periodontal ligament in the rat. *Archs Oral Biol.* **19**, 17–22.

Roberts, W. E., and Morey, E. R. (1985). Proliferation and differ- entiation sequence of osteoblast histogenesis under physio- logical conditions in rat periodontal ligament. *Amer. J. Anat.* **174**, 105–118.

Roberts, W. E., Mozsary, P. G., and Klingler, E. (1982). Nuclear size as a cell-kinetic marker for osteoblast differentiation. *Amer. J. Anat.* **165**, 373–384.

Robertson, J. C., and Kelley, D. B. (1996). Thyroid hormone controls the onset of androgen sensitivity in the developing larynx of *Xenopus laevis*. *Devel. Biol.* **176**, 108–123.

Robertson, K. E., Tickle, C., and Darling, S. M. (1997). *Shh, Fgf4,* and *Hoxd* gene expression in the mouse limb mutant hypo- dactyly. *Int. J. Devel. Biol.* **41**, 733–736.

Robertson, P. B., and Miller, E. J. (1972). Cartilage collagen: Inability to serve as a substrate for collagenases active against skin and bovine collagen. *Biochim. Biophys. Acta* **289**, 247–250.

Robey, P. G., Young, M. F., Flanders, K. C., Roche, N. S., Kondaiah, P., Reddi, A. H., Termine, J. D., Sporn, M. B., and Roberts, A. B. (1987). Osteoblasts synthesize and respond to transforming growth factor-type β (TGF-β) *in vitro*. *J. Cell Biol.* **105**, 457–464.

Robin, Ch. (1849). Sur l'existence de deux epcèces nouvelles d'éléments anatomiques qui se trouvent dans le canal médul- laire des os. *C.R. Séanc. Soc. Biol.* **1**, 149–150.

Robinson, H., and Allenby, K. (1974). The effect of nerve growth factor on hindlimb regeneration in *Xenopus laevis* froglets. *J. Exp. Zool.* **189**, 215–226.

Robinson, P. D., and Poswillo, D. E. (1994). Temporomandibular joint development in the marmoset – a mirror of man. *J. Craniofac. Genet. Devel. Biol.* **14**, 245–251.

Robling, A.-G., and Stout, S. D. (1999). Morphology of the drift- ing osteon. *Cells Tissues & Organs* **164**, 192–204.

Robotti, E, Zimbler, A. G., Kenna, D., and Grossman, J. A. (1999). The effect of pulsed electromagnetic fields on flexor tendon healing in chickens. *J. Hand Surg Brit.* **24**, 56–58.

Robson, P., Wright, G. M., and Keeley, F. W. (2000). Distinct non- collagen based cartilages comprising the endoskeleton of the Atlantic hagfish, *Myxine glutinoda*. *Anat. Embryol.* **202**, 281–290.

Robson, P., Wright, G. M., Sitarz, E., Maiti, A., Rawat, M., Youson, J. H., and Keeley, F. W. (1993). Characterization of lamprin, an unusual matrix protein from lamprey cartilage: implications for evolution, structure and assembly of elastin and other fibrillar proteins. *J. Biol. Chem.* **268**, 1440–1447.

Robson, P., Wright, G. M., Youson, J. H., and Keeley, F. W. (1997). A family of non-collagen-based cartilages in the skeleton of the sea lamprey, *Petromyzon marinus*. *Comp. Biochem. Physiol. B* **118**, 71–78.

Roček, Z. (1996). Skull of the neotenic salamandrid amphibian *Triturus alpestris* and abbreviated development in the Tertiary Salamandridae. *J. Morphol.* **230**, 187–197.

Roček, Z., and Vesely, M. (1989). Development of the ethmoidal structures of the endocranium in the anuran *Pipa pipa*. *J. Morphol.* **200**, 301–320.

Rodan, G. A., Bourret, L. A., and Cutler, L. S. (1977). Membrane changes during cartilage maturation. Increase in 5′-nucleotides and decrease in adenosine inhibition of adenylate cyclase. *J. Cell Biol.* **72**, 493–501.

Rodan, G. A., Bourret, L. A., Harvey, A., and Mensi, T. (1975b). Cyclic AMP and cyclic GMP: Mediators of the mechanical effects of bone remodeling. *Science* **189**, 467–469.

Rodan, G. A., Bourret, L. A., and Norton, L. A. (1978). DNA synthesis in cartilage cells is stimulated by oscillating electric fields. *Science* **199**, 690–692.

Rodan, G. A., and Harada, S.-i. (1997). The missing bone. *Cell* **89**, 677–680.

Rodan, G. A., and Martin, T. J. (1981). Role of osteoblasts in hormonal control of bone resorption – a hypothesis. *Calcif. Tissue Int.* **33**, 349–352.

Rodan, G. A., and Martin, T. J. (2000). Therapeutic approaches to bone diseases. *Science* **289**, 1508–1514.

Rodan, G. A., Mensi, T., and Harvey, A. (1975a). A quantitative method for the application of compressive forces to bone in tissue culture. *Calcif. Tissue Res.* **18**, 125–132.

Rodan, S. B., and Rodan, G. A. (1974). The effect of parathyroid hormone and thyrocalcitonin on the accumulation of cyclic adenosine 3′,5′ monophosphate in freshly isolated bone cells. *J. Biol. Chem.* **249**, 3068–3074.

Rodbard, S. (1970). Negative feedback mechanisms in the architecture and function of the connective and cardiovascular tissues. *Perspect. Biol. Med.* **13**, 507–527.

Rodgers, B. J., Kulyk, W. M., and Kosher, R. A. (1989). Stimulation of limb cartilage differentiation by cyclic AMP is dependent on cell density. *Cell Differ. Devel.* **29**, 179–188.

Rodrigo, I., Hill, R. E., Balling, R., Münsterberg, A., and Imai, K. (2003). Pax1 and Pax9 activate *Bapx1* to induce chondrogenic differentiation in the sclerotome. *Development* **130**, 473–482.

Rodriguez, J. I., Delgardo, E., and Paniagua, K. (1985). Changes in young rat radius following excision of the perichondrial ring. *Calcif. Tissue Int.* **37**, 677–683.

Rodriguez-Esteban, C., Schwabe, J. W. R., De La Pena, J., Foys, B., Eshelman, B., and Izpisúa-Belmonte, J. C. (1997). Radical fringe positions the apical ectodermal ridge at the dorsoven- tral boundary of the vertebrate limb. *Nature* **398**, 814–818.

Rodriguez-Esteban, C., Tsukui, T., Yonei, S., Magallon, J., Tamura, K., and Izpisúa-Belmonte, J. C. (1999). The T-box genes *Tbx4* and *Tbx5* regulate limb outgrowth and identity. *Nature* **398**, 814–818.

Rodriguez-Valquez, J. F., Mérida Velasco, J. R., Arráez-Aybar, L. A., and Jiménez Collade, J. (1997b). A duplicated Meckel's cartilage in a human fetus. *Anat. Embryol.* **195**, 497–502.

Rodriguez-Valquez, J. F., Mérida Velasco, J. R., and Jiménez Collade, J. (1991). A study of the os goniale in man. *Acta Anat.* **142**, 188–192.

Rodriguez-Valquez, J. F., Mérida Velasco, J. R., Mérida Velasco, J. R., Sanchez-Montesinos, I., Espinferra, J., and Jiménez Collade, J. (1997a). Development of Meckel's cartilage in the symphyseal region in man. *Anat. Rec.* **149**, 249–254.

Roesler, H. (1981). Some historical remarks on the theory of can- cellous bone structure (Wolff's law). In *Mechanical Properties of Bone* (S. C. Cowin, ed.), pp. 27–42. ASME Publications, New York, NY.

Rogers, M. J. (2000). Statins: lower lipids *and* better bones? *Nature Med.* **6**, 21–23.

Rogina, B., Coelho, C. N. D., Kosher, R. A., and Upholt, W. R. (1992). The pattern of expression of the chicken homolog of Hox 1i in the developing limb suggests a possible role in the ectodermal inhibition of chondrogenesis. *Devel. Dynam.* **193**, 92–101.

Rojas, M., and Montenegro, M. A. (1995). An anatomical and embryological study of the clavicle in cats (*Felis domesticus*) and sheep (*Ovis aries*) during the prenatal period. *Acta Anat.* **154**, 128–134.

Rojas, M., Posada, J., and Montenegro, M. A. (1996). Comparative study of the ontogeny of mandibular cartilage (Meckel) in sheep (*Ovis aries*) and cat (*Felis domestica*). *Int. J. Devel. Biol.* Suppl. 1, S243–244.

Romanes, G. J. (1901). *Darwin and After Darwin. An Exposition of the Darwinian Theory and a Discussion of Post-Darwinian Questions. Volume I. The Darwinian Theory.* Third Edn. The Open Court Publishing Company, Chicago, IL.

Romer, A. S. (1942). Cartilage an embryonic adaptation. *Amer. Nat.* **76**, 394–404.

Romer, A. S. (1963). The ancient history of bone. *Ann. NY. Acad. Sci.* **109**, 168–176.

Romer, A. S. (1969). Vertebrate paleontology and zoology. *The Biologist* **51**, 49–53.

Rong, P. M., Teillet, M.-A., Ziller, C., and Le Douarin, N. M. (1992). The neural tube/notochord complex is necessary for vertebral but not limb and body wall striated muscle differentiation. *Development* **115**, 657–672.

Rönning, O. (1966). Observations on the intracerebral transplantation of the mandibular condyle. *Acta Odontol. Scand.* **24**, 443–457.

Rönning, O. (1995). Basicranial synchondroses and the mandibular condyle in craniofacial growth. *Acta Odontol. Scand.* **53**, 162–166.

Rönning, O., and Kantomaa, T. (1988). The growth pattern of the clavicle in the rat. *J. Anat.* **159**, 173–179.

Rönning, O., and Kylämarkula, S. (1982). Morphogenetic potential of rat growth cartilages as isogenic transplants in the interparietal suture area. *Archs Oral Biol.* **27**, 581–588.

Rönning, O., Kylämarkula, S., and Peltomäki, T. (1991a). Growth potential of primary and secondary cartilage: interosseal transplantation of cartilaginous structures. In *Fundamentals of Bone Growth: Methodology and Applications* (A. D. Dixon, B. G. Sarnat and D. A. N. Hoyte, eds), pp. 471–477. CRC Press, Boca Raton, FL.

Rönning, O., Rintala, M., Odont, L., and Kantomaa, T. (1991b). Growth potential of the rat clavicle. *J. Oral Maxillofac. Surg.* **49**, 1176–1180.

Rönning, O., Salo, L. A., Larmas, M., and Nieminen, M. (1990). Ossification of the antler in the Lapland reindeer (*Rangifer tarandus tarandus*). *Acta Anat.* **137**, 359–362.

Rooij, P. P. de., Siebrecht, M. A. N., Tägil, M., and Aspenberg, P. (2001). The fate of mechanically induced cartilage in an unloaded environment. *J. Biomech.* **34**, 961–966.

Rooney, P. (1994). Intratendinous ossification. In *Bone, Volume 8: Mechanisms of Bone Development and Growth* (B. K. Hall, ed.), pp. 47–83. CRC Press, Boca Raton, FL.

Rooney, P., Grant, M. E., and McClure, J. (1992). Endochondral ossification and *de novo* collagen synthesis during repair of the rat Achilles tendon. *Matrix* **12**, 274–281.

Rooney, P., Walker, D., Grant, M. E., and McClure, J. (1993). Cartilage and bone formation in repairing Achilles tendons within diffusion chambers: Evidence for tendon-cartilage and cartilage-bone conversion *in vivo*. *J. Pathol.* **169**, 375–381.

Ros, M. A., Dahn, R. D., Fernandez-Teran, M., Rashka, K., Caruccio, N. C., Hasso, S. M., Bitgood, J. J., Lancman, J. J., and Fallon, J. F. (2003). The chick *oligozeugodactyly* (*ozd*) mutant lacks sonic hedgehog function in the limb. *Development* **130**, 527–537.

Ros, M. A., Lyons, G., Kosher, R. A., Upholt, W. B., Coelho, C. N. D., and Fallon, J. F. (1992). Apical ridge dependent and independent mesodermal domains of GHox-7 and Ghox-8 expression in chick limb buds. *Development* **116**, 811–818.

Ros, M. A., Macias, D., Fallon, J. F., and Hurlé, J. M. (1994). Formation of extra digits in the interdigital spaces of the chick leg bud is not preceded by changes in the expression of the *Msx* and *Hoxd* genes. *Anat. Embryol.* **190**, 375–382.

Ros, M. A., Piedra, M. E., Fallon, J. F., and Hurlé, J. M. (1997a). Morphogenetic potential of the chick leg interdigital mesoderm when diverted from the cell death program. *Devel. Dynam.* **208**, 406–419.

Ros, M. A., Sefton, M., and Nieto, M. A. (1997b). *Slug*, a zinc finger gene previously implicated in the early patterning of the mesoderm and the neural crest, is also involved in chick limb development. *Development* **124**, 1821–1829.

Rosa-Molinar, E., Hendricks, S. E., Rodriguez-Sierra, J. F., and Fritzsch, B. (1994). Development of the anal fin appendicular support in the western mosquitofish *Gambusia affinis affinis* (Baird and Girard, 1854): A reinvestigation and reinterpretation. *Acta Anat.* **151**, 20–35.

Rose, C. S. (1995a). Skeletal morphogenesis of the urodele skull. I. Postembryonic development in the Hemidactyliini (Amphibia: Plethodontidae). *J. Morphol.* **223**, 125–148.

Rose, C. S. (1995b). Skeletal morphogenesis of the urodele skull. II. Effect of developmental stage in TH-induced remodelling. *J. Morphol.* **223**, 149–166.

Rose, C. S. (1995c). Skeletal morphogenesis of the urodele skull. III. Effect of hormone dosage in TH-induced remodelling. *J. Morphol.* **223**, 243–261.

Rose, C. S. (1995d). Intraspecific variation in ceratobranchial number in *Hemidactylium scutatum* (Amphibia: Plethodontidae): developmental an systematic implications. *Copeia* **1995**, 228–232.

Rose, C. S. (1996). An endocrine-based model for developmental and morphogenetic diversification in metamorphic and paedomorphic urodeles. *J. Zool. Lond.* **239**, 253–284.

Rose, C. S. (2004). Thyroid hormone-mediated development in vertebrates: What makes frogs unique. In *Environment, Development and Evolution: Toward a Synthesis* (B. K. Hall, R. D. Pearson and G. B. Müller, eds), pp. 197–237. The MIT Press, Cambridge, MA.

Rose, R. W. (1989). Embryonic growth rates of marsupials with a note on monotremes. *J. Zool.* (London) **218**, 11–16.

Rose, S. M. (1948). The role of nerves in amphibian regeneration. *Ann. N. Y. Acad. Sci.* **49**, 818–833.

Rosen, V., and Clark, N. B. (1982). Effects of 25-hydroxyvitamin D_3 and 1,24-dihydroxyvitamin D_3 on embryonic chick bone in organ culture. *J. Exp. Zool.* **224**, 97–102.

Rosen, V., and Thies, R. S. (1992). The BMP proteins in bone formation and repair. *Trends Genet.* **8**, 97–102.

Rosen, V., Thies, R. S., and Lyons, K. (1996). Signaling pathways in skeletal formation: A role for BMP receptors. *Ann. N. Y. Acad. Sci.* **785**, 59–69.

Rosen, V., Wozney, J. M., Wang, E. A., Cordes, P., Celeste, A., McQuaid, D., and Kurtzberg, L. (1989). Purification and molecular cloning of a novel group of BMP's and localization of BMP mRNA in developing bone. *Conn. Tissue Res.* **20**, 313–320.

Rosenberg, G. D., Campbell, S. C., and Simmons, D. J. (1984). The effects of spaceflight on the mineralization of rat incisor dentin. *Proc. Soc. Exp. Biol. Med.* **175**, 429–437.

Rosenberg, H. I., and Richardson, K. C. (1995). Cephalic morphology of the Honey Possum, *Tarsipes rostratus* (Marsupialia: Tarsipedidae); an obligate nectarivore. *J. Morphol.* **223**, 303–323.

Rosenberg, M. J., and Caplan, A. I. (1974). Nicotinamide-adenine dinucleotide levels in cells of developing chick limbs: Possible control of muscle and cartilage development. *Devel. Biol.* **38**, 157–164.

Ross, G. T. (1976). Glycosaminoglycan Synthesis and the initiation of Secondary Chondrogenesis in the Embryo Chick. Ph.D. Thesis, 90 pp. Dalhousie University, Halifax, Nova Scotia.

Ross, R. (1975). Connective tissue cells, cell proliferation and synthesis of extracellular matrix – a review. *Phil. Trans. R. Soc. London, Ser. B.* **271**, 247–260.

Rossant, J., and Frels, W. I. (1980). Interspecific chimeras in mammals: Successful production of live chimeras between *Mus musculus* and *Mus caroli. Science* **208**, 419–421.

Rossel, M., and Capecchi, M. R. (1999). Mice mutant for both *Hoxa1* and *Hoxb1* show extensive remodeling of the hindbrain and effects in craniofacial development. *Development* **126**, 5027–5040.

Rossi, F., MacLean, H. E., Yuan, W., Francis, R. O., Semenova, E., Lin, C. S., Kronenberg, H. M., and Cobrinik, D. (2002). *Devel. Biol.* **247**, 271–285.

Roth, E., and Taylor, H. B. (1966). Heterotopic cartilage in the uterus. *Obstet. Gynecol.* **27**, 838–844.

Roth, M. (1973). The relative osteo-neural growth: A concept of normal and pathological (teratogenic) skeletal morphogenesis. *Morphol. Jahrb.* **119**, 250–274.

Roth, S., Müller, K., Fischer, D.-C., and Dannhauer, K.-H. (1997). Specific properties of the extracellular chondroitin sulphate proteoglycans in the mandibular condylar growth centre in pigs. *Archs Oral Biol.* **42**, 63–76.

Rouleau, M. F., Mitchell, J., and Goltzman, D. (1988). *In vivo* distribution of parathyroid hormone receptors n bone: Evidence that a predominant osseous target cell is not the mature osteoblast. *Endocrinology* **123**, 187–191.

Roux, W. (1885). Beiträge zur Morphologie der funktionellen anpassung. 3. Beschreibung und Erläuterung einer knöchernen Kniegelenkankylose. *Arch. Anat. Physiol. Anat. Abt.* **9**, 120–158.

Rowe, D. A., and Fallon, J. F. (1982). The proximodistal determination of skeletal parts in the developing chick leg. *J. Embryol. Exp. Morphol.* **68**, 1–7.

Roy, P. K., Witten, P. E., Hall, B. K., and Lall, S. P. (2004). Effects of dietary phosphorous on bone growth and mineralization of vertebrae in haddock (*Melanogrammus aeglefinus* L.). *Fish Physiol. Biochem.* **27**, 35–48.

Roy, S., and Gardiner, D. M. (2002). Cyclopamine induces digit loss in regenerating axolotl limbs. *J. Exp. Zool.* **293**, 186–190.

Royal, P. D., and Goetinck, P. F. (1977). *In vitro* chondrogenesis in mouse limb mesenchymal cells; changes in ultrastructure and proteoglycan synthesis. *J. Embryol. Exp. Morphol.* **39**, 79–95.

Ruano-Gil, D., Nardi-Vilardaga, J., and Teixidor-Johé, A. (1985). Embryonal hypermobility and articular development. *Acta Anat.* **123**, 90–92.

Ruben, J. A. (1983). Mineralized tissues and exercise physiology of snakes. *Amer. Zool.* **23**, 377–381.

Ruben, J. A. (1989). Activity physiology and evolution of the vertebrate skeleton. *Amer. Zool.* **29**, 195–203.

Ruben, J. A., and Bennett, A. A. (1980). Antiquity of the vertebrate pattern of activity metabolism and its possible relation to vertebrate origins. *Nature* **286**, 886–888.

Ruben, J. A., and Bennett, A. A. (1987). The evolution of bone. *Evolution* **41**, 1187–1197.

Ruberte, E., Dolle, P., Krust, A., Zelent, A., Morriss-Kay, G. M., and Chambon, P. (1990). Specific spatial and temporal distribution of retinoic acid receptor γ transcripts during mouse embryogenesis. *Development* **108**, 213–222.

Ruberte, E., Friederich, V., Morriss-Kay, G. M., and Chambon, P. (1992). Differential distribution patterns of CRABP-I and CRABP-II transcripts during mouse embryogenesis. *Development* **115**, 973–987.

Rubin, C. T., and Lanyon, L. E. (1987). Kappa Delta award paper: Osteoregulatory nature of mechanical stimuli: function as a determinate for adaptive remodeling in bone. *J. Orthop. Res.* **5**, 300–310.

Rubin, C. T., McLeod, K. J., and Lanyon, L. E. (1989). Prevention of osteoporosis by pulsed electromagnetic fields. *J. Bone Joint Surg. Am. Vol.* **71**, 411–416.

Rubin, L., and Saunders, J. W., Jr. (1972). Ectodermal-mesodermal interactions in the growth of limb buds in the chick embryo: Constancy and temporal limits of the ectodermal induction. *Devel. Biol.* **28**, 94–112.

Rudert, M. (2002). Histological evaluation of osteochondral defects: consideration of animal models with emphasis on the rabbit, experimental set up, follow-up and applied methods. *Cells Tissues Organs* **171**, 229–240.

Rudert, M., Hirschmann, F., Schulze, M., and Wirth, C. J. (2000). Bioartificial cartilage. *Cells Tissues Organs* **167**, 95–105.

Rudnick, D. (1945a). Limb-forming potencies of the chick blastoderm: Including notes on associated trunk structures. *Trans Conn. Acad. Arts Sci.* **36**, 353–377.

Rudnick, D. (1945b). Differentiation of the prospective limb material from creeper chick embryos in coelomic grafts. *J. Exp. Zool.* **100**, 1–14.

Rudnicki, J. A., and Brown, A. M. C. (1997). Inhibition of chondrogenesis by *Wnt* gene expression *in vivo* and *in vitro*. *Devel. Biol.* **185**, 104–118.

Ruf, S., and Pancherz, H. (1998). Temporomandibular joint growth adaptation in Herbst treatment – a prospective magnetic resonance imaging and cephalometric roentgenographic study. *Eur. J. Orthodont.* **20**, 375–388.

Rufai, A., Benjamin, M., and Ralphs, J. R. (1992). Development and ageing of phenotypically distinct fibrocartilages associated with the rabbit Achilles tendon. *Anat. Embryol.* **186**, 611–618.

Rufai, A., Benjamin, M., and Ralphs, J. R. (1995). The development of the fibrocartilage in the rat intervertebral disc. *Anat. Embryol.* **192**, 53–62.

Rufai, A., Ralphs, J. R., and Benjamin, M. (1996). Ultrastructure of fibrocartilage at the insertion of the rat Achilles tendon. *J. Anat.* **189**, 185–191.

Ruggeri, A. (1972). Ultrastructural, histochemical and autoradiographic studies on the developing chick notochord. *Z. Anat. Entwickl.-Gesch.* **138**, 20–33.

Ruhin, B., Creuzet, S., Vincent, C., Benouaiche, L., Le Douarin, N. M., and Couly, G. (2003). Patterning of the hyoid cartilage depends upon signals arising from the ventral foregut endoderm. *Devel. Dynam.* **228**, 239–246.

Ruibal, R., and Shoemaker, V. (1984). Osteoderms in anurans. *J. Herpetol.* **18**, 313–328.

Runner, M. N. (1988). Caffeine, an exquisitely specific inhibitor of osteogenic differentiation. *Teratology* **38**, 599–604.

Ruoslahti, E., and Yamaguchi, Y. (1991). Proteoglycans as modulators of growth factor activities. *Cell* **64**, 867–869.

Russell, A. P., and Bauer, A. M. (1988). Paraphalangeal elements of Gekkonid lizards: A comparative study. *J. Morphol.* **197**, 221–240.

Russell, A. P., and Joffe, D. J. (1985). The early development of the quail (*Coturnix c. japonica*) furcula reconsidered. *J. Zool. London A* **206**, 69–81.

Russell, A. P., and Rewcastle, S. C. (1979). Digital reduction in *Sitana* (Reptilia: Agamidae) and the dual roles of the fifth metatarsal in lizards. *Can. J. Zool.* **57**, 1129–1135.

Russell, E. S. (1916). *Form and Function. A Contribution to the History of Animal Morphology.* John Murray, London. (Reprinted 1982 by The University of Chicago Press, Chicago, IL, with a new introduction by G. V. Lauder).

Russell, J. E., Walker, W. V., and Simmons, D. J. (1984). Adrenal/parathyroid regulation of DNA, collagen and protein synthesis in rat epiphyseal cartilage and bone. *J. Endocrinol.* **103**, 49–57.

Russell, R. G. G., Caswell, A. M., Hearn, P. R., and Shakrard, R. M. (1986). Calcium in mineralized tissues and pathological calcification. *Brit. Med. Bull.* **42**, 435–446.

Russell, R. G. G., and Kanis, J. A. (1984). Ectopic calcification and ossification. In *Metabolic Bone and Stone Disease* (B. E. C. Nordin, ed.), 2nd Edn, pp. 344–365. Churchill Livingstone, London.

Ruth, E. B. (1932). A study of the development of the mammalian pelvis. *Anat. Rec.* **53**, 207–225.

Ruth, E. B. (1934). The os priapi: A study in bone development. *Anat. Rec.* **60**, 231–249.

Ruth, E. B. (1935). Metamorphosis of the pubic symphysis. I. The white rat (*Mus norvegicus albinus*). *Anat. Rec.* **64**, 1–5.

Ruth, E. B. (1936a). Metamorphosis of the pubic symphysis. II. The guinea pig. *Anat. Rec.* **67**, 69–79.

Ruth, E. B. (1936b). Metamorphosis of the pubic symphysis. III. Histological changes in the symphysis of the pregnant guinea pig. *Anat. Rec.* **67**, 409–419.

Rüther, U., Garber, C., Komitowski, D., Muller, R., and Wagner, E. F. (1987). Deregulated *c-fos* expression interferes with normal bone development in transgenic mice. *Nature* **325**, 412–416.

Rutledge, J. C., Shourbaji, A. G., Hughes, L. A., Polifka, J. E., Cruz, Y. P., Bishop, J. B., and Generoso, W. M. (1994). Limb and lower-body duplication induced by retinoic acid in mice. *Proc. Natl Acad. U.S.A.* **91**, 5436–5440.

Rutledge, J. J., Eisen, E. J., and Legates, J. E. (1974). Correlated response in skeletal traits and replicate variation in selected lines of mice. *Theoret. Applied Genet.* **45**, 26–31.

Ruvinsky, I., and Gibson-Brown, J. J. (2000). Genetic and developmental bases of serial homology in vertebrate limb evolution. *Development* **127**, 5233–5244.

Ruvinsky, I., Silver, L. M., and Gibson-Brown, J. J. (2000). Phylogenetic analysis of T-Box genes demoinstrates the importance of amphioxus for understanding evolution of the vertebrate skeleton. *Genetics* **156**, 1249–1257.

Ruvinsky, I., Silver, L. M., and Ho, R. K. (1998). Characterization of the zebrafish *tbx16* gene and evolution of the vertebrate T-box family. *Devel. Genes Evol.* **208**, 94–99.

Ryan, G. B., Cliff, W. J., Gabbiani, G., Irle, C., Statkou, P. R., and Majno, G. (1973). Myofibroblasts in a vascular fibrous tissue. *Lab. Invest.* **29**, 197–206.

Ryder, M. I., Jenkins, S. D., and Horton, J. E. (1981). The adherence to bone by cytoplasmic elements of osteoclasts. *J. Dent. Res.* **60**, 1349–1355.

Ryg, M., and Langvatn, R. (1982). Seasonal changes in weight gain, growth hormone, and thyroid hormone in male red deer (*Cervus elaphus atlanticus*).

Ryou, H. M., Hoffmann, H. M., Beumer, T., Frenkel, B., Towler, D. A., Stein, G. S., Stein, J. L., Van Wijner, A. J., and Lian, J. B. (1997). Stage-specific expression of Dlx-5 during osteoblast differentiation: involvement in regulation of osteocalcin gene expression. *Mol. Endocrinol.* **11**, 1681–1694.

Rzehak, K., and Singer, M. (1966). Limb regeneration and nerve fiber number in *Rana sylvatica* and *Xenopus laevis*. *J. Exp. Zool.* **162**, 15–22.

S

Saber, G. M., Parker, S. B., and Minkoff, R. (1989). Influence of epithelial-mesenchymal interaction on the viability of facial mesenchyme *in vitro*. *Anat. Rec.* **225**, 56–66.

Sabet, S. J., Tarbet, K. J., Lemke, B. N., Smith, M. E., and Albert, D. M. (2001). Subperiosteal hematoma of the orbit with osteoneogenesis. *Arch. Ophthalmol.* **119**, 301–303.

Sadaghiani, B., and Thiébaud, C. H. (1987). Neural crest development in the *Xenopus laevis* embryo, studied by interspecific transplantation and scanning electron microscopy. *Devel. Biol.* **124**, 91–110.

Sadaghiani, B., and Vielkind, J. R. (1989). Neural crest development in *Xiphophorus* fishes: Scanning electron and light microscopic studies. *Development* **105**, 487–504.

Sadler, J., Pankhurst, P. M., and King, H. R. (2001). High prevalence of skeletal deformity and reduced gill surface area in triploid Atlantic salmon (*Salmo salar*). *Aquaculture* **198**, 369–386.

Saga, Y., Yagi, T., Ikawa, Y., Sakakura, T., and Aizama, S. (1992). Mice develop normally without tenascin. *Genes & Devel.* **6**, 1821–1831.

Sagai, T., Masuya, H., Tamura, M., Shimizu, K., Yada, Y., Wakana, S., Gondo, Y., Noda, T., and Shiroishi, T. (2004). Phylogenetic conservation of a limb-specific, *cis*-acting regulator of Sonic hedgehog (*Shh*). *Mammalian Genome Genes & Phenotypes* **15**, 23–34.

Sage, H., and Gray, W. R. (1979). Studies on the evolution of elastin – I. Phylogenetic distribution. *Comp. Biochem. Physiol.* **64B**, 313–328.

Sage, H., and Gray, W. R. (1980). Studies on the evolution of elastin – II. Histology. *Comp. Biochem. Physiol.* **66B**, 13–22.

Sahni, M., Ambrosetti, D.-C., Mansukhani, A., Gertner, R., Levy, D., and Basilico, C. (1999). FGF signaling inhibits chondrocyte proliferation and regulates bone development through the STAT-1 pathway. *Genes & Devel.* **13**, 1361–1366.

Sai Htay, W., Nonaka, K., Sasaki, Y., and Nakata, M. (1997). A longitudinal study of the postnatal maternal effect on the craniofacial growth of mouse offspring by cross-nursing. *J. Craniofac. Genet. Devel. Biol.* **17**, 148–159.

Saint-Paul, U., and Bernardhino, G. (1988). Behavioural and ecomorphological responses of the Neotropical pacu *Piaractus mesopotamicus* (Teleostei; Serrasalmidae) to oxygen-deficient waters. *Exp. Biol.* (*Berlin*) **48**, 19–26.

Saito, D., Yonai-Tamura, S., Kano, K., Ide, H., and Tamura, K. (2002). Specification and determination of limb identity: evidence for inhibitory regulation of *Tbx* gene expression. *Development* **129**, 211–229.

Saitoh, S., Takahashi, I., Mizoguchi, I., Sasano, Y., Kagayama, M., and Mitani, H. (2000). Compressive force promotes chondrogenic differentiation and hypertrophy in midpalatal suture cartilage in growing rats. *Anat. Rec.* **260**, 392–401.

Saklatvala, J., and Dingle, J. T. (1980). Identification of catabolin, a protein from synovium which induces degradation of cartilage in organ culture. *Biochem. Biophys. Res. Commun.* **96**, 1225–1231.

Saklatvala, J., Sarsfield, S. J., and Pilsworth, L. M. C. (1983). Characterization of proteins from human synovium and mononuclear leucocytes that induce resorption of cartilage proteoglycan *in vitro*. *Biochem. J.* **209**, 337–344.

Salas-Vidal, E., Valencia, C., and Covarrubias, L. (2001). Differential tisue growth and patterns of cell death in mouse limb autopod morphogenesis. *Devel. Dynam.* **220**, 295–306.

Salomon, D. S., and Pratt, R. M. (1976). Glucocorticoid receptors in murine embryonic facial mesenchyme cells. *Nature* **264**, 174–177.

Salter, R. B. (1975). Constant motion helps joints heal. *The Toronto Globe & Mail*, 10 April, p. W7.

Salter, R. B., Simmons, D. F., Malcolm, B. W., Rumble, E. J., MacMichael, D., and Clements, N. D. (1980). The biological effect of continuous passive motion on the healing of full-thickness defects in articular cartilage. An experimental investigation in the rabbit. *J. Bone Joint Surg. Am. Vol.* **62**, 1232–1251.

Salzgeber, B., and Guénet, J.-L. (1984). Studies on 'repeated epilation' mouse mutant embryos. II. Development of limb, tail, and skin defects. I. *J. Craniofac. Genet. Devel. Biol.* **4**, 95–114.

Sampath, T. K., Coughlin, J. E., Whetstone, R. M., Banavch, D., Corbett, C., Ridge, R. J., Ozkaynak, E., Oppermann, H., and Rueger, D. C. (1990). Bovine osteogenic protein is composed of dimers of OP-1 and BMP-2A, two members of the transforming growth factor-β superfamily. *J. Biol. Chem.* **265**, 13198–13205.

Sampath, T. K., Maliakal, J. C., Hauschke, P. V., Jones, W. K., Sasak, H., Tucker, R. F., White, K. H., Coughlin, J. E., Tucker, M. M., Pang, R. H. L., Corbett, C., Ozkaynak, E., Oppermann, H., and Rueger, D. C. (1992). Recombinant human osteogenic protein-1 (Hop-1) induces new bone formation *in vivo* with a specific activity comparable with natural bovine osteogenic protein and stimulates osteoblast proliferation and differentiation *in vitro*. *J. Biol. Chem.* **267**, 20352–20362.

Sampath, T. K., Muthukumaran, N., and Reddi, A. H. (1987). Isolation of osteogenin, an extracellular matrix-associated, bone-inductive protein, by heparin affinity chromatography. *Proc. Natl Acad. Sci. U.S.A.* **84**, 7109–7113.

Sampath, T. K., Rashka, K. E., Doctor, J. S., Tucker, R. F., and Hofmann, F. M. (1993). *Drosophila* transforming growth factor β superfamily proteins induce endochondral bone formation in mammals. *Proc. Natl Acad. Sci. U.S.A.* **90**, 6004–6008.

Sánchez, M. R., Gemballa, S., Nummela, S., Smith, K. K., and Maier, W. (2002). Ontogenetic and phylogenetic transformations of the ear ossicles in marsupial mammals. *J. Morphol.* **251**, 219–238.

Sánchez-Villagra, M. R., (2002). Comparative patterns of postcranial ontogeny in therian mammals: An analysis of relative timing of ossification events. *J. Exp. Zool. (Mol. Devel. Evol.)*, **294**, 264–273.

Sánchez-Villagra, M. R., and Dottling, M. (2003). Carpal ontogeny in *Dasyurus viverrinus* and notes on carpal evolution in the Dasyuromorphia among the Marsupialia. *Mammal Biol.* **68**, 329–340.

Sánchez-Villagra, M. R, and Maier, W. (2002). Ontogenetic data and the evolutionary origin of the mammalian scapula. *Naturwissenschaften* **89**, 459–461.

Sánchez-Villagra, M. R, and Maier, W. (2003). Ontogenesis of the scaular in marsupial mammals, with special emphasis on perinatal stages of Didelphids and remarks on the origin of the therian scapular. *J. Morphol.* **258**, 115–129.

Sandberg, M., Aro, H., Multimaki, P., Aho, H., and Vuorio, E. (1989). *In situ* localization of collagen production by chondrocytes and osteoblasts in fractures. *J. Bone Joint Surg.* **71A**, 69–77.

Sandell, L. J., Morris, N., Robbins, J. R., and Goldring, M. B. (1991). Alternatively spliced type II procollagen mRNA define distinct populations of cells during vertebral development: Differential expression of the amino-propeptide. *J. Cell Biol.* **114**, 1307–1319.

Sander, P. M. (2000). Longbone histology of the Tendaguru sauropods: implications for growth and biology. *Paleobiology* **26**, 466–488.

Sanders, E. J., Prasad, S., and Hu, N. (1993). The involvement of TGFβ1 in early avian development: gastrulation and chondrogenesis. *Anat. Embryol.* **187**, 573–581.

Sandhu, H. S., Kwong-Hing, A., Herskovits, M. S., and Singh, I. J. (1990). The early effects of surgical sympathectomy on bone resorption in the rat incisor pocket. *Archs Oral. Biol.* **35**, 1003–1007.

Sanerkin, N. G. (1980). Definitions of osteosarcoma, chondrosarcoma and fibrosarcoma of bone. *Cancer* **46**, 178–185.

Sanford, L. P., Ormsby, I., Gittensberger-de Groot, A. C., Sariota, H., Friedman, R., Boivin, G. P., Cardell, E. L., and Doetschman, T. (1997). TGFβ2 knockout mice have multiple developmental defects that are non-overlapping with other TGFβ knockout phenotypes. *Development* **124**, 2659–2670.

Sannasgala, S. S. M. M. M. K., and Johnson, D. R. (1990). Kinetic parameters in the growth plate of normal and chondroplastic (cn/cn) mice. *J. Anat.* **172**, 245–258.

Sansom, I. J., Smith, M. P., Armstrong, H. A., and Smith, M. M. (1992). Presence of the earliest vertebrate hard tissues in conodonts. *Science* **256**, 1308–1311.

Sansom, I. J., Smith, M. P., and Smith, M. M. (1994). Dentine in conodonts. *Nature* **368**, 591.

Santamaria, J. A., and Becerra, J. (1991). Tail fin regeneration in teleosts: cell-extracellular matrix interactions in blastemal differentiation. *J. Anat.* **176**, 9–21.

Santamaria, J. A., Mari-Beffa, M., and Becerra, J. (1992). Interactions of the lepidotrichial matrix components during tail fin regeneration in teleosts: *Differentiation* **49**, 143–150.

Sanz-Ezquerro, J., and Tickle, C. (2000). Autoregulation of *Shh* expression and *Shh* induction of cell death suggests a mechanism for modulating polarizing activity during chick limb development. *Development* **127**, 4811–4823.

Sarin, V. K., Erickson, G. M., Giori, N. J., Bergman, A. G., and Carter, D. R. (1999). Coincident development of sesamoid bones and clues to their evolution. *Anat. Rec. (New Anat.)* **257**, 174–180.

Sarkar, S., Petiot, A., Copp, A., Ferretti, P., and Thorogood, P. (2001). FGF2 promotes skeletogenic differentiation of cranial neural crest cells. *Development* **128**, 2143–2152.

Sarmiento, E. E., Stiner, E., and Mowbray, K. (2002). Morphology-based systematics (MBS) and problems with fossil hominoid and hominid systematics. *Anat. Rec. (New Anat.)* **269**, 50–66.

Sarnat, B. G. (ed.) (1964). *The Temporomandibular Joint*, 2nd Edn. C. C. Thomas, Springfield, IL.

Sarnat, B. G. (1968). Growth of bones as revealed by implant markers in animals. *Amer. J. Phys. Anthrop.* **29**, 255–286.

Sarnat, B. G. (1986). Growth pattern of the mandible: some reflections. *Amer. J. Orthodont. Dentofac. Orthopedics* **90**, 221–233.

Sarnat, B. G. (1992). Cranio-maxillofacial growth at joints. In *Bone, Volume 6: Bone Growth – A.* (B. K. Hall, ed.), pp. 249–273. CRC Press, Boca Raton, FL.

Sarnat, B. G. (2001). Effects and non effects of personal environmental experimentation on postnatal craniofacial growth. *J. Craniofac. Surg.* **12**, 205–217.

Sarnat, B. G., and Muchnic, H. (1971). Facial skeletal changes after mandibular condylectomy in the adult monkey. *J. Anat.* **108**, 323–338.

Sarnat, B. G., and Schour, I. (1944). Effect of experimental fracture of bone, dentin and enamel: Study of the mandible and incisor in the rat. *Archs Surg.* **49**, 23–38.

Sarnat, B. G., and Selman, A. J. (1982). Growth pattern of the rabbit nasal bone region. *Rhinology* **20**, 93–105.

Sarnat, H. B., and Flores-Sarnat, L. (2004). Integrative classification of morphology and molecular genetics in central nervous system malformations. *Am. J. Med. Genet.* **126A**, 386–392.

Sarras, M. P., Jr. (1996). BMP-1 and the astacin family of metalloproteinases: a potential link between the extracellular matrix, growth factors and pattern formation. *BioEssays* **18**, 439–442.

Sasagawa, I. (2002). Mineralization patterns in elasmobranch fish. *Microsc. Res. Tech.* **59**, 396–407.

Sasaki, Y., Nonaka, K., Nakata, M. (1994a). The effect of embryo transfer on the intrauterine growth of the mandible in mouse fetuses. *J. Craniofac. Genet. Devel. Biol.* **14**, 111–117.

Sasaki, Y., Nonaka, K., Nakata, M. (1994b). The effect of four strains of recipients on the intrauterine growth of the mandible in mouse fetuses. *J. Craniofac. Genet. Devel. Biol.* **14**, 118–123.

Sasaki, Y., Nonaka, K., Nakata, M. (1995). The strain effect of Dam on intrauterine incisal growth in mouse fetuses. *J. Craniofac. Genet. Devel. Biol.* **15**, 140–145.

Sasano, Y., Furusawa, M., Ohtani, H., Mizoguchi, I., Takahashi, I., and Kayogama, M. (1996). Chondrocytes synthesise type I collagen and accumulate the protein in the matrix during development of rat tibial articular cartilage. *Anat. Embryol.* **194**, 247–252.

Sasano, Y., Kamakura, S., Homma, H., Suzuki, O., Mizoguchi, I., and Kagayama, M. (1999). Implanted octacalcium phosphate (OCP) stimulates osteogenesis by osteoblastic cells and/or committed osteoprogenitors in rat calvarial periosteum. *Anat. Rec.* **256**, 1–6.

Sasano, Y., Li, H. C., Zhu, J. X., Imanaka-Yoshida, K., Mizoguchi, I., and Kagayama, M. (2000). Immunohistochemical localization of type I collagen, fibronectin and tenascin C during embryonic osteogenesis in the dentary of mandibles and tibias in rats. *Histochem. J.* **32**, 591–598.

Sasano, Y., Mizoguchi, I., Furusawa, M., Aiba, N., Ohtani, E., Iwamatsu, Y., and Kagayama, M. (1993a). The process of calcification during development of the rat tracheal cartilage characterized by distribution of alkaline phosphatase activity and immunolocalization of types I and II collagens and glycosaminoglycans of proteoglycans. *Anat. Embryol.* **188**, 31–39.

Sasano, Y., Mizoguchi, I., Kagayama, M., Shum, L., Bringas, P., Jr., and Slavkin, H. C. (1992). Distribution of type I collagen, type II collagen and PNA binding glycoconjugates during chondrogenesis of three distinct embryonic cartilages. *Anat. Embryol.* **186**, 205–213.

Sasano, Y., Mizoguchi, I., Takahashi, I., Kagayama, M., Saito, T., and Kuboki, Y. (1997). BMPs induce endochondral ossification in rats when implanted ectopically within a carrier made of fibrous glass membrane. *Anat. Rec.* **247**, 472–478.

Sasano, Y., Ohtani, E., Narita, K., Kagayama, M., Murata, M., Saito, T., Shigenobu, K., Takita, H., Mizuno, M., and Kuboki, Y. (1993b). BMPs induce direct bone formation in ectopic sites independent of the endochondral ossification *in vivo*. *Anat. Rec.* **236**, 373–380.

Sassoon, D., Segil, N., and Kelley, D. (1986). Androgen-induced myogenesis and chondrogenesis in the larynx of *Xenopus laevis*. *Devel. Biol.* **113**, 135–140.

Satchell, P. G., Anderton, X., Ryu, O. H., Luan, X., Ortega, A. J., Opamen, R., Berman, B. J., Witherspoon, D. E., Gutmann, J. L., Yamane, A., Zeichner-David, M., Simmer, J. P., Shuler, C. F., and Diekwisch, T. G. H. (2002). Conservation and variation in enamel protein distribution during vertebrate tooth development. *J. Exp. Zool. (Mol. Devel. Evol.)* **284**, 91–106.

Sato, I., Sunohara, M., and Sato, T. (1999). Quantitative analysis of extracellular matrix proteins in hypertrophic layers of the mandibular condyle and temporal bone during human fetal development. *Cells Tissues & Organs* **165**, 81–90.

Sato, K., Miura, T., and Iwata, H. (1988). Cartilaginous transdifferentiation of rat tenosynovial cells under the influence of bone morphogenetic protein in tissue culture. *Clin. Orthop. Rel. Res.* **236**, 233–239.

Satokata, I., and Maas, R. (1994). *Msx1* deficient mice exhibit cleft palate and abnormalities of craniofacial and tooth development. *Nat. Genet.* **6**, 348–356.

Satomura, K., and Nagayama, M. (1991). Ultrastructure of mineralized nodules formed in rat bone marrow stromal cell cultures *in vitro*. *Acta Anat.* **142**, 97–104.

Saunders, J. W., Jr. (1948). The proximo-distal sequence of origin of the parts of the chick wing and the role of the ectoderm. *J. Exp. Zool.* **108**, 363–404.

Saunders, J. W., Jr. (1966). Death in embryonic systems. *Science* **154**, 604–612.

Saunders, J. W., Jr. (1977). The experimental analysis of chick limb bud development. In *Vertebrate Limb and Somite Morphogenesis* (D. A. Ede, J. R. Hinchliffe, and M. Balls, eds), pp. 1–24. Cambridge University Press, Cambridge.

Saunders, J. W., Jr. (2002). Is the progress zone model a victim of progress? *Cell* **110**, 541–543.

Saunders, J. W., Jr., Cairns, J. M., and Gasseling, M. T. (1957). The role of the apical ridge of ectoderm in the differentiation of the morphological structure and inductive specificity of limb parts in the chick. *J. Morphol.* **101**, 57–87.

Saunders, J. W., Jr., and Fallon, J. F. (1967). Cell death in morphogenesis. *Symp. Soc. Devel. Biol.* **25**, 289–314.

Saunders, J. W., Jr., and Gasseling, M. T. (1963). Transfilter propagation of apical ectoderm maintenance factor in the chick embryo wing bud. *Devel. Biol.* **7**, 64–78.

Saunders, J. W., Jr., and Gasseling, M. T. (1968). Ectodermal-mesenchymal interactions in the origin of limb symmetry. In *Epithelial-Mesenchymal Interactions* (R. Fleischmajer and R. E. Billingham, eds), pp. 78–97. Williams & Wilkins, Baltimore, MD.

Saunders, J. W., Jr., Gasseling, M. T., and Cairns, J. M. (1955). Effect of implantation site on the development of an implant in the chick embryo. *Nature* **175**, 673.

Saunders, J. W., Jr., Gasseling, M. T., and Cairns, J. M. (1959). The differentiation of prospective thigh mesoderm grafted beneath the apical ectodermal ridge of the wing bud in the chick embryo. *Devel. Biol.* **1**, 281–301.

Saunders, J. W., Jr., Gasseling, M. T., and Errick, J. E. (1976). Inductive activity and enduring cellular constitution if a supernumerary apical ectodermal ridge grafted to the limb bud of the chick embryo. *Devel. Biol.* **50**, 16–25.

Saunders, J. W., Jr., Gasseling, M. T., and Gfeller, M. D. (1958). Interactions of ectoderm and mesoderm in the origin of axial relationships in the wing of the fowl. *J. Exp. Zool.* **137**, 39–74.

Saunders, J. W., Jr., Gasseling, M. T., and Saunders, L. C. (1962). Cellular death in morphogenesis of the avian wing. *Devel. Biol.* **5**, 147–178.

Saunders, J. W., Jr., and Reuss, C. (1974). Inductive and axial properties of prospective wing-bud mesoderm in the chick embryo. *Devel. Biol.* **38**, 41–50.

Savagner, P., Karavanova, I., Perantoni, A., Thiery, J. P., and Yamada, K. M. (1998). Slug mRNA is expressed by specific mesodermal derivatives during rodent organogenesis. *Devel. Dynam.* **213**, 182–187.

Savostin-Asling, I., and Asling, C. W. (1973). Resorption of calcified cartilage as seen in Meckel's cartilage in rats. *Anat. Rec.* **176**, 345–360.

Savostin-Asling, I., and Asling, C. W. (1975). Transmission and scanning electron microscope studies of calcified cartilage resorption. *Anat. Rec.* **183**, 373–392.

Sawai, S., Shimono, A., Wakamatsu, Y., Palmes, C., Hanaoka, K., and Kohdoh, H. (1993). Defects of embryonic organogenesis resulting from targeted disruption of the *N-myc* gene in the mouse. *Development* **117**, 1445–1455.

Sawyer, L. M. (1982). Fine structural analysis of limb development in the wingless mutant chick embryo. *J. Embryol. Exp. Morphol.* **68**, 69–86.

Sawyer, L. M., and Goetinck, P. F. (1981). Chondrogenesis in the mutant nanomelia. Changes in the fine structure and proteoglycan synthesis in high density limb bud cell cultures. *J. Exp. Zool.* **216**, 121–131.

Saxén, L. (1976). Review article. Mechanisms of teratogenesis. *J. Embryol. Exp. Morphol.* **26**, 1–12.

Sayegh, F. S., Solomon, G. C., and Davis, R. W. (1974). Ultrastructure of intracellular mineralization in the deer's antler. *Clin. Orthop. Rel. Res.* **99**, 267–284.

Scaal, M., Bonafede, A., Dathe, U., Sachs, M., Cann, G., Christ, B., and Brand-Saberi, B. (1999). SF/HGF is a mediator between limb patterning and muscle development. *Development* **126**, 4885–4893.

Scadding, S. R. (1977). Phylogenetic distribution of limb regeneration capacity in adult Amphibia. *J. Exp. Zool.* **202**, 57–68.

Scadding, S. R. (1981). Limb regeneration in adult amphibia. *Can. J. Zool.* **59**, 34–46.

Scadding, S. R. (1982). Can differences in limb regeneration ability between amphibian species be explained by differences in quantity of innervation? *J. Exp. Zool.* **219**, 81–85.

Scadding, S. R. (1983). Can differences in limb regeneration ability between individuals within certain amphibian species be explained by differences in quantity of innervation? *J. Exp. Zool.* **226**, 75–80.

Schaberg, S. J., Liboff, A. R., and Falk, M. C. (1985). Wire-induced osteogenesis in marrow. *J. Biomed. Materials Res.* **19**, 673–684.

Schaedler, J. M., Krook, L., Wootton, J. A. M., Hover, B., Brodsky, B., Naresh, M. D., Gillette, D. D., Madsen, D. B., Horne, R. H., and Minor, R. R. (1992). Studies of collagen in bone and dentin matrix of a Columbian mammoth (late Pleistocene) of Central Utah. *Matrix* **12**, 297–307.

Schaefer, S. A., and Buitrago-Svárez, U. A. (2002). Odontode morphology and skin surface features of Andean astroblepid catfishes (Siluriformes, Astroblepidae). *J. Morphol.* **254**, 139–148.

Schaeffer, B. (1961). Differential ossification in the fishes. *Trans. NY. Acad. Sci.* [2] **23**, 501–505.

Schaeffer, B. (1977). The dermal skeleton in fishes. In *Problems in Vertebrate Evolution* (S. M. Andrews, R. S. Miles and A. D. Walker, eds), Linnean Soc. Symposium No. 4, pp. 25–52. Academic Press, London.

Schäfer, E. A. (1878). Note on the structure and development of osseous tissue. *Q. J. Microsc. Sci.* **18**, 132–141 (including a Postscript by W. Sharpey, pp. 142–144).

Schaffer, J. (1888). Die Verknöcherung des Unterkeifers und die Metaplasiefrage ein Beitrag zur Lehre von der osteogenese. *Arch. Mikr. Anat.* **32**, 266–377.

Schaffer, J. (1930). Die Stutzbeweβe. In *Handbuch der Mikroskopischen Anatomie des Menschen* (W. von Mallendorff, ed.), Volume II, Part 2, pp. 338–350. Springer-Verlag, Berlin.

Schaller, S. A., and Muneoka, K. (2001). Inhibition of polarizing activity in the anterior limb bud is regulated by extracellular factors. *Devel. Biol.* **240**, 443–457.

Schatzker, J., Waddell, J., and Stoll, J. E. (1989). The effects of motion on the healing of cancellous bone. *Clin. Orthop. Rel. Res.* **245**, 282–287.

Schendel, S. A., and Delaire, J. (1982). Familial osteodysplasia. *Head & Neck Surg.* **4**, 335–343.

Schenk, R. K. (1973). Fracture repair – overview. In *Proceedings of the Ninth European Symposium on Calcified Tissues, 1972* (H. Czitober and J. Eschberger, eds), pp. 13–22. Facta-Publication, Vienna.

Schenk, R. K., Spiro, D., and Wiener, J. (1967). Cartilage resorption in the tibial epiphyseal plate of growing rats. *J. Cell Biol.* **34**, 275–291.

Schenk, R. K., Wiener, J., and Spiro, D. (1968). Fine structural aspects of vascular invasion of the tibial epiphyseal plate of growing rats. *Acta Anat.* **69**, 1–17.

Scherft, J. P. (1978). The lamina limitans of the organic bone matrix: formation *in vitro*. *J. Ultrastruct. Res.* **64**, 173–181.

Scherft, J. P., and Daems, W. T. H. (1967). Single cilia in chondrocytes. *J. Ultrastruct. Res.* **19**, 546–555.

Scheven, B. A. A., and Hamilton, N. J. (1991). Longitudinal bone growth *in vitro*: effects of insulin-like growth factor I and growth hormone. *Acta Endocrinol.* **124**, 602–607.

Schiebinger, L. (1986). Skeletons in the closet: the first illustrations of the female skeleton in eighteenth century anatomy. *Representations* **14**, 42–82.

Schiebinger, L. (2003). Skelettestreit. *Isis* **94**, 307–313.

Schilling, T. F., and Kimmel, C.B. (1997). Musculoskeletal patterning in the pharyngeal segments of the zebrafish embryo. *Development* **124**, 2945–2960.

Schilling, T. F., Piotrowski, T., Grandel, H., Brand, M., Heisenberg, C.-P., Jiang, Y.-J., Beuchle, D., Hammerschmidt, M., Kane, D. A., Mullins, M. C., Eeden, F. J. M., Kelsh, R. N., Furutani-Seiki, M., Granato, M., Haffter, P., Odenthal, J., Warga, R. M., Trowe, T., and Nüsslein-Volhard, C. (1996b). Jaw and branchial arch mutants in zebrafish. I: branchial arches. *Development* **1213**, 329–344.

Schilling, T. F., Walker, C., and Kimmel, C. B. (1996a). The Chinless mutation and neural crest cell-interactions in zebrafish jaw development. *Development* **122**, 1417–1426.

Schimmang, T., Lemaistre, M., Vortkamp, A., and Rüther, U. (1992). Expression of the zinc finger gene *Gli3* is affected in the morphogenetic mouse mutant *extra toes (xt)*. *Development* **116**, 799–804.

Schindler, F. H., Ose, M. A., and Solursh, M. (1975). Synthesis of cartilage collagen by rabbit and human chondrocytes in primary cell culture. *In Vitro* **12**, 44–47.

Schipani, E., Lanske, B., Hunzelman, J., Luz, A., Kovacs, C. S., Lee, K., Pirro, A., Kronenberg, H. M., and Jüppner, H. (1997). Targeted expression of constitutively active receptors for parathyroid hormone and parathyroid hormone-related peptide delays endochondral bone formation and rescues mice that lack parathyroid hormone-related peptide. *Proc. Natl Acad. Sci. U.S.A.* **94**, 13689–13694.

Schirrmacher, K., Schmitz, I., Winterhager, E., Traub, O., Brümmer, F., Jones, D., and Bingmann, D. (1992). Characterization of gap junctions between osteoblast-like cells in culture. *Calcif. Tissue Int.* **51**, 285–290.

Scholmbs, K., Wagner, T., and Scheel, J. (2003). Site-1 protease is required for cartilage development in zebrafish. *Proc. Natl Acad. Sci. U.S.A.* **100**, 14024–14029.

Schlosser, G. (2002a). Development and evolution of lateral line placodes in amphibians. I. Development. *Zoology* **105**, 119–146.

Schlosser, G. (2002b). Development and evolution of lateral line placodes in amphibians. II. Evolutionary diversification. *Zoology* **105**, 177–193.

Schlosser, G., Kintner, C., and Northcutt, R. G. (1999). Loss of ectodermal competence for lateral line placode formation in the direct developing frog *Eleutherodactylus coqui*. *Devel. Biol.* **213**, 354–369.

Schmalhausen, O. I. (1939). Role of the olfactory sac in the development of the cartilage of the olfactory organ in Urodela. *C.R. Acad. Sci. U.S.S.R.* **23**, 395–398.

Schmid, P., Cox, D., Bilbe, G., Maier, R., and McMaster, G. K. (1991). Differential expression of TGFβ1, β2, and β3 genes during mouse embryogenesis. *Development* **111**, 117–130.

Schmid, T. M., and Linsenmayer, T. F. (1985a). Developmental acquisition of type X collagen in the embryonic chick tibiotarsus. *Devel. Biol.* **107**, 373–381.

Schmid, T. M., and Linsenmayer, T. F. (1985b). Immunohistochemical localization of short chain cartilage collagen (type X) in avian tissues. *J. Cell Biol.* **100**, 598–605.

Schmidmaier, G., Wildemann, B., Ostapowicz, D., Kandziora, F., Stange, R., Haas, N. P., and Raschke, M. (2004). Long-term effects of local growth factor (IGF-1 and TGF-β1) treatment on fracture healing. A safety study for using growth factors. *J. Orthop. Res.* **22**, 514–519.

Schmitz, J. P., Schwartz, Z., Hollinger, J. O., and Boyan, B. D. (1990). Characterization of rat calvarial nonunion defects. *Acta Anat.* **138**, 185–192.

Schneider, B. F., and Norton, S. (1979). Equivalent ages in rat, mouse and chick embryos. *Teratology* **19**, 273–278.

Schneider, G. B. (1985). Cellular specificity of the cure for osteopetrosis: Isolation of and treatment with pluripotent hemopoietic stem cells. *Bone* **6**, 241–248.

Schneider, G. B., and Byrnes, J. E. (1983). Cellular specificity of the cure for neonatal osteopetrosis in the *ia* rat. *Exp. Cell Biol.* **51**, 44–50.

Schneider, G. B., Relfson, M., and Nicolas, J. (1986). Pluripotent hemopoietic stem cells give rise to osteoclasts. *Amer. J. Anat.* **177**, 505–512.

Schneider, R. A., and Helms, J. A. (2003). The cellular and molecular origins of beak morphology. *Science* **299**, 565–568.

Schneider, R. A., Hu, D., and Helms, J. A. (1999). From head to toe: conservation of molecular signals regulating limb and craniofacial morphogenesis. *Cell Tissue Res.* **296**, 103–109.

Schneider, R. A. (1999). Neural crest can form cartilage normally derived from mesoderm during development of the avian head skeleton. *Devel. Biol.* **208**, 441–455.

Schneider, R. A., Hu, D., and Helms, J. A. (1999). From head to toe: Conservation of molecular signals regulating limb and craniofacial morphogenesis. *Cell Tissue Res.* **296**, 103–109.

Schneider, R. A., Hu, D., Rubenstein, J. L. R., Maden, M., and Helms, J. A. (2001). Local retinoid signaling coordinates forebrain and facial morphogenesis by maintaining FGF8 and SHH. *Development* **128**, 2755–2767.

Schoch, R. R. (2002). The early formation of the skull in extant and Paleozoic amphibians. *Paleobiology* **28**, 278–296.

Schoch, R. R. (2004). Skeleton formation in the Branchiosauridae: A case study in comparing ontogenetic trajectories. *J. Vert. Paleont.* **24**, 309–319.

Schoch, R. R., and Carroll, R. L. (2003). Ontogenetic evidence for the Paleozoic ancestry of salamanders. *Evol. & Devel.* **5**, 314–324.

Schofield, J. N., and Wolpert, L. (1990). Effect of TGF-beta-1, TGF-beta-2, and bFGF on chick cartilage and muscle cell differentiation. *Exp. Cell Res.* **191**, 144–148.

Schorle, H., Meier, P., Buchert, M., Jaenisch, R., and Mitchell, P. J. (1996). Transcription factor AP-2 essential for cranial closure and craniofacial development. *Nature* **381**, 235–238.

Schour, I. (1936). Measurements of bone growth by alizarine injection. *Proc. Soc. Exp. Biol.* **34**, 140–141.

Schour, I., Hoffman, M. M., Sarnat, B. G., and Engel, N. B. (1941). Vital staining of growing bones and teeth with alizarine red 'S'. *J. Dent. Res.* **20**, 411–418.

Schowing, J. (1961). Influence inductrice de l'encéphale et de la chorde sur la morphogenèse du squelette cranien chez l'embryon de poulet. *J. Embryol. Exp. Morphol.* **9**, 326–334.

Schowing, J. (1968a). Influence inductrice de l'encéphale embryonnaire sur le développement du crâne chez le poulet. I. Influence de l'excision des territoires nerveux antérieurs sur le développement crânien. *J. Embryol. Exp. Morphol.* **19**, 9–22.

Schowing, J. (1968b). Influence inductrice de l'encéphale embryonnaire sur le développement du crâne chez le poulet. II. Influence de l'excision de la chorde et des territoires encéphaliques moyen et postérieur sur le développement crânien. *J. Embryol. Exp. Morphol.* **19**, 23–32.

Schowing, J. (1968c). Influence inductrice de l'encéphale embryonnaire sur le développement du crâne chez le poulet. III. Mise en évidence du rôle inducteur de l'encéphale dans l'ostéogenèse du crâne embryonnaire du poulet. *J. Embryol. Exp. Morphol.* **19**, 83–94.

Schowing, J. (1974). Role morphogenèse de l'encéphale embryonnaire dans l'organogenèse du crâne chez l'oiseau. *Annee Biol.* **13**, 69–76.

Schowing, J., and Robadey, M. (1971). Substitution à l'encéphale embryonnaire de poulet (*Gallus gallus*) d'un encéphale embryonnaire de caille (*Coturnix coturnix japonica*) de même stade. *C.R. Hebd. Seances Acad. Sci.* **271**, 2382–2384.

Schrarer, H. (1970). *Biological Calcification: Cellular and Molecular Aspects.* Appleton, New York, NY.

Schrarer, H., and Hunter, S. J. (1985). The development of medullary bone: A model for osteogenesis. *Comp. Biochem. Physiol. A. Comp. Physiol.* **82**, 13–18.

Schubert, D., and Lacorbiere, M. (1976). Phenotypic transformation of clonal myogenic cells to cells resembling chondrocytes. *Proc. Natl Acad. Sci. U.S.A.* **73**, 1989–1993.

Schubert, M., Holland, L. Z., Stokes, M. D., and Holland, N. D. (2001). Three amphioxus *Wnt* genes (*AmphiWnt3*, *AmphiWnt5*, and *AmphiWnt6*) associated with the tail bud: the evolution of somitogenesis in chordates. *Devel. Biol.* **140**, 262–273.

Schultz, A., Donath, K., and Delling, G. (1974). Ultrastructure and development of cortical osteocytes – experimental investigation in the rat. *Virchows Arch. A Pathol.* **364**, 347–356.

Schumacher, B., Albrechtsen, J., Keller, J., Flyvbjerg, A., and Hvid, I. (1996). Periosteal insulin-like growth factor I and bone formation – Changes during tibial lengthening in rabbits. *Acta Orthop. Scand.* **67**, 237–241.

Schuster, G. S., Dirksen, T. R., and Harms, W. S. (1975). Effect of exogenous lipid on lipid synthesis by bone and bone cell cultures. *J. Dent. Res.* **54**, 131–139.

Schusterman, L., Bernard, G., and Junge, D. (1974). Metabolic components of the transmembrane potential of osteoblasts in culture. *J. Dent. Res.* **53**, 245.

Schwab, W., and Funk, R. H. W. (1998). Innervation pattern of different cartilaginous tissues in the rat. *Acta Anat.* **163**, 184–190.

Schwartz, E. R., and Adamy, L. (1976). Effect of ascorbic acid on arylsulfatase A and B activities in human chondrocyte cultures. *Connect. Tissue Res.* **4**, 211–218.

Schwartz, N. B., and Dorfman, A. (1975). Stimulation of chondroitin sulfate proteoglycan production by chondrocytes in monolayer. *Connect. Tissue Res.* **3**, 115–122.

Schwarz, M., Harbers, K., and Kratochvil, K. (1990). Transcription of a mutant collagen I gene is a cell type and stage-specific marker for odontoblast and osteoblast differentiation. *Development* **108**, 717–726.

Schweitzer, R., Chyung, J. H., Murtaugh, L. C., Brent, A. E. Rosen, V., Olson, E. N., Lassar, A., and Tabin, C. J. (2001). Analysis of the tenson cell fate using scleraxis, a specific marker for tendons and ligaments. *Development* **128**, 3855–3866.

Scopelliti, R. (1975). Sur des effets provoque's *in vivo* par quelques hormones sexuelles sur la croissance des tibias des embryons de poulet normaux et chondrodystrophiques (creeper). *C.R. Hebd. Seances Acad. Sci.* **200**, 901–904.

Scott, B. L. (1965). Fine structure and intercellular relationships of osteogenic cell types. *J. Ultrastruct. Res.* **13**, 560.

Scott, B. L. (1967). Thymidine-³H electron microscope radioautoradiography of osteogenic cells in the fetal rat. *J. Cell Sci.* **35**, 115–126.

Scott, B. L. (1969). Thymidine-³H study of developing tooth germs and osteogenic tissue. *J. Dent. Res.* **48**, Suppl., 753–760.

Scott, B. L., and Glimcher, M. J. (1971). Distribution of glycogen in osteoblasts of the fetal rat. *J. Ultrastruct. Res.* **36**, 565–586.

Scott, B. L., and Pease, D. C. (1956). Electron microscopy of the epiphyseal apparatus. *Anat. Rec.* **126**, 465–495.

Scott, C. K., Bain, S. D., and Hightower, J. A. (1994). Intramembranous bone matrix is osteoinductive. *Anat. Rec.* **238**, 23–30.

Scott, C. K., and Hightower, J. A. (1991). The matrix of endochondral bone differs from the matrix of intramembranous bone. *Calcif. Tissue Int.* **49**, 349–354.

Scott, I. C., Blitz, I. L., Pappano, W. N., Imamura, Y., Clark, T. G., Steiglitz, B. M., Thomas, C. L., Maas, S. A., Takahara, K., Cho, K. W. Y., and Greenspan, D. S. (1999). Mammalian BMP-1/tolloid-related metalloproteinases, including novel family member mammalian tolloid-like 2, have differential enzymatic activities and distributions of expression relevant to patterning and skeletogenesis. *Devel. Biol.* **213**, 283–290.

Scott, I. C., Steiglitz, B. M., Clark, T. G., Pappano, W. N., and Greenspan, D. S. (2000). Spatiotemporal expression patterns of mammalian chordin during postgastrulation embryogenesis and in postnatal brain. *Devel. Dynam.* **217**, 449–456.

Scott, J. E. (1975). Physiological function and chemical composition of pericellular proteoglycan (an evolutionary view). *Philos. Trans. R. Soc. London, Ser. B* **271**, 235–242.

Scott, J. E., and Hughes, E. W. (1981). Chondroitin sulphate from fossilized antlers. *Nature* **291**, 580–581.

Scott, J. H. (1951). The development of joints concerned with early jaw movements in the sheep. *J. Anat.* **85**, 36–43.

Scott, J. H. (1954). The growth of the human face. *Proc. R. Soc. Med.* **47**, 91–100.

Scott, J. H., and Symons, N. B. B. (1974). *Introduction to Dental Anatomy.* 7th Edn. Churchill, London.

Scott, J. P. (1937). The embryology of the guinea pig. III. The development of the polydactylous monster. A case of growth accelerated at a particular period by a semi-dominant lethal gene. *J. Exp. Zool.* **77**, 123–157.

Scott, J. P. (1938). The embryology of the guinea pig. II. The polydactylous monster. A new teras produced by the genes Px Px. *J. Morphol.* **62**, 299–321.

Scott, J. P. (1986). Critical periods in organizational processes. In *Human Growth. A Comprehensive Treatise. 2nd Edn, Volume 1. Developmental Biology, Prenatal Growth* (F. Falkner and J. M. Tanner, eds), pp. 181–196. Plenum Press, New York, NY.

Scott, P. G., Nahano, T., and Dodd, C. M. (1995). Small proteoglycans from different region of the fibrocartilaginous temporomandibular joint disc. *BBA Gen. Subj.* **1244**, 121–128.

Scott, W. J., Ritter, E. J., and Wilson, J. G. (1977). Delayed appearance of ectodermal cell death as a mechanism of polydactyly induction. *J. Embryol. Exp. Morphol.* **43**, 93–104.

Scott-Savage, P., and Hall, B. K. (1979). The timing of the onset of osteogenesis in the tibia of the embryonic chick. *J. Morphol.* **162**, 453–464.

Scott-Savage, P., and Hall, B. K. (1980). Differentiative ability of the tibial periosteum from the embryonic chick. *Acta Anat.* **106**, 129–140.

Scriven, P. N., and Bauchau, V. (1992). The effect of hybridization on mandible morphology in an island population of the house mouse. *J. Zool. Lond.* **226**, 573–583.

Searls, R. L. (1965a). An autoradiographic study of the uptake of S³⁵-sulfate during the differentiation of limb bud cartilage. *Devel. Biol.* **11**, 155–168.

Searls, R. L. (1965b). Isolation of mucopolysaccharide from the pre-cartilaginous chick limb bud. *Proc. Soc. Exp. Biol. Med.* **118**, 1172–1176.

Searls, R. L. (1967). The role of cell migration in the development of the embryonic chick limb bud. *J. Exp. Zool.* **166**, 39–50.

Searls, R. L. (1968). Development of the embryonic chick limb bud in avascular culture. *Devel. Biol.* **17**, 382–399.

Searls, R. L. (1971). Segregation of cells that differentiate without cell movement from a single precursor population. *Exp. Cell Res.* **64**, 163–169.

Searls, R. L. (1972). Cellular segregation: A 'late' differentiative characteristic of chick limb bud cartilage cells. *Exp. Cell Res.* **73**, 57–64.

Searls, R. L. (1973a). Newer knowledge of chondrogenesis. *Clin. Orthop. Rel. Res.* **96**, 327–344.

Searls, R. L. (1973b). Chondrogenesis. In *Developmental Regulation. Aspects of Cell Differentiation* (S. J. Coward, ed.), pp. 219–251. Academic Press, New York, NY.

Searls, R. L. (1976). Effect of dorsal and ventral limb ectoderm on the development of the limb of the embryonic chick. *J. Embryol. Exp. Morphol.* **35**, 369–381.

Searls, R. L., Hilfer, S. R., and Mirow, S. M. (1972). An ultrastructural study of early chondrogenesis in the chick wing bud. *Devel. Biol.* **28**, 123–137.

Searls, R. L., and Janners, M. Y. (1969). The stabilization of cartilage properties in the cartilage-forming mesenchyme of the embryonic chick limb. *Exp. Cell Res.* **170**, 365–376.

Searls, R. L., and Janners, M. Y. (1971). The initiation of limb bud outgrowth in the embryonic chick. *Devel. Biol.* **24**, 198–213.

Searls, R. L., and Zwilling, E. (1964). Regeneration of the apical ectodermal ridge of the chick limb bud. *Devel. Biol.* **9**, 38–55.

Sêbek, J., Skalova, J., and Hert, J. (1972). Reaction of bone to mechanical stimuli: Part 8. Local differences in structure and strength of periosteum. *Folia Morphol.* **20**, 29–37.

Sedmera, D., Misek, I., and Klima, M. (1997a). On the development of cetacean extremities. I. Hind limb rudimentation in the spotted dolphin (*Stenella attenuata*). *Eur. J. Morphol.* **35**, 25–30.

Sedmera, D., Misek, I., and Klima, M. (1997b). On the development of cetacean extremities. II. Morphogenesis and histogenesis of the flippers of the spotted dolphin (*Stenella attenuata*). *Eur. J. Morphol.* **35**, 117–123.

Seegmiller, R. E. (1977). Time of onset and selective response of chondrogenic core of 5-day chick limb after treatment with 6-aminonicotinamide. *Devel. Biol.* **58**, 164–173.

Seegmiller, R. E., Brown, K., and Chandrasekhar, S. (1988). Histochemical, immunofluorescence, and ultrastructural differences in fetal cartilage among three genetically distinct chondrodystrophic mice. *Teratology* **38**, 579–592.

Seegmiller, R. E., Ferguson, C. C., and Sheldon, H. (1972b). Studies on cartilage. VI. A genetically determined defect in tracheal cartilage. *J. Ultrastruct. Res.* **38**, 288–301.

Seegmiller, R. E., and Fraser, F. C. (1977). Mandibular growth retardation as a cause of cleft palate in mice homozygous for the chondrodysplastic gene. *J. Embryol. Exp. Morphol.* **38**, 227–238.

Seegmiller, R. E., Fraser, F. C., and Sheldon, H. (1971). A new chondrodystrophic mutant in mice. Electron microscopy of normal and abnormal chondrogenesis. *J. Cell Biol.* **48**, 580–593.

Seegmiller, R. E., Overman, D. O., and Runner, M. N. (1972a). Histological and fine structural changes during chondrogenesis in micromelia induced by 6-aminonicotinamide. *Devel. Biol.* **28**, 555–572.

Seegmiller, R. E., and Runner, M. N. (1974). Normal incorporation rates and precursors of collagen and mucopolysaccharide during expression of micromelia induced by 6-aminonicotinamide. *J. Embryol. Exp. Morphol.* **31**, 305–312.

Sela, J., Amir, D., Schwartz, Z., and Weinberg, H. (1987). Ultrastructural tissue morphometry of the distribution of extracellular matrix vesicles in remodeling rat tibial bone six days after injury. *Acta Anat.* **128**, 295–300.

Selleck, S. B. (2000). Proteoglycans and pattern formation: Sugar biochemistry meets developmental genetics. *Trends Genet.* **16**, 206–212.

Selye, H. (1934). On the mechanism controlling the growth in length of the long bones. *J. Anat.* **68**, 289–292.

Selz, T., Caverzasio, J., and Bonjour, J. P. (1991). Fluoride selectively stimulates Na-dependent phosphate transport in osteoblast-like cells. *Amer. J. Physiol.* **260**, E833–E838.

Semba, I., Nonaka, K., Takahashi, I., Takahashi, K., Dashner, R., Shum, L., Nuckolls, G. H., and Slavkin, H. C. (2000). Positionally-dependent chondrogenesis induced by BMP4 is co-regulated by *Sox9* and *Msx2*. *Devel. Dynam.* **217**, 404–414.

Senn, N. (1889). On the healing of aseptic bone cavities by implantation of antiseptic decalcified bone. *Amer. J. Med. Sci.* **98**, 219–243.

Seno, T., and Büyüközer, I. (1958). Cartilage formation in somite grafts of chick blastoderms. *Proc. Natl Acad. Sci. U.S.A.* **44**, 1274–1284.

Serafini-Fracassini, A., and Smith, J. W. (1974). *The Structure and Biochemistry of Cartilage*. Churchill-Livingstone, London.

Serra, R., and Chang, C. (2003). TGF-β signaling in human skeletal and patterning disorders. *Birth Defects Res.* (*Part C*) **69**, 333–351.

Serra, R., Johnson, M., Filvaroff, E. H., LaBorde, J., Sheehan, D. M., Derynck, R., and Moses, H. L. (1997). Expression of a truncated kinase-defective TGF-β type II receptor in mouse skeletal tissue promotes terminal chondrocyte differentiation and osteoarthritis. *J. Cell Biol.* **139**, 541–552.

Serra, R., Karaplis, A., and Sohn, P. (1999). Parathyroid hormone-related peptide (PTHrP)-dependent- and independent effects of transforming growth factor β (TGF-β) on endochondral bone formation. *J. Cell Biol.* **145**, 783–794.

Service, R. F. (2000). Tissue engineers build new bone. *Science* **289**, 1498–1500.

Servoss, J. M. (1973). An *in vivo* and *in vitro* autoradiographic investigation of growth in synchondrosal cartilage. *Amer. J. Anat.* **136**, 479–486.

Seufert, D. W., and Hall, B. K. (1990). Tissue interactions involving cranial neural crest in cartilage formation in *Xenopus laevis* (Daudin). *Cell Differ. Devel.* **32**, 153–166.

Seufert, D. W., Hanken, J., and Klymkowsky, M. W. (1994). Type II collagen distribution during cranial development in *Xenopus laevis*. *Anat. Embryol.* **189**, 81–89.

Sewertzoff, A. N. (1931). Morphologische Gesetzmassigkeiten der Evolution. Studien über die Reduktion der Organs der Wirbeltiere. *Zool. Jahrb. Abt. F. Anat.* **53**, 611–700.

Seyedin, S. M., Thompson, A. Y., Bentz, H., Rosen, D. M., McPherson, J. M., Conti, A., Siegel, N. R., Galluppi, G. R., and Piez, K. A. (1986). Cartilage-inducing factor-A. Apparent identity to transforming growth factor-β. *J. Biol. Chem.* **261**, 5693–5695.

Shalhoub, V., Jackson, M. E., Lian, J. B., Stein, G. S., and Marks, S. C., Jr. (1991). Gene expression during skeletal development in three osteopetrotic rat mutations. Evidence for osteoblast abnormalities. *J. Biol. Chem.* **266**, 9487–9856.

Shambaugh, J., and Elmer, W. A. (1980). Analysis of glycosaminoglycans during chondrogenesis of normal and brachypod mouse limb mesenchyme. *J. Embryol. Exp. Morphol.* **56**, 225–238.

Shanfeld, J. L., Shapiro, I., and Davidovitch, Z. (1975). The measurement of adenosine 3',5'-monophosphate in bone. *Anal. Biochem.* **66**, 450–459.

Shapiro, F. (1992). Vertebral development of the chick embryo during days 3–29 of incubation. *J. Morphol.* **213**, 317–333.

Shapiro, F., Holtrop, M. E., and Glimcher, M. J. (1977). Organization and cellular biology of the perichondrial ossification groove of Ranvier. *J. Bone Joint Surg. Am. Vol.* **59**, 703–723.

Shapiro, M. D. (2002). Developmental morphology of limb reduction in *Hemiergis* (Squamata: Scincidae): chondrogenesis, osteogenesis, and heterochrony. *J. Morphol.* **254**, 211–231.

Shapiro, M. D., Hanken, J., and Rosenthal, N. (2003). Developmental basis of evolutionary digit loss in the Australian lizard *Hemiergis. J. Exp. Zool. (Mol. Dev. Evol.)* **297B**, 48–56.

Shapiro, M. D., Marks, M. E., Peichel, C. L., Blackman, B. K., Nereng, K. S., Jonsson, B., Schluter, D., and Kingsley, D. M. (2004). Genetic and developmental basis of evolutionary pelvic reduction in threespine sticklebacks. *Nature* **428**, 717–723.

Sharpe, J., Nonchev, S., Gould, A., Whiting, J., and Krumlauf, R. (1998). Selectivity, sharing and competitive interactions in the regulation of *Hoxbx* genes. *EMBO J.* **17**, 1788–1798.

Sharpe, P. T. (2001). Fish scale development: hair today, teeth and scales yesterday? *Curr. Biol.* **11**, R751–R752.

Sharpe, W. D. (1979). Age changes in human bone: An overview. *Bull. NY. Acad. Med.* **55**, 757–773.

Sharpey, W. (1848). In *Elements of Anatomy* (J. Quain, ed.), 5th Edn, Volume 1. Taylor, Walton & Maberley, London.

Shashikant, C. S., Kim, C. B., Borbély, M. A., Wang, W. C. H., and Ruddle, F. H. (1998). Comparative studies on mammalian *Hoxc8* early enhancer sequence reveal a baleen whale-specific deletion of a cis-acting element. *Proc. Natl Acad. Sci. U.S.A.* **95**, 15446–15451.

Shaw, J. L., and Bassett, C. A. L. (1967). The effects of varying oxygen concentration on osteogenesis and embryonic cartilage *in vitro. J. Bone Joint Surg. Br. Vol.* **49**, 73–80.

Shaw, S. R., Vailas, A. C., Grindeland, R. E., and Zernicke, R. F. (1988). Effects of a 1-wk spaceflight on morphological and mechanical properties of growing bone. *Amer. J. Physiol.* **254**, R78–R83.

Shea, B. T. (1983a). Size and diet in the evolution of African ape craniodental form. *Folia Primatol.* **40**, 32–68.

Shea, B. T. (1983b). Allometry and heterochrony in the African apes. *Amer. J. Phys. Anthropol.* **62**, 275–289.

Shea, B. T. (1983c). Paedomorphosis and neoteny in the pygmy chimpanzee. *Science* **222**, 521–522.

Shea, B. T. (1985). On aspects of skull form in African apes and orangutans with implications for Hominoid evolution. *Amer. J. Phys. Anthropol.* **68**, 329–342.

Shea, B. T. (2002). Are some heterochronic transformations likelier than others? In *Human Evolution Through Developmental Change* (N. Minugh-Purves and K. J. McNamara, eds), pp. 79–101. Johns Hopkins University Press, Baltimore, MD.

Shea, B. T., Hammer, R. E., Brinster, R. L., and Ravosa, M. R. (1990). Relative growth of the skull and postcranium in giant transgenic mice. *Genet. Res.* **56**, 21–34.

Sheldon, H. (1964a). Studies on cartilage. IV. On the fine structure of the elastic fiber in elastic cartilage. *Z. Zellforsch.* **62**, 526–530.

Sheldon, H. (1964b). Cartilage. In *Electron Microscopic Anatomy* (S. M. Kurtz, ed.), pp. 295–314. Academic Press, New York, NY.

Sheldon, H. (1983). Transmission electron microscopy of cartilage. In *Cartilage, Volume 1, Structure, Function and Biochemistry* (B. K. Hall, ed.), pp. 87–104. Academic Press, New York, NY.

Sheldon, H., and Kimball, F. B. (1962). Studies on cartilage. III. The occurrence of collagen within vacuoles of the Golgi apparatus. *J. Cell Biol.* **12**, 599–610.

Shellis, R. P., and Miles, A. E. W. (1974). Autoradiographic study of the formation of enameloid and dentine matrices in teleost fishes using tritiated amino acids. *Proc. R. Soc. London, Ser. B* **185**, 51–72.

Shen, H., Wilke, T., Ashique, A. M., Narvey, M., Zerucha, T., Savino, E., Williams, T., and Richman, J. M. (1997). Chicken transcription factor AP-2: cloning, expression and its role in outgrowth of facial prominences and limb buds. *Devel. Biol.* **188**, 248–266.

Shepard, N., and Mitchell, N. (1977). The localization of articular cartilage proteoglycan by electron microscopy. *Anat. Rec.* **187**, 463–476.

Shepard, N., and Mitchell, N. (1985). Ultrastructural modifications of proteoglycans coincident with mineralization in local regions of rat growth plate. *J. Bone Joint Surg. Am. Vol.* **67**, 455–464. *Anat. Rec.* **187**, 463–476.

Shepard, T. H. (1971). Organ-culture studies of achondroplastic rabbit cartilage: Evidence for a metabolic defect in glucose utilization. *J. Embryol. Exp. Morphol.* **25**, 347–363.

Shephard, T. H., Fry, L. R., and Moffett, B. C., Jr. (1969). Microscopic studies of achondroplastic rabbit cartilage. *Teratology* **2**, 13–22.

Shi, Y.-B. (2000). *Amphibian Metamorphosis.* Wiley-Liss, New York, NY.

Shibata, N., Tatsumi, N., Tanaka, K., Okamura, Y., and Senda, N. (1972). A contractile protein possessing Ca^{2+} sensitivity (natural actomyosin) from leucocytes. Its extraction and some of its properties. *Biochim. Biophys. Acta* **256**, 565–576.

Shibata, S., Baba, O., Ohsako, M., Shikano, S., Terashima, T., Yamashita, Y., and Ichijo, T. (1991). Histological observation of large light cells that seem to be surviving hypertrophic chondrocytes in the rat mandibular condyle. *Archs Oral Biol.* **36**, 541–544.

Shibata, S., Baba, O., Sakamoto, Y., Ohsako, M., Yamashita, Y., Shikano, S., and Ichijo, T. (1993a). An ultrastructural study of the mitotic preosteoblasts in the primary spongiosa of the rat mandibular condyle. *Bone* **14**, 35–40.

Shibata, S., Fukada, K., Susuki, S., and Yamashita, Y. (1997b). Imunohistochemistry of collagen types II and X, and enzyme-histochemistry of alkaline phosphatase in the developing condylar cartilage of the fetal mouse mandible. *J. Anat.* **191**, 561–570.

Shibata, S., Suda, N., Yamazaki, K., Kuroda, T., Beck, F., Senior, P. V., and Hammond, V. E. (2000). Mandibular deformaties in parathyroid hormone-related protein (PTHrP) deficient mice: Possible involvement of masseter muscle. *Anat. Embryol.* **202**, 85–93.

Shibata, S., Susuki, S., Tengan, T., Ishii, M., and Kuroda, T. (1996). A histological study of the developing condylar cartilage of the fetal mouse mandible using coronal sections. *Archs Oral Biol.* **41**, 47–54.

Shibata, S., Susuki, S., and Yamashita, Y. (1997a). An ultrastructural study of cartilage resorption at the site of initial endochondral bone formation in the fetal mouse mandibular condyle. *J. Anat.* **191**, 65–76.

Shibata, S., Susuki, S., Yamashita, Y., and Ichijo, T. (1993b). A comparative ultrastructural study of the mitotic chondrogenic cells in the mandibular condyle and tibial growth plate of the rat. *Archs Oral Biol.* **38**, 845–851.

Shibuya, T., Fujio, Y., and Kondon, K. (1972). Studies on the action of creeper gene in Japanese chicken. *Jap. J. Genet.* **47**, 23–32.

Shibuya, T., and Kuroda, Y. (1973). Studies on growth and differentiation of cartilage cells from creeper chick embryos in culture. *Jap. J. Genet.* **48**, 197–206.

Shigeno, S., Kidokoro, H., Goto, T., Tsuchiya, K., and Segawa, S. (2001). Early ontogeny of the Japanese common squid *Todarodes pacificus* (Cephalopoda, Ommastrephidae) with special reference to its characteristic morphology and ecological significance. *Zool. Sci.* **18**, 1011–1026.

Shigetani, Y., Nobusada, Y., and Kuratani, S. (2000). Ectodermally derived FGF8 defines the maxillomandibular region in the early chick embryo: epithelial-mesenchymal interactions in the specification of the craniofacial ectomesenchyme. *Devel. Biol.* **228**, 73–85.

Shigetani, Y., Sugahara, F., Kawakami, Y., Murakimi, Y., Hirano, S., and Kuratani, S. (2002). Heterotopic shift of epithelial-mesenchymal interactions in vertebrate jaw evolution. *Science* **296**, 1316–1319.

Shih, M. S., and Norrdin, R. W. (1985). Regional acceleration of remodeling during healing of bone defects in beagles of various ages. *Bone* **6**, 377–379.

Shimeld, S. (2003). Evolutionary aspects of vertebrate patterning. In *Patterning in Vertebrate Development* (C. Tickle, ed.), pp. 214–232. Oxford University Press, Oxford.

Shimizu, T., Sasano, Y., Nakajo, S., Kagayama, M., and Shimaucho, H. (2001). Osteoblastic differentiation of periosteum-derived cells is promoted by the physical contact with the bone matrix *in vivo*. *Anat. Rec.* **264**, 72–81.

Shimo, T., Gentili, C., Iwamoto, M., Wu, C., Koyama, E., and Pacifici, M. (2004). Indian hedgehog and syndecans-3 coregulate chondrocyte proliferation and function during chick limb skeletogenesis. *Devel. Dyn.* **229**, 607–617.

Shimomura, Y., Wezeman, F. H., and Ray, R. D. (1973). The growth cartilage plate of the rat rib: Cellular differentiation. *Clin. Orthop. Rel. Res.* **90**, 246–254.

Shimomura, Y., Yoneda, T., and Suzuki, F. (1975). Osteogenesis by chondrocytes from growth cartilage of rat rib. *Calcif. Tissue Res.* **19**, 179–188.

Shin, V., Zebboudj, A. F., and Bostrom, K. (2004). Endothelial cells modulate osteogenesis in calcifying vascular cells. *J. Vascul. Res.* **41**, 193–201.

Shine, R. (2000). Vertebral numbers in male and female snakes: the roles of natural, sexual and fecundity selection. *J. Evol. Biol.* **13**, 455–465.

Shipman, P. (2001). *The Man Who Found the Missing Link. Eugène Dubois and His Lifelong Quest to Prove Darwin Right.* Simon & Schuster, New York, NY.

Shipman, P., Walker, A., and Bichell, D. (1986). *The Human Skeleton.* Harvard University Press, Cambridge, MA.

Shu, D., Morris, S. C., Zhang, Z. F., Liu, J. N., Han, J., Chen, L., Zhang, X. L., and Li, Y. (2003a). A new species of Yunnanozoan with implications for deuterostome evolution. *Science* **299**, 1380–1384.

Shu, D.-G., Conway Morris, S., Han, J., Zhang, Z,-F., Yasui, K., Janvier, P., Chen, L., Zhang, X,-L., Liu, J.-N., and Lio, H.-Q. (2003b). Head and backbone of the early Cambrian vertebrate *Haikouichthys*. *Nature* 421, 526–529.

Shubin, N. H. (1991). The implications of 'the Bauplan' for development and evolution of the tetrapod limb. In *Developmental Patterns of the Vertebrate Limb* (J. R. Hinchliffe, J. M. Hurlé and D. Summerbell, eds), pp. 411–421. Plenum Press, New York, NY.

Shubin, N. H. (1995). The evolution of paired fins and the origin of tetrapod limbs: Phylogenetic and transformational approaches. *Evol. Biol.* **28**, 39–86.

Shubin, N. H., and Alberch, P. (1986). A morphogenetic approach to the origin and basic organization of the tetrapod limb. *Evol. Biol.* **20**, 319–387.

Shubin, N. H., and Marshall, C. R. (2000). Fossils, genes and the origin of novelty. *Paleobiology*, Suppl. **26**(4), 324–340.

Shubin, N. H., Tabin, C., and Carroll, S. (1997). Fossils, genes and the evolution of animal limbs. *Nature* **388**, 639–648.

Shubin, N. H., Wake, D. B., and Crawford, A. J. (1995). Morphological variation in the limbs of *Taricha granulosa* (Caudata: Salamandridae): Evolutionary and phylogenetic implications. *Evolution* **49**, 874–884.

Shuey, D. L., Sadler, T. W., and Lauder, J. M. (1992). Serotonin as a regulator of craniofacial morphogenesis – site specific malformations following exposure to serotonin uptake inhibitors. *Teratology* **46**, 367–378.

Shuey, D. L., Sadler, T. W., Tamir, H., and Lauder, J. M. (1993). Serotonin and morphogenesis. Transient expression of serotonin uptake and binding protein during craniofacial morphogenesis in the mouse. *Anat. Embryol.* **187**, 75–85.

Shukunami, C., Iyama, K.-I., Inoue, H., and Hiraki, Y. (1999). Spatiotemporal pattern of the mouse *chondromodulin-1* gene expression and its regulatory role in vascular invasion into cartilage during endochondral bone formation. *Int. J. Devel. Biol.* **43**, 39–49.

Shukunami, C., Ohta, Y., Sakuda, M., and Hiraki, Y. (1998). Sequential progression of the differentiation program by bone morphogenetic protein 2 in chondrogenic cell line ATDC5. *Exp. Cell Res.* **241**, 1–11.

Shulman, H. J., and Opler, A. (1974). The stimulatory effect of calcium on the synthesis of cartilage proteoglycan. *Biochem. Biophys. Res. Commun.* **59**, 914–919.

Shum, L., Sakakura, Y., Bringas, P., Jr., Luo, W., Snead, M. L., Mayo, M., Crohin, C., Millar, S., Werb, Z., Buckley, S., Hall, F. L., Warburton, D., and Slavkin, H. C. (1993). EGF abrogation-induced *fusilli*-form dysmorphogenesis of Meckel's cartilage during embryonic mouse mandibular *in vitro*. *Development* **118**, 903–917.

Shum, L., Wang, X., Kane, A. A., and Nuckolls, G. H. (2003). BMP4 promotes chondrocyte proliferation and hypertrophy in the endochondral cranial base. *Int. J. Devel. Biol.* **47**, 423–431.

Shyng, Y. C., Devlin, H., Riccardi, D., and Sloan, P. (1999). Expression of cartilage-derived retinoic acid-sensitive protein during healing of the rat tooth-extraction socket. *Archs Oral Biol.* **44**, 751–757.

Siddhanti, S. R., and Quarles, L. D. (1994). Molecular to pharmacological control of osteoblast proliferation and differentiation. *J. Cellular Biochem.* **55**, 310–320.

Siebenmann, R. E. (1977). Die ohrknorpelverknöcherung beim Morbus Addison. *Schweiz. Med. Wochenschr.* **107**, 468–474.

Siegal, T., Segal, S., Nevo, Z., Lev-El, A., Altaratz, C., Katznelson, A., and Nebel, L. (1977). Replacement of massive bone loss by fetal bone transplantation: Biochemical and immunological aspects. *Transpl. Proc.* **9**, 351–354.

Silau, A. M., Fischer Hansen, B., and Kjaer, I. (1995). Normal prenatal development of the human parietal bone and interparietal suture. *J. Craniofac. Genet. Devel. Biol.* **15**, 81–86.

Silberberg, R., Hasler, M., and Lesker, P. (1976). Ultrastructure of articular cartilage of achondroplastic mice. *Acta Anat.* **96**, 162–175.

Silberberg, R., Hasler, M., and Silberberg, M. (1966). Articular cartilage of dwarf mice: Light and electron microscopic studies. *Acta Anat.* **65**, 275–298.

Silberberg, R., and Lesker, P. (1975). Skeletal growth and development of achondroplastic mice. *Growth* **39**, 17–34.

Silbermann, M., and Frommer, J. (1972a). The nature of endochondral ossification in the mandibular condyle of the mouse. *Anat. Rec.* **172**, 659–668.

Silbermann, M., and Frommer, J. (1972b). Further evidence for the vitality of chondrocytes in the mandibular condyle as revealed by ^{35}S-sulfate autoradiography. *Anat. Rec.* **174**, 503–512.

Silbermann, M., and Frommer, J. (1972c). Vitality of chondrocytes in the mandibular condyle as revealed by collagen formation. An autoradiographic study with ^{3}H-proline. *Amer. J. Anat.* **135**, 359–370.

Silbermann, M., and Frommer, J. (1973a). Dynamic changes in acid mucopolysaccharides during mineralization of the mandibular condylar cartilage. *Histochemie* **36**, 185–192.

Silbermann, M., and Frommer, J. (1973b). Ultrastructural localization of acid phosphatase in cartilage of young mandibular condyles. *Histochemie* **37**, 365–372.

Silbermann, M., and Frommer, J. (1974). Ultrastructure of developing cartilage in the mandibular condyle of the mouse. *Acta Anat.* **90**, 330–346.

Silbermann, M., and Kadar, T. (1977). Age-related changes in the cellular population of the growth plate of normal mouse. *Acta Anat.* **97**, 459–468.

Silbermann, M., Lewinson, D., Gonen, H., Lizarbe, M. A., and von der Mark, K. (1983). *In vitro* transformation of chondroprogenitor cells into osteoblasts and the formation of new membrane bone. *Anat. Rec.* **206**, 373–383.

Silbermann, M., Reddi, A. H., Hand, A. R., Leapman, R., von der Mark, K., and Franzen, A. (1987a). Chondroid bone arises from mesenchymal stem cells in organ culture of mandibular condyles. *J. Craniofac. Genet. Devel. Biol.* **7**, 59–80.

Silbermann, M., Tenenbaum, H., Livne, E., Leapman, R., von der Mark, K., and Reddi, A. H. (1987b). The *in vitro* behavior of fetal condylar cartilage in serum-free hormone-supplemented medium. *Bone* **8**, 117–126.

Silbermann, M., and von der Mark, K. (1990). An immunohistochemical study of the distribution of matricial proteins in the mandibular condyle of neonatal mice. I. Collagens. *J. Anat.* **170**, 11–22.

Silbermann, M., von der Mark, K., and Heinegard, D. (1990). An immunohistochemical study of the distribution of matricial proteins in the mandibular condyle of neonatal mice. II. Noncollagenous proteins. *J. Anat.* **170**, 23–31.

Silberzahn, N. (1968). Histocompatibilite du tissu cartilagineux. Étude en culture *in vitro* *C.R. Hebd. Seances Acad. Sci.* **267**, 352–355.

Siman, C. M., Gittenberger-de Groot, A. C., Wisse, B., and Eriksson, U. J. (2000). Malformations in offspring of diabetic rats: morphometric analysis of neural crest-derived organs and effects of maternal vitamin E treatment. *Teratology* **61**, 355–367.

Simon, H.-G., and Tabin, C. J. (1993). Analysis of *Hox-4.5* and *Hox-3.6* expression during newt limb regeneration: differtential regulation of paralogous *Hox* genes suggest different roles for members of different *Hox* clusters. *Development* **117**, 1397–1407.

Simon, T. M., Van Sickle, D. C., Kunishima, D. H., and Jackson, D. W. (2003). Cambium cell stimulation from surgical release of the periosteum. *J. Orthopaed. Res.* **21**, 470–480.

Simmons, D. J. (1962). Diurnal periodicity in epiphyseal growth cartilage. *Nature* **195**, 82–83.

Simmons, D. J. (1971). Calcium and skeletal tissue physiology in teleost fishes. *Clin. Orthop. Rel. Res.* **76**, 244–280.

Simmons, D. J. (1980). Fracture Healing. In *Fundamental and Clinical Bone Physiology* (M. R. Urist, ed.), pp. 283–330. J. B. Lippincott Co., Philadelphia, PA.

Simmons, D. J. (1985). Fracture healing perspectives. *Clin. Orthop. Rel. Res.* **200**, 100–113.

Simmons, D. J. (1992). Circadian aspects of bone growth. In *Bone, Volume 6: Bone Growth – A.* (B. K. Hall, ed), pp. 91–128. CRC Press, Boca Raton, FL.

Simmons, D. J., Arsenis, C., Whitson, S. W., Kahn, S. E., Boskey, A. L., and Gollub, N. (1983b). Mineralization of rat epiphyseal cartilage: A circadian rhythm. *Min. Electrolyte Metab.* **9**, 28–37.

Simmons, D. J., and Cohen, M. (1980). Postfracture linear bone growth in rats: A diurnal rhythm. *Clin. Orthop. Rel. Res.* **149**, 240–248.

Simmons, D. J., Ellsasser, J. C., Cummins, H., and Lesker, P. (1973). The bone inductive potential of a composite bone allograft – marrow autograft in rabbits. *Clin. Orthop. Rel. Res.* **97**, 237–247.

Simmons, D. J., and Kahn, A. J. (1979). Cell lineage in fracture healing in chimeric bone grafts. *Calcif. Tissue Intern.* **27**, 247–253.

Simmons, D. J., and Kunin, A. S. (1979). *Skeletal Research: An Experimental Approach. Volume 1.* Academic Press, New York, NY.

Simmons, D. J., Loeffelman, K., Frier, C., McCoy, R., Friedman, B., Melville, S., and Kahn, A. J. (1984). Circadian changes in the osteogenic competence of marrow stromal cells. In *Chronobiology* (E. Haus and H. F. Kabat, eds), pp. 37–42. S. Karger AG, Basel.

Simmons, D. J., Menton, D. N., Miller, S., and Lozano, R. (1993). Periosteal attachment fibers in the rat calvarium. *Calcif. Tissue Int.* **53**, 424–427.

Simmons, D. J., Russell, J. E., Winter, F., van Tran, P., Vignery, A., Baron, R., Rosenberg, G. D., and Walker, W. V. (1983a). Effect of spaceflight on the non-weight-bearing bones of rat skeleton. *Amer. J. Physiol.* **244**, R319–R326.

Simmons, D. J., Seitz, P., Kidder, L., Klein, G. L., Waedtz, M., Gundberg, C. M., Tabuchi, C., Yang, C., and Zhang, R. W. (1991). Partial characterization of rat marrow stromal cells. *Calcif. Tissue Int.* **48**, 326–334.

Simmons, D. J., Sherman, N. E., and Lesker, P. A. (1974). Allograft induced osteoinduction in rats: A circadian rhythm. *Clin. Orthop. Rel. Res.* **103**, 252–261.

Simmons, D. J., Whiteside, L. A., and Whitson, W. (1979). Biorhythmic profiles in the rat skeleton. *Metab. Bone Dis. Rel. Res.* **2**, 49–64.

Simmons, N. B., and Marshall, J. H. (1970). The uptake of calcium^{45} in the acellular-boned toadfish. *Calcif. Tissue Res.* **5**, 206–221.

Simon, H.-G., Kittappa, R., Khan, P. A., Tsilidis, C., Liversage, R. A., and Openheimer, S. (1997). A novel family of T-box genes in urodele amphibian limb development and regeneration: candidate genes involved in vertebrate forelimb/hindlimb patterning. *Development* **124**, 1355–1366.

Simon, M. R. (1977). The role of compressive forces in the normal migration of the condylar cartilage in the rat. *Acta Anat.* **97**, 351–360.

Simon, W. H., Richardson, S., Herman, W., Parsons, J. R., and Lane, J. (1976). Long-term effects of chondrocyte death on rabbit articular cartilage *in vitro*. *J. Bone Joint Surg. Am. Vol.* **58**, 517–525.

Simons, E. V. (1974). The effects of experimental unilateral anotia on skull development in the chick embryo. I. Introduction, techniques and preliminary results. *Acta Morphol. Neerl. Scand.* **12**, 331–344.

Simons, E. V. (1979). *Control Mechanisms of Skull Morphogenesis. A Study on Unilaterally Anotic Chick Embryos.* Swets & Zeitlinger BV, Lisse.

Singer, M., Weckesser, E. C., Géraudie, J., Maier, C. E., and Singer, J. (1987). Open finger tip healing and replacement after distal amputation in Rhesus monkeys with comparison to limb regeneration in lower vertebrates. *Anat. Embryol.* **177**, 29–36.

Singh, G., Supp, D. M., Schreiner, C., McNeish, J., Merker, H.-J., Copeland, N. G., Jenkins, N. A., Potter, S. S., and Scott, W. (1991). *Legless* insertional mutation: morphological, molecular, and genetic characterization. *Genes & Devel.* **5**, 2245–2255.

Singh, I. J., Herskovits, M. S., Chiego, D. J., Jr., and Klein, R. M. (1982). Modulation of osteoblastic activity by sensory and autonomic innervation of bone. In *Factors and Mechanisms Influencing Bone Growth* (A. D. Dixon and B. G. Sarnat, eds), pp. 535–551. Alan R. Liss Inc., New York, NY.

Singh, I. J., Klein, R. M., and Herskovits, M. S. (1981). Autoradiographic assessment of ^{3}H-proline uptake by osteoblasts following guanethidine-induced sympathectomy in the rat. *Cell Tissue Res.* **216**, 215–220.

Singh, I. J., Sandhu, H. S., and Herskovits, M. S. (1991). Bone vascularity. In *Bone, Volume 3: Bone Matrix and Bone Specific Products* (B. K. Hall, ed.), pp. 141–164. CRC Press, Boca Raton, FL.

Singh, I. J., Tonna, E. A., and Gandel, C. P. (1974). A comparative histological study of mammalian bone. *J. Morphol.* **144**, 421–436.

Sire, J.-Y. (1990). From ganoid to elasmoid scales in the actinopterygian fishes. *Neth. J. Zool.* **40**, 75–92.

Sire, J.-Y. (1993). Development and fine structure of the bony scales in *Corydoras arcuatus* (Siluriformes, Callichthyidae). *J. Morphol.* **215**, 225–244.

Sire, J.-Y. (2001). Teeth outside the mouth in teleost fishes: how to benefit from a developmental accident. *Evol. & Devel.* **3**, 104–108.

Sire, J.-Y., and Akimenko, M.-A. (2004). Scale development in fish: a review, with description of *sonic hedgehog* (*shh*) expression in the zebrafish (*Danio rerio*). *Int. J. Devel. Biol.* **48**, 233–247.

Sire, J.-Y., Allizard, F., Babier, O., Bourguignon, J., and Quilhac, A. (1997b). Scale development in zebrafish (*Danio rerio*). *J. Anat.* **190**, 545–561.

Sire, J.-Y., and Arnulf, I. (1990). The development of squamation in four teleostean fishes with a survey of the literature. *Jap. J. Ichthyol.* **37**, 133–143.

Sire, J.-Y., and Arnulf, I. (2000). Structure and development of the ctenoid spines on the scales of a teleost fish, the cichlid, *Cichlasoma nigrofasciatum*. *Acta Zool.* (*Stockh.*), **81**, 139–158.

Sire, J.-Y., Boulekbache, H., and Joly, C. (1990a). Epidermal-dermal and fibronectin cell-interactions during fish scale regeneration: immunofluorescence and TEM studies. *Biol. Cell* **68**, 147–158.

Sire, J.-Y., and Huysseune, A. (1993). Fine structure of the developing frontal bones and scales of the cranial vault in the cichlid fish *Hemichromis bimaculatus* (Teleostei, Perciformes). *Cell Tissue Res.* **273**, 511–524.

Sire, J.-Y., and Huysseune, A. (1996). Structure and development of the odontodes in an armoured catfish, *Corydoras aeneus* (Siluriformes, Callichthyidae). *Acta Zool.* **77**, 51–72.

Sire, J.-Y., and Huysseune, A. (2003). Formation of skeletal and dental tissues in fish: A comparative and evolutionary approach. *Biol. Rev. Camb. Philos. Soc.* **78**, 219–249.

Sire, J.-Y., Huysseune, A., and Meunier, F. J. (1990b). Osteoclasts in teleost fish – light microscopical and electron microscopical observations. *Cell Tissue Res.* **260**, 85–94.

Sire, J.-Y., Marin, S., and Allizard, F. (1998). Comparison of teeth and dermal denticles (odontodes) in the teleost *Denticeps clupeoides* (Clupeomorpha). *J. Morphol.* **237**, 237–255.

Sire, J.-Y., and Meunier, F. J. (1994). The canaliculi of Williamson in Holostean bone (Osteichthyes, Actinopterygii): a structural and ultrastructural study. *Acta Zool.* **75**, 235–247.

Sire, J.-Y., Quilhac, A., Bourguignon, J., and Allizard, F. (1997a). Evidence for participation of the epidermis in the deposition of superficial layer of scales in zebrafish (*Danio rerio*) – a SEM and TEM study. *J. Morphol.* **231**, 161–174.

Sissons, H. A. (1971). The growth of bone. In *The Biochemistry and Physiology of Bone* (G. H. Bourne, ed.), volume 1, pp. 145–180. Academic Press, New York, NY.

Sivakumar, P., and Chandrakasan, G. (1998). Occurrence of a novel collagen with three distinct chains in the cranial cartilage of the squid *Sepia officinalis*: comparison with shark cartilage collagen. *Biochem. Biophys. Acta* **138**, 161–169.

Skoog, V., Widenfal, B., Ohlsen, L., and Wasteson, A. (1990). The effect of growth factors and synovial fluid on chondrogenesis in perichondrium. *Scand. J. Plast. Surg.* **24**, 89–95.

Skreb, N., and Svajger, A. (1973). Histogenic capacity of rat and mouse embryonic shields cultivated *in vitro*. *W. Roux. Arch. EntwickMech.* **173**, 228–234.

Skreb, N., and Svajger, A. (1975). Experimental teratomas in rats. In *Teratomas and Differentiation* (M. Sherman and D. Solter, eds), pp. 83–97. Academic Press. New York, NY.

Skreb, N., Svajger, A., and Levak-Svajger, B. (1976). Developmental potentialities of the germ layer in mammals. In *Embryogenesis in Mammals*, CIBA Foundation Symposium #40 (new series), pp. 27–45. Elsevier-North-Holland, Amsterdam.

Slack, J. M. W. (1997). We have a morphogen! *Nature* **327**, 553–554.

Slack, J. M. W., and Savage, S. (1978). Regeneration of reduplicated limbs in contravention of the complete circle rule. *Nature* **271**, 760–761.

Slavkin, H. C. (ed.) (1972). *The Comparative Molecular Biology of Extracellular Matrices.* Academic Press, New York, NY.

Slavkin, H. C., and Greulich, R. C. (eds) (1975). *Extracellular Matrix Influences on Gene Expression.* Academic Press, New York, NY.

Slavkin, H. C., Shum, L., Bringas, P., Jr., Sakakura, Y., Chai, Y., Mayo, M., Santos, V., and Werb, Z. (1992). EGF regulation of Meckel's cartilage morphogenesis during mandibular morphogenesis in serumless, chemically-defined medium *in vitro*. In *Chemistry and Biology of Mineralized Tissues* (H. C. Slavkin and P. Price, eds), pp. 361–367. Elsevier Science Publishers BV, Amsterdam.

Sledge, C. B. (1973). Growth hormone and articular cartilage. *Fed. Proc., Fed. Amer. Soc. Exp. Biol.* **32**, 1503–1505.

Sledge, C. B. (1981). Developmental anatomy of joints. In *Diagnosis of Bone and Joint Disorders* (D. Resnick and G. Niwayama, eds), Volume 1, pp. 2–20. W. B. Saunders Co., Philadelphia, PA.

Sledge, C. B., and Dingle, J. T. (1965). Oxygen-induced resorption of cartilage in organ culture. *Nature* **205**, 140.

Slijper, E. J. (1979). *Whales*. Translated by A. J. Pomerans, 2nd Edn with a new foreword, concluding chapter and bibliography by R. J. Harrison. Cornell University Press, Ithaca, NY.

Slootweg, M. C., van Buuloffers, S. C., Herrmann-Erlee, M. P. M., van der Meer, J. M., and Duursma, S. A. (1988). Growth hormone is mitogenic for fetal mouse osteoblasts but not for undifferentiated bone cells. *J. Endocrinol.* **116**, R11–R13.

Small, K. M., and Potter, S. S. (1993). Homeotic transformations and limb defects in *Hox A11* mutant mice. *Genes & Devel.* **7**, 2318–2328.

Smetana, K., Jr., and Holub, M. (1990). Ossification in nude mice. I. Macroscopical study. *APMIS* **98**, 729–734.

Smirnov, S. V. (1990). Evidence of neoteny: a paedomorphic morphology and retarded development in *Bombina orientalis* (Anura, Discoglossidae). *Zool. Anz.* **225**, 324–332.

Smirnov, S. V. (1991). The anuran middle ear: developmental heterochronies and adult morphology diversification. *Belg. J. Zool.* **121**, 99–110.

Smirnov, S. V. (1992). The influence and variation in larval period on adult cranial diversity in *Pelobates fuscus* (Anura: Pelobatidae). *J. Zool. Lond.* **226**, 601–612.

Smirnov, S. V. (1994). Postmaturation skull development in *Xenopus laevis* (Anura; Pipidae): Late-appearing bones and their bearing on the pipid ancestral morphology. *Russ. J. Herpetol.* **1**, 21–29.

Smirnov, S. V. (1995). Extra bones in the *Pelobates* skull as evidence of the paedomorphic origin in the anurans. *Zh. Obsh. Biol.* **56**, 317–328.

Smirnov, S. V. (1997). Additional dermal ossifications in the anuran skull: morphological novelties or archaic elements? *Russ. J. Herpetol.* **4**, 17–27.

Smith, A. A., Searls, R. L., and Hilfer, S. R. (1975). Differential accumulation of extracellular materials beneath the ectoderm during development of the embryonic chick limb bud and flank regions. *Devel. Biol.* **46**, 222–226.

Smith, A. R., and Crawley, A. M. (1977). The pattern of cell division during growth of the blastema of regenerating newt forelimbs. *J. Embryol. Exp. Morphol.* **37**, 33–48.

Smith, A. R., Lewis, J. H., Crawley, A., and Wolpert, L. (1974). A quantitative study of blastemal growth and bone regression during limb regeneration in *Triturus cristatus*. *J. Embryol. Exp. Morphol.* **32**, 375–390.

Smith, B. H., and Taylor, H. B. (1969). The occurrence of bone and cartilage in mammary tumours. *Amer. J. Clin. Pathol.* **51**, 610–618.

Smith, D. M., Johnston, C. C., Jr., Severson, A. R., and Bell, N. (1973). Studies of the metabolism of separated bone cells. I. Techniques of separation and identification. *Calcif. Tissue Res.* **11**, 56–69.

Smith, G. N., Jr., Toole, B. P., and Gross, J. (1975). Hyaluronidase activity and glycosaminoglycan synthesis in the amputated newt limb: Comparison of denervated nonregenerating limbs with regenerates. *Devel. Biol.* **43**, 221–232.

Smith, H. G., and McKeown, M. (1974). Experimental alteration of the coronal sutural area; A histological and quantitative microscopic assessment. *J. Anat.* **118**, 543–561.

Smith, H. M. (1947). Classification of bone. *Turtox News* **25**, 234–236.

Smith, J. (1999). T-Box genes. What they do and how they do it. *Trends Genet.* **15**, 154–158.

Smith, J. D., and Abramson, M. (1974). Membranous vs. endochondral bone autografts. *Arch. Otolarnygol.* **99**, 203–205.

Smith, J. W., and Serafini-Fracassini, A. (1967). The distribution of the protein polysaccharide complex and collagen in bovine aricular cartilage. *J. Cell Sci.* **2**, 129–136.

Smith, K. K. (1996). Integration of craniofacial structures during development in mammals. *Amer. Zool.* **36**, 70–79.

Smith, K. K. (1997). Comparative patterns of craniofacial development in eutherian and metatherian mammals. *Evolution* **51**, 1663–1678.

Smith, K. K. (2003). Time's arrow: heterochrony and the evolution of development. *Int. J. Devel. Biol.* **47**, 613–621.

Smith, K. K., and Schneider, R. A. (1998). Have gene knockouts caused evolutionary reversals in the mammalian first arch? *BioEssays* **20**, 245–255.

Smith, K. K., and Van Nievelt, A. F. H. (1997). Comparative rates of development in *Monodelphis* and *Didelphis*. *Science* **275**, 683–684.

Smith, L., and Thorogood, P. V. (1983). Transfilter studies on the mechanism of epithelio-mesenchymal interaction leading to chondrogenic differentiation of neural crest cells. *J. Embryol. Exp. Morphol.* **75**, 165–188.

Smith, M. M. (1979). Scanning electron microscopy of odontodes in the scales of a coelacanth embryo, *Latimeria chalumnae* Smith. *Archs Oral Biol.* **24**, 179–183.

Smith, M. M. (1991). Putative skeletal neural crest cells in early Late Ordovician vertebrates from Colorado. *Science* **251**, 301–303.

Smith, M. M. (2003). Vertebrate dentitions at the origin of jaws: when and how pattern evolved. *Evol. & Devel.* **5**, 394–413.

Smith, M. M., and Coates, M. I. (1998). Evolutionary origins of vertebrate teeth: oropharyngeal phylogenetic dental patterns and developmental evolution. *Eur. J. Oral Biol.* **106** (S1), 482–500.

Smith, M. M., and Hall, B. K. (1990). Developmental and evolutionary origins of vertebrate skeletogenic and odontogenic tissues. *Biol. Rev. Camb. Philos. Soc.* **65**, 277–374.

Smith, M. M., and Hall, B. K. (1993). A developmental model for evolution of the vertebrate exoskeleton and teeth: the role of cranial and trunk neural crest. *Evol. Biol.* **27**, 387–448.

Smith, M. M., Hickman, A., Amanzee, D., Lumsden, A., and Thorogood, P. (1994). Trunk neural crest origin of caudal fin mesenchyme in the zebrafish *Brachydanio rerio*. *Phil. Trans R. Soc. Lond. B* **256**, 137–145.

Smith, M. M., and Johanson, Z. (2003a). Separate evolutionary origins of teeth from evidence in fossil jawed vertebrates. *Science* **299**, 1235–1236.

Smith, M. M., and Johanson, Z. (2003b). Response to Comment on Separate evolutionary origins of teeth from evidence in fossil jawed vertebrates *Science* **300**, 1661.

Smith, M. M., and Krupina, N. I. (2001). Conserved developmental processes constrain evolution of lungfish dentition. *J. Anat.* **199**, 161–168.

Smith, M. M., Krupina, N. I., and Joss, J. (2002). Developmental constraints conserve evolutionary pattern in an osteichthyan dentition. *Connect. Tissue Res.* **43**, 113–119.

Smith, M. M., and Sansom, I. J. (1997). Exoskeletal micro-remains of an Ordovician fish from the Harding Sandstone of Colorado. *Palaeontology* **40**, 645–658.

Smith, M. M., and Sansom, I. J. (2000). Evolutionary origins of dentine in the fossil record of early vertebrates: diversity, development and function. In *Development, Function and Evolution of Teeth* (M. F. Teaford, M. M. Smith, and M. W. J. Ferguson, eds), pp. 65–81. Cambridge University Press, Cambridge.

Smith, M. P. (1992). Vertebrate homeobox gene nomenclature. *Cell* **71**, 551–553.

Smith, M. P., Sansom, I. J., and Repetski, J. E. (1996). Histology of the first fish. *Nature* **380**, 702–704.

Smith, P. H. (1972). Autoradiographic evidence for the concurrent synthesis of collagen and chondroitin sulfates by chick sternal chondrocytes. *Connect. Tissue Res.* **1**, 181–188.

Smith, R., and Triffitt, J. T. (1986). Bones in muscles – the problems of soft tissue ossification. *Q. J. Med.* **61**, 985–990.

Smith, R. L., and Nagel, D. A. (1983). Effect of pulsing electromagnetic fields on bone growth and articular cartilage. *Clin. Orthop. Rel. Res.* **181**, 277–283.

Smith, S. C., Graveson, A. C., and Hall, B. K. (1994). Evidence for a developmental and evolutionary link between placodal ectoderm and neural crest. *J. Exp. Zool.* **270**, 292–301.

Smith, S. D. (1967). Induction of partial limb regeneration in *Rana pipiens* by galvanic stimulation. *Anat. Rec.* **158**, 89–98.

Smith, S. M., and Eichele, G. (1991). Temporal and regional differences in the expression pattern of distinct retinoic acid receptor-β transcripts in the chick embryo. *Development* **111**, 245–252.

Smithberg, M. (1954). The origin and development of the tail of the frog, *Rana pipiens. J. Exp. Zool.* **127**, 397–425.

Smith-Vaniz, W. F., Kaufman, L. S., and Glowacki, J. (1995). Species-specific patterns of hyperostosis in marine teleost fishes. *Marine Biol.* **121**, 573–580.

Smits, P., Dy, P., Mitra, S., and Lefebvre, V. (2004). Sox5 and Sox6 are needed to develop and maintain source, columnar, and hypertrophic chondrocytes in the cartilage growth plate. *J. Cell Biol.* **164**, 747–758.

Smits, P., and Lefebvre, V. (2003). Sox5 and Sox6 are required for notochord extracellular matrix sheath formation, notochord cell survival and development of the nucleus pulposus in intervertebral discs. *Development* **130**, 1135–1148.

Snow, M. H. L. (1986). Control of embryonic growth rate and fetal size in mammals. In *Human Growth. A Comprehensive Treatise. 2nd Edn, Volume 1. Developmental Biology, Prenatal Growth* (Falkner, F., and J. M. Tanner, eds), pp. 67–82. Plenum Press, New York, NY.

Sofaer, J. A. (1975). Interaction between tooth germs and the adjacent dental lamina in the mouse. *Archs Oral Biol.* **20**, 57–62.

Sofaer, J. A. (1985). Developmental stability in the mouse vertebral column. *J. Anat.* **140**, 131–141.

Sohal, G. S., Ali, A. A., and Ali, M. M. (1998b). Ventral neural tube cells differentiate into craniofacial skeletal muscles. *Biochem. Biophys. Res. Commun.* **252**, 675–678.

Sohal, G. S., Ali, M. M., Ali, A. A., and Bockman, D. E. (1999a). Ventral neural tube cells differentiate into hepatocytes in the chick embryo. *Cell Mol. Life Sci.* **55**, 128–130.

Sohal, G. S., Ali, M. M., Ali, A. A., and Dai, D. (1999b). Ventrally emigrating neural tube cells contribute to the formation of Meckel's and quadrate cartilage. *Devel. Dynam.* **216**, 37–44.

Sohal, G. S., Ali, M. M., Galileo, D. S., and Ali, A. A. (1998a). Emigration of neuroepithelial cells from the hindbrain neural tube in the chick embryo. *Int. J. Devel. NeuroSci.* **16**, 477–481.

Sohal, G. S., Bockman, D. E., Ali, M. M., and Tsai, N. T. (1996). Di labelling and homeobox gene Islet-1 expression reveal the contribution of ventral neural tube cells to the formation of the avian trigeminal ganglion. *Int. J. Devel. NeuroSci.* **14**, 419–427.

Sokol, O. M. (1981). The larval chondrocranium of *Peoldytes punctatus* with a review of tadpole chondrocrania. *J. Morphol.* **169**, 161–183.

Sokoloff, L. (1974). Cell biology and the repair of articular cartilage. *J. Rheumatol.* **1**, 9–16.

Sokoloff, L. (ed.) (1978). *The Joints and Synovial Fluid.* Volume 1. Academic Press, New York, NY.

Sokoloff, L. (ed.) (1980). *The Joints and Synovial Fluid.* Volume 2. Academic Press, New York, NY.

Sokoloff, L., Malemud, C. J., and Green, W. T., Jr. (1970). Sulfate incorporation by articular chondrocytes in monolayer culture. *Arthritis Rheum.* **13**, 118–124.

Sokoloff, L., Malemud, C. J., and Srivastava, V. M. L. (1973). *In vitro* culture of articular chondrocytes. *Fed. Proc., Fed. Amer. Soc. Exp. Biol.* **32**, 1499–1502.

Soler-Gijon, R. (1999). Occipital spine of *Orthacanthus* (Xenacanthidae, Elasmobranchii): Structure and growth. *J. Morphol.* **242**, 1–45.

Solomon, K. S., Kwak, S.-J., and Fritz, A. (2004). Genetic interactions underlying otic placode induction and formation. *Devel. Dynam.* **230**, 419–433.

Solursh, M. (1984). Ectoderm as a determinant of early tissue pattern in the limb bud. *Cell Differentiation* **15**, 17–24.

Solursh, M., Ahrens, P. B., and Reiter, R. S. (1978). A tissue culture analysis of the steps in limb chondrogenesis. *In Vitro* **14**, 51–61.

Solursh, M., Jensen, K. L., Reiter, R. S., Schmid, T. M., and Linsenmayer, T. F. (1986). Environmental regulation of type X collagen production by cultures of limb mesenchyme, mesectoderm, and sternal chondrocytes. *Devel. Biol.* **117**, 90–101.

Solursh, M., Jensen, K. L., Zanetti, N. C., Linsenmayer, T. F., and Reiter, R. S. (1984). Extracellular matrix mediates epithelial effects on chondrogenesis *in vitro. Devel. Biol.* **105**, 451–457.

Solursh, M., and Karp, G. C. (1975). An effect of accumulated matrix on sulfation among cells in a cartilage colony: An autoradiographic study. *J. Exp. Zool.* **191**, 73–84.

Solursh, M., and Meier, S. (1972). The requirement for RNA synthesis in the differentiation of cultured chick embryo chondrocytes. *J. Exp. Zool.* **181**, 253–262.

Solursh, M., and Meier, S. (1973). A conditioned medium (CM) factor provided by chondrocytes that promotes their own differentiation. *Devel. Biol.* **30**, 279–289.

Solursh, M., and Meier, S. (1974). Effects of cell density on the expression of differentiation by chick embryo chondrocytes. *J. Exp. Zool.* **187**, 311–322.

Solursh, M., Meier, S., and Vaerewyck, S. (1973). Modulation of extracellular matrix production by conditioned medium. *Amer. Zool.* **13**, 1051–1066.

Solursh, M., and Reiter, R. S. (1988). Inhibitory and stimulatory effects of limb ectoderm on *in vitro* chondrogenesis. *J. Exp. Zool.* **248**, 147–154.

Solursh, M., Reiter, R. S., Jensen, K. L., Kato, M., and Bernfield, M. (1990). Transient expression of a cell surface heparan sulfate proteoglycan (syndecan) during limb development. *Devel. Biol.* **140**, 83–92.

Solursh, M., Singley, C. T., and Reiter, R. S. (1981). The influence of epithelia on cartilage and loose connective tissue formation by limb mesenchyme cultures. *Devel. Biol.* **86**, 471–482.

Solursh, M., Vaerewyck, S., and Reiter, R. S. (1974). Depression by hyaluronic acid of glycosaminoglycan synthesis by cultured chick embryo chondrocytes. *Devel. Biol.* **41**, 233–244.

Somerman, M. J., Nathanson, M. A., Sauk, J. J., and Manson, B. (1987). Human dentin matrix induces cartilage formation *in vitro* by mesenchymal cells derived from embryonic muscle. *J. Dent. Res.* **66**, 1551–1558.

Somers, G. F (1962). Thalidomide and congenital abnormalities. *Lancet* **1**, 912–913.

Somjen, D., Binderman, I., Berger, E., and Harell, A. (1980). Bone remodelling induced by physical stress is prostaglandin E$_2$ mediated. *Biochem. Biophys. Acta* **627**, 91–100.

Sommerfeldt, D. W., Zhi, J., Rubin, C. T., and Hadjiargyroo, M. (2002). Proline-rich transcript of the brain (*prtb*) is a serum-responsive gene in osteoblasts and upregulated during adhesion. *J. Cell. BioChem.* **84**, 301–308.

Soni, N. N., and Malloy, R. B. (1974). Effect of removal of the temporal muscle on the coronoid process in guinea pigs: Quantitative triple flurochrome study. *J. Dent. Res.* **53**, 474–480.

Soni, N. N., and Malloy, R. B. (1977). Mandibular condylectomy in the guinea pig: Quantitative triple fluorochrome study. *J. Dent. Res.* **55**, 848–853.

Sordino, P., and Duboule, D. (1996). A molecular approach to the evolution of vertebrate paired appendages. *Trends Ecol. Evol.* **11**, 114–119.

Sordino, P., van der Hoeven, F., and Duboule, D. (1995). *Hox* gene expression in teleost fins and the origin of vertebrate digits. *Nature* **375**, 678–681.

Sorgente, N., Kuettner, K. E., Soble, L. W., and Eisenstein, R. (1975). The resistance of certain tissues to invasion. II. Evidence for extractable factors in cartilage which inhibit invasion by vascularized mesenchyme *Lab. Investig.* **32**, 217–222.

Sorrell, J. M., and Caterson, B. (1989). Detection of age-related changes in the distribution of keratan sulfates and chondroitin sulfates in developing chick limbs: an immunocytochemical study. *Development* **106**, 657–663.

Sorrell, J. M., and Weiss, L. (1980). A light and electron microscopic study of the region of cartilage resorption in the embryonic chick femur. *Anat. Rec.* **198**, 513–530.

Sorrell, J. M., and Weiss, L. (1982). The cellular organization of fibroblastic cells and macrophages at regions of uncalcified cartilage resorption in the embryonic chick femur as revealed by alkaline and acid phosphatase histochemistry. *Anat. Rec.* **202**, 491–499.

Soussi-Yanicostas, N., Barbet, J. P., Laurent-Winter, C., Barton, P., and Butler-Browne, G. S. (1990). Transition of myosin isozymes during development of human masseter muscle. Persistence of developmental isoforms during postnatal stage. *Development* **108**, 239–249.

Spadaro, J. A. (1977). Electrically stimulated bone growth in animals and man. Review of the literature. *Clin. Orthop. Rel. Res.* **122**, 325–332.

Spadaro, J. A. (1982). Electrically enhanced osteogenesis at various metal cathodes. *J. Biomed. Materials Res.* **16**, 861–874.

Spadaro, J. A., Mino, D. E., Chase, S. E., Werner, F. W., and Murray, D. G. (1986). Mechanical factors in electrode-induced osteogenesis. *J. Orthop. Res.* **4**, 37–44.

Spencer-Dene, N., Thorogood, P., Nair, S., Kenny, A. J., Harris, M., and Henderson, B. (1994). Distribution of, and putative role for, the cell surface neutral metallo-endopeptidases during mammalian craniofacial development. *Development* **120**, 3213–3226.

Sperber, G. H. (2001). *Craniofacial Development*. B. C. Decker, Hamilton, Ontario.

Spicer, A. P., and Tien, J. Y. L. (2004). Hyaluronan and morphogenesis. *Birth Defects Res. (Part C)*, **72**, 89–108.

Spigelius, A. (1631). *De Formato Foetuliber Singularis opera posthuma studio Liberalis cremae tarvisni ed. ta.* Francofurti M. Merianus.

Spinage, C. A. (1968). Horns and other bony structures of the skull of the giraffe, and their functional significance. *East Afr. Wildlife J.* **6**, 53–61.

Spitz, F., and Duboule, D. (2001). The art of making a joint. *Science* **291**, 1713–1714.

Sporn, M. B., and Roberts, A. B. (1990). TGF-β – problems and prospects. *Cell Regulation* **1**, 875–882.

Sporn, M. B., and Roberts, A. B. (1992). Transforming growth factor-β: recent progress and new challenges. *J. Cell Biol.* **119**, 1017–1021.

Spyropoulos, M. N. (1977). The morphogenetic relationship of the temporal muscle to the coronoid process in human embryos and fetuses. *Amer. J. Anat.* **150**, 395–410.

Srivastava, H. C. (1992). Ossification of the membranous portion of the squamous part of the occipital bone in man. *J. Anat.* **180**, 219–224.

Srivastava, V. M. L., Malemud, C. J., Hough, A. J., Bland, J. H., and Sokoloff, L. (1974b). Preliminary expression with cell culture of human articular chondrocytes. *Arthritis Rheum.* **17**, 165–169.

Srivastava, V. M. L., Malemud, C. J., and Sokoloff, L. (1974a). Chondroid expression by lapine articular chondrocytes in spinner culture following monolayer culture. *Connect. Tissue Res.* **2**, 127–136.

Stadler, H. S., Higgins, K. M., and Capecchi, M. R. (2001). Loss of *Eph-receptor* expression correlates with loss of cell adhesion and chondrogenic capacity in *Hoxa13* mutant limbs. *Development* **128**, 4177–4188.

Stanescu, V., Chaminade, F., and Do Pham, T. (1991). Immunological detection of the EGF-like domain in the core proteins of large proteoglycans from human and baboon cartilage. *Conn. Tissue Res.* **26**, 283–294.

Stanka, P. (1975). Occurrence of cell junctions and microfilaments in osteoblasts. *Cell Tissue Res.* **159**, 413–422.

Stanka, P., and Bargsten, G. (1983). Experimental study on the haematogenous origin of multinucleate osteoclasts in the rat. *Cell Tissue Res.* **233**, 125–132.

Stanka, P., Bargsten, G., and Herrmann, G. (1981). Woher kommen die vielkernigen osteoklasten? *Verh. Anat. Ges.* **75**, 237–238.

Starck, D. (1975). The development of the chondrocranium in primates. In *Phylogeny of the Primates* (W. P. Luckett and F. S. Szalay, eds), pp. 127–155. Plenum Press, New York, NY.

Starck, D. (1979). *Vergleichende Anatomie der Wirbeltiere auf Evolutionsbiologischer Grundlage. Band 2. Das Skeletsystem allgemeines Skelettsubstanzen, Skelet der Wirbeltiere einschliesslich Lokomotionstypen.* Springer-Verlag, Berlin.

Starck, J. M. (1993). Evolution of avian ontogenies. *Curr. Ornithol.* **10**, 275–366.

Starck, J. M. (1994). Quantitative design of the skeleton in bird hatchlings: does tissue compartmentalization limit posthatching growth rates? *J. Morphol.* **222**, 113–131.

Starck, J. M. (1996). Comparative morphology and cytokinetics of skeletal growth in hatchlings of altricial and precocial birds. *Zool. Anz.* **235**, 53–75.

Starck, J. M., and Ricklefs, R. E. (eds) (1998). *Avian Growth and Development: Evolution within the Altricial-Precocial Spectrum.* Oxford University Press, Oxford.

Stark, M., Miller, E. J., and Kühn, K. (1972). Comparative electron microscope studies on the collagens extracted from cartilage, bone, and skin. *Eur. J. BioChem.* **27**, 192–196.

Stark, R. J., and Searls, R. L. (1973). A description of chick wing bud development and a model of limb morphogenesis. *Devel. Biol.* **33**, 138–153.

Stark, R. J., and Searls, R. L. (1974). The establishment of the cartilage pattern in the embryonic chick wing, and evidence for a role of the dorsal and ventral ectoderm in normal wing development. *Devel. Biol.* **38**, 51–63.

Stearley, R. F. (1992). Historical ecology of Salmoninae, with special reference to *Oncorhynchus*. In *Systematics, Historical Ecology, and North American Freshwater Fishes* (R. L. Mayden, ed.), pp. 622–658. Stanford University Press, Stanford.

Steen. T. P. (1968). Stability of chondrocyte differentiation and contribution of muscle to cartilage during limb regeneration in the axolotl (*Siredon mexicanum*). *J. Exp. Zool.* **167**, 49–78.

Steen, T. P. (1970). Origin and differentiative capacities of cells in the blastema of the regenerating salamander limb. *Amer. Zool.* **10**, 119–132.

Steen, T. P. (1973). The role of muscle cells in *Xenopus* limb regeneration. *Amer. Zool.* **13**, 1340.

Steen, T. P., and Thornton, C. S. (1963). Tissue interactions in amputated aneurogenic limbs of *Ambystoma* larvae. *J. Exp. Zool.* **154**, 207–221.

Stefansson, K., Wollmann, R. L., Moore, B. W., and Arnason, B. G. W. (1982). S-100 protein in human chondrocytes. *Nature* **295**, 63–64.

Stein, B. R. (1989). Bone density and adaptation in semiaquatic mammals. *J. Mammal.* **70**, 467–476.

Stein, G. S., and Lian, J. B. (1993a). Molecular mechanisms mediating proliferation/differentiation interrelationships during progressive development of the osteoblast phenotype. *Endocrinol. Rev.* **14**, 424–442.

Stein, G. S., and Lian, J. B. (1993b). Molecular mechanisms mediating development and hormone-regulated expression of genes in osteoblasts: an integrated relationship of cell growth and differentiation. In *Cellular and Molecular Biology of Bone* (M. Noda, ed.), pp. 47–95. Academic Press, San Diego, CA.

Stein, G. S., Lian, J. B., Gerstenfeld, L. G., Shalhoub, V., Aronow, M., Owen, T., and Markose, E. (1989). The onset and progression of osteoblast differentiation is functionally related to cellular proliferation. *Conn. Tissue Res.* **20**, 3–123.

Stein, G. S., Lian, J. B., and Owen, T. A. (1990). Bone cell differentiation: a functionally coupled relationship between expression of cell-growth- and tissue-specific genes. *Curr. Opin. Cell Biol.* **2**, 1018–1027.

Stein, G. S., Lian, J. B., Stein, J. L., Van Wijnen, A. J., and Montecino, M. (1996). Transcriptional control of osteoblast growth and differentiation. *Physiol. Rev.* **76**, 593–619.

Stein, S., Fritsch, R., Lemaire, L., and Kessel, M. (1996). Checklist: vertebrate homeobox genes. *Mech. Devel.* **55**, 91–108.

Steinberg, B., Singh, I. J., and Mitchell, O. G. (1981). The effects of cold-stress, hibernation and prolonged inactivity on bone dynamics in the Golden Hamster, *Mesodricetus auratus*. *J. Morphol.* **167**, 43–51.

Steinetz, B. G., and Beach, V. L. (1963). Hormonal requirements for interpubic ligament formation in hypophysectomized mice. *Endocrinology* **72**, 771–776.

Steinetz, B. G., Manning, J. P., Butler, M., and Beach, V. L. (1965). Relationships of growth hormone, steroids, and relaxin in the transformation of pubic joint cartilage to ligament in hypophysectomized mice. *Endocrinology* **76**, 876–882.

Steinetz, B. G., Matthews, J. R., Butler, M., and Thompson, S. W. (1973). Inhibition by thyrocalcitonin of estrogen-induced bone resorption in the mouse pubic symphysis. *Amer. J. Pathol.* **73**, 735–746.

Stenner, D. D., Tracy, R. P., Riggs, B. L., and Mann, K. G. (1986). Human platelets contain and secrete osteonectin, a major protein of mineralized bone. *Proc. Natl Acad. Sci. U.S.A.* **83**, 6892–6896.

Stensio, E. A. (1947). The sensory lines and dermal bones of the cheek in fishes and amphibians. *Kungl. Svenska Veten. Handl. Tredje Ser.* **24**, 1–195.

Stephan, P. (1900). Recherches histologiques sur la structure du tissu osseux des poissons. *Bull. Sci. Fr. Belg.* **33**, 281–429.

Stephens, T. D., and Seegmiller, R. E. (1976). Normal production of cartilage glycosaminoglycans in mice homozygous for the chondrodysplasia gene. *Teratology* **13**, 317–326.

Stephenson, N. G., and Tomkins, J. K. N. (1964). Transplantation of embryonic cartilage and bone on to the chorioallantois of the chick. *J. Embryol. Exp. Morphol.* **12**, 825–839.

Stêrba, O., Klima, M., and Schildger, B. (2000). Embryology of dolphins: staging and ageing of embryos and fetuses of some cetaceans. *Adv. Anat. Embryol. Cell Biol.* **157**, 1–133.

Stern, C. D., and Holland, P. W. H. (eds) (1993). *Essential Developmental Biology. A Practical Approach.* IRL Press at Oxford University Press, Oxford.

Stern, P. H. (1980). The vitamins-D and bone. *Pharmacol. Rev.* **32**, 47–80.

Steven, F. S. (1972). Current concepts of collagen structure. *Clin. Orthop. Rel. Res.* **85**, 257–274.

Stevens, L. C. (1968). The development of teratomas from intratesticular grafts of tubal mouse eggs. *J. Embryol. Exp. Morphol.* **20**, 329–341.

Stevenson, S., Hunziker, E. B., Herrmann, W., and Schenk, R. K. (1990). Is longitudinal bone growth influenced by diurnal variation in the mitotic activity of chondrocytes of the growth plate? *J. Orthop. Res.* **8**, 132–135.

Stewart, P. A., and McCallion, D. J. (1975). Establishment of the scleral cartilage in the chick. *Devel. Biol.* **46**, 383–389.

Stiassny, M. L. J. (1992). Atavisms, phylogenetic character reversals, and the origin of evolutionary novelties. *Neth. J. Zool.* **42**, 260–276.

Stickens, D., Brown, D., and Evans, G. A. (2000). EXT genes are differentially expressed in bone and cartilage during mouse embryogenesis. *Devel. Dynam.* **218**, 452–464.

Stickney, H. L., Barresi, M. J. F., and Devoto, S. H. (2000). Somite development in zebrafish. *Devel. Dynam.* **219**, 287–303.

Stieda, L. (1872). *Die Bildung des Knochengewebes. Festschrift des Naturforschervereins zu Riga zur Feier des fünfzigjährigen Bestehens der Gesellschaft practischer Aertze zu Riga.* Engelmann, Leipzig.

Stirpe, N. S., Dickerson, K. T., and Goetinck, P. F. (1990). The chicken embryonic mesonephros synthesizes link protein, an extracellular matrix molecule usually found in cartilage. *Devel. Biol.* **137**, 419–424.

Stirpe, N. S., and Goetinck, P. F. (1989). Gene regulation during cartilage differentiation: temporal and spatial expression of link protein and cartilage matrix protein in the developing limb. *Development* **107**, 23–33.

St-Jacques, B., Hammerschmidt, M., and McMahon, A. P. (1999). Indian hedgehog signalling regulates proliferation and differentiation of chondrocytes and is essential for bone formation. *Genes & Devel.* **13**, 2072–2086.

Stockard, C. R. (1921). Developmental rate and structural expression: An experimental study of twins, 'double monsters' and single deformities, and the interactions among embryonic organs during their origin and development. *Amer. J. Anat.* **28**, 115–277.

Stockard, C. R. (1930). The presence of a factorial basis for characters lost in evolution. The atavistic reappearance of digits in mammals. *Amer. J. Anat.* **45**, 345–378.

Stockdale, F., Holtzer, H., and Lash, J. W. (1961). An experimental analysis of the development of the spinal column. VII. Response of dissociated somite cells. *Acta Embryol. Morphol. Exp.* **4**, 40–46.

Stockdale, F. E., Nikovits, W., Jr., and Christ, B. (2000). Molecular and cellular biology of avian somite development. *Devel. Dynam.* **219**, 304–321.

Stockwell, R. A. (1967). The cell density of human articular and costal cartilage. *J. Anat.* **101**, 753–764.

Stockwell, R. A. (1979). *Biology of Cartilage Cells.* Cambridge University Press, Cambridge.

Stockwell, R. A. (1983). Metabolism of cartilage. In *Cartilage, Volume 1. Structure, Function and Biochemistry* (B. K. Hall, ed.), pp. 253–280. Academic Press, New York, NY.

Stockwell, R. A., and Scott, J. E. (1997). Possums, articular cartilage and oxygen: A comment on the papers by Archer *et al.* (1996) and Morrison *et al.* (1996). *J. Anat.* **190**, 623–627.

Stocum, D. L. (1968). The urodele limb regeneration blastema: A self-organizing system. I. Differentiation *in vitro. Devel. Biol.* **18**, 441–456.

Stocum, D. L. (1975a). Regulation after proximal or distal transposition of limb regeneration blastemas and determination of the proximal boundary of the regenerate. *Devel. Biol.* **45**, 112–136.

Stocum, D. L. (1975b). Outgrowth and pattern formation during limb ontogeny and regeneration. *Differentiation* **3**, 167–182.

Stocum, D. L., and Dearlove, G. E. (1972). Epidermal–mesodermal interaction during morphogenesis of the limb regeneration blastema in larval salamanders. *J. Exp. Zool.* **181**, 49–62.

Stokstad, E. (2003). Primitive jawed fishes had teeth of their own design. *Science* **299**, 1164.

Stolberg, M. (2003). A woman down to her bones. The anatomy of sexual difference in the sixteenth and early seventeenth centuries. *Isis* **94**, 274–299.

Stoleson, S. H., and Beissinger, S. R. (1995). Hatching asynchrony and the onset of incubation in birds, revisited. When is the critical period? *Curr. Ornithol.* **12**, 191–270.

Stone, C. A. (2000). Unravelling the secrets of foetal wound healing: an insight into fracture repair in the mouse fetus and perspectives for clinical application. *Brit. J. Plast. Surg.* **53**, 337–341.

Stone, J. R., and Hall, B. K. (2004). Latent homologues for the neural crest as an evolutionary novelty. *Evol. Devel.* **6**, 123–129.

Stopper, G. F., Hecker, L., Franssen, R. A., and Sessions, S. K. (2002). How trematodes cause limb deformities in amphibians. *J. Exp. Zool. (Mol. Dev. Evol.)* **294**, 252–263.

Storey, E. (1972). Growth and remodeling of bone and bones. *Amer. J. Orthod.* **62**, 142–165.

Storey, E. (1973). Tissue response to the movement of bones. *Amer. J. Orthod.* **64**, 229–247.

Storey, E., and Feik, S. A. (1982). Remodelling of bone and bones. Effects of altered mechanical stress on anlages. *Brit. J. Exp. Pathol.* **63**, 184–193.

Storey, E., and Feik, S. A. (1985). Remodelling of bone and bones. Effects of altered mechanical stress on caudal vertebrae. *J. Anat.* **140**, 37–48.

Storey, E., Feik, S. A., and Ellender, G. (1992). Vertebral growth. In *Bone, Volume 6: Bone Growth – A.* (B. K. Hall, ed.), pp. 209–247. CRC Press, Boca Raton, FL.

Storm, E. E., Huynh, T. V., Copeland, N. G., Jenkins, N. A., Kingsley, D. M., and Lee, S.-J. (1994). Limb alteration in *brachypodism* mice due to mutations in a new member of the TGF-β-superfamily. *Nature* **368**, 639–643.

Storm, E. E., and Kingsley, D. M. (1996). Joint patterning defects caused by single and double mutations in members of the bone morphogenetic protein (BMP) family. *Development* **122**, 3969–3979.

Storm, E. E., and Kingsley, D. M. (1999). GDF5 coordinates bone and joint formation during digit development. *Devel. Biol.* **209**, 11–27.

Storti, R. V., and Ruch, A. (1976). Chick cytoplasmic actin and muscle actin have different structural genes. *Proc. Natl Acad. Sci. U.S.A.* **73**, 2346–2350.

Stott, D., Kispert, A., and Herrmann, B. G. (1993). Rescue of the tail defect of *Brachyury* mice. *Genes & Devel.* **7**, 197–203.

Stott, N. S., and Chuong, C.-M. (1997). Dual action of sonic hedgehog on chondrocyte hypertrophy: retrovirus mediated ectopic sonic hedgehog expression in limb bud micromass culture induces novel cartilage nodules that are positive for alkaline phosphatase and type X collagen. *J. Cell Sci.* **110**, 2691–2701.

Stott, N. S., Jiang, T. X., and Chuong, C.-M. (1999). Successive formative stages of precartilaginous mesenchymal condensations *in vitro*: modulation of cell adhesion by wnt-7a and BMP-2. *J. Cell Physiol.* **180**, 314–324.

Stottmann, R. W., Anderson, R. M., and Klingensmith, J. (2001). The BMP antagonists chordin and noggin have essential but redundant roles in mouse mandibular outgrowth. *Devel. Biol.* **240**, 457–473.

Straney, D. O. (1984). The nasal bones of *Chiroderma* (Phyllostomidae). *J. Mammal.* **65**, 163–165.

Stratford, T. H., Kostakopoulou, K., and Maden, M. (1997). *Hoxb-8* has a role in establishing early anterior-posterior polarity in chick forelimb but not hindlimb. *Development* **124**, 4225–4234.

Straus, W. L., Jr., and Rawles, M. E. (1953). An experimental study of the origin of the trunk musculature and ribs in the chick. *Amer. J. Anat.* **92**, 471–509.

Strauss, P. G., Closs, E. I., Schmidt, J., and Erfle, V. (1990). Gene expression during osteogenic differentiation in mandibular condyles *in vitro. J. Cell Biol.* **110**, 1369–1378.

Strauss, R. E. (1990). Heterochronic variation in the developmental timing of cranial ossification in poeciliid fishes (Cyprinodontiformes). *Evolution* **44**, 1558–1567.

Strauss, R. E. (1991). Developmental variability and heterochronic evolution in Poeciliid fishes (Cyprinodontiformes). In *Systematics, Historical Ecology and North American Freshwater Fishes* (R. L. Mayden, ed.), pp. 429–514. Stanford University Press, Stanford, CA.

Streck, R. D., Wood, T. L., Hsu, H.-S., and Pintar, J. E. (1992). Insulin-like growth factor I and II and insulin-like growth factor binding protein-2 RNAs are expressed in adjacent tissues within rat embryonic and fetal limbs. *Devel. Biol.* **151**, 586–596.

Streicher, J. (1991). Plasticity in skeletal development: knee-joint morphology in fibula-deficient chick embryos. In *Developmental Patterns of the Vertebrate Limb* (J. R. Hinchliffe, J. M. Hurlé, and D. Summerbell, eds), pp. 407–409. Plenum Press, New York, NY.

Streicher, J., and Müller, G. (1992). Natural and experimental reduction of the avian fibula: developmental thresholds and evolutionary constraint. *J. Morphol.* **214**, 269–285.

Stricker, S., Fundele, R., Vortkamp, A., and Mundlos, S. (2002). Role of Runx genes in chondrocyte differentiation. *Devel. Biol.* **245**, 95–108.

Stricker, S. A., and Reed, C. G. (1985). Development of the pedicle in the articulate brachiopod *Terebratalia transversa* (Brachiopoda, Terebratulida). *Zoomorphology* **105**, 253–264.

Strickler, A. G., Famuditimi, K., and Jeffery, W. R. (2002). Retinol homeobox genes and the role of cell proliferation in cavefish eye degeneration. *Int. J. Devel. Biol.* **46**, 285–294.

Stringa, E., and Tuan, R. S. (1996). Chondrogenic cell subpopulation of chick embryonic calvarium: isolation by peanut agglutinin affinity chromatography and *in vitro* characterization. *Anat. Embryol.* **194**, 427–437.

Strom, C. M., and Dorfman, A. (1976a). Distribution of 5-bromodeoxyuridine and thymidine in the DNA of developing chick cartilage. *Proc. Natl Acad. Sci. U.S.A.* **73**, 1019–1023.

Strom, C. M., and Dorfman, A. (1976b). Amplification of moderately repetitive DNA sequences during chick cartilage differentiation. *Proc. Natl Acad. Sci. U.S.A.* **73**, 3428–3432.

Strong, D. D., Beachler, A. L., Wergedal, J. E., and Linkhart, T. A. (1991). Insulin-like growth factor II and transforming growth factor β regulate collagen expression in human osteoblastlike cells *in vitro*. *J. Bone Min. Res.* **6**, 15–23.

Strong, R. M. (1925). The order, time and rate of ossification of the albino rat (*Mus norvegicus albinus*) skeleton. *Amer. J. Anat.* **36**, 313–355.

Strudel, G. (1953a). Conséquences de l'excision de tronçons du tube nerveux sur la morphogenèse de l'embryo de poulet et sur la différenciation de ses organes: Contribution à la genèse de l'orthosympathique. *Ann. Sci. Nat. Zool. Biol. Anim.* [11] **15**, 251–329.

Strudel, G. (1953b). Influence morphogène du tube nerveux et de la chorde sur la différenciation de la colonee vertébrale. *C.R. Seances Soc. Biol. Ser. Fil.* **47**, 132–133.

Strudel, G. (1955). L'action morphogène du tube nerveux et de la corde sur la différenciation des vertèbres et des muscles vertébraux chez l'embryon de poulet. *Arch. Anat. Microsc. Morphol. Exp.* **44**, 209–235.

Strudel, G. (1962). Induction de cartilage *in vitro* par l'extrait de tube nerveux et de chorde de l'embryon de poulet. *Devel. Biol.* **4**, 67–86.

Strudel, G. (1963). Autodifférenciation et induction de cartilage à partir de mésenchyme somitique de poulet cultivé *in vitro*. *J. Embryol. Exp. Morphol.* **11**, 399–412.

Strudel, G. (1967). Some aspects of organogenesis of the chick spinal column. In *Experimental Biology and Medicine. Morphological and Biochemical Aspects of Cytodifferentiation* (E. Hagen, W. Wechsler, and F. Zilliken, eds), Volume 1, pp. 183–198. S. Karger, Basel.

Strudel, G. (1971). Matériel extracellulaire et chondrogenèse vertébrale. *C.R. Hebd. Seances Acad. Sci.* **272**, 473–476.

Strudel, G. (1972). Differenciation d'ebauches chondrogènes d'embryons de poulet cultivées *in vitro* sur differents milieux. *C.R. Hebd. Seances Acad. Sci.* **274**, 112–115.

Strudel, G. (1973a). Etude de la differenciation du cartilage vertebral. *Lyon Med.* **229**, 29–42.

Strudel, G. (1973b). Relationship between the chick periaxial metachromatic extracellular material and vertebral chondrogenesis. In *Biology of Fibroblasts* (E. Kulonen and J. Pikkarairen, eds), pp. 93–101. Academic Press, New York, NY.

Strudel, G. (1973c). Matériel extracellulaire périaxial et chondrogenèse vertébrale. *Ann. Biol.* **12**, 401–416.

Strudel, G. (1975a). Effect of an L-proline analogue, the L-azetidine-2-carboxylic acid, on the phenotypic differentiation of the chick somitic mesenchyme *C.R. Hebd. Seances Acad. Sci.* **280**, 1007–1010.

Strudel, G. (1975b). Periaxial extracellular material and vertebral chondrogenesis. In *Protides of the Biological Fluids, 22nd Colloquium*, pp. 51–58. Pergamon Press, Oxford.

Strudel, G. (1975c). Control of the phenotypic vertebral cartilage differentiation by the periaxial extracellular material. In *Extracellular Matrix Influences on Gene Expression* (H. C. Slavkin and R. C. Greulich, eds), pp. 655–670. Academic Press, New York, NY.

Strudel, G., and Gateau, G. (1971). Etude de l'action tératogène du sulfate de nicotine sur les stades jeunes de l'embryon de poulet. *C.R. Hebd. Seances Acad. Sci.* **272**, 2480–2483.

Strudel, G., and Pinot, M. (1965). Differentiation en culture *in vitro* du mesonephros de l'embryon de poulet. *Devel. Biol.* **11**, 284–299.

Stump, C. W. (1925). The histogenesis of bone. *J. Anat.* **59**, 136–154.

Sturdee, A. P., and Connock, M. (1975). The embryonic limb bud of the urodeles: Morphological studies of the apex. *Differentiation* **3**, 43–50.

Stutzmann, J., and Petrovic, A. (1970). Particularités de croissance de la suture palatine sagittale de jeune rat. *Bull. Assoc. Anat.* **54**, 552–562.

Stutzmann, J., and Petrovic, A. (1974). Effect de la resection du muscle pterygoiden externe sur la croissance du cartilage condylien de jeune rat. *Bull. Assoc. Anat.* **58**, 1–8.

Stutzmann, J., and Petrovic, A. (1975a). Nature et aptitudes evolutives des cellules du compartiment mitotique des cartilages secondaires de la mandible et du maxillaire de jeune rat. Experiences de culture cytotypique et d'homotransplantation. *Bull. Assoc. Anat.* **59**, 523–534.

Stutzmann, J., and Petrovic, A. (1975b). Régulation intrinseque de la croissance du cartilage condylien de la mandibule: Inhibition de la prolifération préchondroblartique par les chondroblastes. *C.R. Hebd. Seances Acad. Sci.* **281**, 175–178.

Stutzmann, J., and Petrovic, A. (1989). Responsiveness of alveolar bone to orthodontic treatment in adult patients. In *Orthodontics in an Aging Society* (D. S. Carlson, ed.), pp. 181–199. Center for Human Growth and Development, University of Michigan Ann Arbor, MI.

Stutzmann, J., Petrovic, A., and George, D. (1991). Nature and number of dividing cells and distribution of collagens type I, Ii and X in the medial cartilage of the clavicle during spontaneous and biomechanically modulated postnatal; growth. In *Fundamentals of Bone Growth: Methodology and Applications*

(A. D. Dixon, B. G. Sarnat, and D. A. N. Hoyte, eds), pp. 95–111, CRC Press, Boca Raton, FL.

Stutzmann, J., Petrovic, A., and Malan, A. (1981). Seasonal variation of the human alveolar bone turnover. A quantitative evaluation in organ culture. *J. Interdiscipl. Cycle Res.* **12**, 177–180.

Styrud, J., and Ericksson, U. J. (1990). Effects of D-glucose and β-hydroxybutyric acid on the *in vitro* development of (pre)chondrocytes from embryos of normal and diabetic rats. *Acta Endocrinol. (Copenh.)* **122**, 487–498.

Styrud, J., and Ericksson, U. J. (1991). *In vitro* effects of glucose and growth factors on limb bud and mandibular arch chondrocytes maintained at various serum concentrations. *Teratology* **44**, 65–76.

Subramanian, V., Meyer, B. I., and Gruss, P. (1995). Disruption of the murine homeobox gene *Cdx1* affects axial skeletal identities by altering the mesodermal expression domains of *Hox* genes. *Cell* **83**, 641–653.

Sucheson, M. E. (1993). Drugs and bone growth. In *Bone, Volume 7: Bone Growth – B* (B. K. Hall, ed.), pp. 179–244. CRC Press, Boca Raton, FL.

Suda, T., Testa, N. G., Allen, T. D., Onions, D., and Jarrett, O. (1983). Effect of hydrocortisone on osteoclasts generated in cat bone marrow cultures. *Calcif. Tissue Int.* **35**, 82–86.

Sudo, H., Kodama, H.-A., Amagai, Y., Yamamoto, S., and Kasai, S. (1983). In vitro differentiation and calcification in a new clonal osteogenic cell line derived from newborn mouse calvaria. *J. Cell Biol.* **96**, 191–198.

Sudo, H., Takahashi, Y., Tonegawa, A., Arase, Y., Aoyama, H., Mizutani, K. Y., Moriya, H., Wilting, J., Christ, B., and Koseki, H. (2001). Inductive signals from the somatopleure mediated by bone morphogenetic proteins are essential for the formation of the sternal components of avian ribs. *Devel. Biol.* **232**, 284–300.

Sugahara, K., Ho, P.-L., and Dorfman, A. (1981). Chemical and immunological characterization of proteoglycans of embryonic chick calvaria. *Devel. Biol.* **85**, 180–189.

Sugahara, K., Tanaka, Y., Yamada, S., Seno, N., Kitagawa, H., Haslam, S. M., Morris, H. R., and Dell, A. (1996). Novel sulfated oligosaccharide containing 3-o-sulfated glucuronic acid from King Crab cartilage chondroitin sulfate K. *J. Biol. Chem.* **271**, 26745–26754.

Sugiura, T., Taniguchi, Y., Tazaki, A., Ueno, N., Watanabe, K., and Mochii, M. (2004). Differential gene expression between the embryonic tail bud and regenerating larval tail in *Xenopus laevis*. *Devel. Growth Differ.* **46**, 97–105.

Sullivan, G. E. (1966). Prolonged paralysis of the chick embryo, with special reference to effects on the vertebral column. *Aust. J. Zool.* **14**, 1–17.

Sullivan, G. E. (1967). Abnormalities of the muscular anatomy in the shoulder region of paralyzed chick embryos. *Aust. J. Zool.* **15**, 911–940.

Sullivan, G. E. (1974). Skeletal abnormalities in chick embryos paralyzed with decamethonium. *Aust. J. Zool.* **22**, 429–438.

Sullivan, G. E. (1975). Paralysis and skeletal abnormalities in chick embryos treated with physostigmine. *Aust. J. Zool.* **23**, 1–8.

Summerbell, D. (1974a). Interaction between the proximo-distal and antero-posterior coordinates of positional value during the specification of positional information in the early development of the chick limb bud. *J. Embryol. Exp. Morphol.* **32**, 227–238.

Summerbell, D. (1974b). A quantitative analysis of the effect of excision of the AER from the chick limb-bud. *J. Embryol. Exp. Morphol.* **32**, 651–660.

Summerbell, D. (1976). A descriptive study of the rate of elongation and differentiation of the skeleton of the developing chick wing. *J. Embryol. Exp. Morphol.* **35**, 241–260.

Summerbell, D. (1977a). Reduction of the rate of outgrowth, cell density and cell division following removal of the apical ectodermal ridge of the chick limb-bud. *J. Embryol. Exp. Morphol.* **40**, 1–21.

Summerbell, D. (1977b). Regulation of deficiencies along the proximal distal axis of the chick wing-bud: A quantitative analysis. *J. Embryol. Exp. Morphol.* **41**, 137–159.

Summerbell, D. (1981). Evidence for regulation of growth, size and pattern in the developing chick limb bud. *J. Embryol. Exp. Morphol.* **65** (Suppl.), 129–150.

Summerbell, D., and Lewis, J. H. (1975). Time, place and positional; value in the chick limb-bud. *J. Embryol. Exp. Morphol.* **33**, 621–643.

Summerbell, D., Lewis, J. H., and Wolpert, L. (1973). Positional information in chick limb morphogenesis. *Nature New Biol.* **244**, 492–496.

Summerbell, D., and Wolpert, L. (1972). Cell density and cell division in the early morphogenesis of the chick wing. *Nature New Biol.* **239**, 24–26.

Summerbell, D., and Wolpert, L. (1973). Precision of development in chick limb morphogenesis. *Nature* **244**, 228–230.

Summers, A. P. (2000). Stiffening the stingray skeleton – an investigation of durophagy in Myliobatid stingrays (Chondrichthyes, Batoidea, Myliobatidae). *J. Morphol.* **243**, 113–126.

Summers, A. P., Ketcham, R. A., and Rowe, T. (2004). Structure and function of the horn shark (*Heterodontus francisci*) cranium through ontogeny: Development of a hard prey specialist. *J. Morphol.* **260**, 1–12.

Summers, A. P., and Koob, T. J. (2002). The evolution of tendon – morphology and material properties. *Comp. Biochem. Physiol. Part A* **133**, 1159–1170.

Summers, A. P., Koob, T. J., and Brainerd, E. L. (1998). Stingray jaws strut their stuff. *Nature* **395**, 450–451.

Summers, A. P., Koob-Emunds, M. M., Kajimura, S. M., and Koob, T J. (2003). A novel fibrocartilaginous tendon from an elasmobranch fish (*Rhinoptera bonasus*). *Cell Tissue Res.* **312**, 221–227.

Sun, X., Mariani, F. V., and Martin, G. R. (2002). Functions of FGF signalling from the apical ectodermal ridge in limb development. *Nature* **418**, 501–508.

Sutfin, L. V., Holtrop, M. E., and Ogilvie, R. E. (1971). Microanalysis of individual mitochondrial granules with diameters less than 1000 angstroms. *Science* **174**, 947–949.

Suttie, J. M. (1980). The effect of antler removal on dominance and fighting behaviour in farmed red deer stags. *J. Zool. (London)* **190**, 217–224.

Suzuki, A., Nishimatsu, S.-I., Shoda, A., Takebayashi, K., Murakami, K., and Ueno, N. (1993). Biochemical properties of amphibian bone morphogenetic protein-4 expressed in CHO cells. *Biochem. J.* **291**, 413–417.

Suzuki, H. K. (1963). Studies on the osseous system in the slider turtle. *Ann. NY. Acad. Sci.* **109**, 351–410.

Suzuki, N., Suzuki, T., and Kurokawa, T. (2000a). Suppression of osteoclastic activity by calcitonin in the scales of goldfish (freshwater teleost) and nibbler fish (sea water teleost). *Peptides* **21**, 115–124.

Suzuki, N., Suzuki, T., and Kurokawa, T. (2000b). Cloning of a calcitonin gene-related peptide receptor and a novel calcitonin receptor-like receptor from the gill of a flounder, *Paralichthys olivaceus*. *Gene* **244**, 81–88.

Suzuki, T., and Kurokawa, T. (1996). Functional analyses of FGF during pharyngeal cartilage development in flounder (*Paralichthys olivaceus*) embryo. *Zool. Sci.* **13**, 883–891.

Suzuki, T., Kurokawa, T., and Srivastava, A. S. (2001). Induction of bent cartilaginous skeletons and undulating notochord in flounder embryos by disulfiram and α, α-dypyridyl. *Zool. Sci.* **18**, 345–351.

Suzuki, T., Oohara, I., and Kurokawa, T. (1998). *Hoxd4* expression during pharyngeal arch development in flounder (*Paralichthys olivaceus*) embryos and effects of retinoic acid on expression. *Zool. Sci.* **15**, 57–67.

Suzuki, T., Oohara, I., and Kurokawa, T. (1999a). Retinoic acid given at late embryonic stage depresses sonic hedgehog and Hoxd-4 expression in the pharyngeal area and induces skeletal malformations in flounder (*Paralichthys olivaceus*) embryos. *Devel. Growth Differ.* **41**, 143–152.

Suzuki, T., Srivastava, A. S., and Kurokawa, T. (1999b). *Hoxb-5* is expressed in gill arch 5 during pharyngeal arch development of flounder *Paralichthys olivaceus* embryos. *Int. J. Devel. Biol.* **43**, 357–359.

Suzuki, T., Srivastava, A. S., and Kurokawa, T. (2000). Experimental induction of jaw, gill and pectoral fin malformations in Japanese flounder, *Paralichthys olivaceus* larvae. *Aquaculture* **185**, 175–187.

Svajger, A. (1970). Chondrogenesis in the external ear of the rat. *Z. Anat. Entwick. Gesch.* **131**, 236–242.

Svajger, A. (1971). Differentiation of rat auricular cartilage in organ culture. *Period. Biol.* **73**, 39–43.

Svajger, A., and Levak-Svajger, B. (1971). Differentiation of rat auricular cartilage in the anterior chamber of the eye. *Periodontol. Biol.* **73**, 33–37.

Svajger, A., and Levak-Svajger, B. (1974). Regional developmental capacities of the rat embryonic endoderm at the head-fold stage. *J. Embryol. Exp. Morphol.* **32**, 461–467.

Svajger, A., and Levak-Svajger, B. (1975). Technique of separation of germ layers in rat embryonic shields. *W. Roux. Arch. EntwickMech.* **178**, 303–308.

Svajger, A., and Levak-Svajger, B. (1976). Differentiation in renal homografts of isolated parts of rat embryonic ectoderm. *Experientia* **32**, 378–379.

Swain, D. P. (1992a). The functional basis of natural selection for vertebral traits of larvae in the stickleback *Gasterosteus aculeatus*. *Evolution* **46**, 987–997.

Swain, D. P. (1992b). Selective predation for vertebral phenotype in *Gasterosteus aculeatus*: reversal in the direction of selection at different larval sizes. *Evolution* **46**, 998–1013.

Swain, D. P., and Lindsey, C. C. (1984). Selective predation for vertebral number of young sticklebacks, *Gasterosteus aculeatus*. *Can. J. Fish. Aquat. Sci.* **41**, 1231–1233.

Swain, D. P., and Lindsey, C. C. (1986a). Meristic variation in a clone of cyprinodont fish *Rivulus marmoratus* related to temperature history of the parents and of the embryo. *Can. J. Zool.* **64**, 1444–1455.

Swain, D. P., and Lindsey, C. C. (1986b). Influence of reproductive history of parents on meristic variation in offspring in the cyprinodont fish *Rivulus marmoratus*. *Can. J. Zool.* **64**, 1456–1459.

Swart, C. C., and De Sá, R. O. (1999). The chondrocranium of the Mexican burrowing toad, *Rhinophrynus dorsalis*. *J. Herpetol.* **33**, 23–28.

Sweeney, R. M., and Watterson, R. L. (1969). Rib development in chick embryos analyzed by means of tantalum foil barriers. *Amer. J. Anat.* **126**, 127–150. *J. Paleontol.* **59**, 485–494.

Sweet, W. C. (1985). Conodonts: Those fascinating little whatzits. *J. Paleont.* **59**, 485–494.

Sweet, W. C., and Donoghue, P. C. J. (2001). Conodonts: past, present, future. *J. Paleont.* **75**, 1174–1184.

Swett, F. H. (1937). Determination of limb axes. *Q. Rev. Biol.* **12**, 322–339.

Swiderski, R. E., and Solursh, M. (1992). Differential co-expression of long and short form type IX collagen transcripts during avian limb chondrogenesis *in vitro*. *Development* **115**, 169–179.

Swinson, D. R., Tam, C. S., Reed, R., Hoffman, D., Little, A. H., and Cruickshank, B. (1975). Bone growth kinetics. IV: A preliminary investigation on a biorhythm in human osteogenesis. *J. Pathol.* **116**, 13–16.

Swinyard, C. A. (1969). *Limb Development and Deformity: Problems of Evaluation and Rehabilitation*. C. C. Thomas, Springfield, IL.

Syftestad, G., and Urist, M. R. (1982). Bone aging. *Clin. Orthop. Rel. Res.* **162**, 288–297.

Syftestad, G., Weitzhandler, M., and Caplan, A. I. (1985). Isolation and characterization of osteogenic cells derived from first bone of the embryonic tibia. *Devel. Biol.* **110**, 275–283.

Syme, J. (1840). On the power of the periosteum to form new bone. *Trans. R. Soc. Edin.* **14**, 158–163.

Symons, N. B. B. (1951). Studies on the growth and form of the mandible. *Dent. Rec.* **71**, 42–53.

Symons, N. B. B. (1952). The development of the human mandibular joint. *J. Anat.* **86**, 326–332.

Szabo, P., Moitra, J., Rencendorj, A., Rakhely, G., Rauch, T., and Kiss, I. (1995). Identification of a nuclear Factor-I family protein binding site in the silencer region of the cartilage matrix protein gene. *J. Biol. Chem.* **270**, 10212–10221.

Szalay, F. S. (1994). *Evolutionary History of the Marsupials and an Analysis of Osteological Characters*. Cambridge University Press, Cambridge.

Szebenyi, G., and Fallon, J. F. (1999). Fibroblast growth factors as multifunctional signaling factors. *Int. Rev. Cytol.* **185**, 45–106.

Szuwart, T., Gath, U., Althoff, J., and Höhling, H. J. (1994a). Biochemical and histological study of the ossification in the early developing pedicle of the fallow deer (*Dama dama*). *Cell Tissue. Res.* **277**, 123–129.

Szuwart, T., Kierdorf, H., Kierdorf, U., Althoff, J., and Clemen, G. (1994b). Tissue differentiation and correlated changes in enzymatic activities during primary antler development in fallow deer (*Dama dama*). *Anat. Rec.* **243**, 413–420.

T

Tabas, J. A., Zasloff, M., Wasmuth, J. J., Emanuel, B. S., Altherr, M. R., McPherson, J. D., Wozney, J. M., and Kaplan, F. S. (1991). Bone morphogenetic protein: chromosomal localization of human genes for BMP1, BMP2A, and BMP3. *Genomics* **9**, 283–289.

Tabata, T., and Takei, Y. (2004). Morphogens: their identification and regulation. *Development* **131**, 703–712.

Tabin, C. J. (1990). Why we have (only) five fingers per hand: Hox genes and the evolution of paired limbs. *Development* **116**, 289–296.

Tabin, C. J. (1998). A developmental model for thalidomide defects. *Nature* **396**, 322–323.

Tabin, C. J., and Laufer, E. (1993). Hox genes and serial homology. *Nature* **361**, 692–693.

Tacchetti, C., Quarto, R., Campanile, G., and Cancedda, R. (1989). Calcification *in vitro* of developed hypertrophic cartilage. *Devel. Biol.* **132**, 442–447.

Tacchetti, C., Quarto, R., Nitsch, L., and Hartmann, D. J. (1987). In vitro morphogenesis of chick embryo hypertrophic cartilage. *J. Cell Biol.* **105**, 999–1006.

Tacchetti, C., Tavella, S., Dozin, B., Quarto, R., Robino, G., and Cancedda, R. (1992). Cell condensation in chondrogenic differentiation. *Exp. Cell Res.* **200**, 26–33.

Tague, R. G. (1997). Variability of a vestigial structure: First metacarpal in *Colobus quereza* and *Ateles geoffroyi. Evolution* **51**, 595–605.

Tague, R. G. (2002). Variability of metapodials in primates with rudimentary digits: *Ateles geoffroyi, Colobus guereza,* and *Perodicticus potto. Amer. J. Phys. Anthropol.* **117**, 195–208.

Tajbakhsh, S., and Spörle, R. (1998). Somite development: constructing the vertebrate body. *Cell* **92**, 9–16.

Takada, J., Baylink, D. J., and Lau, K. H. W. (1995). Pretreatment with low doses of Norethindrone potentiates the osteogenic effects of fluoride on human osteosarcoma cells. *J. Bone Min. Res.* **10**, 1512–1522.

Takada, J., Chevalley, T., Baylink, D. J., and Lau, K. H. W. (1996). Dexamethasone enhances the osteogenic effects of fluoride in human Te 85 osteosarcoma cells in vitro. *Calcif. Tissue Int.* **58**, 355–361.

Takagi, Y., and Kaneko, T. (1995). Developmental sequence of bone-resorbing cells induced by intramuscular implantation of mineral-containing bone particles into rainbow trout, *Oncorhynchus mykiss. Cell Tissue Res.* **280**, 153–158.

Takahashi, H., and Frost, H. M. (1965). A tetracycline-based evaluation of the relative prevalence and incidence of formation of secondary osteons in human cortical bone. *Can. J. Physiol. Pharmacol.* **43**, 783–791.

Takahashi, H., and Frost, H. M. (1966). Age and sex related changes in the amount of cortex in normal human ribs. *Acta Orthop. Scand.* **37**, 122–130.

Takahashi, I., Mizoguchi, I., Nakamura, M., Kagayama, M., and Mitani, H. (1995). Effects of lateral pterygoid muscle hyperactivity on differentiation of mandibular condyles in rats. *Anat. Rec.* **241**, 328–336.

Takahashi, I., Mizoguchi, I., Nakamura, M., Sasano, Y., Saitoh, S., Kagayama, M., and Mitani, H. (1996a). Effects of expansive force on the differentiation of midpalatal suture cartilage in rats. *Bone* **18**, 341–348.

Takahashi, I., Mizoguchi, I., Sasano, Y., Saitoh, S., Ishida, M., Kagayama, M., and Mitani, H. (1996b). Age-related changes in the localization of glycosaminoglycans in condylar cartilage of the mandible in rats. *Anat. Embryol.* **194**, 489–500.

Takahashi, I., Nuckolls, G. H., Takahashi, K., Tanaka, O., Semba, I., Dashner, R., Shum, L., and Slavkin, H. C. (1998). Compressive force promotes Sox9, type II collagen and aggregan and inhibits IL-1-β expression resulting in chondrogenesis in mouse embryonic limb bud mesenchymal cells. *J. Cell Sci.* **111**, 2067–2076.

Takahashi, I., Onodera, K., Sasano, Y., Mizoguchi, I., Bae, J.-W., Mitani, H., Kagayama, M., and Mitani, H. (2003). Effects of stretching on gene expression of β1 integrin and focal adhesion kinase and on chondrogenesis through cell–extracellular matrix interactions. *Eur. J. Cell Biol.* **82**, 182–192.

Takahashi, K., Nuckolls, G. H., Takahashi, I., Nonaka, K., Nagata, M., Ikura, T., Slavkin, H. C., and Shum, L. (2001). Msx2 is a repressor of chondrogenic differentiation in migratory cranial neural crest cells. *Devel. Dynam.* **222**, 252–262.

Takahashi, M. M., and Noumura, T. (1993). Laterally asymmetric development of duck syrinx: Inhibition of growth and chondrogenesis of the right syringeal half by the left half. *Devel. Growth Differ.* **35**, 583–592.

Takahashi, Y., Bontoux, M., and Le Douarin, N. (1991). Epitheliomesenchymal interactions are critical for Quox-7 expression and membrane bone differentiation in the neural crest derived mandibular mesenchyme. *EMBO J.* **10**, 2387–2393.

Takahashi, Y., and Le Douarin, N. (1990). cDNA cloning of a quail homeobox gene and its expression in neural crest-derived mesenchyme and lateral plate mesoderm. *Proc. Natl Acad. Sci. U.S.A.* **87**, 7482–7486.

Takahashi, Y., Monsoro-Burq, A.-H., Bontoux, M., and Le Douarin, N. (1992). A role for Quox-7 in the establishment of the dorsoventral pattern during vertebrate development. *Proc. Natl Acad. Sci. U.S.A.* **89**, 10237–10241.

Takaoka, K., Koesuka, M., and Nakahara, H. (1991). Telopeptide-derived bovine skin collagen as a carrier for bone morphogenetic protein. *J. Orthop. Res.* **9**, 902–907.

Takeda, K., Oida, S., Ichijo, H., Iimura, T., Maruoka, Y., Amagasa, T., and Sasaki, S. (1994). Molecular cloning of rat bone morphogenetic protein (BMP) type IA receptor and its expression during ectopic bone formation induced by BMP. *Biochem. Biophys. Res. Commun.* **204**, 203–209.

Takeda, M., Iwata, H., Suzuki, S., Brown, K. S., and Kimata, K. (1986). Correction of abnormal matrix formed by *cmd/cmd* chondrocytes in culture by exogenously added cartilage proteoglycan. *J. Cell Biol.* **103**, 1605–1614.

Takeichi, M. (1973). The factor affecting the spreading of chondrogenesis upon inorganic substrate. *J. Cell Sci.* **13**, 193–204.

Takeuchi, J. K., Koshiba-Takeuchi, K., Suzuki, T., Kamimura, M., Ogura, K., and Ogura, T. (2003). Tbx5 and Tbx4 trigger limb initiation through activation of the Wnt/Fgf signaling cascade. *Development* **130**, 2729–2739.

Takigawa, M., Okada, M., Takano, T., Ohmae, H., Sakuda, M., and Susuki, F. (1984b). Studies on chondrocytes from mandibular condylar cartilage, nasal septal cartilage and spheno-occipital synchondrosis in culture. 1. Morphology, growth, glycosaminoglycan synthesis and responsiveness to bovine parathyroid hormone (1–34). *J. Dent. Res.* **63**, 19–22.

Takigawa, M., Takano, T., Shirai, E., and Susuki, F. (1984a). Cytoskeleton and differentiation: Effects of cytochalasin B and colchicine on expression of the differentiated phenotype of rabbit costal chondrocytes in culture. *Cell Differ.* **14**, 187–204.

Takio, Y., Pasqualetti, M., Kuraku, S., Hirano, S., Rijli, F. R., and Kuratani, S. (2004). *Lamprey* Hox genes and the evolution of jaws. *Nature* **429**, 386–387.

Takuma, S. (1962). Electron microscopy of cartilage resorption by chondroclasts. *J. Dent. Res.* **41**, 883–892.

Takuwa, Y., Ohse, C., Wang, E. A., Wozney, J. M., and Yamashita, K. (1991). Bone morphogenetic protein-2 stimulates alkaline phosphatase activity and collagen synthesis in cultured osteoblastic cells, MC3T3-E1. *Biochem. Biophys. Res. Commun.* **174**, 96–101.

Tallquist, M. D., Weismann, K. E., Hellström, M., and Soriano, P. (2000). Early myotome specification regulates PDGFA expression and axial skeleton development. *Development* **127**, 5059–5070.

Talmage, R. V. (1947). Changes produced in the symphysis pubis of the guinea pig by the sex steroids and relaxin. *Anat. Rec.* **99**, 91–113.

Tam, C. S., Reed, R., Little, A. H., and Cruikshank, B. (1974). Bone growth kinetics. III. A biorhythm in bone growth in the rabbit. *J. Pathol.* **114**, 127–134.

Tanaka, M., Cohn, M. J., Ashby, P., Davey, M., Martin, P., and Tickle, C. (2000). Distribution of polarizing activity and potential for limb formation in mouse and chick embryos and possible relationships to polydactyly. *Development* **127**, 4011–4021.

Tanaka, S. (1990). Age and growth studies on the calcified structures of newborn sharks in laboratory aquaria using tetracycline. In *Elasmobranchs as Living Resources: Advances in the Biology, Ecology, Systematics, and the Status of the Fisheries* (H. L. Pratt, Jr., S. H. Gruber, and T. Taniuchi, eds), pp. 189–202. NOAA Technical Report NMFS90, U.S. Department of Commerce.

Tanaka, T., Taniguchi, Y., Gotoh, K., Satoh, R., Inazu, M., and Ozawa, H. (1993). Morphological study of recombinant human transforming growth factor β1-induced intramembranous ossification in neonatal rat parietal bone. *Bone* **14**, 117–123.

Tanaka, T., and Uhthoff, H. K. (1981). Significance of resegmentation in the pathogenesis of vertebral body malformations. *Acta Orthop. Scand.* **52**, 331–338.

Tang, M. K., Leung, A. K. C., Kwong, W. H., Chow, P. H., Chan, J. Y. H., Ngo-Muller, V., Li, M., and Lee, K. K. H. (2000). BMP-4 requires the presence of the digits to initiate programmed cell death in limb interdigital tissues. *Devel. Biol.* **218**, 89–98.

Taniguchi, Y., Tanaka, T., Gotoh, K., Satoh, R., and Inazu, M. (1993). Transforming growth factor β1-induced cellular heterogeneity in the periosteum of rat parietal bone. *Calcif. Tissue Int.* **53**, 122–126.

Tanner, W. E. (1946). *Sir W. Arbuthnot Lane, Bart. C.B., M.S., F.R.C.S. His Life and Work.* Baillière, Tindall and Cox, London.

Tapp, E. (1966). The effects of hormones on bone in growing rats. *J. Bone Joint Surg. Br. Vol.* **48**, 526–531.

Tardieu, C. (1997). Femur ontogeny in humans and great apes: heterochronic implications for hominid evolution. *C. R. Acad. Sci. Paris. Earth & Planetary Sciences* **325**, 899–904.

Tardieu, C. (1998). Short adolescence in early hominids: infantile and adolescent growth of the human femur. *Amer. J. Phys. Anthropol.* **107**, 163–178.

Tardieu, C., Blanchard, O., Tabary, J.-C., and Le Lous, M. (1983). Tendon adaptation to bone shortening. *Conn. Tissue Res.* **11**, 35–44.

Tarin, D., and Sturdee, A. P. (1971). Early limb development in *Xenopus laevis. J. Embryol. Exp. Morphol.* **26**, 169–179.

Tarin, D., and Sturdee, A. P. (1974). Ultrastructural features of ectodermal-mesenchymal relationships in the developing limb of *Xenopus laevis. J. Embryol. Exp. Morphol.* **31**, 287–303.

Tarlo, L. B. H. (1964). The origin of bone. In *Bone and Teeth* (H. J. J. Blackwood, ed.), pp. 3–17. Pergamon Press, Oxford.

Tassava, R. A. (2004). Forelimb spike regeneration in *Xenopus laevis*: testing for adaptiveness. *J. Exp. Zool.* **301A**, 150–159.

Tassava, R. A., Bennett, L. L., and Zitnik, G. D. (1974). DNA synthesis without mitosis in amputated denervated forelimbs of larval axolotls. *J. Exp. Zool.* **190**, 111–116.

Tassava, R. A., and Olsen, C. L. (1982). Higher vertebrates do not regenerate digits and legs because the wound epidermis is not functional. A hypothesis. *Differentiation* **22**, 151–155.

Tassava, R. A., and Olsen-Winner, C. L. (2003). Responses to amputation of denervated *Ambystoma* limbs containing aneurogenic limb grafts. *J. Exp. Zool.* **297A**, 64–79.

Tavassoli, M., and Crosby, W. H. (1971). Bone formation in heterotopic implants of kidney tissue. *Proc. Soc. Exp. Biol. Med.* **137**, 641–644.

Tavella, S., Raffo, P., Tacchetti, C., Cancedda, R., and Castagnola, P. (1994). N-CAM and N-cadherin expression during *in vitro* chondrogenesis. *Exp. Cell Res.* **215**, 354–362.

Tawk, M., Tuil, D., Torrente, Y., Vriz, S., and Paulin, D. (2002). High-efficiency gene transfer into adult fish: a new tool to study fin regeneration. *Genesis* **32**, 27–31.

Taylor, A. B. (2002). Masticatory form and function in the African apes. *Am. J. Phys. Anthropol.* **117**, 133–156.

Taylor, G. P., Anderson, R., Reginelli, A. D., and Muneoka, K. (1994). FGF-2 induces regeneration of the chick limb bud. *Devel. Biol.* **163**, 282–284.

Taylor, J. F., Warrell, E., and Evans, R. A. (1987). The response of the rat tibial growth plates to distal periosteal division. *J. Anat.* **151**, 221–232.

Taylor, J. J., and Yeager, V. L. (1966). The fine structure of elastic fibers in the fibrous periosteum of the rat femur. *Anat. Rec.* **156**, 129–141.

Taylor, L. H., Hall, B. K., and Cone, D. (1993). Experimental infections of yellow perch (*Perca flavescens*) with *Apophallus brevis* (Digenia, Heterophyidae): parasite invasion, encystment and bony ossicle development. *Can. J. Zool.* **71**, 1886–1894.

Taylor, L. H., Hall, B. K., Miyake, T., and Cone, D. K. (1994). Ectopic ossicles associated with metacercaria of *Apophallus brevis* (Trematoda) in yellow perch, *Perca flavescens* (Teleostei): development and identification of bone and chondroid bone. *Anat. Embryol.* **190**, 29–46.

Tchernavin, V. (1937a). Preliminary account of the breeding changes in the skulls of *Salmo* and *Oncorhynchus. Proc. Linn. Soc. Lond.* **149**, 11–19.

Tchernavin, V. (1937b). Skulls of salmon and trout. A brief study of their differences and breeding changes. *Salmon & Trout Mag.* **88**, 235–242.

Tchernavin, V. (1938a). Notes on the chondrocranium and branchial skeleton of *Salmo. Proc. Zool. Soc. Lond. B.* **108**, 347–364.

Tchernavin, V. (1938b). The absorption of bones in the skull of salmon during their migration to rivers. *Fisheries, Scotland, Salmon Fish* No. VI, 1–4.

Tchernavin, V. (1938c). Notes on the chondrocranium and branchial skeleton of *Salmo*. II. Changes in the salmon skull. *Trans Zool. Soc. Lond.* **24**, 104–185.

Tchernavin, V. (1938d). The mystery of a salmon kype. *Salmon & Trout Mag.* **90**, 37–44.

Tchernavin, V. (1939). Ripe salmon Parr: A summary of research. *Proc. R. Physiol. Soc. Edinb.* **23**, 73–78.

Tchernavin, V. (1944). The breeding characters of salmon in relation to their size. *Proc. Zool. Soc. Lond. B.* **113**, 206–232.

Tchernov, E., Rieppel, O., Zaher, H., Polcyn, M. J., and Jacobs, L. L. (2000). A fossil snake with limbs. *Science* **287**, 2010–2012.

Teaford, M. F., Smith, M. M., and Ferguson, M. W. J. (eds) (2000). *Development, Function and Evolution of Teeth.* Cambridge University Press, Cambridge.

Teaford, M. F., and Walker, A. (1983). Prenatal jaw movements in the guinea pig, *Cavia porcellus*: Evidence from pattern of tooth wear. *J. Mammal.* **64**, 534–536.

Teillet, M.-A., Watanabe, Y., Jeffs, P., Duprez, D., Lapointe, F., and Le Douarin, N. M. (1998). Sonic hedgehog is required for survival of both myogenic and chondrogenic somitic lineages. *Development* **125**, 2019–2030.

Teitelbaum, S. L. (2000). Bone resorption by osteoclasts. *Science* **289**, 1504–1508.

Telhag, H. (1972). Mitosis of chondrocytes in experimental 'osteoarthritis' in rabbits. *Clin. Orthop. Rel. Res.* **86**, 224–229.

Telhag, H. (1973). DNA-synthesis in degenerated and normal joint cartilage in full-grown rabbits. *Acta Orthop. Scand.* **44**, 604–610.

Telhag, H. (1976). Nucleic acid in human normal and osteoarthritic articular cartilage. *Acta Orthop. Scand.* **47**, 585–587.

Telhag, H., and Havdrup, T. (1975). Nucleic acid in articular cartilage from rabbits of different ages. *Acta Orthop. Scand.* **46**, 185–189.

ten Berge, D., Brouwer, A., Bahi, S. E., Guénet, J.-L., Robert, B., and Meijlink, F. (1998b). Mouse *Alx3*: an *aristaless*-like homeobox gene expressed during embryogenesis in ectomesenchyme and lateral plate mesoderm. *Devel. Biol.* **199**, 11–25.

ten Berge, D., Brouwer, A., Korving, J., Martin, J. F., and Meijlink, F. (1998a). *Prx1* and *Prx2* in skeletogenesis: Roles in the craniofacial region, inner ear and limbs *Development* **125**, 3831–3842.

ten Berge, D., Brouwer, A., Korving, J., Reijnen, M. J., van Raaij, E. J., Verbeek, F., Gaffield, W., and Meijlink, F. (2001). *Prx1* and *Prx2* are upstream regulators of sonic hedgehog and control cell proliferation during mandibular arch morphogenesis. *Development* **128**, 2929–2938.

Ten Cate, A. R. (1972). Morphological studies of fibrocytes in connective tissue undergoing rapid remodelling. *J. Anat.* **112**, 401–414.

Ten Cate, A. R. (1975). Formation of supporting bone in association with periodontal ligament organization in the mouse. *Archs Oral Biol.* **20**, 137–138.

Ten Cate, A. R., and Deporter, D. A. (1975). The degradative role of the fibroblast in the remodelling and turnover of collagen in soft connective tissue. *Anat. Rec.* **182**, 1–14.

Ten Cate, A. R., Deporter, D. A., and Freeman, E. (1976). The role of fibroblasts in the remodelling of periodontal ligament during physiological tooth movement. *Amer. J. Orthod.* **69**, 155–168.

Ten Cate, A. R., and Freeman, E. (1974). Collagen remodelling by fibroblasts in wound repair. Preliminary observations. *Anat. Rec.* **179**, 543–546.

Ten Cate, A. R., Freeman, E., and Dickinson, J. B. (1974). The origin of alveolar bone. *J. Dental Res.* **53**, 245.

Ten Cate, A. R., Freeman, E., and Dickinson, J. B. (1977). Sutural development: Structure and its response to rapid expansion. *Amer. J. Orthod.* **71**, 622–636.

Ten Cate, A. R., and Mills, S. C. (1972). Development of the periodontium – Origin of alveolar bone. *Anat. Rec.* **173**, 69–78.

Ten Cate, A. R., and Syrbu, S. (1974). A relationship between alkaline phosphatase activity and the phagocytosis and degradation of collagen by the fibroblast. *J. Anat.* **117**, 351–360.

Tencer, A. F., Allen, B. L. Jr., Woodard, P. L., Self, J., L'Ilfeureux, A., Calhoun, J. H., and Brown, K. L. (1989). The effect of local controlled release of sodium fluoride on the stimulation of bone growth. *J. Biomed. Mat. Res.* **23**, 571–590.

ten Dijke, P., Yamashita, H., Sampath, T. K., Reddi, A. H., Estevez, M., Riddle, D. L., Ichijo, H., Heldin, C.-H., and Miyazono, K. (1994). Identification of type I receptors for osteogenic protein-1 and bone morphogenetic protein-4. *J. Biol. Chem.* **269**, 16985–16988.

Tenenbaum, H. C., and Heersche, J. N. M. (1982). Differentiation of osteoblasts and formation of mineralized bone in vitro. *Calcif. Tissue Int.* **34**, 76–79.

Tenenbaum, H. C., and Heersche, J. N. M. (1986). Differentiation of osteoid-producing cells *in vitro*: possible evidence for the requirement of a microenvironment. *Calcif. Tissue Int.* **38**, 262–267.

Tenenbaum, H. C., Limeback, H., McCulloch, C. A. G., Mamujee, H., Sukhu, B., and Torontali, M. (1992a). Osteogenic phase-specific co-regulation of collagen synthesis and mineralization by β-glycerophosphate in chick periosteal cultures. *Bone* **13**, 129–138.

Tenenbaum, H. C., McCulloch, C. A. G., Fair, C., and Birek, C. (1989). The regulatory effects of phosphates on bone metabolism *in vitro*. *Cell Tissue Res.* **257**, 555–563.

Tenenbaum, H. C., Palangio, K. G., Holmyard, D. P., and Pritzker, K. P. H. (1986). An ultrastructural study of osteogenesis in chick periosteum *in vitro*. *Bone* **7**, 295–302.

Tenenbaum, H. C., Richards, J., Holmyard, D., Mamujee, H., and Grynpas, M. D. (1992c). The effect of fluoride on osteogenesis *in vitro*. *Cells & Materials* **1**, 317–327.

Tenenbaum, H., Thiebold, J., and Bolender, C. (1976). Développement de la mandibule foetale de rat en transplantation dur la membrane chorio-allantoide de l'embryon de poulet. *J. Biol. Buccale* **4**, 261–274.

Tenenbaum, H. C., Torontali, M., and Sukhu, B. (1992b). Effects of bisphosphonates and inorganic pyrophosphate in osteogenesis *in vitro*. *Bone* **13**, 249–255.

Teng, S., and Herring, S. W. (1995). A stereological study of trabecular architecture in the mandibular condyle of the pig. *Archs Oral Biol.* **40**, 299–310.

Teng, S., and Herring, S. W. (1998). Compressive loading on bone surfaces from muscular contraction: an in vivo study in the miniature pig, *Sus scrofa*. *J. Morphol.* **238**, 71–80.

Tengan, T. (1990). Histogenesis and three-dimensional observation on condylar cartilage in prenatal mice. *J. Stomatol. Soc. Japan* **57**, 32–57 (in Japanese).

Terashima, Y., and Urist, M. R. (1975). Differentiation of cartilage from calvarial bone under influence of bone matrix gelatin *in vitro*. *Clin. Orthop. Rel. Res.* **113**, 168–177.

Termine, J. D., Kleinman, H. K., Whitson, S. W., Conn, K. M., McGarvery, M. L., and Martin, G. R. (1981). Osteonectin, a bone-specific protein linking mineral to collagen. *Cell* **26**, 99–106.

Thaller, C., and Eichele, G. (1987). Identification and spatial distribution of retinoids in the developing chick limb bud. *Nature* **327**, 625–628.

Thesingh, C. W., Groot, C. G., and Wassenaar, A. M. (1991). Transdifferentiation of hypertrophic chondrocytes into osteoblasts in murine fetal metatarsal bones, induced by co-cultured cerebrum. *Bone Min.* **12**, 25–40.

Thesingh, C. W., and Scherft, J. P. (1985). Fusion disability of embryonic osteoclast precursor cells and macrophages in the microphthalmic osteopetrotic mouse. *Bone* **6**, 43–52.

Thewissen, J. G. M., and Fish, F. E. (1997). Locomotor evolution in the earliest cetaceans: functional model, modern analogues, and paleontological evidence. *Paleobiology* **23**, 482–490.

Thewissen, J. G. M., and Hussain, S. T. (1993). Origin of underwater hearing in whales. *Nature* **361**, 444–445.

Thewissen, J. G. M., Hussain, S. T., and Arif, M. (1994). Fossil evidence for the origin of aquatic locomotion in Archaeocete whales. *Science* **263**, 210–212.

Thewissen, J. G. M., and Williams, E. M. (2002). The early radiations of Cetacea (Mammalia): evolutionary pattern and developmental constraints. *Annu. Rev. Ecol. Syst.* **33**, 73–90.

Thewissen, J. G. M., Williams, E. M., Roe, L. J., and Hussain, S. T. (2001). Skeletons of terrestrial cetaceans and the relationship of whales to artiodactyls. *Nature* **413**, 277–281.

Thies, R. S., Bauduy, M., Ashton, B. A., Kurtzberg, L., Wozney, J. M., and Rosen, V. (1992). Recombinant human bone morphogenetic protein-2 induces osteoblastic differentiation in W20-17 (W2017) stromal cells. *Endocrinology* **130**, 1318–1324.

Thomas, A. B., Hashimoto, H., Baylink, D. J., and Lau, K. H. W. (1996). Fluoride at mitogenic concentrations increases the steady state phosphotyrosyl phosphorylation level of cellular proteins in human bone cells. *J. Clin. Endocrinol. Metabol.* **81**, 2570–2578.

Thomas, B. L., Liu, J. K., Rudenstein, J. L. R., and Sharpe, P. T. (2000). Independent regulation of *Dlx2* expression in the epithelium and mesenchyme of the first branchial arch. *Development* **127**, 217–224.

Thomas, J. T., Kilpatrick, M. W., Lin, K., Ehlacher, L., Lembessis, P., Costa, T., Tsipouras, P., and Luyten, F. P. (1997). Disruption of human limb morphogenesis by a dominant negative mutation of Cdmp1. *Nature Genet.* **17**, 58–64.

Thomas, R. D. K., Shearman, R. M., and Stewart, G. W. (2000). Evolutionary exploitation of design options by the first animals with hard skeletons. *Science* **288**, 1239–1242.

Thomas, T., Kurihara, H., Yamagishi, H., Kurihara, Y., Yazaki, Y., Olson, E. N., and Srivastava, D. (1998). A signaling cascade involving endothelin-1, dHAND and Msx1 regulates development of neural crest-derived branchial arch mesenchyme. *Development* **125**, 3005–3014.

Thomason, J. J. (1985). Estimation of locomotory forces and stresses in the limb bones of recent and extinct equids. *Paleobiology* **11**, 209–220.

Thompson, A. Y., Piez, K. A., and Seyedin, S. M. (1985). Chondrogenesis in agarose gel culture. A model for chondrogenic induction, proliferation and differentiation. *Exp. Cell Res.* **157**, 483–494.

Thompson, D'A. W. (1917). *On Growth and Form.* Cambridge University Press, Cambridge.

Thompson, T. J., Owens, P. D. A., and Wilson, D. J. (1989). Intramembranous osteogenesis and angiogenesis in the chick embryo. *J. Anat.* **166**, 55–65.

Thomson, D. A. R. (1986). Meckel's cartilage in *Xenopus laevis* during metamorphosis: A light and electron microscope study. *J. Anat.* **149**, 77–87.

Thomson, D. A. R. (1987). A quantitative analysis of cellular and matrix changes in Meckel's cartilage in *Xenopus laevis. J. Anat.* **151**, 249–254.

Thomson, D. A. R. (1989). A preliminary investigation into the effects of thyroid hormones on the metamorphic changes in Meckel's cartilage in *Xenopus laevis. J. Anat.* **162**, 149–155.

Thomson, K. S. (1966). The evolution of the tetrapod middle ear in the Rhipidistian–Amphibian transition. *Amer. Zool.* **6**, 379–397.

Thomson, K. S. (1991). Parallelism and convergence in the horse limb: The internal–external dichotomy. In *New Perspectives on Evolution* (L. Warren and H. Koprowski, eds), pp. 101–122. Wiley-Liss, Inc., New York, NY.

Thomson, K. S. (1993). The origin of the tetrapods. *Amer. J. Sci.* **293-A**, 33–62.

Thonard, J. C., and Wiebkin, O. W. (1973). Localization of mucopolysaccharides in epithelial-like cells cultured *in vitro. J. Periodontol. Res.* **8**, 101–105.

Thorngren, K.-G., and Hansson, L. I. (1973a). Cell kinetics of the growth plate in hypophysectomized rats treated with growth hormone and thyroxine. *Z. Zellforsch. Mikrosk. Anat.* **142**, 431–442.

Thorngren, K.-G., and Hansson, L. I. (1973b). Cell kinetics and morphology of the growth plate in the normal and hypophysectomized rat. *Calcif. Tissue Res.* **13**, 113–129.

Thornton, C. S. (1938). The histogenesis of the regenerating fore limb of larval *Amblystoma* after exarticulation of the humerus. *J. Morphol.* **62**, 219–235.

Thornton, C. S. (1942). Studies on the origin of the regeneration blastema in *Triturus viridescens. J. Exp. Zool.* **89**, 375–390.

Thorogood, P. V. (1979). *In vitro* studies on skeletogenic potential of membrane bone periosteal cells. *J. Embryol. Exp. Morphol.* **54**, 185–207.

Thorogood, P. V. (1983). Morphogenesis of cartilage. In *Cartilage* (B. K. Hall, ed.), *Volume 2: Development, Differentiation and Growth*, pp. 223–254. Academic Press, New York, NY.

Thorogood, P. V. (1991). The development of the teleost fin and implications for our understanding of tetrapod limb evolution. In *Developmental Patterning of the Vertebrate Limb* (J. R. Hinchliffe, J. M. Hurlé, and D. Summerbell, eds), pp. 347–354. Plenum Press, New York, NY.

Thorogood, P. V., Bee, J., and von der Mark, K. (1986). Transient expression of collagen type II at epithelio-mesenchymal interfaces during morphogenesis of the cartilaginous neurocranium. *Devel. Biol.* **116**, 497–509.

Thorogood, P. V., and Craig Gray, J. (1975). The cellular changes during osteogenesis in bone and bone marrow composite autografts. *J. Anat.* **120**, 27–47.

Thorogood, P. V., and Hall, B. K. (1976). The use of lactate/malic dehydrogenase ratios as a technique to distinguish between progenitor cells of cartilage and bone. *J. Embryol. Exp. Morphol.* **36**, 305–313.

Thorogood, P. V., and Hall, B. K. (1977). Analysis of variable lactate/malic dehydrogenase ratios to distinguish between progenitor cells of cartilage and bone in the embryonic chick. *Calcif. Tissue Res.* **22**, Suppl., 314–317.

Thorogood, P. V., and Hanken, J. (1992). Body building exercises. *Curr. Biol.* **2**, 83–85.

Thorogood, P. V., and Hinchliffe, J. R. (1975). An analysis of the condensation process during chondrogenesis in the embryonic chick hind limb. *J. Embryol. Exp. Morphol.* **33**, 581–606.

Thorogood, P. V., Sarkar, S., and Moore, R. (1998). Skeletogenesis in the head. In *Oral Biology at the Turn of the Century* (B. Guggenheim and S. Shapiro, eds), pp. 93–100. Karger, Basle.

Thorp, B. H. (1990). Absence of cartilage canals in the long bone extremities of four species of skeletally immature marsupials. *Anat. Rec.* **226**, 440–446.

Thorp, B. H., Anderson, I., and Jakowlew, S. B. (1992). Transforming growth factor-β1, -β2 and β3 in cartilage and bone cells during endochondral ossification in the chick. *Development* **114**, 907–911.

Thorp, B. H., and Dixon, J. M. (1991). Cartilaginous bone extremities of growing monotremes appear unique. *Anat. Rec.* **229**, 447–452.

Thorpe, F. F., and Dorfman, A. (1967). Differentiation of connective tissues. *Curr. Top. Devel. Biol.* **2**, 151–190.

Threlkeld, A. J., and Smith, S. D. (1988). Unilateral hindpaw amputation causes bilateral articular cartilage remodeling of the rat hip joint. *Anat. Rec.* **221**, 576–583.

Thulborn, R. A. (1982). Liassic plesiosaur embryos reinterpreted as shrimp burrows. *Palaeontology* **25**, 351–359.

Thulborn, R. A. (1985). Birds as neotenous dinosaurs. *Records N. Z. Geol. Survey* **9**, 90–92.

Thulborn, R. A., and Hamley, T. L. (1982). The reptilian relationships of *Archeopteryx. Aust. J. Zool.* **30**, 611–642.

Thurston, M. N., Johnson, D. R., and Kember, N. F. (1985a). Cell kinetics of growth cartilage of achondroplastic (cn) mice. *J. Anat.* **140**, 425–434.

Thurston, M. N., Johnson, D. R., and Kember, N. F. (1985b). Cell kinetics of growth cartilage in spondylo-metaphyseal chondrodysplasia (smc) mice. *J. Anat.* **140**, 435–445.

Thurston, M. N., Johnson, D. R., Kember, N. F., and Moore, W. J. (1983). Cell kinetics of growth cartilage in stumpy: A new chondrodystrophic mutant in the mouse. *J. Anat.* **136**, 407–415.

Thyberg, J. (1977). Electron microscopy of cartilage proteoglycans. *Histochemie*, **9**, 259–266.

Thyberg, J., and Friberg, U. (1971). Ultrastructure of the epiphyseal plate of the normal guinea pig. *Z. Zellforsch. Mikrosk. Anat.* **122**, 254–272.

Thyberg, J., Nilsson, S., and Friberg, U. (1973). Electron microscopic study on guinea pig rib cartilage. Structural heterogeneity and effects of extraction with guanidine-HCl. *Z. Zellforsch. Mikrosk. Anat.* **146**, 83–102.

Tibone, K. W., and Bernard, G. W. (1982). A new in vitro model of intramembranous osteogenesis from adult bone marrow stem cells. In *Factors and Mechanisms Influencing Bone Growth* (A. D. Dixon and B. G. Sarnat, eds), pp. 107–123. Alan R. Liss Inc., New York, NY.

Tickle, C. (1996). Vertebrate limb development. *Sem. Cell Devel. Biol.* **7**, 137–143.

Tickle, C. (2000). Limb development: an international model for vertebrate pattern formation. *Int. J. Devel. Biol.* **44**, Sp. Issue, 101–108.

Tickle, C. (2002). The early history of the polarizing region: from classical embryology to molecular biology. *Int. J. Devel. Biol.* **46**, 847–852.

Tickle, C., Shellswell, G., Crawley, A. M., and Wolpert, L. (1976). Positional signalling by mouse limb polarizing region in the chick wing bud. *Nature* **259**, 396–397.

Tickle, C., Summerbell, D., and Wolpert, L. (1975). Positional signalling and specification of digits in chick limb morphogenesis. *Nature* **254**, 199–202.

Tilney, L. G., and Mooseker, M. (1971). Actin in the brush-border of epithelial cells of the chicken intestine. *Proc. Natl Acad. Sci. U.S.A.* **68**, 2611–2615.

Timmons, P. M., Wallin, J., Rigby, P. W. J., and Balling, R. (1994). Expression and function of *Pax-1* during development of the pectoral girdle. *Development* **120**, 2773–2785.

Tink, K., Petropull, L. A., Iwatsuki, M., and Nishimura, J. (1993). Altered cartilage phenotype expressed during intramembranous bone formation. *J. Bone Min. Res.* **8**, 1377–1387.

Tinsley, R. C., and Kobel, H. R. (eds) (1996). *The Biology of Xenopus.* Academic Press, London.

Tintut, Y., Parhami, F., Bostrom, K., Jackson, S. M., and Demer, L. L. (1998). cAMP stimulates osteoblast-like differentiation of calcifying vascular cells – Potential signaling pathway for vascular calcification. *J. Biol. Chem.* **273**, 7547–7553.

Tissier-Seta, J.-P., Mucchielli, M.-L., Mark, M., Mattei, M.-G., Goridis, C., and Brunet, J.-F. (1995). *Barx1*, a new mouse homeodomain transcription factor expressed in cranio-facial ectomesenchyme and the stomach. *Mech. Devel.* **51**, 3–15.

Todd, E., and Bowman, J. A. (1845). *The Physiological Anatomy and Physiology of Man.* Blanchard and Lea, Philadelphia, PA.

Toerien, M. J. (1965). An experimental approach to the development of the ear capsule in the turtle, *Chelydra serpentina. J. Embryol. Exp. Morphol.* **13**, 141–149.

Tokimasa, C., Kawata, T., Fujita, T., Kaku, M., Kawasoko, S., Kohno, S., and Tanne, K. (2000). Effects of insulin-like growth factor-I on nasopremaxillary growth under different masticatory loadings in growing mice. *Archs Oral Biol.* **45**, 871–878.

Toma, C. D., Ashkar, S., Gray, M. L., Schaffer, J. L., and Gerstenfeld, L. C. (1997a). Signal transduction of mechanical stimuli is dependent on microfilament integrity: Identification of osteopontin as a mechanically induced gene in osteoblasts. *J. Bone Min. Res.* **12**, 1626–1636.

Toma, C. D., Schaffer, J. L., Meazzini, M. C., Zurakowski, D., Nah, H. D., and Gerstenfeld, L. C. (1997b). Developmental restriction of embryonic calvarial cell populations as characterized by their in vitro potential for chondrogenic differentiation. *J. Bone Min. Res.* **12**, 2024–2039.

Tomes, J. (1839). Osseous tissue. In *Todd's Cyclopaedia of Anatomy and Physiology* **3**, 847.

Tomes, J., and De Morgan, C. (1853). Observations on the structure and development of bone. *Phil. Trans R. Soc. London* **143**, 109–139.

Tomlinson, B. L., Tomlinson, D. E., and Tassava, R. A. (1985). Pattern-deficient forelimb regeneration in adult bullfrogs. *J. Exp. Zool.* **236**, 313–326.

Tomsa, J. M., and Langeland, J. A. (1999). *Otx* expression during lamprey embryogenesis provides insights into the evolution of the vertebrate head and jaw. *Devel. Biol.* **207**, 26–37.

Tonegawa, Y. (1973). Inductive tissue interactions in the beak of a chick embryo. *Devel., Growth & Differ.* **15**, 57–72.

Tonna, E. A. (1960). Osteoclasts and the aging skeleton: A cytological, cytochemical and autoradiographic study. *Anat. Rec.* **137**, 251–270.

Tonna, E. A. (1961). The cellular complement of the skeletal system studied autoradiographically with tritiated thymidine (H^3TDr) during growth and aging, *J. Biophys. Biochem. Cytol.* **9**, 813–824.

Tonna, E. A. (1965). Skeletal cell aging and its effects on the osteogenetic potential. *Clin. Orthop. Rel. Res.* **40**, 57–80.

Tonna, E. A. (1966). H^3-histidine and H^3-thymidine autoradiographic studies of the possibility of osteoclast aging. *J. Lab. Invest.* **15**, 435–448.

Tonna, E. A. (1974). Hormonal influence on skeletal growth and regeneration. In *Humoral Control of Growth and Differentiation* (L. LoBue and A. S. Gordon, ed.), Volume 1, pp. 275–360. Academic Press, New York, NY.

Tonna, E. A., and Cronkite, E. P. (1962). An autoradiographic study of periosteal cell proliferation with tritiated thymidine. *Lab. Invest.* **11**, 455–462.

Tonna, E. A., and Pentel, L. (1972). Chondrogenic cell formation via osteogenic cell progeny transformation. *Lab. Invest.* **27**, 418–426.

Tonna, E.A., Singh, I. J., and Sandhu, H. S. (1987). Autoradiographic investigation of circadian rhythms in alveolar bone periosteum and cementum in young rats. *Histol. Histopathol.* **2**, 129–133.

Toole, B. P. (1969). Solubility of collagen fibrils formed *in vitro* in the presence of sulphated acid mucopolysaccharide-protein. *Nature* **222**, 872–873.

Toole, B. P. (1972). Hyaluronate turnover during chondrogenesis in the developing chick limb and axial skeleton. *Devel. Biol.* **29**, 321–329.

Toole, B. P. (1973a). Hyaluronate inhibition of chondrogenesis: Antagonism of thyroxine, growth hormone, and calcitonin. *Science* **180**, 302–303.

Toole, B. P. (1973b). Hyaluronate and hyaluronidase in morphogenesis and differentiation. *Amer. Zool.* **13**, 1061–1067.

Toole, B. P. (1990). Hyaluronan and its binding proteins, the hyaladherins. *Curr. Opin. Cell Biol.* **2**, 839–844.

Toole, B. P. (2001). Hyaluronan in morphogenesis. *Sem. Cell Devel. Biol.* **12**, 79–87.

Toole, B. P., and Gross, J. (1971). The extracellular matrix of the regenerating newt limb: Synthesis and removal of hyaluronate prior to differentiation. *Devel. Biol.* **25**, 57–77.

Toole, B. P., Jackson, G., and Gross, J. (1972a). Hyaluronate in morphogenesis: Inhibition of chondrogenesis *in vitro*. *Proc. Natl Acad. Sci. U.S.A.* **69**, 1384–1386.

Toole, B. P., Kang, A. H., Trelstad, R. L., and Gross, J. (1972b). Collagen heterogeneity within different growth regions of long bones of rachitic and non-rachitic chicks. *Biochem. J.* **127**, 715–720.

Toole, B. P., and Lowther, D. A. (1968a). Dermatan sulphate-protein: Isolation from and interaction with collagen. *Archs Biochem. Biophys.* **128**, 567–578.

Toole, B. P., and Lowther, D. A. (1968b). The effect of chondroitin sulphate-protein on the formation of collagen fibrils *in vitro*. *Biochem. J.* **109**, 857–866.

Toole, B. P., and Trelstad, R. L. (1971). Hyaluronate production and removal during corneal development in the chick. *Devel. Biol.* **26**, 28–35.

Toriumi, D. M., Kotler, H.-S., Luxenberg, D. P., Holtrop, M. E., and Wang, E. A. (1991). Mandibular reconstruction with a recombinant bone-inductive factor. *Arch. Otolaryngol. Head Neck Surg.* **117**, 1101–1112.

Torres, A. de L. (1917). On the cartilaginous tissue of the heart of Ophidia. *Anat. Rec.* **13**, 443–445.

Townsend, D. S., and Stewart, M. M. (1985). Direct development in *Eleutherodactylus coqui*: A staging table. *Copeia* **1985**, 423–436.

Tran, S., and Hall, B. K. (1989). Growth of the clavicle and development of clavicular secondary cartilage in the embryonic mouse. *Acta Anat.* **135**, 200–207.

Travis, D. F. (1968). Comparative ultrastructure and organization of inorganic crystals and organic matrices of mineralized tissues. In *Biology of the Mouth* (P. Person, ed.), Publ. No. 89, pp. 237–298. Amer. Assoc. Adv. Sci., Washington, DC.

Travis, D. F., François, C. J., Bonar, L. C., and Glimcher, M. J. (1967). Comparative studies of the organic matrices of invertebrate mineralized tissues. *J. Ultrastruct. Res.* **18**, 519–550.

Treharne, R. W. (1981). Review of Wolff's Law and its proposed means of operation. *Orthop. Rev.* **10**, 35–47.

Treharne, R. W., Brighton, C. T., Korostoff, E., and Pollack, S. R. (1980). An *in vitro* study of electrical osteogenesis using direct and pulsating currents. *Clin. Orthop. Rel. Res.* **145**, 300–306.

Trelstad, R. L. (1973). The developmental biology of vertebrate collagens. *J. Histochem. Cytochem.* **21**, 521–528.

Trelstad, R. L. (1977). Mesenchymal cell polarity and morphogenesis of chick cartilage. *Devel. Biol.* **59**, 153–163.

Trelstad, R. L., Kang, A. H., Cohen, A. M., and Hay, E. D. (1973). Collagen synthesis *in vitro* by embryonic spinal cord epithelium. *Science* **179**, 295–297.

Tremaine, R., and Hall, B. K. (1979). Retention during embryonic life of the ability of avian spinal cord to induce somitic chondrogenesis *in vitro*. *Acta Anat.* **105**, 78–85.

Trevisan, R. A., and Scapino, R. P. (1976a). Secondary cartilages in growth and development of the symphysis menti in the hamster. *Acta Anat.* **94**, 40–58.

Trevisan, R. A., and Scapino, R. P. (1976b). The symphyseal cartilage and growth of the symphysis menti in the hamster. *Acta Anat.* **96**, 335–355.

Triboli, C., and Lufkin, T. (1999). The murine *Bapx1* homeobox gene plays a critical role in embryonic development of the axial skeleton and spleen. *Development* **126**, 5699–5711.

Trivett, M. K., Officer, R. A., Clement, J. G., Walker, T. I., Joss, J. M., Ingleton, P. M., Martin, T. J., and Danks, J. A. (1999). Parathyroid hormone-related protein (PTHrp) in cartilaginous and bony fish tissues. *J. Exp. Zool.* **284**, 541–548.

Trock, D. H., Bollet, A. J., Dyer, R. H., Jr., Fielding, L. P., Miner, W. K., and Markoll, R. A. (1993). A double-blind trial of the clinical effects of pulsed electromagnetic fields in osteoarthritis. *J. Rheumatol.* **20**, 456–460.

Troitzky, W. (1932). Zur Frage der Formbildung des Schädeldaches. (Experimentelle Untersuchung der Schädeldachnähte und der damit verbundenen Erscheinungen). *Z. Morphol. Anthrop.* **30**, 504–534.

Trollope, J. (1999). *Other Peoples Children*. McArthur and Company, Toronto.

Trout, J. J., Buckwalter, J. A., and Moore, K. C. (1982). Ultrastructure of the human intervertebral disc. II. Cells of the nucleus pulposus. *Anat. Rec.* **204**, 307–314.

Trueb, L. (1985). A summary of osteocranial development in anurans with notes on the sequence of cranial ossification in *Rhinophrynus dorsalis* (Anura, Pipoidea, Rhinophrynidae). *S. Afr. J. Sci.* **81**, 181–185.

Trueb, L. (1996). Historical constraints and morphological novelties in the evolution of the skeletal system of pipid frogs (Anura: Pipidae). In *The Biology of* Xenopus (R. C. Tinsley and H. R. Kobel, eds), pp. 349–377. Academic Press, London.

Trueb, L., and Alberch, P. (1983). Miniaturization and the anuran skull: A case study of heterochrony. In *Functional Morphology of Vertebrates* (H. R. Duncker and G. Fleischer, eds), pp. 113–121. Gustav Fischer Verlag, Stuttgart

Trueb, L., and Hanken, J. (1992). Skeletal development in *Xenopus laevis* (Anura: Pipidae). *J. Morphol.* **214**, 1–42.

Trueb, L., Púgener, L. A., and Maglia, A. M. (2000). Ontogeny of the bizarre: an osteological description of *Pipa pipa* (Anura: Pipidae), with an account of skeletal development in the species. *J. Morphol.* **243**, 75–104.

Trueta, J. (1968). *Studies of the Development and Decay of the Human Frame*. William Heinemann Medical Books, London.

Trueta, J., and Little, K. (1960). The vascular contribution to osteogenesis. II. Studies with the electron microscope. *J. Bone Joint Surg. Br. Vol.* **42**, 267–276.

Trumpp, A., Depew, M. J., Rubenstein, J. L. R., Bishop, J. M., and Martin, G. R. (1999). Cre-mediated gene inactivation demonstrates that FGF8 is required for cell survival and patterning of the first branchial arch. *Genes & Devel.* **13**, 3136–3148.

Tsaltas, T. T. (1962). Metaplasia of aortic connective tissue to cartilage and bone induced by the intravenous injection of papain. *Nature* **196**, 1006–1007.

Tschanz, K. (1988). Allometry and heterochrony in the growth of the neck of Triassic prolacertiform reptiles. *Palaeontology* **31**, 997–1011.

Tschumi, P. A. (1957). Growth of the hind limb bud of *Xenopus laevis* and its dependence upon the epidermis. *J. Anat.* **91**, 149–173.

Tsilemov, A., Giannicopolou, P., and Vynios, D. H. (1998). Identification of a protein in squid cranial cartilage with link protein properties. *Biochemie* **80**, 591–594.

Tsilfidis, C., and Liversage, R. A. (1991). In vitro effects of implanted spinal ganglia on cartilage cells in Xenopus laevis forelimb regenerates. Can. J. Zool. **69**, 1546–1549.

Tsubai, T., Higashi, Y., and Scott, J. E. (2000). The effect of epidermal growth factor on the fetal rabbit mandibular condyle and isolated condylar fibroblasts. Archs Oral Biol. **45**, 507–515.

Tsukahara, J., and Hall, B. K. (1994). Transmembrane signaling in bone cell differentiation. In Bone, Volume 9. Differentiation and Morphogenesis of Bone (B. K. Hall, ed.), pp. 109–135. CRC Press, Boca Raton, FL.

Tsukahara, T., Okamura, M., Suzuki, S., Iwata, H., Miura, T., and Kimata, K. (1991). Enhanced expression of fibronectin by cmd/cmd chondrocytes and its modulation by exogenously added proteoglycan. J. Cell Sci. **100**, 387–395.

Tsumaki, N., Kimura, T., Matsui, Y., Nakata, K., and Ochi, T. (1996). Separable cis-regulatory elements that contribute to tissue- and site-specific α2(XI) collagen gene expression in the embryonic mouse cartilage. J. Cell Biol. **134**, 1573–1582.

Tsuruga, E., Takita, E., Itoh, H., Wakisaka, Y., and Kuboki, Y. (1997). Pore size of porous hydroxyapatite as the cell-substratum controls BMP-induced osteogenesis. J. Biochem. **121**, 317–324.

Tuan, R. S., and Lynch, M. H. (1983). Effect of experimentally induced calcium deficiency on the developmental expression of collagen types in chick embryonic skeleton. Devel. Biol. **100**, 374–386.

Tuan, R. S., Ono, T., Akins, R. E., and Koide, M. (1991). Experimental studies on cultured, shell-less fowl embryos; calcium transport, skeletal development, and cardio-vascular functions. In Egg Incubation: Its Effects on Embryonic Development in Birds and Reptiles (D. C. Deeming and M. W. J. Ferguson, eds), pp. 419–433. Cambridge University Press, Cambridge.

Tucker, A. S., Al Khamis, A., Ferguson, C. A., Bach, I., Rosenfeld, M. G., and Sharpe, P. T. (1998b). Conserved regulation of mesenchymal gene expression by fgf-8 in face and limb development. Development **126**, 221–228.

Tucker, A. S., and Lumsden, A. (2004). Neural crest cells provide species-specific patterning information in the developing branchial skeleton. Evol. & Devel. **6**, 32–40.

Tucker, A. S., Matthews, K. L., and Sharpe, P. T. (1998c). Transformation of tooth type induced by inhibition of BMP signaling. Science **282**, 1136–1138.

Tucker, A. S., and Slack, J. M. W. (1995a). Tail bud determination in the vertebrate embryo. Curr. Biol. **5**, 807–813.

Tucker, A. S., and Slack, J. M. W. (1995b). The Xenopus laevis tail-forming region. Development **121**, 249–262.

Tucker, A. S., Watson, R. P., Lettice, L. A., Yamada, G., and Hill, R. E. (2004). Bapx1 regulates patterning in the middle ear: altered regulatory role in the transition from the proximal jaw during vertebrate evolution. Development **131**, 1235–1245.

Tucker, A. S., Yamada, G., Grigoriou, M., Pachnis, V., and Sharpe, P. T. (1998a). FGF-8 determines rostral–caudal polarity in the first branchial arch. Development **126**, 51–61.

Tucker, R. P. (1993). The in situ localization of tenascin splice variants and thrombospondin 2 mRNA in the avian embryo. Development **117**, 347–358.

Tucker, R. P., Adams, J. C., and Lawler, J. (1995). Thrombospondin-4 is expressed by early osteogenic tissues in the chick embryo. Devel. Dynam. **203**, 477–490.

Tucker, R. P., Chiquet-Ehrismann, R., Chevron, M P., Martin, D., Hall, R. J., and Rubin, B. P. (2001) Teneurin-2 is expressed in tissues that regulate limb and somite pattern formation and is induced in virtro and in situ by FGF8. Devel. Dynam. **220**, 27–39.

Tucker, R. P., Hammarback, J. A., Jenrath, D. A., Mackie, E. J., and Xu, Y. (1993). Tenascin expression in the mouse: In situ localization and induction in vitro by bFGF. J. Cell Sci. **104**, 69–76.

Tucker, R. P., Spring, J., Baumgartner, S., Martin, D., Hagios, C., Poss, P. M., and Chiquet, E. R. (1994). Novel tenascin variants with a distinctive pattern of expression in the avian embryo. Development **120**, 637–647.

Tuckermann, J. P., Vallon, R., Gack, S., Grigoriadis, A. E., Porte, D., Lutz, A., Wagner, E. F., Schmidt, J., and Angel, P. (2001). Expression of collagenase-3 (MMP-13) in c-Fos-induced osteosarcome and chondrosarcomas is restricted to a subset of cells of the osteo-chondrogenic lineage. Differentiation **69**, 49–57.

Tufan, A. C., and Tuan, R. S. (2001). Wnt regulation of limb mesenchymal chondrogenesis is accompanied by altered N-cadherin-related functions. FASEB J. **15**, 1436–1438.

Tumlinson, R., and McDaniel, V. R. (1984). A description of the baculum of the bobcat (Felis rufus), with comments on its development and taxonomic implications. Can. J. Zool. **62**, 1172–1176.

Tuominen, M., Kantomaa, T., Pirttiniemi, P., and Poikela, A. (1996). Growth and type-II collagen expression in the glenoid fossa of the temporomandibular joint during altered loading – A study in the rat. Eur. J. Orthodont. **18**, 3–9.

Turksen, K., and Aubin, J. E. (1991). Positive and negative immunoselection for enrichment of two classes of osteoprogenitor cells. J. Cell Biol. **114**, 373–384.

Turksen, K., Bhargava, U., Moe, H. K., and Aubin, J. E, (1992). Isolation of monoclonal antibodies recognizing rat bone-associated molecules in vitro and in vivo. J. Histochem. Cytochem. **40**, 1339–1352.

Turner, C. H., Garetto, L. P., Dunipace, A. J., Zhang, W., Wilson, M. E., Grynpas, M. D., Chachra, D., McClintock, R., Peacock, M., and Stookey, G. K. (1997). Fluoride treatment increases serum IGF-1, bone turnover, and bone mass, but not bone strength in rabbits. Calcif. Tissue Int. **61**, 77–83.

Turner, R. S., and Burger, M. M. (1973). Involvement of a carbohydrate group in the active site for surface guided reassociation of animal cells. Nature **244**, 509–510.

Turner, R. T., Backup, P., Sherman, P. J., Hill, E., Evans, G. L., and Spelsberg, T.-C. (1992). Mechanism of action of estrogen on intramembranous bone formation: regulation of osteoblast differentiation and activity. Endocrinology **131**, 883–889.

Turner, R. T., Bell, N. H., Duvall, P., Bobyn, J. D., Spector, M., Holton, E. M., and Baylink, D. J. (1985). Spaceflight results in formation of defective bone. Proc. Soc. Exp. Biol. Med. **180**, 54–549.

Turner, R. T., Farley, J., Van der Steenhoven, J. J., Epstein, S., Bell, N. H., and Baylink, D. J. (1988). Demonstration of reduced mitogenic and osteoinductive activities in demineralized allogeneic bone matrix from vitamin-D-deficient rats. J. Clin. Invest. **82**, 212–217.

Tykoski, R. S., Forster, C. A., Rowe, T., Sampson, S. D., and Munyikwa, D. (2002). A furcula in the coelophysid theropod Syntarsus. J. Vert. Paleont. **22**, 728–733.

Tykoski, R. S., Rowe, T., Ketcham, R. A., and Colbert, M. W. (2002). Calsoyasuchus valliceps, a new crocodyliform from the

Early Jurassic Kayenta formation of Arizona. *J. Vert. Paleont.* **22**, 593–611.

Tyler, M. S. (1978). Epithelial influences on membrane bone formation in the maxilla of the embryonic chick. *Anat. Rec.* **192**, 225–234.

Tyler, M. S. (1983). Development of the frontal bone and cranial meninges in the embryonic chick: An experimental study of tissue interactions. *Anat. Rec.* **206**, 61–70.

Tyler, M. S., and Hall, B. K. (1977). Epithelial influences on skeletogenesis in the mandible of the embryonic chick. *Anat. Rec.* **188**, 229–240.

Tyler, M. S., and Koch, W. E. (1977a). *In vitro* development of palatal tissues from embryonic mice. II. Tissue isolation and recombination studies. *J. Embryol. Exp. Morphol.* **38**, 19–36.

Tyler, M. S., and Koch, W. E. (1977b). *In vitro* development of palatal tissues from embryonic mice. III. Interactions between palatal epithelium and heterotypic oral mesenchyme. *J. Embryol. Exp. Morphol.* **38**, 37–48.

U

Ueoka, C., Nadanaka, S., Seno, N., Khoo, K.-H., and Sugahara, K. (1999). Structural determination of novel tetra- and hexa-saccharide sequences isolated from chondroitin sulfate H (oversulfated dermatan sulfate) of hagfish notochord. *Glycoconjugate J.* **16**, 291–305.

Uesugi, Y., Taguchi, O., Noumura, T., and Iguchi, T. (1992). Effects of sex steroids on the development of sexual dimorphism in mouse innominate bones. *Anat. Rec.* **234**, 541–548.

Ueta, C., Iwamoto, M., Kanatani, N., Yoshida, C., Liu, Y., Enomoto, I. M., Ohmori, T., Enomoto, H., Nakata, K., Takada, K., Kusuri, K., and Komori, T. (2001). Skeletal malformations caused by overexpression of cbfa1 or its dominant negative form in chondrocytes. *J. Cell Biol.* **153**, 87–89.

Uhen, M. D., and Gingerich, P. D. (2001). New genus of dorudontine archaeocyte (Cetacea) from the middle-to-late Eocene of South Carolina. *Marine Mamm. Sci.* **17**, 1–34.

Uhthoff, H. K., and Germain, J.-P. (1977). The reversal of tissue differentiation around screws. *Clin. Orthop. Rel. Res.* **123**, 248–252.

Uitto, J., Hoffmann, H.-P., and Prockop, D. J. (1977). Purification and partial characterization of type II procollagen synthesized by embryonic cartilage cells. *Archs Biochem. Biophys.* **179**, 654–662.

Ulrich, M. M. W., Perizonius, W. R. K., Spoor, C. F., Sandberg, P., and Vermeer, C. (1987). Extraction of osteocalcin from fossil bones and teeth. *Biochem. Biophys. Res. Commun.* **149**, 712–719.

Umansky, R. (1966). The effect of cell population density on the developmental fate of reaggregating mouse limb-bud mesenchyme. *Devel. Biol.* **13**, 31–56.

Underhill, T. M., Sampaio, A. V., and Weston, A. D. (2001). Retinoid signaling and skeletal development. In *The Molecular Basis of Skeletogenesis* (G. Cardew and J. Goode, eds; B. K. Hall, chair), Novartis Foundation Symposium 232, pp. 171–188. John Wiley & Sons, Chichester.

Underhill, T. M., and Weston, A. D. (1998). Retinoids and their receptors in skeletal development. *Microsc. Res. Tech.* **43**, 137–155.

Underwood, J. L., and DeLuca, H. F. (1984). Vitamin D is not directly necessary for bone growth and mineralization. *Amer. J. Physiol.* **246**, E493–E498.

Uneno, S., Yamamoto, I., Jamamuro, T., Okumura, H., Ohta, S., Lee, K., Kasai, R., and Konishi, J. (1989). Transforming growth factor β modulates proliferation of osteoblastic cells: Relation to its effect on receptor levels for epidermal growth factor. *J. Bone Min. Res.* **4**, 165–172.

Unger, E., and Eriksson, U. J. (1992). Regionally disturbed production of cartilage proteoglycans in malformed fetuses from diabetic rats. *Diabetology* **35**, 517–521.

Unwin, D. M. (2003). Smart-winged pterosaurs. *Nature* **425**, 910–911.

Unwin, D. M., Frey, E., Martill, D. M., Clarke, J. B., and Riess, J. (1996). On the nature of the pteroid in pterosaurs. *Proc. R. Soc. Lond. B.* **263**, 45–52.

Upton, L. G. (1972). Bone induction studies using an improved diffusion chamber. *J. Oral Surg.* **30**, 486–490.

Urho, L. (2002). Characters of larvae – what are they? *Folia Zool.* **51**, 161–186.

Urist, M. R. (1953). The physiological basis of bone graft surgery, with special reference to the theory of induction. *Clin. Orthop. Rel. Res.* **1**, 207–216.

Urist, M. R. (1962). The bone–body fluid continuum: Calcium and phosphorus in the skeleton and blood of extinct and living vertebrates. *Perspect. Biol. Med.* **6**, 75–115.

Urist, M. R. (1965). Bone: Formation by autoinduction. *Science* **150**, 893–899 [reprinted in 1997, *J. NIH Res.* **9**, 43–50].

Urist, M. R. (1970). Induction and differentiation of cartilage and bone cells. In *Cell Differentiation* (O. A. Schjeide and J. de Vellis, eds), pp. 504–528. Van Nostrand-Reinhold, Princeton, NJ.

Urist, M. R. (1972). Growth hormone and skeletal tissue metabolism. In *The Biochemistry and Physiology of Bone* (G. H. Bourne, ed.), 2nd Edn, Volume 2, pp. 155–195. Academic Press, New York, NY.

Urist, M. R. (1980). *Fundamental and Clinical Bone Physiology.* J. B. Lippincott Co., Philadelphia, PA.

Urist, M. R. (1991). Emerging concepts of bone morphogenetic protein. In *Fundamentals of Bone Growth: Methodology and Applications* (A. D. Dixon, B. G. Sarnat, and D. A. N. Hoyte, eds), pp. 189–198. CRC Press, Boca Raton, FL.

Urist, M. R., DeLange, R. J., and Finerman, G. A. M. (1983). Bone cell differentiation and growth factors. *Science* **220**, 680–686.

Urist, M. R., and Dowell, T. A. (1968). The inductive substratum for osteogenesis in pellets of particulate bone matrix. *Clin. Orthop. Rel. Res.* **61**, 61–78.

Urist, M. R., Hay, P. H., Dubuc, F., and Buring, K. (1969). Osteogenetic competence. *Clin. Orthop. Rel. Res.* **64**, 194–220.

Urist, M. R., Huo, Y. K., Brownell, A. G., Hohl, W. M., Buyske, J., Lietze, A., Tempst, P., Hunkapiller, M., and DeLange, R. J. (1984). Purification of bovine bone morphogenetic protein by hydroxyapatite chromatography. *Proc. Natl Acad. Sci. U.S.A.* **81**, 371–375.

Urist, M. R., and MacLean, F. C. (1952). Osteogenic potency and new bone formation by induction in transplants to the anterior chamber of the eye. *J. Bone Joint Surg. Amer. Vol.* **34**, 443–446.

Urist, M. R., Maeda, H., Shamie, A. W., and Teplica, D. (1998). Endogenous bone morphogenetic protein expression in transplants of urinary bladder. *Plast. Reconstruct. Surg.* **101**, 408–415.

Urist, M. R., Moss, M. J., and Adams, J. M., Jr. (1964). Calcification of tendon. A triphasic local mechanism. *Archs Pathol.* **77**, 594–608.

Urist, M. R., Nillsson, O. S., Hudak, R., Huo, Y. K., Rasmussen, J., Hirota, Q., and Lietze, A. (1985). Immunologic evidence of a bone morphogenetic protein in the Milieu-Interieur. *Ann. Biol. Clin.* **43**, 755–766.

Urist, M. R., Raskin, K., Goltz, D., and Merickel, K. (1997). Endogenous bone morphogenetic protein – immunohistochemical localization in repair of a punch hole in the rabbit's ear. *Plast. Reconstruct. Surg.* **99**, 1382–1389.

Ushiki, T. (2002). Collagen fibers, reticular fibers and elastic fibers. A comprehensive understanding from a morphological viewpoint. *Arch. Histol. Cytol.* **65**, 109–126.

Uwa, H. (1969). Changes in RNA-, DNA- and protein-synthetic activity during the formation of anal-fin processes in ethisterone-treated females of *Oryzias latipes*. *Devel., Growth & Differ.* **11**, 77–87.

Uwa, H. (1974). Ultrastructural study on the scleroblast of *Oryzias latipes* during ethisterone-induced anal-fin process formation. *Devel., Growth & Differ.* **16**, 41–54.

Uwa, H., and Nagata, T. (1976). Cell population kinetics of the scleroblast during ethisterone-induced anal-fin process formation in adult females of the medaka, *Oryzias latipes*. *Devel., Growth & Differ.* **18**, 279–288.

V

Väänänen, H. K., Morris, D. C., and Anderson, H. C. (1983). Calcification of cartilage matrix in chondrocyte cultures derived from rachitic rat growth plate cartilage. *Metab. Bone Dis. Rel. Res.* **5**, 87–92.

Vaes, G. (1988). Cellular biology and biochemical mechanism of bone resorption. A review of recent developments on the formation, activation and mode of action of osteoclasts. *Clin. Orthop. Rel. Re.* **231**, 239–271.

Vaes, G. N., and Nichols, G. (1962). Oxygen tension and the control of bone cell metabolism. *Nature* **193**, 379–380.

Vaessen, M.-J., Meijers, J. H. C., Bootsma, D., and Van Kessel, A. G. (1990). The cellular retinoic-acid-binding protein is expressed in tissues associated with retinoic-acid-induced malformations. *Development* **110**, 371–378.

Vaglia, J. V., Babcock, S. K., and Harris, R. N. (1997). Tail development and regeneration throughout the life cycle of the four-toed salamander *Hemidactylium scutatum*. *J. Morphol.* **233**, 15–29.

Vaglia, J. L., and Hall, B. K. (1999). Regulation of neural crest cell populations in vertebrates: Occurrence, distribution and underlying mechanisms. *Int. J. Devel. Biol.* **43**, 95–110.

Vakeva, L., Mackie, E., Kantomaa, T., and Thesleff, I. (1990). Comparison of the distribution patterns of tenascin and alkaline phosphatase in developing teeth, cartilage and bone of rats and mice. *Anat. Rec.* **228**, 69–76.

Valcourt, U., Ronziere, M. C., Winkler, P., Rosen, V., Herbage, D., and Mallein-Gerin, F. (1999). Different effects of bone morphogenetic proteins 2, 4, 12, and 13 on the expression of cartilage and bone markers in the MC615 chondrocyte cell line. *Exp. Cell Res.* **251**, 264–274.

van de Kamp, N. (1968). Fine structural analysis of the conjunctival papillae in the chick embryo: A reassessment of their morphogenesis and developmental significance. *J Exp. Zool.* **169**, 447–462.

Van den Akker, E., Fromental-Ramain, C., de Graaff, W., Le Mouellic, H., Brûlet, P., Chambon, P., and Deschamps, J. (2001). Axial skeletal patterning in mice lacking all paralogous group 8 Hox genes. *Development* **128**, 1911–1921.

van der Brugghen, W., and Janvier, P. (1992). Denticles in thelodonts. *Nature* **364**, 107.

Van der Hoeven, F., Schimmang, T., Volkmann, A., Mattei, M. G., Kyewski, B., and Ruether, U. (1994). Programmed cell death is affected in the novel mouse mutant *Fused toes* (*Ft*). *Development* **120**, 2601–2607.

van der Meulen, M. C. H., and Carter, D. R. (1995). Developmental mechanics determine long bone allometry. *J Theoret. Biol.* **172**, 323–327.

Van der Plas, A., and Nijweide, P. J. (1988). Cell–cell interactions in the osteogenic compartment of bone. *Bone* **9**, 107–111.

Van der Rest, M., and Mayne, R. (1988). Type IX collagen proteoglycan from cartilage is covalently cross-linked to type II collagen. *J. Biol. Chem.* **263**, 1615–1618.

Vandersea, M. W., Fleming, P., McCarthy, R. A., and Smith, D. G. (1998b). Fin duplications and deletions induced by disruption of retinoic acid signaling. *Devel. Genes Evol.* **208**, 61–68.

Vandersea, M. W., McCarthy, R. A., Fleming, P., and Smith, D. G. (1998a). Exogenous retinoic acid during gastrulation induces cartilaginous and other craniofacial defects in *Fundulus heteroclitus*. *Biol. Bull.* **194**, 281–296.

Van der Steenhoven, J. J., DeLustro, F. A., Bell, N. H., and Turner, R. T. (1988). Osteoinduction by implants of demineralized allogeneic bone matrix is diminished in vitamin D-deficient rats. *Calcif. Tissue Int.* **42**, 39–45.

Van der Stricht, O. (1890). Recherches sur le cartilage articulaire des oiseaux. *Arch. Biol.* **10**, 1–41.

Van de Velde, J. P., Vermeiden, J. P. W., and Bloot, A. M. (1985). Medullary bone matrix formation, mineralization, and remodeling related to the daily egg-laying cycle of Japanese quail: A histological and radiological study. *Bone* **6**, 321–327.

Vandewalle, P., Chikov, A., Layèyé, P., Parmentier, E., Huriaux, F., and Focant, B. (1999). Early development of the chondrocranium in *Chrysichthys auratus*. *J. Fish Biol.* **55**, 795–808.

Vandewalle, P., Focant, B., Huriaux, F., and Chardon, M. (1992). Early development of the cephalic skeleton of *Barbus barbus* (Teleostei, Cyprinidae). *J. Fish Biol.* **41**, 43–62.

Van de Water, T. (1980). Developmental significance of epithelial/mesenchymal interactions in labyrinthine organogenesis. In *Current Research Trends in Prenatal Craniofacial Development* (R. M. Pratt and R. L. Christiansen, eds), p. 451. Elsevier/ North-Holland, New York, NY.

Van de Water, T. R., and Galinovic-Schwartz, V. (1986). Dysmorphogenesis of the inner ear: disruption of extracellular matrix (ECM) formation by an L-proline analog in otic explants. *J. Craniofac. Genet. Devel. Biol.* **6**, 113–120.

Van de Water, T. R., Li, C. W., Ruben, R. J., and Shea, C. A. (1980a). Ontogenic aspects of mammalian inner ear development. *Birth Defects Orig. Artic. Ser.* **16**, 5–45.

Van de Water, T. R., Maderson, P. F. A., and Jaskoll, T. F. (1980b). The morphogenesis of the middle and external ear. In *Morphogenesis and Malformation of the Ear* (R. Gorlin, ed.), pp. 147–180. Alan R. Liss, New York, NY.

Van de Wijngaert, F. P., Schipper, C. A., Tas, M. C., and Burger, E. H. (1988). Role of mineralizing cartilage in osteoclast and osteoblast recruitment. *Bone* **9**, 81–88.

Van Exan, R. J., and Hall, B. K. (1983). Epithelial induction of osteogenesis in embryonic chick mandibular mesenchyme: A possible role of basal lamina. *Can. J. Biochem. Cell Biol.* **61**, 967–979.

Van Exan, R. J., and Hall, B. K. (1984). Epithelial induction of osteogenesis in embryonic chick mandibular mesenchyme studied by transfilter tissue recombinations. *J. Embryol. Exp. Morphol.* **79**, 225–242.

Van Kampen, G. P. J., Velduijzen, J. P., Kuijer, R., Van der Stadt, R. J., and Schipper, C. A. (1985). Cartilage response to mechanical force in high-density chondrocyte cultures. *Arthrit. Rheum.* **28**, 419–424.

van Limborgh, J. (1970). A new view on the control of the morphogenesis of the skull. *Acta Morphol. Neerl. Scand.* **8**, 143–160.

van Limborgh, J. (1972). The role of genetic and local environmental factors in the control of postnatal craniofacial morphogenesis. *Acta Morphol. Neerl. Scand.* **10**, 37–47.

van Limborgh, J. (1982). Factors controlling skeletal morphogenesis. In *Factors and Mechanisms Influencing Bone Growth* (A. D. Dixon and B. G. Sarnat, eds), pp. 1–17. Alan. R. Liss Inc., New York, NY.

Van Vlasselaer, P., Borremans, B., van Gorp, U., Dasch, J. R., and De Waal-Malefyt, R. (1994). Interleukin-10 inhibits transforming growth factor-β (TGF-β) synthesis required for osteogenic commitment of mouse bone marrow cells. *J. Cell Biol.* **124**, 569–577.

Van Vliet, G., Styne, D. M., Kaplan, S. L., and Grumbach, M. M. (1983). Growth hormone treatment for short stature. *New Engl. J. Med.* **309**, 1016–1022.

Vargesson, N., Clarke, J. D. W., Vincent, K., Coles, C., Wolpert, L., and Tickle, C. (1997). Cell fate in the chick limb bud and relationship to gene expression. *Development* **124**, 1909–1918.

Vasan, N. S. (1981). Analysis of perinotochordal materials. 1. Studies on proteoglycan synthesis. *J. Exp. Zool.* **215**, 229–233.

Vasan, N. S. (1983). Analysis of perinotochordal materials. 2. Studies on the influence of proteoglycan in somite chondrogenesis. *J. Embryol. Exp. Morphol.* **73**, 263–274.

Vasan, N. S. (1987). Somite chondrogenesis: The role of the microenvironment. *Cell Differ.* **21**, 147–159.

Vasan, N. S., Lamb, K. M., and Manna, O. La. (1986a). Somite chondrogenesis *in vitro*. 1. Alterations in proteoglycan synthesis. *Cell Differ.* **18**, 79–90.

Vasan, N. S., Lamb, K. M., and Manna, O. La. (1986b). Somite chondrogenesis *in vitro*. 2. Changes in the hyaluronic acid synthesis. *Cell Differ.* **18**, 91–99.

Vasan, N. S., and Lash, J. W. (1977). Heterogeneity of proteoglycans in developing chick limb cartilage. *Biochem. J.* **164**, 179–184.

Vasan, N. S., and Miller, E. (1985). Somite chondrogenesis *in vitro*: Differential inductions by modified matrix – a biochemical and morphological study. *Devel. Growth Differ.* **27**, 405–418.

Vasse, J. (1974). Etudes expérimentales sur le rôle des somites au cours des premiers stades du développement du membre antérieur chez l'embryon du chélonien, *Emys orbicularis* L. *J. Embryol. Exp. Morphol.* **32**, 417–430.

Vasse, J. (1977). Etudes expérimentales sur les premiers stades du développement du membre antérieur chez l'embryon du chélonien, *Emys orbicularis* L: Détermination en mosaïque régulation. *J. Embryol. Exp. Morphol.* **42**, 135–148.

Vaughan, J. M. (1975). *The Physiology of Bone*, 2nd Edn. Oxford University Press, Oxford.

Veldhuijzen, J. P., Bourret, L. A., and Rodan, G. A. (1979). *In vitro* studies of the effect of intermittent compressive forces on cartilage cell proliferation. *J. Cell Physiol.* **98**, 299–306.

Verbout, A. J. (1976). A critical review of the 'Neugliederung' concept in relation to the development of the vertebral column. *Acta Biotheor.* **25**, 219–258.

Verbout, A. J. (1985). The development of the vertebral column. *Adv. Anat. Embryol. Cell Biol.* **90**, 1–122.

Veis, A. (1987). Report on nomenclature and standards for bone proteins and growth factors. *Connect. Tissue Res.* **16**, 109–110.

Verraes, W., and Ismail, M. H. (1980). Developmental and functional aspects of the frontal bones in relation to some other bony and cartilaginous parts of the head roof in *Haplochromis elegans* Trewavas, 1933 (Teleostei, Cichlidae). *Neth. J. Zool.* **30**, 450–472.

Verraes, W., Ismail, M., and Huysseune, A. (1979). Developmental aspects of the pharyngeal jaws and the pharyngo-branchial neurocranial apophysis in *Haplochromis elegans* (Teleostei: Cichlidae). *Amer. Zool.* **19**, 1013.

Verwoerd, C. D. A., Urbanus, N. A. M., and Mastenbroek, G. J. (1980). The influence of partial resections of the nasal septal cartilage on the growth of the upper jaw and the nose: An experimental study in rabbits. *Clin. Otolaryngol.* **5**, 291–302.

Vetter, U., Pirsig, W., and Heinze, E. (1983). Growth activity in human septal cartilage: Age-dependent incorporation of labeled sulfate in different anatomic locations. *Plast. Reconstruct. Surg.* **71**, 167–170.

Vetter, U., Pirsig, W., and Heinze, E. (1984). Postnatal growth of the human septal cartilage. Preliminary report. *Acta Otolaryngol. (Stock.)* **97**, 131–136.

Vickaryous, M., Fedak, T. J., Franz-Odendaal, T., Hall, B. K., and Stone, J. (2003). Dusting off bone ontologies: considering skeletons in the palaeo-closet. *Palaeont. Assoc. Newsletter* **53**, 48–51.

Vickaryous, M., Fedak, T. J., W. Olson, W., J. Stone, J., and Hall, B. K. (2002). Ontogeny & morphology: past and present. *Palaeont. Assoc. Newsletter* **50**, 29–32.

Vickaryous, M. K., Russell, A. P., and Currie, P. J. (2001). Cranial ornamentation of ankylosaurs (Ornithischia: Thyreophora): Reappraisal of developmental hypothesesi. In *The Armored Dinosaurs* (K. Carpenter, ed.), pp. 318–340. Indiana University Press, Bloomington, IN.

Vico, L., Chappard, D., Alexandre, C., Palle, S., Minaire, P., Riffat, G., Novikov, V. E., and Bakulin. A. V. (1987). Effects of weightlessness on bone mass and osteoclast number in pregnant rats after a five-day spaceflight (Cosmos 1514). *Bone* **8**, 95–103.

Vidinov, N., and Vasilev, V. (1985). Cilia in rat articular chondrocytes. *Anat. Anz.* **158**, 51–56.

Vielle-Grosjean, I., Hunt, P., Gulisano, M., Boncinelli, E., and Thorogood, P. (1997). Branchial Hox gene expression and human craniofacial development. *Devel. Biol.* **183**, 49–60.

Vig, K. W. L. (1990). Orthodontic considerations applied to cranial dysmorphology. *Cleft Palate J.* **27**, 141–145.

Vig, K. W. L., and Burdi, A. R. (eds) (1988). *Craniofacial Morphogenesis and Dysmorphogenesis*. Center for Human Growth and development, University of Michigan, Ann Arbor, MI.

Vignery, A., and Baron, R. (1980). Dynamic histomorphometry of alveolar bone remodeling in the adult rat. *Anat. Rec.* **196**, 191–200.

Villanueva, J. E., and Nimni, M. E. (1990). Modulation of osteogenesis by isolated calvarial cells – a model for tissue interactions. *Biomaterials* **11**, 19–21.

Villiger, P. M., and Lotz, M. (1992). Differential expression of TGFβ isoforms by human articular chondrocytes in response to growth factors. *J. Cell Physiol.* **151**, 318–325.

Vilmann, A., and Vilmann, H. (1983). Os penis of the rat. IV. The proximal growth cartilage. *Acta Anat.* **117**, 136–144.

Vilmann, H. (1982a). The mandibular angular cartilage in the rat. *Acta Anat.* **113**, 61–68.

Vilmann, H. (1982b). Os penis of the rat. III. Formation and growth of the bone. *Acta Morphol. Neerl-Scand.* **20**, 309–318.

Vilmann, H., Juhl, M., and Kirkeby, S. (1985). The skeleton of the muscular dystrophic mouse. *Fortschr. Zool.* **30**, 103–106.

Vilmann, H., and Vilmann, A. (1979). Os penis of the rat. II. Morphology of the mature bone. *Anat. Anz.*, **146**, 483–93.

Vincent, N. A., and Hall, B. K. (2004). Mechanisms of development of the neural crest-derived skeleton in *Hymenochirus boettgeri*. *J. Morphol.* **260**, 338.

Vinkka, H. (1982). Secondary cartilages in the facial skeleton of the rat. *Proc. Finn. Dent. Soc.* **78** (Suppl. 7), 1–137.

Vinkka-Puhakka, H. (1991a). Craniofacial secondary cartilages in the hamster and rat. In *Fundamentals of Bone Growth: Methodology and Applications* (A. D. Dixon, B. G. Sarnat, and D. A. N. Hoyte, eds), pp. 131–137, CRC Press, Boca Raton, FL.

Vinkka-Puhakka, H. (1991b). Secondary cartilages in the auditory bulla of the hamster. *Proc. Finn. Dent. Soc.* **87**, 99–107.

Vinkka-Puhakka, H., and Thesleff, I. (1993). Initiation of secondary cartilage in the mandible of the Syrian hamster in the absence of muscle function. *Archs Oral Biol.* **38**, 49–54.

Vinson, W. C., and Seyer, J. M. (1974). Synthesis of type III collagen by embryonic chick skin. *Biochem. Biophys. Res. Commun.* **58**, 58–65.

Vinyard, C. J., and Ravosa, M. J. (1998). Ontogeny, function and scaling of the mandibular symphysis in papioin primates. *J. Morphol.* **235**, 157–175.

Virchow, R. L. K. (1851). Die Identität von Knochen-Korpel und Bindergewebskorper suchen sowie über Schleingewebe. *Verhandl. Physik-med Gesellsch.* **2**, 150–187.

Virchow, R. L. K. (1853). Das Normale Knochenwachsthum und die Rachitische Störung Desselben. *Virchows Arch. Pathol. Anat. Physiol.* **5**, 409.

Viriot, L., Lesot, H., Vonesch, J.-L., Ruch, J.-V., Peterka, M., and Peterkova, R. (2000). The presence of rudimentary odontogenic structures in the mouse embryonic mandible requires reinterpretation of developmental control of first lower molar histomorphogenesis. *Int. J. Devel. Biol.* **44**, 233–240.

Visconti, C. S., Kavalkovich, K., Wu, J.-J., and Niyibizi, C. (1996). Biochemical analysis of collagens of the ligament–bone interface reveals presence of cartilage-specific collagens. *Archs Biochem. Biophys.* **328**, 134–142.

Visnapuu, V., Peltomäki, T., Isotupa, K., Kantomaa, T., and Helenius, H. (2000). Distribution and characterization of proliferative cells in the rat mandibular condyle during growth. *Eur. J. Orthod.* **22**, 631–638.

Visnapuu, V., Peltomäki, T., Säämänen, A.-M., and Rönning, O. (2000). Collagen I and II mRNA distribution in the rat temporomandibular joint region during growth. *J. Craniofac. Genet. Devel. Biol.* **20**, 144–149.

Vivian, J. L., Olson, E. N., and Klein, W. H. (2000). Thoracic skeletal defects in myogenin- and MRF4-deficient mice correlate with early defects in myotome and intercostal musculature. *Devel. Biol.* **224**, 29–41.

Vivien, D., Galera, P., Loyau, G., and Pujol, J.-P. (1991). Differential response of cultured rabbit articular chondrocytes (RA) to transforming growth factor β (TGF-β) – evidence for a role of serum factors. *Eur. J. Cell Biol.* **54**, 217–223.

Vizcaino, S. F., and Milne, N. (2002). Structure and function in armadillolimbs (Mammalia: Zenarthra: Dasypodidae). *J. Zool. Lond.* **257**, 117–127.

Vleminckx, K., and Kemler, R. (1999). Cadherins and tissue formation: Integrating adhesion and signaling. *BioEssays* **21**, 211–220.

Vogel, A., and Tickle, C. (1993). FGF-4 maintains polarizing activity of posterior limb bud cells *in vivo* and *in vitro*. *Development* **119**, 199–206.

Vogel, K. G., and Kelley, R. O. (1977). Cell surface glycosaminoglycans: Identification and organization in cultured human embryo fibroblasts. *J. Cell Physiol.* **92**, 469–480.

Vogl, C., Atchley, W. R., Cowley, D. E., Crenshaw, P., Murray, J. D., and Pomp, D. (1993). The epigenetic influence of growth hormone on skeletal development. *Growth Devel. Aging* **57**, 163–182.

Vogl, C., Atchley, W. R., and Xu, S. (1994). The ontogeny of morphological differences in the mandible in two inbred strains of mice. *J. Craniofac. Genet. Devel. Biol.* **14**, 97–110.

Vogt, C. (1842). *Untersuchungen über die Entwicklungsgeschichte der Geburtshelferkroete (Alytes obstetricans)*. Jent und Gassmann, Solothurn.

Vogt, T. F., and Duboule, D. (1999). Antagonists go out on a limb. *Cell* **99**, 563–566.

Volek-Smith, H., and Urist, M. R. (1996). Recombinant human bone morphogenetic protein (rhBMP) induced heterotopic bone development *in vivo* and *in vitro*. *Proc. Soc. Exp. Biol. Med.* **211**, 265–272.

Volk, S. W., and Leboy, P. S. (1999). Regulating the regulators of chondrocyte hypertrophy. *J. Bone Min. Res.* **14**, 483–486.

Vonau, R. L., Bostrom, M. P. G., Aspenberg, P., and Sams, A. E. (2001). Combination of growth factors inhibits ingrowth in the bone harvest chamber. *Clin. Orthop. Rel. Res.* **386**, 243–251.

von der Mark, H., von der Mark, K., and Gay, S. (1976). Study of differential collagen synthesis during development of the chick embryo by immunofluorescence. I. Preparation of collagen type I and type II specific antibodies and their application to early stages of the chick embryo. *Devel. Biol.* **48**, 237–249.

von der Mark, K. (1980). Immunological studies on collagen type transition in chondrogenesis. *Curr. Topics Devel. Biol.* **14**, 199–225.

von der Mark, K., and Bornstein, P. (1973). Characterization of the proα1 chain of procollagen: Isolation of a sequence unique to the precursor chain. *J. Biol. Chem.* **248**, 2285–2289.

von der Mark, K., Gauss, V., von der Mark, H., and Müller, P. (1977a). Relationship between cell shape and type of collagen synthesized as chondrocytes lose their cartilage phenotype in culture. *Nature* **267**, 531–532.

von der Mark, K., and von der Mark, H. (1977a). The role of three genetically distinct collagen types in endochondral ossification and calcification of cartilage. *J. Bone Joint Surg. Br. Vol.* **59**, 458–464.

von der Mark, K., and von der Mark, H. (1977b). Immunological and biochemical studies of collagen type transition during *in vitro* chondrogenesis of chick limb mesodermal cells. *J. Cell Biol.* **73**, 736–747.

von der Mark, K., von der Mark, H., and Gay, S. (1976). Study of differential collagen synthesis during development of the chick embryo by immunofluorescence. II. Localization of type I and type II collagen during long bone development. *Devel. Biol.* **53**, 153–170.

von der Mark, K., von der Mark, H., Timple, R., and Trelstad, R. L. (1977b). Immunofluorescent localization of collagen types I, II, and III in the embryonic chick eye. *Devel. Biol.* **59**, 75–85.

von Meyer, H. (1867). Die Architecteur der Spongiosa. *Arch. Anat. Physiol., Physiol. Abt.* **34**, 615–642.

Vorobyeva, E., and Hinchliffe, J. R. (1996). From fins to limbs: Developmental perspectives on paleontological and morphological evidence. *Evol. Biol.* **29**, 263–311.

Vorster, W. (1989). The development of the chondrocranium of *Gallus gallus*. *Adv. Anat. Embryol. Cell Biol.* **113**, 1–77.

Vortkamp, A. (1997). Defining the skeletal elements. *Curr. Biol.* **7**, R104–R107.

Vortkamp, A., Lee, K., Lanske, B., Segre, G. V., Kronenberg, H. M., and Tabin, C. J. (1996). Regulation of rate of cartilage differentiation by Indian hedgehog and PTH-related protein. *Science* **273**, 613–622.

Vukicevic, S., Helder, M. N., and Luyten, F. P. (1994b). Developing human lung and kidney are major sites for synthesis of bone morphogenetic protein-3 (osteogenin). *J. Histochem. Cytochem.* **42**, 869–875.

Vukicevic, S., Kleinman, H. K., Luyten, F. P., Roberts, A. B., Roche, N. S., and Reddi, A. H. (1992). Identification of multiple active growth factors in basement membrane Matrigel suggests caution in interpretation of cellular activity related to extracellular matrix components. *Exp. Cell Res.* **202**, 1–8.

Vukicevic, S., Latin, V., Chen, P., Batorsky, R., Reddi, A. H., and Sampath, T. K. (1994a). Localization of osteogenic protein 1 (bone morphogenetic protein-7) during human embryonic development: high affinity binding to basement membranes. *Biochem. Biophys. Res. Commun.* **198**, 693–700.

Vukicevic, S., Luyten, F. P., Kleinman, H. K., and Reddi, A. H. (1990b). Differentiation of canalicular cell processes in bone cells by basement membrane matrix components: regulation by discrete domains of laminin. *Cell* **63**, 437–445.

Vukicevic, S., Luyten, F. P., and Reddi, A. H. (1989). Stimulation of the expression of osteogenic and chondrogenic phenotypes *in vitro* by osteogenin. *Proc. Natl Acad. Sci. U.S.A.* **86**, 8793–8797.

Vukicevic, S., Paralkar, V. M., Cunningham, N. S., Gutkind, J. S., and Reddi, A. H. (1990a). Autoradiographic localization of osteogenin binding sites in cartilage and bone during rat embryonic development. *Devel. Biol.* **140**, 209–214.

Vukicevic, S., Paralkar, V. M., and Reddi, A. H. (1993). Extracellular matrix and bone morphogenetic proteins in cartilage and bone development and repair. *Adv. Mol. Cell Biol.* **6**, 207–224.

Vuust, J., Abildsten, D., and Lund, T. (1983). Control of type I collagen synthesis; Evidence for pretranslational coordination of proα1(I) and proα2(I) chain synthesis in embryonic chick bone. *Conn. Tissue Res.* **11**, 185–192.

Vynios, D. H., Aletras, A., Tsiganos, C. P., Tsegenidis, T., Antonopoulos, C. A., Hjerpe, A., and Engfeldt, B. (1985). Proteoglycans from squid cartilage: Extraction and characterization. *Comp. Biochem. Physiol. B.* **80**, 761–766.

Vynios, D. H., and Tsiganos, C. P. (1990). Squid proteoglycans: Isolation and characterization of three populations from cranial cartilage. *Biochim. Biophys. Acta* **1033**, 139–147.

W

Wachtler, F., Christ, B., and Jacob, H. J. (1981). On the determination of mesodermal tissues in the avian embryonic wing bud. *Anat. Embryol.* **161**, 283–290.

Wachtler, F., Christ, B., and Jacob, H. J. (1982). Grafting experiments on determination and migratory behaviour of presomitic, somitic and somatopleural cells in avian embryos. *Anat. Embryol.* **164**, 369–378.

Wada, N., Kawakami, Y., Ladher, R., Francis-West, P. H., and Nohno, T. (1999). Involvement of *Frzb-1* in mesenchymal condensation and cartilage differentiation in the chick limb bud. *Int. J. Devel. Biol.* **43**, 495–500.

Wada, N., Tanaka, H., Ide, H., and Nohno, T. (2003). Ephrin-A2 regulates position-specific cell affinity and is involved in cartilage morphogenesis in the chick limb bud. *Devel. Biol.* **264**, 550–563.

Wagemans, F., and Vandewalle, P. (1999). Development of the cartilaginous skull in *Solea solea*: trends in Pleuronectiformes. *Ann. Sci. Nat-Zool. Biol. Anim.* **20**, 39–52.

Wagemans, P. A. H. M., Van De Velde, J. P., and Kuijpers-Jagtman, A. M. (1988). Sutures and forces: A review. *Amer. J. Orthod.* **94**, 129–141.

Wagner, G. P., Booth, G., and Bagheri-Chaichian, H. (1997). A population genetic theory of canalization. *Evolution* **51**, 329–347.

Wagner, G. P., and Gauthier, J. A. (1999). 1, 2, 3 = 2, 3, 4: A solution to the problem of the homology of the digits in the avian hand. *Proc. Natl Acad. Sci. U.S.A.* **96**, 5111–5116.

Wagner, G. P., and Misof, B. Y. (1992). Evolutionary modification of regenerative capability in vertebrates: a comparative study on teleost pectoral fin regeneration. *J. Exp. Zool.* **261**, 62–78.

Wahba, G. M., Hostikka, S. L., and Carpenter, E. M. (2001). The paralogous Hox genes Hoxa10 and Hoxd10 interact to pattern the mouse hindlimb peripheral nervous system and skeleton. *Devel. Biol.* **231**, 87–102.

Wahl, L. M. (1971). Collagenolytic activity in pubic symphysis. *Anat. Rec.* **169**, 448.

Wahl, L. M., Wahl, S. M., Mergenhagen, S. E., and Martin, G. R. (1975). Collagenase production by lymphokine-activated macrophage. *Science* **187**, 261–263.

Wahl, M., Shukunami, C., Heinzmann, U., Hamajima, K., Hiraki, Y., and Imai, K. J. (2004). Transcriptome analysis of early chondrogenesis in ATDC5 cells induced by bone morphogenetic protein 4. *Genomics* **83**, 45–58.

Wake, D. B. (1976). On the correct scientific names of urodeles. *Differentiation* **6**, 195.

Wake, D. B. (1980). Evidence of heterochronic evolution: A nasal bone in the Olympic salamander, *Rhyacotriton olympicus*. *J. Herpetol.* **14**, 292–295.

Wake, D. B. (1997). Incipient species formation in salamanders of the *Ensatina* complex. *Proc. Natl Acad. Sci. U.S.A.* **94**, 7761–7767.

Wake, D. B., and Jockusch, E. L. (2000). Detecting species borders using diverse data sets: Examples from plethodontid salamanders in California. In *The Biology of Plethodontid Salamanders* (R. C. Bruce, R. G. Jaeger and L. D. Houck, eds), pp. 95–119. Kluwer Academic Publishers, New York, NY.

Wake, D. B., and Lawson, R. (1973). Developmental and adult morphology of the vertebral column in the plethodontid salamander *Eurycea bislineata*, with comments on vertebral evolution in the amphibia. *J. Morphol.* **139**, 251–300.

Wake, D. B., and Shubin, N. H. (1998). Limb development in the Pacific giant salamanders, *Dicamptodon* (Amphibia, Caudata, Dicamptodontidae). *Can. J. Zool.* **76**, 2058–2066.

Wake, D. B., and Wake, M. H. (1986). On the development of vertebrae in gymnophione amphibians. *Mém. Soc. Zool. France* **43**, 67–70.

Wake, M. H. (1980). Morphometrics of the skeleton of *Dermophis mexicanus* (Amphibia: Gymnophiona). Part 1. The vertebrae, with comparisons to other species. *J. Morphol.* **165**, 117–130.

Wake, M. H. (1985). The comparative morphology and evolution of the eyes of caecilians (Amphibia: Gymnophiona). *Zoomorphol.* **105**, 277–295.

Wake, M. H. (1989). Metamorphosis of the hyobranchial apparatus in *Epicrinops* (Amphibia:Gymnophiona: Rhinatrematidae): Replacement of bone by cartilage. *Ann. Sci. Nat. Zool.* **10**, 171–182.

Wake, M. H. (1998). Cartilage in the cloaca: Phallodeal spicules in caecilians (Amphibia: Gymnophiona). *J. Morphol.* **237**, 177–186.

Wake, M. H., Exbrayat, J.-M., and Delsol, M. (1985). The development of the chondrocranium of *Typhlonectes compressicaudus* (Gymnophiona), with comparison to other species. *J. Herpetol.* **19**, 68–77.

Wake, M. H., and Hanken, J. (1982). Development of the skull in *Dermophis mexicanus* (Amphibia: Gymnophiona), with comments on skull kinesis and amphibian relationships. *J. Morphol.* **173**, 203–223.

Wake, T. A., Wake, D. B., and Wake, M. H. (1983). The ossification sequence of *Aneides lugubris*, with comments on heterochrony. *J. Herpetol.* **17**, 10–22.

Walder, D. N. (1976). Aseptic necrosis of bone. In *Diving Medicine* (R. H. Strauss, ed.), pp. 97–108. Grune & Stratton, New York, NY.

Walker, D. G. (1972). Enzymatic and electron microscopic analysis of isolated osteoclasts. *Calcif. Tissue Res.* **9**, 296–309.

Walker, D. G. (1975a). Bone resorption restored in osteopetrotic mice by transplants of normal bone marrow and spleen cells. *Science* **190**, 784–785.

Walker, D. G. (1975b). Spleen cells transmit osteopetrosis. *Science* **190**, 785–787.

Walker, K. V. R., and Kember, N. F. (1972). Cell kinetics of growth cartilage in the rat tibia. I. Measurement in young male rats. *Cell Tissue Kinet.* **5**, 401–408.

Wall, N. A., Blessing, M., Wright, C. V. E., and Hogan, B. L. M. (1993). Biosynthesis and *in vivo* localization of the decapenta-plegic-Vg-related protein, DVR-6 (bone morphogenetic protein-6). *J. Cell Biol.* **120**, 493–502.

Wall, N. A., and Hogan, B. L. M. (1995). Expression of bone morphogenetic protein-4 (BMP-4), bone morphogenetic protein-7 (BMP-7), fibroblast growth factor-8 (FGF-8) and sonic hedgehog (SHH) during branchial arch development in the chick. *Mech. Devel.* **53**, 383–392.

Wallace, H. (1981). *Vertebrate Limb Regeneration*. John Wiley, Chichester.

Wallace, H., Clyman, R. I., and Pierro, L. P. (1969). Nucleic acid deficiency in the prothanic homozygous mutant of creeper fowl. *J. Exp. Zool.* **172**, 245–252.

Wallace, H., Maden, M., and Wallace, B. M. (1974). Participation of cartilage grafts in amphibian limb regeneration. *J. Embryol. Exp. Morphol.* **32**, 391–404.

Wallin, J., Wilting, J., Koseki, H., Fritsch, R., Christ, B., and Balling, R. (1994). The role of *Pax-1* in axial skeleton development. *Development* **120**, 1109–1121.

Wallis, G. A. (1993). Bone growth: Here today, bone tomorrow. *Curr. Biol.* **3**, 687–689.

Wallis, G. A. (1996). Bone growth: Coordinating chondrocyte differentiation. *Curr. Biol.* **6**, 1577–1580.

Walls, G. (1942). *The Vertebrate Eye and its Adaptive Radiation*. Hafner Publ. Co., New York, NY.

Walsh, A. J. L., Bradford, D. S., and Lotz, J. C. (2004). *In vivo* growth factor treatment of degenerated intervertebral discs. *Spine* **29**, 156–163.

Walsh, A. J. L., and Lotz, J. C. (2004). Biological response of the intervertebral disc to dynamic loading. *J. Biomech.* **37**, 329–337.

Walshe, J., and Mason, I. (2003). Fgf signalling is required for formation of cartilage in the head. *Devel. Biol.* **264**, 522–536.

Walter, L. R. (1985). The formation of secondary centers of ossification in kannemeyeriid dicynodonts. *J. Paleont.* **59**, 1486–1488.

Walter, Ph. von (1821). Wiedereinheilung der bei der Trapanation ausgebohrten Knochenscheibe. *J. Chir. Augen Heilkunde* **2**, 571.

Walton, A. G. (1974). Molecular aspects of calcified tissue. *J. Biomed. Mater. Res.* **8**, 409–426.

Wang, B., Fallon, J. F., and Beachy, P. A. (2000). Hedgehog-regulated processing of Gli3 produces an anterior/posterior repressor gradient in the developing vertebrate limb. *Cell* **100**, 423–434.

Wang, E. A. (1993). Bone morphogenetic proteins (BMPs): Therapeutic potential in healing bone defects. *Trends Biotechnol.* **11**, 379–383.

Wang, E. A., Gerhart, T. N., and Toriumi, D. M. (1992). Bone morphogenetic proteins and development. In *Chemistry and Biology of Mineralized Tissues* (H. C. Slavkin and P. Price, eds), pp. 351–359. Elsevier Science Publishers, BV, Amsterdam.

Wang, E. A., Israel, D. I., Kelly, S., and Luxenberg, D. P. (1993). Bone morphogenetic protein-2 causes commitment and differentiation of C3H10T1/2 [C3H10T12] and 3T3 cells. *Growth Factors* **9**, 57–71.

Wang, E. A., Rosen, V., D'Alessandro, J. S., Bauduy, M., Cordes, P., Harada, T., Israel, D. I., Hewick, R. M., Kerns, K. M., LaPan, P., Luxenberg, D. P., McQuaid, D., Moutsatsos, I. K., Nove, J., and Wozney, J. M. (1990). Recombinant human bone morphogenetic protein induces bone formation. *Proc. Natl Acad. Sci. U.S.A.* **87**, 2220–2224.

Wang, N., Butler, J. P., and Ingber, D. E. (1993). Mechanotransduction across the cell surface and through the cytoskeleton. *Science* **260**, 1124–1127.

Wang, Q., Green, R. P., Zhao, G., and Ornitz, D. M. (2001). Differential regulation of endochondral bone growth and joint development by FGFR1 and FGFR3 tyrosine kinase domains. *Development* **128**, 3867–3876.

Wang, Y., Hu, Y., Meng, J., and Li, C. (2001). An ossified Meckel's cartilage in two Cretaceous mammals and origin of the mammalian middle ear. *Science* **294**, 357–361.

Wang, Y., and Sassoon, D. (1995). Ectoderm–mesenchyme and mesenchyme–mesenchyme interactions regulate Msx-1 expression and cellular differentiation in the murine limb bud. *Devel. Biol.* **168**, 374–382.

Wang, Y., Spatz, M. K., Kannan, K., Hayk, H., Avivi, A., Gorivodsky, M., Pines, M., Yayon, A., Lonai, P., and Givol, D. (1999). A mouse model for achondroplasia produced by targeting fibroblast growth factor receptor 3. *Proc. Natl Acad. Sci. U.S.A.* **96**, 4455–4460.

Wang, Y.-H., Rutherford, B., Upholt, W. B., and Mina, M. (1999). Effects of BMP-7 on mouse tooth mesenchyme and chick mandibular mesenchyme. *Devel. Dynam.* **216**, 320–335.

Wang, Z.-Q., Grigoriadis, A. E., Möhle-Steinlein, U., and Wagner, E. F. (1991). A novel target cell for *c-fos*-induced oncogenesis: development of chondrogenic tumours in embryonic stem cell chimeras. *EMBO J.* **10**, 2437–2450.

Wang, Z.-Q., Ovitt, C., Grigoriadis, A. E., Möhle-Steinlein, U., Rüther, U., and Wagner, E. F. (1992). Bone and haematopoietic defects in mice lacking *c-fos*. *Nature* **360**, 741–745.

Ward, F. O. (1838). *Outlines of Human Osteology*. Henry Renslaw, London.

Warrell, E., and Taylor, J. F. (1979). The role of periosteal tension in the growth of long bones. *J. Anat.* **128**, 179–184.

Warheit, K. I., Good, D. A., and de Queiroz, K. (1989). Variation in numbers of scleral ossicles and their phylogenetic transformations within the Pelicaniformes. *The Auk* **106**, 383–388.

Warshafsky, B., Aubin, J. E., and Heersche, J. N. M. (1985). Cytoskeletal rearrangements during calcitonin-induced changes in osteoclast motility in vitro. *Bone* **6**, 179–185.

Warshawsky, H., Goltzman, D., Rouleau, M. F., and Bergeron, J. J. M. (1980). Direct in vivo demonstration by radioautography of specific binding sites for calcitonin in skeletal and renal tissues of the rat. *J. Cell Biol.* **85**, 682–694.

Washburn, S. L. (1947). The relation of the temporal muscle to the form of the skull. *Anat. Rec.* **99**, 239–248.

Watabe, N. (1965). Studies on shell formation. XI. Crystal-matrix relationships in the inner layers of Mollusk shells. *J. Ultrastruct. Res.* **12**, 351–370.

Watanabe, K., Bruder, S. P., and Caplan, A. I. (1994). Transient expression of type II collagen and tissue mobilization during development of the scleral ossicle, a membranous bone, in the chick embryo. *Devel. Dynam.* **200**, 212–226.

Watanabe, K., and Imura, K. (1983). Significance of the egg shell in the development of the chick embryo: A study using shell-less culture. *Zool. Mag.* **92**, 64–72.

Watanabe, K., Yagi, K., Ohya, Y., and Kimata, K. (1992). Scleral fibroblasts of the chick embryo differentiate into chondrocytes in soft-agar culture. *In Vitro Cell. Devel. Biol.* **28**, 603–608.

Watanabe, M., and Okada, T. S. (1975). The relationship between cell-substrate adhesiveness and cell growth. A study on chondrocytes cultured *in vitro* with conditioned medium. *Devel., Growth & Differ.* **17**, 51–60.

Watanabe, Y., Duprez, D., Mansoro-Burq, A.-H., Vincent, C., and Le Douarin, N. M. (1998a). Two domains in vertebral development: antagonistic regulation by SHH and BMP4 proteins. *Development* **125**, 2631–2639.

Watanabe, Y., and Le Douarin, N. M. (1996). A role for BMP-4 in the development of subcutaneous cartilage. *Mech. Devel.* **57**, 69–78.

Watanabe, Y., Nonaka, K., Sasaki, Y., and Nakata, M. (1998b). A longitudinal observation of the postnatal craniofacial growth in artificial monozygotic twin mice. *J. Craniofac. Genet. Devel. Biol.* **18**, 107–118.

Watson, A. G., and Bonde, R. K. (1986). Congenital malformations of the flipper in three West Indian manatees, *Trichechus manatus*, and a proposed mechanism for development of ectrodactyly and cleft hand in mammals. *Clin. Orthop. Rel. Res.* **202**, 294–301.

Watson, J., DeHaas, W. G., and Hauser, S. S. (1975). Effect of electric fields on growth rate of embryonic chick tibiae in vitro. *Nature* **254**, 331–332.

Watterson, R. L. (1952). Neural tube extirpation in *Fundulus heteroclitus* and resultant neural arch defects. *Biol. Bull.* **103**, 310.

Watterson, R. L., Fowler, I., and Fowler, B. J. (1954). The role of the neural tube and notochord in development of the axial skeleton of the chick. *Amer. J. Anat.* **95**, 337–399.

Watts, E. S. (1990a). A comparative study of neonatal skeletal development in *Cebus* and other primates. *Folia Primatol.* **54**, 217–224.

Watts, E. S. (1990b). Evolutionary trends in primate growth and development. In *Primate Life History and Evolution* (C. J. de Rousseau, ed.), pp. 89–104, Wiley-Liss, New York, NY.

Wawersik, S., Purcell, P., Rauchman, M., Dudley, A. T., Robertson, E. J., and Maas, R. (1999). BMP7 acts in murine lens placode development. *Devel. Biol.* **207**, 176–188.

Wayne, R. K. (1986a). Cranial morphology of domestic and wild canids: The influence of development on morphological change. *Evolution* **40**, 243–261.

Wayne, R. K. (1986b). Limb morphology of domestic and wild canids: The influence of development on morphological change. *J. Morphol.* **187**, 301–319.

Wayne, R. K., and Ruff, C. B. (1993). Domestication and bone growth. In *Bone, Volume 7: Bone Growth – B* (B. K. Hall, ed.), pp. 101–131. CRC Press, Boca Raton, FL.

Weatherbee, S. D., and Carroll, S. B. (1999). Selector genes and limb identity in arthropods and vertebrates. *Cell* **97**, 283–286.

Weaver, V. M., and Roskelley, C. D. (1997). Extracellular matrix: the central regulator of cell and tissue homeostasis. *Trends Cell Biol.* **7**, 40–42.

Webb, J. F. (1989a). Developmental constraints and evolution of the lateral line system in teleost fishes. In *The Mechanosensory Lateral Line: Neurobiology and Evolution* (S. Coombs, P. Görner, and H. Münz, eds), pp. 79–97. Springer-Verlag, New York, NY.

Webb. J. F. (1989b). Gross morphology and evolution of the mechanoreceptive lateral-line system in teleost fishes. *Brain, Behav. Evol.* **33**, 34–53.

Webb. J. F. (1989c). Neuromast morphology and lateral line trunk canal ontogeny in two species of cichlids: An SEM study. *J. Morphol.* **202**, 53–68.

Webb, J. F., and Noden, D. M. (1993). Ectodermal placodes: Contributions to the development of the vertebrate head. *Amer. Zool.* **33**, 434–447.

Webb, J. F., and Shirey, J. E. (2003). Postembryonic development of the cranial lateral line canals and neuromasts in zebrafish. *Devel. Dynam.* **228**, 370–385.

Webber, R. J., Malemud, C. J., and Sokoloff, L. (1977). Species differences in cell culture of mammalian articular chondrocytes. *Calcif. Tissue Res.* **23**, 61–66.

Webster, D. A., and Gross, J. (1970). Studies on possible mechanisms of programmed cell death in the chick embryo. *Devel. Biol.* **22**, 157–184.

Webster, M. K., and Donoghue, D. J. (1997). FGFR activation in skeletal disorders: too much of a good thing. *Trends Genet.* **13**, 178–182.

Wedden, S. E., Lewin-Smith, M. R., and Tickle, C. (1986). The patterns of chondrogenesis of cells from facial primordia of chick embryos in micromass culture. *Devel. Biol.* **117**, 71–82.

Wedlock, W. E., and McCallion, D. J. (1969). Induction of scleral cartilage in the chorioallantois of the chick embryo. *Can. J. Zool.* **47**, 142–143.

Wegner, G. (1872). Myeloplaxen und Knochenresorption. *Virchows Arch. Anat. Pathol.* **56**, 523.

Wei, X., and Messner, K. (1996). The postnatal development of the insertions of the medial collateral ligament in the rat knee. *Anat. Embryol.* **193**, 53–59.

Weijs, W. A., and Hillen, B. (1984). Relationships between masticatory muscle cross-section and skull shape. *J. Dent. Res.* **63**, 1154–1157.

Weijs, W. A., and Hillen, B. (1986). Correlations between the cross-sectional area of the jaw muscles and craniofacial size and shape. *Amer. J. Phys. Anthropol.* **70**, 423–432.

Weiner, S., and Traub, W. (1992). Bone structure: From ångstrom to microns. *FASEB J.* **6**, 879–885.

Weinger, J. M., and Holtrop, M. E. (1974). An ultrastructural study of bone cells: The occurrence of microtubules, microfilaments and tight junctions. *Calcif. Tissue Res.* **14**, 15–30.

Weinreb, M., Shinar, D., and Rodan, G. A. (1990). Different pattern of alkaline phosphatase, osteopontin, and osteocalcin expression in developing rat bone visualized by *in situ* hybridization. *J. Bone Min. Res.* **5**, 831–842.

Weis, J. S. (1972). The effect of nerve growth factor on fin regeneration in the goldfish, *Carassius auratus*. *Growth* **36**, 155–162.

Weisel, G. F. (1967). Early ossification in the skeleton of the sucker (*Catostomus macrocheilus*) and the guppy (*Poecilia reticulata*). *J. Morphol.* **121**, 1–18.

Weiss, A., Livne, E., von der Mark, K., Heinegard, D., and Silbermann, M. (1988). Growth and repair of cartilage: Organ culture system utilizing chondroprogenitor cells of condylar cartilage in newborn mice. *J. Bone Min. Res.* **3**, 93–100.

Weiss, A., von der Mark, K., and Silbermann, M. (1986). A tissue culture system supporting cartilage cell differentiation, extracellular mineralization, and subsequent bone formation, using mouse condylar progenitor cells. *Cell Differ.* **19**, 103–113.

Weiss, K. M., and Buchanan, A. V. (2004). *Genetics and the Logic of Evolution*. John Wiley and Sons, Hoboken, NJ.

Weiss, P. (1925). Unabhängigkeit der Extremitäten-regeneration vom Skellet (bei *Triton cristatus*). *Wilhelm Roux Arch. Entwicklungsmech. Org.* **104**, 359–394.

Weiss, P., and Amprino, R. (1940). The effect of mechanical stress on the differentiation of scleral cartilage *in vitro* and in the embryo. *Growth* **4**, 245–258.

Weiss, P., and Moscona, A. A. (1958). Type-specific morphogenesis of cartilages developed from dissociated limb and scleral mesenchyme *in vitro*. *J. Embryol. Exp. Morphol.* **6**, 238–246.

Weiss, R. E., and Reddi, A. H. (1980). Synthesis and localization of fibronectin during collagenous matrix-mesenchymal cell interaction and differentiation of cartilage and bone *in vivo*. *Proc. Natl Acad. Sci. U.S.A.* **77**, 2074–2078.

Weiss, R. E., and Reddi, A. H. (1981a). Appearance of fibronectin during the differentiation of cartilage, bone and bone marrow. *J. Cell Biol.* **88**, 630–636.

Weiss, R. E., and Reddi, A. H. (1981b). Role of fibronectin in collagenous matrix-induced mesenchymal cell proliferation and differentiation *in vivo*. *Exp. Cell Res.* **133**, 247–254.

Weiss, R. E., Reddi, A. H., and Nimni, M. E. (1981). Somatostatin can locally inhibit proliferation and differentiation of cartilage and bone precursor cells. *Calcif. Tissue int.* **33**, 425–430.

Weiss, R. E., and Watabe, N. (1978a). Studies on the biology of fish bone. I. Bone resorption after scale removal. *Comp. Biochem. Physiol. (A)* **60**, 207–211.

Weiss, R. E., and Watabe, N. (1978b). Studies on the biology of fish bone. II. Bone matrix changes during resorption. *Comp. Biochem. Physiol. (A)* **61**, 245–252.

Weiss, R. E., and Watabe, N. (1979). Studies on the biology of fish bone. III. Ultrastructure of osteogenesis and resorption in osteocytic (cellular) and anosteocytic (acellular) bones. *Calcif. Tissue Int.* **28**, 43–56.

Wellik, D. M., and Capecchi, M. R. (2003). *Hox10* and *Hox11* genes are required to globally pattern the mammalian skeleton. *Science* **301**, 363–367.

Wells, C. (1973). The paleopathology of bone disease. *Practitioner* **210**, 384–391.

Wendling, O., Dennefeld, C., Chambon, P., and Mark, M. (2000). Retinoid signaling is essential for patterning the endoderm of the third and fourth pharyngeal arches. *Development* **127**, 1553–1562.

Werb, Z., and Chin, J. R. (1998). Extracellular matrix remodeling during morphogenesis. *Ann. N. Y. Acad. Sci.* **857**, 110–118.

Wergedal, J. E., Lau, K. H. M., and Baylink, D. J. (1988). Fluoride and bovine bone extract influence cell proliferation and phosphatase activities in human bone cell cultures. *Clin. Orthop. Rel. Res.* **233**, 274–282.

Wergedal, J. E., Mohan, S., Lundy, M., and Baylink, D. J. (1990). Skeletal growth factor and other growth factors known to be present in bone matrix stimulate proliferation and protein synthesis in human bone cells. *J. Bone Min. Res.* **5**, 179–186.

Werner, Y. L. (1967). Regeneration of the caudal axial skeleton in a Gekkonid lizard (*Hemidactlyus*) with particular reference to the 'latent' period. *Acta Zool. (Stockholm)* **48**, 103–126.

Wertheim, M. G. (1847). Memoire sur l'elasticité et la Cohesion des Principaux Tissues du Corps Humain. *Ann. Chim. Phys.* **21**, 385.

West, C. M., Lanza, R., Rosenbloom, J., Lowe, M., Holtzer, H., and Audalovic, N. (1979). Fibronectin alters the phenotypic properties of cultured chick embryo chondroblasts. *Cell* **17**, 491–502.

Westbroek, P., and Marin, F. (1998). A marriage of bone and nacre. *Nature* **392**, 861–862.

Westoll, T. S. (1941). Latero-sensory canals and dermal bones. *Nature* **148**, 168.

Weston, A. D., Rosen, V., Chandraratna, R. A. S., and Underhill, T. M. (2000). Regulation of skeletal progenitor differentiation by the BMP and retinoid signaling pathways. *J. Cell Biol.* **148**, 679–690.

Weston, E. M. (2003). Evolution of ontogeny in the hippopotamus skull; using allometry to dissect developmental change. *Biol. J. Linn. Soc.* **80**, 625–638.

Weston, J. A. (1963). A radioautographic analysis of the migration and localization of trunk neural crest cells in the chick. *Devel. Biol.* **6**, 279–310.

Weston, J. A. (1970). The migration and differentiation of neural crest cells. *Adv. Morphog.* **8**, 41–114.

Weston, J. A., and Butler, S. L. (1966). Temporal factors affecting localization of neural crest cells in the chicken embryo. *Devel. Biol.* **14**, 246–266.

Wezeman, F. H. (1998). Morphological foundations of precartilage development in mesenchyme. *Microsc. Res. Tech.* **43**, 91–101.

Wezeman, F. H., and Childs, G. V. (1982). Ultrastructural immunohistochemical localization of anti-invasion factor (AIF) in bovine cartilage matrix. *J. Histochem. Cytochem.* **30**, 524–531.

Wharton, K. A., Thomsen, G. H., and Gelbart, W. M. (1991). *Drosophila* 60A gene, another transforming growth factor β family member, is closely related to human bone morphogenetic proteins. *Proc. Natl Acad. Sci. U.S.A.* **88**, 9214–9218.

Wheatley, S. C., Isacke, C. M., and Crossley, P. H. (1993). Restricted expression of the hyaluronan receptor, CD44, during postimplantation mouse embryogenesis suggests key roles in tissue formation and patterning. *Development* **119**, 295–306.

Whedon, G. D. (1986). The consequences of space flight. *Bone. Clin. Biochem. News Rep.* **3**, 6–7.

Whedon, G. D., and Heaney, R. P. (1993). Effects of physical inactivity, paralysis, and weightlessness on bone growth. In *Bone, Volume 7: Bone Growth – B* (B. K. Hall, ed.), pp. 1–36. CRC Press, Boca Raton, FL.

Whedon, G. D., Lutwak, L., Rambaut, P., Whittle, M., Leach, C., Reid, J., and Smith, M. (1976). Effect of weightlessness on mineral metabolism; metabolic studies on Skylab orbital space flights. *Calcif. Tissue Res.* **21** (Suppl), 423–430.

White, A. A., III. (1975). Fracture treatment. The still unsolved problem. *Clin. Orthop. Rel. Res.* **106**, 279–284.

White, D. G., Hershey, H. P., Moss, J. A., Daniels, H., Tuan, R. S., and Bennett, V. D. (2003). Functional analysis of fibronectin isoforms in chondrogenesis: Full-length recombinant mesenchymal fibronectin reduces spreading and promotes condensation and chondrogenesis of limb mesenchymal cells. *Differentiation* **71**, 251–261.

White, J. A., Boffa, M. B., Jones, B., and Petkovich, M. (1994). A zebrafish retinoic acid receptor expressed in the regenerating caudal tail. *Development* **120**, 1861–1872.

White, P. H., Farkas, D. R., McFadden, E. E., and Chapman, D. L. (2003). Defective somite patterning in mouse embryos with reduced levels of *Tbx6*. *Development* **130**, 1681–1690.

Whitehead, P. E., and McEwan, E. H. (1973). Seasonal variation in the plasma testosterone concentration of reindeer and caribou. *Can. J. Zool.* **51**, 651–658.

Whitfield, G. K., Dang, H. T. J., Schulter, S. F., Berenstein, R. M., Bunag, T., Manzon, L. A., Hsieh, G., Encinas-Dominguez, C., Youson, J. H., Haussler, M. R., and Marchalonis, J. J. (2003). Cloning of a functional vitamin D receptor from the lamprey (*Petromyzon marinus*), an ancient vertebrate lacking a calcified skeleton and teeth. *Endocrynology* **144**, 2704–2716.

Whitfield, J. F., Rixon, R. H., MacManus, J. P., and Balk, S. D. (1973). Calcium, cyclic adenosine 3',5'-monophosphate, and the control of cell proliferation: A review. *In Vitro* **8**, 257–278.

Whitten, J. M. (1969). Cell death during early morphogenesis: Parallels between insect limb and vertebrate limb development. *Science* **163**, 1456–1457.

Whyte, J. R., González, L., Cisneros, A. I., Yus, C., Torres, A., and Sarrat, R. (2002). Fetal development of the human tympanic ossicular chair articulations. *Cells Tissues & Organs* **171**, 241–249.

Wible, J. R., and Bhatnagar, K. P. (1996). Chiropteran vomeronasal complex and the interfamilial relationships of bats. *J. Mammal. Evol.* **3**, 285–314.

Wible, J. R., and Novacek, M. J. (1988). Cranial evidence for the monophyletic origin of bats. *Amer. Mus. Novitatis* # **2911**, 1–19.

Widelitz, R. B., Jiang, T.-X., Murray, B. A., and Chuong, C.-M. (1993). Adhesion molecules in skeletogenesis. II. Neural cell adhesion molecules mediate precartilaginous mesenchymal condensations and enhance chondrogenesis. *J. Cell Physiol.* **156**, 399–411.

Wiebkin, O. W., and Muir, H. (1973). Factors affecting the biosynthesis of sulphated glycosaminoglycans by chondrocytes in short-term maintenance culture isolated from adult tissue. In *Biology of Fibroblast* (E. Kulonen and J. Pikkarairen, eds), pp. 231–236. Academic Press, New York, NY.

Wiebkin, O. W., and Muir, H. (1975). The effect of hyaluronic acid on proteoglycan synthesis and secretion by chondrocytes of adult cartilage. *Philos. Trans. R. Soc. London, Ser. B* **271**, 283–292.

Wiebkin, O. W., and Muir, H. (1977). Synthesis of cartilage-specific proteoglycan by suspension cultures of adult chondrocytes. *Biochem. J.* **164**, 269–272.

Wiens, J. J. (1989). Ontogeny of the skeleton of *Spea bombifrons* (Anura: Pelobatidae). *J. Morphol.* **202**, 29–51.

Wiens, J. J., and Slingluff, J. L. (2001). How lizards turn into snakes: A phylogenetic analysis of body-form evolution in anguid lizards. *Evolution* **55**, 2303–2318.

Wientroub, S., and Reddi, A. H. (1988). Influence of irradiation on the osteoinductive potential of demineralized bone matrix. *Calcif. Tissue Int.* **42**, 255–260.

Wientroub, S., Wahl, L. M., Feverstein, N., Winter, C. C., and Reddi, A. H. (1983). Changes in tissue concentration of prostaglandins during endochondral bone differentiation. *Biochem. Biophys. Res. Commun.* **117**, 746–750.

Wieslander, J., and Heinegard, D. (1981). Immunochemical analysis of cartilage proteoglycans. Cross-reactivity of molecules isolated from different species. *Biochem. J.* **199**, 81–87.

Wiffen, J., de Buffrénil, V., de Ricqlès, A. J., and Maxin, J.-M. (1995). Ontogenetic evolution of bone structure in Late Cretaceous Plesiosauria from New Zealand. *Geobios* **28**, 625–640.

Wight, P. A. L., and Duff, S. R. I. (1985). Ectopic pulmonary cartilage and bone in domestic fowl. *Res. Vet. Sci.* **39**, 188–195.

Wight, T. N., and Mecham, R. P. (eds) (1987). *Biology of Proteoglycans*. Academic Press, Orlando, FL.

Wigley, C. B. (1975). Differentiated cells *in vitro*. *Differentiation* **4**, 25–55.

Wika, M. (1980). On growth of reindeer antlers. In *Proceedings of the Second International Reindeer/Caribou Symposium, Køros, Norway, 1979* (E. Reimers, E Gaare., and S. Skjenneberg, eds), pp. 416–421. Direktoratet for Vilt og Ferskvannsfish, Trondheim.

Wika, M. (1982a). Antlers – a mineral source in *Rangifer*. *Acta Zool.* (*Stockholm*) **63**, 7–10.

Wika, M. (1982b). Foetal stages of antler development. *Acta Zool.* (*Stockholm*) **63**, 187–189.

Wika, M., and Krog, J. (1980). Antler 'disposable vascular bed.' In *Proceedings of the Second International Reindeer/Caribou Symposium, Køros, Norway, 1979* (E. Reimers, E. Gaare., and S. Skjenneberg, eds), pp. 422–424. Direktoratet for Vilt og Ferskvannsfish, Trondheim.

Wikström, B., Gay, R., Hjerpe, A., Mengarelli, S., Reinholt, F. P., and Engfeldt, B. (1984a). Morphological studies of the epiphyseal growth zone in the brachymorphic (*bm/bm*) mouse. *Virchows Arch. B Cell Pathol.* **47**, 167–176.

Wikström, B., Hjerpe, A., Hultenby, K., Reinholt, F. P., and Engfeldt, B. (1984b). Stereological analysis of the epiphyseal growth cartilage in the brachymorphic (*bm/bm*) mouse, with special reference to the distribution of matrix vesicles. *Virchows Arch. B Cell Pathol.* **47**, 199–218.

Wiktor-Jedrzejczak, W., Ahmed, A., Szczlik, C., and Skelly, R. R. (1982). Hematological characterization of congenital osteopetrosis in *op/op* mouse. Possible mechanism for abnormal macrophage differentiation. *J. Exp. Med.* **156**, 1516–1527.

Wilby, O. K., and Ede, D. A. (1975). A model generating the pattern of cartilage skeletal elements in the embryonic chick limb. *J. Theor. Biol.* **52**, 199–218.

Wild, E. R. (1997). Description of the adult skeleton and developmental osteology of the hyperossified horned frog, *Ceratophrys cornuta* (Anura, Leptodactylidae). *J. Morphol.* **232**, 169–206.

Wild, E. R. (1999). Description of the chondrocranium and osteogenesis of the Chacoan burrowing frog, *Chacophrys pierotti* (Anura, Leptodactylidae). *J. Morphol.* **242**, 29–246.

Wild, R. (1973). Die Triasfauna der Tessiner Kalkalpen. XXIII. *Tanystropheus longobardicus* (Bassani) Neue Ergebnisse. *Schweiz. Paläont. Abh.* **95**, 1–160.

Wilde, S. M., Wedden, S. E., and Tickle, C. (1987). Retinoids programme pre-bud mesenchyme to give changes in limb pattern. *Development* **100**, 723–733.

Wildermann, B., Schmidmaier, G., Ordel, S., Strange, R., Haas, N. P., and Raschke, M. (2003). Cell proliferation and differentiation during fracture healing are influenced by locally applied IGF-I and TGF-β1; Comparison of two proliferation markers, PCNA and BrdU. *J. Biomed. Mater. Res. Part B. Appl. Biomater.* **65B**, 150–156.

Wilkie, A. O. M., and Morriss-Kay, G. M. (2001). Genetics of craniofacial development and malformation. *Nature Revs* **2**, 458–467.

Wilkie, A. O. M., Oldridge, M., Tang, Z., and Maxson, R. E., Jr. (2001). Craniosynostosis and related limb anomalies. In *The Molecular Basis of Skeletogenesis* (G. Cardew and J. Goode, eds; B. K. Hall, chair), Novartis Foundation Symposium 232, pp. 122–145. John Wiley & Sons, Chichester.

Wilkins, A. S. (2002). *The Evolution of Developmental Pathways.* Sinauer Associates, Inc., Sunderland, MA.

Wilkinson, R. G. (1975). Techniques of ancient skull surgery. *Nat. Hist.* **84** (8), 94–101.

Williams, J. L. (1942). The development of cervical vertebrae in the chick under normal and experimental conditions. *Amer. J. Anat.* **71**, 153–177.

Williams, L. W. (1910). The somites of the chick. *Amer. J. Anat.* **11**, 55–100.

Williams-Ashman, H. G., and Reddi, A. H. (1991). Differentiation of mesenchymal tissues during phallic morphogenesis with emphasis on the os penis: Roles of androgens and other regulatory agents. *J. Steroid Biochem. Mol. Biol.* **39**, 873–881.

Williams-Boyce, P. K., and Daniel, J. C., Jr. (1980). Regeneration of rabbit ear tissue. *J. Exp. Zool.* **212**, 243–253.

Williams-Boyce, P. K., and Daniel, J. C., Jr. (1986). Comparison of ear tissue regeneration in mammals. *J. Anat.* **149**, 55–63.

Willis, R. A. (1962). *The Borderland of Embryology and Pathology.* Butterworth, London.

Willmer, E. N. (1935). *Tissue Culture.* Methuen & Co., London.

Willmer, E. N. (1951). Some aspects of evolutionary cytology. In *Cytology and Cell Physiology* (G. H. Bourne, ed.), 2nd Edn. Oxford University Press, Oxford.

Willmer, E. N. (ed.). (1965). *Cells and Tissues in Cultures. Methods, Biology and Physiology.* Two vols. Academic Press, New York, NY.

Willmer, E. N. (1970). *Cytology and Evolution,* 2nd Edn. Academic Press, New York, NY.

Wilsman, N. J., Farnum, C. E., Leiferman, E. M., Fry, M., and Barreto, C. (1996). Differential growth by growth plates as a function of multiple paramaters of chondrocytic kinetics. *J. Orthop. Res.* **14**, 927–936.

Wilsman, N. J., and Fletcher, T. F. (1978). Cilia of neonatal articular chondrocytes. Incidence and morphology. *Anat. Rec.* **190**, 871–890.

Wilson, D. B., and Wyatt, D. P. (1995). Alterations in cranial morphogenesis in the *Lp* mutant mouse. *J. Craniofac. Genet. Devel. Biol.* **15**, 182–189.

Wilson, D. J., and Hinchliffe, R. J. (1985). Experimental analysis of the role of the ZPA in the development of the wing buds of wingless (*ws*) mutant embryos. *J. Embryol. Exp. Morphol.* **85**, 271–283.

Wilson, G., and Geikie, A. (1861). *Memoir of Edward Forbes, F.R.S.* Macmillan, Cambridge and London; Edmonston and Douglas, Edinburgh.

Wilson, M. V. H., and Caldwell, M. W. (1993). New Silurian and Devonian forktailed 'thelodonts' are jawless vertebrates with stomachs and deep bodies. *Nature* **361**, 442–444.

Wilting, J., Christ, B., and Bokeloh, M. (1991). A modified chorioallantoic membrane (CAM) assay for qualitative and quantitative study of growth factors. *Anat. Embryol.* **183**, 259–271.

Wilting, J., Ebensperger, C., Müller, T. S., Koseki, H., Wallin, J., and Christ, B. (1995). *Pax-1* in the development of the cervicooccipital transitional zone. *Anat. Embryol.* **192**, 221–227.

Winnard, R. G., Gerstenfeld, L. C., Toma, C. D., and Franceschi, R. T. (1995). Fibronectin gene expression, synthesis, and accumulation during in vitro differentiation of chicken osteoblasts. *J. Bone Min. Res.* **10**, 1969–1977.

Winograd, J., Reilly, M. P., Roe, R., Lutz, J., Laughner, E., Xu, X., Hu, L., Asakura, T., Vanderkolk, C., Strandberg, J. D., and Semenza, G. L. (1997). Perinatal lethality and multiple craniofacial malformations in Msx2 transgenic mice. *Human Mol. Genet.* **6**, 369–379.

Winter, G. D. (1971). Heterotopic bone induced by synthetic sponge implants. *Israel J. Med. Sci.* **7**, 433.

Winter, G. D., and Simpson, B. J. (1969). Heterotopic bone formation in a synthetic sponge in the skin of young pigs. *Nature* **223**, 88–90.

Winterburn, P. J., and Phelps, C. F. (1970). Relevance of feedback inhibition applied to the biosynthesis of hexosamines. *Nature* **228**, 1311–1312.

Wislocki, G. B., Weatherford, H. L., and Singer, M. (1947). Osteogenesis of antlers investigated by histological and histochemical methods. *Anat. Rec.* **99**, 265–284.

Witten, P. E. (1997). Enzyme histochemical characteristics of osteoblasts and mononucleated osteoclasts in a teleost fish with acellular bone (*Oreochromis niloticus*, Cichlidae). *Cell Tissue Res.* **287**, 591–599.

Witten, P. E., Gil-Martens, L., Hall, B. K., Huysseune, A., and Obach, A. (2005). Compressed vertebrae in Atlantic salmon (*Salmo salar*) Evidence for metaplastic chondrogenesis as a skeletogenic response late in ontogeny. *Diseases Aquat. Organisms* (in press).

Witten, P. E., and Hall, B. K. (2001). The salmon Kype. *Atlantic Salmon J.* **50**, 36–39.

Witten, P. E., and Hall, B. K. (2002). Differentiation and growth of the kype skeletal tissue in anadromous male Atlantic salmon (*Salmo salar*). *Int. J. Devel. Biol.* **46**, 719–730.

Witten, P. E., and Hall, B. K. (2003). Seasonal change in the lower jaw skeleton in male Atlantic salmon (*Salmo salar* L.): remodelling and regression of the kype after spawning. *J. Anat.* **203**, 435–450.

Witten, P. E., Holliday, L. S., Delling, G., and Hall, B. K. (1999b). Immunohistochemical identification of a vacuolar proton pump (V-ATPase) in bone-resorbing cells of an advanced teleost species (*Oreochromis niloticus*, Teleostei, Cichlidae). *J. Fish Biol.* **55**, 1258–1272.

Witten, P. E., Huysseune, A., Franz-Odendaal, T., Fedak, T. J., Vickaryous, M., Cole, A. G., and Hall, B. K. (2004). Acellular teleost bone: dead or alive, primitive or derived? *Palaeont. Assoc. Newsletter* **55**, 37–41.

Witten, P. E., and Villwock, W. (1997). Growth requires bone resorption at particular skeletal elements in a teleost fish with acellular bone (*Oreochromis niloticus*, Teleostei, Cichlidae). *J. Appl. Ichthyol.* **13**, 149–158.

Witten, P. E., Villwock, W., Peter, H., and Hall, B. K. (2000). Bone resorbing and bone remodeling cells in juvenile carp (*Cyprinus carpio*). *J. Appl. Ichthyol.* **16**, 254–261.

Witten, P. E., Villwock, W., and Watermann, B. (1999a). About the use of radiological, histological, and enzyme histochemical procedures to visualize bone resorption at developing and at pathological altered skeletons of advanced teleosts. In *Krankheiten der Aquatischen Organismen* (ed. H. Wedekind), pp. 188–200. Inst. für Binnenfischerei Postdam, Sacrow e.v., Groβ Glimicke.

Wlodarski, K. H. (1969). The inductive properties of epithelial established cell lines. *Exp. Cell Res.* **57**, 446–448.

Wlodarski, K.H. (1985). Orthotopic and ectopic chondrogenesis and osteogenesis by neoplastic cells. *Clin. Orthop. Rel. Res.* **200**, 248–265.

Wlodarski, K. H. (1989). Normal and heterotopic periosteum. *Clin. Orthop. Rel. Res.* **241**, 265–277.

Wlodarski, K. H. (1991). Bone formation in soft tissues. In *Bone, Volume 5 Fracture Repair and Regeneration* (B. K. Hall, ed.), pp. 313–337. CRC Press, Boca Raton, FL.

Wlodarski, K. H., Hinek, A., and Ostrowski, K. (1970). Investigations on cartilage and bone induction in mice grafted with FL and WISH line human amniotic cells. *Calcif. Tissue Res.* **5**, 70–79.

Wlodarski, K.H., and Jakobisiak, M. (1981). Heterotopic induction of osteogenesis in mice lethally irradiated and repopulated with syngeneic bone marrow cells. *Archs Immunol. Therap. Exp.* **29**, 509–514.

Wlodarski, K. H., Kobus, M., Luczak, M., and Dolowy, R. (1981). Virally induced periosteal osteogenesis in mice. *Calcif. Tissue Int.* **33**, 135–142.

Wlodarski, K. H., Moskalewski, S., Skarzynska, S., Poltorak, A., and Ostrowski, K. (1971b). Irradiation and the bone induction properties of epithelial cells. *Bull. Acad. Pol. Sci., Ser. Sci. Biol.* **19**, 821–825.

Wlodarski, K. H., Ostrowski, K., Chtopkiewicz, B., and Koziorowska, J. (1974). Correlation between the agglutin-ability of living cells by concanavalin A and their ability to induce cartilage and bone formation. *Calcif. Tissue Res.* **16**, 251–256.

Wlodarski, K. H., Poltorak, A., and Koziorowska, J. (1971a). Species specificity of osteogenesis induced by WISH cell line and bone induction by *Vaccinia* virus transformed human fibroblasts. *Calcif. Tissue Res.* **7**, 345–352.

Wlodarski, K. H., and Reddi, A. H. (1987a). Heterotopically induced bone does not develop functional periosteal membrane. *Arch. Immunol. Therap. Exp.* **34**, 583–594.

Wlodarski, K. H., and Reddi, A. H. (1986). Alkaline phosphatase as a marker of osteoinductive cells. *Calcif. Tissue Int.* **39**, 382–385.

Wlodarski, K. H., and Reddi, A. H. (1987b). Tumor cells stimulate *in vivo* periosteal bone formation. *Bone Mineral* **2**, 185–192.

Wlodarski, K. H., and Thyberg, J. (1984). Demonstration of virus particles in Moloney murine sarcoma virus-induced periosteal bone in mice. *Virchows Arch. Cell Pathol. B.* **46**, 109–118.

Wlodarski, P. K., Wlodarski, K. H., Galus, K., Skarpetowski, A. C., Kowalski, M., Luczak, M., and Marciniak, M. M. (1995). *In vivo* exposure to sodium fluoride does not modify the yield of viral tumor-induced periosteal bone nor of heterotopic bone induced by human tumor KD cells in mice. *Folia Biol.* **41**, 88–96.

Woessner, J. F. (1973). Mammalian collagenases. *Clin. Orthop. Rel. Res.* **96**, 310–326.

Wolff, E., and Hampé, A. (1954). Sur la régulation de la patte du poulet après résection d'un segment intermediare du bourgéon de patte. *C.R. Soc. Biol. Paris* **148**, 154–156.

Wolff, E., and Kieny, M. (1957). Mise en evidence d'une action inhibitrice de l'extrait d'embryons de la race de poules 'Courte-pattes' Creeper sur la croissance des tibias cultivée *in vitro. C.R. Hebd. Seances Acad. Sci.* **244**, 1089–1091.

Wolff, E., and Kieny, M. (1963). Recherches sur la nature d'un facteur inhibiteur de la croissance des os longs dans la race de poule creeper. *Devel. Biol.* **7**, 324–341.

Wolff, J. (1870). Über die innere Architecture des Knochen und ihre Bedeutung für die Frage von Knochenwachstum. *Virchows Arch. Pathol. Anat.* **50**, 389–450.

Wolff, J. (1885). Markirversuche am Scheitelstirn- und Nasenbein der Kaninchen. *Arch. Path. Anat. Physiol. Klin. Med.* **101**, 572–630.

Wolff, J. (1892). *Das Gesetz der Transformation der Knochen.* Berlin.

Wolpert, L. (1971). Positional information and pattern formation. *Curr. Top. Devel. Biol.* **6**, 183–224.

Wolpert, L. (2002). Limb patterning: reports of model's death exaggerated. *Curr. Biol.* **12**, 628–630.

Wong, G. L. (1982). Skeletal actions of parathyroid hormone. *Min. Electrol. Metab.* **8**, 188–198.

Wong, G. L. (1984). Paracrine interactions in bone-secreted products of osteoblasts permit osteoclasts to respond to parathyroid hormone. *J. Biol. Chem.* **259**, 4019–4022.

Wong, G. L., and Cohn, D. V. (1974). Separation of parathyroid hormone and calcitonin-sensitive cells from non-responsive bone cells. *Nature* **252**, 713–714.

Wong, G. L., and Cohn, D. V. (1975). Target cells in bone for parathormone and calcitonin are different: Enrichment for each cell type by sequential digestion of mouse calvaria and selective adhesion to polymeric surfaces. *Proc. Natl Acad. Sci. U.S.A.* **72**, 3167–3171.

Wong, G. L., and Kocour, B. A. (1983). Differential sensitivity of osteoclasts and osteoblasts suggests that prostaglandin E_1 effects on bone may be mediated primarily through the osteoclasts. *Arch. Biochem. Biophys.* **224**, 29–35.

Wong, G. L., Luben, R. A., and Cohn, D. V. (1977). 1,25-dihydroxycholcalciferol and parathormone: Effects on isolated osteoclast-like and osteoblast-like cells. *Science* **197**, 663–665.

Wong, M., and Carter, D. R. (1990). A theoretical model of endochondral ossification and bone architectural construction in long bone ontogeny. *Anat. Embryol.* **181**, 523–532.

Wong, M., Lawton, T., Goetinck, P. F., Kuhn, J. L., Goldstein, S. A., and Bonadio, J. (1992). Aggrecan core protein is expressed in membranous bone of the chick embryo. Molecular and biochemical studies of normal and nanomelic embryos. *J. Biol. Chem.* **267**, 5592–5598.

Wong, M., and Tuan, R. S. (1995). Interactive cellular modulation of chondrogenic differentiation *in vitro* by subpopulations of chick embryonic calvarial cells. *Devel. Biol.* **167**, 130–147.

Wong, Y. Ch., and Buck, R. C. (1972). The development of sites of metaplastic change in regenerating tendon. *Z. Zellforsch. Mikrosk. Anat.* **134**, 175–182.

Woo, S. L.-Y., Matthews, J. V. Åkeson, W. H., Amiel, D., and Convery, F. R. (1975). Connective tissue response to immobility. *Arthritis Rheum.* **18**, 257–264.

Wood, A. (1982). Early pectoral fin development and morphogenesis of the apical ectodermal ridge in the killifish, *Aphyosemion scheeli. Anat. Rec.* **204**, 349–356.

Wood, A., Ashhurst, D. E., Corbett, A., and Thorogood, P. (1991). The transient expression of type II collagen at tissue interfaces during mammalian craniofacial development. *Development* **111**, 955–968.

Wooley, D. E., Crossley, M. J., and Evanson, J. M. (1977). Collagenase at sites of cartilage erosion in the rheumatoid joint. *Arthritis Rheum.* **20**, 1231–1239.

Wooley, D. E., Glanville, R. W., Lindberg, K. A., Bailey, A. J., and Evanson, J. M. (1973). Action of human skin collagenase on cartilage collagen. *FEBS Lett.* **34**, 267–269.

Worthington, R. D. (1974). High incidence of anomalies in a natural population of spotted salamanders, *Ambystoma maculatum. Herpetologica* **30**, 216–220.

Wozney, J. M. (1992). The bone morphogenetic protein family and osteogenesis. *Mol. Reprod. Devel.* **32**, 160–167.

Wozney, J. M., Capparella, J., and Rosen, V. (1993). The bone morphogenetic proteins in cartilage and bone development. In *Molecular Basis of Morphogenesis* (M. Bernfield, ed.), 51st Symp. Soc. Devel. Biol. pp. 221–230. Wiley-Liss, New York, NY.

Wozney, J. M., Rosen, V., Celeste, A. J., Mitsock, L. M., Whitters, M. J., Kriz, R. W., Hewick, R. M., and Wang, E. A. (1988). Novel regulators of bone formation: Molecular clones and activities. *Science* **242**, 1528–1534.

Wright, B. A., and Cohen, M. M., Jr. (1983). Tumors of cartilage. In *Cartilage, Volume 3: Biomedical Aspects* (B. K. Hall, ed.), pp. 143–164. Academic Press, New York, NY.

Wright, D. M., and Moffett, B. C., Jr. (1974). The post-natal development of the human temporomandibular joint. *Amer. J. Anat.* **141**, 235–250.

Wright, G. C., Jr., Miller, F., and Sokoloff, L. (1985). Induction of bone by xenografts of rabbit growth plate chondrocytes in the nude mouse. *Calcif. Tissue Int.* **37**, 250–256.

Wright, G. M., Armstrong, L. A., Jacques, A. M., and Youson, J. H. (1988). Trabecular, nasal, branchial, and pericardial cartilages in the sea lamprey, *Petromyzon marinus*: Fine structure and immunohistochemical detection of elastin. *Amer. J. Anat.* **182**, 1–15.

Wright, G. M., Keely, F. W., and Robson, P. (2001). The unusual cartilaginous tissues of jawless craniates, cephalochordates and invertebrates. *Cell Tissue Res.* **304**, 165–174.

Wright, G. M., Kelley, F. W., and Youson, J. H. (1983). Lamprin: A new vertebrate protein comprising the major structural protein of adult lamprey cartilage. *Experientia* **39**, 495–496.

Wright, G. M., Kelley, F. W., Youson, J. H., and Babineau, D. L. (1984). Cartilage in the Atlantic hagfish, *Myxine glutinosa. Amer. J. Anat.* **169**, 407–424.

Wright, G. M., and Leblond, C. P. (1981). Immunohistochemical localization of procollagens. III. Procollagen antigenicity in osteoblasts and prebone (osteoid). *J. Histochem. Cytochem.* **29**, 791–804.

Wright, G. M., and Youson, J. H. (1982). Ultrastructure of muco-cartilage in the larval sea lamprey, *Petromyzon marinus. Amer. J. Anat.* **165**, 39–51.

Wright, G. M., and Youson, J. H. (1983). Ultrastructure of cartilage from young adult sea lamprey, *Petromyzon marinus* L: A new type of vertebrate cartilage. *Amer. J. Anat.* **167**, 59–70.

Wright, H. V., Asling, C. W., Dougherty, H. L., Nelson, M. M., and Evans, H. M. (1958). Prenatal development of the skeleton in Long-Evans rats. *Anat. Rec.* **130**, 659–672.

Wright, M. O., Stockwell, R. A., and Nuki, G. (1992). Response of plasma membrane to applied hydrostatic-pressure in chondrocytes and fibroblasts. *Connect. Tissue Res.* **28**, 49–70.

Wright, S. (1934a). Polydactylous guinea-pigs. Two types respectively heterozygous and homozygous in the same mutant gene. *J. Hered.* **25**, 359–362.

Wright, S. (1934b). An analysis of variability in number of digits in an inbred strain of guinea pigs. *Genetics* **19**, 506–536.

Wright, S. (1934c). The results of crosses between inbred strains of guinea pigs, differing in number of digits. *Genetics* **19**, 537–551.

Wright, S. (1935). A mutation in the guinea pig, tending to restore the pentadactyly foot when heterozygous, producing a monstrosity when homozygous. *Genetics* **20**, 84–107.

Wright, T. J., and Mansour, S. L. (2003). Fgf3 and Fgf10 are required for mouse otic placode induction. *Development* **130**, 3379–3390.

Wroblewski, J., and Edwall, C. (1992). PDGF BB stimulates proliferation and differentiation in cultured chondrocytes from rat rib growth plate. *Cell Biol. Intern. Rep.* **16**, 133–144.

Wu, K.-C. (1994). Entwicklung, Stimulation und Paralyse der embryonalen Motorik. *Diss. Med Fac. Univ. Wien* pp. 1–83.

Wu, K.-C. (1996). Entwicklung, Stimulation und Paralyse der embryonalen Motorik. *Winer Klin. Wochensch.* **108**, 303–305.

Wu, K.-C., Streicher, J., Lee, M. L., Hall, B. K., and Müller, G. B. (2001). Role of motility in embryonic development. I: Embryo movement and amnion contractions in the chick and the influence of illumination. *J. Exp. Zool.* **291**, 186–194.

Wu, L. N. Y., Sauer, G. R., Genge, B. R., and Wuthier, R. E. (1989). Induction of mineral deposition by primary cultures of chicken growth plate chondrocytes in ascorbate-containing media. Evidence of an association between matrix vesicles and collagen. *J. Biol. Chem.* **264**, 21346–21355.

Wu, Q.-q., Zhang, Y., and Chen, Q. (2001). Indian hedgehog is an essential component of mechanotransduction complex to stimulate chondrocyte proliferation. *J. Biol. Chem.* **276**, 35290–35296.

Wu, X.-C., Li, Z., Zhou, B.-C., and Dong, Z.-M. (2003). A polydactylous amniote from the Triassic period. *Nature* **426**, 516.

Wurst, W., Auerbach, A. B., and Joyner, A. L. (1994). Multiple developmental defects in *Engrailed-1* mutant mice: an early mid-hindbrain deletion and patterning defects in forebrain and sternum. *Development* **120**, 2065–2075.

Wurtz, T., Krüger, A., Christersson, C., and Lundmark, C. (2001). A new protein expressed in bone marrow cells and osteoblasts with implications in osteoblast recruitment. *Exp. Cell Res.* **263**, 236–242.

Wyss, A. (1990). Clues to the origin of whales. *Nature* **347**, 428–429.

X

Xu, Z., Parker, S. B., and Minkoff, R. (1990a). Influence of epithelial-mesenchymal interactions on the viability of facial mesenchyme. II. Synthesis of basement-membrane components during tissue recombination. *Anat. Rec.* **228**, 58–68.

Xu, Z., Parker, S. B., and Minkoff, R. (1990b). Distribution of type IV collagen, laminin, and fibronectin during maxillary process formation in the chick embryo. *Amer. J. Anat.* **187**, 232–246.

Xue, Z.-G., Gehring, W. J., and Le Douarin, N. M. (1991). Quox-1, a quail homeobox gene expressed in the embryonic central nervous system, including the forebrain. *Proc. Natl Acad. Sci. U.S.A.* **88**, 2427–2437.

Xue, Z.-G., Xue, X. J., and Le Douarin, N. M. (1993). Quox-1, an *Antp*-like homeobox gene of the avian embryo: a developmental study using a Quox-1 specific antiserum. *Mech. Devel.* **43**, 149–158.

Y

Yabe, H., and Hanaoka, H. (1985). Investigation of the origin of the osteoclast by use of transplantation on chick chorioallantoic membrane. *Clin. Orthop. Rel. Res.* **197**, 255–265.

Yablokov, A. V. (1974). *Variability of Mammals. Revised by the author.* Published by the Smithsonian Institution and the National Science Foundation, Washington, D.C., by Amerind Publishing Co., New Delhi.

Yagami, K., Suh, J.-Y., Enomoto-Iwamoto, M., Koyama, E., Abrams, W. R., Shapiro, I. M., Pacifici, M., and Iwamoto, M. (1999). Matrix GLA protein is a developmental regulator of chondrocyte mineralization and, when constituitively expressed, blocks endochondral and intramembranous ossification in the limb. *J. Cell Biol.* **147**, 1097–1108.

Yagiela, J. A., and Woodbury, D. M. (1977). Enzymatic isolation of osteoblasts from fetal rat calvaria. *Anat. Rec.* **188**, 287–306.

Yajima, H., Hara, K., Ide, H., and Tamura, K. (2002). Cell adhesiveness and affinity for limb pattern formation. *Int. J. Devel. Biol.* **46**, 897–904.

Yajima, H., Yonei-Tamura, S., Watanabe, N., Tamura, K., and Ide, H. (1999). Role of N-cadherin in the sorting-out of mesenchymal cells and in the positional identity along the proximodistal axis of the chick limb bud. *Devel. Dynam.* **216**, 274–284.

Yalamanchi, N., Klein, M.B., Pham, H. M, Longaker, M. T., and Chang, J. (2004). Flexor tendon wound healing *in vitro*: Lactate up-regulation of TGF-beta expression and functional activity. *Plast. Reconst. Surg.* 625–632.

Yamada, G., Mansouri, A., Torres, M., Stuart, E. T., Blum, M., Schultz, M., de Robertis, E. M., and Gruss, P. (1995). Targeted mutation of the mouse *goosecoid* gene results in craniofacial defects and neonatal death. *Development* **121**, 2917–2922.

Yamada, J. (1971). A fine structural aspect of the development of scales in the chum salmon fry. *Bull. Jap. Soc. Scient. Fish.* **37**, 18–29.

Yamada, K., Shimizu, S., Brown, K. S., and Kimata, K. (1984). The histochemistry of complex carbohydrates in certain organs of homozygous brachymorphic (*bm/bm*) mice. *Histochem. J.* **16**, 587–600.

Yamada, M., Fujimori, K., Takeuchi, H., Yamamoto, K., and Takakusu, A. (1977). Hematopoiesis in bovine heart bone. *Cell Struct. Funct.* **2**, 353–360.

Yamada, T. (1991). Selective staining methods for cartilage of rat fetal specimens previously treated with alizarin-red-S. *Teratology* **43**, 615–620.

Yamada, T., Kurihara, A., Nishiyama, T., and Uchida, H. (1993). Application of the Bromophenol blue (BPB) staining method to rat fetal cartilage previously stained with alizarin red S. *Exp. Anim.* **42**, 457–461.

Yamagata, M., Shinomura, T., and Kimata, K. (1993). Tissue variation of two large chondroitin sulfate proteoglycans (PG-M/versican and PG-H/aggrecan) in chick embryos. *Anat. Embryol.* **187**, 433–444.

Yamagishi, T., Nishimatsu, S.-i., Nomura, S., Asashima, M., Murakami, K., and Ueno, N. (1995). Expression of BMP-2,4 genes during early development in *Xenopus. Zool. Sci.* **12**, 355–358.

Yamaguchi, A. (1995). Regulation of differentiation pathway of skeletal mesenchymal cells in cell lines by transforming growth factor-β superfamily. *Sem. Cell Biol.* **6**, 165–173.

Yamaguchi, A., and Kahn, A. J. (1991). Clonal osteogenic cell lines express myogenic and adipocytic developmental potential. *Calcif. Tissue Int.* **49**, 221–225.

Yamaguchi, A., Katagiri, T., Ikeda, T., Wozney, J. M., Rosen, V., Wang, E. A., Kahn, A. J., Suda, T., and Yoshiki, S. (1991). Recombinant human bone morphogenetic protein-2 stimulates osteoblastic maturation and inhibits myogenic differentiation *in vitro. J. Cell Biol.* **113**, 681–687.

Yamaguchi, A., Yamanouchi, M., and Yoshiki, S. (1985). Osteoblastic and osteoclastic differentiation of mononuclear cells facing the resorbing surface of uncalcified cartilage in the tibia of embryonic chick. *Cell Tissue Res.* **240**, 425–431.

Yamaguchi, T. P., Bradley, A., McMahon, A. P., and Jones, S. (1999). A *Wnt5a* pathway underlies outgrowth of multiple structures in the vertebrate embryo. *Development* **126**, 1211–1223.

Yamaguchi, Y., Mann, D. M., and Ruoslahti, E. (1990). Negative regulation of transforming growth factor-β by the proteoglycan decorin. *Nature* **346**, 281–284.

Yamamoto, M. (1989). An electron microscopic study of the distal segment of the os penis of the rat. *Arch. Histol. Cytol.* **52**, 529–541.

Yamamoto, T., Ecarot, B., and Glorieux, F. H. (1991). *In vivo* osteogenic activity of isolated human bone cells. *J. Bone Min. Res.* **6**, 45–51.

Yamamoto, Y., Espinasa, L., Stock., D. W., and Jeffrey, W. R. (2003). Development and evolution of craniofacial patterning is mediated by eye-dependent and -independent processes in the cavefish *Astyanax. Evol. & Devel.* **5**, 435–446.

Yamashita, K., Horisaka, Y., Okamoto, Y., Yoshimura, Y., Matsumoto, N., Kawada, J., and Takagi, T. (1991). Architecture of implanted bone matrix gelatin influences heterotopic calcification and new bone formation. *Proc. Soc. Exp. Biol. Med.* **197**, 342–347.

Yamashita, K., and Takagi, T. (1992a). Ultrastructural observation of calcification preceding new bone formation induced by demineralized bone matrix gelatin. *Acta Anat.* **143**, 261–267.

Yamashita, K., and Takagi, T. (1992b). Appearance of adipose cells in connective tissue at the implantation site of bone matrix gelatin and bupivacaine injection. *Acta Anat.* **145**, 406–411.

Yamashita, K., and Takagi, T. (1992c). Calcification preceding new bone formation induced by demineralized bone matrix gelatin. *Arch. Histol. Cytol.* **55**, 31–43.

Yamaza, H., Matzuo, K., Kiyoshima, T., Shigemura, N., Kobayashi, I., Wada, H., Akamine, A., and Sakai, H. (2001). Detection of differentially expressed genes in the early developmental stage of the mouse mandible. *Int. J. Devel. Biol.* **45**, 675–680.

Yamazaki, K., and Eyden, B. P. (1995). A study of intercellular relationships between trabecular bone and marrow stronal cells in the murine femoral metaphysis. *Anat. Embryol.* **192**, 9–20.

Yamazaki, K., Suda, N., and Kuroda, T. (1997). Immunochemical localization of parathyroid hormone-related protein in developing mouse Meckel's cartilage and mandible. *Archs Oral Biol.* **42**, 787–794.

Yamazaki, K., Suda, N., and Kuroda, T. (1999). Distribution of parathyroid hormone-related protein (PTHrp) and type I parathyroid hormone (PTH) PthrP receptor in developing mouse mandibular condylar cartilage. *Archs Oral Biol.* **44**, 853–860.

Yan, Y.-L., Hatta, K., Riggleman, B., and Postelthwait, J. H. (1995). Expression of a type II collagen gene in the zebrafish embryonic axis. *Devel. Dynam.* **203**, 363–376.

Yan, Y.-L., Jowett, T., and Postelthwait, J. H. (1998). Ectopic expression of *hoxb2* after retinoic acid treatment or mRNA injection: Disruption of hindbrain and craniofacial morphogenesis in zebrafish embryos. *Devel. Dynam.* **213**, 3730–365.

Yanagisawa, H., Clouthier, D. E., Richardson, J. A., Charité, J., and Olson, E.N. (2003). Targeted deletion of a branchial arch-specific enhancer reveals a role of *dHAND* in craniofacial development. *Development* **130**, 1069–1078.

Yaneza, M., Gilthorpe, J. D., Lumsden, A., and Tucker, A. S. (2002). No evidence for ventrally migrating neural tube cells from the mid- and hindbrain. *Devel. Dynam.* **223**, 163–167

Yang, B. B., Zhang, Y., Cao, L., and Yang, B. L. (1998). Aggrecan and link protein affect cell adhesion to culture plates and to type II collagen. *Matrix Biol.* **16**, 541–561.

Yang, X., Chen, L., Xu, X., Li, C., Huang, C., and Deng, C.-X. (2001). TGFβ/Smad3 signals repress chondrocyte hypertrophic differentiation and are rewired for maintaining articular cartilage. *J. Cell Biol.* **153**, 35–46.

Yang, Y. (2003). Wnts and wing: Wnt signaling in vertebrate limb development and musculoskeletal morphogenesis. *Birth Defects Res.* (*Part C*), **69**, 305–317.

Yang, Y., Buillot, P., Boyd, Y., Lyon, M. F., and McMahon, A. (1998). Evidence that preaxial polydactyly in the Doublefoot mutant is due to ectopic Indian hedgehog signaling. *Development* **125**, 3123–3132.

Yang, Y., Drossopoulou, G., Chuang, P.-T., Duprez, D., Marti, E., Bumcrot, D., Vargesson, N., Clarke, J., Niswander, L., McMahon, A., and Tickle, C. (1997). Relationship between dose, distance and time in *Sonic hedgehog*-mediated regulation of anteroposterior polarity in the chick limb. *Development* **124**, 4393–4404.

Yasuda, M. (1975). Pathogenesis of preaxial polydactyly of the hand in human embryos. *J. Embryol. Exp. Morphol.* **33**, 745–756.

Yasuda, Y. (1973). Differentiation of human limb bones *in vitro*. *Anat. Rec.* **175**, 561–578.

Yasui, N., Benya, P. D., and Nimni, M. E. (1986). Coordinate regulation of type IX and type II collagen synthesis during growth of chick chondrocytes in retinoic acid and of 5-bromo-2'-deoxyuridine. *J. Biol. Chem.* **261**, 7997–8001.

Yasui, N., Ono, K., Konomi, H., and Nagai, Y. (1984). Transitions in collagen types during endochondral ossification in human growth cartilage. *Clin. Orthop. Rel. Res.* **183**, 215–218.

Yasutake, J., Inohaya, K., and Kudo, A. (2004). *Twist* functions in vertebral column formation in medaka, *Oryzias latipes*. *Mech. Devel.* **121**, 883–894.

Ye, X. J., Terato, K., Nakatani, H., Cremer, M. A., and Yoo, T. J. (1991). Monoclonal antibodies against bovine type IX collagen (LMW fragment): Production, characterization and use for immunohistochemical localization studies. *J. Histochem. Cytochem.* **39**, 265–271.

Yeager, V. L., Chiemchanya, S., and Chaiseri, P. (1975). Changes in size of lacunae during the life of osteocytes in osteons of compact bone. *J. Gerontol.* **30**, 9–14.

Yeh, J. (2002). The effect of miniaturized body size on skeletal morphology in frogs. *Evolution* **53**, 628–641.

Yelon, D., Ticho, B., Halpern, M. E., Ruvinsky, I., Ho, R. K., Silver, L. M., and Stainier, D. Y. R. (2000). The bHLH transcription factor Hand2 plays parallel roles in zebrafish heart and pectoral fin development. *Development* **127**, 2573–2582.

Yeung, H.-Y., Lee, K.-M., Fung, K.-P., and Leung, K.-S. (2002). Sustained expression of transforming growth factor-β1 by distraction during distraction osteogenesis. *Life Sci.* **71**, 67–79.

Yi, S. E., Daluiski, A., Pederson, R., Rosen, V., and Lyons, K. M. (2000). The type 1 BMP receptor BMPR1B is required for chondrogenesis in the mouse limb. *Development* **127**, 621–630.

Yih, W. H., Zysset, M., and Merrill, R. G. (1992). Histologic study of the fate of autogenous auricular cartilage grafts in the human temporomandibular joint. *J. Oral Maxillofac. Surg.* **50**, 964–967.

Yin, M., and Pacifici, M. (2001). Vascular regression is required for mesenchymal condensation and chondrogenesis in the developing limb. *Devel. Dynam.* **222**, 522–533.

Yokouchi, Y., Nakazato, S., Yamamoto, M., Goto, Y., Kameda, T., Iba, H., and Kuroiwa, A. (1995). Misexpression of *Hoxa-13* induces cartilage homeotic transformation and changes cell adhesiveness in chick limb buds. *Genes & Devel.* **9**, 2509–2522.

Yokouchi, Y., Ohsugi, K., Sasaki, H., and Kuroiwa, A. (1991a). Chicken homeobox gene *Msx-1*: Structure, expression in limb buds and effect of retinoic acid. *Development* **113**, 431–444.

Yokouchi, Y., Sasaki, H., and Kuroiwa, A. (1991b). Homeobox gene expression correlated with the bifurcation process of limb cartilage development. *Nature* **353**, 443–445.

Yokoyama, H., Endo, T., Tamura, K., Yajima, H., and Ide, H. (1998). Multiple digit formation in *Xenopus* limb bud recombinants. *Devel. Biol.* **196**, 1–10.

Yokoyama, H., Yonei-Tamura, S., Endo, T., Izpisúa-Belmonte, J. CV., Tamura, K., and Ide, H. (2000). Mesenchyme with *fgf-10* expression is responsible for regenerative capacity in *Xenopus* limb buds. *Devel. Biol.* **219**, 18–29.

Yonei-Tamura, S., Endo, T., Yajima, H., Ohuchi, H., Ide, H., and Tamura, K. (1999). FGF7 and FGF10 directly induce the apical ectodermal ridge in chick embryos. *Devel. Biol.* **211**, 133–143.

Yoon, K., Buenaga, R., and Rodan, G. A. (1987). Tissue specificity and developmental expression of rat osteopontin. *Biochem. Biophys. Res. Commun.* **148**, 1129–1136.

Yoshida, Y., Tanaka, S., Umemori, H., Minowa, O., Usui, M., Ikematsu, N., Hosoda, E., Imamura, T., Kuno, J., Yamashita, T., Miyazomo, K., Noda, M., Noda, T., and Yamamoto, T. (2000). Negative regulation of BMP/Smad signaling by Tob in osteoblasts. *Cell* **103**, 1085–1097.

Yoshikawa, T., Ohgushi, H., Okumura, M., Tamai, S., Dohi, Y., and Moriyama, T. (1992). Biochemical and histological sequences of membranous ossification in ectopic sites. *Calcif. Tissue Int.* **50**, 184–188.

Yoshioka, C., and Yagi, T. (1988). Electron microscopic observations on the fate of hypertrophic chondrocytes in condylar cartilage of rat mandible. *J. Craniofac. Genet. Devel. Biol.* **8**, 253–264.

Young, B. A. (1994). Cartilago cordis in Serpents. *Anat. Rec.* **240**, 243–247.

Young, H. E., Ceballos, E. M., Smith, J. C., Lucas, P. A., and Morrison, D. C. (1992). Isolation of embryonic chick myosattelite and pluripotential stem cells. *J. Tissue Cult. Meth.* **14**, 85–92.

Young, H. E., Ceballos, E. M., Smith, J. C., Mancini, M. L., Wright, R. P., Ragan, B. L., Bushell, I., and Lucas, P. A. (1993).

Pluripotential mesenchymal stem cells reside within avian connective tissue matrices. *In Vitro Cell Devel. Biol. Anim.* **29**, 723–736.

Young, N. M. (2003). A reassessment of living hominoid postcranial variability: implications for ape evolution. *J. Human Evol.* **45**, 441–464.

Young, N. M. (2004). Modularity and integration in the hominoid scapula. *J. Exp. Zool.* (*Mol. Dev. Evol.*) **302B**, 226–240.

Young, R. W. (1962). Cell proliferation and specialization during endochondral osteogenesis in young rats. *J. Cell Biol.* **14**, 357–370.

Young, R. W. (1963). Nucleic acids, protein synthesis and bone. *Clin. Orthop. Rel. Res.* **26**, 147–160.

Young, R. W. (1964). Specialization of bone cells. In *Bone Biodynamics* (H. M. Frost, ed.), pp. 117–139. Little, Brown, Boston, MA.

Yousen, S. A., Dauber, J. H., and Griffith, B. P. (1990). Bronchial cartilage alterations in lung transplantation. *Chest* **98**, 1121–1124.

Youson, J. H. (2004). The impact of environmental and hormonal cues on the evolution of fish metamorphosis. In *Environment, Development and Evolution: Toward a Synthesis* (B. K. Hall, R. D. Pearson and G. B. Müller, eds), pp. 239–237. The MIT Press, Cambridge, MA.

Youssef, E. H. (1969). Development of the membrane bones and ossification of the chondrocranium of the albino rat. *Acta Anat.* **72**, 603–623.

Yu, J. C., McClintock, J. S., Gannon, F., Gao, X. X., Mobasser, J. P., and Sharawy, M. (1997). Regional differences of dura osteoinduction – squamous dura induces osteogenesis, sutural dura induces chondrogenesis and osteogenesis. *Plast. Reconstruct. Surg.* **100**, 23–31.

Yu, K., Xu, J., Liu, Z., Sosic, D., Shao, J., Olson, E. N., Towler, D. A., and Ornitz, D. M. (2003). Conditional inactivation of FGF receptor 2 reveals an essential role for FGF signaling in the regulation osteoblast function and bone growth. *Development* **130**, 3063–3074.

Yu, Y. M., Becvar, R., Yamada, Y., and Reddi, A. H. (1991). Changes in the gene expression of collagens, fibronectin, integrin and proteoglycans during matrix-induced bone morphogenesis. *Biochem. Biophys. Res. Commun.* **177**, 427–432.

Yueh, Y. G., Gardner, D. P., and Kappen, C. (1998). Evidence for regulation of cartilage differentiation by the homeobox gene *Hoxc-8*. *Proc. Natl Acad. Sci. U.S.A.* **95**, 9956–9961.

Yumoto, T., Poel, W. E., Kodama, T., and Dmochowski, L. (1970). Studies on FBJ virus-induced bone tumors in mice. *Texas Rep. Biol. Med.* **28**, 145–165.

Yuodelis, R. A. (1966). The morphogenesis of the human temporomandibular joint and its associated structures. *J. Dent. Res.* **45**, 182–191.

Z

Zákány, J., and Duboule, D. (1996). Synpolydactyly in mice with a targeted deficiency in the *HoxD* complex. *Nature* **384**, 69–71.

Zákány, J., Fromenthal-Ramain, C., Warot, X., and Duboule, D. (1997). Regulation of number and size of digits by posterior *Hox* genes: a dose-dependent mechanism with potential evolutionary implications. *Proc. Natl Acad. Sci. U.S.A.* **94**, 13695–13700.

Zaman, G., Dallas, S. L., and Lanyon, L. E. (1992). Cultured embryonic bone shafts show osteogenic responses to mechanical loading. *Calcif. Tissue Int.* **51**, 132–136.

Zambonin Zallone, A., and Teti, A. (1981). The osteoclast of hen medullary bone under hypocalcaemic conditions. *Anat. Embryol.* **162**, 379–392.

Zambonin Zallone, A., and Teti, A. (1985). Autoradiographic demonstration of in vitro fusion of blood monocytes with osteoclasts. *Basic Appl. Histochem.* **29**, 45–48.

Zambonin Zallone, A., Teti, A., Primavera, M. V., and Pace, G. (1983). Mature osteocytes behaviour in a repletion period: The occurrence of osteoplastic activity. *Basic Appl. Histochem.* **27**, 191–204.

Zanetti, N. C., Dress, V. M., and Solursh, M. (1990). Comparison between ectoderm-conditioned medium and fibronectin in their effects on chondrogenesis by limb bud mesenchymal cells. *Devel. Biol.* **139**, 383–395.

Zanetti, N. C., and Solursh, M. (1984). Induction of chondrogenesis in limb mesenchymal cultures by disruption of the actin cytoskeleton. *J. Cell Biol.* **99**, 115–123.

Zardoya, R., and Meyer, A. (1998). Complete mitochondrial genome suggests diapsid affinities of turtles. *Proc. Natl Acad. Sci. U.S.A.* **95**, 14266–14231.

Zauner, H., Begemann, G., Marí-Beffa, M., and Meyer, A. (2003). Differential regulation of *msx* genes in the development of the gonopodium, an intromittent organ, and of the 'sword,' a sexually selected trait of swordtail fishes (*Xiphophorus*). *Evol. & Devel.* **5**, 466–477.

Zelditch, M. L. (1987). Evaluating models of developmental integration in the laboratory rat using confirmatory factor analysis. *Syst. Zool.* **36**, 368–380.

Zelditch, M. L. (1988). Ontogenetic variation in patterns of phenotypic integration in the laboratory rat. *Evolution* **42**, 28–41.

Zelditch, M. L. (ed.) (2001). *Beyond Heterochrony: the Evolution of Development*. Wiley-Liss, New York, NY.

Zelditch, M. L., Bookstein, F. L., and Lundrigan, B. L. (1992). Ontogeny of integrated skull growth in the cotton rat *Sigmodon fulviventer*. *Evolution* (*Lawrence, Kansas*), **46**, 1164–1180.

Zelditch, M. L., Bookstein, F. L., and Lundrigan, B. L. (1993). The ontogenetic complexity of developmental constraints. *J. Evol. Biol.* **6**, 621–641.

Zelditch, M. L., and Carmichael, A. C. (1989a). Ontogenetic variation in patterns of developmental and functional integration in skulls of *Sigmodon fulviventer*. *Evolution* **43**, 814–824.

Zelditch, M. L., and Carmichael, A. C. (1989b). Growth and intensity of integration through postnatal growth in the skull of *Sigmodon fulviventer*. *J. Mammal.* **70**, 477–484.

Zelditch, M. L., and Fink, W. L. (1996). Heterochrony and heterotopy: Stability and innovation in the evolution of form. *Paleobiology* **22**, 241–254.

Zelditch, M. L., Lundrigan, B. L., and Garland, T. (2004). Developmental regulation of skull morphology. I. Ontogenetic dynamics of variance. *Evol. & Devel.* **6**, 194–206.

Zelditch, M. L., Straney, D. O., Swiderski, D. L., and Carmichael, A. C. (1990). Variation in developmental constraints in *Sigmodon*. *Evolution* (*Lawrence, Kansas*), **44**, 1738–1747.

Zeller, R., and Duboule, D. (1997). Dorso-ventral limb polarity and origin of the ridge: on the fringe of independence? *BioEssays* **19**, 541–546.

Zeller, R., Jackson-Grusby, L., and Leder, P. (1989). The limb deformity gene is required for apical ectodermal ridge differentiation and anteroposterior limb pattern formation. *Genes & Devel.* **3**, 1481–1492.

Zeller, U. (1988). *Die Entwicklung und Morphologie des Schädels von Ornithorhynchus anatinus* (*Monotremata, Protheria,*

Mammalia). Habilitationsschrift, Georg August Universität, Göttingen.

Zelzer, E., McLean, W., Na, Y.-S., Fukai, N., Reginato, A. M., Lovejoy, S., S'Amore, P. A., and Olsen, B. R. (2002). Skeletal defects in VEGF120/120 mice reveal multiple roles for VEGF in skeletogenesis. *Development* **129**, 1893–1904.

Zelzer, E., and Olsen, B. R. (2003). The genetic basis for skeletal diseases. *Nature* **423**, 343–348.

Zernik, J., Twarog, K., and Upholt, W. B. (1990). Regulation of alkaline phosphatase and alpha 2(I) procollagen synthesis during early intramembranous bone formation in the rat mandible. *Differentiation* **44**, 207–215.

Zhang, J., Hagopian-Donaldson, S., Serbedzija, G., Elsemore, J., Plehn-Dujowich, D., McMahon, A. P., Flavell, R. A., and Williams, T. (1996). Neural tube, skeletal and body wall defects in mice lacking transcription factor AP-2. *Nature* **381**, 238–241.

Zhang, J., and Williams, T. (2003). Identification and regulation of tissue-specific cis-acting elements associated with the human AP-2α gene. *Devel. Dynam.* **228**, 194–207.

Zhang, Q., Carr, D. W., Lerea, K. M., Scott, J. D., and Newman, S. A. (1996). Nuclear localization of type II cAMP-dependent protein kinase during limb cartilage differentiation is associated with a novel developmentally regulated A-kinase anchoring protein. *Devel. Biol.* **176**, 51–61.

Zhang, Y., Cao, L., Kiaii, C. G., Yang, B. L., and Yang, B. B. (1998). The G3 domain of versican inhibits mesenchymal chondrogenesis via the epidermal growth factor like motifs. *J. Biol. Chem.* **273**, 33054–33063.

Zhang, Y., Wu, Y., Cao, L., Lee, V., Chen, K., Lin, Z., Kiani, C., Adamas, M. E., and Yang, B. B. (2001). Versican modulates embryonic chondrocyte morphology via the epidermal growth factor-like motifs in G3. *Exp. Cell Res.* **263**, 33–42.

Zhao, G.-Q., Eberspaecher, H., Seldin, M. F., and de Crombrugghe, B. (1994). The gene for the homeodomain-containing protein CART-1 is expressed in cells that have a chondrogenic potential during embryonic development. *Mech. Devel.* **48**, 245–254.

Zhao, W. G., and Neufeld, D. A. (1995). Bone regrowth in young mice stimulated by nail organ. *J. Exp. Zool.* **271**, 155–159.

Zhao, W. G., and Neufeld, D. A. (1996). Bone regeneration after amputation stimulated by basic fibroblast growth factor in vitro. *In Vitro Cell Devel. Biol.* **32**, 63–65.

Zhao, X., Zhang, Z., Song, Y., Zhang, X., Zhang, Y., Hu, Y., Fromm., S. H., and Chen, Y. (2000). Transgenically ectopic expression of BMP4 to the Msx1 mutant dental mesenchyme restores downstream gene expression by represses Shh and BMP2 in the enamel knot of wild type tooth germ. *Mech. Devel.* **99**, 29–38.

Zhao, Z., Stock, D. W., Buchanan, A. V., and Weiss, K. M. (2000). Expression of Dlx genes during the development of the murine dentition. *Devel. Genes Evol.* **210**, 270–275.

Zhi, J., Sommerfeldt, D. W., Rubin, C. T., and Hadjiargyrou, M. (2001). Differential expression of neuroleukin in osseous tissues and its involvement in muneralization during osteoblast differentiation. *J. Bone Min. Res.* **16**, 1994–2004.

Zhou, H., Chernecky, R., and Davies, J. E. (1993). Scanning electron microscopy of the osteoclast-bone interface in vivo. *Cells & Matrix* **3**, 141–150.

Zhou, X.-N., and Van de Water, T. R. (1987). The effect of target tissues on survival and differentiation of mammalian statoacoustic ganglion neurons in organ culture. *Acta Otolaryngol. (Stockh.)* **104**, 90–98.

Zhu, C. C., Yamada, G., Nakamura, S., Terashi, T., Schweickert, A., and Blum, M. (1998). Malformation of trachea and pelvic region in goosecoid mutant mice. *Devel. Dynam.* **211**, 371–381.

Zhu, Y., McAlinder, A., and Sandell, L. J. (2001). Type IIA procollagen in development of the human intervertebral disc: regulated expression of the NH2-propeptide by enzymic processing reveals a unique developmental pathway. *Devel. Dynam.* **220**, 307–322.

Zika, J. M., and Klein, L. (1975). Comparison of whole calvarial bones and long bones during early growth in rats. Histology and collagen composition. *Calcif. Tissue Res.* **18**, 101–111.

Zilliken, F. (1967). Notochord induced cartilage formation in chick somites: Intact tissue versus extracts. *Exp. Biol. Med.* **1**, 199–212.

Zimmermann, A., Bachman, M., Stocker, F., and Weber, J. (1979). Knorpelbildung im Bereuch des Linken Ventrikels und des Hisschen Bündels bei einem Komplexen angeborenen Herzvitium. *Klin. Pädiat.* **191**, 584–588.

Zimmermann, B. (1984). Assembly and disassembly of gap junctions during mesenchymal cell condensation and early chondrogenesis in limb buds of mouse embryos. *J. Anat.* **138**, 351–364.

Zimmermann, B. (1992). Degeneration of osteoblasts involved in intramembranous ossification of fetal rat calvaria. *Cell Tissue Res.* **267**, 75–84.

Zimmermann, B., and Cristea, R. (1993). Dexamethasone induces chondrogenesis in organoid culture of cell mixtures from mouse embryos. *Anat. Embryol.* **187**, 67–73.

Zimmermann, B., Moegelin, A., de Souza, P., and Bier, J. (1998). Morphology of the development of the sagittal suture of mice. *Anat. Embryol.* **197**, 155–165.

Zimmermann, B., Scharlach, E., and Kaatz, R. (1982). Cell contact and surface coat alterations of limb-bud mesenchymal cells during differentiation. *J. Embryol. Exp. Morphol.* **72**, 1–18.

Zimmermann, B., Wachtel, H. C., and Somogyi, H. (1990). Endochondral mineralization in cartilage organoid culture. *Cell Differ. Devel.* **31**, 11–22.

Zimny, M. L., and Redler, I. (1969). An ultrastructural study of patellar chondromalacia in humans. *J. Bone Joint Surg. Am. Vol.* **51**, 1179–1190.

Zimny, M. L., and Redler, I. (1972). An ultrastructural study of chondromalacia patellae. *Clin. Orthop. Rel. Res.* **82**, 37–44.

Zollikofer, C. P. E., and Ponce de León, M. S. (2004). Kinematics of cranial ontogeny: Heterotopy, heterochrony, and geometric morphometric analysis of growth models. *J. Exp. Zool. (Mol. Dev. Evol.)*, **302B**, 322–340.

Zou, H., and Niswander, L. (1996). Requirement for BMP signaling in interdigital apoptosis and scale formation. *Science* **272**, 738–741.

Zou, H., Wieser, R., Massagué, J., and Niswander, L. (1997). Distinct roles of type I bone morphogenetic protein receptors in the formation and differentiation of cartilage. *Genes & Devel.* **11**, 2191–2203.

Zschäbitz, A. (1998). Glycoconjugate expression and cartilage development of the cranial skeleton. *Acta Anat.* **161**, 254–274.

Zschäbitz, A., Gabius, H.-J., Krahn, V., Michiels, I., Schmidt, W., Koepp, H., and Stofft, E. (1995a). Distribution patterns of neoglycoprotein-binding sites (endogenous lectins) and lectin-reactive glycoconjugates during cartilage and bone formation in human finger. *Acta Anat.* **154**, 272–282.

Zschäbitz, A., Krahn, V., Gabius, H.-J., Weiser, H., Khaw, A., Biesalski, H. K., and Stofft, E. (1995b). Glycoconjugate expression of chondrocytes and perichondrium during hyaline cartilage development in the rat. *J. Anat.* **187**, 67–83.

Zusi, R. L., and Warheit, K. I. (1992). On the evolution of intraramal mandibular joints in *Pseudodontorus* (Aves: odontopterygia). In *Papers in Avian Paleontology Honoring Pierce Brodkorb* (K. E. Campbell, Jr., ed.), pp. 351–360. No 36, Sci. Ser. Nat. Hist. Mus. L. A. County, CA.

Zuzarte-Luís, V., and Hurlé, J. M. (2002). Programmed cell death in the developing limb. *Int. J. Devel. Biol.* **46**, 871–876.

Zwilling, E. (1949). The role of epithelial components in the developmental origin of the 'wingless' syndrome of chick embryos. *J. Exp. Zool.* **111**, 175–188.

Zwilling, E. (1955). Ectoderm-mesoderm relationships in the development of the chick embryo limb bud. *J. Exp. Zool.* **128**, 423–441.

Zwilling, E. (1956a). Interaction between limb bud ectoderm and mesoderm in the chick embryo. II. Experimental limb duplication. *J. Exp. Zool.* **132**, 173–187.

Zwilling, E. (1956b). Interaction between limb bud ectoderm and mesoderm in the chick embryo. IV. Experiments with a wingless mutant. *J. Exp. Zool.* **132**, 241–253.

Zwilling, E. (1959). Micromelia as a direct effect of insulin – evidence from an *in vitro* and *in vivo* experiment. *J. Morphol.* **103**, 159–179.

Zwilling, E. (1964). Development of fragmented and of dissociated limb bud mesoderm. *Devel. Biol.* **9**, 20–37.

Zwilling, E. (1966). Cartilage formation from so-called myogenic tissues of chick embryo limb bud. *Ann. Med. Exp. Biol. Fenn.* **44**, 134–139.

Zwilling, E. (1968). Morphogenetic phases in development. *Devel. Biol., Suppl.* **2**, 184–207.

Zwilling, E. (1972). Limb morphogenesis. *Devel. Biol.* **28**, 12–17.

Zwilling, E. (1974). Effects of contact between mutant (wingless) limb buds and those of genetically normal chick embryos: Confirmation of a hypothesis. *Devel. Biol.* **38**, 37–48.

Zwilling, E., and Hansborough, L. (1956). Interaction between limb bud ectoderm and mesoderm in the chick embryo. III. Experiments with polydactylous limbs. *J. Exp. Zool.* **132**, 219–239.

Zylberberg, L., and Castanet, J. (1985). New data on the structure and the growth of the osteoderms in the reptile *Anguis fragilis* L (Anguidae, Squamata). *J. Morphol.* **186**, 327–342.

Zylberberg, L., Castanet, J., and de Ricqlès, A. J. (1980). Structure of the dermal scales in Gymnophiona (Amphibia). *J. Morphol.* **165**, 41–54.

Zylberberg, L., and Wake, M. H. (1990). Structure of the scales of *Dermophis* and *Microcaecilia* (Amphibia: Gymnophiona) and a comparison to dermal ossifications of other vertebrates. *J. Morphol.* **206**, 25–44.

Index

This single index serves as a subject index and as a taxonomic index to common and species names. Numbers in parentheses refer to endnotes. In general, abbreviations are not listed, exceptions being gene families with multiple members where readers may want to seek information on an individual family member; Bmp-2 or Fgf-8 for example. Abbreviations are listed in the annotated list of abbreviations that follows the table of contents and which contains older names for genes (Hox 1.1 for *Hoxa-7*, for example). These older names are not duplicated in the index.